HANDBOOK OF ENGINEERING MECHANICS

McGRAW-HILL HANDBOOKS

Abbott and Stetka · National Electrical Code Handbook, 10th ed.
Aljian · Purchasing Handbook
American Institute of Physics · American Institute of Physics Handbook
American Society of Mechanical Engineers · ASME Handbooks:
    Engineering Tables
    Metals Engineering—Processes
    Metals Engineering—Design
    Metals Properties
American Society of Tool and Manufacturing Engineers:
    Die Design Handbook
    Tool Engineers Handbook, 2d ed.
    Handbook of Fixture Design
Beeman · Industrial Power Systems Handbook
Berry, Bollay, and Beers · Handbook of Meteorology
Blatz · Radiation Hygiene Handbook
Brady · Materials Handbook, 8th ed.
Burington · Handbook of Mathematical Tables and Formulas, 3d ed.
Burington and May · Handbook of Probability and Statistics with Tables
Carroll · Industrial Instrument Servicing Handbook
Cockrell · Industrial Electronics Handbook
Condon and Odishaw · Handbook of Physics
Considine · Process Instruments and Controls Handbook
Crocker · Piping Handbook, 4th ed.
Croft and Carr · American Electricians' Handbook, 8th ed.
Davis · Handbook of Applied Hydraulics, 2d ed.
Dudley · Gear Handbook
Etherington · Nuclear Engineering Handbook
Factory Mutual Engineering Division · Handbook of Industrial Loss Prevention
Fink · Television Engineering Handbook
Flügge · Handbook of Engineering Mechanics
Frick · Petroleum Production Handbook, 2 vols.
Guthrie · Petroleum Products Handbook
Harris · Handbook of Noise Control
Harris and Crede · Shock and Vibration Handbook, 3 vols.
Henney · Radio Engineering Handbook, 5th ed.
Hunter · Handbook of Semiconductor Electronics, 2d ed.
Huskey and Korn · Computer Handbook
Jasik · Antenna Engineering Handbook
Juran · Quality-control Handbook

KALLEN · Handbook of Instrumentation and Controls
KETCHUM · Structural Engineers' Handbook, 3d ed.
KING · Handbook of Hydraulics, 4th ed.
KNOWLTON · Standard Handbook for Electrical Engineers, 9th ed.
KOELLE · Handbook of Astronautical Engineering
KORN AND KORN · Mathematical Handbook for Scientists and Engineers
KURTZ · The Lineman's Handbook, 3d ed.
LA LONDE AND JANES · Concrete Engineering Handbook
LANDEE, DAVIS, AND ALBRECHT · Electronic Designers' Handbook
LANGE · Handbook of Chemistry, 10th ed.
LAUGHNER AND HARGAN · Handbook of Fastening and Joining of Metal Parts
LE GRAND · The New American Machinist's Handbook
LIDDELL · Handbook of Nonferrous Metallurgy, 2d ed.
MAGILL, HOLDEN, AND ACKLEY · Air Pollution Handbook
MANAS · National Plumbing Code Handbook
MANTELL · Engineering Materials Handbook
MARKS AND BAUMEISTER · Mechanical Engineers' Handbook, 6th ed.
MARKUS · Handbook of Electronic Control Circuits
MARKUS AND ZELUFF · Handbook of Industrial Electronic Circuits
MARKUS AND ZELUFF · Handbook of Industrial Electronic Control Circuits
MAYNARD · Industrial Engineering Handbook
MERRITT · Building Construction Handbook
MOODY · Petroleum Exploration Handbook
MORROW · Maintenance Engineering Handbook
PERRY · Chemical Business Handbook
PERRY · Chemical Engineers' Handbook, 3d ed.
SHAND · Glass Engineering Handbook, 2d ed.
STANIAR · Plant Engineering Handbook, 2d ed.
STREETER · Handbook of Fluid Dynamics
STUBBS · Handbook of Heavy Construction
TERMAN · Radio Engineers' Handbook
TRUXAL · Control Engineers' Handbook
URQUHART · Civil Engineering Handbook, 4th ed.
WALKER · NAB Engineering Handbook, 5th ed.
WOODS · Highway Engineering Handbook
YODER, HENEMAN, TURNBULL, AND STONE · Handbook of Personnel Management and Labor Relations

HANDBOOK OF ENGINEERING MECHANICS

Edited by

W. FLÜGGE, Dr.-Ing.

Professor of Engineering Mechanics
Stanford University

FIRST EDITION

New York Toronto London

McGRAW-HILL BOOK COMPANY, INC.

1962

HANDBOOK OF ENGINEERING MECHANICS

Library of Congress Catalog Card Number: 61-13165

21392

THE MAPLE PRESS COMPANY, YORK, PA.

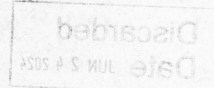

LIST OF CONTRIBUTORS

D. N. de G. Allen, Professor of Applied Mathematics, The University, Sheffield, England. *Relaxation Method.*

F. R. Arnold, Ph.D., Associate Professor of Mechanical Engineering and Director, Mechanical Engineering Laboratories, Stanford University. *Algebra.*

J. S. Aronofsky, Ph.D., Manager, Electronic Computer Center, Socony Mobil Oil Company, Inc. *Flow through Porous Media.*

D. G. Ashwell, Ph.D., Senior Lecturer, University College, Cardiff, United Kingdom. *Nonlinear Problems in Elasticity.*

R. S. Ayre, Ph.D., Professor and Chairman of Civil Engineering, Yale University. *Kinematics of Vibrations.*

D. C. Baxter, Ph.D., Assistant Research Officer, National Research Council of Canada. *Dimensionless Parameters.*

P. W. Berg, Ph.D., Associate Professor of Mathematics, Stanford University. *Calculus of Variations.*

A. Bronwell, D.Engr.(Hon.), President, Worcester Polytechnic Institute. *Calculus.*

C. E. Brown, Chief, Theoretical Mechanics Division, Langley Research Center, National Aeronautics and Space Administration. *Basic Concepts of Fluid Mechanics.*

A. E. Bryson, Jr., Ph.D., Associate Professor of Mechanical Engineering, Division of Engineering and Applied Physics, Harvard University. *Slender-body Theory.*

I. D. Chang, Ph.D., Assistant Professor, Department of Aeronautics and Astronautics, Stanford University. *Flow at Low Reynolds Numbers.*

E. Chwalla, Dr.techn., Dr.-Ing.h.c., Late Professor of Theory of Structures, Technische Hochschule, Graz, Austria. *Second-order Theory of Structures.*

L. Collatz, Dr., Dr.h.c., Professor, University of Hamburg. *Eigenvalue Problems.*

S. H. Crandall, Ph.D., Professor of Mechanical Engineering, Massachusetts Institute of Technology. *Rotating and Reciprocating Machines.*

F. R. E. Crossley, D.Eng., Associate Professor of Mechanical Engineering, Massachusetts Institute of Technology. *Systems of Several Degrees of Freedom.*

D. C. Drucker, Ph.D., Professor of Engineering, Brown University. *Basic Concepts of Plasticity and Viscoelasticity.*

J. E. Duberg, Ph.D., Research Engineer, Langley Research Center, National Aeronautics and Space Administration. *Inelastic Buckling.*

A. J. Eggers, Jr., Ph.D., Chief, Vehicle-environment Division, Ames Research Center, National Aeronautics and Space Administration. *Hypersonic Flow.*

A. C. Eringen, Ph.D., Professor, School of Aeronautical and Engineering Sciences, Purdue University. *Stochastic Loads.*

W. Flügge, Dr.-Ing., Professor of Engineering Mechanics, Stanford University. *Influence Diagrams. Shells.*

I. Flügge-Lotz, Dr.-Ing., Professor of Aeronautical Engineering, Stanford University. *Three-dimensional Ideal Fluid Flow.*

W. F. Freiberger, Ph.D., Associate Professor of Applied Mathematics, Brown University. *Plastic Torsion.*

Y. C. Fung, Ph.D., Professor of Aeronautics, Guggenheim Aeronautical Laboratory, California Institute of Technology. *Flutter.*

I. E. Garrick, Hunsaker Professor of Aeronautical Engineering, 1956–1957, Massachusetts Institute of Technology; Chief, Dynamic Loads Division, Langley Research Center, National Aeronautics and Space Administration. *Flutter.*

J. M. Gere, Ph.D., Associate Professor of Civil Engineering, Stanford University. *Statically Determinate Structures. Statically Indeterminate Structures.*

J. N. Goodier, Ph.D., Sc.D., Professor of Engineering Mechanics, Stanford University. *Elastic Torsion.*

A. P. Green, Assistant Director, Tube Investments Research Laboratories, Hinxton Hall, England. *Two-dimensional Problems of Plasticity.*

H. V. Hahne, Ph.D., Associate Professor of Civil Engineering, University of Santa Clara. *Moments of Inertia. Elastic Bending of Beams.*

M. A. Heaslet, Ph.D., Chief, Theoretical Branch, Ames Research Center, National Aeronautics and Space Administration. *Integral Equations.*

M. Hetényi, Ph.D., Walter P. Murphy Professor of Engineering Science, The Technological Institute, Northwestern University. *Beams on Elastic Foundation.*

P. G. Hodge, Jr., Ph.D., Professor of Mechanics, Illinois Institute of Technology. *Plastic Plates and Shells.*

G. W. Housner, Ph.D., Professor of Engineering, California Institute of Technology. *Dynamics.*

Ch. Jaeger, Dr. ès Sc.techn., Special Lecturer, Imperial College of Science and Technology, and Consulting Engineer, English Electric Company, Ltd. *Surge Tanks.*

J. H. B. Kemperman, Ph.D., Professor of Mathematics, University of Rochester. *Probability.*

J. Kempner, Ph.D., Professor of Aeronautical Engineering, Department of Aerospace Engineering and Applied Mechanics, Polytechnic Institute of Brooklyn. *Viscoelastic Buckling.*

J. Kestin, Ph.D.(Eng.), Professor of Engineering, Brown University. *Incompressible Boundary Layers.*

K. Klotter, Dr.-Ing., Professor of Engineering Mechanics, Technische Hochschule, Darmstadt, Germany; formerly Professor of Engineering Mechanics, Stanford University. *Nonlinear Vibrations.*

P. Kuhn, Assistant Chief, Structures Research Division, Langley Research Center, National Aeronautics and Space Administration. *Torsion-box Analysis.*

P. A. Lagerstrom, Ph.D., Professor of Aeronautics, Guggenheim Aeronautical Laboratory, California Institute of Technology. *Flow at Low Reynolds Numbers.*

C. Lanczos, Ph.D., Senior Professor, Dublin Institute for Advanced Studies. *Variational Principles of Mechanics.*

H. Lass, Ph.D., Research Specialist, Jet Propulsion Laboratory, California Institute of Technology. *Complex Variables.*

G. E. Latta, Ph.D., Associate Professor of Mathematics, Stanford University. *Ordinary Differential Equations.*

E. H. Lee, Ph.D., Professor of Applied Mathematics, Brown University. *Viscoelasticity.*

C. Libove, Instructor of Mechanical Engineering, Syracuse University; formerly Associate Professor of Aeronautical Engineering, Tri-State College. *Elastic Stability.*

J. L. Lubkin, Ph.D., Senior Research Engineer, Research and Development Division, American Machine & Foundry Company. *Contact Problems.*

K. Marguerre, Dr.-Ing., Professor of Applied Mechanics, Technische Hochschule, Darmstadt, Germany. *Basic Concepts of Elasticity.*

Ch. Massonet, Professor of Engineering Mechanics and Theory of Structures, University of Liège, Belgium. *Two-dimensional Problems of Elasticity.*

O. K. Mawardi, Ph.D., Professor of Engineering, Case Institute of Technology; formerly Professor of Electrical Engineering, Massachusetts Institute of Technology. *Acoustics.*

E. Mettler, Dr.rer.techn., Professor of Mechanics, Technische Hochschule, Karlsruhe, Germany. *Dynamic Buckling.*

H. E. Newell, Ph.D., Deputy Director, Space Flight Program, National Aeronautics and Space Administration; formerly Superintendent, Atmosphere and Astrophysics Division, U.S. Naval Research Laboratory. *Vector Analysis.*

R. H. Niemann, Ph.D., Associate Professor of Mathematics, Colorado State University; formerly Associate Professor of Mathematics, Worcester Polytechnic Institute. *Calculus.*

F. Oberhettinger, Dr.rer.nat., Dr.phil.habil., Professor of Mathematics, Oregon State University. *Special Functions.*

G. G. O'Brien, Ph.D., Director, Operations Research, Division of Management Sciences, Touche, Ross, Baily and Smart. *Numerical Solution of Hyperbolic Equations.*

K. Oswatitsch, Dr. phil., Professor of Fluid Mechanics, Technische Hochschule, Vienna, Austria, and Director of the Institute for Theorectical Gas Dynamics, Deutsche Versuchsanstalt für Luftfahrt (DVL), Aachen, Germany. *General Thermodynamics.*

S. I. Pai, Ph.D., Research Professor, Institute for Fluid Dynamics and Applied Mathematics, University of Maryland. *Turbulence.*

H. Parkus, Dr.techn., Professor of Mechanics, University of Technology, Vienna, Austria. *Thermal Stresses.*

B. Perry, Ph.D., Assistant Professor, Stanford University. *Cavitation.*

A. Phillips, Dr.-Ing., Professor of Civil Engineering, Yale University. *Plastic Bending of Beams.*

M. S. Plesset, Ph.D., Professor of Applied Mechanics, California Institute of Technology. *Cavitation.*

H. Poritsky, Ph.D., Consulting Engineer, General Electric Laboratory, General Electric Company. *Lubrication.*

J. R. M. Radok, Dr.-Ing., Editorial Consultant to Interscience Publishers, Inc. *Complex-variable Approach.*

F. H. Raven, Ph.D., Assistant Professor, University of Notre Dame. *Kinematics.*

E. A. Ripperger, Ph.D., Professor of Engineering Mechanics, University of Texas. *Geometry.*

A. Robinson, Ph.D., D.Sc., Professor of Mathematics, The Hebrew University, Jerusalem. *Airfoil Theory.*

A. Roshko, Ph.D., Associate Professor of Aeronautics, Guggenheim Aeronautical Laboratory, California Institute of Technology. *Thermodynamics of Gas Flow.*

N. Rott, Dr.sc.techn., Professor of Engineering, University of California, Los Angeles. *Subsonic Flow.*

M. G. Salvadori, Dr.Ing.Civ., Dr.Mat.Pura, Professor of Civil Engineering and Architecture, Columbia University. *Numerical Algebra.*

J. B. Scarborough, Ph.D., Professor Emeritus of Mathematics, U.S. Naval Academy. *Numerical Integration.*

A. Schild, Ph.D., Professor of Mathematics, University of Texas. *Tensor Analysis.*

E. J. Scott, Ph.D., Associate Professor of Mathematics, University of Illinois. *Laplace Transformation.*

G. A. Smith, Ph.D., Research Engineer, Ames Research Center, National Aeronautics and Space Administration, and Lecturer in Electrical Engineering, Stanford University. *Servomechanisms, Automatic Control.*

I. N. Sneddon, D.Sc., Simson Professor of Mathematics, The University of Glasgow. *Partial Differential Equations.*

J. R. Spreiter, Ph.D., Aeronautical Research Scientist, Theoretical Branch, Ames Research Center, National Aeronautics and Space Administration, and Lecturer, Department of Aeronautical Engineering, Stanford University. *Transonic Flow.*

K. Stewartson, Ph.D., Professor of Applied Mathematics, The Durham Colleges in the University of Durham, England. *Compressible Boundary Layers.*

V. L. Streeter, Sc.D., Professor of Hydraulics, University of Michigan. *Two-dimensional Ideal Fluid Flow.*

P. S. Symonds, Ph.D., Chairman, Division of Engineering, and Professor of Engineering, Brown University. *Limit Analysis.*

C. A. Syvertson, Chief, 3 by 5 Foot Hypersonic Wind Tunnel Branch, Ames Research Center, National Aeronautics and Space Administration. *Hypersonic Flow.*

W. T. Thomson, Ph.D., Professor of Engineering, University of California, Los Angeles, and Consultant in Dynamics, Space Technology Laboratories. *Systems of One Degree of Freedom.*

M. D. Van Dyke, Ph.D., Professor of Aeronautical Engineering, Stanford University. *Supersonic Flow.*

C.-K. Wang, Ph.D., Professor of Civil Engineering, University of Wisconsin. *Statically Indeterminate Structures.*

S. Way, Sc.D., Consultant in Mechanics, Westinghouse Research Laboratories, and Lecturer in Fluid Dynamics, University of Pittsburgh. *Elastic Plates.*

H. J. Weiss, D.Sc., Associate Professor of Mathematics, Iowa State University, and Mathematical Consultant, Iowa State Engineering Experiment Station. *Basic Equations of Elasticity in Tensor Notation.*

T. Y. Wu, Ph.D., Associate Professor of Applied Mechanics, California Institute of Technology. *Surface Waves.*

D. Young, Ph.D., Professor of Civil Engineering, Yale University. *Continuous Systems.*

D. H. Young, Sc.D., Professor of Engineering Mechanics, Stanford University. *Statics.*

Y. Y. Yu, Ph.D., Professor, Department of Mechanical Engineering, Polytechnic Institute of Brooklyn. *Bodies of Revolution.*

E. E. Zajac, Ph.D., Member of the Technical Staff, Bell Telephone Laboratories, Inc. *Propagation of Elastic Waves.*

H. Ziegler, Dr.Sc.math., Professor of Mechanics, Federal Institute of Technology, Zurich, Switzerland. *Gyroscopes.*

PREFACE

There is no lack of good textbooks and monographs on the many subjects which, together, are called engineering mechanics or applied mechanics. There is also a handful of books which attempt to present the field in its entity and, in particular, the common foundations of all its parts, but, so far, engineering mechanics has not had its own handbook. The present book has been written to fill this gap.

As in any handbook, emphasis has been placed on ready-to-use material. The book contains many formulas and tables which give immediate answers to questions arising from practical work. However, mechanics is a complicated subject, and more than facts and formulas is needed to apply it properly. Therefore, each chapter of this book contains a brief description of the essential lines of thought of its subject, thus enabling the reader to understand the logical background of the ready results and to think beyond them. In some chapters this presentation turned out to become the main part, in particular in those which lie on the boundaries between applied mechanics and neighboring fields. Lengthy derivations have, of course, always been omitted; they may be found in the books listed as references at the end of the chapters.

Six of the seven parts of the book cover the various fields of engineering mechanics. The subdivision into these parts is necessarily somewhat arbitrary, since the many subjects of the individual chapters are cross-linked in many ways and cannot be arranged in a one-dimensional sequence without preference being given to some of these links at the expense of others.

Each chapter is essentially concerned with the theoretical side of its subject. Where a comparison with experiments helps to throw light on a subject or to increase confidence in a theoretical argument, such comparison has been made, but detailed experimental data should be sought in books and papers on specific fields of engineering design rather than in a handbook on mechanics in general.

In a book that uses mathematics to the extent that has been done here, it is imperative that the mathematical tools themselves be properly presented for ready reference. The first part of the handbook is dedicated to this task. As the table of contents shows, the emphasis is on advanced subjects; however, some frequently needed information of an elementary character has been included for convenience. It is a matter of course that in this field all proofs had to be omitted.

Most of the chapters of this book deal with well-established subjects, but others are devoted to subjects which are relatively new, some of them having emerged within the last decade. It is clear that between these categories the presentation varies. In the older subjects that material could be chosen which has proved its usefulness in many applications, while the chapters on the youngest subjects are more or less a description of the state of the art at the time when they were written, i.e., some time between 1957 and 1960. A particular difficulty presented itself in those subjects which have mushroomed during the years while this book was being written, edited, and printed. In all these a last-minute effort has been made to add the latest significant publications to the list of references.

This handbook has not been written for readers, but for users, and its place is not on the bookshelf, but on the desk. There are several groups of users for which this book has been prepared in particular. First among them is the expert in any one of the fields covered, who wants to use the book in his work. To him it offers data, formulas, and ideas, and indicates sources which he may consult for further details.—Then there is the expert in one field who wants to be informed about thoughts, methods, and results in an adjacent field, with which he is not thoroughly familiar. He may study the corresponding chapter as an introduction to or a survey of the subject, and he may use the ready results as far as he sees fit.—Finally, the book also is aimed at the student of engineering mechanics or of a field of engineering in which mechanics plays an important part. He may consider the entire book as a survey of engineering mechanics, browse in it and find the topics on which he wants to concentrate his attention. He may use it as an outline of any of the special subjects, to be studied before he approaches a textbook.

It is expected that many of the users of this book will feel that their particular subject has not been given enough attention. The editor is well aware of the fact that much more could be written, and so are the authors. With their competence and their enthusiasm a book of twice the present size could easily have been filled, but a compromise had to be found between what we all would have liked to see printed and what could find room in a book of a marketable size.

The chapters of this handbook have been written by 88 authors, all experts in their fields. The editor has taken great pains to integrate their work in such a way that it forms a coherent book rather than a collection of loosely connected individual articles. He has tried to establish a uniform system of notations, which necessitated here and there deviations from notations that had developed without regard to the needs of neighbor fields. He has also made an effort to reduce duplication of subject matter to a small amount, and to watch that no important subject has been overlooked. All this turned out to be an iterative process, and the editor wishes to take this opportunity to thank all the contributors who helped to make this process converge, for their cooperation and their patience. He also wishes to thank his wife for her patience and understanding on the many evenings and weekends devoted to work on the handbook.

W. FLÜGGE

CONTENTS

PART 2. MECHANICS OF RIGID BODIES

PART 6. VIBRATIONS

COMMONLY USED NOTATIONS

The following symbols have been used as consistently as possible in this book. Because of the limited number of letters available, the same symbols have, of course, also been used for other purposes, and occasionally other symbols than those listed here have been used to denote the same quantities.

Mathematical Symbols

x, y, z	rectilinear coordinates (preferably, but not necessarily a right-handed system)
r, θ	plane polar coordinates
r', θ, z	cylindrical coordinates, see Fig. 4.12 (r and ρ are also used instead of r')
r, θ, ϕ	spherical coordinates, see Fig. 4.12
s	arc length
f_x	$= \partial f/\partial x$
f'	$= df/dx$, possibly also the derivative of f with respect to some other independent variable
D	differential operator, $= d/dx$ or $= d/dt$
∇^2	Laplace operator in three or two dimensions, see Eqs. (11.43*a–c*), (11.64), (6.23), (6.24)
Δx	finite increment of x
i	imaginary unit
$\mathcal{R}(), \mathcal{I}()$	real, imaginary part of ()
$\|z\|$	absolute value of a real or complex quantity z
$\|\mathbf{z}\|, z$	absolute value of a vector $\mathbf{z}$
$\exp x$	$= e^x$
Sinh x, Cosh x	hyperbolic sine and cosine of x
Tanh x, Coth x	Hyperbolic tangent and cotangent of x
$\ln x$	natural logarithm of x
$\log x$	decimal logarithm of x
$\Delta(x)$	unit step function, see Eq. (19.4)
$\delta(x)$	Dirac delta function, see Eq. (19.7)
$O()$	order of
sgn x	signum of x, i.e., $+1$ if $x > 0$, -1 if $x < 0$
$\mathbf{i}_x$	unit vector in x direction
det A	determinant of the matrix A
tr A	trace of the matrix A, i.e., the sum of the elements in the principal diagonal, see p. 7–6
δ_{ij}	Kronecker delta, $\delta_{ii} = 1$, $\delta_{ij} = 0$ if $i \neq j$
$\approx$	approximately equal
$\sim$	approximately
$\propto$	proportional

Boldface type indicates vectors. Subscripts are used to indicate components:

$$\mathbf{r} = r_x\mathbf{i}_x + r_y\mathbf{i}_y + r_z\mathbf{i}_z = (r_x, r_y, r_z)$$

Statics and Dynamics, including Vibrations

dim ϕ	dimension of ϕ
L, T, F	length, time, force (in dimensional analysis)
t	time
$\dot{q}$	$= dq/dt$ or $= \partial q/\partial t$
v	velocity
u, v, w	components of velocity in x, y, z directions
a	linear acceleration
g	acceleration of gravity
ω	angular velocity, circular frequency
F, P	forces
R	resultant or reaction
W	weight
q	intensity of a distributed force, force per unit of length
M	moment of a force or of a couple
μ	coefficient of dry (Coulomb) friction
m	mass
ρ	mass density (mass per unit volume)
μ	mass per unit length (in strings, bars) or per unit area (in membranes, plates, shells)
I_x	mass moment of inertia with respect to the x axis
H	moment of momentum, angular momentum
T	kinetic energy
V	potential energy
L	$= T - V$
q_n	normal coordinate
Q_n	generalized force

Elasticity, Plasticity, Structures

x	coordinate measured along the axis of a bar or beam
$y, z; r, \theta$	coordinates in the plane of the cross section of a bar or a beam
$x, y; r, \theta$	coordinates in the middle surface of a plate
z	distance from the middle surface of a plate or a shell
u, v, w	displacements in directions x, y, z
$u_r, u_\theta, \ldots$	displacements in the directions indicated by the subscripts
ϵ	tensile strain
γ	shear strain
κ	change of curvature of a beam due to deformation
θ	twist of a bar subjected to torsion
κ_x, κ_{xy}	change of curvature and twist of a plate or shell
σ	normal stress
τ	shear stress
N	normal (axial) force in bars and beams
V	shear force
M	bending moment
M_T	torque
T	shear flow
N_x, N_{xy}	normal and shear forces in plates and shells
M_x, M_{xy}	bending moment and twisting moment in plates and shells
Q_x, Q_y	transverse (shear) forces in plates and shells
P	concentrated load
q, p	load per unit area

A	area of a cross section
A_S	shear area (modified cross-sectional area which determines the shear deformation of a beam)
I	area moment of inertia (subscript indicates axis if necessary)
I_{yz}	area product of inertia for the (orthogonal) axes y, z
i_x	radius of gyration, $i_x^2 = I_x/A$
I_p	polar moment of inertia
J	torsion constant of cross section
Γ	warping integral
t	thickness of plate or shell
E	elastic (Young's) modulus
G	shear modulus
ν	Poisson's ratio
EI	flexural rigidity of a beam
GJ	torsional rigidity of a bar
D	$= Et/(1 - \nu^2)$, extensional rigidity of a plate or a shell
K	$= Et^3/12(1 - \nu^2)$, flexural rigidity of a plate or a shell
U	strain energy
T	temperature (measured from any convenient reference temperature)

Special Notations, Theory of Structures

θ	slope of the deflection line
δ	deflection
δ_{ij}	deflection (or slope, or relative displacement, or relative rotation) at section i, caused by a unit load (force, couple, pair of forces or of couples) at j, coefficient of X_j in the ith equation
X_j	jth redundant in a statically indeterminate system

Special Notations, Plasticity

σ_e, σ_0	yield stress in uniaxial tension of a perfectly plastic material
k	yield stress in shear
ϵ^e, γ^e	elastic parts of the strains ϵ, γ
ϵ^p, γ^p	plastic parts of the strains ϵ, γ
s_{ij}	stress deviation component, $= \sigma_{ij} - \sigma_m\delta_{ij}$
σ_m	$= \frac{1}{3}(\sigma_x + \sigma_y + \sigma_z)$, hydrostatic stress

Fluid Mechanics

$D\phi/Dt$	rate of change of ϕ, observed when following a moving particle, see Eq. (68.15)
$\partial\phi/\partial t$	rate of change of ϕ, observed at a fixed point x, y, z
u, v, w	velocity components in directions x, y, z
w	$= u - iv$, complex velocity
$\mathbf{q}$	velocity vector
q, v	absolute value of the velocity
U, V, W	components of velocity at infinity or of a moving body
ω_x, ω_y, ω_z	components of the rotation, see Eqs. (68.2), or components of the vorticity, see Eqs. (71.3)
Γ	$= \oint \mathbf{q} \cdot d\mathbf{s}$, circulation, see Eq. (70.21)
ϕ	velocity potential, see Eqs. (70.10)
ψ	stream function, see pp. 70–6, 71–3
Φ	$= \phi + i\psi$, complex potential
Æ	aspect ratio of a wing
p	pressure

p_t	total pressure, isentropic stagnation pressure
C_p, C_L	pressure coefficient, lift coefficient, etc.
a	velocity of sound
M	Mach number
μ	viscosity, see Eqs. (68.6)
ν	kinematic viscosity, see Eq. (69.2)
ρ	mass density, specific mass
γ	specific weight
γ	$= c_p/c_v$
c_p, c_v	specific heats at constant pressure and at constant volume
R	gas constant
T	absolute temperature
E, u	internal energy, see p. 68–10
h	enthalpy
s,S	entropy

HANDBOOK OF ENGINEERING MECHANICS

Part 1

MATHEMATICS

CHAPTER 1

ALGEBRA

BY

F. R. ARNOLD, Ph.D., Stanford, Calif.

1.1. DETERMINANTS, MATRICES, AND LINEAR ALGEBRAIC EQUATIONS

Determinants

A determinant of the nth order is a square array of n^2 quantities such as

$$D = \begin{vmatrix} a_{11} & a_{12} & a_{13} & \cdots & a_{1n} \\ a_{21} & a_{22} & a_{23} & \cdots & a_{2n} \\ a_{31} & a_{32} & a_{33} & \cdots & a_{3n} \\ \cdots & \cdots & \cdots & \cdots & \cdots \\ a_{n1} & a_{n2} & a_{n3} & \cdots & a_{nn} \end{vmatrix} = \det a_{ij} \tag{1.1}$$

with elements a_{ij} arranged in horizontal rows and vertical columns. The first subscript i indicates the row in which the element may be found. The second subscript j indicates the column.

The complementary minor, or minor M_{ij}, of the element a_{ij} is the determinant of order $(n - 1)$ which remains after striking out the elements in the ith row and jth column of the original determinant.

The cofactor C_{ij} of the element a_{ij} is defined as $C_{ij} = (-1)^{i+j}M_{ij}$.

The value of the determinant D is defined as

$$\begin{aligned} D &= \sum_{i=1}^{n} \sum_{j=1}^{n} \cdots \sum_{p=1}^{n} \epsilon_{ij\cdots p} a_{1i} a_{2j} \cdots a_{np} \\ &= \sum_{i=1}^{n} \sum_{j=1}^{n} \cdots \sum_{p=1}^{n} \epsilon_{ij\cdots p} a_{i1} a_{j2} \cdots a_{pn} \end{aligned} \tag{1.2}$$

where the permutation operator $\epsilon_{ij\cdots p}$ is defined as $\epsilon_{ij\cdots p} = +1$ if its subscripts are in natural order or in such an order that an even number of interchanges will make them so; otherwise $\epsilon_{ij\cdots p} = -1$ if an odd number of interchanges is needed and $\epsilon_{ij\cdots p} = 0$ if any two or more of the subscripts are equal.

The value of a determinant may as well be found by expanding it in terms of a row or a column:

$$D = \sum_{j=1}^{n} a_{ij}C_{ij} = \sum_{j=1}^{n} a_{ji}C_{ji} \tag{1.3}$$

Here the subscript i is arbitrary but fixed. The cofactors are $(n - 1)$st-order determinants and may be evaluated in a similar way.

Special Properties of Determinants. The value of a determinant becomes zero when:

1. Each element of one row or one column is zero, or
2. The elements of a row (column) are multiples of the corresponding elements of another row (column).

The value of a determinant is **unchanged** if:

1. Columns and rows are interchanged, or
2. A linear combination of any rows (columns) is added to any row (column).

If all the elements of a row or column are multiplied by a factor, then the value of the determinant is multiplied by the same factor.

If two rows or two columns are interchanged, the value of the determinant changes sign.

A determinant may be written as the algebraic sum of two or more determinants of the same order if each element of any one row appears as the algebraic sum of two or more quantities.

Example 1

$$\begin{vmatrix} a_{11} \pm b_{11} & a_{12} & a_{13} \pm b_{13} \pm c_{13} \\ a_{21} & a_{22} & a_{23} \\ a_{31} & a_{32} & a_{33} \end{vmatrix} = \begin{vmatrix} a_{11} & a_{12} & a_{13} \\ a_{21} & a_{22} & a_{23} \\ a_{31} & a_{32} & a_{33} \end{vmatrix} \pm \begin{vmatrix} b_{11} & 0 & b_{13} \\ a_{21} & a_{22} & a_{23} \\ a_{31} & a_{32} & a_{33} \end{vmatrix} \pm \begin{vmatrix} 0 & 0 & c_{13} \\ a_{21} & a_{22} & a_{23} \\ a_{31} & a_{32} & a_{33} \end{vmatrix} \tag{1.4}$$

As a result of the special properties of determinants, a determinant of higher order may be reduced to one of lower order by judiciously subtracting some multiple of the elements of a row (or column) from corresponding elements of another row (or column) until all the elements but one in the latter row (or column) have been reduced to *zero*.

Example 2

$$\begin{vmatrix} 1 & 2 & 3 & 5 \\ 4 & 1 & 1 & 2 \\ 3 & 0 & 4 & 0 \\ 2 & 6 & 1 & 5 \end{vmatrix} = \begin{vmatrix} 1 & 2 & 3 & 5 \\ 4 & 1 & 1 & 2 \\ 3 & 0 & 4 & 0 \\ 0 & 2 & -5 & -5 \end{vmatrix} \quad \text{as a result of subtracting twice the first row from the bottom row}$$

$$= \begin{vmatrix} 1 & 2 & 3 & 5 \\ 4 & 1 & 1 & 2 \\ 0 & -6 & -5 & -15 \\ 0 & 2 & -5 & -5 \end{vmatrix} \quad \text{as a result of subtracting three times the first row from the third}$$

$$= \begin{vmatrix} 1 & 2 & 3 & 5 \\ 0 & -7 & -11 & -18 \\ 0 & -6 & -5 & -15 \\ 0 & 2 & -5 & -5 \end{vmatrix} \quad \text{by subtracting four times the first row from the second}$$

$$= 1 \begin{vmatrix} -7 & -11 & -18 \\ -6 & -5 & -15 \\ 2 & -5 & -5 \end{vmatrix} \quad \text{by expanding in terms of elements of the first column and their cofactors}$$

Solution of a System of Simultaneous Linear Equations (Cramer's Rule)

Let a set of n linear algebraic equations in n unknowns of type x_j be indicated by the following:

$$\begin{aligned} a_{11}x_1 + a_{12}x_2 + \cdots a_{1n}x_n &= c_1 \\ a_{21}x_1 + a_{22}x_2 + \cdots a_{2n}x_n &= c_2 \\ a_{n1}x_1 + a_{n2}x_2 + \cdots a_{nn}x_n &= c_n \end{aligned} \tag{1.5}$$

where the coefficients a_{ij} and the right-hand terms c_j are supposed to be known. The unknowns x_j may be found as quotients of two determinants each. The elements of the denominator determinant are the coefficients a_{ij}. The numerator determinant is similarly constituted except that, in place of the column containing the coefficients of the particular x_j being determined, one introduces a column made up of the terms $c_1, c_2, \ldots, c_n$.

Example 3

$$x_j = \frac{D_j}{D} \quad \text{where} \quad D = \begin{vmatrix} a_{11} & a_{12} & \cdots & a_{1j} & \cdots & a_{1n} \\ a_{21} & a_{22} & \cdots & a_{2j} & \cdots & a_{2n} \\ \cdots & \cdots & \cdots & \cdots & \cdots & \cdots \\ a_{n1} & a_{n2} & \cdots & a_{nj} & \cdots & a_{nn} \end{vmatrix}$$

$$D_j = \begin{vmatrix} a_{11} & a_{12} & \cdots & c_1 & \cdots & a_{1n} \\ a_{21} & a_{22} & \cdots & c_2 & \cdots & a_{2n} \\ \cdots & \cdots & \cdots & \cdots & \cdots & \cdots \\ a_{n1} & a_{n2} & \cdots & c_n & \cdots & a_{nn} \end{vmatrix} \tag{1.6}$$

For other methods of solving linear equations, see p. 2–5.

Matrices

A rectangular array of numbers (objects) arranged in m rows and n columns is called an m-by-n matrix.

Example 4

$$A = \begin{bmatrix} a_{11} & a_{12} & \cdots & a_{1n} \\ a_{21} & a_{22} & \cdots & a_{2n} \\ \cdots & \cdots & \cdots & \cdots \\ a_{m1} & a_{m2} & \cdots & a_{mn} \end{bmatrix} = [a_{ij}] \tag{1.7}$$

Unlike determinants, a matrix may be rectangular as well as square. A matrix consisting of but one row or one column is called a row vector or column vector accordingly. Two matrices A and B are of the same type if they have the same number of rows and the same number of columns. Two matrices are conformable if, when they are placed side by side as for multiplication, the number of columns of the first or left-hand matrix is equal to the number of rows of the second or right-hand matrix.

A square matrix has the same number of rows as columns; that is, $m = n$. The order of a square matrix is the number of rows or columns.

A symmetric matrix is a square matrix with the elements related in that $a_{ij} = a_{ji}$. An antimetric matrix or skew-symmetric matrix is a square matrix in which $a_{ij} = -a_{ji}$ and $a_{ii} = 0$.

A diagonal matrix is a square matrix having all the elements except those on a diagonal equal to *zero;* that is, $a_{ij} = 0$ if $i \neq j$ and $a_{ii} \neq 0$.

A unit matrix is a diagonal matrix with each diagonal element $a_{ii} = 1$.

Example 5. Third-order Unit Matrix

$$I = \begin{bmatrix} 1 & 0 & 0 \\ 0 & 1 & 0 \\ 0 & 0 & 1 \end{bmatrix} = [\delta_{ij}] \tag{1.8}$$

where $$\delta_{ij} = \begin{cases} 1 & \text{if } i = j \\ 0 & \text{if } i \neq j \end{cases} \quad \text{(Kronecker } \delta) \tag{1.9}$$

A null matrix is a square matrix having each element equal to *zero.*

Matrix Manipulations

EQUALITY: $A = B$ if $a_{ij} = b_{ij}$
ADDITION: $C = A + B$ if $c_{ij} = a_{ij} + b_{ij}$
MULTIPLICATION BY A SCALAR: $C = kA$ if $c_{ij} = ka_{ij}$
ASSOCIATIVE LAW: $(A + B) + C = A + (B + C)$

MATRIX MULTIPLICATION: Matrix multiplication is possible only when two matrices concerned are conformable. Then

$$C = AB \quad \text{where} \quad c_{ij} = \sum_k a_{ik}b_{kj}$$

and $k = 1, 2, \ldots, n \qquad i = 1, 2, \ldots, m \qquad j = 1, 2, \ldots, p$

Since $AB \neq BA$ in general, it is important that the order of the matrices be definitely prescribed. In the product AB, the matrix A is the premultiplier or prefactor in the premultiplication of B by A. Correspondingly the matrix B is the postmultiplier or postfactor in the postmultiplication of A by B. Where the product is unaffected by the order of the matrices being multiplied, the latter are said to commute. Both the unit matrix I and the null matrix 0 commute with any conformable square matrix A, so that $IA = AI = A$ and $0A = A0 = 0$.

ASSOCIATIVE LAW: $ABC = (AB)C = A(BC)$
DISTRIBUTIVE LAW:

$$C(A + B) = CA + CB \qquad (A + B)C = AC + BC$$

The premultiplication of a column vector by a row vector having the same number of elements yields a matrix with but one element, a scalar quantity. The premultiplication of a row vector by a column vector yields a square matrix with a number of rows or columns equal to the number of elements in each of the related vectors, and its rows are proportionally related as are its columns. An $n \times p$ matrix B premultiplied by an $m \times n$ matrix A will yield an $m \times p$ matrix C.

Example 6. Matrix Multiplication

$$AB = \begin{bmatrix} a_{11} & a_{12} & a_{13} \\ a_{21} & a_{22} & a_{23} \\ a_{31} & a_{32} & a_{33} \end{bmatrix} \begin{bmatrix} b_{11} & b_{12} & b_{13} \\ b_{21} & b_{22} & b_{23} \\ b_{31} & b_{32} & b_{33} \end{bmatrix} = C \tag{1.10}$$

$$= \begin{bmatrix} (a_{11}b_{11} + a_{12}b_{21} + a_{13}b_{31}) & (a_{11}b_{12} + a_{12}b_{22} + a_{13}b_{32}) & (a_{11}b_{13} + a_{12}b_{23} + a_{13}b_{33}) \\ (a_{21}b_{11} + a_{22}b_{21} + a_{23}b_{31}) & (a_{21}b_{12} + a_{22}b_{22} + a_{23}b_{32}) & (a_{21}b_{13} + a_{22}b_{23} + a_{23}b_{33}) \\ (a_{31}b_{11} + a_{32}b_{21} + a_{33}b_{31}) & (a_{31}b_{12} + a_{32}b_{22} + a_{33}b_{32}) & (a_{31}b_{13} + a_{32}b_{23} + a_{33}b_{33}) \end{bmatrix}$$

where the terms in parentheses are the elements of the product matrix.

Derived Matrices. The transpose A' of a matrix A is formed by interchanging its rows with its columns: $a'_{ij} = a_{ji}$. $(AB)' = B'A'$.

The complex conjugate A^* of A is formed by replacing any complex elements of the original matrix A with their complex conjugates. $(AB)^* = A^*B^*$.

The hermitian conjugate, hermitian adjoint, or associate matrix A^+ of A is defined by $a^+_{ij} = a^*_{ji}$. $A^+ = (A^*)' = (A')^*$; $(AB)^+ = B^+A^+$.

The reciprocal or inverse A^{-1} of A is defined such that $AA^{-1} = A^{-1}A = I$. Let α_{ij} represent the ijth element of the inverse matrix A^{-1}. Then

$$\alpha_{ij} = \frac{C_{ji}}{\det A} = \frac{(-1)^{i+j}M_{ji}}{\det A} \tag{1.11}$$

where C_{ji} and M_{ji} are the cofactor and the minor of a_{ji}, respectively, and

$$\det A = \det a_{ij}$$

is the determinant formed with the elements of the matrix A.

Only square matrices have reciprocals. If $\det A = 0$, the inverse of A does not exist and A is said to be a singular matrix. $(AB)^{-1} = B^{-1}A^{-1}$.

Special Square Matrices. If a square matrix A remains unchanged after a type of conversion described under Derived Matrices, it is given a special descriptive name depending on the nature of the conversion. A matrix A is:

1. Unitary if $A = (A^+)^{-1}$, $A^+ = A^{-1}$, $A^+A = I$, $AA^+ = I$
2. Symmetric if $A = A'$, $(a_{ij} = a_{ji})$
3. Skew-symmetric if $A = -A'$, $(a_{ij} = -a_{ji})$
4. Orthogonal if $AA' = I$, $A' = A^{-1}$, $A'A = I$
5. Real if $A = A^*$
6. Pure imaginary if $A = -A^*$
7. Hermitian if $A = A^+$, $(a_{ij} = a^*_{ji})$
8. Skew-hermitian if $A = -A^+$, $(a_{ij} = -a^*_{ji})$

Matrices and Simultaneous Linear Equations

The linear equations (1.5) may be represented by the one matrix equation

$$Ax = c \tag{1.12}$$

where

$$A = \begin{bmatrix} a_{11} & a_{12} & \cdots & a_{1n} \\ a_{21} & a_{22} & \cdots & a_{2n} \\ \cdots & \cdots & \cdots & \cdots \\ a_{n1} & a_{n2} & \cdots & a_{nn} \end{bmatrix} \qquad x = \begin{bmatrix} x_1 \\ x_2 \\ \cdot \\ \cdot \\ \cdot \\ x_n \end{bmatrix} \qquad c = \begin{bmatrix} c_1 \\ c_2 \\ \cdot \\ \cdot \\ \cdot \\ c_n \end{bmatrix} \tag{1.13}$$

Premultiplication of both sides of Eq. (1.12) by A^{-1} yields

$$A^{-1}Ax = Ix = x = A^{-1}c \tag{1.14}$$

Consequently, the elements x_j of the matrix x may be found as sums of products as soon as the reciprocal A^{-1} of the coefficient matrix A is known.

When every $c_i = 0$ $(i = 1, 2, \ldots, n)$, it follows from Eq. (1.14) that also all the $x_i = 0$ unless $\det A = 0$.

When $\det A = 0$, the reciprocal A^{-1} does not exist and Eq. (1.14) is not applicable. In this case a nonvanishing solution x is possible even if every $c_i = 0$. If not all elements of the column c are zero, there may or may not be a finite solution.

1.2. QUADRATIC EQUATIONS AND GENERAL PROPERTIES OF INTEGRAL FUNCTIONS OF ONE VARIABLE

Quadratic Equation Involving One Unknown

The quadratic equation

$$ax^2 + bx + c = 0 \tag{1.15}$$

has the two solutions

$$x_{1,2} = \frac{1}{2a}\left(-b \pm \sqrt{b^2 - 4ac}\right) \tag{1.16}$$

which, for real values for a, b, and c, will be real and different if $b^2 - 4ac > 0$, equal if $b^2 - 4ac = 0$, and conjugate complex if $b^2 - 4ac < 0$.

Gauss' Fundamental Theorem of Algebra

The algebraic equation of the nth degree

$$a_0x^n + a_1x^{n-1} + a_2x^{n-2} + \cdots + a_{n-1}x + a_n = 0 \tag{1.17}$$

where n is an integer, each a_i is a given real, complex, or imaginary number, and $a_0 \neq 0$, has n roots, not necessarily different; that is, there are exactly n numbers, $r_1, r_2, \ldots, r_n$, real, complex, or imaginary, such that

$$a_0x^n + a_1x^{n-1} + \cdots + a_{n-1}x + a_n = a_0(x - r_1)(x - r_2) \cdots (x - r_n) \tag{1.18}$$

for all values of x.

Relations between Roots and Coefficients

$$\begin{aligned} r_1 + r_2 + r_3 + \cdots + r_n &= -\frac{a_1}{a_0} \\ r_1r_2 + r_1r_3 + \cdots + r_{n-1}r_n &= \frac{a_2}{a_0} \\ \cdots\cdots\cdots&\cdots\cdots\cdots \\ r_1 \cdot r_2 \cdot r_3 \cdots\cdots r_n &= (-1)^n \frac{a_n}{a_0} \end{aligned} \tag{1.19}$$

Descartes's Rule of Signs

An algebraic equation $f(x) = 0$ cannot have more positive real roots than it has variations in the signs of its terms. It cannot have more negative real roots than there are variations in signs of the terms of $f(-x) = 0$. The minimum number of complex roots is found by subtracting from the degree of the equation the maximum number of real roots except that, if the number so obtained is odd, the next higher even number should be used as the minimum number of complex roots.

Further Rules

If one of two conjugate complex numbers is a root of an algebraic equation having real coefficients, its conjugate is also a root.

Every algebraic equation of odd degree and with real coefficients has at least one real root, and if it has more than one it has an odd number.

1.3. SERIES

Definitions

A sequence is a succession of terms each of which is an expression formed in accordance with a specified rule.

A series is the sum of the terms of a sequence, and may be finite or infinite depending on the number of terms to be summed.

The general term or nth term of a sequence or series is the expression indicating the law of formation of the terms.

Example 7. With nth term

Sequence: $$1, x, \frac{x^2}{1 \cdot 2}, \frac{x^3}{1 \cdot 2 \cdot 3}, \ldots, \frac{nx^{n-1}}{n!}, \ldots$$

Infinite series: $$1 + x + \frac{x^2}{1 \cdot 2} + \frac{x^3}{1 \cdot 2 \cdot 3} + \cdots + \frac{nx^{n-1}}{n!} + \cdots$$

A series is convergent if the sum S approaches a limit as the number of terms increases without limit. A nonconvergent series is divergent.

A series which neither increases indefinitely nor approaches a limit as the number of terms increases without limit is called an oscillating series and is divergent.

Example 8. $S = a - a + a - a + a \cdots$ is oscillating.

Where a series is made up of expressions containing a variable which may assume or be assigned different values, the range of values of the variable for which the series is convergent is called the interval of convergence.

An alternating series is an infinite series in which terms are alternately positive and negative.

An infinite series having both positive and negative terms, not necessarily alternating, is absolutely convergent if a series consisting of the absolute values of its terms is convergent. If a series having positive and negative terms is convergent but not absolutely convergent, it is conditionally convergent.

General Theorems

Theorem 1. If $S_1, S_2, \ldots, S_n, \ldots$ is an infinite sequence of numbers which increase as n increases but never exceed a finite fixed number A, then the limit of S_n as $n \to \infty$ is $S_\infty \leqq A$.

Theorem 2. If $S_1, S_2, \ldots, S_n, \ldots$ is an infinite sequence of numbers which decrease as n increases but are never less than a fixed finite number B, then the limit of S_n as $n \to \infty$ is $S_\infty \geqq B$.

Tests for Convergence

Fundamental Test by Cauchy's Criterion for Convergence. The necessary and sufficient condition for convergence of an infinite series is, that for any given positive number ϵ, however small, a number n exists such that, for every positive value of k,

$$|u_{n+1} + u_{n+2} + \cdots + u_{n+k}| < \epsilon$$

A necessary but not sufficient condition for convergence is that, for a convergent infinite series,

$$S = u_1 + u_2 + u_3 + \cdots + u_n + \cdots$$

the terms must approach zero as n increases; that is,

$$\lim_{n\to\infty} u_n = 0$$

Comparison Test for Convergence and Divergence. If one series consisting of all positive terms is known to be convergent, and a second series of positive terms exists such that after some term all the following terms become and remain as small as or smaller than the remaining terms of the first series, then the second series is convergent as well.

If one series consisting of all positive terms is known to be divergent, and a second series exists such that after some term all the following terms become and remain at least as large as the terms of the first series, then the second series is divergent as well.

Ratio Test. One of the simplest tests to apply is the ratio test, which is shown in the following theorem:

Let $S = u_1 + u_2 + \cdots + u_n + \cdots$ be an infinite series of terms which may be of different signs. Then, if

(1) $$\lim_{n\to\infty} \left|\frac{u_{n+1}}{u_n}\right| < 1 \qquad \text{the series converges}$$

(2) $$\lim_{n\to\infty} \left|\frac{u_{n+1}}{u_n}\right| > 1 \qquad \text{the series diverges}$$

(3) $$\lim_{n\to\infty} \left|\frac{u_{n+1}}{u_n}\right| = 1 \qquad \text{no information is provided}$$

Cauchy's Integral Test. Let

$$u_1 + u_2 + \cdots + u_n + \cdots = \sum_{n=1}^{\infty} u_n$$

be a series having positive terms with $u_{n+1} < u_n$. If a positive but decreasing function $f(n)$ for $n > 1$ exists such that $f(n) = u_n$, then the series converges if the integral

$$\int_1^{\infty} f(n)\, dn \quad \text{exists}$$

If the integral does not exist, the series is divergent.

Alternating Series. For an alternating series

$$S = u_1 - u_2 + u_3 - u_4 + - \cdots$$

where each term is numerically less than or equal to the one it follows, then, if

$$\lim_{n \to \infty} u_n = 0 \qquad \text{the series is convergent}$$

A useful property of alternating series is the size of the error made by terminating the series at any particular term. Such error does not exceed numerically the value of the first term disregarded.

Power Series. A series of the form

$$S = a_0 + a_1x + a_2x^2 + \cdots + a_nx^n + \cdots$$

is called a power series in x. Such a series may converge for some ranges of values of x and diverge for others. The Cauchy integral test and the Cauchy ratio test are convenient for establishing the interval of convergence.

Theorem 3. Two power series may be added term by term, and the result will be convergent for those values of x for which the component series are each convergent.

Theorem 4. Two power series may be multiplied to form a product series which converges for the interval of convergence common to the two series.

Theorem 5. A series formed as the quotient of two given series has its own intervals of convergence, if any, separate from those of the given series.

Theorem 6. If the series

$$z = a_0 + a_1y + a_2y^2 + \cdots + a_ny^n + \cdots$$

converges for $|y| < r_1$, and the series

$$y = b_0 + b_1x + b_2x^2 + \cdots + b_nx^n + \cdots$$

converges for $|x| < r_2$, then, if $|b_0| < r_1$, one may introduce in place of y in the first series its value in terms of a series expansion of the second series to obtain z as a power series in x. The resulting series will converge for x sufficiently small.

Theorem 7. A convergent power series may be differentiated term by term, and the resulting series has the same interval of convergence.

Theorem 8. A convergent power series may be integrated term by term, and the resulting series is convergent for the same interval of convergence.

Theorem 9. The coefficients of like powers of the variable in two equal power series must be equal.

Series of Functions. A series of the form

$$S(x) = u_0(x) + u_1(x) + \cdots + u_n(x) + \cdots$$

where each term is a continuous function of the variable x in an interval $a \leq x \leq b$, and which converges for every value of x in that interval, is said to be uniformly convergent in that interval if, corresponding to any preassigned positive number δ, a positive integer N, independent of x, can be found such that the absolute value of the remainder R_n of the given series, where

$$R_n = u_{n+1}(x) + u_{n+2}(x) + \cdots$$

is less than δ for every value of $n \geq N$ and for every value of x in the interval (a,b).

The Weierstrass M Test for Uniform Convergence. Let

$$S(x) = u_0(x) + u_1(x) + \cdots + u_n(x) + \cdots$$

be a series with each term a continuous function of x in an interval $a \leq x \leq b$. If a convergent series of the form

$$M = M_0 + M_1 + M_2 + \cdots + M_n + \cdots$$

where each term is a positive constant, can be found such that $|u_n| \leq M_n$ for all values of x and of n in the interval (a,b), then the series $S(x)$ converges uniformly in that interval.

Theorem 10. Any uniformly convergent series of functions may be integrated term by term provided the limits of integration are finite and lie in the interval of convergence of the integrand series.

Theorem 11. Any uniformly convergent series may be differentiated term by term if the resulting series converges uniformly.

Useful Series

The following series are convergent for all values of the variable unless otherwise indicated by inclusion of the interval of convergence following the series.

Binomial Series

$$(a+b)^n = a^n + \binom{n}{1} a^{n-1}b + \binom{n}{2} a^{n-2}b^2 + \cdots + \binom{n}{k} a^{n-k}b^k + \cdots \qquad b^2 < a^2$$

where the binomial coefficient

$$\binom{n}{k} = (n)_k = \frac{n!}{(n-k)!k!} \tag{1.20}$$

$$(1 \pm x)^n = 1 \pm nx + \frac{n(n-1)}{2!} x^2 \pm \frac{n(n-1)(n-2)}{3!} x^3 + \cdots + \frac{(\pm 1)^k n! x^k}{(n-k)!k!} + \cdots \qquad x^2 < 1$$

$$(1 \pm x)^{-n} = 1 \mp nx + \frac{n(n+1)}{2!} x^2 \mp \frac{n(n+1)(n+2)}{3!} x^3 + \cdots + (\mp 1)^k \frac{(n+k-1)! x^k}{(n-1)!k!} + \cdots \qquad x^2 < 1$$

$$(1 \pm x)^{-1} = 1 \mp x + x^2 \mp x^3 + x^4 \mp x^5 + \cdots \qquad x^2 < 1$$

$$(1 \pm x)^{1/2} = 1 \pm \frac{1}{2}x - \frac{1}{2 \cdot 4}x^2 \pm \frac{1 \cdot 3}{2 \cdot 4 \cdot 6}x^3 - \frac{1 \cdot 3 \cdot 5}{2 \cdot 4 \cdot 6 \cdot 8}x^4 \pm \cdots \qquad x^2 < 1$$

$$(1 \pm x)^{-1/2} = 1 \mp \frac{1}{2}x + \frac{1 \cdot 3}{2 \cdot 4}x^2 \mp \frac{1 \cdot 3 \cdot 5}{2 \cdot 4 \cdot 6}x^3 + \frac{1 \cdot 3 \cdot 5 \cdot 7}{2 \cdot 4 \cdot 6 \cdot 8}x^4 \mp \cdots \qquad x^2 < 1$$

Bernoulli Numbers and Euler Numbers. The Bernoulli numbers B_n are defined as the coefficients in the power-series expansion

$$\frac{x(e^x+1)}{2(e^x-1)} = -\frac{B_0}{0!}x^0 + \frac{B_1}{2!}x^2 - \frac{B_2}{4!}x^4 + - \cdots$$

They may be calculated by actually expanding this function or from the recurrence formula

$$\binom{2n+2}{2}B_1 - \binom{2n+2}{4}B_2 + \binom{2n+2}{6}B_3 - + \cdots + (-1)^{n-1}\binom{2n+2}{2n}B_n = n \qquad (1.21)$$

which is valid for $n = 1, 2, 3, \ldots$. Various notations are in use for the Bernoulli numbers. In the notation used here the first few are

$$B_0 = -1, B_1 = 1/6, B_2 = 1/30, B_3 = 1/42, B_4 = 1/30, B_5 = 5/66, B_6 = 691/2730, \ldots$$

The Euler numbers E_n are defined as the coefficients in the power-series expansion of sec x (see below). With $E_0 = 1$ they may be calculated from the recurrence formula

$$E_0 - \binom{2n}{2}E_1 + \binom{2n}{4}E_2 - + \cdots + (-1)^n E_n = 0 \qquad (1.22)$$

The first few of them are

$$E_0 = 1, E_1 = 1, E_2 = 5, E_3 = 61, E_4 = 1385, \ldots$$

Various Series

$$e^x = 1 + x + \frac{x^2}{2!} + \frac{x^3}{3!} + \cdots$$

$$a^x = 1 + x \ln a + \frac{(x \ln a)^2}{2!} + \frac{(x \ln a)^3}{3!} + \cdots$$

$$\sin x = x - \frac{x^3}{3!} + \frac{x^5}{5!} - \frac{x^7}{7!} + \cdots$$

$$\cos x = 1 - \frac{x^2}{2!} + \frac{x^4}{4!} - \frac{x^6}{6!} + - \cdots$$

$$\tan x = x + \frac{x^3}{3} + \frac{2x^5}{15} + \frac{17x^7}{315} + \frac{62x^9}{2835} + \cdots$$

$$= \sum_{n=1}^{\infty} \frac{4^n(4^n-1)}{(2n)!} B_n x^{2n-1} \qquad x^2 < \frac{\pi^2}{4}$$

$$\cot x = \frac{1}{x} - \frac{x}{3} - \frac{x^3}{45} - \frac{2x^5}{945} - \frac{x^7}{4725} - \cdots = -\sum_{n=0}^{\infty} \frac{4^n}{(2n)!} B_n x^{2n-1} \qquad x^2 < \pi^2$$

$$= \frac{1}{x} + \frac{1}{x-\pi} + \frac{1}{x+\pi} + \frac{1}{x-2\pi} + \frac{1}{x+2\pi} + \cdots \qquad x^2 < \pi^2$$

$$\sec x = 1 + \frac{x^2}{2!} + \frac{5x^4}{4!} + \frac{61x^6}{6!} + \cdots = \sum_{n=0}^{\infty} \frac{1}{(2n)!} E_n x^{2n} \qquad x^2 < \frac{\pi^2}{4}$$

$$\csc x = \frac{1}{x} + \frac{x}{6} + \frac{7x^3}{360} + \frac{31x^5}{15{,}120} + \cdots = \sum_{n=0}^{\infty} \frac{4^n - 2}{(2n)!} B_n x^{2n-1} \qquad x^2 < \pi^2$$

$$\sin^{-1} x = x + \frac{x^3}{6} + \frac{1 \cdot 3}{2 \cdot 4} \cdot \frac{x^5}{5} + \frac{1 \cdot 3 \cdot 5}{2 \cdot 4 \cdot 6} \cdot \frac{x^7}{7} + \cdots = \frac{\pi}{2} - \cos^{-1} x \qquad x^2 < 1$$

$$\tan^{-1} x = x - \frac{x^3}{3} + \frac{x^5}{5} - \frac{x^7}{7} + \cdots = \frac{\pi}{2} - \cot^{-1} x \qquad x^2 < 1$$

$$\tan^{-1} x = \frac{\pi}{2} - \frac{1}{x} + \frac{1}{3x^3} - \frac{1}{5x^5} + \cdots \qquad x^2 > 1$$

$$\sec^{-1} x = \frac{\pi}{2} - \frac{1}{x} - \frac{1}{6x^3} - \frac{1 \cdot 3}{2 \cdot 4 \cdot 5 \cdot x^5} - \frac{1 \cdot 3 \cdot 5}{2 \cdot 4 \cdot 6 \cdot 7 \cdot x^7} - \cdots$$

$$= \frac{\pi}{2} - \csc^{-1} x \qquad x^2 > 1$$

$$\text{Sinh}\, x = x + \frac{x^3}{3!} + \frac{x^5}{5!} + \frac{x^7}{7!} + \cdots$$

$$\text{Cosh}\, x = 1 + \frac{x^2}{2!} + \frac{x^4}{4!} + \frac{x^6}{6!} + \frac{x^8}{8!} + \cdots$$

$$\text{Tanh}\, x = x - \frac{x^3}{3} + \frac{2x^5}{15} - \frac{17x^7}{315} + \cdots$$

$$= \sum_{n=1}^{\infty} (-1)^{n-1} \frac{4^n(4^n-1)}{(2n)!} B_n x^{2n-1} \qquad x^2 < \frac{\pi^2}{4}$$

$$\text{Coth}\, x = \frac{1}{x} + \frac{x}{3} - \frac{x^3}{45} + \frac{2x^5}{945} - \frac{x^7}{4725} + - \cdots$$

$$= -\sum_{n=0}^{\infty} (-1)^n \frac{4^n}{(2n)!} B_n x^{2n-1} \qquad x^2 < \pi^2$$

$$\text{Sech}\, x = 1 - \frac{x^2}{2!} + \frac{5x^4}{4!} - \frac{61x^6}{6!} + \frac{1385x^8}{8!} - + \cdots$$

$$= \sum_{n=0}^{\infty} (-1)^n \frac{1}{(2n)!} E_n x^{2n} \qquad x^2 < \frac{\pi^2}{4}$$

$$\text{Csch}\, x = \frac{1}{x} - \frac{x}{6} + \frac{7x^3}{360} - \frac{31x^5}{15{,}120} + - \cdots$$

$$= \sum_{n=0}^{\infty} (-1)^n \frac{4^n-2}{(2n)!} B_n x^{2n-1} \qquad x^2 < \pi^2$$

$$\text{Sinh}^{-1} x = \ln 2x + \frac{1}{2 \cdot 2 \cdot x^2} - \frac{3}{2 \cdot 4 \cdot 4 \cdot x^4} + \frac{3 \cdot 5}{2 \cdot 4 \cdot 6 \cdot 6 \cdot x^6} + \cdots \qquad x > 1$$

$$\text{Sinh}^{-1} x = x - \frac{x^3}{2 \cdot 3} + \frac{3x^5}{2 \cdot 4 \cdot 5} - \frac{3 \cdot 5 \cdot x^7}{2 \cdot 4 \cdot 6 \cdot 7} + \cdots \qquad x^2 < 1$$

$$\text{Cosh}^{-1} x = \pm \left(\ln 2x - \frac{1}{2 \cdot 2 \cdot x^2} - \frac{1 \cdot 3}{2 \cdot 4 \cdot 4 \cdot x^4} - \frac{1 \cdot 3 \cdot 5}{2 \cdot 4 \cdot 6 \cdot 6 \cdot x^6} \cdots \right) \qquad x > 1$$

$$\text{Tanh}^{-1} x = x + \frac{x^3}{3} + \frac{x^5}{5} + \frac{x^7}{7} \cdots \qquad x^2 < 1$$

$$\text{Coth}^{-1} x = \frac{1}{x} + \frac{1}{3x^3} + \frac{1}{5x^5} + \cdots \qquad x^2 > 1$$

$$\text{Sech}^{-1} x = \pm \left(\ln \frac{2}{x} - \frac{1}{2 \cdot 2} x^2 - \frac{1 \cdot 3}{2 \cdot 4 \cdot 4} x^4 - \frac{1 \cdot 3 \cdot 5}{2 \cdot 4 \cdot 6 \cdot 6} x^6 - \cdots \right) \qquad 0 < x < 1$$

$$\text{Csch}^{-1} x = \frac{1}{x} - \frac{1}{2 \cdot 3 \cdot x^3} + \frac{3}{2 \cdot 4 \cdot 5 \cdot x^5} - \frac{3 \cdot 5}{2 \cdot 4 \cdot 6 \cdot 7 \cdot x^7} + \cdots \qquad x^2 < 1$$

$$1 + \frac{1}{2^{2n}} + \frac{1}{3^{2n}} + \frac{1}{4^{2n}} + \cdots = \frac{4^n \pi^{2n}}{2(2n)!} B_n$$

$$1 - \frac{1}{2^{2n}} + \frac{1}{3^{2n}} - \frac{1}{4^{2n}} + - \cdots = \frac{(4^n - 2)\pi^{2n}}{2(2n)!} B_n$$

$$1 + \frac{1}{3^{2n}} + \frac{1}{5^{2n}} + \frac{1}{7^{2n}} + \cdots = \frac{(4^n - 1)\pi^{2n}}{2(2n)!} B_n$$

$$1 - \frac{1}{3^{2n+1}} + \frac{1}{5^{2n+1}} - \frac{1}{7^{2n+1}} + - \cdots = \frac{\pi^{2n+1}}{4^{n+1}(2n)!} E_n$$

Infinite Products and Logarithmic Series

$$\sin x = x\left(1 - \frac{x^2}{\pi^2}\right)\left(1 - \frac{x^2}{4\pi^2}\right)\left(1 - \frac{x^2}{9\pi^2}\right) \cdots = x \prod_{n=1}^{\infty} \left(1 - \frac{x^2}{n^2\pi^2}\right)$$

$$\cos x = \left(1 - \frac{4x^2}{\pi^2}\right)\left(1 - \frac{4x^2}{9\pi^2}\right)\left(1 - \frac{4x^2}{25\pi^2}\right) \cdots = \prod_{n=1}^{\infty} \left[1 - \frac{4x^2}{(2n-1)^2\pi^2}\right]$$

$$\text{Sinh } x = x\left(1+\frac{x^2}{\pi^2}\right)\left(1+\frac{x^2}{4\pi^2}\right)\left(1+\frac{x^2}{9\pi^2}\right)\cdots = x\prod_{n=1}^{\infty}\left(1+\frac{x^2}{n^2\pi^2}\right)$$

$$\text{Cosh } x = \left(1+\frac{4x^2}{\pi^2}\right)\left(1+\frac{4x^2}{9\pi^2}\right)\left(1+\frac{4x^2}{25\pi^2}\right)\cdots = \prod_{n=1}^{\infty}\left[1+\frac{4x^2}{(2n-1)^2\pi^2}\right]$$

$$\ln x = (x-1) - \frac{(x-1)^2}{2} + \frac{(x-1)^3}{3} - \cdots \qquad 2 \geq x > 0$$

$$\ln x = \frac{x-1}{x} + \frac{1}{2}\left(\frac{x-1}{x}\right)^2 + \frac{1}{3}\left(\frac{x-1}{x}\right)^3 + \cdots \qquad x > \tfrac{1}{2}$$

$$\ln x = 2\left[\frac{x-1}{x+1} + \frac{1}{3}\left(\frac{x-1}{x+1}\right)^3 + \frac{1}{5}\left(\frac{x-1}{x+1}\right)^5 + \cdots\right] \qquad x > 0$$

$$\ln(1+x) = x - \frac{x^2}{2} + \frac{x^3}{3} - \frac{x^4}{4} + \cdots \qquad x^2 < 1$$

$$\ln\left(\frac{1+x}{1-x}\right) = 2\left(x + \frac{x^3}{3} + \frac{x^5}{5} + \frac{x^7}{7} + \cdots\right) \qquad x^2 < 1$$

$$\ln\left(\frac{x+1}{x-1}\right) = 2\left(\frac{1}{x} + \frac{1}{3x^3} + \frac{1}{5x^5} + \cdots\right) \qquad x^2 > 1$$

$$\ln \sin x = \ln x - \frac{x^2}{6} - \frac{x^4}{180} - \frac{x^6}{2835} - \cdots$$

$$= \ln x - \sum_{n=1}^{\infty}\frac{4^n}{2n(2n)!}B_n x^{2n} \qquad x^2 < \pi^2$$

$$\ln \cos x = -\frac{x^2}{2} - \frac{x^4}{12} - \frac{x^6}{45} - \frac{17x^8}{2520} - \cdots$$

$$= -\sum_{n=1}^{\infty}\frac{4^n(4^n-1)}{2n(2n)!}B_n x^{2n} \qquad x^2 < \frac{\pi^2}{4}$$

$$\ln \tan x = \ln x + \frac{x^2}{3} + \frac{7x^4}{90} + \frac{62x^6}{2835} + \cdots$$

$$= \ln x + \sum_{n=1}^{\infty}\frac{4^n(4^n-2)}{2n(2n)!}B_n x^{2n} \qquad x^2 < \frac{\pi^2}{4}$$

Series for logarithms to any arbitrary base may be found by means of the following relations:

The logarithm of a number to a given base is the power to which the base must be raised to yield the number; i.e., if $b^x = N$, then $\log_b N = x$. $\log_a N = \ln N \cdot \log_a e$, where $\ln N = \log_e N$ (natural or Napierian); $\ln N = \log_a N \cdot \ln a$.

1.4. REFERENCES

[1] C. Atkinson: Polynomial root solving on the electronic differential analyser, *Math. Tables and Other Aids to Computation*, **9** (1955), 139–143.
[2] M. Bôcher: "Introduction to Higher Algebra," Macmillan, New York, 1924.
[3] A. Fletcher, J. C. P. Miller, L. Rosenhead: "An Index of Mathematical Tables," Scientific Computing Service, London, 1946.
[4] T. Fort: "Infinite Series," Oxford Univ. Press, New York, 1930.
[5] J. M. Hyslop: "Infinite Series," Interscience, New York, 1945.
[6] K. Knopp: "Theorie und Anwendung der unendlichen Reihen," 4th ed., Springer, Berlin, 1947.
[7] D. E. Littlewood: "A University Algebra," Heinemann, London, 1950.
[8] A. D. Michel: "Matrix and Tensor Calculus with Applications to Mechanics, Elasticity, and Aeronautics," Wiley, New York, 1947.
[9] H. Margenau, G. M. Murphy: "The Mathematics of Physics and Chemistry," 2d ed., Van Nostrand, Princeton, N.J., 1956.
[10] B. O. Peirce: "A Short Table of Integrals," Ginn, Boston, 1920.
[11] L. A. Pipes: "Applied Mathematics for Engineers and Physicists," 2d ed., McGraw-Hill, New York, 1958.

CHAPTER 2

NUMERICAL ALGEBRA

BY

M. G. SALVADORI, Dr. Ing. Civ., Dr. Mat. Pura, New York

2.1. SOLUTION OF ALGEBRAIC EQUATIONS

The *n*th-degree Algebraic Equation

$$y(x) \equiv a_n x^n + a_{n-1}x^{n-1} + \cdot \cdot \cdot + a_1 x + a_0 = 0 \tag{2.1}$$

has n roots $x_1 \leq x_2 \leq x_3 \leq \cdot \cdot \cdot \leq x_n$, of which some may be real, either separate or repeated, and some couples of complex conjugate numbers with different or identical moduli. The real roots are evaluated first.

The number of positive roots is equal to the number of sign changes in the coefficients or less than that by an even number; the number of negative roots is equal to the number of sign constancies or less than that by an even number. Zero coefficients are to be ignored in the count. (Descartes's rule of signs.)

To scan for the real roots of Eq. (2.1) less than 1 in modulus, evaluate $y(x)$ by synthetic division for $-1 \leq x \leq 1$ in steps of 0.2; to scan for the roots larger than 1 in modulus, evaluate similarly the equation

$$\bar{y}(\xi) \equiv a_0 \xi^n + a_1 \xi^{n-1} + \cdot \cdot \cdot + a_{n-1}\xi + a_n = 0 \tag{2.1a}$$

whose roots are $\xi_i = 1/x_i$

If n is odd, then Eqs. (2.1) and (2.1*a*) have at least one real root of sign opposite to the sign of a_0/a_n.

The roots of Eq. (2.1) may be checked by means of Newton's relations:

$$\sum_{i=1}^{n} x_i = -\frac{a_{n-1}}{a_n}, \quad \prod_{i,j=1}^{n} x_i x_j = \frac{a_{n-2}}{a_n}, \quad \ldots \; , \quad x_1 \cdot x_2 \cdot \cdot \cdot x_n = (-1)^n \frac{a_0}{a_n} \tag{2.2}$$

Evaluation of Real Separate Roots (Horner's Method)

Approximate values of the smallest and largest roots of Eq. (2.1) are given respectively by

$$x_1 \approx -\frac{a_0}{a_1} \qquad x_n \approx -\frac{a_{n-1}}{a_n}$$

(provided these roots are not repeated).

Successive, rapidly converging approximations $x^{(m)}$ to a root x are given by Newton's first-order formula,

$$x^{(m+1)} = x^{(m)} - \frac{y^{(m)}}{y'^{(m)}} \tag{2.3}$$

or by Newton's second-order formula,

$$x^{(m+1)} = x^{(m)} + h_m \qquad \frac{1}{h_m} = -\frac{y'^{(m)}}{y^{(m)}} + \frac{\tfrac{1}{2}y''^{(m)}}{y'^{(m)}} \tag{2.4}$$

where $\quad y^{(m)} = y(x^{(m)}) \qquad y'^{(m)} = \left[\dfrac{dy}{dx}\right]_{x=x^{(m)}} \qquad y''^{(m)} = \left[\dfrac{d^2y}{dx^2}\right]_{x=x^{(m)}}$

The values of y, y', and y'' at $x = x^{(m)}$ are best obtained by synthetic division according to the following scheme, in which zero coefficients must appear explicitly.

Horner's Scheme (Synthetic Division)

$\cdots$ | $x^{(m)}$ | 1 | $\rightarrow$

a_n	a_{n-1} $b_n x^{(m)}$	a_{n-2} $b_{n-1}x^{(m)}$	$\cdots$ $\cdots$	a_2 $b_3x^{(m)}$	a_1 $b_2x^{(m)}$	a_0 $b_1x^{(m)}$
$b_n \equiv a_n$	b_{n-1} $c_n x^{(m)}$	b_{n-2} $c_{n-1}x^{(m)}$	$\cdots$ $\cdots$	b_2 $c_3x^{(m)}$	b_1 $c_2x^{(m)}$	$b_0 \equiv y^{(m)}$
$c_n \equiv b_n$	c_{n-1} $d_n x^{(m)}$	c_{n-2} $d_{n-1}x^{(m)}$	$\cdots$ $\cdots$	c_2 $d_3x^{(m)}$	$c_1 \equiv y'^{(m)}$	
$d_n \equiv c_n$	d_{n-1}	d_{n-2}	$\cdots$	$d_2 \equiv \tfrac{1}{2}y''^{(m)}$		

Here $\quad b_{n-1} = b_n x^{(m)} + a_{n-1}, \quad b_{n-2} = b_{n-1}x^{(m)} + a_{n-2}, \quad \ldots,$
$\quad c_{n-1} = c_n x^{(m)} + b_{n-1}, \quad$ etc.

The process is stopped when the remainder b_0 equals zero within the number of significant figures used in the computations. The reduced equation or quotient

$$Q(x) = b_n x^{n-1} + b_{n-1}x^{n-2} + \cdots + b_2 x + b_1 = 0$$

contains the remaining $(n - 1)$ roots of Eq. (2.1) and is solved in analogous manner for the next real root.

The process is continued until all the real roots are evaluated. The roots are checked by the first and last of relations (2.2).

Repeated Real Roots

A root repeated k times is a root of the k equations: $y(x) = 0$, $y'(x) = 0$, $y''(x) = 0$, $\ldots$, $y^{(k-1)}(x) = 0$, and is detected by synthetic division. Newton's method cannot be applied to a repeated root x_1 since $y'(x_1) = 0$.

Two almost repeated roots (very near roots), x_1 and x_2, are indicated by small values of y and y', and are better evaluated by obtaining first the root $x = a$ of the equation $y'(x) = 0$ lying between them, and then using the equation

$$x_{1,2} = a \pm \sqrt{\frac{-y(a)}{\tfrac{1}{2}y''(a)}} \tag{2.5}$$

Complex Conjugate Roots with Different Moduli

Two complex conjugate roots $x = a \pm bi$ of Eq. (2.1) are contained in the quadratic factor

$$[x - (a + bi)][x - (a - bi)] \equiv x^2 + px + q = q\left(1 + \frac{p}{q}x + \frac{1}{q}x^2\right) \tag{2.6}$$

which can be isolated using the following schemes of synthetic division in descending and ascending powers of x, provided the couples of conjugate roots have different moduli.

Division in Descending Powers of x

$\cdots$	$-q$	$-p$	1	$\rightarrow$

a_n	a_{n-1}	a_{n-2}	$\cdots$	a_2	a_1	a_0
	$-pb_n$	$-pb_{n-1}$	$\cdots$	$-pb_3$	$-pb_2$	—
	—	$-qb_n$	$\cdots$	$-qb_4$	$-qb_3$	$-qb_2$
$b_n \equiv a_n$	b_{n-1}	b_{n-2}	$\cdots$	b_2	b_1	b_0

Division in Ascending Powers of x

$\leftarrow$	1	$-p/q$	$-1/q$	$\cdots$

a_n	a_{n-1}	a_{n-2}	$\cdots$	a_2	a_1	a_0
—	$-\frac{p}{q}b'_{n-2}$	$\frac{p}{q}b'_{n-3}$	$\cdots$	$\frac{p}{q}b'_1$	$-\frac{p}{q}b'_0$	
$-\frac{1}{q}b'_{n-2}$	$-\frac{1}{q}b'_{n-3}$	$-\frac{1}{q}b'_{n-4}$	$\cdots$	$-\frac{1}{q}b'_0$	—	
b'_n	b'_{n-1}	b'_{n-2}	$\cdots$	b'_2	b'_1	$b'_0 \equiv a_0$

The quadratic factor is isolated when the remainder $b_1x + b_0 = 0$, that is, when $b_1 = b_0 = 0$. The reduced equation, or quotient,

$$Q(x) \equiv b_nx^{n-2} + b_{n-1}x^{n-3} + \cdots + b_3x + b_2 = 0$$

contains the remaining complex roots of Eq. (2.1).

Successive approximations of a quadratic factor are obtained by (1) dividing the original equation by a trial value of the smallest quadratic factor in ascending powers of x; (2) dividing the original equation by the quotient $Q(x)$ of the first division in descending powers of x; (3) continuing these alternate divisions until convergence is obtained for the quadratic factor.

The successive alternate divisions can be combined in a single scheme, in which a_{n-2} is repeated twice and convergence is indicated by the equality of the starred coefficients b_{n-2} and b'_{n-2}, without carrying either division all the way through.

Initial values of the largest and smallest quadratic factors are given respectively by

$$x^2 + \frac{a_{n-1}}{a_n}x + \frac{a_{n-2}}{a_n} \qquad x^2 + \frac{a_1}{a_2}x + \frac{a_0}{a_2}$$

Example 1

$$x^6 - 16x^5 + 128x^4 - 504x^3 + 1156x^2 - 1360x + 800 = 0$$

(1) Division by trial largest factor $x^2 - 16x + 128 = 128(1 - 0.125x + 0.008x^2)$ in ascending powers of x; (2) division by quotient

$$Q(x) = 74x^4 - 370x^3 + 993x^2 + \cdots = 74x^2(x^2 - 5x + 13.4)$$

in descending powers of x; (3) repetition of alternate divisions (see table).

Repetition of Alternate Divisions

1	−16		128	128	−504	1156		−1360	800	
−13.4 |	5 |	1	|→				←| 1 |	0.125 |		−0.008
	5		−55	−46	124	−157		100		
			−13	−8	10	−6				
1	−11		60	74	−370	993		−1260	800	
−16.8 |	5.76 |	1	|→				←| 1 |	0.18 |		−0.017
	6		−58	−57	166	−219		144		
			−17	−16	21	−14				
1	−10		53	55	−317	923		−1216	800	
−17.5 |	+5.92 |	1	|→				←| 1 |	0.19 |		−0.019
	6		−59	−59	173	−230		152		
			−18	−17	23	−15				
1	−10		51*	52*	−308	911		−1208	800	

$$x^2 - 10x + 51 = 0 \qquad 52(x^4 - 5.92x^3 + 17.52x^2 - 23.23x + 15.38) = 0$$

True factors:

$$x^2 - 10x + 50 = 0 \qquad x^4 - 6x^3 + 18x^2 - 24x + 16 = 0$$

Method for Quartic Equation

Given the 4th-degree equation

$$x^4 + a_3x^3 + a_2x^2 + a_1x + a_0 = 0 \tag{2.7}$$

1. Evaluate:

$$b_1 = a_3a_1 - 4a_0 \qquad b_0 = a_0(4a_2 - a_3^2) - a_1^2 \tag{2.8}$$

2. Evaluate the algebraically largest real root z_3 of the cubic equation:

$$z^3 - a_2z^2 + b_1z + b_0 = 0 \tag{2.9}$$

3. Evaluate:

$$c_{1,2} = \frac{a_3}{2} \pm \sqrt{\left(\frac{a_3}{2}\right)^2 - a_2 + z_3} \qquad d_{1,2} = \frac{z_3}{2} \pm \sqrt{\left(\frac{z_3}{2}\right)^2 - a_0} \tag{2.10}$$

and check which of the two values of d is d_1 and which is d_2 by the equation

$$c_1d_2 + c_2d_1 = a_1 \tag{2.11}$$

4. The two quadratic factors of Eq. (2.7) are

$$x^2 + c_1x + d_1 = 0 \qquad x^2 + c_2x + d_2 = 0 \tag{2.12}$$

Complex Conjugate Roots with Equal Moduli

Isolating a quadratic factor according to Eq. (2.6) becomes a poorly convergent procedure when there are pairs of complex roots with almost equal moduli. In this case Graeffe's root-squaring method is helpful. When the given equation is $y(x) = 0$, then an equation of the same degree for $-x^2$ has coefficients computed in the following scheme, in which a_n are the coefficients defined by Eq. (2.1) and b_n the coefficients of the equation

$$b_n(-x^2)^n + b_{n-1}(-x^2)^{n-1} + \cdot\cdot\cdot + b_1(-x^2) + b_0 = 0$$

Graeffe's Scheme

a_n	a_{n-1}	a_{n-2}	a_{n-3}	$\cdots$	a_2	a_1	a_0
a_n^2	a^2_{n-1}	a^2_{n-2}	a^2_{n-3}	$\cdots$	a_2^2	a_1^2	a_0^2
	$-2a_na_{n-2}$	$-2a_{n-1}a_{n-3}$	$-2a_{n-2}a_{n-4}$	$\cdots$	$-2a_3a_1$	$-2a_2a_0$	
		$2a_na_{n-4}$	$2a_{n-1}a_{n-5}$	$\cdots$	$2a_4a_0$		
				$\vdots$			
b_n	b_{n-1}	b_{n-2}	b_{n-3}	$\cdots$	b_2	b_1	b_0

Repeated application of this scheme produces nth-degree equations for x^4, x^8, ... When after a few steps the double products do not become negligible in comparison with the squares, the given equation has complex roots; k pairs of conjugate roots with identical moduli are indicated in the mth row by $(2k - 1)$ coefficients with random values flanked by coefficients which are pure squares. The ratio of the right flanking coefficient divided by the left flanking coefficient gives the $2km$th power of the identical moduli. Once their modulus is known, the roots are obtained by Newton's relations (2.2).

2.2. SOLUTION OF TRANSCENDENTAL EQUATIONS

Real Roots

Newton's methods [Eqs. (2.3) and (2.4)] can be applied to the location of real roots of transcendental equations as soon as initial values are computed by trial and error, graphically, or by power-series expansion.

Complex Roots

Complex roots of transcendental equations are located by letting $x = a + bi$ in the equation and by solving simultaneously for a and b the two equations for the real and the imaginary parts of the original equation. No general methods of solution exist for simultaneous transcendental equations. Trial and error and Newton's method in two dimensions are the best available techniques.

2.3. SOLUTION OF SIMULTANEOUS LINEAR EQUATIONS

The nth-order System of n Linear Equations

$$\begin{aligned} a_{11}x_1 + a_{12}x_2 + &\cdots + a_{1n}x_n - c_1 = 0 \\ a_{21}x_1 + a_{22}x_2 + &\cdots + a_{2n}x_n - c_2 = 0 \\ &\cdots\cdots\cdots\cdots \\ a_{n1}x_1 + a_{n2}x_2 + &\cdots + a_{nn}x_n - c_n = 0 \end{aligned} \tag{2.13}$$

has a unique set of roots x_j ($j = 1, 2, \ldots, n$) if and only if the determinant Δ of its coefficients is different from zero.

Solution by Determinants

This is convenient only if one or a few of the n roots are to be evaluated, and n is not larger than 5 or 6. The roots x_j are given by

$$x_j = \frac{\Delta_j}{\Delta}$$

where

$$\Delta = \begin{vmatrix} a_{11} & a_{12} & \cdots & a_{1n} \\ a_{21} & a_{22} & \cdots & a_{2n} \\ \cdots & \cdots & \cdots & \cdots \\ a_{n1} & a_{n2} & \cdots & a_{nn} \end{vmatrix} \qquad \Delta_j = \begin{vmatrix} a_{11} & a_{12} & \cdots & a_{1,j-1} & c_1 & a_{1,j+1} & \cdots & a_{1n} \\ a_{21} & a_{22} & \cdots & a_{2,j-1} & c_2 & a_{2,j+1} & \cdots & a_{2n} \\ \cdots & \cdots & \cdots & \cdots & \cdots & \cdots & \cdots & \cdots \\ a_{n1} & a_{n2} & \cdots & a_{n,j-1} & c_n & a_{n,j+1} & \cdots & a_{nn} \end{vmatrix}$$

The most practical method for evaluating determinants is *pivotal condensation*, according to the following scheme:

$$\Delta = \frac{1}{a_{11}{}^{n-2}} \begin{vmatrix} \begin{vmatrix} a_{11} & a_{12} \\ a_{21} & a_{22} \end{vmatrix} & \begin{vmatrix} a_{11} & a_{13} \\ a_{21} & a_{23} \end{vmatrix} & \cdots & \begin{vmatrix} a_{11} & a_{1n} \\ a_{21} & a_{2n} \end{vmatrix} \\ \begin{vmatrix} a_{11} & a_{12} \\ a_{31} & a_{32} \end{vmatrix} & \begin{vmatrix} a_{11} & a_{13} \\ a_{31} & a_{33} \end{vmatrix} & \cdots & \begin{vmatrix} a_{11} & a_{1n} \\ a_{31} & a_{3n} \end{vmatrix} \\ \cdots & \cdots & \cdots & \cdots \\ \begin{vmatrix} a_{11} & a_{12} \\ a_{n1} & a_{n2} \end{vmatrix} & \begin{vmatrix} a_{11} & a_{13} \\ a_{n1} & a_{n3} \end{vmatrix} & \cdots & \begin{vmatrix} a_{11} & a_{1n} \\ a_{n1} & a_{nn} \end{vmatrix} \end{vmatrix} \tag{2.14}$$

Solution by Elimination (Gauss' Scheme)

One of the unknowns is eliminated at a time, by multiplication of the equations by suitable coefficients and subtraction. An additional *check column s* is carried in the computations; it contains the sum of all the coefficients and the constant of each equation, and is operated upon as any coefficient of that equation.

Example 2

Row	x_1	x_2	x_3	c	s	Explanation
1	2.12	2.00	4.51	19.40	28.03	Eq. I
2	1.20	3.14	2.05	13.65	20.04	Eq. II
3	1.25	3.20	4.15	19.50	28.10	Eq. III
4	1.20	1.13	2.55	10.98	15.86	(1.20/2.12) × eq. I
5	0	2.01	−0.50	2.67	4.18	Row 2 − row 4
6	1.25	1.18	2.66	11.45	16.54	(1.25/2.12) × eq. I
7	0	2.02	1.49	8.05	11.56	Row 3 − row 6
8		2.02	−0.50	2.68	4.20	(2.02/2.01) × row 5
9		0	1.99	5.37	7.36	Row 7 − row 8

$x_3 = 5.37/1.99 = 2.70$ from row 9
$x_2 = (2.68 + 0.5 \times 2.70)/2.02 = 2.00$ from row 8
$x_1 = (19.40 - 4.51 \times 2.70 - 2.00 \times 2.00)/2.12 = 1.52$ from row 1

Roots must always be checked by substitution in the original equations.

Solution by Crout's Scheme (Cholewsky's Method)

The coefficient matrix A of the system (2.13) is equated to the product of a lower triangular matrix L and an upper triangular matrix T ($A = LT$), with

$$A = \begin{bmatrix} a_{11} & a_{12} & \cdots & a_{1n} & c_1 \\ a_{21} & a_{22} & \cdots & a_{2n} & c_2 \\ a_{n1} & a_{n2} & \cdots & a_{nn} & c_n \end{bmatrix} \qquad L = \begin{bmatrix} l_{11} & 0 & 0 & \cdots & 0 \\ l_{21} & l_{22} & 0 & \cdots & 0 \\ l_{n1} & l_{n2} & l_{n3} & \cdots & l_{nn} \end{bmatrix}$$

$$T = \begin{bmatrix} 1 & t_{12} & t_{13} & \cdots & t_{1n} & k_1 \\ 0 & 1 & t_{23} & \cdots & t_{2n} & k_2 \\ \cdots & \cdots & \cdots & \cdots & \cdots & \cdots \\ 0 & 0 & 0 & \cdots & 1 & k_n \end{bmatrix} \tag{2.15}$$

The unknown elements l_{ij}, t_{ij} are evaluated by equating to a_{ij} the product of the ith row of L by the jth column of T, proceeding by rows, and starting with l_{11}; for example:

$$a_{11} = l_{11} \times 1 + 0 \qquad l_{11} = a_{11} \qquad a_{12} = l_{11}t_{12} + 0 \qquad t_{12} = \frac{a_{12}}{l_{11}}$$
$$a_{22} = l_{21}t_{12} + l_{22}1 + 0 \qquad l_{22} = a_{22} - l_{21}t_{12}$$

Notice that $l_{i1} = a_{i1}$ and $t_{1j} = a_{1j}/a_{11}$.

Check sums can be carried in A and T. The unknowns are obtained by backward substitution from the triangular system T.

Solution by Iteration (Gauss-Seidel Method)

This converges only if, in each equation, a different unknown has a coefficient, the absolute value of which is larger than or equal to the sum of the absolute values of the other coefficients. Each equation is solved for the large-coefficient unknown; initial values of the other unknowns are set in the right-hand member of the equations; the left-hand-member unknowns are evaluated and their new values are set in the right-hand members, continuing until convergence is obtained.

The method is ideal when only a few unknowns appear in each equation of large systems.

Solution by Relaxation

This permits the solution of special types of large systems by pencil and paper. Indicate by R_i (residuals) the value of the left-hand member of the ith equation of system (2.13) for given initial value x_j° of the unknowns:

$$\begin{aligned} a_{11}x_1^\circ + a_{12}x_2^\circ + \cdots + a_{1n}x_n^\circ - c_1 &= R_1 \\ a_{21}x_1^\circ + a_{22}x_2^\circ + \cdots + a_{2n}x_n^\circ - c_2 &= R_2 \\ \cdots\cdots\cdots\cdots\cdots\cdots\cdots\cdots \\ a_{n1}x_1^\circ + a_{n2}x_2^\circ + \cdots + a_{nn}x_n^\circ - c_n &= R_n \end{aligned} \tag{2.16}$$

A change from x_j to $x_j + \Delta x_j$ changes R_j by $a_{jj}\,\Delta x_j$, and all other R_i by $a_{ij}\,\Delta x_j$: hence the operations table giving the changes in the residuals for unit changes in the unknowns is the transpose of the coefficient matrix. Relaxation consists in reducing to zero the residuals by successive changes in the unknowns.

The simplest relaxation technique consists in reducing to zero at each step the largest residual, say R_n, by the change

$$\Delta x_m = -\frac{R_m}{a_{mm}} \tag{2.17}$$

which increases the R_i by

$$\Delta R_i = -a_{im}\frac{R_m}{a_{mm}} \tag{2.18}$$

The most efficient use of relaxation consists in changing the values of the unknowns on the basis of a physical knowledge of the problem at hand.

Example 3

x_1	x_2	x_3	$-c$
10	−2	−2	−6
−1	10	−2	−7
−1	−1	10	−8

Initial values: $x_1^{(0)} = 0$, $x_2^{(0)} = 0$, $x_3^{(0)} = 0$
Initial residuals: $R_1 = -6$, $R_2 = -7$, $R_3 = 8$ (Appear multiplied by 100 in solution scheme to avoid decimal point.)

Operations Table

	x_1	x_2	x_3
R_1	10	−1	−1
R_2	−2	10	−1
R_3	−2	−2	10

The largest residual $R_3 = -800$ is reduced to zero by a change

$$\Delta x_3 = -\frac{-800}{a_{33}} = \frac{800}{10} = 80,$$

which introduces changes $\Delta R_2 = -2\Delta x_3 = -160$; $\Delta R_1 = -2\Delta x_3 = -160$. The next largest residual $R_2 = -860$ is reduced to zero by a change

$$\Delta x_2 = -\frac{-860}{a_{22}} = \frac{860}{10} = 86,$$

introducing changes $\Delta R_1 = -2\Delta x_2 = -172$; $\Delta R_3 = -\Delta x_2 = -86$. The process is terminated when the residuals are negligible or zero.

The relaxation table is conveniently shortened by omitting ΔR_i and writing directly $R_i + \Delta R_i$.

Shortened Relaxation Table for Example 3

x_1	R_1	x_2	R_2	x_3	R_3
0	−600	0	−700	0	−800
	−760	86	−860	80	−86
93	−932		−93	18	−179
	−2	13	−129		1
	−38		1		−12
6	−64		−5		−18
	−4	1	−9	2	2
	−8		1		1
1	−10				
100	0	100	0	100	0

Convergence is often accelerated by "overrelaxing" or "underrelaxing," i.e., by introducing changes larger or smaller than those required to reduce residuals to zero, and thus obtaining residuals of opposite signs in the equations.

Error Equations

All the preceding methods may be used to obtain additional significant figures in the unknowns by solving *error equations*, in which the coefficients are those of the original system and the constants are the negative of the residuals (computed by means of the available values of the unknowns). The roots of the error equations, i.e., the corrections, are then added to the values of the corresponding unknowns. Elimination and Crout's method are well suited to this process, since only one extra column of constants must be evaluated to obtain the corrections.

In view of the loss of accuracy characteristic of the solution of simultaneous equations, the number N of significant or decimal figures to be carried in the computations must be greater than the number N' of correct figures needed in the roots. At least two extra figures must be carried for each 10 equations, even if the coefficients are not given with an accuracy of more than N' figures.

2.4. REFERENCES

Graeffe's Method

[1] J. B. Scarborough: "Numerical Mathematical Analysis," 3d ed., Johns Hopkins Press, Baltimore, 1955.

Simultaneous Nonlinear Equations

[2] W. E. Milne: "Numerical Calculus," Princeton Univ. Press, Princeton, N.J., 1949.

Linear Equations by Matrix Partition

[3] R. A. Frazer, W. J. Duncan, A. R. Collar: "Elementary Matrices," Cambridge Univ. Press, New York, 1955.

Solution of Algebraic and Simultaneous Linear Equations

[4] M. G. Salvadori, M. L. Baron: "Numerical Methods in Engineering," Prentice-Hall, Englewood Cliffs, N.J., 1961.

CHAPTER 3

PROBABILITY

BY

J. H. B. KEMPERMAN, Ph.D., Rochester, N.Y.

3.1. PROBABILITY

Let $\mathcal{E}$ be a given experiment; by an event A we shall mean the occurrence of an outcome having a prescribed quality, e.g., "no collision," "average weight greater than 20." If A is an event, then $\bar{A}$ will denote the event that A does not happen. The union $A = A_1 \cup A_2 \cdots$ of the events $A_1, A_2, \ldots$ is defined as the event that at least one of the A_i happens, (read $\cup$ as "or"). Similarly, the intersection $A = A_1A_2A_3 \cdots$ denotes the event that each A_i happens. Finally, two events A_1 and A_2 are said to be disjoint if they cannot happen simultaneously.

A probabilistic model of a "random" experiment $\mathcal{E}$ is defined as follows. Let $\mathcal{A}$ be a class of a priori possible events, such that $\bar{A}_i$, $A_1 \cup A_2 \cup \cdots$, and $A_1A_2 \cdots$ are in $\mathcal{A}$ whenever the A_i $(i = 1, 2, \ldots)$ are in $\mathcal{A}$. Furthermore, let to each A in $\mathcal{A}$ correspond a non-negative number Prob (A), such that Prob (A) + Prob $(\bar{A}) = 1$ and that Prob $(A_1 \cup A_2 \cup \cdots)$ = Prob (A_1) + Prob $(A_2) + \cdots$, whenever the A_i $(i = 1, 2, \ldots)$ are pairwise disjoint. Here, Prob (A) is called the probability that the event A will occur; its heuristic interpretation is as follows. Let m denote the number of times that A happens in N repetitions of the experiment $\mathcal{E}$. Then, for N large enough, it is "practically certain" that the frequency m/N differs only very slightly from the number Prob (A).

As a simple illustration, let the experiment $\mathcal{E}$ consist of n tossings of a fair coin. By an elementary event we shall mean the event that the outcome is a specific sequence HHTHT $\cdots$ of heads and tails. To each of the 2^n elementary events we assign the probability 2^{-n}. Furthermore, with each class A^* of elementary events there corresponds an event A, such that A happens if and only if the actual outcome of $\mathcal{E}$ belongs to the class A^*. Finally, if A^* consists of m_A elementary events, let Prob $(A) = m_A 2^{-n}$. For example,

$$\text{Prob}\,(A) = \binom{n}{k} 2^{-n}$$

if A denotes the event that head occurs exactly k times. Here,

$$\binom{n}{k} = \frac{n!}{k!(n-k)!}$$

The *conditional probability* Prob $(A|B)$ of the event A, given that B happened, is defined as

$$\text{Prob}\,(A|B) = \text{Prob}\,(AB)/\text{Prob}\,(B) \tag{3.1}$$

[if Prob $(B) = 0$, we let Prob $(A|B)$ = Prob (A)]. In the above example, Prob $(A|B)$ is equal to the ratio m_{AB}/m_B—the number m_{AB} of elementary events belonging to both A^* and B^*, to the number m_B of elementary events in the class B^*.

If Prob $(A|B)$ = Prob (A), we say that the events A and B are *independent*. In heuristic terms: The knowledge that B happened provides no information whatsoever concerning the occurrence of A. In general, the events $A_1, \ldots, A_n$ are said to be completely independent if

$$\text{Prob}\,(A_{i_1}A_{i_2} \cdots A_{i_k}) = \text{Prob}\,(A_{i_1}) \cdots \text{Prob}\,(A_{i_k}) \tag{3.2}$$

for each subset $(A_{i_1}, \ldots, A_{i_k})$ of $(A_1, \ldots, A_n)$.

3.2. RANDOM VARIABLES

Consider an experiment $\mathcal{E}$ with random real-valued outcomes $X_1, X_2, \ldots$. Suppose that the probabilistic model of $\mathcal{E}$ is such that, for each choice of the continuous functions $g_j(x_1, \ldots, x_n)$ and real numbers c_j $(j = 1, \ldots, m)$, a probability Prob (A) is assigned to the event that, simultaneously, $Y_1 \leq c_1, \ldots, Y_m \leq c_m$, where $Y_j = g_j(X_1, \ldots, X_n)$. Then the X_i and also the Y_j are called random variables. Furthermore,

$$\text{Prob}\,(A) = \text{Prob}\,(Y_1 \leq c_1, \ldots, Y_m \leq c_m) = F_{Y_1,\ldots,Y_m}(c_1, \ldots, c_m)$$

say, is called the *joint distribution function* (d.f.) of the random variables $Y_1, \ldots, Y_m$.

The random variables $Y_1, \ldots, Y_m$ are said to be *independent* if, for each choice of the c_j,

$$F_{Y_1,\ldots,Y_m}(c_1, \ldots, c_m) = F_{Y_1}(c_1) \cdots F_{Y_m}(c_m) \tag{3.3}$$

in other words, if the m events $\{Y_1 \leq c_1\}, \ldots, \{Y_m \leq c_m\}$ are always completely independent. In heuristic terms: Knowing some of the actual outcomes Y_j provides no information whatsoever concerning the remaining outcomes Y_k.

As a simple illustration, let $\mathcal{E}$ consist of n independent tossings of a coin which might be biased, and let X_i denote the number of heads in the ith tossing $(X_i = 0 \text{ or } 1)$. In the probabilistic model for $\mathcal{E}$, $X_1, \ldots, X_n$ are independent random variables, such that Prob $(X_i = 1) = p$ $(i = 1, \ldots, n)$, p denoting a constant $(0 \leq p \leq 1)$. The random variable $Y = X_1 + \cdots + X_n$ denotes the total number of heads. It has the d.f.

$$F_Y(c) = \text{Prob}\,(Y \leq c) = \sum_{k=0}^{c'} \binom{n}{k} p^k(1-p)^{n-k} \tag{3.4}$$

(c' denoting the largest integer $\leq c$ and $\leq n$). Any random variable Y having the latter d.f. will be called a *binomial* random variable (with parameters p and n).

3.3. EXPECTATION VALUES

Let Y be a random variable having the d.f. $F_Y(c) = F(c)$. Then its *mean* or *expectation value* $E\{Y\}$ is defined by the Stieltjes integral

$$E\{Y\} = \int_{-\infty}^{\infty} y\, dF(y) \tag{3.5}$$

The mean $E\{Y\}$ is said to exist if the latter integral converges absolutely. Suppose $F(y)$ admits a continuous derivative; then

$$f(y) = F'(y) \tag{3.6}$$

is called the *probability density function* (p.d.f.) of Y. In this case, the above definition becomes

$$E\{Y\} = \int_{-\infty}^{\infty} yf(y)\, dy \tag{3.7}$$

On the other hand, if Y can only assume certain discrete values $y_1, y_2, \ldots$, then the d.f. $F(c)$ is a nondecreasing step function having at y_j a jump of size Prob $(Y = y_j)$. In this case,

$$E\{Y\} = \sum_j y_j \text{ Prob } (Y = y_j) \tag{3.8}$$

For example, if a lottery ticket of \$2 offers a chance of 5% to win \$10 and a chance of 5% to win \$25, the expectation value of the gain Y is equal to

$$E\{Y\} = -2 \times 0.90 + 8 \times 0.05 + 23 \times 0.05 = -0.25$$

We mention that always

$$E\{a_1Y_1 + \cdots + a_mY_m\} = a_1E\{Y_1\} + \cdots + a_mE\{Y_m\} \tag{3.9}$$

if a_j is constant and $E\{Y_j\}$ exists $(j = 1, \ldots, m)$. Moreover,

$$E\{Y_1 \cdots Y_m\} = E\{Y_1\}E\{Y_2\} \cdots E\{Y_m\} \tag{3.10}$$

provided that $Y_1, \ldots, Y_m$ are independent and the $E\{Y_j\}$ exist.

3.4. MOMENTS

Let X be a random variable. Its kth-order moment $(k = 0, 1, \ldots)$ is defined by

$$\nu_k = E\{X^k\} = \int_{-\infty}^{\infty} x^k \, dF_X(x) \tag{3.11}$$

It is said to exist if the latter integral converges absolutely. For convenience, we shall assume that all second-order moments, considered in the sequel, do exist.

The first moment $\nu_1 = E\{X\}$ is usually called the mean of X. Furthermore, the so-called *variance* σ^2 of X is defined by

$$\sigma^2 = E\{(X - \nu_1)^2\} = \nu_2 - \nu_1^2 \tag{3.12}$$

its square root $\sigma \geq 0$ is called the *standard deviation of X*. The latter is a measure of the dispersion of X from its mean ν_1, as may also be seen from the Bienaymé-Tchebycheff inequality which states that, for each $\lambda \geq 1$,

$$\text{Prob } (|X - \nu_1| > \lambda\sigma) < \lambda^{-2} \tag{3.13}$$

Especially, if $\sigma = 0$, then $X = \nu_1 =$ const (with probability 1).

Suppose that $X_1, \ldots, X_n$ are independent random variables, all having the same mean ν_1 and standard deviation σ. Then their sum $Y = X_1 + \cdots + X_n$ is a random variable with mean $n\nu_1$ and variance $n\sigma^2$ (thus, their average has a mean ν_1 and a variance σ^2/n). For example, the binomial random variable (cf. Sec. 3.2) with parameters n and p can be regarded as the sum Y of n independent random variables $X_1, \ldots, X_n$, each assuming only the values 0 and 1, Prob $(X_i = 1) = p$ $(i = 1, \ldots, n)$. Now, $E\{X_i\} = 0 \cdot (1 - p) + 1 \cdot p = p$; thus, each X_i has the variance

$$\sigma^2 = E\{(X_i - p)^2\} = p^2(1 - p) + (1 - p)^2p = p(1 - p)$$

Consequently, the binomial variable Y has a mean np and a variance $np(1 - p)$.

3.5. CORRELATION

Let $X_1, \ldots, X_n$ be random variables (not necessarily independent), and let μ_i denote the mean of X_i. We mention that the point $(\mu_1, \ldots, \mu_n)$ is precisely the center of gravity of the mass distribution in the n-dimensional space R_n, for which each subset A of R_n contains a mass equal to the probability that the random point $(X_1, \ldots, X_n)$ will fall in A.

The so-called *covariance* of X_i and X_k is defined by

$$\mu_{ik} = E\{(X_i - \mu_i)(X_k - \mu_k)\} = E\{X_iX_k\} - \mu_i\mu_k \tag{3.14}$$

It is equal to 0 when X_i and X_k are independent (but *not vice versa*). Note that $\mu_{ii} = \sigma_i^2$ is precisely the variance of X_i. If $\sigma_i > 0$, $\sigma_k > 0$, then

$$\rho_{ik} = \mu_{ik}/\sigma_i\sigma_k \tag{3.15}$$

defines the so-called *correlation coefficient* of X_i and X_k, where $-1 \leq \rho_{ik} \leq 1$. It measures the degree of *linear* dependence between X_i and X_k. More precisely, if $\rho_{ik} \neq 0$, then part of the variance of X_i may be removed by subtracting a linear function $aX_k + b$ of X_k (a and b constants). The best result is obtained by choosing

$$a = \rho_{ik}\sigma_i/\sigma_k \qquad b = \mu_i - a\mu_k \qquad \eta_{ik} = X_i - aX_k - b \tag{3.16}$$

This so-called *residual* η_{ik} satisfies

$$E\{\eta_{ik}\} = 0, \; E\{\eta_{ik}^2\} = (1 - \rho_{ik}^2)\sigma_i^2 < \sigma_i^2$$

(thus, if $\rho_{ik} = 1$ then $E\{\eta_{ik}^2\} = 0$; hence, $X_i = aX_k + b$). On the other hand, if $\rho_{ik} = 0$, such a reduction is not possible and X_i, X_k are said to be *uncorrelated.*

Suppose that ρ_{ik} is substantially different from zero, but that the correlation coefficient $\rho_{ij\cdot k}$ (say) between η_{ik} and η_{jk} is nearly equal to 0. This may indicate that the correlation between X_i and X_j is mainly due to the fact that both depend on X_k. This coefficient of partial correlation of X_i and X_j (after removal of the part of the variance due to their dependence on X_k) is given by

$$\rho_{ij\cdot k} = (\rho_{ij} - \rho_{ik}\rho_{jk})(1 - \rho_{ik}^2)^{-1/2}(1 - \rho_{jk}^2)^{-1/2} \tag{3.17}$$

3.6. THE NORMAL DISTRIBUTION

Let

$$\Phi(a) = \frac{1}{2\pi}\int_{-\infty}^{a} e^{-x^2/2}\,dx \tag{3.18}$$

Tables of the function $\Phi(a)$ can be found in any book on mathematical statistics; one has

$$\Phi(0) = 0.50 \qquad \Phi(1.65) = 0.95 \qquad \Phi(1.96) = 0.975$$

A random variable X is said to be normally distributed with mean μ and standard deviation $\sigma > 0$, [normal (μ,σ) for shortness] if it has the d.f.

$$F_X(x) = \Phi\left(\frac{x - \mu}{\sigma}\right)$$

that is, if it has the p.d.f.

$$p_X(x) = F'_X(x) = \frac{1}{\sigma\sqrt{2\pi}}\exp\left[-\frac{1}{2}\left(\frac{x - \mu}{\sigma}\right)^2\right] \tag{3.19}$$

Furthermore, if $X = \mu = \text{const}$, then X is said to have the singular normal distribution $(\mu,0)$.

Let $X_1, \ldots, X_n$ be independent random variables, and $a_1, \ldots, a_n$ nonzero constants. It can be shown that the linear combination $a_1X_1 + \cdots + a_nX_n$ is normally distributed if and only if each X_i is normally distributed.

Dropping the independence assumption, we shall say that the random variables $X_1, \ldots, X_n$ have a *normal joint distribution* if, for each choice of the constants $c_1, \ldots, c_n$, the linear combination $Z = c_1X_1 + \cdots + c_nX_n$ is normally distributed. The corresponding d.f., $F_{X_1,\ldots,X_n}(x_1, \ldots, x_n)$, is then called an n-dimensional normal d.f.

It may be shown that, for arbitrary random variables $Y_1, \ldots, Y_n$, there exists a *unique* n-dimensional normal d.f. having the same means and covariances as (the joint d.f. of) $Y_1, \ldots, Y_n$. Another important feature is: If Y_1, Y_2 have normal joint distribution, then Y_1 and Y_2 are independent *if* and only if they are uncorrelated.

3.7. CORRELATION CONTINUED. REGRESSION

Let X and Y be random variables having the correlation coefficient ρ. In practice, the true value of ρ is usually unknown. Assuming that X and Y have a normal joint distribution, an estimate r of ρ may be obtained as follows. In N independent repetitions of the experiment $\mathcal{E}$, let x_i, y_i denote the observed outcomes of X, Y in the ith experiment ($i = 1, \ldots, N$). Furthermore, let

$$N\bar{x} = x_1 + \cdots + x_N \qquad Ns_1^2 = (x_1 - \bar{x})^2 + \cdots + (x_N - \bar{x})^2$$

similarly, $\bar{y}$ and Ns_2^2 ($s_1 \geq 0$, $s_2 \geq 0$). Finally, let

$$Nm_{11} = \sum_i (x_i - \bar{x})(y_i - \bar{y}) = \sum_i x_i y_i - N\bar{x}\bar{y}$$

and

$$r = m_{11}/s_1 s_2$$

Actually, r is itself a random variable having a mean ρ; its standard deviation is about equal to $(1 - \rho^2)N^{-\frac{1}{2}}$. If N is large and $|r|\sqrt{N} > 1.96$, one can be 95% sure that X and Y are dependent; more precisely, if actually $\rho = 0$ (thus, X and Y independent), there would be only a 5% probability that the event $|r|\sqrt{N} < 1.96$ will occur. For a test of independence, also holding for small N, cf. Art. 3.13*a*.

Suppose we want to know the best linear combination $aX + b$ to Y in the sense that $E\{(Y - aX - b)^2\}$ assumes its minimum value. From Sec. 3.5, $a = \rho\sigma_2/\sigma_1$,

$$b = \nu_2 - a\nu_1$$

if μ_1, σ_1 and μ_2, σ_2 denote the mean and standard deviation of X and Y, respectively. However, the latter quantities are usually unknown. Therefore, one adopts the approximation $a^*X + b^*$ to Y which minimizes the sum

$$\sum_i (y_i - a^*x_i - b^*)^2$$

the so-called *method of least squares*. One has the explicit formulas

$$a^* = rs_2/s_1 = m_{11}/s_1^2 \qquad b^* = \bar{y} - a^*\bar{x}$$

Here, the estimate a^* of a is (a priori) itself a random variable. However, we have the useful inequality

$$|a - a^*| < 2s_2/s_1\sqrt{(1 - r^2)/(N - 2)} \tag{3.20}$$

holding with a 95% probability, provided that $N > 20$ and that X, Y have a normal joint distribution.

3.8. THE CENTRAL-LIMIT THEOREM

The following result is an example chosen from a collection of theorems, usually referred to as *the* central-limit theorem. Let X be a random variable with mean μ and variance σ^2. Furthermore, let $X_1, \ldots, X_n$ be independent random variables with the same d.f. as X, and let $S_n = X_1 + \cdots + X_n$. Then, for n sufficiently large, in good approximation,

$$\text{Prob } (S_n \leq s) \approx \Phi\left(\frac{s - n\mu}{\sigma\sqrt{n}}\right) \tag{3.21}$$

As an application, consider n independent trials, each with a probability p of success; let X_i denote the number of successes in the ith trial, $X_i = 0, 1$; Prob $(X_i = 1) = p$; and let $S_n = X_1 + \cdots + X_n$ denote the total number of successes in the first n trials. All the X_i have the same d.f., a mean p, and a variance $p(1 - p)$ (cf. Sec. 3.4). Consequently, for n sufficiently large, in good approximation,

$$\text{Prob}\,(S_n \leq s) \approx \Phi(a) \quad \text{where} \quad a = \frac{s - pn + \frac{1}{2}}{\sqrt{p(1-p)n}} \tag{3.22}$$

(the extra term ½ will result in a somewhat better approximation if s is an integer). For $n = 400$, $p = \frac{1}{2}$, $s = 223$ we find $a = 2.35$, $\Phi(a) = 0.9906$. Hence, the probability that in 400 tossings with a fair coin the number of heads will not exceed 223 is equal to 99%.

In many cases, much stronger assertions can be made. In fact, there exists an *absolute* constant c, such that the following is true. Let $X_1, \ldots, X_n$ be independent random variables, not necessarily having the same d.f., and let μ_i and σ_i denote the mean and the standard deviation of X_i $(i = 1, \ldots, n)$. Finally, let

$$S_n = (X_1 - \mu_1) + \cdots + (X_n - \mu_n)$$

Then

$$\left| \text{Prob}\,(S_n \leq s) - \Phi\left(\frac{s}{\sigma}\right) \right| < c\sigma^{-3} \sum_{i=1}^{n} E\{|X_i - \mu_i|^3\} \tag{3.23}$$

Here, σ denotes the standard deviation of S_n, more precisely, $\sigma^2 = \sigma_1^2 + \cdots + \sigma_n^2$. If we assume that, moreover, each X_i can deviate at most an amount ϵ from its mean μ_i, then the right-hand side of the above inequality is at most equal to $cn\left(\frac{\epsilon}{\sigma}\right)^3$. Hence, if ϵ^3 is small with respect to $\frac{\sigma^3}{n}$, the sum S_n is approximately normally distributed. Roughly speaking, the sum of a large number of independent small disturbances tends to be normally distributed. This explains why so many random variables occurring in nature seem to be normally distributed, e.g., the error in the outcome of a physical experiment, the position of a particle describing a Brownian motion.

3.9. STOCHASTIC PROCESSES

Any collection $X_t = X(t)$ of random variables, t running through a set T of real numbers, will be called a stochastic process. We may think of X_t as the position at time t of a particle describing some random motion along a straight line. An important example is the so-called *Markov chain:* a sequence X_t $(t = 0, 1, \ldots)$ of integer-valued random variables, such that

$$\text{Prob}\,(X_{t_0} = c_0 | X_{t_1} = c_1, \ldots, X_{t_n} = c_n) = \text{Prob}\,(X_{t_0} = c_0 | X_{t_1} = c_1) \tag{3.24}$$

whenever $t_0 > t_1 > \cdots > t_n \geq 0$. In heuristic terms: If the present position X_{t_1} is known, the future $(t > t_1)$ is independent of the past $(t < t_1)$.

As a simple application, consider a so-called Markov chain of independent increments $Y_t = X_t - X_{t-1}$, where $Y_1, Y_2, \ldots$ are independent random variables. Suppose, moreover, that $X_0 = 0$, $Y_t = \pm 1$, Prob $(Y_t = 1) = {}^{18}\!/_{37}$ $(t = 1, 2, \ldots)$. Then X_t can be interpreted as the gain after t trials for a gambler playing roulette by betting each time one dollar on red. Assuming that the bank's capital is infinite, the number N of trials leading to the gambler's ruin is on the average equal to $E\{N\} = 37a$, a denoting the player's initial capital. For further details, see [1], chap. 14.

3.10. SPECTRAL ANALYSIS OF A STATIONARY PROCESS

In this section, $X_t = X(t)$ denotes a so-called stationary stochastic process, that is, a process for which

$$E\{X_{t+\tau}X_t\} = \rho(\tau) \tag{3.25}$$

is independent of t, t running through all real numbers. In order to simplify the exposition, let us assume that $E\{X_t\} = 0$ for all t [thus, $\rho(\tau)$ denotes the covariance of X_t and $X_{t+\tau}$], that $\rho(\tau)$ is continuous and $\int_{-\infty}^{\infty} |\rho(\tau)|\, d\tau < \infty$. Then

$$g(\zeta) = \frac{1}{2\pi} \int_{-\infty}^{\infty} e^{-i\zeta\tau} \rho(\tau)\, d\tau \tag{3.26}$$

defines the so-called *spectral density* of our stationary process; $g(\zeta)$ is an even, nonnegative, and continuous function. Moreover, $X(t)$ admits the representation

$$X(t) = \int_{-\infty}^{\infty} \sqrt{g(\zeta)}\, e^{i\zeta t}\, dY(\zeta) \qquad -\infty < t < +\infty \tag{3.27}$$

(the integral existing in a certain sense; see [2], p. 429). Here, $Y(\zeta)$ denotes an essentially unique complex-valued stochastic process satisfying:

(1) $E\{Y(\zeta)\} = 0$
(2) if $\zeta_1 < \zeta_2 \leq \zeta_3 < \zeta_4$, $U = Y(\zeta_2) - Y(\zeta_1)$, $V = Y(\zeta_4) - Y(\zeta_3)$, then $E\{U\bar{V}\} = 0$, $E\{|U|^2\} = \zeta_2 - \zeta_1$

Roughly speaking, this amounts to a decomposition of $X(t)$ into a sum of simple vibrations $e^{i\zeta t}\, dA(\zeta)$, where the complex amplitude $dA(\zeta) = \sqrt{g(\zeta)}\, dY(\zeta)$ has zero mean and a variance $g(\zeta)\, d\zeta$ while different amplitudes are uncorrelated random variables.

Consider a linear filter transforming the simple input vibration $e^{i\zeta t}$ into an output $C(\zeta)e^{i\zeta t}$, $C(\zeta)$ denoting a complex constant. Then the input $X(t)$ will result in the output

$$Z(t) = \int_{-\infty}^{\infty} \sqrt{g(\zeta)}\, C(\zeta) e^{i\zeta t}\, dY(\zeta) \tag{3.28}$$

The latter integral has meaning if

$$\int_{-\infty}^{\infty} g(\zeta)|C(\zeta)|^2\, d\zeta < \infty$$

For the special case $C(\zeta) = i\zeta$, the process $Z(t)$ becomes the so-called mean-square derivative of $X(t)$:

$$\lim_{h \to 0} E\left\{\left(\frac{X_{t+h} - X_t}{h} - Z(t)\right)^2\right\} = 0 \tag{3.29}$$

In fact, $X(t)$ has a kth mean-square derivative if and only if the covariance function $\rho(\tau)$ possesses a continuous $2k$th derivative.

Suppose that X_t is also gaussian; that is, each finite set of X_ts has a normal joint distribution. Then, for each continuous function $g(u_0, u_1, \ldots, u_k)$ and each choice of the real constants $a_0, \ldots, a_k$,

$$\lim_{T\to\infty} (2T)^{-1} \int_{-T}^{T} g(X_{a_0+t}, \ldots, X_{a_k+t})\, dt = E\{g(X_{a_0}, \ldots, X_{a_k})\} \tag{3.30}$$

(with probability 1, in a certain technical sense), assuming that the latter expectation value exists. Especially,

$$\lim_{T\to\infty} \frac{1}{2T} \int_{-T}^{T} X(t+\tau)X(t)\,dt = \rho(\tau) \tag{3.31}$$

a formula which is sometimes taken as the definition of $\rho(\tau)$.

3.11. STATISTICAL METHODS

Consider a random experiment $\mathcal{E}$ resulting in the real-valued outcomes $X^{(1)}, \ldots, X^{(h)}$ which, in the corresponding probabilistic model, are treated as random variables. Further, let $F = F(c_1, \ldots, c_h)$ denote the joint d.f. of $X^{(1)}, \ldots, X^{(h)}$; in other words, $F(c_1, \ldots, c_h)$ denotes the a priori probability that the outcomes of $\mathcal{E}$ will be such that, simultaneously, $X^{(1)} \leq c_1, \ldots, X^{(h)} \leq c_h$. Finally, in n independent repetitions of the experiment $\mathcal{E}$ (under identical circumstances), let $x_i^{(1)}, \ldots, x_i^{(h)}$ denote the actually observed outcomes of the ith experiment, a so-called *observation* on the h-dimensional random variable $(X^{(1)}, \ldots, X^{(h)})$.

Suppose that the actual d.f. F is unknown, as is true in most cases. Suppose, further, that we are willing to assume that F belongs to an explicitly given class H of a priori possible joint d.f.'s for $X^{(1)}, \ldots, X^{(h)}$. Finally, let H_0 be a given subclass of H.

Question: How often shall we repeat the experiment $\mathcal{E}$, and how shall we use the resulting observations $x_i^{(r)}$ $(i = 1, 2, \ldots; r = 1, \ldots, h)$ in deciding whether or not F belongs to H_0? In general, the answer to this question will also depend on the possible "damage" caused by a wrong decision, the cost of performing the experiment $\mathcal{E}$ and the cost of measuring the $x_i^{(r)}$, etc.

Without further justification we shall now list a number of important procedures for answering certain predetermined questions of this type.

Let $Z_0, \ldots, Z_n$ be independent random variables, each normal (0,1). Then the random variable $\chi^2[n] = Z_1^2 + \cdots + Z_n^2$ has a so-called χ^2 distribution with n degrees of freedom. Furthermore, the random variable $t[n] = Z_0(\chi^2[n]/n)^{-1/2}$ has a so-called t distribution with n degrees of freedom. Finally, Z_0 itself has the d.f. $\Phi(z)$. In the sequel, z_α, $t_\alpha[n]$, $\chi_\alpha^2[n]$ will denote positive numbers, such that

$$\text{Prob}\,(|Z_0| > z_\alpha) = \text{Prob}\,(|t[n]| > t_\alpha[n]) = \text{Prob}\,(\chi^2[n] > \chi_\alpha^2[n]) = \alpha$$

Here, α usually is taken as one of the values 0.01, 0.05, 0.10. Tables of the functions z_α, $t_\alpha[n]$, and $\chi_\alpha^2[n]$ can be found in any book on mathematical statistics. For example, we have $z_{0.95} = 0.0627$, $z_{0.10} = 1.6649$, $z_{0.05} = 1.9600$. Furthermore, for n large,

$$t_\alpha[n] \approx z_\alpha \qquad \text{and} \qquad \chi_\alpha^2[n] \approx (z_{2\alpha} + \sqrt{2n-1})^2/2$$

Also $t_{0.10}[10] = 1.812$, $t_{0.05}[10] = 2.228$, $\chi^2_{0.95}[20] = 10.851$, $\chi^2_{0.05}[20] = 31.410$.

3.12. TESTING WHETHER TWO RANDOM VARIABLES HAVE THE SAME d.f.

a. The Normal Case

Let X and Y be random variables and suppose that we are willing to assume: (1) that X and Y are independent; (2) that there exists *unknown* numbers μ_1, μ_2, and σ, such that X is normal (μ_1,σ) and Y is normal (μ_2,σ). *Problem:* To decide whether or not $\mu_1 = \mu_2$. Here, the following procedure is available (where 0.05 can be replaced by any fixed $\alpha < 1$).

Take a set of n_1 observations, $x_1, \ldots, x_{n_1}$ on X (a so-called sample of size n_1 on X), and compute the sample mean $\bar{x}$ and the sample variance s_1^2 from

$$n_1\bar{x} = x_1 + \cdots + x_{n_1} \qquad n_1 s_1^2 = (x_1 - \bar{x})^2 + \cdots + (x_{n_1} - \bar{x})^2$$

Similarly, take a sample $y_1, \ldots, y_{n_2}$ on Y, and compute the corresponding numbers $\bar{y}$ and s_2^2. Furthermore, compute Δ from

$$\Delta = \frac{(n_1 s_1^2 + n_2 s_2^2)(n_1 + n_2)}{n_1 n_2 (n_1 + n_2 - 2)}$$

Finally, let $t = t_{0.05}[n_1 + n_2 - 1]$ (for example, $t = 2.23$ if $n_1 = n_2 = 6$). *Rule:* If and only if

$$|\bar{x} - \bar{y}| > t\sqrt{\Delta} \tag{3.32}$$

we shall conclude that $\mu_1 \neq \mu_2$. *Justification:* If actually μ_1 would be equal to μ_2, then this procedure is such that the probability of arriving at the (wrong) decision that $\mu_1 \neq \mu_2$ is precisely equal to 0.05.

Example. The breaking force of a certain material in six trials was 172, 128, 126, 57, 242, 301. A second material yielded in six trials the breaking forces 298, 262, 120, 335, 205, 308. We have $\bar{x} = 171$, $\bar{y} = 255$, $n_1 s_1^2 = 38{,}812$, $n_2 s_2^2 = 31{,}052$, $\sqrt{\Delta} = 48.3$, $t\sqrt{\Delta} = 107.6$. But $|\bar{x} - \bar{y}| = 84 < 107$; thus, at the 5% level, we cannot yet conclude that the two materials have different average breaking forces (each breaking force assumed to be normally distributed).

Given the above mentioned assumptions (1) and (2), this so-called t test is slightly better (in the probability of concluding that the μ_i are equal when they are actually different) than the following so-called Wilcoxon test, which, however, is valid under much weaker assumptions.

b. The Continuous Case

Here, we assume only that X and Y are independent, each having a continuous d.f. *Problem:* To decide whether or not X and Y have the same d.f. [that is, $F_X(c) = F_Y(c)$ for all c]. Take a sample $x_1, \ldots, x_{n_1}$ on X and a sample $y_1, \ldots, y_{n_2}$ on Y. Furthermore, let U denote the number of pairs (x_i,y_j) for which $y_j < x_i$ (thus, a priori, U can asume any of the values $0, 1, \ldots, n_1n_2$). *Rule:* If and only if

$$|U - \tfrac{1}{2}n_1n_2| \geq z_\alpha \sqrt{\tfrac{1}{12}n_1n_2(n_1 + n_2 - 1)} \tag{3.33}$$

we shall conclude that the d.f.'s of X and Y are different. In case they are actually the same, the probability of deciding that they are different is equal to α (at least in good approximation for n_1 and n_2 not too small).

c. The Discrete Case

Assume: (1) that X and Y are independent random variables; (2) that both X and Y can take only the finitely many values $a_1, \ldots, a_r$. *Problem:* To decide whether or not Prob $(X = a_j) =$ Prob $(Y = a_j)$ for $j = 1, \ldots, r$. Take a sample $x_1, \ldots, x_{n_1}$ on X, a sample $y_1, \ldots, y_{n_2}$ on Y. Let k_j (or m_j) denote the number of x_i (or y_i, respectively) equal to a_j $(j = 1, \ldots, r)$. Furthermore, let

$$U = n_1n_2 \sum_{j=1}^{r} (k_j + m_j)^{-1} \left(\frac{k_j}{n_1} - \frac{m_j}{n_2}\right)^2$$

and $t = \chi_\alpha^2[r - 1]$ (for $r = 2$ we have $t = z_\alpha^2$). *Rule:* If and only if $U > t$, we shall decide that X and Y have a different distribution. Assuming that they are actually the same, the probability of arriving at the conclusion that they are different is equal to α (at least in good approximation for n_1, n_2 large).

Letting $n_2 \to \infty$, $m_j/n_2 \to p_j$, we obtain a useful test for deciding whether or not

Prob $(X = a_j) = p_j$ for $j = 1, \ldots, r$, the p_j denoting given numbers. Here, the decision depends on whether or not we have

$$\frac{1}{n_1} \sum_{j=1}^{r} \frac{(k_j - n_1 p_j)^2}{p_j} > \chi_\alpha{}^2[r - 1] \tag{3.34}$$

3.13. TESTING WHETHER TWO RANDOM VARIABLES ARE INDEPENDENT

a. The Normal Case

Assume that X and Y are random variables having a normal joint distribution. *Problem:* To determine whether or not X and Y are independent (that is, cf. Sec. 3.6, whether or not X and Y are uncorrelated). Take a sample of n observations (x_i, y_i) on (X, Y), and compute the corresponding sample correlation coefficient r, as indicated in Sec. 3.7. Furthermore, let $t = t_\alpha[n - 2]$. If and only if $|r| \geq (n - 2 + t)^{-1/2} t$, we shall conclude that X and Y are dependent. If they are actually independent, then the probability of arriving at the (wrong) decision that they are dependent is precisely equal to α. *Example:* Let $n = 20$, $\alpha = 0.05$; then $t = t_{0.05}[18] = 2.101$. If and only if the numerical value of the sample correlation coefficient exceeds 0.444, one may conclude, at the 5% level, that X and Y are dependent.

b. The Continuous Case

Assume that X and Y have an (unknown) continuous joint d.f. $F_{X,Y}(x,y)$. *Problem:* To decide whether or not X and Y are independent. Take n pairs of observations (x_i, y_i) $(i = 1, \ldots, n)$ on (X, Y). Let S denote the number of pairs x_i, y_j with $i < j$ and $(x_j - x_i)(y_j - y_i) < 0$ (thus, a priori, S can take any of the values $0, 1, \ldots, n(n - 1)/2$). If and only if the inequality

$$|S - \tfrac{1}{4}n(n - 1)| \geq \tfrac{1}{2} + z_\alpha \sqrt{\tfrac{1}{72}n(n - 1)(2n + 5)} \tag{3.35}$$

holds true, we shall conclude that X and Y are dependent. If X and Y are actually independent, the probability of concluding that they are dependent is equal to α (at least in good approximation for n sufficiently large).

c. The Discrete Case

Let X and Y be random variables, such that X takes only the values $a_1, \ldots, a_r$ and Y takes only the values $b_1, \ldots, b_s$. *Problem:* To determine whether or not X and Y are dependent. Take n pairs of observations (x_k, y_k) on (X, Y). Let $\nu_{ij} = np_{ij}$ denote the number of pairs (x_k, y_k) with $x_k = a_i$ and $y_k = b_j$. Furthermore, let np_i denote the number of pairs (x_k, y_k) with $x_k = a_i$ and, similarly, nq_j the number of pairs with $y_k = b_j$. Finally, let

$$U = n \sum_{i=1}^{r} \sum_{j=1}^{s} \frac{(p_{ij} - p_i q_j)^2}{p_i q_j} \tag{3.36}$$

and let $t = \chi_\alpha{}^2[(r - 1)(s - 1)]$. If and only if $U > t$, we shall conclude that X and Y are dependent. If they are actually independent then, for n sufficiently large, the probability of deciding that they are dependent is approximately equal to α.

3.14. REFERENCES

[1] W. Feller: "An Introduction to Probability Theory and Its Applications," vol. 1, 2d ed., Wiley, New York, 1957.
[2] J. L. Doob: "Stochastic Processes," Wiley, New York, 1953.
[3] H. Cramér: "Mathematical Methods of Statistics," Princeton Univ. Press, Princeton, N.J., 1946.

CHAPTER 4

GEOMETRY

BY

E. A. RIPPERGER, Ph.D., Austin, Tex.

4.1. COORDINATE SYSTEMS

Cartesian Coordinates

In a two-dimensional cartesian coordinate system (Fig. 4.1), a point P is located by its coordinates x, y. The slope $m_1 = dy/dx$ of the straight line P_1P_2 is given by

$$m_1 = \frac{y_2 - y_1}{x_2 - x_1} \tag{4.1}$$

Lines perpendicular to P_1P_2 have a slope $m_2 = -1/m_1$. The length of the line P_1P_2 is

$$l = \sqrt{(y_2 - y_1)^2 + (x_2 - x_1)^2} \tag{4.2}$$

If P_1P_2 is to be divided in the ratio of r_1/r_2, the coordinates of the point of division P_0 are

$$x_0 = \frac{r_2x_1 + r_1x_2}{r_1 + r_2} \qquad y_0 = \frac{r_2y_1 + r_1y_2}{r_1 + r_2}$$

Equations of a Straight Line in Cartesian Coordinates

Slope form $\quad y = mx + b$

Point-slope form $\quad y - y_1 = m(x - x_1)$

Intercept form $\quad x/a + y/b = 1$

Two-point form $\quad \dfrac{y - y_1}{x - x_1} = \dfrac{y_2 - y_1}{x_2 - x_1}$

Normal form

$$x \cos \beta + y \sin \beta - p = 0 \tag{4.3}$$

p is the perpendicular from the origin to the line, and β is the angle which p makes with the x axis.

General form

$$Ax + By + C = 0 \tag{4.4}$$

The Intersection of Two Straight Lines. Let the equations of two lines be

$$A_1x + B_1y + C_1 = 0 \qquad A_2x + B_2y + C_2 = 0$$

Then

$$x_1 = \frac{B_2C_1 - B_1C_2}{A_2B_1 - A_1B_2} \qquad y_1 = \frac{A_1C_2 - A_2C_1}{A_2B_1 - A_1B_2} \tag{4.5}$$

are the coordinates of the point of intersection if the two lines intersect. If $A_1/A_2 \neq B_1/B_2$ the lines intersect in one point; if $A_1/A_2 = B_1/B_2 \neq C_1/C_2$, the lines are parallel; if $A_1/A_2 = B_1/B_2 = C_1/C_2$, the lines coincide.

The distance from a line to a point is (Fig. 4.2)

$$d = \frac{A_2x_0 + B_2y_0 + C_2}{\pm \sqrt{A_2^2 + B_2^2}} \tag{4.6}$$

The sign of the denominator is opposite to the sign of C_2.

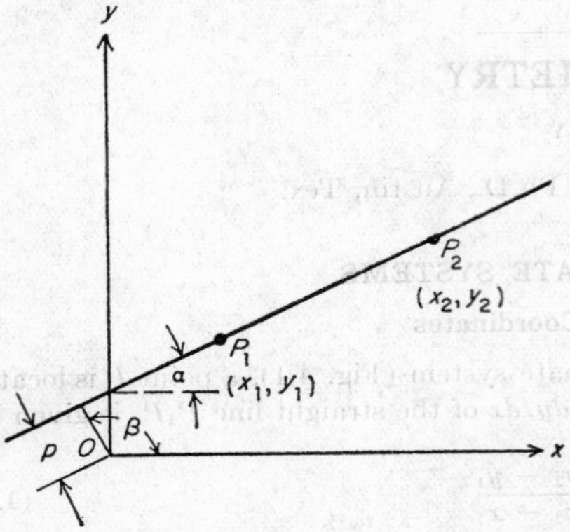

FIG. 4.1

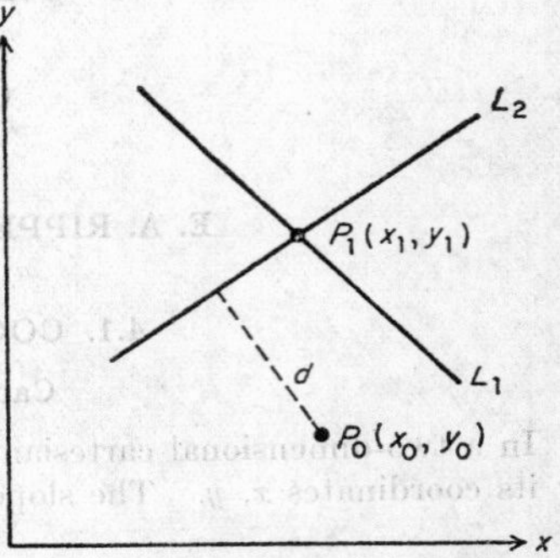

FIG. 4.2

Transformation of Coordinates

For a rotation of axes

$$x' = x \cos \alpha + y \sin \alpha \qquad y' = -x \sin \alpha + y \cos \alpha \tag{4.7}$$

$$x = x' \cos \alpha - y' \sin \alpha \qquad y = x' \sin \alpha + y' \cos \alpha \tag{4.8}$$

Simplification of a Second-degree Equation by Rotation. Consider the equation

$$Ax^2 + 2Bxy + Cy^2 + Dx + Ey + F = 0 \tag{4.9}$$

If the axes are rotated so that

$$\cot 2\alpha = \frac{A - C}{2B}$$

the equation becomes

$$A'x'^2 + C'y'^2 + D'x' + E'y' + F' = 0$$

This is the transformation of a conic on principal axes. $AC - B^2$ is invariant with respect to a rotation. The degree of an equation is invariant with respect to a transformation by rotation or translation.

Oblique Coordinates

The x coordinate of a point is defined as the distance measured from Oy to the point, parallel to the x axis. A similar definition is given for the y coordinate. The equations for transformation from rectangular to oblique axes as shown in Fig. 4.4 are

$$x = x' - y' \cot \beta \qquad y = y' \csc \beta \tag{4.10}$$

Since these transformation equations are all of the first degree, the transformation of an equation from rectangular to oblique coordinates cannot raise or lower its degree.

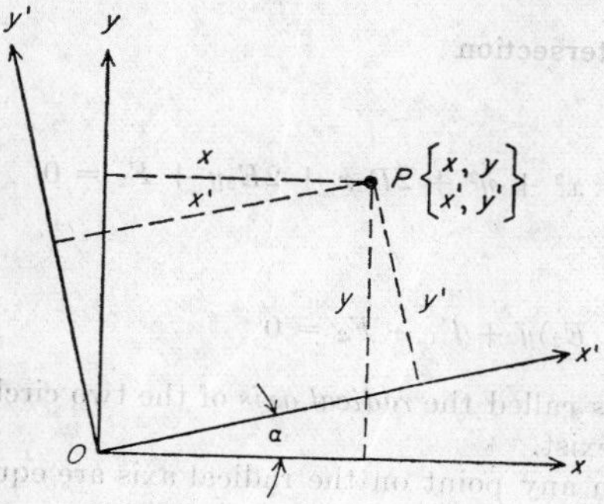

FIG. 4.3

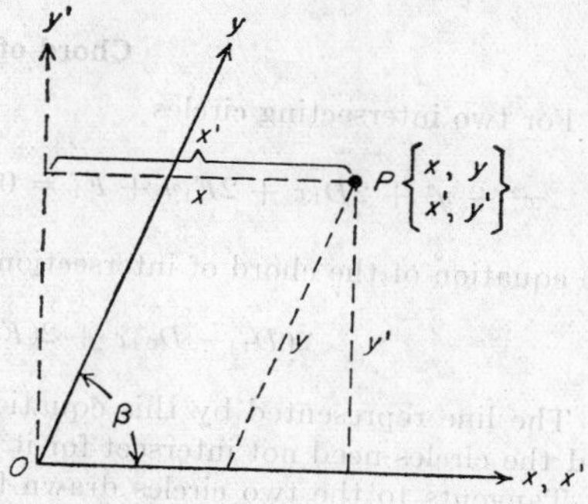

FIG. 4.4

Polar Coordinates

The polar coordinates of a point P are r, θ. Point O is called the pole, and Ox is called the polar axis. If the rectangular coordinates of the point are x, y, then

$$x = r\cos\theta \qquad y = r\sin\theta \tag{4.11}$$

$$\theta = \tan^{-1} y/x \qquad r = \sqrt{x^2 + y^2} \tag{4.12}$$

The distance between two points (r_1,θ_1) and (r_2,θ_2) is given by

$$d = \sqrt{r_1{}^2 + r_2{}^2 - 2r_1r_2\cos(\theta_1 - \theta_2)} \tag{4.13}$$

Equations in Polar Coordinates. Straight Line (Fig. 4.1)

$$r\sin(\theta - \alpha) = p \tag{4.14}$$

Spirals

Logarithmic	$r = a^\theta$	
Archimedes	$r = a\theta$	(4.15a–c)
Hyperbolic	$r = a/\theta$	

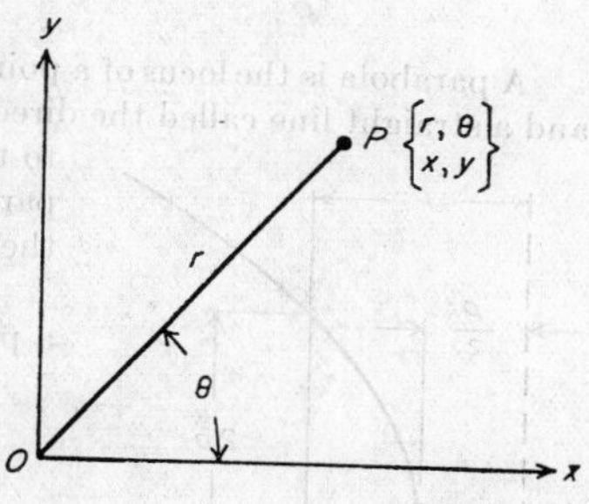

FIG. 4.5

4.2. CIRCLE

Equations of a Circle

Standard form	$(x - h)^2 + (y - k)^2 = a^2$	(4.16)
Normal form	$x^2 + y^2 + 2Dx + 2Ey + F = 0$	(4.17)
Parameter form	$x = h + a\cos\theta \qquad y = k + a\sin\theta$	(4.18)

Tangents

A line of slope m and tangent to a circle (4.16) has the equation

$$y - k = m(x - h) + a\sqrt{m^2 + 1} \tag{4.19}$$

The length t of a tangent, from a given point $P(x',y')$ outside a given circle to the circle, is

$$t^2 = (x' - h)^2 + (y' - k)^2 - a^2 \tag{4.20a}$$

In terms of the coefficients in the normal equation (4.17)

$$t^2 = x'^2 + y'^2 + 2Dx' + 2Ey' + F \tag{4.20b}$$

Chord of Intersection

For two intersecting circles

$$x^2 + y^2 + 2D_1x + 2E_1y + F_1 = 0 \qquad x^2 + y^2 + 2D_2x + 2E_2y + F_2 = 0$$

the equation of the chord of intersection is

$$2(D_1 - D_2)x + 2(E_1 - E_2)y + F_1 - F_2 = 0$$

The line represented by this equation is called the *radical axis* of the two circles, and the circles need not intersect for it to exist.

Tangents to the two circles drawn from any point on the radical axis are equal.

Orthogonal Circles

Two given circles intersect orthogonally if

$$2D_1D_2 + 2E_1E_2 = F_1F_2$$

4.3. PARABOLA

Equations

A parabola is the locus of a point equally distant from a fixed point, called the focus, and a straight line called the directrix. The line through the focus and perpendicular to the directrix is the principal axis. The vertex of the parabola is a point halfway between the directrix and the focus on the principal axis.

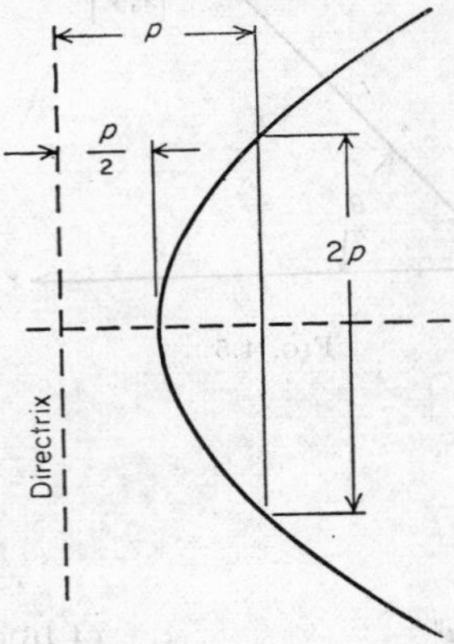

Fig. 4.6

The second-degree equation (4.9) is the equation of a parabola if

$$AC - B^2 = 0$$

If the principal axis is parallel to either the x or the y axis, the equation reduces to

$$Cy^2 + Ey + Dx + F = 0$$
$$Ax^2 + Dx + Ey + F = 0$$

For a parabola with the y axis as the principal axis,

$$x^2 = 2py \tag{4.21}$$

where p is the distance between the focus and the directrix. This is called the *standard form* of the equation for a parabola. For this form, the focus is at $(0,p/2)$, and the directrix is $y = -p/2$.

The polar equation with the pole at the focus and polar axis coinciding with the principal axis of the parabola is

$$r = \frac{p}{1 - \cos\theta} \tag{4.22}$$

If the pole is at the vertex, $r = 2p \cot\theta \csc\theta$.

A parabola with its principal axis parallel to the y axis can be represented in terms of a variable t as follows:

$$x = at \qquad y = bt + ct^2 \tag{4.23}$$

Tangents

The line $y = mx + p/2m$ is tangent to the parabola $y^2 = 2px$ for any value of m other than zero.

For the tangent which passes through a point $P'(x',y')$ on the parabola,

$$m = \frac{2y' \pm 2\sqrt{y'^2 - 2px'}}{4x'}$$

4.4. ELLIPSE

An ellipse is the locus of a point whose distance from a certain fixed point, the *focus*, is in a constant ratio $e < 1$ to its distance from a fixed straight line, the *directrix*. The ratio e is called the *eccentricity*.

Equations

The second-degree equation (4.9) represents the general equation of the ellipse if $AC - B^2 > 0$.

If $B > 0$, and A, C, and $(D^2/4A + E^2/4C - F)$ are of the same sign, the ellipse has its major and minor axes parallel to the coordinate axes.

The equation

$$\frac{x^2}{a^2} + \frac{y^2}{b^2} = 1 \tag{4.24}$$

represents an ellipse in the standard position with major and minor axes coincident with the x, y axes. The length of the major axis is $2a$, and that of the minor axis $2b$.

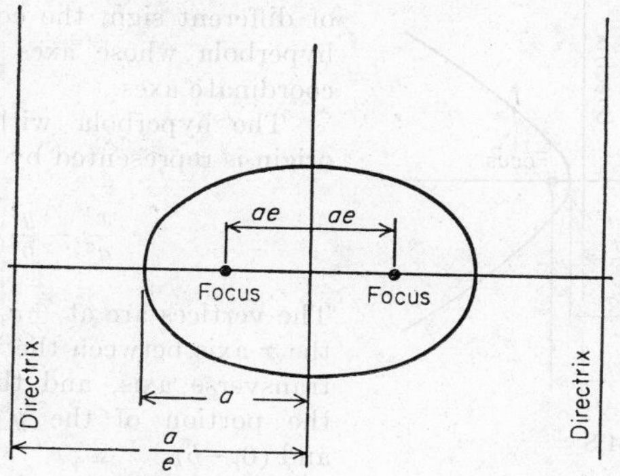

FIG. 4.7

The foci are located at $(\pm ae, 0)$ and the directrices are $x = \pm a/e$. In terms of the major and minor axes, the eccentricity is

$$e = \sqrt{1 - b^2/a^2}$$

The polar equation of the ellipse is

$$r = \frac{a(1 - e^2)}{1 - e \cos \theta} \tag{4.25}$$

with the pole at a focus and the polar axis coinciding with the principal axis through the focus. If the pole is at the center of the ellipse,

$$r = \frac{ab}{\sqrt{b^2 \cos^2 \theta + a^2 \sin^2 \theta}}$$

The ellipse may be represented in terms of a parameter θ by the equations

$$x = a \cos \theta \qquad y = b \sin \theta \tag{4.26}$$

Tangents

A tangent to a standard ellipse at the point $P(x_1, y_1)$ has the equation

$$b^2 x_1 x + a^2 y y_1 = a^2 b^2 \tag{4.27}$$

and the lines

$$y = mx \pm \sqrt{a^2 m^2 + b^2} \tag{4.28}$$

are tangent to the ellipse for any value of m.

Conjugate Diameters

If one diameter of an ellipse bisects the chords parallel to a second diameter, the second diameter bisects all the chords which are parallel to the first diameter. The two diameters are said to be *conjugate diameters.*

4.5. HYPERBOLA

Equations

The hyperbola is defined in the same way as the ellipse, but the eccentricity $e > 1$. The curve has two foci located at distances $\pm ae$ from the center and two directrices at distances $\pm a/e$ from the center. Equation (4.9) represents the hyperbola if $AC - B^2 < 0$. If $B = 0$ and A and C are of different sign, the equation represents a hyperbola whose axes are parallel to the coordinate axes.

The hyperbola with its center at the origin is represented by

$$\frac{x^2}{a^2} - \frac{y^2}{b^2} = 1 \tag{4.29}$$

The vertices are at $\pm a$, 0. The portion of the x axis between the vertices is called the transverse axis, and the conjugate axis is the portion of the y axis between $(0,b)$ and $(0,-b)$.

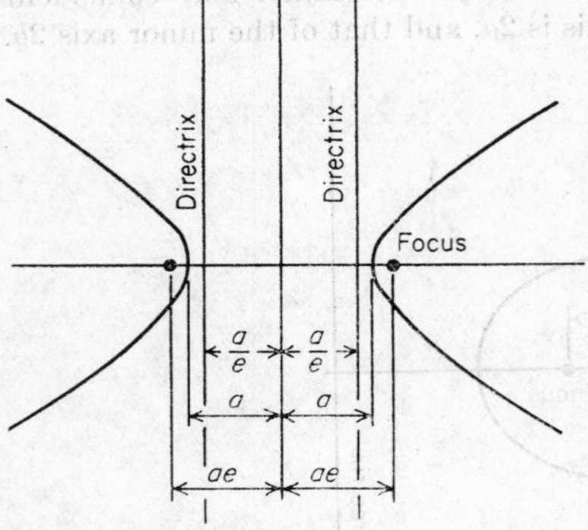

FIG. 4.8

In polar coordinates, the equation of a hyperbola is

$$r = \frac{a(e^2 - 1)}{1 - e \cos \theta} \tag{4.30}$$

In parametric form, the equations for the hyperbola are

$$x = a \operatorname{Cosh} u \qquad y = b \operatorname{Sinh} u \tag{4.31}$$

where u is A/ab. A is twice the area bounded by one branch of the hyperbola, the x axis, and a line drawn from the origin to a point $P(x,y)$ on the curve. The curve has the same relation to the hyperbolic functions as a circle does to trigonometric functions.

The hyperbola may also be represented by

$$x = a \sec v \qquad y = b \tan v \tag{4.32}$$

where v is an angle of no special geometric significance.

Tangents

The equation of the tangent at x_1y_1 is

$$b^2x_1x - a^2y_1y = a^2b^2 \tag{4.33}$$

The lines

$$y = mx \pm \sqrt{a^2m^2 - b^2} \tag{4.34}$$

are tangent to the hyperbola (4.29) for any value of m.

The slope of the tangent to the hyperbola from a given point $P(x',y')$ is

$$m = \frac{x'y' \pm \sqrt{a^2b^2 - (b^2x'^2 - a^2y'^2}}{x'^2 - a^2} \tag{4.35}$$

Rectangular Hyperbolas

A hyperbola is said to be rectangular or equilateral if the asymptotes are perpendicular to each other. Thus

$$x^2 - y^2 = a^2$$

represents a rectangular hyperbola. If the axes are rotated 45°, the equation becomes

$$x'y' = a^2/2$$

Conjugate Hyperbolas

Two hyperbolas are conjugate if the transverse axis of each is the conjugate axis of the other.

4.6. OTHER CURVES

The Catenary

This is the form assumed by a perfectly flexible chain of uniform density hanging from two supports not in the same vertical line (Fig. 4.9). The curve has the equation

$$y = a \operatorname{Cosh} \frac{x}{a} \tag{4.36}$$

For small x/a, a parabola is a close approximation of the catenary.

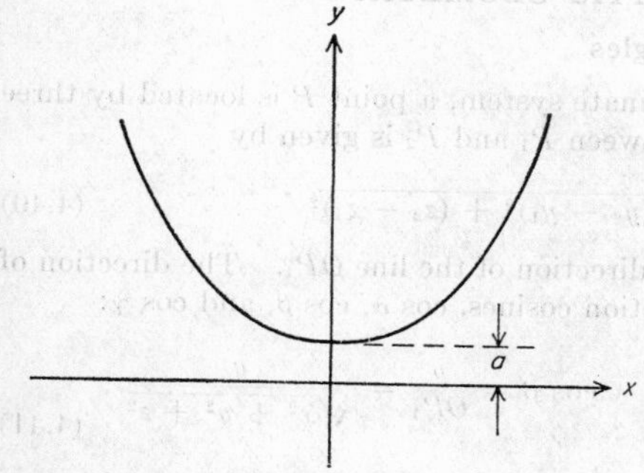

FIG. 4.9

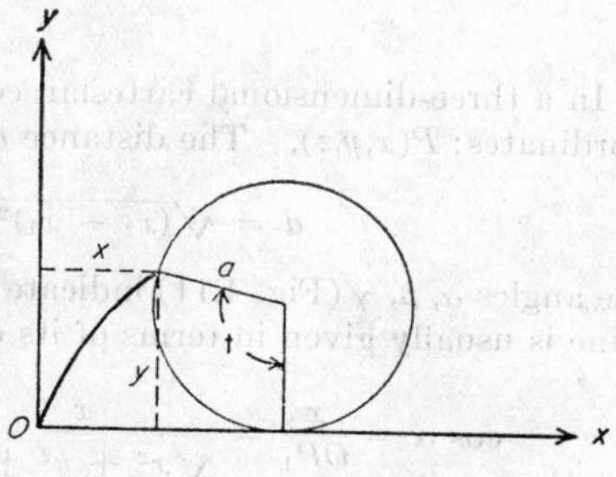

FIG. 4.10

The Cycloid

This is the path of a point of a circle rolling upon a fixed line (Fig. 4.10). The curve is most conveniently represented in the parametric form

$$x = a(t - \sin t) \qquad y = a(1 - \cos t) \tag{4.37}$$

where a is the radius of the generating circle and t is the angle through which a radius of the circle turns in generating a segment of the cycloid.

4.7. CURVILINEAR ORTHOGONAL COORDINATES

A point can be located by the intersection of two curves

$$f_1(x,y) = \alpha \qquad f_2(x,y) = \beta$$

when these curves are suitably chosen. Different values of α and β give different curves and different points of intersection. Hence, each point in the x,y plane will be characterized by a definite pair of values α and β which make the two curves pass through the point. Thus, α and β may be regarded as the coordinates of the point, and, since they locate the point with two intersecting curves, they are called *curvilinear* coordinates. The most convenient systems of curvilinear coordinates are those which cut each other at right angles everywhere. The condition for orthogonality is

$$\frac{\partial \alpha}{\partial x}\frac{\partial \beta}{\partial x} + \frac{\partial \alpha}{\partial y}\frac{\partial \beta}{\partial y} = 0 \tag{4.38}$$

Functions of a complex variable can be used for conveniently constructing systems of orthogonal coordinates.

Let $f(x + iy) = f_1(x,y) + if_2(x,y)$

If these functions satisfy the Cauchy-Riemann equations (9.14), then

$$f_1(x,y) = \alpha \qquad f_2(x,y) = \beta \quad \text{are orthogonal}$$

As an example, a system of elliptical coordinates can be constructed by taking

$$\begin{aligned} x + iy &= C \operatorname{Cosh} (\alpha + i\beta) \\ &= C \cos \beta \operatorname{Cosh} \alpha + iC \sin \beta \operatorname{Sinh} \alpha \end{aligned}$$

Thus

$$x = C \cos \beta \operatorname{Cosh} \alpha \qquad y = C \sin \beta \operatorname{Sinh} \alpha \tag{4.39}$$

These are the parametric equations of ellipses or of hyperbolas, depending on whether α or β is constant. By taking different values of α and β in these equations, an orthogonal set of ellipses and hyperbolas is obtained.

4.8. SOLID ANALYTIC GEOMETRY

Angles

In a three-dimensional cartesian coordinate system, a point P is located by three coordinates: $P(x,y,z)$. The distance d between P_1 and P_2 is given by

$$d = \sqrt{(x_2 - x_1)^2 + (y_2 - y_1)^2 + (z_2 - z_1)^2} \tag{4.40}$$

The angles α, β, γ (Fig. 4.11) indicate the direction of the line OP_1. The direction of a line is usually given in terms of its direction cosines, $\cos \alpha$, $\cos \beta$, and $\cos \gamma$:

$$\cos \alpha = \frac{x}{OP_1} = \frac{x}{\sqrt{x^2 + y^2 + z^2}} \qquad \cos \beta = \frac{y}{OP_1} = \frac{y}{\sqrt{x^2 + y^2 + z^2}}$$
$$\cos \gamma = \frac{z}{OP_1} = \frac{z}{\sqrt{x^2 + y^2 + z^2}} \tag{4.41}$$

$$\cos^2 \alpha + \cos^2 \beta + \cos^2 \gamma = 1 \tag{4.42}$$

The angle ω between any two lines having direction angles α_1, β_1, γ_1 and α_2, β_2, γ_2 is given by

$$\cos \omega = \cos \alpha_1 \cos \alpha_2 + \cos \beta_1 \cos \beta_2 + \cos \gamma_1 \cos \gamma_2 \tag{4.43}$$

Surface of Revolution

A surface of revolution is generated by the revolution of a plane curve about an axis that lies in the plane of the curve. In general, if the axis of revolution is the z

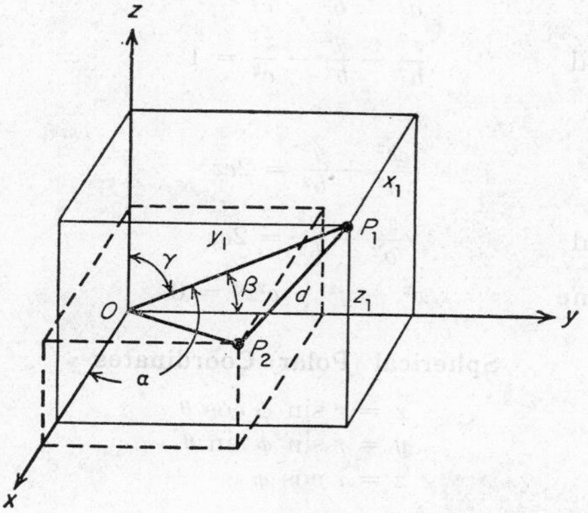

FIG. 4.11

axis and the generating curve is in the (x,z) plane, the equation of the surface is obtained by replacing x in the equation of the generating curve with $\sqrt{x^2 + y^2}$.

Plane

The general equation of a plane is

$$Ax + By + Cz + D = 0 \tag{4.44}$$

When this equation is written

$$x \cos \alpha + y \cos \beta + z \cos \gamma - p = 0 \tag{4.45}$$

it is called the *normal form,* and $\cos \alpha$, $\cos \beta$, $\cos \gamma$ are the direction cosines of the normal from the origin to the plane.

$$\frac{x}{a} + \frac{y}{b} + \frac{z}{c} = 1$$

is called the intercept form of the equation of a plane.

The distance to the plane from a point $P(x_1,y_1,z_1)$ is

$$d = \frac{Ax_1 + By_1 + Cz_1 + D}{\pm\sqrt{A^2 + B^2 + C^2}} \tag{4.46}$$

with sign of the radical opposite to that of D if the distance is taken as positive when measured in the same direction as the normal to the plane from the origin.

Surface

The surface defined by an equation of the second degree in x, y, and z is called a quadric surface.

Ellipsoid

$$\frac{x^2}{a^2} + \frac{y^2}{b^2} + \frac{z^2}{c^2} = 1 \tag{4.47}$$

If $a \neq b$, but $b = c$, the figure is called a spheroid. If $a < b$, it is an oblate spheroid, and, if $a > b$, it is a prolate spheroid.

Hyperboloids

One-sheet hyperboloid $$\frac{x^2}{a^2} + \frac{y^2}{b^2} - \frac{z^2}{c^2} = 1 \tag{4.48}$$

Two-sheet hyperboloid $$\frac{x^2}{a^2} - \frac{y^2}{b^2} - \frac{z^2}{c^2} = 1 \tag{4.49}$$

Paraboloids

Elliptic paraboloid $$\frac{x^2}{a^2} + \frac{y^2}{b^2} = 2cz \tag{4.50}$$

Hyperbolic paraboloid $$\frac{x^2}{a^2} - \frac{y^2}{b^2} = 2cz \tag{4.51}$$

Right Circular Cone $$x^2 + y^2 - c^2z^2 = 0 \tag{4.52}$$

Spherical (Polar) Coordinates

$$\begin{aligned} x &= r \sin \phi \cos \theta \\ y &= r \sin \phi \sin \theta \\ z &= r \cos \phi \end{aligned} \tag{4.53}$$

Cylindrical Coordinates

The point P in Fig. 4.12 can also be determined by r', θ, z, the cylindrical coordinates of the point.[1]

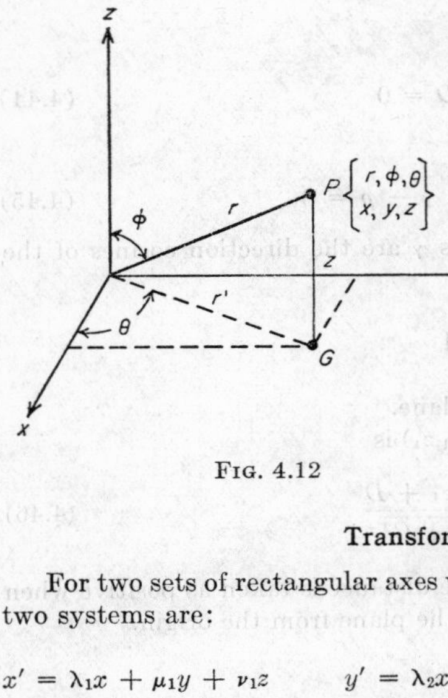

FIG. 4.12

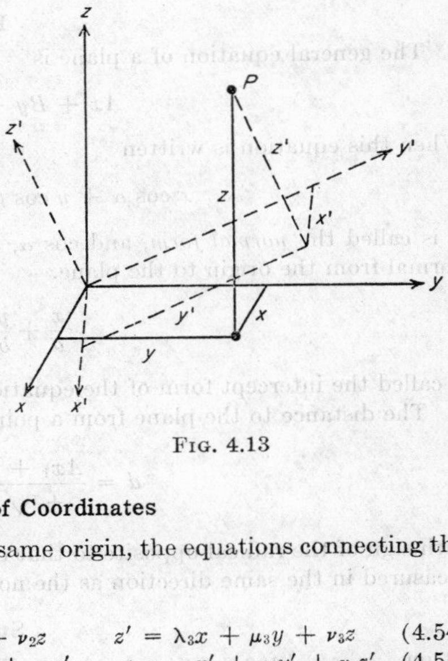

FIG. 4.13

Transformation of Coordinates

For two sets of rectangular axes with the same origin, the equations connecting the two systems are:

$$x' = \lambda_1 x + \mu_1 y + \nu_1 z \qquad y' = \lambda_2 x + \mu_2 y + \nu_2 z \qquad z' = \lambda_3 x + \mu_3 y + \nu_3 z \tag{4.54}$$
$$x = \lambda_1 x' + \lambda_2 y' + \lambda_3 z' \qquad y = \mu_1 x' + \mu_2 y' + \mu_3 z' \qquad z = \nu_1 x' + \nu_2 y' + \nu_3 z' \tag{4.55}$$

where $(\lambda_1,\lambda_2,\lambda_3)$, (μ_1,μ_2,μ_3), and (ν_1,ν_2,ν_3) are the direction cosines of Ox, Oy, and Oz with respect to Ox', Oy', and Oz'.

[1] Where no mistake is possible, r will be used instead of r' in this handbook.

4.9. DIFFERENTIAL GEOMETRY

Plane Curves—Curvature

The rate of change of direction of a curve at a given point is called the curvature κ at the point.

In rectangular coordinates,

$$\kappa = \frac{d^2y/dx^2}{[1 + (dy/dx)^2]^{3/2}} \tag{4.56}$$

In polar coordinates,

$$\kappa = \frac{r^2 + 2r'^2 - rr''}{(r^2 + r'^2)^{3/2}} \tag{4.57}$$

where r' and r'' are the first and second derivatives with respect to θ.

The *radius of curvature* ρ, at a given point on a curve, is the reciprocal of curvature at that point: $\rho = 1/\kappa$.

Circle of Curvature or Osculating Circle

If a circle be drawn through three neighboring points P_0, P_1, and P_2 on a curve, and if P_1 and P_2 be made to approach P_0 along the curve as a limiting position, the limiting circle is called the osculating circle and is identical with the circle of curvature. The radius of this circle is the radius of curvature of the curve at P_0. The center of this circle is the *center of curvature* of the curve at P_0 and has the coordinates x_c, y_c, such that

$$x_c = x_0 - \frac{y_0'(1 + y_0'^2)}{y_0''} \qquad y_c = y_0 + \frac{(1 + y_0'^2)}{y_0''} \tag{4.58}$$

Space Curves

If PP' is a tangent to a given twisted curve, and P'' is a point on the curve near P, the plane which contains the tangent and the point P'' becomes the osculating plane at P in the limit as P'' approaches P along the curve.

The normal to the curve which lies in the osculating plane of the point where the normal meets the curve is called the *principal normal* at that point.

The normal to the curve which is perpendicular to the osculating plane at the point where the normal meets the curve is called the *binormal* at that point.

The curvature κ at a point P is the limit of the angle between the tangents at adjacent points P and P', divided by their distance PP', as P' approaches P. The curvature is the reciprocal of the radius of the circle of curvature which is lying in the osculating plane and tangent to the curve at P. The torsion τ at a point P is the limit of the angle between the osculating planes at adjacent points P and P', divided by their distance PP', as P' approaches P.

Surfaces

The general equation of a surface expressed in rectangular coordinates is

$$F(x,y,z) = 0$$

In parametric form, the equations are

$$x = f_1(u,v) \qquad y = f_2(u,v) \qquad z = f_3(u,v)$$

A curve on the surface may be defined by

$$v = v(u)$$

If a point P on the surface is an ordinary point, all the tangents to all of the curves on the surface through P lie in a plane called the *tangent plane.*

If a plane, which includes the normal to a surface at P is passed through the surface, it produces a section called the normal section. The curvature of the normal section is called the normal curvature. If the normal curvature is the same for all sections through the point, the point is called an umbilical point. (Every point on a spherical surface is an umbilical point.) If P is not an umbilical point, there will be two sections at P such that one section has a minimum and the other the maximum curvature. These are called the principal curvatures, and the tangents to the sections are called principal directions. The principal directions are perpendicular to each other. The curves on the surface which, at every point, are tangent to one of the principal directions, are called the lines of curvature.

The sum of the maximum and minimum normal curvatures is called the average curvature, and their product is called the gaussian curvature.

CHAPTER 5

MOMENTS OF INERTIA

BY

H. V. HAHNE, Ph.D., Los Altos, Calif.

The *moment of inertia* of a plane figure (Fig. 5.1) about an axis x located in the plane of the figure is defined as the integral

$$I_x = \int_A y^2 \, dA \tag{5.1}$$

where y is the distance of the area element dA from the x axis, and the integration is meant to extend over the entire area A of the figure. The subscript of I indicates the axis about which the moment of inertia is taken. Accordingly, the moment of inertia of the figure about the y axis is

$$I_y = \int_A x^2 \, dA \tag{5.2}$$

where x is the distance of the area element dA from this axis.

The quantities i_x, i_y, defined by the equations

$$i_x{}^2 A = I_x \qquad i_y{}^2 A = I_y \tag{5.3a,b}$$

are called the *radii of gyration* of the figure about the respective axes.

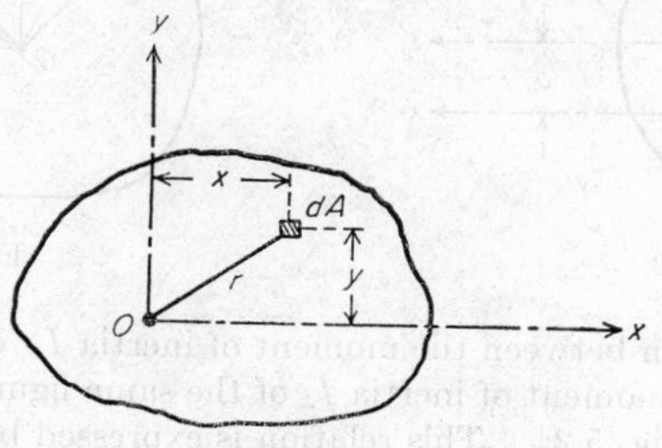

FIG. 5.1

The moment of inertia of a plane figure (Fig. 5.1) about a point O in its plane is called the *polar moment of inertia* of the figure with respect to this point. It is defined as the integral

$$I_p = \int_A r^2 \, dA \tag{5.4}$$

where r is the distance of the area element dA from the point O. Since $r^2 = x^2 + y^2$,

$$I_p = \int_A (x^2 + y^2) \, dA = I_x + I_y \tag{5.5}$$

Thus, the polar moment of inertia of a plane figure about a point in its plane is equal to the sum of moments of inertia of the figure about two orthogonal axes which pass through that point and are located in the plane of the figure.

The *product of inertia* of a plane figure (Fig. 5.1) about an orthogonal system of axes x and y is defined as the integral

$$I_{xy} = \int_A xy\,dA \tag{5.6}$$

The two subscripts of I indicate the axes about which the product of inertia is taken.

Moments of inertia are always positive, but a product of inertia can assume negative values.

Whenever the outlines of a figure can be expressed analytically, the integrals in Eqs. (5.1), (5.2), (5.4), and (5.6) can be evaluated, and an analytical expression for the respective quantities can be obtained. More complex figures often may be subdivided into portions whose moments of inertia are known, and the total moment of inertia of the figure is obtained by adding the individual moments of inertia. The same method may be applied to obtain the products of inertia of such figures.

If the outlines of a figure are completely irregular, its moment and product of inertia can be obtained by evaluating the respective integrals numerically or by using a graphical method. The moments of inertia and radii of gyration of commonly encountered figures are listed in Table 20.1. Corresponding quantities for commercially available shapes (cross sections of steel and aluminum bars) can be found in engineering handbooks and design manuals.

Of special importance are moments of inertia about centroidal axes, i.e., axes which pass through the centroid of the figure.

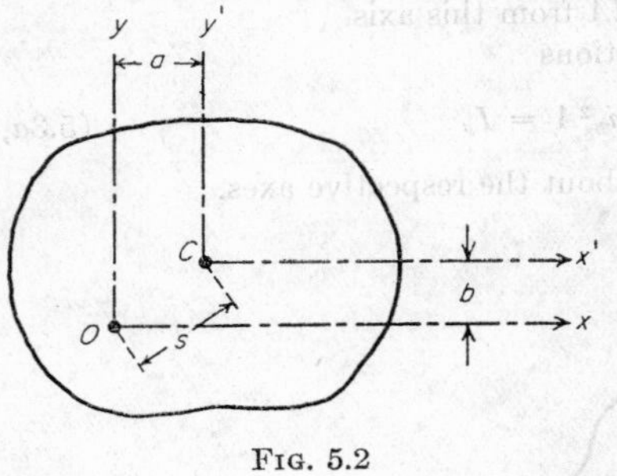

FIG. 5.2

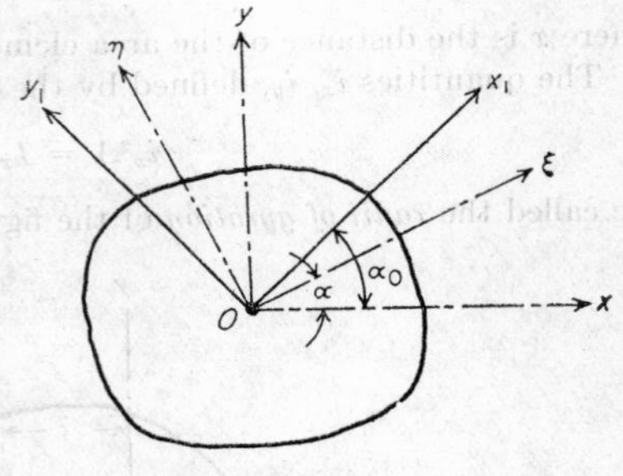

FIG. 5.3

There exists a relation between the moment of inertia $I_{x'}$ of a plane figure about a centroidal axis and the moment of inertia I_x of the same figure about an axis parallel to the centroidal axis (Fig. 5.2). This relation is expressed by the equation

$$I_x = I_{x'} + Ab^2 \tag{5.7}$$

where both axes are assumed to be in the plane of the figure, A is the area of the figure, and b is the distance between the two axes. Equation (5.7) is known as the parallel-axes theorem.

An analogous relation exists between the polar moment of inertia I_{pC} of a plane figure about the centroid and the polar moment of inertia I_p of the same figure about any other point O in its plane. If A denotes the area of the figure and s the distance between the centroid and the point O (Fig. 5.2), then

$$I_p = I_{pC} + As^2 \tag{5.8}$$

A similar relation can also be established for the products of inertia. Let $I_{x'y'}$ denote the product of inertia of a plane figure about a system of orthogonal centroidal axes x', y' located in the plane of the figure (Fig. 5.2). The product of inertia I_{xy} of the same figure about a system of orthogonal axes x, y located in its plane and oriented parallel to the x', y' axes is then given by the equation

$$I_{xy} = I_{x'y'} + Aab \tag{5.9}$$

If a system of orthogonal coordinate axes is rotated about its point of origin, the moments and the product of inertia about these axes become functions of the angle of rotation. Let I_x and I_y denote the moments of inertia, and I_{xy} denote the product of inertia of a plane figure about a set of orthogonal axes x, y in its plane (Fig. 5.3). Let I_ξ, I_η, and $I_{\xi\eta}$ denote the corresponding quantities for a set of orthogonal axes ξ, η also located in the plane of the figure and having a common origin with the (x,y) system. If α denotes the angle by which the ξ, η axes are rotated with respect to the (x,y) system, then I_ξ, I_η, and $I_{\xi\eta}$ can be expressed in terms of α, I_x, I_y, and I_{xy} as follows:

$$\begin{aligned} I_\xi &= I_x \cos^2 \alpha + I_y \sin^2 \alpha - 2I_{xy} \sin \alpha \cos \alpha \\ I_\eta &= I_x \sin^2 \alpha + I_y \cos^2 \alpha + 2I_{xy} \sin \alpha \cos \alpha \\ I_{\xi\eta} &= \tfrac{1}{2}(I_x - I_y) \sin 2\alpha + I_{xy} \cos 2\alpha \end{aligned} \tag{5.10a–c}$$

The moments of inertia I_ξ and I_η are functions of the angle α, and it is possible to determine values of α for which I_ξ and I_η assume extreme values. This value of α denoted by α_0 is given by the equation

$$\tan 2\alpha_0 = \frac{2I_{xy}}{I_y - I_x} \tag{5.11}$$

It is found that, whenever I_ξ is a maximum, I_η is a minimum, and vice versa. It is also found that, when $\alpha = \alpha_0$, $I_{\xi\eta} = 0$.

Any two orthogonal axes ξ, η about which the product of inertia vanishes and the moments of inertia assume maximum and minimum values are called *principal axes of inertia* of the figure. If their origin is in the centroid of the figure, they are called centroidal principal axes. The orientation of the principal axes with respect to a reference coordinate system (x,y) is given by Eq. (5.11). In a symmetrical figure, the axis of symmetry always coincides with a centroidal principal axis of inertia. The moments of inertia about the principal axes are called the *principal moments of inertia*, and their values are given by the equations

$$\begin{aligned} I_{\max} &= \tfrac{1}{2}(I_x + I_y) + \sqrt{\tfrac{1}{4}(I_x - I_y)^2 + I_{xy}^2} \\ I_{\min} &= \tfrac{1}{2}(I_x + I_y) - \sqrt{\tfrac{1}{4}(I_x - I_y)^2 + I_{xy}^2} \end{aligned} \tag{5.12a,b}$$

For commercially available shapes, the orientation of the principal axes and the values of the principal moments of inertia are found in design manuals and tables published by the manufacturers.

It can be shown that

$$I_{\max} + I_{\min} = I_x + I_y = I_\xi + I_\eta$$

or, generally, that the sum of moments of inertia of a plane figure about systems of orthogonal axes in its plane with a common origin is independent of the orientation of the axes.

CHAPTER 6

VECTOR ANALYSIS

BY

H. E. NEWELL, Ph.D., Washington, D.C.

6.1. VECTOR ALGEBRA

A *scalar* is a quantity that can be specified by a single number. Examples are: π, e, temperature, pressure, density, voltage, energy. A region throughout which a scalar function is defined is known as a *scalar field.*

A vector is a quantity that possesses both a scalar magnitude and a characteristic direction, and also obeys the familiar parallelogram rule. By the last is meant that the equivalent, or *resultant* **R**, of two vectors[1] **A** and **B** of similar physical meaning is given by the diagonal of the parallelogram constructed from **A** and **B**, as shown in Fig. 6.1. In constructing the parallelogram, the vectors **A** and **B** are represented as arrows of suitable length and appropriate direction, whereupon the diagonal **R** correctly gives both the magnitude and direction of the resultant vector. It is plain

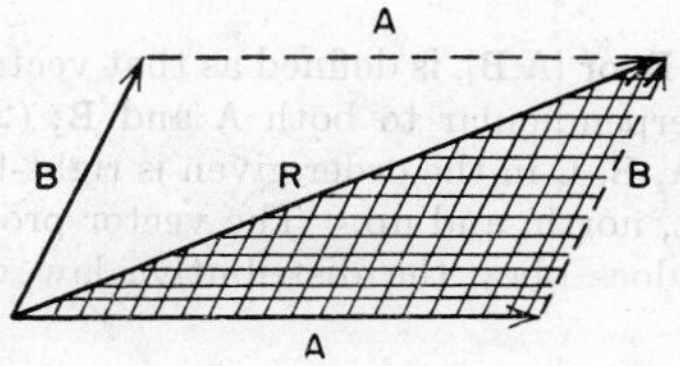

FIG. 6.1

from the crosshatched portion of Fig. 6.1 that the resultant of **A** and **B** can also be obtained by a simple triangle rule.

Examples of vector quantities are: position in space relative to a fixed origin, velocity, acceleration, force, current density, electric field intensity, magnetic flux density.

A region throughout which a vector function of position is defined is called a *vector field.*

The combination of two vectors **A** and **B** by either the parallelogram rule or the equivalent triangle law is called addition, and is denoted by $\mathbf{A} + \mathbf{B}$. The *difference* $\mathbf{A} - \mathbf{B}$ is defined by the relation

$$(\mathbf{A} - \mathbf{B}) + \mathbf{B} = \mathbf{A}$$

A vector of zero length is called the *null vector,* and is designated by 0. The vector $-\mathbf{A}$ has the same length as **A** but the opposite direction. The vector $m\mathbf{A}$, where m is a positive scalar, has the same direction as **A**, but is m times as long.

[1] A vector will be denoted by a boldface letter, and its magnitude either by the same letter in italic form or by means of absolute value signs.

With the above definitions and simple geometry, one can now prove the following relations:

$$\begin{aligned} \mathbf{A} + \mathbf{B} &= \mathbf{B} + \mathbf{A} && \text{(commutative law)} \\ (\mathbf{A} + \mathbf{B}) + \mathbf{C} &= \mathbf{A} + (\mathbf{B} + \mathbf{C}) && \text{(associative law)} \\ \mathbf{A} + (-\mathbf{A}) &= 0 \\ \mathbf{A} - \mathbf{B} &= \mathbf{A} + (-\mathbf{B}) \\ m\mathbf{A} + n\mathbf{A} &= (m + n)\mathbf{A} \\ m\mathbf{A} + m\mathbf{B} &= m(\mathbf{A} + \mathbf{B}) \end{aligned} \tag{6.1a–f}$$

where in the last two relations m and n are scalars. As far as addition is concerned, the algebra of vectors is identical with that of ordinary real numbers.

The *scalar product* $\mathbf{A} \cdot \mathbf{B}$, or $(\mathbf{A},\mathbf{B})$, is defined as

$$\mathbf{A} \cdot \mathbf{B} = AB \cos \theta \tag{6.2}$$

where θ is the angle between **A** and **B**. The notation $\mathbf{A}^2$ is often used to denote $\mathbf{A} \cdot \mathbf{A}$.

The scalar product obeys the following laws:

$$\begin{aligned} \mathbf{A} \cdot \mathbf{B} &= \mathbf{B} \cdot \mathbf{A} && \text{(commutative law)} \\ \mathbf{A} \cdot (\mathbf{B} + \mathbf{C}) &= \mathbf{A} \cdot \mathbf{B} + \mathbf{A} \cdot \mathbf{C} && \text{(distributive law with respect to addition)} \\ \mathbf{A} \cdot \mathbf{B} &= 0 \text{ if } \mathbf{A} \perp \mathbf{B} \\ \mathbf{A}^2 &= A^2 \end{aligned} \tag{6.3a–d}$$

A *unit vector* is a vector of unit length. Let **i** be a unit vector. Then $\mathbf{A} \cdot \mathbf{i}$ is the orthogonal projection of the length of **A** onto the direction of **i**. The vector $\mathbf{A} \cdot \mathbf{i}\, \mathbf{i}$ is called the *component* of **A** in the direction of **i**. When a vector is written as the sum of three mutually perpendicular vectors, it is said to have been resolved into orthogonal components.

The *vector product* $\mathbf{A} \times \mathbf{B}$, or $[\mathbf{A},\mathbf{B}]$, is defined as that vector **C** having all the following properties: (1) **C** is perpendicular to both **A** and **B**; (2) the magnitude of **C** is $AB \sin \theta$; and (3) the set **A**, **B**, **C** in the order given is right-handed, i.e., has the same relative orientation as east, north, and up. The vector product is neither commutative nor associative, but does obey the distributive law with respect to addition. One has

$$\begin{aligned} \mathbf{A} \times \mathbf{B} &= -\mathbf{B} \times \mathbf{A} \\ \mathbf{A} \times (\mathbf{B} + \mathbf{C}) &= \mathbf{A} \times \mathbf{B} + \mathbf{A} \times \mathbf{C} \end{aligned} \tag{6.4a,b}$$

The following laws are especially useful in vector calculations.

$$\mathbf{A} \times (\mathbf{B} \times \mathbf{C}) = (\mathbf{A} \cdot \mathbf{C})\mathbf{B} - (\mathbf{A} \cdot \mathbf{B})\mathbf{C} \tag{6.5}$$

The *triple scalar product* (**ABC**), defined as $\mathbf{A} \cdot \mathbf{B} \times \mathbf{C}$, is positive or negative according as **A**, **B**, **C** are, in the order given, right-handed or left-handed, and is numerically equal to the volume of the parallelepiped constructed from **A**, **B**, **C** as coterminous edges.

$$(\mathbf{A} \times \mathbf{B}) \cdot (\mathbf{C} \times \mathbf{D}) = (\mathbf{A} \cdot \mathbf{C})(\mathbf{B} \cdot \mathbf{D}) - (\mathbf{A} \cdot \mathbf{D})(\mathbf{B} \cdot \mathbf{C}) \tag{6.6}$$

This last relation is known as the generalized identity of Lagrange; the following more familiar identity of Lagrange being a special case:

$$(\mathbf{A} \times \mathbf{B})^2 = \mathbf{A}^2\mathbf{B}^2 - (\mathbf{A} \cdot \mathbf{B})^2 \tag{6.7}$$

Any vector can be expressed uniquely as a linear combination of three given noncoplanar vectors. Such a triplet of vectors can, therefore, be used as a *basis* for representing vectors in space, and the various systems of vector triplets that can be used give rise to different coordinate systems. The simplest coordinate system is the cartesian system in which a fixed right-handed set $\mathbf{i}_x$, $\mathbf{i}_y$, $\mathbf{i}_z$ of mutually perpendicular unit vectors is used as a basis. In this case:

$$\mathbf{A} = A_1\mathbf{i}_x + A_2\mathbf{i}_y + A_3\mathbf{i}_z \qquad A^2 = A_1^2 + A_2^2 + A_3^2$$
$$\mathbf{A} \pm \mathbf{B} = (A_1 \pm B_1)\mathbf{i}_x + (A_2 \pm B_2)\mathbf{i}_y + (A_2 \pm B_2)\mathbf{i}_z \tag{6.8a–d}$$
$$\mathbf{A} \cdot \mathbf{B} = A_1B_2 + A_2B_2 + A_3B_3$$

$$\mathbf{A} \times \mathbf{B} = \begin{vmatrix} \mathbf{i}_x & \mathbf{i}_y & \mathbf{i}_z \\ A_1 & A_2 & A_3 \\ B_1 & B_2 & B_3 \end{vmatrix} \qquad (\mathbf{ABC}) = \begin{vmatrix} A_1 & A_2 & A_3 \\ B_1 & B_2 & B_3 \\ C_1 & C_2 & C_3 \end{vmatrix} \tag{6.8e,f}$$

6.2. VECTOR CALCULUS

The definitions of limit, continuity, and derivative for vector functions are formally identical with the corresponding definitions for scalar functions. Thus, letting $\mathbf{A}(u)$ depend on the scalar variable u,

$$\lim_{u \to u_0} \mathbf{A}(u) = \mathbf{L}$$

if and only if

$$\lim_{u \to u_0} |\mathbf{A}(u) - \mathbf{L}| = 0$$

The vector $\mathbf{A}(u)$ is continuous at u_0 if and only if

$$\lim_{u \to u_0} \mathbf{A}(u) = \mathbf{A}(u_0)$$

The derivative $d\mathbf{A}/du$, or $\mathbf{A}'(u)$, is defined as

$$\frac{d\mathbf{A}}{du} = \lim_{u \to u_0} \frac{\mathbf{A}(u) - \mathbf{A}(u_0)}{u - u_0}$$

provided the limit exists.

The limit of a sum, difference, scalar product, or vector product of two vectors is the sum, difference, scalar product, or vector product of the individual limits of the vectors. Similarly, the sum, difference, scalar product, or vector product of two continuous vector functions is itself a continuous vector function. Also, the usual rules of differentiation, familiar in the scalar calculus, apply:

$$(\mathbf{A} \pm \mathbf{B})' = \mathbf{A}' \pm \mathbf{B}' \qquad (\mathbf{A} \cdot \mathbf{B})' = \mathbf{A}' \cdot \mathbf{B} + \mathbf{A} \cdot \mathbf{B}' \tag{6.9a,b}$$
$$(\mathbf{A} \times \mathbf{B})' = \mathbf{A}' \times \mathbf{B} + \mathbf{A} \times \mathbf{B}' \qquad \frac{d}{dw}[\mathbf{A}(u(w))] = \frac{d\mathbf{A}}{du}\frac{du}{dw} \tag{6.9c,d}$$

Finally, partial derivatives of vector functions of two or more independent variables are obtained by differentiating with respect to one of the variables while holding the others fixed.

Let ϕ be a differentiable scalar function of position, and fix attention upon a specific point P. Let $d\mathbf{r}$ be an elemental vector displacement from P to an adjacent point P'. The *gradient of* ϕ at P, written grad ϕ, is that vector which satisfies the relation

$$\text{grad } \phi \cdot d\mathbf{r} = d\phi \tag{6.10}$$

where

$$d\phi = \phi(P') - \phi(P)$$

The vector grad $\phi(P)$ is normal to the surface $\phi = \phi(P)$ at the point P.

Let $\mathbf{A}$ be a continuous function of position in a closed region R of space. Divide R into n elemental volumes ΔV, and let P_ν be a point in the νth element $\Delta_\nu V$. Form the sum over all of R,

$$\sum_\nu \mathbf{A}(P_\nu)\Delta_\nu V$$

The limit of this sum, as the number of subdivisions of R is increased indefinitely and all the $\Delta_\nu V$ are made to vanish, exists and is called the definite integral of **A** over the region R, written

$$\lim \sum \mathbf{A}(P_\nu)\Delta_\nu V = \int_R \mathbf{A}\, dV \qquad \text{as } \Delta_\nu V \to 0,\ n \to \infty$$

The definition is precisely analogous to that of an integral in the scalar calculus. Line and surface integrals are similarly defined. In the case of line integrals, the element of arc is often used in its vector form $d\mathbf{r}$, giving rise to such integrals as

$$\int_C \mathbf{A} \cdot d\mathbf{r} \qquad \text{and} \qquad \int_C \mathbf{A} \times d\mathbf{r}$$

where C is the path of integration. Surface integrals of the form

$$\int_S \mathbf{A} \cdot d\mathbf{S} \qquad \text{and} \qquad \int_S \mathbf{A} \times d\mathbf{S}$$

are often encountered, where S is the surface of integration, and where the elemental vector $d\mathbf{S}$ is the element of area dS times the unit vector **n** normal to the area dS.

Integration over a closed surface or around a closed path is indicated by a small circle through the integral sign, thus:

$$\oint_C \mathbf{A} \cdot d\mathbf{r}$$

Let **A** have continuous first derivatives in a region R of space, and fix attention upon a specific point P. Let ΔV be a small volume containing P in its interior. Then the *divergence* of **A** at P, written div **A**, is defined as

$$\operatorname{div} \mathbf{A}(P) = \lim_{\Delta V \to 0} \frac{1}{\Delta V} \oint_{\substack{\text{bdy} \\ \text{of } \Delta V}} \mathbf{A} \cdot d\mathbf{S} \tag{6.11}$$

where $d\mathbf{S}$ has the direction of the outer normal to the boundary of ΔV. The divergence is a measure of the intensity at P of sources or sinks for the field **A**.

Let C be a closed curve in a region R where **A** has continuous first derivatives. The *circulation* of **A** about C is defined as

$$\oint_C \mathbf{A} \cdot d\mathbf{r}$$

Now consider a point P in R, and let ΔS be an elemental area that contains P, has the boundary C, and is normal to the fixed unit vector **n**. The *circulation intensity* L_n of the field **A** at P normal to **n** is given by

$$L_n = \lim \frac{1}{\Delta S} \oint_C \mathbf{A} \cdot d\mathbf{r} \qquad \text{as } \Delta S \to 0,\ C \to P,\ \mathbf{n} \text{ fixed} \tag{6.12a}$$

where the sense of integration around C is counterclockwise when viewed from the tip of **n**. The *curl* of **A** at P, written curl **A**, is that vector which has the property that

$$\operatorname{curl} \mathbf{A}(P) \cdot \mathbf{n} = L_n(P) \tag{6.12b}$$

for all unit vectors **n**.

The operator $\boldsymbol{\nabla}$, called *del* or *nabla*, is defined as follows:

$$\boldsymbol{\nabla} = \lim_{\Delta V \to 0} \frac{1}{\Delta V} \oint_{\substack{\text{bdy} \\ \text{of } \Delta V}} d\mathbf{S} \tag{6.13}$$

where $d\mathbf{S}$ points outwardly from ΔV. The operator is always applied from the left, and, when used at point P, the volume ΔV is so chosen as to contain P in its interior.

The following general relationships hold:

$$\text{curl}\,(\text{grad}\,\phi) = 0 \qquad \text{div}\,(\text{curl}\,\mathbf{A}) = 0 \qquad \text{div}\,(\text{grad}\,\phi \times \text{grad}\,\psi) = 0 \tag{6.14a–c}$$

$$\int_V \text{div}\,\mathbf{A}\,dV = \int_{\substack{\text{bdy}\\ \text{of } V}} \mathbf{A} \cdot d\mathbf{S} \qquad \text{(Gauss' theorem)} \tag{6.15}$$

$$\int_S \text{curl}\,\mathbf{A} \cdot d\mathbf{S} = \int_{\substack{\text{bdy}\\ \text{of } S}} \mathbf{A} \cdot d\mathbf{r} \qquad \text{(Stokes' theorem)} \tag{6.16}$$

$$\text{div}\,(\phi\mathbf{A}) = \phi\,\text{div}\,\mathbf{A} + \mathbf{A} \cdot \text{grad}\,\phi \qquad \text{curl}\,(\phi\mathbf{A}) = \phi\,\text{curl}\,\mathbf{A} + \text{grad}\,\phi \times \mathbf{A}$$
$$\text{div}\,(\mathbf{A} \times \mathbf{B}) = \mathbf{B} \cdot \text{curl}\,\mathbf{A} - \mathbf{A} \cdot \text{curl}\,\mathbf{B} \tag{6.17a–c}$$

$$\int_V \text{grad}\,\phi\,dV = \oint_{\substack{\text{bdy}\\ \text{of } V}} \phi\,d\mathbf{S} \qquad \oint_S d\mathbf{S} = 0$$
$$\int_V \text{curl}\,\mathbf{A}\,dV = \oint_{\substack{\text{bdy}\\ \text{of } V}} d\mathbf{S} \times \mathbf{A} \qquad \int_S d\mathbf{S} \times \text{grad}\,\phi = \oint_{\substack{\text{bdy}\\ \text{of } S}} \phi\,d\mathbf{r} \tag{6.18a–d}$$

$$\int_V [\phi\,\text{div grad}\,\psi + \text{grad}\,\phi \cdot \text{grad}\,\psi]\,dV = \oint_{\substack{\text{bdy}\\ \text{of } V}} \phi\,\text{grad}\,\psi \cdot d\mathbf{S} \qquad \text{(Green's first theorem)} \tag{6.19}$$

$$\int_V [\phi\,\text{div grad}\,\psi - \psi\,\text{div grad}\,\phi]\,dV = \oint_{\substack{\text{bdy}\\ \text{of } V}} [\phi\,\text{grad}\,\psi - \psi\,\text{grad}\,\phi] \cdot d\mathbf{S} \quad \text{(Green's second theorem)} \tag{6.20}$$

In all of Eqs. (6.15), (6.16), (6.18), (6.19), and (6.20), $d\mathbf{S}$ has the direction of the outer normal to the boundary of V. In Eqs. (6.16) and (6.18d), a positive side is designated for S, all $d\mathbf{S}$ are taken as pointing toward the positive side, and, as viewed from that side, the sense of integration around the boundary of S is counterclockwise. Continuing:

$$\nabla\phi = \text{grad}\,\phi \qquad \nabla \cdot \mathbf{A} = \text{div}\,\mathbf{A} \qquad \nabla \times \mathbf{A} = \text{curl}\,\mathbf{A} \tag{6.21a–c}$$

Also, if $\mathbf{r}$ is the radius vector from the origin,

$$\text{div}\,\mathbf{r} = 3 \qquad \text{curl}\,\mathbf{r} = 0$$

In rectangular coordinates x, y, z one has:

$$\text{grad}\,\phi = \frac{\partial\phi}{\partial x}\mathbf{i}_x + \frac{\partial\phi}{\partial y}\mathbf{i}_y + \frac{\partial\phi}{\partial z}\mathbf{i}_z$$
$$\text{div}\,\mathbf{A} = \frac{\partial A_1}{\partial x} + \frac{\partial A_2}{\partial y} + \frac{\partial A_3}{\partial z}$$
$$\text{curl}\,\mathbf{A} = \begin{vmatrix} \mathbf{i}_x & \mathbf{i}_y & \mathbf{i}_z \\ \dfrac{\partial}{\partial x} & \dfrac{\partial}{\partial y} & \dfrac{\partial}{\partial z} \\ A_1 & A_2 & A_3 \end{vmatrix} \tag{6.22a–e}$$
$$\text{div grad}\,\phi = \frac{\partial^2\phi}{\partial x^2} + \frac{\partial^2\phi}{\partial y^2} + \frac{\partial^2\phi}{\partial z^2}$$
$$\nabla = \mathbf{i}_x\frac{\partial}{\partial x} + \mathbf{i}_y\frac{\partial}{\partial y} + \mathbf{i}_z\frac{\partial}{\partial z}$$

The *Laplacian operator* ∇^2, or Δ, is defined in general by the relations:

$$\nabla^2\phi = \operatorname{div}(\operatorname{grad}\phi) \qquad \nabla^2\mathbf{A} = \operatorname{grad}(\operatorname{div}\mathbf{A}) - \operatorname{curl}(\operatorname{curl}\mathbf{A}) \qquad \operatorname{curl}\nabla^2\mathbf{A} = \nabla^2\operatorname{curl}\mathbf{A} \tag{6.23a–c}$$

It then follows that, in rectangular coordinates,

$$\nabla^2 = \nabla\cdot\nabla = \frac{\partial^2}{\partial x^2} + \frac{\partial^2}{\partial y^2} + \frac{\partial^2}{\partial z^2} \tag{6.24}$$

The operator $\mathbf{A}\cdot\nabla$ is defined only for rectangular coordinates as

$$\mathbf{A}\cdot\nabla = A_1\frac{\partial}{\partial x} + A_2\frac{\partial}{\partial y} + A_3\frac{\partial}{\partial z} \tag{6.25}$$

With this operator the following relations hold in rectangular coordinates:

$$\begin{aligned}\nabla\times(\mathbf{A}\times\mathbf{B}) &= (\mathbf{A}\nabla\cdot - \mathbf{A}\cdot\nabla)\mathbf{B} - (\mathbf{B}\nabla\cdot - \mathbf{B}\cdot\nabla)\mathbf{A}\\ \nabla(\mathbf{A}\cdot\mathbf{B}) &= \mathbf{A}\cdot\nabla\mathbf{B} + \mathbf{B}\cdot\nabla\mathbf{A} + \mathbf{A}\times(\nabla\times\mathbf{B}) + \mathbf{B}\times(\nabla\times\mathbf{A})\end{aligned} \tag{6.26a,b}$$

6.3. GENERAL COORDINATES

If a one-to-one continuous correspondence exists between a set of number triples (u_1,u_2,u_3) and the points of a region R of space, then the number triples can be used as coordinates of the points of R. The u_α can be expressed as continuous functions of rectangular coordinates x, y, z:

$$u_\alpha = u_\alpha(x,y,z) \qquad \alpha = 1, 2, 3$$

The most practical coordinate systems are those for which the u_α have continuous first and second partial derivatives.

In general, the coordinate surfaces, u_α = const, are not planes, and are not necessarily orthogonal. Likewise, the coordinate lines, which are the intersections of pairs of coordinate surfaces, are often curved, and are not necessarily orthogonal. At each point, two sets of vectors recommend themselves as bases for representing the general vector. First is the set $\mathbf{i}_\alpha$ of unit vectors tangent to the coordinate lines, and directed toward increasing u_α. Second, there is the set $\mathbf{e}_\alpha$ of unit vectors perpendicular to the coordinate surfaces, also pointing toward increasing u_α. Generally both sets of vectors vary from point to point, and this must be kept in mind in calculating with general coordinates. If and only if the coordinates are orthogonal, then the $\mathbf{i}_\alpha$ coincide with the $\mathbf{e}_\alpha$.

Let $\mathbf{r}$ be the radius vector from the origin of coordinates. Then

$$\mathbf{i}_\alpha \parallel \partial\mathbf{r}/\partial u_\alpha \qquad \alpha = 1, 2, 3$$

Also

$$\mathbf{e}_\alpha \parallel \nabla u_\alpha \qquad \alpha = 1, 2, 3$$

Moreover,

$$(\partial\mathbf{r}/\partial u_\alpha)\cdot\nabla u_\beta = \begin{cases}0 & \alpha\neq\beta\\ 1 & \alpha=\beta\end{cases} \tag{6.27}$$

Two sets of vectors related in this fashion are referred to as *reciprocal sets* of vectors.

Let

$$g_{\alpha\beta} = (\partial\mathbf{r}/\partial u_\alpha)\cdot(\partial\mathbf{r}/\partial u_\beta) \qquad G_{\alpha\beta} = \nabla u_\alpha\cdot\nabla u_\beta \tag{6.28a,b}$$

$$g = \begin{vmatrix} g_{11} & g_{12} & g_{13}\\ g_{21} & g_{22} & g_{23}\\ g_{31} & g_{32} & g_{33}\end{vmatrix} \tag{6.29}$$

Then the general vector $\mathbf{A}$ can be represented as

$$\mathbf{A} = \sum_{\alpha=1}^{3}\sqrt{g_{\alpha\alpha}G_{\alpha\alpha}}\,(\mathbf{A}\cdot\mathbf{e}_\alpha)\mathbf{i}_\alpha = \sum_{\alpha=1}^{3}\sqrt{g_{\alpha\alpha}G_{\alpha\alpha}}\,(\mathbf{A}\cdot\mathbf{i}_\alpha)\mathbf{e}_\alpha \tag{6.30}$$

The differential element of volume is

$$dV = \sqrt{g}\, du_1\, du_2\, du_3 \tag{6.31}$$

The differential element of arc length is given by

$$ds^2 = \sum_{\alpha,\beta=1}^{3} g_{\alpha\beta}\, du_\alpha\, du_\beta \tag{6.32}$$

One also has:

$$\text{grad } \Phi = \sum_{\alpha=1}^{3} \sqrt{G_{\alpha\alpha}} \frac{\partial \Phi}{\partial u_\alpha} \mathbf{e}_\alpha \qquad \text{div } \mathbf{A} = \frac{1}{\sqrt{g}} \sum_{\alpha=1}^{3} \frac{\partial}{\partial u_\alpha} (\sqrt{g G_{\alpha\alpha}}\, \mathbf{A} \cdot \mathbf{e}_\alpha) \tag{6.33a,b}$$

$$\text{curl } \mathbf{A} = \frac{1}{\sqrt{g}} \begin{vmatrix} \sqrt{g_{11}}\, \mathbf{i}_1 & \sqrt{g_{22}}\, \mathbf{i}_2 & \sqrt{g_{33}}\, \mathbf{i}_3 \\ \dfrac{\partial}{\partial u_1} & \dfrac{\partial}{\partial u_2} & \dfrac{\partial}{\partial u_3} \\ \sqrt{g_{11}}\, (\mathbf{A} \cdot \mathbf{i}_1) & \sqrt{g_{22}}\, (\mathbf{A} \cdot \mathbf{i}_2) & \sqrt{g_{33}}\, (\mathbf{A} \cdot \mathbf{i}_3) \end{vmatrix} \tag{6.33c}$$

$$\text{div grad } \Phi = \frac{1}{\sqrt{g}} \sum_{\alpha=1}^{3} \frac{\partial}{\partial u_\alpha} \left(\sqrt{g} \sum_{\beta=1}^{3} G_{\alpha\beta} \frac{\partial \Phi}{\partial u_\beta} \right) \tag{6.33d}$$

For orthogonal coordinates, the $g_{\alpha\beta}$ and $G_{\alpha\beta}$ all vanish if $\alpha \neq \beta$. Setting $h_\alpha = \sqrt{g_{\alpha\alpha}}$, and remembering that the $\mathbf{e}_\alpha$ coincide with the $\mathbf{i}_\alpha$, one gets:

$$\begin{aligned} \mathbf{A} &= \sum_{\alpha=1}^{3} (\mathbf{A} \cdot \mathbf{i}_\alpha) \mathbf{i}_\alpha \\ dV &= h_1 h_2 h_3\, du_1\, du_2\, du_3 \\ ds^2 &= h_1^2\, du_1^2 + h_2^2\, du_2^2 + h_3^2\, du_3^2 \\ \text{div } \mathbf{A} &= \frac{1}{h_1 h_2 h_3} \sum_{\alpha=1}^{3} \frac{\partial}{\partial u_\alpha} \left(\frac{h_1 h_2 h_3}{h_\alpha} \mathbf{A} \cdot \mathbf{i}_\alpha \right) \\ \text{curl } \mathbf{A} &= \frac{1}{h_1 h_2 h_3} \begin{vmatrix} h_1 \mathbf{i}_1 & h_2 \mathbf{i}_2 & h_3 \mathbf{i}_3 \\ \dfrac{\partial}{\partial u_1} & \dfrac{\partial}{\partial u_2} & \dfrac{\partial}{\partial u_3} \\ h_1 \mathbf{A} \cdot \mathbf{i}_1 & h_2 \mathbf{A} \cdot \mathbf{i}_2 & h_3 \mathbf{A} \cdot \mathbf{i}_3 \end{vmatrix} \\ \text{div grad } \Phi &= \frac{1}{h_1 h_2 h_3} \sum_{\alpha=1}^{3} \frac{\partial}{\partial u_\alpha} \left[\frac{h_1 h_2 h_3}{h_\alpha^2} \frac{\partial \Phi}{\partial u_\alpha} \right] \end{aligned} \tag{6.34a–f}$$

For cylindrical coordinates r, θ, z (see Fig. 4.12, where r' stands for r), one has

$$\begin{aligned} dV &= r\, dr\, d\theta\, dz \qquad ds^2 = dr^2 + r^2\, d\theta^2 + dz^2 \\ \text{grad } \Phi &= \frac{\partial \Phi}{\partial r} \mathbf{i}_r + \frac{1}{r} \frac{\partial \Phi}{\partial \theta} \mathbf{i}_\theta + \frac{\partial \Phi}{\partial z} \mathbf{i}_z \\ \text{div } \mathbf{A} &= \frac{1}{r} \frac{\partial}{\partial r} (r \mathbf{A} \cdot \mathbf{i}_r) + \frac{1}{r} \frac{\partial}{\partial \theta} (\mathbf{A} \cdot \mathbf{i}_\theta) + \frac{\partial}{\partial z} (\mathbf{A} \cdot \mathbf{i}_z) \\ \text{curl } \mathbf{A} &= \frac{1}{r} \begin{vmatrix} \mathbf{i}_r & r \mathbf{i}_\theta & \mathbf{i}_z \\ \dfrac{\partial}{\partial r} & \dfrac{\partial}{\partial \theta} & \dfrac{\partial}{\partial z} \\ \mathbf{A} \cdot \mathbf{i}_r & r \mathbf{A} \cdot \mathbf{i}_\theta & \mathbf{A} \cdot \mathbf{i}_z \end{vmatrix} \\ \text{div grad } \Phi &= \frac{1}{r} \frac{\partial}{\partial r} \left(r \frac{\partial \Phi}{\partial r} \right) + \frac{1}{r^2} \frac{\partial^2 \Phi}{\partial \theta^2} + \frac{\partial^2 \Phi}{\partial z^2} \end{aligned} \tag{6.35a–f}$$

For spherical polar coordinates r, ϕ, θ (see Fig. 4.12), one has

$$dV = r^2 \sin\phi \, dr \, d\phi \, d\theta \qquad ds^2 = dr^2 + r^2 \, d\phi^2 + r^2 \sin^2\phi \, d\theta^2 \qquad (6.36a,b)$$

$$\text{grad } \Phi = \frac{\partial \Phi}{\partial r} \mathbf{i}_r + \frac{1}{r}\frac{\partial \Phi}{\partial \phi} \mathbf{i}_\phi + \frac{1}{r \sin\phi}\frac{\partial \Phi}{\partial \theta} \mathbf{i}_\theta$$

$$\text{div } \mathbf{A} = \frac{1}{r^2}\frac{\partial}{\partial r}(r^2 \mathbf{A} \cdot \mathbf{i}_r) + \frac{1}{r \sin\phi}\frac{\partial}{\partial \phi}(\sin\phi \, \mathbf{A} \cdot \mathbf{i}_\phi) + \frac{1}{r \sin\phi}\frac{\partial}{\partial \theta}(\mathbf{A} \cdot \mathbf{i}_\theta)$$

$$\text{curl } \mathbf{A} = \frac{1}{r^2 \sin\phi} \begin{vmatrix} \mathbf{i}_r & r\,\mathbf{i}_\phi & r \sin\phi \, \mathbf{i}_\theta \\ \dfrac{\partial}{\partial r} & \dfrac{\partial}{\partial \phi} & \dfrac{\partial}{\partial \theta} \\ \mathbf{A} \cdot \mathbf{i}_r & r\mathbf{A} \cdot \mathbf{i}_\phi & r \sin\phi \, \mathbf{A} \cdot \mathbf{i}_\theta \end{vmatrix} \qquad (6.36c\text{–}f)$$

$$\text{div grad } \Phi = \frac{1}{r^2}\frac{\partial}{\partial r}\left(r^2 \frac{\partial \Phi}{\partial r}\right) + \frac{1}{r^2 \sin\phi}\frac{\partial}{\partial \phi}\left(\sin\phi \frac{\partial \Phi}{\partial \phi}\right) + \frac{1}{r^2 \sin^2\phi}\frac{\partial^2 \Phi}{\partial \theta^2}$$

6.4. POTENTIAL THEORY

If the vector curl **A** vanishes everywhere throughout a region R, then **A** is said to be *irrotational* within R. Suppose that **A** has continuous first derivatives, and that R is simply connected. Then each of the following is a necessary and sufficient condition that **A** be irrotational in R:

1. curl $\mathbf{A} \equiv 0$.
2. **A** is of the form grad ϕ.
3. For every simple closed curve C in R, $\oint_C \mathbf{A} \cdot d\mathbf{r}$ vanishes.

The function ϕ is said to be a *scalar potential* of **A**, and is unique except for an additive constant.

Suppose now that, in addition to having continuous first derivatives, **A** is $O(1/r^2)$ at ∞. Set

$$\text{div } \mathbf{A} = a \qquad (6.37)$$

The points at which a does not vanish are called *sources* ($a > 0$) and *sinks* ($a < 0$) of **A**. It can be shown that **A** is given by

$$\mathbf{A}(P') = \text{grad}' \, \phi(P')$$

where

$$\phi(P') = -\frac{1}{4\pi} \int_{\substack{\text{all} \\ \text{space}}} \frac{a(P)}{R(P,P')} \, dV(P) \qquad (6.38)$$

where $R(P,P')$ is the distance from P' to the variable point P of integration, and where grad′ signifies differentiation with respect to P'. The function $\phi(P')$ is $O(1/r)$ at ∞ and is said to be a *regular* potential of **A**.

If div **A** vanishes everywhere throughout a region R, **A** is said to be *solenoidal* within R. Suppose that **A** has continuous first derivatives. Then a necessary and sufficient condition that **A** be solenoidal is that it be of the form

$$\mathbf{A} = \text{curl } \mathbf{B} \qquad (6.39)$$

The vector **B** is said to be a *vector potential* of **A**. Any two vector potentials of **A** differ at most by the gradient of some scalar function.

Let **A** be solenoidal everywhere, have continuous second derivatives, and be $O(1/r^2)$ at ∞. Set

$$\text{curl } \mathbf{A} = \mathbf{b}$$

The points at which $\mathbf{b} \neq 0$ are called *vortices* of $\mathbf{A}$. Under the conditions specified, the function

$$\mathbf{B}(P') = \frac{1}{4\pi} \int_V \frac{\mathbf{b}(P)}{R(P,P')} \, dV(P) \tag{6.40}$$

is a vector potential of $\mathbf{A}(P')$, so that

$$\mathbf{A}(P') = \text{curl}' \, \mathbf{B}(P')$$

where curl′ denotes differentiation with respect to P'.

Finally, suppose that the vector field $\mathbf{A}$ possesses continuous second derivatives everywhere and is $O(1/r^2)$ at ∞. Then $\mathbf{A}$ can be resolved in one and only one way into the sum of an irrotational field $\mathbf{B}$ plus a solenoidal field $\mathbf{C}$. Specifically,

$$\mathbf{A}(P') = \mathbf{B}(P') + \mathbf{C}(P') \tag{6.41}$$

where

$$\mathbf{B}(P') = \text{grad}' \left[-\frac{1}{4\pi} \int_{\substack{\text{all} \\ \text{sources} \\ \text{of } \mathbf{A}}} \frac{a(P)}{R(P,P')} \, dV(P) \right] \qquad a = \text{div } \mathbf{A} \tag{6.42}$$

and

$$\mathbf{C}(P') = \text{curl}' \left[\frac{1}{4\pi} \int_{\substack{\text{all} \\ \text{vortices} \\ \text{of } \mathbf{A}}} \frac{\mathbf{b}(P)}{R(P,P')} \, dV(P) \right] \qquad \mathbf{b} = \text{curl } \mathbf{A} \tag{6.43}$$

6.5. REFERENCES

[1] H. E. Newell, Jr.: "Vector Analysis," McGraw-Hill, New York, 1955.
[2] H. B. Phillips: "Vector Analysis," Wiley, New York, 1933.

CHAPTER 7

TENSOR ANALYSIS

BY

A. SCHILD, Ph.D., Austin, Tex.

7.1. INTRODUCTION

Some geometrical or physical objects can be represented immediately by a number (in suitable units), e.g., the length of a line segment, the mass or temperature of a body. Others are more complicated and cannot be represented by a single number, e.g., velocity, acceleration, force, stress, or strain. However, if a coordinate system is first introduced, then each of the latter objects can be described by a set of numbers, the set of *components* of the object.

There is a great deal of arbitrariness in the choice of a coordinate system. In geometry and physics we are interested in relations or laws which are independent of this arbitrary choice of coordinates. Equations between geometrical or physical objects which have the same form in all coordinate systems are called *invariant*. There are two ways of dealing with invariant relations.

The first is essentially the method of *vector analysis*. Here no coordinates are introduced at all, so that the problem of invariance is simply sidestepped. Objects such as force or displacement, which have magnitude and direction, are immediately represented by symbols $\mathbf{F}$ or $\mathbf{s}$. Operations are defined in terms of the objects themselves. Thus the scalar product $\mathbf{F} \cdot \mathbf{s}$ is defined as the product of the two magnitudes multiplied by the cosine of the angle between the directions of $\mathbf{F}$ and $\mathbf{s}$. Clearly an equation such as $\mathbf{F} \cdot \mathbf{s} = 5$ is invariant. Another example is the divergence of a vector $\mathbf{T}$ given at all points of space; it can be defined invariantly by

$$\operatorname{div} \mathbf{T} = \lim_{\Delta V \to P} \oint_{(S)} \mathbf{T} \cdot \mathbf{n}\, dS$$

where S is the closed surface bounding the infinitesimal volume ΔV which shrinks to the point P, and $\mathbf{n}$ is the unit outward normal to S.

The method of vector analysis becomes clumsy if we deal with (1) objects more complicated than force, e.g., stress or strain; (2) special problems with a complicated geometry which is naturally related to a curvilinear coordinate system, e.g., toroidal coordinates for a problem on the heat conduction in a ring; (3) spaces of dimension higher than three; or (4) curved non-Euclidean spaces, e.g., a shell [1]. In these cases the second method is preferable.

This is the method of *tensor analysis*. Here all possible coordinate systems are considered. By studying how the components of an object transform under an arbitrary change of coordinates, it is possible to single out the invariant relations which remain unchanged.

Tensor analysis has its historical origin in the work on differential geometry by Gauss, Riemann, Beltrami, Christoffel, and others. As a formal system it was first developed and published by Ricci and Levi-Civita [2].

7.2. COORDINATES. TRANSFORMATIONS

In the following, we shall deal throughout with ordinary three-dimensional Euclidean space. A general set of *coordinates* for a point in space is denoted by x^1, x^2, x^3, or x^m ($m = 1, 2, 3$). Thus x^1, x^2, x^3 may represent rectangular cartesians x, y, z; cylindrical polar coordinates r, θ, z; spherical polar coordinates r, θ, ϕ; etc. Where several coordinate systems are considered, they are distinguished by primes: x^m, x'^m, x''^m, etc. A *coordinate transformation* from any one coordinate system to any other is given by general one-to-one functional relations:

$$x'^m = x'^m(x^1,x^2,x^3);\ x^m = x^m(x'^1,x'^2,x'^3) \qquad m = 1, 2, 3 \tag{7.1}$$

The functions are assumed to be well-behaved analytically, i.e., differentiable, etc., and the *Jacobian* of the transformation

$$\left|\frac{\partial x'}{\partial x}\right| \equiv \begin{vmatrix} \dfrac{\partial x'^1}{\partial x^1} & \dfrac{\partial x'^1}{\partial x^2} & \dfrac{\partial x'^1}{\partial x^3} \\ \dfrac{\partial x'^2}{\partial x^1} & \dfrac{\partial x'^2}{\partial x^2} & \dfrac{\partial x'^2}{\partial x^3} \\ \dfrac{\partial x'^3}{\partial x^1} & \dfrac{\partial x'^3}{\partial x^2} & \dfrac{\partial x'^3}{\partial x^3} \end{vmatrix} \tag{7.2}$$

is assumed not to be zero or infinite.

From here on, we shall restrict ourselves to right-handed coordinate systems, i.e., to right-handed rectangular cartesians and to curvilinear coordinates which can be obtained from them by a coordinate transformation with positive Jacobian. It follows from (7.4) below that all coordinate transformations considered will have positive Jacobians: $|\partial x'/\partial x| > 0$. The restriction is not serious, since any coordinate system can be made right-handed by relabeling the coordinates x^1, x^2, x^3 in a suitable order.

For a succession of two coordinate transformations, the chain rule of partial differentiation gives

$$\sum_{n=1}^{3} \frac{\partial x''^m}{\partial x'^n}\frac{\partial x'^n}{\partial x^s} = \frac{\partial x''^m}{\partial x^s} \qquad m, s = 1, 2, 3 \tag{7.3}$$

From the theorem that the determinant of a matrix product equals the product of the determinants of the factor matrices, it follows that

$$\left|\frac{\partial x''}{\partial x'}\right| \cdot \left|\frac{\partial x'}{\partial x}\right| = \left|\frac{\partial x''}{\partial x}\right| \tag{7.4}$$

We shall now introduce two notational conventions due to A. Einstein:

1. *Range convention:* When a small Latin suffix (subscript or superscript) occurs unrepeated in a term, it is understood to take all the values 1, 2, 3.

2. *Summation convention:* When a small Latin suffix is repeated in a term, summation with respect to this suffix is understood over the range 1, 2, 3. Thus, the nine equations (7.3) can be written simply as

$$\frac{\partial x''^m}{\partial x'^n}\frac{\partial x'^n}{\partial x^s} = \frac{\partial x''^m}{\partial x^s} \tag{7.3'}$$

A repeated suffix or a pair of *dummies* can be changed in a single term because of the implied summation; thus $a_{rm}b^m \equiv a_{rn}b^n$. To avoid being led to false conclusions by the conventions, the same suffix must not be used more than twice in any one term; thus $\left(\sum_{m=1}^{3} a_m x^m\right)^2$ will be written $a_m a_n x^m x^n$, but not $a_m a_m x^m x^m$.

The *Kronecker delta* δ_n^m is defined by

$$\delta_n^m = \begin{cases} 1 & \text{if } m = n \\ 0 & \text{if } m \neq n \end{cases} \tag{7.5}$$

It is also known as the *substitution symbol* because of $\delta_n^m a_m = a_n$.

When the coordinate systems x^m and x''^m coincide, (7.3) and (7.4) become

$$\frac{\partial x^m}{\partial x'^n}\frac{\partial x'^n}{\partial x^s} = \delta_s^m \tag{7.6}$$

$$\left|\frac{\partial x}{\partial x'}\right| \cdot \left|\frac{\partial x'}{\partial x}\right| = 1 \tag{7.7}$$

since obviously $\partial x^m/\partial x^s = \delta_s^m$.

7.3. RELATIVE TENSORS. EXAMPLES

The most general objects we shall consider are *relative tensors.* They are defined as follows:

A set of quantities $T_m{}^{rs}{}_n{}^{\cdots}_{\cdots}$, associated with a coordinate system x^m and with a point P, is said to be the components of a relative tensor of weight W if they transform, on change to any other coordinate system x'^m, according to the equation

$$T'_m{}^{rs}{}_n{}^{\cdots}_{\cdots} = \left|\frac{\partial x}{\partial x'}\right|^W \frac{\partial x^a}{\partial x'^m}\frac{\partial x^b}{\partial x'^n} \cdots \frac{\partial x'^r}{\partial x^c}\frac{\partial x'^s}{\partial x^d} \cdots T_a{}^{cd}{}_b{}^{\cdots}_{\cdots} \tag{7.8}$$

where the partial derivatives are evaluated at P.

Relative tensors of weight zero are called *absolute tensors* or simply *tensors,* since they are the most common. Relative tensors of weight one are also called *tensor densities.* The total number of free suffixes is called the *order* of the relative tensor. Suffixes, such as r or s in (7.8), which correspond to a factor of the type $\partial x'^r/\partial x^c$ in the transformation equation are called *contravariant* suffixes and are always written as superscripts. Suffixes, such as m or n in (7.8), which correspond to a factor of the type $\partial x^a/\partial x'^m$, are called *covariant* suffixes and are always written as subscripts. Relative tensors with only contravariant suffixes are called *contravariant relative tensors;* those with only covariant suffixes are *covariant relative tensors;* others are *mixed relative tensors.* First-order relative tensors are also called (contravariant or covariant) *relative vectors.* Zeroth-order relative tensors are also called *relative invariants* or *relative scalars.*

A relative tensor defined at all points of a portion of space or throughout all space is called a *relative tensor field.*

The tensor transformation law is (1) reflexive, (2) symmetric, and (3) transitive. By this we mean that (1) the tensor components remain unchanged under the identity transformation $x'^m = x^m$ [follows from (7.8)]; (2) under the inverse transformation from x'^m to x^m, the tensor components $T'_m{}^{rs}{}_n{}^{\cdots}_{\cdots}$ are transformed to their original values $T_m{}^{rs}{}_n{}^{\cdots}_{\cdots}$ [follows from (7.8) and (7.6), (7.7)]; and (3) the sequence of transformations from x^m to x'^m and from x'^m to x''^m gives the same final components $T''_m{}^{rs}{}_n{}^{\cdots}_{\cdots}$ as does the direct transformation from x^m to x''^m [follows from (7.8) and (7 3), (7.4)].

The most important property of the tensor transformation law is that it 's *linear* and *homogeneous;* i.e., each new component, $T'_m{}^{rs}{}_n{}^{\cdots}_{\cdots}$, is a linear homogeneous combination of all the old components, $T_1{}^{11}{}_1{}^{\cdots}_{\cdots}$, $T_2{}^{11}{}_1{}^{\cdots}_{\cdots}$, etc. As a result, *if all the components of a relative tensor vanish in one coordinate system, then they vanish in all coordinate systems;* that is, $T_m{}^{rs}{}_n{}^{\cdots}_{\cdots} = 0$ implies $T'_m{}^{rs}{}_n{}^{\cdots}_{\cdots} = 0$. This is the first example of an invariant or *tensor equation.* In fact, all tensor equations are essentially equivalent to the vanishing of a relative tensor and are therefore invariant.

We shall now give some examples of relative tensors:

1. Ordinary numbers (-2.5, π, etc.) are invariants. In Newtonian mechanics the mass of a body is an invariant. The temperature distribution in a room or a solid is an *invariant field,* and transforms according to the simple equation

$$T' = T \tag{7.9}$$

2. The coordinate differences dx^m between two neighboring points $P \equiv (x^1,x^2,x^3)$ and $Q \equiv (x^1 + dx^1,\ x^2 + dx^2,\ x^3 + dx^3)$ are called the components of an *infinitesimal displacement.* They transform according to the equation (formula for total differential)

$$dx'^m = \frac{\partial x'^m}{\partial x^n}\, dx^n \tag{7.10}$$

and therefore form a *contravariant vector.* Similarly, the velocity of a particle

$$v^m = dx^m/dt$$

is a contravariant vector, and the velocities of the particles of a fluid constitute a contravariant vector field. If $x^m = x^m(u)$ are the parametric equations of a curve in space, then dx^m/du is a contravariant vector, the *tangent* vector to the curve.

3. Let T be an invariant field. Then the partial derivatives $\partial T/\partial x^m$ are called the components of the *gradient* of T. They transform according to the equation (chain rule for partial differentiation)

$$\frac{\partial T'}{\partial x'^m} = \frac{\partial x^n}{\partial x'^m}\frac{\partial T}{\partial x^n} \tag{7.11}$$

and therefore form a *covariant vector field.*

4. The Kronecker delta is a mixed second-order tensor. This follows from

$$\frac{\partial x'^m}{\partial x^a}\frac{\partial x^b}{\partial x'^n}\,\delta^a_b = \frac{\partial x'^m}{\partial x^a}\frac{\partial x^a}{\partial x'^n} = \delta^m_n = \delta'^m_n$$

5. The *permutation symbol* ϵ_{mns} or ϵ^{mns} is defined in all coordinate systems by

$$\epsilon_{mns} = \epsilon^{mns} = \begin{cases} 0 & \text{if } m, n, s \text{ are not all distinct} \\ 1 & \text{if } m, n, s \text{ is an even permutation of } 1, 2, 3 \\ -1 & \text{if } m, n, s \text{ is an odd permutation of } 1, 2, 3 \end{cases} \tag{7.12}$$

Consider the expression $\epsilon_{abc}(\partial x^a/\partial x'^m)(\partial x^b/\partial x'^n)(\partial x^c/\partial x'^s)$. By the definition of ϵ_{abc} and of a determinant, this is equal to

$$\begin{vmatrix} \dfrac{\partial x^1}{\partial x'^m} & \dfrac{\partial x^1}{\partial x'^n} & \dfrac{\partial x^1}{\partial x'^s} \\ \dfrac{\partial x^2}{\partial x'^m} & \dfrac{\partial x^2}{\partial x'^n} & \dfrac{\partial x^2}{\partial x'^s} \\ \dfrac{\partial x^3}{\partial x'^m} & \dfrac{\partial x^3}{\partial x'^n} & \dfrac{\partial x^3}{\partial x'^s} \end{vmatrix}$$

If m, n, s are not all distinct, this determinant vanishes, since two or more columns are equal. If m, n, s are distinct, the determinant is plus or minus the Jacobian $|\partial x/\partial x'|$, depending on whether m, n, s is an even or odd permutation of 1, 2, 3. Thus, in all cases the original expression equals $|\partial x/\partial x'|\ \epsilon_{mns}$ or, equivalently, $|\partial x/\partial x'|\ \epsilon'_{mns}$, so that

$$\epsilon'_{mns} = \left|\frac{\partial x}{\partial x'}\right|^{-1}\frac{\partial x^a}{\partial x'^m}\frac{\partial x^b}{\partial x'^n}\frac{\partial x^c}{\partial x'^s}\,\epsilon_{abc} \tag{7.13}$$

This shows that the permutation symbol is a *third-order covariant relative tensor of weight* -1. Similarly, by considering $\epsilon^{abc}\,(\partial x'^m/\partial x^a)(\partial x'^n/\partial x^b)(\partial x'^s/\partial x^c)$, it is shown that the permutation symbol is also a *third-order contravariant tensor density.*

Important identities involving permutation symbols are

$$\epsilon_{mns} = \epsilon_{nsm} = \epsilon_{smn} = -\epsilon_{nms} = -\epsilon_{msn} = -\epsilon_{snm} \tag{7.14}$$

$$\tfrac{1}{6}\epsilon_{mns}\epsilon^{mns} = 1 \qquad \tfrac{1}{2}\epsilon_{mns}\epsilon^{mnr} = \delta^r_s \qquad \epsilon_{mrs}\epsilon^{mab} = \delta^a_r\delta^b_s - \delta^b_r\delta^a_s \tag{7.15}$$

The first set is obvious; the second set can be checked by noting that both sides are equal for all choices of the free suffixes.

6. Let a_{mn} be a second-order covariant tensor. Its determinant is denoted by $a \equiv |a_{mn}|$. Then

$$a'_{mn} = \frac{\partial x^a}{\partial x'^m}\frac{\partial x^b}{\partial x'^n}a_{ab}$$

Taking determinants on both sides gives

$$a' = \left|\frac{\partial x}{\partial x'}\right|^2 a \qquad \sqrt{a'} = \left|\frac{\partial x}{\partial x'}\right|\sqrt{a} \tag{7.16}$$

Thus, a is a relative invariant of weight 2, and $\sqrt{a}$ is an invariant density.

7.4. TENSOR ALGEBRA

Addition and Subtraction

Two relative tensors of the same weight, order, and type and associated with the same point in space may be added or subtracted to give another tensor of the same weight, order, and type and at the same point. Thus, for example, if $A^m{}_{ns}$ and $B_s{}^m{}_n$ are both relative tensors of weight W, then $S^m{}_{ns}$ and $D^m{}_{ns}$ defined by

$$S^m{}_{ns} = A^m{}_{ns} + B_s{}^m{}_n \qquad D^m{}_{ns} = A^m{}_{ns} - B_s{}^m{}_n \tag{7.17}$$

are also relative tensors of weight W. This follows immediately from the linearity and homogeneity of the tensor transformation law (7.8).

Symmetry Properties

The *symmetry* of a relative tensor with respect to two suffixes at the same level, for example, $S_{mn}{}^s = S_{nm}{}^s$, is conserved under coordinate transformations; and so is *skew-symmetry* or *antimetry;* for example, $A^{mn} = -A^{nm}$. This follows from the fact that $S_{mn}{}^s - S_{nm}{}^s$ and $A^{mn} + A^{nm}$ are themselves relative tensors, and so vanish in all coordinate systems if they vanish in one. Similarly, more complicated symmetries, such as the cyclic symmetry given by $C_{mns} + C_{nsm} + C_{smn} = 0$, are tensorial. However, symmetries with respect to suffixes at different levels, for example, $D_m{}^n = D_n{}^m$, are *not* tensorial and are of no interest, since they may hold in one coordinate system and not hold in another.

Any covariant (or contravariant) second-order relative tensor can be written as the sum of a symmetric and a skew-symmetric relative tensor:

$$T_{mn} = S_{mn} + A_{mn} \qquad S_{mn} = \tfrac{1}{2}(T_{mn} + T_{nm}) \qquad A_{mn} = \tfrac{1}{2}(T_{mn} - T_{nm}) \tag{7.18}$$

If a_{mn} is symmetric and b^{mn} is skew-symmetric, then

$$a_{mn}b^{mn} = 0 \tag{7.19}$$

This can be shown by writing the left side out in full, or else by noting that

$$a_{mn}b^{mn} = a_{nm}b^{nm}$$

(by change of dummies), $= a_{mn}b^{nm}$ (by symmetry of a_{mn}), $= -a_{mn}b^{mn}$ (by skew-symmetry of b^{mn}).

Outer Multiplication

Any two relative tensors of the same or of different types and with different literal suffixes, but associated with the same point in space, may simply be written in juxtaposition. The resulting product will be a relative tensor whose weight and order are respectively equal to the sums of the weights and orders of the factors. Thus, if $A^m{}_{ns}$ and B^{rt} are relative tensors of weights W_1 and W_2 and of the types indicated, then

$$C^m{}_{ns}{}^{rt} = A^m{}_{ns}B^{rt} \tag{7.20}$$

is a fifth-order relative tensor of weight $(W_1 + W_2)$. The proof follows from the structure of (7.8).

Contraction

If in a mixed relative tensor a contravariant and a covariant suffix are replaced by a pair of dummies, then there results a new relative tensor of the same weight and of the type indicated by the remaining free suffixes, so that the order is reduced by 2. This may easily be proved using Eqs. (7.6) and (7.8). An example is the *trace* of a tensor $T^m{}_n$ defined as tr $T^m{}_n = T^m{}_m$. Note that contraction must not be performed on the same level; e.g., if $P_m{}^n$ and Q_{mn} are tensors, then $P_m{}^m$ is an invariant, but Q_{mm} is *not* an invariant.

Inner Multiplication

This is a combination of outer multiplication and contraction. Thus, if A_{mn} and $B_r{}^s$ are relative tensors, contraction of the outer product $A_{mn}B_r{}^s$ gives the inner products $A_{mn}B_r{}^m$, $A_{mn}B_r{}^n$. The inner product of a covariant vector A_m and of a contravariant vector B^m is the invariant A_mB^m, which is also called the *scalar product* of the two vectors.

Tensor Equation

The different tensor operations may, of course, be combined to give new relative tensors. An equation in which each term is a relative tensor of the same type and weight is a tensor equation which is invariant under coordinate transformations, since it is equivalent to the vanishing of a relative tensor. As an example, consider the tensor equation

$$A_m{}^n = C_{rm}D^{nr} + E_mF^n \tag{7.21}$$

where $A_m{}^n$, C_{rm}, D^{nr}, E_m, F^n are relative tensors of the type indicated and of respective weights W_1, W_2, W_3, W_4, W_5, such that $W_1 = W_2 + W_3 = W_4 + W_5$. This equation is valid in all coordinate systems if it holds in one, since it is equivalent to the vanishing of all components of the mixed second-order relative tensor of weight W_1: $T_m{}^n \equiv A_m{}^n - C_{rm}D^{nr} - E_mF^n$.

Conjugate Second-order Tensor

If a_{mn} is a covariant second-order tensor of nonzero determinant ($a \neq 0$), then a conjugate set of components A^{mn} is defined uniquely by the equations

$$a_{mn}A^{mr} = \delta^r_n \tag{7.22}$$

Let Δ^{mn} be the cofactor of a_{mn} in the determinant a. Cramér's rule for the solution of linear equations shows that the A^{mn} are given by the explicit expressions

$$A^{mn} = \frac{\Delta^{mn}}{a} \tag{7.23}$$

It will now be shown that the A^{mn} form a contravariant second-order tensor. Define

$\bar{A}^{mn}$ as being given in any one coordinate system x^m by (7.22) or (7.23) and as being obtained in any other coordinate system x'^m by the transformation equations for a contravariant tensor. Then $\bar{A}^{mn}$ is a tensor, and $a_{mn}\bar{A}^{mr} = \delta_n^r$ is a tensor equation which, being valid by definition in the original coordinates x^m, holds in all other coordinate systems. Thus $a'_{mn}\bar{A}'^{mr} = \delta_n^r$. The quantities A'^{mr} are defined by equations analogous to (7.22): $a'_{mn}A'^{mr} = \delta_n^r$. Since the solution of these linear equations is unique, $A'^{mn} = \bar{A}'^{mn}$ in an arbitrary coordinate system x'^m, and thus, since $\bar{A}^{mn}$ is a tensor, so is A^{mn}.

The relation between a_{mn} and A^{mn} is a reciprocal one; a_{mn} and A^{mn} are called conjugate tensors. The equation (7.22) can also be used to define conjugate relative tensors whose weights are negatives of each other. If a_{mn} is symmetric, then so are the cofactors Δ^{mn} and hence, by (7.23), A^{mn} is symmetric.

Differentiation

From here on, it will be convenient to denote partial differentiation with respect to coordinates by subscripts following a comma; thus $T_{,m} \equiv \partial T/\partial x^m$, $T'_{,m} \equiv \partial T'/\partial x'^m$, $T^r{}_{m,st} \equiv \partial^2 T^r{}_m/\partial x^s\, \partial x^t$, etc. Since it is immaterial in what order partial differentiations of a well-behaved function are performed, expressions are always symmetric in all suffixes which follow a comma, for example, $T^r{}_{m,st} = T^r{}_{m,ts}$.

On p. 7–4 it was shown that the partial derivatives $T_{,m}$ of an invariant form a tensor. This property does not generalize to other relative tensor fields. The reason is that such differentiation with respect to position in space essentially involves the subtraction of two relative tensors at *different* (though neighboring) points, and this is not a tensor operation. As an example, consider a covariant vector field T_m, whose transformation law is

$$T'_m = \frac{\partial x^a}{\partial x'^m} T_a \tag{7.24}$$

Differentiating with respect to x'^n, and remembering that $\partial/\partial x'^n \equiv (\partial x^b/\partial x'^n)(\partial/\partial x^b)$, this becomes

$$T'_{m,n} = \frac{\partial x^a}{\partial x'^m}\frac{\partial x^b}{\partial x'^n} T_{a,b} + \frac{\partial^2 x^a}{\partial x'^m \partial x'^n} T_a \tag{7.25}$$

The first term on the right-hand side is just the one giving the transformation law for a second-order covariant tensor; the second term, which contains the second derivatives of the coordinate transformation and which does not contain $T_{a,b}$, spoils the tensor character of the transformation equation (7.25). This illustrates that partial differentiation is, in general, *not* a tensor operation. A suitable general substitute for common differentiation will be discussed in Sec. 7.6.

In spite of the statements above, certain special combinations of partial derivatives are tensorial. Thus, in the transformation law for the skew-symmetric combination $T_{m,n} - T_{n,m}$, the last term in (7.25) cancels because of symmetry so that $T_{m,n} - T_{n,m}$ is a second-order covariant tensor. Similarly, if $F_{mn} = -F_{nm}$ is a skew-symmetric covariant tensor, then $F_{mn,r} + F_{nr,m} + F_{rm,n}$ is a third-order covariant tensor; if D^m is a contravariant vector density, then $D^m{}_{,m}$ is an invariant density; if $S^{mn} = -S^{nm}$ is a skew-symmetric contravariant tensor density, then $S^{mn}{}_{,n}$ is a contravariant vector density.

7.5. METRIC TENSOR. LINE ELEMENT

Let x'^m be a rectangular cartesian coordinate system. By Pythagoras' theorem, the square of the distance ds between two neighboring points with infinitesimal displacement dx'^m is

$$ds^2 = (dx'^1)^2 + (dx'^2)^2 + (dx'^3)^2 \tag{7.26}$$

Define g'_{mn} as being given by $g'_{11} = g'_{22} = g'_{33} = 1$, $g'_{mn} = 0$ $(m \neq n)$, in the special rectangular cartesian coordinate system x'^m and as being obtained in any other coordinate system by the transformation equations for a second-order covariant tensor. Then

$$ds^2 = g_{mn}\, dx^m\, dx^n \tag{7.27}$$

is an invariant and thus, since (7.27) agrees with (7.26) in the special coordinates x'^m, it gives the square of the distance between neighboring points in an arbitrary coordinate system.

The tensor field g_{mn} and its conjugate g^{mn}, given by

$$g_{mn}g^{mr} = \delta^r_n \tag{7.28}$$

are called the *metric tensors; ds^2* is called the *metric form*, the *fundamental form*, or the square of the *line element*. Clearly, the metric tensors are symmetric

$$g_{mn} = g_{nm} \qquad g^{mn} = g^{nm} \tag{7.29}$$

The determinant $g = |g_{mn}|$ is always positive, since in rectangular cartesians $g' = 1$, and since $g = |\partial x'/\partial x|^2 g'$, $|\partial x'/\partial x| \neq 0$. In fact, the quadratic form, i.e., homogeneous quadratic polynomial, (7.27) is *positive-definite;* this means that $ds^2 > 0$ if $dx^m \neq 0$, and this follows from the invariance of ds^2 and from the special form (7.26).

If the line element is given in some coordinate system, then the components of the metric tensor can be read off immediately; this is so because identical polynomials have equal coefficients. Thus, if $(x^1,x^2,x^3) \equiv (r,\phi,\theta)$ are spherical polars (see Fig. 4.12), it follows from elementary geometry that

$$ds^2 = (dx^1)^2 + (x^1)^2(dx^2)^2 + (x^1)^2 \sin^2 x^2 (dx^3)^2$$

hence, $g_{11} = 1$, $g_{22} = (x^1)^2$, $g_{33} = (x^1)^2 \sin^2 x^2$, $g_{mn} = 0$ $(m \neq n)$. In more complicated coordinates, the metric tensor is obtained from the formula

$$g_{mn} = \frac{\partial x'^1}{\partial x^m}\frac{\partial x'^1}{\partial x^n} + \frac{\partial x'^2}{\partial x^m}\frac{\partial x'^2}{\partial x^n} + \frac{\partial x'^3}{\partial x^m}\frac{\partial x'^3}{\partial x^n} \tag{7.30}$$

where x'^m are rectangular cartesians.

The metric tensors are used in the operations of *raising* and *lowering suffixes.* To raise a subscript, form an inner product with the contravariant metric tensor; to lower a superscript, form an inner product with the covariant metric tensor. For example,

$$g^{mn}T_{nr}{}^{s} = T^{m}{}_{r}{}^{s} \qquad g_{mn}S^n = S_m \tag{7.31}$$

These operations change the type of a relative tensor as indicated, but leave the order and weight unchanged. The raising and lowering of a suffix are inverse operations; i.e., if a suffix is first raised and then lowered (or vice versa), the original relative tensor is regained; this follows from (7.28) and (7.29). For example, $g_{nm}T^{m}{}_{r}{}^{s} = T_{nr}{}^{s}$, the original relative tensor of (7.31). Relative tensors obtained from each other by raising and lowering suffixes are denoted by the same principal letter [as in (7.31)] and are regarded as the "same relative tensor" in a different representation. This is perhaps loose language, but experience shows that it can be used safely and effectively. Thus the covariant velocity vector v_m of a particle is automatically understood to mean $g_{mn}v^n = g_{mn}\, dx^n/dt$. It is to avoid confusion when carrying out the operations of raising and lowering suffixes that we never write two suffixes one underneath the other; the only exception is the Kronecker delta. It is easily seen by using (7.28) that the metric tensors g_{mn} and g^{mn} can be obtained from each other by raising or lowering two suffixes; the mixed metric tensor is $g^{m}{}_{n} = \delta^m_n$.

The determinant of the metric tensor

$$g = |g_{mn}| \tag{7.32}$$

is a relative invariant of weight 2 [cf. (7.16)]. It can be used to *change the weight* of a relative tensor by multiplication. In particular, to any relative tensor $T_r{}^s{}^{\cdots}_{\cdots}$ of weight W, there corresponds the absolute tensor

$$\tilde{T}_r{}^s{}^{\cdots}_{\cdots} = g^{-\frac{1}{2}W} T_r{}^s{}^{\cdots}_{\cdots} \tag{7.33}$$

Thus

$$\tilde{\epsilon}_{mns} = \sqrt{g}\,\epsilon_{mns} \qquad \tilde{\epsilon}^{mns} = \frac{1}{\sqrt{g}}\,\epsilon^{mns} \tag{7.34}$$

are absolute skew-symmetric tensors. It is easily seen that $\tilde{\epsilon}_{mns}$ and $\tilde{\epsilon}^{mns}$ can be obtained from each other by raising or lowering three suffixes; this is *not* true of ϵ_{mns} and ϵ^{mns}.

The $\tilde{\epsilon}$s are used to define *duality of skew-symmetric tensors.* If A, B_m, C^m, $D_{mn}(= -D_{nm})$, $E^{mn}(= -E^{nm})$, $F_{mns}(= -F_{nms} = -F_{msn})$, $G^{mns}(= -G^{nms} = -G^{msn})$ are skew-symmetric tensors of the type indicated, then their duals are the skew-symmetric tensors defined by

$$\begin{aligned} \hat{A}^{mns} &= \tilde{\epsilon}^{mns}A & \hat{A}_{mns} &= \tilde{\epsilon}_{mns}A \\ \hat{B}^{mn} &= \tilde{\epsilon}^{mns}B_s & \hat{C}_{mn} &= \tilde{\epsilon}_{mns}C^s \\ \hat{D}^m &= \tfrac{1}{2}\tilde{\epsilon}^{mns}D_{ns} & \hat{E}_m &= \tfrac{1}{2}\tilde{\epsilon}_{mns}E^{ns} \\ \hat{F} &= \tfrac{1}{6}\tilde{\epsilon}^{mns}F_{mns} & \hat{G} &= \tfrac{1}{6}\tilde{\epsilon}_{mns}G^{mns} \end{aligned} \tag{7.35}$$

From (7.15) it follows that duality is a reciprocal relationship; i.e., the dual of $\hat{B}^{mn}$ is $\hat{\hat{B}}_s = B_s$, the dual of $\hat{A}_{mns}$ is $\hat{\hat{A}} = A$, etc. Duality can also be extended to skew-symmetric relative tensors, a pair of duals having the same weight.

The ordinary operations of vector algebra can now be performed in the tensor notation. The square of the *magnitude* T of a vector T^m is

$$(T)^2 = T^m T_m = g_{mn}T^m T^n \tag{7.36}$$

The scalar product $S = \mathbf{A} \cdot \mathbf{B}$ of vectors A^m, B^m, whose directions form an angle θ, is

$$S = A^m B_m = g_{mn}A^m B^n = AB\cos\theta \tag{7.37}$$

$$\cos\theta = \frac{A^m B_m}{\sqrt{A^r A_r}\cdot\sqrt{B^s B_s}} \tag{7.38}$$

The dual of S is

$$\hat{S}^{mns} = A^m\hat{B}^{ns} + A^n\hat{B}^{sm} + A^s\hat{B}^{mn} \tag{7.39}$$

The vector product $\mathbf{V} = \mathbf{A} \times \mathbf{B}$ of vectors A_m, B_m is represented by

$$V^m = \tilde{\epsilon}^{mns}A_n B_s = -\hat{A}^{ms}B_s = -A_n\hat{B}^{nm} \tag{7.40}$$

or by its dual

$$\hat{V}_{mn} = A_m B_n - A_n B_m \tag{7.41}$$

$$V = AB\sin\theta \tag{7.42}$$

and V^m is perpendicular to both A_m and B_m. All the above follows from the fact that the equations are tensor equations, valid in rectangular cartesians, and therefore true generally.

The usual formulas of vector algebra follow immediately, and do not require any commitment to memory except for the one identity

$$\tilde{\epsilon}^{mrs}\tilde{\epsilon}_{mab} = \delta^r_a\,\delta^s_b - \delta^r_b\,\delta^s_a \tag{7.43}$$

which is equivalent to (7.15). Thus **A × A** is

$$\tilde{\epsilon}^{mns}A_nA_s = 0 \tag{7.44}$$

since $\tilde{\epsilon}^{mns}$ is skew-symmetric in n,s and A_nA_s is symmetric [cf. (7.19)]. The identity **A · B × C = A × B · C** for triple scalar products is due to the fact that either side can be written as

$$\tilde{\epsilon}_{mns}A^mB^nC^s \tag{7.45}$$

Equation (7.43) yields the identity **A × (B × C) = A · CB − A · BC** for triple vector products:

$$\begin{aligned}\tilde{\epsilon}^{mns}A_n(\tilde{\epsilon}_{sab}B^aC^b) &= \tilde{\epsilon}^{smn}\tilde{\epsilon}_{sab}A_nB^aC^b \\ &= (\delta^m_a\delta^n_b - \delta^m_b\delta^n_a)A_nB^aC^b \\ &= A_nC^nB^m - A_nB^nC^m\end{aligned} \tag{7.46}$$

7.6. COVARIANT AND ABSOLUTE DERIVATIVES

The *Christoffel symbol of the first kind* is defined as the following combination of the partial derivatives of the metric tensor:

$$\Gamma_{mnr} = \tfrac{1}{2}(g_{nr,m} + g_{mr,n} - g_{mn,r}) \tag{7.47}$$

The more important *Christoffel symbol of the second kind* is obtained by the formal operation of raising the last suffix of Γ_{mnr}:

$$\Gamma^s_{mn} = g^{sr}\Gamma_{mnr} \qquad \Gamma_{mnr} = g_{rs}\Gamma^s_{mn} \tag{7.48}$$

The following identities hold:

$$\Gamma_{mnr} = \Gamma_{nmr} \qquad \Gamma^s_{mn} = \Gamma^s_{nm} \tag{7.49}$$

$$\Gamma_{rmn} + \Gamma_{rnm} = g_{mn,r} \tag{7.50}$$

$$\Gamma^n_{rn} = \frac{1}{\sqrt{g}}(\sqrt{g})_{,r} = (\ln\sqrt{g})_{,r} \tag{7.51}$$

All are obvious except the last. To prove (7.51), ignore for the moment the symmetry of g_{mn}, and regard g as a function of nine independent components g_{mn}, so that $\partial g/\partial g_{mn} = \Delta^{mn} = gg^{mn}$, where Δ^{mn} is the cofactor of g_{mn} in the determinant g. Then

$$\Gamma^n_{rn} = \tfrac{1}{2}g^{mn}g_{mn,r} = \frac{1}{2g}\frac{\partial g}{\partial g_{mn}}g_{mn,r} = \frac{1}{2g}g_{,r}$$

Equation (7.51) follows.

A somewhat lengthy but completely straightforward calculation shows that the transformation equation for Christoffel symbols is

$$\Gamma'^r_{mn} = \frac{\partial x'^r}{\partial x^c}\frac{\partial x^a}{\partial x'^m}\frac{\partial x^b}{\partial x'^n}\Gamma^c_{ab} + \frac{\partial x'^r}{\partial x^a}\frac{\partial^2 x^a}{\partial x'^m\partial x'^n} \tag{7.52}$$

As in (7.25), the first term on the right-hand side corresponds to a tensor transformation law, but the second term spoils the tensor character. If (7.52) is multiplied by a covariant vector T'_r on the left, and by the equivalent expression $(\partial x^s/\partial x'^r)T_s$ on the right, it is seen that the last term becomes the same as the last term in (7.25). Subtraction eliminates this unwanted term to give

$$T'_{m,n} - \Gamma'^r_{mn}T'_r = \frac{\partial x^a}{\partial x'^m}\frac{\partial x^b}{\partial x'^n}(T_{a,b} - \Gamma^c_{ab}T_c) \tag{7.53}$$

This shows that

$$T_m|_n = T_{m,n} - \Gamma^r_{mn}T_r \tag{7.54}$$

is a second-order covariant tensor. It is called the *covariant derivative* of T_m with respect to x^n.

In a similar way it can be shown, for a general relative tensor field $T_m{}^{rs}{}_n{}^{\cdots}_{\cdots}$ of weight W, that the *covariant derivative*, defined by

$$T_m{}^{rs}{}_n{}^{\cdots}_{\cdots}|_p = T_m{}^{rs}{}_n{}^{\cdots}_{\cdots,p} - W\Gamma^a_{ap}T_m{}^{rs}{}_n{}^{\cdots}_{\cdots} - \Gamma^a_{mp}T_a{}^{rs}{}_n{}^{\cdots}_{\cdots} - \Gamma^a_{np}T_m{}^{rs}{}_a{}^{\cdots}_{\cdots} - \cdots + \Gamma^r_{ap}T_m{}^{as}{}_n{}^{\cdots}_{\cdots} + \Gamma^s_{ap}T_m{}^{ra}{}_n{}^{\cdots}_{\cdots} + \cdots \tag{7.55}$$

is a relative tensor of the same weight W and of order higher by one, p being an additional covariant suffix. Covariant differentiation is the tensorial substitute for partial differentiation. It will always be denoted by suffixes following a vertical stroke; for example, $T^r{}_s|_{pq} \equiv (T^r{}_s|_p)|_q$ is the second covariant derivative of $T^r{}_s$ with respect to x^p and x^q.

If a relative tensor field $T_m{}^r\cdots$ is defined along a curve $x^r(u)$, so that its components are functions of the parameter u, then the ordinary derivative $dT_m{}^r{}^{\cdots}_{\cdots}/du$ is *not* tensorial. The tensorial substitute for the ordinary derivative is easily shown to be the *absolute derivative*

$$\frac{\delta T_m{}^r{}^{\cdots}_{\cdots}}{\delta u} = \frac{dT_m{}^r{}^{\cdots}_{\cdots}}{du} - W\Gamma^a_{ap}T_m{}^r{}^{\cdots}_{\cdots}\frac{dx^p}{du} - \Gamma^a_{mp}T_a{}^r{}^{\cdots}_{\cdots}\frac{dx^p}{du} - \cdots + \Gamma^r_{ap}T_m{}^a{}^{\cdots}_{\cdots}\frac{dx^p}{du} + \cdots \tag{7.56}$$

$\delta T_m{}^r{}^{\cdots}_{\cdots}/\delta u$ is a relative tensor of the same weight, order, and type as $T_m{}^r{}^{\cdots}_{\cdots}$. If the tensor field is, in fact, defined throughout space, then clearly

$$\frac{\delta T_m{}^r{}^{\cdots}_{\cdots}}{\delta u} = T_m{}^r{}^{\cdots}_{\cdots}|_p \frac{dx^p}{du} \tag{7.57}$$

We shall now state properties of covariant and absolute differentiation.

1. The metric tensors, the Kronecker delta, the ϵs, and $\tilde{\epsilon}$s all behave like constants under covariant or absolute differentiation; i.e., they have zero derivatives:

$$g_{mn}|_p = 0 \qquad g^{mn}|_p = 0 \qquad g|_p = 0 \qquad \delta^n_m|_p = 0$$
$$\epsilon_{mns}|_p = 0 \qquad \epsilon^{mns}|_p = 0 \qquad \tilde{\epsilon}_{mns}|_p = 0 \qquad \tilde{\epsilon}^{mns}|_p = 0 \tag{7.58}$$

2. All the usual rules of the calculus apply to covariant and absolute derivatives; e.g.,

$$\frac{\delta T_m{}^r{}^{\cdots}_{\cdots}}{\delta u} = \frac{dv}{du}\frac{\delta T_m{}^r{}^{\cdots}_{\cdots}}{\delta v} \tag{7.59}$$

$$(T_m{}^r{}^{\cdots}_{\cdots} + S_m{}^r{}^{\cdots}_{\cdots})|_p = T_m{}^r{}^{\cdots}_{\cdots}|_p + S_m{}^r{}^{\cdots}_{\cdots}|_p \tag{7.60}$$
$$(A_m{}^r{}^{\cdots}_{\cdots}B_{ab}{}^{\cdots}_{\cdots})|_p = A_m{}^r{}^{\cdots}_{\cdots}|_p\, B_{ab}{}^{\cdots}_{\cdots} + A_m{}^r{}^{\cdots}_{\cdots}B_{ab}{}^{\cdots}_{\cdots}|_p \tag{7.61}$$
$$T_m{}^r{}^{\cdots}_{\cdots}|_{pq} = T_m{}^r{}^{\cdots}_{\cdots}|_{qp} \tag{7.62}$$

Equations (7.58) to (7.62) could be checked directly from the definitions. However, a simpler and much more instructive proof can be given:

In rectangular cartesian coordinates, g_{mn} are constants (in fact, 0 or 1) and therefore, by (7.47) and (7.48), all Christoffel symbols vanish. It then follows from (7.55) and (7.56) that covariant and absolute differentiation reduces to partial and ordinary differentiation in rectangular cartesians. Now equations (7.58) to (7.62) are tensor equations. They hold in rectangular cartesians, since they reduce to well-known facts of the ordinary calculus. Hence, they are generally valid.

A tensor field $T_r{}^s{}^{\cdots}_{\cdots}$ is called *parallel* or *constant* along a curve or throughout space if its rectangular cartesian components are constants. It is clear that the conditions for parallelism in general coordinates are respectively

$$\frac{\delta T_r{}^{s\cdots}_{\cdots}}{\delta u} = 0 \tag{7.63}$$

$$T_r{}^{s\cdots}_{\cdots}|_p = 0 \tag{7.64}$$

A curve $x^r(s)$, s being the arc length, is a straight line if its unit tangent vectors dx^r/ds are parallel. Thus, in an arbitrary coordinate system, the necessary and sufficient conditions for a *straight line* are

$$\frac{\delta}{\delta s}\frac{dx^r}{ds} = \frac{d^2x^r}{ds^2} + \Gamma^r_{mn}\frac{dx^m}{ds}\frac{dx^n}{ds} = 0 \tag{7.65}$$

These are also easily seen to be the Euler-Lagrange equations of the variational principle $\delta\int ds = 0$. It may be mentioned in passing that often the quickest way of finding the Christoffel symbols is to first obtain the differential equations of $\delta\int ds = 0$ and then, by comparison with (7.65), to read off all nonvanishing Γ^r_{mn}.

In certain covariant derivatives, the Christoffel symbols cancel. If S is an invariant, T_m a covariant vector, $F_{mn} = -F_{nm}$ a skew-symmetric covariant tensor, D^m a contravariant vector density, and $S^{mn} = -S^{nm}$ a skew-symmetric contravariant tensor density, then (7.55) shows that

$$S|_p = S_{,p} \tag{7.66}$$

$$T_m|_n - T_n|_m = T_{m,n} - T_{n,m} \tag{7.67}$$

$$F_{mn}|_r + F_{nr}|_m + F_{rm}|_n = F_{mn,r} + F_{nr,m} + F_{rm,n} \tag{7.67'}$$

$$D^m|_m = D^m{}_{,m} \tag{7.68}$$

$$S^{mn}|_n = S^{mn}{}_{,n} \tag{7.69}$$

This provides an alternate proof of the statements at the end of Sec. 7.4.

The differential operations of ordinary vector analysis can now be discussed. It should be kept in mind that there are many additional operations on tensors, for example, $T_r{}^{mn}|_{mn}$, which have no counterpart in elementary vector analysis.

The *gradient* of a relative tensor field is simply its first covariant derivative; for an invariant T, (7.66) shows that the gradient is $T_{,m}$. The *divergence* of a contravariant vector field T^m is

$$T^m|_m = \frac{1}{\sqrt{g}}(\sqrt{g}\,T^m)_{,m} \tag{7.70}$$

the second expression following from (7.68) and the facts that $\sqrt{g}\,T^m$ is a vector density and that $(\sqrt{g})|_m = 0$. Combining divergence and gradient, the Laplacian ∇^2 is defined as div grad or $\nabla^2 T^{\cdot\cdot}_{\cdot\cdot} = g^{mn}T^{\cdot\cdot}_{\cdot\cdot}|_{mn}$, $T^{\cdot\cdot}_{\cdot\cdot}$ being any relative tensor field. For an invariant T, this gives the simple expression

$$\nabla^2 T = g^{mn}T|_{mn} = \frac{1}{\sqrt{g}}(\sqrt{g}\,g^{mn}T_{,n})_{,m} \tag{7.71}$$

The curl C^r and its dual $\hat{C}_{mn}$, of a covariant vector field T_m, are defined by

$$\hat{C}_{mn} = T_n|_m - T_m|_n = T_{n,m} - T_{m,n} \qquad C^r = \tfrac{1}{2}\epsilon^{rmn}\hat{C}_{mn} = \epsilon^{rmn}T_{n,m} \tag{7.72}$$

Corresponding definitions for the divergence of a covariant vector field and for the curl of a contravariant vector field are obtained from the above by lowering or raising suffixes.

The usual formulas of vector analysis follow immediately or by use of (7.43). Thus, the curl grad of an invariant T is $\epsilon^{rmn}T_{,nm} = 0$, by the skew-symmetry of

$\bar{\epsilon}^{rmn}$ and the symmetry of $T_{,nm}$. The relation curl curl = div grad − ∇^2 is obtained as follows for a vector field T_m:

$$\begin{aligned}\bar{\epsilon}_{rps}g^{pt}(\bar{\epsilon}^{smn}T_{n}|_{m})|_{t} &= g^{pt}\bar{\epsilon}_{srp}\bar{\epsilon}^{smn}T_{n|mt} \\ &= (\delta_r^m g^{nt} - \delta_r^n g^{mt})T_{n|mt} \\ &= T^t|_{rt} - g^{mt}T_{r|mt} \\ &= (T^t|_t)_{,r} - g^{mt}T_{r|mt}\end{aligned} \tag{7.73}$$

This relation holds in any (curvilinear) coordinates in Euclidean space; the occasional reference to the contrary [13] is only meant to stress the fact that the Laplacian of a vector field T_r is, of course, not given by (7.71).

The simplest integral theorems are

$$\int_{A(C)}^{B} T_{,m}\,dx^m = T(B) - T(A) \qquad \oint_{(C)} T_{,m}\,dx^m = 0 \tag{7.74}$$

$$\int_{(S)} \bar{\epsilon}^{mns}T_{s,n}n_m\,dS = \oint_{(C)} T_m dx^m \qquad \oint_{(S)} \bar{\epsilon}^{mns}T_{s,n}n_m\,dS = 0 \tag{7.75}$$

$$\int_{(V)} T^m|_m\,dV = \int_{(V)} \frac{1}{\sqrt{g}}(\sqrt{g}\,T^m)_{,m}\,dV = \oint_{(S)} T^m n_m\,dS \tag{7.76}$$

The notation is as follows: T is an invariant field, T_m and T^m are vector fields; $\oint$ denotes that the integral is taken over a closed curve or surface; in (7.74), C is a curve with end points A and B (or a closed curve in the second equation); in (7.75), S is a surface with (closed) boundary curve C (or a closed surface in the second equation), dS is a surface element with unit normal n_m, and the sense of the normal and of the boundary curve are related by the usual right-hand rule; in (7.76), V is a volume bounded by the (closed) surface S, and n_m is the outward normal to S. The integral theorem (7.74) is obvious. The other two follow from the fact that they are equations between invariants and that they reduce to the usual theorems of Stokes and of Gauss or Green in rectangular cartesians (cf. Sec. 7.7). More general integral theorems can be stated, [3][11]. This will not be done here for curvilinear coordinates; the special case of rectangular cartesians is taken up in the following paragraphs.

7.7. CARTESIAN TENSORS

If we restrict ourselves to (right-handed) rectangular cartesian coordinates, then tensor analysis becomes much simpler.

Transformations between rectangular cartesians are called *orthogonal transformations;* if the coordinates are both right-handed, they are called *proper* orthogonal transformations. For an orthogonal transformation,

$$\frac{\partial x'^r}{\partial x^s} = \frac{\partial x^s}{\partial x'^r} = \alpha_{rs} \tag{7.77}$$

a set of constants, since both $\partial x'^r/\partial x^s$ and $\partial x^s/\partial x'^r$ are easily seen to be cos (x'^r,x^s), the cosine of the angle between the x'^r axis and the x^s axis. An orthogonal transformation is a linear transformation

$$x'^r = \alpha_{rs}x^s + \alpha_r \tag{7.78}$$

where the *translation coefficients* α_r are arbitrary constants and the *rotation coefficients* α_{rs}, by (7.6) and (7.77), satisfy the relations

$$\alpha_{nm}\alpha_{ns} = \delta_{ms} \tag{7.79}$$

and, hence, $\alpha = |\alpha_{nm}| = \pm 1$ [cf. (7.7)]. Since any right-handed cartesian coordinate system can be obtained from another such right-handed system by a continuous motion, the corresponding orthogonal transformation must be such that it can be changed continuously into the identity transformation, and, hence, the determinant α must have the same value $+1$ as for the identity transformation. Thus, the Jacobian of a proper orthogonal transformation is

$$\left|\frac{\partial x'}{\partial x}\right| = |\alpha_{nm}| = 1 \tag{7.80}$$

The metric tensor in rectangular cartesians is, by (7.26),

$$g_{mn} = \delta_{mn} \tag{7.81}$$

In both (7.79) and (7.81), the Kronecker delta has been written with two subscripts.

Because of (7.77) the distinction between contravariant and covariant suffixes disappears, and *cartesian tensors* will always be written with subscripts alone (coordinates being denoted by x_r, x_r', etc.). Because of (7.80) the distinction between relative and absolute tensors disappears, so that all cartesian tensors are absolute. The transformation law for a general cartesian tensor is

$$T'_{rs\ldots} = \alpha_{ra}\alpha_{sb} \cdots T_{ab\ldots} \tag{7.82}$$

Contraction at the same level, for example, T_{mmr}, is now a tensor operation. By (7.81), and its obvious consequence $g^{mn} = \delta_{mn}$, the raising and lowering of suffixes leave the components of a cartesian tensor unchanged; these operations therefore become superfluous. The ϵs and $\bar{\epsilon}$s all reduce to the one symbol ϵ_{mns}.

Since all Christoffel symbols vanish in cartesian coordinates, the distinction disappears between covariant and partial differentiation and between absolute and ordinary differentiation. The comma and the straight derivative sign d/du will always be used to denote these operations; for example, $T_{m,m}$ and dT_{rs}/du.

Finally, cartesian tensors at *different* points in space can be added, subtracted, or multiplied to yield new cartesian tensors. This is because the coefficients $\alpha_{ra}\alpha_{sb} \cdots$ of the tensor transformation law are constants, independent of position in space, for orthogonal transformations. It follows that an integral (limit of a sum) of a cartesian tensor field is a cartesian tensor, for example,

$$\int_{(V)} T_{rs}\, dV \quad \text{or} \quad \oint_{(S)} T_r n_m\, dS$$

It is important to realize that such integrals are *not* tensorial under general coordinate transformations, the one exception being integrals of invariants.

For cartesian coordinates, the general integral theorems in 3-space will be written down. Using the same notation as in Sec. 7.6, but with $T\ldots$ representing a general cartesian tensor field, they are

$$\int_{A(C)}^{B} T_{\ldots,n}\, dx_n = T_{\ldots}(B) - T_{\ldots}(A) \tag{7.83}$$

$$\int_{(S)} \epsilon_{mnr} n_m T_{\ldots,n}\, dS = \oint_{(C)} T_{\ldots}\, dx_r \tag{7.84}$$

$$\int_{(V)} T_{\ldots,n}\, dV = \oint_{(S)} T_{\ldots} n_n\, dS \tag{7.85}$$

The first of these is obvious. For $T_{\ldots} = T_r$, (7.84) reduces to the usual Stokes theorem; for $T_{\ldots} = T_n$, (7.85) reduces to the usual divergence theorem of Gauss and Green. All these theorems essentially correspond to the simple fact that the integral of the derivative of a function can be evaluated immediately in terms of contributions of the function over the boundary. The proof of the generalized Stokes and Gauss-Green theorems will now be sketched.

Let u,v be a two-parameter network over the surface S. Then for the infinitesimal parallelogram of Fig. 7.1, $n_m\,dS$ is the vector product of $(\partial x^m/\partial u)\,du$ and $(\partial x^m/\partial v)\,dv$: $n_m\,dS = \epsilon_{mpq}(\partial x_p/\partial u)(\partial x_q/\partial v)\,du\,dv$. Thus,

$$\begin{aligned}\int_{(S)} \epsilon_{mnr} n_m T_{\ldots,n}\,dS &= \int_{(S)} \epsilon_{mnr}\epsilon_{mpq} \frac{\partial x_p}{\partial u}\frac{\partial x_q}{\partial v} T_{\ldots,n}\,du\,dv \\ &= \int_{(S)} \left(\frac{\partial x_n}{\partial u} T_{\ldots,n} \frac{\partial x_r}{\partial v} - \frac{\partial x_n}{\partial v} T_{\ldots,n}\frac{\partial x_r}{\partial u}\right) du\,dv \\ &= \int_{(S)} \left(\frac{\partial T_{\ldots}}{\partial u}\frac{\partial x_r}{\partial v} - \frac{\partial T_{\ldots}}{\partial v}\frac{\partial x_r}{\partial u}\right) du\,dv \\ &= \int_{(S)} \left[\frac{\partial}{\partial u}\left(T_{\ldots}\frac{\partial x_r}{\partial v}\right) - \frac{\partial}{\partial v}\left(T_{\ldots}\frac{\partial x_r}{\partial u}\right)\right] du\,dv \\ &= \oint_{(C)} T_{\ldots}\left(\frac{\partial x_r}{\partial v}\,dv + \frac{\partial x_r}{\partial u}\,du\right) \\ &= \oint_{(C)} T_{\ldots}\,dx_r\end{aligned}$$

where the last two integrals over (C) are taken in the sense indicated in Fig. 7.1. This proves (7.84).

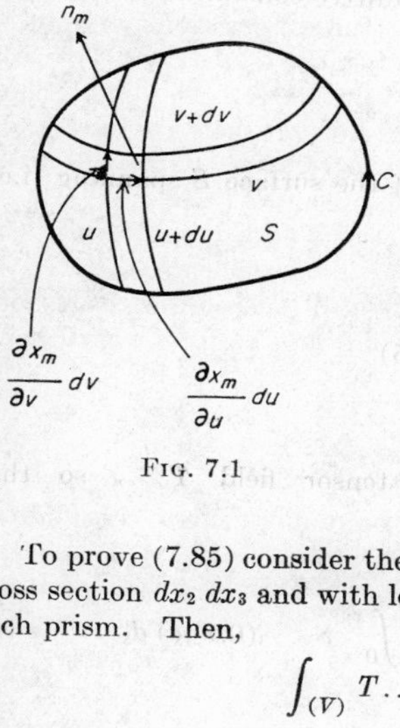

Fig. 7.1

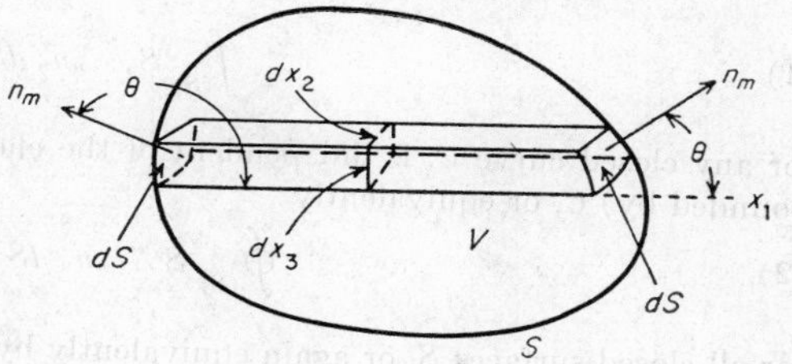

Fig. 7.2

To prove (7.85) consider the volume V divided into prisms, each with infinitesimal cross section $dx_2\,dx_3$ and with long sides parallel to the x_1 axis. Figure 7.2 shows one such prism. Then,

$$\begin{aligned}\int_{(V)} T_{\ldots,1}\,dV &= \int_{(V)} T_{\ldots,1}\,dx_1\,dx_2\,dx_3 \\ &= \oint_{(S)} T_{\ldots}(\pm dx_2\,dx_3) \\ &= \oint_{(S)} T_{\ldots}\cos\theta\,dS \\ &= \oint_{(S)} T_{\ldots}\,n_1\,dS\end{aligned}$$

This establishes (7.85) for $n = 1$. Similarly (7.85) can be proved for $n = 2$ and 3.

It is clear that the integral theorems (7.83) to (7.85) are valid, irrespective of the transformation properties of $T_{\ldots}$. However, if $T_{\ldots}$ is a cartesian tensor field, then each of the two sides of an integral theorem is a cartesian tensor, and the integral theorem is a cartesian tensor equation.

If a cartesian tensor field $S_{\ldots n}$ is such that

(1) $$\int_{A(C)}^{B} S_{\ldots n}\, dx_n$$

for any two points A and B, is independent of the choice of the curve C joining them or equivalently

(2) $$\oint_{(C)} S_{\ldots n}\, dx_n = 0$$

for all closed curves C, or again equivalently by (7.84)

(3) $$\epsilon_{rmn} S_{\ldots n,m} = 0$$

throughout space, then there exists a cartesian tensor field $T_{\ldots}$ so that $S_{\ldots n} = T_{\ldots,n}$. This is shown constructively by choosing a fixed point O, by defining

$$T_{\ldots}(P) = \int_{O}^{P} S_{\ldots n}\, dx_n$$

and by checking that the field $T_{\ldots}$ so defined has the required property $T_{\ldots,n} = S_{\ldots n}$. Clearly $S_{\ldots n}$ determines $T_{\ldots}$ to within an arbitrary additive constant $C_{\ldots}$.

If a cartesian tensor field $S_{\ldots n}$ is such that

(1) $$\int_{(S)} S_{\ldots n} n_n\, dS$$

for any closed curve C, is independent of the choice of the surface S spanning (i.e., bounded by) C, or equivalently

(2) $$\oint_{(S)} S_{\ldots n} n_n\, dS = 0$$

for all closed surfaces S, or again equivalently by (7.85)

(3) $$S_{\ldots n,n} = 0$$

throughout space, then there exists a cartesian tensor field $T_{\ldots n}$ so that $S_{\ldots n} = \epsilon_{nrs} T_{\ldots s,r}$. This is shown by defining

$$T_{\ldots 1}(x_1,x_2,x_3) = 0$$
$$T_{\ldots 2}(x_1,x_2,x_3) = \int_0^{x_1} S_{\ldots 3}(\xi,x_2,x_3)\, d\xi - \int_0^{x_1} S_{\ldots 1}(0,x_2,\zeta)\, d\zeta$$
$$T_{\ldots 3}(x_1,x_2,x_3) = -\int_0^{x_1} S_{\ldots 2}(\xi,x_2,x_3)\, d\xi$$

and by checking that the field $T_{\ldots n}$ so defined has the required property

$$\epsilon_{nrs} T_{\ldots s,r} = S_{\ldots n}$$

Clearly $S_{\ldots n}$ determines $T_{\ldots n}$ to within an arbitrary additive gradient $V_{\ldots,n}$.

Tensors whose components have the same values in all rectangular cartesian coordinate systems are called *isotropic* or *numerical* tensors. They are important because tensors which describe physical properties of a material medium, e.g., dielectric constants or elastic constants in a generalized Hooke's law, are isotropic tensors whenever the medium is isotropic (noncrystalline). It can be shown (see [6], chap. VII) that there are no isotropic vectors, and that the most general isotropic cartesian tensors of second, third, and fourth orders are $\lambda\delta_{mn}$, $\lambda\epsilon_{mns}$, $\lambda\delta_{mn}\delta_{rs} + \mu(\delta_{mr}\delta_{ns} + \delta_{ms}\delta_{nr}) + \nu(\delta_{mr}\delta_{ns} - \delta_{ms}\delta_{nr})$, where λ, μ, ν are invariants. The corresponding isotropic tensors in curvilinear coordinates, written here as covariant absolute tensors, are λg_{mn}, $\lambda\bar{\epsilon}_{mns}$, $\lambda g_{mn}g_{rs} + \mu(g_{mr}g_{ns} + g_{ms}g_{nr}) + \nu(g_{mr}g_{ns} - g_{ms}g_{nr})$. Obviously any invariant or relative invariant is isotropic.

7.8. TENSOR FORM OF PHYSICAL LAWS

A general method for obtaining the differential equations of mathematical physics in tensor form, valid in arbitrary (curvilinear) coordinates, can be illustrated by means of a special example.

In the usual vector notation, the hydrodynamic equations of a perfect fluid are

$$\frac{\partial \rho}{\partial t} + \operatorname{div}(\rho \mathbf{v}) = 0 \tag{7.86}$$

$$\frac{\partial \mathbf{v}}{\partial t} + \mathbf{v} \cdot \nabla \mathbf{v} = \mathbf{X} - \frac{1}{\rho} \operatorname{grad} p \tag{7.87}$$

where ρ is the density of the fluid, v the velocity, p the pressure, and X the body force per unit mass. The first equation expresses the law of conservation of mass, the second equation the law of linear momentum.

Transcribe these equations into cartesian tensor notation:

$$\frac{\partial \rho}{\partial t} + (\rho v_m)_{,m} = 0 \tag{7.88}$$

$$\frac{\partial v_m}{\partial t} + v_r v_{m,r} = X_m - \frac{1}{\rho} p_{,m} \tag{7.89}$$

Raise or lower suffixes, replace ordinary or partial differentiation by absolute or covariant differentiation, until the form of a general tensor equation is obtained:

$$\frac{\partial \rho}{\partial t} + (\rho v^m)|_m = 0 \tag{7.90}$$

$$\frac{\partial v^m}{\partial t} + v^r v^m|_r = X^m - \frac{1}{\rho} g^{mr} p_{,r} \tag{7.91}$$

These are the required tensor equations, valid in all coordinates.

The justification is simple. In a rectangular cartesian coordinate system, (7.90) and (7.91) are identical with (7.88) and (7.89), and therefore valid. But (7.90) and (7.91) are tensor equations and thus, being valid in one coordinate system, are valid in all.

7.9. ORTHOGONAL COORDINATES. PHYSICAL COMPONENTS

In this section the summation convention is suspended. All summations will be indicated explicitly.

The most common curvilinear coordinate systems have the special property of being *orthogonal*. By this is meant that the square of the line element is free of cross terms:

$$ds^2 = \sum_{m=1}^{3} (h_m\, dx^m)^2 = (h_1)^2(dx^1)^2 + (h_2)^2(dx^2)^2 + (h_3)^2(dx^3)^2 \tag{7.92}$$

or that the metric tensor is diagonal:

$$g_{mm} = (h_m)^2 \qquad g_{mn} = 0 \qquad m \neq n \tag{7.93}$$

It immediately follows that

$$g^{mm} = \frac{1}{(h_m)^2} \qquad g^{mn} = 0 \qquad m \neq n \tag{7.94}$$

$$g = (h_1h_2h_3)^2 \qquad \sqrt{g} = h_1h_2h_3 \tag{7.95}$$

and the Christoffel symbols are easily calculated to be

$$\Gamma^r_{mn} = 0 \qquad (r,\ m,\ n \text{ all distinct})$$

$$\Gamma^r_{rm} = \frac{h_{r,m}}{h_r} \qquad \Gamma^m_{rr} = -\frac{h_r h_{r,m}}{(h_m)^2} \qquad r \neq m \tag{7.96}$$

$$\Gamma^r_{rr} = \frac{h_{r,r}}{h_r}$$

The formulas for raising and lowering suffixes, for the scalar and vector products of two vectors, for the divergence and curl of a vector field, and for the Laplacian of an invariant simplify to:

$$T_1 = (h_1)^2T^1 \qquad T_2 = (h_2)^2T^2 \qquad T_3 = (h_3)^2T^3 \tag{7.97}$$

$$\mathbf{T}\cdot\mathbf{S} = (h_1)^2T^1S^1 + (h_2)^2T^2S^2 + (h_3)^2T^3S^3 \tag{7.98}$$

$$(\mathbf{T}\times\mathbf{S})^1 = \frac{1}{h_1h_2h_3}(T_2S_3 - T_3S_2) \qquad (\mathbf{T}\times\mathbf{S})^2 = \frac{1}{h_1h_2h_3}(T_3S_1 - T_1S_3)$$

$$(\mathbf{T}\times\mathbf{S})^3 = \frac{1}{h_1h_2h_3}(T_1S_2 - T_2S_1) \tag{7.99}$$

$$\text{div}\ \mathbf{T} = \frac{1}{h_1h_2h_3}[(h_1h_2h_3T^1)_{,1} + (h_1h_2h_3T^2)_{,2} + (h_1h_2h_3T^3)_{,3}] \tag{7.100}$$

$$(\text{curl}\ \mathbf{T})^1 = \frac{1}{h_1h_2h_3}(T_{3,2} - T_{2,3}) \qquad (\text{curl}\ \mathbf{T})^2 = \frac{1}{h_1h_2h_3}(T_{1,3} - T_{3,1})$$

$$(\text{curl}\ \mathbf{T})^3 = \frac{1}{h_1h_2h_3}(T_{2,1} - T_{1,2}), \tag{7.101}$$

$$\nabla^2T = \frac{1}{h_1h_2h_3}\left[\left(\frac{h_2h_3}{h_1}T_{,1}\right)_{,1} + \left(\frac{h_3h_1}{h_2}T_{,2}\right)_{,2} + \left(\frac{h_1h_2}{h_3}T_{,3}\right)_{,3}\right] \tag{7.102}$$

The most common orthogonal coordinates will now be enumerated, with the transformations which link them to rectangular cartesians x, y, z, and with the h_m which determine the metric. Unless indicated otherwise, coordinates denoted by Latin letters have the range $-\infty$ to $+\infty$, coordinates denoted by Greek letters are angles and have the range 0 to 2π, with 0 and 2π regarded as the same angle; a, b, c are positive constants.

Cylindrical Systems

Cylindrical Polar Coordinates

$$x^m \equiv (r,\theta,z) \qquad r \geq 0$$

$$x = r\cos\theta \qquad y = r\sin\theta \qquad z = z \tag{7.103}$$

$$h_1 = 1 \qquad h_2 = r \qquad h_3 = 1$$

Elliptic Coordinates

$$x^m \equiv (u,\eta,z) \qquad u \geq 0$$

$$x = a\ \text{Cosh}\ u\cos\eta \qquad y = a\ \text{Sinh}\ u\sin\eta \qquad z = z \tag{7.104}$$

$$h_1 = h_2 = a(\text{Cosh}^2 u - \cos^2\eta)^{1/2} \qquad h_3 = 1$$

Parabolic Coordinates

$$x^m \equiv (u,v,z) \qquad v \geq 0$$

$$x = \tfrac{1}{2}(u^2 - v^2) \qquad y = uv \qquad z = z \tag{7.105}$$

$$h_1 = h_2 = (u^2 + v^2)^{1/2} \qquad h_3 = 1$$

Bipolar Coordinates

$$x^m \equiv (\xi, v, z)$$

$$x = \frac{a \operatorname{Sinh} v}{\operatorname{Cosh} v - \cos \xi} \qquad y = \frac{a \sin \xi}{\operatorname{Cosh} v - \cos \xi} \qquad z = z \tag{7.106}$$

$$h_1 = h_2 = \frac{a}{\operatorname{Cosh} v - \cos \xi} \qquad h_3 = 1$$

Rotational Systems

Spherical Polar Coordinates

$$x^m \equiv (r, \theta, \phi) \qquad r \geq 0 \qquad 0 \leq \theta \leq \pi$$

$$x = r \sin \theta \cos \phi \qquad y = r \sin \theta \sin \phi \qquad z = r \cos \theta \tag{7.107}$$

$$h_1 = 1 \qquad h_2 = r \qquad h_3 = r \sin \theta$$

Prolate Spheroidal Coordinates

$$x^m \equiv (u, \eta, \phi) \qquad u \geq 0 \qquad 0 \leq \eta \leq \pi$$

$$x = a \operatorname{Sinh} u \sin \eta \cos \phi \qquad y = a \operatorname{Sinh} u \sin \eta \sin \phi$$

$$z = a \operatorname{Cosh} u \cos \eta \tag{7.108}$$

$$h_1 = h_2 = a(\operatorname{Cosh}^2 u - \cos^2 \eta)^{1/2} \qquad h_3 = a \operatorname{Sinh} u \sin \eta$$

Oblate Spheroidal Coordinates

$$x^m \equiv (u, \eta, \phi) \qquad u \geq 0 \qquad -\frac{\pi}{2} \leq \eta \leq \frac{\pi}{2}$$

$$x = a \operatorname{Cosh} u \cos \eta \cos \phi \qquad y = a \operatorname{Cosh} u \cos \eta \sin \phi$$

$$z = -a \operatorname{Sinh} u \sin \eta \tag{7.109}$$

$$h_1 = h_2 = a(\operatorname{Cosh}^2 u - \cos^2 \eta)^{1/2} \qquad h_3 = a \operatorname{Cosh} u \cos \eta$$

Paraboloidal Coordinates

$$x^m \equiv (u, v, \phi) \qquad u \geq 0 \qquad v \geq 0$$

$$x = uv \cos \phi \qquad y = uv \sin \phi \qquad z = \tfrac{1}{2}(u^2 - v^2) \tag{7.110}$$

$$h_1 = h_2 = [u^2 + v^2]^{1/2} \qquad h_3 = uv$$

Toroidal Coordinates

$$x^m \equiv (\xi, v, \phi) \qquad v \geq 0$$

$$x = \frac{a \operatorname{Sinh} v}{\operatorname{Cosh} v - \cos \xi} \cos \phi \qquad y = \frac{a \operatorname{Sinh} v}{\operatorname{Cosh} v - \cos \xi} \sin \phi$$

$$z = -\frac{a \sin \xi}{\operatorname{Cosh} v - \cos \xi} \tag{7.111}$$

$$h_1 = h_2 = \frac{a}{\operatorname{Cosh} v - \cos \xi} \qquad h_3 = \frac{a \operatorname{Sinh} v}{\operatorname{Cosh} v - \cos \xi}$$

Ellipsoidal Coordinates

$$x^m \equiv \lambda_m \qquad -a^2 < \lambda_1 < -b^2 < \lambda_2 < -c^2 < \lambda_3$$

λ_m are the roots λ, lying in the ranges indicated above, of the cubic equation

$$\frac{x^2}{a^2 + \lambda} + \frac{y^2}{b^2 + \lambda} + \frac{z^2}{c^2 + \lambda} = 1 \tag{7.112}$$

$$h_1 = \frac{1}{2} \left[\frac{(\lambda_1 - \lambda_2)(\lambda_1 - \lambda_3)}{(\lambda_1 + a^2)(\lambda_1 + b^2)(\lambda_1 + c^2)} \right]^{1/2}$$

and h_2, h_3 are obtained from h_1 by cyclic permutation of the subscripts 1, 2, 3.

Additional, but only trivially different, orthogonal coordinates can be obtained from any of the above systems by transformations of the form

$$x'^1 = x'^1(x^1) \qquad x'^2 = x'^2(x^2) \qquad x'^3 = x'^3(x^3) \tag{7.113}$$

$$h_1 = h_1' \frac{dx'^1}{dx^1} \qquad h_2 = h_2' \frac{dx'^2}{dx^2} \qquad h_3 = h_3' \frac{dx'^3}{dx^3} \tag{7.114}$$

Some authors associate the names used above with coordinate systems obtained from ours by some such transformation, e.g., *elliptic coordinates* for v, w, z, where $v = \text{Cosh } u$, $w = \cos \eta$.

It may also be remarked that from any cylindrical coordinate system (u,v,z) such that

$$\begin{aligned} x &= f(u,v) \qquad & y &= g(u,v) \qquad & z &= z \\ h_1 &= \alpha(u,v) \qquad & h_2 &= \beta(u,v) \qquad & h_3 &= 1 \end{aligned} \tag{7.115}$$

two rotational coordinate systems (u,v,ϕ) can be obtained:

$$\begin{aligned} x &= g(u,v) \cos \phi \qquad & y &= g(u,v) \sin \phi \qquad & z &= f(u,v) \\ h_1 &= \alpha(u,v) \qquad & h_2 &= \beta(u,v) \qquad & h_3 &= g(u,v) \end{aligned} \tag{7.116}$$

and

$$\begin{aligned} x &= f(u,v) \cos \phi \qquad & y &= f(u,v) \sin \phi \qquad & z &= -g(u,v) \\ h_1 &= \alpha(u,v), \qquad & h_2 &= \beta(u,v) \qquad & h_3 &= f(u,v) \end{aligned} \tag{7.117}$$

where the ranges of u and v are restricted to cover only a suitable half of the u,v plane. The rotational systems above were obtained in this manner from the cylindrical systems.

7.10. REFERENCES

[1] W. Z. Chien: The intrinsic theory of thin shells and plates, *Quart. Appl. Math.*, **1** (1944)' 297; **2** (1944), 43, 120.
[2] G. Ricci, T. Levi-Civita: Méthodes du calcul différentiel absolu et leurs applications, *Math. Ann.*, **54** (1901), 125–201.
[3] J. A. Schouten: "Der Ricci-Kalkül," Berlin, 1924.
[4] T. Levi-Civitá: "The Absolute Differential Calculus" (Calculus of Tensors), London, 1927.
[5] A. Duschek, W. Mayer: "Lehrbuch der Differentialgeometrie," Leipzig and Berlin, 1930.
[6] H. Jeffreys: "Cartesian Tensors," Cambridge, 1931.
[7] A. J. McConnell: "Applications of the Absolute Differential Calculus," London, 1931.
[8] L. Brillouin: "Les Tenseurs en mécanique et en élasticité," Paris, 1938; New York, 1946.
[9] H. V. Craig: "Vector and Tensor Analysis," McGraw-Hill, New York, 1943.
[10] L. Brand: "Vector and Tensor Analysis," Wiley, New York, 1947.
[11] J. L. Synge, A. Schild: "Tensor Calculus," Toronto, 1949.
[12] B. Spain: "Tensor Calculus," Edinburgh and London, 1953.
[13] J. A. Stratton: "Electromagnetic Theory," pp. 25, 50, McGraw-Hill, New York, 1941

CHAPTER 8

CALCULUS

BY

A. BRONWELL, D.Engr. (Hon.), Worcester, Mass.
R. H. NIEMANN, Ph.D., Fort Collins, Colo.

8.1. FUNCTION, LIMIT, CONTINUITY

Function of One Variable

When for each value of a variable x in a certain domain there is associated exactly one value of a variable y, we say that y is a function of x. The domain of definition may be closed ($a \leqq x \leqq b$) or open ($a < x < b$), or it may at one side or both tend to infinity. Some symbols frequently used for functions of x are $f(x)$ read f of x, $g(x)$, $\phi(x)$, etc. The value of $f(x)$ when $x = a$ is designated by $f(a)$. The variables considered in this section are assumed to be real.

Limit

A variable x approaches a limit c (written symbolically as $x \to c$) if the difference $|x - c|$ becomes and remains less than any positive preassigned constant no matter how small. The symbol $x \to \infty$ means x is increasing without bound and $x \to -\infty$ means x is decreasing without bound. The *left-hand limit* (written symbolically as $x \to c^-$) means x approaches c through values less than c. The *right-hand limit* (written symbolically as $x \to c^+$) means x approaches c through values greater than c.

A function $f(x)$ approaches a limit b as x approaches c if, given any arbitrarily small $\epsilon > 0$, there exists a $\delta > 0$ such that $|f(x) - b| < \epsilon$ for $0 < |x - c| < \delta$. Intuitively, this means that, for x close to c, $f(x)$ is close to b. This is written as

$$\lim_{x \to c} f(x) = b$$

Note that the limit is independent of the value of $f(c)$.

Continuity

The function $f(x)$ is continuous at $x = c$, if

$$\lim_{x \to c} f(x) = f(c)$$

that is, if given any arbitrarily small $\epsilon > 0$ there exists a $\delta > 0$ such that

$$|f(x) - f(c)| < \epsilon$$

for $|x - c| < \delta$. This definition implies three things: (1) $f(x)$ is defined at $x = c$, (2) $\lim_{x \to c} f(x)$ exists, and (3) this limit is $f(c)$. Intuitively, this means that the curve does not have a break in it at the point. At a point $x = c$ where a function $f(x)$ is discontinuous, the following possibilities exist: (1) $f(x)$ has a finite jump; (2) $\lim_{x \to c} f(x)$

becomes infinite; (3) $f(c)$ is not defined; (4) $f(c)$ is defined but $\lim_{x\to c} f(x) \neq f(c)$. If a function is continuous at every point in an interval, it is said to be *continuous in the interval.*

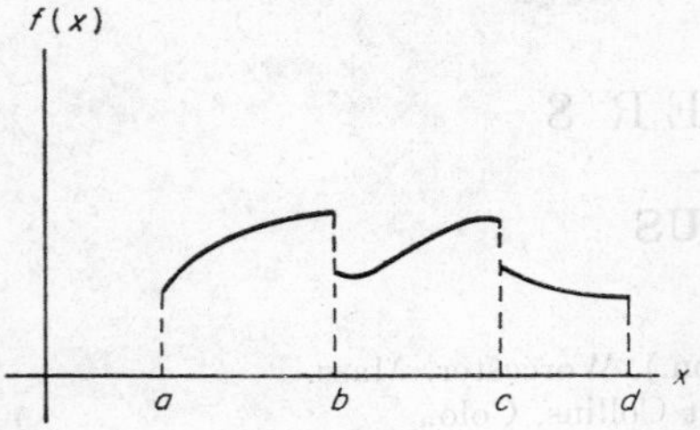

FIG. 8.1

Example 1. The function

$$f(x) = 1/(x - 3)$$

is continuous in every interval which does not contain the point $x = 3$ and is discontinuous at $x = 3$.

Another type of continuity which occurs is sectional or piecewise continuity. A function $f(x)$ is said to be *sectionally continuous* in the interval $a < x < b$ if the interval can be subdivided into a finite number of subintervals such that $f(x)$ is continuous in each of the subintervals and the right- and left-hand limits exist at each of the points of subdivision. An example is shown in Fig. 8.1. The function is defined in the interval (a,d) and is continuous in the subintervals (a,b), (b,c), and (c,d).

8.2. DIFFERENTIATION

Let $y = f(x)$ be a function of x; then the derivative $y' = f'(x)$ of y with respect to x is defined as

$$y' = \lim_{\Delta x\to 0} \frac{f(x + \Delta x) - f(x)}{\Delta x} = \lim_{\Delta x\to 0} \frac{\Delta y}{\Delta x} = \frac{dy}{dx} \tag{8.1}$$

if this limit exists. The symbol Δx is the change or increment of the independent variable x, which may be positive or negative. If the limit does not exist at any point, the function is not differentiable; if it exists at only certain points, the derivative is defined at only those points; if it exists at every point in an interval, then the function is *differentiable in the interval.* The derivative at a point is a number, and it may be interpreted geometrically as the slope of the tangent to the graph of the curve $y = f(x)$ at the point (x,y). If a function is differentiable, it is necessarily continuous, but a continuous function is not necessarily differentiable.

In the definition of the derivative, it was assumed that Δx could be positive or negative; however, if Δx is restricted to negative values only, the limit, if it exists, is called the *left-hand derivative* and is denoted by $f'_-(x)$; that is,

$$f'_-(x) = \lim_{\Delta x\to 0^-} \frac{f(x + \Delta x) - f(x)}{\Delta x} \tag{8.2}$$

Likewise, for $\Delta x > 0$, the *right-hand derivative* is

$$f'_+(x) = \lim_{\Delta x\to 0^+} \frac{f(x + \Delta x) - f(x)}{\Delta x} \tag{8.3}$$

The derivative $f'(x)$ is a function of x, and, as such, it may be possible to differentiate this new function. If

$$\lim_{\Delta x\to 0} \frac{f'(x + \Delta x) - f'(x)}{\Delta x}$$

exists, it is called the *second derivative* of $f(x)$ and is denoted by $f''(x)$ or d^2y/dx^2. Higher-order derivatives are defined in a similar manner. Derivatives are usually not found by using the definition (8.1) but rather by using the formulas given in Table 8.1 at the end of this chapter combined with the chain rule (8.6).

Example 2. Let $y = x^3$; then $y' = 3x^2$ and $y'' = 6x$.

The variables x and y may be given in terms of a third variable t, which is called a parameter, and the equations are called *parametric equations*. If $x = x(t)$ and $y = y(t)$ are functions of t, and if the derivatives $\dot{x}(t) \equiv dx/dt$ and $\dot{y}(t) \equiv dy/dt$ exist and $\dot{x}(t) \neq 0$, then the derivative of y with respect to x is given by the formula

$$\frac{dy}{dx} = \frac{\dot{y}(t)}{\dot{x}(t)} \tag{8.4}$$

The *differentials* of x and y are denoted by dx and dy, respectively, and are defined by the equations $dx = \Delta x$ and $dy = f'(x)\,dx$, where $y = f(x)$. Figure 8.2 illustrates

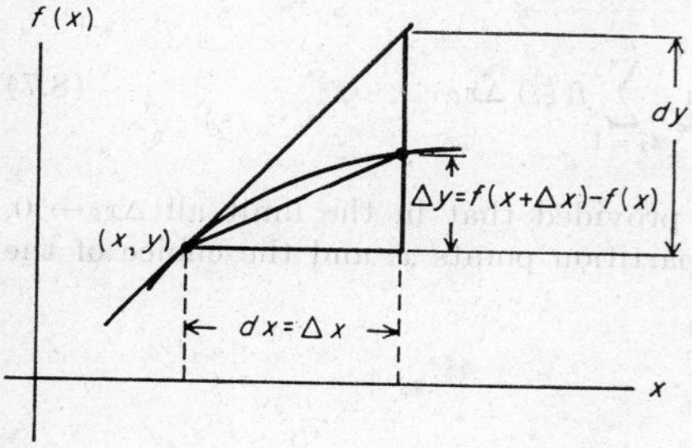

FIG. 8.2

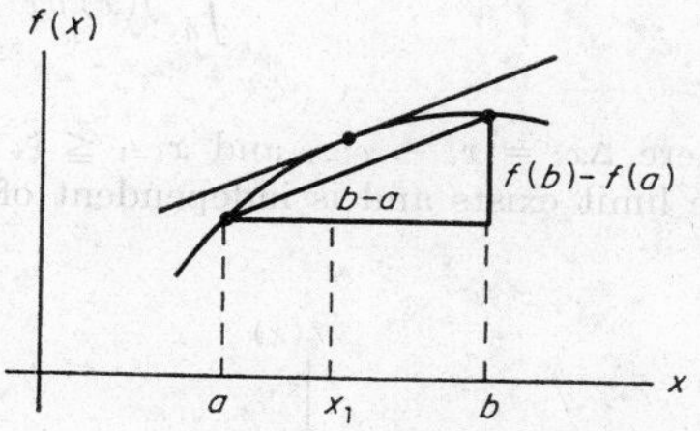

FIG. 8.3

the relationship between dy and $\Delta y = f(x + \Delta x) - f(x)$. One application of differentials is the approximate error problem as illustrated by the following example.

Example 3. If the radius of a circle is $r = 20 \pm 2$, the approximate error in the area may be found as follows: $A = \pi r^2$. So $dA = 2\pi r\,dr$, and the approximate error dA is $\pm 80\pi$.

The quotient $[f(x + \Delta x) - f(x)]/\Delta x$ is often called the difference quotient. A relationship between this quotient and the derivative is known as the *mean-value theorem*, which may be stated as follows: If $f(x)$ is single-valued and continuous in the closed interval $a \leqq x \leqq b$, and differentiable in the open interval $a < x < b$, then there exists a value x_1, where $a < x_1 < b$, such that

$$f(b) - f(a) = (b - a)f'(x_1) \tag{8.5}$$

Another form of this theorem may be obtained by taking $a = x$, $b = x + \Delta x$, and $x_1 = x + \theta\,\Delta x$. Then

$$f(x + \Delta x) - f(x) = f'(x + \theta\,\Delta x)\,\Delta x \qquad 0 < \theta < 1$$

The mean-value theorem may be used for purposes of estimation.

Example 4. Estimate the value of $\ln 1001 - \ln 1000$. Choose $f(x) = \ln x$, $a = 1000$, and $b = 1001$. $f'(x) = 1/x$. So $\ln 1001 - \ln 1000 = (1001 - 1000)/x_1$, $1000 < x_1 < 1001$. Hence $1/1001 < \ln 1001 - \ln 1000 < 1/1000$.

Differentiation of a Function of a Function (Chain Rule)

If $y = f(u)$ is a differentiable function of u, and $u = g(x)$ is a differentiable function of x, then $y = f[g(x)]$ is a differentiable function of x, and

$$\frac{dy}{dx} = \frac{dy}{du}\frac{du}{dx} \tag{8.6}$$

Example 5. $y = u^3$ and $u = 1 - 2x$; then

$$\frac{dy}{dx} = 3u^2(-2) = -6(1 - 2x)^2$$

8.3. INTEGRATION

Definite Integral

The definite integral of $f(x)$ over the finite interval $a \leqq x \leqq b$, denoted by $\int_a^b f(x)\,dx$, is defined as follows: Let the interval from a to b be divided by points $x = x_i$ $(i = 1, 2, \ldots, n - 1)$ as shown in Fig. 8.4; then

$$\int_a^b f(x)\,dx = \lim_{n\to\infty} \sum_{i=1}^{n} f(\xi_i)\,\Delta x_i \tag{8.7}$$

where $\Delta x_i = x_i - x_{i-1}$ and $x_{i-1} \leqq \xi_i \leqq x_i$, provided that in the limit all $\Delta x_i \to 0$, the limit exists and is independent of the partition points x_i and the choice of the

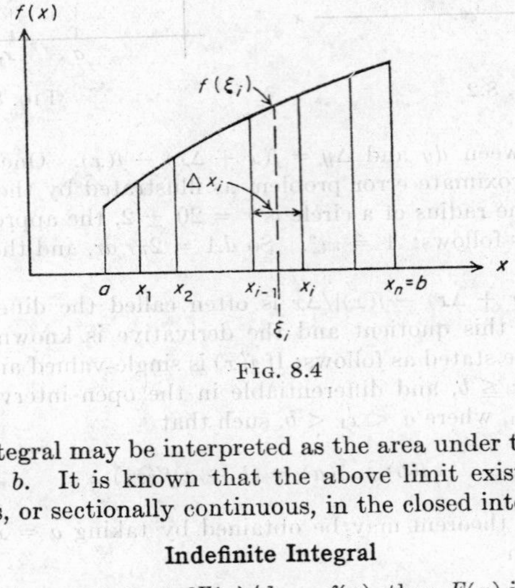

FIG. 8.4

ξ_i. The definite integral may be interpreted as the area under the graph of the curve $y = f(x)$ from a to b. It is known that the above limit exists for every function which is continuous, or sectionally continuous, in the closed interval from a to b.

Indefinite Integral

If $F(x)$ is a function such that $dF(x)/dx = f(x)$, then $F(x)$ is called the indefinite integral (antiderivative), and is written

$$\int f(x)\,dx = F(x) \tag{8.8}$$

If $F(x)$ satisfies Eq. (8.8), then also $F(x) + C$ with an arbitrary constant C satisfies this equation; i.e., the indefinite integral of a given function $f(x)$ is determined except for an additive constant.

The relationship between the definite and indefinite integrals is exhibited in the *fundamental theorem of integral calculus:* If $F(x)$ is a function such that $dF(x)/dx = f(x)$, then

$$\int_a^b f(x)\,dx = F(b) - F(a) \tag{8.9}$$

Methods of Integration

Integral tables (see Table 20.2) may be used to evaluate a number of different types of integrands. However, not all functions appear in the tables. Some integrands which do not appear in the tables may be converted into forms which do appear in the tables by the methods of partial fractions, substitutions, or integration by parts.

Partial Fractions. Theoretically any rational function (quotient of two polynomials) can be integrated by splitting the integrand into partial fractions.

If $g(x)$ is a polynomial with real coefficients, then it is possible to write

$$g(x) = a_0(x - a_1)^{n_1}(x - a_2)^{n_2} \cdots (x^2 + b_1x + c_1)^{m_1}(x^2 + b_2x + c_2)^{m_2} \cdots \tag{8.10}$$

where a_r, b_r, c_r are real quantities and $b_r^2 < 4c_r$. If $h(x)$ is another polynomial of degree less than $g(x)$, then the quotient h/g may be split in a sum of fractions:

$$f(x) \equiv \frac{h(x)}{g(x)} = \sum_{r=1}^{n_1} \frac{A_r}{(x - a_1)^r} + \sum_{r=1}^{n_2} \frac{B_r}{(x - a_2)^r} + \cdots$$
$$+ \sum_{r=1}^{m_1} \frac{C_rx + D_r}{(x^2 + b_1x + c_1)^r} + \sum_{r=1}^{m_2} \frac{E_rx + F_r}{(x^2 + b_2x + c_2)^r} + \cdots \tag{8.11}$$

The A_r, B_r, . . . are unknown real constants and are to be determined such that the equation above is an identity. One method of finding these constants is to clear the equation of fractions and equate coefficients of like powers of x. If the numerator of $f(x)$ is of equal or higher degree than the denominator, carrying out the division will split it in a polynomial and a fraction whose numerator is of lower degree than $g(x)$. In any case the problem of integrating $f(x)$ is converted into the problem of integrating individual fractions.

Integration by Substitution. Difficult integration problems may sometimes be simplified by introducing a new variable. A theorem that applies is: If $u = \psi(x)$ is a monotonic differentiable function, $\psi'(x) \neq 0$ in the interval, and $x = \phi(u)$ is the inverse function of $\psi(x)$, then

$$\int f(x)\,dx = \int f[\phi(u)]\phi'(u)\,du \tag{8.12}$$

After the integration is performed, the variable u is replaced by $\psi(x)$. For definite integrals, the formula is

$$\int_a^b f(x)\,dx = \int_{\psi(a)}^{\psi(b)} f[\phi(u)]\phi'(u)\,du \tag{8.12'}$$

The following list indicates some suggested substitutions for various integrands· $R(x,y)$ denotes a rational function of x and y.

Integrand	*Substitution*	
$R(x, \sqrt{a^2 - x^2})$	$x = a \sin u$	
$R(x, \sqrt{x^2 - a^2})$	$x = a \sec u$	
$R(x, \sqrt{x^2 + a^2})$	$x = a \tan u$	
$R(\sin x, \cos x)$	$u = \tan x/2$	
$R(\text{Sinh } x, \text{Cosh } x)$	$u = \text{Tanh } (x/2)$	
$R(x, \sqrt{ax^2 + bx + c})$	$u = (2ax + b)/\sqrt{4ac - b^2}$	if $b^2 - 4ac < 0$
	$u = (2ax + b)/\sqrt{b^2 - 4ac}$	if $b^2 - 4ac > 0$
$R(x, \sqrt[n]{ax + b})$	$u^n = ax + b$	

Integration by Parts. The integration by parts formula

$$\int u\,dv = uv - \int v\,du \tag{8.13}$$

may be used to reduce one integration problem to another one whose solution may be known or easier to find. The formula holds when any two of the three terms exist.

Mean-value Theorem

The mean-value theorem for integral calculus is

$$\int_a^b f(x)\,dx = (b-a)f(\xi)$$

The formula holds if $f(x)$ is continuous in the closed interval $a \leqq x \leqq b$ and ξ is some fixed value between a and b. It is the basis of numerical integration methods.

8.4. FUNCTION OF TWO VARIABLES

Definition and Continuity

When to each pair of values of the independent variables x and y in a certain domain exactly one value of a dependent variable z is assigned, we say that z is a function of x and y and write $z = f(x,y)$. The function may be defined for all values of x and y or only inside a certain domain of the x,y plane.

The function $f(x,y)$ is *continuous* at the point (a,b) if given an arbitrarily small $\epsilon > 0$ there exists a $\delta > 0$ such that $|f(x,y) - f(a,b)| \leqq \epsilon$ for $(x-a)^2 + (y-b)^2 \leqq \delta^2$. Intuitively, this means that for the point (x,y) close to the point (a,b) the function $f(x,y)$ is close to $f(a,b)$.

Partial Derivatives

If one of the independent variables, say x, is held constant, then z is a function of only one variable y. If the derivative of z with respect to y exists, it is called the *partial derivative* of z with respect to y, and is written as $\partial z/\partial y$ or f_y. This means that

$$f_y \equiv \frac{\partial z}{\partial y} = \lim_{\Delta y \to 0} \frac{f(x, y + \Delta y) - f(x,y)}{\Delta y} \tag{8.14}$$

If y is held constant, the partial derivative of z with respect to x is obtained as the limit

$$f_x \equiv \frac{\partial z}{\partial x} = \lim_{\Delta x \to 0} \frac{f(x + \Delta x, y) - f(x,y)}{\Delta x} \tag{8.15}$$

The rules for finding partial derivatives are the same as for ordinary derivatives. A function $f(x,y)$ is called *differentiable* if f_x and f_y exist and are continuous. The notation $f_x(a,b)$ means that the partial derivative is evaluated at the point (a,b).

Partial derivatives are, in general, functions of both x and y, and so they may be differentiated (if the partial derivatives exist) to obtain higher-order partial derivatives. The following possibilities occur for the second partial derivatives:

$$\begin{aligned} \frac{\partial}{\partial x}\left(\frac{\partial f}{\partial y}\right) &= \frac{\partial^2 f}{\partial x\,\partial y} = f_{yx} \qquad \frac{\partial}{\partial y}\left(\frac{\partial f}{\partial x}\right) = \frac{\partial^2 f}{\partial y\,\partial x} = f_{xy} \\ \frac{\partial}{\partial x}\left(\frac{\partial f}{\partial x}\right) &= \frac{\partial^2 f}{\partial x^2} = f_{xx} \qquad \frac{\partial}{\partial y}\left(\frac{\partial f}{\partial y}\right) = \frac{\partial^2 f}{\partial y^2} = f_{yy} \end{aligned} \tag{8.16}$$

Two of these four, namely, f_{yx} and f_{xy}, are equal if f_{xy} and f_{yx} are continuous. The higher-order partial derivatives are formed in a similar manner.

The partial derivative has a geometric interpretation much the same as the derivative of a function of one variable. If $z = f(x,y)$ is a function of the two independent variables x and y, then the equation represents a surface in three dimensions, and $\partial f/\partial x$ gives the slope of the tangent line to the surface in the plane y = const and at the point $[x,y,f(x,y)]$. The other partial derivatives have similar interpretations.

Example 6. $f(x,y) = xy^2 + x^2 - 2x + 1$, $f_x = y^2 + 2x - 2$, and $f_y = 2xy$; $f_x(1,2) = 4$, which is the slope of the tangent line to the surface in the plane $y = 2$ and at the point (1,2,4). Also $f_{xy} = f_{yx} = 2y$.

Total Differential

If $f(x,y)$ is a function of two variables x and y and f_x and f_y are continuous, then the total differential df is defined by the equation

$$df = f_x\,dx + f_y\,dy \tag{8.17}$$

This is often a good approximation to Δf, where

$$\Delta f = f(x + \Delta x, y + \Delta y) - f(x,y) \tag{8.18}$$

Example 7. Determine the possible error in the area of a rectangle if the sides are 10 ± 1 and 15 ± 2. The approximate error dA is used instead of the exact error ΔA because it is easier to compute. Let $A = xy$ denote the area; then

$$dA = x\,dy + y\,dx = 15(\pm 1) + 10(\pm 2)$$

The possible error in the area is ± 35.

Chain Rule

If $u = g(x,y)$ and $v = h(x,y)$ are differentiable functions of x and y at the point (x,y), and if $f(u,v)$ is a differentiable function in some rectangle containing the point $[g(x,y),h(x,y)]$, then the compound function $z = f[g(x,y),h(x,y)] = F(x,y)$ is also a differentiable function of x and y at the point (x,y), and the partial derivatives are given by the formulas

$$f_x = f_u u_x + f_v v_x \qquad f_y = f_u u_y + f_v v_y \tag{8.19}$$

Example 8. $f(x,y) = \exp\,(x^2 \sin^2 y + xy)$. Let $u = x^2 \sin^2 y$ and $v = xy$; then $f(x,y) = e^{(u+v)}$, and $f_x = (2x \sin^2 y + y) \exp\,(x^2 \sin^2 y + xy)$.

Implicit Function

Sometimes a function y of a variable x is given by an equation of the form $F(x,y) = 0$. This equation is said to define y as an implicit function of x or x as an implicit function of y. It may be possible to solve the equation for y, for example, and obtain y as an explicit function of x. In case it is impossible or undesirable to find such an explicit function, it may still be possible to find the derivative y'. If $F(x,y)$ has continuous partial derivatives, F_x and F_y, and if $F_y \neq 0$, then

$$\frac{dy}{dx} = -\frac{F_x}{F_y} \tag{8.20}$$

Higher-order derivatives may be formed by differentiating Eq. (8.20).

8.5. INDETERMINATE FORMS

Let $f(x)$, $g(x)$ be two differentiable functions of x, and $\phi(x)$ any of the functions f/g, fg, $f - g$, f^g. As $x \to c$, it may happen that $\phi(x)$ takes one of the forms $0/0$, ∞/∞, $0 \times \infty$, $\infty - \infty$, 0^0, 1^∞, ∞^0. All these expressions are void of a definite meaning. However, it is often sensible to define $\lim_{x\to c} \phi$ as the value of $\phi(c)$. This limit is found in the following way:

The Case 0/0. If $\lim_{x\to c} f(x) = f(c) = \lim_{x\to c} g(x) = g(c) = 0$

then
$$\lim_{x\to c} \frac{f(x)}{g(x)} = \lim_{x\to c} \frac{f'(x)}{g'(x)} \tag{8.21}$$

This limit may be zero, finite, or infinite. If it assumes the form 0/0, then Eq. (8.21) may be applied again:

$$\lim_{x\to c} \frac{f'(x)}{g'(x)} = \lim_{x\to c} \frac{f''(x)}{g''(x)}$$

This process may be continued until either the numerator or the denominator or both are different from zero. Equation (8.21) is known as L'Hospital's rule.

Example 9. $\displaystyle \lim_{x\to 0} \frac{x^2}{\cos 2x - 1} = \lim_{x\to 0} \frac{2x}{-2 \sin 2x} = \lim_{x\to 0} \frac{-1}{2 \cos 2x} = -\frac{1}{2}$

The Case ∞/∞. If $\lim_{x\to c} |f(x)| = \lim_{x\to c} |g(x)| = \infty$, then L'Hospital's rule may be applied as in the previous case.

The forms $0 \times \infty$, $\infty - \infty$, 0^0, 1^∞, and ∞^0 may often be evaluated by transforming them into forms such that L'Hospital's rule may be applied.

The Case 0 × ∞. If $\lim_{x\to c} f(x) = 0$ and $\lim_{x\to c} g(x) = \infty$

let $h(x) = 1/g(x)$; then

$$\lim_{x\to c} f(x)g(x) = \lim_{x\to c} \frac{f(x)}{h(x)}$$

and L'Hospital's rule applies. The replacement $k(x) = 1/f(x)$ could have been used instead.

The Case ∞ − ∞. If $\lim_{x\to c} f(x) = \lim_{x\to c} g(x) = \infty$

let $h(x) = 1/g(x)$ and $k(x) = 1/f(x)$; then

$$\lim_{x\to c} [f(x) - g(x)] = \lim_{x\to c} \frac{h(x) - k(x)}{h(x)k(x)}$$

and L'Hospital's rule applies.

The Cases 0^0, 1^∞, ∞^0. The limit of $f(x)^{g(x)}$ may be converted from one of the forms 0^0, 1^∞, ∞^0 to the form 0/0 or ∞/∞ by using logarithms.

Example 10. Let $y = f(x)^{g(x)}$; then

$$\ln y = g(x) \ln f(x) \text{ and } \lim_{x\to c} \ln y = \lim_{x\to c} [g(x) \ln f(x)]$$

This takes the form $0 \times \infty$ and the rules for that form may be applied. If

$$\lim_{x\to c} \ln f(x)^{g(x)} = L \qquad \text{then} \qquad \lim_{x\to c} f(x)^{g(x)} = e^L$$

Example 11. To find $\lim_{x\to 0} x^x$, put $y = x^x$; then

$$\ln y = x \ln x \qquad \text{and} \qquad \lim_{x\to 0} \ln y = \lim_{x\to 0} \frac{\ln x}{1/x} = 0$$

hence $\displaystyle \lim_{x\to 0} x^x = e^0 = 1$

8.6. MAXIMA AND MINIMA

One Variable

The number $f(c)$ is called a maximum value of the function $f(x)$ if in some interval about the point $x = c$, where c is not an end point of the interval, $f(x) \leq f(c)$ for all $x \neq c$. Likewise, the number $f(c)$ is called a minimum value of the function $f(x)$ if in some interval about the point $x = c$, where c is not an end point of the interval, $f(x) \geq f(c)$ for all $x \neq c$. Both maxima and minima are called extrema.

If the function $y = f(x)$ is differentiable, then the tangent to the graph of this function must be parallel to the x axis at an extreme value; hence, the derivative at an extreme value must be zero. This is a necessary condition for an extreme value at an interior point but not a sufficient condition. The first derivative must also change sign at the point in order to ensure that the point is an extreme value.

The *procedure for finding the extreme values* of a differentiable function $f(x)$ is to set $f'(x) = 0$ and solve the equation for x. The values of x that satisfy this equation are the values that may make $f(x)$ take on an extreme value. There are two methods of determining whether or not a value of x thus obtained gives an extreme value for $f(x)$, and whether it is a maximum or minimum value. (1) If c is a solution of the equation $f'(x) = 0$, and $f'(x)$ changes from positive to negative as x increases from $x < c$ to $x > c$, then $f(c)$ is a maximum. Similarly, if $f'(x)$ changes from negative to positive as x increases from $x < c$ to $x > c$, then $f(c)$ is a minimum. If $f'(x)$ does not change sign, then $f(c)$ is neither a maximum nor a minimum, but the point $(c,f(c))$ is a point of inflection with a horizontal tangent. A *point of inflection* is a point where the curve changes from concave upward to concave downward or vice versa. (2) If c is a solution of the equation $f'(x) = 0$ and $f''(c) < 0$, then $f(c)$ is a maximum. If $f'(c) = 0$ and $f''(c) > 0$, then $f(c)$ is a minimum. If $f''(c) = 0$, use method 1. Figure 8.5 illustrates a graph with a point of inflection with a horizontal tangent at a, a maximum at b, and a minimum at c.

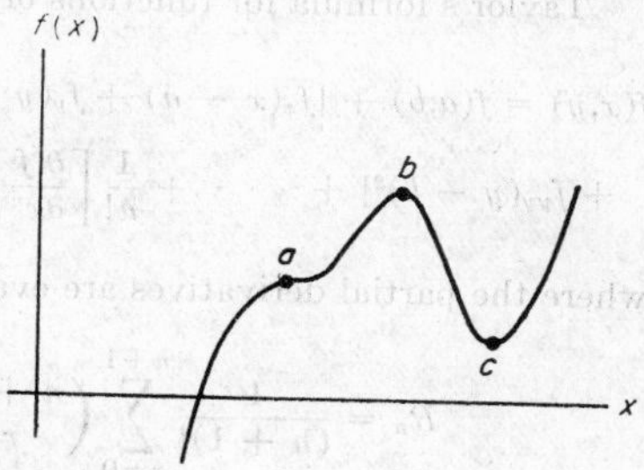

FIG. 8.5

If $f(x)$ is not differentiable, then the function may have an extreme value for which $f'(x) \neq 0$.

Example 12. The function $f(x) = (x - 2)^{2/3}(x - 9)^{5/3}$ has a maximum at $x = 2$, but $f'(2)$ does not exist. It has a minimum at $x = 4$ and a point of inflection with a horizontal tangent at $x = 9$.

Two or More Variables

The number $f(a,b)$ is called the maximum value of the function $f(x,y)$ in a region containing (a,b) in its interior if $f(a,b) \geqq f(x,y)$ for every (x,y) in the region. The minimum value may be defined in a similar manner. Geometrically this means that the tangent plane to the surface represented by the differentiable function $z = f(x,y)$ is parallel to the x,y plane at extreme values.

If $f(x,y)$ is a differentiable function, then the procedure for finding extreme values is to set $f_x = 0$ and $f_y = 0$. If (a,b) is a solution for this simultaneous system of equations and $A = f_{yx}{}^2(a,b) - f_{xx}(a,b)f_{yy}(a,b)$; $B = f_{xx}(a,b) + f_{yy}(a,b)$; then $f(a,b)$ is a maximum if $A < 0$ and $B < 0$. $f(a,b)$ is a minimum if $A < 0$ and $B > 0$. The extreme value is a saddle point if $A > 0$. The nature of the extreme value cannot be determined by this analysis if $A = 0$. In the latter case, f_x and f_y may be studied for values near (a,b).

8.7. TAYLOR'S SERIES AND FORMULA

If $f(x)$ has derivatives of all orders in any interval containing the number a, and if the following series converges, then $f(x)$ may be represented by the power series

$$f(x) = f(a) + \frac{f'(a)}{1!}(x-a) + \cdots + \frac{f^{(n)}(a)}{n!}(x-a)^n + \cdots \tag{8.22}$$

This is known as Taylor's series; in the special case $a = 0$ the series is known as Maclaurin's series. Here $f^{(n)}(a)$ means the nth derivative of $f(x)$ evaluated at $x = a$. Series (8.22) may be expressed with a finite number of terms and a remainder, in which case it is called Taylor's formula. If $f(x)$ has continuous derivatives up to the $(n+1)$st order, then

$$f(x) = f(a) + \frac{f'(a)}{1!}(x-a) + \cdots + \frac{f^{(n)}(a)}{n!}(x-a)^n + R_n \tag{8.23}$$

where R_n may be written in several different forms. One form (Lagrange's) is

$$R_n = f^{(n+1)}(x^*)\frac{(x-a)^{n+1}}{(n+1)!} \qquad a < x^* < x \tag{8.24}$$

Here x^* is some fixed value between a and x, but, in general, it is not known exactly. Taylor's formula reduces to the mean-value theorem if $n = 0$.

Taylor's formula for functions of two variables is

$$f(x,y) = f(a,b) + [f_x(x-a) + f_y(y-b)] + \frac{1}{2!}[f_{xx}(x-a)^2 + 2f_{xy}(x-a)(y-b)$$

$$+ f_{yy}(y-b)^2] + \cdots + \frac{1}{n!}\left[\frac{\partial^n f}{\partial x^n}(x-a)^n + \cdots + \frac{\partial^n f}{\partial y^n}(y-b)^n\right] + R_n \tag{8.25}$$

where the partial derivatives are evaluated at (a,b) and

$$R_n = \frac{1}{(n+1)!}\sum_{r=0}^{n+1}\binom{n+1}{r}\frac{\partial^{n+1}f(x^*,y^*)}{\partial x^r\,\partial y^{n+1-r}}(x-a)^r(y-b)^{n+1-r} \tag{8.26}$$

For the definition of the binomial coefficient $\binom{n+1}{r}$ see Eq. (1.20). (x^*,y^*) denotes a certain fixed point on the line segment joining the points (a,b) and (x,y).

8.8. IMPROPER INTEGRALS

An integral is called improper if the integrand becomes infinite in the closed interval of integration, if the range of integration is infinite, or if both of these occur.

Integrand Infinite

If the function $f(x)$ is continuous in the interval $a \leqq x \leqq b$ except at the point b, then the integral is defined as the limit

$$\int_a^b f(x)\,dx = \lim_{t \to b^-}\int_a^t f(x)\,dx \tag{8.27}$$

If this limit exists, it is the value of the integral, and the integral is said to converge. If it does not exist, the integral diverges. The integral converges if there exists a positive number M independent of x and an $m < 1$ such that $|f(x)| \leqq M(b-x)^{-m}$. It diverges if there exists a number $M > 0$ and an $m \geqq 1$ such that $|f(x)| \geqq M(b-x)^{-m}$. A similar test applies if $f(a)$ is infinite. If the infinite discontinuity occurs between a

and b, break the interval up so that the discontinuity occurs at the end of the subinterval, and then apply the above test to each part separately.

Example 13. $\int_0^2 (2 - x)^{-1/2}\,dx$ converges, since m may be chosen as ½ and M as 1.

Example 14. $\int_0^2 (2 - x)^{-1}\,dx$ diverges. Choose $m = 1$, $M = 1$.

Infinite Interval of Integration

If the limit

$$\int_a^\infty f(x)\,dx = \lim_{t\to\infty} \int_a^t f(x)\,dx \tag{8.28}$$

exists, then the integral $\int_a^\infty f(x)\,dx$ is said to converge. If the limit does not exist the integral diverges. A criterion for testing a given integral is the following: The integral converges if there exists a number $M > 0$ and an $m > 1$ such that $|f(x)| \leqq Mx^{-m}$ for all x no matter how large. The integral diverges if $|f(x)| \geqq Mx^{-1}$ for x sufficiently large.

Example 15. $\int_0^\infty (x + 1)^{-2}\,dx$ converges since $(x + 1)^{-2} \leqq x^{-2}$ for $x > 0$.

Example 16. $\int_0^\infty (x + 1)^{-1/2}\,dx$ diverges since $|(x + 1)^{-1/2}| > x^{-1}$ for $x > 3$.

8.9. DEFINITE INTEGRALS CONTAINING ARBITRARY PARAMETERS

Definition

The integral $\int_a^b f(x,y)\,dx$ is said to contain y as a parameter; i.e., the integration may be carried out with respect to x for a fixed value of y. Different values of y in general give different values of the integral, hence the integral is a function of y. It may be written

$$F(y) = \int_a^b f(x,y)\,dx \tag{8.29}$$

The limits may also be functions of the parameter y, and it may be written

$$F(y) = \int_{g(y)}^{h(y)} f(x,y)\,dx \tag{8.30}$$

If the integral $\int_a^\infty f(x,y)\,dx$ converges to $F(y)$ for each y ($\alpha \leqq y \leqq \beta$), then the integral *converges uniformly* to $F(y)$ in the interval. A very useful test for determining uniform convergence is the *Weierstrass M test:* If $f(x,y)$ is continuous in $a \leqq x < \infty$, $\alpha \leqq y \leqq \beta$, $M(x)$ is a continuous function in $a \leqq x < \infty$, $|f(x,y)| < M(x)$ for $a \leqq x < \infty$ and $\alpha \leqq y \leqq \beta$, and $\int_a^\infty M(x)\,dx$ exists; then $\int_a^\infty f(x,y)\,dx$ converges uniformly in $\alpha \leqq y \leqq \beta$.

Example 17. $\int_0^\infty e^{-xy}\,dx$ converges uniformly in any closed interval for $y \geqq 1$. Choose $M(x) = 2e^{-x}$; then

$$\int_0^\infty 2e^{-x}\,dx = 2$$

$f(x,y)$ and $M(x)$ are continuous and $|f(x,y)| < M(x)$; so the theorem applies.

Differentiation under the Integral Sign

It is often desirable to be able to differentiate an integral containing a parameter, in which case the following theorem applies. Let $F(y)$ be defined as in (8.30). If $g(y)$ and $h(y)$ are differentiable in the interval $\alpha \leqq y \leqq \beta$ and $f(x,y)$ and f_y are continuous in the region $\alpha \leqq y \leqq \beta$, $g(y) \leqq x \leqq h(y)$, then

$$\frac{dF}{dy} = \int_{g(y)}^{h(y)} f_y\,dx - f(g,y)\frac{dg}{dy} + f(h,y)\frac{dh}{dy} \tag{8.31}$$

The notation $f(g,y)$ means that $g(y)$ is computed for a given y; then f is evaluated. If $g(y)$ and $h(y)$ are constant, the last two terms of (8.31) are zero.

If one of the limits is infinite and the other a constant, a useful theorem is: If $f(x,y)$ and f_y are continuous in the region $a \leqq x \leqq b$ (b arbitrary), $\alpha \leqq y \leqq \beta$; if $\int_a^\infty f(x,y)\,dx$ converges to $F(y)$; and if $\int_a^\infty f_y\,dx$ converges uniformly in $\alpha \leqq y \leqq \beta$; then

$$\frac{dF}{dy} = \int_a^\infty f_y\,dx \qquad \alpha \leqq y \leqq \beta \tag{8.32}$$

Integration under the Integral Sign

The integral in (8.29) defining $F(y)$ is a function of y, and as such it may be necessary to integrate $F(y)$. A theorem which permits the order of integration to be changed is: If $f(x,y)$ is continuous for $a \leqq x \leqq b$, $\alpha \leqq y \leqq \beta$, then

$$\int_{y_1}^{y_2} F(y)\,dy = \int_{y_1}^{y_2}\left[\int_a^b f(x,y)\,dx\right]dy = \int_a^b\left[\int_{y_1}^{y_2} f(x,y)\,dy\right]dx \tag{8.33}$$

y_1 and y_2 are any two points in the interval (α,β). In the integral on the right of the equal signs, the integration with respect to y is to be performed first, and then the integration with respect x. The order is the reverse for the term between the equal signs.

For infinite limits, the same formula applies if the integral $\int_a^\infty f(x,y)\,dx$ converges uniformly in $\alpha \leqq y \leqq \beta$.

8.10. DOUBLE INTEGRALS

A double integral is the analog, for functions of two independent variables, of the definite integral of one variable, and it is defined in a similar manner. Let $f(x,y)$ be a function of x and y which is defined at all points of a region R in the x,y plane. If the region R is divided into n subregions ΔR_i, then the double integral is

$$\iint_R f(x,y)\,dR = \lim \sum_{i=0}^{n} f(x_i,y_i)\,\Delta R_i \tag{8.34}$$

where the limit is taken as $n \to \infty$ and the maximum diagonal of the subregions $\Delta R_i \to 0$.

The double integral exists if $f(x,y)$ is continuous; however, it is changed to an iterated (repeated) integral for purposes of evaluation. The theorem which permits this change is: If $f(x,y)$ has at most jump discontinuities on a finite number of sectionally continuous curves, then the double integral is equal to the iterated integral; i.e.,

$$\iint_R f(x,y)\,dR = \int_c^d\left[\int_{X_1}^{X_2} f(x,y)\,dx\right]dy = \int_a^b\left[\int_{Y_1}^{Y_2} f(x,y)\,dy\right]dx \tag{8.35}$$

Here $x = X_1(y)$ is the equation of the arc DAB in Fig. 8.6, and $x = X_2(y)$ is the equation of the arc BCD; $y = Y_1(x)$ is the equation of the arc ABC, and $y = Y_2(x)$ is the equation of the arc CDA; and a, b, c, and d are constants.

If $f(x,y) = 1$ at every point in R, then the double integral over R gives the area of the region R. If $f(x,y) \geqq 0$, then the value of the integral is the volume below the surface represented by the equation $z = f(x,y)$ and above the region R.

Example 18. The volume in the first octant bounded by the surface

$$z = 6 - 3x - 2y$$

is given by the integral

$$V = \iint_R (6 - 3x - 2y)\,dR = \int_0^3 \int_0^{(6-2y)/3} (6 - 3x - 2y)\,dx\,dy$$

$$= \int_0^3 \left[6x - \frac{3}{2}x^2 - 2xy \right]_0^{(6-2y)/3} dy = 6$$

The notion of double integrals may be generalized to integrals of functions of more than two variables.

Example 19. A Triple Integral. The mass of a solid in the first octant bounded by the plane $3x + 2y + z = 6$ and above the plane $z = 2$ where the density varies as the distance from the plane $z = 0$, is given by

$$\text{Mass} = \iiint_V kz\,dV = \int_2^6 \int_0^{(6-z)/2} \int_0^{(6-z-2y)/3} kz\,dx\,dy\,dz$$

V is the region bounded by two coordinate planes, the plane $z = 2$ and the given plane.

$$\text{Mass} = \frac{k}{3} \int_2^6 \int_0^{(6-z)/2} z(6 - z - 2y)\,dy\,dz = \frac{16k}{3}$$

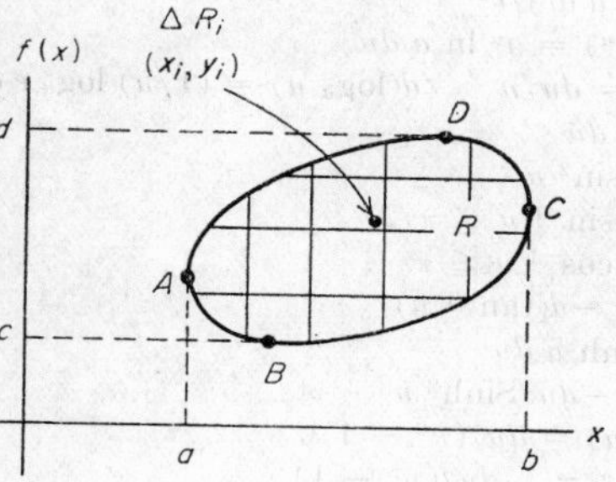

FIG. 8.6

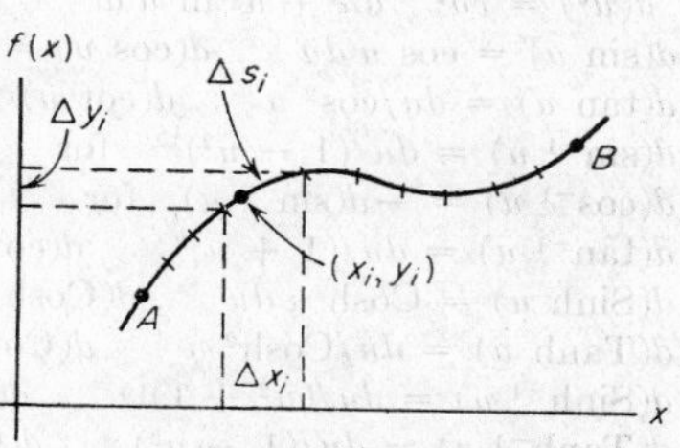

FIG. 8.7

8.11. LINE INTEGRAL

A line integral (curvilinear integral in the plane) is an extension of the ordinary definite integral. It has many applications in such fields as vector analysis and complex variables. The path of integration for the ordinary integral is along the x axis from a to b; however, for a line integral the path of integration may be a curve between two points. The definition follows: Let C be a sectionally continuous curve joining the points A and B in the x,y plane, and $f(x,y)$ a continuous function defined at every point of the curve C. If the curve C is divided into n segments of length

Δs_i, and (x_i,y_i) is any point on the segment Δs_i, then the line integral is

$$\int_C f(x,y)\,ds = \lim \sum_{i=0}^{n} f(x_i,y_i)\,\Delta s_i \tag{8.36}$$

The limit is taken as $n \to \infty$, and each of the $\Delta s_i \to 0$. Two other types of line integrals are defined by the sums

$$\int_C f(x,y)\,dx = \lim \sum_{i=0}^{n} f(x_i,y_i)\,\Delta x_i \qquad \int_C f(x,y)\,dy = \lim \sum_{i=0}^{n} f(x_i,y_i)\,\Delta y_i \tag{8.37}$$

Δx_i and Δy_i are the projections of Δs_i on the x and y axes, respectively. Line integrals are also denoted by $\int_A^B f(x,y)\,ds$ and $\int_{A,B} f(x,y)\,ds$. Line integrals usually occur in the form $\int_C [P(x,y)\,dx + Q(x,y)\,dy]$

Example 20. Evaluate $\int_{(0,1)}^{(2,3)} [(2xy - 1)\,dx + (x^2 + 1)\,dy]$ along the path $y = x + 1$.

$$\int_{(0,1)}^{(2,3)} [(2xy - 1)\,dx + (x^2 + 1)\,dy] = \int_0^2 [2x(x+1) - 1]\,dx$$

$$+ \int_1^3 [(y-1)^2 + 1]\,dy = 12$$

The line integral may be generalized to the case where C is not a plane curve but rather a more general curve in space.

Table 8.1

In this table a is a constant and u, v, w are functions of x.

$d(au) = a\,du \qquad d(u + v + w + \cdots) = du + dv + dw + \cdots$

$d(uv) = u\,dv + v\,du \qquad d(u/v) = (v\,du - u\,dv)/v^2$

$d(u^a) = au^{a-1}\,du \qquad d(e^u) = e^u\,du \qquad d(a^u) = a^u \ln a\,du$

$d(u^v) = vu^{v-1}\,du + u^v \ln u\,dv \qquad d(\ln u) = du/u \qquad d(\log_a u) = (1/u)\log_a e\,du$

$d(\sin u) = \cos u\,du \qquad d(\cos u) = -\sin u\,du$

$d(\tan u) = du/\cos^2 u \qquad d(\cot u) = -du/\sin^2 u$

$d(\sin^{-1} u) = du/(1 - u^2)^{1/2}$ for $-\pi/2 \leqq \sin^{-1} u \leqq \pi/2$

$d(\cos^{-1} u) = -d(\sin^{-1} u)$ for $0 \leqq \cos^{-1} u \leqq \pi$

$d(\tan^{-1} u) = du/(1 + u)^2 \qquad d(\cot^{-1} u) = -d(\tan^{-1} u)$

$d(\text{Sinh}\,u) = \text{Cosh}\,u\,du \qquad d(\text{Cosh}\,u) = \text{Sinh}\,u\,du$

$d(\text{Tanh}\,u) = du/\text{Cosh}^2 u \qquad d(\text{Coth}\,u) = -du/\text{Sinh}^2 u$

$d(\text{Sinh}^{-1} u) = du/(u^2 + 1)^{1/2} \qquad d(\text{Cosh}^{-1} u) = du/(u^2 - 1)^{1/2}$

$d(\text{Tanh}^{-1} u) = du/(1 - u^2) \qquad d(\text{Coth}^{-1} u) = -du/(u^2 - 1)$

For a table of integrals see p. 20–7.

8.12. REFERENCES

[1] A. B. Bronwell: "Advanced Mathematics in Physics and Engineering," McGraw-Hill, New York, 1953.
[2] R. Courant: "Differential and Integral Calculus," vol. 1, trans. by E. J. McShane, Interscience, New York, 1937.
[3] P. Franklin: "Methods of Advanced Calculus," McGraw-Hill, New York, 1944.
[4] C. E. Love: "Differential and Integral Calculus," Macmillan, New York, 1934.
[5] L. L. Smail: "Calculus," Appleton-Century-Crofts, New York, 1949.
[6] G. B. Thomas: "Calculus and Analytic Geometry," Addison-Wesley, Reading, Mass., 1953.

CHAPTER 9

COMPLEX VARIABLES

BY

H. LASS, Ph.D., Pasadena, Calif.

9.1. INTRODUCTION

In order to attach a solution to the equation $x^2 + 1 = 0$, one is forced to invent a new number, i, such that $i^2 + 1 = 0$. One says that $i = \sqrt{-1}$ is an imaginary number in order to distinguish it from the elements of the real-number system. A solution of the quadratic equation $az^2 + bz + c = 0$ $(a \neq 0)$ requires a discussion of complex numbers of the form $z = x + yi$, x and y real. Complex numbers obey the following rules with respect to the operations of addition and multiplication:

$$(a + bi) \pm (c + di) = (a \pm c) + (b \pm d)i \tag{9.1}$$

$$(a + bi)(c + di) = (ac - bd) + (bc + ad)i \tag{9.2}$$

$$\frac{a + bi}{c + di} = \frac{(a + bi)(c - di)}{(c + di)(c - di)} = \frac{ac + bd}{c^2 + d^2} + \frac{bc - ad}{c^2 + d^2}\, i \tag{9.3}$$

$$\begin{aligned} &1 = 1 + 0i \qquad 1i = i \qquad 1 \cdot z = z \cdot 1 = z \\ &0 = 0 + 0i \qquad 0 + z = z + 0 = z \qquad 0 \cdot z = z \cdot 0 = 0 \\ &z_1 + z_2 = z_2 + z_1 \qquad z_1 + (z_2 + z_3) = (z_1 + z_2) + z_3 \\ &z_1 z_2 = z_2 z_1 \qquad z_1(z_2 z_3) = (z_1 z_2) z_3 \\ &z_1(z_2 + z_3) = z_1 z_2 + z_1 z_3 \end{aligned} \tag{9.4}$$

The definitions of the unit element and the zero element as well as the associative, commutative, and distributive laws for the complex-number field are found in Eqs. (9.4).

9.2. GRAPHICAL REPRESENTATION OF COMPLEX NUMBERS. THE ARGAND PLANE

The complex number $z = x + yi$ admits of a very simple geometric representation. We may consider z to be a vector, with its origin at the origin of the Euclidean x,y plane of analytic geometry, and its terminus at the point with coordinates (x,y) (see Fig. 9.1). The projection of z on the x axis is called the real part of z, and the projection of z on the y axis is called the imaginary part of z. Thus, for $z = x + yi$ we have $x = \Re z$, $y = \Im z$. The length of $z = x + yi$ is designated by $|z| = r = (x^2 + y^2)^{1/2} = \text{mod } z$, and is called the *modulus* or *amplitude* of z. The argument of z is the angle between the positive x axis and the vector z, measured in the counterclockwise sense. The argument of z is not single-valued, for, if $\theta = \arg z$, then $\theta \pm 2\pi n$ is also an argument of z for any integer n. The principal value of arg z is defined by $-\pi < \arg z \leq \pi$.

Every point $P(x,y)$ in the Euclidean plane determines a complex number $z = x + yi$, and, conversely, every complex number $z = x + yi$ determines a unique point $P(x,y)$ in the Euclidean plane. This correspondence between points of the Euclidean plane and the totality of complex numbers yields the Argand plane, usually denoted as the z plane.

Addition of complex numbers obeys the parallelogram law for addition of vectors (see Fig. 9.2).

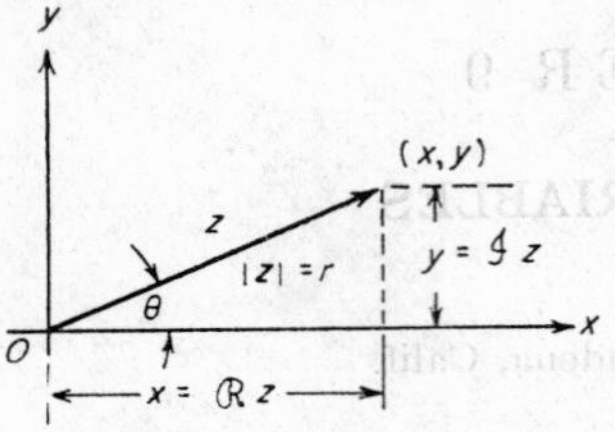

FIG. 9.1

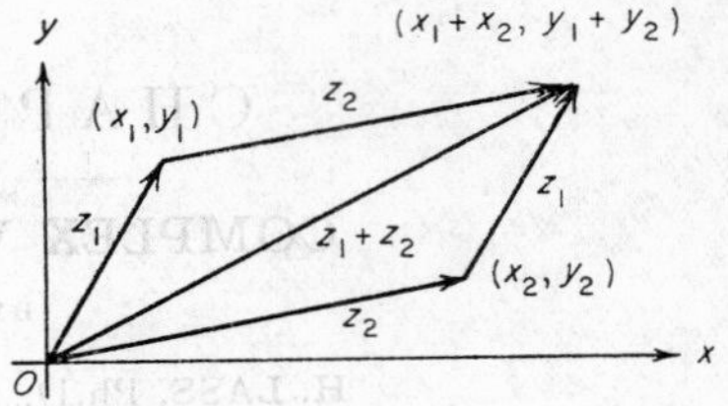

FIG. 9.2

The polar coordinate form of a complex number is at times most useful. From Fig. 9.1 one has $x = r\cos\theta$, $y = r\sin\theta$, so that

$$z = x + yi = r\cos\theta + ir\sin\theta = r(\cos\theta + i\sin\theta) \tag{9.5}$$

If $z_1 = r_1(\cos\theta_1 + i\sin\theta_1)$, $z_2 = r_2(\cos\theta_2 + i\sin\theta_2)$, then

$$\begin{aligned} z_1z_2 &= r_1r_2\,[\cos(\theta_1+\theta_2) + i\sin(\theta_1+\theta_2)] \\ \frac{z_1}{z_2} &= \frac{r_1}{r_2}\,[\cos(\theta_1-\theta_2) + i\sin(\theta_1-\theta_2)] \end{aligned} \tag{9.6}$$

Thus the product of two complex numbers z_1 and z_2 is a new complex number, whose modulus is the product of the moduli, and whose argument is the sum of the arguments of z_1 and z_2.

The complex conjugate of $z = x + yi$ is defined by $\bar{z} = x - yi$. Thus $\Re z = \Re\bar{z}$, $\Im z = -\Im\bar{z}$. It is a very simple matter to deduce the following results concerning complex numbers.

$$\begin{aligned} &\Re(z_1+z_2) = \Re z_1 + \Re z_2 \qquad \Im(z_1+z_2) = \Im z_1 + \Im z_2 \\ &\qquad\qquad |z_1z_2| = |z_1|\,|z_2| \\ &-|z| \leqq \Re z \leqq |z| \qquad -|z| \leqq \Im z \leqq |z| \\ &\overline{z_1+z_2} = \bar{z}_1 + \bar{z}_2 \qquad \overline{z_1z_2} = \bar{z}_1\bar{z}_2 \\ &z + \bar{z} = 2\Re z \qquad z - \bar{z} = 2i\Im z \\ &z\bar{z} = |z|^2 \qquad |z_1+z_2| \leqq |z_1| + |z_2| \end{aligned} \tag{9.7}$$

De Moivre's Formula

From (9.6) it follows that $z^n = \cos n\theta + i\sin n\theta$ for $z = \cos\theta + i\sin\theta$, so that

$$(\cos\theta + i\sin\theta)^n = \cos n\theta + i\sin n\theta \tag{9.8}$$

Equation (9.8) is De Moivre's formula. For $n = 3$, one has

$$\begin{aligned} (\cos\theta + i\sin\theta)^3 &= (\cos^3\theta - 3\cos\theta\sin^2\theta) + i(3\cos^2\theta\sin\theta - \sin^3\theta) \\ &= \cos 3\theta + i\sin 3\theta \end{aligned}$$

so that

$$\begin{aligned} \cos 3\theta &= \cos^3\theta - 3\cos\theta\sin^2\theta \\ \sin 3\theta &= 3\cos^2\theta\sin\theta - \sin^3\theta \end{aligned}$$

To determine the roots of $z^k = 1$, k a positive integer, let $z = r(\cos\theta + i\sin\theta)$, so that

$$r^k(\cos k\theta + i\sin k\theta) = 1 = 1(\cos 2\pi n + i\sin 2\pi n) \qquad n = 0, 1, 2, \ldots, k-1$$

Hence $r = 1$, $k\theta = 0, 2\pi, 4\pi, \ldots, 2\pi(k-1)$, and $\theta = 0, 2\pi/k, \ldots, 2\pi(k-1)/k$. The kth roots of unity are:

$$z_0 = 1 \qquad z_3 = \cos\frac{6\pi}{k} + i\sin\frac{6\pi}{k}$$

$$z_1 = \cos\frac{2\pi}{k} + i\sin\frac{2\pi}{k} \qquad \vdots$$

$$z_2 = \cos\frac{4\pi}{k} + i\sin\frac{4\pi}{k} \qquad z_{k-1} = \cos\frac{2\pi(k-1)}{k} + i\sin\frac{2\pi(k-1)}{k}$$

9.3. DEFINITION OF A COMPLEX FUNCTION, CONTINUITY

Let $Z = \{z\}$ be a domain of complex numbers in the Argand z plane. Now assume that by some rule or set of rules one can set up a correspondence such that, for every number z of Z, there exists a unique complex number w which determines a totality of complex numbers $W = \{w\}$ in an Argand w plane represented by $w = u + vi$. This correspondence or mapping, designated by $Z \to W$, defines a complex function of z over the set Z. The correspondence between Z and W is usually denoted by

$$w = f(z) = u(x,y) + iv(x,y) \tag{9.9}$$

since, given z, we are given x and y, which in turn determine the real and imaginary parts of w, designated by $u(x,y)$ and $v(x,y)$, respectively. Since a unique w exists for each z of Z, one says that w is a single-valued complex function of z.

Example 1. Let Z be the set of complex numbers $\{z\}$, such that $|z| < 2$. The correspondence $z \to w$ such that $w = z + |z| + i$ defines a complex function of z for the domain of definition of z. In this case,

$$u(x,y) = x + (x^2 + y^2)^{1/2} \qquad v(x,y) = y + 1 \qquad \text{for } x^2 + y^2 < 4$$

If $w = f(z)$ represents a single-valued function of z over its domain of definition, one can define continuity of $f(z)$ at $z = z_0$, z_0 in Z, as follows: If, for every sequence of points of Z given by $z_1, z_2, \ldots, z_n, \ldots$, with

$$\lim_{n\to\infty} z_n = z_0 \qquad \text{one has} \qquad \lim_{n\to\infty} f(z_n) = f(z_0)$$

then $f(z)$ is said to be continuous at $z = z_0$. From a simple point of view, continuity guarantees that points sufficiently close to z_0 will map into points near $w_0 = f(z_0)$.

As in the calculus it can be shown that, if $f(z)$ and $g(z)$ are continuous at $z = z_0$, then $f(z) \pm g(z)$, $f(z)g(z)$, $f(z)/g(z)$ with $g(z_0) \neq 0$ are continuous at $z = z_0$. Since it is easily seen that $w = \text{const}$, and $w = z$ are continuous for all z, it follows that any polynomial in z is continuous for all z. A rational function of z defined as the quotient of two polynomials is also continuous at all points for which the denominator is different from zero.

9.4. DIFFERENTIABILITY. THE CAUCHY-RIEMANN EQUATIONS

Let $w = f(z)$ be defined for all points in an immediate neighborhood of $z = z_0$. If

$$\lim_{\Delta z\to 0} \frac{f(z_0 + \Delta z) - f(z_0)}{\Delta z} \tag{9.10}$$

exists independent of the approach to zero of Δz, then $f(z)$ is said to be differentiable at z_0. This definition of differentiability is a simple generalization of the corresponding definition encountered in the calculus of real-variable theory. There is one essential point of difference. Since z is complex, there is an infinity of ways for Δz to approach zero. Recall that $\Delta z = \Delta x + i\,\Delta y$, with Δx and Δy independent of each other.

The subscript in z_0 will be omitted in the following considerations. Let

$$f(z) = u(x,y) + iv(x,y)$$
$$f(z + \Delta z) = u(x + \Delta x,\ y + \Delta y) + iv(x + \Delta x,\ y + \Delta y)$$

so that

$$\frac{f(z+\Delta z) - f(z)}{\Delta z} = \frac{u(x+\Delta x,\ y+\Delta y) - u(x,y) + i[v(x+\Delta x,\ y+\Delta y) - v(x,y)]}{\Delta x + i\,\Delta y} \tag{9.11}$$

One simple way in which Δz can approach zero is to hold y constant ($\Delta y = 0$) so that (9.11) yields

$$\lim_{\Delta z\to 0} \frac{f(z+\Delta z) - f(z)}{\Delta z} = \lim_{\Delta x\to 0} \frac{u(x+\Delta x,\ y) - u(x,y)}{\Delta x} + i \lim_{\Delta x\to 0} \frac{v(x+\Delta x,\ y) - v(x,y)}{\Delta x}$$
$$= \frac{\partial u}{\partial x} + i\frac{\partial v}{\partial x} \tag{9.12}$$

provided $\partial u/\partial x$ and $\partial v/\partial x$ exist. Another simple way to allow Δz to approach zero is to hold x constant ($\Delta x = 0$) so that (9.11) yields

$$\lim_{\Delta z\to 0} \frac{f(z+\Delta z) - f(z)}{\Delta z} = \frac{1}{i}\frac{\partial u}{\partial y} + \frac{\partial v}{\partial y} = \frac{\partial v}{\partial y} - i\frac{\partial u}{\partial y} \tag{9.13}$$

Hence, if $f(z)$ is differentiable, then of necessity

$$\frac{\partial u}{\partial x} + i\frac{\partial v}{\partial x} = \frac{\partial v}{\partial y} - i\frac{\partial u}{\partial y}$$

so that

$$\frac{\partial u}{\partial x} = \frac{\partial v}{\partial y} \qquad \frac{\partial u}{\partial y} = -\frac{\partial v}{\partial x} \tag{9.14}$$

The Cauchy-Riemann equations (9.14) are necessary conditions if $f(z)$ is to be differentiable. Now it is an important fact, stated without proof, that, if the Cauchy-Riemann equations hold, and if the four partial derivatives of (9.14) are continuous, then $f(z)$ is differentiable, and

$$\frac{dw}{dz} = f'(z) = \lim_{\Delta z\to 0} \frac{f(z+\Delta z) - f(z)}{\Delta z} = \frac{\partial u}{\partial x} + i\frac{\partial v}{\partial x} = \frac{\partial v}{\partial y} - i\frac{\partial u}{\partial y} \tag{9.15}$$

The limit process yields a unique value independent of the approach to zero of Δz. The notation $f'(z)$ represents the first derivative of $f(z)$ with respect to z.

From (9.14) we obtain

$$\frac{\partial^2 u}{\partial x^2} = \frac{\partial^2 v}{\partial x\,\partial y} = -\frac{\partial^2 u}{\partial y^2}$$

as well as

$$\frac{\partial^2 v}{\partial y^2} = \frac{\partial^2 u}{\partial y\,\partial x} = -\frac{\partial^2 v}{\partial x^2}$$

providing the second derivatives exist and are continuous. Thus, if $f(z) = u + vi$ is differentiable, then u and v satisfy Laplace's equation:

$$\nabla^2 u \equiv \frac{\partial^2 u}{\partial x^2} + \frac{\partial^2 u}{\partial y^2} = 0 \qquad \nabla^2 v \equiv \frac{\partial^2 v}{\partial x^2} + \frac{\partial^2 v}{\partial y^2} = 0 \tag{9.16}$$

As in the calculus, it follows that, if $f'(z)$ and $g'(z)$ exist, then

$$\frac{d}{dz}[f(z) \pm g(z)] = f'(z) \pm g'(z)$$
$$\frac{d}{dz}[f(z)g(z)] = f(z)g'(z) + f'(z)g(z) \qquad (9.17a\text{–}c)$$
$$\frac{d}{dz}\left(\frac{f(z)}{g(z)}\right) = \frac{g(z)f'(z) - f(z)g'(z)}{[g(z)]^2} \qquad g(z) \neq 0$$

Example 2. Let $f(z) = z^2 = (x^2 - y^2) + 2xyi$, so that $u = x^2 - y^2$, $v = 2xy$. Thus

$$\frac{\partial u}{\partial x} = 2x = \frac{\partial v}{\partial y} \qquad \frac{\partial u}{\partial y} = -2y = -\frac{\partial v}{\partial x}$$

Since the Cauchy-Riemann equations hold, and since the partial derivatives are continuous, then of necessity $f(z) = z^2$ is differentiable. From (9.15),

$$f'(z) = 2x + 2yi = 2z$$

This result can be obtained also from

$$f'(z) = \lim_{\Delta z \to 0} \frac{(z + \Delta z)^2 - z^2}{\Delta z} = \lim_{\Delta z \to 0} \frac{2z\,\Delta z + (\Delta z)^2}{\Delta z} = 2z$$

Example 3. The function $f(z) = \bar{z} = x - yi$ is nowhere differentiable. From $u = x$, $v = -y$, it follows that

$$\frac{\partial u}{\partial x} = 1 \neq \frac{\partial v}{\partial y} = -1$$

so that the Cauchy-Riemann equations do not hold.

Example 4. Let us consider

$$f(z) = e^x \cos y + ie^x \sin y = e^x(\cos y + i \sin y)$$

It is a simple matter to show that $f(z)$ is differentiable everywhere. It is to be noted that $f(z)$ reduces to e^x for z real, $z = x$, $y = 0$. It can be shown that $f(z)$ is the only function which is differentiable for all z and which reduces to e^x for z real. We write

$$e^z = e^{x+yi} = e^x(\cos y + i \sin y)$$

For $x = 0$, we obtain Euler's formula $e^{iy} = \cos y + i \sin y$.

Similarly

$$\sin z = \sin x \operatorname{Cosh} y + i \cos x \operatorname{Sinh} y$$
$$\cos z = \cos x \operatorname{Cosh} y - i \sin x \operatorname{Sinh} y$$

are differentiable everywhere and reduce to $\sin x$ and $\cos x$, respectively, for z real. It can easily be shown that

$$\frac{d}{dz}(\sin z) = \cos z \qquad \frac{d}{dz}(\cos z) = -\sin z \qquad \sin^2 z + \cos^2 z = 1$$

The elementary functions of real-variable theory obey the same differentiation formulas in complex-variable theory.

Example 5. Let $f(z)$ and $g(z)$ be differentiable at z_0, with $f(z_0) = g(z_0) = 0$. We have

$$\lim_{z \to z_0} \frac{f(z)}{g(z)} = \lim_{z \to z_0} \frac{[f(z) - f(z_0)]/(z - z_0)}{[g(z) - g(z_0)]/(z - z_0)} = \frac{f'(z_0)}{g'(z_0)} \qquad g'(z_0) \neq 0$$

This is L'Hospital's rule for complex-variable theory.

Definition. Let $f(z)$ be differentiable at $z = z_0$. If, furthermore, there exists a circle with center at z_0 such that $f(z)$ is differentiable at every interior point of this circle, then $f(z)$ is regular, or analytic, at z_0. If $f(z)$ is not analytic at z_0, then z_0 is a singular point of $f(z)$.

9.5. CONFORMAL MAPPING

We have seen that every point $P(x,y)$ in the Euclidean plane determines a complex number $z = x + yi$, and, conversely, every complex number $z = x + yi$ determines a unique point $P(x,y)$ in the Euclidean plane. Let us now examine the set of complex numbers

$$\begin{aligned} w = z^2 &= (x + yi)^2 \\ &= (x^2 - y^2) + 2xyi = u + vi \end{aligned}$$

with $u = x^2 - y^2$, $v = 2xy$. It is highly beneficial to construct a new Argand plane, called the w plane, with $w = u + iv$, such that u and v are rectangular coordinates in the same sense that x and y are rectangular coordinates. The equation $w = z^2$ may be looked upon as a mapping of points in the z plane onto points of the w plane. The point $P(x,y)$ in the z plane maps onto the point $Q(u = x^2 - y^2, v = 2xy)$ in the w plane. A curve in the z plane will, in general, map onto a curve in the w plane. In the above example, the hyperbolas $x^2 - y^2 = \text{const}$ map onto the straight lines $u = \text{const}$ in the w plane. Similarly, the hyperbolas $2xy = \text{const}$ map onto the straight lines $v = \text{const}$.

Let us consider now the relationship $w = f(z) = u(x,y) + iv(x,y)$ with $f(z)$ differentiable. The curve $u(x,y) = \text{const}$ maps onto the straight line $u = \text{const}$ in the w plane. The slope of the curve $u(x,y) = \text{const}$ is given by

$$m_1 = -\frac{\partial u/\partial x}{\partial u/\partial y}$$

Similarly, the slope of the curve $v(x,y) = \text{const}$ is given by

$$m_2 = -\frac{\partial v/\partial x}{\partial v/\partial y} \qquad \text{so that} \qquad m_1 m_2 = \frac{(\partial u/\partial x)\,(\partial v/\partial x)}{(\partial u/\partial y)\,(\partial v/\partial y)} = -1$$

from the Cauchy-Riemann equations, provided $f'(z) \neq 0$. Thus the curves $u(x,y) = \text{const}$ intersect the curves $v(x,y) = \text{const}$ at right angles. Their images in the w plane, namely, the curves $u = \text{const}$, $v = \text{const}$, also intersect at right angles in the w plane. More generally, let two curves in the z plane intersect at a point z_0, and let dz_1 and dz_2 be two tangent vectors to these curves at z_0. These curves will map onto two curves in the w plane, which intersect at the point $w_0 = f(z_0)$. If dw_1 and dw_2 are corresponding tangent vectors to these curves at w_0, and if $f(z)$ is differentiable at z_0, then $dw_1/dz_1 \approx f'(z_0)$, $dw_2/dz_2 \approx f'(z_0)$, so that $dw_2/dw_1 = dz_2/dz_1$, provided $f'(z_0) \neq 0$. Thus

$$\arg\frac{dw_2}{dw_1} = \arg dw_2 - \arg dw_1 = \arg\frac{dz_2}{dz_1} = \arg dz_2 - \arg dz_1$$

which shows that angles are preserved (conformal mapping) under the mapping $w = f(z)$.

Example 6. Let us examine the transformation

$$w = z + \frac{1}{z} \qquad z \neq 0 \tag{9.18}$$

It is useful to express z in its polar form, $z = r(\cos\theta + i\sin\theta)$, so that

$$w = u + vi = r(\cos\theta + i\sin\theta) + \frac{1}{r}(\cos\theta - i\sin\theta)$$

$$= \left(r + \frac{1}{r}\right)\cos\theta + i\left(r - \frac{1}{r}\right)\sin\theta$$

Thus $u = (r + 1/r)\cos\theta$, $v = (r - 1/r)\sin\theta$, and so from $\cos^2\theta + \sin^2\theta = 1$ it follows that

$$\frac{u^2}{(r + 1/r)^2} + \frac{v^2}{(r - 1/r)^2} = 1 \qquad r \neq 0, 1$$

The circles $r = \text{const} \neq 1$ of the z plane map onto ellipses of the w plane.

Example 7. Let us examine the important bilinear transformation

$$w = \frac{az + b}{cz + d} \qquad ad - bc \neq 0 \tag{9.19}$$

(If $ad - bc = 0$, then $w = b/d = a/c$ for all z.) The constants a, b, c, d are complex numbers. Upon division, (9.19) can be written in the form

$$w = \frac{a}{c} + \frac{bc - ad}{c(cz + d)} \qquad c \neq 0 \tag{9.20}$$

(For $c = 0$, (9.19) reduces to the simpler form $w = (a/d)z + b/d$.)

If we define $z_1 = cz + d$, $z_2 = 1/z_1$, then (9.20) can be written as

$$w = \frac{a}{c} + \frac{bc - ad}{c} z_2$$

Thus, to analyze (9.19) we need only consider the simpler transformations

$$w = Ez + F \qquad w = \frac{1}{z} \tag{9.21}$$

The equation of a circle of radius r with center at z_0 can be represented by

$$z = z_0 + r(\cos\theta + i\sin\theta)$$

where $0 \leq \theta < 2\pi$. Under the transformation $w = Ez + F$, with

$$E = |E|(\cos\alpha + i\sin\alpha)$$

the circle will map onto the points given by

$$w = (Ez_0 + F) + |E|r[\cos(\theta + \alpha) + i\sin(\theta + \alpha)] \qquad 0 \leq \theta < 2\pi \tag{9.22}$$

Equation (9.22) represents a circle in the w plane with center at $w_0 = Ez_0 + F$ and radius $|E|r$. Thus circles map onto circles under the transformation $w = Ez + F$.

Let us now turn our attention to the transformation $w = 1/z$. Expressing z and w in polar coordinate form yields $\rho(\cos\phi + i\sin\phi) = (1/r)(\cos\theta - i\sin\theta)$, so that $\rho = 1/r$, $\phi = -\theta$. The general equation of a circle in the z plane is given by $x^2 + y^2 + Ax + By + C = 0$; A, B, C being real constants. In polar coordinate form this becomes $r^2 + r(A\cos\theta + B\sin\theta) + C = 0$. Under the transformation $w = 1/z$, the circle maps onto the curve

$$C\rho^2 + \rho(A\cos\phi - B\sin\phi) + 1 = 0 \tag{9.23}$$

Now (9.23) also represents a circle in the w plane for $C \neq 0$. If $C = 0$, then (9.23) represents a straight line (a circle of infinite radius). The transformation $w = 1/z$ maps circles onto circles. Those circles in the z plane which pass through the origin map onto straight lines. Hence, the bilinear transformation (9.19) maps circles onto circles.

Since not all the constants of (9.19) are zero, let us assume that $d \neq 0$, so that $w = (Az + B)/(Cz + 1)$. Any circle in the z plane is determined from three points z_1, z_2, z_3 on this circle. Similarly, any circle in the w plane is determined from three points w_1, w_2, w_3. Setting $w_n = (Az_n + B)/(Cz_n + 1)$ $(n = 1, 2, 3)$ yields three linear equations for the three unknown constants A, B, C. Hence any circle in the z plane can be mapped onto any circle in the w plane through a bilinear transformation.

Example 8. We examine the transformation $w = \sqrt{z}$. Let $z = re^{i\theta}$, so that $w = \sqrt{r}\, e^{i\theta/2}$, with $\sqrt{r}$ the positive square root of r. If we begin at a point z $(|z| = R \neq 0)$, and traverse the circle $|z| = R$ once in a counterclockwise manner until the point z is reached again, then the angle θ will have increased by an amount 2π. Thus, the value of $w = \sqrt{z}$ will be changed by a factor -1 since θ is replaced by $\theta + 2\pi$, so that $w = \sqrt{r}\, e^{\frac{1}{2}i(\theta+2\pi)} = -\sqrt{r}\, e^{i\theta/2}$, if we recall that $e^{\pi i} = -1$. A further rotation of 2π will bring us back to the original value of the function. Thus the function $w = \sqrt{z}$ is not single-valued. If we make a cut in the z plane along the negative real axis from $z = 0$ to $z = -\infty$ (a branch line), and restrict the passage of a point across this branch line, the function $f(z) = \sqrt{z}$ will be single-valued. The placing of the cut along the negative real axis is purely arbitrary. The points $z = 0$ and $z = -\infty$ (the ends of the branch line) are called *branch points* of the function. These points are not arbitrary. Two single-valued functions $f_1(z) = \sqrt{z}$ and $f_2(z) = \sqrt{z}$ are obtained by choosing $f_1(1) = 1$, $f_2(1) = -1$.

We can proceed as follows to construct a Riemann surface which exhibits the above peculiarities such that $f(z) = \sqrt{z}$ will be single-valued on the entire Riemann surface so constructed. Consider two z planes, both with cuts along the negative real axis, and placed one on top of the other with their branch points and branch lines coinciding. For simplicity, the top z plane is such that $\sqrt{1} = 1$, while the bottom z plane is such that $\sqrt{1} = -1$. Now imagine that the upper edge of the cut in the top plane is connected to the lower edge of the cut in the bottom plane, and that, at the same time, the lower edge of the cut in the top plane is connected to the upper edge of the cut in the bottom plane. (One cannot construct a paper model of this connection.) The Riemann surface consisting of the two planes connected in this peculiar way has the property that, as a point moves around the origin, it first slides into one plane, and then into the other.

More generally, if n is a positive integer, $w = \sqrt[n]{z - a}$ has branch points of order $(n - 1)$ at $z = a$ and $z = \infty$, and the Riemann surface is composed of n interconnected planes. The Riemann surface for the function $\ln z$ defined on p. 9–12 requires an infinite number of interconnected planes for its construction.

9.6. THE DEFINITE INTEGRAL

Integration in the complex plane can be reduced to the evaluation of line integrals of real-variable theory. First we consider line integrals of the type

$$\int_\Gamma f(x,y)\, dx \qquad \int_\Gamma f(x,y)\, dy \tag{9.24}$$

where $x = g(t)$, $y = h(t)$, $t_0 \leqq t \leqq t_1$, is a parametric representation of a curve Γ in the x,y plane. If $g(t)$ and $h(t)$ are differentiable, then $dx = g'(t)\, dt$, $dy = h'(t)\, dt$, and the integrals above reduce to ordinary Riemann integrals

$$\int_{t_0}^{t_1} f(g(t),h(t))g'(t)\, dt \qquad \int_{t_0}^{t_1} f(g(t),h(t))h'(t)\, dt$$

Now let us consider a curve Γ in the z plane. If $f(z)$ is continuous at every point of Γ, we define the line integral of $f(z)$ over Γ by

$$\begin{aligned}\int_\Gamma f(z)\,dz &= \int_\Gamma [u(x,y) + iv(x,y)](dx + i\,dy)\\ &= \left[\int_\Gamma u(x,y)\,dx - \int_\Gamma v(x,y)\,dy\right] + i\left[\int_\Gamma u(x,y)\,dy + \int_\Gamma v(x,y)\,dx\right]\end{aligned} \tag{9.25}$$

The integrals of (9.25) are of the types given by (9.24).

Example 9. Let $f(z) = z$, and let Γ be the parabola $y = x^2$ $(0 \leqq x \leqq 1)$, so that $x = t$, $y = t^2$ $(0 \leqq t \leqq 1)$ represents Γ. Along Γ we have

$$z = x + yi = t + t^2 i \qquad dz = (1 + 2ti)\,dt$$

$$f(z) = z = t + t^2 i \qquad f(z)\,dz = [(t - 2t^3) + 3t^2 i]\,dt$$

$$\int_\Gamma f(z)\,dz = \int_0^1 (t - 2t^3 + 3t^2 i)\,dt = i = \frac{(1+i)^2}{2}$$

The same result is obtained if we consider the straight-line path $x = t$, $y = t$ $(0 \leqq t \leqq 1)$. It will be shown later that $\int_{z_0}^{z_1} z\,dz$ is independent of the path joining z_0 and z_1 since $f(z) = z$ is analytic everywhere.

Example 10. Let $f(z) = 1/z$, and let Γ be the closed circle of radius a with center at the origin. Along Γ we have $z = a(\cos\theta + i\sin\theta)$, $dz = a(-\sin\theta + i\cos\theta)\,d\theta$, $f(z) = (1/a)(\cos\theta - i\sin\theta)$, and

$$\oint_\Gamma \frac{dz}{z} = i\int_0^{2\pi} d\theta = 2\pi i$$

The notation $\oint$ represents a line integral around a closed path.

The following statements are easily verified:

$$\int_\Gamma f(z)\,dz + \int_\Gamma g(z)\,dz = \int_\Gamma [f(z) + g(z)]\,dz$$

$$\int_\Gamma cf(z)\,dz = c\int_\Gamma f(z)\,dz \qquad \text{for} \qquad c = \text{const}$$

$$\underset{(\Gamma)}{\int_\alpha^\beta} f(z)\,dz = -\underset{(\Gamma)}{\int_\beta^\alpha} f(z)\,dz \tag{9.26a–e}$$

$$\left|\int_\Gamma f(z)\,dz\right| \leqq ML \qquad \text{if } |f(z)| < M,\ L = \text{length of } \Gamma$$

$$\left|\int_\Gamma f(z)\,dz\right| \leqq \int_\Gamma |f(z)|\,|dz|$$

9.7. CAUCHY'S INTEGRAL THEOREM

If $f(z)$ is analytic in a simple region R of the Argand plane, and if Γ is any simple closed path in R, then

$$\oint_\Gamma f(z)\,dz = 0 \tag{9.27}$$

We omit proof of this fundamental result due to Cauchy.

Some immediate consequences of Cauchy's integral theorem are as follows:

1. Let $f(z)$ be analytic in a simply connected region R. For $z = a$, $z = b$ in R,

$$\int_a^b f(z)\,dz$$

is independent of the path Γ lying in R joining a to b. From Fig. 9.3 we have

$$\int_{a\,(\Gamma_1)}^{b} f(z)\,dz + \int_{b\,(\Gamma_2)}^{a} f(z)\,dz = 0$$

so that

$$\int_{a\,(\Gamma_1)}^{b} f(z)\,dz = -\int_{b\,(\Gamma_2)}^{a} f(z)\,dz = \int_{a\,(\Gamma_2)}^{b} f(z)\,dz$$

2. Let C_1 and C_2 be simple closed paths lying in a region R for which $f(z)$ is analytic, with C_2 interior to C_1. Then

$$\oint_{C_1} f(z)\,dz = \oint_{C_2} f(z)\,dz$$

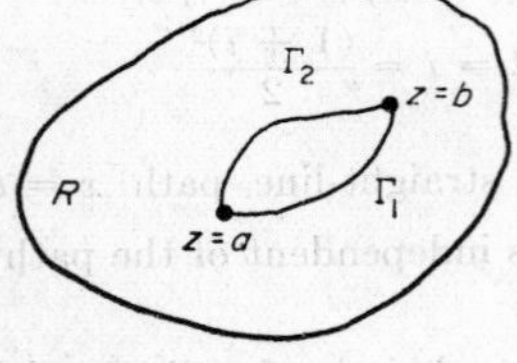

FIG. 9.3

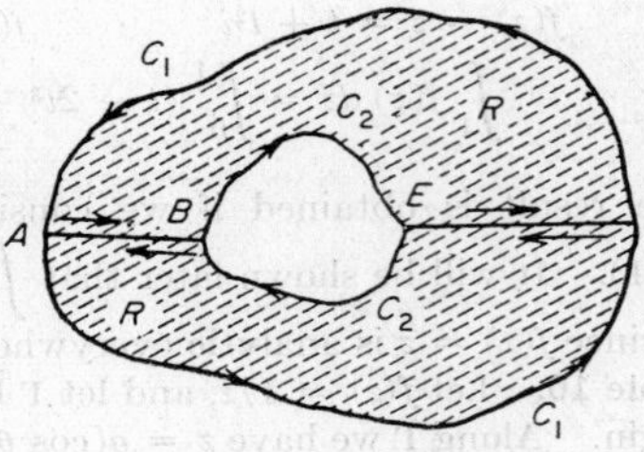

FIG. 9.4

From Fig. 9.4 and Cauchy's theorem, we have

$$\oint_{ABC_2EFC_1A} f(z)\,dz = 0 \qquad \oint_{AC_1FEC_2BA} f(z)\,dz = 0$$

Adding yields

$$\oint_{C_1} f(z)\,dz + \oint_{C_2} f(z)\,dz = 0 \qquad \text{and} \qquad \oint_{C_1} f(z)\,dz = \oint_{C_2} f(z)\,dz$$

Notice that nothing need be known about $f(z)$ outside C_1 or inside C_2.

3. Let $f(z)$ be analytic inside a simply connected region R, and consider

$$F(z) = \int_{z_0}^{z} f(t)\,dt \tag{9.28}$$

where z_0 and z are in R. The path of integration is omitted since the integral is independent of the path (see consequence 1). The letter t is the complex variable of integration. We state without proof that

$$F'(z) = \frac{d}{dz}\int_{z_0}^{z} f(t)\,dt = f(z) \tag{9.29}$$

4. *The Fundamental Theorem of the Integral Calculus.* Let $G(z)$ be any function whose derivative is $f(z)$, $f(z)$ analytic. Then

$$G'(z) - F'(z) = 0$$

from (9.29), so that $F(z) = G(z) + C$. From (9.28)

$$G(z) + C = \int_{z_0}^{z} f(t)\,dt \qquad G(z_0) + C = 0$$

so that

$$G(z) - G(z_0) = \int_{z_0}^{z} f(t)\,dt \tag{9.30}$$

Example 11. We have seen that

$$\oint_\Gamma \frac{dz}{z} = 2\pi i$$

where Γ is any circle with center at the origin. Note that $1/z$ is not analytic at the origin, an interior point of Γ. From paragraph 2 of this section we note that, if C is any simple closed path containing the origin in its interior, then

$$\oint_C \frac{dz}{z} = 2\pi i$$

9.8. CAUCHY'S INTEGRAL FORMULA

From Cauchy's integral theorem we derive the important integral formula due also to Cauchy. Let $f(z)$ be analytic in a simply connected region R, and let Γ be any simple closed path in R. Cauchy's integral formula states that

$$f(a) = \frac{1}{2\pi i}\oint_\Gamma \frac{f(z)}{z-a}\,dz \tag{9.31}$$

if $z = a$ is an interior point of Γ. The proof proceeds as follows: From paragraph 2 of Sec. 9.7, we can replace the curve of integration Γ by a circle C with center at $z = a$ of radius $b > 0$. On C we have

$$z = a + b(\cos\theta + i\sin\theta) \qquad 0 \le \theta < 2\pi$$
$$dz = b(-\sin\theta + i\cos\theta)\,d\theta \qquad f(z) = f(a) + \eta$$

where $\eta \to 0$ as $b \to 0$, since $f(z)$ is continuous at $z = a$. Hence

$$\begin{aligned}\frac{1}{2\pi i}\oint_\Gamma \frac{f(z)}{z-a}\,dz &= \frac{1}{2\pi}\int_0^{2\pi}[f(a)+\eta]\,d\theta \\ &= f(a) + \frac{1}{2\pi}\int_0^{2\pi}\eta\,d\theta\end{aligned} \tag{9.32}$$

The left-hand side of (9.32) is independent of b, and the right-hand side of (9.32) approaches $f(a)$ as $b \to 0$, since $\eta \to 0$ as $b \to 0$. Thus (9.31) is seen to be true. The importance of analyticity is seen when one examines (9.31). The value of $f(z)$ at $z = a$ can be obtained if one knows the values of $f(z)$ on Γ.

It is not difficult to show that successive derivatives of $f(z)$ exist at $z = a$ by performing the differentiations with respect to a inside those integrals obtained in this manner from (9.31). Thus

$$f'(a) = \frac{1}{2\pi i}\oint_\Gamma \frac{f(z)}{(z-a)^2}\,dz \qquad f^{(n)}(a) = \frac{n!}{2\pi i}\oint_\Gamma \frac{f(z)\,dz}{(z-a)^{n+1}} \tag{9.33}$$

If a function is analytic at $z = a$, all its derivatives exist at $z = a$. In real-variable theory the existence of $f'(x)$ in a neighborhood of $x = a$ in no way yields any information about the existence of further derivatives of $f(x)$ at $x = a$.

9.9. TAYLOR SERIES EXPANSION

Let $f(z)$ be analytic in a simply connected region R. Let z_0 be any interior point of R, and let C be the largest circle with center at z_0 such that $f(z)$ is analytic inside C. Finally, let Γ be any circle with center at z_0 containing an arbitrary point z of R in its interior (see Fig. 9.5).

Let ζ be any point on Γ. From Cauchy's integral formula,

$$f(z) = \frac{1}{2\pi i}\oint \frac{f(\zeta)}{\zeta - z}\,d\zeta = \frac{1}{2\pi i}\oint \frac{f(\zeta)\,d\zeta}{(\zeta - z_0) - (z - z_0)}$$
$$= \frac{1}{2\pi i}\oint \frac{f(\zeta)\,d\zeta}{(\zeta - z_0)(1 - (z - z_0)/(\zeta - z_0))} \tag{9.34}$$

Since $\left|\dfrac{z - z_0}{\zeta - z_0}\right| < \kappa < 1$, we have

$$\frac{1}{1 - (z - z_0)/(\zeta - z_0)} = \sum_{n=0}^{\infty} \frac{(z - z_0)^n}{(\zeta - z_0)^n}$$

Substituting into (9.34) yields

$$f(z) = \frac{1}{2\pi i}\oint \sum_{n=0}^{\infty} \frac{(z - z_0)^n f(\zeta)\,d\zeta}{(\zeta - z_0)^{n+1}} \tag{9.35}$$

We can justify interchanging the order of integration and summation, so that

$$f(z) = \sum_{n=0}^{\infty} \left(\frac{1}{2\pi i}\oint \frac{f(\zeta)\,d\zeta}{(\zeta - z_0)^{n+1}}\right)(z - z_0)^n$$
$$= \sum_{n=0}^{\infty} \frac{f^{(n)}(z_0)}{n!}(z - z_0)^n \tag{9.36}$$

by making use of (9.33). Equation (9.36) is the Taylor series expansion of $f(z)$ about $z = z_0$. This series converges for any point z inside C of Fig. 9.5. The proof is omitted that the series of (9.36), representing $f(z)$, is unique.

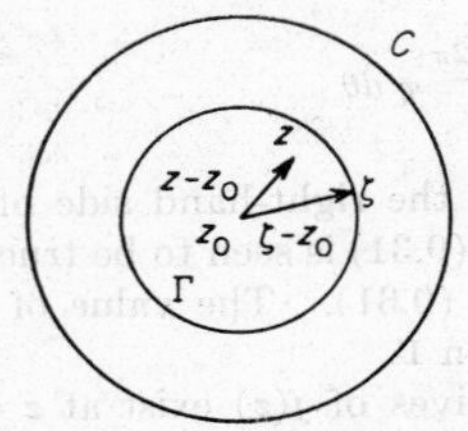

FIG. 9.5

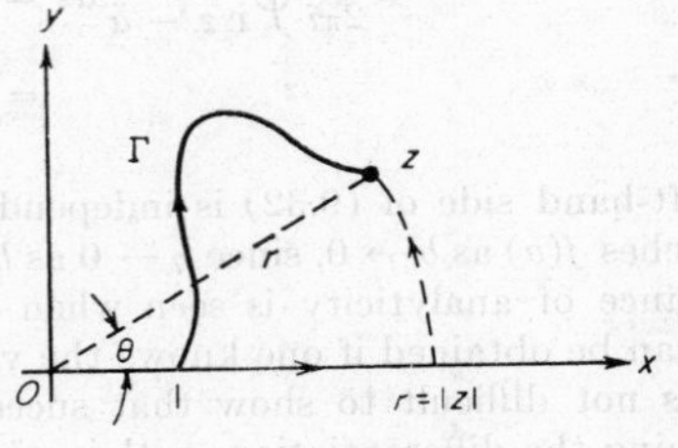

FIG. 9.6

Example 12. The function e^z defined in Sec. 9.4 has the property that $d^n e^z/dz^n = e^z$ for all integers n. Since $e^0 = 1$, we have $f^{(n)}(0) = 1$. From (9.36), we obtain

$$e^z = \sum_{n=0}^{\infty} \frac{z^n}{n!} \tag{9.37}$$

by choosing $z_0 = 0$.

Example 13. Let us define ln z as follows: Let z be any complex number other that $z = 0$, $z = r(\cos\theta + i\sin\theta) = re^{i\theta}$ $(-\pi < \theta \leq \pi)$. We evaluate

$$\ln z = \int_1^z \frac{dt}{t} \tag{9.38}$$

along any curve Γ not passing through the origin or crossing the negative x axis (see Fig. 9.6). We can replace Γ by the straight-line path from $x = 1$ to $x = |z|$, followed by the arc of the circle with radius $|z|$ and center at the origin, until we reach z. This yields

$$\ln z = \int_1^{|z|} \frac{dx}{x} + \int_0^{\theta} \frac{ire^{i\phi}}{re^{i\phi}}\, d\phi = \ln |z| + i \arg z$$

For z real and positive, $\arg z = 0$ and $\ln x = \ln |x|$. From (9.38) it follows that $d \ln z/dz = 1/z, \ldots, d^n \ln z/dz^n = (-1)^{n+1}(n-1)!/z^n$, so that, for $z_0 = 1$, (9.36) yields

$$\ln z = \sum_{n=1}^{\infty} \frac{(-1)^{n+1}}{n} (z-1)^n \tag{9.39}$$

Example 14. The function $f(z) = 1/(1-z)$, $z \neq 1$, has for its only singularity the point $z = 1$. Since $f(z)$ is analytic at $z = 0$, we can obtain the Taylor series expansion for $f(z)$ about $z = 0$. This series will converge for only those values of z such that $|z| < 1$, since the distance from $z = 0$ to the nearest singularity of $f(z)$, namely, $z = 1$, is one. It is easy to show that

$$\frac{1}{1-z} = \sum_{n=0}^{\infty} z^n \tag{9.40}$$

Now $1/(1-z)$ is analytic at $z = i/2$. Since the distance from $z = i/2$ to the nearest singularity $z = 1$ is $\sqrt{5}/2$, the Taylor series expansion of $1/(1-z)$ about $z = i/2$ will converge for all z such that $|z - i/2| < \sqrt{5}/2$. We write

$$\frac{1}{1-z} = \frac{1}{1 - i/2 - (z - i/2)} = \frac{1}{1-i/2} \, \frac{1}{1 - \dfrac{z-i/2}{1-i/2}}$$

$$= \frac{1}{1-i/2} \sum_{n=0}^{\infty} \left(\frac{z-i/2}{1-i/2}\right)^n \tag{9.41}$$

The interiors of the circles of convergence, $|z| < 1$ and $|z - i/2| < \sqrt{5}/2$, overlap. In the common region of overlapping, the series of (9.40) and (9.41) converge to the same value of $f(z) = 1/(1-z)$. Each series, however, is a valid representation of $1/(1-z)$ for those z in their respective circles of convergence. We say that (9.40) and (9.41) are analytic continuations of each other since they represent the same function in different regions of the z plane.

In the example above, the form of $f(z)$ was given explicitly for $z \neq 1$ (a global definition), so that, at any point $z \neq 1$, we can form the Taylor series expansion

$$f_{z_0}(z) = \sum_{n=0}^{\infty} \frac{f^{(n)}(z_0)}{n!} (z - z_0)^n$$

The radius of convergence of this series is $R = |z_0 - 1|$. The totality of all such function elements, $\{f_{z_0}(z)\}$, $z_0 \neq 1$, represents the analytic function $f(z) = 1/(1-z)$. The totality of all the function elements can be obtained from any given function element by a process known as *analytic continuation* described below.

Suppose we wish to examine a function given in the form of a power series with a finite nonzero radius of convergence, say

$$f_0(z) = \sum_{n=0}^{\infty} a_n z^n \qquad |z| < R$$

This regular function is thus defined locally. It can be shown that $f_0(z)$ has at least one singular point on the circumference of the circle of convergence, for otherwise one could extend the radius of convergence. Now let $z = b$ be any interior point of the circle $|z| = R$ (see Fig. 9.7). Since $f_0(z)$ is regular at $z = b$, one can compute $f_0^{(n)}(b)$ $(n = 0, 1, 2, \ldots)$, from which the function element

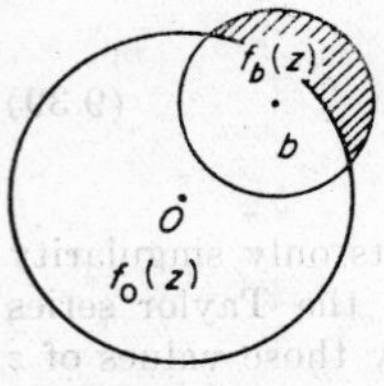

FIG. 9.7

$$f_b(z) = \sum_{n=0}^{\infty} \frac{f_0^{(n)}(b)}{n!} (z - b)^n$$

can be formed. The radius of convergence $R(b)$ of this new function element is at least $R - |b|$. If $R(b) = R - |b|$, the circles of convergence of $f_0(z)$ and $f_b(z)$ have only one point in common, and this point is a singularity of $f_0(z)$. If, however, $R(b) > R - |b|$, then a crescent-shaped region will exist outside the circle $|z| = R$ for which the function element $f_b(z)$ represents an analytic continuation of $f_0(z)$.

In the common region of intersection, $f_0(z) \equiv f_b(z)$. We can now consider a point $z = c$ in the crescent-shaped region of Fig. 9.7, and can obtain the function element $f_c(z)$ in exactly the same manner as described above. This process can be continued until a singular point is reached. More generally, it can be shown that, if C is any continuous curve starting at the origin, one can continue $f_0(z)$ analytically along C (the centers of the circles of convergence lie on C); this process of analytic continuation extends up to a point of singularity which lies on C. The totality of all function elements obtained by analytic continuation by considering all continuous curves passing through the origin yields a global definition of the analytic function $f(z)$, originally represented by $f_0(z)$ for $|z| < R$.

9.10. LAURENT'S EXPANSION

In a manner analogous to the development of the Taylor series expansion, Laurent showed that, if $f(z)$ is analytic in an annular region given by $r_1 < |z - z_0| < r_2$, then

$$f(z) = \sum_{n=-\infty}^{\infty} a_n (z - z_0)^n$$
$$a_n = \frac{1}{2\pi i} \oint_\Gamma \frac{f(\zeta)\, d\zeta}{(\zeta - z_0)^{n+1}} \qquad n = 0, \pm 1, \pm 2, \ldots \qquad (9.42)$$

where Γ is any path lying in the annular region described above.

If we express (9.42) in the form

$$f(z) = \sum_{n=0}^{\infty} a_n (z - z_0)^n + \sum_{n=1}^{\infty} \frac{a_{-n}}{(z - z_0)^n} \qquad (9.43)$$

the first series converges for $|z - z_0| < r_2$, and the second series converges for $|z - z_0| > r_1$. Both series converge in the annular region described above. The second series in (9.43) is known as the principal part of the Laurent expansion of $f(z)$.

Example 15. The function

$$f(z) = \frac{1}{(z-1)(z-2)} \tag{9.44}$$

is analytic in the annular region given by $1 < |z| < 2$. Of course, $f(z)$ is analytic everywhere except at $z = 1$ and $z = 2$. Now we write

$$f(z) = \frac{1}{z-2} - \frac{1}{z-1}$$

$$= -\frac{1}{2}\frac{1}{1-z/2} - \frac{1}{z}\frac{1}{1-1/z} = -\sum_{0}^{\infty} \frac{z^n}{2^{n+1}} - \sum_{1}^{\infty} z^{-n} \tag{9.45}$$

The first series of (9.45) converges for $|z| < 2$, and the second series converges for $|z| > 1$. Hence (9.45) represents $f(z)$ in the annular region $1 < |z| < 2$.

9.11. SINGULAR POINTS. RESIDUES

We have stated previously that z_0 is a singular point of $f(z)$ if $f(z)$ is not analytic at z_0. We say that z_0 is an isolated singular point of $f(z)$ if $f(z)$ is analytic for some neighborhood of z_0 with the exception of z_0. In this case, the Laurent expansion of $f(z)$ given by (9.45) converges for $0 < |z - z_0| < r$, where r is the distance from z_0 to the nearest singular point of $f(z)$ other than z_0 itself. We distinguish now among three types of functions with isolated singular points.

Case 1. The series

$$f(z) = \sum_{n=0}^{\infty} a_n(z - z_0)^n$$

is such that $a_{-n} = 0$ for $n \geqq 1$. If $f(z_0) \neq a_0$, we need only redefine $f(z)$ at z_0 to have the value a_0 in order for $f(z)$ to be analytic at $z = z_0$. A singularity of this type is said to be a removable singularity.

Case 2. All but a finite number of the a_n vanish, so that

$$f(z) = \frac{a_{-n}}{(z-z_0)^n} + \cdots + \frac{a_{-1}}{z-z_0} + \sum_{n=0}^{\infty} a_n(z-z_0)^n \qquad a_{-n} \neq 0$$

For this case we say that z_0 is a pole (or a regular singular point) of order n of $f(z)$. The constant a_{-1} is called the residue of $f(z)$ at $z = z_0$, written as $a_{-1} = \text{res } f(z_0)$. Its importance will appear in the next section. Note that

$$(n-1)!a_{-1} = \left\{\frac{d^{n-1}}{dz^{n-1}}[(z-z_0)^n f(z)]\right\}_{z=z_0} \tag{9.46}$$

A simple pole exists if $a_{-1} \neq 0$, $a_{-n} = 0$ for $n > 1$.

Case 3. An infinite number of the a_{-n} do not vanish. In this case we say that z_0 is an essential singular point. For example,

$$e^{1/z} = 1 + \frac{1}{z} + \frac{1}{2!}\frac{1}{z^2} + \cdots + \frac{1}{n!}\frac{1}{z^n} + \cdots$$

obtained from e^z by replacing z by $1/z$ yields an essential singularity at $z = 0$.

The point at infinity plays an important role in the development of differential equations in the complex domain. The point $z = \infty$ is an isolated singular point of $f(z)$ if $z = 0$ is an isolated singular point of $f(1/z)$. The nature of $f(1/z)$ at $z = 0$, by definition, yields the type of singularity of $f(z)$ at $z = \infty$. Thus $f(z) = z - 1/z$ yields $f(1/z) = 1/z - z$ which has a simple pole at $z = 0$ with residue 1, so that $f(z) = z - 1/z$ has a simple pole at $z = \infty$ with residue 1.

9.12. THE RESIDUE THEOREM. CONTOUR INTEGRATION

If $z = z_0$ is an isolated singular point of $f(z)$, then

$$f(z) = \sum_{n=-\infty}^{\infty} a_n(z - z_0)^n$$

for $0 < |z - z_0| < r$. Now let Γ be a simple closed path encircling $z = z_0$ and lying in the region $|z - z_0| < r$. Then

$$\oint f(z)\,dz = \oint \sum_{n=-\infty}^{\infty} a_n(z - z_0)^n\,dz = \sum_{n=-\infty}^{\infty} a_n \oint (z - z_0)^n\,dz$$

We omit proof of the interchange of integration and summation. Now

$$\oint_\Gamma (z - z_0)^n\,dz = 0 \qquad \text{except for } n = -1 \qquad \text{and} \qquad \oint \frac{dz}{z - z_0} = 2\pi i$$

so that

$$\oint_\Gamma f(z)\,dz = 2\pi i a_{-1} = 2\pi i \operatorname{res} f(z_0)$$

If $f(z)$ contains a finite number of singularities inside Γ, then

$$\oint_\Gamma f(z)\,dz = 2\pi i \sum \operatorname{res} f(z_k) \tag{9.47}$$

where the sum is extended over all points z_k inside Γ.

Equation (9.47) is the important residue theorem which is most useful in evaluating certain integrals.

Example 16. Let us evaluate

$$\int_{-\infty}^{\infty} \frac{dx}{1 + x^2}$$

by use of the residue theorem. We deal naturally with $f(z) = 1/(1 + z^2)$, which has simple poles at $z = \pm i$. Now we look for a path of integration which will eventually embrace the real axis and will contain a singularity of $f(z)$. Let us choose for Γ the straight-line path from $x = -R$ to $x = R$, and the upper semicircle $|z| = R > 1$, (see Fig. 9.8).

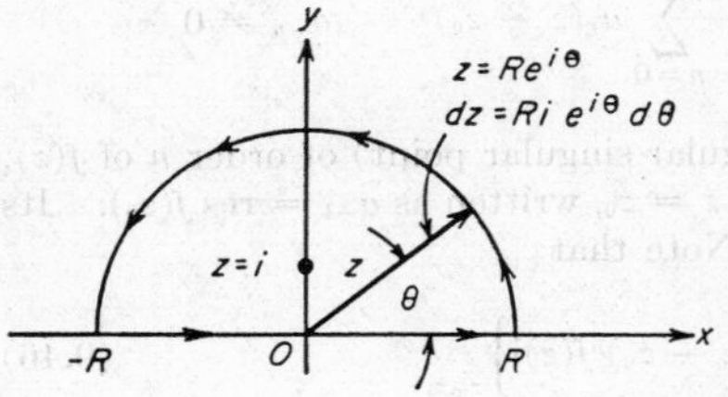

FIG. 9.8

From (9.46), the residue of $f(z) = 1/(1 + z^2)$ at $z = i$ is

$$a_{-1} = \lim_{z\to i} \frac{z - i}{(z - i)(z + i)} = \frac{1}{2i}$$

so that

$$\int_{-R}^{R} \frac{dx}{1 + x^2} + \int_0^\pi \frac{Rie^{i\theta}\,d\theta}{1 + R^2e^{2i\theta}} = 2\pi i \frac{1}{2i} = \pi \tag{9.48}$$

Now

$$\left| \int_0^\pi \frac{Rie^{i\theta}\,d\theta}{1 + R^2e^{2i\theta}} \right| \le \frac{R}{R^2 - 1} \int_0^\pi d\theta = \frac{R\pi}{R^2 - 1}$$

since $|1 + R^2 e^{2i\theta}| \geqq R^2 - 1$ for $R > 1$. As $R \to \infty$, (9.48) becomes

$$\lim_{R\to\infty} \int_{-R}^{R} \frac{dx}{1+x^2} = \int_{-\infty}^{\infty} \frac{dx}{1+x^2} = \pi$$

Example 17. Let us evaluate $\int_{-\infty}^{\infty} \frac{x \sin x \, dx}{x^2 + c^2}$, c positive. Instead of dealing with $f(z) = (z \sin z)/(z^2 + c^2)$, it is much simpler to consider $f(z) = ze^{iz}/(z^2 + c^2)$. The function $e^{iz} = \cos z + i \sin z$ introduces the term $\sin z$. Now $f(z)$ has simple poles at $z = \pm ci$. However, the contour of Fig. 9.8 yields some difficulties. The choice of the contour of integration is not apparent in a great many cases. For this example, the contour in the shape of the rectangle with vertices at $z = -R, R, R + Ri, -R + Ri$ yields a significant result. The residue of $f(z)$ at $z = ci$ is

$$a_{-1} = \lim_{z\to ci} \frac{ze^{iz}(z - ci)}{z^2 + c^2} = \tfrac{1}{2}e^{-c}$$

As $R \to \infty$, three of the four integrals tend to zero, so that

$$\int_{-\infty}^{\infty} \frac{xe^{ix}\, dx}{x^2 + c^2} = 2\pi i\,(\tfrac{1}{2}e^{-c}) = \pi i e^{-c}$$

Replacing e^{ix} by $\cos x + i \sin x$, and equating imaginary parts yields

$$\int_{-\infty}^{\infty} \frac{x \sin x}{x^2 + c^2}\, dx = \pi e^{-c} \tag{9.49}$$

Example 18. By use of the residue theorem it can be shown that, if $f(z)$ has N zeros and P poles inside Γ, then

$$\oint_{\Gamma} \frac{f'(z)}{f(z)}\, dz = 2\pi i(N - P) \tag{9.50}$$

This result is intimately connected with the Nyquist criterion encountered in servomechanism theory (see p. 59–5).

9.13. THE SCHWARTZ-CHRISTOFFEL TRANSFORMATION

Let us consider a closed polygon in the w plane. It is assumed that the polygon does not intersect itself (see Fig. 9.9).

Now we wish to find a function $z = F(w)$ or $w = f(z)$ which maps the polygon into the real axis of the z plane. Let P_1 be mapped into $(x_1,0)$; P_2 into $(x_2,0)$; etc. As one

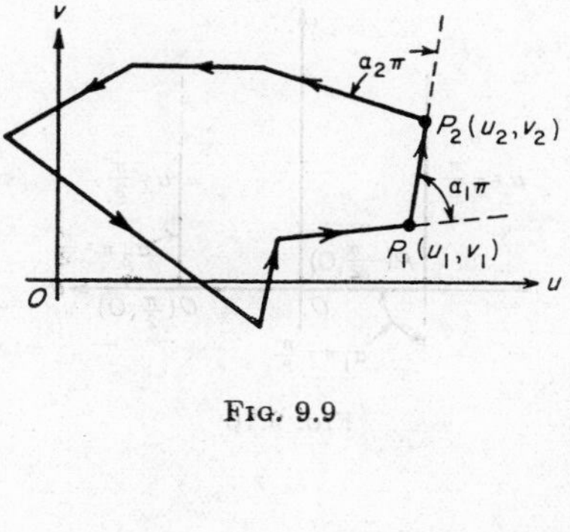

Fig. 9.9

moves along the polygon in the w plane, one will move to the right along the x axis in the z plane. If the transformation $w = f(z)$ exists, then, from $dw/dz = f'(z)$, $dw \approx f'(z)\,dz$, we obtain

$$\arg dw = \arg f'(z) + \arg dz$$

Since $\arg dz = \arg dx = 0$, it follows that

$$\arg dw = \arg f'(z)$$

along the polygon. If $\Delta \arg dw$ represents an abrupt change in the direction of dw (this occurs at the vertices of the polygon), then

$$\Delta \arg dw = \Delta \arg f'(z)$$

so that $\arg f'(z)$ has a discontinuity at $z = x_1, x_2, \ldots$. Thus, at $z = x_k$ we have $\Delta \arg f'(x_k) = \alpha_k\pi (k = 1, 2, 3, \ldots)$. Now consider $f'(z) = (z - x_k)^{-\alpha_k}$, so that $\arg f'(z) = -\alpha_k \arg (z - x_k)$. For $z < x_k$ we have $\arg (z - x_k) = -\pi$, and for $z > x_k$ we have $\arg (z - x_k) = 0$. Thus $\Delta \arg (z - x_k) = \pi$. This suggests that the transformation we are looking for satisfies

$$\frac{dw}{dz} = f'(z) = A(z - x_1)^{-\alpha_1}(z - x_2)^{-\alpha_2} \cdots (z - x_m)^{-\alpha_m}$$

It can be shown that

$$w = f(z) = A \int_0^z (z - x_1)^{-\alpha_1} \cdots (z - x_m)^{-\alpha_m}\,dz + B \tag{9.51}$$

is the required transformation. A and B are arbitrary constants in the *Schwarz-Christoffel transformation* given by (9.51).

Example 19. We consider the polygon of Fig. 9.10. Two of the vertices of the polygon are at $(-\pi/2,0)$, $(\pi/2,0)$, and the other two vertices are at $w = \pm\pi/2 + i\infty$. Let us map P into $(-a,0)$ and Q into $(a,0)$. At P we have $\alpha_1 = \frac{1}{2}$, and at Q we have $\alpha_2 = \frac{1}{2}$. Thus (9.51) yields

$$w = A \int_0^z (z + a)^{-1/2}(z - a)^{-1/2}\,dz + B$$

$$= A \int_0^z \frac{dz}{\sqrt{z^2 - a^2}} + B = \frac{A}{i} \sin^{-1}\frac{z}{a} + B$$

From the conditions $z = -a$, $w = -\pi/2$, and $z = a$, $w = \pi/2$, we obtain

$$w = \sin^{-1}\frac{z}{a} \qquad z = a \sin w \tag{9.52}$$

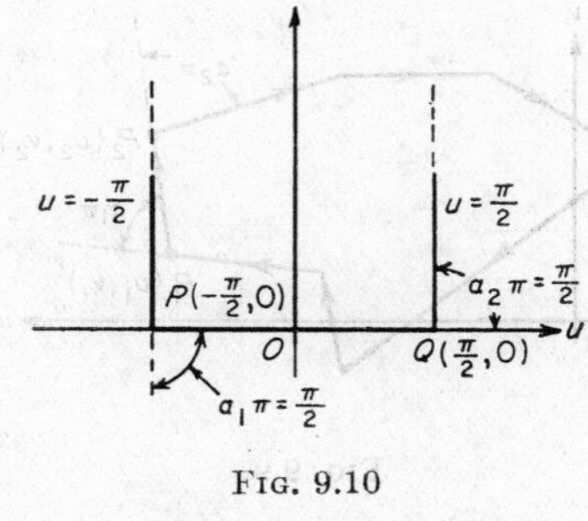

FIG. 9.10

If we let $w = \pi/2 + Ri$, then $z = a \cos Ri = a \operatorname{Cosh} R$, so that the point $w = \pi/2 + i\infty$ maps into the point $x = \infty$. For $w = -\pi/2 + Ri$ we have $z = -a \operatorname{Cosh} R$, so that the point $-\pi/2 + i\infty$ maps into $x = -\infty$. The transformation (9.52) unfolds the polygon.

For $z = x + yi$, $w = u + vi$, (9.52) yields

$$x = a \sin u \operatorname{Cosh} v \qquad y = a \cos u \operatorname{Sinh} v$$

so that

$$\frac{x^2}{\operatorname{Cosh}^2 v} + \frac{y^2}{\operatorname{Sinh}^2 v} = a^2$$

Hence the straight lines $v = \text{const} \neq 0$ map into ellipses in the z plane. Similarly, the straight lines $u = \text{const} \neq 0, \pi/2$, map into the hyperbolas in the z plane.

9.14. REFERENCES

[1] L. V. Ahlfors: "Complex Analysis," McGraw-Hill, New York, 1953.
[2] A. Betz: "Konforme Abbildung," Springer, Berlin, 1948.
[3] E. T. Copson: "Theory of Functions of a Complex Variable," Oxford Univ. Press, New York, 1935.
[4] K. Knopp: "Theory of Functions," Dover, New York, 1945.
[5] H. Kober: "Dictionary of Conformal Representations," Dover, New York, 1957.
[6] T. M. McRobert: "Functions of a Complex Variable," St. Martin's, New York, 1938.
[7] E. C. Titchmarsh: "The Theory of Functions," Oxford Univ. Press, New York, 1932.
[8] E. T. Whittaker, G. N. Watson: "A Course of Modern Analysis," Macmillan, New York, 1944.

CHAPTER 10

ORDINARY DIFFERENTIAL EQUATIONS

BY

G. E. LATTA, Ph.D., Stanford, Calif.

10.1. GENERAL REMARKS ABOUT ORDINARY DIFFERENTIAL EQUATIONS

An ordinary differential equation is a relation connecting an independent variable x and a number of derivatives of a dependent variable $y(x)$. Such equations are classified as to *order*, which is the order n of the highest derivative appearing in the given relation, and to *degree*, which is the algebraic degree of the highest derivative, it being assumed in the latter case that the equation is algebraic in the derivatives of $y(x)$. A differential equation is *linear* if the dependent variable and all its derivatives appear to be first degree only, and no two such terms are multiplied together; that is, the coefficients of y, or any of its derivatives, are functions of the independent variable alone.

Thus $f(x,y,y', \ldots, y^{(n)}) = 0$ is an nth-order equation, while

$$y^{(n)} = F(x,y,y', \ldots, y^{(n-1)})$$

is of nth order and first degree, if F is algebraic in $y', \ldots, y^{(n-1)}$. The most general linear equation of the nth order is

$$p_0(x)y^{(n)} + p_1(x)y^{(n-1)} + \cdots + p_n(x)y = r(x)$$

By a *solution* of a differential equation is meant any expression, free of derivatives, and involving x and y, such as $g(x,y,c) = 0$, $y = h(x)$, etc., which is compatible with the differential equation. A solution of an equation of order n which involves n arbitrary constants, and which cannot be written in terms of fewer constants, is called the *general solution* of the equation. Solutions obtained by specializing these constants are called *particular integrals*, or particular solutions. Relations obtained from an nth-order equation which involve only derivatives of order less than n, and which are compatible with the original equation, are called *intermediate integrals*.

Singular solutions are solutions of the differential equation which are not particular integrals; they occur, for example, in some nonlinear equations of first order whose general solution is a one-parameter family of curves having an envelope. The envelope is a singular solution. If $f(x,y,p) = 0$, $p = y'$, is the differential equation, then $\partial f/\partial p = 0$, together with $f(x,y,p) = 0$, contains any possible envelope, and hence any possible singular solution.

The general solution of an nth-order equation is an n parameter family of curves, and n subsidiary conditions, called boundary conditions, are necessary to single out one particular member of the family. For first-order equations, a precise statement is the

Existence and Uniqueness Theorem

If $f(x,y)$ is continuous in x,y, and if $|f(x,y_2) - f(x,y_1)| \leqq M|y_2 - y_1|$ for all x,y_1,y_2 close to the preassigned point (x_0,y_0), then $y' = f(x,y)$ has a unique solution which has the value y_0 when $x = x_0$.

The condition $|f(x,y_2) - f(x,y_1)| \leqq M|y_2 - y_1|$ is called a Lipschitz condition, and is automatically satisfied under the more restrictive situation when $\partial f/\partial y$ is continuous near (x_0,y_0). An immediate corollary is that the Lipschitz condition holds for linear equations $y' + p(x)y = r(x)$ with continuous $p(x)$, since $\partial f/\partial y = -p(x)$.

As a generalization of the above existence theorem, we consider the simultaneous solution of a system of equations

$$\frac{dy_i(x)}{dx} = f_i(x,y_1, \ldots, y_n) \qquad i = 1, \ldots, n$$

which can be written in matrix form

$$\frac{dY}{dx} = F(x,Y) \tag{10.1}$$

where Y stands for the column matrix of ys, and $F(x,Y)$ is the column matrix of fs. The existence and uniqueness theorem for (10.1) can be stated

$$\frac{dY}{dx} = F(x,Y) \qquad \text{subject to} \quad Y(x_0) = Y_0 \quad \text{an arbitrary constant vector}$$

has a unique solution if $F(x,Y)$ is continuous in x,Y, and if $F(x,Y)$ satisfies a Lipschitz condition in Y; we can write this as $|F(x,Y) - F(x,Z)| \leqq M|Z - Y|$, which stands for

$$|f_i(x,y_1, \ldots, y_n) - f_i(x,z_1, \ldots, z_n)| \leqq M\{|z_1 - y_1| + |z_2 - y_2| + \cdots + |z_n - y_n|\} \quad \text{for each } i = 1, \ldots, n$$

The connection between such first-order systems and single equations of higher order is that an equation of order n can be transformed into such a system by the following device.

Let

$$y^{(n)} = f(x,y,y', \ldots, y^{(n-1)}) \tag{10.2}$$

be the nth-order equation in question. Define

$$y_1 = y, \quad y_2 = y', \quad y_3 = y'', \ldots, \quad y_n = y^{(n-1)}$$

Then (10.2) is equivalent to the system

$$y_1' = y_2, \quad y_2' = y_3, \ldots, \quad y_{n-1}' = y_n, \quad y_n' = f(x,y_1, \ldots, y_n)$$

If, now, the Lipschitz condition is satisfied, we can prescribe an arbitrary vector Y_0 at x_0, and obtain a unique solution. In other words, we can prescribe $y(x_0)$, $y'(x_0)$, . . ., $y^{(n-1)}(x_0)$ in advance. Since these n conditions are all at the same point x_0, this problem is referred to as the *initial-value problem*. Again, linear equations automatically fulfill these conditions, so that the initial-value problem for linear equations possesses a unique solution.

REFERENCES: [1], chaps. 1, 3; [2], chap. 1

10.2. ELEMENTARY METHODS OF INTEGRATION

First Order and First Degree

$$\frac{dy}{dx} = f(x,y) \qquad dy - f(x,y)\,dx = 0$$

Since either x or y can be considered as independent in this case, we consider the equation in the more flexible form

$$M(x,y)\,dx + N(x,y)\,dy = 0 \tag{10.3}$$

If this equation falls into one of the following categories, the general solution is "reduced to quadratures"; that is, the solution is given in terms of a number of indefinite integrals.

(1) Variables Separable. If each of $M(x,y)$, $N(x,y)$ is a product of a function of x alone, and a function of y alone, then, by a rearrangement of terms, (10.3) becomes

$$P(x)\,dx + Q(y)\,dy = 0$$

giving

$$\int^x P(x')\,dx' + \int^y Q(y')\,dy' = c$$

which is in quadrature form.

(2) Exact Equations. Since $dF(x,y) = (\partial F/\partial x)\,dx + (\partial F/\partial y)\,dy$ and $dF(x,y) = 0$ implies $F(x,y) = c$, the general solution of $F_x\,dx + F_y\,dy = 0$ is $F(x,y) = c$. If it happens that $M(x,y) = F_x$ and $N(x,y) = F_y$, then (10.3) is said to be exact, and the solution is $F(x,y) = c$. A necessary and sufficient condition for exactness, which does not involve the solution of the problem, is $P_y = Q_x$. A practical method of integration is the following:

$$P = F_x \qquad F(x,y) = \int^x P(x',y)\,dx'$$

where y is constant for the integration, and the constant of integration is an arbitrary function of y. Then

$$F_y = Q = \frac{\partial}{\partial y}\int^x P(x',y)\,dx'$$

leads to the determination of $F(x,y)$.

(3) Nonexact Equations. If (10.3) is not exact, but satisfies the conditions of the existence theorem of Sec. 10.1, then there exist functions $\theta(x,y)$, called integrating factors, so that

$$\theta M\,dx + \theta N\,dy = 0$$

is exact. The condition on $\theta(x,y)$ is

$$(\theta M)_y = (\theta N)_x$$

which involves two independent variables, and is a partial differential equation. The problem of finding an integrating factor is as difficult as solving the original equation, in general. It can be shown that the quotient θ_1/θ_2 of two distinct integrating factors is an $F(x,y)$ so that $\theta_1/\theta_2 = c$ is a general solution of (10.3).

(4) Homogeneous Functions. If $M(x,y)$, $N(x,y)$ are homogeneous functions of the same degree k, then $y = vx$ yields an equation in x and v for which the variables are separable. $M(x,y)$ is homogeneous of degree k if $M(\lambda x,\lambda y) \equiv \lambda^k M(x,y)$. In this case $M(x,y) = x^k M(1,y/x)$, so that (10.3) becomes

$$[M(1,v) + vN(1,v)]\,dx + xN(1,v)\,dv = 0$$

(5) Linear Equations. The first-order linear equation

$$y' + p(x)y = r(x)$$

has $e^{\int p\,dx}$ as an integrating factor, and solution

$$y = e^{-\int p\,dx}\int e^{\int p\,dt} r(t)\,dt + ce^{-\int p\,dx} \tag{10.4}$$

(6) Bernoulli's Equation

$$y' + p(x)y = q(x)y^a \qquad a \neq 1$$

becomes linear in v where $v = y^{1-a}$.

(7) Ricatti's Equation

$$y' = p(x) + q(x)y + r(x)y^2$$

This equation cannot, in general, be reduced to quadratures. If any particular solution is known, say $y_1(x)$, the substitution $y = y_1(x) + c/v(x)$ leads to a linear equation for $v(x)$.

One of the most important properties of the Ricatti equation is its connection with linear second-order equations. The substitution

$$y(x) = -\frac{1}{r(x)}\frac{z'(x)}{z(x)}$$

transforms the equation into

$$z'' + \left(\frac{r'}{r} - q\right) z' + rpz = 0$$

First Order, Higher Degree

$$f(x,y,y') \equiv (y')^n + p_1(x,y)(y')^{n-1} + \cdots + p_{n-1}(x,y)y' + p_n(x,y) = 0$$

Theoretically, the polynomial in y' can be factored, so that

$$(y' - q_1)(y' - q_2) \cdots (y' - q_n) = 0 \qquad \text{with } q_i = q_i(x,y) \tag{10.5}$$

Since a product vanishes if, and only if, one of its factors vanishes, then $y(x)$ is a solution to (10.5) if, and only if, $dy/dx = q_i(x,y)$ for some particular $i = 1, \ldots, n$. Thus, if $\phi_i(x,y,c) = 0$ is the general solution of $y' = q_i(x,y)$, then the general solution of (10.5) is

$$\phi_1(x,y,c) \cdot \phi_2(x,y,c) \cdots \phi_n(x,y,c) = 0$$

In general, if y is a solution of $f(x,y,y') = 0$, it is also a solution of

$$\frac{d}{dx} f(x,y,y') = 0 = f_x(x,y,y') + f_y(x,y,y')y' + f_{y'}(x,y,y')\frac{dy'}{dx} \tag{10.6}$$

It may happen that the simultaneous consideration of (10.6) and $f(x,y,y') = 0$ leads to a solution. For example, a first integral of (10.6) which is not equivalent to $f(x,y,y') = 0$ gives a parametric representation of the solution in the form

$$\phi(x,y,p,c) = 0 \quad \text{(first integral)} \qquad p = y'$$
$$f(x,y,p) = 0$$

Another example is Clairaut's equation

$$xy' - y + f(y') = 0 \tag{10.7}$$

Here (10.6) becomes $y''(x + f'(y')) = 0$, yielding $y'' = 0$, $y' = c$, so that

$$xc - y + f(c) = 0$$

is the general solution. The equation $x + f'(y') = 0$ leads to the singular solution.

Similar results hold if $f(x,y,y') = 0$ is differentiated with respect to y.

Higher Order

If x is missing from the equation, then

$$f(y,y',y'', \ldots, y^{(n)}) = 0$$

can be reduced to an equation of order $(n-1)$ by taking y as a new independent variable; $y' = v$, $y'' = v\,dv/dy$, etc.

If y is missing, $f(x,y',y'', \ldots ,y^{(n)}) = 0$ is formally an $(n-1)$st-order equation in $v = y'$, $v' = y''$, etc.

These techniques are most useful for low-order equations, and may lead to intermediate integrals.

A last example is

$$f(y,xy',x^2y'', \ldots ,x^ny^{(n)}) = 0$$

The change of variable $x = e^t$ gives

$$x^k \frac{d^k}{dx^k} = D(D-1) \cdots (D-k+1) \qquad D = \frac{d}{dt}$$

transforming the equation to a new form

$$F\left(y, \frac{dy}{dt}, \ldots, \frac{d^ny}{dt^n}\right) = 0$$

in which t is missing.

REFERENCE: [1], chap. 2

10.3. LINEAR DIFFERENTIAL EQUATIONS

a. General Theory of Linear Equations

The equation

$$Ly \equiv p_0(x)y^{(n)} + p_1(x)y^{(n-1)} + \cdots + p_n(x)y = r(x) \tag{10.8}$$

is the general linear equation of order n. We assume $p_i(x)$, $r(x)$ continuous for all x under consideration. The initial-value problem has a unique solution provided $p_i(x)/p_0(x)$ is continuous, or $p_0(x) \neq 0$. Those x for which $p_0(x) \neq 0$ are called ordinary points of (10.8), while those x for which $p_0(x) = 0$ are called singular points (later classified as regular singular points and irregular points; see p. 10–11).

If $r(x) \equiv 0$, the equation is homogeneous; otherwise nonhomogeneous. From the linearity of Ly, it follows that $L(c_1y_1 + c_2y_2 + \cdots + c_ky_k) = c_1Ly_1 + c_2Ly_2 + \cdots + c_kLy_k$, where the c_i are arbitrary constants, so that, if $y_1, \ldots, y_k$ are solutions of the homogeneous equation, $Ly_i = 0$, then $c_1y_1 + \cdots + c_ky_k$ is also a solution.

The functions $y_1, \ldots, y_k$ are linearly independent if $c_1y_1 + c_2y_2 + \cdots + c_ky_k \equiv 0$ implies each of the constants $c_1, \ldots, c_k = 0$; otherwise, they are linearly dependent. In this section we refer only to *linear* dependence and independence, and so drop the term *linear*.

If $y_1, \ldots, y_k$ are dependent, then constants $c_1, c_2, \ldots, c_k \not\equiv 0$ exist so that

$$c_1y_1 + c_2y_2 + \cdots + c_ky_k = 0$$

In turn, this implies

$$\begin{aligned} c_1y_1' + c_2y_2' + \cdots + c_ky_k' &= 0 \\ \cdots\cdots\cdots\cdots\cdots\cdots& \\ c_1y_1^{(k-1)} + \cdots + c_ky_k^{(k-1)} &= 0 \end{aligned}$$

It now follows that the determinant of the coefficients of c_i must vanish; i.e.,

$$W(y_1,y_2, \ldots ,y_k) \equiv \begin{vmatrix} y_1 & \cdots & y_k \\ \cdots & \cdots & \cdots \\ y_1^{(k-1)} & \cdots & y_k^{(k-1)} \end{vmatrix} = 0$$

Conversely, if $W(y_1, \ldots ,y_k) \neq 0$, then $c_1 = c_2 = \cdots = c_k = 0$, so that a necessary and sufficient condition for the independence of $y_1, \ldots, y_k$ is that the determinant $W(y_1, \ldots, y_k) \neq 0$. This quantity is called the Wronskian of $y_1, \ldots, y_k$.

For the homogeneous equation, the Wronskian of n solutions satisfies $W' = -[p_1(x)/p_0(x)]W$, and

$$W(x) = W(x_0) \exp \left[-\int_{x_0}^{x} \frac{p_1(t)}{p_0(t)}\, dt \right]$$

This is Abel's identity, and shows that the Wronskian [provided $p_0(x) \neq 0$] never vanishes if it is nonzero at any initial point.

If $y_p(x)$ is any *particular integral* of (10.8), then $y = y_p + v$ leads to

$$Ly = L(y_p + v) = Ly_p + Lv = r(x) + Lv = r(x)$$

and $Lv = 0$. The general solution of the homogeneous equation is called the *complementary function*, and we see that the general solution of (10.8) is any particular integral plus the complementary function.

If we try to construct an integrating factor $w(x)$ for the differential expression

$$Ly = p_0(x)y'' + p_1(x)y' + p_2(x)y$$

that is,

$$wLy = w(p_0y'' + p_1y' + p_2y)$$

is to be an exact derivative, integration by parts yields

$$wLy = \frac{d}{dx}[p_0wy' + p_1wy - (p_0w)'y] + y[(p_0w)'' - (p_1w)' + p_2w]$$

which we write as

$$wLy = \frac{d}{dx} P(y,w) + y\bar{L}w \tag{10.9}$$

The expression $\bar{L}w = (p_0w)'' - (p_1w)' + p_2w$ is the adjoint operator to Ly, and the quantity $P(y,w) = p_0wy' - (p_0w)'y + p_1wy$, which is linear in y,y', and also in w,w', is referred to as the bilinear concomitant.

From the identity (10.9) it is clear that $w(x)$ is an integrating factor for Ly provided $\bar{L}w = 0$, or, in other words, w is a solution of the adjoint equation. Conversely, from the symmetry of (10.9), $y(x)$ is an integrating factor of $\bar{L}w$ if $Ly = 0$. In particular, then, Ly is the adjoint of $\bar{L}w$.

A particularly important case occurs when the adjoint of Ly is identical with Ly; in this case Ly is said to be self-adjoint. The condition for self-adjointness in this case is that $p_1 = p_0'$, and $Ly = (p_0y')' + p_2y$. The identity (10.9) now takes the form

$$wLy - yLw = \frac{d}{dx}[p_0(x)(wy' - yw')]$$

known as Lagrange's identity.

While the majority of second-order linear differential equations one obtains from the descriptions of physical laws are self-adjoint, there are many which are not, and it is important in certain applications to deal with self-adjoint equations. To this end, any such equation, $y'' + Py' + Qy = 0$, which is not self-adjoint for $P \neq 0$, can be made self-adjoint by multiplying the equation through by $g(x)$, so chosen that $gP = g'$ (the condition for self-adjointness). This gives

$$g(x) = \exp\left(\int_a^x P\, dt\right)$$

Another important consequence of Eq. (10.9) is that, if all solutions of $Ly = 0$ are known, then the general solution of $\bar{L}w = 0$ is known, and conversely. To see this, let y_1, y_2 be two independent solutions of $Ly = 0$. Then (10.9) gives

$$y_i \bar{L} w = -\frac{d}{dx} P(y_i, w) \qquad i = 1, 2$$

To find solutions of $\bar{L}w = 0$, we merely have to solve $P(y_i, w) = c_i$ for arbitrary constants c_i. Since $P(y,w)$ is linear in w, w', we are actually solving a system of linear algebraic equations for w, w'; the solution for $w(x)$ contains two arbitrary constants, and so is the general solution.

For equations of nth order, the above discussion remains valid, except that it is no longer possible, in general, to make a non-self-adjoint equation self-adjoint by multiplying by a factor $g(x)$.

Particular Integrals. If the general solution of the homogeneous equation (or its adjoint) is known, a particular integral can be found by quadrature. There are several methods, more or less equivalent, for doing this. First, consider

$$Ly = r(x)$$

where the general solution of $\bar{L}v = 0$ is known $v_1, v_2, \ldots, v_n$ being a fundamental system, that is, n independent solutions. Then, from the identities

$$v_i L y = \frac{d}{dx} P\{y, v_i\} + y \bar{L} v_i \qquad i = 1, \ldots, n$$

and, since $\bar{L}v_i = 0$, we have

$$P\{y, v_i\} = \int_{x_0}^{x} v_i r \, dt + c_i \qquad i = 1, \ldots, n \tag{10.10}$$

which are n equations, linear in $y, y', \ldots, y^{(n-1)}$, so that $y(x)$ can be expressed as a quotient of determinants. This form is particularly applicable if the solution of the initial-value problem is desired, as each c_i is then determined at once. For a particular integral alone, we may take each $c_i = 0$.

A second method, due to Bernoulli and Lagrange, is known as *variation of parameters*. If $y_1, \ldots, y_n$ are independent solutions of $Ly = 0$, we try to represent a particular integral of $Ly = r(x)$ as $y = c_1(x)y_1 + c_2(x)y_2 + \cdots + c_n(x)y_n$. We can impose n conditions on the n functions $c_i(x)$, and, moreover, we can do this in such a way as to reduce their determination to quadrature. This is accomplished by requiring that

$$y^{(k)} = c_1(x)y_1^{(k)} + c_2(x)y_2^{(k)} + \cdots + c_n(x)y_n^{(k)} \qquad \text{for } k = 1, 2, 3, \ldots, n-1$$

and that y satisfy $Ly = r(x)$. These equations are equivalent to

$$\begin{aligned}
c_1'y_1 + c_2'y_2 + \cdots + c_n'y_n &= 0 \\
c_1'y_1' + c_2'y_2' + \cdots + c_n'y_n' &= 0 \\
\cdots\cdots\cdots\cdots\cdots\cdots\cdots\cdots\cdots \\
c_1'y_1^{(n-1)} + c_2'y_2^{(n-1)} + \cdots + c_n'y_n^{(n-1)} &= r(x)
\end{aligned}$$

where we assume $p_0(x) = 1$, for simplicity. Since the determinant of the unknown coefficients c_i' is the Wronskian $W(y_1, \ldots, y_n) \neq 0$, $y_1, \ldots, y_n$ being independent, each c_i' is obtained explicitly, and thus each c_i is determined.

Lowering of Order, Removal of Terms. For the linear equation

$$Ly = p_0 y^{(n)} + p_1 y^{(n-1)} + \cdots + p_n y = 0$$

the substitution $y = uv$ can be used to remove various terms from the equation. To illustrate, let $n = 2$,

$$Ly = p_0y'' + p_1y' + p_2y = 0$$
$$L(uv) = p_0(u''v + 2u'v' + uv'') + p_1(u'v + uv') + p_2uv$$
$$= vLu + up_0v'' + (2p_0u' + p_1u)v'$$

If, now, $u(x)$ is a solution of $Lu = 0$, we have a first-order linear equation in v',

$$up_0(v')' + (2p_0u' + p_1u)v' = 0$$

so that

$$v' = \frac{c_1}{u^2}\exp\int^x \frac{p_1}{p_0}\,dt$$

$$y = c_1u\int^x\left(\frac{1}{u^2}\exp\int^x\frac{p_1}{p_0}\,dt\right)dx + c_2u \tag{10.11}$$

gives the general solution.

Alternatively, we can form an equation with no first derivative term by choosing $u(x)$ to satisfy $2p_0u' + p_1u = 0$, or

$$u = \exp\left(-\int^x\frac{p_1}{2p_0}\,dt\right)$$

yielding $up_0v'' + vLu = 0$. Similar procedures hold for higher-order equations.

b. Constant-coefficient Equations

As standard form, we take

$$Ly = y^{(n)} + a_1y^{(n-1)} + \cdots + a_ny = r(x) \tag{10.12}$$

with $a_1, \ldots, a_n$ real constants, and $r(x)$ an arbitrary function of x. The general solution is $y = y_p + y_c$, where y_p is any particular integral, and y_c is the complementary function, or the general solution of the homogeneous equation.

The Complementary Function

$$L(e^{\lambda x}) = (\lambda^n + a_1\lambda^{n-1} + \cdots + a_n)e^{\lambda x} = f(\lambda)e^{\lambda x}$$

so that $e^{\lambda x}$ is a solution of $Ly = 0$ if $f(\lambda) = 0$. This is called the *auxiliary equation.* If the n roots of $f(\lambda) = 0$ are all real and distinct, then $e^{\lambda_i x}$ $(i = 1, \ldots, n)$ are n independent solutions of $Ly = 0$, and so the general solution is $c_1e^{\lambda_1x} + c_2e^{\lambda_2x} + \cdots + c_ne^{\lambda_nx}$. If some of the n distinct roots are complex, they occur in conjugate pairs $\lambda_1 = \mu + i\nu$, $\lambda_2 = \mu - i\nu$; then two solutions in real form corresponding to λ_1, λ_2 are $e^{\mu x}(c_1\cos\nu x + c_2\sin\nu x)$. This follows from Euler's formula,

$$e^{i\theta} = \cos\theta + i\sin\theta$$

Finally, if $f(\lambda) = 0$ has repeated roots, say $f(\lambda) = (\lambda - \lambda_1)^p\phi(\lambda)$ where $\phi(\lambda_1) \neq 0$, we can find p independent solutions pertaining to the exponent λ_1 by the following device, due to d'Alembert,

$$Le^{\lambda x} = (\lambda - \lambda_1)^p\phi(\lambda)e^{\lambda x}$$

and, because of the factor $(\lambda - \lambda_1)^p$, the right-hand side, together with all its derivatives up to order $(p - 1)$, vanishes for $\lambda = \lambda_1$. This gives

$$Le^{\lambda x}, \frac{\partial}{\partial\lambda}Le^{\lambda x}, \frac{\partial^2}{\partial\lambda^2}Le^{\lambda x}, \ldots, \frac{\partial^{p-1}}{\partial\lambda^{p-1}}Le^{\lambda x}$$

all vanishing for $\lambda = \lambda_1$. In turn, then

$$Le^{\lambda x}, L(xe^{\lambda x}), \ldots, L(x^{p-1}e^{\lambda x})$$

vanish for $\lambda = \lambda_1$, so that

$$e^{\lambda_1x}, xe^{\lambda_1x}, x^2e^{\lambda_1x}, \ldots, x^{p-1}e^{\lambda_1x}$$

are all solutions, and are independent. In this way, to each root of multiplicity p we obtain p independent solutions, the general form being

$$e^{\lambda_1 x}(c_1 + c_2 x + \cdots + c_p x^{p-1})$$

An alternate procedure which is very powerful depends on the algebraic properties of the differentiation operator $D \equiv d/dx$. We have $D^2 = d/dx(d/dx) = d^2/dx^2$, $D^n = d^n/dx^n$, and $Ly = 0$ becomes

$$(D^n + a_1 D^{n-1} + \cdots + a_n)y \equiv f(D)y = 0$$

In this form, the *zeros* of the polynomial $f(D)$ are the roots of the auxiliary equation. The identities

$$D^k(e^{\lambda x}v) = e^{\lambda x}(D + \lambda)^k v \qquad f(D)(e^{\lambda x}v) = e^{\lambda x}f(D + \lambda)v$$

follow immediately, and we can obtain the independent solutions corresponding to multiple roots of the auxiliary equation as follows. If $(D - \lambda)^p y = 0$, let $y = e^{\lambda x}v$; then $(D - \lambda)^p e^{\lambda x}v = e^{\lambda x}D^p v = 0$ implies $d^p v/dx^p = 0$, or $v = c_1 + c_2 x + \cdots + c_p x^{p-1}$, as previously.

The Particular Integral. As well as the methods outlined in Art. 10.3*a* for obtaining the particular integral, there are various special methods for constant coefficient equations. One of these is the method of *undetermined coefficients*. If $r(x)$ itself satisfies a constant coefficient homogeneous equation, $g(D)r(x) = 0$, then $f(D)y = r(x)$ and $g(D)f(D)y = 0$. The auxiliary equation for $g(D)f(D)y = 0$ is $g(\lambda)f(\lambda) = 0$, and the general solution of $g(D)f(D)y = 0$ is $y = c_1y_1 + c_2y_2 + \cdots + c_Ny_N$.

Thus the desired particular integral of $f(D)y = r(x)$ is of the form $c_1y_1 + c_2y_2 + \cdots + c_Ny_N$. If we delete from this general expression the complementary function of $f(D)y = r(x)$, we have the form of $y_p = c_{n+1}y_{n+1} + \cdots + c_Ny_N$, and the *undetermined coefficients* $c_{n+1}, \ldots, c_N$ can be found by substitution in $Ly_p = r(x)$. From a practical point of view, it is desirable to choose the lowest-order equation possible for $g(D)r(x) = 0$. The computations can frequently be simplified by noting that, if $r(x)$ is a sum of simpler terms, $r_1 + r_2 + \cdots$, we can form y_p by adding the particular integrals corresponding to $r_1, r_2, \ldots$.

Another method for finding particular integrals depends on the operational calculus of $D = d/dx$. Noting that the solution of the first-order equation $(D - a)y = r(x)$ is

$$y = e^{+ax}\int^x e^{-at}r(t)\,dt + ce^{+ax}$$

we can represent a particular integral

$$y_p = \frac{1}{D - a}r(x) = e^{ax}\int^x e^{-at}r\,dt \tag{10.13}$$

Now, by induction, we can handle

$$f(D)y = r(x) = (D - \lambda_1)f_1(D)y$$

call $f_1(D)y = v$; then $(D - \lambda_1)v = r(x)$

$$v = \frac{1}{D - \lambda_1}r(x) = e^{\lambda_1 x}\int^x e^{-\lambda_1 t}r\,dt$$

Then

$$f_1(D)y = e^{\lambda_1 x}\int^x e^{-\lambda_1 t}r\,dt$$

is an equation of order $(n - 1)$, which can be successively reduced to yield a particular integral y_p. A slight modification of this procedure is as follows:

$$f(D)y = r(x) \qquad y_p = \frac{1}{f(D)} r(x)$$

If we perform the partial fractions decomposition of $1/f(D)$, then y_p is a sum of terms, the most general of which is $[1/(D - \lambda)^p]r(x)$, and this is the particular integral of $(D - \lambda)^p v = r(x)$. A straightforward way to find this is as follows. Let $v = e^{\lambda x}w$. Then

$$(D - \lambda)^p e^{\lambda x} w = e^{\lambda x} D^p w = r(x)$$

or

$$D^p w = \frac{d^p w}{dx^p} = e^{-\lambda x} r(x)$$

A particular integral for w is then the p-fold integral of $e^{-\lambda x}r(x)$, or

$$\frac{1}{\Gamma(p)} \int^x (x - t)^{p-1} e^{-\lambda t} r(t)\, dt$$

and hence

$$v_p = \frac{e^{\lambda x}}{\Gamma(p)} \int^x (x - t)^{p-1} e^{-\lambda t} r\, dt = \frac{1}{(D - \lambda)^p} r(x)$$

A number of special formulas can be derived to expedite these computations. For example,

$$\frac{1}{f(D)} e^{ax} = \frac{1}{f(a)} e^{ax} \qquad \text{if } f(a) \neq 0$$

$$\frac{1}{(D - a)^p f_1(D)} e^{ax} = \frac{1}{f_1(a)} e^{ax} \frac{x^p}{p!}$$

$$\frac{1}{f(D^2)} \begin{matrix}\cos\\ \sin\end{matrix}\, ax = \frac{1}{f(-a^2)} \begin{matrix}\cos\\ \sin\end{matrix}\, ax \qquad \text{if } f(-a^2) \neq 0$$

$$\frac{1}{f(D)} e^{ax} v = e^{ax} \frac{1}{f(D + a)} v$$

To evaluate $1/f(D)x^p$, expand $1/f(D)$ in a series of ascending powers of D,

$$\frac{1}{f(D)} x^p = (\alpha_0 + \alpha_1 D + \alpha_2 D^2 + \cdots + \alpha_n D^n + \cdots) x^p$$

and all the terms for $n > p$ vanish.

REFERENCES: [1], chaps. 5, 6; [2], chap. 3

c. Equations with Variable Coefficients

As mentioned on p. 10–5, the special equation

$$Ly \equiv x^n y^{(n)} + a_1 x^{n-1} y^{(n-1)} + \cdots + a_n y = 0$$

with a_i = const can be simplified by the change of variable $x = e^t$. If $D = d/dt$, then $x\,d/dx = D$, $x^2 d^2/dx^2 = D(D - 1)$, $x^3 d^3/dx^3 = D(D - 1)(D - 2)$, and so on, leading to a constant coefficient equation for $y(t)$. The solutions being $e^{\lambda t} = (e^t)^\lambda = x^\lambda$, we can form an auxiliary equation for Ly as follows:

$$Lx^\lambda = [\lambda(\lambda - 1) \cdots (\lambda - n + 1) + a_1 \lambda(\lambda - 1) \cdots (\lambda - n + 2) + \cdots + a_n] x^\lambda$$

the zeros of the polynomial in brackets being the required exponents λ, so that x^λ is a solution.

More generally, let

$$Ly = p_0(x)y^{(n)} + p_1(x)y^{(n-1)} + \cdot \cdot \cdot + p_n(x)y = 0$$

We make the assumption that each of the coefficients $p_i(x)$ can be expanded into a convergent power series about some point x_0, which without loss of generality we may take to be zero. Thus

$$p_i(x) = \sum_{j=0}^{\infty} p_{ij}x^j$$

If $p_0(0) \neq 0$, the initial-value problem for $Ly = 0$ has a unique solution in a neighborhood of $x = 0$, and $x = 0$ is called an *ordinary point* of the equation. Dividing by $p_0(x)$, we have $L_1y = y^{(n)} + P_1(x)y^{(n-1)} + \cdot \cdot \cdot + P_n(x)y = 0$, and the series for each $P_i(x)$ converges for $|x| < R_i$. One general method for obtaining solutions of $L_1y = 0$ is by means of power series. Let

$$y = \sum_0^{\infty} a_n x^n$$

and form L_1y. Equating the coefficient of each power of x in L_1y to zero determines each a_i in terms of $a_0, a_1, \ldots, a_{n-1}$, which remain arbitrary. The series so found converges at least for $|x| < \min(R_1, \ldots, R_n)$, and contains n arbitrary constants. This gives us the required general solution.

If $p_0(0) = 0$, then the above analysis fails, $x = 0$ is a singular point of the equation, and the initial-value problem can no longer be solved at $x = 0$. A special case which can still be handled occurs if the equation can be brought into the form

$$x^n y^{(n)} + x^{n-1}P_1(x)y^{(n-1)} + \cdot \cdot \cdot + P_n(x)y = 0 \qquad (10.14)$$

where each $P_i(x)$ can be expanded in a series

$$P_i(x) = \sum_{j=0}^{\infty} P_{ij}x^j \qquad \text{converging for } |x| < R_i$$

In this case, $x = 0$ is called a *regular singular point,* and the following method, due to Frobenius, again yields solutions in the form of power series.

Guided by the form of the solutions in the two preceding cases, it is reasonable to expect solutions of (10.14) of the form

$$y = x^{\lambda} \sum_0^{\infty} a_n x^n$$

The method outlined below is quite general, but, for simplicity, we consider the second-order equation

$$Ly = x^2y'' + xP(x)y' + Q(x)y = 0 \qquad (10.15)$$

$$P(x) = \sum_0^{\infty} p_n x^n \qquad Q(x) = \sum_0^{\infty} q_n x^n$$

Then

$$L\left(\sum_0^\infty a_n x^{n+\lambda}\right) = \sum_0^\infty (n+\lambda)(n+\lambda-1)a_n x^{n+\lambda} + \sum_0^\infty (n+\lambda)a_n x^{n+\lambda} \sum_0^\infty p_n x^n + \sum_0^\infty q_n x^n \sum_0^\infty a_n x^{n+\lambda}$$

$$= x^\lambda \sum_0^\infty \left[(n+\lambda)(n+\lambda-1)a_n + \sum_0^n (r+\lambda)a_r p_{n-r} + \sum_0^n a_r q_{n-r} \right] x^n \tag{10.16}$$

The first term on the right of (10.16) is

$$[\lambda(\lambda-1) + \lambda p_0 + q_0]a_0 x^\lambda = f(\lambda)a_0 x^\lambda$$

and the nth term is

$$x^{n+\lambda}[f(\lambda+n)a_n + \text{terms linear in } a_0,a_1, \ldots ,a_{n-1}] = x^{n+\lambda}[f(\lambda+n)a_n + F_n(a_0,a_1, \ldots ,a_{n-1})]$$

For $n \geq 1$, we choose $f(\lambda+n)a_n + F_n(a_0,a_1, \ldots ,a_{n-1}) = 0$. This leaves

$$L\left(\sum_0^\infty a_n x^{n+\lambda}\right) = a_0 x^\lambda f(\lambda) \qquad \text{with } a_n = a_n(\lambda)$$

In order that the above process be legitimate, and that

$$y = \sum_0^\infty a_n x^{n+\lambda}$$

be a solution of $Ly = 0$, we must choose λ so that $f(\lambda) = 0$, while $f(\lambda+n) \neq 0$ $(n \geqq 1)$. (We assume that $a_0 \neq 0$.) The equation $f(\lambda) = \lambda(\lambda-1) + \lambda p_0 + q_0 = 0$ is called the *indicial equation,* and has roots λ_1, λ_2. If $\lambda_1 \geq \lambda_2$, or $\mathcal{R}\,\lambda_1 \geq \mathcal{R}\,\lambda_2$, then $f(\lambda_1 + n) \neq 0$ for all $n \geq 1$, and we have a formal solution

$$y_1 = \sum_0^\infty a_n(\lambda_1)x^{n+\lambda_1} \tag{10.17}$$

In fact, this is an actual solution, the series converging for $|x| < R$, where each of the series for $P(x)$, $Q(x)$ converges for $|x| < R$.

If $\lambda_1 - \lambda_2 \neq$ integer, then $f(\lambda_2 + n) \neq 0$ for all $n \geq 1$, and a second solution is

$$y_2 = \sum_0^\infty a_n(\lambda_2)x^{n+\lambda_2}$$

which is independent of y_1. If $\lambda_1 - \lambda_2 =$ integer, then either $\lambda_1 = \lambda_2$, or $f(\lambda_2 + n) = 0$ for some value of n, causing difficulty in satisfying the condition

$$f(\lambda+n)a_n + F_n(a_0,a_1, \ldots ,a_{n-1}) = 0 \tag{10.18}$$

There are three cases. First of all, consider $\lambda_1 = \lambda_2$. Then $f(\lambda) = (\lambda - \lambda_1)^2$, and

$$L\left(\sum_0^\infty a_n(\lambda)x^{n+\lambda}\right) = a_0x^\lambda(\lambda - \lambda_1)^2 \tag{10.19}$$

Because of the factor $(\lambda - \lambda_1)^2$ on the right of (10.19), both $a_0x^\lambda(\lambda - \lambda_1)^2$ and $(\partial/\partial\lambda)[a_0x^\lambda(\lambda - \lambda_1)^2]$ vanish at $\lambda = \lambda_1$. This gives

$$\left[\frac{\partial}{\partial\lambda} L\left(\sum_0^\infty a_n(\lambda)x^{n+\lambda}\right)\right]_{\lambda=\lambda_1} = \left\{L\left[\sum_0^\infty a_n'(\lambda)x^{n+\lambda} + \ln x \sum_0^\infty a_n(\lambda)x^{n+\lambda}\right]\right\}_{\lambda=\lambda_1} = 0$$

and hence a second solution

$$y_2(x) = \sum_0^\infty a_n'(\lambda_1)x^{n+\lambda_1} + y_1(x) \ln x$$

which is independent of $y_1(x)$. Again, these series converge, so that $y_2(x)$ is an actual solution.

If $\lambda_1 - \lambda_2 = N$, then $f(\lambda_2 + N) = 0$, and (10.18)

$$f(\lambda_2 + N)a_N + F_N(a_0,a_1, \ldots ,a_{N-1}) = 0$$

cannot be solved for a_N; however, it may happen that $F_N(a_0,a_1, \ldots ,a_{N-1}) = 0$ for $\lambda = \lambda_2$. This leaves a_N completely arbitrary, and thus

$$y_2 = \sum_0^\infty a_n(\lambda_2)x^{n+\lambda_2}$$

contains two arbitrary constants, a_0 and a_N, and is the general solution.

Finally, if $F_N(a_0,a_1, \ldots ,a_{N-1}) \neq 0$ for $\lambda = \lambda_2$, we can construct two independent solutions by the artifice $a_0 = a_0^*(\lambda - \lambda_2)$, $a_0^* \neq 0$; then

$$L\left(\sum_0^\infty a_n^*(\lambda)x^{n+\lambda}\right) = a_0^*(\lambda - \lambda_1)(\lambda - \lambda_2)^2x^\lambda$$

together with the λ derivative, vanishes for $\lambda = \lambda_2$. This yields two independent solutions

$$y_1 = \sum_0^\infty a_n^*(\lambda_2)x^{n+\lambda_2} \qquad y_2 = \sum_0^\infty \left[\frac{da_n^*(\lambda)}{d\lambda}\right]_{\lambda_2} x^{n+\lambda_2} + y_1 \ln x$$

REFERENCE: [1], chap. 7

d. Definite Integral Representations

An important method of representing the solutions of certain differential equations is by means of definite integrals. The general theory of this technique is as follows. Let

$$y = \int_a^b K(x,t)F(t)\,dt \tag{10.20}$$

represent a solution of the linear equation

$$Ly = p_0y^{(n)} + p_1y^{(n-1)} + \cdots + p_ny = 0$$

The function $K(x,t)$ is the kernel of the integral, or transformation from y to F, and a, b are fixed numbers, finite or infinite. All the following operations are assumed valid.

$$Ly = \int_a^b L_x K(x,t)F(t)\,dt$$

If $L_x K(x,t) = M_t K(x,t)$, where L_x is the given differential operator in terms of x, and M_t is another differential operator in terms of t, then

$$\begin{aligned} Ly &= \int_a^b M_t K(x,t)F(t)\,dt \\ &= [P\{K(x,t),F(t)\}]_a^b + \int_a^b K(x,t)\bar{M}_t F\,dt \end{aligned}$$

where $\bar{M}_t$ is the adjoint of M_t. Thus (10.20) will be a solution of $L_x y = 0$ if we can satisfy both $\bar{M}_t F(t) = 0$ and

$$[P\{K(x,t),F(t)\}]_a^b = 0$$

This procedure admits of generalization. All we need is two functions $K(x,t)$, $k(x,t)$ so that $L_x K(x,t) = M_t k(x,t)$ for the method to work, a and b being chosen as previously.

Perhaps the most important example of this procedure is for equations $Ly = 0$ whose coefficients are polynomials in x, with $K(x,t) = e^{xt}$. The integral representations so obtained are called Laplace integrals. In particular, if the coefficients are linear in x, the equation $\bar{M}_t F = 0$ is a first-order linear equation, and so can be integrated in general. Thus, for a second-order equation,

$$\begin{aligned} L_x y &= (a_1 + a_2 x)y'' + (b_1 + b_2 x)y' + (c_1 + c_2 x)y = 0 \\ L_x e^{xt} &= [(a_1 + a_2 x)t^2 + (b_1 + b_2 x)t + (c_1 + c_2 x)]e^{xt} \\ &= (a_1 t^2 + b_1 t + c_1)e^{xt} + (a_2 t^2 + b_2 t + c_2)\frac{d}{dt}e^{xt} = M_t e^{xt} \\ M_t &= (a_2 t^2 + b_2 t + c_2)\frac{d}{dt} + (a_1 t^2 + b_1 t + c_1) \\ \bar{M}_t F &= -\frac{d}{dt}(a_2 t^2 + b_2 t + c_2)F + (a_1 t^2 + b_1 t + c_1)F \end{aligned}$$

For example, with $a_1 = b_2 = c_1 = 0 \qquad a_2 = b_1 = 1 \qquad c_2 = -1$

$$L_x y = xy'' + y' - xy = 0 \qquad \bar{M}_t F = +\frac{d}{dt}(1 - t^2)F + tF = 0$$

$$F(t) = \frac{1}{(1 - t^2)^{1/2}} \qquad [P\{K,F\}]_a^b = -[(1 - t^2)^{1/2} e^{xt}]_a^b$$

so that $a = -1$, $b = 1$ leads to

$$y(x) = \int_{-1}^{1} \frac{e^{xt}\,dt}{(1 - t^2)^{1/2}}$$

Another choice is

$$F = \frac{1}{(t^2 - 1)^{1/2}} \qquad a = -\infty,\ b = -1$$

with

$$y(x) = \int_{-\infty}^{-1} \frac{e^{xt}\,dt}{(t^2 - 1)^{1/2}} \qquad \text{valid for } x > 0$$

Other kernels of common application are of the form $K(x - t)$, $K(xt)$.

REFERENCE: [1], chap. 8

e. Green's Functions

The ideas of this subsection are applicable to linear equations of all orders, but we restrict ourselves to discussions of second-order equations.

Let $Ly = p_0y'' + p_1y' + p_2y$, and consider $Ly = r(x)$ $(a < x < b)$ with some homogeneous boundary-value problem, such as $y(a) = y(b) = 0$. Boundary conditions are called homogeneous if, when satisfied by $y(x)$, they are satisfied for $ky(x)$ for arbitrary constant k. Now consider the identity

$$\int_a^b (vL_\xi y - y\bar{L}_\xi v)\, d\xi = [P\{v,y\}]_a^b = [p_0(vy' - yv') + (p_1 - p_0')yv]_a^b \qquad (10.21)$$

and choose v, a solution of $\bar{L}v = 0$, so that the right side becomes $y(x)$. In order to achieve this, we take $v(\xi,x)$ to be one solution of $\bar{L}_\xi v = 0$ in $a < \xi < x$, and a second solution in $x < \xi < b$. Properly speaking, $\int_a^b (vLy - y\bar{L}v)\, d\xi$ should be written $\int_a^x (vLy - y\bar{L}v)\, d\xi + \int_x^b (vLy - y\bar{L}v)\, d\xi$ and the right side $[P\{y,v\}]_a^x + [P\{y,v\}]_x^b$. In this way, with $v(a) = v(b) = 0$, $v(\xi)$ continuous at x, and

$$p_0(x)v'(x + 0) - p_0(x)v'(x - 0) = 1$$

we have

$$\int_a^b (vLy - y\bar{L}v)\, d\xi = y(x) \qquad (10.22)$$

The function $v = G(x,\xi)$ satisfying these conditions

$\bar{L}_\xi G(x,\xi) = 0$, $G(x,a) = G(x,b) = 0$
$G(x,\xi)$ continuous at $\xi = x$
$G_\xi(x,\xi)$ having an upward jump of $1/p_0(x)$ at $\xi = x$

is called the *Green's function* for the differential expression Ly in $a < x < b$, and the homogeneous boundary conditions $y(a) = y(b) = 0$. In terms of the Green's function, the solution of the boundary-value problem is

$$y(x) = \int_a^b G(x,\xi)r(\xi)\, d\xi$$

If $H(x,\xi)$ is the Green's function for the adjoint $\bar{L}y$ in $a < x < b$, with the same end conditions, then an application of (10.21) with $G(x_1,\xi)$, $H(x_2,\xi)$ shows

$$H(x_2,x_1) = G(x_1,x_2)$$

and, in particular, if Ly is self-adjoint,

$$G(x,\xi) = G(\xi,x)$$

REFERENCE: [1] chap. 11

f. Systems of Linear Equations

As we saw in Sec. 10.1, every nth-order equation can be transformed to a system of n first-order equations. We consider the most general linear case,

$$\frac{dx_i}{dt} = \dot{x}_i = \sum_{j=1}^{n} a_{ij}(t)x_j + h_i(t) \qquad i = 1, \ldots, n$$

where the $a_{ij}(t)$, $h_i(t)$ are assumed continuous. In matrix notation, this system becomes

$$\dot{x} = A(t)x + h(t) \qquad (10.23)$$

From the existence theorem of Sec. 10.1, the system $\dot\theta = A(t)\theta$, θ being an $n \times n$ matrix, with $\theta(t_0) = I$, the identity matrix, possesses a unique solution. [If A is a constant matrix, $\theta(t) = e^{A(t-t_0)} = I + A(t - t_0) + \cdots + A^n(t - t_0)^n/n! + \cdots$.] Let $W(t) = \det \theta(t)$. Then

$$\begin{aligned} W(t+h) &= \det \theta(t+h) = \det [\theta(t) + h\dot\theta(t) + O(h^2)] \\ &= \det [\theta(t) + hA(t)\theta(t) + O(h^2)] \\ &= W(t)[1 + h \operatorname{tr} A(t) + O(h^2)] \qquad \text{where tr } A = \text{trace of } A \\ \frac{W(t+h) - W(t)}{h} &= W(t) \operatorname{tr} A(t) + O(h) \end{aligned}$$

Hence
$$\dot W(t) = W(t) \operatorname{tr} A(t)$$

and
$$W(t) = W(t_0) \exp \int_{t_0}^{t} \operatorname{tr} A(t)\, dt = \exp \int_{t_0}^{t} \operatorname{tr} A(t)\, dt$$

since $W(t_0) = \det I = 1$. This shows that $\theta(t)$ is nonsingular for all t, and so possesses a continuous inverse $\theta^{-1}(t)$.

The solution of $\dot x = A(t)x$, $x(t_0) = x_0$ is $x(t) = \theta(t)x_0$. To solve the general equation

$$\dot x = A(t)x + h(t) \qquad x(t_0) = x_0$$

set $x(t) = \theta(t)x_0 + y(t)$. Then $\dot y = A(t)y + h(t)$ with $y(t_0) = 0$. This is solved by variation of parameters, $y = \theta(t)w$, giving

$$\dot y = \dot\theta w + \theta\dot w = A\theta w + \theta\dot w = A(t)\theta w + h(t)$$

or $\theta\dot w = h(t)$, and thus $\dot w = \theta^{-1}(t)h(t)$. This gives

$$w(t) = \int_{t_0}^{t} \theta^{-1}(t)h(t)\, dt \qquad \text{and} \qquad y(t) = \theta(t) \int_{t_0}^{t} \theta^{-1}(t)h(t)\, dt$$

is a particular integral.

If $\phi(t)$ is any matrix solution of $\dot\phi(t) = A(t)\phi(t)$, then, with $\phi = \theta\psi$, we find $\dot\theta\psi + \theta\dot\psi = A(\theta\psi) = A\theta\psi + \theta\dot\psi$ or $\theta\dot\psi = 0$, and, since θ is nonsingular, $\dot\psi = 0$, giving $\psi = \text{const}$. Thus, every solution of $\dot\phi = A\phi$ is given by $\phi = \theta C$, where C is a constant matrix.

Equations with Periodic Coefficients; the Floquet Theory. In the equation $\dot x = A(t)x$, let $A(t)$ be periodic, with period ω, and consider $\theta(t)$ as defined above. From $\dot\theta(t) = A(t)\theta(t)$, and $A(t + \omega) = A(t)$, it follows that $\theta(t + \omega)$ satisfies the same equation as $\theta(t)$, and, hence, $\theta(t + \omega) = \theta(t)P$; P is a constant matrix, and is nonsingular, since $\theta(t)$ is nonsingular.

We can now investigate the periodic solutions, if any, of $\dot x = A(t)x$. Since P is nonsingular, it has a logarithm, and we can write $P = e^{\omega B}$. Next, we define the matrix $\Gamma(t)$ by the relation $\theta(t) = \Gamma(t)e^{tB}$, and then $\Gamma(t)$ is periodic, with period ω. This follows from $\theta(t + \omega) = \Gamma(t + \omega)e^{(t+\omega)B} = \Gamma(t)e^{tB}P = \Gamma(t)e^{(t+\omega)B}$, so that $\Gamma(t + \omega) = \Gamma(t)$. Thus, in order to investigate periodic solutions of $\dot x = A(t)x$, we use $x(t) = \theta(t)x_0 = \Gamma(t)e^{tB}x_0$; if now $x(t)$ is periodic, with period ω, then $x(t + \omega) = x(t)$, implying $\Gamma(t + \omega)e^{(t+\omega)B}x_0 = \Gamma(t)e^{tB}x_0 = \Gamma(t)e^{(t+\omega)B}x_0$, or $e^{\omega B}x_0 = x_0$. This can hold, for $x_0 \neq 0$, only if $\det (e^{\omega B} - I) = 0$, which is a relation involving only the constant matrix B (or P). In other words, the determination of periodic solutions of $\dot x = A(t)x$, with $A(t)$ periodic, can be reduced to the study of an equivalent system of equations with constant coefficients, namely, $\dot y = By$. The characteristic roots of B are called *characteristic exponents*, while those of P are referred to as *characteristic multipliers*.

Other periods than ω are frequently of importance for $x(t)$. Thus, if $e^{\omega B}$ has -1 as a characteristic root, and x_0 is a solution of $(e^{\omega B} + I)x_0 = 0$, then $x(t) = \theta(t)x_0$ has period 2ω, and so on.

Closely related to the periodicity of solutions of equations with periodic coefficients is the stability of such systems. Thus, if the characteristic multipliers are all distinct, and of absolute value unity, every solution of $\dot{x} = A(t)x$ is uniformly bounded in $-\infty < t < \infty$. Unbounded solutions occur (for $t > 0$, say), if multiple roots occur, or if one of the multipliers is greater than one in magnitude. It is frequently possible to determine relations between the coefficients of $A(t)$ leading to such stable solutions, even though the explicit evaluation of B, or P, is not feasible.

REFERENCE: [2], chap. 3

g. Dependence on Parameters

If a system of equations has coefficients which depend on a parameter $\dot{x} = A(t,\epsilon)x$, then for $A(t,\epsilon)$ continuous in ϵ, as well as in t, the solutions of the system depend continuously on the parameter. A particularly important case occurs when $A(t,\epsilon) = A_0(t) + \epsilon A_1(t)$. In this case, we can expand the solution vector $x(t,\epsilon)$ as $x(t,\epsilon) = x_0(t) + \epsilon x_1(t) + \cdots + \epsilon^n x_n(t) + \cdots$ where $\dot{x}_0 = A_0 x_0$,

$$\dot{x}_n = A_0 x_n + A_1 x_{n-1}$$

The series, in general, converges for all finite ϵ.

When boundary-value problems permit such an expansion as above, they are usually called ordinary, or *regular perturbation problems.*

Another class of perturbation problems of considerable importance can be written as $\epsilon\dot{x} = A(t,\epsilon)x$, the solution of some boundary-value problem being desired as $\epsilon \to 0$. The difficulty here is that the limit problem, with $\epsilon = 0$, is of lower order than the original problem, and is referred to as a *singular perturbation problem.* The expansion of the solution is no longer a power series in ϵ, but contains terms typified by $e^{-t/\epsilon}$ ($x > 0$, $\epsilon > 0$) which exhibit nonuniform convergence in that, as $\epsilon \to 0$ then $t \to 0$, the result is zero, while, if $t \to 0$, then $\epsilon \to 0$, the result is unity. The regions in which these nonuniformities occur are referred to as *boundary layers,* or *shocks.*

REFERENCE: [2], chap. 6

h. Linear Equations in the Complex Domain

The methods of solution outlined in Art. 10.3*c* properly belong to a discussion of equations in the complex domain. We illustrate the type of result obtainable with a discussion of second-order equations

$$Lw = w'' + p(z)w' + q(z)w = 0 \tag{10.24}$$

The coefficients $p(z)$, $q(z)$ are supposed to be single-valued analytic functions with, at worst, isolated singular points. If $p(z)$, $q(z)$ are analytic at $z = 0$, and, hence, in a neighborhood of $z = 0$, say $|z| < R$, then the point $z = 0$ is an *ordinary point* of the equation, and the classical existence theorem guarantees a unique solution to the initial-value problem at $z = 0$. The solution is automatically single-valued and analytic in $|z| < R$, and so can be expanded in a convergent power series about $z = 0$.

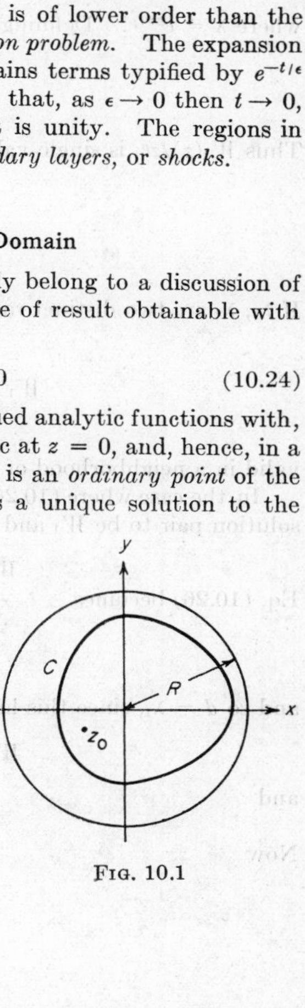

FIG. 10.1

If $p(z)$, $q(z)$ are analytic in $|z| < R$, except possibly for the point $z = 0$, we can construct a fundamental solution pair w_1, w_2 in a neighborhood of any fixed point z_0 in $|z| < R$. These functions exist in $|z| < R$ except possibly at $z = 0$.

Now let C denote a simple closed curve encircling $z = 0$, and let w^* be the functional element of $w(z)$ obtained by transporting $w(z)$ around C once in the positive direction. Then the fundamental solution pair w_1, w_2 become w_1^*, w_2^*, which again are independent. Hence, for some constants a, b, c, d, we have

$$w_1^* = aw_1 + bw_2 \qquad w_2^* = cw_1 + dw_2 \tag{10.25}$$

and the determinant

$$\begin{vmatrix} a & b \\ c & d \end{vmatrix} \neq 0$$

It is now possible to select a particular solution $W(z) = \alpha w_1 + \beta w_2$ so that

$$W^*(z) = \lambda W(z)$$

This is equivalent to

$$[\alpha(a - \lambda) + \beta c]w_1 + [\alpha b + \beta(d - \lambda)]w_2 = 0$$

and hence,

$$\begin{vmatrix} a - \lambda & c \\ b & d - \lambda \end{vmatrix} = 0 \tag{10.26}$$

This equation is independent of the fundamental solution pair chosen.

If λ_1 is a root of (10.26), then $W_1^* = \lambda_1 W_1$. On the other hand, the elementary function z^ρ has the same sort of multivalued behavior on being carried around C. Thus

$$(z^\rho)^* = (e^{\rho \ln z})^* = e^{\rho(\ln z + 2\pi i)} = e^{2\pi i \rho} z^\rho = \lambda z^\rho$$

where $\lambda = e^{2\pi i \rho}$. Defining $\rho_1 = (1/2\pi i) \ln \lambda_1$, we see that

$$\left(\frac{W_1}{z^{\rho_1}}\right)^* = \frac{\lambda_1 W_1}{e^{2\pi i \rho_1} z^{\rho_1}} = \frac{W_1}{z^{\rho_1}}$$

Thus $W_1(z)/z^{\rho_1}$ is single-valued in $|z| < R$, and so, by Laurent's theorem (p. 9–14),

$$W_1(z) = z^{\rho_1} \sum_{-\infty}^{\infty} a_n z^n$$

If λ_1, λ_2 are two distinct roots of (10.26), then we have a fundamental solution pair

$$W_1 = z^{\rho_1} \sum_{-\infty}^{\infty} a_n z^n \qquad W_2 = z^{\rho_2} \sum_{-\infty}^{\infty} b_n z^n$$

valid in a neighborhood of $z = 0$.

In the case where (10.26) has a double root ($\lambda = \lambda_1$), by choosing our fundamental solution pair to be W_1 and W_2, for which

$$W_1^* = \lambda_1 W_1 \qquad W_2^* = cW_1 + dW_2$$

Eq. (10.26) becomes

$$\begin{vmatrix} \lambda_1 - \lambda & c \\ 0 & d - \lambda \end{vmatrix} = 0$$

and so $d = \lambda_1$, since this has a double root. Thus

$$W_1^* = \lambda_1 W_1 \qquad W_2^* = cW_1 + \lambda_1 W_2$$

and

$$\left(\frac{W_2}{W_1}\right)^* = \frac{W_2}{W_1} + \frac{c}{\lambda_1}$$

Now

$$\left(\frac{c}{\lambda_1 2\pi i} \ln z\right)^* = \frac{c}{\lambda_1 2\pi i} \ln z + \frac{c}{\lambda_1}$$

so that $W_2/W_1 - (c/\lambda_1 2\pi i) \ln z$ is single-valued in $|z| < R$, so that Laurent's theorem yields a representation for

$$W_2(z) = z^{\rho_1}\left(A \ln z \sum_{-\infty}^{\infty} a_n z^n + \sum_{-\infty}^{\infty} b_n z^n\right)$$

This discussion remains valid in the neighborhood of each finite isolated singular point of the coefficients $p(z)$, $q(z)$. To handle the case $z = \infty$, set $z = 1/t$, and then (10.24) becomes

$$\frac{d^2w}{dt^2} + \left[\frac{2}{t} - \frac{1}{t^2}\, p\left(\frac{1}{t}\right)\right]\frac{dw}{dt} + \frac{1}{t^4}\, q\left(\frac{1}{t}\right) w = 0$$

and the nature of these coefficients at $t = 0$ describes the situation for $z = \infty$.

If the Laurent series in the above expansions have only a finite principal part, the singular point $z = 0$ is called a regular singular point. We have

Fuchs' Theorem. A necessary and sufficient condition that $z = a$ be a regular singular point of $w'' + p(z)w' + q(z)w = 0$ is that $p(z)$ have (at worst) a pole of order 1, and $q(z)$ have (at worst) a pole of order 2, at $z = a$.

In this case, the equation can be written (for $a = 0$) $z^2w'' + zP(z)w' + Q(z)w = 0$, where $P(z)$, $Q(z)$ are analytic at $z = 0$, and we see that the method of Frobenius is guaranteed to yield the desired solutions.

If $z = 0$ (or $z = a$) is a singular point, and the conditions of Fuchs' theorem are not fulfilled, then $z = 0$ is called an irregular singular point, and the solutions, in general, possess an infinite principal part. In this case, the indicial equation is of degree 1 or 0. If it is of degree 1, there is still the possibility of a series solution with a finite principal part. Even if the series is so obtainable, it diverges, in general, and is the asymptotic development, in the sense of Poincaré, of one of the solutions of the equation.

If the indicial equation is of degree zero, it may happen that the substitution $w = v \exp P(1/z)$, where $P(1/z)$ is a polynomial in $1/z$, yields an equation for v which has a first-degree indicial equation. The series solution thus obtained is called a *normal integral* at $z = 0$. *Subnormal integrals* of the form $w = v \exp\, P(1/z^\alpha)$ may exist for some rational value of α.

REFERENCE: [1], chaps. 15, 16

10.4. NONLINEAR EQUATIONS

Practically no general analytical methods exist for solving nonlinear equations. Given a specific equation, we can try techniques which sometimes work. For example, if $y^{(n)} = f(x,y', \ldots, y^{(n-1)})$, and $f(z_1,z_2, \ldots, z_n)$ is analytic in $z_1, \ldots, z_n$, we might try to expand $y(x)$ as a power series

$$y = \sum_0^\infty a_n x^n$$

One of the major difficulties in the nonlinear case is the existence of singularities of the solutions which are not fixed by the equation, but depend on the boundary conditions.

Another method of practical importance is iteration. The successful application of such techniques frequently depends on a fortunate choice of iteration scheme. For example, given $y' = f(x,y)$, the scheme $y'_{n+1} = f(x,y_n)$ converges for suitable $f(x,y)$, but the convergence may be improved by a modification such as

$$y'_{n+1} = f_1(x,y_n,y_{n+1})$$

If the equation contains a parameter, we might be able to expand the solution in a series in terms of this parameter. In order for this to be a practical method, we should be able to solve the separate equations for the various powers of the parameter.

Another device in common use is linearization of an equation. For example, if $y' = f(x,y)$, and $y = y_0(x)$ is a known solution, then we can obtain information about nearby solutions by setting $y = y_0(x) + \epsilon v$. Then

$$y' = y_0' + \epsilon v' = f(x,y_0 + \epsilon v) = f(x,y_0) + \epsilon \frac{\partial f}{\partial y}(x,y_0)v + \cdots$$

and vanishing of the coefficient of ϵ yields a linear equation in v, $v' = (\partial f/\partial y)(x,y_0)v$. If we know that $|v|$ is small, so that v^2 can be neglected in favor of v, the solution of the linear equation gives a first-order correction to $y_0(x)$. Similar results hold for higher-order equations and systems.

A comprehensive list of solvable nonlinear equations appears in [3].

Special Properties of Two-dimensional Systems

Although it may not, in general, be feasible to obtain the solutions of nonlinear differential equations in closed analytical form, it is frequently possible to infer many important properties of their solution. In particular, we consider the general first-order first-degree equation $y' = f(x,y)$, or equivalently, the two-(space)dimensional system

$$\dot{x} = dx/dt = P(x,y) \qquad \dot{y} = dy/dt = Q(x,y) \tag{10.27}$$

where we have introduced an auxiliary parameter t, which we may interpret as time. This system is *autonomous* (not involving time explicitly) and involves only two space dimensions, x and y.

In the following discussion, we assume sufficient continuity and differentiability of the functions $P(x,y)$ and $Q(x,y)$ to guarantee existence and uniqueness of solutions of these equations, and investigate the *global* behavior of solutions, i.e., as $t \to \pm\infty$.

Any point (x_0,y_0) for which $P(x_0,y_0) = Q(x_0,y_0) = 0$ is called a critical point of the system (10.27). Clearly $x = x_0$, $y = y_0$ is then a solution of (10.27) for all t. The nature of the solutions near a critical point, to a large extent, governs the behavior of the solution curves, called characteristics, for large values of t. To study this behavior, we assume the critical point to be isolated, and by a translation bring any specific one to the origin. Accordingly, we assume the MacLaurin's expansion of $P(x,y)$, $Q(x,y)$ to take the form

$$P(x,y) = ax + by + r^2P_2(x,y) \qquad Q(x,y) = cx + dy + r^2Q_2(x,y) \tag{10.28}$$
$$r^2 = x^2 + y^2$$

with P_2, Q_2 continuous in some circle centered at $x = y = 0$.

If $ad - bc \neq 0$, the behavior of the characteristics near the origin is determined by the behavior of the solutions of the corresponding linear system obtained by setting $P_2 = Q_2 = 0$. There are two exceptions to this result, and these will be mentioned later. If no first-order terms are present at a critical point, the behavior is much more complicated, and will not be considered here. We now turn to the linear problem

$$\dot{x} = ax + by \qquad \dot{y} = cx + dy \tag{10.29}$$

or, in matrix notation,

$$\dot{\xi} = A\xi \qquad \text{where } \xi = \begin{bmatrix} x \\ y \end{bmatrix}, A = \begin{bmatrix} a & b \\ c & d \end{bmatrix}$$

A number of cases arise, depending on the nature of the characteristic roots of the matrix A, and it is convenient to make a preliminary change of variables of the form $\xi = P\eta$, where P is a nonsingular matrix, so that we can solve for $\eta = P^{-1}\xi$. Our linear system (10.29) now becomes

$$\dot{\eta} = P^{-1}AP\eta = A_1\eta \tag{10.30}$$

Depending on the nature of A, we can choose P so that A_1 takes one of the following forms:

$$(1)\ A_1 = \begin{bmatrix} \lambda_1 & 0 \\ 0 & \lambda_2 \end{bmatrix} \qquad (2)\ A_1 = \begin{bmatrix} \lambda_1 & 0 \\ 1 & \lambda_1 \end{bmatrix} \qquad (3)\ A_1 = \begin{bmatrix} \alpha & \beta \\ -\beta & \alpha \end{bmatrix}$$

where all the quantities indicated are real. λ_1, λ_2 are the characteristic roots of A, real for (1) and (2) [and not necessarily distinct in (1)], while in (3), $\alpha \pm i\beta$ are the characteristic roots. We denote the new coordinates η by $\begin{bmatrix} u \\ v \end{bmatrix}$.

Case 1.

$$\dot{\eta} = \begin{bmatrix} \lambda_1 & 0 \\ 0 & \lambda_2 \end{bmatrix} \eta$$

or $\dot{u} = \lambda_1 u$, $\dot{v} = \lambda_2 v$, so that

$$u = ce^{\lambda_1 t} \qquad v = de^{\lambda_2 t}$$

with c, d arbitrary constants.

Now, if λ_1, λ_2 are of the same sign, we have, for $\lambda_2 < \lambda_1 < 0$, both $u, v \to 0$ as $t \to \infty$, and for $c \neq 0$, $v/u = (d/c)e^{(\lambda_2 - \lambda_1)t} \to 0$, while for $\lambda_1 = \lambda_2 < 0$, $v/u = d/c$. These situations are illustrated in Figs. 10.2 and 10.3. All solutions tend to the

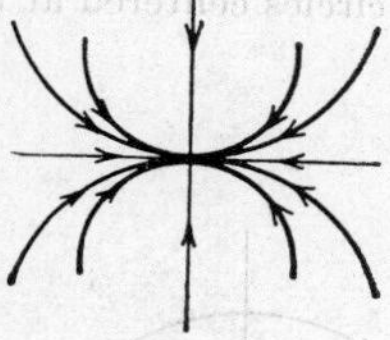

FIG. 10.2

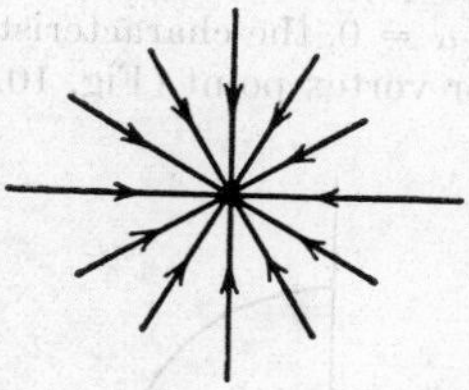

FIG. 10.3

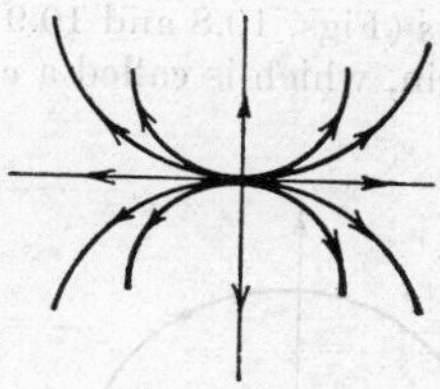

FIG. 10.4

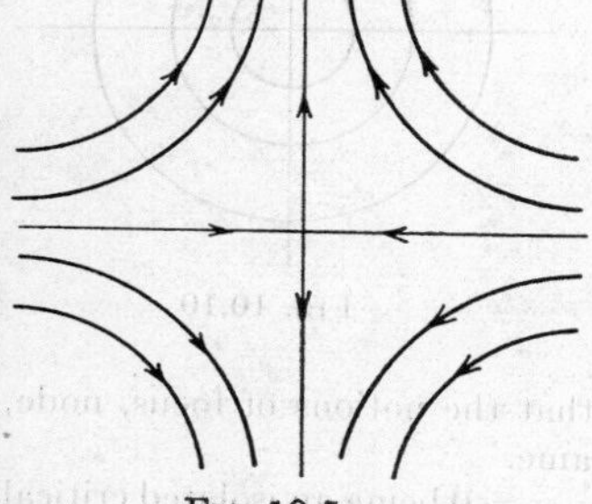

FIG. 10.5

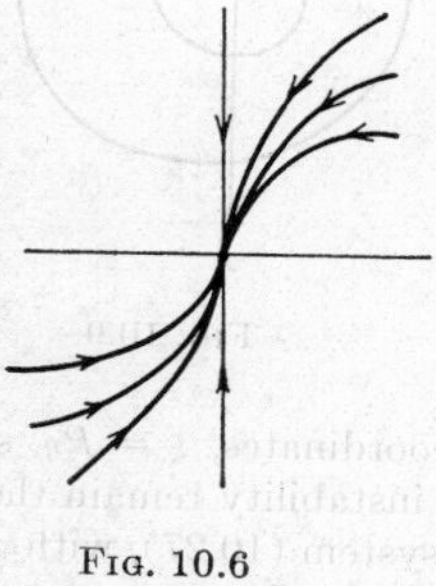

FIG. 10.6

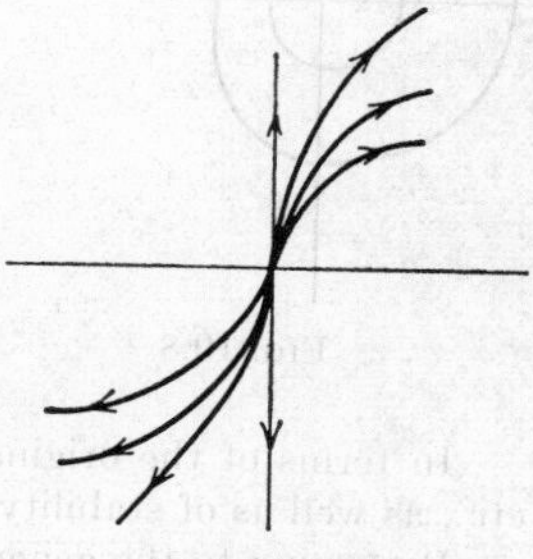

FIG. 10.7

origin as $t \to \infty$, the coordinate axes for $c = 0$, $d = 0$ being special characteristics, and the critical point is called a *stable node.* If we consider $t \to -\infty$, the characteristics leave the origin, and the arrows are reversed on the figures (see Fig. 10.4). For both roots positive, the whole picture is reversed; as $t \to -\infty$, all solutions tend to the critical point, while, as $t \to \infty$, the solutions leave the critical point, which is an *unstable node.*

Finally, if $\lambda_1 < 0 < \lambda_2$, $u \to 0$ and $v \to \infty$ as $t \to \infty$; the axes are characteristics corresponding to $c = 0$, $d = 0$, while, for $cd \neq 0$, the general form of the characteristics is as shown in Fig. 10.5. The origin is called a *saddle point,* and again, as $t \to -\infty$, the arrows are reversed, while the over-all nature of the solutions remains unchanged. We observe that two particular characteristics approach the critical point while all others leave.

Case 2.

$$\dot{\eta} = \begin{bmatrix} \lambda_1 & 0 \\ 1 & \lambda_1 \end{bmatrix} \eta$$

or $\dot{u} = \lambda_1 u$, $\dot{v} = u + \lambda_1 v$, giving

$$u = ce^{\lambda_1 t} \qquad v = (ct + d)e^{\lambda_1 t}$$

If $\lambda_1 < 0$; then $u, v \to 0$ as $t \to \infty$ and we have a *stable node.* If $t \to -\infty$ we have an *unstable node,* while for $\lambda_1 > 0$, the same situation arises with a stable node as $t \to -\infty$, unstable as $t \to \infty$. These situations are shown in Figs. 10.6 and 10.7. We see that all the characteristics enter (or leave) the critical point, and along the same direction.

Case 3

$$\dot{\eta} = \begin{bmatrix} \alpha & \beta \\ -\beta & \alpha \end{bmatrix} \eta$$

or

$$\dot{u} = \alpha u + \beta v \qquad \dot{v} = -\beta u + \alpha v$$

This is the case of complex roots, and is most easily described by the introduction of polar coordinates $u = r \cos \theta, v = r \sin \theta$. Then $\dot{r} = \alpha r$, $\dot{\theta} = -\beta$ with $r = ce^{\alpha t}$, $\theta = -\beta t + d$.

If $\alpha \neq 0$, this curve is a logarithmic spiral. For $\alpha < 0$, $t \to \infty$, the curve spirals into the origin which is a *stable focus,* while, for $\alpha > 0$, $t \to \infty$, we have an *unstable focus* (Figs. 10.8 and 10.9). For $\alpha = 0$, the characteristics are circles centered at the origin, which is called a center, or vortex point (Fig. 10.10).

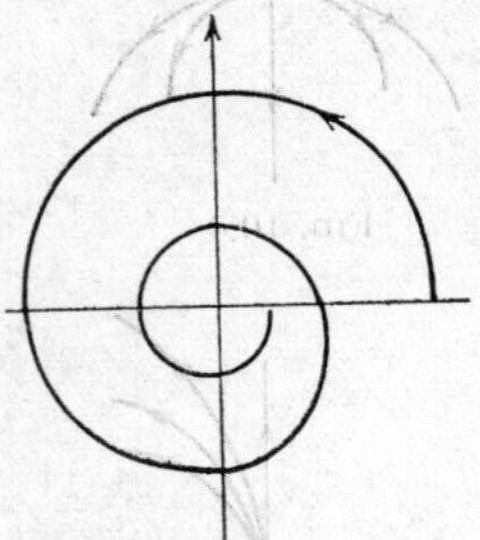

FIG. 10.8

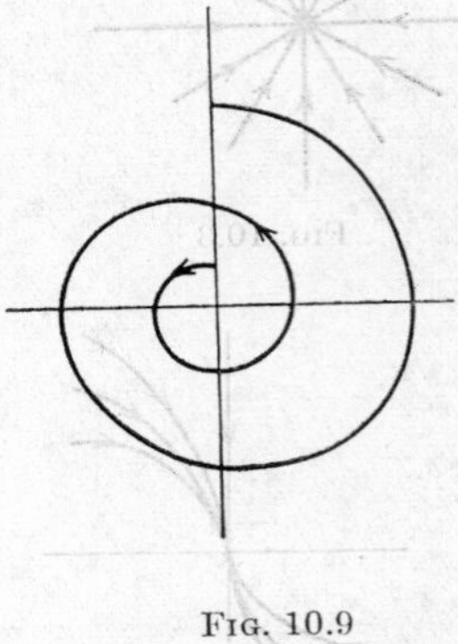

FIG. 10.9

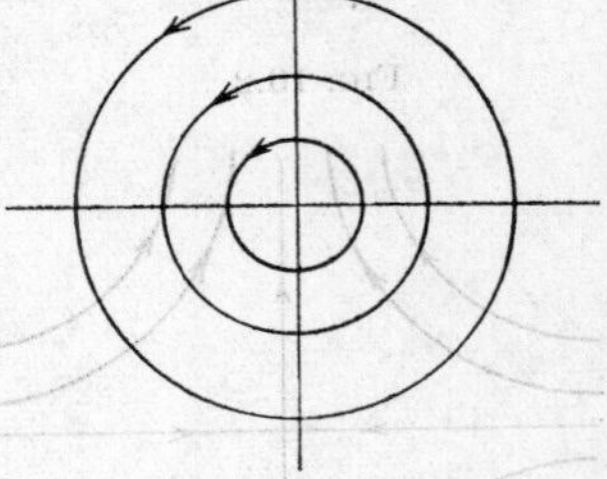

FIG. 10.10

In terms of the original coordinates, $\xi = P\eta$, so that the notions of focus, node, etc., as well as of stability or instability remain the same.

Returning to the general system (10.27), with $x = y = 0$ being an isolated critical point, and P, Q satisfying (10.28), we can now state the following.

Theorem 1. If $ad - bc \neq 0$, the behavior of the characteristics near the critical point $x = y = 0$ is the same as for the linear approximation (10.29), with two exceptions, noted below. This means, in particular, that, if $x, y \to 0$ as $t \to \infty$, for the linear case, then any solution entering a sufficiently small circle about the origin, in the general case, remains there, and $\to (0,0)$ as $t \to \infty$. The exceptional cases are (1) pure imaginary roots may lead to either a focus or a center, and (2) in the case of equal roots (case 1), the direction of a characteristic need not approach a constant value as $t \to \infty$.

One of the most important applications of the study of the behavior of the characteristics near a critical point is the global structure of the characteristics and, in particular, the existence of periodic solutions. (The existence and uniqueness theorems are purely local in nature, and, in themselves, give little or no information about the solutions as $t \to \infty$.) Typical results along these lines are, for example, (1) characteristics, in general, begin and end at critical points ($t \to \mp \infty$), (2) either a closed characteristic begins and ends on the same critical point, or else it is a periodic orbit, and then must enclose at least one critical point.

In connection with these results we have

Theorem 2 (Poincaré-Bendixson). If, as $t \to \infty$, a characteristic does not approach a critical point, and remains in a bounded region of the x,y plane, then either it is a periodic orbit or else it "spirals" around a periodic orbit. In the latter case, the periodic orbit is called a limit cycle. These ideas are illustrated by the example

$$\dot{x} = y \qquad \dot{y} = -x + (1 - x^2 - y^2)y$$

In polar coordinates $x = r \cos \theta$, $y = r \sin \theta$.

$$\dot{r} = r(1 - r^2) \sin^2 \theta \qquad \dot{\theta} = -1 - (1 - r^2) \sin \theta \cos \theta$$

Every characteristic starting inside $r = 1$ at $t = 0$ has $\dot{\theta} < 0$, $\dot{r} > 0$ so that it spirals ever closer to $r = 1$, but never reaches it. $r = 1$ is then a limit cycle for such trajectories, and is a periodic orbit. Characteristics starting outside spiral in toward $r = 1$. There is only one critical point, $x = y = 0$, and this lies inside the limit cycle, and is approached by those characteristics inside the circle $r = 1$ as $t \to -\infty$, making it a focus.

Reference: [2], chaps. 15, 16

10.5. REFERENCES

[1] E. L. Ince: "Ordinary Differential Equations," Dover, New York, 1956.
[2] E. A. Coddington, N. Levinson: "Theory of Ordinary Differential Equations," McGraw-Hill, New York, 1955.
[3] E. Kamke: "Differentialgleichungen, Lösungsmethoden und Lösungen," 3d ed., Chelsea, New York, 1948.

CHAPTER 11

PARTIAL DIFFERENTIAL EQUATIONS

BY

I. N. SNEDDON, D.Sc., Glasgow, Scotland

11.1. PARTIAL DIFFERENTIAL EQUATIONS: THEIR NATURE AND OCCURRENCE

If a function $u(x_1,x_2, \ldots ,x_r)$ is known to depend upon the r independent variables $x_1, x_2, \ldots , x_r$, then any relation between its partial derivatives

$$f\left(x_1,x_2, \ldots ,x_r, \frac{\partial u}{\partial x_1}, \ldots , \frac{\partial^2 u}{\partial x_1 \, \partial x_r}, \ldots \right) = 0 \qquad (11.1)$$

is called a *partial differential equation.* The *order* of the differential equation is defined to be the order of the derivative of highest order occurring in the equation. For example, the equation

$$\frac{\partial^4 z}{\partial x^4} = \frac{\partial^2 z}{\partial y^2}$$

is a fourth-order equation in the independent variables x and y.

If we have a relation of the type

$$F(x,y,z,a,b) = 0 \qquad (11.2)$$

connecting the independent variables x, and y, the dependent variable z, and two arbitrary constants a and b, then, regarding z as a function of x and y, this relation becomes an identity which may be differentiated with respect to x and y. In this way we may obtain the pair of equations[1] $F_x + F_z p = 0$, $F_y + F_z q = 0$. Eliminating a and b from this pair of equations and Eq. (11.2), we obtain a relation of the type

$$f(x,y,z,p,q) = 0 \qquad (11.3)$$

In other words, the relation (11.2), involving two arbitrary parameters, gives rise to a partial differential equation of the first order in two independent variables.

Suppose now that a function z is given in terms of the independent variables x and y by a relation of the type

$$z = \theta(u) + \psi(v) + w \qquad (11.4)$$

in which the functions θ and ψ are arbitrary functions of u and v, respectively, and u, v, and w are prescribed functions of x and y. If we differentiate both sides of this equation with respect to x and y, respectively, we obtain the pair

$$p = \theta'(u)u_x + \psi'(v)v_x + w_x \qquad q = \theta'(u)u_y + \psi'(v)v_y + w_y$$

Differentiating the first of these equations with respect to x, the second with respect to y, and the first with respect to y, we obtain three relations involving θ', θ'', ψ', ψ''

[1] Here we use the conventional notation: $F_x = \partial F/\partial x$, etc.; $p = \partial z/\partial x$, $q = \partial z/\partial y$.

and the second-order derivatives $r = \partial^2 z/\partial x^2$, $s = \partial^2 z/\partial x\, \partial y$, $t = \partial^2 z/\partial y^2$. In this way we obtain five equations involving the four arbitrary quantities θ', θ'', ψ', ψ''; eliminating these four quantities from the five equations, we obtain a relation of the type

$$Rr + Ss + Tt + Pp + Qq = W \tag{11.5}$$

where R, S, T, P, Q, W are known functions of x and y. In other words, the relation (11.4) leads to a second-order partial differential equation. It is readily shown that any relation of the type

$$z = \sum_{r=1}^{n} \theta_r(v_r)$$

where the functions $\theta_1, \ldots, \theta_n$ are arbitrary and the functions v_r are prescribed, leads to a partial differential equation of the nth order.

A partial differential equation is said to be *linear* if only first powers of the highest-order derivatives occur in it. For example, the equation

$$z\frac{\partial^2 z}{\partial x^2} = \left(\frac{\partial z}{\partial y}\right)^2$$

is linear, but

$$\left(\frac{\partial^2 z}{\partial x^2}\right)^2 = \frac{\partial z}{\partial y}$$

is not.

11.2. FIRST-ORDER PARTIAL DIFFERENTIAL EQUATIONS

We shall discuss first-order equations only very briefly since they do not arise frequently in engineering mechanics. We saw in the last section that relations of the type $F(x,y,z,a,b) = 0$ led to partial differential equations of the first order. A relation of this kind, involving two arbitrary constants (a and b) satisfying a given partial differential equation, is said to be a *complete solution* of that equation. On the other hand, a relation of the type $F(u,v) = 0$, involving an arbitrary function F connecting two known functions u and v, is called a *general solution* of the equation it satisfies. It would appear that a general solution provides a much broader set of solutions of any partial differential equation than does a complete solution; it can be shown, however, that this is not true—in the sense that it is possible to derive a general integral of the equation once a complete integral has been found. (See [15], chap. II, sec. 12.)

Linear Equations of the First Order

The general form of the linear equation of the first order in n variables is

$$P_1\frac{\partial z}{\partial x_1} + P_2\frac{\partial z}{\partial x_2} + \cdots + P_n\frac{\partial z}{\partial x_n} = R \tag{11.6}$$

where $P_1, \ldots, P_n$, R are prescribed functions of the independent variables and the dependent variable.

It was shown by Lagrange that, if $u_i(x_1,x_2, \ldots ,x_n,z) = c_i$ ($i = 1,2, \ldots ,n$; the c_is constant) are independent solutions of the simultaneous ordinary differential equations

$$\frac{dx_1}{P_1} = \frac{dx_2}{P_2} = \cdots = \frac{dx_n}{P_n} = \frac{dz}{R} \tag{11.7}$$

then the relation

$$\Phi(u_1,u_2, \ldots ,u_n) = 0 \tag{11.8}$$

in which the function Φ is arbitrary, is a general solution of the partial differential equation (11.6).

As an example, consider the linear equation

$$x_1 \frac{\partial z}{\partial x_1} + x_2 \frac{\partial z}{\partial x_2} + x_3 \frac{\partial z}{\partial x_3} = nz \tag{11.9}$$

For this equation the set corresponding to (11.7) is obviously

$$\frac{dx_1}{x_1} = \frac{dx_2}{x_2} = \frac{dx_3}{x_3} = \frac{dz}{nz}$$

The ordinary equation $dx_1/x_1 = dx_2/x_2$ of this set has the solution $x_2/x_1 = c_1$, and, similarly, we can derive a solution $x_3/x_1 = c_2$. The equation $dx_1/x_1 = dz/nz$ has the solution $z/x_1^n = c_3$. We have thus been able to find three independent solutions of the relevant set of simultaneous ordinary differential equations. Using Lagrange's theorem, we find that a general solution of Eq. (11.9) is

$$\Phi\left(\frac{x_2}{x_1}, \frac{x_3}{x_1}, \frac{z}{x_1^n}\right) = 0$$

where the function $\Phi(\xi,\eta,\zeta)$ is an arbitrary function of ξ, η, and ζ. Another way of writing the same result would be to say that

$$z = x_1^n F\left(\frac{x_2}{x_1}, \frac{x_3}{x_1}\right)$$

is a general solution with $F(\xi,\eta)$ an arbitrary function of ξ and η.

Nonlinear Equations of the First Order

Finding solutions of the partial differential equation

$$f(x,y,z,p,q) = 0 \tag{11.10}$$

when F is not linear in p and q, is a much more difficult business. It can be shown quite generally that any partial differential equation of the type (11.10) has solutions of the type:

1. $f(x,y,z,a,b) = 0$ involving two arbitrary constants a and b; these are called *complete integrals.*

2. If we take any one-parameter subsystem $f\{x,y,z,a,\Phi(a)\} = 0$ of system 1 and form its envelope, we obtain a solution of Eq. (11.10); it is called a *general integral* of Eq. (11.10).

3. If the envelope of the two-parameter system $f(x,y,z,a,b) = 0$ exists, it is also a solution of the partial differential equation; it is called the *singular integral* of Eq. (11.10).

The simplest method of solving nonlinear first-order equations is that due to Charpit. Charpit's method consists basically of finding a solution of the subsidiary equations

$$\frac{dx}{f_p} = \frac{dy}{f_q} = \frac{dz}{pf_p + qf_q} = \frac{dp}{-(f_x + pf_z)} = \frac{dq}{-(f_y + qf_z)} \tag{11.11}$$

involving an arbitrary constant a, say. If this solution is

$$g(x,y,z,p,q,a) = 0 \tag{11.12}$$

we can solve Eqs. (11.10) and (11.12) to obtain $p = p(x,y,z,a)$, $q = q(x,y,z,a)$. If we now integrate the equation

$$dz = p(x,y,z,a)\,dx + q(x,y,z,a)\,dy \tag{11.13}$$

we obtain a complete integral of Eq. (11.10).

For example, if we wish to solve the equation

$$q = (z + px)^2$$

we form the auxiliary equations

$$\frac{dx}{2x(z + px)} = \frac{dy}{-1} = \frac{dz}{2xp(z + px) - q} = \frac{dp}{-4p(z + px)} = \frac{dq}{-2q(z + px)}$$

according to scheme (11.11) above. From this set of equations we pick out the equation

$$2\frac{dx}{x} + \frac{dp}{p} = 0$$

which is readily seen to have solution $x^2p = a$, where a is a constant of integration. Using this result and the original partial differential equation, we find that

$$p = \frac{a}{x^2} \qquad q = \left(z + \frac{a}{x}\right)^2$$

Substituting these values into Eq. (11.13), we obtain the equation

$$dz = \frac{a}{x^2}\,dx + \left(z + \frac{a}{x}\right)^2 dy$$

which may be written in the form

$$\frac{dz - (a/x^2)\,dx}{(z + a/x)^2} = dy$$

from which it follows immediately that it has the solution

$$-\frac{1}{z + a/x} = y - b$$

where b is a constant of integration. In this way we obtain the complete integral

$$z = \frac{1}{b - y} - \frac{a}{x}$$

of the given equation.

Special Types of First-order Equations

Certain special kinds of first-order partial differential equations arise so frequently that it is worth listing the solutions obtained by Charpit's method.

Equations of the Type $f(p,q) = 0$. If $q = Q(a)$ is the solution of the equation $f(a,q) = 0$, then the solution of the partial differential equation $f(p,q) = 0$ is

$$z = ax + Q(a)y + b$$

Equations of the Type $f(z,p,q) = 0$. In this case p and q can be found from the pair of equations $p = aq$, $f(z,p,q) = 0$, and the solution found by integrating

$$dz = p\,dx + q\,dy$$

Equations of the Type $f(x,p) = g(y,q)$. We determine p and q from the relations $f(x,p) = a$, $g(y,q) = a$ and then proceed as in the general theory.

Clairaut Equations: $z = px + qy + f(p,q)$. Equations of this type have a complete integral of the form $z = ax + by + f(a,b)$.

11.3. CLASSIFICATION OF SECOND-ORDER LINEAR EQUATIONS

A great many second-order linear equations can be written in the form

$$Lz + f(x,y,z,p,q) = 0 \tag{11.14}$$

in which z is the dependent variable, x and y are the independent variables, $p = \partial z/\partial x$, $q = \partial z/\partial y$, and L denotes the operator

$$L = R\frac{\partial^2}{\partial x^2} + S\frac{\partial^2}{\partial x\,\partial y} + T\frac{\partial^2}{\partial y^2} \tag{11.15}$$

in which R, S, and T are continuous functions of x and y, possessing continuous partial derivatives with respect to both x and y.

If we change the independent variables from x and y to $\xi = \xi(x,y)$ and $\eta = \eta(x,y)$ and write $z(x,y) = \zeta(\xi,\eta)$, then (11.14) assumes the form

$$A(\xi_x,\xi_y)\frac{\partial^2\zeta}{\partial\xi^2} + 2B(\xi_x,\xi_y;\eta_x,\eta_y)\frac{\partial^2\zeta}{\partial\xi\,\partial\eta} + A(\eta_x,\eta_y)\frac{\partial^2\zeta}{\partial\eta^2} = F(\xi,\eta,\zeta,\zeta_\xi,\zeta_\eta) \tag{11.16}$$

where $A(u,v) = Ru^2 + Suv + Tv^2$, $B(u,v;u'v') = Ruu' + \frac{1}{2}S(uv' + u'v) + Tvv'$ and the function F is easily derived from the function f. The problem of reducing Eq. (11.16) to a simple form by a suitable choice of ξ and η is straightforward when the quantity $S^2 - 4RT$ is *everywhere* either positive, negative, or zero.

Case 1. $S^2 - 4RT > 0$. In this case we choose $\xi = f_1(x,y)$, $\eta = f_2(x,y)$, where $f_1 = c_1$ and $f_2 = c_2$ are the solutions of the first-order ordinary differential equations $y' + \lambda_1(x,y) = 0$, $y' + \lambda_2(x,y) = 0$, in which λ_1 and λ_2 are the roots of the quadratic equation $R\lambda^2 + S\lambda + T = 0$. When we make these substitutions, we find that Eq. (11.14) reduces to the form

$$\frac{\partial^2\zeta}{\partial\xi\,\partial\eta} = \phi(\xi,\eta,\zeta,\zeta_\xi,\zeta_\eta) \tag{11.17}$$

Case 2. $S^2 - 4RT = 0$. In this case the roots of the quadratic equation $R\lambda^2 + S\lambda + T = 0$ are equal. If we define ξ precisely as in case 1 above, and take η to be any function of x and y which is independent of ξ, we find that Eq. (11.14) reduces to the form

$$\frac{\partial^2\zeta}{\partial\eta^2} = \phi(\xi,\eta,\zeta,\zeta_\xi,\zeta_\eta) \tag{11.18}$$

Case 3. $S^2 - 4RT < 0$. If we go through the procedure of case 1 to obtain ξ and η, and then write $\alpha = \frac{1}{2}(\xi + \eta)$, $\beta = \frac{1}{2}i(\eta - \xi)$, we find that Eq. (11.14) becomes

$$\frac{\partial^2\zeta}{\partial\alpha^2} + \frac{\partial^2\zeta}{\partial\beta^2} = \psi(\alpha,\beta,\zeta,\zeta_\alpha,\zeta_\beta) \tag{11.19}$$

In this way we may classify second-order linear equations of the type (11.14). We say that an equation of this type is

1. *Hyperbolic* if $S^2 - 4RT > 0$ (e.g., the wave equation $z_{xx} = z_{yy}$)
2. *Parabolic* if $S^2 - 4RT = 0$ (e.g., the diffusion equation $z_{xx} = z_y$)
3. *Elliptic* if $S^2 - 4RT < 0$ (e.g., the potential equation $z_{xx} + z_{yy} = 0$)

The procedure is not quite so simple in the case of second-order linear equations in three independent variables. If we write x_1, x_2, x_3 for the independent variables and u for the dependent variable, and denote $\partial u/\partial x_i$ by p_i, $\partial^2 u/\partial x_i\,\partial x_j$ by p_{ij}, we may write a typical equation in the form

$$Lu = \sum_{i,j=1}^{3} a_{ij}p_{ij} + \sum_{i=1}^{3} b_i p_i + cu = 0 \tag{11.20}$$

To characterize these equations, we introduce the quadratic algebraic form

$$\Phi = \sum_{i,j=1}^{3} a_{ij}x_i x_j \tag{11.21}$$

1. If Φ is positive-definite in the x_i, we say that Eq. (11.20) is *elliptic*.
2. If Φ is indefinite, we say that the equation is *hyperbolic*.
3. If the determinant $|a_{ij}|$ of the form Φ vanishes, we say that the equation is *parabolic*.

This classification is in line with the one for equations in two independent variables and has the advantage that it is readily generalized to equations in n independent variables.

It should be observed here that, when a partial differential equation of the second order has to be solved numerically, the relaxation method (see Chap. 13) works very well for boundary-value problems relating to elliptic equations. With parabolic and hyperbolic equations, the boundary conditions are seldom specified on a closed boundary, and relaxation methods are generally ineffective. There are, however, satisfactory step-by-step methods suitable for the solution of such equations (see Chap. 14).

11.4. METHODS OF SOLUTION OF SECOND-ORDER PARTIAL DIFFERENTIAL EQUATIONS

a. Separation of Variables

Suppose that we have a second-order equation in three independent variables ξ, η, and ζ, say, and that we write the equation in the form

$$\mathsf{L}(u) = 0 \tag{11.22}$$

where u denotes the dependent variable, and L denotes a differential operator of the second order. In certain circumstances if we make the substitution

$$u = u_1(\xi)u_2(\eta)u_3(\zeta), \tag{11.23}$$

in which u_1, u_2, and u_3 are respectively (unknown) functions of ξ, η, and ζ, we find that Eq. (11.22) can be written in the form

$$\frac{\mathsf{L}_1 u_1}{u_1} + \frac{\mathsf{L}_2 u_2}{u_2} + \frac{\mathsf{L}_3 u_3}{u_3} = 0 \tag{11.24}$$

in which the operator L_1 contains only operations relevant to the variable ξ, etc. In these circumstances it is obvious that, if u_1, u_2, and u_3 satisfy the set of ordinary differential equations

$$\mathsf{L}_1 u_1 - a u_1 = 0 \qquad \mathsf{L}_2 u_2 - b u_2 = 0 \qquad \mathsf{L}_3 u_3 + (a + b)u_3 = 0$$

in which a and b are constants, then the product $u_1u_2u_3$ is a solution of the original partial differential equation. When this procedure works, we say that the given equation is *separable* in the coordinates ξ, η, ζ.

For example, consider the equation

$$\frac{\partial^2 u}{\partial x^2} + \frac{\partial^2 u}{\partial y^2} + \frac{\partial^2 u}{\partial z^2} = 0$$

If we let $u = X(x)Y(y)Z(z)$ in this equation, we find that it can be written in the form

$$\frac{X''(x)}{X} + \frac{Y''(y)}{Y} + \frac{Z''(z)}{Z} = 0$$

where $X''(x)$ denotes d^2X/dx^2, etc. Hence, the form chosen will be a solution of the partial differential equation if we choose X, Y, and Z so that

$$X''(x) - aX = 0 \qquad Y''(y) - bY = 0 \qquad Z''(z) + (a + b)Z = 0$$

where a and b are arbitrary constants. In this way the solution of the partial differential equation is reduced to that of a system of ordinary differential equations.

The method of separation of variables is often combined with the use of a fundamental property of linear partial differential equations. If the equation in question is linear, and if u and v are two independent solutions of the equation, then the linear combination $\lambda u + \mu v$, where λ and μ are constants, is also a solution of the equation.

To illustrate the use of this procedure, take $a = -m^2$, $b = -n^2$ in the last example, and assume that m and n are integers. Then we shall have

$$X = \cos(mx + \delta_m) \qquad Y = \cos(ny + \epsilon_n) \qquad Z = A \exp[-(m^2 + n^2)^{1/2}z] + B \exp[+(m^2 + n^2)^{1/2}z]$$

where δ_m, ϵ_n, A, and B are constants. It follow that

$$u = \{A \exp[-(m^2 + n^2)^{1/2}z] + B \exp[+(m^2 + n^2)^{1/2}z]\} \cos(mx + \delta_m) \cos(ny + \epsilon_n)$$

is a solution of the partial differential equation. From what we have said about the combination of solutions, it follows that

$$u = \sum_{m=0}^{\infty} \sum_{n=0}^{\infty} \{A_{mn} \exp[-(m^2 + n^2)^{1/2}z] + B_{mn} \exp[+(m^2 + n^2)^{1/2}z]\} \cos(mx + \delta_m) \cos(ny + \epsilon_n)$$

is also a solution for arbitrary values of the constants A_{mn}, B_{mn}, δ_m, ϵ_n (subject to conditions ensuring the convergence of the series).

b. The Method of the Green's Function

It is difficult to describe the method of the Green's function in the general case, and so we shall illustrate it by describing a particular example. Suppose that we wish to find a solution of Laplace's equation $\nabla^2 u = 0$ in the interior of a solid bounded by the simple closed surface S, with the additional property that the solution will take the value f on the surface (where f is a prescribed function of position). By a systematic use of Green's theorem relating surface to volume integrals, we can show that, if we can find a function $G(\mathbf{r},\mathbf{r}')$ satisfying the conditions

(1) $$G(\mathbf{r},\mathbf{r}') = H(\mathbf{r},\mathbf{r}') + \frac{1}{|\mathbf{r} - \mathbf{r}'|}$$

(2) $$\left(\frac{\partial^2}{\partial x'^2} + \frac{\partial^2}{\partial y'^2} + \frac{\partial^2}{\partial z'^2}\right) H(\mathbf{r},\mathbf{r}') = 0$$

(3) $$G(\mathbf{r},\mathbf{r}') = 0 \qquad \text{if } \mathbf{r}' \text{ is on } S$$

then the solution of our problem will be given by the integral expression

$$u(\mathbf{r}) = -\frac{1}{4\pi}\int_S f(\mathbf{r}')\frac{\partial G(\mathbf{r},\mathbf{r}')}{\partial n}\,dS'$$

The function $G(\mathbf{r},\mathbf{r}')$, which is uniquely determined by the three conditions (1) to (3) above is called the Green's function of this particular problem. The determination of a Green's function is often a very difficult problem, and great ingenuity is sometimes required in the process.

c. The Method of Characteristics for Hyperbolic Equations

There exists a useful general tool for the solution of hyperbolic equations of the second order in two independent variables.[1] Consider the second-order partial differential equation

$$ar + bs + ct = f \tag{11.25}$$

where a, b, c, and f are functions of $p = \partial z/\partial x$, and $q = \partial z/\partial y$, and, in the usual notation, $r = \partial^2 z/\partial x^2$, $s = \partial^2 z/\partial x\,\partial y$, $t = \partial^2 z/\partial y^2$. It is assumed that the equation is hyperbolic so that, throughout the domain in which we are interested, $b^2 > 4ac$.

Suppose that we are given a curve in the x,y plane and that values of z, p and q are assigned at each point of the curve. To ensure that p and q are actually derivatives of z, the prescribed values must satisfy the relation

$$dz = p\,dx + q\,dy \tag{11.26}$$

The problem we now pose is to discover whether the values of z, p, and q so prescribed, and the knowledge that r, s, and t satisfy the partial differential equation (11.25), are sufficient to determine the values of r, s, and t at each point of the curve. For this to be so we must have

$$dp = r\,dx + s\,dy \qquad dq = s\,dx + t\,dy \tag{11.27}$$

The three equations (11.25) and (11.27) have, in general, the unique solution given by the equations

$$\frac{r}{f\,dy^2 + c(dx\,dp - dy\,dq) - b\,dy\,dp} = \frac{-s}{f\,dx\,dy - a\,dp\,dy - c\,dq\,dx}$$
$$= \frac{t}{f\,dx^2 - a\,dx\,dp - b\,dx\,dq + a\,dy\,dq} = \frac{1}{a\,dy^2 - b\,dx\,dy + c\,dx^2}$$

so that unique values of the second derivatives are determined at each point of the curve. If, however,

$$a\,dy^2 - b\,dx\,dy + c\,dx^2 = 0 \tag{11.28}$$

then the equations have no solution unless also[2]

$$f\,dx\,dy - a\,dp\,dy - c\,dq\,dx = 0 \tag{11.29}$$

If this is so, then we know from the theory of linear algebraic equations that there is an infinite number of solutions.

Since the equation we are dealing with is hyperbolic, $b^2 > 4ac$, so that Eq. (11.28) has two distinct real roots, λ and μ, giving

$$\frac{dy}{dx} = \lambda \qquad \frac{dy}{dx} = \mu \tag{11.30}$$

[1] The theory of characteristics has recently been extended to equations in three or more independent variables, but the method lacks the simplicity of the two variable cases.

[2] It is easily shown that, if (11.28) and (11.29) hold, the denominators of r and t are also zero.

A curve satisfying either of the differential equations (11.30) is called a *characteristic curve* of Eq. (11.25). The entity consisting of a characteristic curve and the set of values of z, p, q prescribed at each point on it is called a *characteristic strip*. The relation (11.29) must be satisfied along a characteristic.

We shall now consider the solution of Eq. (11.25) by characteristics in the case where values of z, p, and q are prescribed at every point of a curve Γ in the x, y plane (Fig. 11.1). It is assumed that Γ is not a characteristic curve of the equation. If

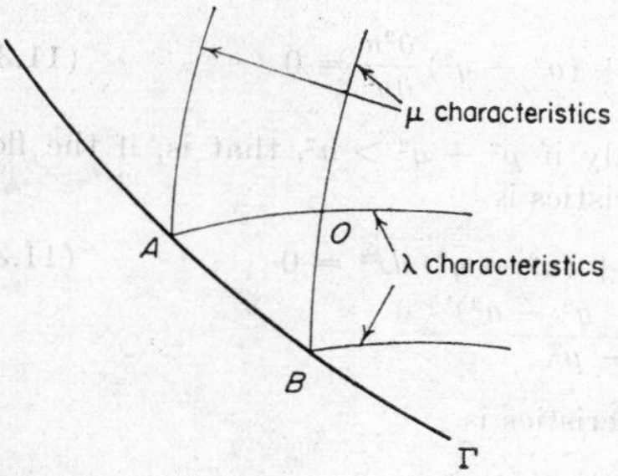

FIG. 11.1

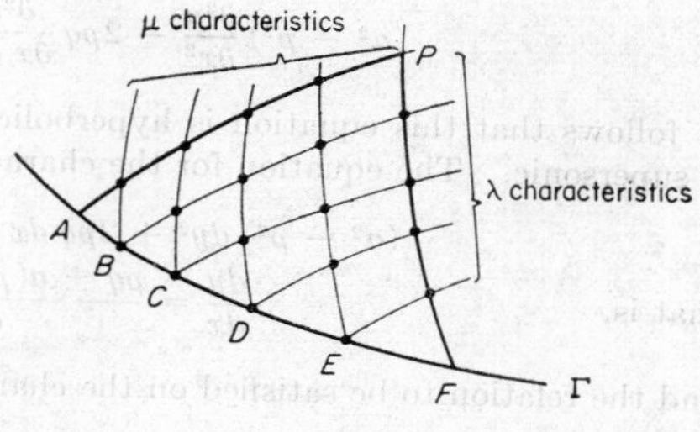

FIG. 11.2

we take two neighboring points A and B on the curve Γ, then there are two characteristic curves through A, and two through B. The λ characteristic through A will meet the μ characteristic through B in a point O, say, and, if A is close to B, the point O will be close to both of them, and we may regard the segments AO, BO of the two characteristics as being approximately straight. If we denote the value of λ at A by λ_A, etc., then the equations of the segments AO and BO are respectively

$$y_O - y_A = \lambda_A(x_O - x_A) \qquad y_O - y_B = \mu_B(x_O - x_B)$$

From these equations we can easily determine approximate values for the coordinates x_O, y_O of the point of intersection of the two characteristics. Using these approximations in Eq. (11.29), we see that:

$$\begin{aligned} &\text{On } AO: \quad \lambda_A a_A p_O + c_A q_O = \lambda_A f_A(x_O - x_A) + \lambda_A a_A p_A + c_A q_A \\ &\text{On } BO: \quad \mu_B a_B p_O + c_B q_O = \mu_B f_B(x_O - x_B) + \mu_B a_B p_B + c_B q_B \end{aligned} \tag{11.31}$$

The pair of linear equations (11.31) gives us two approximate values for the partial derivatives p_O and q_O. If we denote the mean values of p and q along AO by $\bar{p}$ and $\bar{q}$, then since, to this degree of approximation,

$$dz = \bar{p}\,dx + \bar{q}\,dy$$

we have

$$z_O = z_A + \tfrac{1}{2}(p_A + p_O)(x_O - x_A) + \tfrac{1}{2}(q_A + q_O)(y_O - y_A) \tag{11.32}$$

From these values of z_O, p_O, q_O we can now calculate λ_O and μ_O.

We can improve the values so obtained by a method of successive approximations by now replacing λ_A, μ_B, etc., in the above equations by $\tfrac{1}{2}(\lambda_A + \lambda_O)$, $\tfrac{1}{2}(\lambda_B + \lambda_O)$, etc. The process is repeated until two successive approximations to x_O, y_O, p_O, q_O, z_O agree to the desired accuracy.

We now select a number of conveniently spaced points $A, B, C, \ldots$ on our initial curve Γ and, by the process outlined above, obtain values of x, y, z, p, q at all points of the mesh shown in Fig. 11.2. It is obvious from this diagram that, if we wish to obtain the solution of the partial differential equation at the point P, then we need only compute the values of x, y, z, p, q, at points of the mesh which lie within the curvilinear triangle PAF formed by the initial curve and the two characteristic curves through P. For that reason this region is known as the *domain of dependence* of the point P.

As an example of the determination of the characteristics of a partial differential equation, we shall consider an equation which arises in the theory of compressible flow. If we denote the component of the velocity of the fluid in the x direction by $-p = -\partial u/\partial x$, and the component in the y direction by $-q = -\partial u/\partial y$, then the equation governing the flow is

$$(a^2 - p^2)\frac{\partial^2 u}{\partial x^2} - 2pq\frac{\partial^2 u}{\partial x\,\partial y} + (a^2 - q^2)\frac{\partial^2 u}{\partial y^2} = 0 \tag{11.33}$$

It follows that this equation is hyperbolic only if $p^2 + q^2 > a^2$, that is, if the flow is supersonic. The equation for the characteristics is

$$(a^2 - p^2)\,dy^2 + 2pq\,dx\,dy + (a^2 - q^2)\,dx^2 = 0 \tag{11.34}$$

that is,

$$\frac{dy}{dx} = \frac{pq \pm a(p^2 + q^2 - a^2)^{1/2}}{a^2 - p^2}$$

and the relation to be satisfied on the characteristics is

$$(a^2 - p^2)dp\,dy + (a^2 - q^2)\,dq\,dx = 0$$

which, because of (11.34), can be written in the form

$$(a^2 - q^2)\left(\frac{dq}{dp}\right)^2 - 2pq\frac{dq}{dp} + (a^2 - p^2) = 0$$

The main application of the method of characteristics is to the solution of problems in compressible flow. For a full account of the method, the reader is referred to R. E. Meyer [11]. In most problems, the various equations have to be integrated numerically. The mechanism by which errors are propagated through a computation by the method of characteristics has been analyzed and the resultant effect of the various types of error determined by Hall [6]. Hall also outlines a procedure for planning such computations to achieve specified accuracy with minimum labor.

d. The Use of Integral Transforms

In recent years the theory of integral transforms has been applied extensively to the solution of partial differential equations [14]. If we are given a function $K(\xi x)$ which is integrable over the range (α,β), then, provided the integral is convergent, we can define a function by the relation

$$\bar{f}(\xi) = \int_\alpha^\beta f(x)K(\xi x)\,dx \tag{11.35}$$

This function is called the *integral transform* of $f(x)$ with respect to the *kernel* $K(\xi x)$. We may regard (11.35) as an integral equation determining $f(x)$ when $\bar{f}(\xi)$ is known. In some circumstances this integral equation has a solution of the type

$$f(x) = \int_\gamma^\delta \bar{f}(\xi)H(\xi x)\,d\xi \tag{11.36}$$

A result of this type is known as an *inversion theorem* for the integral transform defined by Eq. (11.35). If it happens that $\gamma = \alpha$, $\delta = \beta$, and $H(\xi x) = K(\xi x)$ so that the relation between a function and its integral transform is completely symmetrical, we say that $K(\xi x)$ is a *Fourier kernel*. For instance, when $(\alpha,\beta) = (0,\infty)$, the kernels

$$\sqrt{\frac{2}{\pi}}\cos\xi x \qquad \sqrt{\frac{2}{\pi}}\sin\xi x \qquad (\xi x)^{1/2}J_\nu(\xi x) \quad \nu \geq -\tfrac{1}{2}$$

are Fourier kernels. They define the Fourier cosine, the Fourier sine, and the Hankel transforms, respectively. The Fourier exponential transform defined by $(\alpha,\beta) = (-\infty,\infty)$, $K(\xi x) = (2\pi)^{-1/2}e^{i\xi x}$ has $(\gamma,\delta) = (-\infty,\infty)$ and $H(\xi x) = (2\pi)^{-1/2}e^{-i\xi x}$. In the Laplace transform $(\alpha,\beta) = (0,\infty)$, $K(\xi x) = e^{-\xi x}$, where $\mathcal{R}(\xi) > c$, and $(\gamma,\delta) = (\epsilon - i\infty, \epsilon + i\infty)$ with $\epsilon > c$, $H(\xi x) = (2\pi i)^{-1}e^{\xi x}$.

Multiple integral transforms may be built up by combining these simple kernels; thus, from the kernels $K_1(\xi x)$, $K_2(\xi x)$ we might form the double transform

$$\bar{f}(\xi_1,\xi_2) = \int_{\alpha_1}^{\beta_1}\int_{\alpha_2}^{\beta_2} f(x,y)K_1(\xi_1 x)K_2(\xi_2 y)\,dx\,dy$$

Suppose now that we have a partial differential equation of the form

$$\mathbf{L}u + \mathbf{M}u = f(x_1,x_2, \ldots, x_n) \tag{11.37}$$

where $\mathbf{L}$ is a linear differential operator involving derivatives with respect to $x_1, \ldots, x_r$, and $\mathbf{M}$ is a linear differential operator involving derivatives with respect to $x_{r+1}, \ldots, x_n$. If we introduce a multiple transform

$$\bar{u}(\xi_1,\xi_2, \ldots, \xi_r, x_{r+1}, \ldots, x_n) = \int\int \cdots \int u(x_1,x_2, \ldots, x_n)K(\xi_1 x_1, \ldots, \xi_r x_r)\,dx_1 \cdots dx_r$$

and find that

$$\int\int \cdots \int (\mathbf{L}u)K\,dx_1 \cdots dx_r = \lambda(\xi_1,\xi_2, \ldots, \xi_r)\bar{u}$$

where λ is a function of $\xi_1, \ldots, \xi_r$, then Eq. (11.37) can be transformed to

$$\mathbf{M}\bar{u} = \bar{f} - \lambda\bar{u} \tag{11.38}$$

which is now a partial differential equation in $(n - r)$ variables. If this equation can be solved to give the form of $\bar{u}$, then the form of u can be obtained by using the appropriate inversion theorem. In practice, we should hope to use transforms in such a way that Eq. (11.38) was either a purely algebraic equation or an ordinary differential equation, the solution of which is always easier than that of a partial differential equation.

As an example, consider the solution $u(x,y)$ of

$$\frac{\partial^2 u}{\partial x^2} = \frac{\partial u}{\partial y} \qquad x > 0,\ y > 0$$

satisfying $u(x,0) = 0$, $u(0,y) = k$, a constant. If we use a Fourier sine transform and observe that (as a result of integrating by parts)

$$\sqrt{\frac{2}{\pi}}\int_0^\infty \frac{\partial^2 u}{\partial x^2}\sin \xi x\,dx = \sqrt{\frac{2}{\pi}}\,k\xi - \xi^2\bar{u}$$

we see that the Fourier sine transform

$$\bar{u} = \frac{2}{\pi}\int_0^\infty u(x,y)\sin \xi x\,dx$$

satisfies the ordinary differential equation

$$\frac{d\bar{u}}{dy} + \xi^2\bar{u} = \sqrt{\frac{2}{\pi}}\,k\xi \qquad \bar{u}(\xi,0) = 0$$

which has the solution

$$\bar{u} = \sqrt{\frac{2}{\pi}}\,k\,\frac{1 - e^{-\xi^2 y}}{\xi}$$

Using the inversion theorem for Fourier sine transforms, we see that the required solution is

$$u(x,y) = \sqrt{\frac{2}{\pi}} \int_0^\infty \bar{u}(\xi,y) \sin \xi x \, d\xi = \frac{2}{\pi} k \int_0^\infty (1 - e^{-\xi^2 y}) \frac{\sin \xi x}{\xi} \, d\xi$$

e. Reduction of Partial Differential Equations to Integral Equations

In certain conditions, the solution of a partial differential equation, satisfying certain boundary conditions, can be reduced to that of an integral equation by means of the theory of integral transforms or by the method of the Green's function. We shall illustrate the procedure by considering the mixed boundary-value problem:

$$\frac{\partial^2 f}{\partial x^2} + \frac{\partial^2 f}{\partial y^2} + k^2 f = 0 \qquad -\infty < x < \infty \qquad y \geq 0$$

when $y = 0$

$$f = \Phi(x) \qquad x > 0$$

$$\frac{\partial f}{\partial y} = \psi(x) \qquad x < 0$$

If we use the Fourier exponential transform

$$F(\alpha,y) = \frac{1}{\sqrt{2\pi}} \int_{-\infty}^\infty f(x,y) e^{i\alpha x} \, dx \tag{11.39}$$

of $f(x)$, then it is easily seen that, if $f \to 0$ as $y \to \infty$, then

$$F(\alpha,y) = F(\alpha) e^{-\gamma y} \tag{11.40}$$

where $\gamma^2 = \alpha^2 - k^2$. If we extend $\psi(x)$ to denote the (unknown) value of $\partial f/\partial y$ on $y = 0$, $x > 0$, then $F(\alpha) = -\Psi(\alpha)/\gamma$, and we obtain the solution

$$f(x,y) = -\frac{1}{2\pi} \int_{-\infty}^\infty \gamma^{-1} e^{-i\alpha x - \gamma y} \, d\alpha \int_{-\infty}^\infty \psi(\xi) e^{i\alpha \xi} \, d\xi$$

Interchanging the order of integration, letting $y \to 0$, and splitting the range of integration in ξ into two semi-infinite stretches, we obtain the integral equation

$$\int_0^\infty \psi(\xi) K(x - \xi) \, d\xi = \chi(x) \tag{11.41}$$

where $K(x - \xi)$ denotes the kernel defined by the equation

$$K(x - \xi) = -\frac{1}{2\pi} \int_{-\infty}^\infty \gamma^{-1} e^{i\alpha(\xi - x)} \, d\alpha$$

and $\chi(x)$ denotes the known function

$$\phi(x) - \int_{-\infty}^0 \psi(\xi) K(x - \xi) \, d\xi$$

Such integral equations can be solved by the Wiener-Hopf technique ([12], pp. 978–990).

The same boundary-value problem can be formulated in terms of dual integral equations. If we insert from (11.40) into (11.39) and invert by means of Fourier's theorem, we find that

$$f(x,y) = \frac{1}{\sqrt{2\pi}} \int_{-\infty}^\infty F(\alpha) e^{-\gamma y - i\alpha x} \, d\alpha \tag{11.42}$$

If we now insert the boundary conditions, we find that (11.42) is a solution of the problem provided that $F(\alpha)$ is a solution of the dual integral equations:

$$\frac{1}{\sqrt{2\pi}}\int_{-\infty}^{\infty} F(\alpha)e^{-i\alpha x}\,d\alpha = \phi(x) \qquad x > 0$$

$$\frac{1}{\sqrt{2\pi}}\int_{-\infty}^{\infty} \gamma F(\alpha)e^{-i\alpha x}\,d\alpha = \psi(x) \qquad x < 0$$

11.5. ELLIPTIC EQUATIONS

We shall now consider in a little more detail the partial differential equations of elliptic type which arise frequently in physical problems. The prototype of such equations is Laplace's equation $\nabla^2\psi = 0$, where, in cartesian coordinates x, y, z,

$$\nabla^2 = \frac{\partial^2}{\partial x^2} + \frac{\partial^2}{\partial y^2} + \frac{\partial^2}{\partial z^2} \tag{11.43a}$$

in cylindrical polar coordinates ρ, θ, z,

$$\nabla^2 = \frac{\partial^2}{\partial \rho^2} + \frac{1}{\rho}\frac{\partial}{\partial \rho} + \frac{\partial^2}{\partial z^2} + \frac{1}{\rho^2}\frac{\partial^2}{\partial \theta^2} \tag{11.43b}$$

and in spherical polar coordinates r, θ, ϕ (see Fig. 4.12),

$$\nabla^2 = \frac{1}{r^2}\frac{\partial}{\partial r}\left(r^2\frac{\partial}{\partial r}\right) + \frac{1}{r^2 \sin\phi}\frac{\partial}{\partial \phi}\left(\sin\phi\frac{\partial}{\partial \phi}\right) + \frac{1}{r^2\sin^2\phi}\frac{\partial^2}{\partial \theta^2} \tag{11.43c}$$

Most of the results stated for elliptic equations will refer to Laplace's equation, but the same methods of solution apply to other elliptic equations of the second order.

Laplace's Equation $\nabla^2\psi = 0$

Introductory Remarks. A great deal of theoretical work in electricity and hydrodynamics rests upon the fact that

$$\psi = \frac{1}{|\mathbf{r} - \mathbf{r}'|} \tag{11.44}$$

where $|\mathbf{r} - \mathbf{r}'|$ denotes the distance between the points with position vectors $\mathbf{r}$ and $\mathbf{r}'$ ($\mathbf{r}'$ being fixed and $\mathbf{r}$ "current"), is a solution of Laplace's equation everywhere except at the point $\mathbf{r} = \mathbf{r}'$. By differentiating this expression and also by constructing integral expressions with it as their core, it is possible to build up a variety of solutions of Laplace's equation with physical significance such as "point charges," "dipoles," "quadrupoles," "surface layers," etc. The nature of such solutions is discussed in most modern textbooks on electricity.

Another useful theorem for constructing solutions out of more elementary ones is due to Lord Kelvin. It states that if $\psi_0(\mathbf{r})$ is a solution of Laplace's equation, then so are the functions

$$\frac{a}{r}\psi_0\left(\frac{a^2\mathbf{r}}{r^2}\right) \qquad \frac{a}{r}\int_0^1 f(u)\psi_0\left(\frac{u^2a^2\mathbf{r}}{r^2}\right)$$

where f is an arbitrary function such that the integral exists. For example, if $\psi(\mathbf{r}) \sim \psi_0(\mathbf{r})$ as $r \to \infty$ and $\psi = 0$ on the sphere $r = a$, then, for $r > a$,

$$\psi(\mathbf{r}) = \psi_0(\mathbf{r}) - \frac{a}{r}\psi_0\left(\frac{a^2}{r^2}\mathbf{r}\right) \tag{11.45a}$$

Similarly, if $\psi(\mathbf{r}) \to \psi_0(\mathbf{r})$ as $r \to \infty$ and $\partial\psi/\partial r = 0$ on $r = a$, then, for $r > a$,

$$\psi(\mathbf{r}) = \psi_0(\mathbf{r}) + \frac{a}{r}\psi_0\left(\frac{a^2}{r^2}\mathbf{r}\right) - \frac{2}{ar}\int_0^a u\psi_0\left(\frac{u^2}{a^2}\mathbf{r}\right)du \tag{11.45b}$$

In practical applications, we usually have to find a function which satisfies not only Laplace's equation in a certain region but also certain conditions on the boundary of the region. Any problem in which we have to determine such a function is called a boundary-value problem for Laplace's equation. If the value of the function ψ is uniquely specified at every point of the boundary of the region, we have what is called a *Dirichlet problem;* if it is the normal derivative $\partial\psi/\partial n$ which is specified at all points of the boundary, then we have a *Neumann problem;* if ψ is prescribed at certain parts of the boundary, and $\partial\psi/\partial n$ is prescribed over the remainder, then we are said to have a *mixed* boundary-value problem. It will be noted that the solution of the Dirichlet problem by the method of the Green's function has been given on p. 11–7.

Solutions in Cartesian Coordinates. It is easily shown by the method of separation of variables (or directly) that the function $\psi = \exp(i\alpha x + i\beta y + \gamma z)$ is a solution of Laplace's equation if $\alpha^2 + \beta^2 = \gamma^2$.

SEMI-INFINITE SOLID. If $-\infty < x,y < \infty$, $z > 0$, and $\psi = f(x,y)$ on $z = 0$, then

$$\psi(x,y,z) = \frac{1}{2\pi}\int_{-\infty}^{\infty}\int_{-\infty}^{\infty} \bar{f}(\xi,\eta)\exp\left[-i\xi x - i\eta y - (\xi^2+\eta^2)^{1/2}z\right] d\xi\, d\eta$$

where

$$\bar{f}(\xi,\eta) = \frac{1}{2\pi}\int_{-\infty}^{\infty}\int_{-\infty}^{\infty} f(x,y)e^{i\xi x + i\eta y}\,dx\,dy$$

is the double Fourier transform of $f(x,y)$. This expression can be simplified, using the theory of Fourier transforms, to give the solution

$$\psi(x,y,z) = \frac{z}{2\pi}\int_{-\infty}^{\infty}\int_{-\infty}^{\infty}\frac{f(x',y')\,dx'\,dy'}{[(x-x')^2 + (y-y')^2 + z^2]^{3/2}} \tag{11.46}$$

which can also be derived by means of the method of the Green's function.

SEMI-INFINITE PARALLELEPIPED. If $\psi = 0$ on $x = 0$, a and on $y = 0$, b and if $\psi = f(x,y)$ on $z = 0$, while $\psi \to 0$ as $z \to \infty$, then

$$\psi(x,y,z) = \sum_{m=1}^{\infty}\sum_{n=1}^{\infty} f_{mn}\sin\frac{m\pi x}{a}\sin\frac{n\pi y}{b}\,e^{-\gamma_{mn}z} \tag{11.47}$$

where $\gamma_{mn}{}^2 = \pi^2(m^2/a^2 + n^2/b^2)$ and

$$f_{mn} = \frac{4}{ab}\int_0^a\int_0^b f(x,y)\sin\frac{m\pi x}{a}\sin\frac{n\pi y}{b}\,dx\,dy$$

If $\partial\psi/\partial n$ vanishes instead of ψ on the stated faces, then we replace the sines by cosines throughout.

CUBOID. If $\psi = 0$ on $x = 0$, a and on $y = 0$, b, then

$$\psi(x,y,z) = \sum_{m=1}^{\infty}\sum_{n=1}^{\infty} g_{mn}\sin\frac{m\pi x}{a}\sin\frac{n\pi y}{b}\,\text{Sinh}(\gamma_{mn}z + \epsilon_{mn}) \tag{11.48}$$

where g_{mn} and ϵ_{mn} are arbitrary constants and γ_{mn} is given by the value in the last section. For example, if $\psi = 0$ on $z = c$ and $\psi = f(x,y)$ on $z = 0$, then, in the notation of the last problem,

$$\psi(x,y,z) = \sum_{m=1}^{\infty}\sum_{n=1}^{\infty} f_{mn}\sin\frac{m\pi x}{a}\sin\frac{n\pi y}{a}\,\text{Sinh}\,[\gamma_{mn}(c-z)]\,\text{Csch}\,\gamma_{mn}c \tag{11.49}$$

If $\partial\psi/\partial n$ vanishes on $x = 0$, a and $y = 0$, b, we merely replace the sines by cosines.

Solutions in Cylindrical Coordinates. It is easily seen by the method of separation of variables that the double series

$$\psi = \sum_m \sum_n (A_{mn}e^{mz} + B_{mn}e^{-mz})[C_{mn}J_n(m\rho) + D_{mn}Y_n(m\rho)] \sin (n\theta + \epsilon_{mn})$$

$$\psi = \sum_m \sum_n (A_{mn} \cos mz + B_{mn} \sin mz)[C_{mn}I_n(m\rho) + D_{mn}K_n(m\rho)] \sin (n\theta + \epsilon_{mn}) \tag{11.50a,b}$$

are solutions of Laplace's equation; J_n and Y_n denote Bessel functions, and I_n and K_n denote modified Bessel functions.

If there is symmetry about the z axis, these solutions reduce to

$$\psi = \sum_m (A_m e^{mz} + B_m e^{-mz})[J_0(m\rho) + C_m Y_0(m\rho)]$$

$$\psi = \sum_m (A_m \cos mz + B_m \sin mz)[I_0(m\rho) + C_m K_0(m\rho)] \tag{11.51a,b}$$

In a region containing the z axis and in problems for which ψ is finite on the z axis, we have the solutions

$$\psi = \sum_m \sum_n (A_{mn}e^{mz} + B_{mn}e^{-mz})J_n(m\rho) \sin (n\theta + \epsilon_{mn})$$

$$\psi = \sum_m \sum_n (A_{mn} \cos mz + B_{mn} \sin mz)I_n(m\rho) \sin (n\theta + \epsilon_{mn}) \tag{11.52a,b}$$

SEMI-INFINITE SOLID WITH AXIAL SYMMETRY. If $\psi \to 0$ as $z \to \infty$, $\rho \to \infty$ and $\psi = f(\rho)$ on $z = 0$, then

$$\psi = \int_0^\infty \xi \bar{f}(\xi) e^{-\xi z} J_0(\xi\rho)\, d\xi \qquad \text{where} \qquad \bar{f}(\xi) = \int_0^\infty u f(u) J_0(\xi u)\, du \tag{11.53}$$

SEMI-INFINITE CYLINDER OF FINITE RADIUS. If $\psi = 0$ on $\rho = a$, $\psi \to 0$ as $z \to \infty$, $\psi = f(\rho)$ on $z = 0$, then

$$\psi = \sum_s A_s J_0\left(\frac{\lambda_s \rho}{a}\right) e^{-\lambda_s z/a} \tag{11.54}$$

where the sum is taken over all the positive roots of the equation $J_0(\lambda) = 0$ and

$$A_s = \frac{2}{[aJ_1(\lambda_s)]^2} \int_0^a u f(u) J_0(u\lambda_s/a)\, du$$

FINITE CYLINDER OF HEIGHT h AND RADIUS a

1. If $\psi = 0$ on $\rho = a$ and on $z = h$, and if $\psi = f(\rho)$ on $z = 0$, then

$$\psi = \sum_s A_s J_0\left(\frac{\lambda_s \rho}{a}\right) \operatorname{Sinh} \frac{\lambda_s(h - z)}{a} \operatorname{Csch} \frac{\lambda_s h}{a} \tag{11.55}$$

where A_s is given by the same expression as in the last problem.

2. If $\psi = 0$ on $z = 0$, $z = h$, and if $\psi = f(z)$ on $\rho = a$, then

$$\psi = \sum_{m=1}^{\infty} f_m \sin \frac{m\pi z}{h} \frac{I_0(m\pi\rho/h)}{I_0(m\pi a/h)} \tag{11.56}$$

where
$$f_m = \frac{2}{h} \int_0^h f(u) \sin \frac{m\pi u}{h}\, du$$

Electrified Disk. If

$$\psi = \sum_n G_n(\rho)\cos(n\theta + \epsilon_n) \quad \text{on } z = 0,\ 0 < \rho < 1 \qquad \partial\psi/\partial z = 0 \quad \text{on } z = 0,\ \rho > 1$$

then

$$\psi = \sum_n \cos(n\theta + \epsilon_n)\int_0^\infty f_n(t)e^{-t|z|}J_n(\rho t)\,dt \tag{11.57}$$

where

$$f_n(t) = \sqrt{\frac{2}{\pi}}\left[t^{1/2}J_{n-1/2}(t)\int_0^1 \frac{y^{n+1}G(y)\,dy}{\sqrt{1-y^2}} + \int_0^1 \frac{u^{n+1}\,du}{\sqrt{1-u^2}}\int_0^1 G(yu)(ty)J_{n+1/2}(ty)\,dy\right]$$

Spherical Polar Coordinates. A general solution of Laplace's equation in spherical polars is

$$\psi = \sum_{n=0}^{\infty}\sum_{m=1}^{n}(A_nr^n + B_nr^{-n-1})(A_{mn}\cos m\theta + B_{mn}\sin m\theta)[P_n{}^m(\cos\phi) + C_{mn}Q_n{}^m(\cos\phi)] \tag{11.58}$$

Spherical Region. If $0 < r < a$ and ψ is finite at $r = 0$ and along $\phi = 0, \pi$, then

$$\psi = \sum_{n=0}^{\infty}\sum_{m=1}^{n}(A_{mn}\cos m\theta + B_{mn}\sin m\theta)\left(\frac{r}{a}\right)^n P_n{}^m(\cos\phi) \tag{11.59}$$

If, in addition, $\psi(a,\theta,\phi) = f(\theta,\phi)$, then

$$\psi = \frac{a(r^2 - a^2)}{4\pi}\int_0^{2\pi} d\theta'\int_0^\pi \frac{f(\theta',\phi')\sin\phi'\,d\phi'}{(a^2 + r^2 - 2ar\cos\Phi)^{3/2}}, \tag{11.60}$$

where $\cos\Phi = \cos\phi\cos\phi' + \sin\phi\sin\phi'\cos(\theta - \theta')$.

Infinite Solid with Spherical Hollow. If $r > a$ and $\psi \to 0$ as $r \to \infty$ and ψ is finite along $\phi = 0, \pi$, then

$$\psi = \sum_{n=0}^{\infty}\sum_{m=1}^{n}\left(\frac{a}{r}\right)^{n+1}(A_{mn}\cos m\theta + B_{mn}\sin m\theta)\,P_n{}^m(\cos\phi) \tag{11.61}$$

Region Bounded by Two Cones. If $\psi = 0$ on the cone $\phi = \alpha$ and $\psi = \Sigma\psi_nr^n$ on the cone $\phi = \beta$, then, if $\alpha < \phi < \beta$,

$$\psi = \sum_n \psi_n\left[\frac{Q_n(\cos\alpha)P_n(\cos\phi) - P_n(\cos\alpha)Q_n(\cos\phi)}{Q_n(\cos\alpha)P_n(\cos\beta) - P_n(\cos\alpha)Q_n(\cos\beta)}\right]r^n \tag{11.62}$$

Axial Symmetry. If along the axis $\phi = 0$

$$\psi = \psi(z,0,0) = \sum_n (a_nz^n + b_nz^{-n-1})$$

then at a general point

$$\psi(r,\theta,\phi) = \sum_n (a_nr^n + b_nr^{-n-1})P_n(\cos\phi) \tag{11.63}$$

The Two-dimensional Laplace Equation

We shall now consider briefly the two-dimensional Laplace equation $\nabla_1{}^2\psi = 0$, where $\nabla_1{}^2$ denotes the two-dimensional Laplacian operator, which, in plane cartesian coordinates x, y and polar coordinates r, θ, is

$$\nabla_1^2 = \frac{\partial^2}{\partial x^2} + \frac{\partial^2}{\partial y^2} = \frac{\partial^2}{\partial r^2} + \frac{1}{r}\frac{\partial}{\partial r} + \frac{1}{r^2}\frac{\partial^2}{\partial \theta^2} \tag{11.64}$$

If we write $\mathbf{r} = (x,y)$, $\mathbf{r}' = (x',y')$ and denote the distance between the fixed point r' and the current point r by $|\mathbf{r} - \mathbf{r}'|$, then it is readily shown that

$$\psi = A \ln |\mathbf{r} - \mathbf{r}'|, \tag{11.65}$$

with A a constant, is a solution of the equation $\nabla_1^2\psi = 0$. From this basic solution we can construct, by summation or differentiation, other solutions of the two-dimensional Laplace equation. Because of the central part which this basic solution plays in the theory, a solution of the two-dimensional Laplace equation is often referred to as a *logarithmic potential.*

A valuable source of solutions of this equation arises from the fact that, if we write $x + iy = z$, the real and imaginary parts of an analytic function of the complex variable z are logarithmic potentials.

For example, the real and imaginary parts of the analytic function $w = \text{Cosh } z$ are $u = \text{Cosh } x \cos y$ and $v = \text{Sinh } x \sin y$, and it is easily verified that both u and v satisfy $\nabla_1^2\psi = 0$.

The real power of the complex-variable method rests in the fact that, if $\zeta = f(z)$, that is, $\xi + i\eta = f(x + iy)$ is a conformal mapping taking $\psi(x,y)$ into $\Psi(\xi,\eta)$, then

$$\frac{\partial^2\Psi}{\partial\xi^2} + \frac{\partial^2\Psi}{\partial\eta^2} = \left|\frac{dz}{d\zeta}\right|^2 \left(\frac{\partial^2\psi}{\partial x^2} + \frac{\partial^2\psi}{\partial y^2}\right) \tag{11.66}$$

so that, if ψ is a logarithmic potential in x and y, Ψ is one in the variables ξ and η. Also any curve in the x,y plane along which ψ is constant is mapped into a curve in the ξ,η plane along which Ψ is constant. For a complete account of the theory of conformal mapping and its application to the theory of the logarithmic potential, the reader is referred to Nehari [18].

The theory of the logarithmic potential is also closely linked to the theory of integral equations. Suppose we wish to find a function ψ satisfying $\nabla_1^2\psi = 0$ in the area S enclosed by the curve C and such that $\psi = f$ on C. Then, if we write

$$\psi = \int_C \mu(s') \frac{\cos(\mathbf{n},\mathbf{r} - \mathbf{r}')}{|\mathbf{r} - \mathbf{r}'|}\, ds' \tag{11.67}$$

it is easily shown that $\mu(s)$ is a solution of the nonhomogeneous integral equation

$$\mu(s) + g(s) = \int \mu(s')K(s,s')\, ds' \tag{11.68}$$

where the function $g(s) = f(s)/\pi$ is known, and $K(s,s')$ denotes the kernel

$$\frac{1}{\pi}\frac{\cos(\mathbf{n},\mathbf{r} - \mathbf{r}')}{|\mathbf{r} - \mathbf{r}'|}$$

It is possible to develop a method of the Green's function for the two-dimensional equation. For the Dirichlet problem considered in the last paragraph we can show that, if we can find a Green's function $G(x,y;x',y') = w(x,y;x',y') - \ln |\mathbf{r} - \mathbf{r}'|$ where $w(x,y;x'y') = \ln |\mathbf{r} - \mathbf{r}'|$ on C and

$$\left(\frac{\partial^2}{\partial x'^2} + \frac{\partial^2}{\partial y'^2}\right) w(x,y;x',y') = 0$$

then

$$\psi(x,y) = -\frac{1}{2\pi}\int_C \psi(x',y') \frac{\partial G(x,y;x',y')}{\partial n}\, ds' \tag{11.69}$$

where n is the outward-drawn normal to C.

Solutions in Cartesian Coordinates. We can construct, by the method of separation of variables, solutions of the kind

$$\psi(x,y) = \sum_n (A_n e^{nx} + B_n e^{-nx}) \cos(ny + \epsilon_n) \tag{11.70}$$

where A_n, B_n, and ϵ_n are constants, depending only on the parameter n (which need not be an integer).

HALF-PLANE: $x \geq 0, \quad -\infty < y < \infty$. If $\psi \to 0$ as $x \to \infty$ and $\psi = f(y)$ on $x = 0$, then

$$\psi(x,y) = \frac{1}{2\pi} \int_{-\infty}^{\infty} \bar{f}(\xi) e^{-\xi x - i\xi y}\, d\xi \tag{11.71}$$

where

$$\bar{f}(\xi) = \frac{1}{2\pi} \int_{-\infty}^{\infty} f(y) e^{i\xi y}\, dy$$

Using the theory of Fourier transforms (or, alternatively, the method of the Green's function), we can show that this solution is equivalent to

$$\psi = \frac{x}{\pi} \int_{-\infty}^{\infty} \frac{f(u)\, du}{x^2 + (y-u)^2} \tag{11.72}$$

SEMI-INFINITE STRIP: $x \geq 0$, $0 < y < b$: If $\psi \to 0$ as $x \to \infty$, $\psi = f(y)$ on $x = 0$ and $\psi = 0$ on $y = 0$ and $y = b$, then

$$\psi = \sum_{n=1}^{\infty} f_n e^{-n\pi x/b} \sin \frac{n\pi y}{b} \tag{11.73}$$

where

$$f_n = \frac{2}{b} \int_0^b f(u) \sin \frac{n\pi u}{b}\, du$$

RECTANGLE: $0 \leq x \leq a$, $0 \leq y \leq b$. If $\psi = f(y)$ when $x = 0$, $\psi = 0$ when $x = a$, $y = b$, $y = 0$, then

$$\psi = \sum_{n=1}^{\infty} f_n \sin \frac{n\pi y}{b} \operatorname{Sinh} \frac{n\pi(a-x)}{b} \operatorname{Csch} \frac{n\pi a}{b} \tag{11.74}$$

where f_n is defined as in the previous problem.

Polar Coordinates. A general solution in polar coordinates is

$$\psi = \sum_n (A_n r^n + B_n r^{-n}) \cos(n\theta + \epsilon_n) \tag{11.75}$$

where A_n, B_n, and ϵ_n are constants.

CIRCULAR REGION: $0 \leq r \leq a$. If ψ remains finite as $r \to 0$, then

$$\psi = \Sigma A_n r^n \cos(n\theta + \epsilon_n) \tag{11.76}$$

In particular, if $\psi = f(\theta)$ on $r = a$, then

$$\psi = \frac{a^2 - r^2}{2\pi} \int_0^{2\pi} \frac{f(\theta')\, d\theta'}{a^2 - 2ar\cos(\theta' - \theta) + r^2} \tag{11.77}$$

This equation is known as Poisson's integral.

INFINITE REGION WITH A CIRCULAR HOLE: $r \geq a$. If ψ tends to zero as $r \to \infty$, then

$$\psi = \sum_{n=1}^{\infty} \frac{a^n A_n}{r^n} \cos(n\theta + \epsilon_n) \tag{11.78}$$

If $\psi = f(\theta)$ on $r = a$, then

$$A_0 \cos \epsilon_0 = \frac{1}{2\pi} \int_0^{2\pi} f(\theta)\, d\theta \qquad A_n \cos \epsilon_n = \frac{1}{\pi} \int_0^{2\pi} f(\theta) \cos n\theta\, d\theta$$

$$A_n \sin \epsilon_n = \frac{1}{\pi} \int_0^{2\pi} f(\theta) \sin n\theta\, d\theta$$

Poisson's Equation

Another elliptic equation which occurs frequently in applications is Poisson's equation

$$\nabla^2 \psi + 4\pi\rho(\mathbf{r}) = 0 \tag{11.79}$$

in which the function $\rho(\mathbf{r})$ is a prescribed function of position. It is readily seen that, if ψ_1 is any solution of Poisson's equation, and ψ_2 is a solution of Laplace's equation, then $\psi_1 + \psi_2$ is a solution of Poisson's equation. By means of this result, we can combine general solutions of Laplace's equation with particular integrals of Poisson's equation to achieve general solutions of Poisson's equation. If $\rho(\mathbf{r})$ is defined through a region V, then, at any point $\mathbf{r}$ within the region, the function

$$\psi(r) = \int_V \frac{\rho(\mathbf{r}')\, d\tau'}{|\mathbf{r} - \mathbf{r}'|} \tag{11.80}$$

is a solution of Poisson's equation.

In constructing particular integrals of Poisson's equation, it is useful to note that

$$\psi = \frac{r^{m+2} P_n(\cos \theta)}{(m - n + 2)(m + n + 3)} \tag{11.81}$$

is a particular integral of the equation

$$\nabla^2 \psi = r^m P_n(\cos \theta) \tag{11.82}$$

Similarly

$$\psi = \frac{r^{m+2} \cos (n\theta + \epsilon)}{(m - n + 2)(m + n + 2)} \tag{11.81$'$}$$

is a solution of the two-dimensional equation

$$\nabla_1{}^2 \psi = r^m \cos (n\theta + \epsilon) \tag{11.82$'$}$$

11.6. HYPERBOLIC EQUATIONS

The most frequently occurring hyperbolic equations of the second order are the wave equation and the nonhomogeneous wave equation, and we shall restrict our attention to these two.

The One-dimensional Wave Equation

A general solution of the one-dimensional wave equation

$$\frac{\partial^2 \psi}{\partial x^2} = \frac{1}{c^2} \frac{\partial^2 \psi}{\partial t^2} \tag{11.83}$$

is

$$\psi = f(x + ct) + g(x - ct) \tag{11.84}$$

where the functions f and g are arbitrary.

Infinite Region: $-\infty < x < \infty$, $t > 0$. If $\psi = p(x)$, $\partial\psi/\partial t = v(x)$, when $t = 0$, then (*d'Alembert's solution*)

$$\psi = \tfrac{1}{2}[p(x + ct) + p(x - ct)] + \frac{1}{2c} \int_{x-ct}^{x+ct} v(\xi)\, d\xi \tag{11.85}$$

Semi-infinite Region: $x \geq 0,\ t > 0$. If $\psi = P(x)$, $\partial\psi/\partial t = V(x)$, when $t = 0$, then the solution is given by (11.85) where

$$p(u) = \begin{cases} P(u) & u \geq 0 \\ -P(-u) & u < 0 \end{cases} \qquad v(u) = \begin{cases} V(u) & u \geq 0 \\ -V(-u) & u < 0 \end{cases}$$

Finite Region: $0 \leq x \leq l,\ t > 0$.

SOLUTION BY FOURIER SERIES. If $\psi = 0$, $\partial\psi/\partial t = 0$ at $x = 0$ and $x = l$, and $\psi = p(x)$, $\partial\psi/\partial t = v(x)$ at $t = 0$, then

$$\psi = \sum_{m=1}^{\infty} \left(p_m \cos \frac{m\pi ct}{l} + \frac{lv_m}{\pi cm} \sin \frac{m\pi ct}{l} \right) \sin \frac{m\pi x}{l} \tag{11.86}$$

where $$(p_m, v_m) = \frac{2}{l} \int_0^l (p,v) \sin \frac{m\pi x}{l}\, dx$$

are the Fourier sine coefficients of the functions p and v.

SOLUTION BY THE METHOD OF CHARACTERISTICS. To illustrate the use of the method of characteristics we shall solve the wave equation

$$\frac{\partial^2 z}{\partial x^2} = \frac{\partial^2 z}{\partial y^2} \qquad 0 \leq x \leq 1,\ y \geq 0 \tag{11.87}$$

subject to the conditions that z, p, and q (in the notation of Sec. 11.4*c*) are prescribed on the line $y = 0$ and both z and q are zero on the lines $x = 0$, $x = 1$ (Fig. 11.3). It will be observed that the boundary-value problem of the last section can be put into this form by a trivial change of the independent variables. For this equation the

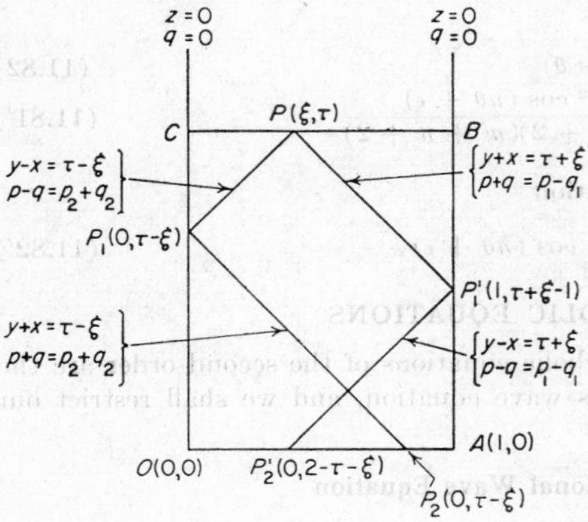

FIG. 11.3

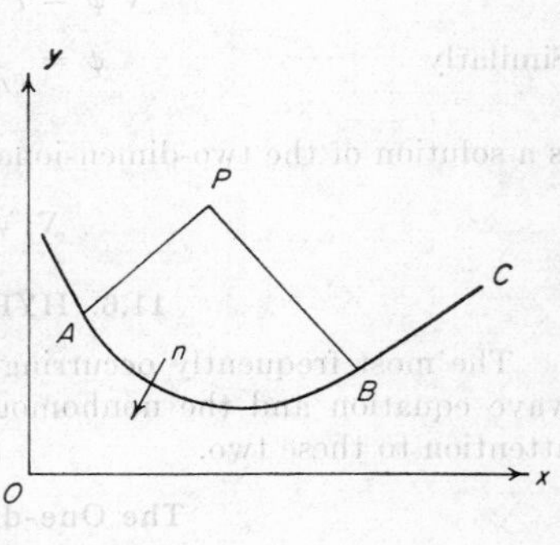

FIG. 11.4

characteristics have equations $x \pm y = $ const, and along these lines the first derivatives satisfy $p \pm q = $ const. Suppose that we wish to find the value of z at "time" $y = \tau$. We draw the line $y = \tau$ to cut the lines $x = 0$, $x = 1$ in the points C and B, respectively. The two characteristics through a typical point $P(\xi,\eta)$ will be PP_1 and PP_1' with equations $y - x = \tau - \xi$ and $y + x = \tau + \xi$, respectively. The other characteristic through P_1 will meet $y = 0$ in P_2 (for the particular choices of ξ and τ which are shown in Fig. 11.3), and P_1P_2 will have equation $y + x = \tau - \xi$ so that P_2 will have coordinates $(0, \tau - \xi)$; since the values of p and q are prescribed at each point of OA, we know the values of p and q at $P_2 - p_2, q_2$ say. In a similar way, we can see that the other characteristic through P_1' has equation

$$y - x = \tau + \xi - 2$$

and that it meets OA in $P_2'(0, 2 - \tau - \xi)$ where the values of p and q are p_1 and q_1, respectively.

Hence, along $P_2'P_1'$ we have $p - q = p_1 - q_1$, so that, at the point P_1', we have $p = p_1 - q_1$, $q = 0$; similarly we find that, at P_1, $p = p_2 + q_2$, $q = 0$. Hence, along P_1P we have $p - q = p_2 + q_2$, and along $P_1'P$ we have $p + q = p_1 - q_1$, so that at P we have

$$p = \tfrac{1}{2}(p_1 - q_1 + p_2 + q_2) \qquad q = \tfrac{1}{2}(p_1 - q_1 - p_2 - q_2)$$

In this way we find the values of p and q at all points of CB. Integrating p with respect to x, starting at $x = 0$ where $z = 0$, we obtain the expression

$$z = \int_0^x p\,dx$$

for the value of z at the points of BC. As a check on the accuracy of the integrations (which might have to be performed numerically), we have

$$\int_0^1 p\,dx = 0$$

(since z vanishes when $x = 1$).

THE RIEMANN-VOLTERRA SOLUTION. If we are given the values of z and $\partial z/\partial n$ at each point of a curve C, then the solution of the wave equation

$$\frac{\partial^2 z}{\partial x^2} = \frac{\partial^2 z}{\partial y^2} \tag{11.88}$$

at the point P in the x,y plane is given by the expression

$$z_P = \tfrac{1}{2}(z_A + z_B) - \frac{1}{2}\int_{AB}\left(\frac{\partial z}{\partial y}\,dy - \frac{\partial z}{\partial x}\,dx\right) \tag{11.89}$$

where the characteristics through P meet the curve C in the points A and B (Fig. 11.4).

The Two-dimensional Wave Equation

The two-dimensional wave equation

$$\nabla_1^2\psi = \frac{1}{c^2}\frac{\partial^2\psi}{\partial t^2} \tag{11.90}$$

has solutions of the type

$$\psi = A_{pq}\exp\left[\pm i(px + qy + kct)\right] \tag{11.91}$$

where $k^2 = p^2 + q^2$. In plane polar coordinates r and θ, we have solutions of the type

$$\psi = J_m(k_{mn}r)e^{\pm im\theta \pm ik_{mn}ct} \tag{11.92}$$

If $\psi = 0$ on the circle $r = a$, the constants k_{mn} must be chosen to be the roots of the transcendental equation $J_m(ka) = 0$.

Infinite Region: $r > 0$. If $\psi = f(x,y)$, $\partial\psi/\partial t = 0$ at $t = 0$ and $\psi \to 0$ as $r \to \infty$, then

$$\psi = \frac{1}{2\pi}\int_{-\infty}^{\infty}\int_{-\infty}^{\infty} F(u,v)\cos\left[c(u^2 + v^2)^{1/2}t\right]e^{-i(ux+vy)}\,du\,dv \tag{11.93}$$

where
$$F(u,v) = \frac{1}{2\pi}\int_{-\infty}^{\infty}\int_{-\infty}^{\infty} f(x,y)e^{i(ux+vy)}\,dx\,dy$$

This solution can be written in the form

$$\psi = -\frac{1}{2\pi c}\int_{-\infty}^{\infty}\int_{-\infty}^{\infty}\frac{\partial}{\partial t}\frac{f(\alpha,\beta)\,d\alpha\,d\beta}{[c^2t^2 - (x-\alpha)^2 - (y-\beta)^2]^{1/2}} \tag{11.94}$$

Rectangle: $0 \le x \le a$, $0 \le y \le b$. If $\psi = 0$ on the bounding lines and $\psi = f(x,y)$, $\partial\psi/\partial t = g(x,y)$ when $t = 0$, then

$$\psi = \sum_{m=1}^{\infty}\sum_{n=1}^{\infty}\left(f_{mn}\cos k_{mn}ct + \frac{g_{mn}}{k_{mn}c}\sin k_{mn}ct\right)\sin\frac{m\pi x}{a}\sin\frac{n\pi y}{b} \tag{11.95}$$

where $(f_{mn}, g_{mn}) = \dfrac{4}{ab}\displaystyle\int_0^a dx\int_0^b (f,g)\sin\frac{m\pi x}{a}\sin\frac{n\pi y}{b}\,dy$

Circular Region: $0 \le r \le a$. If $\psi = 0$ on $r = a$, $\psi = f(r)$, $\partial\psi/\partial t = 0$ at $t = 0$, then

$$\psi = \sum_n f_n J_0(k_n r)\cos k_n ct \tag{11.96}$$

where $k_1, k_2, \ldots, k_n, \ldots$ are the positive zeros of $J_0(ka)$ and

$$f_n = \frac{2}{a^2 J_1{}^2(k_n a)}\int_0^a rf(r)J_0(k_n r)\,dr$$

Three-dimensional Wave Equation

In cartesian coordinates, the equation

$$\nabla^2\psi = \frac{1}{c^2}\frac{\partial^2\psi}{\partial t^2} \tag{11.97}$$

has the solution

$$\psi = \exp\,[\pm i(lx + my + nz + kct)] \tag{11.98}$$

provided that $k^2 = l^2 + m^2 + n^2$.

In polar coordinates r, θ, ϕ, we have solutions of the form

$$\psi = J_{\pm(n+1/2)}(kr)P_n{}^m(\cos\phi)e^{\pm im\theta\pm ikct} \tag{11.99}$$

important special cases of which are

$$\psi = \frac{1}{r}\sin kr\, e^{\pm ikct} \quad\text{and}\quad \psi = \frac{1}{r}\left(\frac{\sin kr}{kr} - \cos kr\right)\cos\phi\, e^{\pm ikct} \tag{11.100a,b}$$

If there is spherical symmetry, then the general solution of the wave equation is

$$\psi = \frac{1}{r}[f(r - ct) + g(r + ct)] \tag{11.101}$$

where the functions f and g are arbitrary.

In cylindrical coordinates ρ, θ, z, we have solutions of the type

$$\psi = J_m(\omega\rho)e^{ikct - i\gamma z\pm im\theta} \tag{11.102}$$

where $\omega^2 = k^2 - \gamma^2$. If we replace J_m by a Bessel function of the same order, we get another solution; e.g., we could replace $J_m(\omega\rho)$ by $H_m{}^{(1)}(\omega\rho)$ or $H_m{}^{(2)}(\omega\rho)$ and obtain alternative solutions. The simplest solutions of these types are

$$\psi = H_0{}^{(1)}(\omega\rho)e^{ikct - i\gamma z} \qquad \psi = H_0{}^{(1)}(k\rho)e^{ikct} \tag{11.103a,b}$$

representing outgoing cylindrical waves and

$$\psi = H_0^{(2)}(\omega\rho)e^{ikct-i\gamma z} \qquad \psi = H_0^{(2)}(k\rho)e^{ikct} \tag{11.104a,b}$$

representing incoming cylindrical waves.

11.7. PARABOLIC EQUATIONS

The Diffusion Equation

We shall now consider solutions of the diffusion equation

$$k\,\nabla^2\psi = \frac{\partial\psi}{\partial t} \tag{11.105}$$

An important general result for extending known solutions is provided by *Duhamel's theorem* which states that, if $\phi(\mathbf{r},t,t')$ is the solution of the boundary-value problem (in which t' is a parameter),

$$k\,\nabla^2\phi = \frac{\partial\phi}{\partial t} \qquad \mathsf{L}\phi = G(\mathbf{r}',t') \qquad \phi(\mathbf{r},0,t') = J(\mathbf{r}) \tag{11.106}$$

where $\mathbf{r}'$ denotes a point on the boundary of the region under consideration and L is a boundary differential operator (assumed linear), then the solution of Eq. (11.105) satisfying the conditions

$$\mathsf{L}\psi = G(\mathbf{r}',t) \quad \text{on } S, \qquad \psi(\mathbf{r},0) = J(\mathbf{r})$$

is given by the equation

$$\psi(\mathbf{r},t) = \frac{\partial}{\partial t}\int_0^t \phi(\mathbf{r},t-t',t')\,dt' \tag{11.107}$$

The One-dimensional Equation. INFINITE REGION: $-\infty < x < \infty$, $t \geq 0$. The Poisson integral

$$\psi(x,t) = \frac{1}{2\sqrt{\pi kt}}\int_{-\infty}^{\infty}\phi(\xi)\exp\left[-\frac{(x-\xi)^2}{4kt}\right]d\xi = \frac{1}{\sqrt{\pi}}\int_{-\infty}^{\infty}\phi(x+2u\sqrt{kt})e^{-u^2}\,du \tag{11.108}$$

satisfies the boundary-value problem:

$$k\psi_{xx} = \psi_t \qquad \psi(x,0) = \phi(x) \tag{11.109}$$

SEMI-INFINITE REGION: $x \geq 0$, $t \geq 0$. The solution satisfying $\psi(x,0) = f(x)$, $\psi(0,t) = 0$ is

$$\psi(x,t) = \frac{1}{2\sqrt{\pi kt}}\int_0^{\infty} f(u)(e^{-(x-u)^2/4kt} - e^{-(x+u)^2/4kt})\,du \tag{11.110}$$

FINITE SLAB: $0 \leq x \leq a$, $t \geq 0$. If $\psi(0,t) = \psi(a,t) = 0$, $\psi(x,0) = f(x)$, then

$$\psi(x,t) = \frac{2}{a}\sum_{n=1}^{\infty} e^{-n^2\pi^2a^{-2}kt}\sin\frac{n\pi x}{a}\int_0^a f(u)\sin\frac{n\pi u}{a}\,du \tag{11.111}$$

The Three-dimensional Equation. If we put $\psi(\mathbf{r},t) = \Psi(\mathbf{r})e^{-k\lambda^2 t}$ in the diffusion equation [with $f(\mathbf{r},t) = 0$], we find that Ψ satisfies Helmholtz's equation

$$(\nabla^2 + \lambda^2)\Psi = 0$$

In cylindrical coordinates we have solutions of the type

$$\sum_{\lambda,\mu,\gamma} A_{\lambda\mu\gamma} J_{\pm\gamma}(\rho\sqrt{\lambda^2+\mu^2})e^{\pm\mu z \pm i\gamma\theta - \lambda^2 kt}$$

For example, for the infinite cylinder $0 < \rho < a$, we find that, if $\psi(\rho,0) = f(\rho)$, $\psi(a,t) = 0$, then

$$\psi(\rho,t) = \Sigma A_n J_0(\rho\xi_n)e^{-kt\xi_n^2} \tag{11.112}$$

where $\xi_1, \xi_2, \ldots$ are the roots of $J_0(\xi a) = 0$ and

$$A_n = \frac{2}{[aJ_1(\xi_n a)]^2}\int_0^a uf(u)J_0(\xi_n u)\,du$$

In spherical polar coordinates r, θ, ϕ, we obtain solutions of the type

$$\sum_{m,n,\lambda} C_{mn\lambda}(\lambda r)^{-1/2}J_{n+1/2}(\lambda r)P_n^m(\cos\phi)e^{\pm im\theta - \lambda^2 kt}$$

For instance, the function $\psi(r,\phi,t)$ satisfying the diffusion equation and the conditions $\psi(r,\phi,0) = f(r,\phi)$, $\psi(a,\phi,t) = 0$ is

$$\psi(r,\phi,t) = \sum_{n=1}^{\infty}\sum_{i=1}^{\infty} C_{ni}(\lambda_{ni}r)^{-1/2}J_{n+1/2}(\lambda_{ni}r)P_n(\cos\phi)e^{-\lambda_{ni}^2 kt} \tag{11.113}$$

where $\lambda_{n1}, \lambda_{n2}, \ldots$ are the positive roots of the equations $J_{n+1/2}(\lambda a) = 0$, and

$$C_{ni} = \frac{(2n+1)\lambda_{ni}^{1/2}}{[aJ'_{n+1/2}(\lambda_{ni}a)]^2}\int_0^a r^{3/2}J_{n+1/2}(r\lambda_{ni})\,dr\int_0^\pi f(r,\phi)P_n(\cos\phi)\sin\phi\,d\phi$$

The Diffusion Equation with Sources

If we wish to solve the boundary-value problem for $\psi(r,t)$:

$$\frac{\partial\psi}{\partial t} = k\nabla^2\psi + \chi(\mathbf{r},t) \qquad \psi(\mathbf{r},0) = f(\mathbf{r}) \qquad \psi(\mathbf{r},t) = \phi(\mathbf{r},t) \qquad \text{for } \mathbf{r}\epsilon S \tag{11.114}$$

then it is necessary only to find a solution ψ_1 of the diffusion equation (without the χ) satisfying $\psi_1(\mathbf{r},0) = f(\mathbf{r})$, $\psi_1(\mathbf{r},t) = \phi(\mathbf{r},t)$ for $\mathbf{r}\epsilon S$, and a particular integral $\psi_2(\mathbf{r},t)$ of the nonhomogeneous equation satisfying $\psi_2(\mathbf{r},0) = 0$, $\psi_2(\mathbf{r},t) = 0$ for $\mathbf{r}\epsilon S$, for the required function is then $\psi_1 + \psi_2$.

The methods available for the solution of the diffusion equation are also available for the solution of the nonhomogeneous diffusion equation. For instance, if the method of separation of variables has been applied to determine the function $\psi_1(\mathbf{r},t)$, the same type of expansion may be employed in the determination of $\psi_2(\mathbf{r},t)$, or, if a particular kind of integral transform has been used to find $\psi_1(\mathbf{r},t)$, it may also be used to determine $\psi_2(\mathbf{r},t)$.

11.8. EQUATIONS OF HIGHER ORDER THAN THE SECOND

Symbolic Methods of Solution

We shall consider here the solution of the equation

$$F(D,D')z = f(x,y) \tag{11.115}$$

where $D = \partial/\partial x$, $D' = \partial/\partial y$, and $F(D,D')$ is a polynomial with constant coefficients in D and D'. $F(D,D')$ is said to be *reducible* if it can be written as a product of linear factors of the form $(D + aD' + b)$, where a, b are constants; otherwise it is irreducible.

If $F(D,D')$ is reducible, then the order in which the linear factors occur is unimportant. If u is a sloution of $F(D,D')u = 0$, and z_1 is any solution of Eq. (11.115), then $u + z_1$ is a solution of (11.115). u is called a complementary function and z_1 a particular integral of Eq. (11.115). The difficult part is the construction of suitable complementary functions.

If $F(D,D')$ is reducible and can be written in the form

$$F(D,D') = \prod_{r=1}^{n} (\alpha_r D + \beta_r D' + \gamma_r)^{m_r} \tag{11.116}$$

in which none of the α_rs is zero, then the corresponding complementary function is

$$u = \sum_{r=1}^{n} \exp\left(-\frac{\gamma_r x}{\alpha_r}\right) \sum_{s=1}^{m_r} x^{s-1}\phi_{rs}(\beta_r x - \alpha_r y) \tag{11.117}$$

where the functions ϕ_{rs} ($s = 1, \ldots, m_r$; $r = 1, \ldots, n$) are arbitrary. If some of the αs are zero, the corresponding terms in the series are replaced by terms of the kind

$$\exp\left(-\frac{\gamma_r y}{\beta_r}\right) \sum_{s=1}^{m_r} x^{s-1}\phi_{rs}(\beta_r x)$$

When the operator $F(D,D')$ is irreducible, it is not always possible to find a solution with the full number of arbitrary functions, but it is possible to construct solutions containing as many arbitrary constants as we wish. In doing this we often make use of the results

$$F(D,D')e^{ax+by} = F(a,b)e^{ax+by} \qquad F(D,D')[e^{ax+by}g(x,y)] = e^{ax+by}F(D + a, D' + b)g(x,y) \tag{11.118}$$

For instance, $z = c_r \exp(a_r x + b_r y)$ is a solution of $F(D,D')z = 0$ if the constants a_r, b_r are connected through the relation $F(a_r,b_r) = 0$. To find the particular integral of (11.115) we write it symbolically as

$$z = \frac{1}{F(D,D')} f(x,y)$$

We expand F^{-1} by means of the binomial theorem and then interpret D^{-1}, D'^{-1} as integrations.

For example, consider the equation $(D^2 - D')z = y(y - x^2)$. Since

$$(D^2 - D')e^{ax+by} = (a^2 - b)e^{ax+by}$$

it follows, by taking $a = \pm in$, $b = -n^2$ that a suitable complementary function is

$$u = \sum_n a_n e^{\pm inx - n^2 y}$$

For the particular integral we write

$$z = \frac{1}{D^2 - D'}(y^2 - x^2 y) = -\left(\frac{1}{D'} + \frac{D^2}{D'^2}\right)(y^2 - x^2 y) = -(\tfrac{1}{3}y^3 - \tfrac{1}{2}x^2y^2 - \tfrac{1}{3}y^3) = \tfrac{1}{2}x^2y^2$$

Hence, we obtain the solution

$$z = \tfrac{1}{2}x^2y^2 + \sum_n a_n e^{-n^2 y} \cos(nx + \epsilon_n)$$

The Biharmonic Equation

We can derive solutions of the biharmonic equation from the theory of Laplace's equation and Poisson's equation by writing $\nabla^2\psi = \chi$, $\nabla^2\chi = 0$. In spherical coordinates, we obtain the axially symmetrical solution

$$\psi = \sum_{n=0}^{\infty} (A_n r^n + B_n r^{n+2} + C_n r^{-n-1} + D_n r^{-n+1}) P_n(\cos\phi) \tag{11.119}$$

If we are concerned with the region $r \leq a$ and with problems in which ψ remains finite as $r \to 0$, then we take

$$\psi = \sum_{n=0}^{\infty} \left[A_n \left(\frac{r}{a}\right)^n + C_n \left(\frac{r}{a}\right)^{n+2} \right] P_n(\cos\phi) \tag{11.120}$$

For example, if $\psi = f(\cos\phi)$, $\partial\psi/\partial r = 0$ on $r = a$, then

$$\psi = \frac{1}{2} \sum_{n=0}^{\infty} \left[(n+2)\left(\frac{r}{a}\right)^n - n\left(\frac{r}{a}\right)^{n+2} \right] f_n P_n(\cos\phi)$$

$$\text{where} \quad f_n = \left(n + \frac{1}{2}\right) \int_{-1}^{1} f(\mu) P_n(\mu)\, d\mu \tag{11.121}$$

An example which arises in hydrodynamics is the solution satisfying the conditions $\partial\psi/\partial\phi = 0$, $\partial\psi/\partial r = 0$, on $r = a$, $\psi \sim \frac{1}{2} U y^2$ as $r \to \infty$; we find that

$$\psi = \tfrac{1}{2} U \left(r^2 - \tfrac{3}{2} ar + \frac{1}{2}\frac{a^3}{r} \right) \sin^2\phi \tag{11.122}$$

In cylindrical coordinates we have solutions of the type

$$\psi = \sum_n (A + Bnz)(C \operatorname{Cosh} nz + D \operatorname{Sinh} nz) B_0(n\rho) \tag{11.123}$$

in which n is a constant, not necessarily an integer, and $B_0(n\rho)$ is any solution of Bessel's equation of zero order. For instance, if ψ remains finite as $\rho \to 0$ and $\psi \to 0$ as $z \to \infty$, then

$$\psi(\rho,z) = \sum_n (A + Bnz) e^{-nz} J_0(n\rho) \tag{11.124}$$

If we have a semi-infinite cylinder $0 \leq \rho \leq a$, $z \geq 0$, and if $\psi(a,z) = 0$, $\psi(\rho,0) = f(\rho)$, $\partial\psi(\rho,0)/\partial z = 0$, then

$$\psi(\rho,z) = \Sigma f_n (1 + \lambda_n z) e^{-\lambda_n z} J_0(\lambda_n \rho) \tag{11.125}$$

where the sum is taken over all the positive roots of the equation $J_0(\lambda a) = 0$, and

$$f_n = \frac{2}{[aJ_1(\lambda_n a)]^2} \int_0^a u f(u) J_0(\lambda_n u)\, du$$

For the semi-infinite solid $z \geq 0$, with $\psi(\rho,0) = f(\rho)$, $\partial\psi(\rho,0)/\partial z = 0$, $\psi \to 0$ as $\rho \to \infty$, and as $z \to \infty$, we have

$$\psi(\rho,z) = \int_0^\infty u\bar{f}(u)(1 + uz) e^{-uz} J_0(u\rho)\, du \qquad \text{with} \qquad \bar{f}(u) = \int_0^\infty \rho f(\rho) J_0(\rho u)\, d\rho \tag{11.126}$$

The two-dimensional form of the biharmonic equation $\nabla_1^4\psi = 0$ also arises in practical problems. In cartesian coordinates x and y, it has solutions of the type

$$(A + Bny)(C \operatorname{Cosh} ny + D \operatorname{Sinh} ny)e^{\pm inx} \tag{11.127}$$

It can also be shown that, if ψ_1, ψ_2, ψ_3 are two-dimensional harmonic functions (i.e., satisfy $\nabla_1^2\psi = 0$), then

$$\psi = x\psi_1 + y\psi_2 + \psi_3 \tag{11.128}$$

is a biharmonic function.

This has a complex-variable form. If the functions $\phi_1(z)$, $\phi_2(\bar{z})$, $\phi_3(z)$, $\phi_4(\bar{z})$ are arbitrary functions of the variables stated, z being $x + iy$ and $\bar{z}$ being $x - iy$, then the function

$$\psi = \phi_1(z) + \phi_2(\bar{z}) + \bar{z}\phi_3(z) + z\phi_4(\bar{z}) \tag{11.129}$$

is a biharmonic function. If ψ is real, then we must take $\phi_2(\bar{z}) = \overline{\phi_1(z)}, = \overline{g(z)}$ say, $\phi_4(\bar{z}) = \overline{\phi_3(z)} = \overline{f(z)}$, and

$$\psi = \bar{z}f(z) + z\overline{f(z)} + g(z) + \overline{g(z)} \tag{11.130}$$

where f and g are arbitrary.

For the semi-infinite region $y \geq 0$, with $\nabla_1^4\psi = 0$, $\psi \to 0$ as $y \to \infty$, $\partial\psi/\partial x = f(x)$, $\partial\psi/\partial y = g(x)$ on $y = 0$, we have

$$\psi(x,y) = \frac{1}{\pi}\int_{-\infty}^{\infty} f(u)\left[\tan^{-1}\frac{x-u}{y} + \frac{y(x-u)}{y^2+(x-u)^2}\right]du + \frac{y^2}{\pi}\int_{-\infty}^{\infty}\frac{g(u)\,du}{(x-u)^2+y^2} \tag{11.131}$$

In plane polar coordinates r and θ, we have solutions of the type

$$\psi = \sum_n (A_nr^n + B_nr^{-n} + C_nr^{n+2} + D_nr^{-n+2})\cos(n\theta + \epsilon_n) \tag{11.132}$$

For instance, if $r \leq a$, and $\psi = f(\theta)$, an *odd* function of θ, on $r = a$, and $\partial\psi/\partial r = 0$ on $r = a$, then

$$\psi = \frac{1}{2}\sum_{n=1}^{\infty}\left[(n+2)\left(\frac{r}{a}\right)^n - n\left(\frac{r}{a}\right)^{n+2}\right]f_n \sin n\theta \tag{11.133}$$

where $$f_n = \frac{2}{\pi}\int_0^{\pi} f(\theta)\sin n\theta\,d\theta$$

A general result is that, if $\nabla_1^2\psi = 0$, then $\nabla_1^4(r^2\psi) = 0$. If ψ depends on r alone, then $\psi = c_1r^2\ln r + c_2\ln r + c_3r^2 + c_4$, where the cs are constants.

Equations of Elastic Vibrations

An equation which arises frequently in engineering problems is the equation

$$EI\frac{\partial^4 y}{\partial x^4} + \mu\frac{\partial^2 y}{\partial t^2} = 0$$

governing the transverse vibrations of an elastic beam. If we let $y = ue^{i\omega t}$ and write $n^4 = \mu\omega^2/EI$, we find that u satisfies the ordinary differential equation

$$\frac{d^4u}{dx^4} - n^4u = 0 \tag{11.134}$$

which has solution

$$u_n = A\cos nx + B\sin nx + C\operatorname{Cosh} nx + D\operatorname{Sinh} nx \tag{11.135}$$

We can, therefore, build up solutions of the form $\sum_n u_n e^{i\omega t}$. The actual form of u_n will depend on the particular boundary conditions imposed on y. For example, if $y = 0$ and $\partial y/\partial x = 0$ at $x = 0$ and at $x = a$, then the n must be chosen so that $\cos na \operatorname{Cosh} na = 1$, and we get solutions of the type

$$y = \sum_i C_i[(\operatorname{Sinh} n_i x - \sin n_i x)(\operatorname{Cosh} n_i a - \cos n_i a) - (\operatorname{Cosh} n_i x - \cos n_i x)(\operatorname{Sinh} n_i a - \sin n_i a)] \sin (\omega_i t + \epsilon_i) \tag{11.136}$$

where $n_1, n_2, \ldots$ are the roots of the transcendental equation $\cos na \operatorname{Cosh} na = 1$ and $\omega_i^2 = n_i^4 EI/\mu$.

The equation

$$D\,\nabla_1^4 w + 2\rho h \frac{\partial^2 w}{\partial t^2} = Z(x,y,t) \tag{11.137}$$

governing the transverse vibrations of a flat elastic plate, can be treated in a similar manner. In the case in which $Z = 0$, if we let $w = W(x,y)e^{i\omega t}$ and put $2\rho h\omega^2/D = k^4$, we find that

$$(\nabla_1^2 - k^2)(\nabla_1^2 + k^2)W = 0 \tag{11.138}$$

so that W can be made up of the sum of solutions of $(\nabla_1^2 - k^2)W = 0$ and

$$(\nabla_1^2 + k^2)W = 0$$

In an infinite region $r > 0$, if at $t = 0$, $w = f(r)$, $\partial w/\partial t = 0$, then $(Z = 0)$

$$w(r,t) = \frac{1}{2bt}\int_0^\infty uf(u)J_0\left(\frac{ur}{2bt}\right) \sin \frac{u^2 + r^2}{4bt}\, du \tag{11.139}$$

If $Z(r,t) = 16hbf(r)\psi'(t)$, and $w = \partial w/\partial t = 0$, when $t = 0$, then

$$w(r,t) = 4\int_{-\infty}^{t} \frac{\psi(u)\, du}{t - u} \int_0^\infty vf(v) \sin \frac{v^2 + r^2}{4b(t - u)}\, J_0\left(\frac{vr}{2b(t - u)}\right) dv \tag{11.140}$$

11.9. REFERENCES

[1] R. d'Adhémar: "Les équations aux dérivées partielles à charactéristiques réelles," Gauthier-Villars, Paris, 1907.

[2] H. Bateman: "Partial Differential Equations of Mathematical Physics," Dover, New York, 1944.

[3] R. V. Churchill: "Fourier Series and Boundary Value Problems," McGraw-Hill, New York, 1941.

[4] R. Courant, D. Hilbert: "Methoden der mathematischen Physik," vol. 1, 2d ed., 1931; vol. 2, 1937, Springer, Berlin.

[5] J. Crank: "The Mathematics of Diffusion," Oxford Univ. Press, New York, 1956.

[6] M. G. Hall: The accuracy of the method of characteristics for plane supersonic flow, *Quart. J. Mech. Appl. Math.*, **9** (1956), 320–333.

[7] L. Hopf: "Introduction to the Differential Equations of Physics," transl. by W. Nef, Dover, New York, 1948.

[8] E. Kamke: "Differentialgleichungen, Lösungsmethoden und Lösungen, part 1: Gewöhnliche Differentialgleichungen," 5th ed., Akad.-Verlag., Leipzig, 1956.

[9] E. Kamke: "Differentialgleichungen, Lösungsmethoden und Lösungen, part 2, Akad.-Verlag., Leipzig, 1944.

[10] O. D. Kellogg: "Foundations of Potential Theory," Springer, Berlin, 1929.

[11] R. E. Meyer: The method of characteristics, in: L. Howarth (ed.), "Modern Developments in Fluid Mechanics: High Speed Flow," vol. 1, pp. 71–104, Oxford Univ. Press, New York, 1953.

[12] P. M. Morse, H. Feshbach: "Methods of Theoretical Physics," vols. 1 and 2, McGraw-Hill, New York, 1953.

[13] I. G. Petrovsky: "Lectures on Partial Differential Equations," transl. by A. Shenitzer, Interscience, New York, 1954.
[14] I. N. Sneddon: "Fourier Transforms," McGraw-Hill, New York, 1951.
[15] I. N. Sneddon: "Elements of Partial Differential Equations," McGraw-Hill, New York, 1957.
[16] A. Sommerfeld: "Partial Differential Equations in Physics," transl. by E. G. Strauss, Academic Press, New York, 1949.
[17] A. G. Webster: "Partial Differential Equations of Mathematical Physics," Dover, New York, 1955.
[18] Z. Nehari: "Conformal Mapping," McGraw-Hill, New York, 1952.

CHAPTER 12

NUMERICAL INTEGRATION

BY

J. B. SCARBOROUGH, Ph.D., Annapolis, Md.

12.1. DIFFERENCES AND INTERPOLATION

Differences

Let $y_0, y_1, y_2, \ldots, y_n$ denote a set of values of a function $y = f(x)$. Then $y_1 - y_0$, $y_2 - y_1$, etc., are called the *first differences* of y and are written with the notation

$$\Delta y_n = y_{n+1} - y_n$$

The differences of these first differences are called *second differences*, so that

$$\Delta^2 y_n = \Delta y_{n+1} - \Delta y_n = y_{n+2} - 2y_{n+1} + y_n$$

Likewise,

$$\Delta^3 y_n = \Delta^2 y_{n+1} - \Delta^2 y_n = y_{n+3} - 3y_{n+2} + 3y_{n+1} - y_n$$

A short table of these differences is Table 12.1. This is called a *diagonal* difference table. For some purposes it is written in the more compact form of a *horizontal* difference table shown in Table 12.2. Note the difference in notation in the two tables.

The relation between diagonal and horizontal differences is expressed by the equation

$$\Delta^m y_k = \Delta_m y_{k+m}$$

The relation between differences and derivatives is

$$\Delta^n f(x) = (\Delta x)^n f^{(n)}(x + \theta n\, \Delta x) \qquad 0 < \theta < 1$$

Consequently,

$$\lim_{\Delta x \to 0} \frac{\Delta^n f(x)}{(\Delta x)^n} = f^{(n)}(x)$$

Interpolation

The process of interpolation is used not only for finding in-between values of a function from a numerical table, but also for replacing complicated functions by simpler ones which can be made to coincide with them as closely as desired over a given interval. Let the function $y = f(x)$ be given by the values $y_0, y_1, y_2, \ldots, y_n$ which it takes for the corresponding values $x_0, x_1, x_2, \ldots, x_n$ of x. Then, if $f(x)$ is replaced in a given interval by a simpler function $\phi(x)$ whose form has been chosen in advance, the process constitutes interpolation in the broad sense of the term and $\phi(x)$ is called an interpolating function. The function $\phi(x)$ may take a variety of forms; but, since polynomials are the simplest functions, interpolating functions are usually

In the preparation of this chapter the author has drawn freely on his "Numerical Mathematical Analysis" [7].

polynomials, the coefficients of whose terms are expressed in differences of ascending order.

The most important and useful of the polynomial interpolation formulas are listed below, with brief instructions concerning their use. In these formulas the values of

Table 12.1. Diagonal Difference Table

x	y	Δy	$\Delta^2 y$	$\Delta^3 y$	$\Delta^4 y$	$\Delta^5 y$
x_{-3}	y_{-3}					
		Δy_{-3}				
x_{-2}	y_{-2}		$\Delta^2 y_{-3}$			
		Δy_{-2}		$\Delta^3 y_{-3}$		
x_{-1}	y_{-1}		$\Delta^2 y_{-2}$		$\Delta^4 y_{-3}$	
		Δy_{-1}		$\Delta^3 y_{-2}$		$\Delta^5 y_{-3}$
x_0	y_0		$\Delta^2 y_{-1}$		$\Delta^4 y_{-2}$	
		Δy_0		$\Delta^3 y_{-1}$		$\Delta^5 y_{-2}$
x_1	y_1		$\Delta^2 y_0$		$\Delta^4 y_{-1}$	
		Δy_1		$\Delta^3 y_0$		
x_2	y_2		$\Delta^2 y_1$			
		Δy_2				
x_3	y_3					

Table 12.2. Horizontal Difference Table

x	y	$\Delta_1 y$	$\Delta_2 y$	$\Delta_3 y$	$\Delta_4 y$	$\Delta_5 y$
x_{-3}	y_{-3}					
x_{-2}	y_{-2}	$\Delta_1 y_{-2}$				
x_{-1}	y_{-1}	$\Delta_1 y_{-1}$	$\Delta_2 y_{-1}$			
x_0	y_0	$\Delta_1 y_0$	$\Delta_2 y_0$	$\Delta_3 y_0$		
x_1	y_1	$\Delta_1 y_1$	$\Delta_2 y_1$	$\Delta_3 y_1$	$\Delta_4 y_1$	
x_2	y_2	$\Delta_1 y_2$	$\Delta_2 y_2$	$\Delta_3 y_2$	$\Delta_4 y_2$	$\Delta_5 y_2$
x_3	y_3	$\Delta_1 y_3$	$\Delta_2 y_3$	$\Delta_3 y_3$	$\Delta_4 y_3$	$\Delta_5 y_3$

the function are assumed to be given at equidistant intervals h of the independent variable.

Newton's Formula for Forward Interpolation

$$y = y_0 + u\,\Delta y_0 + \frac{u(u-1)}{2!}\,\Delta^2 y_0 + \frac{u(u-1)(u-2)}{3!}\,\Delta^3 y_0 + \frac{u(u-1)(u-2)(u-3)}{4!}\,\Delta^4 y_0 + \cdots + \frac{u(u-1)(u-2)\cdots(u-n+1)}{n!}\,\Delta^n y_0 \qquad (12.1)$$

where $u = (x - x_0)/h$, and it is understood that y denotes any value of the interpolating polynomial $\phi(x)$.

Equation (12.1) is called Newton's formula for *forward* interpolation, because it employs ys from y_0 forward to the right. It is used mainly for interpolating near the beginning of a table. The value y_0 may, of course, be any tabular value, but the formula will contain only values of y which come *after* the value chosen as the starting point.

Newton's Formula for Backward Interpolation

$$y = y_n + u\,\Delta_1 y_n + \frac{u(u+1)}{2!}\Delta_2 y_n + \frac{u(u+1)(u+2)}{3!}\Delta_3 y_n + \frac{u(u+1)(u+2)(u+3)}{4!}\Delta_4 y_n + \cdots + \frac{u(u+1)(u+2)\cdots(u+n-1)}{n!}\Delta_n y_n \qquad (12.2)$$

where $u = (x - x_n)/h$. This formula is used mainly for interpolation near the end of a table and for extrapolating one step beyond y_n, because it employs functional values which precede y_n.

Note that Eq. (12.2) employs horizontal differences, whereas (12.1) employs diagonal differences.

Stirling's Formula of Interpolation

$$y = y_0 + u\frac{\Delta y_{-1} + \Delta y_0}{2} + \frac{u^2}{2!}\Delta^2 y_{-1} + \frac{u(u^2-1)}{3!}\frac{\Delta^3 y_{-2} + \Delta^3 y_{-1}}{2} + \frac{u^2(u^2-1)}{4!}\Delta^4 y_{-2} + \frac{u(u^2-1)(u^2-2^2)}{5!}\frac{\Delta^5 y_{-3} + \Delta^5 y_{-2}}{2} + \frac{u^2(u^2-1)(u^2-2^2)}{6!}\Delta^6 y_{-3} + \cdots \qquad (12.3)$$

where $u = (x - x_0)/h$. This formula employs differences lying on and adjacent to a horizontal line through y_0 in the diagonal difference table.

Stirling's formula converges most rapidly for small values of u, and should be used when $-0.25 < u < 0.25$.

Bessel's Interpolation Formula. The symmetrical form of Bessel's formula is

$$y = \frac{y_0 + y_1}{2} + v\,\Delta y_0 + \frac{(v^2 - \frac{1}{4})}{2}\frac{\Delta^2 y_{-1} + \Delta^2 y_0}{2} + \frac{v(v^2 - \frac{1}{4})}{3!}\Delta^3 y_{-1} + \frac{(v^2 - \frac{1}{4})(v^2 - \frac{9}{4})}{4!}\frac{\Delta^4 y_{-2} + \Delta^4 y_{-1}}{2} + \frac{v(v^2 - \frac{1}{4})(v^2 - \frac{9}{4})}{5!}\Delta^5 y_{-2} + \frac{(v^2 - \frac{1}{4})(v^2 - \frac{9}{4})(v^2 - \frac{25}{4})}{6!}\frac{\Delta^6 y_{-3} + \Delta^6 y_{-2}}{2} + \cdots \qquad (12.4)$$

where $v = u - \frac{1}{2} = (x - x_0)/h - \frac{1}{2}$.

The differences occurring in this formula lie on and adjacent to the horizontal line through Δy_0 in the diagonal difference table.

Bessel's formula converges most rapidly when v is small (thus making u near $\frac{1}{2}$), and should be used when $-0.25 < v < 0.25$, that is, when $0.25 < u < 0.75$.

Because Stirling's and Bessel's formulas employ differences near the middle of a table, they are called *central-difference* formulas.

Bessel's Formula for Interpolating to Halves. Upon putting $v = 0$ in Eq. (12.4), a formula is obtained for finding the value of y exactly halfway between any two given values y_0 and y_1:

$$y = \frac{y_0 + y_1}{2} - \frac{1}{8}\frac{\Delta^2 y_{-1} + \Delta^2 y_0}{2} + \frac{3}{128}\frac{\Delta^4 y_{-2} + \Delta^4 y_{-1}}{2} - \frac{5}{1024}\frac{\Delta^6 y_{-3} + \Delta^6 y_{-2}}{2} + \cdots \qquad (12.5)$$

In terms of horizontal differences, this becomes

$$y = \frac{y_0 + y_1}{2} - \frac{1}{8}\frac{\Delta_2 y_1 + \Delta_2 y_2}{2} + \frac{3}{128}\frac{\Delta_4 y_2 + \Delta_4 y_3}{2} - \frac{5}{1024}\frac{\Delta_6 y_3 + \Delta_6 y_4}{2} + \cdots \qquad (12.6)$$

Inverse Interpolation

Inverse interpolation is the process of finding the value of the independent variable corresponding to a given value of the function when the latter is between two tabular values. The problem of inverse interpolation can be solved by several methods, but the method of successive approximations by iteration is satisfactory and sufficient in most cases.

To illustrate the procedure, we take Eq. (12.1) and write it in the form

$$u = \frac{y - y_0}{\Delta y_0} - \frac{1}{\Delta y_0}\left[\frac{u(u-1)}{2}\Delta^2 y_0 + \frac{u(u-1)(u-2)}{3!}\Delta^3 y_0 + \frac{u(u-1)(u-2)(u-3)}{4!}\Delta^4 y_0 + \cdots\right] \tag{12.7}$$

$$= \phi(u) \qquad \text{say}$$

To get a first approximation to u, neglect all differences beyond the second in (12.7), and solve the resulting quadratic equation for u, thereby obtaining two values of u. Use the positive value which lies between 0 and 1. Call this $u^{(1)}$. Substitute $u^{(1)}$ into the right-hand member of (12.7) and thereby get a second approximation

$$u^{(2)} = \phi(u^{(1)})$$

The succeeding approximations, as far as needed, will be $u^{(3)} = \phi(u^{(2)})$, $u^{(4)} = \phi(u^{(3)})$, etc. Continue the iteration as long as it gives any improvement in u.

After finding u, find x from the relation $x = x_0 + hu$.

The above iteration process converges to the true value of u when $|\phi'(u)| < 1$ for all values of u in the neighborhood of the value sought.

12.2. NUMERICAL DIFFERENTIATION AND INTEGRATION

Numerical Differentiation

The derivative of a tabulated function can be found at any point by representing the function by an appropriate interpolation formula and then differentiating the formula with respect to u. Then the derivative is given by the relation

$$\frac{dy}{dx} = \frac{1}{h}\frac{dy}{du}$$

From Eq. (12.3), for example, we have

$$\frac{dy}{dx} = \frac{1}{h}\left(\frac{\Delta y_{-1} + \Delta y_0}{2} + u\Delta^2 y_{-1} + \frac{3u^2 - 1}{3!}\frac{\Delta^3 y_{-2} + \Delta^3 y_{-1}}{2} + \frac{4u^3 - 2u}{4}\Delta^4 y_{-2} + \cdots\right)$$

$$\frac{d^2 y}{dx^2} = \frac{1}{h^2}\left(\Delta^2 y_{-1} + u\frac{\Delta^3 y_{-2} + \Delta^3 y_{-1}}{2} + \frac{12u^2 - 2}{4!}\Delta^4 y_{-2} + \cdots\right) \qquad \text{etc.}$$

Since $u = 0$ when $x = x_0$, we have

$$\left(\frac{dy}{dx}\right)_0 = \frac{1}{h}\left(\frac{\Delta y_{-1} + \Delta y_0}{2} - \frac{1}{3!}\frac{\Delta^3 y_{-2} + \Delta^3 y_{-1}}{2} + \cdots\right) \qquad \text{etc.}$$

To find the maximum or minimum value of a tabulated function, find several orders of differences, and substitute them into an appropriate interpolation formula, equate to zero the first derivative of the formula, and solve for u.

Numerical Evaluation of Definite Integrals

The numerical value of a definite integral can be found, to any desired degree of accuracy, by means of any of several quadrature formulas which express the integral as a linear function of given values of the integrand. In any problem, the definite integral should be set up before a quadrature formula is applied.

When the values of the integrand are given at equidistant intervals of width h of the independent variable, various quadrature formulas can be derived by integrating any of the standard interpolation formulas between given limits. By integrating (12.1), for example, and retaining differences of successively higher orders, we can derive the following formulas:

The Trapezoidal Rule

$$\int_a^b y\,dx = \int_a^b f(x)\,dx = h(\tfrac{1}{2}y_0 + y_1 + y_2 + \cdots + y_{n-2} + y_{n-1} + \tfrac{1}{2}y_n) \qquad (12.8)$$

where $h = (b - a)/n$, and n is either even or odd.

$$\text{Correction term} = \text{error} = -\frac{nh^3}{12} f''(\xi) \qquad a \le \xi \le b$$

Equation (12.8) is the simplest of the quadrature formulas, but is also the least accurate. The accuracy can be increased by decreasing the interval h.

Simpson's Rule

$$\int_a^b y\,dx = \int_a^b f(x)\,dx = \frac{h}{3}(y_0 + 4y_1 + 2y_2 + 4y_3 + 2y_4 + \cdots + 2y_{n-2} + 4y_{n-1} + y_n)$$

$$= \frac{h}{3}\sum_0^n cy \qquad \text{where } c = 1, 4, 2, \ldots, 2, 4, 1 \qquad (12.9)$$

$$\text{Error} = -\frac{nh^5}{180} f^{\text{iv}}(\xi) \qquad a \le \xi \le b$$

$$= -\frac{(b-a)}{180} h^4 f^{\text{iv}}(\xi) \approx -\frac{h}{90}\sum \Delta^4 y$$

In (12.9), n must be *even*.

When the number of given values $y_0, y_1, y_2, \ldots$ is fairly large, the computation should be arranged in tabular form.

Inasmuch as Simpson's rule is based on the assumption that the graph of $y = f(x)$ has been replaced by either $n/2$ arcs of vertical parabolas (parabolas with vertical axes) or $n/2$ arcs of polynomials of the third degree, none of which arcs are anywhere vertical, then it cannot give a reliable result in regions where the graph of $y = f(x)$ is vertical. In such cases, the graph is replaced by a parabola with horizontal axis. Then the value of the integral $\int y\,dx$ in such a region is given to a close approximation by the formula for the area of a parabolic segment, which is two-thirds of the area of the circumscribing rectangle.

Because of its simplicity, flexibility, and relatively high accuracy, Simpson's rule is the most useful of all the quadrature formulas.

Weddle's Rule

$$\int_a^b y\,dx = \int_a^b f(x)\,dx = \frac{3h}{10}[(y_0 + 5y_1 + y_2 + 6y_3 + y_4 + 5y_5 + y_6) + (y_6 + 5y_7 + y_8 + 6y_9 + y_{10} + 5y_{11} + y_{12}) + \cdots + (y_{n-6} + 5y_{n-5} + y_{n-4} + 6y_{n-3} + y_{n-2} + 5y_{n-1} + y_n)] \qquad (12.10)$$

Here n must be a multiple of 6, and, for the sake of clearness, the terms in brackets are written in parenthetical groups of seven ordinates each. If the terms were not grouped, the coefficients of $y_6, y_{12}, \ldots, y_{n-6}$ would be 2.

$$\text{Error in each group} = -\frac{h^7}{140} f^{\text{vi}}(x_m) = -\frac{h}{140}\Delta^6 y_k \quad \text{(approximately)}$$

where x_m denotes the mid-point of a group interval, and y_k denotes the first ordinate of a group.

Although Weddle's rule is more accurate, in general, than Simpson's, it has the disadvantage of requiring that an interval of integration be subdivided into 6, 12, 18, etc., subintervals of width h.

Euler's Quadrature Formula

$$\int_a^b y\,dx = h[\tfrac{1}{2}y_0 + y_1 + y_2 + \cdots + y_{n-1} + \tfrac{1}{2}y_n] - \frac{h^2}{12}[y_n' - y_0'] + \frac{h^4}{720}[y_n''' - y_0'''] - \cdots \quad (12.11)$$

The right-hand member of (12.11) is an asymptotic series in which the terms decrease for a time and then begin to increase. The error committed by stopping at any term is less then twice the first neglected term. Hence, the most accurate result is obtained by stopping with the term just before the smallest.

The first term in the right-hand member of (12.11) constitutes the trapezoidal rule, to which the further terms may be regarded as corrections. The first two bracketed terms will give a result more accurate than Simpson's rule for the same value of h, the error being only one-fourth as large as in Simpson's rule.

Gauss' Quadrature Formula

For the same number of functional values, Gauss' formula will give a more accurate result than any of the formulas presented up to this point. The subdivision points of the interval of integration are not equidistant, but they are symmetrically placed with respect to the mid-point of the interval.

To evaluate the integral $\int_a^b y\,dx$ by Gauss' formula, we first change the independent variable from x to u by the substitution

$$x = (b-a)u + \frac{a+b}{2} \quad (12.12)$$

Then $$y = f(x) = f\left[(b-a)u + \frac{a+b}{2}\right] = \phi(u) \quad \text{say} \quad (12.13)$$

and, since $dx = (b-a)\,du$, the given integral becomes

$$\int_a^b y\,dx = (b-a)\int_{-1/2}^{1/2} \phi(u)\,du \quad (12.14)$$

Gauss' formula for the integral in the right-hand member of (12.14) is

$$\int_{-1/2}^{1/2} \phi(u)\,du = R_1\phi(u_1) + R_2\phi(u_2) + \cdots + R_n\phi(u_n) \quad (12.15)$$

where $u_1, u_2, \ldots, u_n$ are the subdivision points in the interval $u = -\tfrac{1}{2}$ to $u = \tfrac{1}{2}$ and $R_1, R_2, \ldots, R_n$ are the corresponding weighting coefficients. The corresponding values of x are $x_1 = (b-a)u_1 + (a+b)/2$, $x_2 = (b-a)u_2 + (a+b)/2$, etc.

A formula for the correction term or error in Gauss' formula is

$$E_n = \frac{f^{(2n)}(\xi)}{(2n)!} \frac{(b-a)^{2n+1}}{(2n+1)(C_n{}^{2n})^2} \qquad a \leq \xi \leq b \tag{12.16}$$

where $C_n{}^{2n}$ denotes the combination of $2n$ things taken n at a time.

The factor $f^{(2n)}(\xi)$ in (12.16) shows that the error is zero if $f(x)$ is a polynomial of any degree up to and including $(2n-1)$.

The numerical values of the us and corresponding Rs for $n = 2$ to $n = 7$ are given in Table 12.3. More extensive tables can be found in [5] and [7].

Table 12.3

$n = 2$	$u_{\pm 1} = 0.2886751346$	$R = \frac{1}{2}$
$n = 3$	$u_0 = 0$	$R = \frac{4}{9}$
	$u_{\pm 1} = 0.3872983346$	$R = \frac{5}{18}$
$n = 4$	$u_{\pm 1} = 0.1699905218$	$R = 0.3260725774$
	$u_{\pm 2} = 0.4305681558$	$R = 0.1739274226$
$n = 5$	$u_0 = 0$	$R = 0.2844444444$
	$u_{\pm 1} = 0.2692346551$	$R = 0.2393143352$
	$u_{\pm 2} = 0.4530899230$	$R = 0.1184634425$
$n = 6$	$u_{\pm 1} = 0.1193095930$	$R = 0.2339569673$
	$u_{\pm 2} = 0.3306046932$	$R = 0.1803807865$
	$u_{\pm 3} = 0.4662347571$	$R = 0.08566224619$
$n = 7$	$u_0 = 0$	$R = 0.2089795918$
	$u_{\pm 1} = 0.2029225757$	$R = 0.1909150253$
	$u_{\pm 2} = 0.3707655928$	$R = 0.1398526957$
	$u_{\pm 3} = 0.4745539562$	$R = 0.06474248308$

Since the values of u are symmetrical with respect to the mid-point of the interval $u = -\frac{1}{2}$ to $u = \frac{1}{2}$, the pairs of numerically equal us are denoted by $u_{\pm}$, the plus subscript referring to the positive u, and the negative subscript to the negative u. The corresponding Rs apply to both values of u, and are always positive.

12.3. NUMERICAL SOLUTION OF ORDINARY DIFFERENTIAL EQUATIONS

The numerical solution of an ordinary differential equation consists of a table of numerical values of the function corresponding to assigned numerical values of the independent variable. Starting with given initial values x_0 and y_0, the solution is constructed step by step from that point onward as far as desired.

Several methods have been devised for finding additional values $y_1, y_2, y_3, \ldots$ corresponding to the values $x_1, x_2, x_3, \ldots$ of the independent variable, the most important of which will be outlined below. The first is a rough-and-ready method of limited accuracy, but has the advantage of being applicable to any differential equation of the first order.

a. The Average-slope Method of Solution

Any differential equation of the first order in the variables x and y can be written in the form

$$\frac{dy}{dx} = f(x,y) \tag{12.17}$$

The slope of the integral curve at the point (x_0,y_0) is therefore

$$\left(\frac{dy}{dx}\right)_0 = f(x_0,y_0)$$

If $x_1 = x_0 + h$, the corresponding value of y is approximately

$$y_1^{(1)} = y_0 + h\left(\frac{dy}{dx}\right)_0 \tag{12.18}$$

Then, by (12.17), the slope of the integral curve at (x_1,y_1) is approximately

$$\left(\frac{dy}{dx}\right)_1^{(1)} = f(x_1,y_1^{(1)})$$

A more accurate value of the increment in y is obtained by taking h times the *mean* of the slopes at (x_0,y_0) and (x_1,y_1), giving

$$y_1^{(2)} = y_0 + \frac{h}{2}\left[\left(\frac{dy}{dx}\right)_0 + \left(\frac{dy}{dx}\right)_1^{(1)}\right] \tag{12.19}$$

This $y_1^{(2)}$ is then substituted into (12.17) to get a better value of the slope at (x_1,y_1), which in turn will give a better value of y_1 by the relation

$$y_1^{(3)} = y_0 + \frac{h}{2}\left[\left(\frac{dy}{dx}\right)_0 + \left(\frac{dy}{dx}\right)_1^{(2)}\right] \tag{12.20}$$

The cycle of operations is continued until no improvement is made in the value of y_1. The step interval h must be short in this method of finding the successive ys. First approximations to $y_2, y_3 \ldots$ are found by means of the formula

$$y_{n+1}^{(1)} = y_{n-1} + 2h\left(\frac{dy}{dx}\right)_n \tag{12.21}$$

Then these first approximations are improved by the cycle of substitutions into (12.17) and the formula

$$y_{n+1}^{(2)} = y_n + \frac{h}{2}\left[\left(\frac{dy}{dx}\right)_n + \left(\frac{dy}{dx}\right)_{n+1}^{(1)}\right] \tag{12.22}$$

The reason for using (12.21) for finding first approximations after y_1 has been found is that (12.21) is more accurate than a formula analogous to (12.18).

The process outlined above is convergent when

$$\left|\frac{h}{2}\frac{\partial f(x,y)}{\partial y}\right| < 1$$

If the successive derivatives of (12.17) are easily found, the starting values y_1 and y_2, or preferably y_{-1} and y_1, for substitution in (12.21) should be found from Taylor's series, as explained in Art. 12.3*c*.

b. Methods Based on Interpolating Polynomials

Although the method of the preceding article is applicable to any differential equation of the first order in two variables, and will give results of slide-rule accuracy for several steps, it is neither accurate enough nor fast enough for general use in solving differential equations numerically. More accurate and more rapid methods will now be outlined.

Formulas in Terms of Differences. After four values of y, in addition to the initial y_0, have been found by one of the methods explained in Art. 12.3*c* the solution is continued by means of the following formulas:

$$\bar{y}_{n+1} = y_n + h(y_n' + \tfrac{1}{2}\Delta_1 y_n' + \tfrac{5}{12}\Delta_2 y_n' + \tfrac{3}{8}\Delta_3 y_n' + \tfrac{1}{3}\Delta_4 y_n') \tag{12.23a}$$

for finding a first approximation to the next y ahead;

$$y_{n+1} = y_n + h(\bar{y}'_{n+1} - \tfrac{1}{2}\Delta_1\bar{y}'_{n+1} - \tfrac{1}{12}\Delta_2\bar{y}'_{n+1} - \tfrac{1}{24}\Delta_3\bar{y}'_{n+1} - \tfrac{1}{38}\Delta_4\bar{y}'_{n+1}) \qquad (12.23b)$$

for checking and correcting the new y found by (12.23a).

Here y' denotes dy/dx and is found from (12.17). Note that the first approximation to the new y is written with a bar over it. Equation (12.23a) is used only for integrating ahead by extrapolation to find the first approximation to the new y. Equation (12.23b) is used in conjunction with (12.17) as many times as it will produce improvement in y_{n+1}.

The cycle of operations in computing y_{n+1} is therefore as follows:

1. Use (12.23a) to get the first approximation $\bar{y}_{n+1}$.
2. Substitute $\bar{y}_{n+1}$ into (12.17) to get the first approximation to y'_{n+1}. Call this $\bar{y}'_{n+1}$.
3. Substitute this $\bar{y}'_{n+1}$ into (12.23b) to check and improve the $\bar{y}_{n+1}$ found by (12.23a).
4. If the y_{n+1} found from (12.23b) is not the same as the $\bar{y}_{n+1}$ found from (12.23a), repeat steps 2 and 3 one or more times until no improvement is made in y_{n+1}.

After the final values of y_{n+1} and y'_{n+1} have been found, compute the differences $\Delta_1 y'$, $\Delta_2 y'$, etc., for that line of the table.

The interval h should be short enough to make $\tfrac{1}{3}\Delta_4 y'$ negligibly small.

Note that horizontal differences are used in (12.23a,b).

Another formula for integrating ahead by extrapolation is

$$\bar{y}_{n+1} = y_{n-1} + 2h[y'_n + \tfrac{1}{6}(\Delta_2 y'_n + \Delta_3 y'_n + \Delta_4 y'_n) - \tfrac{1}{180}\Delta_4 y'_n] \qquad (12.24)$$

This formula is simpler than (12.23a) and may be used in its place if desired. The term $-\tfrac{1}{180}\Delta_4 y'_n$ should be small enough to be neglected.

Formulas of Quadrature Type in Terms of Equidistant Functional Values. Use

$$\bar{y}_{n+1} = y_{n-3} + \frac{4h}{3}(2y'_{n-2} - y'_{n-1} + 2y'_n) \qquad (12.25a)$$

for integrating ahead to find a first approximation to y_{n+1}, and

$$y_{n+1} = y_{n-1} + \frac{h}{3}(y'_{n-1} + 4y'_n + \bar{y}'_{n+1}) \qquad (12.25b)$$

for checking and correcting the first approximation found by (12.25a). The $\bar{y}'_{n+1}$ in (12.25b) was found by substituting $\bar{y}_{n+1}$ into (12.17). Note that (12.25a) requires four consecutive ys as starting values.

Equations (12.25a,b) are used in exactly the same way as Eqs. (12.23a,b) or as (12.24) and (12.23b) after four starting values are available.

If y_{n+1} denotes the final value obtained by applications of (12.25b) and (12.17), then the principal part of the error in each step is

$$E = \frac{\bar{y}_{n+1} - y_{n+1}}{29} = \frac{D}{29} \quad \text{say} \qquad (12.26)$$

The computation should carry a column for D, so that the error in each step can be estimated at a glance. If the Ds become erratic at any stage of the computation, look for an arithmetical mistake. If E becomes large enough to affect, by more than one unit, the last digit to be retained in the computation, then the interval h should be reduced to half its value.

c. Methods of Starting the Solution

In the numerical solution of differential equations, it is of the utmost importance to have correct starting values. The first four or five values of y and y' should be determined accurately to the number of significant figures desired in the final results. Sufficiently accurate starting values can be found by one of the following methods:

1. If successive derivatives of $y' = f(x,y)$ can be easily found, the best and quickest method of starting a solution is to find y_{-2}, y_{-1}, y_1, y_2 from the Taylor series

$$y_k = y_0 + y_0'(kh) + \frac{y_0''(kh)^2}{2} + \frac{y_0'''(kh)^3}{3!} + \frac{y^{\text{iv}}(kh)^4}{4!} \qquad (12.27)$$

with $k = -2, -1, 1, 2$ in succession and using small values of h. Then, by substituting these ys into the given equation, the values of y_{-2}', y_{-1}', y_1', y_2' can be found.

2. If successive derivatives of $y' = f(x,y)$ cannot be easily found, but y'' can be found without difficulty, starting values can be found by the following procedure:

(a) Find first approximations to y_1' and y_{-1}' from the formulas

$$y_1' = y_0' + hy_0'' \qquad y_{-1}' = y_0' - hy_0'' \qquad (12.28)$$

(b) Substitute these approximations into the formulas

$$y_1 = y_0 + \frac{h}{24}(y_{-1}' + 16y_0' + 7y_1') + \frac{h^2y_0''}{4}$$
$$y_{-1} = y_0 - \frac{h}{24}(7y_{-1}' + 16y_0' + y_1') + \frac{h^2y_0''}{4} \qquad (12.29a,b)$$

to get approximations to y_1 and y_{-1}.

(c) Substitute these y_1 and y_{-1} into the given equation $y' = f(x,y)$ to get improved values of y_1' and y_{-1}'.

(d) Substitute these y_1' and y_{-1}' into (12.29a,b) to get improved values of y_1 and y_{-1}.

(e) Continue the iteration of steps c and d until no change is produced in y_1 and y_{-1}.

The following additional steps are required for finding y_2 and y_{-2}:

(f) Substitute the final y_1' and y_{-1}' into the formulas

$$y_2 = y_0 + \frac{2h}{3}(5y_1' - y_0' - y_{-1}') - 2h^2y_0''$$
$$y_{-2} = y_0 - \frac{2h}{3}(5y_{-1}' - y_0' - y_1') \quad 2h^2y_0'' \qquad (12.30a,b)$$

to get first approximations to y_2 and y_{-2}.

(g) Substitute these y_2 and y_{-2} into the given equation $y' = f(x,y)$ to get first approximations to y_2' and y_{-2}'.

(h) Substitute these y_2' and y_{-2}' into the formulas

$$y_2 = y_0 + \frac{h}{3}(y_0' + 4y_1' + y_2') \qquad y_{-2} = y_0 - \frac{h}{3}(y_{-2}' + 4y_{-1}' + y_0') \qquad (12.31a,b)$$

to check and improve the y_2 and y_{-2} found from (12.30a,b).

(i) If the y_2 and y_{-2} found from (12.31) do not agree with those found from (12.30), repeat the iterative steps g and h as long as any improvement is made in y_2 and y_{-2}.

3. If methods 1 and 2 cannot be used to obtain correct starting values, use the average-slope method of Art. 12.3a with small values of h.

d. Changing the Length of the Step Interval

Soon after the integration is started, it will usually be advisable to double the value of h that was used in obtaining the starting values. Also, if the second differences become small at any stage of the computation, it will be advisable to double the value of h.

To double the length of interval, when using Eqs. (12.23*a,b*) and (12.24), make a table of differences for intervals of width $2h$, and then proceed with the use of the formulas just mentioned, but with a doubled value of h.

If, at some stage of the computation, fourth differences of y' should become large, or if $D/29$ should become large enough to affect the last digit by more than one unit, then the value of h should be reduced to half its value. To do this, compute the value of y' at two consecutive mid-interval points by Eq. (12.6) expressed in terms of y', namely

$$y'_{\frac{1}{2}} = \frac{y'_0 + y'_1}{2} - \frac{1}{8}\,\frac{\Delta_2 y'_1 + \Delta_2 y'_2}{2} + \frac{3}{128}\,\frac{\Delta_4 y'_2 + \Delta_4 y'_3}{2} \qquad (12.32)$$

where $y'_{\frac{1}{2}}$ is the value of y' halfway between y'_0 and y'_1. This procedure will give five consecutive values of y' at distances $h/2$ apart.

Since changing the interval is equivalent to starting a new computation, the five consecutive values of y' must be accurate to the number of digits used throughout the computation.

Illustrative Example. Table 12.4 shows the result of the first seven steps in the numerical solution of

$$\frac{dy}{dx} + \cos y = x$$

with the initial conditions $x_0 = 0$, $y_0 = 0$.

The values of y_{-2}, y_{-1}, y_1, y_2 were found from Taylor's series, and the corresponding values of y' were found by substituting these ys into the given equation. The last two lines of the table were found by application of Eqs. (12.23*a,b*) in connection with the given equation.

The last two lines can also be found by means of Eqs. (12.25*a,b*).

Table 12.4

x	y	Δy	y'	$\Delta_1 y'$	$\Delta_2 y'$	$\Delta_3 y'$	$\Delta_4 y'$
−0.2	0.2185		−1.1762				
−0.1	0.1048	−0.1137	−1.0945	0.0817			
0.0	0.0000	−0.1048	−1.0000	0.0945	0.0128		
0.1	−0.0948	−0.0948	−0.8955	0.1045	0.0100	−0.0028	
0.2	−0.1789	−0.0841	−0.7840	0.1115	0.0070	−0.0030	−0.0002
0.3	−0.2515	−0.0726	−0.6685	0.1155	0.0040	−0.0030	0.0000
0.4	−0.3125	−0.0610	−0.5516	0.1169	0.0014	−0.0026	0.0004

e. Equations of the Second and Higher Orders

The General Equation of the Second Order. The substitution of $dy/dx = y'$ reduces a second-order equation of the form

$$\frac{d^2y}{dx^2} + P\frac{dy}{dx} + Qy = f(x)$$

to the two first-order equations

$$\frac{dy}{dx} = y' \qquad y'' = f(x) - Py' - Qy \tag{12.33}$$

A similar procedure will reduce an equation of any order to a system of first-order equations, and these can be solved numerically by the methods already given and by those soon to be given.

After five starting values have been found by Taylor's series or otherwise, the solution of Eqs. (12.33) is continued by the following means: Eq. (12.23*a*) is applied, replacing there y' by y'' and y by y'. It yields a trial value of y'_{n+1}. This value is entered in the difference table, and then Eq. (12.23*b*) is used for finding the first value of y_{n+1}. From Eq. (12.33) y''_{n+1} is found, and then Eq. (12.23*b*) is used to find improved values for y'_{n+1} and y_{n+1}. Repeated use of the same equation is needed until the result shows no further improvement.

A convenient and accurate check formula is

$$y_{n+1} = 2y_n - y_{n-1} + h^2(y''_n + 1/12\ \Delta_2 y''_{n+1}) \tag{12.34}$$

Eqs. (12.25*a*,*b*) may be used instead of Eqs. (12.23*a*,*b*).

Second-order Equations with First Derivative Absent. Many second-order differential equations are of the form

$$y'' = f(x,y) \tag{12.35}$$

and can be solved more simply and easily than those of the general type considered above.

After starting values have been found by the methods of Art. 12.3*c*, then the solution is continued by means of the formulas

$$\bar{y}_{n+1} = 2y_n - y_{n-1} + h^2[y''_n + 1/12(\Delta_2 y''_n + \Delta_3 y''_n + \Delta_4 y''_n)] \tag{12.36a}$$

for finding the first approximation to y_{n+1};

$$y_{n+1} = 2y_n - y_{n-1} + h^2(y''_n + 1/12\ \Delta_2 \bar{y}''_{n+1}) \tag{12.36b}$$

for checking and correcting the $\bar{y}_{n+1}$ found from (12.36*a*). The $\bar{y}''_{n+1}$ in (12.36*b*) was obtained by substituting $\bar{y}_{n+1}$ into (12.35).

Equations (12.35) and (12.36*a*,*b*) are applied in the following order: (12.36*a*), (12.35), (12.36*b*); (12.35), (12.36*b*); (12.35), (12.36*b*) until no improvement is made in y_{n+1}.

f. Systems of Simultaneous Equations

Systems of simultaneous differential equations can be solved numerically by the methods already outlined. After finding starting values by Taylor's series or otherwise, solve the given equations for the highest derivatives in terms of the variables and the lower derivatives, as was done in the first half of Art. 12.3*e*. Then, continue the solution by either of the methods outlined in Art. 12.3*b*. Further details and illustrative examples can be found in [4] and [7].

In systems of second-order equations in which first derivatives are absent, find the first approximation to each dependent variable at the next step ahead by means of (12.36*a*). Then substitute these approximate values into the given equations to get $\bar{y}''_{n+1}$, $\bar{z}''_{n+1}$, etc. Substitute these latter values into (12.36*b*) to check and correct the $\bar{y}_{n+1}$, $\bar{z}_{n+1}$, etc., found from (12.36*a*). Continue the use of (12.36*b*) and the given equations as long as they effect improvement in y_{n+1}, z_{n+1}, etc. Further details will be found in [4] and [7].

12.4. REFERENCES

[1] A. A. Bennett, W. E. Milne, H. Bateman: "Numerical Integration of Differential Equations," Washington, 1933; reprint, Dover, New York, 1956.
[2] L. Collatz: "Numerische Behandlung von Differentialgleichungen," 2d ed., Springer, Berlin, 1955.
[3] W. E. Milne: "Numerical Calculus," Princeton Univ. Press, Princeton, N.J., 1949.
[4] W. E. Milne: "Numerical Solution of Differential Equations," Wiley, New York, 1953.
[5] B. P. Moors: "Valeur approximative d'une intégrale définie," Gauthier-Villars, Paris, 1905.
[6] C. Runge, H. König: "Numerisches Rechnen," Springer, Berlin, 1924.
[7] J. B. Scarborough: "Numerical Mathematical Analysis," 3d ed., Johns Hopkins Press, Baltimore, 1955.
[8] E. T. Whittaker, G. Robinson: "The Calculus of Observations," Blackie, Glasgow, 1924.

CHAPTER 13

RELAXATION METHOD

BY

D. N. de G. ALLEN, Sheffield, England

Relaxation is a numerical method of analyzing engineering problems whose theoretical formulation presents, for solution, either a set of algebraic equations or one or more differential equations. Its most extensive application is to find particular solutions to linear differential equations, either ordinary (in one dimension) or partial (in two or three dimensions). Such a solution, obtained by relaxation, comprises a number of values (of the wanted functions or unknowns) situated at a finite number of points regularly distributed over the domain covered by the problem. Thus, in one dimension, values are obtained at points uniformly spaced along the range of integration of the differential equation; in two dimensions, values are found at the nodal

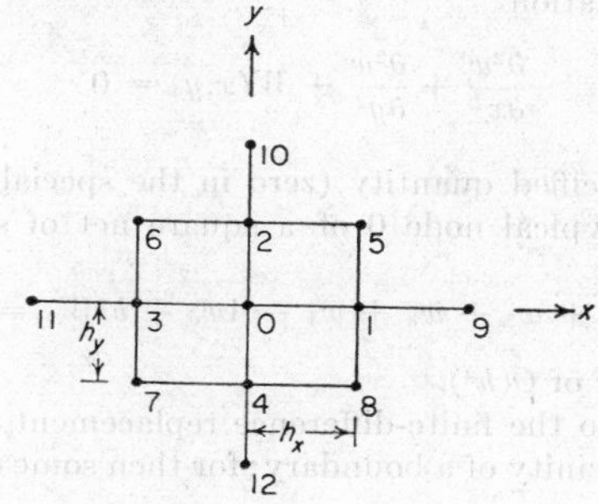

FIG. 13.1

points of a regular net, usually square but sometimes rectangular, covering the area of integration of the governing differential equation; in three dimensions, values are determined similarly at the intersections of a regular lattice, usually cubical but possibly rectangular.

To determine a unique, particular solution to a given differential equation requires that appropriate end or boundary conditions be also stated. The relaxation method utilizes finite differences, in order to approximate both to the differential equation and to the associated boundary conditions. In two dimensions, a typical group of points on a rectangular net are designated as shown in Fig. 13.1, the mesh lengths of the net being denoted by h_x, h_y in the x and y directions, respectively. We list here the simpler and more frequently required finite-difference replacements which approximate to derivatives of a dependent function w, at a typical node 0, in terms of values of w at 0 itself and at neighboring nodes; the order of the error in these approximations is also listed:

Finite-difference approximation — *Error*

$$h_x\left(\frac{\partial w}{\partial x}\right)_0 = \tfrac{1}{2}(w_1 - w_3) \qquad O(h^3)$$

$$h_y\left(\frac{\partial w}{\partial y}\right)_0 = \tfrac{1}{2}(w_2 - w_4) \qquad O(h^3) \qquad (13.1a,b)$$

$$h_x^2\left(\frac{\partial^2 w}{\partial x^2}\right)_0 = w_1 + w_3 - 2w_0 \qquad O(h^4)$$

$$h_y^2\left(\frac{\partial^2 w}{\partial y^2}\right)_0 = w_2 + w_4 - 2w_0 \qquad O(h^4) \qquad (13.2a\text{–}c)$$

$$h_x h_y\left(\frac{\partial^2 w}{\partial x\,\partial y}\right)_0 = \tfrac{1}{4}(w_5 - w_6 + w_7 - w_8) \qquad O(h^4)$$

$$h_x^3\left(\frac{\partial^3 w}{\partial x^3}\right)_0 = \tfrac{1}{2}(w_9 - 2w_1 + 2w_3 - w_{11}) \qquad O(h^5)$$

$$h_y^3\left(\frac{\partial^3 w}{\partial y^3}\right)_0 = \tfrac{1}{2}(w_{10} - 2w_2 + 2w_4 - w_{12}) \qquad O(h^5) \qquad (13.3a,b)$$

$$h_x^4\left(\frac{\partial^4 w}{\partial x^4}\right)_0 = w_9 - 4w_1 + 6w_0 - 4w_3 + w_{11} \qquad O(h^6)$$

$$h_y^4\left(\frac{\partial^4 w}{\partial y^4}\right)_0 = w_{10} - 4w_2 + 6w_0 - 4w_4 + w_{12} \qquad O(h^6) \qquad (13.4a\text{–}c)$$

$$h_x^2 h_y^2\left(\frac{\partial^4 w}{\partial x^2\,\partial y^2}\right)_0 = w_5 + w_6 + w_7 + w_8 + 4w_0 - 2(w_1 + w_2 + w_3 + w_4) \qquad O(h^6)$$

These formulas enable finite-difference approximations to be written down for most linear differential equations which arise in engineering mechanics. Thus, as an illustration, Poisson's equation

$$\frac{\partial^2 w}{\partial x^2} + \frac{\partial^2 w}{\partial y^2} + W(x,y) = 0$$

in which $W(x,y)$ is a specified quantity (zero in the special case of Laplace's equation), is replaced at a typical node 0 of a square net of side $h_x = h_y = h$ by the approximation

$$w_1 + w_2 + w_3 + w_4 - 4w_0 + h^2 W_0 = 0 \tag{13.5}$$

in which there is an error of $O(h^4)$.

Special adjustments to the finite-difference replacement must be made when the typical node 0 is in the vicinity of a boundary; for then some of the neighboring nodes,

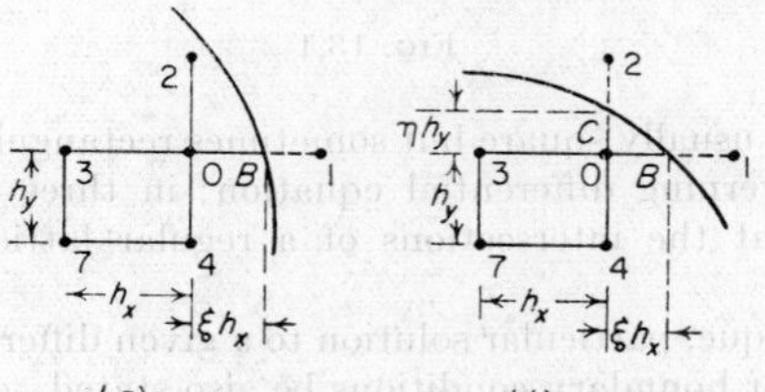

FIG. 13.2

1, 2, 3, etc., may lie outside the domain of the problem. In such a case, those nodes are described as fictitious, and no values of the unknown function w really exist at them. However, corresponding fictitious values of w may appear in the finite-difference approximation to the differential equation at nodes near to but within the boundary. Figure 13.2 illustrates two possibilities: (a) where for a real node, 0, one of the neighboring nodes, 1, is fictitious, the mesh line 01 of the net cutting across the boundary, at the point B where $0B = \xi h_x$, (b) where two of the neighboring nodes, 1 and 2, are fictitious, the mesh line 02 also cutting across the boundary, at a point C where $0C = \eta h_y$. These two illustrations sufficiently exemplify the kinds of situation which may arise in the neighborhood of a real node 0 near to the boundary. Provided a sufficiently fine net is being used, the situation should never arise that more than two of the mesh lines radiating from any node 0 cut across the boundary.

In order to ensure satisfaction of the finite-difference approximation at the node 0, some means must be provided of attaching values to w at the fictitious nodes such as 1 in Fig. 13.2a, 1 and 2 in Fig. 13.2b—for these fictitious values enter into the finite-difference relation at 0. Formulas for so attaching values to w are derived from the boundary conditions (which must be available to ensure a solution) holding at the point B in Fig. 13.2a, the points B and C in Fig. 13.2b. We now list these formulas for the most frequently met types of boundary condition:

Boundary Values of w Specified

Suppose w is specified at B (Fig. 13.2a) to be equal to K_B. Then

$$w_1 = \frac{1-\xi}{1+\xi} w_3 - \frac{2(1-\xi)}{\xi} w_0 + \frac{2K_B}{\xi(1+\xi)} \tag{13.6}$$

If, in addition (Fig. 13.2b), w is specified at C to be equal to K_C, then

$$w_2 = \frac{1-\eta}{1+\eta} w_4 - \frac{2(1-\eta)}{\eta} w_0 + \frac{2K_C}{\eta(1+\eta)} \tag{13.7}$$

The above two formulas can be used together, but are independent of each other.

Normal Gradients of w Specified

Suppose $\partial w/\partial \nu$ is specified at B (Fig. 13.2a) to be equal to k_B (ν denoting the direction of the outward-drawn normal). Then

$$w_1 = \frac{1 - 2\xi[1-(h_x/h_y)\tan\beta]}{1+2\xi} w_3 + \frac{2\xi[2-(h_x/h_y)\tan\beta]}{1+2\xi} w_0 + \frac{2k_B h_x \sec\beta}{1+2\xi} - \frac{(h_x/h_y)\tan\beta}{1+2\xi} w_2 + \left(\frac{h_x}{h_y}\tan\beta\right) w_4 - \frac{2\xi(h_x/h_y)\tan\beta}{1+2\xi} w_7 \tag{13.8}$$

where β is the angle between the normal (ν) direction at B and the mesh line 01.

If, in addition (Fig. 13.2b), $\partial w/\partial \nu$ is specified at C to be equal to k_C, then

$$w_2 = \frac{1 - 2\eta[1-(h_y/h_x)\tan\gamma]}{1+2\eta} w_4 + \frac{2\eta[2-(h_y/h_x)\tan\gamma]}{1+2\eta} w_0 + \frac{2k_C h_y \sec\gamma}{1+2\eta} - \frac{(h_y/h_x)\tan\gamma}{1+2\eta} w_1 + \left(\frac{h_y}{h_x}\tan\gamma\right) w_3 - \frac{2\eta(h_y/h_x)\tan\gamma}{1+2\eta} w_7 \tag{13.9}$$

where γ is the angle between the normal (ν) direction at C and the mesh line 02.

The above two formulas cannot both be used independently of each other, since the first expresses w_1 in a form involving w_2 and, vice versa, the second expresses w_2 in a form involving w_1; when both formulas are needed (i.e., when both points 1 and 2 are fictitious as in Fig. 13.2b), then these formulas must be solved as a simultaneous pair for the two fictitious values w_1 and w_2.

Boundary Values and Normal Gradients of w Specified

In the solution of second-order differential equations, only one boundary condition can be specified, but, in the solution of a fourth-order equation (e.g., the biharmonic equation), two conditions must be specified. Suppose (Fig. 13.2a) that, at B,

w is given to be equal to K_B, and that $\partial w/\partial \nu$ is also specified. It is clear that then $\partial w/\partial x$ at B is able to be determined (since knowledge of w *along* the boundary implies knowledge of $\partial w/\partial s$, s being the direction of the tangent at B); suppose, then, that $\partial w/\partial x$ is found to be equal to k_B. In these circumstances the two boundary conditions between them yield two formulas, one expressing the fictitious value w_1, the other expressing the real value w_0. *For fourth-order equation points like* 0 *in Fig.* 13.2, *less than one mesh length distant from, but within, the boundary, are treated in the same manner as the fictitious points.* The formulas are:

$$\begin{aligned} w_1 &= \left(\frac{1-\xi}{1+\xi}\right)^2 w_3 + \frac{4\xi K_B}{(1+\xi)^2} + 2\frac{1-\xi}{1+\xi}k_B h_x \\ w_0 &= \left(\frac{\xi}{1+\xi}\right)^2 w_3 + \frac{(1+2\xi)K_B}{(1+\xi)^2} - \frac{\xi}{1+\xi}k_B h_x \end{aligned} \qquad (13.10a,b)$$

If, in addition (Fig. 13.2b), w is specified at C to be equal to K_C, and $\partial w/\partial y$ at C is found to be equal to k_C, then

$$\begin{aligned} w_2 &= \left(\frac{1-\eta}{1+\eta}\right)^2 w_4 + \frac{4\eta K_C}{(1+\eta)^2} + 2\frac{1-\eta}{1+\eta}k_C h_y \\ w_0 &= \left(\frac{\eta}{1+\eta}\right)^2 w_4 + \frac{(1+2\eta)K_C}{(1+\eta)^2} - \frac{\eta}{1+\eta}k_C h_y \end{aligned} \qquad (13.11a,b)$$

When (Fig. 13.2b) both points 1 and 2 are fictitious, we thus derive two alternative formulas expressing w_0. In the computation, if these formulas lead to differing numerical values for w_0, their mean result is taken. In practice, and particularly as the calculations approach convergence, the two values for w_0 should not differ significantly.

It may be pointed out here that the circumstances requiring special formulas, in order to attach values of w to fictitious points in a problem in two dimensions, have no parallel in the solution of a (one-dimensional) ordinary differential equation. For in one dimension the mesh length h can always be chosen (and should be chosen) so that the whole range of integration comprises a whole number of complete meshes.

The relaxation computations proceed, once the several finite-difference formulas have been prepared in the manner described above. We begin by making a guess (preferably, and when possible, an inspired guess) as to the answer, by attaching numerical values to w at all nodes of the net within the boundary (except, in the case of a fourth-order equation, at those nodes less than one mesh length within, which, as already remarked, are treated in the same manner as the fictitious nodes). We then test, by the trial of direct substitution, the worth of this guess. As an illustration, we have already shown that, in any solution of Poisson's equation on a square net, the finite-difference relation (13.5) should be satisfied by the w values relative to any typical node 0. At each node, in turn, the guessed values of w are inserted in the expression

$$F_0 \equiv w_1 + w_2 + w_3 + w_4 - 4w_0 + h^2 W_0 \qquad (13.12)$$

and so, node by node, quantities F_0 are calculated. At nodes in the immediate vicinity of the boundary, the expression for F_0 will have been modified by prior algebraic substitution of formulas which eliminate any fictitious w values from the general expression. If the guessed values of w are, in fact, those of the solution sought, then the satisfaction of the finite-difference approximation to the governing differential equation will be demonstrated by the evanescence of every F_0. More usually, the F values will be found not to be zero, and their magnitude is a measure of the incorrectness of the guess as to the w values. The quantities F_0 are known as residual errors or, more shortly, as *residuals*. The aim of the relaxation process is systematically to apply corrections to the guessed w values, with the purpose gradually, but eventually, of persuading all the residuals (one at each node, corresponding there to an originally guessed w value) to assume zero or negligible proportions.

This aim is achieved by systematic and repeated use of the computational operation of adding appropriate increments to the individual w values. The effect on the residuals of a unit increment, added to any typical w value, can be recorded in a relaxation pattern. For Poisson's equation, it is easily seen, by inspection of the general formula which defines the residuals' values, that an increment $\Delta w_0 = 1$, applied at any node, changes the residual *at that node* by -4, the residuals at the four immediately surrounding nodes each by $+1$. The pattern of these changes is displayed in Fig. 13.3. It should, of course, be pointed out that the numerical elements in this pattern suffer modification at nodes in the vicinity of the boundary, where the effect of the elimination of fictitious values from the residual formulas causes local individual variations.

The computation is carried out on a large plan of the field of integration, drawn to such a scale that the mesh sides of the net are not less than 1 in. in both x and y directions. At each node, the original w value is recorded to the left, and the corresponding calculated F value to the right. At any node, the value of the residual indicates the magnitude of the increment Δw_0 which should be added to the original w value in order to reduce that residual to insignificance. Thus, with the pattern (Fig. 13.3) corresponding to Poisson's equation, a residual of 45 at any node is reduced to 1 by the addition of an increment to w there of $+11$. At the same time, the pattern shows that the residuals at the four neighboring nodes are each increased by 11.

	+1	
+1	−4	+1
	+1	

Fig. 13.3

	+1	+1	+1	
+1	−3	−2	−3	+1
	+1	+1	+1	

Fig. 13.4

The calculation should record, node by node, all w increments which the computer applies together with the running totals, node by node, of the residuals. In general, the computer should concentrate his attention successively on the largest residuals, relaxing them to insignificant values, one at a time, by repeated use of his operational tool, the appropriate relaxation pattern. He should remember that any residual which has once been made small may become significant again later, by virtue of operations performed at neighboring nodes. Thus, a series of w increments may, in time, come to be applied at the same node.

With experience, and only with experience, the computer will learn tricks of the trade which shorten his labors—mathematically, which accelerate the convergence. He will realize, for instance, that a residual at one node may sometimes better be overrelaxed rather than just relaxed, in anticipation of residual error being, in part, reflected back to that node when neighboring nodes are dealt with (overrelaxation consists, of course, in applying w increments larger than the pattern would seem to demand). The computer also learns to construct block patterns so that, in one operation, he can apply w increments (of uniform magnitude) simultaneously at a group of nodes. Figure 13.4 displays a simple block pattern for Poisson's equation, showing the effect on the residuals' values of a unit increment applied to the w values at a group of three collinear nodes. These patterns are constructed by superposition of a corresponding number of basic point patterns, each of the form of that in Fig. 13.3. Any device or trick which he can contrive is justified by its success, and individual computers can be expected to develop their own techniques in much the same way that any craftsman develops his own personal methods.

There remains to mention briefly the question of the accuracy of the results of a relaxation calculation. On a chosen net, the relaxation can be regarded as completed only when the final w values yield residuals which, on checking, are so small that they could not be made any smaller without increasing the number of significant figures in the w values. But the resulting answer may be inaccurate by virtue of the inherent error involved in replacing derivatives by their finite-difference approximations. The only satisfactory test of whether a solution found on a net of side h is sufficiently accurate for the purpose for which that solution is required is to find the solution again on a finer net of side, say $h/2$. (The first solution enables a good guess to be made of the second, and so much extra work should not be involved.) A comparison of the two solutions provides evidence as to the effect of the finite-difference approximations. It may indicate that the net of side h was, in fact, fine enough; or it may suggest that a solution is desirable on yet a finer net, of side $h/4$. Once again, experience is the best asset in choosing the size of net for a particular problem: It should be just small enough to provide the requisite accuracy with minimum time and labor of computation involved.

REFERENCES

[1] R. V. Southwell: "Relaxation Methods in Theoretical Physics," vol. 1, Oxford Univ. Press, New York, 1946.

[2] D. N. de G. Allen: "Relaxation Methods," McGraw-Hill, New York, 1954.

[3] R. V. Southwell: "Relaxation Methods in Theoretical Physics," vol. 2, Oxford Univ. Press, New York, 1956.

CHAPTER 14

NUMERICAL SOLUTION OF HYPERBOLIC EQUATIONS

BY

G. G. O'BRIEN, Ph.D., New York

14.1. INTRODUCTION

Equations of the form

$$A\frac{\partial^2 T}{\partial x^2} + B\frac{\partial^2 T}{\partial x\,\partial t} + C\frac{\partial^2 T}{\partial t^2} + D\frac{\partial T}{\partial x} + E\frac{\partial T}{\partial t} + FT = G$$

where T is the dependent variable and x and t are independent variables, and where A, B, C, D, E, F, and G are functions of x and t, represent a very important class of linear partial differential equations of the second order. If the relation

$$B^2 - 4AC > 0$$

is satisfied, the equations are called hyperbolic, and physically represent various types of wave motion.

While the next few pages are devoted chiefly to the numerical solution of hyperbolic equations, the methods are equally applicable to many other types of partial differential equations.

14.2. DIFFERENCES

A partial derivative of a function $T(x,t)$ is defined as

$$T_x \equiv \frac{\partial T(x,t)}{\partial x} \equiv \lim_{h\to 0}\frac{T(x+h,\,t) - T(x,t)}{h} \tag{14.1}$$

In the numerical solution of differential equations, the differential quotients are approximated by the quotients of finite differences used for their definition. There are various ways of approximating the same derivative, e.g.,

$$T_x \approx \frac{1}{h}[T(x+h,\,t) - T(x,t)] \approx \frac{1}{h}[T(x,t) - T(x-h,\,t)]$$

$$\approx \frac{1}{2h}[T(x+h,\,t) - T(x-h,\,t)]$$

or $$T_{xx} \approx \frac{1}{h^2}[T(x+h,\,t) - 2T(x,t) + T(x-h,\,t)]$$

$$\approx \frac{1}{3h^2}[T(x+2h,\,t) - T(x+h,\,t) - T(x-h,\,t) + T(x-2h,\,t)]$$

Similarly, differences can be taken with respect to t, using the same increment, h, or a different one, k. In like manner, one way of representing T_{xt} by differences is

$$T_{xt} \approx \frac{1}{hk}[T(x+h,\,t+k) - T(x,\,t+k) - T(x+h,\,t) + T(x,t)]$$

The judicious choice between these finite-difference approximations of the partial derivatives is one of the major problems of the numerical solution of a partial differential equation.

Consider now the simple hyperbolic equation

$$\frac{\partial^2 T}{\partial t^2} = \frac{\partial^2 T}{\partial x^2} \tag{14.2}$$

where $T = T(x,t)$. This equation may be approximated by

$$\frac{1}{k^2}[U(x, t + k) - 2U(x,t) + U(x, t - k)] = \frac{1}{h^2}[U(x + h, t) - 2U(x,t) + U(x - h, t)] \tag{14.3}$$

It is often convenient to rearrange the terms of (14.3):

$$U(x, t + k) = 2U(x,t) - U(x, t - k) + \frac{k^2}{h^2}[U(x + h, t) - 2U(x,t) + U(x - h, t)] \tag{14.4}$$

Difference quotients can be represented pictorially by the lattice points of rectangular cartesian coordinates. Let the x,t plane be marked off by the parallel lines $x = mh$ and the parallel lines $t = nk$, as shown in Fig. 14.1. Any mesh point (x,t) has a

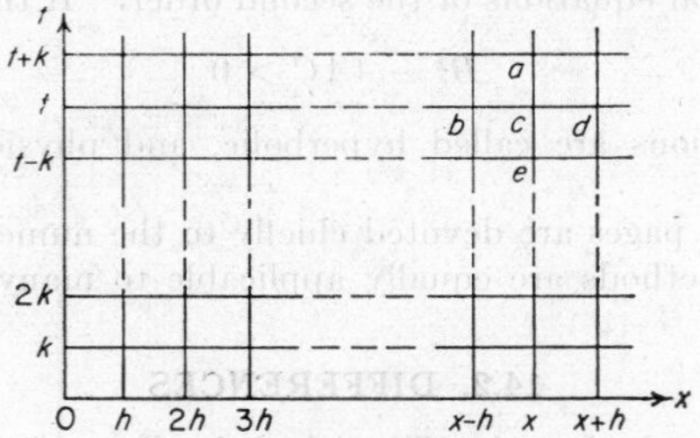

FIG. 14.1

value associated with it, namely, $T(x,t)$, the magnitude of which can be determined from the value of the function at certain points nearby. The particular points to be used, and the manner in which we use them, depend upon the difference equation. For example, if Eq. (14.2) is approximated by the difference equation (14.4), then the value of the function at point a (Fig. 14.1) can be computed from the values of the function at points b, c, d, and e. Thus, if the values of the function are known at each point on the first two horizontal lines, and along the two vertical boundaries, then the values can be computed, line by line, until they are known at every mesh point.

14.3. THE NUMERICAL SOLUTION

The numerical solution (or numerical approximation to the solution) is accomplished by means of applying directly the partial difference equation with the accompanying initial and boundary conditions. Thus, in explicit difference equations, such as Eq. (14.4), the solution is determined on the mesh points, line by line in the time direction, and with the aid of a high-speed computer if the solution is to be carried very far. After the values of the numerical solution are determined on the mesh points of a given line, the formula then uses these values to aid in computing the solution on the mesh points of the succeeding line. The larger the values of h and k, the less computing is necessary to determine the solution at a given point (x,t).

However, the larger we choose h and k, the greater the error (usually) in our result.

What is required in a numerical solution $u(mh,nk)$ is that

$$|u(mh,nk) - T(mh,nk)| \leqq \epsilon$$

for all mesh points (mh,nk) within our region of interest, where ϵ is an acceptable error, and $T(mh,nk)$ is the exact solution of the differential equation. In fact, we wish to satisfy

$$|u(x,t) - T(x,t)| \leqq \epsilon$$

for all points within this region, where $u(x,t)$ is determined by interpolation if it is not a mesh point.

14.4. ERROR ANALYSIS

When a partial differential equation is approximated by a difference equation, there is introduced an error called the *truncation error*. When the difference equation is solved by numerical methods there are additional errors (assuming faultless computing) due to round-offs; the accumulation of these is called the *numerical error*. When thousands of numerical steps are necessary for the solution of a problem, these errors may accumulate to such a magnitude that the results are worthless. To find the conditions under which the truncation error can be satisfactorily controlled is the problem of *convergence;* to find the conditions under which the numerical errors remain sufficiently small is the problem of *stability*. Unless the numerical representation is both convergent and stable, the results derived from its use are generally not good approximations to the true values.

14.5. THE TRUNCATION ERROR

The magnitude of the error involved in replacing a derivative by a difference can be determined by means of a Taylor expansion. To illustrate, consider the approximation

$$\frac{\partial^2 U(x,t)}{\partial x^2} \approx \frac{1}{h^2}[U(x+h,\,t) - 2U(x,t) + U(x-h,\,t)] \tag{14.5}$$

$$U(x \pm h,\,t) = U(x,t) \pm \frac{\partial U(x,t)}{\partial x}\frac{h}{1!} + \frac{\partial^2 U(x,t)}{\partial x^2}\frac{h^2}{2!} \pm \frac{\partial^3 U(x,t)}{\partial x^3}\frac{h^3}{3!} + \frac{\partial^4 U(x,t)}{\partial x^4}\frac{h^4}{4!} + \cdots \tag{14.6}$$

Substitute the values from (14.6) into (14.5) and simplify:

$$\frac{\partial^2 U(x,t)}{\partial x^2} \approx \frac{\partial^2 U(x,t)}{\partial x^2} + \frac{\partial^4 U(x,t)}{\partial x^4}\frac{h^2}{12} + O(h^4) \tag{14.7}$$

The approximation (14.7) indicates that the truncation error involved in (14.5) is of the order of magnitude of h^2. The term $O(h^4)$ indicates the presence of additional terms in h of the fourth or higher power. It is assumed that h is chosen small enough to ensure the convergence of the series.

We have already seen that a derivative may be represented by more than one type of difference. An analysis of the nature of the one above will demonstrate that the magnitude of the truncation error will vary considerably according to the difference representation chosen. However, what is gained in choosing a more rapidly converging difference expression may be more than lost in having a more complicated numerical iteration to perform, or it may be that the difference equation is not stable, in spite of the relatively small truncation error. Some higher-powered formulas are troublesome to use, because they lead to implicit equations, or because their point pattern presents difficulties near the boundaries because they demand the values of points outside the boundaries.

14.6. THE NUMERICAL ERROR

It bears repeating that a numerical solution based upon a difference equation that will converge to the differential equation, that is, such that the difference between the differential and difference equations can be made arbitrarily small by choosing h and k sufficiently small, is only part of the solution to the numerical problem. No matter how small this difference may be, round-off errors, introduced at each step in the iteration, and accumulated and modified with each succeeding step, may yield a result far from the true value. To control this error is the problem of stability.

It is easier to study the behavior of a single line of errors than to study the accumulation and compounding of errors that occur as the numerical iteration is continued. For the wave equations, as well as the heat-flow and other equations, the use of an explicit difference equation satisfying the simpler criterion ensures uniform boundedness in the solution of the equation. Hence, we use the term *stability* in this sense, and proceed to a method of determining the conditions under which stability will be realized.

When dealing with an implicit equation, the appropriate assumption may depend upon the method employed to solve the system of equations. If a relaxational solution is used, then we might assume that the residuals are uniformally small.

14.7. VON NEUMANN'S METHOD

The method of von Neumann can easily be explained and understood by applying it to particular equations. Hence, we shall proceed to determine the conditions under which the simplest heat equation and wave equation will be stable.

14.8. STABILITY OF THE HEAT EQUATION

Let

$$\frac{\partial T}{\partial t} = \frac{\partial^2 T}{\partial x^2} \tag{14.8}$$

represent heat flow along a bar (one direction), where $T = T(x,t)$ represents the temperature, t the time, and x the distance measured along the bar. Let (14.8) be approximated by

$$U(x,\, t + k) = U(x,t) + r[U(x + h,\, t) - 2U(x,t) + U(x - h,\, t)] \tag{14.9}$$

where k and h are the time and space increments, respectively, and the ratio $r = k/h^2$. Let l be the length of the domain, and let $Mh = l$.

Let $U(x,t)$ be the exact solution of a set of finite-difference equations (14.9). Owing to accumulated round-off errors, the values actually used in calculating are $u(x,t) = U(x,t) + \epsilon(x,t)$, where $\epsilon(x,t)$ is the accumulated error. Under the assumption that no additional error is introduced when using Eq. (14.9) to calculate $U(x,\, t + k)$, then a value $u(x,\, t + k) = U(x,\, t + k) + \epsilon(x,\, t + k)$ will be found which contains the error

$$\epsilon(x,\, t + k) = \epsilon(x,t) + r[\epsilon(x + h,\, t) - 2\epsilon(x,t) + \epsilon(x - h,\, t)] \tag{14.10}$$

The form of (14.10) is the same as that of (14.9) because of the fact that (14.9) is linear.

The errors (round-offs) $\epsilon(x,0)$ at each of the $(M - 1)$ interior mesh points on the line $t = 0$ are arbitrary quantities and small. They may be represented by a Fourier series of complex exponentials of the form

$$\epsilon(x,0) = \sum_{n=1}^{M-1} A_n e^{ib_n x} \tag{14.11}$$

It is easily seen that Eq. (14.10) is satisfied by

$$\epsilon(x,t) = \sum_{n=1}^{M-1} A_n e^{a_n t} e^{ib_n x} \qquad \text{with} \qquad a_n = \frac{1}{k} \ln \left(1 - 4r \sin^2 \frac{b_n h}{2}\right) \tag{14.12}$$

and that this solution reduces to (14.11) when $t = 0$. Any term in the sum (14.12) is a solution of (14.10); let such a term be represented by

$$Ae^{at}e^{ibx}$$

The value of b depends on M and on the length of the rod l. Here $a = a(b)$ is generally complex. The ratio of error growth for one integration step is e^{ak}, so that a sufficient condition for stability is

$$|e^{ak}| = \left| 1 - 4r \sin^2 \frac{bh}{2} \right| \leqq 1 \tag{14.13}$$

This requires that

$$-1 \leqq 1 - 4r \sin^2 \frac{bh}{2} \leqq 1 \tag{14.14}$$

The right side of the inequality is evidently true; the left inequality holds for all b if and only if $r \leqq \frac{1}{2}$. This must be made to hold for all b because, in a numerical solution, all frequencies are possible, owing to the small errors always present. Thus, in this problem, the value $r = \frac{1}{2}$ separates the region of stability, where errors damp out, from the region of instability, where some errors grow. Thus the *ratio* r is the determining factor in the stability of a difference representation.

14.9. STABILITY OF THE WAVE EQUATION

In a similar manner we shall obtain the conditions determining stability in the hyperbolic equation (14.2), using the difference equation (14.4) to approximate it.

Replacing $U(x,t)$ by $e^{at}e^{ibx}$, setting $r \equiv k/h$ and $y \equiv e^{ak}$, we get

$$y + \frac{1}{y} - 2 + 4r^2 \sin^2 \frac{bh}{2} = 0 \tag{14.15}$$

or

$$y^2 - 2Hy + 1 = 0 \qquad \text{where} \qquad H = 1 - 2r^2 \sin^2 \frac{bh}{2}$$

The quadratic has solutions

$$y_1 = H + \sqrt{H^2 - 1} \qquad y_2 = H - \sqrt{H^2 - 1} = \frac{1}{y_1} \tag{14.16}$$

If $H > 1$, $|y_1| > 1$ and if $H < -1$, $|y_2| > 1$. However, if $|H| \leqq 1$, then $|y_1| = |y_2| = 1$. Thus for stability,

$$-1 \leqq 1 - 2r^2 \sin^2 \frac{bh}{2} \leqq 1$$

The right side of the inequality is always satisfied; the left side is satisfied for all b if and only if $r \leqq 1$. Thus, in the wave equation being considered, we find that $k/h \leqq 1$ ensures stability.

14.10. SOME COMMENTS ON THE STABILITY CRITERIA

If the differential equation had not been linear, or if the coefficients had not been constants, we would still get a linear error equation but with variable coefficients. The method employed to determine stability assumes constant coefficients. This difficulty can be overcome, laboriously, by applying the method successively to a sequence of small overlapping regions, the regions being taken so small that we may, with negligible error, assume the coefficients of the error equation to be constant throughout the region.

If the differential equation contains terms of lower order, then the same conditions for stability will hold as before. For example, the equation

$$\frac{\partial T}{\partial t} = \frac{\partial^2 T}{\partial x^2} + p\frac{\partial T}{\partial x} + qT$$

where p and q are constants, will be stable under the same conditions as Eq. (14.8).

The values $r \equiv k/h^2 \leqq \frac{1}{2}$ in Eq. (14.9) and $r \equiv k/h \leqq 1$ in (14.4) are also necessary and sufficient to ensure convergence, except under very special initial conditions where convergence can be obtained for all values of r.

14.11. AN UNSTABLE DIFFERENCE EQUATION

Let the heat equation (14.8) be represented by a slightly different difference equation:

$$\frac{1}{2k}[U(x, t+k) - U(x, t-k)] = \frac{1}{h^2}[U(x+h, t) - 2U(x,t) + U(x-h, t)] \qquad (14.17)$$

Again let the error be $e^{at}e^{ibx}$, and the ratio $r \equiv k/h^2$. Then substitution into the variational equation gives

$$e^{ak} - e^{-ak} = 2r(e^{ibh} - 2 + e^{-ibh}) \qquad (14.18)$$

Let $y = e^{ak}$, and simplify (14.18) to the form

$$y^2 + \left(8r\sin^2\frac{bh}{2}\right)y - 1 = 0$$

then $$y = -4r\sin^2\frac{bh}{2} \pm \sqrt{16r^2\sin^4\frac{bh}{2} + 1} \qquad (14.19)$$

For stability we must have $-1 \leqq y \leqq 1$ for all values of b. Three of the four inequalities are satisfied for all b; the fourth,

$$-1 \leqq -4r\sin^2\frac{bh}{2} - \sqrt{16r^2\sin^4\frac{bh}{2} + 1}$$

is generally false. Thus stability is impossible, regardless of the choice of r.

This is surprising, in view of the fact that (14.17) is so much like the difference equation previously considered, where stability depended on choosing $r \leqq \frac{1}{2}$. It is more surprising still if we determine the truncation error in each case, for the truncation error in (14.17) is the smaller of the two.

14.12. IMPLICIT DIFFERENCE EQUATIONS

To satisfy the stability criterion of the difference equation being used will often force us to choose k, the time increment, inconveniently small. This factor becomes less important as our access to electronic digital computers of greater speed increases, but it still may often remain a serious problem. This problem can be avoided by the use of implicit difference equations, for which stability exists for all values of r.

We may replace Eq. (14.8) by the implicit relation

$$U(x, t+k) - U(x,t) = \frac{r}{2}[U(x+h, t+k) - 2U(x, t+k) + U(x-h, t+k) + U(x+h, t) - 2U(x,t) + U(x-h, t)] \qquad (14.20)$$

As before, let $U(x,t)$ be replaced by e^{at+ibx}, and let $y = e^{ak}$; then (14.20) becomes, when simplified,

$$y = 1 - \frac{r}{2}\left[(y+1)4\sin^2\frac{bh}{2}\right]$$

Thus

$$y = \frac{1 - 2r\sin^2(bh/2)}{1 + 2r\sin^2(bh/2)}$$

and is seen to satisfy $|y| \leqq 1$ for all b, regardless of the value of r.

Similarly, the wave equation (14.2) may be replaced by the implicit equation

$$\begin{aligned} U(x, t+k) - 2U(x,t) + U(x, t-k) &= r^2\{c[U(x+h, t+k) - 2U(x, t+k) \\ &\quad + U(x-h, t+k)] + (1-2c)[U(x+h, t) - 2U(x,t) + U(x-h, t)] \\ &\quad + c[U(x+h, t-k) - 2U(x, t-k) + U(x-h, t-k)]\} \end{aligned} \tag{14.21}$$

where $c \geqq 0$ is a real parameter and at our disposal.

Again let $y = e^{ak}$, and let $H = r^2 \sin^2(bh/2)$, to obtain, upon simplification,

$$y + \frac{1}{y} = 2 - \frac{4H}{1 + 4cH}$$

The stipulation that $|y| \leqq 1$ leads to the inequality

$$-1 \leqq 1 - \frac{2H}{1 + 4cH} \leqq 1$$

The right inequality is evidently true for all permissible c and H. The left inequality depends upon c for its form:

$$H \leqq \frac{1}{1-4c}, c \leqq \frac{1}{4}; \qquad H > \frac{1}{1-4c}, c > \frac{1}{4} \tag{14.22}$$

Since $H \geqq 0$, the second part of (14.22) is always true. Using max H in the first part of (14.22), we have

$$r \leqq \frac{1}{\sqrt{1-4c}} \qquad c \leqq 1/4$$

Thus, if $c > 1/4$, we have satisfied our stability requirement, regardless of the size of r, so that r is no longer a factor in the choice of k.

The point patterns for the implicit equations (14.20) and (14.21) are respectively

```
. . .        . . .
. . .        . . .
             . . .
```

One seldom gets much for nothing: The implicit forms are not exceptions. In using them we are faced with the problem of solving many (the number depending primarily on the number of subdivisions in the x direction) algebraic equations simultaneously at each step in the time direction.

14.13. REFERENCES

[1] W. G. Bickley: Finite difference formulae for the square lattice, *Quart. J. Mech. Appl. Math.*, **1** (1948), 35–42.

[2] G. Blanch: On the numerical solution of parabolic partial differential equations, Natl. Bur. Standards, Los Angeles, prepublication copy, pp. 343–356, 1951.

[3] G. H. Bruce, D. W. Peaceman, H. H. Rachford, J. D. Rice: Calculations of unsteady-state gas flow through porous media, *J. Petrol. Technol.*, **198** (1953), 79–91.

[4] R. Courant, E. Isaacson, M. Rees: On the solution of nonlinear hyperbolic differential equations by finite differences, *Commun. Pure Appl. Math.*, **5** (1952), 243–255.

[5] R. Courant, K. Friedrichs, H. Lewy: Über die partiellen Differenzengleichungen der mathematischen Physik, *Math. Ann.*, **100** (1928), 32–74.

[6] S. Crandall: On a stability criterion for partial difference equations, *J. Math. Phys.*, **32** (1953), 80–81.

[7] J. Douglas: On the numerical integration of $U_{xx} + U_{yy} = U_t$ by implicit methods, *J. Soc. Ind. Appl. Math.*, **3** (1955), 42–65.

[8] J. Douglas: On the relation between stability and convergence in the numerical solution of linear parabolic and hyperbolic differential equations, *J. Soc. Ind. Appl. Math.*, **4** (1956), 20–37.

[9] J. Douglas, T. Gallie: On the numerical integration of a parabolic differential equation subject to a moving boundary condition, *Duke Math. J.*, **22** (1955), 557–566.

[10] J. Douglas, H. Rockford: On the numerical solution of heat-conduction problems in two and three space variables, *Trans. Am. Math. Soc.*, **82** (1956), 421–433.

[11] E. du Fort, S. Frankel: Stability conditions in the numerical treatment of parabolic differential equations, *Math. Tables and Other Aids to Computation*, **7** (1953), 135–152.

[12] H. W. Emmons: The numerical solution of partial differential equations, *Quart. Appl. Math.*, **2** (1944), 173–195.

[13] G. Evans, R. Brousseau, R. Kleirstock: Stability considerations for various difference equations derived for the linear heat-conduction equation, *J. Math. Phys.*, **34** (1956), 267–285.

[14] S. Frankel: Convergence rates of iterative treatments of partial differential equations, *Math. Tables and Other Aids to Computation*, **4** (1950), 65–75.

[15] F. B. Hildebrand: On the convergence of numerical solutions of the heat flow equation, *J. Math. Phys.*, **31** (1952), 35–41.

[16] M. A. Hyman: Non-iterative numerical solutions of boundary value problems, *Appl. Sci. Research*, **B**,2 (1952), 321–335.

[17] F. John: On integration of parabolic equations by difference methods, *Commun. Pure Appl. Math.*, **5** (1952), 155–211.

[18] W. Leutert, G. G. O'Brien: On the convergence of approximate solutions of the wave equation to the exact solution, *J. Math. Phys.*, **30** (1952), 252–256.

[19] G. O'Brien, M. A. Hyman, S. Kaplan: A study of the numerical solution of partial differential equations, *J. Math. Phys.*, **29** (1951), 223–251.

[20] J. Todd: A direct approach to the problem of stability in the numerical solution of partial differential equations, *Trans. Symposium Partial Differential Equations* (Univ. of California, 1955), Interscience, New York, 299–314.

CHAPTER 15

SPECIAL FUNCTIONS

BY

F. OBERHETTINGER, Dr.rer.nat., Dr.phil.habil., Corvallis, Oreg.

15.1. TRIGONOMETRIC AND HYPERBOLIC FUNCTIONS

The trigonometric functions can be defined by means of the exponential functions as

$$\sin x = -\frac{i}{2}(e^{ix} - e^{-ix}) \qquad \cos x = \tfrac{1}{2}(e^{ix} + e^{-ix})$$
$$\tan x = \frac{\sin x}{\cos x} \qquad \cot x = \frac{\cos x}{\sin x} \tag{15.1a–d}$$

The hyperbolic functions are defined as

$$\text{Sinh } x = \tfrac{1}{2}(e^{x} - e^{-x}) \qquad \text{Cosh } x = \tfrac{1}{2}(e^{x} + e^{-x})$$
$$\text{Tanh } x = \frac{\text{Sinh } x}{\text{Cosh } x} \qquad \text{Coth } x = \frac{\text{Cosh } x}{\text{Sinh } x} \tag{15.2a–d}$$

The corresponding inverse functions can be expressed as:

$$\sin^{-1} x \equiv \arcsin x = -i \ln (ix + \sqrt{1 - x^2})$$
$$\cos^{-1} x \equiv \arccos x = -i \ln (x + i\sqrt{1 - x^2})$$
$$\tan^{-1} x \equiv \arctan x = -\frac{i}{2} \ln \frac{1 + ix}{1 - ix}$$
$$\cot^{-1} x \equiv \text{arccot } x = -\frac{i}{2} \ln \frac{ix - 1}{ix +} \tag{15.3a–h}$$
$$\text{Sinh}^{-1} x = \ln (x + \sqrt{x^2 + 1}) \qquad \text{Cosh}^{-1} x = \ln (x + \sqrt{x^2 - 1})$$
$$\text{Tanh}^{-1} x = \tfrac{1}{2} \ln \frac{1 + x}{1 - x} \qquad \text{Coth}^{-1} x = \tfrac{1}{2} \ln \frac{x + 1}{x - 1}$$

Relations between the Functions

Trigonometric and hyperbolic functions are connected by the relations

$$\sin ix = i \text{ Sinh } x \qquad \cos ix = \text{Cosh } x \qquad \tan ix = i \text{ Tanh } x$$
$$\text{Sinh } ix = i \sin x \qquad \text{Cosh } ix = \cos x \qquad \text{Tanh } ix = i \tan x \tag{15.4a–f}$$

The following relations are frequently needed:

$$\sin (x \pm y) = \sin x \cos y \pm \cos x \sin y$$
$$\cos (x \pm y) = \cos x \cos y \mp \sin x \sin y \tag{15.5a–c}$$
$$\tan (x \pm y) = \frac{\tan x \pm \tan y}{1 \mp \tan x \tan y}$$

$$\text{Sinh } (x \pm y) = \text{Sinh } x \text{ Cosh } y \pm \text{Cosh } x \text{ Sinh } y$$
$$\text{Cosh } (x \pm y) = \text{Cosh } x \text{ Cosh } y \pm \text{Sinh } x \text{ Sinh } y \tag{15.5d–f}$$
$$\text{Tanh } (x \pm y) = \frac{\text{Tanh } x \pm \text{Tanh } y}{1 \pm \text{Tanh } x \text{ Tanh } y}$$

$$\sin 2x = 2 \sin x \cos x \qquad \cos 2x = \cos^2 x - \sin^2 x$$
$$\tan 2x = \frac{2 \tan x}{1 - \tan^2 x} \tag{15.6a–c}$$

$$\sin \frac{x}{2} = \sqrt{\tfrac{1}{2}(1 - \cos x)} \qquad \cos \frac{x}{2} = \sqrt{\tfrac{1}{2}(1 + \cos x)}$$
$$\tan \frac{x}{2} = \frac{1 - \cos x}{\sin x} = \frac{\sin x}{1 + \cos x} \tag{15.7a–c}$$

$$\text{Sinh } 2x = 2 \text{ Sinh } x \text{ Cosh } x \qquad \text{Cosh } 2x = \text{Cosh}^2 x + \text{Sinh}^2 x$$
$$\text{Tanh } 2x = \frac{2 \text{ Tanh } x}{1 + \text{Tanh}^2 x} \tag{15.8a–c}$$

$$\text{Sinh} \frac{x}{2} = \sqrt{\tfrac{1}{2}(\text{Cosh } x - 1)} \qquad \text{Cosh} \frac{x}{2} = \sqrt{\tfrac{1}{2}(\text{Cosh } x + 1)}$$
$$\text{Tanh} \frac{x}{2} = \frac{\text{Cosh } x - 1}{\text{Sinh } x} = \frac{\text{Sinh } x}{\text{Cosh } x + 1} \tag{15.9a–c}$$

$$\sin x \pm \sin y = 2 \sin \frac{x \pm y}{2} \cos \frac{x \mp y}{2}$$
$$\cos x + \cos y = 2 \cos \frac{x + y}{2} \cos \frac{x - y}{2} \tag{15.10a–c}$$
$$\cos x - \cos y = -2 \sin \frac{x + y}{2} \sin \frac{x - y}{2}$$

$$\text{Sinh } x \pm \text{Sinh } y = 2 \text{ Sinh} \frac{x \pm y}{2} \text{ Cosh} \frac{x \mp y}{2}$$
$$\text{Cosh } x + \text{Cosh } y = 2 \text{ Cosh} \frac{x + y}{2} \text{ Cosh} \frac{x - y}{2} \tag{15.11a–c}$$
$$\text{Cosh } x - \text{Cosh } y = 2 \text{ Sinh} \frac{x + y}{2} \text{ Sinh} \frac{x - y}{2}$$

$$2 \sin x \sin y = \cos (x - y) - \cos (x + y)$$
$$2 \cos x \cos y = \cos (x + y) + \cos (x - y) \tag{15.12a–c}$$
$$2 \sin x \cos y = \sin (x + y) + \sin (x - y)$$

$$2 \text{ Sinh } x \text{ Sinh } y = \text{Cosh } (x + y) - \text{Cosh } (x - y)$$
$$2 \text{ Cosh } x \text{ Cosh } y = \text{Cosh } (x + y) + \text{Cosh } (x - y) \tag{15.13a–c}$$
$$2 \text{ Sinh } x \text{ Cosh } y = \text{Sinh } (x + y) + \text{Sinh } (x - y)$$

Orthogonality

The system of functions $u_n(x) = \sin (n\pi x/l)$ and $v_m(x) = \cos (m\pi x/l)$ $(n, m = 0, 1, 2, \ldots)$ is orthogonal (see p. 17–2) in the interval $(0,2l)$; that is,

$$\int_0^{2l} u_n(x)u_m(x)\,dx = \int_0^{2l} v_n(x)v_m(x)\,dx = \begin{cases} 0 & \text{if } m \neq n \\ l & \text{if } m = n \neq 0 \end{cases}$$
$$\int_0^{2l} v_0{}^2(x)\,dx = 2l \qquad \int_0^{2l} u_n(x)v_m(x)\,dx = 0 \tag{15.14}$$

Fourier's Theorem

A consequence of these orthogonality relations is Fourier's theorem which admits (under certain conditions) the expansion of a function $f(x)$ into a trigonometric series in the form

$$f(x) = \sum_{n=0}^{\infty} \left(A_n \cos \frac{n\pi x}{l} + B_n \sin \frac{n\pi x}{l} \right) \tag{15.15}$$

where $A_0 = \dfrac{1}{2l} \displaystyle\int_0^{2l} f(x)\,dx \qquad A_n = \dfrac{1}{l} \displaystyle\int_0^{2l} f(x) \cos \dfrac{n\pi x}{l}\,dx \qquad n = 1, 2, \ldots$

$$B_n = \frac{1}{l} \int_0^{2l} f(x) \sin \frac{n\pi x}{l}\,dx \tag{15.16}$$

The above expansion is valid in the interval $(0,2l)$.

Examples of such expansions are

$$\sum_{n=1}^{\infty} \frac{1}{n} \sin nx = \frac{\pi - x}{2} \qquad \sum_{n=1}^{\infty} \frac{1}{n} \cos nx = -\ln\left(2 \sin \frac{x}{2}\right) \qquad 0 < x < 2\pi$$

$$\sum_{n=0}^{\infty} t^n \cos nx = \frac{1 - t \cos x}{1 - 2t \cos x + t^2} \qquad |t| < 1$$

$$\sum_{n=1}^{\infty} t^n \sin nx = \frac{t \sin x}{1 - 2t \cos x + t^2} \qquad |t| < 1$$

$$\sum_{n=1}^{\infty} \frac{t^n}{n} \cos nx = -\tfrac{1}{2} \ln (1 - 2t \cos x + t^2) \qquad |t| \leq 1 \qquad (15.17a\text{–}h)$$

$$\sum_{n=1}^{\infty} \frac{t^n}{n} \sin nx = \arctan \frac{t \sin x}{1 - t \cos x} \qquad |t| \leq 1$$

$$\cos ax = \frac{1}{\pi} \sin a\pi \left[2a \sum_{n=1}^{\infty} (-1)^n \frac{\cos nx}{a^2 - n^2} + \frac{1}{a}\right] \qquad -\pi \leq x \leq \pi$$

$$\sin ax = \frac{2}{\pi} \sin a\pi \sum_{n=1}^{\infty} \frac{(-1)^n n \sin nx}{a^2 - n^2} \qquad -\pi < x < \pi$$

For further expansions, see Table 20.3.

15.2. ORTHOGONAL POLYNOMIALS

Definition

A set of polynomials $p_n(x)$ ($n = 0, 1, 2, \ldots$) of degree n in x is orthogonal in the interval (a,b) with respect to a weight function $w(x)$ if

$$\int_a^b w(x) p_n(x) p_m(x)\, dx = \begin{cases} 0 & \text{for } m \neq n \\ a_n & \text{for } m = n \end{cases} \qquad m, n = 0, 1, 2, \ldots \qquad (15.18)$$

This relation admits (under certain conditions) the representation of a function $f(x)$ in the form

$$f(x) = \sum_{n=0}^{\infty} c_n p_n(x) \qquad \text{with} \qquad c_n = \frac{1}{a_n} \int_a^b f(x) w(x) p_n(x)\, dx \qquad (15.19)$$

Classical Polynomials

Of particular interest are the "classical" (Tchebycheff, Legendre, Gegenbauer, Jacobi, Hermite, Laguerre) polynomials. Table 15.1 gives the interval of orthogonality (a,b), the weight function $w(x)$, and the coefficients a_n as defined before for these polynomials. For the function $\Gamma(x)$, see Sec. 15.3. In the following,

$$(c)_n = c(c + 1)(c + 2) \cdots (c + n - 1) \qquad (15.20)$$

means Pochhammer's symbol.

Table 15.1

Polynomial $p_n(x)$	Interval (a,b)	Weight function $w(x)$	$a_n = \int_a^b w(x) p_n^2(x)\,dx$
Tchebycheff $T_n(x)$	$(-1,1)$	$(1-x^2)^{-1/2}$	$a_0 = \pi,\ a_n = \frac{1}{2}\pi,\ n \geqq 1$
Legendre $P_n(x)$	$(-1,1)$	1	$(n+\frac{1}{2})^{-1}$
Gegenbauer $C_n^{\nu}(x)$	$(-1,1)$	$(1-x^2)^{\nu-1/2}$	$\frac{2^{1-2\nu}\pi\Gamma(n+2\nu)[\Gamma(\nu)]^{-2}}{n!(\nu+n)}$
Jacobi $P_n^{(\alpha,\beta)}(x)$	$(-1,1)$	$(1-x)^{\alpha}(1+x)^{\beta}$	$\frac{2^{\alpha+\beta+1}\Gamma(n+\alpha+1)\Gamma(n+\beta+1)}{n!(2n+\alpha+\beta+1)\Gamma(n+\alpha+\beta+1)}$
Laguerre $L_n^{\alpha}(x)$	$(0,\infty)$	$x^{\alpha}e^{-x}$	$\Gamma(1+\alpha+n)/n!$
Hermite $\mathrm{He}_n(x)$	$(-\infty,\infty)$	$e^{-\frac{1}{2}x^2}$	$(2\pi)^{1/2}n!$

Explicit Representation of the Classical Polynomials

The first four polynomials listed in Table 15.5 can be expressed as terminating hypergeometric series (see Sec. 15.4).

$$T_n(x) = F(-n, n; \tfrac{1}{2}; \tfrac{1}{2} - \tfrac{1}{2}x) \qquad P_n(x) = F(-n, n+1; 1; \tfrac{1}{2} - \tfrac{1}{2}x)$$

$$C_n^{\nu}(x) = \frac{(2\nu)_n}{n!} F(-n, n+2\nu; \nu+\tfrac{1}{2}; \tfrac{1}{2} - \tfrac{1}{2}x) \tag{15.21a–d}$$

$$P_n^{(\alpha,\beta)}(x) = \frac{(\alpha+1)_n}{n!} F(-n, \alpha+\beta+n+1; \alpha+1; \tfrac{1}{2} - \tfrac{1}{2}x)$$

Hence, it follows that

$$T_n(x) = \cos(n\cos^{-1} x) = \frac{n}{2}\lim_{\nu\to 0}\Gamma(\nu)C_n^{\nu}(x) = \frac{(2^n n!)^2}{(2n)!}P_n^{(-1/2,1/2)}(x) \tag{15.22a–c}$$

$$P_n(x) = C_n^{1/2}(x) = P_n^{(0,0)}(x) \qquad C_n^{\nu}(x) = \frac{(2\nu)_n}{(\nu+\frac{1}{2})_n}P_n^{(\nu-1/2,\nu-1/2)}(x)$$

The Hermite and Laguerre polynomials can be given in the form of a terminating Kummer series (see Sec. 15.5).

$$\mathrm{He}_{2n}(x) = \frac{(2n)!}{(-2)^n n!}\,{}_1F_1(-n; \tfrac{1}{2}; \tfrac{1}{2}x^2)$$

$$\mathrm{He}_{2n+1}(x) = \frac{(2n+1)!}{(-2)^n n!}\,x\,{}_1F_1(-n; \tfrac{3}{2}; \tfrac{1}{2}x^2) \tag{15.23a–d}$$

$$L_n^{\alpha}(x) = \frac{(\alpha+1)_n}{n!}\,{}_1F_1(-n; \alpha+1; x) \qquad L_n^0(x) = L_n(x)$$

Hence it follows that

$$\mathrm{He}_{2n}(x) = (-2)^n n! L_n^{-1/2}(\tfrac{1}{2}x^2) \qquad \mathrm{He}_{2n+1}(x) = (-2)^n n! x L_n^{1/2}(\tfrac{1}{2}x^2) \tag{15.24a,b}$$

From these hypergeometric series follow the explicit expressions:

$$T_n(x) = \tfrac{1}{2}n \sum_{m=0}^{[n/2]} \frac{(-1)^m(n-m-1)!}{m!(n-2m)!}(2x)^{n-2m}$$

$$P_n(x) = 2^{-n} \sum_{m=0}^{[n/2]} (-1)^m \binom{n}{m}\binom{2n-2m}{n} x^{n-2m}$$

$$C_n{}^{\nu}(x) = \sum_{m=0}^{[n/2]} \frac{(-1)^m(\nu)_{n-m}}{m!(n-2m)!}(2x)^{n-2m} \qquad (15.25a\text{–}f)$$

$$P_n{}^{(\alpha,\beta)}(x) = 2^{-n} \sum_{m=0}^{n} \binom{n+\alpha}{m}\binom{n+\beta}{n-m}(x-1)^{n-m}(x+1)^m$$

$$\mathrm{He}_n(x) = n! \sum_{m=0}^{[n/2]} \frac{(-2)^{-m}(x)^{n-2m}}{m!(n-2m)!}$$

$$L_n{}^{\alpha}(x) = \sum_{m=0}^{n} \binom{n+\alpha}{n-m}\frac{(-x)^m}{m!}$$

Here $[n/2]$ is the largest integer $\leqq n/2$,

$$\binom{\alpha}{m} = \frac{\alpha(\alpha-1)\cdots(\alpha-m+1)}{m!} = \frac{\Gamma(\alpha+1)}{m!\Gamma(\alpha+1-m)}$$

Special Values of x

Of particular interest are the values of the above polynomials for $x = -1, 0, 1$. It follows immediately that

$$T_n(1) = (-1)^n T_n(-1) = (-1)^n T_{2n}(0) = 1 \qquad T_{2n+1}(0) = 0$$

$$P_n(1) = (-1)^n P_n(-1) = 1 \qquad P_{2n}(0) = (-1)^n 2^{-2n}(2n)!(n!)^{-2} \qquad P_{2n+1}(0) = 0$$

$$C_n{}^{\nu}(1) = (-1)^n C_n{}^{\nu}(-1) = \frac{(2\nu)_n}{n!} \qquad C_{2n}{}^{\nu}(0) = \frac{(-1)^n(\nu)_n}{n!} \qquad C_{2n+1}^{\nu}(0) = 0 \qquad (15.26)$$

$$P_n{}^{(\alpha,\beta)}(1) = \frac{(\alpha+1)_n}{n!} \qquad P_n{}^{(\alpha,\beta)}(-1) = \frac{(-1)^n(\beta+1)_n}{n!}$$

$$\mathrm{He}_{2n}(0) = \frac{(-2)^{-n}(2n)!}{n!} \qquad \mathrm{He}_{2n+1}(0) = 0$$

$$L_n{}^{\alpha}(0) = \frac{(\alpha+1)_n}{n!}$$

Representations by Means of a Generating Function and as a Differentiation Formula

The polynomials considered here can also be expressed by means of a generating function

$$\sum_{n=0}^{\infty} b_n p_n(x) z^n = f(x,z)$$

or by a differentiation formula:

$$p_n(x) = [K_n w(x)]^{-1} \frac{d^n}{dx^n}[w(x)X^n]$$

Here $w(x)$ is the weight function, K_n is a constant, and X is a polynomial in x whose coefficients are independent of n.

$$\frac{1 - z^2}{1 - 2zx + z^2} = 1 + 2 \sum_{n=1}^{\infty} T_n(x) z^n \qquad |z| < 1,\ -1 < x < 1$$

$$T_n(x) = (-2)^n \frac{n!}{(2n)!} (1 - x^2)^{1/2} \frac{d^n}{dx^n} [(1 - x^2)^{n - 1/2}] \tag{15.27a,b}$$

$$(1 - 2zx + z^2)^{-1/2} = \sum_{n=0}^{\infty} P_n(x) z^n \qquad |z| < \min |x \pm \sqrt{x^2 - 1}|$$

$$P_n(x) = \frac{1}{2^n n!} \frac{d^n}{dx^n} [(x^2 - 1)^n] \tag{15.28a,b}$$

$$(1 - 2zx + z^2)^{-\nu} = \sum_{n=0}^{\infty} C_n^{\nu}(x) z^n \qquad |z| < \min |x \pm \sqrt{x^2 - 1}|$$

$$C_n^{\nu}(x) = \frac{(-2)^n (\nu)_n}{n!(2\nu + n)_n} (1 - x^2)^{1/2 - \nu} \frac{d^n}{dx^n} [(1 - x^2)^{n - 1/2 + \nu}] \tag{15.29a,b}$$

$$2^{\alpha+\beta} R^{-1} (1 - z + R)^{-\alpha} (1 + z + R)^{-\beta} = \sum_{n=0}^{\infty} P_n^{(\alpha,\beta)}(x) z^n \qquad |z| < 1$$

$$R = (1 - 2xz + z^2)^{1/2} \tag{15.30a,b}$$

$$P_n^{(\alpha,\beta)}(x) = \frac{(-2)^{-n}}{n!} (1 - x)^{-\alpha} (1 + x)^{-\beta} \frac{d^n}{dx^n} [(1 - x)^{n+\alpha} (1 + x)^{n+\beta}]$$

$$e^{zx - 1/2 z^2} = \sum_{n=0}^{\infty} \mathrm{He}_n(x) \frac{z^n}{n!} \qquad \mathrm{He}_n(x) = (-1)^n e^{1/2 x^2} \frac{d^n}{dx^n} e^{-1/2 x^2} \tag{15.31a,b}$$

$$(1 - z)^{-\alpha-1} e^{-xz/(1-z)} = \sum_{n=0}^{\infty} L_n^{\alpha}(x) z^n \qquad |z| < 1$$

$$L_n^{\alpha}(x) = \frac{e^x x^{-\alpha}}{n!} \frac{d^n}{dx^n} (e^{-x} x^{n+\alpha}) \tag{15.32a,b}$$

Recurrence Relations, Differential Quotients, and Differential Equations

Recurrence relations:

$$\begin{gathered}
T_{n+1}(x) - 2xT_n(x) + T_{n-1}(x) = 0 \\
nP_n(x) - (2n - 1)xP_{n-1}(x) + (n - 1)P_{n-2}(x) = 0 \\
(n + 2)C_{n+2}^{\nu}(x) = 2(\nu + n + 1)xC_{n+1}^{\nu}(x) - (2\nu + n)C_n^{\nu}(x)
\end{gathered}$$

$$\begin{aligned}
&2(n + 1)(n + \alpha + \beta + 1)(2n + \alpha + \beta)P_{n+1}^{(\alpha,\beta)}(x) \\
&\quad = (2n + \alpha + \beta + 1)[(2n + \alpha + \beta)(2n + \alpha + \beta + 2)x + \alpha^2 - \beta^2]P_n^{(\alpha,\beta)}(x) \\
&\qquad - 2(n + \alpha)(n + \beta)(2n + \alpha + \beta + 2)P_{n-1}^{(\alpha,\beta)}(x)
\end{aligned} \tag{15.33}$$

$$\begin{gathered}
\mathrm{He}_{n+1}(x) = x\mathrm{He}_n(x) - n\mathrm{He}_{n-1}(x) \\
nL_n^{\alpha}(x) = (2n - x + \alpha - 1)L_{n-1}^{\alpha}(x) - (n + \alpha - 1)L_{n-2}^{\alpha}(x)
\end{gathered}$$

Differential quotients:

$$\begin{aligned}
(1 - x^2)T_n'(x) &= n[T_{n-1}(x) - xT_n(x)] \\
(1 - x^2)P_n'(x) &= n[P_{n-1}(x) - xP_n(x)] \\
C_n^{\nu\prime}(x) &= 2\nu C_{n-1}^{\nu+1}(x) \\
(2n + \alpha + \beta)(1 - x^2)P_n^{(\alpha,\beta)\prime}(x) &= n[\alpha - \beta - (2n + \alpha + \beta)x]P_n^{(\alpha,\beta)}(x) \\
&\quad + 2(n + \alpha)(n + \beta)P_{n-1}^{(\alpha,\beta)}(x) \\
\mathrm{He}_n'(x) = n\mathrm{He}_{n-1}(x) \qquad xL_n^{\alpha\prime}(x) &= nL_n^{\alpha}(x) - (n + \alpha)L_{n-1}^{\alpha}(x)
\end{aligned} \tag{15.34}$$

Differential equations:

$$\begin{aligned}
&y = T_n(x) \text{ satisfies } (1 - x^2)y'' - xy' + n^2y = 0\\
&y = P_n(x) \text{ satisfies } (1 - x^2)y'' - 2xy' + n(n+1)y = 0\\
&y = C_n^{\nu}(x) \text{ satisfies } (1 - x^2)y'' - (2\nu + 1)xy' + n(n + 2\nu)y = 0\\
&y = P_n^{(\alpha,\beta)}(x) \text{ satisfies } (1 - x^2)y'' + [\beta - \alpha - (\alpha + \beta + 2)x]y'\\
&\qquad\qquad + n(n + \alpha + \beta + 1)y = 0 \qquad (15.35a\text{–}f)\\
&y = He_n(x) \text{ satisfies } y'' - xy' + ny = 0\\
&y = L_n^{\alpha}(x) \text{ satisfies } xy'' + (\alpha + 1 - x)y' + ny = 0\\
&\qquad\qquad n = 0, 1, 2, \ldots
\end{aligned}$$

15.3. THE GAMMA FUNCTION

Definitions

The function $\Gamma(z)$ can be defined by an integral expression which is valid for complex values of the argument z provided $\Re z > 0$ (Euler):

$$\Gamma(z) = \int_0^\infty e^{-t}t^{z-1}\,dt \qquad \Re z > 0 \tag{15.36}$$

If z is a positive integer, $z = n + 1$ $(n = 0, 1, 2, \ldots)$, it follows by integration by parts from the formula above that

$$\Gamma(n + 1) = \int_0^\infty e^{-t}t^n\,dt = n! \tag{15.37}$$

Frequently used and equivalent with the integral expression are the following representations by an infinite product:

$$\Gamma(z) = z^{-1}\prod_{n=1}^{\infty}\left[\left(1 + \frac{1}{n}\right)^z\left(1 + \frac{z}{n}\right)^{-1}\right] \qquad \text{(Gauss)}$$

$$[\Gamma(z)]^{-1} = ze^{\gamma z}\prod_{n=1}^{\infty}\left[\left(1 + \frac{z}{n}\right)e^{-z/n}\right] \qquad \text{(Weierstrass)} \tag{15.38a,b}$$

Here γ is Euler's constant defined by

$$\gamma = \lim_{m\to\infty}\left(\sum_{n=1}^{m}\frac{1}{n} - \ln m\right) = 0.57721\ldots \tag{15.39}$$

It is evident that the product representations are valid in the whole complex z plane with the exception of the points $z = 0, -1, -2, \ldots$ for the first product. These points are the only singular points of $\Gamma(z)$.

Functional Equations

The gamma function satisfies a number of functional equations, which can easily be derived from one of the definitions above:

$$\Gamma(1 + z) = z\Gamma(z)$$

and, hence,

$$\begin{aligned}
\Gamma(z + n) &= z(z + 1)(z + 2)\cdots(z + n - 1)\Gamma(z)\\
\Gamma(z)\Gamma(1 - z) &= \pi/\sin \pi z\\
\Gamma(\tfrac{1}{2} - z)\Gamma(\tfrac{1}{2} + z) &= \pi/\cos \pi z \qquad (15.40)\\
\Gamma(2z) &= \pi^{-1/2}2^{2z-1}\Gamma(z)\Gamma(\tfrac{1}{2} + z) \qquad \text{(duplication formula)}
\end{aligned}$$

Special Values of the Argument

The function $\Gamma(z)$ becomes elementary for $z = \frac{1}{2} \pm n \quad (n = 0, 1, 2, \ldots)$.

$$\Gamma(n+1) = n! \qquad \Gamma(\tfrac{1}{2} + n) = \sqrt{\pi}\,\frac{(2n)!}{4^n n!} \qquad \Gamma(\tfrac{1}{2} - n) = (-1)^n \sqrt{\pi}\,\frac{4^n n!}{(2n)!} \tag{15.41a–c}$$

Definite Integrals Related to the Gamma Function

A number of definite integrals with algebraic or trigonometric integrand can be expressed in terms of the gamma function. In the following formulas x, y are complex parameters with $\Re\,(x,y) > 0$ if not otherwise stated.

Algebraic integrand:

$$\int_0^1 t^{x-1}(1 - t^c)^{y-1}\,dt = c^{-1}\frac{\Gamma(x/c)\Gamma(y)}{\Gamma(y + x/c)} \qquad c > 0$$

$$\int_0^1 t^{x-1}(1-t)^{y-1}(1+bt)^{-x-y}\,dt = (1+b)^{-x}\frac{\Gamma(x)\Gamma(y)}{\Gamma(x+y)} \qquad b > -1$$

$$\int_{-1}^1 (1+t)^{2x-1}(1-t)^{2y-1}(1+t^2)^{-x-y}\,dt = 2^{x+y-2}\frac{\Gamma(x)\Gamma(y)}{\Gamma(x+y)} \tag{15.42a–f}$$

$$\int_b^a (t-b)^{x-1}(a-t)^{y-1}\,dt = (a-b)^{x+y-1}\frac{\Gamma(x)\Gamma(y)}{\Gamma(x+y)} \qquad a > b > 0$$

$$\int_b^a (t-b)^{x-1}(a-t)^{y-1}(t-c)^{-x-y}\,dt = \frac{(a-b)^{x+y-1}}{(a-c)^x(b-c)^y}\frac{\Gamma(x)\Gamma(y)}{\Gamma(x+y)} \qquad a > b > c > 0$$

$$\int_b^a (t-b)^{x-1}(a-t)^{y-1}(c-t)^{-x-y}\,dt = \frac{(a-b)^{x+y-1}}{(c-a)^x(c-b)^y}\frac{\Gamma(x)\Gamma(y)}{\Gamma(x+y)} \qquad c > a > b > 0$$

Trigonometric integrand:

$$\int_0^{\pi/2} (\sin t)^{2x-1}(\cos t)^{2y-1}\,dt = \frac{\Gamma(x)\Gamma(y)}{2\Gamma(x+y)}$$

$$\int_0^{\pi/2} (1 + a\sin^2 t)^{-x-y}(\sin t)^{2x-1}(\cos t)^{2y-1}\,dt = \frac{1}{2(1+a)^x}\frac{\Gamma(x)\Gamma(y)}{\Gamma(x+y)} \qquad a > -1$$

$$\int_0^{\pi} \sin^x t \cos yt\,dt = \frac{\pi}{2^x}\cos\frac{\pi y}{2}\,\frac{\Gamma(1+x)}{\Gamma(1 + x/2 + y/2)\Gamma(1 + x/2 - y/2)} \tag{15.43a–e}$$

$$\int_0^{\pi} \sin^x t \sin yt\,dt = \frac{\pi}{2^x}\sin\frac{\pi y}{2}\,\frac{\Gamma(1+x)}{\Gamma(1 + x/2 + y/2)\Gamma(1 + x/2 - y/2)}$$

$$\int_0^{\frac{1}{2}\pi} \cos^x t \cos yt\,dt = \frac{\pi}{2^{x+1}}\,\frac{\Gamma(1+x)}{\Gamma(1 + x/2 + y/2)\Gamma(1 + x/2 - y/2)}$$

Equations (43c–e) are valid for $\Re\,x > -1$ and y real.

$$\int_0^{\infty} \cos\,(at^c)t^{x-1}\,dt = \frac{1}{ca^{x/c}}\cos\frac{\pi x}{2c}\,\Gamma\left(\frac{x}{c}\right) \qquad 0 < \Re\,x < c,\ c > 0$$

$$\int_0^{\infty} \sin\,(at^c)t^{x-1}\,dt = \frac{1}{ca^{x/c}}\sin\frac{\pi x}{2c}\,\Gamma\left(\frac{x}{c}\right) \qquad -c < \Re\,x < c,\ c > 0 \tag{15.43f,g}$$

Asymptotic Expansions for $\Gamma(z)$

The function $\Gamma(z)$ can be approximated by an asymptotic expansion for large values of the modulus of z in the range $-\pi < \arg z < \pi$. This expansion is known as Stirling's formula:

$$\Gamma(z) \sim \sqrt{\frac{2\pi}{z}}\,e^{z(\ln z - 1)}\left(1 + \frac{1}{12z} + \frac{1}{288z^2} + \cdots\right) \tag{15.44}$$

A consequence of this formula is

$$\frac{\Gamma(z+a)}{\Gamma(z+b)} z^{a-b} \sim [1 + \tfrac{1}{2}z^{-1}(a-b)(a+b-1) + \cdots] \tag{15.45}$$

for large $|z|$ if $-\pi < \arg z < \pi$ and a, b arbitrary.

The Function $\psi(z)$

The logarithmic derivative of $\Gamma(z)$ is denoted by $\psi(z)$:

$$\psi(z) = \frac{d}{dz} \ln \Gamma(z) = \frac{\Gamma'(z)}{\Gamma(z)} \tag{15.46}$$

Explicit expressions for $\psi(z)$ are obtained by means of this definition under consideration of the infinite product representation for $\Gamma(z)$, such as

$$\psi(z) = \ln z + \sum_{n=0}^{\infty} \left[\ln\left(1 + \frac{1}{z+n}\right) - \frac{1}{z+n} \right]$$

$$\psi(z) = -\gamma + \sum_{n=0}^{\infty} \left(\frac{1}{n+1} - \frac{1}{z+n} \right) \tag{15.47a,b}$$

The functional equations for $\psi(z)$ follow from the functional equations for $\Gamma(z)$:

$$\psi(1+z) = \frac{1}{z} + \psi(z) \qquad \psi(z+n) = \frac{1}{z} + \frac{1}{z+1} + \cdots + \frac{1}{z+n-1} + \psi(z)$$

$$n = 2, 3, \ldots \tag{15.48a,b}$$

$$\psi(z) - \psi(1-z) = -\pi \cot \pi z \qquad \psi(\tfrac{1}{2} + z) - \psi(\tfrac{1}{2} - z) = \pi \tan \pi z$$

Special Values of the Argument. The function $\psi(z)$ becomes elementary for $z = n + 1$ and $z = \tfrac{1}{2} \pm n$ ($n = 0, 1, 2, \ldots$):

$$\psi(1) = -\gamma \qquad \psi(n+1) = 1 + \tfrac{1}{2} + \cdots + \frac{1}{n} - \gamma \qquad n = 1, 2, 3, \ldots$$

$$\psi(\tfrac{1}{2}) = -\gamma - 2\ln 2 \qquad \psi(-\tfrac{1}{2}) = 2 - \gamma - 2\ln 2 \tag{15.49}$$

$$\psi(\tfrac{1}{2} + n) = \psi(\tfrac{1}{2} - n) = 2\left(1 + \tfrac{1}{3} + \cdots + \frac{1}{2n-1} - \ln 2\right) - \gamma$$

$$n = 1, 2, \ldots$$

15.4. THE HYPERGEOMETRIC FUNCTION

Definition

A solution of the hypergeometric differential equation

$$z(1-z)y'' - [(a+b+1)z - c]y' - aby = 0 \tag{15.50}$$

in the form of a power series in z is Gauss' hypergeometric series

$$y = F(a,b;c;z) = \sum_{n=0}^{\infty} \frac{(a)_n (b_n)}{(c)_n} \frac{z^n}{n!} \tag{15.51}$$

$|z| < 1 \qquad c \neq 0, -1, -2, \ldots,$

$$(k)_n = k(k+1)(k+2) \cdots (k+n-1) \qquad k_0 = 1$$

Equivalent with this series is the integral representation

$$F(a,b;c;z) = \frac{\Gamma(c)}{\Gamma(a)\Gamma(c-b)} \int_0^1 t^{b-1}(1-t)^{c-b-1}(1-tz)^{-a}\,dt \qquad (15.52)$$

valid for $\Re\, c > \Re\, b > 0$, $0 < \arg(z-1) < 2\pi$.

Special Cases

Obviously the series for F reduces to a polynomial of degree n in z when a or b is equal to $-n$ ($n = 0, 1, \ldots$) (these cases are treated in Sec. 15.2). A number of elementary functions are special cases of the hypergeometric function such as

$$\begin{aligned} &F(a,b;b;z) = (1-z)^{-a} \qquad F(1,1;2;z) = z^{-1}\ln(1-z) \\ &F(\tfrac{1}{2},1;\tfrac{3}{2};z^2) = \tfrac{1}{2}z^{-1}\ln\frac{1+z}{1-z} \qquad F(\tfrac{1}{2},1;\tfrac{3}{2};-z^2) = z^{-1}\tan^{-1} z \\ &F(\tfrac{1}{2},\tfrac{1}{2};\tfrac{3}{2};z^2) = z^{-1}\sin^{-1} z \qquad F(\tfrac{1}{2}a, \tfrac{1}{2}+\tfrac{1}{2}a; 1+a; z) = 2^a(1+\sqrt{1-z})^{-a} \\ &F(a, a+\tfrac{1}{2}; \tfrac{1}{2}; z^2) = \tfrac{1}{2}[(1+z)^{-2a} + (1-z)^{-2a}] \end{aligned} \qquad (15.53)$$

Transformation Formulas

The function $F(a,b;c;z)$ as defined before admits a number of linear transforms. These are:

$$\begin{aligned} F(a,b;c;z) &= (1-z)^{-a}F\left(a, c-b; c; \frac{z}{z-1}\right) \\ &= (1-z)^{-b}F\left(b, c-a; c; \frac{z}{z-1}\right) \\ &= (1-z)^{c-a-b}F(c-a, c-b; c; z) \end{aligned} \qquad (15.54)$$

$$\begin{aligned} F(a,b;c;z) = (1-z)^{-a}&\frac{\Gamma(c)\Gamma(b-a)}{\Gamma(b)\Gamma(c-a)}F\left(a, c-b; a-b+1; \frac{1}{1-z}\right) \\ &+ (1-z)^{-b}\frac{\Gamma(c)\Gamma(a-b)}{\Gamma(a)\Gamma(c-b)}F\left(b, c-a; b-a+1; \frac{1}{1-z}\right) \end{aligned} \qquad (15.55)$$

$$\begin{aligned} F(a,b;c;z) = &\frac{\Gamma(c)\Gamma(c-a-b)}{\Gamma(c-a)\Gamma(c-b)}F(a, b; a+b-c+1; 1-z) \\ &+ (1-z)^{c-a-b}\frac{\Gamma(c)\Gamma(a+b-c)}{\Gamma(a)\Gamma(b)}F(c-a, c-b; c-a-b+1; 1-z) \end{aligned} \qquad (15.56)$$

$$\begin{aligned} F(a,b;c;z) = &\frac{\Gamma(c)\Gamma(b-a)}{\Gamma(b)\Gamma(c-a)}(-z)^{-a}F\left(a, 1-c+a; 1-b+a; \frac{1}{z}\right) \\ &+ \frac{\Gamma(c)\Gamma(a-b)}{\Gamma(a)\Gamma(c-b)}(-z)^{-b}F\left(b, 1-c+b; 1-a+b; \frac{1}{z}\right) \end{aligned} \qquad (15.57)$$

$$\begin{aligned} F(a,b;c;z) = &\frac{\Gamma(c)\Gamma(c-a-b)}{\Gamma(c-a)\Gamma(c-b)}z^{-a}F\left(a, a-c+1; a+b-c+1; 1-\frac{1}{z}\right) \\ &+ \frac{\Gamma(c)\Gamma(a+b-c)}{\Gamma(a)\Gamma(b)}(1-z)^{c-a-b}z^{a-c} \\ &\times F\left(c-a, 1-a; c-a-b+1; 1-\frac{1}{z}\right) \end{aligned} \qquad (15.58)$$

If, and only if the numbers $\pm(1-c)$, $\pm(a-b)$, $\pm(a+b-c)$ have the property that one of them is equal to $\tfrac{1}{2}$ or that two of them are equal to each other, then there exists a quadratic transform.

Transformation formulas of this type have been given by Kummer:

$$\begin{aligned} &F(a,b;2b;z) = (1-\tfrac{1}{2}z)^{-a}F\left[\tfrac{1}{2}a, \tfrac{1}{2}a+\tfrac{1}{2}; \tfrac{1}{2}+b; \left(1-\frac{2}{z}\right)^{-2}\right] \\ &F(a, a+\tfrac{1}{2}-b; b+\tfrac{1}{2}; z^2) = (1+z)^{-2a}F[a, b; 2b; 4z(1+z)^{-2}] \\ &F(2a, 2a+1-c; c; z) = (1+z)^{-2a}F[a, a+\tfrac{1}{2}; c; 4z(1+z)^{-2}] \\ &F[a, b; a+b+\tfrac{1}{2}; 4z(1-z)] = F(2a, 2b; a+b+\tfrac{1}{2}; z) \end{aligned} \qquad (15.59a\text{–}d)$$

Special Values of the Argument z

Under certain conditions $F(a,b;c;z)$ can be expressed in terms of the gamma function (see Sec. 15.3) when the argument has one of the values ± 1, $\frac{1}{2}$. Particularly

$$F(a,b;c;1) = \frac{\Gamma(c)\Gamma(c-a-b)}{\Gamma(c-a)\Gamma(c-b)} \qquad c \neq 0, -1, -2, \ldots \qquad \mathcal{R}(a+b-c) < 0$$

$$F(a,\,b;\,1+a-b;\,-1) = \frac{\sqrt{\pi}}{2^a}\,\frac{\Gamma(1+a-b)}{\Gamma(1-b+\frac{1}{2}a)\Gamma(\frac{1}{2}+\frac{1}{2}a)} \qquad 1+a-b \neq 0, -1, -2, \ldots \qquad (15.60a\text{–}e)$$

$$F(1,\,a;\,a+1;\,-1) = \tfrac{1}{2}a[\psi(\tfrac{1}{2}+\tfrac{1}{2}a) - \psi(\tfrac{1}{2}a)]$$

$$F(2a,\,2b;\,a+b+\tfrac{1}{2};\,\tfrac{1}{2}) = \sqrt{\pi}\,\frac{\Gamma(a+b+\frac{1}{2})}{\Gamma(a+\frac{1}{2})\Gamma(b+\frac{1}{2})} \qquad \tfrac{1}{2}+a+b \neq 0, -1, -2, \ldots$$

$$F(a,\,1-a;\,b;\,\tfrac{1}{2}) = 2^{1-b}\sqrt{\pi}\,\frac{\Gamma(b)}{\Gamma(\frac{1}{2}a+\frac{1}{2}b)\Gamma(\frac{1}{2}+\frac{1}{2}b-\frac{1}{2}a)} \qquad b \neq 0, -1, -2, \ldots$$

The differential quotient is

$$\frac{dF}{dz} = \frac{ab}{c}F(a+1,\,b+1;\,c+1;\,z)$$

and

$$\frac{d^nF}{dz^n} = \frac{(a)_n(b)_n}{(c)_n}F(a+n,\,b+n;\,c+n;\,z) \qquad (15.61a,b)$$

$$n = 2, 3, 4, \ldots \qquad (a)_n = a(a+1)\cdots(a+n-1)$$

15.5. THE CONFLUENT HYPERGEOMETRIC FUNCTION

Kummer's Function

A solution of Kummer's differential equation

$$zy'' - (z-c)y' - ay = 0 \qquad (15.62)$$

in the form of a series in ascending powers of z is Kummer's series

$$y = {}_1F_1(a;c;z) = 1 + \frac{a}{c} + \frac{a(a+1)}{c(c+1)}z + \frac{a(a+1)(a+2)}{c(c+1)(c+2)}\frac{z^2}{2!} + \cdots \qquad c \neq 0, -1, -2, \ldots \qquad (15.63)$$

Equivalent with this series is the integral representation

$$ {}_1F_1(a;c;z) = \frac{2^{1-c}\Gamma(c)}{\Gamma(a)\Gamma(c-a)}e^{\frac{1}{2}z}\int_{-1}^{1} e^{\frac{1}{2}zt}(1-t)^{c-a-1}(1+t)^{a-1}\,dt \qquad 0 < \mathcal{R}\,a < \mathcal{R}\,c \qquad (15.64)$$

From these formulas there follow a number of functional relations:

$$\frac{d}{dz}\,{}_1F_1(a;c;z) = \frac{a}{c}\,{}_1F_1(a+1;\,c+1;\,z)$$

$$a\,{}_1F_1(a+1;\,c+1;\,z) = (a-c)\,{}_1F_1(a;\,c+1;\,z) + c\,{}_1F_1(a;c;z) \qquad (15.65a\text{–}d)$$

$$a\,{}_1F_1(a+1;\,c;\,z) = (z+2a-c)\,{}_1F_1(a;c;z) + (c-a)\,{}_1F_1(a-1;\,c;\,z)$$

$$ {}_1F_1(a;c;z) = e^z\,{}_1F_1(c-a;\,c;\,-z)$$

Whittaker's Functions

The substitution $y = x^{-\frac{1}{2}c}e^{\frac{1}{2}z}u$, $a = \frac{1}{2} - k + \mu$, $c = 1 + 2\mu$ transforms Kummer's equation into Whittaker's equation:

$$u'' - \left(\tfrac{1}{4} - \frac{k}{z} - \frac{\frac{1}{4}-\mu^2}{z^2}\right)u = 0 \qquad (15.66)$$

A system of two linearly independent solutions are the two Whittaker functions $M_{k,\mu}(z)$ and $W_{k,\mu}(z)$ defined by

$$M_{k,\mu}(z) = z^{\mu+\frac{1}{2}}e^{-\frac{1}{2}z}{}_1F_1(\tfrac{1}{2} - k + \mu; 1 + 2\mu; z)$$
$$W_{k,\mu}(z) = \frac{\Gamma(-2\mu)}{\Gamma(\frac{1}{2} - k - \mu)} M_{k,\mu}(z) + \frac{\Gamma(2\mu)}{\Gamma(\frac{1}{2} - k + \mu)} M_{k,-\mu}(z) \qquad (15.67)$$

Equivalent with this are the integral representations

$$M_{k,\mu}(z) = \frac{2^{-2\mu}\Gamma(2\mu + 1)z^{\frac{1}{2}+\mu}}{\Gamma(\frac{1}{2} - k + \mu)\Gamma(\frac{1}{2} + k + \mu)} \int_{-1}^{1} e^{\frac{1}{2}zt}(1 - t)^{\mu+k-\frac{1}{2}}(1 + t)^{\mu-k-\frac{1}{2}}\, dt$$
$$\Re(\mu \pm k) > -\tfrac{1}{2} \qquad (15.68)$$
$$W_{k,\mu}(z) = \frac{z^k e^{-\frac{1}{2}z}}{\Gamma(\frac{1}{2} - k + \mu)} \int_0^\infty t^{\mu-k-\frac{1}{2}}e^{-t}\left(1 + \frac{t}{z}\right)^{k-\frac{1}{2}+\mu} dt$$
$$\Re(\mu - k) > -\tfrac{1}{2}, \quad -\pi < \arg z < \pi$$

15.6. ERROR FUNCTIONS, FRESNEL'S INTEGRALS, EXPONENTIAL INTEGRAL, SINE AND COSINE INTEGRALS

Error Functions

The error functions are defined by

$$\operatorname{erf} z = \frac{2}{\sqrt{\pi}} \int_0^z e^{-t^2}\, dt = \frac{2}{\sqrt{\pi z}} M_{-\frac{1}{4},\frac{1}{4}}(z^2)e^{-\frac{1}{2}z^2}$$
$$\operatorname{erfc} z = \frac{2}{\sqrt{\pi}} \int_z^\infty e^{-t^2}\, dt = 1 - \operatorname{erf} z = \frac{1}{\sqrt{\pi z}} e^{-\frac{1}{2}z^2} W_{-\frac{1}{4},\frac{1}{4}}(z^2) \qquad (15.69a,b)$$

They are special cases of the previously defined Whittaker functions. A series expansion for erf z is

$$\operatorname{erf} z = \frac{2}{\sqrt{\pi}}\left(z - \frac{z^3}{1!\cdot 3} + \frac{z^5}{2!\cdot 5} - \frac{z^7}{3!\cdot 7} + \cdots\right) \qquad (15.70)$$
$$= 1 - \operatorname{erfc} z$$

An asymptotic expansion for erfc z is

$$\operatorname{erfc} z \sim \frac{e^{-z^2}}{z\sqrt{\pi}}\left[1 - \frac{1}{2z^2} + \frac{3}{(2z^2)^2} - \frac{3\cdot 5}{(2z^2)^3} + \cdots\right] \qquad |z| \gg 1, \quad \frac{-3\pi}{4} < \arg z < \frac{3\pi}{4} \qquad (15.71)$$

Fresnel's Integrals

Fresnel's integrals

$$C(x) = (2\pi)^{-\frac{1}{2}} \int_0^x t^{-\frac{1}{2}} \cos t\, dt \qquad S(x) = (2\pi)^{-\frac{1}{2}} \int_0^x t^{-\frac{1}{2}} \sin t\, dt \qquad (15.72a,b)$$

are connected with the error function by the relations

$$(\mp \tfrac{1}{2}i)^{\frac{1}{2}} \operatorname{erf} \sqrt{\pm ix} = C(x) \mp iS(x)$$
$$2C(x) = (\tfrac{1}{2}i)^{\frac{1}{2}} \operatorname{erf} \sqrt{-ix} + (-\tfrac{1}{2}i)^{\frac{1}{2}} \operatorname{erf} \sqrt{ix} \qquad (15.73a\text{–}c)$$
$$2iS(x) = (\tfrac{1}{2}i)^{\frac{1}{2}} \operatorname{erf} \sqrt{-ix} - (-\tfrac{1}{2}i)^{\frac{1}{2}} \operatorname{erf} \sqrt{ix}$$

The Fresnel integrals are asymptotically represented by

$$C(x) \sim \tfrac{1}{2} + (2\pi x)^{-\frac{1}{2}} \sin x \qquad S(x) \sim \tfrac{1}{2} - (2\pi x)^{-\frac{1}{2}} \cos x \qquad (15.74a,b)$$

Exponential Integral, Sine and Cosine Integrals

The exponential integral is defined by

$$\operatorname{Ei}(-z) = -\int_z^\infty e^{-t}\frac{dt}{t} = -z^{-\frac{1}{2}}e^{-\frac{1}{2}z}W_{-\frac{1}{2},0}(z) \qquad -\pi < \arg z < \pi \qquad (15.75)$$

The definition of the sine and cosine integrals is

$$\begin{aligned} \mathrm{si}\,(x) &= -\int_x^\infty \sin t\,\frac{dt}{t} = -\tfrac{1}{2}i[\mathrm{Ei}\,(ix) - \mathrm{Ei}\,(-ix)] \\ \mathrm{Ci}\,(x) &= -\int_x^\infty \cos t\,\frac{dt}{t} = \tfrac{1}{2}[\mathrm{Ei}\,(ix) + \mathrm{Ei}\,(-ix)] \end{aligned} \qquad x > 0 \quad (15.76a,b)$$

These functions have the following power-series expansions:

$$\begin{aligned} \mathrm{Ei}\,(-z) - \gamma - \ln z &= \sum_{n=1}^{\infty} \frac{(-z)^n}{nn!} \\ \mathrm{si}\,(x) + \tfrac{1}{2}\pi &= \sum_{n=0}^{\infty} \frac{(-1)^n x^{2n+1}}{(2n+1)(2n+1)!} \\ \mathrm{Ci}\,(x) - \gamma - \ln x &= \sum_{n=1}^{\infty} \frac{(-1)^n x^{2n}}{2n(2n)!} \end{aligned} \qquad (15.77a\text{–}c)$$

[For γ see Eq. (15.39).]

The asymptotic expansions of these functions are:

$$\begin{aligned} &\mathrm{Ei}\,(-z) \sim -z^{-1}e^{-z}(1 - z^{-1} + 2!z^{-2} - 3!z^{-3} + \cdots) \\ &\qquad\qquad |z| \gg 1 \qquad -3\pi/2 < \arg z < 3\pi/2 \qquad (15.78a,b) \\ &\mathrm{Ci}\,(x) + i\,\mathrm{si}\,(x) \sim -ix^{-1}e^{ix}[1 + (ix)^{-1} + 2!(ix)^{-2} + 3!(ix)^{-3} + \cdots] \end{aligned}$$

A function related to $\mathrm{Ei}\,(-z)$ is

$$\overline{\mathrm{Ei}}\,(x) = -\int_{-x}^{\infty} e^{-t}\,\frac{dt}{t} \qquad (15.79)$$

Here x is positive, and the integral is a Cauchy principal value. The connection of $\overline{\mathrm{Ei}}\,(x)$ with $\mathrm{Ei}\,(-x)$ is

$$\overline{\mathrm{Ei}}\,(x) = \mathrm{Ei}\,(-xe^{\pm i\pi}) \mp i\pi \qquad (15.80)$$

Furthermore,

$$\begin{aligned} \overline{\mathrm{Ei}}\,(x) - \gamma - \ln x &= \sum_{n=1}^{\infty} \frac{x^n}{nn!} \qquad (15.81a,b) \\ \overline{\mathrm{Ei}}\,(x) &\sim x^{-1}e^{x}(1 + x^{-1} + 2!x^{-2} + 3!x^{-3} + \cdots) \qquad x \gg 1 \end{aligned}$$

15.7. BESSEL FUNCTIONS

Bessel Functions of General Order

Bessel functions are solutions of Bessel's differential equation

$$z^2y'' + zy' + (z^2 - \nu^2)y = 0 \qquad (15.82)$$

where z and ν are unrestricted. A solution of the differential equation in the form of a power series is

$$y = J_\nu(z) = (\tfrac{1}{2}z)^\nu \sum_{m=0}^{\infty} \frac{(-1)^m(\tfrac{1}{2}z)^{2m}}{m!\Gamma(\nu + m + 1)} \qquad (15.83)$$

This series is convergent in the whole complex z plane. [For the function $\Gamma(z)$ see Sec. 15.3.]

With $J_\nu(z)$ the following linear combinations are likewise solutions of Bessel's differential equation:

$$\begin{aligned} Y_\nu(z) &= [\sin(\pi\nu)]^{-1}[J_\nu(z)\cos(\nu\pi) - J_{-\nu}(z)] \\ H_\nu^{(1)}(z) &= J_\nu(z) + iY_\nu(z) = (i\sin\pi\nu)^{-1}[J_{-\nu}(z) - J_\nu(z)e^{-i\nu\pi}] \\ H_\nu^{(2)}(z) &= J_\nu(z) - iY_\nu(z) = (i\sin\pi\nu)^{-1}[J_\nu(z)e^{i\nu\pi} - J_{-\nu}(z)] \\ &\qquad \nu \neq 0, \pm 1, \pm 2, \ldots \end{aligned} \tag{15.84a–c}$$

The functions $J_\nu(z)$, $Y_\nu(z)$, $H_\nu^{(1)}(z)$, $H_\nu^{(2)}(z)$ denote Bessel's function, Neumann's function, the first and second Hankel functions, all of the order ν, respectively.

The modified Bessel differential equation is obtained by replacing z by iz in Bessel's differential equation:

$$z^2y'' + zy' - (z^2 + \nu^2)y = 0 \tag{15.85}$$

Solutions of this equation are called modified Bessel functions. Two kinds of modified Bessel functions are usually distinguished:

$$\begin{aligned} I_\nu(z) &= (\tfrac{1}{2}z)^\nu \sum_{m=0}^{\infty} \frac{(\tfrac{1}{2}z)^{2m}}{m!\Gamma(\nu + m + 1)} \\ K_\nu(z) &= \frac{\pi}{2\sin\pi\nu}[I_{-\nu}(z) - I_\nu(z)] \qquad \nu \neq 0, \pm 1, \pm 2, \ldots \end{aligned} \tag{15.86a,b}$$

$I_\nu(z)$ and $K_\nu(z)$ denote the modified Bessel and the modified Hankel functions of the order ν. There exist the relations

$$I_\nu(z) = e^{-\frac{1}{2}i\nu\pi}J_\nu(iz) \qquad K_\nu(z) = \tfrac{1}{2}i\pi e^{i\frac{1}{2}\nu\pi}H_\nu^{(1)}(iz) \tag{15.87a,b}$$

Bessel Functions of the Orders $\nu = \pm n$ and $\nu = \pm(n + \frac{1}{2})$

If $n = 0, 1, 2, \ldots$, then

$$\begin{aligned} J_n(z) &= (-1)^n J_{-n}(z) = (\tfrac{1}{2}z)^n \sum_{m=0}^{\infty} \frac{(-1)^m(\tfrac{1}{2}z)^{2m}}{m!(m+n)!} \\ I_n(z) &= I_{-n}(z) = (\tfrac{1}{2}z)^n \sum_{m=0}^{\infty} \frac{(\tfrac{1}{2}z)^{2m}}{m!(m+n)!} \\ I_n(-z) &= (-1)^n I_n(z) \qquad J_n(-z) = (-1)^n J_n(z) \\ Y_0(z) &= \frac{2}{\pi}\left(\gamma + \ln\frac{z}{2}\right)J_0(z) - \frac{2}{\pi}\sum_{m=0}^{\infty}\frac{(-1)^m}{m!^2}h_m\left(\frac{z}{2}\right)^{2m} \\ K_0(z) &= -\left(\gamma + \ln\frac{z}{2}\right)J_0(z) + \sum_{m=0}^{\infty}\frac{h_m}{m!^2}\left(\frac{z}{2}\right)^{2m} \\ Y_n(z) &= (-1)^n Y_{-n}(z) = \frac{2}{\pi}\left(\gamma + \ln\frac{z}{2}\right)J_n(z) \\ &\quad - \frac{1}{\pi}\sum_{m=0}^{n-1}\frac{(n-m-1)!}{m!}\left(\frac{z}{2}\right)^{2m-n} - \frac{1}{\pi}\sum_{m=0}^{\infty}\frac{(-1)^m(h_{m+n} + h_m)}{m!(n+m)!}\left(\frac{z}{2}\right)^{n+2m} \\ K_n(z) &= K_{-n}(z) = (-1)^{n+1}\left(\gamma + \ln\frac{z}{2}\right)I_n(z) \\ &\quad + \frac{1}{2}\sum_{m=0}^{n-1}\frac{(-1)^m(n-m-1)!}{m!}\left(\frac{z}{2}\right)^{2m-n} + \frac{(-1)^n}{2}\sum_{m=0}^{\infty}\frac{h_{n+m} + h_m}{m!(n+m)!}\left(\frac{z}{2}\right)^{n+2m} \end{aligned} \tag{15.88}$$

In these formulas, $n = 1, 2, 3, \ldots, \gamma = 0.57721 \ldots, h_k = 1 + \frac{1}{2} + \frac{1}{3} + \cdots + 1/k$, $h_0 = 0$. The function $J_n(z)$ can also be represented by means of a generating function

$$\exp\left[\frac{1}{2} z\left(t - \frac{a^2}{t}\right)\right] = \sum_{n=-\infty}^{\infty} \left(\frac{t}{a}\right)^n J_n(az) \tag{15.89}$$

A consequence of this relation is the Jacobi-Anger formulas:

$$e^{iz \sin t} = \sum_{n=-\infty}^{\infty} e^{int} J_n(z) \qquad e^{iz \cos t} = \sum_{n=-\infty}^{\infty} i^n e^{int} J_n(z) \tag{15.90a,b}$$

$$\begin{aligned} \cos(z \sin t) &= J_0(z) + 2 \sum_{n=1}^{\infty} J_{2n}(z) \cos 2nt \\ \sin(z \sin t) &= 2 \sum_{n=1}^{\infty} J_{2n-1}(z) \sin(2n-1)t \\ \cos(z \cos t) &= J_0(z) + 2 \sum_{n=1}^{\infty} (-1)^n J_{2n}(z) \cos 2nt \\ \sin(z \cos t) &= -2 \sum_{n=1}^{\infty} (-1)^n J_{2n-1}(z) \cos(2n-1)t \end{aligned} \tag{15.90c–f}$$

The Bessel functions become elementary when the order $\nu = \pm(n + \frac{1}{2})$, where $n = 0, 1, 2, \ldots$. It follows from the power series for $J_\nu(z)$ that

$$\begin{aligned} &J_{\frac{1}{2}}(z) = Y_{-\frac{1}{2}}(z) = (\tfrac{1}{2}\pi z)^{-\frac{1}{2}} \sin z \qquad Y_{\frac{1}{2}}(z) = -J_{-\frac{1}{2}}(z) = -(\tfrac{1}{2}\pi z)^{-\frac{1}{2}} \cos z \\ &H_{\frac{1}{2}}^{(1)}(z) = -iH_{-\frac{1}{2}}^{(1)}(z) = -i(\tfrac{1}{2}\pi z)^{-\frac{1}{2}} e^{iz} \qquad H_{\frac{1}{2}}^{(2)}(z) = iH_{-\frac{1}{2}}^{(2)}(z) = i(\tfrac{1}{2}\pi z)^{-\frac{1}{2}} e^{-iz} \\ &I_{\frac{1}{2}}(z) = (\tfrac{1}{2}\pi z)^{-\frac{1}{2}} \operatorname{Sinh} z \qquad I_{-\frac{1}{2}}(z) = (\tfrac{1}{2}\pi z)^{-\frac{1}{2}} \operatorname{Cosh} z \\ &K_{\frac{1}{2}}(z) = K_{-\frac{1}{2}}(z) = \left(\frac{\pi}{2z}\right)^{\frac{1}{2}} e^{-z} \end{aligned} \tag{15.91}$$

or, generally,

$$\begin{aligned} J_{n+\frac{1}{2}}(z) &= (-1)^n Y_{-n-\frac{1}{2}}(z) = (-1)^n (\tfrac{1}{2}\pi z)^{-\frac{1}{2}} z^{n+1} \left(\frac{1}{z}\frac{d}{dz}\right)^n \frac{\sin z}{z} \\ Y_{n+\frac{1}{2}}(z) &= -(-1)^n J_{-n-\frac{1}{2}}(z) = -(-1)^n (\tfrac{1}{2}\pi z)^{-\frac{1}{2}} z^{n+1} \left(\frac{1}{z}\frac{d}{dz}\right)^n \frac{\cos z}{z} \\ I_{n+\frac{1}{2}}(z) &= (\tfrac{1}{2}\pi z)^{-\frac{1}{2}} z^{n+1} \left(\frac{1}{z}\frac{d}{dz}\right)^n \frac{\operatorname{Sinh} z}{z} \\ K_{n+\frac{1}{2}}(z) &= K_{-n-\frac{1}{2}}(z) = (-1)^n z^{n+1} \left(\frac{\pi}{2z}\right)^{\frac{1}{2}} \left(\frac{1}{z}\frac{d}{dz}\right)^n \frac{e^{-z}}{z} \end{aligned} \tag{15.92}$$

Differentiation Formulas, Recurrence Relations, Integrals

From the power series for $J_\nu(z)$ follow

$$\frac{d}{dz}[z^\nu J_\nu(z)] = z^\nu J_{\nu-1}(z) \qquad \frac{d}{dz}[z^{-\nu} J_\nu(z)] = -z^{-\nu} J_{\nu+1}(z) \tag{15.93a,b}$$

and, hence,

$$J_{\nu-1}(z) + J_{\nu+1}(z) = \frac{2\nu}{z} J_\nu(z) \qquad J_{\nu-1}(z) - J_{\nu+1}(z) = 2J'_\nu(z) \tag{15.93c,d}$$

The same relations are valid for $Y_\nu(z)$ and for $H_\nu^{(1),(2)}(z)$. The corresponding relations for the modified Bessel functions are:

$$\frac{d}{dz}[z^\nu I_\nu(z)] = z^\nu I_{\nu-1}(z) \qquad \frac{d}{dz}[z^{-\nu} I_\nu(z)] = z^{-\nu} I_{\nu+1}(z)$$
$$\frac{d}{dz}[z^\nu K_\nu(z)] = -z^\nu K_{\nu-1}(z) \qquad \frac{d}{dz}[z^{-\nu} K_\nu(z)] = -z^{-\nu} K_{\nu+1}(z) \tag{15.94a–h}$$
$$I_{\nu-1}(z) - I_{\nu+1}(z) = \frac{2\nu}{z} I_\nu(z) \qquad I_{\nu-1}(z) + I_{\nu+1}(z) = 2I'_\nu(z)$$
$$K_{\nu-1}(z) - K_{\nu+1}(z) = \frac{-2\nu}{z} K_\nu(z) \qquad K_{\nu-1}(z) + K_{\nu+1}(z) = -2K'_\nu(z)$$

From those formulas follow

$$\int z^{\nu+1} J_\nu(z)\, dz = z^{\nu+1} J_{\nu+1}(z) \qquad \int z^{1-\nu} J_\nu(z)\, dz = -z^{1-\nu} J_{\nu-1}(z) \tag{15.95a,b}$$

The same formulas are valid for the Neumann and for the Hankel functions. Similarly,

$$\int z^{\nu+1} I_\nu(z)\, dz = z^{\nu+1} I_{\nu+1}(z) \qquad \int z^{1-\nu} I_\nu(z)\, dz = z^{1-\nu} I_{\nu-1}(z)$$
$$\int z^{\nu+1} K_\nu(z)\, dz = -z^{\nu+1} K_{\nu+1}(z) \qquad \int z^{1-\nu} K_\nu(z)\, dz = -z^{1-\nu} K_{\nu-1}(z) \tag{15.95c–f}$$

Differential Equations

Let $Z_\nu(z)$ be a solution of Bessel's differential equation. Then a solution of the differential equation

$$z^2 y'' + (1 - 2a)zy' + [(bz^c)^2 + a^2 - (\nu c)^2]y = 0$$

is

$$y = z^a Z_\nu\left(\frac{bz^c}{c}\right) \tag{15.96a}$$

Special cases are:

$$y'' + c^2 z^{-2+2/b} y = 0 \qquad \text{solution } y = z^{1/2} Z_{1/2b}(bcz^{1/b})$$
$$y'' + (e^{2z} - \nu^2)y = 0 \qquad \text{solution } y = Z_\nu(e^z) \tag{15.96b–d}$$
$$y'' + z^{-4}(e^{2/z} - \nu^2)y = 0 \qquad \text{solution } y = zZ_\nu(e^{1/z})$$

The functions $J_\nu(z)$, $Y_\nu(z)$, $I_\nu(z)$, and $K_\nu(z)$ are linearly independent solutions of the following differential equation of the fourth order:

$$y'''' + 2z^{-1}y''' - (1 + 2\nu^2)z^{-2}y'' + (1 + 2\nu^2)z^{-3}y' - (z^4 + 4\nu^2 - \nu^4)z^{-4}y = 0 \tag{15.97}$$

Kelvin's Functions

Kelvin's functions $\mathrm{ber}_\nu(z)$, $\mathrm{bei}_\nu(z)$, $\mathrm{ker}_\nu(z)$, $\mathrm{kei}_\nu(z)$ are defined by

$$\mathrm{ber}_\nu(z) \pm i\,\mathrm{bei}_\nu(z) = J_\nu(ze^{\pm i3\pi/4}) \qquad \mathrm{ker}_\nu(z) \pm i\,\mathrm{kei}_\nu(z) = e^{\mp i\frac{1}{2}\nu\pi} K_\nu(ze^{\pm i\frac{1}{4}\pi}) \tag{15.98a,b}$$

The order ν and the argument z are unrestricted. We write, for $\nu = 0$, $\mathrm{ber}_0(z) = \mathrm{ber}\, z$, $\mathrm{bei}_0(z) = \mathrm{bei}\, z$, $\mathrm{ker}_0(z) = \mathrm{ker}\, z$, $\mathrm{kei}_0(z) = \mathrm{kei}\, z$.

With the notation $(d/dx)\,\mathrm{ber}\, x = \mathrm{ber}'\, x$, etc., the following formulas for the second derivatives hold:

$$\mathrm{ber}''\, x = -\mathrm{bei}\, x - x^{-1}\,\mathrm{ber}'\, x \qquad \mathrm{bei}''\, x = \mathrm{ber}\, x - x^{-1}\,\mathrm{bei}'\, x$$
$$\mathrm{ker}''\, x = -\mathrm{kei}\, x - x^{-1}\,\mathrm{ker}'\, x \qquad \mathrm{kei}''\, x = \mathrm{ker}\, x - x^{-1}\,\mathrm{kei}'\, x \tag{15.99a–d}$$

Integral Representations

All the Bessel functions as defined before can be expressed as definite integrals. The integral representations by Hansen and Bessel are valid for integer order

$$n = 0, 1, 2, \ldots$$

$$\pi J_n(z) = i^{-n} \int_0^\pi e^{iz\cos t} \cos nt\, dt = \int_0^\pi \cos(z \sin t - nt)\, dt$$

$$\pi J_{2n}(z) = 2\int_0^{\frac{1}{2}\pi} \cos(z \sin t)\cos 2nt\, dt \qquad (15.100a\text{–}c)$$

$$\pi J_{2n+1}(z) = 2\int_0^{\frac{1}{2}\pi} \sin(z \sin t)\sin(2n+1)t\, dt$$

The argument z is unrestricted. The integral representations by Poisson are valid for complex order:

$$\Gamma(\nu + \tfrac{1}{2})J_\nu(z) = 2\pi^{-\frac{1}{2}}(\tfrac{1}{2}z)^\nu \int_0^1 (1-t^2)^{\nu-\frac{1}{2}} \cos zt\, dt \qquad \Re\, \nu > -\tfrac{1}{2}$$

$$\Gamma(\nu + \tfrac{1}{2})Y_\nu(z) = 2\pi^{-\frac{1}{2}}(\tfrac{1}{2}z)^\nu \left[\int_0^1 (1-t^2)^{\nu-\frac{1}{2}} \sin zt\, dt - \int_0^\infty e^{-zt}(1+t^2)^{\nu-\frac{1}{2}}\, dt\right] \qquad \Re\, z > 0,\ \Re\, \nu > -\tfrac{1}{2} \qquad (15.101a\text{–}c)$$

$$\Gamma(\nu + \tfrac{1}{2})K_\nu(z) = \pi^{\frac{1}{2}}(\tfrac{1}{2}z)^\nu \int_1^\infty e^{-zt}(t^2-1)^{\nu-\frac{1}{2}}\, dt \qquad \Re\, z > 0,\ \Re\, \nu > -\tfrac{1}{2}$$

Heine's formulas are:

$$\pi J_\nu(z) = e^{i\frac{1}{2}\nu\pi}\left(\int_0^\pi e^{-iz\cos t}\cos \nu t\, dt - \sin \nu\pi \int_0^\infty e^{iz \operatorname{Cosh} t - \nu t}\, dt\right) \qquad 0 < \arg z < \pi$$

$$= e^{-i\frac{1}{2}\nu\pi}\left(\int_0^\pi e^{iz\cos t}\cos \nu t\, dt - \sin \nu\pi \int_0^\infty e^{-iz \operatorname{Cosh} t - \nu t}\, dt\right) \qquad -\pi < \arg z < 0 \qquad (15.102)$$

$$K_\nu(z) = \int_0^\infty e^{-z \operatorname{Cosh} t} \operatorname{Cosh} \nu t\, dt \qquad \Re\, z > 0$$

Schläfli's formulas are:

$$\pi J_\nu(z) = \int_0^\pi \cos(z \sin t - \nu t)\, dt - \sin \nu\pi \int_0^\infty e^{-z \operatorname{Sinh} t - \nu t}\, dt$$

$$\pi Y_\nu(z) = \int_0^\pi \sin(z \sin t - \nu t)\, dt - \int_0^\infty e^{-z \operatorname{Sinh} t}(e^{\nu t} + \cos \nu\pi e^{-\nu t})\, dt \qquad (15.103a,b)$$

Both formulas are valid for $\Re\, z > 0$.

Mehler-Sonine's formulas are:

$$\Gamma(\tfrac{1}{2} - \nu)J_\nu(x) = 2\pi^{-\frac{1}{2}}(\tfrac{1}{2}x)^{-\nu} \int_1^\infty (t^2-1)^{-\nu-\frac{1}{2}} \sin xt\, dt$$

$$\Gamma(\tfrac{1}{2} - \nu)Y_\nu(x) = -2\pi^{-\frac{1}{2}}(\tfrac{1}{2}x)^{-\nu} \int_1^\infty (t^2-1)^{-\nu-\frac{1}{2}} \cos xt\, dt \qquad (15.104a,b)$$

Both formulas are valid for $x > 0$ and $-\frac{1}{2} < R\, \nu < \frac{1}{2}$.

$$\pi J_\nu(x) = 2\int_0^\infty \sin(x \operatorname{Cosh} t - \tfrac{1}{2}\nu\pi) \operatorname{Cosh} \nu t\, dt$$

$$\pi Y_\nu(x) = -2\int_0^\infty \cos(x \operatorname{Cosh} t - \tfrac{1}{2}\nu\pi) \operatorname{Cosh} \nu t\, dt$$

$$\cos \tfrac{1}{2}\nu\pi\, K_\nu(x) = \int_0^\infty \cos(x \operatorname{Sinh} t) \operatorname{Cosh} \nu t\, dt \qquad (15.105a\text{–}d)$$

$$\sin \tfrac{1}{2}\nu\pi\, K_\nu(x) = \int_0^\infty \sin(x \operatorname{Sinh} t) \operatorname{Sinh} \nu t\, dt$$

The last four expressions are valid for $x > 0$ and $-1 < \Re\, \nu < 1$. Generalizations of the representations listed above are:

$$\pi \left(\frac{z+Z}{z-Z}\right)^{\frac{1}{2}\nu} J_\nu(\sqrt{z^2 - Z^2}) = \int_0^\pi e^{Z \cos t} \cos (z \sin t - \nu t)\, dt$$
$$- \sin \nu\pi \int_0^\infty e^{-z \operatorname{Sinh} t - Z \operatorname{Cosh} t - \nu t}\, dt \qquad \Re(z + Z) > 0 \tag{15.106a}$$

$$\left(\frac{z+Z}{z-Z}\right)^{\frac{1}{2}\nu} K_\nu(\sqrt{z^2 - Z^2}) = \tfrac{1}{2} \int_{-\infty}^\infty e^{-z \operatorname{Cosh} t - Z \operatorname{Sinh} t - \nu t}\, dt \qquad \Re(z \pm Z) > 0 \tag{15.106b}$$

Addition Theorems

There are two kinds of expansions of Bessel functions which are known as addition theorems: Gegenbauer's type and Graf's type. They become identical when the order ν is zero. Let $Z_\nu(z)$ be any of the Bessel functions $J_\nu(z)$, $Y_\nu(z)$, $H_\nu^{(1),(2)}(z)$; and $\bar{Z}_\nu(z)$ any of the modified Bessel functions $I_\nu(z), K_\nu(z)$. The addition theorem for the order zero is:

$$Z_0(\sqrt{a^2 + b^2 - 2ab \cos v}) = \sum_{n=0}^\infty \epsilon_n J_n(a) Z_n(b) \cos nv$$
$$\bar{Z}_0(\sqrt{a^2 + b^2 - 2ab \cos v}) = \sum_{n=0}^\infty \epsilon_n I_n(a) \bar{Z}_n(b) \cos nv \tag{15.107a,b}$$

Here $a < b$, $\epsilon_0 = 1$, $\epsilon_n = 2$ for $n = 1, 2, 3, \ldots$.

For arbitrary order ν, Gegenbauer's addition theorem is

$$(a^2 + b^2 - 2ab \cos v)^{-\frac{1}{2}\nu} Z_\nu(\sqrt{a^2 + b^2 - 2ab \cos v})$$
$$= (\tfrac{1}{2}ab)^{-\nu}\Gamma(\nu) \sum_{n=0}^\infty (\nu + n) C_n^\nu(\cos v) J_{\nu+n}(a) Z_{\nu+n}(b) \qquad a < b$$

for $C_n^\nu(z)$ see Table 15.1 (15.107c)

$$(a^2 + b^2 - 2ab \cos v)^{-\frac{1}{2}\nu} \bar{Z}_\nu(\sqrt{a^2 + b^2 - 2ab \cos v})$$
$$= (\tfrac{1}{2}ab)^{-\nu}\Gamma(\nu) \sum_{n=0}^\infty (\nu + n) C_n^\nu(\cos v) I_{\nu+n}(a) \bar{Z}_{\nu+n}(b) \qquad a < b \tag{15.107d}$$

Graf's addition theorem is

$$\left(\frac{a - be^{-iv}}{a - be^{iv}}\right)^{\nu/2} Z_\nu(\sqrt{a^2 + b^2 - 2ab \cos v}) = \sum_{n=-\infty}^\infty J_n(a) Z_{\nu+n}(b) e^{inv} \qquad a < b \tag{15.108a}$$

For the modified functions,

$$\left(\frac{a - be^{-iv}}{a - be^{iv}}\right)^{\nu/2} I_\nu(\sqrt{a^2 + b^2 - 2ab \cos v}) = \sum_{n=-\infty}^\infty (-1)^n I_n(a) I_{\nu+n}(b) e^{inv}$$

$$\left(\frac{a - be^{-iv}}{a - be^{iv}}\right)^{\nu/2} K_\nu(\sqrt{a^2 + b^2 - 2ab \cos v}) = \sum_{n=-\infty}^\infty I_n(a) K_{\nu+n}(b) e^{inv} \qquad a < b \tag{15.108b,c}$$

Special cases of the above formulas are:

$$(a^2 + b^2 - 2ab \cos v)^{-1/2} \exp [i(a^2 + b^2 - 2ab \cos v)^{1/2}]$$

$$= i\pi(ab)^{-1/2} \sum_{n=0}^{\infty} (n + \tfrac{1}{2}) J_{n+1/2}(a) H^{(1)}_{n+1/2}(b) P_n(\cos v)$$

$$a < b \qquad \text{for } P_n(z) \text{ see Table 15.1} \qquad (15.109a)$$

$$J_0(2z \sin \tfrac{1}{2}v) = [J_0(z)]^2 + 2 \sum_{n=1}^{\infty} [J_n(z)]^2 \cos nv$$

$$e^{iz \cos v} = \sqrt{\frac{\pi}{2z}} \sum_{m=0}^{\infty} i^m (2m + 1) J_{m+1/2}(z) P_m(\cos v) \qquad (15.109b,c)$$

Asymptotic Expansions

Asymptotic expansions of the modified Bessel functions and of the Bessel functions for large values of the argument have been given by Hankel. The abbreviation (Hankel's symbol)

$$(\nu,m) = 2^{-2m} \frac{(4\nu^2 - 1)(4\nu^2 - 3^2) \cdots [4\nu^2 - (2m - 1)^2]}{m!}$$

$$(\nu,0) = 1, \quad m = 1, 2, 3, \ldots \qquad (15.110)$$

is used. The following asymptotic expansions are valid for $|z| \gg 1$, $|z| \gg \nu$:

$$(\tfrac{1}{2}\pi z)^{1/2} J_\nu(z) \approx \cos (z - \tfrac{1}{2}\nu\pi - \tfrac{1}{4}\pi) \sum_{m=0}^{\infty} (-1)^m (\nu,2m)(2z)^{-2m}$$

$$- \sin (z - \tfrac{1}{2}\nu\pi - \tfrac{1}{4}\pi) \sum_{m=0}^{\infty} (-1)^m (\nu,2m + 1)(2z)^{-2m-1} \qquad (15.111a,b)$$

$$(\tfrac{1}{2}\pi z)^{1/2} Y_\nu(z) \approx \sin (z - \tfrac{1}{2}\nu\pi - \tfrac{1}{4}\pi) \sum_{m=0}^{\infty} (-1)^m (\nu,2m)(2z)^{-2m}$$

$$+ \cos (z - \tfrac{1}{2}\nu\pi - \tfrac{1}{4}\pi) \sum_{m=0}^{\infty} (-1)^m (\nu,2m + 1)(2z)^{-2m-1}$$

Both formulas are valid in the region $-\pi < \arg z < \pi$.

$$(\tfrac{1}{2}\pi z)^{1/2} H_\nu^{(1)}(z) \approx \exp [i(z - \tfrac{1}{2}\nu\pi - \tfrac{1}{4}\pi)] \sum_{m=0}^{\infty} (\nu,m)(-2iz)^{-m}$$

$$-\pi < \arg z < 2\pi$$

$$(\tfrac{1}{2}\pi z)^{1/2} H_\nu^{(2)}(z) \approx \exp [-i(z - \tfrac{1}{2}\nu\pi - \tfrac{1}{4}\pi)] \sum_{m=0}^{\infty} (\nu,m)(2iz)^{-m} \qquad (15.112a\text{–}c)$$

$$-2\pi < \arg z < \pi$$

$$K_\nu(z) \approx \left(\frac{\tfrac{1}{2}\pi}{z}\right)^{1/2} e^{-z} \sum_{m=0}^{\infty} (\nu,m)(2z)^{-m} \qquad -\frac{3\pi}{2} < \arg z < \frac{3\pi}{2}$$

Asymptotic expansions for large values of the order ν have been given by Debye and Langer. The formulas below represent the leading terms of asymptotic expansions, valid for large positive real values of the order ν and arbitrary positive real values of the argument x (Langer's formulas).

$$\begin{aligned} J_\nu(x) &\sim (w\nu/z)^{-1/2}[J_{1/3}(z)\cos(\pi/6) - Y_{1/3}(z)\sin(\pi/6)] \\ Y_\nu(x) &\sim (w\nu/z)^{-1/2}[J_{1/3}(z)\sin(\pi/6) + Y_{1/3}(z)\cos(\pi/6)] \end{aligned} \qquad (15.113a,b)$$

Here $\nu < x$, $w = (x^2\nu^{-2} - 1)^{1/2}$, $z = \nu\,(w - \arctan w)$.

$$J_\nu(x) \sim \pi^{-1}(w\nu/z)^{-1/2}K_{1/3}(z) \qquad Y_\nu(x) \sim -(w\nu/z)^{-1/2}[I_{1/3}(z) + I_{-1/3}(z)] \qquad (15.113c,d)$$

Here $\nu > x$, $w = (1 - x^2\nu^{-2})^{1/2}$, $z = \nu(\mathrm{Tanh}^{-1}\, w - w)$.

Airy's Integrals and Some Other Integral Formulas

Airy's integrals are defined by

$$\begin{aligned} \mathrm{Ai}\,(x) &= \pi^{-1}\int_0^\infty \cos\left(xt + \frac{t^3}{3}\right) dt \\ \mathrm{Bi}\,(x) &= \pi^{-1}\int_0^\infty \left[\exp\left(xt - \frac{t^3}{3}\right) + \sin\left(xt + \frac{t^3}{3}\right)\right] dt \end{aligned} \qquad (15.114a,b)$$

The functions Ai $(\pm x)$, Bi $(\pm x)$ can be expressed in terms of the Bessel or modified Bessel functions of the order $\frac{1}{3}$:

$$\begin{aligned} \mathrm{Ai}\,(x) &= \pi^{-1}\left(\frac{x}{3}\right)^{1/2} K_{1/3}(z) \\ \mathrm{Ai}\,(-x) &= \tfrac{1}{3}x^{1/2}[J_{1/3}(z) + J_{-1/3}(z)] \\ \mathrm{Bi}\,(x) &= \left(\frac{x}{3}\right)^{1/2} [I_{1/3}(z) + I_{-1/3}(z)] \\ \mathrm{Bi}\,(-x) &= \left(\frac{x}{3}\right)^{1/2} [J_{-1/3}(z) - J_{1/3}(z)] \qquad z = \tfrac{2}{3}x^{3/2} \end{aligned} \qquad (15.115a\text{–}d)$$

The Fourier cosine and the Fourier sine transforms of the Bessel functions are:

$$\int_0^\infty J_\nu(ax)\cos xy\,dx = \begin{cases} \dfrac{\cos[\nu\sin^{-1}(y/a)]}{(a^2 - y^2)^{1/2}} & \text{for } 0 < y < a \\ -\dfrac{a^\nu \sin\frac{1}{2}\nu\pi}{(y^2 - a^2)^{1/2}[y + (y^2 - a^2)^{1/2}]^\nu} & \text{for } y > a,\ \Re\,\nu > -1 \end{cases} \qquad (15.116a)$$

$$\int_0^\infty J_\nu(ax)\sin xy\,dx = \begin{cases} \dfrac{\sin[\nu\sin^{-1}(y/a)]}{(a^2 - y^2)^{1/2}} & \text{for } 0 < y < a \\ \dfrac{a^\nu \cos\frac{1}{2}\nu\pi}{(y^2 - a^2)^{1/2}[y + (y^2 - a^2)^{1/2}]^\nu} & \text{for } y > a,\ \Re\,\nu > -2 \end{cases} \qquad (15.116b)$$

$$\int_0^\infty Y_0(ax)\cos xy\,dx = \begin{cases} 0 & \text{for } 0 < y < a \\ -(y^2 - a^2)^{-1/2} & \text{for } y > a \end{cases} \qquad (15.116c)$$

$$\int_0^\infty Y_0(ax)\sin xy\,dx = \begin{cases} \dfrac{2\sin^{-1}(y/a)}{\pi(a^2 - y^2)^{1/2}} & \text{for } 0 < y < a \\ \dfrac{2\ln\{y/a - [(y/a)^2 - 1]^{1/2}\}}{\pi(y^2 - a^2)^{1/2}} & \text{for } y > a \end{cases} \qquad (15.116d)$$

$$\begin{aligned} \int_0^\infty K_\nu(ax)\cos xy\,dx = \frac{\pi\sec\frac{1}{2}\nu\pi}{4(a^2 + y^2)^{1/2}} &\{a^{-\nu}[y + (a^2 + y^2)^{1/2}]^\nu \\ &+ a^\nu[y + (a^2 + y^2)^{1/2}]^{-\nu}\} \qquad -1 < \Re\,\nu < 1 \end{aligned} \qquad (15.116e)$$

$$\int_0^\infty K_\nu(ax) \sin xy\, dx = \frac{\pi \csc \frac{1}{2}\nu\pi}{4a^\nu(a^2 + y^2)^{1/2}} \{[(a^2 + y^2)^{1/2} + y]^\nu - [(a^2 + y^2)^{1/2} - y]^\nu\} \qquad -2 < \Re\, \nu < 2 \quad (15.116f)$$

$$\int_0^\infty K_0(ax) \sin xy\, dx = \frac{\ln \{y/a + [(y/a)^2 + 1]^{1/2}\}}{(a^2 + y^2)^{1/2}} \quad (15.116g)$$

The Laplace transforms of these functions are:

$$\int_0^\infty J_\nu(at)e^{-pt}\, dt = \frac{[(a^2 + p^2)^{1/2} - p]^\nu}{a^\nu(a^2 + p^2)^{1/2}} \qquad \Re\, \nu > -1$$

$$\int_0^\infty Y_0(at)e^{-pt}\, dt = -\frac{2 \ln \{p/a + [1 + (p/a)^2]^{1/2}\}}{\pi(a^2 + p^2)^{1/2}} \quad (15.117a,b)$$

$$\int_0^\infty K_0(at)e^{-pt}\, dt = \begin{cases} \dfrac{\cos^{-1}(p/a)}{(a^2 - p^2)^{1/2}} & \text{for } 0 < p < a \\ \dfrac{\ln \{p/a + [(p/a)^2 - 1]^{1/2}\}}{(p^2 - a^2)^{1/2}} & \text{for } p > a \end{cases} \quad (15.117c)$$

15.8. ELLIPTIC INTEGRALS

Definitions

The elliptic integrals, $F(k,\phi)$ of the first kind, $E(k,\phi)$ of the second kind, and $\Pi(k,\nu,\phi)$ of the third kind, of the modulus k in Legendre's normal form are given by

$$F(k,\phi) = \int_0^\phi \frac{dt}{\sqrt{1 - k^2 \sin^2 t}} \qquad E(k,\phi) = \int_0^\phi \sqrt{1 - k^2 \sin^2 t}\, dt$$

$$\Pi(k,\nu,\phi) = \int_0^\phi \frac{dt}{(1 + \nu \sin^2 t)\sqrt{1 - k^2 \sin^2 t}} \quad (15.118a\text{–}c)$$

Clearly $F(k,\phi) = \Pi(k,0,\phi)$.

For $\phi = \frac{1}{2}\pi$, they are the complete elliptic integrals of the first, second, and third kind, respectively. These are denoted by

$$F(k,\tfrac{1}{2}\pi) = K(k) \qquad E(k,\tfrac{1}{2}\pi) = E(k) \qquad \Pi(k,\nu,\tfrac{1}{2}\pi) = \Pi(k,\nu)$$

If $k' = (1 - k^2)^{1/2}$ denotes the modulus complementary to k, then it is customary to write

$$K(k') = K'(k) \qquad E(k') = E'(k)$$

Between K, K', E, E' there exists Legendre's relation

$$EK' + E'K - KK' = \tfrac{1}{2}\pi \quad (15.119)$$

$K(k)$ and $E(k)$ are special cases of the hypergeometric function (see p. 15–9):

$$K(k) = \tfrac{1}{2}\pi F(\tfrac{1}{2},\tfrac{1}{2};1;k^2) \qquad E(k) = \tfrac{1}{2}\pi F(-\tfrac{1}{2};\tfrac{1}{2};1;k^2)$$

These relations define K and E for complex values of the modulus k.

Transformation Formulas

Let the modulus k in $K(k)$ and $E(k)$ be replaced by a new quantity $k_1 = k_1(k)$. Then $K(k_1)$ and $E(k_1)$ can be expressed in terms of $K(k)$ and $F(k)$ for a proper choice of the transformation $k_1(k)$. Table 15.2 lists these cases.

Table 15.2

k_1	$K(k_1)$	$K'(k_1)$	$E(k_1)$	$E'(k_1)$
$\frac{1}{k}$	$k(K + iK')$	kK'	$\frac{E + iE' - k'^2K - ik^2K'}{k}$	$\frac{E'}{k}$
k'	K'	K	E'	E
$\frac{1}{k'}$	$k'(K' + iK)$	$k'K$	$\frac{E' + iE - k^2K' - ik'^2K}{k'}$	$\frac{E}{k'}$
$\frac{ik}{k'}$	$k'K$	$k'(K' - iK)$	$\frac{E}{k'}$	$\frac{E' + iE - k^2K' - ik'^2K}{k'}$
$\frac{k'}{ik}$	kK'	$k(K + iK')$	$\frac{E'}{k}$	$\frac{E - iE' - k'^2K + ik^2K'}{k}$
$\frac{1 - k'}{1 + k'}$	$(\frac{1}{2} + \frac{1}{2}k')K$	$(1 + k')K'$	$\frac{E + k'K}{1 + k'}$	$\frac{2E' - k^2K'}{1 + k'}$
$\frac{2k^{1/2}}{1 + k}$	$(1 + k)K$	$(\frac{1}{2} + \frac{1}{2}k)K'$	$\frac{2E - k'^2K}{1 + k}$	$\frac{E' + kK'}{1 + k}$

Similarly, when k and ϕ in $F(k,\phi)$ and $E(k,\phi)$ are suitably replaced by $k_1 = k_1(k)$ and $\phi_1 = \phi_1(\phi)$, then $F(k_1,\phi_1)$ and $E(k_1,\phi_1)$ can be expressed in terms of $F(k,\phi)$ and $E(k,\phi)$. These transformation formulas are listed in Table 15.3. The abbreviation

$$\Delta = (1 - k^2 \sin^2 \phi)^{1/2} \tag{15.120}$$

is used in this table.

Table 15.3

k_1	$\sin \phi_1$	$\cos \phi_1$	$F(k_1,\phi_1)$	$E(k_1,\phi_1)$
$\frac{1}{k}$	$k \sin \phi$	Δ	$kF(k,\phi)$	$\frac{E(k,\phi) - k'^2F(k,\phi)}{k}$
k'	$-i \tan \phi$	$\frac{1}{\cos \phi}$	$-iF(k,\phi)$	$i[E(k,\phi) - F(k,\phi) - \Delta \tan \phi]$
$\frac{1}{k'}$	$-ik' \tan \phi$	$\frac{\Delta}{\cos \phi}$	$-ik'F(k,\phi)$	$\frac{i[E(k,\phi) - k'^2F(k,\phi) - \Delta \tan \phi]}{k'}$
$\frac{ik}{k'}$	$\frac{k' \sin \phi}{\Delta}$	$\frac{\cos \phi}{\Delta}$	$k'F(k,\phi)$	$\frac{E(k,\phi) - \frac{1}{2}k^2 \Delta^{-1} \sin 2\phi}{k'}$
$\frac{k'}{ik}$	$\frac{-ik \sin \phi}{\Delta}$	$\frac{1}{\Delta}$	$-ikF(k,\phi)$	$\frac{i[E(k,\phi) - F(k,\phi) - \frac{1}{2}k^2 \Delta^{-1} \sin 2\phi]}{k}$
$\frac{1 - k'}{1 + k'}$	$\frac{(\frac{1}{2} + \frac{1}{2}k') \sin 2\phi}{\Delta}$	$\frac{\cos^2 \phi - k' \sin^2 \phi}{\Delta}$	$(1 + k')F(k,\phi)$	$\frac{2}{1 + k'}\left[E(k,\phi) + k'F(k,\phi) - (\frac{1}{2} - \frac{1}{2}k)\frac{\sin 2\phi}{\Delta}\right]$

Reduction of Some Elliptic Integrals to Legendre's Form

Numerical tables for $F(k,\phi)$ and $E(k,\phi)$ as defined before are available [22]. A large number of integrals with algebraic integrand can be reduced to these tabulated functions. Table 15.4 gives such examples. The integrals listed in the first column are equal to $AF(k,\phi)$, where A, k, ϕ are available from the remaining columns.

Table 15.4

$AF(k,\phi)$	A	k	$\cos\phi$
$\int_x^\infty (t^3-1)^{-1/2}\,dt$	$3^{-1/4}$	$\sin\frac{\pi}{12}$	$\frac{x-1-3^{1/2}}{x-1+3^{1/2}}$
$\int_1^x (t^3-1)^{-1/2}\,dt$	$3^{-1/4}$	$\sin\frac{\pi}{12}$	$\frac{3^{1/2}+1-x}{3^{1/2}-1+x}$
$\int_x^1 (1-t^3)^{-1/2}\,dt$	$3^{-1/4}$	$\sin\frac{5\pi}{12}$	$\frac{3^{1/2}-1+x}{3^{1/2}+1-x}$
$\int_{-\infty}^x (1-t^3)^{-1/2}\,dt$	$3^{-1/4}$	$\sin\frac{5\pi}{12}$	$\frac{1-x-3^{1/2}}{1-x+3^{1/2}}$
$\int_x^1 (1+t^4)^{-1/2}\,dt$	$\frac{1}{2}$	$2^{-1/2}$	$\frac{x2^{1/2}}{(1+x^4)^{1/2}}$
$\int_0^x (1+t^4)^{-1/2}\,dt$	$\frac{1}{2}$	$2^{-1/2}$	$\frac{1-x^2}{1+x^2}$
$\int_x^\infty (1+t^4)^{-1/2}\,dt$	$\frac{1}{2}$	$2^{-1/2}$	$\frac{x^2-1}{x^2+1}$
$\int_0^x [(a^2-t^2)(b^2-t^2)]^{-1/2}\,dt$	$\frac{1}{a}$	$\frac{b}{a}$	$\left[1-\left(\frac{x}{b}\right)^2\right]^{1/2}$
$\int_x^b [(a^2-t^2)(b^2-t^2)]^{-1/2}\,dt$	$\frac{1}{a}$	$\frac{b}{a}$	$\left[\frac{(a/b)^2-1}{(a/x)^2-1}\right]^{1/2}$
$\int_b^x [(a^2-t^2)(t^2-b^2)]^{-1/2}\,dt$	$\frac{1}{a}$	$\left[1-\left(\frac{b}{a}\right)^2\right]^{1/2}$	$\left[\frac{(a/x)^2-1}{(a/b)^2-1}\right]^{1/2}$
$\int_x^a [(a^2-t^2)(t^2-b^2)]^{-1/2}\,dt$	$\frac{1}{a}$	$\left[1-\left(\frac{b}{a}\right)^2\right]^{1/2}$	$\left[\frac{(x/b)^2-1}{(a/b)^2-1}\right]^{1/2}$
$\int_a^x [(t^2-a^2)(t^2-b^2)]^{-1/2}\,dt$	$\frac{1}{a}$	$\frac{b}{a}$	$\left[\frac{(b/a)^2-1}{(x/a)^2-1}\right]^{1/2}$
$\int_x^\infty [(t^2-a^2)(t^2-b^2)]^{-1/2}\,dt$	$\frac{1}{a}$	$\frac{b}{a}$	$\left[1-\left(\frac{a}{x}\right)^2\right]^{1/2}$
$\int_0^x [(a^2+t^2)(b^2+t^2)]^{-1/2}\,dt$	$\frac{1}{a}$	$\left[1-\left(\frac{b}{a}\right)^2\right]^{1/2}$	$\left[1+\left(\frac{x}{b}\right)^2\right]^{-1/2}$
$\int_x^\infty [(a^2+t^2)(b^2+t^2)]^{-1/2}\,dt$	$\frac{1}{a}$	$\left[1-\left(\frac{b}{a}\right)^2\right]^{1/2}$	$\left[1+\left(\frac{a}{x}\right)^2\right]^{-1/2}$
$\int_0^x [(a^2-t^2)(b^2+t^2)]^{-1/2}\,dt$	$(a^2+b^2)^{-1/2}$	$a(a^2+b^2)^{-1/2}$	$\left[\frac{1-(x/a)^2}{1+(x/b)^2}\right]^{1/2}$
$\int_x^a [(a^2-t^2)(b^2+t^2)]^{-1/2}\,dt$	$(a^2+b^2)^{-1/2}$	$a(a^2+b^2)^{-1/2}$	$\frac{x}{a}$
$\int_b^x [(a^2+t^2)(t^2-b^2)]^{-1/2}\,dt$	$(a^2+b^2)^{-1/2}$	$a(a^2+b^2)^{-1/2}$	$\frac{b}{x}$
$\int_x^\infty [(a^2+t^2)(t^2-b^2)]^{-1/2}\,dt$	$(a^2+b^2)^{-1/2}$	$a(a^2+b^2)^{-1/2}$	$\left[\frac{1-(b/x)^2}{1+(a/x)^2}\right]^{1/2}$

Similar reductions to the elliptic integral of the second kind $E(k,\phi)$ in Legendre's normal form are listed in Table 15.5.

Table 15.5

$AE(k,\phi)$	A	k	$\cos\phi$
$\int_0^x (a^2-t^2)^{1/2}(b^2-t^2)^{-1/2}\,dt$	a	$\dfrac{b}{a}$	$\left[1-\left(\dfrac{x}{b}\right)^2\right]^{1/2}$
$\int_x^a (b^2+t^2)^{1/2}(a^2-t^2)^{-1/2}\,dt$	$(a^2+b^2)^{1/2}$	$\left[1+\left(\dfrac{b}{a}\right)^2\right]^{-1/2}$	$\dfrac{x}{a}$
$\int_b^x t^{-2}(t^2+a^2)^{1/2}(t^2-b^2)^{-1/2}\,dt$	$\dfrac{(a^2+b^2)^{1/2}}{b^2}$	$\left[1+\left(\dfrac{b}{a}\right)^2\right]^{-1/2}$	$\dfrac{b}{x}$
$\int_b^x t^2(t^2+a^2)^{1/2}(t^2-b^2)^{-1/2}\,dt$	$(a^2+b^2)^{1/2}$	$\left[1+\left(\dfrac{b}{a}\right)^2\right]^{-1/2}$	$\left[\dfrac{1-(b/x)^2}{1+(a/x)^2}\right]^{1/2}$
$\int_0^x (a^2+t^2)^{1/2}(b^2+t^2)^{-3/2}\,dt$	$\dfrac{a}{b^2}$	$\left[1-\left(\dfrac{b}{a}\right)^2\right]^{1/2}$	$\left[1+\left(\dfrac{x}{b}\right)^2\right]^{-1/2}$
$\int_0^x t^{-2}[(t^2-b^2)(a^2-t^2)]^{-1/2}\,dt$	$(ab^2)^{-1}$	$\left[1-\left(\dfrac{b}{a}\right)^2\right]^{1/2}$	$\left[\dfrac{(a/x)^2-1}{(a/b)^2-1}\right]^{1/2}$

The complete elliptic integral of the third kind $\Pi(k,\nu,\tfrac{1}{2}\pi)$ can be expressed in terms of incomplete elliptic integrals of the first and second kind. Furthermore, all indefinite integrals

$$\int R[t,(a_0t^4+a_1t^3+a_2t^2+a_3t+a_4)^{1/2}]\,dt$$

where R is a rational function of its arguments with constants a_0, a_1, a_2, a_3, a_4, can be expressed linearly in terms of elementary functions and elliptic integrals of the first, second, and third kinds.

15.9. ELLIPTIC FUNCTIONS

Theta Functions

The four theta functions ϑ_0, ϑ_1, ϑ_2, ϑ_3 of the variable v and the parameter τ as introduced by Jacobi are defined by

$$\vartheta_0(v,\tau) = (-i\tau)^{-1/2}\sum_{n=-\infty}^{\infty}\exp\left[-\frac{i\pi}{\tau}(v+n-\tfrac{1}{2})^2\right]$$

$$= \sum_{n=0}^{\infty}(-1)^n\epsilon_n e^{i\pi\tau n^2}\cos 2n\pi v \qquad (15.121a,b)$$

$$\vartheta_1(v,\tau) = (-i\tau)^{-1/2}\sum_{n=-\infty}^{\infty}(-1)^n\exp\left[-\frac{i\pi}{\tau}(v+n-\tfrac{1}{2})^2\right]$$

$$= 2\sum_{n=0}^{\infty}(-1)^n\exp[i\pi\tau(n+\tfrac{1}{2})^2]\sin[(2n+1)\pi v]$$

$$\vartheta_2(v,\tau) = (-i\tau)^{-\frac{1}{2}} \sum_{n=-\infty}^{\infty} (-1)^n \exp\left[-\frac{i\pi}{\tau}(v+n)^2\right]$$

$$= 2 \sum_{n=0} \exp\,[i\pi\tau(n+\tfrac{1}{2})^2] \cos\,[(2n+1)\pi v] \qquad (15.121c,d)$$

$$\vartheta_3(v,\tau) = (-i\tau)^{-\frac{1}{2}} \sum_{n=-\infty}^{\infty} \exp\left[-\frac{i\pi}{\tau}(v+n)^2\right] = \sum_{n=0}^{\infty} \epsilon_n e^{i\pi\tau n^2} \cos 2n\pi v$$

Here $\epsilon_0 = 1$, $\epsilon_n = 2$ for $n \geq 1$. The variable v can be arbitrarily complex while $\mathscr{I}\,\tau > 0$. It is customary to write $\vartheta_k(v,\tau) \equiv \vartheta_k(v)$. When the argument v of one of the theta functions is increased by $\frac{1}{2}(m + n\tau)$ where m, n are integers, then the result is a theta function of the original argument, but generally of a different kind, multiplied by an exponential factor. This property is displayed in Table 15.6.

Table 15.6

Increase of the argument	$\vartheta_0(v,\tau)$	$\vartheta_1(v,\tau)$	$\vartheta_2(v,\tau)$	$\vartheta_3(v,\tau)$	Exponential factor
$m + n\tau$	$(-1)^n\vartheta_0(v,\tau)$	$(-1)^{m+n}\vartheta_1(v,\tau)$	$(-1)^m\vartheta_2(v,\tau)$	$\vartheta_3(v,\tau)$	$e^{-n\pi i(2v+n\tau)}$
$m - \frac{1}{2} + n\tau$	$\vartheta_3(v,\tau)$	$(-1)^{m+n}\vartheta_2(v,\tau)$	$(-1)^{m+n}\vartheta_1(v,\tau)$	$(-1)^n\vartheta_0(v,\tau)$	
$m + (n + \frac{1}{2})\tau$	$(-1)^n i\vartheta_1(v,\tau)$	$(-1)^{m+n} i\vartheta_0(v,\tau)$	$(-1)^m\vartheta_3(v,\tau)$	$\vartheta_2(v,\tau)$	$e^{-(n+\frac{1}{2})\pi i[2v+(n+\frac{1}{2})\tau]}$
$m - \frac{1}{2} + (n + \frac{1}{2})\tau$	$\vartheta_2(v,\tau)$	$\vartheta_3(v,\tau)$	$(-1)^{m+n} i\vartheta_0(v,\tau)$	$(-1)^n i\vartheta_1(v,\tau)$	

The zeros of $\vartheta_k(v,\tau)$ are at $v = n + m\tau + \frac{1}{2}\tau$, $n + m\tau$, $n + m\tau + \frac{1}{2}$, and $n + m\tau + \frac{1}{2} + \frac{1}{2}\tau$ (n, m integers) for $k = 0$, 1, 2, and 3, respectively.

Jacobian Elliptic Functions

The Jacobian elliptic functions sn (v,k), cn (v,k), dn (v,k) of the (arbitrarily complex) variable v and the modulus k can be defined by means of the four theta functions:

$$\text{sn}\,(v,k) = \frac{1}{\sqrt{k}}\frac{\vartheta_1(z)}{\vartheta_0(z)} \qquad \text{cn}\,(v,k) = \sqrt{\frac{k'}{k}}\frac{\vartheta_2(z)}{\vartheta_0(z)} \qquad \text{dn}\,(v,k) = \sqrt{k'}\frac{\vartheta_3(z)}{\vartheta_0(z)} \qquad (15.122a\text{–}c)$$

Here $z = v/2K(k)$, $k' = (1 - k^2)^{\frac{1}{2}}$. The parameter τ of the theta functions is $\tau = iK'/K$.

The functions sn v, cn v, dn v are doubly periodic functions of v. For $k = 0$ the following relations hold:

$$\text{sn}\,(v,0) = \sin v \qquad \text{cn}\,(v,0) = \cos v \qquad \text{dn}\,(v,0) = 1 \qquad (15.123)$$

The power expansions of the elliptic functions are:

$$\text{sn}\,(v,k) = v - \frac{1}{3!}(1+k^2)v^3 + \frac{1}{5!}(1+14k^2+k^4)v^5 - \cdots$$

$$\text{cn}\,(v,k) = 1 - \frac{1}{2!}v^2 + \frac{1}{4!}(1+4k^2)v^4 - \frac{1}{6!}(1+44k^2+16k^4)v^6 + \cdots \qquad (15.124a\text{–}c)$$

$$\text{dn}\,(v,k) = 1 - \frac{1}{2!}k^2v^2 + \frac{1}{4!}k^2(4+k^2)v^4 - \frac{1}{6!}k^2(16+44k^2+k^4)v^6 + \cdots$$

There exist the relations

$$\operatorname{sn}^2(v,k) + \operatorname{cn}^2(v,k) = 1 \qquad \operatorname{dn}^2(v,k) + k^2 \operatorname{sn}^2(v,k) = 1 \tag{15.125}$$

The primitive periods, the zeros, and the poles of these functions are shown in Table 15.7, where $m, n = 0, \pm 1, \pm 2, \ldots$.

Table 15.7

Function	Primitive periods	Zeros	Poles
sn (v,k)	$4K$ $2iK'$	$2mK + 2niK'$	$2mK + (2n+1)iK'$
cn (v,k)	$4K$ $2K + 2iK'$	$(2m+1)K + 2niK'$	$2mK + (2n+1)iK'$
dn (v,k)	$2K$ $4iK'$	$(2m+1)K + (2n+1)iK'$	$2mK + (2n+1)iK'$

The Jacobian elliptic functions have an addition theorem:

$$\operatorname{sn}(u+v) = \frac{\operatorname{sn} u \operatorname{cn} v \operatorname{dn} v + \operatorname{sn} v \operatorname{cn} u \operatorname{dn} u}{1 - k^2 \operatorname{sn}^2 u \operatorname{sn}^2 v}$$

$$\operatorname{cn}(u+v) = \frac{\operatorname{cn} u \operatorname{cn} v - \operatorname{sn} u \operatorname{dn} u \operatorname{sn} v \operatorname{dn} v}{1 - k^2 \operatorname{sn}^2 u \operatorname{sn}^2 v} \tag{15.126a–c}$$

$$\operatorname{dn}(u+v) = \frac{\operatorname{dn} u \operatorname{dn} v - k^2 \operatorname{sn} u \operatorname{cn} u \operatorname{sn} v \operatorname{cn} v}{1 - k^2 \operatorname{sn}^2 u \operatorname{sn}^2 v}$$

The Jacobian elliptic functions are often defined by means of the inverse function of Legendre's normal form of the elliptic integral of the first kind:

$$u(k,\phi) \equiv F(k,\phi) = \int_0^\phi (1 - k^2 \sin^2 t)^{1/2}\, dt$$

When the inverse function is denoted by am (u,k)

$$\phi = \operatorname{am}(u,k)$$

then

$$\operatorname{sn}(u,k) = \sin\phi = \sin \operatorname{am}(u,k)$$

$$\operatorname{cn}(u,k) = \cos\phi = \cos \operatorname{am}(u,k) \tag{15.127a–c}$$

$$\operatorname{dn}(u,k) = (1 - k^2 \sin^2\phi)^{1/2} = [1 - k^2 \operatorname{sn}^2(v,k)]^{1/2}.$$

15.10. REFERENCES

Books

[1] A. Erdélyi: "Higher Transcendental Functions," 3 vols., McGraw-Hill, New York, 1953, 1955.
[2] P. Byrd, M. Friedman: "Handbook of Elliptic Integrals for Engineers and Physicists," Springer, Berlin, 1954.
[3] H. Hancock: "Elliptic Integrals," Wiley, New York, 1917.
[4] H. Hancock: "Lectures on the Theory of Elliptic Functions," Wiley, New York, 1910.
[5] E. W. Hobson: "The Theory of Spherical and Ellipsoidal Harmonics," Cambridge Univ. Press, New York, 1931.
[6] F. Klein: "Vorlesungen über die hypergeometrische Funktion," Springer, Berlin, 1933.
[7] J. Lense: "Kugelfunktionen," Akad.-Verlag., Leipzig, 1950.
[8] T. M. McRobert: "Spherical Harmonics," Methuen, London, 1947.
[9] E. H. Neville: "Jacobian Elliptic Functions," Oxford Univ. Press, New York, 1951.
[10] C. Snow: Hypergeometric and Legendre Functions with Applications to Integral Equations of Potential Theory, *Natl. Bur. Standards Appl. Math. Ser.*, **19** (1952).
[11] G. Szegö: Orthogonal Polynomials, *Colloq. Publ.*, **23**, Am. Math. Soc., Providence, R.I., 1959.

[12] G. N. Watson: "A Treatise on the Theory of Bessel Functions," Cambridge Univ. Press, New York, 1944.

[13] E. T. Whittaker, G. N. Watson: "A Course of Modern Analysis, Macmillan, New York, 1944.

Tables

[14] L. J. Comrie: "Chamber's Six Figure Mathematical Tables," Van Nostrand, Princeton, N.J., 1949.

[15] H. Davis: "Tables of Higher Mathematical Functions," Principia Press, Bloomington, Ind., 1935.

[16] W. Flügge: "Four-place Tables of Transcendental Functions," Pergamon, London, 1954.

[17] E. Jahnke, F. Emde: "Tables of Functions," Dover, New York, 1945.

[18] L. M. Milne-Thomson: "Jacobian Elliptic Function Tables," Dover, New York, 1950.

[19] *Natl. Bur. Standards Appl. Math. Ser.*, U.S. Government Printing Office.

1: Tables of the Bessel Functions $Y_0(x)$, $Y_1(x)$, $K_0(x)$, $K_1(x)$ (1948).

3: Tables of the Confluent Hypergeometric Function $F(n/2;\frac{1}{2};x)$ and Related Polynomials (1948).

9: Tables of Chebycheff Polynomials $S_n(x)$, $C_n(x)$ (1952).

17: Tables of Coulomb Wave Functions (1952).

23: Tables of Normal Probability Functions (1953).

32: Tables of Sine and Cosine Integrals (1954).

34: Tables of the Gamma Function for Complex Argument (1954).

41: Tables of the Error Function and its Derivative (1954).

51: Tables of the Exponential Integral for Complex Argument (1958).

[20] British Association for the Advancement of Sciences, "Tables," Cambridge Univ. Press, New York.

Vol. 6: "Bessel Functions I," 1937.

Vol. 7: "The Probability Integral," 1939.

Vol. 10: "Bessel Functions II," 1952.

[21] M. Schuler, H. Gebelein: "Five Place Tables of Elliptic Functions," Springer, Berlin 1955.

[22] A. M. Legendre: "Exercices de calcul intégral," vol. 3, table IX, Paris, 1816. Recent reprints of this table: F. Emde: "A. M. Legendre Tafeln der elliptischen Normalintegrale," Wittwer, Stuttgart, 1931; K. Pearson: "Tables of the Complete and Incomplete Elliptic Integrals," Univ. College, London, 1934.

CHAPTER 16

CALCULUS OF VARIATIONS

BY

P. W. BERG, Ph.D., Stanford, Calif.

16.1. FUNCTIONALS AND THE FIRST VARIATION

A *functional* is an operation which assigns a number to each of a certain class of functions. A typical form of functional is

$$I[u] = \int_{x_1}^{x_2} f(x,u,u')\,dx \tag{16.1}$$

where u is a function of x and u' is the derivative of u. We will consider primarily functionals of this form and generalizations to include higher-order derivatives, several independent, and several dependent variables.

The central problem of the calculus of variations is to find those functions in a prescribed class of functions for which a given functional has extreme (maximum or minimum) values.

The class of functions in which the extreme values are sought is called the class of *admissible functions*. For example, we might seek among all functions u for which

$$u(x_1) = u_1 \qquad u(x_2) = u_2 \tag{16.2}$$

where u_1 and u_2 are given numbers, those functions which extremize the functional (16.1). The class of admissible functions is here specifed by a *boundary condition*. Other conditions, *subsidiary conditions*, which involve the behavior of the function in the whole interval and not just at the end points, may also be prescribed.

We can immediately derive a necessary condition for an admissible function u to furnish an extreme value of a functional I. A function δu will be called an *admissible variation* if, for all sufficiently small ϵ, $u + \epsilon\,\delta u$ is an admissible function. If δu is an admissible variation, then $I[u + \epsilon\,\delta u]$ is a function of ϵ which has an extreme value when $\epsilon = 0$. From the differential calculus then, we know that the derivative of this function must vanish when $\epsilon = 0$, that is,

$$\delta I[u,\delta u] = \frac{d}{d\epsilon} I(u + \epsilon\,\delta u)\Big|_{\epsilon=0} = 0 \tag{16.3}$$

The functional δI, which depends on δu as well as u, is called the *first variation* of the functional I. This result may be stated: A necessary condition that the function u extremize the functional I is that the first variation, δI, of I at u vanish for all admissible variations δu.

If, for example, I is the functional (16.1), then substituting $u + \epsilon\,\delta u$, differentiating with respect to ϵ under the integral sign, and setting $\epsilon = 0$, we obtain

$$\delta I = \int_{x_1}^{x_2} \left[\delta u \frac{\partial f}{\partial u}(x,u,u') + \delta u' \frac{\partial f}{\partial u'}(x,u,u') \right] dx \tag{16.4}$$

where $\delta u'$ is the derivative of δu. Thus, a necessary condition that u extremize (16.1) is that (16.4) vanish for all admissible variations δu.

The first variation of a functional is the exact analog of the differential of a function of several variables—with the admissible variations of the function playing the role of the differential of the independent variables. The necessity of the vanishing of the first variation at an extremum is analogous to the necessity of the vanishing of the differential at an extremum. The precautions that must be observed in employing these necessary conditions are the same in both cases. Thus, although a necessary condition, the vanishing of the first variation is by no means sufficient. Furthermore, even when a function does furnish an extreme value, it is only a relative extreme value, that is, extreme compared to the values of the functional for all sufficiently close functions. Finally, a functional may have an absolute extreme value which is not a relative extreme value.

The term *extremal* of a functional is generally applied to all functions for which the first variation of the functional vanishes, whether or not the corresponding value of the functional is extreme. The value of a functional for an extremal is generally called a *stationary* value.

16.2. THE EULER EQUATION

In many cases the necessary condition of the preceding section can be transformed into a differential equation, integral equation, or similar operator equation for the extremal functions. In the following discussion, we assume that all functions and derivatives which appear are continuous.

Consider first the problem of extremizing (16.1) when the class of admissible functions is defined by (16.2). If u is an extremal, then, from (16.3) and (16.4), u satisfies

$$\int_{x_1}^{x_2} \left(\delta u \frac{\partial f}{\partial u} + \delta u' \frac{\partial f}{\partial u'} \right) dx = 0 \tag{16.5}$$

for all admissible variations δu. From (16.2), a function δu is an admissible variation if it satisfies

$$\delta u(x_1) = 0 \qquad \delta u(x_2) = 0 \tag{16.6}$$

We transform (16.5) by integrating the second term by parts; that is, we write

$$\delta u' \frac{\partial f}{\partial u'} = \frac{d}{dx} (\delta u) \frac{\partial f}{\partial u'} = \frac{d}{dx} \left(\delta u \frac{\partial f}{\partial u'} \right) - \delta u \frac{d}{dx} \left(\frac{\partial f}{\partial u'} \right) \tag{16.7}$$

and employ the theorem

$$\int_{x_1}^{x_2} \frac{d}{dx} G(x)\, dx = G(x) \Big|_{x_1}^{x_2} = G(x_2) - G(x_1) \tag{16.8}$$

Equation (16.5) then becomes

$$\int_{x_1}^{x_2} \delta u \left[\frac{\partial f}{\partial u} - \frac{d}{dx} \left(\frac{\partial f}{\partial u'} \right) \right] dx + \left[\delta u \frac{\partial f}{\partial u'} \right]_{x_1}^{x_2} = 0 \tag{16.9}$$

and, since the second term vanishes because of (16.6),

$$\int_{x_1}^{x_2} \delta u \left[\frac{\partial f}{\partial u} - \frac{d}{dx} \left(\frac{\partial f}{\partial u'} \right) \right] dx = 0 \tag{16.10}$$

From (16.10) we infer that u satisfies

$$\frac{d}{dx} \left(\frac{\partial f}{\partial u'} \right) - \frac{\partial f}{\partial u} = 0 \tag{16.11}$$

in virtue of a *fundamental lemma* which states: If

$$\int_{x_1}^{x_2} \delta u \, F \, dx = 0$$

for all functions δu which vanish at x_1 and x_2, then F vanishes identically.

Equation (16.11) is called the *Euler equation* for the functional (16.1). It is a second-order, in general nonlinear, ordinary differential equation.

The analogous problem for functions u of two independent variables x and y is to extremize the functional

$$I[u] = \iint_R f(x,y,u,u_x,u_y) \, dx \, dy \tag{16.12}$$

when the class of admissible functions consists of functions which assume given values on the boundary curve C of R. A discussion similar to that above, except that instead of (16.8) we employ the divergence theorem

$$\iint_R \left(\frac{\partial G}{\partial x} + \frac{\partial H}{\partial y} \right) dx \, dy = \int_C (G \, dy - H \, dx) \tag{16.13}$$

shows that the extremal functions u satisfy the second-order partial differential equation

$$\frac{\partial}{\partial x}\left(\frac{\partial f}{\partial u_x}\right) + \frac{\partial}{\partial y}\left(\frac{\partial f}{\partial u_y}\right) - \frac{\partial f}{\partial u} = 0 \tag{16.14}$$

Equation (16.14) is the Euler equation for the functional (16.12).

If the integrands in (16.1) or (16.12) contain derivatives up to the nth order, then the Euler equations are differential equations of order $2n$. Thus, for example, the Euler equations of the functionals

$$I[u] = \int_{x_1}^{x_2} f(x,u,u',u'') \, dx \tag{16.15}$$

$$I[u] = \iint_R f(x,y,u,u_x,u_y,u_{xx},u_{xy},u_{yy}) \, dx \, dy \tag{16.16}$$

are

$$\frac{d^2}{dx^2}\left(\frac{\partial f}{\partial u''}\right) - \frac{d}{dx}\left(\frac{\partial f}{\partial u'}\right) + \frac{\partial f}{\partial u} = 0 \tag{16.17}$$

$$\frac{\partial^2}{\partial x^2}\left(\frac{\partial f}{\partial u_{xx}}\right) + \frac{\partial^2}{\partial x \, \partial y}\left(\frac{\partial f}{\partial u_{xy}}\right) + \frac{\partial^2}{\partial y^2}\left(\frac{\partial f}{\partial u_{yy}}\right) - \frac{\partial}{\partial x}\left(\frac{\partial f}{\partial u_x}\right) - \frac{\partial}{\partial y}\left(\frac{\partial f}{\partial u_y}\right) + \frac{\partial f}{\partial u} = 0 \tag{16.18}$$

respectively.

When a functional depends on several functions, then, by considering all but one of the functions fixed and varying the remaining one, we see that a set of extremal functions satisfies the Euler equation with respect to each of the functions; that is, the extremal functions satisfy a *system* of differential equations. In the case

$$I[u] = \int_{x_1}^{x_2} f(x,u_1,u_2, \ldots ,u_n,u_1',u_2', \ldots ,u_n') \, dx \tag{16.19}$$

for example, a set of functions $u_1, \ldots, u_n$ which extremizes the functional satisfies the system

$$\frac{d}{dx}\left(\frac{\partial f}{\partial u_i'}\right) - \frac{\partial f}{\partial u_i} = 0 \qquad i = 1, 2, \ldots, n \tag{16.20}$$

The fact that a differential equation is the Euler equation for some functional is of considerable consequence for the discussion of the equation. One simple but useful property is the *covariance* of the Euler equation with respect to coordinate changes—the form of the Euler equation is unchanged when the coordinates are changed. If, for example, in (16.1) we introduce a new independent variable ξ,

$$I[u] = \int_{x_1}^{x_2} f\left(x,u,\frac{du}{dx}\right) dx = \int_{\xi_1}^{\xi_2} f\left(x,u,\frac{du/d\xi}{dx/d\xi}\right)\frac{dx}{d\xi}\,d\xi$$
$$= \int_{\xi_1}^{\xi_2} \phi\left(\xi,u,\frac{du}{d\xi}\right) d\xi \qquad (16.21)$$

then

$$\frac{d}{dx}\left(\frac{\partial f}{\partial u_x}\right) - \frac{\partial f}{\partial u} = \frac{1}{dx/d\xi}\left[\frac{d}{d\xi}\left(\frac{\partial \phi}{\partial u_\xi}\right) - \frac{\partial \phi}{\partial u}\right] \qquad (16.22)$$

Correspondingly, if ξ, η are introduced as new independent variables in (16.12),

$$\frac{\partial}{\partial x}\left(\frac{\partial f}{\partial u_x}\right) + \frac{\partial}{\partial y}\left(\frac{\partial f}{\partial u_y}\right) - \frac{\partial f}{\partial u} = \frac{1}{\partial(x,y)/\partial(\xi,\eta)}\left[\frac{\partial}{\partial \xi}\left(\frac{\partial \phi}{\partial u_\xi}\right) + \frac{\partial}{\partial \eta}\left(\frac{\partial \phi}{\partial u_\eta}\right) - \frac{\partial \phi}{\partial u}\right] \qquad (16.23)$$

These relations permit a change of variables in the Euler equation to be made by merely changing variables in the integrand of the functional.

The methods we have indicated are applicable to functionals of more general form than those discussed above, and yield equations from which the extremal functions can be determined. These equations need not be differential equations, however. For example, the Euler equation for the functional

$$I[u] = \int_a^b \int_a^b K(x,y)u(x)u(y)\,dx\,dy + \int_a^b u^2(y)\,dy - 2\int_a^b f(y)u(y)\,dy \qquad (16.24)$$

where $K(x,y)$ and $f(y)$ are given functions and $K(x,y) = K(y,x)$, is the integral equation

$$f(x) = u(x) + \int_a^b K(x,y)u(y)\,dy \qquad (16.25)$$

16.3. THE SECOND VARIATION AND SUFFICIENT CONDITIONS

Let I be a functional, u an admissible function, and δu an admissible variation. Then $I[u + \epsilon\,\delta u]$ is a function of ϵ. If we expand this function by Taylor's theorem, the coefficient of ϵ will be the first variation $\delta I[u,\delta u]$. The coefficient of $\epsilon^2/2$ is called the *second variation* of I and is denoted by $\delta^2 I[u,\delta u]$. We have

$$I[u + \epsilon\,\delta u] = I[u] + \epsilon\,\delta I[u,\delta u] + \frac{\epsilon^2}{2}\,\delta^2 I[u,\delta u] + \cdots \qquad (16.26)$$

The second variation can also be defined by

$$\delta^2 I[u,\delta u] = \frac{d^2}{d\epsilon^2} I[u + \epsilon\,\delta u]\Big|_{\epsilon=0} \qquad (16.27)$$

Just as the first variation is the analog of the differential of a function, the second variation is the analog of the second differential.

Now suppose that I has a minimum at u. Then $\delta I[u,\delta u] = 0$ and, from (16.26), we obtain

$$\tfrac{1}{2}\delta^2 I[u,\delta u] + \cdots = \frac{I[u + \epsilon\delta u] - I[u]}{\epsilon^2} \geq 0 \qquad (16.28)$$

where the omitted terms on the left vanish when $\epsilon = 0$. Letting ϵ tend to zero, we conclude that

$$\delta^2 I[u,\delta u] \geq 0 \qquad (16.29)$$

for all admissible variations δu. Thus: A necessary condition that the extremal u make I a minimum is that the second variation of I at u, $\delta^2 I[u, \delta u]$, be non-negative for all admissible variations δu. Correspondingly, a necessary condition for a maximum is that the second variation be nonpositive for all admissible variations.

One might suspect, by analogy with ordinary minimum problems, that strict inequality in (16.29) would be a sufficient condition for a minimum. This, however, is not correct. To find a sufficient condition we expand $I[u + \epsilon\, \delta u]$ by Taylor's theorem with a remainder after two terms. We obtain

$$I[u + \epsilon\, \delta u] = I[u] + \epsilon\, \delta I[u, \delta u] + \frac{\epsilon^2}{2}\, \delta^2 I[u + \bar{\epsilon}\, \delta u, \delta u] \tag{16.30}$$

where $\bar{\epsilon}$ is a number between 0 and ϵ. If u is an extremal, then

$$I[u + \epsilon\, \delta u] - I[u] = \frac{\epsilon^2}{2}\, \delta^2 I[u + \bar{\epsilon}\, \delta u, \delta u] \tag{16.31}$$

Now suppose that inequality (16.29) holds not just for the extremal u, but for all admissible functions u. Then we may set $\epsilon = 1$ in (16.31) to obtain

$$I[u + \delta u] - I[u] = \delta^2 I[u + \bar{\epsilon}\, \delta u, \delta u] \geq 0 \tag{16.32}$$

which shows that u actually minimizes I.

A functional for which (16.29) holds for all admissible functions u and admissible variations δu is called *convex*. If strict inequality holds unless $\delta u = 0$, the functional is called *strictly convex*. We have shown that: A sufficient condition that an extremal u minimize a functional I is that I be convex. The minimum in this case is the absolute minimum of the functional. If the functional is strictly convex, then we can also conclude from (16.32) that the minimizing function is the only extremal of the functional.

Let I be the functional (16.1). Setting $u + \epsilon\, \delta u$ in the integrand, and expanding the integrand by Taylor's theorem, we find

$$\delta^2 I[u, \delta u] = \int_{x_1}^{x_2} \left[\frac{\partial^2 f}{\partial u^2} (\delta u)^2 + 2 \frac{\partial^2 f}{\partial u\, \partial u'}\, \delta u\, \delta u' + \frac{\partial^2 f}{\partial u'^2} (\delta u')^2 \right] dx \tag{16.33}$$

From this it can be shown, with a suitable choice of the variation δu, that a necessary condition that an extremal u of I minimize I is that *Legendre's condition*

$$\frac{\partial^2 f}{\partial u'^2} \geq 0 \tag{16.34}$$

hold along the extremal. A sufficient condition for a minimum is that the quadratic form under the integral sign in (16.33) be positive semidefinite for all admissible u and δu, that is,

$$\frac{\partial^2 f}{\partial u'^2} \frac{\partial^2 f}{\partial u^2} - \left(\frac{\partial^2 f}{\partial u\, \partial u'} \right)^2 \geq 0 \tag{16.35}$$

for all admissible u. If strict inequality holds in (16.35), then the variational problem has a unique solution.

Similar statements can be made in the cases of several independent or several dependent variables. Examples will be given in the next section.

16.4. QUADRATIC FUNCTIONALS

Functionals in which the integrand is quadratic in u and its derivatives are called *quadratic functionals*, and are of particular interest because their Euler equations are linear. If the functionals (16.1), (16.12) are quadratic, they may be written in the forms

$$Q[u] = \int_{x_1}^{x_2} [a(u')^2 + bu^2]\, dx \tag{16.36}$$

$$Q[u] = \iint_R [au_x^2 + 2bu_xu_y + cu_y^2 + eu^2]\, dx\, dy \tag{16.37}$$

where $a, b, \ldots$ are functions of x or x and y, respectively. The corresponding Euler equations are:

$$\frac{d}{dx}(au') - bu = 0 \tag{16.38}$$

$$\frac{\partial}{\partial x}(au_x + bu_y) + \frac{\partial}{\partial y}(cu_y + bu_x) - eu = 0 \tag{16.39}$$

From (16.38), (16.39) and the corresponding equations for functionals involving higher-order derivatives, we see that a necessary and sufficient condition that a linear differential equation be the Euler equation of a quadratic functional is that it be *self-adjoint*. Thus, for example, the potential equation and the wave equation can be derived from variational problems, but the heat equation cannot.

The case of second-order ordinary differential equations is special. Every such equation can be put into self-adjoint form. If the equation is first written in the form

$$u'' + pu' + qu = 0 \tag{16.40}$$

then, multiplying by $e^{\int p\,dx}$, we obtain

$$\frac{d}{dx}(e^{\int p\,dx}u') + qe^{\int p\,dx}u = 0 \tag{16.41}$$

which is of the form (16.38). The corresponding possibility does not exist for partial differential equations or for higher-order ordinary differential equations.

If Q is a quadratic functional, the second variation of Q is easily found to be

$$\delta^2 Q[u,\delta u] = 2Q[\delta u] \tag{16.42}$$

and is independent of u. Thus, for quadratic functionals the condition

$$Q[\delta u] \geq 0 \tag{16.43}$$

for all admissible variations δu is both necessary and sufficient for an extremal to minimize the functional. If strict inequality holds in (16.43), then there is at most one extremal.

From (16.43) we infer that

$$a \geq 0 \qquad b \geq 0 \tag{16.44}$$

is a sufficient condition for a minimum of (16.36), and the first of these inequalities is also necessary. A sufficient condition for a minimum of (16.37) is that the quadratic form

$$a\xi^2 + 2b\xi\eta + c\eta^2 + e\zeta^2 \tag{16.45}$$

be positive-semidefinite, that is,

$$a \geq 0 \qquad ac - b^2 \geq 0 \qquad e \geq 0 \tag{16.46}$$

The first two of these inequalities are also necessary. They express the requirement that the differential equation (16.39) be of *elliptic* (or elliptic-parabolic) type.

If strict inequality holds in (16.44) or (16.46), then the solution of the boundary-value problem for the corresponding Euler equation is unique.

16.5. BOUNDARY CONDITIONS

We have seen that, when the boundary conditions are those of (16.2), the boundary-value problem for the differential equation (16.11) is equivalent to a variational problem. To investigate the possibility of formulating an equivalent variational problem when the boundary conditions are more general, we first consider the problem of extremizing (16.1) when the admissible functions are not restricted by any condition.

In this case, the requirement of the vanishing of the first variation leads, just as in Sec. 16.2, to Eq. (16.9). Now, however, every variation δu is admissible. Since, in particular, those variations which vanish at x_1 and x_2 are admissible, we see just as before that an extremal u satisfies the Euler equation. Thus, the first term in (16.9) vanishes for all variations δu, and we have

$$\left[\delta u \frac{\partial f}{\partial u'}\right]_{x_1}^{x_2} = \delta u(x_2) \frac{\partial f}{\partial u'}(x_2,u_2,u_2') - \delta u(x_1) \frac{\partial f}{\partial u'}(x_1,u_1,u_1') = 0 \tag{16.47}$$

where

$$u_1 = u(x_1) \qquad u_2 = u(x_2) \qquad u_1' = u'(x_1) \qquad u_2' = u'(x_2) \tag{16.48}$$

Since (16.47) must hold for all variations δu, we conclude that an extremal u must also satisfy the *natural boundary conditions:*

$$\frac{\partial f}{\partial u'}(x_1,u_1,u_1') = 0 \qquad \frac{\partial f}{\partial u'}(x_2,u_2,u_2') = 0 \tag{16.49}$$

Thus, when the boundary conditions are given by (16.49), the boundary-value problem for (16.11) is equivalent to the problem of extremizing (16.1) when the class of admissible functions is unrestricted.

The importance of the natural boundary conditions lies in the fact that, by adding suitable boundary terms to a functional, it is possible to alter the natural boundary conditions without changing the Euler equation. If to the functional (16.1) we add $\phi(u_1) + \psi(u_2)$, where ϕ and ψ are arbitrary functions, we obtain a functional

$$\int_{x_1}^{x_2} f(x,u,u')\,dx + \phi(u_1) + \psi u_2) \tag{16.50}$$

whose first variation is

$$\int_{x_1}^{x_2} \delta u \left[\frac{\partial f}{\partial u} - \frac{d}{dx}\frac{\partial f}{\partial u'}\right] dx + \delta u \left(\frac{\partial f}{\partial u'} + \psi'\right)\Big|_{x_2} - \delta u \left(\frac{\partial f}{\partial u'} - \phi'\right)\Big|_{x_1} \tag{16.51}$$

so that the Euler equation is still (16.11), while the natural boundary conditions are

$$\frac{\partial f}{\partial u'}(x_1,u_1,u_1') - \phi'(u_1) = 0 \qquad \frac{\partial f}{\partial u'}(x_2,u_2,u_2') + \psi'(u_2) = 0 \tag{16.52}$$

We apply these results to the quadratic functional (16.36). If the coefficient $a(x)$ does not vanish at x_1 or x_2, then a function which extremizes (16.36) when the class of admissible functions is unrestricted satisfies the boundary conditions

$$u'(x_1) = 0 \qquad u'(x_2) = 0 \tag{16.53}$$

and is thus a solution of the *homogeneous second boundary-value problem* for the differential equation (16.38). If we modify (16.36) by adding the terms $a(x_2)k_2u^2(x_2) - a(x_1)k_1u^2(x_1)$, where k_1 and k_2 are arbitrary constants, then the boundary conditions become those of the *homogeneous third boundary-value problem,*

$$u'(x_1) + k_1u(x_1) = 0 \qquad u'(x_2) + k_2u(x_2) = 0 \tag{16.54}$$

It is worth observing that the conditions of the *homogeneous first boundary-value problem,*

$$u(x_1) = 0 \qquad u(x_2) = 0 \tag{16.55}$$

cannot be obtained as natural boundary conditions, but may be regarded as the limiting case of (16.54) as k_1 and k_2 become infinite.

The *inhomogeneous* boundary conditions,

$$u'(x_1) = l_1 \qquad u'(x_2) = l_2 \tag{16.56}$$

$$u'(x_1) + k_1u(x_1) = l_1 \qquad u'(x_2) + k_2u(x_2) = l_2 \tag{16.57}$$

corresponding to (16.53), (16.54), can be obtained by adding to the functional the further terms $2a(x_1)l_1u(x_1) - 2a(x_2)l_2u(x_2)$.

A similar discussion holds for problems involving several independent variables. The natural boundary condition for the functional (16.12) is

$$\frac{\partial f}{\partial u_x}\frac{dy}{ds} - \frac{\partial f}{\partial u_y}\frac{dx}{ds} = 0 \tag{16.58}$$

where $x = x(s)$, $y = y(s)$ are the parametric equations, in terms of the arc length s, of the boundary C of the region R. If to (16.12) we add a *boundary integral,*

$$\int_C \phi(s,u)\, ds \tag{16.59}$$

then the natural boundary condition becomes

$$\frac{\partial f}{\partial u_x}\frac{dy}{ds} - \frac{\partial f}{\partial u_y}\frac{dx}{ds} + \frac{\partial \phi}{\partial u} = 0 \tag{16.60}$$

For the important special case of the functional

$$\iint\limits_R (u_x^2 + u_y^2)\, dx\, dy \tag{16.61}$$

whose Euler equation is Laplace's equation, the natural boundary condition is

$$u_x\frac{dy}{ds} - u_y\frac{dx}{ds} = \frac{\partial u}{\partial n} = 0 \tag{16.62}$$

where $\partial u/\partial n$ is the derivative of u in the direction of the outer normal to the boundary. If $k(s)$ is an arbitrary function, the functional

$$\iint\limits_R (u_x^2 + u_y^2)\, dx\, dy + \int_C k(s)u^2\, ds \tag{16.63}$$

has the same Euler equation as (16.61), but the associated natural boundary condition is

$$\frac{\partial u}{\partial n} + k(s)u = 0 \tag{16.64}$$

Finally, if a functional depends on several functions, each of which is unrestricted by boundary conditions, then the natural boundary conditions with respect to each function must be satisfied, and we obtain a system of boundary conditions.

16.6. SUBSIDIARY CONDITIONS

Consider the problem of extremizing (16.1) when the admissible functions are required to satisfy, in addition to (16.2), the *subsidiary condition*,

$$J[u] = \int_{x_1}^{x_2} g(x,u,u')\,dx = c \tag{16.65}$$

where c is some constant. Such a problem is called an *isoperimetric problem.*

Let u be a solution of this problem, and let δu_1 and δu_2 be arbitrary variations satisfying (16.6). Then $I[u + \epsilon_1\,\delta u_1 + \epsilon_2\,\delta u_2]$ and $J[u + \epsilon_1\,\delta u_1 + \epsilon_2\,\delta u_2]$ are functions of the variables ϵ_1 and ϵ_2, and $\epsilon_1 = \epsilon_2 = 0$ is a solution of the problem of extremizing the first function subject to the condition that the second function have the value c. From the differential calculus, then, we know that there is a constant λ, called a *Lagrange multiplier*, such that

$$\begin{aligned} \left\{\frac{\partial}{\partial \epsilon_1}(I[u + \epsilon_1\,\delta u_1 + \epsilon_2\,\delta u_2] + \lambda J[u + \epsilon_1\,\delta u_1 + \epsilon_2\,\delta u_2])\right\}_{\epsilon_1=\epsilon_2=0} &= 0 \\ \left\{\frac{\partial}{\partial \epsilon_2}(I[u + \epsilon_1\,\delta u_1 + \epsilon_2\,\delta u_2] + \lambda J[u + \epsilon_1\,\delta u_1 + \epsilon_2\,\delta u_2])\right\}_{\epsilon_1=\epsilon_2=0} &= 0 \end{aligned} \tag{16.66}$$

Either of these equations may be written

$$\delta(I + \lambda J) = 0 \tag{16.67}$$

and, since in (16.67) the variation of u is arbitrary except for the condition (16.6), we see that u satisfies the Euler equation,

$$\frac{d}{dx}\frac{\partial}{\partial u'}(f + \lambda g) - \frac{\partial}{\partial u}(f + \lambda g) = 0 \tag{16.68}$$

Equations (16.68) and (16.65), together with the boundary conditions, determine the extremal u and the multiplier λ.

This result, which is known as the *multiplier rule,* is easily extended to the case of any number of subsidiary conditions of the form (16.65). The general statement is: If u extremizes the functional

$$\int_{x_1}^{x_2} f(x,u,u')\,dx$$

subject to the subsidiary conditions,

$$\int_{x_1}^{x_2} g_i(x,u,u')\,dx = c_i \qquad i = 1, 2, \ldots, n$$

then u satisfies the Euler equation

$$\frac{d}{dx}\frac{\partial}{\partial u'}(f + \lambda_1 g_1 + \cdots + \lambda_n g_n) - \frac{\partial}{\partial u}(f + \lambda_1 g_1 + \cdots + \lambda_n g_n) = 0 \tag{16.69}$$

where the multipliers $\lambda_1, \ldots, \lambda_n$ are constants.

In the case of functionals depending on several functions, the multiplier rule continues to hold. Thus, if u and v extremize

$$\int_{x_1}^{x_2} f(x,u,v,u',v')\,dx \tag{16.70}$$

subject to the condition

$$\int_{x_1}^{x_2} g(x,u,v,u',v')\,dx = c \tag{16.71}$$

then u and v satisfy the system

$$\begin{aligned}\frac{d}{dx}\frac{\partial}{\partial u'}(f + \lambda g) - \frac{\partial}{\partial u}(f + \lambda g) &= 0\\ \frac{d}{dx}\frac{\partial}{\partial v'}(f + \lambda g) - \frac{\partial}{\partial v}(f + \lambda g) &= 0\end{aligned} \tag{16.72}$$

where the multiplier λ is a constant.

The subsidiary conditions in an isoperimetric problem are *functional equations.* Subsidiary conditions may, however, also have the form of *operator equations,* such as differential or integral equations. For such subsidiary conditions, a multiplier rule holds which is formally similar to that for isoperimetric problems, with the important difference that the multiplier is not a constant but a function of the independent variable. For example, if u and v extremize (16.70) in the class of function pairs which satisfy the differential equation

$$g(x,u,v,u',v') = 0 \tag{16.73}$$

then u and v satisfy the system (16.72), where now λ is a function of x. The three functions u, v, and λ are determined by the system of three differential equations (16.72) and (16.73) and the boundary conditions.

For functionals involving several independent variables statements corresponding exactly to those above hold.

16.7. EIGENVALUE PROBLEMS

The multiplier rule of the last section permits us to give a variational formulation to the problem of determining the eigenvalues and eigenfunctions of a self-adjoint differential operator.

Let $\lambda_1 < \lambda_2 < \cdots < \lambda_n < \cdots$ be the eigenvalues and $u_1, u_2, \ldots, u_n, \ldots$ the associated eigenfunctions of the eigenvalue problem:

$$\frac{d}{dx}(au') - bu + \lambda ru = 0 \qquad 0 \le x \le l \tag{16.74}$$

$$u(0) = 0 \qquad u(l) = 0 \tag{16.75}$$

Here a, b, and r are functions of x. The eigenfunctions are determined only up to a constant factor, which we assume so chosen that

$$\int_0^l r u_i^2\,dx = 1 \qquad i = 1, 2, \ldots, n, \ldots \tag{16.76}$$

We shall use the relation

$$\int_0^l [a(u_i')^2 + b u_i^2]\,dx = \lambda_i \tag{16.77}$$

which is obtained by multiplying (16.74) by u and then integrating by parts.

Now consider the problem of minimizing

$$\int_0^l [a(u')^2 + bu^2]\,dx \tag{16.78}$$

subject to the boundary conditions (16.75) and the subsidiary condition

$$\int_0^l ru^2\,dx = 1 \tag{16.79}$$

Using the multiplier rule, writing $-\lambda$ instead of λ for the multiplier, we find that the Euler equation for this problem is exactly (16.74). Thus, the solution of the problem is an eigenfunction, and, from (16.77), the corresponding minimum value is an eigenvalue, which is clearly the smallest eigenvalue. We have obtained a variational characterization of λ_1 and u_1.

If u_1 has been determined, we can determine u_2 as the solution of the problem of minimizing (16.78) in the class of functions satisfying (16.75), (16,79) and the additional subsidiary condition

$$\int_0^l ruu_1\,dx = 0 \tag{16.80}$$

For the Euler equation of this problem will be

$$\frac{d}{dx}(au') - bu + \lambda ru + \mu ru_1 = 0 \tag{16.81}$$

where λ and μ are multipliers. Writing

$$\frac{d}{dx}(au_1') - bu_1 + \lambda_1 ru_1 = 0 \tag{16.82}$$

multiplying (16.81) by u_1 and (16.82) by u, subtracting one equation from the other, and integrating by parts, we obtain, using (16.80),

$$\mu \int_0^l ru_1^2\,dx = 0 \tag{16.83}$$

so that $\mu = 0$.

Continuing in this fashion, we establish the following *minimum principle:* The nth eigenfunction u_n is the solution of the problem of minimizing

$$\int_0^l [a(u')^2 + bu^2]\,dx$$

subject to the boundary conditions and the subsidiary conditions

$$\int_0^l ru^2\,dx = 1 \qquad \int_0^l ruu_i\,dx = 0 \qquad i = 1, 2, \ldots, n-1$$

The corresponding minimum value is λ_n, the nth eigenvalue. It is useful to remark that, since the integrands in (16.78) and (16.79) are both homogeneous of the second degree, the subsidiary condition (16.79) can be dropped if the functional (16.78) is replaced by the quotient

$$\frac{\int_0^l [a(u')^2 + bu^2]\,dx}{\int_0^l ru^2\,dx} \tag{16.84}$$

A principal consequence of a variational formulation of a problem is the possibility of stating *comparison theorems*. These are based on the following simple propositions: If the class of admissible functions is the same in two variational problems, but, for each admissible function, the value of the first functional is smaller than the value of the second, then the minimum in the first problem is not larger than the minimum in the second; if the functional is the same in two variational problems, but the class of admissible functions in the first problem is larger than (that is, includes) that in the second, then the minimum in the frst problem is not larger than that in the second. With these considerations and the minimum principle for eigenvalues, we see easily that an increase in the coefficients a or b causes an increase in the first eigenvalue.

Similarly, using (16.84), an increase in the coefficient r causes a decrease in the first eigenvalue. Finally, an increase in the length l of the interval of integration causes a decrease in the first eigenvalue, since every function in the smaller interval can be regarded as defined in the larger interval if we take its value to be zero wherever it is not already defined.

These relations hold also for higher-order eigenvalues. They cannot be established on the basis of the minimum principle, however, for, since the first eigenfunctions in two different problems will not be the same, the classes of admissible functions in the problems for the second- and higher-order eigenvalues will be different. What is needed is a variational characterization of the nth eigenvalue which does not require the prior determination of the first $(n-1)$ eigenvalues and eigenfunctions. This is furnished by the *maximinimum principle:* Let $\phi_1, \phi_2, \ldots, \phi_{n-1}$ be an arbitrary set of functions. Denote by $\lambda(\phi_1,\phi_2, \ldots,\phi_{n-1})$ the minimum of

$$\int_0^l [a(u')^2 + bu^2]\,dx$$

in the class of functions satisfying the boundary conditions and the subsidiary conditions,

$$\int_0^l ru^2\,dx = 1 \qquad \int_0^l ru\phi_i\,dx = 0 \qquad i = 1, 2, \ldots, n-1$$

Then the nth eigenvalue λ_n is the maximum of $\lambda(\phi_1,\phi_2, \ldots,\phi_{n-1})$ when all possible choices of the set $\phi_1, \phi_2, \ldots, \phi_{n-1}$ are admitted. Using this principle, and comparing maximinima instead of minima, the comparison theorems for higher-order eigenvalues are readily proved.

For the sake of simplicity we have discussed the eigenvalue problem for a second-order ordinary differential operator with the homogeneous first boundary condition. Our conclusions hold equally well for partial differential operators, higher-order operators, and other boundary conditions. It should be emphasized, however, that the operator and boundary conditions must be self-adjoint.

16.8. APPROXIMATION METHODS

A variational formulation of a boundary-value problem is particularly useful for the approximate computation of the solution. The most widely used approximating procedures are the Ritz method and the closely related Galerkin method.

The *Ritz method* is applicable to those variational problems which satisfy the sufficiency conditions of Secs. 16.3 and 16.4. The central idea is that of a *minimizing sequence.* A sequence $\bar{u}_1, \bar{u}_2, \ldots, \bar{u}_n, \ldots$ of admissible functions is called a minimizing sequence for the functional I if $I[\bar{u}_n]$ converges to the minimum value of I. Let $\bar{u}$ be the admissible function for which I has its minimum value. Then, since $I[\bar{u}_n]$ converges to $I[\bar{u}]$, we may expect that $\bar{u}_n$ will converge to $\bar{u}$. In many cases this expectation is correct.

The answer to the question whether every minimizing sequence converges to the minimizing function depends on the number of independent variables and the order of the derivatives in the integrand of the functional. For problems in one independent variable, it is affirmative in every case; for problems in two or three independent variables, the integrand must contain derivatives of second or higher order. However, even when the convergence of every minimizing sequence cannot be guaranteed, suitably chosen minimizing sequences will be convergent.

To construct a minimizing sequence, we first select a set $\phi_1, \phi_2, \ldots, \phi_n, \ldots$ of admissible functions such that each admissible function can be approximated arbitrarily closely, together with its derivatives, by a suitable linear combination with constant coefficients

$$c_1\phi_1 + c_2\phi_2 + \cdots + c_n\phi_n \tag{16.85}$$

and its derivatives. Such a set of admissible functions is called *complete.*

Now consider $I[c_1\phi_1 + c_2\phi_2 + \cdots + c_n\phi_n]$. This is a function of the n variables $c_1, c_2, \ldots, c_n$. We minimize this function by solving the algebraic system of equations,

$$\frac{\partial}{\partial c_i} I[c_1\phi_1 + c_2\phi_2 + \cdots + c_n\phi_n] = 0 \qquad i = 1, 2, \ldots, n \tag{16.86}$$

Let $\bar{c}_1, \bar{c}_2, \ldots, \bar{c}_n$ be the solution of (16.86), and set

$$\bar{u}_n = \bar{c}_1\phi_1 + \bar{c}_2\phi_2 + \cdots + \bar{c}_n\phi_n \tag{16.87}$$

Clearly

$$I[\bar{u}_1] \geq I[\bar{u}_2] \geq \cdots \geq I[\bar{u}_n] \geq \cdots \geq I[\bar{u}] \tag{16.88}$$

The sequence $\bar{u}_1, \bar{u}_2, \ldots, \bar{u}_n, \ldots$ is a minimizing sequence. For there are functions u_n of the form (16.85) which are arbitrarily close to $\bar{u}$, and whose derivatives are arbitrarily close to those of $\bar{u}$. Then $I[u_n]$ will be arbitrarily close to $I[\bar{u}]$, and from

$$I[\bar{u}] \leq I[\bar{u}_n] \leq I[u_n] \tag{16.89}$$

it follows that $I[\bar{u}_n]$ will be still closer to $I[\bar{u}]$.

When the class of admissible functions is the class of all functions, the complete sets of functions most commonly used are the powers x^n ($n = 0, 1, 2, \ldots$) in one-variable problems, and the power products x^ny^m ($n = 0, 1, 2, \ldots; m = 0, 1, 2, \ldots$) in two-variable problems. When the admissible functions are those satisfying the homogeneous first boundary condition, complete sets of functions can be obtained by multiplying each of the powers or power products by some simple function which vanishes on the boundary. For example, $x(l - x)x^n$ ($n = 0, 1, 2, \ldots$) is a complete set of functions vanishing on the boundary of the interval $0 \leq x \leq l$, and $xy(l - x)(l - y)x^ny^m$ ($n = 0, 1, 2, \ldots; m = 0, 1, 2, \ldots$) is a complete set of functions vanishing on the boundary of the square $0 \leq x \leq l$, $0 \leq y \leq l$. Other complete sets of functions in these cases are $\sin(n\pi x/l)$ ($n = 1, 2, \ldots$) and $\sin(n\pi x/l)\sin(m\pi y/l)$ ($n = 1, 2, \ldots; m = 1, 2, \ldots$). If the boundary conditions are inhomogeneous, the problem should first be transformed into one with homogeneous boundary conditions by subtracting from the dependent variable some simple function which satisfies the boundary conditions.

As indicated above, the Ritz procedure will certainly be convergent in all one-variable problems, and in all two- or three-variable problems in which the order of the Euler differential equation is at least four. In other cases, sufficient conditions for convergence can be stated, and usually are satisfied, although they are complicated to verify. As a practical matter, it is found that good approximations are obtained with as few as two or three terms.

The Ritz procedure is particularly simple and effective for linear boundary-value problems, since, in this case, the algebraic problem to which it leads is a system of linear equations. Consider, for example, the problem

$$\frac{d}{dx}(au') - bu - f = 0 \qquad u(x_1) = 0,\ u(x_2) = 0 \tag{16.90}$$

The differential equation (16.90) is the Euler equation of the functional

$$I[u] = \int_{x_1}^{x_2} [a(u')^2 + bu^2 + 2fu]\,dx \tag{16.91}$$

If in (16.91) we substitute

$$u_n = \sum_{i=1}^{n} c_i \phi_i \tag{16.92}$$

differentiate with respect to c_i, and equate to zero, we obtain the system

$$\int_{x_1}^{x_2} [a u_n' \phi_i' + b u_n \phi_i + f \phi_i] \, dx = 0 \qquad i = 1, 2, \ldots, n \tag{16.93}$$

which may also be written as

$$\sum_{j=1}^{n} \alpha_{ij} c_j + \beta_i = 0 \qquad i = 1, 2, \ldots, n \tag{16.94}$$

with

$$\alpha_{ij} = \int_{x_1}^{x_2} [a \phi_j' \phi_i' + b \phi_j \phi_i] \, dx \qquad \beta_i = \int_{x_1}^{x_2} f \phi_i \, dx \tag{16.95}$$

The Ritz method can also be employed for the approximate determination of eigenvalues and eigenfunctions. To approximate the solutions to the problem (16.74), (16.75), we substitute (16.92) in (16.78) and (16.79), and use the multiplier rule for ordinary minimum problems. We thus obtain the system

$$\int_{x_1}^{x_2} [a u_n' \phi_i' + b u_n \phi_i - \lambda u_n \phi_i] \, dx = 0 \qquad i = 1, 2, \ldots, n \tag{16.96}$$

or

$$\sum_{j=1}^{n} (\alpha_{ij} - \lambda \gamma_{ij}) c_j = 0 \qquad i = 1, 2, \ldots, n \tag{16.97}$$

with

$$\gamma_{ij} = \int_{x_1}^{x_2} r \phi_i \phi_j \, dx \tag{16.98}$$

The system (16.97) will have a nontrivial solution if and only if its determinant vanishes,

$$\begin{vmatrix} \alpha_{11} - \lambda\gamma_{11} & \cdots & \alpha_{1n} - \lambda\gamma_{1n} \\ \cdots & \cdots & \cdots \\ \alpha_{n1} - \lambda\gamma_{n1} & \cdots & \alpha_{nn} - \lambda\gamma_{nn} \end{vmatrix} = 0 \tag{16.99}$$

This is an nth-degree equation for λ, and its n roots are approximations to the first n eigenvalues of the problem. The corresponding nontrivial solutions substituted in (16.92) furnish approximations to the corresponding eigenfunctions.

The *Galerkin method* is a reformulation of the Ritz method; that is, the approximating functions have the same form, and the equations for the coefficients are the same, although written in a different fashion. We arrive at the Galerkin formulation for the problem (16.90) by transforming equations (16.93) by integration by parts of the first terms. We obtain

$$\int_{x_1}^{x_2} \left[\frac{d}{dx} (a u_n') - b u_n - f \right] \phi_i \, dx = 0 \qquad i = 1, 2, \ldots, n \tag{16.100}$$

Observe that the coefficient of ϕ_i is just the differential operator of the problem applied to u_n. In this form, the equations are more easily remembered and written down. The principal advantage of the Galerkin formulation, however, is that it is also applicable to equations which are not the Euler equations of proper variational problems, for example, to non-self-adjoint equations.

There is a variant of the Ritz method for two-dimensional problems which is useful. When the boundary condition is the homogeneous first boundary condition, we seek an approximating solution of the form

$$u_n(x,y) = w(x,y)[v_0(x) + y v_1(x) + \cdots + y^n v_n(x)] \tag{16.101}$$

where $w(x,y)$ is some function which vanishes on the boundary, and $v_0, v_1, \ldots, v_n$ are unknown functions of x which are to be determined. If we substitute (16.101) in the functional and seek the functions $v_0, \ldots, v_n$ which minimize it, we are led to a system of $(n + 1)$ ordinary differential equations. This method furnishes better approximations than the Ritz method. It is particularly useful for partial differential equations with constant coefficients in rectangular domains, since then the system of ordinary differential equations to be solved has constant coefficients.

16.9. REFERENCES

[1] R. Courant, D. Hilbert: "Methods of Mathematical Physics," vol. 1, Interscience, New York, 1953.

[2] L. V. Kantorovich, V. I. Krylov: "Approximate Methods of Higher Analysis," transl. by C. D. Benster, Noordhoff, Groningen, Netherlands, 1958.

[3] H. Lewy: "Aspects of the Calculus of Variations," Univ. of California Press, Berkeley, Calif., 1939.

CHAPTER 17

INTEGRAL EQUATIONS

BY

M. A. HEASLET, Ph.D., Los Altos, Calif.

17.1. INTRODUCTION

An integral equation is an equation in which an unknown function appears under an integral sign. Many theoretical analyses in engineering can be formulated alternatively through the use of differential or integral equations (e.g., eigenvalue problems in the theory of elasticity); in the latter case, the boundary conditions are built into the problem itself. On the other hand, if certain phenomena involving averaging or statistical effects are to be studied, then the use of integrals and integral equations is of paramount utility in the analytical characterization.

Linear integral equations occur when the unknown function appears linearly, and the most far-reaching theory stems from consideration of the two standard linear types:

Fredholm's integral equation of the first kind

$$g(x) = \int_a^b K(x,\xi)f(\xi)\,d\xi \tag{17.1}$$

Fredholm's integral equation of the second kind

$$f(x) = g(x) + \lambda \int_a^b K(x,\xi)f(\xi)\,d\xi \tag{17.2}$$

The functions $K(x,\xi)$ and $g(x)$, the constant limits a and b, and the parameter λ in Eq. (17.2) are known. The function $f(x)$ is to be determined so as to satisfy the given equation in the region (a,b), that is, for $a \leqq x \leqq b$. $K(x,\xi)$ is called the *kernel* of the equation.

An important class of equations appears when $K(x,\xi) \equiv 0$ for $x < \xi$. The upper limit on the integral sign is then x, and Eqs. (17.1) and (17.2) are called respectively *Volterrs's integral equations of the first and second kind.*

Singular integral equations occur when the region of integration $(a \leqq \xi \leqq b)$ becomes infinite or when the kernel possesses a singularity within the region. Such equations require special consideration, and the following sections treat, first, nonsingular equations where $g(x)$ is continuous within the region of integration and the kernel is piecewise-continuous. The restriction on $K(x,\xi)$ can, in fact, be relaxed further [1] to include kernels for which the integrals

$$\int_a^b \int_a^b [K(x,\xi)]^2\,dx\,d\xi \qquad \int_a^b [K(x,\xi)]^2\,d\xi \qquad \int_a^b [K(x,\xi)]^2\,dx$$

exist and for which the two latter integrals are bounded functions of x and ξ, respectively.

When $g(x) \equiv 0$, then Eq. (17.2) becomes the homogeneous Fredholm equation

$$f(x) = \lambda \int_a^b K(x,\xi) f(\xi)\, d\xi \tag{17.3}$$

Linear Independence. A set of functions $h_1(x)$, $h_2(x)$, . . . , $h_n(x)$ is said to be *linearly independent* if *no* constants a_1, a_2, . . . , a_n exist (at least one of which is not zero) for which

$$a_1 h_1(x) + a_2 h_2(x) + \cdots + a_n h_n(x) \equiv 0 \tag{17.4}$$

If such a set of constants does exist, the functions are *linearly dependent.* Suppose that m and only m of the functions are linearly independent ($m < n$); it follows that, by means of Eq. (17.4), each member of the original set can be expressed as a linear combination of the m linearly independent functions.

Orthogonal Functions. Two functions (fx) and $g(x)$ are said to be *orthogonal* over the region (a,b) if the definite integral of their product vanishes:

$$\int_a^b f(x) g(x)\, dx = 0 \tag{17.5}$$

If, moreover, a function $f(x)$ satisfies the condition

$$\int_a^b f^2(x)\, dx = 1 \tag{17.6}$$

it is said to be *normalized.* If all the functions of a set satisfy Eqs. (17.5) and (17.6), then an *orthonormal* set results. An orthogonal set of functions is always linearly independent.

If n linearly independent functions $h_1(x)$, $h_2(x)$, . . . , $h_n(x)$ are given, it is possible to form n linear combinations of them into the orthogonal set $\phi_1(x)$, $\phi_2(x)$, . . . , $\phi_n(x)$. This is accomplished by means of the following *Schmidt orthogonalization procedure.* Start by letting $\phi_1 = h_1$. Next, determine the constant a_1' so that $\phi_2 = h_2 - a_1'\phi_1$ is orthogonal to ϕ_1. Thus, we have

$$a_1' = \frac{\int_a^b \phi_1 h_2\, dx}{\int_a^b \phi_1^2\, dx}$$

Next set $\phi_3 = h_3 - a_1''\phi_1 - a_2''\phi_2$ and determine a_1'', a_2'' so that ϕ_3 is orthogonal to ϕ_1 and ϕ_2; that is, set

$$a_1'' = \frac{\int_a^b \phi_1 h_3\, dx}{\int_a^b \phi_1^2\, dx} \qquad a_2'' = \frac{\int_a^b \phi_2 h_3\, dx}{\int_a^b \phi_2^2\, dx}$$

The process can be continued in this way until the desired orthogonal set of n functions has been calculated. The normalization follows directly if each function is divided by a constant chosen so that Eq. (17.6) is satisfied. Thus, the set ϕ_n/N_n becomes orthonormal if

$$N_n^2 = \int_a^b \phi_n^2\, dx$$

Example 1. Take the set 1, x, x^2, x^3, . . . in the region $(-1,1)$. The orthogonalization procedure yields functions proportional to Legendre's polynomials $P_n(x)$ $(n = 0, 1, 2, \ldots)$. The first five of these polynomials are:

$$P_0(x) = 1 \qquad P_1(x) = x \qquad P_2(x) = \tfrac{3}{2}x^2 - \tfrac{1}{2}$$
$$P_3(x) = \tfrac{5}{2}x^3 - \tfrac{3}{2}x \qquad P_4(x) = \tfrac{35}{8}x^4 - \tfrac{15}{4}x^2 + \tfrac{3}{8}$$

The normalization factor can be found from the relation

$$\int_{-1}^{1} P_n{}^2(x)\,dx = \frac{2}{2n+1}$$

Linear Differential and Integral Equations. The solution of the most general form of a linear, nth-order, inhomogeneous differential equation with homogeneous boundary conditions is equivalent to the solution of a Fredholm integral equation of the second kind if a *Green's function* for the problem exists. [Homogeneous boundary conditions arise when linear functions of the unknown and its first $(n-1)$ derivatives vanish at the end points of the region.] Specific results for second-order equations follow. Consider

$$\frac{d}{dx}\left(p\frac{dy}{dx}\right) + (q - \lambda r)y = s \tag{17.7}$$

where $p(x) > 0$, $q(x)$, $r(x)$, and $s(x)$ are functions of the independent variable x defined in the region (a,b). The boundary conditions are given in the form (the prime denoting differentiation)

$$\alpha_0 y(a) + \beta_0 y'(a) = 0 \qquad \alpha_1 y(b) + \beta_1 y'(b) = 0 \tag{17.8}$$

where α_0, β_0, as well as α_1, β_1, are constants that are not both zero. The integral equation associated with this Sturm-Liouville system [Eqs. (17.7) and (17.8)] is

$$y(x) = h(x) + \lambda \int_a^b G(x,\xi)r(\xi)y(\xi)\,d\xi \tag{17.9}$$

where
$$h(x) = \int_a^b G(x,\xi)s(\xi)\,d\xi$$

and $G(x,\xi)$ is called the Green's function or *influence function* of the problem. When $s \equiv 0$, Eq. (17.9) becomes a homogeneous integral equation.

The function $G(x,\xi)$ is the solution of the differential equation

$$L(y) \equiv py'' + p'y' + qy = \delta(x - \xi) \tag{17.10}$$

for the given boundary conditions and where $\delta(x)$ is the Dirac pulse function, which has the properties that it vanishes everywhere except at $x = 0$ and

$$\int_{-\infty}^{\infty} \delta(x)\,dx = 1$$

In more physical terms, the equation $L(y) = f(x)$ can be considered as the condition for the equilibrium of a physical system subjected to a force distribution $f(x)$. The influence function $G(x,\xi)$ gives the displacement of the system under limiting conditions for which $f(x)$ represents a unit point force exerted at $x = \xi$; in some cases, it may be determined experimentally and thus be known numerically or graphically. Mathematically, its existence implies that $G(x,\xi)$ can be determined subject to the conditions:

1. $G(x,\xi)$ is continuous over the region (a,b).
2. $G(x,\xi)$ satisfies the differential equation

$$p\frac{\partial^2 G}{\partial x^2} + p'\frac{\partial G}{\partial x} + qG = 0 \qquad \text{when } x < \xi,\ x > \xi$$

3. $G(x,\xi)$ satisfies the boundary conditions of Eqs. (17.8).
4. The first derivative has a discontinuity equal to $1/p(\xi)$ at $x = \xi$, that is,

$$\lim_{\epsilon\to 0}\left[\frac{\partial G(x,\xi)}{\partial x}\bigg|_{x=\xi+\epsilon} - \frac{\partial G(x,\xi)}{\partial x}\bigg|_{x=\xi-\epsilon}\right] = \frac{1}{p(\xi)}$$

In Eqs. (17.7) and (17.10), the differential expression $L(y)$ is self-adjoint (see p. 10–6), from which it follows that Green's function is symmetric; that is,

$$G(x,\xi) = G(\xi,x)$$

This symmetry corresponds to a reciprocity principle common to many physical problems. In the present case, it implies that the effect at point x produced by a force acting at point ξ is equal to the effect at point ξ that would be produced by the same force acting at point x.

Table 17.1 lists Green's function $G(x,\xi)$ for a few particular differential equations $L(y) - \lambda ry = s$. In each case, the influence function is symmetric, and can be used directly in Eq. (17.9). The right column gives $G(x,\xi)$ for $x \leqq \xi$; when $x \geqq \xi$, Green's function is $G(\xi,x)$. (Additional cases are to be found in [1] and table IV of [2].) It should be noted that differences in the over-all sign of Green's function appear often in the literature as a result of the particular way it is defined.

Table 17.1. Green's Function $G(x,\xi)$

Boundary conditions	$G(x,\xi)$ for $x \leqq \xi$
$L(y) \equiv y''$	
$y(0) = y(l) = 0$	$\frac{x}{l}(\xi - l)$
$y(0) = y'(l) = 0$	$-x$
$y(-1) = y(1) = 0$	$\frac{1}{2}[\lvert\xi - x\rvert + x\xi - 1]$
$L(y) \equiv y'' + k^2y$	
$y(0) = y(l) = 0$	$-\left[l \sin k\frac{x}{l} \sin k\left(1 - \frac{\xi}{l}\right)\right] \Big/ k \sin k$
$L(y) \equiv y'' - k^2y$	
$y(0) = y(l) = 0$	$-\left[l \operatorname{Sinh} k\frac{x}{l} \operatorname{Sinh} k\left(1 - \frac{\xi}{l}\right)\right] \Big/ k \operatorname{Sinh} k$
$L(y) \equiv y^{\text{iv}}$	
$y(0) = y(l) = y'(0) = y'(l) = 0$	$\frac{1}{6l^3}x^2(l - \xi)^2(3l\xi - 2x\xi - lx)$
$y(0) = y(l) = y''(0) = y''(l) = 0$	$\frac{1}{6l}x(\xi - l)(x^2 - 2l\xi + \xi^2)$
$y(0) = y'(0) = y''(l) = y'''(l) = 0$	$\frac{1}{6}x^2(3\xi - x)$

Example 2. The differential equation governing the natural oscillation of a string of length l and under uniform tension T is $y'' = -\lambda\rho(x)y$ where $y(x)\sin\omega t$ gives the deflection of the string in time and space ($\omega^2 = \lambda T$), and $\rho(x)$ is mass per unit of length at x. The boundary conditions for fixed ends are $y(0) = y(l) = 0$. The equivalent integral equation is thus

$$y(x) = -\lambda \int_0^l \rho(\xi)G(x,\xi)y(\xi)\,d\xi$$

where $G(x,\xi)$ is the first entry in the table of Green's functions. If $\rho(x) = \rho_0 = \text{const}$, the differential equation yields immediately the possible solutions

$$y_n(x) = \sin\,(\lambda\rho_0)^{1/2}x \qquad \lambda = n^2\pi^2/l^2\rho_0 \qquad n = 1, 2, \ldots, \infty$$

17.2. INTEGRAL EQUATIONS OF THE SECOND KIND

Separable Kernels

When $K(x,\xi)$ is a finite sum of N products of functions of x and ξ, respectively,

$$K(x,\xi) = \sum_{n=1}^{N} \alpha_n(x)\beta_n(\xi) \tag{17.11}$$

it is said to be a *separable* (or *degenerate*) kernel. In all such cases, the summation may be written so that the functions $\alpha_n(x)$ (and $\beta_n(\xi)$) are linearly independent. Equation (17.2) then becomes

$$f(x) = g(x) + \lambda \sum_{n=1}^{N} b_n\alpha_n(x) \quad \text{where} \quad b_n = \int_a^b \beta_n(\xi)f(\xi)\,d\xi \tag{17.12}$$

If we multiply Eq. (17.12) successively by $\beta_1(x)$, $\beta_2(x)$, . . . , $\beta_N(x)$ and integrate over the region (a,b), the N simultaneous linear equations

$$b_m - \lambda \sum_{n=1}^{N} c_{mn}b_n = g_m \qquad m = 1, 2, \ldots, N \tag{17.13}$$

in the unknowns b_1, b_2, . . . , b_N result where

$$c_{mn} = \int_a^b \beta_m(x)\alpha_n(x)\,dx \qquad g_m = \int_a^b g(x)\beta_m(x)\,dx$$

Once the values of b_m are determined by solving Eqs. (17.13), they may be substituted into Eq. (17.12) to give the desired solution of the integral equation. The possibilities arising in the solution of Eqs. (17.13) follow directly from the theory of algebraic linear equations and lead to the basic results established by Fredholm. Consider, first, the homogeneous integral equation, $g(x) \equiv 0$. The trivial solution $f(x) = 0$ is the only solution unless λ has a value for which the determinant of the coefficients in the left member of Eq. (17.13) vanishes. In this latter case, λ is called an *eigenvalue* (or *characteristic value*), and the homogeneous integral equation possesses a finite number of linearly independent solutions $\phi_1(x)$, $\phi_2(x)$, . . . , $\phi_k(x)$ called *eigenfunctions* (or *characteristic functions*). If $g(x) \not\equiv 0$ and λ is equal to an eigenvalue, a solution of Eq. (17.2) exists if and only if $g(x)$ satisfies the orthogonality condition,

$$\int_a^b g(x)\phi_i(x)\,dx = 0 \qquad i = 1, 2, \ldots, k \tag{17.14}$$

Under these latter conditions, the solution is composed of a particular solution plus the linear combination $c_1\phi_1 + c_2\phi_2 + \cdots + c_k\phi_k$, the coefficients c_i being arbitrary constants.

Successive Approximations and Series Solutions

Let $f(x)$ be approximated by the arbitrary function $p_0(x)$, and calculate the sequence of functions $p_0(x), p_1(x), \ldots, p_n(x), \ldots$ by means of the recursion formula,

$$p_n(x) = g(x) + \lambda \int_a^b K(x,\xi)p_{n-1}(\xi)\, d\xi \tag{17.15}$$

It can be shown that $p_n(x)$ approaches the solution of Eq. (17.2) provided λ satisfies the inequality

$$|\lambda| \leqq \left\{\int_a^b \int_a^b [K(x,\xi)]^2\, dx\, d\xi\right\}^{-1/2} \tag{17.16}$$

By means of this iterative procedure, it is possible to express the solution as a power series in λ. The formal development yields the so-called Liouville-Neumann series

$$f(x) = g(x) + \sum_{n=1}^{\infty} \lambda^n \int_a^b K^{(n)}(x,\xi)g(\xi)\, d\xi \tag{17.17}$$

where $K^{(n)}(x,\xi)$ is the nth iterated kernel of $K(x,\xi)$:

$$K^{(1)}(x,\xi) = K(x,\xi) \qquad K^{(n)}(x,\xi) = \int_a^b K^{(n-1)}(x,\xi_1)K(\xi_1,\xi)\, d\xi_1 \qquad n \geqq 2$$

The series converges for values of λ satisfying the inequality (17.16). Improvement of the convergence is discussed in [3].

For Volterra-type integral equations, series (17.17) converges for any value of λ; when $g(x) \equiv 0$, the only solution is $f(x) \equiv 0$.

The relation

$$L(x,\xi;\lambda) = \sum_{n=1}^{\infty} \lambda^{n-1}K^{(n)}(x,\xi) \tag{17.18}$$

may be used to express $L(x,\xi;\lambda)$, the *resolvent kernel*, and from Eq. (17.17) the solution of Eq. (17.2) becomes symbolically

$$f(x) = g(x) + \lambda \int_a^b L(x,\xi;\lambda)g(\xi)\, d\xi \tag{17.19}$$

The Liouville-Neumann series is valid only for a restricted range of values of λ. The reason for this becomes apparent in Fredholm's general solution, in which case the resolvent kernel is expressed in terms of the ratio of two series in λ. The general solution can be calculated as follows:

$$f(x) = g(x) + \lambda \int_a^b \frac{D(x,\xi;\lambda)}{\Delta(\lambda)}\, g(\xi)\, d\xi \tag{17.20}$$

where

$$D(x,\xi;\lambda) = \sum_{n=0}^{\infty} \frac{(-\lambda)^n}{n!} D_n(x,\xi) \qquad \Delta(\lambda) = \sum_{n=0}^{\infty} \frac{(-\lambda)^n}{n!} C_n$$

and $D_n(x,\xi)$ and C_n are calculated by means of the recursion relations:

$$D_0(x,\xi) = K(x,\xi) \qquad C_0 = 1 \qquad C_n = \int_a^b D_{n-1}(x,x)\,dx$$

$$D_n(x,\xi) = C_nK(x,\xi) - n\int_a^b K(x,\xi_1)D_{n-1}(\xi_1,\xi)\,d\xi_1$$

Equation (17.17) is, in fact, the expansion of Eq. (17.20) into a power series in λ, and its region of convergence extends to the smallest root of $\Delta(\lambda) = 0$. This root is the smallest eigenvalue of the homogeneous equation (17.3), and a lower bound for its magnitude is given by the right side of Eq. (17.16). When the kernel is separable, as in Eq. (17.1), $D(x,\xi;\lambda)$ and $\Delta(\lambda)$ become polynomials of degree N in λ, and, at most, N different eigenvalues are determined from $\Delta(\lambda) = 0$.

Example 3. If the kernel separates simply into the product of two functions of x and ξ alone, we have

$$K(x,\xi) = \alpha(x)\beta(\xi) \qquad A = \int_a^b \alpha(x)\beta(x)\,dx$$

and the explicit solution of the integral equation is

$$f(x) = g(x) + \frac{\lambda}{1 - A\lambda}\int_a^b K(x,\xi)g(\xi)\,d\xi$$

It is obvious that here the single eigenvalue $\lambda = 1/A$ occurs.

Function approximations often lead to adequate orders of accuracy in practice. Let $\bar{K}(x,\xi)$ and $G(x)$ approximate $K(x,\xi)$ and $g(x)$ such that

$$K(x,\xi) = \bar{K}(x,\xi) + \delta(x) \qquad g(x) = G(x) + \eta(x) \tag{17.21}$$

where $\delta(x)$ and $\eta(x)$ are known error terms. If

$$f(x) = F(x) + \epsilon(x) \tag{17.22}$$

and $F(x)$ is the solution of the equation

$$F(x) = G(x) + \lambda\int_a^b \bar{K}(x,\xi)F(\xi)\,d\xi \tag{17.23}$$

then the deviation $\epsilon(x)$ from the exact solution satisfies the equation [4]

$$\epsilon(x) = [\eta(x) + \lambda\int_a^b \delta(\xi)F(\xi)\,d\xi] + \lambda\int_a^b K(x,\xi)\epsilon(\xi)\,d\xi \tag{17.24}$$

If necessary, the approximation can be repeated.

Hilbert-Schmidt Theory for Symmetric Kernels

In many fields of application, the kernel satisfies the relations

$$K(x,\xi) = K(\xi,x), \quad \int_a^b\int_a^b K(x,\xi)q(x)q(\xi)\,dx\,d\xi > 0 \tag{17.25}$$

where $q(x)$ is any piecewise-continuous function in the interval (a,b). When the first condition is satisfied, the kernel is said to be *symmetric*; in the latter case, it is said to be *positive-definite*. That both conditions may hold is often intuitively obvious in practice when the kernel is identified with a physical influence function; in a nondissipative vibrating system with one degree of freedom, for example, the symmetry of the influence function follows directly from the law of conservation of energy, and the double integral is twice the value of the positive work done by the arbitrary force distribution $q(x)$ in displacing the system.

When the kernel is symmetric, the following results hold:

1. Eigenvalues and eigenfunctions always exist, and their number is denumerably infinite if and only if the kernel is nonseparable, and finite if and only if the kernel is separable.
2. All eigenvalues of a real kernel are real.
3. If there are infinitely many eigenvalues, their magnitudes increase indefinitely.
4. The eigenfunctions of different eigenvalues are orthogonal.
5. The eigenvalues are all positive if and only if the kernel is positive-definite. (The eigenvalues are always negative if the integral of Eq. (17.25) is always negative; the kernel is then said to be *negative-definite*.)

In many cases of practical interest, each eigenvalue has a unique eigenfunction. (This is true for the Sturm-Liouville-type problems defined in Eqs. (17.7) and (17.8), and, as shown in Example 2, for a uniform string. The eigenvalues, which in this case were associated with natural frequencies of oscillation, corresponded to unique eigenfunctions, which corresponded to natural mode shapes.) It was apparent, however, in the development under Separable Kernels, that an eigenvalue may have a finite number of linearly independent eigenfunctions. (This happens, for example, when a natural frequency corresponds to more than one mode shape.) Complications of notation can be avoided by first modifying these linearly independent functions so that they are replaced by an equal number of orthonormal functions (see Schmidt procedure, p. 17–2). The homogeneous integral equation thus has the eigenvalues $\lambda_1 \leqq \lambda_2 \leqq \lambda_3 \leqq \cdots$ and their respective orthonormal eigenfunctions $\phi_1(x)$, $\phi_2(x)$, $\phi_3(x)$, ...

If λ is not an eigenvalue, the unique solution of Eq. (17.2) is

$$f(x) = g(x) + \lambda \sum_n \frac{g_n}{\lambda_n - \lambda} \phi_n(x) \qquad \lambda \neq \lambda_n \tag{17.26}$$

where $$g_n = \int_a^b g(x)\phi_n(x)\,dx$$

If λ is an eigenvalue, and $g(x)$ is orthogonal to all its eigenfunctions, then Eq. (17.26) is still a particular solution, to which can be added the linear combination, with arbitrary coefficients, of eigenfunctions corresponding to the value λ.

If $K(x,\xi)$ is a definite, continuous, symmetric kernel or if it has only a finite number of eigenvalues of one sign, then $K(x,\xi)$ and the resolvent kernel [see Eq. (17.19)] are given [5] by

$$K(x,\xi) = \sum_n \frac{\phi_n(x)\phi_n(\xi)}{\lambda_n} \qquad L(x,\xi;\lambda) = \sum_n \frac{\phi_n(x)\phi_n(\xi)}{\lambda_n - \lambda} \tag{17.27}$$

Example 4. If $|\lambda - \lambda_0| \ll 1$, where λ_0 is an eigenvalue with a single eigenfunction, then the resolvent kernel is approximated by

$$L(x,\xi;\lambda) \approx \frac{\phi_0(x)\phi_0(\xi)}{\lambda_0 - \lambda}$$

The approximate solution of Eq. (17.2) then takes the form

$$f(x) \approx g(x) + \frac{\lambda}{\lambda_0 - \lambda} \phi_0(x) \int_a^b g(x)\phi_0(x)\,dx$$

Eigenvalues and eigenfunctions can be calculated by means of a method of successive approximations. One starts with a normalized arbitrary function $\psi_0(x)$ and calculates by recursion

$$\psi_n(x) = \lambda_n' \int_a^b K(x,\xi)\psi_{n-1}(\xi)\, d\xi \tag{17.28}$$

where λ_n' is so determined that $\psi_n(x)$ is normalized. Then $\psi_n(x)$ and λ_n' converge respectively to an eigenfunction and an eigenvalue. A suitable starting function is $\psi_0(x) \sim K(x,c)$ where c is any constant for which $K(x,c)$ is not identically zero. In subsequent calculations of other eigenfunctions, the starting function is chosen orthogonal to the eigenfunctions previously determined.

Computational aids are provided by the following results. Symmetrical kernels yield, in general,

$$\sum_n \lambda_n^{-2} = \int_a^b \int_a^b [K(x,\xi)]^2\, dx\, d\xi = c$$

As a consequence, we get the inequality $\lambda_2 \geqq (c - \lambda_1^{-2})^{-1/2}$. Under the same conditions imposed to get Eq. (17.27),

$$\sum_n \lambda_n^{-1} = \int_a^b K(x,x)\, dx$$

(see [5]). Let $\psi(x)$ be a continuous function. If the function

$$G(x) = \frac{\psi(x)}{\int_a^b K(x,\xi)\psi(\xi)\, d\xi}$$

does not change sign in the region (a,b) and lies between the finite limits $G_{\min}$ and $G_{\max}$, then an eigenvalue λ exists [6] which satisfies the inequality

$$G_{\min} \leqq \lambda \leqq G_{\max}$$

Polar Kernels. When the kernel is of the form

$$K(x,\xi) = \sigma(\xi)H(x,\xi) \tag{17.29}$$

where $\sigma(\xi)$ is a continuous function that does not change sign in the region of integration and $H(x,\xi)$ is symmetric, then the transformations

$$F(x) = [\sigma(x)]^{1/2}f(x) \qquad G(x) = [\sigma(x)]^{1/2}g(x)$$

lead to the equation

$$F(x) = G(x) + \lambda \int_a^b [\sigma(x)]^{1/2}H(x,\xi)[\sigma(\xi)]^{1/2}F(\xi)\, d\xi \tag{17.30}$$

where the new kernel $[\sigma(x)\sigma(\xi)]^{1/2}H(x,\xi)$ is now symmetric.

The Hilbert-Schmidt methods can be extended if orthogonality is redefined to include the weighting function $\sigma(x)$. If $\phi_m(x)$ and $\phi_n(x)$ are two eigenfunctions corresponding to the eigenvalues λ_m and λ_n, then they satisfy the orthogonality conditions

$$\int_a^b \sigma(x)\phi_m(x)\phi_n(x)\, dx = 0 \qquad m \neq n \tag{17.31}$$

The function $\phi_n(x)$ is normalized with respect to the weighting function $\sigma(x)$ if

$$\int_a^b \sigma_n(x)\phi_n^2(x)\, dx = 1 \tag{17.32}$$

Equation (17.26) is still the solution of Eq. (17.2) where

$$g_n = \int_a^b \sigma(x)g(x)\phi_n(x)\,dx \qquad n = 1, 2, \ldots$$

The eigenvalues and eigenfunctions for the polar kernel can also be determined by the method of successive approximations if the weighting function $\sigma(x)$ is a factor of the integrand in Eq. (17.28) and the calculated functions are normalized according to Eq. (17.32).

Numerous applications of these methods to the study of elastic vibrations are given in [7]. A discussion of numerical methods of solution is given in [8], chap. VIII.

Approximation Methods

When the kernel is known only in graphical or numerical form or is considered analytically intractable, the integral in Eq. (17.2) may be approximated by a finite sum of terms, and the problem thereby reduced to the solution of simultaneous linear equations. Let the region (a,b) be divided into N intervals with lengths $(\delta x)_1$, $(\delta x)_2, \ldots, (\delta x)_N$, and let x_j be chosen within the interval $(\delta x)_j$. Equation (17.2) is thus replaced by

$$f(x) = g(x) + \lambda \sum_{j=1}^{N} a_j K(x,x_j) f(x_j) \tag{17.33}$$

where the coefficients a_j depend on the integration rule used. (Either Simpson's rule (12.9) or the trapezoidal rule with equal intervals (12.8) will be the most obvious choice. Further refinements are discussed in [8].) If Eq. (17.33) is to be satisfied at each of the chosen points, we get N simultaneous linear equations,

$$f(x_i) = g(x_i) + \lambda \sum_{j=1}^{N} a_j K(x_i,x_j) f(x_j) \qquad i = 1, 2, \ldots, N \tag{17.34}$$

in the unknowns $f(x_i)$ $(i = 1, 2, \ldots, N)$. The numerical solution provides N ordinates on the graph of the unknown function $f(x)$. In practice, estimates of the order of accuracy could be made by increasing the number of intervals.

Another method that is widely used to obtain approximate solutions starts by replacing $f(x)$ by a linear combination of known linearly independent functions $\psi_1(x), \psi_2(x), \ldots, \psi_N(x)$. (These functions could be successive powers of x or, perhaps, a sequence of orthogonal functions.) Thus, set

$$f(x) \approx \sum_{i=1}^{N} c_i \psi_i(x) \tag{17.35}$$

where the constant coefficients c_i are to be determined. If $f(x)$ is substituted into Eq. (17.2) and the notation

$$\gamma_i(x) = \psi_i(x) - \lambda \int_a^b K(x,\xi)\psi_i(\xi)\,d\xi$$

is used, we get the single approximate relation

$$\sum_{i=1}^{N} c_i \gamma_i(x) - g(x) \approx 0 \tag{17.36}$$

In the *method of collocation*, the left member of Eq. (17.36) is made to vanish at N arbitrary points within the region (a,b). After these N simultaneous equations are solved for the coefficients c_i, then Eq. (17.35) yields the approximate solution.

Refinements of technique lead to the two following processes. First, product integrals involving the left member of Eq. (17.36) and the functions $\psi_j(x)$ are set equal to zero. Thus, the N equations

$$\int_a^b \left[\sum_{i=1}^N c_i \gamma_i(x) - g(x) \right] \psi_j(x)\, dx = 0 \qquad j = 1, 2, \ldots, N \tag{17.37}$$

are solved for the coefficients c_i. When the functions $\psi_j(x)$ form the sequence $1, x, x^2, \ldots$, the first $(N - 1)$ moments of the error are thereby forced to vanish. Second, by the *method of least squares* the minimization of the expression

$$\int_a^b \left[\sum_{i=1}^N c_i \gamma_i(x) - g(x) \right]^2 dx$$

leads to the N equations,

$$\int_a^b \left[\sum_{i=1}^N c_i \gamma_i(x) - g(x) \right] \gamma_j(x)\, dx = 0 \qquad j = 1, 2, \ldots, N$$

In explicit terms, the equations to be solved are

$$\sum_{i=1}^N c_i \int_a^b \left[\psi_i(x) - \lambda \int_a^b K(x,\xi)\psi_i(\xi)\, d\xi \right] \left[\psi_j(x) - \lambda \int_a^b K(x,\xi)\psi_j(\xi)\, d\xi \right] dx$$

$$= \int_a^b \left[\psi_j(x) - \lambda \int_a^b K(x,\xi)\psi_j(\xi)\, d\xi \right] g(x)\, dx \qquad j = 1, 2, \ldots, N \tag{17.38}$$

For discussion and evaluation of error bounds, see [9].

17.3. INTEGRAL EQUATIONS OF THE FIRST KIND

General Methods

Any function $g(x)$ that can be represented by the relation

$$g(x) = \int_a^b K(x,\xi) f(\xi)\, d\xi \tag{17.39}$$

is called an *integral transform* of $f(x)$. If the kernel is continuous, and $f(\xi)$ is piecewise-continuous, then $g(x)$ is a continuous function. Thus, in general, the totality of piecewise-continuous functions transforms into a more restricted class of functions, and there is no assurance that a continuous solution $f(x)$ can be found corresponding to a continuous function $g(x)$. When more specific information is at hand, however, the situation is improved, and the theory of integral equations of the first kind deals, to a large extent, with special types. In practice, singular equations appear often.

Symmetric Kernel. Every continuous function $g(x)$ which is represented as in Eq. (17.39), with $K(x,\xi)$ symmetric and $f(\xi)$ piecewise-continuous, can be expanded into a series of the eigenfunctions of $K(x,\xi)$, that is, a series of solutions of Eq. (17.3). Let $\phi_i(x)$ $(i = 1, 2, \ldots)$ be the normalized eigenfunctions with eigenvalues λ_i, and assume that $g(x)$ can be expanded in the series

$$g(x) = \sum_n A_n \phi_n(x) \qquad A_n = \int_a^b \phi_n(x) g(x)\, dx$$

A solution of Eq. (17.39) is then

$$f(x) = \sum_n \lambda_n A_n \phi_n(x) \tag{17.40}$$

provided the series converges.

When the number of eigenfunctions is finite, as happens in the case of a separable kernel, it follows that $g(x)$ must be limited to the class of functions represented by a linear combination of the eigenfunctions. It also follows that Eq. (17.40) may not possibly represent a complete solution, since to the right member can be added any function $F(x)$ that has a null integral transform, that is,

$$0 = \int_a^b K(x,\xi)F(\xi)\,d\xi \tag{17.41}$$

It is obvious that this relation can be satisfied if $F(x)$ is orthogonal to any eigenfunction. For most practical cases, an infinite number of eigenfunctions of $K(x,\xi)$ exist, and they form a complete set, in the sense that no null integral transform exists. An intuitive grasp of the physical problem is valuable in settling such questions; if, for example, $K(x,\xi)$ is the influence function giving, say, the deflection of a physical system corresponding to a unit point force, then Eq. (17.41) will hold only if a force distribution $F(\xi)$ exists that produced no deflection.

Volterra's Equation. The problem of solving the equation

$$g(x) = \int_a^x K(x,\xi)f(\xi)\,d\xi \tag{17.42}$$

can sometimes be reduced to solving a Volterra's equation of the second kind. If the kernel is differentiable, then Eq. (17.42) yields

$$g'(x) = K(x,x)f(x) + \int_a^x \frac{\partial K(x,\xi)}{\partial x} f(\xi)\,d\xi$$

and, if $K(x,x)$ never vanishes, the relation becomes

$$f(x) = \frac{g'(x)}{K(x,x)} + \int_a^x \left[\frac{-1}{K(x,x)} \frac{\partial K(x,\xi)}{\partial x}\right] f(\xi)\,d\xi \tag{17.43}$$

Equation (17.43) is of the form discussed previously, and its solution is expressible either as a series or by means of successive substitutions.

Singular Integral Equations

The general results relevant to eigenfunctions and eigenvalues in the theory of nonsingular integral equations no longer apply when singularities occur in the kernel or when the range of integration is unlimited. A single eigenvalue may possess an infinite number of eigenfunctions, and, in other cases, the eigenvalues may form a continuous spectrum of values, as opposed to the discrete spacing for nonsingular equations. The importance of singular integral equations in practice can, however, hardly be overestimated. Particularly, the use of transform theory and operational calculus, which often introduces singular integrals, rests upon the possibility of inverting an integral equation of the first kind. Examples of common integral transforms and their inversions are listed below.

The piecewise continuity of $f(x)$ and the convergence of the integral $\int_{-\infty}^{\infty} |f(x)|\,dx$ are sufficient to establish the following transforms and inversions:

Fourier Cosine Transform

$$g(x) = \left(\frac{2}{\pi}\right)^{1/2} \int_0^\infty f(\xi) \cos(x\xi)\, d\xi \qquad f(\xi) = \left(\frac{2}{\pi}\right)^{1/2} \int_0^\infty g(x) \cos(\xi x)\, dx \qquad (17.44)$$

Fourier Sine Transform

$$g(x) = \left(\frac{2}{\pi}\right)^{1/2} \int_0^\infty f(\xi) \sin(x\xi)\, d\xi \qquad f(\xi) = \left(\frac{2}{\pi}\right)^{1/2} \int_0^\infty g(x) \sin(\xi x)\, dx \qquad (17.45)$$

Fourier Transform

$$g(x) = \frac{1}{(2\pi)^{1/2}} \int_{-\infty}^{\infty} f(\xi) e^{ix\xi}\, d\xi \qquad f(\xi) = \frac{1}{(2\pi)^{1/2}} \int_{-\infty}^{\infty} g(x) e^{-i\xi x}\, dx \qquad (17.46)$$

Hankel Transform

$$g(x) = \int_0^\infty \xi J_\nu(x\xi) f(\xi)\, d\xi \qquad f(\xi) = \int_0^\infty x J_\nu(\xi x) g(x)\, dx \qquad (17.47)$$

where $J_\nu(x)$ denotes the Bessel function of the first kind of order $\nu \geqq -\frac{1}{2}$.

If $f(x)$ and $f'(x)$ are piecewise-continuous and, for some k,

$$\lim_{x \to \infty} e^{kx} f(x) = \text{const}$$

then for $c > k$ we get the

Laplace Transform

$$g(s) = \int_0^\infty f(\xi) e^{-s\xi}\, d\xi \qquad f(x) = \frac{1}{2\pi i} \int_{c-i\infty}^{c+i\infty} g(s) e^{sx}\, ds \qquad (17.48)$$

If, for some $k > 0$, the integral $\int_0^\infty \xi^{k-1} |f(\xi)|\, d\xi$ is bounded and $c > k$, we get the

Mellin Transform

$$g(x) = \int_0^\infty f(\xi) \xi^{x-1}\, d\xi \qquad f(\xi) = \frac{1}{2\pi i} \int_{c-i\infty}^{c+i\infty} g(x) \xi^{-x}\, dx \qquad (17.49)$$

For specified functional forms of $g(x)$, it is very often possible to find $f(\xi)$ merely by referring to existing tables (see, e.g., [10]–[12] and the section on the Laplace transformation in this handbook).

If $\int_{-\infty}^{\infty} |f(x)|^\alpha\, dx$ converges for $\alpha > 1$, we get the

Hilbert Transform

$$g(x) = \frac{1}{\pi} \int_{-\infty}^{\infty} \frac{f(\xi)\, d\xi}{x - \xi} \qquad f(x) = \frac{-1}{\pi} \int_{-\infty}^{\infty} \frac{g(\xi)}{\xi - x}\, dx \qquad (17.50)$$

If $g(x)$ is a continuously differentiable function and $0 < \alpha < 1$, we get the

Abel Integral Equation

$$g(x) = \int_a^x \frac{f(\xi)\, d\xi}{(x - \xi)^\alpha} \qquad f(x) = \frac{\sin \pi\alpha}{\pi} \frac{d}{dx} \int_a^x \frac{g(\xi)\, d\xi}{(x - \xi)^{1-\alpha}} \qquad (17.51)$$

Example 5. Consider the free vibrations of a thin elastic plate of infinite radius and thickness $2h$ [13]. Let $w(r,t)$ be the displacement at time t of the central plane, and let $w = f(r)$, $\partial w/\partial t = 0$ represent initial conditions at $t = 0$. If we wish to determine the initial displacement that would produce the motion $w(0,t) = g(t)$ of the center of the plate, then the integral equation

$$g(t) = \frac{1}{2\beta t} \int_0^\infty \tau f(\tau) \sin\left(\frac{\tau^2}{4\beta t}\right) d\tau$$

results where $\beta^2 = Eh^2/[3\rho(1 - \nu^2)]$, E is Young's modulus, ν is Poisson's ratio, and ρ is the uniform density of the plate. The substitutions $\tau^2 = \xi$, $1/(4\beta t) = x$ reduce the equation to the Fourier sine transform

$$\frac{1}{x}\, g\left(\frac{1}{4\beta x}\right) = \int_0^\infty f(\xi^{1/2}) \sin x\xi \, d\xi$$

In terms of the original variables, the general solution is

$$f(r) = \frac{2}{\pi} \int_0^\infty \frac{g(t)}{t} \sin \frac{r^2}{4\beta t} \, dt$$

Example 6. A point particle of mass m starts from rest at time $t = 0$, and slides under the influence of gravity along a smooth curve $y = f(x)$ in a vertical plane (the x axis to be directed upward). Let $T(x)$ be the time required for the particle to descend from the height x to the lowest point $x = 0$. Then

$$T(x) = \int_0^x \frac{s'(\xi)\, d\xi}{[2g(x - \xi)]^{1/2}}$$

where $s(\xi)$ is distance measured along the curve from $\xi = 0$. If $T(x)$ is specified, the inversion of the Abel equation yields

$$s'(x) = \frac{(2g)^{1/2}}{\pi} \int_0^x \frac{T(\xi)\, d\xi}{(x - \xi)^{1/2}}$$

The equation of the curve is then given by

$$y = \int_0^x [s'(x)^2 - 1]^{1/2} \, dx$$

If the integrals $\int_{-a}^{a} g(x)\, dx$ and $\int_{-a}^{a} g^2(x)(a^2 - x^2)^{1/2}\, dx$ converge and $g(x)$ has, at most, a finite number of discontinuities, we have the

Airfoil Equation

$$g(x) = \frac{1}{2\pi} \int_{-a}^{a} \frac{f(\xi)\, d\xi}{x - \xi} \qquad f(x) = \frac{1}{\pi(a^2 - x^2)^{1/2}} \left[\Gamma - 2\int_{-a}^{a} \frac{g(\xi)(a^2 - \xi^2)^{1/2}}{x - \xi}\, d\xi\right] \tag{17.52}$$

$$\Gamma \text{ (arbitrary)} = \int_{-a}^{a} f(\xi)\, d\xi$$

The arbitrariness in the value of Γ is due to the fact that $(a^2 - x^2)^{-1/2}$ has a null integral transform for the given kernel. An infinity of answers is thus possible, and, if a unique solution for a physical problem is sought, some additional condition must be available in order to fix the answer.

Example 7. Consider a two-dimensional thin cambered plate of length $2a$, and assume that the local inclination of its surface deviates only slightly from its direction of flight. In low-speed aerodynamic theory, Eqs. (17.52) relate the local geometry and the aerodynamic loading on such a surface. If the flight direction is along the negative x axis, $g(x)$ is proportional to local surface slope and $f(x)$ is proportional to the local loading. The Kutta-Joukowski principle states, moreover, that loading is zero at the trailing edge; that is, $f(a) = 0$. This condition fixes Γ in Eq. (17.52), and the inversion takes the form

$$f(x) = \frac{2}{\pi}\left(\frac{a-x}{a+x}\right)^{1/2}\int_{-a}^{a}\frac{g(x_1)\,dx_1}{(a^2-x_1{}^2)^{1/2}} - \frac{2}{\pi}(a^2-x^2)^{1/2}\int_{-a}^{a}\frac{g(x_1)\,dx_1}{(x-x_1)(a^2-x_1{}^2)^{1/2}}$$

If the surface is a flat plate at a small angle of attack to the flight direction, we have $g(x) \sim \alpha$. The second integral in the last equation vanishes for $-a < x < a$, and the local aerodynamic loading becomes

$$f(x) \sim 2\alpha\left(\frac{a-x}{a+x}\right)^{1/2}$$

Convolution Integrals

Let the Laplace transform and inverse transform of a function be written respectively as follows:

$$\mathfrak{L}[f(x)]_s = \int_0^\infty f(\xi)e^{-s\xi}\,d\xi = F_l(s)$$

$$\mathfrak{L}^{-1}[F_l(s)]_x = \frac{1}{2\pi i}\int_{c-i\infty}^{c+i\infty} F_l(s)e^{xs}\,ds = f(x) \qquad (17.53a,b)$$

Similarly, denote the Fourier transform and inverse transform by

$$\mathfrak{F}[f(x)]_\omega = \frac{1}{(2\pi)^{1/2}}\int_{-\infty}^{\infty} f(\xi)e^{i\omega\xi}\,d\xi = F_f(\omega)$$

$$\mathfrak{F}^{-1}[F_f(\omega)]_x = \frac{1}{(2\pi)^{1/2}}\int_{-\infty}^{\infty} F_f(\omega)e^{-ix\omega}\,d\omega = f(x) \qquad (17.54a,b)$$

This operational notation, together with the use of the convolution theorems given below, is especially adapted to the solution of integral equations with kernels of the form $K(x - \xi)$. Tables of transforms enable one, in many cases, to pass directly to the required inversion.

Convolution Theorem for Laplace Transforms. Let $F_l(s)$ and $G_l(s)$ be the Laplace transforms, respectively, of the functions $f(x)$ and $g(x)$; then the product $F_l(s)G_l(s)$ is the Laplace transform of a convolution (Faltung) integral, that is,

$$F_l(s)G_l(s) = \mathfrak{L}\left[\int_0^x g(\xi)f(x-\xi)\,d\xi\right]_s \qquad (17.55)$$

Convolution Theorem for Fourier Transforms. Let $F_f(\omega)$ and $G_f(\omega)$ be the Fourier transforms, respectively, of the functions $f(x)$ and $g(x)$; the product $(2\pi)^{1/2}F_f(\omega)G_f(\omega)$ is then the Fourier transform of a convolution (Faltung) integral; that is,

$$(2\pi)^{1/2}F_f(\omega)G_f(\omega) = \mathfrak{F}\left[\int_{-\infty}^{\infty} g(\xi_1)f(x-\xi_1)\,d\xi_1\right]_\omega \qquad (17.56)$$

Equation (17.55) can be used to solve the following Volterra equations of the first and second kind:

$$f(x) = g(x) + \int_0^x K(x-\xi)f(\xi)\,d\xi$$

$$g(x) = \int_0^x K(x-\xi)f(\xi)\,d\xi \qquad (17.57a,b)$$

If we apply the Laplace operator $\mathfrak{L}[\ ]$, the symbolic solutions are

$$F_l(s) = \frac{G(s)}{1 - \bar{K}(s)} \qquad F_l(s) = \frac{G(s)}{\bar{K}(s)} \tag{17.58a,b}$$

where $\bar{K}(s) = \mathfrak{L}[K]$. The inversion then yields another convolution integral. Fourier transforms can be applied similarly to Fredholm equations for which the region of integration is $(-\infty, \infty)$; for further discussion, see [14], pp. 286–287.

Example 8. Consider the Abel equation (17.51), and for simplicity set $a = 0$. From a table of transforms we get $\mathfrak{L}[x^{-\alpha}] = s^{\alpha-1}\Gamma(1-\alpha)$, where Γ denotes the gamma function. From Eq. (17.58*b*), $F(s) = G(s)s^{\alpha-1}/\Gamma(1-\alpha) = sG(s)s^{-\alpha}/\Gamma(1-\alpha)$. Thus

$$f(x) = \frac{1}{\Gamma(1-\alpha)}\frac{d}{dx}\mathfrak{L}^{-1}[G(s)s^{-\alpha}] = \frac{1}{\Gamma(\alpha)\Gamma(1-\alpha)}\frac{d}{dx}\int_0^x g(\xi)(x-\xi)^{\alpha-1}\,d\xi$$

and, since $\Gamma(\alpha)\Gamma(1-\alpha) = \pi \csc \pi\alpha$, the inversion agrees with Eq. (17.51).

17.4. REFERENCES

[1] R. Courant, D. Hilbert: "Methods of Mathematical Physics," vol. 1, Interscience, New York, 1953.
[2] L. Collatz: "Eigenwertprobleme," Chelsea, New York, 1948.
[3] C. Wagner: On the numerical evaluation of Fredholm integral equations with the aid of the Liouville-Neumann series, *J. Math. Phys.*, **30** (1951), 232–234.
[4] P. Moon, D. E. Spencer: Errors in the solution of integral equations, *J. Franklin Inst.*, **264** (1954), 29–41.
[5] I. Mercer: Functions of positive and negative type and their connection with the theory of integral equations, *Phil. Trans. Roy. Soc. London*, **A,209** (1909), 415–446.
[6] L. Collatz: Einschliessungssatz für die Eigenwerte, *Math. Z.*, **47** (1941), 395–398.
[7] G. Temple, W. G. Bickley: "Rayleigh's Principle and Its Application to Engineering," Dover, New York, 1956.
[8] Z. Kopal: "Numerical Analysis," Wiley, New York, 1956.
[9] A. T. Lonseth: Approximate solutions of Fredholm type integral equations, *Am. Math. Soc. Bull.*, **60** (1954), 415–430.
[10] G. A. Campbell, R. M. Foster: "Fourier Integrals for Practical Application," Van Nostrand, Princeton, N.J., 1948.
[11] I. N. Sneddon: "Fourier Transforms," McGraw-Hill, New York, 1951.
[12] A. Erdélyi (ed.): "Tables of Integral Transforms," vols. 1 and 2, Bateman Manuscript Project, McGraw-Hill, New York, 1954.
[13] I. N. Sneddon: The symmetrical vibrations of a thin elastic plate, *Proc. Cambridge Phil. Soc.*, **41** (1945), 27–43.
[14] I. N. Sneddon: Functional Analysis, in: S. Flügge (ed.), "Encyclopedia of Physics," vol. 2, Springer, Berlin, 1955.
[15] H. Buckner: "Die praktische Behandlung von Integral-Gleichungen," Springer, Berlin, 1952.
[16] F. B. Hildebrand: "Methods of Applied Mathematics," Prentice-Hall, Englewood Cliffs, N.J., 1952.
[17] W. V. Lovitt: "Linear Integral Equations," Dover, New York, 1950.
[18] N. I. Muskhelishvili: "Singular Integral Equations," transl. by J. R. M. Radok, Noordhoff, Groningen, Netherlands, 1953.
[19] W. Schmeidler: "Integralgleichungen mit Anwendungen in Physik und Technik," Akad.-Verlag, Leipzig, 1950.
[20] E. T. Whittaker, G. N. Watson: "A Course of Modern Analysis," Macmillan, New York, 1944.

CHAPTER 18

EIGENVALUE PROBLEMS

BY

L. COLLATZ, Dr., Dr. h.c., Hamburg, Germany

The present section contains a discussion of those types of eigenvalue problems which are of particular importance in engineering, namely, the eigenvalue problems associated with finite matrices, with ordinary and partial differential equations for finite domains, and with integral equations. The eigenvalue problems associated with infinite matrices and differential equations for infinite domains (which play a large role in physics) are not treated.

A uniform notation, Eq. (18.1), has been chosen for all eigenvalue problems. Many formulas which otherwise must be written separately for matrices, for differential equations, and for integral equations thus need to be written down only once. At the same time the unity of the various kinds of eigenvalue problems is seen more clearly, and the presentation is closer to the ideas of modern functional analysis. Of course, in order to accomplish this, it was necessary to deviate slightly from commonly used notations. Here the *eigenelement* (eigenvector, eigenfunction) is always denoted by y, indeed not the usual notation with partial differential equations. Moreover, much use is made of the notion of the scalar product (y,z) of two elements y and z. For vectors this means the ordinary scalar product, except that when the components are complex, the first vector must be changed to its complex conjugate [Eq. (18.6)]. For functions the scalar product means the integral of the product of the two functions over the basic domain [Eqs. (18.18) and (18.28)].

18.1. CLASSIFICATION OF MATRIX EIGENVALUE PROBLEMS

a. General and Special Matrix Eigenvalue Problems

Free undamped vibrations of mechanical or electrical systems with a finite number of degrees of freedom lead to problems of the form

$$Ny = \kappa My \qquad \text{or} \qquad \lambda Ny = My \tag{18.1a,b}$$

Here $y = (y_1, y_2, \ . \ . \ . \ , y_n)$ is a vector whose elements y_j represent the amplitudes of the vibrations corresponding to the individual degrees of freedom, $N = [n_{jk}]$ and $M = [m_{jk}]$ are given square n-rowed matrices which depend on the data of the system, and κ is a *characteristic number,* $\lambda = 1/\kappa$ the *eigenvalue* of the system. One seeks those values of λ or κ for which Eq. (18.1) has a *nontrivial* solution, i.e., an *eigenvector* or *eigenelement* y whose components y_j do not all vanish simultaneously. In vibration problems, κ usually is the square of the natural frequency. Also, κ is a root of the frequency equation

$$\det (N - \kappa M) = \begin{vmatrix} n_{11} - \kappa m_{11} & \cdots & n_{1n} - \kappa m_{1n} \\ \cdots & \cdots & \cdots \\ n_{n1} - \kappa m_{n1} & \cdots & n_{nn} - \kappa m_{nn} \end{vmatrix} = 0 \tag{18.2}$$

Translated from the German by R. L. Causey, Los Altos Calif.

Table 18.1. Examples of the Occurrence of the Different Matrix Eigenvalue Problems

System or method	Special problem $Ny = \kappa y$ Matrix N is: Triple diagonal	Special problem $Ny = \kappa y$ Matrix N is: Full	Intermediate problem $Ny = \kappa Dy$ Matrix N is: Triple diagonal	Intermediate problem $Ny = \kappa Dy$ Matrix N is: Full	General problem $Ny = \kappa My$
Examples of Mechanical Systems					
Spring-mass system			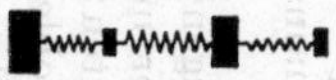	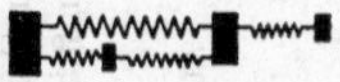	
Torsional vibrations	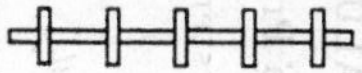		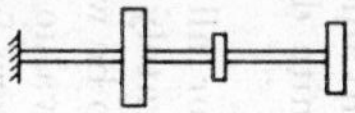	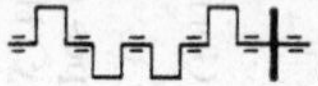	
Beams					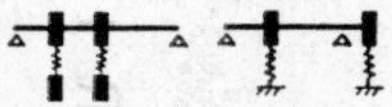
Pendulums					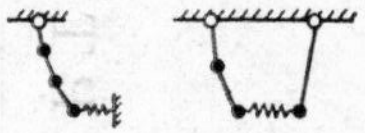
Other systems			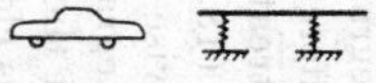		
Examples of Computation Methods					
Finite-difference methods for $y'' + \lambda p(x) y = 0$, $y(a) = y(b) = 0$	$p = \text{const}$; ordinary difference method	$p = \text{const}$; method of higher approximation	$p \neq \text{const}$; ordinary difference method	$p \neq \text{const}$; method of higher approximation	Hermitian method
Finite-difference methods for $\nabla^2 u + \lambda u = 0$, $u = 0$ on the boundary (possibly $u = \partial u / \partial \nu$ on part of the boundary)		Ordinary difference method		Part $u = 0$, part $u = \partial u / \partial \nu$: Ordinary difference method	Hermitian method

In general, this equation has n roots $\kappa_1, \ldots, \kappa_n$, which can be ordered according to their moduli, viz.,

$$|\kappa_1| \geq |\kappa_2| \geq \cdots \geq |\kappa_n| \tag{18.3}$$

There may be less than n roots κ_j, the κ_j may be complex, and some of them may be repeated.

We now take up the following classification of matrix eigenvalue problems:

1. Neither M nor N is a *diagonal matrix* (by definition, all the nonzero elements of a diagonal matrix are on the principal diagonal). Then (18.1) is a *general eigenvalue problem.*
2. M (or N) is a diagonal matrix. This is an *intermediate problem.*
3. M (or N) is the identity matrix I. This is a *special eigenvalue problem.*

If $M = I$, then the roots κ of (18.2) are the *characteristic numbers* of the matrix N and are, therefore, the zeros of the *characteristic polynomial* $\phi(\kappa)$:

$$\phi(\kappa) = \det (N - \kappa I) \tag{18.4}$$

An eigenvector y is a nontrivial solution of $(N - \kappa I)y = 0$. If one substitutes the matrix N in place of κ in the characteristic polynomial $\phi(\kappa)$, the result is the null (or zero) matrix. This is the Cayley-Hamilton theorem: Every matrix satisfies its characteristic equation $\phi(\kappa) = 0$. Thus

$$\phi(N) = 0 \tag{18.5}$$

Table 18.1 gives some illustrations and examples of the occurrence of various kinds of matrix eigenvalue problems. Here a matrix N is called respectively a triple diagonal matrix or a full matrix, according to whether $n_{jk} = 0$ for $|j - k| > 1$ holds or not. In connection with the examples, it is important to remark that most of them can be arranged in other forms by transformation of equations. In the table they are arranged in a form in which they can occur and, by use of the most natural coordinates, normally will occur.

b. Some Elementary Concepts Useful in Matrix Theory

The *transpose* of a matrix $N = [n_{jk}]$ is denoted by $N^T = [n_{kj}]$.

The *conjugate* $\bar{N}$ of N is the matrix obtained from N by replacing every element by its complex conjugate: $\bar{N} = [\bar{n}_{jk}]$. In many engineering problems, the elements n_{jk} are real numbers; and in this case the bar can be omitted, since $\bar{n}_{jk} = n_{jk}$.

If $N = N^T$, we call N *symmetric.*

If $N = \bar{N}^T$, then N is called *hermitian.* A real symmetric matrix is always hermitian.

A matrix N is called *orthogonal* if $NN^T = I$, that is, if $N^T = N^{-1}$.

A matrix N is called *unitary* if $N\bar{N}^T = I$, that is, if $\bar{N}^T = N^{-1}$. In the real case the notion of an orthogonal matrix and that of a unitary matrix coincide.

A matrix N is called *normal* if it commutes with its own conjugate transpose, i.e., if $N\bar{N}^T = \bar{N}^T N$. Every hermitian and every unitary matrix is normal.

The *scalar product* of two vectors y and z with components y_j and z_j is defined by

$$(y,z) = \sum_{j=1}^{n} \bar{y}_j z_j \tag{18.6}$$

This is the ordinary scalar product whenever all the components are real. The vectors y and z are said to be *orthogonal to each other* if $(y,z) = 0$.

A vector whose components do not all vanish will be called a *test vector* or *test element.*

A matrix M is called respectively *positive-definite, negative-definite,* or *positive-semidefinite* if the inequality

$$(z,Mz) > 0 \qquad (z,Mz) < 0 \qquad \text{or} \qquad (z,Mz) \geq 0 \tag{18.7}$$

holds for all test vectors z. The eigenvalue problem (18.1) is called *half-definite* if N is positive-definite and *completely definite* if both M and N are positive-definite.

The eigenvalue problem (18.1) is called *hermitian* if the matrices M and N are both hermitian.

If in a hermitian eigenvalue problem M is (positive or negative) definite, or M and N commute (i.e., $MN = NM$), then all existing eigenvalues are real.

c. The Three Principal Classes of Matrices

One can now divide square matrices N into the following three principal classes:

1. *Normal matrices* N are precisely those matrices which have a system of n linearly independent, mutually orthogonal eigenvectors y_j, and they are also precisely those matrices which can be brought into diagonal form by a unitary transformation; i.e., there is a unitary matrix U such that $U^{-1}NU = D$ is a diagonal matrix. The diagonal elements of D are the characteristic numbers κ_j, and U can be chosen so that

$$D = \begin{bmatrix} \kappa_1 & & & & 0 \\ & \kappa_2 & & & \\ & & \cdot & & \\ & & & \cdot & \\ 0 & & & & \kappa_n \end{bmatrix} \tag{18.8}$$

In the special case of a hermitian matrix (or a real symmetric matrix) the characteristic numbers are all real.

2. *Normalizable matrices*[1] are the matrices N which have a system of n linearly independent eigenvectors y_j or, equivalently, those matrices which can be brought into diagonal form by a similarity transformation. In other words there is a nonsingular matrix P ($\det P \neq 0$) such that $P^{-1}NP = D$ is a diagonal matrix (cf., e.g., [15]). The y_j can therefore serve as an (oblique-angled) coordinate system; i.e., any vector z can be expressed as a linear combination of the y_j.

3. *Non-normalizable matrices* are the matrices N for which the properties of class 2 no longer hold. Here transformation into diagonal form is impossible, but instead we may transform into the so-called Jordan canonical form (in which directly beside the principal diagonal elements there stands in certain places a 1 instead of a 0, while all other elements outside the principal diagonal vanish, cf. [15], p. 78). There are less than n linearly independent eigenvectors. When one is led by a physical system to such a matrix, then this means that one can no longer describe the general motion as a superposition of eigenmotions, since these do not exist in sufficient numbers. In lieu of the missing eigenmotions there are secular motions, and the role of the missing eigenvectors is taken over by the principal vectors. A vector $h \neq 0$ is called a *principal vector* of the matrix N if an integer $r \geq 1$ and a complex number κ exist so that

$$(N - \kappa I)^r h = 0 \tag{18.9}$$

Indeed, h is called a principal vector of degree r if r is the smallest positive integer for which (18.9) holds. The eigenvectors are principal vectors of degree 1. The numbers κ can be none other than the characteristic numbers of N. The representation theorem states that any vector z can be expressed as a linear combination of principal

[1] What is here called a normalizable matrix is also sometimes called a diagonalizable or a diagonable matrix [Trans.].

vectors. In the classical theory, the nondiagonable matrices are those which have nonlinear elementary divisors.

Example 1. Consider a simple mechanical system (Fig. 18.1), consisting of two masses m_1, m_2 (displacements x_1, x_2), and containing a linear spring and two linear dampers as shown. The equations of motion are:

$$m_1\ddot{x}_1 = k(x_2 - x_1) - c_1\dot{x}_1 + c_2(\dot{x}_2 - \dot{x}_1)$$
$$m_2\ddot{x}_2 = k(x_1 - x_2) \qquad\qquad + c_2(\dot{x}_1 - \dot{x}_2)$$

If for simplicity we set $m_1 = m_2$, $c_1 = c_2$, $k/m_1 = 1$, $c_1/m_1 = 1$, then we get the system of differential equations $\dot{y} = Ny$, where y is the vector with components x_1, x_2, $\dot{x}_1$, $\dot{x}_2$ and

$$N = \begin{bmatrix} 0 & 0 & 1 & 0 \\ 0 & 0 & 0 & 1 \\ -1 & 1 & -2 & 1 \\ 1 & -1 & 1 & -1 \end{bmatrix}$$

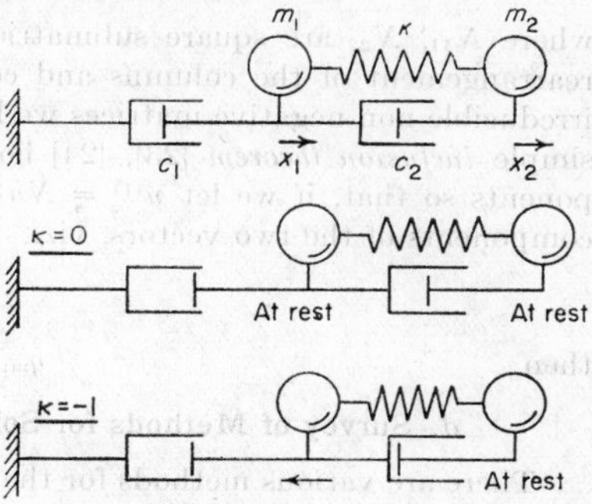

FIG. 18.1

Here the polynomial (18.4) is $\phi(\kappa) \equiv \kappa(\kappa + 1)^3$. For the characteristic number $\kappa = 0$, there is the eigenvector $y^T = \gamma'[1,1,0,0]$ and, for $\kappa = -1$, there is only a single eigenvector, namely, $y^T = \gamma''[1,0,-1,0]$; the numbers γ' and γ'' are arbitrary constants. In Fig. 18.1 the *eigenmotions* corresponding to these two eigenvectors have been indicated ($\kappa = 0$: displacement, no velocity; and $\kappa = -1$: m_1 experiences aperiodic motion while m_2 remains stationary). These two eigenmotions are not sufficient to produce the general state of motion by superposition. Here there are principal vectors h_2, h_3 of the second and third degree, respectively. Thus one gets, e.g.:

$$[N - (-1)I]^3 = \begin{bmatrix} 1 & 0 & 1 & 1 \\ 1 & 0 & 1 & 1 \\ 0 & 0 & 0 & 0 \\ 0 & 0 & 0 & 0 \end{bmatrix}$$

The solution of $(N + I)^3 h_3 = 0$ is the vectors $h_3{}^T = [\alpha, \beta, \gamma, -(\alpha + \beta + \gamma)]$ where α, β, γ are arbitrary constants. If, however, $\delta = \alpha - \beta + \gamma = 0$ holds, then h_3 reduces to a principal vector of second or first degree. From the theory of systems of differential equations, it follows that

$$\left(h_3 + \frac{t}{1!} h_2 + \frac{t^2}{2!} h_1 \right) e^{\kappa t}$$

is a solution of the equations of motion, where $h_2 = (N - \kappa I)h_3$, $h_1 = (N - \kappa I)h_2$. Therefore, the general solution of $\dot{y} = Ny$ (with four arbitrary constants α, β, δ, ϵ) is given by

$$y = \begin{bmatrix} \alpha + t(\beta + \delta) - \frac{1}{2}t^2\delta \\ \beta - t\delta \\ \beta + \delta - \alpha - t(2\delta + \beta) + \frac{1}{2}t^2\delta \\ -\beta - \delta + t\delta \end{bmatrix} e^{-t} + \begin{bmatrix} \epsilon \\ \epsilon \\ 0 \\ 0 \end{bmatrix}$$

There is another class of matrices which occurs frequently in the applications: the matrices $N \not\equiv 0$ with non-negative elements or, more briefly, the non-negative matrices. Every such matrix has a characteristic number $\kappa = r$ which is non-negative and whose absolute value is exceeded by no other characteristic number. Consequently, r is called the *dominant* or *maximal root.* A matrix is called *irreducible* if it is not of the form

$$N = \begin{bmatrix} N_{11} & N_{12} \\ 0 & N_{22} \end{bmatrix}$$

where N_{11}, N_{22} are square submatrices, and cannot be brought into this form by rearrangement of the columns and corresponding rearrangement of the rows. For irreducible non-negative matrices we have $r > 0$, and for such matrices the following simple *inclusion theorem* [23], [24] holds: There is a vector $u^{(0)}$ with positive components so that, if we let $u^{(1)} = Nu^{(0)}$ and form the quotients q_j of the individual components of the two vectors, viz.,

$$q_j = \frac{u_j^{(1)}}{u_j^{(0)}} \tag{18.10}$$

then

$$q_{\min} \leq r \leq q_{\max} \tag{18.11}$$

d. Survey of Methods for Solving Matrix Eigenvalue Problems

There are various methods for the numerical calculation of eigenvalues and eigenvectors of the problem (18.1).

(1) Direct Calculation of the Characteristic Polynomial. We can then determine the roots κ of the characteristic equation (18.2) by known approximate methods and, for each such root κ_j, compute an eigenvector by the solution of the corresponding system of equations (18.1). In the special eigenvalue problems, the algebraic equation $\phi(\kappa) = 0$ corresponding to (18.4) is formed by special methods, e.g., the method of Hessenberg ([22], p. 136).

(2) Iteration Methods for Special Eigenvalue Problems. Here we avoid the computation of Eq. (18.2). We start from a vector $u^{(0)}$, which already, as nearly as possible, is parallel to the eigenvector corresponding to the characteristic number κ_1 of largest absolute value, and perform the iteration $u^{(s+1)} = Nu^{(s)}$ [cf. Eq. (18.59)]. If $|\kappa_1| > |\kappa_2|$ [the κ_j are ordered according to (18.3)], then the vectors

$$\frac{u^{(s)}}{|u^{(s)}|}$$

converge to the eigenvector y_1 which corresponds to κ_1. The symbol in the denominator in the last expression stands for the length of the vector $u^{(s)}$. However, if $|\kappa_1| = |\kappa_2|$, then, in general, there is no convergence, but, in special cases, one can assert something more [25].

The iteration method converges rapidly only if $|\kappa_1|$ is very much larger than $|\kappa_2|$; it yields directly only the eigenvector y_1 belonging to κ_1. The characteristic number κ_n of smallest modulus can be obtained by application of the iteration to the inverse matrix. For real symmetric matrices, additional characteristic numbers can be obtained. Here we replace the vectors $u^{(s)}$ in the iteration by the vectors $v^{(s)}$, which are orthogonal to the eigenvectors belonging to the characteristic numbers κ_j of larger modulus (we subtract from $u^{(s)}$ its components corresponding to these κ_j). Of course, the method then becomes somewhat unwieldy and inaccurate. There is another variant in which one employs a suitable polynomial in N; see, e.g., Hartree [10], Householder [11].

(3) Inclusion Theorems. If we have an approximate eigenvector, then for normal matrices the theorem of Art. 18.4*e*(2) yields bounds for arbitrary eigenvalue problems. For positive-definite problems, the "main formula" (18.66) is also applicable.

(4) Iterative Methods for General Eigenvalue Problems. Here the iteration formula is given by (18.58), and we have to solve a system of linear equations at each iteration step. For completely definite problems, Eq. (18.66) is again applicable.

(5) Jacobi's Method for Special Eigenvalue Problems. This method uses the fact that, for any matrix M, the sum of squares of the moduli of the elements

$$[M] = \sum_{j,k=1}^{n} |m_{jk}|^2$$

remains invariant under a unitary transformation U: $[M] = [UMU^{-1}]$.

Now let N be a hermitian matrix. Let n_{jk} be an off-diagonal element of largest modulus. Then let U be a unitary transformation which transforms a vector y into z such that all components except the jth and kth remain unaltered and such that the two-row matrix

$$\begin{bmatrix} n_{jj} & n_{jk} \\ n_{kj} & n_{kk} \end{bmatrix}$$

is transformed to diagonal form. In that way the part of $[N]$ stemming from the elements outside the principal diagonal is reduced. We now find a new off-diagonal element of largest modulus, and continue the process until the characteristic values of N stand on the principal diagonal to required accuracy. The convergence of the method can be proved (see [17], p. 143). A suitable form for the arrangement of the computation can be found in a paper by Givens [26].

(6) Influence of Inaccuracies. If the elements n_{jk} of the matrix N are not known exactly (perhaps originating out of measurements or having been subjected to round-off errors), then the calculated characteristic numbers of an inexact matrix $\tilde{N}$ can differ considerably from those of N. As an example, consider the matrices N and $\tilde{N}$ with the elements

$$n_{jk} = \begin{cases} 10 & \text{for } j = k + 1 \\ 0 & \text{otherwise} \end{cases} \qquad \tilde{n}_{jk} = \begin{cases} 10^{-n+1} & \text{for } j = 1,\ k = n \\ n_{jk} & \text{otherwise} \end{cases}$$

N has the characteristic numbers $\kappa_j = 0$, but $\tilde{N}$ has the characteristic equation $\kappa^n = 1$, so that its characteristic numbers are the nth roots of unity. It is, therefore, important to know estimates for the influence of small changes in the matrix elements. If we use as a norm of a matrix the quantity $||N|| = [N]^{1/2}$, and if we have two normal matrices N and $\tilde{N}$ with characteristic numbers κ_j and $\tilde{\kappa}_j$, respectively, then we can number the κ_j so that

$$\sum_{j=1}^{n} |\kappa_j - \tilde{\kappa}_j|^2 \leq ||N - \tilde{N}||$$

18.2. CLASSIFICATION OF EIGENVALUE PROBLEMS FOR ORDINARY DIFFERENTIAL EQUATIONS

a. Differential Equations and Boundary Conditions

Consider the homogeneous linear differential equation

$$\sum_{\nu=0}^{p} a_\nu(x,\lambda) y^{(\nu)}(x) = 0 \tag{18.12}$$

with p linear homogeneous boundary conditions:

$$U_\rho(y,\lambda) = 0 \qquad \rho = 1, \ldots, p \tag{18.13}$$

The $a_\nu(x,\lambda)$ are given functions of x and λ with $a_p \not\equiv 0$. The variable x runs over an interval $\langle a,b \rangle$, which can extend to infinity, and λ can assume any (real or complex) values. The U_ρ are linear homogeneous in the values of $y(x)$ and possibly in some derivatives of $y(x)$ at given places. The eigenvalue problem consists in finding the *eigenvalues;* i.e., those values of λ for which Eqs. (18.12) and (18.13) have a nontrivial (i.e., not identically vanishing in $\langle a,b \rangle$) solution $y(x)$, which then is called an *eigen-unction* or *eigenele ment.*

Many investigations are concerned with a very specialized class of problems. However, this class is still so comprehensive that numerous eigenvalue problems originating from engineering problems are included. In this class, the differential equation is linear in λ, and has the form

$$My = \lambda Ny \tag{18.14}$$

where

$$My = \sum_{\nu=0}^{m} (-1)^\nu [f_\nu(x) y^{(\nu)}(x)]^{(\nu)}$$
$$Ny = \sum_{\nu=0}^{n} (-1)^\nu [g_\nu(x) y^{(\nu)}(x)]^{(\nu)} \tag{18.15a,b}$$

The $f_\nu(x)$ and $g_\nu(x)$ are in the interval $\langle a,b \rangle$ given ν times continuously differentiable functions with

$$f_m(x) \neq 0 \qquad g_n(x) \neq 0 \qquad m > n \geq 0 \tag{18.16}$$

The $2m$ linearly independent boundary conditions are assumed to be of the form

$$U_\mu y \equiv \sum_{\nu=0}^{2m-1} [\alpha_\nu y^{(\nu)}(a) + \beta_\nu y^{(\nu)}(b)] = 0 \tag{18.17}$$

where the α_ν, β_ν are given real constants not all zero (or are sometimes given functions of λ).

Example 2. A rod of length l with variable flexural rigidity $\alpha(x) = EI(x)$, which is clamped at the end $x = 0$, and simply supported at the other end $x = l$, is loaded by an axial force P. The Euler buckling load $\lambda = P$ is calculated from the eigenvalue problem

$$\frac{d^2}{dx^2}(\alpha(x)y'') = -\lambda y'' \qquad y(0) = y'(0) = y(l) = y''(l) = 0$$

b. Some Important Concepts Connected with Eigenvalue Problems for Ordinary Differential Equations [28]

It is useful to distinguish between *essential* and *suppressible* boundary conditions. They are defined by the following process. We try to eliminate the derivatives of mth and higher order from as many as possible of the $2m$ original boundary conditions (18.17) by forming linear combinations of these conditions. The k boundary conditions thus obtained are the essential conditions. There are $(2m - k)$ conditions left over, out of which no more essential condition can be extracted. These are the suppressible conditions.

Example 3. $-y'' = \lambda y$, $y(0) - y'(0) = 0$, $2y(1) + y'(0) = 0$; here we have one essential boundary condition, namely, $y(0) + 2y(1) = 0$, and one suppressible boundary condition, $y(0) - y'(0) = 0$.

A function which is m times continuously differentiable and which satisfies the essential boundary conditions is called an *admissible function.* A function $y(x) \not\equiv 0$ which is $2m$ times continuously differentiable and which satisfies all the boundary conditions is called a *test function* or a *test element.*

The *scalar product* (y,z) of two real functions y, z, which are integrable in the interval $\langle a,b\rangle$, is defined as

$$(y,z) = \int_a^b y(x)z(x)\,dx \tag{18.18}$$

If $(y,z) = 0$, y and z are called orthogonal.

The eigenvalue problem (18.14) through (18.17) is called *self-adjoint* if, for two arbitrary test functions $y(x)$ and $z(x)$,

$$(y,Mz) = (z,My) \qquad \text{and} \qquad (y,Nz) = (z,Ny) \tag{18.19}$$

Ordinarily, we ascertain whether an eigenvalue problem is self-adjoint or not by using integration by parts; e.g., the problem

$$-y'' = \lambda y \qquad y(0) = 0 \qquad y'(1) + cy(1) = 0 \tag{18.20}$$

is self-adjoint, for, if $y(x)$ and $z(x)$ satisfy the boundary conditions, then

$$\int_0^1 [y(-z'') - z(-y'')]\,dx = [-yz' + zy']_0^1 + \int_0^1 [y'z' - z'y']\,dx = 0$$

The self-adjointness of differential operators corresponds to the symmetry of matrices.

The eigenvalue problem (18.14) through (18.17) is called completely definite if, for each test function $z(x)$,

$$(z,Mz) > 0 \qquad (z,Nz) > 0 \tag{18.21}$$

The complete definiteness is also most easily investigated with the help of integration by parts. For example, the above problem (18.20) is completely definite for $c > 0$. To see this, we note that

$$J = \int_0^1 z(-z'')\,dx = [-zz']_0^1 + \int_0^1 z'^2\,dx$$
$$= c[z(1)]^2 + \int_0^1 z'^2\,dx \geq 0$$

J can be equal to 0 only for $z(1) = 0$, $z' \equiv 0$; that is, $z = \text{const} = 0$. There is, however, no test function $z \equiv 0$; that is $J > 0$ for every test function.

Example 4. The following eigenvalue problems are self-adjoint and completely definite:

1. Differential equation:

$$-(f_0(x)y')' = \lambda g_0(x)y \qquad f_0 > 0,\ g_0 > 0 \quad (f_0' \text{ and } g_0 \text{ continuous})$$

Boundary conditions:

$$y(a) - c_1y'(a) = 0 \qquad y(b) + c_2y'(b) = 0 \qquad \text{with } c_1 \geq 0,\ c_2 \geq 0$$

or $\quad y(a) = y'(b) = 0 \quad$ or $\quad y(a) - y(b) = y'(a) - y'(b) = 0$

2. Differential equation:

$$(f_0(x)y'')'' - (f_1(x)y')' + f_2(x)y = \lambda g_0(x)y \qquad f_j > 0,\ g_0 > 0 \quad (f_0'', f_1', f_2, g_0 \text{ continuous})$$

Boundary conditions:

$$y(a) = y(b) = y^{(i)}(a) = y^{(j)}(b) = 0 \qquad \text{for } i = 1 \text{ or } 2,\ j = 1 \text{ or } 2$$

or

$$y(a) = y'(a) = y''(b) = y'''(b) = 0$$

The eigenvalue problem is called *half-definite* if, for each test function $z(x)$,

$$(z,Nz) > 0 \tag{18.22}$$

We now define the following three classes of eigenvalue problems.

1. For the eigenvalue problem (18.14) through (18.17) we have $n > 0$, and in (18.15*b*) at least two of the functions $g_\nu(x)$ are not identically zero. Then we have a *general eigenvalue problem.*

2. The sum (18.15*b*) contains only a single term:

$$Ny = (-1)^n[g_n(x)y^{(n)}]^{(n)} \tag{18.23}$$

and $n > 0$. Moreover, for two test functions $y(x)$, $z(x)$

$$(y,Nz) = (z^{(n)},g_n y^{(n)}) \tag{18.24}$$

The eigenvalue problem then belongs to the *one-term class.*

3. If Ny has the form

$$Ny = g_0(x)y(x) \tag{18.25}$$

then we have a *special eigenvalue problem.*

c. The Expansion Theorem

For a test function $u(x)$, we can calculate its *Fourier coefficients* a_j relative to an orthonormal system of eigenfunctions y_i according to the formula

$$a_j = (u,Ny_j)$$

and formally write down the *Fourier series* $\Sigma a_j y_j(x)$.

If the eigenvalue problem (18.14) through (18.17) is self-adjoint, then $\sum_{j=1}^{\infty} a_j^2$ and (even) $\sum_{j=1}^{\infty} \lambda_j a_j^2$ converge and Parseval's equation and Bessel's inequality hold:

$$\sum_{j=1}^{\infty} a_j^2 = (u,Nu) \qquad \sum_{j=1}^{\infty} \lambda_j a_j^2 \le (u,Mu)$$

The Fourier series then converges, even absolutely and uniformly, but whether its sum

$$v(x) = \sum_{j=1}^{\infty} a_j y_j(x)$$

coincides with $u(x)$ is a rather deep question. However, if we have a problem of the one-term class, and if among the boundary conditions the equations

$$y(a) = y'(a) = \cdots = y^{(n-1)}(a) = y(b) = y'(b) = \cdots = y^{(n-1)}(b) = 0$$

occur (this condition arises in special eigenvalue problems), then

$$u(x) = \sum_{j=1}^{\infty} a_j y_j(x)$$

and this equation may be differentiated term by term $(m - 1)$ times.

d. Survey of Methods for the Treatment of Eigenvalue Problems for Ordinary Differential Equations

There are many methods at our disposal for the numerical treatment.

1. If the differential equation (18.14) has constant coefficients, then we can express its general solution with $2m$ parameters α_ν in closed form, and obtain from the bound-

Table 18.2. Equations for the Eigenvalues of some Simple Problems

Differential equation, abbreviations	Boundary conditions	Equation for the eigenvalues
$-y'' = \lambda y$ $\lambda = k^2$ $kl = \zeta$ $y_0 = y(0)$ $y'_l = y'(l)$ etc.	$ay'_0 + by_0 = 0$ $cy'_l + dy_l = 0$	$\tan \zeta = \dfrac{(ad - bc)l\zeta}{bdl^2 + ac\zeta^2}$
	$y_0 = y_l = 0$ $y_0 = y'_l = 0$ $y'_0 = y'_l = 0$	$\sin \zeta = 0$ $\cos \zeta = 0$ $\sin \zeta = 0$
	$y_0 = y'_l + dy_l = 0$ $y'_0 = y'_l + dy_l = 0$ $y_0 = y_l,\ y'_0 = y'_l$	$dl \tan \zeta = -\zeta$ $\zeta \tan \zeta = dl$ $\cos \zeta = 1$
$y^{IV} = \lambda y$ $\lambda = k^4,\ kl = \zeta$ $A = \operatorname{Cosh} \zeta \sin \zeta + \operatorname{Sinh} \zeta \cos \zeta$ $B = \operatorname{Cosh} \zeta \sin \zeta - \operatorname{Sinh} \zeta \cos \zeta$ $C = 2 \operatorname{Cosh} \zeta \cos \zeta$ $S = 2 \operatorname{Sinh} \zeta \sin \zeta$ $D = \operatorname{Cosh} \zeta \cos \zeta - 1$ $E = \operatorname{Cosh} \zeta \cos \zeta + 1$	$y_0 = y'_0 = 0$ $ay''_l + by'_l = 0$ $cy'''_l + dy_l = 0$	$ack^4E + bck^3A - adkB + bdD = 0$
	$y_0 = y'_0 = y_l = y'_l = 0$ $y_0 = y'_0 = y''_l = y'''_l = 0$ $y_0 = y'_0 = y'_l = y'''_l = 0$	$D = 0$ $E = 0$ $\tan \zeta + \operatorname{Tanh} \zeta = 0$
	$y_0 = y''_0 = 0$ $ay''_l + by'_l = 0$ $cy'''_l + dy_l = 0$	$(ack^4 + bd)B - bck^3C + adkS = 0$
	$y_0 = y''_0 = y_l = y''_l = 0$ $y_0 = y''_0 = y_l = y'_l = 0$ $y_0 = y''_0 = y''_l = y'''_l = 0$	$\sin \zeta = 0$ $\tan \zeta = \operatorname{Tanh} \zeta$ $\tan \zeta = \operatorname{Tanh} \zeta$
	$y_0 = y_l = 0$ $ay''_0 + by'_0 = 0$ $cy''_l + dy'_l = 0$	$ack^2S + bdD + k(ad - bc)B = 0$
	$y''_0 = y'''_0 = y''_l = y'''_l = 0$	$D = 0$
$y^{IV} = -\lambda y''$ $\lambda = k^2$ $kl = \zeta$	$y_0 = y'_0 = 0$ $ay''_l + by'_l = 0$ $cy'''_l + dy_l = 0$	$k^4ac + [k^3bc + kd(bl - a)] \sin \zeta$ $+ k^2lad \cos \zeta - 2bd(1 - \cos \zeta) = 0$
	$y_0 = y'_0 = y_l = y'_l = 0$	$\sin \zeta/2 = 0$ and $\tan \zeta/2 = \zeta/2$
	$y_0 = y''_0 = 0$ $ay''_l + by'_l = 0$ $cy'''_l + dy_l = 0$	$b(k^3c + kld) \cos \zeta = d(k^2la + b) \sin \zeta$
	$y_0 = y_l = 0$ $ay''_0 + by'_0 = 0$ $cy''_l + dy'_l = 0$	$aclk^3 \sin \zeta$ $+ bd(\zeta \sin \zeta - 2 + 2 \cos \zeta)$ $+ k(bc - ad)(\zeta \cos \zeta - \sin \zeta) = 0$

ary conditions $2m$ linear homogeneous equations for the α_ν. If we set the determinant of the coefficients equal to zero, we get a condition (ordinarily a transcendental equation) for the eigenvalues λ. Table 18.2 gives a compilation of the transcendental equations for several frequently occurring simple cases.

For ordinary differential equations with nonconstant coefficients and for partial differential equations, it is rarely possible to solve the eigenvalue problem analytically in closed form. We are then led to numerical or graphical methods of approximation.

2. If we need only general information concerning the location of the eigenvalues, then we can often compare the problem with a similar one which has constant coefficients. For the one-term class, we can use the inclusion theorem of Art. 18.4*e*, p. 18–21. Here we can frequently improve the bounds by suitable choice of accompanying free parameters.

3. An easily feasible method which is generally applicable also to partial differential equations is the ordinary finite-difference method of Art. 18.5*c*, p. 18–26. For large mesh width (and therewith a moderate amount of computational work) it yields information concerning the location of the lower eigenvalues and the behavior of the accompanying eigenfunctions. If we desire results of greater accuracy, then we do not calculate with a smaller mesh width, but use instead one of the improved methods of Art. 18.5*d*. For differential equations of simple structure, the hermitian method of Art. 18.5*d*(2) has proved very helpful. For more complicated differential equations, the method of higher approximation of Art. 18.5*d*(1) is more suitable. For both improved methods it should be noted that, according to the greater accuracy of the procedure, it is necessary to use correspondingly accurate expressions for the boundary conditions.

4. If we are particularly interested in the first eigenvalue (e.g., in buckling problems), then the iteration method of Art. 18.4*c* is very suitable. When the additional hypotheses of Art. 18.4*d* are satisfied, the main formula (18.66) generally yields very useful upper and lower bounds for the first eigenvalue. The iteration method is, because of the nature of this type of problem, computationally or in many cases also graphically feasible. We can also use the iteration method for higher eigenvalues; for that, however, one must note what has been said for matrices in No. 18.1*d*(2).

5. For the simultaneous determination of several eigenvalues, the Ritz method in Art. 18.5*a* and its variants in Art. 18.5*b* are very suitable. For self-adjoint completely definite problems, it furnishes (by the use of a p-termed linear combination of test functions) upper bounds for the first p eigenvalues. Corresponding lower bounds can often be obtained with the help of the Temple quotients (Art. 18.4*d*). The quality of Ritz approximate values (particularly the higher ones) depends strongly on the number and choice of the test functions selected.

6. Moreover, there is a large number of additional methods which can be very suitable for the preceding cases. Some of these methods are mentioned in Art. 18.5*e*.

18.3. EIGENVALUE PROBLEMS FOR PARTIAL DIFFERENTIAL EQUATIONS AND INTEGRAL EQUATIONS

a. Some Fundamental Concepts for Partial Differential Equations

The following examples may illustrate the general problem. A homogeneous membrane covers a region B with boundary Γ in the x_1,x_2 plane and is clamped along this boundary. Its transverse vibrations satisfy the differential equation

$$\nabla^2 y \equiv \frac{\partial^2 y}{\partial x_1^2} + \frac{\partial^2 y}{\partial x_2^2} = -\lambda y \text{ in } B \qquad y = 0 \text{ on } \Gamma$$

The problems of plate vibrations and of plate buckling lead to partial differential equations of the fourth order. For the shear buckling of a plate, the differential equation is

$$\nabla^2\nabla^2 y = -\lambda \frac{\partial^2 y}{\partial x_1\, \partial x_2}$$

and, for the buckling of a circular cylindrical shell of large radius, we have the differential equation ([1], vol. 2, p. 506):

$$\frac{\partial^8 y}{\partial x_1^8} + c_1 \frac{\partial^4 y}{\partial x_2^4} = -\lambda \frac{\partial^6 y}{\partial x_1^6}$$

In the general problem we have a differential equation

$$My = \lambda Ny \text{ in } B \tag{18.26}$$

for a function $y(x_1,x_2, \ldots ,x_n)$, where M, N are given (possibly real) linear homogeneous partial differential operators. The boundary conditions are of the form

$$U_\mu y = 0 \text{ on } \Gamma_\mu \qquad \mu = 1, 2, \ldots , k \tag{18.27}$$

B is a bounded closed domain in $(x_1, \ldots ,x_n)$ space and Γ_μ is a sufficiently smooth hypersurface (ordinarily part of the boundary of B). The $U_\mu y$ are given linear homogeneous expressions in y and the partial derivatives of y. A function $y(x_1, \ldots ,x_n)$, which possesses continuous partial derivatives of orders as high as occur in the differential equations, and fulfils the boundary conditions, is called a *test function* or a *test element*. Using the scalar product of two functions y and z (integrable in B)

$$(y,z) = \int_B y(s)z(s)\,ds \tag{18.28}$$

(where s stands for the totality of the independent variables $x_1, \ldots ,x_n$), we can introduce the concepts of self-adjoint, completely definite, and half-definite by the same formulas (18.19), (18.21), and (18.22) as for ordinary differential equations.

Example 5. The eigenvalue problem with the elliptic differential equation

$$My = -\frac{\partial}{\partial x_1}\left(p\frac{\partial y}{\partial x_1}\right) - \frac{\partial}{\partial x_2}\left(p\frac{\partial y}{\partial x_2}\right) + qy = \lambda g_0 y \tag{18.29}$$

[where $p(x_1,x_2)$, $q(x_1,x_2)$, $g_0(x_1,x_2)$ are given continuous functions in B, p with continuous partial derivatives of first order, $p > 0$, $g_0 > 0$ in B] and with the boundary condition on the boundary Γ of B

$$\alpha y + \beta\frac{\partial y}{\partial \nu} = 0 \tag{18.30}$$

(where ν denotes the inner normal, and α and β are given position functions which do not vanish simultaneously on the boundary) is self-adjoint, as can easily be verified with the help of Gauss' integral theorem. For $\beta = 0$, $q \geq 0$, it is also completely definite.

b. Eigenvalue Problems for Integral Equations

Certain formulations of engineering problems lead directly to linear integral equations. However, the particular importance of integral equations stems from the fact that, for a certain class of integral equations, a well-rounded theory has been established, and that many boundary-value and eigenvalue problems of ordinary and partial differential equations lead to integral equations of this class, so that the theory can be immediately applied.

Consider the integral equation

$$y(s) = \lambda \int_B K(s,t)y(t)\,dt \tag{18.31}$$

where s stands for finitely many variables $x_1, \ldots , x_n$, and B is a bounded measurable domain in $(x_1, \ldots ,x_n)$ space. In the simplest case, s is a real variable and B is a finite interval $\langle a,b\rangle$ in which s lies. By means of the integral, an *integral transformation* is given

$$Ny(s) = \int_B K(s,t)y(t)\,dt \tag{18.32}$$

which relates the function $y(s)$ to the function $Ny(s)$. If M is the identity operator, i.e., every function $y(s)$ remains unchanged: $My(s) = y(s)$, then (18.31) assumes the same form as (18.1*b*)

$$My = \lambda Ny$$

As before, those values of λ, the *eigenvalues*, are sought for which the integral equation has a nonidentically vanishing square integrable solution $y(s)$, an *eigenfunction* or *eigenelement*. The eigenvalues of (18.31) are also called the eigenvalues of the *kernel* K. The *scalar product* of two test elements, i.e., square integrable functions $y(s)$, $z(s)$, is again introduced as

$$(y,z) = \int_B y(s)z(s)\, ds \tag{18.33}$$

We assume the kernel $K(s,t)$ to be symmetric: $K(s,t) = K(t,s)$, and we then call the operator N or the integral equation symmetric. We consider here only real variables. The theory can be carried through in exactly the same way for complex valued functions, if we assume the kernel to be hermitian; $K(s,t) = \bar{K}(t,s)$ but this case seldom appears in engineering applications. In any event, a real symmetric kernel is also hermitian. Furthermore, we assume the kernel to be *square integrable* and of *mean continuity;* i.e., the integrals

$$\int_B K(s,t)\, dt \qquad \int_B K^2(s,t)\, dt \qquad \iint_B K(s,t)\, ds\, dt \qquad \iint_B K^2(s,t)\, ds\, dt$$

exist and are bounded, and, for any fixed s_1,

$$\lim_{s \to s_1} \int_B [K(s,t) - K(s_1,t)]^2\, dt = 0$$

Of course, for a continuous kernel $K(s,t)$, both these assumptions are satisfied (see [9], p. 68).

While an arbitrary nonsymmetric matrix always has at least one characteristic number, a nonsymmetric kernel may have no eigenvalue. However, a kernel which satisfies the above assumptions has at least one eigenvalue. Among these kernels, those that have only finitely many eigenvalues are precisely the *degenerate* kernels. A kernel is called degenerate if it can be represented as a finite sum of the form

$$K(s,t) = \sum_{k=1}^{r} a_k(s)b_k(t) \tag{18.34}$$

For a degenerate kernel we can immediately reduce (18.31) to a matrix eigenvalue problem by setting

$$y(s) = \sum_{j=1}^{r} c_j a_j(s)$$

where the c_j are undetermined constants.

If $\tilde{N} = [\tilde{n}_{jk}]$ is the matrix of the numbers

$$\tilde{n}_{jk} = \int_B b_j(t)a_k(t)\, dt$$

and if we interpret the c_j as components of a vector c then

$$c = \lambda \tilde{N} c \tag{18.35}$$

holds. That is, the eigenvalues of (18.30) are here also the eigenvalues of the matrix problem (18.35). A possibility of treating an integral equation (18.31) with a nondegenerate kernel exists if the kernel can be approximated by degenerate kernels whose eigenvalues can be easily calculated.

c. Results from the Theory of Integral Equations

Under the assumptions of Art. 18.3*b*, the following statement holds concerning the kernel: The integral equation (18.31) has, if the kernel is nondegenerate, a denumerable infinity of eigenvalues, which we can, therefore, order in a sequence. If eigenvalues appear with both signs (half-definite problem), then we assign positive and negative indices to the eigenvalues which we then order in the following way:

$$\cdots \lambda_{-3} \leq \lambda_{-2} \leq \lambda_{-1} < 0 < \lambda_1 \leq \lambda_2 \leq \lambda_3 \cdots \tag{18.36}$$

We denote the corresponding eigenfunctions by

$$\ldots y_{-3}, y_{-2}, y_{-1}, y_1, y_2, y_3, \ldots$$

Of course, it can happen that only positive eigenvalues occur (positive-definite problem) or only negative eigenvalues (negative-definite problem). We then say that the kernel is definite. We can also characterize the definiteness as follows: The kernel $K(s,t)$ is positive-definite if, for all square integrable functions $y(s) \not\equiv 0$, we have

$$(y,Ny) = \iint_B K(s,t)y(s)y(t)\,ds\,dt > 0 \tag{18.37}$$

The eigenvalues cannot have a finite accumulation point, and they can have only a finite multiplicity (for the definition of multiplicity, see p. 18–17).

A function $h(s)$ is called *representable* by a source distribution if there exists a square integrable function $g(s)$ so that

$$h(s) = \int_B K(s,t)g(t)\,dt \tag{18.38}$$

(The operator N maps the test elements into the functions representable by a source distribution.)

Expansion Theorem. The integral equation has, under the assumptions given above, an orthogonal system of eigenfunctions y_j with

$$(y_j,y_k) = \begin{cases} 0 & \text{for } j \neq k \\ 1 & \text{for } j = k \end{cases} \tag{18.39}$$

Every function $h(s)$ which can be represented in the form (18.38) is expandable in an absolutely and uniformly convergent series of these eigenfunctions:

$$h(s) = \sum_j h_j y_j(s) \tag{18.40}$$

where the expansion coefficients h_j are given by $h_j = (h,y_j)$.

If the kernel $K(s,t)$ is continuous, and has only finitely many negative eigenvalues, then it can be expanded in the absolutely and uniformly convergent series:

$$K(s,t) = \sum_j \frac{\phi_j(s)\phi_j(t)}{\lambda_j} \tag{18.41}$$

and, therefore,

$$\int_B K(s,s)\,ds = \sum_j \frac{1}{\lambda_j} \tag{18.42}$$

If a kernel $K(s,t)$ satisfies the assumptions of Art. 18.3*b* but is otherwise arbitrary, then the first iterated kernel

$$K^{(2)}(s,t) = \int_B K(s,v)K(v,t)\,dv \tag{18.43}$$

can be expanded in an absolutely and uniformly convergent series

$$K^{(2)}(s,t) = \sum_j \frac{y_j(s)y_j(t)}{\lambda_j^2} \tag{18.44}$$

$K^{(2)}$ is always positive-definite, and the following formula may be used for eigenvalue estimates

$$\int_B K^{(2)}(s,s)\,ds = \iint_B [K(s,t)]^2\,ds\,dt = \sum_j \frac{1}{\lambda_j^2} \tag{18.45}$$

Inclusion Theorem. If $h(s)$ is a continuous function and the function

$$\phi(s) = \frac{v(s)}{Nv(s)} \tag{18.46}$$

has a fixed sign and lies in B between finite limits $\phi_{\min}$ and $\phi_{\max}$, then at least one eigenvalue λ_k lies between these limits:

$$\phi_{\min} \leq \lambda_k \leq \phi_{\max} \tag{18.47}$$

***d.* Application of Integral Equations to Eigenvalue Problems for Differential Equations**

An inhomogeneous boundary-value problem for a function y with given position function r

$$My = r$$

and with homogeneous boundary conditions $U_\mu y = 0$ ($\mu = 1, 2, \ldots, k$) can often be solved with the help of the Green's function (influence function) in the form

$$y(s) = \int_B G(s,t)r(t)\,dt \tag{18.48}$$

Here s stands for the totality of the independent variables $x_1, \ldots, x_n$.

Example 6. $My \equiv -y'' = r(x)$, $y(0) = y'(1) = 0$. Here, if we write x instead of s, and ξ instead of t, then the Green's function reads:

$$G(x,\xi) = \begin{cases} x & \text{for } x \leq \xi \\ \xi & \text{for } x \geq \xi \end{cases}$$

so that the solution of the boundary-value problem reads

$$y(x) = \int_0^1 G(x,\xi)r(\xi)\,d\xi$$

For self-adjoint boundary-value problems, the Green's function is symmetric; that is, $G(s,t) = G(t,s)$. If the Green's function exists, the eigenvalue problem (18.14), (18.17) goes over into the integral equation

$$y(s) = \lambda \int_B G(s,t)Ny(t)\,dt \tag{18.49}$$

For special eigenvalue problems with

$$Ny = g_0(s)y(s) \tag{18.50}$$

Eq. (18.49) is an ordinary integral equation of the form (18.31). For ordinary differential equations (as long as $\lambda = 0$ is not an eigenvalue), problems of the one-term class (18.14), (18.23) can be converted into integral equations of the form (18.31) with a continuous, real symmetric kernel, so that the theory of integral equations is immediately applicable.

Likewise, for partial differential equations, we can prove the existence of a Green's function in many cases; see, e.g., [5], [6], [13]. If, for a partial differential equation, N has the form (18.50) and a Green's function exists, then the theory of integral equations is applicable. We obtain the expansion theorem, the extremal theorem for the Rayleigh quotient (18.55), and there is the inclusion theorem (18.73) of Art. 18.4*e*, where ϕ has the form

$$\phi = \frac{Mu^{[1]}}{g_0 u^{[1]}} \tag{18.51}$$

and $u^{[1]}$ is a test function.

18.4. EXTREMAL PROPERTIES OF THE EIGENVALUES AND FURTHER INCLUSION THEOREMS

In the following, the eigenvalue problems are assumed to be half-definite and, moreover, for differential equations, to be self-adjoint, and for matrices and integral equations, to be hermitian (or, if they are real, symmetric).

Then all existing eigenvalues are real. For two distinct eigenvalues λ_i, λ_j, the corresponding eigenelements are orthogonal to each other "in the general sense"; i.e.,

$$\begin{aligned}(y_i,Ny_j) &= 0 \qquad \text{for } \lambda_i \neq \lambda_j \\ (y_i,My_j) &= 0 \qquad \text{for } \lambda_i \neq \lambda_j\end{aligned} \tag{18.52}$$

If the eigenvalue problem is completely definite, then all the eigenvalues are positive. Often the eigenelements are "normalized" by making

$$(y_i,Ny_i) = 1 \tag{18.53}$$

Then we have

$$(y_i,My_i) = \lambda_i \tag{18.54}$$

If, for an eigenvalue λ, there are exactly p linearly independent eigenelements, then λ is called a p-fold eigenvalue, or we say that λ has multiplicity p.

a. Minimum Properties of the Eigenvalues for Completely Definite Problems

An important role is played by the Rayleigh quotient

$$R[u] = \frac{(u,Mu)}{(u,Nu)} \tag{18.55}$$

which can be formed for each test element. If $u = y_j$ is the eigenelement corresponding to the eigenvalue λ_j, then evidently

$$R[y_j] = \lambda_j \tag{18.56}$$

For differential equations we can often bring the Rayleigh quotient also into another form [see Eq. (18.80)].

We have the following theorems:

Theorem 1. Minimum Property of λ_1. If the eigenvalue problem is self-adjoint (hermitian) and completely definite, then the first eigenelement y_1 solves the extremal problem in which the Rayleigh quotient $R[u]$ is to be made a minimum. In fact, the corresponding minimum of R, as u runs through the range of all test elements, is equal to λ_1. Therefore,

$$R[u] \geq \lambda_1 \tag{18.57}$$

Theorem 2. Recursive Minimum Property of the Higher Eigenvalues. Under the assumptions of Theorem 1, there are for matrices finitely many (namely, n) and for differential equations infinitely many real and positive eigen values λ_1, $\lambda_2, \ldots$. The $(k + 1)$st eigenvalue is equal to the minimum value which the Rayleigh quotient $R[u]$ can assume as u runs through the range of all test elements which are (in the general sense) orthogonal to the first k eigenelements, i.e., for which

$$(u,Ny_j) = 0 \qquad \text{for } j = 1, 2, \ldots, k$$

Here multiple eigenvalues are to be counted several times according to their multiplicity.

Theorem 3. Courant's Independent Minimum Property of the Higher Eigenvalues. Under the assumptions of Theorem 1, let $F(v_1,v_2, \ldots ,v_k)$ be the greatest lower bound of the Rayleigh quotient $R[u]$, as u runs through the range of all test elements which are orthogonal to k given linearly independent elements $v_1, \ldots, v_k$; that is, $(u,v_j) = 0$ for $j = 1, 2, \ldots, k$. Then the $(k + 1)$st eigenvalue λ_{k+1} is equal to the largest value which $F(v_1, \ldots ,v_k)$ can assume when the $v_1, \ldots, v_k$ run through the range of all admissible systems of elements.

b. Extremal Properties for Half-definite Eigenvalue Problems

If the eigenvalue problem is not completely definite, but only half-definite, then the eigenvalues can appear with both signs, even infinitely many of each. E. Kamke ([28], vol. 45) has developed a theory of the *polar* eigenvalue problems; however, they do not directly enter here since they seldom occur in engineering problems. We give here only theorems for matrix eigenvalue problems.

Theorem 4. In (18.1*a*), let M, N be hermitian, and let M be positive-definite. Then there are n characteristic numbers κ_j with $\kappa_1 \geq \kappa_2 \geq \cdots \geq \kappa_n$. κ_1 is the maximum and κ_n the minimum of $1/R[u]$ as u runs through all test vectors. κ_{k+1} is the maximum of $1/R[u]$ as u runs through all test vectors with $(u,My_j) = 0$ $(j = 1,2, \ldots ,k)$, and the maximum is assumed when $u = y_{k+1}$, an eigenvector for the characteristic number κ_{k+1}. If we normalize these eigenvectors by $(y_j,My_j) = 1$, then, when the column matrices are placed side by side, there results a matrix

$$Y = [y_1,y_2, \ldots ,y_n]$$

which, by $u = Yv$, transforms simultaneously the two hermitian forms (u,Nu) and (u,Mu) to diagonal form:

$$\sum_{j=1}^{n} k_j|v_j|^2 \qquad \text{and} \qquad \sum_{j=1}^{n} |v_j|^2$$

respectively (the v_j are the components of v).

Principal Coordinates. The theorem just mentioned plays a role in conservative mechanical systems with finitely many degrees of freedom. The system is described by $-M\ddot{y} = Ny$ with real symmetric positive-definite matrices M, N. By splitting off a time factor $\cos \omega t$ from y, we obtain the equation $\kappa My = Ny$, where $\kappa = \omega^2$. M^{-1} and N^{-1} exist, and so the system can be written in the acceleration-coupled form $\kappa N^{-1}My = y$ or in the displacement-coupled form $\kappa y = M^{-1}Ny$. The theorem asserts that, by a suitable linear transformation, i.e., by the introduction of new coordinates, the principal coordinates, the same system can be brought into the form $\kappa D_1y = D_2y$, where D_1, D_2 are diagonal matrices. That is, the system can be written in completely uncoupled form.

Example 7. Motor-vehicle Vibrations. A motor vehicle may be idealized as a rigid body resting on two springs. If we introduce the displacements from the equilibrium position as coordinates, then the equations of motion turn out to be coupled. For each fundamental and higher vibration, there is on the body (or on its extension) one nodal point. If we introduce the displacements at these two nodal points as new (principal) coordinates, then the equations of motion turn out to be uncoupled.

The theorems of Art. 18.4*a* are transferable to half-definite eigenvalue problems for integral equations whose kernels fulfil the assumptions required in Art. 18.3*b*. If we work with the reciprocal value of the Rayleigh quotient and the eigenvalues, then, e.g., Theorem 2 of Art. 18.4*a* (for integral equations with $Mu \equiv u$) corresponds to:

Theorem 2′. The reciprocal value of the kth positive eigenvalue λ_k, if it exists, is equal to the maximum of the reciprocal Rayleigh quotient

$$\frac{1}{\lambda_k} = \max \frac{(u,Nu)}{(u,u)}$$

as u runs through the range of all test elements which are orthogonal to the eigenelements y_j which belong to the first $(k-1)$ positive eigenvalues: $(u,y_j) = 0$ for $j = 1, 2, \ldots, k-1$.

[Theorem 2 has $(u,Ny_j) = 0$ which, because $\lambda_j \neq 0$, is equivalent to $(u,y_j) = 0$]. A corresponding statement holds for the negative eigenvalues.

c. The Iteration Method for Eigenvalue Problems

If an eigenvalue problem is linear in the eigenvalue λ, then the following iterative procedure may be used. In all terms containing λ, y is replaced by $u^{[k]}$, and in all other terms by $u^{[k+1]}$, and λ is replaced by 1. Then, starting from an arbitrary element $u^{[0]}$, the sequence $u^{[0]}, u^{[1]}, \ldots$ is calculated. Its elements converge to an eigenelement.

Example 8. Matrices. For Eq. (18.1*b*), the formula for the determination of a new vector $u^{[k+1]}$ reads

$$Mu^{[k+1]} = Nu^{[k]} \qquad k = 0, 1, \ldots \tag{18.58}$$

If M has an inverse, then this becomes

$$u^{[k]} = Cu^{[k-1]} = C^k u^{[0]} \qquad k = 1, 2, \ldots$$

where $C = M^{-1}N$. In the special case $M = I$, we have

$$u^{[k]} = Nu^{[k-1]} = N^k u^{[0]} \tag{18.59}$$

Remarks concerning this have already been made in Art. 18.1*d*(2) and (4).

Example 9. Ordinary Differential Equations. Here we have to determine $u^{[k+1]}$ by solving a boundary-value problem. In particular, if λ does not appear in the boundary conditions, then the formula reads:

$$\begin{aligned} Mu^{[k+1]} &= Nu^{[k]} & k &= 0, 1, 2, \ldots \\ U_\mu u^{[k+1]} &= 0 & \mu &= 1, \ldots, 2m \end{aligned} \tag{18.60}$$

Example 10. Integral Equations. The formula here is again (18.58) or, since $My \equiv y$, simply

$$u^{[k+1]}(s) = \int_B K(s,t)u^{[k]}(t)\,dt \qquad k = 0, 1, 2, \ldots \tag{18.61}$$

With the help of the iterated kernel

$$K^{(p)}(s,t) = \int_B K(s,v)K^{(p-1)}(v,t)\,dv \qquad p = 2, 3, \ldots \tag{18.62}$$

and $K^{(1)} = K$, we can express $u^{[k]}$ directly in terms of $u^{[0]}$:

$$u^{[k]}(s) = \int_B K^{(k)}(s,t)u^{[0]}(t)\,dt \tag{18.63}$$

Occasionally the convergence can be improved by writing $u^{[k+1]} - \theta u^{[k]}$ in place of $u^{[k+1]}$ in the iteration formula, for example,

$$Mu^{[k+1]} = Nu^{[k]} + \theta M u^{[k]} \tag{18.64}$$

in place of Eq. (18.58). Here θ is a suitable real number.

In the application of the iteration method, one often uses Schwartz's constants a_k and Schwartz's quotients μ_k:

$$\begin{aligned} a_{2k} &= (u^{[k]},Nu^{[k]}) \\ a_{2k+1} &= (u^{[k+1]},Nu^{[k]}) \qquad k = 0, 1, 2, \ldots \\ \mu_{k+1} &= \frac{a_k}{a_{k+1}} \end{aligned} \tag{18.65}$$

d. Main Formula and Temple's Quotient for Differential Equations

For iteration methods applied to ordinary differential equations, we have:

Theorem 5. Suppose that the eigenvalue problem (18.14) through (18.17) is self-adjoint and completely definite, that the eigenvalue λ does not appear in the boundary conditions, and that the smallest eigenvalue λ_1 is simple. Let the iteration be started from a $2n$-times continuously differentiable function $u^{[0]} \not\equiv 0$, which satisfies so many of the given boundary conditions that, for a test function z, there is $(u^{[0]},Nz) = (z,Nu^{[0]})$ (for special eigenvalue problems, this is automatically satisfied).

If we have carried out the method so far that we know two Schwartz quotients μ_k, μ_{k+1}, and if l_2 is a lower bound for the second eigenvalue λ_2 (l_2 must also be larger than μ_{k+1}), then we can include the first eigenvalue λ_1 between the bounds (*main formula*):

$$\mu_{k+1} - \frac{\mu_k - \mu_{k+1}}{l_2/\mu_{k+1} - 1} \leq \lambda_1 \leq \mu_{k+1} \qquad k = 1, 2, \ldots \tag{18.66}$$

The μ_k form a monotonically nonincreasing sequence:

$$\mu_1 \geq \mu_2 \geq \cdots \geq \lambda_1$$

Also, for half-definite problems, there are formulas (naturally somewhat complicated) corresponding to (18.66); see L. Collatz [3], p. 300. (For matrices, the a_k have a different meaning from here!)

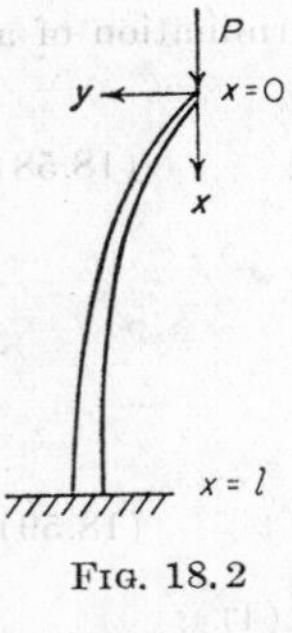

FIG. 18.2

Example 11. We want to find the Euler buckling load P for a column of variable flexural rigidity EI, supported as shown in Fig. 18.2. This amounts to calculating the smallest eigenvalue λ_1 of

$$-y'' = \lambda p(x) y \qquad y(0) = 0,\ y'(l) = 0 \tag{18.67}$$

with $1/EI = cp(x) = c(1 + \cos x)$, $l = \pi/2$.

From λ_1 we find $P = \lambda_1/c$. There are two functions $u^{[0]}$, $u^{[1]}$ to be found with $-(u^{[1]})'' = p(x)u^{[0]}$, and $u^{[1]}$ must satisfy the boundary conditions. If we choose $u^{[1]} = \sin x$, then

$$u^{[0]} = \frac{\sin x}{1 + \cos x} = \tan \frac{x}{2}$$

Frequently one starts from $u^{[0]}$ and determines $u^{[1]}$ by integration. If the iteration method is carried out graphically (method of Vianello), then one always proceeds so that one graphically only integrates, never differentiates. In the present case there follows immediately

$$a_0 = \int_0^{\pi/2} p(u^{[0]})^2\,dx = 2\int_0^{\pi/2} \sin^2 \frac{x}{2}\,dx = 0.57080$$

$$a_1 = \int_0^{\pi/2} pu^{[0]}u^{[1]}\,dx = \int_0^{\pi/2} \sin^2 x\,dx = 0.785398$$

$$a_2 = \int_0^{\pi/2} p(u^{[1]})^2\,dx = \int_0^{\pi/2} (1 + \cos x)\sin^2 x\,dx = 1.118732$$

$$\mu_1 = \frac{a_0}{a_1} = 0.72676 \qquad \mu_2 = \frac{a_1}{a_2} = 0.70204$$

For the utilization of (18.66) we need a rough lower bound for the second eigenvalue λ_2; we obtain it here from the comparison with the eigenvalue problem with constant coefficients $-y'' = \bar{\lambda}2y$, $y(0) = 0$, $y'(\pi/2) = 0$. Here $\bar{\lambda}_1 = 1/2$, $\bar{\lambda}_2 = 9/2 = l_2$. With this value of l_2 we obtain bounds which are sufficiently exact for engineering purposes:

$$0.69748 \leq \lambda_1 \leq 0.70204$$

The mean value 0.69976 has an error of, at most, 0.33%.

For self-adjoint (hermitian) completely definite eigenvalue problems, the main formula (18.66) can also be carried over to the higher eigenvalues. Also it often yields the possibility of an error estimate for the Ritz method. For differential equations it is assumed that the function $u^{[0]}$ can be expanded in a uniformly convergent series of the normalized eigenfunctions y_j:

$$u^{[0]} = \sum_{j=1}^{\infty} c_j y_j \tag{18.68}$$

and that the series can be differentiated term by term a sufficient number of times. $u^{[0]}$, $u^{[1]}$, a_0, a_1, a_2, μ_1, μ_2 are assumed known. We now form, from the Schwartz constants a_ν, the Temple quotient,

$$T(t) = \frac{a_0 - ta_1}{a_1 - ta_2} \tag{18.69}$$

Let the eigenvalues λ be ordered according to their magnitudes. If λ_s is a (single or multiple) eigenvalue, then let λ_{s-} and λ_{s+} be the two neighboring eigenvalues, $\lambda_{s-} < \lambda_s < \lambda_{s+}$. If rough bounds L_{s-} and l_{s+} for the neighboring eigenvalues are known, and if μ_2 lies between these bounds, i.e., if

$$\lambda_{s-} \leq L_{s-} < \mu_2 < l_{s+} \leq \lambda_{s+} \tag{18.70}$$

then for λ_s we have the bounds

$$T(l_{s+}) \leq \lambda_s \leq T(L_{s-}) \tag{18.71}$$

e. Inclusion Theorem for the One-term Class and for Matrices

(1) For the One-term Class. This theorem holds for self-adjoint completely definite eigenvalue problems of the one-term class [(18.14), (18.23)]. Let $u^{[1]}$ be a test function and $u^{[0]}$ be a $2n$-times continuously differentiable function with $Mu^{[1]} = Nu^{[0]}$. Let the function (quotient of the nth derivatives of $u^{[0]}$ and $u^{[1]}$)

$$\phi(x) = \frac{u^{[0](n)}}{u^{[1](n)}} \tag{18.72}$$

remain, in the interval $\langle a,b \rangle$, between finite positive limits $\phi_{\min}$ and $\phi_{\max}$; then at least one eigenvalue λ_s lies between these limits:

$$\phi_{\min} \leq \lambda_s \leq \phi_{\max} \tag{18.73}$$

Example 12. For the same example as in (18.67) we put $u^{[1]} = f(x) \sin [g(x)]$ and choose $n = 0$. In order to have a factor $\sin [g(x)]$ cancel out in numerator and denominator of (18.72), it is necessary that $g'f^2 = \text{const.}$ This condition is satisfied if we assume

$$u^{[1]} = \sqrt{c^2 + x^2} \sin \left(\alpha \tan^{-1} \frac{x}{c} \right)$$

where α, c are free parameters. We have

$$\phi = \frac{c^2(\alpha^2 - 1)}{D(x)} \qquad \text{where} \qquad D(x) = p(x)(c^2 + x^2)^2$$

c is chosen so that $D(x)$ approximates a constant as well as possible. We have $D(0) = D(\pi/2)$ for $c^2 = \frac{1}{4}\pi^2(1 + \sqrt{2}) = 5.9568$ and $D_{\min} = 70.96$ and $D_{\max} = 74.68$. Now α is determined so that $(u^{[1]})'(\pi/2) = 0$. With $\rho = \tan^{-1} (\pi/2c)$ and $\alpha\rho = \xi$, we obtain for ξ the equation $\tan \xi = -\xi(2c/\pi\rho) = -2.7171\xi$. The table contains the first three zeros of this equation and the eigenvalue bounds resulting therefrom. By computation of additional zeros, we would obtain equally good bounds (from a percentage standpoint) for arbitrarily many eigenvalues.

ξ	$b = c^2\left[\left(\frac{\xi}{\rho}\right)^2 - 1\right]$	Bounds for	Lower bound $b/D_{\max}$	Upper bound $b/D_{\min}$
1.7752	51.45	λ_1	0.6889	0.7250
4.789	411.9	λ_2	5.515	5.804
7.901	1131	λ_3	15.15	15.94

(2) For Matrices: Let N be a normal matrix, and let u and v be two vectors with components u_j and v_j such that $v = Nu$. Now form the quotients $q_j = v_j/u_j$. If a circular region K in the complex plane (which in the limiting case may be a half plane) contains all quotients q_j for which u_j and v_j do not vanish simultaneously, then K contains also at least one characteristic number of the matrix N [23],[29],[30].

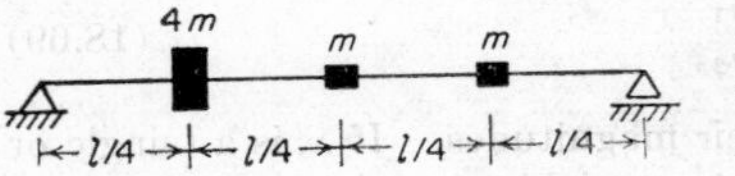

FIG. 18.3

Special Case. All quantities are considered to be real and N is symmetric. Then at least one characteristic number of N lies in the interval $\langle q_{\min}, q_{\max} \rangle$. A corresponding inclusion theorem [31] holds also for normalizable matrices N.

Example 13. A simply supported beam has three masses attached to it, as shown in Fig. 18.3. For its free vibrations, the following equation holds

$$\omega^2 CDz = z$$

where

$$C = \beta \begin{bmatrix} 9 & 11 & 7 \\ 11 & 16 & 11 \\ 7 & 11 & 9 \end{bmatrix} \qquad D = \begin{bmatrix} 4 & 0 & 0 \\ 0 & 1 & 0 \\ 0 & 0 & 1 \end{bmatrix}$$

$$D^{1/2} = \begin{bmatrix} 2 & 0 & 0 \\ 0 & 1 & 0 \\ 0 & 0 & 1 \end{bmatrix} \qquad z = \begin{bmatrix} z_1 \\ z_2 \\ z_3 \end{bmatrix} \qquad \beta = \frac{l^3}{768EI}$$

and where the z_j ($j = 1, 2, 3$) are the displacement amplitudes of the three masses. With

$$y = D^{1/2}z \qquad \omega^2\beta\kappa = 1 \qquad \beta N = D^{1/2}CD^{1/2}$$

we have $Ny = \kappa y$, where

$$N = \begin{bmatrix} 36 & 22 & 14 \\ 22 & 16 & 11 \\ 14 & 11 & 9 \end{bmatrix}$$

In the following table $u^T = [6,4,3]$ is used first in $v = Nu$, and the quotients q_i and the "spread" $\sigma = q_{\max} - q_{\min}$ are calculated. It is easily seen that, by reduction of the third component $u_3 = 3$ to 2.7, the spread can be considerably reduced. The table also contains estimates for the higher modes and corrections according to a similar relaxation, which has the advantages and disadvantages of relaxation methods used in other fields. The short calculation yields, for the three characteristic numbers, bounds, which can easily be improved by continuation of the method:

$$56.425 \le \kappa_1 \le 56.967 \qquad 3.680 \le \kappa_2 \le 3.742 \qquad 0.52 \le \kappa_3 \le 0.60$$

u_j			v_j			q_i			σ
6	4	3	346	229	155	57.7	57.25	51.7	6.0
6	4	2.7	341.8	225.7	152.3	56.967	56.425	56.5	0.542
5	−4	−5	22	−9	−19	4.4	2.25	3.8	2.15
5	−3.4	−6.2	18.4	−12.6	−23.2	3.68	3.706	3.742	0.062
2	−6	5	10	3	7	5	−0.5	1.4	5.5
2	−6.4	5	1.2	−3.4	2.6	0.6	0.53	0.52	0.08

The bounds can now be easily (and significantly) improved with the help of the Temple quotients. If we use, for example, $u^{[0]T} = [6,4,2.7]$ and $u^{[1]} = Nu^{[0]}$, then by the fact that $\kappa_2 \le 3.75$, we find

$$56.75299 \le \kappa_1 \le 56.75309$$

18.5. FURTHER METHODS FOR DIFFERENTIAL EQUATIONS

a. The Ritz Method

In the Ritz method we seek to come as close as possible to an extremal value λ of the Rayleigh quotient (18.55) of a self-adjoint (or hermitian) and half-definite eigenvalue problem. We assume u to have the form

$$u = \sum_{\nu=1}^{p} a_\nu v_\nu \tag{18.74}$$

so that R becomes a quotient of two quadratic forms in the a_ν, namely, $Q_1 = (u,Mu)$ and $Q_2 = (u,Nu)$. The extremum Λ of this quotient $R = \phi(a_1, \ldots, a_p)$ is determined from

$$\frac{\partial R}{\partial a_j} = 0 \qquad j = 1, \ldots, p \tag{18.75}$$

The v_ν are fixed chosen test elements. The equations (18.75) have the form

$$\frac{\partial Q_1}{\partial a_j} - \Lambda \frac{\partial Q_2}{\partial a_j} = 0 \qquad j = 1, \ldots, p \tag{18.76}$$

and are linear equations in the a_j (Galerkin equations):

$$\sum_{j=1}^{p} (\bar{m}_{ij} - \Lambda \bar{n}_{ij}) a_j = 0 \qquad i = 1, \ldots, p \tag{18.77}$$

where

$$\bar{m}_{ij} = (v_i, Mv_j) \qquad \text{and} \qquad \bar{n}_{ij} = (v_i, Nv_j) \tag{18.78}$$

Setting the determinant of coefficients of (18.77) equal to zero results in an equation of the pth degree in Λ, whose roots turn out to be real. The roots can be ordered as

$$\Lambda_1 \leq \Lambda_2 \leq \cdot \cdot \cdot \leq \Lambda_p$$

and are regarded as approximate values for the correspondingly ordered eigenvalues $\lambda_1, \ldots, \lambda_p$. For self-adjoint, completely definite eigenvalue problems for ordinary differential equations, all Λ_j are upper bounds for the corresponding eigenvalues:

$$\Lambda_j \geq \lambda_j \qquad j = 1, 2, \ldots, p \tag{18.79}$$

The quality of the Ritz approximate values often depends critically on the proper choice of the functions v_ν used in (18.74).

Example 14. For the problem (18.67), the functions $v_1 = \sin x$ and $v_2 = \sin 3x$ are used. Then

$$m_{11} = \int_0^{\pi/2} \sin^2 x \, dx = \frac{\pi}{4} \qquad m_{12} = \int_0^{\pi/2} \sin x \sin 3x \, dx = 0 \qquad \text{etc.}$$

and a short calculation yields the determinant

$$\begin{vmatrix} \frac{\pi}{4} - \left(\frac{\pi}{4} + \frac{1}{3}\right)\Lambda & -\frac{1}{5}\Lambda \\ -\frac{1}{5}\Lambda & \frac{9\pi}{4} - \left(\frac{\pi}{4} + \frac{17}{35}\right)\Lambda \end{vmatrix} = 0$$

with the roots $\Lambda_1 = 0.699215$, $\Lambda_2 = 5.7451$. By (18.79) these are upper bounds for the eigenvalues λ_1 and λ_2.

b. Additional Minimum Principles for Differential Equations

Minimum Principle of Kamke (see [28], vol. 48). For the eigenvalue problem (18.14) through (18.17), we can bring the Rayleigh quotient (by using integration by parts in the numerator and denominator) into the form

$$R[u] = \frac{\sum_{\nu=0}^{m} (f_\nu u^{(\nu)}, u^{(\nu)}) + M_0 u}{\sum_{\nu=0}^{n} (g_\nu u^{(\nu)}, u^{(\nu)}) + N_0 u} \tag{18.80}$$

Here the *Dirichlet boundary parts,* M_0u, N_0u, are quadratic forms in the quantities $u, u', u'', \ldots$, evaluated at the end points a and b. The eigenvalue problem is called *K-definite* if these two quadratic forms M_0u, N_0u are positive-definite in the appearing variables and if $f_\nu(x) \geq 0$, $g_\nu(x) \geq 0$ for all ν. A K-definite eigenvalue problem is also completely definite. The K-definiteness is very easily verified. All theorems given in the preceding number hold here, and we need only require the v_j to be admissible functions. (In Art. 18.5*a* the v_j must be test functions.)

Schwartz Quotients. For special eigenvalue problems with $Ny = g_0 y$, there can also be used the minimum principle in which the Schwartz quotients

$$\mu_1[u] = \frac{(g_0^{-1}Mu, Mu)}{(u, Mu)} \tag{18.81}$$

are minimized [32]. With u from (18.74) we now arrive at Grammel's equations [33]:

$$\sum_{j=1}^{p} (m_{ij}^* - \Lambda^* n_{ij}^*)a_j = 0 \qquad i = 1, \ldots, p \tag{18.82}$$

The equation obtained by setting the determinant of coefficients equal to zero yields the approximate values $\Lambda_1^*, \ldots, \Lambda_p^*$.

Energy Method. Consider an undamped mechanical system capable of vibrating, e.g., an elastic body which can vibrate about a position of equilibrium. For a certain vibration mode, the displacement η depends on the coordinates $x_1, \ldots, x_n$ and the time t in the form

$$\eta(x_1, \ldots, x_n, t) = y(x_1, \ldots, x_n) \sin \omega t \tag{18.83}$$

Let ρ denote the density, and dx_j the volume element, in the domain B covered by the body. Then the kinetic energy T is given by

$$T(t) = \tfrac{1}{2} \int_B \rho \left(\frac{\partial \eta}{\partial t}\right)^2 dx_j = T^* \omega^2 \cos^2 \omega t \tag{18.84}$$

where

$$T^* = \tfrac{1}{2} \int_B \rho [y(x_1, \ldots, x_n)]^2 \, dx_j \tag{18.85}$$

The potential energy $U(t)$, that is, the energy needed to produce the displacement η (work done against external forces, strain energy, etc.) is a quadratic function of y and, hence, has the form $U(t) = U_{\max} \sin^2 \omega t$. According to the energy theorem, the total energy $E = T(t) + U(t)$ is a constant in time. When the distribution of the amplitudes $y(x_1, \ldots, x_n)$ is known, both energies may be calculated, and from

$$E = T_{\max} = \omega^2 T^* = U_{\max} \tag{18.86}$$

the frequency ω may be found. Now Rayleigh's principle states the following: If in

$$\hat{R}[y] = \frac{U_{\max}}{T^*} \tag{18.87}$$

one uses an estimated distribution $u(x_1, \ldots, x_n)$ in place of the true, in general unknown amplitude distribution $y(x_1, \ldots, x_n)$, then one obtains a value which is greater than or equal to ω^2:

$$\hat{R}[u] \geq \omega^2 \tag{18.88}$$

In a large number of the cases arising in engineering, e.g., if one is led to the cases in Art. 18.5*a*, the Rayleigh principle can be rigorously proved. However, it is also used in many other cases.

When (18.74) is introduced into (18.88), a system of equations

$$\frac{\partial U_{\max}}{\partial a_j} - \Lambda \frac{\partial T^*}{\partial a_j} = 0 \qquad j = 1, \ldots, p \tag{18.89}$$

is obtained, which is analogous to the system (18.76), and which yields approximate values for ω^2.

In many cases the expressions for T^* and $U_{\max}$ are easily found, e.g., for the lateral vibrations of beams, frames, trusses, etc. In the simplest case, when axial displacements, shear deformation, rotatory inertia are neglected, we have

$$U_{\max} = \tfrac{1}{2} \int_B \alpha (u'')^2 \, dx \qquad T^* = \tfrac{1}{2} \int_B \rho A u^2 \, dx \tag{18.90}$$

FIG. 18.4

Here ρ denotes the density, A the cross section, $\alpha = EI$ the flexural rigidity, x the arc length measured in the axis of the system, and $u(x)$ the lateral displacement. Also lumped masses, torsional vibrations of shafts with added lumped masses (wheels), three-dimensional systems, etc., can be included. For example, the expressions for a beam clamped at one end and supporting a concentrated mass M at the other end $x = l$ (see Fig. 18.4) read

$$U_{\max} = \tfrac{1}{2}\int_0^l \alpha(u'')^2\,dx \qquad T^* = \tfrac{1}{2}\int_0^l \rho A u^2\,dx + \tfrac{1}{2}Mu^2(1)$$

c. Ordinary Finite-difference Methods

(1) Ordinary Differential Equations. The interval $\langle a,b\rangle$ of the eigenvalue problem is subdivided into n equal parts of length h with

$$x_j = x_0 + jh \qquad j = 1, 2, \ldots, n \qquad x_0 = a \qquad x_n = b$$

Let Y_j be an approximation for the value $y(x_j)$ of an eigenfunction at the point x_j. We now set up *finite equations*, corresponding to the differential equations and to the boundary conditions. These are linear equations for the Y_j, in which the differential quotients are replaced by difference quotients, and λ is replaced by an approximate value Λ which is to be calculated. A derivative $y^{(k)}(x_j)$ of even order k is replaced by the kth difference quotient

$$\frac{1}{h^k}\,\Delta^k Y_{j-k/2} \tag{18.91}$$

and the same derivative for odd k by

$$\frac{1}{2h^k}\,(\Delta^k Y_{j-(k+1)/2} + \Delta^k Y_{j-(k-1)/2}) \tag{18.92}$$

Here Δ is the difference operator,

$$\Delta f(x) = f(x+h) - f(x) \qquad \Delta Y_j = Y_{j+1} - Y_j \tag{18.93}$$

For an eigenvalue problem of order $k = 2m$, and $2m$ boundary conditions at the points a and b, we replace the differential equation at each of the $(n + 1)$ points x for $j = 0, 1, \ldots, n$ and each boundary condition by a finite equation, and thus obtain $(n + 1 + 2m)$ linear homogeneous equations for the same number of unknowns: $Y_{-m}, Y_{-m+1}, \ldots, Y_{n+m}$. These equations have a nontrivial solution if and only if the determinant of the coefficients vanishes. Certain elements contain the quantity Λ. Thus we obtain an algebraic equation in Λ whose roots serve as approximate values for the eigenvalues λ.

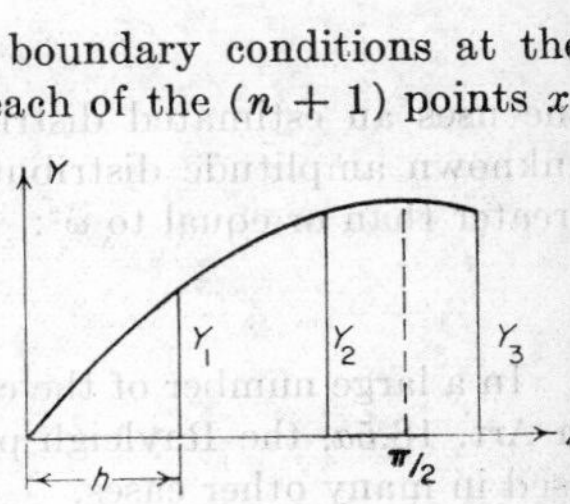

FIG. 18.5

Example 15. For the problem (18.67), a mesh width $h = \pi/5$ is chosen (only for the illustration of the method; for accurate results we would, of course, choose smaller h); then the boundary conditions yield $Y_0 = 0$ and $Y_2 = Y_3$ (see Fig. 18.5). The differential equation corresponds [with $\Lambda h^2 = \mu$ and $p_j = p(jh)$] to the equation

$$Y_2 - 2Y_1 + Y_0 + \mu p_1 Y_1 = 0 \qquad Y_3 - 2Y_2 + Y_1 + \mu p_2 Y_2 = 0$$

After Y_0 and Y_3 have been eliminated with the help of the boundary conditions, these are two equations for Y_1, Y_2 with the coefficient determinant

$$\begin{vmatrix} -2 + \mu p_1 & 1 \\ 1 & -1 + \mu p_2 \end{vmatrix} = 0$$

This yields the approximate values Λ_j for the λ_j:

$$\Lambda_1 = 0.6658 \qquad \Lambda_2 = 4.070$$

(2) Partial Differential Equations.[1] For convenience, the case of only two independent variables will be considered, although the method is applicable in exactly the same way for more variables. In the x,y plane, a rectangular net with mesh widths h and l and with "lattice points" having the coordinates

$$\begin{aligned} x_j &= x_0 + jh \\ y_k &= y_0 + kl \end{aligned} \qquad j, k = 0, \pm 1, \pm 2, \ldots \tag{18.94}$$

is used. A lattice with $h = l$ is called square. Function values at the lattice points (x_j, y_k) are designated by the subscripts j, k. Thus U_{jk} is an approximate value for the value u_{jk} of a function $u(x,y)$ at the point (x_j,y_k). As with ordinary differential equations, the differential equations and boundary conditions are replaced by finite equations, in which all differential quotients are replaced by difference quotients.

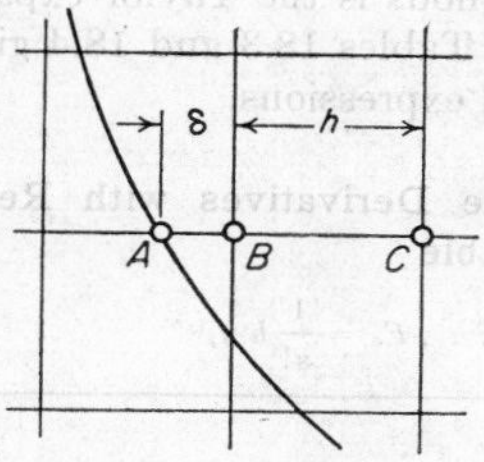

FIG. 18.6

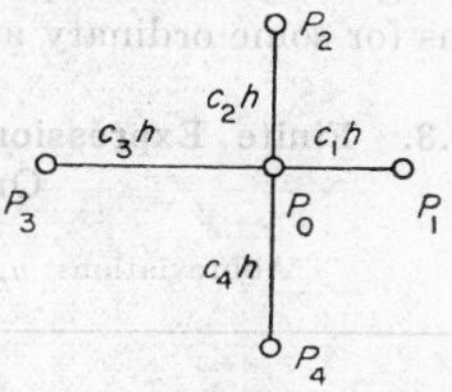

FIG. 18.7

Of course, this may often be done in various ways. For example, the Laplacian $\nabla^2 u = \partial^2 u/\partial x^2 + \partial^2 u/\partial y^2$ for a square lattice corresponds to the finite expression

$$\frac{1}{h^2}(U_{j+1,k} + U_{j,k+1} + U_{j-1,k} + U_{j,k-1} - 4U_{j,k}) \tag{18.95}$$

at the point (x_j,y_k). Table 18.4 contains some "stars" of finite expressions.

Particular care is often required in dealing with the boundary conditions. For curvilinear boundaries, if we want to calculate very roughly, we can apply linear interpolation. For example, the value $U(A)$ of U at a boundary point (Fig. 18.6) is connected with those at two neighboring points by the relation

$$(\delta + h)U(B) = hU(A) + \delta U(C) \tag{18.96}$$

A higher accuracy can be obtained by using the Taylor expansion of the unknown function $u(x,y)$. For example, for the differential equation

$$\nabla^2 u = -\lambda g(x,y)u$$

the following formula may be established (see Fig. 18.7):

$$\frac{u(P_1)}{c_1(c_1+c_3)} + \frac{u(P_2)}{c_2(c_2+c_4)} + \frac{u(P_3)}{c_3(c_3+c_1)} + \frac{u(P_4)}{c_4(c_4+c_2)} - \left(\frac{1}{c_1c_3} + \frac{1}{c_2c_4}\right)u(P_0)$$
$$= \frac{h^2}{2}\left(\frac{\partial^2 u}{\partial x^2} + \frac{\partial^2 u}{\partial y^2}\right)(P_0) + \text{a remainder of third order} \tag{18.97}$$

[1] In the following discussion, in conformity with the usual notation for partial differential equations, we use u (instead of y) to denote the function.

When we omit the remainder term, write the approximation U in place of u, and replace $\nabla^2 u(P_0)$ by $-\lambda g(P_0)U(P_0)$, we have a finite equation for the $U(P_j)$.

d. Sharpening of the Finite-difference Method

(1) Method of Higher Approximation. Here the individual derivatives in the differential equation and in the boundary conditions are replaced not by difference quotients, but by finite expressions, which approximate them more closely than the difference quotients. For example, the deviation of the expression

$$\frac{1}{12h}\left(-y_{j+2} + 8y_{j+1} - 8y_{j-1} + y_{j-2}\right)$$

from $y'(x_j)$ amounts to, at most, $(h^4/18)|y^{(5)}|_{\max}$ in $\langle x_{j-2}, x_{j+2}\rangle$ for a five-times continuously differentiable function, while the deviation of the ordinary difference quotient $(1/h)\ (y_{j+1} - y_j)$ from $y'(x_j)$ can be as large as $(h/2)|y''|_{\max}$ in $\langle x_j, x_{j+1}\rangle$. This can be important, particularly for small values of h. The general tool for the formation of formulas for all types of difference methods is the Taylor expansion of all appearing terms and comparison of coefficients. Tables 18.3 and 18.4 give finite expressions for some ordinary and partial differential expressions.

Table 18.3. Finite Expressions Approximating the Derivatives with Respect to One Independent Variable

Abbreviations: $y_j = y(jh),\ y'_j = y'(jh),\ \ldots,\ F_s = \dfrac{1}{s!}h^s y_j^{(s)}$

Finite expressions	First term neglected
$hy'_j = \frac{1}{2}(-y_{j-1} + y_{j+1}) + \cdots$	$-F_3$
$12hy'_j = y_{j-2} - 8y_{j-1} + 8y_{j+1} - y_{j+2} + \cdots$	$+48F_5$
$2hy'_j = -3y_j + 4y_{j+1} - y_{j+2} + \cdots$	$+4F_3$
$h(y'_{j-1} + 4y'_j + y'_{j+1}) + 3(y_{j-1} - y_{j+1}) = 0 + \cdots$	$+4F_5$
$h(y'_j + y'_{j+1}) + 2(y_j - y_{j+1}) = 0 + \cdots$	$+F_3$
$h^2y''_j = y_{j-1} - 2y_j + y_{j+1} + \cdots$	$-2F_4$
$12h^2y''_j = -y_{j-2} + 16y_{j-1} - 30y_j + 16y_{j+1} - y_{j+2} + \cdots$	$+96F_6$
$h^2y''_j = 2y_j - 5y_{j+1} + 4y_{j+2} - y_{j+3} + \cdots$	$+22F_4$
$h^2(y''_{j-1} + 10y''_j + y''_{j+1}) - 12(y_{j-1} - 2y_j + y_{j+1}) = 0 + \cdots$	$+36F_6$
$h^2(y''_{j-1} - 8y''_j + y''_{j+1}) + 9h(y'_{j-1} - y'_{j+1}) + 24(y_{j-1} - 2y_j + y_{j+1}) = 0 + \cdots$	$+16F_8$
$2h^3y'''_j = -y_{j-2} + 2y_{j-1} - 2y_{j+1} + y_{j+2} + \cdots$	$-60F_5$
$h^3(y'''_{j-1} + 2y'''_j + y'''_{j+1}) + 2(y_{j-2} - 2y_{j-1} + 2y_{j+1} - y_{j+2}) = 0 + \cdots$	$-84F_7$
$h^4y^{\text{iv}}_j = y_{j-2} - 4y_{j-1} + 6y_j - 4y_{j+1} + y_{j+2} + \cdots$	$-120F_6$
$h^4(y^{\text{iv}}_{j-1} + 4y^{\text{iv}}_j + y^{\text{iv}}_{j+1}) - 6(y_{j-2} - 4y_{j-1} + 6y_j - 4y_{j+1} + y_{j+2}) = 0 + \cdots$	$+336F_8$

(2) Hermitian Method.[1] For differential equations of simple structure, the following method often leads to good results, whereas, for complicated differential equations, it can be quite involved. Here, for each individual finite equation, the differential equation is approximated at several points, and thereby an improvement is achieved over the usual finite-difference method. To this end one sets up once

[1] The German term for this is Mehrstellenverfahren. The method is not due to Hermite, but the formulas used are related to Hermite's interpolation formula. Hence the term *hermitian method* [Trans.].

Table 18.4. Finite Expressions Approximating $\nabla^2 u$ and $\nabla^2\nabla^2 u$

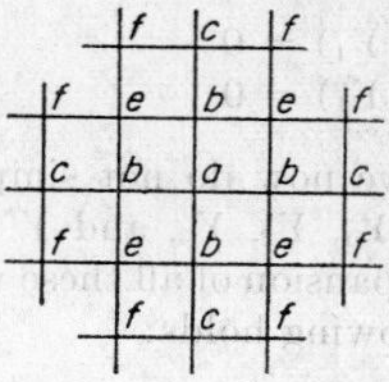

Quadratic net, mesh width h, u_a is the value of u at the point a; Σu_b is the sum of the values of u at all the points marked b in the figure. R_q = remainder term containing partial derivatives of u of order q.

Method	Finite expressions	Remainder
Ordinary difference method	$h^2\nabla^2 u_a = -4u_a + \Sigma u_b + \cdots$	R_4
	$h^2\nabla^2\nabla^2 u_a = 20u_a - 8\Sigma u_b + \Sigma u_c + 2\Sigma u_e + \cdots$	R_6
Higher approximation	$12h^2\nabla^2 u_a = -60u_a + 16\Sigma u_b - \Sigma u_c + \cdots$	R_6
Hermitian method	$h^2(8\nabla^2 u_a + \Sigma\nabla^2 u_b) + 40u_a - 8\Sigma u_b - 2\Sigma u_e = 0 + \cdots$	R_6
	$\frac{1}{2}h^4(2\nabla^2\nabla^2 u_a + \Sigma\nabla^2\nabla^2 u_b) - 36u_a + 10\Sigma u_b - \Sigma u_c + 2\Sigma u_e - \Sigma u_f = 0 + \cdots$	R_8
	$\frac{3}{10}h^4\nabla^2\nabla^2 u_a + \frac{1}{15}h^2(82\nabla^2 u_a + \Sigma\nabla^2 u_b + \Sigma\nabla^2 u_e) + 20u_a - 4\Sigma u_b - \Sigma u_e = 0 + \cdots$	R_8

and for all expressions of the type:

$$P = \sum_{\nu=-p}^{p} (a_\nu{}^{(k)} y_{j+\nu} + A_\nu{}^{(k)} y_{j+\nu}^{(k)}) \tag{18.98}$$

P includes, therefore, the values of y and of a fixed chosen derivative $y^{(k)}$ at certain division points $x_{j+\nu}$. The coefficients $a_\nu{}^{(k)}$, $A_\nu{}^{(k)}$ are so determined that the Taylor expansion of the expression P vanishes up to a remainder of highest possible order. For example, for $k = 2$,

$$-12(y_{j-1} - 2y_j + y_{j+1}) + h^2(y''_{j-1} + 10y''_j + y''_{j+1}) = Ch^6 \tag{18.99}$$

holds in $\langle x_{j-1}, x_{j+1}\rangle$ where $|C| \leq (\frac{1}{20})|y^{(6)}|_{\max}$. Tables 18.3 and 18.4 contain hermitian expressions for some ordinary and partial differential expressions.

Let us describe the method for an ordinary differential equation of the kth order. We put

$$\sum_{\nu=-p}^{p} (a_\nu{}^{(k)} Y_{j+\nu} + A_\nu{}^{(k)} Y_{j+\nu}^{(k)}) = 0 \tag{18.100}$$

for all interior points $x_j (a < x_j < b)$, and, by using the differential equation, replace $Y_{j+\nu}^{(k)}$ by the lower derivatives $Y_{j+\nu}^{(k-1)}, \ldots, Y_{j+\nu}$. These $Y_{j+\nu}^{(q)}$ are additional unknowns, and, for all the derivatives which appear in the differential equation, we write the corresponding hermitian formulas for all interior points. Near the boundary we must always be particularly careful, in order that the accuracy gain of the hermitian formulas should not be lost again by inaccurate formulas. At the boundary we are sometimes compelled to use *one-sided* formulas.

Example 16. For comparison with the ordinary difference method, the same example (18.67) is chosen as in Art. 18.5c(1), and with the same mesh width $h = \pi/5$ (Fig. 18.5). The differential equation corresponds to the following hermitian equations, where we set $Y_j'' = -\Lambda p_j Y_j$ and $\Lambda h^2 = 12\nu$:

$$Y_2 - 2Y_1 + \nu(p_2Y_2 + 10p_1Y_1) = 0$$
$$Y_3 - 2Y_2 + Y_1 + \nu(p_3Y_3 + 10p_2Y_2 + p_1Y_1) = 0$$

When dealing with the boundary condition $y'(\pi/2) = 0$, we now do not simply set $Y_2 = Y_3$, but we use a relation connecting the values Y_0, Y_1, Y_2, Y_3, and Y' at the point $\pi/2$ with one another. It is obtained by the Taylor expansion of all these values: For a sufficiently often differentiable function $f(x)$, the following holds:

$$48hf'(0) - f(-5h) + 3f(-3h) + 21f(-h) - 23f(h) = 16h^4f^{(4)}(0) + \text{terms with higher derivatives}$$

Therefore, we put here

$$3Y_1 + 21Y_2 - 23Y_3 = 0$$

Now Y_3 can be eliminated, and we obtain without difficulty, as in Art. 18.5*c*, a quadratic equation in ν and the approximate values which are much better than in Art. 18.5*c*:

$$\Lambda_1 = 0.7087 \qquad \Lambda_2 = 5.536$$

e. Other Methods

Collocation. This is a primitive method, very simple to handle, which provides a rough survey and sometimes furnishes surprisingly good results. However, it is not always reliable.

As an approximation, y is assumed in the form (18.74) where the v_ν must satisfy the boundary conditions for the differential equation, and this expression for u is substituted for y in the differential equation or integral equation. There arises a *defect* ϵ

$$\epsilon = Mu - \Lambda Nu \tag{18.101}$$

(Λ is written as an approximation for λ). We now choose p points $P_1, \ldots, P_p$, distributed as uniformly as possible in the basic domain, and require that ϵ vanish at these points:

$$\epsilon(P_j) = 0 \qquad j = 1, \ldots, p \tag{18.102}$$

These are p linear homogeneous equations in the a_j whose determinant of coefficients set equal to zero is an algebraic equation for the determination of the approximate values of Λ.

Example 17. For the problem (18.67), we put

$$y \approx u = a_1 \sin x + a_2 \sin 3x$$

If we choose $x = \pi/6$ and $\pi/3$ as *collocation points*, then we obtain from

$$\begin{vmatrix} \tfrac{1}{2} - \Lambda(1 + \tfrac{1}{2}\sqrt{3})\tfrac{1}{2} & 9 - \Lambda(1 + \tfrac{1}{2}\sqrt{3}) \\ \tfrac{1}{2}\sqrt{3} - \Lambda\tfrac{3}{2}\,\tfrac{1}{2}\sqrt{3} & 0 \end{vmatrix} = 0$$

the approximations $\Lambda_1 = 0.667$, $\Lambda_2 = 5.28$, which, for so short a calculation, are reasonably good.

Formulas for Composite Systems. For the problem (18.14) through (18.17), let Ny be the sum

$$Ny = \sum_{\rho=1}^{r} N_\rho y \tag{18.103}$$

Suppose that each partial problem

$$My = \lambda^{(\rho)}N_\rho y \qquad U_\mu y = 0 \tag{18.104}$$

is self-adjoint and completely definite, and let its smallest eigenvalue $\lambda_1^{(\rho)}$ be known. For the original problem, the formula (Dunkerley's formula)

$$\lambda_1 \sum_{\rho=1}^{r} \frac{1}{\lambda_1^{(\rho)}} \geq 1 \tag{18.105}$$

holds. If, correspondingly,

$$My = \sum_{\rho=1}^{r} M_\rho y \qquad \text{and if} \qquad M_\rho y = \lambda^{(\rho)} Ny$$

then $U_\mu y = 0$ is self-adjoint and completely definite, and $\lambda_1^{(\rho)}$ is the accompanying smallest eigenvalue; then we have, for the original problem, Southwell's formula:

$$\lambda_1 \geq \sum_{\rho=1}^{r} \lambda_1^{(\rho)} \tag{18.106}$$

Other Methods. There are still various other methods which cannot be discussed here, e.g., the method of perturbation [34],[35],[36],[37]; methods of smallest mean-square error; etc. Also series expansions can be very suitable in certain cases.

18.6. REFERENCES

Books

[1] C. B. Biezeno, R. Grammel: "Engineering Dynamics," vols. 1 and 2, transl. by M. L. Meyer from the 2d German ed., Blackie, Glasgow, 1955, 1956.

[2] E. A. Coddington, N. Levinson: "Theory of Ordinary Differential Equations," McGraw-Hill, New York, 1955.

[3] L. Collatz: "Eigenwertaufgaben mit Technischen Anwendungen," Akad.-Verlag, Leipzig, 1949.

[4] L. Collatz: "The Numerical Treatment of Differential Equations," 3d ed., Springer, Berlin, 1959.

[5] R. Courant, D. Hilbert: "Methoden der mathematischen Physik," vol. 1, 2d ed., 1931; vol. 2, 1937, Springer, Berlin.

[6] G. F. D. Duff: "Partial Differential Equations," Univ. of Toronto Press, Toronto, 1956.

[7] P. Frank, R. V. Mises: "Die Differential- und Integralgleichungen der Mechanik und Physik," 2d ed., Rosenberg, New York, 1943.

[8] B. Friedman: "Principles and Techniques of Applied Mathematics," Wiley, New York, 1956.

[9] G. Hamel: "Integralgleichungen," 2d ed., Springer, Berlin, 1949.

[10] D. R. Hartree: "Numerical Analysis," 2d ed., Oxford Univ. Press, New York, 1958.

[11] A. S. Householder: "Principles of Numerical Analysis," McGraw-Hill, New York, 1953.

[12] E. L. Ince: "Ordinary Differential Equations," Dover, New York, 1956.

[13] I. G. Petrovsky: "Lectures on Partial Differential Equations," transl. by A. Shenitzer, Interscience, New York, 1954.

[14] J. Ratzersdorfer: "Die Knicksicherheit von Stäben und Stabwerken," Springer, Vienna, 1936.

[15] W. Schmeidler: "Vorträge über Determinanten und Matrizen mit Anwendungen in Physik und Technik," Akad.-Verlag, Leipzig, 1949.

[16] W. Schmeidler: "Integralgleichungen mit Anwendungen in Physik und Technik," Akad.-Verlag., Leipzig, 1950.

[17] E. Sperner: "Einführung in die analytische Geometrie und Algebra," part 2, 2d ed., Vandenhoeck u. Ruprecht, Göttingen, 1957.

[18] I. Szabó: "Höhere Technische Mechanik," 2d ed., Springer, Berlin, 1958.

[19] G. Temple, W. G. Bickley: "Rayleigh's Principle and Its Application to Engineering," Dover, New York, 1956.

[20] S. Timoshenko, D. H. Young: "Vibration Problems in Engineering," 3d ed., Van Nostrand, Princeton, N.J., 1955.

[21] R. Zurmühl: "Matrizen," 2d ed., Springer, Berlin, 1958.

[22] R. Zurmühl: "Praktische Mathematik für Ingenieure und Physiker," 3d ed., Springer, Berlin, 1961.

Papers

[23] L. Collatz: Einschliessungssatz für die charakteristischen Zahlen von Matrizen, *Math. Z.*, **48** (1942), 221–226.

[24] H. Wielandt: Unzerlegbare, nichtnegative Matrizen, *Math. Z.*, **52** (1950), 642–648.

[25] H. Wielandt: Das Iterationsverfahren bei nicht selbstadjungierten Eigenwertaufgaben, *Math. Z.*, **50** (1944), 93–143.

[26] W. Givens: Numerical computation of the characteristic values of a real symmetric matrix, Oak Ridge National Laboratory, 1954.

[27] A. J. Hoffmann, H. Wielandt: The variation of the spectrum of a normal matrix, *Duke Math. J.*, **20** (1953), 37–39.

[28] E. Kamke: Über die definiten selbstadjungierten Eigenwertaufgaben bei gewöhnlichen linearen Differentialgleichungen, *Math. Z.*, **45** (1939), 759–787; **46** (1940), 231–250, 251–286; **48** (1942), 67–100.

[29] A. G. Walker, J. D. Weston: Inclusion theorems for the eigenvalues of a normal matrix, *J. London Math. Soc.*, **24** (1949), 28–31.

[30] H. Wielandt: Inclusion theorems for eigenvalues, *Natl. Bur. Standards Appl. Math. Ser.*, **29** (1953), 75–78.

[31] H. Wielandt: Ein Einschliessungssatz für charakteristische Wurzeln normaler Matrizen, *Arch. Math.*, **1** (1949), 348–352.

[32] L. Collatz: Genäherte Berechnung von Eigenwerten, *Z. angew. Math. Mech.*, **19** (1939), 228–249.

[33] R. Grammel: Ein neues Verfahren zur Lösung technischer Eigenwertprobleme, *Ing.-Arch.*, **10** (1939), 35–46.

[34] W. Meyer zur Capellen: Methode zur angenäherten Lösung von Eigenwertproblemen mit Anwendungen auf Schwingungsprobleme, *Ann. Physik*. **V,8** (1931), 297–352.

[35] W. Meyer zur Capellen: Genäherte Berechnung von Eigenwerten, *Ing.-Arch.*, **10** (1939), 167–174.

[36] F. Rellich: Störungstheorie der Spektralzerlegung, *Math. Ann.*, **113** (1937), 600–619; **116** (1939), 555–570; **117** (1940), 356–382.

[37] J. Schröder: Fehlerabschätzung zur Störungsrechnung für lineare Eigenwertprobleme bei gewöhnlichen Differentialgleichungen, *Z. angew. Math. Mech.*, **34** (1954), 140–149.

CHAPTER 19

LAPLACE TRANSFORMATION

BY

E. J. SCOTT, Ph.D., Urbana, Ill.

19.1. DEFINITION AND EXISTENCE OF THE LAPLACE TRANSFORM

Let $f(t)$ be a real function of the positive real variable t. Then the Laplace transform of $f(t)$ is defined by the integral

$$\bar{f}(s) = L\{f(t)\} = \int_0^\infty e^{-st} f(t)\, dt \qquad (19.1)$$

where $s = x + iy$. Thus, $\bar{f}(s)$ is a function of the complex variable s for those values of s for which the integral exists. In what follows we shall consistently use the bar notation for the transform of a function.

A function $f(t)$ is said to be *sectionally continuous* over a finite interval if that interval can be divided into a finite number of subintervals, over each of which $f(t)$ is continuous and possesses finite limits as t tends to either end of the subinterval from the interior.

A function $f(t)$ is said to be of *exponential order* as $t \to \infty$ if a constant c exists such that $e^{-ct}|f(t)|$ is bounded for all $t > T$, where T is a finite number. In other words, if M is a bound, then

$$|f(t)| < Me^{ct} \qquad \text{for } t > T$$

When this inequality is satisfied, we often say that $f(t)$ is of the order of e^{ct}, and write briefly $f(t) = O(e^{ct})$.

The function $f(t) = \text{const}$ is obviously of exponential order with $c = 0$. So also are functions such as t^n $(n \geq 0)$ with $c > 0$, $e^{at} \cos b(t - t_0)$ with $c > a$, and $t^3 \sin 3t$ with $c > 0$. It is to be noted that *any bounded* function is of exponential order with $c = 0$. The function $e^{(t-1)^2}$ is an example of a function which is *not* of exponential order.

If we now assume that $f(t)$ is sectionally continuous in every finite interval for which $t \geq 0$, and of exponential order as t tends to infinity, then the Laplace transform of $f(t)$ exists. For we have

$$\int_0^\infty |e^{-st} f(t)|\, dt < \int_0^T e^{-xt} |f(t)|\, dt + M \int_T^\infty e^{-(x-c)t}\, dt$$

Now, the first integral on the right exists because $f(t)$ is sectionally continuous, whereas the second integral exists if $x = \mathcal{R}\, s > c$. Hence, the Laplace transform not only converges, but converges *absolutely* when $\mathcal{R}\, s > c$.

Although these conditions can be relaxed somewhat by allowing $f(t)$ to have certain infinite discontinuities at $t = 0$, for most practical purposes they suffice.

The following example illustrates the direct calculation of a Laplace transform:

$$L\{\text{Sinh } at\} = \int_0^\infty e^{-st} \text{ Sinh } at\, dt = -\left[\frac{e^{-st}(s \text{ Sinh } at + a \text{ Cosh } at)}{s^2 - a^2}\right]_0^\infty$$
$$= \frac{a}{s^2 - a^2} \qquad \text{provided } \Re\, s > |a| \tag{19.2}$$

If $L\{f(t)\} = \bar{f}(s)$, then $f(t)$ is said to be the *inverse* Laplace transform of $\bar{f}(s)$, and we write

$$f(t) = L^{-1}\{\bar{f}(s)\}$$

As a direct consequence of the definition of the transformation, we have

$$L\left[\sum_{i=1}^{n} a_i f_i(t)\right] = \sum_{i=1}^{n} a_i L\{f_i(t)\} \tag{19.3}$$

where $a_1, a_2, \ldots, a_n$ are constants. This equation states that the Laplace transform is *linear*.

19.2. THE UNIT STEP FUNCTION, THE UNIT IMPULSE, AND THE UNIT DOUBLET

Before considering further properties of the Laplace transform we shall discuss three functions which play an extremely useful role in the solution of physical problems

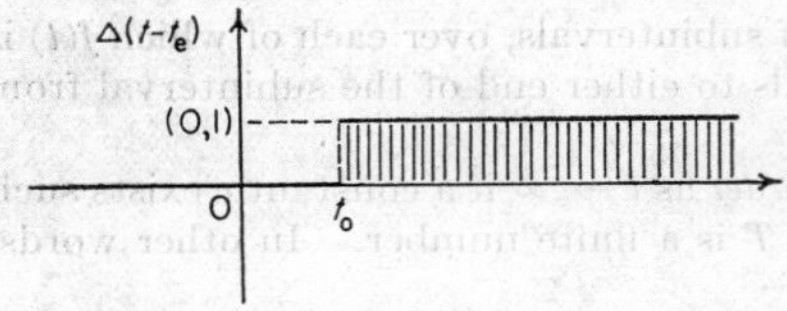

FIG. 19.1

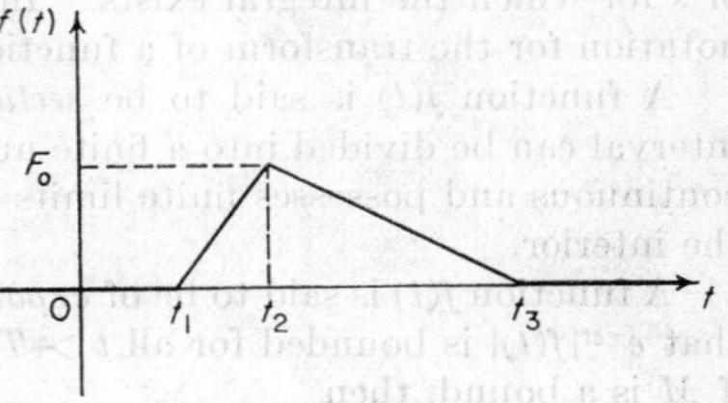

FIG. 19.2

by the operational method. The first of these, which is known as the unit step function, is defined as follows:

$$\Delta(t - t_0) \begin{cases} = 0 & t < t_0 \\ = 1 & t \geq t_0 \end{cases} \tag{19.4}$$

Its graph is shown in Fig. 19.1.

The Laplace transform of this function is

$$L\{\Delta(t - t_0)\} = \int_{t_0}^\infty e^{-st}\, dt = \frac{e^{-st_0}}{s} \qquad \Re\, s > 0 \tag{19.5}$$

In particular, if $t_0 = 0$, we have

$$L\{\Delta(t)\} = 1/s$$

By combining unit step functions appropriately, many functions can be expressed analytically in terms of them. For example, the triangular pulse (Fig. 19.2) is given by the equation

$$f(t) = F_0(t - t_1)\frac{\Delta(t - t_1) - \Delta(t - t_2)}{t_2 - t_1} - F_0(t - t_3)\frac{\Delta(t - t_2) - \Delta(t - t_3)}{t_3 - t_2}$$
$$= F_0 \frac{[(t_3 - t_2)(t - t_1)\,\Delta(t - t_1) - (t_3 - t_1)(t - t_2)\,\Delta(t - t_2) + (t_2 - t_1)(t - t_3)\,\Delta(t - t_3)]}{(t_2 - t_1)(t_3 - t_2)}$$

An expression which has played an increasingly important role in operational methods ever since the time of Heaviside is the so-called *unit impulse* (Fig. 19.3), which is defined as follows:

$$\Delta'(t - t_0) = \lim_{\epsilon \to 0} \frac{\Delta(t - t_0) - \Delta(t - t_0 - \epsilon)}{\epsilon} \tag{19.6}$$

The function $\Delta'(t)$ is identical with the Dirac delta function, defined as

$$\delta(t) = 0 \quad \text{for } t \neq 0 \qquad \delta(0) = \infty \qquad \int_{-\epsilon}^{+\epsilon} \delta(t)\, dt = 1 \tag{19.7}$$

The following examples will serve to indicate the usefulness of the unit impulse function. Suppose that a mechanical system is subjected to an impulse of magnitude

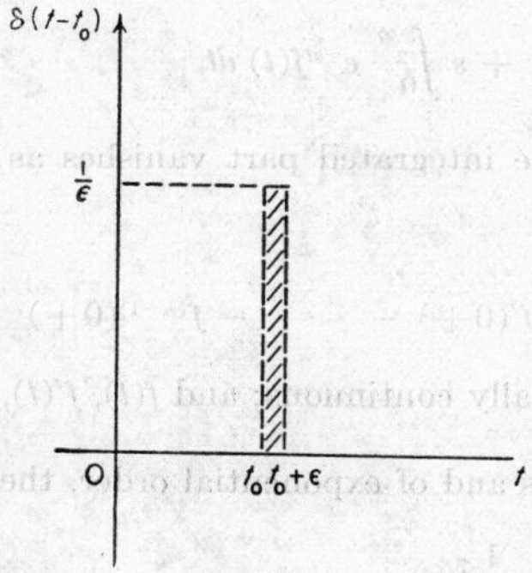

FIG. 19.3

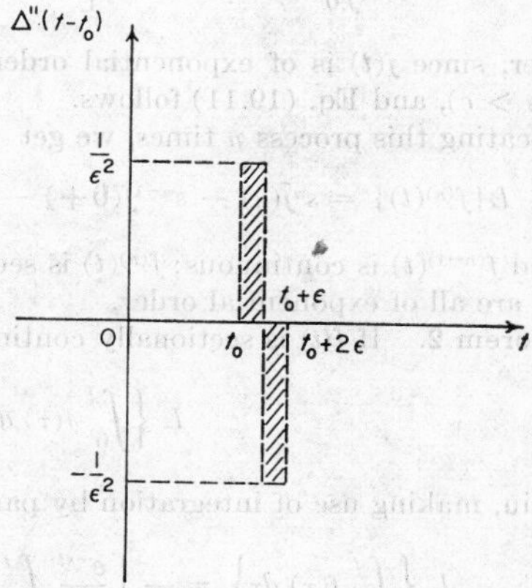

FIG. 19.4

I_0 at time $t = t_0$. Then the impulsive force F can be represented by $I_0\,\Delta'(t - t_0)$. Similarly, a concentrated force P_0, acting on a beam at $x = x_0$, can be considered as being replaced by the limit of the distributed load P_0/ϵ, acting over the distance ϵ; that is, by $P_0\,\Delta'(x - x_0)$. In the first example, $\Delta'(t - t_0)$ has the dimensions of $(\text{time})^{-1}$ and, in the second example, $\Delta'(x - x_0)$ has the dimensions of $(\text{length})^{-1}$.

The transform of the unit impulse is

$$L\{\Delta'(t - t_0)\} = \lim_{\epsilon \to 0} \frac{e^{-st_0} - e^{-s(t_0+\epsilon)}}{s\epsilon} = e^{-t_0 s} \tag{19.8}$$

after using L'Hospital's rule to evaluate the limit.

The unit doublet (Fig. 19.4) is defined as follows:

$$\begin{aligned} \Delta''(t - t_0) &= \lim_{\epsilon \to 0} \frac{\Delta'(t - t_0) - \Delta'(t - t_0 - \epsilon)}{\epsilon} \\ &= \lim_{\epsilon \to 0} \frac{\Delta(t - t_0) - 2\Delta(t - t_0 - \epsilon) + \Delta(t - t_0 - 2\epsilon)}{\epsilon^2} \end{aligned} \tag{19.9}$$

Its transform is easily shown to be

$$L\{\Delta''(t - t_0)\} = se^{-t_0 s} \tag{19.10}$$

The importance of the unit doublet in mechanics arises from the fact that a couple of moment M_0, applied at the point $x = x_0$, can be represented by $M_0\,\Delta''(x - x_0)$. Here $\Delta''(x - x_0)$ has the dimensions of (length)$^{-2}$.

19.3. FUNDAMENTAL PROPERTIES OF LAPLACE TRANSFORMS

In deriving some of the following fundamental operational properties of the Laplace transformation, we shall proceed formally. For rigorous derivations in these cases, we refer the reader to the treatises mentioned in the bibliography.

Theorem 1. Let $f(t)$ be continuous and $f'(t)$ sectionally continuous in every finite interval $(0,T)$. If $f(t)$ is also of exponential order, then

$$L\{f'(t)\} = s\bar{f}(s) - f(0+) \tag{19.11}$$

Applying the formula of integration by parts, we obtain

$$\int_0^\infty e^{-st}f'(t)\,dt = \Big[e^{-st}f(t)\Big]_0^\infty + s\int_0^\infty e^{-st}f(t)\,dt$$

However, since $f(t)$ is of exponential order, the integrated part vanishes as $t \to \infty$ (for $\Re\, s > c$), and Eq. (19.11) follows.

Repeating this process n times, we get

$$L\{f^{(n)}(t)\} = s^n\bar{f}(s) - s^{n-1}f(0+) - s^{n-2}f'(0+) - \cdots - f^{(n-1)}(0+) \tag{19.12}$$

provided $f^{(n-1)}(t)$ is continuous; $f^{(n)}(t)$ is sectionally continuous; and $f(t), f'(t), \ldots, f^{(n-1)}(t)$ are all of exponential order.

Theorem 2. If $f(t)$ is sectionally continuous and of exponential order, then

$$L\left\{\int_0^t f(\tau)\,d\tau\right\} = \frac{1}{s}\bar{f}(s) \tag{19.13}$$

Again, making use of integration by parts, we have

$$L\left\{\int_0^t f(\tau)\,d\tau\right\} = -\left[\frac{e^{-st}}{s}\int_0^t f(\tau)\,d\tau\right]_0^\infty + \frac{1}{s}\int_0^\infty e^{-st}f(\tau)\,d\tau$$

Since $f(t)$ is of exponential order, and, hence, $\int_0^t f(\tau)\,d\tau$ also, the integrated part vanishes at the upper limit if $\Re\, s$ is sufficiently large, and Eq. (19.13) results.

Theorem 3 (Shifting Theorem). For any real constant, $a \geq 0$,

$$L\{f(t-a)\,\Delta(t-a)\} = e^{-as}\bar{f}(s) \tag{19.14}$$

By definition,

$$L\{f(t-a)\,\Delta(t-a)\} = \int_0^\infty e^{-st}f(t-a)\,\Delta(t-a)\,dt = \int_a^\infty e^{-st}f(t-a)\,dt$$

Making the substitutions $t - a = \tau$, $dt = d\tau$, we have

$$L\{f(t-a)\,\Delta(t-a)\} = e^{-as}\int_0^\infty e^{-s\tau}f(\tau)\,d\tau = e^{-as}\bar{f}(s) \qquad \text{(q.e.d.)}$$

We may interpret this result as stating that the multiplication of the transform of $f(t)$ by e^{-as} *translates* the function $f(t)$ by a units to the right (Figs. 19.5 and 19.6).

Theorem 4 (Shifting Theorem). For any constant a,

$$L\{e^{-at}f(t)\} = \bar{f}(s+a) \qquad \Re\,(s+a) > c \tag{19.15}$$

Let $L\{f(t)\} = \bar{f}(s)$, $\Re\, s > c$. Then,

$$L\{e^{-at}f(t)\} = \int_0^\infty e^{-(s+a)t}f(t)\,dt = \bar{f}(s+a)$$

provided $\Re\,(s+a) > c$ (q.e.d.).

Here, we see that multiplication of $f(t)$ by e^{-at} is reflected by a translation of the origin in the complex plane.

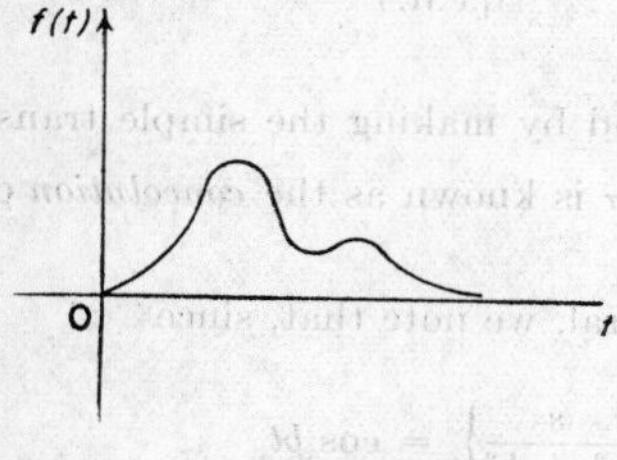

FIG. 19.5

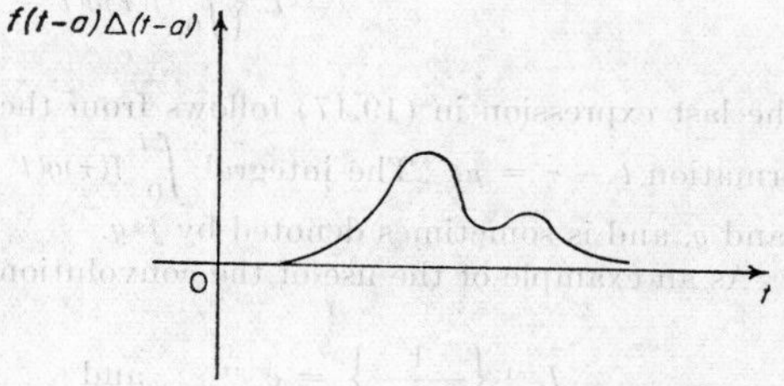

FIG. 19.6

As an illustration, since, by (19.2),

$$L^{-1}\left\{\frac{a}{s^2 - a^2}\right\} = \text{Sinh}\, at \qquad \Re\, s > |a|$$

then
$$L^{-1}\left\{\frac{a}{(s+b)^2 - a^2}\right\} = e^{-bt}\,\text{Sinh}\, at \qquad \Re\,(s+b) > |a|$$

Theorem 5. If $f(t)$ is sectionally continuous and of the order of e^{ct}, then, when $\Re\, s > c$,

$$L\{(-t)^n f(t)\} = \bar{f}^{(n)}(s) \qquad n = 1, 2, \ldots \tag{19.16}$$

For the type of function being considered here, differentiating

$$\bar{f}(s) = \int_0^\infty e^{-st}f(t)\,dt$$

with respect to s under the integral sign is a valid operation. Therefore, we have

$$\bar{f}'(s) = \int_0^\infty -te^{-st}f(t)\,dt = L\{-tf(t)\}$$

Repeated differentiation gives (19.16) (q.e.d.).

This theorem is useful in the solution of differential equations whose coefficients are polynomials in t.

Theorem 6 (Convolution Theorem). If $f(t)$ and $g(t)$ are sectionally continuous and of the order of e^{ct}, then

$$\bar{f}(s)\bar{g}(s) = L\left\{\int_0^t f(t-\tau)g(\tau)\,d\tau\right\} = L\left\{\int_0^t f(\tau)g(t-\tau)\,d\tau\right\} \tag{19.17}$$

where $\bar{f}(s) = L\{f(t)\}$ and $\bar{g}(s) = L\{g(t)\}$.

We have, in view of the absolute convergence of the integrals defining $\bar{f}(s)$ and $\bar{g}(s)$,

$$\begin{aligned}\bar{f}(s)\bar{g}(s) &= \int_0^\infty e^{-s\tau}f(\tau)\,d\tau \int_0^\infty e^{-sv}g(v)\,dv \\ &= \int_0^\infty \int_0^\infty e^{-s(\tau+v)}f(\tau)g(v)\,dv\,d\tau\end{aligned}$$

Upon making the substitutions $\tau + v = t$, $dv = dt$ in the last integral and interchanging order of integration, we find that

$$\begin{aligned}\bar{f}(s)\bar{g}(s) &= \int_0^\infty \left\{\int_0^\infty e^{-st} f(\tau)g(t-\tau)\,dt\right\} d\tau \\ &= \int_0^\infty e^{-st} \left\{\int_0^t f(\tau)g(t-\tau)\,d\tau\right\} dt \\ &= L\left\{\int_0^t f(\tau)g(t-\tau)\,d\tau\right\} \qquad \text{(q.e.d.)}\end{aligned}$$

The last expression in (19.17) follows from the second by making the simple transformation $t - \tau = \mu$. The integral $\int_0^t f(\tau)g(t-\tau)\,d\tau$ is known as the *convolution* of f and g, and is sometimes denoted by $f*g$.

As an example of the use of the convolution integral, we note that, since

$$L^{-1}\left\{\frac{1}{s+a}\right\} = e^{-at} \qquad \text{and} \qquad L^{-1}\left\{\frac{s}{s^2+b^2}\right\} = \cos bt$$

then $$L^{-1}\left\{\frac{s}{(s+a)(s^2+b^2)}\right\} = \int_0^t e^{-a(t-\tau)} \cos b\tau\,d\tau = \frac{a\cos bt + b\sin bt - ae^{-at}}{a^2+b^2}$$

Theorem 7. If $f(t)$ is a periodic function of period T, then

$$L\{f(t)\} = \int_0^T \frac{f(t)e^{-st}\,dt}{1-e^{-sT}} \tag{19.18}$$

By definition,

$$L\{f(t)\} = \int_0^\infty e^{-st}f(t)\,dt = \sum_{n=0}^{\infty} \int_{nT}^{(n+1)T} e^{-st}f(t)\,dt$$

If we make the substitution $t = \tau + nT$, then, because of the periodicity of $f(t)$, $f(\tau + nT) = f(\tau)$, and we obtain

$$L\{f(t)\} = \left(\sum_{n=0}^{\infty} e^{-nsT}\right)\int_0^T e^{-s\tau}f(\tau)\,d\tau$$

But $$\sum_{n=0}^{\infty} e^{-nsT} = \frac{1}{1-e^{-sT}}$$

and Theorem 7 follows.

The transforms of the following three frequently occurring periodic functions can be easily verified by an application of the preceding theorem.

1. Rectangular wave

$$L\{f(t)\} = \frac{1}{s}\,\text{Tanh}\,\frac{sT}{2}$$

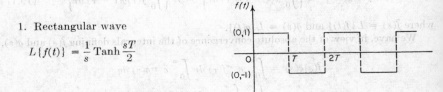

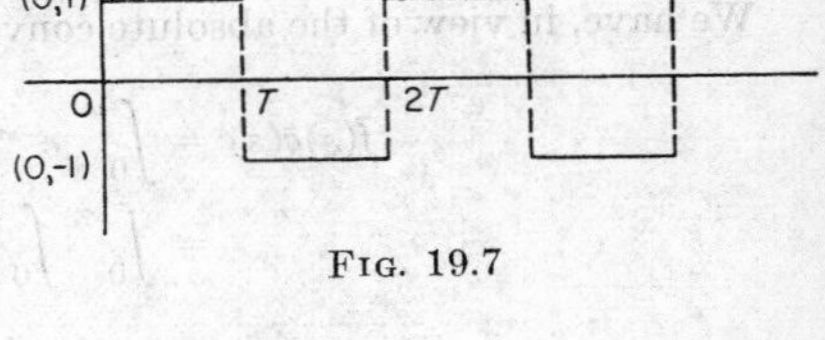

Fig. 19.7

2. Triangular wave

$$L\{f(t)\} = \frac{1}{s^2}\operatorname{Tanh}\frac{sT}{2}$$

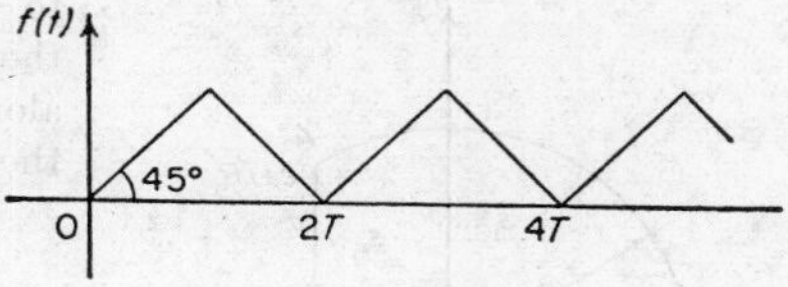

FIG. 19.8

3. Sawtooth wave

$$L\{f(t)\} = \frac{1 + sT}{s^2} - \frac{T}{s(1 - e^{-sT})}$$

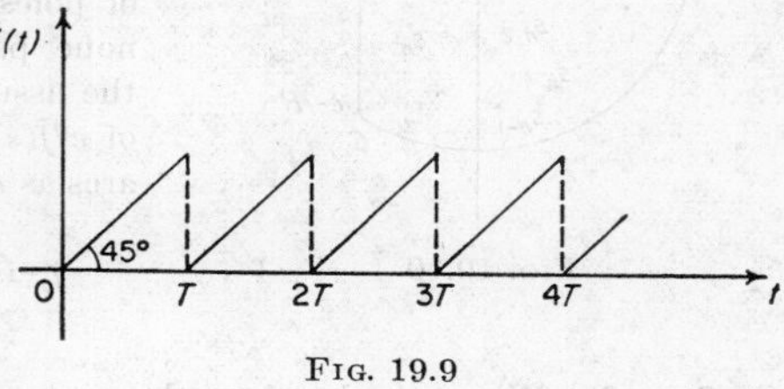

FIG. 19.9

19.4. THE COMPLEX INVERSION THEOREM

The success of the Laplace transform method is largely due to the fact that, if $f(t) = 0(e^{c_0t})$ for $t \geq 0$, then $\bar{f}(s)$ is an analytic function for $\Re\, s > c_0$. Thus, the powerful tools of analytic function theory, and especially contour integration, can be brought to bear on the problem of finding inverses of transforms.

We now derive the so-called *complex inversion formula* which is basic for the solution of differential equations. To do this, we make use of the complex form of the Fourier integral theorem:

$$\phi(t) = \frac{1}{2\pi}\lim_{R\to\infty}\int_{-R}^{R} e^{iyt}\left\{\int_{-\infty}^{\infty} e^{-iy\tau}\phi(\tau)\,d\tau\right\}dy \tag{19.19}$$

Substituting in this formula $e^{-ct}f(t)$ for $\phi(t)$, where $f(t)$ vanishes for $t < 0$, we get

$$e^{-ct}f(t) = \frac{1}{2\pi}\lim_{R\to\infty}\int_{-R}^{R} e^{iyt}\left\{\int_{0}^{\infty} e^{-(c+iy)\tau}f(\tau)\,d\tau\right\}dy \tag{19.20}$$

The conditions for (19.19) to hold are satisfied if $f(t)$ and $f'(t)$ are sectionally continuous for $t \geq 0$, $f(t) = 0(e^{c_0t})$, and $c > c_0$. Making the substitution $s = c + iy$ and dividing by e^{-ct}, we obtain

$$f(t) = \frac{1}{2\pi i}\lim_{R\to\infty}\int_{c-iR}^{c+iR} e^{st}\left\{\int_{0}^{\infty} e^{-s\tau}f(\tau)\,d\tau\right\}ds \tag{19.21}$$

Whence, upon replacing the integral within the braces by $\bar{f}(s)$,

$$f(t) = \frac{1}{2\pi i}\lim_{R\to\infty}\int_{c-iR}^{c+iR} e^{st}\bar{f}(s)\,ds \tag{19.22}$$

The preceding integral taken along the infinite line $x = c$ can be evaluated, in general, by contour integration. The modus operandi is as follows:

Case 1. Suppose that $\bar{f}(s)$ has a finite number of poles in the complex s plane. As has been observed, they will lie to the left of the line $x = c$, since $\bar{f}(s)$ is analytic for $x > c$. Enclose these singularities by the contour shown in Fig. 19.10. If $\bar{f}(s)$ tends

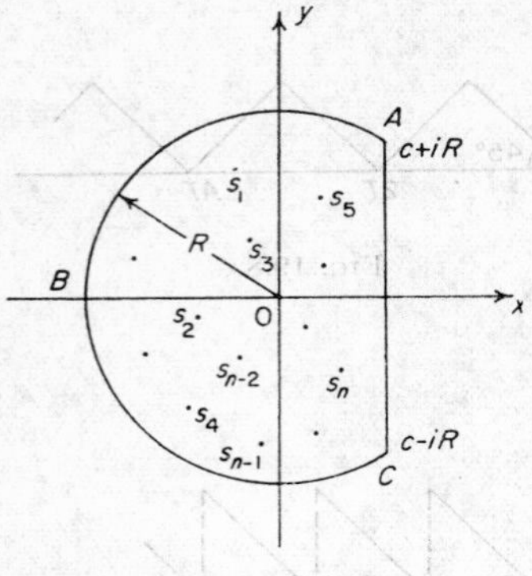

FIG. 19.10

to zero uniformly along the circular arc ABC, then it can be shown that the integral of $e^{st}\bar{f}(s)$ along this arc vanishes, and, by Cauchy's residue theorem [Eq. (9.47)],

$$L^{-1}\{\bar{f}(s)\} = f(t) = \sum_{k=1}^{n} \text{res}\,[e^{s_k t}\bar{f}(s_k)] \qquad (19.23)$$

Case 2. If $\bar{f}(s)$ possesses an infinite number of poles, a sequence of circular arcs of radii R_n, none passing through any pole, is taken. On the assumption that $\bar{f}(s)$ is such that the integral of $e^{st}\bar{f}(s)$ vanishes along this sequence of circular arcs as $n \to \infty$, we find that

$$L^{-1}\{\bar{f}(s)\} = f(t) = \sum_{k=1}^{\infty} \text{res}\,[e^{s_k t}\bar{f}(s_k)] \qquad (19.24)$$

Case 3. When $\bar{f}(s)$ has branch points and poles, suitable cuts must be made in order to render the function single-valued. We proceed as in the previous cases, except that now $f(t)$ will be given as a combination of residues and real infinite integrals taken along the cuts.

In the applications, $\bar{f}(s)$ is usually encountered in fractional form; i.e.,

$$\bar{f}(s) = \frac{A(s)}{B(s)} \qquad (19.25)$$

where $A(s)$ and $B(s)$ are analytic at $s = s_k$ and $A(s_k) \neq 0$. If $s = s_k$ is a simple pole of $\bar{f}(s)$, the residue of $e^{st}\bar{f}(s)$ is given by

$$\text{res}\,[e^{s_k t}\bar{f}(s_k)] = \frac{A(s_k)}{B'(s_k)}\, e^{s_k t} \qquad (19.26)$$

Hence, when $\bar{f}(s)$ has an infinite number of simple poles *only*, Eq. (19.24) yields the result

$$f(t) = L^{-1}\left\{\frac{A(s)}{B(s)}\right\} = \sum_{k=1}^{\infty} \frac{A(s_k)}{B'(s_k)}\, e^{s_k t} \qquad (19.27)$$

In particular, when $A(s)$ and $B(s)$ are polynomials, where the degree of $B(s)$ is greater than that of $A(s)$, there are only a finite number of poles, and (19.27) becomes

$$f(t) = \sum_{k=1}^{n} \frac{A(s_k)}{B'(s_k)}\, e^{s_k t} \qquad (19.28)$$

which is known as the *Heaviside expansion formula.*

If $s = s_r$ is a pole of order m, the residue of $e^{st}\bar{f}(s)$ at this pole is given by

$$\text{res}\,[e^{s_r t}\bar{f}(s_r)] = \frac{1}{(m-1)!} \lim_{s \to s_r} \frac{d^{m-1}}{ds^{m-1}}\left[(s - s_r)^m e^{st}\, \frac{(As)}{B(s)}\right] \qquad (19.29)$$

The use of (19.29) may involve a considerable amount of algebra, especially when $A(s)$ and $B(s)$ are not polynomials. In such cases, it is usually best to appeal directly to a Laurent series expansion.

When $m = 1$, Eq. (19.29) reduces to

$$\text{res}\,[e^{s_r t}\bar{f}(s_r)] = \lim_{s \to s_r} \left[(s - s_r)e^{st} \frac{A(s)}{B(s)} \right] \tag{19.30}$$

which, in some instances, is more convenient to use than (19.26).

Example 1. Determine $L^{-1}\left\{\dfrac{s+2}{s(s^2+2s+2)}\right\}$. We have, $A(s) = s + 2$, $B(s) = s(s^2 + 2s + 2)$ and $B'(s) = 3s^2 + 4s + 2$, with simple poles occurring at $s_1 = 0$, $s_2 = -1 + i$, $s_3 = -1 - i$. Furthermore,

$$\begin{aligned} A(s_1) &= 2, & B'(s_1) &= 2 \\ A(s_2) &= 1 + i & B'(s_2) &= -2(1 + i) \\ A(s_3) &= 1 - i & B'(s_3) &= -2(1 - i) \end{aligned}$$

Therefore, by (19.28),

$$L^{-1}\left\{\frac{s+2}{s(s^2+2s+2)}\right\} = 1 - \tfrac{1}{2}e^{(-1+i)t} - \tfrac{1}{2}e^{(-1-i)t} = 1 - e^{-t}\cos t$$

Example 2. Find $L^{-1}\left\{\dfrac{s+5}{(s+1)(s-3)^2}\right\}$. By inspection, $s = -1$ is a simple pole, and $s = 3$ is a double pole. Using Eqs. (19.26) and (19.29) to compute the residues at -1 and 3, respectively, we obtain $\frac{1}{4}e^{-t}$ and $2(t - \frac{1}{8})e^{3t}$. Whence,

$$L^{-1}\left\{\frac{s+5}{(s+1)(s-3)^2}\right\} = \tfrac{1}{4}e^{-t} + 2(t - \tfrac{1}{8})e^{3t}$$

Example 3. Obtain $L^{-1}\left\{\dfrac{\text{Sinh}\, a\sqrt{s}}{s\, \text{Sinh}\, b\sqrt{s}}\right\}$. An examination of the function within the braces reveals that there is a simple pole at $s = 0$. Moreover, there are simple poles when $\text{Sinh}\, b\sqrt{s} = 0$; that is, $b\sqrt{s} = \pm n\pi i$, or $s = -n^2\pi^2/b^2$ $(n = 1, 2, 3, \ldots)$. The residue at $s = 0$ is

$$\lim_{s\to 0} \frac{e^{st}\,\text{Sinh}\, a\sqrt{s}}{\text{Sinh}\, b\sqrt{s}} = \lim_{s\to 0} \frac{ae^{st}[1 + (a\sqrt{s})^2/3! + (a\sqrt{s})^4/5! + \cdots]}{b[1 + (b\sqrt{s})^2/3! + (b\sqrt{s})^4/5! + \cdots]} = \frac{a}{b}$$

whereas the residue at $s = -n^2\pi^2/b^2$ is

$$\left[\frac{e^{st}\,\text{Sinh}\, a\sqrt{s}}{s\, d(\text{Sinh}\, b\sqrt{s})/ds}\right]_{s=-n^2\pi^2/b^2} = (-1)^n \frac{2}{n\pi} e^{-n^2\pi^2 t/b^2} \sin\frac{n\pi a}{b} \qquad n = 1, 2, 3, \ldots$$

Therefore,

$$L^{-1}\left\{\frac{\text{Sinh}\, a\sqrt{s}}{s\,\text{Sinh}\, b\sqrt{s}}\right\} = \frac{a}{b} + \frac{2}{\pi}\sum_{n=1}^{\infty}(-1)^n \frac{1}{n} e^{-n^2\pi^2 t/b^2} \sin\frac{n\pi a}{b} \tag{19.31}$$

Example 4. Obtain $L^{-1}\left\{\dfrac{\sqrt{s}}{s^2+a^2}\right\}$. The function here is multiple-valued with a *branch point* at $s_0 = 0$ and simple poles at $s_1 = ai$ and $s_2 = -ai$. In order for the Cauchy residue theorem to apply, we must render the function single-valued by introducing a suitable branch cut such as that shown in Fig. 19.11. Thus, we have

$$\int_{AB} \phi(s)\,ds + \int_{BC} \phi(s)\,ds + \int_{CDE} \phi(s)\,ds + \int_{EF} \phi(s)\,ds + \int_{FG} \phi(s)\,ds + \int_{c-iR}^{c+iR} \phi(s)\,ds = 2\pi i \sum_{k=1}^{2} \text{res}\,\phi(s_k) \tag{19.32}$$

where $\phi(s) = e^{st}\sqrt{s}/(s^2 + a^2)$.

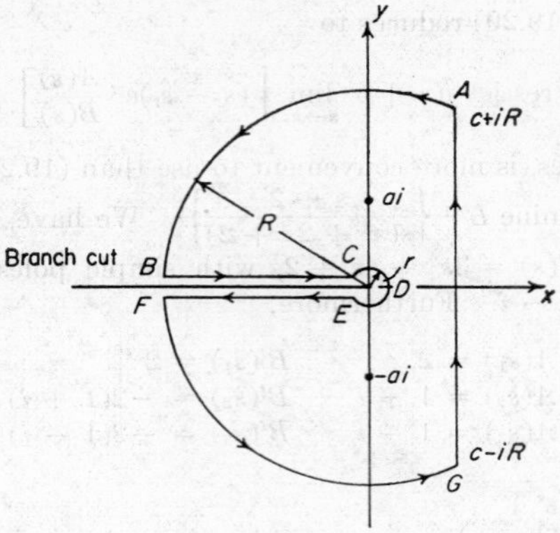

FIG. 19.11

As $R \to \infty$, it can be shown that

$$\int_{AB} \phi(s)\, ds \to 0 \quad \text{and} \quad \int_{FG} \phi(s)\, ds \to 0$$

Also, as $r \to 0$,

$$\int_{CDE} \phi(s)\, ds \to 0$$

On the line BC, $s = xe^{i\pi}$ in the limit, so that, as $R \to \infty$,

$$\int_{BC} \phi(s)\, ds = -i \int_{\infty}^{0} \frac{e^{-xt}\sqrt{x}\, dx}{x^2 + a^2}$$

Similarly, on the line EF, $s = xe^{-i\pi}$ in the limit. Hence, as $R \to \infty$,

$$\int_{EF} \phi(s)\, ds = i \int_{0}^{\infty} \frac{e^{-xt}\sqrt{x}\, dx}{x^2 + a^2}$$

Therefore, from (19.32), after dividing both sides by $2\pi i$, we have

$$\frac{1}{2\pi i} \lim_{R\to\infty} \int_{c-iR}^{c+iR} \frac{e^{st}\sqrt{s}}{s^2 + a^2}\, ds + \frac{1}{\pi}\int_0^\infty \frac{e^{-xt}\sqrt{x}}{x^2 + a^2}\, dx = \sum_{k=1}^{2} \operatorname{res} \phi(s_k) \tag{19.33}$$

Now, $$\operatorname{res} \phi(ai) = \left[\frac{e^{st}\sqrt{s}}{2s}\right]_{s=ai} = \frac{e^{ait}\sqrt{ai}}{2ai}$$

and $$\operatorname{res} \phi(-ai) = \left[\frac{e^{st}\sqrt{s}}{2s}\right]_{s=-ai} = \frac{e^{-ait}\sqrt{-ai}}{-2ai}$$

The sum of these residues is

$$\frac{1}{\sqrt{a}} \sin\left(at + \frac{\pi}{4}\right)$$

Therefore, solving (19.33) for the first term, we obtain

$$L^{-1}\left\{\frac{\sqrt{s}}{s^2 + a^2}\right\} = \frac{1}{\sqrt{a}} \sin\left(at + \frac{\pi}{4}\right) - \frac{1}{\pi}\int_0^\infty \frac{e^{-xt}\sqrt{x}}{x^2 + a^2}\, dx$$

19.5. ORDINARY DIFFERENTIAL EQUATIONS

One of the most frequently occurring differential equations in dynamics is the second-order differential equation

$$y''(t) + 2ay'(t) + (a^2 + b^2)y(t) = f(t) \tag{19.34}$$

where $f(t)$ is a given function. We wish to find the general solution of (19.34).

Taking the Laplace transform of both sides of (19.34), we find that

$$\bar{y}(s) = \frac{ay(0) + y'(0)}{(s+a)^2 + b^2} + \frac{(s+a)y(0)}{(s+a)^2 + b^2} + \frac{\bar{f}(s)}{(s+a)^2 + b^2} \tag{19.35}$$

Applying formulas 37 and 38 from Table 19.2 and the convolution theorem, we obtain immediately

$$y(t) = [ay(0) + y'(0)]e^{-at}\frac{1}{b}\sin bt + y(0)e^{-at}\cos bt + \frac{1}{b}\int_0^t f(t-\tau)e^{-a\tau}\sin b\tau\, d\tau \tag{19.36}$$

As a second example, let us consider the angular motion of a disk of moment of inertia I attached to a shaft of torsional stiffness k (Fig. 19.12) and initially at rest when it is subjected to the triangular torque pulse of Fig. 19.13.

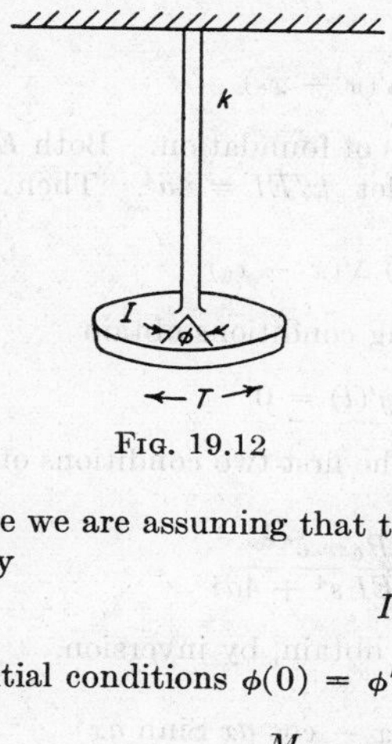

FIG. 19.12

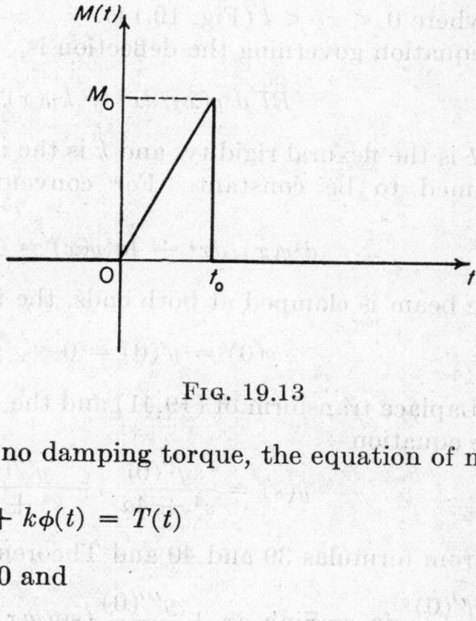

FIG. 19.13

Since we are assuming that there is no damping torque, the equation of motion is given by

$$I\phi''(t) + k\phi(t) = T(t) \tag{19.37}$$

with initial conditions $\phi(0) = \phi'(0) = 0$ and

$$M(t) = \frac{M_0}{t_0}[t\,\Delta(t) - (t - t_0)\,\Delta(t - t_0) - t_0\,\Delta(t - t_0)] \tag{19.38}$$

The Laplace transform of (19.37) yields the equation

$$\bar{\phi}(s) = \frac{M_0}{t_0 I}\,\frac{1 - e^{-t_0 s} - st_0e^{-t_0 s}}{s^2(s^2 + k/I)} \tag{19.39}$$

Using formulas 27 and 28 from Table 19.2 and Theorem 3, we readily find that

$$\phi(t) = \frac{M_0}{t_0 I\beta^3}\{(\beta t - \sin\beta t) - [\beta t - \sin\beta(t - t_0) - \beta t_0\cos\beta(t - t_0)]\,\Delta(t - t_0)\}$$

where $\beta = (k/I)^{1/2}$.

In the preceding two problems, the values of the dependent variable and its derivative were specified at *one* point. Many engineering problems, however, lead to

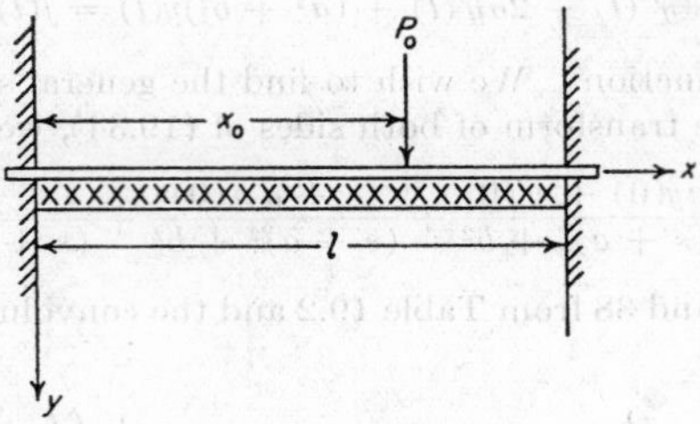

FIG. 19.14

differential equations in which boundary conditions are specified at *two* or *more* points. We shall illustrate the modus operandi in the latter case by solving the following problem:

Determine the deflection of a uniform beam of length l, clamped at both ends, which is attached to an elastic foundation and carries a concentrated load P_0 at $x = x_0$, where $0 < x_0 < l$ (Fig. 19.14).

The equation governing the deflection is

$$EI\, d^4y(x)/dx^4 + ky(x) = P_0\, \Delta'(x - x_0) \tag{19.40}$$

where EI is the flexural rigidity, and k is the modulus of foundation. Both EI and k are assumed to be constant. For convenience, let $k/EI = 4a^4$. Then (19.40) becomes

$$d^4y(x)/dx^4 + 4a^4y(x) = (P_0/EI)\, \Delta'(x - x_0) \tag{19.41}$$

Since the beam is clamped at both ends, the following conditions obtain

$$y(0) = y'(0) = 0 \qquad y(l) = y'(l) = 0 \tag{19.42}$$

The Laplace transform of (19.41) and the use of the first two conditions of (19.42) yield the equation

$$\bar{y}(s) = \frac{sy''(0)}{s^4 + 4a^4} + \frac{y'''(0)}{s^4 + 4a^4} + \frac{P_0}{EI}\frac{e^{-x_0 s}}{s^4 + 4a^4} \tag{19.43}$$

Hence, from formulas 39 and 40 and Theorem 3, we obtain, by inversion,

$$\begin{aligned} y(x) = {} & \frac{y''(0)}{2a^2} \sin ax \operatorname{Sinh} ax + \frac{y'''(0)}{4a^3} (\sin ax \operatorname{Cosh} ax - \cos ax \operatorname{Sinh} ax) \\ & + \frac{P_0}{4a^3EI} [\sin a(x - x_0) \operatorname{Cosh} a(x - x_0) - \cos a(x - x_0) \operatorname{Sinh} a(x - x_0)]\, \Delta(x - x_0) \end{aligned} \tag{19.44}$$

The last two conditions of (19.42), namely, $y(l) = y'(l) = 0$, yield the simultaneous equations:

$$\begin{aligned} & ay''(0) \sin al \operatorname{Sinh} al + y'''(0)(\sin al \operatorname{Cosh} al - \cos al \operatorname{Sinh} al) \\ & \qquad = \frac{P_0}{2EI} [\cos a(l - x_0) \operatorname{Sinh} a(l - x_0) - \sin a(l - x_0) \operatorname{Cosh} a(l - x_0)] \\ & ay''(0)(\sin al \operatorname{Cosh} al + \cos al \operatorname{Sinh} al) + y'''(0)(\sin al \operatorname{Sinh} al) \\ & \qquad = -\frac{P_0}{EI} [\sin a(l - x_0) \operatorname{Sinh} a(l - x_0)] \end{aligned}$$

from which $y''(0)$ and $y'''(0)$ can be determined. These values, when substituted in (19.44), give the desired solution.

As a last illustrative example of this section, we consider the solution of the system of ordinary equations,

$$\frac{dx(t)}{dt} = x(t) - 4y(t) \qquad \frac{dy(t)}{dt} = 2x(t) + y(t) \tag{19.45}$$

subject to the conditions $x(0) = y(0) = 3$.

The Laplace transforms of (19.45) reduce the differential equations to the following simultaneous equations:

$$(s - 1)\bar{x}(s) + 4\bar{y}(s) = 3 \qquad -2\bar{x}(s) + (s - 1)\bar{y}(s) = 3$$

which when solved yield the results

$$\bar{x}(s) = \frac{3(s-5)}{(s-1)^2+8} = 3\frac{s-1}{(s-1)^2+8} - 12\frac{1}{(s-1)^2+8}$$
$$\bar{y}(s) = \frac{3(s+1)}{(s-1)^2+8} = 3\frac{s-1}{(s-1)^2+8} + 6\frac{1}{(s-1)^2+8}$$

Upon taking inverse transforms of both sides, we find that

$$x(t) = 3L^{-1}\left\{\frac{s-1}{(s-1)^2+8}\right\} - 12L^{-1}\left\{\frac{1}{(s-1)^2+8}\right\}$$
$$y(t) = 3L^{-1}\left\{\frac{s-1}{(s-1)^2+8}\right\} + 6L^{-1}\left\{\frac{1}{(s-1)^2+8}\right\}$$

whence, from formulas 37 and 38 in Table 19.2, we obtain

$$x(t) = 3e^t[\cos(2\sqrt{2}\,t) - \sqrt{2}\sin(2\sqrt{2}\,t)]$$
$$y(t) = 3e^t[\cos(2\sqrt{2}\,t) + (\sqrt{2}/2)\sin(2\sqrt{2}\,t)]$$

19.6. INTEGRODIFFERENTIAL AND INTEGRAL EQUATIONS

In some cases, the statements of physical laws lead directly to integral or integro-differential equations.

An important case is the integral equation of the convolution type, namely,

$$y(t) = f(t) + \int_0^t k(t - \tau)y(\tau)\,d\tau \tag{19.46}$$

where $k(t - \tau)$, the *kernel* of the equation, and $f(t)$ are *known* functions, and $y(t)$ is the *unknown* function. Such an equation is solved directly by Laplace transforms. For, upon taking transforms of both sides of (19.46), we obtain

$$\bar{y}(s) = \bar{f}(s) + \bar{k}(s)\bar{y}(s)$$

whence

$$\bar{y}(s) = \frac{\bar{f}(s)}{1 - \bar{k}(s)}$$

and, upon inversion,

$$y(t) = L^{-1}\left\{\frac{\bar{f}(s)}{1 - \bar{k}(s)}\right\} \tag{19.47}$$

This result may be used as a formula for the solution of an integral equation of the convolution type.

Suppose, for example, that it is required to solve the integral equation

$$y(t) = \sin t + \int_0^t (t - \tau)y(\tau)\,d\tau$$

By comparison with (19.46), we have

$$\bar{f}(s) = L\{\sin t\} = \frac{1}{s^2 + 1} \qquad \bar{k}(s) = L\{t\} = \frac{1}{s^2}$$

Hence, from (19.47) and formula 45 in Table 19.2, we obtain

$$y(t) = L^{-1}\left\{\frac{s^2}{(s^2 + 1)(s^2 - 1)}\right\} = \tfrac{1}{2}(\sin t + \text{Sinh } t)$$

19.7. PARTIAL DIFFERENTIAL EQUATIONS

The Laplace transform is quite effective in the solution of boundary-value problems involving certain types of partial differential equations. As illustrations, we shall consider two boundary-value problems, the first having to do with the so-called one-dimensional heat equation,

$$u_t(x,t) = ku_{xx}(x,t) \tag{19.48}$$

where k is the diffusivity constant, and the second involving the one-dimensional wave equation,

$$u_{tt}(x,t) = a^2u_{xx}(x,t) \tag{19.49}$$

where a^2 is a constant.

Consider a slab bounded by the planes $x = 0$ and $x = l$, but otherwise infinite in extent. If the entire slab is initially at temperature T_0, and faces $x = 0$ and $x = l$ are maintained at temperatures T_1 and T_2, respectively, what is the temperature of the slab for $t > 0$?

The boundary-value problem to be solved is

$$u_t(x,t) = ku_{xx}(x,t) \qquad 0 < x < l,\ t > 0 \tag{19.50}$$

$$u(x,0) = T_0 \qquad 0 < x < l \tag{19.51}$$

$$u(0,t) = T_1 \qquad u(l,t) = T_2 \qquad t > 0 \tag{19.52}$$

Transforming the relations (19.50), (19.52) with respect to t and making use of (19.51), we reduce the preceding problem in partial differential equations to one in ordinary differential equations, namely,

$$s\bar{u}(x,s) - T_0 = k\frac{d^2\bar{u}(x,s)}{dx^2} \qquad x > 0$$

$$\bar{u}(0,s) = \frac{T_1}{s} \qquad \bar{u}(l,s) = \frac{T_2}{s}$$

the solution of which is

$$\bar{u}(x,s) = \frac{T_0}{s} + (T_1 - T_0)\frac{\text{Sinh }(l - x)\sqrt{s/k}}{s\text{ Sinh }(l\sqrt{s/k})} + (T_2 - T_0)\frac{\text{Sinh }(x\sqrt{s/k})}{s\text{ Sinh }(l\sqrt{s/k})} \tag{19.53}$$

Making use of (19.31) for the last two terms, we have, upon inversing and simplifying,

$$u(x,t) = T_1 + (T_2 - T_1)\frac{x}{l} + \frac{2}{\pi}\sum_{n=1}^{\infty}\frac{1}{n}[(-1)^n(T_2 - T_0) + (T_0 - T_1)]e^{-n^2\pi^2kt/l^2}\sin\frac{n\pi x}{l} \tag{19.54}$$

A useful alternative solution of the preceding problem can be obtained in terms of the complementary error function. We have

$$\frac{\operatorname{Sinh}(l-x)\sqrt{s/k}}{s \operatorname{Sinh}(l\sqrt{s/k})} = \frac{e^{(l-x)\sqrt{s/k}} - e^{-(l-x)\sqrt{s/k}}}{s(1-e^{-2l\sqrt{s/k}})} e^{-l\sqrt{s/k}}$$

$$= \frac{e^{(l-x)\sqrt{s/k}} - e^{-(l-x)\sqrt{s/k}}}{s} \sum_{n=0}^{\infty} e^{-(2n+1)l\sqrt{s/k}}$$

$$= \frac{1}{s} \sum_{n=0}^{\infty} (e^{-(2nl+x)\sqrt{s/k}} - e^{-[(2n+1)l-x]\sqrt{s/k}})$$

and, similarly,

$$\frac{\operatorname{Sinh}(x\sqrt{s/k})}{s \operatorname{Sinh}(l\sqrt{s/k})} = \frac{1}{s} \sum_{n=0}^{\infty} (e^{-[(2n+1)l-x]\sqrt{s/k}} - e^{-[(2n+1)l+x]\sqrt{s/k}})$$

Therefore, substituting these results in (19.53) and inversing, we obtain

$$u(x,t) = T_0 + (T_1 - T_0) \sum_{n=0}^{\infty} \left[\operatorname{erfc} \frac{2nl + x}{2\sqrt{kt}} - \operatorname{erfc} \frac{2(n+1)l - x}{2\sqrt{kt}} \right]$$

$$+ (T_2 - T_0) \sum_{n=0}^{\infty} \left[\operatorname{erfc} \frac{(2n+1)l - x}{2\sqrt{kt}} - \operatorname{erfc} \frac{(2n+1)l + x}{2\sqrt{kt}} \right] \tag{19.55}$$

The one-dimensional wave equation (19.49) arises in a number of physical problems, e.g., the vibrating string, the dissipationless transmission line, sound waves in a tube, and the longitudinal displacements of an elastic bar. As an illustration of the last case,

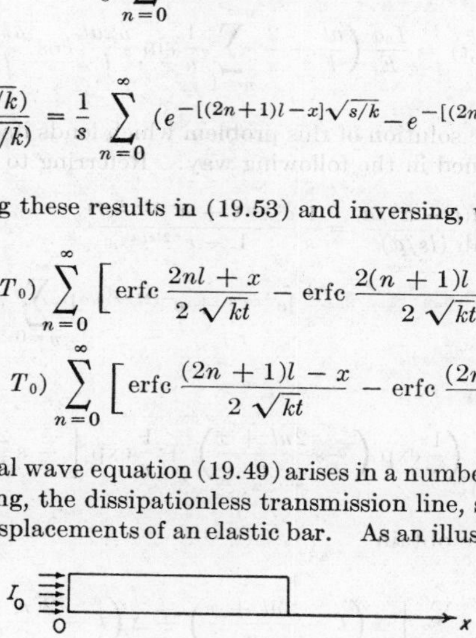

Fig. 19.15

let us suppose that a uniform bar of length l lying on a horizontal frictionless plane is initially at rest and unstrained. At $t = 0$, the end $x = 0$ is subjected to an axial impact of impulse I_0 per unit cross-sectional area (Fig. 19.15). Determine the longitudinal displacement $u(x,t)$ for $t > 0$.

The boundary-value problem to be solved is:

$$u_{tt}(x,t) = a^2 u_{xx}(x,t) \qquad 0 < x < l,\ t > 0 \tag{19.56}$$

$$u(x,0) = 0 \qquad u_t(x,0) = 0 \qquad 0 \le x \le l \tag{19.57}$$

$$Eu_x(0,t) = -I_0\,\Delta'(t) \qquad u_x(l,t) = 0 \qquad t > 0 \tag{19.58}$$

When transformed with respect to t, this problem becomes

$$\frac{d^2\bar{u}(x,s)}{dx^2} - \frac{s^2}{a^2}\bar{u}(x,s) = 0 \tag{19.59}$$

$$E\frac{d\bar{u}(0,s)}{dx} = -I_0 \qquad \frac{d\bar{u}(l,s)}{dx} = 0 \tag{19.60}$$

whose solution is

$$\bar{u}(x,s) = \frac{I_0 a}{E} \frac{\operatorname{Cosh}[(l-x)s/a]}{s \operatorname{Sinh}(ls/a)} \tag{19.61}$$

As is readily seen, the preceding function has a *double* pole at $s = 0$, and simple poles at $s = \pm in\pi a/l$ $(n = 1, 2, 3, \ldots)$. Now,

$$\text{res } \bar{u}(x,0) = \lim_{s \to 0} \frac{d}{ds} \left[\frac{e^{st} \text{ Cosh } (l - x)s/a}{s \text{ Sinh } (ls/a)} \right] = \frac{at}{l} \tag{19.62}$$

and

$$\text{res } \bar{u}\left(x, \pm \frac{in\pi}{l}\right) = \left[\frac{e^{st} \text{ Cosh } (l - x)s/a}{s \dfrac{d}{ds}\left(\text{Sinh } \dfrac{ls}{a}\right)} \right]_{s = \pm in a\pi/l} = \pm \frac{e^{\pm ina\pi t/l} \cos (n\pi x/l)}{in\pi}$$

Therefore, by the complex inversion theorem,

$$u(x,t) = \frac{I_0 a}{E} \left(\frac{at}{l} + \frac{2}{\pi} \sum_{n=1}^{\infty} \frac{1}{n} \sin \frac{n\pi a t}{l} \cos \frac{n\pi x}{l} \right) \tag{19.63}$$

Another form of the solution of this problem which lends itself to a vivid graphical representation is obtained in the following way. Referring to (19.61), we write

$$\frac{\text{Cosh } s(l - x)/a}{s \text{ Sinh } (ls/a)} = \frac{1}{s} \frac{e^{-xs/a} + e^{-(2l-x)s/a}}{1 - e^{-2ls/a}}$$

$$= \frac{1}{s} [e^{-xs/a} + e^{-(2l-x)s/a}] \sum_{n=0}^{\infty} e^{-2nls/a}$$

Consequently,

$$\bar{u}(x,s) = \frac{I_0 a}{E} \sum_{n=0}^{\infty} \left(\frac{1}{s} \exp\left(-s \frac{2nl + x}{a} \right) + \frac{1}{s} \exp\left[-s \frac{2(n+1)l - x}{a} \right] \right)$$

and, upon inversing, we obtain

$$u(x,t) = \frac{I_0 a}{E} \sum_{n=0}^{\infty} \left[\Delta\left(t - \frac{2nl + x}{a}\right) + \Delta\left(t - \frac{2(n+1)l - x}{a}\right) \right] \tag{19.64}$$

The graph of this function (Fig. 19.16) shows the displacement of an arbitrary section at any time t.

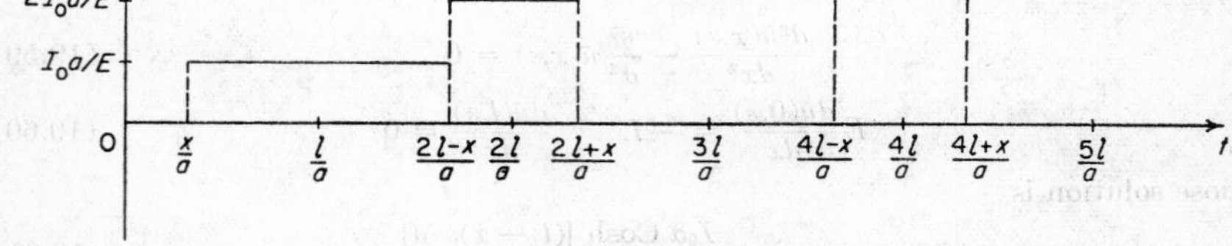

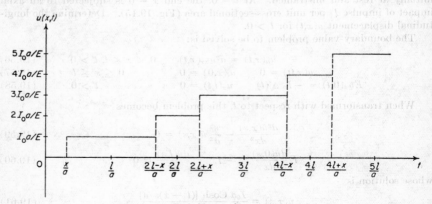

FIG. 19.16

19.8. DIFFERENCE EQUATIONS

Many of the problems which arise in mechanics, statistics, probability, theory of structures, and network theory give rise to *difference equations*. This is due to the fact that the problems leading to them involve, in general, structures or entities having many identical component parts. Examples of such structures are: (1) equal masses on an elastic string, (2) electrical filters, (3) continuous beams with equally spaced supports, and (4) crankshafts of multicylinder engines. In what follows, we shall confine ourselves to a difference equation of the second order, referring the reader to works in the bibliography for more extensive treatments.

Thus, let it be required to solve the *second-order difference equation*,

$$ay_t + by_{t+1} + cy_{t+2} = f_t \qquad t = 0, 1, 2, \ldots \tag{19.65}$$

where a, b, and c are constants and $f_0, f_1, f_2, \ldots$ are a given set of numbers. This means that we are to find a sequence of numbers $y_0, y_1, y_2, \ldots$ which satisfy (19.65) for all non-negative integral values of t. Graphically, the solution is a set of discrete points, as shown in Fig. 19.17.

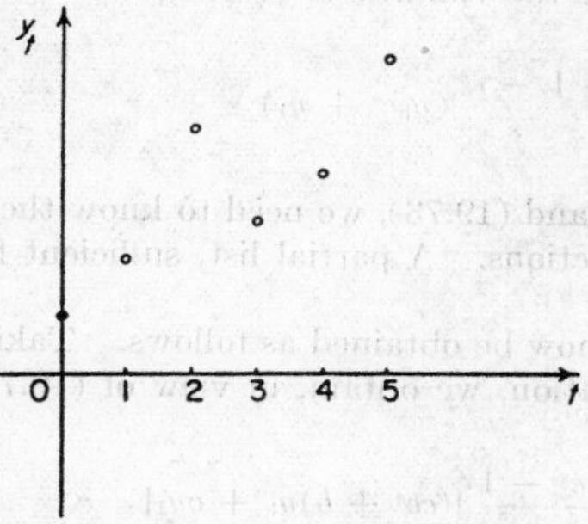

FIG. 19.17

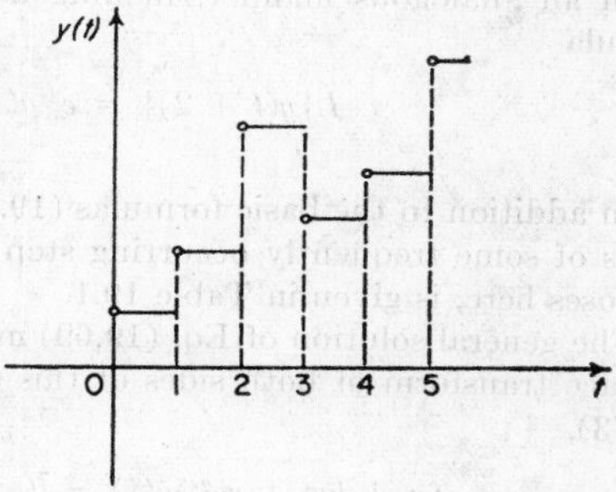

FIG. 19.18

To make use of the Laplace transform in solving difference equations, we construct from the sequence $y_0, y_1, y_2, \ldots$ a *step function* $y(t)$ defined as follows

$$y(t) = y_{[t]} \qquad t \geq 0 \tag{19.66}$$

where the symbol $[t]$, read the *greatest integer in* t, is the integer such that

$$[t] \leq t < [t] + 1 \tag{19.67}$$

For example, $[\frac{1}{4}] = 0$, $[-3\frac{1}{2}] = -4$, $[2] = 2$, and $[\sqrt{5}] = 2$. From (19.67), we also have

$$t - 1 < [t] \leq t \tag{19.68}$$

Thus, the graph of Fig. 19.17 becomes the step function of Fig. 19.18. Similarly, a step function $f(t)$ is constructed from the sequence $f_0, f_1, f_2, \ldots$. Therefore, in terms of these step functions, Eq. (19.65) is replaced by

$$ay(t) + by(t + 1) + cy(t + 2) = f(t) \qquad t \geq 0 \tag{19.69}$$

In order to solve (19.69), we must know the transforms of $y(t + 1)$ and $y(t + 2)$. For the transform of $y(t + 1)$, we have

$$L\{y(t + 1)\} = \int_0^\infty e^{-st} y(t + 1)\, dt \tag{19.70}$$

Making the substitutions $t + 1 = \tau$, $dt = d\tau$, we obtain

$$\int_0^\infty e^{-st}y(t+1)\,dt = e^s \int_1^\infty e^{-s\tau}y(\tau)\,d\tau$$
$$= e^s \int_0^\infty e^{-s\tau}y(\tau)\,d\tau - e^s \int_0^1 e^{-s\tau}y(\tau)\,d\tau$$

If the initial conditions stipulate that $y(t)$ is such that

$$\begin{aligned} y(t) &= y_0 \qquad 0 \le t < 1 \\ y(t) &= y_1 \qquad 1 \le t < 2 \end{aligned} \tag{19.71a,b}$$

then
$$\int_0^1 e^{-s\tau}y(\tau)\,d\tau = \frac{1 - e^{-s}}{s}\,y_0$$

and (19.70) becomes

$$L\{y(t+1)\} = e^s\bar{y}(s) + \frac{1 - e^s}{s}\,y_0 \tag{19.72}$$

In an analogous manner, making use of the conditions (19.71), we derive the formula

$$L\{y(t+2)\} = e^{2s}\bar{y}(s) + \frac{1 - e^s}{s}(y_0e^s + y_1) \tag{19.73}$$

In addition to the basic formulas (19.72) and (19.73), we need to know the transforms of some frequently occurring step functions. A partial list, sufficient for our purposes here, is given in Table 19.1.

The general solution of Eq. (19.69) may now be obtained as follows. Taking the Laplace transform of both sides of this equation, we obtain, in view of (19.72) and (19.73),

$$(a + be^s + ce^{2s})\bar{y}(s) = \bar{f}(s) + \frac{e^s - 1}{s}[(ce^s + b)y_0 + cy_1]$$

whence

$$\bar{y}(s) = \frac{\bar{f}(s)}{a + be^s + ce^{2s}} + \frac{(e^s - 1)[(ce^s + b)y_0 + cy_1]}{s(a + be^s + ce^{2s})} \tag{19.74}$$

and, by inversion,

$$y(t) = L^{-1}\left\{\frac{\bar{f}(s)}{a + be^s + ce^{2s}}\right\} + L^{-1}\left\{\frac{(e^s - 1)[(ce^s + b)y_0 + cy_1]}{s(a + be^s + ce^{2s})}\right\} \tag{19.75}$$

Equation (19.75) may now be used to solve a given second-order difference equation. We shall give two examples to illustrate the method of procedure (see p. 19–21).

Table 19.1

	$\bar{y}(s)$	$y(t)$
1	$1/s$	1
2	$1/s(e^s - 1)$	$[t]$
3	$(e^s + 1)/s(e^s - 1)^2$	$[t]^2$
4	$(e^s - 1)/s(e^s - a)$	$a^{[t]}$
5	$a(e^s - 1)/s(e^s - a)^2$	$[t]a^{[t]}$

For the explanation of the symbol $[t]$, see Eq. (19.67).

Table 19.2. Laplace Transforms

	$\bar{f}(s)$	$f(t)$
1	$\bar{f}(s) = \int_0^\infty e^{-st}f(t)\,dt$	$f(t)$
2	$a_1\bar{f}_1(s) + a_2\bar{f}_2(s) + \cdots + a_n\bar{f}_n(s)$	$a_1f_1(t) + a_2f_2(t) + \cdots + a_nf_n(t)$
3	$s\bar{f}(s) - f(0)$	$f'(t)$
4	$s^2\bar{f}(s) - sf(0) - f'(0)$	$f''(t)$
5	$s^n\bar{f}(s) - \sum_{k=1}^{n} s^{n-k}f^{(k-1)}(0)$	$f^{(n)}(t)$
6	$s^{-n}\bar{f}(s)$	$\int_0^t \cdots \int_0^t f(t)\,(dt)^n$
7	$(-1)^n\bar{f}^{(n)}(s)$	$t^nf(t)$
8	$\int_s^\infty \cdots \int_s^\infty \bar{f}(s)\,(ds)^n$	$t^{-n}f(t)$
9	$\bar{f}(as - b)$	$e^{bt/a}f(t/a)/a$
10	$\bar{f}(s - a)$	$e^{at}f(t)$
11	$e^{-as}\bar{f}(s) \quad a \geq 0$	$f(t - a)\,\Delta(t - a)$
12	$\bar{f}_1(s)\bar{f}_2(s)$	$\int_0^t f_1(\tau)f_2(t - \tau)\,d\tau = \int_0^t f_1(t - \tau)f_2(\tau)\,d\tau$
13	$1/s$	1
14	$1/s^n \quad n = 1, 2, 3, \ldots$	$t^{n-1}/(n - 1)!$
15	$1/s^k \quad k > 0$	$t^{k-1}/\Gamma(k)$
16	$1/(s - a)^n \quad n = 1, 2, 3, \ldots$	$t^{n-1}e^{at}/(n - 1)!$
17	$1/(s - a)^k \quad k > 0$	$t^{k-1}e^{at}/\Gamma(k)$
18	$e^{-as}/s \quad a \geq 0$	$\Delta(t - a)$
19	$e^{-as} \quad a \geq 0$	$\Delta'(t - a)$
20	$se^{-as} \quad a \geq 0$	$\Delta''(t - a)$
21	$\dfrac{1}{(s - a)(s - b)}$	$\dfrac{1}{a - b}(e^{at} - e^{bt})$
22	$\dfrac{s}{(s - a)(s - b)}$	$\dfrac{1}{a - b}(ae^{at} - be^{bt})$
23	$1/(s^2 + a^2)$	$(1/a)\sin at$
24	$s/(s^2 + a^2)$	$\cos at$
25	$1/(s^2 - a^2)$	$(1/a)\,\text{Sinh}\,at$
26	$s/(s^2 - a^2)$	$\text{Cosh}\,at$
27	$1/s(s^2 + a^2)$	$(1 - \cos at)/a^2$
28	$1/s^2(s^2 + a^2)$	$(at - \sin at)/a^3$
29	$1/(s^2 + a^2)^2$	$\sin at - at\cos at$
30	$s/(s^2 + a^2)^2$	$(t/2a)\sin at$
31	$(s^2 - a^2)/(s^2 + a^2)^2$	$t\cos at$
32	$1/(s^2 - a^2)^2$	$(at\,\text{Cosh}\,at - \text{Sinh}\,at)/2a^3$
33	$s/(s^2 - a^2)^2$	$t\,\text{Sinh}\,at$
34	$(s^2 + a^2)/(s^2 - a^2)^2$	$t\,\text{Cosh}\,at$
35	$\dfrac{1}{(s^2 + a^2)^k} \quad k > 0$	$\dfrac{\sqrt{\pi}}{\Gamma(k)}\left(\dfrac{t}{2a}\right)^{k-\frac{1}{2}} J_{k-\frac{1}{2}}(at)$
36	$\dfrac{1}{(s^2 - a^2)^k} \quad k > 0$	$\dfrac{\sqrt{\pi}}{\Gamma(k)}\left(\dfrac{t}{2a}\right)^{k-\frac{1}{2}} I_{k-\frac{1}{2}}(at)$
37	$\dfrac{1}{(s + a)^2 + b^2}$	$\dfrac{1}{b}e^{-at}\sin bt$
38	$\dfrac{s + a}{(s + a)^2 + b^2}$	$e^{-at}\cos bt$
39	$1/(s^4 + 4a^4)$	$(\sin at\,\text{Cosh}\,at - \cos at\,\text{Sinh}\,at)/4a^3$

Table 19.2. Laplace Transforms (*Continued*)

	$\bar{f}(s)$		$f(t)$
40	$s/(s^4 + 4a^4)$		$(\sin at \,\mathrm{Sinh}\, at)/2a^2$
41	$s^2/(s^4 + 4a^4)$		$(\cos at \,\mathrm{Sinh}\, at + \sin at \,\mathrm{Cosh}\, at)/2a$
42	$s^3/(s^4 + 4a^4)$		$\cos at \,\mathrm{Cosh}\, at$
43	$1/(s^4 - a^4)$		$(\mathrm{Sinh}\, at - \sin at)/2a^3$
44	$s/(s^4 - a^4)$		$(\mathrm{Cosh}\, at - \cos at)/2a^2$
45	$s^2/(s^4 - a^4)$		$(\mathrm{Sinh}\, at + \sin at)/2a$
46	$s^3/(s^4 - a^4)$		$(\mathrm{Cosh}\, at + \cos at)/2$
47	$\dfrac{1}{\sqrt{s}\,(1 + a\sqrt{s})}$		$\dfrac{1}{a} e^{t/a^2} \,\mathrm{erfc}\, \dfrac{\sqrt{t}}{a}$
48	$1/(a + \sqrt{s})$		$1/\sqrt{\pi t} - a e^{a^2 t} \,\mathrm{erfc}\, a\sqrt{t}$
49	$1/s(a + \sqrt{s})$		$1 - e^{a^2 t} \,\mathrm{erfc}(a\sqrt{t})/a$
50	$\sqrt{s + a}/(s + b)$		$e^{-at}/\sqrt{\pi t} + \sqrt{a - b}\, e^{-bt} \,\mathrm{erf}\, \sqrt{(a - b)t}$
51	$1/(s + a)\sqrt{s + b}$		$(e^{-at}/\sqrt{b - a}) \,\mathrm{erf}\, \sqrt{(b - a)t}$
52	$\sqrt{s - a} - \sqrt{s - b}$		$\tfrac{1}{2}(e^{bt} - e^{at})/\sqrt{\pi t^3}$
53	$\dfrac{(\sqrt{s^2 + a^2} - s)^k}{\sqrt{s^2 + a^2}}$	$k > -1$	$a^k J_k(at)$
54	$(\sqrt{s^2 + a^2} - s)^k$	$k > 0$	$k a^k J_k(at)/t$
55	$\dfrac{(s - \sqrt{s^2 - a^2})^k}{\sqrt{s^2 - a^2}}$	$k > -1$	$a^k I_k(at)$
56	$(s - \sqrt{s^2 - a^2})^k$	$k > 0$	$k a^k I_k(at)/t$
57	$\dfrac{1}{(s + a)^k (s + b)^k}$	$k > 0$	$\dfrac{\sqrt{\pi}}{\Gamma(k)} \left(\dfrac{t}{a - b}\right)^{k - \frac{1}{2}} e^{-\frac{1}{2}(a+b)t} I_{k-\frac{1}{2}} \left(\dfrac{a - b}{2} t\right)$
58	$\dfrac{1}{(s + a)^{1/2}(s + b)^{3/2}}$		$t e^{-[(a+b)/2]t} \left[I_0\left(\dfrac{a - b}{2} t\right) + I_1\left(\dfrac{a - b}{2} t\right)\right]$
59	$\dfrac{(\sqrt{s + a} - \sqrt{s})^k}{\sqrt{s + a} + \sqrt{s}}$	$k > 0$	$e^{-at/2} I_k\left(\dfrac{at}{2}\right)$
60	$\dfrac{\sqrt{s}}{\sqrt{s + a}\,(\sqrt{s} + \sqrt{s + a})^{2k}}$	$k > 0$	$\dfrac{e^{-at/2}}{4a^{k-1}} \left[I_{k-1}\left(\dfrac{at}{2}\right) - 2I_k\left(\dfrac{at}{2}\right) + I_{k+1}\left(\dfrac{at}{2}\right)\right]$
61	$(\sqrt{s + 2a} - \sqrt{s})/(\sqrt{s + 2a} + \sqrt{s})$		$e^{-at} I_1(at)/t$
62	$\dfrac{(a - b)^k}{(\sqrt{s + a} + \sqrt{s + b})^{2k}}$	$k > 0$	$k e^{-[(a+b)/2]t} I_k\left(\dfrac{a - b}{2} t\right)$
63	$e^{-a\sqrt{s}}$	$a > 0$	$\dfrac{a/2}{\sqrt{\pi t^3}} \exp\left(-\dfrac{a^2}{4t}\right)$
64	$\dfrac{e^{-a\sqrt{s}}}{\sqrt{s}}$	$a \geq 0$	$\dfrac{1}{\sqrt{\pi t}} \exp\left(-\dfrac{a^2}{4t}\right)$
65	$\dfrac{e^{-a\sqrt{s}}}{s}$	$a \geq 0$	$\mathrm{erfc}\, \dfrac{a}{2\sqrt{t}}$
66	$\dfrac{e^{-a\sqrt{s}}}{s^{3/2}}$	$a \geq 0$	$2\dfrac{\sqrt{t}}{\pi} \exp\left(\dfrac{a^2}{4t}\right) - a \,\mathrm{erfc}\, \dfrac{a}{2\sqrt{t}}$
67	$\dfrac{e^{-a/s}}{s^{k+1}}$	$k > -1$	$\left(\dfrac{t}{a}\right)^{k/2} J_k(2\sqrt{at})$
68	$\dfrac{e^{a/s}}{s^{k+1}}$	$k \geq -1$	$\left(\dfrac{t}{a}\right)^{k/2} I_k(2\sqrt{at})$

Table 19.2. Laplace Transforms (*Continued*)

	$\bar{f}(s)$		$f(t)$
69	$\dfrac{(\sqrt{s+a}-\sqrt{s})^{2k}}{\sqrt{s(s+a)}}$	$k > -1$	$a^k e^{-(a/2)t} I_k\left(\dfrac{at}{2}\right)$
70	$\dfrac{(\sqrt{s}-\sqrt{s-a})^{2k}}{\sqrt{s(s-a)}}$	$k > -1$	$a^k e^{at/2} I_k\left(\dfrac{at}{2}\right)$
71	$e^{-bs} - e^{-b\sqrt{s^2+a^2}}$		$\dfrac{ab}{\sqrt{t^2-b^2}} J_1(a\sqrt{t^2-b^2})\,\Delta(t-b)$
72	$e^{-b\sqrt{s^2-a^2}} - e^{-bs}$		$\dfrac{ab}{\sqrt{t^2-b^2}} I_1(a\sqrt{t^2-b^2})\,\Delta(t-b)$
73	$\dfrac{e^{-b\sqrt{s^2+a^2}}}{\sqrt{s^2+a^2}}\left(\dfrac{a}{s+\sqrt{s^2+a^2}}\right)^k$	$k > -1$	$\left(\dfrac{t-b}{t+b}\right)^{k/2} J_k(a\sqrt{t^2-b^2})\,\Delta(t-b)$
74	$\dfrac{e^{-b(\sqrt{s^2-a^2})}}{\sqrt{s^2-a^2}}\left(\dfrac{a}{s+\sqrt{s^2+a^2}}\right)^k$	$k > -1$	$\left(\dfrac{t-b}{t+b}\right)^{k/2} I_k(a\sqrt{t^2-b^2})\,\Delta(t-b)$
75	$\dfrac{e^{-b(\sqrt{s^2+a^2}-s)}}{\sqrt{s^2+a^2}}\left(\dfrac{a}{s+\sqrt{s^2+a^2}}\right)^k$	$k > -1$	$\dfrac{t^{k/2}}{(t+2b)^{k/2}} J_k(a\sqrt{t^2+2bt})$
76	$e^{a^2s^2}\,\mathrm{erfc}\,as$	$a > 0$	$\dfrac{1}{a\sqrt{\pi}}\exp\left(-\dfrac{t^2}{4a^2}\right)$
77	$\dfrac{1}{s}e^{a^2s^2}\,\mathrm{erfc}\,as$	$a > 0$	$\mathrm{erf}\,\dfrac{t}{2a}$
78	$e^{as}\,\mathrm{erfc}\,\sqrt{as}$	$a > 0$	$\sqrt{a}/\pi(t+a)\sqrt{t}$
79	$\dfrac{1}{\sqrt{s}}\,\mathrm{erfc}\,\sqrt{as}$	$a \geq 0$	$(\pi t)^{-1/2}\,\Delta(t-a)$
80	$\dfrac{1}{\sqrt{s}}e^{as}\,\mathrm{erfc}\,\sqrt{as}$	$a > 0$	$\dfrac{1}{\sqrt{\pi(t+a)}}$

Example 5. Solve $6y(t) + 5y(t+1) + y(t+2) = 0$, when $y_0 = 0$, $y_1 = 1$. From (19.75), we have

$$y(t) = L^{-1}\left\{\frac{e^s - 1}{s(e^{2s} + 5e^s + 6)}\right\}$$

$$= L^{-1}\left\{\frac{e^s - 1}{s(e^s + 2)} - \frac{e^s - 1}{s(e^s + 3)}\right\} = (-2)^{[t]} - (-3)^{[t]}$$

The solution of the corresponding equation,

$$6y_t + 5y_{t+1} + y_{t+2} = 0$$

is, therefore,

$$y_t = (-2)^t - (-3)^t \qquad t = 0, 1, 2, \ldots$$

Example 6. Find the solution of

$$10y(t) + 3y(t+1) - y(t+2) = 3^{[t]}$$

subject to the conditions $y_0 = y_1 = 0$.

Equation (19.75) now becomes

$$\begin{aligned} y(t) &= L^{-1}\left\{\frac{-(e^s-1)}{s(e^s-3)(e^{2s}-3e^s-10)}\right\} \\ &= \tfrac{1}{10}L^{-1}\left\{\frac{e^s-1}{s(e^s-3)}\right\} - \tfrac{5}{70}L^{-1}\left\{\frac{e^s-1}{s(e^s-5)}\right\} - \tfrac{2}{70}L^{-1}\left\{\frac{e^s-1}{s(e^s+2)}\right\} \\ &= \tfrac{1}{10}(3)^{[t]} - \tfrac{5}{70}(5)^{[t]} - \tfrac{2}{70}(-2)^{[t]} \end{aligned}$$

The solution of

$$10y_t + 3y_{t+1} - y_{t+2} = 3^t$$

is, consequently,

$$y_t = \tfrac{1}{10}\,3^t - \tfrac{5}{70}\,5^t - \tfrac{2}{70}(-2)^t \qquad t = 0, 1, 2, \ldots$$

19.9. REFERENCES

Theoretical Aspects of Integral Transforms

[1] S. Bochner, K. Chandrasekharen: "Fourier Transforms," Princeton Univ. Press, Princeton, N.J., 1949.
[2] E. C. Titchmarsh: "Introduction to the Theory of Fourier Integrals," Oxford Univ. Press, New York, 1948.
[3] D. V. Widder: "The Laplace Transform," Princeton Univ. Press, Princeton, N.J., 1941.
[4] N. Wiener: "The Fourier Integral and Certain of Its Applications," Cambridge Univ. Press, New York, 1933.

Books Emphasizing the Application of Integral Transforms

[5] H. S. Carslaw, J. C. Jaeger: "Operational Methods in Applied Mathematics," Oxford Univ. Press, New York, 1947.
[6] R. V. Churchill: "Operational Mathematics," 2d ed., McGraw-Hill, New York, 1958.
[7] G. Doetsch: "Theorie und Anwendung der Laplace-Transformation," Springer, Berlin, 1937.
[8] N. W. McLachlan: "Complex Variable Theory and Transform Calculus," Cambridge Univ. Press, New York, 1953.
[9] E. J. Scott: "Transform Calculus with an Introduction to Complex Variables," Harper, New York, 1955.
[10] I. N. Sneddon: "Fourier Transforms," McGraw-Hill, New York, 1951.
[11] C. J. Tranter: "Integral Transforms in Mathematical Physics," Methuen, London, 1956.

Tables of Laplace Transforms

[12] G. A. Campbell, R. M. Foster: "Fourier Integrals for Practical Applications," Van Nostrand, Princeton, N.J., 1948.
[13] W. Magnus, F. Oberhettinger: "Formeln und Sätze für die speziellen Funktionen der mathematischen Physik," pp. 122–136, Springer, Berlin, 1943.
[14] G. Doetsch: "Tabellen zur Laplace-Transformation," Springer, Berlin, 1947.

CHAPTER 20

TABLES

Table 20.1. Arcs, Areas, Volumes, Centroids, Moments of Inertia

NOTATION. In this table the following notation is used:

l = arc length

A = area of a plane figure

area of the *curved* part of the surface of a solid

V = volume

I_x = moment of inertia of a curve, a plane figure, or a solid with respect to the x axis.

All moments of inertia are geometric moments of inertia. To obtain mass moments of inertia, multiply by mass per unit length, mass per unit area, or mass density. All angles are understood to be measured in radians.

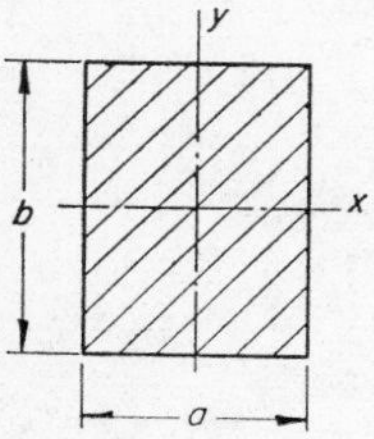

FIG. 20.1

Rectangle

$$A = ab \qquad I_x = \tfrac{1}{12}ab^3 \qquad I_y = \tfrac{1}{12}a^3b$$

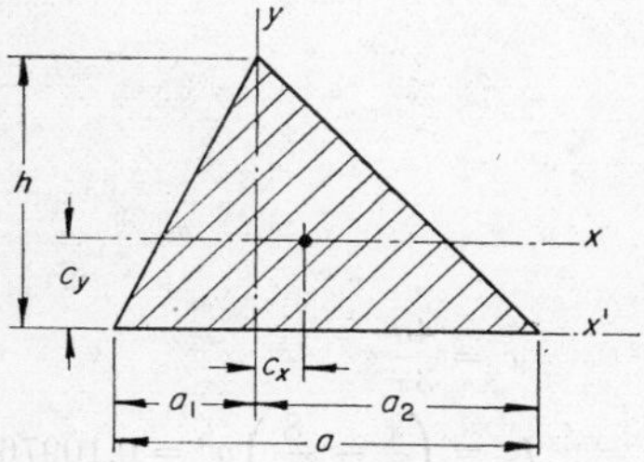

FIG. 20.2

Triangle

$$A = \tfrac{1}{2}ah \qquad c_x = \tfrac{1}{3}(a_2 - a_1) \qquad c_y = \tfrac{1}{3}h$$

$$I_x = \frac{ah^3}{36} \qquad I_{x'} = \frac{ah^3}{12}$$

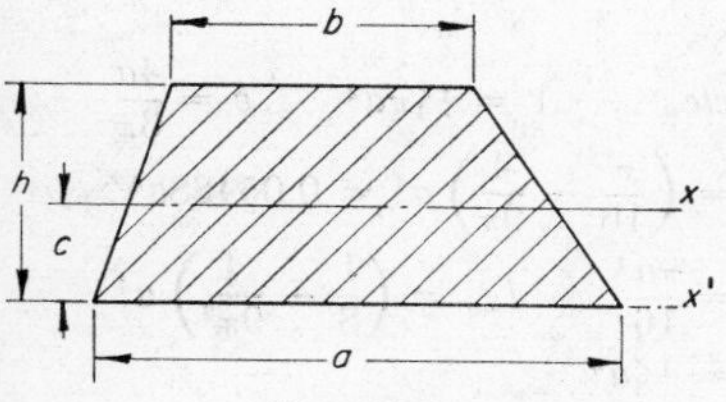

FIG. 20.3

Trapezoid

$$A = \tfrac{1}{2}(a + b)h \qquad c = \frac{h\ (a + 2b)}{3\ (a + b\ )}$$

$$I_x = \frac{(a^2 + 4ab + b^2)h^3}{36(a + b)} \qquad I_{x'} = \tfrac{1}{12}(a + 3b)h^3$$

Table 20.1. Arcs, Areas, Volumes, Centroids, Moments of Inertia (*Continued*)

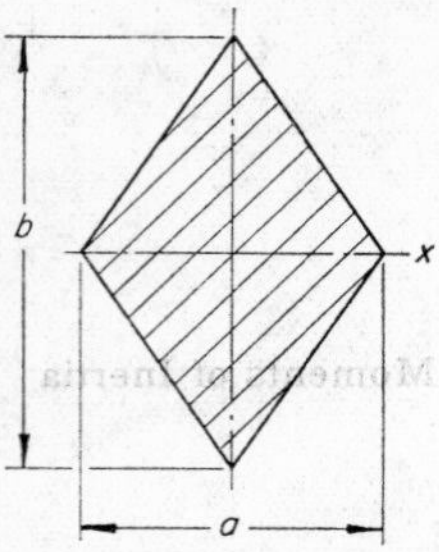

FIG. 20.4

Rhombus $A = \frac{1}{2}ab$ $I_x = \frac{ab^3}{48}$

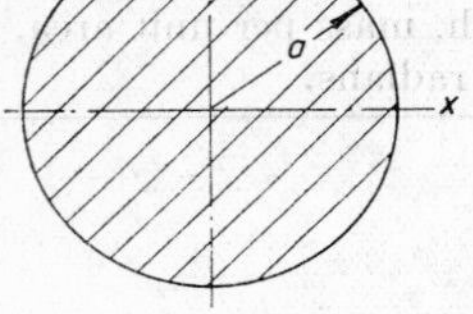

FIG. 20.5

Circle *Arc:* $l = 2\pi a$ $I_x = \pi a^3$

Area: $A = \pi a^2$ $I_x = \frac{1}{4}\pi a^4$

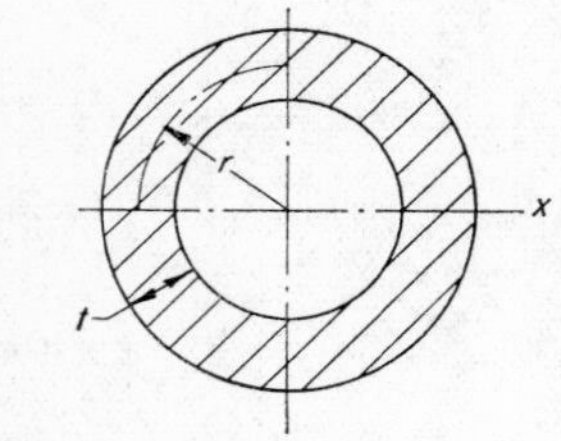

FIG. 20.6

Annulus $A = 2\pi rt$ $I_x = \pi r^3 t\left(1 + \frac{t^2}{4r^2}\right)$

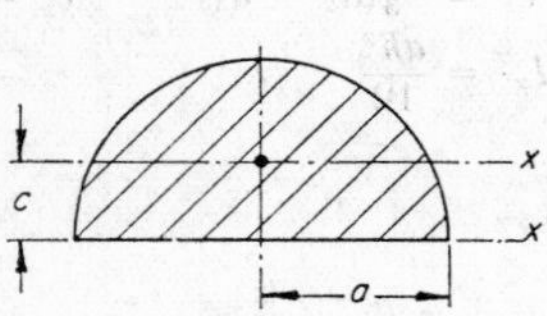

FIG. 20.7

Semicircle

$A = \frac{1}{2}\pi a^2$ $c = \frac{4a}{3\pi}$

$I_{x'} = \frac{\pi a^4}{8}$ $I_x = \left(\frac{\pi}{8} - \frac{8}{9\pi}\right)a^4 = 0.10976a^4$

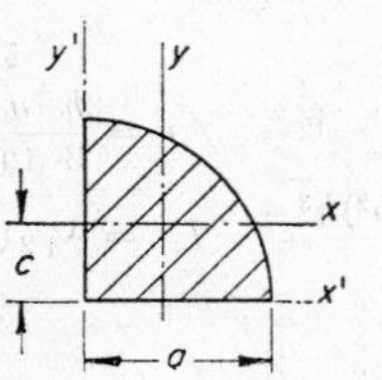

FIG. 20.8

Quarter Circle $A = \frac{1}{4}\pi a^2$ $c = \frac{4a}{3\pi}$

$I_x = \left(\frac{\pi}{16} - \frac{4}{9\pi}\right)a^4 = 0.05488a^4$

$I_{x'} = \frac{\pi a^4}{16}$ $I_{xy} = \left(\frac{1}{8} - \frac{4}{9\pi^2}\right)a^4$

$I_{x'y'} = \frac{1}{8}a^4$

Table 20.1. Arcs, Areas, Volumes, Centroids, Moments of Inertia *(Continued)*

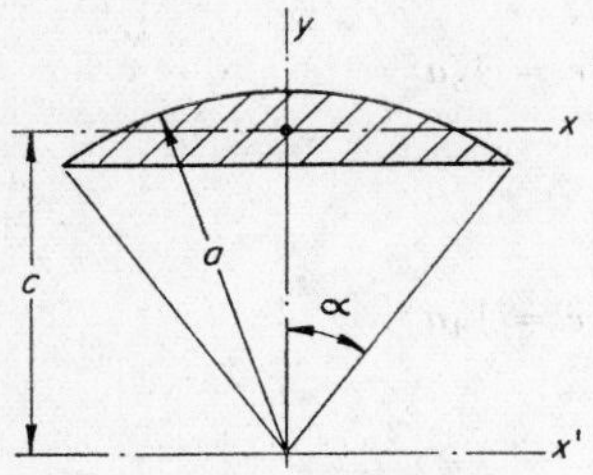

FIG. 20.9

Segment of a Circle

Arc: $l = 2\alpha a \qquad c = a\,\dfrac{\sin\alpha}{\alpha}$

$$I_x = a^3\left(\alpha + \tfrac{1}{2}\sin 2\alpha - 2\,\frac{\sin^2\alpha}{\alpha}\right)$$

$$I_{x'} = a^3(\alpha + \tfrac{1}{2}\sin 2\alpha)$$

$$I_y = a^3(\alpha - \tfrac{1}{2}\sin 2\alpha)$$

Area:

$$A = \tfrac{1}{2}a^2(2\alpha - \sin 2\alpha) \qquad c = \tfrac{4}{3}a\,\frac{\sin^3\alpha}{2\alpha - \sin 2\alpha}$$

$$I_{x'} = \frac{a^4}{24}\,[6\alpha - \sin 2\alpha(4 - \cos 2\alpha)]$$

$$I_y = \tfrac{1}{4}a^4(\alpha - \tfrac{1}{2}\sin 2\alpha - \tfrac{1}{3}\sin 2\alpha\,\sin^2\alpha)$$

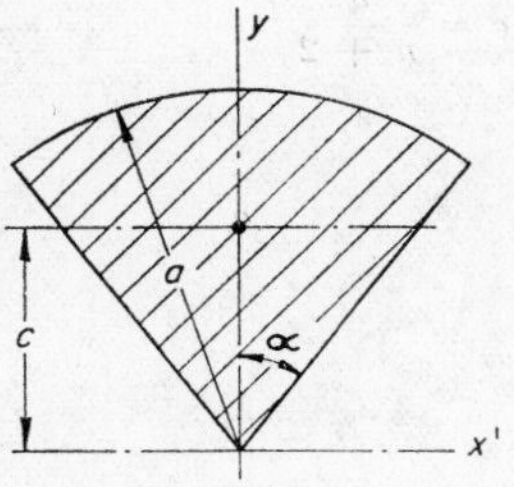

FIG. 20.10

Sector of a Circle $\quad A = \alpha a^2 \qquad c = \dfrac{2a}{3}\,\dfrac{\sin\alpha}{\alpha}$

$$I_{x'} = \tfrac{1}{4}a^4(\alpha + \tfrac{1}{2}\sin 2\alpha)$$

$$I_y = \tfrac{1}{4}a^4(\alpha - \tfrac{1}{2}\sin 2\alpha)$$

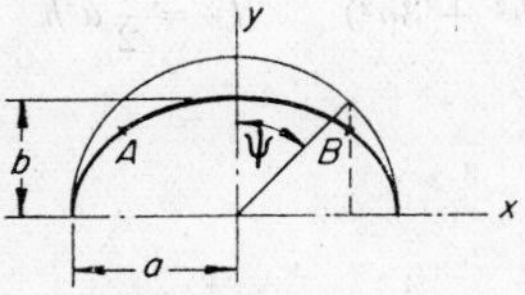

FIG. 20.11

Ellipse

Arc AB: $l = 2aE(k,\psi) \qquad k = e/a,$

$e^2 = a^2 - b^2 \qquad e =$ focal distance

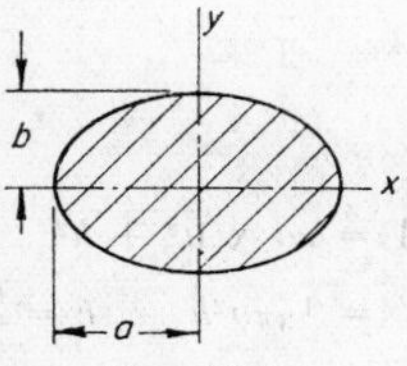

FIG. 20.12

Area: $A = \pi ab \qquad I_x = \tfrac{1}{4}\pi ab^3 \qquad I_y = \tfrac{1}{4}\pi a^3 b$

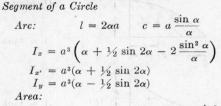

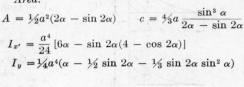

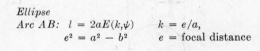

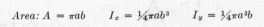

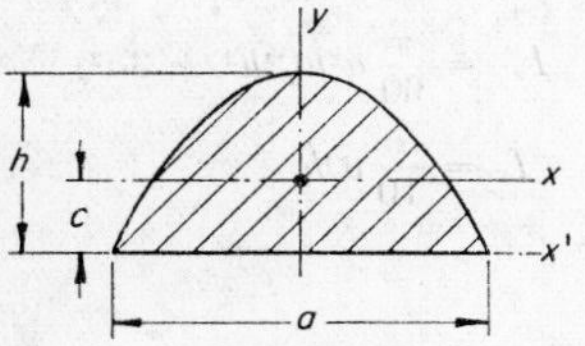

FIG. 20.13

Parabola

Arc: $l = \tfrac{1}{2}\sqrt{a^2 + 4h^2} + \dfrac{a^2}{4h}\ln\left(\dfrac{2h}{a} + \dfrac{1}{a}\sqrt{a^2 + 4h^2}\right)$

Area: $A = \tfrac{2}{3}ah \qquad c = \tfrac{2}{5}h$

$I_x = \tfrac{8}{175}ah^3 \qquad I_{x'} = \tfrac{16}{105}ah^3$

$I_y = \tfrac{1}{30}a^3h$

Table 20.1. Arcs, Areas, Volumes, Centroids, Moments of Inertia (*Continued*)

Quadratic Parabolas

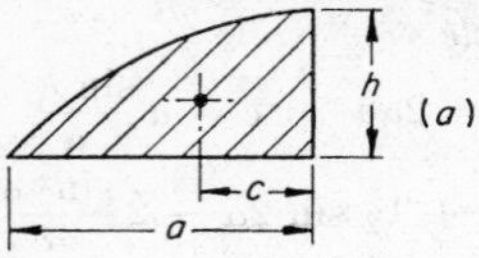

$A = \frac{2}{3}ah \qquad c = \frac{3}{8}a$

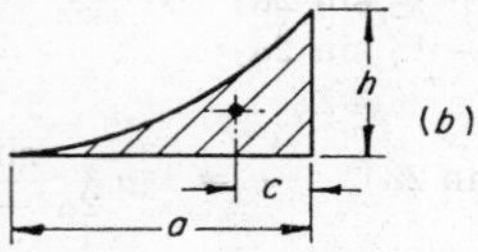

$A = \frac{1}{3}ah \qquad c = \frac{1}{4}a$

Parabola of the nth Degree

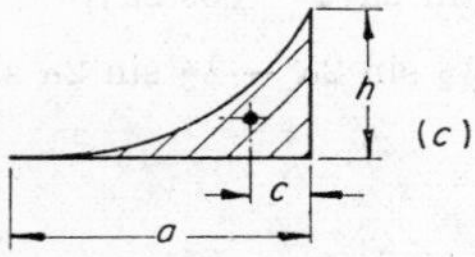

$$A = \frac{ah}{n+1} \qquad c = \frac{a}{n+2}$$

Fig. 20.14

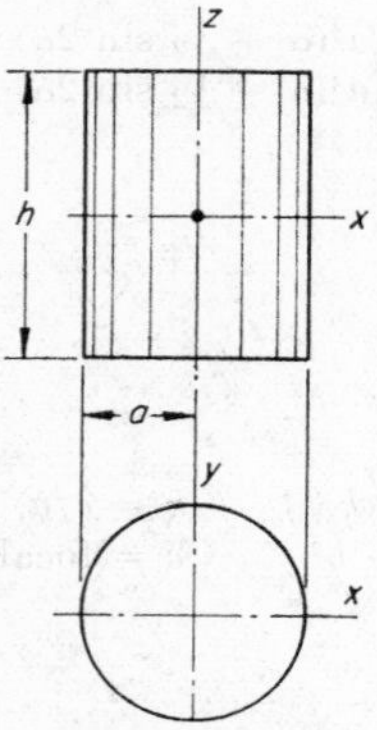

Fig. 20.15

Circular Cylinder $\quad A = 2\pi ah \qquad V = \pi a^2 h$

$$I_x = \frac{\pi}{12} a^2 h(h^2 + 3a^2) \qquad I_z = \frac{\pi}{2} a^4 h$$

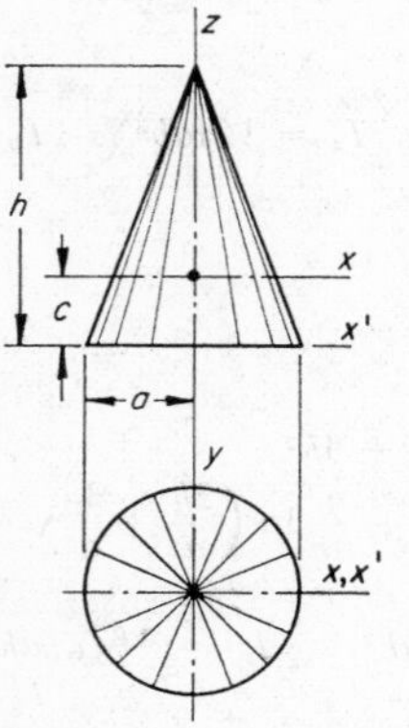

Fig. 20.16

Circular Cone $\quad A = \pi a \sqrt{a^2 + h^2}$

$$V = \frac{1}{3}\pi a^2 h \qquad c = \frac{h}{4}$$

$$I_x = \frac{\pi}{80} a^2 h(h^2 + 4a^2)$$

$$I_{x'} = \frac{\pi}{60} a^2 h(2h^2 + 3a^2)$$

$$I_z = \frac{\pi}{10} a^4 h$$

Table 20.1. Arcs, Areas, Volumes, Centroids, Moments of Inertia *(Continued)*

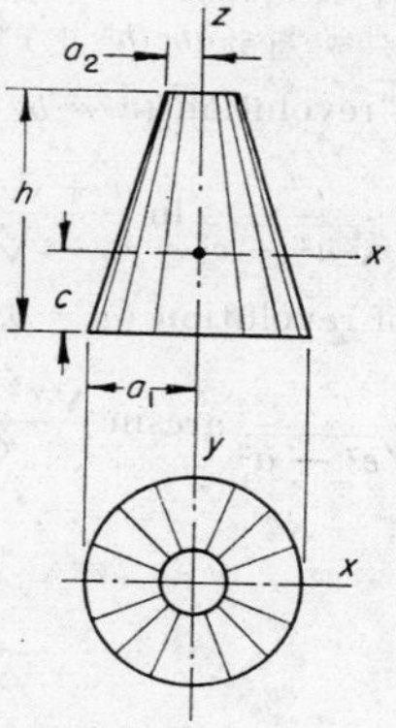

FIG. 20.17

Truncated Cone

$$A = \pi(a_1 + a_2)\sqrt{(a_1 - a_2)^2 + h^2}$$

$$V = \frac{\pi}{3} h(a_1^2 + a_1 a_2 + a_2^2)$$

$$c = \frac{h}{4} \frac{a_1^2 + 2a_1 a_2 + 3a_2^2}{a_1^2 + a_1 a_2 + a_2^2}$$

$$I_z = \frac{\pi}{10} h \frac{a_1^5 - a_2^5}{a_1 - a_2}$$

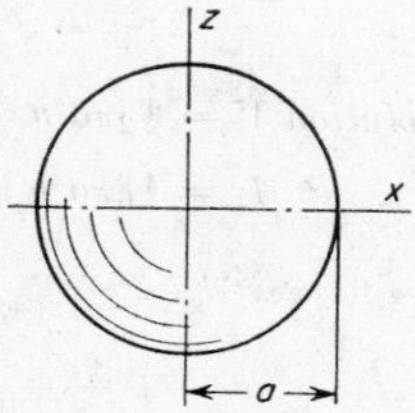

FIG. 20.18

Sphere $A = 4\pi a^2$ $V = \tfrac{4}{3}\pi a^3$

$$I_x = \frac{8\pi a^5}{15}$$

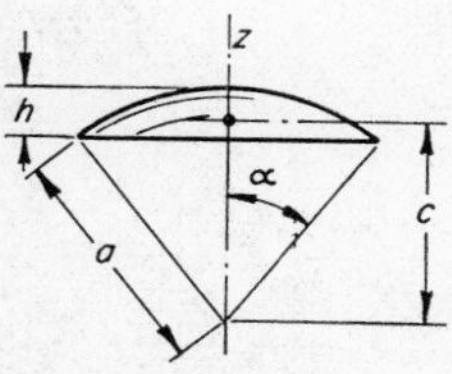

FIG. 20.19

Spherical Cap

$$A = 2\pi a h$$

$$V = \tfrac{1}{3}\pi h^2(3a - h)$$

$$c = \frac{3(2a - h)^2}{4(3a - h)}$$

$$I_z = \frac{\pi a^5}{60}\left[8 - (8 + 4\sin^2\alpha + 3\sin^4\alpha)\cos\alpha\right]$$

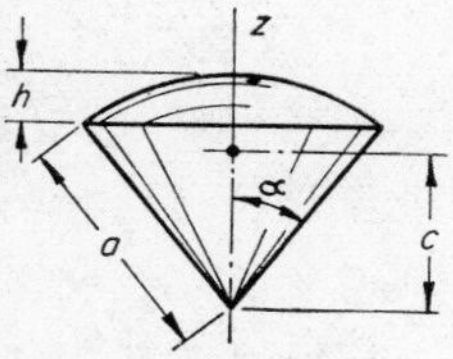

FIG. 20.20

Spherical Sector

$$V = \tfrac{2}{3}\pi a^3(1 - \cos\alpha) = \tfrac{2}{3}\pi a^2 h$$

$$c = \tfrac{3}{8}a(1 + \cos\alpha)$$

$$I_z = \frac{\pi a^5}{60}\left[8 - (8 + 4\sin^2\alpha - 3\sin^4\alpha)\cos\alpha\right]$$

Table 20.1. Arcs, Areas, Volumes, Centroids, Moments of Inertia *(Continued)*

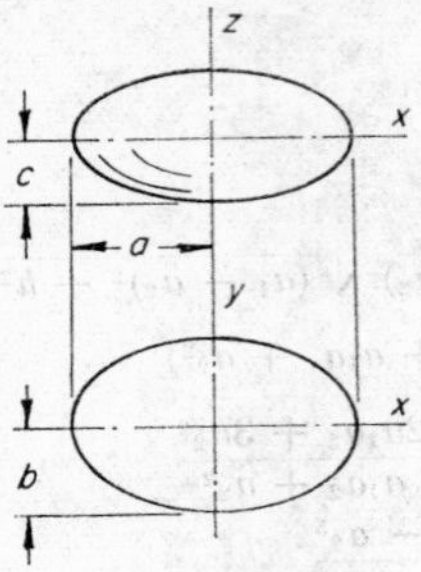

Fig. 20.21

Ellipsoid $V = \frac{4}{3}\pi abc$

$I_x = \frac{4}{15}\pi abc(b^2 + c^2)$

Oblate ellipsoid of revolution ($a = b$, $c < a$):

$$A = 2\pi a\left(a + \frac{c^2}{2\sqrt{a^2 - c^2}} \ln \frac{a + \sqrt{a^2 + c^2}}{a - \sqrt{a^2 - c^2}}\right)$$

Prolate ellipsoid of revolution ($a = b$, $c > a$):

$$A = 2\pi a\left(a + \frac{c^2}{\sqrt{c^2 - a^2}} \arcsin \frac{\sqrt{c^2 - a^2}}{c}\right)$$

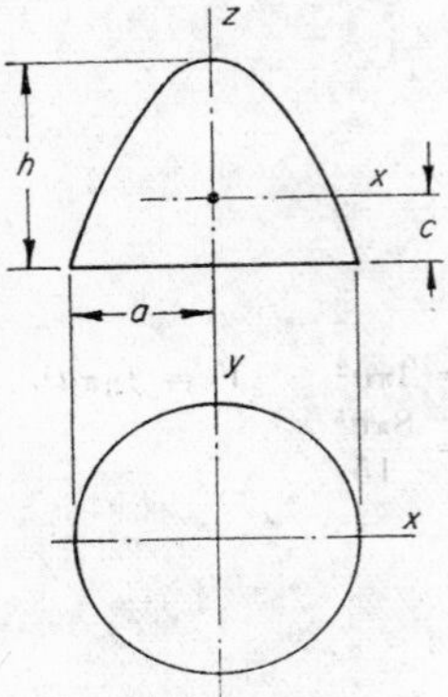

Fig. 20.22

Paraboloid of Revolution $V = \frac{1}{2}\pi a^2 h$ $\quad c = \dfrac{h}{3}$

$I_z = \frac{1}{6}\pi a^4 h$

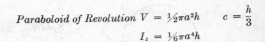

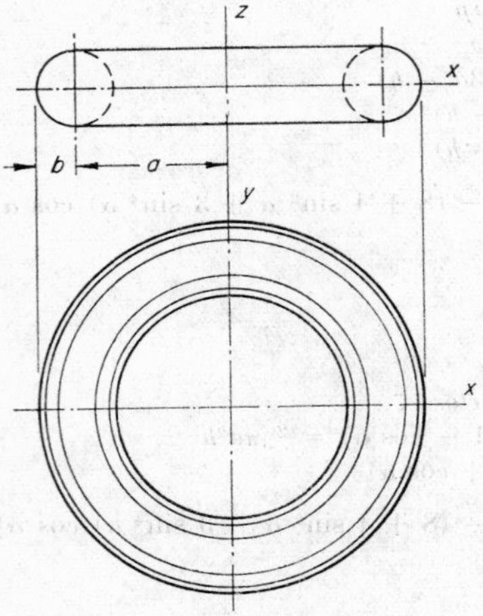

Fig. 20.23

Tore $A = 4\pi^2 ab$ $\quad V = 2\pi^2 ab^2$

$I_x = \frac{1}{4}\pi^2 ab^2(4a^2 + 5b^2)$

$I_z = \frac{1}{2}\pi^2 ab^2(4a^2 + 3b^2)$

Table 20.2. Integrals

$$\int x^n\,dx = \frac{1}{n+1}\,x^{n+1} \qquad n \neq -1 \tag{20.1}$$

$$\int \frac{dx}{x} = \ln x \tag{20.2}$$

$$\int \frac{dx}{a+bx^2} = \frac{1}{\sqrt{ab}}\tan^{-1}\left(\sqrt{\frac{b}{a}}\,x\right) \qquad a>0,\ b>0 \tag{20.3}$$

$$\int \frac{dx}{a-bx^2} = \frac{1}{2\sqrt{ab}}\ln\frac{\sqrt{a}+x\sqrt{b}}{\sqrt{a}-x\sqrt{b}} \qquad a>0,\ b>0 \tag{20.4}$$

$$\int \frac{dx}{(a+bx^2)^2} = \frac{1}{2a}\left(\frac{x}{a+bx^2}+\int\frac{dx}{a+bx^2}\right) \tag{20.5}$$

$$\begin{aligned}\int \frac{dx}{a+2bx+cx^2} &= \frac{1}{\sqrt{ac-b^2}}\tan^{-1}\frac{b+cx}{\sqrt{ac-b^2}} && ac-b^2>0\\ &= \frac{1}{2\sqrt{b^2-ac}}\ln\frac{\sqrt{b^2-ac}-b-cx}{\sqrt{b^2-ac}+b+cx} && ac-b^2<0\\ &= -\frac{1}{b+cx} && ac-b^2=0\end{aligned} \tag{20.6}$$

$$\int \frac{(\alpha+\beta x)\,dx}{a+2bx+cx^2} = \frac{\beta}{2c}\ln(a+2bx+cx^2) + \frac{\alpha c-\beta b}{c}\int\frac{dx}{a+2bx+cx^2} \tag{20.7}$$

$$\begin{aligned}\int \frac{dx}{(a+2bx+cx^2)^n} &= \frac{1}{2(n-1)(ac-b^2)}\,\frac{b+cx}{(a+2bx+cx^2)^{n-1}}\\ &\quad + \frac{(2n-3)c}{2(n-1)(ac-b^2)}\int\frac{dx}{(a+2bx+cx^2)^{n-1}}\end{aligned} \tag{20.8}$$

$$\begin{aligned}\int \frac{(\alpha+\beta x)dx}{(a+2bx+cx^2)^n} &= -\frac{\beta}{2(n-1)c}\,\frac{1}{(a+2bx+cx^2)^{n-1}}\\ &\quad + \frac{\alpha c-\beta b}{c}\int\frac{dx}{(a+2bx+cx^2)^n}\end{aligned} \tag{20.9}$$

$$\int \sqrt{a^2+x^2}\,dx = \frac{x}{2}\sqrt{a^2+x^2} + \frac{a^2}{2}\ln\left(x+\sqrt{a^2+x^2}\right) \tag{20.10}$$

$$\int \sqrt{x^2-a^2}\,dx = \frac{x}{2}\sqrt{x^2-a^2} - \frac{a^2}{2}\ln\left(x+\sqrt{x^2-a^2}\right) \tag{20.11}$$

$$\int \sqrt{a^2-x^2}\,dx = \frac{x}{2}\sqrt{a^2-x^2} + \frac{a^2}{2}\sin^{-1}\frac{x}{a} \tag{20.12}$$

$$\begin{aligned}\int \sqrt{a+2bx+cx^2}\,dx &= \frac{b+cx}{2c}\sqrt{a+2bx+cx^2}\\ &\quad + \frac{ac-b^2}{2c}\int\frac{dx}{\sqrt{a+2bx+cx^2}}\end{aligned} \tag{20.13}$$

$$\int \frac{(\alpha+\beta x)\,dx}{\sqrt{a+bx}} = \frac{2}{3b^2}(3\alpha b-2\beta a+\beta bx)\sqrt{a+bx} \tag{20.14}$$

$$\int \frac{dx}{x\sqrt{a+bx}} = \frac{1}{\sqrt{a}}\ln\frac{\sqrt{a+bx}-\sqrt{a}}{\sqrt{a+bx}+\sqrt{a}} \tag{20.15}$$

$$\int \frac{dx}{\sqrt{a^2+x^2}} = \ln\left(x+\sqrt{a^2+x^2}\right) \tag{20.16}$$

$$\int \frac{dx}{\sqrt{x^2-a^2}} = \ln\left(x+\sqrt{x^2-a^2}\right) \tag{20.17}$$

$$\int \frac{dx}{\sqrt{a^2-x^2}} = \sin^{-1}\frac{x}{a} \tag{20.18}$$

Table 20.2. Integrals (*Continued*)

In Eqs. (20.19) through (20.22) the abbreviation

$$X = \sqrt{a + 2bx + cx^2}$$

has been used.

$$\int \frac{dx}{X} = \frac{1}{\sqrt{c}} \ln (b + cx + \sqrt{c}\, X) \qquad c > 0$$
$$= -\frac{1}{\sqrt{-c}} \sin^{-1} \frac{b + cx}{\sqrt{b^2 - ac}} \qquad c < 0 \tag{20.19}$$

$$\int \frac{(\alpha + \beta x)\, dx}{X} = \frac{\beta X}{c} + \frac{\alpha c - \beta b}{c} \int \frac{dx}{X} \tag{20.20}$$

$$\int \frac{x^n\, dx}{X} = \frac{x^{n-1} X}{nc} - \frac{(2n-1)b}{nc} \int \frac{x^{n-1}\, dx}{X} - \frac{(n-1)a}{nc} \int \frac{x^{n-2}\, dx}{X} \tag{20.21}$$

$$\int \frac{dx}{xX} = -\frac{1}{\sqrt{a}} \ln \left(\frac{X + \sqrt{a}}{x\sqrt{c}} + \frac{b}{\sqrt{ac}} \right) \tag{20.22}$$

$$\int a^x\, dx = \frac{a^x}{\ln a} \tag{20.23}$$

$$\int x^n e^{\alpha x}\, dx = \frac{1}{\alpha} e^{\alpha x} \left[x^n - \frac{n}{\alpha} x^{n-1} + \frac{n(n-1)}{\alpha^2} x^{n-2} - + \cdots \pm \frac{n!}{\alpha^n} \right] \tag{20.24}$$

$$\int \ln x\, dx = x \ln x - x \tag{20.25}$$

$$\int x^n \ln x\, dx = \frac{x^{n+1}}{n+1} \ln x - \frac{x^{n+1}}{(n+1)^2} \qquad n \neq -1 \tag{20.26}$$

$$\int x^{-1} \ln x\, dx = \tfrac{1}{2}(\ln x)^2 \tag{20.27}$$

$$\int \frac{(\ln x)^n}{x}\, dx = \frac{1}{n+1} (\ln x)^{n+1} \tag{20.28}$$

$$\int \sin x\, dx = -\cos x \qquad \int \cos x\, dx = \sin x \tag{20.29a,b}$$

$$\int \tan x\, dx = -\ln \cos x \qquad \int \cot x\, dx = \ln \sin x \tag{20.30a,b}$$

$$\int \sin^2 x\, dx = -\tfrac{1}{4} \sin 2x + \tfrac{1}{2} x$$
$$\int \cos^2 x\, dx = \tfrac{1}{4} \sin 2x + \tfrac{1}{2} x \tag{20.31a,b}$$

$$\int \tan^2 x\, dx = \tan x - x \qquad \int \cot^2 x\, dx = -\cot x - x \tag{20.32a,b}$$

$$\int \sin^n x\, dx = -\frac{1}{n} \sin^{n-1} x \cos x + \frac{n-1}{n} \int \sin^{n-2} x\, dx$$
$$\int \cos^n x\, dx = \frac{1}{n} \cos^{n-1} x \sin x + \frac{n-1}{n} \int \cos^{n-2} x\, dx \tag{20.33a,b}$$

$$\int \tan^n x\, dx = \frac{1}{n-1} \tan^{n-1} x - \int \tan^{n-2} x\, dx$$
$$\int \cot^n x\, dx = -\frac{1}{n-1} \cot^{n-1} x - \int \cot^{n-2} x\, dx \tag{20.34a,b}$$

$$\int \frac{dx}{\sin x} = \ln \tan \frac{x}{2} \qquad \int \frac{dx}{\cos x} = \ln \tan \left(\frac{x}{2} + \frac{\pi}{4} \right) \tag{20.35a,b}$$

$$\int \frac{dx}{\sin^2 x} = -\cot x \qquad \int \frac{dx}{\cos^2 x} = \tan x \qquad \int \frac{dx}{\sin x \cos x} = \ln \tan x \tag{20.36a–c}$$

$$\int \frac{dx}{\sin^n x} = -\frac{\cos x}{(n-1)\sin^{n-1} x} + \frac{n-2}{n-1} \int \frac{dx}{\sin^{n-2} x}$$
$$\int \frac{dx}{\cos^n x} = \frac{\sin x}{(n-1)\cos^{n-1} x} + \frac{n-2}{n-1} \int \frac{dx}{\cos^{n-2} x} \qquad n \neq 1 \tag{20.37a,b}$$

Table 20.2. Integrals (*Continued*)

$$\int \sin^m x \cos^n x \, dx = \frac{\sin^{m+1} x \cos^{n-1} x}{m+n} + \frac{n-1}{m+n} \int \sin^m x \cos^{n-2} x \, dx$$

$$= -\frac{\sin^{m-1} x \cos^{n+1} x}{m+n} + \frac{m-1}{m+n} \int \sin^{m-2} x \cos^n x \, dx \qquad (20.38)$$

$$\int \frac{\cos^n x}{\sin^m x} dx = -\frac{\cos^{n+1} x}{(m-1) \sin^{m-1} x} + \frac{m-n-2}{m-1} \int \frac{\cos^n x}{\sin^{m-2} x} dx$$

$$\int \frac{\sin^m x}{\cos^n x} dx = \frac{\sin^{m+1} x}{(n-1) \cos^{n-1} x} + \frac{n-m-2}{n-1} \int \frac{\sin^m x}{\cos^{n-2} x} dx \qquad (20.39a,b)$$

$$\int x \sin x \, dx = \sin x - x \cos x$$

$$\int x \cos x \, dx = \cos x + x \sin x \qquad (20.40a,b)$$

$$\int x^2 \sin x \, dx = 2x \sin x - (x^2 - 2) \cos x$$

$$\int x^2 \cos x \, dx = 2x \cos x + (x^2 - 2) \sin x \qquad (20.41a,b)$$

$$\int x^3 \sin x \, dx = 3(x^2 - 2) \sin x - x(x^2 - 6) \cos x$$

$$\int x^3 \cos x \, dx = 3(x^2 - 2) \cos x + x(x^2 - 6) \sin x \qquad (20.42a,b)$$

$$\int x^4 \sin x \, dx = 4x(x^2 - 6) \sin x - (x^4 - 12x^2 + 24) \cos x$$

$$\int x^4 \cos x \, dx = 4x(x^2 - 6) \cos x + (x^4 - 12x^2 + 24) \sin x \qquad (20.43a,b)$$

$$\int x^{-1} \sin x \, dx = \operatorname{Si} x = \frac{\pi}{2} + \operatorname{si} x$$

(see p. 15-13)

$$\int x^{-1} \cos x \, dx = \operatorname{Ci} x \qquad (20.44a,b)$$

$$\int \frac{dx}{1 + \sin x} = \tan\left(\frac{x}{2} - \frac{\pi}{4}\right) \qquad \int \frac{dx}{1 - \sin x} = -\cot\left(\frac{x}{2} - \frac{\pi}{4}\right) \qquad (20.45a,b)$$

$$\int \frac{dx}{1 + \cos x} = \tan \frac{x}{2} \qquad \int \frac{dx}{1 - \cos x} = -\cot \frac{x}{2} \qquad (20.46a,b)$$

$$\int \frac{dx}{a + b \cos x} = \frac{2}{\sqrt{a^2 - b^2}} \tan^{-1}\left(\sqrt{\frac{a-b}{a+b}} \tan \frac{x}{2}\right) \qquad a^2 > b^2$$

$$= \frac{1}{\sqrt{a^2 - b^2}} \tan^{-1} \frac{\sqrt{a^2 - b^2} \sin x}{b + a \cos x} \qquad a^2 > b^2 \qquad (20.47)$$

$$= \frac{1}{\sqrt{b^2 - a^2}} \ln \frac{b + a \cos x + \sqrt{b^2 - a^2} \sin x}{a + b \cos x} \qquad a^2 < b^2$$

$$\int \frac{\cos x \, dx}{a + b \cos x} = \frac{x}{b} - \frac{a}{b} \int \frac{dx}{a + b \cos x} \qquad (20.48)$$

$$\int \cos mx \cos nx \, dx = \frac{\sin (m-n)x}{2(m-n)} + \frac{\sin (m+n)x}{2(m+n)}$$

$$\int \sin mx \sin nx \, dx = \frac{\sin (m-n)x}{2(m-n)} - \frac{\sin (m+n)x}{2(m+n)} \qquad (20.49a\text{–}c)$$

$$\int \cos mx \sin nx \, dx = \frac{\cos (m-n)x}{2(m-n)} - \frac{\cos (m+n)x}{2(m+n)}$$

$$\int e^{mx} \sin nx \, dx = \frac{m \sin nx - n \cos nx}{m^2 + n^2} e^{mx}$$

$$\int e^{mx} \cos nx \, dx = \frac{m \cos nx + n \sin nx}{m^2 + n^2} e^{mx} \qquad (20.50a,b)$$

Table 20.3. Fourier Series

In the following diagrams the shaded part represents the period of length $2l$ of a function $y = f(x)$. The coefficients A_n and B_n are defined by the Fourier expansion

$$f(x) = \sum_{n=0}^{\infty} A_n \cos \frac{n\pi x}{l} + \sum_{n=1}^{\infty} B_n \sin \frac{n\pi x}{l}$$

For further details see p. 15-2.

A. Piecewise Constant Functions

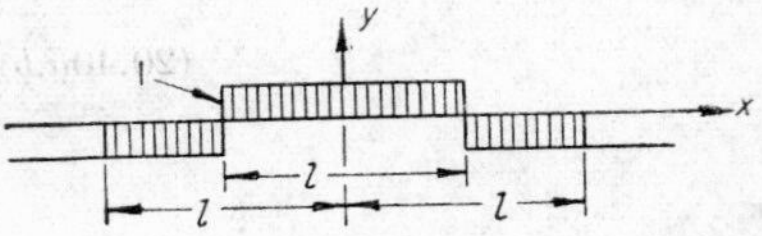

FIG. 20.24

$$A_n = 0 \qquad n = 0, 2, 4, 6, \ldots$$
$$A_n = +\frac{4}{\pi n} \qquad n = 1, 5, 9, \ldots$$
$$A_n = -\frac{4}{\pi n} \qquad n = 3, 7, 11, \ldots$$
$$B_n = 0$$

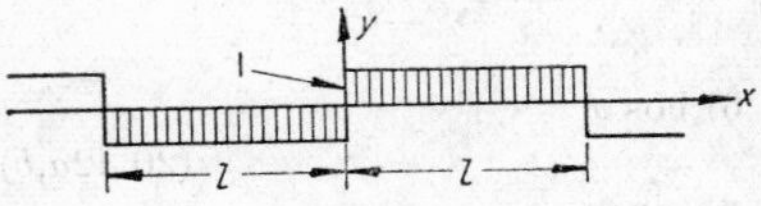

FIG. 20.25

$$A_n = 0$$
$$B_n = 0 \qquad n = \text{even}$$
$$B_n = \frac{4}{\pi n} \qquad n = \text{odd}$$

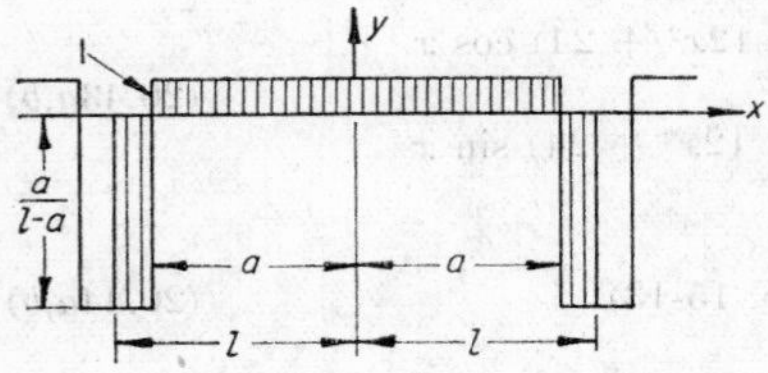

FIG. 20.26

$$A_0 = 0$$
$$A_n = \frac{2l}{(l-a)\pi n} \sin \frac{n\pi a}{l} \qquad n = 1, 2, 3, \ldots$$
$$B_n = 0$$

Limiting case[1] $a = l$:

$$A_0 = 0 \qquad A_n = +2 \qquad n = 1, 3, 5, \ldots$$
$$A_n = -2 \qquad n = 2, 4, 6, \ldots$$

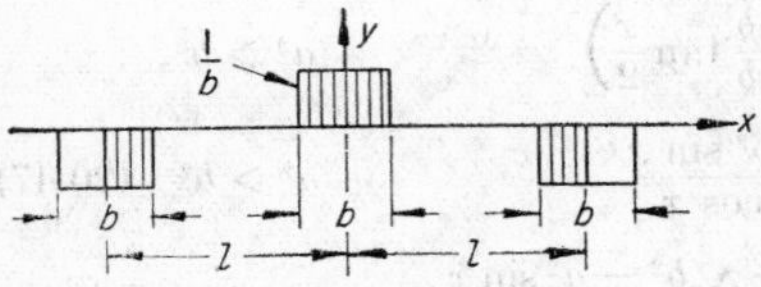

FIG. 20.27

$$A_n = 0 \qquad n = \text{even}$$
$$A_n = \frac{4}{\pi b n} \sin \frac{n\pi b}{2l} \qquad n = \text{odd}$$
$$B_n = 0$$

Limiting case[1] $b = 0$:
$$A_n = \frac{2}{l} \qquad n = \text{odd}$$

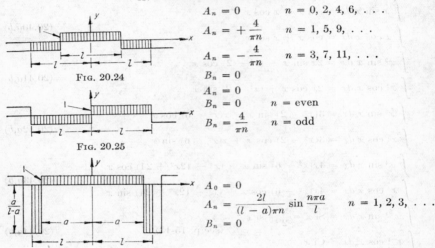

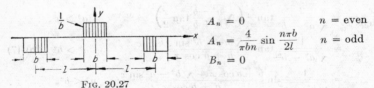

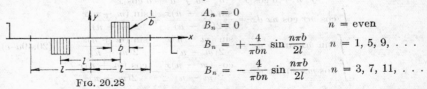

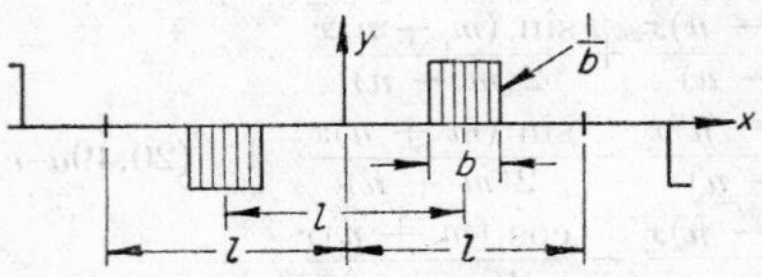

FIG. 20.28

$$A_n = 0$$
$$B_n = 0 \qquad n = \text{even}$$
$$B_n = +\frac{4}{\pi b n} \sin \frac{n\pi b}{2l} \qquad n = 1, 5, 9, \ldots$$
$$B_n = -\frac{4}{\pi b n} \sin \frac{n\pi b}{2l} \qquad n = 3, 7, 11, \ldots$$

---

[1] In this limiting case the series represents a singular function. The series is divergent, but summable.

Table 20.3. Fourier Series (*Continued*)

Limiting case[1] $b = 0$: $B_n = +\frac{2}{l}$ $n = 1, 5, 9, \ldots$

$B_n = -\frac{2}{l}$ $n = 3, 7, 11, \ldots$

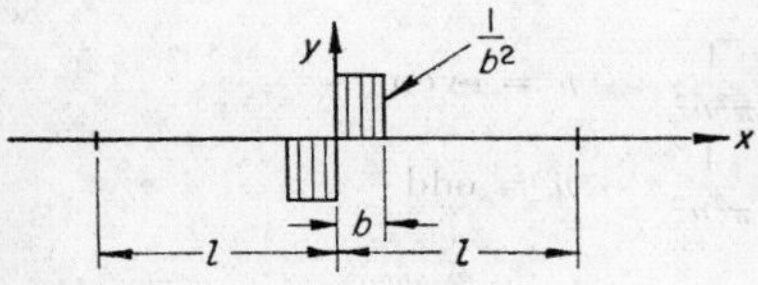

FIG. 20.29

$A_n = 0 \qquad B_n = \frac{2}{\pi b^2 n}\left(1 - \cos\frac{n\pi b}{l}\right)$

Limiting case[1] $b = 0$:

$B_n = \frac{\pi n}{l^2}$

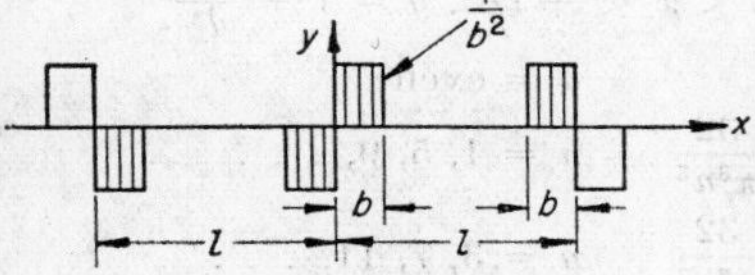

FIG. 20.30

$A_n = 0$

$B_n = 0$ $n = \text{even}$

$B_n = \frac{4}{\pi b^2 n}\left(1 - \cos\frac{n\pi b}{l}\right)$ $n = \text{odd}$

Limiting case[1] $b = 0$: $B_n = \frac{2\pi n}{l}$ $n = \text{odd}$

B. Sawtooth Curves

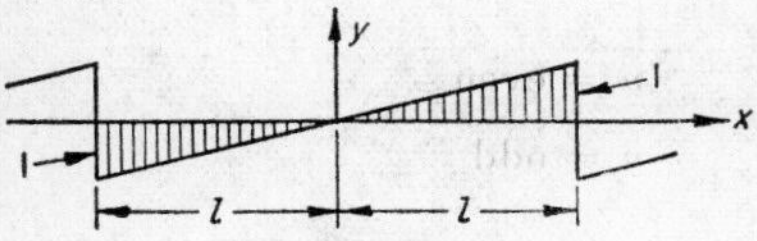

FIG. 20.31

$A_n = 0$

$B_n = +\frac{2}{\pi n}$ $n = \text{odd}$

$B_n = -\frac{2}{\pi n}$ $n = \text{even}$

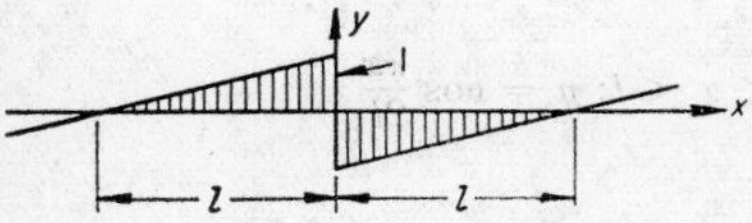

FIG. 20.32

$A_n = 0 \qquad B_n = -\frac{2}{\pi n}$

C. Zigzag Curves

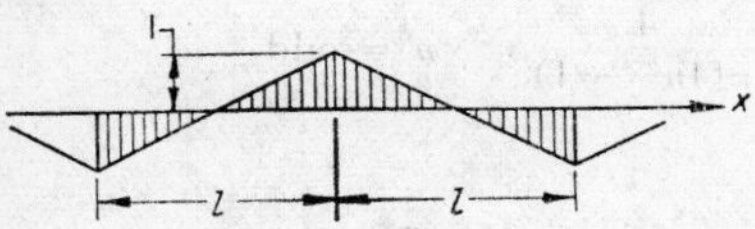

FIG. 20.33

$A_n = 0$ $n = \text{even}$

$A_n = \frac{8}{\pi^2 n^2}$ $n = \text{odd}$

$B_n = 0$

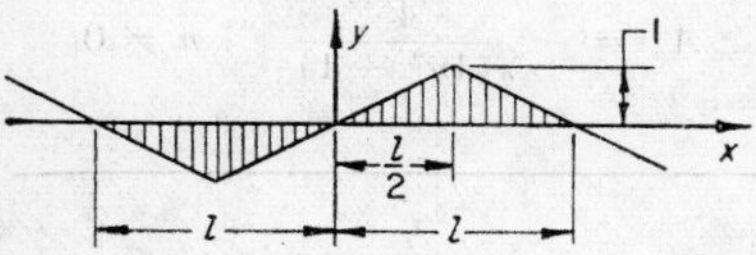

FIG. 20.34

$A_n = 0$

$B_n = +\frac{8}{\pi^2 n^2}$ $n = 1, 5, 9, \ldots$

$B_n = -\frac{8}{\pi^2 n^2}$ $n = 3, 7, 11, \ldots$

$B_n = 0$ $n = \text{even}$

[1] In this limiting case the series represents a singular function. The series is divergent, but summable.

Table 20.3. Fourier Series (*Continued*)

D. Quadratic Parabolas

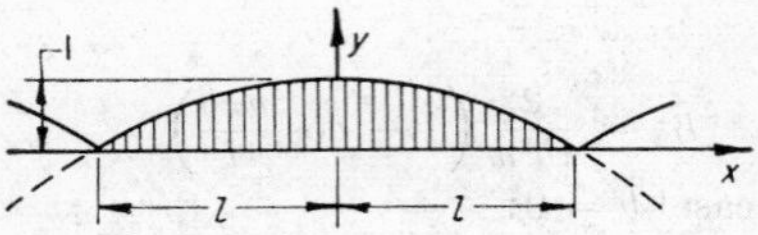

FIG. 20.35

For $-l < x < l$: $y = 1 - \frac{x^2}{l^2}$

$A_0 = \frac{2}{3}$

$A_n = -\frac{4}{\pi^2 n^2}$ n = even

$A_n = +\frac{4}{\pi^2 n^2}$ n = odd

$B_n = 0$

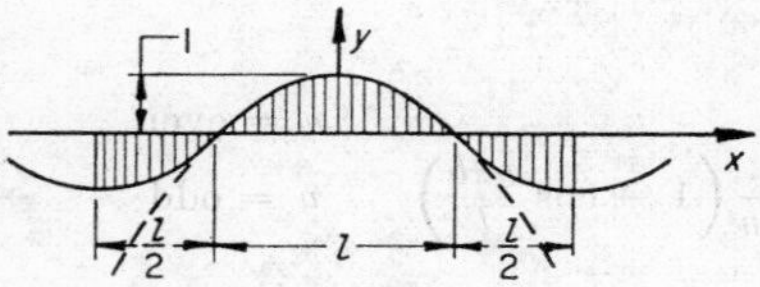

FIG. 20.36

For $-l/2 < x < +l/2$: $y = 1 - \frac{4x^2}{l^2}$

$A_n = 0$ n = even

$A_n = +\frac{32}{\pi^3 n^3}$ $n = 1, 5, 9, \ldots$

$A_n = -\frac{32}{\pi^3 n^3}$ $n = 3, 7, 11, \ldots$

$B_n = 0$

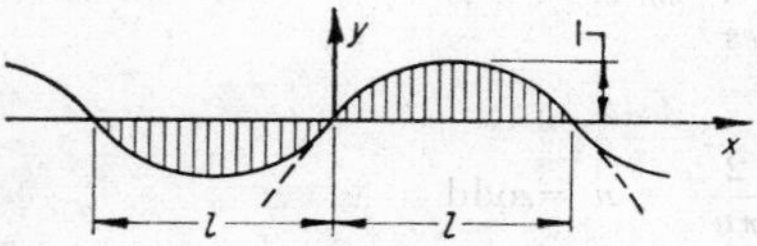

FIG. 20.37

For $0 < x < l$: $y = \frac{4}{l^2} x(l - x)$

$A_n = 0$

$B_n = 0$ n = even

$B_n = \frac{32}{\pi^3 n^3}$ n = odd

E. Sine Curves

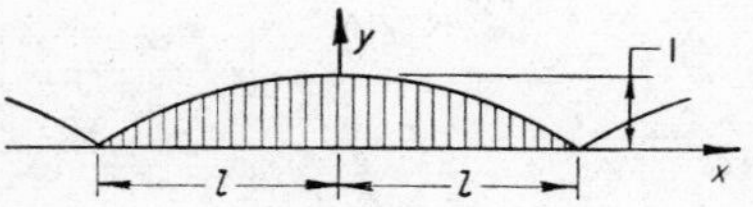

FIG. 20.38

For $-l < x < l$: $y = \cos \frac{\pi x}{2l}$

$A_0 = \frac{2}{\pi}$

$A_n = -\frac{4}{\pi(4n^2 - 1)}$ n = even

$A_n = +\frac{4}{\pi(4n^2 - 1)}$ n = odd

$B_n = 0$

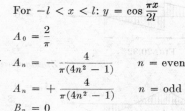

FIG. 20.39

For $0 < x < 2l$: $y = \sin \frac{\pi x}{2l}$

$A_0 = \frac{2}{\pi}$ $A_n = -\frac{4}{\pi(4n^2 - 1)}$ $n \neq 0$

$B_n = 0$

Part 2

MECHANICS OF RIGID BODIES

CHAPTER 21

STATICS

BY

D. H. YOUNG, Sc.D., Palo Alto, Calif.

21.1. FUNDAMENTALS OF PLANE STATICS

Statics. This subject deals with the equilibrium of rigid bodies under the action of applied forces. *Force* is defined as any action that tends to change the state of rest of a body to which it is applied. A force is defined by its *magnitude, direction,* and *point of application.* These properties can be represented by a directed line segment such as AB or AC in Fig. 21.1*a*, where the length of the line represents the magnitude; the arrow, the direction; and the end point A, the point of application.

If several forces are applied to a body at various points and acting in various directions, we have a *system of forces.* Force systems are usually classified as being *coplanar* or *noncoplanar.* Noncoplanar systems will be discussed separately in Sec. 21.4. The general problem of statics consists in reducing a given system of forces to its simplest equivalent, i.e., to its *resultant.* The conditions under which this resultant vanishes are of special interest since they represent the *conditions of equilibrium* of the system.

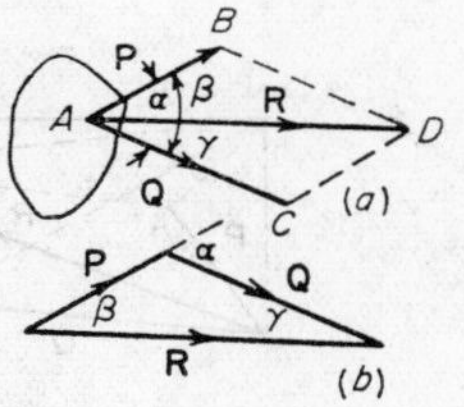

FIG. 21.1

Parallelogram Law. If two forces **P** and Q, represented by directed line segments AB and AC, acting under an angle α (Fig. 21.1*a*) are applied to a body at point A, their resultant **R** is represented by the directed line segment AD, obtained as the diagonal of the parallelogram constructed on AB and AC as shown. Thus, forces may be considered as *vectors* and the rules of vector algebra are applicable to them.

The resultant **R** can also be obtained by geometric addition of the *vectors* **P** and **Q** as shown in Fig. 21.1*b*. The forces **P** and **Q** are called *components* of the force **R**. The resultant **R** in Fig. 21.1 can be defined analytically as follows:

$$\mathbf{R} = \sqrt{P^2 + Q^2 + 2PQ\cos\alpha}$$
$$\sin\beta = \frac{Q}{R}\sin\alpha \qquad \sin\gamma = \frac{P}{R}\sin\alpha \tag{21.1}$$

Equilibrium Law. If two forces **F** applied to a body at point A are equal in magnitude, collinear in action, and opposite in sense, then the diagonal of their parallelogram vanishes, that is, $\mathbf{R} = 0$, and the forces are said to be in equilibrium.

Superposition Law. Two equal, opposite, collinear forces, being in equilibrium, can be superimposed on, or removed from, any given force system without altering the action of that system on any rigid body.

Transmissibility of Force. The point of application of a force may be moved along its line of action without altering the effect of the force on any rigid body to which it is applied.

Projection of a Force. The *projection* of a force **F** on any axis in its plane of action is defined as the product of the magnitude F of the force by the cosine of the angle α between its line of action and the chosen axis. Thus, in Fig. 21.2, the projection of the force **F** on the x axis will be $X = F \cos \alpha$.

Moment of a Force. The *moment* of a force **F** with respect to any point in its plane of action is defined as the product of the magnitude F of the force and the distance d from the chosen point to the line of action of the force. Thus, in Fig. 21.2, the moment of the force **F** with respect to the origin O will be $M_O = Fd$. This moment has the dimension of *force* × *length* and is considered positive when tending to produce counterclockwise rotation around point O as shown. We see from Fig. 21.2 that the moment of the force **F** about point O is represented by the doubled area of the triangle OAB obtained by joining the ends A and B of the given force vector with the moment center O.

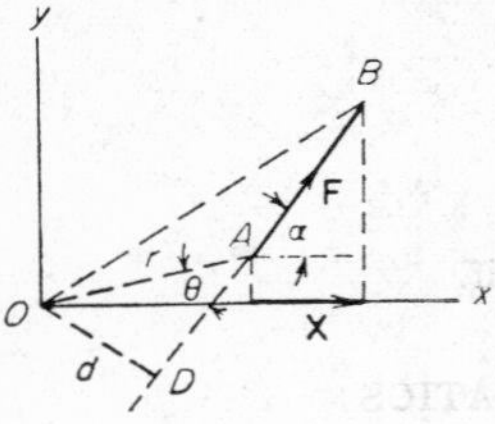

FIG. 21.2

We can also see from Fig. 21.2 that the moment of the force **F** about point O may be defined as the *vector product* of the radius vector OA and the force vector AB. Thus, in vector notation, $\mathbf{M} = \mathbf{r} \times \mathbf{F}$.

Theorem of Varignon. The algebraic sum of moments of any two intersecting forces **P** and **Q** about a chosen moment center O in their plane of action is equal to the moment of their resultant **R** about the same center. This is *Varignon's theorem*, which may be proved as follows: In Fig. 21.3, let $ABCD$ be the given parallelogram of forces, and O the chosen moment center. Through O we draw OX perpendicular to OA, and let B', C', D' be the projections on OX of B, C, D, respectively. Then

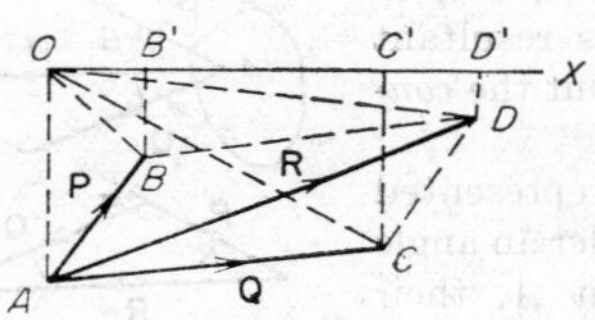

FIG. 21.3

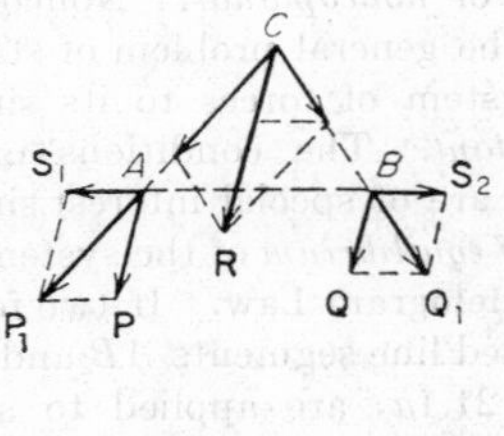

FIG. 21.4

$$\text{Moment of } \mathbf{P} = 2(\text{area } \triangle OAB) = OA \times OB'$$
$$\text{Moment of } \mathbf{Q} = 2(\text{area } \triangle OAC) = OA \times OC'$$

Adding these expressions, we get

$$\text{Moment of } \mathbf{P} + \text{moment of } \mathbf{Q} = OA(OB' + OC') = OA(OB' + B'D')$$

This, being double the area of $\triangle OAD$, is the moment of **R** about point O, and the theorem is proved.

Parallel Forces. To find the resultant of two parallel forces **P** and **Q** applied to a body at points A and B (Fig. 21.4), we proceed as follows: In accordance with the law of superposition, we add to the system at A and B two equal, opposite, collinear forces $\mathbf{S}_1$ and $\mathbf{S}_2$ of arbitrary magnitude as shown. Then, by the parallelogram law, we replace **P** and $\mathbf{S}_1$ by their resultant $\mathbf{P}_1$, and also **Q** and $\mathbf{S}_2$ by their resultant $\mathbf{Q}_1$. Transmitting $\mathbf{P}_1$ and $\mathbf{Q}_1$ to point C, we find their resultant **R**, as shown, which is the resultant of the given parallel forces **P** and **Q**. We see from the geometry of the figure that this resultant is parallel to **P** and **Q** and equal to their algebraic sum.

We locate the position of the line of action of **R** in Fig. 21.4 by using the theorem of Varignon. For any chosen moment center O in the plane of action of the forces:

$$\begin{aligned}\text{Moment of } \mathbf{P}_1 &= \text{moment of } \mathbf{P} + \text{moment of } \mathbf{S}_1\\ \text{Moment of } \mathbf{Q}_1 &= \text{moment of } \mathbf{Q} + \text{moment of } \mathbf{S}_2\\ \text{Moment of } \mathbf{R} &= \text{moment of } \mathbf{P}_1 + \text{moment of } \mathbf{Q}_1\end{aligned}$$

Adding the first two expressions and substituting into the third, we get

$$\text{Moment of } \mathbf{R} = \text{moment of } \mathbf{P} + \text{moment of } \mathbf{Q}$$

since moment of $\mathbf{S}_1$ + moment of $\mathbf{S}_2 = 0$. Thus the theorem of moments holds also for parallel forces.

The Force Couple. Two equal, opposite, noncollinear forces **P** applied to a body at points A and B (Fig. 21.5) cannot be reduced to a single resultant force. Such a pair of forces is called a *couple*. The magnitude P of either force is called the force of the couple; the distance d between their lines of action is called the arm of the couple.

The sum of moments of the two forces of a couple is the same about any moment center O in their plane of action. Thus, in Fig. 21.5,

$$M_O = P(d + x) - Px = Pd = M$$

This product of force times arm is called the *moment* of the couple; it is considered positive when the forces tend to produce counterclockwise rotation of the body, as shown.

Properties of a Couple. A given couple may be moved about in its plane of action without changing its effect on a rigid body. Both the force and the arm of a couple may be changed if their product, i.e., the moment M of the couple, is unchanged. Couples in the same plane may be added simply by adding their moments algebraically. In short, a couple is completely defined by (1) the magnitude of its moment, (2) its plane of action, and (3) its sense, i.e., clockwise or counterclockwise.

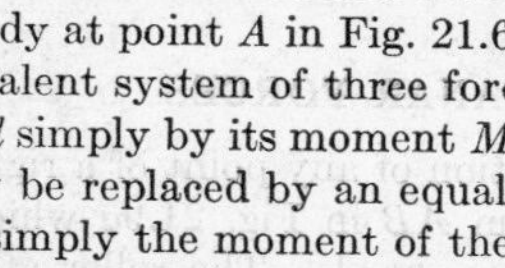

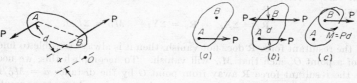

FIG. 21.5

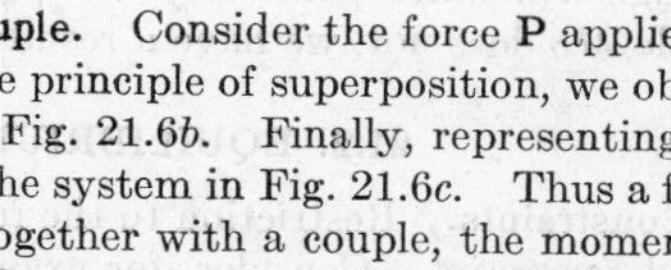

FIG. 21.6

Resolution of a Force into a Force and a Couple. Consider the force **P** applied to a rigid body at point A in Fig. 21.6a. Using the principle of superposition, we obtain the equivalent system of three forces shown in Fig. 21.6b. Finally, representing the couple Pd simply by its moment M, we obtain the system in Fig. 21.6c. Thus a force at A may be replaced by an equal force at B together with a couple, the moment of which is simply the moment of the given force **P** at A about the chosen point B.

Concurrent, Coplanar Forces. Any number of concurrent forces $\mathbf{F}_1$, $\mathbf{F}_2$, . . . , $\mathbf{F}_n$ applied at point O and acting in one plane (Fig. 21.7a) may be reduced to a single resultant force **R** applied at O. To define this resultant analytically, we first replace each force $\mathbf{F}_i$ by its rectangular components,

$$X_i = F_i \cos \alpha_i \qquad Y_i = F_i \cos \beta_i$$

as shown. We then have two systems of collinear forces acting along the x and y axes, and may add forces algebraically, obtaining

$$\mathbf{R}_x = \Sigma \mathbf{X}_i \qquad \mathbf{R}_y = \Sigma \mathbf{Y}_i \tag{21.2}$$

These rectangular components completely define the resultant **R** but, if desired, its magnitude and inclination to the x axis may be computed as follows:

$$R = \sqrt{R_x^2 + R_y^2} \qquad \tan \theta_x = R_y/R_x$$

Nonconcurrent, Coplanar Forces. Consider now the general case of any system of coplanar forces $\mathbf{F}_1, \mathbf{F}_2, \ldots, \mathbf{F}_n$, as shown in Fig. 21.8*a*. To define analytically the resultant of such a system, we first resolve each force $\mathbf{F}_i$ into a force $\mathbf{F}_i'$ at O and a couple F_id_i as shown. In this way, we obtain a system of forces, $\mathbf{F}_1', \mathbf{F}_2', \ldots, \mathbf{F}_n'$, concurrent at O, and a system of coplanar couples, the moments of which are the moments of the forces in their original positions about the chosen point O. Then the forces concurrent at O have the resultant already defined by Eqs. (21.2), and the

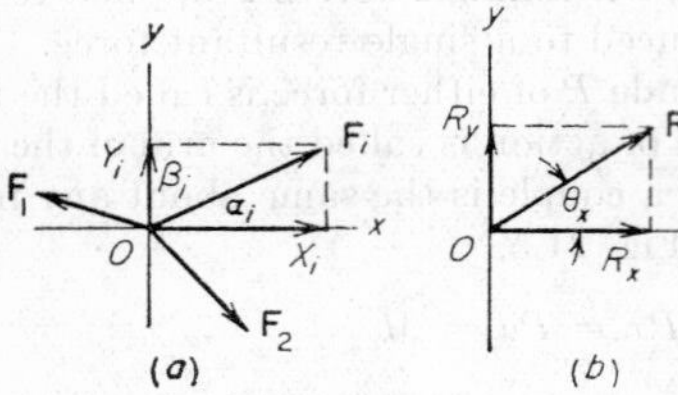

FIG. 21.7

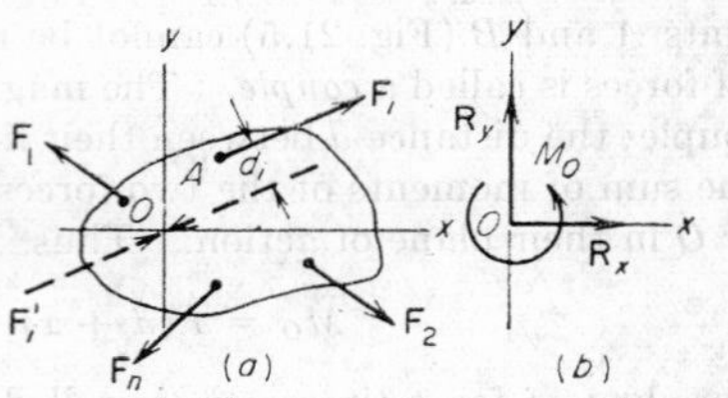

FIG. 21.8

couples have the resultant $M_O = \Sigma F_id_i$, as shown in Fig. 21.8*b*. Thus, in general, we may reduce a system of coplanar forces to a resultant force **R** applied at a chosen point O in the plane of action of the forces and a resultant couple M_O. We see that the resultant force **R** is independent of the position of the chosen point O, but the couple M_O, equal to the algebraic sum of moments of the given forces with respect to point O, does depend upon its position.

For further reference, we define the resultant force and couple in Fig. 21.8*b* as follows:

$$\mathbf{R}_x = \Sigma \mathbf{X}_i \qquad \mathbf{R}_y = \Sigma \mathbf{Y}_i \qquad M_O = \Sigma F_id_i \tag{21.3}$$

If the resultant force **R** does not vanish, then it is always possible to find the position of a point O' such that $M_{O'}$ will vanish. To accomplish this, we need only to move the resultant force **R** away from point O by the distance $d = M_O/R$ so as to counteract the couple M_O. In such case, the system reduces to a resultant force through a definite point O'. If the resultant force **R** vanishes, but the resultant couple M_O does not, we have a resultant couple.

21.2. EQUILIBRIUM OF COPLANAR FORCES

Constraints. Restriction to the freedom of motion of any point of a rigid body is called *constraint*. Consider, for example, the beam AB in Fig. 21.9*a* which is supported in a plane by a *roller* at B and a fixed *hinge* at A. The roller at B allows freedom of motion in the horizontal direction, but restricts motion vertically downward. Thus it is said to constitute *one degree of constraint.* The fixed hinge at A, on the other hand, allows neither horizontal nor vertical motion of point A. Thus it is said to constitute *two degrees of constraint.*

Law of Action and Reaction. In engineering machines and structures, various types of constraint are used in supporting rigid bodies so that they may resist the action of applied forces without moving. Whenever a body, not entirely free to move, is acted upon by applied forces such as **P** and **Q** in Fig. 21.9, it, in turn, exerts forces on its supports. These actions of the constrained body on its supports induce *reactions* from the supports, such that action and reaction at each point are equal, opposite, collinear forces. This is the law of action and reaction.

Free-body Diagrams. To investigate the forces (reactions) arising at the points of support of a rigid body, we imagine that each support is removed, and replace it by the reaction that it exerts on the body. Proceeding in this manner with the beam in Fig. 21.9*a*, for example, we replace the roller at B by a vertical force $\mathbf{R}_B$, since the roller offers constraint only in the vertical direction. Similarly, we replace the pin at A by horizontal and vertical reactions $\mathbf{X}_A$ and $\mathbf{Y}_A$, since this pin offers two degrees of constraint. The diagram in Fig. 21.9*b*, in which the beam is shown completely isolated from its supports, and on which all forces are shown by vectors, is called a *free-body diagram.*

We may now state briefly the general problem of statics. We have a body, either partially or completely constrained, which remains at rest under the action of applied loads. We isolate the body from its supports, and show all forces acting upon it, both active and reactive. We then consider what conditions this system of forces must fulfill to be in equilibrium, i.e., in order that their resultant will completely vanish.

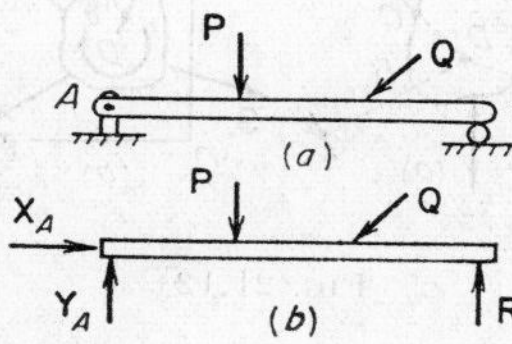

FIG. 21.9

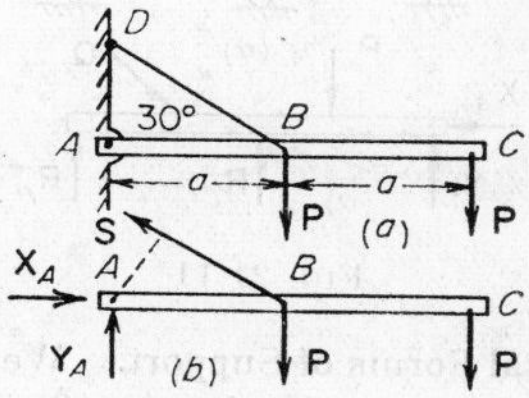

FIG. 21.10

Equations of Equilibrium. Referring to Eqs. (21.3), defining the resultant of any coplanar system of forces, we see that this resultant vanishes completely only if

$$\Sigma X_i = 0 \qquad \Sigma Y_i = 0 \qquad \Sigma F_i d_i = 0 \tag{21.4}$$

Thus we have three independent conditions of equilibrium from which, in general, three reaction components can be evaluated.

To illustrate, let us find the reactions at A and B for the beam supported and loaded as shown in Fig. 21.10*a*. The corresponding free-body diagram of the beam is shown in Fig. 21.10*b*. Here the pull of the cable BD on the bar has been replaced by a reactive force $\mathbf{S}$ applied at B and directed along the axis of the cable; the pin at A has been replaced by horizontal and vertical reactions $\mathbf{X}_A$ and $\mathbf{Y}_A$ as shown. This system of five coplanar forces is in equilibrium, and their resultant must vanish completely. Applying Eqs. (21.4), with point A as a moment center, gives

$$X_A - 0.866S = 0 \qquad Y_A + 0.500S - 2P = 0 \qquad Sa/2 - Pa - 2Pa = 0$$

from which $S = 6P$, $X_A = 5.20P$, $Y_A = -P$. The negative value for Y_A simply means that this reaction component acts downward instead of upward as assumed in the free-body diagram.

Statically Determinate Systems. Reconsidering the set of equations of equilibrium (21.4), we observe that three independent equations are not only necessary but also sufficient to guarantee that the resultant of a coplanar system of forces vanishes. This means that, in dealing with constrained bodies where unknown reactions are to be evaluated, we will not be able to determine more than three unknown elements. For this reason, a system of physical constraints of a rigid body in a plane that gives rise to just three unknown reactive elements, such as X_A, Y_A, S, in the example above, is said to be *statically determinate*. Since a rigid body in a plane has just three degrees of freedom, we conclude that three degrees of constraint, properly arranged, will be both necessary and sufficient to prevent motion of the body in its plane. Thus, the minimum requirements for complete constraint of a rigid body in a plane constitutes at the same time a statically determinate system of supports.

Statically Indeterminate Systems. If a system of supports of a rigid body in a plane comprises more than three degrees of constraint, there will be set up under the action of applied loads a set of reactions involving more than three unknown elements. Consequently, the three equations of equilibrium (21.4) will be inadequate. Such systems of constraints are said to be *statically indeterminate.* The beam supported as shown in Fig. 21.11*a* is an example. In this case, the free-body diagram will be as shown in Fig. 21.11*b*, but the three equations of equilibrium will not determine the four unknowns X_A, Y_A, R_B, R_C.

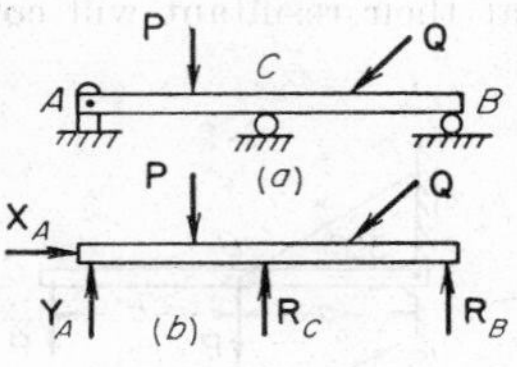

FIG. 21.11

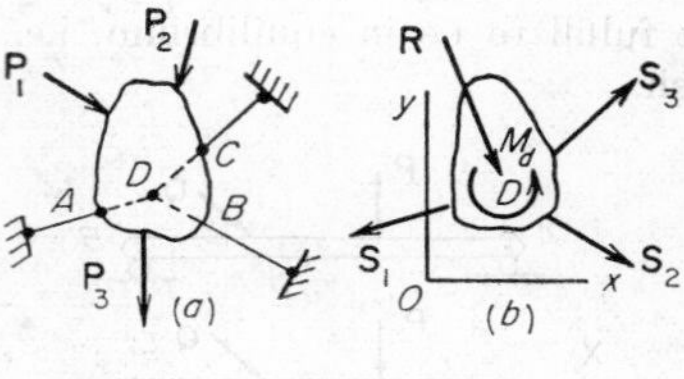

FIG. 21.12

Critical Forms of Support. We have stated above that three degrees of constraint, when properly arranged, are sufficient for the complete constraint of a rigid body in a plane. A few further remarks in this connection may be of interest. Consider, for example, the rigid body supported in a plane by three simple struts hinged at their ends, as shown in Fig. 21.12*a*. Under the action of forces $\mathbf{P}_1$, $\mathbf{P}_2$, $\mathbf{P}_3$, applied to the body, each strut will be subjected to some tension or compression, and will exert a reaction on the body acting along the axis of the strut. Thus, we obtain the free-body diagram shown in Fig. 21.12*b*. Normally, such a set of bars furnishes a statically determinate system of supports, and there will be no difficulty in finding the reactions $\mathbf{S}_1$, $\mathbf{S}_2$, $\mathbf{S}_3$.

Suppose, however, that, by accident, the axes of the three struts happen to meet in one point D as shown. Then let the resultant of the active forces P_i be represented by a force $\mathbf{R}$ at D, together with a couple M_D, as shown in Fig. 21.12*b*. Clearly, the reactive forces $\mathbf{S}_1$, $\mathbf{S}_2$, $\mathbf{S}_3$ concurrent at D cannot develop a moment around point D to balance the moment M_D, unless they are of infinite magnitude. Physical struts, of course, cannot produce such reactions. Even if the moment M_D vanishes for a particular state of loading, we are still left with an indeterminate problem, inasmuch as we then have a system of forces concurrent at point D, and only two equations of equilibrium (21.2) with which to find three unknowns. Thus the system of supports shown in Fig. 21.12*a* is said to be of *critical form.* In general, we exclude such critical forms of support from our discussion.

Before dismissing the subject, it may be of interest to point out what does happen with the equations of equilibrium (21.4) if they are written for a system having a critical form of support. In Fig. 21.12*b*, let a_1, a_2, a_3 be the moment arms of $\mathbf{S}_1$, $\mathbf{S}_2$, $\mathbf{S}_3$, with respect to the origin O, and let $\alpha_1, \beta_1; \alpha_2, \beta_2; \alpha_3, \beta_3$, be the direction cosines of their lines of action, with reference to coordinate axes x and y through point O. Then Eqs. (21.4) become:

$$\alpha_1 S_1 + \alpha_2 S_2 + \alpha_3 S_3 = R_x \qquad \beta_1 S_1 + \beta_2 S_2 + \beta_3 S_3 = R_y \qquad a_1 S_1 + a_2 S_2 + a_3 S_3 = M_D$$

This system of linear simultaneous algebraic equations defines a definite set of values for S_1, S_2, S_3, only so long as their coefficient determinant

$$\begin{vmatrix} \alpha_1 & \alpha_2 & \alpha_3 \\ \beta_1 & \beta_2 & \beta_3 \\ a_1 & a_2 & a_3 \end{vmatrix}$$

is different from zero [see Eq. (1.6)]. If the system of supports has a critical form, then the above determinant will vanish, and the equations of equilibrium refuse to yield a solution.

Friction. Whenever the surfaces of two bodies are in contact, there will be a limited amount of resistance to sliding between them, which is called *friction.* Such friction often influences the reactions at points of support of a body, and must be taken into account in discussing equilibrium problems.

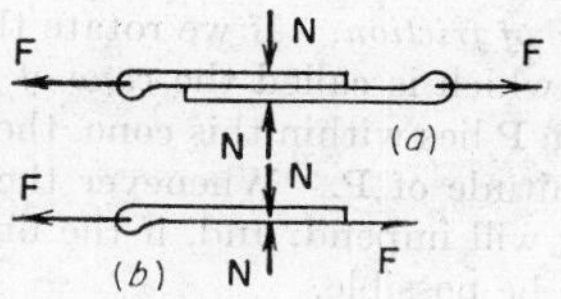

FIG. 21.13

Laws of Friction. Consider two plates with clean dry surfaces, which are pressed together by normal forces **N**, as shown in Fig. 21.13*a*. To overcome friction and cause sliding between the plates, certain forces **F** acting in the plane of contact will be required. The physical laws governing this friction were stated by Coulomb in 1781 and are as follows:

1. The total friction that can be developed is independent of the magnitude of the area of contact.
2. The total friction that can be developed is proportional to the normal force.
3. For low velocities of sliding, the total friction that can be developed is practically independent of the velocity.

The above laws of friction may be expressed by the formula

$$F = \mu N \tag{21.5}$$

where μ is called the *coefficient of friction.* If F is taken as the force necessary to start sliding, μ is called the coefficient of *static friction.* If F is taken as the somewhat smaller force necessary to maintain sliding, once it has been started, μ is called the coefficient of *kinetic friction.* Both static and kinetic coefficients of friction vary widely for different materials and for different conditions of their surfaces, as can be seen from the following table:

Coefficients of Friction

Materials	Static	Kinetic
Steel on steel	0.2–0.3	0.15–0.25
Wood on wood	0.4–0.7	0.3 –0.5
Leather on wood	0.5–0.6	0.3 –0.5
Hemp rope on wood	0.5–0.8	0.5 –0.6
Rubber tire on concrete	0.6–0.8	0.5 –0.7

The Friction Cone. Consider a block resting on a plane and acted upon by a force **P** inclined to the normal to the plane of contact by an angle α, as shown in Fig. 21.14. Clearly, the reaction exerted on the block by the supporting plane must be a force **R**, equal, opposite, and collinear with the applied force **P** if the block remains in equilibrium. Resolving this force into rectangular components as shown in Fig. 21.14*b*, we have

$$F = N \tan \alpha \tag{21.6}$$

where F represents the friction force resisting sliding, and N the normal force. Comparing Eq. (21.6) with Eq. (21.5) above, we see that, so long as $\tan \alpha < \mu$, the condition of equilibrium (21.6) will govern the relation between F and N, but, when α reaches a certain limiting value ϕ such that

$$\tan \phi = \mu \tag{21.7}$$

slipping will impend, and equilibrium cannot exist for larger values of α. This limiting value ϕ of the angle α that the line of action of **R** in Fig. 21.14*a* can make with the normal is called the *angle of friction*. If we rotate this line about the normal, we generate a cone of angle 2ϕ which is called the *cone of friction*. So long as the line of action of the applied force **P** lies within this cone, the block will remain in equilibrium, regardless of the magnitude of **P**. Whenever the line of action of **P** lies in the surface of this cone, slipping will impend, and, if the line of action of **P** falls outside the cone, no equilibrium will be possible.

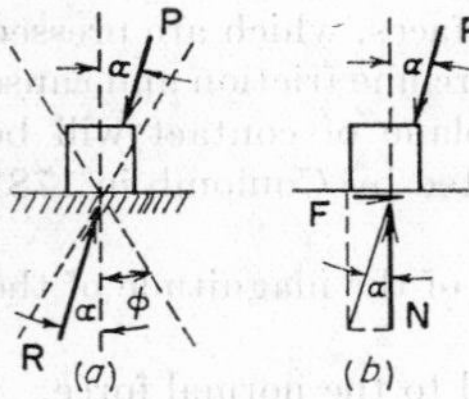

FIG. 21.14

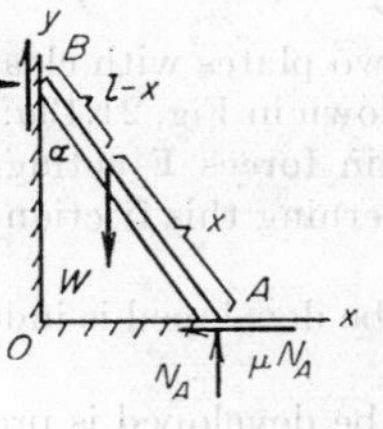

FIG. 21.15

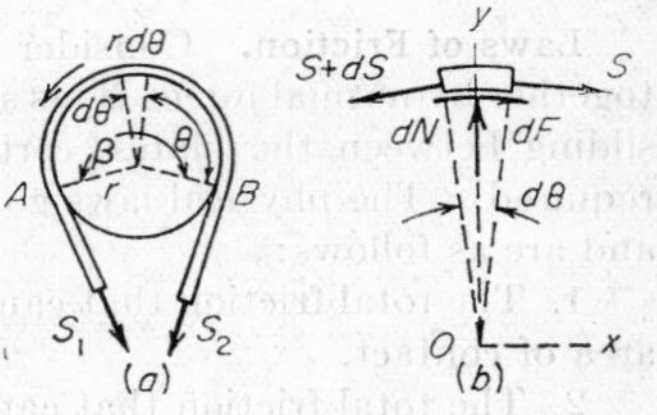

FIG. 21.16

If we assume an ideal smooth surface for which the coefficient of friction is zero, then the angle of friction $\phi = 0$, and the total reaction **R** is normal to the surface of contact. Thus an *ideal smooth surface* offers only one degree of constraint to a rigid body that it supports, but a rough surface offers, to a limited extent, two degrees of constraint. We see that a system of supports of a rigid body in a plane that involves rough surfaces may be statically indeterminate, statically determinate, or insufficient, depending on whether the condition of impending slip is not reached, just realized, or exceeded.

Example. A ladder of negligible weight is supported in a vertical plane by a horizontal floor and a vertical wall, as shown in Fig. 21.15. It is required to find the value of x, defining the position of a man of weight W, for which slipping at A and B will impend, if the coefficient of friction at both points of support is μ. The ladder makes the angle α with the plane of the wall.

We begin with a free-body diagram of the ladder, assuming that slip impends for the configuration shown. Otherwise, the relations between the friction and normal components of the reactions at A and B are ambiguous, and the problem is statically indeterminate. If we choose O as a moment center, the equations of equilibrium become

$$N_B - \mu N_A = 0 \qquad N_A + \mu N_B - W = 0$$
$$N_A l \sin \alpha - N_B l \cos \alpha - W(l - x) \sin \alpha = 0$$

Eliminating N_A and N_B and solving for x gives

$$x = \frac{\mu l}{1 + \mu^2}(\mu + \cot \alpha)$$

For $\alpha < \phi = \tan^{-1} \mu$, this gives values of $x > l$. Thus, we conclude that, for safety, the ladder should not incline to the wall by an angle greater than the angle of friction.

Belt Friction. Another example of interest is possible slipping between the rim of a pulley and a flexible belt or rope overrunning it. In Fig. 21.16*a* we have a pulley driven at uniform speed (equilibrium), by virtue of the friction developed between its rim and the encircling belt AB. It is required to find the limiting value of the ratio S_1/S_2 of tensions in the two branches of the belt when all possible power is being developed and slip impends.

To investigate this problem, let us consider the conditions of equilibrium of any element of the belt of length $ds = r\,d\theta$ located at the angle θ from the point of tangency B. A free-body diagram of this element showing all forces acting upon it is given in Fig. 21.16*b*. Treating this as a concurrent coplanar system, the equations of equilibrium (21.4) become

$$S + dF - (S + dS) = 0 \qquad dN - S\frac{d\theta}{2} - (S + dS)\frac{d\theta}{2} = 0$$

Neglecting small quantities of second order, these two equations reduce to

$$dF = dS \qquad dN = S\,d\theta$$

Now, when slipping impends, we have $dF = \mu\,dN$, which, if we use the two equations of statics, gives

$$dS = \mu S\,d\theta \qquad \text{or} \qquad \frac{dS}{S} = \mu\,d\theta \tag{21.8}$$

Integrating Eq. (21.8) over the entire line of contact BA, that is, from $\theta = 0$ to $\theta = \beta$, we obtain

$$\ln \frac{S_1}{S_2} = \mu\beta \qquad \text{or} \qquad \frac{S_1}{S_2} = e^{\mu\beta} \tag{21.9}$$

21.3. FUNDAMENTALS OF GRAPHIC STATICS

Concurrent, Coplanar Forces. If several coplanar forces $\mathbf{F}_1, \mathbf{F}_2, \ldots, \mathbf{F}_n$ act at a single point O (Fig. 21.17*a*), they can always be reduced to a resultant force $\mathbf{R}$, which also acts through point O. This resultant force can be found graphically by successive applications of the parallelogram law as illustrated in Fig. 21.17*a*, or as the closing side of the polygon of forces constructed as shown in Fig. 21.17*b*. Thus, the resultant $\mathbf{R}$ of any system of concurrent coplanar forces is obtained as the *geometric sum* of the free vectors representing the given forces.

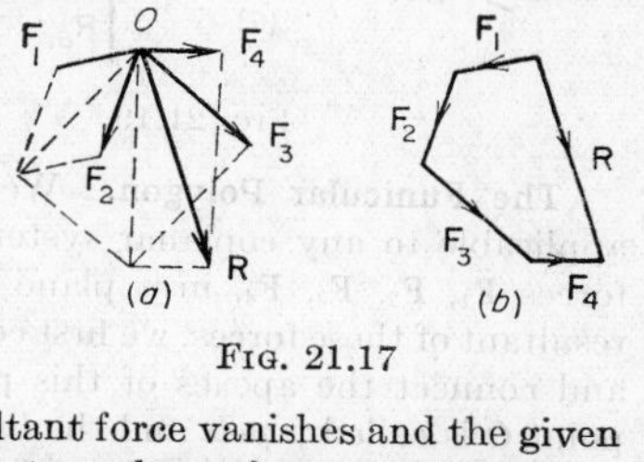

FIG. 21.17

If the polygon of forces happens to close, the resultant force vanishes and the given forces are in equilibrium. Thus, if several concurrent, coplanar forces are known to be in equilibrium, their free vectors must build a closed polygon. This graphical condition of equilibrium is sometimes more useful than the analytic conditions discussed in Sec. 21.2.

For example, in Fig. 21.18*a*, we have two rollers of weights **P** and **Q** supported by two mutually perpendicular smooth inclined planes and connected by a string as shown. It is required to find the angle ϕ that the string makes with the horizontal when the system is in equilibrium. Free-body diagrams for the two rollers are shown in Fig. 21.18*a*. The three forces **P**, $\mathbf{R}_A$, **S** acting on the roller *A* are in equilibrium and must build a closed triangle. Likewise, the three forces **Q**, $\mathbf{R}_B$, **S**, acting on the roller *B* are in equilibrium and must build a closed triangle. By setting these two triangles of forces in juxtaposition as shown in Fig. 21.18*b*, we get a very simple graphical solution of the problem and can measure the angle ϕ with a protractor, or it can also be found from the trigonometry of the force triangles as follows:

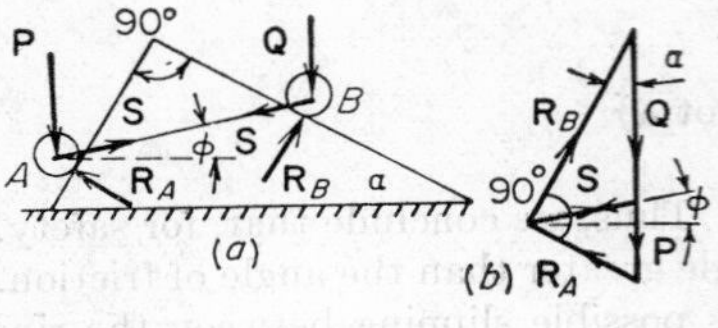

FIG. 21.18

$$R_B = (P + Q)\cos\alpha \qquad R_A = (P + Q)\sin\alpha$$

$$\tan\phi = \frac{R_B\cos\alpha - Q}{R_A\cos\alpha} = \cot\alpha - \frac{2Q}{P+Q}\csc 2\alpha$$

Theorem of Three Forces. Three nonparallel forces acting in one plane can be in equilibrium only if their lines of action intersect in one point. This follows from the fact that one force can equilibrate the resultant of the other two only if it is equal, opposite, and collinear with their resultant. Thus, it must act through the intersection of their lines of action. This is called the *theorem of three forces;* it is sometimes very useful in studying the equilibrium of coplanar force systems.

Consider, for example, a beam *AB*, supported as shown in Fig. 21.19*a* and subjected to the action of a force **P** applied at *C* as shown. What reactions are induced at the supports? To answer this question entirely on the basis of graphical constructions, we intersect the known lines of action of **P** and $\mathbf{R}_B$ at point *D*. Then, since $\mathbf{R}_A$ is the third of three forces in equilibrium, and since its point of application *A* is known, we obtain its line of action *AD* by the theorem of three forces. Thereafter it is a simple matter to construct the closed triangle of forces in Fig. 21.19*b* and find graphically the magnitudes of $\mathbf{R}_A$ and $\mathbf{R}_B$.

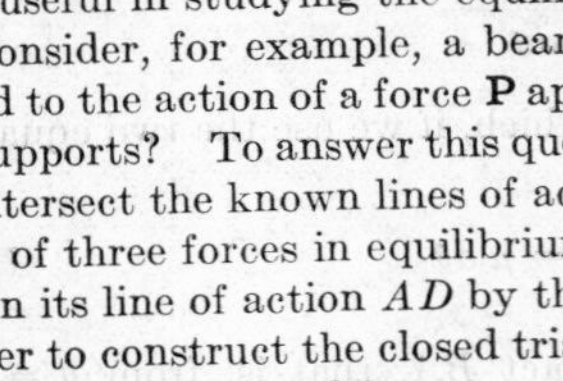

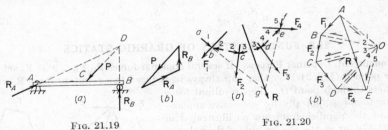

FIG. 21.19

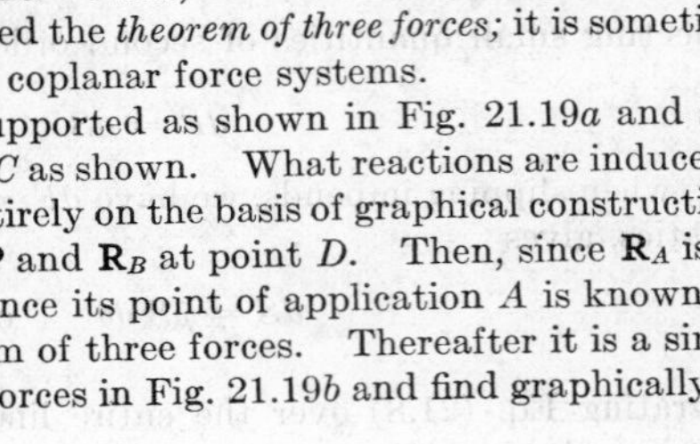

FIG. 21.20

The Funicular Polygon. We shall now develop a general graphical treatment applicable to any coplanar system of forces. Consider, for example, the system of forces $\mathbf{F}_1$, $\mathbf{F}_2$, $\mathbf{F}_3$, $\mathbf{F}_4$, in a plane as shown in Fig. 21.20*a*. To find graphically the resultant of these forces, we first construct the polygon of forces *ABCDE* in Fig. 21.20*b*, and connect the apexes of this polygon with an arbitrary point *O* as shown. The point *O* is called a *pole* and the lines 1, 2, 3, 4, 5, are called *rays*. The length of each ray represents a magnitude of force and we see that, by the described construction, we have resolved the given forces $\mathbf{F}_1$, $\mathbf{F}_2$, . . . into components represented by the rays. For example, forces represented by the vectors *AO* and *OB* have for their resultant the force $\mathbf{F}_1$, etc.

Now to proceed further, we construct in the plane of action of the forces (Fig. 21.20*a*) a polygon *abcdef*, the apexes of which fall on the lines of action of the given forces and the sides of which are parallel to the rays of the force polygon (Fig. 21.20*b*). Specifically, the two sides of this polygon which intersect at *b* on the line of action of $\mathbf{F}_1$ are parallel to the two rays which straddle $\mathbf{F}_1$ in the force polygon, etc. The polygon *abcdef* obtained in this way is called a *funicular polygon.*

Having completed the above described constructions, we finally replace $\mathbf{F}_1$ at *b* by its components 1 and 2 acting along *ab* and *cb*, as shown in Fig. 21.20*a*. In the same way, we replace $\mathbf{F}_2$ by its components 2 and 3 intersecting at *c*, etc. Thus the given forces $\mathbf{F}_1$, $\mathbf{F}_2$, . . . are transformed into the statically equivalent system of forces acting along the sides of the funicular polygon. Of these, however, the pairs (2,2), (3,3), and (4,4) cancel as equal, opposite, collinear forces. Thus, only 1 and 5 remain, and their resultant **R** is given by the vector AE representing their geometric sum in the force polygon, and acts through the intersection *g* of the first and last sides of the funicular polygon.

If we imagine a weightless string going along the sides of the funicular polygon *abcdef* and fixed at *a* and *f*, this string will be in equilibrium under the action of the forces $\mathbf{F}_1$, $\mathbf{F}_2$, $\mathbf{F}_3$, $\mathbf{F}_4$, applied to it at points *b, c, d, e*. The tensions in the branches *ab* and *bc* of this string will be represented by the rays AO and OB, and their action on the knot *b* will balance the force $\mathbf{F}_1$. In the same way, the tensions in the portions *bc* and *cd* of the string balance the force $\mathbf{F}_2$, etc. This explains the origin of the name *funicular polygon*, i.e., string polygon.

If the polygon of forces is closed, the possibility of a resultant force vanishes. In such case, the first and last rays coincide; hence, the first and last sides of the funicular polygon become parallel or coincide. If they are parallel, the two equal and opposite forces acting along them represent a *resultant couple.* If they coincide, the two equal, opposite forces acting along them are balanced, and there is no resultant; the forces are in equilibrium.

The first of the above two special cases is illustrated in Fig. 21.21. To obtain the moment of the resultant couple represented by the equal and opposite forces 1 and 5 acting along *ab* and *fe*, we scale the magnitude of force from the ray OA in Fig. 21.21*b* and multiply by the arm, measured to the scale of length in Fig. 21.21*a*.

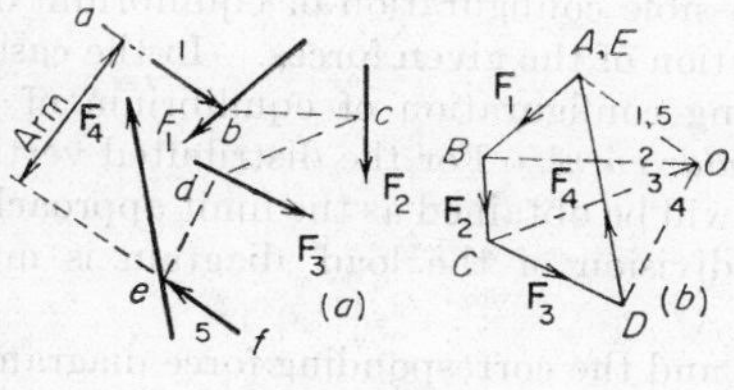

FIG. 21.21

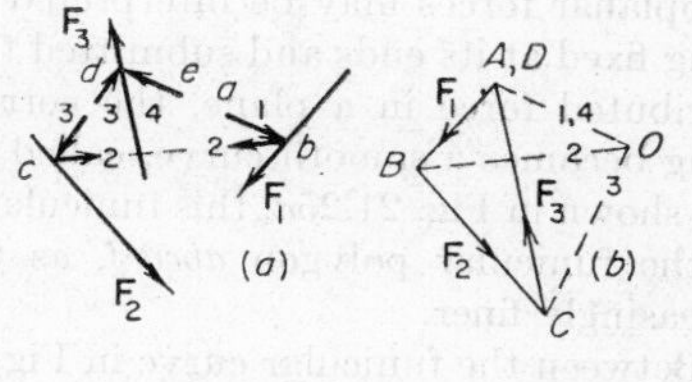

FIG. 21.22

The second special case, where the two forces are in equilibrium, is illustrated in Fig. 21.22. In this case, the first and last sides *ab* and *de* coincide; hence, the components of force 1 and 4, being collinear, cancel each other, and all possibility of either a resultant force or couple vanishes.

From the above discussion, we see that we may take, as graphical conditions of equilibrium for any coplanar force system, the following statements:

1. The force polygon must close.
2. The funicular polygon must close.

These graphical conditions are equivalent to the analytic conditions of equilibrium represented by Eqs. (21.4).

To illustrate the use of the funicular polygon in finding reactions, let us consider the beam supported in a vertical plane by three simple struts and loaded as shown in Fig. 21.23*a*. Replacing each strut by a coaxial reactive force, we obtain a system of five coplanar forces in equilibrium, three of which, S_1, S_2, S_3, are of unknown magnitude.

To find these three quantities graphically, we begin with the force polygon shown in Fig. 21.23*b*. By assembling the free vectors in the order shown, we contrive to determine the magnitude of S_2 at once, leaving only the apex D of the force polygon undetermined. Choosing a pole O, drawing rays 1, 2, 3, 4, in Fig. 21.23*b* and the corresponding portion *abcde* of the funicular polygon in Fig. 21.23*a*, we establish the *closing side ea* as shown. Then a ray 5 drawn parallel to *ea* through the pole O in Fig. 21.23*b* determines the apex D of the force polygon and, correspondingly, the magnitudes S_3 and S_1 of the strut reactions.

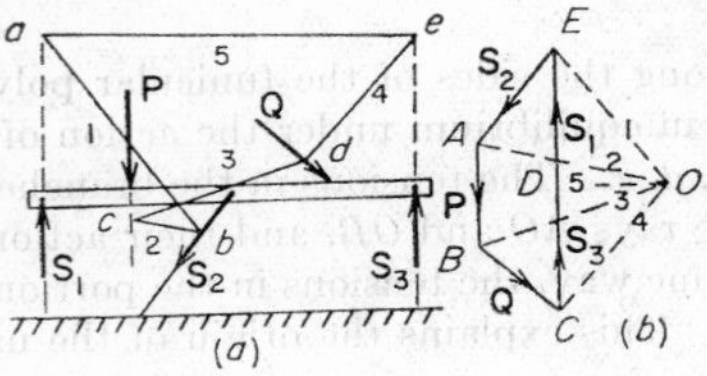

FIG. 21.23

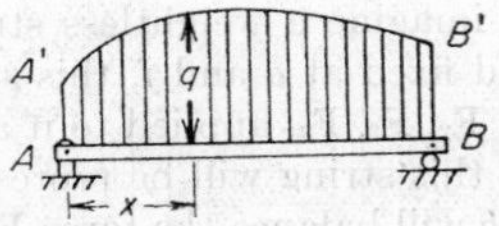

FIG. 21.24

Distributed Force. In previous discussions, we have dealt entirely with concentrated forces. Sometimes, we need to consider the equilibrium of bodies under the action of *distributed force*. Consider, for example, the beam AB in Fig. 21.24, which is loaded continuously along its length by distributed vertical load. This type of loading can be represented by a diagram $AA'BB'$, the ordinate q of which at any point along the beam represents the intensity of load at that point, i.e., the force per unit length of AB. Such distributed vertical load can also be defined analytically by specifying $q = f(x)$.

Funicular Curves. We have already seen that any funicular polygon for a system of coplanar forces may be interpreted as a possible configuration of equilibrium of a string fixed at its ends and submitted to the action of the given forces. In the case of distributed force in a plane, the corresponding configuration of equilibrium of the string becomes a smooth curve, called a *funicular curve.* For the distributed vertical load shown in Fig. 21.25*a*, this funicular curve will be obtained as the limit approached by the funicular polygon *abcdef,* as the subdivision of the load diagram is made increasingly finer.

Between the funicular curve in Fig. 21.25*a* and the corresponding force diagram in Fig. 21.25*b* there are the following relationships: (1) For every tangent to the funicular curve there is a corresponding parallel ray in the force diagram, the length of which represents the tension in the string at the point of tangency. (2) The shortest ray, called the *pole distance* H, represents the tension at the vertex of the funicular curve and, likewise, the horizontal projection of tension at any other point on the string. (3) For any two points on the funicular curve, rays parallel to the corresponding tangents cut from the load line AD that portion of the total load which acts between those two points.

Differential Equation of the Funicular Curve. It may be desirable to develop a funicular curve for a given distributed load analytically, instead of graphically. Referring to Fig. 21.26*a*, let $MM'NN'$ be a given distribution of vertical load, and let *abcd* be a funicular curve corresponding to the pole O in Fig. 21.26*b*. On this funicular curve, we choose two adjacent points b and c as defined by the coordinates x and $x + dx$. The corresponding slopes are dy/dx and $dy/dx + (d^2y/dx^2)\,dx$, respectively. Then from the relationships discussed in connection with Fig. 21.25, we may write

$$\frac{dy}{dx} = \tan BOE = \frac{BE}{OE} \qquad \frac{dy}{dx} + \frac{d^2y}{dx^2}\,dx = \tan COE = \frac{CE}{OE}$$

Subtracting the first of those expressions from the second, we obtain

$$\frac{d^2y}{dx^2}\,dx = \frac{BC}{OE} = \frac{q\,dx}{H}$$

which reduces to

$$\frac{d^2y}{dx^2} = \frac{q}{H} \tag{21.10}$$

This is the differential equation of a family of funicular curves corresponding to different values of the arbitrary pole distance H. For any particular load distribution,

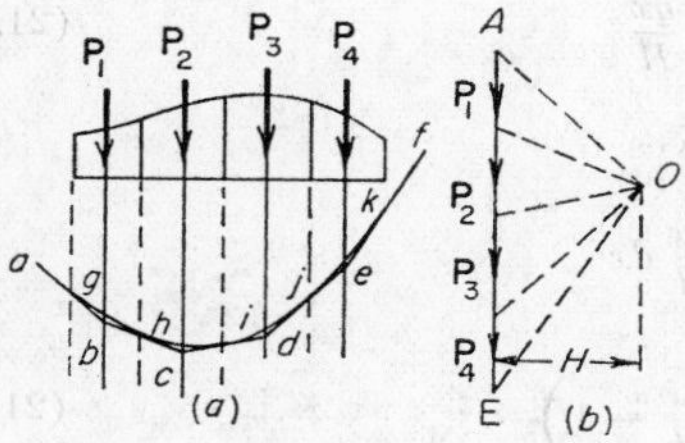

FIG. 21.25

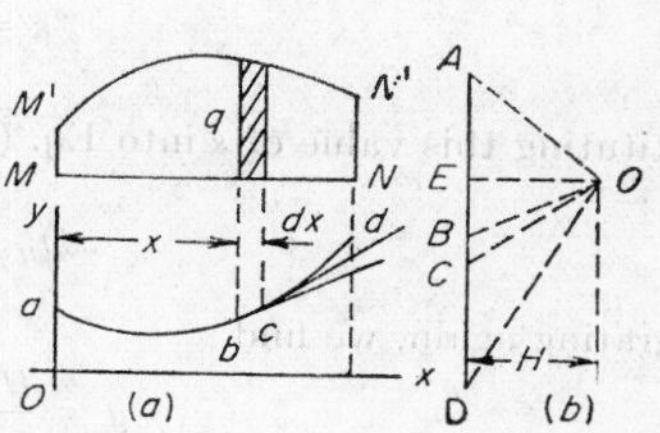

FIG. 21.26

as defined by the function $q(x)$, the corresponding family of curves can be obtained by integration of this differential equation.

The Parabolic Cable. For the particular case of uniformly distributed vertical load, such as we have in the cables of a suspension bridge, $q = q_0 =$ const and the solution of Eq. (21.10) becomes

$$y = \frac{q_0x^2}{2H} + C_1x + C_2$$

Taking the vertex of this parabola as the origin of coordinates, we have $y = 0$ and $dy/dx = 0$ when $x = 0$. Thus $C_1 = C_2 = 0$, and we obtain

$$y = \frac{q_0x^2}{2H} \tag{21.10a}$$

This defines the configuration of equilibrium of a flexible cable subjected to a uniformly distributed load of intensity q_0 along the horizontal span of the cable.

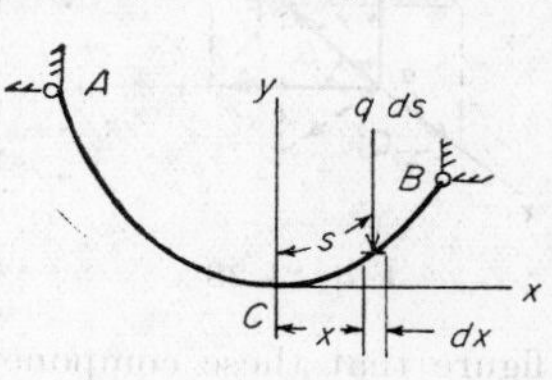

FIG. 21.27

The Catenary Cable. A flexible cable or chain fastened at its two ends and hanging freely under its own weight in a vertical plane presents a funicular curve of special interest. In this case (Fig. 21.27), the vertical load is uniformly distributed along the axis of the cable itself, and Eq. (21.10) becomes

$$\frac{d^2y}{dx^2} = \frac{q}{H}\sqrt{1 + \left(\frac{dy}{dx}\right)^2} \tag{21.11}$$

where q is the uniform weight of the cable per unit length. To integrate this equation, we observe that

$$ds = dx\sqrt{1 + \left(\frac{dy}{dx}\right)^2} \tag{21.12}$$

Substituting (21.12) into (21.11) and integrating, we obtain

$$\frac{dy}{dx} = \frac{qs}{H} \tag{21.13}$$

so that expression (21.12) may be written in the form

$$ds = \sqrt{1 + \left(\frac{qs}{H}\right)^2}\, dx$$

This may now be integrated, and we obtain, for the coordinate axes shown in Fig. 21.27,

$$s = \frac{H}{q} \operatorname{Sinh} \frac{qx}{H} \tag{21.10b}$$

Substituting this value of s into Eq. (21.13) gives

$$dy = \operatorname{Sinh} \frac{qx}{H}\, dx$$

Integrating again, we find

$$y = \frac{H}{q}\left(\operatorname{Cosh} \frac{qx}{H} - 1\right) \tag{21.10c}$$

Equations (21.10*b*,*c*) define the configuration of equilibrium of a freely suspended cable or chain under the action of its own weight. Such a funicular curve is called a *catenary*.

21.4. SPACE STATICS

Resolution of a Force into Three Orthogonal Components. Consider the force $\mathbf{F}_i$ applied at point O in Fig. 21.28 and making angles α_i, β_i, γ_i with rectangular coordinate axes x, y, z, respectively, as shown. Using the parallelogram law, we may resolve this force into three rectangular components as follows: First, we pass a plane through the line of action of the force $\mathbf{F}_i$ and the z axis. In this plane, we resolve the force into components Z_i and Q_i as shown. Then, again by the parallelogram law, we resolve the force $\mathbf{Q}_i$ in the x,y plane into rectangular components X_i and Y_i, as shown. Thus the three rectangular components X_i, Y_i, Z_i, are statically equivalent to the original force $\mathbf{F}_i$. We see also from the geometry of the figure that these components are simply the projections of the given force on the chosen axes. Thus,

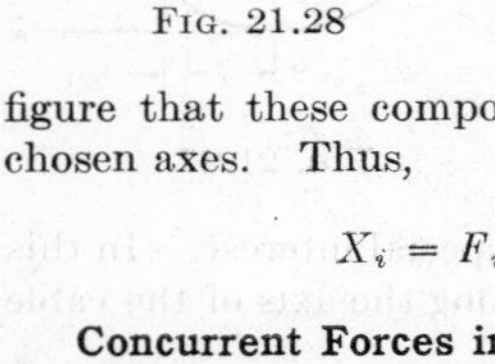

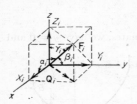

FIG. 21.28

$$X_i = F_i \cos \alpha_i \qquad Y_i = F_i \cos \beta_i \qquad Z_i = F_i \cos \gamma_i$$

Concurrent Forces in Space. If we have several concurrent forces in space, we replace each force by its rectangular components as discussed above, sum these collinear forces along each axis, algebraically, and obtain for the three rectangular components of the resultant force the following expressions:

$$R_x = \Sigma X_i \qquad R_y = \Sigma Y_i \qquad R_z = \Sigma Z_i$$

Then the complete resultant force will have the magnitude and direction cosines, as follows:

$$R = \sqrt{R_x^2 + R_y^2 + R_z^2}$$

$$\cos \theta_x = R_x/R \qquad \cos \theta_y = R_y/R \qquad \cos \theta_z = R_z/R \tag{21.14}$$

Moment of a Force about an Axis. In dealing with force systems in space, the concept of moment of force about an axis is very helpful. In Fig. 21.29, the moment of the force $\mathbf{F}_i$ applied at A with respect to the axis OZ is defined as follows: Through point A, pass a plane normal to the axis OZ. Project the force $\mathbf{F}_i$ onto this plane as shown. Then the moment of this projection about the point O in its plane is defined as the moment of the given force $\mathbf{F}_i$ about the axis OZ, that is,

$$M_{zi} = (F_i \cos \alpha)d$$

This quantity has the dimension of *force* × *length* and is considered positive when the rotation is in the direction indicated by the fingers of the right hand when the thumb

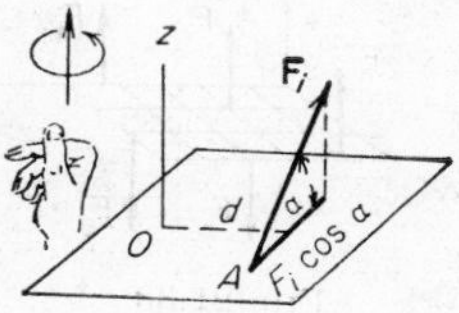

FIG. 21.29

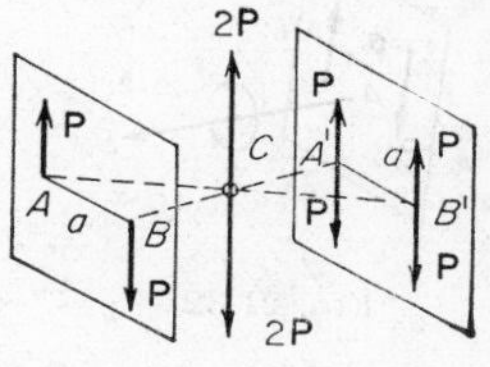

FIG. 21.30

is pointed in the positive direction of the z axis. The moment of a force about an axis vanishes if the line of action of the force is parallel to, or intersects, the axis. In the first case the projection $F_i \cos \alpha$ vanishes, and in the second case the arm $d = 0$.

Couples in Parallel Planes. In Sec. 21.1, it was shown that a given couple is completely defined by its plane of action and the magnitude and sign of its moment. We will now see that the plane of the couple may be moved parallel to itself without altering the action of the couple on a rigid body. In Fig. 21.30, we begin with the couple $M = Pa$ in the plane AB. Then, in a parallel plane $A'B'$, we take two equal and opposite couples, each of the same moment Pa as the given couple. This superposition of a self-equilibrated system does not alter the action of the original couple. We now replace the force $\mathbf{P}$ at B and the downward force $\mathbf{P}$ at A' by their downward resultant $2\mathbf{P}$ at C as shown, and the force $\mathbf{P}$ at A and the upward force $\mathbf{P}$ at B' by their upward resultant $2\mathbf{P}$, also at C. Finally, removing the equilibrated forces at C, we are left with only a couple Pa in the plane $A'B'$, which proves the theorem.

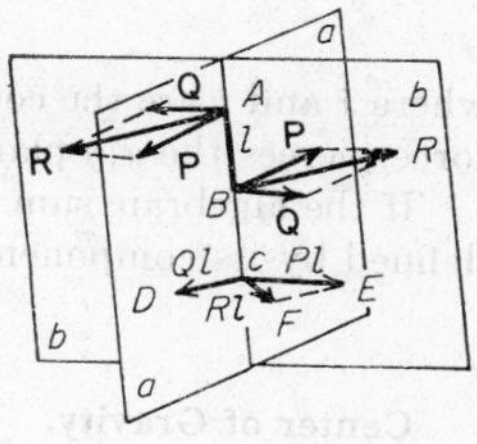

FIG. 21.31

Couples in Intersecting Planes. In Fig. 21.31, consider the couple Pl in the plane aa, and the couple Ql in the intersecting plane bb. Adding the forces $\mathbf{P}$ and $\mathbf{Q}$ at A by the parallelogram law, we obtain a single force $\mathbf{R}$ at A as shown. Similarly, the forces $\mathbf{P}$ and $\mathbf{Q}$ at B produce an equal parallelogram of forces, giving an equal, opposite, and parallel resultant force $\mathbf{R}$ at B as shown. Thus the given couples Pl and Ql in intersecting planes reduce to a resultant couple Rl in an intermediate plane.

If, in Fig. 21.31, we construct, at any point C on the line of intersection AB of the two planes aa and bb, line segments CD and CE proportional in length to the moments of the couples Pl and Ql, respectively, and normal to their planes of action, we may construct thereon a parallelogram, the diagonal CF of which will be proportional to the resultant couple Rl and normal to its plane of action. This follows from the fact that the parallelogram $CDFE$ is geometrically similar to either of the parallelograms of forces above it and rotated by 90° about the axis ABC.

From the above discussion we may conclude that the three properties of a couple, (1) magnitude of moment, (2) aspect of plane, and (3) sense of rotation, can be represented by a directed line segment and that such line segments obey the parallelogram law of addition. Thus, couples are *vector quantities* and may be represented by vectors as shown in Fig. 21.32.

Since the position of a couple in its plane is of no consequence, and since the plane of the couple can be displaced parallel to itself without altering the action of the couple, it follows that any system of couples in space can be represented by a set of moment vectors concurrent at any chosen point O. Then by analogy to the case of concurrent forces in space, it follows that the resultant couple can be defined analyti-

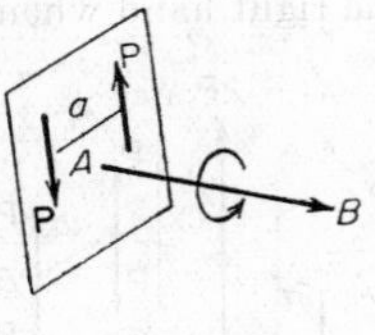

FIG. 21.32

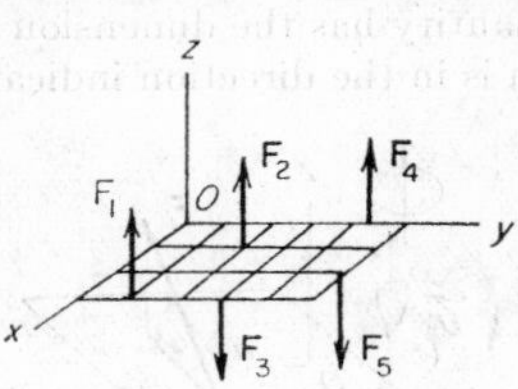

FIG. 21.33

cally by expressions analogous to Eqs. (21.14) defining resultant force. Thus, the projections of the resultant couple are:

$$M_x = \Sigma M_{xi} \qquad M_y = \Sigma M_{yi} \qquad M_z = \Sigma M_{zi}$$

while the couple itself is defined as

$$M = \sqrt{M_x^2 + M_y^2 + M_z^2}$$
$$\cos\theta_x = M_x/M \qquad \cos\theta_y = M_y/M \qquad \cos\theta_z = M_z/M \tag{21.15}$$

Parallel Forces in Space. Consider a system of parallel forces in space, as shown in Fig. 21.33. If the algebraic sum of these forces is different from zero, they reduce to a resultant force defined as follows:

$$R = \sum F_i \qquad \bar{x} = \frac{\Sigma F_i x_i}{\Sigma F_i} \qquad \bar{y} = \frac{\Sigma F_i y_i}{\Sigma F_i} \tag{21.16a}$$

where $\bar{x}$ and $\bar{y}$ are the coordinates of the point where the line of action of the resultant force pierces the x,y plane.

If the algebraic sum of the forces is zero, the system reduces to a resultant couple defined by its components

$$M_x = \Sigma F_i y_i \qquad M_y = -\Sigma F_i x_i \tag{21.16b}$$

Center of Gravity. Equations (21.16a) are very useful in defining the position of the *center of gravity* of a body, i.e., the point of application of the resultant of all parallel gravity forces acting upon its various elements. Assume, for example, that a body is subdivided into many particles, each of weight w_i and having coordinates x_i, y_i, z_i. Then, by reference to Eqs. (21.16a), we conclude that the position of the center of gravity C is defined by the coordinates

$$x_c = \frac{\Sigma w_i x_i}{\Sigma w_i} \qquad y_c = \frac{\Sigma w_i y_i}{\Sigma w_i} \qquad z_c = \frac{\Sigma w_i z_i}{\Sigma w_i} \tag{21.17a}$$

If the material of the body is homogeneous, the weight w_i of each element is proportional to its volume dV, and Eqs. (21.17a) take the form

$$x_c = \frac{\int x\,dV}{\int dV} \qquad y_c = \frac{\int y\,dV}{\int dV} \qquad z_c = \frac{\int z\,dV}{\int dV} \tag{21.17b}$$

where, of course, the integration is to be extended over the entire volume of the body.

Center of Pressure. As an example of another application of Eqs. (21.16a), we consider the case of water pressure against the vertical bulkhead of a horizontal prismatic trough of irregular cross section, as shown in Fig. 21.34. In such case, the elemental forces $d\mathbf{P}$ exerted by the water on various elements of area dA of the bulkhead represent a system of parallel forces in space. Therefore, Eqs. (21.16a) can be used to find their resultant $\mathbf{P}$ and the coordinates $\bar{x}$, $\bar{y}$, defining its point of application D, which is called the *center of pressure*. Replacing F_i in Eqs. (21.16a) by dP, they become

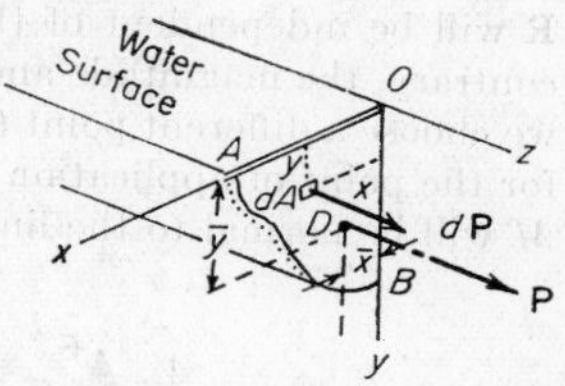

FIG. 21.34

$$\mathbf{P} = \int d\mathbf{P} \qquad \bar{x} = \frac{\int x\,dP}{\int dP} \qquad \bar{y} = \frac{\int y\,dP}{\int dP}$$

where the integration must cover the entire area of the bulkhead.

Denoting by γ the specific weight of water and recalling that water pressure is proportional to depth, we have

$$dP = \gamma y\,dA$$

so that the above equations give

$$\mathbf{P} = \gamma \int y\,dA = \gamma A y_c \qquad \bar{x} = \frac{\gamma \int xy\,dA}{\gamma \int y\,dA} = \frac{I_{xy}}{A y_c} \qquad \bar{y} = \frac{\gamma \int y^2\,dA}{\gamma \int y\,dA} = \frac{I_x}{A y_c} \tag{21.18}$$

These equations show that the total force $\mathbf{P}$ exerted on the bulkhead is obtained simply by multiplying the area A by the intensity of pressure γy_c at the centroid of the area, but it does not act at the centroid. The position of the center of pressure D, as defined by $\bar{x}$ and $\bar{y}$, depends on the product of inertia I_{xy} and the moment of inertia I_x of the area with respect to the coordinate axes taken as shown in the figure.

If, for example, the bulkhead OAB in Fig. 21.34 is a quarter circle of radius r, we have

$$I_{xy} = \frac{r^4}{8} \qquad I_x = \frac{\pi r^4}{16} \qquad A = \frac{\pi r^2}{4} \qquad y_c = \frac{4r}{3\pi}$$

Substituting these values into Eqs. (21.18), we find

$$\mathbf{P} = \frac{\gamma r^3}{3} \qquad \bar{x} = \frac{3r}{8} \qquad \bar{y} = \frac{3\pi r}{16}$$

Nonconcurrent, Nonparallel Forces in Space. We consider now the general case of any system of forces $\mathbf{F}_1, \mathbf{F}_2, \mathbf{F}_3, \ldots$ applied to a rigid body at points $A, B, C, \ldots$ as shown in Fig. 21.35. To define the resultant of such a system, we first use the idea of resolution of a force into a force and a couple as discussed in Sec. 21.1. Thus, we replace each force $\mathbf{F}_i$ by an equal force $\mathbf{F}_i'$ applied at an arbitrary point O, together with a couple M_i, the moment of which is the moment of the given force $\mathbf{F}_i$ with respect to the point O. The plane of this couple is defined by the original line of action of the force $\mathbf{F}_i$ and the point O, as shown by the shaded triangles in Fig. 21.35. In this way we obtain a system of forces $\mathbf{F}'_1, \mathbf{F}'_2, \mathbf{F}'_3, \ldots$ concurrent at point O and a system of couples $M_1, M_2, M_3, \ldots$ in space. The concurrent forces at O reduce to a resultant force $\mathbf{R}$ already defined by Eqs. (21.14), and the system of couples reduces to a resultant couple $\mathbf{M}$ as previously defined by Eqs. (21.15). Thus, in general, the complete resultant of a system of forces in space is a *resultant force* and a *resultant couple.*

Proceeding as above, we see that the magnitude and direction of the resultant force $\mathbf{R}$ will be independent of the position of its chosen point of application O. On the contrary, the magnitude and plane of the resultant couple M will be altered each time we choose a different point O. This suggests that we can perhaps find such a point O_1 for the point of application of $\mathbf{R}$ that the plane of the corresponding resultant couple M will be normal to the line of action of $\mathbf{R}$.

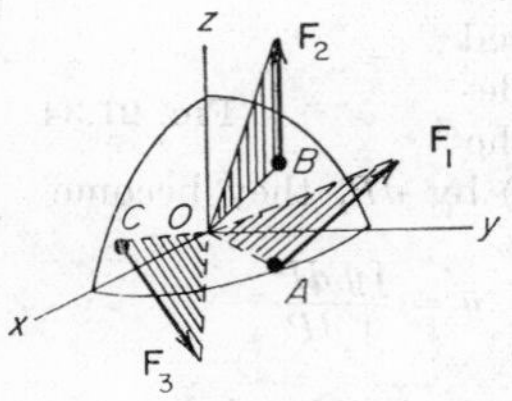

Fig. 21.35

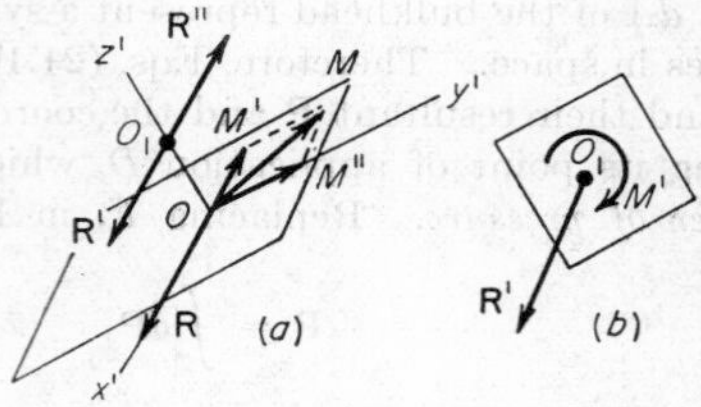

Fig. 21.36

To accomplish this objective, we note that the resultant force vector $\mathbf{R}$ and the resultant moment vector $\mathbf{M}$ obtained as explained before always define a plane through the chosen point O. We now take a new set of coordinate axes x', y', z', through O such that the plane defined by the vectors $\mathbf{R}$ and $\mathbf{M}$ becomes the x',y' plane as shown in Fig. 21.36. In this plane we resolve the moment vector $\mathbf{M}$ into rectangular components M' and M'' such that M' is collinear with $\mathbf{R}$. Now on the axis z' we choose a point O_1 such that $OO_1 = M''/R$, and at this point we superimpose two oppositely directed forces $\mathbf{R}'$ and $\mathbf{R}''$, each equal and parallel to $\mathbf{R}$. The force $\mathbf{R}''$ at O_1 and the force $\mathbf{R}$ at O now constitute a couple which balances the couple M'', and all this may be removed from the system. All that remains is the force $\mathbf{R}'$ at O_1 and the couple M', the plane of which is normal to the line of action of $\mathbf{R}'$. Thus, we obtain the simplified representation of the total resultant, as shown in Fig. 21.36*b*. This simplest representation of a system of forces in space is sometimes called a *wrench.*

Equations of Equilibrium. If a system of forces in space is in equilibrium, both the resultant force and the resultant couple must vanish. Referring to Eqs. (21.14) and Eqs. (21.15) we see that this requires

$$\left.\begin{array}{lll} \Sigma X_i = 0 & \Sigma Y_i = 0 & \Sigma Z_i = 0, \\ \Sigma M_{xi} = 0 & \Sigma M_{yi} = 0 & \Sigma M_{zi} = 0 \end{array}\right\} \tag{21.19}$$

These are the equations of equilibrium for any system of forces. They are much used in finding reactions at the points of support of a rigid body subjected to the actions of applied loads. In writing these conditions of equilibrium, however, it is not necessary to adhere to rectangular axes x, y, z for projections and moments. If the forces are known to be in equilibrium, then the algebraic sum of their projections on any axis must be zero, and the algebraic sum of their moments with respect to any axis must be zero.

Consider, for example, an equilateral triangular slab ABC with sides of length a, supported by six simple struts, arranged as shown in Fig. 21.37*a* To calculate the axial force in each strut due to the action of a couple M in the plane of the slab, we consider the free-body diagram in Fig. 21.37*b*, where each strut has been replaced by its reaction on the slab. Noting that all the unknown forces except $\mathbf{S}_5$ either are parallel to or intersect the z axis, we write $\Sigma M_{zi} = 0$ and obtain

$$S_5 \cos \beta \frac{\sqrt{3}}{2} a - M = 0$$

from which $S_5 = 2M/\sqrt{3}\, a \cos \beta$. Similarly, by writing $\Sigma M_{xi} = 0$, we have

$$S_5 \sin \beta \frac{\sqrt{3}}{2} a + S_2 \frac{\sqrt{3}a}{2} = 0$$

from which

$$S_2 = -S_5 \sin \beta = -2M \tan \beta / \sqrt{3}a$$

Statically Determinate Systems. We discuss now the problem of how to support a rigid body in space so that it cannot move under the action of any system of applied forces. Referring to the body in Fig. 21.38, we first attach a point A to the foundation

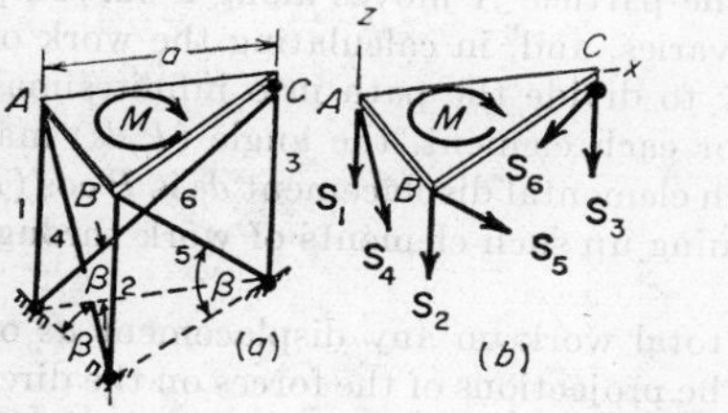

FIG. 21.37

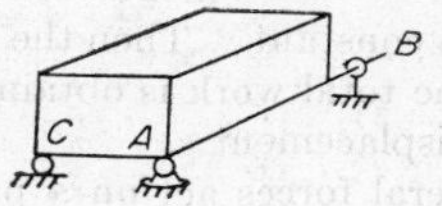

FIG. 21.38

by a *spherical hinge*. This completely constrains point A in space, but the body is still free to rotate about A. In so doing, another point B is constrained to remain on the surface of a sphere of radius AB and centered at A. Hence, a *ring guide* at B, as shown, completes the constraint of this point. The body can now rotate only about the fixed axis AB, and a third point C must follow the arc of a circle normal to AB. Hence, a *ball bearing* at C completes the constraint of the body. We see that the spherical hinge at A comprises three degrees of constraint; the ring guide at B, two degrees of constraint; and the ball bearing at C, one degree of constraint. Thus, in all, *six degrees of constraint* are required for completely fixing a rigid body in space. These need not be arranged exactly as shown in Fig. 21.38. The six simple struts in Fig. 21.37*a*, for example, represent another possible arrangement. We conclude, in general, that six degrees of constraint are necessary, and, when properly arranged, sufficient for the support of a rigid body in space.

Since we have six equations of equilibrium [Eqs. (21.19)], we conclude further that such systems of support as described above will, in general, be statically determinate. That is, we can always find the magnitudes of the reactive forces induced at such supports under the action of applied forces.

There is always the possibility that six constraints may be misarranged so that they allow some freedom of motion of the body; i.e., they constitute a system of supports of *critical form*, as already discussed in Sec. 21.2. In this event, the determinant of the six equations of equilibrium will vanish, and the equations refuse to yield a solution.

21.5. PRINCIPLE OF VIRTUAL DISPLACEMENTS

Work. Referring to Fig. 21.39, imagine that a *particle* A, acted upon by a force $\mathbf{P}$ and free to move in any direction, is displaced from position A to A_1. Then the vector AA_1 is called a *displacement* of the particle, and its length is denoted by s.

When the particle undergoes this displacement s, the force **P** is said to produce *work*. This work U is defined as the product of the displacement and the projection of the force on the directions of the displacement. Thus,

$$U = P \cos (P,s) \cdot s \tag{21.20}$$

which has the dimension of *force* × *length*. We see from this definition that work is represented by the scalar product of the force vector **P** and the displacement vector **s**:

$$\mathbf{U} = \mathbf{P} \cdot \mathbf{s} \tag{21.20'}$$

If the direction of the displacement is perpendicular to that of the force, then $\cos (P,s) = 0$ and the work (21.20) vanishes. For instance, if a particle moves on a smooth surface, the reaction exerted by this surface is normal to the displacement of its point of application and does no work.

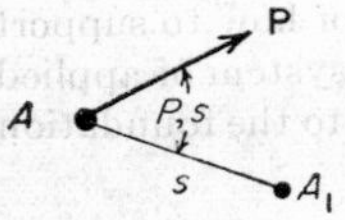

FIG. 21.39

If, in Fig. 21.39, the particle A moves along a curved path, then the angle (P,s) varies, and, in calculating the work of the force **P**, it is necessary to divide the path into infinitesimal elements ds such that, for each element, the angle (P,ds) may be treated as constant. Then the work on each elemental displacement ds is $P \cos (P,ds)\, ds$, and the total work is obtained by summing up such elements of work throughout the full displacement s.

If several forces act on a particle, the total work on any displacement ds of the particle is obtained as the algebraic sum of the projections of the forces on the direction of the displacement multiplied by the displacement. Since the sum of these projections of the forces is equal to the corresponding projection of their resultant, we conclude that the work done by a system of concurrent forces acting on a particle, on any displacement of the particle, is equal to the work of their resultant on the same displacement.

Virtual Work. Let us consider now a particle free to move in any direction, and acted upon by a system of concurrent forces, the resultant of which we define by its three rectangular components R_x, R_y, R_z, parallel, respectively, to arbitrarily chosen coordinate axes, x, y, z. We imagine now that this particle undergoes an arbitrary infinitesimal displacement δs, which we call a *virtual displacement* to distinguish it from any actual displacement that the particle may suffer if the resultant of the applied forces is not zero. Then the work done by the acting forces on such a virtual displacement δs is

$$\delta U = \mathbf{R} \cdot \delta \mathbf{s} = R_x\, \delta x + R_y\, \delta y + R_z\, \delta z \tag{21.21}$$

where δx, δy, δz, are components of $\delta \mathbf{s}$ in the x, y, z directions, respectively. Such work is called *virtual work*. If the forces are in equilibrium, we have

$$R_x = 0 \qquad R_y = 0 \qquad R_z = 0$$

and conclude from Eq. (21.21) that, in such case, the work done by the acting forces on any virtual displacement of the particle is equal to zero, i.e.,

$$\delta U = R_x\, \delta x + R_y\, \delta y + R_z\, \delta z = 0 \tag{21.22}$$

This represents the principle of virtual work for a system of concurrent forces. Since we may assume any direction for the virtual displacement δs, its rectangular components δx, δy, δz, are entirely arbitrary infinitesimal quantities. Thus, by taking only one component of virtual displacement at a time different from zero, we can use

Eq. (21.22) three times and obtain three independent conditions of equilibrium based on the principle of virtual work.

We consider now a particle A not entirely free to move, but partially constrained by a weightless rigid bar OA connecting it to a fixed point O, as shown in Fig. 21.40. In such case, the particle can move only on the surface of a sphere of radius OA centered at O. Consequently, a virtual displacement δs of the particle A, compatible with its constraints, must lie in a plane tangent to the sphere at A, and can be defined by two components δs_1 and δs_2 in this plane as shown. We say that this particle has only two degrees of freedom, and only two independent possibilities of virtual displacement.

If we denote by R_1, R_2, R_3 the rectangular components of the resultant of the active forces applied to the particle, and by **S** the axial force in the bar OA, the equation of virtual work becomes

$$R_1\,\delta s_1 + R_2\,\delta s_2 = 0 \tag{21.23}$$

Neither the component R_3 of the active forces nor the reactive force **S** normal to the tangent plane at A enters into Eq. (21.23) because there can be no virtual displacement δs_3 in this direction. We see that, in this case, the work done by the active forces alone on any virtual displacement of the particle must be zero. In calculating this virtual work, it is not necessary to consider the reactive force **S**, since its work is automatically zero.

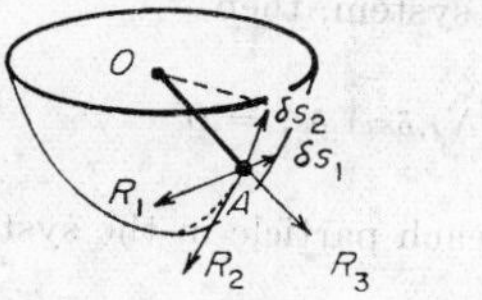

FIG. 21.40

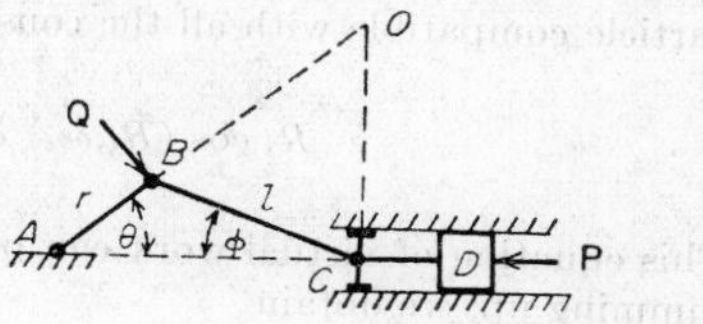

FIG. 21.41

In a similar manner, the problem of the equilibrium of a particle constrained to remain on a smooth surface, or a bead constrained to slide along a smooth wire, can be discussed. In all such cases, we come to the conclusion that the necessary and sufficient conditions of equilibrium are that the work of the active forces alone for each virtual displacement of the particle compatible with its constraints must be zero. Thus, the principle of virtual work is particularly useful in the solution of those problems of statics in which we are interested, not in the magnitudes of reactive forces, but only in the conditions that the active forces must satisfy in order to have equilibrium.

Ideal Systems. If several particles are interconnected in such a manner that they are not entirely free in their relative motion, but are subject to some constraint, we have a *system of particles*. Take, for example, the familiar system of a crank, connecting rod, and piston, as shown in Fig. 21 41. In discussing the equilibrium of this system under the action of applied forces **P** and **Q**, we regard it as consisting of four particles, A, B, C, and D, interconnected by weightless rigid bars AB, BC, and CD. These bars we consider simply as internal constraints between the particles of the system. We have also several external constraints, since point A cannot move at all, and the particles C and D have to follow the axis of the guiding cylinder.

In all further discussion, we assume no friction on the sliding surfaces of the crosshead C and piston D. Thus the reactive forces exerted by the guides act normal to the direction of any possible displacement of their points of application, and they do not produce work on such displacements. It is assumed further that the bars joining the several particles of the system are absolutely rigid. Thus, they can undergo no changes in length, and the tensile or compressive forces which they carry cannot produce work. Such systems of particles interconnected by inextensional bars and guided by frictionless surfaces are called *ideal systems*. Any rigid body may be considered as an ideal system of many particles.

Considering possible displacements of the ideal system shown in Fig. 21.41, we see that the positions of all particles are defined by the magnitude of the angle θ that the crank AB makes with the fixed axis AC. This angle is called the coordinate of the system, and, because one coordinate completely specifies the configuration, the system is said to have *one degree of freedom*. Thus, its several particles can have only one set of related virtual displacements. Giving to the angle θ an infinitesimal increase $\delta\theta$, the corresponding virtual displacement of each particle in the system will be defined. For example, the particle B will have the displacement $r\ \delta\theta$ in the direction perpendicular to AB. Similarly the piston D will have the displacement $\delta x = (OC/OB)r\ \delta\theta$ where O is the instantaneous center (see p. 22-6) for the connecting rod BC.

Principle of Virtual Work. Let us consider now any ideal system of particles like the one shown in Fig. 21.41 and subjected to a system of applied active forces. If such a system of particles is in equilibrium, then each individual particle by itself is in equilibrium. Hence, the work of all forces (both active and reactive) acting on any one particle for any displacement of that particle must be zero. If $\mathbf{R}_i$ is the resultant of the active forces applied to a particle i of the system, and $\mathbf{N}_i$ is the resultant of all reactive forces on the same particle, while δs_i is the virtual displacement of that particle compatible with all the constraints of the system, then

$$R_i \cos (R_i,\delta s_i)\ \delta s_i + N_i \cos (N_i,\delta s_i)\ \delta s_i = 0$$

This equation of virtual work can be written for each particle in the system. Then, summing up, we obtain

$$\Sigma R_i \cos (R_i,\delta s_i)\ \delta s_i + \Sigma N_i \cos (N_i,\delta s_i)\ \delta s_i = 0 \tag{21.24}$$

Now the second sum in this expression represents the work of all reactive and internal forces in the system, and, as we have seen above, this net work is always zero for an ideal system. Hence, Eq. (21.24) reduces to

$$\Sigma R_i \cos (R_i,\delta s_i)\ \delta s_i = 0 \tag{21.25}$$

which expresses algebraically the principle of virtual work for any ideal system of particles. If such a system is in equilibrium, the total work of all active forces on any virtual displacement of the system compatible with its constraints is equal to zero. The reactive forces due to constraints do not produce work, and do not enter into Eq. (21.25).

To illustrate the application of the principle of virtual work to a practical problem, let us consider the screw press shown in Fig. 21.42, and inquire what turning moment M applied to the wheel of the press is required to maintain a specified compressive force $\mathbf{Q}$ on the block, if friction in the screw thread is neglected. To answer this question, we replace the block by forces $\mathbf{Q}$ as shown, and obtain an ideal system with one degree of freedom. Defining a virtual displacement of the system by an angle of rotation $\delta\theta$ of the wheel, we have, for the corresponding displacement of the screw, $h\ \delta\theta/2\pi$, where h is the pitch of the screw thread. Hence, the equation of virtual work becomes

$$M\ \delta\theta - \frac{Qh\ \delta\theta}{2\pi} = 0 \qquad \text{from which} \qquad M = \frac{Qh}{2\pi}$$

We see that, by using the principle of virtual work here, we get a simple answer without the necessity of considering the somewhat complicated reactive forces exerted on the screw by the frame of the press.

As a second example, let us consider the compound beam $ABCDEF$ supported and loaded as shown in Fig. 21.43. To find one of the reactions, say $\mathbf{R}_b$, without taking the system all apart, we can use the principle of virtual work. To do this, we replace the roller at B by its reaction $\mathbf{R}_b$, and thereby obtain a movable system with one degree of freedom. Defining a virtual displacement of this system by a small rota-

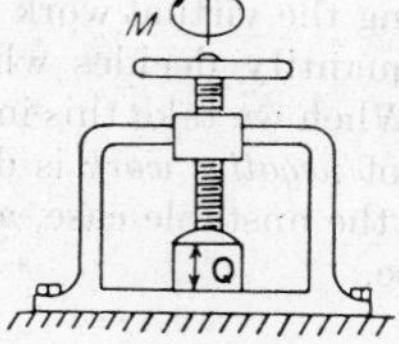

FIG. 21.42

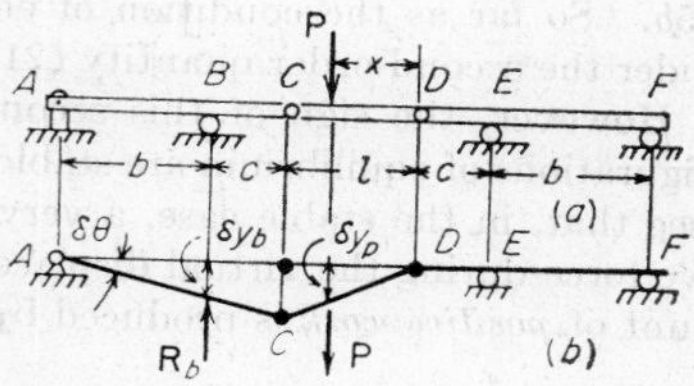

FIG. 21.43

tion $\delta\theta$ of AC about the fixed pin at A, we see that the corresponding virtual displacements of the points of application of $\mathbf{R}_b$ and $\mathbf{P}$ are respectively

$$\delta y_b = b\,\delta\theta \qquad \text{and} \qquad \delta y_p = \frac{b+c}{l}\,x\,\delta\theta$$

Hence, the equation of virtual work becomes

$$-R_b b\,\delta\theta + P\frac{b+c}{l}\,x\,\delta\theta \qquad \text{from which} \qquad R_b = \frac{Px}{l}\left(1+\frac{c}{b}\right)$$

Stable and Unstable Equilibrium. The principle of virtual work is useful, not only in establishing configurations of equilibrium, but also in answering the question of

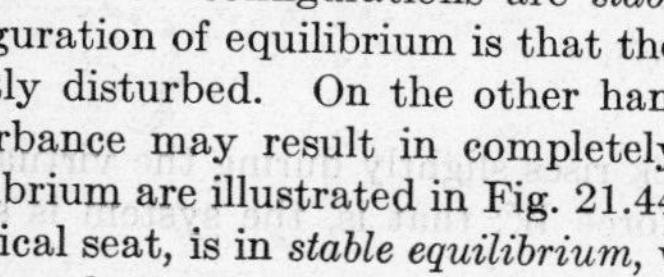

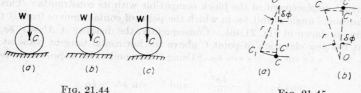

FIG. 21.44

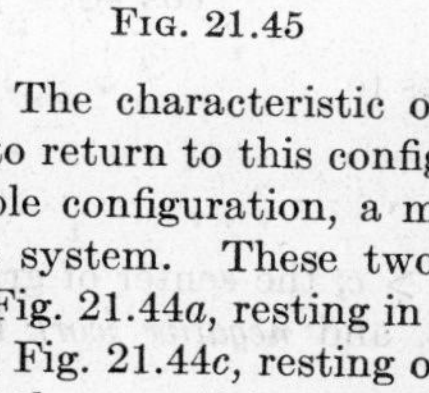

FIG. 21.45

whether such configurations are *stable* or *unstable.* The characteristic of a stable configuration of equilibrium is that the system tends to return to this configuration if slightly disturbed. On the other hand, in an unstable configuration, a momentary disturbance may result in completely upsetting the system. These two types of equilibrium are illustrated in Fig. 21.44. The ball in Fig. 21.44*a*, resting in a concave spherical seat, is in *stable equilibrium,* while the ball in Fig. 21.44*c*, resting on top of a convex spherical surface, is in *unstable equilibrium.* If the supporting surface is plane as in Fig. 21.44*b*, the equilibrium of the ball is said to be *neutral.*

In each case illustrated in Fig. 21.44, a virtual displacement of the ball will be an infinitesimal displacement, in which the ball remains in contact with its supporting surface. For such a small displacement, the center C of the ball moves essentially in a horizontal plane, and the work of the active gravity force W is zero. Hence, in each case the ball is in a condition of equilibrium.

To decide the question of stability, however, a more refined calculation of the virtual work in each case is required. For the ball in Fig. 21.44*a*, the path of point C during a virtual displacement is actually a very flat circular arc CC_1, as shown in Fig. 21.45*a*, and point C rises by the amount

$$CC_1' = r(1 - \cos\delta\phi) \approx \frac{r}{2}\,\delta\phi^2 \tag{21.26}$$

while the center C of the ball in Fig. 21.44*c* is lowered by this amount, as shown in Fig. 21.45*b*. So far as the condition of equilibrium is concerned, it is not necessary to consider the second-order quantity (21.26) in calculating the virtual work of the force W. However, the sign of this second-order small quantity decides whenever the configurations of equilibrium are stable or unstable. When we take this into account, we see that, in the stable case, a very small amount of *negative work* is done by the active force during the virtual displacement, while, in the unstable case, a very small amount of *positive work* is produced by the active force.

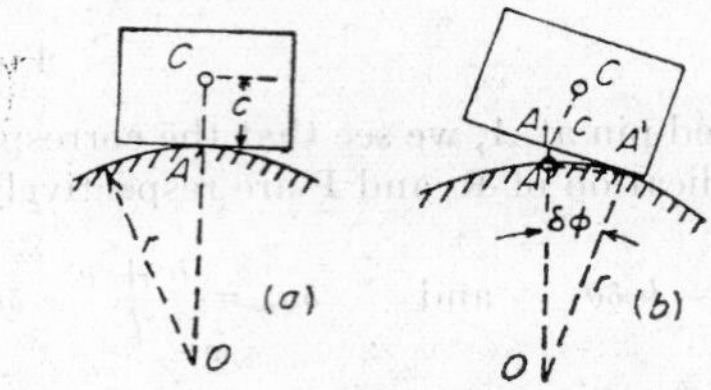

FIG. 21.46

As an example of a study of the stability of a configuration of equilibrium of an ideal system, let us consider a block with center of gravity at C, and resting with one plane face on top of a cylindrical surface of radius r, as shown in Fig. 21.46*a*. Clearly this is a configuration of equilibrium in which the weight W and the reaction at A are two equal, opposite, collinear forces. To investigate the problem of stability, we make a virtual displacement of the block compatible with its constraints. This will be an infinitesimal angle of roll $\delta\phi$, in which the point of contact moves from A to A' without slip, as shown in Fig. 21.46*b*. Consequently, the distance $A_1A' = r\,\delta\phi$, and we see that the new elevation of point C above the horizontal tangent plane at A is $r(\cos\delta\phi - 1) + r\,\delta\phi \sin\delta\phi + c\cos\delta\phi$. Using the second-order approximations

$$\cos\delta\phi \approx 1 - \frac{\delta\phi^2}{2} \qquad \text{and} \qquad \sin\delta\phi \approx \delta\phi$$

this reduces to

$$c + \frac{r - c}{2}\,\delta\phi^2$$

Thus, if $r > c$, the center of gravity of the block rises slightly during the virtual displacement, and *negative work* is done by the force W; that is, the system is stable if $r > c$.

CHAPTER 22

KINEMATICS

BY

F. H. RAVEN, Ph.D., Notre Dame, Ind.

Kinematics is concerned with the motion of particles and of rigid bodies, without regard to the forces which cause the motion.

22.1. MOTION OF A PARTICLE

The linear displacement of a particle or a point is the vector change in position. As a point moves along the path shown in Fig. 22.1 from P to P', its displacement is the vector $\Delta\mathbf{r}$. The scalar quantity Δs is the distance measured along the path, and the average linear velocity $\mathbf{v}_{av}$ is the displacement $\Delta\mathbf{r}$ divided by the time Δt required for the particle to move from position P to P'. The instantaneous linear velocity $\mathbf{v}$ of a particle at any position along its path is the instantaneous time rate of displacement:

$$\mathbf{v} = \lim_{\Delta t \to 0} \frac{\Delta \mathbf{r}}{\Delta t} = \frac{d\mathbf{r}}{dt} = \frac{ds}{dt}\frac{d\mathbf{r}}{ds} \qquad (22.1)$$

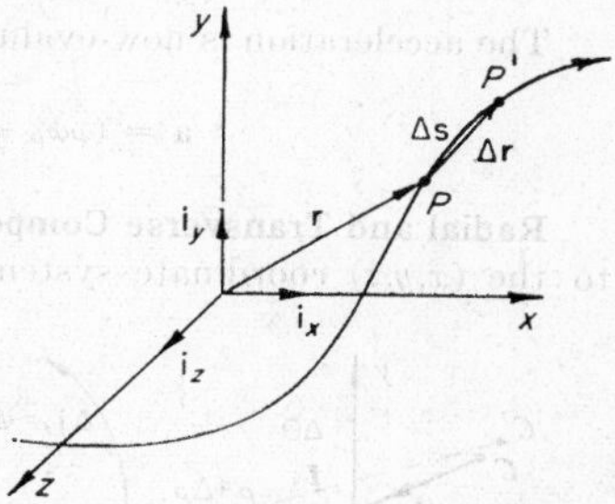

FIG. 22.1

The scalar quantity ds/dt is the instantaneous speed or magnitude of the velocity, and the term $d\mathbf{r}/ds$ is a unit vector which indicates that the direction of the instantaneous velocity is always tangent to the path.

The instantaneous linear acceleration is the time rate of change of the velocity:

$$\mathbf{a} = \lim_{\Delta t \to 0} \frac{\Delta \mathbf{v}}{\Delta t} = \frac{d\mathbf{v}}{dt} = \frac{d^2\mathbf{r}}{dt^2} \qquad (22.2)$$

Equations (22.1) and (22.2) for velocity and acceleration are in a general form, and are usually evaluated from components. The particular components which are obtained for the velocity and acceleration of a particle will depend on the method chosen to specify the position of the particle.

Rectangular Components. The path of motion of point P with respect to the (x,y,z) coordinate system is shown in Fig. 22.1. The unit vectors $\mathbf{i}_x$, $\mathbf{i}_y$, $\mathbf{i}_z$ are fixed to the x, y, and z axes, respectively. The position of point P relative to this coordinate system is

$$\mathbf{r} = x\mathbf{i}_x + y\mathbf{i}_y + z\mathbf{i}_z \qquad (22.3)$$

The velocity is the first derivative of Eq. (22.3) with respect to time, and the acceleration is the second derivative. Thus,

$$\mathbf{v} = \frac{d\mathbf{r}}{dt} = \frac{dx}{dt}\mathbf{i}_x + \frac{dy}{dt}\mathbf{i}_y + \frac{dz}{dt}\mathbf{i}_z = v_x\mathbf{i}_x + v_y\mathbf{i}_y + v_z\mathbf{i}_z \qquad (22.4)$$

$$\mathbf{a} = \frac{d\mathbf{v}}{dt} = \frac{d^2\mathbf{r}}{dt^2} = \frac{d^2x}{dt^2}\mathbf{i}_x + \frac{d^2y}{dt^2}\mathbf{i}_y + \frac{d^2z}{dt^2}\mathbf{i}_z = a_x\mathbf{i}_x + a_y\mathbf{i}_y + a_z\mathbf{i}_z \qquad (22.5)$$

The direction of the resultant velocity must be tangent to the path of motion, but the acceleration may have any direction, and is obtained by evaluating Eq. (22.5).

Tangential and Normal Components. When the position of a point P is specified with respect to the center of curvature of its path of motion, then the acceleration will be expressed in terms of components which are tangential and normal to the path. The velocity, of course, can have only a tangential component. As shown in Fig. 22.2, the position vector of point P with respect to its center of curvature C is $\boldsymbol{\rho} = \rho \mathbf{i}_n$, where ρ is the radius of curvature of the path, and $\mathbf{i}_n$ is a unit vector which is normal to the path.

The instantaneous linear velocity is evaluated as

$$\mathbf{v} = v\mathbf{i}_t = \lim_{\Delta t \to 0} \frac{\Delta s}{\Delta t}\mathbf{i}_t = \lim_{\Delta t \to 0} \rho \frac{\Delta \theta}{\Delta t}\mathbf{i}_t = \rho\omega_\rho \mathbf{i}_t \tag{22.6}$$

where ω_ρ is the magnitude of the instantaneous angular velocity of the radius of curvature, and the vector $\mathbf{i}_t$ is a unit vector which is tangent to the path at the point P.

The acceleration is obtained by differentiation of Eq. (22.6):

$$\mathbf{a} = (\rho\dot{\omega}_\rho + \dot{\rho}\omega_\rho)\mathbf{i}_t + \rho\omega_\rho(\mathbf{i}_t)^{\cdot}$$

where $$(\mathbf{i}_t)^{\cdot} = \lim_{\Delta t \to 0} \frac{\Delta \mathbf{i}_t}{\Delta t} = -\lim_{\Delta t \to 0} \frac{\Delta \theta}{\Delta t}\mathbf{i}_n = -\omega_\rho \mathbf{i}_n$$

The acceleration is now evaluated as

$$\mathbf{a} = (\rho\dot{\omega}_\rho + \dot{\rho}\omega_\rho)\mathbf{i}_t - \rho\omega_\rho{}^2\mathbf{i}_n = \dot{v}\mathbf{i}_t - \frac{v^2}{\rho}\mathbf{i}_n \tag{22.7}$$

Radial and Transverse Components. Let the position of the point P with respect to the (x,y,z) coordinate system shown in Fig. 22.3 be designated by the vector

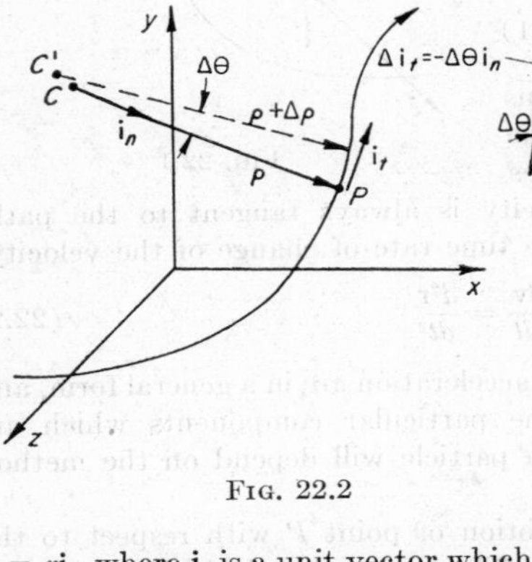

FIG. 22.2

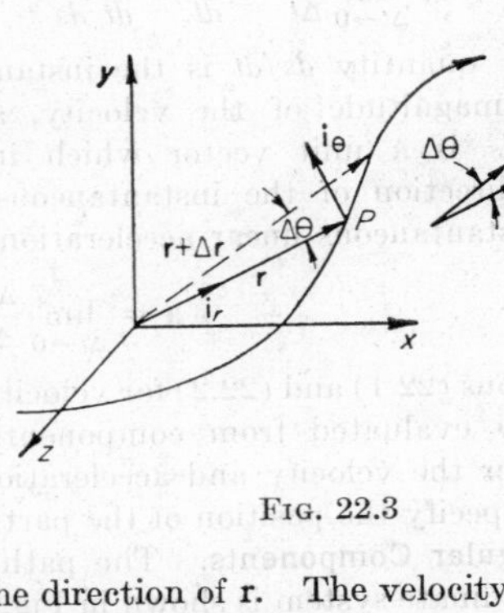

FIG. 22.3

$\mathbf{r} = r\mathbf{i}_r$, where $\mathbf{i}_r$ is a unit vector which has the direction of $\mathbf{r}$. The velocity of point P is obtained by differentiating this position vector:

$$\mathbf{v} = \dot{r}\mathbf{i}_r + r(\mathbf{i}_r)^{\cdot}$$

where $$(\mathbf{i}_r)^{\cdot} = \lim_{\Delta t \to 0} \frac{\Delta \mathbf{i}_r}{\Delta t} = \lim_{\Delta t \to 0} \frac{\Delta \theta}{\Delta t}\mathbf{i}_\theta = \omega_r \mathbf{i}_\theta$$

and $\mathbf{i}_\theta$ is a unit vector in the transverse direction and lies in the plane determined by $\mathbf{r}$ and $(\mathbf{r} + \Delta\mathbf{r})$. The term ω_r is the magnitude of the angular velocity of the vector $\mathbf{r}$. Substituting the value of $(\mathbf{i}_r)^{\cdot}$ into the equation for the velocity $\mathbf{v}$ yields

$$\mathbf{v} = \dot{r}\mathbf{i}_r + r\omega_r \mathbf{i}_\theta \tag{22.8}$$

The acceleration is

$$\mathbf{a} = \dot{r}(\mathbf{i}_r)^{\cdot} + \ddot{r}\mathbf{i}_r + (r\dot{\omega}_r + \dot{r}\omega_r)\mathbf{i}_\theta + r\omega_r(\mathbf{i}_\theta)^{\cdot}$$

where $(\mathbf{i}_\theta)^{\cdot} = -\omega_r\mathbf{i}_r$. The acceleration in terms of radial and transverse components is

$$\mathbf{a} = (\ddot{r} - r\omega_r^2)\mathbf{i}_r + (r\dot{\omega}_r + 2\dot{r}\omega_r)\mathbf{i}_\theta \tag{22.9}$$

Cylindrical Coordinates. In Fig. 22.4, the position of point P is given by $\mathbf{R} = r\mathbf{i}_r + z\mathbf{i}_z$. It is seen that $r = R \sin \phi$ and $z = R \cos \phi$. The unit vector $\mathbf{i}_r$ is parallel to the vector $\mathbf{r}$, the unit vector $\mathbf{i}_z$ is parallel to the z axis, and $\mathbf{i}_\theta$ is perpendicular

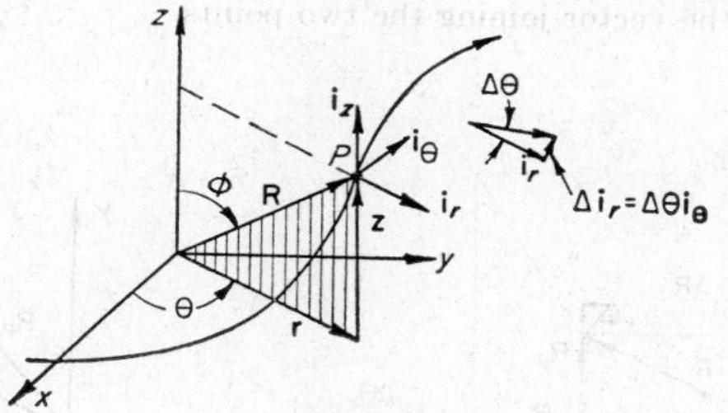

FIG. 22.4

to both $\mathbf{i}_r$ and $\mathbf{i}_z$. The velocity of point P is

$$\mathbf{v} = \frac{d\mathbf{R}}{dt} = \dot{r}\mathbf{i}_r + r(\mathbf{i}_r)^{\cdot} + \dot{z}\mathbf{i}_z$$

The substitution of $(\mathbf{i}_r)^{\cdot} = (d\theta/dt)\,\mathbf{i}_\theta = \dot{\theta}\mathbf{i}_\theta$ into the preceding expression yields

$$\mathbf{v} = \dot{r}\mathbf{i}_r + r\dot{\theta}\mathbf{i}_\theta + \dot{z}\mathbf{i}_z \tag{22.10}$$

Differentiation of Eq. (22.10) yields the following equation for the acceleration of point P:

$$\mathbf{a} = \ddot{r}\mathbf{i}_r + \dot{r}(\mathbf{i}_r)^{\cdot} + (\dot{r}\dot{\theta} + r\ddot{\theta})\mathbf{i}_\theta + r\dot{\theta}(\mathbf{i}_\theta)^{\cdot} + \ddot{z}\mathbf{i}_z$$

Noting that $(\mathbf{i}_\theta)^{\cdot} = -\dot{\theta}\mathbf{i}_r$, then

$$\mathbf{a} = (\ddot{r} - r\dot{\theta}^2)\mathbf{i}_r + (2\dot{r}\dot{\theta} + r\ddot{\theta})\mathbf{i}_\theta + \ddot{z}\mathbf{i}_z \tag{22.11}$$

Angular Velocity. In Fig. 22.5 is shown a fixed point P in the (x,y,z) coordinate system. Its position vector is $\mathbf{R} = R_x\mathbf{i}_x + R_y\mathbf{i}_y + R_z\mathbf{i}_z$. As is seen from Fig. 22.5, if the coordinate system rotates about the z axis an angle $\Delta\theta_z$, then the motion $\Delta\mathbf{R}_3$ is

$$\Delta\mathbf{R}_3 = R_x\,\Delta\theta_z\mathbf{i}_y - R_y\,\Delta\theta_z\mathbf{i}_x$$

Similarly, for a rotation $\Delta\theta_y$ or $\Delta\theta_x$ about the y or x axis, respectively, $\mathbf{R}$ changes by

$$\Delta\mathbf{R}_2 = R_z\,\Delta\theta_y\mathbf{i}_x - R_x\,\Delta\theta_y\mathbf{i}_z \qquad \Delta\mathbf{R}_1 = R_y\,\Delta\theta_x\mathbf{i}_z - R_z\,\Delta\theta_x\mathbf{i}_y$$

The total change in the position of point P is

$$\Delta\mathbf{R} = \Delta\mathbf{R}_1 + \Delta\mathbf{R}_2 + \Delta\mathbf{R}_3$$

The velocity of point P relative to the origin of the coordinate system is

$$\mathbf{v} = \lim_{\Delta t \to 0} \frac{\Delta\mathbf{R}}{\Delta t} = \begin{vmatrix} \mathbf{i}_x & \mathbf{i}_y & \mathbf{i}_z \\ \omega_x & \omega_y & \omega_z \\ R_x & R_y & R_z \end{vmatrix} = \boldsymbol{\omega} \times \mathbf{R} \tag{22.12}$$

where $\omega_x = \lim (\Delta\theta_x/\Delta t)$, . . . ; $\boldsymbol{\omega} = \omega_x\mathbf{i}_x + \omega_y\mathbf{i}_y + \omega_z\mathbf{i}_z$ is the angular velocity of the coordinate system; and $\mathbf{R}$ is the vector from the origin to the fixed point P. The vector $\boldsymbol{\omega}$ is the total angular velocity of the coordinate system. Because the vector $\mathbf{R}$ is fixed in this system, then $\mathbf{R}$ is also moving with angular velocity $\boldsymbol{\omega}$. The cross product $\boldsymbol{\omega} \times \mathbf{R}$ is the linear velocity of point P relative to the origin of the vector $\mathbf{R}$.

For any two points P_1 and P_2 fixed in the (x,y,z) coordinate system and located at positions $\mathbf{R}_1$ and $\mathbf{R}_2$, respectively, from the origin of the coordinate system, the linear velocity of point P_1 with respect to P_2 is obtained by application of Eq. (22.12).

$$\mathbf{v}_{P_1} - \mathbf{v}_{P_2} = \boldsymbol{\omega} \times \mathbf{R}_1 - \boldsymbol{\omega} \times \mathbf{R}_2 = \boldsymbol{\omega} \times (\mathbf{R}_1 - \mathbf{R}_2)$$

where $(\mathbf{R}_1 - \mathbf{R}_2)$ is the vector joining the two points.

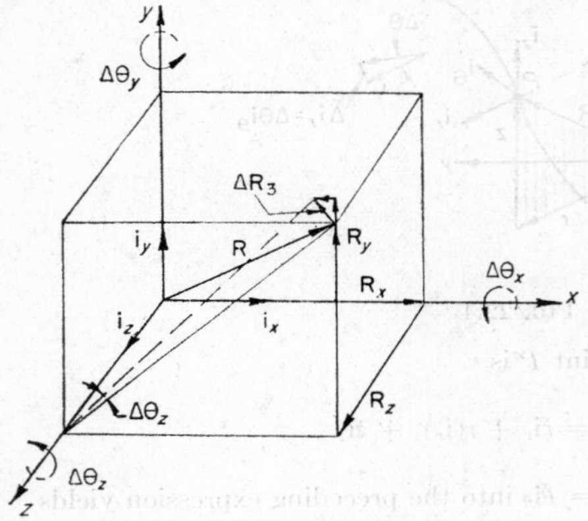

FIG. 22.5

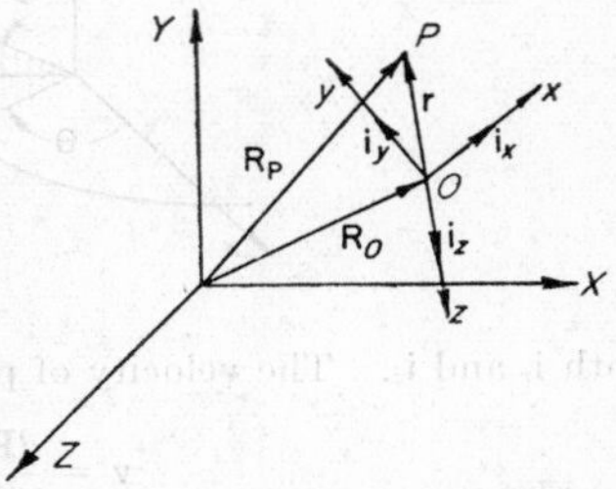

FIG. 22.6

If the fixed point P is selected at the tip of the arrowhead of one of the unit vectors $\mathbf{i}_x$, $\mathbf{i}_y$, $\mathbf{i}_z$ in Fig. 22.6, then application of Eq. (22.12) yields

$$(\mathbf{i}_x)^{\cdot} = \boldsymbol{\omega} \times \mathbf{i}_x \qquad (\mathbf{i}_y)^{\cdot} = \boldsymbol{\omega} \times \mathbf{i}_y \qquad (\mathbf{i}_z)^{\cdot} = \boldsymbol{\omega} \times \mathbf{i}_z$$

Relative Motion. In the preceding analyses, the motion of the particle was described with respect to a reference set of coordinate axes. When the coordinate axes are fixed to the earth, then the motion is called absolute motion. The term relative motion generally designates that the motion is relative to something which is not fixed to the earth.

Let points P and O in Fig. 22.6 be any two points moving in the fixed (X,Y,Z) coordinate system. The position, velocity, and acceleration of point P are given by

$$\mathbf{R}_P = \mathbf{R}_O + \mathbf{r} \qquad \mathbf{v}_P = \mathbf{v}_O + \mathbf{v} \qquad \mathbf{a}_P = \mathbf{a}_O + \mathbf{a} \tag{22.13}$$

Thus, the motion of any point P is seen to be the vector sum of the motion of any point O plus the motion of point P with respect to point O.

Moving Coordinate System. Let the motion of point P in Fig. 22.6 be known with respect to the (x,y,z) coordinate system, which is moving relative to the (X,Y,Z) coordinate system. It is seen that the position of point P is given by the equation

$$\mathbf{R}_P = \mathbf{R}_O + \mathbf{r} \tag{22.14}$$

The unit vectors $\mathbf{i}_x$, $\mathbf{i}_y$, and $\mathbf{i}_z$ are fixed to the x, y, and z axes, respectively, of the moving coordinate system, so that

$$\mathbf{r} = x\mathbf{i}_x + y\mathbf{i}_y + z\mathbf{i}_z$$

Differentiating this equation with respect to time yields

$$\dot{\mathbf{r}} = (\dot{x}\mathbf{i}_x + \dot{y}\mathbf{i}_y + \dot{z}\mathbf{i}_z) + [x(\mathbf{i}_x)^{\cdot} + y(\mathbf{i}_y)^{\cdot} + z(\mathbf{i}_z)^{\cdot}]$$

Since $(\mathbf{i}_x)^{\cdot} = \boldsymbol{\omega} \times \mathbf{i}_x$, etc., where $\boldsymbol{\omega}$ is the angular velocity of the moving coordinate system relative to the fixed system, the expression for $\dot{\mathbf{r}}$ becomes

$$\dot{\mathbf{r}} = \mathbf{v}_m + \boldsymbol{\omega} \times \mathbf{r}$$

where $\mathbf{v}_m = \dot{x}\mathbf{i}_x + \dot{y}\mathbf{i}_y + \dot{z}\mathbf{i}_z$ is the velocity of P with respect to the moving coordinate system.

Differentiating Eq. (22.14) yields the following expression for the velocity of P with respect to the (X,Y,Z) coordinate system:

$$\dot{\mathbf{R}}_P = \dot{\mathbf{R}}_O + \dot{\mathbf{r}} = \dot{\mathbf{R}}_O + \mathbf{v}_m + \boldsymbol{\omega} \times \mathbf{r} \tag{22.15}$$

The acceleration of the point P with respect to the fixed (X,Y,Z) coordinate system is obtained from Eq. (22.15) as follows:

$$\begin{aligned}\ddot{\mathbf{R}}_P &= \ddot{\mathbf{R}}_O + (\ddot{x}\mathbf{i}_x + \ddot{y}\mathbf{i}_y + \ddot{z}\mathbf{i}_z) + [\dot{x}(\mathbf{i}_x)^{\cdot} + \dot{y}(\mathbf{i}_y)^{\cdot} + \dot{z}(\mathbf{i}_z)^{\cdot}] + \dot{\boldsymbol{\omega}} \times \mathbf{r} + \boldsymbol{\omega} \times \dot{\mathbf{r}} \\ &= \ddot{\mathbf{R}}_O + \mathbf{a}_m + \boldsymbol{\omega} \times \mathbf{v}_m + \dot{\boldsymbol{\omega}} \times \mathbf{r} + \boldsymbol{\omega} \times \mathbf{v}_m + \boldsymbol{\omega} \times (\boldsymbol{\omega} \times \mathbf{r}) \\ \ddot{\mathbf{R}}_P &= \ddot{\mathbf{R}}_O + \mathbf{a}_m + 2\boldsymbol{\omega} \times \mathbf{v}_m + \dot{\boldsymbol{\omega}} \times \mathbf{r} + \boldsymbol{\omega} \times (\boldsymbol{\omega} \times \mathbf{r})\end{aligned} \tag{22.16}$$

where $\mathbf{a}_m$ is the acceleration of point P relative to the moving coordinate system, and the term $2\boldsymbol{\omega} \times \mathbf{v}_m$ is known as the Coriolis component of acceleration.

22.2. MOTION OF A RIGID BODY

Absolute Motion. Let the (x,y,z) coordinate system shown in Fig. 22.7 be fixed to the rigid body, which is moving with respect to the fixed (X,Y,Z) coordinate system.

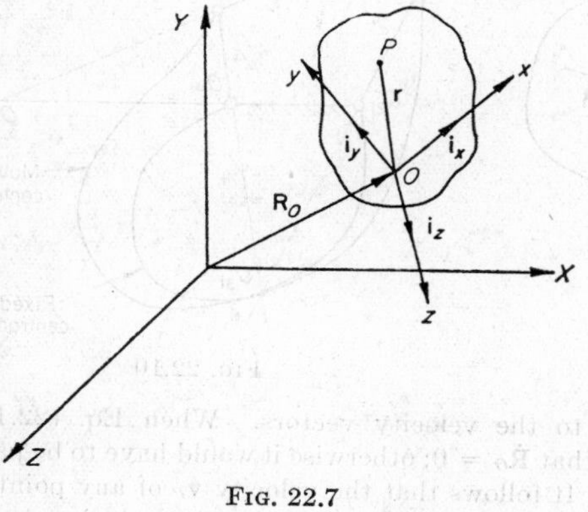

FIG. 22.7

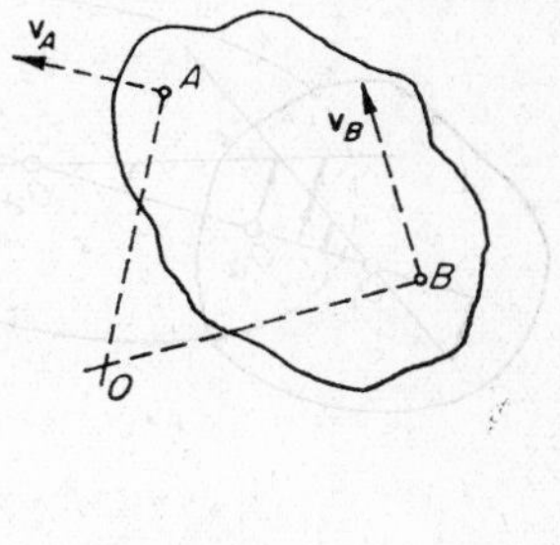

FIG. 22.8

The point P is a fixed point in the rigid body, and the unit vectors, $\mathbf{i}_x$, $\mathbf{i}_y$, and $\mathbf{i}_z$, are attached to the x, y, and z axes, respectively. The absolute velocity and acceleration of point P are obtained directly from Eqs. (22.15) and (22.16) by noting that the terms $\mathbf{v}_m$ and $\mathbf{a}_m$ are both zero, because point P has no motion with respect to the (x,y,z) coordinate system. Thus,

$$\mathbf{v}_P = \dot{\mathbf{R}}_P = \dot{\mathbf{R}}_O + \boldsymbol{\omega} \times \mathbf{r} \tag{22.17}$$

$$\mathbf{a}_P = \ddot{\mathbf{R}}_P = \ddot{\mathbf{R}}_O + \dot{\boldsymbol{\omega}} \times \mathbf{r} + \boldsymbol{\omega} \times (\boldsymbol{\omega} \times \mathbf{r}) \tag{22.18}$$

where $\boldsymbol{\omega}$ is the angular velocity of the rigid body with respect to the fixed (X,Y,Z) system, and, similarly, $\dot{\boldsymbol{\omega}}$ is the angular acceleration of the rigid body.

Relative Motion. The velocity of any fixed point P in the rigid body with respect to any other fixed point O is simply

$$\mathbf{v}_{P/O} = \dot{\mathbf{R}}_P - \dot{\mathbf{R}}_O = \boldsymbol{\omega} \times \mathbf{r} \tag{22.19}$$

Thus, the velocity of point P relative to point O is the same as though the body were rotating with an angular velocity $\boldsymbol{\omega}$ about the point O considered to be fixed.

Similarly, the acceleration of point P with respect to point O is seen to be

$$\mathbf{a}_{P/O} = \ddot{\mathbf{R}}_P - \ddot{\mathbf{R}}_O = \dot{\boldsymbol{\omega}} \times \mathbf{r} + \boldsymbol{\omega} \times (\boldsymbol{\omega} \times \mathbf{r}) \tag{22.20}$$

This is the same result that would be obtained if the motion of the body were purely a rotation about a fixed point O. Therefore, both the velocity and the acceleration of any fixed point in a rigid body, relative to another fixed point, are computed by considering the body as rotating about the second point.

Translation and Rotation. From Eq. (22.17) it is seen that the instantaneous velocity of any fixed point P in a rigid body may be considered to be the vector sum of a translational velocity $\dot{\mathbf{R}}_O$ plus a velocity due to the angular velocity $\boldsymbol{\omega}$ of the body about any point O. Similarly, for finite motions, it can be shown that the motion of any point in a rigid body may be regarded as the sum of a translation plus a rotation about any point. When $\boldsymbol{\omega} = 0$, then every point in the body has the same velocity. For this case, the motion of the body is entirely translational.

Plane Motion, Instant Center. Suppose that, for a rigid body in plane motion (Fig. 22.8), the velocities of two points, A and B, are known. The point O is the

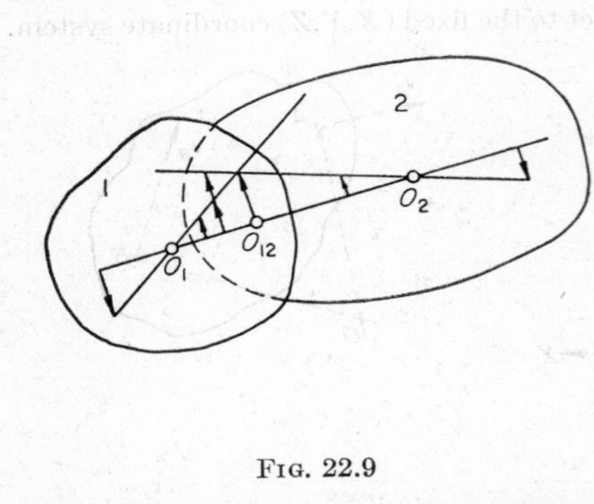

FIG. 22.9

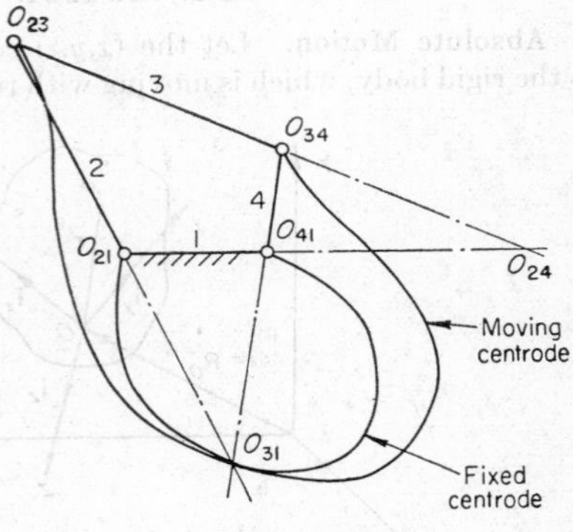

FIG. 22.10

intersection of the perpendiculars to the velocity vectors. When Eq. (22.17) is applied to points A and B, it is seen that $\dot{\mathbf{R}}_O = 0$; otherwise it would have to be parallel to $\mathbf{v}_A$ and $\mathbf{v}_B$, which is impossible. It follows that the velocity $\mathbf{v}_P$ of any point P of this body can be derived from a pure rotation about O with angular velocity ω. This point O is called the instant center.

For the case of general three-dimensional motion, there is no instant center of velocity. For example, the motion of a screw is such that all points have translation in addition to the rotation, and thus no point is at rest. However, a three-dimensional motion can always be described as the sum of a rotation about a certain axis and a translation in the direction of this axis.

Kennedy's Theorem. Figure 22.9 shows two rigid bodies in plane motion. Their instant centers are O_1 and O_2. For all points on the straight line O_1O_2 (including its extensions to both sides), the velocity is perpendicular to this line, no matter to which of the two bodies the point is attached. Somewhere on this line there is a point O_{12}, for which the velocity of both bodies is the same. The relative motion of either body with respect to the other is a rotation about O_{12}. This point is called the instant center of relative motion, and it is seen that, for any two rigid bodies, the instant centers O_1, O_2, O_{12} of absolute and relative motion lie on a straight line. Evidently, also, the more general theorem holds, that the instant centers O_{12}, O_{23}, O_{31} of the relative motion of three rigid bodies lie on a straight line (Kennedy's theorem).

Instant Center of Acceleration. For a rigid body having plane motion, the point which has no acceleration with respect to the (X,Y) coordinate system is called the instant center of acceleration. This will be the one point O for which $\ddot{\mathbf{R}}_O$ is zero. From Eq. (22.18), it is seen that the acceleration of any point P is computed by regarding the motion of the body as a pure rotation about its instant center of acceleration. In general, the instant center of velocity and the instant center of acceleration will not be the same point in the body. The term *instant center* when unmodified means the instant center of velocity.

Centrodes. The locus of the path traced by an instant center on a body or an extension of it is called a centrode. When this path is described on a fixed body, it is further distinguished by the term *fixed centrode,* and the locus of an instant center on a moving body is called a *moving centrode.*

For the four-bar linkage shown in Fig. 22.10, the path traced by point O_{31} relative to link 3 extended is the moving centrode $O_{23}O_{31}O_{34}$. Similarly, the locus of O_{31} with respect to link 1 is the fixed centrode $O_{21}O_{31}O_{41}$. Because there is no relative motion between bodies 1 and 3 at point O_{31}, it follows that, if the moving centrode moved with pure rolling motion relative to the fixed centrode, then the motion of link 3 with respect to link 1 would be identical with that given by the original mechanism.

22.3. METHODS FOR MOTION ANALYSIS OF MECHANISMS

The preceding principles governing the motion of particles and of rigid bodies form the basis for several methods of velocity and acceleration analysis of mechanisms.

Velocity Analysis by Composition and Resolution. The method of composition and resolution is most suited for the determination of the velocity of a point which has a known path of motion. The basis of this method is the fact that the component of velocity along the line connecting two points in a rigid body must be the same for the two points. This is illustrated for the slider crank mechanism shown in Fig. 22.11, where the velocity $\mathbf{v}_A$ is known and it is desired to determine the velocity of point B. Because the component $\mathbf{v}_{A3}$ along link 3 of the known velocity $\mathbf{v}_A$ must be equal to the component $\mathbf{v}_{B3}$ of the velocity of point B, then, knowing $\mathbf{v}_{B3}$ and the direction of motion of point B, the velocity $\mathbf{v}_B$ may be determined.

This method may be extended to members in direct contact by noting that, at the point of contact, the adjacent points on the two members must have the same component of velocity along the common normal to the surfaces.

Velocity Analysis by Instant Centers. Velocity analysis by means of instant centers is based upon the following characteristics of instant centers. (1) The magnitude of the velocity of any point in one body relative to another body is proportional to the distance from the point to the instant center, and the direction of this velocity is perpendicular to the line connecting the instant center of motion and the point. (2) This instant center is the point at which there is no relative motion between the two bodies, and, therefore, at this center, the two bodies have the same velocity relative to a third body.

The proportionality property makes it possible to determine the velocity of any point on a link if the velocity of one point is known. The characteristic of no relative motion between two bodies at their common instant center enables one to transfer the velocity of the instant center from one link to another. When a velocity is transferred via the common instant center, this center is termed a *transfer point.*

In Fig. 22.12, let the absolute velocity (i.e., relative to link 1) of point A, which is coincident with O_{23}, be known, and let it be desired to determine the absolute velocity of point B, which is coincident with O_{34}. Since points A and B may both be considered to be on link 3, the proportionality property is used to determine the magnitude v_B. In Fig. 22.12, it is seen that triangle $AO_{31}B$ is similar to triangle $A'O_{31}B'$, and thus $v_B/v_A = BB'/AA'$. The direction of $\mathbf{v}_B$ is perpendicular to the line joining point B and the center O_{31}. The magnitude of the angular velocity of link 3 is

$$\omega_3 = \frac{v_A}{AO_{31}} = \frac{v_B}{BO_{31}}$$

and the direction is seen to be clockwise.

Another approach to this problem is to regard points A and B as being on links 2 and 4, respectively, and use point O_{24} as a transfer point. Considering point O_{24} as a point on body 2, its absolute velocity is determined by using the proportionality prop-

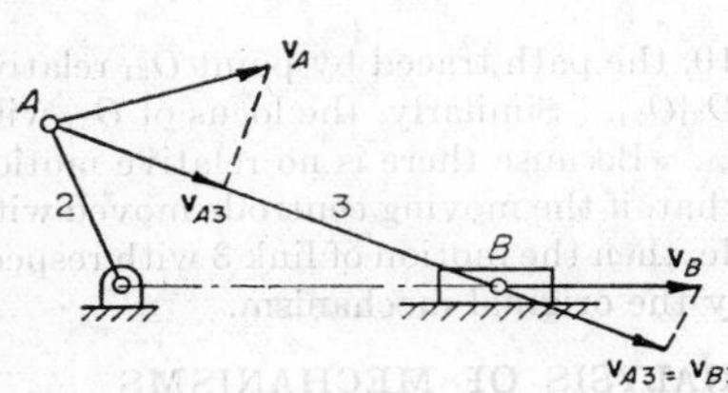

FIG. 22.11

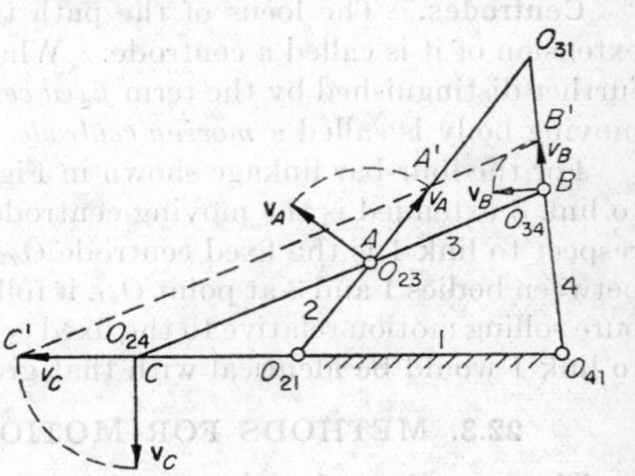

FIG. 22.12

erty. Because triangle $AO_{21}C$ is similar to triangle $A'O_{21}C'$, then $v_C/v_A = CC'/AA'$. By considering point O_{24} as a point on body 4, it is now possible to determine the velocity of any other point on body 4.

Velocity and Acceleration Analysis by Vector Diagrams. In this method, the equations of relative motion are solved by drawing vector diagrams. Images of the links in the original mechanism appear in the vector diagrams, and the velocity or acceleration of any point in a link is represented by the vector drawn from the origin of the vector diagram to the corresponding point in the vector image.

The application of this method to a slider crank mechanism is illustrated in Fig. 22.13. The known velocity of point A is drawn to scale. The direction of $\mathbf{v}_{B/A}$ must be perpendicular to line AB, and the direction of $\mathbf{v}_B$ is also known. Therefore, the equation $\mathbf{v}_B = \mathbf{v}_A + \mathbf{v}_{B/A}$ may be solved vectorially as shown. Since $\mathbf{v}_{B/A}$, $\mathbf{v}_{A/C}$, and $\mathbf{v}_{B/C}$ are perpendicular to the lines AB, AC, and BC, respectively, triangle ABC is similar to triangle $A'B'C'$ and rotated by 90°. The magnitude of the angular velocity of link 3 is $\omega_3 = v_{B/A}/AB = v_{A/C}/AC = v_{B/C}/BC$, and the direction is seen to be counterclockwise. The velocity image of link 2 is seen to be the line OA', and the image of the slider is the point B', as all points in the slider have the same velocity. The acceleration vector diagram is similarly obtained by solving the acceleration equations for the mechanism.

With this method, care must be taken to use the equations which properly describe the motion of one point relative to another. This is of particular importance in constructing acceleration diagrams which may involve Coriolis components of acceleration, in addition to the usual normal and tangential components.

Method of Independent Position Equations. A position equation is a vector equation which expresses the position of a point of a mechanism as a function of the geometry of the mechanism. To write a position equation, we start from a point whose position is known (e.g., the hinge O_4 in Fig. 22.14), and follow a "path" from one link to the next. A loop equation is obtained when the path is closed, i.e., when it leads back to the starting point. Depending on the geometry of the mechanism, there are one or more independent loop equations. They are algebraic equations from which unknown position quantities may be calculated. The first and second time derivatives of the loop equation yield velocities and accelerations of the links.

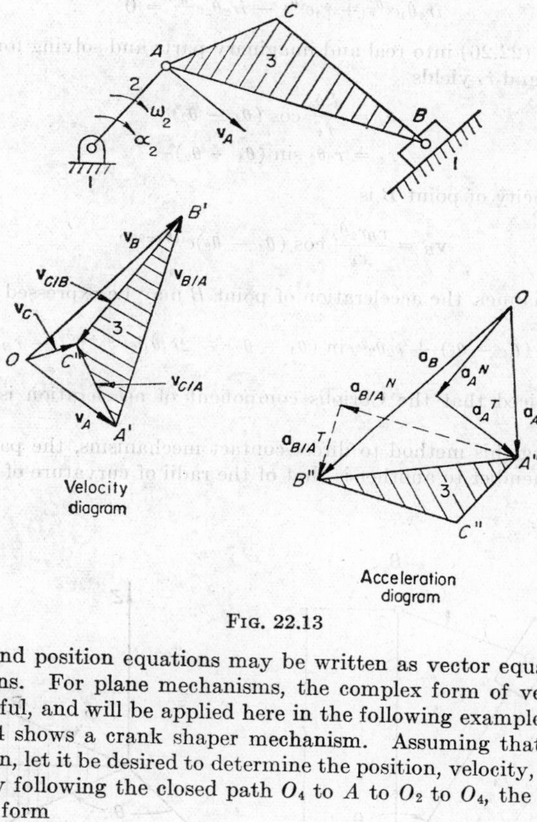

FIG. 22.13

The loop and position equations may be written as vector equations or as component equations. For plane mechanisms, the complex form of vectors $\mathbf{r} = re^{i\theta}$ is particularly useful, and will be applied here in the following example.

Figure 22.14 shows a crank shaper mechanism. Assuming that r_2, r_B, l, θ_2, $\dot{\theta}_2$, and $\ddot{\theta}_2$ are known, let it be desired to determine the position, velocity, and acceleration of point B. By following the closed path O_4 to A to O_2 to O_4, the loop equation is obtained in the form

$$r_4e^{i\theta_4} - r_2e^{i\theta_2} - il = 0$$

where $e^{i\theta}$ is the complex-plane designation for a unit vector which is rotated an angle θ counterclockwise from the real axis, as is shown in Fig. 22.14.

The two unknown position terms θ_4 and r_4 may be evaluated from the preceding equation. Thus,

$$\tan \theta_4 = \frac{r_2 \sin \theta_2 + l}{r_2 \cos \theta_2} \tag{22.21}$$

and

$$r_4{}^2 = r_2{}^2 + 2r_2l \sin \theta_2 + l^2 \tag{22.22}$$

The simplest position equation for point B is seen to be

$$\mathbf{r}_B = r_Be^{i\theta_4} \tag{22.23}$$

Successive differentiation of Eq. (22.23) yields, for the velocity and acceleration of point B:

$$\mathbf{v}_B = \frac{d\mathbf{r}_B}{dt} = ir_B\dot{\theta}_4 e^{i\theta_4} = r_B\dot{\theta}_4 e^{i(\theta_4+90°)} \tag{22.24}$$

$$\mathbf{a}_B = \frac{d\mathbf{v}_B}{dt} = r_B(i\ddot{\theta}_4 - \dot{\theta}_4^2)e^{i\theta_4} \tag{22.25}$$

The first time derivative of the loop equation is

$$ir_4\dot{\theta}_4 e^{i\theta_4} + \dot{r}_4 e^{i\theta_4} - ir_2\dot{\theta}_2 e^{i\theta_2} = 0 \tag{22.26}$$

Separation of Eq. (22.26) into real and imaginary parts and solving for the unknown velocity terms $\dot{\theta}_4$ and $\dot{r}_4$ yields

$$\dot{\theta}_4 = \frac{r_2\dot{\theta}_2}{r_4}\cos(\theta_4 - \theta_2) \tag{22.27}$$

and

$$\dot{r}_4 = r_2\dot{\theta}_2 \sin(\theta_4 - \theta_2) \tag{22.28}$$

Thus, the velocity of point B is

$$\mathbf{v}_B = \frac{r_B r_2 \dot{\theta}_2}{r_4}\cos(\theta_4 - \theta_2)e^{i(\theta_4+90°)} \tag{22.29}$$

In a similar manner, the acceleration of point B may be expressed as

$$\mathbf{a}_B = \frac{r_B}{r_4}[r_2\ddot{\theta}_2\cos(\theta_4 - \theta_2) + r_2\dot{\theta}_2^2\sin(\theta_4 - \theta_2) - 2\dot{r}_4\dot{\theta}_4]e^{i(\theta_4+90°)} - r_B\dot{\theta}_4^2 e^{i\theta_4} \tag{22.30}$$

It should be noticed that the Coriolis component of acceleration is automatically included.

When applying this method to direct-contact mechanisms, the path to follow in going from one member to another is that of the radii of curvature of the surfaces at

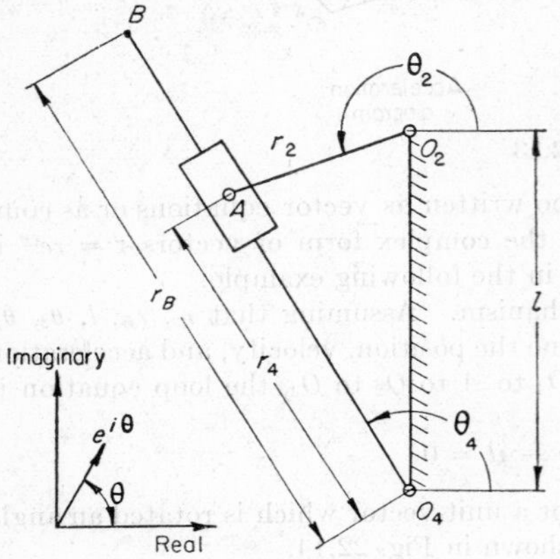

FIG. 22.14

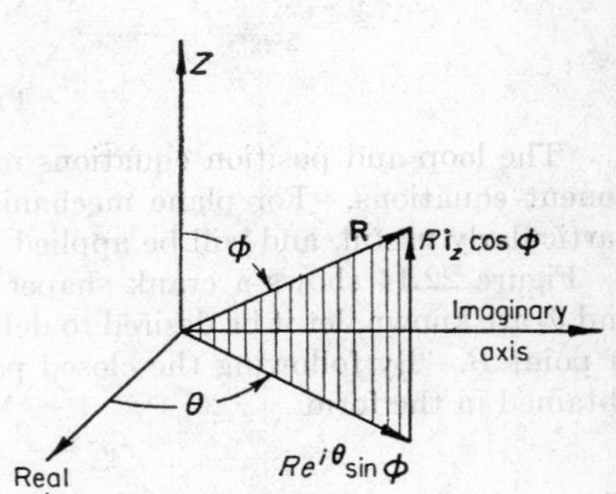

FIG. 22.15

the point of contact. For position, velocity, and acceleration analysis, these radii of curvature may be considered to remain constant for an incremental motion of the members.

This same method is readily applied to spatial mechanisms by using the complex-plane representation shown in Fig. 22.15 for designating position vectors in space. The projection of $\mathbf{R}$ on the complex plane is $e^{i\theta}R\sin\phi$, where $e^{i\theta}$ is a unit vector at an angle θ from the real axis. The vertical component of **R** is $\mathbf{i}_z R \cos \phi$. Thus

$$\mathbf{R} = R(e^{i\theta} \sin \phi + \mathbf{i}_z \cos \phi)$$

The advantage of using this notation, as opposed to the cylindrical notation of Fig. 22.4, is that the exponential form of the unit vector may be differentiated directly, whereas differentiation of the Gibbs-type vector $\mathbf{i}_r$ requires a physical interpretation of the derivative. Because $\mathbf{i}_z$ is always vertical, its time derivatives are zero.

22.4. REFERENCES

[1] L. Brand: "Vectorial Mechanics," Wiley, New York, 1930.
[2] C. W. Ham, E. J. Crane, W. L. Rogers: "Mechanics of Machinery," 4th ed., McGraw-Hill, New York, 1958.
[3] G. W. Housner, D. E. Hudson: "Applied Mechanics—Dynamics," Van Nostrand, Princeton, N.J., 1950.
[4] H. H. Mabie, F. W. Ocvirk: "Mechanisms and Dynamics of Machinery," Wiley, New York, 1957.
[5] R. T. Hinkle: "Kinematics of Machines," 2d ed., Prentice-Hall, Englewood Cliffs, N.J., 1960.
[6] F. H. Raven: Velocity and acceleration analysis of plane and space mechanisms by means of independent position equations, *J. Appl. Mech.*, **25** (1958), 1–6.
[7] F. H. Raven: Position, velocity, and acceleration analysis and kinematic synthesis of plane and space mechanisms by independent position equations, Diss., Cornell Univ., 1958.

CHAPTER 23

DYNAMICS

BY

G. W. HOUSNER, Ph.D., Pasadena, Calif.

23.1. INTRODUCTION

Newton's Equations

A study of dynamics presumes the existence of a basic frame of reference, with respect to which the motion is measured, and a reference event, with respect to which time is measured. It will be assumed that these concepts are sufficiently understood so that they may be left undefined. A satisfactory basic reference frame for engineering problems will vary according to the problem under consideration, and it is understood to be one in which Newton's second law is sufficiently accurate when the acceleration is measured with respect to the reference frame.

Although preferred definitions of force and mass are still being debated, it is customary, in engineering, to define force by means of a physical process, such as the force of gravity, or the force exerted by a spring balance. With a definition of force, it then follows that the mass of a body is defined by Newton's second law as the ratio of the force to the acceleration produced by it. A common statement of Newton's laws is the following:

1. Every body persists in a state of rest or of uniform motion in a straight line, except insofar as it may be compelled by force to change that state.

2. The time rate of change of momentum is equal to the force producing it, and the change takes place in the direction in which the force is acting,

$$F = \frac{d}{dt}(mv) \tag{23.1}$$

3 To every action there is an equal and opposite reaction, or the mutual actions of any two bodies are always equal and oppositely directed.

These statements are interpreted as summing up the results of experimental investigations on a class of physical processes (classical mechanics), and their validity rests on the fact that observations on these processes subsequent to Newton's time are in agreement with them. The v in the second law refers to the velocity of the point coinciding with the center of mass of the body. If the body is of infinitesimal size, having mass Δm, it is said to be a *particle*.

Dynamical Units

By international agreement, the *unit of mass* is defined as the mass of a particular platinum-iridium cylinder, called the international prototype kilogram, which is in the possession of the International Committee of Weights and Measures at Sèvres, France. In the United States, the *pound mass avoirdupois* has been defined legally by Congress as the 1/2.2046 part of the international kilogram. The *pound force* is defined as the gravitational force exerted on a standard pound mass when the acceleration of gravity has the "standard" value of 32.174 ft/sec². An empirical formula for gravitational acceleration at a latitude ϕ and an elevation h ft is

$$g = 32.089(1 + 0.00524 \sin^2 \phi)(1 - 0.000000096h) \text{ ft/sec}^2 \tag{23.2}$$

In engineering the pound force is taken as the unit of force, and it then follows from Newton's second law that the engineering unit of mass is that mass which is given an acceleration of 1 ft/sec² by a force of 1 lb force. This unit of mass is called a *slug* and is equal to 32.174 lb mass.

The unit of time is the *second*, which is defined as the 1/86,400 part of a mean solar day. The *mean solar day* is the yearly average of the time intervals between successive transits of the sun past a meridian of the earth.

The international standard of length is the *standard meter*, which is defined as the distance, at 0°C, between two lines on a platinum-iridium bar in the possession of the International Committee of Weights and Measures. The United States *yard* is defined legally by Congress as the 3600/3937 part of the standard meter, and the *foot* is defined as one-third of a yard.

Displacement, Velocity, and Acceleration

The displacement of a point P from the origin of a coordinate system is described by the magnitude and direction of the radius vector $\mathbf{r}$, which extends from the origin to the point. During a change $\Delta\mathbf{r}$ in the displacement of the point in time Δt, the average change of $\mathbf{r}$ per unit time is $\Delta\mathbf{r}/\Delta t$, and the velocity of the point at time t is the vector

$$\mathbf{v} = \lim_{\Delta t \to 0} \frac{\Delta \mathbf{r}}{\Delta t} = \frac{d\mathbf{r}}{dt} \tag{23.3}$$

The change of velocity with time is illustrated by a diagram in which $\mathbf{v}$ is drawn as a radius vector. The curve traced by the end point of $\mathbf{v}$ in this diagram is called the *hodograph* of the motion. For a change of velocity $\Delta\mathbf{v}$ in time Δt, the average change per unit time is $\Delta\mathbf{v}/\Delta t$, and the acceleration of P at time t is the vector

$$\mathbf{a} = \lim_{\Delta t \to 0} \frac{\Delta \mathbf{v}}{\Delta t} = \frac{d^2\mathbf{r}}{dt^2} \tag{23.4}$$

FIG. 23.1

The hodograph in Fig. 23.1 illustrates the time derivative of a vector. It shows that the derivative of any vector $\mathbf{a}$ is the velocity of the end point of $\mathbf{a}$ plotted in hodograph form.

In vector notation, the equation of motion is written

$$\mathbf{F} = m\frac{d\mathbf{r}^2}{dt^2} = m\ddot{\mathbf{r}} \tag{23.5}$$

The velocity and acceleration of a point, when expressed in terms of three commonly used sets of coordinates, are:

Rectangular x, y, z coordinates

$$\mathbf{r} = x\mathbf{i}_x + y\mathbf{i}_y + z\mathbf{i}_z \qquad \mathbf{v} = \dot{x}\mathbf{i}_x + \dot{y}\mathbf{i}_y + \dot{z}\mathbf{i}_z \qquad \mathbf{a} = \ddot{x}\mathbf{i}_x + \ddot{y}\mathbf{i}_y + \ddot{z}\mathbf{i}_z \tag{23.6}$$

Cylindrical r, θ, z coordinates (Fig. 23.2*a*)

$$\mathbf{v} = \dot{r}\mathbf{i}_r + r\dot{\theta}\mathbf{i}_\theta + \dot{z}\mathbf{i}_z \qquad \mathbf{a} = (\ddot{r} - r\dot{\theta}^2)\mathbf{i}_r + (r\ddot{\theta} + 2\dot{r}\dot{\theta})\mathbf{i}_\theta + \ddot{z}\mathbf{i}_z \tag{23.7}$$

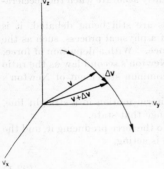

Spherical r, θ, ϕ coordinates (Fig. 23.2b)

$$\mathbf{v} = \dot{r}\mathbf{i}_r + r\dot{\phi}\mathbf{i}_\phi + r\dot{\theta}\,(\sin\phi)\mathbf{i}_\theta$$
$$\mathbf{a} = (\ddot{r} - r\dot{\phi}^2 - r\dot{\theta}^2\sin^2\phi)\mathbf{i}_r + (2\dot{r}\dot{\phi} + r\ddot{\phi} - r\dot{\theta}^2\sin\phi\cos\phi)\mathbf{i}_\phi + (2\dot{r}\dot{\theta}\sin\phi + r\ddot{\theta}\sin\phi + 2r\dot{\theta}\dot{\phi}\cos\phi)\mathbf{i}_\theta \quad (23.8)$$

In terms of the arc length s, measured along the path of motion of P (see Fig. 23.2c), and the principal radius of curvature ρ at the point P,

$$\mathbf{v} = \dot{s}\mathbf{i}_t \qquad \mathbf{a} = \ddot{s}\mathbf{i}_t - \frac{\dot{s}^2}{\rho}\mathbf{i}_\rho \quad (23.9)$$

If a rigid body is rotating about a fixed axis that passes through the origin of the coordinate system, then the relation between the angular velocity $\boldsymbol{\omega}$ of the body and

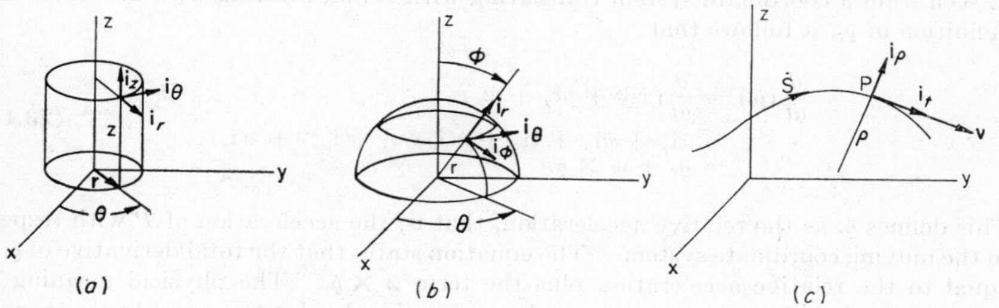

FIG. 23.2

the linear velocity $\mathbf{v}_i$ of a point in the body having displacement $\mathbf{r}_i$ from the origin is given by

$$\mathbf{v}_i = \boldsymbol{\omega} \times \mathbf{r}_i \quad (23.10)$$

Motion Referred to a Moving Coordinate System

Let the displacement, $\boldsymbol{\rho}$, of a point P be determined with respect to a moving (x,y,z) coordinate system whose origin has a translational velocity $\dot{\mathbf{R}}$ and an angular

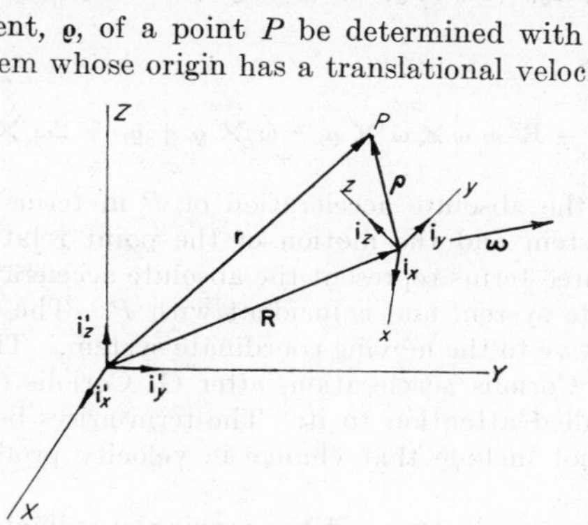

FIG. 23.3

velocity $\boldsymbol{\omega}$ with respect to a basic (X,Y,Z) coordinate system, as shown in Fig. 23.3. The motion of P in the (X,Y,Z) system will be called the *absolute motion,* and it is given by

$$\mathbf{r} = X\mathbf{i}_x' + Y\mathbf{i}_y' + Z\mathbf{i}_z' \qquad \dot{\mathbf{r}} = \dot{X}\mathbf{i}_x' + \dot{Y}\mathbf{i}_y' + \dot{Z}\mathbf{i}_z' \qquad \ddot{\mathbf{r}} = \ddot{X}\mathbf{i}_x' + \ddot{Y}\mathbf{i}_y' + \ddot{Z}\mathbf{i}_z' \quad (23.11)$$

The motion of P as measured in the moving (x,y,z) system will be called the *relative motion.* The absolute motion is the consequence of the relative motion plus the motion of the (x,y,z) system; that is,

$$\begin{aligned} \mathbf{r} &= \mathbf{R} + \boldsymbol{\rho} = \mathbf{R} + x\mathbf{i}_x + y\mathbf{i}_y + z\mathbf{i}_z \\ \dot{\mathbf{r}} &= \dot{\mathbf{R}} + \dot{\boldsymbol{\rho}} = \dot{\mathbf{R}} + \dot{x}\mathbf{i}_x + \dot{y}\mathbf{i}_y + \dot{z}\mathbf{i}_z + x(\mathbf{i}_x)^{\cdot} + y(\mathbf{i}_y)^{\cdot} + z(\mathbf{i}_z)^{\cdot} \end{aligned} \tag{23.12}$$

From the definition of the derivative of a vector, $(\mathbf{i})^{\cdot} = \boldsymbol{\omega} \times \mathbf{i}$, etc.; so $\dot{\mathbf{r}}$ may be written

$$\begin{aligned} \dot{\mathbf{r}} &= \dot{\mathbf{R}} + \dot{x}\mathbf{i}_x + \dot{y}\mathbf{i}_y + \dot{z}\mathbf{i}_z + \boldsymbol{\omega} \times (x\mathbf{i}_x + y\mathbf{i}_y + z\mathbf{i}_z) \\ \dot{\mathbf{r}} &= \dot{\mathbf{R}} + \dot{\boldsymbol{\rho}}_r + \boldsymbol{\omega} \times \boldsymbol{\rho} \end{aligned} \tag{23.13}$$

The last two equations define $\dot{\boldsymbol{\rho}}_r$ as the relative velocity of P. It should be noted that $\dot{\boldsymbol{\rho}} = \dot{\boldsymbol{\rho}}_r + \boldsymbol{\omega} \times \boldsymbol{\rho}$, and that $\dot{\boldsymbol{\rho}}$ is the apparent velocity of P as seen from a coordinate system translating with $\dot{\mathbf{R}}$ but not rotating, whereas $\dot{\boldsymbol{\rho}}_r$ is the apparent velocity of P as seen from a coordinate system translating with $\dot{\mathbf{R}}$ and rotating with $\boldsymbol{\omega}$. From the definition of $\dot{\boldsymbol{\rho}}_r$, it follows that

$$\begin{aligned} \frac{d}{dt}(\dot{\boldsymbol{\rho}}_r) &= \frac{d}{dt}(\dot{x}\mathbf{i}_x + \dot{y}\mathbf{i}_y + \dot{z}\mathbf{i}_z) \\ &= \ddot{x}\mathbf{i}_x + \ddot{y}\mathbf{i}_y + \ddot{z}\mathbf{i}_z + \dot{x}(\mathbf{i}_x)^{\cdot} + \dot{y}(\mathbf{i}_y)^{\cdot} + \dot{z}(\mathbf{i}_z)^{\cdot} \\ &= \ddot{\boldsymbol{\rho}}_r + \boldsymbol{\omega} \times \dot{\boldsymbol{\rho}}_r \end{aligned} \tag{23.14}$$

This defines $\ddot{\boldsymbol{\rho}}_r$ as the relative acceleration, that is, the acceleration of P with respect to the moving coordinate system. The equation states that the total derivative of $\dot{\boldsymbol{\rho}}_r$ is equal to the relative acceleration plus the term $\boldsymbol{\omega} \times \dot{\boldsymbol{\rho}}_r$. The physical meaning of this term is that the velocity vector $\dot{\boldsymbol{\rho}}_r$ has a rotational velocity $\boldsymbol{\omega}$, and hence there is an acceleration even if $\ddot{\boldsymbol{\rho}}_r = 0$.

The absolute acceleration of the point P is

$$\begin{aligned} \ddot{\mathbf{r}} &= \frac{d\dot{\mathbf{r}}}{dt} = \frac{d}{dt}(\dot{\mathbf{R}} + \dot{\boldsymbol{\rho}}_r + \boldsymbol{\omega} \times \boldsymbol{\rho}) \\ &= \ddot{\mathbf{R}} + (\ddot{\boldsymbol{\rho}}_r + \boldsymbol{\omega} \times \dot{\boldsymbol{\rho}}_r) + [\dot{\boldsymbol{\omega}} \times \boldsymbol{\rho} + \boldsymbol{\omega} \times (\dot{\boldsymbol{\rho}}_r + \boldsymbol{\omega} \times \boldsymbol{\rho})] \end{aligned}$$

Collecting terms gives

$$\ddot{\mathbf{r}} = \ddot{\mathbf{R}} + \boldsymbol{\omega} \times \boldsymbol{\omega} \times \boldsymbol{\rho} + \dot{\boldsymbol{\omega}} \times \boldsymbol{\rho} + \ddot{\boldsymbol{\rho}}_r + 2\boldsymbol{\omega} \times \dot{\boldsymbol{\rho}}_r \tag{23.15}$$

This equation gives the absolute acceleration of P in terms of the motion of the (x,y,z) coordinate system and the motion of the point relative to this coordinate system. The first three terms represent the absolute acceleration of a point fixed in the moving coordinate system and coincident with P. The term $\ddot{\boldsymbol{\rho}}_r$ represents the acceleration of P relative to the moving coordinate system. The last term, $(2\boldsymbol{\omega} \times \dot{\boldsymbol{\rho}}_r)$, is usually called the Coriolis acceleration, after G. Coriolis (1792–1843), a French engineer who first called attention to it. The term arises because the other terms in the equation do not include that change in velocity produced by $\boldsymbol{\omega}$ when $\boldsymbol{\rho}_r$ is changing.

The equation of motion in terms of the moving coordinate system may thus be written

$$\mathbf{F} = m\ddot{\mathbf{R}} + m\boldsymbol{\omega} \times \boldsymbol{\omega} \times \boldsymbol{\rho} + m\dot{\boldsymbol{\omega}} \times \boldsymbol{\rho} + m\ddot{\boldsymbol{\rho}}_r + 2m\boldsymbol{\omega} \times \dot{\boldsymbol{\rho}}_r \tag{23.16}$$

if Newton's law is valid in the basic coordinate system. In most engineering problems (excluding ballistics), if the moving coordinate system is one fixed to the surface of the earth, then the term $m\ddot{\boldsymbol{\rho}}_r$ is dominant, and the other terms, being very small compared to it, may be neglected; hence such a coordinate system may be considered a basic frame of reference.

23.2. PARTICLE DYNAMICS

Solution of Problems

In many engineering problems involving the motion of a particle, or the motion of a mass center, the magnitude and direction of the force are specified, and the solution of the problem is given by the integration of the component differential equations

$$f_x = m\ddot{x} \qquad f_y = m\ddot{y} \qquad f_z = m\ddot{z} \qquad \text{etc.}$$

The differential equations and the solutions of some commonly discussed problems in particle dynamics are given below.

1. Body of mass m under a constant gravitational acceleration g:

$$m\ddot{y} = mg \qquad m\ddot{x} = 0$$
$$\dot{y} = gt + \dot{y}_0 \qquad \dot{x} = \dot{x}_0$$
$$y = \tfrac{1}{2}gt^2 + \dot{y}_0 t + y_0 \qquad x = \dot{x}_0 t + x_0$$
$$y - y_0 = \tfrac{1}{2}g\left(\frac{x - x_0}{\dot{x}_0}\right)^2 + \dot{y}_0 \frac{x - x_0}{\dot{x}_0}$$

The details of the motion depend on the initial values ($t = 0$) of the displacements and velocities. For the particular case $y_0 = \dot{y}_0 = x_0 = \dot{x}_0 = 0$:

$$\dot{y} = gt \qquad y = \tfrac{1}{2}gt^2 \qquad \dot{y} = \sqrt{2gy}$$

2 Body falling under constant gravitational force and a drag force proportional to the velocity ($f_d = -kv$):

$$m\ddot{y} = mg - k\dot{y} \qquad \frac{\dot{y}k}{mg} = 1 - \left(1 - \frac{\dot{y}_0 k}{mg}\right)e^{-kt/m}$$

$$\frac{yk^2}{m^2 g} = \frac{y_0 k^2}{m^2 g} + \frac{kt}{m} - \left(1 - \frac{\dot{y}_0 k}{mg}\right)(1 - e^{-kt/m}) \qquad \dot{y}_{\max} = \frac{mg}{k} = \text{terminal velocity}$$

For a sphere falling in a fluid medium, Stokes' law states that $f_d = -3\pi d\mu v$, where μ = dynamic viscosity, d = diameter, and $\rho vd/\mu < 1.0$, ρ being the density of the medium.

3. Body falling with drag force proportional to velocity squared ($f_d = -kv^2$), and $y_0 = \dot{y}_0 = 0$:

$$m\ddot{y} = mg - k\dot{y}^2 \qquad \dot{y}\left(\frac{k}{mg}\right)^{1/2} = \operatorname{Tanh}\left(\frac{kgt^2}{m}\right)^{1/2} \qquad \frac{yk}{m} = \ln \operatorname{Cosh}\left(\frac{kgt^2}{m}\right)^{1/2}$$

4. Projectile retarded by resisting force $f_d = -a - bv^2$, and $x_0 = \dot{y}_0 = 0$ (Poncelet's penetration problem):

$$m\ddot{x} = -a - b\dot{x}^2 \qquad m\ddot{y} = 0$$
$$\tan^{-1}\dot{x}\left(\frac{b}{a}\right)^{1/2} = (ab)^{1/2}\frac{t}{m} + \tan^{-1}\dot{x}_0\left(\frac{b}{a}\right)^{1/2}$$
$$\frac{bx}{m} = \tfrac{1}{2}\ln\frac{(b/a)\dot{x}_0^2 + 1}{(b/a)\dot{x}^2 + 1}$$

5. Projectile with drag force $f_d = -kv^2$, and $x_0 = y_0 = 0$; approximate solution for flat trajectory:

$$m\ddot{x} = -kv^2\cos\phi = -k\dot{x}^2\left(1 + \left(\frac{\dot{y}}{\dot{x}}\right)^2\right)^{1/2} = -k\dot{x}^2$$
$$m\ddot{y} = -mg - kv^2\sin\phi = -mg - k\dot{x}\dot{y}\left(1 + \left(\frac{\dot{y}}{\dot{x}}\right)^2\right)^{1/2} = -mg - k\dot{x}\dot{y}$$
$$\frac{\dot{x}}{\dot{x}_0} = \left(1 + \frac{k\dot{x}_0}{m}t\right)^{-1} \qquad \frac{2k\dot{x}_0\dot{y}}{mg} = \left(1 + \frac{2k\dot{x}_0\dot{y}_0}{mg}\right)\left(1 + \frac{k\dot{x}_0 t}{m}\right)^{-1} - \left(1 + \frac{k\dot{x}_0 t}{m}\right)$$

6. Projectile with drag force $f_d = -kv$, and $x_0 = y_0 = 0$:

$$m\ddot{x} = -k\dot{x} \qquad m\ddot{y} = -k\dot{y} - mg$$

$$\frac{xk}{\dot{x}_0 m} = 1 - e^{-kt/m} \qquad \frac{yk^2}{gm^2} = \left(1 + \frac{\dot{y}_0 k}{gm}\right)(1 - e^{-kt/m}) - \frac{k}{m}t$$

7. Mass m performing forced vibrations under action of force F:

$$m\ddot{x} + c\dot{x} + kx = F(t)$$

$$x = \frac{T}{2\pi m}\int_0^t F(\tau)e^{-c(t-\tau)/2m} \sin\frac{2\pi}{T}(t-\tau)\,d\tau + \frac{\dot{x}_0 T}{2\pi}\sin\frac{2\pi}{T}t + x_0\cos\frac{2\pi}{T}t$$

$$T = \frac{2\pi}{\sqrt{k/m - (c/2m)^2}}$$

8. Planetary motion of a particle of mass m about a fixed point with inverse-square attraction. For plane motion (polar coordinates):

$$m(2\dot{r}\dot{\theta} + r\ddot{\theta}) = 0 \qquad m(\ddot{r} - r\dot{\theta}^2) = -\frac{k}{r^2}$$

$$mr^2\dot{\theta} = h = \text{constant angular momentum} \qquad \tfrac{1}{2}m(\dot{r}^2 + r^2\dot{\theta}^2) - \frac{k}{r} = E = \text{constant energy}$$

$$r = \frac{(h/m)^2}{E(1 + e\cos\theta)} \quad \text{(orbit)} \qquad e^2 = 1 + \frac{2Eh^2}{mk^2}$$

The orbital equation is that of a conic having a focus at the origin. If $e < 1$, the orbit is an ellipse; if $e = 1$, it is a parabola; and, if $e > 1$, it is a hyperbola; e depends on the magnitudes of the initial E and h but not on the direction of initial velocity.

Equation of Impulse and Momentum

The two first integrals of the equation of motion lead to the concepts of momentum and energy. If the equation of motion is integrated with respect to time, there is obtained:

$$\int_{t_1}^{t_2} \mathbf{F}\,dt = \int_{t_1}^{t_2} m\dot{\mathbf{v}}\,dt = [m\mathbf{v}]_{t_1}^{t_2}$$

$$\int_{t_1}^{t_2} \mathbf{F}\,dt = m\mathbf{v}_2 - m\mathbf{v}_1 \tag{23.17}$$

The term on the left side of this equation is called the impulse of the force $\mathbf{F}$, and the term $m\mathbf{v}$ is called the momentum of the particle. Equation (23.17) states that the impulse is equal to the change in momentum.

Moment of Momentum

If the vector product with the radius vector $\mathbf{r}$ is performed on both sides of the equation of motion, and the result is integrated with respect to time, there is obtained:

$$\int_{t_1}^{t_2} (\mathbf{r} \times \mathbf{F})\,dt = \int_{t_1}^{t_2} (\mathbf{r} \times m\dot{\mathbf{v}})\,dt = \int_{t_1}^{t_2} \frac{d}{dt}(\mathbf{r} \times m\mathbf{v})\,dt = [\mathbf{r} \times m\mathbf{v}]_{t_1}^{t_2}$$

$$\int_{t_1}^{t_2} (\mathbf{r} \times \mathbf{F})\,dt = \mathbf{r}_2 \times m\mathbf{v}_2 - \mathbf{r}_1 \times m\mathbf{v}_1 \tag{23.18}$$

The term on the left side of this equation is called the moment of impulse ($\mathbf{r} \times \mathbf{F}$ is the moment of the force), and the term $\mathbf{r} \times m\mathbf{v}$ is called the moment of momentum, or angular momentum, of the particle. Equation (23.18) states that the moment of impulse is equal to the change in moment of momentum.

Conservation of Momentum

If in a period of time (t_1,t_2) the impulse on a particle is zero, it follows from Eq. (23.17) that the momentum of the particle at t_2 is the same as it was at t_1; that is, the momentum is conserved. Furthermore, if two or more particles are interacting, it follows from Newton's third law that the mutual impulses exerted on one another by the particles are equal and opposite, and, hence, the sum of all the impulses is zero, and the total momentum is conserved. This may be stated as a principle of conservation of momentum: *If in a system of particles only mutual interactions are involved, the momentum of the system is conserved.* A corresponding statement may be made for the moment of momentum of a system of particles.

Equation of Work and Energy

If the equation of motion is integrated with respect to the displacement, there is obtained:

$$\int_{r_1}^{r_2} \mathbf{F} \cdot d\mathbf{r} = \int_{r_1}^{r_2} m\dot{\mathbf{v}} \cdot d\mathbf{r} = \int_{t_1}^{t_2} m\dot{\mathbf{v}} \cdot \mathbf{v}\, dt$$

$$= \int_{t_1}^{t_2} \tfrac{1}{2} \frac{d}{dt}(m\mathbf{v} \cdot \mathbf{v})\, dt = \int_{t_1}^{t_2} \frac{d}{dt}(\tfrac{1}{2}mv^2)\, dt$$

$$\int_{r_1}^{r_2} \mathbf{F} \cdot d\mathbf{r} = \tfrac{1}{2}mv_2^2 - \tfrac{1}{2}mv_1^2 \tag{23.19}$$

The integral on the left side of this equation is called the work done by the force **F**, and term $\frac{1}{2}mv^2$ is called the kinetic energy of the particle. The equation states that the work done by the force is equal to the change in kinetic energy of the particle.

The vector displacement $d\mathbf{r}$ is tangent to the path of motion of the particle, and so the scalar product $\mathbf{F} \cdot d\mathbf{r}$ represents the component of the force in the direction of the displacement multiplied by the displacement. The work done by a force whose point of application moves along a path from point A to point B is given by the line integral over the path:

$$W = \int_A^B \mathbf{F} \cdot d\mathbf{r}$$

The rate of doing work $(\mathbf{F} \cdot d\mathbf{r}/dt = \mathbf{F} \cdot \mathbf{v})$ is called the power. Work, power, and kinetic energy are scalar quantities.

Potential Energy

If the magnitude and direction of a force **F** are known functions of position only (not velocities, etc.), and are derivable from a potential, as follows:

$$\mathbf{F} = F_x\mathbf{i}_x + F_y\mathbf{i}_y + F_z\mathbf{i}_z = \frac{\partial \phi}{\partial x}\mathbf{i}_x + \frac{\partial \phi}{\partial y}\mathbf{i}_y + \frac{\partial \phi}{\partial z}\mathbf{i}_z = \text{grad } \phi$$

then the left side of Eq. (23.19) can be integrated directly to give

$$\int_1^2 \mathbf{F} \cdot d\mathbf{r} = \int_1^2 \left(\frac{\partial \phi}{\partial x} dx + \frac{\partial \phi}{\partial y} dy + \frac{\partial \phi}{\partial z} dz \right) = \int_1^2 d\phi = \phi_2 - \phi_1$$

The quantity $(\phi_2 - \phi_1)$ is the work done by **F** as it moves from point 1 to point 2, and it is independent of the path followed between these points. If the notation $V = -\phi$ is introduced,

$$V_2 - V_1 = -\int_1^2 \mathbf{F} \cdot d\mathbf{r} \tag{23.20}$$

The quantity $(V_2 - V_1)$ is called the change in potential energy of the force as it moves from point 1 to point 2, and it is equal to the negative of the work done during this process. In general, the potential energy of the force is

$$V = -\int \mathbf{F} \cdot d\mathbf{r} + C_0 \tag{23.21}$$

where C_0 is a constant of integration. The value assigned to C_0 determines the zero value (datum) for the potential energy.

Conservation of Energy

When a particle is acted upon by a force that has a potential energy, the work-energy equation (23.19) may be written

$$V_1 + \tfrac{1}{2}mv_1^2 = V_2 + \tfrac{1}{2}mv_2^2 \tag{23.22}$$

This equation states that the sum of the potential energy and the kinetic energy remains a constant, so that

$$V + T = C \tag{23.23}$$

where T is the kinetic energy. This equation implies a principle of conservation of mechanical energy: *If all the forces acting on the particles of a system are derivable from potential functions, the total potential energy plus kinetic energy remains a constant.* Such a system of particles is said to be a conservative system, and the forces are said to be conservative forces. It should be noted that being a function of position only does not ensure the force to be conservative; the force must also be derivable from a potential which requires (see p. 6–8)

$$\text{curl } \mathbf{F} \equiv \left(\frac{\partial F_z}{\partial y} - \frac{\partial F_y}{\partial z}\right)\mathbf{i}_x + \left(\frac{\partial F_x}{\partial z} - \frac{\partial F_z}{\partial x}\right)\mathbf{i}_y + \left(\frac{\partial F_y}{\partial x} - \frac{\partial F_x}{\partial y}\right)\mathbf{i}_z = 0$$

Energy-dissipating forces arising from dry friction, viscous or magnetic damping, etc., are not conservative.

Central Impact

An impact problem is here defined to be the collision of two bodies with the following special features: The contact force has a known direction, and any energy loss that occurs during impact is known. In a central impact, the line of action of the contact force passes through the mass center of each body; hence, a central impact has no effect on the rotation of either body. The problem of noncentral impact is discussed in the section on rigid bodies.

Let a central impact occur between bodies of mass m_1 and m_2, and let $\mathbf{v}_1$, $\mathbf{v}_2$ and $\mathbf{V}_1$, $\mathbf{V}_2$ be the velocity before and after impact, respectively. Let $\mathbf{I}$ be the impulse exerted on body 1 by body 2, and $-\mathbf{I}$ the impulse exerted on 2 by 1. Note that $\mathbf{I} = I(l_x\mathbf{i}_x + l_y\mathbf{i}_y + l_z\mathbf{i}_z)$, where I is the magnitude of impulse (unknown), and l_x, l_y, l_z are the direction cosines of $\mathbf{I}$ with respect to the x, y, z axes (known). The equations of impulse-momentum and conservation of energy are:

$$m_1(\mathbf{V}_1 - \mathbf{v}_1) = \mathbf{I} \qquad m_2(\mathbf{V}_2 - \mathbf{v}_2) = -\mathbf{I} \tag{23.24}$$
$$\tfrac{1}{2}m_1v_1^2 + \tfrac{1}{2}m_2v_2^2 = \tfrac{1}{2}m_1V_1^2 + \tfrac{1}{2}m_2V_2^2 + E$$

where E is the energy loss during impact (known). The unknown quantities are $\mathbf{V}_1$, $\mathbf{V}_2$, I, which comprise seven unknown scalar components. The two vector equations of impulse-momentum plus the energy equation furnish seven scalar equations, which determine the unknowns. It should be noted that, if the two momentum equations are added together, there results the equation of conservation of momentum, but this equation contains less information than the two momentum equations (except for collinear impact).

If the coordinate axes are oriented so that the x axis coincides with the line of action of **I**, the change of motion is described by

$$m_1(V_{1x} - v_{1x}) = I \qquad m_2(V_{2x} - v_{2x}) = -I$$
$$\tfrac{1}{2}m_1 V_{1x}^2 + \tfrac{1}{2}m_2 v_{2x}^2 = \tfrac{1}{2}m_1 V_{1x}^2 + \tfrac{1}{2}m_2 V_{2x}^2 + E \tag{23.25}$$

If there is no energy loss, $E = 0$, and, by using the first two equations, the last equation may be put in the form $V_{1x} - V_{2x} = -(v_{1x} - v_{2x})$, which states that the relative rebound velocity is equal and opposite to the relative velocity of approach. Energy loss may thus be expressed by

$$V_{1x} - V_{2x} = -e(v_{1x} - v_{2x}) \tag{23.26}$$

where the coefficient of restitution e in actual problems may have values $0 \lesseqgtr e < 1$. The values of e are determined experimentally, or by analysis of stress wave propagation, and they will depend on the materials of which the bodies are made, and on the shapes of the bodies, and may also depend on the intensities of impact stresses.

23.3. SYSTEMS OF PARTICLES

In most dynamics problems, it is not possible to approximate the system by a single particle, but it must be treated as a collection of particles, each of which obeys Newton's laws. A multiplicity of particles introduces algebraic complexities that are best coped with by certain general dynamical principles. The following principles are completely general, and apply to all systems of masses.

Motion of the Mass Center

The center of mass of a system of particles is defined as the point located by the vector $\mathbf{r}_c$, where

$$\mathbf{r}_c = \frac{\Sigma m_i \mathbf{r}_i}{\Sigma m_i} = \frac{\Sigma m_i x_i}{\Sigma m_i}\mathbf{i}_x + \frac{\Sigma m_i y_i}{\Sigma m_i}\mathbf{i}_y + \frac{\Sigma m_i z_i}{\Sigma m_i}\mathbf{i}_z \tag{23.27}$$

The subscript i refers to the ith particle, and the summation is over all the particles in the system. If the mass of the system has a continuous distribution, the total mass is

$$m = \int dm \qquad \text{and} \qquad \mathbf{r}_c = \frac{\int \mathbf{r}\, dm}{m}$$

Let Newton's equation for the ith particle of the system be written in the form

$$m_i \ddot{\mathbf{r}}_i = \mathbf{F}_i + \mathbf{f}_i$$

where $\mathbf{f}_i$ represents the component of force arising from the mutual interaction between particles of the system (internal forces), and $\mathbf{F}_i$ represents the component originating from outside the system (external forces). Newton's third law states that the internal forces occur in equal and opposite pairs, so that, for the complete system, $\Sigma \mathbf{f}_i = 0$ and $\Sigma \mathbf{F}_i = \mathbf{F}$, and, therefore,

$$\Sigma m_i \ddot{\mathbf{r}}_i = \mathbf{F}$$

From the definition of the center of mass, the left side of the equation is just $m\ddot{\mathbf{r}}_c$, so that the mass center of a system of particles moves according to the equation

$$\mathbf{F} = m\ddot{\mathbf{r}}_c \tag{23.28}$$

where **F** is the resultant force exerted on the system, and m is the total mass of the system. This equation holds for all systems of particles, such as solids, liquids, and powders. The mass center behaves like a particle of mass m, acted on by the force **F**.

Momentum and Energy of the Mass Center

All the principles of particle dynamics apply to the mass center, and, from this analogy, $m\dot{\mathbf{r}}_c$ is called the momentum of the mass center and $\frac{1}{2}m\dot{r}_c^2$ is called the kinetic energy of the mass center. The equations of impulse-momentum and work-energy apply:

$$\int_1^2 \mathbf{F}\,dt = [m\dot{\mathbf{r}}_c]_1^2 \tag{23.29}$$

$$\int_1^2 \mathbf{F}\cdot d\mathbf{r}_c = [\tfrac{1}{2}m\dot{r}_c^2]_1^2 \tag{23.30}$$

The moment-of-impulse–moment-of-momentum equation is

$$\int_1^2 \mathbf{r}_c \times \mathbf{F}\,dt = [\mathbf{r}_c \times m\dot{\mathbf{r}}_c]_1^2 \tag{23.31}$$

The principles of conservation of momentum, energy, and moment of momentum apply, as for a particle.

Kinetic Energy of a System of Particles

The total kinetic energy of a system of particles, $T = \Sigma\frac{1}{2}m_i\dot{r}_i^2$, may be expressed in a convenient form by introducing the displacement of the mass center $\mathbf{r}_c$ and expressing the displacement of the ith particle as $\mathbf{r}_i = \mathbf{r}_c + \boldsymbol{\varrho}_i$, where $\boldsymbol{\varrho}_i$ is the displacement of the ith particle with respect to the mass center ($\Sigma m_i\boldsymbol{\varrho}_i = 0$). In terms of $\mathbf{r}_c$ and $\boldsymbol{\varrho}_i$,

$$\dot{r}_i^2 = \dot{\mathbf{r}}_i \cdot \dot{\mathbf{r}}_i = (\dot{\mathbf{r}}_c + \dot{\boldsymbol{\varrho}}_i)\cdot(\dot{\mathbf{r}}_c + \dot{\boldsymbol{\varrho}}_i) = \dot{r}_c^2 + 2\dot{\mathbf{r}}_c\cdot\dot{\boldsymbol{\varrho}}_i + \dot{\rho}_i^2$$

and $$T = \Sigma\tfrac{1}{2}m_i\dot{r}_c^2 + \dot{\mathbf{r}}_c\cdot\Sigma m_i\dot{\boldsymbol{\varrho}}_i + \Sigma\tfrac{1}{2}m_i\dot{\rho}_i^2$$

The kinetic energy may therefore be written

$$T = \tfrac{1}{2}m\dot{r}_c^2 + \Sigma\tfrac{1}{2}m_i\dot{\rho}_i^2 \tag{23.32}$$

The last term in this equation is the kinetic energy of motion relative to the mass center; that is, it is the kinetic energy that would be inferred from the particle velocities as seen from a nonrotating coordinate system attached to the center of mass. The total kinetic energy of a system of particles is, thus, the sum of the kinetic energy of the mass center and the kinetic energy of motion relative to the mass center.

The total work of all the forces of the system is

$$\int_1^2 \sum(\mathbf{F}_i + \mathbf{f}_i)\cdot d\mathbf{r}_i = \int_1^2 \sum(\mathbf{F}_i + \mathbf{f}_i)\cdot(d\mathbf{r}_c + d\boldsymbol{\varrho}_i)$$

and, since $\Sigma\mathbf{f}_i = 0$, the equation of work-energy for the system may be written

$$\int_1^2 \mathbf{F}\cdot d\mathbf{r}_c + \int_1^2 \sum(\mathbf{F}_i + \mathbf{f}_i)\cdot d\boldsymbol{\varrho}_i = [\tfrac{1}{2}m\dot{r}_c^2]_1^2 + \left[\sum \tfrac{1}{2}m_i\dot{\rho}_i^2\right]_1^2$$

The first term on each side is the corresponding term in the work-energy equation for the mass center, and so the equation may be written in two parts:

$$\int_1^2 \mathbf{F}\cdot d\mathbf{r}_c = [\tfrac{1}{2}m\dot{r}_c^2]_1^2$$

$$\int_1^2 \sum(\mathbf{F}_i + \mathbf{f}_i)\cdot d\boldsymbol{\varrho}_i = \left[\sum \tfrac{1}{2}m_i\dot{\rho}_i^2\right]_1^2 \tag{23.33}$$

The second equation is exactly the work-energy equation that would be written for the particle velocities as seen from a nonrotating coordinate system attached to the center of mass. The work and energy of a system of particles may thus be separated into two parts: the work-energy relations for the mass center, and the work-energy relations for the motion relative to the mass center. Under the appropriate conditions, the equation of conservation of energy may be written for each of these motions.

The Equation of Moment of Momentum

The total moment of momentum, **H**, of a system of particles about a fixed point is $\Sigma \mathbf{r}_i \times m_i\dot{\mathbf{r}}_i$, and the time derivative of this is $\Sigma \mathbf{r}_i \times m_i\ddot{\mathbf{r}}_i$. Operating on the equation of motion of a particle as follows:

$$m_i\ddot{\mathbf{r}}_i = \mathbf{F}_i + \mathbf{f}_i \qquad \Sigma \mathbf{r}_i \times m_i\ddot{\mathbf{r}}_i = \Sigma \mathbf{r}_i \times (\mathbf{F}_i + \mathbf{f}_i) = \Sigma \mathbf{r}_i \times \mathbf{F}_i$$

it is seen that the the left side of the preceding equation is $\dot{\mathbf{H}}$, and the right side is the resultant moment, **M**, of the external forces acting on the system. The application of Newton's equation to the particles of a system thus leads to the equation

$$\dot{\mathbf{H}} = \mathbf{M} \tag{23.34}$$

or the time rate of change of the moment of momentum about a fixed point is equal to the resultant moment of the external forces about the point.

The time integral of Eq. (23.34) is

$$\int_1^2 \mathbf{M}\,dt = [\mathbf{H}]_1^2 \tag{23.35}$$

which states that the moment of impulse is equal to the change in moment of momentum. If the moment of impulse is zero, the equation reduces to

$$\mathbf{H} = \text{const}$$

which expresses a principle of conservation of moment of momentum: *If there is no moment of external force about a fixed point, the moment of momentum about that point remains constant.*

The moment-of-momentum equation can be separated into two parts, as was done for the work-energy equation, by substituting $\mathbf{r}_i = \mathbf{r}_c + \boldsymbol{\varrho}_i$, where $\mathbf{r}_c$ is the displacement of the mass center. Making use of $\Sigma m_i \boldsymbol{\varrho}_i = 0$ and $\Sigma \mathbf{F}_i = \mathbf{F}$, this substitution gives

$$\Sigma \mathbf{r}_i \times m_i\ddot{\mathbf{r}}_i = \Sigma \mathbf{r}_i \times \mathbf{F}_i \qquad \mathbf{r}_c \times m\ddot{\mathbf{r}}_c + \Sigma \boldsymbol{\varrho}_i \times m_i\ddot{\boldsymbol{\varrho}}_i = \mathbf{r}_c \times \mathbf{F} + \Sigma \boldsymbol{\varrho}_i \times \mathbf{F}_i$$

The first term on each side of the last equation is the corresponding term in the equation of moment of momentum of the mass center; so two separate equations may be written

$$\mathbf{r}_c \times m\ddot{\mathbf{r}}_c = \mathbf{r}_c \times \mathbf{F} \tag{23.36}$$

$$\Sigma \boldsymbol{\varrho}_i \times m_i\ddot{\boldsymbol{\varrho}}_i = \Sigma \boldsymbol{\varrho}_i \times \mathbf{F}_i \tag{23.37}$$

The first equation states that $\dot{\mathbf{H}}_o$, the rate of change of the moment of momentum of the mass center about the fixed point O, is equal to $\mathbf{M}_o$, the moment about O of the force **F** taken as acting through the mass center. The second equation states that $\dot{\mathbf{H}}_c$, the rate of change of the moment of momentum about the mass center, is equal to $\mathbf{M}_c$, the moment of force about the mass center. The moment-of-momentum equations may be summed up as follows:

$$\begin{aligned} \dot{\mathbf{H}} &= \mathbf{M} & \int_1^2 \mathbf{M}\,dt &= [\mathbf{H}]_1^2 \\ \dot{\mathbf{H}}_o &= \mathbf{M}_o & \int_1^2 \mathbf{M}_o\,dt &= [\mathbf{H}_o]_1^2 \\ \dot{\mathbf{H}}_c &= \mathbf{M}_c & \int_1^2 \mathbf{M}_c\,dt &= [\mathbf{H}_c]_1^2 \end{aligned} \tag{23.38a–c}$$

The equations may be expressed in rectangular coordinates by means of the following relations for a single particle:

$$\mathbf{H} = \mathbf{r} \times m\dot{\mathbf{r}} = m(y\dot{z} - z\dot{y})\mathbf{i}_x + m(z\dot{x} - x\dot{z})\mathbf{i}_y + m(x\dot{y} - y\dot{x})\mathbf{i}_z$$
$$\mathbf{M} = \mathbf{r} \times \mathbf{F} = (yF_z - zF_y)\mathbf{i}_x + (zF_x - xF_z)\mathbf{i}_y + m(xF_y - yF_x)\mathbf{i}_z$$

Moments and Products of Inertia

For a system of mass m, the moments of inertia I_{xx}, I_{yy}, I_{zz} and the products of inertia I_{xy}, I_{yz}, I_{zx}, with respect to an orthogonal (x,y,z) coordinate system, are defined by

$$I_{xx} = \int (y^2 + z^2)\, dm \qquad I_{xy} = -\int xy\, dm$$
$$I_{yy} = \int (z^2 + x^2)\, dm \qquad I_{yz} = -\int yz\, dm$$
$$I_{zz} = \int (x^2 + y^2)\, dm \qquad I_{zx} = -\int zx\, dm$$

The integrals are taken over all the mass, m, of the system. It should be noted that the products of inertia are here defined to be the negatives of the integrals, whereas, in many engineering books and tables, they are defined as the integrals without the minus signs.[1] The rotary inertia of a body is specified by the matrix

$$\begin{bmatrix} I_{xx} & I_{yx} & I_{zx} \\ I_{xy} & I_{yy} & I_{zy} \\ I_{xz} & I_{yz} & I_{zz} \end{bmatrix}$$

The radius of gyration about the origin of the coordinate system is defined as

$$\mathbf{i}_g = i_x\mathbf{i}_x + i_y\mathbf{i}_y + i_z\mathbf{i}_z = \sqrt{\frac{I_{xx}}{m}}\,\mathbf{i}_x + \sqrt{\frac{I_{yy}}{m}}\,\mathbf{i}_y + \sqrt{\frac{I_{zz}}{m}}\,\mathbf{i}_z \tag{23.39}$$

The i_x, i_y, i_z are respectively the radii of gyration about the x, y, z axes, while $\mathbf{i}_x$, $\mathbf{i}_y$, $\mathbf{i}_z$ as usual, are the unit vectors.

A general expression that represents all moments and products of inertia is

$$I_{\alpha\beta} = \int (r^2\mathbf{i}_\alpha \cdot \mathbf{i}_\beta - \alpha\beta)\, dm \tag{23.40}$$

where α, β may be x, y, or z, and $r^2 = x^2 + y^2 + z^2$, and $\mathbf{i}_\alpha$, $\mathbf{i}_\beta$ are unit vectors in the α, β directions ($\mathbf{i}_\alpha \cdot \mathbf{i}_\beta = 1$ if $\alpha = \beta$; $= 0$ if $\alpha \neq \beta$). This form is particularly useful for deriving the equations for transformations of coordinates.

Let (x',y',z') be a coordinate system with origin at the center of mass, and let (x,y,z) be a parallel system such that, in the (x,y,z) system, the coordinates of the mass center are $\mathbf{r}_c = x_c\mathbf{i}_x + y_c\mathbf{i}_y + z_c\mathbf{i}_z$. For a transformation from the (x',y',z') to the (x,y,z) system, $x = x' + x_c$, or, in general, $\alpha = \alpha' + \alpha_c$, $\beta = \beta' + \beta_c$, from which

$$I_{\alpha\beta} = I_{\alpha'\beta'} + (mr_c^2\mathbf{i}_\alpha \cdot \mathbf{i}_\beta - m\alpha_c\beta_c) \tag{23.41}$$

This equation is sometimes called the *parallel-axis theorem.*

Let (x,y,z) and (x',y',z') be two nonparallel coordinate systems whose origins are coincident, and let the direction cosines between the primed and unprimed axes be:

	x'	y'	z'
x	$l_{xx'}$	$l_{xy'}$	$l_{xz'}$
y	$l_{yx'}$	$l_{yy'}$	$l_{yz'}$
z	$l_{zx'}$	$l_{zy'}$	$l_{zz'}$

Then

$$x = l_{xx'}x' + l_{xy'}y' + l_{xz'}z' = \sum_{\alpha'} l_{x\alpha'}\alpha' \qquad \text{etc.}$$

[1] Note that, in other parts of this handbook, the products of inertia are understood to be

$$I_{xy} = +\int xy\, dm \qquad \text{etc.}$$

or, in general,

$$\alpha = \sum_{\alpha'} l_{\alpha\alpha'}\alpha' \qquad \beta = \sum_{\beta'} l_{\beta\beta'}\beta'$$

and, similarly,

$$\mathbf{i}_\beta = \sum_{\beta'} l_{\beta\beta'}\mathbf{i}_{\beta'} \qquad \mathbf{i}_\alpha = \sum_{\alpha'} l_{\alpha\alpha'}\mathbf{i}_{\alpha'}$$

The transformation from the (x',y',z') to the (x,y,z) systems is deduced as follows:

$$\begin{aligned} I_{\alpha\beta} &= \int (r^2\mathbf{i}_\alpha \cdot \mathbf{i}_\beta - \alpha\beta)\, dm \\ &= \int \Big(r^2 \sum_{\alpha'}\sum_{\beta'} l_{\alpha\alpha'}l_{\beta\beta'}\mathbf{i}_{\alpha'} \cdot \mathbf{i}_{\beta'} - \sum_{\alpha'}\sum_{\beta'} l_{\alpha\alpha'}l_{\beta\beta'}\alpha'\beta'\Big)\, dm \\ &= \sum_{\alpha'}\sum_{\beta'} l_{\alpha\alpha'}l_{\beta\beta'} \int (r^2\mathbf{i}_{\alpha'} \cdot \mathbf{i}_{\beta'} - \alpha'\beta')\, dm \\ I_{\alpha\beta} &= \sum_{\alpha'}\sum_{\beta'} l_{\alpha\alpha'}l_{\beta\beta'}I_{\alpha'\beta'} \end{aligned} \tag{23.42}$$

In expanded form, this equation gives expressions such as

$$\begin{aligned} I_{xx} = &+ l_{xx'}l_{xx'}I_{x'x'} + l_{xy'}l_{xx'}I_{y'x'} + l_{xz'}l_{xx'}I_{z'x'} \\ &+ l_{xx'}l_{xy'}I_{x'y'} + l_{xy'}l_{xy'}I_{y'y'} + l_{xz'}l_{xy'}I_{z'y'} \\ &+ l_{xx'}l_{xz'}I_{x'z'} + l_{xy'}l_{xz'}I_{y'z} + l_{xz'}l_{xz'}I_{z'z'} \\ I_{xy} = &+ l_{xx'}l_{yx'}I_{x'x'} + l_{xy'}l_{yx'}I_{y'x'} + l_{xz'}l_{yx'}I_{z'x'} \\ &+ l_{xx'}l_{yy'}I_{x'y'} + l_{xy'}l_{yy'}I_{y'y'} + l_{xz'}l_{yy'}I_{z'y'} \\ &+ l_{xx'}l_{yz'}I_{x'z'} + l_{xy'}l_{yz'}I_{y'z'} + l_{xz'}l_{yz'}I_{z'z'} \end{aligned}$$

It is always possible to find an orientation of axes x', y', z' such that the products of inertia are identically zero and the inertia matrix has the form

$$\begin{bmatrix} I_{x'x'} & 0 & 0 \\ 0 & I_{y'y'} & 0 \\ 0 & 0 & I_{z'z'} \end{bmatrix}$$

The axes for which the products of inertia are all zero are called the *principal axes.* The moments of inertia listed in tables are usually with respect to principal axes, and, in this case, the equations for transformation to rotated axes are:

$$I_{xx} = l_{xx'}^2 I_{x'x'} + l_{xy'}^2 I_{y'y'} + l_{xz'}^2 I_{z'z'} \tag{23.43}$$

$$I_{xy} = l_{xx'}l_{yx'}I_{x'x'} + l_{xy'}l_{yy'}I_{y'y'} + l_{xz'}l_{yz'}I_{z'z'} \tag{23.44}$$

In general, the three moments of inertia are not equal, and, for the principal axes, the inequality is an extremum; that is, the largest principal moment of inertia is the maximum, and the smallest principal moment of inertia is the minimum, for any orientation. In particular, for centroidal axes, the smallest principal moment of inertia is the smallest possible for any axis.

23.4. DYNAMICS OF RIGID BODIES

A rigid body is defined to be one for which the motion of any point P is described by the equations

$$\begin{aligned} \mathbf{r} &= \mathbf{r}_c + \boldsymbol{\rho} \\ \dot{\mathbf{r}} &= \dot{\mathbf{r}}_c + \dot{\boldsymbol{\rho}} = \mathbf{r}_c + \boldsymbol{\omega} \times \boldsymbol{\rho} \\ \ddot{\mathbf{r}} &= \ddot{\mathbf{r}}_c + \boldsymbol{\omega} \times \boldsymbol{\omega} \times \boldsymbol{\rho} + \dot{\boldsymbol{\omega}} \times \boldsymbol{\rho} \end{aligned} \tag{23.45}$$

where **r** is the displacement of the point P, and $\mathbf{r}_c$ is the displacement of some particular point c in the body, and $\boldsymbol{\omega}$ is the angular velocity of the body, as shown in Fig. 23.4.

In the particular case of plane motion ($\dot{\mathbf{r}}$ parallel to $\boldsymbol{\omega} \times \boldsymbol{\rho}$), a point c can be found for which $\dot{\mathbf{r}}_c = 0$. All points lying on the line passing through c parallel to $\boldsymbol{\omega}$ will have zero velocity. This line is called the *instantaneous axis of rotation.* At any instant, the velocity of any point in the body is as though the body were rotating with angular velocity $\boldsymbol{\omega}$ about a fixed axis coinciding with the instantaneous axis of rotation.

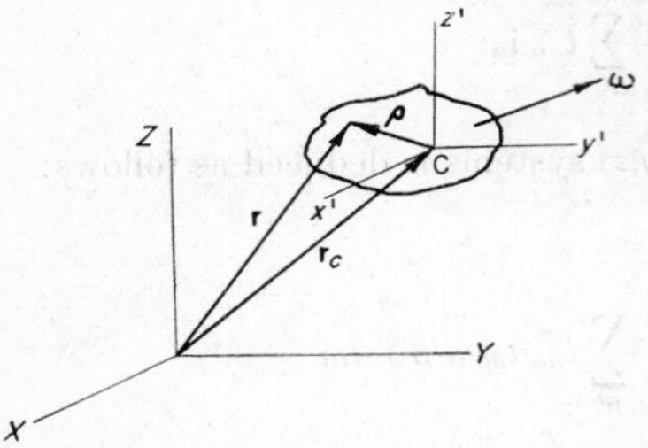

FIG. 23.4

In the more general case (nonplanar), the motion of the body may be represented as a rotation $\boldsymbol{\omega}$ about the instantaneous axis of rotation, plus a translational velocity $\mathbf{v}_c$ along the axis, which is the same as the motion of a screw with angular velocity $\boldsymbol{\omega}$ and pitch v_c/ω. This is analogous to the composition of a force system into a force **F** along the central axis, plus a moment **M** about the axis (see p. 21–20).

Moment of Momentum of a Rigid Body

The moment of momentum, **H**, about a fixed point O is

$$\mathbf{H} \equiv \int \mathbf{r} \times \dot{\mathbf{r}}\, dm$$

For a rigid body, **r** and $\dot{\mathbf{r}}$ are as given above, and

$$\mathbf{H} \equiv \int \mathbf{r}_c \times \dot{\mathbf{r}}_c\, dm + \int \boldsymbol{\rho} \times \dot{\boldsymbol{\rho}}\, dm + \int \mathbf{r}_c \times \dot{\boldsymbol{\rho}}\, dm + \int \boldsymbol{\rho} \times \dot{\mathbf{r}}_c\, dm$$

If the point c coincides with the mass center, then **H** reduces to

$$\mathbf{H} \equiv \mathbf{r}_c \times m\dot{\mathbf{r}}_c + \int \boldsymbol{\rho} \times \dot{\boldsymbol{\rho}}\, dm = \mathbf{H}_o + \mathbf{H}_c \tag{23.46}$$

The moment of momentum about the mass center $\mathbf{H}_c$, when expressed in terms of a centroidal coordinate system (x',y',z'), is

$$\begin{aligned} \mathbf{H}_c &= H_{x'}\mathbf{i}_{x'} + H_{y'}\mathbf{i}_{y'} + H_{z'}\mathbf{i}_{z'} \\ H_{x'} &= I_{x'x'}\omega_{x'} + I_{x'y'}\omega_{y'} + I_{x'z'}\omega_{z'} \\ H_{y'} &= I_{y'x'}\omega_{x'} + I_{y'y'}\omega_{y'} + I_{y'z'}\omega_{z'} \\ H_{z'} &= I_{z'x'}\omega_{x'} + I_{z'y'}\omega_{y'} + I_{z'z'}\omega_{z'} \end{aligned} \tag{23.47}$$

where $I_{x'y'} = -\int x'y'\, dm$, etc. The only restriction on the (x',y',z') coordinate system is that its origin be at the center of mass. For most problems, it is advantageous to take the coordinate system fixed in the body and rotating with it, in which case $I_{x'x'}$, $I_{x'y'}$, etc., are constants (not varying with time). In some cases, the axes chosen are not fixed in the body, but these are usually special problems, such as gyroscopes, in which symmetry ensures that the moments of inertia are constant, despite the fact that the body is rotating with respect to the axes.

General Equations of Motion for a Rigid Body

The equation of motion of the mass center of the body is

$$\mathbf{F} = m\ddot{\mathbf{r}}_c$$

and the moment-of-momentum equation of the mass center is

$$\mathbf{M}_o = \dot{\mathbf{H}}_o$$

where $\mathbf{M}_o = \mathbf{r}_c \times \mathbf{F}$, $\mathbf{H}_o = \mathbf{r}_c \times m\dot{\mathbf{r}}_c$, m = mass of body, $\mathbf{r}_c$ = displacement of mass center from origin O, and F = resultant force on the body.

The equation of motion about the mass center is $\mathbf{M}_c = \dot{\mathbf{H}}_c$, where $\mathbf{M}_c$ is the resultant moment of force about the mass center, and $\mathbf{H}_c$ is the moment of momentum about the mass center. In particular,

$$\mathbf{M}_c = \frac{d}{dt}(H_{x'}\mathbf{i}_{x'} + H_{y'}\mathbf{i}_{y'} + H_{z'}\mathbf{i}_{z'}) \tag{23.48a}$$

The x', y', z' axes are centroidal, and, if in addition they are fixed in the body (rotating with it), the derivatives of the unit vectors are $(\mathbf{i}_{x'})^{\cdot} = \boldsymbol{\omega} \times \mathbf{i}_{x'} = \omega_{z'}\mathbf{i}_{y'} - \omega_{y'}\mathbf{i}_{z'}$, etc., and the equation of motion is

$$\mathbf{M}_c = (\dot{H}_{x'} - \omega_{z'}H_{y'} + \omega_{y'}H_{z'})\mathbf{i}_{x'} + (\dot{H}_{y'} - \omega_{x'}H_{z'} + \omega_{z'}H_{x'})\mathbf{i}_{y'} + (\dot{H}_{z'} - \omega_{y'}H_{x'} + \omega_{x'}H_{y'})\mathbf{i}_{z'} \tag{23.48b}$$

The three scalar equations of motion are:

$$M_{x'} = \dot{H}_{x'} - \omega_{z'}H_{y'} + \omega_{y'}H_{z'} \qquad M_{y'} = \dot{H}_{y'} - \omega_{x'}H_{z'} + \omega_{z'}H_{x'}$$
$$M_{z'} = \dot{H}_{z'} - \omega_{y'}H_{x'} + \omega_{x'}H_{y'} \tag{23.48c}$$

It is customary to make the x', y', z' axes the principal axes of the body, in which case $H_{x'} = I_{x'x'}\omega_{x'}$, $H_{y'} = I_{y'y'}\omega_{y'}$, $H_{z'} = I_{z'z'}\omega_{z'}$, and the equations of motion are:

$$\begin{aligned} M_{x'} &= I_{x'x'}\dot{\omega}_{x'} + (I_{z'z'} - I_{y'y'})\omega_{y'}\omega_{z'} \\ M_{y'} &= I_{y'y'}\dot{\omega}_{y'} + (I_{x'x'} - I_{z'z'})\omega_{z'}\omega_{x'} \\ M_{z'} &= I_{z'z'}\dot{\omega}_{z'} + (I_{y'y'} - I_{x'x'})\omega_{x'}\omega_{y'} \end{aligned} \tag{23.49}$$

These are called Euler's equations of motion of a rigid body.

Equations of Impulse-Momentum

The equation of impulse-momentum for the mass center is

$$\int_1^2 \mathbf{F}\,dt = [m\mathbf{v}_c]_1^2 \tag{23.50}$$

and the moment-of-momentum equation is

$$\int_1^2 \mathbf{M}_o\,dt = [\mathbf{H}_o]_1^2 \tag{23.51}$$

The motion about the mass center is described by the equation

$$\int_1^2 \mathbf{M}_c\,dt = [\mathbf{H}_c]_1^2 \tag{23.52}$$

which states that the moment of impulse about the mass center is equal to the change in moment of momentum about the mass center.

The Equation of Work-Energy

The equation of work-energy for the mass center is

$$\int_1^2 \mathbf{F}\cdot d\mathbf{r}_c = \tfrac{1}{2}[m\dot{r}_c{}^2]_1^2 = \tfrac{1}{2}[\mathbf{v}_c\cdot m\mathbf{v}_c]_1^2 \tag{23.53}$$

The work-energy equation for motion about the mass center is derived as follows:

$$\mathbf{M}_c = \dot{\mathbf{H}}_c$$

$$\int_1^2 \boldsymbol{\omega} \cdot \mathbf{M}_c \, dt = \int_1^2 \boldsymbol{\omega} \cdot \dot{\mathbf{H}}_c \, dt$$

$$= \int_1^2 \frac{1}{2} \frac{d}{dt} (\boldsymbol{\omega} \cdot \mathbf{H}_c) \, dt$$

$$\int_1^2 \boldsymbol{\omega} \cdot \mathbf{M}_c \, dt = \tfrac{1}{2}[\boldsymbol{\omega} \cdot \mathbf{H}_c]_1^2 \tag{23.54}$$

In expanded form the kinetic energy of motion about the mass center is

$$\tfrac{1}{2}\boldsymbol{\omega} \cdot \mathbf{H}_c = \tfrac{1}{2}(I_{x'x'}\omega_{x'}^2 + I_{y'y'}\omega_{y'}^2 + I_{z'z'}\omega_{z'}^2 + 2I_{x'y'}\omega_{x'}\omega_{y'} + 2I_{y'z'}\omega_{y'}\omega_{z'} + 2I_{z'x'}\omega_{z'}\omega_{x'}) \tag{23.55}$$

where $I_{x'y'} = -\int x'y' \, dm$, etc.

The total kinetic energy of a rigid body is

$$T = \tfrac{1}{2}\mathbf{v}_c \cdot m\mathbf{v}_c + \tfrac{1}{2}\boldsymbol{\omega} \cdot \mathbf{H}_c \tag{23.56}$$

If there is no moment of force about the mass center, the kinetic energy of rotation is conserved,

$$\tfrac{1}{2}\boldsymbol{\omega} \cdot \mathbf{H}_c = \text{const}$$

and the moment of momentum is also conserved:

$$\mathbf{H}_c = \text{const}$$

It is seen from these two equations that, in this case, $\boldsymbol{\omega}$ is not necessarily a constant, for, if it is resolved into a component $\boldsymbol{\omega}_h$ parallel to $\mathbf{H}_c$ and a component $\boldsymbol{\omega}_n$ normal to $\mathbf{H}_c$, the equations describing the conservation of energy and moment of momentum are

$$\tfrac{1}{2}\omega_h H_c = \text{const} \qquad \mathbf{H}_c = \text{const}$$

from which it is inferred that, although $\boldsymbol{\omega}_h$ is a constant, $\boldsymbol{\omega}_n$ may vary in such a way that $\mathbf{H}_c$ remains constant. For example, $\boldsymbol{\omega}_n$ describes the wobbling of a freely spinning disk.

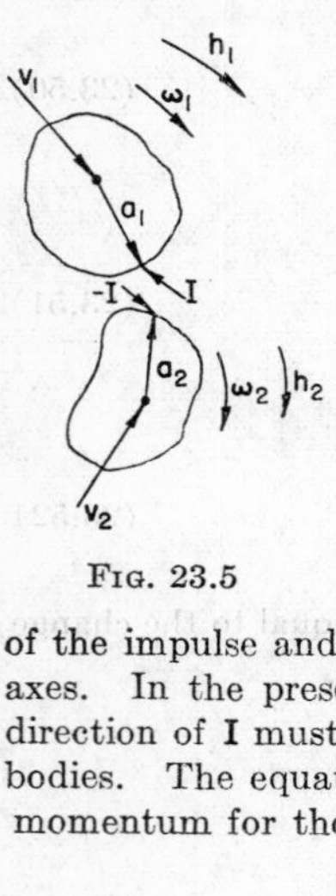

FIG. 23.5

Impact between Two Bodies

A general impact between two bodies involves the quantities shown in Fig. 23.5, where $\mathbf{h}_1$, $\mathbf{h}_2$ are the moments of momentum about the mass centers of bodies 1 and 2 before impact; $\mathbf{v}_1$, $\mathbf{v}_2$ are the velocities of the mass centers before impact; $\boldsymbol{\omega}_1$, $\boldsymbol{\omega}_2$ are the angular velocities before impact. $\mathbf{H}_1$, $\mathbf{H}_2$, $\mathbf{V}_1$, $\mathbf{V}_2$, $\boldsymbol{\Omega}_1$, $\boldsymbol{\Omega}_2$ are the corresponding quantities after impact. $\mathbf{I}$ is the impulse exerted upon body 1 by body 2, and $-\mathbf{I}$ is the impulse exerted on 2 by 1; $\mathbf{a}_1$ and $\mathbf{a}_2$ are the vectors from the mass centers to the points of application of I and $-I$, respectively. It should be noted that $\mathbf{I} = I(l_x\mathbf{i}_x + l_y\mathbf{i}_y + l_z\mathbf{i}_z)$, where I is the magnitude of the impulse and l_x, l_y, l_z are the direction cosines of $\mathbf{I}$ with respect to the x, y, z axes. In the present analysis, the direction cosines are known quantities, for the direction of $\mathbf{I}$ must be determined from a prior analysis of contact pressures between bodies. The equations of impulse-momentum and moment of impulse–moment of momentum for the two bodies are:

$$\begin{aligned} m_1(\mathbf{V}_1 - \mathbf{v}_1) &= \mathbf{I} & \mathbf{H}_1 - \mathbf{h}_1 &= \mathbf{a}_1 \times \mathbf{I} \\ m_2(\mathbf{V}_2 - \mathbf{v}_2) &= -\mathbf{I} & \mathbf{H}_2 - \mathbf{h}_2 &= \mathbf{a}_2 \times (-\mathbf{I}) \end{aligned} \tag{23.57}$$

The equation of conservation of energy is

$$\tfrac{1}{2}m_1v_1^2 + \tfrac{1}{2}m_2v_2^2 + \tfrac{1}{2}\boldsymbol{\omega}_1 \cdot \mathbf{h}_1 + \tfrac{1}{2}\boldsymbol{\omega}_2 \cdot \mathbf{h}_2 = \tfrac{1}{2}m_1V_1^2 + \tfrac{1}{2}m_2V_2^2 + \tfrac{1}{2}\boldsymbol{\Omega}_1 \cdot \mathbf{H}_1 + \tfrac{1}{2}\boldsymbol{\Omega}_2 \cdot \mathbf{H}_2 + E \quad (23.58)$$

where E is the energy lost during impact (presumed to be known). The unknown quantities are $\mathbf{V}_1$, $\mathbf{V}_2$, $\mathbf{H}_1$, $\mathbf{H}_2$, I, which comprise 13 unknown scalar components. The four vector equations plus the energy equation give 13 scalar equations which suffice to determine the unknowns. When solving such a problem, it will be advantageous to express $\mathbf{v}$, etc., in a coordinate system having an axis parallel to $\mathbf{I}$, and to express the moments of momentum with respect to principal axes.

Work Done by Impulsive Forces

Let a body of mass m be moving with velocity of mass center $\mathbf{v}_c$, angular velocity $\boldsymbol{\omega}$, and moment of momentum about the mass center $\mathbf{H}_c$. At time t, there are applied impulsive forces whose impulse can be written $\mathbf{F}\,\Delta t$ and whose moment of impulse about the mass center can be written $M_c\,\Delta t$. During time Δt, there is a change in motion represented by

$$m\,\Delta\mathbf{v}_c = \mathbf{F}\,\Delta t \qquad \Delta\mathbf{H}_c = \mathbf{M}_c\,\Delta t$$

The kinetic energy will change by an amount

$$\Delta T = \Delta(\tfrac{1}{2}m\mathbf{v}_c \cdot \mathbf{v}_c + \tfrac{1}{2}\boldsymbol{\omega} \cdot \mathbf{H}_c) = \left(\mathbf{v}_c + \frac{\Delta\mathbf{v}_c}{2}\right) \cdot m\,\Delta\mathbf{v}_c + \left(\boldsymbol{\omega} + \frac{\Delta\boldsymbol{\omega}}{2}\right) \cdot \Delta\mathbf{H}_c$$

The work done may thus be expressed in terms of the impulse by

$$\Delta W = (\mathbf{F}\,\Delta t) \cdot \left(\mathbf{v}_c + \frac{\Delta\mathbf{v}_c}{2}\right) + (\mathbf{M}_c\,\Delta t) \cdot \left(\boldsymbol{\omega} + \frac{\Delta\boldsymbol{\omega}}{2}\right) \quad (23.59)$$

Equations of Motion for a Translating Body

The motion of a body whose moment of momentum about the mass center is a constant ($\dot{\mathbf{H}}_c = 0$) is described by the equations

$$\mathbf{F} = m\ddot{\mathbf{r}}_c \qquad \mathbf{M}_c = 0 \quad (23.60)$$

where $\mathbf{F}$ is the resultant force on the body, m is the mass, $\mathbf{r}_c$ is the displacement of the mass center, and $\mathbf{M}_c$ is the resultant moment of force about the mass center. In rectangular coordinates, the equations are:

$$\begin{aligned} F_x &= m\ddot{x}_c & M_{cx} &= 0 \\ F_y &= m\ddot{y}_c & M_{cy} &= 0 \\ F_z &= m\ddot{z}_c & M_{cz} &= 0 \end{aligned} \quad (23.61)$$

The equations of impulse-momentum and work-energy are

$$\int_1^2 \mathbf{F}\,dt = [m\dot{\mathbf{r}}_c]_1^2 \qquad \int_1^2 \mathbf{F} \cdot d\mathbf{r}_c = \tfrac{1}{2}[m\dot{\mathbf{r}}_c{}^2]_1^2 \quad (23.62)$$

In rectangular coordinates, the component equations may be written

$$\int_1^2 F_x\,dt = [m\dot{x}_c]_1^2 \qquad \int_1^2 F_x\,dx_c = \tfrac{1}{2}[m\dot{x}_c{}^2]_1^2 \qquad \text{etc.}$$

Rotation about a Fixed Axis

Let the (x,y,z) coordinate system be fixed in the body with the z axis coinciding with the axis of rotation. The motion is described by the equation $\mathbf{M} = \dot{\mathbf{H}}$, where

M and **H** are the moment of force and the moment of momentum, respectively, about the origin of the coordinate system. In expanded form, the equation is:

$$M_x = -I_{yz}\omega_z^2 + I_{xz}\dot{\omega}_z \qquad M_y = I_{xz}\omega_z^2 + I_{yz}\dot{\omega}_z \qquad M_z = I_{zz}\dot{\omega}_z \tag{23.63}$$

If x', y', z' are a parallel set of axes fixed in the body and rotating with it, with the z' axis passing through the center of mass, the motion is described by

$$M_{x'} = -I_{y'z'}\omega_z^2 + I_{x'z'}\dot{\omega}_z \qquad M_{y'} = I_{x'z'}\omega_z^2 + I_{y'z'}\dot{\omega}_z \qquad M_{z'} = I_{z'z'}\dot{\omega}_z \tag{23.64}$$

where $I_{x'y'} = -\int x'y'\,dm$. In addition, there is the equation of motion of the mass center $\mathbf{F} = m\ddot{\mathbf{r}}_c$.

The impulse-momentum and work-energy equations are

$$\int_1^2 M_z\,dt = [I_{zz}\omega_z]_1^2 \qquad \int_1^2 M_z\omega_z\,dt = [\tfrac{1}{2}I_{zz}\omega_z^2]_1^2 \tag{23.65}$$

or, in alternate forms,

$$\int_1^2 M_{z'}\,dt = [I_{z'z'}\omega_z]_1^2 \qquad \int_1^2 M_{z'}\omega_z\,dt = [\tfrac{1}{2}I_{z'z'}\omega_z^2]_1^2 \tag{23.66}$$

$$\int_1^2 \mathbf{r}_c \times \mathbf{F}\omega_z\,dt = [mr_c^2\omega_z]_1^2 \qquad \int_1^2 \mathbf{F}\cdot d\mathbf{r}_c = [\tfrac{1}{2}mr_c^2\omega_z^2]_1^2 \tag{23.67}$$

For the balancing of rotors (no applied forces, zero gravity force), it should be noted that M_x and M_y are the total moments exerted by the rotor supports; $M_{x'}$ and $M_{y'}$ are the support moments due to the dynamic unbalance (nonzero $I_{x'z'}$ and $I_{y'z'}$); and **F** is the support force due to static unbalance (nonzero $\mathbf{r}_c$).

Centers of Oscillation and Percussion

A compound pendulum (Fig. 23.6) is a body, free to rotate about a fixed horizontal axis, that oscillates under the action of gravity. Let $\mathbf{r}_c$ be the displacement of the

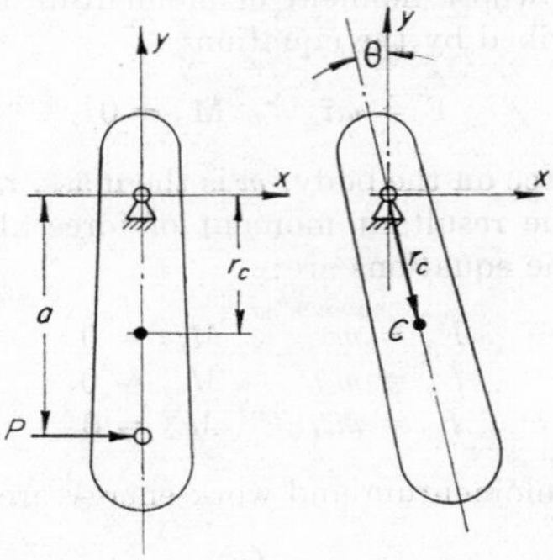

Fig. 23.6

mass center from the center of suspension, θ the angular deviation of $\mathbf{r}_c$ from the vertical, and **R** the force exerted on the body by the support. The equation of moment of momentum about the axis of rotation is

$$-r_c mg \sin\theta = (mr_c^2 + I_c)\ddot{\theta} \qquad \text{or} \qquad \ddot{\theta} + \frac{g}{r_c(1 + I_c/mr_c^2)}\sin\theta = 0$$

The quantity $l = r_c(1 + I_c/mr_c^2)$ is called the equivalent length of the compound pendulum (it is the length of a simple pendulum, $I_c = 0$). If the body is hanging at rest, and is acted on by an impulse $P\,\Delta t$, where P is normal to r_c and to the axis, and is at a distance a from the axis, the equations of impulse–momentum and moment of impulse–moment of momentum are:

$$(R_x + P)\,\Delta t = mr_c\dot{\theta} \qquad (P(a - r_c) - R_x r_c)\,\Delta t = I_c\dot{\theta}$$

The condition for zero horizontal reaction at the support ($R_x = 0$) is

$$a = r_c\left(1 + \frac{I_c}{mr_c^2}\right)$$

This is just the equivalent length of the pendulum, and the point lying on the extension of r_c at this distance from the axis is called the center of percussion. The center of oscillation (rotation) and the center of percussion are interchangeable, for, if the pendulum is suspended from a parallel axis through the center of percussion, the period of oscillation ($2\pi\sqrt{l/g}$) for small motion is unchanged, and the original center of oscillation is the new center of percussion.

Plane Motion of a Rigid Body

Let all velocity components lie in the x,y plane; the angular velocity of the body is then $\omega_z\mathbf{i}_z$, and the velocity of the mass center is $\dot{\mathbf{r}}_c = \dot{x}_c\mathbf{i}_x + \dot{y}_c\mathbf{i}_y$. Let the x', y', z' axes be fixed in the body and rotating with it, with the origin at the mass center. The motion of the body is described by the equations

$$\mathbf{F} = m\ddot{\mathbf{r}}_c \qquad \mathbf{M}_c = \dot{\mathbf{H}}_c \tag{23.68}$$

where $\mathbf{F}$ is the resultant force acting on the body and $\mathbf{M}_c$ is the resultant moment about the mass center. In rectangular coordinates, the equations are:

$$\begin{aligned} F_x &= m\ddot{x}_c & M_{x'} &= -I_{y'z'}\omega_z^2 + I_{x'z'}\dot{\omega}_z \\ F_y &= m\ddot{y}_c & M_{y'} &= I_{x'z'}\omega_z^2 + I_{y'z'}\dot{\omega}_z \\ F_z &= 0 & M_{z'} &= I_{z'z'}\dot{\omega}_z \end{aligned} \tag{23.69}$$

The impulse-momentum and work-energy equations are:

$$\int_1^2 M_{z'}\,dt = [I_{z'z'}\omega_z]_1^2 \qquad \int_1^2 M_{z'}\omega_z\,dt = [\tfrac{1}{2}I_{z'z'}\omega_z^2]_1^2 \tag{23.70}$$

$$\int_1^2 \mathbf{r}_c \times \mathbf{F}\omega_z\,dt = [mr_c^2\omega_z]_1^2 \qquad \int_1^2 \mathbf{F}\cdot d\mathbf{r}_c = [\tfrac{1}{2}mr_c^2\omega_z^2]_1^2 \tag{23.71}$$

Rotation about a Fixed Point

It is customary to express the equations of motion in terms of the so-called *Eulerian angles* which are explained in Figs. 23.7*a* and *b*. The motion of a body such as a spinning top is expressed naturally in terms of the angles shown in Fig. 23.7*a*, where θ measures the declination of the spin axis from the vertical, ψ measures the rotation of the spin axis or the rotation of plane *abo* about the z_0 axis, and the angle ϕ measures the rotation of the body relative to the *abo* plane. Figure 23.7*b* shows the coordinate systems used to represent the motion in terms of θ, ψ, ϕ. The x', y', z' axes are fixed in the body and rotating with it, and so the motion of x', y', z' is the motion of the body. The x, y, z axes are fixed to the *abo* plane and rotating with it, and so the motion of x, y, z is the motion of the *abo* plane. The angular velocities $\dot{\theta}$, $\dot{\psi}$, $\dot{\phi}$ are the same in Fig. 23.7*b* as in Fig. 23.7*a*.

The angular velocity of the (x,y,z) system is

$$\boldsymbol{\omega} = \omega_x\mathbf{i}_x + \omega_y\mathbf{i}_y + \omega_z\mathbf{i}_z = \dot{\theta}\mathbf{i}_x + \dot{\psi}\,(\sin\theta)\mathbf{i}_y + \dot{\psi}\,(\cos\theta)\mathbf{i}_z \tag{23.72}$$

where ω_x, ω_y, ω_z are the components in the (x,y,z) system. The angular velocity of the body is

$$\boldsymbol{\omega}_b = \omega_x \mathbf{i}_x + \omega_y \mathbf{i}_y + \Omega \mathbf{i}_z \tag{23.73}$$

where $\Omega = \omega_z + \dot{\phi}$ is the total spin velocity of the body.

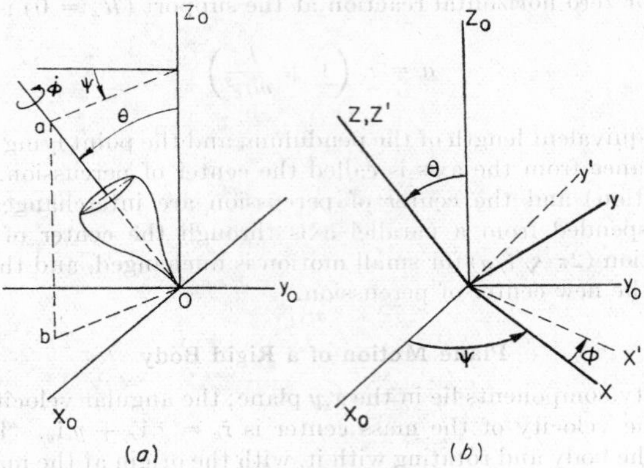

FIG. 23.7

The equation of motion of the body is $\mathbf{M} = \dot{\mathbf{H}}$, where $\mathbf{M}$ and $\dot{\mathbf{H}}$ are the resultant moment and the total moment of momentum about the point O. In the (x,y,z) system, this equation is expressed by

$$M_x = \dot{H}_x - H_y\omega_z + H_z\omega_y \qquad M_y = \dot{H}_y - H_z\omega_x + H_x\omega_z$$
$$M_z = \dot{H}_z - H_x\omega_y + H_y\omega_x \tag{23.74}$$

The moment of momentum of the body is:

$$H_x = I_{xx}\omega_x + I_{xy}\omega_y + I_{xz}\Omega \qquad H_y = I_{xy}\omega_x + I_{yy}\omega_y + I_{yz}\Omega$$
$$H_z = I_{zx}\omega_x + I_{zy}\omega_y + I_{zz}\Omega \tag{23.75}$$

where $I_{xy} = -\int xy\, dm$, etc.

If the body is symmetrical about the spin axis ($I_x = I_y = I$), the equations reduce to:

$$M_x = I(\dot{\omega}_x - \omega_y\omega_z) + I_z\Omega\omega_y \qquad M_y = I(\dot{\omega}_y + \omega_z\omega_x) - I_z\Omega\omega_x \qquad M_z = I_z\dot{\Omega} \tag{23.76}$$

If H_z is large compared to H_x and H_y, the body is said to be a *gyroscope*, and its properties of motion are quite different from the case when H_z is of the order of H_x, H_y. Reference should be made to Chap. 25 for details of this motion.

23.5. ALTERNATE FORMS OF THE EQUATIONS OF MOTION

In the preceding sections, the quantities force and momentum, and moment of force and moment of momentum, played the key roles, and there is no doubt that this approach gives, in general, the best physical insight into the properties of motion. There are alternate forms of presenting the equations of motion that have advantages for certain purposes; and, in engineering, d'Alembert's principle, Lagrange's equations, and Hamilton's principle are often used. The chief advantage afforded by the use of these methods in engineering dynamics is that, in a certain sense, they minimize the thought and effort required to derive the correct differential equations of motion.

D'Alembert's Principle

The engineering form of d'Alembert's principle consists of the statement that the equations of motion

$$\mathbf{F} = m\ddot{\mathbf{r}}_c \qquad \mathbf{M}_c = \dot{\mathbf{H}}_c$$

where **F** is the resultant force acting on the body (through the mass center), and $\mathbf{M}_c$ is the resultant moment about the mass center, may be written in the form

$$\mathbf{F} + (-m\ddot{\mathbf{r}}_c) = 0 \qquad \mathbf{M}_c + (-\dot{\mathbf{H}}_c) = 0 \tag{23.77}$$

and these can be interpreted as equations of static equilibrium if $-m\ddot{\mathbf{r}}_c$ is considered to be a force applied to the body through the mass center, and $-\dot{\mathbf{H}}_c$ is considered to be an applied moment about the mass center. These terms are called the *inertia force* and the *inertia moment*, respectively. D'Alembert's principle is thus equivalent to the statement that, at any instant, the body is in static equilibrium under the action of the applied forces and moments and the inertia force and moment.

Persons more conversant with statics problems than with dynamics problems may find the use of d'Alembert's principle advantageous. This is particularly true for

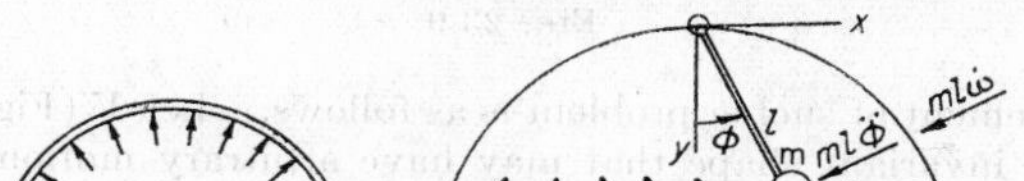

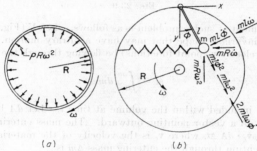

FIG. 23.8

those problems where interest lies in the internal stresses and forces during motion. This may be illustrated by the problem of a thin, circular hoop of radius R and mass per unit length μ that is spinning with constant angular velocity ω. By d'Alembert's principle, this problem is equivalent to that of a hoop subjected to an outward pressure of $\mu R\omega^2$ per unit length, μ being the mass of an element of unit length, and $R\omega^2$ directed inwardly being the acceleration of the element. As shown in Fig. 23.8*a*, this is the well-known problem of hoop stress under an internal pressure, and the circumferential force in the hoop is $\mu R^2\omega^2$.

A second example is a simple pendulum of mass m, restrained by a spring of stiffness k, pivoted on the rim of a wheel rotating with angular velocity ω and angular acceleration $\dot{\omega}$. By d'Alembert's principle, the equation of equilibrium of the mass m is

$$\mathbf{F} - m\ddot{\mathbf{R}} - m\boldsymbol{\omega} \times (\boldsymbol{\omega} \times \boldsymbol{\varrho}) - m\dot{\boldsymbol{\omega}} \times \boldsymbol{\varrho} - m\ddot{\boldsymbol{\varrho}}_r - 2m\boldsymbol{\omega} \times \dot{\boldsymbol{\varrho}}_r = 0$$

where **R** is the displacement of the pivot point from the center of rotation, and $\boldsymbol{\varrho}$ is the displacement of m from the pivot point (see section on Motion Referred to a Moving Coordinate System, p. 23-3). In terms of r, l, ϕ, ω, the equation is

$$-kl\,(\sin\phi)\mathbf{i}_x - (mR\dot{\omega}\mathbf{i}_x + mR\omega^2\mathbf{i}_y) + ml\omega^2\mathbf{i}_\rho + ml\dot{\omega}\mathbf{i}_\phi + (ml\dot{\phi}^2\mathbf{i}_\rho - ml\ddot{\phi}\mathbf{i}_\phi) - 2ml\omega\dot{\phi}\mathbf{i}_\rho = 0$$

The inertia forces are shown in Fig. 23.8*b*, from which the strain in the spring and the force in the pendulum rod can be determined by the methods of statics.

It should be noted that, in addition to the engineering form of d'Alembert's principle, there is another form (see p. 24-5) called *Lagrange's form* or the variational form of d'Alembert's principle, and this plays a basic role in the theory of mechanics.

Equations of Motion for a System with Changing Mass

There is a class of problems that involves a system whose mass is changing. Examples are a rocket whose mass is decreasing during combustion, and a fluid or a particulate matter such as sand or grain that is flowing through some designated volume whose enclosed mass is thus formed of different particles at subsequent times.

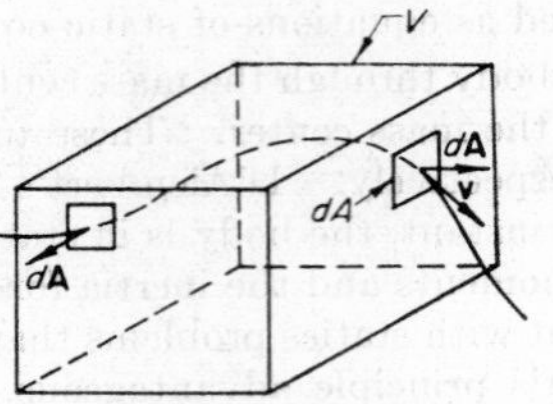

FIG. 23.9

A particular statement of such a problem is as follows. Let V (Fig. 23.9) be a designated volume of invariant shape that may have arbitrary motion. Let a material of density ρ and velocity $\mathbf{v}$ (point functions) be flowing through the volume. Let

$$m = \int_0^m dm$$

be the total mass included within the volume at time t. Let dA be the element of surface area, and $d\mathbf{A}$ a vector pointing outward. The mass entering through $d\mathbf{A}$ in the time Δt is $-\rho \mathbf{v}_r \cdot d\mathbf{A}\, \Delta t$, where $\mathbf{v}_r$ is the velocity of the material relative to $d\mathbf{A}$. The gain in momentum through the entering mass Δm is

$$-\int_A \rho \mathbf{v}(\mathbf{v}_r \cdot d\mathbf{A})\, \Delta t$$

The total momentum at time t is

$$m\, \mathbf{v}_c = \int_0^m \mathbf{v}\, dm$$

and at time $t + \Delta t$ it is

$$m\mathbf{v}_c + \Delta(m\mathbf{v}_c) = \int_0^m (\mathbf{v} + \dot{\mathbf{v}}\, \Delta t)\, dm - \int_A \rho \mathbf{v}(\mathbf{v}_r \cdot d\mathbf{A})\, \Delta t$$

hence

$$\frac{\Delta(m\mathbf{v}_c)}{\Delta t} = \int_0^m \dot{\mathbf{v}}\, dm - \int_A \rho \mathbf{v}(\mathbf{v}_r \cdot d\mathbf{A})$$

The first integral on the right-hand side is the force $\mathbf{F}$ that must be applied to impart accelerations $\dot{\mathbf{v}}$ to the mass elements dm. For $\Delta t \to 0$, the quotient on the left-hand side becomes $d(m\mathbf{v}_c)/dt$, that is, the rate of change of the total momentum included in V. It follows that

$$\mathbf{F} = \frac{d(m\mathbf{v}_c)}{dt} + \int_A \rho \mathbf{v}(\mathbf{v}_r \cdot d\mathbf{A}) \tag{23.78}$$

Similarly,

$$\mathbf{M} = \frac{d\mathbf{H}}{dt} + \int_A \mathbf{r} \times \rho \mathbf{v}(\mathbf{v}_r \cdot d\mathbf{A}) \tag{23.79}$$

where $\mathbf{H}$ is the total moment of momentum with respect to the point from which the radius vector $\mathbf{r}$ is measured.

As an example, consider a rocket with total mass $m = m_0 - kt$, including shell; velocity of mass center $\mathbf{v}_c$; velocity of rocket shell $\mathbf{u}$; and exit velocity of gas relative to nozzle $\mathbf{v}_e$; and let V just enclose the rocket. The equation of motion is

$$\mathbf{F} = \frac{\partial(m\mathbf{v}_c)}{\partial t} + \int_A \rho(\mathbf{u} + \mathbf{v}_e)(\mathbf{v}_e \cdot d\mathbf{A})$$

Letting $F = pA_n$ (pressure times nozzle area), and approximating $\mathbf{v}_c$ by $\mathbf{u}$, the equation reduces to

$$pA_n = m\dot{u} - \rho v_e^2 A_n$$

Generalized Coordinates

A system of n particles m_i will have a kinetic energy

$$T = \sum_{i=1}^{n} \tfrac{1}{2} m_i(\dot{x}_i^2 + \dot{y}_i^2 + \dot{z}_i^2)$$

The displacement of a particle may be described by quantities other than x_i, y_i, z_i, and these may be lengths, angles, or any other convenient parameters. Let such a general set of quantities be $q_1, q_2, q_3, \ldots, q_k$. These are called the generalized coordinates of the system. The q_i have the following properties: The displacement of the system is described completely by the q_i; they are independent, so that q_j may be varied while the remaining q_i are held constant, and, thus, each q_i corresponds to a degree of freedom of the system; the q_i are related to the x_i, y_i, z_i by relations of the form

$$x_i = \phi_i(q_1,q_2, \ldots) \qquad y_i = \psi_i(q_1,q_2, \ldots) \qquad z_i = \theta_i(q_1,q_2, \ldots)$$

which do not contain the time t explicitly. For problems in which it is desired to use q_i that are not independent, or for problems requiring more q_i than there are degrees of freedom (nonholonomic), or for problems in which the coordinate relations involve t explicitly, reference should be made to an advanced textbook on dynamics.

The kinetic energy of the system is expressed in terms of the generalized coordinates by

$$T = \sum_i \frac{m_i}{2}\left[\sum_j \left(\frac{\partial \phi_i}{\partial q_j}\dot{q}_j\right)^2 + \sum_j \left(\frac{\partial \psi_i}{\partial q_j}\dot{q}_j\right)^2 + \sum_j \left(\frac{\partial \theta_i}{\partial q_j}\dot{q}_j\right)^2\right] \tag{23.80}$$

Mechanical systems involve two types of elements: those that may be treated as rigid bodies, and those that must be treated as elastic bodies (springs, beams, etc.). In the most general case, a rigid body has six degrees of freedom, and its motion is described completely by six generalized coordinates. An elastic body requires an infinite set of coordinates to describe its motion completely.

As the system moves through displacements dq_i, an increment of work will be done by the forces of the system, which will be of the form

$$dW = \sum_j Q_j\, dq_j \tag{23.81}$$

The Q_j are called the generalized forces corresponding to the generalized coordinates q_j.

The determination of the generalized forces is a straightforward process for systems composed of rigid bodies. In the case of elastic systems, the procedure is illustrated by the following example. A taut, massless cord supporting six equal masses, equally spaced, is vibrating freely in the second mode, as shown in Fig. 23.10.

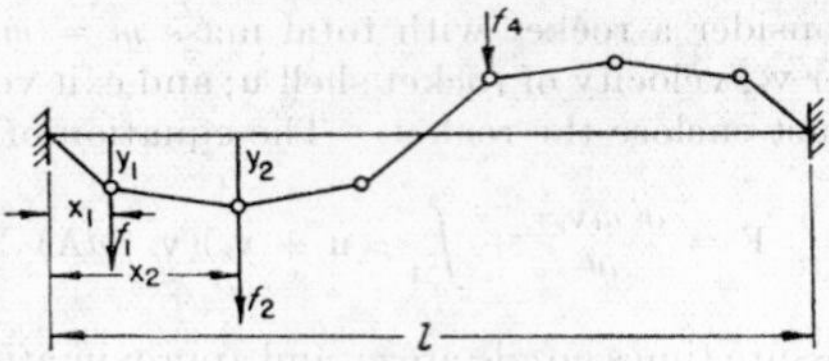

FIG. 23.10

The displacements of the m_i are given by

$$y_i = C \sin \frac{2\pi x_i}{l} \sin \omega_2 t$$

A generalized coordinate for this mode is any quantity that specifies the displacement. Let the generalized coordinate q_2 be the displacement at $x = x_1$:

$$q_2 = C \sin \frac{2\pi x_1}{l} \sin \omega_2 t \qquad \text{and hence} \qquad y_i = q_2 \frac{\sin (2\pi x_i/l)}{\sin (2\pi x_1/l)}$$

The quantity $\frac{\sin (2\pi x_i/l)}{\sin (2\pi x_1/l)}$ is the mode shape ($= 1$ at x_1). The work done during displacement dq_2 is

$$dW = \sum f_i \, dy_i = \left(\sum f_i \frac{\sin (2\pi x_i/l)}{\sin (2\pi x_1/l)} \right) dq_2 = Q_2 \, dq_2$$

Q_2 is the generalized force corresponding to the generalized coordinate q_2.

Lagrange's Equation

In complex engineering problems, Lagrange's equations are a most useful tool for deriving the equations of motion. They are based on the use of generalized coordinates. The kinetic energy of a system of particles, expressed in terms of the generalized coordinates, is a homogeneous, quadratic function of the $\dot{q}_j$, and, hence, it follows from Euler's theorem for homogeneous functions that

$$\sum_j \frac{\partial T}{\partial \dot{q}_j} \dot{q}_j = 2T$$

The rate at which work is done by the forces of the system is

$$\dot{W} = \sum_j Q_j \dot{q}_j$$

and the rate of change in kinetic energy is

$$\dot{T} = \sum_j \left(\frac{\partial T}{\partial \dot{q}_j} \ddot{q}_j + \frac{\partial T}{\partial q_j} \dot{q}_j \right)$$

$$= \sum_j \left[\frac{d}{dt} \left(\frac{\partial T}{\partial \dot{q}_j} \dot{q}_j \right) - \frac{d}{dt} \left(\frac{\partial T}{\partial \dot{q}_j} \right) \dot{q}_j + \frac{\partial T}{\partial q_j} \dot{q}_j \right]$$

$$= 2\dot{T} - \sum_j \left[\frac{d}{dt} \left(\frac{\partial T}{\partial \dot{q}_j} \right) \dot{q}_j - \frac{\partial T}{\partial q_j} \dot{q}_j \right]$$

$$\dot{T} = \sum_j \left(\frac{d}{dt} \frac{\partial T}{\partial \dot{q}_j} - \frac{\partial T}{\partial q_j} \right) \dot{q}_j$$

Equating $\dot{T}$ and $\dot{W}$ gives the work-energy equation:

$$\sum_j \left(\frac{d}{dt} \frac{\partial T}{\partial \dot{q}_j} - \frac{\partial T}{\partial q_j} - Q_j \right) \dot{q}_j = 0$$

It may be shown (see p. 24–7) that each term of this equation is equal to zero, and, hence,

$$\frac{d}{dt} \frac{\partial T}{\partial \dot{q}_j} - \frac{\partial T}{\partial q_j} = Q_j \quad j = 1, 2, \ldots \tag{23.82}$$

These are Lagrange's equations, and there are as many equations as there are q_i (degrees of freedom). If the system is conservative, with potential energy V, the generalized forces are $Q_j = -\partial V/\partial q_j$, and, since V is independent of $\dot{q}_j$, it follows that $\partial V/\partial \dot{q}_j = 0$, and Lagrange's equations may be written

$$\frac{d}{dt} \frac{\partial L}{\partial \dot{q}_j} - \frac{\partial L}{\partial q_j} = 0 \tag{23.83}$$

where $L = T - V$ is called the *Lagrangian function* or the *kinetic potential* of the system.

As an example, a simple pendulum of mass m supported on an elastic string of equilibrium length l, total length $(l + x)$, and elastic constant k, has a Lagrangian function for small oscillations,

$$L = \tfrac{1}{2}m(\dot{x}^2 + l^2\dot{\phi}^2) - \tfrac{1}{2}kx^2 + [x \cos \phi - l(1 - \cos \phi)]mg$$

where x and ϕ are the two generalized coordinates. The application of Eq. (23.83) gives

$$m\ddot{x} + kx - mg \cos \phi = 0 \qquad ml^2\ddot{\phi} + mg(l + x) \sin \phi = 0$$

For simple problems, Lagrange's equations afford no particular advantage, but, for complex problems (multiple degrees of freedom), they have the great advantage of producing the correct equations of motion if the kinetic energy and the generalized forces can be written down; and, hence, they obviate most of the geometrical considerations required when Newton's equations are applied directly. Furthermore, reaction forces and forces of constraints which do no work will not appear in an analysis involving Lagrange's equations, although this is not always an advantage in engineering problems.

Lagrange's Equations for Impulsive Forces

If a system is acted upon by impulsive forces (large magnitude and short time duration Δt), the equations of Lagrange may be integrated to give

$$\int_1^2 \left(\frac{d}{dt} \frac{\partial T}{\partial \dot{q}_j} - \frac{\partial T}{\partial q_j} \right) dt = \int_1^2 Q_j \, dt$$

The first term on the left side may be integrated directly, and, if $\Delta t = t_2 - t_1$ is sufficiently small, the integral of the second term is negligible, and, hence,

$$\left[\frac{\partial T}{\partial \dot{q}_j} \right]_1^2 = \int_1^2 Q_j \, dt \tag{23.84}$$

Terms of the form of the left side of this equation are called generalized components of momentum, and the term on the right is called generalized impulse.

Ignorable Coordinates

If for some particular i, say $i = 1$,

$$\frac{\partial L}{\partial q_1} = 0$$

q_1 is called an ignorable coordinate, and, from Lagrange's equation (23.83),

$$\frac{\partial T}{\partial \dot{q}_1} = c_1$$

where c_1 is a constant. If this equation is solved for $\dot{q}_1$ (in terms of c_1 and the other q_i), the function

$$R = L - c_1\dot{q}_1$$

will satisfy the equations (for a conservative system)

$$\frac{d}{dt}\frac{\partial R}{\partial \dot{q}_j} - \frac{\partial R}{\partial q_j} = 0 \qquad j = 2, 3, \ldots \tag{23.85}$$

and q_1 may now be ignored in the solution of the equations. R is called the *Routhian function.*

Derivation of Equations for Complex Continuous Systems

In engineering problems, difficulties are sometimes encountered in deriving the differential equations of motion (or equilibrium) of continuous, elastic systems. If a system is complex, it sometimes happens that significant elements of the problem are overlooked, and incorrect equations are deduced. Furthermore, in the analysis of engineering problems, it is common practice to introduce simplifications at the outset, by assuming specified forms for strain or stress distribution, and this sometimes causes difficulty in deriving the differential equations, or in establishing the boundary conditions. For example, in the theory of elasticity, the equations of equilibrium, when expressed in terms of the displacements, form a set of three partial differential equations of the second order ($3 \times 2 = 6$), whereas, in problems of beams and plates, certain assumptions are made concerning the form of the strain distribution, and this leads to a single equation of equilibrium of the fourth order. The reduction in order means that the boundary conditions are also affected, and that arbitrary boundary conditions on the stresses or strains cannot be imposed.

Problems of the afore-mentioned types are best handled by means of *Hamilton's principle,* which states that, for a conservative system,

$$\int_1^2 \delta(T - V)\,dt = 0 \tag{23.86}$$

If, in addition to the potential energy, nonconservative forces are acting that do increments of work δW during the variation, the equation has the form

$$\int_1^2 [\delta(T - V) + \delta W]\,dt = 0 \tag{23.87}$$

For statics problems, the corresponding equation is

$$\delta V = \delta W \tag{23.88}$$

Reference should be made to p. 24-5 or to textbooks on dynamics and calculus of variations for the derivation of Hamilton's principle, and for information on the variational procedures required in the application of the above equations. The use of Eq. (23.86) is illustrated by the following example.

A pipeline above ground, supported as a continuous beam, and containing flowing fluid, has a total mass (pipe plus fluid) per unit length m, moment of inertia of pipe I, and modulus of elasticity E. The fluid has a mass per unit length μ and a constant velocity v relative to the pipe. The coordinate x is measured along the pipe, and the vertical displacement of the pipe center line is z. It is desired to find the equation of motion for free vibrations in the vertical plane and the appropriate boundary conditions. The velocity components of the fluid are

$$v_x = v \qquad v_z = \frac{\partial z}{\partial t} + v\frac{\partial z}{\partial x} = \dot{z} + vz'$$

The kinetic energy of pipe shell plus fluid is

$$T = \int_{x_1}^{x_2} \left[\frac{1}{2}(m-\mu)\dot{z}^2 + \frac{1}{2}\mu v^2 + \frac{1}{2}\rho(\dot{z} + vz')^2\right] dx$$

The potential energy of bending is

$$V = \int_{x_1}^{x_1} \frac{1}{2} EI(z'')^2\, dx$$

It should be noted that T and V embody simplifying assumptions concerning the velocity profile of the fluid and the strain distribution in the pipe wall, and the derived differential equation and boundary conditions will be appropriate only to a system described by this set of T and V. Hamilton's principle gives

$$\int_{t_1}^{t_2}\int_{x_1}^{x_2} [(m-\mu)\dot{z}\delta\dot{z} + \mu(\dot{z} + vz')(\delta\dot{z} + v\delta z') - EIz''\delta z'']\, dx\, dt = 0 \qquad (23.89)$$

Integrating by parts, after using the relations

$$\delta\dot{z} = \frac{\partial}{\partial t}\delta z \qquad \delta z' = \frac{\partial}{\partial x}\delta z \qquad \delta z'' = \frac{\partial^2}{\partial x^2}\delta z$$

gives, after collecting terms according to the δz, $\delta z'$, etc.,

$$\int_{t_1}^{t_2}\int_{x_1}^{x_2} (m\ddot{z} + 2\mu v\dot{z}' + \mu v^2 z'' + EIz^{\text{iv}})\delta z\, dx\, dt + \int_{t_1}^{t_2} (EIz''' + \mu v(\dot{z} + vz')\delta z \Big|_{x_1}^{x_2} dt - \int_{t_1}^{t_2} (EIz'')\delta z' \Big|_{x_1}^{x_2} dt + \int_{x_1}^{x_2} (m\dot{z} + \mu vz')\delta z \Big|_{t_1}^{t_2} dx = 0 \qquad (23.90)$$

To satisfy this equation, each of the terms must be zero because of the independence of the variations δz, etc., appearing in them. Each term is the product of two parts, either of which may be zero and thus cause the term to vanish. In the first term, $\delta z = 0$ represents the case where the pipe is at rest. If $\delta z \neq 0$, the term in the parentheses must be zero, and this gives the differential equation of motion,

$$EI\frac{\partial^4 z}{\partial x^4} + \mu v^2\frac{\partial^2 z}{\partial x^2} + 2\mu v\frac{\partial^2 z}{\partial x\,\partial t} + m\frac{\partial^2 z}{\partial t^2} = 0 \qquad (23.91)$$

The remaining terms specify the appropriate boundary and initial conditions. The second term states that, at the ends x_1, x_2, there may be imposed specified displacements ($\delta z = 0$), or that the vertical shear force is zero $[EIz''' + \mu v(\dot{z} + vz') = 0]$. It should be noted that EIz''' is the beam shear, and $\mu v(\dot{z} + vz')$ is the vertical component of the momentum flux which must be included in the net shear. The third term states that the slope must be specified ($\delta z' = 0$), or the bending moment must be zero ($EIz'' = 0$). The last term states that, at the initial time and at the final time, either the displacement must be specified ($\delta z = 0$), or the vertical momentum must be zero ($m\dot{z} + \mu v z' = 0$), and these are equivalent to specifying the initial velocity and displacement. These conditions show that, if $v = 0$, it is possible to find initial conditions such that, at time t_2, the pipe will have a specified displacement and zero velocity, and all the energy is in the form of strain energy; if $v \neq 0$, this is not possible, and it may be concluded that the fluid flow has a fundamental effect on the character of the motion.

If in the above problem the restraints are modified, for example, if beam-shear strain is included, the strain energy will be modified, and the differential equation and the boundary conditions will also be modified. If forces, energy-storing devices, etc., are acting on the boundaries, they must be included in the formulation of Hamilton's principle and they will also affect the derived boundary conditions.

23.6. REFERENCES

[1] R. Dugas: "A History of Mechanics," Central Book Co., New York, 1955. A historical presentation of the development of mechanics from Aristotle to quantum mechanics.

[2] E. Mach: "The Science of Mechanics," Open Court Publ. Co., Chicago, 1902. A philosophical examination of the development of classical mechanics.

[3] S. Timoshenko, D. H. Young: "Engineering Mechanics," vol. 2, "Dynamics," 4th ed., McGraw-Hill, New York, 1956. A treatment of elementary mechanics from the engineering point of view.

[4] G. W. Housner, D. E. Hudson: "Applied Mechanics—Dynamics," 2d ed., Van Nostrand, Princeton, 1959. A treatment of engineering dynamics requiring somewhat more mathematical background than the typical engineering text.

[5] S. Timoshenko, D. H. Young: "Advanced Dynamics," McGraw-Hill, New York, 1948. Covers engineering dynamics through vibration theory, Lagrange's equations, Hamilton's principle, and gyroscopes. Contains many illustrative examples.

[6] J. L. Synge, B. A. Griffith: "Principles of Mechanics," 3d ed., McGraw-Hill, New York, 1959. A nonengineering treatment noteworthy for its logical development of the subject.

[7] A. Sommerfeld: "Mechanics," Academic Press, New York, 1952. An elementary treatment of dynamics by a famous European physicist.

[8] J. C. Slater, N. H. Frank: "Mechanics," McGraw-Hill, New York, 1947. Dynamics presented from a physicist's point of view.

[9] H. Goldstein: "Classical Mechanics," Addison-Wesley, Reading, Mass., 1951. Advanced dynamics from the physicist's point of view.

[10] W. E. Byerly: "Generalized Coordinates," Ginn, Boston, 1916. A detailed discussion of generalized coordinates and their use.

[11] E. J. Routh: "Dynamics of a Particle," Stechert, New York, 1945. A rich collection of problems in particle dynamics.

[12] E. J. Routh: "Dynamics of Rigid Bodies," vol. 1, Macmillan, New York, 1897. A nineteenth century treatment with numerous examples.

[13] E. J. Routh: "Dynamics of Rigid Bodies," vol. 2, Dover, New York, 1955. The advanced part of the subject with many illustrative examples.

[14] A. G. Webster: "Dynamics," Teubner, Stuttgart, 1925. A classical treatment with a nineteenth century flavor.

[15] E. T. Whittaker: "Analytical Dynamics," Dover, New York, 1944. Advanced dynamics from the mathematician's point of view.

CHAPTER 24

VARIATIONAL PRINCIPLES OF MECHANICS

BY

C. LANCZOS, Ph.D., Dublin, Ireland

24.1. THE NATURE OF VARIATIONAL PRINCIPLES

The characteristic feature of the variational treatment of mechanics (statics and dynamics) is that we dispense with vectors and the parallelogram of forces and moments, and derive our results solely on the basis of two fundamental scalars, the *potential energy* V and the *kinetic energy* T. Even these two scalars are combined into one single function, called the *Lagrangian function* $L = T - V$. It seems astonishing that an arbitrary number of equations should be derivable on the basis of one single quantity. Yet this is the case and comes about by a certain manner of reasoning, which is best explained by examining a simple but characteristic example:

Example 1. Find the equilibrium of a chain of rigid bars of given masses and lengths, loosely jointed at their ends and suspended at the two free ends. The motion is restricted to the x,z plane.

In the Newtonian treatment of this problem, we would analyze the forces which act on each link of the system, establishing the conditions of equilibrium for each link. In the variational treatment we are confronted with a mechanism which has well-defined geometrical properties and a certain kinematic mobility. This purely geometrical and kinematical problem is transformed into a physical problem by the presence of the force of gravity, which causes the chain to move in a certain way and come in equilibrium in a very definite geometrical configuration. What distinguishing feature characterizes this particular configuration?

We know that the force of gravity possesses a potential energy V, which can be evaluated for any of the kinematically possible configurations. It is given by the expression

$$V = g\Sigma m_i z_i \tag{24.1}$$

where g is the acceleration due to gravity, m_i the mass of the ith bar, and z_i the vertical distance of its mid-point above an arbitrarily chosen horizontal plane. (The potential energy of the forces which maintain the rigidity of the bars is not taken into account because, under the given circumstances, it adds a mere constant to V; see Sec. 24.11.)

We now try out every kinematically permissible configuration of the system, and record in every case the value of the sum (24.1). Looking over this infinite collection of values, we find one particular value $V = V_0$ which is smaller than any of the other values. We check this particular value and look up the particular configuration to which it belongs. It is this configuration which yields the equilibrium position of the given system of bars. The mathematical formulation of this procedure takes the following form. *The equilibrium state of an arbitrary mechanical system is that particular configuration in which the potential energy assumes its minimum.*

In actual fact, the condition of an *absolute* minimum is unnecessarily restrictive. It suffices that V shall have a *local* minimum. This means that $V = V_0$ is smaller than any other value $V = V_1$ associated with a configuration which is arbitrarily near to the configuration which belongs to $V = V_0$. Still less restrictive is the condition of a *stationary value,* in which the *arbitrarily small neighborhood* of the distinguished configuration associated with V_0 is changed into the *infinitesimal neighborhood* of that configuration (cf. Sec. 24.3).

While we have demonstrated here the operation of a minimum principle for the establishment of an equilibrium position, the general procedure remains the same for the establishment of the equations which govern the motion of arbitrary mechanical systems, although the quantity to be minimized is no longer the potential energy, but an integral with respect to the time t, called *action:*

$$A = \int_{t_1}^{t_2} L\,dt \tag{24.2}$$

The mathematical discipline which deals with the minimum of such integrals is called the *calculus of variations.*

Such integrals, extended over given space domains, occur frequently, even in equilibrium problems, as the following examples demonstrate.

Example 2. The equilibrium position of a completely flexible uniform chain, suspended at the two ends, requires the minimization of the integral

$$V = \int_0^l z(x)\sqrt{1 + (z')^2}\,dx \qquad z(0) = z(l) = 0 \tag{24.3}$$

Example 3. The equilibrium of an elastic bar of the length l, bent by the amount $z(x)$ due to the action of the load per unit length $p(x)$, but supported at the two end points $x = 0$ and $x = l$, requires the minimum of the integral

$$V = \int_0^l [\tfrac{1}{2}EI(x)\,(z'')^2 - p(x)\,z(x)]\,dx \tag{24.4}$$

where E is Young's modulus, and $I(x)$ the moment of inertia of a cross section, taken around a horizontal axis which goes through the centroid of that cross section.

The variational treatment of the laws of mechanics became a guiding light in all the later developments of theoretical physics. It guides our attention from the bewildering variety of individual events to the discovery of the *basic Lagrangian,* which lies hidden in the depth. But, apart from the philosophical value of these principles, the mathematical techniques associated with them are often of eminent practical value in the actual solution of advanced engineering problems. The fundamental feature of these methods, to be applicable to any set of coordinates, is of excellent help in simplifying the equations of motion of complicated systems, by formulating them in an adequately chosen frame of reference. Moreover, direct minimizing procedures are well adapted to the operation of electronic computers, which may thus be entrusted with the task of generating a direct numerical solution of a Lagrangian problem, without going into any analytical details.

24.2. THE CONFIGURATION SPACE

In the problem of the jointed bars, the z_i of the various links cannot be prescribed freely, in view of the rigidity conditions. If our chain is composed of four links, for example, we may prescribe z_1 and z_3 freely (within certain limits), but then z_2 is already determined by the geometry of the problem. Generally the position of an arbitrary mechanical system may be characterized by a certain set of independent parameters, say $q_1, q_2, \ldots, q_n$, called the *generalized coordinates* of the system. The number n of independent parameters determines the *degree of freedom* of the system. Our example of four jointed bars, for instance, has two degrees of freedom. Or let us consider a rigid body, freely movable in space. This body may be composed of an arbitrary (even infinite) number of particles. Nevertheless, its position in space can be fixed by giving six data. We may prescribe, for example, the position of the center of mass, by giving its three rectangular coordinates x, y, z in some cartesian system. To this we must add the rotation of the body around this point, which requires three additional data. The name *generalized coordinates* refers to the fact that we are not obliged to choose our coordinates according to some preconceived scheme (e.g., we might have chosen polar coordinates r, θ, ϕ instead of x, y, z, or any other set of three suitable variables). The only condition is that the coordinates x_i, y_i, z_i of an arbitrary point of the rigid body shall be expressible as continuous and differentiable functions of the given six parameters. This freedom in the choice of coordinates is one of the most important advantages of the variational treatment of dynamical problems. The solution of many otherwise difficult problems is accomplished, or at least facilitated, by formulating the problem in particularly suitable coordinates which lead to a partial or complete integration of the equations of motion.

We now construct an imaginary space of n dimensions, called *configuration space* (or briefly C-space), by plotting the coordinates $q_1, q_2, \ldots, q_n$ as rectangular coordinates of a Euclidean space; (a more adequate definition of the geometry of this space will be discussed in Sec. 24.8). A definite point of this space characterizes a definite position of the mechanical system. For all our mathematical discussions an abstract geometrical point takes the place of the complicated physical entity, called *mechanical system.* An *arbitrary configuration* of the mechanical system is replaced by a geometrical point P which roams freely within a well-circumscribed space of n dimensions, defined by the given variability of the coordinates q_i. The equilibrium position belongs to a definite point P_0 which may be either in the inside or on the boundary of the C-space. We explore the *immediate neighborhood* of this point P_0. This means that we construct a little solid *sphere* around P_0, defined by

$$|P_1 - P_0| = \sqrt{\sum_{i=1}^{n} (q_i' - q_i^\circ)^2} \leq \epsilon \tag{24.5}$$

where ϵ, the radius of the sphere, may be chosen as small as we wish. (This sphere is full if P_0 is an inside point, but only partly full if P_0 is a boundary point.) We compare the value of the potential energy $V = V_1$ at every point P_1 within this sphere, with the value $V = V_0$ at the center P_0 of the sphere. The necessary and sufficient condition of a stable equilibrium is that, for all points P_1, we shall find $V_1 > V_0$. (If for some or all points $V_1 = V_0$, the equilibrium is called *neutral.*)

24.3. THE STATIONARY VALUE OF V

The coordinates q_i' of the point P_1 differ from the coordinates q_i° of the center by small amounts δq_i, called the *variations* of q_i°. We will choose our point P_1 on the circumference of the sphere. This imposes the condition

$$\sum_{i=1}^{n} \delta q_i^2 = \epsilon^2 \tag{24.6}$$

We want to investigate the change of the potential energy ΔV along the radius P_0P_1. This means

$$\Delta V = V(q_i^\circ + \tau\, \delta q_i) - V(q_i^\circ) \qquad 0 \leq \tau \leq 1 \tag{24.7}$$

Since V is a continuous and even differentiable function of the coordinates q_i, the same holds of ΔV, considered as a function of the single variable τ. We will now forgo a full investigation of the difference coefficient $\Delta V/\Delta\tau$, and restrict ourselves to the investigation of its *limit,* approached as $\Delta\tau$ converges to zero. This quantity is denoted by δV and is called the *first variation* of V:

$$\delta V = \lim_{\Delta\tau \to 0} \frac{\Delta V}{\Delta\tau} = \left(\frac{dV}{d\tau}\right)_0 = \sum_{i=1}^{n} \left(\frac{\partial V}{\partial q_i}\right)_0 \delta q_i \qquad (24.8)$$

Now the limit of a quantity which is always positive cannot be negative, but only positive or possibly zero. Let us assume that the point P_0 is an *inside* point of the C-space. Then, picking out one definite subscript, say $i = k$, the choice $\delta q_k = \pm\epsilon$, all other $\delta q_i = 0$ is a permissible choice for the position of P_1. For this choice,

$$\delta V = \pm\left(\frac{\partial V}{\partial q_k}\right)_0 \epsilon \qquad (24.9)$$

Since, however, negative values have to be excluded, we get the conditions

$$\left(\frac{\partial V}{\partial q_k}\right)_0 = 0 \qquad k = 1, 2, \ldots, n \qquad (24.10)$$

as necessary conditions of a minimum, if that minimum occurs at an inside point of the space C. But the fulfillment of (24.10) does not guarantee the minimum of V. If (24.10) holds, the difference (24.7) may become positive throughout the sphere (24.5), local minimum, but it may also become negative throughout, local maximum, or positive in some directions and negative in others, in which case no extremum exists even in the local sense. However, the conditions (24.10), although insufficient for a minimum, have a significance of their own. If we substitute Eqs. (24.10) in (24.8), we see that δV becomes zero for any arbitrary choice of the δq_i. And vice versa, the condition $\delta V = 0$ for arbitrary δq_i leads to (24.10). Equations (24.10) can thus be conceived as the *necessary and sufficient conditions that the first variation δV shall vanish for arbitrary variations δq_i of the coordinates.* We then say that V has a *stationary value* at the point P_0.

Example 4. Consider a ball rolling inside of a spherical bowl of larger radius. The equilibrium is attained if the ball touches the spherical bowl at its lowest point. At this point, an infinitesimal displacement of the ball means horizontal motion, and thus the potential energy of the force of gravity (proportional to the height of the center of mass of the ball) remains stationary. But, if the ball is raised inside the bowl, the quantity δV is no longer zero but positive. The condition of a stationary value is thus violated. But here we are not inside but on the boundary of the C-space, which in this example is composed of the inner portion of the bowl, but terminated by the surface of the bowl. The condition of a stationary value holds only in all the reversible directions, but not otherwise.

24.4. THE PRINCIPLE OF VIRTUAL WORK

It is sometimes more convenient to operate with the forces rather than the potential energy of the forces. The negative change of the potential energy is equal to the work of the forces during that particular displacement. Let us assume that we know all the active forces F_i which operate on the system at the given points $Q_1, Q_2, \ldots, Q_m$. We can now replace the principle of a stationary value of V by another condition called the *principle of virtual work:*

$$\sum_{i=1}^{m} F_i \cdot \delta R_i = 0 \qquad (24.11)$$

where δR_i denote arbitrarily small virtual displacements of the points Q_i. This condition is more general than the condition (24.10) of the previous section, since it includes forces which are not derivable from a potential energy (e.g., forces due to friction, viscosity, electrical resistance, generally forces associated with the generation of heat).

The condition (24.11) does not lead to $F_i = 0$ since the virtual displacements δR_i are usually not freely choosable but bound to the given kinematic conditions. They have to be expressed in terms of the $\delta q_1, \delta q_2, \ldots, \delta q_n$, which can be chosen freely.

The words *active forces* refer to the following: The forces which maintain certain given kinematic conditions, e.g., the rigidity of the links in Example 1, do not come into evidence if we honor the kinematic conditions during the virtual displacements of the system. These forces will thus not contribute anything to the virtual work, and we can completely ignore their existence (cf. Sec. 24.11).

24.5. D'ALEMBERT'S PRINCIPLE

The principle of virtual work can be extended from statics to dynamics by adding to the active forces an additional *dynamical force*, called the *inertia force*, and defined as the *negative time derivative of the momentum:*

$$K_i = -\frac{d}{dt}(m_i\dot{R}_i) \tag{24.12}$$

Hence d'Alembert's principle assumes the form

$$\Sigma(F_i + K_i) \cdot \delta R_i = 0 \tag{24.13}$$

While the active forces F_i may be concentrated in certain isolated points Q_i of the system, the forces of inertia are present wherever masses are in motion.

24.6. HAMILTON'S PRINCIPLE

The inertia forces do not possess a potential energy. Hence, it seems that we will not succeed with the program of reducing the problem of dynamics to a mere scalar, as we succeeded with the problem of statics in the presence of *conservative forces*, which could be reduced to the single scalar function $V(q_i)$. However, the virtual work of the inertial forces allows the following transformation:

$$-\sum_i \frac{d}{dt}(m_i\dot{R}_i) \cdot \delta R_i = -\frac{d}{dt}\sum_i (m_i\dot{R}_i \cdot \delta R_i) + \sum_i m_i\dot{R}_i \cdot \delta\dot{R}_i \tag{24.14}$$

The second term on the right side can be written as the variation δT of a new scalar, called *kinetic energy* and denoted by T:

$$T = \tfrac{1}{2}\sum_i m_i\dot{R}_i{}^2 = \tfrac{1}{2}\sum_i m_i v_i{}^2 \tag{24.15}$$

Hence d'Alembert's principle, for the case of conservative forces, may be written as follows:

$$\delta(T - V) - \frac{d}{dt}\sum_i (m_i\dot{R}_i) \cdot \delta R_i) = 0 \tag{24.16}$$

We will now integrate with respect to the time t, between two definite time limits t_1 and t_2, which may be chosen in any way we like. The integration in the last term gives a mere boundary term. We wish to agree, however, that the variations δR_1 shall vanish at the two time limits t_1 and t_2 (we call that *variation between definite limits*). In this case the boundary term drops out, and we obtain *Hamilton's principle:*

$$\delta\int_{t_1}^{t_2} (T - V)\, dt = 0 \tag{24.17}$$

This principle is very similar in its character to the principle (24.10), which demanded the stationary value of the potential energy V. Once more we have a stationary value principle,

$$\delta A = 0 \tag{24.18}$$

but the quantity which has to be made stationary is no longer the potential energy V, but the time integral of a certain scalar L, called the Lagrangian function,

$$A = \int_{t_1}^{t_2} L\,dt \tag{24.19}$$

We will call this quantity *action.*

The *principle of least action* refers to the program of minimizing this integral, or at least making it stationary. In Hamilton's principle, the function L is defined as the difference between the kinetic and the potential energy:

$$L = T - V \tag{24.20}$$

We will not insist, however, on this definition of L since various formulations of the principle of least action can be given. Moreover, in relativistic mechanics—which takes the place of ordinary mechanics in the case of fast-moving particles—or if the motion of an electrically charged particle in an external electromagnetic field is involved, the Lagrangian function does not have the form $(T - V)$, and yet the principle of the stationary value of an integral of the general form (24.19) remains valid. We will thus leave the definition of L free, except for the general assumption that it is some given function of the coordinates q_i and their time derivatives $\dot{q}_i$; for the sake of generality, we will even allow an explicit dependence of L on the time t:

$$L = L(q_1, \ldots, q_n; \dot{q}_1, \ldots, \dot{q}_n; t) \tag{24.21}$$

Afterward we can still specify L in any form we like.

24.7. THE EULER-LAGRANGE EQUATIONS

The problem of minimizing a definite integral can be solved in full analogy to the procedure of Sec. 24.3. Once more we construct a little sphere of radius ϵ around the (inside) point P_0, and once more we investigate the behavior of the difference quotient $\Delta L/\Delta\tau$, as function of the single variable τ. Once more we restrict ourselves to the investigation of the limit $\Delta\tau \to 0$. The only difference is that now the $\dot{q}_i$ are added to the q_i. Moreover, all these quantities are functions of the time t (continuous and even differentiable), and an integration with respect to t is demanded. The two steps of the procedure are, thus,

$$\frac{\Delta A}{\Delta \tau} = \int_{t_1}^{t_2} \frac{\Delta L}{\Delta \tau}\,dt \tag{24.22}$$

and, passing over to the limit $\Delta\tau \to 0$,

$$\delta A = \left(\frac{dA}{d\tau}\right)_0 = \int_{t_1}^{t_2} \sum_{i=1}^{n} \left(\frac{\partial L}{\partial q_i}\,\delta q_i + \frac{\partial L}{\partial \dot{q}_i}\,\delta \dot{q}_i\right) dt \tag{24.23}$$

Before we can draw any further conclusions, an integration by parts has to be carried out in the second term:

$$\int_{t_1}^{t_2} \frac{\partial L}{\partial \dot{q}_i}\,\delta\dot{q}_i\,dt = \int_{t_1}^{t_2} \frac{d}{dt}\left(\frac{\partial L}{\partial \dot{q}_i}\,\delta q_i\right) dt - \int_{t_1}^{t_2} \left(\frac{d}{dt}\frac{\partial L}{\partial \dot{q}_i}\right) \delta q_i\,dt \tag{24.24}$$

The first term on the right side yields a mere boundary term, which drops out, however, since we have agreed that all variations become zero at the two end points of integration (variation between definite limits). The remaining term can be combined with the first term on the right side of (24.23), and we obtain

$$\delta A = \int_{t_1}^{t_2} \sum_{i=1}^{n} \left(\frac{\partial L}{\partial q_i} - \frac{d}{dt} \frac{\partial L}{\partial \dot{q}_i} \right) \delta q_i \, dt \tag{24.25}$$

As before in Sec. 24.3, we make the choice $\delta q_k = \pm \epsilon(t)$, all other $\delta q_i = 0$. But now δq_k is a function of t, and we can dispose of this function freely, except for the condition of continuity and differentiability. We will choose $\epsilon(t)$ as zero everywhere, except in the small time interval $t = t_0 \pm \alpha$, where t_0 is some definite time moment between t_1 and t_2. The integration on the right side of (24.25) is now reduced to the small time interval $(t_0 - \alpha)$ to $(t_0 + \alpha)$. By the mean-value theorem of integral calculus, we obtain

$$\delta A = \pm \left(\frac{\partial L}{\partial q_k} - \frac{d}{dt} \frac{\partial L}{\partial \dot{q}_k} \right) (t_0 + \theta \alpha) \int_{t_0 - \alpha}^{t_0 + \alpha} \epsilon(t) \, dt \qquad -1 \leq \theta \leq 1 \tag{24.26}$$

Since negative values of δA contradict the existence of a minimum, we obtain, passing to the limit $\alpha \to 0$,

$$\frac{\partial L}{\partial q_k} - \frac{d}{dt} \frac{\partial L}{\partial \dot{q}_k} = 0 \qquad k = 1, 2, \ldots, n \tag{24.27}$$

Since t_0 may be chosen as any time moment between t_1 and t_2, these equations, called the *Euler-Lagrange equations*, have to hold throughout the time interval $\langle t_1, t_2 \rangle$. Moreover, if these equations—which are necessary but not sufficient conditions of a minimum of A—are satisfied, the substitution in (24.25) shows that

$$\delta A = 0 \tag{24.28}$$

for any arbitrary choice of the δq_i. In full analogy to Eqs. (24.10), Eqs. (24.27) have a significance of their own, quite apart from the existence of a minimum. They can be conceived as the necessary and sufficient conditions for the vanishing of the first variation δA, thus guaranteeing a *stationary value* of the integral A.

According to Hamilton's principle (cf. Sec. 24.6), Eqs. (24.27) yield (for $L = T - V$) the dynamical equations of motion, in the presence of conservative forces, in arbitrary coordinates. We can conceive them as generalizations of Newton's equation of motion: "The time rate of change of the momentum is equal to the moving force." Accordingly, we will define

$$\frac{\partial L}{\partial \dot{q}_i} = p_i \tag{24.29}$$

as *components of the generalized momentum*, and

$$\frac{\partial L}{\partial q_i} = F_i \tag{24.30}$$

as *components of the generalized force.* The Lagrangian equations of motion then appear in the Newtonian form:

$$\frac{dp_i}{dt} = F_i \tag{24.31}$$

If some of the forces present are nonconservative (i.e., not derivable from a potential energy), the right side of (24.31) has to be augmented by F_i', where F_i' is defined by the virtual work of the nonconservative forces, which is given by $\sum_{i=1}^{n} F_i' \, \delta q_i$.

Example 5. Write down the Lagrangian equations of motion for a particle of mass m in a field of central force.

(a) RECTANGULAR COORDINATES x, y (NEWTONIAN TREATMENT)

$$L = \tfrac{1}{2}m(\dot{x}^2 + \dot{y}^2) - V(\sqrt{x^2 + y^2}) \qquad m\ddot{x} = -V'(\sqrt{x^2 + y^2})\frac{x}{\sqrt{x^2 + y^2}}$$

$$m\ddot{y} = -V'(\sqrt{x^2 + y^2})\frac{y}{\sqrt{x^2 + y^2}} \tag{24.32}$$

(b) POLAR COORDINATES r, θ

$$L = \tfrac{1}{2}m(\dot{r}^2 + r^2\dot{\theta}^2) - V(r) \qquad m\ddot{r} = mr\dot{\theta}^2 - V'(r) \qquad m\frac{d}{dt}(r^2\dot{\theta}) = 0 \tag{24.33}$$

24.8. THE GEOMETRY OF THE CONFIGURATION SPACE

In Sec. 24.2 we have constructed the configuration space by plotting the coordinates $q_1, q_2, \ldots, q_n$ as rectangular coordinates of a Euclidean space. The infinitesimal distance ds (the *line element*) between two neighboring points is thus given by the Pythagorean expression

$$ds^2 = dq_1^2 + dq_2^2 + \cdots + dq_n^2 \tag{24.34}$$

In actual fact, the geometry of the configuration space can be defined in a much more adequate way.

One of the fundamental scalars of analytical mechanics is the kinetic energy T. It is defined as $\Sigma\tfrac{1}{2}mv^2$, to be extended over all the masses of the system. If the system is composed of rigid links which are in some given kinematical relation to each other, the number n of coordinates $q_1, \ldots, q_n$ necessary and sufficient for the characterization of the position of the system will be usually much smaller than the number of masses involved. For example, in the case of a rigid body, the coordinates x_i, y_i, z_i of each particle will be expressible in the six coordinates which characterize the position of the mechanical system. Generally the kinetic energy of an arbitrary system, expressed in terms of the position coordinates $q_1, \ldots, q_n$, will come out in the following form:

$$T = \sum \tfrac{1}{2}m_i(\dot{x}_i^2 + \dot{y}_i^2 + \dot{z}_i^2) = \frac{1}{2}\sum_{i,k=1}^{n} a_{ik}\dot{q}_i\dot{q}_k \tag{24.35}$$

where the coefficients a_{ik} are some given functions of the coordinates q_i. We will now introduce a definite metrical structure in the C-space by assuming that the infinitesimal distance between the points $Q = (q_1, \ldots, q_n)$ and

$$Q + dR = (q_1 + dq_1, \ldots, q_n + dq_n)$$

is not given by the Pythagorean expression (24.34), but by the following expression,

$$ds^2 = \Sigma a_{ik}\, dq_i\, dq_k \tag{24.36}$$

The great advantage of this definition is that now the total kinetic energy of the mechanical system becomes

$$T = \frac{1}{2}\left(\frac{ds}{dt}\right)^2 = \tfrac{1}{2}v^2 \tag{24.37}$$

This means that the point P_0, whose motion in the C-space represents the motion of the entire mechanical system, behaves with respect to the kinetic energy like a single particle of the mass 1.

24.9. ELIMINATION OF ALGEBRAIC VARIABLES

Let L contain some variables $u_1, u_2, \ldots, u_m$ which enter L in themselves but not with their derivatives:

$$L = L(q_1, \ldots, q_n; \dot{q}_1, \ldots, \dot{q}_n; u_1, \ldots, u_m; t) \tag{24.38}$$

According to (24.23), we obtain for δA, in consequence of varying the u_i,

$$\delta A = \int_{t_1}^{t_2} \sum_{i=1}^{m} \frac{\partial L}{\partial u_i} \delta u_i \, dt \tag{24.39}$$

and the condition $\delta A = 0$ yields directly, without any integration by parts, the conditions

$$\frac{\partial L}{\partial u_i} = 0 \qquad i = 1, 2, \ldots, m \tag{24.40}$$

We can conceive these m equations as determining equations of the u_i, expressing them as functions of the nonalgebraic variables and their velocities (provided that this elimination is actually possible):

$$u_i = f_i(q_1, \ldots, q_n; \dot{q}_1, \ldots, \dot{q}_n; t) \tag{24.41}$$

We substitute these expressions in (24.38). The result is that we have eliminated the algebraic variables from our problem in advance, and start with a Lagrangian problem in which only the nonalgebraic variables q_i occur.

It is not self-evident that this procedure is justified. Equations (24.40) are demanded only for the *actual* motion. To eliminate the u_i with the help of these equations means that we assume their validity for the varied motion as well. But then we see from (24.41) that the condition of "varying between definite limits" will generally not hold, since the variations of u_i will not be necessarily zero at the two time moments t_1 and t_2. On the other hand, we see from (24.39) that δA is zero in consequence of (24.40), *without* any further conditions. Hence, the vanishing of the δu_i at the two end points is not demanded, and the elimination of algebraic variables in the case that Eqs. (24.40) are actually solvable for the u_i is always justified.

24.10. THE LAGRANGIAN MULTIPLIER METHOD

Consider a variational problem with the following Lagrangian:

$$L' = L + \lambda f(q_1, \ldots, q_n, t) \tag{24.42}$$

The dynamical variables of the problem are the q_i and λ. We will transform our coordinates by solving the equation

$$u = f(q_1, \ldots, q_n, t) \tag{24.43}$$

for one of the q_i, say q_n, and consider as our new coordinates the variables $q_1, \ldots, q_{n-1}, u$. We now have

$$L' = L + \lambda u \tag{24.44}$$

Variation with respect to λ gives

$$u = 0 \tag{24.45}$$

On the other hand, the variations with respect to the q_i do not involve the last term but only L. In these variations, the u and $\dot{u}$ do not participate actively, and we lose nothing if we substitute in L for u and $\dot{u}$ the values zero in advance. Consider the variational problem with the Lagrangian L and the added auxiliary condition

$$u = 0 \tag{24.46}$$

The only equation we lose in this fashion is the equation obtained by varying (24.44) with respect to u:

$$\frac{\partial L}{\partial u} - \frac{d}{dt}\frac{\partial L}{\partial \dot{u}} + \lambda = 0 \tag{24.47}$$

But this equation can be conceived as the determination of the multiplier λ. It does not add anything new to the problem of motion as far as the other variables are concerned.

Returning to our original variables, we can express the result of our discussion as follows. The free variational problem (24.42), without any auxiliary condition, and the variational problem, with the Lagrangian L and the added auxiliary condition

$$f(q_1, \ldots, q_n, t) = 0 \tag{24.48}$$

are entirely equivalent problems as far as the equations of motion are concerned. The only equation we lose in the second formulation is the equation which determines the Lagrangian multiplier λ.

In a completely analogous fashion, we can demonstrate the equivalence of the following two variational problems.

1. Given the Lagrangian function L with the added auxiliary conditions

$$\begin{aligned} f_1(q_1, \ldots, q_n, t) &= 0 \\ \cdots\cdots\cdots & \\ f_m(q_1, \ldots, q_n, t) &= 0 \end{aligned} \tag{24.49}$$

(We assume that these m statements are independent of each other and free of contradictions.)

2. Given the modified Lagrangian

$$L' = L + \lambda_1 f_1 + \lambda_2 f_2 + \cdots + \lambda_m f_m \tag{24.50}$$

without any auxiliary conditions, but containing the additional variables $\lambda_1, \lambda_2, \ldots, \lambda_m$.

The merit of the second formulation of our problem is that the equations of motion plus the auxiliary conditions appear as the consequence of *one unified principle.* Moreover, we need not eliminate any variables from our problem but can handle all variables on equal footing. In formulation 1, the imposition of the conditions (24.49) in advance of any variation would demand that we eliminate m properly chosen q_i with the help of Eqs. (24.49), and substitute these expressions in the given L. This often destroys the natural symmetry of the system, apart from the clumsy algebraic operations necessitated by the process of elimination.

Consider, for example, the motion of two particles which are kept at a fixed distance l from each other. The position of each particle is given by three coordinates, e.g., the three rectangular coordinates x_1, y_1, z_1, and x_2, y_2, z_2. Hence the configuration space is six-dimensional, but restricted by the auxiliary condition

$$(x_1 - x_2)^2 + (y_1 - y_2)^2 + (z_1 - z_2)^2 - l^2 = 0 \tag{24.51}$$

This means that the motion is restricted to a given "surface" of the configuration space. We could eliminate one of the coordinates, and thus reduce the configuration space from the beginning to only five dimensions. It is much more symmetric, however, to leave the configuration space as six-dimensional and modify the given

Lagrangian L to

$$L' = L + \lambda[(x_1 - x_2)^2 + (y_1 - y_2)^2 + (z_1 - z_2)^2 - l^2] \tag{24.52}$$

Instead of reducing the number of variables by one, we have actually increased their number by one, since we have added the new variable λ. But the merit of the extended Lagrangian is that now the equations of motion, *together* with the given auxiliary condition, appear as the consequence of one *single and unrestricted variational principle.*

The Lagrangian multiplier method brings about an admirable flexibility in the formulation of variational problems, and is often of great value in the actual solution of dynamical problems, by adding or subtracting dynamical variables of a suitable type.

In kinematic conditions brought about by physical circumstances, the velocities $\dot{q}_i$ seldom occur. But, for many mathematical applications of the λ method, it is important to know that the multiplier method can be extended without any change to auxiliary conditions involving velocities (cf. Secs. 24.13 through 24.15).

24.11. PHYSICAL SIGNIFICANCE OF THE LAGRANGIAN MULTIPLIER METHOD

The physical significance of the multiplier method, in the case of given kinematic conditions, is often misunderstood, because not enough attention is paid to the difference between purely *mathematical* auxiliary conditions, which are absolute and inviolable, and auxiliary conditions of a *physical* type, which are not absolute but maintained by strong (although not infinite) forces.

It occurs frequently that the potential energy of a certain system of forces takes the following form,

$$V_1 = \frac{1}{2}\frac{f^2}{\epsilon g} \tag{24.53}$$

where ϵ is a small positive constant, while f and g are given functions of the q_i and possibly the time t. The function g has the property that it remains positive throughout the domain of the variables q_i. We then speak of a *potential well*, and the small value of ϵ has the consequence that this parabolic well increases very steeply. Consequently, any appreciable deviation from the equilibrium position $f = 0$ excites very strong forces.

Such a potential well appears in macroscopic relations as though the condition $f = 0$ were *imposed* on the system. The potential energy (24.53) is mathematically equivalent to the potential energy

$$V_2 = -\left(\lambda f + \frac{\epsilon}{2}\lambda^2 g\right) \tag{24.54}$$

where λ is an added mechanical variable. Indeed, since this variable is purely algebraic, we can eliminate it in advance (cf. Sec. 24.9) by putting the partial derivative with respect to λ equal to zero,

$$f + \epsilon\lambda g = 0 \tag{24.55}$$

which gives

$$\lambda = -\frac{f}{\epsilon g} \tag{24.56}$$

Substitution in (24.54) brings us back to (24.53).

Now, if the remaining system of forces has the potential energy V and the Lagrangian $L = T - V$, the complete Lagrangian L' becomes

$$L' = L + \lambda f + \frac{\epsilon}{2}\lambda^2 g \tag{24.57}$$

but the last term remains *macroscopically latent,* in view of the smallness of ϵ. What we have at our disposal is then

$$L' = L + \lambda f \tag{24.58}$$

and this L' can be interpreted as the Lagrangian L with the added auxiliary condition $f = 0$. The *force of reaction* associated with this auxiliary condition becomes, according to (24.30),

$$F_i = \lambda \frac{\partial f}{\partial q_i} \tag{24.59}$$

while the actual potential energy (24.54) gives the force

$$F_i = \lambda \frac{\partial f}{\partial q_i} + \frac{\epsilon}{2} \lambda^2 \frac{\partial g}{\partial q_i} \tag{24.60}$$

The difference is too small to be detected, in view of the smallness of ϵ. However, while from the mathematical viewpoint the force (24.59) is interpreted as the force demanded for the *maintenance* of the auxiliary condition $f = 0$, in actual fact the (practically equivalent) force (24.60) has been excited by the microscopic *violation* of the condition $f = 0$. In the exact equilibrium position $f = 0$, no force is excited, but a large force is excited by the small deviation $f = -\epsilon\lambda g$ which exists [according to (24.56)] from the state of equilibrium $f = 0$. By going to the limit $\epsilon = 0$, we obtain a mathematical description of a macroscopic phenomenon in which $f = 0$ appears as an *imposed auxiliary condition,* although in actual fact the condition $f = 0$ is only *practically* fulfilled, in view of the large (but nevertheless finite) forces which are excited by any small deviation from the equilibrium position $f = 0$.

This analysis shows that the *physical* interpretation of the Lagrangian multiplier method is essentially different from its *mathematical* interpretation, which comes about by going to the limit $\epsilon = 0$. Physically, this limit process is meaningless, because it would make the forces which maintain the kinematic condition $f = 0$ infinitely strong.

Example 6. Consider a ball rolling down on a given curved surface $f = 0$. If the ball did not make a microscopic indention on the surface, the force of reaction which keeps the ball on the surface would not be exerted. The ball could stay on the surface $f = 0$ *exactly,* only if the force of reaction were provided by some other source, e.g., the centrifugal force. But then the auxiliary condition could be dropped altogether, since it is satisfied anyway. The *mathematical* solution avoids the indention by going to the limit $\epsilon = 0$ which, however, would make the elastic constant of the surface *infinitely large.*

24.12. ELIMINATION OF KINOSTHENIC VARIABLES

A certain counterpart of the algebraic variables of Sec. 24.9 are the *kinosthenic* variables (also called *ignorable*) which do not participate in L in themselves but only with their *derivatives:*

$$L = L(q_1, \ldots, q_n; \dot{q}_1, \ldots, \dot{q}_n; \dot{u}_1, \ldots, \dot{u}_m; t) \tag{24.61}$$

We can replace the $\dot{u}_k$ by the algebraic variables w_k, provided that we add the auxiliary conditions

$$\dot{u}_k - w_k = 0 \qquad k = 1, 2, \ldots, m \tag{24.62}$$

to our problem. Making use of the Lagrangian multiplier method, we replace the given L by the new

$$L' = L(q_1, \ldots, q_n; \dot{q}_1, \ldots, \dot{q}_n; w_1, \ldots, w_m; t) + \sum_{k=1}^{m} \lambda_k(\dot{u}_k - w_k) \tag{24.63}$$

with the $(n + 3m)$ dynamical variables

$$q_i \ (i = 1, 2, \ldots, n); \ u_k, \lambda_k, w_k \ (k = 1, 2, \ldots, m)$$

Now the Euler-Lagrange equations (24.27) associated with the u_k give

$$\lambda_k = \text{const} = \alpha_k \tag{24.64}$$

If we replace the λ_k by these constants α_k, we deprive ourselves of the variations with respect to the λ_k. We thus lose the equations $w_k = \dot{u}_k$, but this loss is compensated for by keeping in mind that the algebraic variables w_k are actually the original $\dot{u}_k$.

Now the term $\Sigma\alpha_k\dot{u}_k$, if integrated, gives a mere boundary term, which for the variation behaves like a constant and can be omitted. Hence, our final Lagrangian becomes

$$L' = L - \sum_{k=1}^{m} \alpha_k w_k \tag{24.65}$$

But now we can eliminate the purely algebraic variables w_k, according to Eqs. (24.40), which in our case become

$$\frac{\partial L}{\partial w_k} - \alpha_k = 0 \tag{24.66}$$

to be solved for the w_k. We substitute these expressions of w_k in L' and have thus eliminated the kinosthenic variables, reducing our problem to the nonkinosthenic variables q_i.

The final procedure can be expressed as follows. The elimination of the kinosthenic variables $\dot{u}_1, \dot{u}_2, \ldots, \dot{u}_m$ occurs in two steps.

1. Eliminate the $\dot{u}_k$ from the equations

$$\frac{\partial L}{\partial \dot{u}_k} = \alpha_k \qquad k = 1, 2, \ldots, m \tag{24.67}$$

2. Substitute these expressions of $\dot{u}_k$ into the modified Lagrangian

$$L' = L - \sum_{k=1}^{m} \alpha_k \dot{u}_k \tag{24.68}$$

Example 7. In Example 5b (p. 24–8), the angle variable θ is kinosthenic. Hence

$$mr^2\,\dot{\theta} = \alpha = mA$$

(Kepler's *area law*), from which

$$\dot{\theta} = \frac{A}{r^2}$$

The modified Lagrangian $L' = L - \alpha\dot{\theta}$, after the elimination of $\dot{\theta}$, becomes

$$L' = \tfrac{1}{2}m\dot{r}^2 - \tfrac{1}{2}m\frac{A^2}{r^2} - V(r)$$

After solving the reduced problem for the q_i, we still have to obtain the u_k. Since, however, we have expressed the $\dot{u}_k$ in terms of the q_i, these q_i are now given as explicit functions of the time t. Hence, all the $\dot{u}_k$ are now explicit functions of t, and the u_k are obtainable by mere quadratures.

24.13. JACOBI'S PRINCIPLE

The configuration space of Sec. 24.2 served the purpose of representing the position of a mechanical system by one single point P_0 of this space. This point P_0 changes its position during the motion of the system. We get a still more adequate geometrical picture of the mechanical problem if we operate with an "extended configuration space" of $(n + 1)$ dimensions, in which the time t is plotted as an additional coordinate. A point of this space has now the coordinates $q_1, q_2, \ldots, q_n, t$, and the motion of the mechanical system is represented by a definite curve S of this space. We can conceive the equations $q_i = q_i(t)$ as the equations of this curve. A more adequate representation of this curve will be obtained, however, if we give the q_i and t as functions of some unspecified parameter τ, which can later be fixed in any way we like. In this new *parametric* form of the curve S, we have added the time t to the dynamical variables and have a Lagrangian problem in the $(n + 1)$ variables $q_i (i = 1, 2, \ldots, n)$, t, to be considered as functions of τ. We will express the Lagrangian L in these new variables, denoting the derivatives with respect to τ by a prime. We have to replace the velocities $\dot{q}_i$ by

$$\dot{q}_i = \frac{q_i'}{t'} \tag{24.69}$$

The action integral appears in the new variable τ as

$$A = \int_{\tau_1}^{\tau_2} L\left(q_i, \frac{q_i'}{t'}; t\right) t' \, d\tau \tag{24.70}$$

and thus our new Lagrangian becomes

$$L_1 = L\left(q_i, \frac{q_i'}{t'}; t\right) t' \tag{24.71}$$

Now it is true that, for the sake of generality, we have allowed that L shall explicitly depend on t. But, in most dynamical problems, neither T nor V depends explicitly on t, and thus also $L = T - V$ is a function of q_i and $\dot{q}_i$ *alone*, without the addition of t. In this case, (24.71) can be replaced by

$$L_1 = L\left(q_i, \frac{q_i'}{t'}\right) t' \tag{24.72}$$

and we observe that now the time t becomes a *kinosthenic variable* which can be eliminated, according to the procedure of Sec. 24.12. For this purpose we have to eliminate t' with the help of Eq. (24.67), which in our case becomes

$$L - \sum_{i=1}^{m} p_i \frac{q_i'}{t'} = -E \tag{24.73}$$

[replacing the constant α by the more suitable notation $-E$; the *momenta* p_i are defined according to (24.29)]. Now, in ordinary mechanics, $L = T - V$, and V is a function of the q_i only, while T is of the form (24.35). But then

$$p_i = \sum_{k=1}^{n} a_{ik} \dot{q}_k \tag{24.74}$$

and Eq. (24.73) assumes the form $(T - V) - 2T = -E$, or

$$T + V = E \tag{24.75}$$

which is the energy theorem of conservative systems (the name *conservative* refers to the fact that the *total energy* $T + V$ is conserved during the motion). We have to eliminate t' from this equation, but, at the same time, we have to modify (24.72) to

$$
\begin{aligned}
L_2 &= Lt' + Et' = [(T - V) + (T + V)]t' \\
&= 2Tt' = 2(E - V)t'
\end{aligned}
\tag{24.76}
$$

Now in (24.36) we have introduced a definite *line element* ds in configuration space due to which T could be written in the form (24.37). In the new variable τ, the energy theorem (24.75) has to be written as follows:

$$\frac{1}{2}\frac{s'^2}{t'^2} + V = E \tag{24.77}$$

The elimination of t' gives

$$t' = \frac{s'}{\sqrt{2(E - V)}} \tag{24.78}$$

Substituting in (24.76), we obtain the new Lagrangian L_2, from which the time is completely eliminated:

$$L_2 = \sqrt{2(E - V)}\, s' \tag{24.79}$$

The associated action integral becomes

$$A = \int_{\tau_1}^{\tau_2} \sqrt{2(E - V)}\, ds \tag{24.80}$$

It determines the *path* of the point P_0 in C-space, without any reference to the time t associated with the motion. The principle of minimizing (24.80) is called *Jacobi's principle.* It can be formulated in striking geometrical terms by saying that the curve described by the point P_0 in C-space is a *geodesic* or *shortest line* of a certain Riemanian manifold (of n dimensions) whose line element is defined by

$$d\sigma = \sqrt{E - V}\, ds \tag{24.81}$$

After solving the problem of the path, the associated time is available by quadrature, on the basis of (24.78):

$$t = \int_{\tau_1}^{\tau} \frac{ds}{\sqrt{2(E - V)}} + \text{const} \tag{24.82}$$

Example 8. The path of a particle under the action of a central force (cf. Example 5*b*, p. 24–8), is described by the action integral

$$A = \int_{r_1}^{r_2} \sqrt{E - V(r)}\, \sqrt{1 + r^2\theta'^2}\, dr$$

considering r as the parameter τ. Since θ is kinosthenic, it can be eliminated,

$$\sqrt{E - V}\, \frac{r^2\theta'}{\sqrt{1 + r^2\theta'^2}} = A$$

from which $\theta(r)$ is obtainable by quadrature:

$$\theta = A \int \frac{dr}{r^2 \sqrt{E - V - A^2/r^2}}$$

24.14. THE PRINCIPLE OF EULER-LAGRANGE

Jacobi's principle is a modification of Hamilton's principle (24.17), obtained by eliminating the time t for conservative systems. Jacobi's action integral has [according to (24.76)] the following significance:

$$A = \int_{\tau_1}^{\tau_2} 2T \frac{dt}{d\tau} d\tau = \int_{\tau_1}^{\tau_2} 2T\, dt = \int_{\tau_1}^{\tau_2} \left(\frac{ds}{dt}\right)^2 dt$$
$$= \int_{\tau_1}^{\tau_2} \frac{ds}{dt} ds = \int_{\tau_1}^{\tau_2} v\, ds \tag{24.83}$$

Since the *mass* of the particle moving in C-space was normalized to 1, [see (24.37)], we can interpret the last form of the integral as the line integral of the momentum mv. It is this quantity which the early masters of the variational treatment of mechanics, Euler and Lagrange, (following some vague leads of Maupertuis), defined as *action.* The *principle of least action,* as formulated by Euler and Lagrange, uses, thus, the same action integral as the principle of Jacobi. Yet the actual solution of the given variational problem occurred according to different mathematical procedures. Jacobi eliminated the kinosthenic variable from the energy equation (24.77). Euler and Lagrange, on the other hand, used no elimination, but considered the energy theorem (24.77) as an *auxiliary condition* of the variational problem.

Now it makes no difference whether we say that we should minimize the integral

$$A = \int_{\tau_1}^{\tau_2} 2T\, dt = \int_{\tau_1}^{\tau_2} \frac{s'^2}{w} d\tau \tag{24.84}$$

eliminating w (which stands for t') from the energy relation

$$\frac{s'^2}{2w^2} + V = E \tag{24.85}$$

or that we should minimize (24.84) under the auxiliary condition (24.85). The elimination of w from (24.85) merely recognizes the existence of the auxiliary condition (24.85), and removes it from the scene in advance. But we can equally well *maintain* the auxiliary condition (24.85), and treat it by the Lagrangian multiplier method. We will then have the modified Lagrangian,

$$L_1 = \frac{s'^2}{w} + \lambda\left(\frac{s'^2}{2w^2} + V - E\right) \tag{24.86}$$

with no auxiliary condition but the added variables w and λ. Variation with respect to w gives the relation

$$\lambda = -w \tag{24.87}$$

and substituting this (algebraic) value of λ in our L_1, we now have

$$L_1 = \left(\frac{1}{2}\frac{s'^2}{w^2} - V + E\right) w \tag{24.88}$$

and

$$A_1 = \int_{\tau_1}^{\tau_2} \left(\frac{1}{2}\frac{s'^2}{w^2} - V + E\right) w\, d\tau \tag{24.89}$$

But then, introducing a new independent variable t by the definition

$$w\, d\tau = dt \tag{24.90}$$

we can write A_1 in the form

$$A_1 = \int_{t_1}^{t_2} \left[\frac{1}{2}\left(\frac{ds}{dt}\right)^2 - V + E\right] dt \tag{24.91}$$

The last term, if integrated out, gives a mere constant which may be omitted. But then we are back to Hamilton's principle (24.17).

This explains how Euler and Lagrange, although starting with a variational principle which is substantially equivalent to Jacobi's principle, derived their equations of motion in a form which follows more directly from Hamilton's principle.

24.15. THE CANONICAL EQUATIONS OF HAMILTON

A very great advance in the methods of analytical dynamics was achieved by Hamilton, who succeeded in formulating the equations of motion of an arbitrary dynamical system in a remarkably simplified form. His procedure can be conceived as an application of the Lagrangian multiplier method.

The velocities $\dot{q}_i$ which appear in the Lagrangian function $L(q_i,\dot{q}_i,t)$ are not independent of the q_i, and yet play variationally the role of a second set of variables. We can make the $\dot{q}_i$ completely independent by replacing them with a new set of variables u_i:

$$L = L(q_1, \ldots, q_n; u_1, \ldots, u_n; t) \tag{24.92}$$

provided that we add the n auxiliary conditions

$$\dot{q}_i - u_i = 0 \tag{24.93}$$

We now make use of the Lagrangian multiplier method, drop the auxiliary conditions (24.93), and replace L by the modified

$$L' = L(q_i,u_i,t) + \sum_{i=1}^{n} p_i(\dot{q}_i - u_i) \tag{24.94}$$

(we have denoted the Lagrangian multipliers λ_i by p_i). The $3n$ variables of the new variational problem are the q_i, p_i, and u_i.

The new Lagrangian appears in the following remarkable form,

$$L' = \sum_{i=1}^{n} p_i\dot{q}_i - H \tag{24.95}$$

where

$$H = \sum_{i=1}^{n} p_i u_i - L(q_i,u_i,t) \tag{24.96}$$

H is free of all derivatives. The derivatives $\dot{q}_i$ appear solely in the *first term* of the expression (24.95), and are present in a *linear* and *normalized* combination, while the original $L(q_i,\dot{q}_i,t)$ may have contained the $\dot{q}_i$ in arbitrarily complicated relations.

The variables u_i appear solely in H. They are algebraic and can thus be eliminated, according to the procedure of Sec. 24.9. Equations (24.40) now become

$$p_i = \frac{\partial L}{\partial u_i} \tag{24.97}$$

They have to be solved for the u_i and the resulting expressions substituted in H. Since in the final H, called the *Hamiltonian function*, the u_i are no longer present, we can replace the u_i by $\dot{q}_i$, and describe the construction of the Hamiltonian function in the following two steps:

1. Define

$$H = \sum_{i=1}^{n} p_i\dot{q}_i - L(q_i,\dot{q}_i,t) \tag{24.98}$$

2. Eliminate the $\dot{q}_i$ from the equations

$$p_i = \frac{\partial L}{\partial \dot{q}_i} \tag{24.99}$$

expressing them as functions of q_i, p_i, t. Then substitute these expressions in (24.98), thus obtaining H as a function of q_i, p_i, t:

$$H = H(q_i, p_i, t) \tag{24.100}$$

We now have the Lagrangian problem

$$L' = \sum_{i=1}^{n} p_i \dot{q}_i - H(q_i, p_i, t) \tag{24.101}$$

for the $2n$ dynamical variables q_i, p_i. The Euler-Lagrange equations (24.27), applied to the present problem, give

$$\dot{q}_i = \frac{\partial H}{\partial p_i} \qquad \dot{p}_i = -\frac{\partial H}{\partial q_i} \tag{24.102}$$

These are the celebrated *canonical equations* of Hamilton, which replace the n Lagrangian equations of second order by a system of $2n$ equations of first order, characterized by a remarkably high degree of symmetry.

Example 9. In Example 5*b* (p. 24–8), we get

$$p_1 = m\dot{r} \qquad \dot{r} = \frac{p_1}{m}$$

$$p_2 = mr^2\dot{\theta} \qquad \dot{\theta} = \frac{p_2}{mr^2}$$

$$H = p_1\dot{r} + p_2\dot{\theta} - \frac{m}{2}(\dot{r}^2 + r^2\dot{\theta}^2) + V(r)$$

$$H = \frac{1}{2m}\left(p_1^2 + \frac{p_2^2}{r^2}\right) + V(r)$$

The elimination of kinosthenic variables is greatly simplified in the Hamiltonian system. The application of the procedure of Sec. 24.12 to the present problem shows that, if H is free of a certain q_i, we can immediately replace the associated (called *conjugate*) momentum p_i by a constant α_i, and at the same time drop the term $p_i \dot{q}_i$ from the canonical Lagrangian (24.101). In the previous example, for instance, H is independent of q_2. Hence, we can replace p_2 by a constant,

$$H = \frac{1}{2m}\left(p_1^2 + \frac{\alpha^2}{r^2}\right) + V(r) \tag{24.103}$$

and our problem is reduced to a *single pair* of canonical equations.

24.16. THE PHASE SPACE AND THE PHASE FLUID

The canonical equations (24.102) are well suited for a step-by-step numerical integration of the mechanical problem, starting from a definite initial position,

$$q_i(t_1) = q_i^\circ \qquad p_i(t_1) = p_i^\circ \tag{24.104}$$

and continuing in small increments Δt of the independent variable. In the end, the coordinates q_i, p_i of the mechanical path will appear as definite functions of the time t:

$$q_i = q_i(t) \qquad p_i = p_i(t) \tag{23.105}$$

These equations can be conceived as the parametric equations of a curve, constructed in a space of $2n$ dimensions whose coordinates are the q_i and the p_i. The motion of a particle in ordinary space is described by giving the three coordinates x, y, z as functions of the time t. Now we have obtained the description of the motion of an arbitrary mechanical system in an entirely similar manner, but have replaced the three space coordinates by the $2n$ coordinates q_i, p_i of an abstract space of $2n$ dimensions, called the *phase space*. Compared with the Lagrangian "configuration space" of Sec. 24.2 (which included all the q_i but without the p_i), we have added the momenta p_i as additional dimensions. Hence, the configuration space has n and the phase space $2n$ dimensions.

We can now think of the totality of mechanical paths which are possible under arbitrary initial conditions (24.104). Then we see our phase space populated by an infinity of curves, and we arrive at the concept of a *fluid*, called *phase fluid*, which fills out the phase space. The various particles of the phase fluid move along a family of *streamlines* which describe the motion of the mechanical system as it develops from one definite initial position. Moreover, we see a definite *velocity field* in our phase space, established by the velocity vector

$$\mathbf{U} = (\dot{q}_i, \dot{p}_i) = \left(\frac{\partial H}{\partial p_i}, -\frac{\partial H}{\partial q_i}\right) \tag{24.106}$$

If the Hamiltonian function H contains the time explicitly, the velocity field will change from time moment to time moment, and we have to distinguish between the streamlines, obtained by following the velocity vector from point to point at a definite time moment, and the pathlines, obtained by following the history of a definite particle. If H is independent of t, (conservative systems), the two types of lines coincide, and the phase fluid is in a state of *steady motion*.

A very fundamental property of the motion of the phase fluid can be derived by forming the divergence of the velocity vector $\mathbf{U}$:

$$\operatorname{div} \mathbf{U} = \sum_{i=1}^{n} \left(\frac{\partial U_i}{\partial q_i} + \frac{\partial U_{n+i}}{\partial p_i}\right) = \sum_{i=1}^{n} \left(\frac{\partial^2 H}{\partial q_i\, \partial p_i} - \frac{\partial^2 H}{\partial p_i\, \partial q_i}\right) = 0 \tag{24.107}$$

The law that the divergence of the velocity field vanishes has in hydrodynamics the significance that the fluid is *incompressible*, in consequence of which an arbitrary volume composed of fluid particles maintains its volume during the motion. We thus arrive at the theorem of Liouville, according to which the motion of the phase fluid is an incompressible motion.

24.17. THE EXTENDED PHASE SPACE

The geometrical picture associated with the canonical equations becomes still more adequate if we include the time t among the dynamical variables. We can write the action integral of the canonical equations in the following form,

$$A = \int_{t_1}^{t_2} \left(\sum_{i=1}^{n} p_i \dot{q}_i - H\right) dt = \int_{t_1}^{t_2} \left(\sum_{i=1}^{n} p_i\, dq_i - H\, dt\right) = \int_{t_1}^{t_2} \sum_{i=1}^{n+1} p_i\, dq_i \tag{24.108}$$

if we put

$$t = q_{n+1} \qquad -H = p_{n+1} \tag{24.109}$$

Here the time t joins the mechanical variables on equal footing, and we can assume that we describe the motion of the phase fluid in terms of some parameter τ instead of t. The action integral of this generalized form of the canonical equations becomes

$$A = \int_{\tau_1}^{\tau_2} \sum_{i=1}^{n+1} p_i q_i'\, d\tau \tag{24.110}$$

where the prime refers to differentiation with respect to τ. Equations (24.109) can be conceived as an auxiliary condition of the variation

$$p_{n+1} + H(q_1, \ldots, q_{n+1}; p_1, \ldots, p_n) = 0 \tag{24.111}$$

which may be put in the more symmetric form,

$$K(q_1, \ldots, q_{n+1}; p_1, \ldots, p_{n+1}) = 0 \tag{24.112}$$

If this equation is solved for p_{n+1}, we come back to the more specified auxiliary condition (24.111).

Our phase space has now $(2n + 2)$ dimensions, and the auxiliary condition (24.111) means that the phase fluid has to remain on a definite *surface* of this space. The canonical equations of motion now become

$$q_i' = \frac{\partial K}{\partial p_i} \qquad p_i' = -\frac{\partial K}{\partial q_i} \tag{24.113}$$

Thus the function K takes the place of the previous Hamiltonian function H. If the first set of equations is multiplied by p_i', the second set by $-q_i'$, and the sum formed, we obtain

$$\sum_{i=1}^{n+1} \left(\frac{\partial K}{\partial q_i} q_i' + \frac{\partial K}{\partial p_i} p_i' \right) = \frac{dK}{d\tau} = 0 \tag{24.114}$$

which gives

$$K = \text{const} \tag{24.115}$$

This means that, if K is zero at the initial value $\tau = 0$, it will remain permanently zero, and the auxiliary condition (24.112) is fulfilled. If K is of the form (24.111), we obtain

$$q_{n+1}' = 1 \tag{24.116}$$

Then the parameter τ can be identified with q_{n+1}, that is, the time t, and we return to the original form (24.102) of the canonical equations. But now the last equation of the system (24.113) adds the relation

$$p_{n+1}' = -\frac{\partial H}{\partial t} \tag{24.117}$$

Since $p_{n+1} = -H$ has the significance of the negative total energy, we obtain the fundamental result that the total energy is conserved only if H is independent of t, whereas, if H is an explicit function of t, the time rate of change of the total energy is equal to $\partial H/\partial t$.

Another application of the extended phase space arises if H is independent of t. Then the time t becomes a kinosthenic variable which can be dropped from the canonical integral [cf. (24.103)], replacing p_{n+1} by the negative energy constant $-E$. We now have the canonical action integral

$$A = \int_{\tau_1}^{\tau_2} \sum_{i=1}^{n} p_i q_i' \, d\tau \tag{24.118}$$

with the auxiliary condition

$$H(q_1, \ldots, q_n; p_1, \ldots, p_n) = E \tag{24.119}$$

The general parameter τ of Eqs. (24.113) is now replaceable by the time t.

24.18. CANONICAL TRANSFORMATIONS

In Sec. 24.2 we emphasized the great advantages to be derived from the flexibility of the coordinates, which is a characteristic feature of the Lagrangian equations. The solution of many a mechanical problem becomes greatly simplified by setting up the problem in the proper frame of reference. For example, the planetary motion is by far simpler if described in polar rather than rectangular coordinates. In fact, we can say quite generally that, since we do not possess any universal analytical procedure for the integration of the equations of motion, we should look for a proper transformation of the coordinates by which the given set of differential equations becomes either partially or fully solved.

In the Hamiltonian formulation of mechanics, we have a much more extensive group of transformations at our disposal, since we now move in a space of $2n$ instead of n dimensions. On the other hand, it would be a mistake to believe that an *arbitrary* transformation of the (q_i,p_i) into a new set of variables $(\bar{q}_i,\bar{p}_i)$ is at our disposal. We do not want to sacrifice the canonical form of the equations, and thus we will restrict our transformation possibilities to a group of transformations which preserve the canonical form (24.110) [or, in the case of conservative systems, the form (24.118)] of the action integral. Such transformations are called *canonical.*

The preservation of the canonical integral (24.110) is equivalent to the demand that the differential form $\Sigma p_i\, dq_i$ shall be preserved. If we succeed in finding a transformation of the (q_i,p_i) into a new system $(\bar{q}_i,\bar{p}_i)$ which satisfies the condition

$$\sum_{i=1}^{n+1} p_i\, dq_i = \sum_{i=1}^{n+1} \bar{p}_i\, d\bar{q}_i \tag{24.120}$$

the canonical equations will carry over from the old to the new system. In actual fact, a wider group of transformations is permissible. If we put on the right side of (24.120) the perfect differential of a function S, the integral of this term will contribute to the action integral a mere *boundary term,* which has no effect on the resulting differential equations. We can thus define a general canonical transformation by the condition

$$\sum_{i=1}^{n+1} (p_i\, dq_i - \bar{p}_i\, d\bar{q}_i) = dS \tag{24.121}$$

Let us now assume that the function S is given as a function of the variables q_i and $\bar{q}_i$:

$$S = S(q_1, \ldots, q_{n+1}; \bar{q}_1, \ldots, \bar{q}_{n+1}) \tag{24.122}$$

Then

$$dS = \sum_{i=1}^{n+1} \left(\frac{\partial S}{\partial q_i}\, dq_i + \frac{\partial S}{\partial \bar{q}_i}\, d\bar{q}_i \right) \tag{24.123}$$

and the comparison of the left and right sides of (24.121) yields

$$p_i = \frac{\partial S}{\partial q_i} \qquad \bar{p}_i = -\frac{\partial S}{\partial \bar{q}_i} \tag{24.124}$$

The function S is called the *generating function* of a canonical transformation. Equations (24.124) do not give the desired transformation in explicit form, inasmuch as we did not express the old (q_i,p_i) in terms of the new $(\bar{q}_i,\bar{p}_i)$, or vice versa, but the old and new *momenta* are expressed in terms of the old and new *position coordinates.* If our aim is to express the old variables in terms of the new ones, we have to solve the second set of equations (24.124) for the q_i, and then substitute these expressions in the first set. Then the q_i and p_i will appear as explicit functions of the $(\bar{q}_i,\bar{p}_i)$.

Equations (24.124) do not include all possible canonical transformations. Indeed, even the simple identity transformation

$$q_i = \bar{q}_i \qquad p_i = \bar{p}_i \tag{24.125}$$

is not included in the relation (24.124). The reason is that, in deriving the relation, we have considered the q_i and $\bar{q}_i$ as *independent variables*, while the transformation law (24.125) establishes the $(n+1)$ relations $q_i = \bar{q}_i$ between the position coordinates. More generally, an arbitrary point transformation

$$q_i = f_i(\bar{q}_1, \bar{q}_2, \ldots, \bar{q}_{n+1}) \tag{24.126}$$

establishes $(n+1)$ relations between the q_i and the $\bar{q}_i$. This again can be conceived as the special case $m = n + 1$ of the more general situation in which m functional relations are given between the q_i and the $\bar{q}_i$:

$$F_k(q_1, \ldots, q_{n+1}; \bar{q}_1, \ldots, \bar{q}_{n+1}) = 0 \qquad k = 1, 2, \ldots, m \tag{24.127}$$

These equations establish m auxiliary conditions between the differentials dq_i and $d\bar{q}_i$. If these conditions are handled by the Lagrangian multiplier method, we obtain the most general implicit form of a canonical transformation by the following equations:

$$p_i = \frac{\partial S}{\partial q_i} + \lambda_1 \frac{\partial F_1}{\partial q_i} + \cdots + \lambda_m \frac{\partial F_m}{\partial q_i} \qquad -\bar{p}_i = \frac{\partial S}{\partial \bar{q}_i} + \lambda_1 \frac{\partial F}{\partial \bar{q}_i} + \cdots + \lambda_m \frac{\partial F_m}{\partial \bar{q}_i} \tag{24.128}$$

Two special cases are of particular interest:

1. The group of Lagrangian *point transformations;* characterized by Eqs. (24.126) and omitting the function S
2. The group of transformations characterized by the function S alone, without any additional conditions

24.19. THE PARTIAL DIFFERENTIAL EQUATION OF HAMILTON-JACOBI

We can conceive the goal of the transformation theory to obtain a canonical transformation which simplifies the auxiliary condition (24.112) as much as possible. We may choose, for example, the function K itself as one of the new variables, let us say $\bar{q}_{n+1}$:

$$\bar{q}_{n+1} = K(q_1, \ldots, q_{n+1}; p_1, \ldots, p_{n+1}) \tag{24.129}$$

Then, in the new frame of reference, K has the simple form $\bar{q}_{n+1}$, and the canonical system (24.113) becomes immediately integrable:

$$\begin{aligned} \bar{q}_i &= \text{const} = \gamma_i & i &= 1, 2, \ldots, n \\ \bar{p}_i &= \text{const} = -\beta_i & i &= 1, 2, \ldots, n \\ \bar{q}_{n+1} &= 0 & \bar{p}_{n+1} &= \tau_0 - \tau \end{aligned} \tag{24.130}$$

Hence, it suffices to find an arbitrary canonical transformation which will satisfy the single condition (24.129). Now we have seen that we can characterize a canonical transformation by the generating function S which is a function of the q_i and the $\bar{q}_i$. In view of the relations (24.124), we must replace in (24.129) the p_i by $\partial S/\partial q_i$. This yields a definite partial differential equation of the first order for the determination of the function S, called the *partial differential equation of Hamilton-Jacobi:*

$$K\left(q_1, \ldots, q_{n+1}; \frac{\partial S}{\partial q_1}, \ldots, \frac{\partial S}{\partial q_{n+1}}\right) = \bar{q}_{n+1} \tag{24.131}$$

It is necessary, however, that S shall come out as a function of the q_i and the $\bar{q}_i$. These $\bar{q}_i$ do not participate actively in the formation of the differential equation (24.131), whose variables are solely the q_i. They remain latent as variables, and play the role of constants of integration. We must find a solution of the partial differential equation (24.131) which contains n undetermined constants of integration (apart from the constant $\bar{q}_{n+1}$ which appears on the right side of the equation):

$$S = S(q_1, \ldots, q_{n+1}; \alpha_1, \ldots, \alpha_n, \bar{q}_{n+1}) \tag{24.132}$$

The n constants $\alpha_1, \alpha_2, \ldots, \alpha_n$ can now be identified with the n variables $\bar{q}_1, \ldots, \bar{q}_n$. But now the integration of the canonical equations in the new reference system shows [cf. (24.130)] that the $\bar{q}_i$ become all constants. Hence, it is unnecessary to convert the α_i into variables and then make them constants again. We can leave them as constants of integration. Moreover, the auxiliary condition (24.112) demands that the value of the constant $\bar{q}_{n+1}$ become zero. We do not lose anything essential if we make $\bar{q}_{n+1}$ zero in advance, i.e., if we solve the partial differential equation

$$K\left(q_1, \ldots, q_{n+1}; \frac{\partial S}{\partial q_1}, \ldots, \frac{\partial S}{\partial q_{n+1}}\right) = 0 \tag{24.133}$$

obtaining K as a function of the $(n + 1)$ variables q_i and the n constants of integration $\alpha_1, \ldots, \alpha_n$. The last equation of the system (24.130) may be dropped since the first $2n$ equations contain already the complete solution of the problem of motion. If we form the n partial derivatives of S with respect to the α_i, we obtain, according to (24.124), the $-\bar{p}_i$, which, however, in view of (24.130), assume the constant values β_i. This gives the n equations,

$$\frac{\partial S}{\partial \alpha_i} = \beta_i \qquad i = 1, \ldots, n \tag{24.134}$$

and, if these n equations are solved for the q_i $(i = 1, \ldots, n)$, we obtain the position coordinates q_i in terms of the variable q_{n+1}, which is in fact the time t, and the $2n$ constants α_i and β_i. These constants can be adjusted to arbitrary initial conditions (24.104).

We have thus, by transforming to a suitably moving frame of reference, *reduced the motion to rest* and solved the entire motion problem in terms of *constants*. The proper transformation was generated by a function S, which had to satisfy a single partial differential equation of the first order. In particular, if the auxiliary condition (24.112) is given in the usual form (24.111), the differential equation of Hamilton-Jacobi appears in the following form:

$$\frac{\partial S}{\partial t} + H\left(q_1, \ldots, q_n, t; \frac{\partial S}{\partial q_1}, \ldots, \frac{\partial S}{\partial q_n}\right) = 0 \tag{24.135}$$

Although we have no general method of integrating the partial differential equation of Hamilton-Jacobi, yet, in many important problems of physics and astronomy, an integration becomes possible by the method of separation of variables. We try a solution as a sum of functions of one single variable:

$$S = S_1(q_1) + S_2(q_2) + \cdots + S_{n+1}(t) \tag{24.136}$$

If the separation is possible at all, it leads to the right number of essential constants.

24.20. REFERENCES

[1] J. S. Ames, F. D. Murnaghan: "Theoretical Mechanics," Ginn, Boston, 1929.
[2] H. Goldstein: "Classical Mechanics," Addison-Wesley, Reading, Mass., 1951.
[3] C. Lanczos: "The Variational Principles of Mechanics," Univ. Press, Toronto, 1949.
[4] E. T. Whittaker: "A Treatise on the Analytical Dynamics of Particles and Rigid Bodies," 4th ed., Cambridge Univ. Press, New York, 1937.

CHAPTER 25

GYROSCOPES

BY

H. ZIEGLER, Dr. Sc. math., Zürich, Switzerland

25.1. BASIC CONCEPTS

A gyro (gyroscope) is an arbitrary rigid body free to rotate about a fixed point O. Convenient coordinates are Euler's angles, ψ, θ, ϕ, defined by Fig. 25.1. Here, (x,y,z) is a coordinate system at rest with origin O; (ξ,η,ζ) is the system of principal axes for O, moving with the gyro. The nodal axis κ is the intersection of the planes (ξ,η), (x,y), and the transverse axis λ is the intersection of (ξ,η) and (z,ζ).

The motion may be conceived as the result of three rotations: precession ψ, nodding motion θ, and spin ϕ. The state of motion is a rotation with an instantaneous angular velocity $\boldsymbol{\omega}$, the resultant of the vectors of $\dot\psi$, $\dot\theta$, $\dot\phi$. From Fig. 25.1 follow the relations

$$\omega_\kappa = \dot\theta \qquad \omega_\lambda = \dot\psi \sin\theta \qquad \omega_\zeta = \dot\psi\cos\theta + \dot\phi \tag{25.1}$$

and $\quad p = \omega_\xi = \omega_\kappa \cos\phi + \omega_\lambda \sin\phi \qquad q = \omega_\eta = -\omega_\kappa \sin\phi + \omega_\lambda \cos\phi$

$$r = \omega_\zeta \tag{25.2}$$

and thus the kinematic relations

$$p = \dot\psi \sin\theta \sin\phi + \dot\theta\cos\phi \qquad q = \dot\psi\sin\theta\cos\phi - \dot\theta\sin\phi \qquad r = \dot\psi\cos\theta + \dot\phi \tag{25.3}$$

Let I denote the moment of inertia for an arbitrary axis passing through O, and let I_ξ, I_η, I_ζ be the principal moments of inertia (with $I_\xi + I_\eta > I_\zeta, \ldots, \ldots$). Plotting $I^{-1/2}$ on every axis through O, we obtain the ellipsoid of inertia with the equation

$$I_\xi\xi^2 + I_\eta\eta^2 + I_\zeta\zeta^2 = 1 \tag{25.4}$$

and the semiaxes $I_\xi^{-1/2}, \ldots, \ldots$.

The angular momentum for O is defined by

$$\mathbf{H} = \int \mathbf{r} \times \mathbf{v}\, dm \qquad \text{with} \qquad \mathbf{v} = \boldsymbol{\omega} \times \mathbf{r} \tag{25.5}$$

Evaluation in components yields

$$H_\xi = I_\xi p \qquad H_\eta = I_\eta q \qquad H_\zeta = I_\zeta r \tag{25.6}$$

in the system (ξ,η,ζ), and

$$H_i = I_i\omega \tag{25.7}$$

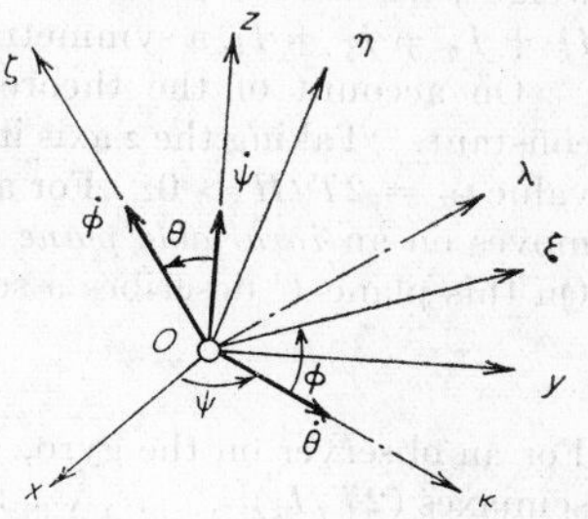

FIG. 25.1. Euler's angles.

for the component along the instantaneous axis i. Similarly, the kinetic energy, defined by

$$T = \tfrac{1}{2}\int v^2\, dm \tag{25.8}$$

takes either one of the forms

$$T = \tfrac{1}{2}(I_\xi p^2 + I_\eta q^2 + I_\zeta r^2) \qquad \text{or} \qquad T = \tfrac{1}{2}I_i\omega^2 \tag{25.9}$$

Comparison of (25.6) and (25.9) yields

$$T = \tfrac{1}{2}\mathbf{H}\cdot\boldsymbol{\omega} \geqq 0 \tag{25.10}$$

Since T is positive-definite, the angle between $\mathbf{H}$ and $\boldsymbol{\omega}$ is never obtuse.

The point of intersection Q of the instantaneous axis i and the ellipsoid of inertia (Fig. 25.2) has the coordinates $I_i^{-1/2}p/\omega, \ldots, \ldots$. The tangential plane E at this point is given by $I_\xi I_i^{-1/2}(p/\omega)\xi + \cdots + \cdots = 1$ or, on account of (25.6) and (25.9), by

$$\frac{H_\xi}{H}\xi + \frac{H_\eta}{H}\eta + \frac{H_\zeta}{H}\zeta = \frac{I_i^{1/2}\omega}{H} = \frac{(2T)^{1/2}}{H} \tag{25.11}$$

From (25.11) follows immediately $\mathbf{H} \perp E$. According to (25.10), $\mathbf{H}$ is directed toward E. If $(I')^{-1/2}$ denotes the distance OE in Fig. 25.2, then Eq. (25.11) yields

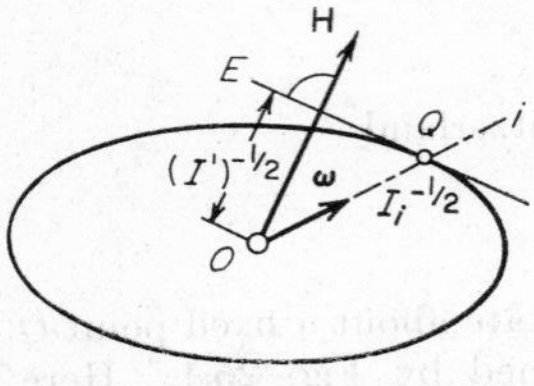

FIG. 25.2. Ellipsoid of inertia and angular momentum.

$$H = (I_i I')^{1/2}\omega \tag{25.12}$$

Thus, $\mathbf{H}$ is determined geometrically by the ellipsoid and by $\boldsymbol{\omega}$.

The external forces acting on the gyro (loads and reactions) reduce to a single force $\mathbf{R}$ and a couple of moment $\mathbf{M}$ at O. In the absence of friction, $\mathbf{M}$ is (a load and therefore) known. Thus, the motion can be obtained by the theorem of angular momentum. For an observer in the system (ξ,η,ζ), this theorem reads

$$\frac{d'\mathbf{H}}{dt} + \boldsymbol{\omega} \times \mathbf{H} = \mathbf{M} \tag{25.13}$$

where $d'\mathbf{H}/dt$ is the velocity of the end point of $\mathbf{H}$ for the observer on the gyro. Decomposing (25.13) in the system (ξ,η,ζ), we obtain Euler's equations:

$$I_\xi\dot{p} + (I_\zeta - I_\eta)qr = M_\xi \qquad I_\eta\dot{q} + (I_\xi - I_\zeta)rp = M_\eta \qquad I_\zeta\dot{r} + (I_\eta - I_\xi)pq = M_\zeta \tag{25.14}$$

25.2. FREE GYROS

A gyro with $M = 0$ is called a free gyro (*example:* a gyro under the sole action of its weight, suspended at its centroid S). It is referred to as an asymmetric gyro if $I_\xi \neq I_\eta \neq I_\zeta \neq I_\xi$, a symmetric gyro if $I_\xi = I_\eta \neq I_\zeta$, a spherical gyro if $I_\xi = I_\eta = I_\zeta$.

On account of the theorems of energy and angular momentum, T and $\mathbf{H}$ are constant. Taking the z axis in the direction of $\mathbf{H}$, we obtain from (25.10) the constant value $\omega_z = 2T/H > 0$. For an observer at rest, the end point U of $\boldsymbol{\omega}$ (Fig. 25.3) thus moves on an *invariable plane* E' normal to $\mathbf{H}$ and seen from O in the direction of $\mathbf{H}$. On this plane U describes a so-called *herpolhode*. From (25.9) follows

$$I_\xi p^2 + I_\eta q^2 + I_\zeta r^2 = 2T \tag{25.15}$$

For an observer on the gyro, U moves on the Poinsot or energy ellipsoid (25.15) with semiaxes $(2T/I_\xi)^{1/2}, \ldots, \ldots$, similar to the ellipsoid of inertia. On this ellipsoid, U describes a so-called *polhode*. The tangential plane of the ellipsoid at U has the equation

$$\frac{H_\xi}{H}\xi + \frac{H_\eta}{H}\eta + \frac{H_\zeta}{H}\zeta = \frac{2T}{H} = \omega_z \tag{25.16}$$

and thus is identical with E'. During the whole motion, the ellipsoid keeps touching E'. Since the point of contact U lies on the instantaneous axis, the ellipsoid rolls on the invariable plane (without gliding). This is the so-called *Poinsot motion.*

Polhodes and herpolhodes are easily obtained; they offer a basis for a qualitative discussion of the motion. A rigorous treatment is possible as well [7], since the homogeneous system (25.14) (with $M_\xi = M_\eta = M_\zeta = 0$) can be integrated by means of Jacobi's elliptic functions.

If the gyro is symmetric ($I_\xi = I_\eta \neq I_\zeta$), then ζ is its figure axis, and (ξ,η) the equatorial plane. I_ζ is the axial and I_ξ the equatorial moment of inertia. The energy ellipsoid (25.15) has rotational symmetry. Depending on its shape, the gyro is called *elongated* or *oblate*. The polhodes are circles about the axis ζ, and the herpolhodes are circles about M (Fig. 25.3).

For the analytical treatment, we might start from (25.14). Owing to the symmetry, however, (κ,λ,ζ) is a system of principal axes, taking part in the precession and the nodding motion, but not in the spin of the gyro. In this system, $\mathbf{H} = (I_\xi\omega_\kappa, I_\xi\omega_\lambda, I_\zeta r)$, and the system itself rotates with $\boldsymbol{\omega}' = (\omega_\kappa, \omega_\lambda, r - \omega_s)$, where $\omega_s = \dot\phi$ is the angular velocity of spin. Thus, the theorem of angular momentum for

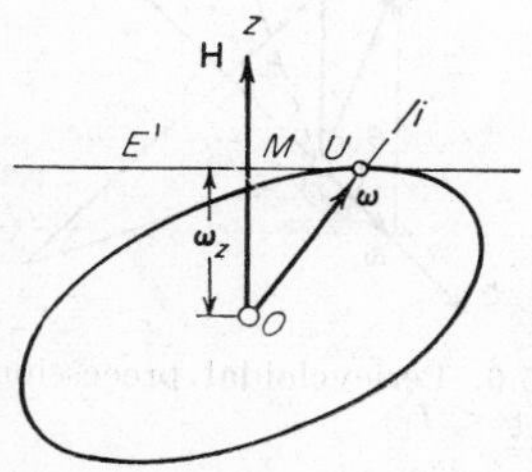

FIG. 25.3. Poinsot motion.

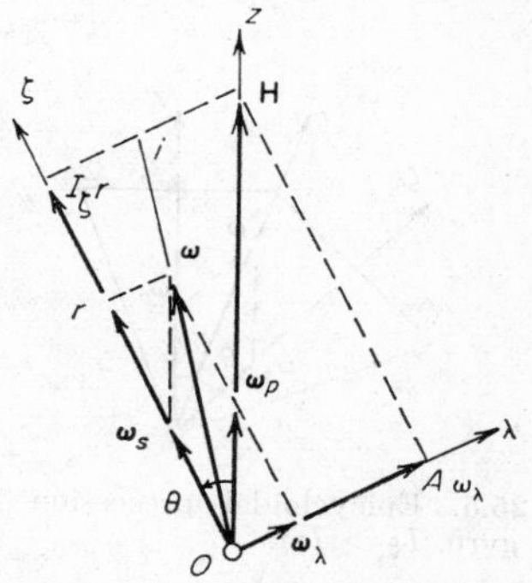

FIG. 25.4. Regular precession of a free symmetric gyro.

an observer in the system (κ,λ,ζ),

$$\frac{d'\mathbf{H}}{dt} + \boldsymbol{\omega}' \times \mathbf{H} = \mathbf{M} \tag{25.17}$$

yields

$$\left.\begin{aligned} I_\xi\dot\omega_\kappa + [I_\xi\omega_s + (I_\zeta - I_\xi)r]\omega_\lambda &= M_\kappa \\ I_\xi\dot\omega_\lambda - [I_\xi\omega_s + (I_\zeta - I_\xi)r]\omega_\kappa &= M_\lambda \\ I_\zeta\dot r &= M_\zeta \end{aligned}\right\} \tag{25.18a–c}$$

the so-called modified Euler equations, which may be completed by the kinematic relation

$$\omega_s = r - \omega_\lambda \cot\theta \tag{25.19}$$

following from (25.1) and (25.2).

For a free gyro ($M_\kappa = M_\lambda = M_\zeta = 0$), Eq. (25.18*c*) yields $r =$ const. Since $\mathbf{H}$ lies in the z axis, $\omega_\kappa = 0$: there is no nodding motion, and θ, according to (25.1), is constant. The figure axis describes a circular precession cone about the precession axis z. From Eq. (25.18*b*) follows $\omega_\lambda =$ const. Thus, $\boldsymbol{\omega}$ is fixed in the precession plane (z,ζ) (Fig. 25.4), and rotates with it about the z axis. Decomposing $\boldsymbol{\omega}$ into $\boldsymbol{\omega}_s$ and the angular velocity of precession, $\omega_p = \dot\psi$, we finally obtain

$$\theta = \text{const} \qquad \omega_p = \dot\psi = \text{const} \qquad \omega_s = \dot\phi = \text{const} \tag{25.20}$$

This motion is called a regular precession. It consists of two uniform rotations. Precession plane and figure axis rotate with ω_p about z, the figure axis describing the precession cone. The gyro rotates with ω_s about the figure axis.

The regular precession is entirely determined by the direction of $\mathbf{H}$ and the quantities θ, ω_p, ω_s. However, on account of Eq. (25.18*a*), these quantities are connected by the relation

$$[I_\xi\omega_s + (I_\zeta - I_\xi)r]\omega_\lambda = 0 \tag{25.21}$$

A first solution of (25.21) is $\omega_\lambda = 0$. This implies, according to Fig. 25.4, $\theta = 0$: the motion is a permanent rotation about the figure axis. A second solution, representing a more general motion, is given by $I_\xi\omega_s + (I_\zeta - I_\xi)r = 0$ or, on account of (25.1) and (25.2), by

$$I_\zeta\omega_s + (I_\zeta - I_\xi)\omega_p \cos\theta = 0 \tag{25.22}$$

It includes as special cases ($\theta = \pi/2{:}\ \omega_s = 0$) the permanent rotations about equatorial axes.

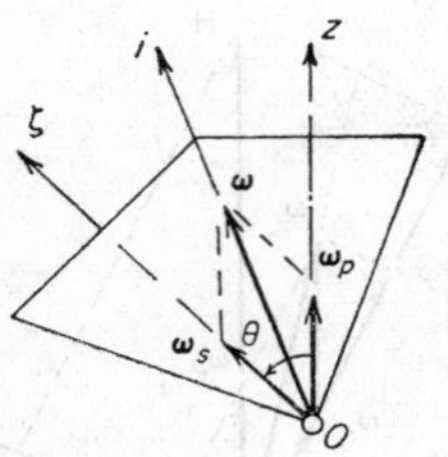

FIG. 25.5. Epicycloidal precession (elongated gyro, $I_\xi > I_\zeta$).

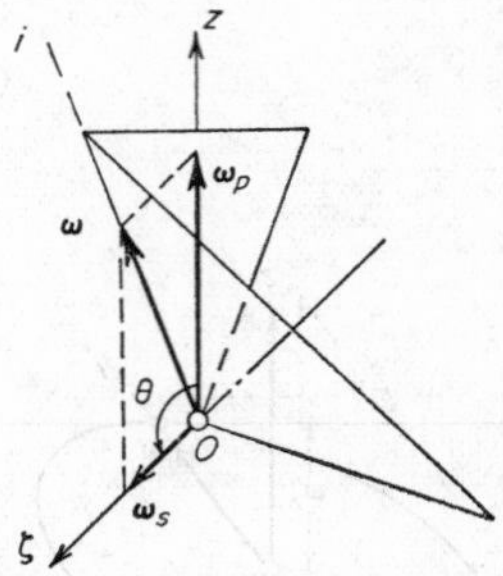

FIG. 25.6. Pericycloidal precession (oblate gyro, $I_\xi < I_\zeta$).

A gyroscopic motion can also be represented by a body cone rolling on a fixed cone. It follows from Fig. 25.4 that, for a regular precession, both cones are circular, and the motion is uniform. On account of (25.22), the precession is epicycloidal (Fig. 25.5) if the gyro is elongated, and pericycloidal (Fig. 25.6) if it is oblate.

25.3. CONSTRAINED PRECESSION

A constrained precession is a regular precession enforced by suitable loads or constraints. Considering first a symmetric gyro, we admit $S \neq O$, but require (as part of the definition of the symmetric gyro) that the mass center S lies on the figure axis ζ.

The precession is determined by the constant quantities $\boldsymbol{\omega}_p$, $\boldsymbol{\omega}_s$, θ, where the direction of $\boldsymbol{\omega}_p$ defines the precession axis (and the direction of $\boldsymbol{\omega}_s$ the positive figure axis). As long as **M** is arbitrary, these quantities may be chosen independently; there is no such relation as (25.22) between them. Hence, the precession may be epicycloidal (Fig. 25.5), pericycloidal (Fig. 25.6), or hypocycloidal (Fig. 25.7). [In a free gyro, the last motion is impossible on account of (25.10).]

According to (25.1),

$$\omega_\kappa = 0 \qquad \omega_\lambda = \omega_p \sin\theta = \text{const} \qquad r = \omega_p \cos\theta + \omega_s = \text{const} \tag{25.23}$$

Inserting these values into (25.18), we obtain

$$M_\kappa = [I_\xi\omega_s + (I_\zeta - I_\xi)r]\omega_\lambda \qquad M_\lambda = M_\zeta = 0 \tag{25.24}$$

or, on account of (25.23),

$$M = M_\kappa = [I_\zeta\omega_s + (I_\zeta - I_\xi)\omega_p \cos\theta]\omega_p \sin\theta \tag{25.25}$$

Thus, the regular precession is sustained by a moment of constant magnitude, lying in the nodal axis, and rotating with it about the precession axis. This result is illustrated in Fig. 25.8: **H** is fixed in the precession plane, rotating with $\boldsymbol{\omega}_p$. Thus

$\dot{\mathbf{H}} = \mathbf{M}$ has the direction of the nodal axis. Because of the rotation with ω_p, **H** remains perpendicular to **M** and, hence, does no work.

Equation (25.25) explains almost everything which seems paradoxical in the behavior of a gyroscope: **M** does not increase θ as one might expect intuitively. The gyro reacts, in accordance with the theorem of angular momentum, with a motion

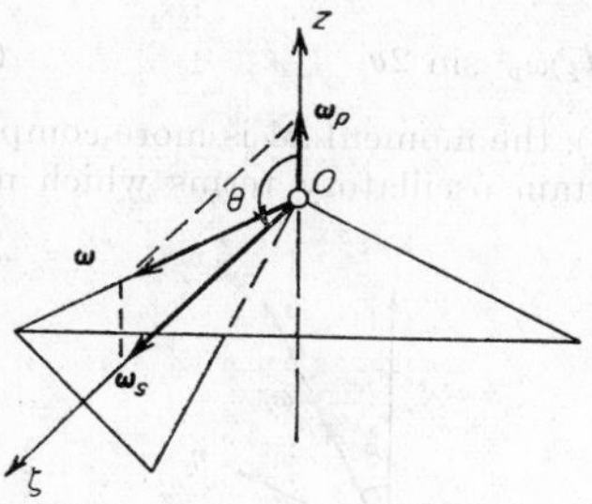

FIG. 25.7. Hypocycloidal precession.

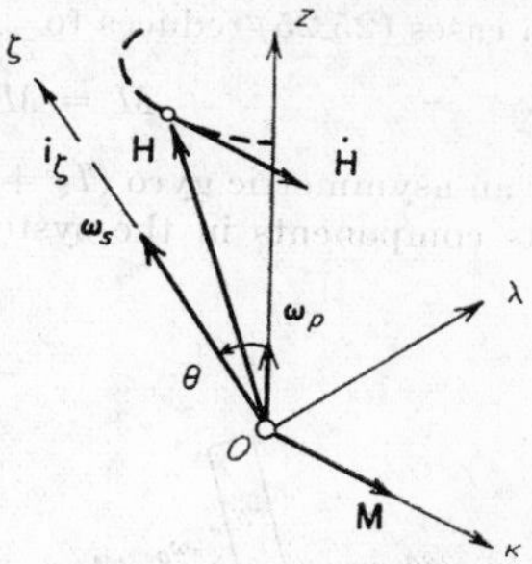

FIG. 25.8. Constrained precession.

tending to bring **H** into the direction of **M**. However, since **M** also rotates, **H** remains normal to **M**.

If $\mathbf{i}_\zeta$ is the unit vector in the figure axis (Fig. 25.8), we may write

$$\mathbf{M} = [I_\zeta \omega_s + (I_\zeta - I_\xi)\omega_p \cos\theta]\omega_p \times \mathbf{i}_\zeta \tag{25.26}$$

instead of (25.25). Note that this vector, on account of (25.22), vanishes in the case of a Poinsot motion (the limiting cases in the form of permanent rotations included).

Usually the external forces reduce to a moment **M** and a single force **R** in O. The force **R** follows from the theorem of momentum and is independent of the spin. If $\mathbf{R} \neq 0$, the resultant moment for S is different from the moment (25.26) for O. It is given by (25.26) if I_ξ and I_ζ denote the moments of inertia for S.

The reaction of the gyro on its surroundings consists in the gyroscopic moment $\mathbf{M}' = -\mathbf{M}$ and the centrifugal force $\mathbf{R}' = -\mathbf{R}$. On account of d'Alembert's principle, $\mathbf{M}'$ and $\mathbf{R}'$ may also be interpreted as the result of a reduction of the inertia forces.

The crusher m (Fig. 25.9) rolling on the horizontal plane E is a gyro with fixed point O. According to (25.26), **M** is directed backward and of magnitude

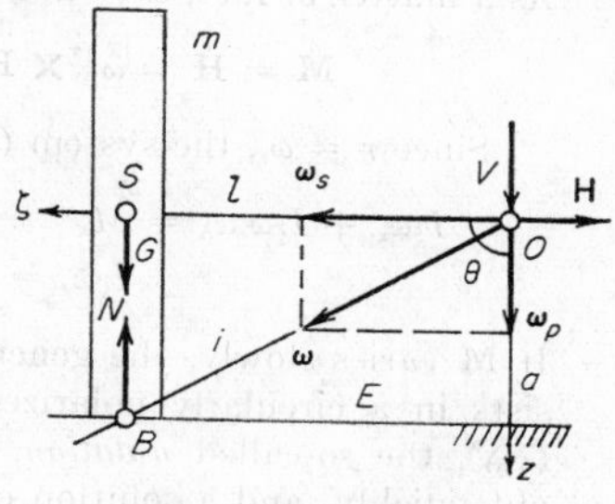

FIG. 25.9. Crusher.

$M = I_\zeta \omega_s \omega_p = I_\zeta (l/a)\omega_p^2$, while **R** is directed to the right and has the magnitude $R = ml\omega_p^2$. Hence,

$$ml\omega_p^2 = H \qquad 0 = N - G - V \qquad I_\zeta \frac{l}{a}\omega_p^2 = (N - G)l \tag{25.27}$$

and, thus,

$$H = ml\omega_p^2 \qquad V = \frac{I_\zeta}{a}\omega_p^2 \qquad N = G + \frac{I_\zeta}{a}\omega_p^2 \tag{25.28}$$

The terms containing I_ζ represent the gyroscopic effect.

In certain cases, e.g., if a rotor is mounted slightly oblique (Fig. 25.10), $\omega_s = 0$. In such cases (25.25) reduces to

$$M = M_\kappa = \tfrac{1}{2}(I_\zeta - I_\xi)\omega_p^2 \sin 2\theta \tag{25.29}$$

For an asymmetric gyro ($I_\xi \neq I_\eta \neq I_\zeta \neq I_\xi$), the moment **M** is more complicated [7]. Its components in the system (κ,λ,ζ) contain oscillatory terms which may be

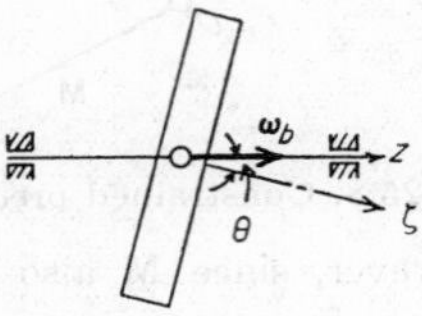

FIG. 25.10. Wheel oblique to its axle.

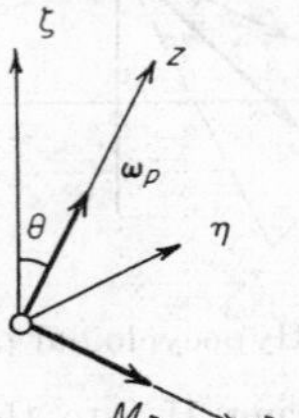

FIG. 25.11. Precession without spin.

devastating. If $\omega_s = 0$, the system (κ,λ,ζ) can be made to coincide with (ξ,η,ζ), and, if the precession axis z lies in the plane (η,ζ) (Fig. 25.11), the only nonvanishing component of **M** is

$$M_\xi = \tfrac{1}{2}(I_\zeta - I_\eta)\omega_p^2 \sin 2\theta \tag{25.30}$$

as for a symmetric gyro.

25.4. FAST SYMMETRIC GYROS

If r is large compared with ω_κ, ω_λ, the gyro is called fast. The vectors $\boldsymbol{\omega}$ and **H** then nearly lie in the figure axis, and it can be shown that the gyro remains fast, provided **M** is small compared with T and varies slowly.

For a constrained precession, $\omega_s \gg \omega_p$, and (25.26) reduces to

$$\mathbf{M} = I_\zeta \boldsymbol{\omega}_p \times \boldsymbol{\omega}_s \tag{25.31}$$

As a matter of fact, $\boldsymbol{\omega} \approx \boldsymbol{\omega}_s$ and $\mathbf{H} \approx I_\zeta \boldsymbol{\omega}_s$; thus

$$\mathbf{M} \approx \dot{\mathbf{H}} \approx \boldsymbol{\omega}_p \times \mathbf{H} = I_\zeta \boldsymbol{\omega}_p \times \boldsymbol{\omega}_s$$

Since $r \approx \omega_s$, the system (25.18) reduces to

$$I_\xi\dot\omega_\kappa + I_\zeta\omega_s\omega_\lambda = M_\kappa \qquad I_\xi\dot\omega_\lambda - I_\zeta\omega_s\omega_\kappa = M_\lambda$$
$$I_\zeta\dot\omega_s = M_\zeta \tag{25.32}$$

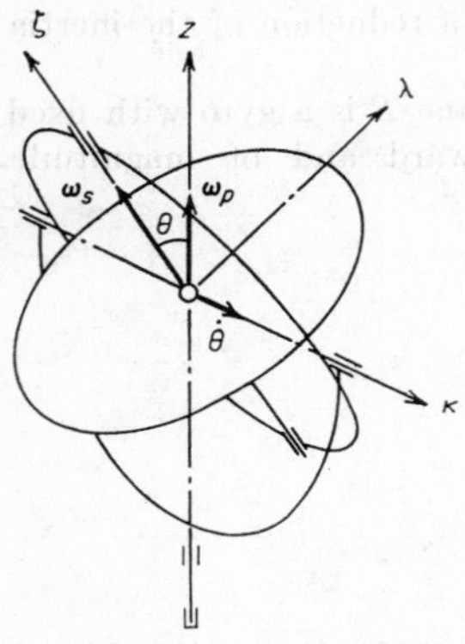

FIG. 25.12. Cardan-suspended gyro.

If **M** varies slowly, the general solution of (25.32) consists in a circularly polarized oscillation in the plane (κ,λ), the so-called *nutation*, which usually is damped out quickly, and a solution of

$$I_\zeta\omega_s\omega_\lambda = M_\kappa \qquad -I_\zeta\omega_s\omega_\kappa = M_\lambda \qquad I_\zeta\dot\omega_s = M_\zeta \tag{25.33}$$

supplying the secular terms. According to (25.1), Eq. (25.33) is equivalent to the system

$$I_\zeta\omega_s\omega_p \sin\theta = M_\kappa \qquad -I_\zeta\omega_s\dot\theta = M_\lambda \qquad I_\zeta\dot\omega_s = M_\zeta \tag{25.34}$$

which is of particular use in Cardan suspended gyros (Fig. 25.12) where ω_p, $\dot{\theta}$, ω_s denote rotations about material axles.

25.5. THE HEAVY SYMMETRIC GYRO

A symmetric gyro ($I_\xi = I_\eta \neq I_\zeta$, $S \neq O$ on ζ) is called heavy if it is subject to the sole influence of its weight W. Taking the figure axis (Fig. 25.13) in the direction OS, and z vertically upward, we have $s > 0$, and the angular velocities ω_p, $\dot{\theta}$, ω_s become algebraic quantities.

The moment **M** lies in the κ axis; its components are

$$M_\kappa = Ws \sin\theta \geqq 0, \; M_\lambda = M_\zeta = 0 \tag{25.35}$$

The simplest motion is a regular precession about the z axis. Comparing (25.35) with (25.25), and excluding the case $\sin\theta = 0$ of permanent rotation about z, we obtain, for this case, the condition

$$[I_\zeta\omega_s + (I_\zeta - I_\xi)\omega_p \cos\theta]\omega_p = Ws \tag{25.36}$$

between the parameters of the precession. Since (25.36) is quadratic in ω_p, the number of real solutions ω_p for given values of θ and ω_s is two, one, or zero, depending

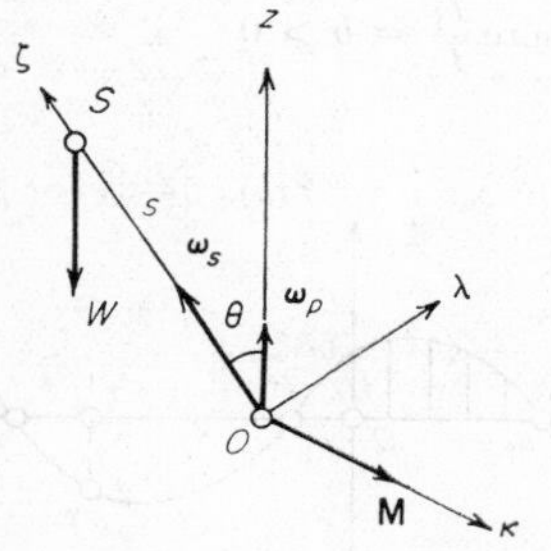

FIG. 25.13. Heavy symmetric gyro.

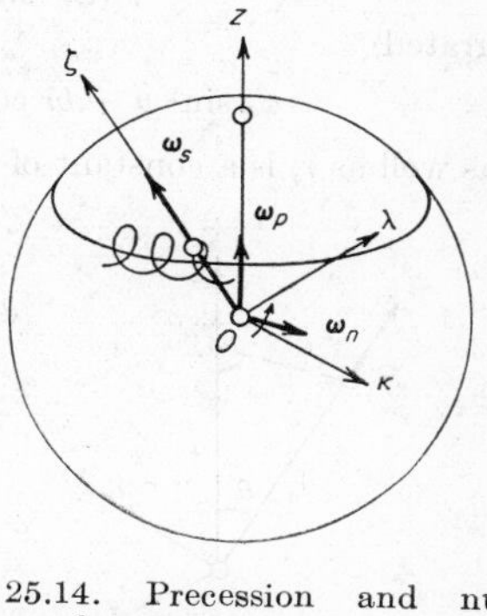

FIG. 25.14. Precession and nutation: pseudoregular precession.

on the values of the remaining constants. If two solutions exist, they are referred to as the *fast* and the *slow* precession.

The regular precession is a particular solution, appearing only under special initial conditions. However, it also represents an approximation for an arbitrary motion of the fast gyro. As a matter of fact, (25.34) with (25.35) reduces to the system

$$I_\zeta\omega_s\omega_p = Ws \qquad I_\zeta\omega_s\dot{\theta} = 0 \qquad I_\zeta\dot{\omega}_s = 0 \tag{25.37}$$

from which follows immediately

$$\omega_s = \text{const} \qquad \theta = \text{const} \qquad \omega_p = \frac{Ws}{I_\zeta\omega_s} = \text{const} \tag{25.38}$$

in agreement with the linearized relation (25.36). Incidentally, the precession, on account of the last equation, has always the same sense as the spin.

Actually, this motion is superposed by a nutation, i.e., by a third angular velocity ω_n performing a circularly polarized oscillation of circular frequency $\sigma = (I_\zeta/I_\xi)\omega_s$ in the plane (κ,λ). Thus, the rotation of the figure axis on its precession cone is modified by a fast whirling motion, as illustrated by Fig. 25.14.

The discussion of this pseudoregular precession, as well as the exact solution, may be based on the system (25.18) with the right-hand sides (25.35). The last equation, together with (25.1) and (25.2), immediately yields

$$\omega_p \cos\theta + \omega_s = r = \text{const} \tag{25.39}$$

On account of (25.39) and (25.1), the first two equations can be written

$$\begin{aligned} I_\xi\ddot{\theta} + (I_\zeta r - I_\xi\omega_p\cos\theta)\omega_p\sin\theta &= Ws\sin\theta \\ I_\xi(\omega_p\sin\theta)^{\cdot} - (I_\zeta r - I_\xi\omega_p\cos\theta)\dot{\theta} &= 0 \end{aligned} \tag{25.40a,b}$$

Multiplying these equations by $\dot{\theta}$ and $\omega_p \sin\theta$, respectively, and adding them, we obtain

$$\tfrac{1}{2}I_\xi(\dot{\theta}^2 + \omega_p^2\sin^2\theta)^{\cdot} + Ws(\cos\theta)^{\cdot} = 0 \tag{25.41}$$

or, integrated,

$$\dot{\theta}^2 + \omega_p^2\sin^2\theta + a\cos\theta = \alpha \qquad \text{where } a = \frac{2Ws}{I_\xi} > 0 \tag{25.42}$$

and α is a constant of integration. Equation (25.40*b*) is equivalent to

$$I_\xi\dot{\omega}_p\sin\theta + 2I_\xi\omega_p(\cos\theta)\dot{\theta} - I_\zeta r\dot{\theta} = 0 \tag{25.43}$$

Multiplying by $\sin\theta$, we obtain

$$I_\xi(\omega_p\sin^2\theta)^{\cdot} + I_\zeta r(\cos\theta)^{\cdot} = 0 \tag{25.44}$$

or, integrated,

$$\omega_p\sin^2\theta + br\cos\theta = \beta \qquad \text{where } \frac{I_\zeta}{I_\xi} = b > 0 \tag{25.45}$$

and β, as well as r, is a constant of integration.

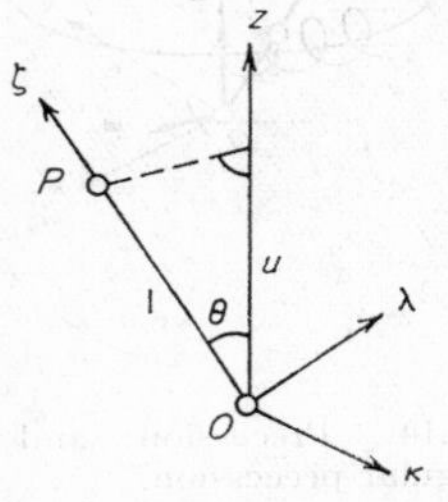

FIG. 25.15. P and u.

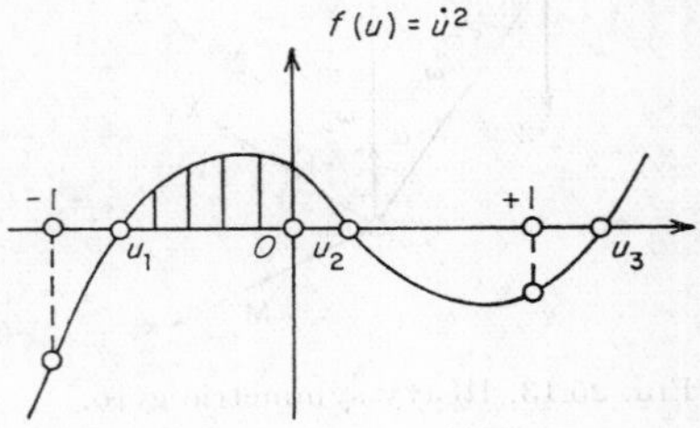

FIG. 25.16. Gyro function.

Elimination of ω_p from (25.42) and (25.45) yields

$$(\sin^2\theta)\dot{\theta}^2 = (\alpha - a\cos\theta)\sin^2\theta - (\beta - br\cos\theta)^2 \tag{25.46}$$

Introducing the point P (Fig. 25.15) with $OP = 1$ on the positive figure axis, and denoting its ordinate by u, we have

$$u = \cos\theta \qquad \dot{u} = -(\sin\theta)\dot{\theta} \tag{25.47}$$

and thus, instead of (25.46),

$$\dot{u}^2 = f(u) = (\alpha - au)(1 - u^2) - (\beta - bru)^2 \tag{25.48}$$

This is a differential equation for $u(t)$. The integral

$$t = \int_{u_0}^{u} \frac{dv}{\sqrt{f(v)}} \tag{25.49}$$

can be evaluated in terms of elliptic functions [7]. As soon as $u(t)$, that is, the nodding motion $\theta(t)$, is obtained, (25.45) and (25.47) yield

$$\omega_p = \dot{\psi} = \frac{\beta - bru}{1 - u^2} \tag{25.50}$$

and from (25.39), (25.47) follows

$$\omega_s = \dot{\phi} = r - u\omega_p \tag{25.51}$$

Thus, two simple integrations supply the precession and the spin.

A qualitative picture of the motion can be obtained without integration, by a mere discussion of the so-called *gyro function* $f(u)$. In the general case, i.e., if $\beta \neq \pm br$, $f(u)$ has the general aspect given in Fig. 25.16. There are two real roots, u_1, u_2 (possibly coinciding) in the interval $-1 < u < 1$, and, since $\dot{u}^2 = f(u)$, only the domain $u_1 \leqq u \leqq u_2$ has a physical meaning. P is restricted (Fig. 25.17) to the region between the parallels u_1, u_2 on the unit sphere. Since

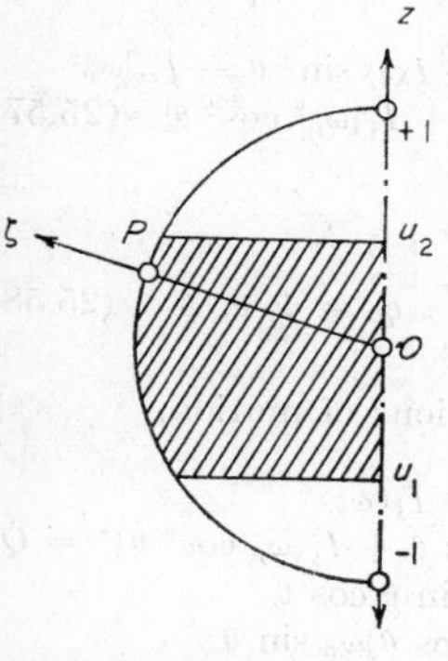

Fig. 25.17. Strip on the unit sphere.

$$\dot{u} = \pm\sqrt{f(u)} \tag{25.52}$$

P travels across the whole strip, and, since

$$\ddot{u} = \frac{1}{2}\frac{df}{du} \tag{25.53}$$

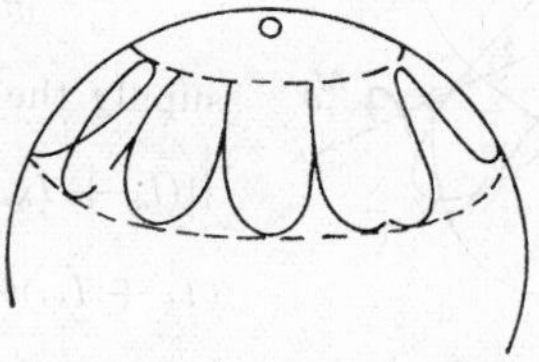

Fig. 25.18. Possible paths of P.

P oscillates between the limiting parallels. This motion is periodic with the period

$$T = 2\int_{u_1}^{u_2} \frac{du}{\sqrt{f(u)}} \tag{25.54}$$

and, on account of (25.50) and (25.51), ω_p and ω_s (but not ψ and ϕ) have the same period. The trajectory of P on the unit sphere is likewise periodic, consisting of congruent arcs. The trajectory, however, is usually not closed, corresponding to the fact that the entire motion is not periodic.

It is clear that the motion is a generalization of the pseudoregular precession. As a matter of fact, the path of P is essentially as illustrated in Fig. 25.14, with or without loops or vertices at the upper parallel, but without loops and vertices at the lower one (Fig. 25.18).

If $\beta = br$ and $\alpha = a$, the gyrofunction (25.48) reduces to

$$f(u) = (1 - u)^2[a(1 + u) - b^2r^2] \tag{25.55}$$

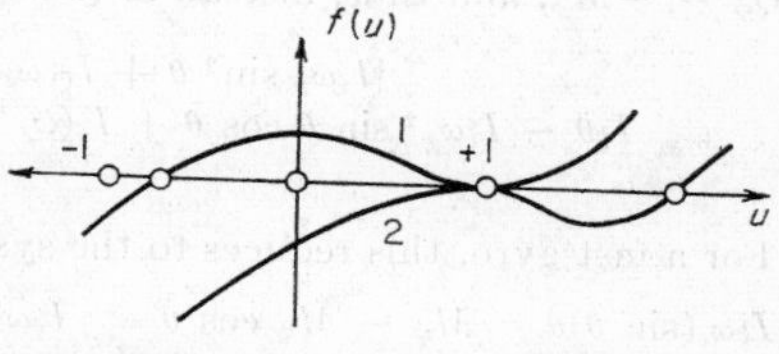

Fig. 25.19. Permanent rotation of the upright gyro.

Since $u = +1$ is a double root, the permanent rotation of the upright gyro ($\theta = 0$) is a solution. Comparing the two possible aspects of $f(u)$ in Fig. 25.19, and admitting the presence of small perturbations which slightly displace the curves, we immediately conclude that the rotation is unstable in case 1 and stable in case 2. In the second case, the third root is greater than 1; hence, the stability condition, according to (25.55), is $b^2r^2 > 2a$ or, on account of (25.42) and (25.45),

$$r > \frac{2}{I_\zeta}\sqrt{I_\xi W s} \tag{25.56}$$

The last result remains valid for a more general force field with rotational symmetry about the z axis, for instance, for missiles with an air drag W acting at a distance s ahead of the centroid. In this and similar cases, stability is obtained by means of a sufficient spin.

25.6. CARDAN GIMBALS AND FRICTION

Denoting the moments of inertia of the internal gimbal (Fig. 25.20) with respect to the axes κ, λ, ζ by $I_{\kappa i}$, $I_{\lambda i}$, $I_{\zeta i}$, and those of the external gimbal with respect to κ, μ, z by $I_{\kappa e}$, $I_{\mu e}$, I_{ze}, we obtain, for the total kinetic energy of the Cardan-suspended gyro, the expression

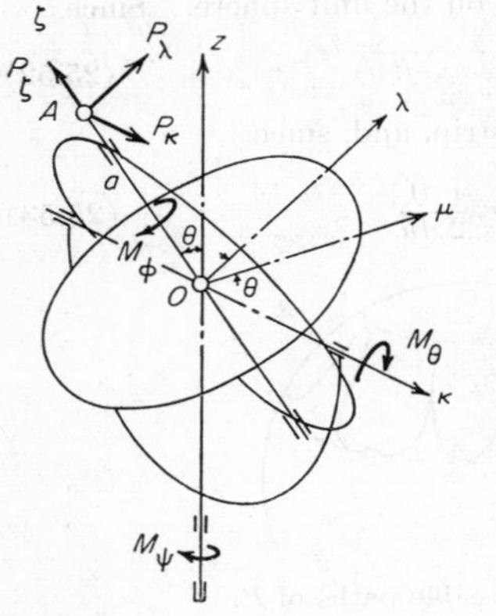

FIG. 25.20. Cardan-suspended gyro.

$$T = \tfrac{1}{2}[(I_\xi + I_{\kappa i})\dot{\theta}^2 + [(I_\xi + I_{\lambda i}) \sin^2 \theta + I_{ze}]\omega_p^2 + I_\zeta(\omega_s + \omega_p \cos \theta)^2 + I_{\zeta i}\omega_p^2 \cos^2 \theta] \qquad (25.57)$$

Lagrange's relations,

$$\frac{d}{dt}\frac{\partial T}{\partial \dot{q}_k} - \frac{\partial T}{\partial q_k} = Q_k \qquad q_k = \psi, \theta, \phi \qquad (25.58)$$

supply the differential equations of motion,

$$\begin{aligned} \{[(I_\xi + I_{\lambda i}) \sin^2 \theta + I_{ze}]\omega_p + I_\zeta(\omega_s &+ \omega_p \cos \theta) \cos \theta + I_{\zeta i}\omega_p \cos^2 \theta\}^{\cdot} = Q_\psi \\ (I_\xi + I_{\kappa i})\ddot{\theta} - (I_\xi + I_{\lambda i})\omega_p^2 \sin \theta \cos \theta &+ I_\zeta (\omega_s + \omega_p \cos \theta)\omega_p \sin \theta \\ &+ I_{\zeta i}\omega_p^2 \cos \theta \sin \theta = Q_\theta \\ I_\zeta(\omega_s + \omega_p \cos \theta)^{\cdot} &= Q_\phi \end{aligned} \qquad (25.59a\text{–}c)$$

where the right-hand sides are the generalized forces.

System (25.59) takes care of the influence of the gimbals on the motion of the actual gyro. In the case of a regular precession about the z axis, for example, ω_p, θ, and ω_s are constant; (25.59) yields $Q_\psi = Q_\phi = 0$; and

$$Q_\theta = [I_\zeta\omega_s + (I_\zeta - I_\xi - I_{\lambda i} + I_{\zeta i})\omega_p \cos \theta]\omega_p \sin \theta \qquad (25.60)$$

Q_θ is the component M_κ of the external moment; thus, (25.60) is a simple generalization of (25.25).

Neglecting again the gimbals, and assuming instead frictional moments M_ψ, M_θ, M_ϕ in the bearings of a heavy Cardan gyro, we obtain $Q_\psi = -M_\psi$, $Q_\theta = W_s \sin \theta - M_\theta$, $Q_\phi = -M_\phi$, and thus, instead of (25.59),

$$\begin{aligned} [I_\xi\omega_p \sin^2 \theta + I_\zeta(\omega_s + \omega_p \cos \theta) \cos \theta]^{\cdot} &= -M_\psi \\ I_\xi\ddot{\theta} - I_\xi\omega_p^2 \sin \theta \cos \theta + I_\zeta(\omega_s + \omega_p \cos \theta)\omega_p \sin \theta &= Ws \sin \theta - M_\theta \\ I_\zeta(\omega_s + \omega_p \cos \theta)^{\cdot} &= -M_\phi \end{aligned} \qquad (25.61)$$

For a fast gyro, this reduces to the system

$$I_\zeta\omega_s(\sin \theta)\dot{\theta} = M_\psi - M_\phi \cos \theta \qquad I_\zeta\omega_s\omega_p \sin \theta = Ws \sin \theta - M_\theta \qquad I_\zeta\dot{\omega}_s = -M_\phi \qquad (25.62)$$

which can be integrated for various laws of friction.

Returning to system (25.59), and assuming that the gyro is acted upon by a force $\mathbf{P} = (P_\kappa, P_\lambda, P_\zeta)$ at the point A in the distance a from O on the figure axis, we obtain, if M_ψ is the only moment of friction, the generalized forces,

$$Q_\psi = P_\kappa a \sin \theta - M_\psi \qquad Q_\theta = -P_\lambda a \qquad Q_\phi = 0 \qquad (25.63)$$

Inserting (25.63) into (25.59), we conclude at once from the last equation that

$$\omega_s + \omega_p \cos \theta = r = \text{const} \qquad (25.64)$$

Thus, the remaining equations become

$$\{I_\zeta r \cos\theta + [(I_\xi + I_{\lambda i})\sin^2\theta + I_{\zeta i}\cos^2\theta + I_{ze}]\omega_p\}^{\cdot} = P_\kappa a \sin\theta - M_\psi$$
$$(I_\xi + I_{\kappa i})\ddot{\theta} + [I_\zeta r - (I_\xi + I_{\lambda i} - I_{\zeta i})\omega_p \cos\theta]\omega_p \sin\theta = -P_\lambda a \qquad (25.65a,b)$$

If the figure axis of a gyro is not free, but compelled to rotate about a fixed axis, the gyro has only two degrees of freedom. *Examples:* a Cardan-suspended gyro with the external gimbal blocked or with the internal gimbal blocked with respect to the external one. The first case follows from the case just considered by increasing M_ψ so that ω_p becomes zero. Equation (25.65*b*) then yields the motion

$$(I_\xi + I_{\xi i})\ddot{\theta} = -P_\lambda a \qquad (25.66)$$

while the first one supplies the necessary moment of friction

$$M_\psi = (P_\kappa a + I_\kappa r\dot{\theta}) \sin\theta \qquad (25.67)$$

The spin affects M_ψ but not the motion of the figure axis. The axis ζ is in equilibrium when $P_\lambda = 0$, that is, when the plane (κ,ζ) contains the force **P**. If **P** is constant, the plane (κ,ζ) oscillates about the plane determined by κ and the direction of **P** as though the gyro had no spin.

The last statement is based on the assumption that the third degree of freedom is completely absent. Actually, it cannot be suppressed completely, because of the elasticity of rotor and gimbals. Hence, the oscillation about the equilibrium is usually considerably slower than in the absence of spin.

25.7. FAST SYMMETRIC GYROS ON VEHICLES

If a gyro G (Fig. 25.21) is mounted on a vehicle V, its motion with respect to the vehicle depends, not only on the real forces, but also on the fictitious forces of relative

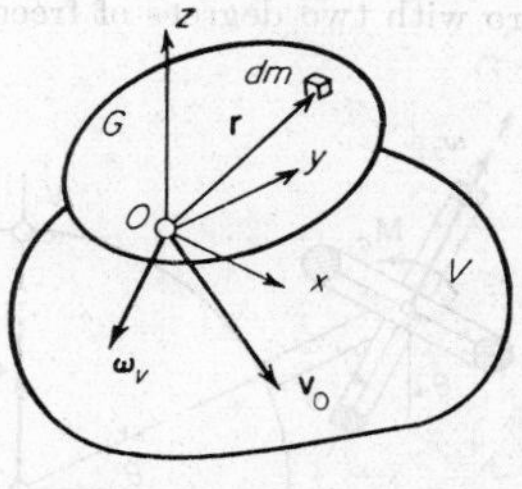

FIG. 25.21. Gyro on vehicle.

motion: the centrifugal and Coriolis forces,

$$d\mathbf{Z} = -\mathbf{a}_v\, dm \qquad d\mathbf{C} = 2\mathbf{v}_r \times \boldsymbol{\omega}_v\, dm \qquad (25.68)$$

acting in the mass elements dm. Here, $\mathbf{v}_r$ is the velocity of dm relative to the vehicle; $\boldsymbol{\omega}_v = (\omega_{v\kappa}, \omega_{v\lambda}, r_v)$ is the angular velocity of the vehicle; and $\mathbf{a}_v$ is the acceleration of dm under the assumption that dm is rigidly connected with the vehicle.

Restricting ourselves to fast symmetric gyros, representing the state of motion of the vehicle by $\mathbf{v}_o$ and $\boldsymbol{\omega}_v$ (Fig. 25.21), and reducing the forces (25.68) to the point O, we obtain the fictitious moments,

$$\mathbf{M}_{z1} = -m\mathbf{r}_s \times \mathbf{a}_o \qquad \mathbf{M}_{z2} = -(I_\xi \dot{\omega}_{v\kappa}, I_\xi \dot{\omega}_{v\lambda}, I_\zeta \dot{r}_v) \qquad (25.69)$$

where $\mathbf{r}_s$ is the radius vector of S, and $\mathbf{a}_o$ the acceleration of O, and

$$\mathbf{M}_c = I_\zeta \omega_s \times \boldsymbol{\omega}_v \qquad (25.70)$$

The moments (25.69) are independent of the spin; they represent the centrifugal forces due to the translation and the rotation of the vehicle. The moment (25.70) is due to the Coriolis forces.

If the rotation of the vehicle is uniform, the second moment (25.69) is zero. Very often r_s and $\mathbf{a}_o$ are small, so that also the first moment (25.69) can be dropped. The only fictitious moment then is the Coriolis moment (25.70).

In the case of a free gyro, the motion for an observer on the vehicle is given, according to (25.31) and (25.70), by

$$I_\zeta \boldsymbol{\omega}_p \times \boldsymbol{\omega}_s = I_\zeta \boldsymbol{\omega}_s \times \boldsymbol{\omega}_v \tag{25.71}$$

Thus, $(\boldsymbol{\omega}_p + \boldsymbol{\omega}_v)$ has the direction of $\boldsymbol{\omega}_s$: the relative precession is such that the figure axis remains absolutely at rest. In spite of this property, free gyros are poor compasses, on account of errors due to eccentricities and friction.

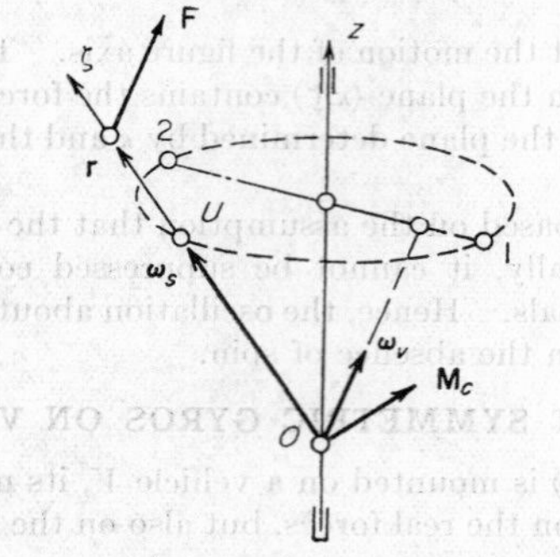

FIG. 25.22. Gyro with two degrees of freedom on vehicle.

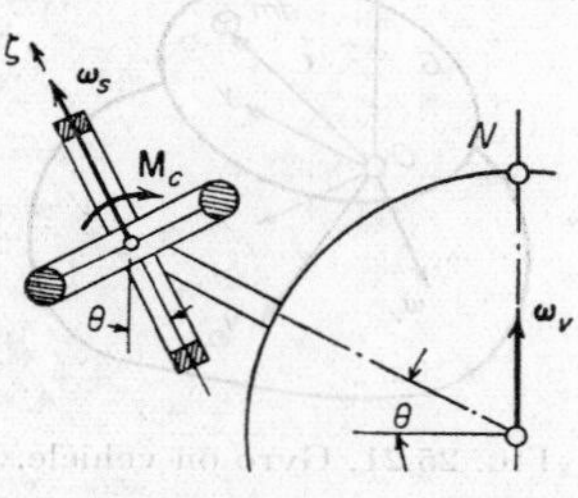

FIG. 25.23. Inclination gyro.

For a gyro with two degrees of freedom (Fig. 25.22), the Coriolis moment (25.70) rotates the figure axis ζ about the precession axis z. There are two equilibrium positions (1 stable and 2 unstable) where the figure axis lies in the plane determined by z and $\boldsymbol{\omega}_v$. As far as the motion (not the reactions) is concerned, the moment $\mathbf{M}_c$ might be replaced by a force $\mathbf{F}$ in the direction of $\boldsymbol{\omega}_v$, acting at the positive figure axis (Fig. 25.22), and of magnitude

$$F = \frac{I_\zeta}{r}\,\omega_s\omega_v \tag{25.72}$$

As a matter of fact, the moment of this force with respect to O is

$$\mathbf{M} = \mathbf{r} \times \frac{I_\zeta}{r}\,\boldsymbol{\omega}_s\omega_v = I_\zeta \boldsymbol{\omega}_s \times \boldsymbol{\omega}_v = \mathbf{M}_c \tag{25.73}$$

Thus, the rotation of the vehicle has the same effect on the gyro as such a force $\mathbf{F}$.

The gyro of Fig. 25.23 with external gimbal fixed to the earth has two degrees of freedom. The moment (25.70) is of magnitude

$$M_c = I_\zeta \omega_s \omega_v \sin \theta \tag{25.74}$$

It tends to render the figure axis parallel to the axis of the earth. Hence, a gyro of this type indicates the inclination of the point of observation.

If the gyro is mounted according to Fig. 25.24, its Coriolis moment may be written

$$\mathbf{M}_c = I_\zeta \boldsymbol{\omega}_s \times \boldsymbol{\omega}_1 + I_\zeta \boldsymbol{\omega}_s \times \boldsymbol{\omega}_2 \tag{25.75}$$

The second moment is normal to the precession axis, and thus without influence on the motion. The first one has the magnitude

$$M_c = I_\zeta \omega_s \omega_1 \sin \theta = I_\zeta \omega_s \omega_v \cos \phi \sin \theta \tag{25.76}$$

and it tends to rotate the figure axis into the horizontal northern direction.

The last instrument represents the basic idea of the modern gyrocompass. Its essential advantage compared with a free gyro is the presence of a directional moment.

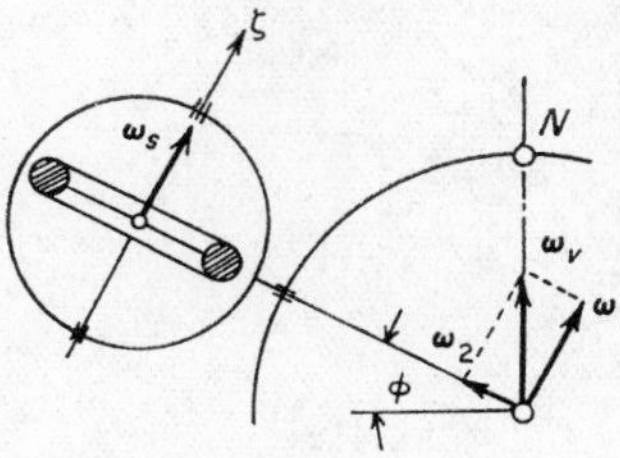

FIG. 25.24. Foucault's compass.

The modern compass is an elaborate apparatus. Its development required essentially the following improvements of Foucault's gyro (Fig. 25.24): (1) an increase of the moment (25.70), (2) elimination of dry friction, and (3) a rapid damping of oscillations. These problems were solved by using fast electromotors as gyros, by new types of suspension (Sperry and Anschütz, introducing again a third degree of freedom), and by artificial damping devices.

25.8. REFERENCES

[1] H. Crabtree: "An Elementary Treatment of the Theory of Spinning Tops and Gyroscopic Motion, Longmans, London, 1923.
[2] R. F. Deimel: "Mechanics of the Gyroscope," Dover, New York, 1950.
[3] M. Davidson: "The Gyroscope and Its Applications," Hutchinson, London, New York, 1947.
[4] F. Klein, A. Sommerfeld: "Über die Theorie des Kreisels," Teubner, Leipzig, 1897–1910.
[5] F. Klein: "The Mathematical Theory of the Top," Princeton Lectures, Scribner, New York, 1897.
[6] E. S. Ferry: "Applied Gyrodynamics," Wiley, New York, 1932.
[7] R. Grammel: "Der Kreisel, seine Theorie und seine Anwendungen," Springer, Berlin, 1950.
[8] G. Greenhill: "Report on Gyroscopic Theory," H. M. Stationery Off., London, 1914.
[9] A. N. Krylov, U. A. Krutkov: "General Theory of Gyroscopics" (in Russian), Leningrad, 1932.
[10] A. Lucas: "Des Phénomènes gyroscopiques et leurs principales applications à la navigation," Challamel, Paris, 1918.
[11] A. Lucas: "Théorie élémentaire du compas gyroscopique à l'usage des marins et des aviateurs," Soc. d'Éditions géographiques etc., Paris, 1940.
[12] J. B. Scarborough: "The Gyroscope, Theory and Applications," Interscience, New York, 1958.

Part 3

THEORY OF STRUCTURES

CHAPTER 26

STATICALLY DETERMINATE STRUCTURES

BY

J. M. GERE, Ph.D., Palo Alto, Calif.

26.1. INTRODUCTION

Structures which lie in a single plane and which are subjected to forces acting in the same plane are called *planar structures.* Many common structures of the beam, truss, and rigid-frame types are in this category. *Space structures* are three-dimensional structures, and plane structures subjected to loads which do not act exclusively in the plane of the structure. Many space structures can be decomposed into planar structures for purposes of analysis.

Types of Supports for Planar Structures[1]

The supports of a planar structure are classified according to the type of restraint which they provide. A *roller support* or *simple support* (represented by any one of the symbols in Fig. 26.1*a*) is free to rotate and to translate in the direction parallel to the plane on which the roller rests. It provides restraint against deflection in the direction normal to the plane of support, and thus may develop only one reaction component R.

A *pin support* (Fig. 26.1*b*) is free to rotate, but is restrained against translation in any direction. Thus, the pin support may develop two reaction components, R_1 and R_2. The *fixed* or *clamped support* shown in Fig. 26.1*c* completely restrains the end of the structure against rotation and translation, and thus may develop three reaction components, M_1, R_1, and R_2.

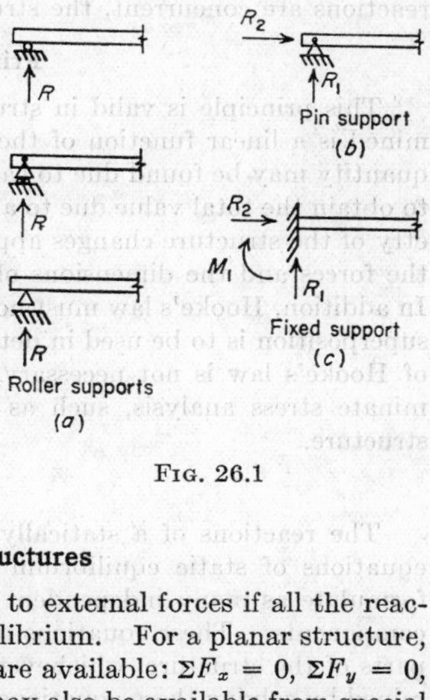

FIG. 26.1

Statically Determinate Structures

A structure is statically determinate with regard to external forces if all the reactions can be determined from equations of static equilibrium. For a planar structure, three equations for the equilibrium of the structure are available: $\Sigma F_x = 0$, $\Sigma F_y = 0$, $\Sigma M = 0$. Additional static-equilibrium equations may also be available from special conditions of construction of the structure. For example, a three-hinged arch with a pin at the center has one additional equation of statics, since the moment at the pin is zero. If there are a greater number of reaction components than equations of equilibrium, the structure is statically indeterminate externally. The degree of

[1] Supports for space structures are discussed in Sec. 26.6, p. 26–25.

indeterminateness is the excess of reaction components over the number of available equations.

A structure is internally statically determinate if all the internal forces in the structure (axial forces, bending moments, etc.) may be found from equilibrium conditions, assuming the reactions and external forces to be known. The total degree of indeterminateness of a structure is the sum of the degrees of external and internal indeterminateness.

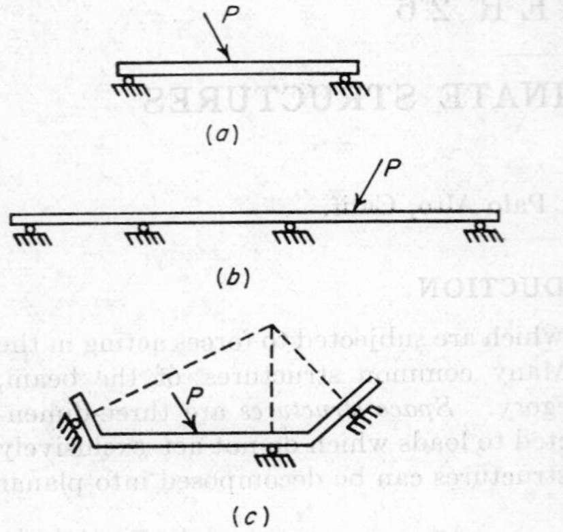

FIG. 26.2

Conditions for Mobility of Planar Structures

The reactions of a structure must be adequate in number and in geometrical arrangement to ensure that the structure will not move when loads are applied. Let r represent the number of external-reaction components and e the total number of equations of static equilibrium available for finding reactions. If $r < e$, the structure will be mobile. This is illustrated in Fig. 26.2*a*, which shows a beam on two roller supports ($r = 2$, $e = 3$). If $r = e$ or if $r > e$, the structure is immovable provided the reactions are neither parallel nor concurrent. If the reactions are parallel (Fig. 26.2*b*) the structure is mobile even though $r > e$. The structure shown in Fig. 26.2*c* has three reaction components, but, since these reactions are concurrent, the structure will move under the action of the load P.

Principle of Superposition

This principle is valid in structural analysis whenever the quantity to be determined is a linear function of the applied loads. Under such conditions, the desired quantity may be found due to each load acting separately, and the results superposed to obtain the total value due to all loads. Superposition will not be valid if the geometry of the structure changes appreciably during loading, so that the lines of action of the forces and the dimensions of the structure are altered from the original values. In addition, Hooke's law must hold for the material of the structure if the principle of superposition is to be used in determining stresses and deflections. The requirement of Hooke's law is not necessary, however, for quantities found by statically determinate stress analysis, such as reactions and bending moments in a determinate structure.

26.2. REACTIONS

The reactions of a statically determinate structure are found either by solving equations of static equilibrium or by graphical constructions. It is necessary to formulate as many independent equations of statics as there are unknown reaction components. These equations may be written for the structure in its entirety or for parts of the structure, whichever may be convenient. After the reactions are determined by solving these equations, additional equations of equilibrium may be used as a check.

A sign convention for the directions of forces and moments must be followed consistently in formulating the equations of static equilibrium. In this article, positive signs will be given to clockwise moments and to forces acting in the positive directions of the axes. When summing forces in the horizontal and vertical directions, forces to the right and upward, respectively, will be taken positive.

As an example, consider the problem of determining the reactions for the beam in Fig. 26.3*a*. The beam has a pin at *B*, and carries a uniformly varying load of maximum intensity q_0 and a concentrated load *P*. The reactions are assumed to act in the directions shown in Fig. 26.3*a*, and free-body diagrams of *BC* and *AB* are given in Fig. 26.3*b* and *c*. In Fig. 26.3*b*, equilibrium of moments about point *B* gives

$$R_c = +P/2$$

and about point *C* gives $V_b = +P/2$. The equation $\Sigma H = 0$ requires that $H_b = 0$. A similar analysis of *AB* (Fig. 26.3*c*) gives

$$M_a = -\frac{q_0 l^2}{24} - \frac{Pl}{4}$$

$$R_{av} = \frac{q_0 l}{4} + \frac{P}{2}$$

$$R_{ah} = 0$$

The negative sign for M_a means that the moment acts in the opposite direction to that assumed in Fig. 26.3. Finally, the equilibrium of the entire structure *ABC* may be used as a check on the results.

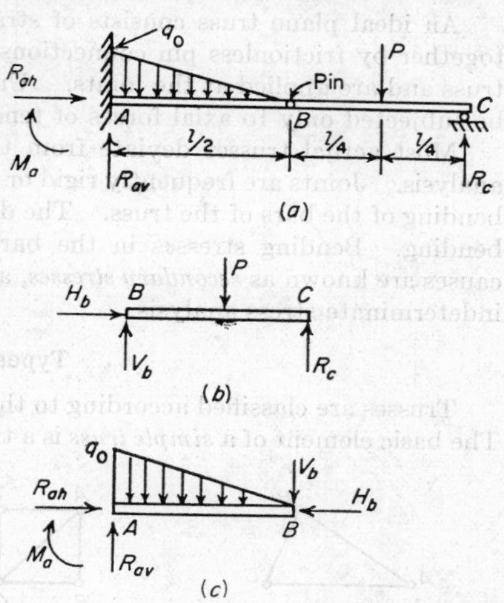

FIG. 26.3

Graphical Determination of Reactions

If the forces acting on a statically determinate structure are not all parallel, the reactions may be found graphically by the *three-force method,* as illustrated in Fig. 26.4. The force *P* represents the resultant of all the applied loads acting on the beam *AB*, and intersects the reaction R_b at point *O*. Since three nonparallel forces must be

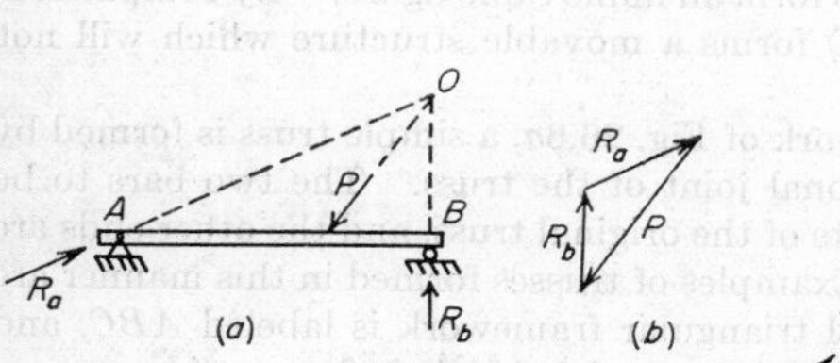

FIG. 26.4

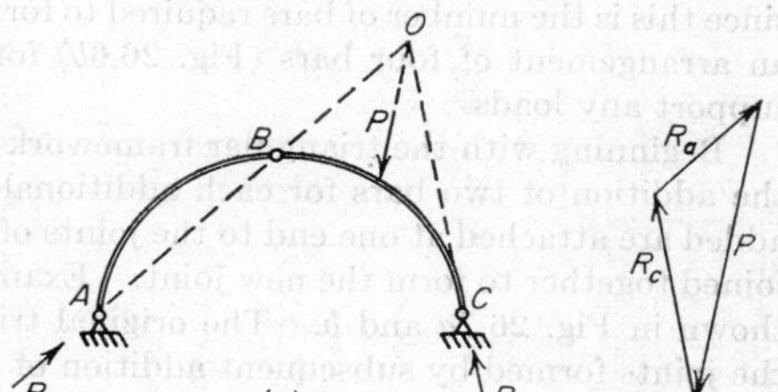

FIG. 26.5

concurrent if they are in equilibrium, the line of action of R_a must also pass through *O*, and thus the force polygon (Fig. 26.4*b*) can be constructed, and the magnitudes of the forces determined.

The reactions of a three-hinged arch may be found in a similar manner (Fig. 26.5). The force *P* represents the resultant of all forces acting on *BC*, and the force polygon (Fig. 26.5*b*) yields the reactions due to these forces. The reactions due to loads acting on *AB* are found in the same way, and then are superposed on the reactions due to *P*, to give the total reactions at *A* and *C*.

If the forces acting on the structure are all parallel, it is necessary to construct the funicular polygon as well as the force polygon (see, for example, [1], p. 17).

26.3. PLANE TRUSSES

An ideal plane truss consists of straight bars lying in a single plane and joined together by frictionless pin connections. All external forces act in the plane of the truss and are applied at the joints. Under these conditions each bar of the truss will be subjected only to axial forces of tension or compression, known as *primary forces.*

Most actual trusses deviate from the conditions assumed in the primary stress analysis. Joints are frequently rigid or semirigid instead of pinned, and this results in bending of the bars of the truss. The dead weight of the members also contributes to bending. Bending stresses in the bars of a truss resulting from these and other causes are known as *secondary stresses,* and may be found by the methods of statically indeterminate stress analysis.

Types of Trusses

Trusses are classified according to the arrangement of the bars forming the trusses. The basic element of a *simple truss* is a triangular framework of three bars (Fig. 26.6*a*),

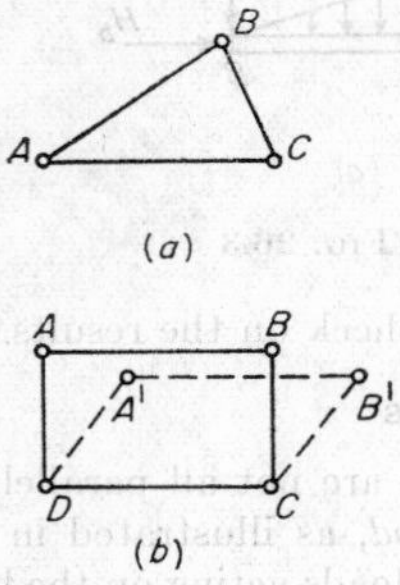

FIG. 26.6

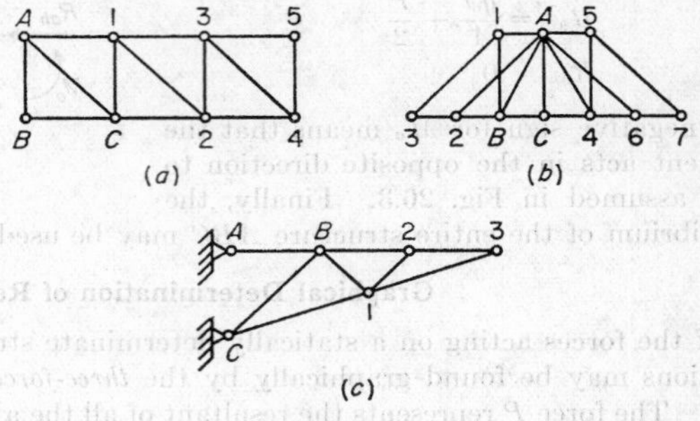

FIG. 26.7

since this is the number of bars required to form an immovable figure. By comparison, an arrangement of four bars (Fig. 26.6*b*) forms a movable structure which will not support any loads.

Beginning with the triangular framework of Fig. 26.6*a*, a simple truss is formed by the addition of two bars for each additional joint of the truss. The two bars to be added are attached at one end to the joints of the original truss, and the other ends are joined together to form the new joint. Examples of trusses formed in this manner are shown in Fig. 26.7*a* and *b*. The original triangular framework is labeled *ABC,* and the joints formed by subsequent addition of bars are labeled 1, 2, 3, . . . in consecutive order.

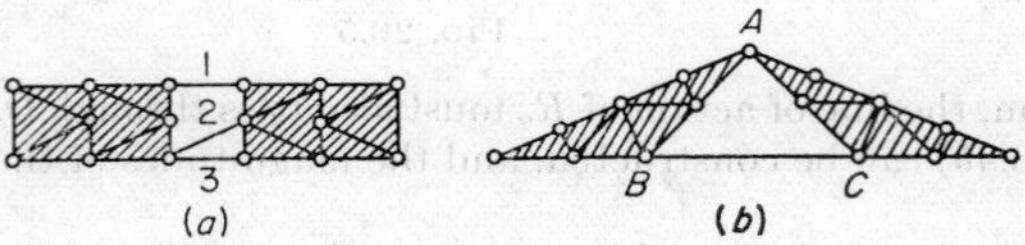

FIG. 26.8

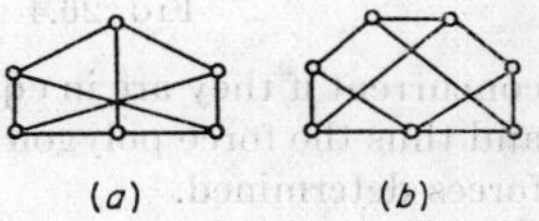

FIG. 26.9

Simple trusses may also be formed by beginning with two bars pinned to a support. An example is shown in Fig. 26.7*c*, where *ABC* represents the original triangle consisting of two bars plus the foundation, and subsequent joints are numbered in order. Trusses which form a rigid and self-contained structure (see Fig. 26.7*a* and *b*) without being attached to a foundation are called *free trusses.*

Two or more simple trusses may be joined together to form a *compound truss*, as illustrated in Fig. 26.8, where the simple trusses are shown shaded.

Trusses which cannot be classified as either simple or compound, and yet are statically determinate, are called *complex trusses*. Examples are shown in Fig. 26.9. These trusses are not formed using the triangle as the basic element, but the bars are so arranged that each truss forms a rigid structure.

Conditions for Mobility and Statical Determinateness

A free truss is statically determinate with regard to the internal structure of the truss if

$$b = 2j - 3$$

where b is the number of bars of the truss, and j the number of joints. If $b > 2j - 3$, the truss is internally indeterminate, and, if $b < 2j - 3$, the truss is mobile, and does not form a rigid structure. The equation represents a necessary condition, but is not sufficient, since a truss may satisfy the equation, and yet be mobile and indeterminate at the same time. A free simple truss will be immobile if it can be constructed by beginning with a triangle and adding two bars for each new joint, as previously described (Fig. 26.7*a*). A free compound truss will be immobile if it is constructed by joining simple trusses in the manner illustrated in Fig. 26.8. For a free complex truss, no simple rules for checking mobility can be given. A stress analysis of the truss can always be attempted, and, if the truss is mobile, the results will be inconsistent, indeterminate, or infinite.

If a simple truss is formed by beginning with two bars pinned to a foundation (Fig. 26.7*c*), the equation for statical determinateness is

$$b = 2j$$

provided the points of attachment to the foundation are not counted as joints. The truss is immovable if it is formed by the addition of two bars for each new joint, as previously described.

The external conditions for mobility and determinateness are given in Sec. 26.1. Thus, if a free truss which is determinate and immobile is supported by three reaction components which are neither parallel nor concurrent, the entire structure will be determinate and immovable. A free mobile truss must be supported by more than three bars, one more for each degree of mobility, in order to have an immovable structure.

If a truss is to be statically determinate in its entirety, that is, both externally and internally, a necessary condition is that

$$r + b = 2j$$

In this equation, r represents the total number of reaction components, b the number of bars, and j the total number of joints, including points of attachment to the supports. The equation is not a sufficient condition since the truss must also be immovable. If $r + b$ is less than $2j$, the truss is mobile, and, if greater, the truss is indeterminate.

Examples illustrating the preceding rules are given in Fig. 26.10. The simple truss in Fig. 26.10*a* satisfies the equation $r + b = 2j$, since $r = 3$, $b = 13$, $j = 8$, and is, therefore, determinate provided it is also immovable. The three reaction components are neither parallel nor concurrent, and the truss can be constructed by beginning with a triangle and adding two bars for each new joint. Therefore, the truss is immobile and determinate under any system of applied loads. The truss in Fig. 26.10*b* also satisfies the equation $r + b = 2j$, but the arrangement of bars is unsatisfactory, and the truss will collapse under general loading conditions.

The compound truss in Fig. 26.10*c* satisfies the relation $r + b = 2j$, and is constructed by joining two simple trusses by three bars (1, 2, and 3) which are neither parallel nor concurrent. Hence, the truss is statically determinate and immovable. The truss in Fig. 26.10*d* also satisfies the equation, but is movable because of the arrangement of bars. This truss can be considered as a compound truss formed by joining two simple trusses by three members. These members are the bars 1 and 2 and the foundation, which may be considered as a third member connecting points *A* and *B*. Since these three members are parallel, the truss is movable. The truss in Fig. 26.10*e* is determinate and immovable, since the bars 1 and 2 and the foundation are equivalent to three nonparallel, nonconcurrent bars.

In Fig. 26.10*f* is shown a truss for which $r = 4$, $b = 10$, $j = 6$, and therefore the truss is statically indeterminate to the second degree. The conditions of support and the arrangement of bars are such that the truss is immobile.

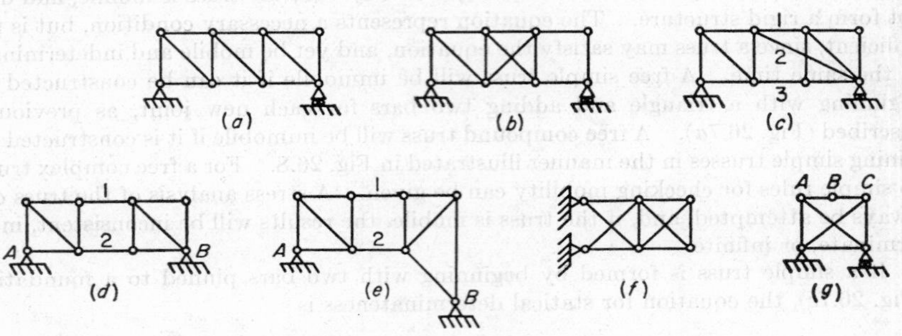

FIG. 26.10

The truss in Fig. 26.10*g* satisfies the equation $r + b = 2j$, but, since the two bars *AB* and *BC* are collinear, the truss is movable. This truss is said to be of *critical form*, since it is movable in spite of the fact that it is formed by beginning with a triangle and adding two bars for each new joint. A critical form will always be evident when a stress analysis is attempted since indeterminate or infinite results are obtained.

Analysis of Trusses

The forces in the bars of a statically determinate truss may be found either analytically (using equations of static equilibrium) or graphically. In using the analytic approach, the first step is to determine the external reactions for the truss. Then a portion of the truss is isolated as a free body, and equations of equilibrium are used to determine the unknown bar forces. If the individual joints of the truss are isolated as free bodies, the analysis is sometimes called the *method of joints*. Likewise, if a section is cut through the truss, and a portion of the entire truss isolated, it is called the *method of sections*. Usually, both methods are used in making a complete analysis of a truss. The choice of method is based solely on convenience and efficiency in carrying out the computations. Because of the great variety of arrangements of simple and compound trusses, no general rules can be given for the most efficient method of solution. For complex trusses, the method of joints and sections usually requires the solution of numerous simultaneous equations. This difficulty can be avoided by Henneberg's method (see [1], p. 82).

Tensile forces in the members of a truss are usually considered positive, and compressive forces negative.

As an example, consider the simply supported truss shown in Fig. 26.11*a*. The forces F_1 and F_2 in bars 1 and 2 can be found by using the method of joints, and solving the two equations of equilibrium of joint A (Fig. 26.11b). To find the forces in bars 3, 4, and 5, a section is cut vertically through the truss, intersecting all three bars, and a free-body diagram of the left-hand part of the truss is drawn (Fig. 26.11c). The force F_3 is found from an equation of moments about joint C. For convenience in writing this equation, F_3 is resolved into components at joint D. The force in bar 4 may be found by considering the vertical equilibrium of the left-hand part of the truss, or from an equation of moments about point E. Finally, the force in bar 5 may be

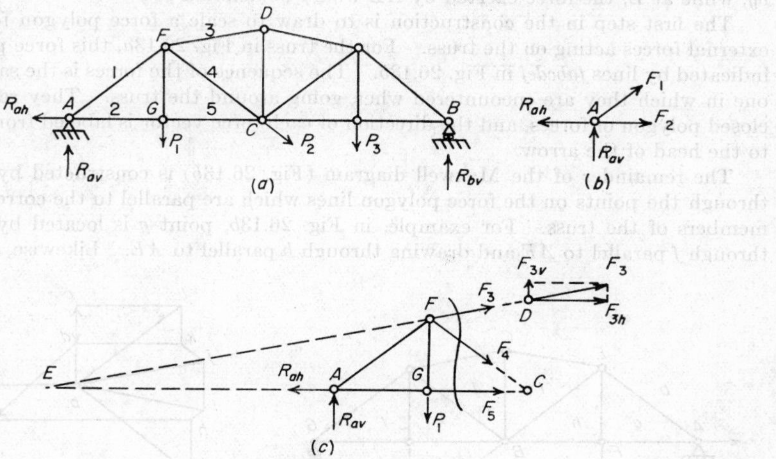

FIG. 26.11

determined by taking moments about joint F (Fig. 26.11c), or by using the method of joints at G.

Three-hinged Trussed Arch

The forces in the bars of a three-hinged trussed arch (see Fig. 26.12) may be found by the same methods as those used for the analysis of simple and compound trusses.

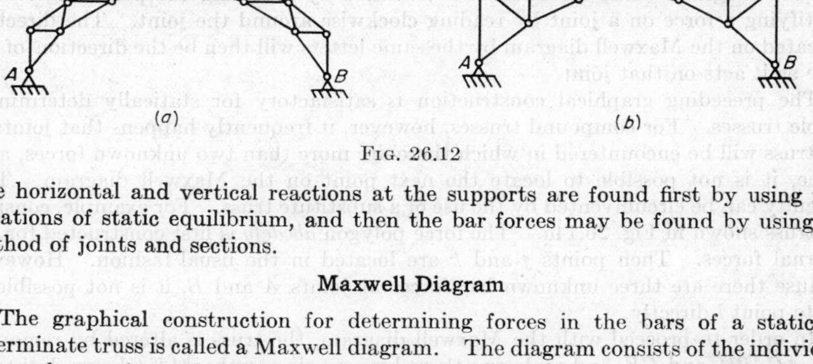

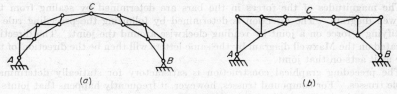

FIG. 26.12

The horizontal and vertical reactions at the supports are found first by using four equations of static equilibrium, and then the bar forces may be found by using the method of joints and sections.

Maxwell Diagram

The graphical construction for determining forces in the bars of a statically determinate truss is called a Maxwell diagram. The diagram consists of the individual force polygons for each joint of the truss superimposed to form one complete figure.

In order to systemize the identification of forces, *Bow's notation* will be used. In this notation, the spaces between the lines of action of the forces are designated by letters (or numbers, if preferred). Then each force is denoted by the letters of the spaces on either side of the force. For example, in the truss shown in Fig. 26.13*a*, the spaces between the external forces are labeled *a, b, c, d, e, f* and the spaces between the internal bar forces are labeled *g, h, i, j, k, l*. The forces acting at any joint of the truss are denoted by naming the spaces around the joint in the clockwise direction. Thus the horizontal reactive force at *A* is denoted *ab*, the vertical reaction at *A* is *fa*, and the inclined load at *B* is *de*. This same scheme is followed for the internal bar forces. For example, the force exerted on joint *A* by the force in bar *AE* is designated *bg*, while at *E*, the force exerted by *AE* would be denoted *gb*.

The first step in the construction is to draw to scale a force polygon for all the external forces acting on the truss. For the truss in Fig. 26.13*a*, this force polygon is indicated by lines *fabcdef* in Fig. 26.13*b*. The sequence of the forces is the same as the one in which they are encountered when going around the truss. They represent a closed polygon of forces, and the direction of each force vector is labeled from the tail to the head of the arrow.

The remainder of the Maxwell diagram (Fig. 26.13*b*) is constructed by drawing through the points on the force polygon lines which are parallel to the corresponding members of the truss. For example, in Fig. 26.13*b*, point *g* is located by drawing through *f* parallel to *AF* and drawing through *b* parallel to *AE*. Likewise, point *h* is

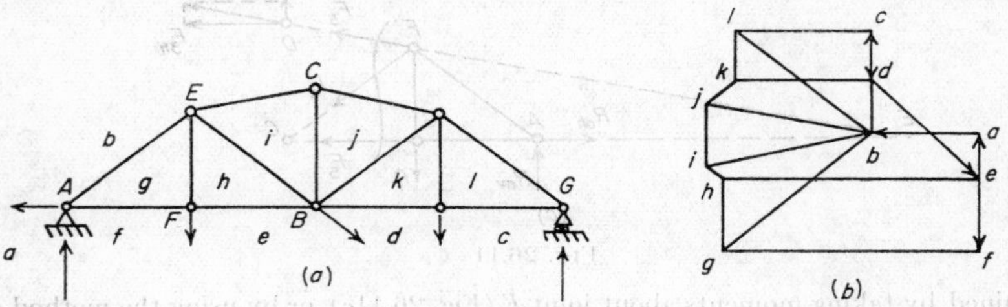

FIG. 26.13

located by drawing parallel to *FB* and *EF* through points *e* and *g*, respectively. This process is continued until the last point *l* is located. A closure is obtained at the end of the construction, since *l* may be located by the intersection of three lines.

The magnitudes of the forces in the bars are determined by scaling from the Maxwell diagram. The direction is determined by following the preceding rule of identifying a force on a joint by reading clockwise around the joint. The direction indicated on the Maxwell diagram by the same letters will then be the direction of the force as it acts on that joint.

The preceding graphical construction is satisfactory for statically determinate simple trusses. For compound trusses, however, it frequently happens that joints of the truss will be encountered in which there are more than two unknown forces, and, hence, it is not possible to locate the next point on the Maxwell diagram. This difficulty can be circumvented by the use of a substitute truss. For example, consider the truss shown in Fig. 26.14*a*. The force polygon *abcdefg* is first constructed for the external forces. Then points *j* and *k* are located in the usual fashion. However, because there are three unknown bar forces at joints *A* and *B*, it is not possible to locate point *l* directly.

In order to proceed with the Maxwell diagram, the truss is altered by removing bars *AC, BC,* and *CE*, and replacing them by a single member *AE* (shown dashed in Fig. 26.14*a*). The space within triangle *ABE* of this new or substitute truss will be denoted *x*, and the space *ADE* is *y*. Points *x* and *y* can readily be located on the Maxwell diagram and the construction continued to point *o*. Point *o* is observed to be a true point; that is, its location on the diagram for the substitute truss is the same as for the original truss. This can be seen since the force in member DF is not affected by the arrangement of bars in panel $ABED$ of the truss.

Since point o is a true point, the diagram for the original truss can now be completed by working back from point o to points n, m, and l. It should be noticed that points y and n coincide, inasmuch as it is also true that the force in AD is not affected by the substitution of bar AE for bars AC, BC, and CE. The remainder of the diagram

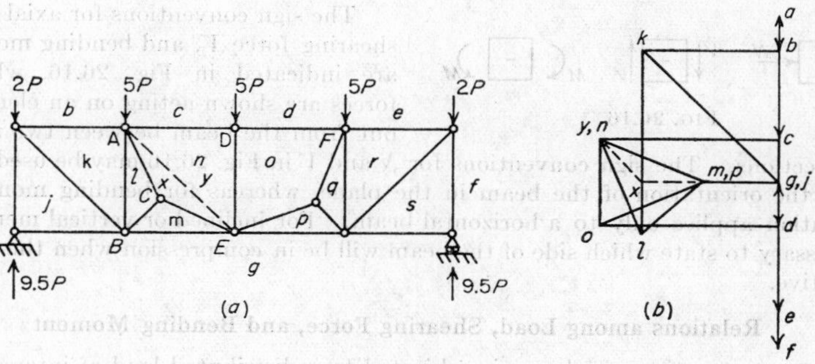

FIG. 26.14

may now be constructed without further difficulty. Because of the symmetry of the truss, it is not necessary to construct the complete diagram, however. Graphical checks on the construction are obtained by noting that points n and o are symmetrically located on the Maxwell diagram, and that points m and p coincide.

26.4. BEAMS

Types of Beams

The primary function of a beam is to support lateral loads, although axial loads may also be present. Several common types of beams, classified according to the

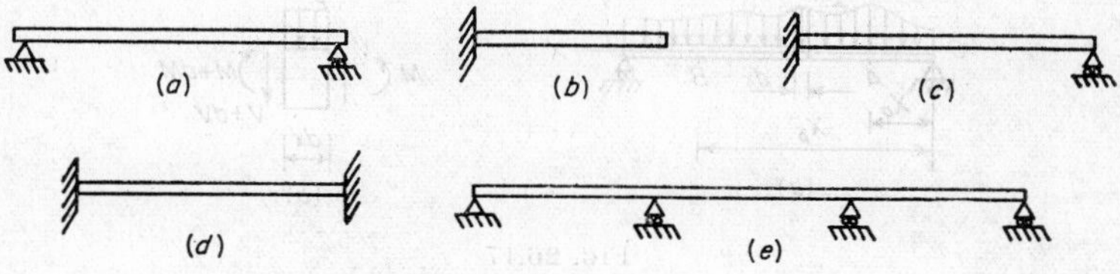

FIG. 26.15

conditions of support, are shown in Fig. 26.15. The *simple beam* in Fig. 26.15*a* and the *cantilever beam* in Fig. 26.15*b* are statically determinate since each has three reaction components. The *propped cantilever*, or fixed-simple beam, of Fig. 26.15*c* is indeterminate to the first degree, while the *fixed-end beam* (or clamped beam) of Fig. 26.15*d* is indeterminate to the third degree. A *continuous beam* on four supports is shown in Fig. 26.15*e* and is indeterminate to the second degree.

Axial Force, Shearing Force, and Bending Moment

If a beam is cut transversely at a point on its axis, stresses will be exerted by the two parts of the beam on each other. The resultant of these stresses may be defined by two force components and a moment, called the axial force, shearing force, and bending moment, respectively. The axial force is defined as the resultant force component acting normal to the plane of the cross section, and the shearing force is the resultant in the transverse direction (normal to the axial force). The resultant moment about the centroid of the cross section is called the bending moment. In a statically determinate beam, all three of these *stress resultants* may be found from equations of equilibrium of the part of the beam to one side or the other of the cross section.

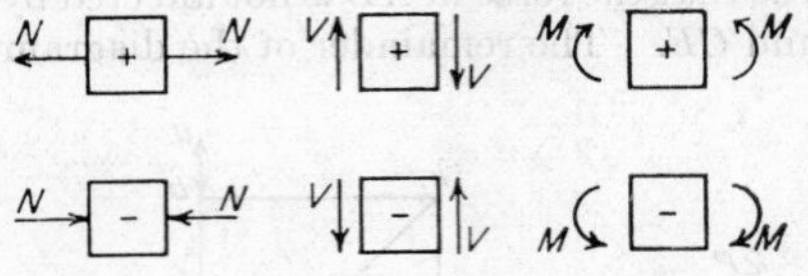

FIG. 26.16

The sign conventions for axial force N, shearing force V, and bending moment M are indicated in Fig. 26.16, where the forces are shown acting on an element cut out from the beam between two adjacent cross sections. The sign conventions for N and V in Fig. 26.16 may be used regardless of the orientation of the beam in the plane, whereas for bending moment the convention applies only to a horizontal beam. For inclined or vertical members, it is necessary to state which side of the beam will be in compression when the moment is positive.

Relations among Load, Shearing Force, and Bending Moment

When any portion of a beam is subjected to a distributed load of intensity $q(x)$, assumed positive downward (Fig. 26.17), the following relations hold

$$V = \frac{dM}{dx} \tag{26.1}$$

$$q = -\frac{dV}{dx} = -\frac{d^2M}{dx^2} \tag{26.2}$$

The first equation expresses the moment equilibrium of an element of the beam (Fig. 26.17*b*), and the second is obtained from its vertical equilibrium. From Eq. (26.1) it is concluded that the bending moment will be maximum or minimum at sections of the beam where the shearing force vanishes.

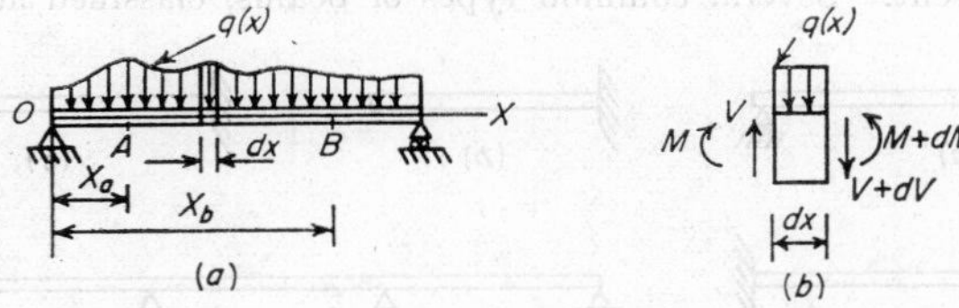

FIG. 26.17

The difference in shearing forces between any two cross sections A and B of the beam is

$$V_b - V_a = -\int_{x_a}^{x_b} q\,dx - \sum_a^b P_i \tag{26.3}$$

where the integral on the right-hand side represents the effect of distributed loads, and ΣP_i represents the sum of all vertical concentrated loads between A and B. The difference in bending moments between A and B is

$$M_b - M_a = \int_{x_a}^{x_b} V\,dx + \sum_a^b M_i \tag{26.4}$$

where ΣM_i represents the algebraic sum of all applied external couples between A and B, clockwise couples being considered as positive.

Axial-force, Shearing-force, and Bending-moment Diagrams

Diagrams or graphs showing the variation of axial force, shearing force, and bending moment with distance along the axis may be constructed for any beam. The general features of the diagrams can be obtained from Eqs. (26.1) and (26.2), and the ordinates to the diagrams can be calculated from equations of static equilibrium and from Eqs. (26.3) and (26.4).

As an example, consider the beam shown in Fig. 26.18. Evidently, at every section the axial force N is positive, and equals the load H applied at the right end. The corresponding diagram is shown directly under the beam. In the span AB, the shearing force is found to be $V = -R_a$, while in the cantilever span it begins at a value $V = R_b - R_a$ and decreases linearly toward $V = 0$ at the free end.

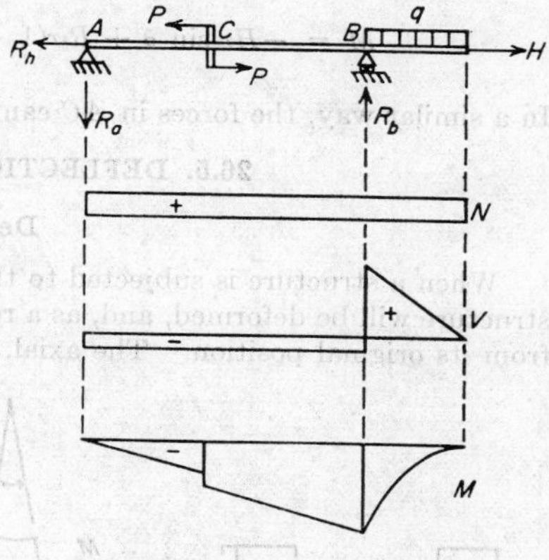

FIG. 26.18

The bending-moment diagram is constructed following a similar procedure. The bending moment is negative everywhere, and its absolute value first increases proportional to the distance from A. The external couple acting at C causes a step in the M diagram. At each section, the slope of the moment diagram is equal to the shearing force, and the maximum of $|M|$ occurs at a section where V changes sign.

Three-hinged Arch

The three-hinged arch is a statically determinate structure and the axial force, shearing force, and bending moment at any section of the arch rib may be found in the same manner as for beams. A free-body diagram is constructed for the portion of the

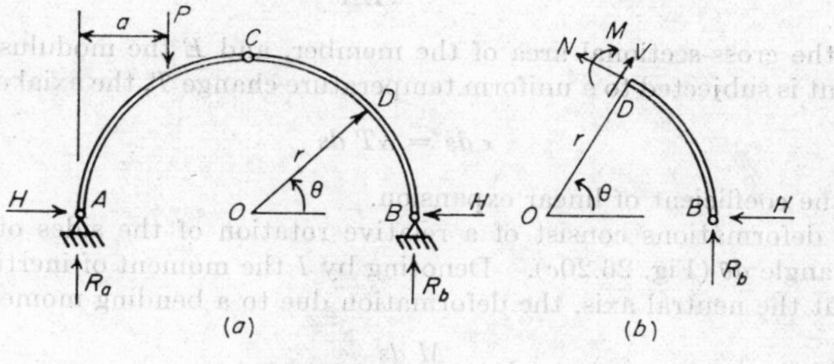

FIG. 26.19

arch to one side or the other of the section under consideration, and then equations of static equilibrium are solved for the unknown forces and moments.

A circular arch of radius r, supporting a vertical load P, is shown in Fig. 26.19a. The reactions may be found from equations of static equilibrium as

$$R_b = \frac{Pa}{2r} \qquad R_a = P\left(1 - \frac{a}{2r}\right) \qquad H = \frac{Pa}{2r}$$

For the right-hand portion BC, the forces at any cross section D are determined with the aid of the free-body diagram in Fig. 26.19*b*. The three equations of static equilibrium yield

$$N = -H \sin\theta - R_b \cos\theta = -\frac{Pa}{2r}(\sin\theta + \cos\theta)$$

$$V = H\cos\theta - R_b \sin\theta = \frac{Pa}{2r}(\cos\theta - \sin\theta)$$

$$M = -Hr\sin\theta + R_b r(1 - \cos\theta) = \frac{Pa}{2}(1 - \cos\theta - \sin\theta)$$

In a similar way, the forces in AC can be determined.

26.5. DEFLECTIONS OF STRUCTURES

Deformations

When a structure is subjected to the action of applied loads, each element of the structure will be deformed, and, as a result, the axis of the structure will be deflected from its original position. The axial, bending, shearing, and torsional deformations

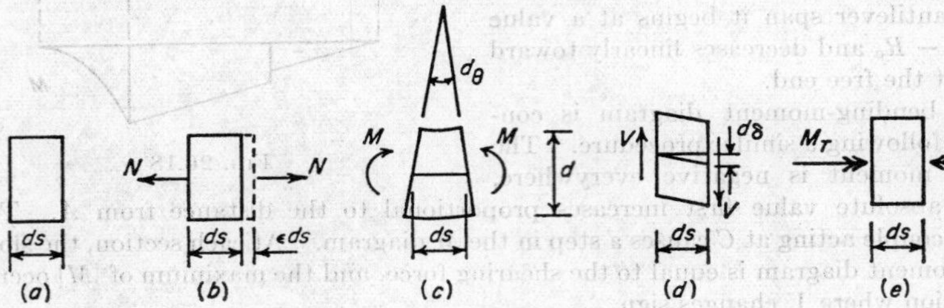

FIG. 26.20

of an element of length ds are shown in Fig. 26.20. If the element is subjected to a centrally applied axial force N (Fig. 26.20*b*) and if Hooke's law holds, the axial deformation will be

$$\epsilon\, ds = \frac{N\, ds}{EA} \tag{26.5}$$

where A is the cross-sectional area of the member, and E the modulus of elasticity. If the element is subjected to a uniform temperature change T, the axial deformation is

$$\epsilon\, ds = \alpha T\, ds \tag{26.6}$$

where α is the coefficient of linear expansion.

Bending deformations consist of a relative rotation of the sides of the element through an angle $d\theta$ (Fig. 26.20*c*). Denoting by I the moment of inertia of the cross section about the neutral axis, the deformation due to a bending moment M is

$$d\theta = \frac{M\, ds}{EI} = \kappa\, ds \tag{26 7}$$

where $\kappa = M/EI$ represents the elastic curvature of the axis of the member (see p. 35–7). If the element is subjected to a temperature change which varies linearly from T_1 at the top to T_2 at the bottom, the angle $d\theta$ becomes

$$d\theta = \frac{\alpha(T_2 - T_1)\, ds}{d} \tag{26.8}$$

where d is the depth of the member.

The deformation due to a shearing force V consists of a displacement $d\delta$ of one side of the element with respect to the other (Fig. 26.20*d*). It may be written as

$$d\delta = \frac{V\,ds}{GA_s} \tag{26.9}$$

where G is the shear modulus, and $A_s \neq A$ because the shearing stresses are not uniformly distributed over the cross section. For I beams including the wide-flange profiles A_s may be taken as the web area. For rectangular sections $A/A_s = 1.2$; for a circle $A/A_s = 1.11$.

Torsional deformations of the element are given by the angle of twist $d\phi$, which represents the difference in the angles of rotation about a longitudinal axis of the two sides of the element. If twisting couples M_T act on the element (Fig. 26.20*e*), the angle of twist is

$$d\phi = \frac{M_T\,ds}{GJ} = \theta\,ds \tag{26.10}$$

where J is the torsion constant, and $\theta = M_T/GJ$ is the angle of twist per unit length. For a circular cross section, the torsion constant J becomes the polar moment of inertia (see p. 36–2).

Dummy-load Method

The dummy-load method is extremely general in application, and may be used to determine deflections caused by axial, bending, shearing, and torsional deformations, as well as deflections caused by temperature changes, misfit of parts, etc. The equations of the dummy-load method are usually derived from the principle of virtual work, and, hence, it is sometimes called the *method of virtual work*. It is also known as the *Maxwell-Mohr method.*

Two systems of loading must be considered in the dummy-load method. The first system consists of the structure in its actual condition, that is, subjected to the actual loads, temperature changes, etc. The second system consists of the same structure subjected to a dummy unit load corresponding to the desired deflection of the actual structure. By a *corresponding load* is meant a load at the particular point of the structure where the deflection is to be determined, and acting in the direction of that deflection. Load and deflection are to be considered in the *generalized* sense. Thus, if the dummy load is a single force, the corresponding deflection is the component of displacement in the direction of the force; if the dummy load consists of two collinear and oppositely directed forces applied at two different points of the structure, the corresponding displacement is the relative displacement of the two points along the line joining them; if the dummy load is a couple, the displacement is an angle of rotation (or slope, since the deflections are small); and, if the dummy load consists of two oppositely directed couples, the corresponding deflection is a relative rotation.

The forces in the structure due to the dummy unit load (second system of loading) constitute a force system in equilibrium. According to the principle of virtual work (see p. 21–24), if this system is given a small virtual displacement, the total work of all the forces will be zero. The forces caused by the unit load in the second system will be represented by the symbols $\bar{N}, \bar{M}, \bar{V}, \bar{M}_T$, denoting the axial force, bending moment, shearing force, and twisting moment, respectively, acting at any cross section of the structure. It should be noted that they have the dimensions of force or moment per unit of the dummy load, and must be measured in the corresponding units. The virtual displacements of the second system are taken the same as the actual displacements of the first system, and the corresponding deformations of any element (see Fig. 26.20) will be denoted as $\epsilon\,ds$, $d\theta$, $d\delta$, and $d\phi$, given by Eqs. (26.5) to (26.10). Taking the work done by the forces in the second system acting through the displacements of the first system, the following equation is obtained from the principle of virtual work:

$$1 \cdot \delta = \int \bar{N}\epsilon \, ds + \int \bar{M} \, d\theta + \int \bar{V} \, d\delta + \int \bar{M}_T \, d\phi \tag{26.11}$$

where the integrations are carried out for the entire structure. The left-hand side of this equation represents the work of the dummy unit load, acting through the corresponding displacement δ in the first system of loading. Thus, Eq. (26.11) gives the deflection δ of the original structure, in terms of forces caused by the dummy load corresponding to δ, and in terms of the actual deformations, which may be due to any cause whatsoever. Usually only one or two of the terms in Eq. (26.11) must be considered, depending on the type of structure. For example, in finding deflections of beams, only the deflection due to bending moment is usually important. Equation (26.11) is valid for a structure of any material, whether Hooke's law holds or not. However, in the equations which follow it is assumed that Hooke's law is valid.

In calculating the deflections of an ideal pin-connected truss, only the first term of Eq. (26.11) must be considered. Furthermore, if the deflection δ is caused only by loads applied at the joints, and if the dummy unit load is also applied at a joint, the axial strain ϵ and the force $\bar{N}$ will be constant throughout the length of each member. In this case, the dummy-load equation becomes

$$\delta = \sum \frac{F\bar{F}l}{EA} \tag{26.12}$$

where the summation is carried out for all bars of the truss, F = bar forces due to actual loads, and $\bar{F}$ = bar forces due to unit load. For simplicity, the factor 1 has been dropped from the left-hand side of this equation.

If the bending, shearing, and torsional deformations in the actual structure are caused by applied loads only, the resulting deflection is given by the formula

$$\delta = \int \frac{M\bar{M}\,ds}{EI} + \int \frac{V\bar{V}\,ds}{GA_s} + \int \frac{M_T\bar{M}_T\,ds}{GJ} \tag{26.13}$$

This equation is obtained by combining Eqs. (26.7), (26.9), and (26.10) with the general formula (26.11). For prismatic members, each of the integrals appearing in Eq. (26.13) is of the form

$$\int_0^l M\bar{M}\,dx$$

Values of this integral are given in Table 26.1 for various functions M and $\bar{M}$.

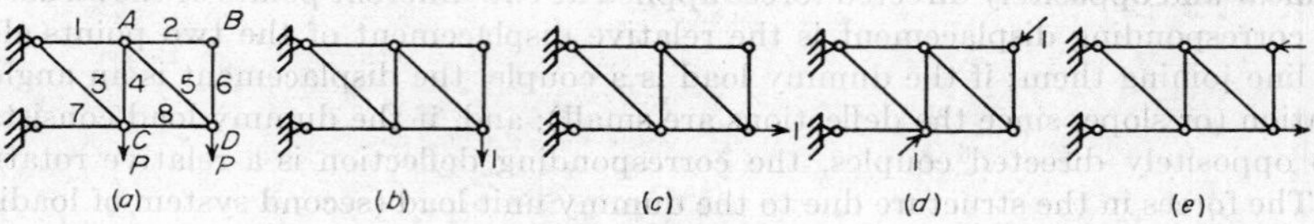

FIG. 26.21

Example 1. Figure 26.21*a* shows a truss carrying two loads P. They cause forces F in the bars. Figure 26.21*b*,*c* shows the dummy loads needed to find the vertical or horizontal displacement of the joint D, caused by the loads P. When the forces caused by either of these dummy loads are introduced as $\bar{F}$ in Eq. (26.12), δ will be the corresponding displacement component.

The dummy-load system shown in Fig. 26.21*d* would be used to find the relative displacement of joints B and C along the line joining them. Likewise, if the angle of rotation of member BD is desired, the dummy load would be taken as the unit couple shown in Fig. 26.21*e*.

Example 2. Assume that the same truss as in Example 1 carries no load, but that the length of the bar AC is $(l + b)$ instead of l, where $b \ll l$. The excess length b may be caused by faulty fabrication or by local heating. It is desired to find the vertical displacement of joint D resulting from this change in length of the bar AC.

Table 26.1. Values of $\int_0^l M\bar{M}\,dx$

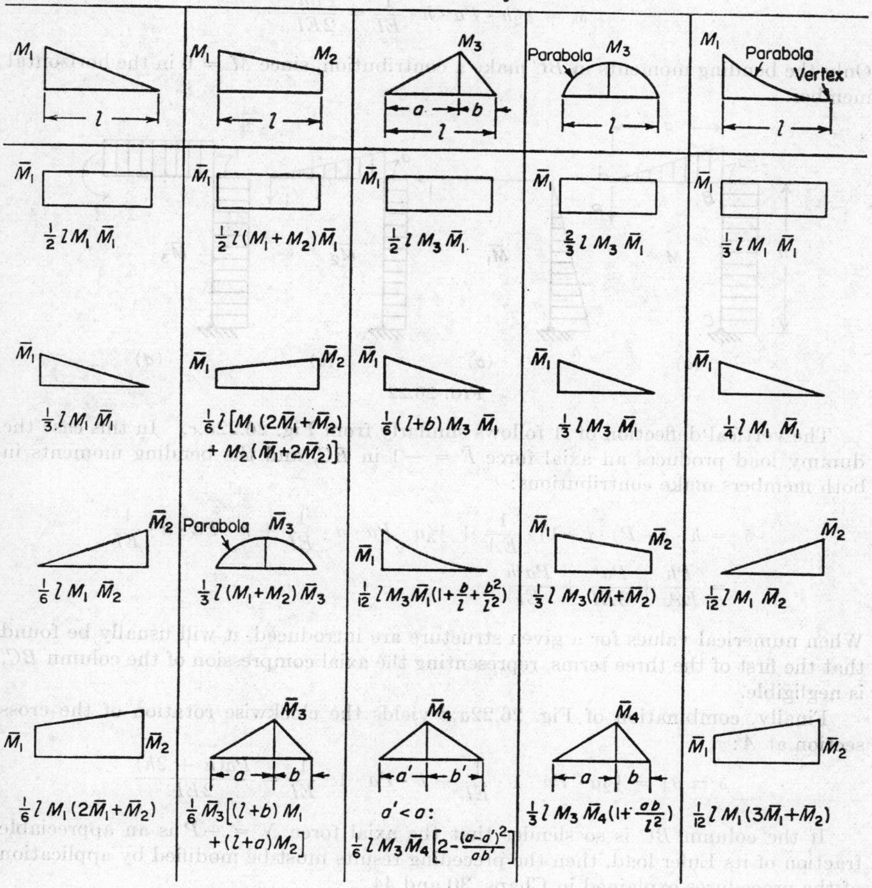

$\frac{1}{2} l M_1 \bar{M}_1$	$\frac{1}{2} l (M_1 + M_2) \bar{M}_1$	$\frac{1}{2} l M_3 \bar{M}_1$	$\frac{2}{3} l M_3 \bar{M}_1$	$\frac{1}{3} l M_1 \bar{M}_1$
$\frac{1}{3} l M_1 \bar{M}_1$	$\frac{1}{6} l [M_1 (2\bar{M}_1 + \bar{M}_2) + M_2 (\bar{M}_1 + 2\bar{M}_2)]$	$\frac{1}{6} (l+b) M_3 \bar{M}_1$	$\frac{1}{3} l M_3 \bar{M}_1$	$\frac{1}{4} l M_1 \bar{M}_1$
$\frac{1}{6} l M_1 \bar{M}_2$	$\frac{1}{3} l (M_1 + M_2) \bar{M}_3$	$\frac{1}{12} l M_3 \bar{M}_1 (1 + \frac{b}{l} + \frac{b^2}{l^2})$	$\frac{1}{3} l M_3 (\bar{M}_1 + \bar{M}_2)$	$\frac{1}{12} l M_1 \bar{M}_2$
$\frac{1}{6} l M_1 (2\bar{M}_1 + \bar{M}_2)$	$\frac{1}{6} \bar{M}_3 [(l+b) M_1 + (l+a) M_2]$	$a' < a$: $\frac{1}{6} l M_3 \bar{M}_4 \left[2 - \frac{(a-a')^2}{ab'}\right]$	$\frac{1}{3} l M_3 \bar{M}_4 (1 + \frac{ab}{l^2})$	$\frac{1}{12} l M_1 (3\bar{M}_1 + \bar{M}_2)$

The dummy load is shown in Fig. 26.21b. For this case, Eq. (26.11) becomes

$$\delta = \int \bar{F} \epsilon \, ds = \Sigma \bar{F}\, \Delta l \tag{26.14}$$

The only bar which has undergone a change in length is the bar AC itself, for which $\Delta l = b$ and $\bar{F} = -1$. Thus, the deflection is

$$\delta_v = \Sigma \bar{F}\, \Delta l = -b$$

and joint D deflects upward by an amount b.

Example 3. Determine the horizontal and vertical deflections and the angle of rotation at point A of the cantilever bracket in Fig. 26.22a. The members AB and BC have flexural rigidity EI and axial rigidity EA.

The bending moments M caused by the load P are shown in Fig. 26.22*a*, plotted on the tension side of the bars. There is an axial force $N = -P$ in the member BC. Figure 26.22*b* through *d* shows the dummy loads needed and the corresponding dummy moments $\bar{M}$. The horizontal deflection of A (positive when directed toward the right) is found from Eq. (26.11), using the first two terms only and applying Table 26.1:

$$\delta_h = \tfrac{1}{2}h \cdot Pa \cdot h \cdot \frac{1}{EI} = \frac{Pah^2}{2EI}$$

Only the bending moments in BC make a contribution, since $\bar{M} = 0$ in the horizontal member.

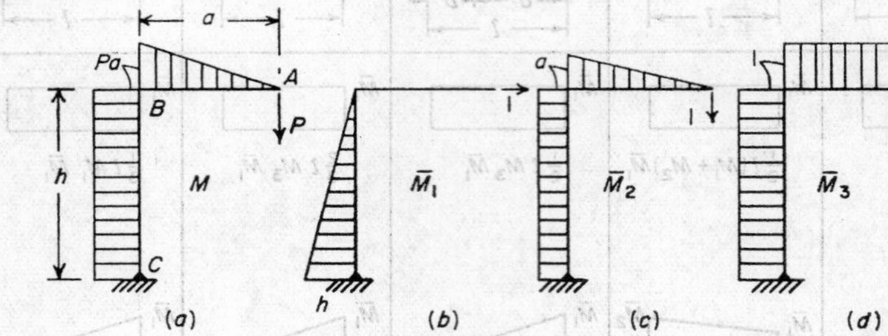

FIG. 26.22

The vertical deflection of A follows similarly from Fig. 26.22*a,c*. In this case the dummy load produces an axial force $\bar{F} = -1$ in BC, and the bending moments in both members make contributions:

$$\delta_v = h \cdot (-P) \cdot (-1) \cdot \frac{1}{EA} + \tfrac{1}{3}a \cdot Pa \cdot a \cdot \frac{1}{EI} + h \cdot Pa \cdot a \cdot \frac{1}{EI}$$
$$= \frac{Ph}{EA} + \frac{Pa^3}{3EI} + \frac{Pa^2h}{EI}$$

When numerical values for a given structure are introduced, it will usually be found that the first of the three terms, representing the axial compression of the column BC, is negligible.

Finally, combination of Fig. 26.22*a,d* yields the clockwise rotation of the cross section at A:

$$\delta = \theta_A = \tfrac{1}{2}a \cdot Pa \cdot 1 \cdot \frac{1}{EI} + h \cdot Pa \cdot 1 \cdot \frac{1}{EI} = \frac{Pa(a + 2h)}{2EI}$$

If the column BC is so slender that the axial force $N = -P$ is an appreciable fraction of its Euler load, then the preceding results must be modified by application of the procedures explained in Chaps. 30 and 44.

Castigliano's Theorem

This theorem is valid for structures composed of materials which are elastic and follow Hooke's law. In addition, any change in geometry of the structure under loading must be so small that the action of the forces is not affected. Under these conditions, the theorem states that the partial derivative of the strain energy U of the structure with respect to any external force P is equal to the displacement δ in the structure corresponding to that force:

$$\delta = \frac{\partial U}{\partial P} \tag{26.15}$$

The terms force and displacement are used here in the generalized sense.

In using Castigliano's theorem, the strain energy must be expressed as a function of statically independent applied forces. Reactions which are determined by statics from the applied loads are not statically independent of those loads, and, hence, cannot appear in the strain-energy expression. Also, if the deflection is desired at a point of the structure where there is no load, it is necessary to place a load at that point, proceed with the calculations in the usual way, and then, finally, set the load equal to zero.

The expression for the strain energy of a bar is

$$U = \int \frac{N^2\,ds}{2EA} + \int \frac{M^2\,ds}{2EI} + \int \frac{V^2\,ds}{2GA_s} + \int \frac{M_T^2\,ds}{2GJ} \tag{26.16}$$

where the terms on the right-hand side represent the strain energy due to axial forces, bending moments, shearing forces, and twisting moments, respectively. The deflection calculations are somewhat simplified if the differentiation [see Eq. (26.15)] is performed before the integration required by Eq. (26.16). If this is done, the equation for the deflection becomes

$$\delta = \frac{\partial U}{\partial P} = \int \frac{N(\partial N/\partial P)\,ds}{EA} + \int \frac{M(\partial M/\partial P)\,ds}{EI} + \int \frac{V(\partial V/\partial P)\,ds}{GA_s} + \int \frac{M_T(\partial M_T/\partial P)\,ds}{GJ} \tag{26.17}$$

The derivatives which appear in the numerators of the four integrands may be interpreted physically as the rate of change of N, M, V, and M_T with respect to P. Thus these derivatives are equal to the values of N, M, V, and M_T as caused by a unit load ($P = 1$), so that

$$\frac{\partial N}{\partial P} = \bar{N} \qquad \frac{\partial M}{\partial P} = \bar{M} \qquad \frac{\partial V}{\partial P} = \bar{V} \qquad \frac{\partial M_T}{\partial P} = \bar{M}_T \tag{26.18}$$

It is seen that the deflection calculations using Castigliano's theorem are the same as those with the dummy-load method.

Complementary Energy

If the load-deflection characteristics of a structure are nonlinear, the displacement δ corresponding to a force P is given by the equation

$$\delta = \frac{\partial U_c}{\partial P} \tag{26.19}$$

where U_c represents the complementary energy of the structure expressed as a function of the applied loads. As before, the deflection δ and the corresponding force P are generalized displacement and force.

The complementary energy is given by the expression

$$U_c = \sum_{i=1}^{n} \int \delta_i\,dP_i \tag{26.20}$$

where n represents the total number of applied forces P. The general expression for the strain energy of the structure is

$$U = \sum_{i=1}^{n} \int P_i\,d\delta_i \tag{26.21}$$

and the strain energy becomes the same as the complementary energy if the structure has a linear load-deflection relationship.

Reciprocal Theorem

The reciprocal theorem applies to structures for which the principle of superposition is valid in determining deflections. If the structure is subjected to two states of loading, then the theorem states that the sum of the products formed by multiplying the forces in the first state by the corresponding displacements of the second state is equal to the sum of the products of the forces in the second state multiplied by the corresponding displacements of the first state. In this form the theorem is also known as Betti's law. Forces and displacements are considered in the generalized sense.

As a simple illustration of the reciprocal theorem, consider the beam of length l shown in Fig. 26.23. In the first state of loading (Fig. 26.23*a*), the beam is subjected to a moment M_1 at the left end and a concentrated force P_1 at the center. The second state of loading is shown in Fig. 26.23*b*, and consists of two equal loads P_2 applied at the quarter points. Denoting the deflections of the first beam at the quarter points by

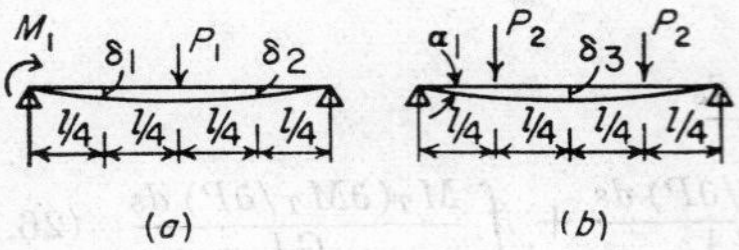

FIG. 26.23

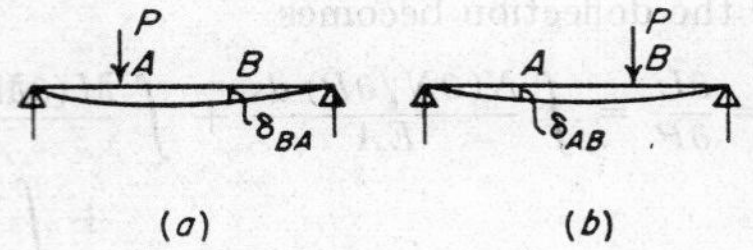

FIG. 26.24

δ_1 and δ_2, the rotation of the left end of the second beam by α_1, and the center deflection of the second beam by δ_3, the following relation is obtained from the reciprocal theorem:

$$M_1\alpha_1 + P_1\delta_3 = P_2\delta_1 + P_2\delta_2$$

The validity of this equation can be checked readily for a beam having known values of the deflections. Note that each term in the equation has units of work. The reciprocal theorem is not limited to beams, as in this example, but is valid also for structures in which the displacements are caused by axial forces, shearing forces, and twisting moments.

Maxwell's theorem is a special form of the reciprocal theorem, applying to a structure carrying a single load. If the load is applied first at point A and second at point B, Maxwell's theorem states that the deflection at B due to the load at A is equal to the deflection at A due to the load at B. Thus, in Fig. 26.24, Maxwell's theorem gives the relation $\delta_{AB} = \delta_{BA}$. If the applied load is a couple, the deflections must be the corresponding angles of rotation of the member.

Beam Deflections by Successive Integration

The differential equation for the deflection curve of a beam is

$$EI\,\frac{d^2w}{dx^2} = -M \tag{26.22}$$

where x denotes the coordinate distance along the axis of the beam and w represents lateral deflections (see Fig. 26.25). The equation gives the deflection due to bending moment only, neglecting the effect of shear, and assumes that Hooke's law is valid and that the deflections are small. From Eqs. (26.1) and (26.2), the following additional equations are obtained:

$$EI\,\frac{d^3w}{dx^3} = -V \tag{26.23}$$

$$EI\,\frac{d^4w}{dx^4} = q \tag{26.24}$$

By successively integrating these equations and evaluating the constants of integration from the known boundary conditions of the beam, deflections and slopes at any point of the beam may be determined. There will always be a sufficient number of boundary conditions to determine the constants of integration. At a free end of a beam, the bending moment and shearing force are zero, and the boundary conditions are

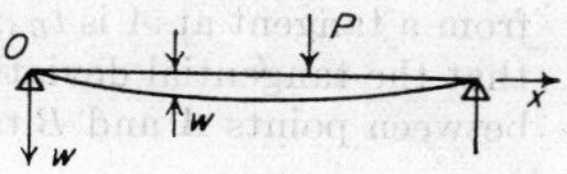

FIG. 26.25

$$\frac{d^2w}{dx^2} = 0 \qquad \frac{d^3w}{dx^3} = 0$$

For a simply supported end, the deflection and the bending moment are zero, so that

$$w = 0 \qquad \frac{d^2w}{dx^2} = 0$$

and, for a fixed or clamped end, the conditions are

$$w = 0 \qquad \frac{dw}{dx} = 0$$

If the integration is being carried out separately for various regions along the axis of the beam, because of sudden changes in the loading, there will be additional conditions expressing the continuity of the deflection curve between the regions.

For statically determinate beams, the expression for the bending moment M can be obtained readily, and, hence, deflections can be found by integrating Eq. (26.22) twice. This method is sometimes called the *method of double integration.* For statically indeterminate beams, it is frequently preferable to begin with one of the later equations, involving V or q, and then carry out the necessary steps of successive integration. Because of the numerous constants to be evaluated, this procedure may be lengthy if the load q is not continuous, or if there are several concentrated loads.

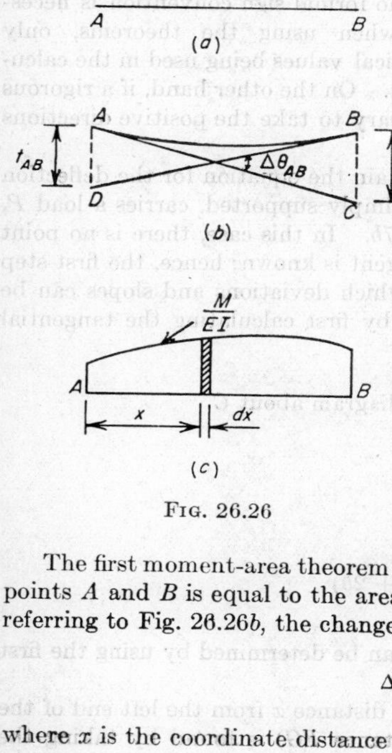

FIG. 26.26

Moment-Area Theorems

These theorems are used to determine deflections of beams and usually offer a more rapid solution than the method of successive integration in cases where the deflection of only one point is desired. Let AB in Fig. 26.26a represent a portion of the axis of an initially straight beam. After the beam is loaded, the portion AB takes a bent shape as shown in Fig. 26.26b, where line AC is a tangent to the elastic curve at A, and line BD is a tangent at B. The bending-moment diagram divided by the flexural rigidity EI for the portion AB of the beam is shown in Fig. 26.26c.

The first moment-area theorem states that the change in slope of the beam between points A and B is equal to the area of the M/EI diagram between A and B. Thus, referring to Fig. 26.26b, the change in slope $\Delta\theta_{AB}$ between A and B is given by

$$\Delta\theta_{AB} = \int_A^B \frac{M\,dx}{EI} \qquad (26.25)$$

where x is the coordinate distance along the axis of the beam.

The second theorem provides a means of determining the deflection or deviation of a point on the axis of the beam from a tangent to some other point. In Fig. 26.26*b*, the distance AD represents the deviation of point A from a tangent drawn at B, and is called the *tangential deviation* t_{AB}. Likewise, the tangential deviation of B from a tangent at A is t_{BA}, or distance BC. The second moment-area theorem states that the tangential deviation t_{AB} is equal to the static moment of the M/EI diagram between points A and B taken about a vertical through point A or

$$t_{AB} = \int_A^B \frac{Mx\,dx}{EI} \tag{26.26}$$

In a similar way, the tangential deviation t_{BA} is equal to the static moment of the M/EI diagram taken about a vertical through point B. Equations (26.25) and (26.26) are derived from Eq. (26.22), and are, therefore, valid only for small deflections of the beam.

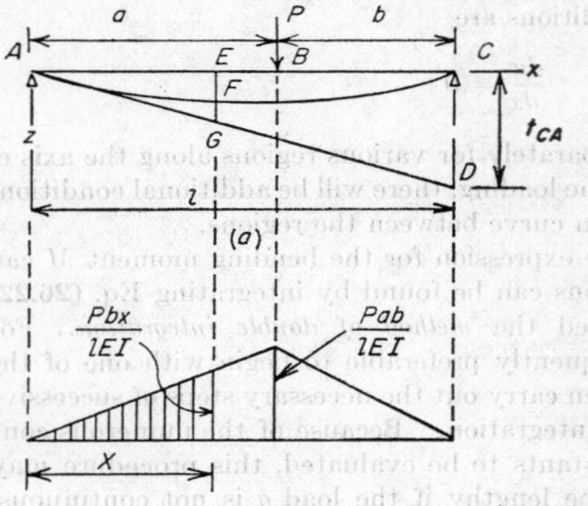

FIG. 26.27

In order to calculate deflections and slopes in a beam, they must be related first to slope changes and to tangential deviations, since these are the quantities found from the theorems. This is accomplished by considering the geometry of the deflected beam, as illustrated in the example which follows. Usually the directions of the slopes and deflections can be determined by inspection, and then no formal sign convention is necessary when using the theorems, only numerical values being used in the calculations. On the other hand, if a rigorous sign convention is to be followed, it is only necessary to take the positive directions of $\Delta\theta$, t, and M/EI, as shown in Fig. 26.26.

As an example, assume that it is desired to obtain the equation for the deflection curve of the beam in Fig. 26.27*a*. The beam is simply supported, carries a load P, and has an M/EI diagram as shown in Fig. 26.27*b*. In this case, there is no point along the beam for which the direction of the tangent is known; hence, the first step is to determine the direction of a tangent from which deviations and slopes can be measured. The tangent AD can be determined by first calculating the tangential deviation t_{CA} as follows:

$$t_{CA} = \text{static moment of } M/EI \text{ diagram about } C$$
$$= \frac{Pab}{6EI}(a + 2b)$$

The slope at the left end of the beam is equal to

$$\theta_A = \frac{t_{CA}}{l} = \frac{Pab}{6lEI}(a + 2b)$$

and from this value the slope at any other point can be determined by using the first moment-area theorem.

In order to obtain the deflection w at point E, distance x from the left end of the beam, the tangential deviation t_{FA} (equal to distance FG) is found by taking the static moment of the shaded part of the M/EI diagram (Fig. 26.27b) about point F:

$$t_{FA} = \overline{FG} = \frac{Pbx^3}{6lEI}$$

The deflection w is then

$$w = \overline{EG} - t_{FA} = t_{CA}\frac{x}{l} - t_{FA} = \frac{Pbx}{6lEI}(l^2 - b^2 - x^2)$$

This equation is valid for $0 \leq x \leq a$. The deflection of the right-hand part of the beam can be found in a similar manner.

Williot-Mohr Diagram

The Williot-Mohr diagram is a graphical method for determining deflections of ideal plane trusses. It is a method which gives the complete deflection of every joint of the truss by means of a single graphical construction, in contrast to the dummy-load method, which gives only one component of deflection of one joint for each calculation. The diagram is constructed in two parts. The Williot diagram is drawn first, and gives a set of relative joint displacements of the truss, based on an assumed direction of one member of the truss. Then the Mohr diagram is drawn, in order to correct the Williot diagram by the addition of a rigid-body rotation of the truss. This rigid-body rotation returns the truss to its true position from the arbitrarily assumed position in the Williot construction. If the final direction of a member of the truss is known in advance, however, then the direction of that particular member may be used as a reference direction in the Williot diagram. The displacements obtained from the Williot diagram will then be true displacements, and no Mohr correction diagram is needed.

The Williot diagram is based upon the following construction. Assume that the displacements of joints A and B of a truss are known, and that it is desired to find the displacement of joint C (Fig. 26.28a). The Williot diagram (Fig. 26.28b) is constructed by drawing only the relative displacements of the joints, to a suitable scale. Starting with point A,B, representing the original positions of joints A and B, the displacements δ_A and δ_B are drawn, in order to locate the displaced positions A' and B'. To determine the final location of joint C, consider that the bars AC and BC are temporarily cut at C. If bar AC elongates by Δ_{AC}, joint C will move away from A along the line AC by that amount. Since the bar may then rotate about A, and since the deflections are small, joint C will move along a perpendicular to AC. In the Williot diagram, Δ_{AC} is drawn parallel to AC, and a perpendicular is constructed. In the same way, joint C may be located by drawing Δ_{BC} parallel to BC and constructing a perpendicular. The final location C' of joint C is at the intersection of the two perpendiculars. If Δ_{BC} represents a shortening of member BC, it would be drawn in the opposite direction from B'. The deflection δ_C of joint C is measured from point A,B, representing the undeflected position, to point C'. For a truss, the deflections of the joints are determined by carrying out this construction for each consecutive joint, as illustrated in the following example.

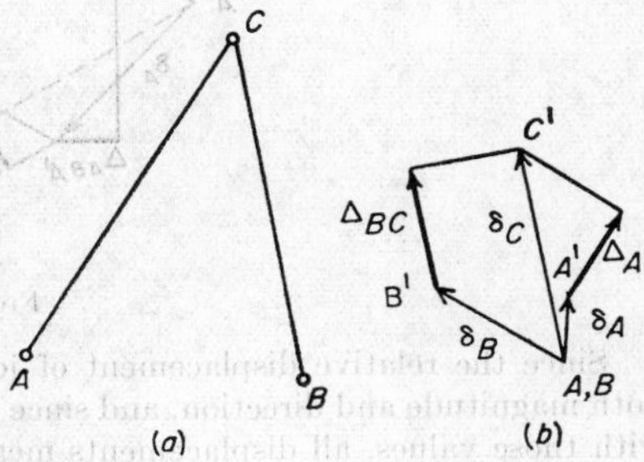

FIG. 26.28

Example 4. Determine the joint displacements for the truss shown in Fig. 26.29a. The changes in length (in some suitable units) of the members are shown on the sketch.

Because of the symmetry of the truss, the member AB will not rotate when the loads are applied, and, hence, its direction can be taken as a reference direction in plotting the Williot diagram. This solution is shown in Fig. 26.29*b*, which begins with point A' as a starting point. Joint B on the truss moves to the left a distance of 40 units with respect to joint A, inasmuch as member AB shortens by that amount. On the Williot diagram, this displacement is shown by a heavy line and labeled Δ_{AB}. With points A' and B' located, point D' can next be determined by following the construction shown in Fig. 26.28. Then C' is located from D' and A', and finally E' from B' and D'.

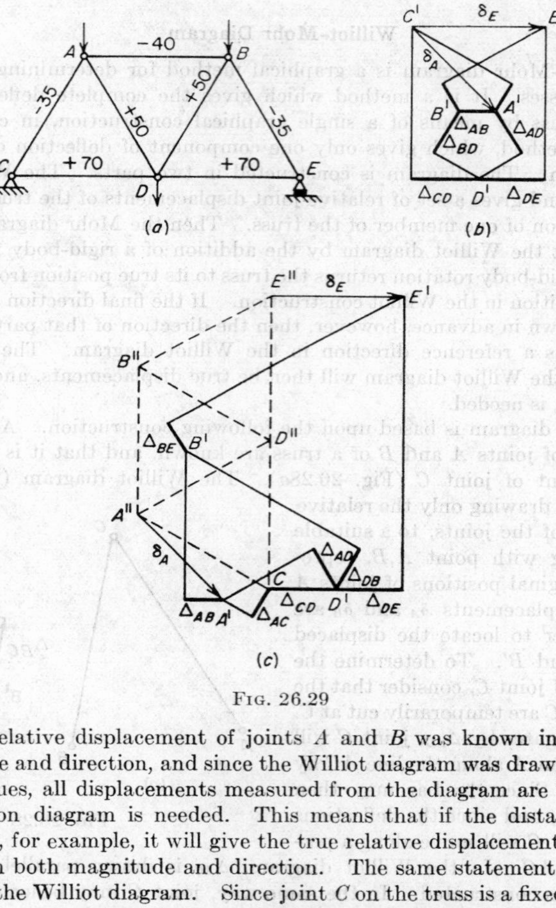

FIG. 26.29

Since the relative displacement of joints A and B was known in advance as to both magnitude and direction, and since the Williot diagram was drawn in accordance with those values, all displacements measured from the diagram are correct, and no Mohr correction diagram is needed. This means that if the distance from A' to D' is measured, for example, it will give the true relative displacement of those joints on the truss, in both magnitude and direction. The same statement is true for any two points on the Williot diagram. Since joint C on the truss is a fixed point in space and does not move, the absolute displacement of any joint can be found by measuring from point C' on the Williot diagram to the corresponding point. For example, to determine the displacement of joint E on the truss, the distance δ_E from point C' to point E' on the Williot diagram is measured. This displacement is seen to be horizontal, to the right, and equal to 140 units. Likewise, the displacement δ_A of joint A is measured from C' to A', and similarly for joints B and D.

In the preceding construction, the direction of AB was known in advance. In an unsymmetrical truss, this would not be the case, and a starting direction must be assumed in order to construct the Williot diagram. This is illustrated in Fig. 26.29*c*, where the Williot diagram for the same truss is drawn, assuming member *CD* fixed in direction. Beginning the construction at point *C*, points D', A', B', and E' are located consecutively. The truss must now be given a rigid-body rotation, to rotate member *CD* from its assumed horizontal direction to its true position. The rotation will take place with *C* as a center, since *C* is a fixed point in space. During this rotation, each joint of the truss will move in a direction perpendicular to a line from joint *C* to that joint, and will move by an amount proportional to the distance of the joint from *C*. This gives rise to the Mohr correction diagram as a figure which is similar to the drawing of the truss itself (Fig. 26.29*a*) but rotated by 90°. The Mohr diagram (Fig. 26.29*c*) is drawn by starting with point *C* as a fixed point and locating point E'' on the Mohr diagram from the known condition that joint *E* of the truss moves only horizontally. Thus point E'' must be located on a horizontal line through point E' on the Williot diagram. With line $E''C$ established, the remainder of the Mohr diagram is drawn readily, by constructing a figure similar to the shape of the truss (dashed in Fig. 26.29*c*).

The final deflection of each joint of the truss is obtained by measuring from the corresponding point on the Mohr diagram to the corresponding point on the Williot diagram. For example, the deflection δ_E of joint *E* is measured from E'' to E'; the deflection δ_A is from A'' to A'; etc.

The validity of the Mohr correction diagram can be seen by considering the displacement of joint *E*. The vector from *C* to E' on the Williot diagram represents the displacement of *E*, assuming *CD* fixed in direction. The vector from E'' to *C* on the Mohr diagram represents the displacement of *E* during a rigid-body rotation of the truss about joint *C*. This latter displacement is perpendicular to a line drawn from joint *C* to joint *E* in the original truss, and is proportional to that distance. The resultant of these two separate displacements is δ_E, the final displacement of joint *E*. Similarly, the displacement of *A* is made up of two parts: the displacement represented by the line from *C* to A', and the displacement during rotation equal to the vector from A'' to *C*. Note that the latter displacement is perpendicular to the line from *C* to *A* in the original truss, and is proportional to that distance. The sum of the two displacements $\overline{CA'}$ and $\overline{A''C}$ is δ_A. Similar reasoning gives the general conclusion that the displacement of any joint is found by measuring from the Mohr diagram to the Williot diagram.

The Williot-Mohr diagram becomes very large and awkward to construct if the initially assumed direction is very inaccurate. Hence, it is advantageous to begin with a bar which rotates only a small amount. For simply supported trusses, this means beginning with a bar near the center of the truss rather than one near the ends.

26.6. SPACE TRUSSES

A space truss is a three-dimensional structure in which the members are assumed to be joined by ideal pin connections and the loads are applied only at the joints. Under these conditions the bars of the truss are subjected to axial forces of tension or compression only. If the truss is statically determinate, it is analyzed by solving equations of static equilibrium, which may be written for the structure as a whole, for a single joint, or for part of the structure. The solution of these equations will give the reactions and the unknown bar forces.

At each point of support of the truss, there may be one or more reaction components, depending on the construction. If the joint consists of a pin connection, there will be reaction components in the *x*, *y*, and *z* directions. On the other hand, the joint may behave in the manner of a roller in two directions, and offer support in the third direction only, thus permitting displacement in the first two directions.

Still another possibility is that the support will behave as a roller in one direction, and offer restraint in the other two directions.

The total number of unknown forces to be determined is equal to the total number of reaction components plus the number of bar forces. At each joint of the truss, both internal joints and reaction points, there are three equations of static equilibrium available for determining forces. Thus, a necessary condition for statical determinateness of the structure in its entirety is that

$$r + b = 3j$$

where r = number of reaction components, b = number of bars, and j = total number of joints. This condition is necessary, but not sufficient, since the bars and reactions of the truss must also be suitably arranged so that the structure is immovable. If $r + b < 3j$, the structure will be movable, and, if $r + b > 3j$, the structure is statically indeterminate.

With regard to external forces only, the necessary condition for statical determinateness is usually that the number of reaction components r must equal six, since

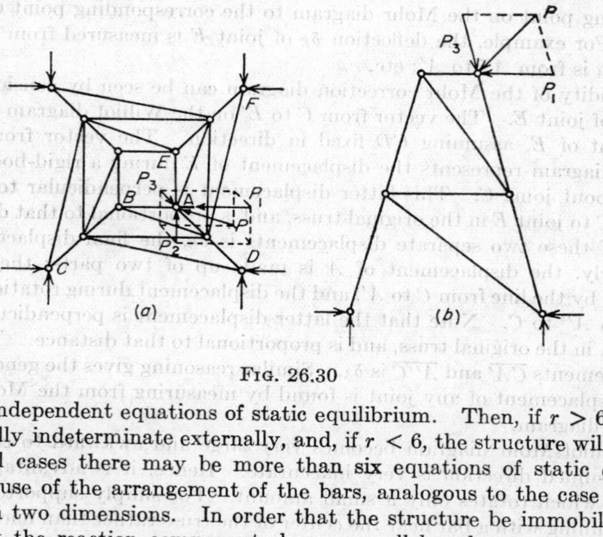

FIG. 26.30

there are six independent equations of static equilibrium. Then, if $r > 6$, the structure is statically indeterminate externally, and, if $r < 6$, the structure will be mobile. In exceptional cases there may be more than six equations of static equilibrium available because of the arrangement of the bars, analogous to the case of a three-hinged arch in two dimensions. In order that the structure be immobile, it is also necessary that the reaction components be nonparallel and nonconcurrent, and so arranged that they are capable of resisting translation of the structure in any direction, and rotation of the structure about any axis.

In some cases, the analysis of a space truss can be resolved into the analysis of component plane trusses, as illustrated by the tower shown in Fig. 26.30. This truss is statically determinate since $r + b = 36 = 3j$, and the arrangement of reactions and bars is sufficient to ensure an immovable structure. Assume that the tower is subjected to a load P at joint A. The force P can be resolved into three components: P_1, P_2, and P_3. The component P_1 is along the line of member AB, and hence lies in the plane of the truss $ABCD$ constituting the front side of the tower. This component causes stresses in truss $ABCD$ only, and the bar forces can be found on the basis of plane truss analysis. The second component, P_2, is along the line of AE, lies in the plane of truss $AEFD$, and causes stresses in that truss only. Finally, the third component, P_3, is along the line of the tower leg AD. This component causes axial forces in the tower leg AD, but causes no stress in any other member. Thus, the analysis of the space truss in Fig. 26.30, for any joint loading, is reduced to the analysis of plane trusses.

For many space structures, the preceding simplification is not possible, and it becomes necessary to use equations of static equilibrium in three dimensions in order to obtain reactions and bar forces. In writing these equations, the components of the forces, rather than the bar forces themselves, will usually appear in the equations. These components can be related to one another and to the total bar force by using the dimensions of the member, since the ratios between the force components in a particular bar will be the same as the ratios between the projected lengths of that bar. Before beginning the force calculations, therefore, it is advantageous to list in a table the total length and the projected lengths on the x, y, and z axes of each member of the truss. Since most space trusses are complex, simultaneous solution of some of the equations of static equilibrium is usually required.

26.7. REFERENCES

[1] S. P. Timoshenko, D. H. Young: "Theory of Structures," McGraw-Hill, New York, 1945.
[2] C. H. Norris, J. B. Wilbur: "Elementary Structural Analysis," 2d ed., McGraw-Hill, New York, 1960.
[3] N. J. Hoff: "Analysis of Structures," Wiley, New York, 1956.
[4] S. P. Timoshenko: "Strength of Materials," 2 vols., Van Nostrand, Princeton, N.J., 1955, 1956.
[5] C. O. Harris: "Introduction to Stress Analysis," Macmillan, New York, 1959.
[6] E. P. Popov: "Mechanics of Materials," Prentice-Hall, Englewood Cliffs, N.J., 1952.

CHAPTER 27

STATICALLY INDETERMINATE STRUCTURES

BY

J. M. GERE, Ph.D., Palo Alto, Calif.

C.-K. WANG, Ph.D., Madison, Wisc.

27.1. INTRODUCTION

Definitions

A structure is statically indeterminate if the total number of forces and moments, both internal and external, is greater than the number which can be found from equations of static equilibrium. The *degree of indeterminateness* is the number of these quantities in excess of the number of equilibrium equations. The excess forces or reactions may be called *redundants.* When some or all of the redundants are removed, the structure which remains is called the *primary structure* or *base structure.* Usually the primary structure is selected so as to be statically determinate, but this is not always necessary.

It is assumed in this discussion that the arrangement of the members and supports of the structure is such that it is immovable or stable (see Sec. 26.1). The redundants must be selected in such a way that the primary structure is immovable also.

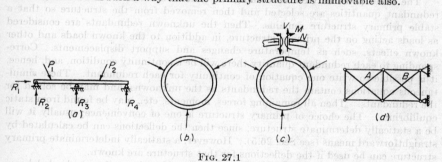

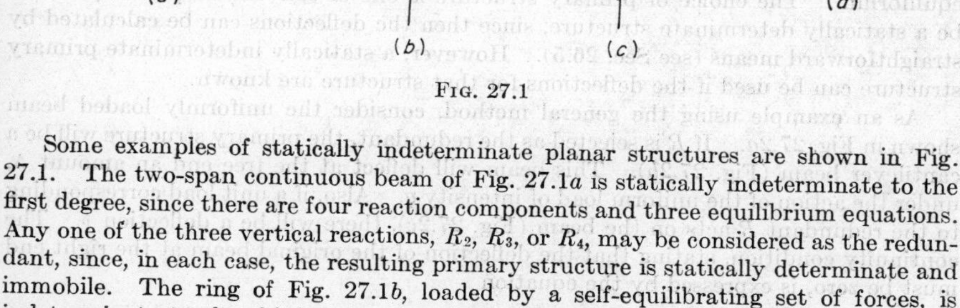

FIG. 27.1

Some examples of statically indeterminate planar structures are shown in Fig. 27.1. The two-span continuous beam of Fig. 27.1*a* is statically indeterminate to the first degree, since there are four reaction components and three equilibrium equations. Any one of the three vertical reactions, R_2, R_3, or R_4, may be considered as the redundant, since, in each case, the resulting primary structure is statically determinate and immobile. The ring of Fig. 27.1*b*, loaded by a self-equilibrating set of forces, is indeterminate to the third degree, and the axial force, shearing force, and bending moment at any section (see Fig. 27.1*c*) may be considered as the three redundant quantities. The pin-connected truss of Fig. 27.1*d* is indeterminate to the second degree, since, if the two bars A and B are removed, a statically determinate truss will remain.

Continuity Conditions

In the determination of reactions and internal forces in a statically indeterminate structure, it is necessary to satisfy not only all conditions of static equilibrium, but also conditions of continuity.[1] Continuity conditions must be satisfied by the deformations or displacements of the structure, and are determined by the manner of support and the way in which the parts of the structure are connected. For example, the vertical displacements at the supports of the beam of Fig. 27.1*a* must be zero, and the elastic curve of the beam must be continuous over the middle support. In Fig. 27.1*c* the relative displacements (horizontal, vertical, and rotational) of the cut ends of the ring must be zero since the ring is actually continuous (Fig. 27.1*b*).

The satisfaction of the continuity conditions is the basis for the most general method of analysis (see Sec. 27.2). The particular continuity conditions which are to be used in the analysis depend on the choice of redundants.

27.2. THE GENERAL METHOD

The general method of analysis of statically indeterminate structures is known by several names, such as the method of consistent deformations, the method of superposition equations, and Maxwell's method. It is the most fundamental method of analysis, and can be used to analyze any type of statically indeterminate structure, including beams, rigid frames, trusses, and composite structures. The only requirement is that the principle of superposition be valid, which means that Hooke's law must hold and the deflections must be small (see Sec. 26.1). In some cases, such as composite structures in which some members are subjected primarily to axial forces and others to bending loads, it is the only method which can be used conveniently, whereas, in other structures, such as continuous beams, more specialized methods (e.g., moment distribution) may be preferred to the general method.

The usual point of view concerning the general method is the following. The redundant quantities are selected and then removed from the structure so that a stable primary structure remains. Then the unknown redundants are considered as loads acting on the primary structure, in addition to the known loads and other known effects, such as temperature changes and support displacements. Corresponding to each redundant quantity there will be a continuity condition, and, hence, it is possible to write one equation of continuity for each redundant. These simultaneous equations contain the redundants as the unknowns, and may be solved for the redundants. Then all remaining forces, moments, etc., may be found from static equilibrium. The choice of primary structure is one of convenience; usually it will be a statically determinate structure, since then the deflections can be calculated by straightforward means (see Sec. 26.5). However, a statically indeterminate primary structure can be used if the deflections for that structure are known.

As an example using the general method, consider the uniformly loaded beam shown in Fig. 27.2*a*. If R is selected as the redundant, the primary structure will be a cantilever beam (Fig. 27.2*b*). This beam will deflect at the free end an amount Δ, under the action of the uniform load of intensity q. Also, if a unit load corresponding to the redundant R acts on the beam (Fig. 27.2*c*), there will be a deflection δ. The continuity condition, stating that the deflection of the original beam at the right end must be zero, is expressed by the equation

$$\Delta - R\delta = 0 \quad \text{or} \quad R = \Delta/\delta \tag{27.1}$$

Note that the deflection Δ has units of length, and δ has units of length per unit force. The determination of these deflections constitutes a separate problem, and any suitable

[1] These also are referred to as conditions of geometry, compatibility, or consistent deformations.

method may be used (see Sec. 26.5); in this case, $\Delta = ql^4/8EI$, $\delta = l^3/3EI$, and therefore $R = 3ql/8$.

The same beam may be analyzed as shown in Fig. 27.3, by taking the reactive moment M_1 at the fixed end as the redundant. Since the angle of rotation at the left end of the beam is zero, the continuity condition is

$$\Theta - M_1\theta = 0 \qquad \text{or} \qquad M_1 = \Theta/\theta \tag{27.2}$$

where Θ is the angle of rotation at the left end due to the uniform load q, and θ is due to the unit moment. These values are $\Theta = ql^3/24EI$, $\theta = l/3EI$, and hence $M_1 = ql^2/8$. Equations (27.1) and (27.2) are based upon continuity conditions which correspond to the selection of redundant. The choice of redundant usually is based on the ease with which the corresponding deflections can be found, as well as the complexity of the resulting equations (see, for example, Sec. 27.4).

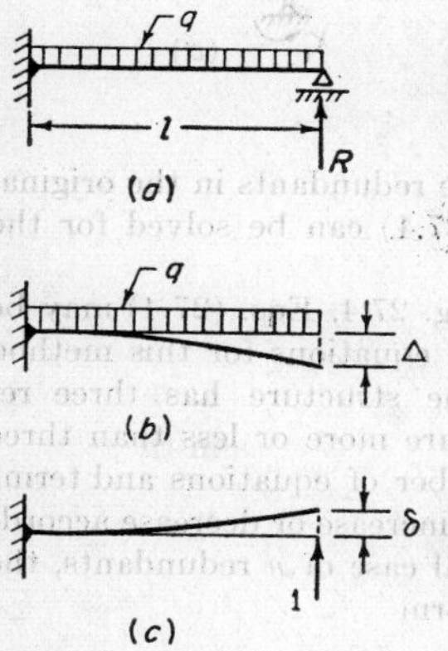

FIG. 27.2

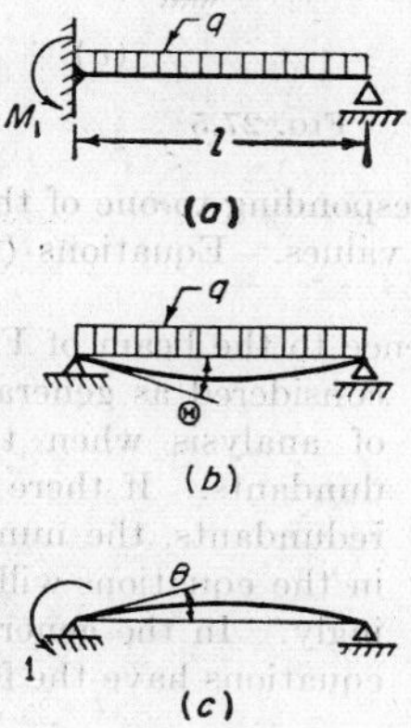

FIG. 27.3

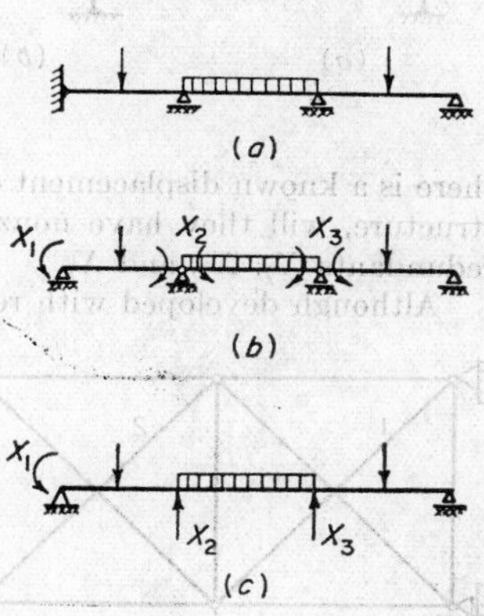

FIG. 27.4

If the right end of the beam in Fig. 27.2*a* undergoes a downward displacement Δ_1, then the equation of continuity [Eq. (27.1)] becomes

$$\Delta - R\delta = \Delta_1 \qquad \text{or} \qquad R = \frac{\Delta - \Delta_1}{\delta} \tag{27.3}$$

An analogous change in the equation of continuity (27.2) can be made if the left end of the beam in Fig. 27.3*a* rotates through a known angle.

An example of a beam which is statically indeterminate to the third degree is shown in Fig. 27.4*a*. There are many possibilities for the selection of three redundants X_1, X_2, X_3 for this beam; two such choices are shown in Fig. 27.4*b* and *c*. Let δ_{1P}, δ_{2P}, and δ_{3P} represent the deflections in the primary structure caused by the applied loads, as well as any other effects such as temperature changes, fabrication errors, and support displacements. These deflections correspond to the redundants in the generalized sense (see p. 26–15); that is, δ_{1P} is the generalized deflection corresponding to X_1, etc. In general, the quantity δ_{iP} is defined as the deflection corresponding to the redundant X_i due to the effect of actual loads, temperature changes, etc., acting on the primary structure.

It is also necessary to determine the deflections of the primary structure due to unit values of the redundants. The quantity δ_{ij} will be defined as the deflection corresponding to the redundant X_i due to a unit value of the redundant X_j. Thus, δ_{12} represents the deflection in the primary structure corresponding to X_1 caused by a load $X_2 = 1$. Then the conditions of continuity for the beam of Fig. 27.4*a* give the following equations:

$$\begin{aligned} \delta_{11}X_1 + \delta_{12}X_2 + \delta_{13}X_3 + \delta_{1P} &= \delta_{10} \\ \delta_{21}X_1 + \delta_{22}X_2 + \delta_{23}X_3 + \delta_{2P} &= \delta_{20} \\ \delta_{31}X_1 + \delta_{32}X_2 + \delta_{33}X_3 + \delta_{3P} &= \delta_{30} \end{aligned} \qquad (27.4a\text{–}c)$$

where δ_{10}, δ_{20}, and δ_{30} represent the deflections in the original structure corresponding to the redundants X_1, X_2, and X_3. In Fig. 27.4 these deflections will be zero; only if

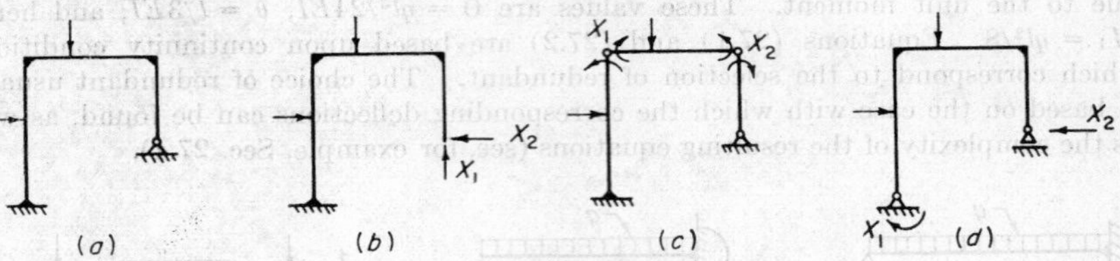

FIG. 27.5

there is a known displacement corresponding to one of the redundants in the original structure, will they have nonzero values. Equations (27.4) can be solved for the redundants X_1, X_2, and X_3.

Although developed with reference to the beam of Fig. 27.4, Eqs. (27.4) may be considered as general equations for this method of analysis when the structure has three redundants. If there are more or less than three redundants, the number of equations and terms in the equations will increase or decrease accordingly. In the general case of n redundants, the equations have the form

$$\sum_{j=1}^{n} \delta_{ij}X_j + \delta_{iP} = \delta_{i0} \qquad i = 1, 2, \ldots, n \qquad (27.5)$$

The calculation of the deflections appearing in Eqs. (27.4) and (27.5) is facilitated by use of Maxwell's reciprocal theorem [see Eqs. (33.85)], which gives

$$\delta_{ij} = \delta_{ji} \qquad (27.6)$$

Thus it is seen that the coefficients of the redundants X_j in Eqs. (27.5) form a symmetrical array.

Equations (27.5) are applicable to the analysis of all types of structures, of which two additional examples are shown in Figs. 27.5 and 27.6. The rigid frame of Fig. 27.5*a* can be analyzed by selecting either reaction components or bending moments as redundants; three possibilities are shown in Fig. 27.5*b*, *c*, and *d*. Two unknown bar forces X_1 and X_2 are selected as redundants for the truss of Fig. 27.6*a*. If the bars are removed from the truss, leaving the primary structure shown in Fig. 27.6*b*, the corresponding deflec-

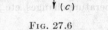

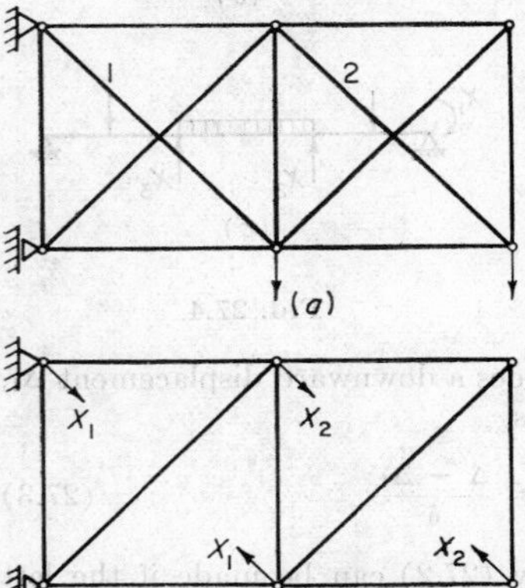

FIG. 27.6

tions δ_{10} and δ_{20} are the actual shortenings of bars 1 and 2 in the original truss, for example, $-X_1l_1/A_1E_1$ and $-X_2l_2/A_2E_2$. On the other hand, if the redundant bars are cut but not removed from the structure (Fig. 27.6c), the deflections corresponding to X_1 and X_2 in the original truss will be zero.

To verify completely the results of an analysis of a statically indeterminate structure, it is necessary to check that both equilibrium conditions and continuity conditions are satisfied. One reliable way of doing this is to take a new primary structure and, using the results of the first analysis, determine whether equilibrium and continuity are satisfied for the new structure.

27.3. METHOD OF LEAST WORK

The method of least work can be used to analyze structures in which the members are subjected to bending, axial, shear, and torsional deformations. The strain energy of the structure is expressed in terms of the redundant forces or moments, and then Castigliano's theorem (see Sec. 26.5) is used to evaluate the deflections corresponding to the redundants. If these deflections are zero, Castigliano's theorem becomes a statement that the partial derivative of the strain energy with respect to each redundant is zero. Hence, the *method of least work* is based on the principle that the redundants have such values that the strain energy of the structure is a minimum, assuming no support movement and no temperature change. This can be expressed in the form

$$\frac{\partial U}{\partial X_1} = 0, \qquad \frac{\partial U}{\partial X_2} = 0, \qquad \cdots \tag{27.7}$$

where U is the strain energy of the structure [see Eq. (26.16)], and X_1, X_2, . . . are the redundants.

In the particular case of the prismatic beam illustrated in Fig. 27.2a, the reaction R is the redundant, and the theorem of least work gives

$$\frac{dU}{dR} = 0$$

Upon substituting the expression for strain energy of bending [see Eq. (26.16)], this expression becomes

$$\frac{dU}{dR} = \frac{d}{dR}\int \frac{M^2\,dx}{2EI} = \int \frac{M(dM/dR)\,dx}{EI} = 0$$

Since the bending moment at distance x from the right end of the beam is $M = Rx - qx^2/2$, the above equation yields

$$\frac{1}{EI}\int_0^l \left(Rx - \frac{qx^2}{2}\right) x\,dx = 0 \qquad \text{or} \qquad \frac{Rl^3}{3EI} - \frac{ql^4}{8EI} = 0$$

from which $R = 3ql/8$. Similarly, if the reactive moment M_1 at the left support is taken as redundant (Fig. 27.3), the expressions are

$$M = \left(\frac{ql}{2} - \frac{M_1}{l}\right)x - \frac{qx^2}{2} \qquad \frac{dM}{dM_1} = -\frac{x}{l}$$

and $M_1 = ql^2/8$.

In a more general case, the method of least work leads to a set of simultaneous equations, with the redundant forces and moments as unknowns. The equations will be essentially the same as obtained by the general method if the same choice of redundants is made. This is evident from the fact that each of Eqs. (27.7) is a statement that a particular deflection in the actual structure is zero. This same condition on the deflections is used in the general method; the only difference in the two methods is the point of view which is adopted for calculating the deflections.

27.4. THREE-MOMENT EQUATIONS

The three-moment equations provide a means for analyzing continuous beams subjected to lateral loading and settlement of supports. The bending moments in the beam at the points of support are taken as the redundant quantities, and, after they are determined, all other force quantities can be found from static equilibrium.

Let AB and BC (Fig. 27.7*a*) be two adjacent spans of a continuous beam in which, owing to unequal settlements, the supports A and C are at higher elevations than support B by the amounts h_A and h_C, respectively. The span lengths and moments of

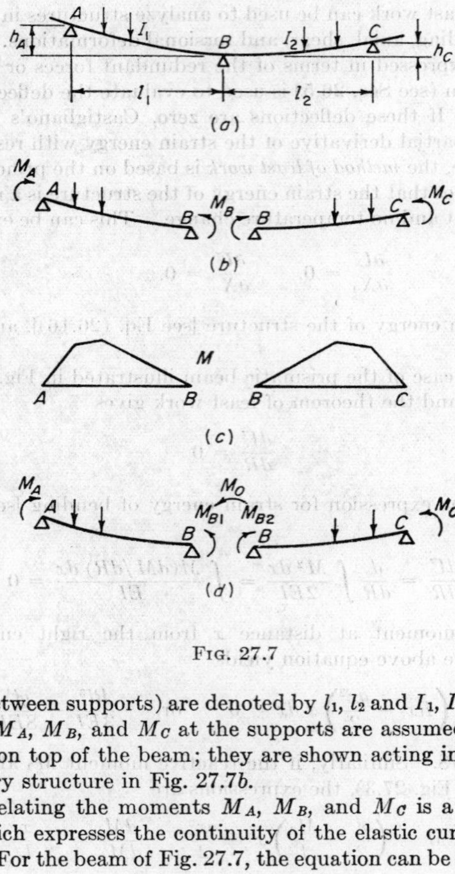

FIG. 27.7

inertia (constant between supports) are denoted by l_1, l_2 and I_1, I_2, respectively. The bending moments M_A, M_B, and M_C at the supports are assumed positive when they cause compression on top of the beam; they are shown acting in their positive directions on the primary structure in Fig. 27.7*b*.

The equation relating the moments M_A, M_B, and M_C is a continuity equation [see Eq. (27.5)] which expresses the continuity of the elastic curve of the continuous beam at point B. For the beam of Fig. 27.7, the equation can be expressed in the form

$$\delta_{BA}M_A + \delta_{BB}M_B + \delta_{BC}M_C + \delta_{BP} = 0 \qquad (27.8)$$

in which δ_{BB} is the generalized deflection corresponding to M_B, that is, the relative rotation of the end tangents of the beams joined at B, due to $M_B = 1$,

$$\delta_{BB} = \frac{l_1}{3EI_1} + \frac{l_2}{3EI_2}$$

δ_{BA} is the same relative rotation, but due to $M_A = 1$,

$$\delta_{BA} = \frac{l_1}{6EI_1}$$

δ_{BC} is the same relative rotation, but due to $M_C = 1$,

$$\delta_{BC} = \frac{l_2}{6EI_2}$$

and δ_{BP} is the same relative rotation, but due to lateral loads and support displacements.

The deflection δ_{BP} can be evaluated in two parts. First, the deflection due to the actual loads can be found from the moment-area theorems (see Sec. 26.5). The simple-beam bending-moment diagrams for spans AB and BC, due to the lateral loads, are shown in Fig. 27.7*c*. The angle of rotation at the right end of beam AB (Fig. 27.7*b*) is equal to the static moment Q_{AB} of the area of the bending-moment diagram, taken about point A, divided by the flexural rigidity and the length of the span; that is, Q_{AB}/EI_1l_1. Similarly, the angle of rotation at the left end of beam BC due to lateral loads is Q_{CB}/EI_2l_2, where Q_{CB} is the static moment of the area of the bending-moment diagram about point C. The second part of the deflection δ_{BP} is the effect of support displacement, which is $-h_A/l_1 - h_C/l_2$. Thus, the expression for the deflection δ_{BP} becomes

$$\delta_{BP} = \frac{Q_{AB}}{EI_1l_1} + \frac{Q_{CB}}{EI_2l_2} - \frac{h_A}{l_1} - \frac{h_C}{l_2}$$

After substituting the above expressions for the deflections into Eq. (27.8) and simplifying, the three-moment equation is obtained:

$$M_A\frac{l_1}{I_1} + 2M_B\left(\frac{l_1}{I_1} + \frac{l_2}{I_2}\right) + M_C\frac{l_2}{I_2} = -6\left(\frac{Q_{AB}}{I_1l_1} + \frac{Q_{CB}}{I_2l_2} - \frac{Eh_A}{l_1} - \frac{Eh_C}{l_2}\right) \quad (27.9)$$

As an example of the use of the three-moment equation, consider the continuous beam of Fig. 27.8. The three redundant bending moments at supports B, C, and D

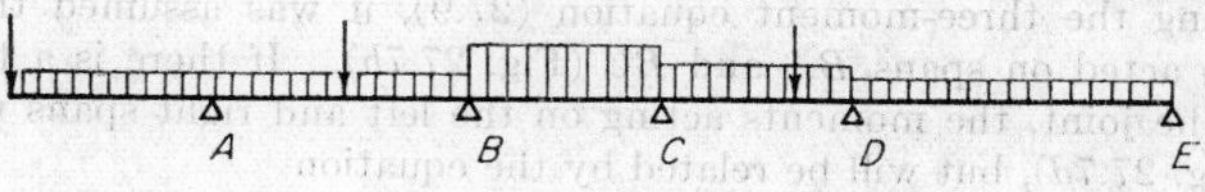

FIG. 27.8

can be found by writing the three-moment equation for each of these supports. The first equation, expressing the condition of continuity at B, will contain M_B and M_C as unknowns, since the moment M_A is statically determinate. The second equation (for support C) will contain M_B, M_C, and M_D as unknowns, and the third equation will contain M_C and M_D as unknowns. After these equations are solved for the moments, it is possible to find all other force quantities from equilibrium.

If the continuous beam has a fixed end (see Fig. 27.9*a*), its effect may be represented by adding an imaginary span A_0A (Fig. 27.9*b*) of any length l_0 and having an infinitely large moment of inertia. Thus the three-moment equation, when applied to joint A (Fig. 27.9*b*), becomes

$$2M_A\frac{l_2}{I_2} + M_B\frac{l_2}{I_2} = -6\left(\frac{Q_{BA}}{I_2l_2} - \frac{Eh_B}{l_2}\right)$$

Two additional equations, expressing the conditions of continuity at supports B and C, can be written also. Then the three equations can be solved for the three redundant bending moments.

One of the principal advantages of using three-moment equations, which are merely special forms of the general equations (27.5), is that the resulting simultaneous equations form a simpler pattern than when other quantities, such as reactions, are taken

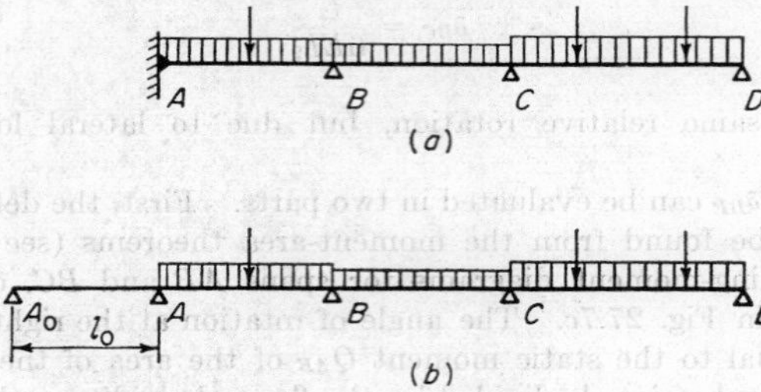

FIG. 27.9

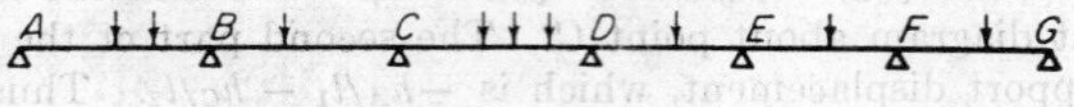

FIG. 27.10

as redundants. For example, the pattern of the three-moment equations for the six-span beam of Fig. 27.10 would be:

$$
\begin{aligned}
\delta_{11}M_B + \delta_{12}M_C \qquad\qquad\qquad\qquad\qquad\qquad &= \delta_{10}\\
\delta_{21}M_B + \delta_{22}M_C + \delta_{23}M_D \qquad\qquad\qquad\qquad &= \delta_{20}\\
\delta_{32}M_C + \delta_{33}M_D + \delta_{34}M_E \qquad\qquad &= \delta_{30}\\
\delta_{43}M_D + \delta_{44}M_E + \delta_{45}M_F &= \delta_{40}\\
\delta_{54}M_E + \delta_{55}M_F &= \delta_{50}
\end{aligned}
$$

where the δ_{ij} and δ_{i0} are constants. On the other hand, if five of the vertical reactions are selected as redundants, there will be obtained five simultaneous equations, each containing all five unknowns.

In deriving the three-moment equation (27.9), it was assumed that the same moment M_B acted on spans BA and BC (Fig. 27.7*b*). If there is a third member meeting at the joint, the moments acting on the left and right spans will no longer be equal (Fig. 27.7*d*), but will be related by the equation

$$M_{B1} - M_{B2} = M_O$$

where M_O is the external couple acting on joint B. Under these conditions, the equation of continuity at B becomes

$$M_A \frac{l_1}{I_1} + 2M_{B1}\frac{l_1}{I_1} + 2M_{B2}\frac{l_2}{I_2} + M_C\frac{l_2}{I_2} = -6\left(\frac{Q_{AB}}{I_1 l_1} + \frac{Q_{CB}}{I_2 l_2} - \frac{Eh_A}{l_1} - \frac{Eh_C}{l_2}\right) \qquad (27.10)$$

called the *four-moment equation*.

27.5. THE SLOPE-DEFLECTION METHOD

The slope-deflection method may be used to analyze statically indeterminate beams and rigid frames in which the effect of bending deformations only is considered. The method differs from those described previously in that angles of rotation and other displacement quantities are taken as the unknowns, rather than forces and moments. All moments acting at the ends of the members of the frame can be expressed in terms of the unknown angles of rotation and displacements; these are known as the *slope-deflection equations.* Then we can write equations of static equilibrium containing the end moments. When the expressions for the moments, in terms of the unknown quantities, are substituted into the equilibrium equations, the resulting set of simultaneous equations can be solved for the angles and displacements. From these, the end moments can be found.

In order to derive the slope-deflection equations, let us consider the beam AB of Fig. 27.11*a*, which represents a member removed from a rigid-frame or continuous-beam structure. The beam is subjected to lateral loading and to end moments M_{AB} and M_{BA}, assumed positive in the clockwise direction. The angle between the original axis AB' and the chord AB joining the ends of the beam in the final deflected position is called the *chord-rotation angle* ψ. This angle is equal to Δ/l, where Δ is the relative transverse displacement of one end of the beam with respect to the other, and l is the length. The angle ψ and the angles of rotation θ_A and θ_B at the ends of the beam are assumed positive in the same direction as positive end moments, that is, clockwise. The beam loaded and deflected as shown in Fig. 27.11*a* can be obtained by the superposition of the loading patterns shown in Fig. 27.11*b*, *c*, and *d*. In Fig. 27.11*b*, the ends of the beam are fixed against rotation, and the lateral loads are applied; the resulting moments $M_{AB}{}^F$ and $M_{BA}{}^F$ are termed *fixed-end moments.*[1] Figure 27.11*c* shows the effect of rotating the chord AB through the angle ψ while the ends of the beam are fixed against rotation. The moments developed at the ends of the beam are

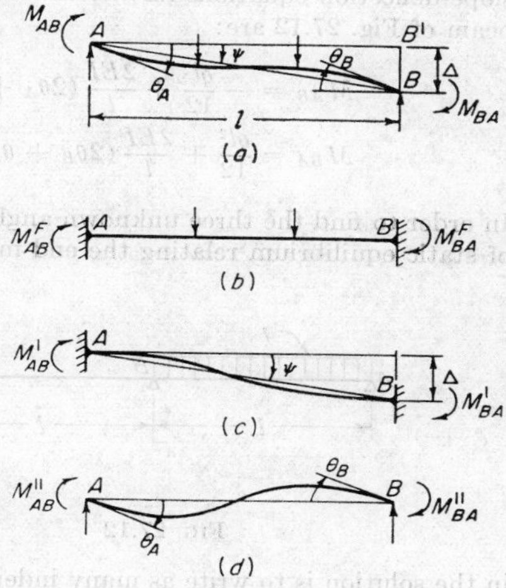

FIG. 27.11

$$M'_{AB} = M'_{BA} = -\frac{6EI\psi}{l} \tag{27.11}$$

as can be derived readily by the methods described in Sec. 26.5. Finally, in Fig. 27.11*d* is shown the effect of rotating the ends of the member through the angles θ_A and θ_B. These angles of rotation are equal to

$$\theta_A = \frac{M''_{AB}l}{3EI} - \frac{M''_{BA}l}{6EI} \qquad \theta_B = -\frac{M''_{AB}l}{6EI} + \frac{M''_{BA}l}{3EI}$$

and, hence,

$$M''_{AB} = \frac{2EI}{l}(2\theta_A + \theta_B) \tag{27.12a}$$

$$M''_{BA} = \frac{2EI}{l}(2\theta_B + \theta_A) \tag{27.12b}$$

The end moments in the beam of Fig. 27.11*a* are obtained by superposition,

$$M_{AB} = M_{AB}{}^F + M'_{AB} + M''_{AB}$$

[1] Formulas for fixed-end moments are found in Table 32.1, Figs. 32.13 to 32.22. A more extensive table is given in [9].

and similarly for M_{BA}. Thus the slope-deflection equations become

$$M_{AB} = M_{AB}^F + \frac{2EI}{l}(2\theta_A + \theta_B - 3\psi) \qquad (27.13a)$$

$$M_{BA} = M_{BA}^F + \frac{2EI}{l}(2\theta_B + \theta_A - 3\psi) \qquad (27.13b)$$

These equations express the end moments in terms of the angles of rotation at the ends, the chord-rotation angle, and the applied loads.

The first step to be used in analyzing a continuous beam consists in writing the slope-deflection equations for the end moments. For example, the equations for the beam of Fig. 27.12 are:

$$M_{AB} = -\frac{ql^2}{12} + \frac{2EI}{l}(2\theta_A + \theta_B) \qquad M_{BC} = \frac{2EI}{l}(2\theta_B + \theta_C)$$

$$M_{BA} = \frac{ql^2}{12} + \frac{2EI}{l}(2\theta_B + \theta_A) \qquad M_{CB} = \frac{2EI}{l}(2\theta_C + \theta_B) \qquad (27.14)$$

In order to find the three unknown angles of rotation, θ_A, θ_B, and θ_C, three equations of static equilibrium relating the end moments are needed. Hence, the second step

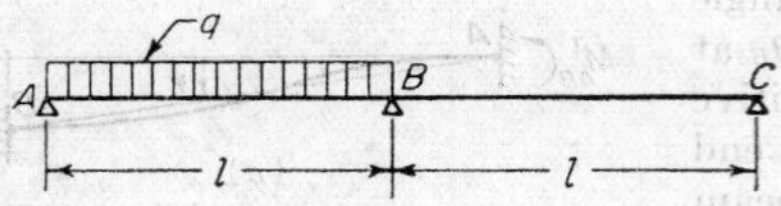

FIG. 27.12

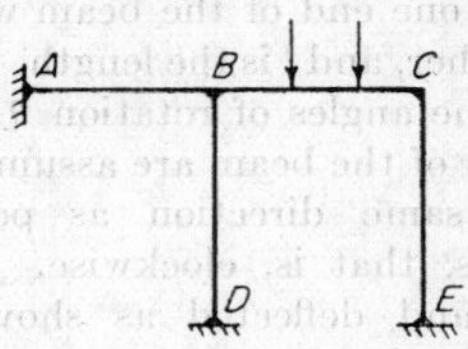

FIG. 27.13

in the solution is to write as many independent equations of equilibrium as there are unknowns. For the beam of Fig. 27.12 the equations of moment equilibrium for the joints are

$$M_{AB} = 0 \qquad M_{BA} + M_{BC} = 0 \qquad M_{CB} = 0 \qquad (27.15)$$

When expressions (27.14) are substituted into Eqs. (27.15), three simultaneous equations are obtained:

$$\begin{aligned} 2\theta_A + \theta_B \quad &= \frac{ql^3}{24EI} \\ \theta_A + 4\theta_B + \theta_C &= -\frac{ql^3}{24EI} \\ \theta_B + 2\theta_C &= 0 \end{aligned} \qquad (27.16)$$

Next, these equations are solved for the angles of rotation, yielding

$$\theta_A = \frac{ql^3}{32EI} \qquad \theta_B = -\frac{ql^3}{48EI} \qquad \theta_C = \frac{ql^3}{96EI}$$

and, after these results are substituted into Eqs. (27.14), the moments are found to be

$$M_{AB} = 0 \qquad M_{BA} = -M_{BC} = \frac{ql^2}{16} \qquad M_{CB} = 0$$

After the end moments are found in this way, all reactions, etc., can be determined from equilibrium equations.

The effect of a known vertical displacement of any support of the beam in Fig. 27.12 can be included in the analysis without difficulty, since it means only the addition of a numerical value for ψ in Eqs. (27.14).

The rigid frame of Fig. 27.13 can be analyzed by the slope-deflection method, using only two simultaneous equations, since there are only two unknown angles of rotation, θ_B and θ_C. In contrast, the use of the general method with forces or moments as redundants would require the solution of six simultaneous equations, since there are six degrees of indeterminateness. This particular example is one in which the use of deformations as unknowns offers considerable advantage over the methods in which forces are taken as unknowns. The opposite situation exists in the case of the beam of Fig. 27.12, which is indeterminate to the first degree. There are three unknown angles of rotation, requiring three equations for a solution by slope deflection, whereas one equation is sufficient when a force or moment is used as the redundant.

In cases where the joints of a frame are free to deflect, as in Fig. 27.14, the joint displacements will appear also as unknown quantities in the slope-deflection equations. For example, the equations for the frame of Fig. 27.14 will contain θ_B, θ_C, and Δ as unknowns. The moment-equilibrium equations at joints B and C will supply two equations; the third is obtained by expressing the horizontal reactions at A and D in

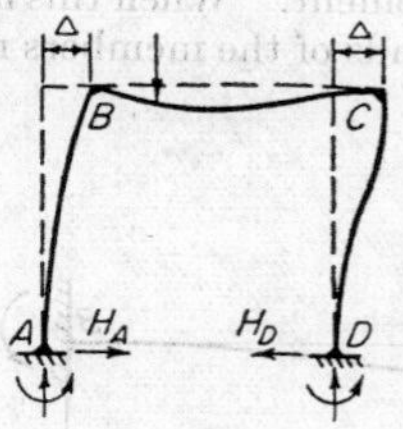

FIG. 27.14

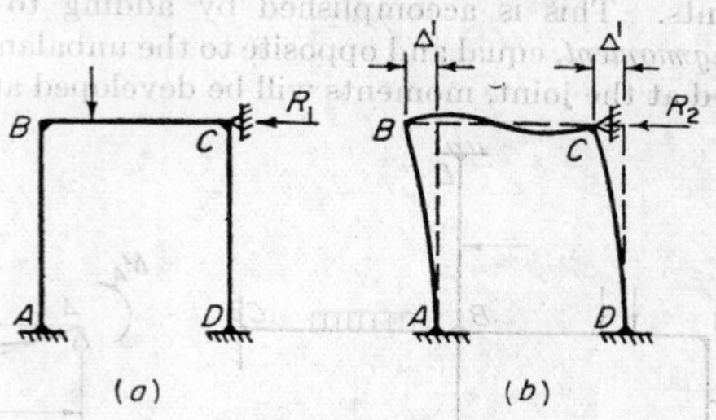

FIG. 27.15

terms of the end moments M_{AB}, M_{BA}, M_{DC}, M_{CD}, and then equating the sum of all horizontal forces to zero. These three equations can be solved for the three unknowns, and then the moments can be found from the slope-deflection equations.

In order to reduce the number of simultaneous equations which must be solved, the frame of Fig. 27.14 can also be analyzed by means of a two-stage solution. The frame is supported at joint C (Fig. 27.15*a*), in which case there are two unknown angles, θ_B and θ_C, to be found. After the angles are found, the moments at the ends of the members and the reaction R_1 are calculated. Next, the frame is displaced to the left by an assumed amount Δ'; again there are two unknown angles, and the moments and the reaction R_2 can be found. The loads in the frames of Fig. 27.15*a* and *b* can be superposed to give those in the original frame (Fig. 27.14), provided the following equation is satisfied,

$$R_1 + bR_2 = 0 \tag{27.17}$$

from which $b = -R_1/R_2$, where b is a constant of proportionality. Thus the moments in the frame of Fig. 27.15*a* are added to b times the moments in the frame of Fig. 27.15*b*. The sums will be the moments in the original frame. This scheme replaces the solution of one frame with three unknowns by the solution of two frames with two unknowns.

The method described above can be extended to more complicated rigid-frame and continuous-beam structures. As the number of unknown quantities increases, there is a corresponding increase in the number of simultaneous equations which must be solved.

27.6. THE MOMENT-DISTRIBUTION METHOD

Moment distribution is a method of structural analysis which is applicable to continuous beams and rigid frames composed of members which act primarily in bending, although, with modifications, the method can be applied to structures in which other effects, such as axial load, are important. Since the principle of superposition is utilized, it is assumed that the deflections are small, and the material follows Hooke's law. In each case, the object of the analysis is to determine the bending moments at the joints of the structure; when these are known, the reactions, etc., can be found from equilibrium.

The method of moment distribution can be illustrated by a discussion of the rigid-frame structure shown in Fig. 27.16. The first step in the process consists of the introduction at joints A and B of external constraints which prevent the joints from rotating. Then, when the loads are applied to the structure, each member is in the condition of a fixed-end beam, and moments will be developed at the joints, called *fixed-end moments* (see Sec. 27.5). At each locked joint of the structure, the external constraint will exert a moment, called the *unbalanced moment,* on the structure. The unbalanced moment is equal to the sum of the fixed-end moments at the joint.

The next step in the moment-distribution process is the removal of the external constraints. This is accomplished by adding to the joint a moment, called the *balancing moment,* equal and opposite to the unbalanced moment. When this moment is applied at the joint, moments will be developed at the ends of the members meeting

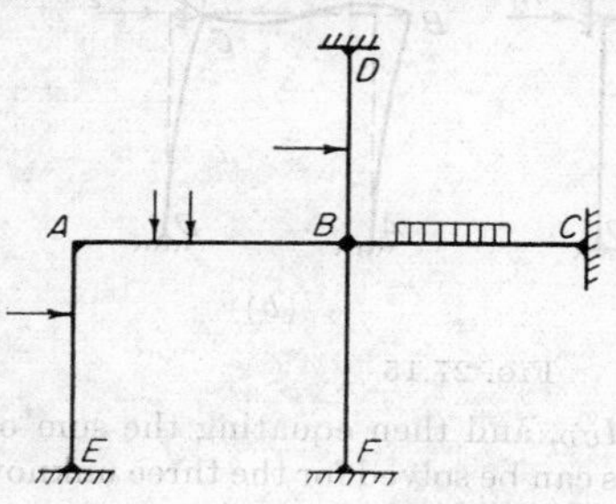

FIG. 27.16

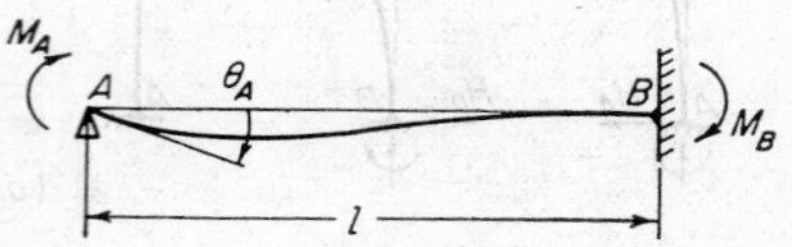

FIG. 27.17

at the joint, called *distributed moments.* For example, when the balancing moment is applied at joint A (Fig. 27.16), moments will be developed in members AB and AE. Since all the members meeting at the joint undergo the same amount of rotation when the balancing moment is applied, the proportion of the balancing moment which is distributed to each member (called the *distribution factor* for the member) is dependent upon the *rotational stiffness* of the member. For example, if member AB (Fig. 27.16) has twice the rotational stiffness of AE, the moment distributed to AB will be two-thirds of the balancing moment, and the other one-third will act on AE. The distribution of a moment to the *near end* of a member, that is, the end of the member at the joint which is being balanced, results in a *carry-over moment* being developed at the other, or *far end,* of the member. The ratio of the carry-over moment to the distributed moment is the *carry-over factor.* Thus, in Fig. 27.16, the distribution of moments at joint A to members AB and AE results in carry-over moments in the same members at joints B and E.

After a joint of the structure has been balanced, no external constraint at that joint is required. However, there may be other joints in the structure which are still constrained by artificial means; hence, the next step is to relock the joint which has just been balanced, and then proceed to unlock another joint in the structure. Thus, after the balancing of joint A (Fig. 27.16) has been completed, it is necessary to relock that joint while joint B is unlocked and then balanced. The distribution of moments at B results in a carry-over moment to A through member AB. This means that joint A is again unbalanced; hence, joint B must be relocked and the process of balancing joint A begun again. The moment-distribution process is continued until the unbalanced moments at the joints are negligible. The process of locking and unlocking each joint once only is called a *cycle* of moment distribution.

As in the preceding section, moments acting on the beams in the clockwise direction will be considered positive.

The rotational stiffness will be defined as the moment required to rotate the near end of the member through a unit angle. Thus the stiffness of beam AB in Fig. 27 17 is

$$K_{AB} = \frac{M_A}{\theta_A} \tag{27.18}$$

where θ_A is the angle of rotation caused by the couple M_A. The stiffness depends on the conditions of support at end B and on the shape of the beam. Stiffness formulas

Table 27.1. Stiffness Formulas for Prismatic Beams

Support conditions	Rotational stiffness K_{AB}	Carry-over factor C_{AB}
1. Far end fixed 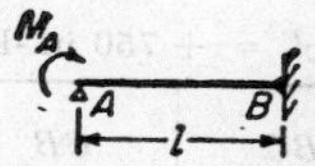	$\frac{4EI}{l}$	$\frac{1}{2}$
2. Far end simply supported 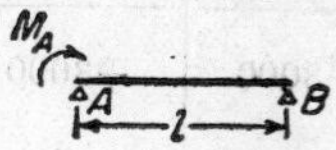	$\frac{3EI}{l}$	0
3. Symmetric moment at far end 	$\frac{2EI}{l}$	0
4. Antimetric moment at far end 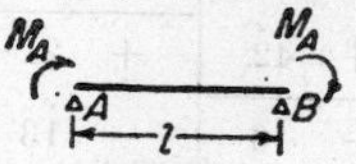	$\frac{6EI}{l}$	0

for prismatic beams are given in Table 27.1. Conditions 1 and 2 are for beams with the far ends fixed and simply supported, respectively. Condition 3 represents a beam in which the far end is subjected to a moment which is opposite in sign to the distributed moment, and condition 4 is a beam with a moment of the same sign at the far end. The latter two conditions arise in the analysis of symmetrical structures carrying symmetric and antimetric loads.

The carry-over factor is defined as the ratio (see Fig. 27.17)

$$C_{AB} = \frac{M_B}{M_A} \tag{27.19}$$

and is given in Table 27.1 also.

For nonprismatic beams, values of rotational stiffness, carry-over factors, and fixed-end moments can be obtained by making use of the standard methods of analysis, such as the moment-area method (see p. 26–21). For most cases of practical interest,

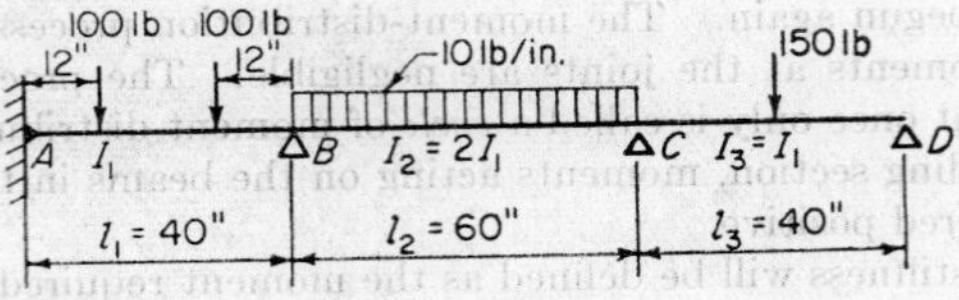

$$K_{BA} = \frac{4EI_1}{l_1} \propto 12 \qquad D_{BA} = \frac{12}{12+16} = 0.428$$

$$K_{BC} = \frac{4EI_2}{l_2} \propto 16 \qquad D_{BC} = \frac{16}{12+16} = 0.572$$

$$K_{CB} = \frac{4EI_2}{l_2} \propto 16 \qquad D_{CB} = \frac{16}{16+9} = 0.640$$

$$K_{CD} = \frac{3EI_3}{l_3} \propto 9 \qquad D_{CD} = \frac{9}{16+9} = 0.360$$

$$M_{AB}{}^F = -\frac{100(12)(28)^2}{(40)^2} - \frac{100(28)(12)^2}{(40)^2} = -840 \text{ in.-kips}$$

$$M_{BA}{}^F = +840 \text{ in.-kips}$$

$$M_{BC}{}^F = -\frac{10(60)^2}{12} = -3000 \text{ in.-kips}$$

$$M_{CB}{}^F = +3000 \text{ in.-kips}$$

$$M_{CD}{}^F = -750 \text{ in.-kips} \qquad M_{DC}{}^F = +750 \text{ in.-kips}$$

Member. . . . . D	AB 0	BA 0.428	BC 0.572	CB 0.640	CD 0.360	DC
M^F, kip-in.	−840	+ 840	−3000	+3000	− 750 − 375	+750 −750
	+463	+ 926	+1234	+ 617		
			− 798	−1595	− 897	
	+171	+ 342	+ 456	+ 228		
			− 73	− 146	− 82	
	+ 16	+ 31	+ 42	+ 21		
			− 6	− 13	− 8	
	+ 2	+ 3	+ 3	+ 2		
				− 1	− 1	
M, kip-in.	−188	+2142	−2142	+2113	−2113	0

FIG. 27.18

tables and graphs have been prepared from which the factors can be obtained directly (see [7]–[9]).

The distribution factor for a member at a joint of a structure is the ratio of the rotational stiffness of the member to the sum of the stiffnesses of all members meeting at the joint; that is,

$$D = \frac{K}{\Sigma K} \tag{27.20}$$

For example, the distribution factor for member AB at joint A (Fig. 27.16) is

$$D_{AB} = \frac{K_{AB}}{K_{AB} + K_{AE}} \tag{27.21}$$

Obviously, the sum of the distribution factors for all members meeting at a joint is unity. Also, it should be observed that relative values of stiffness can be used in calculating the distribution factor.

An example illustrating the method is shown in Fig. 27.18. The calculations for the stiffness and distribution factors are carried out as shown, and the fixed-end moments are determined on the assumption that all joints are fixed against rotation. The first step in the moment-distribution solution is to unlock joint D, resulting in a carry-over moment to C, after which joint D is left in the simply supported condition. Joint B is unlocked next, then relocked, and the same steps are followed at C. This process is repeated until the desired accuracy in the moments is obtained, after which the final moments are obtained by summation.

If a structure is symmetrical and has either symmetric or antimetric loading, it may be possible to use the stiffness and carry-over factors from cases 3 and 4 of Table 27.1. Then the moments are balanced for one-half the structure only, as illustrated in Fig. 27.19.

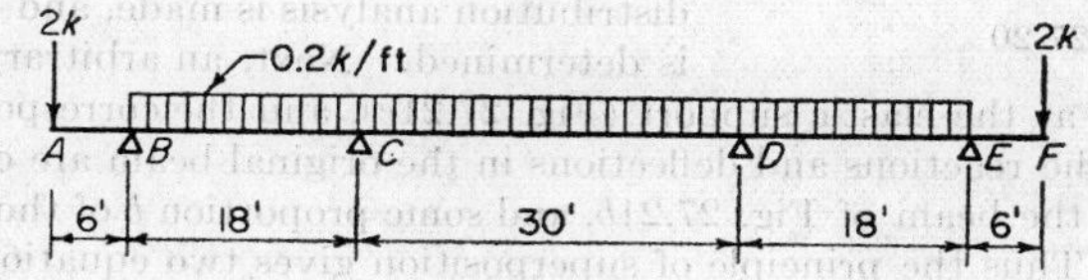

Constant EI

$$K_{CB} = \frac{3EI}{18} \propto 5 \qquad D_{CB} = 5/7$$

$$K_{CD} = \frac{2EI}{30} \propto 2 \qquad D_{CD} = 2/7$$

Member........	BA	BC	CB	CD
D	0	1	5/7	2/7
M^F, kip-ft	+12.0	− 5.4	+ 5.4	− 15.0
	0	− 6.6	− 3.3	
			+ 9.2	+ 3.7
Final moments	+12.0	− 12.0	+11.3	− 11.3

FIG. 27.19

When the supports of the structure deflect or rotate by a known amount, the analysis can be carried out by locking the joints against rotation, and then displacing them to their final (deflected and rotated) positions. The fixed-end moments at the joints can be obtained from the formulas in Fig. 27.20, after which the analysis is carried out in the same manner as though the fixed-end moments were caused by loads.

The effects of temperature changes are handled in a similar way. The deflections and rotations at the supports are allowed to occur as though there were no constraints, after which the supports are locked and returned to their actual positions. The resulting fixed-end moments can be found also by the formulas of Fig. 27.20.

The effects of axial changes in lengths can be taken into account by the same technique. An initial estimate of the changes in length is made, thereby determining the joint displacements. The joints of the structure are then displaced to the positions so determined, resulting in a set of fixed-end moments which can be superimposed on those due to the loads. The moment distribution is now carried out, and, from the results, a second approximation to the axial changes in length can be obtained. If these changes in length are significantly different from the first set, the entire process can be repeated. This procedure is frequently used in the analysis of secondary stresses in rigid-joint trusses, the initial estimate being made on the basis that all joints are pinned (primary analysis). The nonlinear effect of axial forces on stiffness, carry-over, etc. (beam-column effect), is discussed in Chap. 30.

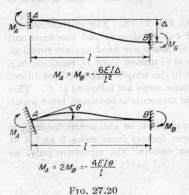

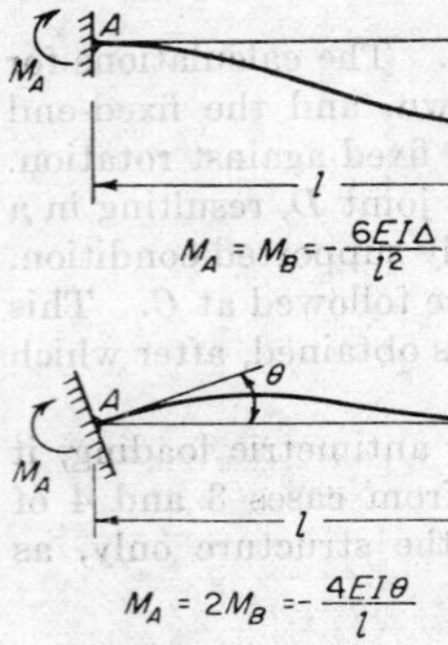

FIG. 27.20

If the structure has an elastic support, as illustrated by the beam of Fig. 27.21*a*, the analysis requires two steps. The structure is first supported so that no deflection can occur at the elastic support (Fig. 27.21*b*), a moment-distribution analysis is made, and the reaction R' is determined. Next, an arbitrary displacement Δ'' is introduced at the elastic support (Fig. 27.21*c*), and the corresponding reaction R'' is found. The reactions and deflections in the original beam are obtained as the sum of those in the beam of Fig. 27.21*b*, and some proportion b of those in the beam of Fig. 27.21*c*. Thus the principle of superposition gives two equations,

$$R = R' + bR'' \qquad \Delta = b\Delta''$$

and, since $R = k\Delta$, where k is the stiffness of the elastic support, this gives

$$b = \frac{R'}{k\Delta'' - R''} \tag{27.22}$$

With b determined, the reactions, bending moments, etc., in the original beam can be determined by superposition. This technique can be generalized to include the effect of any number of elastic supports; if, for example, there are three elastic supports, there will be four subsidiary moment-distribution analyses and three simultaneous equations of superposition to determine three constants of proportionality.

A rigid-frame structure in which joint translation occurs can be analyzed by a similar procedure. Consider, for example, the structure illustrated in Fig. 27.22*a*. The first moment-distribution analysis is carried out as in (*b*), where the structure is artificially supported so that no translation of the joints (or sidesway) may occur. The reactions R_1 and R_2 are calculated from equilibrium after the moments have been found. Next an arbitrary displacement Δ_1 is introduced at the support corresponding to R_1 (Fig. 27.22*c*), and the corresponding reactions R_1' and R_2' are found from a moment-distribution analysis. Finally, the structure is analyzed for a displacement Δ_2 corresponding to R_2 (Fig. 27.22*d*). The loading on the original frame can be considered as the sum of (*b*), plus a times (*c*), plus b times (*d*); that is,

$$\begin{aligned} R_1 + aR_1' + bR_1'' &= 0 \\ R_2 + aR_2' + bR_2'' &= 0 \end{aligned} \tag{27.23}$$

from which the constants a and b can be found. Then the moments in the original frame are obtained as the sum of those from analysis (b) plus a times those from (c), plus b times the moments in (d). The number of simultaneous equations to be solved is the same as the number of degrees of freedom of joint translation of the structure.

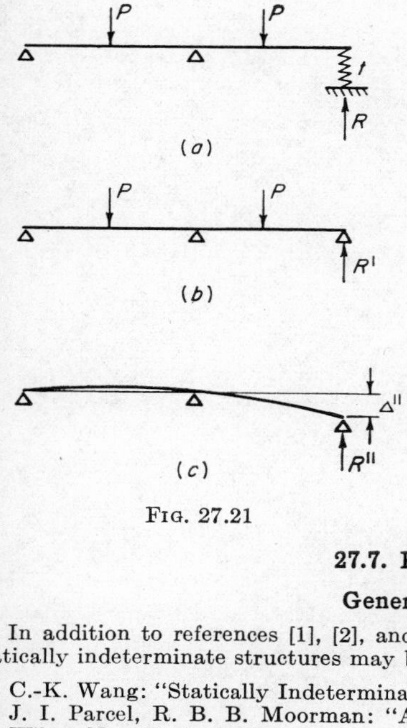

FIG. 27.21

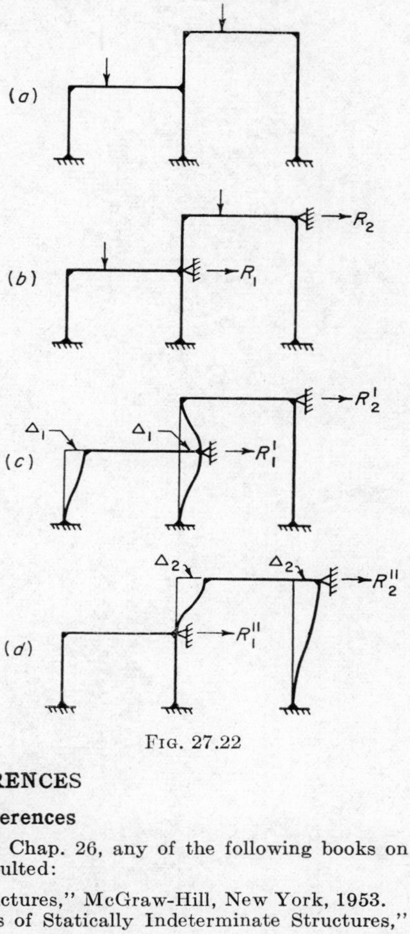

FIG. 27.22

27.7. REFERENCES

General References

In addition to references [1], [2], and [3] of Chap. 26, any of the following books on statically indeterminate structures may be consulted:

[1] C.-K. Wang: "Statically Indeterminate Structures," McGraw-Hill, New York, 1953.
[2] J. I. Parcel, R. B. B. Moorman: "Analysis of Statically Indeterminate Structures," Wiley, New York, 1955.
[3] S. F. Borg, J. J. Gennaro: "Advanced Structural Analysis," Van Nostrand, Princeton, N.J., 1959.
[4] J. S. Kinney: "Indeterminate Structural Analysis," Addison-Wesley, Reading, Mass., 1957.
[5] S. T. Carpenter: "Structural Mechanics," Wiley, New York, 1960.
[6] P. B. Morice: "Linear Structural Analysis," Ronald Press, New York, 1959. This book uses matrix algebra.

Specific References

[7] "Handbook of Frame Constants," Portland Cement Association, Chicago, 1958.
[8] J. M. Gere: "Moment Distribution Factors for Beams of Tapered I-section," American Institute of Steel Construction, New York, 1958.
[9] J. M. Gere: "Moment Distribution," Van Nostrand, Princeton, N.J., to be published in 1962.

CHAPTER 28

INFLUENCE DIAGRAMS

BY

W. FLÜGGE, Dr.-Ing., Los Altos, Calif.

28.1. DEFINITION

Let M be the bending moment at point x of a beam (Fig. 28.1), caused by a load P at point ξ. From Fig. 32.1

$$M(\xi;x) = \frac{P}{l}(l - \xi)x \qquad \text{for } x < \xi$$

$$M(\xi;x) = \frac{P}{l}\xi(l - x) \qquad \text{for } x > \xi \tag{28.1}$$

Figure 28.2 shows a plot of this function. The common moment diagram is a section ξ = const through the three-dimensional diagram. A section x = const shows how the moment in a fixed cross section varies when the load is moved. This diagram is called the influence line of the bending moment M at x. Similar influence lines can be drawn for any other quantity that depends on the position of the load, e.g., the shear force V, the deflection w, any reaction, stress, strain, etc.

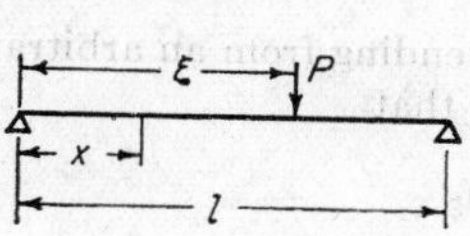

Fig. 28.1

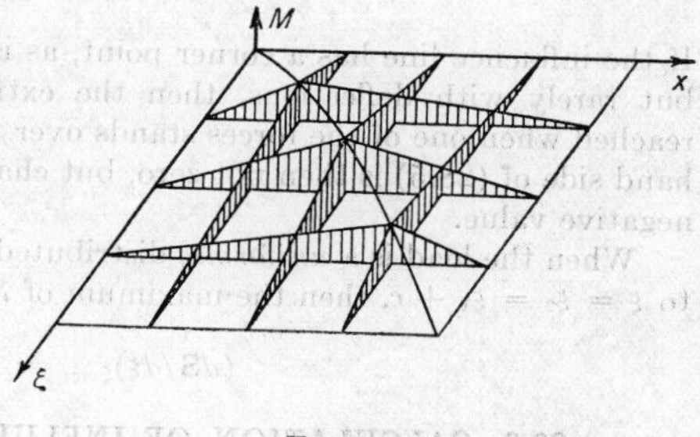

Fig. 28.2

Since influence lines are most useful when linear superposition is applicable, they are usually drawn for a unit load, e.g., for $\mathbf{M} = M/P$, $\mathbf{V} = V/P$, $\mathbf{w} = w/P$, etc. It should be noted that, because of the division by P, the dimension of $\mathbf{M}$ is $FL/F = L$, etc. The functions $\mathbf{M}(\xi)$, $\mathbf{V}(\xi)$, $\mathbf{w}(\xi)$, etc., are called *influence functions.* In what follows, the symbol $\mathbf{S}(\xi)$ or $\mathbf{S}(\xi;x)$ will be used to denote any of them. A plot of $\mathbf{S}(\xi)$ versus ξ is called an *influence line.*

28.2. USE OF INFLUENCE LINES

Figure 28.3 shows the influence line $\mathbf{S}(\xi)$ for some quantity (in this case, the bending moment M at A). The beam carries a distributed load $q(\xi)$ and concentrated forces P_n (n = 1, 2). It follows from linear superposition that the value of S (that is, the bending moment M at A) under this

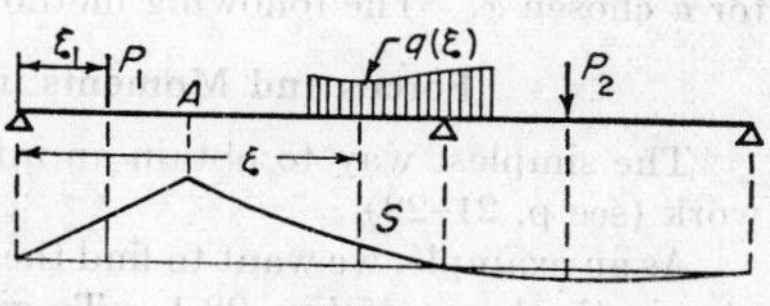

Fig. 28.3

load is

$$S = \sum_n P_n \mathbf{S}(\xi_n) + \int q(\xi)\mathbf{S}(\xi)\,d\xi \tag{28.2}$$

With the help of the Dirac function [see Eq. (19.7)], the concentrated forces P_n may be written as a special kind of distributed load:

$$q_n(\xi) = P_n\delta(\xi - \xi_n) \tag{28.3}$$

In this way these forces may be made part of the load $q(\xi)$, and Eq. (28.2) may be simplified to read

$$S = \int q(\xi)\mathbf{S}(\xi)\,d\xi \tag{28.4}$$

The right-hand side is called the *influence integral.*

A problem frequently encountered in bridge design and elsewhere is the following: A certain load configuration (e.g., a group of forces representing the wheel loads of a vehicle or of a railroad train, or a uniform load extended over a specified length c) can be moved along a beam. In which position will it cause the quantity S to assume its maximum value?

When the load consists of a single force, it should, evidently, be placed where the influence function $\mathbf{S}(\xi)$ has its maximum.

When the load consists of several forces P_n, the correct position is the one for which the sum in Eq. (28.2) has a stationary value, i.e., the one for which

$$\frac{dS}{d\xi} = \sum_n P_n \frac{d\mathbf{S}(\xi_n)}{d\xi} = 0 \tag{28.5}$$

If the influence line has a corner point, as is always the case with bending moments, but rarely with deflections, then the extremum is mostly (but not necessarily!) reached when one of the forces stands over this corner point. The sum on the right-hand side of (28.5) is then not zero, but changes discontinuously from a positive to a negative value.

When the load is a *uniformly* distributed load, extending from an arbitrary $\xi = \xi_1$ to $\xi = \xi_2 = \xi_1 + c$, then the maximum of S requires that

$$(d\mathbf{S}/d\xi)_{\xi=\xi_1} = -(d\mathbf{S}/d\xi)_{\xi=\xi_2} \tag{28.6}$$

28.3. CALCULATION OF INFLUENCE FUNCTIONS OF BEAMS

In simple cases, like the one represented by Fig. 28.1 and Eq. (28.1), it is possible to write closed-form expressions for an influence function $\mathbf{S}(\xi)$, usually containing some point of discontinuity. However, in many cases numerical work intervenes, and then it is impracticable to find a general expression for $\mathbf{S}(\xi;x)$ and to specialize it for x = const; it is preferable to work directly toward the influence function $\mathbf{S}(\xi)$ for a chosen x. The following methods are available.

Forces and Moments in Statically Determinate Systems

The simplest way to obtain an influence line is the use of the principle of virtual work (see p. 21–22).

As an example, we want to find the influence function $\mathbf{M}(\xi)$ for the bending moment at x in the beam of Fig. 28.1. To give the bending moment M at x a chance to do virtual work, we insert at this point an ideal hinge, and apply external couples M to both sides (Fig. 28.4a). Since the beam was statically determinate, it is now a mechanism with one degree of freedom. An (infinitesimal) virtual displacement is shown in Fig. 28.4*b*. The virtual work done by the moments M is $M\psi$. The work done by the load P is negative, namely, $-P\,\delta(\xi)$. The moment needed for equilibrium must satisfy the condition that

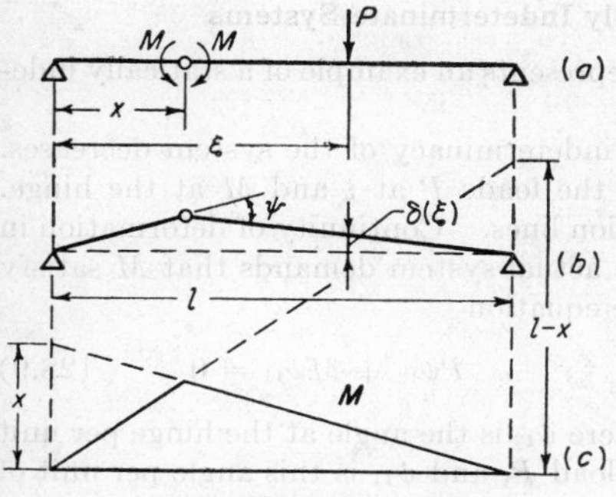

FIG. 28.4

$$M\psi - P\,\delta(\xi) = 0 \qquad (28.7)$$

whence

$$\mathbf{M} = \frac{M}{P} = \frac{\delta(\xi)}{\psi} \qquad (28.8)$$

In Fig. 28.4*c* these values of $\mathbf{M}$ have been plotted. This diagram is the influence line for the bending moment at x.

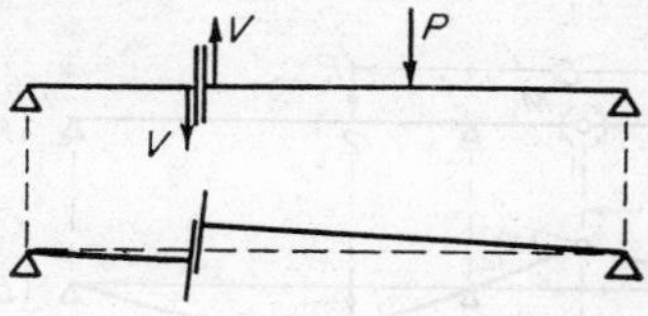

FIG. 28.5

To find the influence line for a shear force, we must replace the hinge by a connection which allows a pair of forces V to do virtual work (see Fig. 28.5).

To find the influence line for a reaction, we simply remove the corresponding support.

Deflections and Slopes

Figure 28.6 shows a beam with a load P placed in two different positions. From Maxwell's reciprocity law [Eq. (33.85*a*)], it follows that $w_{12} = w_{21}$. Now, for $x = \text{const}$ and variable ξ,

$$\frac{1}{P}\,w_{12}(\xi;x) = \mathbf{w}(\xi;x)$$

represents the influence line of the deflection of point 1, while w_{21}/P is the deflection of all the points 2 under a unit load applied at 1. It follows that the influence line

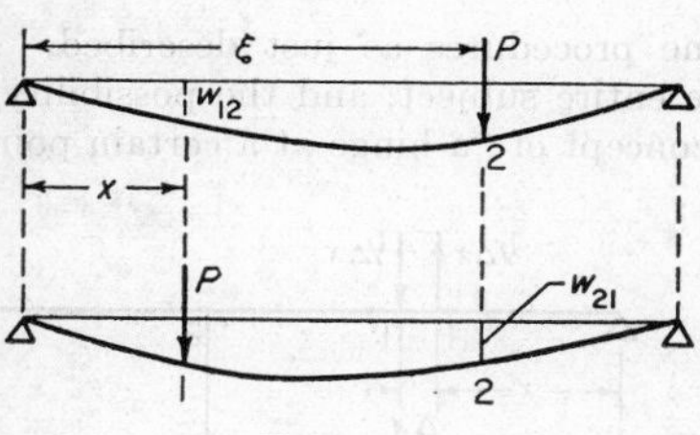

FIG. 28.6

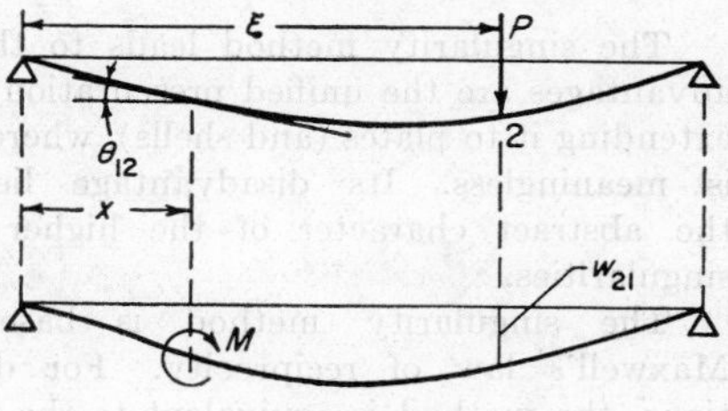

FIG. 28.7

for the deflection of a point is obtained by applying a unit load at this very point and drawing the deflection line.

This argument can easily be extended to slopes (Fig. 28.7). In this case, the reciprocity relation reads

$$M\theta_{12} = Pw_{21}$$

and the influence function

$$\boldsymbol{\theta}(\xi;x) = \frac{1}{P}\,\theta_{12} = \frac{1}{M}\,w_{21}$$

is found as the deflection at ξ produced by a unit couple acting at x. Note that the dimension of a deflection per unit of a moment is $L/FL = F^{-1}$, which is precisely the dimension of a slope per unit of force.

The preceding statements, which are based on Maxwell's reciprocity law, apply to statically determinate and indeterminate systems.

Forces and Moments in Statically Indeterminate Systems

If we disregard the hinge at x, Fig. 28.8a represents an example of a statically indeterminate beam. We want to find $\mathbf{M}(\xi;x)$.

When an ideal hinge is inserted at x, the indeterminacy of the system decreases. To this altered system, we apply separately the loads P at ξ and M at the hinge. Figure 28.8b,c shows the corresponding deflection lines. Continuity of deformation in the actual system demands that M satisfy the equation

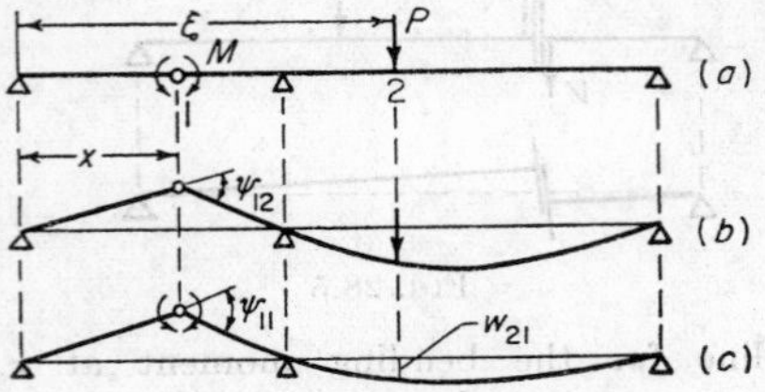

FIG. 28.8

$$P\psi_{12} + M\psi_{11} = 0 \qquad (28.9)$$

where ψ_{12} is the angle at the hinge per unit of load P, and ψ_{11} is this angle per unit of M. From (28.9) then follows that

$$\mathbf{M}(\xi;x) = \frac{M}{P} = -\frac{\psi_{12}}{\psi_{11}}$$

From the reciprocity law, we conclude that $\psi_{12} = w_{21}$, which is the deflection at ξ produced by $M = 1$, whence

$$\mathbf{M}(\xi;x) = -\frac{w_{21}}{\psi_{11}} \qquad (28.10)$$

This equation may be interpreted in two ways: (1) To find M, insert a hinge at x, apply unit couples to both sides of the hinge, determine the deflection line $w(\xi)$, and divide its ordinates by the relative rotation ψ_{11} found on this line; change signs. (2) To find M, insert a hinge as before; enforce there a relative rotation $\psi = -1$. The corresponding deflection $\mathbf{w}(\xi)$ is the influence ordinate $\mathbf{M}(\xi)$.

The same procedure applies to influence lines for the shear force V or a reaction R.

Singularity Method

The singularity method leads to the same procedures as just described. Its advantages are the unified presentation of the entire subject, and the possibility of extending it to plates (and shells), where the concept of "a hinge at a certain point" is meaningless. Its disadvantage lies in the abstract character of the higher load singularities.

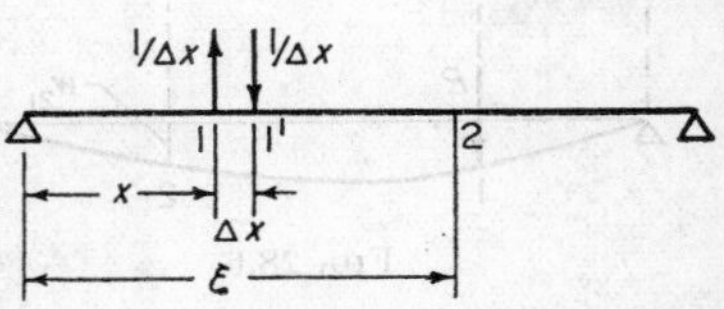

FIG. 28.9

The singularity method is based on Maxwell's law of reciprocity. For deflections, the method is equivalent to the statement made on p. 28–4: The influence line for the deflection $\mathbf{w}(\xi;x)$ is the deflection line for a unit load applied at x.

We next consider the slope $\boldsymbol{\theta}(\xi;x)$ at point 1 caused by a unit load at point 2 (see Fig. 28.9). If $w(\xi;x)$ is the corresponding deflection, then the definition of the differential quotient yields

$$\boldsymbol{\theta}(\xi;x) = \frac{\partial w(\xi;x)}{\partial x} = \lim_{\Delta x \to 0} \frac{w(\xi;x + \Delta x) - w(\xi;x)}{\Delta x}$$

Now, from the reciprocity law, $w(\xi;x) = w(x;\xi)$; hence

$$\boldsymbol{\theta}(\xi;x) = \lim_{\Delta x \to 0} \left[\frac{1}{\Delta x} w(x + \Delta x;\xi) + \left(-\frac{1}{\Delta x}\right) w(x;\xi) \right] \qquad (28.11)$$

i.e., this slope can be obtained as the limiting case ($\Delta x \to 0$) of the deflection at ξ, when the beam is loaded with an upward force $P = 1/\Delta x$ at point 1, and a downward force of the same magnitude at the neighbor point 1′. In the limit, this load is an external unit couple applied in clockwise direction at point 1. This leads to the same conclusion as on p. 28-3.

To obtain the influence line for the bending moment M, we start from the relation

$$\mathbf{M}(\xi;x) = -EI \frac{\partial^2 w(\xi;x)}{\partial x^2}$$

where EI is the flexural rigidity at the point x, and $w(\xi;x) \equiv \mathbf{w}(\xi;x)$ is the deflection at x caused by a unit load at ξ. The definition of the second derivative and the reciprocity law yield

$$\begin{aligned}\mathbf{M}(\xi;x) &= -EI \lim_{\Delta x \to 0} \frac{w(\xi;x+\Delta x) - 2w(\xi;x) + w(\xi;x-\Delta x)}{\Delta x^2} \\ &= -EI \lim_{\Delta x \to 0} \left[\frac{1}{\Delta x^2} w(x+\Delta x;\xi) + \left(\frac{-2}{\Delta x^2}\right) w(x;\xi) + \frac{1}{\Delta x^2} w(x-\Delta x;\xi)\right]\end{aligned} \tag{28.12}$$

i.e., the influence function $-\mathbf{M}/EI$ is found as the deflection of a beam to which, at three adjacent points, the loads $1/\Delta x^2$, $-2/\Delta x^2$, $1/\Delta x^2$ are applied. Figure 28.10*a* shows this load group. It has neither a resultant nor a resultant moment and, therefore, is not encountered when dealing with the equilibrium of *rigid* bodies. As is seen here, it has a significance as a load applied to elastic bodies. We call it a *unit pinch.*

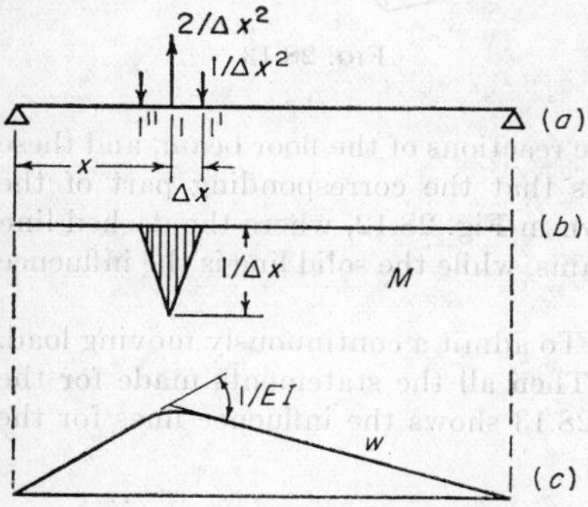

FIG. 28.10

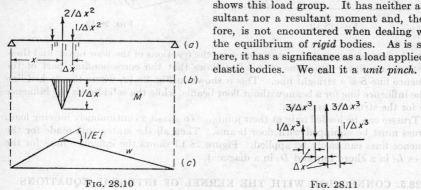

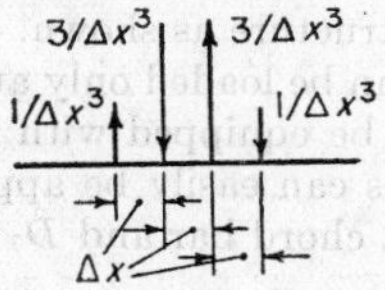

FIG. 28.11

Figure 28.10*b* shows the bending moments which a finite approximation of a unit pinch produces in a statically determinate beam. The total area of the moment diagram is equal to unity. It then follows from the first moment-area theorem (p. 26-21), that the slope of the deflection line changes between points 1″ and 1′ by $1/EI$. Figure 28.10*c* shows this deflection line. For the real unit pinch, i.e., for $\Delta x = 0$, the moment diagram is a Dirac δ function, and the deflection line consists of two straight parts. In statically indeterminate beams there will be nonzero reactions and, therefore, finite bending moments everywhere, but the deflection line will still have the same discontinuity of its tangent.

The unit pinch may be understood as a highly localized, self-equilibrating load, of such a kind that it produces a discontinuity $1/EI$ in the slope of the deflection line.

The interpretation of Eq. (28.12) leads to the interpretation 2 of Eq. (28.10) and a similar interpretation of Eq. (28.8), but it avoids the need for a hinge.

Continuing along the line followed so far, it may easily be seen that the influence line for a shear force is generated by the load singularity obtained in the limit $\Delta x \to 0$ from the load group shown in Fig. 28.11.

28.4. INDIRECT LOADING, TRUSSES

Bridges with floor beams present the problem illustrated by Fig. 28.12. The load moves along a set of simple beams (usually many more than shown in the figure), and an influence line for some stress or deformation quantity S in the main girder is wanted.

When the load stands over one of the crossbeams which connect the floor-beam chain with the main girder, then S is the same as though there were no floor beams at all. The corresponding point of the influence line **S** may be taken from the influence line for a simple beam. When the load is moved from one end of a floor beam to the

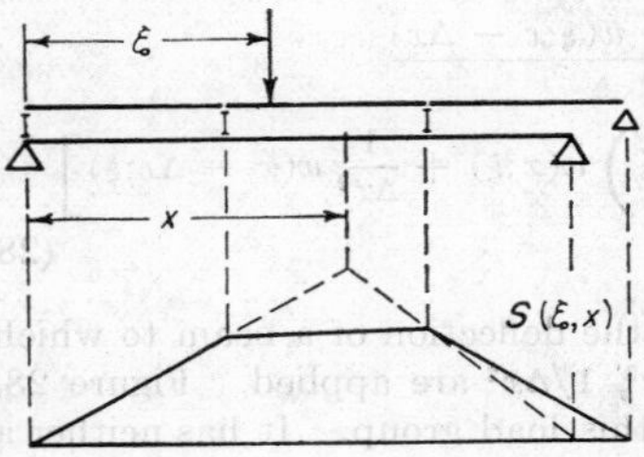

FIG. 28.12

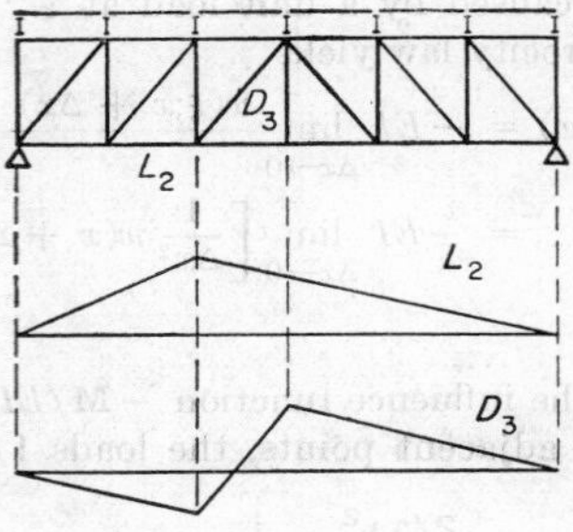

FIG. 28.13

next, the forces acting on the main girder are the reactions of the floor beam, and these reactions are linear functions of ξ. It follows that the corresponding part of the influence line **S** is a straight line. This is shown in Fig. 28.12, where the dashed line is the influence line for a beam without floor beams, while the solid line is the influence line for the structure as shown.

Trusses can be loaded only at their joints. To admit a continuously moving load, a truss must be equipped with floor beams. Then all the statements made for the influence lines can easily be applied. Figure 28.13 shows the influence lines for the forces L_2 in a chord bar and D_3 in a diagonal.

28.5. CONNECTION WITH THE KERNEL OF INTEGRAL EQUATIONS

Forced Deflection

A slender bar is put between two matching rigid dies (Fig. 28.14). When the upper die is lowered, a certain deflection $w(x)$ of the bar is enforced. We ask for the distribution of the load $q(x)$ acting between the dies and the bar.

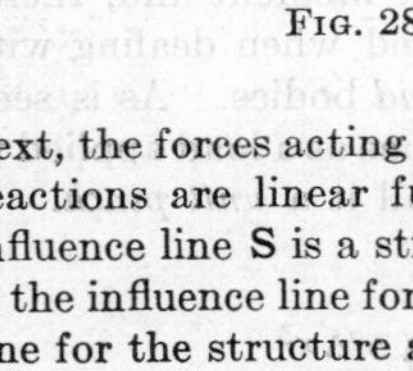

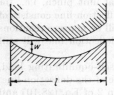

FIG. 28.14

The deflection at any point x is the sum of the influences of all the forces $q(\xi)\, d\xi$ acting on every line element $d\xi$ of the bar, which is supported at both ends. This influence is expressed by the influence integral (28.4):

$$w(x) = \int_0^l \mathbf{w}(\xi;x)q(\xi)\, d\xi \tag{28.13}$$

This is an integral equation of the form (17.1) with $f(\xi) = q(\xi)$, $g(x) = w(x)$, $K(x,\xi) = \mathbf{w}(\xi;x)$; that is, it is an inhomogeneous Fredholm equation of the first kind for $q(\xi)$.

Vibrations of a Beam

Let a beam perform free lateral vibrations. If μ is its mass per unit length, then an inertia load

$$q(\xi) = -\mu \frac{\partial^2 w(\xi,t)}{\partial t^2}$$

is acting on it, and the deflection at any point x is described by the influence integral (28.4):

$$w(x,t) = -\int_0^l \mathbf{w}(\xi;x)\mu(\xi) \frac{\partial^2 w(\xi,t)}{\partial t^2}\, d\xi \tag{28.14}$$

If the vibration is harmonic, $w(x,t) = W(x) \sin \omega t$, then

$$W(x) = \omega^2 \int_0^l \mathbf{w}(\xi;x)\mu(\xi) W(\xi)\, d\xi \tag{28.15}$$

This is an integral equation of the form (17.2) with $f(\xi) = W(\xi)$, $g(x) = 0$, $K(x,\xi) = \mathbf{w}(\xi;x)\mu(\xi)$; that is, it is a homogeneous Fredholm equation of the second kind. It determines an infinite number of discrete eigenvalues ω^2 and the corresponding eigenfunctions $W(\xi)$. Its kernel is not symmetric, but may be symmetrized by using $\mu^{1/2} W$ as the unknown function.

Buckling of Columns

Figure 28.15 shows an example of a straight prismatic bar subjected to the action of an axial compressive force P. For certain values of P, the critical or buckling loads, a lateral deflection can be maintained without applying a lateral load. Such a deflection must satisfy the differential equation (44.3). For constant EI and P, it reads

$$EI \frac{d^4 w(\xi)}{d\xi^4} = -P \frac{d^2 w(\xi)}{d\xi^2} \tag{28.16}$$

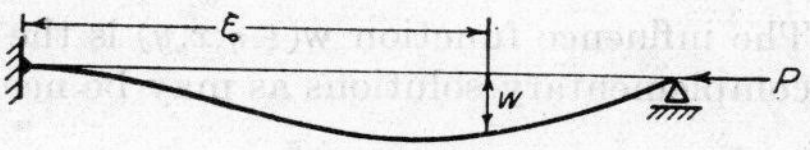

FIG. 28.15

When this equation is compared with Eq. (35.44), it is seen that its right-hand side can be interpreted as a fictitious load

$$q(\xi) = -P \frac{d^2 w(\xi)}{d\xi^2} \tag{28.17}$$

It is then possible to apply the influence integral (28.4) and to write the deflection $w(x)$ in terms of the load $q(\xi)$:

$$w(x) = -P \int_0^l \mathbf{w}(\xi;x) \frac{d^2 w(\xi)}{d\xi^2}\, d\xi \tag{28.18}$$

It should be noted that this equation does not apply to columns whose ends are not laterally supported. In such cases, further terms appear, which complicate the situation substantially.

When the right-hand side of Eq. (28.18) is subjected to repeated integration by parts, it turns out that the boundary terms are zero, and the equation assumes the form

$$w(x) = -P \int_0^l \frac{\partial^2 \mathbf{w}(\xi;x)}{\partial \xi^2} w(\xi)\, d\xi \tag{28.19}$$

and this is a Fredholm integral equation of the second kind. The critical values of P are its eigenvalues, and its eigenfunctions $w(x)$ are the buckling modes.

28.6. INFLUENCE DIAGRAMS FOR PLATES

Deflection

We consider a plate of arbitrary shape, and we admit any kind of linear boundary conditions to be imposed along its edge. At point 2, a concentrated force P is applied. When we keep point 2 fixed, calculate the deflections w_{12} of all points 1 of the plate, and plot them as ordinates at these points, we have the deflection surface pertaining to the load P. When we keep point 1 (the observation point) fixed, let point 2 (the load point) assume all possible positions on the plate, and plot the deflections w_{12} at points 2, we have the influence surface for the deflection of point 1 and for a moving load P. When we divide by the load, we obtain the influence function $\mathbf{w}(\xi,\eta;x,y) = w_{12}/P$. It depends on the coordinates ξ, η of the load point and on the parameters x, y, which are the coordinates of the observation point.

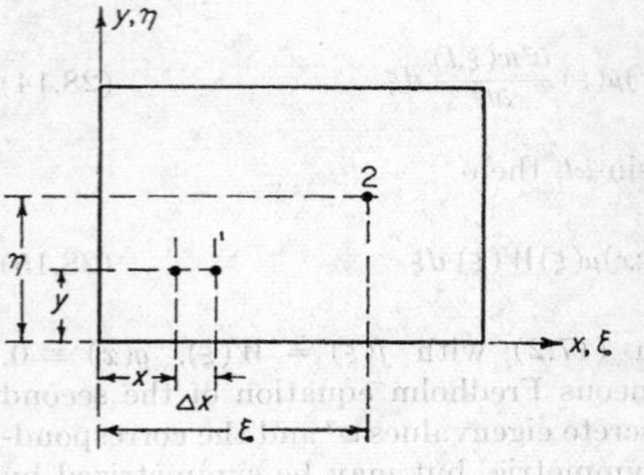

FIG. 28.16

According to Maxwell's reciprocity law, we have $w_{12} = w_{21}$; that is, this influence surface is identical with the deflection surface produced by a unit load at point 1.

The singular solution of the plate equation (39.10′), which corresponds to the presence of a force P at (x,y), is

$$w(\xi,\eta;x,y) = \frac{P}{16\pi K}[(\xi - x)^2 + (\eta - y)^2] \ln [(\xi - x)^2 + (\eta - y)^2] \qquad (28.20)$$

The influence function $\mathbf{w}(\xi,\eta;x,y)$ is the sum of this solution with $P = 1$ and of such complementary solutions as may be needed to satisfy the boundary conditions.

Slope

The slope $\partial w/\partial x$ at point 1, caused by a unit load at 2, is defined as

$$(w_x)_1 \equiv \left(\frac{\partial w}{\partial x}\right)_1 = \lim_{\Delta x \to 0} \frac{w_{1'2} - w_{12}}{\Delta x}$$

The same procedure which led to Eq. (28.11) shows that the influence function $\mathbf{w}_x(\xi,\eta;x,y)$ can be obtained as the deflection surface produced by a unit couple applied at point (x,y), and acting as shown in Fig. 28.17.

The singular solution of the plate equation, which corresponds to the action of an external couple M as shown, is obtained by writing Eq. (28.20) for forces $-P$ and P at 1 and 1′, respectively; adding both expressions; and then going to the limit $\Delta x \to 0$ with $P \cdot \Delta x = M$. This process is identical with replacing P by M in (28.20) and differentiating partially with respect to x. In either way we obtain, after dropping an irrelevant regular term, the following singular plate solution:

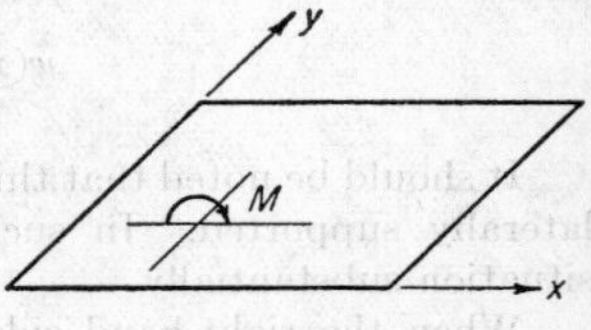

FIG. 28.17

$$w(\xi,\eta;x,y) = \frac{M}{8\pi K}(x - \xi) \ln [(x - \xi)^2 + (y - \eta)^2] \qquad (28.21)$$

The influence function $\mathbf{w}(\xi,\eta;x,y)$ consists of this solution with $M = 1$ plus the appropriate complementary solutions, as demanded by the boundary conditions.

Bending Moment, Twisting Moment, Shear

According to Eq. (39.3), the bending moment M_x depends on the second derivatives of $\mathbf{w}(\xi,\eta;x,y)$ with respect to x and y. Applying the same argument as used with Eq. (28.12), it can easily be shown that the influence function for the second derivative

$$\mathbf{w}_{xx}(\xi,\eta;x,y) \equiv \frac{\partial^2 w(\xi,\eta;x,y)}{\partial x^2}$$

can be obtained as the deflection of the point (ξ,η) under the action of a unit pinch applied at (x,y), its three component forces being lined up along a line y = const, as shown in Fig. 28.18.

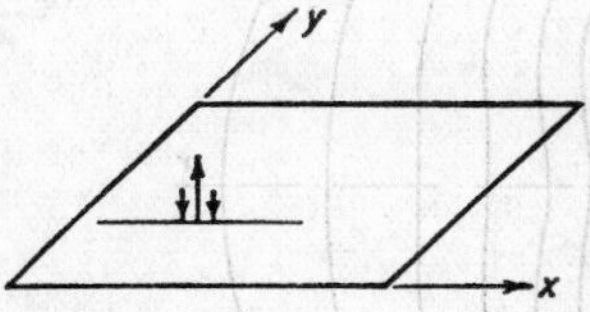

FIG. 28.18

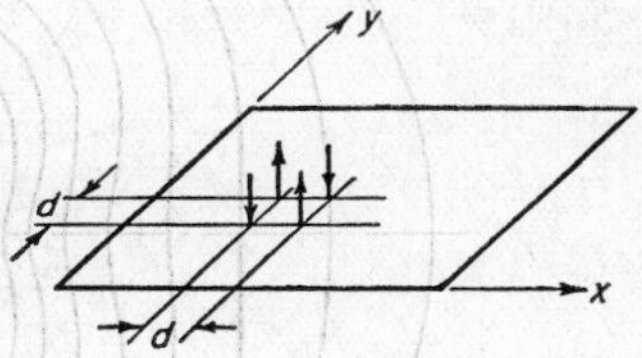

FIG. 28.19

The generating singularity for this load can be found from Eq. (28.20). It is

$$w(\xi,\eta;x,y) = \frac{1}{8\pi K}\left\{\ln\left[(x-\xi)^2 + (y-\eta)^2\right] + \frac{2(x-\xi)^2}{(x-\xi)^2+(y-\eta)^2}\right\} \tag{28.22}$$

For the twisting moment, a load singularity is needed, which consists of four forces of magnitude $1/d^2$, arranged as shown in Fig. 28.19. They may be interpreted as two couples $M = 1/d$, acting in opposite sense in two parallel planes at distance d. This singularity, which is akin to the quadrupole of potential fields, is connected with the singular solution

$$w(\xi,\eta;x,y) = \frac{1}{4\pi K}\frac{(x-\xi)(y-\eta)}{(x-\xi)^2+(y-\eta)^2} \tag{28.23}$$

According to Eq. (39.8a), the shearing force Q_x is

$$Q_x = -K\left(\frac{\partial^3 w}{\partial x^3} + \frac{\partial^3 w}{\partial x\,\partial y^2}\right)$$

Each of the third-order derivatives requires a load singularity consisting of two pinches of magnitude $1/d$, applied at adjacent points and with opposite signs. Figure 28.20a,b shows the two parts separately, and Fig. 28.20c shows the complete load singularity. The corresponding singular solution of Eq. (39.10′) is

$$w(\xi,\eta;x,y) = \frac{1}{2\pi K}\frac{x-\xi}{(x-\xi)^2+(y-\eta)^2} \tag{28.24}$$

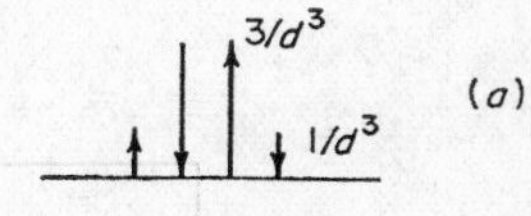

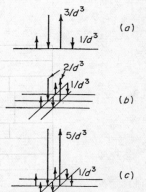

FIG. 28.20

Graphic Representation of Influence Functions

There exist three methods of graphic representation of influence functions which depend on two coordinates, ξ and η. One of them consists in drawing lines $\mathbf{S}$ = const in the ξ,η plane. Figure 28.21 shows as an example the influence field $\mathbf{M}_x$ at the center of a simply supported circular plate ($\nu = 0$). At the center, the influence ordinate is infinite, corresponding to the fact that, at the point of application of a concentrated force, the plate has infinite bending moments.

A second way of presentation is the *influence pavement*. An example is shown in Fig. 28.22 for a rectangular plate. The plate is divided into a number of strips, and each of them is subdivided into rectangles of such length, that, for each of them,

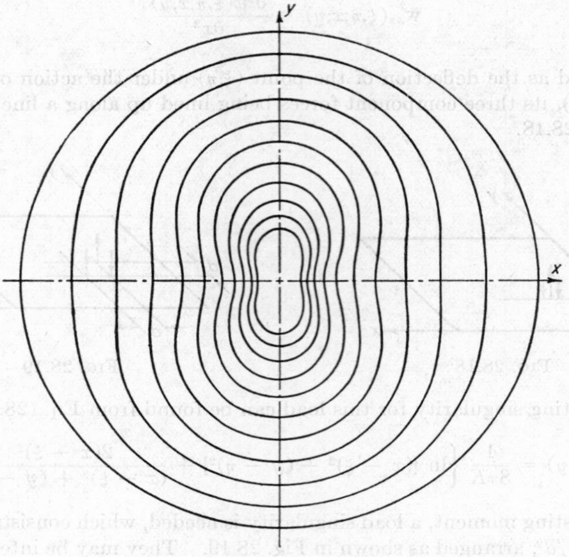

FIG. 28.21

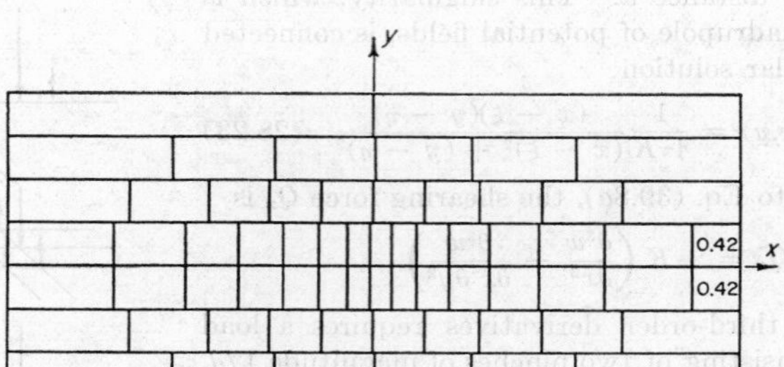

FIG. 28.22

$\int S \, d\xi \, d\eta$ has the same constant value. At the end of each strip there is, of course, a remainder rectangle, for which the integral is only a fraction of the standard value.

A third method of presentation consists in plotting a number of profiles through the three-dimensional influence surface $S(\xi,\eta)$, for example, along a number of lines $y =$ const in Fig. 28.22.

Each of these methods has its advantages and its disadvantages, both in producing and in using the resulting diagrams.

28.7. ROTATING LOAD

Quite different from all the influence functions described so far is the one for a load which is always applied at the same point, but in varying directions. Such a load, as shown in Fig. 28.23, can always be resolved into components

$$V = P \cos \theta \qquad H = P \sin \theta$$

Let $\mathbf{S}(\theta)$ be the influence function, i.e., some displacement, bending moment, or the like, per unit of P. Then $\mathbf{S}$ can be represented as the sum of the influences of V and H.

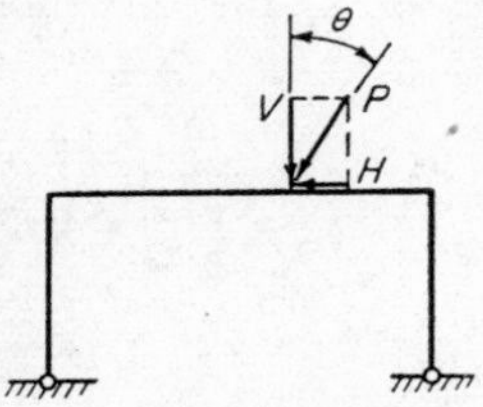

Fig. 28.23

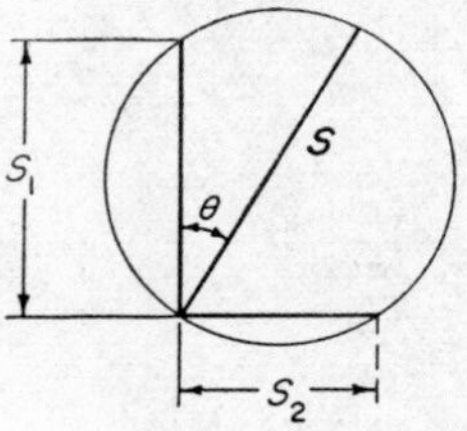

Fig. 28.24

Let S_1 be the value of S per unit of V, S_2 the value per unit of H, then

$$\mathbf{S}(\theta) = S_1 \cos \theta + S_2 \sin \theta \tag{28.25}$$

It can easily be seen that, in this case, the influence line, plotted as a polar diagram, is a circle like the one shown in Fig. 28.24.

28.8. REFERENCES

Influence lines for beams and trusses may be found in most of the textbooks listed in Chaps. 26 and 27. The following papers and books deal with the singularity method and with influence diagrams for plates.

[1] P. Neményi: Über die Singularitätenmethode in der Elastizitätstheorie, *Z. angew. Math. Mech.*, **10** (1930), 383–399.
[2] A. Pucher: Über die Singularitätenmethode an elastischen Platten, *Ing.-Arch.*, **12** (1941), 76–100.
[3] E. Bittner: "Momententafeln und Einflussflächen für kreuzweis bewehrte Eisenbetonplatten," Springer, Vienna, 1938.
[4] H. Olsen, F. "Reinitzhuber: Die zweiseitig gelagerte Platte," vols, 1 and 2, Ernst, Berlin, 1950, 1951.
[5] A. Pucher: "Einflussfelder elastischer Platten," Springer, Vienna, 1951.
[6] G. Hoeland: "Stützmomenten-Einflussfelder durchlaufender Platten," Springer, Berlin, 1957.

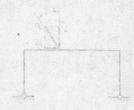

CHAPTER 29

TORSION-BOX ANALYSIS

BY

P. KUHN, Newport News, Va.

29.1. INTRODUCTION

A torsion box is an elongated four-wall shell, designed to transmit torques along its length. Along the four edges, distinct corner flanges are usually provided. At the ends of the box and at intermediate stations, bulkheads (diaphragms) are provided.

Chiefly as a device for simplifying the presentation, the concept of an idealized torsion box is used. The theory of the idealized box is followed by a discussion of the analysis of real boxes, which uses this theory as an intermediate step.

The theory presented here is the simplest theory possible for cases beyond the scope of the Bredt-Batho formula (constant-section tube loaded at the ends and free to warp); see Eqs. (36.44) and (36.45). It is considered to be adequate for most civil engineering structures. The theory was formulated by H. Reissner [1], and developed by H. Ebner [2]; a detailed presentation is given in [3].

29.2. IDEALIZED BOX OF RECTANGULAR DOUBLY SYMMETRICAL CROSS SECTION

In an idealized torsion box, the walls are assumed to carry only shear stresses. At the four edges of the box, concentrated flanges of cross-sectional area A are provided. Figure 29.1 shows the cross section of such a box; the convention of showing the flanges as circles is used to symbolize that the box is idealized.

The box is divided into bays (Fig. 29.2a) by bulkheads (diaphragms), which are assumed to be rigid in their own planes, but to offer no resistance to warping out of their planes. Torques ΔM are assumed to be applied only at bulkheads. At one bulkhead, supports are provided to furnish the

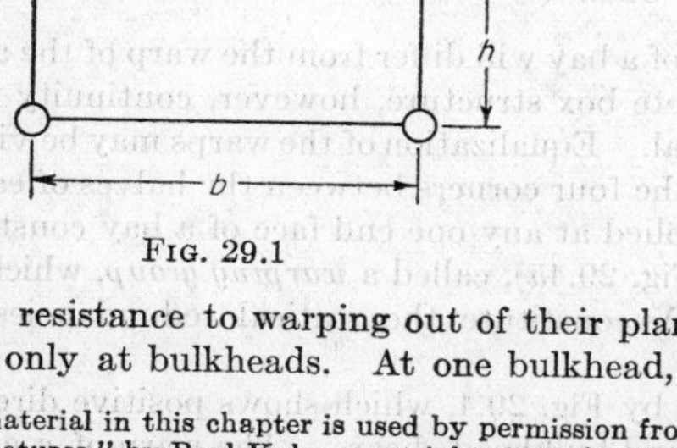

Fig. 29.1

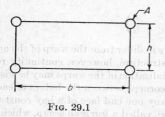

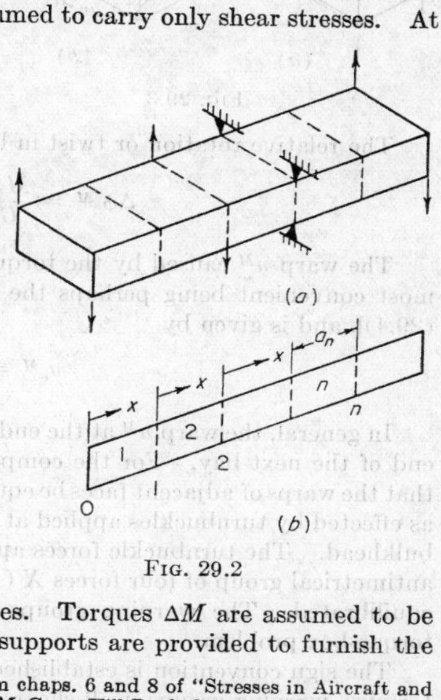

Fig. 29.2

The material in this chapter is used by permission from chaps. 6 and 8 of "Stresses in Aircraft and Shell Structures," by Paul Kuhn, copyright, 1956, by the McGraw-Hill Book Company, Inc.

reaction torque. Starting at one end, the bays and bulkheads are numbered as indicated in Fig. 29.2*b*.

Imagine now that each bulkhead is split on its median plane, and that the two halves are separated and reconnected by a short torque tube (Fig. 29.3*a*). This change converts the torsion box into a statically determinate structure. Each bay is a simple torsion tube, and the shear flow $T_n = (\tau t)_n$ in each wall of bay n is given by the Bredt formula (36.44):

$$T_{bn}{}^M = T_{hn}{}^M = \frac{M_n}{2bh} \tag{29.1}$$

where

$$M_n = \sum_0^{n-1} \Delta M$$

Superscripts are used to denote the force (here, the torque) causing the stress or deformation under consideration (here, the shear flow). The subscripts b and h refer to the horizontal and vertical walls, respectively.

Under the action of the torque M_n, the nth bay undergoes two types of deformations in the statically determinate configuration:

1. One end bulkhead rotates with respect to the other end bulkhead through the angle $\Delta\phi_n$.

2. Each bulkhead warps out of its original plane. The amount of this warp, measured at one corner, is denoted by u (Fig. 29.3*b*). Because of the double symmetry postulated, the (absolute) magnitudes of u are the same at the four corners.

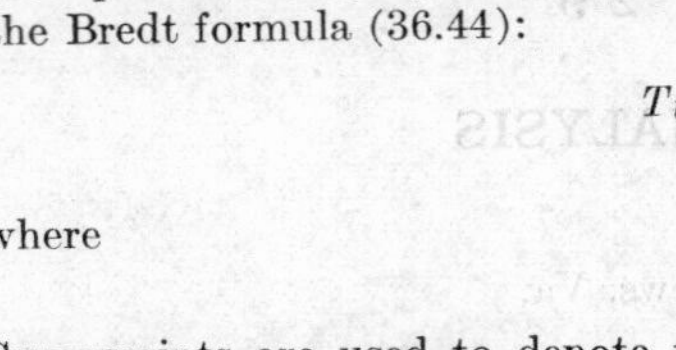

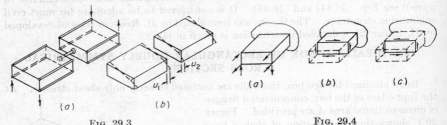

Fig. 29.3

Fig. 29.4

The relative rotation or twist in bay n is given by Eq. (36.46):

$$\Delta\phi_n{}^M = \frac{M_n a_n}{GJ} = \frac{M_n a}{2Gb^2h^2}\left(\frac{b}{t_b} + \frac{h}{t_h}\right) \tag{29.2}$$

The warp u^M caused by the torque can be calculated readily in several ways, the most convenient being perhaps the method of dummy unit loading [utilizing Eqs. (29.4)], and is given by

$$u_n{}^M = \frac{M_n}{8Gbh}\left(\frac{b}{t_b} - \frac{h}{t_h}\right) \tag{29.3}$$

In general, the warp u^M at the end of a bay will differ from the warp of the adjacent end of the next bay. For the complete box structure, however, continuity requires that the warps of adjacent faces be equal. Equalization of the warps may be visualized as effected by turnbuckles applied at the four corners between the halves of each split bulkhead. The turnbuckle forces applied at any one end face of a bay constitute an antimetrical group of four forces X (Fig. 29.4*a*), called a *warping group*, which is self-equilibrated. The warping groups X constitute the statical redundancies of the torque-box problem.

The sign convention is established by Fig. 29.4, which shows positive direction of torques, warping groups, wall shears and bulkhead shears. The warp of a cross section is considered positive when it is in the direction of a positive warping group applied at the lower-numbered ("outboard") end of a bay.

The shear flows caused in the walls of an individual ("free") bay by the application of a warping group X on one end can be calculated readily from equilibrium equations, and the force in each flange varies linearly from X at the end of application to zero at the other end. (The sign of the stress, tension, or compression is established by inspection with the aid of Fig. 29.4*a*.) The total stress system in the nth bay of the complete torque box under the simultaneous action of the torque and the warping group at each end of the bay is given by the formulas

$$\sigma_n = \frac{(a_n - x)X_{n-1}}{a_n A_n} + \frac{xX_n}{a_n A_n} \tag{29.4a}$$

$$T_{bn} = \frac{M_n}{2bh} - \frac{X_n - X_{n-1}}{2a_n} \tag{29.4b}$$

$$T_{hn} = \frac{M_n}{2bh} + \frac{X_n - X_{n-1}}{2a_n} \tag{29.4c}$$

and the shear flow in bulkhead n is given by

$$T_{Bn} = \frac{X_n - X_{n-1}}{2a_n} - \frac{X_{n+1} - X_n}{2a_{n+1}} \tag{29.4d}$$

A warping group applied at one end of an individual free bay causes the cross sections to warp in a similar manner as a torque. Denote by p the warp at the outboard end with a warping group of unit magnitude applied at this end, and by q the warp at the inboard end with a unit warping group applied at the outboard end. Expressions for p and q can be established readily by the method of dummy unit loading [utilizing Eqs. (29.4)], and are:

$$p = \frac{a_n}{3EA} + \frac{1}{8Ga_n}\left(\frac{b}{t_b} + \frac{h}{t_h}\right) \qquad q = -\frac{a_n}{6EA} + \frac{1}{8Ga_n}\left(\frac{b}{t_b} + \frac{h}{t_h}\right) \tag{29.5}$$

Utilizing these unit warps p and q, the total warp $(u^M + u^X)$ of the nth bay at bulkhead n can be written as

$$u(n)_n = u_n{}^M + q_n X_{n-1} - p_n X_n$$

and the total warp of bay $(n + 1)$ at bulkhead n can be written as

$$u(n + 1)_n = u^M_{n+1} + p_{n+1} X_n - q_{n+1} X_{n+1}$$

Equating the two warps as required by the condition of continuity of the box at bulkhead n and rearranging terms gives the recurrence formula,

$$q_n X_{n-1} - (p_n + p_{n+1})X_n + q_{n+1}X_{n+1} = -u_n{}^M + u^M_{n+1} \tag{29.6}$$

Substitution of successive values of n (1,2, . . .) into this formula yields a set of equations from which the warping groups X can be calculated. Equations (29.4) can then be used to calculate the flange stresses and the shear flows in the walls and in the bulkheads.

The warping group X_0 at the near end of the box is zero for a case such as shown in Fig. 29.2*a*, and similarly at the far end. In some cases, an external tip structure exists which applies a warping group X_0 of known magnitude to the end bulkhead. If an end of the box is attached to a rigid abutment, the abutment is considered as an additional bay having $p = q = u^M = 0$. (It should be noted that a practical end bulkhead, even if very heavy by usual engineering standards, as a rule does not have anywhere near sufficient stiffness against warping out of its plane to justify the assumption that it is rigid against warping. The stiffness of such an end plate should be estimated and used in the calculation. The condition of zero warping exists in practical structures only as a result of symmetry.)

In practical box beams, a support similar to that shown in Fig. 29.2*a* is usually provided at another bulkhead, in order to enable the box to carry bending due to its own weight and to other loads. The second torque reaction is then an external redundancy, and must be found by an appropriate calculation.

The two warping groups acting at the ends of the *n*th bay produce a twist of the bay, which is given by the formula

$$\Delta\phi_n{}^X = -\frac{1}{2bhG}\left(\frac{b}{t_b} - \frac{h}{t_h}\right)(X_n - X_{n-1}) \tag{29.7}$$

which reduces the twist $\Delta\phi_n{}^M$ caused directly by the torque in the usual case where $b/t_b > h/t_h$ and $X_n > X_{n-1}$. The reduction in twist can be very substantial near supports and near large torques.

29.3. THE REAL BOX OF RECTANGULAR CROSS SECTION

The action of a real box (Fig. 29.5*a*) is quite similar to that of an idealized box; interference between the warping of adjacent bays produces longitudinal stresses and

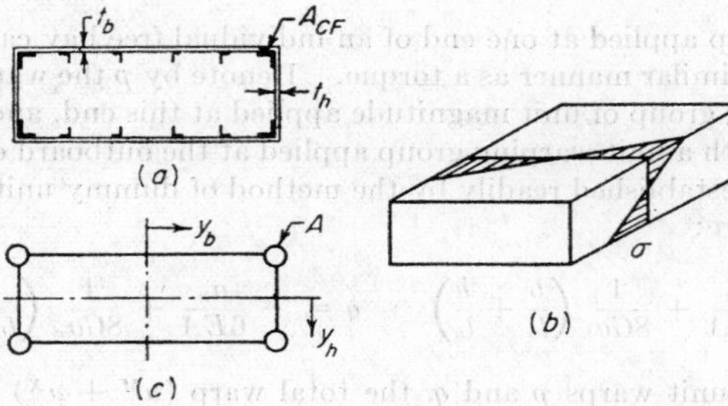

FIG. 29.5

accompanying changes in the shear flows. The longitudinal stresses are tensile along one edge of any one wall, and vary gradually to compressive values along the next edge; in other words, each wall is subjected to bending in its own plane. The difference between the real box and the idealized box lies chiefly in the fact that each wall of the real box develops longitudinal stresses over its entire width instead of only at the edges; since these longitudinal stresses vary over the width of the wall, the shear flow also varies.

In first approximation, it may be assumed that the longitudinal stresses in each wall are distributed linearly across each wall (Fig. 29.5*b*), in conformance with the engineering theory of bending. From this theory, the contribution of the web of a plate girder to the bending-moment resistance of the girder can be expressed by adding one-sixth of the web area to each flange (and considering the web then as shear element only). In the same manner, the contribution of the walls of a torsion box to the system of longitudinal stresses can be expressed by using an idealized box (Fig. 29.5*c*) having corner flanges of an area

$$A = A_{CF} + \tfrac{1}{6}b\bar{t}_b + \tfrac{1}{6}h\bar{t}_h \tag{29.8}$$

where A_{CF} is the area of the real corner flange, $\bar{t}_b$ the effective thickness of a horizontal wall, and $\bar{t}_h$ the effective thickness of a vertical wall. The effective thickness of a horizontal wall is defined by

$$\bar{t}_b = t_b + \frac{A_{ST}}{b}$$

where t_b is the actual thickness of the plate, and A_{ST} is the total area of all the (closely spaced) stringers attached to the plate. The actual thickness t governs the shear stiffness of each wall and is, therefore, used in Eqs. (29.2), (29.3), (29.5), and (29.7); it is also used to compute shear stresses from shear flows.

To summarize, then, the analysis of a real torque box is effected by substituting idealized cross sections (Fig. 29.5*c*) for real ones (Fig. 29.5*a*), and analyzing the idealized box by the formulas of the preceding section. The flange stresses obtained by Eq. (29.4*a*) represent the corner stresses in the real box; the longitudinal stresses in the walls follow directly (Fig. 29.5*b*). The bulkhead shear flows T_B represent directly those in the real box. The wall shear flows T_b and T_h obtained by Eqs. (29.4*a*,*b*), however, represent only average values, because that portion of these shear flows which is associated with the contribution of the walls to the bending action is distributed parabolically over the cross section, in conformance with Eq. (35.31) of the engineering theory of bending. To obtain the detailed distribution, the second terms in Eqs. (29.4*b*,*c*) must be modified to read:

$$T_b = -\frac{X_n - X_{n-1}}{2a}\left(1 + \frac{b\bar{t}_b}{6A} - \frac{2\bar{t}_b}{bA}\,y_b^2\right) \qquad (29.9a)$$

$$T_h = \frac{X_n - X_{n-1}}{2a}\left(1 + \frac{h\bar{t}_h}{6A} - \frac{2t_h}{hA}\,y_h^2\right) \qquad (29.9b)$$

To be consistent with this distribution, the expressions (29.5) for the unit warps p and q should be modified by changing the last term to read

$$\left(\frac{b}{t_b} + \frac{h}{t_h} + \frac{b^3\bar{t}_b^2\bar{t}_h + h^3\bar{t}_h^2\bar{t}_b}{45A^2\bar{t}_b\bar{t}_h}\right) \qquad (29.10)$$

The engineering theory of bending of a beam becomes inaccurate when the warping of the cross section due to shear is interfered with; the theory of the torsion box presented here, consequently, suffers from the same defect. A more exact theory would require specific attention to the manner in which the edges of the bulkheads deform; such a theory does not appear to have been published. Some empirical rules may be given, however, for estimating the distribution of the shear flows as follows:

1. At a station where the warping is zero (rigid abutment or plane of symmetry), the shear flow may be taken as distributed uniformly over the width of each wall.

2. At an infinitesimal distance from the plane of application of a concentrated torque (not near either end of the box), the distribution of the shear due to this torque may be taken as halfway between a uniform distribution and the parabolic one defined by Eqs. (29.9*a*,*b*).

3. At a distance greater than one-half the width of the wall from a rigid root or a concentrated torque, the parabolic distributions (29.9*a*,*b*) may be used. For intermediate distances, linear interpolation may be used.

Modifications of the flange stresses corresponding to these empirical modifications of the shear flows are believed to be unimportant when the corner flange is not too small (say $A_{CF} > 0.2A$). When the corner flange is very small or absent, the accuracy of the entire theory may be low.

The assumption of all bulkheads being rigid against shear is probably adequate for most structures encountered in civil engineering. For certain types of aircraft structures, it may be desirable or necessary to use a theory which takes shear deformation of the bulkheads into account [2],[3].

29.4. BOX TAPERED IN WIDTH AND DEPTH

A box which has a slow taper in width and depth can be analyzed with adequate accuracy by the usual device of using the formulas for an untapered box and representing each quantity by the average value for each bay in turn. Thus, with linear taper, the unit warps p and q [Eqs. (29.5)] and the warp due to torque [Eqs. (29.3)] are computed by inserting for b and h the values for the middle of the bay. The shear flows due to the warping group [Eqs. (29.4*b,c*), second terms] may be considered as constant from bulkhead to bulkhead. The shear flow due to torque, however, varies more markedly within each bay and should be computed for each end of a bay by Eqs. (29.1) where necessary.

A somewhat more elaborate theory of the tapered box has been developed [2], [3]. Comparative calculations show [3] that the results obtained by this theory and by the approximate method outlined above differ very little for a (total) taper angle as high as 14°, and thus justify the use of the approximate method in this range. For large taper angles, drastic modifications of the torsion theory are necessary, according to Southwell [4], and it is not known whether the idealizing concept of cover plates carrying only shear is still useful for such configurations.

29.5. CURVILINEAR DOUBLE SYMMETRICAL CROSS SECTIONS

Curvilinear cross sections (Fig. 29.6) with double symmetry can be treated by slight extensions of the formulas for rectangular cross sections, with the aid of an auxiliary constant

$$\alpha = \frac{F_h - F_b}{F} \tag{29.11}$$

where F_h and F_b are the areas defined in the figure, and F is the total enclosed area of the cross section. The shear flows caused by an outboard warping group acting on a bay can be written in the form

$$T_b = \frac{X}{2a}(1 + \alpha) \qquad T_h = -\frac{X}{2a}(1 - \alpha) \tag{29.12}$$

and, consequently, Eqs. (29.4*a,b*) are modified simply by adding a $(1 + \alpha)$ or $(1 - \alpha)$ term, respectively, to the X terms. The equations for the warps take the modified form:

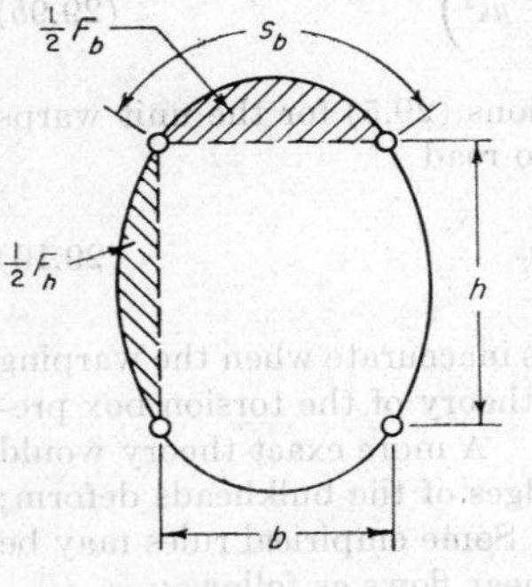

Fig. 29.6

$$u^M = \frac{M}{8GF}\left[\frac{s_b}{t_b}(1 + \alpha) - \frac{s_h}{t_h}(1 - \alpha)\right] \tag{29.13}$$

$$p = \frac{a}{3EA} + \frac{1}{8aG}\left[\frac{s_b}{t_b}(1 + \alpha)^2 + \frac{s_h}{t_h}(1 - \alpha)^2\right]$$

$$q = -\frac{a}{6EA} + \frac{1}{8aG}\left[\frac{s_b}{t_b}(1 + \alpha)^2 + \frac{s_h}{t_h}(1 - \alpha)^2\right] \tag{29.14a,b}$$

where s_b and s_h are *developed* widths (Fig. 29.6).

29.6. ARBITRARY FOUR-FLANGE SECTIONS

Arbitrary four-flange sections are treated best numerically from the outset. The procedure will be outlined for the example shown in Fig. 29.7.

The four forces comprising the warping group may be written in the form $k_i X$ (Fig. 29.7*b*). One of the flanges is chosen as *reference flange;* with the flanges numbered as in Fig. 29.7a, either flange 1 or flange 3 should be chosen as reference, in order to retain the sign convention used previously. Here, flange 1 will be chosen as reference; the coefficient k_1 is then assigned unit value. The coefficients k_2, k_3, and k_4 are calculated from the equilibrium conditions for a warping group, in this case (omitting the factor X and assigning temporarily positive signs to tensile forces),

$$1 + k_2 + k_3 + k_4 = 0$$
$$30k_3 + 30k_4 = 0$$
$$10k_2 + 5k_3 = 0$$

and thus, finally,

$$k_1 = 1 \qquad k_2 = -1 \qquad k_3 = 2 \qquad k_4 = -2$$

The warping group defined by these coefficients is the *standard* for the case considered.

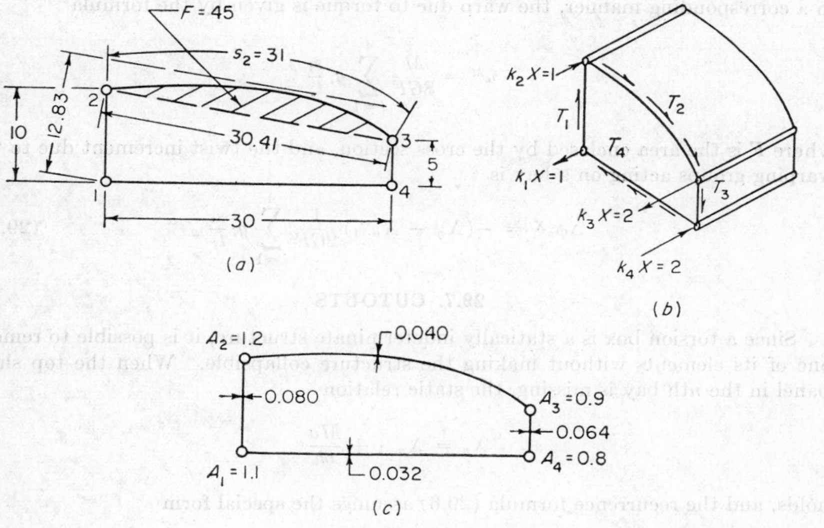

FIG. 29.7

The shear flows T_1 to T_4 caused in an isolated bay by the application of a standard warping group are calculated by equilibrium considerations. The torque equation for the cross section, and the equation of longitudinal equilibrium for each of three flanges may be used. To facilitate application of the torque equation, Fig. 29.7a shows the line of action of the resultant of the shear flow in the curved wall and its moment arm about flange 1. The equilibrium equations used are

$$T_2 \times 30.41 \times 12.83 + T_3 \times 5 \times 30 = 0$$
$$X + T_1 - T_4 = 0 \qquad X + T_1 - T_2 = 0 \qquad 2X + T_3 - T_2 = 0$$

with the solution

$$T_1 = -\frac{4}{9}\frac{X}{a} = g_1\frac{X}{a} \qquad T_2 = \frac{5}{9}\frac{X}{a} = g_2\frac{X}{a}$$
$$T_3 = -\frac{13}{9}\frac{X}{a} = g_3\frac{X}{a} \qquad T_4 = \frac{5}{9}\frac{X}{a} = g_4\frac{X}{a}$$

where new coefficients g_i have been introduced for convenience in the next step.

The warping of the end face of a bay may be defined as the displacement of the end face of the reference flange (here, flange 1) with respect to a plane passed through the end faces of the remaining three flanges. One-fourth of this warping may be regarded as representing the average warp of the cross section, and this *average warp* will be used in order to permit direct comparison with the warps for rectangular sections, as defined previously in Eq. (29.5). The formula for the warp p due to $X = 1$ becomes

$$p = \frac{a}{12E} \sum_{i=1}^{4} \frac{k_i^2}{A_i} + \frac{1}{4aG} \sum_{i=1}^{4} g_i^2 \frac{s_i}{t_i} \tag{29.15}$$

where the coefficients g_i are those used to define the shear flows due to a warping group, and s_i is the developed width of wall i, as in Eqs. (29.13) and (29.14). The formula for the unit warp q is obtained from Eq. (29.15) by substituting $-1/24$ for $1/12$. Defined in a corresponding manner, the warp due to torque is given by the formula

$$u^M = \frac{M}{8GF} \sum_{i=1}^{4} g_i \frac{s_i}{t_i}$$

where F is the area enclosed by the cross section, and the twist increment due to the warping groups acting on a bay is

$$\Delta\phi_n{}^X = -(X_n - X_{n-1}) \frac{1}{2GF} \sum_{i=1}^{4} g_i \frac{s_i}{t_i} \tag{29.16}$$

29.7. CUTOUTS

Since a torsion box is a statically indeterminate structure, it is possible to remove one of its elements without making the structure collapsible. When the top shear panel in the nth bay is missing, the static relation

$$X_n = X_{n-1} + \frac{Ma}{bh}$$

holds, and the recurrence formula (29.6) assumes the special form

$$q_{n-1}X_{n-2} - \left(p_{n-1} + p_{n+1} + \frac{a}{EA}\right) X_{n-1} + q_{n+1}X_{n+1} = -u_{n-1}^M + u_{n+1}^M + \frac{Ma}{bh}\left(p_{n+1} + \frac{a}{2EA}\right) \tag{29.17}$$

Here A is the cross section of the corner flange in the nth bay, and the quantities a, b, h and M are also those pertinent to this bay.

29.8. REFERENCES

[1] H. Reissner: Neuere Probleme der Flugzeugstatik, *Z. Flugtechn. Motorl.*, **17** (1926), 384–393; **18** (1927), 153–158.
[2] H. Ebner: Die Beanspruchung dünnwandiger Kastenträger auf Drillung bei behinderter Querschnittswölbung, *Z. Flugtechn. Motorl.*, **24** (1933), 645–655, 684–692.
[3] Paul Kuhn: "Stresses in Aircraft and Shell Structures," McGraw-Hill, New York, 1956.
[4] R. V. Southwell: On the torsion of conical shells, *Proc. Roy. Soc., London*, **163** (1937), 337–355.

CHAPTER 30

SECOND-ORDER THEORY

BY

E. CHWALLA,† Dr. techn., Dr.-Ing. h.c., Graz, Austria

30.1. GENERAL CONSIDERATIONS

a. Introduction

The classical theory of structures (the first-order theory, Chaps. 26 to 28) neglects the deformation of the structure when formulating the equilibrium conditions. For statically determinate structures, the resulting equations are linear in the loads and in the internal forces, and linear superposition applies to the internal forces caused by different loads or load groups. If Hooke's law is assumed to be valid, linear superposition applies also to the displacements and, hence, to the stresses and displacements of statically indeterminate structures.

In the second-order theory, the equilibrium conditions are written for the geometry of the deformed structure. Since in this theory the coefficients of the equilibrium equations depend on the displacements and, hence, on the internal forces, these equations are no longer linear, and linear superposition is not applicable.

In trusses and other structures of relatively great stiffness, the first-order theory gives reasonably accurate results, since the elastic displacements remain rather small. The first-order theory is also sufficient for the analysis of slender beams if the beams are only loaded transversely, since the change which the lever arms of the transverse forces undergo owing to beam deflection is relatively small. However, when a slender beam is loaded, not only with transverse loads, end moments, or unsymmetrical heating, but also with axial end loads N, then the moment arm of these axial loads can be influenced appreciably by the beam deflection $w(x)$. In such a case (Fig. 30.1), the first-order theory is not sufficient. The bending moment M_0, which is produced by the transverse loading, is increased by the amount Nw because of a compressive force N, and decreased by an equal amount if N is a tensile force. Hence, in the deflected equilibrium condition, the bending moment is

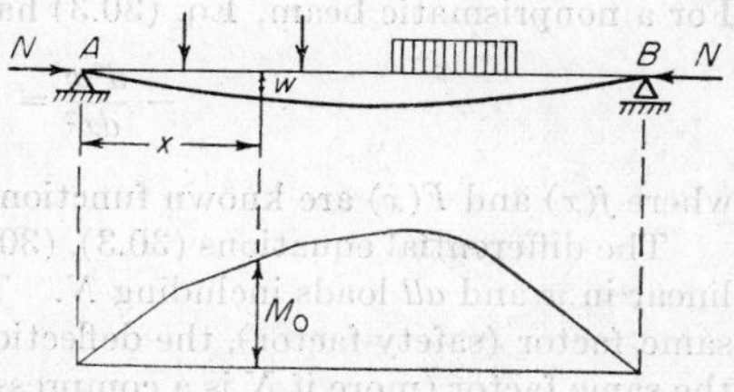

FIG. 30.1

$$M = M_0 \pm Nw \tag{30.1}$$

In the following discussion, only such structures are considered as behave similarly to the slender beam loaded with simultaneous transverse and axial loads, so that the effect of the second-order theory becomes important. We neglect the small influence of the shear force on the deflection, and make two approximate assumptions: (1) Only

Translated from the German by W. R. Blythe.
† Deceased.

the change of the lever arms of the axial end loads is taken into account, (2) the tangential slope $dw/dx \ll 1$, so that the exact expression for the curvature of the beam [Eq. (35.40)] may be replaced by

$$\kappa = -d^2w/dx^2 \tag{30.2}$$

With these two assumptions, the bending moment M_0 (produced by the transverse loads) is calculated according to the first-order theory, which results in the basic equation of the problem being a linear differential equation. As long as the value of N is held unchanged, the principle of superposition applies to the bending moments and the deflections. If we consider a beam with extremely small bending rigidity, assumption 2 is no longer admissible, and the basic equations are no longer linear differential equations (third-order theory).

The curvature κ consists of an elastic part $\kappa_e = M/EI$ and a thermal part κ_t (see p. 30–6). Therefore, Eqs. (30.1) and (30.2) may be combined into the following differential equation

$$-\frac{d^2w}{dx^2} = \frac{M_0}{EI} \pm \frac{Nw}{EI} + \kappa_t \tag{30.3}$$

For a prismatic beam (EI = const) the equation becomes

$$\frac{d^2w}{dx^2} \pm \lambda^2 w + F(x) = 0 \tag{30.4}$$

where λ is a constant. In most cases, $F(x)$ is a polynomial of not more than fifth degree. If N is a compressive force, Eq. (30.4) has the general solution

$$w = C_1 \sin \lambda x + C_2 \cos \lambda x - \frac{1}{\lambda^2} F(x) + \frac{1}{\lambda^4}\frac{d^2F}{dx^2} - \frac{1}{\lambda^6}\frac{d^4F}{dx^4} \tag{30.5}$$

and, when N is a tensile force, the general solution is

$$w = C_1 \operatorname{Sinh} \lambda x + C_2 \operatorname{Cosh} \lambda x + \frac{1}{\lambda^2} F(x) + \frac{1}{\lambda^4}\frac{d^2F}{dx^2} + \frac{1}{\lambda^6}\frac{d^4F}{dx^4} \tag{30.6}$$

For a nonprismatic beam, Eq. (30.3) has the form

$$-\frac{d^2w}{dx^2} = \pm wf(x) + F(x) \tag{30.7}$$

where $f(x)$ and $F(x)$ are known functions.

The differential equations (30.3), (30.4), (30.7) are linear in w, M_0 (and κ_t), but not linear in w and *all* loads including N. Therefore, when all loads are increased by the same factor (safety factor), the deflection and the bending stresses do not increase by the same factor (more if N is a compression, less if it is a tension). It follows that, in the second-order theory, a safety factor must be applied, not to the allowable stresses, but to the loads, and the analysis must be made for the design loads multiplied by the appropriate safety factor.

b. Basic Equations of a Prismatic Bar

Consider the bar AB deformed elastically as shown in Fig. 30.2. Since, in this case, $F(x)$ is a linear function, it may be eliminated from Eq. (30.4) by differentiating this equation twice:

$$\frac{d^4w}{dx^4} \pm \lambda^2 \frac{d^2w}{dx^2} = 0$$

The boundary conditions are:

at $x = 0$:	$w = 0$	$dw/dx = \theta_A$
at $x = l_{AB}$:	$w = \psi_{AB} l_{AB}$	$dw/dx = -\theta_B$

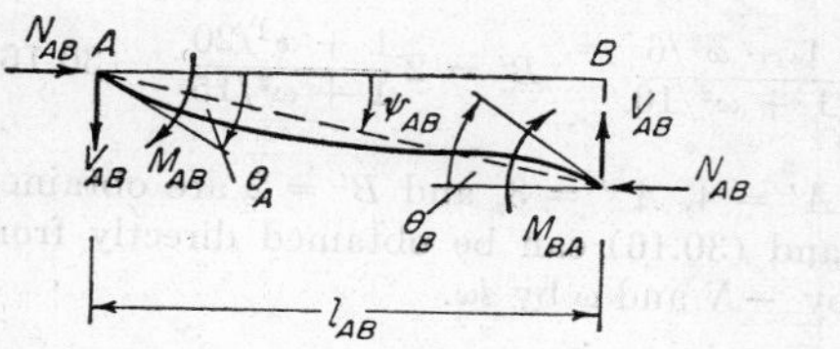

FIG. 30.2

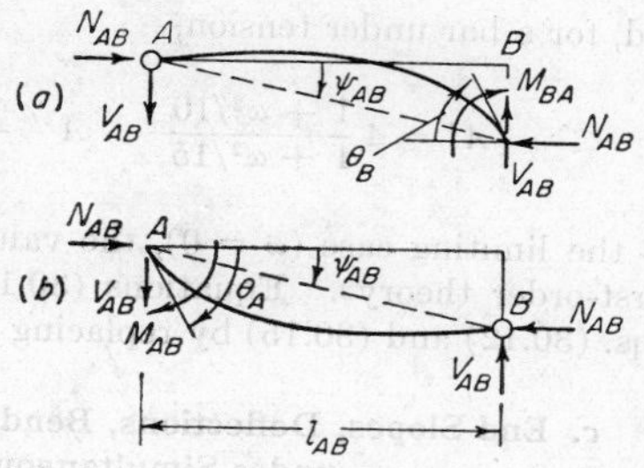

FIG. 30.3

When the solution (30.5) is introduced, the following relations may be derived:

$$\begin{aligned} M_{AB} &= A_{AB}\theta_A + B_{AB}\theta_B - (A_{AB} + B_{AB})\psi_{AB} \\ M_{BA} &= A_{AB}\theta_B + B_{AB}\theta_A - (A_{AB} + B_{AB})\psi_{AB} \\ V_{AB} &= N_{AB}\psi_{AB} + (A_{AB} + B_{AB})(\theta_A + \theta_B - 2\psi_{AB})/l_{AB} \end{aligned} \tag{30.8a–c}$$

If the bar has a pin joint at A (Fig. 30.3a), Eqs. (30.8) must be replaced by

$$M_{BA} = A^\circ_{AB}(\theta_B - \psi_{AB}) \qquad V_{AB} = N_{AB}\psi_{AB} + A^\circ_{AB}(\theta_B - \psi_{AB})/l_{AB} \tag{30.9a,b}$$

and, for the bar shown in Fig. 30.3b, we have

$$M_{AB} = A^\circ_{AB}(\theta_A - \psi_{AB}) \qquad V_{AB} = N_{AB}\psi_{AB} + A^\circ_{AB}(\theta_A - \psi_{AB})/l_{AB} \tag{30.9'a,b}$$

The sign of N_{AB} changes in the formulas for V_{AB} when the axial force is tension rather than compression. In Eqs. (30.8) through (30.9′), we define

$$A_{AB} = \frac{E_{AB}I_{AB}}{l_{AB}}A'_{AB} \qquad A^\circ_{AB} = \frac{E_{AB}I_{AB}}{l_{AB}}A^{\circ\prime}_{AB} \qquad B_{AB} = \frac{E_{AB}I_{AB}}{l_{AB}}B'_{AB} \tag{30.10a–c}$$

where the factors A'_{AB}, $A^{\circ\prime}_{AB}$, B'_{AB} are tabulated [1] transcendental functions of the parameter

$$\omega_{AB} = l_{AB}\sqrt{\frac{N_{AB}}{E_{AB}I_{AB}}} \tag{30.11}$$

When N is a compressive force, we have

$$A' = \frac{\omega \sin\omega - \omega^2\cos\omega}{2(1-\cos\omega) - \omega\sin\omega} \qquad A^{\circ\prime} = \frac{\omega^2\sin\omega}{\sin\omega - \omega\cos\omega}$$
$$B' = \frac{\omega^2 - \omega\sin\omega}{2(1-\cos\omega) - \omega\sin\omega} \tag{30.12a–c}$$

and, when N is a tensile force,

$$A' = \frac{\omega\,\mathrm{Sinh}\,\omega - \omega^2\,\mathrm{Cosh}\,\omega}{2(\mathrm{Cosh}\,\omega - 1) - \omega\,\mathrm{Sinh}\,\omega} \qquad A^{\circ\prime} = \frac{\omega^2\,\mathrm{Sinh}\,\omega}{\omega\,\mathrm{Cosh}\,\omega - \mathrm{Sinh}\,\omega}$$
$$B' = \frac{\omega^2 - \omega\,\mathrm{Sinh}\,\omega}{2(\mathrm{Cosh}\,\omega - 1) - \omega\,\mathrm{Sinh}\,\omega} \tag{30.13a–c}$$

and, in general,

$$A^{\circ\prime} = (A'^2 - B'^2)/A' \tag{30.14}$$

For small values of the parameter ω, a power-series expansion gives, to the first approximation, for a compressed bar,

$$A' = 4\,\frac{1-\omega^2/10}{1-\omega^2/15} \qquad A^{\circ\prime} = 3\,\frac{1-\omega^2/6}{1-\omega^2/10} \qquad B' = 2\,\frac{1-\omega^2/20}{1-\omega^2/15} \tag{30.15}$$

and, for a bar under tension,

$$A' = 4\frac{1+\omega^2/10}{1+\omega^2/15} \qquad A^{\circ\prime} = 3\frac{1+\omega^2/6}{1+\omega^2/10} \qquad B' = 2\frac{1+\omega^2/20}{1+\omega^2/15} \tag{30.16}$$

In the limiting case ($\omega = 0$), the values $A' = 4$, $A^{\circ\prime} = 3$, and $B' = 2$ are obtained (first-order theory). Equations (30.13) and (30.16) can be obtained directly from Eqs. (30.12) and (30.15) by replacing N by $-N$ and ω by $i\omega$.

c. End Slopes, Deflections, Bending Moments, and Shear Forces of Beams under Simultaneous Axial and Lateral Loading

Consider the prismatic beam AB shown in Fig. 30.4, acted upon by an axial load N and an end moment M_{AB}. We set, in Eq. (30.3), $M_0 = M_{AB}(l - x)/l$ and $\kappa_t = 0$,

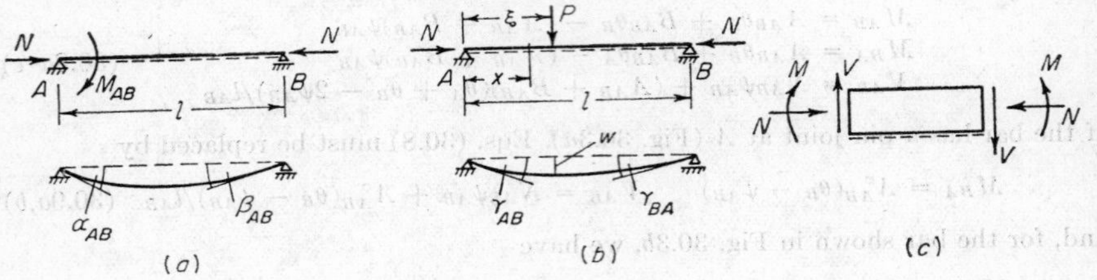

FIG. 30.4

so that, in Eq. (30.11),

$$\lambda = \sqrt{N/EI} \tag{30.17}$$

and $F(x) = M_{AB}(l - x)/lEI$. When the two boundary conditions are considered (for $x = 0$, $w = 0$, and for $x = l$, $w = 0$) the general solution [Eq. (30.5)] leads to the end slopes,

$$\alpha_{AB} = \alpha'_{AB}M_{AB}l/EI \qquad \beta_{AB} = \beta'_{AB}M_{AB}l/EI \tag{30.18}$$

where α'_{AB}, β'_{AB} are tabulated transcendental functions[1] of the parameter ω. If N is a compressive force, we have

$$\alpha' = \frac{\sin\omega - \omega\cos\omega}{\omega^2\sin\omega} \qquad \beta' = \frac{1}{\omega^2}\left(\frac{\omega}{\sin\omega} - 1\right) \tag{30.19}$$

and, for very small values of ω, a power-series expansion leads to

$$\alpha' = \frac{1}{3}\frac{1-\omega^2/10}{1-\omega^2/6} \qquad \beta' = \frac{1}{6}\frac{1-\omega^2/20}{1-\omega^2/6} \tag{30.20}$$

If N is a tensile force, we have instead

$$\alpha' = \frac{\omega\,\mathrm{Cosh}\,\omega - \mathrm{Sinh}\,\omega}{\omega^2\,\mathrm{Sinh}\,\omega} \qquad \beta' = \frac{1}{\omega^2}\left(1 - \frac{\omega}{\mathrm{Sinh}\,\omega}\right) \tag{30.21}$$

and approximately

$$\alpha' = \frac{1}{3}\frac{1+\omega^2/10}{1+\omega^2/6} \qquad \beta' = \frac{1}{6}\frac{1+\omega^2/20}{1+\omega^2/6} \tag{30.22}$$

[1] For compression bars, see [2], table 13, pp. 204–207, and [3], appendix, pp. 499–505. Reference [3] gives values of $3\alpha'$ and $6\beta'$. For tension bars, see [4], table 14, p. 153.

For $\omega = 0$ (first-order theory), we obtain the well-known values $\alpha' = 1/3$ and $\beta' = 1/6$. Equations (30.21) and (30.22) can be obtained directly from Eqs. (30.19) and (30.20) by replacing N by $-N$ and ω by $i\omega$. Among the coefficients α', β', A', $A°$, and B' there exist the relationships:

$$\alpha' = \frac{1}{A°'} = \frac{A'}{A'^2 - B'^2} \qquad \beta' = \frac{\alpha' B'}{A'} = c\alpha' = \frac{B'}{A'^2 - B'^2}$$
$$A' = \frac{\alpha'}{\alpha'^2 - \beta'^2} \qquad B' = \frac{\beta'}{\alpha'^2 - \beta'^2} \tag{30.23}$$
$$\alpha'^2 - \beta'^2 = \frac{1}{A'^2 - B'^2} \qquad A' - B' = \frac{1}{\alpha' + \beta'}$$

The signs of M_{AB}, α_{AB}, and β_{AB} are not determined by the right-hand screw rule, but rather by the reference-fiber rule. In Fig. 30.4, an additional dashed line is drawn, which is defined as the reference fiber for the beam (here the bottom of the beam is chosen). The values M_{AB}, α_{AB}, β_{AB} are positive if they act with reference to the dashed line as shown in Fig. 30.4.

We investigate now a prismatic beam with an axial load N and a concentrated transverse load P at a distance ξ from end A, as shown in Fig. 30.4*b*. Integration of Eq. (30.4) must be done separately for the range $x \leqq \xi$ with $F(x) = Px(l - \xi)/lEI$, and for the range $x \geqq \xi$ with $F(x) = P(l - x)\xi/lEI$; and, in addition to the two boundary conditions ($w = 0$ for $x = 0$ and $x = l$), there are also the two continuity conditions for $x = \xi$ (continuity of w and dw/dx) to consider. The following expressions for the end slopes are obtained:

$$\gamma_{AB} = \frac{l}{EI}\frac{Pl}{\omega^2}\left[\frac{\sin\lambda(l-\xi)}{\sin\omega} - \frac{l-\xi}{l}\right] = \frac{l}{EI}\gamma'_{AB}$$
$$\gamma_{BA} = \frac{l}{EI}\frac{Pl}{\omega^2}\left[\frac{\sin\lambda\xi}{\sin\omega} - \frac{\xi}{l}\right] = \frac{l}{EI}\gamma'_{BA} \tag{30.24}$$

with λ from Eq. (30.17) and $\omega = l\lambda$. The deflection at a point x is:

$$\text{For } x \leqq \xi: \quad w = \frac{Pl^3}{\omega^2 EI}\left[\frac{\sin\lambda(l-\xi)\sin\lambda x}{\omega\sin\omega} - \frac{x}{l}\left(1 - \frac{\xi}{l}\right)\right]$$
$$\text{For } x \geqq \xi: \quad w = \frac{Pl^3}{\omega^2 EI}\left[\frac{\sin\lambda\xi\sin\lambda(l-x)}{\omega\sin\omega} - \frac{\xi}{l}\left(1 - \frac{x}{l}\right)\right] \tag{30.25}$$

and the bending moment at point x is:

$$\text{For } x \leqq \xi: \quad M = \frac{Pl}{\omega\sin\omega}\sin\lambda(l-\xi)\sin\lambda x$$
$$\text{For } x \geqq \xi: \quad M = \frac{Pl}{\omega\sin\omega}\sin\lambda\xi\sin\lambda(l-x) \tag{30.26}$$

The shear force follows from the relation $V = dM/dx$. The values of P, γ_{AB}, γ_{BA}, M, V, and w are positive if, with respect to the dashed line, they are directed as shown in Fig. 30.4. If the beam is not compressed, but instead a tensile load is applied, we must replace N by $-N$, λ by $i\lambda$, and ω by $i\omega$.

With $P = 1$ and ξ as a variable, Eqs. (30.24), (30.25), and (30.26) give the influence lines for the end slope, the deflection, and the bending moment at a point x. With the help of these equations, we can compute the values γ_{AB}, γ_{BA}, w, and M for an arbitrarily distributed transverse load $p(\xi)$, by integrating the right-hand sides of these equations with respect to ξ. For example, we find for the right-hand end slope of the prismatic beam shown in Fig. 30.5, when we replace, in Eq. (30.24), P by $p\,d\xi$, λ by $i\lambda$, and ω by $i\omega$, the formula

$$\gamma_{BA} = \frac{l^2}{\omega^2 EI}\int_{l_1}^{l_2}\left(\frac{\xi}{l} - \frac{\text{Sinh}\,\lambda\xi}{\text{Sinh}\,\omega}\right)p\,d\xi$$
$$= \frac{l}{EI}\frac{pl^2}{\omega^2}\left(\frac{l_2^2 - l_1^2}{2l^2} - \frac{\text{Cosh}\,\lambda l_2 - \text{Cosh}\,\lambda l_1}{\omega\,\text{Sinh}\,\omega}\right) \tag{30.27}$$

For a full load ($l_1 = 0$, $l_2 = l$), there follows[1]

$$\gamma_{BA} = \frac{l}{EI}\frac{pl^2}{\omega^3}\frac{(\omega/2)\,\text{Cosh}\,(\omega/2) - \text{Sinh}\,(\omega/2)}{\text{Cosh}\,(\omega/2)} = \frac{l}{EI}\gamma'_{BA} = \frac{pl^3}{24\,EI}\gamma''_{BA} \tag{30.28}$$

and, for small values of the parameter ω, the power-series expansion yields

$$\gamma_{BA} = \frac{pl^3}{24EI}\frac{1 + \omega^2/40}{1 + \omega^2/8} \tag{30.29}$$

For $\omega = 0$ (first-order theory), we obtain the well-known formula $\gamma_{BA} = pl^3/24EI$.

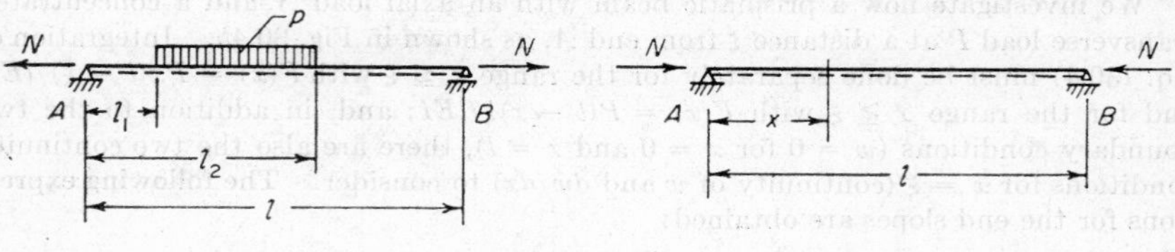

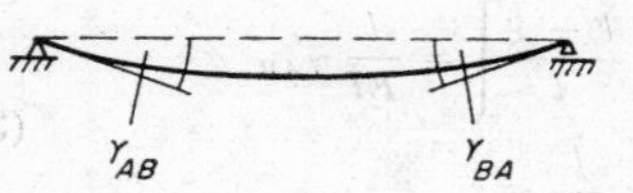

FIG. 30.5

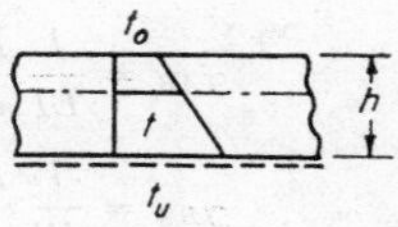

FIG. 30.6

d. End Slope for Nonuniformly Heated Beams under Axial Loading

In Fig. 30.6 is shown a prismatic beam carrying an axial load and subjected to nonuniform heating. It is assumed that the temperature is linearly distributed over the depth h of the cross section, so that the thermal curvature of the beam is

$$\kappa_t = \alpha_t\frac{t_u - t_o}{h} \tag{30.30}$$

The differential equation for $w(x)$ is obtained from Eq. (30.3) for $M_0 = 0$. For a prismatic beam, it can be written in the form

$$\frac{d^2w}{dx^2} \pm \lambda^2 w + \frac{M_t}{EI} = 0 \tag{30.31}$$

where λ is determined by Eq. (30.17), and

$$M_t = EI\kappa_t \tag{30.32}$$

is a fictitious bending moment. If κ_t = const, and N is a compression force, there results from Eq. (30.5), considering the boundary conditions ($w = 0$ for $x = 0$ and $x = l$), the deflection

$$w = \frac{\kappa_t}{\lambda^2}\left(\frac{1 - \cos\omega}{\sin\omega}\sin\lambda x + \cos\lambda x - 1\right) \tag{30.33}$$

[1] The factor γ''_{BA} for the compressed beam with uniform loading has been tabulated; see [3], pp. 499–505.

as well as the end slope

$$\gamma_{AB} = \left[\frac{dw}{dx}\right]_{x=0} = \frac{l}{EI} M_t \frac{1 - \cos\omega}{\omega \sin\omega} = \frac{l}{EI}\gamma'_{AB} \tag{30.34}$$

and the bending moment

$$M = Nw = M_t\left(\frac{1 - \cos\omega}{\sin\omega}\sin\lambda x + \cos\lambda x - 1\right) \tag{30.35}$$

where $\omega = l\lambda$.

e. Three-moment Equations

In the three-moment equation (27.8), the coefficients δ_{BA}, δ_{BB}, δ_{BC} are defined on p. 27-6, but the formulas given there are those of the first-order theory. In the second-order theory, they must be replaced by the following formulas:

$$\delta_{BA} = \beta_{AB} = \frac{\beta'_{AB} l_1}{EI_1}$$

$$\delta_{BB} = \alpha_{BA} + \alpha_{BC} = \frac{\alpha'_{BA} l_1}{EI_1} + \frac{\alpha'_{BC} l_2}{EI_2} \tag{30.36}$$

$$\delta_{BC} = \beta_{BC} = \frac{\beta'_{BC} l_2}{EI_2}$$

which may be read from Eqs. (30.18) with the notations of Figs. 27.7*a* and 30.4.

The quantity δ_{BP} in Eq. (27.8) is

$$\delta_{BP} = \gamma_{BA} + \gamma_{BC} - \psi_{AB} + \psi_{BC} \tag{30.37}$$

and the angles γ are calculated from equations like (30.24) and (30.27) to (30.29).

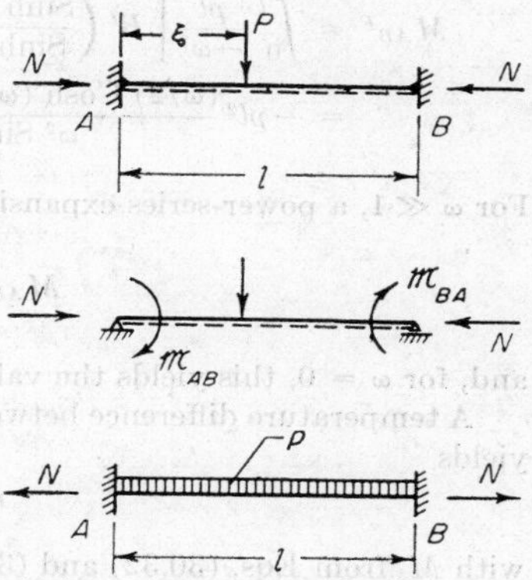

FIG. 30.7

The extension to the four-moment equation (27.10) is obvious.

f. Fixed-end Moments

We consider a prismatic beam clamped at both ends (Fig. 30.7). Its clamping moments are the fixed-end moments $M_{AB}{}^F$ and $M_{BA}{}^F$ of the moment-distribution method (see p. 27-9). These moments, counted positive when clockwise, may be calculated from two equations stating that the ends of the beam do not rotate. They are

$$\begin{aligned} M_{AB}{}^F\alpha'_{AB} - M_{BA}{}^F\beta'_{AB} + \gamma'_{AB} &= 0 \\ M_{AB}{}^F\beta'_{BA} - M_{BA}{}^F\alpha'_{BA} + \gamma'_{BA} &= 0 \end{aligned} \tag{30.38}$$

where not only $\beta'_{BA} = \beta'_{AB}$, but also $\alpha'_{BA} = \alpha'_{AB}$. The solution is

$$M_{AB}{}^F = \frac{\beta'_{AB}\gamma'_{BA} - \alpha'_{AB}\gamma'_{AB}}{\alpha'^2_{AB} - \beta'^2_{AB}} \qquad M_{BA}{}^F = \frac{\alpha'_{AB}\gamma'_{BA} - \beta'_{AB}\gamma'_{AB}}{\alpha'^2_{AB} - \beta'^2_{AB}} \tag{30.39}$$

which, with the help of Eqs. (30.23), may be written as

$$M_{AB}{}^F = B'\gamma'_{BA} - A'\gamma'_{AB} \qquad M_{BA}{}^F = A'\gamma'_{BA} - B'\gamma'_{AB} \tag{30.40}$$

In the special case of a concentrated load P, we obtain, from Eqs. (30.40) and (30.24),

$$\begin{aligned} M_{AB}{}^F &= \frac{P}{\omega^2}\left[B'\left(\frac{\sin\lambda\xi}{\sin\omega} - \frac{\xi}{l}\right) - A'\left(\frac{\sin\lambda(l-\xi)}{\sin\omega} - \frac{l-\xi}{l}\right)\right] \\ M_{BA}{}^F &= \frac{P}{\omega^2}\left[A'\left(\frac{\sin\lambda\xi}{\sin\omega} - \frac{\xi}{l}\right) - B'\left(\frac{\sin\lambda(l-\xi)}{\sin\omega} - \frac{l-\xi}{l}\right)\right] \end{aligned} \tag{30.41}$$

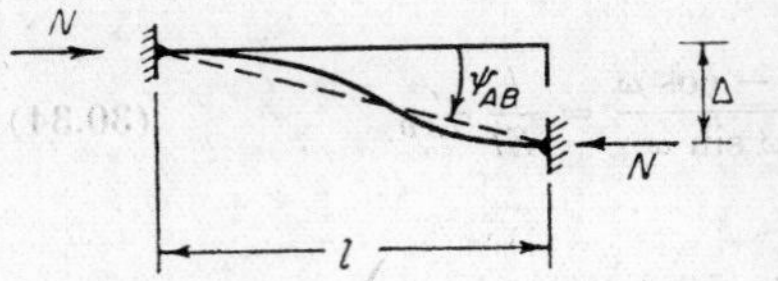

FIG. 30.8

In these equations, A', B' are defined by Eqs. (30.12a,c), with $\omega = \lambda l$ and λ from (30.17). If N is a tensile force, λ must be replaced by $i\lambda$, ω by $i\omega$, and A', B' must be taken from Eqs. (30.13a,c).

With $P = 1$ and ξ as a variable, Eqs. (30.41) give the influence lines for the fixed-end moments, and we can find these moments for any lateral load $p(\xi)$ by the same integration process as used in Eq. (30.27). For example, for a beam acted on by a uniform load p and a tensile force N (Fig. 30.7), we have

$$M_{AB}{}^F = \int_0^l \frac{pl}{-\omega^2}\left[B'\left(\frac{\text{Sinh }\lambda\xi}{\text{Sinh }\omega} - \frac{\xi}{l}\right) - A'\left(\frac{\text{Sinh }\lambda(l-\xi)}{\text{Sinh }\omega} - 1 + \frac{\xi}{l}\right)\right] d\xi$$
$$= -pl^2\,\frac{(\omega/2)\text{ Cosh }(\omega/2) - \text{Sinh }(\omega/2)}{\omega^2\text{ Sinh }(\omega/2)} \tag{30.42}$$

For $\omega \ll 1$, a power-series expansion yields approximately

$$M_{AB}{}^F = -\frac{pl^2}{12}\,\frac{1 + \omega^2/40}{1 + \omega^2/24} \tag{30.43}$$

and, for $\omega = 0$, this yields the value $M_{AB}{}^F = -pl^2/12$ of the first-order theory.

A temperature difference between the upper and lower sides of the beam (Fig. 30.6) yields

$$-M_{AB}{}^F = M_{BA}{}^F = M_t \tag{30.44}$$

with M_t from Eqs. (30.32) and (30.30).

For a lateral displacement Δ of one end of the beam (Fig. 30.8), the fixed-end moments are

$$M_{AB}{}^F = M_{BA}{}^F = -\frac{EI}{l^2}(A' + B')\,\Delta \tag{30.45}$$

g. Carry-over Factors

Consider the prismatic beam shown in Fig. 30.9, simply supported at A and elastically clamped at B. The reactive moment M_{BA} is proportional to the slope $\theta_B = -M_{BA}/k$, where k is the spring constant of the support. The unknowns θ_A and M_{BA} follow from Eqs. (30.8a,b) when we put there $M_{AB} = M_A$, $\psi_{AB} = 0$, and $\theta_B = -M_{BA}/k$. Making use of Eq. (30.14), we find

$$\frac{M_A}{\theta_A} = A_{AB}\,\frac{A^\circ_{AB} + k}{A_{AB} + k} \qquad \frac{M_{BA}}{M_A} = \frac{B'_{AB}}{A'_{AB}}\,\frac{k}{A^\circ_{AB} + k} \tag{30.46a,b}$$

The left-hand sides of these equations are the stiffness K_{AB} and the carry-over factor C_{AB} defined by Eqs. (27.17) and (27.18):

$$K_{AB} = A_{AB}\,\frac{A^\circ_{AB} + k}{A_{AB} + k} \qquad C_{AB} = \frac{B'_{AB}}{A'_{AB}}\,\frac{k}{A^\circ_{AB} + k} \tag{30.47a,b}$$

If the beam is rigidly clamped at B, then $1/k = 0$; hence,

$$K_{AB} = A_{AB} = \frac{E_{AB}I_{AB}}{l_{AB}}A'_{AB} \qquad C_{AB} = \frac{B'_{AB}}{A'_{AB}} \tag{30.48a,b}$$

If the end B is simply supported, $k = 0$, and

$$K_{AB} = A^\circ_{AB} = \frac{E_{AB} I_{AB}}{l_{AB}} A^{\circ\prime}_{AB} \qquad C_{AB} = 0 \tag{30.49a,b}$$

Numerical values of $\frac{1}{4}A'$, $\frac{1}{4}A^{\circ\prime}$, and C have been tabulated [1].

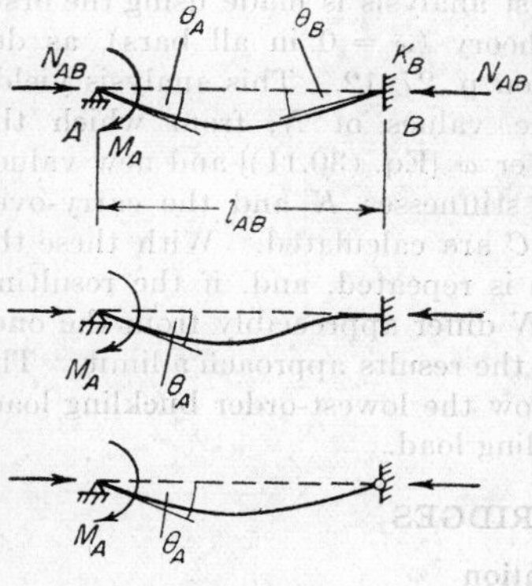

FIG. 30.9

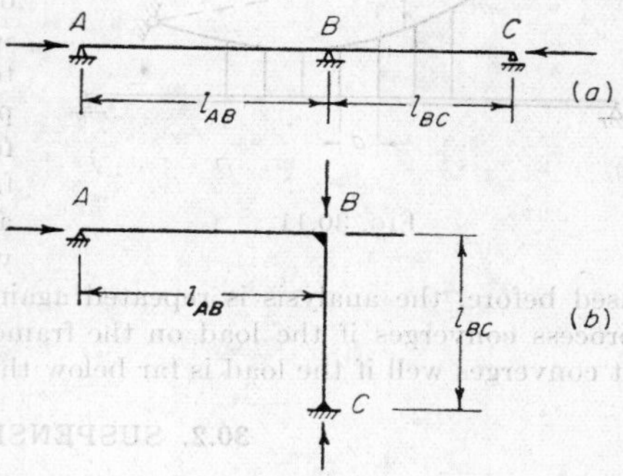

FIG. 30.10

Equations (30.47) may be used to calculate the stiffness and the carry-over factor for the bar AB in Figs. 30.10*a*,*b*. At its right end, this bar is connected rigidly with another bar BC, and, therefore, elastically clamped. The clamping constant k is identical with the stiffness K_{BC}; hence,

$$K_{AB} = A_{AB} \frac{A^\circ_{AB} + K_{BC}}{A_{AB} + K_{BC}} \qquad C_{AB} = \frac{B'_{AB}}{A'_{AB}} \frac{K_{BC}}{A^\circ_{AB} + K_{BC}} \tag{30.50a,b}$$

In the continuous beam (Fig. 30.10*a*), the bar BC has a hinged end C; hence,

$$K_{BC} = A^\circ_{BC}$$

In the frame (Fig. 30.10*b*), the column BC is clamped at C; hence, $K_{BC} = A_{BC}$. Through a step-by-step application of Eqs. (30.50), it is possible to find values K and C for the spans of a continuous beam that may be connected to the left end of the bar AB (see [5],[6]).

h. Beams and Bars of Variable Cross Section

In the case of the prismatic bar, we have the differential equation (30.4) with the general solutions (30.5) and (30.6). If the cross-sectional variation of a bar is small, we can closely approximate the bar by a prismatic bar, by setting the average flexural rigidity EI = const, and assign to this substitute bar a parameter ω, according to Eq. (30.11). If the flexural rigidity varies severely, however, we have to take care of this fact and must go to the differential equation (30.7). In general, numerical integration will be needed to solve this equation. It will yield all the quantities A, A°, B, α, β, γ, K, C which have been defined for the prismatic beam. When the EI distribution is not symmetric with respect to the center of the span, then $A_{AB} \neq A_{BA}$, $\alpha_{AB} \neq \alpha_{BA}$, $C_{AB} \neq C_{BA}$. However, $B_{AB} = B_{BA}$ and $\beta_{AB} = \beta_{BA}$, since these relations are derived from Maxwell's reciprocity theorem. For tables of the stiffness K and the carry-over factor C of bars whose ends are stiffened by gusset plates, see [7] and [8], Art. 2.4.

i. Moment Distribution in Frames

In frame structures, the axial forces N in the bars depend on the distribution of the bending moments, and are not known at the beginning of the analysis. Therefore, the following iterative procedure must be applied.

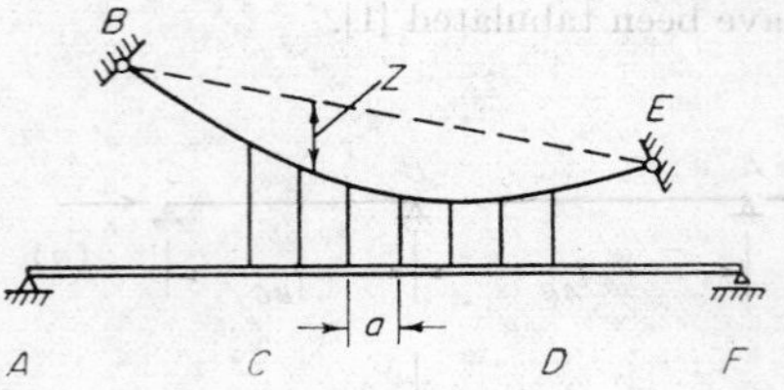

FIG. 30.11

A first analysis is made using the first-order theory ($\omega = 0$ in all bars), as described on p. 27–12. This analysis yields tentative values of N, from which the parameter ω [Eq. (30.11)] and new values for the stiffnesses K and the carry-over factors C are calculated. With these the analysis is repeated, and, if the resulting values N differ appreciably from the ones used before, the analysis is repeated again, until the results approach a limit. The process converges if the load on the frame is below the lowest-order buckling load. It converges well if the load is far below the buckling load.

30.2. SUSPENSION BRIDGES

a. Differential Equation

Consider the suspension bridge shown in Figs. 30.11 and 30.12. In the erection of such bridges, short sections of the stiffening girder are hung from the cable, and are combined with one another, so that a part $g(x)$ of the dead load is carried by the

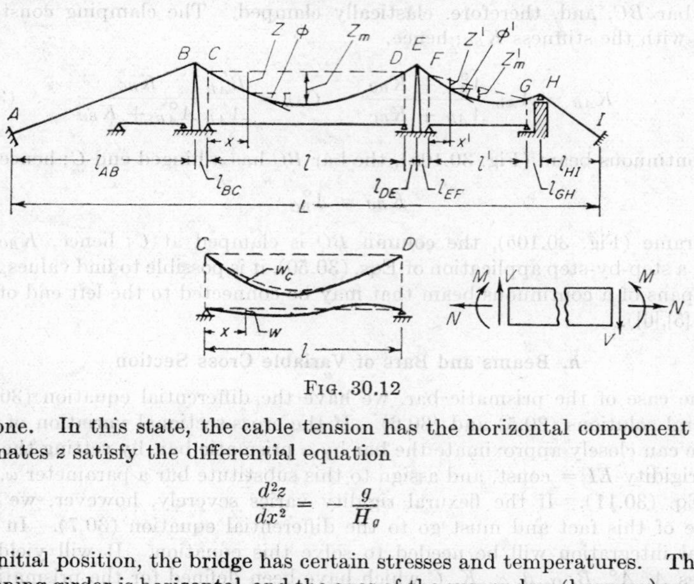

FIG. 30.12

cable alone. In this state, the cable tension has the horizontal component H_g, and the ordinates z satisfy the differential equation

$$\frac{d^2z}{dx^2} = \frac{g}{H_g} \tag{30.51}$$

In this initial position, the bridge has certain stresses and temperatures. The axis of the stiffening girder is horizontal, and the axes of the suspenders (hangers) are vertical. The cable axis has a determined form $z(x)$, and, at a point x, there is a known slope ϕ. If $g =$ const, it follows that

$$z = 4z_m x \frac{l - x}{l^2} \qquad \tan\phi = \frac{dz}{dx} = 4z_m \frac{l - 2x}{l^2} \qquad \frac{d^2z}{dx^2} = -\frac{8z_m}{l^2} = \text{const} \tag{30.52a–c}$$

Now we bring the load $p(x)$ (the rest of the dead load and the entire live load) onto the stiffening girder, and change the temperature of the cables, the hangers, and the stiffening girder. Because of this, there is developed in the cable the *additional* horizontal component H_p of the cable tension and the deflection $w_c(x)$, while the stiffening girder undergoes the deflection $w(x)$. If the hanger stretches the amount Δs (elastic and thermal elongation), we have $w_c = w - \Delta s$. In the classic suspension-bridge theory, the following assumptions are made for the calculation of these quantities: (1) Hooke's law holds; (2) the cables are assumed to be perfectly flexible; (3) the cable slides over the tops of the towers, so that H_g and H_p are the same in all spans; (4) the hanger forces are considered as distributed loads; (5) the hangers are assumed to remain vertical during the deformation of the bridge; (6) the effect of the change in slope ϕ of the cable under the load $p(x)$ is neglected; (7) the deflection of the stiffening girder due to shear and the effect of tower shortening are neglected.

The value of H_p is, for the present, unknown, and is set equal to an assumed value H_a [for example, the value which it has according to the first-order theory due to the load $p(x)$]. Of the load $p(x)$, the component $p_c(x)$ acts on the cable, and the rest $(p - p_c)$ acts on the stiffening girder. Since the cable with ordinates $(z + w_c)$ is the funicular curve for the load $(g + p_c)$, we have

$$g + p_c = -(H_g + H_a)\frac{d^2(z + w_c)}{dx^2} = -(H_g + H_a)\left(\frac{d^2w}{dx^2} - \frac{g}{H_g} - \frac{d^2\,\Delta s}{dx^2}\right) \qquad (30.53)$$

and, for the elastic curve of the stiffening girder,

$$p - p_c = \frac{d^2}{dx^2}\left(EI\,\frac{d^2w}{dx^2} + M_t\right) \qquad (30.54)$$

where M_t results from the nonuniform heating of the stiffening girder, and is determined by Eq. (30.32). Elimination of p_c from Eqs. (30.53) and (30.54) results in the differential equation,

$$\frac{d^2}{dx^2}\left(EI\,\frac{d^2w}{dx^2}\right) - (H_g + H_a)\frac{d^2w}{dx^2} - p(x) + \frac{H_a}{H_g}\,g(x) + \frac{d^2M_t}{dx^2} + (H_g + H_a)\frac{d^2\,\Delta s}{dx^2} = 0 \qquad (30.55)$$

This equation applies only to the sections of the stiffening girder where the cable is attached; thus, in Fig. 30.11 only for section CD, and in Fig. 30.12 only for sections CD and FG. In the remaining sections, Eq. (30.54) applies with $p_c = 0$. In what follows we shall assume that $\Delta s = 0$. We shall also assume that there are hangers everywhere in the span of the stiffening girder, as shown in Fig. 30.12. If we measure z vertically downward from the cable chord CD or FG, we have, for the bending moment M of the stiffening girder CD, the relation

$$M = M_0 - H_a(z + w) - H_g w = M_0 - H_a z - (H_g + H_a)w \qquad (30.56)$$

where M_0 is a fictitious bending moment, which would arise from the load $p(x)$ on the stiffening girder if we cut the cable. We consider this separated stiffening girder as a beam analog for which $d^2M_0/dx^2 = -p(x)$. Introducing M from Eq. (30.56) into the general beam equation,

$$-\frac{d^2w}{dx^2} = \frac{M + M_t}{EI} \qquad (30.57)$$

we obtain the fundamental equation in the form

$$\frac{d^2w}{dx^2} - \frac{H_g + H_a}{EI}\,w + \frac{H_a}{EI}\left(\frac{M_0 + M_t}{H_a} - z\right) = 0 \qquad (30.58)$$

When we multiply this equation by EI, and then differentiate twice with respect to x, we arrive at Eq. (30.55) with the influence of Δs neglected. For a prismatic stiffening girder (EI = const), Eq. (30.58) can be written in the form of Eq. (30.4), by the introduction of the quantity

$$\lambda^2 = \frac{H_g + H_a}{EI} = \text{const} \tag{30.59}$$

and the general solution [Eq. (30.6)] becomes

$$w = \frac{H_a}{H_g + H_a}\left[C_1 \operatorname{Sinh} \lambda x + C_2 \operatorname{Cosh} \lambda x + F_1(x) + \frac{1}{\lambda^2}\frac{d^2F_1}{dx^2} + \frac{1}{\lambda^4}\frac{d^4F_1}{dx^4}\right] \tag{30.60}$$

where

$$F_1(x) = \frac{M_0 + M_t}{H_a} - z$$

and it is assumed that $d^6F_1/dx^6 = 0$.

b. Relation of the Suspension-bridge Problem to the Problem of a Beam with an Additional Axial Load

The differential equation (30.55) of the suspension bridge has the same form as the general differential equation of a simple beam which is simultaneously loaded with an axial end force and a transverse load (beam analog):

$$\frac{d^2}{dx^2}\left(EI\frac{d^2w}{dx^2}\right) - N\frac{d^2w}{dx^2} - p(x) = 0 \tag{30.61}$$

Both equations become identical if we make $N = H_g + H_a$, and replace $p(x)$ in Eq. (30.61) by the fictitious load,

$$\bar{p}(x) = p(x) + \left[H_a\frac{d^2z}{dx^2} - \frac{d^2M_t}{dx^2} - (H_g + H_a)\frac{d^2\,\Delta s}{dx^2}\right] = p(x) + q(x) \tag{30.62}$$

Since Eq. (30.55) is valid only for those sections of the stiffening girder where the cable is attached, the analogy applies only for the portions CD and FG in Fig. 30.12. The axial tensile force ($H_g + H_a$) is fictitious here and, therefore, produces no tensile stresses in the stiffening girder. The same analogy is seen to exist between Eqs. (30.58) and (30.3) when the former is written as

$$\frac{d^2w}{dx^2} - \frac{H_g + H_a}{EI}w + \frac{M_I}{EI} + \kappa_t = 0 \tag{30.63}$$

where $M_I = M_0 - H_az$, and is the bending moment in the stiffening girder according to the first-order theory. Evidently, we must assign to the beam analog the bending moment $M_0 = M_I$ and the axial tension $N = H_g + H_a$.

If the stiffening girder is prismatic, the normal-force parameter becomes

$$\omega = l\lambda = l\sqrt{\frac{H_g + H_a}{EI}} = \text{const} \tag{30.64}$$

and all the quantities given in Sec. 30.1 for the beam with an axial tensile load (A, $A°$, B, α, β, γ, K, and C) and the methods (three-moment equations, four-moment equations, moment distribution) can be applied to the suspension-bridge problems. The principle of superposition is valid as long as the value of H_a (temporarily assumed for the unknown H_p) remains constant. If we wish to superimpose two loadings, we must, therefore, calculate each of these cases under the assumed value H_a which results from the entire superimposed loading.

c. Examples

We assume that the suspension bridge shown in Fig. 30.12 has a prismatic stiffening girder, and that the cable in its initial position is parabolic [Eq. (30.52)]. We place, on the portion CD of the stiffening girder, a concentrated transverse load P at the distance ξ from the left support, and allow heating of the stiffening girder with κ_t = const (hence, M_t = const). The influence of the elongation of the hangers is ignored, and a fixed value H_a is assumed for the unknown H_p.

$$\text{Then for } x \leqq \xi: \quad M_{01} = P\frac{l-\xi}{l}x \qquad\qquad \text{and for } x \geqq \xi: \quad M_{02} = P\frac{\xi}{l}(l-x) \tag{30.65}$$

and from Eq. (30.60) there follows, when we fulfill the two boundary conditions ($w = 0$ for $x = 0$ and $x = l$),

$$\text{For } x \leqq \xi: \quad w_1 = \frac{H_a}{H_g + H_a}\left[C_1 \operatorname{Sinh} \lambda x - c(\operatorname{Cosh} \lambda x - 1) + \frac{M_{01}}{H_a} - z\right]$$

$$\text{For } x \geqq \xi: \quad w_2 = \frac{H_a}{H_g + H_a}\left[C_2(\operatorname{Sinh} \lambda x - \operatorname{Tanh} \omega \operatorname{Cosh} \lambda x) - c\left(\frac{\operatorname{Cosh} \lambda x}{\operatorname{Cosh} \omega} - 1\right) + \frac{M_{02}}{H_a} - z\right] \tag{30.66}$$

$$\text{where} \qquad c = \frac{8z_m}{\omega^2} + \frac{M_t}{H_a} = \frac{EI}{H_a}\left(\frac{8z_m}{l^2}\frac{H_a}{H_g + H_a} + \kappa_t\right) = \text{const} \tag{30.67}$$

Application of the two continuity conditions at $x = \xi$ ($w_1 = w_2$, $dw_1/dx = dw_2/dx$) determines the two integration constants,

$$C_1 = c\frac{\operatorname{Cosh} \omega - 1}{\operatorname{Sinh} \omega} - \frac{Pl}{\omega H_a}\frac{\operatorname{Sinh} \lambda(l-\xi)}{\operatorname{Sinh} \omega} \qquad C_2 = c\frac{\operatorname{Cosh} \omega - 1}{\operatorname{Sinh} \omega} + \frac{Pl}{\omega H_a}\frac{\operatorname{Sinh} \lambda\xi}{\operatorname{Tanh} \omega} \tag{30.68}$$

which appear in Eq. (30.66), and leads to the solution:

$$\begin{aligned} \text{For } x \leqq \xi: \quad w_1 &= \frac{Pl}{H_g + H_a}\left[\frac{(l-\xi)x}{l^2} - \frac{\operatorname{Sinh} \lambda(l-\xi) \operatorname{Sinh} \lambda x}{\omega \operatorname{Sinh} \omega}\right] \\ &\quad + \frac{H_a}{H_g + H_a}\left[c\left(\frac{\operatorname{Cosh} \omega - 1}{\operatorname{Sinh} \omega} \operatorname{Sinh} \lambda x - \operatorname{Cosh} \lambda x + 1\right) - z\right] \\ &= w_1'(\xi) + w'' \\ \text{For } x \geqq \xi: \quad w_2 &= \frac{Pl}{H_g + H_a}\left[\frac{\xi(l-x)}{l^2} - \frac{\operatorname{Sinh} \lambda\xi \operatorname{Sinh} \lambda(l-x)}{\omega \operatorname{Sinh} \omega}\right] \\ &\quad + \frac{H_a}{H_g + H_a}\left[c\left(\frac{\operatorname{Cosh} \omega - 1}{\operatorname{Sinh} \omega} \operatorname{Sinh} \lambda x - \operatorname{Cosh} \lambda x + 1\right) - z\right] \\ &= w_2'(\xi) + w'' \end{aligned} \tag{30.69}$$

Equation (30.56) gives the bending moments:

$$\begin{aligned} \text{For } x \leqq \xi: \quad M_1 &= M_{01} - H_a z - (H_g + H_a)w_1 = Pl\frac{\operatorname{Sinh} \lambda(l-\xi) \operatorname{Sinh} \lambda x}{\omega \operatorname{Sinh} \omega} \\ &\quad - H_a c\left(\frac{\operatorname{Cosh} \omega - 1}{\operatorname{Sinh} \omega} \operatorname{Sinh} \lambda x - \operatorname{Cosh} \lambda x + 1\right) = M_1'(\xi) + M'' \\ \text{For } x \geqq \xi: \quad M_2 &= M_{02} - H_a z - (H_g + H_a)w_2 = Pl\frac{\operatorname{Sinh} \lambda\xi \operatorname{Sinh} \lambda(l-x)}{\omega \operatorname{Sinh} \omega} \\ &\quad - H_a c\left(\frac{\operatorname{Cosh} \omega - 1}{\operatorname{Sinh} \omega} \operatorname{Sinh} \lambda x - \operatorname{Cosh} \lambda x + 1\right) = M_2'(\xi) + M'' \end{aligned} \tag{30.70}$$

and the shear force is:

$$\text{For } x \leqq \xi: \qquad V_1 = \frac{dM_1}{dx} = V_1'(\xi) + V''$$
$$\text{For } x \geqq \xi: \qquad V_2 = \frac{dM_2}{dx} = V_2'(\xi) + V'' \tag{30.71}$$

We see that the quantities w, M, and V are composed of two parts each. The first part, w', or M', or V', depends directly on the given transverse load, and is obtained from the beam analog loaded according to Fig. 30.13*a*. For a beam in compression, the proper solution is given in Eqs. (30.25) and (30.26). We have to replace there N by $-(H_g + H_a)$ to obtain w' and M' in Eqs. (30.69) and (30.70). The second part, w'', M'', or V'', is obtained if we subject the beam analog to the given loading $q(x)$ [Eq. (30.62)], and the given nonuniform heating, and again allow the axial tensile force $(H_g + H_a)$ to act. Since M_t = const, and the influence of Δs is neglected, the loading $q(x)$ becomes simply a uniform loading,

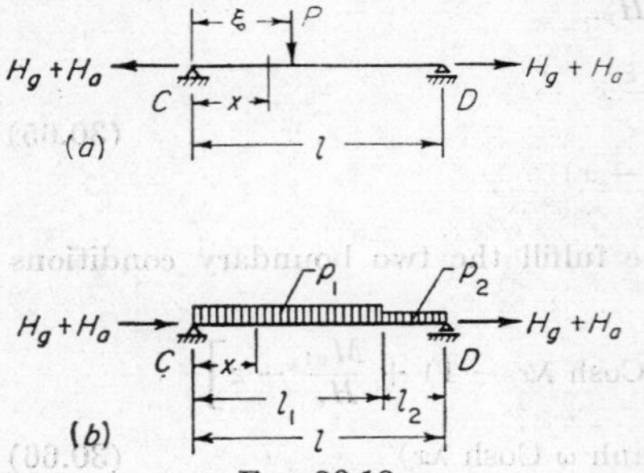

FIG. 30.13

$$q(x) = H_a \frac{d^2z}{dx^2} = \frac{8z_m}{l^2} H_a = \text{const} \tag{30.72}$$

for which

$$w = \frac{H_a}{H_g + H_a}\left[\frac{8z_m}{\omega^2}\left(\frac{\text{Cosh }\omega - 1}{\text{Sinh }\omega}\text{ Sinh }\lambda x - \text{Cosh }\lambda x + 1\right) - z\right]$$
$$M = M_0 - (H_g + H_a)w = -H_a \frac{8z_m}{\omega^2}\left(\frac{\text{Cosh }\omega - 1}{\text{Sinh }\omega}\text{ Sinh }\lambda x - \text{Cosh }\lambda x + 1\right) \tag{30.73}$$

For nonuniform heating of the beam analog, we obtain, from Eqs. (30.33) and (30.35), when we replace N by $-(H_g + H_a)$,

$$w = \frac{M_t}{H_g + H_a}\left(\frac{\text{Cosh }\omega - 1}{\text{Sinh }\omega}\text{ Sinh }\lambda x - \text{Cosh }\lambda x + 1\right)$$
$$M = -(H_g + H_a)w = -M_t\left(\frac{\text{Cosh }\omega - 1}{\text{Sinh }\omega}\text{ Sinh }\lambda x - \text{Cosh }\lambda x + 1\right) \tag{30.74}$$

When we superimpose Eqs. (30.73) and (30.74), we obtain indeed the expressions w'' and M'' of Eqs. (30.69) and (30.70).

Setting the load $P = 1$ in the expressions for w' and M' [Eqs. (30.69) and (30.70)], and taking ξ as a variable, we obtain the influence lines for deflection and bending moment in this section. If we wish to calculate, for example, the portion of the bending moment M_1' which appears at a point $x \leqq l_1$, due to the load on the stiffening girder shown in Fig. 30.13*b*, we must replace, in Eq. (30.70), the load P by $p\,d\xi$ and integrate:

$$\begin{aligned} M_1' &= \frac{p_1 \text{ Sinh }\lambda(l - x)}{\omega \text{ Sinh }\omega}\int_0^x \text{Sinh }\lambda\xi\,d\xi \\ &\quad + \frac{p_1 l \text{ Sinh }\lambda x}{\omega \text{ Sinh }\omega}\int_x^{l_1}\text{Sinh }\lambda(l - \xi)\,d\xi + \frac{p_2 l \text{ Sinh }\lambda x}{\omega \text{ Sinh }\omega}\int_{l_1}^{l}\text{Sinh }\lambda(l - \xi)\,d\xi \\ &= \frac{p_1 l^2}{\omega^2}\left(\frac{\text{Sinh }\lambda x}{\text{Tanh }\omega} - \text{Cosh }\lambda x + 1\right) - \frac{p_2 l^2}{\omega^2}\frac{\text{Sinh }\lambda x}{\text{Sinh }\omega} \\ &\qquad - \frac{(p_1 - p_2)l^2}{\omega^2}\frac{\text{Sinh }\lambda x \text{ Cosh }\lambda l_2}{\text{Sinh }\omega} \end{aligned} \tag{30.75}$$

In a similar way, the expression for M_2' for a point $x \geqq l_1$ or the expression for a deflection w' can be found. If the expressions for M'' and w'' from Eq. (30.69) and (30.70) are added to the expressions obtained above, the solution for the loading $p(x)$ on the stiffening girder in Fig. 30.13*b* is obtained:

$$\text{For } x \leqq l_1: \quad M_1 = -H_a[C_1' \operatorname{Sinh} \lambda x - c_1(\operatorname{Cosh} \lambda x - 1)] \qquad V_1 = \frac{dM_1}{dx}$$

$$\text{For } x \geqq l_1: \quad M_2 = -H_a\{C_2' \operatorname{Sinh} \lambda(l - x) - c_2[\operatorname{Cosh} \lambda(l - x) - 1]\} \qquad V_2 = \frac{dM_2}{dx} \tag{30.76}$$

and

$$\text{For } x \leqq l_1: \quad w_1 = \frac{H_a}{H_g + H_a}\left[C_1' \operatorname{Sinh} \lambda x - c_1(\operatorname{Cosh} \lambda x - 1) + \frac{M_{01}}{H_a} - z\right]$$

$$\text{For } x \geqq l_1: \quad w_2 = \frac{H_a}{H_g + H_a}\left\{C_2' \operatorname{Sinh} \lambda(l - x) - c_2[\operatorname{Cosh} \lambda(l - x) - 1] + \frac{M_{02}}{H_a} - z\right\} \tag{30.77}$$

where

$$\text{For } x \leqq l_1: \quad M_{01} = x\left[\frac{p_2 l}{2} - (p_1 - p_2)\left(1 - \frac{l_1}{2l}\right) l_1\right] - \frac{p_1 x^2}{2}$$

$$\text{For } x \geqq l_1: \quad M_{02} = (l - x)\left[\frac{p_1 l}{2} - (p_1 - p_2)\left(1 - \frac{l_2}{2l}\right) l_2\right] - \frac{p_2 (l - x)^2}{2} \tag{30.78}$$

and

$$c_{1,2} = \frac{8 z_m}{\omega^2} + \frac{M_t}{H_a} - \frac{p_{1,2} l^2}{H_a \omega^2}$$

$$C_{1,2}' = \frac{1}{\operatorname{Sinh} \omega}\left[c_{1,2} \operatorname{Cosh} \omega - c_{2,1} \pm \frac{(p_1 - p_2) l^2}{H_a \omega^2} \operatorname{Cosh} \frac{\omega l_{2,1}}{l}\right] \tag{30.79}$$

d. The Basic Equation for H_p (Cable Condition)

Consider that the cable in Fig. 30.12 is cut through at the center of the middle span (where $\phi = 0$ and $z = z_m$), and that there a force $H_p = 1$ is applied on either side of the cut. Because of assumption 3, the horizontal component 1 of the cable tension exists in the whole cable (over the entire horizontal distance L between the anchorages). Consider, further, that the whole suspension bridge is a rigid body, so that we may employ the first-order theory. The tension in the cable is $\bar{N} = 1/\cos \phi$, and in the hangers it is $\bar{S} = \tan \phi_1 - \tan \phi_2$. In the stiffening girder, there exists the bending moment $\bar{M} = -z$ and the shear $\bar{V} = -\tan \phi$. To this fictitious state of equilibrium we apply the principle of virtual work, using as the virtual deformations those which actually occur in the second-order theory, and setting equal to zero the virtual work of the force H_p. In this way, we express the fact that, for the cable with elastic and thermal deformations, the distance L (Fig. 30.12) is unchanged.

According to the second-order theory, an element of the cable undergoes the elongation

$$\epsilon_e + \epsilon_t = \frac{N}{E_c A_c} + \epsilon_t = \frac{H_p}{E_c A_c \cos(\phi + \Delta\phi)} + \epsilon_t \tag{30.80}$$

and an element of the hanger undergoes an extension

$$\Delta s = \left(\frac{a p_c}{E_c A_c} + \epsilon_t\right) l_s \tag{30.81}$$

An element of the stiffening girder has the curvatures $\kappa_e = M/EI$ and κ_t, according to Art. 30.1*d*, as well as the shear strain $\gamma_e = V/GA_s$. The principle of virtual work gives, with reference to Fig. 30.12,

$$\int_l \bar{M}(\kappa_e + \kappa_t)\,dx + \int_{l'} \bar{M}'(\kappa_e' + \kappa_t')\,dx' + \int_l \bar{V}\gamma_e\,dx + \int_{l'} \bar{V}'\gamma_e'\,dx' + \int_L \bar{N}(\epsilon_e + \epsilon_t)\frac{dx}{\cos(\phi + \Delta\phi)} + \Sigma\bar{S}\,\Delta s + \Delta\bar{S}'\,\Delta s' = 0 \quad (30.82)$$

which simplifies, if we neglect the influence of $\Delta\phi$ and γ_e, according to assumptions (6) and (7). If we, in addition, ignore the influence of the hanger elongation Δs, and consider ϵ_t = const, Eq. (30.82) takes the form

$$\int_l z\frac{d^2w}{dx^2}\,dx + \int_{l'} z'\frac{d^2w'}{dx'^2}\,dx' + \frac{H_p}{E_cA_c}\int_L \sec^3\phi\,dx + \epsilon_t\int_L \sec^2\phi\,dx = 0 \quad (30.83)$$

For an eyebar cable (assuming A_c varying with the slope), we have $A_c = A_0 \sec\phi$, and, therefore, the basic equation (30.82) becomes

$$\int_l z\frac{d^2w}{dx^2}\,dx + \int_{l'} z'\frac{d^2w'}{dx^2}\,dx' + \left(\frac{H_p}{E_cA_0} + \epsilon_t\right)\int_L \sec^2\phi\,dx = 0 \quad (30.84)$$

If the parts *CD* and *FG* are initially parabolic (g = const, and equal forces in all suspenders), then

$$\int_l \sec^2\phi\,dx = l\left(1 + \frac{16}{3}\frac{z_m^2}{l^2}\right) = l_i$$
$$\int_l \sec^3\phi\,dx = \int_l (1 + \tfrac{3}{2}\tan^2\phi + \tfrac{3}{8}\tan^4\phi - \cdots)\,dx \approx l\left(1 + \frac{8z_m^2}{l^2} + \frac{96}{5}\frac{z_m^4}{l^4}\right) = l_i^* \quad (30.85a,b)$$

and, for the cable *FG*,

$$l_i' = l'\left(\sec^2\phi_{FG} + \frac{16}{3}\frac{z_m'^2}{l'^2}\right)$$
$$l_i'^* = l'\left(\sec^3\phi_{FG} + \frac{8z_m'^2}{l'^2} + \frac{96}{5}\frac{z_m'^4}{l'^4}\right) \quad (30.85c,d)$$

and, hence,

$$\int_L \sec^2\phi\,dx = l_i + l_i' + \Sigma l_{NM}\sec^2\phi_{NM}$$
$$\int_L \sec^3\phi\,dx = l_i^* + l_i'^* + \Sigma l_{NM}\sec^3\phi_{NM} \quad (30.86)$$

The sums in these equations refer to the straight parts of *AB*, *BC*, *DE*, *EF*, *GH*, and *HI* of the cable, which are devoid of hangers. The first and second terms in Eq. (30.83) or (30.84) can be transformed by integration by parts,

$$\int_l z\frac{d^2w}{dx^2}\,dx = \left[z\frac{dw}{dx}\right]_0^l - \left[\frac{dz}{dx}w\right]_0^l + \int_l w\frac{d^2z}{dx^2}\,dx = \int_l w\frac{d^2z}{dx^2}\,dx \quad (30.87)$$

since $w = 0$ and $z = 0$ for $x = 0$ and $x = l$. For the parabolic cable [Eq. (30.52)], this becomes

$$\int_l w\frac{d^2z}{dx^2}\,dx = -\frac{g}{H_g}\int_l w\,dx \quad (30.88)$$

and, therefore, the cable condition (30.83) takes the form

$$-\frac{g}{H_g}\int_l w\,dx - \frac{g}{H_g}\int_{l'} w'\,dx' + \frac{H_p}{E_c A_c}(l_i^* + l_i'^* + \Sigma l_{NM}\sec^3\phi_{NM}$$
$$+ \epsilon_t(l_i + l_i' + \Sigma l_{NM}\sec^2\phi_{NM}) = 0 \quad (30.89)$$

When we introduce here Eq. (30.69), which refers to the load P indicated in Fig. 30.13a, then

$$J \equiv \int_l w\,dx = \int_0^\xi w_1'\,dx + \int_\xi^l w_2'\,dx + \int_0^l w''\,dx = \frac{P}{H_g + H_p}\left[\frac{\xi(l-\xi)}{2}\right.$$
$$\left.-\frac{\text{Cosh}\,\omega - 1}{\lambda^2\,\text{Sinh}\,\omega}\,\text{Sinh}\,\lambda\xi + \frac{\text{Cosh}\,\lambda\xi - 1}{\lambda^2}\right] + \frac{H_p}{H_g + H_p}\left[cl\left(1 - 2\,\frac{\text{Cosh}\,\omega - 1}{\omega\,\text{Sinh}\,\omega}\right)\right.$$
$$\left. - \tfrac{2}{3}lz_m\right] \equiv f' + f'' \quad (30.90)$$

where c is determined by Eq. (30.67), and

$$\omega = l\lambda = l\sqrt{\frac{H_g + H_p}{EI}} = \text{const} \quad (30.91)$$

If we want the value of the integral J for the distributed load of Fig. 30.13b, we must, in Eq. (30.90), replace the load P in the expression for f' by $p\,d\xi$, and integrate with respect to ξ. In this way, we obtain

$$f' = \frac{1}{H_g + H_p}\left\{\int_0^l M_0\,dx - \frac{l^2}{\omega^2}(p_1 l_1 + p_2 l_2) + \frac{p_1 l^3}{\omega^3}\left[\text{Sinh}\,\lambda l_1\right.\right.$$
$$\left.\left. - \frac{\text{Cosh}\,\omega - 1}{\text{Sinh}\,\omega}(\text{Cosh}\,\lambda l_1 - 1)\right] + \frac{p_2 l^3}{\omega^3}\left[\text{Sinh}\,\lambda l_2 - \frac{\text{Cosh}\,\omega - 1}{\text{Sinh}\,\omega}(\text{Cosh}\,\lambda l_2 - 1)\right]\right\} \quad (30.92)$$

where M_0 is the bending moment of the simple beam given in Eq. (30.78). In order to obtain J, we must include, with the expression f', the expression f'', which is independent of ξ. Making use of the definitions (30.79), we find ultimately

$$J = \int_l w\,dx = \frac{H_p}{H_g + H_p}\left[\frac{1}{H_p}\int_0^l M_0\,dx - \frac{2}{3}lz_m + \frac{C_1'}{\lambda}(\text{Cosh}\,\lambda l_1 - 1)\right.$$
$$\left. + \frac{C_2'}{\lambda}(\text{Cosh}\,\lambda l_2 - 1) - \frac{c_1}{\lambda}\,\text{Sinh}\,\lambda l_1 - \frac{c_2}{\lambda}\,\text{Sinh}\,\lambda l_2 + c_1 l_1 + c_2 l_2\right] \quad (30.93)$$

In a similar manner, the integral value for the side span EH in Fig. 30.12 is obtained, and introduced in Eq. (30.89). If we substitute, for the still unknown $(H_g + H_p)$, a fictitious value $(H_g + H_a)$, we obtain the cable condition in the form $H_pF_1 + F_2 = 0$, similar to the first-order theory. The calculation must be repeated until the value H_a in $(H_g + H_a)$ agrees with the value of H_p. In practice, it is sufficient to make two different estimates of $(H_g + H_a)$, and to obtain H_p by linear interpolation. When, for the given load $p(x)$, the value H_p has been found, it can be put into Eqs. (30.76) through (30.79) (in place of H_a), and the bending moment, shear, and deflection determined.

e. Additional Notes

The influence of the horizontal displacement component u of the cable (assumption 5) is considered in [21] and [22]. The influence of the slope change $\Delta\phi$ (assumption 6) is discussed in [16] and [17]. The influence of the finite distance of the hangers (assumption 4) has been treated in [19], and that of different arrangements of the towers is discussed in [21]. The nonprismatic stiffening girder is discussed in [18] through [22]. For the influence of static wind forces, see [14],[15],[18], and [22]. For prismatic beams with small normal-force parameters ω, the functions Sinh ω and Cosh ω can be replaced by the power-series expansion as in Eqs. (30.16), (30.22), and (30.29), and, for large values of ω, we can write Sinh $\omega \approx$ Cosh $\omega \approx \frac{1}{2}e^{\omega}$.

30.3. WIDE-SPAN ARCHES

a. Simplified Differential Equation

Consider the two-hinged arch shown in Fig. 30.14. In the initial position, the arch carries the dead load $g(x)$, and has a certain temperature and a known form $z(x)$. The horizontal component of the thrust of the arch has the known value H_g. There are no bending moments. We now apply the load $p(x)$ to the arch (the rest of the dead load and the entire live load), and heat the arch nonuniformly. There occurs, on account of this, an additional horizontal component H_p of the thrust, and a point on the axis undergoes the vertical displacement $w(x)$ and the horizontal displacement $u(x)$. In the simplified theory, the following assumptions are made: (1) Hooke's law holds; (2) the influence of the horizontal component of the deflection is neglected; (3) the change in slope of the axis of the arch under the load $p(x)$ is neglected; (4) the load $p(x)$ is assumed to act directly on the arch; (5) the radius of curvature is large in comparison with the depth of the cross section; (6) the effect of shearing forces on the deformation is negligible.

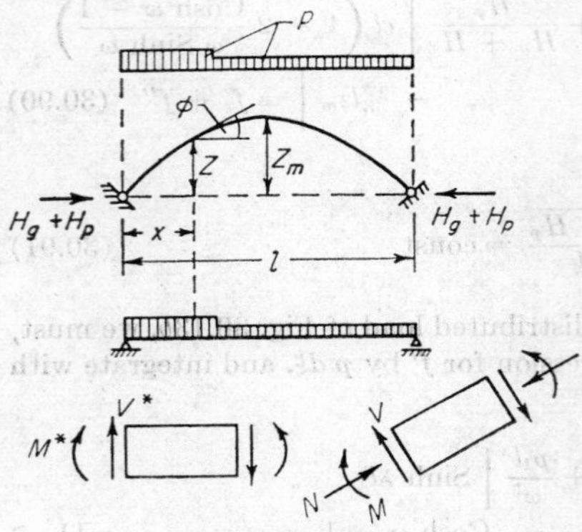

FIG. 30.14

For the analysis of the arch, we use as auxiliary quantities the bending moment M_0 and the shear V_0 which would occur under the given load $p(x)$ in a simple beam having the same span as the arch. In the arch, there is the normal force (positive for compression)

$$N = V_0 \sin \phi + H_p \cos \phi \tag{30.94}$$

Because of assumption 3, this is also the value of N in the second-order theory. The bending moment is, because of assumption 2,

$$M = M_0 - H_p(z - w) + H_g w = M_0 - H_p z + (H_g + H_p)w \tag{30.95}$$

and the differential equation for the vertical deflection is

$$\frac{d^2w}{dx^2} + \frac{M + M_t}{EI \cos \phi} + \frac{d}{dx}\left(\epsilon_t \tan \phi - \frac{N}{EA} \tan \phi\right) = 0 \tag{30.96}$$

where $M_t = EI\kappa_t$, and is calculated according to Art. 30.1*d*. ϵ_t is the given temperature elongation of the arch axis, and $\tan \phi = dz/dx$. For the unknown thrust H_p, we first assume a quantity H_a (possibly determined from the first-order theory), which is held constant during our investigation, so that the principle of superposition is applicable. If we substitute Eqs. (30.94) and (30.95) into (30.96), we obtain the differential equation

$$\frac{d^2w}{dx^2} + \frac{H_g + H_a}{EI \cos \phi} w + \frac{M_0 + M_t}{EI \cos \phi} - \frac{H_a z}{EI \cos \phi} + \frac{d}{dx}\left(\epsilon_t - \frac{N}{EA}\right)\frac{dz}{dx} + \left(\epsilon_t - \frac{N}{EA}\right)\frac{d^2z}{dx^2} = 0 \tag{30.97}$$

After two differentiations with respect to x, we have, because of $d^2M_0/dx^2 = -p$, the form

$$\frac{d^2}{dx^2}\left(EI\cos\phi\,\frac{d^2w}{dx^2}\right) + (H_g + H_a)\frac{d^2w}{dx^2} - \bar{p}(x) = 0 \tag{30.98}$$

where

$$\bar{p}(x) = p(x) + \left[H_a\frac{d^2z}{dx^2} - \frac{d^2M_t}{dx^2} + \frac{d^2}{dx^2}\left(EI\cos\phi\,\frac{d}{dx}\left(\frac{N}{EA}\tan\phi - \epsilon_t\tan\phi\right)\right)\right] \tag{30.99}$$

We compare Eq. (30.98) with the beam-column equation (30.61). Because of the changed sign convention for N, we must now write it in the form

$$\frac{d^2}{dx^2}\left(EI\,\frac{d^2w}{dx^2}\right) + N\,\frac{d^2w}{dx^2} - p(x) = 0 \tag{30.100}$$

Evidently, this is identical with the arch equation (30.98) when we identify the quantities EI, N, and p of the beam column with the quantities $EI\cos\phi$, $H_g + H_a$, and $\bar{p}$ of the arch. Whereas the beam analog for the suspension bridge is a beam with simultaneous transverse load and axial tension (see Art. 30.2*b*), the beam analog for the arch is a beam under simultaneous transverse load and axial compression. If $EI\cos\phi = EI_c = \text{const}$, we can assign to this beam analog a normal-force parameter

$$\omega = l\lambda = l\sqrt{\frac{H_g + H_a}{EI_c}} = \text{const} \tag{30.101}$$

Equation (30.97) then takes the form of Eq. (30.4), with the general solution (30.5).

b. Examples

We consider a two-hinged arch, for which $EI\cos\phi = EI_c = \text{const}$, and which in the initial position is parabolic according to Eq. (30.52). The distributed vertical load $p(x)$ is shown in Fig. 30.15. Nonuniform heating is admitted to the extent that $N/EA - \epsilon_t = \text{const}$ and $M_t = EI\kappa_t = \text{const}$. Equation (30.97) then simply reads

$$\frac{d^2w}{dx^2} + \lambda^2 w + \frac{H_a}{EI_c}\left(\frac{M_0 + M_t}{H_a} - z\right) + \frac{8z_m}{l^2}\left(\frac{N}{EA} - \epsilon_t\right) = 0 \tag{30.102}$$

where M_0 is determined from Eq. (30.78), and λ from Eq. (30.101). The solution, analogous to Eq. (30.5), is

$$\text{For } x \le l_1\text{:}\quad w_1 = \frac{H_a}{H_g + H_a}\left[C_1'\sin\lambda x + c_1(1 - \cos\lambda x) - \left(\frac{M_{01}}{H_a} - z\right)\right]$$

$$\text{For } x \ge l_1\text{:}\quad w_2 = \frac{H_a}{H_g + H_a}\left[C_2'\sin\lambda(l - x) + c_2[1 - \cos\lambda(l - x)] - \left(\frac{M_{02}}{H_a} - z\right)\right] \tag{30.103}$$

where

$$c_{1,2} = \frac{8z_m}{\omega^2} - \frac{p_{1,2}l^2}{\omega^2 H_a} - \frac{M_t}{H_a} + \frac{EI_c}{H_a}\frac{8z_m}{l^2}\left(\epsilon_t - \frac{N}{EA}\right) \tag{30.104}$$

The solution already satisfies the two boundary conditions, which require $w = 0$ for $x = 0$ and $x = l$. The continuity conditions require $w_1 = w_2$ and $dw_1/dx = dw_2/dx$ for $x = l_1$, and yield the integration constants,

$$C_{1,2}' = \frac{1}{\sin\omega}\left[c_{1,2}\cos\omega - c_{2,1} \mp (c_1 - c_2)\cos\lambda l_{2,1}\right] \tag{30.105}$$

The bending moment and the shear at a point x are given by the following expressions:

$$\text{For } x \leqq l_1:\quad M_1 = M_{01} - H_a z + (H_g + H_a)w_1 = H_a[c_1' \sin \lambda x + c_1(1 - \cos \lambda x)] \qquad V_1 = \frac{dM_1}{dx}$$

$$\text{For } x \geqq l_1:\quad M_2 = M_{02} - H_a z + (H_g + H_a)w_2 = H_a\{C_2' \sin \lambda(l - x) + c_2[1 - \cos \lambda(l - x)]\} \qquad V_2 = \frac{dM_2}{dx} \tag{30.106}$$

As a second example, we consider the two-hinged arch, with parabolic axis and loaded by a vertical load P at some distance ξ from the left support, as shown in Fig.

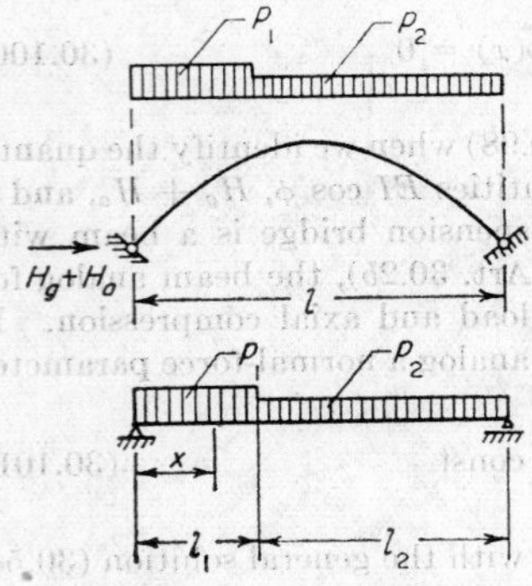

FIG. 30.15

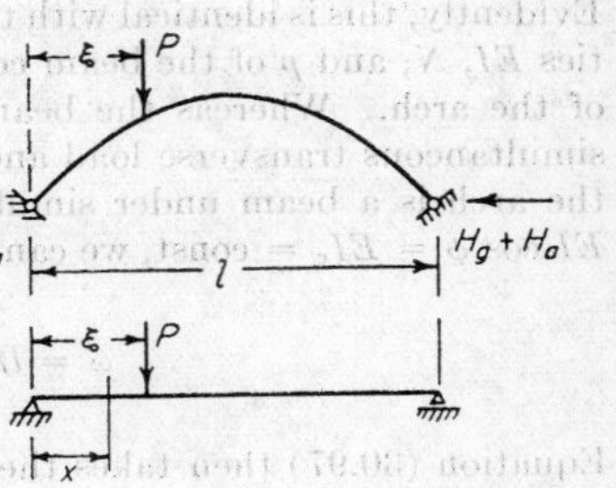

FIG. 30.16

30.16. Let A = const, $\epsilon_t = 0$, $M_t = 0$, and $EI \cos \phi = EI_c$ = const. Then we have

$$\text{For } x \leqq \xi:\quad M_{01} = \frac{P(l - \xi)}{l}x \qquad V_{01} = \frac{P(l - \xi)}{l}$$

$$\text{For } x \geqq \xi:\quad M_{02} = \frac{P\xi}{l}(l - x) \qquad V_{02} = -\frac{P\xi}{l} \tag{30.107}$$

With the help of Eq. (30.52*b*), $\sin \phi$ and $\cos \phi$ appearing in Eq. (30.94) can be expanded in series,

$$\sin \phi = a_0 + a_1(x/l) + a_2(x/l)^2 + \cdots$$
$$\cos \phi = b_0 + b_1(x/l) + b_2(x/l)^2 + \cdots \tag{30.108}$$

where the coefficients a and b depend only on z_m/l. If these series are introduced in Eq. (30.94), the basic equation (30.97) takes the form

$$\frac{d^2w}{dx^2} + \lambda^2 w + \frac{H_a}{EI_c}\left(\frac{M_0}{H_a} - z\right) + \frac{f_N}{EA} = 0 \tag{30.109}$$

where the expression

$$f_N = -\frac{4z_m}{l}\left(1 - \frac{2x}{l}\right)\frac{dN}{dx} + \frac{8z_m}{l^2}N \tag{30.110}$$

contains the forces H_a, V_0, and may also be expanded in a power series in x/l with coefficients depending only on z_m/l. When integrating Eq. (30.109), we distinguish between the two regions $x \leqq \xi$ and $x \geqq \xi$ and, consequently, between $w_1(x)$ and $w_2(x)$, and satisfy the boundary conditions ($w_1 = 0$ for $x = 0$ and $w_2 = 0$ for $x = l$), as well as the continuity conditions ($w_1 = w_2$ and $dw_1/dx = dw_2/dx$ for $x = \xi$). From $w_1(x)$ and $w_2(x)$ there follows, with the help of Eq. (30.95), the bending moments

$M_1(x)$, $M_2(x)$, and, from these, the shears $V_1 = dM_1/dx$ and $V_2 = dM_2/dx$. If the arch is loaded by a distributed vertical load $p(\xi)$, we can replace P in this solution by $p\,d\xi$ and integrate with respect to ξ. In all these equations, the *assumed* thrust H_a is used for the actual value H_p, which is still unknown.

c. The Basic Equation for H_p (Thrust Condition)

In the two-hinged arch shown in Fig. 30.14, we replace the right hinge support by a horizontally movable roller support, and apply there the horizontal force $H_p = 1$. The arch is considered as a rigid body, so that the first-order theory may be employed For a point x, we obtain the values

$$\bar{M} = -z \qquad \bar{N} = \cos\phi \qquad \bar{V} = -\sin\phi \tag{30.111}$$

To this fictitious state of equilibrium, we apply the principle of virtual work and, for virtual displacements, use those deformations which actually appear in the second-order theory. We set the virtual work of the force H_p equal to zero, and thus obtain an equation for the value of H_p which acts in the two-hinged arch. An arch element experiences, according to the second-order theory, the axial strain and the curvature

$$-\epsilon_e - \epsilon_t = \frac{N}{EA} - \epsilon_t \qquad \kappa_e + \kappa_t = \frac{M + M_t}{EI} \tag{30.112}$$

and, therefore, the principle of virtual work gives, with due regard to assumptions 3 and 6,

$$\int_l \bar{M}(\kappa_e + \kappa_t)\frac{dx}{\cos\phi} + \int_l \bar{N}(-\epsilon_e - \epsilon_t)\frac{dx}{\cos\phi} = 0 \tag{30.113}$$

Upon introduction of Eqs. (30.111) and (30.112), this equation assumes the simple form

$$-\int_l z\frac{M + M_t}{EI\cos\phi}\,dx + \int_l\left(\frac{N}{EA} - \epsilon_t\right)dx = 0 \tag{30.114}$$

where, according to Eq. (30.96),

$$\frac{M + M_t}{EI\cos\phi} = -\frac{d^2w}{dx^2} - \frac{d}{dx}\left(\epsilon_t\tan\phi - \frac{N}{EA}\tan\phi\right) \tag{30.115}$$

Since, for $x = 0$ and for $x = l$, $w = 0$ as well as $z = 0$ (Fig. 30.14), we can reuse Eq. (30.87) and have additionally the relation

$$\int_l z\frac{d}{dx}\left(\epsilon_t\tan\phi - \frac{N}{EA}\tan\phi\right)dx = \left[\left(\epsilon_t\tan\phi - \frac{N}{EA}\tan\phi\right)x\right]_0^l - \int_l\left(\epsilon_t\tan\phi - \frac{N}{EA}\tan\phi\right)\frac{dz}{dx}\,dx = \int_l\left(\frac{N}{EA} - \epsilon_t\right)\tan^2\phi\,dx \tag{30.116}$$

so that Eq. (30.114) reduces to

$$\int_l w\frac{d^2z}{dx^2}\,dx + \int_l\left(\frac{N}{EA} - \epsilon_t\right)\sec^2\phi\,dx = 0 \tag{30.117}$$

We shall formulate this equation for the first of the examples treated in Art. 30.3*b*, involving the load shown in Fig. 30.15. Application of Eqs. (30.52*c*) and (30.85*a*) to Eq. (30.117) yields

$$-\frac{8z_m}{l^2}\int_0^l w\,dx + \left(\frac{N}{EA} - \epsilon_t\right)l_i = 0 \tag{30.118}$$

and, upon introduction of w from Eq. (30.103), this yields

$$\int_0^l w\,dx = \int_0^{l_1} w_1\,dx + \int_{l_1}^l w_2\,dx = \frac{H_p}{H_g + H_p}\left(C_1'l\frac{1-\cos\lambda l_1}{\omega} + C_2'l\frac{1-\cos\lambda l_2}{\omega} - \frac{1}{\omega}c_1 l\sin\lambda l_1 - \frac{1}{\omega}c_2 l\sin\lambda l_2 + c_1l_1 + c_2l_2 + \tfrac{2}{3}z_m l + \frac{1}{H_p}\int_l M_0\,dx\right) \qquad (30.119)$$

The values of c_1 and c_2 are given by Eq. (30.104), those of C_1' and C_2' by Eq. (30.105), M_0 by Eqs. (30.78), and λ and ω by Eq. (30.101). If we introduce for $(H_g + H_p)$ a fictitious value $(H_g + H_a)$, we obtain the thrust condition in the form $H_pF_1 + F_2 = 0$, similar to the first-order theory. The calculation is repreated until the value of H_a in $(H_g + H_a)$ agrees with the calculated value H_p. H_p is then substituted into Eqs. (30.103) through (30.106) in place of H_a to obtain w, M, and V. The normal force $N(x)$ follows from Eq. (30.94).

d. Additional Notes

The second-order theory of wide-span hingeless arches is treated in [23] through [26], under the assumptions which are described in Art. 30.3*a*. In addition to the horizontal thrust H_p due to line load, there are also the unknown bending moments at the left and right supports. The arch with three joints or with one joint is considered in [24] and [26]. In [24] and [25], the influence of a large curvature of the arch is considered. The second-order theory of pin-jointed (e.g., eyebar) arches with stiffening girder is described in [28]. For Coriolis arches with constant compressive stress (or the Coriolis eyebar cable with constant tensile stress) in the problem of the limiting span of bridges, see [29]. Of the assumptions described in Art. 30.3*a*, assumption 2 is no longer justified for arches with large z_m/l. The influence of the additional bending moment, which arises on account of the horizontal component of deflection $u(x)$ is determined by step-by-step approximation. For the initial position, we have

$$\frac{du}{dx} = \frac{dw}{dx}\tan\phi + (\epsilon_e + \epsilon_t)\sec^2\phi \qquad (30.120)$$

For the influence of $u(x)$, see [6] and [22], pp. 231–240. Assumption 4 is also, in most practical cases, hard to justify.

30.4. REFERENCES

[1] E. E. Lundquist, W. D. Kroll: Tables of stiffness and carry-over factors for structural members under axial load, *NACA*, *Tech. Note* 652 (1938).
[2] F. Bleich: "Buckling Strength of Metal Structures," McGraw-Hill, New York, 1952.
[3] S. P. Timoshenko, J. M. Gere: "Theory of Elastic Stability," 2d ed., McGraw-Hill, New York, in press.
[4] C. F. Kollbrunner, M. Meister: "Knicken," Springer, Berlin, 1955.
[5] E. E. Lundquist: Stability of structural members under axial load, *NACA Tech. Note* 617 (1937).
[6] D. B. Steinman: Moment distribution by linked rigidities, *Eng. News-Record*, **132** (1944), 802–808.
[7] N. J. Hoff, B. A. Boley, S. V. Nardo, S. Kaufman: Buckling of rigid-jointed plane trusses, *Trans. ASCE*, **116** (1951), 958–986.
[8] N. J. Hoff: "Analysis of Structures," Wiley, New York, 1956.
[9] D. B. Steinman: "Suspension Bridges," 2d ed., Wiley, New York, 1929.
[10] S. P. Timoshenko: The stiffness of suspension bridges, *Trans. ASCE*, **94** (1930), 377–405.
[11] D. B. Steinman: Deflection theory for continuous suspension bridges, *Intern. Assoc. Bridge Struct. Eng. Publs.*, **2** (1934), 400–451.

[12] S. P. Timoshenko, S. Way: Suspension bridges with a continuous stiffening truss, *Intern. Assoc. Bridge Struct. Eng. Publs.*, **2** (1934), 452–466.
[13] A. A. Jakkula: The theory of the suspension bridge, *Intern. Assoc. Bridge Struct. Eng. Publs.*, **4** (1936), 333–358.
[14] A. Selberg: Calculation of lateral truss in suspension bridges, *Intern. Assoc. Bridge Struct. Eng. Publs.*, **7** (1943/44), 311–325.
[15] A. Aas-Jakobsen: Calculating anchored suspension bridges for vertical and horizontal loading (in German), *Intern. Assoc. Bridge Struct. Eng. Publs.*, **7** (1943/44), 15–48.
[16] S. O. Asplund: Deflection theory analysis of suspension bridges, *Intern. Assoc. Bridge Struct. Eng. Publs.*, **9** (1949), 1–33.
[17] E. Egerváry: Bases of a general theory of suspension bridges using a matricial method of calculation (in German), *Intern. Assoc. Bridge Struct. Eng. Publs.*, **16** (1956), 149–184.
[18] H. H. Bleich: "Die Berechnung verankerter Hängebrücken," Springer, Vienna, 1935.
[19] H. Neukirch: Berechnung der Hängebrücken bei Berücksichtigung der Verformung des Kabels, *Ing.-Arch.*, **7** (1936), 140–155.
[20] K. H. Lie: Praktische Berechnung von Hängebrücken nach der Theorie zweiter Ordnung, *Stahlbau*, **14** (1941), 65–69, 78–84.
[21] K. Klöppel, K. H. Lie: Nebeneinflüsse bei der Berechnung von Hängebrücken nach der Theorie zweiter Ordnung, *Forschgh. Stahlbau*, **5** (1942).
[22] F. Stüssi: "Baustatik," vol. 2, Birkhäuser, Basel, Switzerland, 1954.
[23] J. Melan: Der biegsame eingespannte Bogen, *Bauing.*, **6** (1925), 143–147.
[24] B. Fritz: "Theorie und Berechnung vollwandiger Bogenträger bei Berücksichtigung des Einflusses der Systemverformung," Springer, Berlin, 1934.
[25] A. Freudenthal: Deflection theory for arches, *Intern. Assoc. Bridge Struct. Eng. Publs.*, **3** (1935), 100–116.
[26] F. Dischinger: Untersuchungen über die Knicksicherheit, die elastischen Verformungen und das Kriechen des Betons bei Bogenbrücken, *Bauing.*, **18** (1937), 487–520.
[27] S. G. Bergström: On vertical stability of bridge arches (in Swedish), Thesis, Roy. Inst. Technology, Stockholm, 1951.
[28] F. Stüssi: Der Formänderungseinfluss beim versteiften Stabbogen, *Schweiz. Bauztg.*, **108** (1936), 57–59.
[29] H. Ziegler: Zum Problem der grossen Spannweiten. *Österr. Ing.-Arch.*, **9** (1955) 250–262.
[30] Deutscher Stahlbau-Verband: "Hilfstafeln zur Berechnung von Spannungsproblemen der Theorie zweiter Ordnung und von Knickproblemen." Stahlbau-Verlag, Köln, 1959.

CHAPTER 31

BEAMS ON ELASTIC FOUNDATION

BY

M. HETÉNYI, Ph.D., Evanston, Ill.

31.1. BASIC ASSUMPTIONS, DIFFERENTIAL EQUATION, GENERAL SOLUTION

Let us consider a beam supported along its entire length on an elastic medium (Fig. 31.1). With regard to the nature of this supporting medium, we assume that it will exert a resistance p, proportional at every point to the deflection w of the beam at that point:

$$p = kw \tag{31.1}$$

This is by far the simplest assumption we can make regarding the behavior of an elastic foundation, and, fortunately, it also happens to represent, with a greater or

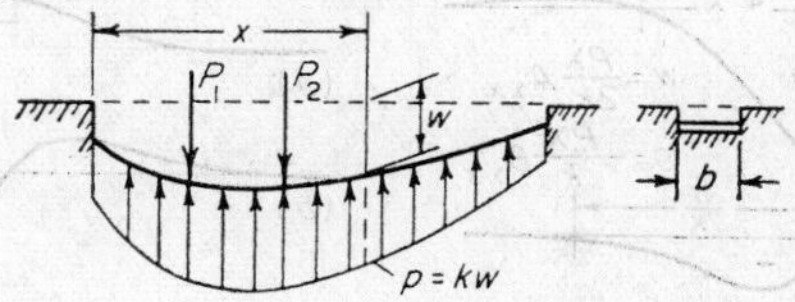

FIG. 31.1

lesser precision, the nature of elastic supports encountered in a large variety of technical problems. It may be visualized as a foundation composed of closely spaced independent spring elements. The elastic resistance of such a foundation can be defined by the amount of pressure (per unit area) required to produce a unit of deflection, k_0, which is called the *modulus of the foundation.* On this basis, the proportionality factor k in Eq. (31.1) will be equivalent to $k = k_0 b$, where b is the width of the beam under consideration. Hence, the dimensions of p will be force per unit length, representing the intensity of the distributed elastic reactions along the length of the deflected beam.

If the beam is subjected to concentrated forces, as indicated in Fig. 31.1, the only distributed loading on the beam will be the foundation pressure p. Hence, we can write, on basis of the elementary bending theory,

$$p = -EI \frac{d^4w}{dx^4} \tag{31.2}$$

where EI represents the flexural stiffness of the beam in the vertical plane of bending, and the minus sign on the right side indicates that the distributed reaction of the foundation always opposes the deflection of the beam.

Combining Eqs. (31.1) and (31.2), we obtain the differential equation of a beam on elastic foundation,

$$\frac{d^4w}{dx^4} + \frac{k}{EI}\, w = 0 \tag{31.3}$$

the general solution of which can be written in the form

$$w = C_1 e^{\lambda x} \cos \lambda x + C_2 e^{\lambda x} \sin \lambda x + C_3 e^{-\lambda x} \cos \lambda x + C_4 e^{-\lambda x} \sin \lambda x \tag{31.4}$$

where

$$\lambda = \sqrt[4]{\frac{k}{4EI}} \tag{31.5}$$

The integration constants in this general solution are defined, in any particular case, by the prescribed end conditions of the beam under consideration.

31.2. THE INFINITELY LONG BEAM

The solution for the deflection curve takes a particularly simple form if the beam is of unlimited length. This will be shown in the following, for various types of transverse loads.

Concentrated Force

Let us assume that the infinitely long beam is loaded by a single concentrated force P (Fig. 31.2). Measuring the x coordinate along the length of beam from the

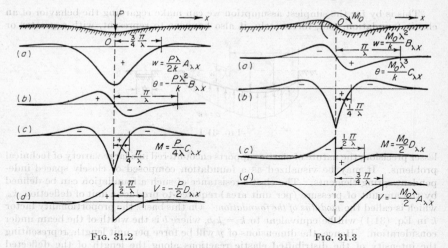

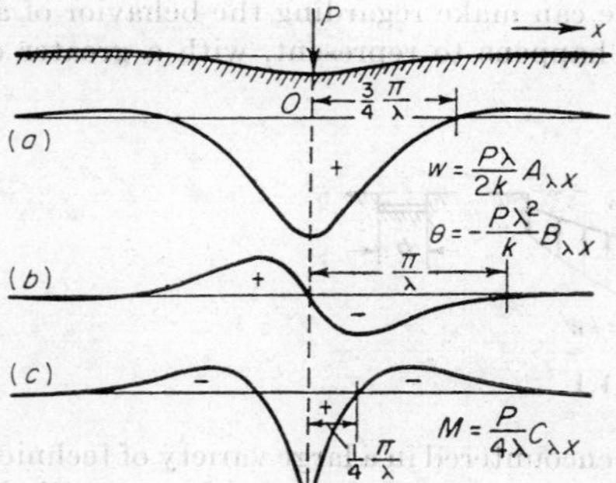

FIG. 31.2

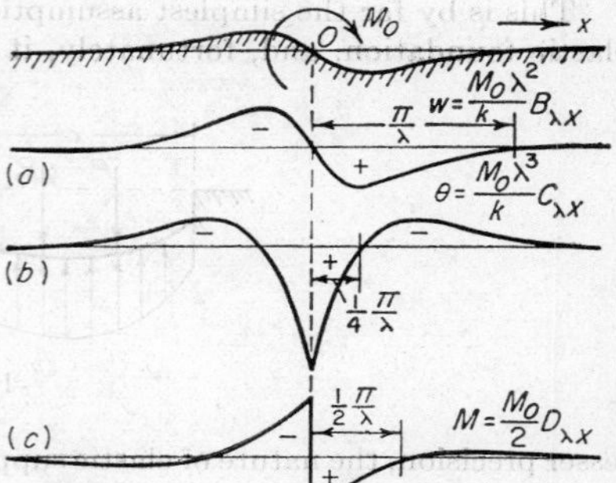

FIG. 31.3

point of application of the load, it is obvious that, as $x \to \infty$, the deflection w and all its derivatives with respect to x should vanish. This condition can be fulfilled only if we put $C_1 = C_2 = 0$ in the general solution (31.4). The remaining two integration constants have to satisfy the first requirement that, on account of symmetry, the slope $\theta = dw/dx = 0$ at $x = 0$, or $C_3 = C_4 = C$. The second requirement is that, to maintain equilibrium, the total pressure in the foundation must balance the load, or

$$2 \int_0^\infty kw\, dx = P$$

which gives $C = P\lambda/2k$.

The equations of the resulting deflection curve $w = w(x)$ and its successive derivatives, defining the slope $\theta = dw/dx$, bending moment $M = -EI\, d^2w/dx^2$, and shearing force $V = -EI\, d^3w/dx^3$, are the following, for $x \geq 0$:

$$w = \frac{P\lambda}{2k} e^{-\lambda x}(\cos \lambda x + \sin \lambda x) \qquad \theta = -\frac{P\lambda^2}{k} e^{-\lambda x} \sin \lambda x$$
$$M = \frac{P}{4\lambda} e^{-\lambda x}(\cos \lambda x - \sin \lambda x) \qquad V = -\frac{P}{2} e^{-\lambda x} \cos \lambda x \tag{31.6a–d}$$

For $x \leq 0$, x must be replaced by $|x|$ in these equations, and the signs of θ and V must be reversed. Positive quantities of w, M, and V are defined in Figs. 35.27 and 35.6.

The curves shown in Fig. 31.2 have several interesting properties. For instance, they not only represent the w, θ, M, and V distributions due to a concentrated load on the beam, but also are influence lines for the same quantities, if the signs of θ and V are reversed. Hence, on the basis of these curves alone, it is feasible to establish a complete solution for the beam under any combination of vertical loading. Furthermore, since all these curves have the character of damped waves, with ordinates of alternating signs, it is possible to load the beam by a number of suitably spaced concentrated forces, and still have smaller values for the resultant maximum deflection or bending moment than the corresponding quantities which would have been produced by one of those forces alone. As indicated by the governing differential equation (31.3), these curves will also periodically repeat themselves in further differentiation, the fourth derivative of w being proportional to w itself, the fifth derivative proportional to the first one, and so on.

Since the expressions for these damped waves will be used frequently in the subsequent discussion, we will now introduce for them the following abbreviated notations:

$$e^{-\lambda x}(\cos \lambda x + \sin \lambda x) = f_1(\lambda x) \qquad e^{-\lambda x} \sin \lambda x = f_2(\lambda x)$$
$$e^{-\lambda x}(\cos \lambda x - \sin \lambda x) = f_3(\lambda x) \qquad e^{-\lambda x} \cos \lambda x = f_4(\lambda x) \tag{31.7}$$

These functions are connected by the relations

$$df_1(\lambda x)/dx = -2\lambda f_2(\lambda x) \qquad df_2(\lambda x)/dx = \lambda f_3(\lambda x)$$
$$df_3(\lambda x)/dx = -2\lambda f_4(\lambda x) \qquad df_4(\lambda x)/dx = -\lambda f_1(\lambda x) \tag{31.8}$$

Concentrated Moment

Having recognized that the curve in Fig. 31.2*b* is an influence line for angular deflection at point O, on the basis of the reciprocity theorem we can conclude that the same curve will also represent the deflection line due to a concentrated moment M_O applied at point O, if we only replace P by M_O in Eq. (31.6*b*). This leads to the set of formulas given in Eqs. (31.9*a–d*) which are also represented graphically in Fig. 31.3*a–d*.

$$w = \frac{M_O\lambda^2}{k} f_2(\lambda x) \qquad \theta = \frac{M_O\lambda^3}{k} f_3(\lambda x)$$
$$M = \tfrac{1}{2} M_O f_4(\lambda x) \qquad V = -\tfrac{1}{2} M_O \lambda f_1(\lambda x) \tag{31.9a–d}$$

It is to be noted that these equations define the w, θ, M, and V curves in accordance with the previously introduced sign convention.

Distributed Loading

An element of a distributed loading may be regarded in itself as an infinitesimal concentrated force. Consequently, the deflection curve of a beam subjected to a distributed load between points A and B (Fig. 31.4) can be derived by substituting $q\,d\xi$ instead of P in Eq. (31.6*a*), and then integrating with respect to ξ between the limits of a and b.

If the point is outside and to the left of the loaded region, as shown in Fig. 31.4, we get the following formulas for w, θ, M, and V at C, due to a constant q loading:

$$w_C = \frac{q}{2k}[f_4(\lambda a) - f_4(\lambda b)] \qquad \theta_C = \frac{q\lambda}{2k}[f_1(\lambda a) - f_1(\lambda b)]$$
$$M_C = -\frac{q}{4\lambda^2}[f_2(\lambda a) - f_2(\lambda b)] \qquad V_C = \frac{q}{4\lambda}[f_3(\lambda a) - f_3(\lambda b)] \tag{31.10a–d}$$

If point C is within the loaded region, the equation for the deflection curve becomes

$$w_C = \frac{q}{2k}[2 - f_4(\lambda a) - f_4(\lambda b)] \tag{31.11}$$

while the formulas for θ_C, M_C, and V_C remain the same as before, with the exception of the sign for M_C being reversed, if a and b are interpreted the same way as before,

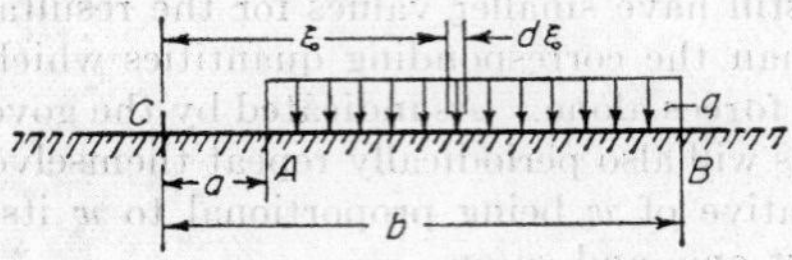

Fig. 31.4

namely, as denoting the distances of point C from the A and B ends of the loaded region.

By this same method of integration, we can derive the solution for any other type of distributed loading, when q is not constant but is any specified function of the variable ξ.

31.3. THE METHOD OF END CONDITIONING

For beams of finite length, the standard procedure is to determine the four integration constants in the general solution [Eq. (31.4)], from the four end conditions prescribed at the two ends of the beam in any particular case. This approach requires the solution of four simultaneous equations and leads, in general, to rather involved expressions, as may be found in [1].

Instead of this procedure, another method of solution will be outlined now, which is based on fulfilling the required end conditions by means of a pair of forces and moments, superimposed on the infinite beam, together with the given loading, and applied immediately on the outside of points A and B, where respective ends of a finite beam are to be produced. It is obvious that, by the simultaneous application of the given load and these end-conditioning forces and moments, the behavior of the infinite beam between, and including, points A and B will be identical with that of the required finite beam. The particular advantage of this method is that it employs only formulas developed for the infinite beam, which are, as we have seen, always of a simple form.

The use of this method will be illustrated now in connection with semi-infinite beams and beams of finite length.

Semi-infinite Beams

Let us assume now that an infinite beam is subjected to a given combination of transverse loading, and we wish to fulfill two prescribed end conditions at point A, by means of an end-conditioning force P_{OA} and moment M_{OB}, applied directly to the left of point A (Fig. 31.5a).

If we wish to produce at A a free end, the corresponding values of P_{OA} and M_{OA} can be determined from the conditions that, under the simultaneous action of these and the given loading, the moment and the shearing force at A should vanish. This means that, on the basis of Eqs. (31.6c,d) and (31.9c,d),

$$M_A + \frac{P_{OA}}{4\lambda} + \frac{M_{OA}}{2} = 0 \qquad \text{and} \qquad V_A - \frac{P_{OA}}{2} - \frac{M_{OA}\lambda}{2} = 0$$

where M_A and V_A denote the bending moment and shearing force produced by the given loading at point A on the infinite beam. Hence, we have

$$P_{OA} = 4(\lambda M_A + V_A) \qquad M_{OA} = -\frac{2}{\lambda}(2\lambda M_A + V_A) \tag{31.12}$$

The same procedure can be followed if at A we are required to produce the conditions of a simply supported end ($w = 0$, $M = 0$), or a fixed end ($w = 0$, $\theta = 0$). In

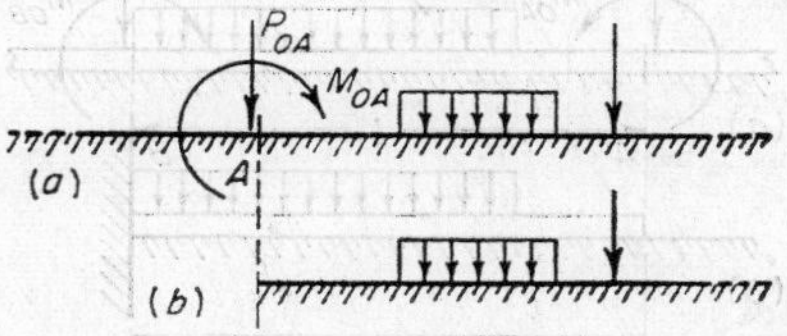

FIG. 31.5

the latter case, the solution takes a particularly simple form. Since a concentrated force produces no rotation on the infinite beam at its point of application, and, conversely, a concentrated moment produces no deflection at the same point, the fulfillment of the end conditions will not necessitate now the solution of two simultaneous equations, but P_{OA} and M_{OA} can be determined separately.

In the following, we give a few of the basic formulas which may be derived in the above manner for semi-infinite beams.

Semi-infinite beam subjected to a concentrated force P at its end

$$w = \frac{2P\lambda}{k} f_4(\lambda x) \tag{31.13}$$

Semi-infinite beam subjected to a concentrated clockwise moment M_O at its end

$$w = -\frac{2M_O\lambda^2}{k} f_3(\lambda x) \tag{31.14}$$

Semi-infinite beam subjected to a prescribed displacement w_O and rotation θ_O at its end

$$w = w_O f_1(\lambda x) + \frac{1}{\lambda}\theta_O f_2(\lambda x) \tag{31.15}$$

Beams of Finite Length

The application of the end-conditioning idea will be illustrated now by means of an example (Fig. 31.6), where the beam is subjected to partially distributed loading, and it is required that there should be a free end at A and a fixed end at B. The corresponding end conditions, under the simultaneous action of the given loading q and the end-conditioning forces P_{OA}, P_{OB} and moments M_{OA}, M_{OB}, all applied on the infinite beam, can be written by means of Eqs. (31.6) and (31.9), as follows:

$$
\begin{aligned}
M_A + \frac{P_{OA}}{4\lambda} + \frac{P_{OB}}{4\lambda} f_3(\lambda l) + \frac{M_{OA}}{2} + \frac{M_{OB}}{2} f_4(\lambda l) &= 0 \\
V_A - \frac{P_{OA}}{2} + \frac{P_{OB}}{2} f_4(\lambda l) - \frac{\lambda M_{OA}}{2} + \frac{\lambda M_{OB}}{2} f_1(\lambda l) &= 0 \\
w_B + \frac{\lambda P_{OA}}{2k} f_1(\lambda l) + \frac{\lambda P_{OB}}{2k} + \frac{\lambda^2 M_{OA}}{k} f_2(\lambda l) &= 0 \\
\theta_B - \frac{\lambda^2 P_{OA}}{k} f_2(\lambda l) + \frac{\lambda^3 M_{OA}}{k} f_3(\lambda l) - \frac{\lambda^3 M_{OB}}{k} &= 0
\end{aligned}
\tag{31.16}
$$

Here M_A, V_A and w_B, θ_B denote the bending moment and shearing force at A, and the deflection and slope at B, respectively, produced on the infinite beam by the loading q according to Eq. (31.10).

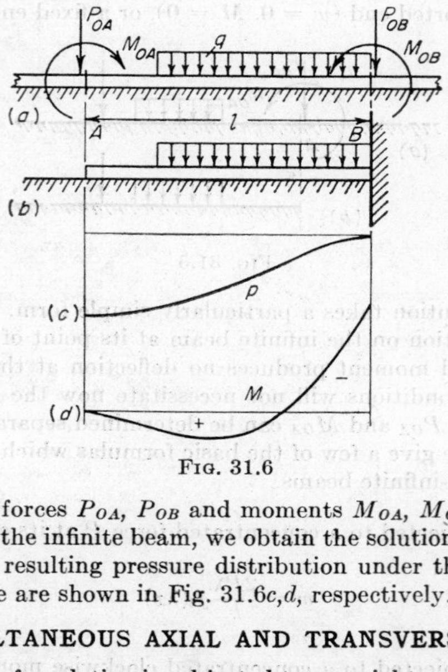

FIG. 31.6

By applying these forces P_{OA}, P_{OB} and moments M_{OA}, M_{OB} simultaneously with the given loading q on the infinite beam, we obtain the solution for the required beam of finite length. The resulting pressure distribution under the beam ($p = kw$) and bending-moment curve are shown in Fig. 31.6c,d, respectively.

31.4. SIMULTANEOUS AXIAL AND TRANSVERSE FORCES

If a beam on elastic foundation is subjected to an axial tensile force N, in addition to transverse loads, it can be shown [1] that the resulting differential equation of the elastic line is

$$
EI \frac{d^4 w}{dx^4} - N \frac{d^2 w}{dx^2} + kw = 0 \tag{31.17}
$$

The general solution of this equation may be written in the form

$$
w = (C_1 e^{\alpha x} + C_2 e^{-\alpha x}) \cos \beta x + (C_3 e^{\alpha x} + C_4 e^{-\alpha x}) \sin \beta x \tag{31.18}
$$

where
$$
\alpha = \sqrt{\lambda^2 + \frac{N}{4EI}} \qquad \beta = \sqrt{\lambda^2 - \frac{N}{4EI}} \tag{31.19}
$$

and, as before,
$$
\lambda = \sqrt[4]{\frac{k}{4EI}}
$$

For beams of unlimited length, the expressions for the deflection lines will be again of relatively simple form. We have, for instance, the following solutions;

Infinite beam loaded by a concentrated force P at O $(x \geq 0)$

$$w = \frac{P}{2k} \frac{\lambda^2}{\alpha\beta} e^{-\alpha x}(\beta \cos \beta x + \alpha \sin \beta x) \tag{31.20}$$

Semi-infinite beam loaded by a concentrated force P at O

$$w = \frac{P}{k} \frac{2\lambda^2}{\beta(3\alpha^2 - \beta^2)} e^{-\alpha x} [2\alpha\beta \cos \beta x + (\alpha^2 - \beta^2) \sin \beta x] \tag{31.21}$$

The above formulas indicate the general rule that, even if axial force is present, the deflection line and all its derivatives will be always proportional to the transverse loading (but not to the axial force; see also p. 30–2). This means that the method of superposition and, consequently, the concept of end conditioning can be again applied to derive solutions for beams of finite length.

If the axial force is compression, we obtain the corresponding solutions by simply reversing the sign of N in the expressions for α and β, given in Eq. (31.19). We can also obtain directly from these deflection formulas the respective buckling loads. Since, by definition, the buckling load produces transverse deflection without the presence of transverse loading, the buckling criteria will be defined by equating the denominator to zero in the corresponding deflection formulas.

This gives, for the infinite beam, when $\alpha\beta = 0$, the value of $N_{cr} = 2\sqrt{kEI}$, while, for the semi-infinite beam, by putting $3\alpha^2 - \beta^2 = 0$, we have the corresponding buckling load as $N_{cr} = \sqrt{kEI}$. It is seen that the buckling load for the semi-infinite beam, with unrestrained end, is exactly one-half that for the infinite beam.

31.5. SOLUTION BY TRIGONOMETRIC SERIES

Simply Supported Ends

Let us assume, for a beam of finite length l, the deflection curve $[w = w(x)]$ in the form of a sine series:

$$w = \sum_{n=1}^{\infty} a_n \sin \frac{n\pi x}{l} \tag{31.22}$$

Because this is an orthogonal series, and because every term of it satisfies the end conditions of a simply supported beam ($w = 0$ and $M = -EI\, d^2w/dx^2 = 0$), the resulting formulas will be particularly simple. In the derivation, the principle of virtual work will be employed, in minimizing the total energy of the system.

Consider that the beam is subjected to a concentrated transverse load P, placed at an arbitrary distance c from the left support, and also to an axial tensile force N, as shown in Fig. 31.7. If the expression in Eq. (31.22) is to represent the deflected state

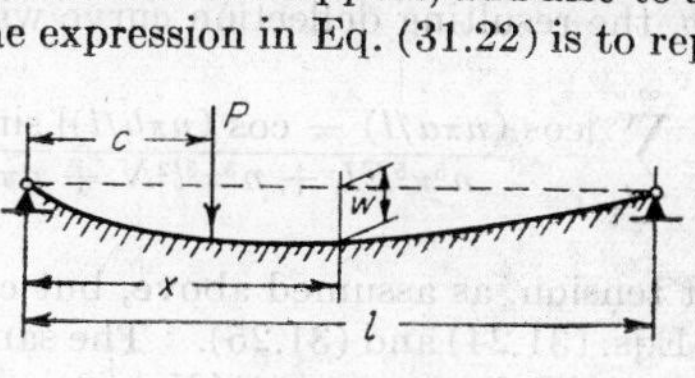

FIG. 31.7

of equilibrium of the beam, any small deviation δa_n from this state should result in a zero change in the total energy of the system.

The total energy will be composed now of four principal sources. Stating each part by means of the expression for w in Eq. (31.22), we have

Strain energy of bending in the beam

$$U_1 = \tfrac{1}{2}EI \int_0^l \left(\frac{d^2w}{dx^2}\right)^2 dx = \frac{\pi^4 EI}{4l^3} \sum_{n=1}^{\infty} n^4 a_n{}^2 \tag{31.23a}$$

Work done in deforming the foundation

$$U_2 = \tfrac{1}{2}k \int_0^l w^2\,dx = \frac{kl}{4} \sum_{n=1}^{\infty} a_n{}^2 \tag{31.23b}$$

Work done against the axial force, due to shortening of span

$$U_3 = N\,\Delta l = \tfrac{1}{2}N \int_0^l \left(\frac{dw}{dx}\right)^2 dx = \frac{\pi^2 N}{4l} \sum_{n=1}^{\infty} n^2 a_n{}^2 \tag{31.23c}$$

Work done by the force P

$$U_4 = Pw_C = P \sum_{n=1}^{\infty} a_n \sin \frac{n\pi c}{l} \tag{31.23d}$$

Taking now the work of the force P positive, this being an input, and the others negative, since they represent stored energies, the application of the principle of virtual work leads to the following equation:

$$P \frac{\partial w_C}{\partial a_n}\,da_n - \left(\frac{\partial U_1}{\partial a_n} + \frac{\partial U_2}{\partial a_n} + N \frac{\partial \Delta l}{\partial a_n}\right) da_n = 0$$

Substituting here the corresponding expressions from Eqs. (31.23*a*–*d*), we have

$$P \sin \frac{n\pi c}{l} - \frac{\pi^4 EI}{2l^3} n^4 a_n - \frac{kl}{2} a_n - N \frac{\pi^2}{2l} n^2 a_n = 0$$

Hence, it is possible to determine directly and independently any amplitude a_n in the formula for w. Expressing a_n from the above equation, and substituting it into Eq. (31.22), we have the deflection curve for the case shown in Fig. 31.7, for any value of the abscissa $0 \le x \le l$, as

$$w = 2Pl^3 \sum_{n=1}^{\infty} \frac{\sin (n\pi c/l) \sin (n\pi x/l)}{n^4\pi^4 EI + n^2\pi^2 l^2 N + kl^4} \tag{31.24}$$

If the beam is subjected to a distributed load q of constant intensity between the points at $x = a$ and $x = b$, the resulting deflection curve will be

$$w = 4ql^4 \sum_{n=1}^{\infty} \frac{[\cos (n\pi a/l) - \cos (n\pi b/l)] \sin (n\pi x/l)}{n^5\pi^5 EI + n^3\pi^3 l^2 N + n\pi k l^4} \tag{31.25}$$

If the axial force is not tension, as assumed above, but compression, we need only to reverse the sign of N in Eqs. (31.24) and (31.25). The same formulas are, of course, valid also when there is no axial force present ($N = 0$), or when the foundation is absent ($k = 0$) and the beam is merely supported at its ends.

Other Types of End Conditions

The above-outlined method can also be used in the analysis of beams whose ends are subject to other types of end conditions than those of simple supports.

For a beam with free ends, we will have to include, in addition to the expansion of the elastic line by means of Eq. (31.22), a constant rigid-body displacement w_0 and rotation α, getting, thus,

$$w = w_0 + \alpha\left(\frac{l}{2} - x\right) + \sum_{n=1}^{\infty} a_n \sin\frac{n\pi x}{l} \tag{31.26}$$

For beams with both ends built in ($w = 0$, $\theta = 0$), the following expression may be used:

$$w = \sum_{n=1}^{\infty} \frac{1}{2} a_n \left(1 - \cos\frac{2n\pi x}{l}\right) \tag{31.27}$$

It is to be noted, however, that both the above expressions have limited fields of application. First of all, they do not form orthogonal series, and, therefore, the determination of n coefficients a_n will require the solution of n simultaneous equations. In addition to this, the expression in Eq. (31.26) may not fully satisfy the condition of zero shearing force at the free ends, and Eq. (31.27) can be used only for loads symmetrical with respect to the center line, $x = l/2$, of the beam.

31.6. FIELDS OF APPLICATION. FOUNDATIONS OF VARIOUS TYPES

The fundamental assumption, $p = kw$, which formed the basis for the solutions in this section, presupposes two characteristics, namely, that the foundation is elastic, and that it is essentially discontinuous, the pressure at any point being independent of the deflections produced at other points. This theory is, therefore, applicable to soil foundations only in exceptional cases, where the above two conditions are satisfactorily fulfilled. For this reason, the most fruitful fields of application of this theory are in connection with technical problems where the foundation is not soil.

The first technical use of this theory was made in the analysis of railroad tracks [2]. Here, first the crossties themselves may be regarded as beams on elastic foundation. These, in turn, support the rails elastically. Though this takes place at separate points, in the analysis it was found permissible to replace the effect of these separate elastic reactions by an equivalent continuously distributed foundation of the $p = kw$ type. The same type of replacement is justifiable in any other case where a beam rests on a large number of elastic supports. This made it possible to analyze by means of this theory all types of structures composed of intersecting beams, such as beam grillages.

There are two instances where the conditions implied in the assumption $p = kw$ are exactly fulfilled. One involves the axially symmetrical deformation of thin circular tubes (see p. 40-21), which, in its extension, provided a method for the approximate calculation of any rotationally symmetrical thin-shell problem [3]. The other instance is of a floating beam, which may be the entire hull of a ship, where the forces of buoyancy are proportional at any point to the depth of submersion.

The extreme counterpart of the independent spring foundation is the case where the supporting medium is itself a continuous elastic solid. There are several solutions for problems of this type [4],[5], but, on account of inherent difficulties, the development has not been extensive in this direction.

In order to bridge the gap between the cases corresponding to the completely continuous (elastic-solid) and completely discontinuous (elastic-spring) foundations, we can devise a mechanical model which is capable of representing partial continuity in the foundation. Such a model may be constructed by embedding a beam into an elastic foundation of the spring type. Since we may have different spring constants above and below this embedded beam, these, together with the flexural stiffness of the beam itself, provide three free parameters, which may be selected, in any particular case, in such a manner as to represent the behavior of any given, partially continuous, elastic foundation [6].

31.7. REFERENCES

[1] M. Hetényi: "Beams on Elastic Foundation," Univ. of Michigan Press, Ann Arbor, Mich., 1946.

[2] H. Zimmermann: "Die Berechnung des Eisenbahnoberbaues," Ernst u. Korn, Berlin, 1888.

[3] J. W. Geckeler: Über die Festigkeit achsensymmetrischer Schalen, *Forsch. Gebiete Ingenieurw.*, **276** (1926).

[4] M. A. Biot: Bending of an infinite beam on an elastic foundation, *J. Appl. Mech.*, **4** (1937), 1–7.

[5] G. Bosson: The flexure of an infinite strip on an elastic foundation, *Phil. Mag.*, [7] **27** (1939), 37–50.

[6] M. Hetényi: A general solution for the bending of beams on an elastic foundation of arbitrary continuity, *J. Appl. Phys.*, **21** (1950), 55–58.

CHAPTER 32

TABLE OF REACTIONS, BENDING MOMENTS, AND DEFLECTION OF BEAMS

In the formulas, the following notation has been used:

l = span
R_1, R_2 = reactions at left and right ends, respectively, positive up
M_1, M_2 = reactive moments, positive when they correspond to a positive bending moment
M = bending moment
w = deflection $\theta = dw/dx$
x = distance from left end of beam
$\xi = x/l \qquad \alpha = a/l \qquad \beta = b/l$

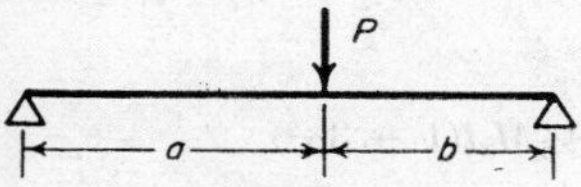

$$R_1 = P\beta \qquad R_2 = P\alpha$$

FIG. 32.1

For $\xi \le \alpha$: $M = Pl\beta\xi$ For $\xi \ge \alpha$: $M = Pl\alpha(1 - \xi)$ For $\xi = \alpha$: $M_{\max} = Pl\alpha\beta$

For $\xi \le \alpha$: $EIw = \dfrac{Pl^3}{6}\beta\xi[\alpha(1 + \beta) - \xi^2]$

For $\xi \ge \alpha$: $EIw = \dfrac{Pl^3}{6}\alpha(1 - \xi)[\xi(2 - \xi) - \alpha^2]$

$$EI\theta_1 = \frac{Pl^2}{6}\alpha\beta(1 + \beta) \qquad EI\theta_2 = -\frac{Pl^2}{6}\alpha\beta(1 + \alpha)$$

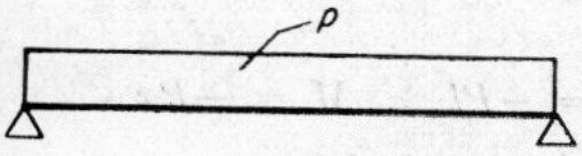

$$R_1 = R_2 = \tfrac{1}{2}pl \qquad M = \tfrac{1}{2}pl^2\xi(1 - \xi)$$

FIG. 32.2 At $x = l/2$: $M_{\max} = pl^2/8$

$$EIw = \frac{pl^4}{24}(\xi - 2\xi^3 + \xi^4) \qquad \text{At } x = \frac{l}{2}\text{: } EIw_{\max} = \frac{5pl^4}{384}$$

$$EI\theta_1 = -EI\theta_2 = pl^3/24$$

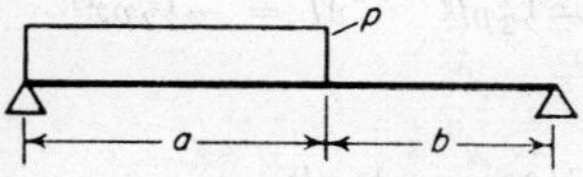

$$R_1 = \tfrac{1}{2}pl\alpha(1 + \beta) \qquad R_2 = \tfrac{1}{2}pl\alpha^2$$

FIG. 32.3

For $\xi \le \alpha$: $M = \tfrac{1}{2}pl^2\xi[\alpha(1 + \beta) - \xi]$
For $\xi \ge \alpha$: $M = \tfrac{1}{2}pl^2\alpha^2(1 - \xi)$

For $\xi = \tfrac{1}{2}l\alpha(1 + \beta)$: $M_{\max} = \dfrac{pl^2}{8}\alpha^2(1 + \beta)^2$

For $\xi \le \alpha$: $EIw = \dfrac{pl^4}{24}\xi[\alpha^2(1 + \beta)^2 - 2\alpha(1 + \beta)\xi^2 + \xi^3]$

For $\xi \ge \alpha$: $EIw = \dfrac{pl^4}{24}\alpha^2(1 - \xi)[-\alpha^2 + 4\xi - 2\xi^2]$

$$EI\theta_1 = \frac{pl^3}{24}\alpha^2(1 + \beta)^2 \qquad EI\theta_2 = -\frac{pl^3}{24}\alpha^2(2 - \alpha^2)$$

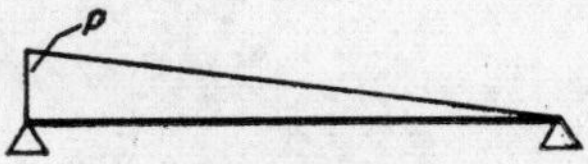

$$R_1 = \tfrac{1}{3}pl \qquad R_2 = \tfrac{1}{6}pl$$

Fig. 32.4

$$M = \tfrac{1}{6}pl^2\xi(1-\xi)(2-\xi)$$

At $\xi = 1 - 1/\sqrt{3} = 0.4226$: $\qquad M_{\max} = \dfrac{pl^2}{27}\sqrt{3}$

$$EIw = \frac{pl^4}{360}\,\xi(1-\xi)(2-\xi)(4+6\xi-3\xi^2)$$

$$EI\theta_1 = pl^3/45 \qquad EI\theta_2 = -\tfrac{7}{360}pl^3$$

$$R_1 = -R_2 = -\frac{M_1}{l}$$

Fig. 32.5

$$M = M_1(1-\xi) \qquad M_{\max} = M_1$$

$$EIw = \tfrac{1}{6}M_1l^2\xi(1-\xi)(2-\xi)$$

$$EI\theta_1 = \tfrac{1}{3}M_1l \qquad EI\theta_2 = -\tfrac{1}{6}M_1l$$

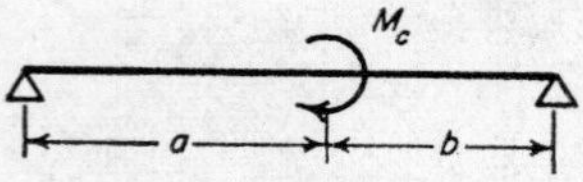

$$R_1 = -R_2 = -M_c/l$$

Fig. 32.6

For $\xi \le \alpha$: $M = -M_c\xi$ $\qquad$ For $\xi \ge \alpha$: $M = M_c(1-\xi)$

For $\xi \le \alpha$: $EIw = -\tfrac{1}{6}M_cl^2\xi(1-3\beta^2-\xi^2)$

For $\xi \ge \alpha$: $EIw = \tfrac{1}{6}M_cl^2(1-\xi)(2\xi-\xi^2-3\alpha^2)$

$$EI\theta_1 = -\tfrac{1}{6}M_cl(1-3\beta^2) \qquad EI\theta_2 = -\tfrac{1}{6}M_cl(1-3\alpha^2)$$

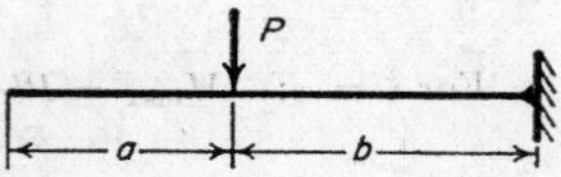

$$R = P \qquad M_2 = -Pb$$

Fig. 32.7

For $\xi \le \alpha$: $M = 0$ $\qquad$ For $\xi \ge \alpha$: $M = -Pl(\xi - \alpha)$

For $\xi \le \alpha$: $EIw = \tfrac{1}{6}Pl^3\beta^2(2+\alpha-3\xi)$

For $\xi \ge \alpha$: $EIw = \tfrac{1}{6}Pl^3(1-\xi)^2(2\beta-\alpha+\xi)$

$$EI\theta_1 = -\tfrac{1}{2}Pb^2$$

$$R = P \qquad M_2 = -Pl \qquad M = -Px$$

Fig. 32.8

$$EIw = \tfrac{1}{6}Pl^3(1-\xi)^2(2+\xi) \qquad EI\theta_1 = -\tfrac{1}{2}Pl^2$$

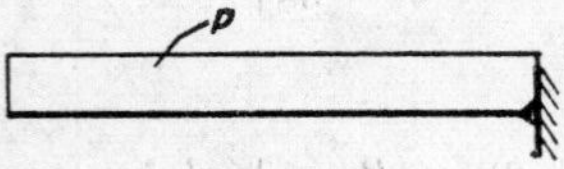

$$R = pl \qquad M_2 = -\tfrac{1}{2}pl^2 \qquad M = -\tfrac{1}{2}px^2$$

Fig. 32.9

$$EIw = \frac{pl^4}{24}(1-\xi)^2(3+2\xi+\xi^2) \qquad EI\theta_1 = -\tfrac{1}{6}pl^3$$

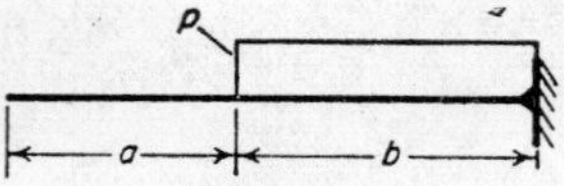

$$R = pb \qquad M_2 = -\tfrac{1}{2}pb^2$$

Fig. 32.10

For $\xi \le \alpha$: $M = 0$ $\qquad$ For $\xi \ge \alpha$: $M = -\tfrac{1}{2}pl^2(\xi-\alpha)^2$

For $\xi \le \alpha$: $EIw = \dfrac{pl^4}{24}(3+\alpha-4\xi)$

For $\xi \ge \alpha$: $EIw = \dfrac{pl^4}{24}(1-\xi)^2[6\beta^2 - 4\beta(1-\xi) + (1-\xi)^2]$

$$EI\theta_1 = -\frac{pb^3}{6}$$

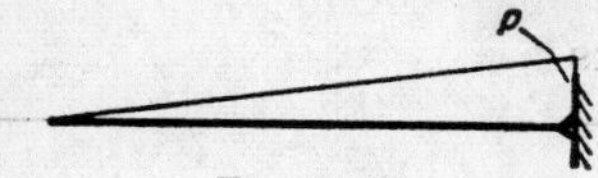

FIG. 32.11

$$R = \tfrac{1}{2}pl \qquad M_2 = -\tfrac{1}{6}pl^2 \qquad M = -\tfrac{1}{6}pl^2\xi^3$$

$$EIw = \frac{pl^4}{120}(4 - 5\xi + \xi^5) \qquad EI\theta_1 = -\frac{pl^3}{24}$$

FIG. 32.12

$$R = 0 \qquad M_2 = M_1 \qquad M = M_1$$

$$EIw = -\tfrac{1}{2}M_1l^2(1 - \xi)^2 \qquad EI\theta_1 = M_1l$$

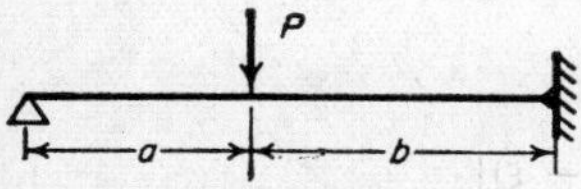

FIG. 32.13

$$R_1 = \tfrac{1}{2}P\beta^2(2 + \alpha) \qquad R_2 = \tfrac{1}{2}P\alpha(2 + \beta)^2$$
$$M_2 = -\tfrac{1}{2}Pl\alpha\beta(1 + \alpha)$$

For $\xi \leq \alpha$: $M = \tfrac{1}{2}Pl\beta^2(2 + \alpha)\xi$

For $\xi \geq \alpha$: $M = \tfrac{1}{2}Pl\alpha[2 - (2 + \beta + \alpha\beta)\xi]$

For $\xi \leq \alpha$: $EIw = \dfrac{Pl^3}{12}\beta^2\xi[3\alpha - (2 + \alpha)\xi^2]$

For $\xi \geq \alpha$: $EIw = \dfrac{Pl^3}{12}\alpha(1 - \xi)^2[-2\alpha^2 + (2 + \beta + \alpha\beta)\xi]$

$$EI\theta_1 = \tfrac{1}{4}Pl^2\alpha\beta^2$$

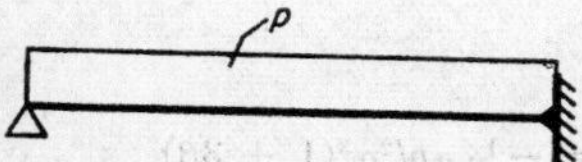

FIG. 32.14

$$R_1 = \tfrac{3}{8}pl \qquad R_2 = \tfrac{5}{8}pl \qquad M_2 = -\tfrac{1}{8}pl^2$$
$$M = \tfrac{1}{8}pl^2\xi(3 - 4\xi)$$

For $\xi = \tfrac{3}{8}$: $M_{max} = \tfrac{9}{128}pl^2$

$$EIw = \frac{pl^4}{48}\xi(1 - \xi)^2(1 + 2\xi) \qquad EI\theta_1 = \frac{pl^3}{48}$$

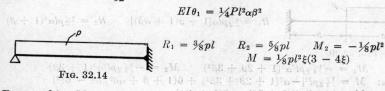

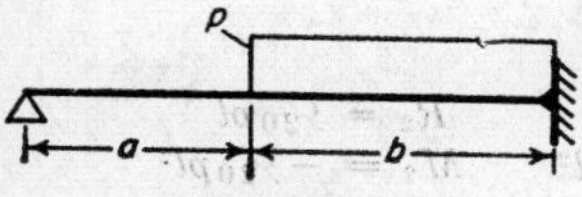

FIG. 32.15

$$R_1 = \tfrac{1}{8}pl\beta^3(3 + \alpha) \qquad R_2 = \tfrac{1}{8}pl\beta(1 + \alpha)(5 - \alpha^2)$$
$$M_2 = -\tfrac{1}{8}pl^2\beta^2(1 + \alpha)^2$$

For $\xi \leq \alpha$: $M = \tfrac{1}{8}pl^2\beta^3(3 + \alpha)\xi$

For $\xi \geq \alpha$: $M = \tfrac{1}{8}pl^2[(3 + \alpha)\beta^3\xi - 4(\xi - \alpha)^2]$

For $\xi \leq \alpha$: $EIw = \dfrac{pl^4}{48}\beta^3\xi[(1 + 3\alpha) - (3 + \alpha)\xi^2]$

For $\xi \geq \alpha$: $EIw = \dfrac{pl^4}{48}(1 - \xi)^2[2\alpha^4 + (1 - 6\alpha^2 + \alpha^4)\xi + 2\xi^2]$

$$EI\theta_1 = \frac{pl^3}{48}\beta^3(1 + 3\alpha)$$

$$R_1 = \tfrac{1}{10}pl \qquad R_2 = \tfrac{2}{5}pl \qquad M_2 = -\tfrac{1}{15}pl^2$$
$$M = \frac{pl^2}{30}\xi(3 - 5\xi^2)$$

FIG. 32.16

$$EIw = \frac{pl^4}{120}\xi(1 - \xi^2)^2 \qquad EI\theta_1 = \frac{pl^3}{120}$$

$$R_1 = -R_2 = -\frac{3M_1}{2l} \qquad M_2 = -\tfrac{1}{2}M_1$$

$$M = \tfrac{1}{2}M_1(2 - 3\xi)$$

FIG. 32.17

$$EIw = \tfrac{1}{4}M_1 l^2 \xi(1 - \xi)^2 \qquad EI\theta_1 = \tfrac{1}{4}M_1 l$$

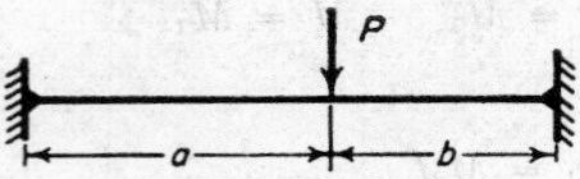

$$R_1 = P\beta^2(1 + 2\alpha) \qquad R_2 = P\alpha^2(1 + 2\beta)$$

$$M_1 = -Pl\alpha\beta^2 \qquad M_2 = -Pl\alpha^2\beta$$

FIG. 32.18

For $\xi \leq \alpha$: $M = Pl\beta^2[(1 + 2\alpha)\xi - \alpha]$
For $\xi \geq \alpha$: $M = Pl\alpha^2[(1 + \beta) - (1 + 2\beta)\xi]$
For $\xi \leq \alpha$: $EIw = \tfrac{1}{6}Pl^3\beta^2\xi^2[3\alpha - (1 + 2\alpha)\xi]$
For $\xi \geq \alpha$: $EIw = \tfrac{1}{6}Pl^3\alpha^2(1 - \xi)^2[3\beta - (1 + 2\beta)(1 - \xi)]$

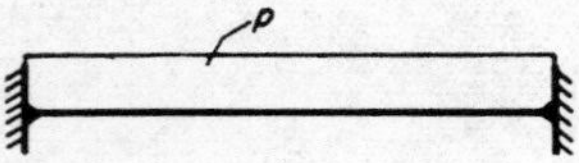

$$R_1 = R_2 = \tfrac{1}{2}pl \qquad M_1 = M_2 = -\frac{pl^2}{12}$$

$$M = -\frac{pl^2}{12}(1 - 6\xi + 6\xi^2)$$

FIG. 32.19

$$EIw = \frac{pl^4}{24}\xi^2(1 - \xi)^2$$

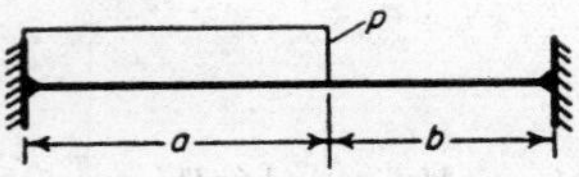

$$R_1 = \tfrac{1}{2}pl\alpha[1 + \beta(1 + \alpha\beta)] \qquad R_2 = \tfrac{1}{2}pl\alpha^2(1 + \beta)$$

FIG. 32.20

$$M_1 = -\tfrac{1}{12}pl^2\alpha^2(1 + 2\beta + 3\beta^2) \qquad M_2 = -\tfrac{1}{12}pl^2\alpha^3(1 + 3\beta)$$

For $\xi \leq \alpha$: $M = \tfrac{1}{12}pl^2[-\alpha^2(1 + 2\beta + 3\beta^2) + 6(1 + \beta + \alpha\beta^2)\alpha\xi - 6\xi^2]$
For $\xi \geq \alpha$: $M = \tfrac{1}{12}pl^2\alpha^3[(5 + 3\beta) - 6(1 + \beta)\xi]$
For $\xi \leq \alpha$: $EIw = \dfrac{pl^4}{24}\xi^2[\alpha^2(1 + 2\beta + 3\beta^2) - 2\alpha\xi(1 + \beta + \alpha\beta^2) + \xi^2]$
For $\xi \geq \alpha$: $EIw = \dfrac{pl^4}{24}\alpha^3(1 - \xi)^2[-\alpha + 2(1 + \beta)\xi]$

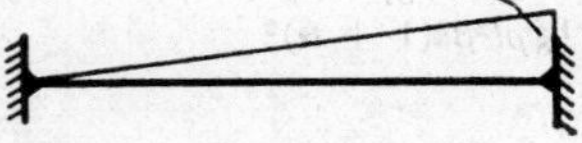

$$R_1 = \tfrac{3}{20}pl \qquad R_2 = \tfrac{7}{20}pl$$

$$M_1 = -\tfrac{1}{30}pl^2 \qquad M_2 = -\tfrac{1}{20}pl^2$$

FIG. 32.21

$$M = -\frac{pl^2}{60}(2 - 9\xi + 10\xi^3) \qquad EIw = \frac{pl^4}{120}\xi^2(1 - \xi)^2(2 + \xi)$$

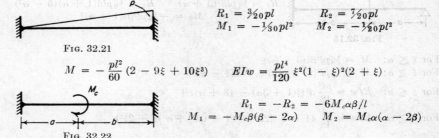

$$R_1 = -R_2 = -6M_c\alpha\beta/l$$

$$M_1 = -M_c\beta(\beta - 2\alpha) \qquad M_2 = M_c\alpha(\alpha - 2\beta)$$

FIG. 32.22

For $\xi \leq \alpha$: $M = M_c\beta[(2\alpha - \beta) - 6\alpha\xi]$
For $\xi \geq \alpha$: $M = M_c\alpha[(\alpha - 2\beta) + 6\beta(1 - \xi)]$
For $\xi \leq \alpha$: $EIw = \tfrac{1}{2}M_c l^2\beta\xi^2[(\beta + 2\alpha) + 2\alpha\xi]$
For $\xi \geq \alpha$: $EIw = -\tfrac{1}{2}M_c l^2\alpha(1 + \xi)^2[(\alpha - 2\beta) + 2\beta(1 - \xi)]$

Part 4

ELASTICITY

CHAPTER 33

BASIC CONCEPTS

BY

K. MARGUERRE, Dr.-Ing., Darmstadt, Germany

33.1. THE CONCEPTS OF STRESS AND STRAIN

Consider a solid body under the action of a system of forces in equilibrium. When we cut this body into two pieces, passing an arbitrary surface (section) through it, each of these pieces will be in equilibrium if an additional force (or possibly a couple) is applied to it. In the uncut body this force must be transmitted through the imaginary section, and it is plausible to assume that each area element dA makes its contribution $d\mathbf{F}$. The limit of the quotient $d\mathbf{F}/dA$ for vanishing dA is called the stress

$$\mathbf{s} = \lim \frac{d\mathbf{F}}{dA} \tag{33.1}$$

in the element. It is a vector and may be resolved into components normal to dA and in dA, the normal stress σ and the shear stress τ. Obviously, τ can once more be resolved into two components in two arbitrary directions in the area element dA.

When different sections are passed through the same point, different stresses are found in areas dA of different orientation. The entity of all stresses thus associated with one point is no longer a vector, but a dyadic. It may be written in different forms, e.g., as a stress matrix [see Eq. (33.32)] or as a tensor [see Eq. (34.31)]. The entity of all stresses in a body constitutes a stress field.

Since the stresses occurring in a solid body are responsible for permanent deformations or fracture, their calculation is an important problem.

When a solid body is subjected to a stress field, it undergoes a certain deformation. Two quite different sets of quantities can be used to describe this deformation. Either the displacements of all points may be specified, or the change of size and shape of every volume element. The displacements constitute a vector field. The deformation of a volume element may be described in various ways, depending on its shape. If the undeformed element is a small sphere, the deformed element is an ellipsoid, and, to describe the deformation, the orientation of its axes may be indicated and their lengths compared before and after deformation. For each axis (and for any other line element) we define here the quotient of its increase in length divided by its original length as its extensional or tensile strain ϵ.

If the volume element is a small cube, its deformation can be described by the tensile strains ϵ of its three sides and by the changes γ of its originally right angles, the shear strains.

The strain components ϵ, γ at a point define a strain dyadic, a strain matrix, and a strain tensor, similar to the corresponding quantities for the stress. The entity of all strains occurring in a solid body forms a strain field.

Stress and strain are related by a physical law, which may have many forms. Depending on this law, the theories of elastic, plastic, viscoelastic, and other bodies are distinguished from one another.

The relations between the strain field and the displacement field are independent of the stress-strain law and are called the kinematic relations of continuum mechanics.

33.2. STRAIN AND COMPATIBILITY

Definition of Strain for Small Displacements

Figure 33.1 shows a rectangle $ABCD$ and the quadrilateral $AB'C'D'$ into which it has been deformed. When the sides AB and AD decrease in length, $AB'C'D'$ approaches a parallelogram and may be considered as such for the purpose of the following analysis.

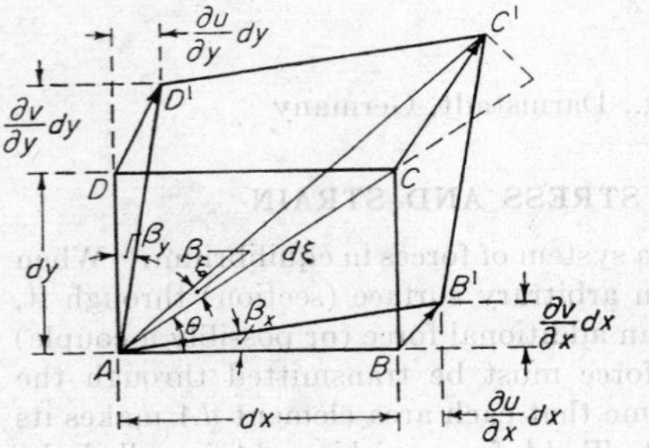

FIG. 33.1

During the deformation the point A undergoes a displacement with the components u, v in x, y directions. In Fig. 33.1 the deformed element has been moved back so that its lower left corner is in the position which it had before deformation. Consequently, the vector **BB′** has the components $(\partial u/\partial x)\,dx$ and $(\partial v/\partial x)\,dx$. When the partial derivatives of the displacements are small compared with 1, the elongation of the line element AB is $AB' - AB = (\partial u/\partial x)\,dx$, and the strain ϵ_x of this line element is this quantity divided by the original length $AB = dx$. Similarly, the strains ϵ_y and ϵ_z of line elements dy and dz are obtained:

$$\epsilon_x = \frac{\partial u}{\partial x} \qquad \epsilon_y = \frac{\partial v}{\partial y} \qquad \epsilon_z = \frac{\partial w}{\partial z} \tag{33.2a–c}$$

In two dimensions there is only one further strain component, the decrease of the right angle between the sides AB and AD of the element:

$$\gamma_{xy} = \beta_x + \beta_y = \frac{\partial v}{\partial x} + \frac{\partial u}{\partial y}$$

The difference

$$2\omega_z = \frac{\partial v}{\partial x} - \frac{\partial u}{\partial y} \tag{33.3}$$

defines the average rotation ω_z of the element and is not a strain component. In three dimensions there are three shear strains:

$$\gamma_{xy} = \frac{\partial v}{\partial x} + \frac{\partial u}{\partial y} \qquad \gamma_{yz} = \frac{\partial w}{\partial y} + \frac{\partial v}{\partial z} \qquad \gamma_{zx} = \frac{\partial u}{\partial z} + \frac{\partial w}{\partial x} \tag{33.2d–f}$$

Equations (33.2), which connect the strains with the displacements, are known as the kinematic relations. Because of the assumption that all the displacement derivatives are small compared with unity, Eqs. (33.2) and (33.3) are linear relations between these derivatives and the strains.

Transformation Law in Two Dimensions

By means of the four quantities ϵ_x, ϵ_y, γ_{xy}, ω_z the dimensions and the angular position of the deformed element $dx\,dy$ are determined; it must, therefore, be possible to find from them the elongation and the rotation of the diagonal $d\xi$ and of any line element which may be made a diagonal by a suitable choice of dx/dy.

The vector **CC′** has the x component

$$\frac{\partial u}{\partial x}\,dx + \frac{\partial u}{\partial y}\,dy = \left(\frac{\partial u}{\partial x} + \frac{\partial u}{\partial y}\tan\theta\right)dx$$

and the y component

$$\left(\frac{\partial v}{\partial x} + \frac{\partial v}{\partial y}\tan\theta\right)dx$$

Its projection on the diagonal $AC = d\xi = dx/\cos\theta$ is the elongation of this line element and yields the strain

$$\epsilon_\xi = \left[\left(\frac{\partial u}{\partial x} + \frac{\partial u}{\partial y}\tan\theta\right)\cos\theta + \left(\frac{\partial v}{\partial x} + \frac{\partial v}{\partial y}\tan\theta\right)\sin\theta\right]\cos\theta$$

while the projection on a normal to AC yields, after division by $d\xi$, the counterclockwise rotation of AC:

$$\beta_\xi = \left[-\left(\frac{\partial u}{\partial x} + \frac{\partial u}{\partial y}\tan\theta\right)\sin\theta + \left(\frac{\partial v}{\partial x} + \frac{\partial v}{\partial y}\tan\theta\right)\cos\theta\right]\cos\theta$$

Similar considerations for a line element $d\eta$ normal to $d\xi$ yield its strain ϵ_η and clockwise rotation β_η. From all these we find with Eqs. (33.2) and $\gamma_{\xi\eta} = \beta_\xi + \beta_\eta$ the strains for a pair of orthogonal directions (ξ,η):

$$\begin{aligned}\epsilon_\xi &= \epsilon_x\cos^2\theta + \epsilon_y\sin^2\theta + \gamma_{xy}\cos\theta\sin\theta\\ \epsilon_\eta &= \epsilon_x\sin^2\theta + \epsilon_y\cos^2\theta - \gamma_{xy}\cos\theta\sin\theta\\ \gamma_{\xi\eta} &= 2(\epsilon_y - \epsilon_x)\cos\theta\sin\theta + \gamma_{xy}(\cos^2\theta - \sin^2\theta)\end{aligned} \qquad (33.4a–c)$$

These equations are the transformation law for the strains in two dimensions. They may be written in the alternate form:

$$\begin{aligned}\epsilon_\xi &= \tfrac{1}{2}(\epsilon_x + \epsilon_y) + \tfrac{1}{2}(\epsilon_x - \epsilon_y)\cos 2\theta + \tfrac{1}{2}\gamma_{xy}\sin 2\theta\\ \epsilon_\eta &= \tfrac{1}{2}(\epsilon_x + \epsilon_y) - \tfrac{1}{2}(\epsilon_x - \epsilon_y)\cos 2\theta - \tfrac{1}{2}\gamma_{xy}\sin 2\theta\\ \gamma_{\xi\eta} &= (\epsilon_y - \epsilon_x)\sin 2\theta + \gamma_{xy}\cos 2\theta\end{aligned} \qquad (33.5a–c)$$

Note that $\epsilon_x + \epsilon_y = \epsilon_\xi + \epsilon_\eta$ for any two pairs of orthogonal directions. With the abbreviations

$$\begin{aligned}\bar\epsilon &= \tfrac{1}{2}(\epsilon_x + \epsilon_y) &\quad \delta_{xy} &= \tfrac{1}{2}(\epsilon_x - \epsilon_y) &\quad \epsilon_{xy} &= \tfrac{1}{2}\gamma_{xy}\\ &= \tfrac{1}{2}(\epsilon_\xi + \epsilon_\eta) & \delta_{\xi\eta} &= \tfrac{1}{2}(\epsilon_\xi - \epsilon_\eta) & \epsilon_{\xi\eta} &= \tfrac{1}{2}\gamma_{\xi\eta}\end{aligned} \qquad (33.6)$$

the transformation law assumes the simple form

$$\begin{aligned}\epsilon_\xi &= \bar\epsilon + \delta_{\xi\eta} &\quad \epsilon_\eta &= \bar\epsilon - \delta_{\xi\eta}\\ \delta_{\xi\eta} &= \delta_{xy}\cos 2\theta + \epsilon_{xy}\sin 2\theta & \epsilon_{\xi\eta} &= -\delta_{xy}\sin 2\theta + \epsilon_{xy}\cos 2\theta\end{aligned} \qquad (33.7)$$

Definition of Strain for Large Displacements

Equations (33.2*a–f*) are contingent upon the assumption that the displacement derivatives, and hence the strains, are small. The following strain definitions are valid without any such restriction.

In an (x,y,z) coordinate system with unit vectors $\mathbf{i}_x$, $\mathbf{i}_y$, $\mathbf{i}_z$ we consider two points, A and B, with position vectors $\mathbf{OA} = \mathbf{r}$ and $\mathbf{OB} = \mathbf{r} + d\mathbf{r}$. Let $\mathbf{u} = \mathbf{AA'}$ be the displacement vector of A, and $\mathbf{u} + d\mathbf{u}$ that of B. Then

$$\begin{aligned}\mathbf{r} &= \mathbf{i}_x x + \mathbf{i}_y y + \mathbf{i}_z z\\ \mathbf{r}' &= \mathbf{r} + \mathbf{u} = \mathbf{i}_x(x + u) + \mathbf{i}_y(y + v) + \mathbf{i}_z(z + w)\\ d\mathbf{r} &= \mathbf{i}_x\,dx + \mathbf{i}_y\,dy + \mathbf{i}_z\,dz\\ d\mathbf{r}' &= d\mathbf{r} + d\mathbf{u} = \mathbf{g}_x\,dx + \mathbf{g}_y\,dy + \mathbf{g}_z\,dz\end{aligned} \qquad (33.8a,b)$$

with

$$\mathbf{g}_x = \frac{\partial \mathbf{r}'}{\partial x} = \mathbf{i}_x + \frac{\partial \mathbf{u}}{\partial x} = \mathbf{i}_x\left(1 + \frac{\partial u}{\partial x}\right) + \mathbf{i}_y\frac{\partial v}{\partial x} + \mathbf{i}_z\frac{\partial w}{\partial x}, \quad \ldots \qquad (33.9)$$

The vectors $\mathbf{g}_x$, $\mathbf{g}_y$, $\mathbf{g}_z$ into which the unit vectors $\mathbf{i}_x$, $\mathbf{i}_y$, $\mathbf{i}_z$ are transformed by the deformation are called the base vectors of the deformed body.

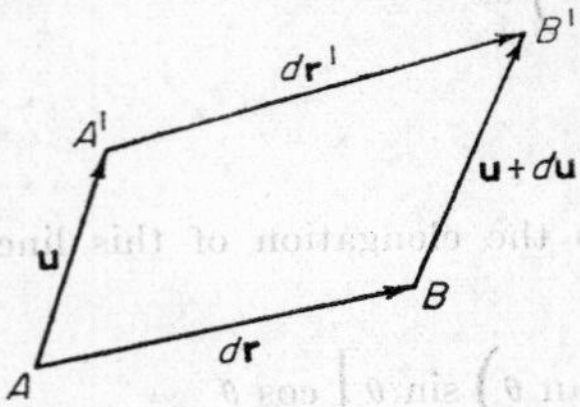

FIG. 33.2

From Eqs. (33.8) we find the squares of the line elements $d\mathbf{r}$ and $d\mathbf{r}'$ and their difference

$$dr'^2 - dr^2 = (\mathbf{g}_x{}^2 - 1)\,dx^2 + (\mathbf{g}_y{}^2 - 1)\,dy^2 + 2\mathbf{g}_x \cdot \mathbf{g}_y\,dx\,dy + \cdots \quad (33.10)$$

Here, as in many of the following formulas, only the terms needed for the two-dimensional case have been written in full; the missing ones may easily be added when needed. The coefficients in Eq. (33.10) describe the change in length of an arbitrary line element with the components dx, dy, dz and hence are strain components. Introducing the base vectors from Eqs. (33.9), we find

$$\begin{aligned}
\gamma_{xx} &= \mathbf{g}_x{}^2 - 1 = 2\frac{\partial u}{\partial x} + \left(\frac{\partial u}{\partial x}\right)^2 + \left(\frac{\partial v}{\partial x}\right)^2 + \left(\frac{\partial w}{\partial x}\right)^2 \\
\gamma_{yy} &= \mathbf{g}_y{}^2 - 1 = 2\frac{\partial v}{\partial y} + \left(\frac{\partial u}{\partial y}\right)^2 + \left(\frac{\partial v}{\partial y}\right)^2 + \left(\frac{\partial w}{\partial y}\right)^2 \\
\gamma_{zz} &= \mathbf{g}_z{}^2 - 1 = 2\frac{\partial w}{\partial z} + \left(\frac{\partial u}{\partial z}\right)^2 + \left(\frac{\partial v}{\partial z}\right)^2 + \left(\frac{\partial w}{\partial z}\right)^2 \\
\gamma_{xy} &= \mathbf{g}_x \cdot \mathbf{g}_y = \frac{\partial v}{\partial x} + \frac{\partial u}{\partial y} + \frac{\partial u}{\partial x}\frac{\partial u}{\partial y} + \frac{\partial v}{\partial x}\frac{\partial v}{\partial y} + \frac{\partial w}{\partial x}\frac{\partial w}{\partial y} \\
\gamma_{yz} &= \mathbf{g}_y \cdot \mathbf{g}_z = \frac{\partial w}{\partial y} + \frac{\partial v}{\partial z} + \frac{\partial u}{\partial y}\frac{\partial u}{\partial z} + \frac{\partial v}{\partial y}\frac{\partial v}{\partial z} + \frac{\partial w}{\partial y}\frac{\partial w}{\partial z} \\
\gamma_{zx} &= \mathbf{g}_z \cdot \mathbf{g}_x = \frac{\partial u}{\partial z} + \frac{\partial w}{\partial x} + \frac{\partial u}{\partial z}\frac{\partial u}{\partial x} + \frac{\partial v}{\partial z}\frac{\partial v}{\partial x} + \frac{\partial w}{\partial z}\frac{\partial w}{\partial x}
\end{aligned} \quad (33.11a\text{–}f)$$

Since these strain quantities have been derived from the squares of the line element, they are not identical with those defined in Eqs. (33.2), even for small strains. The strain γ_{xx} of Eq. (33.11a) is connected with ϵ_x of Eq. (33.2a) by the relation

$$\mathbf{g}_x{}^2 = 1 + \gamma_{xx} = (1 + \epsilon_x)^2 \quad (33.12a)$$

and similar relations hold for γ_{yy}, γ_{zz}. The strains γ_{xy} of Eqs. (33.11d) and (33.2d) are obviously not identical. If the angle between $\mathbf{g}_x$ and $\mathbf{g}_y$ is denoted by $(\pi/2 - \psi_{xy})$, then

$$\mathbf{g}_x \cdot \mathbf{g}_y = \gamma_{xy} = (1 + \epsilon_x)(1 + \epsilon_y)\sin\psi_{xy} \quad (33.12b)$$

When ϵ_x, $\gamma_{xy} \ll 1$, Eqs. (33.12) assume the simple form

$$\gamma_{xx} = 2\epsilon_x \qquad \gamma_{xy} = \psi_{xy} \quad (33.13a,b)$$

Transformation of Strain in Three Dimensions

Besides (x,y,z) we consider another set of orthogonal axes (ξ,η,ζ) with the base vectors $\mathbf{g}_\xi$, $\mathbf{g}_\eta$, $\mathbf{g}_\zeta$ given by

$$\mathbf{g}_\xi = \frac{\partial \mathbf{r}'}{\partial \xi} = \frac{\partial \mathbf{r}'}{\partial x}\cos(x,\xi) + \frac{\partial \mathbf{r}'}{\partial y}\cos(y,\xi) + \frac{\partial \mathbf{r}'}{\partial z}\cos(z,\xi)$$

and two similar relations. With Eqs. (33.9) and (33.11), equations of the following type may be derived:

$$\begin{aligned}
\gamma_{\xi\xi} &= \gamma_{xx}\cos^2(x,\xi) + \gamma_{yy}\cos^2(y,\xi) + 2\gamma_{xy}\cos(x,\xi)\cos(y,\xi) + \cdots \\
\gamma_{\xi\eta} &= \gamma_{xx}\cos(x,\xi)\cos(x,\eta) + \gamma_{yy}\cos(y,\xi)\cos(y,\eta) \\
&\quad + \gamma_{xy}[\cos(x,\xi)\cos(y,\eta) + \cos(x,\eta)\cos(y,\xi)] + \cdots
\end{aligned} \quad (33.14)$$

They are the transformation law of the strains γ in three dimensions. When the axes z and ζ coincide, these equations become identical with Eqs. (33.4).

From Eqs. (33.14) it is seen that the γ_{ik} are the components of a tensor Γ. In the linear case the tensor $\mathcal{E} = \frac{1}{2}\Gamma$ is, in matrix notation,

$$\mathcal{E} = \begin{bmatrix} \epsilon_x & \epsilon_{xy} & \epsilon_{xz} \\ \epsilon_{yx} & \epsilon_y & \epsilon_{yz} \\ \epsilon_{zx} & \epsilon_{zy} & \epsilon_z \end{bmatrix} = \begin{bmatrix} \dfrac{\partial u}{\partial x} & \dfrac{1}{2}\left(\dfrac{\partial v}{\partial x} + \dfrac{\partial u}{\partial y}\right) & \dfrac{1}{2}\left(\dfrac{\partial w}{\partial x} + \dfrac{\partial u}{\partial z}\right) \\ \dfrac{1}{2}\left(\dfrac{\partial u}{\partial y} + \dfrac{\partial v}{\partial x}\right) & \dfrac{\partial v}{\partial y} & \dfrac{1}{2}\left(\dfrac{\partial w}{\partial y} + \dfrac{\partial v}{\partial z}\right) \\ \dfrac{1}{2}\left(\dfrac{\partial u}{\partial z} + \dfrac{\partial w}{\partial x}\right) & \dfrac{1}{2}\left(\dfrac{\partial v}{\partial z} + \dfrac{\partial w}{\partial y}\right) & \dfrac{\partial w}{\partial z} \end{bmatrix} \tag{33.15}$$

It is the symmetric part of the matrix of the nine displacement derivatives:

$$\begin{bmatrix} \dfrac{\partial u}{\partial x} & \dfrac{\partial v}{\partial x} & \dfrac{\partial w}{\partial x} \\ \dfrac{\partial u}{\partial y} & \dfrac{\partial v}{\partial y} & \dfrac{\partial w}{\partial y} \\ \dfrac{\partial u}{\partial z} & \dfrac{\partial v}{\partial z} & \dfrac{\partial w}{\partial z} \end{bmatrix}$$

The antimetric part of this matrix,

$$\begin{bmatrix} 0 & \omega_z & -\omega_y \\ -\omega_z & 0 & \omega_x \\ \omega_y & -\omega_x & 0 \end{bmatrix} = \begin{bmatrix} 0 & \dfrac{1}{2}\left(\dfrac{\partial v}{\partial x} - \dfrac{\partial u}{\partial y}\right) & \dfrac{1}{2}\left(\dfrac{\partial w}{\partial x} - \dfrac{\partial u}{\partial z}\right) \\ \dfrac{1}{2}\left(\dfrac{\partial u}{\partial y} - \dfrac{\partial v}{\partial x}\right) & 0 & \dfrac{1}{2}\left(\dfrac{\partial w}{\partial y} - \dfrac{\partial v}{\partial z}\right) \\ \dfrac{1}{2}\left(\dfrac{\partial u}{\partial z} - \dfrac{\partial w}{\partial x}\right) & \dfrac{1}{2}\left(\dfrac{\partial v}{\partial z} - \dfrac{\partial w}{\partial y}\right) & 0 \end{bmatrix} \tag{33.16}$$

is equivalent to a vector and defines a rigid-body rotation of the volume element [see the remark following Eq. (33.3)].

Through the deformation, the volume of the element $dx\,dy\,dz$ is changed to

$$(1 + \epsilon_x)(1 + \epsilon_y)(1 + \epsilon_z)\,dx\,dy\,dz = [1 + (\epsilon_x + \epsilon_y + \epsilon_z) + \text{higher terms}]\,dx\,dy\,dz$$

This is obviously true when x, y, z are the principal directions of the strain tensor, and, since

$$\operatorname{tr} \mathcal{E} = e = \epsilon_x + \epsilon_y + \epsilon_z \tag{33.17}$$

is invariant against rotations of the reference axes, it must be true for any set of orthogonal axes (x,y,z). The quantity e is called the *dilatation*.

When we split $\mathcal{E}$ into two parts,

$$\mathcal{E} = \frac{1}{3}\begin{bmatrix} e & 0 & 0 \\ 0 & e & 0 \\ 0 & 0 & e \end{bmatrix} + \begin{bmatrix} \epsilon_x' & \epsilon_{xy}' & \epsilon_{xz}' \\ \epsilon_{yx}' & \epsilon_y' & \epsilon_{yz}' \\ \epsilon_{zx}' & \epsilon_{zy}' & \epsilon_z' \end{bmatrix} \tag{33.18}$$

then the first matrix represents a pure dilatation without change of shape, and the second matrix represents a deformation without change of volume and is called the deviator of $\mathcal{E}$.

In curvilinear coordinates the matrix (33.15) contains additional terms. We list the components of $\mathcal{E}$ in cylindrical and in spherical coordinates:

Cylindrical coordinates (see Fig. 4.12)

$$\epsilon_r = \frac{\partial u_r}{\partial r} \qquad \epsilon_\theta = \frac{u_r}{r} + \frac{1}{r}\frac{\partial u_\theta}{\partial \theta} \qquad \epsilon_z = \frac{\partial u_z}{\partial z}$$

$$\gamma_{zr} = \frac{\partial u_r}{\partial z} + \frac{\partial u_z}{\partial r} \qquad \gamma_{\theta r} = \frac{\partial u_\theta}{\partial r} - \frac{u_\theta}{r} + \frac{1}{r}\frac{\partial u_r}{\partial \theta} \qquad \gamma_{z\theta} = \frac{1}{r}\frac{\partial u_z}{\partial \theta} + \frac{\partial u_\theta}{\partial z} \qquad (33.19a\text{–}f)$$

Spherical coordinates (see Fig. 4.12)

$$\epsilon_r = \frac{\partial u_r}{\partial r} \qquad \epsilon_\theta = \frac{1}{r \sin\phi}\frac{\partial u_\theta}{\partial \theta} + \frac{u_r}{r} + \frac{u_\phi}{r}\cot\phi$$

$$\epsilon_\phi = \frac{1}{r}\frac{\partial u_\phi}{\partial \phi} + \frac{u_r}{r} \qquad \gamma_{r\theta} = \frac{1}{r \sin\phi}\frac{\partial u_r}{\partial \theta} + \frac{\partial u_\theta}{\partial r} - \frac{u_\theta}{r} \qquad (33.20a\text{–}f)$$

$$\gamma_{\phi\theta} = \frac{1}{r}\frac{\partial u_\theta}{\partial \phi} - \frac{u_\theta}{r}\cot\phi + \frac{1}{r \sin\phi}\frac{\partial u_\phi}{\partial \theta} \qquad \gamma_{r\theta} = \frac{\partial u_\phi}{\partial r} - \frac{u_\phi}{r} + \frac{1}{r}\frac{\partial u_r}{\partial \phi}$$

In these equations u_r, u_z, u_ϕ, u_θ are the displacement components in the directions indicated by the subscripts.

Compatibility

The kinematic relations (33.2) permit the six strains, $\epsilon_x, \ldots, \gamma_{zx}$, to be derived from only three displacements, u, v, w (in two dimensions: three strains from two displacements). Mathematically speaking, this means that the strains cannot all be prescribed arbitrarily as functions of the coordinates: they are connected by a corresponding number of equations. Physically speaking, it means that the strains must be *compatible:* the deformed elements must fit together.

In two dimensions there exists one compatibility equation. It is found by eliminating u and v from Eqs. (33.2*a*,*b*,*d*):

$$\frac{\partial^2 \epsilon_x}{\partial y^2} + \frac{\partial^2 \epsilon_y}{\partial x^2} - \frac{\partial^2 \gamma_{xy}}{\partial x\,\partial y} = 0 \qquad (33.21a)$$

In three dimensions there are two more equations of this type:

$$\frac{\partial^2 \epsilon_y}{\partial z^2} + \frac{\partial^2 \epsilon_z}{\partial y^2} - \frac{\partial^2 \gamma_{yz}}{\partial y\,\partial z} = 0 \qquad \frac{\partial^2 \epsilon_z}{\partial x^2} + \frac{\partial^2 \epsilon_x}{\partial z^2} - \frac{\partial^2 \gamma_{zx}}{\partial z\,\partial x} = 0 \qquad (33.21b,c)$$

By conducting the elimination of u, v, and w differently, we may as well obtain the following three compatibility conditions:

$$2\frac{\partial^2 \epsilon_z}{\partial x\,\partial y} = \frac{\partial}{\partial z}\left(\frac{\partial \gamma_{yz}}{\partial x} + \frac{\partial \gamma_{zx}}{\partial y} - \frac{\partial \gamma_{xy}}{\partial z}\right) \qquad 2\frac{\partial^2 \epsilon_x}{\partial y\,\partial z} = \frac{\partial}{\partial x}\left(\frac{\partial \gamma_{zx}}{\partial y} + \frac{\partial \gamma_{xy}}{\partial z} - \frac{\partial \gamma_{yz}}{\partial x}\right)$$

$$2\frac{\partial^2 \epsilon_y}{\partial z\,\partial x} = \frac{\partial}{\partial y}\left(\frac{\partial \gamma_{xy}}{\partial z} + \frac{\partial \gamma_{yz}}{\partial x} - \frac{\partial \gamma_{zx}}{\partial y}\right) \qquad (33.21d\text{–}f)$$

Using the notation (33.13*a*), the six compatibility equations may be written in the following form:

$$c_{11} \equiv \frac{\partial}{\partial y}\left(\frac{\partial \gamma_{zz}}{\partial y} - \frac{\partial \gamma_{yz}}{\partial z}\right) - \frac{\partial}{\partial z}\left(\frac{\partial \gamma_{zy}}{\partial y} - \frac{\partial \gamma_{yy}}{\partial z}\right) = 0$$

$$c_{22} = 0 \qquad c_{33} = 0 \qquad (33.22a\text{–}f)$$

$$c_{32} = c_{23} \equiv \frac{\partial}{\partial z}\left(\frac{\partial \gamma_{yx}}{\partial x} - \frac{\partial \gamma_{xx}}{\partial y}\right) - \frac{\partial}{\partial x}\left(\frac{\partial \gamma_{yz}}{\partial x} - \frac{\partial \gamma_{xz}}{\partial y}\right) = 0$$

$$c_{13} = c_{31} = 0 \qquad c_{21} = c_{12} = 0$$

The six equations (33.22) cannot all be independent since, if they were, they would be sufficient to determine the six strain components, which, of course, is impossible.

It is easily verified that the following identities hold:

$$\frac{\partial c_{11}}{\partial x} + \frac{\partial c_{12}}{\partial y} + \frac{\partial c_{13}}{\partial z} = 0 \qquad \frac{\partial c_{21}}{\partial x} + \frac{\partial c_{22}}{\partial y} + \frac{\partial c_{23}}{\partial z} = 0 \qquad \frac{\partial c_{31}}{\partial x} + \frac{\partial c_{32}}{\partial y} + \frac{\partial c_{33}}{\partial z} = 0 \quad (33.23)$$

They are *Bianchi's identities.*

33.3. STRESS AND EQUILIBRIUM

Definition of Stress for Small Deformations

Consider a volume element $dx\,dy\,dz$ (Fig. 33.3). Among its six faces we distinguish three front faces and three rear faces. The front faces are those where the outer normal points in the direction of a positive coordinate axis.

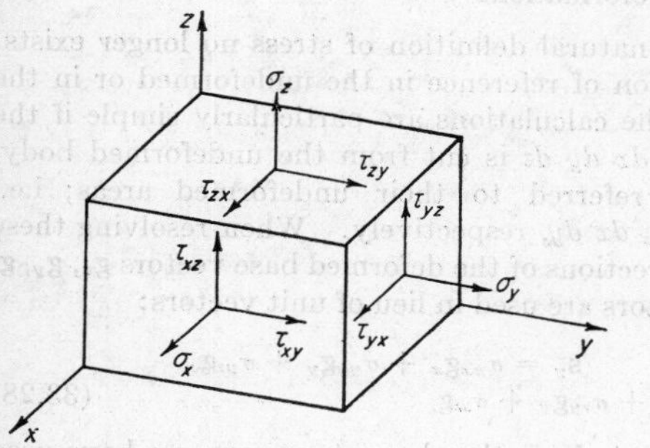

FIG. 33.3

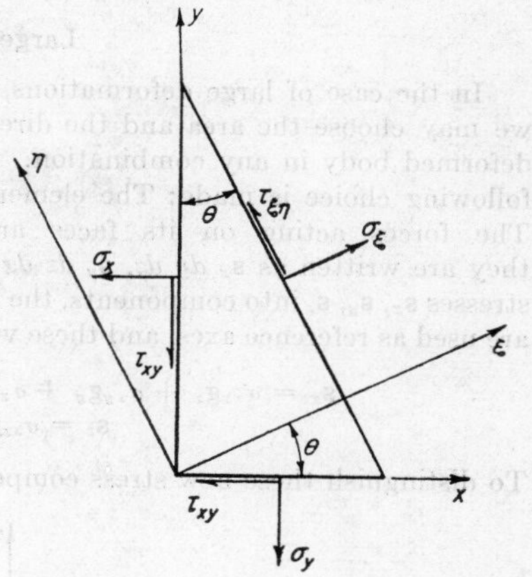

FIG. 33.4

The stress acting on the front side (x = const) has been resolved into its rectangular components, the normal stress σ_x and the shear stresses τ_{xy}, τ_{xz}. For these and the other stresses shown in the figure, the following notation applies:

The subscript of a normal stress indicates the section in which this stress acts (σ_x in the section x = const), and the coordinate axis to which the stress is parallel. The first subscript of the shear stress τ indicates the *section* in which the stress acts, and the second subscript indicates the *direction* of the stress. On a front (rear) side, a stress is positive when it points in the direction of a positive (negative) coordinate axis.

The moment equilibrium of the forces acting on the volume element $dx\,dy\,dz$ yields three relations among the nine stresses just defined. When writing the moments with respect to a line connecting the centers of the faces $dx\,dy$, it is found that only the stresses τ_{xy} and τ_{yx} make contributions, and that between them the first of the following equations must hold:

$$\tau_{xy} = \tau_{yx} \qquad \tau_{yz} = \tau_{zy} \qquad \tau_{zx} = \tau_{xz} \qquad (33.24)$$

Transformation in Two Dimensions

Consider a small triangular prism as shown in Fig. 33.4. Let dA be the area of the inclined face. Then the areas of the other faces are $dA \cos\theta$ and $dA \sin\theta$, and the equilibrium of forces in directions ξ and η yields two of the following equations:

$$\begin{aligned} \sigma_\xi &= \sigma_x \cos^2\theta + \sigma_y \sin^2\theta + 2\tau_{xy}\cos\theta\sin\theta \\ \tau_{\eta\xi} = \tau_{\xi\eta} &= (\sigma_y - \sigma_x)\cos\theta\sin\theta + \tau_{xy}(\cos^2\theta - \sin^2\theta) \\ \sigma_\eta &= \sigma_x \sin^2\theta + \sigma_y \cos^2\theta - 2\tau_{xy}\cos\theta\sin\theta \end{aligned} \qquad (33.25a\text{–}c)$$

Equation (33.25*c*) is derived from a prism containing a face η = const.

With the abbreviations

$$\begin{aligned} \bar\sigma &= \tfrac{1}{2}(\sigma_x + \sigma_y) = \tfrac{1}{2}(\sigma_\xi + \sigma_\eta) \\ \rho_{xy} &= \tfrac{1}{2}(\sigma_x - \sigma_y) \qquad \rho_{\xi\eta} = \tfrac{1}{2}(\sigma_\xi - \sigma_\eta) \end{aligned} \qquad (33.26)$$

Eqs. (33.25) may be written in the compact form

$$\begin{aligned} &\sigma_\xi = \bar\sigma + \rho_{\xi\eta} \qquad \sigma_\eta = \bar\sigma - \rho_{\xi\eta} \\ &\rho_{\xi\eta} = \rho_{xy}\cos 2\theta + \tau_{xy}\sin 2\theta \qquad \tau_{\xi\eta} = -\rho_{xy}\sin 2\theta + \tau_{xy}\cos 2\theta \end{aligned} \qquad (33.27a\text{–}d)$$

Large Deformations

In the case of large deformations, a natural definition of stress no longer exists; we may choose the area and the direction of reference in the undeformed or in the deformed body in any combination. The calculations are particularly simple if the following choice is made: The element $dx\ dy\ dz$ is cut from the undeformed body. The forces acting on its faces are referred to their undeformed areas; i.e., they are written as $\mathbf{s}_x\ dy\ dz$, $\mathbf{s}_y\ dz\ dx$, $\mathbf{s}_z\ dx\ dy$, respectively. When resolving these stresses $\mathbf{s}_x$, $\mathbf{s}_y$, $\mathbf{s}_z$ into components, the directions of the deformed base vectors $\mathbf{g}_x$, $\mathbf{g}_y$, $\mathbf{g}_z$ are used as reference axes, and these vectors are used in lieu of unit vectors:

$$\begin{aligned} \mathbf{s}_x &= \sigma_{xx}\mathbf{g}_x + \sigma_{xy}\mathbf{g}_y + \sigma_{xz}\mathbf{g}_z \qquad \mathbf{s}_y = \sigma_{yx}\mathbf{g}_x + \sigma_{yy}\mathbf{g}_y + \sigma_{yz}\mathbf{g}_z \\ \mathbf{s}_z &= \sigma_{zx}\mathbf{g}_x + \sigma_{zy}\mathbf{g}_y + \sigma_{zz}\mathbf{g}_z \end{aligned} \qquad (33.28)$$

To distinguish these new stress components from the elementary ones, we have used

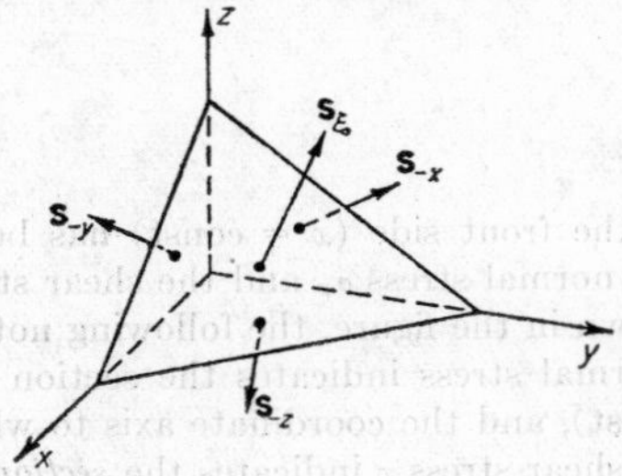

FIG. 33.5

the letter σ and double subscripts everywhere. The definition (33.28) has the advantage that the moment equilibrium yields the symmetry relations

$$\sigma_{yx} = \sigma_{xy} \qquad \sigma_{zy} = \sigma_{yz} \qquad \sigma_{xz} = \sigma_{zx} \qquad (33.29)$$

Transformation in Three Dimensions

For the stresses defined by Eqs. (33.28), the transformation formulas are rather simple and will be derived here. They may be specialized to small deformations by letting

$$\sigma_{xx} = \sigma_x, \ \ldots, \ \sigma_{xy} = \tau_{xy}, \ \ldots$$

The equilibrium of a tetrahedron (Fig. 33.5) yields the vector equation

$$\mathbf{s}_\xi + \mathbf{s}_{-x}\cos(x,\xi) + \mathbf{s}_{-y}\cos(y,\xi) + \mathbf{s}_{-z}\cos(z,\xi) = 0 \qquad (33.30)$$

in which the cosines stand for the ratios of the areas of the four triangular faces. Upon introduction of components according to (33.28), this equation assumes the form

$$\sigma_{\xi\xi}\mathbf{g}_\xi + \sigma_{\xi\eta}\mathbf{g}_\eta + \cdots = (\sigma_{xx}\mathbf{g}_x + \sigma_{xy}\mathbf{g}_y + \cdots)\cos(x,\xi) + (\sigma_{yx}\mathbf{g}_x + \sigma_{yy}\mathbf{g}_y + \cdots)\cos(y,\xi) + \cdots$$

Now, since

$$\mathbf{g}_x = \mathbf{g}_\xi \cos(x,\xi) + \mathbf{g}_\eta \cos(x,\eta) + \cdots$$
$$\mathbf{g}_y = \mathbf{g}_\xi \cos(y,\xi) + \mathbf{g}_\eta \cos(y,\eta) + \cdots \qquad \mathbf{g}_z = \cdots$$

the preceding equation may be split into the following component equations

$$\sigma_{\xi\xi} = \sigma_{xx}\cos^2(x,\xi) + 2\sigma_{xy}\cos(x,\xi)\cos(y,\xi) + \sigma_{yy}\cos^2(y,\xi) + \cdots$$
$$\sigma_{\xi\eta} = \sigma_{xx}\cos(x,\xi)\cos(x,\eta) + \sigma_{xy}[\cos(x,\xi)\cos(y,\eta) + \cos(x,\eta)\cos(y,\xi)] + \sigma_{yy}\cos(y,\xi)\cos(y,\eta) + \cdots \qquad (33.31a,b)$$

When specialized for two dimensions and small deformations, these equations reduce to Eqs. (33.25).

Because the stress components are transformed according to Eqs. (33.25) and (33.31), the stress has tensor character.

In analogy to the strain tensor, the stress tensor

$$\mathcal{S} = \begin{bmatrix} \sigma_x & \tau_{xy} & \tau_{xz} \\ \tau_{yx} & \sigma_y & \tau_{yz} \\ \tau_{zx} & \tau_{zy} & \sigma_z \end{bmatrix}$$

may be split in two parts

$$\mathcal{S} = \begin{bmatrix} \bar\sigma & 0 & 0 \\ 0 & \bar\sigma & 0 \\ 0 & 0 & \bar\sigma \end{bmatrix} + \begin{bmatrix} \sigma'_x & \tau'_{xy} & \tau'_{xz} \\ \tau'_{yx} & \sigma'_y & \tau'_{yz} \\ \tau'_{zx} & \tau'_{zy} & \sigma'_z \end{bmatrix} \qquad (33.32)$$

with $\sigma'_x = \sigma_x - \bar\sigma, \ldots$, and $\tau'_{xy} = \tau_{xy}, \ldots$. The first part represents a *hydrostatic* stress system, and the second is the stress deviator.

Equilibrium Equations

Between the components of the stress tensor there exists, besides the symmetry relations (33.24) or (33.29), a set of three differential conditions of equilibrium. They state that the infinitesimal differences between stresses on opposite faces of a volume element must have a vanishing resultant. In the notation of Eqs. (33.28) and for the element of Fig. 33.3, the difference between the forces on the two faces x = const of a volume element is $(\partial\mathbf{s}_x/\partial x)\,dx\,dy\,dz$. Therefore, the equilibrium condition, even for large displacements is simply

$$\frac{\partial\mathbf{s}_x}{\partial x} + \frac{\partial\mathbf{s}_y}{\partial y} + \frac{\partial\mathbf{s}_z}{\partial z} + \mathbf{F} = 0 \qquad (33.33)$$

where $\mathbf{F} = X\mathbf{g}_x + Y\mathbf{g}_y + Z\mathbf{g}_z$ is the body force per unit of the undeformed volume. Introducing Eqs (33.9) and (33.28) yields

$$\frac{\partial}{\partial x}\left[\sigma_{xx}\left(1 + \frac{\partial u}{\partial x}\right) + \sigma_{xy}\frac{\partial u}{\partial y} + \cdots\right] + \frac{\partial}{\partial y}\left[\sigma_{yx}\left(1 + \frac{\partial u}{\partial x}\right) + \sigma_{yy}\frac{\partial u}{\partial y} + \cdots\right] + \cdots + X = 0$$

$$\frac{\partial}{\partial x}\left[\sigma_{xx}\frac{\partial v}{\partial x} + \sigma_{xy}\left(1 + \frac{\partial v}{\partial y}\right) + \cdots\right] + \frac{\partial}{\partial y}\left[\sigma_{yx}\frac{\partial v}{\partial x} + \sigma_{yy}\left(1 + \frac{\partial v}{\partial y}\right) + \cdots\right] + \cdots + Y = 0 \qquad (33.34a\text{–}c)$$

$$\cdots + \cdots + \cdots = 0$$

In Eqs. (33.34), as before, only those terms have been written explicitly which remain in the case of plane elasticity. Equations (33.34) must be used, not only when dealing with materials which, like rubber, are capable of large deformation, but also as a starting point for the formulation of stability problems. In this case, however, many of the terms of the type $\sigma_{xx}\partial u/\partial x$ become small of a higher order.

In linear elasticity, Eqs. (33.34) reduce to the set

$$\frac{\partial \sigma_x}{\partial x} + \frac{\partial \tau_{yx}}{\partial y} + \frac{\partial \tau_{zx}}{\partial z} + X = 0 \qquad \frac{\partial \tau_{xy}}{\partial x} + \frac{\partial \sigma_y}{\partial y} + \frac{\partial \tau_{zy}}{\partial z} + Y = 0$$
$$\frac{\partial \tau_{xz}}{\partial x} + \frac{\partial \tau_{yz}}{\partial y} + \frac{\partial \sigma_z}{\partial z} + Z = 0 \qquad (33.35a\text{–}c)$$

and, in two dimensions, they are:

$$\frac{\partial \sigma_x}{\partial x} + \frac{\partial \tau_{xy}}{\partial y} + X = 0 \qquad \frac{\partial \tau_{xy}}{\partial x} + \frac{\partial \sigma_y}{\partial y} + Y = 0 \qquad (33.36a,b)$$

They are three equations for six unknowns, and two equations for three unknowns, respectively. Therefore, the stress problem of an elastic body is internally statically indeterminate.

At the surface of the body, the surface tractions may be prescribed. Let p_x, p_y, p_z be the components of the external force per unit area acting on a surface element with the outer normal **n**. Then the equations of equilibrium

$$\begin{aligned} \sigma_x \cos(x,n) + \tau_{yx} \cos(y,n) + \tau_{zx} \cos(z,n) &= p_x \\ \tau_{xy} \cos(x,n) + \sigma_y \cos(y,n) + \tau_{zy} \cos(z,n) &= p_y \\ \tau_{xz} \cos(x,n) + \tau_{yz} \cos(y,n) + \sigma_z \cos(z,n) &= p_z \end{aligned} \qquad (33.37a\text{–}c)$$

must hold.

In cylindrical coordinates r, θ, z (see Fig. 4.12, but replace r' by r), the equilibrium equations (33.35) assume the form

$$\begin{aligned} \frac{\partial \sigma_r}{\partial r} + \frac{1}{r}\frac{\partial \tau_{r\theta}}{\partial \theta} + \frac{\partial \tau_{zr}}{\partial z} + \frac{\sigma_r - \sigma_\theta}{r} + F_r &= 0 \\ \frac{\partial \tau_{r\theta}}{\partial r} + \frac{1}{r}\frac{\partial \sigma_\theta}{\partial \theta} + \frac{\partial \tau_{z\theta}}{\partial z} + 2\frac{\tau_{r\theta}}{r} + F_\theta &= 0 \\ \frac{\partial \tau_{rz}}{\partial r} + \frac{1}{r}\frac{\partial \tau_{z\theta}}{\partial \theta} + \frac{\partial \sigma_z}{\partial z} + \frac{\tau_{rz}}{r} + F_z &= 0 \end{aligned} \qquad (33.38a\text{–}c)$$

where F_r, F_θ, F_z are the components of the body force. For the special case of axial symmetry and absence of body forces, these equations reduce to

$$\frac{\partial \sigma_r}{\partial r} + \frac{\partial \tau_{zr}}{\partial z} + \frac{\sigma_r - \sigma_\theta}{r} = 0 \qquad \frac{\partial \tau_{rz}}{\partial r} + \frac{\partial \sigma_z}{\partial z} + \frac{\tau_{rz}}{r} = 0 \qquad (33.39a,b)$$

In spherical coordinates r, θ, ϕ (Fig. 4.12), the equilibrium equations are:

$$\begin{aligned} &\frac{\partial \sigma_r}{\partial r} + \frac{1}{r \sin \phi}\frac{\partial \tau_{r\theta}}{\partial \theta} + \frac{1}{r}\frac{\partial \tau_{r\phi}}{\partial \phi} + \frac{1}{r}(2\sigma_r - \sigma_\theta - \sigma_\phi + \tau_{r\phi} \cot \phi) + F_r = 0 \\ &\frac{\partial \tau_{r\theta}}{\partial r} + \frac{1}{r \sin \phi}\frac{\partial \sigma_\theta}{\partial \theta} + \frac{1}{r}\frac{\partial \tau_{\phi\theta}}{\partial \phi} + 3\frac{\tau_{r\theta}}{r} + 2\frac{\tau_{\phi\theta}}{r}\cot \phi + F_\theta = 0 \\ &\frac{\partial \tau_{r\phi}}{\partial r} + \frac{1}{r \sin \phi}\frac{\partial \tau_{\phi\theta}}{\partial \theta} + \frac{1}{r}\frac{\partial \sigma_\phi}{\partial \phi} + \frac{\sigma_\phi - \sigma_\theta}{r}\cot \phi + 3\frac{\tau_{r\phi}}{r} + F_\phi = 0 \end{aligned} \qquad (33.40a\text{–}c)$$

33.4. MOHR'S CIRCLE

Equations (33.4) and (33.25), as well as the more general equations (33.14) and (33.31), have identical form; they all describe the transformation which the components of a tensor undergo when the coordinate system is rotated. Because of the formal likeness of these equations, the idea of Mohr's circle, described here for the stresses, may as well be applied to the strains.

Mohr's Circle in Two Dimensions

Squaring and adding Eqs. (33.27*c*,*d*) shows that

$$\rho_{\xi\eta}^2 + \tau_{\xi\eta}^2 = \rho_{xy}^2 + \tau_{xy}^2 = \rho_*^2 \tag{33.41}$$

is another stress invariant besides $2\bar{\sigma} = \sigma_x + \sigma_y$. Equation (33.41) can be interpreted as the equation of a circle (Mohr's circle) in a coordinate system (ρ,τ) or (σ,τ) (Fig. 33.6), each point of its circumference representing a pair of values (σ,τ) for some reference frame (ξ,η). Equations (33.27*c*,*d*) are a parameter form of the equation of this circle with $\measuredangle DCE = 2\theta$ and, hence, $\measuredangle DBE = \theta$. In Fig. 33.4 the angle θ is measured

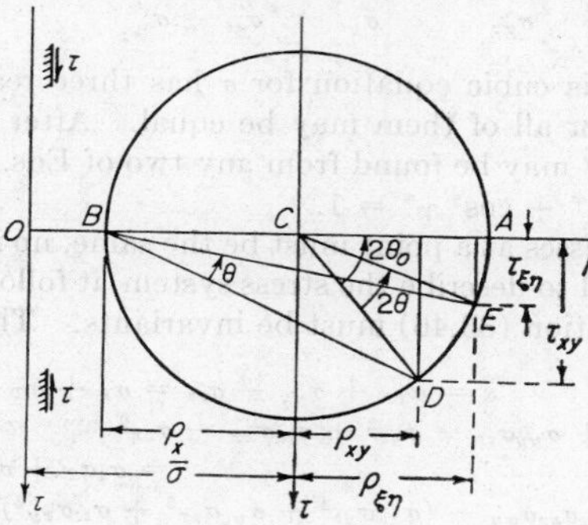

FIG. 33.6

counterclockwise from the x axis to the ξ axis. In order to have θ and 2θ in Fig. 33.6 also appear as counterclockwise angles, the τ axis had to be pointed downward.

At A and B, $\tau = 0$ and σ has extreme values ($\sigma_{\max} = \bar{\sigma} + \rho_*$, $\sigma_{\min} = \bar{\sigma} - \rho_*$). It may be seen from Fig. 33.6 that

$$\tan 2\theta_0 = \frac{\tau_{xy}}{\rho_{xy}} = \frac{2\tau_{xy}}{\sigma_x - \sigma_y} \tag{33.42}$$

This equation has, for any choice of σ_x, σ_y, τ_{xy}, two solutions θ_0 differing by 90°, corresponding to points A and B. The directions thus determined are the principal directions of the stress tensor, and the corresponding value of the normal stress,

$$\sigma_{\substack{\max\\ \min}} = \bar{\sigma} \pm \rho_* = \frac{\sigma_x + \sigma_y}{2} \pm \sqrt{\left(\frac{\sigma_x - \sigma_y}{2}\right)^2 + \tau_{xy}^2} \tag{33.43}$$

are the principal stresses. Lines which have at every point a principal direction as a tangent are the stress trajectories.

Principal Directions in Three Dimensions

In three dimensions there exists a set of three principal directions, which are perpendicular to one another. They can be found from Eq. (33.30), multiplying it in turn by the unit vectors $\mathbf{i}_x$, $\mathbf{i}_y$, $\mathbf{i}_z$:

$$\begin{aligned} \mathbf{s}_\xi \cdot \mathbf{i}_x &= \sigma_{xx} \cos(x,\xi) + \sigma_{yx} \cos(y,\xi) + \sigma_{zx} \cos(z,\xi) \\ \mathbf{s}_\xi \cdot \mathbf{i}_y &= \sigma_{xy} \cos(x,\xi) + \sigma_{yy} \cos(y,\xi) + \sigma_{zy} \cos(z,\xi) \\ \mathbf{s}_\xi \cdot \mathbf{i}_z &= \sigma_{xz} \cos(x,\xi) + \sigma_{yz} \cos(y,\xi) + \sigma_{zz} \cos(z,\xi) \end{aligned} \tag{33.44}$$

If, in a section ξ = const, there exists only a normal stress σ and no shear, then

$$\mathbf{s}_\xi \cdot \mathbf{i}_x = \sigma \cos \alpha^* \qquad \mathbf{s}_\xi \cdot \mathbf{i}_y = \sigma \cos \beta^* \qquad \mathbf{s}_\xi \cdot \mathbf{i}_z = \sigma \cos \gamma^*$$

where α^*, β^*, γ^* are the angles between the principal direction ξ and the x, y, z axes. It follows that

$$\begin{aligned}(\sigma_{xx} - \sigma)\cos\alpha^* + \sigma_{yx}\cos\beta^* + \sigma_{zx}\cos\gamma^* &= 0\\ \sigma_{xy}\cos\alpha^* + (\sigma_{yy} - \sigma)\cos\beta^* + \sigma_{zy}\cos\gamma^* &= 0\\ \sigma_{xz}\cos\alpha^* + \sigma_{yz}\cos\beta^* + (\sigma_{zz} - \sigma)\cos\gamma^* &= 0\end{aligned} \qquad (33.45)$$

These are three linear, homogeneous equations for the direction cosines. To admit a nonzero solution, they must have a vanishing coefficient determinant:

$$\begin{vmatrix} \sigma_{xx} - \sigma & \sigma_{yx} & \sigma_{zx} \\ \sigma_{xy} & \sigma_{yy} - \sigma & \sigma_{zy} \\ \sigma_{xz} & \sigma_{yz} & \sigma_{zz} - \sigma \end{vmatrix} = 0 \qquad (33.46)$$

It may be shown that this cubic equation for σ has three real roots, σ_1, σ_2, σ_3, the principal stresses. Two or all of them may be equal. After they have been calculated, the angles α^*, β^*, γ^* may be found from any two of Eqs. (33.45) and the condition that $\cos^2\alpha^* + \cos^2\beta^* + \cos^2\gamma^* = 1$.

Since the principal stresses at a point must be the same, no matter which reference frame (x,y,z) has been used to describe the stress system, it follows that the coefficients of the determinantal equation (33.46) must be invariants. They are:

$$\begin{aligned} s = \sigma_{xx} + \sigma_{yy} + \sigma_{zz} &= \sigma_1 + \sigma_2 + \sigma_3 \\ \sigma_{xx}\sigma_{yy} - \sigma_{xy}{}^2 + \sigma_{yy}\sigma_{zz} - \sigma_{yz}{}^2 + \sigma_{zz}\sigma_{xx} - \sigma_{zx}{}^2 & \\ &= \sigma_1\sigma_2 + \sigma_2\sigma_3 + \sigma_3\sigma_1 \qquad (33.47a\text{–}c) \\ \sigma_{xx}\sigma_{yy}\sigma_{zz} + 2\sigma_{yz}\sigma_{zx}\sigma_{xy} - (\sigma_{xx}\sigma_{yz}{}^2 + \sigma_{yy}\sigma_{zx}{}^2 + \sigma_{zz}\sigma_{xy}{}^2) & \\ &= \sigma_1\sigma_2\sigma_3 \end{aligned}$$

and can be expressed in terms of the principal stresses as indicated.

33.5. HOOKE'S LAW AND THE FUNDAMENTAL EQUATIONS

Hooke's Law in Three Dimensions

The preceding discussion of stress and strain is concerned with kinematics and statics, and the relations obtained are valid for any deformable medium, solid or fluid. However, the kinematic relations (33.2) and the equilibrium conditions (33.35) are only 9 equations for 15 unknown displacements, strains, and stresses and, hence, are insufficient for determining these quantities. The third set of equations needed consists of relations between stress and strain, which must be found from experiment and which depend highly on the material. According to their stress-strain laws, we distinguish, as mentioned above, among elastic, plastic, viscoelastic solids, viscous and ideal fluids, etc.

The theory of elasticity is built on Hooke's law, which involves the following properties of the material: (1) proportionality of stress and strain, (2) elasticity, i.e., the capability of storing energy which can be recovered without loss, (3) isotropy, i.e., independence of the elastic properties of the direction of the stress, (4) homogeneity, i.e., independence of these properties of the coordinates of the point considered.

It may easily be verified that the most general law which corresponds to the properties (1) and (3) is:

$$E\epsilon_1 = \sigma_1 - \nu(\sigma_2 + \sigma_3) \quad E\epsilon_2 = \sigma_2 - \nu(\sigma_3 + \sigma_1) \quad E\epsilon_3 = \sigma_3 - \nu(\sigma_1 + \sigma_2) \qquad (33.48a\text{–}c)$$

The modulus of elasticity (Young's modulus) E and Poisson's ratio ν must be found from experiment. E has the dimension of a stress, and varies widely between different materials. Poisson's ratio is confined to lie between the limits 0 and ½. According to (2), E and ν are independent of time; according to (4), they are independent of x, y, z.

Equations (33.48) are valid only for small strain; but there is no need that the displacement derivatives be small.

With the help of the transformation equations (33.14) and (33.31) and the relations

$$\cos^2(\xi,x) + \cos^2(\xi,y) + \cos^2(\xi,z) = 1$$
$$\cos(\xi,x)\cos(\eta,x) + \cos(\xi,y)\cos(\eta,y) + \cos(\xi,z)\cos(\eta,z) = 0$$

Hooke's law may be transformed to an arbitrary orthogonal reference frame (x,y,z). With $s = 3\bar{\sigma}$ from Eq. (33.47*a*), there follows that

$$\begin{aligned}\epsilon_x &= \epsilon_1\cos^2(1,x) + \epsilon_2\cos^2(2,x) + \epsilon_3\cos^2(3,x)\\ &= \frac{1+\nu}{E}[\sigma_1\cos^2(1,x) + \sigma_2\cos^2(2,x) + \sigma_3\cos^2(3,x)] - \frac{\nu}{E}s\\ &= \frac{\sigma_x}{E}(1+\nu) - \frac{\nu}{E}s\end{aligned}$$

and

$$\begin{aligned}\gamma_{xy} &= 2\frac{1+\nu}{E}[\sigma_1\cos(1,x)\cos(1,y) + \sigma_2\cos(2,x)\cos(2,y) + \sigma_3\cos(3,x)\cos(3,y)]\\ &= \frac{2(1+\nu)}{E}\tau_{xy}\end{aligned}$$

and four similar equations for ϵ_y, ϵ_z, γ_{yz}, γ_{zx}. Introducing the shear modulus

$$G = \frac{E}{2(1+\nu)} \tag{33.49}$$

we may bring these equations into the form

$$\begin{aligned}\epsilon_x &= \frac{1}{E}(\sigma_x - \nu\sigma_y - \nu\sigma_z) = \frac{1}{2G}\left(\sigma_x - \frac{\nu}{1+\nu}s\right)\\ \epsilon_y &= \frac{1}{E}(\sigma_y - \nu\sigma_z - \nu\sigma_x) = \frac{1}{2G}\left(\sigma_y - \frac{\nu}{1+\nu}s\right)\\ \epsilon_z &= \frac{1}{E}(\sigma_z - \nu\sigma_x - \nu\sigma_y) = \frac{1}{2G}\left(\sigma_z - \frac{\nu}{1+\nu}s\right)\\ \gamma_{xy} &= \frac{\tau_{xy}}{G} \qquad \gamma_{yz} = \frac{\tau_{yz}}{G} \qquad \gamma_{zx} = \frac{\tau_{zx}}{G}\end{aligned} \tag{33.50a–f}$$

When solving the first three of these equations for the stresses, we obtain

$$\sigma_x = 2G\left(\epsilon_x + \frac{\nu}{1-2\nu}e\right) = \frac{E}{(1+\nu)(1-2\nu)}[(1-\nu)\epsilon_x + \nu(\epsilon_y + \epsilon_z)] \tag{33.51}$$

and two similar equations, where e is the cubic dilatation defined by Eq. (33.17).

Adding Eqs. (33.50*a–c*) yields the relation

$$e = \frac{3(1-2\nu)}{E}\bar{\sigma} = \frac{\bar{\sigma}}{K} \tag{33.52}$$

K as defined by this equation is the bulk modulus.

When stress and strain tensors are split according to Eqs. (33.32) and (33.18), Hooke's law may be written in the particularly simple form

$$\bar{\sigma} = Ke \qquad \sigma'_{ik} = 2G\epsilon'_{ik} \qquad i, k = x, y, z \tag{33.53}$$

Hooke's Law in Two Dimensions

There are two ways of specializing Eqs. (33.50) and (33.51) to two dimensions, letting either $\sigma_z = 0$ (plane stress) or $\epsilon_z = 0$ (plane strain). The equations relating shear stress and shear strain reduce, in both cases, to (33.50*d*), but the equations for normal stress and strain are as follows:

Plane stress

$$\sigma_z = 0 \qquad \epsilon_x = \frac{1}{E}(\sigma_x - \nu\sigma_y) \qquad \sigma_x = \frac{E}{1-\nu^2}(\epsilon_x + \nu\epsilon_y) \qquad (33.54a,b)$$

Plain strain

$$\epsilon_z = 0 \qquad \epsilon_x = \frac{1-\nu^2}{E}\left(\sigma_x - \frac{\nu}{1-\nu}\sigma_y\right) \qquad \sigma_x = \frac{E}{(1+\nu)(1-2\nu)}[(1-\nu)\epsilon_x + \nu\epsilon_y] \qquad (33.55a,b)$$

Fundamental Equations

If the displacements are small, we have 15 linear field equations for 15 unknowns (6 stresses, 6 strains, 3 displacements), namely:

1. Three equilibrium conditions (33.35)
2. Six kinematic relations (33.2)
3. Six stress-strain equations (33.50)

In addition to the field equations, a set of boundary conditions is needed: either the displacements u, v, w or the surface tractions p_x, p_y, p_z must be prescribed on the surface of the body. The latter ones are related to the stresses through Eqs. (33.37).

A reduction of the 15 field equations can be accomplished in various ways, two of which will be mentioned here.

The first of the reduced systems is obtained, when we introduce relations 2 and 3 into conditions 1. This yields

$$2\frac{\partial^2 u}{\partial x^2} + \frac{\partial}{\partial y}\left(\frac{\partial v}{\partial x} + \frac{\partial u}{\partial y}\right) + \frac{\partial}{\partial z}\left(\frac{\partial u}{\partial z} + \frac{\partial w}{\partial x}\right) + \frac{2\nu}{1-2\nu}\frac{\partial e}{\partial x} + \frac{X}{G} = 0$$

whence

$$\nabla^2 u + \frac{1}{1-2\nu}\frac{\partial e}{\partial x} + \frac{X}{G} = 0$$

$$\nabla^2 v + \frac{1}{1-2\nu}\frac{\partial e}{\partial y} + \frac{Y}{G} = 0 \qquad (33.56a\text{–}c)$$

$$\nabla^2 w + \frac{1}{1-2\nu}\frac{\partial e}{\partial z} + \frac{Z}{G} = 0$$

where ∇^2 is the three-dimensional Laplacian operator. Equations (33.56) are often referred to as the fundamental equations of the theory of elasticity.

When Eqs. (33.56a–c) are differentiated with respect to x, y, z, respectively, and then added, it is seen that

$$\nabla^2 e = -\frac{1-2\nu}{2(1-\nu)G}\left(\frac{\partial X}{\partial x} + \frac{\partial Y}{\partial y} + \frac{\partial Z}{\partial z}\right) \qquad (33.57)$$

whence, on account of (33.52),

$$\nabla^2 s = 3\nabla^2\bar{\sigma} = -\frac{1+\nu}{1-\nu}\left(\frac{\partial X}{\partial x} + \frac{\partial Y}{\partial y} + \frac{\partial Z}{\partial z}\right) \qquad (33.58)$$

that is, for vanishing body forces the dilatation e and the average stress $\bar{\sigma}$ are potential functions. In the special case of cartesian coordinates, we may further conclude that, for $X = Y = Z = 0$, the displacements obey the equations

$$\nabla^2\nabla^2 u = 0 \qquad \nabla^2\nabla^2 v = 0 \qquad \nabla^2\nabla^2 w = 0$$

but, different from the invariant statements (33.57) and (33.58), these equations have little practical value as they are not valid in curvilinear coordinates.

Through a different grouping of terms, Eqs. (33.56) may be brought into the following form:

$$\begin{aligned}
\frac{2(1-\nu)}{1-2\nu}\frac{\partial e}{\partial x} + \frac{\partial}{\partial z}\left(\frac{\partial u}{\partial z} - \frac{\partial w}{\partial x}\right) - \frac{\partial}{\partial y}\left(\frac{\partial v}{\partial x} - \frac{\partial u}{\partial y}\right) + \frac{X}{G} &= 0 \\
\frac{2(1-\nu)}{1-2\nu}\frac{\partial e}{\partial y} + \frac{\partial}{\partial x}\left(\frac{\partial v}{\partial x} - \frac{\partial u}{\partial y}\right) - \frac{\partial}{\partial z}\left(\frac{\partial w}{\partial y} - \frac{\partial v}{\partial z}\right) + \frac{Y}{G} &= 0 \qquad (33.59) \\
\frac{2(1-\nu)}{1-2\nu}\frac{\partial e}{\partial z} + \frac{\partial}{\partial y}\left(\frac{\partial w}{\partial y} - \frac{\partial v}{\partial z}\right) - \frac{\partial}{\partial x}\left(\frac{\partial u}{\partial z} - \frac{\partial w}{\partial x}\right) + \frac{Z}{G} &= 0
\end{aligned}$$

The second system of equations to be presented here are the six equations of Beltrami for the stresses. To obtain them, we start from the compatibility conditions (33.21*a–f*), that is, from equations of group 2, introduce the stresses with the aid of group 3, and transform with group 1.

Adding Eqs. (33.21*a,c*) yields

$$\left(\frac{\partial^2}{\partial y^2} + \frac{\partial^2}{\partial z^2}\right)\epsilon_x + \frac{\partial^2}{\partial x^2}(\epsilon_y + \epsilon_z) - \frac{\partial}{\partial x}\left(\frac{\partial \gamma_{xy}}{\partial y} + \frac{\partial \gamma_{xz}}{\partial z}\right) = 0$$

When Hooke's law (33.50) is used to express the strains in terms of the stresses, it follows that

$$\left(\frac{\partial^2}{\partial y^2} + \frac{\partial^2}{\partial z^2}\right)\sigma_x + \frac{\partial^2}{\partial x^2}(\sigma_y + \sigma_z) - \frac{\nu}{1+\nu}\left(\frac{\partial^2 s}{\partial x^2} + \nabla^2 s\right) - 2\frac{\partial}{\partial x}\left(\frac{\partial \tau_{xy}}{\partial y} + \frac{\partial \tau_{xz}}{\partial z}\right) = 0$$

Applying Eq. (33.35*a*) to the last term and making use of Eq. (33.58), we find the first of the following equations:

$$\begin{aligned}
\nabla^2\sigma_x + \frac{1}{1+\nu}\frac{\partial^2 s}{\partial x^2} + \frac{\nu}{1-\nu}\left(\frac{\partial X}{\partial x} + \frac{\partial Y}{\partial y} + \frac{\partial Z}{\partial z}\right) + 2\frac{\partial X}{\partial x} &= 0 \\
\nabla^2\sigma_y + \frac{1}{1+\nu}\frac{\partial^2 s}{\partial y^2} + \frac{\nu}{1-\nu}\left(\frac{\partial X}{\partial x} + \frac{\partial Y}{\partial y} + \frac{\partial Z}{\partial z}\right) + 2\frac{\partial Y}{\partial y} &= 0 \qquad (33.60a\text{–}c) \\
\nabla^2\sigma_z + \frac{1}{1+\nu}\frac{\partial^2 s}{\partial z^2} + \frac{\nu}{1-\nu}\left(\frac{\partial X}{\partial x} + \frac{\partial Y}{\partial y} + \frac{\partial Z}{\partial z}\right) + 2\frac{\partial Z}{\partial z} &= 0
\end{aligned}$$

The three equations for the shear stresses are obtained by a quite similar procedure. Combining Eq. (33.21*d*) with Hooke's law (33.50) yields

$$\frac{\partial^2\sigma_z}{\partial x\,\partial y} + \frac{\partial^2\tau_{xy}}{\partial z^2} - \frac{\partial^2\tau_{xz}}{\partial y\,\partial z} - \frac{\partial^2\tau_{yz}}{\partial x\,\partial z} - \frac{\nu}{1+\nu}\frac{\partial^2 s}{\partial x\,\partial y} = 0$$

Putting $\sigma_z = s - (\sigma_x + \sigma_y)$ and using Eqs. (33.35*a,b*), we find the first of the following three equations:

$$\begin{aligned}
\nabla^2\tau_{xy} + \frac{1}{1+\nu}\frac{\partial^2 s}{\partial x\,\partial y} + \frac{\partial X}{\partial y} + \frac{\partial Y}{\partial x} &= 0 \\
\nabla^2\tau_{yz} + \frac{1}{1+\nu}\frac{\partial^2 s}{\partial y\,\partial z} + \frac{\partial Y}{\partial z} + \frac{\partial Z}{\partial y} &= 0 \qquad (33.60d\text{–}f) \\
\nabla^2\tau_{zx} + \frac{1}{1+\nu}\frac{\partial^2 s}{\partial z\,\partial x} + \frac{\partial Z}{\partial x} + \frac{\partial X}{\partial z} &= 0
\end{aligned}$$

Equations (33.60*a–f*) are six independent equations for the six stresses, arising from the conditions of compatibility and of equilibrium. These six equations of the second order require six boundary conditions. Three of them are Eqs. (33.37), and the other three are the first-order differential equations (33.35), which appear in (33.60) but in differentiated form.

33.6. ENERGY PRINCIPLES

Strain Energy

Consider an element $dx\,dy\,dz$ (Fig. 33.7) whose sides have the directions of the principal stresses σ_1, σ_2, σ_3. An incremental deformation consists of elongations

$d\delta u = \delta\epsilon_1\, dx$, $d\delta v = \delta\epsilon_2\, dy$, $d\delta w = \delta\epsilon_3\, dz$ of the sides of this element. During this deformation, the stresses do the work:

$$\delta W = \sigma_1\, dy\, dz\, \delta\epsilon_1\, dx + \cdot\cdot\cdot = (\sigma_1\,\delta\epsilon_1 + \sigma_2\,\delta\epsilon_2 + \sigma_3\,\delta\epsilon_3)\, dx\, dy\, dz$$
$$= \delta\bar{U}_i\, dx\, dy\, dz$$

Here

$$\delta\bar{U}_i = \sigma_1\,\delta\epsilon_1 + \sigma_2\,\delta\epsilon_2 + \sigma_3\,\delta\epsilon_3 \tag{33.61}$$

is the incremental strain energy per unit volume. When the stresses and strains grow slowly from zero to their end values σ_1, σ_2, σ_3 and ϵ_1, ϵ_2, ϵ_3, then the total strain energy per unit volume is

$$\bar{U}_i = \int \delta\bar{U}_i$$

With the help of Hooke's law, this may be shown to be

$$\bar{U}_i = \tfrac{1}{2}(\sigma_1\epsilon_1 + \sigma_2\epsilon_2 + \sigma_3\epsilon_3) \tag{33.62}$$

Equations (33.61) and (33.62) are valid also for large displacements as long as the linear stress-strain law holds.

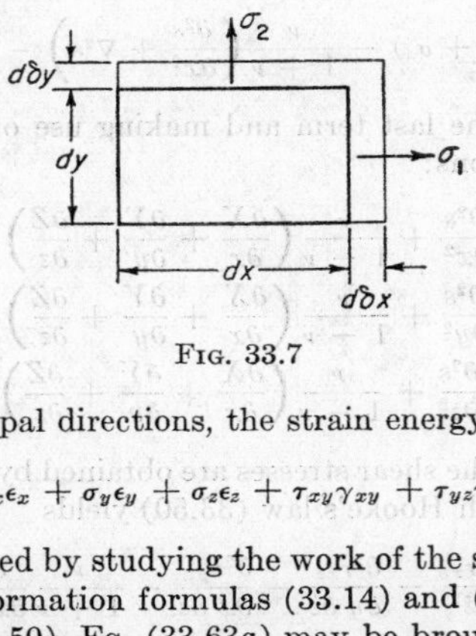

FIG. 33.7

If x, y, z are not principal directions, the strain energy is

$$\bar{U}_i = \tfrac{1}{2}(\sigma_x\epsilon_x + \sigma_y\epsilon_y + \sigma_z\epsilon_z + \tau_{xy}\gamma_{xy} + \tau_{yz}\gamma_{yz} + \tau_{zx}\gamma_{zx}) \tag{33.63a}$$

This formula may be derived by studying the work of the stresses at a volume element or by applying the transformation formulas (33.14) and (33.31) to Eq. (33.62). By means of Hooke's law (33.50), Eq. (33.63a) may be brought in any of the following forms:

$$\begin{aligned}\bar{U}_i &= \frac{1}{4G}\left[\sigma_x^2 + \sigma_y^2 + \sigma_z^2 - \frac{\nu}{1+\nu}s^2 + 2(\tau_{xy}^2 + \tau_{yz}^2 + \tau_{zx}^2)\right]\\ &= \frac{1}{4G}\left[\frac{1}{1+\nu}s^2 - 2(\sigma_x\sigma_y + \sigma_y\sigma_z + \sigma_z\sigma_x - \tau_{xy}^2 - \tau_{yz}^2 - \tau_{zx}^2)\right]\\ &= G\left[\epsilon_x^2 + \epsilon_y^2 + \epsilon_z^2 + \frac{\nu}{1-2\nu}e^2 + \tfrac{1}{2}(\gamma_{xy}^2 + \gamma_{yz}^2 + \gamma_{zx}^2)\right]\\ &= G\left[\frac{1-\nu}{1-2\nu}e^2 - 2(\epsilon_x\epsilon_y + \epsilon_y\epsilon_z + \epsilon_z\epsilon_x) + \tfrac{1}{2}(\gamma_{xy}^2 + \gamma_{yz}^2 + \gamma_{zx}^2)\right]\end{aligned} \tag{33.63b–e}$$

Corresponding formulas in two dimensions are with

$$s_2 = \sigma_x + \sigma_y \qquad e_2 = \epsilon_x + \epsilon_y$$

for plane stress ($\sigma_z = 0$):

$$\begin{aligned}\bar{U}_i &= \frac{1}{4G}\left[\frac{1}{1+\nu}s_2^2 - 2(\sigma_x\sigma_y - \tau_{xy}^2)\right]\\ &= G\left[\frac{1}{1-\nu}e_2^2 - 2\epsilon_x\epsilon_y + \tfrac{1}{2}\gamma_{xy}^2\right]\end{aligned} \tag{33.64a,b}$$

for plane strain ($\epsilon_z = 0$):

$$\bar{U}_i = \frac{1}{4G}\left[(1-\nu)s_2{}^2 - 2(\sigma_x\sigma_y - \tau_{xy}{}^2)\right]$$
$$= G\left[\frac{1-\nu}{1-2\nu}e_2{}^2 - 2\epsilon_x\epsilon_y + \tfrac{1}{2}\gamma_{xy}{}^2\right] \qquad (33.64c,d)$$

When stress and strain tensors are split into a hydrostatic part and a deviator [see Eqs. (33.18) and (33.32)], the strain energy splits into dilatation energy and distortion energy:

$$\bar{U}_i = \tfrac{1}{2}\bar{\sigma}e + \tfrac{1}{2}(\sigma'_x\epsilon'_x + \sigma'_y\epsilon'_y + \sigma'_z\epsilon'_z + \tau_{xy}\gamma_{xy} + \tau_{yz}\gamma_{yz} + \tau_{zx}\gamma_{zx}) \qquad (33.65)$$

Alternate forms are given here in principal stresses only:

$$\bar{U}_i = G[(\epsilon'_1)^2 + (\epsilon'_2)^2 + (\epsilon'_3)^2] + \tfrac{1}{2}Ke^2$$
$$= \frac{1}{4G}[(\sigma'_1)^2 + (\sigma'_2)^2 + (\sigma'_3)^2] + \frac{\bar{\sigma}^2}{2K}$$
$$= \frac{1}{3G}\left[\left(\frac{\sigma_1-\sigma_2}{2}\right)^2 + \left(\frac{\sigma_2-\sigma_3}{2}\right)^2 + \left(\frac{\sigma_3-\sigma_1}{2}\right)^2\right] + \frac{\bar{\sigma}^2}{2K}$$
$$= \tfrac{1}{6}[(\sigma'_1-\sigma'_2)(\epsilon'_1-\epsilon'_2) + (\sigma'_2-\sigma'_3)(\epsilon'_2-\epsilon'_3) + (\sigma'_3-\sigma'_1)(\epsilon'_3-\epsilon'_1)] + \tfrac{1}{2}\bar{\sigma}e \qquad (33.66a\text{–}d)$$

The last two of these expressions may be written without the primes, since $\sigma'_1 - \sigma'_2 = \sigma_1 - \sigma_2$, etc.

The Principle of Virtual Work

Consider an elastic body subjected to body forces X, Y, Z and surface tractions p_x, p_y, p_z, as defined on pp. 33–11 and 33–12. Let this body undergo a small additional deformation described by virtual displacements δu, δv, $\delta w(x,y,z)$. The virtual work done by the external forces during this deformation is

$$\delta W = \int(X\,\delta u + Y\,\delta v + Z\,\delta w)\,dV + \int(p_x\,\delta u + p_y\,\delta v + p_z\,\delta w)\,dA \qquad (33.67)$$

where the first integral is extended over the volume V of the body, and the second integral over its surface A.

The volume integral may easily be written in vector notation, and, by means of Eq. (33.33), it may be brought into the form

$$\int \mathbf{F}\cdot\delta\mathbf{u}\,dV = -\int\left(\frac{\partial \mathbf{s}_x}{\partial x} + \frac{\partial \mathbf{s}_y}{\partial y} + \frac{\partial \mathbf{s}_z}{\partial z}\right)\cdot\delta\mathbf{u}\,dV \qquad (33.68)$$

The surface integral in Eq. (33.67) may be transformed as follows. Application of Eq. (33.37a), with the large-deflection notation σ_{xy} for τ_{xy}, yields

$$\int p_x\,\delta u\,dA = \int[\sigma_{xx}\,\delta u\cos(x,n) + \sigma_{xy}\,\delta u\cos(y,n) + \sigma_{xz}\,\delta u\cos(z,n)]\,dA$$

The integrand is now the scalar product of the vectors $\mathbf{i}_x\sigma_{xx}\,\delta u + \mathbf{i}_y\sigma_{xy}\,\delta u + \mathbf{i}_z\sigma_{xz}\,\delta u$ and $\mathbf{n}\,dA$ (where $\mathbf{n}$ is the outer normal), and, according to Gauss' theorem (6.15), the integral may be transformed into a volume integral:

$$\int p_x\,\delta u\,dA = \int\left[\frac{\partial}{\partial x}(\sigma_{xx}\,\delta u) + \frac{\partial}{\partial y}(\sigma_{xy}\,\delta u) + \frac{\partial}{\partial z}(\sigma_{xz}\,\delta u)\right]dV$$

When the same transformation is applied to the other parts of the surface integral in Eq. (33.67), this integral may be brought into the form

$$\int(p_x\,\delta u + p_y\,\delta v + p_z\,\delta w)\,dA = \int\left[\frac{\partial}{\partial x}(\mathbf{s}_x\cdot\delta\mathbf{u}) + \frac{\partial}{\partial y}(\mathbf{s}_y\cdot\delta\mathbf{u}) + \frac{\partial}{\partial z}(\mathbf{s}_z\cdot\delta\mathbf{u})\right]dV \qquad (33.69)$$

and it is seen that

$$\delta W = \int \left(\mathbf{s}_x \cdot \frac{\partial \delta \mathbf{u}}{\partial x} + \mathbf{s}_y \cdot \frac{\partial \delta \mathbf{u}}{\partial y} + \mathbf{s}_z \cdot \frac{\partial \delta \mathbf{u}}{\partial z} \right) dV \tag{33.70}$$

that is, the virtual work of the external forces can be written as a volume integral.

By means of Eqs. (33.28) the stress vectors $\mathbf{s}_x$, $\mathbf{s}_y$, $\mathbf{s}_z$ can be replaced by their components, and the derivatives of $\delta\mathbf{u}$ can be expressed in terms of the strains. Indeed, from Eqs. (33.9) there follows that $\partial \delta \mathbf{u}/\partial x = \delta \mathbf{g}_x$, etc., and from Eqs. (33.11) we see that

$$\mathbf{g}_x \cdot \delta \mathbf{g}_x = \tfrac{1}{2}\,\delta\gamma_{xx} \qquad \mathbf{g}_x \cdot \delta \mathbf{g}_y + \mathbf{g}_y \cdot \delta \mathbf{g}_x = \delta\gamma_{xy} \qquad \text{etc.}$$

Using all these relations, we may bring δW into the following forms:

$$\begin{aligned} \delta W &= \int (\mathbf{s}_x \cdot \delta \mathbf{g}_x + \mathbf{s}_y \cdot \delta \mathbf{g}_y + \mathbf{s}_z \cdot \delta \mathbf{g}_z)\, dV \\ &= \int (\tfrac{1}{2}\sigma_{xx}\, \delta\gamma_{xx} + \tfrac{1}{2}\sigma_{yy}\, \delta\gamma_{yy} + \tfrac{1}{2}\sigma_{zz}\, \delta\gamma_{zz} \\ &\qquad + \sigma_{xy}\, \delta\gamma_{xy} + \sigma_{yz}\, \delta\gamma_{yz} + \sigma_{zx}\, \delta\gamma_{zx})\, dV \end{aligned} \tag{33.71}$$

For small strains (but not necessarily small displacement derivatives), this can be written as

$$\delta W = \int (\sigma_x\, \delta\epsilon_x + \sigma_y\, \delta\epsilon_y + \sigma_z\, \delta\epsilon_z + \tau_{xy}\, \delta\gamma_{xy} + \tau_{yz}\, \delta\gamma_{yz} + \tau_{zx}\, \delta\gamma_{zx})\, dV \tag{33.72}$$

When comparing the right-hand side of this equation with Eq. (33.63*a*), we see that it represents the increment of the strain energy $U_i = \int \bar{U}_i\, dV$ connected with the virtual displacement. A similar interpretation is possible for Eq. (33.71), and both equations state that, during a virtual deformation, the work of the external forces equals the increase of the strain energy,

$$\delta W = \delta U_i \tag{33.73}$$

Since any small actual change of the deformation qualifies as a virtual deformation, the actual deformation may be broken down into a sequence of increments to which Eq. (33.73) applies. By integration there follows that

$$W = U_i \tag{33.74}$$

The derivation of Eqs. (33.73) and (33.74) does not depend on Hooke's law, and is valid for unelastic bodies as well. However, only for elastic bodies can the energy U_i be recovered upon unloading and thus represents a potential energy. In the special case that the external forces can be derived from a potential U_e, we have $\delta W = -\delta U_e$, and hence

$$\delta U_i + \delta U_e = \delta U = 0 \tag{33.75}$$

$U = U_i + U_e$ is called the *total potential*. For the exact as well as the approximate solution of elasticity problems, Eq. (33.75) proves to be very useful.

We may go backward through the preceding derivation, showing that the sum of the right-hand sides of Eqs. (33.68) and (33.69) is the strain energy U_i. On the other hand, the sum of the left-hand sides of these equations is the virtual work defined by Eq. (33.67), and it is seen that (33.73) holds for arbitrary virtual displacements $\delta\mathbf{u}$ if and only if the equilibrium conditions (33.33) and (33.37) are satisfied. This statement is the principle of virtual work for elastic bodies.

Castigliano's Principles

In the linear theory of elasticity, a second principle can be introduced, which is a dual counterpart of the principle of virtual work, and which is known as Castigliano's second principle. To make the analogy clear, both principles will be developed here in common for the two-dimensional case.

We begin with the equilibrium conditions (33.36). Multiplying them by the virtual displacements δu, δv and adding and integrating over the appropriate domain yields

$$\iint \left[\left(\frac{\partial \sigma_x}{\partial x} + \frac{\partial \tau_{xy}}{\partial y} \right) \delta u + \left(\frac{\partial \tau_{xy}}{\partial x} + \frac{\partial \sigma_y}{\partial y} \right) \delta v \right] dx\, dy = - \iint (X\, \delta u + Y\, \delta v)\, dx\, dy$$

Integration by parts and use of the two-dimensional form of Eqs. (33.37),

$$\sigma_x \cos \theta + \tau_{xy} \sin \theta = p_x \qquad \tau_{xy} \cos \theta + \sigma_y \sin \theta = p_y \tag{33.76}$$

yields

$$\iint \left[\sigma_x \frac{\partial \delta u}{\partial x} + \tau_{xy} \left(\frac{\partial \delta v}{\partial x} + \frac{\partial \delta u}{\partial y} \right) + \sigma_y \frac{\partial \delta v}{\partial y} \right] dx\, dy = \iint (X\, \delta u + Y\, \delta v)\, dx\, dy + \oint (p_x\, \delta u + p_y\, \delta v)\, ds$$

that is, Eq. (33.73), and this is equivalent to the equilibrium equations (33.36) and (33.76), since the equations follow from each other also in the reverse order.

The second principle replaces the kinematic relations (33.2*a,b,d*) by an energy statement. First, we multiply these relations by virtual stresses $\delta\sigma_x$, $\delta\tau_{xy}$, $\delta\sigma_y$; add them; integrate over $dx\, dy$; and apply integration by parts to the right-hand side of the resulting equation. Second, if the virtual stresses satisfy the equilibrium conditions; in the interior,

$$\frac{\partial \delta\sigma_x}{\partial x} + \frac{\partial \delta\tau_{xy}}{\partial y} = -\delta X \qquad \frac{\partial \delta\tau_{xy}}{\partial x} + \frac{\partial \delta\sigma_y}{\partial y} = -\delta Y \tag{33.77}$$

on the boundary,

$$\delta\sigma_x \cos \theta + \delta\tau_{xy} \sin \theta = \delta p_x \qquad \delta\tau_{xy} \cos \theta + \delta\sigma_y \sin \theta = \delta p_y \tag{33.78}$$

then it follows that

$$\iint (\epsilon_x\, \delta\sigma_x + \gamma_{xy}\, \delta\tau_{xy} + \epsilon_y\, \delta\sigma_y)\, dx\, dy = \iint (u\, \delta X + v\, \delta Y)\, dx\, dy + \oint (u\, \delta p_x + v\, \delta p_y)\, ds \tag{33.79}$$

Both sides of this equation have the dimension of an energy. The right-hand side is the work done by the virtual forces on the actual displacements [which in Eq. (33.2) are supposed to be small]. On the left-hand side there is an energy expression

$$\delta U_i^* = \iint (\epsilon_x\, \delta\sigma_x + \gamma_{xy}\, \delta\tau_{xy} + \epsilon_y\, \delta\sigma_y)\, dx\, dy \tag{33.80}$$

which is called the *complementary energy*, because it complements the strain energy to an integral of terms $\epsilon\sigma$. (For an elastic material obeying Hooke's law, $\delta U_i^* = \delta U_i$.) With these notations Eq. (33.79) may be written as

$$\delta U_i^* = \delta W^* \tag{33.81}$$

and, since we can go from (33.81) back to (33.2), this equation, if satisfied for arbitrary virtual stresses, is equivalent to the kinematic relations. This statement is *Castigliano's second principle.*

When the virtual stresses are so chosen that they are in equilibrium with one single load δP, then $\delta W^* = u\, \delta P$, where u is the component in the direction of δP of the actual displacement of the point where this force is applied. In this case, Eq. (33.81) specializes to $u\, \delta P = \delta U_i^*$, which may also be written as

$$u = \frac{\partial U_i^*}{\partial P} \tag{33.82}$$

This relation is known as *Castigliano's first principle* and is widely used for the evaluation of displacements. For further details see p. 26–18.

Reciprocity Theorems of Betti and Maxwell

Consider two groups of loads applied to an elastic body. The loads $X_1, \ldots,$ $p_{x1}, \ldots$ produce displacements u_1, v_1, w_1, and the loads $X_2, \ldots, p_{x2}, \ldots$ produce displacements u_2, v_2, w_2. The law of superposition states that the simultaneous application of both load groups produces the displacements $(u_1 + u_2)$, $(v_1 + v_2)$, $(w_1 + w_2)$. This law is based on the linearity of all the equations 1, 2, and 3 on p. 33–16, and it is valid where these equations apply.

If $X, \ldots, p_x, \ldots$ are the final values of the external forces obtained by quasi-static loading, and if u, v, w are the corresponding displacements, then the work done during loading is

$$W = \tfrac{1}{2}\int(Xu + Yv + Zw)\,dV + \tfrac{1}{2}\int(p_xu + p_yv + p_zw)\,dA \qquad (33.83)$$

As in Eq. (33.63*a*), the factor $\frac{1}{2}$ is a consequence of the proportionality of load and displacement during loading. For Eq. (33.83), we write symbolically

$$W = \tfrac{1}{2}\mathbf{F}\cdot\mathbf{u} \qquad (33.83')$$

We now consider two load systems, $\mathbf{F}_1$ and $\mathbf{F}_2$. While $\mathbf{F}_1$ is applied, the work $W_{11} = \frac{1}{2}\mathbf{F}_1\cdot\mathbf{u}_1$ is done. When now $\mathbf{F}_2$ is added, work of two kinds is done. The forces $\mathbf{F}_2$ do the work $W_{22} = \frac{1}{2}\mathbf{F}_2\cdot\mathbf{u}_2$, and the forces $\mathbf{F}_1$ do the work $W_{21} = \mathbf{F}_1\cdot\mathbf{u}_2$. The total work done is, therefore,

$$W = W_{11} + W_{22} + W_{21} \qquad (33.84a)$$

Because of the superposition law, this work must be independent of the order in which the loads have been applied; i.e., it must be equal to

$$W = W_{22} + W_{11} + W_{12} \qquad (33.84b)$$

with $W_{12} = \mathbf{F}_2\cdot\mathbf{u}_1$, and, therefore, $W_{12} = W_{21}$. This is *Betti's theorem:* When two systems of forces act on an elastic body, then the work of the forces 1 on the displacements 2 is equal to the work of the forces 2 on the displacements 1.

Maxwell's reciprocity theorem is a special case of Betti's theorem. We define the influence coefficient α_{ik} as the displacement of the point i resulting from a unit force at the point k. Then *Maxwell's reciprocity theorem* states that

$$\alpha_{ik} = \alpha_{ki} \qquad (33.85a)$$

A similar relation applies to beams subjected to external couples. Let β_{ik} be the rotation of a line element i resulting from an external unit couple applied at a line element k; then

$$\beta_{ik} = \beta_{ki} \qquad (33.85b)$$

Finally, if a unit force at i produces an angular displacement (a rotation) γ_{ki} at k, and a unit couple at k produces a deflection δ_{ik} at i, then

$$\delta_{ik} = \gamma_{ki} \qquad (33.85c)$$

Note that this equation is dimensionally correct, since an angle per unit of force applied has the same dimension as a displacement per unit of moment.

33.7. METHODS FOR SOLVING THE BASIC EQUATIONS

As shown on p. 33–16, there exist two methods for reducing the basic 15 equations: the first leading to three equations (33.56) or (33.59) for the displacements, the second leading to six equations (33.60) for the stresses. Here it will be shown how things must be arranged in order to get solutions for practically important problems in terms of a restricted number of elementary solutions. The basic idea is to introduce certain potentials or bipotentials from which the stresses and displacements can be found by differentiation. It is obvious that these potentials must be so defined that their connection with the stresses and displacements is independent of the particular choice of the coordinate system. Since it is usually not difficult to find a particular solution for given loads X, Y, Z, attention will be restricted here to the homogeneous problem, where $X = Y = Z = 0$.

The oldest of the techniques is the *method of Airy*. It is particulary simple, but essentially restricted to plane problems. The procedure is as follows: We satisfy the equations (33.36) of the force equilibrium by introducing two stress functions Φ_1 and Φ_2, putting

$$\sigma_x = \frac{\partial \Phi_1}{\partial y} \qquad \tau_{yx} = -\frac{\partial \Phi_1}{\partial x} \qquad \tau_{xy} = \frac{\partial \Phi_2}{\partial y} \qquad \sigma_y = -\frac{\partial \Phi_2}{\partial x}$$

The moment equation $\tau_{yx} = \tau_{xy}$ yields a connection between Φ_1 and Φ_2; it is satisfied if

$$\Phi_1 = \frac{\partial \Phi}{\partial y} \qquad \Phi_2 = -\frac{\partial \Phi}{\partial x}$$

so that, finally,

$$\sigma_x = \frac{\partial^2 \Phi}{\partial y^2} \qquad \tau_{xy} = \tau_{yx} = -\frac{\partial^2 \Phi}{\partial x\, \partial y} \qquad \sigma_y = \frac{\partial^2 \Phi}{\partial x^2} \tag{33.86}$$

When, by means of Hooke's law, we introduce these equations into the compatibility condition (33.21a), we obtain an equation for Φ:

$$\nabla^2 \nabla^2 \Phi = 0 \tag{33.87}$$

Therefore, any function Φ defines a state of stress which is statically possible, and every bipotential Φ defines a state of stress which, in addition, is kinematically possible in an elastic body.

In three dimensions, we start from the basic equations (33.59), which are equivalent to the vector equation

$$\frac{2G}{1 - 2\nu} \operatorname{grad} \operatorname{div} \mathbf{u} - \frac{G}{1 - \nu} \operatorname{curl} \operatorname{curl} \mathbf{u} = 0 \tag{33.88}$$

According to a well-known theorem of Helmholtz (see p. 6–9), every vector field can be considered as the superposition of an irrotational and a solenoidal field. Consequently, we put

$$2G\mathbf{u} = (1 - 2\nu) \operatorname{grad} \Phi - 2(1 - \nu) \operatorname{curl} \mathbf{\Psi} \tag{33.89}$$

The vector $\mathbf{\Psi}$ is restricted by the condition that $\operatorname{div} \mathbf{\Psi} = 0$, so that only three new scalar functions replace u, v, w.

Upon introduction of (33.89) into (33.88) and application of (6.23b) there follows

$$\nabla^2 (\operatorname{grad} \Phi - \operatorname{curl} \mathbf{\Psi}) = 0 \tag{33.90}$$

This equation states that the expression in parentheses is a vector potential:

$$\operatorname{grad} \Phi - \operatorname{curl} \mathbf{\Psi} = \tfrac{1}{2}\boldsymbol{\phi} = \tfrac{1}{2}(\mathbf{i}_x \phi_1 + \mathbf{i}_y \phi_2 + \mathbf{i}_z \phi_3) \tag{33.91}$$

Taking the divergence of this equation, we obtain the differential equation

$$\nabla^2 \Phi = \tfrac{1}{2} \operatorname{div} \boldsymbol{\phi} \tag{33.92}$$

which has the general solution

$$\Phi = \tfrac{1}{4}\mathbf{r} \cdot \boldsymbol{\phi} + \phi_0 \tag{33.93}$$

Here $\mathbf{r}$ is the position vector, and ϕ_0 is an arbitrary scalar potential. Elimination of $\mathbf{\Psi}$ between Eqs. (33.89) and (33.91) yields

$$2G\mathbf{u} = -\operatorname{grad} \Phi + (1 - \nu)\boldsymbol{\phi} \tag{33.94}$$

where Φ is an abbreviation for the right-hand side of Eq. (33.93). This right-hand side contains four potentials, ϕ_0, ϕ_1, ϕ_2, ϕ_3, one of which can be dropped without loss of generality (see [10], p. 22). The functions ϕ_i are known as the stress functions of Papkovich and Neuber (see [10], p. 24).

From Eqs. (33.94) and (33.50), (33.51) we find expressions for the stresses, for example,

$$\sigma_x = -\frac{\partial^2\Phi}{\partial x^2} + \nu\,\nabla^2\Phi + (1-\nu)\frac{\partial\phi_1}{\partial x}$$
$$\tau_{xy} = -\frac{\partial^2\Phi}{\partial x\,\partial y} + \tfrac{1}{2}(1-\nu)\left(\frac{\partial\phi_1}{\partial y} + \frac{\partial\phi_2}{\partial x}\right) \qquad (33.95)$$

Compared with the approach of Galerkin and Westergaard (see [11], p. 248), to which we are easily led from Eq. (33.90), the approach of Neuber has one great advantage: In Eqs. (33.95) the stresses appear as second and not as third derivatives, a fact which considerably facilitates the transformation to curvilinear coordinates.

Equations (33.88) through (33.95) may easily be specialized for the case of plane strain. The vector $\mathbf{\Psi}$ is then parallel to the z axis, and the condition div $\mathbf{\Psi} = 0$ is trivial. Equations (33.88) through (33.92) remain unchanged, with the understanding that all z components are zero, and Eq. (33.93) reads

$$\Phi = \tfrac{1}{4}(x\phi_1 + y\phi_2) + \phi_0 \qquad (33.96)$$

Now, instead of dropping one of the three functions ϕ_i (which, again, would be possible without loss of generality), it is preferable to connect ϕ_1 and ϕ_2 by choosing them as the real and imaginary parts of a complex function, i.e., by requiring that they satisfy the Cauchy-Riemann equations (9.14):

$$\frac{\partial\phi_1}{\partial x} = \frac{\partial\phi_2}{\partial y} \qquad \frac{\partial\phi_2}{\partial x} = -\frac{\partial\phi_1}{\partial y} \qquad (33.97)$$

Equation (33.92) can then be written in the form

$$\nabla^2\Phi = \tfrac{1}{2}\left(\frac{\partial\phi_1}{\partial x} + \frac{\partial\phi_2}{\partial y}\right) = \frac{\partial\phi_1}{\partial x} = \frac{\partial\phi_2}{\partial y} \qquad (33.98)$$

From this equation ϕ_1 and ϕ_2 can be calculated if the potential $P = \nabla^2\Phi$ is known: $(\phi_1 + i\phi_2)$ is the integral of the complex analytic function $(P + iQ)$ whose real part is P:

$$\phi_1 + i\phi_2 = \int (P + iQ)\, d(x + iy) \qquad (33.99)$$

Indeed, from this equation there follows

$$\frac{\partial\phi_1}{\partial x} = \frac{\partial\phi_2}{\partial y} = P \qquad -\frac{\partial\phi_1}{\partial y} = \frac{\partial\phi_2}{\partial x} = Q$$

Equation (33.94) for the displacement vector does not change. In components, it reads

$$2Gu = -\frac{\partial\Phi}{\partial x} + (1-\nu)\phi_1 \qquad 2Gv = -\frac{\partial\Phi}{\partial y} + (1-\nu)\phi_2 \qquad (33.100)$$

However, when use is made of Eqs. (33.97) and (33.98), the stress equations (33.95) simplify to

$$\sigma_x = \frac{\partial^2\Phi}{\partial y^2} \qquad \tau_{xy} = -\frac{\partial^2\Phi}{\partial x\,\partial y} \qquad \sigma_y = \frac{\partial^2\Phi}{\partial x^2} \qquad (33.101)$$

Comparison with Eqs. (33.86) shows that, in this case, Φ is identical with Airy's stress function.

In cylindrical coordinates r, θ, z, the vector operators in Eqs. (33.88) through (33.92) are represented by Eqs. (6.35). The component representation of $\boldsymbol{\phi}$ in

Eq. (33.91) is, of course, no longer applicable. In the axisymmetric case we write instead

$$\boldsymbol{\phi} = \mathbf{i}_r\rho + \mathbf{i}_z\phi$$

and it is found that ρ and ϕ must satisfy the differential equations

$$\left(\nabla^2 - \frac{1}{r^2}\right)\rho = 0 \qquad \nabla^2\phi = 0 \tag{33.102}$$

with

$$\nabla^2 = \frac{\partial^2}{\partial r^2} + \frac{1}{r}\frac{\partial}{\partial r} + \frac{\partial^2}{\partial z^2}$$

[see Eq. (11.43*b*)]. Equation (33.93) assumes the form

$$\Phi = \tfrac{1}{4}(r\rho + z\phi) + \phi_0 \tag{33.103}$$

As in the plane case, it is advantageous to retain some sort of symmetry between r and z by choosing ρ and ϕ as "conjugate" functions, i.e., such that

$$\frac{\partial\rho}{\partial r} + \frac{\rho}{r} = \frac{\partial\phi}{\partial z} \qquad \frac{\partial\phi}{\partial r} = -\frac{\partial\rho}{\partial z} \tag{33.104}$$

These equations differ from Eqs. (33.97) by the term ρ/r. This lack of complete symmetry is inherent in cylindrical coordinates, and cannot be removed by any means. Obviously, the equation

$$\nabla^2\Phi = \frac{1}{2}\left(\frac{\partial\rho}{\partial r} + \frac{\rho}{r} + \frac{\partial\phi}{\partial z}\right) = \frac{\partial\rho}{\partial r} + \frac{\rho}{r} = \frac{\partial\phi}{\partial z} \tag{33.105}$$

which follows from (33.103) and (33.104), cannot be used as the basis for introducing a complex function of which $\nabla^2\Phi$ would be the real part.

The displacement $\mathbf{u} = \mathbf{i}_r u_r + \mathbf{i}_z u_z$ follows, nevertheless, simply from Eq. (33.94):

$$2Gu_r = -\frac{\partial\Phi}{\partial r} + (1-\nu)\rho \qquad 2Gu_z = -\frac{\partial\Phi}{\partial z} + (1-\nu)\phi \tag{33.106}$$

The shear stress τ_{rz} is easily found from Eqs. (33.50). To find expressions for the normal stresses from Eq. (33.51), it is necessary to calculate first the dilatation e, using Eqs. (33.17), (33.19), and (33.105). We find

$$2Ge = 2G\left(\frac{\partial u_r}{\partial r} + \frac{u_r}{r} + \frac{\partial u_z}{\partial z}\right) = (1-2\nu)\,\nabla^2\Phi$$

and then

$$\sigma_z = \frac{\partial^2\Phi}{\partial r^2} + \frac{1}{r}\frac{\partial\Phi}{\partial r} \qquad \tau_{rz} = -\frac{\partial^2\Phi}{\partial r\,\partial z}$$

$$\sigma_r = \frac{\partial^2\Phi}{\partial x^2} + \frac{1}{r}\frac{\partial\Phi}{\partial r} - \frac{\rho}{r}(1-\nu) \qquad \sigma_\theta = \frac{\partial^2\Phi}{\partial z^2} + \frac{\partial^2\Phi}{\partial r^2} - \frac{\partial\rho}{\partial r}(1-\nu) \tag{33.107a–d}$$

In two of these relations, the function ρ occurs, and it cannot be eliminated with the help of Eq. (33.105). It is seen that Φ is by no means a *stress function*, neither practically, in the sense that the state of stress could be found from Φ alone when the boundary conditions involve only stresses, nor essentially, in the sense that Φ were a function by means of which the equilibrium conditions are identically satisfied. In Eqs. (33.107), Φ is nothing else but an abbreviation for the combination of potentials (33.103).

It is quite obvious that nothing like the Airy stress function of the plane problem can be found in the axisymmetric case, although, in both, the displacement vector has only two components. The difference between the two problems lies in the presence of the strain $\epsilon_\theta = u_r/r$ to which a stress σ_θ is related. A simple count (four strains, two displacements) shows that two independent compatibility conditions must exist, and these, of course, cannot be satisfied by only one function Φ. It is equally obvious that we cannot eliminate four stresses from the two equations (33.39) by introducing only one function. It is, however, possible to go through the entire stress-function theory if, as in the three-dimensional case, we make reference to the tensor of stress functions (which in the plane, by mere accident, reduces to a scalar). Practically this leads again to the same result as Eq. (33.103) (see [11], p. 253).

The system (33.103) through (33.107) is well suited for the solution of axisymmetric problems. These equations are a variation of Boussinesq's classic approach in which Eq. (33.103) is replaced by

$$\Phi^* = \tfrac{1}{2} z\phi^* + \phi_0 \tag{33.108}$$

where ϕ^* and ϕ_0 are potentials, and Φ^* is a bipotential. With (33.108), the displacements are

$$2Gu_r = -\frac{\partial \Phi^*}{\partial r} \qquad 2Gu_z = -\frac{\partial \Phi^*}{\partial z} + 2(1-\nu)\phi^* \tag{33.109}$$

Since

$$\frac{\partial \phi^*}{\partial z} = \nabla^2 \Phi^*$$

we can express the three normal stresses (but not τ_{rz}) by Φ alone; see [11], Eqs. (35′) with $\Phi \to \Phi^*$, $2\beta \to \phi^*$. However, everywhere terms containing ν enter, indicating that also Boussinesq's function Φ^* is not a stress function in the sense of the Airy function.

It is, however, possible to find a set of expressions for the displacements and the stresses in which formally only one function appears explicitly. We have to put

$$\Phi^* = \frac{\partial \chi}{\partial z} \qquad \text{i.e.} \qquad \phi^* = \nabla^2 \chi \tag{33.110}$$

The function χ is Love's function. Like its three-dimensional generalization, Galerkin's vector, it has the important disadvantage that everywhere the order of the derivatives is higher by one, which makes the introduction of curvilinear coordinates more awkward. Nevertheless, Love's function has been used widely in the literature. Displacements and stresses are derived from it by the following equations:

$$2Gu_r = -\frac{\partial^2 \chi}{\partial r\, \partial z} \qquad 2Gu_z = -\frac{\partial^2 \chi}{\partial z^2} + 2(1-\nu)\nabla^2\chi$$

$$\sigma_z = \frac{\partial}{\partial z}\left(-\frac{\partial^2 \chi}{\partial z^2} + (2-\nu)\nabla^2\chi\right) \qquad \sigma_r = \frac{\partial}{\partial z}\left(-\frac{\partial^2 \chi}{\partial r^2} + \nu\nabla^2\chi\right) \tag{33.111a–f}$$

$$\tau_{rz} = \frac{\partial}{\partial r}\left(-\frac{\partial^2 \chi}{\partial z^2} + (1-\nu)\nabla^2\chi\right) \qquad \sigma_\theta = \frac{\partial}{\partial z}\left(-\frac{1}{r}\frac{\partial \chi}{\partial r} + \nu\nabla^2\chi\right).$$

33.8. REFERENCES

[1] A. E. H. Love: "A Treatise on the Mathematical Theory of Elasticity," 4th ed., Dover, New York, 1944.

[2] S. P. Timoshenko, J. N. Goodier: "Theory of Elasticity," 2d ed., McGraw-Hill, New York, 1951.

[3] I. S. Sokolnikoff: "Mathematical Theory of Elasticity," 2d ed., McGraw-Hill, New York, 1956.

[4] A. E. Green, W. Zerna: "Theoretical Elasticity," Oxford Univ. Press, New York, 1954.

[5] C.-T. Wang: "Applied Elasticity," McGraw-Hill, New York, 1953.

[6] C. E. Pearson: "Theoretical Elasticity," Harvard Univ. Press, Cambridge, Mass., 1959.

[7] E. Trefftz: Mechanik der elastischen Körper, in H. Geiger, K. Scheel (eds.), "Handbuch der Physik," vol. 6, Springer, Berlin, 1928.

[8] C. B. Biezeno, R. Grammel: "Engineering Dynamics," vols. 1 and 2, transl. by M. L. Meyer from the 2d German ed., Van Nostrand, Princeton, N.J., 1956.

[9] I. Szabó: "Höhere Technische Mechanik," Springer, Berlin, 1958.

[10] H. Neuber: "Kerbspannungslehre," 2d ed., Springer, Berlin, 1958.

[11] K. Marguerre: Ansätze zur Lösung der Grundgleichungen der Elastizitätstheorie, *Z. angew. Math. Mech.*, **35** (1955), 242–263.

CHAPTER 34

BASIC EQUATIONS IN TENSOR NOTATION

BY

H. J. WEISS, D.Sc., Ames, Iowa

34.1. THE STRAIN TENSOR

Let X, with coordinates $x^i (i = 1, 2, 3)$ be a fixed rectangular cartesian system of axes with origin O and unit base vectors γ_i. At time $t = 0$, consider the undeformed state of a continuous elastic medium referred to an arbitrary curvilinear coordinate system A, with coordinates a^i. The system A is related to the cartesian system X by a continuous single-valued transformation

$$a^i = \bar{a}^i(x^1,x^2,x^3) \tag{34.1}$$

which is assumed to have a continuous single-valued inverse

$$x^i = x_0{}^i(a^1,a^2,a^3)$$

The covariant and contravariant base vectors, α_i and α^i, in the reference frame A are then given by

$$\alpha_i = \frac{\partial x_0{}^r}{\partial a^i}\gamma_r \qquad \alpha^i = \frac{\partial \bar{a}^i}{\partial x^r}\gamma^r \tag{34.2a,b}$$

where the summation convention on repeated subscripts or superscripts is used (see p. 7–2), unless specifically stated otherwise.

Let the medium, or body, be deformed continuously in a one-to-one fashion. At time $t = t$, let the deformed body be referred to an arbitrary curvilinear coordinate system B, with coordinates b^i, related to the frame A by

$$b^i = b^i(a^1,a^2,a^3,t) \tag{34.3a}$$

and, inversely,

$$a^i = a^i(b^1,b^2,b^3,t) \tag{34.3b}$$

where (34.3b) is assumed continuous and single-valued for all values of t. Frame B is related to the cartesian frame X by

$$b^i = \bar{b}^i(x^1,x^2,x^3,t) \qquad x^i = x^i(b^1,b^2,b^3,t) \tag{34.4a,b}$$

As in Eqs. (34.2), the base vectors of B are given by

$$\beta_i = \frac{\partial x^r}{\partial b^i}\gamma_r \qquad \beta^i = \frac{\partial \bar{b}^i}{\partial x^r}\gamma^r \tag{34.5a,b}$$

Consider a point P_0 in the undeformed body, with coordinates a^i. Let the same material point in the deformed medium at time t be P with coordinates b^i referred to B. Let $\mathbf{v}$ and $\mathbf{V}$ be the displacement vectors from P_0 to P referred to frames A and B, respectively. Thus,

$$\mathbf{v} = \mathbf{v}(a^1,a^2,a^3) \qquad \mathbf{V} = \mathbf{V}(b^1,b^2,b^3)$$

and

$$\mathbf{v}(x^1,x^2,x^3) \equiv \mathbf{V}(x^1,x^2,x^3)$$

Then the covariant and contravariant components of $\mathbf{v}$ and $\mathbf{V}$ with respect to their frames of reference are given by

$$\mathbf{v} = v_i \boldsymbol{\alpha}^i = v^i \boldsymbol{\alpha}_i \qquad \mathbf{V} = V_i \boldsymbol{\beta}^i = V^i \boldsymbol{\beta}_i \tag{34.6a,b}$$

The element of arc length in the undeformed state is given by

$$ds_0{}^2 = g_{ij}\, da^i\, da^j \tag{34.7}$$

where $g_{ij} = g_{ij}(a^1,a^2,a^3)$ is the metric tensor in frame A. In the deformed state, the square of the element of arc length is

$$ds^2 = G_{ij}\, db^i\, db^j \tag{34.8}$$

where $G_{ij} = G_{ij}(b^1,b^2,b^3)$ is the metric tensor in frame B. If it is assumed that Eqs. (34.3) are sufficiently continuously differentiable, then either element of arc length can be expressed in terms of either initial coordinates a^i or final coordinates b^i. Thus,

$$ds_0{}^2 = h_{ij}(b^1,b^2,b^3)\, db^i\, db^j \qquad ds^2 = H_{ij}(a^1,a^2,a^3)\, da^i\, da^j \tag{34.9a,b}$$

where

$$h_{ij} = \frac{\partial a^k}{\partial b^i}\frac{\partial a^l}{\partial b^j} g_{kl} \qquad H_{ij} = \frac{\partial b^k}{\partial a^i}\frac{\partial b^l}{\partial a^j} G_{kl} \tag{34.10a,b}$$

The difference $(ds^2 - ds_0{}^2)$ is a measure of the strain produced in the medium by the deformation. From Eqs. (34.7) to (34.10),

$$ds^2 - ds_0{}^2 = 2\eta_{ij}\, da^i\, da^j \tag{34.11a}$$

or

$$ds^2 - ds_0{}^2 = 2\epsilon_{ij}\, db^i\, db^j \tag{34.11b}$$

where

$$\eta_{ij} = \tfrac{1}{2}(H_{ij} - g_{ij}) \tag{34.12a}$$

is the covariant strain tensor referred to initial coordinates a^i, called the *Lagrangian strain tensor;* and

$$\epsilon_{ij} = \tfrac{1}{2}(G_{ij} - h_{ij}) \tag{34.12b}$$

is the covariant strain tensor referred to final coordinates b^i, called the *Eulerian strain tensor.*

The corresponding contravariant and mixed strain tensors can be found from

$$\eta^{ij} = g^{ik}g^{jl}\eta_{kl} \qquad \eta^i_j = g^{ik}\eta_{kj} \tag{34.13a,b}$$

where g^{ij} is the associated metric tensor in the frame A. Similar forms are obtained for ϵ^{ij} and ϵ^i_j with the use of the associated metric tensor G^{ij}.

Let $\mathbf{r}_0$ be the position vector of P_0 with respect to the fixed origin O, and $\mathbf{r}$ the position vector of P with respect to O. Directed line elements $d\mathbf{r}_0$ and $d\mathbf{r}$ can be written as

$$d\mathbf{r}_0 = \boldsymbol{\alpha}_i\, da^i \qquad d\mathbf{r} = \boldsymbol{\beta}_i\, db^i \tag{34.14a,b}$$

Then

$$\begin{aligned} ds_0{}^2 &= d\mathbf{r}_0 \cdot d\mathbf{r}_0 = \boldsymbol{\alpha}_i \cdot \boldsymbol{\alpha}_j\, da^i\, da^j \\ ds^2 &= d\mathbf{r} \cdot d\mathbf{r} = \boldsymbol{\beta}_i \cdot \boldsymbol{\beta}_j\, db^i\, db^j \end{aligned} \tag{34.15a,b}$$

and thus, from (34.7) and (34.8),

$$g_{ij} = \boldsymbol{\alpha}_i \cdot \boldsymbol{\alpha}_j \qquad G_{ij} = \boldsymbol{\beta}_i \cdot \boldsymbol{\beta}_j$$

From Eqs. (34.6),

$$\begin{aligned} d\mathbf{v} &= v^i|_j\, da^j\, \boldsymbol{\alpha}_i = v_i|_j\, da^j\, \boldsymbol{\alpha}^i \\ d\mathbf{V} &= V^i|_j\, db^j\, \boldsymbol{\beta}_i = V_i|_j\, db^j\, \boldsymbol{\beta}^i \end{aligned} \tag{34.16}$$

where the bars denote covariant derivatives with respect to the appropriate arguments [see Eq. (7.54)]. But

$$d\mathbf{v} = d\mathbf{r} - d\mathbf{r}_0$$

Therefore,

$$d\mathbf{r} = d\mathbf{r}_0 + d\mathbf{v} = \boldsymbol{\alpha}_i(da^i + v^i|_j\, da^j) \tag{34.17}$$

Similarly,

$$d\mathbf{V} = d\mathbf{r} - d\mathbf{r}_0$$

and thus

$$d\mathbf{r}_0 = d\mathbf{r} - d\mathbf{V} = \boldsymbol{\beta}_i(db^i - V^i|_j\, db^j) \tag{34.18}$$

Therefore,

$$\begin{aligned} ds_0{}^2 = d\mathbf{r}_0 \cdot d\mathbf{r}_0 &= \boldsymbol{\beta}_i(db^i - V^i|_k\, db^k) \cdot \boldsymbol{\beta}_j(db^j - V^j|_l\, db^l) \\ &= G_{ij}(db^i\, db^j - V^j|_l\, db^i\, db^l - V^i|_k\, db^j\, db^k + V^i|_k V^j|_l\, db^k\, db^l) \end{aligned}$$

Comparison with Eqs. (34.8) and (34.11*b*) gives

$$\epsilon_{ij} = \tfrac{1}{2}(V_i|_j + V_j|_i - V_l|_i V^l|_j) \tag{34.19}$$

Similarly, Eqs. (34.11*a*) and (34.17) give

$$\eta_{ij} = \tfrac{1}{2}(v_i|_j + v_j|_i + v_l|_i v^l|_j) \tag{34.20}$$

Under the assumption of small strains, the nonlinear terms in Eqs. (34.19) and (34.20) are neglected, and usually no distinction is made between ϵ_{ij} and η_{ij}.

The space in which the medium is defined is Euclidean. Thus the Riemann-Christoffel tensors associated with the metrics defined by $ds_0{}^2$ and ds^2 must vanish. The Riemann-Christoffel tensor in the frame B is given by

$$R^B_{ijkl} = \frac{\partial \Gamma_{jli}}{\partial b^k} - \frac{\partial \Gamma_{jki}}{\partial b^l} + \Gamma^p_{jk}\Gamma_{ilp} - \Gamma^p_{jl}\Gamma_{ikp} \tag{34.21}$$

where Γ_{ijk} is the Christoffel symbol of the first kind,

$$\Gamma_{ijk} = \frac{1}{2}\left(\frac{\partial G_{ik}}{\partial b^j} + \frac{\partial G_{jk}}{\partial b^i} - \frac{\partial G_{ij}}{\partial b^k}\right) \tag{34.22}$$

and

$$\Gamma^k_{ij} = G^{kp}\Gamma_{ijp} \tag{34.23}$$

is the Christoffel symbol of the second kind [see Eqs. (7.47) and (7.48)]. Similarly, the Riemann-Christoffel tensor in frame A, R^A_{ijkl}, is given in terms of g_{ij} and its derivatives with respect to a^k. From Eq. (34.12*b*),

$$h_{ij}(b^1,b^2,b^3) = G_{ij} - 2\epsilon_{ij}$$

and, from Eqs. (34.3),

$$\frac{\partial}{\partial a^i} = \frac{\partial b^j}{\partial a^i}\frac{\partial}{\partial b^j}$$

Furthermore, $g_{ij}(a^1,a^2,a^3)$ can be related to h_{ij} by

$$g_{ij} = \frac{\partial b^k}{\partial a^i}\frac{\partial b^l}{\partial a^j} h_{kl}$$

which is the inverse of Eq. (34.10*a*). Thus, R^A_{ijkl} can be written in terms of G_{ij} and ϵ_{ij} in the frame B. The condition that R^A_{ijkl} and R^B_{ijkl} both vanish then gives a set of conditions on the strains ϵ_{ij}, which are the equations of compatibility. For small strains, these equations are

$$\epsilon_{ij}|_{kl} + \epsilon_{kl}|_{ij} - \epsilon_{ik}|_{jl} - \epsilon_{jl}|_{ik} = 0 \tag{34.24}$$

Since there are only six independent nonzero components of the Riemann-Christoffel tensor, there are only six independent equations of the type (34.24).

34.2. PHYSICAL COMPONENTS OF STRAIN

The physical component of tensile strain is defined as the elongation per unit initial length in that direction. Let ϵ_i be the extension in the direction of b^i per unit initial length, of the component in the b^i direction, of a directed line element. The magnitudes of the components in the b^i direction of the initial and final directed line elements are

$$|d\mathbf{r}_0|_i = \sqrt{h_{ii}}\, db^i \qquad |d\mathbf{r}|_i = \sqrt{G_{ii}}\, db^i \qquad (i \text{ not summed}) \tag{34.25a,b}$$

Thus,
$$\epsilon_i = \frac{|d\mathbf{r}|_i - |d\mathbf{r}_0|_i}{|d\mathbf{r}_0|_i} = \sqrt{\frac{G_{ii}}{h_{ii}}} - 1$$

But
$$\epsilon_{ii} = \tfrac{1}{2}(G_{ii} - h_{ii}) \qquad (i \text{ not summed})$$

Thus
$$\epsilon_i = \sqrt{1 + \frac{2\epsilon_{ii}}{h_{ii}}} - 1 \tag{34.26}$$

For small strains,
$$\epsilon_i \approx \epsilon_{ii}/h_{ii}$$

The strain components ϵ_{ij} $(i \neq j)$ are interpreted physically as follows. Let ϕ_{ij} be the angle between two directed line elements, in the deformed medium, in the directions of b^i and b^j, respectively. These are

$$d\mathbf{r}_{(i)} = \boldsymbol{\beta}_i\, db^i \qquad d\mathbf{r}_{(j)} = \boldsymbol{\beta}_j\, db^j \qquad (i, j \text{ not summed})$$

with magnitudes given in Eq. (34.25*b*). Then

$$\cos \phi_{ij} = \frac{d\mathbf{r}_i \cdot d\mathbf{r}_j}{|d\mathbf{r}|_i |d\mathbf{r}|_j} = \frac{G_{ij}}{\sqrt{G_{ii}G_{jj}}} \qquad (i \neq j;\ i, j \text{ not summed})$$

From Eq. (34.12*b*),

$$\cos \phi_{ij} = \frac{h_{ij} + 2\epsilon_{ij}}{\sqrt{h_{ii} + 2\epsilon_{ii}}\, \sqrt{h_{jj} + 2\epsilon_{jj}}} \qquad (i \neq j;\ i, j \text{ not summed})$$

For small strains, $2\epsilon_{ij}$ $(i \neq j)$ may be interpreted as the change in a right angle during the deformation. Analogous relations are obtained for the Lagrangian strain tensor.

34.3. THE STRESS TENSOR

The forces which cause the deformation of the medium are divided into body forces and surface forces. Consider the deformed medium in a state of equilibrium referred to the space frame B with curvilinear coordinates b^i. Let $\mathbf{F} = F^i\boldsymbol{\beta}_i$ be the force per unit volume acting on the mass of the medium. $\mathbf{F}$ is called the body force, and it is assumed that $\mathbf{F}$ is at least a sectionally continuous function of position.

Consider an element of surface area ΔS either within the deformed medium or on its surface. Let $\mathbf{n}$ be the unit normal vector to ΔS. Let the forces acting on ΔS be statically equivalent to a force $\Delta\mathbf{P}$ acting at a point O within ΔS. $\Delta\mathbf{P}$ may be the surface force exerted by one portion of the medium on the other across ΔS, or, if ΔS is on the outer surface of the body, $\Delta\mathbf{P}$ is the surface force applied to the body on ΔS. It is assumed that all such surface forces are at least sectionally continuous. Let $\Delta S \to 0$, always keeping the point O within it. The stress vector $\overset{n}{\mathbf{T}}$ is defined as

$$\overset{n}{\mathbf{T}} = \lim_{\Delta S \to 0} \frac{\Delta\mathbf{P}}{\Delta S}$$

where the superscript n indicates that $\overset{n}{\mathbf{T}}$ is associated with the orientation of the area element ΔS. $\overset{n}{\mathbf{T}}$ is considered positive when it is directed on the same side of ΔS as is the positive normal $\mathbf{n}$. $\overset{n}{\mathbf{T}}$ has components with respect to the coordinate system b^i given by

$$\mathbf{T} = T^i \boldsymbol{\beta}_i$$

It is assumed that $\overset{n}{\mathbf{T}}$ is also at least sectionally continuous.

Consider a small tetrahedral volume element $\Delta\tau$, formed by the coordinate surfaces through a point P, and a neighboring arbitrary surface element ΔS. Let S_i be the elementary surface areas of the coordinate faces of the tetrahedron. Since ΔS is vectorially equivalent to the areas of the surfaces b^i = const, then their magnitudes are related by

$$\Delta S_i = n_i \, \Delta S \tag{34.27}$$

where n_i are the covariant components of $\mathbf{n}$ with respect to the base vectors $\boldsymbol{\beta}^i$. Furthermore,

$$\Delta\tau = h \, \Delta S \tag{34.28}$$

where h is an appropriate linear factor determined by the shape of the tetrahedron.

Let $\overset{i}{\mathbf{T}}$ be the stress vector on the coordinate face b^i = const. The positive normals to these coordinate faces are exterior to $\Delta\tau$. Thus

$$\overset{i}{\mathbf{T}} = \sigma^{ij}(-\boldsymbol{\beta}_j) \tag{34.29}$$

where σ^{ij} are the contravariant components of $\overset{i}{\mathbf{T}}$ in the direction of b^j.

Since the entire strained body is assumed to be in equilibrium, then so is $\Delta\tau$. Thus,

$$(\mathbf{F} + \boldsymbol{\varepsilon}) \, \Delta\tau + (\overset{n}{\mathbf{T}} + \eta) \, \Delta S + \overset{i}{\mathbf{T}} \, \Delta S_i = 0 \tag{34.30}$$

where, by the continuity of $\mathbf{F}$ and $\overset{n}{\mathbf{T}}$, $\boldsymbol{\varepsilon}$ and $\eta \to 0$ as $\Delta\tau$ shrinks to the point P. If now $\Delta\tau \to 0$, keeping the direction of $\mathbf{n}$ fixed, $h \to 0$. Thus,

$$\overset{n}{\mathbf{T}} = \sigma^{ij} n_i \boldsymbol{\beta}_j$$

or

$$T^i = \sigma^{ji} n_j \tag{34.31}$$

Since T^i are the components of a vector, and n_j are the covariant components of an arbitrary vector, then σ^{ji} are the contravariant components of a second-order tensor, the stress tensor. The mixed and covariant stress tensors may be defined in the usual fashion with respect to the metric G_{ij}.

34.4. EQUATIONS OF EQUILIBRIUM

Since the strained medium is in equilibrium, the resultant force and resultant moment in any arbitrary direction must vanish. Thus

$$\int_V \mathbf{F} \cdot \boldsymbol{\lambda} \, d\tau + \int_S \overset{n}{\mathbf{T}} \cdot \boldsymbol{\lambda} \, d\sigma = 0 \tag{34.32}$$

where V and S are respectively the volume and surface of any arbitrary portion of the body, and $\boldsymbol{\lambda}$ is a unit vector in an arbitrarily chosen direction,

$$\boldsymbol{\lambda} = \lambda^i \boldsymbol{\beta}_i = \lambda_i \boldsymbol{\beta}^i$$

Since $\boldsymbol{\lambda}$ is a parallel vector field, $\lambda_i|_j = 0$. Thus, the application of the divergence theorem to (34.32) yields

$$\int_V (F^i + \sigma^{ji}|_j)\lambda_i \, d\tau = 0$$

Since V and λ_i are arbitrary, then

$$F^i + \sigma^{ji}|_j = 0 \tag{34.33}$$

at every point of the medium.

The resultant moment about any point vanishing gives

$$\int_V \mathbf{r} \times \mathbf{F} \cdot \boldsymbol{\lambda} \, d\tau + \int_S \mathbf{r} \times \overset{n}{\mathbf{T}} \cdot \boldsymbol{\lambda} \, d\sigma = 0 \tag{34.34}$$

where $\mathbf{r}$ is the position vector from an arbitrarily chosen fixed point to a generic point, either in V or on S, at which $\mathbf{F}$ and $\overset{n}{\mathbf{T}}$ are given:

$$\mathbf{r} = r^i \boldsymbol{\beta}_i = r_i \boldsymbol{\beta}^i$$

The triple scalar product can be written as [see Eq. (7.45)]

$$\mathbf{r} \times \mathbf{F} \cdot \boldsymbol{\lambda} = \epsilon_{ijk} r^i F^j \lambda^k$$

where ϵ_{ijk} is the skew-symmetric covariant tensor of third order (denoted by $\bar{\epsilon}_{ijk}$ on p. 7–9), the covariant derivatives of which are all zero. When this form, the divergence theorem, and (34.33) are used, Eq. (34.34) becomes

$$\int_V \epsilon_{ijk} r^i|_p \sigma^{pj} \lambda^k \, d\tau = 0$$

But $r^i|_p = \delta^i_p$, and, since V and λ^k are arbitrary, for each choice of k,

$$\epsilon_{ijk}\sigma^{ji} = 0$$

which implies

$$\sigma^{ji} = \sigma^{ij} \tag{34.35}$$

and thus σ^{ij} is symmetric.

34.5. PHYSICAL COMPONENTS OF STRESS

The physical component of stress s_{ij} is defined as the component, in the direction of $\boldsymbol{\beta}_j$, of the stress vector acting on the coordinate surface area element whose normal is in the direction of $\boldsymbol{\beta}_i$. Thus,

$$\overset{i}{\mathbf{T}} = \sum_{j=1}^{3} s_{ij} \frac{\boldsymbol{\beta}_j}{\sqrt{G_{jj}}}$$

where, therefore, $\boldsymbol{\beta}_j/\sqrt{G_{jj}}$ is a unit vector in the direction of b^j. On the other hand,

$$\overset{i}{\mathbf{T}} = \sum_{j=1}^{3} \sigma^{ij} \sqrt{G_{jj}} \frac{\boldsymbol{\beta}_j}{\sqrt{G_{jj}}}$$

Thus,

$$s_{ij} = \sqrt{G_{jj}}\,\sigma^{ij} \qquad \text{(no sum on } j) \tag{34.36}$$

34.6. STRESS-STRAIN RELATIONS

Since the stresses are calculated with respect to the deformed medium, it is convenient to relate σ^{ij} and ϵ_{ij}. For small strains, the usual assumption is that the stress components are linear functions of the strain components and vice versa. This relationship is the generalized Hooke's law,

$$\sigma^{ij} = c^{ijmn}\epsilon_{mn}$$

which may, of course, be written in a number of associated ways.

For an anisotropic elastic medium, c^{ijmn} is a fourth-order tensor function of position, defining the elastic properties of the medium. In general, there are 21 independent c^{ijmn}, due to the symmetry of σ^{ij} and ϵ_{mn}. If the c^{ijmn} are constants, the medium is said to be homogeneous. If, furthermore, the medium is isotropic, the number of independent elastic constants reduces to two. In this latter case, the stress-strain relationship may be written as

$$\sigma_{ij} = \lambda G_{ij}\vartheta + 2\mu\epsilon_{ij} \tag{34.37}$$

where

$$\vartheta = G^{ij}\epsilon_{ij} = \epsilon_i^i \tag{34.38}$$

is the cubic dilatation, and λ and μ are Lamé's constants. Analogously, associated forms of (34.37) may be written in terms of the tensors associated with σ_{ij} and ϵ_{ij}.

If the equations of equilibrium (34.33) are written as

$$G^{jk}\sigma_{ij}|_k + F_i = 0$$

and (34.37) is substituted into these, the result is the Navier equations,

$$(\lambda + \mu)\vartheta|_i + \mu\nabla^2 V_i + F_i = 0 \tag{34.39}$$

where

$$\nabla^2 V_i = G^{jk} V_i|_{jk}$$

is the Laplacian of the ith component of the displacement vector **V**.

34.7. REFERENCES

[1] I. S. Sokolnikoff: "Tensor Analysis: Theory and Applications," Wiley, New York, 1958.
[2] I. S. Sokolnikoff: "Mathematical Theory of Elasticity," 2d ed., McGraw-Hill, New York, 1956.
[3] A. E. Green, W. Zerna: "Theoretical Elasticity," Oxford Univ. Press, New York, 1954.
[4] A. J. McConnell: "Applications of the Absolute Differential Calculus," Blackie, London, 1946.
[5] J. L. Synge, A. Schild: "Tensor Calculus," Univ. Toronto Press, Toronto, 1949.

CHAPTER 35

BENDING OF BEAMS

BY

H. V. HAHNE, Ph.D., Los Altos, Calif.

35.1. STRAIGHT SLENDER BARS IN PURE TENSION OR COMPRESSION

Stresses in Axially Loaded Bars

A straight slender bar of uniform cross section is considered, and the axis of the bar is defined as the line which connects the centroids of its cross sections. If such a bar is subjected to the action of external loads, internal forces arise in the material of the bar. To determine the magnitude and direction of these internal forces at a given point O within the bar, a section is made through this point normal to the axis of the bar, which cuts the bar into two portions (Fig. 35.1). Before the section is made, the bar as a whole, and thus also each of the two portions, is in equilibrium. The loads acting on the upper portion are balanced by the internal forces exerted on it by the lower portion, and vice versa. The direction, magnitude, and location of the resultant of these internal forces can be determined with the help of statics. A state of pure tension or compression is said to exist at a cross section if the external loads are such that the resultant of the internal forces acts at the centroid of the cross section and its direction at this point is collinear with the axis of the bar. The resultant of the internal forces is then called the *normal force* and is denoted by N. A tensile force N is usually considered as positive.

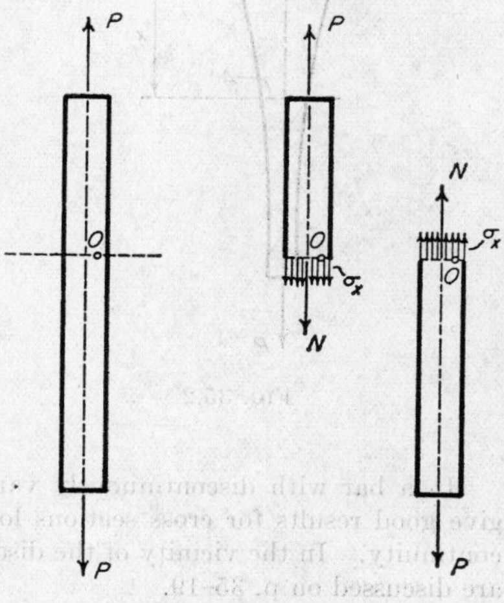

FIG. 35.1

The distribution of the internal forces over the cross section cannot be determined precisely with elementary means. However, for slender bars in pure tension or compression, it can be assumed that, at a cross section located some distance away from the point of application of an external load, the internal forces are distributed uniformly over the cross section, and are all directed parallel to the axis of the bar. On a cross section normal to the axis, therefore, only normal stresses σ_x will be present, and the magnitude of σ_x is given by the equation

$$\sigma_x = \frac{N}{A} \tag{35.1}$$

where A denotes the area of the cross section of the bar.

The sign convention for N determines also the sign of σ_x; if a tensile force N is taken as positive, a tensile stress σ_x is considered to be the positive stress.

With some reservations, Eq. (35.1) can also be used to calculate stresses in axially loaded bars with varying cross-sectional area.

In a bar with a steadily varying cross section (Fig. 35.2), where A is a function of the axial coordinate x, Eq. (35.1) gives fairly accurate results, as long as the rate of change of cross section with x is not excessive.

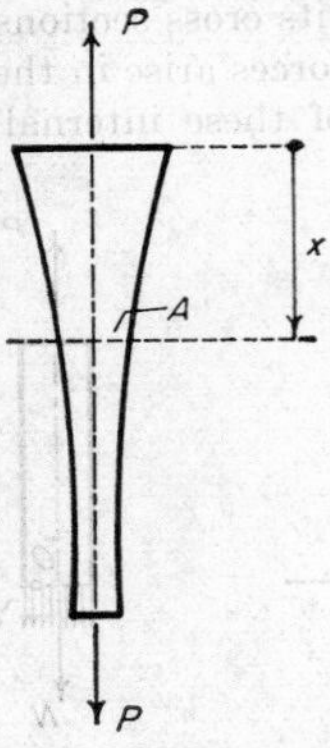

FIG. 35.2

FIG. 35.3

In a bar with discontinuously varying cross section (Fig. 35.3), Eq. (35.1) will give good results for cross sections located some distance away from points of discontinuity. In the vicinity of the discontinuities, stress concentrations arise. These are discussed on p. 35–19.

In a bar subjected to tensile forces, failure will occur if, at some cross section of the bar, the stress exceeds the ultimate tensile stress for the material. However, if the bar is acted upon by compressive forces, the elastic equilibrium of the bar may become unstable, and the bar may fail by buckling, long before the maximum stress in the bar reaches the ultimate value for its material. Methods for determining the critical loads at which a bar in compression fails because of instability are discussed in Chap. 44.

Deformations of Axially Loaded Bars

To investigate the deformations of an axially loaded bar, an element of length dx is isolated from it by means of two sections normal to the axis (Fig. 35.4). On the faces of the element act the uniformly distributed stresses σ_x, which cause the element to elongate by the amount Δdx. The strain in the element then is $\epsilon_x = \Delta dx/dx$. In elastic materials, Hooke's law expresses the relation between stress and strain, and, accordingly,

$$\epsilon_x = \sigma_x/E \qquad (35.2)$$

where E is the modulus of elasticity of the material. Thus, the change in length of the bar element is

$$\Delta dx = \frac{\sigma_x}{E}\,dx$$

The total elongation Δl of a bar of length l is obtained by adding the deformations of the individual elements as follows:

$$\Delta l = \int_0^l \frac{\sigma_x}{E}\,dx = \int_0^l \frac{N}{EA}\,dx \qquad (35.3)$$

To evaluate the integral in Eq. (35.3), N and A must be given as functions of x.

In the foregoing analysis, tensile stress is considered to be the positive stress; as a result, the elongation becomes the positive deformation.

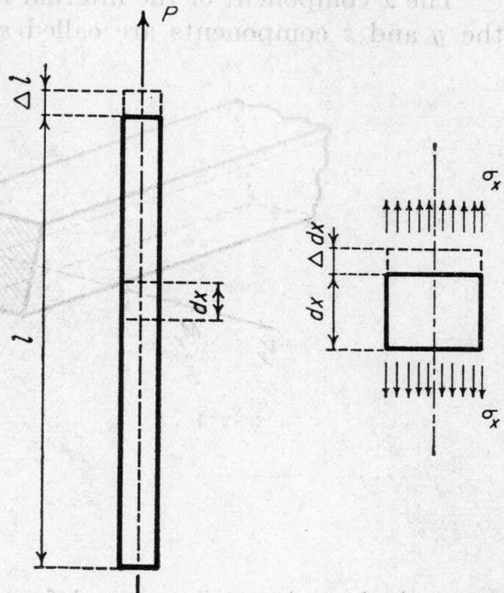

FIG. 35.4

In a bar of length l and uniform cross section A, acted upon by two end loads P (Fig. 35.4), $N = P$ and evaluation of Eq. (35.3) gives

$$\Delta l = \frac{Pl}{EA} \qquad (35.4)$$

Equation (35.3) can be used to calculate the deformations of a bar with continuously varying cross section, as long as the rate of change of the cross section is not excessive. For a bar with a discontinuously varying cross section (Fig. 35.3), the deformations of the individual portions are obtained using Eq. (35.4). The total change of length of the bar is obtained by adding algebraically the deformations of the individual portions. Stress concentrations have but a minor influence on Δl.

35.2. BENDING OF SLENDER BARS

Internal Forces and Moments

In a bar subjected to an external load system of a general type, the normal and shearing stress distributions over any of its cross sections will usually be nonuniform. At any such cross section, the direction and magnitude of the resultant of the internal forces must be such that it balances the external loads applied to the portions of the beam on either side of the cross section. To define this resultant internal force, an orthogonal coordinate system (x,y,z) is assumed at the cross section under consideration, so that the x axis coincides with the axis of the bar (Fig. 35.5). The axis of a bar is defined as the line which connects the centroids of its cross sections. If this axis is curved, the x axis is chosen tangent to the axis, and, hence, it will be changing its direction from cross section to cross section. The directions of the y and z axes can be assumed arbitrarily, and are chosen so as to meet best the needs of the particular problem. A right-handed coordinate system will be used throughout this chapter.

The x component of the internal force is called *normal force* and is denoted by N; the y and z components are called *shearing forces* and are denoted by V_y and V_z,

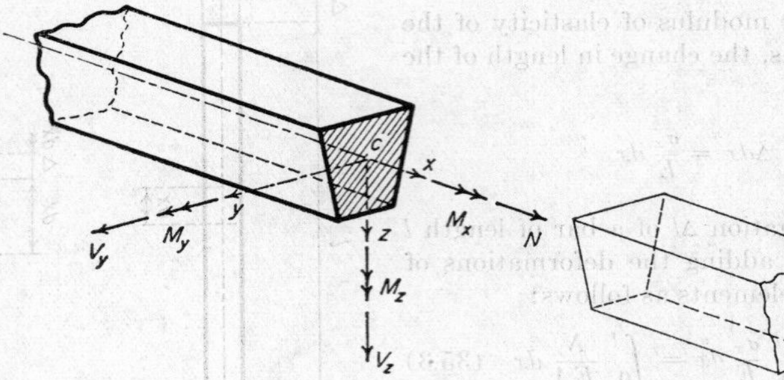

FIG. 35.5

respectively. A tensile normal force generally is considered as positive. For the shearing forces, the following sign convention is adopted: On a cross section whose outward normal is directed in the positive x direction, positive shearing forces act in the direction of the positive y and z axes, respectively.

The line of action of the resultant internal force does not necessarily pass through the origin of the (x,y,z) coordinate system. Its location is determined by the moments of the resultant internal force about these three coordinate axes. The moment about the x axis is denoted by M_x, and is of importance when torsion of the bar is being considered. The moments about the y and z axes are called bending moments and are denoted by M_y and M_z, respectively.

It is often convenient to use vectorial representation for the bending moments. Whenever this is done in this chapter, the right-hand screw rule is used, and the

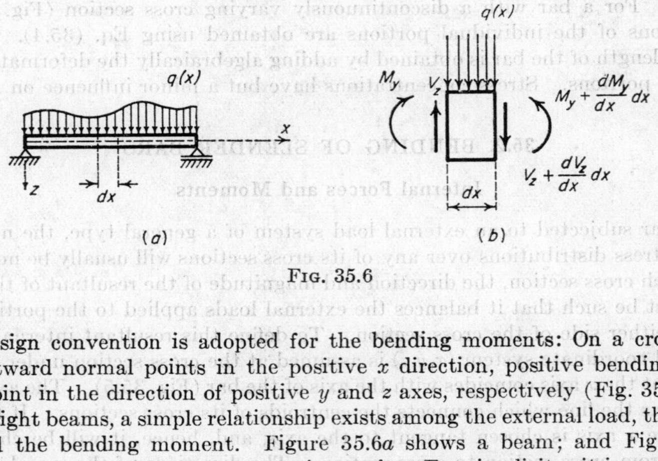

FIG. 35.6

following sign convention is adopted for the bending moments: On a cross section whose outward normal points in the positive x direction, positive bending-moment vectors point in the direction of positive y and z axes, respectively (Fig. 35.5).

In straight beams, a simple relationship exists among the external load, the shearing force, and the bending moment. Figure 35.6a shows a beam; and Fig. 35.6b an element of the beam with the forces acting on it. For simplicity, it is assumed that the loads act in the x,z plane only. The z axis is assumed as positive downward, and loads acting in this direction are taken as positive. If positive directions of moments and shearing forces, as shown in Fig. 35.6*b*, are assumed, the condition of equilibrium of vertical forces on the beam element requires that

$$\frac{dV_z}{dx} = -q \tag{35.5}$$

The condition of moment equilibrium of the beam element yields the equation

$$\frac{dM_y}{dx} = V_z \tag{35.6}$$

Combining Eqs. (35.5) and (35.6) gives the relation

$$\frac{d^2M_y}{dx^2} = -q \tag{35.7}$$

There are several analytical and graphical procedures for determining the bending moments and shearing forces in a beam, as well as standard methods for representing them in a diagram. These are fully discussed in Chap. 26.

Elementary Beam-bending Theory

The elementary theory of slender beams is based on the Navier hypothesis, which assumes that plane cross sections normal to the axis of the beam remain plane and normal to the axis during deformation. This assumption is true only for straight

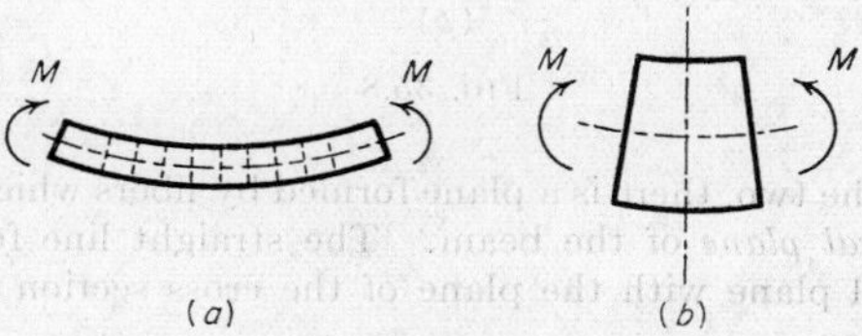

FIG. 35.7

beams of uniform cross section in pure bending, i.e., in the absence of shearing forces (Fig. 35.7*a*). In such a beam, all elements which are located some distance away from the ends will suffer the same deformation, and, consequently, the axis of the beam will be deformed into a circle. Since any diameter of a circle can serve as an axis of symmetry, it may be concluded that plane cross sections, originally normal to the axis, will remain normal to it, and will retain their plane shape (Fig. 35.7*b*).

In the presence of shearing forces, a certain amount of warping of the cross sections will take place. However, for a shearing force which is constant along the axis of the beam, the amount of warping will also be constant along the axis, and will not affect the magnitude and distribution of the normal stresses in the beam. Only a variation of the shearing force along the axis of the beam will produce a disturbance in the pattern of the normal stresses obtained on the basis of the elementary theory. However, for slender beams, the magnitude of this disturbance is rather small, and the elementary bending theory is therefore still applicable.

Normal Stresses in Beams with No Axial Force

Straight Beams with Symmetrical Cross Section. The simplest case of bending is that of a straight beam which has a longitudinal plane of symmetry, and which is subjected to loads acting in this plane only.

Figure 35.8*a* shows an element of such a beam. The z axis is placed in the plane of symmetry, and, since the loads are assumed to act in this plane, only the bending moment M_y is present. Figure 35.8*b* shows the same element in the deformed state. If the beam is assumed to consist of longitudinal fibers, the top fibers of the element near *ab* will contract as deformation takes place, while the bottom fibers near *cd* will

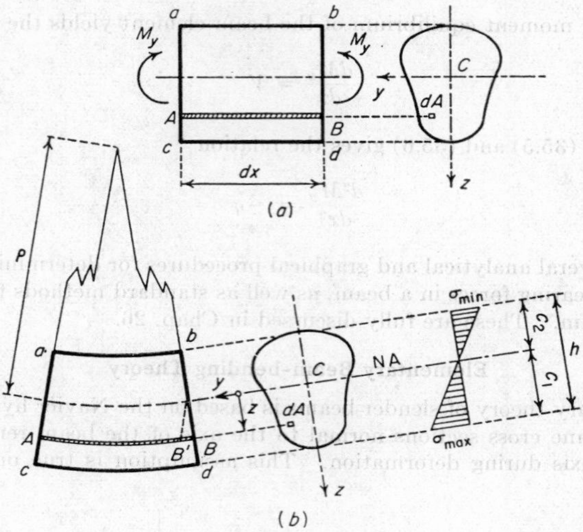

FIG. 35.8

elongate. Between the two, there is a plane formed by fibers which suffer no deformation, called the *neutral plane* of the beam. The straight line formed by the intersection of the neutral plane with the plane of the cross section is called the *neutral axis* of the cross section.

In the undeformed state, an arbitrary fiber, AB, of the beam element has the length dx. As the element deforms, the length of this fiber changes by the amount BB', and the corresponding strain in the fiber is

$$\epsilon_x = \frac{BB'}{dx} = z\kappa \tag{35.8}$$

where $\kappa = 1/\rho$ is the curvature of the deformed beam axis, and z is the distance of the fiber AB from the neutral plane.

In an elastic material, the stress in the fiber AB, according to Hooke's law, is

$$\sigma_x = Ez\kappa \tag{35.9}$$

Conditions of equilibrium require that the resultant of the internal forces on the cross section be zero, and that their moment be equal to the bending moment M_y. Thus,

$$\int_A \sigma_x \, dA = 0 \qquad \text{and} \qquad \int_A \sigma_x z \, dA = M_y$$

with the integral extending over the cross-sectional area A. Substitution of Eq. (35.9) here yields

$$E\kappa \int_A z \, dA = 0 \qquad \text{and} \qquad E\kappa \int_A z^2 \, dA = M_y \tag{35.10a,b}$$

From Eq. (35.10a), it follows that the neutral axis passes through the centroid of the cross section. The integral in Eq. (35.10b) represents the moment of inertia of the cross-sectional area about the neutral axis, and this equation can be written in the form

$$\kappa = \frac{M_y}{EI_y} \tag{35.11}$$

Combination of Eqs. (35.9) and (35.11) yields, for normal stresses,

$$\sigma_x = \frac{M_y}{I_y} z \tag{35.12}$$

The distribution of σ_x is linear over the depth of the beam, and is uniform across its width (Fig. 35.8b). Extreme values of the stress are found at points most remote from the neutral axis. If c_1 and $-c_2$ denote the extreme positive and negative values of z, respectively, the section moduli of the cross section S_1 and S_2 are defined as

$$S_1 = \frac{I_y}{c_1} \qquad \text{and} \qquad S_2 = \frac{I_y}{c_2}$$

With these notations,

$$\sigma_{x\,\max} = \frac{M_y}{S_1} \qquad \text{and} \qquad \sigma_{x\,\min} = -\frac{M_y}{S_2}$$

In the above analysis, it was assumed that the stresses in the beam do not exceed the elastic limit of the material. This assumption requires that the depth of the beam h be small compared to the radius of curvature of the deformed beam ρ, but does not preclude ρ from being of the same order of magnitude as the span of the beam.

As with a bar in compression, a beam in bending can also fail because of lateral instability. Such condition of instability may occur if a beam whose $I_y \gg I_z$ is acted upon by bending moments M_y. Methods for investigating the stability of beams in bending are discussed in Chap. 44.

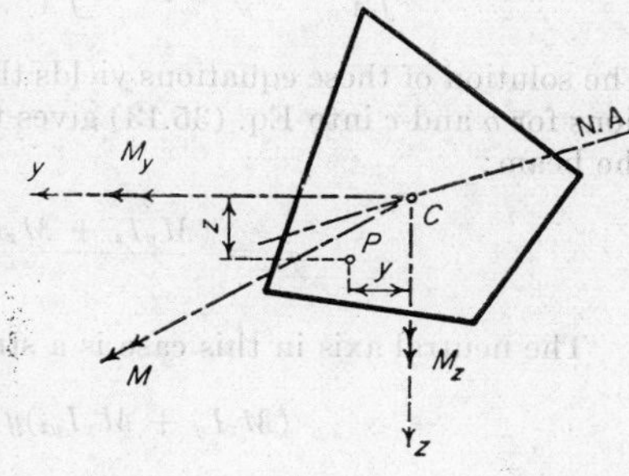

FIG. 35.9

Straight Beams with Unsymmetrical Cross Section. The analysis of bending of beams with unsymmetrical cross section is also based on the assumption that plane cross sections remain plane and normal to the axis of the beam during deformation. The normal stress σ_x at a point P of the cross section is then a linear function of the coordinates y and z of that point (Fig. 35.9), and can be expressed in the general form

$$\sigma_x = a + by + cz \tag{35.13}$$

where a, b, and c are quantities still to be determined.

The conditions of equilibrium for the internal forces acting on the cross section require that the x component of the resultant of the internal forces be zero, and that the moments of the internal forces about the y and z axes be equal to M_y and M_z, respectively.

The first condition yields the equation

$$\int_A \sigma_x \, dA = 0 \tag{35.14}$$

and the latter two conditions yield the equations

$$\int_A \sigma_x z\, dA = M_y \qquad \int_A \sigma_x y\, dA = -M_z \tag{35.15a,b}$$

The negative sign in Eq. (35.15*b*) is due to the adopted sign convention, according to which a positive moment M_z produces a negative (compressive) stress at points with positive y coordinates.

Substitution of Eq. (35.13) into Eqs. (35.14) and (35.15) yields the following three equations for the quantities a, b, and c:

$$a \int_A dA + b \int_A y\, dA + c \int_A z\, dA = 0 \tag{35.16}$$

$$a \int_A z\, dA + b \int_A yz\, dA + c \int_A z^2\, dA = M_y \tag{35.17}$$

$$a \int_A y\, dA + b \int_A y^2\, dA + c \int_A yz\, dA = -M_z \tag{35.18}$$

Since the origin of the y and z coordinates coincides with the centroid of the cross section,

$$\int_A y\, dA = \int_A z\, dA = 0$$

Consequently, $a = 0$, and it follows that the neutral axis passes through the centroid of the cross section. In Eqs. (35.17) and (35.18),

$$\int_A z^2\, dA = I_y \qquad \int_A y^2\, dA = I_z \qquad \int_A yz\, dA = I_{yz}$$

The solution of these equations yields the factors b and c. Substitution of the expressions for b and c into Eq. (35.13) gives the following equation for the normal stress in the beam:

$$\sigma_x = \frac{(M_y I_z + M_z I_{yz})z - (M_z I_y + M_y I_{yz})y}{I_y I_z - I_{yz}^2} \tag{35.19}$$

The neutral axis in this case is a straight line, and its equation is

$$(M_z I_y + M_y I_{yz})y - (M_y I_z + M_z I_{yz})z = 0$$

If the directions of the y and z axes are chosen so that they coincide with the principal axes of inertia of the cross section, η and ζ, then $I_{\eta\zeta} = 0$, and Eq. (35.19) reduces to

$$\sigma_x = \frac{M_\eta}{I_\eta}\zeta - \frac{M_\zeta}{I_\zeta}\eta \tag{35.20}$$

The points of highest stress on the cross section cannot be immediately recognized from Eq. (35.19) or (35.20), and must be found graphically or by calculating the stress at several points of the cross section. There exist several graphical methods for determining the orientation of the neutral axis and the location of points of maximum stress.

Composite Beams. With slight modifications, the theory developed for simple bending of homogeneous beams can also be applied to beams built up of two or more different materials, called composite beams. The present discussion applies only to beams with a symmetrical cross section made of two different materials. However, more complicated cases can be treated similarly.

Both materials of the beam are assumed to obey Hooke's law, and their moduli of elasticity are denoted by E_1 and E_2 (Fig. 35.10). In Eqs. (35.10a,b), the integration must be carried out separately for the areas A_1 and A_2. With $E_2/E_1 = n$, Eq. (35.10a) then assumes the form

$$E_1\kappa\left(\int_{A_1} z\,dA + n\int_{A_2} z\,dA\right) = 0 \qquad (35.21)$$

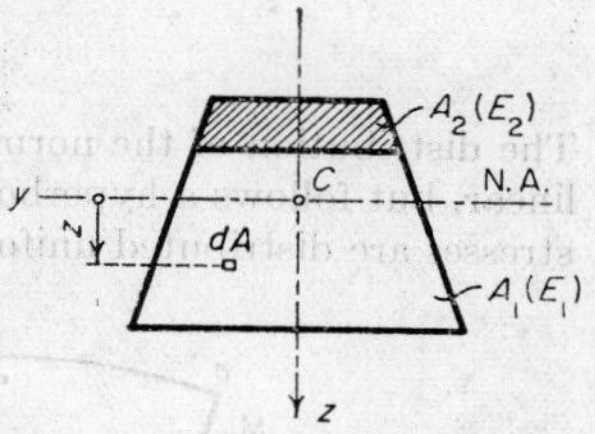

FIG. 35.10

For a given shape of the cross section, Eq. (35.21) determines the location of a fictitious centroid C, through which the neutral axis passes. In Eqs. (35.21) and (35.23), κ denotes the curvature of the deformed beam axis, which is defined as the line connecting the fictitious centroids.

If a fictitious moment of inertia $\bar{I}$ is introduced, where

$$\bar{I} = \int_{A_1} z^2\,dA + n\int_{A_2} z^2\,dA \qquad (35.22)$$

Eq. (35.10b) can be written in the form

$$E_1\kappa\bar{I}_y = M_y \qquad (35.23)$$

Combining Eqs. (35.9) and (35.23) yields the following expressions for the normal stresses in the beam:

In the material with modulus E_1 $\quad \sigma_x = \dfrac{M_y}{\bar{I}_y} z \qquad (35.24a)$

In the material with modulus E_2 $\quad \sigma_x = n\dfrac{M_y}{\bar{I}_y} z \qquad (35.24b)$

Normal Stresses in Curved Beams. Development of the theory of simple bending of beams was based on the assumption that, before deformation, all fibers of the beam element in Fig. 35.8a have the same length dx. If the beam is initially straight, this assumption is true, and it can serve as a good approximation if the curvature of the beam is small. However, for beams with considerable initial curvature, this curvature has to be taken into account.

It will be assumed here that the axis of the beam is a plane curve, and that the beam has a plane of symmetry. It is further assumed that, at the cross section considered, the bending-moment vector is normal to this plane, and that no normal force is present.

The elementary theory of bending of curved beams is again based on the assumption that plane cross sections remain plane during deformation. Possible effects of transverse shearing forces on the deformations of the beam are thus disregarded.

In the course of bending, the cross section bd (Fig. 35.11) will rotate with respect to its original position by the angle $\Delta d\phi$. As a result, the fibers on the concave side are extended, while those on the convex side are compressed. In between lie the fibers which suffer no deformation and form the neutral surface of the beam. The strain ϵ_ϕ in a fiber located a distance ζ from the neutral surface is

$$\epsilon_\phi = \frac{\Delta d\phi}{(r-\zeta)\,d\phi}\zeta \qquad (35.25)$$

In accordance with Hooke's law, the corresponding normal stress is then

$$\sigma_\phi = \frac{E\,\Delta d\phi}{(r - \zeta)\,d\phi}\,\zeta \tag{35.26}$$

The distribution of the normal stress over the depth of the cross section is no longer linear, but follows a hyperbolic law; across the width of the cross section, the normal stresses are distributed uniformly.

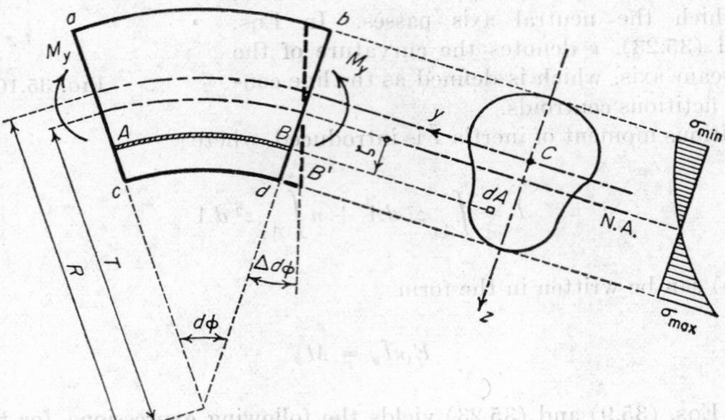

FIG. 35.11

The conditions of equilibrium of internal forces acting on the face of the element require that their resultant force be zero, and that their resultant moment be equal to M_y. This requirement gives the two equations

$$\int_A \sigma_\phi\,dA = \frac{E\,\Delta d\phi}{d\phi}\int_A \frac{\zeta}{r-\zeta}\,dA = 0 \tag{35.27}$$

and
$$\int_A \sigma_\phi \zeta\,dA = \frac{E\,\Delta d\phi}{d\phi}\int_A \frac{\zeta^2}{r-\zeta}\,dA = M_y \tag{35.28}$$

For a given shape of the cross section, Eq. (35.27) gives the value of r, which determines the location of the neutral axis.

If the depth of the beam is small compared to the radius of initial curvature R, then ζ can be neglected against r in Eqs. (35.27) and (35.28). The neutral axis then comes to pass through the centroid of the cross section, and, hence, combination of Eqs. (35.26) and (35.28) yields, for the normal stress, an expression which is identical with that for straight beams. Thus, in curved beams having a small depth compared to the radius of initial curvature, the stresses can be calculated as though the beam were straight.

For pure bending and for several other types of loading of a curved bar, it is possible to obtain exact solutions for the stresses, using a more exact two-dimensional analysis (see [2], p. 55). This analysis shows that the assumption of plane cross sections remaining plane during deformation is fully correct only for the case of pure bending. It also shows that, in a bent curved bar, there are radial stresses which are neglected in the simple theory. These stresses, however, are considerably smaller than the longitudinal stresses σ_ϕ.

For comparison, the maximum and minimum longitudinal stresses in a curved bar of rectangular cross section in pure bending were calculated, using three different theories. The results are presented below.[1] If R_i denotes the inner, and R_a the outer radius of the undeformed bar, the maximum and minimum values of the longitudinal stress can be presented in the form

$$\sigma_\phi = k \frac{M_y}{R_i^2}$$

The table gives the values of the numerical factor k calculated by three different methods for beams of different R_a/R_i ratios. Results of the first column are obtained, neglecting the initial curvature of the bar, which yields a linear stress distribution; the results of the second column are obtained on the basis of a hyperbolic stress distribution; and, in the third column, the results of the two-dimensional analysis are presented. The table shows that there is good agreement between the hyperbolic

R_a/R_i	Linear stress distribution	Hyperbolic stress distribution		Exact solution	
		Max.	Min.	Max.	Min.
1.3	±66.67	+72.98	−61.27	+73.05	−61.35
2.0	±6.00	+7.725	−4.863	+7.755	−4.917
3.0	±1.500	+2.285	−1.095	+2.292	−1.193

stress distribution and the exact theory. For other types of loading, the agreement between the two theories is fairly good but not so good as for pure bending.

Shearing Stresses in Beams

Distribution of Shearing Stresses. In a beam subjected to other than pure bending, transverse shearing forces V_y and V_z, in addition to bending moments, are present on a cross section normal to the axis of the beam. These shearing forces are resultants of the shearing stresses τ_{xy} and τ_{xz}, respectively. In deriving the elementary beam-bending theory, the contribution of the shearing stresses to the deformation of the beam has been neglected. Therefore, these stresses can be found only from equilibrium conditions for a suitably chosen portion of the beam, and the deformations derived from them are in disagreement with the Navier assumption.

Figure 35.12 shows a beam element of symmetrical cross section with the normal stresses acting on it. It is assumed that all loads on the beam act in its vertical plane of symmetry. The shearing force V_z will then act along the vertical axis of symmetry of the cross section, and $V_y = 0$. The bending moment M_y, which acts on the face $ABFD$ of the element, produces there normal stresses $\sigma_x = (M_y/I_y)\zeta$. The bending moment acting on the face $A'B'F'D'$ is $(M_y + dM_y)$, and it produces normal stresses $\sigma_x' = [(M_y + dM_y)/I_y]\zeta$ on this face.

To determine the shearing stresses at a point O, a section $HH'K'K$, normal to the plane $ABDF$, is made through the beam element at this point. The shape and orientation of this section, with respect to the z axis, are chosen to suit the requirements of the particular problem, as discussed later. By means of this section, the portion $DFHKD'F'H'K'$ is separated from the beam element. It is shown separately in Fig. 35.13.

The forces which act on this portion of the element in the x direction are the resultant of the stresses σ_x acting on the area $DFHK$, the resultant of the stresses σ_x'

[1] The table is taken from [2], p. 64.

acting on the area $D'F'H'K'$, and the resultant of the shearing stresses τ_s acting parallel to the beam axis on the area $HH'K'K$. The equation of equilibrium for these forces is

$$\frac{M_y}{I_y}\int_{DFHK}\zeta\,dA - \frac{M_y + dM_y}{I_y}\int_{D'F'H'K'}\zeta\,dA + dx\int_{HK}\tau_s\,ds = 0 \quad (35.29)$$

The integral in the first two terms of this equation is the static moment of the area $DFHK$ about the neutral axis of the cross section, and will be denoted by Q. If the

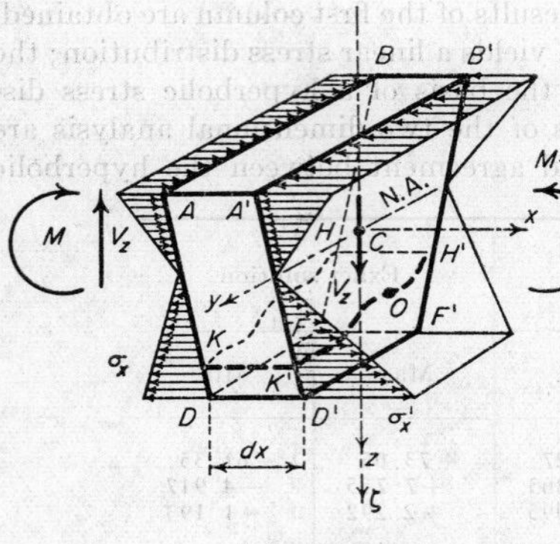

FIG. 35.12

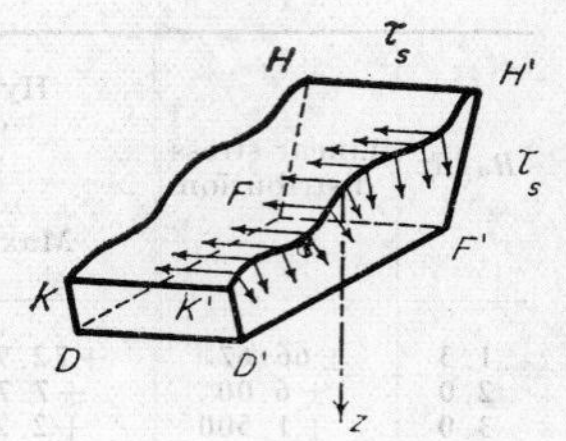

FIG. 35.13

integral in the last term is divided by the length of the line HK, it then represents the average shearing stress along this line, denoted by $\bar{\tau}_s$. Since $dM_y/dx = V_z$, Eq. (35.29) yields

$$\bar{\tau}_s = \frac{V_z Q}{I_y}\,\frac{1}{\overline{HK}} \quad (35.30)$$

To obtain the actual value of the shearing stress at O, the section through this point must be so chosen that an assumption can be made regarding the distribution of the stresses along the line HK.

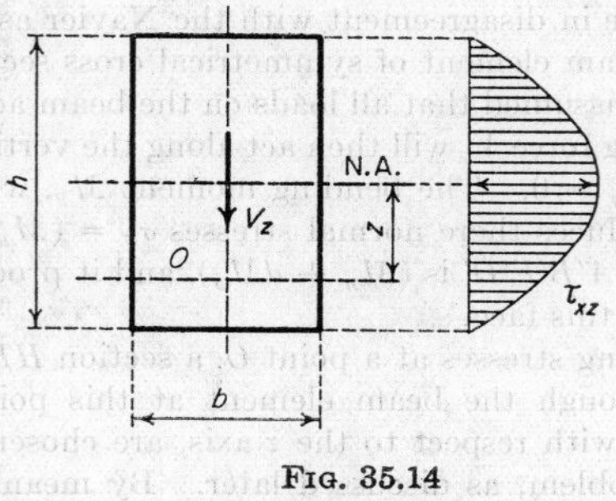

FIG. 35.14

For a beam with a rectangular cross section (Fig. 35.14), this section is made parallel to the neutral plane. Then the stress τ_s becomes identical with τ_{xz}, and it can reasonably be assumed that the shearing stresses are distributed uniformly across the width b of the cross section. A more exact theory shows that, for beams not excessively wide, this assumption is in good agreement with the actual stress distribution (see [2], p. 316). The shearing stress at point O located at a distance z from the neutral axis then is

$$\tau_{xz} = \frac{V_z Q}{I_y b} \tag{35.31}$$

For a beam of depth h,

$$Q = \frac{b}{8}(h^2 - 4z^2) \qquad I_y = \frac{bh^3}{12}$$

and Eq. (35.31) gives

$$\tau_{xz} = \frac{3V_z}{2bh^3}(h^2 - 4z^2)$$

On a rectangular cross section, the shearing stress τ_{xz} thus has a parabolic distribution over the depth of the beam; it has its maximum value at the neutral axis, where $\tau_{xz\,\max} = \frac{3}{2} V_z/bh$, and is zero at the top and the bottom of the beam.

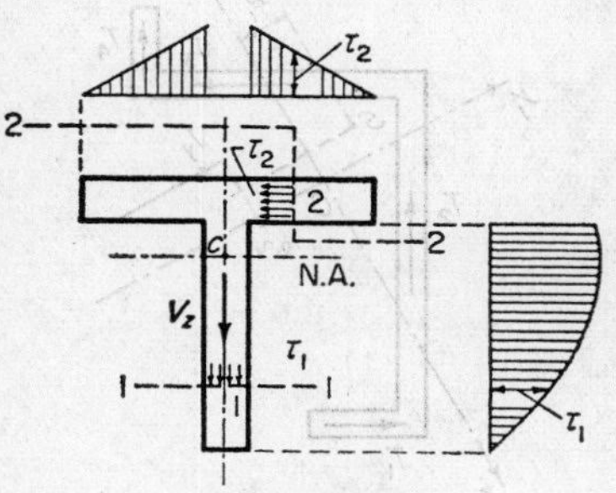

FIG. 35.15

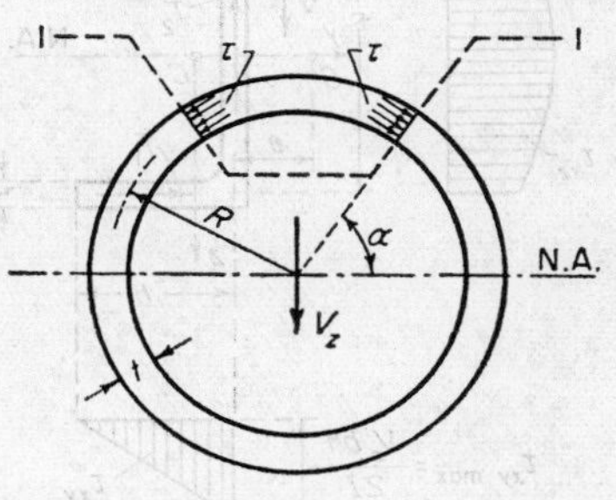

FIG. 35.16

To obtain the vertical shearing stresses τ_1 at point 1 in the web of a T section (Fig. 35.15), a horizontal section 1-1 is made through this point, and it is assumed that the shearing stresses are distributed uniformly across the thickness of the web. To obtain the horizontal shearing stresses τ_2 at point 2 of the flange, the section 2-2 is made, and uniform distribution of shearing stresses over the thickness of the flange is assumed. At the transition between web and flange, the theory is not applicable because no reasonable assumption of stress distribution can be made at this point. The magnitude and distribution of the shearing stresses here depend greatly on the form of the fillet. Also, the values of vertical shearing stresses at point 2 cannot be obtained by means of the present theory, because no reasonable assumption can be made with regard to stress distribution across the width of the flange.

The shearing stresses in a beam of thin tubular cross section can be found by using a section as shown in Fig. 35.16, and by assuming a uniform stress distribution across the thickness of the tube wall. Then, along section 1-1,

$$\tau = \frac{V_z \cos \alpha}{\pi R t}$$

Whenever discontinuities, such as reentrant corners or notches, exist on the outlines of a cross section, concentrations of shearing stresses will occur at these points. An estimate of the value of the shearing stresses at such points can be obtained by using an appropriate stress-concentration factor [4].

In composite beams, the bond stresses τ_b between the different materials are often of importance. Their analysis is based on the same idea as the analysis of shearing stresses in beams of homogeneous material.

Shear Center. For every cross-sectional shape, there exists one point on the cross section through which the resultant of the transverse shearing forces always passes, regardless of the direction of the transverse external forces (see [3], p. 200). This point is called the shear center of the cross section, and the locus of the shear centers of the cross sections of a beam is called the *elastic axis* of the beam. Loads acting on the beam must pass through the elastic axis in order to produce simple bending of the beam with no twisting.

The location of the shear center for a given shape of the cross section is not always readily obtainable. If the cross section has an axis of symmetry, the shear center will be located on this axis. If the cross section has two axes of symmetry, the shear center will be located at their point of intersection, and will thus coincide with the centroid.

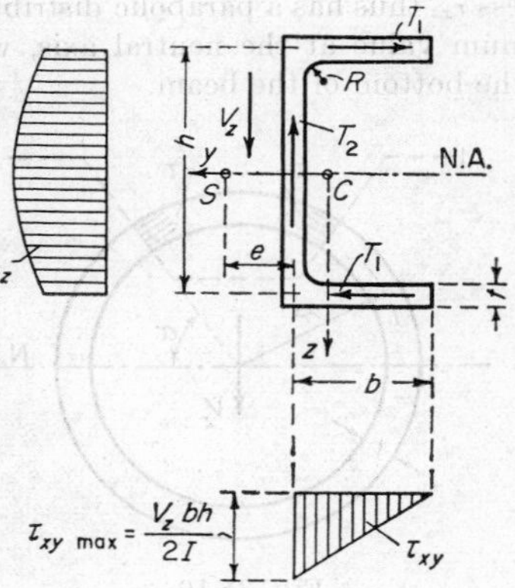

FIG. 35.17

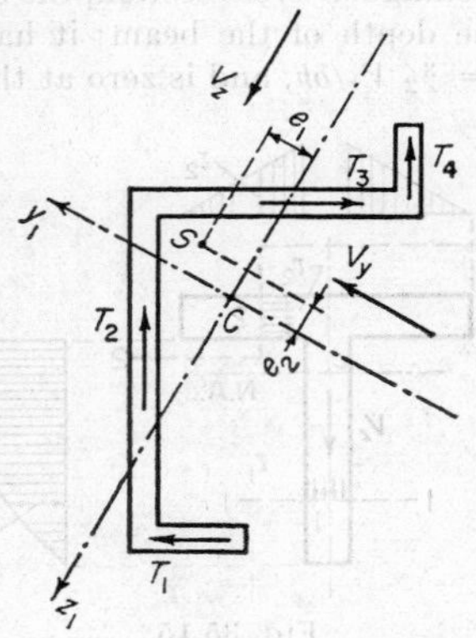

FIG. 35.18

For some cross-sectional shapes, the location of the shear center can be obtained using a two-dimensional analysis.

For thin-walled cross sections, it can be assumed that the shearing stresses are distributed uniformly over the thickness of the walls, and that their direction is always parallel to the boundary of the cross section. On the basis of these assumptions, it is then possible to locate the resultant of the shearing stresses. If this procedure is followed for bending in two different planes, the location of the shear center is found.

To obtain the location of the shear center for a thin-walled section with one axis of symmetry, such as the channel in Fig. 35.17, a shearing force V_z is assumed to act normal to the axis of symmetry. The shearing stresses on the cross section due to V_z are found in the usual manner, and their resultants in the web and the flange portions are determined. The resultant force in the flange is $T_1 = (V_z ht/4I_y)b^2$. The internal force in the web, T_2, can be assumed equal to V_z. By definition, the location of the shear center S must be such that the resultant moment of all internal forces about S is zero. If e denotes the distance between the shear center and the web, this condition is expressed by the equation $T_1 h - T_2 e = 0$. Substitution in this equation of the expressions for T_1 and T_2 yields

$$e = \frac{b^2 h^2 t}{4I_y}$$

For an unsymmetrical thin-walled cross section, such as shown in Fig. 35.18, the principal axes of inertia, y_1 and z_1, first have to be determined. A transverse shearing force V_z is then assumed to act parallel to the z_1 axis. The shearing stresses caused by V_z and their resultants $T_1 \cdots T_4$ for the different portions of the cross section are obtained in the usual manner. Setting the sum of the moments of these forces about S equal to zero yields an equation for the distance e_1. The procedure is repeated for a shearing force V_y parallel to the y_1 axis, which gives the distance e_2.

Combined Bending and Axial Force

Normal Stresses Due to Combined Bending and Axial Force. In a general case of the loading of a bar, a normal force may be present on the cross section under consideration, in addition to the two bending-moment components. Except for a few cases which are noted later, the principle of superposition may be applied to calculate the stresses in such a bar. Figure 35.19 shows a portion of a beam in which the normal force N and the bending moments M_y and M_z act at the cross section $ABDF$. The axes y and z are assumed to be the principal centroidal axes of inertia of the cross-sectional area. The normal stress at any point of the cross section is obtained by superimposing the stress caused by the normal force and the normal stresses caused by the bending moments. The first is given by Eq. (35.1), and the latter by Eq. (35.20). The total stress at a point O with coordinates y and z then is given by

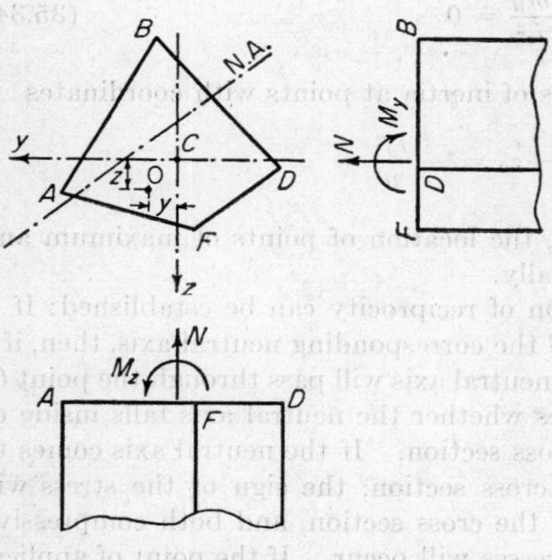

FIG. 35.19

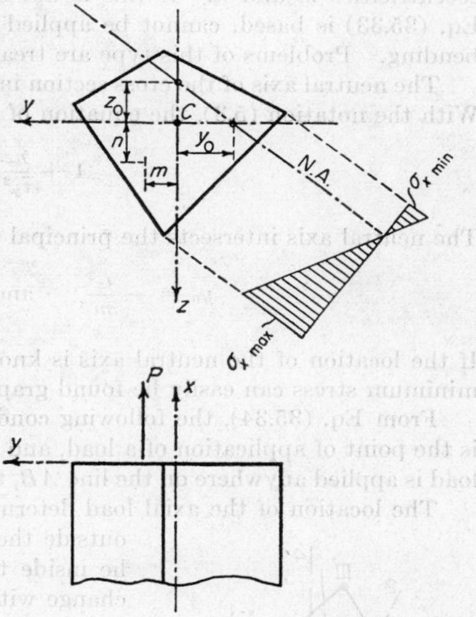

FIG. 35.20

$$\sigma_x = \frac{N}{A} + \frac{M_y}{I_y} z - \frac{M_z}{I_z} y \tag{35.32}$$

The neutral axis for this type of loading is, as before, a straight line, but it does not pass through the centroid of the cross section; in fact, it may fall outside the cross section. In this case, when the normal force is high and the bending moments are relatively low, the normal stresses over the whole cross section have the same sign.

A special case of combined bending and direct stress is a straight, eccentrically loaded bar. Figure 35.20 shows the cross section of such a bar, with y and z being the principal axes of inertia. The line of action of the load P passes at a distance n and m from the y and z axes, respectively; therefore,

$$N = P \qquad M_y = Pn \qquad M_z = -Pm$$

The normal stress at a point of the cross section with coordinates y and z then is

$$\sigma_x = \frac{P}{A} + \frac{Pn}{I_y} z + \frac{Pm}{I_z} y \tag{35.33}$$

If a bar is acted upon by an eccentric compressive force, Eq. (35.33) applies only if the lateral displacements of the bar due to M_y and M_z are small compared to the eccentricities m and n. If this is not so, the principle of superposition, on which Eq. (35.33) is based, cannot be applied and the problem becomes one of nonlinear bending. Problems of this type are treated in Chap. 30.

The neutral axis of the cross section in an eccentrically loaded bar is a straight line. With the notation (5.3), the equation of the neutral axis is

$$1 + \frac{nz}{i_y^2} + \frac{my}{i_z^2} = 0 \tag{35.34}$$

The neutral axis intersects the principal axes of inertia at points with coordinates

$$y_0 = -\frac{i_z^2}{m} \qquad \text{and} \qquad z_0 = -\frac{i_y^2}{n}$$

If the location of the neutral axis is known, the location of points of maximum and minimum stress can easily be found graphically.

From Eq. (35.34), the following condition of reciprocity can be established: If O is the point of application of a load, and AB the corresponding neutral axis, then, if a load is applied anywhere on the line AB, the neutral axis will pass through the point O.

The location of the axial load determines whether the neutral axis falls inside or outside the cross section. If the neutral axis comes to lie inside the cross section, the sign of the stress will change within the cross section, and both compressive and tensile stresses will occur. If the point of application of the axial force falls inside a certain area within the cross section, only stresses of one sign arise on the cross section. This area is called the *kernel* (also core or kern) of the cross section. If the point of application of the axial force lies on the boundary of the kernel, the corresponding neutral axis just touches the boundary of the cross section.

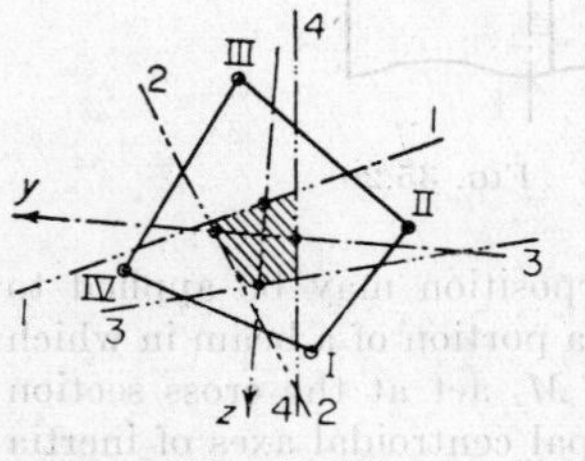

FIG. 35.21

To determine the outlines of the kernel for a given cross section, the rule of reciprocity between the position of the axial load and the corresponding neutral axis can be used to advantage. If the line 1-1 (Fig. 35.21) is the neutral axis for an axial load which passes through point I, then the neutral axis for a load passing anywhere through the line 1-1 will pass through point I. A similar relation exists between line 2-2 and point II, line 3-3 and point III, line 4-4 and point IV. By definition, these lines then form the boundary of the kernel.

Points where the principal axes of inertia intersect the boundaries of the kernel are called *kernel points.*

For regular cross-sectional shapes, the dimensions of the kernel can easily be determined analytically. For a rectangle (Fig. 35.22), the kernel is a rhomboid with diagonals equal to $h/3$ and $b/3$.

For a circular cross section of radius R, the kernel is a circle of radius $r = R/4$.

For an I beam (Fig. 35.23), for instance, the kernel is a rhomboid with $k_1 = 2i_z^2/b$ and $k_2 = 2i_y^2/h$, where i_y and i_z are the radii of gyration about the y and z axes, respectively.

If a beam is acted upon by a bending moment and an axial force, so that bending takes place about one of the principal axes of inertia, it is often useful to employ the

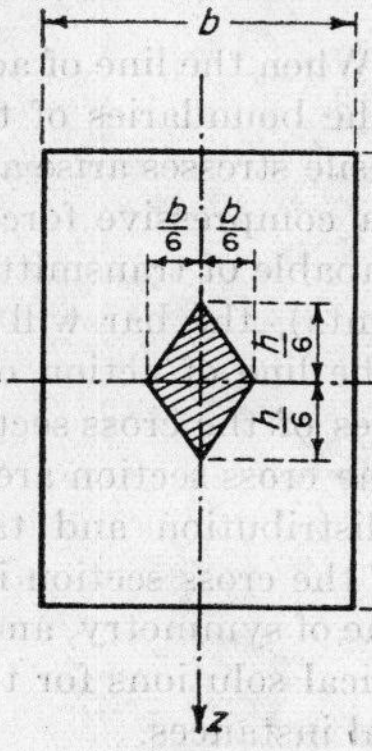

FIG. 35.22

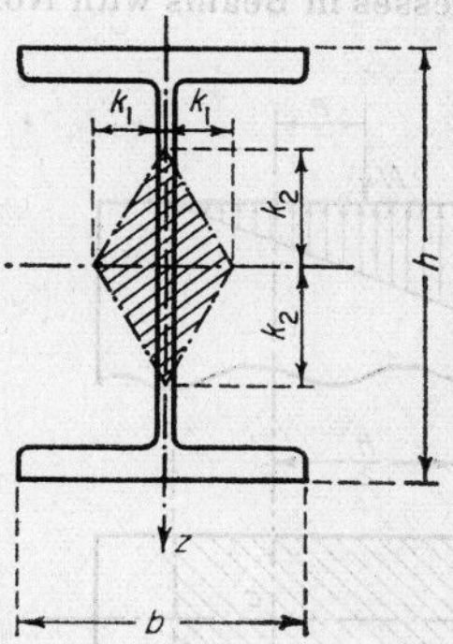

FIG. 35.23

concept of the kernel moment for calculating the maximum and minimum normal stresses. Instead of the normal force being assumed to act at the centroid and the bending moment being calculated around this point, the normal force is assumed to act at either the upper or the lower kernel point, and the moments (called kernel moments) are calculated about these points. To determine the stress in the bottom fibers of the beam shown in Fig. 35.24a, the combination of bending moment M_y and

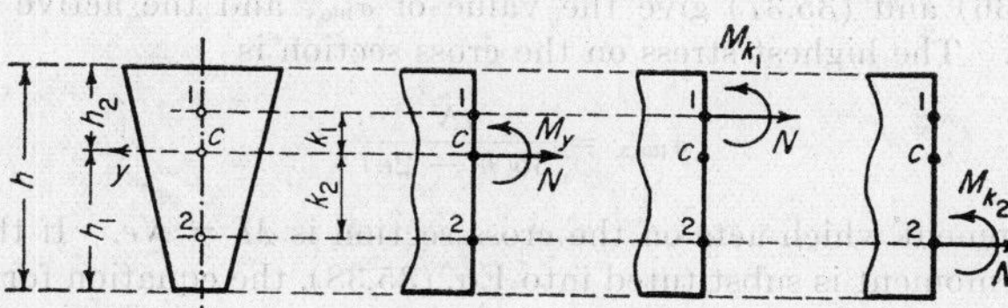

FIG. 35.24

force N in Fig. 35.24b is replaced by the combination of kernel moment M_{k_1} and force N as shown in Fig. 35.24c, where

$$M_{k_1} = M_y + Nk_1$$

A normal force acting at point 1 produces zero stress in the bottom fibers; the normal stress there is due to M_{k_1} only, and thus is given by the equation

$$\sigma_{x1} = \frac{M_{k1}}{I_z} h_1 \qquad (35.35a)$$

To find the maximum stress in the upper fibers, the original internal force and moment system is replaced by that shown in Fig. 35.24d, where

$$M_{k_2} = M_y - Nk_2$$

The normal stress in the upper fiber then is

$$\sigma_{x2} = -\frac{M_{k2}}{I_y} h_2 \qquad (35.35b)$$

In Eqs. (35.35), I_y is the moment of inertia of the cross-sectional area about the centroidal y axis.

Stresses in Beams with Nonexistent Tension Zone. When the line of action of an axial load falls outside the boundaries of the kernel, both compressive and tensile stresses arise at the cross section. If the load is a compressive force, and the material of the bar is incapable of transmitting tensile stresses (e.g., mortar joints), the bar will still be in equilibrium, as long as the line of action of the load falls within the boundaries of the cross section. The stresses which arise on the cross section are obtained, assuming linear stress distribution and taking into account that a portion of the cross section is inactive.

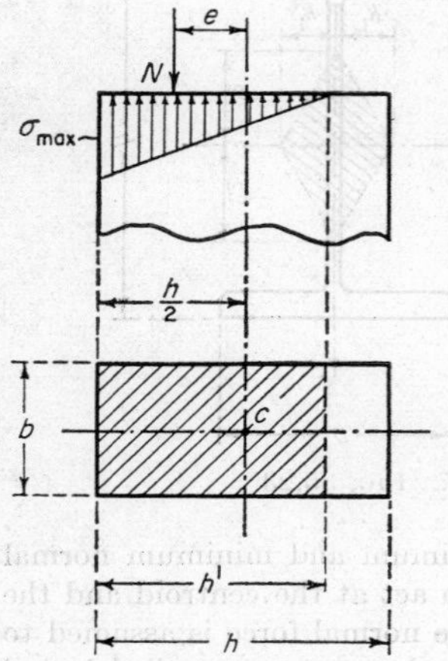

FIG. 35.25

If the beam has a plane of symmetry, and the loads act in this plane, analytical solutions for the stresses can be obtained in several instances.

For a rectangle (Fig. 35.25), the condition of force equilibrium gives the equation

$$N - \tfrac{1}{2}\sigma_{\max} b h' = 0 \qquad (35.36)$$

The condition of moment equilibrium is

$$b\frac{h'}{2}\left(\frac{h}{2} - \frac{h'}{3}\right)\sigma_{\max} - Ne = 0 \qquad (35.37)$$

Equations (35.36) and (35.37) give the value of $\sigma_{\max}$ and the active depth of the cross section h'. The highest stress on the cross section is

$$\sigma_{\max} = \frac{4N}{3b(h - 2e)} \qquad (35.38)$$

The bending moment which acts on the cross section is $M = Ne$. If this expression for the bending moment is substituted into Eq. (35.38), the equation for $\sigma_{\max}$ becomes

$$\sigma_{\max} = \frac{4N^2}{3b(Nh - 2M)} \qquad (35.39)$$

From Eq. (35.39), it can be seen that the stress is not a linear function of the force and moment. This nonlinearity occurs in all cases of inactive tension zone; the principle of superposition, therefore, cannot be applied in these cases.

For a beam of symmetrical cross section of arbitrary shape, loaded in its plane of symmetry, the stress distribution with an inactive tension zone can be obtained graphically.

For rectangular cross sections and an arbitrary location of the line of action of the load, there are tables which help to determine the stress distribution and the location of the neutral axis.

Stress Concentrations

In deriving the equations for normal stresses in bars subjected to an axial load, or bending, or both, it was assumed that the cross section under consideration is located some distance away from points of application of external loads, and that the cross section of the bar is uniform.

The first assumption is made because the stress distribution near points of application of a load cannot be determined using elementary theory.

If the cross section of the bar is not uniform, but changes gradually, equations which were derived for a uniform bar still give fairly accurate results. However, in regions where the cross-sectional shape of the beam changes abruptly, the elementary theory gives stress values which are usually too low. This condition occurs in the vicinity of notches, holes, and fillets. In some instances, the stresses in these regions can be obtained with the help of a more exact two- or three-dimensional analysis; often, however, experimental methods must be employed.

The expressions for shearing stresses in beams subjected to bending have also been derived on the assumption that the cross-sectional shape of the beam is uniform.

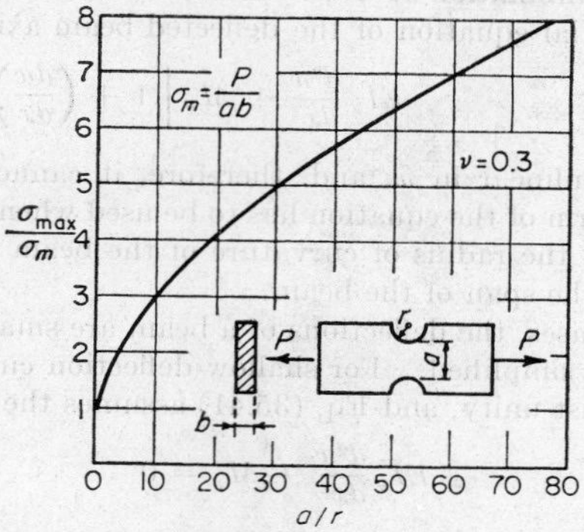

FIG. 35.26. Stress-concentration factor for a flat tension bar with semicircular notches. (*With permission from Ref.* [5], p. 37.)

Any nonuniformity of the cross-sectional shape affects the accuracy of these expressions in its immediate vicinity. In addition, concentrations of shearing stresses occur at points where reentrant corners or other irregularities exist on the outlines of the cross-sectional area. In the channel in Fig. 35.17, for instance, the actual value of the shearing stress at the inside corner will be greater than that predicted by the elementary theory. The value of the shearing stress at this point depends greatly on the magnitude of the radius of the fillet R; it is higher for a small radius.

To give the designer a simple means of estimating the stresses at points where stress concentrations arise, the concept of the stress-concentration factor has been introduced. The stress at such points is first obtained, using elementary theory and disregarding the presence of any irregularity, such as stress σ_m in Fig. 35.26. The result is then multiplied by an appropriate stress-concentration factor, which is taken from a graph or a table [4],[5]. An example of such a graph is given in Fig. 35.26. This graph gives the stress-concentration factor for a bar under axial load with rectangular cross section and semicircular notches.

Deflection of Beams

Effect of Bending Moments and Shearing Forces. The deflections of a loaded beam are analyzed by considering separately the deflections which result from the action of the bending moments, and those which result from the action of the transverse shearing forces. The total deflection is then obtained by adding the two contributions. The more slender the beam, the more predominant is the part of deflection due to the bending moments, and the more negligible is the part due to shearing forces. In many practical applications, therefore, the effect of the shearing forces on the deflection is neglected altogether.

Deflection Caused by Moments. In the following discussion, it will be assumed that the loads on the beam in Fig. 35.27 act in the x,z plane, which contains one of the principal axes of inertia of the beam cross section, and that the bending-moment vector is normal to this plane. In the deflected state, the axis of the beam will then be a plane curve in the x,z plane. The deflection in the positive z direction is denoted by w, and the curvature of the deformed beam axis at any point is denoted by κ. The relation between κ and w is given by the equation

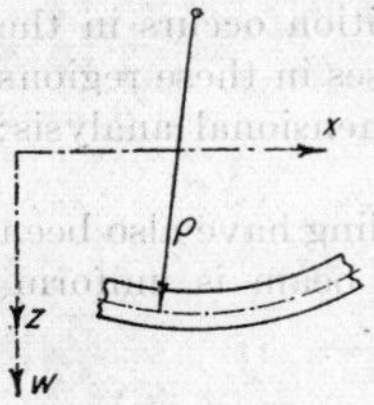

FIG. 35.27

$$\kappa = -\frac{d^2w/dx^2}{[1 + (dw/dx)^2]^{3/2}} \tag{35.40}$$

Combination of Eqs. (35.11) and (35.40) then yields the differential equation of the deflected beam axis as follows:

$$EI_y \frac{d^2w}{dx^2} + M_y \left[1 + \left(\frac{dw}{dx}\right)^2\right]^{3/2} = 0 \tag{35.41}$$

Equation (35.41) is nonlinear in w, and, therefore, it cannot always be solved by simple means. This form of the equation has to be used whenever the deflections of a beam are large, so that the radius of curvature of the beam axis ($\rho = 1/\kappa$) is of the order of magnitude of the span of the beam.

In many practical cases, the deflections of a beam are small compared to its span, and Eq. (35.41) can be simplified. For shallow deflection curves, the term $(dw/dx)^2$ may be neglected against unity, and Eq. (35.41) assumes the simplified form

$$EI_y \frac{d^2w}{dx^2} + M_y = 0 \tag{35.42}$$

If the relation between bending moment and load, as given by Eq. (35.7), is used, Eq. (35.42) can be written in the form

$$\frac{d^2}{dx^2}\left(EI_y \frac{d^2w}{dx^2}\right) = q \tag{35.43}$$

Equations (35.42) and (35.43) are linear in w, and can be easily solved for a given distributed load q, or for a given distribution of the bending moment M_y. These equations can be applied to beams of either uniform or continuously varying cross-sectional area; in the latter case, the equations are valid as long as the assumptions of Navier's hypothesis are satisfied to a fair degree. For beams of constant moment of inertia, Eq. (35.43) can be simplified further and brought into the form

$$EI_y \frac{d^4w}{dx^4} = q \tag{35.44}$$

The solution of Eq. (35.43) or (35.44) contains four integration constants, which have to be determined from the boundary conditions of the beam. These boundary conditions state either the deflections or the slopes of the deformed beam axis at the ends, or they predetermine the bending moments or the transverse shearing forces acting there. In the case of elastic end constraints, the boundary conditions express the relations between the support reactions and movements of the supports.

Various methods for the solution of Eqs. (35.43) and (35.44) have been developed. These methods and their application to beams under different kinds of loading and boundary conditions are discussed in Chap. 26.

In a more general case of loading, a beam may be subjected to loads which act in planes other than those containing the principal axes of inertia of the beam cross section. As long as the deformations are small, the total deflection of the beam can be obtained by resolving the bending-moment vectors into components parallel to the principal axes, calculating separately the deflections in the direction of each of these, and adding the deflections vectorially.

Deflection Caused by Shearing Forces. In developing the elementary beam-bending theory, deformations of the beam which are caused by the shearing stresses have been neglected. Therefore, this theory applies essentially to slender beams, where the effect of the shearing stresses on the total deformation is actually negligible. However, the elementary theory can also be applied, with a fair degree of accuracy, to relatively short and deep beams. The additional deformations due to shearing stresses are then accounted for by means of an approximate analysis.

The theory presented here applies to beams initially straight or of small initial curvature, and, for simplicity, it is assumed that the loads on the beam act in one plane only (the x,z plane).

Since the shearing stresses τ_{xz} in a beam vary with the distance from the neutral plane, the corresponding shearing strain $\gamma_{xz} = \tau_{xz}/G$ also varies over the depth of the beam. For the purpose of further analysis, an average shearing strain $\bar{\gamma}_{xz}$ is introduced, which is determined from the condition that the work $\frac{1}{2}V_z\bar{\gamma}_{xz}$ on a beam element of length dx must be equal to the integrated work of the shearing stresses τ_{xz}. This condition gives the equation

$$\tfrac{1}{2}V_z\bar{\gamma}_{xz} = \frac{1}{2G}\int_A \tau_{xz}^2\,dA$$

If the shearing stress distribution over the cross section is known, the integral in this equation can be evaluated, and the expression for $\bar{\gamma}_{xz}$ can be written in the form

$$\bar{\gamma}_{xz} = \frac{\alpha V_z}{GA} \tag{35.45}$$

where A is the area of the cross section, and α is a numerical factor which depends on the cross-sectional shape (for a rectangle, $\alpha = 1.2$).

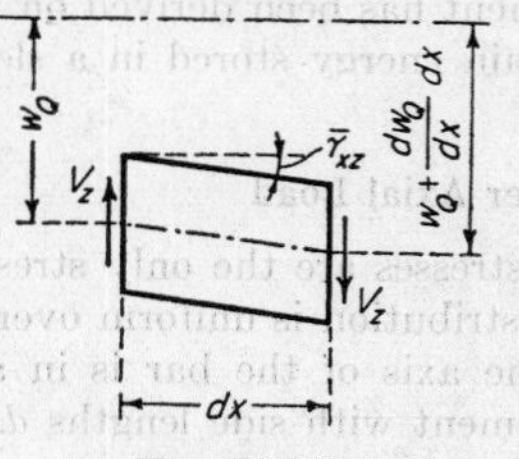

FIG. 35.28

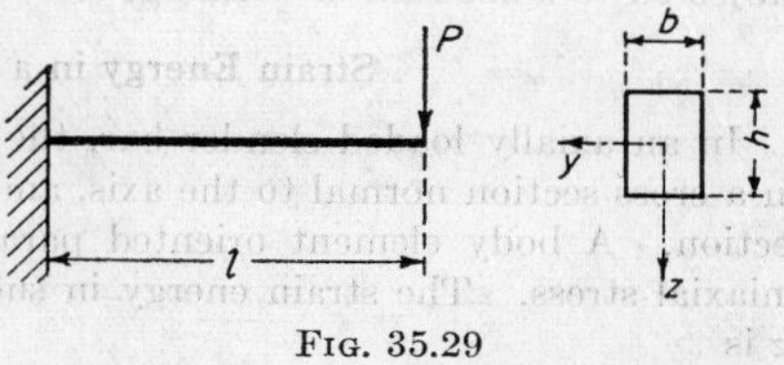

FIG. 35.29

The additional deflection of the beam due to shearing stresses will be denoted by w_s. As Fig. 35.28 shows,

$$\bar{\gamma}_{xz} = \frac{dw_s}{dx} \tag{35.46}$$

Combining Eqs. (35.45) and (35.46) gives the following differential equation for w_s:

$$\frac{dw_s}{dx} - \frac{\alpha V_z}{GA} = 0 \tag{35.47}$$

Integration of Eq. (35.47) gives a solution with one constant of integration, which must be determined from a boundary condition. The more slender the beam, the smaller is the effect of shearing forces on the total deflection of the beam. For instance, the total deflection of the end of the cantilever in Fig. 35.29 is

$$w = \frac{4Pl^3}{Ebh^3}\left[1 + 0.6(1 + \nu)\frac{h^2}{l^2}\right]$$

The second term in the bracket represents the contribution of the shearing force, which is small for $h/l \ll 1$.

For slender curved bars of moderate initial curvature, the effect of shearing forces on deformations may be estimated by neglecting the curvature of the bar and using Eq. (35.47) to calculate deflections. If the initial curvature of the bar is large, the elementary theory provides no means of calculating deflections due to shearing forces. In some cases, solutions may be obtained by means of a two- or three-dimensional analysis.

35.3. STRAIN ENERGY

The Concept of Strain Energy

An elastic body which is acted upon by external loads suffers deformations, and, as a result of these deformations, the points of application of the loads move. If the loads are applied gradually, the kinetic energy imparted to the particles of the body is negligible. All the work done by the loads is transformed into potential energy of strain, and is stored in the body. If the body is perfectly elastic, and the energy losses due to internal friction are neglected, the work done while loading it can be recovered during a gradual unloading. The elastic body thus represents a conservative system, to which the principle of conservation of energy may be applied. This principle proves useful in calculating deformations of elastic bodies and structures; the corresponding methods are discussed in Chap. 26.

In actual physical materials, internal friction causes a small part of the energy to be converted into heat energy; also, the stress-strain relationship during unloading is not the same as during loading. Nevertheless, in the elastic region most elastic bodies can be assumed to represent a conservative system; stresses and deformations obtained on the basis of this assumption prove to be quite accurate.

To apply the concept of strain energy, the amount of strain energy stored in a body subjected to given loads must be known. The general expression for the strain energy stored in an infinitesimal elastic body element has been derived on p. 33–18. It will be used here to derive expressions for strain energy stored in a slender bar, subjected to axial load or bending, or to both.

Strain Energy in a Bar under Axial Load

In an axially loaded slender bar, the normal stresses are the only stresses found on a cross section normal to the axis, and their distribution is uniform over the cross section. A body element oriented parallel to the axis of the bar is in a state of uniaxial stress. The strain energy in such an element with side lengths dx, dy, and dz is

$$dU = \frac{1}{2}\frac{\sigma_x^2}{E}\,dx\,dy\,dz \tag{35.48}$$

The strain energy stored in a portion of the bar of length dx is denoted by $d\bar{U}$ and is obtained by integrating Eq. (35.48) over the cross-sectional area A of the bar. Since σ_x is uniform over the area, and $\sigma_x = N/A$,

$$d\bar{U} = \frac{1}{2}\frac{\sigma_x^2 A}{E}\,dx = \frac{1}{2}\frac{N^2}{EA}\,dx \tag{35.49}$$

where N is the normal force at the cross section under consideration.

The total strain energy U stored in a bar of length l under axial load is obtained by integrating Eq. (35.49) over the length of the bar. This gives

$$U = \frac{1}{2}\int_0^l \frac{N^2}{EA}\,dx \tag{35.50}$$

In a bar of constant area A and of length l loaded by two end loads P (Fig. 35.1), the strain energy is

$$U = \frac{P^2 l}{2EA}$$

Strain Energy in a Beam under Axial Load and Bending

In a beam which is subjected to an axial load and to bending in two planes, an element oriented parallel to the beam axis with side lengths dx, dy, dz will have one normal-stress component, σ_x, and two shearing-stress components, τ_{xy} and τ_{xz}, acting on it. The strain energy stored in such an element, denoted by dU, is given by the equation

$$dU = \left[\frac{\sigma_x^2}{2E} + \frac{1+\nu}{E}(\tau_{xz}^2 + \tau_{xy}^2)\right] dx\,dy\,dz \tag{35.51}$$

The total strain energy in the beam is obtained by integrating Eq. (35.51) over the volume of the beam.

The first term in the bracket in Eq. (35.51) represents strain energy due to normal stresses and for this portion of the energy the stress σ_x can be expressed by using Eq. (35.32). The total strain energy in a beam due to normal stresses then is

$$U_n = \frac{1}{2E}\iiint\left(\frac{N^2}{A^2} + \frac{M_y^2}{I_y^2}z^2 + \frac{M_z^2}{I_z^2}y^2 + 2\frac{N}{A}\frac{M_y}{I_y}z - 2\frac{N}{A}\frac{M_z}{I_z}y - 2\frac{M_yM_z}{I_yI_z}yz\right) dx\,dy\,dz \tag{35.52}$$

If the y and z axes are the centroidal principal axes of inertia, then, upon integration, the last three terms in Eq. (35.52) vanish, and the strain energy due to normal stresses in a beam of length l is

$$U_n = \frac{1}{2E}\int_0^l \frac{N^2}{A}\,dx + \frac{1}{2E}\int_0^l \frac{M_y^2}{I_y}\,dx + \frac{1}{2E}\int_0^l \frac{M_z^2}{I_z}\,dx \tag{35.53}$$

The strain energy in a beam due to normal stresses can thus be obtained by calculating separately and then adding the strain energy produced by the normal-force component and the strain energy produced by the two bending-moment components. This superposition of portions of strain energy is possible because the normal force in a beam does no work on displacements produced by the bending moments, and vice versa.

The second term in Eq. (35.51) represents strain energy due to shearing stresses. For a beam of a given cross-sectional area, the shearing stress at a point of a cross section can be expressed as a function of the shearing force and of the coordinates y and z. For a given external loading, the expression in Eq. (35.51) can then be integrated.

The total strain energy in a beam can thus be found by calculating separately the energy due to normal stresses and the energy due to shearing stresses, and then adding the two contributions. This superposition of strain energies is possible because the normal stresses in a beam do no work during shearing strain, and the shearing stresses do no work during strains produced by the normal stresses.

For slender beams, the portion of strain energy due to shearing stresses is small compared to that due to normal stresses, and, in practical applications, it is often neglected.

35.4. REFERENCES

[1] S. P. Timoshenko: "Strength of Materials," vols. 1 and 2, 3d ed., Van Nostrand, Princeton, N.J., 1955, 1956.
[2] S. P. Timoshenko, J. N. Goodier: "Theory of Elasticity," 2d ed., McGraw-Hill, New York, 1951.
[3] E. E. Sechler: "Elasticity in Engineering," Wiley, New York, 1952.
[4] R. E. Peterson: "Stress Concentration Design Factors," Wiley, New York, 1953.
[5] H. Neuber: "Kerbspannungslehre," 2d ed., Springer, Berlin, 1958.

CHAPTER 36

TORSION

BY

J. N. GOODIER, Ph.D., Sc.D., Stanford, Calif.

36.1. INTRODUCTION: BAR OF CIRCULAR CROSS SECTION

Torsion of a uniform bar of homogeneous isotropic material means a uniform twisting deformation produced by equal and opposite twisting moments (Fig. 36.1*f*) applied to the ends of the bar. Figure 36.1 shows the simplest case—a solid circular cross section. In the general case, the section can be of any shape, with or without holes.

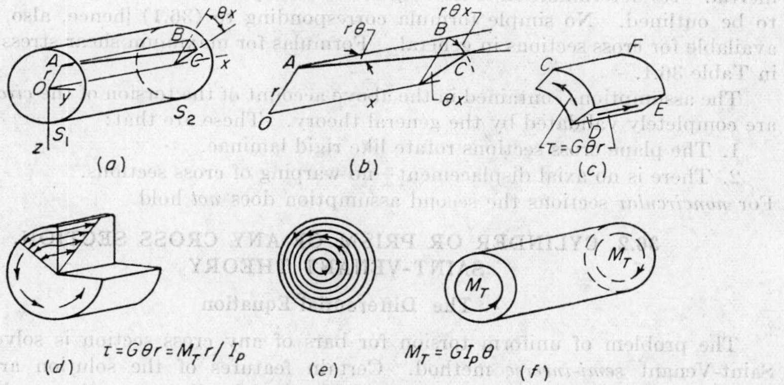

FIG. 36.1

Torsional deformation in the solid circular shaft is illustrated in Fig. 36.1*a* and *b*. In Fig. 36.1*a* the cross section, or end, S_1 does not rotate. The cross section S_2 ($S_1S_2 = x$) rotates by an angle θx about the axis Ox of the shaft, where θ is a constant, the *twist*. The twist and the rotation are positive when right-handed, looking along Ox. A fiber AB, originally parallel to the axis, at a distance r from it, becomes the inclined fiber AC. The arc BC (Fig. 36.1*b*) is $r\theta x$, and the angle CAB is $r\theta$. This angle implies shear strain. In fact, it is the complete shear strain of an element such as $CDEF$ in Fig. 36.1*c*, because a line element CD along the arc remains on the same circle. The shear stress τ (Fig. 36.1*c*) is therefore given by

$$\tau = G\theta r \tag{36.1}$$

It appears as circumferential shear stress on the cross section, for example, on CD in Fig. 36.1*c*, and also as complementary shear stress on an axial section, for example, DE in Fig. 36.1*c*. Figure 36.1*d* shows these two sections, with their shear stresses proportional to r, in accordance with (36.1). On the complete cross section (Fig. 36.1*e*), the shear stresses form a pattern of concentric circles. Their resultant is a counterclockwise twisting moment M_T, given by

$$M_T = \int_0^a \tau \cdot 2\pi r \, dr \cdot r$$

and, from (36.1), this reduces to

$$M_T = GI_p\theta \tag{36.2}$$

where a is the radius of the circular shaft, and I_p is the polar moment of inertia of the cross section about the center, $\frac{1}{2}\pi a^4$. Elimination of θ between (36.1) and (36.2) gives the formula for τ in terms of M_T:

$$\tau = M_T \cdot r/I_p \tag{36.3}$$

Equations (36.1), (36.2), (36.3) are the principal formulas used in design. If the shaft is hollow ($b < r < a$), Eq. (36.1) remains valid. Equations (36.2) and (36.3) remain valid provided I_p is taken to mean the moment of inertia of the hollow cross section, $\frac{1}{2}\pi(a^4 - b^4)$.

When the cross section is noncircular, there is a relation corresponding to (36.2),

$$M_T = GJ\theta \tag{36.4}$$

where J, the *torsional stiffness factor* of the section, differs from its polar moment of inertia. Its determination is no longer elementary, but rests on general theory now to be outlined. No simple formula corresponding to (36.1) [hence, also, (36.3)] is available for cross sections in general. Formulas for maximum shear stress are given in Table 36.1.

The assumptions contained in the above account of the torsion of the *circular* shaft are completely validated by the general theory. These are that:

1. The plane cross sections rotate like rigid laminae.
2. There is no axial displacement—no warping of cross sections.

For *noncircular* sections the second assumption does *not* hold.

36.2. CYLINDER OR PRISM OF ANY CROSS SECTION. SAINT-VENANT THEORY

The Differential Equation

The problem of uniform torsion for bars of any cross section is solved by the Saint-Venant *semi-inverse* method. Certain features of the solution are at first assumed, but the fact that it is then possible to satisfy all the equations and boundary conditions validates the assumptions. According to the uniqueness theorem, there is only one such solution for a given distribution of torsional shear stress on the ends.

It is assumed first that the cross sections rotate about Ox (Fig. 36.2), the (small) angle of rotation being θx, without distortion in their own planes. The displacement components v, w in the plane then correspond to a small rotation θx of a rigid lamina, and are

$$v = -\theta xz \qquad w = \theta xy \tag{36.5}$$

The axial, or warping, displacement u is taken as a function of y and z, and so is the same at all cross sections. It will be written $u = \theta u_1(y,z)$, and the function $u_1(y,z)$ is to be found. The six strain components follow from these displacements as

$$\begin{gathered} \epsilon_x = \frac{\partial u}{\partial x} = 0 \qquad \epsilon_y = \frac{\partial v}{\partial y} = 0 \qquad \epsilon_z = \frac{\partial w}{\partial z} = 0 \\ \gamma_{xy} = \frac{\partial v}{\partial x} + \frac{\partial u}{\partial y} = \theta\left(\frac{\partial u_1}{\partial y} - z\right) \qquad \gamma_{yz} = \frac{\partial w}{\partial y} + \frac{\partial v}{\partial z} = 0 \\ \gamma_{xz} = \frac{\partial u}{\partial z} + \frac{\partial w}{\partial x} = \theta\left(\frac{\partial u_1}{\partial z} + y\right) \end{gathered} \tag{36.6}$$

These give, by Hooke's law, the stress components

$$\sigma_x = \sigma_y = \sigma_z = \tau_{yz} = 0$$

and
$$\tau_{xy} = G\theta\left(\frac{\partial u_1}{\partial y} - z\right) \qquad \tau_{xz} = G\theta\left(\frac{\partial u_1}{\partial z} + y\right) \tag{36.7}$$

These two nonzero components are indicated on the left-hand end of the bar in Fig. 36.2.

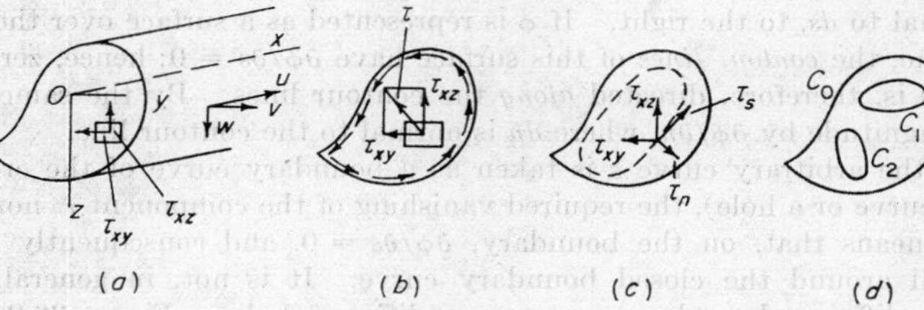

FIG. 36.2

Of the three equations of equilibrium of an element (33.35), two are identically satisfied, and the third reduces to

$$\frac{\partial \tau_{xy}}{\partial y} + \frac{\partial \tau_{xz}}{\partial z} = 0 \tag{36.8}$$

It follows that τ_{xy} and τ_{xz} can be expressed in terms of a single function $\phi(y,z)$, Prandtl's stress function, as

$$\tau_{xy} = \frac{\partial \phi}{\partial z} \qquad \tau_{xz} = -\frac{\partial \phi}{\partial y} \tag{36.9}$$

This function must satisfy a differential equation obtained by eliminating u_1 from the two equations (36.7),

$$\frac{\partial^2 \phi}{\partial y^2} + \frac{\partial^2 \phi}{\partial z^2} = -2G\theta \tag{36.10}$$

and ϕ is uniquely determined when a suitable boundary condition is added.

The Boundary Condition

The three boundary conditions in the general theory [Eqs. (33.37)] of the type

$$l\sigma_x + m\tau_{xy} + n\tau_{xz} = p_x$$

reduce, when applied to the free curved surface (for which $l = 0$), to the single condition

$$m\tau_{xy} + n\tau_{xz} = 0 \tag{36.11}$$

The two others are identically satisfied. This condition is equivalent to the statement that *the* shear stress τ (see Fig. 36.2*b*) at the boundary of the cross section must be directed along the boundary curve (Fig. 36.2*b*). A nonzero component normal to the boundary curve would imply longitudinal drag on the curved surface, violating the free boundary condition.

Figure 36.2*c* shows an element of cross section through which an arbitrary curve passes, with arc length s measured from some fixed point. The shear-stress components τ_{xy}, τ_{xz} can be replaced by a component τ_s along the curve in the direction s-increasing, and a component τ_n normally to the right. Then, y, z, s referring to the arbitrary curve, we have, from Fig. 36.2c,

$$\cos\alpha = \frac{dy}{ds} \qquad \sin\alpha = -\frac{dz}{ds} \qquad \tau_n = -\tau_{xy}\sin\alpha - \tau_{xz}\cos\alpha$$

Using (36.9), this becomes

$$\tau_n = \frac{\partial\phi}{\partial z}\frac{dz}{ds} + \frac{\partial\phi}{\partial y}\frac{dy}{ds} = \frac{\partial\phi}{\partial s} \tag{36.12}$$

In words, the derivative $\partial\phi/\partial s$, for any direction ds, gives the component of shear stress normal to ds, to the right. If ϕ is represented as a surface over the section in the y,z plane, the *contour lines* of this surface have $\partial\phi/\partial s = 0$; hence, zero τ_n. The shear stress is, therefore, directed *along* the contour lines. By the same rule, it is given in magnitude by $\partial\phi/\partial n$, where dn is normal to the contour line.

If now the arbitrary curve s is taken as a boundary curve of the cross section (the outer curve or a hole), the required vanishing of the component τ_n normal to the boundary means that, on the boundary, $\partial\phi/\partial s = 0$, and consequently ϕ must be constant all around the closed boundary curve. It is not, in general, the *same* constant on different boundary curves, e.g., different holes. Figure 36.2*d* shows an outer boundary, C_o, and two holes, C_1 and C_2. Then

$$\phi = c_o \text{ on } C_o \qquad \phi = c_1 \text{ on } C_1 \qquad \phi = c_2 \text{ on } C_2 \tag{36.13}$$

and so on if there are more holes. One of these constants can be arbitrarily assigned as zero, since addition of a constant to ϕ does not affect the stress. But the rest are then fixed by the requirement that u shall be free from discontinuity. This is discussed later.

The torsion problem for a given section with given twist θ is solved by finding the function $\phi(y,z)$ which satisfies the differential equation (36.10) and the boundary condition (36.13). The stress is then obtained from (36.9).

It is easily seen that the stress on the section can have as resultant only a pure twisting couple. There is no normal stress. The shear stress components τ_{xy} might form a resultant force in the y direction. It would be the same at all cross sections. Consequently, equilibrium of a piece between two cross sections would require bending moments. But the zero normal stress means that there are no such moments, and, therefore, the force is zero. A similar argument shows that the force resultant of the component τ_{xz} is also zero. Twisting couple (torque) remains the only possible resultant.

Before proceeding to examples and results, it is advantageous to consider, in general terms, the evaluation of *torque* M_T and *warping displacement* θu_1. In the latter we have to ensure freedom from discontinuity.

Torque in Terms of Twist

It may be seen from Fig. 36.2*a* that the torque M_T (counterclockwise on the left-hand end shown) is given by

$$M_T = \iint(\tau_{xz}y - \tau_{xy}z)\,dy\,dz \tag{36.14}$$

the integral extending over the material cross section (excluding holes).

This will be evaluated in terms of the stress function ϕ, and it will appear that the stress components τ_{xz}, τ_{xy} each contribute one-half the torque, regardless of the shape of the section.

Taking the contribution of τ_{xy} alone, it is convenient to consider first a typical strip of the cross section PQ and RS in Fig. 36.3. Using (36.9), the contribution is

$$-dy\int \tau_{xy}z\,dz = -dy\int \frac{\partial\phi}{\partial z}z\,dz = -dy\left(\int_P^Q \frac{\partial\phi}{\partial z}z\,dz + \int_R^S \frac{\partial\phi}{\partial z}z\,dz\right)$$

With an integration by parts this becomes

$$-dy\left\{[\phi z]_P^Q - \int_P^Q \phi\, dz + [\phi z]_R^S - \int_R^S \phi\, dz\right\} \tag{36.15}$$

The boundary condition on ϕ requires that $\phi_S = \phi_P$ and $\phi_R = \phi_Q$. But one of the boundary constants can be zero. Choosing the outer boundary, we put $\phi_P = 0$. Then (36.15) can be written

$$-dy\left[\phi_Q(z_Q - z_R) - \int_P^Q \phi\, dz - \int_R^S \phi\, dz\right]$$

The term $-dy\, \phi_Q(z_Q - z_R)$ represents ϕ_Q multiplied by the area of the strip QR in the hole. When the contributions of all strips are summed, the sum becomes $\phi_Q A_Q$ where A_Q is the area of the hole. The integral terms become $\iint \phi\, dy\, dz$ over the material cross section M (excluding the hole). This is the volume under the ϕ surface. If the ϕ surface is extended as a flat roof at height ϕ_Q over the hole, the integral $\iint \phi\, dy\, dz$ over the whole area inside the outer boundary represents the complete contribution of τ_{xy} to the torque.

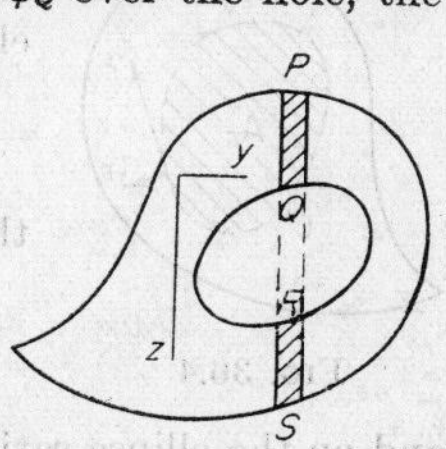

FIG. 36.3

When the contribution of τ_{xz} is worked out in the same way (for horizontal strips), the same result is found. Finally, then, the total torque is given by

$$M_T = 2\iint \phi\, dy\, dz \qquad \phi_P = 0 \tag{36.16}$$

which is twice the volume under the ϕ surface extended over the hole. When there are several holes, the ϕ surface is extended over each, at the constant height belonging to each. When the section is solid, i.e., free from holes, the torque is simply twice the volume under the original ϕ surface.

Warping Displacement. Exclusion of Dislocations. Shear-circulation Theorem

The warping displacement $u = \theta u_1(y,z)$ is important in the extension of the present theory to nonuniform torsion. In the present *uniform* torsion, the correct calculation of stress is not ensured unless the associated warping displacement is shown, or known, to be continuous. The presence of any discontinuity implies that some operation of cutting, and relative movement of the cut surfaces, has been carried out and has contributed to the stress. Such a discontinuity occurs in the *screw dislocation*, as when a tube is slit axially, and one face of the slit is shifted, axially, relative to the other.

The increment of $u_1(y,z)$ corresponding to any arc element $ds(dy,dz)$ anywhere in the material section is

$$du_1 = \frac{\partial u_1}{\partial y}\, dy + \frac{\partial u_1}{\partial z}\, dz$$

When $\partial u_1/\partial y$, $\partial u_1/\partial z$ are obtained from (36.7), we find

$$du_1 = \left(\frac{\tau_{xy}}{G\theta} + z\right) dy + \left(\frac{\tau_{xz}}{G\theta} - y\right) dz \tag{36.17}$$

The total of these increments for all arc elements of a closed curve lying entirely in the material (curve C, Fig. 36.4), represents the change in u_1 as we go round the curve back to the starting point. Continuity is ensured if this change, the line integral $\int du_1$, vanishes for all such closed curves. By (36.17) this means

$$\frac{1}{G\theta}\int_C\left(\tau_{xy}\frac{dy}{ds}+\tau_{xz}\frac{dz}{ds}\right)ds=\int_C y\,dz-\int_C z\,dy$$

The line integral on the left is the same as $-\int_C \tau_s\,ds$. On the right, the first line integral is equal to minus the area A_c enclosed by C, and the second integral with its minus sign is also equal to $-A_c$. The condition for continuous warping displacement becomes

$$\int_C \tau_s\,ds = 2G\theta A_c \tag{36.18}$$

This *shear-circulation theorem* is particularly useful in the analysis of thin-walled tubes of several cells, considered below.

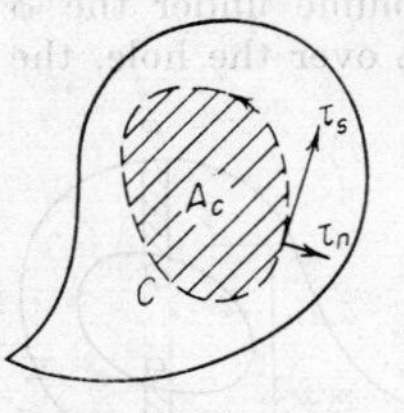

FIG. 36.4

Example 1. The Solid Elliptical Section. The problem is solved by finding the stress function ϕ which satisfies within the ellipse

$$\frac{y^2}{a^2}+\frac{z^2}{b^2}=1$$

the differential equation (36.10), namely,

$$\frac{\partial^2\phi}{\partial y^2}+\frac{\partial^2\phi}{\partial z^2}=-2G\theta$$

and on the ellipse satisfies the boundary condition $\phi = 0$. The function is[1]

$$\phi = m\left(\frac{y^2}{a^2}+\frac{z^2}{b^2}-1\right)\quad\text{where}\quad m=-G\theta\frac{a^2b^2}{a^2+b^2} \tag{36.19}$$

The stress components are, from (36.9),

$$\tau_{xy}=\frac{2m}{b^2}z\qquad \tau_{xz}=-\frac{2m}{a^2}y \tag{36.20}$$

The shear stress τ is the resultant of τ_{xy} and τ_{xz}, and so

$$\tau^2=\tau_{xy}{}^2+\tau_{xz}{}^2 \tag{36.21}$$

In the elliptical section, it follows from (36.20) that the greatest value of τ occurs on the boundary ellipse at the ends of the minor axis, that is, at the points *nearest* the axis of the torsional rotations. This, however, is not the case in all sections. This greatest value, τ_{max}, is given, for the ellipse, by

$$\tau_{max}=G\theta\frac{2a^2b}{a^2+b^2} \tag{36.22}$$

The torque is, from (36.16) and (36.19),

$$M_T=2\iint\phi\,dy\,dz=2m\iint\left(\frac{y^2}{a^2}+\frac{z^2}{b^2}-1\right)dy\,dz$$

and this yields

$$M_T=G\theta\frac{\pi a^3b^3}{a^2+b^2} \tag{36.23}$$

[1] For an account of several particular problems solved by Saint-Venant and others, see, for instance, [1] and [2].

Formulas (36.22) and (36.23) are the principal results so far. They provide $\tau_{\max}$ and M_T when θ, the twist, is given. When it is the torque M_T which is given, θ can, of course, be calculated from (36.23). A formula for $\tau_{\max}$ is obtained by eliminating θ from (36.22) and (36.23). It is

$$\tau_{\max} = \frac{2M_T}{\pi ab^2} \tag{36.24}$$

The formulas (36.22), (36.23), (36.24) are of the type

$$\tau_{\max} = k_3 G\theta \qquad \tau_{\max} = M_T/k_2 \qquad M_T = GJ\theta \tag{36.25}$$

where J, k_2, and k_3 are geometrical constants of the cross section, and $k_3 = J/k_2$. Corresponding constants exist for cross sections of other shapes. A compilation,[1] for a few of the common shapes, is given in Tables 36.1 and 36.2.

Table 36.1. Torsion Constants for Simple Sections
See Eqs. (36.25)

Section	J	k_2
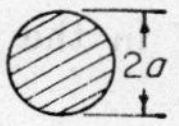	$\frac{1}{2}\pi a^4$	$\frac{1}{2}\pi a^3$
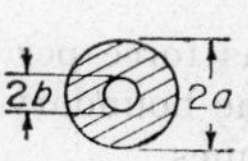	$\frac{1}{2}\pi(a^4 - b^4)$	$\frac{1}{2}\pi \frac{a^4 - b^4}{a}$
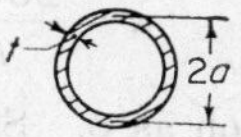	$2\pi a^3 t$	$2\pi a^2 t$
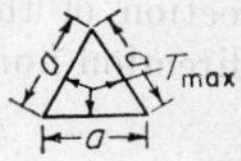	$\frac{\sqrt{3}}{80} a^4$	$\frac{1}{20}a^3$
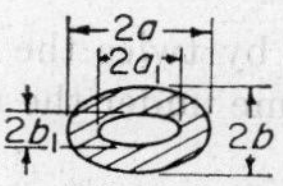 $q = a_1/a = b_1/b$ For solid ellipse put $q = 0$	$\frac{\pi a^3 b^3}{a^2 + b^2}(1 - q^4)$	$\frac{1}{2}\pi ab^2(1 - q^4)$

Table 36.2. Torsion Constants for Rectangular Sections ($b \times a$)
See Eqs. (36.25)

b/a	1.0	1.2	1.5	2.0	2.5	3	4	5	10	∞
J/a^3b	0.1406	0.166	0.196	0.229	0.249	0.263	0.281	0.291	0.312	0.333
k_2/a^2b	0.208	0.219	0.231	0.246	0.258	0.267	0.282	0.291	0.312	0.333

[1] Other shapes are included in the compilation in [8].

Formulas such as those given above for the elliptical section need to be validated by a demonstration that the associated warping displacement $u = \theta u_1$ is free from discontinuity. This displacement is easily found directly, for the special case of the elliptical section. From (36.7), using (36.20), expressions for $\partial u_1/\partial y$, $\partial u_1/\partial z$ are found as

$$\frac{\partial u_1}{\partial y} = \left(1 + \frac{2m}{b^2 G\theta}\right) z \qquad \frac{\partial u_1}{\partial z} = -\left(1 + \frac{2m}{a^2 G\theta}\right) y$$

and from these it follows that

$$u_1 = -\frac{a^2 - b^2}{a^2 + b^2} yz + C$$

where C is an arbitrary constant, representing a rigid-body translation. This is clearly free from discontinuity.

36.3. THE MEMBRANE ANALOGY

The ϕ surface for a given cross section may be realized by making a hole, the shape of the section, in a flat horizontal plate; filling the hole with a membrane of constant tension (e.g., a soap film); and blowing up the membrane slightly by a uniform pressure from below. The membrane adheres to the rim of the hole, and so has zero elevation at this boundary. Its (upward) ordinate U in the interior satisfies the differential equation

$$\frac{\partial^2 U}{\partial y^2} + \frac{\partial^2 U}{\partial z^2} = -\frac{q}{S} \tag{36.26}$$

q being the pressure, and S the membrane tension as force per unit length. The boundary-value problem for U is thus analogous to the boundary-value problem for ϕ. In fact, if (36.26) is multiplied by $2G\theta S/q$, it becomes

$$\left(\frac{\partial^2}{\partial y^2} + \frac{\partial^2}{\partial z^2}\right)(\kappa U) = -2G\theta \qquad \text{where} \qquad \kappa = \frac{2G\theta S}{q} \tag{36.27}$$

showing that κU satisfies the same differential equation as ϕ. Since U satisfies the condition $U = 0$ on the boundary, we have $\phi = \kappa U$ throughout. Consequently, the contour lines of the membrane surface show the direction of the shear stress (see p. 36–4). Its magnitude, given by the slope in the direction normal to the contour line, is

$$\tau_s = \kappa \frac{\partial U}{\partial n} \tag{36.28}$$

The torque M_T is given, for a section without holes, by twice the volume under the ϕ surface [see Eq. (36.16)], and this is twice the volume under the membrane surface multiplied by the factor κ.

When the section has a hole, the ϕ surface is continued (for evaluation of torque) as a flat roof over the hole. Correspondingly, the membrane will be attached to a flat horizontal roof plate having the shape and position of the hole. It is not yet evident at what height this roof plate must be placed. This is decided by the condition that the torsional warping displacement u must be continuous, expressed as the shear-circulation theorem (36.18). Applying (36.18) to a contour line and using (36.28) gives

$$\kappa \int_C \frac{\partial U}{\partial n} ds = 2G\theta A_c \tag{36.29}$$

and, replacing κ by its value from (36.27), this becomes

$$S \int_C \frac{\partial U}{\partial n} ds = qA_c \tag{36.30}$$

But this is the equation of equilibrium of vertical forces acting on the membrane, above the contour line, together with the roof plate (Fig. 36.5), provided that the gas pressure q from below acts on the roof plate, and provided there are no other forces (e.g., weight of plate). The roof plate must, therefore, be weightless (counterbalanced), and it must be guided, without friction, so as to remain horizontal, in order to keep ϕ constant around the hole. The pressure q then raises it automatically to the correct height, and with it the membrane to the correct form. If there are several holes, there is such a plate for each hole. The volume, in the torque evaluation, includes the volume under the roof plates.

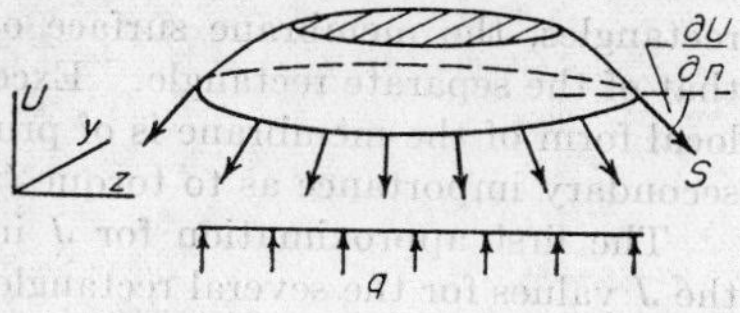

FIG. 36.5

Use of the soap-film or other membrane for actual measurements requires great care [3] in the elimination of numerous sources of error, and sometimes numerical solution of finite-difference equations is preferred.[1] The membrane analogy has an important significance in allowing easy visualization of the ϕ surface. This leads to useful approximate forms of solution, illustrated in the following section.

36.4. APPROXIMATE FORMULAS FOR THIN-WALLED OPEN SECTIONS

a. Thin-walled Open Sections, Uniform Wall Thickness

Formulas for thin-walled open sections such as those appearing in Fig. 36.6 are based on formulas for the thin rectangle (Fig. 36.6*a*). It is easily realized that the membrane surface will have the same shape—a simple arched form—over the whole thin rectangle except near the ends. A typical (parabolic) section is indicated at AA in Fig. 36.6*a*. The slope $\partial U/\partial z$ would then be zero or negligible, and with it $\partial^2 U/\partial z^2$, except near the ends. The equation for the ϕ surface would, therefore, become simply

$$\frac{d^2\phi}{dy^2} = -2G\theta$$

The solution making $\phi = 0$ on the long edges ($y = \pm a/2$) is

$$\phi = -G\theta\left(y^2 - \frac{a^2}{4}\right) \tag{36.31}$$

showing that the section of the surface (AA in Fig. 36.6*a*) is in fact parabolic. The stress formulas are then, from (36.9),

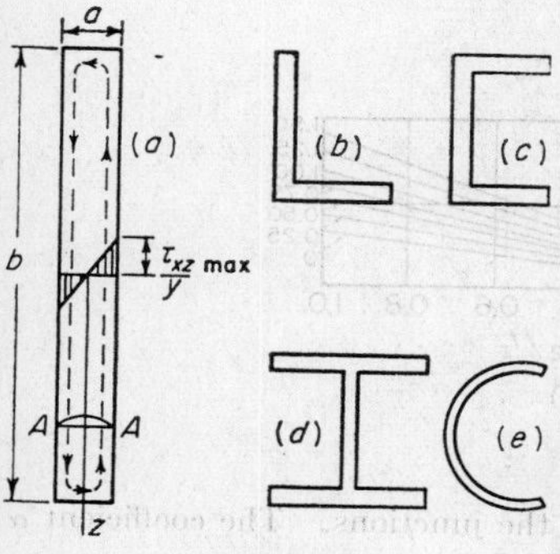

FIG. 36.6

$$\tau_{xy} = 0 \qquad \tau_{xz} = 2G\theta y \qquad \tau_{\max} = G\theta a \tag{36.32}$$

This shear stress, in Fig. 36.6*a*, is vertical. Near the ends, the contour lines (broken line in Fig. 36.6*a*), showing the direction of the shear, turn round as indicated, and here, of course, (36.31) and (36.32) do not apply. The exact theory for the rectangle is required (see [1], p. 275).

The simple parabolic approximate form of the ϕ surface will give a good approximation to the volume, since it differs from the true surface only in small end zones. The approximate torque formula derived from (36.31) is

$$M_T = 2\iint \phi \, dx \, dy = \tfrac{1}{3}a^3 b G\theta \tag{36.33}$$

Here then the factor J in $M_T = GJ\theta$ is $\frac{1}{3}a^3b$. Comparison with Table 36.2 shows that this is correct for $b/a \to \infty$; and, for $b/a > 10$, the error does not exceed 6.5%.

[1] See [1], pp. 289, 301, and appendix.

Combination of (36.32) and (36.33) yields

$$\tau_{\max} = \frac{3M_T}{a^2b} \tag{36.34}$$

For sections as in Fig. 36.6*b* to *d*, regarded as composed of two or three rectangles, the membrane surface over each rectangular part will differ little from that of the separate rectangle. Exception must be made for the junctions, where the local form of the membrane is of primary importance as to stress (slope), but only of secondary importance as to torque (volume).

The first approximation for J in $M_T = GJ\theta$ is accordingly obtained by adding the J values for the several rectangles composing the section. In actual sections, the corners at the junctions will be rounded fillets. Corrected values of J, taking account of the junction with its fillet radius, are expressed for the I section of Fig. 36.7 by [4]

$$J = 2J_1 + \tfrac{1}{3}ht_2^3 + 2\alpha D^4 \tag{36.35}$$

Here $2J_1$ is the contribution from the two flanges, taken for greatest accuracy from Table 36.2. The term $\tfrac{1}{3}ht_2^3$ is the contribution from the web in the approximate

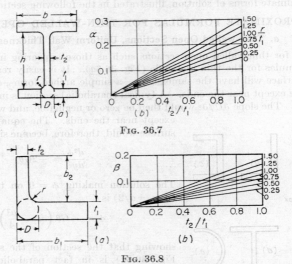

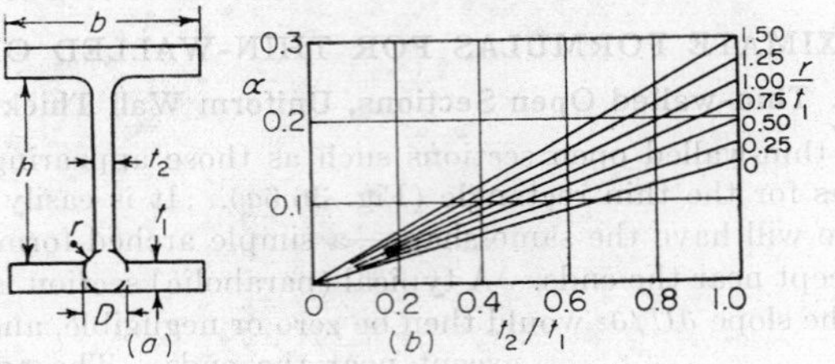

FIG. 36.7

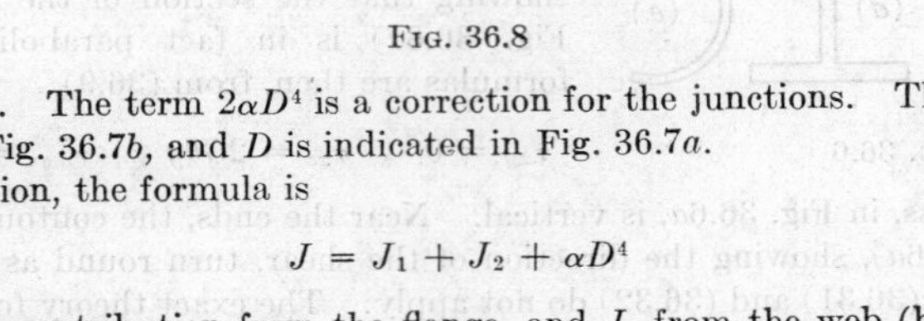

FIG. 36.8

form of (36.33). The term $2\alpha D^4$ is a correction for the junctions. The coefficient α is taken from Fig. 36.7*b*, and D is indicated in Fig. 36.7*a*.

For a T section, the formula is

$$J = J_1 + J_2 + \alpha D^4 \tag{36.36}$$

where J_1 is the contribution from the flange, and J_2 from the web (height h) (from Table 36.2).

For the angle section (Fig. 36.8*a*),

$$J = J_1 + \tfrac{1}{3}b_2t_2^3 + \beta D^4 \tag{36.37}$$

where J_1 is the contribution from the (thicker) horizontal leg as a rectangle ($b_1 \times t_1$), and β is taken from Fig. 36.8*b*.

Estimation of maximum shear stress at fillets is treated in Art. 36.5*b*. It is usually greater than the shear stress in the rectangles away from the fillets.

For a curved thin-walled open section as in Fig. 36.6*e*, of uniform wall thickness, the membrane section will differ little from what it would be if the curved section were straightened out. The rectangle formulas (36.32), (36.33), and (36.34) can be applied, with b denoting the curved length of the section, and a the wall thickness. Better approximations for the stress can be found from Eqs. (36.42) and (36.43) below. A. A. Griffith [7] has found from soap-film measurements that curvature of the section diminishes the torque from that of the straightened section by less than 5%, as long as the least radius of curvature of the boundary is not less than the wall thickness.

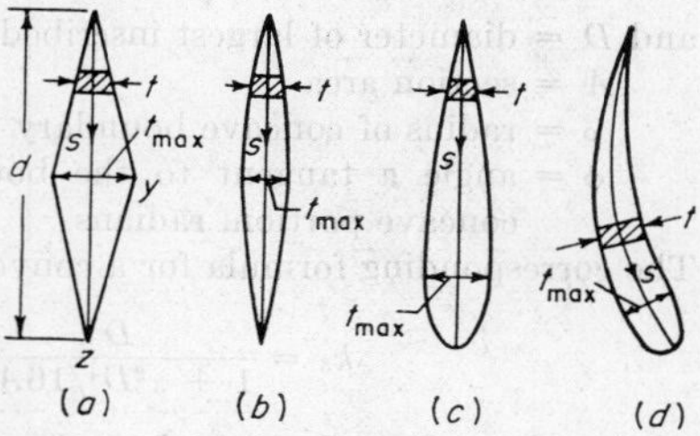

Fig. 36.9

b. Thin-walled Open Sections, Nonuniform Wall Thickness

For sufficiently thin-walled sections, good approximate formulas are obtained by regarding the membrane as parabolic across the thickness t (Fig. 36.9) as in a thin rectangle of the same thickness. Thus the membrane volume over an element $t\,ds$ as in Fig. 36.9*a* to *d* is the same as the membrane volume over the length ds of a rectangle of thickness t, and this will yield a contribution to the torque given by (36.33) when a is replaced by t, and b by ds. The total torque is

$$M_T = \tfrac{1}{3} G\theta \int t^3\, ds \tag{36.38}$$

the integral covering the entire length of the middle line along which s is measured. When the middle line is straight (Fig. 36.9*a* to *c*), this formula becomes

$$M_T = G \cdot 4I_z \cdot \theta \tag{36.39}$$

I_z being the moment of inertia of the section about Oz.

The greatest stress may, accordingly, be estimated from

$$\tau_{\max} = G\theta t_{\max} \tag{36.40}$$

where $t_{\max}$ is the maximum thickness.

The accuracy of Eqs. (36.39) and (36.40) for the diamond section of Fig. 36.9*a* and the symmetrical section of Fig. 36.9*b*, the bounding curves being parabolic, has been examined by comparison with relaxation calculations [6]. For the diamond section, (36.39) gives M_T too high by less than 6%, and (36.40) gives $\tau_{\max}$ too high by less than 7% provided $d/t_{\max} > 7$. For the parabolic section, the corresponding percentages are 3 and 4, respectively.

The greatest stress will be found at the ends of the maximum thickness in sections such as in Fig. 36.7*b* to *d*. In Fig. 36.7*a*, however, these ends are at external corners, and the stress there will be zero. The greatest stress is found some little distance from the corners, to either side. In general it occurs near one of the points of contact of the largest inscribed circle, and, of these, at that one where the boundary curvature is least convex or most concave [7]. Special consideration is needed if the boundary curvature is reentrant and sharp, since localized stress concentration can then occur. The case of a fillet at the vertex of an angle is discussed in Art. 36.5*b*.

In (36.38) the torsional stiffness constant is

$$J = \tfrac{1}{3} \int t^3\, ds$$

A. A. Griffith [7] gives a more accurate formula,

$$J = \frac{4H}{1 + 16H/Ad^2} \qquad \text{where} \qquad H = \tfrac{1}{12} \int t^3\, ds \tag{36.41}$$

and A is the section area. When the middle line of the section is straight (Fig. 36.9a to c), H is the same as I_2. He gives a more accurate formula for the greatest stress at a concave, or reentrant, point in the form

$$\tau_{\max} = k_3 G\theta \quad \text{where} \quad k_3 = \frac{D}{1 + \pi^2 D^4/16A^2} \left\{ 1 + \left[0.118 \ln \left(1 + \frac{D}{2} \right) + 0.238 \frac{D}{2\rho} \right] \operatorname{Tanh} \frac{2\phi}{\pi} \right\} \tag{36.42}$$

and D = diameter of largest inscribed circle
A = section area
ρ = radius of concave boundary curvature at the point (positive)
ϕ = angle a tangent to the boundary rotates through in rolling around the concave portion, radians

The corresponding formula for a convex point is

$$k_3 = \frac{D}{1 + \pi^2 D^4/16A^2} \left[1 + 0.15 \left(\frac{\pi^2 D^4}{16A^2} - \frac{D}{2\rho} \right) \right] \tag{36.43}$$

ρ being the radius of convex boundary curvature (positive).

Other approximation methods are treated in [1] and [8].

c. Thin-walled Closed Sections

The upper part of Fig. 36.10 represents the cross section of a tube of some form. The wall thickness t is small, but need not be uniform. The bounding curves are smooth, in the sense that their smallest radius of curvature is large compared with t. This excludes the local stress concentration which would exist at a sharp bend (see p. 36–15).

In Sec. 36.3, it was shown that the ϕ surface should be extended as a flat horizontal roof over the hole. Such a roof is indicated in Fig. 36.10 at a height h. The membrane connects this roof to the periphery at zero level. As an approximation, the thickness section (two such are shown in the lower part of Fig. 36.10) may be taken as a straight line.[1] This uniform slope across the thickness means that the shear stress τ is uniform across the wall thickness. If the height h refers to the ϕ surface (rather than the membrane surface), $\tau = h/t$, and $\tau t = h$ = const. This constant τt is often called the *shear flow*.

The volume under the ϕ surface with the roof is Ah or $A\tau t$, where A is the area enclosed by the middle (broken) curve. The torque is, therefore,

$$M_T = 2A\tau t$$

This provides, as a formula for τ when M_T is given,

$$\tau = M_T/2At \tag{36.44}$$

and $\tau_{\max}$ is found where t is least.

The shear-circulation theorem (36.18) serves the double purpose of ensuring freedom from displacement discontinuity and providing immediately a formula for the twist θ. Substitution of (36.44) gives

$$\theta = \frac{M_T}{4GA^2} \int \frac{ds}{t} \tag{36.45}$$

A_c in (36.18) being identified with A. The integral is extended all around the middle curve, and is clearly a geometrical constant of the section. For a uniform wall

[1] The line is slightly curved for two reasons: (1) because of the gas pressure q from below, (2) because it would be curved if the height h were maintained and the pressure removed. The curvature can be investigated and its neglect justified for sufficiently thin walls.

thickness, with l as the peripheral length of the middle curve, the formula becomes

$$\theta = \frac{M_T l}{4GA^2 t} \tag{36.46}$$

This is known as *Bredt's second formula*,[1] the first being (36.44). In two-cell sections, as in Fig. 36.11, which have a central web along an axis of symmetry AA, the web will be idle, and the sections may be treated as simple single-cell tubes. This is apparent from the fact that the two flat roofs corresponding to the two holes will be at the same height. Consequently, the membrane joining them, over the web, will be level, and the shear stress in the web will be zero, to the present order of accuracy. The junctions, however, introduce reentrant corners with local stress concentrations (see Art. 36.5*b*).

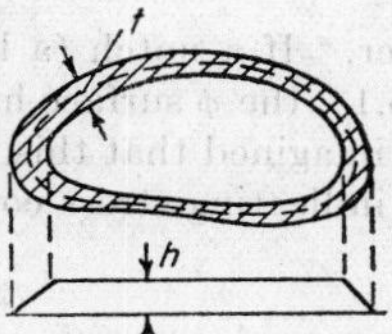

FIG. 36.10

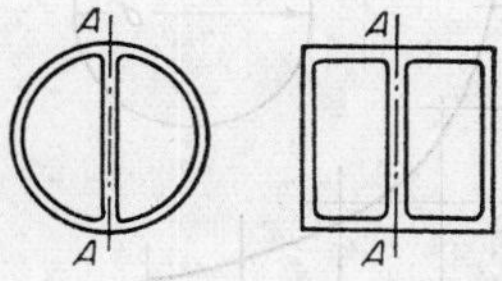

FIG. 36.11

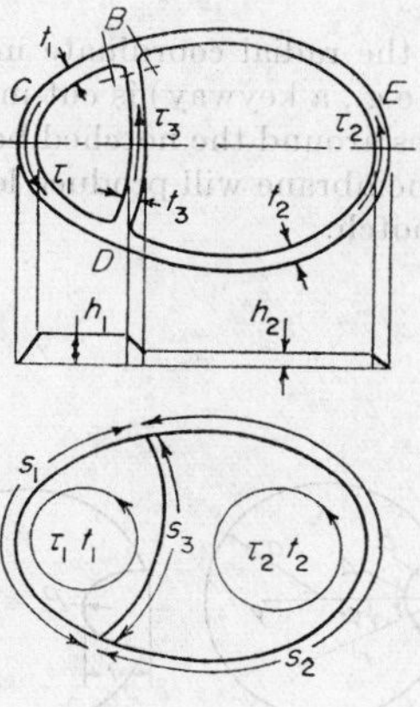

FIG. 36.12

When the symmetry is lacking (Fig. 36.12), the roofs will stand at different heights h_1, h_2. Taking each wall thickness t_1, t_2, t_3 as uniform for simplicity in this illustrative example, the shear flows $\tau_1 t_1$, $\tau_2 t_2$, $\tau_3 t_3$ are constants in the respective parts BCD, DEB, DB.

If the junction at B is cut out by three longitudinal cuts as shown, these shear flows represent *longitudinal* forces (per unit length) on their respective cuts, and longitudinal equilibrium of the junction requires that

$$\tau_3 t_3 = \tau_1 t_1 - \tau_2 t_2$$

This condition is satisfied if, as in Fig. 36.12*b*, the shear flow $\tau_1 t_1$ is regarded as circulating around the left cell, thus acting in the web as well, and if the shear flow $\tau_2 t_2$ similarly circulates in the right cell. There are now two unknowns, τ_1 and τ_2. They may be found by applying the shear-circulation theorem to each cell. With A_1 for the area enclosed by the circuit of the left cell, A_2 for the right cell, the theorem gives

$$\tau_1 s_1 + \frac{\tau_1 t_1 - \tau_2 t_2}{t_3} s_3 = 2G\theta A_1 \qquad \tau_2 s_2 + \frac{\tau_2 t_2 - \tau_1 t_1}{t_3} s_3 = 2G\theta A_2 \tag{36.47}$$

and these determine τ_1, τ_2 when θ is given. The torque is found as

$$M_T = 2(A_1 \tau_1 t_1 + A_2 \tau_2 t_2) \tag{36.48}$$

and this then becomes the torque-twist relation $M_T = GJ\theta$.

[1] Corresponding formulas for a tube with a circular, instead of straight, axis are derived in [16].

This procedure is easily extended to n cells. There are n circulating shear flows to find, and n independent equations, like (36.47), available from the shear-circulation theorem. The results, in terms of θ, are used in the torque formula corresponding to (36.48), having n terms in the parentheses, to obtain the torque-twist relation. A computational procedure for such solutions is given in [5].

36.5. STRESS CONCENTRATION AT NOTCHES, CORNERS, AND KEYWAYS

a. Shaft with Notch or Keyway

The ϕ surface for a solid circular shaft of radius a is a simple paraboloid of revolution given by

$$\phi = -\tfrac{1}{2}G\theta(r^2 - a^2)$$

r being the radial coordinate measured from the center. If a notch (a longitudinal groove, e.g., a keyway) is cut in the shaft, as in Fig. 36.13, the ϕ surface has now zero ordinates around the notched boundary. It is readily imagined that this indentation of the membrane will produce locally increased slopes, indicating stress concentration at the notch.

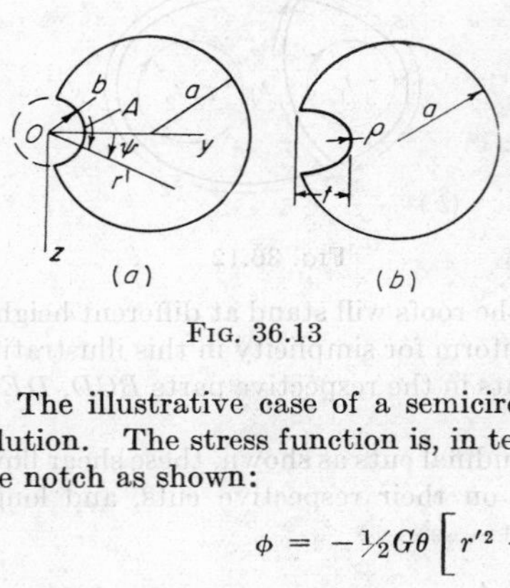

FIG. 36.13

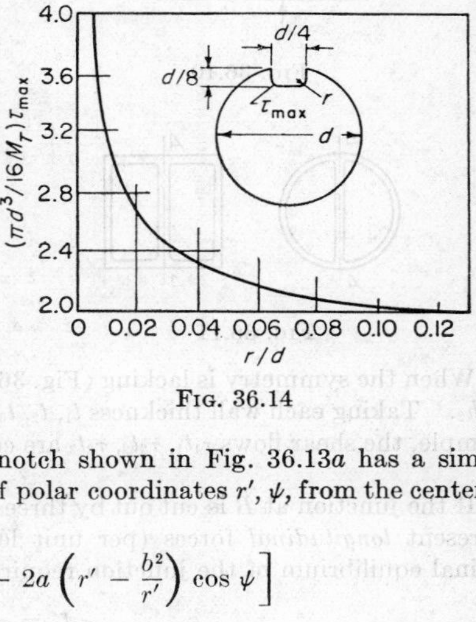

FIG. 36.14

The illustrative case of a semicircular notch shown in Fig. 36.13a has a simple solution. The stress function is, in terms of polar coordinates r', ψ, from the center of the notch as shown:

$$\phi = -\tfrac{1}{2}G\theta\left[r'^2 - b^2 - 2a\left(r' - \frac{b^2}{r'}\right)\cos\psi\right]$$

The maximum shear stress occurs at A, and is

$$\tau_{\max} = G\theta(2a - b) \tag{36.49}$$

Without the notch, $\tau_{\max} = G\theta a$. With a very small notch, (36.49) gives nearly twice this value. Such markedly increased values occur only locally, at and near the bottom of notch A. The torsional stiffness J is little affected by a small notch.

For a small semielliptical notch (Fig. 36.13b), Neuber's formula, derived in [31], is

$$\frac{\tau_{\max}}{\tau_n} = 1 + \sqrt{\frac{t}{\rho}} \tag{36.50}$$

where τ_n, the nominal stress, is the stress in the absence of the notch, at the point corresponding to the bottom of the notch.

For keyways (Fig. 36.14), τ_{max} in the rounded corners may be found from the curve in Fig. 36.14, calculated, and verified by photoelastic tests, by M. M. Leven [15].

An analytical method for multiple notches is given in [10]. For given twist θ, the effect of several notches of the same size and shape is generally less severe than that of a single notch.

***b.* Fillets at Corners of Thin-walled Angle and Channel Sections and Rectangular Tubes**

Approximate formulas for the stress in thin-walled sections have been given in Arts. 36.4*a*, *b*, and *c*, sharp or fairly sharp concave boundary corners being set aside as exceptions to be examined here.

Several (approximate) analytical and experimental methods have been applied to the problem of corner fillets, and a comparison, due originally to Lyse and Johnston, is given in [11]. The most dependable results appear to be those of Huth [11], obtained by relaxation calculations. They are reproduced here in Fig. 36.15. The ordinate gives the ratio of the maximum stress, at the corner, to the edge stress prevailing along the straight sections away from the corners. The results hold for angles with unequal legs, and for rectangular tubes which are not square. They will not hold if the legs or sides are not of equal thickness, or nearly so.

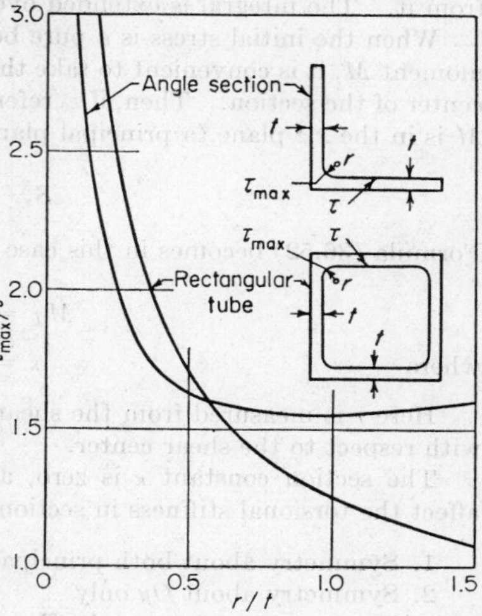

FIG. 36.15

36.6. EFFECT OF INITIAL TENSILE OR BENDING STRESS

In the ordinary torsion of uniform bars as so far considered, the torque M_T is given in terms of the twist θ by $M_T = GJ\theta$, and the torsional stiffness factor J has been given for a number of section shapes. When the bar carries an initial uniform tensile stress σ, this relation is modified, for torsion about the centroidal axis, to

$$M_T = (GJ + \sigma I_p)\theta \tag{36.51}$$

where I_p is the polar moment of inertia of the section about the centroidal axis. The term $\sigma I_p \theta$ represents the torque required to twist a "bar" consisting simply of straight strings connecting two end plates, with tensile stress σ imposed. The torque is applied to the end plates, and resisted by the (inclined) tensions in the strings. In the original elementary theory (cited in [12]), this torque was simply added, for the metal bar, to the ordinary torque $GJ\theta$ to give the formula (36.51). Subsequent analysis and testing, described or cited in [13], have validated the formula. It shows that the torsional stiffness is increased by tension, and decreased by compression. The effect is important in thin-walled open sections on account of their low values of J. The torsional stiffness—the term in parentheses in Eq. (36.51)—can be reduced to zero by compression, and a type of torsional buckling ensues.

Effects of the same nature occur for initial axial stress of other kinds, as bending, or thermal, or residual stress. The general formula, of which (36.51) is a special case, is given in [12] as

$$M_T = \left(GJ + \int S_{zz} r^2 \, dA\right)\theta \tag{36.52}$$

where S_{xx} is the initial axial stress, having any distribution over the section, but not varying in the axial direction x. The axis of torsional rotation is arbitrarily chosen, parallel to the generators, and r is the radial coordinate in the cross section measured from it. The integral is extended over the material cross section.

When the initial stress is a pure bending stress corresponding to uniform bending moment M, it is convenient to take the torsional axis Ox as the axis through the shear center of the section. Then, if z_0 refers to the centroid (Fig. 36.16b), and the moment M is in the x,z plane (a principal plane), the bending (initial) stress is

$$S_{xx} = \frac{M}{I}(z - z_o) \tag{36.53}$$

Formula (36.52) becomes in this case

$$M_T = \left(GJ + \frac{\kappa M}{I}\right)\theta \tag{36.54}$$

where

$$\kappa = \int zr^2\,dA - Ai^2z_o \tag{36.55}$$

Here r is measured from the shear center, and Ai^2 is the polar moment of inertia with respect to the shear center.

The section constant κ is zero, and, consequently, the bending stress does not affect the torsional stiffness in sections having the following types of symmetry:

1. Symmetry about both principal axes
2. Symmetry about Oy only
3. Point symmetry: e.g., the **Z** section

Type 2 is illustrated in Fig. 36.16a, by an angle section.

In Fig. 36.16b and c the bending moment M acts in the plane of symmetry. In Fig. 36.16b the constant κ is, for a thin-walled 90° angle, $tb^4/6\sqrt{2}$. This being positive, the torsional stiffness is increased. For $b/t = 10$, $G = 4 \times 10^6$ lb/in.², and M such as to produce an extreme fiber stress of 2×10^4 lb/in.², the increase is 25%. In Fig. 36.16c, the constant κ has the same magnitude but is negative. The torsional stiffness is accordingly decreased, by 25%. Validating measurements of the bending effect are given in [13].

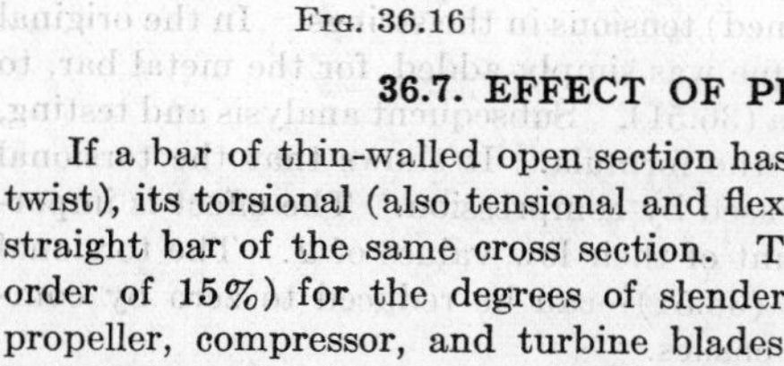

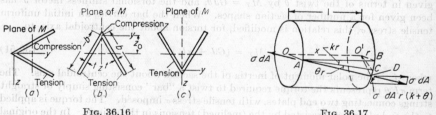

FIG. 36.16

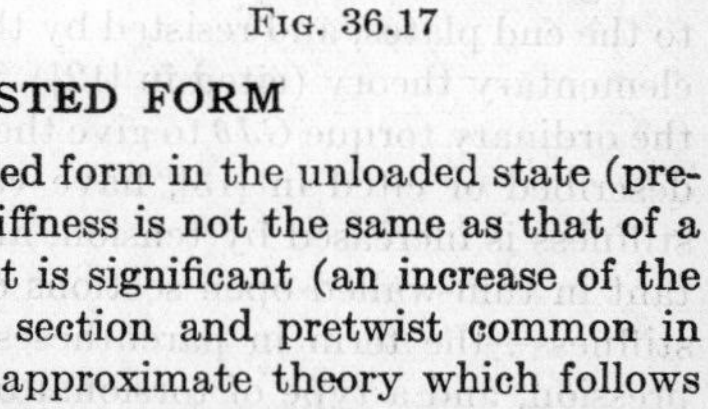

FIG. 36.17

36.7. EFFECT OF PRETWISTED FORM

If a bar of thin-walled open section has a twisted form in the unloaded state (pretwist), its torsional (also tensional and flexural) stiffness is not the same as that of a straight bar of the same cross section. The effect is significant (an increase of the order of 15%) for the degrees of slenderness of section and pretwist common in propeller, compressor, and turbine blades. The approximate theory which follows had its inception in the work of Chen Chu [14], which includes substantiating tests.

A pretwisted bar may be imagined as derived, in a purely geometrical sense, from a straight bar by a twist k, each cross section rotating in its own plane by an angle kx without deformation. Then, in Fig. 36.17, a line AB at a distance r from Ox becomes AC. AC is part of a "fiber" of the pretwisted, unstressed bar. The bar is now loaded by a torque causing an elastic twist θ, and the fiber AC becomes AD. If there is no axial displacement, AD is longer than AC, the extensional strain being

$$\frac{AD - AC}{AC} = \frac{\sqrt{1 + r^2(k + \theta)^2} - \sqrt{1 + r^2k^2}}{\sqrt{1 + r^2k^2}}$$

and if k^2r^2 is neglected in comparison with unity, and only the first power of θ retained, this strain becomes

$$\epsilon_1 = r^2k\theta$$

In general, there will be axial displacement. But it will be the same at corresponding points of all cross sections, and will not alter the strain along AD. This extensional strain by itself would give rise to a nonzero axial force on the cross section. To annul this, each fiber, such as AD, is given a strain β, the same for all, in addition to ϵ_1. The total strain $(\epsilon_1 + \beta)$ corresponds to stress σ *along the fiber*, and, lateral stress being supposed negligible, this is given by

$$\sigma = E(r^2k\theta + \beta) \tag{36.56}$$

These fiber tensions form an axial resultant force (replacing the cosine of the angle DAB by unity)

$$N = EA\beta + EI_pk\theta \tag{36.57}$$

where A is the section area, and I_p the polar moment of inertia of the section about Ox, the axis of the twist and of the pretwist.

The torque M_T is formed by the shear stress on the cross section induced by the twist θ, and by the inclined fiber stresses σ. The shear stress is assumed identical with the Saint-Venant stress for an initially straight bar, forming a torque $GJ\theta$. The total torque is then

$$M_T = GJ\theta + \int \sigma \sin [r(k + \theta)]r\,dA \tag{36.58}$$

and, when terms not of the first degree in θ and β are discarded, this becomes

$$M_T = (GJ + EBk^2)\theta + EI_pk\beta \tag{36.59}$$

where

$$B = \int r^4\,dA \tag{36.60}$$

For pure torsion, β must be chosen so that N vanishes. Then, from (36.57),

$$\beta = -(I_p/A)k\theta$$

which implies that twisting, in the same sense as the pretwist, causes axial shortening. Then the torque is found from (36.59) as

$$M_T = GJ\left[1 + \frac{E}{GJ}\left(B - \frac{I_p{}^2}{A}\right)k^2\right]\theta \tag{36.61}$$

The torsional stiffness factor is different from that of the straight bar, by the factor in square brackets. It can be shown by a Schwarzian inequality that $B - I_p{}^2/A$ is always positive. The torsional stiffness is, therefore, *increased* by the pretwist.

For a thin rectangular section of (small) thickness t and width b, the section constants are

$$A = bt \qquad I_p = \tfrac{1}{12}b^3t \qquad B = \tfrac{1}{80}b^5t \qquad J = \tfrac{1}{3}bt^3$$

Writing J_k for the torsional stiffness factor of the pretwisted bar, in $M_T = GJ_k\theta$, (36.61) gives

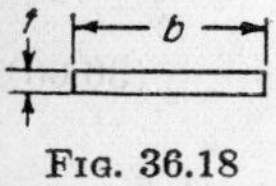

FIG. 36.18

$$\frac{J_k}{J} = 1 + \frac{E}{GJ}\left(B - \frac{I_p{}^2}{A}\right)k^2 = 1 + \frac{1 + \nu}{30}\frac{b^4k^2}{t^2} \tag{36.62}$$

For example, if $b/t = 20$, and $kb = 1/10$ radian (the angle of pretwist rotation for an axial advance b), the value of J_k/J is 1.17. Formula (36.62) is closely verified by the tests of Chen Chu [14].

The corresponding formula for a slender elliptical section of semiaxes a, b (b/a small) is

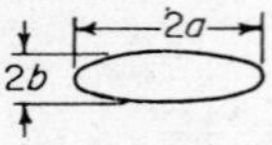

$$\frac{J_k}{J} = 1 + \tfrac{1}{8}(1 + \nu)\frac{a^4k^2}{b^2} \tag{36.63}$$

FIG. 36.19

For a double-wedge or diamond section of semiaxes a, b, the formula is

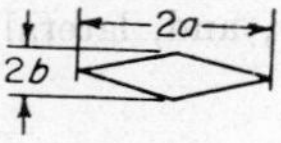

$$\frac{J_k}{J} = 1 + \tfrac{7}{60}(1 + \nu)\frac{a^4k^2}{b^2} \tag{36.64}$$

FIG. 36.20

It may be verified from these formulas that the effect of pretwist is much more significant in slender than in thick sections.

36.8. RESTRAINED WARPING. NONUNIFORM TORSION

a. Nature of the Problem

In the uniform torsion of a uniform bar (Saint-Venant theory, Sec. 36.2), the originally plane cross sections become warped, the warping displacement u being nonzero (unless the section is circular) and such that the cross section becomes a (slightly) curved surface. This warping is the same for all cross sections and is proportional to the twist θ, or, equivalently, to the torque M_T. The Saint-Venant solution prevails only when the warping is free to occur. It needs emendation if the warping is completely or partially prevented, as by an end fitting, or by variation of torque along the length of the bar. If, for instance, a torsional couple is applied at some station along the length, the torque M_T in the bar changes abruptly at that station. The warping displacement of the Saint-Venant theory would show a corresponding abrupt change. In consequence, further stress and strain is induced to preserve the continuity of the material.

Existing theory of such effects deals mainly with thin-walled open and closed sections, and only open sections are considered here. For closed sections, [25] and [26] should be consulted.

b. The Warping Displacement of Uniform Torsion, Thin-walled Open Section

The theory is based on a general, approximate formula for the Saint-Venant warping displacement $u = \theta u_1$ for thin-walled open sections, to be derived here. The increment of u_1 corresponding to any shift ds with components dy, dz is given by (36.17) as

$$du_1 = \frac{1}{G\theta}(\tau_{xy}\,dy + \tau_{xz}\,dz) + (z\,dy - y\,dz) \tag{36.65}$$

The expression in the first pair of parentheses is equal to $-\tau_s\,ds$ (Fig. 36.21). With polar coordinates r, ψ, such that $y = r\cos\psi$, $z = r\sin\psi$, the term in the second pair of parentheses is equal to $-r^2\,d\psi$, and Eq. (36.65) becomes

$$du_1 = -\frac{1}{G\theta}\tau_s\,ds - r^2\,d\psi \tag{36.66}$$

For a thin-walled open section, τ_s along the middle curve is zero, in the approximation adopted in Arts. 36.4*a* and *b*. Along this curve then,

$$du_1 = -r^2\, d\psi \tag{36.67}$$

The expression $\frac{1}{2}r^2\, d\psi$ represents the area swept by the radius r as the middle line (solid curve) is traversed from arc length s to $(s + ds)$. The arc length s is measured from an arbitrarily chosen point. At this point, there is some value of u_1, say u_0. The area $\frac{1}{2}r^2\, d\psi$ is also given by $\frac{1}{2}r_s\, ds$, where r_s (Fig. 36.21) is the perpendicular from O to the tangent from ds. Consequently, (36.67) yields, for points on the middle line,

$$u_1 - u_0 = -\int_0^s r_s\, ds \tag{36.68}$$

Having started from $s = 0$ and traversed the middle line to s, we can move along the normal to the middle line. In this direction, τ_s (in the approximation) is again zero, and (36.67) is again valid. But now $r^2\, d\psi$ is replaceable by $r_n\, dn$, where dn is along the normal, positive when $d\psi$ is positive, and r_n is the perpendicular from O to the normal line (Fig. 36.21). Taking account of this, the value of $(u_1 - u_0)$, when u_1 is evaluated anywhere in the section, is given by

$$u_1 - u_0 = -\int_0^s r_s\, ds - \int_0^n r_n\, dn \tag{36.69}$$

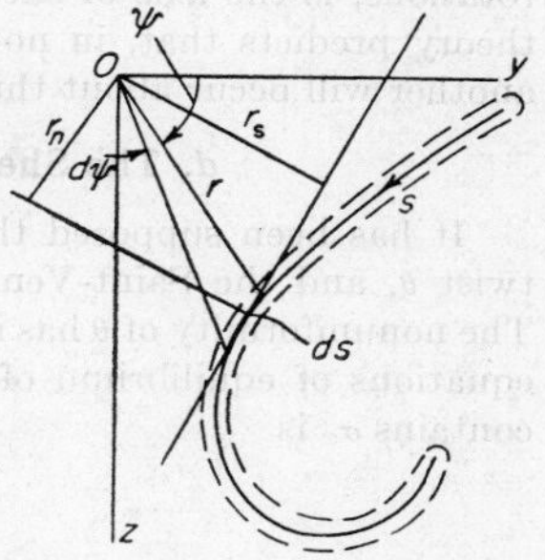

FIG. 36.21

The right-hand side, but for the negative signs, is the area swept by the radius r of a point starting at the reference point $s = 0$ on the middle line, traversing the middle line to a point s, and then moving along the normal direction n to the point of evaluation for u_1. When r_s and r_n are, on the whole, of the same order of magnitude, the second integral in (36.69) will be relatively small because n has a small range (the wall thickness). It will be convenient to denote the right-hand side of (36.69) by u_{10}, so that

$$u_1 - u_0 = u_{10} \tag{36.70}$$

where, for a sufficiently thin-walled section,

$$u_{10} = -\int_0^s r_s\, ds \tag{36.71}$$

c. Nonuniform Torsion. The Axial Stress

The twist θ is now allowed to vary from section to section. The warping displacement $u = \theta u_1$ varies correspondingly, implying axial strain

$$\epsilon_x = \frac{\partial u}{\partial x} = u_1 \frac{\partial \theta}{\partial x}$$

In this approximate theory[1] it is assumed that this is in fact the axial strain, and, further, that the only significant normal stress component is σ_x. Then

$$\sigma_x = E \frac{\partial \theta}{\partial x} u_1 \tag{36.72}$$

This axial stress must, since the loading is purely torsional, form zero-force resultant

[1] A generalization to arbitrary thin-walled open sections, by H. Wagner, of the special case of the I section solved earlier by S. Timoshenko [29]; see [30].

on the cross section, and this requires

$$\int u_1 \, dA = 0 \tag{36.73}$$

We suppose that the reference point $s = 0$ has been (arbitrarily) fixed. Then u_{10} may be calculated [from the right-hand side of (36.69), or by (36.71)]. Since $u_1 = u_0 + u_{10}$, the condition (36.73) is satisfied by choosing u_0 so that

$$u_0 A = -\int u_{10} \, dA \tag{36.74}$$

With (36.73) satisfied, u_1 becomes definite. It is the warping displacement for $\theta = 1$, with rigid-body displacement adjusted to make the mean value zero.

The bending moment of σ_x is zero provided that

$$\int u_1 y \, dA = 0 \qquad \int u_1 z \, dA = 0 \tag{36.75}$$

and it can be shown that these conditions are satisfied when Ox, the axis of torsional rotations, is the axis of shear centers (a proof is given in [27]). Consequently, this theory predicts that, in nonuniform torsion, the rotation of one section relative to another will occur about this axis of shear centers.

d. The Shear Stress and the Torque-Twist Relation

It has been supposed that the slice of the bar between x and $(x + dx)$ has the twist θ, and the Saint-Venant shear stress and torque corresponding to this twist. The nonuniformity of θ has introduced axial stress σ_x (36.72). Of the three differential equations of equilibrium of a volume element in coordinates x, y, z, the one which contains σ_x is

$$\frac{\partial \sigma_x}{\partial x} + \frac{\partial \tau_{xy}}{\partial y} + \frac{\partial \tau_{xz}}{\partial z} = 0$$

Since the Saint-Venant shear stress makes the sum of the second and third terms vanish, this equation can be satisfied only if there is additional shear stress of the types τ_{xy}, τ_{xz}, and it will appear that the additional stress contributes to the torsional moment.

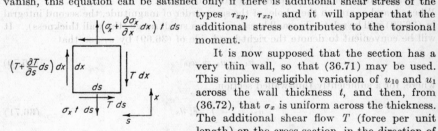

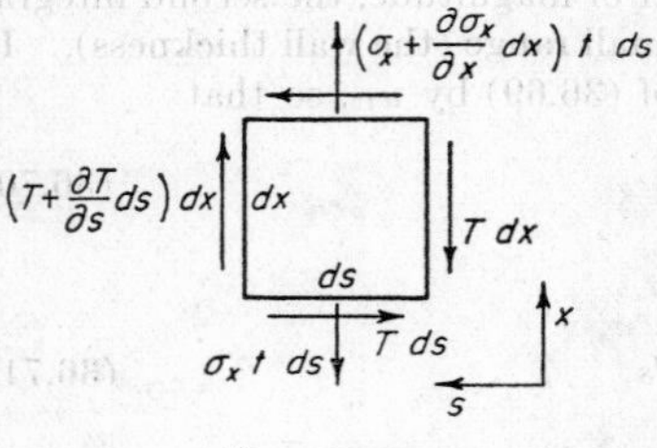

FIG. 36.22

It is now supposed that the section has a very thin wall, so that (36.71) may be used. This implies negligible variation of u_{10} and u_1 across the wall thickness t, and then, from (36.72), that σ_x is uniform across the thickness. The additional shear flow T (force per unit length) on the cross section, in the direction of the middle curve, is determined by the equilibrium of the element in Fig. 36.22 in the vertical (axial) direction. The element has the full wall thickness t normal to the paper, and is seen looking normally to the wall, for instance, from O toward the middle curve in Fig. 36.21. The equilibrium requires that

$$\frac{\partial T}{\partial s} + t \frac{\partial \sigma_x}{\partial x} = 0$$

and this integrates, using (36.72) and (36.70), to

$$T = -E \frac{d^2\theta}{dx^2} \int_{s_1}^{s} u_1' t' \, ds' \tag{36.76}$$

Here the lower limit s_1 is the value of s at the upper end of the middle curve in Fig. 36.21, where T must vanish [it is zero at the other end $s = s_2$ on account of (36.73)].

A running variable s' is used with range s_1 to s, and u_1', t' indicate evaluation of u_1, t, at s'.

The contribution of T to the torsional moment, counterclockwise in Fig. 36.21, is

$$M_{T2} = \int_{s_1}^{s_2} T r_s \, ds \tag{36.77}$$

From (36.71), $du_{10} = -r_s \, ds$, and (36.77) can be written

$$M_{T2} = -\int_{s_1}^{s_2} T \, du_{10}$$

A partial integration yields

$$M_{T2} = -[T u_{10}]_{s_1}^{s_2} + \int_{s_1}^{s_2} u_{10} \, dT$$

The integrated term vanishes because $T = 0$ at both limits. In the integral, dT can be taken from (36.76). Then

$$M_{T2} = -E \frac{d^2\theta}{dx^2} \int_{s_1}^{s_2} u_{10} \cdot u_1 t \, ds = -E \frac{d^2\theta}{dx^2} \int_{s_1}^{s_2} u_1{}^2 t \, ds$$

the last step involving (36.70) and (36.73). The last integral is a section constant. It will be written as Γ, and called the *warping integral*. Thus

$$\Gamma = \int_{s_1}^{s_2} u_1{}^2 t \, ds \tag{36.78}$$

Some values are given in Art. 36.8e.

The torque M_{T2} combines with the Saint-Venant torque $M_{T1} = GJ\theta$ to form the actual torsional moment:

$$M_T = GJ\theta - E\Gamma \frac{d^2\theta}{dx^2} \tag{36.79}$$

For given terminal or distributed torsional loading, M_T is a given function of x, and (36.79) is a differential equation for θ. The end conditions implied when warping is restrained are illustrated in the examples which follow. When θ is known, the stress components are determined by (36.72), (36.76), and the formulas of the Saint-Venant theory.

For applications of this theory to buckling, and its further development, see [28], Chap. III.

Example 2. Warping Prevented at One End: $x = 0$ (Fig. 36.23a). At $x = l$, a torque M_T is applied, and the axial stress σ_x is zero.

Here M_T in (36.79) is constant, and the general solution of (36.79) is

$$\theta = \frac{M_T}{GJ}\left(1 + C_1 \operatorname{Sinh}\frac{x}{a} + C_2 \operatorname{Cosh}\frac{x}{a}\right) \quad \text{where} \quad a^2 = \frac{E\Gamma}{GJ} = 2(1+\nu)\frac{\Gamma}{J} \tag{36.80a,b}$$

The end conditions are $u = 0$ at $x = 0$, and $\sigma_x = 0$ at $x = l$. By (36.72), and the equation $u = \theta u_1$, these require

$$\text{at } x = 0 \quad \theta = 0 \qquad \text{and} \qquad \text{at } x = l \quad d\theta/dx = 0$$

giving

$$C_1 = \operatorname{Tanh}\frac{l}{a} \qquad C_2 = -1$$

The rotation of cross sections is

$$\Theta = \int_0^x \theta \, dx = \frac{M_T}{GJ}\left[x + a \operatorname{Tanh}\frac{l}{a}\left(\operatorname{Cosh}\frac{x}{a} - 1\right) - a \operatorname{Sinh}\frac{x}{a}\right]$$

For a very long bar ($l/a \to \infty$), this becomes

$$\Theta = \frac{M_T}{GJ}(x + ae^{-x/a} - a)$$

or, for large values of x/a,

$$\Theta = \frac{M_T}{GJ}(x - a)$$

If the end $x = 0$ had been free to warp, the term $-a$ in the parentheses would have been absent. It shows the reduction of torsional rotation due to the prevention of warping, effected by the stress σ_x on the end $x = 0$. Since at this end $\theta = 0$, (36.79) shows that the torque M_T is entirely resisted there by the additional shear flow T.

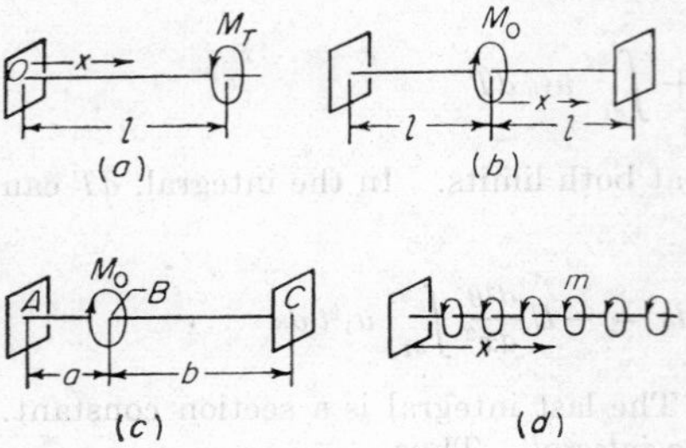

FIG. 36.23

Example 3. Twisting Couple at Midsection (Fig. **36.23***b*). The right-hand half of the bar has a uniform torque $M_T = \frac{1}{2}M_0$. On account of the symmetry about the middle plane $x = 0$, the warping must be zero there, and, therefore, $\theta = 0$. C_1 is available to satisfy one condition at the end $x = l$. If this is taken as $\sigma_x = 0$, the solution is determined, and it is, in fact, the same as in Example 2, when M_T is replaced by $\frac{1}{2}M_0$.

Example 4. Twisting Couple at Any Section (Fig. **36.23***c*). The torque in the left-hand part AB is M_a, that in the right-hand part M_b, and $M_0 = M_b - M_a$. M_a will be negative. Separate solutions of the type (36.80*a*) are required for the two parts, involving four arbitrary constants altogether. These and M_a, M_b constitute six unknowns, determined by (1) continuity of u, σ_x, that is, θ, $d\theta/dx$, at B; (2) continuity of the torsional rotation at B; (3) one condition, for example, $\sigma_x = 0$ at each end A, C; and (4) the equation $M_0 = M_b - M_a$.

Example 5. Uniformly Distributed Torsional Load (Fig. **36.23***d*). The load is m_0 per unit length, and the torque is $M_T = m_0(l - x)$. The end conditions are $u = 0$ at $x = 0$, $\sigma_x = 0$ at $x = l$, giving

$$\theta = 0 \quad \text{at } x = 0 \qquad d\theta/dx = 0 \quad \text{at } x = l$$

The solution of (36.79) satisfying these conditions is

$$\theta = \frac{m_0}{GJ}\left(l - x + C_1 \operatorname{Sinh}\frac{x}{a} + C_2 \operatorname{Cosh}\frac{x}{a}\right)$$

with $\qquad C_1 = a \operatorname{Sech}\frac{l}{a} + l \operatorname{Tanh}\frac{l}{a} \qquad C_2 = -l$

e. Formulas for the Warping Integral Γ

The following formulas are worked out on the basis of Eqs. (36.71), (36.70), (36.74), and (36.78); see also p. 44–28.

I section (Fig. 24*a*)	$\Gamma = \frac{1}{24}b^3h^2t_1$
Z section (Fig. 24*b*)	$\Gamma = \frac{1}{12}b^3h^2t_1 \dfrac{bt_1 + 2ht_2}{2bt_1 + ht_2}$
Channel section (Fig. 24*c*)	$\Gamma = \frac{1}{12}b^3h^2t_1 \dfrac{3bt_1 + 2ht_2}{6bt_1 + ht_2}$
Angle section	$\Gamma = 0$

The zero value holds for all sufficiently thin-walled sections consisting of straight lines intersecting at one point, the + section, for instance.

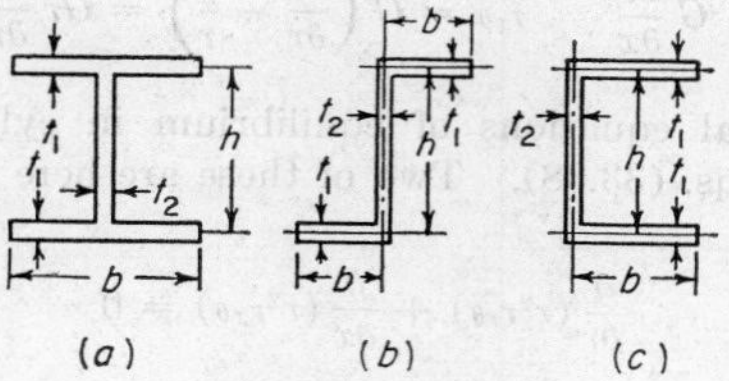

FIG. 36.24

36.9. TORSION OF NONUNIFORM CIRCULAR SHAFTS. MICHELL THEORY

a. Differential Equations

A circular shaft with a shoulder at a change of diameter (Fig. 36.25c) is a special case of a solid of revolution. The general theory of the torsion of such solids, for any profile (generating) curve, is given here.[1] No corresponding theory for noncircular sections has yet been developed.

As in the Saint-Venant theory of uniform bars of noncircular section (Sec. 36.2), certain features of the solution are at first assumed, but the fact that it is then possible to satisfy all the equations and boundary conditions validates the assumptions.

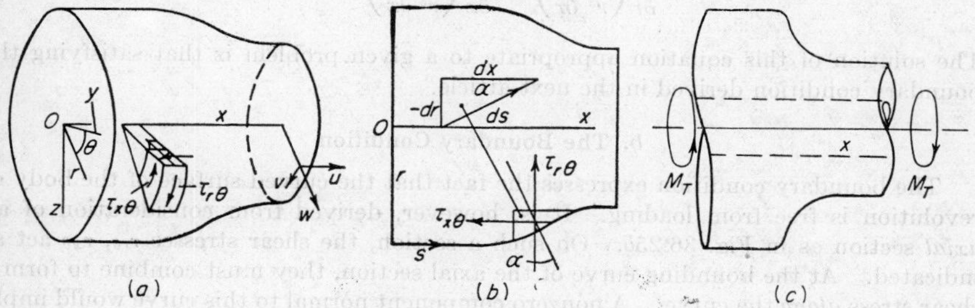

FIG. 36.25

Using cylindrical coordinates x, θ, r (Fig. 36.25a), the displacement components are u, v, w (Fig. 36.25a). Of these, the circumferential component v will evidently represent torsional rotations, but u and w are at once assumed zero. On account of the axial symmetry, v will be independent of θ, but will depend on r (as it does in a cylinder), and also on x. Then

$$u = 0 \qquad v = v(r,x) \qquad w = 0 \tag{36.81}$$

The strain components, in the cylindrical coordinates, are, consequently,

$$\begin{aligned}
&\epsilon_r = \frac{\partial w}{\partial r} = 0 \qquad \epsilon_\theta = \frac{w}{r} + \frac{1}{r}\frac{\partial v}{\partial \theta} = 0 \qquad \epsilon_x = \frac{\partial u}{\partial x} = 0 \\
&\gamma_{r\theta} = \frac{1}{r}\frac{\partial w}{\partial \theta} + \frac{\partial v}{\partial r} - \frac{v}{r} = \frac{\partial v}{\partial r} - \frac{v}{r} \qquad \gamma_{x\theta} = \frac{\partial v}{\partial x} + \frac{1}{r}\frac{\partial w}{\partial \theta} = \frac{\partial v}{\partial x} \\
&\gamma_{xr} = \frac{\partial w}{\partial x} + \frac{\partial u}{\partial r} = 0
\end{aligned} \tag{36.82}$$

[1] Further details, and sources, are given in [1].

It follows that, of the six stress components, $\sigma_r, \sigma_\theta, \sigma_x, \tau_{xr}$ are zero, and $\tau_{x\theta}, \tau_{r\theta}$ are given by

$$\tau_{x\theta} = G\frac{\partial v}{\partial x} \qquad \tau_{r\theta} = G\left(\frac{\partial v}{\partial r} - \frac{v}{r}\right) = Gr\frac{\partial}{\partial r}\left(\frac{v}{r}\right) \tag{36.83a,b}$$

The three differential equations of equilibrium in cylindrical coordinates are given on p. 33–12 as Eqs. (33.38). Two of these are here satisfied identically, and (33.38*b*) reduces to

$$\frac{\partial}{\partial r}(r^2\tau_{r\theta}) + \frac{\partial}{\partial x}(r^2\tau_{x\theta}) = 0 \tag{36.84}$$

This shows that $r^2\tau_{r\theta}$ and $r^2\tau_{x\theta}$ must be expressible in terms of a single function $\phi(r,x)$ by

$$r^2\tau_{r\theta} = \frac{\partial \phi}{\partial x} \qquad r^2\tau_{x\theta} = \frac{\partial \phi}{\partial r} \tag{36.85a,b}$$

To find the differential equation satisfied by ϕ, an equation of compatibility is first found from (36.83*a,b*) by eliminating v. This is

$$\frac{\partial}{\partial r}\left(\frac{1}{r}\tau_{x\theta}\right) = \frac{\partial}{\partial x}\left(\frac{1}{r}\tau_{r\theta}\right)$$

and substitution of $\tau_{r\theta}, \tau_{x\theta}$ from (36.85*a,b*) gives

$$\frac{\partial}{\partial r}\left(\frac{1}{r^3}\frac{\partial \phi}{\partial r}\right) + \frac{\partial}{\partial x}\left(\frac{1}{r^3}\frac{\partial \phi}{\partial x}\right) = 0 \tag{36.86}$$

The solution of this equation appropriate to a given problem is that satisfying the boundary condition derived in the next article.

b. The Boundary Condition

The boundary condition expresses the fact that the curved surface of the body of revolution is free from loading. It is, however, derived from consideration of an *axial* section as in Fig. 36.25*b*. On such a section, the shear stresses $\tau_{r\theta}, \tau_{x\theta}$ act as indicated. At the bounding curve of the axial section, they must combine to form a shear stress *along* the curve. A nonzero component normal to this curve would imply nonzero ringwise shear forces on the curved surface, and the latter would not be free.

The normal direction is shown in Fig. 36.25*b* at an angle α to the r direction. Vanishing of the normal component of shear stress requires that

$$\tau_{x\theta}\sin\alpha + \tau_{r\theta}\cos\alpha = 0 \tag{36.87}$$

If ds is the arc element along the curve, with corresponding dx, dr, the lengths of sides of the elementary triangle in Fig. 36.25*b* are ds, dx, $-dr$, and

$$\cos\alpha = \frac{dx}{ds} \qquad \sin\alpha = -\frac{dr}{ds} \tag{36.88}$$

Now (36.87) becomes, using (36.85*a,b*),

$$-\frac{\partial \phi}{\partial r}\frac{dr}{ds} - \frac{\partial \phi}{\partial x}\frac{dx}{ds} = 0 \quad \text{or} \quad \frac{\partial \phi}{\partial s} = 0 \tag{36.89}$$

Therefore ϕ must be constant along the profile curve in the axial section, the r,x plane. If the shaft is hollow, as indicated by the broken lines in Fig. 36.25*c*, then ϕ must be constant on the inner curve as well, but the two constants will not have the same value. This appears on considering the expression for the torque M_T in terms of ϕ. On a plane cross section, the only stress component appearing is $\tau_{x\theta}$ (Fig. 36.25a). Thus,

$$M_T = \int_b^a \tau_{x\theta} 2\pi r^2\, dr = 2\pi \int_b^a \frac{\partial \phi}{\partial r}\, dr = 2\pi\, [\phi]_{r=b}^{r=a} \tag{36.90}$$

where a is the outer and b the inner radius of the shaft at the particular cross section, each being a given function of x. In a solid shaft, the relation becomes

$$M_T = 2\pi\, [\phi]_{r=0}^{r=a} \tag{36.91}$$

and $\phi_{r=0}$ will be a constant, by (36.85a), for any solution in which $\tau_{r\theta}$ is nonsingular.

c. The Twist

The deformation of the shaft has been specified by v, the circumferential displacement. This is the same for all points on a circle r = const, x = const. The angular displacement of these points is $\psi = v/r$, a function of r and x. It can be found when ϕ is known, from (36.83a,b), which can be written, using (36.85a,b), as

$$G\frac{\partial \psi}{\partial x} = \frac{1}{r^3}\frac{\partial \phi}{\partial r} \qquad G\frac{\partial \psi}{\partial r} = -\frac{1}{r^3}\frac{\partial \phi}{\partial x} \tag{36.92}$$

It follows, by elimination of ϕ, that ψ satisfies the differential equation

$$\frac{\partial}{\partial r}\left(r^3 \frac{\partial \psi}{\partial r}\right) + \frac{\partial}{\partial x}\left(r^3 \frac{\partial \psi}{\partial x}\right) = 0 \tag{36.93}$$

and the boundary condition (36.87) takes the form

$$\frac{\partial \psi}{\partial n} = 0 \tag{36.94}$$

The boundary-value problem (36.93) and (36.94) is the basis of an electrical analogy (described in [1], p. 310).

It is readily verified that the shear stress along the boundary is given by

$$\tau_s = Gr\frac{\partial \psi}{\partial s} = \frac{1}{r^2}\frac{\partial \phi}{\partial n}$$

where n refers to the normal.

The rotation ψ is uniform on surfaces ψ = const, and such surfaces rotate "rigidly." They coincide with the plane cross sections x = const only in the cylindrical shaft.

Example 6. The Conical Shaft. A given problem may be solved by finding either ϕ or ψ. For ϕ, we require a solution of the differential equation (36.86) which is constant on one or both profile curves. It may be verified that (36.86) is satisfied by

$$\phi = C\left(\frac{x}{R} - \frac{1}{3}\frac{x^3}{R^3}\right) \quad \text{where} \quad R^2 = x^2 + r^2 \tag{36.95}$$

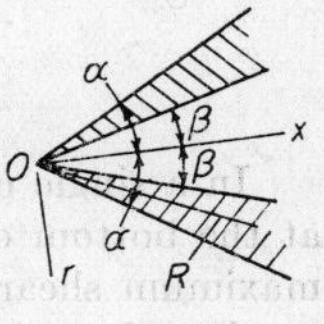

FIG. 36.26

and C is any constant. Since x/R is constant along any ray in the x,r plane through the origin (Fig. 32.26), the function is similarly constant, and, therefore, appropriate for a solid of revolution bounded by any two conical surfaces with vertices at the origin.

For a solid conical shaft of semiangle α, the torque is found from (36.95) and (36.91) as

$$M_T = -2\pi C(2/3 - \cos\alpha + 1/3 \cos^3\alpha) \tag{36.96}$$

and this determines C when M_T is given. The stress components follow from (36.95) and (36.85a,b).

The rotation ψ follows from (36.95) and (36.92) as

$$\psi = \frac{C}{3G}\frac{1}{R^3} \tag{36.97}$$

but for an additive constant representing rigid-body rotation. The surfaces ψ = const which have simple rotation are the spherical surfaces R = const. The relative rotation of two spherical ends of a conical shaft, under given torque M_T, can be found from (36.97) as the difference of the two values of ψ corresponding to the two ends.

For a hollow shaft, the constant C is determined from (36.90) instead of from (36.91). Other solutions may be found in [17] and [18] and in works cited in [1].

d. Shafts with Shoulder Fillets and Circumferential Grooves

In a shoulder fillet (Fig. 36.27), the maximum shear stress occurs in the fillet. The results of measurements by Jacobsen [19], using the electrical analogy mentioned in Art. 36.9*c*, are shown in Fig. 36.27, which also includes strain-gauge measurements by Weigand [20].

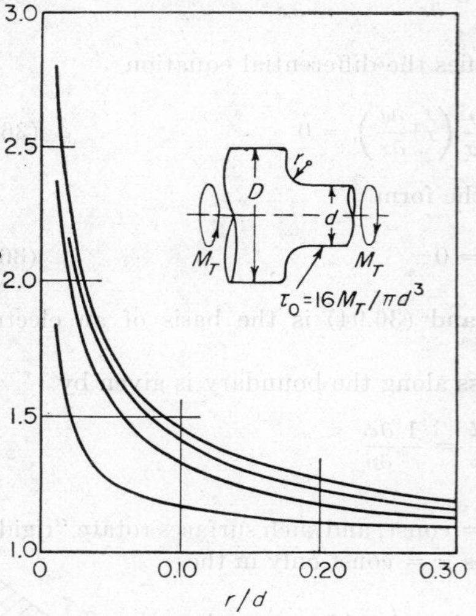

FIG. 36.27

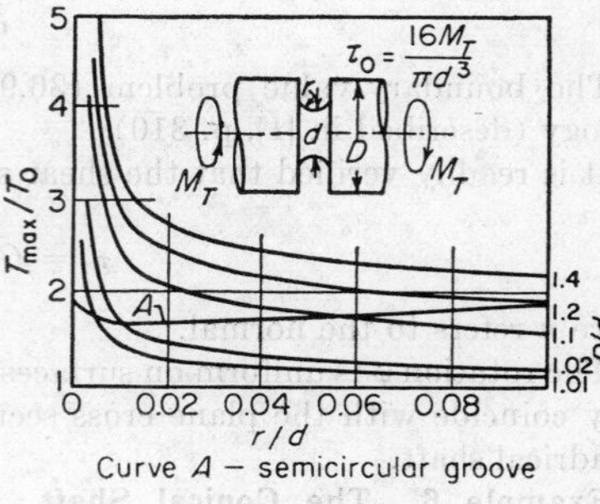

FIG. 36.28

In a single circumferential groove (Fig. 36.28), the maximum shear stress occurs at the bottom of the groove. When the groove is semicircular and very small, the maximum shear stress has twice the value it would have without the groove. The results of strain-gauge measurements by Thum and Bautz [21] are shown in Fig. 36.28. Theoretical results by Ōkubo, with some support from measurements by an electroplating method, are given in [22]. The question of an appropriate multiplier for conversion from a single groove to a regular succession of similar grooves is discussed in [23], and in [24] where further design information and references will be found.

36.10. REFERENCES

[1] S. P. Timoshenko, J. N. Goodier: "Theory of Elasticity," 2d ed., McGraw-Hill, New York, 1951.

[2] I. S. Sokolnikoff: "Mathematical Theory of Elasticity," 2d ed., McGraw-Hill, New York, 1956.

[3] M. Hetényi: "Handbook of Experimental Stress Analysis," Wiley, New York, 1950.

[4] G. W. Trayer, H. W. March: The torsion of members having sections common in aircraft construction, *NACA Rept.* 334 (1930).

[5] F. M. Baron: Torsion of multi-connected thin-walled cylinders, *J. Appl. Mechanics*, **9** (1942), 72–74.

[6] W. J. Carter: Torsion and flexure of slender solid sections, *J. Appl. Mechanics* **25** (1958), 115–121.

[7] A. A. Griffith: The determination of the torsional stiffness and strength of cylindrical bars of any shape, *Brit. Advis. Comm. Aeronaut. Tech. Rept.*, 1917/18, pp. 910–937.

[8] R. J. Roark: "Formulas for Stress and Strain," 3d ed., pp. 174–179, McGraw-Hill, New York, 1954.

[9] H. Ōkubo: On the torsion of a shaft with keyways, *Quart. J. Mech. Appl. Math.*, **3** (1950), 162–172.

[10] H. Ōkubo: Torsion of a circular shaft with a number of longitudinal notches, *J. Appl. Mechanics*, **17** (1950), 359–362.

[11] J. H. Huth: Torsional stress concentration in angle and square tube fillets, *J. Appl. Mechanics*, **17** (1950), 388–390.

[12] J. N. Goodier: Elastic torsion in the presence of initial axial stress, *J. Appl. Mechanics*, **17** (1950), 383–387.

[13] H. L. Engel, J. N. Goodier: Measurements of torsional stiffness changes and instability due to tension, compression and bending, *J. Appl. Mechanics*, **20** (1953), 553–561.

[14] C. Chu: The effect of initial twist on the torsional rigidity of thin prismatical bars and tubular members, *Proc. 1st U.S. Natl. Congr. Appl. Mechanics*, pp. 265–269, Chicago, 1952.

[15] M. M. Leven: Stresses in keyways by photoelastic methods and comparison with numerical solution, *Proc. Soc. Exptl. Stress Anal.*, **7** (1949), 141–154.

[16] E. Reissner: Note on the problem of twisting of a circular ring sector, *Quart. Appl. Math.*, **7** (1949), 342–347.

[17] H. Poritsky: Stress fields of axially symmetric shafts in torsion and related fields, *Proc. Symposia Appl. Math.*, **3** (1950), 163–186.

[18] H. Reissner, G. J. Wennagel: Torsion of non-cylindrical shafts of circular cross section, *J. Appl. Mechanics*, **17** (1950), 275–282.

[19] L. S. Jacobsen: Torsional stress concentrations in shafts of circular cross-section and variable diameter, *Trans. ASME*, **47** (1925), 619–638.

[20] A. Weigand: Determination of the stress concentration factor of a stepped shaft stressed in torsion by means of precision strain gages, *NACA Tech. Mem.* 1179 (1947); transl. from *Luftfahrt-Forsch.*, **20** (1943), 217–219.

[21] A. Thum, W. Bautz: Zur Frage der Formziffer, *Z. VDI*, **79** (1935), 1303–1306.

[22] H. Ōkubo: Stress concentration factors for a circumferential notch in a cylindrical shaft, *Mem. Fac. Eng. Univ. Nagoya*, **6** (1954), 23–29.

[23] H. Ōkubo: Die Formzahlen tordierter Wellen mit mehreren Nuten, *Ing.-Arch.*, **23** (1955), 130–132.

[24] R. B. Heywood: "Designing by Photoelasticity," p. 188ff, Chapman & Hall, London, 1952.

[25] W. Flügge, K. Marguerre: Wölbkräfte in dünnwandigen Profilstäben, *Ing.-Arch.*, **18** (1950), 23–38.

[26] E. Reissner: On torsion with variable twist, *J. Appl. Mechanics*, **23** (1956), 315–316; also *Österr. Ingr.-Arch.*, **9** (1955), 218–224.

[27] J. N. Goodier: The buckling of compressed bars by torsion and flexure, *Cornell Univ. Eng. Expt. Sta. Bull.* 27 (1941).

[28] F. Bleich: "Buckling Strength of Metal Structures," chap. 3, McGraw-Hill, New York, 1952.

[29] S. P. Timoshenko, J. M. Gere: "Theory of Elastic Stability," 2d ed., p. 218, McGraw-Hill, New York, 1961.

[30] S. Timoshenko: Theory of bending, torsion and buckling of thin-walled members of open cross section, *J. Franklin Inst.*, **239** (1945), 201–219, 249–268, 343–361.

[31] H. Neuber: "Kerbspannungslehre," 1st ed., Springer, Berlin, 1937; transl. as "Theory of Notch Stresses," Edwards, Ann Arbor, 1946; 2d ed., 1958.

[32] R. E. Peterson: "Stress Concentration Design Factors," Wiley, New York, 1953.

CHAPTER 37

TWO-DIMENSIONAL PROBLEMS

BY

CH. MASSONNET, Liège, Belgium

37.1. GENERAL PROPERTIES

Plane Stress and Plane Strain

Plane Stress. Consider a thin plate of uniform thickness (Fig. 37.1), made of an elastic material. Assume that all forces applied to its boundary or in its interior are distributed uniformly across the thickness.

Evidently, the stresses σ_z, τ_{xz}, τ_{yz} vanish on both faces of the plate, and it is unlikely that they assume substantial values in the interior. Therefore, their presence may be neglected entirely. Additionally, it is plausible to assume that the other three stresses σ_x, σ_y, τ_{xy} do not depend on z. The stress system described by these assumptions is called *plane stress*.

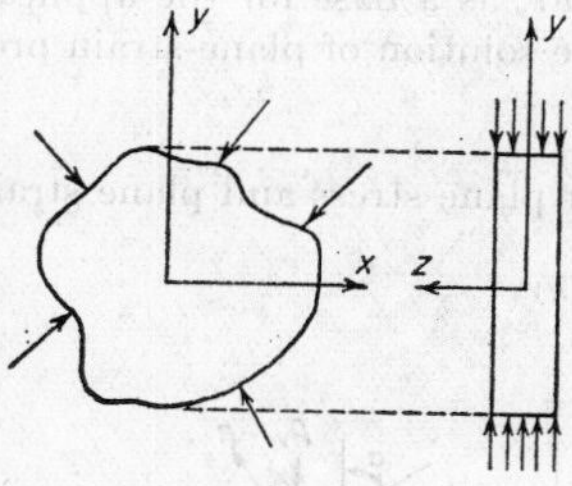

Fig. 37.1

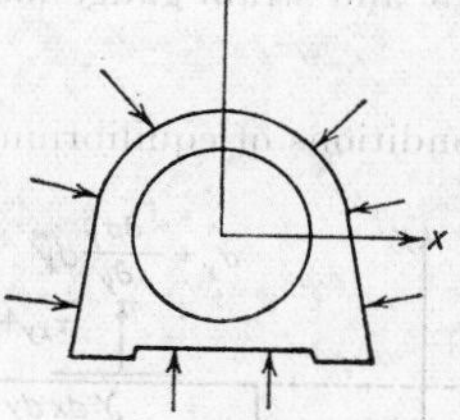

Fig. 37.2

The shear strains γ_{xz}, γ_{yz} are zero, and ϵ_z has no interest. The relations between the other three and the stresses are Eqs. (33.54) and (33.50*d*).

The theory of plane stress is an approximate theory. It yields good results for thin plates when the stresses vary gently as functions of x and y. It will be more or less in error when the stresses vary substantially over distances comparable with the thickness of the plate.

Plane Strain. Consider a very long cylinder or prism with generators parallel to the z axis (Fig. 37.2). Assume that all the surface loads and mass forces acting on this body are perpendicular to the z axis and independent of z. Furthermore, assume that the ends of the cylinder are prevented from moving in the z direction, but are otherwise free. Then the strains ϵ_z, γ_{xz}, γ_{yz} vanish, and the other three are independent of z. Such a body is said to be in a state of *plane strain*

To study and describe the stresses in such a body, it suffices to consider a slice or plate cut from it by two planes z = const. The shear stresses τ_{xz}, τ_{yz} are zero, and, with $\epsilon_z = 0$, Hooke's law [Eq. (33.50c)] yields

$$\sigma_z = \nu(\sigma_x + \sigma_y)$$

The relations between the remaining stresses and strains are Eqs. (33.55) and (33.50d).

The stress σ_z may depend on x and y, but not on z. Consequently, such stresses must act at the terminal cross sections of the cylinder. If they are absent, an additional problem arises which is not covered by plane-strain theory. If the stresses σ_z needed for plane strain are self-equilibrating at each end, their absence only leads to a local disturbance; otherwise, an axial force and a bending moment are imposed on the cylinder. If this end disturbance is neglected, as is often done, then the theory of plane strain is an approximate theory only.

Analytical Connection between Plane Stress and Plane Strain. The problems of plane stress and plane strain, although different in their basic equations, are not independent of each other. When a plane-stress problem has been solved, the solution for the corresponding plane-strain problem may be derived from it by the following procedure: Replace everywhere

$$E \text{ by } \frac{E'}{1-\nu'^2} \quad \text{and} \quad \nu \text{ by } \frac{\nu'}{1-\nu'} \tag{37.1}$$

and then drop the primes. Conversely, if the plane-strain problem has been solved, the solution for plane stress is obtained by this procedure: Replace everywhere

$$E \text{ by } \frac{E'(1+2\nu')}{(1+\nu')^2} \quad \text{and} \quad \nu \text{ by } \frac{\nu'}{1+\nu'} \tag{37.2}$$

and then drop the primes. It is easily seen that neither substitution changes the shear modulus G. The substitutions may be verified by applying them to the elastic law [Eqs. (33.54) and (33.55)]. They are useful to avoid the repetition of lengthy mathematical or numerical work, but, in particular, as a base for the application of photoelastic and strain-gauge measurements to the solution of plane-strain problems.

Equilibrium

The conditions of equilibrium are the same for plane stress and plane strain.

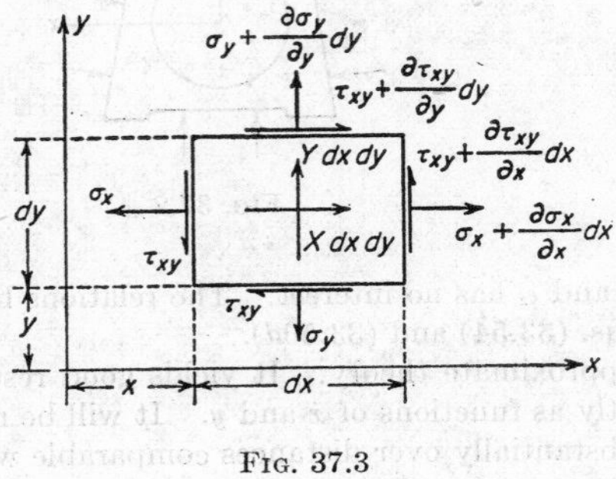

FIG. 37.3

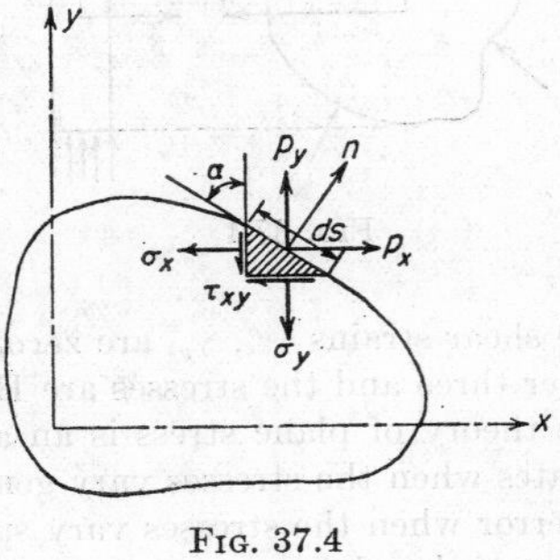

FIG. 37.4

The equilibrium of a plate element (Fig. 37.3) yields the equations [see also Eqs. (33.35)]

$$\frac{\partial \sigma_x}{\partial x} + \frac{\partial \tau_{xy}}{\partial y} + X = 0 \qquad \frac{\partial \tau_{xy}}{\partial x} + \frac{\partial \sigma_y}{\partial y} + Y = 0 \tag{37.3}$$

The equilibrium of a triangular element at the boundary of the plate (Fig. 37.4) requires that

$$\sigma_x \cos\alpha + \tau_{xy} \sin\alpha = p_x \qquad \tau_{xy}\cos\alpha + \sigma_y \sin\alpha = p_y \tag{37.4}$$

If the components p_x, p_y of the stress at the boundary (or the normal and tangential components) are given, these equations constitute the boundary conditions of the stress problem.

Compatibility

Of the six compatibility conditions (33.21), only one is not trivial in the two-dimensional problem:

$$\frac{\partial^2 \epsilon_x}{\partial y^2} + \frac{\partial^2 \epsilon_y}{\partial x^2} = \frac{\partial^2 \gamma_{xy}}{\partial x \, \partial y} \tag{37.5}$$

When we use Eqs. (33.54) to express the strains in terms of the stresses, and when we then use Eqs. (37.3) to eliminate τ_{xy}, we obtain for plane stress

$$\nabla^2(\sigma_x + \sigma_y) = -(1 + \nu)\left(\frac{\partial X}{\partial x} + \frac{\partial Y}{\partial y}\right) \tag{37.6}$$

where ∇^2 is the two-dimensional Laplace operator [Eq. (11.64)]. The corresponding relation for plane strain is obtained from (37.6) with the help of the transformation (37.1):

$$\nabla^2(\sigma_x + \sigma_y) = -\frac{1}{1 - \nu}\left(\frac{\partial X}{\partial x} + \frac{\partial Y}{\partial y}\right) \tag{37.7}$$

Problem Types

Equations (37.3) and (37.6) or (37.7) are the differential equations of the stress field. If the boundary conditions (37.4) are prescribed along the entire boundary, we have the general stress problem. If, instead, the components u, v of the displacement are given on the entire boundary, we have the general displacement problem. In some cases, p_x, p_y are given on part of the boundary; u, v on the remainder. Still more generally, we may choose, at each point of the boundary, two orthogonal directions, and prescribe for each of them independently the component of either the displacement or the external force.

In a simply connected domain, the general stress problem admits only one regular solution. It depends on Poisson's ratio except when the divergence of the external force field vanishes, $\partial X/\partial x + \partial Y/\partial y = 0$. In this case, Eqs. (37.6) and (37.7) are identical, and so are the problems of plane stress and plane strain.

If the domain is multiply connected (plate or cylinder with holes), the solution of the stress problem is no longer unique. The difference of any two solutions is a solution for vanishing load X, Y, p_x, p_y. Such a stress system may be produced by cutting the domain along a curve connecting one of the interior boundaries with the exterior one, then producing a relative displacement (*dislocation*) between points on opposite sides of the cut, and finally reuniting the material (Fig. 37.5). Evidently, the displacement is then discontinuous along the cut. If such a discontinuous displacement is excluded, the solution of the stress problem is unique.

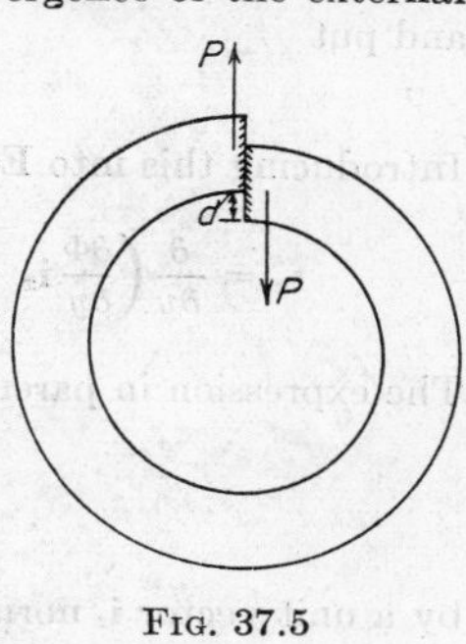

FIG. 37.5

The following methods may be used for solving two-dimensional stress problems: Airy's stress function (see below), the complex variable method (see Chap. 38), the relaxation method applied to the biharmonic equation (37.9) or to the complex-variable approach, the method of singularities (p. 37-28).

Airy's Stress Function

Case of Zero Body Forces. When $X = Y = 0$, Eqs. (37.3), (37.6), (37.7) are homogeneous. Equations (37.3) are identically satisfied when the stresses are expressed as derivatives of Airy's stress function Φ:

$$\sigma_x = \frac{\partial^2 \Phi}{\partial y^2} \qquad \sigma_y = \frac{\partial^2 \Phi}{\partial x^2} \qquad \tau_{xy} = -\frac{\partial^2 \Phi}{\partial x\, \partial y} \tag{37.8}$$

Upon introduction of (37.8) into (37.6) or (37.7), it is found that Φ must satisfy the biharmonic equation,

$$\nabla^2 \nabla^2 \Phi = 0 \tag{37.9}$$

where ∇^2 is the two-dimensional Laplace operator [Eq. (11.64)].

Body Forces Deriving from a Potential. When the body-force field can be described as the gradient of a potential V, that is, when

$$X = \frac{\partial V}{\partial x} \qquad Y = \frac{\partial V}{\partial y} \tag{37.10}$$

the conditions of equilibrium (37.3) may be written as

$$\frac{\partial}{\partial x}(\sigma_x + V) + \frac{\partial \tau_{xy}}{\partial y} = 0 \qquad \frac{\partial \tau_{xy}}{\partial x} + \frac{\partial}{\partial y}(\sigma_y + V) = 0 \tag{37.11}$$

They are satisfied when we put

$$\sigma_x + V = \frac{\partial^2 \Phi}{\partial y^2} \qquad \sigma_y + V = \frac{\partial^2 \Phi}{\partial x^2} \qquad \tau_{xy} = -\frac{\partial^2 \Phi}{\partial x\, \partial y} \tag{37.12}$$

Introducing (37.12) into the compatibility equation (37.6) for plane stress, we find

$$\nabla^2 \nabla^2 \Phi = (1 - \nu) \nabla^2 V \tag{37.13}$$

In the particular case that V is a harmonic function, satisfying the Laplace equation $\nabla^2 V = 0$, Eq. (37.13) as well as its counterpart for plane strain reduces to the biharmonic equation (37.9), but with Eqs. (37.12) for the stresses.

Vectorial Formulation. We introduce unit vectors $\mathbf{i}_x$ and $\mathbf{i}_y$ in the directions x, y and put

$$\mathbf{t}_x = \sigma_x \mathbf{i}_x + \tau_{xy} \mathbf{i}_y \qquad \mathbf{t}_y = \tau_{xy} \mathbf{i}_x + \sigma_y \mathbf{i}_y \tag{37.14}$$

Introducing this into Eqs. (37.12), we obtain

$$\mathbf{t}_x = \frac{\partial}{\partial y}\left(\frac{\partial \Phi}{\partial y}\mathbf{i}_x - \frac{\partial \Phi}{\partial x}\mathbf{i}_y\right) - V\mathbf{i}_x \qquad \mathbf{t}_y = -\frac{\partial}{\partial x}\left(\frac{\partial \Phi}{\partial y}\mathbf{i}_x - \frac{\partial \Phi}{\partial x}\mathbf{i}_y\right) - V\mathbf{i}_y$$

The expression in parentheses is the cross product of

$$\text{grad } \Phi = \frac{\partial \Phi}{\partial x}\mathbf{i}_x + \frac{\partial \Phi}{\partial y}\mathbf{i}_y$$

by a unit vector $\mathbf{i}_z$ normal to the x,y plane; hence,

$$\mathbf{t}_x = \frac{\partial}{\partial y}(\text{grad } \Phi \times \mathbf{i}_z) - V\mathbf{i}_x \qquad \mathbf{t}_y = -\frac{\partial}{\partial x}(\text{grad } \Phi \times \mathbf{i}_z) - V\mathbf{i}_y \tag{37.15}$$

Since grad $\Phi \times \mathbf{i}_z$ is independent of the choice of the axes x, y, Eqs. (37.15) apply to any pair of orthogonal directions r, t:

$$\mathbf{t}_r = \frac{\partial}{\partial t}(\text{grad } \Phi \times \mathbf{i}_z) - V\mathbf{i}_r \qquad \mathbf{t}_t = -\frac{\partial}{\partial r}(\text{grad } \Phi \times \mathbf{i}_z) - V\mathbf{i}_t \tag{37.16}$$

where $\mathbf{i}_r$ and $\mathbf{i}_t$ are unit vectors in directions r and t.

Biot's Theorem. When the body force is the gradient of a potential, the particular solution $\Phi \equiv 0$ of Eq. (37.9) yields, with (37.12), the stresses

$$\sigma_x = \sigma_y = -V \qquad \tau_{xy} = 0 \tag{37.17}$$

They represent the solution of the stress problem if the boundary conditions specify the application of a normal pressure $V(x,y)$ at all points of the boundary, which, consequently, is in equilibrium with the body forces. When forces different from this pressure are prescribed at the boundary, the solution is the sum of (37.17) and a stress system for zero body forces and an edge load equal to the given one plus a normal tension V.

In particular, if the body force is the weight of the body of specific weight γ, then $V = -\gamma y$ (y axis positive upward), and Eqs. (37.17) represent the stresses that result when the body is immersed in a fluid of the same specific weight. In this case, *Biot's theorem* applies: The stresses in a vertical heavy plate can be derived from an auxiliary stress field σ'_x, σ'_y, τ'_{xy} for a weightless plate subjected to the actual edge loads plus a normal stress $-\gamma y$ applied to all points of the boundary; the actual stresses are then

$$\sigma_x = \sigma'_x + \gamma y \qquad \sigma_y = \sigma'_y + \gamma y \qquad \tau_{xy} = \tau'_{xy}$$

Wieghardt's Analogy. Equation (37.9) is identical with Eq. (39.10′) for the deflection w of a plane plate if the distributed load $q \equiv 0$. Consequently, every stress function Φ may be visualized as the deflection w of a plate subjected to edge loads only, and every solution of a two-dimensional stress problem is also a solution of a plate-bending problem, and vice versa.

37.2. PLANE PROBLEMS IN CARTESIAN COORDINATES

Displacements

When the stresses σ_x, σ_y, τ_{xy} are known, the strains ϵ_x, ϵ_y, γ_{xy} may be obtained, using Eqs. (33.50*d*), (33.54), and (33.55). Then the displacement components u, v may be found by integrating the system of partial differential equations:

$$\frac{\partial u}{\partial x} = \epsilon_x \qquad \frac{\partial v}{\partial y} = \epsilon_y \qquad \frac{\partial u}{\partial y} + \frac{\partial v}{\partial x} = \gamma_{xy} \tag{37.18}$$

Integrating the first two of these equations, we obtain

$$u = \int \epsilon_x \, dx + f_1(y) \qquad v = \int \epsilon_y \, dy + f_2(x) \tag{37.19}$$

where $f_1(y)$ and $f_2(x)$ are arbitrary functions. Introduction of these results into the last equation (37.18) leads to an equation for f_1 and f_2, from which it may be found that u and v contain arbitrary linear functions,

$$u_1 = a + by \qquad v_1 = c - bx \tag{37.20}$$

where a, b, and c are constants. These linear functions represent a rigid-body movement consisting of translations a and c and a rotation b about the origin.

Although the stress fields for problems of plane stress and plane strain are identical in a large class of problems, the corresponding displacements differ, because Eqs. (33.54) and (33.55) differ. However, the relations (37.1) and (37.2) may be used to convert a solution of one of the problems into a solution of the other one.

Polynomial Solutions

The most elementary way of finding solutions of Eq. (37.9) is to take homogeneous polynomials of various degrees and adjust their coefficients so that they satisfy Eq. (37.9).

Any quadratic polynomial,

$$\Phi = \Phi_2 = a_2x^2 + b_2xy + c_2y^2 \tag{37.21}$$

is a solution of Eq. (37.9). The corresponding stresses are:

$$\sigma_x = 2c_2 \qquad \sigma_y = 2a_2 \qquad \tau_{xy} = -b_2 \tag{37.22}$$

Any cubic polynomial,

$$\Phi = \Phi_3 = a_3x^3 + b_3x^2y + c_3xy^2 + d_3y^3 \tag{37.23}$$

also satisfies Eq. (37.9). It yields the stresses

$$\sigma_x = 2c_3x + 6d_3y \qquad \sigma_y = 6a_3x + 2b_3y \qquad \tau_{xy} = -2b_3x - 2c_3y \tag{37.24}$$

Polynomials of higher degree are only solutions of Eq. (37.9) if certain relations exist between their coefficients. The number of these relations is $(n - 3)$, if n represents the degree of the polynomial.

For example, the polynomial of fourth degree,

$$\Phi = \Phi_4 = a_4x^4 + b_4x^3y + c_4x^2y^2 + d_4xy^3 + e_4y^4 \tag{37.25}$$

is only a solution if

$$3e_4 = -c_4 - 3a_4$$

The stresses are:

$$\begin{gathered}\sigma_x = 2c_4x^2 + 6d_4xy - (4c_4 + 12a_4)y^2 \qquad \sigma_y = 12a_4x^2 + 6b_4xy + 2c_4y^2 \\ \tau_{xy} = -3b_4x^2 - 4c_4xy - 3d_4y^2\end{gathered} \tag{37.26}$$

The polynomial of the fifth degree,

$$\Phi = \Phi_5 = a_5x^5 + b_5x^4y + c_5x^3y^2 + d_5x^2y^3 + e_5xy^4 + f_5y^5 \tag{37.27}$$

is subjected to the restrictions

$$e_5 = -c_5 - 5a_5 \qquad 5f_5 = -b_5 - d_5$$

The stresses are:

$$\begin{aligned}\sigma_x &= 2c_5x^3 + 6d_5x^2y - 12(c_5 + 5a_5)xy^2 - 4(b_5 + d_5)y^3 \\ \sigma_y &= 20a_5x^3 + 12b_5x^2y + 6c_5xy^2 + 2d_5y^3 \\ \tau_{xy} &= -4b_5x^3 - 6c_5x^2y - 6d_5xy^2 + 4(c_5 + 5a_5)y^3\end{aligned} \tag{37.28}$$

By combining some of these solutions, problems of practical interest may be solved.

Applications of Polynomial Solutions

Elementary Cases. It is evident that solution (37.22) may be adapted to represent simple tension ($c_2 \neq 0$), double tension ($c_2 \neq 0$, $a_2 \neq 0$), or pure shear ($b_2 \neq 0$) of a rectangular plate.

In the same way, solution (37.24) may be adapted to represent pure bending induced by couples applied to vertical sides of the rectangular plate. We have only to put $a_3 = b_3 = c_3 = 0$, $d_3 \neq 0$.

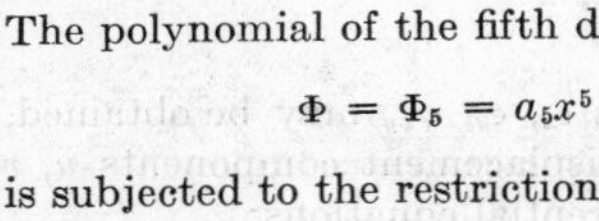

FIG. 37.6

Bending of a Cantilever Loaded at the End. Consider a cantilever having a narrow rectangular section of unit width loaded by a force P applied at the end (Fig. 37.6). In order to find a *simple* solution, it is necessary to leave the precise mode of application of force P undefined, and it can only be asked that, on the vertical side $x = 0$, the normal stresses σ_x must be zero, and the resultant of shearing stresses τ_{xy}

must be equivalent to P, so that

$$\int_{-c}^{+c} \tau_{xy}\, dy = P \tag{37.29}$$

Moreover, the horizontal edges of the cantilever are not loaded so that, at $y = \pm c$,

$$\sigma_y = 0 \qquad \tau_{xy} = 0 \tag{37.30}$$

The problem may be solved by combining solutions Φ_2 and Φ_4 with

$$a_2 = c_2 = a_4 = b_4 = c_4 = 0,$$

so that the stress function reduces to

$$\Phi = b_2 xy + d_4 xy^3$$

The corresponding stress field is

$$\sigma_x = 6d_4xy \qquad \sigma_y = 0 \qquad \tau_{xy} = -b_2 - 3d_4y^2$$

It is seen that it satisfies the first condition (37.30). The second yields $d_4 = -b_2/3c^2$, and, hence,

$$\tau_{xy} = b_2\left(1 - \frac{y^2}{c^2}\right)$$

Introducing this into condition (37.29), we obtain $b_2 = -3P/4c$, and hence the stresses

$$\sigma_x = \frac{3P}{2c^3}xy \qquad \sigma_y = 0 \qquad \tau_{xy} = \frac{3P}{4c}\left(1 - \frac{y^2}{c^2}\right) \tag{37.31}$$

Since $2c^3/3 = I$ is the moment of inertia of the cross section, Eqs. (37.31) may be written as

$$\sigma_x = \frac{Pxy}{I} \qquad \sigma_y = 0 \qquad \tau_{xy} = \frac{P}{2I}(c^2 - y^2) \tag{37.32}$$

They are identical with the corresponding formulas of the elementary beam theory [Eqs. (35.12), (35.31)]. It should be borne in mind, however, that this solution is

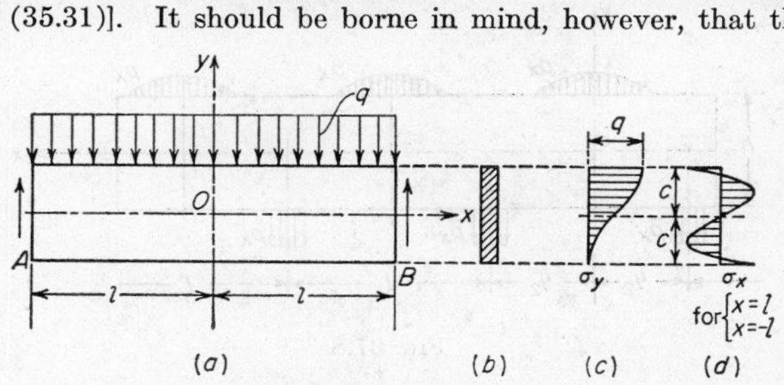

FIG. 37.7

exact only if the force P is distributed over the end section according to Eq. (37.32), and if the constraint at the built-in end of the plate is compatible with the strains accompanying the stresses (37.32).

Beam on Two Supports, Loaded Uniformly. The dimensions and notations are defined by Fig. 37.7. The boundary conditions are:

At $y = -c$:	$\sigma_y = 0$	$\tau_{xy} = 0$
At $y = +c$:	$\sigma_y = -q$	$\tau_{xy} = 0$
At $x = \pm l$:	$\sigma_x = 0$	$\int_{-c}^{+c} \tau_{xy}\, dy = \pm ql$

However, it is impossible to find a simple solution satisfying all these conditions. For this reason, we drop the condition for σ_x at $x = \pm l$, and replace it by the milder conditions

$$\int_{-c}^{+c} \sigma_x \, dy = 0 \qquad \int_{-c}^{+c} \sigma_x y \, dy = 0$$

which prescribe that the normal force and the bending moment shall vanish at the terminal sections of the beam. With this relaxed set of boundary conditions, the problem is solved by the stress function,

$$\Phi = a_2 x^2 + b_3 x^2 y + d_3 y^3 + d_5(x^2 y^3 - \tfrac{1}{5} y^5)$$

After determining the constants from the boundary conditions, the stresses are found to be

$$\sigma_x = -\frac{q}{2I}(l^2 - x^2)y + \frac{q}{2I}(\tfrac{2}{5}c^2 y - \tfrac{2}{3}y^3)$$
$$\sigma_y = \frac{q}{2I}(\tfrac{1}{3}y^3 - c^2 y - \tfrac{2}{3}c^3) \qquad \tau_{xy} = \frac{q}{2I}(c^2 - y^2)x \tag{37.33}$$

with $I = \tfrac{2}{3}c^3$.

It is easily seen that τ_{xy} is the same as in the elementary Navier theory [Eq. (35.31)]. The stress σ_y, which does not exist in the elementary theory, decreases from 0 to $-q$, according to the cubic law shown in Fig. 37.7*c*. The first term in the expression for σ_x coincides with the Navier formula, and the second one is a small correction if the ratio c/l is sufficiently small.

However, $\sigma_x \neq 0$ at the ends of the beam, and presents there a self-equilibrating cubic distribution shown in Fig. 37.7*d*.

Further details of polynomial solutions are found in [1], pp. 39–46.

Fourier Series Solutions

Infinite Strip, Symmetric Edge Loads. Consider an infinite plate strip of height h, subjected to periodic edge loads $p(x)$ at $y = h/2$ and $\bar{p}(x)$ at $y = -h/2$ (Fig. 37.8).

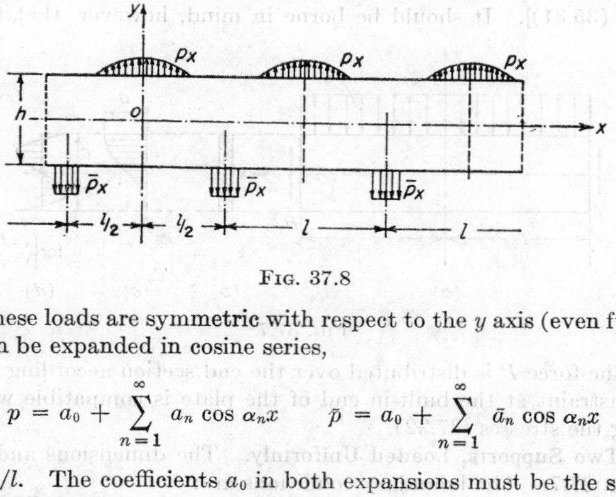

FIG. 37.8

Assume that these loads are symmetric with respect to the y axis (even functions of x). They may then be expanded in cosine series,

$$p = a_0 + \sum_{n=1}^{\infty} a_n \cos \alpha_n x \qquad \bar{p} = a_0 + \sum_{n=1}^{\infty} \bar{a}_n \cos \alpha_n x \tag{37.34}$$

with $\alpha_n = 2n\pi/l$. The coefficients a_0 in both expansions must be the same when the loads are in equilibrium.

The boundary conditions are:

$$\begin{aligned} &\text{At } y = h/2: && \sigma_y = p(x) && \tau_{xy} = 0 \\ &\text{At } y = -h/2: && \sigma_y = \bar{p}(x) && \tau_{xy} = 0 \end{aligned} \tag{37.35}$$

To solve the problem the stress function is also expanded into a Fourier series with coefficients depending on y:

$$\Phi(x,y) = \sum_{n=0}^{\infty} \Phi_n(y) \cos \alpha_n x \tag{37.36}$$

When this is introduced in the differential equation (37.9), an ordinary differential equation for $\Phi_n(y)$ is found:

$$\Phi_n^{\text{iv}} - 2\alpha_n^2 \Phi_n'' + \alpha_n^4 \Phi_n = 0 \tag{37.37}$$

For $n \geq 1$ it has the general solution

$$\Phi_n = \alpha_n^{-2} (A_n \operatorname{Cosh} \alpha_n y + B_n \alpha_n y \operatorname{Cosh} \alpha_n y + C_n \operatorname{Sinh} \alpha_n y + D_n \alpha_n y \operatorname{Sinh} \alpha_n y) \tag{37.38}$$

For $n = 0$, the solution degenerates, and is simply $\Phi_0 = A_0 x^2$. Equations (37.8) yield the stresses corresponding to this solution:

$$\begin{aligned}
\sigma_x &= \sum_{n=1}^{\infty} [(A_n + 2D_n) \operatorname{Cosh} \alpha_n y + B_n \alpha_n y \operatorname{Cosh} \alpha_n y \\
&\qquad + (2B_n + C_n) \operatorname{Sinh} \alpha_n y + D_n \alpha_n y \operatorname{Sinh} \alpha_n y] \cos \alpha_n x \\
\sigma_y &= 2A_0 - \sum_{n=1}^{\infty} (A_n \operatorname{Cosh} \alpha_n y + B_n \alpha_n y \operatorname{Cosh} \alpha_n y \\
&\qquad + C_n \operatorname{Sinh} \alpha_n y + D_n \alpha_n y \operatorname{Sinh} \alpha_n y) \cos \alpha_n x \\
\tau_{xy} &= \sum_{n=1}^{\infty} [(A_n + D_n) \operatorname{Sinh} \alpha_n y + B_n \alpha_n y \operatorname{Sinh} \alpha_n y \\
&\qquad + (B_n + C_n) \operatorname{Cosh} \alpha_n y + D_n \alpha_n y \operatorname{Cosh} \alpha_n y] \sin \alpha_n x
\end{aligned} \tag{37.39}$$

These formulas constitute the general solution for symmetric edge loads (including shear). When they are introduced in the special set of boundary conditions (37.35), it is found that $2A_0 = a_0$ and

$$\begin{aligned}
A_n &= -(a_n + \bar{a}_n) \frac{\operatorname{Sinh} (\alpha_n h/2) + (\alpha_n h/2) \operatorname{Cosh} (\alpha_n h/2)}{\operatorname{Sinh} \alpha_n h + \alpha_n h} \\
B_n &= (a_n - \bar{a}_n) \frac{\operatorname{Cosh} (\alpha_n h/2)}{\operatorname{Sinh} \alpha_n h - \alpha_n h} \\
C_n &= -(a_n - \bar{a}_n) \frac{\operatorname{Cosh} (\alpha_n h/2) + (\alpha_n h/2) \operatorname{Sinh} (\alpha_n h/2)}{\operatorname{Sinh} \alpha_n h - \alpha_n h} \\
D_n &= (a_n + \bar{a}_n) \frac{\operatorname{Sinh} (\alpha_n h/2)}{\operatorname{Sinh} \alpha_n h + \alpha_n h}
\end{aligned} \tag{37.40}$$

Infinite Strip, Antimetric Edge Loads. When the edge loads are antimetric with respect to the y axis, they may be expanded in sine series:

$$p = \sum_{n=1}^{\infty} a_n \sin \alpha_n x \qquad \bar{p} = \sum_{n=1}^{\infty} \bar{a}_n \sin \alpha_n x \tag{37.41}$$

The calculations are similar to those for symmetric load, and the stress function is, in this case,

$$\begin{aligned}
\Phi = \sum_{n=1}^{\infty} \alpha_n^{-2} (A_n \operatorname{Cosh} \alpha_n y + B_n \alpha_n y \operatorname{Cosh} \alpha_n y + C_n \operatorname{Sinh} \alpha_n y \\
+ D_n \alpha_n y \operatorname{Sinh} \alpha_n y) \sin \alpha_n x
\end{aligned} \tag{37.42}$$

Applications. The preceding formulas may be used to find the stresses in the vertical walls of bunkers [13],[14]. These are high walls supported by equidistant columns. We discuss the case shown in Fig. 37.9. The vertical load $\bar{p}$ represents the

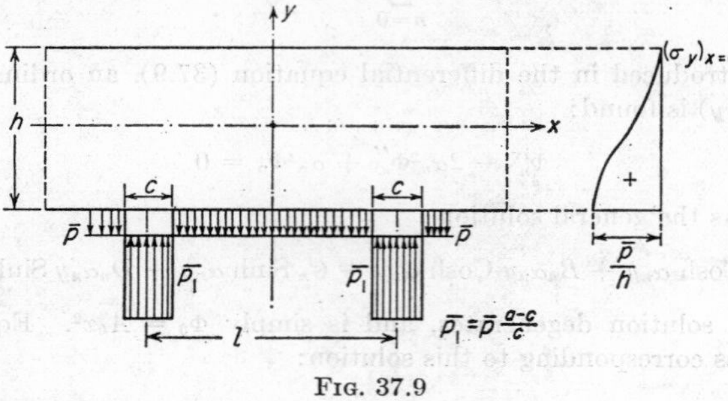

FIG. 37.9

weight of the hoppers, and is in equilibrium with the column reaction $\bar{p}_1$. In this case, $a_n = 0$ for $n \geq 0$, and

$$\bar{a}_n = -\frac{2\bar{p}l}{\pi c}\frac{(-1)^n}{n}\sin\frac{n\pi c}{l} \qquad n \geq 1 \tag{37.43}$$

When this is introduced in Eqs. (37.40), the stresses may be computed. Since usually l/h is far from being a large number, they deviate substantially from those obtained from the elementary beam theory.

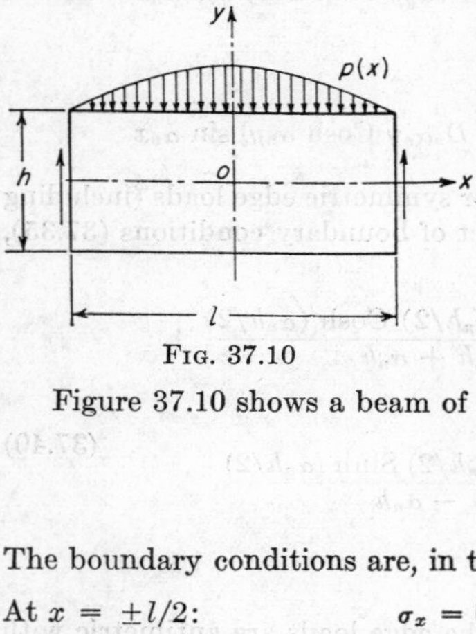

FIG. 37.10

The cases that the load is applied at the upper edge or is the weight of the wall itself may easily be deduced from the preceding solution by adding simple stress fields (see [5], pp. 86–87). For a detailed study of problems of the half space, of infinite strips, and of finite rectangular beams subjected to various loads, see [1], pp. 46–53, and [5], pp. 74–101.

Figure 37.10 shows a beam of span l loaded at its upper edge by a load

$$p(x) = p_1 \cos\frac{\pi x}{l}$$

The boundary conditions are, in this case:

At $x = \pm l/2$: $\sigma_x = 0$

At $y = +h/2$: $\sigma_y = -p_1 \cos\frac{\pi x}{l} \quad \tau_{xy} = 0$

At $y = -h/2$: $\sigma_y = \tau_{xy} = 0$

The solution is given by Eqs. (37.39) if we restrict the sums to the term $n = 1$ and put $\alpha_1 = \pi/l$ and $a_1 = -p_1$. In Fig. 39.11, stress diagrams are shown for two values of the ratio l/h. It may be seen that, for $l/h = 1$, there are important deviations from the linear distribution of σ_x and the parabolic distribution of the shear stress τ_{xy}. They are much less for $l/h = 2$.

The Fourier series solution of the stress problem of the rectangular plate is definitely superior to the polynomial solution, although the latter seems to be the favorite in many textbooks.

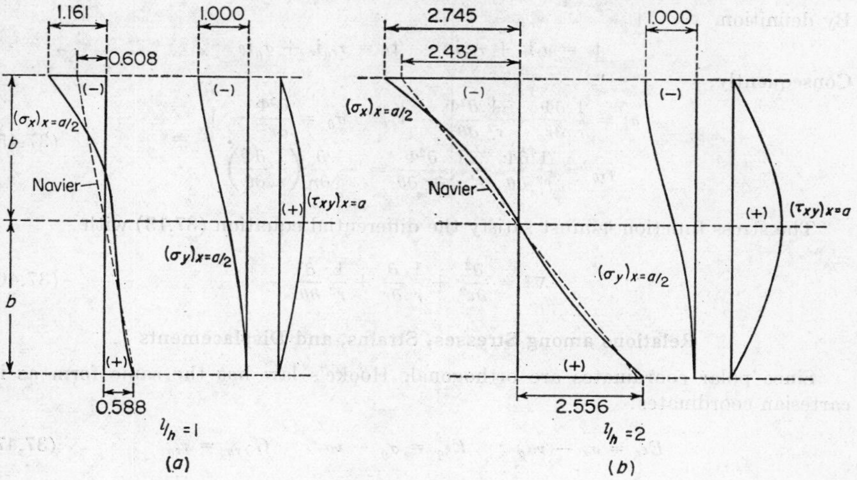

FIG. 37.11

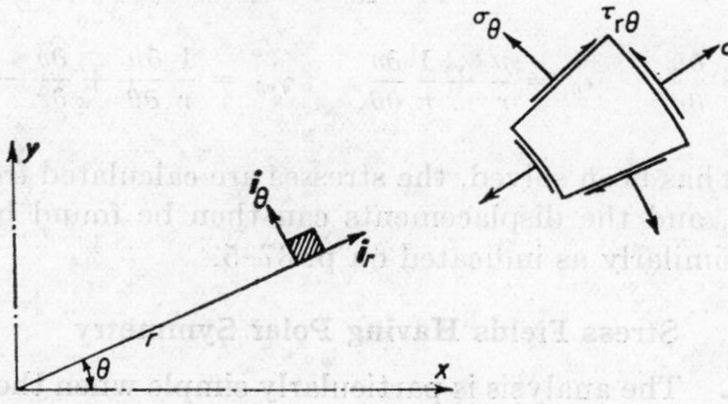

FIG. 37.12

37.3. PLANE PROBLEMS IN POLAR COORDINATES

General Equations

The coordinates r, θ and the stresses σ_r, σ_θ, $\tau_{r\theta}$ are shown in Fig. 37.12. The unit vectors in r and θ directions, respectively, are $\mathbf{i}_r$ and $\mathbf{i}_\theta$. Equations (37.16) yield the stress vectors in terms of Airy's stress function:

$$\mathbf{t}_r = \frac{1}{r}\frac{\partial}{\partial\theta}(\text{grad}\,\Phi \times \mathbf{i}_z) - V\mathbf{i}_r \qquad \mathbf{t}_\theta = -\frac{\partial}{\partial r}(\text{grad}\,\Phi \times \mathbf{i}_z) - V\mathbf{i}_\theta \tag{37.44}$$

Since [see Eq. (6.35c)]

$$\text{grad}\,\Phi \times \mathbf{i}_z = \left(\frac{\partial\Phi}{\partial r}\mathbf{i}_r + \frac{1}{r}\frac{\partial\Phi}{\partial\theta}\mathbf{i}_\theta\right) \times \mathbf{i}_z = -\frac{\partial\Phi}{\partial r}\mathbf{i}_\theta + \frac{1}{r}\frac{\partial\Phi}{\partial\theta}\mathbf{i}_r$$

and $\qquad \partial\mathbf{i}_r/\partial r = \partial\mathbf{i}_\theta/\partial r = 0 \qquad \partial\mathbf{i}_\theta/\partial\theta = -\mathbf{i}_r \qquad \partial\mathbf{i}_r/\partial\theta = \mathbf{i}_\theta$

it follows that

$$\mathbf{t}_r = \frac{1}{r}\left(\frac{\partial\Phi}{\partial r} + \frac{1}{r}\frac{\partial^2\Phi}{\partial\theta^2}\right)\mathbf{i}_r - \frac{1}{r}\left(\frac{\partial^2\Phi}{\partial r\,\partial\theta} - \frac{1}{r}\frac{\partial\Phi}{\partial\theta}\right)\mathbf{i}_\theta - V\mathbf{i}_r$$

$$\mathbf{t}_\theta = \frac{1}{r}\left(\frac{1}{r}\frac{\partial\Phi}{\partial\theta} - \frac{\partial^2\Phi}{\partial r\,\partial\theta}\right)\mathbf{i}_r + \frac{\partial^2\Phi}{\partial r^2}\mathbf{i}_\theta - V\mathbf{i}_\theta$$

By definition,

$$\mathbf{t}_r = \sigma_r \mathbf{i}_r + \tau_{r\theta} \mathbf{i}_\theta \qquad \mathbf{t}_\theta = \tau_{r\theta} \mathbf{i}_r + \sigma_\theta \mathbf{i}_\theta$$

Consequently,

$$\begin{aligned} \sigma_r &= \frac{1}{r}\frac{\partial \Phi}{\partial r} + \frac{1}{r^2}\frac{\partial^2 \Phi}{\partial \theta^2} - V \qquad \sigma_\theta = \frac{\partial^2 \Phi}{\partial r^2} - V \\ \tau_{r\theta} &= \frac{1}{r^2}\frac{\partial \Phi}{\partial \theta} - \frac{1}{r}\frac{\partial^2 \Phi}{\partial r\, \partial \theta} = -\frac{\partial}{\partial r}\left(\frac{1}{r}\frac{\partial \Phi}{\partial \theta}\right) \end{aligned} \tag{37.45}$$

The stress function Φ must satisfy the differential equation (37.13) with

$$\nabla^2 = \frac{\partial^2}{\partial r^2} + \frac{1}{r}\frac{\partial}{\partial r} + \frac{1}{r^2}\frac{\partial^2}{\partial \theta^2} \tag{37.46}$$

Relations among Stresses, Strains, and Displacements

Since polar coordinates are orthogonal, Hooke's law has the same form as in cartesian coordinates:

$$E\epsilon_r = \sigma_r - \nu\sigma_\theta \qquad E\epsilon_\theta = \sigma_\theta - \nu\sigma_r \qquad G\gamma_{r\theta} = \tau_{r\theta} \tag{37.47}$$

The kinematic relations are derived from Eqs. (33.19) by changing the notation from u_r, u_θ, u_z to u, v, w and then dropping w and all z derivatives:

$$\epsilon_r = \frac{\partial u}{\partial r} \qquad \epsilon_\theta = \frac{u}{r} + \frac{1}{r}\frac{\partial v}{\partial \theta} \qquad \gamma_{r\theta} = \frac{1}{r}\frac{\partial u}{\partial \theta} + \frac{\partial v}{\partial r} - \frac{v}{r} \tag{37.48}$$

After Eq. (37.13) has been solved, the stresses are calculated from (37.45) and the strains from (37.47), and the displacements can then be found by integrating Eqs. (37.48), proceeding similarly as indicated on p. 37–5.

Stress Fields Having Polar Symmetry

General Solution. The analysis is particularly simple when the stress distribution is symmetrical about the origin of the coordinates. Then all stresses, strains, displacements, and the stress function Φ depend on r only, and all partial derivatives with respect to θ are zero.

When there are no body forces, $V \equiv 0$ and Φ satisfies the biharmonic equation (37.9), which here reduces to the ordinary differential equation

$$\nabla^2\nabla^2\Phi \equiv \frac{d^4\Phi}{dr^4} + \frac{2}{r}\frac{d^3\Phi}{dr^3} - \frac{1}{r^2}\frac{d^2\Phi}{dr^2} + \frac{1}{r^3}\frac{d\Phi}{dr} = 0 \tag{37.49}$$

It has the general solution

$$\Phi = A \ln r + Br^2 \ln r + Cr^2 + D \tag{37.50}$$

where A, B, C, D are arbitrary constants. The corresponding stresses follow from Eqs. (37.45) with $V = 0$:

$$\begin{aligned} \sigma_r &= \frac{1}{r}\frac{d\Phi}{dr} = \frac{A}{r^2} + B(1 + 2\ln r) + 2C \\ \sigma_\theta &= \frac{d^2\Phi}{dr^2} = -\frac{A}{r^2} + B(3 + 2\ln r) + 2C \end{aligned} \tag{37.51}$$

The shear $\tau_{r\theta}$ vanishes, and the stresses σ_r, σ_θ are principal stresses. The stress trajectories are identical with the circles r = const and the radii θ = const.

The displacements are obtained by integrating Eqs. (37.48). They are

$$u = \frac{1}{E}\,[-(1+\nu)Ar^{-1} + 2(1-\nu)Br\ln r - B(1+\nu)r + 2C(1-\nu)r] + H\sin\theta + K\cos\theta \qquad (37.52)$$

$$v = \frac{4B}{E}\,r\theta + Fr + H\cos\theta - K\sin\theta$$

where F, H, K are arbitrary constants. These displacements do not necessarily have polar symmetry. In particular, the tangential displacement v is multivalued if $B \neq 0$. It increases by $8\pi Br/E$ when θ is increased by 2π. The B term in Eqs. (37.52) represents one of the dislocation solutions described on p. 37–3. If the problem in hand does not admit such a solution, it is necessary to put $B = 0$.

Both the A and B terms in Eqs. (37.51) tend to infinity for $r \to 0$. Therefore, if the plate does not have a hole including this point, the only possible solution is $\sigma_r = \sigma_\theta = 2C$, that is, a state of uniform biaxial tension or compression.

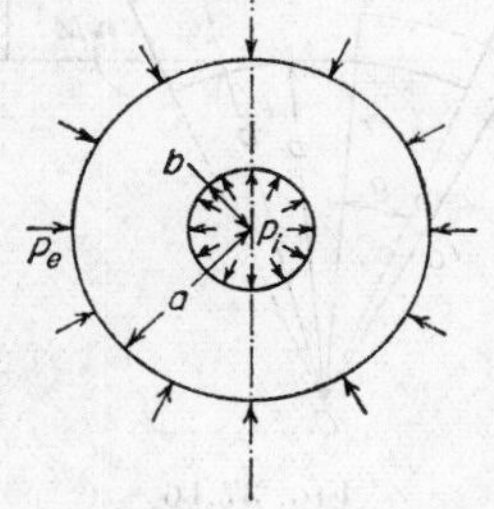

FIG. 37.13

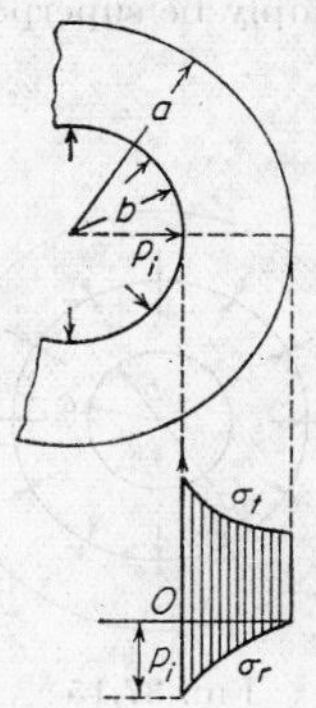

FIG. 37.14

Thick-walled Cylinder. Figure 37.13 shows a cross section of a thick-walled cylinder subjected to an internal pressure p_i and an external pressure p_e. Uniqueness of the displacements requires that $B = 0$. Putting $F = H = K = 0$, we exclude an uninteresting rigid-body displacement, and, finally, A and C may be calculated from the boundary conditions, that, for $r = a$ and $r = b$, the radial stress σ_r assumes the values $-p_e$ and $-p_i$, respectively. The following stresses and displacements result:

$$\sigma_r = \frac{a^2b^2(p_e - p_i)}{a^2 - b^2}\,r^{-2} + \frac{p_ib^2 - p_ea^2}{a^2 - b^2}$$

$$\sigma_\theta = -\frac{a^2b^2(p_e - p_i)}{a^2 - b^2}\,r^{-2} + \frac{p_ib^2 - p_ea^2}{a^2 - b^2} \qquad (37.53)$$

$$u = \frac{1+\nu}{E}\,\frac{a^2b^2(p_i - p_e)}{(a^2 - b^2)r} + \frac{1-\nu}{E}\,\frac{b^2p_i - a^2p_e}{a^2 - b^2}\,r$$

Tube Subjected to Internal Pressure. For $p_e = 0$, Eqs. (37.53) reduce to

$$\sigma_r = \frac{b^2p_i}{a^2 - b^2}\left(1 - \frac{a^2}{r^2}\right) \qquad \sigma_\theta = \frac{b^2p_i}{a^2 - b^2}\left(1 + \frac{a^2}{r^2}\right) \qquad (37.54)$$

Since $a > r$, σ_r is always a compressive and σ_θ a tensile stress (Fig. 37.14). The hoop stress σ_θ is a maximum at the inner surface of the cylinder:

$$\sigma_{\theta\,\max} = \frac{p_i(a^2 + b^2)}{a^2 - b^2}$$

It is always larger than p_i, and approaches this value when a is increased. The minimum of σ_θ occurs at the outer surface of the cylinder. The ratio

$$\frac{\sigma_{\theta\ \max}}{\sigma_{\theta\ \min}} = \frac{a^2 + b^2}{2b^2}$$

increases with the wall thickness. For $a/b \approx 1$, that is, for thin-walled tubes, σ_θ is nearly uniformly distributed in agreement with the assumption made in the elementary theory of thin-walled tubes. The greatest shearing stress, occurring in sections at 45° to the radii, is a maximum at the inner surface: $\tau_{\max} = a^2 p_i/(a^2 - b^2)$.

The preceding results have been derived under the assumption that the cylinder is not stressed axially. If it is closed at the ends, the internal pressure produces a tensile force $p_i \pi b^2$ which, when distributed uniformly, leads to a tensile stress

$$\sigma_z = \frac{b^2 p_i}{a^2 - b^2}$$

which may simply be superposed on the stresses from Eqs. (37.54).

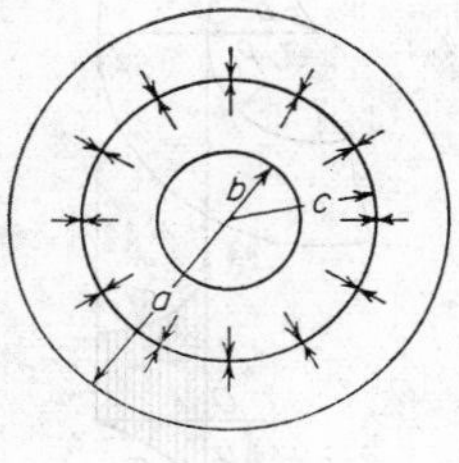

FIG. 37.15

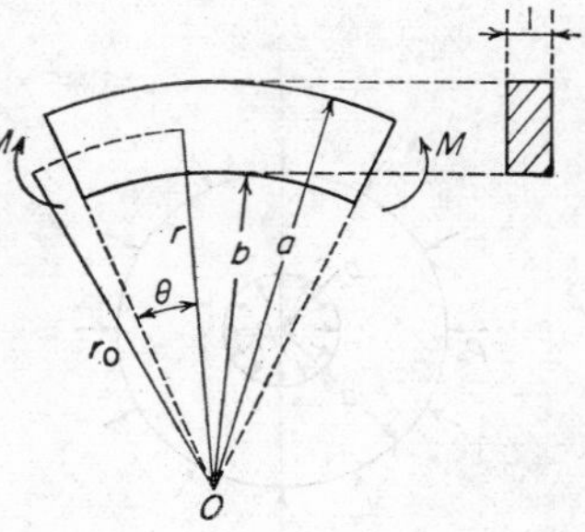

FIG. 37.16

Shrink Fits. In order to reduce the peak value of the hoop stress σ_θ shown in Fig. 37.14, a tube may be built up of two concentric shells (Fig. 37.15). In the unstressed condition, the outer diameter of the inner cylinder is larger than the inner diameter of the outer cylinder, the difference being 2δ. The parts are assembled after heating the outer cylinder. After cooling, there is a shrink-fit pressure p_c between the two faces $r = c$ of the cylinders. It can be calculated by applying the last of Eqs. (37.53), first to the outer cylinder with $p_i = p_c$, and then to the inner cylinder with $p_e = p_c$. The corresponding displacements for $r = c$ are:

Inner face of the outer cylinder

$$u_1 = \frac{cp_c}{E}\left(\frac{a^2 + c^2}{a^2 - c^2} + \nu\right)$$

Outer face of the inner cylinder

$$u_2 = -\frac{cp_c}{E}\left(\frac{b^2 + c^2}{c^2 - b^2} - \nu\right)$$

The pressure p_c is found from the condition that $u_1 - u_2 = \delta$. It is

$$p_c = \frac{E\delta}{c}\,\frac{(a^2 - c^2)(c^2 - b^2)}{2c^2(a^2 - b^2)} \tag{37.55}$$

This pressure produces compressive hoop stresses σ_θ in the inner cylinder and tensile hoop stresses in the outer cylinder. When the built-up cylinder is subjected to loads at its faces $r = a$ and $r = b$, the corresponding stresses may be computed from Eqs.

(37.53) and superposed to the shrink-fit stresses. It is obvious that, for internal pressure, the maximum hoop stress in the built-up cylinder is smaller than in a cylinder without prestress.

Curved Beam Subjected to Pure Bending. Figure 37.16 shows a curved bar under the influence of equal couples M at both ends. The bending moment in any cross section equals M, and we ask for a stress system which is independent of θ. It is represented by Eqs. (37.51). The boundary conditions are as follows: Along the curved edges $r = a$ and $r = b$: $\sigma_r = 0$; along the straight edges and, consequently, for every θ:

$$\int_b^a \sigma_\theta \, dr = 0 \qquad \int_b^a \sigma_\theta r \, dr = -M$$

These conditions are satisfied by the following values for the free constants:

$$A = -\frac{4M}{\Delta} a^2b^2 \ln \frac{a}{b} \qquad B = \frac{2M}{\Delta}(a^2 - b^2)$$
$$C = \frac{M}{\Delta}[a^2 - b^2 + 2(a^2 \ln a - b^2 \ln b)] \tag{37.56}$$
$$\Delta = (a^2 - b^2)^2 - 4a^2b^2\left(\ln \frac{a}{b}\right)^2$$

With these constants Eqs. (37.51) yield the following stresses:

$$\sigma_r = -\frac{4M}{\Delta}\left(\frac{a^2b^2}{r^2} \ln \frac{a}{b} + a^2 \ln \frac{r}{a} + b^2 \ln \frac{b}{r}\right)$$
$$\sigma_\theta = -\frac{4M}{\Delta}\left(-\frac{a^2b^2}{r^2} \ln \frac{a}{b} + a^2 \ln \frac{r}{a} + b^2 \ln \frac{b}{r} + a^2 - b^2\right) \tag{37.57}$$

For a comparison with the elementary theory of curved beams, see p. 35–11.

Rotating Disk of Uniform Thickness

General Remarks. For a disk of uniform thickness and of specific mass ρ, rotating at constant angular velocity ω, the d'Alembert inertia force per unit volume is $R = \rho\omega^2 r$, directed outward. It may be derived from a potential $V = \frac{1}{2}\rho\omega^2 r^2$.

The right-hand side of Eq. (37.13) is, in this case,

$$(1 - \nu)\,\nabla^2 V = (1 - \nu)\left(\frac{d^2V}{dr^2} + \frac{1}{r}\frac{dV}{dr}\right) = 2(1 - \nu)\rho\omega^2$$

and, since $\nabla^2\nabla^2\Phi$ is given by Eq. (37.49), the differential equation

$$\frac{d^4\Phi}{dr^4} + \frac{2}{r}\frac{d^3\Phi}{dr^3} - \frac{1}{r^2}\frac{d^2\Phi}{dr^2} + \frac{1}{r}\frac{d\Phi}{dr} = 2(1 - \nu)\rho\omega^2$$

must be solved.

Obviously,

$$\Phi = \frac{1 - \nu}{32}\rho\omega^2 r^4$$

is a particular solution. Adding the complementary solution (37.50), we have the general solution

$$\Phi = A \ln r + Br^2 \ln r + Cr^2 + D + \frac{1 - \nu}{32}\rho\omega^2 r^4 \tag{37.58}$$

Uniqueness of displacements requires that $B = 0$. What is left yields the stress field,

$$\sigma_r = \frac{A}{r^2} + 2C - \frac{3 + \nu}{8}\rho\omega^2 r^2 \qquad \sigma_\theta = -\frac{A}{r^2} + 2C - \frac{1 + 3\nu}{8}\rho\omega^2 r^2 \tag{37.59}$$

Solid Disk. For a solid disk, regularity of the stress field at $r = 0$ requires that $A = 0$. C is determined from the boundary condition $\sigma_r = 0$ at the edge $r = a$. It yields

$$C = \frac{3 + \nu}{16} \rho\omega^2 a$$

and the stresses are:

$$\sigma_r = \frac{3 + \nu}{8} \rho\omega^2(a^2 - r^2) \qquad \sigma_\theta = \frac{3 + \nu}{8} \rho\omega^2 a^2 - \frac{1 + 3\nu}{8} \rho\omega^2 r^2 \tag{37.60}$$

Both stresses reach their maxima at the center of the disk, where

$$\sigma_r = \sigma_\theta = \frac{3 + \nu}{8} \rho\omega^2 a^2 \tag{37.61}$$

Disk with a Circular Hole. In this case, the boundary conditions are $\sigma_r = 0$ for $r = a$ and $r = b$. They yield the constants

$$A = -\frac{3 + \nu}{8} \rho\omega^2 a^2 b^2 \qquad C = \frac{3 + \nu}{16} \rho\omega^2 (a^2 + b^2)$$

and the stresses

$$\begin{aligned} \sigma_r &= \frac{3 + \nu}{8} \rho\omega^2 \left(a^2 + b^2 - \frac{a^2 b^2}{r^2} - r^2 \right) \\ \sigma_\theta &= \frac{3 + \nu}{8} \rho\omega^2 \left(a^2 + b^2 + \frac{a^2 b^2}{r^2} - \frac{1 + 3\nu}{3 + \nu} r^2 \right) \end{aligned} \tag{37.62}$$

The largest radial stress occurs at $r = \sqrt{ab}$, and is

$$\sigma_{r\,\max} = \frac{3 + \nu}{8} \rho\omega^2 (a - b)^2 \tag{37.63a}$$

The largest circumferential stress is found at the inner edge. Its value

$$\sigma_{\theta\,\max} = \frac{3 + \nu}{4} \rho\omega^2 \left(a^2 + \frac{1 - \nu}{3 + \nu} b^2 \right) \tag{37.63b}$$

is always higher than $\sigma_{r\,\max}$.

When $b \to 0$, the maximum circumferential stress approaches a value twice that obtained from Eq. (37.61) for the solid disk.

Rotating Disk Subjected to Radial Forces at the Edge. Turbine disks are subjected to a radial load at the outer edge, caused by the centrifugal force acting on the blades, and to a shrink-fit pressure at the inner edge. The corresponding stresses may be calculated from Eqs. (37.53).

Rotating Disk of Variable Thickness. In the general case where the thickness t is an arbitrary function of r, the equation of equilibrium in the radial direction is

$$\frac{r}{t} \frac{d}{dr} (t\sigma_r) + \sigma_r - \sigma_\theta + \rho\omega^2 r^2 = 0 \tag{37.64}$$

Calling $u = u(r)$ the radial displacement of any point, the strains are $\epsilon_r = du/dr$ and $\epsilon_\theta = u/r$, so that Hooke's law gives the relations

$$E \frac{du}{dr} = \sigma_r - \nu\sigma_\theta \qquad E \frac{u}{r} = \sigma_\theta - \nu\sigma_r$$

Eliminating u between them, we obtain

$$r \left(\nu \frac{d\sigma_r}{dr} - \frac{d\sigma_\theta}{dr} \right) + (1 + \nu)(\sigma_r - \sigma_\theta) = 0 \tag{37.65}$$

There exists an elementary solution of these equations for the case where the thickness t varies as $t = Cr^n$, where C and n are constants; it is rarely applicable to practical problems. The solution is also known for disks of conical profile. For a complete study of both cases, see [8], pp. 639–645.

Disk of Constant Strength. The lightest disk is that for which each of the stresses σ_r, σ_θ takes a constant value in the whole disk. Equation (37.65) shows that, in this case, σ_r and σ_θ must be equal. Let us call σ their common value. Equation (37.64) reduces then to

$$\frac{1}{t}\frac{dt}{dr} + \frac{\rho\omega^2}{\sigma} r = 0$$

or, integrated,

$$t = t_0 e^{-(\rho\omega^2/2\sigma)r^2} \tag{37.66}$$

These profiles present an inflection point for $r = \sqrt{\sigma/\rho\omega^2}$. They are usable only if the disk has no central hole. For further details on this problem, see [8], pp. 638–639.

Stress Fields without Polar Symmetry

General Solution. The following formulas give stresses and displacements which are regular and single-valued in the entire plane, with the exception of a possible singularity at $r = 0$.

Stresses and displacements independent of θ

$$\Phi = A \ln r + Cr^2 + D\theta \tag{37.67}$$

with the stresses

$$\sigma_r = Ar^{-2} + 2C \qquad \sigma_\theta = -Ar^{-2} + 2C \qquad \tau_{r\theta} = Dr^{-2} \tag{37.68a}$$

and the displacements

$$Eu = -(1+\nu)Ar^{-1} + 2(1-\nu)Cr \qquad Ev = (1+\nu)Dr^{-1} \tag{37.68b}$$

Stresses and displacements proportional cos θ or sin θ

$$\Phi = (Br^3 + Cr^{-1})\cos\theta + D[(1-\nu)r(1-\ln r)\cos\theta + 2r\theta\sin\theta] \tag{37.69}$$

with the stresses

$$\begin{aligned}\sigma_r &= [2Br - 2Cr^{-3} + (3+\nu)Dr^{-1}]\cos\theta \\ \sigma_\theta &= [6Br + 2Cr^{-3} - (1-\nu)Dr^{-1}]\cos\theta \\ \tau_{r\theta} &= [2Br - 2Cr^{-3} - (1-\nu)Dr^{-1}]\sin\theta\end{aligned} \tag{37.70a}$$

and the displacements

$$\begin{aligned}Eu &= [(1-3\nu)Br^2 + (1+\nu)Cr^{-2} + (1+\nu)(3-\nu)D\ln r]\cos\theta \\ Ev &= [(5+\nu)Br^2 + (1+\nu)Cr^{-2} - (1+\nu)D((1+\nu) + (3-\nu)\ln r)]\sin\theta\end{aligned} \tag{37.70b}$$

Stresses and displacements proportional cos $n\theta$ or sin $n\theta$ with $n \geq 2$

$$\Phi = (Ar^n + Br^{n+2} + Cr^{-n} + Dr^{-n+2})\cos n\theta \tag{37.71}$$

with the stresses

$$\begin{aligned}\sigma_r &= -[(n-1)nAr^{n-2} + (n-2)(n+1)Br^n + n(n+1)Cr^{-n-2} \\ &\qquad + (n-1)(n+2)Dr^{-n}]\cos n\theta \\ \sigma_\theta &= [(n-1)nAr^{n-2} + (n+1)(n+2)Br^n + n(n+1)Cr^{-n-2} \\ &\qquad + (n-2)(n-1)Dr^{-n}]\cos n\theta \\ \tau_{r\theta} &= [(n-1)nAr^{n-2} + n(n+1)Br^n - n(n+1)Cr^{-n-2} \\ &\qquad - (n-1)nDr^{-n}]\sin n\theta\end{aligned} \tag{37.72a}$$

and the displacements

$$Eu = -\{n(1+\nu)Ar^{n-1} + [n(1+\nu) - 2(1-\nu)]Br^{n+1} + n(1+\nu)Cr^{-n-1} - [n(1+\nu) + 2(1-\nu)]Dr^{-n+1}\} \cos n\theta$$

$$Ev = \{n(1+\nu)Ar^{n-1} + [n(1+\nu) + 4]Br^{n+1} + n(1+\nu)Cr^{-n-1} + [n(1+\nu) - 4]Dr^{-n+1}\} \sin n\theta \qquad (37.72b)$$

Bending of a Curved Bar Loaded at the End. For the case shown in Fig. 37.17, the bending moment is proportional to sin θ, and we may expect that the same is true for σ_θ. However, Eqs. (37.67) to (37.70) do not solve the problem completely, since they do not contain certain solutions which are not applicable to a complete ring, but are not objectionable when applied to a quarter ring.

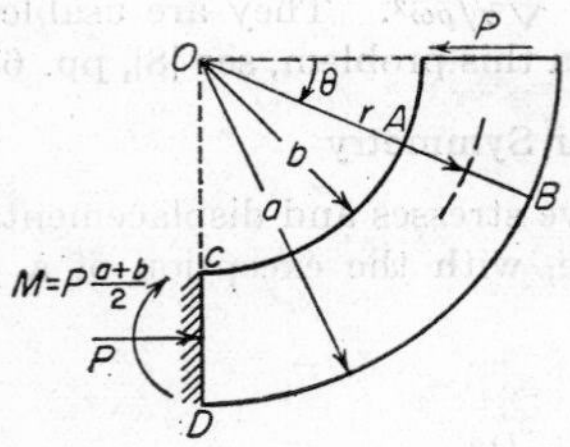

FIG. 37.17

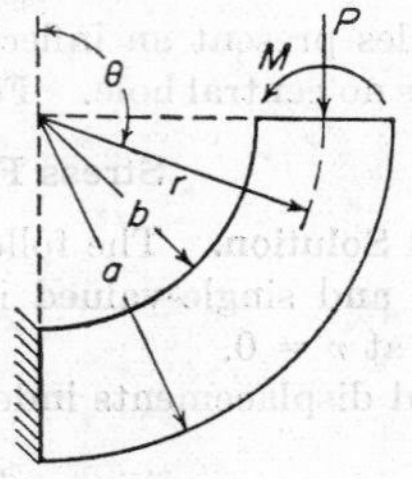

FIG. 37.18

The boundary conditions are, here,

For $r = a$ and $r = b$: $\qquad \sigma_r = \tau_{r\theta} = 0$

For $\theta = 0$: $\qquad \int_b^a \sigma_r \, dr = 0 \qquad \int_b^a \tau_{r\theta} \, dr = P$

To satisfy them, the following stress function may be used:

$$\Phi = (Br^3 + Cr^{-1} + Dr \ln r) \sin \theta$$

It satisfies automatically the condition $\sigma_r = 0$ at $\theta = 0$, and, when the constants B, C, D are chosen to be

$$B = \frac{P}{2\Delta} \qquad C = -\frac{Pa^2b^2}{2\Delta} \qquad D = -\frac{P(a^2+b^2)}{\Delta}$$

with $\qquad \Delta = (a^2 + b^2) \ln \frac{a}{b} - a^2 + b^2$

then also the other conditions are fulfilled, and the stresses are:

$$\sigma_r = \frac{P}{\Delta}\left[r - \frac{a^2+b^2}{r} + \frac{a^2b^2}{r^3}\right] \sin \theta$$

$$\sigma_\theta = \frac{P}{\Delta}\left[3r - \frac{a^2+b^2}{r} - \frac{a^2b^2}{r^3}\right] \sin \theta \qquad (37.73)$$

$$\tau_{r\theta} = -\frac{P}{\Delta}\left[r - \frac{a^2+b^2}{r} + \frac{a^2b^2}{r^3}\right] \cos \theta$$

It follows from the equilibrium of the quarter-ring plate shown in Fig. 37.17 that, at $\theta = 90°$, a compressive force P and a bending moment $M = P(a+b)/2$ are acting. Consequently, the next quarter of the same ring is identical with the curved bar shown in Fig. 37.18, and Eqs. (37.73) are applicable to this case if the angle θ is defined as was done in Fig. 37.18. The solutions pertaining to these two figures and the pure bending solution given on p. 37–15 may be superposed to find the stresses in a curved bar under an arbitrary end load. For further discussion and displacements, see [1], pp. 73–78.

Circular Hole in an Infinite Plate, Uniaxial Tension. Assume an infinite plate subjected to a uniform tensile stress $\sigma_x = \sigma_1$. When a circular hole of radius b is cut in this plate (Fig. 37.19), stresses

$$\sigma_r = \sigma_1 \cos^2 \theta = \tfrac{1}{2}\sigma_1(1 + \cos 2\theta)$$
$$\tau_{r\theta} = -\sigma_1 \sin \theta \cos \theta = -\tfrac{1}{2}\sigma_1 \sin 2\theta \qquad (37.74)$$

must be applied at its edge to maintain the uniform stress field in the plate surrounding it. When we want the edge of the hole free of stresses, we must apply forces opposite and equal to the distribution (37.74), and superpose the resulting stress field to the uniform tension σ_1.

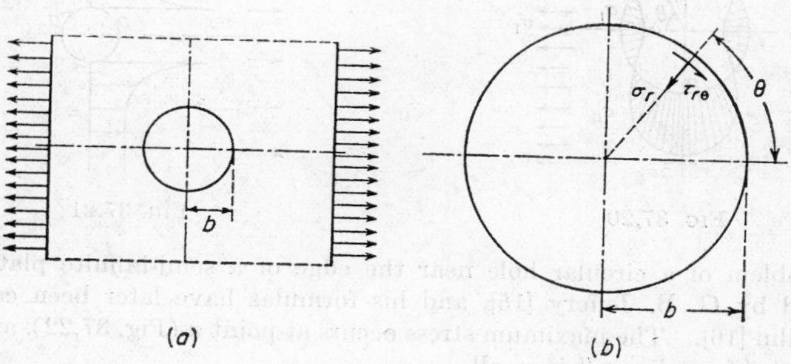

FIG. 37.19

The stress function describing this disturbance of the uniform tension is given by Eqs. (37.51) and (37.72) with $n = 2$. In order to maintain uniform tension at infinity, we must exclude all particular solutions which would yield finite or infinite stresses for $r \to \infty$. This requires that $B = C = 0$ in Eq. (37.51) and $A = B = 0$ in Eqs. (37.72). The other constants follow from the boundary conditions, and lead to the stress function

$$\Phi = -\frac{\sigma_1}{2} b^2 \ln r - \frac{\sigma_1}{4}(b^4 r^{-2} + 2b^2) \cos 2\theta$$

Superposing the corresponding stresses and the uniform tension, we obtain

$$\sigma_r = \frac{\sigma_1}{2}\left(1 - \frac{b^2}{r^2}\right) + \frac{\sigma_1}{2}\left(1 - \frac{b^2}{r^2}\right)\left(1 - \frac{3b^2}{r^2}\right) \cos 2\theta$$
$$\sigma_\theta = \frac{\sigma_1}{2}\left(1 + \frac{b^2}{r^2}\right) - \frac{\sigma_1}{2}\left(1 + \frac{3b^4}{r^4}\right) \cos 2\theta \qquad (37.75)$$
$$\tau_{r\theta} = -\frac{\sigma_1}{2}\left(1 - \frac{b^2}{r^2}\right)\left(1 + \frac{3b^2}{r^2}\right) \sin 2\theta$$

It may be seen that σ_θ is a maximum for $\theta = \pm 90°$, that is, at the ends of the diameter mn perpendicular to the direction of tension (Fig. 37.20); at these points $\sigma_{\theta\,\max} = 3\sigma_1$. This is the largest normal stress occurring in this case.

Like holes, notches in the edge of a plate also introduce stress concentrations. They are usually described by a *stress-concentration factor*, defined as the ratio

$$K = \frac{\text{maximum normal stress at notch}}{\text{nominal stress existing in absence of notch}}$$

It is seen that, in this case, $K = 3$.

At the points p, q, Eqs. (37.75) yield $\sigma_\theta = -\sigma_1$, indicating that there exists a (small) area with compressive stress.

Stresses in a section normal to the direction of the tension are obtained by letting $\theta = 90°$; Eqs. (37.75) yield

$$\sigma_\theta = \frac{\sigma_1}{2}\left(2 + \frac{b^2}{r^2} + \frac{3b^4}{r^4}\right) \qquad \tau_{r\theta} = 0$$

The result is plotted in Fig. 37.21. It shows that the stress disturbance caused by the hole is very localized, and that σ_θ very rapidly approaches σ_1 as r increases.

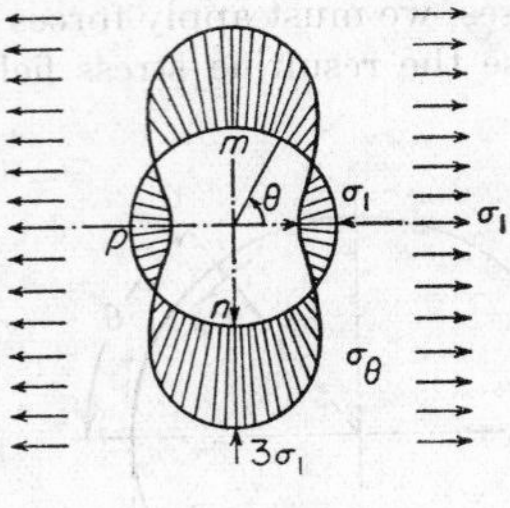

FIG. 37.20

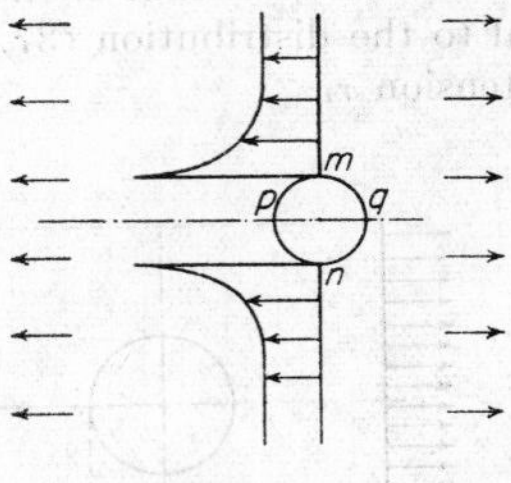

FIG. 37.21

The problem of a circular hole near the edge of a semi-infinite plate has been investigated by G. B. Jeffery [15], and his formulas have later been corrected by R. D. Mindlin [16]. The maximum stress occurs at point n (Fig. 37.22), and is a very large multiple of σ_1 when c/b is small.

Circular Hole in a Plate Strip. The localized character of the stress disturbance around a hole justifies the application of Eqs. (37.75) to plates of finite width in the direction normal to the tension. However, when the width of the plate strip is comparable to the diameter of the hole, Eqs. (37.75) are no longer applicable. Many

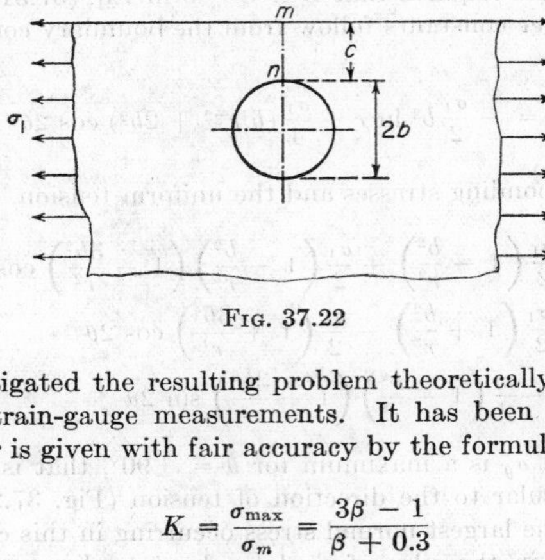

FIG. 37.22

authors have investigated the resulting problem theoretically, photoelastically, and with the help of strain-gauge measurements. It has been found that the stress-concentration factor is given with fair accuracy by the formula

$$K = \frac{\sigma_{\max}}{\sigma_m} = \frac{3\beta - 1}{\beta + 0.3}$$

where β = (width of the strip)/(diameter of the hole), and σ_m = mean stress in the weakened section.

Circular Hole in an Infinite Plate, Biaxial Tension. The stress disturbance caused by a circular hole in a biaxial stress field may easily be derived from Eqs. (37.75). We replace σ_1 by σ_2, and θ by $(\theta + 90°)$, and add the stresses so obtained to those from

Eqs. (37.75). The result is the disturbance in a biaxial stress field with principal stresses σ_1 and σ_2.

If the plate is subjected to a uniform tension in all directions ($\sigma_1 = \sigma_2$), we find at the edge of the hole that $\sigma_{\theta\,\max} = 2\sigma_1$; hence, $K = 2$. If, on the other hand, the plate is subjected to pure shear ($\sigma_1 = -\sigma_2$, Fig. 37.23), then the circumferential stress along the edge of the hole is found to be $\sigma_\theta = -4\sigma_1 \cos 2\theta$. For $\theta = \pm 90°$, that is, at the points m, n in Fig. 37.23, we find $\sigma_\theta = 4\sigma_1$, and, for $\theta = 0°, 180°$, that is, at m_1, n_1, $\sigma_\theta = -4\sigma_1$. Hence, the stress-concentration factor $K = 4$ in this case.[1]

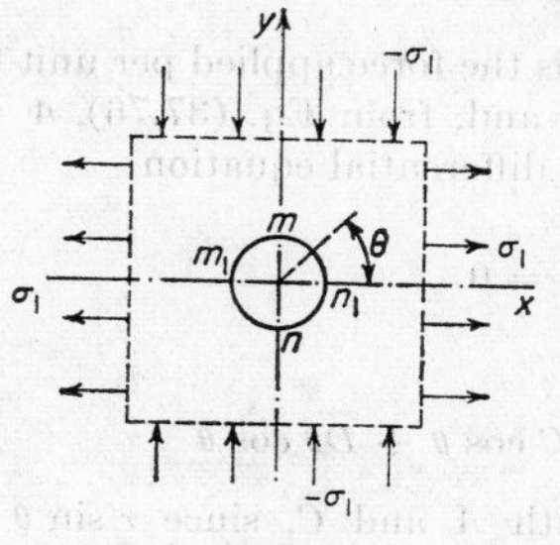

FIG. 37.23

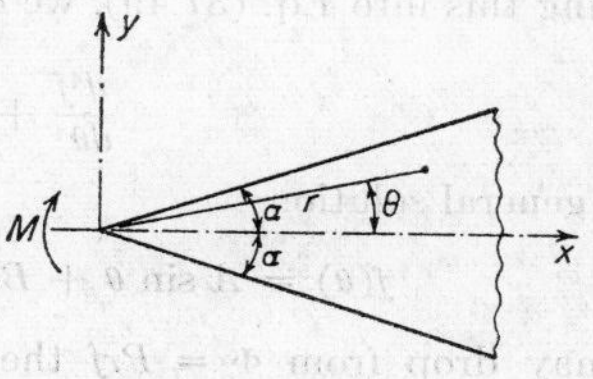

FIG. 37.24

Stresses in Wedges

Dimensional Analysis. Several problems of practical interest pertain to infinite elastic wedges of constant thickness (taken as unity), loaded in different ways. The solution of these problems is greatly facilitated by dimensional analysis.

Let F and L denote the dimensions of force and length, respectively. The stress function has the dimension F since its second derivatives are stresses and have the dimension FL^{-2}. Now, the stress function depends on the coordinates r and θ, and is proportional to a quantity G measuring the applied load, and, hence, must have the form

$$\Phi = Gr^b f(\theta)$$

Let us assume that dim $G = FL^a$; then it is obvious that $b = -a$ to make the preceding equation dimensionally correct. Consequently, Φ must have the form

$$\Phi = Gr^{-a} f(\theta) \tag{37.76}$$

Concentrated Couple at the Apex. The concentrated couple (Fig. 37.24) has the moment M per unit thickness of the plate, and hence the dimension F. Consequently, $a = 0$ and $\Phi = Mf(\theta)$. Upon introducing this in Eq. (37.49), we find

$$\nabla^2\Phi = \frac{M}{r^2}\frac{d^2f(\theta)}{d\theta^2} \qquad \nabla^2\nabla^2\Phi = \frac{M}{r^4}\left(\frac{d^4f}{d\theta^4} + 4\frac{d^2f}{d\theta^2}\right) = 0$$

The last of these relations is an ordinary differential equation for $f(\theta)$, and has the general solution

$$f(\theta) = A\theta + B + C\cos 2\theta + D\sin 2\theta$$

which, except for a missing factor M, is the stress function.

[1] Since plastic deformation does not depend on the largest stress, the stress-concentration factors given here do not apply to the yield load of the plate. Using the Tresca-Mohr yield criterion (see p. 46-7), we find that the greatest difference between two principal stresses is at infinity $\sigma_1 - \sigma_2 = 2\sigma_1$ and at the hole $\sigma_\theta - 0 = 4\sigma_1$; hence, the corresponding stress-concentration factor $= 2$.

The boundary conditions prescribe $\sigma_\theta = \tau_{r\theta} = 0$ for $\theta = \pm\alpha$. After determining the free constants from these conditions, we find

$$\Phi = M \frac{\sin 2\theta - 2\theta \cos 2\alpha}{2(\sin 2\alpha - 2\alpha \cos 2\alpha)} \tag{37.77}$$

and

$$\sigma_r = -\frac{2M \sin 2\theta}{(\sin 2\alpha - 2\alpha \cos 2\alpha)r^2} \qquad \sigma_\theta = 0$$
$$\tau_{r\theta} = \frac{M(\cos 2\theta - \cos 2\alpha)}{(\sin 2\alpha - 2\alpha \cos 2\alpha)r^2} \tag{37.78}$$

Concentrated Force P at the Apex. Since P is the force applied per unit thickness of the plate, we have dim $P = FL^{-1}$, $a = -1$, and, from Eq. (37.76), $\Phi = Prf(\theta)$. Introducing this into Eq. (37.49), we obtain the differential equation

$$\frac{d^4f}{d\theta^4} + 2\frac{d^2f}{d\theta^2} + f = 0$$

with the general solution

$$f(\theta) = A \sin \theta + B\theta \sin \theta + C \cos \theta + D\theta \cos \theta$$

We may drop from $\Phi = Prf$ the terms with A and C, since $r \sin \theta = y$ and $r \cos \theta = x$ have vanishing second derivatives in x and y and, therefore, yield no

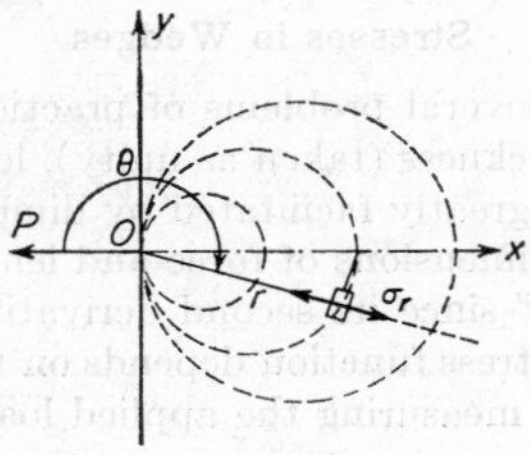

Fig. 37.25

stresses. The remaining terms differ essentially by a phase shift of θ. It is, therefore, sufficient to study the plane-stress system corresponding to the stress function

$$\Phi = Br\theta \sin \theta \tag{37.79}$$

The stresses are

$$\sigma_r = 2B\frac{\cos \theta}{r} \qquad \sigma_\theta = 0 \qquad \tau_{r\theta} = 0 \tag{37.80}$$

Evidently, σ_r and σ_θ are principal stresses, and every point is in a state of uniaxial tension, varying in magnitude and direction. The resultant stress vector for any section at any point has radial direction. Therefore, the stress field (37.80) is called the *simple radial stress field*.

The circles drawn as broken lines in Fig. 37.25 are loci $\sigma_r =$ const. Since $\sigma_\theta \equiv 0$, they are also lines of constant difference between principal stresses, and will show as isochromatics in a photoelastic test.

Since no stresses are transmitted in sections $\theta =$ const, we may cut by such lines arbitrary wedge-shaped parts from the plate, and thus obtain solutions for wedges with different loads applied at the apex. The following two cases are of importance: Pure tension (or compression) (Fig. 37.26)

$$\sigma_r = \frac{2P}{2\alpha + \sin 2\alpha}\frac{\cos \theta}{r} \tag{37.81}$$

Bending and shear (Fig. 37.27)

$$\sigma_r = -\frac{2P}{2\alpha - \sin 2\alpha}\frac{\sin\theta}{r} \tag{37.82}$$

These equations may be applied for any value of $\alpha \leq 180°$. For $\alpha = 180°$, they represent the stresses in a plane with a slit along the line $\theta = 180°$.

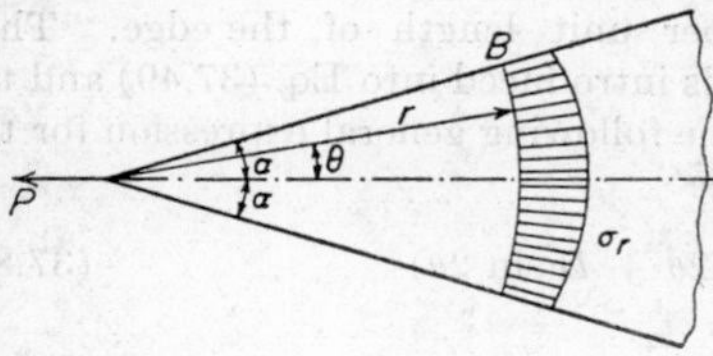

FIG. 37.26

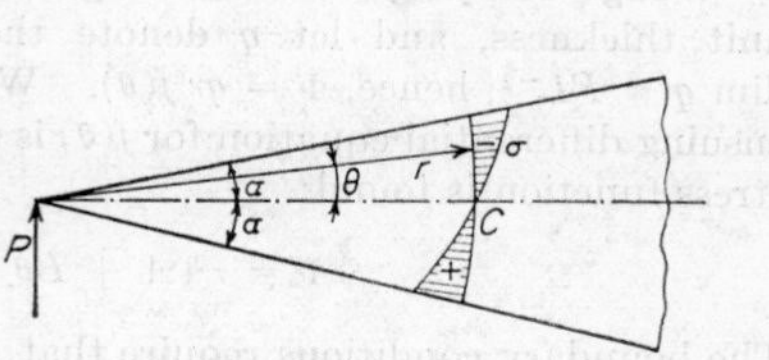

FIG. 37.27

Superposition of the stress fields (37.78), (37.81), and (37.82) yields the solution for the general case of a wedge subjected simultaneously to actions H_0, V_0, and M_0 (Fig. 37.28). Such a solution may be used for obtaining the distribution of stresses in the cross section S of a beam of variable height. By drawing the tangents to the contour of the beam at A and B, we define a wedge of apex O. The normal force N, shear force V, and bending moment M, acting in the section S of the actual beam, are replaced by the statically equivalent loads V_0, H_0, and M_0, acting at the apex. By superposing the corresponding stress fields, we may calculate the stresses in a circular section ACB of radius $r = OA$, which is a good approximation of the stresses in S.

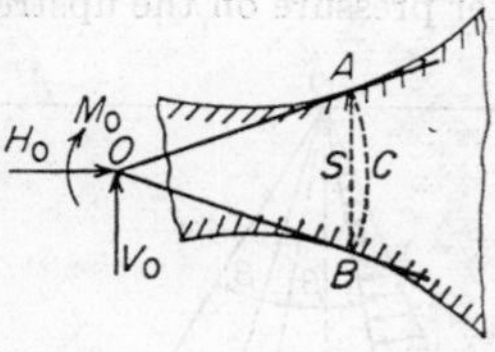

FIG. 37.28

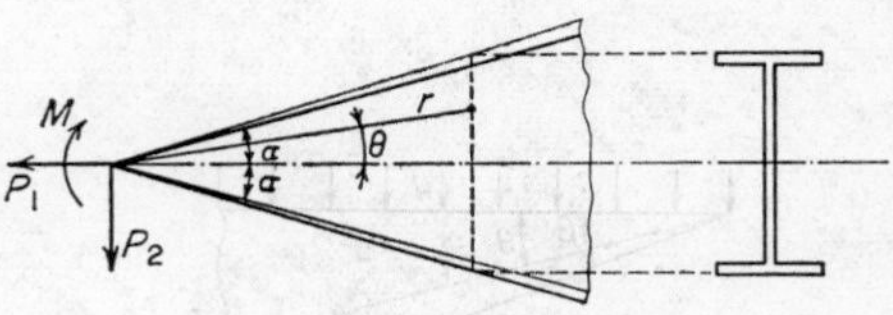

FIG. 37.29

Flanged Wedges. The theory has been generalized by W. R. Osgood [20] for the case of a wedge with equal flanges (Fig. 37.29). Each flange is assumed to be thin, so that its cross-sectional area may be considered concentrated at its centroid. The cross section $A_f = a_f r$ is assumed to increase linearly with the distance from the apex.[1] Moreover, we do no longer consider a plate of unit thickness, but denote the web thickness by t and understand under P, M the total loads, not the loads per unit thickness. Then Eqs. (37.78), (37.81), and (37.82) may be generalized as follows (Fig. 37.29):

Axial load P_1

$$\sigma_r = \frac{P_1}{t(\alpha + \frac{1}{2}\sin 2\alpha) + 2a_f\cos^2\alpha}\frac{\cos\theta}{r} \qquad \sigma_\theta = \tau_{r\theta} = 0 \tag{37.83}$$

Transverse load P_2

$$\sigma_r = \frac{P_2}{t(\alpha - \frac{1}{2}\sin 2\alpha) + 2a_f\cos^2\alpha}\frac{\sin\theta}{r} \qquad \sigma_\theta = \tau_{r\theta} = 0 \tag{37.84}$$

[1] Osgood assumes A_f to be a constant; however, a law of linear variation seems necessary in order to ensure compatibility of strains at the junction web-flange and, in the case of moment loading, to satisfy the equilibrium of shearing and normal forces at the flange element.

Moment load M

$$\sigma_r = -\frac{2M}{t(\sin 2\alpha - 2\alpha \cos 2\alpha) + 4a_f\alpha \sin 2\alpha} \frac{\sin 2\theta}{r^2} \qquad \sigma_\theta = 0$$

$$\tau_{r\theta} = \frac{M[t(\cos 2\theta - \cos 2\alpha) + 2a_f \sin 2\alpha]}{[t(\sin 2\alpha - 2\alpha \cos 2\alpha) + 4a_f\alpha \sin 2\alpha]tr^2} \tag{37.85}$$

Wedge Carrying a Uniform Edge Load. Let Fig. 37.30 again represent a wedge of unit thickness, and let q denote the load per unit length of the edge. Then dim $q = FL^{-2}$; hence, $\Phi = qr^2f(\theta)$. When this is introduced into Eq. (37.49) and the ensuing differential equation for $f(\theta)$ is solved, the following general expression for the stress function is found:

$$\Phi = r^2(A + B\theta + C \cos 2\theta + D \sin 2\theta) \tag{37.86}$$

The boundary conditions require that

$$\text{For } \theta = 0\text{:} \quad \sigma_\theta = -q \quad \tau_{r\theta} = 0 \qquad \text{for } \theta = \beta\text{:} \quad \sigma_r = \tau_{r\theta} = 0$$

They lead to the following stress field:

$$\sigma_r = \frac{q}{\tan \beta - \beta}(\beta - \tfrac{1}{2} \tan \beta - \theta - \tfrac{1}{2} \sin 2\theta + \tfrac{1}{2} \tan \beta \cos 2\theta)$$

$$\sigma_\theta = \frac{q}{\tan \beta - \beta}(\beta - \tfrac{1}{2} \tan \beta - \theta + \tfrac{1}{2} \sin 2\theta - \tfrac{1}{2} \tan \beta \cos 2\theta) \tag{37.87}$$

$$\tau_{r\theta} = \frac{q}{\tan \beta - \beta}(\tfrac{1}{2} - \tfrac{1}{2} \tan \beta \sin 2\theta - \tfrac{1}{2} \cos 2\theta)$$

Triangular Gravity Dam. The solution given here excludes the influence of the water circulating in the pores of the dam. The water pressure on the upstream face

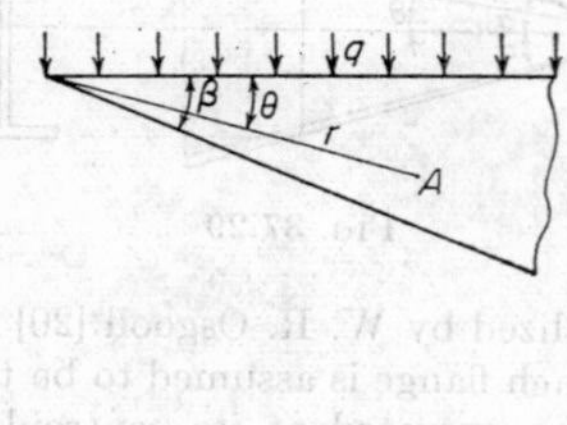

FIG. 37.30

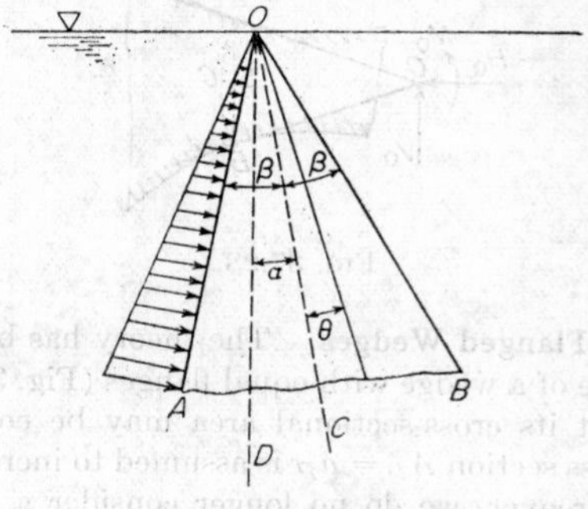

FIG. 37.31

acts as shown in Fig. 37.31. As shown on p. 37–5, the influence of the weight can be studied by immersing the dam in a fictitious fluid having the specific weight γ of the concrete. Consequently, all loading cases may be obtained by superposition of solutions for various hydrostatic loads like the one shown in Fig. 37.31.

The mechanical quantity defining the load is the specific weight γ of the water (or of the concrete); its dimension is FL^{-3}; hence, $\Phi = \gamma r^3 f(\theta)$. With this expression, Eq. (37.49) yields the differential equation

$$\frac{d^4 f}{d\theta^4} + 10\frac{d^2 f}{d\theta^2} + 9f = 0$$

for f. With its general solution, the stress function

$$\Phi = \gamma r^3(A \cos \theta + B \sin \theta + C \cos 3\theta + D \sin 3\theta) \tag{37.88}$$

is formed. It yields the stresses

$$\begin{aligned}\sigma_r &= 2\gamma r(A\cos\theta + B\sin\theta - 3C\cos 3\theta - 3D\sin 3\theta)\\ \sigma_\theta &= 6\gamma r(A\cos\theta + B\sin\theta + C\cos 3\theta + D\sin 3\theta)\\ \tau_{r\theta} &= 2\gamma r(A\sin\theta - B\cos\theta + 3C\sin 3\theta - 3D\cos 3\theta)\end{aligned} \tag{37.89}$$

Now, with α being the angle between the vertical OD and the center line OC of the wedge, the boundary conditions are:

For $\theta = \beta$: $\quad \sigma_\theta = \tau_{r\theta} = 0$
For $\theta = -\beta$: $\quad \sigma_\theta = -\gamma r\cos(\beta - \alpha) \qquad \tau_{r\theta} = 0$

These conditions lead to the following values for the constants:

$$\begin{aligned}A &= \frac{\cos(\beta-\alpha)\sin 3\beta}{4(\cos 3\beta\sin\beta - 3\sin 3\beta\cos\beta)}\\ B &= -\frac{\cos(\beta-\alpha)\cos 3\beta}{4(\sin 3\beta\cos\beta - 3\cos 3\beta\sin\beta)}\\ C &= -\frac{\cos(\beta-\alpha)\sin\beta}{12(\cos 3\beta\sin\beta - 3\sin 3\beta\cos\beta)}\\ D &= \frac{\cos(\beta-\alpha)\cos\beta}{12(\sin 3\beta\cos\beta - 3\cos 3\beta\sin\beta)}\end{aligned} \tag{37.90}$$

Edge Loads on a Half Space or Half Plane

Concentrated Force. Putting $\alpha = 90°$ in Eq. (37.81), and replacing P by $-P$ yields the stress system for the case shown in Fig. 37.32, a half plane loaded by a single compressive force at its edge:

$$\sigma_r = -\frac{2P}{\pi}\frac{\cos\theta}{r} \qquad \sigma_\theta = \tau_{r\theta} = 0 \tag{37.91}$$

The corresponding stress function is

$$\Phi = -\frac{P}{\pi} r\theta\sin\theta \tag{37.92}$$

The same result may be obtained from Eq. (37.82) for a tangential load P if θ is replaced by $\theta - 90°$. This shows that the stress field in an elastic half plane loaded

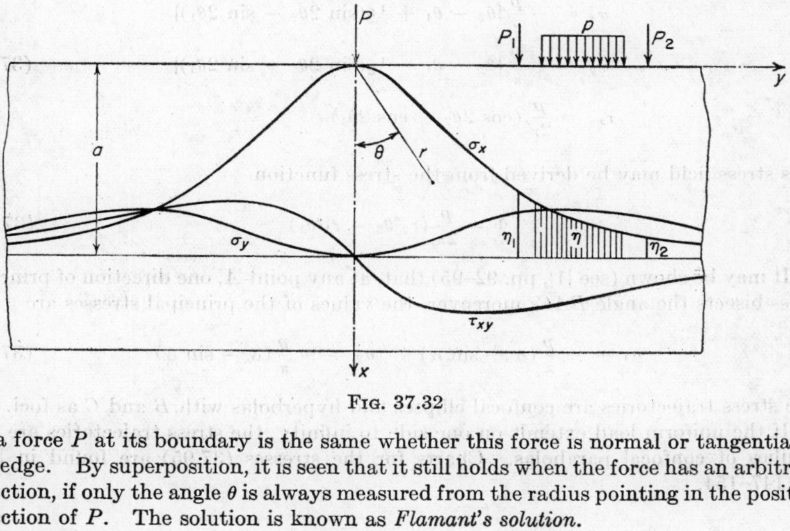

FIG. 37.32

by a force P at its boundary is the same whether this force is normal or tangential to the edge. By superposition, it is seen that it still holds when the force has an arbitrary direction, if only the angle θ is always measured from the radius pointing in the positive direction of P. The solution is known as *Flamant's solution.*

When the solution (37.91) is transformed to rectilinear coordinates x and y (Fig. 37.32), the following stresses are found:

$$\sigma_x = -\frac{2P}{\pi x}\cos^4\theta \qquad \sigma_y = -\frac{2P}{\pi x}\sin^2\theta\cos^2\theta \qquad \tau_{xy} = -\frac{2P}{\pi x}\sin\theta\cos^3\theta \qquad (37.93)$$

The distribution of these stresses is shown in Fig. 37.32. Similar formulas and curves may be established for a tangential edge load.

These curves may be used as influence lines when calculating the stresses produced by an arbitrary edge load. As an example, the vertical stress σ_x at $x = a$, $y = 0$ produced by two concentrated forces P_1 and P_2 and a distributed load p is

$$\sigma_x = P_1\eta_1 + P_2\eta_2 + p\int\eta\,dx \qquad (37.94)$$

where η represents the ordinates of the σ_x curve indicated in Fig. 37.32.

Equations (37.93) show that the stresses may be plotted in dimensionless form as functions of P/x and either θ or y/x. Extended charts of this kind have been given by Magnel [21], pp. 135–144.

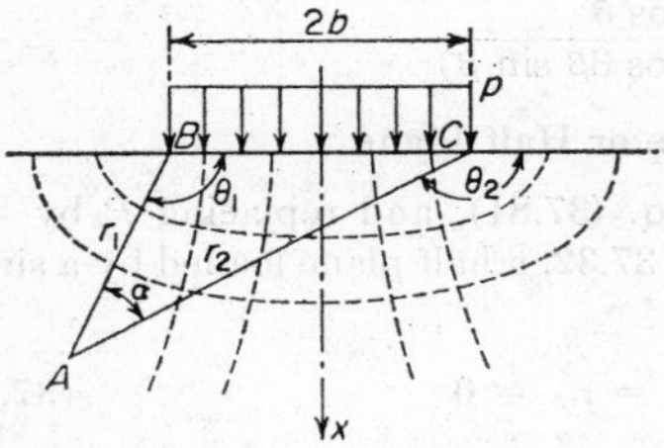

FIG. 37.33

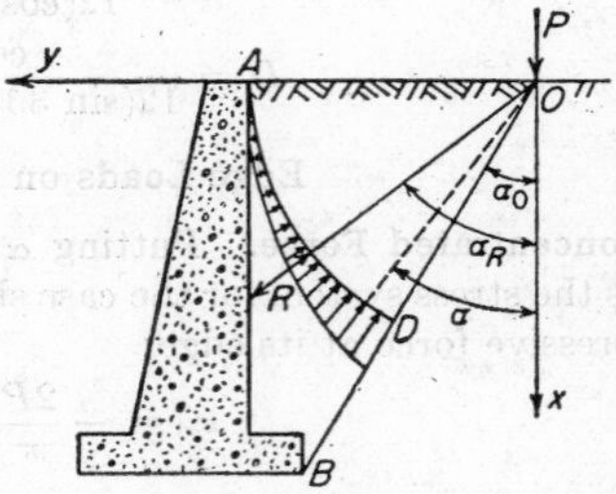

FIG. 37.34

In simple cases the integral in Eq. (37.94) can be evaluated by formal integration. As an example, we give the result for a uniform load p applied to a segment of length $2b$ of the edge, symmetrically to the x axis (Fig. 37.33):

$$\begin{aligned}\sigma_x &= -\frac{p}{\pi}\left[\theta_2 - \theta_1 + \tfrac{1}{2}(\sin 2\theta_2 - \sin 2\theta_1)\right]\\ \sigma_y &= -\frac{p}{\pi}\left[\theta_2 - \theta_1 - \tfrac{1}{2}(\sin 2\theta_2 - \sin 2\theta_1)\right]\\ \tau_{xy} &= \frac{p}{2\pi}(\cos 2\theta_2 - \cos 2\theta_1)\end{aligned} \qquad (37.95)$$

This stress field may be derived from the stress function

$$\Phi = \frac{p}{2\pi}(r_2^2\theta_2 - r_1^2\theta_1) \qquad (37.96)$$

It may be shown (see [1], pp. 92–95) that, at any point A, one direction of principal stress bisects the angle BAC; moreover, the values of the principal stresses are

$$\sigma_1 = -\frac{p}{\pi}(\alpha + \sin\alpha) \qquad \sigma_2 = -\frac{p}{\pi}(\alpha - \sin\alpha) \qquad (37.97)$$

The stress trajectories are confocal ellipses and hyperbolas with B and C as foci.

If the uniform load extends on one side to infinity, the stress trajectories are two families of confocal parabolas. Charts for the stresses (37.95) are found in [21], pp. 147–154.

Similar results may be obtained for tangential loads (see [7], pp. 39–40).

In the design of retaining walls, we sometimes encounter the problem of finding the load exerted on the wall by a live load applied at the surface of the soil. This problem has a simple solution if the soil can be assumed to be elastic. Indeed, since a force P (Fig. 37.34) produces a radial stress system, the part of this force which goes into the angle AOB is independent of the shape of the retaining wall, and can be found by integrating the stresses on the circular arc AD. We find:

$$R_x = \frac{P}{\pi}\left(\frac{\pi}{2} - \alpha_0 - \tfrac{1}{2}\sin 2\alpha_0\right) \qquad R_y = \frac{P}{\pi}\cos^2\alpha_0 \tag{37.98}$$

For further details, see [18], pp. 144–147.

Stresses in Circular Disks

Compressive Forces Applied at the Ends of One Diameter. To solve the problem described by Fig. 37.35, consider first the half plane lying below AA. In it, the upper force P produces a simple radial stress distribution. The same stress field will exist in the plate strip bounded by AA and BB if appropriate edge forces are applied along BB. At point C, there exists a normal stress

$$\sigma_{r1} = -\frac{2P}{\pi}\frac{\cos\theta_1}{r_1}$$

In the same way, the lower force P produces a radial stress field with the stress

$$\sigma_{r2} = -\frac{2P}{\pi}\frac{\cos\theta_2}{r_2}$$

at C. Now, $r_1/\cos\theta_1 = r_2/\cos\theta_2 = 2a$, and, therefore, $\sigma_{r1} = \sigma_{r2}$. Consequently, when we cut from the plate strip a circular plate of radius a, we shall find in it the

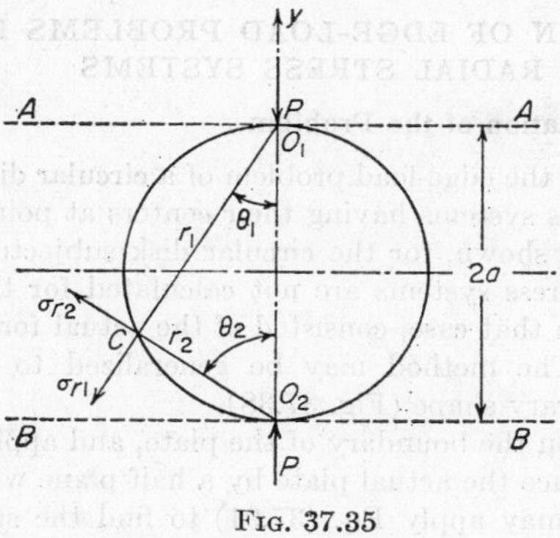

FIG. 37.35

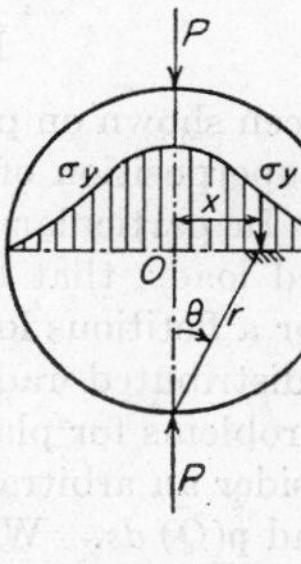

FIG. 37.36

superposition of the two radial stress systems, if we apply to its edge the two forces P and a uniformly distributed radial load of intensity $P/\pi a$, directed inward. In order to get rid of this latter load, we just have to add a field of uniform tensile stress, $\sigma = P/\pi a$. The superposition of this uniform stress and the two radial stress fields represents, therefore, the stresses in a circular disk under the action of two forces P at the ends of its vertical diameter.

In detail, this yields the following stresses: In a section along the vertical diameter, a uniform tensile stress $\sigma_x = P/\pi a$; in a section along the horizontal diameter, the stress

$$\sigma_y = -\frac{4Pa^3}{\pi(a^2+x^2)^2} + \frac{P}{\pi a}$$

This distribution is shown in Fig. 37.36. The stress vanishes at the ends of the diameter.

There are different ways to solve the stress problem of a circular disk subjected to an arbitrary, but self-equilibrating edge load. Let **f** be the intensity of the external load at an arbitrary line element ds (Fig. 37.37). Then the stress field in the disk can be obtained (see [22], pp. 66–72) by superposing the radial stress distributions induced by all such edge loads and a uniform tension of intensity

$$\sigma = \frac{1}{2\pi a} \oint f_n \, ds$$

where f_n is the normal component of **f**. Another solution of the same problem is found in [1], pp. 108–111. A third solution uses the expansion of the surface load **f** in Fourier series (see, for example, [5], pp. 132–135, and [8], pp. 405–430).

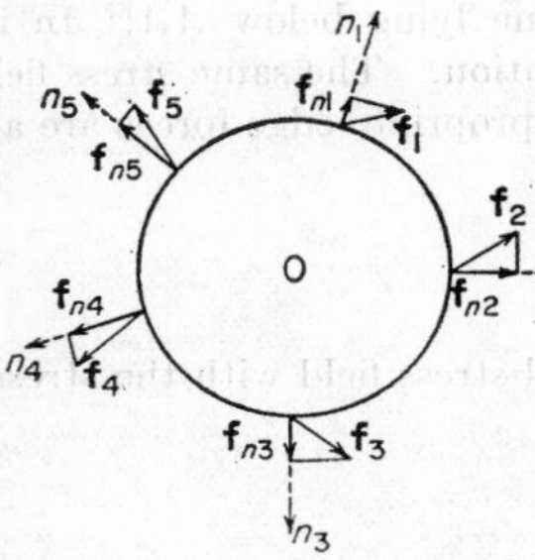

FIG. 37.37

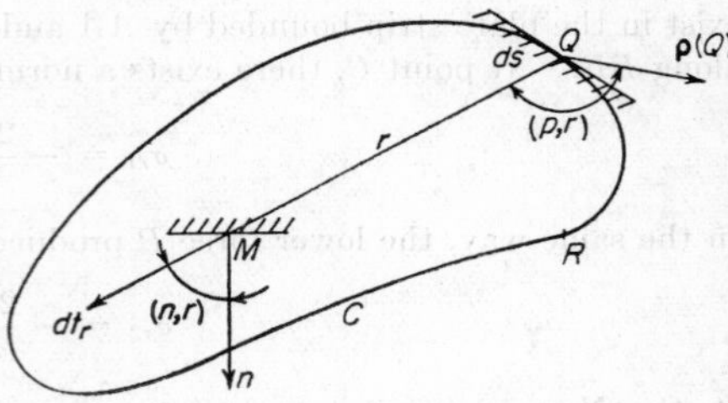

FIG. 37.38

37.4. NUMERICAL SOLUTION OF EDGE-LOAD PROBLEMS BY SUPERPOSITION OF RADIAL STRESS SYSTEMS

Integral Equation of the Problem

It has been shown on p. 37–27 how the edge-load problem of a circular disk can be solved by superposition of radial stress systems having their centers at points of the boundary. In particular, it has been shown, for the circular disk subjected to two concentrated loads, that the radial stress systems are not calculated for the actual load, but for a fictitious load which, in that case, consisted of the actual forces and a uniformly distributed radial load. The method may be generalized to apply to edge-load problems for plates of arbitrary shape (Fig. 37.38).

We consider an arbitrary point Q on the boundary of the plate, and apply there a *fictitious* load $\mathbf{p}(Q)\,ds$. When we replace the actual plate by a half plane whose edge is the tangent to the edge at Q, we may apply Eq. (37.91) to find the stress field produced by this force $\mathbf{p}(Q)\,ds$. At an arbitrary point M in the interior, it consists of a radial stress in a section normal to the radius **r**. In an arbitrary section with exterior normal n, it yields the stress vector

$$d\mathbf{t}_n = -\frac{2}{\pi}\, p(Q) \frac{\cos(p,r)\cos(n,r)}{r^2}\, \mathbf{r}\, ds$$

where p is the absolute value of the vector **p**. Similar loads **p** at other points of the boundary make similar contributions to the stress vector $\mathbf{t}_n$, and they add up to

$$\mathbf{t}_n(M) = -\frac{2}{\pi} \oint p(Q) \frac{\cos(p,r)\cos(n,r)}{r^2}\, \mathbf{r}\, ds \qquad (37.99)$$

When M approaches a point of the boundary, and n becomes the direction of its exterior normal, the integral in (37.99) becomes singular. It can be shown that it may then be replaced by its Cauchy principal value plus the local value of **p**:

$$\mathbf{t}_n(R) = \mathbf{p}(R) - \frac{2}{\pi} \int p(Q) \frac{\cos(p,r) \cos(n,r)}{r^2} \mathbf{r}\, ds \qquad (37.100)$$

This stress must be identical with the *given* boundary load intensity $\mathbf{f}(R)$ at this point:

$$\mathbf{f}(R) = \mathbf{p}(R) - \frac{2}{\pi} \int p(Q) \frac{\cos(p,r) \cos(n,r)}{r^2} \mathbf{r}\, ds \qquad (37.101)$$

This is a vectorial integral equation of the second kind for the unknown fictitious load $\mathbf{p}(R)$. When it has been solved for **p**, the solution of the stress problem is given by Eq. (37.99).

Practical Solution of the Integral Equation

The integral equation (37.101) may be written in the form

$$\mathbf{f} = \mathbf{p} + B(\mathbf{p}) \qquad (37.102)$$

where B is the linear integral operator defined by the last term of Eq. (37.101). It seems most natural to solve this equation by successive approximations as follows:

An arbitrary distribution $\mathbf{p} = \mathbf{p}^{(0)}$ of the fictitious loads is assumed. It may, for example, be identical with the given forces **f**. This distribution is introduced in the integral of Eq. (37.102), and a better approximation is found from

$$\mathbf{p}^{(1)} = \mathbf{f} - B(\mathbf{p}^{(0)}) \qquad (37.103)$$

This procedure may be continued until $\mathbf{p}^{(n+1)} \approx \mathbf{p}^{(n)}$ to the desired degree of accuracy.

However, it may happen that successive approximations $\mathbf{p}^{(n)}$, instead of approaching a limiting distribution, keep oscillating about an average. In such cases, the method must be modified as follows:

Starting from $\mathbf{p}^{(0)} = \mathbf{f}/2$, each approximation is calculated as the average of the preceding one and the outcome of Eq. (37.103); that is,

$$\mathbf{p}^{(n)} = \tfrac{1}{2}[\mathbf{p}^{(n-1)} + \mathbf{f} - B(\mathbf{p}^{(n-1)})] \qquad (37.104)$$

It has been shown (see [19], pp. 129–161) that this modified procedure is always convergent, whatever be the shape of the plate and the given distribution of forces **f**; it has also been proved that the distribution **p** obtained in this way is unique, and forms, like the starting distribution $\mathbf{p}^{(0)}$, a system of vectors in equilibrium.

As soon as the fictitious loads **p** are known, the stress vector $\mathbf{t}_n$ may be calculated everywhere from Eq. (37.99).

The integral $B(\mathbf{p})$ may be evaluated by the usual methods of numerical integration. To facilitate this work, C. Massonnet has designed and built [19] a vectorial summator. With its help a number of practical problems have been solved. As an alternative to its use, the integral may be evaluated on a digital computer. The repetitive nature of the operations to be performed facilitates the programming.

The particular merit of the method described here lies in the fact that the stresses are obtained as integrals, whereas the Airy stress function requires double differentiation which, in numerical work, results in a loss of accuracy.

It is possible to extend the method to other than plane problems, for example, to plate bending and three-dimensional elasticity [21], [22].

37.5. REFERENCES

[1] S. P. Timoshenko, J. N. Goodier: "Theory of Elasticity," 2d ed., McGraw-Hill, New York, 1951.
[2] A. E. H. Love: "A Treatise on the Mathematical Theory of Elasticity," 4th ed., Dover, New York, 1944.
[3] M. Hetényi: "Handbook of Experimental Stress Analysis," Wiley, New York, 1950.
[4] N. I. Muskhelishvili: "Some Basic Problems of the Theory of Elasticity," transl. of 3d Russian ed. by J. R. M. Radok, Noordhoff, Groningen, Netherlands, 1953.
[5] K. Girkmann: "Flächentragwerke," 5th ed., Springer, Vienna, 1959.
[6] S. P. Timoshenko: "Strength of Materials," vol. 2, 3d ed., Van Nostrand, Princeton, N.J., 1956.
[7] L. Föppl: "Drang und Zwang," vol. 3, Leibniz Verlag, Munich, 1947.
[8] C. B. Biezeno, R. Grammel: "Technische Dynamik," 2d ed., Springer, Berlin, 1953.
[9] A. E. Green, W. Zerna: "Theoretical Elasticity," Oxford Univ. Press, New York, 1954.
[10] R. V. Southwell: "Relaxation Methods in Theoretical Physics," vol. 2, Oxford Univ. Press, New York, 1956.
[11] E. G. Coker, L. N. G. Filon: "Photoelasticity," Cambridge Univ. Press, New York, 1931.
[12] M. M. Frocht: "Photoelasticity," vols. 1 and 2, Wiley, New York, 1941, 1948.
[13] F. Dischinger: Beitrag zur Theorie der Halbscheibe und des wandartigen Balkens, *Intern. Assoc. Bridge Struct. Eng.*, Publs., **1** (1932), 69–93; and Die Ermittlung der Eiseneinlagen in wandartigen Trägern, *Beton u. Eisen*, **32** (1933), 237–239.
[14] O. F. Theimer: "Hilfstafeln zur Berechnung wandartiger Stahlbetonträger," Ernst, Berlin, 1956.
[15] G. B. Jeffery: Plane stress and plane strain in bipolar coordinates, *Phil. Trans. Roy. Soc. London*, **A,221** (1921), 265–293.
[16] R. D. Mindlin: Stress distribution around a hole near the edge of a plate under tension, *Proc. Soc. Expt. Stress Anal.*, **5,II** (1948), 56–68.
[17] W. R. Osgood: A theory of flexure for beams with nonparallel extreme fibers, *J. Appl. Mechanics*, **6** (1939), 122–126.
[18] G. Magnel: "Stabilité des constructions," vol. 3, Rombaut-Fecheyr, Ghent, Belgium, 1942.
[19] Ch. Massonnet: Graphomechanical solution of the general problem of plane elasticity (in French), *Bull. CERES Liège*, **4** (1949), 3.
[20] J. H. Michell: On the direct determination of stress in an elastic solid, with application to the theory of plates, *Proc. London Math. Soc.*, **31** (1899), 100.
[21] Ch. Massonnet: Le calcul des pièces à plan moyen, Intern. Assoc. Bridge Struct. Eng., 3d Congr., Liège, 1948, final report, p. 285.
[22] Ch. Massonnet: General solution of the stress problem of three-dimensional elasticity (in French), *Proc. 9th Intern. Congr. Appl. Mechanics*, vol. 5, pp. 168–180, Brussels, Belgium, 1956.
[23] M. L. Williams: Stress singularities resulting from various boundary conditions in angular corners of plates in extension, *J. Appl. Mechanics*, **19** (1952), 526–528.
[24] R. Miche: Le calcul pratique des problèmes élastiques à deux dimensions par la méthode des équations intégrales, *2d Intern. Congr. Appl. Mechanics*, pp. 126–130, Zurich, Switzerland, 1926.
[25] E. Weinel: Die Integralgleichungen des ebenen Spannungszustandes und der Plattentheorie, *Z. ang. Math. Mech.*, **11** (1931), 349–360.

CHAPTER 38

COMPLEX-VARIABLE APPROACH

BY

J. R. M. RADOK, Dr.-Ing., Vienna, Austria

In the field of plane elasticity, complex-variable methods have by now been applied to static and dynamic problems of isotropic and anisotropic plane strain, generalized plane stress and extension, torsion and flexure of homogeneous and nonhomogeneous beams. In all these cases, the basic ideas underlying the introduction of functions of complex variables are the same, and, for this reason, the present chapter, although it deals with isotropic plane strain, may serve as a general introduction. More details of the application of complex methods in other fields and their more advanced aspects may be found in the literature.

38.1. COMPLEX REPRESENTATION OF STRESSES AND DISPLACEMENTS

In the theory of plane strain, it is shown [see Eq. (37.6)] that, in the absence of distributed loads X, Y, the sum of the normal stresses $(\sigma_x + \sigma_y)$ is a harmonic function, i.e., that it satisfies Laplace's equation.

Introducing the complex variable z and its conjugate complex $\bar{z}$, defined by

$$z = x + iy \qquad \bar{z} = x - iy \tag{38.1}$$

we find

$$\frac{\partial}{\partial x} = \left(\frac{\partial}{\partial z} + \frac{\partial}{\partial \bar{z}}\right) \qquad \frac{\partial}{\partial y} = i\left(\frac{\partial}{\partial z} - \frac{\partial}{\partial \bar{z}}\right) \tag{38.2}$$

and, hence,

$$\nabla^2(\sigma_x + \sigma_y) = 4\,\frac{\partial^2(\sigma_x + \sigma_y)}{\partial z\,\partial \bar{z}} = 0 \tag{38.3}$$

Successive integration of this equation with respect to z and $\bar{z}$ leads to

$$\sigma_x + \sigma_y = 2[\Phi(z) + \bar{\Phi}(\bar{z})] \tag{38.4}$$

where the factor 2 has been introduced for the sake of later convenience, and the functions $\Phi(z)$ and $\bar{\Phi}(\bar{z})$ have conjugate complex values, and depend on the conjugate complex variables, z and $\bar{z}$, respectively. The above complex representation of harmonic functions is usually derived by reference to the Cauchy-Riemann equations.

Next, consider the equilibrium equations in the absence of body forces:

$$\frac{\partial \sigma_x}{\partial x} + \frac{\partial \tau_{xy}}{\partial y} = 0 \qquad \frac{\partial \tau_{xy}}{\partial x} + \frac{\partial \sigma_y}{\partial y} = 0 \tag{38.5}$$

Multiplying the second of these equations by i, subtracting it from the first equation, and introducing the complex variables (38.1), we obtain by use of (38.4)

$$\frac{\partial}{\partial \bar{z}}(\sigma_y - \sigma_x + 2i\tau_{xy}) = \frac{\partial}{\partial z}(\sigma_x + \sigma_y) = 2\Phi'(z) \tag{38.6}$$

where

$$\Phi'(z) = \frac{d\Phi}{dz} \tag{38.7}$$

Integration of (38.6) with respect to $\bar{z}$ gives immediately

$$\sigma_y - \sigma_x + 2i\tau_{xy} = 2[\bar{z}\Phi'(z) + \Psi(z)] \tag{38.8}$$

where $2\Psi(z)$ is a second arbitrary function of z. It is thus seen that the stresses σ_x, σ_y, τ_{xy} are represented in terms of two functions of the complex variables z and $\bar{z}$ by the formulas (38.4) and (38.8).

In order to obtain corresponding expressions for the displacements, we must now refer to the stress-strain relations,

$$\sigma_x = \lambda\left(\frac{\partial u}{\partial x} + \frac{\partial v}{\partial y}\right) + 2\mu\frac{\partial u}{\partial x} \qquad \sigma_y = \lambda\left(\frac{\partial u}{\partial x} + \frac{\partial v}{\partial y}\right) + 2\mu\frac{\partial v}{\partial y}$$
$$\tau_{xy} = \mu\left(\frac{\partial v}{\partial x} + \frac{\partial u}{\partial y}\right) \tag{38.9}$$

where λ, μ are Lamé's constants, related to Young's modulus E and Poisson's ratio ν by

$$\lambda = \frac{E\nu}{(1+\nu)(1-2\nu)} \qquad \mu = \frac{E}{2(1+\nu)} \tag{38.10}$$

Using (38.9), (38.2), and (38.8), we find

$$4\mu\frac{\partial}{\partial z}(u - iv) = -(\sigma_y - \sigma_x + 2i\tau_{xy}) = -2[\bar{z}\Phi'(z) + \Psi(z)]$$

and hence, after integrating with respect to z and taking the conjugate complex expression throughout,

$$2\mu(u + iv) = -z\bar{\Phi}(\bar{z}) - \int\bar{\Psi}(\bar{z})\,d\bar{z} + \chi(z) \tag{38.11}$$

where $\chi(z)$ is at present still undetermined. However, compatibility of the displacements demands

$$\sigma_x + \sigma_y = 2(\lambda + \mu)\left(\frac{\partial u}{\partial x} + \frac{\partial v}{\partial y}\right) = 2[\Phi(z) + \bar{\Phi}(\bar{z})]$$

and, hence, we must have

$$\chi(z) = \kappa\int\Phi(z)\,dz$$

where

$$\kappa = \frac{\lambda + 3\mu}{\lambda + \mu} = 3 - 4\nu \tag{38.12}$$

Introducing the two functions $\phi(z)$, $\psi(z)$, defined by

$$\Phi(z) = \phi'(z) \qquad \Psi(z) = \psi'(z) \tag{38.13}$$

the basic formulas (38.4), (38.8), and (38.11) giving the complex representation of the stresses and displacements may be written in various equivalent ways:

$$\begin{aligned} \sigma_x + \sigma_y &= 2[\Phi(z) + \bar{\Phi}(\bar{z})] = 2[\phi'(z) + \bar{\phi}'(\bar{z})] \\ \sigma_y - \sigma_x + 2i\tau_{xy} &= 2[\bar{z}\Phi'(z) + \Psi(z)] = 2[\bar{z}\phi''(z) + \psi'(z)] \\ 2\mu(u + iv) &= \kappa\phi(z) - z\bar{\phi}'(\bar{z}) - \bar{\psi}(\bar{z}) \end{aligned} \tag{38.14}$$

Finally, we obtain from the first two of these formulas, by subtraction,

$$\sigma_x - i\tau_{xy} = \Phi(z) + \bar{\Phi}(\bar{z}) - \bar{z}\Phi'(z) - \Psi(z) \tag{38.15}$$

38.2. TRANSFORMATION OF COORDINATES. BOUNDARY CONDITIONS. THE FUNDAMENTAL BOUNDARY-VALUE PROBLEMS

For plane strain, the stresses σ_x, σ_y, τ_{xy} transform for counterclockwise rotation of the coordinate system by an angle α in the following manner:

$$\begin{aligned}\sigma_{x'} &= \frac{\sigma_x + \sigma_y}{2} + \frac{\sigma_x - \sigma_y}{2}\cos 2\alpha + \tau_{xy}\sin 2\alpha \\ \sigma_{y'} &= \frac{\sigma_x + \sigma_y}{2} - \frac{\sigma_x - \sigma_y}{2}\cos 2\alpha - \tau_{xy}\sin 2\alpha \\ \tau_{x'y'} &= \qquad - \frac{\sigma_x - \sigma_y}{2}\sin 2\alpha + \tau_{xy}\cos 2\alpha\end{aligned} \tag{38.16}$$

The complex representation of these formulas is

$$\begin{aligned}\sigma_{x'} + \sigma_{y'} &= \sigma_x + \sigma_y \\ \sigma_{y'} - \sigma_{x'} + 2i\tau_{x'y'} &= (\sigma_y - \sigma_x + 2i\tau_{xy})e^{2i\alpha}\end{aligned} \tag{38.17}$$

so that, if σ_n, τ_{nt} are the normal and tangential components of the external stress acting on a boundary at a point where the outward normal makes an angle α with the x axis, we have

$$2(\sigma_n - i\tau_{nt}) = \sigma_x + \sigma_y - (\sigma_y - \sigma_x + 2i\tau_{xy})e^{2i\alpha} \tag{38.18}$$

Substituting for the stresses their complex representations (38.14), we find the boundary condition for the first fundamental boundary-value problem of elasticity when the stresses σ_n, τ_{nt} on the boundary L of a body are specified:

$$\sigma_n - i\tau_{nt} = \Phi(z) + \bar{\Phi}(\bar{z}) - [\bar{z}\Phi'(z) + \Psi(z)]e^{2i\alpha} \tag{38.19}$$

for z on L.

This boundary condition may be expressed in a different form, which is often more convenient. It is obtained from (38.19) by taking the integral of the applied stresses along the boundary between some fixed point s_0 and the variable point s,

$$\begin{aligned}f_1 + if_2 = i(X + iY) &= \int_{s_0}^{s} i(\sigma_n + i\tau_{nt})e^{i\alpha}\,ds \\ &= i\int_{s_0}^{s} \{[\Phi(z) + \bar{\Phi}(\bar{z})]e^{i\alpha} - [z\bar{\Phi}'(\bar{z}) + \bar{\Psi}(\bar{z})]e^{-i\alpha}\}\,ds \\ &= \int_{s_0}^{s} \Phi(z)\,dz + \int_{s_0}^{s} [\bar{\Phi}(\bar{z})\,d\bar{z} + z\bar{\Phi}'(\bar{z})\,d\bar{z}] + \int_{s_0}^{s} \bar{\Psi}(\bar{z})\,d\bar{z} \\ &= \phi(z) + z\bar{\phi}'(\bar{z}) + \bar{\psi}(\bar{z}) + \text{const}\end{aligned} \tag{38.20}$$

for z on L, where X and Y are the components of the resultant force, applied to the boundary L between the points s_0 and s.

The boundary condition for the second fundamental boundary-value problem of elasticity, when the displacements u and v on the boundary L are specified, is obtained directly from (38.14):

$$2\mu(u + iv) = 2\mu(g_1 + ig_2) = \kappa\phi(z) - z\bar{\phi}'(\bar{z}) - \bar{\psi}(\bar{z}) \tag{38.21}$$

for z on L, where g_1 and g_2 are the specified displacements.

Thus, the fundamental boundary-value problems of the theory of plane strain have been reduced to boundary-value problems for the two functions of a complex variable $\phi(z)$, $\psi(z)$. The formal similarity of the boundary conditions (38.20) and

(38.21) makes it possible to discuss certain aspects of the analytical solution of these two problems together, by studying the boundary-value problem,

$$K\phi(z) + z\bar{\phi}'(\bar{z}) + \bar{\psi}(\bar{z}) = h_1 + ih_2 \qquad \text{for } z \text{ on } L \tag{38.22}$$

where $K = 1 \quad h_1 = f_1 \quad h_2 = f_2$ for the first problem

$K = -\kappa \quad h_1 = -2\mu g_1 \quad h_2 = -2\mu g_2$ for the second problem

38.3. SOME ELEMENTARY RESULTS. SINGLE-VALUEDNESS OF DISPLACEMENTS

An elementary, but powerful, method for the solution of the boundary-value problem (38.22) assumes the functions $\phi(z)$, $\psi(z)$ in the form of power series. For this purpose, a distinction must be made between problems for which the region bounded by L is finite or infinite, i.e., between regions which contain either the origin or the point at infinity. As will be shown later, use may be made of conformal mapping to reduce the problem for any given region, the boundary L of which satisfies certain conditions, to a corresponding problem for a region having the unit circle as exterior or interior boundary, depending on whether the region is finite or infinite. Another basic region, which is often more convenient, is the half plane, but this region will not be considered here, and the reader will be referred to the literature for details about it.

Before solving the problem (38.22) for the unit circle, it is of interest to investigate separately the roles played by certain terms in the power series for the functions $\phi(z)$, $\psi(z)$. If the region has the unit circle γ as exterior boundary, let

$$\phi(z) = a_0 + a_1 z \qquad \psi(z) = a_0' \tag{38.23}$$

It follows from (38.14) that

$$\sigma_x + \sigma_y = 2(a_1 + \bar{a}_1) \qquad \sigma_y - \sigma_x + 2i\tau_{xy} = 0 \tag{38.24}$$

that is, only the real part of a_1 has an effect on the stress distribution. The corresponding displacements are

$$2\mu(u + iv) = z(a_1\kappa - \bar{a}_1) + a_0\kappa - \bar{a}_0'$$

and, hence, if $a_1 = iC$ so that there occur no stresses,

$$2\mu(u + iv) = ziC(\kappa + 1) + a_0\kappa - \bar{a}_0' \tag{38.25}$$

corresponds to a rigid-body displacement.

Next, consider the infinite region with the unit circle about the origin as the internal boundary. In this case, let

$$\Phi(z) = \phi'(z) = a_0 = B + iC \qquad \Psi(z) = \psi'(z) = a_0' = B' + iC'$$

so that, by (38.14) for $z = \infty$,

$$\sigma_x^\infty + \sigma_y^\infty = 4B \qquad \sigma_y^\infty - \sigma_x^\infty + 2i\tau_{xy}^\infty = 2(B' + iC')$$

or

$$\sigma_x^\infty = 2B - B' \qquad \sigma_y^\infty = 2B + B' \qquad \tau_{xy}^\infty = C' \tag{38.26}$$

It is seen that, in this case, a_0' and the real part of a_0 determine the loading at infinity. In terms of the principal stresses σ_1, σ_2 at infinity, Eqs. (38.26) become

$$\begin{aligned} a_0 &= B = \tfrac{1}{4}(\sigma_1 + \sigma_2) \qquad C = 0 \\ a_0' &= B' + iC' = -\tfrac{1}{2}(\sigma_1 - \sigma_2)e^{-2i\alpha} \end{aligned} \tag{38.27}$$

Finally, for infinite regions, consider

$$\Phi(z) = \frac{a_1}{z} \qquad \Psi(z) = \frac{a_1'}{z} \tag{38.28}$$

or

$$\phi(z) = a_1 \ln z + \text{const} \qquad \psi(z) = a_1' \ln z + \text{const} \tag{38.29}$$

Since

$$z = re^{i\theta} \qquad \bar{z} = re^{-i\theta} \tag{38.30}$$

we find, for the corresponding displacements from (38.14),

$$2\mu(u + iv) = (\kappa a_1 - \bar{a}_1') \ln r - \bar{a}_1 e^{2i\theta} + i(\kappa a_1 + \bar{a}_1')\theta \tag{38.31}$$

Thus the displacements will have a multivalued part unless the condition of single-valuedness

$$\kappa a_1 + \bar{a}_1' = 0 \tag{38.32}$$

is satisfied.

38.4. FOURIER SERIES. THE FUNDAMENTAL PROBLEMS FOR THE UNIT CIRCLE

The basic principle behind the use of power series for the solution of the fundamental problems is the comparison of the coefficients of linearly independent functions, i.e., of the powers of z. For this purpose, when the boundary L is the unit circle, the known functions h_1, h_2 on the right-hand side of (38.22) have to be expanded in power series of $z = e^{i\theta}$, which will be seen to be equivalent to Fourier series. In fact, provided the function $h = h_1 + ih_2$ is continuous for $0 \leqq \theta \leqq 2\pi$, except for finite discontinuities at a finite number of points, we find that

$$h(e^{i\theta}) = h_1(\theta) + ih_2(\theta) = \sum_{-\infty}^{+\infty} A_k e^{ik\theta} \tag{38.33}$$

where

$$A_k = \frac{1}{2\pi} \int_0^{2\pi} h(e^{i\theta}) e^{-ik\theta}\, d\theta \qquad k = 0, \pm 1, \pm 2, \ldots \tag{38.34}$$

This is the complex representation of Fourier series which is a direct consequence of the orthogonal properties of the trigonometric functions.

Consider now the boundary-value problem,

$$K\phi(z) + z\bar{\phi}'(\bar{z}) + \bar{\psi}(\bar{z}) = h_1 + ih_2 = h(z) \qquad z = e^{i\theta} \tag{38.35}$$

for the inside of the unit circle. Substituting in this condition

$$\phi(z) = \sum_{k=0}^{\infty} a_k z^k \qquad \psi(z) = \sum_{k=0}^{\infty} a_k' z^k \qquad z = e^{i\theta} \tag{38.36}$$

and using (38.33), comparison of coefficients of $e^{ik\theta}$ gives

$$\begin{aligned} Ka_0 + 2\bar{a}_2 + \bar{a}_0' &= A_0 \\ Ka_1 + \bar{a}_1 &= A_1 \\ Ka_k &= A_k \qquad k > 1, \\ (k+2)\bar{a}_{k+2} + \bar{a}_k' &= A_{-k} \qquad k > 0 \end{aligned} \tag{38.37}$$

for the determination of the coefficients a_k and a_k'.

As a special case of the first fundamental problem, let there be given the external loading with the components $p_x = p \cos \theta$, $p_y = 0$, or $\sigma_n = \frac{1}{2}p(1 + \cos 2\theta)$, $\tau_{nt} = -\frac{1}{2}p \sin 2\theta$ so that, by (38.20),

$$f = f_1 + if_2 = \frac{p}{2}(e^{i\theta} - e^{-i\theta})$$

Hence, it follows from (38.37) with $K = 1$ that

$$\phi(z) = \tfrac{1}{4}pz \qquad \psi(z) = -\tfrac{1}{2}pz$$

and, from (38.14),

$$\sigma_x = p \qquad \tau_{xy} = 0 \qquad \sigma_y = 0$$

$$u = \frac{p}{8\mu}(\kappa + 1)x \qquad v = \frac{p}{8\mu}(\kappa - 3)y$$

Next, consider the first fundamental problem for the infinite region with a circular hole surrounding the origin. This time, use will be made of the boundary condition (38.19):

$$\Phi(z) + \bar{\Phi}(\bar{z}) - [\bar{z}\Phi'(z) + \Psi(z)]e^{2i\theta} = \sigma_n - i\tau_{nt} \qquad z = e^{i\theta} \tag{38.38}$$

for the outside of the unit circle. Substituting, in (38.38),

$$\Phi(z) = \sum_{k=0}^{\infty} a_k z^{-k} \qquad \Psi(z) = \sum_{k=0}^{\infty} a_k' z^{-k} \tag{38.39}$$

we find, with

$$\sigma_n - i\tau_{nt} = \sum_{-\infty}^{+\infty} A_k e^{ik\theta}$$

by comparing coefficients of $e^{ik\theta}$,

$$\begin{aligned} &a_0 + \bar{a}_0 - a_2' = A_0 \qquad \bar{a}_1 - a_1' = A_1 \qquad \bar{a}_2 - a_0' = A_2 \\ &\bar{a}_k = A_k \quad k \geqq 3 \qquad (1 + k)a_k - a_{k+2}' = A_{-k} \quad k \geqq 1 \end{aligned} \tag{38.40}$$

But, by (38.27),

$$a_0 = \tfrac{1}{4}(\sigma_1 + \sigma_2) \qquad a_0' = -\tfrac{1}{2}(\sigma_1 - \sigma_2)e^{-2i\alpha}$$

may be considered part of the data specifying the problem, since they determine the principal stresses σ_1 and σ_2 at infinity and their orientation α. From (38.40),

$$a_2 = \bar{A}_2 + \bar{a}_0' \qquad a_2' = 2a_0 - A_0$$

since only the real part of a_0 influences the stress distribution. We also have the condition of single-valuedness of the displacements (38.32)

$$\kappa a_1 + \bar{a}_1' = 0$$

which, together with (38.40), gives

$$a_1 = \frac{\bar{A}_1}{1 + \kappa} \qquad a_1' = -\frac{\kappa \bar{A}_1}{1 + \kappa}$$

Since, by (38.34),

$$\begin{aligned} A_1 &= \frac{1}{2\pi}\int_0^{2\pi} (\sigma_n - i\tau_{nt})e^{-i\theta}\, d\theta \\ &= \frac{1}{2\pi}\int_0^{2\pi} [(\sigma_n \cos\theta - \tau_{nt}\sin\theta) - i(\sigma_n \sin\theta + \tau_{nt}\cos\theta)]\, d\theta \end{aligned}$$

it follows that

$$a_1 = -\frac{1}{2\pi}\frac{X + iY}{1 + \kappa} \qquad a_1' = \frac{\kappa}{2\pi}\frac{X - iY}{1 + \kappa} \tag{38.41}$$

where X and Y are the components of the resultant of the external forces acting on the boundary.

38.5. CONFORMAL MAPPING. CURVILINEAR COORDINATES

The special attention given above to regions bounded by the unit circle is justified by Riemann's theorem which ensures the existence of a function

$$z = \omega(\zeta) \tag{38.42}$$

mapping conformally any finite or infinite, simply connected region onto the inside or outside of the unit circle, provided the boundary L of the original region consists of a finite number of continuous curves. Let $\omega(\zeta)$ be such a function, which will be analytic inside or outside the unit circle $\zeta = e^{i\theta}$, depending on whether it maps a finite or an infinite region of the z plane on the inside or outside of the unit circle. The construction of such functions is a separate problem beyond the scope of this chapter. In what follows, consideration will be given to polynomial mapping functions,

$$z = \sum_{k=1}^{m} c_k\zeta^k \tag{38.43}$$

Such functions may be used to construct curvilinear coordinate systems in the z plane, which correspond to the polar coordinate system

$$\rho = \text{const} \qquad \theta = \text{const} \tag{38.44}$$

in the $\zeta = \rho e^{i\theta}$ plane, and which will be orthogonal, since the mapping functions (38.43) are analytic.

Through each point $z = \omega(\rho e^{i\theta})$ of a region mapped by (38.43) on the inside or outside of the unit circle, there will pass a pair of coordinate lines, the tangents to which form a rectilinear orthogonal coordinate system, and the axes of which will be denoted by ρ and θ, respectively. Consider the relationship between the two infinitesimal vectors

$$dz = e^{i\alpha}|dz| \qquad d\zeta = e^{i\theta}|d\zeta| \tag{38.45}$$

corresponding to each other in the z and ζ planes, where α is the angle between the axis ρ and the real axis of the coordinate system in the z plane. It follows from (38.45) and (38.42) that

$$\begin{aligned} e^{i\alpha} &= \frac{dz}{|dz|} = \frac{\omega'(\zeta)\,d\zeta}{|\omega'(\zeta)|\cdot|d\zeta|} = \frac{e^{i\theta}\omega'(\zeta)}{|\omega'(\zeta)|} = \frac{\zeta\omega'(\zeta)}{\rho|\omega'(\zeta)|} \\ e^{-i\alpha} &= \frac{\bar{\zeta}\bar{\omega}'(\bar{\zeta})}{\rho|\omega'(\zeta)|} \end{aligned} \tag{38.46}$$

and, hence, the ρ and θ components of a vector $\mathbf{A}$ in the z plane are given by

$$A_\rho + iA_\theta = \frac{\bar{\zeta}}{\rho}\frac{\bar{\omega}'(\bar{\zeta})}{|\omega'(\zeta)|}(A_x + iA_y) \tag{38.47}$$

This formula will now be used to obtain the transformed expressions for the basic formulas (38.14). Let

$$\phi_1(z),\ \psi_1(z),\ \Phi_1(z),\ \Psi_1(z) \tag{38.48}$$

denote the complex stress functions in the original z plane and

$$\phi(\zeta),\ \psi(\zeta),\ \Phi(\zeta),\ \Psi(\zeta) \tag{38.49}$$

the corresponding functions in the image ζ plane, where (38.42) relates z to ζ. It follows immediately that

$$\Phi(\zeta) = \Phi_1(z) = \frac{d\phi_1}{dz} = \frac{d\phi}{d\zeta}\frac{d\zeta}{dz} = \frac{\phi'(\zeta)}{\omega'(\zeta)} \qquad \Psi(\zeta) = \frac{\psi'(\zeta)}{\omega'(\zeta)} \tag{38.50}$$

The displacements will be considered first. We have

$$\begin{aligned} 2\mu(u + iv) &= \kappa\phi_1(z) - z\bar{\phi}_1'(\bar{z}) - \bar{\psi}_1(\bar{z}) \\ &= \kappa\phi(\zeta) - \frac{\omega(\zeta)}{\bar{\omega}'(\bar{\zeta})}\bar{\phi}'(\bar{\zeta}) - \bar{\psi}(\bar{\zeta}) \end{aligned}$$

and hence, in the curvilinear coordinate system, by (38.46),

$$\begin{aligned} 2\mu(v_\rho + iv_\theta) &= 2\mu(u + iv)e^{-i\alpha} \\ &= \frac{\bar{\zeta}}{\rho}\frac{\bar{\omega}'(\bar{\zeta})}{|\omega'(\zeta)|}\left[\kappa\phi(\zeta) - \frac{\omega(\zeta)}{\bar{\omega}'(\bar{\zeta})}\bar{\phi}'(\bar{\zeta}) - \bar{\psi}(\bar{\zeta})\right] \end{aligned} \tag{38.51}$$

This formula expresses the normal and tangential displacements v_ρ and v_θ in the original plane, in terms of the complex stress functions of the image plane.

The corresponding results for the stresses are easily seen to be

$$\begin{aligned} \sigma_\rho + \sigma_\theta &= 2[\Phi(\zeta) + \bar{\Phi}(\bar{\zeta})] \\ \sigma_\rho - i\tau_{\rho\theta} &= \Phi(\zeta) + \bar{\Phi}(\bar{\zeta}) - [\bar{\omega}(\bar{\zeta})\Phi'(\zeta) + \omega'(\zeta)\Psi(\zeta)]\frac{\zeta^2}{\rho^2}\frac{1}{\bar{\omega}'(\bar{\zeta})} \end{aligned} \tag{38.52}$$

where σ_ρ, σ_θ, $\tau_{\rho\theta}$ are the stress components in the curvilinear system.

The boundary condition (38.22) now becomes

$$K\phi(\sigma) + \frac{\omega(\sigma)}{\bar{\omega}'(\bar{\sigma})}\bar{\phi}'(\bar{\sigma}) + \bar{\psi}(\bar{\sigma}) = h_1 + ih_2 \tag{38.53}$$

where $\sigma = e^{i\theta}$ is a point on the unit circle in the ζ plane, and h_1 and h_2 have the values at the corresponding point of the original boundary $s = \omega(e^{i\theta})$.

As a special example, consider the transformation

$$z = R\left(\zeta + \frac{m}{\zeta}\right) \qquad R > 0,\ m \geqq 0 \tag{38.54}$$

mapping elliptic rings onto circular rings. In fact, we have

$$x = R\left(\rho + \frac{m}{\rho}\right)\cos\theta \qquad y = R\left(\rho - \frac{m}{\rho}\right)\sin\theta \tag{38.55}$$

which, for $\rho = \text{const}$, is the parametric representation of an ellipse with semiaxes

$$a = R\left(\rho + \frac{m}{\rho}\right) \qquad b = R\left(\rho - \frac{m}{\rho}\right) \tag{38.56}$$

The corresponding curvilinear coordinate system in the plane is formed by these ellipses and orthogonal hyperbolas for which $\theta = \text{const}$.

38.6. SOLUTION FOR REGIONS, MAPPED BY POLYNOMIALS

Consider the first fundamental boundary-value problem,

$$\phi(\sigma) + \frac{\omega(\sigma)}{\bar{\omega}'(\bar{\sigma})} \bar{\phi}'(\bar{\sigma}) + \bar{\psi}(\bar{\sigma}) = f_1 + if_2 = \sum_{-\infty}^{+\infty} A_k\sigma^k \tag{38.57}$$

when a finite region is mapped onto the unit circle $\sigma = e^{i\theta}$ by

$$z = \omega(\zeta) = \sum_{j=1}^{n} c_j\zeta^j \tag{38.58}$$

Since $\quad \sigma = e^{i\theta} \qquad \bar{\sigma} = e^{-i\theta} = \frac{1}{\sigma}$

we have

$$\frac{\omega(\sigma)}{\bar{\omega}'(\bar{\sigma})} = \frac{\sum_{j=1}^{n} c_j\sigma^j}{\sum_{j=1}^{n} \bar{c}_j j\bar{\sigma}^{j-1}} = \frac{\sigma^n \sum_{j=1}^{n} c_j\sigma^{j-1}}{\sum_{j=1}^{n} \bar{c}_j j\sigma^{n-j}} = \sum_{j=1}^{n} b_j\sigma^j + R(\sigma) \tag{38.59}$$

where $R(\sigma)$ is a rational function.

Substituting in (38.57)

$$\phi(\zeta) = \sum_{k=1}^{\infty} a_k\zeta^k \qquad \psi(\zeta) = \sum_{k=0}^{\infty} a'_k\zeta^k$$

we find, by comparison of coefficients of σ^m $(m > 0)$,

$$\begin{aligned} a_1 + K_1 &= a_1 + \bar{a}_1b_1 + 2\bar{a}_2b_2 + \cdots + (n-1)\bar{a}_{n-1}b_{n-1} + n\bar{a}_nb_n = A_1 \\ a_2 + K_2 &= a_2 + \bar{a}_1b_2 + 2\bar{a}_2b_2 + \cdots + (n-1)\bar{a}_{n-1}b_n = A_2 \\ &\cdots\cdots\cdots\cdots\cdots\cdots\cdots\cdots \\ a_n + K_n &= a_n + \bar{a}_1b_n = A_n \qquad a_m = A_m,\ m > n \end{aligned} \tag{38.60}$$

so that all the coefficients a_m are determined in terms of the known b_j, A_j. Hence, we may write

$$\phi(\zeta) = \sum_{k=1}^{\infty} a_k\zeta^k = \sum_{k=1}^{\infty} A_k\zeta^k - \sum_{k=1}^{n} K_k\zeta^k \tag{38.61}$$

Instead of determining the coefficients a'_k by comparison of the factors of σ^m $(m \leqq 0)$, it is easier to proceed directly from (38.57), which gives

$$\psi(\sigma) = f_1 - if_2 - \bar{\phi}(\bar{\sigma}) - \frac{\bar{\omega}(\bar{\sigma})}{\omega'(\sigma)}\phi'(\sigma) = \sum_{k=-\infty}^{+\infty} \bar{A}_k\bar{\sigma}^k - \sum_{k=1}^{\infty} \bar{A}_k\bar{\sigma}^k + \sum_{k=1}^{n} K_k\bar{\sigma}^k - \frac{\bar{\omega}(\bar{\sigma})}{\omega'(\sigma)}\phi'(\sigma)$$

whence follows

$$\psi(\zeta) = \sum_{k=0}^{-\infty} \bar{A}_k\zeta^{-k} + \sum_{k=1}^{n} \bar{K}_k\zeta^{-k} - \left[\sum_{l=1}^{n} \bar{b}_l\zeta^{-l} + R\left(\frac{1}{\zeta}\right)\right]\phi'(\zeta) \tag{38.62}$$

The same method applies to the other boundary-value problems for finite and infinite regions mapped by polynomials onto the unit circle.

38.7. OTHER METHODS OF SOLUTION

Consider the boundary condition (38.22) in the form

$$\psi(\sigma) = (h_1 - ih_2) - K\bar{\phi}(\bar{\sigma}) - \bar{\sigma}\phi'(\sigma) \tag{38.63}$$

when the region is the inside of the unit circle $\sigma = e^{i\theta}$. This condition gives the boundary value $\psi(\sigma)$ of a function $\psi(\zeta)$, which may be represented by a power series at any point inside the unit circle. It may be shown, by use of Cauchy's theorem, that a necessary and sufficient condition for this is

$$\frac{1}{2\pi i}\int_\gamma \frac{\bar{\psi}(\bar{\sigma})\,d\sigma}{\sigma - \zeta} = \bar{a} \tag{38.64}$$

for all ζ inside the unit circle γ, where $\bar{a}$ is the value of $\bar{\psi}(\bar{\zeta})$ at $\zeta = 0$.

By application of the calculus of residues, the formula (38.64) may be evaluated and the functions $\phi(\zeta)$, $\psi(\zeta)$ determined in terms of the boundary conditions $(h_1 + ih_2)$. The equivalence of the solutions, obtained by this method of Cauchy integrals, with those obtained by power series is readily established.

Another method of solution of the boundary-value problem (38.63) is based on the relationship between the boundary values of so-called sectionally holomorphic functions, i.e., of functions which can be expanded in power series at any point in the entire z plane except along a curve of discontinuity which may be a closed contour. Such a function may be represented by the Cauchy integral

$$\Phi(z) = \frac{1}{2\pi i}\int_L \frac{f(t)\,dt}{t - z} \tag{38.65}$$

where L is the line of discontinuity. If $\Phi^+(t)$ and $\Phi^-(t)$ are the boundary values of $\Phi(z)$ as z approaches t on L from the left or right side of L, then

$$\Phi^+(t) - \Phi^-(t) = f(t) \tag{38.66}$$

The boundary condition (38.63) may be shown to be expressible in the form (38.66), and, hence, its solution has the form (38.65). It may be of interest to note that the condition (38.66) occurs also in the two-dimensional theory of airfoils, where the solution (38.65) also plays an important part.

38.8. REFERENCES

[1] N. I. Muskhelishvili: "Some Basic Problems of the Theory of Elasticity," transl. of 3d Russian ed., Noordhoff, Groningen, Netherlands, 1953.

[2] A. E. Green, W. Zerna: "Theoretical Elasticity," Oxford Univ. Press, New York, 1954.

[3] I. S. Sokolnikoff: "Mathematical Theory of Elasticity," 2d ed., McGraw-Hill, New York, 1956.

[4] S. G. Lekhnitzky: "Theory of Elasticity of an Anisotropic Body" (in Russian), Moscow, 1950.

[5] S. G. Lekhnitzky: "Anisotropic Plates" (in Russian), Moscow, 1947.

[6] G. N. Savin: "Concentration of Stresses around Openings" (in Russian), Moscow, 1951.

[7] I. Ya. Shtaerman: "The Contact Problems of Elasticity Theory" (in Russian), Moscow-Leningrad, 1949.

CHAPTER 39

PLATES

BY

S. WAY, Sc.D., Pittsburgh, Pa.

39.1. BASIC EQUATIONS OF LATERALLY LOADED PLATES

Underlying Assumptions

For presentation of historical material and more detailed treatments than are possible in this chapter, the reader is referred to [1]–[3].

In the theory of thin plates, it is customary to make the following assumptions: (1) The plate is initially flat. (2) The material is elastic, homogeneous, and isotropic. (3) Thickness is small compared to area dimensions. (4) Slope of the deflection surface is small compared to unity. (5) Deformation is such that straight lines initially normal to the middle surface remain straight and normal to that surface. (Vertical shear strains are neglected.) (6) Strains in the middle surface, arising from the deflection, are neglected compared to strains due to bending. (7) Deflection of the plate occurs by virtue of displacement of points in the mid-surface normal to its initial plane. (8) Direct stress normal to the middle surface is neglected. (9) Near edges and boundaries of loaded areas, stress resultants rather than detailed stress patterns are considered.

Moments and Curvatures

The middle plane of the plate, before deflection, is assumed to lie at $z = 0$ in the coordinate system shown in Fig. 39.1. The plate element has its mid-point at x,y.

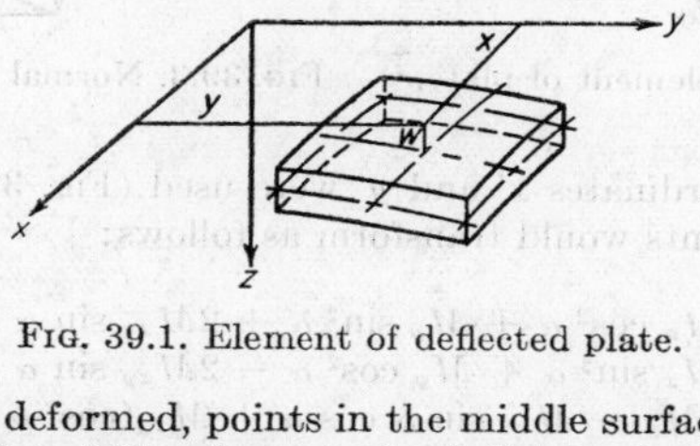

FIG. 39.1. Element of deflected plate.

When the plate is being deformed, points in the middle surface move parallel to the z axis, and, at the mid-point of the element, the deflection is w.

The bending moments will depend on the stresses in the plate, and these, in turn, depend on the extensions. Since the horizontal displacements u and v in the x and y directions depend on the z coordinate and the slope components, we have

$$u = -z\,\frac{\partial w}{\partial x} \qquad v = -z\,\frac{\partial w}{\partial y}$$

and the extensions will be

$$\epsilon_x = -z\,\frac{\partial^2 w}{\partial x^2} \qquad \epsilon_y = -z\,\frac{\partial^2 w}{\partial y^2} \qquad \gamma_{xy} = -2z\,\frac{\partial^2 w}{\partial x\,\partial y}$$

Stresses σ_x, σ_y, τ_{xy} are related to the extensions by Hooke's law for the special case $\sigma_z = 0$. In terms of the plate curvatures, these stresses become

$$\begin{aligned}\sigma_x &= -\frac{Ez}{1-\nu^2}\left(\frac{\partial^2 w}{\partial x^2} + \nu\frac{\partial^2 w}{\partial y^2}\right)\\ \sigma_y &= -\frac{Ez}{1-\nu^2}\left(\frac{\partial^2 w}{\partial y^2} + \nu\frac{\partial^2 w}{\partial x^2}\right)\\ \tau_{xy} &= -\frac{Ez}{1+\nu}\frac{\partial^2 w}{\partial x\,\partial y}\end{aligned} \tag{39.1}$$

By integrating the stress components over the thickness t of the plate, we obtain the bending and twisting moments per unit length:

$$M_x = \int_{-t/2}^{t/2} \sigma_x z\,dz \qquad M_y = \int_{-t/2}^{t/2} \sigma_y z\,dz \qquad M_{xy} = M_{yx} = \int_{-t/2}^{t/2} \tau_{xy} z\,dz \tag{39.2}$$

After substituting the stress expressions (39.1), the moments become

$$M_x = -K\left(\frac{\partial^2 w}{\partial x^2} + \nu\frac{\partial^2 w}{\partial y^2}\right) \qquad M_y = -K\left(\frac{\partial^2 w}{\partial y^2} + \nu\frac{\partial^2 w}{\partial x^2}\right)$$
$$M_{xy} = -K(1-\nu)\frac{\partial^2 w}{\partial x\,\partial y} \tag{39.3}$$

where

$$K = \frac{Et^3}{12(1-\nu^2)} \tag{39.4}$$

is the flexural rigidity or bending stiffness of the plate. Figure 39.2 shows the application of these moments to the plate element.

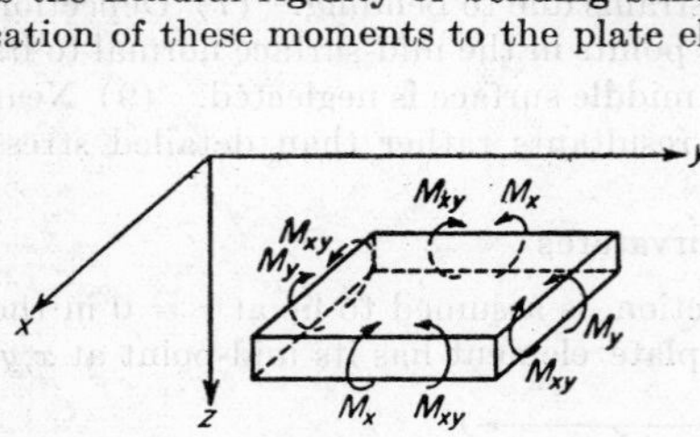

FIG. 39.2. Moments on element of plate.

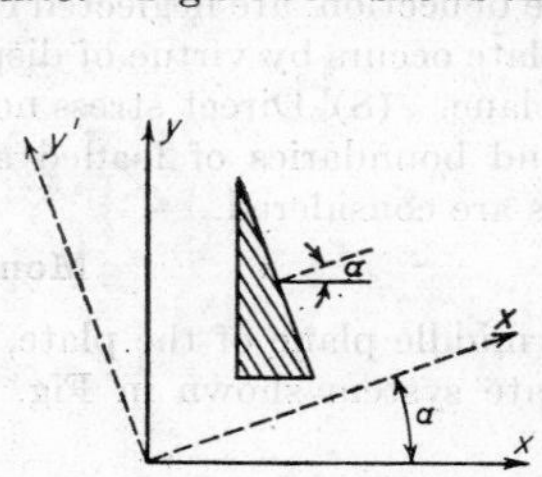

FIG. 39.3. Normal and tangential directions.

If another set of coordinates x' and y' were used (Fig. 39.3), instead of x and y (with $z' = z$), the moments would transform as follows:

$$\begin{aligned}M_{x'} &= M_x \cos^2\alpha + M_y \sin^2\alpha + 2M_{xy}\sin\alpha\cos\alpha\\ M_{y'} &= M_x \sin^2\alpha + M_y \cos^2\alpha - 2M_{xy}\sin\alpha\cos\alpha\\ M_{x'y'} &= (M_y - M_x)\sin\alpha\cos\alpha + M_{xy}(\cos^2\alpha - \sin^2\alpha)\end{aligned} \tag{39.5a–c}$$

These equations may be obtained by resolution of stresses and calculation of the new moments by equations like (39.2), or by consideration of the static equivalence of moments on a triangular plate element.

It should be pointed out that the stresses given by (39.1) are values corresponding to the location x, y, z in the plate in its *undeflected* configuration. For the bending stresses in the outer plate surface ($z = t/2$), combination of (39.1) and (39.3) yields

$$(\sigma_x)_{t/2} = \frac{6M_x}{t^2} \qquad (\sigma_y)_{t/2} = \frac{6M_y}{t^2} \qquad (\tau_{xy})_{t/2} = \frac{6M_{xy}}{t^2} \tag{39.6}$$

Equilibrium Equations

The moments acting on a small square element of the plate will form a system in equilibrium. The moments due to the vertical shear stress resultants Q_x and Q_y, per unit length, must be included, as shown in Fig. 39.4.

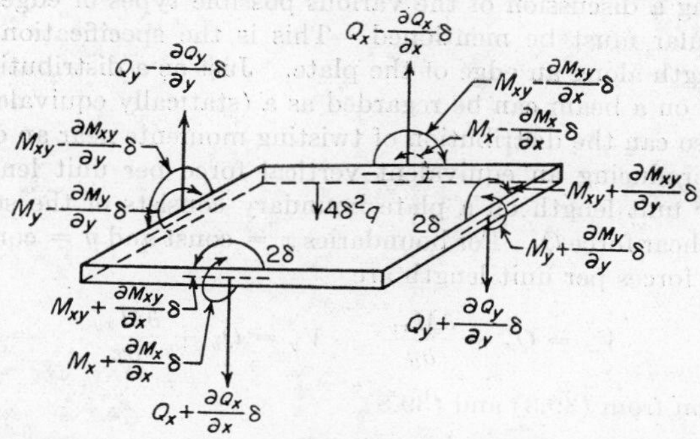

FIG. 39.4. Moments and shears per unit length on sides of infinitesimal element.

Considering equilibrium of moments about the y and x axes, we obtain

$$\frac{\partial M_x}{\partial x} + \frac{\partial M_{xy}}{\partial y} - Q_x = 0 \qquad \frac{\partial M_y}{\partial y} + \frac{\partial M_{xy}}{\partial x} - Q_y = 0 \tag{39.7a,b}$$

These equations, along with (39.3), lead to expressions for the vertical shears,

$$Q_x = -K\frac{\partial}{\partial x}\nabla^2 w \qquad Q_y = -K\frac{\partial}{\partial y}\nabla^2 w \tag{39.8a,b}$$

where operator ∇^2 is the two-dimensional Laplacian

$$\nabla^2 = \frac{\partial^2}{\partial x^2} + \frac{\partial^2}{\partial y^2}$$

Consideration of vertical equilibrium of the plate element gives

$$\frac{\partial Q_x}{\partial x} + \frac{\partial Q_y}{\partial y} + q = 0 \tag{39.9}$$

where q is load per unit area in the z direction. Substitution of expressions (39.8) leads to the following basic differential equation of plate theory:

$$\nabla^2\nabla^2 w \equiv \frac{\partial^4 w}{\partial x^4} + 2\frac{\partial^4 w}{\partial x^2\,\partial y^2} + \frac{\partial^4 w}{\partial y^4} = \frac{q}{K} \tag{39.10}$$

Where there is no lateral load on the plate,

$$\nabla^2\nabla^2 w \equiv \frac{\partial^4 w}{\partial x^4} + 2\frac{\partial^4 w}{\partial x^2\,\partial y^2} + \frac{\partial^4 w}{\partial y^4} = 0 \tag{39.10'}$$

Since the slope of the plate is assumed small compared to unity, it is legitimate to consider Q_x, Q_y, σ_x, σ_y, τ_{xy} as acting on planes normal to the middle surface of the plate in its deflected position.

For the axis system x',y' of Fig. 39.3, we have the shear forces

$$Q_{x'} = Q_x\cos\alpha + Q_y\sin\alpha \qquad Q_{y'} = -Q_x\sin\alpha + Q_y\cos\alpha \tag{39.11}$$

Boundary Conditions

The boundary conditions for a deflected plate, whose deflection satisfies Eq. (39.10), are two in number. They may involve either the deflection and slope or the edge forces and moments, or a combination of these factors.

Before entering a discussion of the various possible types of edge conditions, one matter in particular must be mentioned. This is the specification of the vertical force per unit length along an edge of the plate. Just as a distribution of externally applied moments on a beam can be regarded as a (statically equivalent) distribution of normal loads, so can the distribution of twisting moments near an edge of the plate be regarded as producing an equivalent vertical force per unit length. The total vertical force per unit length on a plate boundary consists of the sum of this force and the vertical shear force Q. For boundaries x = const and y = const, respectively, the vertical edge forces per unit length are

$$V_x = Q_x + \frac{\partial M_{xy}}{\partial y} \qquad V_y = Q_y + \frac{\partial M_{xy}}{\partial x} \tag{39.12}$$

or, by substitution from (39.3) and (39.8),

$$\begin{aligned} V_x &= -K\left[\frac{\partial}{\partial x}\nabla^2 w + (1-\nu)\frac{\partial^3 w}{\partial x\,\partial y^2}\right] \\ V_y &= -K\left[\frac{\partial}{\partial y}\nabla^2 w + (1-\nu)\frac{\partial^3 w}{\partial x^2\,\partial y}\right] \end{aligned} \tag{39.12'}$$

Their positive direction agrees with that of the shearing force Q_x or Q_y.

Wherever the boundary of the plate has a corner, there may be a concentrated force R, positive when downward. At a straight boundary whose outward normal points in the direction of the positive x axis, for the end with the largest y, the concentrated force is $R = -M_{xy}$, and, for the end with the smaller value y, it is $R = +M_{xy}$. Similarly, for a boundary whose outward normal is in the y direction, we have, for larger x, a force $R = -M_{xy}$ and, for smaller x, the force $R = +M_{xy}$. Thus, at a corner at $x = a$, $y = b$, having outward normals in the positive x and y directions, there will be a concentrated force

$$R_{a,b} = -2(M_{xy})_{a,b} \tag{39.13}$$

The various types of boundary conditions may now be mentioned. Subscripts n and t are used to indicate normal and tangential directions.

Clamped Edge. The deflection and slope are zero:

$$w = 0 \qquad \frac{\partial w}{\partial n} = 0 \tag{39.14}$$

Simply Supported Edge. The deflection and normal bending moment are zero:

$$w = 0 \qquad \frac{\partial^2 w}{\partial n^2} + \nu\frac{\partial^2 w}{\partial t^2} = 0 \tag{39.15}$$

If the edge is straight, both normal and tangential bending moments are zero, leading to

$$w = 0 \qquad \nabla^2 w = 0 \tag{39.16}$$

Free Edge. The normal moment and the vertical edge force are zero:

$$\frac{\partial^2 w}{\partial n^2} + \nu\frac{\partial^2 w}{\partial t^2} = 0 \qquad \frac{\partial}{\partial n}\nabla^2 w + (1-\nu)\frac{\partial^3 w}{\partial n\,\partial t^2} = 0 \tag{39.17}$$

In general, the boundary conditions for a plate involve specification of any two of the quantities w, $\partial w/\partial n$ M_n, V_n on every part of the boundary.

Elastic Energy of Thin Plates

In thin-plate theory, only the elastic energy of bending and twisting is considered. If we integrate the stored elastic energy for a rectangular element over the entire plate, we obtain

$$U = \iint \frac{1}{2}\left(- M_x \frac{\partial^2 w}{\partial x^2} - M_y \frac{\partial^2 w}{\partial y^2} - 2M_{xy} \frac{\partial^2 w}{\partial x\, \partial y}\right) dx\, dy \qquad (39.18)$$

When Eqs. (39.3) for the moments are used, we have

$$U = \frac{K}{2} \iint \left[\left(\frac{\partial^2 w}{\partial x^2}\right)^2 + \left(\frac{\partial^2 w}{\partial y^2}\right)^2 + 2\nu \frac{\partial^2 w}{\partial x^2}\frac{\partial^2 w}{\partial y^2} + 2(1-\nu)\left(\frac{\partial^2 w}{\partial x\, \partial y}\right)^2 \right] dx\, dy \qquad (39.19a,b)$$

$$= \frac{K}{2} \iint \left[(\nabla^2 w)^2 - 2(1-\nu)\left\{ \frac{\partial^2 w}{\partial x^2}\frac{\partial^2 w}{\partial y^2} - \left(\frac{\partial^2 w}{\partial x\, \partial y}\right)^2 \right\} \right] dx\, dy$$

The fundamental plate equation (39.10) could have been derived by equating the increase in U during a virtual deflection δw to the corresponding work done by the lateral loads.

The elastic energy expressions (39.19*a*,*b*) are useful in obtaining approximate solutions to plate problems.

39.2. THEORY OF RECTANGULAR PLATES

a. Long Rectangular Plate Bent to a Cylindrical Surface

The plate is assumed to be bounded as shown in Fig. 39.5, and the loading is some function $q(x)$. The differential equation (39.10) now takes the form

$$\frac{d^4 w}{dx^4} = \frac{q}{K} \qquad (39.20)$$

which will be recognized as the same as the equation for flexure of a beam. Consequently, relations for beams developed in Chap. 35 of this handbook can be applied

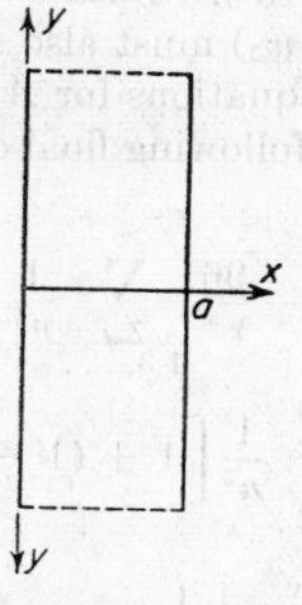

FIG. 39.5. Long rectangular plate.

for the case of the plate bent to a cylindrical surface. We note only that the beam flexural rigidity EI is replaced by K. Also we will have a transverse moment $M_y = \nu M_x$.

In the general case of a nonuniform load $q(x)$, it is often convenient to express it by the Fourier series,

$$q(x) = \sum_{m=1,2,\ldots}^{\infty} q_m \sin \frac{m\pi x}{a} \qquad (39.21)$$

Substitution in (39.20) permits direct solution for w for the case of simply supported edges:

$$w = \frac{a^4}{K} \sum_{1,2,\ldots}^{\infty} \frac{q_m}{(m\pi)^4} \sin \frac{m\pi x}{a} \tag{39.22}$$

b. Semi-infinite Rectangular Plate with Uniform Load

The plate is bounded as shown in Fig. 39.6. The load is uniform ($q = q_0$).

As in many other plate problems, the above case can be solved by superimposing the solution for an infinite strip with a suitable solution of the homogeneous equation (39.10′).

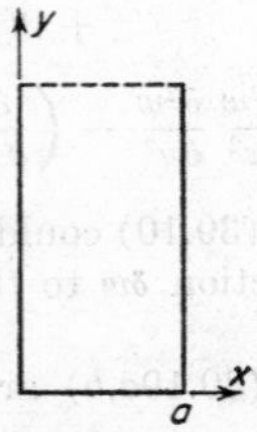

FIG. 39.6. Semi-infinite simply supported rectangular plate.

We assume, in the first case, that the edges are simply supported. Combine the deflection w_1 of an infinite simply supported strip, with loading q_0, with the deflection w_2 needed to complete the boundary conditions:

$$w_1 = \frac{q_0 a^4}{24K}\left[\left(\frac{x}{a}\right)^4 - 2\left(\frac{x}{a}\right)^3 + \frac{x}{a}\right] = \frac{4q_0 a^4}{\pi^5 K} \sum_{1,3,\ldots}^{\infty} \frac{\sin \lambda_n x}{n^5}$$

$$w_2 = \sum_{1,3,\ldots}^{\infty} (A_n + B_n y) e^{-\lambda_n y} \sin \lambda_n x \qquad \lambda_n = \frac{n\pi}{a} \qquad n = 1, 3, \ldots \tag{39.23a,b}$$

The expression w_2 satisfies Eq. (39.10′). Both w_1 and w_2 are symmetric.

The combined deflection $(w_1 + w_2)$ must also satisfy conditions (39.16) at $y = 0$. Substitution gives a pair of linear equations for A_n and B_n, for each n. The calculation of these constants leads to the following final expressions for deflection, moments, and shear forces:

$$w = \frac{q_0 a^4}{24K}\left[\left(\frac{x}{a}\right)^4 - 2\left(\frac{x}{a}\right)^3 + \frac{x}{a} - \frac{96}{\pi^5} \sum_{1,3,\ldots}^{\infty} \frac{1}{n^5}\left(1 + \frac{\lambda_n y}{2}\right) e^{-\lambda_n y} \sin \lambda_n x\right] \tag{39.24}$$

$$M_x = \frac{q_0 a^2}{2}\left\{\frac{x}{a}\left(1 - \frac{x}{a}\right) - \frac{8}{\pi^3} \sum_{1,3,\ldots}^{\infty} \frac{1}{n^3}\left[1 + (1-\nu)\frac{\lambda_n y}{2}\right] e^{-\lambda_n y} \sin \lambda_n x\right\}$$

$$M_y = \frac{q_0 a^2}{2}\left\{\frac{\nu x}{a}\left(1 - \frac{x}{a}\right) - \frac{8}{\pi^3} \sum_{1,3,\ldots}^{\infty} \frac{1}{n^3}\left[\nu - (1-\nu)\frac{\lambda_n y}{2}\right] e^{-\lambda_n y} \sin \lambda_n x\right\} \tag{39.25a–c}$$

$$M_{xy} = -\frac{2q_0 a^2}{\pi^3}(1-\nu) \sum_{1,3,\ldots}^{\infty} \frac{1}{n^3}(1 + \lambda_n y) e^{-\lambda_n y} \cos \lambda_n x$$

$$Q_x = -\frac{2q_0 a^2}{2}\left(1 - \frac{2x}{a} - \frac{8}{\pi^2} \sum_{1,3,\ldots}^{\infty} \frac{1}{n^2} e^{-\lambda_n y} \cos \lambda_n x\right)$$

$$Q_y = \frac{4q_0 a}{\pi^2} \sum_{1,3,\ldots}^{\infty} \frac{1}{n^2} e^{-\lambda_n y} \sin \lambda_n x \tag{39.26a,b}$$

The following cases may be investigated in a similar way [2], using conditions (39.14) and (39.17).

Clamped edge at $y = 0$

$$w = \frac{q_0 a^4}{24K}\left[\left(\frac{x}{a}\right)^4 - 2\left(\frac{x}{a}\right)^3 + \frac{x}{a} - \frac{96}{\pi^5}\sum_{1,3,\ldots}^{\infty}\frac{1}{n^5}(1+\lambda_n y)e^{-\lambda_n y}\sin\lambda_n x\right] \tag{39.27}$$

$$M_x = \frac{q_0 a^2}{2}\left\{\frac{x}{a}\left(1-\frac{x}{a}\right) - \frac{8}{\pi^3}\sum_{1,3,\ldots}^{\infty}\frac{1}{n^3}[1+\nu+(1-\nu)\lambda_n y]e^{-\lambda_n y}\sin\lambda_n x\right\}$$

$$M_y = \frac{q_0 a^2}{2}\left\{\frac{\nu x}{a}\left(1-\frac{x}{a}\right) - \frac{8}{\pi^3}\sum_{1,3,\ldots}^{\infty}\frac{1}{n^3}[1+\nu-(1-\nu)\lambda_n y]e^{-\lambda_n y}\sin\lambda_n x\right\} \tag{39.28a–c}$$

$$M_{xy} = -\frac{4q_0 a y}{\pi^2}(1-\nu)\sum_{1,3,\ldots}^{\infty}\frac{1}{n^2}e^{-\lambda_n y}\cos\lambda_n x$$

Since M_{xy} is zero at the corners, there are no corner reactions.

Free edge at $y = 0$

$$w = \frac{q_0 a^4}{24K}\left[\left(\frac{x}{a}\right)^4 - 2\left(\frac{x}{a}\right)^3 + \frac{x}{a} + \frac{96\nu}{\pi^5(3+\nu)}\sum_{1,3,\ldots}^{\infty}\left(\frac{1+\nu}{1-\nu} - \lambda_n y\right)\frac{e^{-\lambda_n y}}{n^5}\sin\lambda_n x\right] \tag{39.29}$$

$$M_x = \frac{q_0 a^2}{2}\left[\frac{x}{a}\left(1-\frac{x}{a}\right) + \frac{8}{\pi^3}\frac{\nu(1-\nu)}{3+\nu}\sum_{1,3,\ldots}^{\infty}(1-\lambda_n y)\frac{e^{-\lambda_n y}}{n^3}\sin\lambda_n x\right]$$

$$M_y = \frac{q_0 a^2}{2}\left[\frac{\nu x}{a}\left(1-\frac{x}{a}\right) - \frac{8}{\pi^3}\frac{\nu(1-\nu)}{3+\nu}\sum_{1,3,\ldots}^{\infty}\left(\frac{3+\nu}{1-\nu} - \lambda_n y\right)\frac{e^{-\lambda_n y}}{n^3}\sin\lambda_n x\right] \tag{39.30a–c}$$

$$M_{xy} = -\frac{4q_0 a^2}{\pi^3}\frac{\nu(1-\nu)}{3+\nu}\sum_{1,3,\ldots}^{\infty}\left(\frac{2}{1-\nu} - \lambda_n y\right)\frac{e^{-\lambda_n y}}{n^3}\cos\lambda_n x$$

Along the free edge ($y = 0$)

$$w = \frac{q_0 a^4}{24K}\frac{3-\nu}{(3+\nu)(1-\nu)}\left(\frac{x^4}{a^4} - 2\frac{x^3}{a^3} + \frac{x}{a}\right)$$
$$M_x = \frac{q_0 x}{2}(a-x)\frac{(3-\nu)(1+\nu)}{3+\nu} \tag{39.31a,b}$$

We note that the deflection and moment M_x are larger along the free edge than at very large values by constant factors.

c. Line Loading on Infinite and Semi-infinite Strips

Let the plate strip (Fig. 39.5) be loaded by a line load $p(x)$ along the x axis, as shown in Fig. 39.7. Deflections and stresses are symmetrical about the x axis, and so results are given only for $y \geq 0$. Along $y = 0$, we have the boundary conditions [note Eq. (39.8)]

$$\frac{\partial w}{\partial y} = 0 \qquad Q_y = -\frac{p}{2} = -K\frac{\partial}{\partial y}\nabla^2 w \tag{39.32}$$

If the edges are simply supported, an expression satisfying Eq. (39.10′) and the boundary conditions at $x = 0$, $x = a$ has been given in Eqs. (39.23b). To apply this solution to the present problem, we first represent $p(x)$ by the Fourier series

$$p(x) = \sum_{1,2,\ldots}^{\infty} p_n \sin \frac{n\pi x}{a} \tag{39.33}$$

When w is given by (39.23b) and we make use of (39.32) and (39.33), equations can be found for the A_n and B_n in terms of p_n. The resulting deflection is then

$$w = \frac{a^3}{4\pi^3 K} \sum_{n=1,2,\ldots}^{\infty} (1 + \lambda_n y) \frac{p_n}{n^3} e^{-\lambda_n y} \sin \lambda_n x \tag{39.34}$$

where $\lambda_n = n\pi/a$, as before.

Special cases are obtained by appropriate values for the coefficients p_n for various loadings.

For uniform loading $p = p_0$

$$p_n = \frac{4p_0}{\pi n} \quad \text{for } n = 1, 3, 5, \ldots \qquad p_n = 0 \quad \text{for } n = 2, 4, \ldots$$

$$w = \frac{p_0 a^3}{\pi^4 K} \sum_{1,3,\ldots}^{\infty} \frac{1}{n^4} (1 + \lambda_n y) e^{-\lambda_n y} \sin \lambda_n x \tag{39.35}$$

For a concentrated force P *at* $x = \xi$, $y = 0$

$$p_n = \frac{2P}{a} \sin \lambda_n \xi \qquad n = 1, 2, 3, \ldots$$

$$w = \frac{Pa^2}{2\pi^3 K} \sum_{1,2,\ldots}^{\infty} \frac{1}{n^3} (1 + \lambda_n y) e^{-\lambda_n y} \sin \lambda_n \xi \sin \lambda_n x \tag{39.36}$$

Consider next a semi-infinite rectangular plate with loading $p(x)$ on line $y = \eta$, as in Fig. 39.8. Assume that the strip is simply supported on all edges.

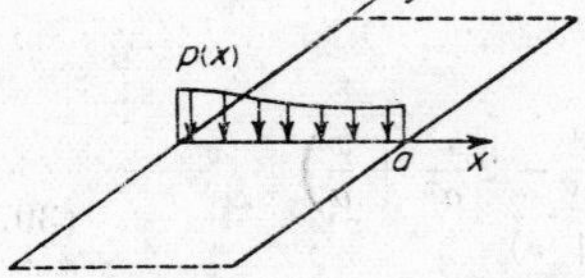

FIG. 39.7. Nonuniform line load on infinite strip.

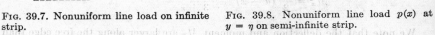

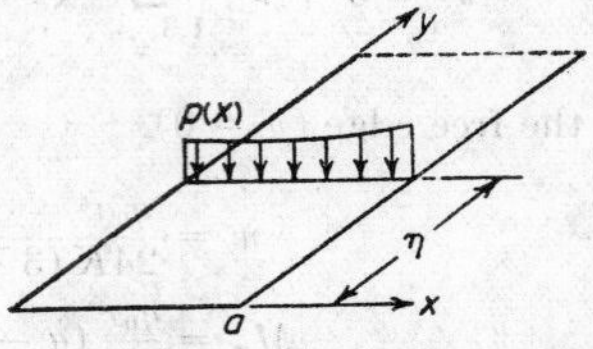

FIG. 39.8. Nonuniform line load $p(x)$ at $y = \eta$ on semi-infinite strip.

The problem is solved by superimposing deflection w_1 for an infinite strip with load $p(x)$ at $y = \eta$, and w_2 for another strip with load $-p(x)$ at $y = -\eta$. This ensures fulfillment of boundary conditions at $y = 0$. We let

$$\begin{aligned} w_1' &= w_1 && \text{for } y \geq \eta \\ w_1'' &= w_1 && \text{for } y \leq \eta \\ w_2' &= w_2 && \text{for } y \geq -\eta \\ w' &= w && \text{for } y \geq \eta \\ w'' &= w && \text{for } 0 \leq y \leq \eta \end{aligned}$$

We are primarily interested in w' and w'', which will be

$$w' = w_1' + w_2' \qquad w'' = w_1'' + w_2'$$

Expressions for w_1', w_1'', and w_2' are easily obtained by transformation from (39.34):

$$w_1' = \frac{a^3}{4\pi^3 K} \sum_{1,2,\ldots}^{\infty} [1 + \lambda_n(y - \eta)] \frac{p_n}{n^3} e^{-\lambda_n(y-\eta)} \sin \lambda_n x$$

$$w_1'' = \frac{a^3}{4\pi^3 K} \sum_{1,2,\ldots}^{\infty} [1 + \lambda_n(\eta - y)] \frac{p_n}{n^3} e^{-\lambda_n(\eta-y)} \sin \lambda_n x$$

$$w_2' = -\frac{a^3}{4\pi^3 K} \sum_{1,2,\ldots}^{\infty} [1 + \lambda_n(y + \eta)] \frac{p_n}{n^3} e^{-\lambda_n(y+\eta)} \sin \lambda_n x$$

We then obtain for w' and w''

$$w' = \frac{a^3}{4\pi^3 K} \sum_{1,2,\ldots}^{\infty} \frac{p_n}{n^3} \{[1 + \lambda_n(y - \eta)]e^{-\lambda_n(y-\eta)} - [1 + \lambda_n(y + \eta)]e^{-\lambda_n(y+\eta)}\} \sin \lambda_n x$$

$$w'' = \frac{a^3}{4\pi^3 K} \sum_{1,2,\ldots}^{\infty} \frac{p_n}{n^3} \{[1 + \lambda_n(\eta - y)]e^{-\lambda_n(\eta-y)} - [1 + \lambda_n(y + \eta)]e^{-\lambda_n(y+\eta)}\} \sin \lambda_n x \qquad (39.37)$$

These formulas represent the solution of the problem in Fig. 39.8 for a general load distribution, Eq. (39.33). We next give results for particular cases.

For a uniform load $p = p_0$ *at* $y = \eta$

$$w' = \frac{p_0 a^3}{\pi^4 K} \sum_{1,3,\ldots}^{\infty} \frac{1}{n^4} \{[1 + \lambda_n(y - \eta)]e^{-\lambda_n(y-\eta)} - [1 + \lambda_n(y + \eta)]e^{-\lambda_n(y+\eta)}\} \sin \lambda_n x$$

$$w'' = \frac{p_0 a^3}{\pi^4 K} \sum_{1,3,\ldots}^{\infty} \frac{1}{n^4} \{[1 + \lambda_n(\eta - y)]e^{-\lambda_n(\eta-y)} - [1 + \lambda_n(y + \eta)]e^{-\lambda_n(y+\eta)}\} \sin \lambda_n x \qquad (39.38)$$

For a concentrated load P *at* $x = \xi$, $y = \eta$

$$w' = \frac{Pa^2}{2\pi^3 K} \sum_{1,2,\ldots}^{\infty} \frac{1}{n^3} \{[1 + \lambda_n(y - \eta)]e^{-\lambda_n(y-\eta)} - [1 + \lambda_n(y + \eta)]e^{-\lambda_n(y+\eta)}\} \sin \lambda_n \xi \sin \lambda_n x$$

$$w'' = \frac{Pa^2}{2\pi^3 K} \sum_{1,2,\ldots}^{\infty} \frac{1}{n^3} \{[1 + \lambda_n(\eta - y)]e^{-\lambda_n(\eta-y)} - [1 + \lambda_n(y + \eta)]e^{-\lambda_n(y+\eta)}\} \sin \lambda_n \xi \sin \lambda_n x \qquad (39.39)$$

For problems involving the concentrated force P the bending-moment expressions calculated from the series given above do not converge rapidly near the point of load application. The solution may be improved in this respect by using the method of analysis of bending moments discussed on p. 39–15.

d. Simply Supported Rectangular Plate

Let the plate be bounded as shown in Fig. 39.9, and assume that all edges are simply supported, so that conditions (39.16) apply.

A solution for various load distributions may be obtained, using double series expressions for the load and the deflection (Navier). We let

$$q(x,y) = \sum_{\substack{m \\ 1,2,\ldots}}^{\infty} \sum_{n}^{\infty} q_{mm} \sin \frac{m\pi x}{a} \sin \frac{n\pi y}{b}$$

$$w(x,y) = \sum_{\substack{m \\ 1,2,\ldots}}^{\infty} \sum_{n}^{\infty} w_{mn} \sin \frac{m\pi x}{a} \sin \frac{n\pi y}{b} \tag{39.40a,b}$$

Coefficients q_{mn} are calculated by usual methods when $q(x,y)$ is known.

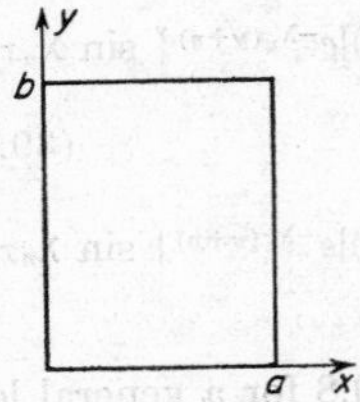

FIG. 39.9. Simply supported plate for Navier solution.

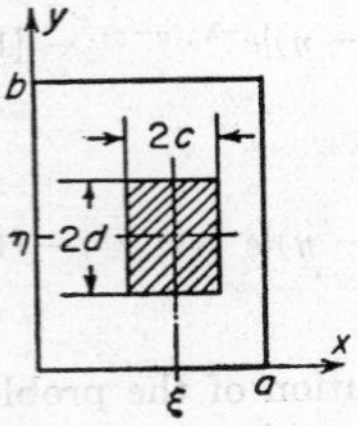

FIG. 39.10. Uniform load on subrectangle.

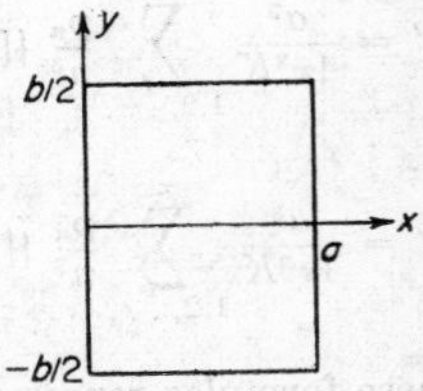

FIG. 39.11. Axis location for single-series solution. Simply supported plate.

Equation (39.40*b*) satisfies the boundary conditions for simple support. When Eqs. (39.40) are substituted in the basic equation (39.10), a relation between constants q_{mn} and w_{mn} is established:

$$w_{mn} = \frac{q_{mn}}{\pi^4 K(m^2/a^2 + n^2/b^2)^2} \tag{39.41}$$

For various particular load distributions, we have the following results:

Uniform load q_0

$$q_{mn} = \frac{16q_0}{\pi^2 mn} \qquad m, n = 1, 3, 5, \ldots \tag{39.42}$$

$$w = \frac{16q_0 a^4}{\pi^6 K} \sum_{\substack{m \\ 1,3,\ldots}}^{\infty} \sum_{n}^{\infty} \frac{\sin (m\pi x/a) \sin (n\pi y/b)}{mn[m^2 + n^2(a^2/b^2)]^2} \tag{39.43}$$

Uniformly loaded rectangular subregion (*Fig.* 39.10)

$$q_{mn} = \frac{16q_0}{\pi^2 mn} \sin \frac{m\pi\xi}{a} \sin \frac{m\pi c}{a} \sin \frac{n\pi\eta}{b} \sin \frac{n\pi d}{b} \qquad m, n = 1, 2, 3, \ldots \tag{39.44}$$

Concentrated load P at ξ, η. For this case, we let c and d in Fig. 39.10 approach zero while $4cdq_0 = P$.

$$q_{mn} = \frac{4P}{ab} \sin \frac{m\pi\xi}{a} \sin \frac{n\pi\eta}{b} \qquad m, n = 1, 2, \ldots \tag{39.45}$$

$$w = \frac{4Pa^3}{\pi^4 bK} \sum_{m}\sum_{n} \frac{\sin (m\pi\xi/a) \sin (n\pi\eta/b)}{[m^2 + n^2(a^2/b^2)]^2} \sin \frac{m\pi x}{a} \sin \frac{n\pi y}{b} \tag{39.46}$$

$$\scriptstyle 1,2,\ldots$$

Calculation of bending moments from the above expressions is not very satisfactory because of slow convergence of the series. An improved solution in this respect has been developed in more recent times by Lévy, Estanave, and Nadai [6]. We proceed to discuss this solution, for the case of uniform load.

Axes are chosen as in Fig. 39.11 with the $b \geq a$. The solution w_1 for an infinite strip of width a is combined with a solution w_2 of the homogeneous equation (39.10′), which is symmetrical in y and satisfies conditions (39.16) for simple support at $x = 0$, $x = a$:

$$w = w_1 + w_2 \qquad w_1 = \frac{q_0 a^4}{24K}\left[\left(\frac{x}{a}\right)^4 - 2\left(\frac{x}{a}\right)^3 + \frac{x}{a}\right]$$

$$w_2 = \sum_{1,3,\ldots}^{\infty} (A_m \operatorname{Cosh} \lambda_m y + C_m \lambda_m y \operatorname{Sinh} \lambda_m y) \sin \lambda_m x \qquad \lambda_m = \frac{m\pi}{a} \tag{39.47}$$

When we satisfy the boundary conditions of simple support at $y = \pm b/2$, the constants A_m and C_m are determined, and the final expression for w is

$$w = \frac{q_0 a^4}{24K}\left[\left(\frac{x}{a}\right)^4 - 2\left(\frac{x}{a}\right)^3 + \frac{x}{a}\right] + 2\frac{q_0 a^4}{\pi^5 K} \sum_{1,3,\ldots}^{\infty} \frac{1}{m^5}\left(-\frac{\alpha_m \operatorname{Tanh} \alpha_m + 2}{\operatorname{Cosh} \alpha_m} \operatorname{Cosh} \lambda_m y + \lambda_m y \frac{\operatorname{Sinh} \lambda_m y}{\operatorname{Cosh} \alpha_m}\right) \sin \lambda_m x \tag{39.48}$$

where

$$\alpha_m = \frac{m\pi b}{2a}$$

The moments and edge forces can be calculated by application of relations (39.3) and (39.12′); for example,

$$M_x = -\tfrac{1}{2} q_0(x^2 - ax) - \frac{2q_0 a^2}{\pi^3} \sum_{1,3,\ldots}^{\infty} \frac{1}{m^3}\left[\frac{2 + (1-\nu)\alpha_m \operatorname{Tanh} \alpha_m}{\operatorname{Cosh} \alpha_m} \operatorname{Cosh} \lambda_m y - \frac{(1-\nu)\lambda_m y \operatorname{Sinh} \lambda_m y}{\operatorname{Cosh} \alpha_m}\right] \sin \lambda_m x$$

$$M_y = -\tfrac{1}{2} q_0 \nu(x^2 - ax) - \frac{2q_0 a^2}{\pi^3} \sum_{1,3,\ldots}^{\infty} \frac{1}{m^3}\left[\frac{2\nu - (1-\nu)\alpha_m \operatorname{Tanh} \alpha_m}{\operatorname{Cosh} \alpha_m} \operatorname{Cosh} \lambda_m y + \frac{(1-\nu)\lambda_m y \operatorname{Sinh} \lambda_m y}{\operatorname{Cosh} \alpha_m}\right] \sin \lambda_m x$$

$$M_{xy} = \frac{2q_0 a^2}{\pi^3}(1-\nu) \sum_{1,3,\ldots}^{\infty} \frac{1}{m^3}\left[\frac{1 + \alpha_m \operatorname{Tanh} \alpha_m}{\operatorname{Cosh} \alpha_m} \operatorname{Sinh} \lambda_m y - \frac{\lambda_m y \operatorname{Cosh} \lambda_m y}{\operatorname{Cosh} \alpha_m}\right] \cos \lambda_m x \tag{39.49a–c}$$

Design calculations are facilitated by tables of numerical coefficients, defined as follows:

$$w_{\max} = \delta_1 \frac{q_0 a^4}{Et^3} \qquad (M_x)_{a/2,0} = \delta_2 q_0 a^2 \qquad (M_y)_{a/2,0} = \delta_3 q_0 a^2$$

$$(V_x)_{0,0} = \delta_4 q_0 a \qquad (V_y)_{a/2,b/2} = -\delta_5 q_0 a \qquad R = \delta_6 q_0 a^2$$

Values of these coefficients are given in Table 39.1 for $\nu = 0.3$. Additional values are given in [1].

Table 39.1. Numerical Coefficients δ_1, δ_2, δ_3, δ_4, δ_5, δ_6 for Uniformly Loaded Simply Supported Plates
($\nu = 0.3$)

b/a	δ_1	δ_2	δ_3	δ_4	δ_5	δ_6
1.0	0.0443	0.0497	0.0479	0.420	0.420	0.065
1.2	0.0616	0.0626	0.0501	0.455	0.453	0.074
1.4	0.0770	0.0753	0.0506	0.478	0.471	0.083
1.6	0.0906	0.0862	0.0493	0.491	0.485	0.086
1.8	0.1017	0.0948	0.0479	0.499	0.491	0.090
2.0	0.1106	0.1017	0.0464	0.503	0.496	0.092
3.0	0.1336	0.1189	0.0404	0.505	0.498	0.093
4.0	0.1400	0.1235	0.0384	0.502	0.500	0.094
∞	0.1422	0.1250	0.0375	0.500	0.500	0.095

We next consider application of the single series solution to cases of nonuniform loading.

Suppose the load is a function of x alone $[q = q(x)]$. It may be expressed in the sine series (39.21). The deflection will consist of one portion w_1 (for the infinite strip), given by (39.22), and another portion w_2 [solution of (39.10′)], identical with (39.47)

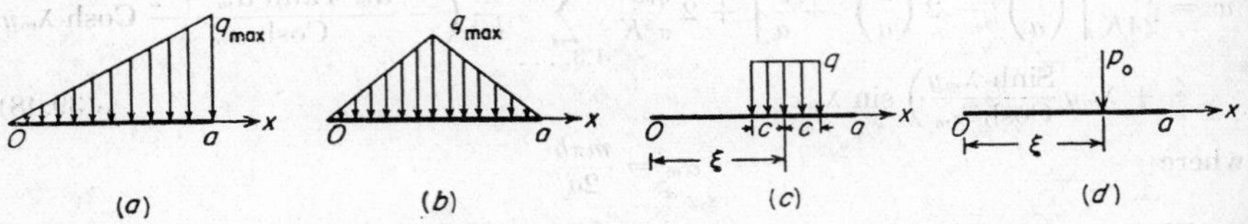

FIG. 39.12. Types of loading $q(x)$ on simply supported rectangular plates.

except that we sum for $m = 1, 2, 3, \ldots$ The constants A_m and C_m are determined from the boundary conditions at $y = \pm b/2$, and the deflection of the plate is then given by

$$w = \frac{a^4}{\pi^4 K} \sum_{1,2,\ldots}^{\infty} \frac{q_m}{m^4}\left(1 - \frac{2 + \alpha_m \operatorname{Tanh} \alpha_m}{2 \operatorname{Cosh} \alpha_m} \operatorname{Cosh} \lambda_m y + \frac{\lambda_m y \operatorname{Sinh} \lambda_m y}{2 \operatorname{Cosh} \alpha_m}\right) \sin \lambda_m x \tag{39.50}$$

For several types of load distributions, as shown in Fig. 39.12, the q_m values are as follows:

Hydrostatic loading ($q = q_{max}x/a$)

$$q_m = \frac{2q_{max}}{m\pi}(-1)^{m+1} \qquad m = 1, 2, \ldots$$

Triangular loading

$$q = 2q_{max}x/a \qquad x \le a/2$$
$$q = 2q_{max}(a - x)/a \qquad x \ge a/2$$
$$q_m = \frac{8q_{max}}{\pi^2 m^2}(-1)^{(m-1)/2} \qquad m = 1, 3, 5, \ldots$$

Uniform load q_0 from $(\xi - c)$ to $(\xi + c)$

$$q_m = \frac{4q_0}{m\pi} \sin \frac{m\pi\xi}{a} \sin \frac{m\pi c}{a} \qquad m = 1, 2, \ldots$$

Line load p_0 at $x = \xi$

$$q_m = \frac{2p_0}{a} \sin \frac{m\pi\xi}{a} \qquad m = 1, 2, \ldots$$

Values of coefficients for design calculations of moments and deflections at specific points have been given by Timoshenko [1]. Let $\gamma_1 \ldots \gamma_8$ be defined as follows for the case of hydrostatic loading (Fig. 39.12*a*):

$$w = \gamma_1 \frac{q_{\max} a^4}{Et^3} \qquad M_x = \gamma_2 q_{\max} a^2 \qquad M_y = \gamma_3 q_{\max} a^2$$

$$(V_x)_{0,0} = \gamma_4 q_{\max} a \qquad (V_x)_{a,0} = -\gamma_5 q_{\max} a \qquad V_y = -\gamma_6 q_{\max} b \tag{39.51}$$

$$R_{a,b/2} = \gamma_7 q_{\max} ab \qquad R_{0,b/2} = \gamma_8 q_{\max} ab$$

Values of these coefficients are given in Tables 39.2 through 39.4.

Table 39.2. Numerical Coefficient γ_1 for Deflection of Simply Supported Plate with Hydrostatic Loading (Fig. 39.12*a*)

$(y = 0,\ \nu = 0.3)$

b/a	$x = 0.25a$	$x = 0.50a$	$x = 0.60a$	$x = 0.75a$
1.0	0.0143	0.0221	0.0220	0.0177
1.2	0.0203	0.0308	0.0305	0.0241
1.4	0.0257	0.0385	0.0380	0.0298
1.6	0.0303	0.0453	0.0444	0.0346
1.8	0.0342	0.0508	0.0497	0.0385
2.0	0.0373	0.0553	0.0539	0.0417
3.0	0.0454	0.0668	0.0647	0.0498
4.0	0.0477	0.0700	0.0679	0.0521
∞	0.0484	0.0711	0.0690	0.0529

Table 39.3. Numerical Coefficients γ_2 and γ_3 for Bending Moments in Simply Supported Rectangular Plate with Hydrostatic Loading (Fig. 39.12*a*)

$(y = 0,\ \nu = 0.3)$

b/a	γ_2				γ_3			
	$x = 0.25a$	$x = 0.50a$	$x = 0.60a$	$x = 0.75a$	$x = 0.25a$	$x = 0.50a$	$x = 0.60a$	$x = 0.75a$
1.0	0.0132	0.0239	0.0264	0.0259	0.0149	0.0239	0.0245	0.0207
1.2	0.0179	0.0313	0.0338	0.0318	0.0158	0.0250	0.0254	0.0213
1.4	0.0221	0.0376	0.0402	0.0367	0.0160	0.0253	0.0254	0.0212
1.6	0.0256	0.0431	0.0454	0.0407	0.0158	0.0246	0.0249	0.0207
1.8	0.0286	0.0474	0.0496	0.0439	0.0153	0.0239	0.0242	0.0202
2.0	0.0309	0.0508	0.0529	0.0463	0.0148	0.0232	0.0234	0.0197
3.0	0.0369	0.0594	0.0611	0.0525	0.0128	0.0202	0.0207	0.0176
4.0	0.0385	0.0617	0.0632	0.0541	0.0120	0.0192	0.0196	0.0168
∞	0.0391	0.0625	0.0640	0.0547	0.0117	0.0187	0.0192	0.0165

Table 39.4. Numerical Coefficients $\gamma_4, \gamma_5, \gamma_6, \gamma_7, \gamma_8$ for Edge and Corner Reactions in Simply Supported Rectangular Plates with Hydrostatic Loading (Fig. 39.12a)

($\nu = 0.3$)

b/a	γ_4	γ_5	$\gamma_6\left(y = \frac{b}{2}\right)$				γ_7	γ_8
			$x = 0.25a$	$x = 0.50a$	$x = 0.60a$	$x = 0.75a$		
1.0	0.126	0.294	0.115	0.210	0.234	0.239	0.039	0.026
1.2	0.144	0.312	0.105	0.189	0.208	0.209	0.037	0.026
1.4	0.155	0.323	0.095	0.169	0.185	0.184	0.035	0.025
1.6	0.162	0.330	0.086	0.151	0.166	0.164	0.032	0.023
1.8	0.166	0.333	0.078	0.136	0.149	0.147	0.029	0.021
2.0	0.168	0.335	0.071	0.124	0.135	0.134	0.026	0.020
3.0	0.169	0.336	0.048	0.083	0.091	0.089	0.018	0.014
4.0	0.168	0.334	0.036	0.063	0.068	0.067	0.014	0.010
∞	0.167	0.333						

e. Bending Moments in Plates Carrying Concentrated Loads

A point load P, as considered in Eqs. (39.36), (39.39), and (39.45), is an idealization of a distributed load $q(x,y)$, which is concentrated in a very small area. At the point of application of a point load, the bending moments M_x and M_y tend to infinity, indicating that, in the vicinity of this point, they depend substantially on the distribution of the load. If the dimensions of the loaded area are of the same order as the thickness of the plate, thin-plate theory even becomes inapplicable [2],[7].

Nevertheless, the solution of point-load problems is of interest, mainly because of its connection with the Green's function of the differential equation (see Chaps. 11 and 28). Since the numerical evaluation of series becomes very tedious as the singular point is approached, there is need for a better solution. It may be found by separating the singularity in closed form and using Fourier series only to satisfy the boundary conditions.

It may easily be verified that

$$w = \frac{P}{16\pi K}(x^2 + y^2)\ln(x^2 + y^2) \tag{39.52}$$

is a solution of Eq. (39.10′), and that the shearing forces Q_x and Q_y derived from it are in equilibrium with a concentrated force P at $x = y = 0$. The solution (39.52) yields the following moments and shear forces:

$$\begin{aligned}
M_x &= -\frac{P}{8\pi}\left[1 + \nu + 2\frac{x^2 + \nu y^2}{x^2 + y^2} + (1 + \nu)\ln(x^2 + y^2)\right] \\
M_y &= -\frac{P}{8\pi}\left[1 + \nu + 2\frac{\nu x^2 + y^2}{x^2 + y^2} + (1 + \nu)\ln(x^2 + y^2)\right] \\
M_{xy} &= -\frac{P}{4\pi}(1 - \nu)\frac{xy}{x^2 + y^2} \\
Q_x &= -\frac{P}{2\pi}\frac{x}{x^2 + y^2} \qquad Q_y = -\frac{P}{2\pi}\frac{y}{x^2 + y^2}
\end{aligned} \tag{39.53a–e}$$

These formulas may appear objectionable, because the argument of the logarithm is not dimensionless. It should, however, be realized that this deficiency may easily be removed by adding obvious solutions of Eq. (39.10′). Since this is done automatically in any application, the form presented here is unobjectionable, and has the advantage of simplicity. Its application is illustrated by the following example.

Consider a simply supported rectangular plate (Fig. 39.13), and assume that it carries a force P at its center. The deflection is composed of three parts, $w = w_1 + w_2 + w_3$, where w_1 is identical with w from Eq. (39.52). The corresponding values w_1 and M_{y1} are evaluated for $y = \pm b/2$ and are expanded in Fourier series. Then the solution w_2 of Eqs. (39.47) is superposed (writing $\cos \lambda_m x$ for $\sin \lambda_m x$ because

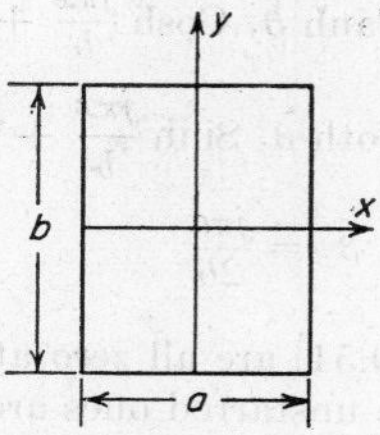

FIG. 39.13. Coordinate axes for rectangular plate.

of the shifting of the coordinate system), and the constants A_m and C_m are so chosen that $(w_1 + w_2)$ satisfies the conditions (39.16) on $y = \pm b/2$. Finally, a solution

$$w_3 = \sum_{1,3,\ldots}^{\infty} (A'_m \operatorname{Cosh} \lambda'_m x + C'_m \lambda'_m x \operatorname{Sinh} \lambda'_m x) \cos \lambda'_m y \qquad \lambda'_m = \frac{m\pi}{b}$$

is added and used to fulfill the boundary conditions on $x = \pm a/2$.

When the load P is not applied at the center, Eqs. (39.47) must be replaced by

$$w_2 = \sum_{1,2,\ldots}^{\infty} (A_m \operatorname{Cosh} \lambda_m y + B_m \operatorname{Sinh} \lambda_m y + C_m \lambda_m y \operatorname{Sinh} \lambda_m y + D_m \lambda_m y \operatorname{Cosh} \lambda_m y) \cos \lambda_m x$$

and the boundary conditions at the edges $y = +b/2$ and $y = -b/2$ yield four linear equations for constants A_m through D_m with every $m = 1, 2, \ldots$. The procedure for w_3 is, of course, similar.

f. Plates with Edge Slopes and Edge Moments Specified

Axes are taken so that the edges of the plate are at $x = \pm a/2$, $y = \pm b/2$, as in Fig. 39.13.

When the plate is loaded by edge moments only, the deflection will satisfy Eq. (39.10′). Solutions of (39.10′) can be found by use of the product form (a function of x times a function of y), and will involve expressions of the type

$$\begin{array}{ll} Y_m \cos \dfrac{m\pi x}{a} & X_m \cos \dfrac{m\pi y}{b} \\ Y_i \sin \dfrac{i\pi x}{a} & X_i \sin \dfrac{i\pi y}{b} \\ Y^*_m \cos \dfrac{m\pi x}{a} & X^*_m \cos \dfrac{m\pi y}{b} \\ Y^*_i \sin \dfrac{i\pi x}{a} & X^*_i \sin \dfrac{i\pi y}{b} \\ m = 1, 3, 5, \ldots & i = 2, 4, 6, \ldots \end{array} \qquad (39.54)$$

where, for $j = 1, 2, 3, \ldots,$

$$Y_j = -\alpha_j \operatorname{Tanh} \alpha_j \operatorname{Cosh} \frac{j\pi y}{a} + \frac{j\pi y}{a} \operatorname{Sinh} \frac{j\pi y}{a}$$

$$Y_j^* = -\alpha_j \operatorname{Coth} \alpha_j \operatorname{Sinh} \frac{j\pi y}{a} + \frac{j\pi y}{a} \operatorname{Cosh} \frac{j\pi y}{a}$$

$$X_j = -\beta_j \operatorname{Tanh} \beta_j \operatorname{Cosh} \frac{j\pi x}{b} + \frac{j\pi x}{b} \operatorname{Sinh} \frac{j\pi x}{b} \tag{39.55}$$

$$X_j^* = -\beta_j \operatorname{Coth} \beta_j \operatorname{Sinh} \frac{j\pi x}{b} + \frac{j\pi x}{b} \operatorname{Cosh} \frac{j\pi x}{b}$$

$$\alpha_j = \frac{j\pi b}{2a} \qquad \beta_j = \frac{j\pi a}{2b}$$

Note that the expressions (39.54) are all zero at $x = \pm a$, $y = \pm b$. The starred functions are antimetric, and the unstarred ones are symmetric.

These expressions are useful in cases of plates loaded by edge couples. In the following discussion, we consider deflections symmetrical about the x axis, but not necessarily symmetrical about the y axis. The nonsymmetric (about the y axis) cases can be treated by resolution into a symmetric part w' and an antimetric part w''.

Case 1. Symmetrically Disposed Moments at $x = \pm a/2$, with Simple Support at $y = \pm b/2$. The deflection is symmetrical about both axes. The loading is pictured in Fig. 39.14a. Since we want $\partial^2 w'/\partial y^2 = 0$ at $y = \pm b/2$, we will have w' in the form

$$w' = \sum_{1,3,\ldots}^{\infty} \frac{af'_m}{m^2 \operatorname{Cosh} \beta_m} X_m \cos \frac{m\pi y}{b} \tag{39.56}$$

where the coefficient has been taken so as to simplify ensuing formulations.

The edge moments are

$$(M'_x)_{x=\pm a/2} = -\frac{2K\pi^2 a}{b^2} \sum_{1,3,\ldots}^{\infty} f'_m \cos \frac{m\pi y}{b} \tag{39.57}$$

If the edge-moment distribution is known, it can be represented by a cosine series, and constants f'_m will be known.

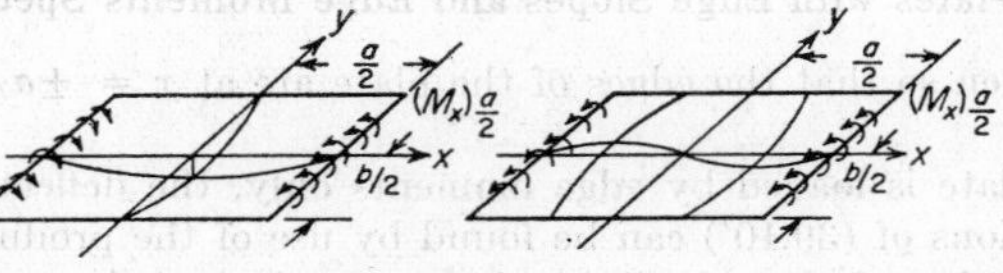

FIG. 39.14. Plates loaded by edge moments M_x at $x = \pm a/2$ and simply supported at $y = \pm b/2$: (a) Symmetric; (b) antimetric.

The edge slopes are:

$$\left(\frac{\partial w'}{\partial x}\right)_{x=\pm a/2} = \pm 2 \sum_{1,3,\ldots}^{\infty} \frac{f'_m \beta_m}{m^2 \operatorname{Cosh} \beta_m} \left(\frac{\beta_m}{\operatorname{Cosh} \beta_m} + \operatorname{Sinh} \beta_m\right) \cos \frac{m\pi y}{b}$$

$$\left(\frac{\partial w'}{\partial y}\right)_{y=\pm b/2} = \mp 2 \sum_{1,3,\ldots}^{\infty} \frac{f'_m \beta_m (-1)^{(m-1)/2}}{m^2 \operatorname{Cosh} \beta_m} X_m \tag{39.58a,b}$$

If the edge slope at $x = \pm a/2$ is specified by

$$\left(\frac{\partial w'}{\partial x}\right)_{x=\pm a/2} = \mp \sum_{1,3,\ldots}^{\infty} \phi_n' \cos \frac{n\pi y}{b}$$

the constants f_m' would be

$$f_m' = \frac{-m^2 \phi_m'}{2\beta_m[(\beta_m/\text{Cosh}^2 \beta_m) + \text{Tanh}\, \beta_m]} \tag{39.59}$$

Case 2. Antimetric Moment Distribution at $x = \pm a/2$ and Simple Support at $y = \pm b/2$ (Fig. 39.14*b*)

$$w'' = \sum_{1,3,\ldots}^{\infty} \frac{af_m''}{m^2 \text{Sinh}\, \beta_m} X_m^* \cos \frac{m\pi y}{b} \tag{39.60}$$

$$(M_x'')_{x=\pm a/2} = \mp 2 \frac{K\pi^2 a}{b} \sum_{1,3,\ldots}^{\infty} f_m'' \cos \frac{m\pi y}{b} \tag{39.61}$$

$$\left(\frac{\partial w''}{\partial x}\right)_{x=\pm a/2} = 2 \sum_{1,3,\ldots}^{\infty} \frac{f_m''\beta_m}{m^2 \text{Sinh}\, \beta_m} \left(\frac{-\beta_m}{\text{Sinh}\, \beta_m} + \text{Cosh}\, \beta_m\right) \cos \frac{m\pi y}{b}$$

$$\left(\frac{\partial w''}{\partial y}\right)_{y=\pm b/2} = \mp 2 \sum_{1,3,\ldots}^{\infty} \frac{f_m''\beta_m}{m^2 \text{Sinh}\, \beta_m} X_m^* (-1)^{(m-1)/2} \tag{39.62a,b}$$

If edge slope at $x = \pm a/2$ is specified as

$$\left(\frac{\partial w''}{\partial x}\right)_{x=\pm a/2} = - \sum_{1,3,\ldots}^{\infty} \phi_n'' \cos \frac{n\pi y}{b}$$

then

$$f_m'' = \frac{-m^2 \phi_m''}{2\beta_m[(-\beta_m/\text{Sinh}^2 \beta_m) + \text{Coth}\, \beta_m]} \tag{39.63}$$

Case 3. Moments Symmetrically Distributed on All Sides (Fig. 39.15*a*). The solution will consist of one contribution due to moments at $x = \pm a/2$, and another due to moments at $y = \pm b/2$:

$$w' = \sum_{1,3,\ldots}^{\infty} \left(\frac{af_m'}{m^2 \text{Cosh}\, \beta_m} X_m \cos \frac{m\pi y}{b} + \frac{bd_m'}{m^2 \text{Cosh}\, \alpha_m} Y_m \cos \frac{m\pi x}{a} \right) \tag{39.64}$$

The edge moments in this case are:

$$(M_x')_{x=\pm a/2} = -2 \frac{K\pi^2 a}{b^2} \sum_{1,3,\ldots}^{\infty} f_m' \cos \frac{m\pi y}{b}$$

$$(M_y')_{y=\pm b/2} = -2 \frac{K\pi^2 b}{a^2} \sum_{1,3,\ldots}^{\infty} d_m' \cos \frac{m\pi x}{a} \tag{39.65a,b}$$

so that specification of moments determines constants f_m' and d_m'.

If edge slopes are specified by the relations

$$\left(\frac{\partial w'}{\partial x}\right)_{x=\pm a/2} = \mp \sum_{1,3,\ldots}^{\infty} \phi_n' \cos\frac{n\pi y}{b} \qquad \left(\frac{\partial w'}{\partial y}\right)_{y=\pm b/2} = \mp \sum_{1,3,\ldots}^{\infty} \psi_n' \cos\frac{n\pi x}{a} \tag{39.66}$$

the equations for determining f_m' and d_m' are obtained by calculating edge slopes from (39.64) and equating them to the series expressions (39.66). In doing this, we have to deal with functions like that in Eq. (39.58*b*). To solve the problem, X_m and Y_m

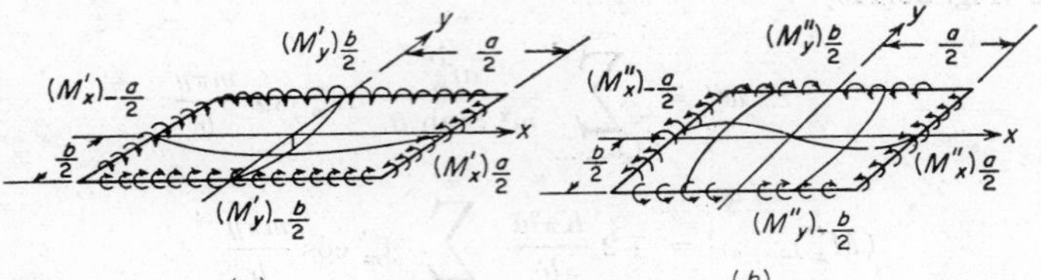

FIG. 39.15. Specified edge slopes or edge moments: (*a*) Symmetric case; (*b*) antimetric case.

must be expanded in cosine series. After carrying out these steps, we obtain the system of equations:

$$-2\frac{f_n'}{n^2}\beta_n\left(\frac{\beta_n}{\text{Cosh}^2\,\beta_n} + \text{Tanh}\,\beta_n\right) - 8\left(\frac{b}{a}\right)^3 n \sum_{m=1,3,\ldots}^{\infty} \frac{d_m'(-1)^{(m+n-2)/2}}{m^3[(b/a)^2 + (n/m)^2]^2} = \phi_n'$$

$$-2\frac{d_n'\alpha_n}{n^2}\left(\frac{\alpha_n}{\text{Cosh}^2\,\alpha_n} + \text{Tanh}\,\alpha_n\right) \tag{39.67a,b}$$

$$-8\left(\frac{a}{b}\right)^3 n \sum_{m=1,3,\ldots}^{\infty} \frac{f_m'(-1)^{(m+n-2)/2}}{m^3[(a/b)^2 + (n/m)^2]^2} = \psi_n'$$

$$n = 1, 3, 5, \ldots$$

For numerical solutions, we may use the first $(n + 1)/2$ equations from each set, and allow m to range from 1 to n.

Case 4. Antimetric Distribution of Edge Moments (Fig. 39.15*b*). The analysis is similar to that for case 3, except that we use the antimetric function X_m^*.

$$w'' = \sum_{1,3,\ldots}^{\infty} \frac{af_m''}{m^2\,\text{Sinh}\,\beta_m} X_m^* \cos\frac{m\pi y}{b} + \sum_{2,4,\ldots}^{\infty} \frac{bd_i''}{i^2\,\text{Cosh}\,\alpha_i} Y_i \sin\frac{i\pi x}{a} \tag{39.68}$$

$$(M_x'')_{x=\pm a/2} = \mp\frac{2\pi^2 Ka}{b^2} \sum_{1,3,\ldots}^{\infty} f_m'' \cos\frac{m\pi y}{b}$$

$$(M_y'')_{y=\pm b/2} = -2\frac{\pi^2 Kb}{a^2} \sum_{2,4,\ldots}^{\infty} d_i'' \sin\frac{i\pi x}{a} \tag{39.69a,b}$$

If the edge slopes are assigned, and given by

$$\left(\frac{\partial w''}{\partial x}\right)_{x=\pm a/2} = -\sum_{1,3,\ldots}^{\infty} \phi_n'' \cos\frac{n\pi y}{b} \qquad \left(\frac{\partial w''}{\partial y}\right)_{y=\pm b/2} = \mp \sum_{2,4,\ldots}^{\infty} \psi_i'' \sin\frac{i\pi x}{a} \tag{39.70}$$

the equations for constants f_n'', d_i'' become

$$-2f_n''\frac{\beta_n}{n^2}\left(\frac{-\beta_n}{\operatorname{Sinh}^2\beta_n} + \operatorname{Coth}\beta_n\right) + 8\left(\frac{b}{a}\right)^3 n \sum_{i=2,4,\ldots}^{\infty} \frac{d_i''(-1)^{(i+n-1)/2}}{i^3[(b/a)^2 + (n/i)^2]^2} = \phi_n''$$

$$i = 2, 4, \ldots \qquad n = 1, 3, \ldots \tag{39.71a,b}$$

$$-2d_i''\frac{\alpha_i}{i^2}\left(\frac{\alpha_i}{\operatorname{Cosh}^2\alpha_i} + \operatorname{Tanh}\alpha_i\right) + 8\left(\frac{a}{b}\right)^3 i \sum_{m=1,3,\ldots}^{\infty} \frac{f_m''(-1)^{(i+m-1)/2}}{m^3[(a/b)^2 + (i/m)^2]^2} = \psi_i''$$

Cases 1 through 4 all involve symmetry about the x axis. The most general type of edge-moment loading can be resolved into four component parts. One of them is symmetric with respect to both coordinate axes; two are symmetric with respect to one axis and antimetric with respect to the other one; and the fourth is antimetric with respect to both axes. Each case may be solved by the appropriate choice among the solutions (39.54).

g. Laterally Loaded Clamped Rectangular Plate

The method here presented is similar to that used by Timoshenko [8]. Any case of loading symmetrical about one of the axes may be treated. The problem of the simply supported plate with the prescribed loading is solved first, giving deflection w_1, and then a deflection w_2 is superimposed, which is that for a plate loaded by edge couples only. The edge couples are so taken that the edge slope for w_2 just cancels that for w_1.

For any loading symmetric about the x axis we will then have

$$w = w_1 + w_2 = w_1 + \sum_{1,3,\ldots}^{\infty}\left(\frac{af_m'}{m^2\operatorname{Cosh}\beta_m}X_m\cos\frac{m\pi y}{b} + \frac{bd_m'}{m^2\operatorname{Cosh}\alpha_m}Y_m\cos\frac{m\pi x}{a} + \frac{af_m''}{m^2\operatorname{Sinh}\beta_m}X_m^*\cos\frac{m\pi y}{b}\right) + \sum_{2,4,\ldots}^{\infty}\frac{bd_i''}{i^2\operatorname{Cosh}\alpha_i}Y_i\sin\frac{i\pi x}{a} \tag{39.72}$$

Deflection w_1 can be calculated by methods described in Art. 39.2*d*. Constants f_m', d_m', f_m'', d_i'' will be determined by Eqs. (39.67) and (39.71), where the ϕ_n', ϕ_n'', ψ_n', ψ_i'' are the values of the slope coefficients for the simply supported plate.

The bending moments within the plate may be calculated by Eqs. (39.3); at the edges they are given by relations (39.65) and (39.69).

Results for special cases will have to do with the particular expressions for w_1 and for the coefficients ϕ_n', ϕ_n'', ψ_n', ψ_i''. Some of these results are summarized below.

Case 1. Uniform Load q_0. With axes taken as in Fig. 39.13, the deflection w_1 for the simply supported plate becomes [transformation of Eq. (39.48)]

$$w_1 = \frac{q_0 a^4}{24K}\left[\left(\frac{x}{a}\right)^4 - \frac{3}{2}\left(\frac{x}{a}\right)^2 + \frac{5}{16}\right] + \frac{2q_0a^4}{\pi^5 K}\sum_{1,3,\ldots}^{\infty}\frac{(-1)^{(m-1)/2}}{m^5 \operatorname{Cosh}\alpha_m}\left(Y_m - 2\operatorname{Cosh}\frac{m\pi y}{a}\right)\cos\frac{m\pi x}{a} \quad (39.73)$$

The edge-slope coefficients to be used in Eq. (39.67) are:

$$\phi_n' = \frac{2q_0b^3}{\pi^4 K}\left[\frac{(-1)^{(n-1)/2}}{n^4}\left(\frac{\beta_n}{\operatorname{Cosh}^2\beta_n} - \operatorname{Tanh}\beta_n\right)\right]$$
$$\psi_n' = \frac{2q_0a^3}{\pi^4 K}\left[\frac{(-1)^{(n-1)/2}}{n^4}\left(\frac{\alpha_n}{\operatorname{Cosh}^2\alpha_n} - \operatorname{Tanh}\alpha_n\right)\right] \quad (39.74)$$

The complete deflection of the uniformly loaded plate is

$$w = w_1 + \sum_{1,3,\ldots}^{\infty}\left(\frac{af_m'}{m^2\operatorname{Cosh}\beta_m}X_m\cos\frac{m\pi y}{b} + \frac{bd_m'}{m^2\operatorname{Cosh}\alpha_m}Y_m\cos\frac{m\pi x}{a}\right) \quad (39.75)$$

where X_m and Y_m are given by Eq. (39.55). Numerical calculations have been carried out by Evans [9], who gives curves for f_m' and d_m' and coefficients for design calculations.[1] Some values for these coefficients are given in Table 39.5.

Table 39.5. Clamped Rectangular Plates with Uniform Load ($\nu = 0.3$)

$w_{\max} = \lambda_1 \dfrac{q_0a^4}{Et^3}$ $\quad (M_x)_{a/2,0} = \lambda_2 q_0 a^2 \quad (M_y)_{0,b/2} = \lambda_3 q_0 a^2$

$(M_x)_{0,0} = \lambda_4 q_0 a^2 \quad (M_y)_{0,0} = \lambda_5 q_0 a^2$

b/a	λ_1	λ_2	λ_3	λ_4	λ_5
1.0	0.0138	−0.0513	−0.0513		
1.2	0.0188	−0.0639	−0.0554	0.0299	0.0228
1.4	0.0226	−0.0726	−0.0568	0.0349	0.0212
1.6	0.0251	−0.0780	−0.0571	0.0381	0.0193
1.8	0.0267	−0.0812	−0.0571	0.0401	0.0174
2.0	0.0277	−0.0829	−0.0571		
∞	0.0285	−0.0833	....	0.0417	0.0125

For a square plate, the solution is simplified since $\alpha_m = \beta_m = m\pi/2$ and $d_m' = f_m'$. In this case,

$$w = w_1 + \sum_{1,3,\ldots}^{\infty}\frac{af_m'}{m^2\operatorname{Cosh}(m\pi/2)}\left(X_m\cos\frac{m\pi y}{a} + Y_m\cos\frac{m\pi x}{a}\right)$$
$$\phi_n' = \frac{2q_0a^3}{\pi^4 K}\frac{(-1)^{(n-1)/2}}{n^4}\left(\frac{n\pi/2}{\operatorname{Cosh}^2(n\pi/2)} - \operatorname{Tanh}\frac{n\pi}{2}\right) \quad (39.76)$$

[1] Evans's $F_m(-1)^{(m-1)/2}/K$ corresponds to our $f_m'\pi^5K/2q_0ab^2$, and Evans's $E_m(-1)^{(m-1)/2}/K$ to our $d_m'\pi^5bK/2a^4q_0$.

The f_n' values determined from a set of four equations are:

$$f_1' = 0.002432 \frac{q_0 a^3}{K} \qquad f_5' = -0.0001163 \frac{q_0 a^3}{K}$$

$$f_3' = 0.0002484 \frac{q_0 a^3}{K} \qquad f_7' = 0.0000563 \frac{q_0 a^3}{K}$$

Computation of $w_{\max}$ for the square plate gives

$$w_{\max} = (0.04436 - 0.03054) \frac{q_0 a^4}{Et^3} = 0.01382 \frac{q_0 a^4}{Et^3}$$

The first term is the contribution from w_1 (simply supported plate) and the second part is that from w_2 (edge-moment loading). The maximum bending moment, occurring at the middle of an edge, is found, using the four terms, to be $-0.0517 q_0 a^2$; calculations with additional terms [8] give the more exact value,

$$M_{\max} = (M_x)_{a/2,0} = -0.0513 q_0 a^2$$

Case 2. Concentrated Load P at Center of Rectangular Plate. Deflection is given, as before, by (39.75), but w_1 in this case is

$$w_1 = \frac{Pa^2}{2K\pi^3} \sum_{1,3,\ldots}^{\infty} \frac{1}{m^3} \cos \frac{m\pi x}{a} \left[\left(\operatorname{Tanh} \alpha_m - \frac{\alpha_m}{\operatorname{Cosh}^2 \alpha_m} \right) \operatorname{Cosh} \frac{m\pi y}{a} - \operatorname{Sinh} \frac{m\pi y}{a} - \operatorname{Tanh} \alpha_m \frac{m\pi y}{a} \operatorname{Sinh} \frac{m\pi y}{a} + \frac{m\pi y}{a} \operatorname{Cosh} \frac{m\pi y}{a} \right] \tag{39.77}$$

The edge-slope coefficients to be used in calculation of f_n', d_n' by (39.67) are

$$\phi_n' = -\frac{Pb}{2K\pi^2} \frac{\beta_n \operatorname{Tanh} \beta_n}{n^2 \operatorname{Cosh} \beta_n} \qquad \psi_n' = -\frac{Pa}{2K\pi^2} \frac{\alpha_n \operatorname{Tanh} \alpha_n}{n^2 \operatorname{Cosh} \alpha_n} \tag{39.78}$$

Detailed numerical calculations have been carried out by Young [10], solving for seven f_n' and seven d_n'. Deflection and moment coefficients obtained by Young are given in Table 39.6.

Table 39.6. Clamped Rectangular Plate, Concentrated Force at Center
($\nu = 0.3$)

b/a	1.0	1.2	1.4	1.6	1.8	2.0
$w_{\max} Et^3/Pa^2$	0.0611	0.0706	0.0755	0.0777	0.0786	0.0788
$(M_x)_{a/2,0}/P$	−0.1257	−0.1490	−0.1604	−0.1651	−0.1677	−0.1674

h. Continuous Plates

Two examples of continuous plates with lateral loading will be discussed. The first is a long rectangular plate with many transverse supports (Fig. 39.16). The other is a plate with uniform load continuous over many column supports, as in Fig. 39.17.

For the first problem, the results of Art. 39.2*f* may be applied. Any panel j will have deflection expressed as $(w_{1j} + w_{2j})$, where w_{1j} is that for a simply supported plate with the prescribed loading, and w_{2j} is a biharmonic function [solution of

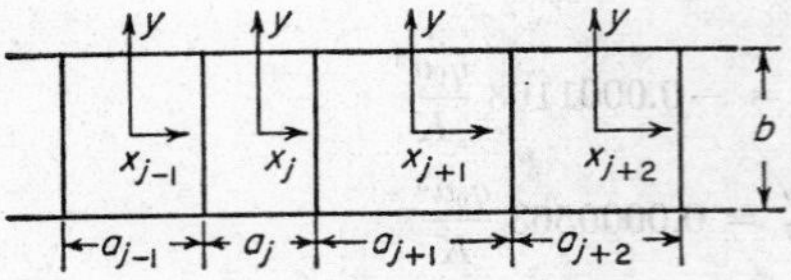

FIG. 39.16. Continuous plate on n bays.

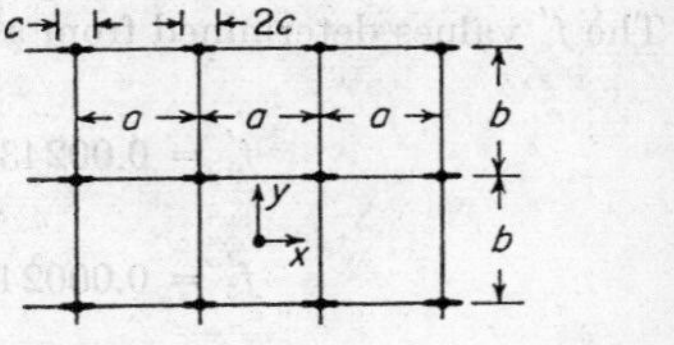

FIG. 39.17. Plate continuous over many column supports.

(39.10′)] satisfying simple support conditions at $y = \pm b/2$ and appropriate matching conditions of slopes and bending moments at the transverse supports at $x_j = \pm a_j/2$. Loading is assumed symmetric in y.

$$w_j = w_{1j} + a \sum_{1,3,\ldots}^{\infty} \left[\frac{f'_{mj}}{m^2 \operatorname{Cosh} \beta_{mj}} \left(-\beta_{mj} \operatorname{Tanh} \beta_{mj} \operatorname{Cosh} \frac{m\pi x_j}{b} + \frac{m\pi x_j}{b} \operatorname{Sinh} \frac{m\pi x_j}{b} \right) + \frac{f''_{mj}}{m^2 \operatorname{Sinh} \beta_{mj}} \left(-\beta_{mj} \operatorname{Coth} \beta_{mj} \operatorname{Sinh} \frac{m\pi x_j}{b} + \frac{m\pi x_j}{b} \operatorname{Cosh} \frac{m\pi x_j}{b} \right) \right] \cos \frac{m\pi y}{b}$$

$$\beta_{mj} = \frac{m\pi a_j}{2b} \tag{39.79}$$

Bending moments at $x_j = \pm a_j/2$ are

$$(M_x)_{x_j=\pm a_j/2} = -2K_j \frac{\pi^2 a_j}{b^2} \sum_{1,3,\ldots}^{\infty} (f'_{mj} \pm f''_{mj}) \cos \frac{m\pi y}{b} \tag{39.80}$$

A similar moment expression could be written for panel $(j + 1)$. Continuity of bending moments then requires that

$$a_j K_j (f'_{mj} + f''_{mj}) = a_{j+1} K_{j+1} (f'_{m,j+1} - f''_{m,j+1}) \tag{39.81}$$

Setting the edge slopes for the panel j at $x_j = a_j/2$ and for the panel $(j + 1)$ at $x_{j+1} = -a_{j+1}/2$ equal to each other leads to

$$f'_{mj} A_{mj} + f''_{mj} B_{mj} + f'_{m,j+1} A_{m,j+1} - f''_{m,j+1} B_{m,j+1} = -(\phi'_{mj} + \phi''_{mj}) - (\phi'_{m,j+1} - \phi''_{m,j+1}) \tag{39.82}$$

where $A_{mj} = \dfrac{2\beta_{mj}}{m^2} \left(\operatorname{Tanh} \beta_{mj} + \dfrac{\beta_{mj}}{\operatorname{Cosh}^2 \beta_{mj}} \right)$ $\quad B_{mj} = \dfrac{2\beta_{mj}}{m^2} \left(\operatorname{Coth} \beta_{mj} - \dfrac{\beta_{mj}}{\operatorname{Sinh}^2 \beta_{mj}} \right)$

and the slope coefficients are specified by

$$\left(\frac{\partial w'_{ij}}{\partial x_j} \right)_{x_j=a_j/2} = \sum_{1,3,\ldots}^{\infty} \phi'_{mj} \cos \frac{m\pi y}{b} \qquad \left(\frac{\partial w''_{ij}}{\partial x_j} \right)_{x_j=a_j/2} = \sum_{1,3,\ldots}^{\infty} \psi''_{mj} \cos \frac{m\pi y}{b} \tag{39.83}$$

Finally, at the ends of the plate (panel 1 and panel n), we have certain conditions; e.g., for simply supported ends the moments vanish, so that

$$f''_{m1} = f'_{m1} \qquad f'_{mn} = -f''_{mn} \tag{39.84}$$

Now for n panels we have $(n-1)$ equations (39.81), $(n-1)$ equations (39.82), and the two equations (39.84), making $2n$ equations for the $2n$ constants f'_{mj}, f''_{mj} $(j = 1, 2, \ldots, n)$.

For the second problem (Fig. 39.17), where the load is uniform, take the axes such that supporting columns are at the corners $(x = \pm a/2, y = \pm b/2)$. The supporting forces are considered, for analytical convenience, as acting over short line segments. Thus, we have a load per unit length $-p_0$ at $y = b/2$ and $a/2 - c < x < a/2 + c$, and similarly for the other support zones. Boundary conditions for the rectangular panel are:

$$\begin{aligned} x &= \pm\frac{a}{2} \qquad -\frac{b}{2} \le y \le \frac{b}{2}: \qquad Q_x = 0 \qquad \frac{\partial w}{\partial x} = 0 \\ y &= \pm\frac{b}{2} \qquad -\frac{a}{2} \le x \le \frac{a}{2}: \qquad Q_y = \pm\frac{p(x)}{2} \qquad \frac{\partial w}{\partial y} = 0 \end{aligned} \tag{39.85}$$

where $p(x)$ is defined as

$$\begin{aligned} -\frac{a}{2} + c \le x \le \frac{a}{2} - c \qquad & p(x) = 0 \\ -\frac{a}{2} \le x < -\frac{a}{2} + c \qquad & \\ & p(x) = -p_0 \\ \frac{a}{2} - c < x \le \frac{a}{2} \qquad & \end{aligned}$$

The deflection may be written as the sum of that for an infinite strip with uniform load and clamped edges plus that for a rectangular plate which satisfies boundary conditions at $x = \pm a/2$:

$$w = \frac{q_0 b^4}{384K}\left(1 - \frac{4y^2}{b^2}\right)^2 + A_0 + \sum_{2,4,\ldots}^{\infty}\left(A_m \operatorname{Cosh}\frac{m\pi y}{a} + B_m \frac{m\pi y}{a} \operatorname{Sinh}\frac{m\pi y}{a}\right)\cos\frac{m\pi x}{a} \tag{39.86}$$

The support force function $p(x)$ can be represented by

$$p(x) = -\frac{2p_0 c}{a} - \frac{4p_0 c}{a} \sum_{2,4,\ldots} (-1)^{m/2}\left(\frac{a}{m\pi c}\right)\sin\frac{m\pi c}{a}\cos\frac{m\pi x}{a} \tag{39.87}$$

We use the conditions at $y = \pm b/2$ with the help of (39.87) and (39.8) to determine constants A_m and B_m. Finally A_0 is found such that $w = 0$ at $x = a/2$, $y = b/2$. The values of the constants are, for $c \ll a$,

$$\begin{aligned} A_m &= -\frac{q_0 b}{K}(-1)^{m/2}\left(\frac{a}{m\pi}\right)^3 \frac{\alpha_m + \operatorname{Tanh}\alpha_m}{\operatorname{Tanh}\alpha_m \operatorname{Sinh}\alpha_m} \\ B_m &= \frac{q_0 b}{K}(-1)^{m/2}\left(\frac{a}{m\pi}\right)^3 \frac{1}{\operatorname{Sinh}\alpha_m} \\ A_0 &= \frac{q_0 b^4}{2\pi^3 K}\left(\frac{a}{b}\right)^3 \sum_{2,4,\ldots}^{\infty} \frac{1}{m^3}\left(\frac{\alpha_m}{\operatorname{Sinh}^2\alpha_m} + \operatorname{Coth}\alpha_m\right) \end{aligned} \tag{39.88}$$

The deflection is given by (39.86).

Having now obtained the deflection, we dispense with the assumption of supporting forces on short line segments, and consider instead that they act on small circular areas. The stresses and moments near a column are then analyzed by treating that portion of the plate as a circular plate, as shown in Fig. 39.18. This procedure is justified as long as the panels are nearly square. Nadai [2] has shown that the appropriate radius for the circular plate is $0.22a$. This plate is loaded by uniform

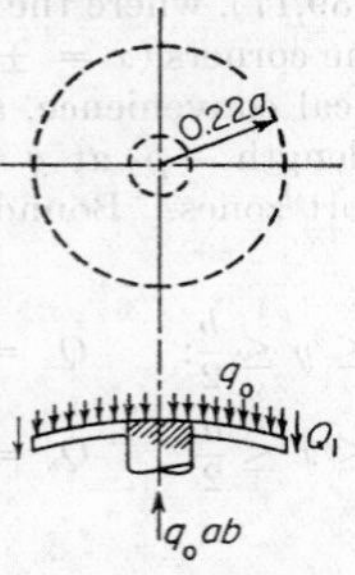

FIG. 39.18. Circular plate equivalent to region around ends of columns supporting a continuous plate.

load q_0, shear forces Q_1 around the periphery, and upward force q_0ab at the inner built-in circular boundary. The shear force Q_1 is

$$Q_1 = 0.723q_0b - 0.11q_0a \tag{39.89}$$

39.3. CIRCULAR PLATES

Plate Equations in Polar Coordinates

When Eq. (39.10), for vertical equilibrium of a plate element, is transformed into polar coordinates r and θ (Fig. 39.19), it has the form

$$\nabla^2\nabla^2 w \equiv \left(\frac{\partial^2}{\partial r^2} + \frac{1}{r}\frac{\partial}{\partial r} + \frac{1}{r^2}\frac{\partial^2}{\partial \theta^2}\right)^2 w = \frac{q}{K} \tag{39.90}$$

Expressions for moments M_r, M_θ, $M_{r\theta}$ can be found by noting that, for $\theta = 0$, these moments coincide, respectively, with M_x, M_y, M_{xy}, and using the operation

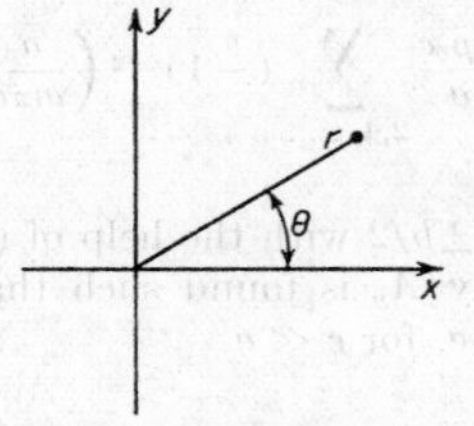

FIG. 39.19. Polar coordinates.

relations

$$\frac{\partial}{\partial x} = \cos\theta\frac{\partial}{\partial r} - \frac{1}{r}\sin\theta\frac{\partial}{\partial \theta} \qquad \frac{\partial}{\partial y} = \sin\theta\frac{\partial}{\partial r} + \frac{1}{r}\cos\theta\frac{\partial}{\partial \theta}$$

In this way we find

$$\begin{aligned} M_r &= -K\left[\frac{\partial^2 w}{\partial r^2} + \nu\left(\frac{1}{r}\frac{\partial w}{\partial r} + \frac{1}{r^2}\frac{\partial^2 w}{\partial \theta^2}\right)\right] \\ M_\theta &= -K\left[\frac{1}{r}\frac{\partial w}{\partial r} + \frac{1}{r^2}\frac{\partial^2 w}{\partial \theta^2} + \nu\frac{\partial^2 w}{\partial r^2}\right] \\ M_{r\theta} &= -(1-\nu)K\left[\frac{1}{r}\frac{\partial^2 w}{\partial r\,\partial \theta} - \frac{1}{r^2}\frac{\partial w}{\partial \theta}\right] \end{aligned} \tag{39.91a–c}$$

Since shear force components depend on the gradient of $\nabla^2 w$,

$$Q_r = -K \frac{\partial}{\partial r} \nabla^2 w \qquad Q_\theta = -\frac{K}{r} \frac{\partial}{\partial \theta} \nabla^2 w \tag{39.91d,e}$$

The edge reactions, for edge with outward normal in the r or θ direction are:

$$\begin{aligned} V_r &= -K \left[\frac{\partial}{\partial r} \nabla^2 w + \frac{1-\nu}{r} \frac{\partial}{\partial \theta} \left(\frac{1}{r} \frac{\partial^2 w}{\partial r\, \partial \theta} - \frac{1}{r^2} \frac{\partial w}{\partial \theta} \right) \right] \\ V_\theta &= -K \left[\frac{1}{r} \frac{\partial}{\partial \theta} \nabla^2 w + (1-\nu) \frac{\partial}{\partial r} \left(\frac{1}{r} \frac{\partial^2 w}{\partial r\, \partial \theta} - \frac{1}{r^2} \frac{\partial w}{\partial \theta} \right) \right] \end{aligned} \tag{39.92}$$

Axially Symmetric Bending of Circular Plates

For radial symmetry, (39.90) reduces to the ordinary equation

$$\left(\frac{d^2}{dr^2} + \frac{1}{r} \frac{d}{dr} \right)^2 w = \frac{q}{K} \tag{39.93}$$

in which $q = q(r)$.

Moments and shear forces simplify to

$$\begin{aligned} M_r &= -K \left(\frac{d^2 w}{dr^2} + \frac{\nu}{r} \frac{dw}{dr} \right) \qquad M_\theta = -K \left(\frac{1}{r} \frac{dw}{dr} + \nu \frac{d^2 w}{dr^2} \right) \\ Q_r &= -K \frac{d}{dr} \left[\frac{1}{r} \frac{d}{dr} \left(r \frac{dw}{dr} \right) \right] \qquad Q_\theta = 0 \qquad V_r = Q_r \end{aligned} \tag{39.94a–e}$$

Problems can often be solved by superposition of elementary cases. In general, the solutions of (39.93) will involve combination of a particular solution and the general solution of the homogeneous equation ($q = 0$). The latter is known to be

$$w = A + Br^2 + C \ln r + Dr^2 \ln r \tag{39.95}$$

Case a. Plate Loaded by Edge Couples M_0 (Fig. 39.20*a*). Equation (39.95) applies, and the terms in C and D will not appear since the deflection and shear are zero at the origin. Thus,

$$w = A + Br^2$$

and evaluation of A and B by the boundary conditions of $w = 0$ and $M_r = M_0$ at $r = a$ leads to

$$w = \frac{M_0 a^2}{2K(1+\nu)} \left(1 - \frac{r^2}{a^2} \right) \tag{39.96a}$$

The bending moments are uniform over the plate:

$$M_r = M_\theta = M_0 \tag{39.96b}$$

Case b. Clamped Plate with Uniform Load (Fig. 39.20*b*). The particular integral may be taken as $q_0 r^4/64K$, and only terms involving A and B from (39.95) apply, as before:

$$w = \frac{q_0 r^4}{64K} + A + Br^2 \tag{39.97}$$

Use of clamped-edge boundary conditions $w = dw/dr = 0$ at $r = a$ leads to

$$w = \frac{q_0 a^4}{64K} \left(1 - \frac{r^2}{a^2} \right)^2 \tag{39.98a}$$

$$\begin{aligned} M_r &= -\frac{q_0 a^2}{16} \left[\frac{r^2}{a^2} (3+\nu) - 1 - \nu \right] \\ M_\theta &= -\frac{q_0 a^2}{16} \left[\frac{r^2}{a^2} (1+3\nu) - 1 - \nu \right] \end{aligned} \tag{39.98b,c}$$

Case c. Simply Supported Plate, Uniform Load (Fig. 39.20c). By superposition of deflections and moments from cases a and b with $M_0 = q_0a^2/8$, we cancel out the edge bending moment and obtain the results for a simply supported plate:

$$w = \frac{q_0a^4}{64K}\left(1 - \frac{r^2}{a^2}\right)\left(\frac{5+\nu}{1+\nu} - \frac{r^2}{a^2}\right) \tag{39.99a–c}$$

$$M_r = \frac{q_0a^2}{16}(3+\nu)\left(1 - \frac{r^2}{a^2}\right) \qquad M_\theta = \frac{q_0a^2}{16}\left[3 + \nu - (1+3\nu)\frac{r^2}{a^2}\right]$$

Case d. Concentrated Load at Center, Clamped Edge (Fig. 39.20d). We use solution (39.95), but only the term with C is omitted. The term with D must be retained, because we now have very high shear forces near the center. Constants A, B, and D are determined by the edge conditions (39.14) plus the fact that the vertical shear Q_r must be

$$Q_r = -\frac{P}{2\pi r}$$

The resulting expressions are:

$$w = \frac{Pa^2}{16\pi K}\left(2\frac{r^2}{a^2}\ln\frac{r}{a} - \frac{r^2}{a^2} + 1\right) \tag{39.100a–c}$$

$$M_r = -\frac{P}{4\pi}\left[1 + (1+\nu)\ln\frac{r}{a}\right] \qquad M_\theta = -\frac{P}{4\pi}\left[\nu + (1+\nu)\ln\frac{r}{a}\right]$$

The moments have a logarithmic singularity at $r = 0$. For its interpretation, see p. 39–14.

Case e. Concentrated Load at Center, Simply Supported Edge (Fig. 39.20e). By superposition of cases a and d with $M_0 = P/4\pi$, we obtain

$$w = \frac{Pa^2}{16\pi K}\left[2\frac{r^2}{a^2}\ln\frac{r}{a} + \frac{3+\nu}{1+\nu}\left(1 - \frac{r^2}{a^2}\right)\right] \tag{39.101a–c}$$

$$M_r = -\frac{P}{4\pi}(1+\nu)\ln\frac{r}{a} \qquad M_\theta = \frac{P}{4\pi}\left[1 - \nu - (1+\nu)\ln\frac{r}{a}\right]$$

Case f. Plates with Load Varying as a Power of the Radius. Let the load be given by

$$q = q_n\left(\frac{r}{a}\right)^n \tag{39.102}$$

It is simplest, in this case, to integrate (39.93) directly;[1] this facilitates the problem of finding the particular integral. Only the terms in A and B from (39.95) appear in the final resulting expression, since the slope and shear force are zero at $r = 0$. For a *clamped edge*, determination of the constants leads to

$$w = \frac{q_0a^4}{K(n+2)^2(n+4)^2}\left[\left(\frac{r}{a}\right)^{n+4} - \left(\frac{r}{a}\right)^2\frac{n+4}{2} + \frac{n+2}{2}\right]$$

$$M_r = \frac{-q_na^2}{(n+2)^2(n+4)}\left[(n+3+\nu)\left(\frac{r}{a}\right)^{n+2} - 1 - \nu\right] \tag{39.103a–c}$$

$$M_\theta = \frac{-q_na^2}{(n+2)^2(n+4)}\left[\{1 + \nu(n+3)\}\left(\frac{r}{a}\right)^2 - 1 - \nu\right]$$

For a simply supported plate, we superimpose the solution for case a with

$$M_o = \frac{q_na^2}{(n+2)(n+4)}$$

[1] Note that $\frac{d^2}{dr^2} + \frac{1}{r}\frac{d}{dr} = \frac{1}{r}\frac{d}{dr}\left(r\frac{d}{dr}\right)$.

For more general types of loading, solutions of the form given above may be superimposed for various values n. In this way, results may be obtained for the case of the loading expressed as a power series.

Case g. Uniform Loading on Outer Portion, Clamped Edge (Fig. 39.20*g*). The loading is assumed to be

$$q = 0 \qquad \text{for} \qquad 0 \le r < b$$
$$q = q_0 \qquad \text{for} \qquad b \le r \le a$$

The central portion ($r \le b$) comprises a plate loaded by edge moments only, namely, the moment M_r at $r = b$. The outer portion is analyzed by direct integration, noting that the shear force Q_r vanishes at $r = b$. In this way we obtain the expressions

$$w = C_1 + C_2 r^2 \qquad r \le b$$
$$w = \frac{q_0 a^4}{64K}\left[\left(\frac{r}{a}\right)^4 - 8\left(\frac{b}{a}\right)^2\left(\frac{r^2}{a^2}\ln\frac{r}{a} - \frac{r^2}{a^2}\right)\right] + C_3\frac{r^2}{a^2} + C_4 \ln\frac{r}{a} + C_5 \qquad r \ge b$$

Constants $C_1 \cdots C_5$ are found from clamped-edge conditions (39.14) at $r = a$ and the three matching relations at $r = b$:

$$(w)_{r\le b} = (w)_{r\ge b} \qquad \left(\frac{\partial w}{\partial r}\right)_{r\le b} = \left(\frac{\partial w}{\partial r}\right)_{r\ge b} \qquad \left(\frac{\partial^2 w}{\partial r^2}\right)_{r\le b} = \left(\frac{\partial^2 w}{\partial r^2}\right)_{r\ge b} \qquad r = b \tag{39.104}$$

The final results for the deflection are:

$$w = \frac{q_0 a^4}{8K}\left\{\left[\frac{1}{8} + \frac{3}{8}\left(\frac{b}{a}\right)^4 - \frac{1}{2}\left(\frac{b}{a}\right)^2 - \frac{1}{2}\left(\frac{b}{a}\right)^4 \ln\frac{b}{a}\right] - \frac{r^2}{a^2}\left[\frac{1}{4} + \left(\frac{b}{a}\right)^2 \ln\frac{b}{a} - \frac{1}{4}\left(\frac{b}{a}\right)^4\right]\right\} \qquad r \le b$$
$$w = \frac{q_0 a^4}{8K}\left\{\left[\frac{1}{8} - \frac{1}{2}\left(\frac{b}{a}\right)^2 - \frac{1}{4}\left(\frac{b}{a}\right)^4\right] - \left(\frac{r}{a}\right)^2\left[\frac{1}{4} - \frac{1}{2}\left(\frac{b}{a}\right)^2 - \frac{1}{4}\left(\frac{b}{a}\right)^4\right] + \frac{1}{8}\left(\frac{r}{a}\right)^4 - \left(\frac{b}{a}\right)^2\left(\frac{r}{a}\right)^2 \ln\frac{r}{a} - \frac{1}{2}\left(\frac{b}{a}\right)^4 \ln\frac{r}{a}\right\} \qquad r \ge b \tag{39.105}$$

Case h. Uniform Loading on Inner Portion, Clamped Edge (Fig. 39.20*h*). We superimpose solutions for cases b and g, with loading $-q_0$ in case g. Results for the deflection are:

$$w = \frac{q_0 a^4}{64K}\left(1 - \frac{r^2}{a^2}\right)^2 + \frac{q_0 a^4}{8K}\left\{-\frac{1}{8} - \frac{3}{8}\left(\frac{b}{a}\right)^4 + \frac{1}{2}\left(\frac{b}{a}\right)^2 + \frac{1}{2}\left(\frac{b}{a}\right)^4 \ln\frac{b}{a} + \frac{r^2}{a^2}\left[\frac{1}{4} + \left(\frac{b}{a}\right)^2 \ln\frac{b}{a} - \frac{1}{4}\left(\frac{b}{a}\right)^4\right]\right\} \qquad r \le b$$
$$w = \frac{q_0 a^4}{64K}\left(1 - \frac{r^2}{a^2}\right)^2 + \frac{q_0 a^4}{8K}\left\{-\frac{1}{8} + \frac{1}{2}\left(\frac{b}{a}\right)^2 + \frac{1}{4}\left(\frac{b}{a}\right)^4 + \frac{r^2}{a^2}\left[\frac{1}{4} - \frac{1}{2}\left(\frac{b}{a}\right)^2 - \frac{1}{4}\left(\frac{b}{a}\right)^4\right] - \frac{1}{8}\left(\frac{r}{a}\right)^4 + \frac{b^2 r^2}{a^4}\ln\frac{r}{a} + \frac{1}{2}\left(\frac{b}{a}\right)^4 \ln\frac{r}{a}\right\} \qquad r \ge b \tag{39.106}$$

The solution for this case converges to that for case 2 when $r \le b$ and $b \to a$, and to that for case d when $r \ge b$, $b \to 0$, and $\pi q_0 b^2 = P$.

Case i. Clamped Edge, Line Load Distributed around a Circle (Fig. 39.20*i*). Let the load p per unit length be applied on the circle of radius b. In both regions $r \le b$ and $r \ge b$, we have solutions of the general form (39.95). For $r \le b$, C and D will not enter, as in case a. Thus, we may write

$$w = C_1 + C_2 r^2 \qquad r \le b$$
$$w = C_3 + C_4 r^2 + C_5 \ln r + C_6 r^2 \ln r \qquad r \ge b$$

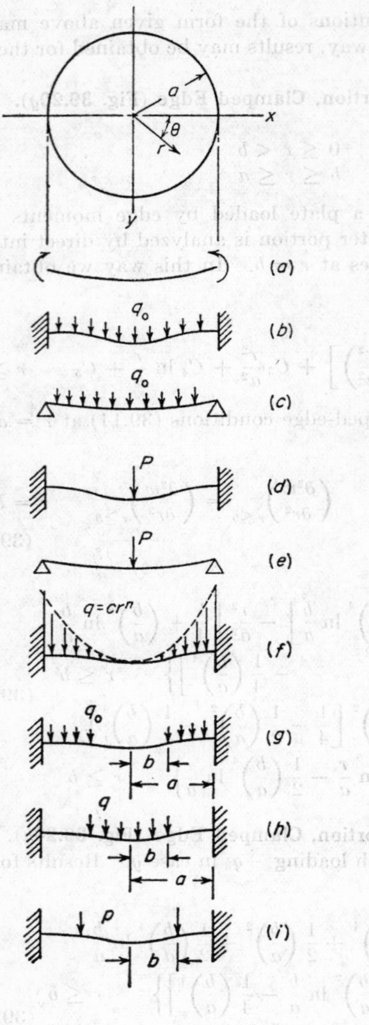

Fig. 39.20. Cases of loading of circular plates.

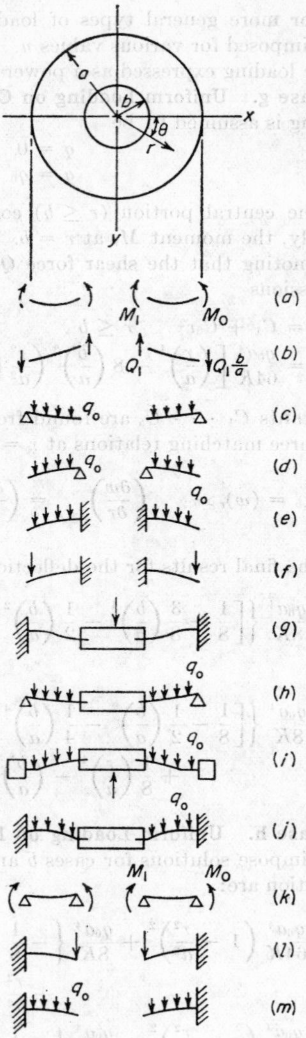

Fig. 39.21. Cases of loading of annular plates.

The six constants are determined by the two edge conditions (39.14) at $r = a$, the three matching conditions (39.104) plus the condition that $Q_r \to -p$ for $r \to b$ while $r \geq b$. We obtain in this way

$$\left.\begin{aligned} w &= \frac{pa^2b}{K}\left[\left(\frac{r^2}{a^2}+\frac{b^2}{a^2}\right)\ln\frac{r}{a}+\frac{1}{2}\left(1+\frac{b^2}{a^2}\right)\left(1-\frac{r^2}{a^2}\right)\right] \\ M_r &= -\frac{pb}{4}\left[2+2(1+\nu)\ln\frac{r}{a}-\frac{b^2}{r^2}(1-\nu)-\frac{b^2}{a^2}(1+\nu)\right] \\ M_\theta &= -\frac{pb}{4}\left[2\nu+2(1+\nu)\ln\frac{r}{a}+\frac{b^2}{r^2}(1-\nu)-\frac{b^2}{a^2}(1+\nu)\right] \end{aligned}\right\} \quad r \geq b$$

$$\left.\begin{aligned} w &= \frac{pa^2b}{K}\left[\left(\frac{r^2}{a^2}+\frac{b^2}{a^2}\right)\ln\frac{b}{a}+\frac{1}{2}\left(1-\frac{b^2}{a^2}\right)\left(1+\frac{r^2}{a^2}\right)\right] \\ M_r &= M_\theta = -(1+\nu)\frac{pb}{2}\left[\ln\frac{b}{a}+\frac{1}{2}\left(1-\frac{b^2}{a^2}\right)\right] \end{aligned}\right\} \quad r \leq b \qquad (39.107a\text{–}e)$$

Axially Symmetric Bending of Annular Plates

The inner boundary is taken at $r = b$, and the outer at $r = a$. Many cases of loading are illustrated in Fig. 39.21.

The process of arriving at solutions for the various cases is either to utilize form (39.95) where applicable or to integrate Eq. (39.93) directly. In the latter case, the solution is expedited by starting with the shear force equation for Q_r [Eq. (39.94c)], and expressing Q_r in terms of r by consideration of vertical equilibrium of the portion of the plate inside radius r. Superposition can be used to good advantage also in many cases. These methods were illustrated in the preceding section, and so we give here only the final results for various cases.

Case a. Annular Plate Loaded by Edge Moments (Fig. 39.21*a*)

$$\begin{aligned} w &= \frac{1}{2}\frac{r^2-a^2}{a^2-b^2}\frac{M_1b^2-M_0a^2}{(1+\nu)K}+\frac{a^2b^2}{a^2-b^2}\frac{M_1-M_0}{(1-\nu)K}\ln\frac{r}{a} \\ M_r &= -\frac{M_1b^2-M_0a^2}{a^2-b^2}+\frac{a^2b^2(M_1-M_0)}{r^2(a^2-b^2)} \\ M_\theta &= -\frac{M_1b^2-M_0a^2}{a^2-b^2}-\frac{a^2b^2(M_1-M_0)}{r^2(a^2-b^2)} \end{aligned} \qquad (39.108a\text{–}c)$$

Case b. Plate Loaded by Shear Force Q_1 at Inner Edge (Fig. 39.21*b*)

$$w = -\frac{Q_1a^2b}{4K}\left\{\left(1-\frac{r^2}{a^2}\right)\left[\frac{3+\nu}{2(1+\nu)}-\frac{b^2}{a^2-b^2}\ln\frac{b}{a}\right]+\frac{r^2}{a^2}\ln\frac{r}{a} + \frac{2b^2}{a^2-b^2}\frac{1+\nu}{1-\nu}\ln\frac{b}{a}\ln\frac{r}{a}\right\} \qquad (39.109)$$

Case c. Uniform Load, Simply Supported Outer Edge and Free Inner Edge (Fig. 39.21*c*)

$$\begin{aligned} w = \frac{q_0a^4}{8K}\Bigg\{&\left[\frac{5+\nu}{8(1+\nu)}-\frac{b^2(3+\nu)}{4a^2(1+\nu)}+\frac{(b/a)^4}{1-(b/a)^2}\ln\frac{b}{a}\right] \\ &-\frac{r^2}{a^2}\left[\frac{3+\nu}{4(1+\nu)}\left(1-\frac{b^2}{a^2}\right)+\frac{(b/a)^4}{1-(b/a)^2}\ln\frac{b}{a}\right]-\ln\frac{r}{a}\left[\frac{3+\nu}{2(1-\nu)}\left(\frac{b}{a}\right)^2 \right. \\ &\left. +\frac{2(1+\nu)}{1-\nu}\frac{(b/a)^4}{1-(b/a)^2}\ln\frac{b}{a}\right]+\frac{1}{8}\left(\frac{r}{a}\right)^4-\frac{r^2b^2}{a^4}\ln\frac{r}{a}\Bigg\} \end{aligned} \qquad (39.110)$$

Cases d through i. The loading configurations are shown in Fig. 39.21. These cases have been treated by Wahl and Lobo [11]. We may define coefficients k and k_1

in terms of maximum stresses and deflections,

$$w_{\max} = k_1 \frac{q_0 a^4}{Et^3} \text{ or } k_1 \frac{Pa^2}{Et^3} \qquad \sigma_{\max} = k \frac{q_0 a^2}{t^2} \text{ or } \frac{kP}{t^2}$$

where P is the total edge force (in cases b, f, and g only).

Values of the coefficients k and k_1 are given in Table 39.7 for cases b through i.

Table 39.7. Stress and Deflection Coefficients of Wahl and Lobo for Annular Plates

$a/b =$	1.25		1.50		2.0		3.0		4.0		5.0	
$b/a =$	0.800		0.667		0.500		0.333		0.250		0.200	
Case	k	k_1	k	k_1	k	k_1	k	k_1	k	k_1	k	k_1
b	1.10	0.341	1.26	0.519	1.48	0.672	1.88	0.734	2.17	0.724	2.34	0.704
c	0.592	0.184	0.976	0.414	1.440	0.664	1.880	0.824	2.08	0.830	2.19	0.813
d	0.66	0.202	1.19	0.491	2.04	0.902	3.34	1.220	4.30	1.300	5.10	1.310
e	0.135	0.00231	0.410	0.0183	1.04	0.0938	2.15	0.293	2.99	0.448	3.96	0.564
f	0.227	0.00510	0.428	0.0249	0.753	0.0877	1.205	0.209	1.514	0.293	1.745	0.350
g	0.115	0.00129	0.220	0.0064	0.405	0.0237	0.703	0.062	0.933	0.092	1.13	0.114
h	0.122	0.00343	0.336	0.0313	0.74	0.1250	1.21	0.291	1.45	0.417	1.59	0.492
i	0.090	0.00077	0.273	0.0062	0.71	0.0329	1.54	0.110	2.23	0.179	2.80	0.234

Cases j through o. These cases, also shown in Fig. 39.21, may be calculated by superposition, as follows:

Case	Superposition	Conditions
j	a, c	Cancel edge slopes
k	a, b	Cancel inner deflection
l	a, $b(M_1 = 0)$	Cancel outer edge slope
m	a, $c(M_1 = 0)$	Cancel outer edge slope
n	b, c	Cancel inner deflection
o	k, n	Cancel both edge slopes

Though these cases can be handled by superposition, it may sometimes be easier to integrate the shear force equation directly, and determine the constants from the boundary conditions.

Nonsymmetrical Deflections of Circular Plates

The deflection at the center of the plate with nonsymmetric loading but symmetric boundary conditions (e.g., a clamped or simply supported plate) can be calculated by certain symmetry and reciprocity relations.

The deflection at the center due to load P at r, θ is the same as the deflection at r (any θ) due to load P at the center. Hence, the deflection at the center due to any distribution of loads on a circle of radius r is equal to the deflection at r with the total of these loads applied at the center. To obtain the deflection at the center with any nonuniform loading $q(r,\theta)$, we therefore may use the formula

$$w_0 = \int_{r=0}^{a} \int_{\theta=0}^{2\pi} q(r,\theta) r f(r)\, dr\, d\theta \tag{39.111}$$

in which $f(r)$ is the deflection at r produced by unit force at the center.

For analysis of deflections and bending moments at points other than the center, we must find appropriate solutions of the basic circular-plate equation (39.90). We let w' be a particular solution, and w'' be a solution of the homogeneous equation ($q \equiv 0$),

$$w = w' + w'' \tag{39.112}$$

The w'' part will contain constants, the values of which are adjusted so that w meets all the boundary conditions.

By conventional methods we arrive at the following biharmonic function w'':

$$w'' = R_0 + \sum_{1,2,\ldots}^{\infty} R_m \cos m\theta + \sum_{1,2,\ldots}^{\infty} R_m^* \sin m\theta \tag{39.113}$$

where

$$\begin{aligned} R_0 &= A_0 + B_0 \ln r + C_0 r^2 + D_0 r^2 \ln r \\ R_1 &= A_1 r + B_1 r^{-1} + C_1 r^3 + D_1 r \ln r \\ R_m &= A_m r^m + B_m r^{-m} + C_m r^{m+2} + D_m r^{-m+2} \qquad m > 1 \end{aligned} \tag{39.114}$$

Similar expressions with constants A_1^*, B_1^*, . . . apply for R_1^* and R_n^*.

Several cases of nonsymmetrical loading may be mentioned, as follows [1], [12]–[15]:

Case 1. Plate Loaded by Nonuniform Symmetric Distribution of Edge Couples. The edge deflection is assumed zero. Let the couples be specified by

$$(M_r)_{r=a} = \sum_{1,2,\ldots} c_m \cos m\theta \tag{39.115}$$

In this case, $w' = 0$ and $w = w''$. Since w, $\partial w/\partial r$, and $\partial^2 w/\partial r^2$ are finite at $r = 0$, we must have

$$B_m = D_m = 0 \qquad m = 0, 1, 2, \ldots$$

Therefore

$$w = \sum_{0,1,\ldots}^{\infty} (A_m r^m + C_m r^{m+2}) \cos m\theta \tag{39.116}$$

Setting $w = 0$ at $r = a$ gives $A_m = -C_m a^2$. Then, computation of the edge moment M_r from (39.116) and utilization of (39.115) determines the C_m and yields finally

$$w = -\frac{1}{2K} \sum_{0,1,\ldots}^{\infty} \frac{c_m (r^{m+2} - a^2 r^m)}{a^m (2m + 1 + \nu)} \cos m\theta \tag{39.117}$$

Case 2. Clamped Plate with Hydrostatic Loading (**Fig. 39.22**). The load is given by

$$q = q_0 + q_1 \frac{r}{a} \cos \theta \tag{39.118}$$

The load varies from a minimum value of $(q_0 - q_1)$ to a maximum value of $(q_0 + q_1)$. Boundary conditions (39.14) apply at $r = a$. An examination shows that the particular solution w' may be taken as

$$w' = \frac{q_1 r^5 \cos \theta}{129aK} + \frac{q_0 r^4}{64K}$$

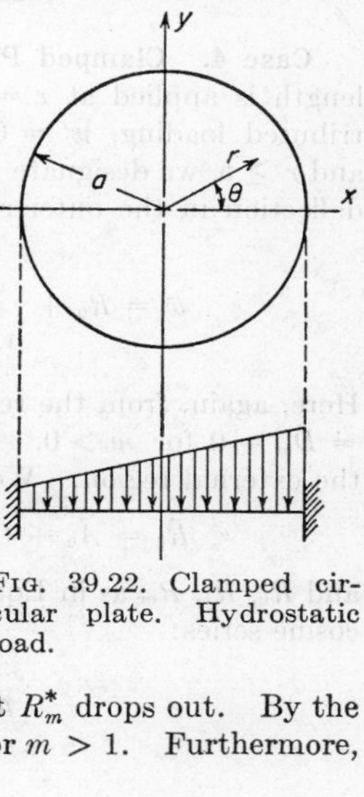

Fig. 39.22. Clamped circular plate. Hydrostatic load.

The portion w'' will be symmetrical in θ, which means that R_m^* drops out. By the nature of the loading, and w', we find we do not need terms for $m > 1$. Furthermore, since w'' is finite with finite derivatives at $r = 0$, we have $B_0 = D_0 = B_1 = D_1 = O$. Consequently,

$$w'' = A_0 + C_0r^2 + (A_1r + C_1r^3)\cos\theta$$

Application of the clamped-edge boundary conditions to $(w' + w'')$ leads to definite values for the constants, and the final deflection expression and bending moments are:

$$w = \frac{q_0}{64K}(a^2 - r^2)^2 + \frac{q_1}{192K}\frac{r}{a}(a^2 - r^2)^2\cos\theta$$

$$M_r = \frac{q_0}{16}[a^2(1+\nu) - r^2(3+\nu)] - \frac{q_1}{48}\left[\frac{r^3}{a}(5+\nu) - ar(3+\nu)\right]\cos\theta$$

(39.119a–d)

$$M_\theta = \frac{q_0}{16}[a^2(1+\nu) - r^2(1+3\nu)] - \frac{q_1}{48}\left[\frac{r^3}{a}(5\nu+1) - ar(3\nu+1)\right]\cos\theta$$

$$M_{r\theta} = -\frac{1-\nu}{48}q_1ra\left(1 - \frac{r^2}{a^2}\right)\sin\theta$$

Case 3. Simply Supported Plate, Hydrostatic Loading. Superimpose on case 2 the solution for case 1 having edge moments equal and opposite to the $(M_r)_{r=a}$ values

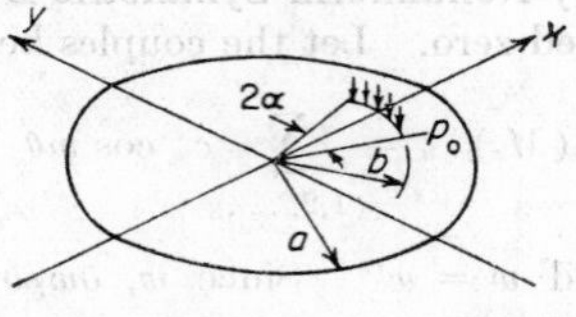

FIG. 39.23. Clamped circular plate loaded along circular arc.

just given in (39.119b). The resulting deflection is [15]

$$w = \frac{q_0a^4}{64K}\left(1 - \frac{r^2}{a^2}\right)\left(\frac{5+\nu}{1+\nu} - \frac{r^2}{a^2}\right) + \frac{q_1a^4}{192K(3+\nu)}\frac{r}{a}\left(1 - \frac{r^2}{a^2}\right)\left[7 + \nu - \frac{r^2}{a^2}(3+\nu)\right]\cos\theta \quad (39.120)$$

Case 4. Clamped Plate, Load on Circular Arc (Fig. 39.23). Load p_0 per unit length is applied at $r = b$ from $\theta = -\alpha$ to $\theta = +\alpha$. Since the plate has no distributed loading, $w' = 0$ and $w = w''$. To distinguish between the regions $r \leq b$ and $r \geq b$, we designate deflection in the inner region by $\bar{w}$, and use symbol w for the deflection in the outer region. Since the antimetric (starred) terms are not needed,

$$\bar{w} = \bar{R}_0 + \sum_{1,2,\ldots}^{\infty} \bar{R}_m \cos m\theta \qquad w = R_0 + \sum_{1,2,\ldots}^{\infty} R_m \cos m\theta \quad (39.121)$$

Here, again, from the required behavior of $\bar{w}$ at $r = 0$, we must have $\bar{B}_0 = \bar{B}_m = \bar{D}_0 = \bar{D}_m = 0$ for $m > 0$. However, all constants must be retained in R_0 and R_m for the external region. We thus have

$$\bar{R}_0 = \bar{A}_0 + \bar{C}_0r^2 \qquad \bar{R}_m = \bar{A}_mr^m + \bar{C}_mr^{m+2} \qquad m = 1, 2, \ldots$$

and R_0, R_1, R_m as in Eqs. (39.114). The line load $p = p(\theta)$ may be expanded into a cosine series:

$$p(\theta) = \frac{\alpha p_0}{\pi} + \frac{2p_0}{\pi}\sum_{1,2,\ldots}^{\infty}\frac{\sin m\alpha}{m}\cos m\theta \quad (39.122)$$

There are six relations that must be fulfilled, pertaining to boundary and matching conditions:

For $r = a$: $\quad w = 0 \quad \dfrac{\partial w}{\partial r} = 0$

For $r = b$: $\quad w = \bar{w} \quad \dfrac{\partial w}{\partial r} = \dfrac{\partial \bar{w}}{\partial r}; \dfrac{\partial^2 w}{\partial r^2} = \dfrac{\partial^2 \bar{w}}{\partial r^2} \quad \bar{Q}_r - Q_r = p(\theta)$

These six relations are sufficient to determine all the constants (six of them with each subscript). When the calculation is carried out, the R functions are found to be

$$R_0 = \frac{\alpha p_0 b a^2}{4\pi K}\left[\frac{1}{2}\left(1 + \frac{b^2}{a^2}\right)\left(1 - \frac{r^2}{a^2}\right) + \frac{b^2 + r^2}{a^2} \ln \frac{r}{a}\right]$$

$$\bar{R}_0 = \frac{\alpha p_0 b a^2}{4\pi K}\left[\frac{1}{2}\left(1 - \frac{b^2}{a^2}\right)\left(1 + \frac{r^2}{a^2}\right) + \frac{b^2 + r^2}{a^2} \ln \frac{b}{a}\right]$$

$$R_1 = \frac{p_0 b^2 a \sin \alpha}{4\pi K}\left[-\frac{2r}{a} \ln \frac{r}{a} - \frac{r}{a}\left(1 - \frac{b^2}{a^2}\right) + \frac{r^3}{2a^3}\left(2 - \frac{b^2}{a^2}\right) - \frac{b^2}{2ar}\right]$$

$$\bar{R}_1 = \frac{p_0 b^2 a \sin \alpha}{4\pi K}\left[-\frac{2r}{a} \ln \frac{b}{a} - \frac{r}{a}\left(1 - \frac{b^2}{a^2}\right) - \frac{a^2}{2b^2}\left(\frac{r}{a}\right)^3\left(1 - \frac{b^2}{a^2}\right)^2\right]$$

$$R_n = \frac{p_0 a^3 \sin n\alpha}{4\pi n^2 K}\left[\left(\frac{r}{a}\right)^n\left\{\left(\frac{b}{a}\right)^{n+3} - \frac{n}{n-1}\left(\frac{b}{a}\right)^{n+1}\right\} + \left(\frac{r}{a}\right)^{n+2}\left\{\left(\frac{b}{a}\right)^{n+1} \right.\right. \tag{39.123}$$

$$\left.\left. - \frac{n}{n+1}\left(\frac{b}{a}\right)^{n+3}\right\} - \left(\frac{r}{a}\right)^{-n} \frac{1}{n+1}\left(\frac{b}{a}\right)^{n+3} + \left(\frac{r}{a}\right)^{-n+2} \frac{1}{n-1}\left(\frac{b}{a}\right)^{n+1}\right]$$

$$\bar{R}_n = \frac{p_0 a^3 \sin n\alpha}{4\pi n^2 K}\left[\left(\frac{r}{a}\right)^n\left\{\frac{1}{n-1}\left(\frac{b}{a}\right)^{-n+3} + \left(\frac{b}{a}\right)^{n+3} - \frac{n}{n-1}\left(\frac{b}{a}\right)^{n+1}\right\}\right.$$

$$\left. + \left(\frac{r}{a}\right)^{n+2}\left\{\left(\frac{b}{a}\right)^{n+1} - \frac{1}{n+1}\left(\frac{b}{a}\right)^{-n+1} - \frac{n}{n+1}\left(\frac{b}{a}\right)^{n+3}\right\}\right]$$

Equations (39.121) then give the deflection, from which moments can also be calculated.

Case 5. Clamped Plate, Concentrated Force P. The force is assumed to be applied at point $r = b$, $\theta = 0$. The solution is obtained from the previous case 4 by allowing α to approach zero while $p_0 \to \infty$, with

$$2b\alpha p_0 = P$$

The expressions for the Rs are:

$$R_0 = \frac{Pa^2}{8\pi K} f_0(r) \qquad \bar{R}_0 = \frac{Pa^2}{8\pi K} \bar{f}_0(r)$$

$$R_1 = \frac{Pab}{8\pi K} f_1(r) \qquad \bar{R}_1 = \frac{Pab}{8\pi K} \bar{f}_1(r) \tag{39.124}$$

$$R_n = \frac{P}{8\pi K} \frac{a^3}{bn} f_n(r) \qquad \bar{R}_n = \frac{P}{8\pi K} \frac{a^3}{bn} \bar{f}_n(r)$$

where the fs stand for the expressions in the brackets in Eqs. (39.123).

The corresponding case of a simply supported circular plate with a concentrated force at a general point has been solved; results are given by Timoshenko [1]. Other solved problems include those of an annular plate clamped at the inner edge and loaded by a concentrated force on the rim [12], a circular plate with several (self-equilibrating) concentrated forces at the edge [13], and a circular plate with elastic edge support loaded over a sector.

39.4. REFERENCES

[1] S. P. Timoshenko, S. Woinowsky-Krieger: "Theory of Plates and Shells," 2d ed., McGraw-Hill, New York, 1959.

[2] A. Nadai: "Die elastischen Platten," Springer, Berlin, 1925.

[3] H. Geiger, K. Scheel: "Handbuch der Physik," vol. 6, Springer, Berlin, 1927.
[4] W. J. Duncan: Galerkin's method in mechanics and differential equations, *Aeronaut. Research Comm. Rept. Mem.* 1798 (1937).
[5] W. J. Duncan: The principles of the Galerkin method, *Aeronaut. Research Comm. Rept. Mem.* 1848 (1938).
[6] A. Nadai: Die Formänderung und Spannungen von rechteckigen elastischen Platten, *VDI Mitt. Forschungsarb.* 170, 171 (1915).
[7] S. Woinowski-Krieger: Der Spannungszustand in dicken elastischen Platten, *Ing.-Arch.*, **4** (1933), 305–331.
[8] S. P. Timoshenko: Bending of rectangular plates with clamped edges, *Proc. 5th. Intern. Congr. Appl. Mechanics*, pp. 40–43, Cambridge, Mass., 1938.
[9] T. H. Evans: Tables of moments and deflections for a rectangular plate fixed on all edges and carrying a uniformly distributed load, *J. Appl. Mechanics*, **6** (1939), 7–10.
[10] D. Young: Clamped rectangular plates with a central concentrated load, *J. Appl. Mechanics*, **6** (1939), 114–116.
[11] A. M. Wahl, G. Lobo: Stresses and deflections in flat circular plates with central holes, *Trans. ASME*, **52** (1930), APM, pp. 29–43.
[12] H. Reissner: Über die unsymmetrische Biegung dünner Kreisringplatten, *Ing.-Arch.*, **1** (1929), 72–83.
[13] A. Nadai: Die Verbiegung in einzelnen Punkten unterstützter kreisförmiger Platten, *Physik. Z.*, **23** (1922), 366–376.
[14] W. A. Bassali, R. H. Dawoud: Bending of an elastically restrained circular plate under normal loading, *J. Appl. Mechanics*, **25** (1958), 37–46.
[15] W. Flügge: Kreisplatten mit linear veränderlichen Belastungen, *Bauing.*, **10** (1929), 221–225.
[16] E. G. Odley: Deflections and moments of a rectangular plate clamped on all edges and under hydrostatic pressure, *J. Appl. Mechanics*, **14** (1947), 289–299.

CHAPTER 40

SHELLS

BY

W. FLÜGGE, Dr.-Ing., Los Altos, Calif.

40.1. DEFINITIONS

A shell is an object whose material is confined to the close vicinity of a curved surface, the middle surface of the shell. In most cases, the shell is made of a solid material filling the space between two parallel or almost parallel surfaces, called its *faces*. Their distance is the shell thickness t, and the middle surface is defined as the surface that halves the shell thickness everywhere. However, a shell may be formed by a liquid (e.g., soap film), a fabric, a combination of several materials (reinforced concrete, sandwich shell), or a combination of a sheet and closely spaced ribs, and even the interface between two liquids displays characteristic features of a shell, the surface tension being identical with a membrane force. The membrane theory described here applies to all these shells, while the bending theory is restricted to homogeneous, elastic, isotropic shells.

To locate a point on the middle surface, we use curvilinear coordinates ξ, η. Figure 40.1 shows a shell element bounded by two pairs of adjacent curves ξ = const and

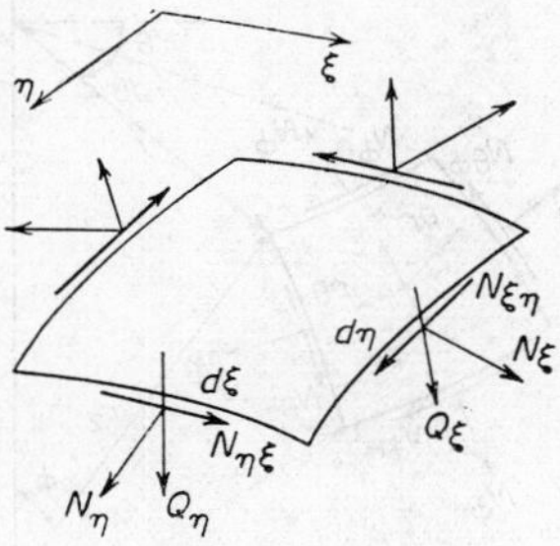

FIG. 40.1

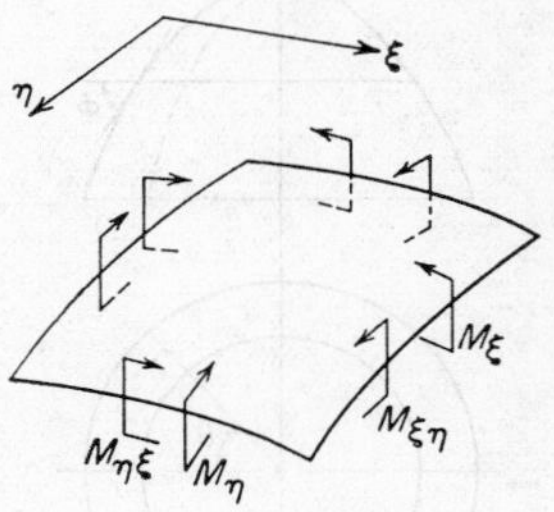

FIG. 40.2

η = const. The force acting on a line element ds_η of a line ξ = const is resolved into these components: a force $N_\xi\, ds_\eta$ tangent to a line η = const, a force $N_{\xi\eta}\, ds_\eta$ tangent to the line element ds_η, and a force $Q_\xi\, ds_\eta$ normal to the shell. Similarly, forces $N_\eta\, ds_\xi$ tangent to a line ξ = const, $N_{\eta\xi}\, ds_\xi$ tangent to a line η = const, and $Q_\eta\, ds_\xi$ normal to the middle surface are transmitted across a side ds_ξ of the shell element. If ξ, η are orthogonal, N_ξ, $N_{\xi\eta}$, Q_ξ are orthogonal; and N_ξ, N_η are called normal forces; $N_{\xi\eta}$, $N_{\eta\xi}$ (tangential) shear forces; and Q_ξ, Q_η transverse (shear) forces. If ξ and η are not orthogonal, then N_ξ and N_η are called fiber forces. All are forces per unit length of a section and are called *internal forces* of the shell.

Normal or fiber forces are positive when tensile. The positive directions of the shear forces depend on those of the coordinates, and are shown in Fig. 40.1. To define the positive directions of the transverse forces, one side of the shell element must be declared its *outside*. Figure 40.1 shows positive transverse forces if the visible side of the element is its outside.

If the shell has no bending rigidity, $Q_\xi = Q_\eta = 0$. If it has bending rigidity, moments similar to the bending and twisting moments in plates (p. 39–2) are acting on the shell element (Fig. 40.2). They are $M_\xi \, ds_\eta$ and $M_{\xi\eta} \, ds_\eta$ on the side ds_η, and $M_\eta \, ds_\xi$ and $M_{\eta\xi} \, ds_\xi$ on the side ds_ξ. We call M_ξ, M_η the bending moments and $M_{\xi\eta}$, $M_{\eta\xi}$ the twisting moments. All are moments per unit length of a section. Both these internal moments and the internal forces are called *stress resultants*. The moments shown in Fig. 40.2 are positive if the visible side of the element is the outside.

40.2. SHELLS OF REVOLUTION, MEMBRANE THEORY

Obviously, the bending and twisting moments of a thin shell must be rather small and, hence, of little influence on the over-all stress picture. The membrane theory is based on the assumption that they are small enough to be neglected. It meets the limits of its applicability when either external moments are applied at the edge of the shell, or the membrane deformations lead to discrepancies which can only be resolved by admitting the existence of small, but finite bending and twisting moments.

If $M_\xi = M_\eta = M_{\xi\eta} = M_{\eta\xi} = 0$, the moment equilibrium of the shell element requires that $N_{\xi\eta} = N_{\eta\xi}$.

Differential Equations

Figure 40.3 shows a surface of revolution and the curvilinear coordinates ϕ, θ. Lines ϕ = const are called parallel circles, lines θ = const are the meridians. We denote by r the radius of the parallel circle, by r_1 the radius of curvature of the meridian, and by $r_2 = r/\sin\phi$ the length of a normal between the surface and the axis of revolution.

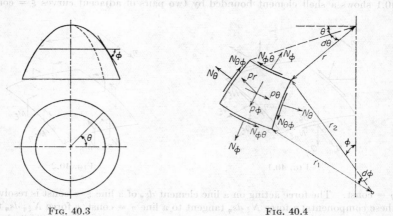

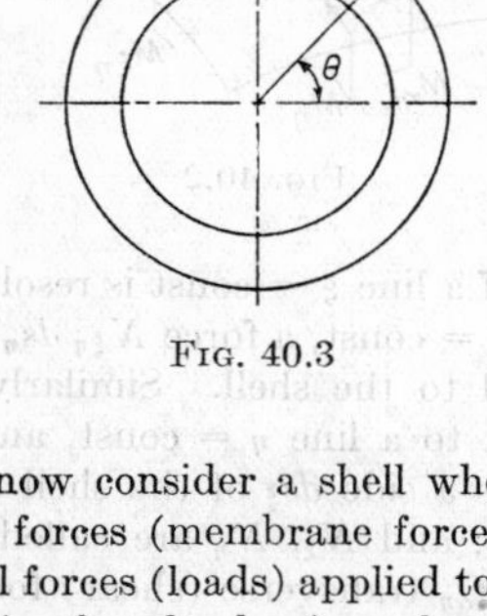

FIG. 40.3

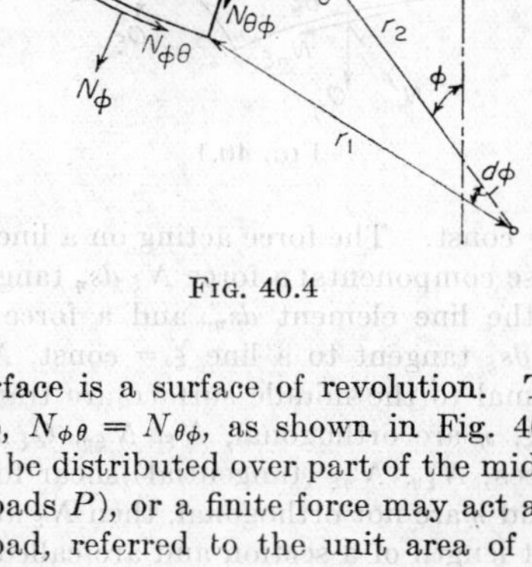

FIG. 40.4

We now consider a shell whose middle surface is a surface of revolution. The internal forces (membrane forces) are N_ϕ, N_θ, $N_{\phi\theta} = N_{\theta\phi}$, as shown in Fig. 40.4. External forces (loads) applied to the shell may be distributed over part of the middle surface (surface loads p) or along a line (line loads P), or a finite force may act at a definite point (point load **P**). The surface load, referred to the unit area of the middle surface is resolved into its components p_ϕ, p_θ, and p_r, as shown in Fig. 40.4. The radial (= normal) component is positive when directed outward.

The equilibrium of the shell element yields the equations

$$\begin{aligned}\frac{\partial}{\partial\phi}(rN_\phi) + r_1\frac{\partial N_{\phi\theta}}{\partial\theta} - r_1N_\theta\cos\phi + p_\phi rr_1 &= 0\\ \frac{\partial}{\partial\phi}(rN_{\phi\theta}) + r_1\frac{\partial N_\theta}{\partial\theta} + r_1N_{\phi\theta}\cos\phi + p_\theta rr_1 &= 0\\ \frac{N_\phi}{r_1} + \frac{N_\theta}{r_2} &= p_r\end{aligned} \qquad (40.1a\text{–}c)$$

They are three equations for three unknown membrane forces. Consequently, the membrane stress problem is internally statically determinate (see p. 26-3), but it may be externally indeterminate. If the loads, including the reactions at the edge, are independent of θ, the stress system is axisymmetric. This requires that $N_{\phi\theta} = 0$, and Eqs. (40.1) reduce to

$$\frac{d}{d\phi}(rN_\phi) - r_1N_\theta\cos\phi + p_\phi rr_1 = 0 \qquad \frac{N_\phi}{r_1} + \frac{N_\theta}{r_2} = p_r \qquad (40.2a,b)$$

Axisymmetric Problem

In the axisymmetric case, Eqs. (40.2) are solved by

$$\begin{aligned}N_\phi &= \frac{1}{r_2\sin^2\phi}\left[\int r_1r_2(p_r\cos\phi - p_\phi\sin\phi)\sin\phi\,d\phi + C\right]\\ N_\theta &= p_rr_2 - N_\phi\frac{r_2}{r_1}\end{aligned} \qquad (40.3)$$

The integral may be solved by formal or numerical integration for any shape and load. If the shell is closed at the apex (Fig. 40.5*a*), the constant C must be determined from a regularity requirement at $\phi = 0$. If the shell is open at the top (Fig. 40.5*b*), the value of N_ϕ may be given at one of its edges. If the shell is closed at the top and the bottom (Fig. 40.5*c*), a regularity requirement at one of these points yields C, and the solution will be regular at the other if the loads are self-equilibrating. If $p_\phi \equiv p_r \equiv 0$, the integral in (40.3) vanishes, and the C term represents a stress system produced by edge loads [homogeneous solution of Eqs. (40.2)].

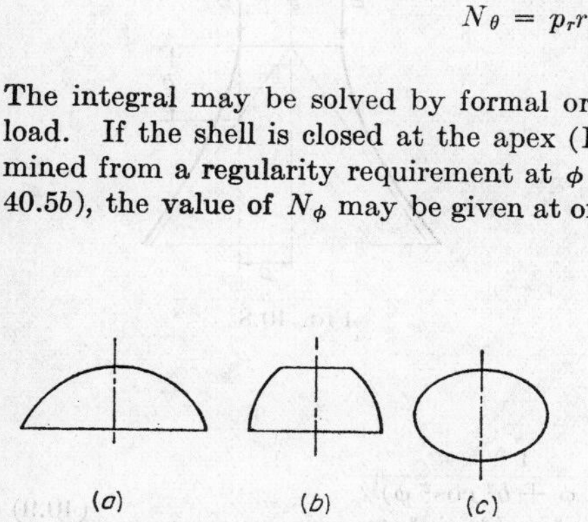

FIG. 40.5

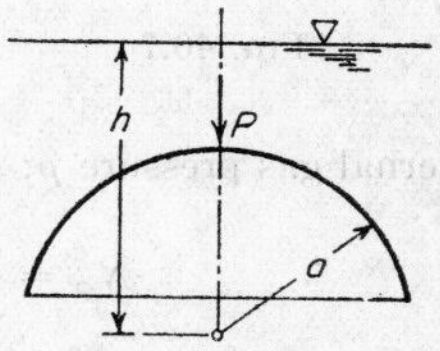

FIG. 40.6

Special Solutions

Sphere (Fig. 40.6). (*a*) Uniform weight p:

$$p_\phi = p\sin\phi \qquad p_r = -p\cos\phi \qquad \text{(same for all shells)}$$

$$N_\phi = -\frac{pa}{1+\cos\phi} \qquad N_\theta = pa\left(\frac{1}{1+\cos\phi} - \cos\phi\right) \qquad (40.4)$$

(*b*) Water pressure (γ = specific weight of water):

$$p_\phi = 0 \qquad p_r = -\gamma(h - a\cos\phi)$$

$$N_\phi = -\frac{\gamma a}{6}\left[3h - \frac{2a}{1+\cos\phi}(1 + \cos\phi + \cos^2\phi)\right]$$
$$N_\theta = -\frac{\gamma a}{6}\left[3h + \frac{2a}{1+\cos\phi}(1 - 2\cos\phi - 2\cos^2\phi)\right] \tag{40.5}$$

(*c*) Concentrated force **P** at apex:

$$N_\phi = -N_\theta = -\frac{\mathbf{P}}{2\pi a \sin^2\phi} \tag{40.6}$$

(*d*) Internal gas pressure p:

$$p_\phi = 0 \qquad p_r = p \qquad N_\phi = N_\theta = \tfrac{1}{2}pa \tag{40.7}$$

Ellipsoid (Fig. 40.7). (*a*) Uniform weight p:

$$N_\phi = -\frac{pb^2}{2}\frac{(a^2\sin^2\phi + b^2\cos^2\phi)^{1/2}}{\sin^2\phi}\left[\frac{1}{b^2} - \frac{\cos\phi}{a^2\sin^2\phi + b^2\cos^2\phi} + \frac{1}{2ac}\ln\frac{(a+c)(a-c\cos\phi)}{(a-c)(a+c\cos\phi)}\right] \tag{40.8}$$

N_θ from Eq. (40.3), $c^2 = a^2 - b^2$

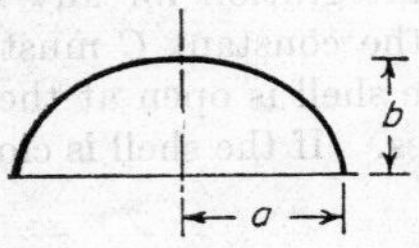

FIG. 40.7

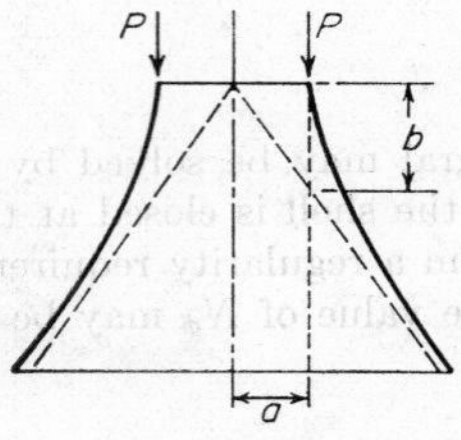

FIG. 40.8

(*b*) Internal gas pressure p:

$$N_\phi = \frac{pa^2}{2}\frac{1}{(a^2\sin^2\phi + b^2\cos^2\phi)^{1/2}}$$
$$N_\theta = \frac{pa^2}{2b^2}\frac{b^2 - (a^2 - b^2)\sin^2\phi}{(a^2\sin^2\phi + b^2\cos^2\phi)^{1/2}} \tag{40.9}$$

Hyperboloid of revolution (Fig. 40.8). (*a*) Uniform weight p:

$$N_\phi = -\frac{pb^2}{2}\frac{(a^2\sin^2\phi - b^2\cos^2\phi)^{1/2}}{\sin^2\phi}\left(\frac{\cos\phi}{a^2\sin^2\phi - b^2\cos^2\phi} + \frac{1}{2ac}\ln\frac{a + c\cos\phi}{a - c\cos\phi}\right) \tag{40.10}$$

N_θ from Eq. (40.3), $c^2 = a^2 + b^2$

(*b*) Vertical line load P at the edge $\phi = 90°$:

$$N_\phi = -\frac{P}{a^2}\frac{(a^2\sin^2\phi - b^2\cos^2\phi)^{1/2}}{\sin^2\phi}$$
$$N_\theta = -\frac{P}{a^2b^2}\frac{(a^2\sin^2\phi - b^2\cos^2\phi)^{3/2}}{\sin^2\phi} \tag{40.11}$$

Toroid of Circular Cross Section (Fig. 40.9). A toroidal shell has either two edges or none. The solutions given below are unique if the shell contains the top circle ($\phi = 0$). They yield finite N_ϕ, N_θ at $\phi = 0$ and, in the case of weight load, require definite reactions at each of the edges. However, if the displacements are calculated, an incompatibility is found at $\phi = 0$, indicating that the shell must have bending stresses in the vicinity of the top circle. If the shell has only points $0 < \phi < 180°$ or

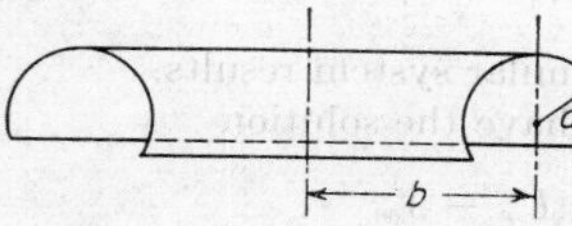

FIG. 40.9

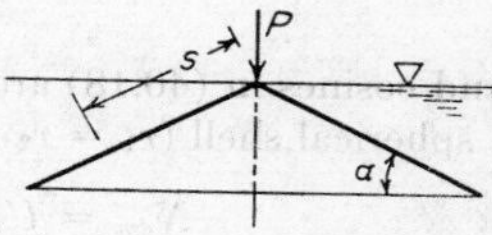

FIG. 40.10

$0 > \phi > -180°$, a homogeneous solution of Eqs. (40.2) may still be added, and neither stresses nor deformations will be singular.

(*a*) Uniform weight p:

$$N_\phi = -\frac{pa}{(b + a\sin\phi)\sin\phi}[b\phi + a(1 - \cos\phi)]$$
$$N_\theta = \frac{pb}{\sin^2\phi}(\phi - \cos\phi\sin\phi) - \frac{pa}{1 + \cos\phi}(\cos\phi - \sin^2\phi) \tag{40.12}$$

(*b*) Internal gas pressure p:

$$N_\phi = \frac{pa}{b + a\sin\phi}\left(b + \frac{a}{2}\sin\phi\right) \qquad N_\theta = \frac{pa}{2} \tag{40.13}$$

Conical Shell. The slope of the meridian $\phi = \alpha$ is a constant and not suitable as a coordinate. We use s as defined in Fig. 40.10.

(*a*) Uniform weight p:

$$N_s = -\frac{ps}{2\sin\alpha} \qquad N_\theta = -ps\cos\alpha\cot\alpha \tag{40.14}$$

(*b*) Water pressure $p = -\gamma s \sin\alpha$:

$$N_s = -\tfrac{1}{3}\gamma s^2\cos\alpha \qquad N_\theta = -\gamma s^2\cos\alpha \tag{40.15}$$

(*c*) Internal gas pressure p:

$$N_s = \frac{ps}{2}\cot\alpha \qquad N_\theta = ps\cot\alpha \tag{40.16}$$

(*d*) Point load **P** at the apex:

$$N_s = -\frac{\mathbf{P}}{\pi\sin 2\alpha}\frac{1}{s} \qquad N_\theta = 0 \tag{40.17}$$

General Problem

When the loads and stress resultants are not independent of θ, they may be written as Fourier series in θ, each coefficient being a function of ϕ. The general term of these series is

$$\begin{aligned} p_\phi &= p_{\phi n}\cos n\theta & p_\theta &= p_{\theta n}\sin n\theta & p_r &= p_{rn}\cos n\theta \\ N_\phi &= N_{\phi n}\cos n\theta & N_\theta &= N_{\theta n}\cos n\theta & N_{\phi\theta} &= N_{\phi\theta n}\sin n\theta \end{aligned} \tag{40.18}$$

Introduction of (40.18) into (40.1) yields three ordinary differential equations for the *amplitudes* of the nth harmonic of the stress resultants:

$$\frac{d}{d\phi}(rN_{\phi n}) + nr_1N_{\phi\theta n} - r_1N_{\theta n}\cos\phi + p_{\phi n}rr_1 = 0$$

$$\frac{d}{d\phi}(rN_{\phi\theta n}) - nr_1N_{\theta n} + r_1N_{\phi\theta n}\cos\phi + p_{\theta n}rr_1 = 0 \qquad \text{(40.19}a\text{–}c\text{)}$$

$$\frac{N_{\phi n}}{r_1} + \frac{N_{\theta n}}{r_2} = p_{rn}$$

If sines and cosines in (40.18) are interchanged, a similar system results.

For a spherical shell ($r_1 = r_2 = a$), Eqs. (40.19) have the solution

$$N_{\phi n} = U_n + V_n \qquad N_{\phi\theta n} = U_n - V_n$$

$$U_n = \frac{\cot^n(\phi/2)}{2\sin^2\phi}\left[C_{1n} - a\int\left(p_{\phi n} + p_{\theta n} - \frac{n+\cos\phi}{\sin\phi}p_{rn}\right)\sin^2\phi\tan^n\frac{\phi}{2}\,d\phi\right] \qquad (40.20)$$

$$V_n = \frac{\tan^n(\phi/2)}{2\sin^2\phi}\left[C_{2n} - a\int\left(p_{\phi n} - p_{\theta n} + \frac{n-\cos\phi}{\sin\phi}p_{rn}\right)\sin^2\phi\cot^n\frac{\phi}{2}\,d\phi\right]$$

$N_{\theta n}$ is found from (40.19*c*).

For other than spherical shells, Eqs. (40.19) may be solved by numerical integration. In some cases, the use of cartesian coordinates z and r in the plane of the meridian is helpful.

Special Solutions

Sphere. (*a*) Uniform weight p (Fig. 40.11):

$$p_\phi = -p\cos\phi\cos\theta \qquad p_\theta = p\sin\theta \qquad p_r = -p\sin\phi\cos\theta$$

$$N_\phi = pa\frac{\sin\phi}{(1+\cos\phi)^2}\cos\theta$$

$$N_\theta = -pa\frac{2+2\cos\phi+\cos^2\phi}{(1+\cos\phi)^2}\sin\phi\cos\theta \qquad (40.21)$$

$$N_{\phi\theta} = -pa\frac{(2+\cos\phi)\sin\phi}{(1+\cos\phi)^2}\sin\theta$$

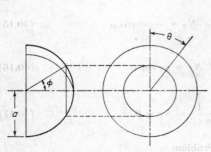

FIG. 40.11

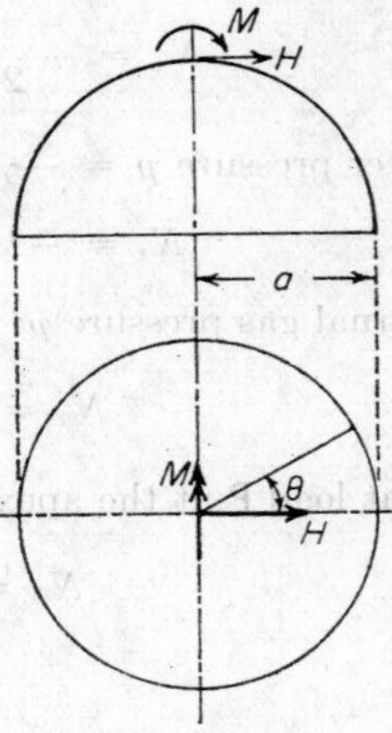

FIG. 40.12

(*b*) Point load **H** at the apex (Fig. 40.12):

$$N_\phi = -N_\theta = -\frac{\mathbf{H}}{\pi a}\frac{\cos\theta}{(1+\cos\phi)\sin\phi} \qquad (40.22)$$

$$N_{\phi\theta} = \frac{\mathbf{H}}{\pi a}\frac{\sin\theta}{(1+\cos\phi)\sin\phi}$$

(*c*) Point couple **M** at the apex (Fig. 40.12):

$$N_\phi = -N_\theta = \frac{\mathbf{M}}{\pi a^2} \frac{\cos\theta}{\sin^3\phi} \qquad N_{\phi\theta} = \frac{\mathbf{M}}{\pi a^2} \frac{\cot\phi}{\sin^2\phi} \sin\theta \tag{40.23}$$

(*d*) Load $P = P_n \cos n\theta$ acting at the edge $\phi = \alpha$ in meridional direction (Fig. 40.13):

$$N_\phi = -N_\theta = P_n \frac{\sin^2\alpha \tan^n(\phi/2)}{\sin^2\phi \tan^n(\alpha/2)} \cos n\theta$$
$$N_{\phi\theta} = -P_n \frac{\sin^2\alpha \tan^n(\phi/2)}{\sin^2\phi \tan^n(\alpha/2)} \sin n\theta \tag{40.24}$$

The value of $N_{\phi\theta}$ for $\phi = \alpha$ defines a tangential edge load that must be supplied simultaneously. If it is absent, a membrane stress system is not possible.

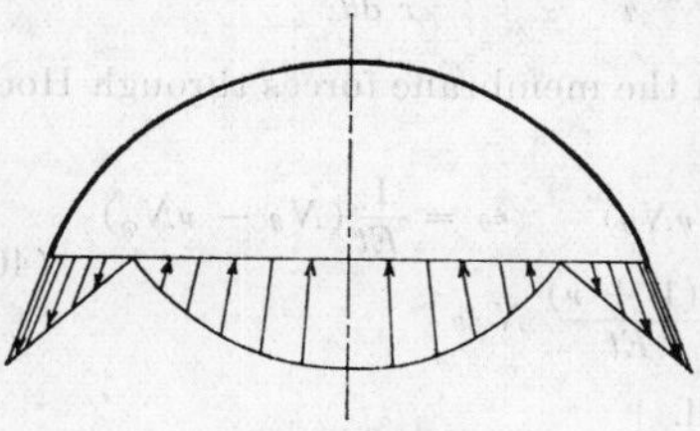

FIG. 40.13

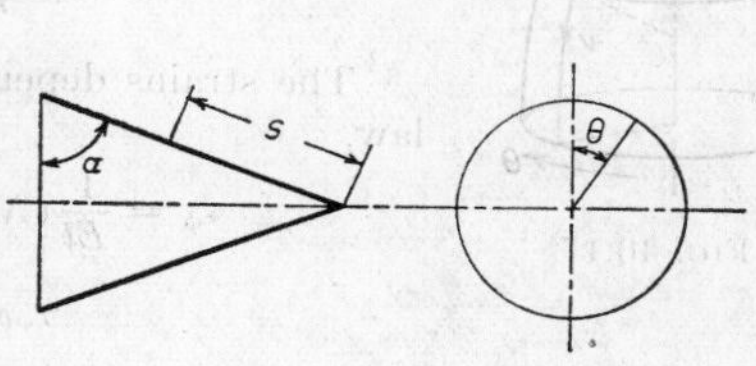

FIG. 40.14

Cone. (*a*) Uniform weight *p* (Fig. 40.14):

$$N_s = \frac{ps\cos\theta}{3\cos\alpha} \qquad N_\theta = -ps\cos\alpha\cos\theta \qquad N_{s\theta} = -\tfrac{2}{3}ps\sin\theta \tag{40.25}$$

(*b*) Point load **H** at the apex (Fig. 40.15):

$$N_s = \frac{\mathbf{H}\cos\theta}{\pi s \cos^2\alpha} \qquad N_\theta = N_{s\theta} = 0 \tag{40.26}$$

(*c*) Point couple **M** at apex (Fig. 40.15):

$$N_s = \frac{\mathbf{M}\cos\theta}{\pi s^2 \cos^2\alpha \sin\alpha} \qquad N_\theta = 0 \qquad N_{s\theta} = \frac{\mathbf{M}\sin\theta}{\pi s^2 \cos\alpha \sin\alpha} \tag{40.27}$$

(*d*) Meridional edge load $P = P_n \cos n\theta$ $(n > 1)$ at $s = l$ (Fig. 40.16):

$$N_s = \frac{P_n l}{s} \cos n\theta \qquad N_\theta = N_{s\theta} = 0 \tag{40.28}$$

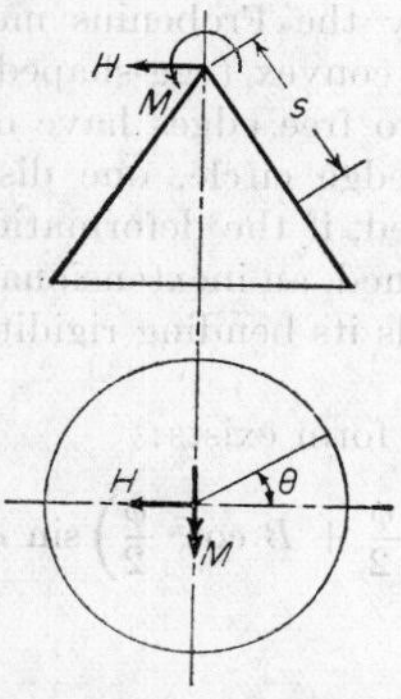

FIG. 40.15

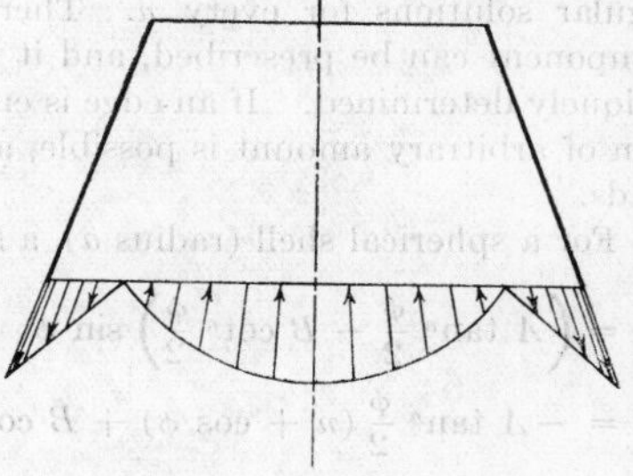

FIG. 40.16

If the shell is truncated, it must be supported at the upper edge, and the reaction is given by the edge value of N_s. If the shell is pointed, the stress system has a singularity at $s = 0$.

Deformations

The deformation of the shell can be described either by the components u, v, w of the displacement of every point of its middle surface (Fig. 40.17) or by the tensile strains ϵ_ϕ, ϵ_θ along meridians and parallel circles and the shear strain $\gamma_{\phi\theta}$. Displacements and strains are connected by the kinematic relations

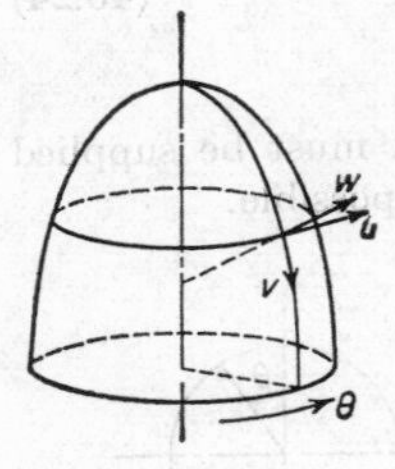

FIG. 40.17

$$\epsilon_\phi = \frac{1}{r_1}\left(\frac{\partial v}{\partial \phi} + w\right) \qquad \epsilon_\theta = \frac{1}{r}\left(\frac{\partial u}{\partial \theta} + v \cos \phi + w \sin \phi\right)$$
$$\gamma_{\phi\theta} = \frac{1}{r_1}\frac{\partial u}{\partial \phi} + \frac{u}{r}\cos \phi + \frac{1}{r}\frac{\partial v}{\partial \theta} \tag{40.29}$$

The strains depend on the membrane forces through Hooke's law,

$$\epsilon_\phi = \frac{1}{Et}(N_\phi - \nu N_\theta) \qquad \epsilon_\theta = \frac{1}{Et}(N_\theta - \nu N_\phi)$$
$$\gamma_{\phi\theta} = \frac{2(1+\nu)}{Et} N_{\phi\theta} \tag{40.30}$$

where t is the thickness of a homogeneous shell.

Inextensional Deformation. When the membrane forces are zero, also

$$\epsilon_\phi = \epsilon_\theta = \gamma_{\phi\theta} = 0$$

and Eqs. (40.29) are a set of three homogeneous partial differential equations for u, v, w. Their solutions are called inextensional deformations. They may be written as

$$u = u_n(\phi) \sin n\theta \qquad v = v_n(\phi) \cos n\theta \qquad w = w_n(\phi) \cos n\theta \tag{40.31}$$

with arbitrary integer n. Upon introducing (40.31) into (40.29) and eliminating u_n and w_n, a second-order equation for v_n may be obtained:

$$\frac{d^2 v_n}{d\phi^2}\frac{r_2}{r_1}\sin^2\phi - \frac{dv_n}{d\phi}\cos\phi\sin\phi + v_n\left(\frac{r_2}{r_1}\sin^2\phi + \cos^2\phi - n^2\right) = 0 \tag{40.32}$$

The substitution $x = 1 - \cos \phi$ makes the coefficients algebraic:

$$\frac{d^2 v_n}{dx^2} + \frac{1-x}{x(2-x)}\frac{r_2 - r_1}{r_2}\frac{dv_n}{dx} + \left(\frac{1}{x(2-x)} + \frac{(1-x)^2 - n^2}{x^2(2-x)^2}\frac{r_1}{r_2}\right) v_n = 0 \tag{40.33}$$

This equation has singular points at $x = 0$ and $x = 2$, corresponding to the apices $\phi = 0$ and $\phi = \pi$ of the shell. It may be solved by the Frobenius method (see p. 10–11) or by numerical integration. For a closed convex (egg-shaped) shell no regular solution exists. Convex shells with one or two free edges have one or two regular solutions for every n. Therefore, at every edge circle, one displacement component can be prescribed, and it must be prescribed, if the deformation is to be uniquely determined. If an edge is entirely unconstrained, an inextensional deformation of arbitrary amount is possible, and the shell needs its bending rigidity to carry loads.

For a spherical shell (radius a), a solution in closed form exists:

$$u_n = \left(A \tan^n \frac{\phi}{2} - B \cot^n \frac{\phi}{2}\right)\sin\phi \qquad v_n = \left(A \tan^n \frac{\phi}{2} + B \cot^n \frac{\phi}{2}\right)\sin\phi \tag{40.34}$$
$$w_n = -A \tan^n \frac{\phi}{2}(n + \cos\phi) + B \cot^n \frac{\phi}{2}(n - \cos\phi)$$

For $n = 0$ and $n = 1$, it degenerates into rigid-body displacements.

Extensional Deformations. When the membrane forces are not identically zero, elimination of the strains between Eqs. (40.29) and (40.30) yields three inhomogeneous partial differential equations for u, v, w. If the membrane forces have the form (40.18), the displacements have the form (40.31), and Eqs. (40.32) and (40.33) have a nonvanishing right-hand side. In most cases, it is a rather complicated function of ϕ, and numerical integration is the easiest means of obtaining a particular solution. In some cases, the Frobenius method is useful.

When the deformations of the shell have been calculated, it is possible to check the adequacy of the membrane theory. This is done by inserting the displacements into an elastic law like Eqs. (40.99), which, however, must include unsymmetric deformations, and by using a general set of equilibrium conditions similar to Eqs. (40.98) to find Q_ϕ and Q_θ. Membrane theory is inadequate if either the moments contribute substantially to the stresses or the transverse shear forces interfere substantially with the force equilibrium of the shell element.

40.3. CYLINDERS, MEMBRANE THEORY

Differential Equations

The coordinates are shown in Fig. 40.18. The angle ϕ is measured between a local tangent plane and a fixed tangent plane, assumed horizontal in the figure. The radius of curvature r of the cross section is a function of ϕ only.

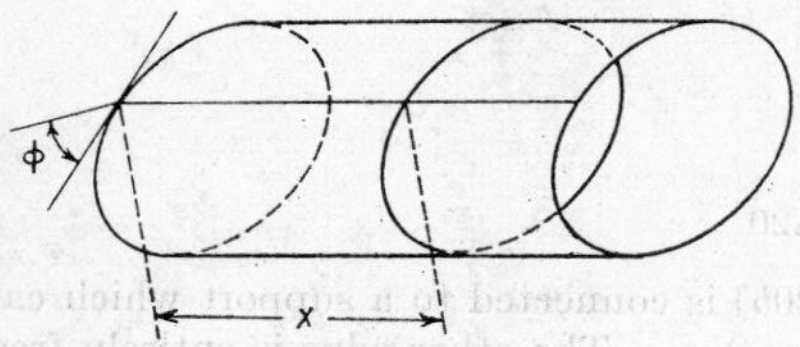

FIG. 40.18

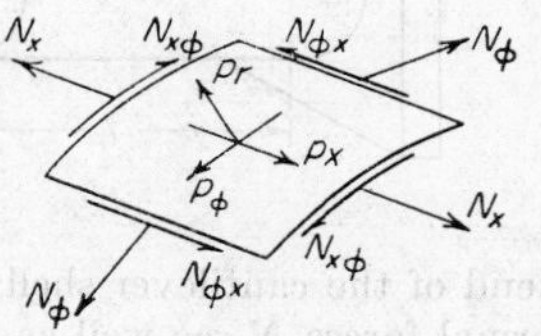

FIG. 40.19

Figure 40.19 shows a shell element. The load components (per unit area) p_x, p_ϕ, p_r are taken in the directions of increasing x and ϕ and normal to the shell. The equilibrium of the element yields the following differential equations:

$$N_\phi = rp_r \qquad \frac{\partial N_{x\phi}}{\partial x} = -p_\phi - \frac{1}{r}\frac{\partial N_\phi}{\partial \phi} \qquad \frac{\partial N_x}{\partial x} = -p_x - \frac{\partial N_{x\phi}}{\partial \phi} \qquad (40.35a\text{–}c)$$

The hoop force N_ϕ depends only on the local load. The other two forces follow from two successive integrations:

$$N_{x\phi} = -\int \left(p_\phi + \frac{1}{r}\frac{\partial N_\phi}{\partial \phi}\right) dx + f_1(\phi) \qquad N_x = -\int \left(p_x + \frac{1}{r}\frac{\partial N_{x\phi}}{\partial \phi}\right) dx + f_2(\phi) \qquad (40.36a,b)$$

In most applications, p_ϕ and p_r are independent of x, and $p_x \equiv 0$. In this case, Eqs. (40.36) can be simplified to

$$N_{x\phi} = -xF(\phi) + f_1(\phi) \qquad N_x = \frac{x^2}{2r}\frac{dF(\phi)}{d\phi} - \frac{x}{r}\frac{df_1(\phi)}{d\phi} + f_2(\phi) \qquad (40.37a,b)$$

with

$$F(\phi) = p_\phi + \frac{1}{r}\frac{dN_\phi}{d\phi} = p_\phi + \frac{dp_r}{d\phi} + \frac{1}{r}p_r\frac{dr}{d\phi}$$

The free functions $f_1(\phi)$ and $f_2(\phi)$ must be determined from two boundary conditions on edges $x =$ const. Two typical cases are the shell supported at both ends by rings or diaphragms (Fig. 40.20a) and the cantilever shell (Fig. 40.20b).

The rings supporting the shell in Fig. 40.20*a* are supposed to offer no resistance to forces in the direction of the shell generators; hence, the boundary conditions are $N_x \equiv 0$ for $x = 0$ and for $x = l$. The stress resultants are then given by (40.35*a*), and

$$N_{x\phi} = \tfrac{1}{2}(l - 2x)F(\phi) + c \qquad N_x = -\frac{x(l - x)}{2r}\frac{dF}{d\phi} \qquad (40.38a,b)$$

The constant *c* represents a kind of torsion of the shell [see Bredt's first formula, Eq. (36.44)]. It can usually be disposed of by a symmetry consideration.

Equation (40.38*a*) yields a nonvanishing shear force $N_{x\phi}$ at the ends of the shell, which is passed on to the rings. These must be strong enough against bending in their own plane to resist this load, and they must be properly supported.

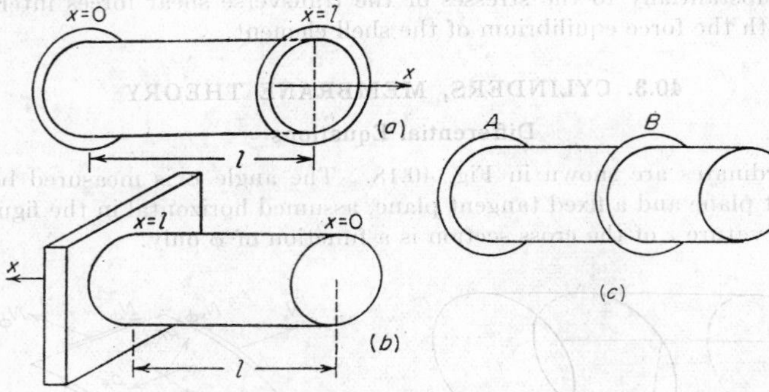

FIG. 40.20

One end of the cantilever shell (Fig. 40.20*b*) is connected to a support which can resist normal forces N_x as well as shear forces $N_{x\phi}$. The other edge is entirely free, and the boundary conditions are $N_x \equiv 0$, $N_{x\phi} \equiv 0$ for $x = 0$. In this case, the stress resultants are given by Eq. (40.35*a*), and

$$N_{x\phi} = -xF(\phi) \qquad N_x = \frac{x^2}{2r}\frac{dF}{d\phi} \qquad (40.39a,b)$$

A shell with an overhanging end (Fig. 40.20*c*) must be treated in two parts. Equations (40.39) apply to the cantilever shell. Matching of the N_x values over the ring *B* yields one of the boundary conditions for the part *AB*.

In the cases described by Fig. 40.20, the spanwise distribution of $N_{x\phi}$ and N_x is that of the shear and the bending moment of a beam supported in the same way. The distribution of N_ϕ, $N_{x\phi}$, N_x in ϕ direction depends on the cross section of the shell.

Special Solutions

Circular Cylinder (Fig. 40.21). (*a*) Uniform weight per unit area *p*:

$$\begin{aligned} p_\phi &= p \sin\phi \qquad p_r = -p\cos\phi \\ N_\phi &= -pa\cos\phi \qquad N_{x\phi} = p(l - 2x)\sin\phi \\ N_x &= -\frac{p}{a}x(l - x)\cos\phi \end{aligned} \qquad (40.40a\text{–}e)$$

If the lower half of the cylinder is cut away, the shell is a barrel-vault roof. Since the shear force $N_{x\phi} \neq 0$ at the straight edges $\phi = 90°$, appropriate forces must be applied as external forces to the shell. They may be furnished as the reaction of an edge member connected to each of these edges (Fig. 40.22). As a consequence of the shear transmitted to it, each edge member carries a tensile force varying between zero at the ends and a maximum in the middle.

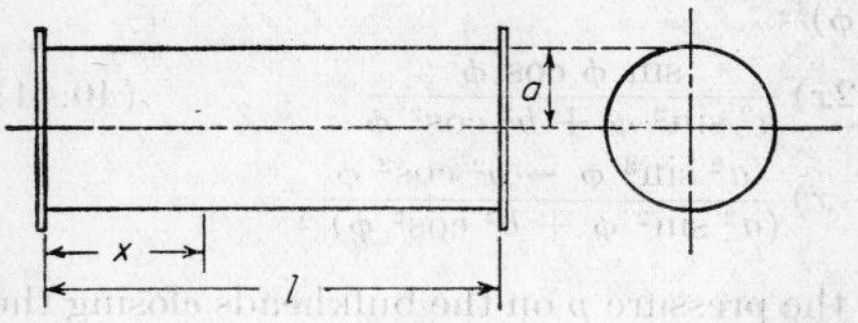

FIG. 40.21

FIG. 40.22

This stress system represents only very roughly the actual stresses occurring in a barrel-vault roof, since these are greatly influenced by the bending rigidity of the shell.

(*b*) Fourier series solution for the same case: the load [Eqs. (40.40*a*,*b*)] may be expanded in a Fourier series:

$$p_\phi = \frac{4p}{\pi} \sin\phi \sum \frac{1}{n} \sin\frac{n\pi x}{l}$$
$$p_r = -\frac{4p}{\pi} \cos\phi \sum \frac{1}{n} \sin\frac{n\pi x}{l} \qquad (40.41a,b)$$

The corresponding stress resultants are

$$N_\phi = -\frac{4pa}{\pi} \cos\phi \sum \frac{1}{n} \sin\frac{n\pi x}{l}$$
$$N_{x\phi} = \frac{8pl}{\pi^2} \sin\phi \sum \frac{1}{n^2} \cos\frac{n\pi x}{l} \qquad (40.41c\text{–}e)$$
$$N_x = -\frac{8pl^2}{\pi^2 a} \cos\phi \sum \frac{1}{n^3} \sin\frac{n\pi x}{l}$$

In these formulas, all sums are to be extended over the values $n = 1, 3, 5, \ldots$.

(*c*) Hydrostatic pressure, zero pressure at the top (γ = specific weight of water):

$$p_\phi = 0 \qquad p_r = \gamma a(1 - \cos\phi)$$
$$N_\phi = \gamma a^2(1 - \cos\phi) \qquad N_{x\phi} = 2\gamma a(l - 2x)\sin\phi \qquad (40.42)$$
$$N_x = -\tfrac{1}{2}\gamma x(l - x)\cos\phi$$

Elliptic cylinder: (*a*) Uniform weight per unit area *p* (Fig. 40.23):

$$N_\phi = -\frac{pa^2b^2\cos\phi}{(a^2\sin^2\phi + b^2\cos^2\phi)^{3/2}}$$
$$N_{x\phi} = \tfrac{1}{2}p(l - 2x)\frac{2a^2 + (a^2 - b^2)\cos^2\phi}{a^2\sin^2\phi + b^2\cos^2\phi}\sin\phi \qquad (40.43)$$
$$N_x = \frac{px(l - x)}{2a^2b^2}\left(1 - 3a^2\frac{b^2 - (a^2 - b^2)\sin^2\phi}{(a^2\sin^2\phi + b^2\cos^2\phi)^2}\right)\frac{\cos\phi}{(a^2\sin^2\phi + b^2\cos^2\phi)^{1/2}}$$

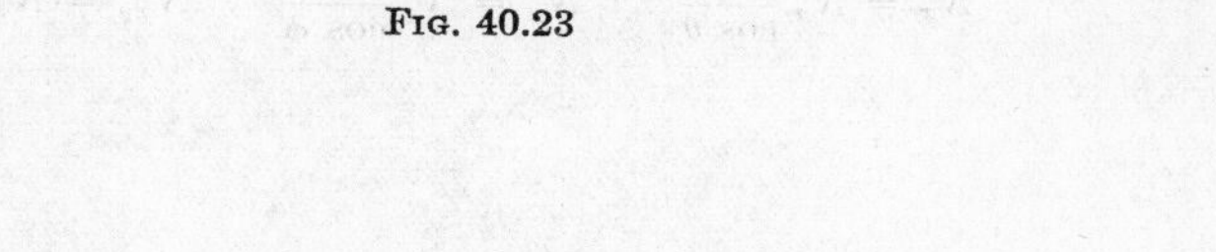

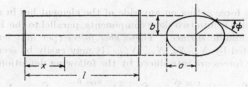

FIG. 40.23

These formulas may also be used for a barrel vault of semielliptic cross section.

(b) Uniform internal pressure p (Fig. 40.23):

$$N_\phi = \frac{pa^2b^2}{(a^2 \sin^2 \phi + b^2 \cos^2 \phi)^{3/2}}$$

$$N_{x\phi} = -3/2 p(a^2 - b^2)(l - 2x) \frac{\sin \phi \cos \phi}{a^2 \sin^2 \phi + b^2 \cos^2 \phi} \tag{40.44}$$

$$N_x = -3/2 p \frac{a^2 - b^2}{a^2b^2} x(l - x) \frac{a^2 \sin^2 \phi - b^2 \cos^2 \phi}{(a^2 \sin^2 \phi + b^2 \cos^2 \phi)^{1/2}}$$

The force N_x does not include the action of the pressure p on the bulkheads closing the cylinder. Equations (40.44) show that a pressure vessel may have a membrane stress system without having circular cross section.

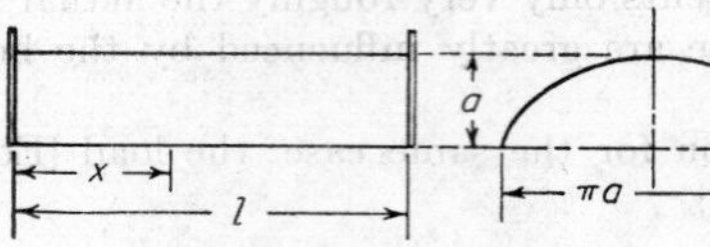

FIG. 40.24

Barrel Vault with a Common Cycloid as its Cross Section (Fig. 40.24). Uniform weight per unit area p:

$$N_\phi = -2pa \cos^3 \phi \qquad N_{x\phi} = 3/2 p(l - 2x) \sin \phi \qquad N_x = -\frac{3p}{4a} x(l - x) \tag{40.45}$$

Like all barrel vaults, this shell needs edge members, which carry large tensile forces.

40.4. GENERAL MEMBRANE THEORY

Differential Equation

Figure 40.25 shows a shell element in a cartesian coordinate system (x,y,z). Its projection into the x,y plane is a rectangle, but the element is not necessarily rectangular. The angle ω between its sides depends, in general, on x and y.

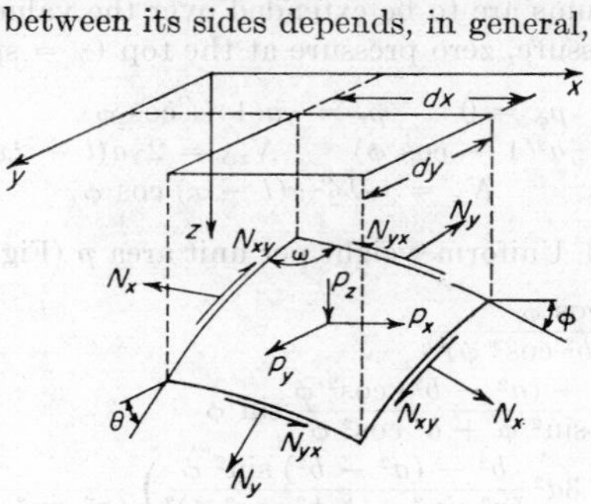

FIG. 40.25

The membrane force acting on any side of the element lies in a tangential plane. It is useful to resolve it into two skew components, parallel to the lines x = const and y = const on the middle surface. These *fiber forces* (per unit length of the actual element) are denoted by N_x, N_y, N_{xy}, N_{yx}. It may easily be seen that $N_{xy} = N_{yx}$.

Projected fiber forces are introduced by the following definitions:

$$\bar{N}_x = N_x \frac{\cos \phi}{\cos \theta} \qquad \bar{N}_y = N_y \frac{\cos \theta}{\cos \phi} \qquad \bar{N}_{xy} = N_{xy} \tag{40.46}$$

They are the horizontal projections of the actual membrane forces, referred to a unit length of the side dx or dy of the projected element.

Conditions of force equilibrium in directions x, y, z yield the following equations, where p_x, p_y, p_z are the load components per unit of projected area $dx \cdot dy$:

$$\frac{\partial \bar{N}_x}{\partial x} + \frac{\partial \bar{N}_{xy}}{\partial y} + p_x = 0 \qquad \frac{\partial \bar{N}_{xy}}{\partial x} + \frac{\partial \bar{N}_y}{\partial y} + p_y = 0$$
$$\frac{\partial}{\partial x}\left(\bar{N}_x \frac{\partial z}{\partial x}\right) + \frac{\partial}{\partial y}\left(\bar{N}_{xy} \frac{\partial z}{\partial x}\right) + \frac{\partial}{\partial x}\left(\bar{N}_{xy} \frac{\partial z}{\partial y}\right) + \frac{\partial}{\partial y}\left(\bar{N}_y \frac{\partial z}{\partial y}\right) + p_z = 0 \tag{40.47a–c}$$

Equations (40.47a,b) are identically satisfied when the projected fiber forces are expressed in terms of a stress function Φ as

$$\bar{N}_x = \frac{\partial^2 \Phi}{\partial y^2} - \int p_x\,dx \qquad \bar{N}_y = \frac{\partial^2 \Phi}{\partial x^2} - \int p_y\,dy \qquad \bar{N}_{xy} = -\frac{\partial^2 \Phi}{\partial x\,\partial y} \tag{40.48}$$

Equation (40.47c) may then be reduced to the form

$$\frac{\partial^2 \Phi}{\partial x^2}\frac{\partial^2 z}{\partial y^2} - 2\frac{\partial^2 \Phi}{\partial x\,\partial y}\frac{\partial^2 z}{\partial x\,\partial y} + \frac{\partial^2 \Phi}{\partial y^2}\frac{\partial^2 z}{\partial x^2} = -p_z + p_x\frac{\partial z}{\partial x} + p_y\frac{\partial z}{\partial y} + \frac{\partial^2 z}{\partial x^2}\int p_x\,dx + \frac{\partial^2 z}{\partial y^2}\int p_y\,dy \tag{40.49}$$

This is a second-order partial differential equation for Φ. It is of the elliptic (hyperbolic) type (see p. 11–5), if the shell has positive (negative) gaussian curvature.

Solutions

An elliptic paraboloid has the equation

$$z = \frac{x^2}{h_1} + \frac{y^2}{h_2} \tag{40.50}$$

A shell formed after this surface is shown in Fig. 40.26. For vertical load, $p_x = p_y = 0$.

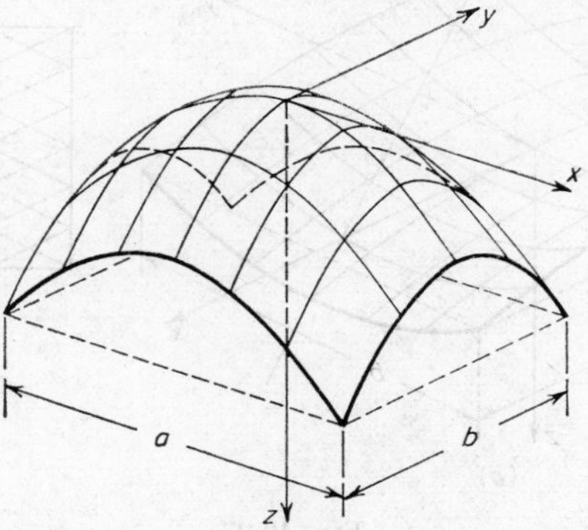

FIG. 40.26

Equation (40.49) simplifies, in this case, to

$$\frac{1}{h_2}\frac{\partial^2 \Phi}{\partial x^2} + \frac{1}{h_1}\frac{\partial^2 \Phi}{\partial y^2} = -\tfrac{1}{2}p_z \tag{40.51}$$

For $p_z = p = \text{const}$, the stress function

$$\Phi = \frac{ph_1}{4}\left[\frac{b^2}{4} - y^2 + \frac{8b^2}{\pi^3}\sum (-1)^{(n+1)/2}\frac{\text{Cosh}\,(n\pi x/c)\cos\,(n\pi y/b)}{n^3\,\text{Cosh}\,(n\pi a/2c)}\right] \tag{40.52}$$

represents a membrane system satisfying the boundary conditions $N_x = 0$ for $x = \pm a/2$ and $N_y = 0$ for $y = \pm b/2$. The shell is supported by shear forces N_{xy} acting on all four edges. To provide these forces, it must be connected to four arches which are subjected to bending in their planes.

The projected membrane forces are:

$$\begin{aligned}\bar{N}_x &= -\tfrac{1}{2}ph_1\left[1 + \frac{4}{\pi}\sum(-1)^{(n+1)/2}\frac{\operatorname{Cosh}(n\pi x/c)\cos(n\pi y/b)}{n\operatorname{Cosh}(n\pi a/2c)}\right]\\ \bar{N}_{xy} &= \frac{2p}{\pi}\sqrt{h_1h_2}\sum(-1)^{(n+1)/2}\frac{\operatorname{Sinh}(n\pi x/c)\sin(n\pi y/b)}{n\operatorname{Cosh}(n\pi a/2c)}\\ \bar{N}_y &= \frac{2ph_2}{\pi}\sum(-1)^{(n+1)/2}\frac{\operatorname{Cosh}(n\pi x/c)\cos(n\pi y/b)}{n\operatorname{Cosh}(n\pi a/2c)}\end{aligned} \tag{40.53a–c}$$

In these formulas $c^2 = b^2h_1/h_2$, and all summations must be extended over odd integers n only. In the interior of the shell, the series converge rapidly. At the corners, the series in Eq. (40.53*b*) diverges, indicating a singularity in the stress system.

For other loads (variable p_z, nonvanishing p_x and p_y), numerical integration of Eq. (40.49) is necessary. The relaxation method is the appropriate tool for this purpose. If possible, the solution Φ should be split into a formal solution Φ_1 like (40.52), containing all singularities, and a regular part $\Phi_2 = \Phi - \Phi_1$, which may be obtained from a relaxation procedure.

A hyperbolic paraboloid has the equation

$$z = \frac{x^2}{h_1} - \frac{y^2}{h_2} \tag{40.54}$$

A shell of this shape is shown in Fig. 40.27*a*. The surface (40.54) contains two sets of straight lines (generators). Their projections into the x,y plane make an angle $\pm\gamma$

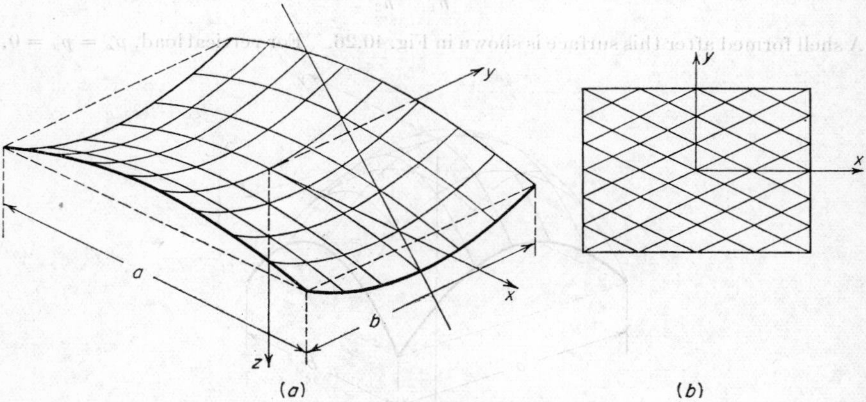

FIG. 40.27

with the x axis, where $\tan^2\gamma = h_2/h_1$. These lines are the characteristics of the differential equation (40.49) which, for the hyperbolic paraboloid, is of the hyperbolic type.

It is always possible to prescribe both components of the membrane force at one edge of the shell (one which is not intersected by characteristics emanating from the corners), and to prescribe one force component on the two adjacent edges. Whether or not a set of more useful boundary conditions can be satisfied depends on the configuration of the characteristics.

In the case illustrated in Fig. 40.28, it is possible to have $N_x = 0$ on the edges $x = \pm a/2$, and $N_y = 0$ on $y = \pm b/2$. The stress system, however, is highly discontinuous. For the load $p_z = p = \text{const}$, it may be written in the form

$$\bar{N}_x = \tfrac{1}{2} p h_1 k_1 \qquad \bar{N}_y = \tfrac{1}{2} p h_2 k_2 \qquad \bar{N}_{xy} = \tfrac{1}{2} p \sqrt{h_1 h_2}\, k_3 \tag{40.55}$$

and the coefficients k_1, k_2, k_3 assume the values indicated in Fig. 40.28.

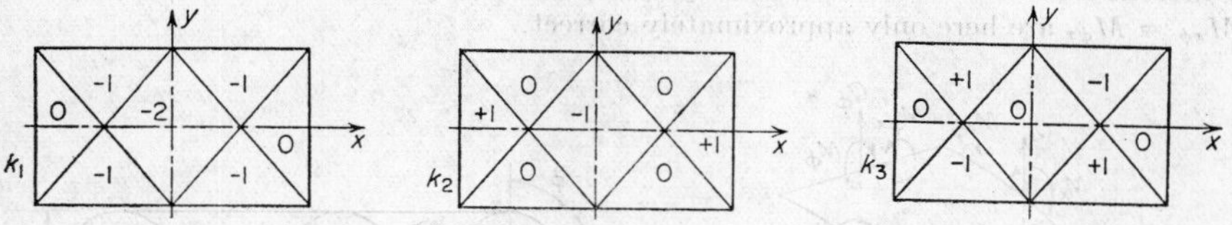

FIG. 40.28

For other arrangements of the corner characteristics (i.e., for different rectangles cut from the same paraboloid), the stress system looks quite different.

The equation

$$z = \frac{xy}{c} \tag{40.56}$$

represents the same paraboloid as Eq. (40.54) with $h_1 = h_2 = 2c$, but in a different position with respect to the coordinate system. In the present case, the straight generators are parallel to the coordinate planes. For $p_x = p_y = 0$, $p_z = p = \text{const}$, Eq. (40.49) assumes the form

$$\frac{2}{c} \frac{\partial^2 \Phi}{\partial x\, \partial y} = p \tag{40.57}$$

and has the solution

$$\Phi = \frac{cp}{2} xy + f_1(x) + f_2(y) \tag{40.58}$$

with two free functions $f_1(x)$ and $f_2(y)$. It yields the membrane force system

$$\bar{N}_x = \frac{d^2 f_2(y)}{dy^2} \qquad \bar{N}_y = \frac{d^2 f_1(x)}{dx^2} \qquad \bar{N}_{xy} = -\tfrac{1}{2} cp \tag{40.59}$$

The shear force cannot be influenced by boundary conditions, and all four edges need edge members to take care of the edge shears. The fiber force N_x is independent of x; that is, a force of this kind applied at one edge $x = \text{const}$ of the shell runs clear through the shell along a generator, and reappears at the opposite edge.

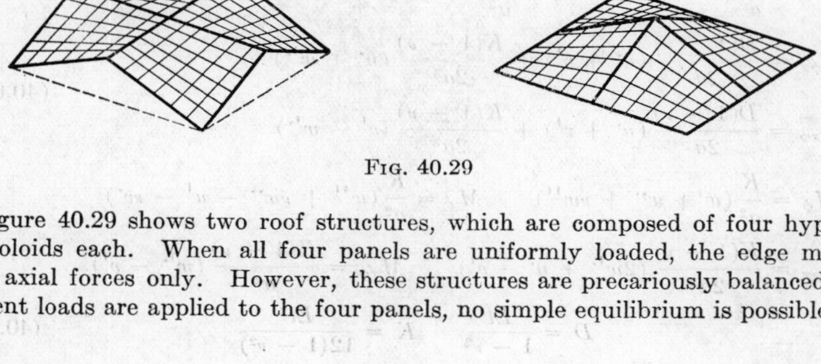

FIG. 40.29

Figure 40.29 shows two roof structures, which are composed of four hyperbolic paraboloids each. When all four panels are uniformly loaded, the edge members carry axial forces only. However, these structures are precariously balanced; when different loads are applied to the four panels, no simple equilibrium is possible.

40.5. CIRCULAR CYLINDER, BENDING THEORY

Differential Equations

The coordinates are the same as in Fig. 40.18. The stress resultants are the membrane forces N_x, N_ϕ, $N_{x\phi}$, $N_{\phi x}$ (Fig. 40.19); the bending and twisting moments M_x, M_ϕ, $M_{x\phi}$, $M_{\phi x}$; and the transverse shear forces Q_x, Q_ϕ of plate theory (Fig. 40.30). Different from membrane shell theory and plate theory the relations $N_{x\phi} = N_{\phi x}$ and $M_{x\phi} = M_{\phi x}$ are here only approximately correct.

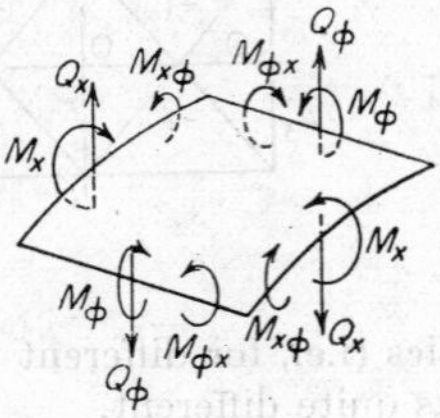

FIG. 40.30

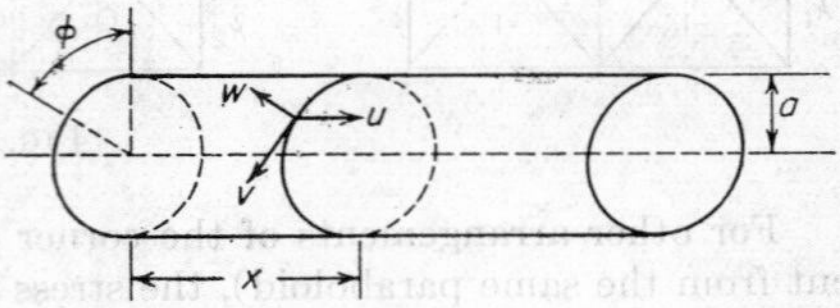

FIG. 40.31

For the derivatives with respect to the dimensionless coordinates ϕ and x/a, a dash-dot notation is in common use:

$$\frac{\partial()}{\partial\phi} = ()^{\cdot} \qquad a\frac{\partial()}{\partial x} = ()^{\prime} \tag{40.60}$$

The equilibrium of forces and moments at a shell element, including the surface loads p_x, p_ϕ, p_r (Fig. 40.19), yields the following equations:

$$\begin{aligned} N_x^{\prime} + N_{\phi x}^{\cdot} &= -p_x a \qquad N_\phi^{\cdot} + N_{x\phi}^{\prime} - Q_\phi = -p_\phi a \\ Q_\phi^{\cdot} &+ Q_x^{\prime} + N_\phi = p_r a \\ M_\phi^{\cdot} + M_{x\phi}^{\prime} &= aQ_\phi \qquad M_x^{\prime} + M_{\phi x}^{\cdot} = aQ_x \\ aN_{x\phi} &- aN_{\phi x} + M_{\phi x} = 0 \end{aligned} \tag{40.61a–f}$$

Since these equations contain 10 unknown stress resultants, the problem is statically indeterminate. Relations between the stress resultants and the displacements u, v, w (Fig. 40.31) may be derived in various ways, starting from the general relations of three-dimensional elasticity or from the assumption of conservation of normals used in plate theory. The latter approach yields the following set:

$$\begin{aligned} N_\phi &= \frac{D}{a}(v^{\cdot} + w + \nu u^{\prime}) + \frac{K}{a^3}(w + w^{\cdot\cdot}) \\ N_x &= \frac{D}{a}(u^{\prime} + \nu v^{\cdot} + \nu w) - \frac{K}{a^3}w^{\prime\prime} \\ N_{\phi x} &= \frac{D(1-\nu)}{2a}(u^{\cdot} + v^{\prime}) + \frac{K(1-\nu)}{2a^3}(u^{\cdot} + w^{\prime\cdot}) \\ N_{x\phi} &= \frac{D(1-\nu)}{2a}(u^{\cdot} + v^{\prime}) + \frac{K(1-\nu)}{2a^3}(v^{\prime} - w^{\prime\cdot}) \\ M_\phi &= \frac{K}{a^2}(w + w^{\cdot\cdot} + \nu w^{\prime\prime}) \qquad M_x = \frac{K}{a^2}(w^{\prime\prime} + \nu w^{\cdot\cdot} - u^{\prime} - \nu v^{\cdot}) \\ M_{\phi x} &= \frac{K(1-\nu)}{2a^2}(2w^{\prime\cdot} + u^{\cdot} - v^{\prime}) \qquad M_{x\phi} = \frac{K(1-\nu)}{a^2}(w^{\prime\cdot} - v^{\prime}) \end{aligned} \tag{40.62a–h}$$

where

$$D = \frac{Et}{1-\nu^2} \qquad K = \frac{Et^3}{12(1-\nu^2)} \tag{40.63a,b}$$

are the extensional and flexural rigidities of the shell. For a nonhomogeneous (but isotropic) shell, other definitions for D and K may be substituted without changing Eqs. (40.62).

The elastic law (40.62) contains rather many terms of minor importance. In every specific application, many of them may be neglected, but not always the same ones. A very radical omission of all but the most important terms leads to Donnell's equations, which appeal by their great simplicity, but have to be used with some caution:

$$N_\phi = \frac{D}{a}(v^{\cdot} + w + \nu u') \qquad N_x = \frac{D}{a}(u' + \nu v^{\cdot} + \nu w)$$

$$N_{\phi x} = N_{x\phi} = \frac{D(1-\nu)}{2a}(u^{\cdot} + v')$$

$$M_\phi = \frac{K}{a^2}(w^{\cdot\cdot} + \nu w'') \qquad M_x = \frac{K}{a^2}(w'' + \nu w^{\cdot\cdot}) \tag{40.64a–f}$$

$$M_{\phi x} = M_{x\phi} = \frac{K(1-\nu)}{a^2} w'^{\cdot}$$

Introduction of the elastic law (40.62) into Eqs. (40.61) yields, after elimination of Q_x and Q_ϕ, the following set of partial differential equations for u, v, w:

$$u'' + \tfrac{1}{2}(1-\nu)u^{\cdot\cdot} + \tfrac{1}{2}(1+\nu)v'^{\cdot} + \nu w' + k[\tfrac{1}{2}(1-\nu)u^{\cdot\cdot} - w''' + \tfrac{1}{2}(1-\nu)w'^{\cdot\cdot}] = -\frac{p_x a^2}{D}$$

$$\tfrac{1}{2}(1+\nu)u'^{\cdot} + v^{\cdot\cdot} + \tfrac{1}{2}(1-\nu)v'' + w^{\cdot} + k[\tfrac{3}{2}(1-\nu)v'' - \tfrac{1}{2}(3-\nu)w''^{\cdot}] = -\frac{p_\phi a^2}{D} \tag{40.65a–c}$$

$$\nu u' + v^{\cdot} + w + k[\tfrac{1}{2}(1-\nu)u'^{\cdot\cdot} - u''' - \tfrac{1}{2}(3-\nu)v''^{\cdot} + w'''' + 2w''^{\cdot\cdot} + w^{\cdot\cdot\cdot\cdot} + 2w^{\cdot\cdot} + w] = \frac{p_r a^2}{D}$$

where

$$k = \frac{K}{Da^2} = \frac{t^2}{12a^2} \tag{40.66}$$

is a small, dimensionless thickness parameter. When Eqs. (40.64) are used instead of (40.62), it is advisable to drop also the Q_ϕ term from Eq. (40.61b). Then the simple equations

$$u'' + \tfrac{1}{2}(1-\nu)u^{\cdot\cdot} + \tfrac{1}{2}(1+\nu)v'^{\cdot} + \nu w' = -\frac{p_x a^2}{D}$$

$$\tfrac{1}{2}(1+\nu)u'^{\cdot} + v^{\cdot\cdot} + \tfrac{1}{2}(1-\nu)v'' + w^{\cdot} = -\frac{p_\phi a^2}{D} \tag{40.67a–c}$$

$$\nu u' + v^{\cdot} + w + k(w'''' + 2w''^{\cdot\cdot} + w^{\cdot\cdot\cdot\cdot}) = \frac{p_r a^2}{D}$$

are obtained. From these equations, u and v may easily be eliminated. The result is an eighth-order equation for w. For the edge-load problem ($p_x = p_\phi = p_r = 0$), it reads

$$k\,\nabla^8 w + (1-\nu^2)w'''' = 0 \tag{40.68}$$

where ∇^8 is $(\nabla^2)^4$ and $\nabla^2() = ()'' + ()^{\cdot\cdot}$.

When the Q_ϕ term in (40.61b) and all the surface loads p are dropped, Eqs.

(40.61a,b) are identical with the equilibrium conditions (37.3) of the plane plate, and may be satisfied by writing the membrane forces as derivatives of a stress function Φ:

$$N_\phi = \Phi'' \qquad N_x = \Phi^{\cdot\cdot} \qquad N_{x\phi} = N_{\phi x} = -\Phi'^{\cdot} \tag{40.69a–c}$$

It is then possible to reduce Eqs. (40.61c–e) and (40.64) to the following pair of fourth-order equations,

$$\nabla^4 w = -\frac{a^3}{K}\Phi'' \qquad \nabla^4\Phi = \frac{D(1-\nu^2)}{a}w'' \tag{40.70a,b}$$

from which ultimately Eq. (40.68) may be derived.

Shell Loaded at a Circular Edge

General Solution. It is assumed that a cylindrical shell is bounded by two circles, $x = 0$ and $x = l$, and that no surface loads are present ($p_x = p_\phi = p_r = 0$). The stresses and deformations of the shell are then exclusively caused by external forces and moments applied to its edges. Such edge forces and moments are necessarily periodic functions of ϕ with the period 2π, and may be expanded in Fourier series. The solution of Eqs. (40.65) or (40.67) corresponding to the general term of these series has the form

$$u = Ae^{\lambda x/a}\cos m\phi \qquad v = Be^{\lambda x/a}\sin m\phi \qquad w = Ce^{\lambda x/a}\cos m\phi \tag{40.71}$$

with integer m. Upon introduction of this solution into Eqs. (40.65), a set of three linear equations for the constants A, B, C is obtained:

$$\begin{aligned}
&[\lambda^2 - \tfrac{1}{2}(1-\nu)m^2(1+k)]A + [\tfrac{1}{2}(1+\nu)\lambda m]B \\
&\qquad\qquad + [\nu\lambda - k(\lambda^3 + \tfrac{1}{2}(1-\nu)\lambda m^2)]C = 0 \\
&[-\tfrac{1}{2}(1+\nu)\lambda m]A + [\tfrac{1}{2}(1-\nu)\lambda^2 - m^2 + \tfrac{3}{2}(1-\nu)k\lambda^2]B \\
&\qquad\qquad + [-m + \tfrac{1}{2}(3-\nu)k\lambda^2 m]C = 0 \\
&[\nu\lambda - k(\lambda^3 + \tfrac{1}{2}(1-\nu)\lambda m^2)]A + [m - \tfrac{1}{2}(3-\nu)k\lambda^2 m]B \\
&\qquad\qquad + [1 + k(\lambda^4 - 2\lambda^2 m^2 + m^4 - 2m^2 + 1)]C = 0
\end{aligned} \tag{40.72}$$

Since they are homogeneous, the determinant of the bracketed coefficients must vanish. This yields an eighth-degree equation for λ:

$$\begin{aligned}
\lambda^8 - 2(2m^2 - \nu)\lambda^6 + [(1-\nu^2)k^{-1} + 6m^2(m^2-1)]\lambda^4 \\
- 2m^2[2m^4 - (4-\nu)m^2 + (2-\nu)]\lambda^2 + m^4(m^2-1)^2 = 0
\end{aligned} \tag{40.73}$$

When the tentative solution (40.71) is introduced into Eqs. (40.67) or into Eq. (40.68), the equation

$$\lambda^8 - 4\lambda^6 m^2 + \lambda^4[(1-\nu^2)k^{-1} + 6m^4] - 4\lambda^2 m^6 + m^8 = 0 \tag{40.74}$$

is obtained instead of (40.73).

To each of the eight solutions λ_j of Eq. (40.73) or (40.74), there corresponds one solution (40.71) with an arbitrary constant C_j. The constants A_j and B_j of this solution depend on C_j through Eqs. (40.72), or through a similar set derived from Eqs. (40.67). All solutions λ_j are complex, and, therefore, A_j and B_j are also complex when a real value has been chosen for C_j.

In Eqs. (40.73) and (40.74), the coefficient of λ^4 is much larger than all the others. Therefore, these equations have four large solutions, $\lambda = \pm\kappa_1 \pm i\mu_1$, and four small solutions, $\lambda = \pm\kappa_2 \pm i\mu_2$. The large ones represent edge disturbances, caused by an edge load and dying out rather rapidly with increasing distance from the edge. The small values λ describe a stress system which penetrates much farther into the shell.

For a semi-infinite shell ($x \geq 0$), only those λ with negative real parts are of interest, and, after separation of real and imaginary parts, the solution may be written as

$$w = \left[e^{-\kappa_1 x/a} \left(C_1 \cos \frac{\mu_1 x}{a} + C_2 \sin \frac{\mu_1 x}{a} \right) + e^{-\kappa_2 x/a} \left(C_3 \cos \frac{\mu_2 x}{a} + C_4 \sin \frac{\mu_2 x}{a} \right) \right] \cos m\phi \quad (40.75)$$

Ready-to-use formulas for the stress resultants may be found in [5], p. 230.

Boundary Conditions. The edge load at $x = 0$, say, consists of forces and moments distributed along this edge. These loads may be either prescribed or the consequence or prescribed displacements of the edge. Since the problem is of the eighth order, four boundary conditions must be specified at each of the two edges x = const.

If displacements are prescribed, it is possible and necessary to prescribe all three components u, v, w and the slope w' (clamped edge). If forces are to be prescribed, it is not possible to prescribe the three components N_x, $N_{x\phi}$, Q_x and the two moments M_x, $M_{x\phi}$. As in the case of the plane plate [see Eqs. (39.12)], Q_x and $M_{x\phi}$ must be combined into an effective transverse shear (Kirchhoff force),

$$S_x = Q_x + \frac{M^{\cdot}_{x\phi}}{a} \quad (40.76a)$$

and, by a similar reasoning, also an effective tangential shear,

$$T_x = N_{x\phi} - \frac{M_{x\phi}}{a} \quad (40.76b)$$

may be established. Prescribing these and N_x, M_x yields an acceptable set of four boundary conditions. It is, of course, possible to prescribe some of the forces and some of the displacements, for example, v, w, N_x, and M_x (simply supported edge).

Shell Loaded at a Straight Edge

General Solution. We consider here a part of a cylindrical shell bounded by two circular arcs (at distance l) and by two generators. When this shell is loaded along the curved edges, this load may be expanded into Fourier series with the actual length of the edge as the half period. Then the solution (40.71) applies with noninteger values of m.

When the shell is loaded along its straight edges, the load is expanded into Fourier series in $\sin (n\pi x/l)$ or $\cos (n\pi x/l)$. The solution then appears also in the form of such Fourier series, with the general term

$$u = Ae^{m\phi} \cos \frac{n\pi x}{l} \qquad v = Be^{m\phi} \sin \frac{n\pi x}{l} \qquad w = Ce^{m\phi} \sin \frac{n\pi x}{l} \quad (40.77)$$

In these formulas, n is an arbitrary integer, whereas m must be found from the differential equations.

Upon introduction of (40.77) into (40.65), a set of three linear equations similar to (40.72) is obtained:

$$\begin{aligned}
&[\lambda^2 - \tfrac{1}{2}(1 - \nu)m^2(1 + k)]A + [-\tfrac{1}{2}(1 + \nu)\lambda m]B \\
&\qquad + [-\nu\lambda - k(\lambda^3 + \tfrac{1}{2}(1 - \nu)\lambda m^2)]C = 0 \\
&[-\tfrac{1}{2}(1 + \nu)\lambda m]A + [m^2 - \tfrac{1}{2}(1 - \nu)\lambda^2 - \tfrac{3}{2}(1 - \nu)k\lambda^2]B \\
&\qquad + [m + \tfrac{1}{2}(3 - \nu)k\lambda^2 m]C = 0 \\
&[-\nu\lambda - k(\lambda^3 + \tfrac{1}{2}(1 - \nu)\lambda m^2)]A + [m + \tfrac{1}{2}(3 - \nu)k\lambda^2 m]B \\
&\qquad + [1 + k(\lambda^4 - 2\lambda^2 m^2 + m^4 + 2m^2 + 1)]C = 0
\end{aligned} \quad (40.78)$$

where $\lambda = n\pi a/l$. The vanishing of the determinant of the bracketed coefficients yields the equation for m:

$$m^8 - 2(2\lambda^2 - 1)m^6 + [6\lambda^4 - 2(4 - \nu)\lambda^2 + 1]m^4 - 2\lambda^2[2\lambda^4 - 3\lambda^2 + (2 - \nu)]m^2 + [(1 - \nu^2)k^{-1}\lambda^4 + \lambda^6(\lambda^2 - 2\nu)] = 0 \qquad (40.79)$$

Its eight roots are all complex, and come in two sets of four: $m = \pm\kappa_1 \pm i\mu_1$ and $m = \pm\kappa_2 \pm i\mu_2$; but, since k^{-1} appears in the *last* term of Eq. (40.79), these roots are not of different orders of magnitude.

To each of the eight roots $m = m_j$, there corresponds a separate solution (40.77), with an arbitrary constant C_j, while the constants A_j and B_j depend on C_j through Eqs. (40.78). After separation of real and imaginary parts, the solution assumes the form

$$w = [e^{-\kappa_1\phi}(C_1 \cos \mu_1\phi + C_2 \sin \mu_1\phi) + e^{-\kappa_2\phi}(C_3 \cos \mu_2\phi + C_4 \sin \mu_2\phi) + e^{\kappa_1\phi}(C_5 \cos \mu_1\phi + C_6 \sin \mu_1\phi) + e^{\kappa_2\phi}(C_7 \cos \mu_2\phi + C_8 \sin \mu_2\phi)] \sin (n\pi x/l) \qquad (40.80)$$

It is, of course, possible and often advisable to combine the exponentials into hyperbolic sines and cosines.

The solution (40.77) satisfies, on the edges $x = 0$ and $x = l$, the boundary conditions of a simply supported (piano-hinge) edge:

$$v = 0 \qquad w = 0 \qquad N_x = 0 \qquad M_x = 0$$

On each edge $\phi =$ const, four boundary conditions may be prescribed, and they determine the constants C_j. These boundary conditions may refer to displacements, to the slope $w^{\cdot}$, and to edge forces and moments. There is again a Kirchhoff force,

$$S_\phi = Q_\phi + \frac{M'_{\phi x}}{a} \qquad (40.81)$$

but the effective tangential shear T_ϕ is identical with $N_{\phi x}$. Ready-to-use formulas for all stress resultants and displacements may be found in [5], Table 5.2.

Barrel Vaults. Barrel vaults are shells of the kind just described. Usually their length l is a good deal larger than the length of the curved edges. In this case, it turns out that M_x, Q_x and both twisting moments are rather small, and that the bending-stress system consists mainly in bending moments M_ϕ and shear forces Q_ϕ. It is then possible to neglect the smaller stress resultants entirely, and to arrive at a differential equation for M_ϕ.

Under the simplifying assumptions just stated, the equilibrium equations (40.61) may be used to express all other stress resultants in terms of M_ϕ:

$$aQ_\phi = M_\phi^{\cdot} \qquad aN_\phi = -M_\phi^{\cdot\cdot} \qquad aN'_{x\phi} = aN'_{\phi x} = M_\phi^{\cdot\cdot\cdot} + M_\phi^{\cdot} \qquad aN''_x = -M_\phi^{\cdot\cdot\cdot\cdot} - M_\phi^{\cdot\cdot} \qquad (40.82a\text{–}d)$$

On the other hand, u, v, w may be eliminated from Eqs. (40.64a–d). When the relation between the stress resultants thus obtained is combined with Eqs. (40.82), the following differential equation results:

$$M_\phi^{\cdot\cdot\cdot\cdot\cdot\cdot\cdot\cdot} + (2 + \nu)M_\phi^{''\cdot\cdot\cdot\cdot\cdot\cdot} + M_\phi^{\cdot\cdot\cdot\cdot\cdot\cdot} + (1 + 2\nu)M_\phi^{''''\cdot\cdot\cdot\cdot} + 2(1 + \nu)M_\phi^{''\cdot\cdot\cdot\cdot} + \nu M_\phi^{''''''\cdot\cdot} + \nu(2 + \nu)M_\phi^{''''\cdot\cdot} + (1 - \nu^2)k^{-1}M_\phi^{''''} = 0 \qquad (40.83)$$

It admits the solution

$$M_\phi = Ce^{m\phi} \sin \frac{\lambda x}{a} \qquad \text{with} \qquad \lambda = \frac{n\pi a}{l} \qquad (40.84)$$

which leads to an eighth-degree equation for m. When, in this equation, only those terms are retained which agree with the corresponding terms of the exact equation (40.79), it assumes the simple form

$$m^8 + (1 - \nu^2)k^{-1}\lambda^4 = 0 \tag{40.85}$$

and has the eight solutions $m = \pm\kappa \pm i\mu$ and $m = \pm\mu \pm i\kappa$, with

$$\kappa = \tfrac{1}{2}\sqrt{2 + \sqrt{2}}\,\zeta\sqrt{n} \qquad \mu = \tfrac{1}{2}\sqrt{2 - \sqrt{2}}\,\zeta\sqrt{n} \tag{40.86}$$

where

$$\zeta^8 = \frac{1 - \nu^2}{k}\frac{\pi^4 a^4}{l^4}$$

When the two edges are at $\phi = \pm\phi_0$ and are loaded symmetrically, the general form of the solution (40.84) is

$$M_\phi = [(A_1 \operatorname{Cosh} \kappa\phi \cos \mu\phi + B_1 \operatorname{Sinh} \kappa\phi \sin \mu\phi) + (A_2 \operatorname{Cosh} \mu\phi \cos \kappa\phi + B_2 \operatorname{Sinh} \mu\phi \sin \kappa\phi)] \sin (n\pi x/l) \tag{40.87}$$

Expressions for the other stress resultants may be derived from it by using Eqs. (40.82). When this is done, the lower-order derivative in Eqs. (40.82*c,d*) should be neglected. The displacements may be found from Eqs. (40.64).

Axisymmetric Case

Axial symmetry of the stress system implies that all derivatives with respect to ϕ vanish, and that there is no load component p_ϕ. Additionally, it will be assumed here that $p_x = 0$, and that the boundary conditions are such that $N_x = 0$. This simplifies the formulas, and excludes only a few trivial cases.

Under the assumptions stated, Eq. (40.67*b*) is identically satisfied, and elimination of u from Eqs. (40.67*a,c*) yields the differential equation of the problem:

$$kw'''' + (1 - \nu^2)w = \frac{p_r a^2}{D} \tag{40.88}$$

Equations (40.64) simplify to the following relations:

$$N_\phi = \frac{D(1 - \nu^2)}{a} w \qquad M_x = \frac{K}{a^2} w'' \qquad M_\phi = \nu M_x \tag{40.89a–c}$$

All other stress resultants are zero.

General Solution. Particular solutions of Eq. (40.88) for gas pressure and hydrostatic pressure are obvious. The homogeneous equation

$$kw'''' + (1 - \nu^2)w = 0 \tag{40.88'}$$

has the general solution

$$w = e^{-\kappa x/a}\left(A_1 \cos \frac{\kappa x}{a} + A_2 \sin \frac{\kappa x}{a}\right) + e^{\kappa x/a}\left(B_1 \cos \frac{\kappa x}{a} + B_2 \sin \frac{\kappa x}{a}\right) \tag{40.90}$$

where

$$\kappa^4 = \frac{1 - \nu^2}{4k} = 3(1 - \nu^2)\frac{a^2}{t^2} \tag{40.91}$$

The A terms describe the stress system produced by loads applied to the edge $x = 0$ of the semi-infinite cylinder $x \geq 0$, while the B terms account for a similar edge disturbance in the semi-infinite shell $x \leq 0$. The A terms of (40.90) may be written in the alternative form

$$w = Ce^{-\kappa x/a} \sin\left(\frac{\kappa x}{a} + \psi\right) \tag{40.92}$$

where C and ψ are the free constants. This form of the solution is preferable when one of the boundary conditions is homogeneous, while (40.90) is advantageous when both boundary conditions specify nonvanishing values of some force or displacement.

The A terms of Eq. (40.90) and the solution (40.92) yield the following expressions for the slope and the stress resultants:

$$
\begin{aligned}
w^{\bullet} &= -\kappa e^{-\kappa x/a}\left[(A_1 - A_2)\cos\frac{\kappa x}{a} + (A_1 + A_2)\sin\frac{\kappa x}{a}\right] \\
&= -\kappa\sqrt{2}\,Ce^{-\kappa x/a}\sin\left(\frac{\kappa x}{a} + \psi - \frac{\pi}{4}\right) \\
M_x &= \frac{2K\kappa^2}{a^2}\,e^{-\kappa x/a}\left(A_1\sin\frac{\kappa x}{a} - A_2\cos\frac{\kappa x}{a}\right) \\
&= -\frac{2K\kappa^2}{a^2}\,Ce^{-\kappa x/a}\cos\left(\frac{\kappa x}{a} + \psi\right) \\
Q_x &= \frac{2K\kappa^3}{a^3}\,e^{-\kappa x/a}\left[(A_1 + A_2)\cos\frac{\kappa x}{a} + (A_2 - A_1)\sin\frac{\kappa x}{a}\right] \\
&= 2\sqrt{2}\,\frac{K\kappa^3}{a^3}\,Ce^{-\kappa x/a}\sin\left(\frac{\kappa x}{a} + \psi + \frac{\pi}{4}\right)
\end{aligned}
\tag{40.93}
$$

Special Solutions

1. Edge load $M_x = M$ with $Q_x = 0$ at $x = 0$ (Fig. 40.32*a*):

$$
M_x = M\sqrt{2}\,e^{-\kappa x/a}\cos\left(\frac{\kappa x}{a} - \frac{\pi}{4}\right) \qquad Q_x = -\frac{2M\kappa}{a}\,e^{-\kappa x/a}\sin\frac{\kappa x}{a}
$$
$$
N_\phi = -2\sqrt{2}\,\frac{M\kappa^2}{a}\,e^{-\kappa x/a}\sin\left(\frac{\kappa x}{a} - \frac{\pi}{4}\right) \tag{40.94a–c}
$$

and, at the edge $x = 0$:

$$
w = \frac{Ma^2}{2K\kappa^2} \qquad w^{\bullet} = -\frac{Ma^2}{K\kappa} \tag{40.94d,e}
$$

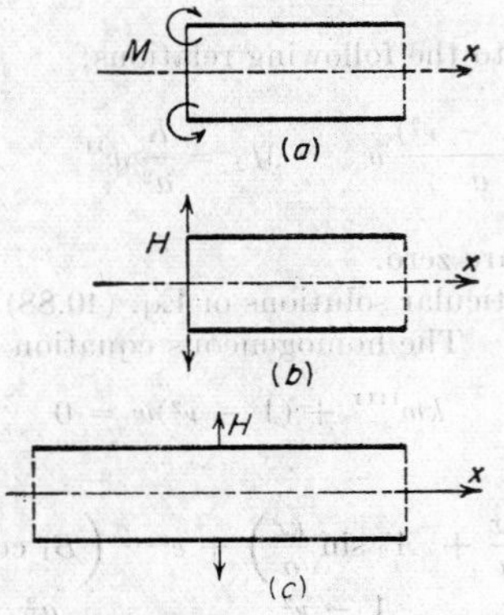

FIG. 40.32

2. Edge load $Q_x = H$ with $M_x = 0$ at $x = 0$ (Fig. 40.32*b*):

$$
M_x = \frac{Ha}{\kappa}\,e^{-\kappa x/a}\sin\frac{\kappa x}{a} \qquad Q_x = \sqrt{2}\,He^{-\kappa x/a}\cos\left(\frac{\kappa x}{a} + \frac{\pi}{4}\right)
$$
$$
N_\phi = 2H\kappa e^{-\kappa x/a}\cos\frac{\kappa x}{a} \tag{40.95a–c}
$$

and, at the edge $x = 0$:

$$w = \frac{Ha^3}{2K\kappa^3} \qquad w' = -\frac{Ha^3}{2K\kappa^2} \tag{40.95d,e}$$

3. Edge load $M_x = M$ with $w = 0$ at $x = 0$ (simply supported edge):

$$M_x = Me^{-\kappa x/a}\cos\frac{\kappa x}{a} \qquad Q_x = -\sqrt{2}\,\frac{M\kappa}{a}\,e^{-\kappa x/a}\sin\left(\frac{\kappa x}{a}+\frac{\pi}{4}\right)$$
$$N_\phi = -\frac{2M\kappa^2}{a}\,e^{-\kappa x/a}\sin\frac{\kappa x}{a} \tag{40.96a–c}$$

and, at the edge $x = 0$:

$$w' = -\frac{Ma^2}{2K\kappa} \tag{40.96d}$$

For a ring of radial forces H (force per unit of circumference) at $x = 0$ (Fig. 40.32c), the stress resultants in the right half of the shell are:

$$M_x = -\sqrt{2}\,\frac{Ha}{\kappa}\,e^{-\kappa x/a}\sin\left(\frac{\kappa x}{a}-\frac{\pi}{4}\right) \qquad Q_x = -2He^{-\kappa x/a}\cos\frac{\kappa x}{a}$$
$$N_\phi = -2\sqrt{2}\,H\kappa e^{-\kappa x/a}\sin\left(\frac{\kappa x}{a}+\frac{\pi}{4}\right) \tag{40.97a–c}$$

The deflection at $x = 0$ is

$$w = -\frac{Ha^3}{2K\kappa^3} \tag{40.97d}$$

40.6. SHELLS OF REVOLUTION, BENDING THEORY

Differential Equations

For most thin shells the membrane solutions may be used as good approximations to particular solutions of the bending equations. Therefore, it suffices to consider here

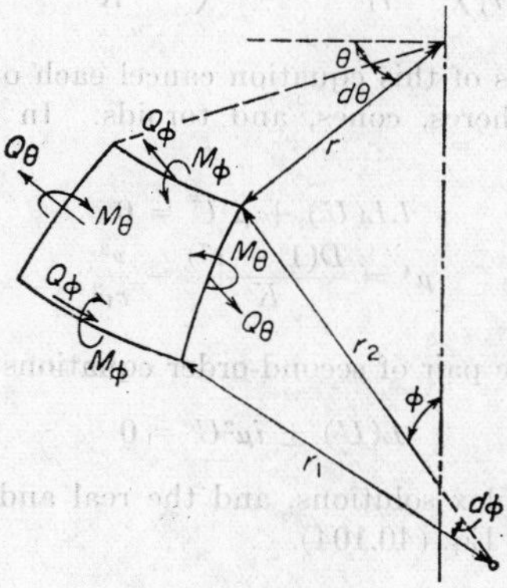

FIG. 40.33

the case that the shell carries no surface loads, its stresses being caused by forces and moments applied to the edges. The discussion will be restricted to axisymmetric loads.

The stress system consists of the normal forces N_ϕ and N_θ shown in Fig. 40.4; the bending moments M_ϕ, M_θ; and the shear force Q_ϕ shown in Fig. 40.33. The equilibrium of forces and moments acting on a shell element yields three conditions of equilibrium. With the abbreviation $\partial()/\partial\phi = ()^{\cdot}$, they are:

$$(rN_\phi)^{\cdot} - r_1 N_\theta \cos\phi - rQ_\phi = 0 \qquad r_1 N_\theta \sin\phi + rN_\phi + (rQ_\phi)^{\cdot} = 0$$
$$(rM_\phi)^{\cdot} - r_1 M_\theta \cos\phi - rr_1 Q_\phi = 0 \tag{40.98}$$

The normal forces and the bending moments depend on the displacements v, w (Fig. 40.17) through the relations

$$N_\phi = D\left(\frac{v^{\cdot} + w}{r_1} + \nu\frac{v\cot\phi + w}{r_2}\right) \qquad N_\theta = D\left(\frac{v\cot\phi + w}{r_2} + \nu\frac{v^{\cdot} + w}{r_1}\right)$$

$$M_\phi = K\left(\frac{\chi^{\cdot}}{r_1} + \nu\frac{\chi}{r_2}\cot\phi\right) \qquad M_\theta = K\left(\frac{\chi}{r_2}\cot\phi + \nu\frac{\chi^{\cdot}}{r_1}\right) \tag{40.99a–d}$$

where
$$\chi = \frac{w^{\cdot} - v}{r_1} \tag{40.100}$$

is the angle through which a line element of the meridian rotates during deformation.

It is possible to reduce Eqs. (40.98) through (40.100) to two second-order differential equations for the variables χ and $U = r_2 Q_\phi$. With the help of the linear operator

$$r_1 L() = \frac{r_2}{r_1}()^{\cdot\cdot} + \left[\left(\frac{r_2}{r_1}\right)^{\cdot} + \frac{r_2}{r_1}\cot\phi\right]()^{\cdot} - \frac{r_1}{r_2}()\cot^2\phi \tag{40.101}$$

they may be written as

$$L(\chi) - \frac{\nu}{r_1}\chi = \frac{U}{K} \qquad L(U) + \frac{\nu}{r_1}U = -D(1-\nu^2)\chi \tag{40.102a,b}$$

When χ is eliminated between these equations, a fourth-order equation for U results:

$$LL(U) + \nu L\left(\frac{U}{r_1}\right) - \frac{\nu}{r_1}L(U) + \left(\frac{D(1-\nu^2)}{K} - \frac{\nu^2}{r_1^2}\right)U = 0 \tag{40.103}$$

The second and third terms of this equation cancel each other when either $\nu = 0$ or $r_1 = \text{const}$, that is, for spheres, cones, and toroids. In these cases, Eq. (40.103) reduces to

$$LL(U) + \mu^4 U = 0 \tag{40.104}$$

with
$$\mu^4 = \frac{D(1-\nu^2)}{K} - \frac{\nu^2}{r_1^2}$$

and it may be split into the pair of second-order equations

$$L(U) \pm i\mu^2 U = 0 \tag{40.105a,b}$$

They have conjugate complex solutions, and the real and imaginary parts of these solutions satisfy separately Eq. (40.104).

Spherical Shell

For a sphere with $r_1 = r_2 = a$, Eq. (40.105a) reads

$$Q_\phi^{\cdot\cdot} + Q_\phi^{\cdot}\cot\phi - Q_\phi\cot^2\phi + 2i\kappa^2 Q_\phi = 0 \tag{40.106}$$

with
$$\kappa^4 = \tfrac{1}{4}a^2\mu^4 = 3(1-\nu^2)a^2/t^2 \tag{40.107}$$

It has singular points at the poles $\phi = 0$ and $\phi = \pi$ of the coordinate system. Transformations like $x = \sin^2\phi$ or $x = \cos^2\phi$ and $Q_\phi = z\sin\phi$ make its coefficients rational and reduce it to a hypergeometric equation. It may be solved by (not tabulated) hypergeometric functions or by numerical integration. For all but very thick shells, the work is rather tedious.

Solution for Large ϕ. When the shell is thin and ϕ is not too small ($\phi > 45°$, say), the first and last terms in (40.106) dominate, and the other two may be neglected. The equation has then solutions of the type

$$Q_\phi = Ce^{\lambda\phi}$$

with $\lambda = \pm(1 \pm i)\kappa$. They are of an oscillatory character, and represent edge disturbances of the shell, caused by edge loads. For a spherical zone $\phi_1 \geq \phi \geq \phi_2$, it is convenient to introduce coordinates

$$\omega_1 = \phi_1 - \phi \qquad \omega_2 = \phi - \phi_2$$

and to write the solution as

$$Q_\phi = e^{-\kappa\omega_1}(A_1 \cos \kappa\omega_1 + B_1 \sin \kappa\omega_1) + e^{-\kappa\omega_2}(A_2 \cos \kappa\omega_2 + B_2 \sin \kappa\omega_2) \tag{40.108a}$$

or as

$$Q_\phi = C_1 e^{-\kappa\omega_1} \sin (\kappa\omega_1 + \psi_1) + C_2 e^{-\kappa\omega_2} \sin (\kappa\omega_2 + \psi_2) \tag{40.108b}$$

In both forms, the first part of the solution describes the stresses near the lower edge $\phi = \phi_1$, and the second part those near the upper edge $\phi = \phi_2$. If $(\phi_1 - \phi_2)$ is not very small, and if the shell is really thin, then a substantial part of the shell is, for all practical purposes, free of any stress when loads are applied to its edges. It should be noted that ϕ_2 must be large enough to keep $\cot \phi_2$ so small that the second and third terms of Eq. (40.106) can safely be neglected. Otherwise, only the first half of the solution may be applicable.

The solution (40.108*a*,*b*) produces the following stress resultants and deformations:

$$\begin{aligned}
N_\theta &= -\kappa e^{-\kappa\omega_1}[(A_1 - B_1) \cos \kappa\omega_1 + (A_1 + B_1) \sin \kappa\omega_1] \\
&\qquad + \kappa e^{-\kappa\omega_2}[(A_2 - B_2) \cos \kappa\omega_2 + (A_2 + B_2) \sin \kappa\omega_2] \\
&= \kappa \sqrt{2}\, [-C_1 e^{-\kappa\omega_1} \sin (\kappa\omega_1 + \psi_1 - \pi/4) \\
&\qquad + C_2 e^{-\kappa\omega_2} \sin (\kappa\omega_2 + \psi_2 - \pi/4)] \\
M_\phi &= \frac{a}{2\kappa} e^{-\kappa\omega_1}[(A_1 + B_1) \cos \kappa\omega_1 - (A_1 - B_1) \sin \kappa\omega_1] \\
&\qquad - \frac{a}{2\kappa} e^{-\kappa\omega_2}[(A_2 + B_2) \cos \kappa\omega_2 - (A_2 - B_2) \sin \kappa\omega_2] \\
&= \frac{a}{2\kappa} \sqrt{2} \left[C_1 e^{-\kappa\omega_1} \sin \left(\kappa\omega_1 + \psi_1 + \frac{\pi}{4}\right) \right. \\
&\qquad \left. - C_2 e^{-\kappa\omega_2} \sin \left(\kappa\omega_2 + \psi_2 + \frac{\pi}{4}\right)\right] \\
D(1 - \nu^2)\chi &= 2\kappa^2 e^{-\kappa\omega_1}(B_1 \cos \kappa\omega_1 - A_1 \sin \kappa\omega_1) \\
&\qquad - 2\kappa^2 e^{-\kappa\omega_2}(B_2 \cos \kappa\omega_2 - A_2 \sin \kappa\omega_2) \\
&= 2\kappa^2[C_1 e^{-\kappa\omega_1} \cos (\kappa\omega_1 + \psi_1) + C_2 e^{-\kappa\omega_2} \cos (\kappa\omega_2 + \psi_2)] \\
N_\phi &= -Q_\phi \cot \phi \qquad M_\theta = \nu M_\phi
\end{aligned} \tag{40.109a–e}$$

From Eqs. (40.109), the solutions for the two cases shown in Fig. 40.34 may be derived. In both cases, $C_2 = 0$. For the horizontal line load of Fig. 40.34*a*, the remaining constants are

$$C_1 = H \sqrt{2} \sin \phi_1 \qquad \psi_1 = -\pi/4 \tag{40.110a}$$

and, for the moment load of Fig. 40.34*b*, they are

$$C_1 = \frac{2\kappa}{a} M \qquad \psi_1 = 0 \tag{40.110b}$$

Solution Near the Apex. Near the apex of a spherical shell, Eq. (40.106) may be simplified in another way, replacing cot ϕ by ϕ^{-1}. It then reads

$$Q_\phi^{\cdot\cdot} + \phi^{-1}Q_\phi^{\cdot} - \phi^{-2}Q_\phi + 2i\kappa^2 Q_\phi = 0 \tag{40.111}$$

and the transformation

$$\xi = x\sqrt{i} = \kappa\phi\sqrt{2i}$$

brings it into the standard form of Bessel's equation

$$\frac{d^2Q_\phi}{d\xi^2} + \frac{1}{\xi}\frac{dQ_\phi}{d\xi} + \left(1 - \frac{1}{\xi^2}\right)Q_\phi = 0$$

Its solutions are Bessel functions of ξ or Kelvin (Thomson) functions of x (see p. 15–16):

$$Q_\phi = A_1 \operatorname{ber}' x + A_2 \operatorname{bei}' x + B_1 \operatorname{ker}' x + B_2 \operatorname{kei}' x \tag{40.112a}$$

The A terms are regular functions, and describe stresses produced by edge loads applied to a shallow spherical cap (Fig. 40.35*a*). The B terms have a singularity at

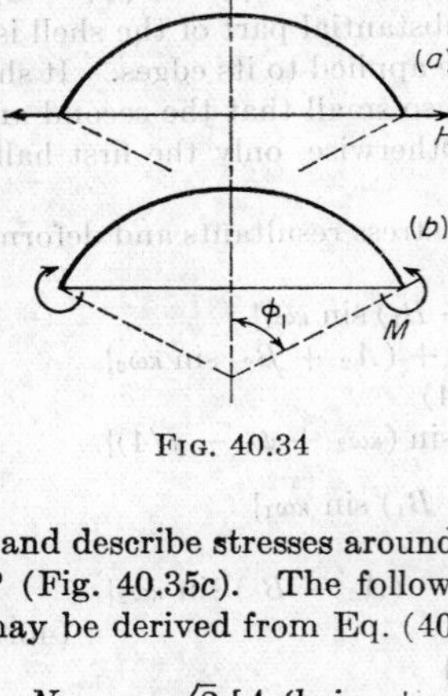

FIG. 40.34

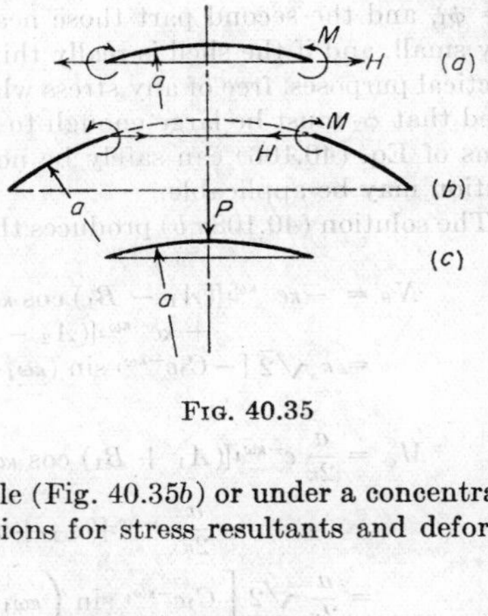

FIG. 40.35

$x = 0$, and describe stresses around a small hole (Fig. 40.35*b*) or under a concentrated force **P** (Fig. 40.35*c*). The following expressions for stress resultants and deformations may be derived from Eq. (40.112*a*):

$$\begin{aligned} N_\theta = \kappa\sqrt{2}\,[&A_1(\operatorname{bei} x + x^{-1}\operatorname{ber}' x) - A_2(\operatorname{ber} x - x^{-1}\operatorname{bei}' x) \\ &+ B_1(\operatorname{kei} x + x^{-1}\operatorname{ker}' x) - B_2(\operatorname{ker} x - x^{-1}\operatorname{kei}' x)] \end{aligned}$$

$$\begin{aligned} M_\phi = \frac{K\kappa\sqrt{2}}{Da(1-\nu^2)}\,\{&A_1[2\kappa^2(\operatorname{ber} x - (1-\nu)x^{-1}\operatorname{bei}' x) \\ &\qquad + \nu(\operatorname{bei} x + (1-\nu)x^{-1}\operatorname{ber}' x)] \\ &+ A_2[2\kappa^2(\operatorname{bei} x + (1-\nu)x^{-1}\operatorname{ber}' x) - \nu(\operatorname{ber} x - (1-\nu)x^{-1}\operatorname{bei}' x)] \\ &+ B_1[2\kappa^2(\operatorname{ker} x - (1-\nu)x^{-1}\operatorname{kei}' x) + \nu(\operatorname{kei} x + (1-\nu)x^{-1}\operatorname{ker}' x)] \\ &+ B_2[2\kappa^2(\operatorname{kei} x + (1-\nu)x^{-1}\operatorname{ker}' x) - \nu(\operatorname{ker} x - (1-\nu)x^{-1}\operatorname{kei}' x)]\} \end{aligned}$$

$$\begin{aligned} M_\theta = \frac{K\kappa\sqrt{2}}{Da(1-\nu^2)}\,\{&A_1[2\kappa^2(\nu\operatorname{ber} x + (1-\nu)x^{-1}\operatorname{bei}' x) \\ &\qquad + \nu(\nu\operatorname{bei} x - (1-\nu)x^{-1}\operatorname{ber}' x)] \\ &+ A_2[2\kappa^2(\nu\operatorname{bei} x - (1-\nu)x^{-1}\operatorname{ber}' x) - \nu(\nu\operatorname{ber} x + (1-\nu)x^{-1}\operatorname{bei}' x)] \\ &+ B_1[2\kappa^2(\nu\operatorname{ker} x + (1-\nu)x^{-1}\operatorname{kei}' x) + \nu(\nu\operatorname{kei} x - (1-\nu)x^{-1}\operatorname{ker}' x)] \\ &+ B_2[2\kappa^2(\nu\operatorname{kei} x - (1-\nu)x^{-1}\operatorname{ker}' x) - \nu(\nu\operatorname{ker} x + (1-\nu)x^{-1}\operatorname{kei}' x)]\} \end{aligned}$$

$$D(1-\nu^2)\chi = A_1(2\kappa^2 \operatorname{bei}' x - \nu \operatorname{ber}' x) - A_2(2\kappa^2 \operatorname{ber}' x + \nu \operatorname{bei}' x) + B_1(2\kappa^2 \operatorname{kei}' x - \nu \operatorname{ker}' x) - B_2(2\kappa^2 \operatorname{ker}' x + \nu \operatorname{kei}' x)$$

$$N_\phi = -\phi^{-1} Q_\phi \qquad (40.112b\text{–}f)$$

Formulas like Eqs. (40.108) covering the cases represented by Fig. 40.35*a,b* are very cumbersome. Numerical values of the constants A and B should rather be derived from Eqs. (40.112) after the numerical data of any specific problem have been inserted.

The case of Fig. 40.35*c* is not quite covered by the solution (40.112), which has been derived under the assumption that the resultant force transmitted through any parallel circle is zero. A simple generalization of Eqs. (40.102) allows inclusion of the case of the point load **P** and leads to the following stress resultants:

$$Q_\phi = \frac{\mathbf{P}\kappa}{\sqrt{2}\,\pi a}\left(\operatorname{ker}' x - \frac{\nu}{2\kappa^2}\operatorname{kei}' x\right) \qquad N_\phi = -\frac{\mathbf{P}\kappa^2 x^{-1}}{\pi a}\left(\operatorname{ker}' x - \frac{\nu}{2\kappa^2}\operatorname{kei}' x + x^{-1}\right)$$

$$N'_\theta = \frac{\mathbf{P}\kappa^2}{\pi a}\left[\operatorname{kei} x + x^{-1}\operatorname{ker}' x + x^{-2} + \frac{\nu}{2\kappa^2}(\operatorname{ker} x - x^{-1}\operatorname{kei}' x)\right] \qquad (40.113a\text{–}f)$$

$$M_\phi = \frac{\mathbf{P}}{2\pi}[\operatorname{ker} x - (1-\nu)x^{-1}\operatorname{kei}' x]$$

$$M_\theta = \frac{\mathbf{P}}{2\pi}[\nu \operatorname{ker} x + (1-\nu)x^{-1}\operatorname{kei}' x]$$

$$D(1-\nu^2)\chi = \frac{\mathbf{P}\sqrt{2}\,\kappa^3}{\pi a}\operatorname{kei}' x$$

The singularity of the bending moments is the same as that known in a circular plate. With increasing ϕ, the moments and Q_ϕ vanish rapidly in damped oscillations, while N_ϕ and N_θ approach the membrane solution (40.6).

Conical Shell

Constant Wall Thickness. With the notations of Fig. 40.10, the basic equation (40.105*a*) may be brought into the form

$$s(sQ_s)'' + (sQ_s)' - s^{-1}(sQ_s) \pm i\lambda^2 sQ_s = 0 \qquad (40.114)$$

where primes indicate differentiation with respect to s, and

$$\lambda^4 = \frac{D(1-\nu^2)}{K}\tan^2\alpha = \frac{12(1-\nu^2)}{t^2}\tan^2\alpha$$

The transformation

$$\eta = y\sqrt{i} = 2\lambda\sqrt{i}\sqrt{s}$$

reduces it to a Bessel equation which has as solutions the Bessel functions of order 2, of η or the following combinations of Kelvin functions (see p. 15–16) of y:

$$sQ_s = A_1(\operatorname{ber} y - 2y^{-1}\operatorname{bei}' y) + A_2(\operatorname{bei} y + 2y^{-1}\operatorname{ber}' y) + B_1(\operatorname{ker} y - 2y^{-1}\operatorname{kei}' y) + B_2(\operatorname{kei} y + 2y^{-1}\operatorname{ker}' y) \qquad (40.115a)$$

From it, the following expressions may be derived:

$$N_\theta = -\frac{\cot\alpha}{2s}[A_1(y\operatorname{ber}' y - 2\operatorname{ber} y + 4y^{-1}\operatorname{bei}' y) + A_2(y\operatorname{bei}' y - 2\operatorname{bei} y - 4y^{-1}\operatorname{ber}' y) + B_1(y\operatorname{ker}' y - 2\operatorname{ker} y + 4y^{-1}\operatorname{kei}' y) + B_2(y\operatorname{kei}' y - 2\operatorname{kei} y - 4y^{-1}\operatorname{ker}' y)]$$

$$M_s = 2y^{-2}\{A_1[y\operatorname{bei}' y - 2(1-\nu)(\operatorname{bei} y + 2y^{-1}\operatorname{ber}' y)] - A_2[y\operatorname{ber}' y - 2(1-\nu)(\operatorname{ber} y - 2y^{-1}\operatorname{bei}' y)] + B_1[y\operatorname{kei}' y - 2(1-\nu)(\operatorname{kei} y + 2y^{-1}\operatorname{ker}' y)] - B_2[y\operatorname{ker}' y - 2(1-\nu)(\operatorname{ker} y - 2y^{-1}\operatorname{kei}' y)]\}$$

$$
\begin{aligned}
M_\theta = 2y^{-2}\{&A_1[\nu y \text{ bei}' y + 2(1-\nu)(\text{bei } y + 2y^{-1}\text{ ber}' y)] \\
&- A_2[\nu y \text{ ber}' y + 2(1-\nu)(\text{ber } y - 2y^{-1}\text{ bei}' y)] \\
&+ B_1[\nu y \text{ kei}' y + 2(1-\nu)(\text{kei } y + 2y^{-1}\text{ ker}' y)] \\
&- B_2[\nu y \text{ ker}' y + 2(1-\nu)(\text{ker } y - 2y^{-1}\text{ kei}' y)]\}
\end{aligned}
$$

$$
\begin{aligned}
\chi = \frac{2\sqrt{3(1-\nu^2)}}{Et^2}\cot\alpha[&A_1(\text{bei } y + 2y^{-1}\text{ ber}' y) - A_2(\text{ber } y - 2y^{-1}\text{ bei}' y) \\
&+ B_1(\text{kei } y + 2y^{-1}\text{ ker}' y) - B_2(\text{ker } y - 2y^{-1}\text{ kei}' y)]
\end{aligned}
$$

$$
N_s = -Q_s \cot\alpha \qquad (40.115b\text{–}f)
$$

Here the subscript s instead of ϕ has been used, with the stress resultants transmitted across a parallel circle.

In Eqs. (40.115), the A terms are associated with loads applied to the lower edge of the shell, while the singular B terms describe the stresses caused by loads applied to the upper edge of a truncated cone.

When the argument y of the Kelvin functions is large ($y > 10$ or > 12), these functions may be replaced by their asymptotic representations. Equations (40.115) assume then the following form:

$$
Q_s = \frac{\exp(y/\sqrt{2})}{s\sqrt{2\pi y}}\left[A_1\cos\left(\frac{y}{\sqrt{2}} - \frac{\pi}{8}\right) + A_2\sin\left(\frac{y}{\sqrt{2}} - \frac{\pi}{8}\right)\right]
$$

$$
N_\theta = -\frac{\cot\alpha}{2\sqrt{2\pi}}\frac{\sqrt{y}\exp(y/\sqrt{2})}{s}\left[A_1\cos\left(\frac{y}{\sqrt{2}} + \frac{\pi}{8}\right) + A_2\sin\left(\frac{y}{\sqrt{2}} + \frac{\pi}{8}\right)\right]
$$

$$
M_s = \frac{2\exp(y/\sqrt{2})}{y\sqrt{2\pi y}}\left[A_1\sin\left(\frac{y}{\sqrt{2}} + \frac{\pi}{8}\right) - A_2\cos\left(\frac{y}{\sqrt{2}} + \frac{\pi}{8}\right)\right] \qquad (40.116a\text{–}f)
$$

$$
\chi = \frac{2\sqrt{3(1-\nu^2)}}{Et^2}\frac{\exp(y/\sqrt{2})}{\sqrt{2\pi y}}\left[A_1\sin\left(\frac{y}{\sqrt{2}} - \frac{\pi}{8}\right) - A_2\cos\left(\frac{y}{\sqrt{2}} - \frac{\pi}{8}\right)\right]
$$

$$
N_s = -Q_s\cot\alpha \qquad M_\theta = \nu M_s
$$

These formulas are particularly useful for truncated cones of almost cylindrical shape.

Variable Wall Thickness. The differential equations (40.102) for U and χ have been derived from Eqs. (40.98) through (40.100) by eliminating all other unknowns. This process requires repeated differentiation with respect to ϕ, and, in this process, it has been assumed that t is independent of ϕ. If this simplifying assumption is dropped, similar but much more complicated equations are obtained. There exists, however, one case in which the differential equations are rather simple, and admit to a solution in elementary functions. This is the conical shell whose wall thickness t is proportional to the distance from the apex:

$$
t = \delta \cdot s \qquad (40.117)
$$

In this case, the pair of differential equations replacing (40.105a,b) reads as follows:

$$
s^2\frac{d^2U}{ds^2} - \left[2(1-\nu) \pm i\sqrt{1-\nu^2}\sqrt{\frac{12}{\delta}\tan^2\alpha - 1}\right]U = 0 \qquad (40.118a,b)
$$

Since, for a thin shell, $\delta \ll 1$, the term -1 under the radical sign may be neglected unless α is extremely small. If this is done, the substitutions

$$
\xi = \tfrac{1}{2}(9 - 8\nu) \qquad \eta = \frac{4}{\delta}\sqrt{3(1-\nu^2)}\tan\alpha
$$

$$
\kappa = \sqrt{\xi + \sqrt{\xi^2 + \eta^2}} \qquad \mu = \sqrt{-\xi + \sqrt{\xi^2 + \eta^2}}
$$

lead to the general solution

$$U = C_1 s^{\frac{1}{2}(1+\kappa-i\mu)} + C_2 s^{\frac{1}{2}(1-\kappa+i\mu)} + C_3 s^{\frac{1}{2}(1+\kappa+i\mu)} + C_4 s^{\frac{1}{2}(1-\kappa-i\mu)}$$

Since the exponents are fractional, the constants C have awkward dimensions. This may be avoided by using a dimensionless coordinate

$$z = \sqrt{s/l}$$

where l is some conveniently chosen reference length. When, finally, the complex-valued powers of z are separated into real and imaginary parts, the following form of the general solution is obtained:

$$U = Q_s s \cot \alpha = z^{\kappa+1}[A_1 \cos (\mu \ln z) + B_1 \sin (\mu \ln z)] + z^{1-\kappa}[A_2 \cos (\mu \ln z) + B_2 \sin (\mu \ln z)] \qquad (40.119a)$$

From it, the following formulas for the stress resultants and the rotation χ may be derived:

$$\begin{aligned} N_\theta = {} & \frac{z^{\kappa-1}}{2l} \{(d_1A_1 + d_2B_1)[(\kappa + 1) \cos (\mu \ln z) - \mu \sin (\mu \ln z)] \\ & + (d_1B_1 - d_2A_1)[(\kappa + 1) \sin (\mu \ln z) + \mu \cos (\mu \ln z)]\} \\ & - \frac{z^{-\kappa-1}}{2l} \{(d_1A_2 - d_2B_2)[(\kappa - 1) \cos (\mu \ln z) + \mu \sin (\mu \ln z)] \\ & + (d_1B_2 + d_2A_2)[(\kappa - 1) \sin (\mu \ln z) - \mu \cos (\mu \ln z)]\} \\ M_s = {} & \frac{\delta z^{\kappa+1}}{24(1 - \nu^2)} \{(d_1A_1 + d_2B_1)[(\kappa - 3 + 2\nu) \cos (\mu \ln z) - \mu \sin (\mu \ln z)] \\ & + (d_1B_1 - d_2A_1)[(\kappa - 3 + 2\nu) \sin (\mu \ln z) + \mu \cos (\mu \ln z)]\} \\ & - \frac{\delta z^{-\kappa+1}}{24(1 - \nu^2)} \{(d_1A_2 - d_2B_2)[(\kappa + 3 - 2\nu) \cos (\mu \ln z) + \mu \sin (\mu \ln z)] \\ & + (d_1B_2 + d_2A_2)[(\kappa + 3 - 2\nu) \sin (\mu \ln z) - \mu \cos (\mu \ln z)]\} \\ E\, \delta^2 s^2 \chi = {} & -z^{\kappa+1}[(d_1A_1 + d_2B_1) \cos (\mu \ln z) + (d_1B_1 - d_2A_1) \sin (\mu \ln z)] \\ & - z^{-\kappa+1}[(d_1A_2 - d_2B_2) \cos (\mu \ln z) + (d_1B_2 + d_2A_2) \sin (\mu \ln z)] \end{aligned} \qquad (40.119b\text{–}d)$$

Here the abbreviations

$$d_1 = (1 - \nu)\, \delta \cot \alpha \qquad d_2 = 2\sqrt{3(1 - \nu^2)}$$

have been used.

Equations (40.119) cover all stress systems produced by axisymmetric, self-equilibrating edge loads acting on a cone or a truncated cone whose wall thickness varies as required by Eq. (40.117). Since, very often, such stress systems extend only over a limited zone near the loaded edge, it may be possible in special cases to use Eqs. (40.119) as an approximate representation of stresses in a shell of constant wall thickness.

40.7. REFERENCES

[1] A. E. H. Love: "A Treatise on the Mathematical Theory of Elasticity," 4th ed., pp. 499–613, Dover, New York, 1944.

[2] A. E. Green, W. Zerna: "Theoretical Elasticity," pp. 375–437, Oxford Univ. Press, New York, 1954.

[3] W. Flügge: "Statik und Dynamik der Schalen," 2d ed., Springer, Berlin, 1957.

[4] S. P. Timoshenko, S. Woinowsky-Krieger: "Theory of Plates and Shells," 2d ed., pp. 429–568, McGraw-Hill, New York, 1959.

[5] W. Flügge: "Stresses in Shells," Springer, Berlin, 1960.

[6] K. Girkmann: "Flächentragwerke," 5th ed., pp. 352–582, Springer, Vienna, 1959.

CHAPTER 41

BODIES OF REVOLUTION

BY

Y. Y. YU, Ph.D., Brooklyn, N.Y.

41.1. BASIC EQUATIONS

There are two ways of solving elasticity problems. The first consists in determining the stresses which satisfy the equilibrium conditions (33.35), the conditions of compatibility (33.60), and the boundary conditions of the problem. After the stresses have been determined, the strains may be obtained from Eqs. (33.50), and the displacements from the kinematic relations (33.2).

A second approach to elasticity problems is based on the *fundamental equations* (33.56), which, together with the proper boundary conditions, define completely the displacements u, v, w. It is not necessary to consider the compatibility conditions, for the latter are simply to guarantee the continuity and single-valuedness of the displacements. After the displacements are known, the strains may be found from Eqs. (33.2), and the stresses, in turn, from Hooke's law.

For a body of revolution deformed symmetrically with respect to the axis of revolution, it is often convenient to use cylindrical coordinates r, θ, z, with the z axis chosen along the axis of symmetry. Because of axial symmetry, the shearing stresses $\tau_{\theta r}$ and $\tau_{\theta z}$ vanish identically, the other stresses are independent of the angle θ, and all the derivatives with respect to θ vanish. It follows that the equilibrium conditions (33.38) reduce to

$$\frac{\partial \sigma_r}{\partial r} + \frac{\partial \tau_{rz}}{\partial z} + \frac{\sigma_r - \sigma_\theta}{r} = 0 \qquad \frac{\partial \tau_{rz}}{\partial r} + \frac{\partial \sigma_z}{\partial z} + \frac{\tau_{rz}}{r} = 0 \tag{41.1a,b}$$

in which body forces are neglected. The compatibility conditions are:

$$\nabla^2 \sigma_r - \frac{2}{r^2}(\sigma_r - \sigma_\theta) + \frac{1}{1+\nu}\frac{\partial^2 s}{\partial r^2} = 0$$

$$\nabla^2 \sigma_\theta + \frac{2}{r^2}(\sigma_r - \sigma_\theta) + \frac{1}{1+\nu}\frac{1}{r}\frac{\partial s}{\partial r} = 0 \tag{41.2a–d}$$

$$\nabla^2 \sigma_z + \frac{1}{1+\nu}\frac{\partial^2 s}{\partial z^2} = 0 \qquad \nabla^2 \tau_{rz} - \frac{1}{r^2}\tau_{rz} + \frac{1}{1+\nu}\frac{\partial^2 s}{\partial r\,\partial z} = 0$$

where

$$\nabla^2 = \frac{\partial^2}{\partial r^2} + \frac{1}{r}\frac{\partial}{\partial r} + \frac{\partial^2}{\partial z^2} \qquad s = \sigma_r + \sigma_\theta + \sigma_z$$

The nonvanishing displacement components, in this case, are u and w in the directions of r and z, and the nonvanishing strain components are

$$\epsilon_r = \frac{\partial u}{\partial r} \qquad \epsilon_\theta = \frac{u}{r} \qquad \epsilon_z = \frac{\partial w}{\partial z} \qquad \gamma_{rz} = \frac{\partial u}{\partial z} + \frac{\partial w}{\partial r} \tag{41.3}$$

which are also related to the stresses by equations similar to Eqs. (33.50). By means of the strain-displacement relations (41.3) and the stress-strain relations, the stress-equilibrium equations (41.1) are transformed into the following displacement-equilibrium equations for the axially symmetrical case:

$$2(1-\nu)\frac{\partial}{\partial r}\left(\frac{\partial u}{\partial r}+\frac{u}{r}+\frac{\partial w}{\partial z}\right)+(1-2\nu)\frac{\partial}{\partial z}\left(\frac{\partial u}{\partial z}-\frac{\partial w}{\partial r}\right)=0$$
$$2(1-\nu)\frac{\partial}{\partial z}\left(\frac{\partial u}{\partial r}+\frac{u}{r}+\frac{\partial w}{\partial z}\right)-(1-2\nu)\frac{1}{r}\frac{\partial}{\partial r}r\left(\frac{\partial u}{\partial z}-\frac{\partial w}{\partial r}\right)=0 \qquad (41.4a,b)$$

41.2. METHODS OF SOLUTION BY THE USE OF STRESS AND DISPLACEMENT FUNCTIONS

Solutions of problems of elasticity may be obtained by solving directly the basic equations. One such example will be given in Sec. 41.3. More often, solutions are obtained through the construction of certain functions whose derivatives define the stresses or displacements. The former are the stress functions, which are chosen in such a manner that the equilibrium equations are identically satisfied. The stress function is then substituted into the compatibility conditions, to yield the governing equation which the function must satisfy. The use of the stress function is therefore associated with the first approach to elasticity problems described in Sec. 41.1. The best-known example is Airy's stress function, used extensively in two-dimensional problems (see p. 37–3). Stress functions have also been suggested in three-dimensional elasticity, such as the Maxwell functions and Morera functions, but their practical use has been very limited, if there is any.

In three-dimensional elasticity, the second approach described in Sec. 41.1 is practically the only one that has been used effectively in obtaining solutions of problems. In connection with this approach, functions are used whose derivatives define the displacements, and the governing equation is deduced from the displacement-equilibrium equations. These functions may properly be referred to as the *displacement functions.* Apparently owing to the fact that the displacements lead directly to the strains, Westergaard [6a] called these the strain functions. Since stresses may naturally be expressed also in terms of these functions, most authors refer to them also as stress functions, making no distinction between the two.

Numerous methods of introducing the displacement function have been devised by various authors. In what follows, we shall discuss only three of them which are considered more important than others and have been used fruitfully to obtain solutions of practical problems. The other methods are all directly or indirectly related to these three, which are also related to one another among themselves, as will be seen presently. A detailed discussion of the various displacement functions, as well as the stress functions, may be found in a recent paper by Marguerre [10].

Love Function

The Love function ϕ [2e] is useful in the investigation of axially symmetrical stress distributions in bodies of revolution. We shall not consider body forces in the general case, although a particular example solved with the aid of the Love function, which involves inertia force, will be given on p. 41–10. The nonvanishing displacements are in the directions of r and z, and are given in terms of ϕ by

$$u=-\frac{1}{2G}\frac{\partial^2\phi}{\partial r\,\partial z} \qquad w=\frac{1-\nu}{G}\nabla^2\phi-\frac{1}{2G}\frac{\partial^2\phi}{\partial z^2} \qquad (41.5a,b)$$

The displacement-equilibrium equation (41.4a) is then identically satisfied, and Eq. (41.4b) becomes

$$\nabla^2\nabla^2\phi=0 \qquad (41.6)$$

which is the governing equation of the Love function. Equation (41.6) is the biharmonic equation, and ϕ is a biharmonic function.

In terms of ϕ, the stresses are:

$$\sigma_r = \frac{\partial}{\partial z}\left(\nu\,\nabla^2\phi - \frac{\partial^2\phi}{\partial r^2}\right) \qquad \sigma_\theta = \frac{\partial}{\partial z}\left(\nu\,\nabla^2\phi - \frac{1}{r}\frac{\partial\phi}{\partial r}\right)$$

$$\sigma_z = \frac{\partial}{\partial z}\left[(2-\nu)\,\nabla^2\phi - \frac{\partial^2\phi}{\partial z^2}\right] \qquad \tau_{rz} = \frac{\partial}{\partial r}\left[(1-\nu)\,\nabla^2\phi - \frac{\partial^2\phi}{\partial z^2}\right] \tag{41.7a–d}$$

Similar to the displacement-equilibrium equations, the stress-equilibrium equation (41.1*a*) is now identically satisfied by the stresses in Eqs. (41.7), and Eq. (41.1*b*) is transformed into Eq. (41.6). Since the Love function is a displacement function, it is not necessary to check the compatibility conditions. However, it may be verified easily that the compatibility conditions (41.2) in terms of stresses are satisfied by the stresses in Eqs. (41.7) when ϕ satisfies Eq. (41.6).

The problem of axially symmetrical stress distribution now reduces to one of determining a Love function which satisfies Eq. (41.6), with the boundary conditions of the problem also fulfilled at the same time. A number of problems will be solved in the next articles by this method.

Galerkin Vector

The Galerkin vector may be considered as a generalization of the Love function for stress distributions which are not necessarily axially symmetrical. The interpretation of a vector function was due to Papkovitch and Westergaard [6*b*], who expressed the displacements and stresses in terms of the vector function, the components of which are the three displacement functions introduced by Galerkin.

The three displacement-equilibrium equations (33.56) may be combined into the following vector form:

$$G\left(\nabla^2\mathbf{u} + \frac{1}{1-2\nu}\nabla\nabla\cdot\mathbf{u}\right) + \mathbf{X} = 0 \tag{41.8}$$

where

$$\nabla = \mathbf{i}_x\frac{\partial}{\partial x} + \mathbf{i}_y\frac{\partial}{\partial y} + \mathbf{i}_z\frac{\partial}{\partial z}$$

and

$$\mathbf{u} = \mathbf{i}_x u + \mathbf{i}_y v + \mathbf{i}_z w \qquad \mathbf{X} = \mathbf{i}_x X + \mathbf{i}_y Y + \mathbf{i}_z Z$$

are the displacement and body-force vectors, respectively. By virtue of Helmholtz's theorem, we resolve the displacement vector into the lamellar and solenoidal components:

$$\mathbf{u} = \nabla\psi + \nabla\times\mathbf{S} \tag{41.9}$$

where ψ is a scalar function, and $\mathbf{S}$ is a vector function such that $\nabla\cdot\mathbf{S} = 0$. Substitution of Eq. (41.9) in (41.8) yields

$$G\,\nabla^2\left(2\,\frac{1-\nu}{1-2\nu}\nabla\psi + \nabla\times\mathbf{S}\right) + \mathbf{X} = 0 \tag{41.10}$$

Since $\mathbf{S}$ is solenoidal, it may be represented in the form

$$\mathbf{S} = -\frac{1-\nu}{G}\nabla\times\mathbf{F} \tag{41.11}$$

The function ψ is independent of $\mathbf{S}$ and may, therefore, be written as

$$\psi = \frac{1-2\nu}{2G}\nabla\cdot\mathbf{F} \tag{41.12}$$

Substituting Eqs. (41.11) and (41.12) in (41.10), and making use of the identity $\nabla \times \nabla \times \equiv \nabla\nabla \cdot - \nabla^2$, we finally obtain

$$\nabla^2\nabla^2\mathbf{F} = -\frac{\mathbf{X}}{1-\nu} \tag{41.13}$$

which is the governing equation of the Galerkin vector: $\mathbf{F} = \mathbf{i}_x F_x + \mathbf{i}_y F_y + \mathbf{i}_z F_z$. When the body force $\mathbf{X} = 0$, Eq. (41.13) reduces to the homogeneous biharmonic equation

$$\nabla^2\nabla^2\mathbf{F} = 0$$

This derivation of the Galerkin vector is due to Mindlin [11], and has the advantage of providing also a proof of its completeness.

Substitution of Eqs. (41.11) and (41.12) in (41.9) yields the following displacements in terms of the Galerkin vector and its component functions:

$$\begin{aligned} u &= \frac{1-\nu}{G}\nabla^2 F_x - \frac{1}{2G}\frac{\partial}{\partial x}\nabla \cdot \mathbf{F} \\ v &= \frac{1-\nu}{G}\nabla^2 F_y - \frac{1}{2G}\frac{\partial}{\partial y}\nabla \cdot \mathbf{F} \\ w &= \frac{1-\nu}{G}\nabla^2 F_z - \frac{1}{2G}\frac{\partial}{\partial z}\nabla \cdot \mathbf{F} \end{aligned} \tag{41.14a–c}$$

The corresponding stresses are:

$$\begin{aligned} \sigma_x &= 2(1-\nu)\frac{\partial}{\partial x}\nabla^2 F_x + \left(\nu\nabla^2 - \frac{\partial^2}{\partial x^2}\right)\nabla \cdot \mathbf{F} \\ \sigma_y &= 2(1-\nu)\frac{\partial}{\partial y}\nabla^2 F_y + \left(\nu\nabla^2 - \frac{\partial^2}{\partial y^2}\right)\nabla \cdot \mathbf{F} \\ \sigma_z &= 2(1-\nu)\frac{\partial}{\partial z}\nabla^2 F_z + \left(\nu\nabla^2 - \frac{\partial^2}{\partial z^2}\right)\nabla \cdot \mathbf{F} \\ \tau_{xy} &= (1-\nu)\left(\frac{\partial}{\partial y}\nabla^2 F_x + \frac{\partial}{\partial x}\nabla^2 F_y\right) - \frac{\partial^2}{\partial x\,\partial y}\nabla \cdot \mathbf{F} \\ \tau_{yz} &= (1-\nu)\left(\frac{\partial}{\partial z}\nabla^2 F_y + \frac{\partial}{\partial y}\nabla^2 F_z\right) - \frac{\partial^2}{\partial y\,\partial z}\nabla \cdot \mathbf{F} \\ \tau_{zx} &= (1-\nu)\left(\frac{\partial}{\partial x}\nabla^2 F_z + \frac{\partial}{\partial z}\nabla^2 F_x\right) - \frac{\partial^2}{\partial z\,\partial x}\nabla \cdot \mathbf{F} \end{aligned} \tag{41.15a–f}$$

The problem is therefore completely solved if a Galerkin vector which satisfies Eq. (41.13) may be determined such that the boundary conditions of the problem are also satisfied.

For axially symmetrical stress distributions in bodies of revolution, the component function F_z of the Galerkin vector is identical with the Love function ϕ. This was pointed out by Westergaard [6c], and may be verified easily by writing the displacements and stresses in cylindrical coordinates in terms of both F_z and ϕ.

Method of Boussinesq, Papkovitch, and Neuber

It is noted that, in the two preceding methods of Love and Galerkin, the stress expressions involve the third derivatives of the displacement function. This often makes it inconvenient to express the stresses in general curvilinear coordinates. In this respect, the method associated with the names Boussinesq, Papkovitch, and Neuber offers some advantage over the other two, and extensive use of the method has been made by recent authors.

As in the derivation of the Galerkin vector, we start again by substituting Eq. (41.9) into the displacement-equilibrium equation (41.8), which then takes the form of

Eq. (41.10). The quantity in the parenthesis in Eq. (41.10) is a vector, which may be denoted by $\mathbf{B} = \mathbf{i}_x B_x + \mathbf{i}_y B_y + \mathbf{i}_z B_z$. We therefore have

$$2\frac{1-\nu}{1-2\nu}\nabla\psi + \nabla \times \mathbf{S} = \mathbf{B} \qquad G\nabla^2\mathbf{B} + \mathbf{X} = 0 \qquad (41.16a,b)$$

Equation (41.16a), operated on by $\nabla\cdot$, yields

$$2\frac{1-\nu}{1-2\nu}\nabla^2\psi = \nabla\cdot\mathbf{B}$$

the complete solution of which is given by

$$4\frac{1-\nu}{1-2\nu}\psi = \mathbf{r}\cdot\mathbf{B} + \beta \qquad (41.17)$$

where $\mathbf{r} = \mathbf{i}_x x + \mathbf{i}_y y + \mathbf{i}_z z$ is the position vector, and β is a scalar function satisfying

$$G\nabla^2\beta = \mathbf{r}\cdot\mathbf{X}$$

Eliminating $\nabla \times \mathbf{S}$ between Eqs. (41.9) and (41.16a), and substituting Eq. (41.17) into the result, we have, finally,

$$\mathbf{u} = \mathbf{B} - \frac{1}{4(1-\nu)}\nabla(\mathbf{r}\cdot\mathbf{B} + \beta) \qquad (41.18)$$

$$G\nabla^2\mathbf{B} = -\mathbf{X} \qquad G\nabla^2\beta = \mathbf{r}\cdot\mathbf{X} \qquad (41.19)$$

The displacement vector is now in terms of the vector function **B** and the scalar function β, which are the Papkovitch functions, whose Laplacians are known if the body force **X** is given. This proof of completeness of the Papkovitch functions is due to Mindlin [13]. Written in nonvector form, Eq. (41.18) gives the displacement components in rectangular coordinates, as follows:

$$\begin{aligned} u &= B_x - \frac{1}{4(1-\nu)}\frac{\partial}{\partial x}(xB_x + yB_y + zB_z + \beta) \\ v &= B_y - \frac{1}{4(1-\nu)}\frac{\partial}{\partial y}(xB_x + yB_y + zB_z + \beta) \\ w &= B_z - \frac{1}{4(1-\nu)}\frac{\partial}{\partial z}(xB_x + yB_y + zB_z + \beta) \end{aligned} \qquad (41.20a\text{–}c)$$

The stresses may likewise be expressed in terms of the Papkovitch functions. To solve the general problem of elasticity, thus, consists of the determination of the Papkovitch functions which satisfy Eqs. (41.19) and which make the boundary conditions of the problem also satisfied.

The Papkovitch functions were shown by Mindlin [11] to be related to the Galerkin vector, in the following manner:

$$G\mathbf{B} = (1-\nu)\nabla^2\mathbf{F} \qquad G\beta = (1-\nu)(2\nabla\cdot\mathbf{F} - \mathbf{r}\cdot\nabla^2\mathbf{F}) \qquad (41.21a,b)$$

Some authors have associated the functions **B** and β with the name of Boussinesq, but, as was pointed out by Mindlin [13], Boussinesq introduced only B_z and β, and used functions of a different type in place of B_x and B_y.

In the absence of body forces, the Papkovitch functions are harmonic: that is,

$$\nabla^2 B_x = \nabla^2 B_y = \nabla^2 B_z = \nabla^2\beta = 0$$

When these functions are now multiplied by $2(1-\nu)/G$, they assume the form introduced independently by Neuber [3]. In this case, the precise circumstances under which the completeness of the solution (41.18), in terms of the four harmonic functions, is established were further examined by Eubanks and Sternberg [7]. These authors also proved that, for any convex body, any one of the four functions may be taken as zero, without loss in completeness of the solution, if the coordinate system is chosen properly, and if 4ν is not an integer.

For the axially symmetrical case, only two of the four functions are needed. There exists no difference in opinion in associating the solution of this special case with the name of Boussinesq [1]. The following formulation of the Boussinesq solution has been used extensively by Sternberg, Eubanks, and Sadowsky [18]:

$$2G[u,v,w] = \nabla(\Phi + z\Psi) - [0,0,4(1-\nu)\Psi] \qquad \nabla^2\Phi(r,z) = 0 \qquad \nabla^2\Psi(r,z) = 0 \tag{41.22}$$

where u, v, w are now in the directions of the cylindrical coordinates r, θ, z. Clearly, Φ and Ψ (which will hereafter be called the *Boussinesq functions*) differ from β and B_z, respectively, by only a nonessential constant coefficient. In terms of Φ and Ψ, results for the displacements and stresses in general curvilinear coordinates are also available [18]. The completeness of the solution of this case, in terms of two harmonic functions, was established by Eubanks and Sternberg [7].

41.3. DIRECT SOLUTION FOR A HOLLOW SPHERE SUBJECTED TO UNIFORM INTERNAL AND EXTERNAL PRESSURES

The solution to this problem will be obtained by solving directly the displacement-equilibrium equation. Because of symmetry with respect to the center of the sphere,

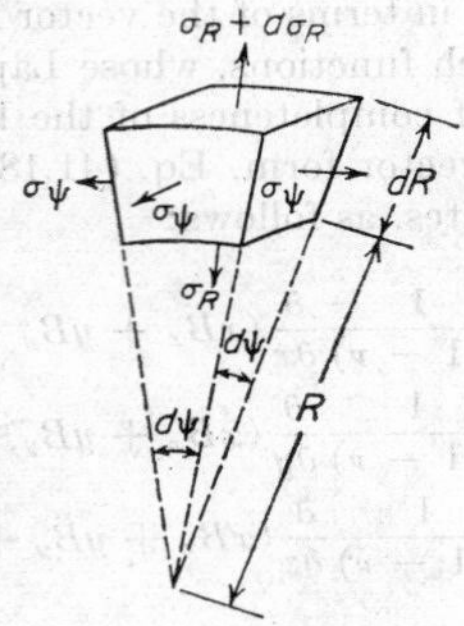

FIG. 41.1

spherical coordinates are used. Consider the element of the sphere shown in Fig. 41.1. The equilibrium of forces on the element in the radial direction requires that

$$\frac{d\sigma_R}{dR} + 2\,\frac{\sigma_R - \sigma_\psi}{R} = 0 \tag{41.23}$$

where σ_R and σ_ψ are normal stresses in the radial and tangential directions. The stress-strain relations are given by Hooke's law:

$$\sigma_R = \frac{E}{1-\nu-2\nu^2}\,[(1-\nu)\epsilon_R + 2\nu\epsilon_\psi] \qquad \sigma_\psi = \frac{E}{1-\nu-2\nu^2}\,(\epsilon_\psi + \nu\epsilon_R) \tag{41.24}$$

In terms of the displacement u in the radial direction, the strains are

$$\epsilon_R = \frac{du}{dR} \qquad \epsilon_\psi = \frac{u}{R} \tag{41.25}$$

Substitution of Eqs. (41.25) into (41.24) and of the result into (41.23) yields the displacement-equilibrium equation,

$$\frac{d^2u}{dR^2} + \frac{2}{R}\frac{du}{dR} - 2\frac{u}{R^2} = 0 \tag{41.26}$$

which may naturally be deduced also from Eq. (41.8).

The general solution of Eq. (41.26) is

$$u = AR + \frac{B}{R^2}$$

from which

$$\sigma_R = \frac{C}{R^3} + D \qquad \sigma_\psi = -\frac{C}{2R^3} + D \tag{41.27}$$

If a and b are the inner and outer radii of the sphere, and p_i and p_o the internal and external pressures, the boundary conditions of the problem are:

$$\text{for } R = a \quad \sigma_R = -p_i \qquad \text{for } R = b \quad \sigma_R = -p_o \tag{41.28}$$

Solving for C and D by virtue of Eqs. (41.28) and substituting the result in Eqs. (41.27), we find

$$\sigma_R = \frac{p_o b^3(R^3 - a^3)}{R^3(a^3 - b^3)} + \frac{p_i a^3(b^3 - R^3)}{R^3(a^3 - b^3)}$$

$$\sigma_\psi = \frac{p_o b^3(2R^3 + a^3)}{2R^3(a^3 - b^3)} - \frac{p_i a^3(2R^3 + b^3)}{2R^3(a^3 - b^3)}$$

These constitute the solution of a thick-walled sphere, which, like that of the thick-walled cylinder, is due to G. Lamé [2a].

41.4. AXIALLY SYMMETRIC STRESS DISTRIBUTIONS IN FINITE-SIZED BODIES

In this section we shall discuss plates, cylinders, and spheres through the use of the Love function and the Boussinesq functions, which are biharmonic and harmonic, respectively. A brief discussion of some useful harmonic and biharmonic functions is therefore given first.

a. Harmonic and Biharmonic Functions

We seek the solutions of the harmonic equation $\nabla^2\phi = 0$ and the biharmonic equation $\nabla^2\nabla^2\phi = 0$ for the axially symmetrical case. In cylindrical coordinates, the

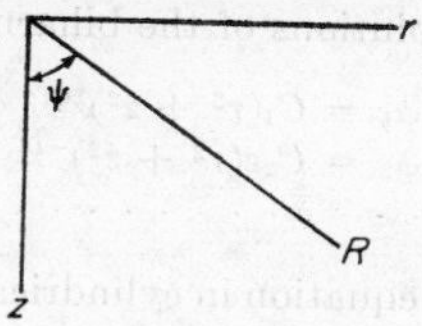

FIG. 41.2

Laplacian in these equations is, therefore, independent of θ, and has the form

$$\nabla^2 = \frac{\partial^2}{\partial r^2} + \frac{1}{r}\frac{\partial}{\partial r} + \frac{\partial^2}{\partial z^2}$$

By means of the transformation (Fig. 41.2)

$$R^2 = r^2 + z^2 \qquad \tan\psi = \frac{r}{z} \tag{41.29}$$

the corresponding form of the Laplacian in spherical coordinates is found to be

$$\nabla^2 = \frac{\partial^2}{\partial R^2} + \frac{2}{R}\frac{\partial}{\partial R} + \frac{1}{R^2}\cot\psi\,\frac{\partial}{\partial\psi} + \frac{1}{R^2}\frac{\partial^2}{\partial\psi^2}$$

Any harmonic function is obviously also a biharmonic function. Furthermore, an additional biharmonic function may be derived by multiplying the harmonic function by $R^2 = r^2 + z^2$.

A solution of the harmonic equation in spherical coordinates has the form

$$\phi_n = A_n R^n P_n(x) \tag{41.30}$$

with

$$x = \cos \psi \tag{41.31}$$

where A_n is a constant coefficient, $P_n(x)$ the Legendre polynomial of degree n, and n a positive or negative integer. For $n = 0, 1, 2, \ldots$, we have

$$\begin{aligned} \phi_0 &= B_0 \\ \phi_1 &= B_1 z \\ \phi_2 &= B_2[z^2 - \tfrac{1}{3}(r^2 + z^2)] \\ &\cdots\cdots\cdots \end{aligned} \tag{41.32}$$

where the functions are now written in cylindrical coordinates by the use of Eqs. (41.29) and (41.31), and B_n are constants. Equations (41.32) are also solutions of the biharmonic equation. By multiplying Eqs. (41.32) by $R^2 = r^2 + z^2$, the following additional biharmonic functions are obtained:

$$\begin{aligned} \phi_2 &= C_2(r^2 + z^2) \\ \phi_3 &= C_3 z(r^2 + z^2) \\ &\cdots\cdots\cdots \end{aligned} \tag{41.33}$$

Another solution of the harmonic equation in spherical coordinates is of the form

$$\phi_n = A_n R^n Q_n(x) \tag{41.34}$$

where $Q_n(x)$ is the Legendre function of the second kind of degree n and, unlike $P_n(x)$, is no longer a polynomial. For $n = -1, -2, \ldots$, we have

$$\begin{aligned} \phi_1 &= B_1(r^2 + z^2)^{-1/2} \\ \phi_2 &= B_2 z(r^2 + z^2)^{-3/2} \\ \phi_3 &= B_3[z^2(r^2 + z^2)^{-5/2} - \tfrac{1}{3}(r^2 + z^2)^{-3/2}] \\ &\cdots\cdots\cdots \end{aligned} \tag{41.35}$$

which are also solutions of the biharmonic equation. By multiplying Eqs. (41.35) by $R^2 = r^2 + z^2$, another series of solutions of the biharmonic equation is obtained:

$$\begin{aligned} \phi_1 &= C_1(r^2 + z^2)^{1/2} \\ \phi_2 &= C_2 z(r^2 + z^2)^{-1/2} \\ &\cdots\cdots\cdots \end{aligned} \tag{41.36}$$

The solution of the harmonic equation in cylindrical coordinates may be in terms of the Bessel function. Thus, it may readily be shown that the equation has the solutions

$$\phi = \frac{\sin}{\cos} kz\,[a_1 I_0(kr) + a_2 K_0(kr)] \tag{41.37}$$

where either sin or cos may be used, k is an arbitrary real number, and $I_0(kr)$ and $K_0(kr)$ are modified Bessel's functions of zero order and of the first and second kinds, respectively. These are, of course, also solutions of the biharmonic equation. In addition, the latter has the following solutions:

$$\phi = \frac{\sin}{\cos} kz\,[a_3 kr I_1(kr) + a_4 kr K_1(kr)] \tag{41.38}$$

where $I_1(kr)$ and $K_1(kr)$ are now of the first order.

Any of the above harmonic and biharmonic functions or their combinations may be taken as a Love function. Similarly, any of the harmonic functions or their combinations may be used as Boussinesq functions. Examples will be given presently.

b. Bending of Circular Plates

By means of the results given in Eqs. (41.32) and (41.33), several problems of axially symmetrical bending of circular plates may be solved. Thus, we take, as a Love function, the polynomial of the third degree

$$\phi = a_3(2z^3 - 3r^2z) + b_3(r^2z + z^3)$$

and obtain, according to Eqs. (41.7),

$$\sigma_r = 6a_3 + (10\nu - 2)b_3 \qquad \sigma_\theta = 6a_3 + (10\nu - 2)b_3$$
$$\sigma_z = -12a_3 + (14 - 10\nu)b_3 \qquad \tau_{rz} = 0$$

The stresses are, therefore, constant throughout the plate. By adjusting the constants a_3 and b_3, a solution may be obtained for any prescribed constant values of the stresses σ_z and σ_r at the surface of the plate.

If, in the Love function,

$$\phi = a_4(8z^4 - 24r^2z^2 + 3r^4) + b_4(2z^4 + r^2z^2 - r^4)$$

which consists of polynomials of the fourth degree, the constants a_4 and b_4 are chosen so that $24a_4 - b_4(8 - 7\nu) = 0$, we then find the corresponding stresses,

$$\sigma_z = \tau_{rz} = 0 \qquad \sigma_r = 28(1 + \nu)b_4z \tag{41.39}$$

This is the solution of pure bending of the plate by a uniform moment along the boundary, z being the distance from the middle plane of the plate.

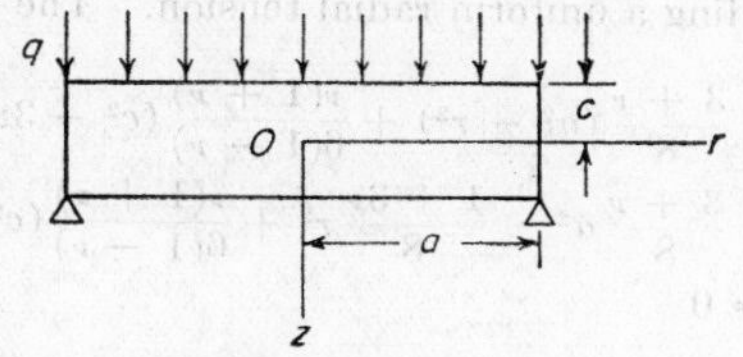

FIG. 41.3

The problem of a uniformly loaded circular plate (Fig. 41.3) may similarly be solved by means of the Love function:

$$\phi = a_3(2z^3 - 3r^2z) + b_3(r^2z + z^3) + a_4(8z^4 - 24r^2z^2 + 3r^4)$$
$$+ a_6(16z^6 - 120z^4r^2 + 90z^2r^4 - 5r^6) + b_6(8z^6 - 16z^4r^2 - 21z^2r^4 + 3r^6)$$

In conjunction with the use of the results in Eqs. (41.39), and by adjusting the constants so that the boundary conditions are satisfied, this yields the following stresses [5a]:

$$\sigma_r = q\left[\frac{2+\nu}{8}\frac{z^3}{c^3} - \frac{3(3+\nu)}{32}\frac{r^2z}{c^3} - \frac{3}{8}\frac{2+\nu}{5}\frac{z}{c} + \frac{3(3+\nu)}{32}\frac{a^2z}{c^3}\right]$$

$$\sigma_z = q\left(-\frac{z^3}{4c^3} + \frac{3}{4}\frac{z}{c} - \frac{1}{2}\right) \qquad \tau_{rz} = \frac{3}{8}\frac{qr}{c^3}(c^2 - z^2)$$

They satisfy exactly the conditions at $z = \pm c$, and yield zero force and moment at $r = a$.

By taking polynomials of a higher degree than the sixth as the Love function, solutions for bending of circular plates by nonuniformly distributed loads may be obtained. Biharmonic functions given in Eqs. (41.35) and (41.36) may be used to investigate circular plates with holes at the center. All the solutions considered here are for small deflections of the plate. When deflections are large compared with the plate thickness, the stretching of the middle plane of the plate must be considered.

c. The Rotating Disk as a Three-dimensional Problem

When the centrifugal force is present, Eq. (41.1*a*) has to include an additional term $\rho\omega^2 r$ on its left-hand side, where ρ is the mass per unit volume, and ω the angular velocity of the disk. Similarly, a term $2\rho\omega^2/(1-\nu)$ is to be added to the left-hand sides of Eqs. (41.2*a*,*b*), and a term $2\nu\rho\omega^2/(1-\nu)$ to the left-hand side of Eq. (41.2*c*). The system of Eqs. (41.1) and (41.2) thus becomes nonhomogeneous, and has the particular integral

$$\begin{aligned} \sigma_r &= -\frac{\rho\omega^2}{3}r^2 - \frac{\rho\omega^2(1+2\nu)(1+\nu)}{6\nu(1-\nu)}z^2 \qquad \sigma_z = \frac{\rho\omega^2(1+3\nu)}{6\nu}r^2 \\ \sigma_\theta &= -\frac{\rho\omega^2(1+2\nu)(1+\nu)}{6\nu(1-\nu)}z^2 \qquad \tau_{rz} = 0 \end{aligned} \tag{41.40}$$

To the stresses given in Eqs. (41.40) are to be added some other stresses derived from a Love function.

For the problem of a rotating circular disk of constant thickness, the Love function needed consists of the fifth-degree polynomials given in Eqs. (41.32) and (41.33):

$$\phi = a_5(8z^5 - 40r^2z^3 + 15r^4z) + b_5(2z^5 - r^2z^3 - 3r^4z)$$

The constants a_5 and b_5 are chosen in such a manner that, when the stresses σ_z and τ_{rz} calculated from ϕ are added to those given in Eqs. (41.40), their resultants vanish. The resultant radial compression over the thickness of the plate at the boundary $r = a$ is next eliminated by adding a uniform radial tension. The final stresses are:

$$\begin{aligned} \sigma_r &= \rho\omega^2\left[\frac{3+\nu}{8}(a^2 - r^2) + \frac{\nu(1+\nu)}{6(1-\nu)}(c^2 - 3z^2)\right] \\ \sigma_\theta &= \rho\omega^2\left[\frac{3+\nu}{8}a^2 - \frac{1+3\nu}{8}r^2 + \frac{\nu(1+\nu)}{6(1-\nu)}(c^2 - 3z^2)\right] \\ \sigma_z &= \tau_{rz} = 0 \end{aligned}$$

The terms involving the factor $(c^2 - 3z^2)$ are not present in the solution when the problem is solved as a two-dimensional one. For thin disks, these terms are small. The nonzero radial stress at the edge should affect only the part of the disk near the edge, according to Saint-Venant's principle.

The present three-dimensional solution of a rotating disk was given by Timoshenko [5*b*].

d. Circular Cylinders of Finite Length

The biharmonic functions in Eqs. (41.37) and (41.38) may be used as Love functions for the investigation of axially symmetrical stress distributions in circular cylinders. For solid cylinders, we need only the modified Bessel functions of the first kind:

$$\phi = \frac{\sin}{\cos} kz\,[a_1 I_0(kr) + a_3 kr I_1(kr)] \tag{41.41}$$

According to Eqs. (41.7), the stresses σ_r and τ_{rz} are

$$\sigma_r = \frac{\cos}{\sin} kz\,[a_1F_1(r) + a_3F_2(r)] \qquad \tau_{rz} = \frac{\sin}{\cos} kz\,[a_1F_3(r) + a_3F_4(r)] \tag{41.42}$$

where $F_1(r), \ldots, F_4(r)$ are functions of r involving $I_0(kr)$ and $I_1(kr)$. For a cylinder of radius a, the values of these stresses at the curved boundary surface $r = a$ must be equal to the external forces acting on the same surface. For instance, when the external force consists of a normal pressure distribution represented by $A_n \cos (n\pi z/l)$, where n is an integer, and l the length of the cylinder, we use the sine term in Eq. (41.41) and obtain, according to Eqs. (41.42),

$$k = \frac{n\pi}{l}$$

$$a_1F_1(a) + a_3F_2(a) = -A_n$$
$$a_1F_3(a) + a_3F_4(a) = 0$$

The last two equations may then be solved for a_1 and a_3. With k, a_1, and a_3 substituted in Eq. (41.41), the Love function is determined, and all the stresses may be calculated. In the general case, both normal and shearing forces may act on the curved surface of the cylinder. The forces may be represented by Fourier series of the form

$$\sum_{n=1}^{\infty} A_n \cos \frac{n\pi z}{l} + \sum_{n=1}^{\infty} B_n \sin \frac{n\pi z}{l}$$

and each term of the series may be treated in the above manner. The solution will give nonzero stresses over the ends of the cylinder, the resultant of which may be made to vanish by adding a simple tension or compression in the axial direction. This, of course, affects only the parts of the cylinder near the ends, according to Saint-Venant's principle. Several examples of symmetrical stress distribution of cylinders were discussed by L. N. G. Filon [2f].

e. A General Solution for the Hollow Sphere

The present solution for the axially symmetrical stress distribution in a hollow sphere of arbitrary uniform thickness is due to Sternberg, Eubanks, and Sadowsky [19]. No body forces are considered, and the solution is based on the Boussinesq approach formulated in Eqs. (41.22), but referred to spherical coordinates. In terms of the Boussinesq functions $\Phi(R,\psi)$ and $\Psi(R,\psi)$, the stresses σ_R and $\tau_{R\psi}$ in spherical coordinates are:

$$\begin{aligned} \sigma_R &= \frac{\partial^2\Phi}{\partial R^2} + Rx\frac{\partial^2\Psi}{\partial R^2} - 2(1-\nu)x\frac{\partial\Psi}{\partial R} - 2\nu\frac{\bar{x}^2}{R}\frac{\partial\Psi}{\partial x} \\ \tau_{R\psi} &= \frac{\bar{x}}{R}\left(\frac{1}{R}\frac{\partial\Phi}{\partial x} - \frac{\partial^2\Phi}{\partial R\,\partial x}\right) - \bar{x}\left[x\frac{\partial^2\Psi}{\partial R\,\partial x} - (1-2\nu)\frac{\partial\Psi}{\partial R} - 2(1-\nu)\frac{1}{R}\frac{\partial\Psi}{\partial x}\right] \end{aligned} \tag{41.43}$$

where $\qquad x = \cos\psi \qquad \bar{x} = \sin\psi$

The harmonic functions in Eqs. (41.30) and (41.34) may be used as Boussinesq functions. The functions $Q_n(x)$ in Eq. (41.34), however, have logarithmic singularities at $x = \pm 1$ and give displacements and stresses which are singular along the z axis. Therefore, only the functions in Eq. (41.30) may be used. The first set of Boussinesq functions to be used in the solution consists of

$$\Phi_n = R^{-n-1}P_n(x) \qquad \Psi_n = 0 \qquad n = 0, \pm 1, \pm 2, \ldots \tag{41.44}$$

from which, according to Eqs. (41.43),

$$\sigma_R = (n+1)(n+2)\frac{P_n(x)}{R^{n+3}} \qquad \tau_{R\psi} = (n+2)\bar{x}\frac{P_n'(x)}{R^{n+3}} \tag{41.45}$$

The second set includes

$$\Phi_n = -(n + 4 - 4\nu)R^{-n}P_{n-1}(x) \qquad n = 0, \pm 1, \pm 2, \ldots \tag{41.46}$$
$$\Psi_n = (2n + 1)R^{-n-1}P_n(x)$$

from which

$$\sigma_R = (n + 1)[(n + 1)(n + 4) - 2\nu] \frac{P_{n+1}(x)}{R^{n+2}}$$
$$\tau_{R\psi} = (n^2 + 2n - 1 + 2\nu)\bar{x} \frac{P'_{n+1}(x)}{R^{n+2}} \tag{41.47}$$

Assume now that the boundary stresses on the two spherical surfaces are given by

$$\sigma_R(R_k,x) = f_k(x) \qquad \tau_{R\psi}(R_k,x) = g_k(x) \qquad k = 1, 2 \quad -1 \leqq x \leqq 1 \tag{41.48}$$

where R_1 and R_2 are the radii of the spherical surfaces, and $0 < R_1 < R_2$. These stresses may be expressed in the following Fourier-Legendre series:

$$f_k(x) = \sum_{n=0}^{\infty} \xi_n^{(k)} P_n(x) \qquad g_k(x) = \bar{x} \sum_{n=1}^{\infty} \eta_n^{(k)} P'_n(x) \qquad k = 1, 2 \quad -1 \leqq x \leqq 1 \tag{41.49}$$

where the coefficients are given by

$$\xi_n^{(k)} = \frac{2n + 1}{2} \int_{-1}^{1} f_k(x)P_n(x)\,dx \qquad n = 0, 1, 2, \ldots$$
$$\eta_n^{(k)} = \frac{2n + 1}{2n(n + 1)} \int_{-1}^{1} \bar{x}g_k(x)P'_n(x)\,dx \qquad n = 1, 2, \ldots \tag{41.50}$$

From the equilibrium of forces in the z direction, it may further be shown that

$$[\xi_1^{(1)} - 2\eta_1^{(1)}]R_1^2 = [\xi_1^{(2)} - 2\eta_1^{(2)}]R_2^2 \tag{41.51}$$

If the solution given in Eqs. (41.44) is denoted symbolically by $[A_n]$, and that in (41.46) by $[B_n]$, then the desired general solution $[S]$ may be taken to be

$$[S] = \sum_{n=-\infty}^{\infty} \{a_n[A_n] + b_n[B_n]\}$$

where a_n and b_n may be determined from Eqs. (41.45) (41.47), and (41.48) to (41.51). Complete expressions for a_n and b_n in terms of the radii R_1, R_2 and the coefficients $\xi_n^{(k)}$, $\eta_n^{(k)}$ are lengthy and may be found in the paper of Sternberg, Eubanks, and Sadowsky [19]. When the Boussinesq functions are thus determined, all displacements and stresses may be calculated.

The problem in which boundary displacements are given instead of stresses may be solved similarly. The technique has also been applied to the investigation of the simpler case of a solid sphere [19]. Calculations were made by Sternberg and Rosenthal [21] for a solid sphere subjected to uniform radial pressure distributions over two spherical caps surrounding the poles of the sphere. The case of a spherical cavity in an infinite body has been treated in a similar manner [19], and a problem of this kind will be solved in Art. 41.7*a* by using another method.

41.5. INTEGRAL SOLUTIONS FOR BODIES EXTENDING TO INFINITY IN CERTAIN DIRECTIONS

For problems involving bodies extending to infinity, the solution may be in integral form. In Art. 41.5*a*, a general solution in integral form is obtained for axially symmetrical stress distributions through the use of the Hankel transform. The solution is then applied to a semi-infinite body in the same article and to a large thick plate in

Art. 41.5*b*. Finally, by the use of a different method, an integral solution is given in Art. 41.5*c* for an infinitely long circular cylinder.

a. Semi-infinite Body Subjected to Surface Loading

When cylindrical coordinates r, z are used, and r ranges from zero to infinity, a general solution of the axially symmetrical problem, as was formulated in terms of the Love function in Eqs. (41.5) to (41.7), may be obtained by means of the Hankel transform. Thus, multiplying the biharmonic equation (41.6) by $rJ_0(kr)$, integrating with respect to r from zero to infinity, and making use of the formula

$$\int_0^\infty r\,\nabla^2\phi\,J_0(kr)\,dr = \left(\frac{d^2}{dz^2} - k^2\right)\bar{\phi} \tag{41.52}$$

we find

$$\left(\frac{d^2}{dz^2} - k^2\right)^2 \bar{\phi}(k,z) = 0 \tag{41.53}$$

In Eqs. (41.52) and (41.53), $\bar{\phi}(k,z)$ is the zero-order Hankel transform,

$$\bar{\phi}(k,z) = \int_0^\infty r\phi(r,z)J_0(kr)\,dr$$

of the Love function $\phi(r,z)$, and Eq. (41.52) gives the zero-order Hankel transform of

$$\nabla^2\phi \equiv \frac{\partial^2\phi}{\partial r^2} + \frac{1}{r}\frac{\partial\phi}{\partial r} + \frac{\partial^2\phi}{\partial z^2}$$

which is obtained by integration by parts, and is valid provided that $r\phi$ and $r\,\partial\phi/\partial r$ tend to zero as r approaches either zero or infinity. The solution of the ordinary differential equation (41.53) is clearly

$$\bar{\phi}(k,z) = (A + Bz)e^{-kz} + (C + Dz)e^{kz} \tag{41.54}$$

where the coefficients A, B, C, and D are functions of k, to be determined by the boundary conditions.

The stresses and displacements are next expressed in integral form in terms of $\bar{\phi}(k,z)$. For instance, multiplying σ_z and τ_{rz} in Eqs. (41.7), respectively, by $rJ_0(kr)$ and $rJ_1(kr)$; integrating over r; and making use of the inversion formula

$$f(r) = \int_0^\infty \bar{f}(k)kJ_n(rk)\,dk$$

where $\bar{f}(k)$ is the Hankel transform of order n of any function $f(r)$; we find

$$\begin{aligned}\sigma_z &= \int_0^\infty k\left[(1-\nu)\frac{\partial^3\bar{\phi}}{\partial z^3} - (2-\nu)k^2\frac{\partial\bar{\phi}}{\partial z}\right]J_0(kr)\,dk\\ \tau_{rz} &= \int_0^\infty k^2\left[\nu\frac{\partial^2\bar{\phi}}{dz^2} + (1-\nu)k^2\bar{\phi}\right]J_1(kr)\,dk\end{aligned} \tag{41.55}$$

The other stresses in Eqs. (41.7) and the displacements in Eqs. (41.5) may be represented in terms of $\bar{\phi}$ and its derivatives by similar expressions. The problem is, therefore, formally solved.

Consider a semi-infinite body bounded by the plane $z = 0$, with the z axis pointing into the body. The surface $z = 0$ is subjected to a normal pressure distribution $p(r)$ symmetrical with respect to the origin, but no shearing force is acting on the surface. The boundary conditions are, thus,

$$\text{at } z = 0: \qquad \sigma_z = -p(r) \qquad \tau_{rz} = 0 \tag{41.56}$$

Since displacements and stresses must tend to zero as r and z increase indefinitely, only the first term on the right side of Eq. (41.54) may be used:

$$\bar{\phi}(k,z) = (A + Bz)e^{-kz} \tag{41.57}$$

We substitute Eq. (41.57) into Eqs. (41.55), and make use of the results in evaluating the following integrals:

$$\int_0^\infty r\sigma_z J_0(kr)\,dr = k^2 e^{-kz}[\nu kA + (1 - 2\nu + \nu kz)B]$$

$$\int_0^\infty r\tau_{rz} J_1(kr)\,dr = k^2 e^{-kz}[kA + (kz - 2\nu)B]$$

which, on putting $z = 0$, become, on account of Eq. (41.56),

$$\frac{1}{k^3} Z(k) = \nu kA + (1 - 2\nu)B \qquad 0 = kA - 2\nu B \tag{41.58}$$

where

$$Z(k) = -k \int_0^\infty rp(r)J_0(kr)\,dr$$

Solving for A and B from Eqs. (41.58), and substituting the result in Eq. (41.57), we find

$$\bar{\phi}(k,z) = \frac{4\nu}{k^4} Z(k)\left(1 + \frac{1}{2\nu} kz\right)e^{-kz} \tag{41.59}$$

which may now be substituted into Eqs. (41.55) and other similar expressions for the stresses and displacements. The solution was first given by Terezawa [24], and the present derivation is due to Sneddon [4a].

For the special case of a total normal force P, distributed uniformly over a circular area of radius a in the plane $z = 0$, we have

$$p(r) = \begin{cases} P/\pi a^2 & 0 \leqq r \leqq a \\ 0 & r > a \end{cases}$$

and

$$Z(k) = -k \int_0^a r\frac{P}{\pi a^2} J_0(kr)\,dr = -\frac{P}{\pi a} J_1(ka) \tag{41.60}$$

Evaluation of the integrals involved in Eqs. (41.55) and other similar expressions in this case is not simple; details of integration may be found in the work of Terezawa [24].

When the force P is concentrated, we use a limit process, in which the radius a of the circle is made to approach zero; thus, Eq. (41.60) becomes

$$Z(k) = -\frac{P}{\pi} \lim_{a\to 0} \frac{J_1(ka)}{a} = -\frac{Pk}{2\pi}$$

With this expression of $Z(k)$ substituted in Eq. (41.59) and the result in Eqs. (41.55) and other similar expressions, we find

$$\begin{aligned}
\sigma_z &= -\frac{3P}{2\pi} z^3(r^2 + z^2)^{-5/2} \qquad \tau_{rz} = -\frac{3P}{2\pi} rz^2(r^2 + z^2)^{-5/2} \\
\sigma_r &= \frac{P}{2\pi}\{(1 - 2\nu)[r^{-2} - zr^{-2}(r^2 + z^2)^{-1/2}] - 3r^2z(r^2 + z^2)^{-5/2}\} \\
\sigma_\theta &= \frac{P}{2\pi}(1 - 2\nu)[-r^{-2} + zr^{-2}(r^2 + z^2)^{-1/2} + z(r^2 + z^2)^{-3/2}] \\
u &= \frac{P}{2\pi Er}(1 - 2\nu)(1 + \nu)\left[z(r^2 + z^2)^{-1/2} - 1 + \frac{1}{1 - 2\nu} r^2z(r^2 + z^2)^{-3/2}\right] \\
w &= \frac{P}{2\pi E}[(1 + \nu)z^2(r^2 + z^2)^{-3/2} + 2(1 - \nu^2)(r^2 + z^2)^{-1/2}]
\end{aligned} \tag{41.61}$$

which constitute the solution of the important problem of Boussinesq [1].

When the concentrated force at the boundary of the semi-infinite body acts in a direction parallel to the boundary, the problem is no longer an axially symmetrical one, and its solution was given by V. Cerruti [2*d*]. Cerruti's problem, as well as Boussinesq's, is a special case of the more general problem of a concentrated force acting at an interior point of a semi-infinite body, which was solved by Mindlin [12], and will be discussed in Art. 41.6*b*.

Integral solutions were given by Yu [24] for a semi-infinite body, the boundary plane of which is subjected to torsional shear loading, including the case of a concentrated torque.

b. Large Thick Plate Subjected to Surface Loading

The above treatment of a semi-infinite body may be extended to the case of a plate of arbitrary thickness bounded by two infinitely large parallel planes. Both terms on the right-hand side of Eq. (41.54) must now be used. For a plate with the boundary planes subjected to normal pressure distributions, the coefficients A, B, C, and D in Eq. (41.54) were determined by Sneddon [4*b*], who then discussed the problem of a plate loaded symmetrically by normal pressures within circular areas, including the special case of a pair of equal and opposite concentrated forces.

Integral solutions were obtained by Yu [24] for a large thick plate, the two boundary planes of which are subjected to torsional shear loading, including the case of a pair of equal and opposite concentrated torques.

c. Infinitely Long Circular Cylinder

We consider the case of a cylinder of radius a subjected to uniform pressure p on one half ($z > 0$) of the curved boundary ($r = a$) and $-p$ on the other half ($z < 0$). Guided by Eqs. (41.37) and (41.38), we formulate the following integral:

$$\phi = \int_0^\infty [AI_0(kr) + BkrI_1(kr)] \cos kz\, dk \tag{41.62}$$

which obviously still satisfies the biharmonic equation and may, therefore, be used as a Love function. The coefficients A and B in the integrand are functions of k, to be determined by the boundary conditions. Substituting ϕ from Eq. (41.62) in (41.7), we find, among the stresses,

$$\begin{aligned} \sigma_r &= \int_0^\infty \left\{ [A + B(1 - 2\nu)]I_0(kr) - \left(\frac{A}{kr} - Bkr\right) I_1(kr) \right\} k^3 \sin kz\, dk \\ \tau_{rz} &= \int_0^\infty \{AI_0'(kr) + B[krI_1'(kr) + I_1(kr) + 2(1-\nu)I_0'(kr)]\} k^3 \cos kz\, dk \end{aligned} \tag{41.63}$$

where primes denote differentiation with respect to kr. By virtue of the definite integral

$$\frac{2p}{\pi} \int_0^\infty \frac{\sin kz}{k}\, dk = \begin{cases} p & z > 0 \\ 0 & z = 0 \\ -p & z < 0 \end{cases}$$

the first of the two boundary conditions at $r = a$

$$\begin{aligned} \sigma_r &= \begin{cases} p & z > 0 \\ -p & z < 0 \end{cases} \\ \tau_{rz} &= 0 \end{aligned} \tag{41.64}$$

may be rewritten as

$$\sigma_r = \frac{2p}{\pi} \int_0^\infty \frac{\sin kz}{k}\, dk \tag{41.65}$$

For the stresses in Eqs. (41.63) to satisfy the boundary conditions (41.64) and (41.65), we must have

$$AI_0'(ka) + B[kaI_1'(ka) + I_1(ka) + 2(1-\nu)I_0'(ka)] = 0$$
$$\left\{[A + B(1-2\nu)]I_0(ka) - \left(A\frac{1}{ka} - Bka\right)I_1(ka)\right\}k^3 = \frac{2p}{\pi k}$$

from which A and B may be calculated. The Love function is thus completely determined, and displacements and stresses may be computed. The present solution was due to A. W. Rankin [5*d*]. By superposition, Rankin further solved the case of a cylinder compressed by a short band of uniform pressure, which simulates the problem of a short collar shrunk on a much longer shaft. Numerical results of the latter problem were also obtained by M. V. Barton [5*d*] by a different method using Fourier series, and another problem similar to the one considered here was solved by C. J. Tranter and J. W. Craggs [4*c*] by means of the Fourier transform.

41.6. PROBLEMS INVOLVING CONCENTRATED FORCES: NUCLEI OF STRAIN

Here we introduce an approach to problems involving concentrated forces, through the use of the nuclei of strain. Mathematically these are singularities which may be located either inside or outside the body under consideration.

a. Concentrated Force at a Point in an Infinite Body

The solution of this important problem associated with the name of Kelvin [2*b*] may be represented by the Love function

$$\phi = C_1(r^2 + z^2)^{1/2} \tag{41.66}$$

which was given in Eq. (41.36), and from which

$$\sigma_r = C_1[(1-2\nu)z(r^2+z^2)^{-3/2} - 3r^2z(r^2+z^2)^{-5/2}]$$
$$\sigma_\theta = C_1(1-2\nu)z(r^2+z^2)^{-3/2}$$
$$\sigma_z = -C_1[(1-2\nu)z(r^2+z^2)^{-3/2} + 3z^3(r^2+z^2)^{-5/2}]$$
$$\tau_{rz} = -C_1[(1-2\nu)r(r^2+z^2)^{-3/2} + 3rz^2(r^2+z^2)^{-5/2}]$$

All these stresses become infinite at the origin, where the concentrated force P is acting in the positive direction of the z axis. The infinite stresses may be avoided if a very small spherical portion of the body surrounding the origin is excluded from consideration. By equating to P the resultant of the forces acting on this spherical surface in the z direction, we find the value of the constant C_1:

$$C_1 = \frac{P}{8\pi(1-\nu)} \tag{41.67}$$

Kelvin's solution in Eq. (41.66), with C_1 given by Eq. (41.67), represents a nucleus of strain. From this and by methods of synthesis and superposition, an unlimited number of other strain nuclei may be obtained. For instance, we may immediately deduce the case of a double force, which consists of a pair of equal and opposite forces acting at points infinitely close to each other. By superposing three mutually perpendicular double forces, we further arrive at the so-called center of compression. From the nuclei of strain, solutions of practically important problems may also be obtained. Thus, the addition of a uniform tension or pressure to the center of compression leads to the solution of the problem of a hollow sphere, discussed in Sec. 41.3. By combining a line of such centers of compression with the Kelvin solution, we may also obtain the solution of Boussinesq's problem given in Art. 41.5*a*. A detailed discussion of the various types of strain nuclei may be found in Love's book [2*c*].

b. Concentrated Force at an Interior Point of a Semi-infinite Body

This problem was solved in two parts by Mindlin [12], through synthesis of nuclei of strain in terms of Galerkin vectors. The semi-infinite body is bounded by the plane $z = 0$, with the z axis pointing into the body. In the first part of the problem a force P acts in the positive z direction normal to the boundary plane, and is applied at a point on the z axis at a distance c from the plane. The required Galerkin vector is also in the z direction and is given by

$$\mathbf{F} = \mathbf{i}_z \frac{P}{8\pi(1-\nu)} \left\{ R_1 + (3-4\nu)R_2 - 2c(z+c)\frac{1}{R_2} - 4(1-2\nu)c \ln (R_2 + z + c) \right.$$
$$\left. + 4(1-\nu)(1-2\nu)[(z+c) \ln (R_2 + z + c) - R_2] + \frac{2c^2}{R_2} \right\} \tag{41.68}$$

which is compounded of six nuclei of strain, and in which

$$R_1 = [r^2 + (z-c)^2]^{1/2} \qquad R_2 = [r^2 + (z+c)^2]^{1/2} \tag{41.69}$$

Since the coefficient of the single symmetrical component of a Galerkin vector in the z direction, such as that in Eq. (41.68), is identical with the Love function, complete results for the displacements and stresses may be derived with the aid of Eqs. (41.5) and (41.7), and were given by Mindlin [12]. When c becomes 0 or ∞, the solution reduces to Boussinesq's and Kelvin's solutions, respectively.

In the second part of Mindlin's problem, the force P at the same point on the z axis is parallel to the boundary. The x axis is chosen in the direction of the force. The Galerkin vector now has a component in the x as well as the z direction:

$$\mathbf{F} = \mathbf{i}_x \frac{P}{8\pi(1-\nu)} \left\{ R_1 + R_2 - \frac{2c^2}{R_2} + 4(1-\nu)(1-2\nu)[(z+c) \ln (R_2 + z + c) - R_2] \right\}$$
$$+ \mathbf{i}_z \frac{P}{8\pi(1-\nu)} \left[\frac{2cx}{R_2} + 2(1-2\nu)x \ln (R_2 + z + c) \right] \tag{41.70}$$

which also consists of six nuclei of strain. Complete expressions of the displacements and stresses in this case may be derived according to Eqs. (41.14) and (41.15), and were given by Mindlin [12]. When c becomes ∞, Eq. (41.70) reduces to Kelvin's solution; when c becomes 0, it reduces to Cerruti's solution, for a force applied tangentially at the boundary of a semi-infinite body.

More recently, Mindlin [13] showed that the above two cases may also be solved directly by means of an application of potential theory. The solutions thus obtained were in the form of Papkovitch functions instead of Galerkin vectors. For the case of a force normal to the boundary, the Papkovitch functions were found to be

$$B_z = \frac{P}{4\pi\nu} \left[\frac{1}{R_1} + \frac{3-4\nu}{R_2} + \frac{2c(z+c)}{R_2{}^3} \right]$$
$$\beta = \frac{P}{4\pi\nu} \left[4(1-\nu)(1-2\nu) \ln (R_2 + z + c) - \frac{c}{R_1} - \frac{(3-4\nu)c}{R_2} \right] \tag{41.71}$$

where R_1 and R_2 have the same meaning as in Eqs. (41.69). For the other case of a force parallel to the boundary, the results are:

$$B_x = \frac{P}{4\pi\nu} \left(\frac{1}{R_1} + \frac{1}{R_2} \right) \qquad B_z = \frac{P}{2\pi\nu} \left[\frac{x(1-2\nu)}{R_2(R_2 + z + c)} - \frac{cx}{R_2{}^3} \right]$$
$$\beta = -\frac{P}{2\pi\nu} \left[\frac{x(1-2\nu)^2}{R_2 + z + c} + \frac{cx(1-2\nu)}{R_2(R_2 + z + c)} \right] \tag{41.72}$$

The Papkovitch functions in Eqs. (41.71) and (41.72) are related to the Galerkin vectors in Eqs. (41.68) and (41.70), respectively, in the manner indicated in Eqs. (41.21).

c. Concentrated Loading on Curved Surfaces

The solution to a problem involving concentrated forces on the surface of a body is not different from the solution to any other type of problem in elasticity, in that it must satisfy both the basic equations of elasticity and the boundary conditions of the problem. Besides, the solution must have singularities at each point of application of a concentrated force, such that the resultant of the stresses on any surface enclosing the point and lying entirely in the body is equal to the force. For a problem involving concentrated forces on a curved surface, Sternberg and his associates [21],[22] showed that the above conditions are not sufficient, and a complete formulation of the problem may be reached only by further considering a modified problem, in which each of the concentrated loads is replaced by an arbitrary distribution of surface forces over finite surface elements surrounding the load points. The solution to the concentrated-force problem is then defined as the limit of the solution to the modified problem, as the surface elements are shrunk to the load points while the resultants of the distributed forces are made to approach the given concentrated forces. This approach was, in fact, used on p. 41–14, to solve Boussinesq's problem, which, of course, deals with only a plane boundary surface.

With the proper formulation of the problem of concentrated forces on curved surfaces, Sternberg and his associates solved the case [21] of a sphere under concentrated forces applied at the ends of a diameter, and that [14] of a body bounded by one sheet of a two-sheeted hyperboloid of revolution, and subjected to an axial concentrated force at the vertex. The former case has application to ball-bearing problems.

The problem of an elastic sphere under two concentrated torques applied at the ends of a diameter was discussed by Huber [9].

41.7. STRESS CONCENTRATIONS IN BODIES OF REVOLUTION

a. Stress Concentration around a Small Spherical Cavity

Let a small spherical cavity of radius a be located in a large body subjected to uniform tension of magnitude σ (Fig. 41.4). Without the cavity, stresses acting on the spherical surface $R = a$ are, in spherical coordinates,

$$\sigma_R = \sigma \cos^2 \psi \qquad \tau_{R\psi} = -\sigma \sin \psi \cos \psi \tag{41.73}$$

When the cavity is present, these stresses at the surface must be annulled, which may be accomplished by adding to Eqs. (41.73) the stresses corresponding to the Love function [5c]

$$\phi = A \cos \psi + B(1 + \cos \psi) \ln R + C \frac{\cos \psi}{R^2} \tag{41.74}$$

The first and second terms of ϕ represent a double force and a center of compression, respectively. Since the stresses σ_R and $\tau_{R\psi}$ are given in terms of the Love function by

$$\sigma_R = \cos \psi \frac{\partial}{\partial R}\left[(2 - \nu)\nabla^2\phi - \frac{\partial^2 \phi}{\partial R^2}\right] + \frac{\sin \psi}{R}\frac{\partial}{\partial \psi}\left[-\nu\nabla^2\phi + \frac{\partial^2\phi}{\partial R^2} - \frac{2}{R}\frac{\partial \phi}{\partial R} + \frac{2\phi}{R^2}\right]$$

$$\tau_{R\psi} = \frac{\cos\psi}{R}\left|\frac{\partial}{\partial\psi}\left[(1-\nu)\nabla^2\phi - \frac{\partial^2\phi}{\partial R^2} + \frac{2}{R}\frac{\partial\phi}{\partial R} - \frac{2\phi}{R^2}\right]\right. + \sin\psi\frac{\partial}{\partial R}\left[-(1-\nu)\nabla^2\phi + \frac{1}{R}\frac{\partial\phi}{\partial R} + \frac{1}{R^2}\frac{\partial^2\phi}{\partial\psi^2}\right]$$

the following results at the spherical surface $R = a$ are obtained from ϕ in Eq. (41.74):

$$\begin{aligned} \sigma_R &= -2(1+\nu)\frac{A}{a^3} + 2(5-\nu)\frac{A}{a^3}\cos^2\psi - \frac{B}{a^3} - \frac{12C}{a^5} + \frac{36C}{a^5}\cos^2\psi \\ \tau_{R\psi} &= 2(1+\nu)\frac{A}{a^3}\sin\psi\cos\psi + \frac{24C}{a^5}\sin\psi\cos\psi \end{aligned} \tag{41.75}$$

When the stresses in Eq. (41.75) are added to those in (41.73), and the sums made to vanish, we find

$$A = -\frac{5\sigma a^3}{2(7-5\nu)} \qquad B = -\frac{\sigma a^3(1-5\nu)}{2(7-5\nu)} \qquad C = \frac{\sigma a^5}{2(7-5\nu)} \tag{41.76}$$

The final solution consists of a uniform tension $\sigma_z = \sigma$ superposed on the system of stresses derived from the function ϕ in Eq. (41.74), with the coefficients A, B, and C given by Eqs. (41.76).

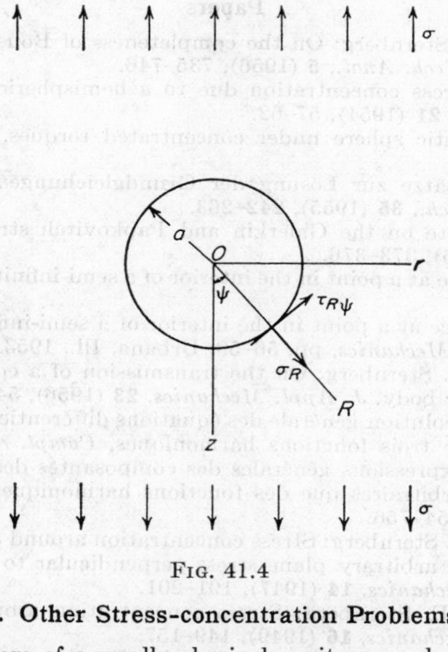

FIG. 41.4

b. Other Stress-concentration Problems

The preceding problem of a small spherical cavity was also solved by Neuber [3], to illustrate his method mentioned on p. 41–5. The method was further applied by him to a large variety of other problems of stress concentration involving hyperboloidal notches and ellipsoidal cavities. Ellipsoidal coordinates were used, and the types of loading included tension, bending, shear, and torsion.

The solution for the small spherical cavity may also be derived in a similar manner as the problem of a hollow sphere was solved in Art. 41.4e, on the basis of the formulation given in Eqs. (41.22). By using this method, Sternberg and Sadowsky [20] solved the axially symmetrical problem of two spherical cavities in an infinite body, in spherical dipolar coordinates, and Eubanks [8] discussed the stress concentration due to a hemispherical pit at a free surface of a semi-infinite body. By generalizing the formulation in Eqs. (41.22) to the nonsymmetrical case, Sadowsky and Sternberg [16],[17] were also able to investigate the stress concentration around an ellipsoidal cavity in an infinite body under arbitrary plane stress perpendicular to the axis of revolution of the cavity, and the stress concentration around a triaxial ellipsoidal cavity.

A comprehensive survey of the literature on three-dimensional stress-concentration problems has been prepared recently by Sternberg [23].

41.8. REFERENCES

Books

[1] J. Boussinesq: "Applications des potentiels à l'étude de l'équilibre et du mouvement des solides élastiques," Gauthier-Villars, Paris, 1885.

[2] A. E. H. Love: "A Treatise on the Mathematical Theory of Elasticity," 4th ed., Dover, New York, 1944: (*a*) p. 142; (*b*) p. 183; (*c*) p. 186; (*d*) p. 241; (*e*) p. 274; (*f*) p. 277.

[3] H. Neuber: "Theory of Notch Stresses," Edwards, Ann Arbor, 1946.

[4] I. N. Sneddon: "Fourier Transforms," McGraw-Hill, New York, 1951: (*a*) p. 469; (*b*) p. 473; (*c*) p. 504.

[5] S. P. Timoshenko, J. N. Goodier: "Theory of Elasticity," 2d ed., McGraw-Hill, New York, 1951: (*a*) p. 351; (*b*) p. 352; (*c*) p. 359; (*d*) p. 388.

[6] H. M. Westergaard: "Theory of Elasticity and Plasticity," Harvard Univ. Press, Cambridge, Mass., 1952: (*a*) p. 27; (*b*) p. 120; (*c*) p. 130.

Papers

[7] R. A. Eubanks, E. Sternberg: On the completeness of Boussinesq-Papkovitch stress functions, *J. Rat. Mech. Anal.*, **5** (1956), 735–746.

[8] R. A. Eubanks: Stress concentration due to a hemispherical pit at a free surface, *J. Appl. Mechanics*, **21** (1954), 57–62.

[9] A. Huber: The elastic sphere under concentrated torques, *Quart. Appl. Math.*, **13** (1955), 98–102.

[10] K. Marguerre: Ansätze zur Lösung der Grundgleichungen der Elastizitätstheorie, *Z. angew. Math. Mech.*, **35** (1955), 242–263.

[11] R. D. Mindlin: Note on the Galerkin and Papkovitch stress functions, *Bull. Am. Math. Soc.*, **42** (1936), 373–376.

[12] R. D. Mindlin: Force at a point in the interior of a semi-infinite solid, *Physics*, **7** (1936), 195–202.

[13] R. D. Mindlin: Force at a point in the interior of a semi-infinite solid, *Proc. 1st Midwestern Conf. Solid Mechanics*, pp. 56–59, Urbana, Ill., 1953.

[14] G. L. Neidhardt, E. Sternberg: On the transmission of a concentrated load into the interior of an elastic body, *J. Appl. Mechanics*, **23** (1956), 541–554.

[15] P. F. Papkovitch: Solution générale des équations différentielles fondamentales d'élasticité, exprimée par trois fonctions harmoniques, *Compt. rend. acad. sci. Paris*, **195** (1932), 513–515; Expressions générales des composantes des tensions, ne renfermant comme fonctions arbitraires que des fonctions harmoniques, *Compt. rend. acad. sci. Paris*, **195** (1932), 754–756.

[16] M. A. Sadowsky, E. Sternberg: Stress concentration around an ellipsoidal cavity in an infinite body under arbitrary plane stress perpendicular to the axis of revolution of cavity, *J. Appl. Mechanics*, **14** (1947), 191–201.

[17] M. A. Sadowsky, E. Sternberg: Stress concentration around a triaxial ellipsoidal cavity, *J. Appl. Mechanics*, **16** (1949), 149–157.

[18] E. Sternberg, R. A. Eubanks, M. A. Sadowsky: On the stress-function approaches of Boussinesq and Timpe to the axisymmetric problem of elasticity theory, *J. Appl. Phys.*, **22** (1951), 1121–1124.

[19] E. Sternberg, R. A. Eubanks, M. A. Sadowsky: On the axisymmetric problem of elasticity theory for a region bounded by two concentric spheres, *Proc. 1st U.S. Natl. Congr. Appl. Mechanics*, pp. 209–215, Chicago, 1952.

[20] E. Sternberg, M. A. Sadowsky: On the axisymmetric problem of the theory of elasticity for an infinite region containing two spherical cavities, *J. Appl. Mechanics*, **19** (1952), 19–27.

[21] E. Sternberg, F. Rosenthal: The elastic sphere under concentrated loads, *J. Appl. Mechanics*, **19** (1952), 413–421.

[22] E. Sternberg, R. A. Eubanks: On the singularity at a concentrated load applied to a curved surface, *Proc. 2d U.S. Natl. Congr. Appl. Mechanics*, pp. 237–245, Ann Arbor, Mich., 1954.

[23] E. Sternberg: Three-dimensional stress concentrations in the theory of elasticity, *Appl. Mech. Rev.*, **11** (1958), 1.

[24] K. Terezawa: On the elastic equilibrium of a semi-infinite solid under given boundary conditions with some applications, *J. Coll. Sci., Imp. Univ. Tokyo*, **37** (1916), art. 7.

[25] Y.-Y. Yu: Torsion of a semi-infinite body and a large thick plate, *Quart. J. Mech. Appl. Math.*, **7** (1954), 287–298.

CHAPTER 42

CONTACT PROBLEMS

BY

J. L. LUBKIN, Ph.D., Springdale, Conn.

42.1. THE HERTZ THEORY OF ELASTIC CONTACT

Assume that two elastic solids are brought into contact at a single point O, as sketched in Fig. 42.1a for the case of spheres. They have a common tangent plane Or and a common normal Oz. If collinear forces N are now applied so as to press the two solids together (Fig. 42.1b), deformation takes place, and we may expect that a small contact area will replace the contact point O of the unloaded state. The first

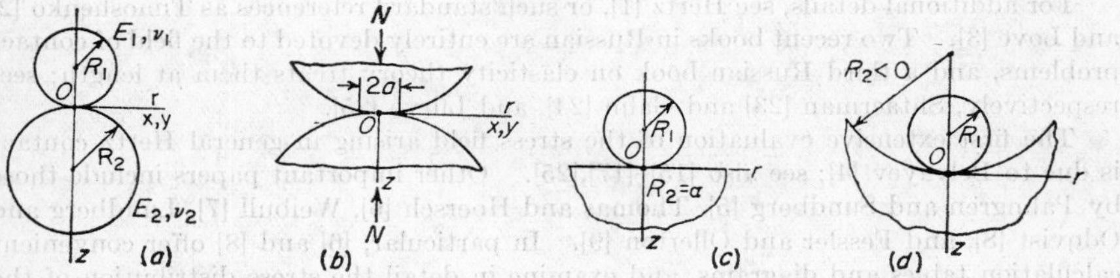

FIG. 42.1. Hertz contact of solids of revolution: (a) Undeformed spheres in contact at point O; (b) formation of a small contact surface under normal forces N. (c) Lower body (plane) has infinite radius of curvature. (d) Lower body has negative radius of curvature.

step is to find the size and shape of this contact area, and the distribution of normal pressure arising on it. Having these, in principle we can calculate the stresses and deformation which the interfacial pressure induces in the contacting members.

a. Assumptions of the Theory

Hertz [1] starts by assuming that the contacting solids are isotropic and linearly elastic,[1] and also that the representative dimensions of the contact area are very small compared to the various radii of curvature of the undeformed bodies, in the vicinity of the contact interface. This leads naturally to two further simplifying assumptions. Near the contact zone, it is considered sufficiently accurate to represent the actual shape of the two nearly flat bodies by general surfaces of the second degree, $z = Ax^2 + By^2 + Hxy$, the cartesian coordinates x and y lying in the common tangent plane. Despite this assumption, Hertz also takes the bodies to be "flat" enough, in the neighborhood of the contact surface, to be treated by the powerful analytical methods available for the semi-infinite solid, or body bounded by a plane. As actually employed by Hertz, these last two assumptions are not mutually inconsistent,

[1] There is a considerable literature dealing with problems of the indentation of an elastic solid by a rigid stamp or punch, which constitutes one type of contact problem. In the Russian literature. the elastic contact problem is often approached from this viewpoint, or simply as a problem of the surface loading of a body bounded by a plane, and the connection to the contact of two elastic solids is taken up last; see, for example, [23],[24].

having different purposes in the analysis. Finally, he assumes that the two solids are perfectly smooth; i.e., only the *normal* pressures which arise during contact are considered. Relative displacements in the plane of x and y, and any shearing tractions which might arise therefrom, are neglected as effects of higher order. Such tractions actually occur only in the contact of bodies with unlike elastic constants.

b. The General Case

Based on the foregoing assumptions, by an application of potential theory, Hertz shows that:

1. In general, the contact area is bounded by an ellipse, of semiaxes a and b, whose positions relative to the two contacting solids can be calculated from their given geometric properties, that is, their principal radii of curvature and the relative spatial orientation of the planes of principal curvature.

2. The normal pressure distribution over this area is $p = p_0[1 - (x/a)^2 - (y/b)^2]^{1/2}$. The maximum pressure p_0 occurs above point O, situated at the center of the ellipse, and is given by $3N/2\pi ab$.

3. The dimensions a and b of the ellipse of contact increase directly as $N^{1/3}$.

4. Regions of the solids which are remote from the contact zone approach each other by an amount α which varies as $N^{2/3}$. Note the nonlinear dependence upon N in these last two items.

For additional details, see Hertz [1], or such standard references as Timoshenko [2] and Love [3]. Two recent books in Russian are entirely devoted to the field of contact problems, and a third Russian book on elasticity theory treats them at length; see, respectively, Shtaerman [23] and Galin [24], and Lurye [25].

The first extensive evaluation of the stress field arising in general Hertz contact is due to Belyayev [4]; see also [15]–[17],[25]. Other important papers include those by Palmgren and Sundberg [5], Thomas and Hoersch [6], Weibull [7], Lundberg and Odqvist [8], and Fessler and Ollerton [9]. In particular, [6] and [8] offer convenient calculation tables and diagrams, and examine in detail the stress distribution of the elliptic contact problem and all its special cases. For the contact stresses which are critical in design for fatigue, see [9],[17], and for static loading see Karas [19].

c. Contact of Spheres and Solids of Revolution

In this case, the problem is axially symmetric about the common normal Oz, and the formulas of the general case simplify considerably. Let R_1 and R_2 be the radii of curvature of the spheres or other solids of revolution at the origin O, before deformation. These are positive for convex bodies, and negative for concave ones. Figure 42.1*a*, *c*, and *d* illustrates a few typical combinations of curvatures. Also, let E_i and ν_i $(i = 1,2)$ be the Young's moduli and Poisson's ratios of the bodies. Then the contact area is circular, and has the radius

$$a = (3NR_0/4E_0)^{1/3} \qquad 1/R_0 = 1/R_1 + 1/R_2$$
$$1/E_0 = (1 - \nu_1^2)/E_1 + (1 - \nu_2^2)/E_2 \tag{42.1}$$

The normal pressure distribution is given by

$$p = p_0(1 - r^2/a^2)^{1/2} \qquad p_0 = 3N/2\pi a^2 \tag{42.2}$$

Points remote from the contact area approach each other by an amount

$$\alpha = R_0^{-1/3}(3N/4E_0)^{2/3} = a^2/R_0 \tag{42.3}$$

The stress distribution on the contact surface ($z = 0$, $r \leqq a$) is [10]:

$$\begin{aligned} \sigma_r/p_0 &= G(\rho) - (1-\rho^2)^{1/2} & \sigma_z/p_0 &= -(1-\rho^2)^{1/2} \\ \sigma_\theta/p_0 &= -G(\rho) - 2\nu(1-\rho^2)^{1/2} & \rho &= r/a \\ G(\rho) &= \frac{1-2\nu}{3\rho^2}\,[1-(1-\rho^2)^{3/2}] \end{aligned} \tag{42.4}$$

On $z = 0$, outside the contact surface ($r > a$):

$$\sigma_r/p_0 = -\sigma_\theta/p_0 = (1-2\nu)/3\rho^2 \tag{42.5}$$

so that the stress field is continuous at $z = 0$, $r = a$. All components of stress not given above are zero in the region concerned. The solution may be invalid for

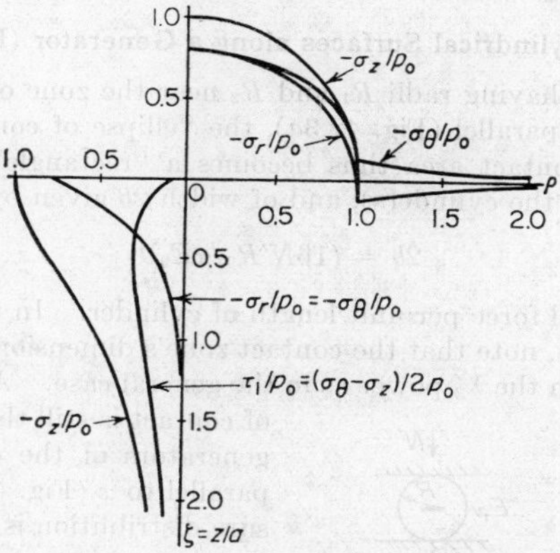

FIG. 42.2. Stress profiles on contact surface $O\rho$ and on axis of symmetry $O\zeta$ for case of circular contact area ($\nu = 0.3$).

$r \gg a$ if the effect of body curvature becomes significant, or wherever the body departs from the Hertz hypothesis of a semi-infinite solid.

On the axis of symmetry Oz, the nonzero stresses are:

$$\begin{aligned} \sigma_r/p_0 = \sigma_\theta/p_0 &= -(1+\nu)[1-\zeta\arctan(1/\zeta)] + \tfrac{1}{2}(1+\zeta^2)^{-1} \\ \sigma_z/p_0 &= -(1+\zeta^2)^{-1} \qquad \zeta = z/a \end{aligned} \tag{42.6}$$

These stresses are also the principal stresses for their respective regions, and the same formulas hold for both solids, using the proper value of ν. Elsewhere in the bodies τ_{rz} is not zero, so that only σ_θ is always a principal stress. Figure 42.2 shows the stress profiles corresponding to Eqs. (42.4) through (42.6), for $\nu = 0.3$; these are symmetrical about the ζ axis.

If σ_i and τ_i ($i = 1, 2, 3$) denote the principal stresses and shears, then

$$\text{At } z = r = 0{:} \quad \begin{aligned} \sigma_1 = \sigma_2 &= -(\nu + \tfrac{1}{2})p_0 \qquad \sigma_3 = -p_0 \\ -\tau_1 = \tau_2 &= (\nu - \tfrac{1}{2})p_0 \qquad \tau_3 = 0 \end{aligned} \tag{42.7}$$

$$\text{At } z = 0,\ r = a{:} \quad \sigma_1 = -\sigma_3 = 2\tau_1 = 2\tau_3 = -\tau_2 = (1-2\nu)p_0/3 \qquad \sigma_2 = 0 \tag{42.8}$$

The largest tensile stress is, thus, $\sigma_1 = \sigma_r$ at $z = 0$, $r = a$, and the largest compressive stress is $\sigma_3 = \sigma_z$ at $z = r = 0$. Further analysis shows that the largest shear stress in the body τ_{max} occurs on the z axis at a point near $\frac{1}{2}a$, whose location varies slightly with ν. For $\nu = 0.3$,

$$\tau_{max} = 0.31p_0 \qquad \text{at } z/a = 0.47 \tag{42.9}$$

Several references, for example, [6] and [8], show that variations of Poisson's ratio in the usual engineering range do not have a strong influence on the important features of the stress pattern. Further discussion of the stress field in this case can be found in papers by Fuchs [11], Morton and Close [12], and also in [2],[6]–[10],[18],[19]. In some cases, the authors' sign conventions bear close scrutiny.

d. Contact of Cylinders Crossed at Right Angles

In the contact of cylindrical bodies having equal radii of curvature near the contact region and perpendicular axes, the area of contact is circular, and the formulas of Art. 42.1*c* still obtain. For cylinders with unequal radii, the problem reverts to the general (elliptical) case.

e. Contact of Cylindrical Surfaces along a Generator (Rolls and Disks)

For long cylinders having radii R_1 and R_2 near the zone of contact, and oriented so that their axes are parallel (Fig. 42.3*a*), the "ellipse of contact" degenerates into parallel lines. The contact area thus becomes a "rectangle" of infinite length (in practice, the length of the cylinders), and of width $2b$ given by

$$2b = (16N'R_0/\pi E_0)^{1/2} \tag{42.10}$$

where N' is the normal force per unit length of cylinder. In this limiting case of the Hertz contact problem, note that the contact zone's dimension varies as the ½ power of the load, rather than the ⅓ power as in the general case. Assuming that the plane of contact is still the x,y plane and that the generators of the cylindrical surfaces are parallel to x (Fig. 42.3*a*), the normal pressure distribution is

$$p = p_0(1 - y^2/b^2)^{1/2} \qquad p_0 = 2N'/\pi b \tag{42.11}$$

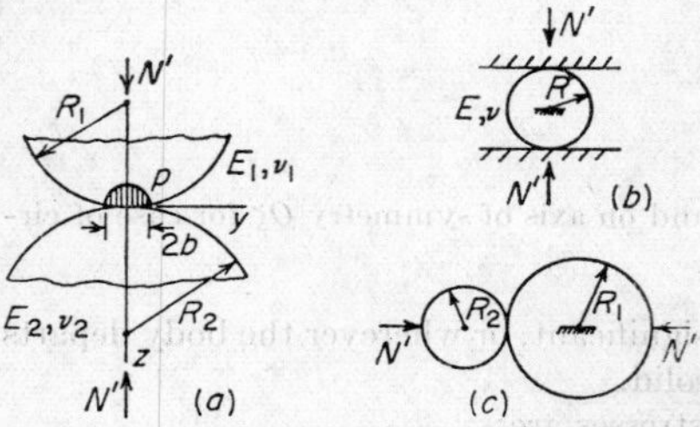

FIG. 42.3. Hertz contact of cylinders along a generator: (*a*) Schematic of the system parameters and normal pressure p; (*b*) roller with fixed center compressed by rigid smooth plates; (*c*) rolls in contact with the axis of one body fixed in space.

No simple expression for the relative approach of all remote points exists for this two-dimensional problem. The general Hertz formulas show an infinite displacement of the contact zone, relative to infinitely removed points in the bodies. This is characteristic of two-dimensional problems of the elastic half space. Hertz [1] interprets this to mean that it is necessary to take account of the actual shapes of the contacting solids, in dealing with two-dimensional contact problems. Thus, if a smooth circular cylinder of radius R is compressed between smooth *rigid* planes (Fig. 42.3*b*), and the center of the cylinder is taken to be fixed in space, Dörr [13] gives, for the displacement of the points of initial contact, the formula

$$\delta = (2N'/\pi)(1 - \nu^2)[\ln(4R/b) - \tfrac{1}{2}]/E \tag{42.12}$$

Hence, if two complete elastic cylinders are in contact, and the axis of one is held fixed (Fig. 42.3*c*), the total displacement of the axis of the other will be

$$\alpha = \delta_1 + \delta_2 \tag{42.13}$$

where the δ_i are given by Eq. (42.12), with appropriate subscripts for E, ν, and R. Similar expressions have been derived in related situations; see Weber [32] and Loo [33].

Huber and Fuchs [14] have obtained the entire stress field for the contact of long cylinders with parallel axes. On the contact surface ($z = 0, |y| \leq b$):

$$\sigma_z/p_0 = \sigma_y/p_0 = -(1 - y^2/b^2)^{1/2} \qquad \sigma_x = 2\nu\sigma_z \tag{42.14}$$

and these are also the principal stresses. Outside the contact zone, all stresses vanish on $z = 0$. In the plane of symmetry ($y = 0$ in Fig. 42.3*a*):

$$\begin{aligned} \sigma_x/p_0 &= -2\nu[(1 + Z^2)^{1/2} - Z] & Z &= z/b \\ \sigma_y/p_0 &= -(1 + Z^2)^{1/2}[2 - (1 + Z^2)^{-1}] + 2Z & \sigma_z/p_0 &= -(1 + Z^2)^{-1/2} \end{aligned} \tag{42.15}$$

These are also the principal stresses on the axis of symmetry. Principal shears are readily calculated from Eqs. (42.14) and (42.15). The shear stress τ_{yz} is zero only on the planes $y = 0$ and $z = 0$; the other shears vanish everywhere. In Fig. 42.4 we show, for $\nu = 0.3$, the stresses of Eqs. (42.14) and (42.15) and the principal shear $\tau = (\sigma_y - \sigma_z)/2$ on $y = 0$. These equations hold for both bodies, if the proper value of ν is used.

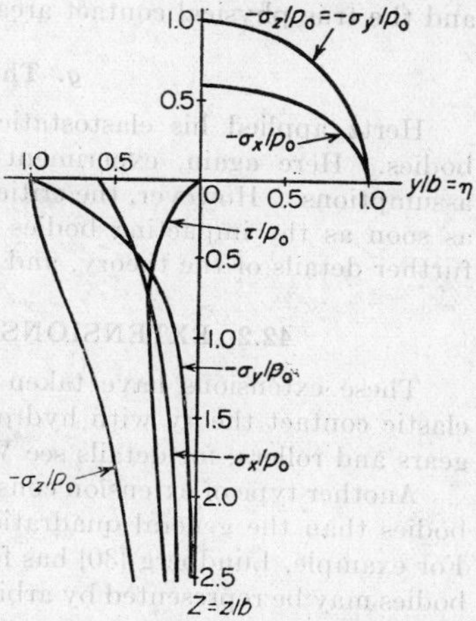

FIG. 42.4. Stress distribution on contact surface $O\eta$ and axis of symmetry OZ for contacting cylinders ($\nu = 0.3$).

The cumbersome cartesian expressions for the stresses at an arbitrary point of each body become quite compact in suitable elliptical coordinates. See [15]–[17], which quote these relations, originally derived by Belyayev. In the contact of cylinders, the principal stress field is entirely compressive, within the valid range of the Hertz analysis [14],[15].

The maximum shear stress occurs on $y = 0$:

$$\tau_{max} = 0.300p_0 \qquad \text{at } z/b = 0.786 \tag{42.16}$$

independent of the value of ν.

If the cylinders in contact are very short instead of very long, as in the contact of thin disks, the problem becomes one of plane stress, and $\sigma_x = 0$ in Eqs. (42.14) and (42.15). The general stress picture is otherwise similar.

Additional discussion of the stress pattern can be found in [6],[8],[9],[14]–[18], and in the literature of roller-bearing, gear, and friction-drive engineering.

f. Experimental Confirmation, Limitations of the Theory

Because of its technological importance, the Hertz theory has been scrutinized in countless experiments. Hertz himself [1] measured the size and shape of the contact area, using soot-covered glass lenses pressed against glass plates, and glass cylinders crossed at various angles. Subsequently, many others have confirmed and extended his findings. In essence, the theory can be checked to within a fraction of a per cent, as long as its hypotheses obtain, i.e., as long as the solids remain linearly elastic and the contact area is small in size compared to the radii of curvature. In particular, the relative approach α of remote points in the bodies, the theoretical size and shape of the contact area, the dependence upon elastic moduli and system geometry, and many features of the stress distribution have been amply confirmed for a variety of materials. Berndt [20] has given a critical evaluation of earlier experimental work, and Palmgren [21] has reviewed the subsequent literature before giving his own exhaustive results.

A recent experimental study of great interest has been carried out by Fessler and

Ollerton [9], who give additional references. They find that the Hertz theory accurately predicts the important shear stresses in elliptical contact problems for cases where the semimajor axis of the ellipse of contact is as large as one-half of the smallest radius of curvature of either of the contacting solids. However, we must not expect the geometric predictions of the theory to be accurate when there is such a considerable departure from the original hypotheses.

Beyond the proportional limit, some of the relations predicted by the theory continue to hold approximately, but the divergence increases with the degree of departure from linear elasticity [20],[21].

The relation between the nominal contact area determined by elasticity theory and the true physical contact area has been discussed by Johnson [26].

g. The Hertz Impact Theory

Hertz applied his elastostatic contact theory to the normal impact of elastic bodies. Here again, experiment shows that the theory is very good, within its assumptions. However, the critical assumption of linear elasticity is quickly violated as soon as the impacting bodies acquire appreciable velocities. Consult [1]–[3] for further details of the theory, and Hunter [22] for a recent evaluation of it.

42.2. EXTENSIONS OF THE BASIC HERTZ THEORY

These extensions have taken a variety of forms, including the combination of elastic contact theory with hydrodynamics. This is important in the lubrication of gears and rollers; for details see Way [15], Lewicki [28], and Petrusevich [29].

Another type of extension considers more complex surface shapes for the contacting bodies than the general quadratic surfaces $z = Ax^2 + By^2 + Hxy$ of Hertz' theory. For example, Lundberg [30] has formulated a general theory of contact, in which the bodies may be represented by arbitrary functions $f_i(x,y)$. Allowance is made for shear tractions on the contact surface ("friction"), in addition to the normal pressures; no problems are solved with friction, however. For the case of the circular contact surface without friction, Lundberg gives the general solution for the potential function which solves the problem, in the form of a single integral. The latter, of course, involves the arbitrary axisymmetric shape functions assumed for the contacting solids. He also considers the contact of cones and truncated cones.

Cattaneo [31] has solved the axisymmetric contact problem for bodies of the form

$$z = a_1r^2 + a_2r^4 \tag{42.17}$$

He still retains the Hertz assumption that each body may be treated as a semi-infinite solid. The normal pressure of Eq. (42.2) is now modified to an expression of the form

$$p = m(1 - r^2/a^2)^{1/2} + n(1 - r^2/a^2)^{3/2} \tag{42.18}$$

where m and n depend on the elastic and geometric parameters of the system, and an algebraic equation of the fifth degree governs the radius a. Shtaerman [23] and Galin [24] consider similar problems. Shtaerman [23] has also discussed the periodic contact problem, which arises, for example, in the contact of a body bounded by a corrugated surface with a body bounded by a plane. Yet another extension of Hertz's theory consists of treating anisotropic solids. Problems of this type have been examined by Conway [34] and Galin [24].

Probably the most important extension of the Hertz contact theory consists of problems involving additional force systems superimposed upon the Hertz normal force. These arise in the contact of wheels and rails, gear teeth, friction drives, etc., and even in the consideration of what happens in the Hertz contact of bodies with unequal elastic moduli. In the latter case, differential displacements occur in the plane of the contact surface, and surface shear stresses must arise unless the coefficient of friction is truly zero (as assumed by Hertz). Schäfer [35] has examined this problem, which is closely related to those discussed next; see also a photoelastic investigation by Löffler, referred to in [16].

The important case shown schematically in Fig. 42.5*a* has been treated by Cattaneo [38], and independently by Mindlin [39], using a different method. Here we have two elastic solids which have first been pressed together with a force N along their common normal Oz A tangential force T_x is then applied to one body in the x direction, with line of action in the plane of contact, tending to slide it relative to the other body. The latter is similarly loaded to preserve equilibrium. By an argument based upon considerations of symmetry, Mindlin [39] shows that the tangential-force problem can

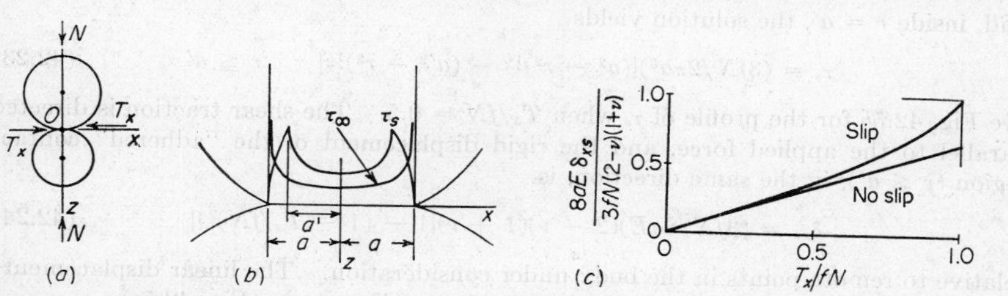

FIG. 42.5. Solids of revolution in Hertz contact with superimposed tangential force system: (*a*) Schematic diagram showing tangential force T_x with line of action in the plane of contact; (*b*) singular shear traction distribution τ_∞ (in absence of slip) and finite traction τ_s (with slip), for $T_x/fN = 0.5$; (*c*) rigid displacement of "adhered zone" of contact surface, with and without slip.

be solved without appreciably affecting the normal pressure distribution of the Hertz problem. The degree of approximation is of the same order as that due to neglecting the small shear tractions which arise in the Hertz contact of materials having unlike elastic properties.

Cattaneo [38] and Mindlin [39] point out that, if the two bodies adhere to each other over the entire contact surface, without relative slip, this interface must translate in the plane of contact as a rigid body. The movement is in the direction of the applied force, and, for the case of the circular contact surface, is of amount

$$\delta_x = (2 - \nu)(1 + \nu)T_x/4aE \qquad (42.19)$$

relative to remote points in the solid under consideration.[1] In all cases, the solution for the tangential traction τ_∞ arising on the contact surface shows that it is everywhere parallel to the applied force T_x. Furthermore, it becomes infinite on the boundary of the ellipse of contact as long as there is no relative slip between the two bodies. For the circular case (Fig. 42.5*b*),

$$\begin{aligned} \tau = \tau_\infty &= (T_x/2\pi a^2)(1 - r^2/a^2)^{-1/2} \qquad r < a \\ \tau = \tau_\infty &= 0 \qquad r > a \end{aligned} \qquad (42.20)$$

Such a distribution is physically untenable. We must assume that the singular surface shear falls to a finite value which the system can physically sustain, and the problem must be reformulated.

Mindlin [39], for the circular case, and Cattaneo [38], for the elliptic case, assume that the shear traction falls to a value equal to the product of a constant coefficient of Coulomb friction f times the Hertz normal pressure of Eq. (42.2). This takes

[1] For the case of the elliptic contact area, this rigid displacement is parallel to the tangential force only if $\nu = 0$ or if the x direction coincides with a principal axis of the ellipse of contact [39].

place in a suitable circular or elliptical annulus, whose size is to be determined. A rigid-body displacement is still assumed to prevail in that part of the contact surface over which relative slip of the two bodies has not yet penetrated.

For the problem thus reformulated, the solution shows [38],[39] that slip takes place over an annulus which, for the circular case, has the radius

$$a' = a(1 - T_x/fN)^{1/3} \tag{42.21}$$

Outside $r = a'$, the shear traction is, by hypothesis,

$$\begin{aligned} \tau &= \tau_s = (3fN/2\pi a^3)(a^2 - r^2)^{1/2} \qquad a' \leqq r \leqq a \\ \tau &= \tau_s = 0 \qquad r \geqq a \end{aligned} \tag{42.22}$$

and, inside $r = a'$, the solution yields

$$\tau_s = (3fN/2\pi a^3)[(a^2 - r^2)^{1/2} - (a'^2 - r^2)^{1/2}] \qquad r \leqq a' \tag{42.23}$$

See Fig. 42.5*b* for the profile of τ_s when $T_x/fN = 0.5$. The shear traction is directed parallel to the applied force, and the rigid displacement of the "adhered" contact region ($r \leqq a'$), in the same direction, is

$$\delta_{xs} = (3fN/8aE)(2 - \nu)(1 + \nu)[1 - (1 - T_x/fN)^{2/3}] \tag{42.24}$$

relative to remote points in the body under consideration. The linear displacement–force relation of Eq. (42.19) has given way to a nonlinear one (Fig. 42.5*c*).

Johnson [26] has extended Mindlin's analysis to calculate the amount of relative slip between the elastic bodies in the annulus $a' \leqq r \leqq a$. He also conducted static and dynamic tests, which furnish convincing confirmation of the important features of the theory, and which explain certain discrepancies appearing at small vibratory amplitudes. It is interesting that a shear force T_x in the x direction induces, in the annulus $a' \leqq r \leqq a$, a small but finite relative slip of the two bodies in the y direction, unless $\nu = 0$. The relative slip in the annulus is not uniform, varying with r. The limiting case where $a' = 0$, or gross sliding impends, has been studied in some detail by Sonntag [45].

The foregoing investigation of the contact of axisymmetric bodies has been extended by Mindlin and Deresiewicz, in an exhaustive study of the situation where both T_x and N are varied, separately or simultaneously. Deresiewicz has more recently further extended the analysis to include the case of elliptical contact. Both references are found in [40]. Cattaneo [50] has also solved the problem corresponding to the body shape of Eq. (42.17) for the case where a tangential force T_x is present, in addition to the normal force N.

Analogous investigations of Hertz contact, with superimposed torsional couple acting about the common normal Oz, have been carried out by Mindlin [39], Lubkin, Cattaneo, and Deresiewicz. Consult [41] for a list of the last three papers, and also Cattaneo [51]. Hetényi and McDonald [42] have treated in some detail, theoretically and experimentally, the limiting case where slip has progressed to the point where gross torsional movement impends.

Mindlin, Deresiewicz, and Duffy have evolved a theory of granular media, based upon these extended contact theories. The literature can be traced with the aid of [43],[44], which also provide an evaluation of related earlier work.

Many authors have carried out two-dimensional studies of the Hertz contact problem with superimposed tangential force. Fromm has conducted an exhaustive investigation of the rolling of elastic disks and cylinders, in a series of papers which are listed in [37]. He shows that a "locked" or "adhered" zone arises in the contact of steadily rolling bodies, just as in the three-dimensional problems discussed above.[1]

[1] This is true for rolling without gross sliding; there is no locked zone if complete slip takes place.

However, he finds that the only physically acceptable position for the nonslip zone is at the *leading edge* of the contact area, where points on the circumference of the rolling bodies first enter the contact region. The zone of relative slip, or *creep*, comprises the rest of the contact surface (see Fig. 42.6). Over the slip zone, the shear traction between the two solids is the largest possible, consistent with the frictional forces present. Under Fromm's hypothesis, which is analogous to that of the Cattaneo-Mindlin theory discussed above, the shear traction in the slip zone is taken to be equal to a constant coefficient of Coulomb friction f times the normal pressure distribution. He observes that the size of the contact surface, and the normal pressure on it, generally remains as in the basic Hertz theory for cylinders with parallel axes.

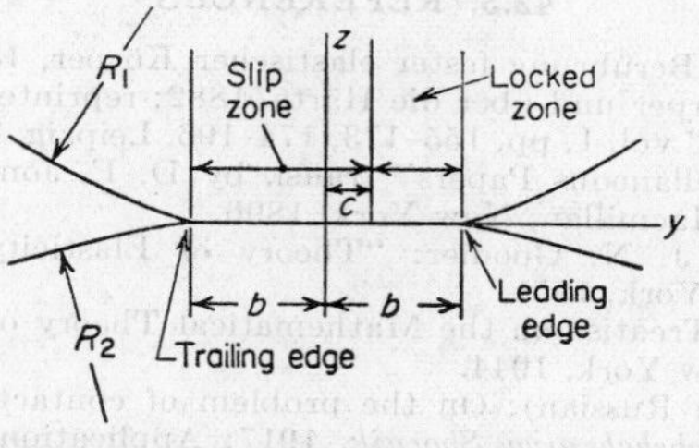

FIG. 42.6. Diagram defining the contact zone configuration in steady rolling of cylinders with friction; points on the cylinders first enter the contact zone at the leading edge $y = b$.

The position of the contact surface is now slightly inclined with respect to the perfectly symmetrical configuration of purely normal contact. Under some circumstances, the Hertz normal pressure distribution can be affected by the addition of the tangential (frictional) force.

Fromm is able to solve the integral equations of the problem when the solids have equal shear moduli, and presents numerous contact-surface distributions of shear traction as a function of the various parameters of the problem. He also obtains the deflected shape of the contact surface and its environs, and the associated stress field. The effect of superimposing a tangential force T'_y (per unit length of cylinder) on the Hertz normal force N' is to increase the stress level considerably over that due to N' alone. The following expressions define the slip zone (see Fig. 42.6) and the shear traction τ_s on the contact surface [37]:

$$c/b = 1 - 2(1 - T'_y/fN')^{1/2} \qquad (42.25)$$

$$\begin{aligned} \tau_s &= (fp_0/b)(b^2 - y^2)^{1/2} && -b \leqq y \leqq c \\ &= (fp_0/b)[(b^2 - y^2)^{1/2} - (b - y)^{1/2}(y - c)^{1/2}] && c \leqq y \leqq b \end{aligned} \qquad (42.26)$$

The values of b and p_0 are calculated from Eqs. (42.10) and (42.11) as usual.

If steady rolling is not involved, all sorts of positions are possible for the locked zone of the contact area, ranging from the perfectly symmetrical (analogous to Fig. 42.5*b* for the axisymmetric case) to the steady-rolling condition (Fig. 42.6), depending on how the final combination of normal and tangential load is applied. Poritsky [46] has carried out an investigation closely related to Fromm's, with quite similar conclusions.

For the limiting case where gross sliding impends under the tangential force T'_y, Karas [27] has given the explicit stress field and calculated the stresses under the combined Hertzian and tangential loadings. This work covers both plane strain (rolls) and plane stress (thin disks). Karas has located the critical point under each of five hypotheses of failure, for values of f from 0 to 0.35, at intervals of 0.05. He also finds the increase in the critical stress due to adding the shear force at the contact surface. M'Ewen [47], Carter [36], and Smith and Liu [48] have also contributed to the problem studied by Karas.

Sonntag [49] has investigated the steady rolling of elastic spheres. Creep or slip takes place here likewise, primarily in a direction parallel to the rolling direction, but a small transverse component is also present. He neglects the effect of this component, and the problem then exhibits a strong resemblance to the two-dimensional case studied by Fromm [37]. As in [36],[37],[46], a locked zone occurs at the leading edge of the contact surface, and the zone of slip comprises the rest of the contact interface. Sonntag finds an approximate solution for the dimensions of the locked zone, and for the shear traction distribution.

42.3. REFERENCES[1]

[1] H. Hertz: Über die Berührung fester elastischer Körper, 1881; Über die Berührung fester elastischer Körper und über die Härte, 1882; reprinted in "Gesammelte Werke von Heinrich Hertz," vol. 1, pp. 155–173, 174–196, Leipzig, 1895, English translations in H. Hertz, "Miscellaneous Papers" transl. by D. E. Jones and G. A. Schott, pp. 146–162, 163–183, Macmillan, New York, 1896.

[2] S. P. Timoshenko, J. N. Goodier: "Theory of Elasticity," 2d ed., pp. 372–384, McGraw-Hill, New York, 1951.

[3] A. E. H. Love: "A Treatise on the Mathematical Theory of Elasticity," 4th ed., pp. 192–200, Dover, New York, 1944.

[4] N. M. Belyayev (in Russian): On the problem of contact stresses, *Petrograd. Inst. Inzhenerov Puteĭ Soobshcheniya Sbornik*, 1917; Application of Hertz' theory to the calculation of local stresses at the contact point of wheels and rails, *Vestnik Inzhenerov i Tekh.*, 2 (1917); Memoirs on the Theory of Structures, Petrograd, 1924; Local stresses in the compression of contacting bodies, *Sbornik Stateĭ Inzhenerov Sooruzhennya i Stroitel. Mekhan. Leningrad*, 1924; On the problem of local stresses in connection with the resistance of rails to distortion, *Leningrad. Inst. Inzhenerov Puteĭ Soobshcheniya Sbornik*, 90 (1929); Computation of the largest calculated stresses in the compression of contacting bodies, *ibid.*, 102 (1929).

[5] A. Palmgren, K. Sundberg: Once again on the subject of ball bearing load capacity (in Swedish), *Tek. Tidskr. (Mekanik)*, **49** (1919), 57–67.

[6] H. R. Thomas, V. A. Hoersch: Stresses due to the pressure of one elastic solid upon another, *Univ. Illinois Eng. Expt. Sta. Bull.* 212 (1930), 54 pp.

[7] W. Weibull: Stresses in the contact of elastic bodies (in Swedish), *Tek. Tidskr. (Mekanik)*, **49** (1919), 160–163.

[8] G. Lundberg, F. K. G. Odqvist: Studies on stress distributions in the vicinity of the contact region between bodies, with applications (in Swedish, long English summary), *Ing. Vetenskaps Akad. Handl.* 116 (1932), 64 pp.

[9] H. Fessler, E. Ollerton: Contact stresses in toroids under radial loads, *Brit. J. Appl. Phys.*, **8:10** (1957), 387–393.

[10] M. T. Huber: Zur Theorie der Berührung fester elastischer Körper, *Ann. Physik*, ser 4, **14** (1904), 153–163.

[11] S. Fuchs: Hauptspannungstrajektorien bei der Berührung einer Kugel mit einer Platte, *Physik. Z.*, **14** (1913), 1282–1285.

[12] W. B. Morton, L. J. Close: Notes on Hertz's theory of the contact of elastic bodies, *Phil. Mag.*, ser 6, **43:254** (1922), 320–329.

[13] J. Dörr: Oberflächenverformungen und Randkräfte bei runden Rollen und Bohrungen, *Stahlbau*, **24** (1955), 202–206.

[14] M. T. Huber, S. Fuchs: Spannungsverteilung bei der Berührung zweier elastischer Zylinder, *Physik. Z.*, **15** (1914), 298–303.

[15] S. Way: Pitting due to rolling contact, *J. Appl. Mechanics*, **2** (1935), 49–58, 110–114.

[16] F. Karas: Werkstoffanstrengung beim Druck achsenparalleler Walzen nach den gebräuchlichen Festigkeitshypothesen, *Forsch. Ingenieurw.*, **11** (1940), 334–339.

[17] E. I. Radzimovsky: Stress distribution and strength condition of two rolling cylinders pressed together, *Univ. Illinois Eng. Expt. Sta. Bull.* 408 (1953), 40 pp.

[18] L. Föppl: Der Spannungszustand und die Anstrengung der Werkstoffe bei der Berührung zweier Körper, *Forsch. Ingenieurw.*, **7** (1936), 209–221.

[19] F. Karas: Der Ort grösster Beanspruchung in Wälzverbindungen mit verschiedenen Druckfiguren, *Forsch. Ingenieurw.*, **12** (1941), 237–243.

[1] References [53] to [64] will lead the reader to many significant studies published recently; [52] contains much of the not easily accessible material in [4].

[20] G. Berndt: Über die Gültigkeit der Hertzschen Formeln zur Berechnung der Abplattung von Messkörpern, *Z. Tech. Physik*, **3** (1922), 14–21, 82–87.
[21] A. Palmgren: Investigations concerning the static load capacity of ball bearings (in Swedish), Diss., Gothenburg, Sweden, 1930, *Stockholm Tek. Högskol. Avhandl.* 7, 48 pp.
[22] S. C. Hunter: Energy absorbed by elastic waves during impact, *J. Mech. Phys. Solids*, **5** (1957), 162–171.
[23] I. Ya. Shtaerman: "The Contact Problem of Elasticity Theory" (in Russian), 270 pp., Moscow-Leningrad 1949.
[24] L. A. Galin: "The Contact Problems of Elasticity Theory" (in Russian), 264 pp., Moscow-Leningrad, 1953.
[25] A. I. Lurye: "Three-dimensional Problems of Elasticity Theory" (in Russian), 491 pp., Moscow-Leningrad, 1955.
[26] K. L. Johnson: Surface interaction between elastically loaded bodies under tangential forces, *Proc. Roy. Soc. (London)*, **A,230** (1955), 531–548.
[27] F. Karas: Die äussere Reibung beim Walzendruck, *Forsch. Ingenieurw.*, **12** (1941), 266–274.
[28] W. Lewicki: Some physical aspects of lubrication in rolling bearings and gears, *Engineer*, **200** (1955), 176–178, 212–215.
[29] A. I. Petrusevich: Basic conclusions from the contact-hydrodynamic theory of lubrication (in Russian), *Izvest. Akad. Nauk SSSR Otdel. Tekh. Nauk*, **2** (1951), 209–223.
[30] G. Lundberg: Elastische Berührung zweier Halbräume, *Forsch. Ingenieurw.*, **10** (1939), 201–211.
[31] C. Cattaneo: Teoria del contatto elastico in seconda approssimazione, *Univ. Rome, Rend. mat. appl.*, **6** (1947), 504–512.
[32] C. Weber: Beitrag zur Berührung gewölbter Oberflächen beim ebenen Formänderungszustand, *Z. angew. Math. Mech.*, **13** (1933), 11–16.
[33] T. T. Loo: Effect of curvature on the Hertz theory for two circular cylinders in contact, *J. Appl. Mechanics*, **25** (1958), 122–124, and subsequent discussion.
[34] H. D. Conway: The pressure distribution between two elastic bodies in contact, *Z. angew. Math. Phys.*, **7** (1956), 460–465.
[35] A. Schäfer: Die Reibung beim Walzendruck bei verschiedenen Elastizitätsmoduln, *Arch. Eisenhüttenw.*, **23** (1952), 253–256.
[36] F. W. Carter: On the action of a locomotive driving wheel, *Proc. Roy. Soc. (London)*, **A,112** (1926), 151–157.
[37] H. Fromm: Zulässige Belastung von Reibungsgetrieben mit zylindrischen oder kegelförmigen Rädern, *Z. VDI*, **73** (1929), 957–962, 1029–1032.
[38] C. Cattaneo: Sul contatto di due corpi elastici, distribuzione locale degli sforzi, *Rend. accad. naz. Lincei*, ser. 6, **27** (1938), 342–348, 434–436, 474–478.
[39] R. D. Mindlin: Compliance of elastic bodies in contact, *J. Appl. Mechanics*, **16** (1949), 259–268.
[40] H. Deresiewicz: Oblique contact of nonspherical elastic bodies, *J. Appl. Mechanics*, **24** (1957), 623–624.
[41] H. Deresiewicz: Contact of elastic spheres under an oscillating torsional couple, *J. Appl. Mechanics*, **21** (1954), 52–56.
[42] M. Hetényi, P. H. McDonald: Contact stresses under combined pressure and twist, *J. Appl. Mechanics*, **25** (1958), 396–401.
[43] H. Deresiewicz: Mechanics of granular matter, *Advances Appl. Mech.*, **5** (1958).
[44] C. W. Thurston, H. Deresiewicz: Analysis of a compression test of a model of a granular medium, *J. Appl. Mechanics*, **26** (1959), 251–258.
[45] G. Sonntag: Halbraum mit halbkugelförmiger Schubbelastung, *Z. angew. Math. Mech.*, **29** (1949), 52–54.
[46] H. Poritsky: Stresses and deflections of cylindrical bodies in contact with application to contact of gears and of locomotive wheels, *J. Appl. Mechanics*, **17** (1950), 191–201, 465–468.
[47] E. M'Ewen: Stresses in elastic cylinders in contact along a generatrix, *Phil. Mag.*, **40** (1949), 454–459.
[48] J. O. Smith, C. K. Liu: Stresses due to tangential and normal loads on an elastic solid with application to some contact stress problems, *J. Appl. Mechanics*, **20** (1953), 157–166.
[49] G. Sonntag: Berechnung des Spannungszustandes und des Gleitens beim Rollen deformierbarer Kugeln, *Z. angew. Math. Mech.*, **30** (1950), 73–83.
[50] C. Cattaneo: Teoria del contatto elastico in seconda approssimazione, compressione obliqua, *Rend. seminar. fac. sci. univ. Cagliari*, **17** (1947), 13–28.
[51] C. Cattaneo: Compressione e torsione nel contatto tra corpi elastici di forma qualunque, *Ann. scuola norm. super. Pisa*, ser. 3, **9** (1955), 23–43.

[52] N. M. Belyayev: "Studies in the Theory of Elasticity and Plasticity," (in Russian), Moscow, 1957.

[53] G. D. Archard, F. C. Gair, W. Hirst: The elasto-hydrodynamic lubrication of rollers, *Proc. Roy. Soc. London*, **A,262** (1961), 51–72.

[54] H. Deresiewicz: A note on second-order Hertz contact, *J. Appl. Mechanics*, **28** (1961) 141–142.

[55] K. L. Johnson: The influence of elastic deformation upon the motion of a ball rolling between two surfaces, Inst. Mech. Engrs. Preprint, 1959, 14 pp.

[56] R. V. Klint: Oscillating tangential forces on cylindrical specimens in contact at displacements within the region of no gross slip, *ASLE Trans.*, **3** (1960), 255–264.

[57] L. E. Goodman: Contact stress analysis of normally loaded rough spheres, to be published in *J. Appl. Mechanics*, **29** (1962).

[58] L. E. Goodman, C. B. Brown: Energy dissipation in contact friction; constant normal and cyclic tangential loading, ASME Winter Annual Meeting, New York, 1961.

[59] S. C. Hunter: The Hertz problem for a rigid spherical indenter and a viscoelastic half-space, *J. Mech. Phys. Solids*, **8** (1960), 219–234.

[60] Y.-H. Pao: Extension of the Hertz theory of impact to the viscoelastic case, *J. Appl. Physics*, **26** (1955), 1083–1088.

[61] M. Pacelli: Contatto con attrito tra due corpi di forma qualunque; compressione e torsione, *Ann. scuola norm. super. Pisa*, ser. 3, **10** (1956), 155–184.

[62] K. L. Johnson: Recent developments in the theory of elastic contact stresses, *Proc. IME Conf. on Lubrication and Wear*, London, 1957.

[63] Rolling Contact Phenomena, *Proc. General Motors Corp. Symposium*, Detroit, Mich., 1960.

[64] S. C. Hunter: The rolling contact of a rigid cylinder with a viscoelastic half-space, paper no. 61-WA-8, ASME Winter Annual Meeting, New York, November, 1961.

CHAPTER 43

THERMAL STRESSES

BY

H. PARKUS, Dr. techn., Vienna, Austria

43.1. GENERAL RELATIONS

If a body is heated or cooled, its elements expand or contract. Thermal stresses are the result of restrictions imposed on these expansions or contractions by the continuity of the body and by conditions prescribed on its boundary. As long as no reactionary forces appear on the boundary, thermal stress systems are self-equilibrating.

The basic equations of thermal stress differ only in the stress-strain relations from those of a body at uniform temperature. They are obtained by superposing, on the strain due to stress, the strain produced by a temperature increase T. Hence, for an isotropic Hookean solid,

$$\epsilon_x = \frac{1}{2G}\left(\sigma_x - \frac{\nu}{1+\nu}s\right) + \alpha T \qquad \gamma_{xy} = \frac{1}{G}\tau_{xy} \qquad \text{etc.} \tag{43.1}$$

where α is the coefficient of linear thermal expansion and $s = \sigma_x + \sigma_y + \sigma_z$. Solving for the stresses yields

$$\sigma_x = 2G\left(\epsilon_x + \frac{\nu}{1-2\nu}e - \frac{1+\nu}{1-2\nu}\alpha T\right) \tag{43.2}$$

and two similar equations, where $e = \epsilon_x + \epsilon_y + \epsilon_z$.

Utilizing Eqs. (43.2), the strain energy U of the body is

$$U = \int_V G\left[\epsilon_x^2 + \epsilon_y^2 + \epsilon_z^2 + \tfrac{1}{2}(\gamma_{xy}^2 + \gamma_{yz}^2 + \gamma_{zx}^2) + \frac{\nu}{1-2\nu}e^2 - \frac{2(1+\nu)}{1-2\nu}\alpha T e\right] dV \tag{43.3}$$

If the principle of complementary energy is employed, U has to be replaced by U^*, where

$$U^* = \int_V \left\{\frac{1}{4G}\left[\sigma_x^2 + \sigma_y^2 + \sigma_z^2 + 2(\tau_{xy}^2 + \tau_{yz}^2 + \tau_{zx}^2) - \frac{\nu}{1+\nu}s^2\right] + \alpha T s\right\} dV \tag{43.4}$$

G and α are, in general, functions of T.

Substituting Eqs. (43.2) in the equations of motion of an elastic solid [i.e., Eqs. (33.35) with $X = -\rho\partial^2 u/\partial t^2$, . . .] and introducing the linearized kinematic relations (33.2), we find the following equations for the displacements u, v, w in a body with small deformations and temperature-independent shear modulus:

$$\begin{aligned}\nabla^2 u + \frac{1}{1-2\nu}\frac{\partial e}{\partial x} - \frac{\rho}{G}\frac{\partial^2 u}{\partial t^2} &= \frac{2(1+\nu)}{1-2\nu}\frac{\partial(\alpha T)}{\partial x}\\ \nabla^2 v + \frac{1}{1-2\nu}\frac{\partial e}{\partial y} - \frac{\rho}{G}\frac{\partial^2 v}{\partial t^2} &= \frac{2(1+\nu)}{1-2\nu}\frac{\partial(\alpha T)}{\partial y}\\ \nabla^2 w + \frac{1}{1-2\nu}\frac{\partial e}{\partial z} - \frac{\rho}{G}\frac{\partial^2 w}{\partial t^2} &= \frac{2(1+\nu)}{1-2\nu}\frac{\partial(\alpha T)}{\partial z}\end{aligned} \tag{43.5}$$

In *cylindrical coordinates* r, θ, z with symmetry about the z axis and displacements u, $v \equiv 0$, w, Eqs. (43.5) read

$$\begin{aligned}\nabla^2 u - \frac{u}{r^2} + \frac{1}{1-2\nu}\frac{\partial e}{\partial r} - \frac{\rho}{G}\frac{\partial^2 u}{\partial t^2} &= \frac{2(1+\nu)}{1-2\nu}\frac{\partial(\alpha T)}{\partial r}\\ \nabla^2 w + \frac{1}{1-2\nu}\frac{\partial e}{\partial z} - \frac{\rho}{G}\frac{\partial^2 w}{\partial t^2} &= \frac{2(1+\nu)}{1-2\nu}\frac{\partial(\alpha T)}{\partial z}\end{aligned} \tag{43.6}$$

where ∇^2 is given by Eq. (11.43*a*). In *spherical coordinates* r, ϕ, θ with symmetry about the origin, Eqs. (43.5) yield one equation for the radial displacement u:

$$\nabla^2 u - \frac{2u}{r^2} - \frac{1-2\nu}{2(1-\nu)}\frac{\rho}{G}\frac{\partial^2 u}{\partial t^2} = \frac{1+\nu}{1-\nu}\frac{\partial(\alpha T)}{\partial r} \tag{43.7}$$

where ∇^2 is given by Eq. (11.43*c*).

A particular solution of Eqs. (43.5) may be obtained by introducing the *thermoelastic potential* Φ. Letting

$$2Gu = \frac{\partial\Phi}{\partial x} \qquad 2Gv = \frac{\partial\Phi}{\partial y} \qquad 2Gw = \frac{\partial\Phi}{\partial z} \tag{43.8}$$

and neglecting the change in density associated with the deformation, the three equations (43.5) reduce to the single equation

$$\nabla^2\Phi - \frac{1-2\nu}{2(1-\nu)}\frac{\rho}{G}\frac{\partial^2\Phi}{\partial t^2} = \frac{E\alpha T}{1-\nu} \tag{43.9}$$

where $E = 2(1+\nu)G$. From Eqs. (33.2) and (43.2), the components of stress corresponding to Φ are:

In *cartesian coordinates*

$$\sigma_x = \frac{\partial^2\Phi}{\partial x^2} - \nabla^2\Phi + \frac{\rho}{2G}\frac{\partial^2\Phi}{\partial t^2} \qquad \tau_{xy} = \frac{\partial^2\Phi}{\partial x\,\partial y} \qquad \text{etc.} \tag{43.10}$$

In *cylindrical coordinates* with axial symmetry

$$\begin{aligned}\sigma_r &= \frac{\partial^2\Phi}{\partial r^2} - \nabla^2\Phi + \frac{\rho}{2G}\frac{\partial^2\Phi}{\partial t^2} & \tau_{rz} &= \frac{\partial^2\Phi}{\partial r\,\partial z}\\ \sigma_\theta &= \frac{1}{r}\frac{\partial\Phi}{\partial r} - \nabla^2\Phi + \frac{\rho}{2G}\frac{\partial^2\Phi}{\partial t^2} & \sigma_z &= \frac{\partial^2\Phi}{\partial z^2} - \nabla^2\Phi + \frac{\rho}{2G}\frac{\partial^2\Phi}{\partial t^2}\end{aligned} \tag{43.11}$$

In *spherical coordinates* with point symmetry

$$\sigma_r = -\frac{2}{r}\frac{\partial\Phi}{\partial r} + \frac{\rho}{2G}\frac{\partial^2\Phi}{\partial t^2} \qquad \sigma_\phi = \sigma_\theta = \frac{1}{r}\frac{\partial\Phi}{\partial r} - \nabla^2\Phi + \frac{\rho}{2G}\frac{\partial^2\Phi}{\partial t^2} \tag{43.12}$$

The general solution of the problem is then obtained by superposing, on the particular solution corresponding to Φ, the general solution of the homogeneous ($T \equiv 0$) equations (43.5).

It has been tacitly assumed, in the preceding discussion, that the temperature field in the body is independent of the deformations of the body. This is not strictly true. Deformations generate or absorb heat, and thus change the temperature field. In general, however, the modifications of the field are very slight, and may be neglected. The temperature distribution can then be considered a given function obtained by solving the problem of heat conduction in the body before the problem of stress distribution.

Table 43.1. Thermal Constants [2]

Material	α °F^{-1} × 10^{-6}	c, Btu/lb. · °F	λ, Btu/in. · hr · °F	κ^2, in.^{2}/hr
Metals				
Aluminum				
Pure	14	0.22	10.7	500
Alloy	13	0.22	7.6	350
Copper				
Pure	9.5	0.09	19	650
Brass (60:40)	10.5	0.09	4.5	163
Bronze (90:10)	10	0.09	2	70
Gold	7.8	0.03	15	730
Iron				
Pure	6.7	0.11	2.8	90
Carbon-steel	6.7	0.11	2.2	71
Steel alloy	6.7	0.11	1 1	36
Cast iron	5.6	0.13	2.6	71
Lead	16	0.03	1.7	136
Magnesium				
Pure	14.5	0.25	7.8	500
Cast alloy	14	0.25	3.9	250
Wrought alloy	14	0.25	2.8	176
Nickel				
Pure	7.2	0.11	2.8	81
Iron-nickel: 5% Ni	6.7	0.11	1.7	53
20	2.8	0.11	0.9	28
36	1.1	0.11	0.6	17
Monel	7.8	0.10	1.1	34
Platinum	5	0.03	3.4	144
Silver..................	11	0.06	20	880
Tin	15	0.05	3.1	230
Bearing alloy	13.5	0.05	1.1	82
Zinc	14.5	0.09	5.6	240
Nonmetals				
Brick wall	1.7	0.20	0.04	2
Ceramics	1.7	0.20	0.06	3.1
Concrete	6.7	0.21	0.05	3
Glass	4.5	0.18	0.04	2.5
Quartz glass	0.2	0.17	0.06	4.2
Ice		0.50	0.11	6.5
Plastics	11	0.37	0.02	0.9
Rocks:				
Sandstone	5	0.19	0.07	4.5
Limestone	4.5	0.19	0.08	4.2
Granite	4.5	0.19	0.15	8.2

NOTE: The values may be used up to temperatures of approximately 400°F. At higher temperatures, all quantities become temperature-dependent.

For a thermally isotropic and homogeneous body with temperature-independent thermal properties, the equation of heat conduction is

$$\kappa^2 \nabla^2 T = \frac{\partial T}{\partial t} \tag{43.13}$$

$\kappa = \lambda/(\rho c)$ is the "diffusivity" of the body, λ its conductivity, ρ its density, and c its specific heat. The constant λ is related to the amount of heat dq crossing a surface element dS per unit of time according to

$$dq = -\lambda \frac{\partial T}{\partial n} dS$$

where $\partial/\partial n$ denotes differentiation in the direction normal to the surface.

The boundary conditions usually associated with Eq. (43.13) are the following: (1) prescribed surface temperature, $T = F(x,y,z,t)$ on the surface of the body, where F is a given function, or (2) heat transfer into the surrounding medium of temperature T_0,

$$\lambda \frac{\partial T}{\partial n} + k(T - T_0) = 0 \tag{43.14}$$

where k is the coefficient of heat transfer, and n is the direction of the outward normal to the surface.

For an exhaustive treatment of problems in heat conduction, the book by Carslaw and Jaeger [1] should be consulted.

The general equations for the Hookean solid as given in Sec. 43.1 simplify considerably for a time-independent temperature field. All time derivatives drop out of the equations. In Secs. 43.2 through 43.10 attention is restricted to such steady-state problems.

43.2. STRESS-FREE TEMPERATURE FIELDS

Three-dimensional Fields

If the body is free of stress, Eqs. (43.1) read $\epsilon_x = \epsilon_y = \epsilon_z = \alpha T$, $\gamma_{xy} = \gamma_{yz} = \gamma_{zx} = 0$. Substituting into the equations of compatibility (33.21), a system of six equations in T is obtained, whose solution is

$$T = a_0 + a_1 x + a_2 y + a_3 z \tag{43.15}$$

with arbitrary coefficients a_i. It follows from (43.15) that $\nabla^2 T = 0$, and hence, from (43.13), that $\partial T/\partial t = 0$. The coefficients a_i are therefore constants.

Condition (43.15) is necessary for a stress-free temperature field. It is also sufficient if there are no external constraints restricting the free expansion on the boundary of the body.

Two-dimensional Fields

For a body in a state of plane strain (p. 43–5) with temperature independent of the coordinate z, the equations of compatibility yield

$$\frac{\partial^2 T}{\partial x^2} + \frac{\partial^2 T}{\partial y^2} = 0 \tag{43.16}$$

as a necessary condition that all stress components except σ_z vanish. σ_z takes on the value

$$\sigma_z = -E\alpha T \tag{43.17}$$

If the body is free to expand along its lateral surface, and if its cross section is simply connected, condition (43.16) is also sufficient. If the cross section is multiply connected (cross section with holes), an additional condition has to be introduced to make the displacements single-valued functions. If there are heat sources located within the holes, this condition will be violated, and the cylinder will not remain free from stress.

43.3. STRAIGHT OR SLIGHTLY CURVED BARS

Let the x axis be in the direction of the axis of the beam, and let the y and z axes coincide with the principal axes at the centroid of the cross section. Assume, further, the axis of the bar to form a plane curve in the x,z plane, and the temperature independent of y: $T = T(x,z)$. Neglecting σ_y and σ_z, we obtain from Eqs. (43.1)

$$\sigma_x = E(\epsilon_x - \alpha T) \tag{43.18}$$

Expanding ϵ_x in a power series in z, and retaining linear terms only, i.e., assuming the cross sections to remain plane,we get

$$\epsilon_x = \epsilon_0 - z\lambda \tag{43.19}$$

ϵ_0, the relative elongation at the centroid, and λ are given in terms of the displacements u and w at the axis (v being identically zero) by

$$\epsilon_0 = \frac{du}{dx} - \kappa w \qquad \lambda = \frac{d(\kappa u)}{dx} + \frac{d^2w}{dx^2} \tag{43.20}$$

$\kappa = 1/r_0$ is the initial curvature of the axis of the beam, with z pointing toward the center of curvature.

Substituting (43.19) into (43.18), assuming all coefficients to be independent of temperature, and integrating over the cross section of the bar yields the resultant axial force

$$N = \int_A \sigma_x \, dA = EA(\epsilon_0 - \alpha T_m) \tag{43.21}$$

T_m is the mean value of T over the cross section:

$$T_m = \frac{1}{A}\int_A T \, dA$$

Multiplying σ_x by z and again integrating gives the bending moment

$$M = \int_A \sigma_x z \, dA = -EI(\lambda + \alpha\Theta) \tag{43.22}$$

I is the moment of inertia of the cross section with respect to the y axis, and Θ is defined by

$$\Theta = \frac{1}{I}\int_A Tz \, dA$$

Substituting for ϵ_0 and λ from Eqs. (43.21) and (43.22) into Eqs. (43.19) and (43.18), the following expression for the axial stress σ_x is obtained:

$$\sigma_x = \frac{N}{A} + \frac{M}{I}z + E\alpha(T_m - T + \Theta z) \tag{43.23}$$

The complementary strain energy U^* of the bar is

$$U^* = \int_L \left(\frac{N^2}{2EA} + \frac{M^2}{2EI} + N\alpha T_m + M\alpha\Theta \right) dx \tag{43.24}$$

This follows from Eq. (43.4) upon substituting σ_x from Eq. (43.23) and dropping all terms independent of N and M.

43.4. PLANE STRAIN

Consider a cylindrical body of arbitrary length, and let the coordinate z be in the direction of the axis of the body. If the temperature field is independent of z, and if the end surfaces z = const are kept fixed, the body is in a state of plane strain (see p. 37–1). Displacements and stresses are then independent of z, and

$$\epsilon_z = 0, \; \tau_{yz} = \tau_{zx} = 0$$

The equations of equilibrium (33.25) are identically satisfied upon putting

$$\sigma_x = \frac{\partial^2 F}{\partial y^2} \qquad \tau_{xy} = -\frac{\partial^2 F}{\partial x\,\partial y} \qquad \sigma_y = \frac{\partial^2 F}{\partial x^2} \tag{43.25}$$

F is Airy's stress function. Hooke's law, together with the equations of compatibility, yields

$$\nabla^2\nabla^2 F = -\frac{E\alpha}{1-\nu}\nabla^2 T \tag{43.26}$$

where $\nabla^2 = \partial^2/\partial x^2 + \partial^2/\partial y^2$.

A particular solution of Eq. (43.26) can be found by using the thermoelastic potential Φ. From Eq. (43.9),

$$\nabla^2\Phi = \frac{E}{1-\nu}\alpha T \tag{43.27}$$

The corresponding stress components follow from Eqs. (43.10). Since these stresses will, in general, not satisfy the boundary conditions on the lateral surface of the cylinder, a second set of stresses that corresponds to the general solution of the homogeneous equation (43.26)

$$\nabla^2\nabla^2 F' = 0 \tag{43.28}$$

has to be superposed. The resulting stresses are then

$$\begin{aligned} \sigma_x &= \frac{\partial^2}{\partial y^2}(F' - \Phi) \qquad \tau_{xy} = -\frac{\partial^2}{\partial x\,\partial y}(F' - \Phi) \\ \sigma_y &= \frac{\partial^2}{\partial x^2}(F' - \Phi) \qquad \sigma_z = \nu\nabla^2 F' - \frac{E}{1-\nu}\alpha T = \nu(\sigma_x + \sigma_y) - E\alpha T \end{aligned} \tag{43.29}$$

Hence F' is subject to the boundary conditions

$$\frac{d}{ds}\frac{\partial F'}{\partial y} = p_x + \frac{d}{ds}\frac{\partial \Phi}{\partial y} \qquad \frac{d}{ds}\frac{\partial F'}{\partial x} = -p_y + \frac{d}{ds}\frac{\partial \Phi}{\partial x} \tag{43.30}$$

where p_x and p_y are the x and y components, respectively, of the external applied stresses on the boundary $x = x(s)$, $y = y(s)$ of the cross section, s being the arc length.

In *polar coordinates*, the stress components are:

$$\begin{aligned} \sigma_r &= \left(\frac{1}{r}\frac{\partial}{\partial r} + \frac{1}{r^2}\frac{\partial^2}{\partial \theta^2}\right)(F' - \Phi) \qquad \sigma_\theta = \frac{\partial^2}{\partial r^2}(F' - \Phi) \\ \tau_{r\theta} &= -\frac{\partial}{\partial r}\left[\frac{1}{r}\frac{\partial}{\partial \theta}(F' - \Phi)\right] \end{aligned} \tag{43.31}$$

43.5. PLANE STRESS

If the faces $z = \pm t/2$ of a thin plate of thickness t are free of applied loads, and if the forces on the lateral surface act in the middle plane, the plate is said to be in a state of plane stress (see p. 37-1). The stress component σ_z is then very small. If it is neglected entirely, Hooke's law (43.2) reads

$$\begin{aligned} \sigma_x &= \frac{2G}{1-\nu}[\epsilon_x + \nu\epsilon_y - (1+\nu)\alpha T] \\ \sigma_y &= \frac{2G}{1-\nu}[\epsilon_y + \nu\epsilon_x - (1+\nu)\alpha T] \\ \tau_{xy} &= G\gamma_{xy} \qquad \sigma_z = 0 \end{aligned} \tag{43.32}$$

and

$$\begin{aligned} E\epsilon_x &= \sigma_x - \nu\sigma_y + E\alpha T \qquad E\epsilon_y = \sigma_y - \nu\sigma_x + E\alpha T \\ (1-\nu)\epsilon_z &= (1+\nu)\alpha T - \nu(\epsilon_x + \epsilon_y) \end{aligned} \tag{43.33}$$

Strictly speaking, strains and stresses in plane stress vary with z. However, if these variations are neglected, or if mean values are taken over the thickness of the plate, Airy's stress function may again be introduced. It is now a solution of the equation

$$\nabla^2\nabla^2 F = -E\alpha\,\nabla^2 T \tag{43.34}$$

while the thermoelastic potential now satisfies the equation

$$\nabla^2\Phi = E\alpha T \tag{43.35}$$

The expressions as given in Eqs. (43.25) and (43.29) for σ_x, τ_{xy}, and σ_y remain unchanged while, for σ_z and ϵ_z, we now have

$$\sigma_z = 0 \qquad \epsilon_z = (1+\nu)\alpha T - \frac{\nu}{E}\nabla^2 F' \tag{43.36}$$

Equations (43.5) have to be replaced by

$$\begin{aligned}(1-\nu)\nabla^2 u + (1+\nu)\frac{\partial e}{\partial x} &= 2(1+\nu)\frac{\partial(\alpha T)}{\partial x}\\ (1-\nu)\,\nabla^2 v + (1+\nu)\frac{\partial e}{\partial y} &= 2(1+\nu)\frac{\partial(\alpha T)}{\partial y} \qquad e = \frac{\partial u}{\partial x} + \frac{\partial v}{\partial y}\end{aligned} \tag{43.37}$$

In the preceding equations, the temperature T is assumed to be constant over the thickness of the plate, $T = T(x,y)$. If this is not the case, the plate will, in general, not remain plane (cf. Sec. 43.6).

43.6. BENDING OF PLATES

If the temperature in a thin plate varies over the thickness of the plate, bending will, in general, occur. σ_z may again be neglected, and the relations (43.32) and (43.33) remain valid. All quantities are, however, now functions of z. Neglecting also the deformations of the plate due to the shearing stresses τ_{xz} and τ_{yz}, expanding the strain components in powers of z, and retaining only terms linear in z, we get

$$\epsilon_x = \epsilon_{x0} - z\frac{\partial^2 w}{\partial x^2} \qquad \epsilon_y = \epsilon_{y0} - z\frac{\partial^2 w}{\partial y^2} \qquad \gamma_{xy} = \gamma_{xy0} - 2z\frac{\partial^2 w}{\partial x\,\partial y} \tag{43.38}$$

where the subscript 0 indicates the strain components in the middle plane $z = 0$ of the plate, and $w(x,y)$ represents the deflection. Substituting ϵ_x, γ_{xy}, ϵ_y from Eqs. (43.38), ϵ_z from (43.33), and $\gamma_{xz} = \gamma_{yz} = 0$ into Eq. (43.3), and integrating with respect to z, assuming temperature-independent coefficients, the following expression for the strain energy of the plate is obtained

$$\begin{aligned}U &= \frac{1}{2}\iint_A D[\epsilon_{x0}^2 + \epsilon_{y0}^2 + 2\nu\epsilon_{x0}\epsilon_{y0} + \tfrac{1}{2}(1-\nu)\gamma_{xy0}^2 - 2(1+\nu)\alpha T_m e_0]\,dA\\ &+ \frac{1}{2}\iint_A K\left\{(\nabla^2 w)^2 + 2(1-\nu)\left[\left(\frac{\partial^2 w}{\partial x\,\partial y}\right)^2 - \frac{\partial^2 w}{\partial x^2}\frac{\partial^2 w}{\partial y^2}\right] + 2(1+\nu)\alpha\Theta\,\nabla^2 w\right\} dA\end{aligned} \tag{43.39}$$

where

$$D = \frac{Et}{1-\nu^2} \qquad K = \frac{Et^3}{12(1-\nu^2)} \qquad T_m = \frac{1}{t}\int_{-t/2}^{+t/2} T\,dz$$
$$\Theta = \frac{12}{t^3}\int_{-t/2}^{+t/2} Tz\,dz \tag{43.40}$$

and $e_0 = \epsilon_{x0} + \epsilon_{y0}$. A is the area of the middle plane of the plate. For finite deflections w, the strain components in the middle plane are related to the displacements of that plane by

$$\epsilon_{x0} = \frac{\partial u}{\partial x} + \frac{1}{2}\left(\frac{\partial w}{\partial x}\right)^2 \qquad \epsilon_{y0} = \frac{\partial v}{\partial y} + \frac{1}{2}\left(\frac{\partial w}{\partial y}\right)^2 \qquad \gamma_{xy0} = \frac{\partial u}{\partial y} + \frac{\partial v}{\partial x} + \frac{\partial w}{\partial x}\frac{\partial w}{\partial y} \tag{43.41}$$

The differential equations of the plate may now be obtained by employing the principle of virtual work (see p. 33–19). Introducing Airy's stress function F for the *stress resultants*,

$$N_x = \int_{-t/2}^{+t/2} \sigma_x \, dz = \frac{\partial^2 F}{\partial y^2} \qquad N_{xy} = \int_{-t/2}^{+t/2} \tau_{xy} \, dz = -\frac{\partial^2 F}{\partial x \, \partial y} \qquad N_y = \int_{-t/2}^{+t/2} \sigma_y \, dz = \frac{\partial^2 F}{\partial x^2} \tag{43.42}$$

the equations are

$$\nabla^2\nabla^2 F = Et\left[\left(\frac{\partial^2 w}{\partial x \, \partial y}\right)^2 - \frac{\partial^2 w}{\partial x^2}\frac{\partial^2 w}{\partial y^2} - \alpha \nabla^2 T_m\right]$$

$$K \nabla^2\nabla^2 w = p + \frac{\partial^2 F}{\partial y^2}\frac{\partial^2 w}{\partial x^2} - 2\frac{\partial^2 F}{\partial x \, \partial y}\frac{\partial^2 w}{\partial x \, \partial y} + \frac{\partial^2 F}{\partial x^2}\frac{\partial^2 w}{\partial y^2} - (1+\nu)K\alpha \nabla^2\Theta \tag{43.43a,b}$$

Here p is the lateral load per unit area of the middle plane. In terms of F and w, bending moments M_x, M_y; twisting moments $M_{xy} = M_{yx}$; and shearing forces Q_x, Q_y—all per unit of arc length in the middle plane—are

$$M_x = -K\left[\frac{\partial^2 w}{\partial x^2} + \nu\frac{\partial^2 w}{\partial y^2} + (1+\nu)\alpha\Theta\right] \qquad M_{xy} = -(1-\nu)K\frac{\partial^2 w}{\partial x \, \partial y} \tag{43.44}$$

$$Q_x + \frac{\partial M_{xy}}{\partial y} = \frac{\partial^2 F}{\partial y^2}\frac{\partial w}{\partial x} - \frac{\partial^2 F}{\partial x \, \partial y}\frac{\partial w}{\partial y} - K\left[\frac{\partial^3 w}{\partial x^3} + (2-\nu)\frac{\partial^3 w}{\partial x \, \partial y^2} + (1+\nu)\alpha\frac{\partial \Theta}{\partial x}\right] \tag{43.45}$$

plus two more equations obtained by interchanging x and y.

After F and w have been determined, the stresses are obtained from the following equations:

$$\sigma_x = \frac{1}{t}N_x + \frac{12z}{t^3}M_x + \frac{E\alpha}{1-\nu}(T_m - T + z\Theta)$$

$$\sigma_y = \frac{1}{t}N_y + \frac{12z}{t^3}M_y + \frac{E\alpha}{1-\nu}(T_m - T + z\Theta) \tag{43.46}$$

$$\tau_{xy} = \frac{1}{t}N_{xy} + \frac{12z}{t^3}M_{xy} \qquad \tau_{xz} = \frac{3Q_x}{2t}\left[1 - \left(\frac{2z}{t}\right)^2\right] \qquad \tau_{yz} = \frac{3Q_y}{2t}\left[1 - \left(\frac{2z}{t}\right)^2\right]$$

The two equations (43.43a,b) in F and w are nonlinear. They are restricted to small displacements u and v but not to small deflections w. If, however, w is also small, the nonlinear terms may be dropped. Stress resultants $N_{\alpha\beta}$ and deflection w are then independent of each other, and Eq. (43.43a) reduces to Eq. (43.34) of plane stress:

$$\nabla^2\nabla^2 F = -Et\alpha \nabla^2 T_m \tag{43.47a}$$

while Eq. (43.43b) simplifies to

$$K \nabla^2\nabla^2 w = p - (1+\nu)K\alpha \nabla^2\Theta \tag{43.47b}$$

Then the terms containing F in Eq. (43.45) are also dropped.

43.7. BODIES OF REVOLUTION

In a body of revolution, an axisymmetric temperature field will produce an axisymmetric stress distribution. Using cylindrical coordinates, a particular solution of the problem may be obtained by solving Eq. (43.9) for the thermoelastic potential Φ. The corresponding stresses follow then from Eqs. (43.11). In order to satisfy the boundary conditions, a second state of stress must, in general, be superposed, which corresponds to the general solution of the homogeneous equations of elasticity. In the present case, these equations can be reduced to a single equation in one function $L(r,z)$, which is known as *Love's displacement function,* satisfying the biharmonic equation

$$\nabla^2\nabla^2 L = 0 \tag{43.48}$$

The corresponding displacements u in the radial and w in the axial direction and the corresponding stress components are:

$$2Gu = -\frac{\partial^2 L}{\partial r\, \partial z} \qquad 2Gw = 2(1-\nu)\,\nabla^2 L - \frac{\partial^2 L}{\partial z^2} \tag{43.49}$$

$$\begin{aligned} \sigma_r &= \frac{\partial}{\partial z}\left(\nu\,\nabla^2 L - \frac{\partial^2 L}{\partial r^2}\right) & \sigma_\theta &= \frac{\partial}{\partial z}\left(\nu\,\nabla^2 L - \frac{1}{r}\frac{\partial L}{\partial r}\right) \\ \sigma_z &= \frac{\partial}{\partial z}\left[(2-\nu)\,\nabla^2 L - \frac{\partial^2 L}{\partial z^2}\right] & \tau_{rz} &= \frac{\partial}{\partial r}\left[(1-\nu)\,\nabla^2 L - \frac{\partial^2 L}{\partial z^2}\right] \end{aligned} \tag{43.50}$$

43.8. THIN SHELLS OF REVOLUTION

The temperature distribution is assumed to be symmetrical about the axis of the shell. The only stress components different from zero will then be σ_θ in the circumferential direction, σ_ϕ in the direction of the tangent of the meridian curve, and shearing stresses $\tau_{\phi z}$, where z is in the direction of the outward normal of the middle surface. As in the theory of thin plates, normal stresses σ_z are neglected.

From these stresses the stress resultants shown in Fig. 40.33 can be formed. They must satisfy the equilibrium conditions (40.98), from which the following relations may be derived:

$$\begin{gathered} N_\phi = -Q_\phi \cot\phi \qquad N_\theta = -\frac{1}{r_1}\frac{d}{d\phi}(Q_\phi r_2) \\ \frac{d}{d\phi}(M_\phi r_2 \sin\phi) - M_\theta r_1 \cos\phi - Q_\phi r_1 r_2 \sin\phi = 0 \end{gathered} \tag{43.51}$$

Since $\sigma_z = 0$, Hooke's law has the form (43.32). The temperature terms of these equations have not been included in the derivation of Eqs. (40.99), which now must be replaced by the following:

$$\begin{aligned} N_\phi &= D\left[\frac{1}{r_1}\left(\frac{dv}{d\phi} + w\right) + \frac{\nu}{r_2}(v\cot\phi + w) - (1+\nu)\alpha T_m\right] \\ N_\theta &= D\left[\frac{1}{r_2}(v\cot\phi + w) + \frac{\nu}{r_1}\left(\frac{dv}{d\phi} + w\right) - (1+\nu)\alpha T_m\right] \\ M_\phi &= K\left[\frac{1}{r_1}\frac{d\chi}{d\phi} + \frac{\nu}{r_2}\chi\cot\phi + (1+\nu)\alpha\Theta\right] \\ M_\theta &= K\left[\frac{1}{r_2}\chi\cot\phi + \frac{\nu}{r_1}\frac{d\chi}{d\phi} + (1+\nu)\alpha\Theta\right] \end{aligned} \tag{43.52a–d}$$

Here χ is again the elastic rotation of a tangent to the meridian, as defined by Eq. (40.100), and D, K, T_m, and Θ are defined in Eqs. (43.40). For constant wall thickness and with $U = r_2 Q_\phi$ as before, the following pair of equations may be derived, which replaces Eqs. (40.102):

$$\begin{aligned} L(\chi) - \frac{\nu}{r_1}\chi &= \frac{U}{K} - (1+\nu)\frac{r_2}{r_1}\alpha\frac{d\Theta}{d\phi} \\ L(U) + \frac{\nu}{r_1}U &= -D(1-\nu^2)\chi + D(1-\nu^2)\frac{r_2}{r_1}\alpha\frac{dT_m}{d\phi} \end{aligned} \tag{43.53a,b}$$

In the following, Eqs. (43.53) are specialized to shells of constant meridian curvature.

Spherical Shell. With $r_1 = r_2 = a$, Eqs. (43.53) simplify to

$$\frac{d^2\chi}{d\phi^2} + \frac{d\chi}{d\phi}\cot\phi - (\cot^2\phi + \nu)\chi = \frac{a}{K}U - (1+\nu)a\alpha\frac{d\Theta}{d\phi}$$
$$\frac{d^2U}{d\phi^2} + \frac{dU}{d\phi}\cot\phi - (\cot^2\phi - \nu)U = -D(1-\nu^2)a\left(\chi - \alpha\frac{dT_m}{d\phi}\right) \qquad (43.54a,b)$$

Conical Shell. We use the notation of Fig. 40.10 with the exception that we call the base angle of the cone β instead of α, to avoid the conflict with the thermal expansion coefficient. Equations (43.53) then assume the following form:

$$s^2\frac{d^2\chi}{ds^2} + s\frac{d\chi}{ds} - \chi = \frac{sU}{K}\tan\beta - (1+\nu)\alpha s^2\frac{d\Theta}{ds}$$
$$s^2\frac{d^2U}{ds^2} + s\frac{dU}{ds} - U = -D(1-\nu^2)\left(s\chi\tan\beta - s^2\alpha\frac{dT_m}{ds}\right) \qquad (43.55a,b)$$

The stress resultants are now

$$N_\phi \equiv N_s = -Q_s\cot\beta \qquad N_\theta = -\frac{d(sQ_s)}{ds}\cot\beta$$
$$M_\phi \equiv M_s = K\left[\frac{d\chi}{ds} + \frac{\nu}{s}\chi + (1+\nu)\alpha\Theta\right] \qquad (43.56a\text{–}d)$$
$$M_\theta = K\left[\frac{\chi}{s} + \nu\frac{d\chi}{ds} + (1+\nu)\alpha\Theta\right]$$

and Eq. (40.100) simplifies to

$$\chi = \frac{dw}{ds} \qquad (43.57)$$

Cylindrical Shell. With $dx = r_1\,d\phi$, $\phi = \pi/2$, $r_1 \to \infty$, $r_2 = r = a$, $Q_\phi = Q_x$, Eqs. (43.53) transform into

$$\frac{d^2\chi}{dx^2} = \frac{Q_x}{K} - (1+\nu)\alpha\frac{d\Theta}{dx} \qquad \frac{d^2Q_x}{dx^2} = -\frac{D(1-\nu^2)}{a}\left(\frac{\chi}{a} - \alpha\frac{dT_m}{dx}\right) \qquad (43.58a,b)$$

The stress resultants are:

$$N_\phi \equiv N_x = 0 \qquad N_\theta = -a\frac{dQ_x}{dx}$$
$$M_\phi \equiv M_x = K\left[\frac{d\chi}{dx} + (1+\nu)\alpha\Theta\right] \qquad M_\theta = K\left[\nu\frac{d\chi}{dx} + (1+\nu)\alpha\Theta\right] \qquad (43.59a\text{–}d)$$

Equation (40.100) now reads

$$\chi = \frac{dw}{dx} \qquad (43.60)$$

and, from Eq. (43.52*a*), there follows

$$\frac{dv}{dx} = (1+\nu)\alpha T_m - \nu\frac{w}{a} \qquad (43.61)$$

43.9. APPLICATIONS

In this section the formulas derived in the preceding sections are applied to special problems.

a. Thin Circular Ring

If the outer and the inner faces of the ring are held at different uniform temperatures, the ring will retain its circular shape. From reasons of equilibrium and symmetry, the axial force N and the axial displacement u will vanish, while the radial displacement w will be constant along the axis. Equations (43.20) to (43.22) then yield, with $\kappa = 1/a$, where a is the initial radius of the axis of the ring,

$$w = \alpha T_m a \qquad \lambda = 0 \qquad M = -EI\alpha\Theta \tag{43.62}$$

The axial stress is, from (43.23), $\sigma_x = E\alpha(T_m - T)$.

b. Thin Rectangular Plate (Length a, Width b, Thickness t)

Case 1. The temperature is assumed to be constant across the thickness (direction of z), and constant along the length (direction of x), $T = T(y)$, where $0 \leq y \leq b$. The plate is in a state of plane stress, and free to expand.

From Eqs. (43.25) and (43.34) with $F = F(y)$,

$$\sigma_x = E(-\alpha T + c_0 + c_1 y) \qquad \tau_{xy} = \sigma_y = 0 \tag{43.63}$$

The constants c_0 and c_1 are determined by making the stress resultants zero along the edges $x = 0$ and $x = a$,

$$\int_0^b \sigma_x \, dy = 0 \qquad \int_0^b \sigma_x y \, dy = 0$$

whence

$$c_0 = \frac{1}{b}\int_0^b T \, dy \qquad c_1 = \frac{12}{b^3}\int_0^b Ty \, dy \tag{43.64}$$

This leaves a self-equilibrating system of stresses at the edges $x = 0$ and $x = a$, whose influence, however, according to Saint-Venant's principle, is confined to the proximity of these edges.

The strain components follow from Eq. (43.33), and, upon integration, yield the following displacement components:

$$\begin{aligned} u &= (c_0 + c_1 y)x + \beta y + \gamma \\ v &= (1 + \nu)\alpha \int_0^y T \, dy - \nu\left(c_0 y + c_1 \frac{y^2}{2}\right) - \beta x - c_1 \frac{x^2}{2} + \delta \end{aligned} \tag{43.65}$$

where β, γ, and δ are arbitrary constants.

Case 2. This is the same as case 1, but expansion of the plate in the x direction is completely prevented. Putting $u = 0$ at $x = a$ yields $c_0 = c_1 = \beta = \gamma = 0$ in the preceding solution.

Case 3. The plate of length $2a$ is at a temperature T_0 along the line $x = 0$, where $-a \leq x \leq a$. T falls rapidly to zero on either side of this line, and it is constant across the thickness (direction of z) and along the width (direction of y).

If the plate is free to expand, a tensile stress $\sigma_x = E\alpha T$ is developed along the two edges parallel to the direction of x, which takes on its maximum $E\alpha T_0$ at $x = 0$. A compressive stress σ_y of equal magnitude occurs at the center of the plate.

c. Thin Plate of Arbitrary Shape

The temperature is assumed to vary across the thickness of the plate, but to be constant in the x and y directions $[T = T(z)]$. The following results are derived from the linearized Eqs. (43.47), and are therefore valid only for sufficiently small deformations.

Free Plate. Upon putting $M_x = M_{xy} = M_y = 0$ and $N_x = N_{xy} = N_y = 0$ in Eqs. (43.46), we get

$$\sigma_x = \sigma_y = \frac{E\alpha}{1 - \nu}(T_m - T + z\Theta) \qquad \tau_{xy} = 0 \tag{43.66}$$

From (43.44), $\partial^2 w/\partial x^2 = \partial^2 w/\partial y^2 = -\alpha\Theta$, and hence $w = -(\alpha\Theta/2)(x^2 + y^2)$. The free plate will remain plane if $\Theta = 0$, that is, if the temperature distribution is symmetric with respect to the middle plane ($z = 0$) of the plate.

Again, a self-equilibrating system of stresses persists along the edges. If the plate is circular, its edge line will, after bending, still lie in a plane. Hence, the results are also valid for the simply supported circular plate.

Plate Clamped along All Edges. The deflection is zero everywhere ($w \equiv 0$). Then, from Eqs. (43.44), $M_x = M_y = -(1+\nu)K\alpha\Theta$, $M_{xy} = 0$, and hence, from Eqs. (43.46),

$$\sigma_x = \frac{N_x}{t} + \frac{E\alpha}{1-\nu}(T_m - T) \qquad \sigma_y = \frac{N_y}{t} + \frac{E\alpha}{1-\nu}(T_m - T) \qquad \tau_{xy} = \frac{N_{xy}}{t} \tag{43.67}$$

The stress resultants $N_{\alpha\beta}$ depend on T_m only [see Eqs. (43.42) and (43.47*a*)]. If the plate is free to expand in its middle plane, $N_x = N_{xy} = N_y = 0$. If, on the other hand, the plate is held fixed along all its edges, $N_x = N_y = -Et\alpha T_m/(1-\nu)$, $N_{xy} = 0$.

d. Circular Disk (Radius *a*, Thickness *t*)

If the temperature is a function of r only, and if the plate is free to expand, the stresses are:

$$\begin{aligned} \sigma_r &= E\alpha\left(\frac{1}{a^2}\int_0^a Tr\,dr - \frac{1}{r^2}\int_0^r Tr\,dr\right) \\ \sigma_\theta &= E\alpha\left(\frac{1}{a^2}\int_0^a Tr\,dr + \frac{1}{r^2}\int_0^r Tr\,dr - T\right) \end{aligned} \qquad \tau_{r\theta} = 0 \tag{43.68}$$

provided the disk is sufficiently thin, and hence can be considered to be in a state of plane stress.

If, in particular, the edge $r = a$ is held at constant temperature T_0, and if there is heat transfer from the two faces of the disk into the surrounding medium of zero temperature, the temperature distribution is given by

$$T = T_0 \frac{I_0(mr)}{I_0(ma)} \tag{43.69}$$

where $m^2 = 2k/\lambda t$ [cf. Eq. (43.14)]. I_0 is the modified Bessel function of order zero. Substituting for T into the stress equations as given and integrating, the following stress distribution is obtained:

$$\begin{aligned} \sigma_r &= \frac{E\alpha T_0}{I_0(ma)}\left[\frac{I_1(ma)}{ma} - \frac{I_1(mr)}{mr}\right] \\ \sigma_\theta &= \frac{E\alpha T_0}{I_0(ma)}\left[\frac{I_1(ma)}{ma} + \frac{I_1(mr)}{mr} - I_0(mr)\right] \end{aligned} \qquad \tau_{r\theta} = 0 \tag{43.70}$$

e. Circular Plate (Radius *a*, Thickness *t*)

If the temperature varies across the thickness and along the radial coordinate r but is independent of the polar angle θ, the general solution of Eq. (43.47*b*) with $p = 0$ is

$$w = (1+\nu)\alpha\left[C_1 + C_2 r^2 + \int_r^a \frac{H(r)}{r}\,dr\right] \tag{43.71}$$

where $\qquad H(r) = \int_0^r \Theta r\,dr$

The bending moments are [cf. Eqs. (43.44)]:

$$\begin{aligned} M_r &= -K\left[\frac{d^2w}{dr^2} + \frac{\nu}{r}\frac{dw}{dr} + (1+\nu)\alpha\Theta\right] = -(1+\nu)K\alpha\left[2(1+\nu)C_2 + \frac{1-\nu}{r^2}H\right] \\ M_\theta &= -K\left[\frac{1}{r}\frac{dw}{dr} + \nu\frac{d^2w}{dr^2} + (1+\nu)\alpha\Theta\right] = -(1+\nu)K\alpha\left[2(1+\nu)C_2 - \frac{1-\nu}{r^2}H + (1-\nu)\Theta\right] \end{aligned} \tag{43.72}$$

Plate Simply Supported. From the boundary conditions $w = 0$ and $M_r = 0$ along $r = a$, it follows that

$$C_1 = -a^2C_2 = \frac{1-\nu}{2(1+\nu)} H(a) \qquad (43.72a)$$

Plate Clamped. The boundary conditions $w = 0$ and $dw/dr = 0$ at $r = a$ yield

$$C_1 = -a^2C_2 = -\tfrac{1}{2}H(a) \qquad (43.72b)$$

f. Long Circular Cylinder (Inner Radius b, Outer Radius a)

The temperature is assumed to be a function of r only. The stresses are, then,

$$\begin{aligned} \sigma_r &= \frac{E\alpha}{1-\nu}\left[\frac{1-b^2/r^2}{1-b^2/a^2}\bar{T}(a) - \bar{T}(r)\right] \\ \sigma_\theta &= \frac{E\alpha}{1-\nu}\left[\frac{1+b^2/r^2}{1-b^2/a^2}\bar{T}(a) + \bar{T}(r) - T(r)\right] \\ \tau_{r\theta} &= 0 \qquad \sigma_z = \sigma_r + \sigma_\theta \end{aligned} \qquad (43.73)$$

where $\bar{T}(r) = \dfrac{1}{r^2}\displaystyle\int_b^r \rho T(\rho)\, d\rho$

The axial stresses σ_z form a self-equilibrating system, making the resultant axial force zero. The solution corresponds, therefore, to a cylinder with free ends. For a cylinder whose axial expansion is completely prevented, σ_z takes on the value $\sigma_z = \nu(\sigma_r + \sigma_\theta) - E\alpha T$ [cf. Eq. (43.29)].

g. Long Circular Tube (Inner Radius b, Outer Radius a)

If the inner surface $r = b$ is kept at uniform temperature T_0, and the outer surface $r = a$ is kept at zero temperature, we obtain, from Eq. (43.13), for the temperature distribution,

$$T = T_0 \frac{\ln (a/r)}{\ln (a/b)}$$

The corresponding stress distribution is then

$$\begin{aligned} \sigma_r &= \frac{-E\alpha T_0}{2(1-\nu)} \frac{1}{\ln (a/b)} \left(\ln \frac{a}{r} - \frac{b^2}{r^2}\frac{a^2 - r^2}{a^2 - b^2} \ln \frac{a}{b}\right) \\ \sigma_\theta &= \frac{-E\alpha T_0}{2(1-\nu)} \frac{1}{\ln (a/b)} \left(\ln \frac{a}{r} - 1 + \frac{b^2}{r^2}\frac{a^2 + r^2}{a^2 - b^2} \ln \frac{a}{b}\right) \\ \tau_{r\theta} &= 0 \qquad \sigma_z = \sigma_r + \sigma_\theta \end{aligned} \qquad (43.74)$$

The resultant of the axial stresses σ_z is zero.

If the temperatures along the inner and the outer surfaces of the tube are not uniform, but vary with the polar angle θ, the stresses may be found by superposition of two parts. The first is calculated as before, by taking for T_0 the difference between the mean inside and the mean outside temperatures. To find the second part, the temperatures T_b along the inside $r = b$ and T_a along the outside $r = a$ are developed in Fourier series:

$$T_b = \sum_{n=1}^{\infty} (A_n \cos n\theta + B_n \sin n\theta) + A_0$$

$$T_a = \sum_{n=1}^{\infty} (A'_n \cos n\theta + B'_n \sin n\theta) + A'_0$$

The stress components of the second part are then

$$\sigma_r = \frac{-E\alpha}{2(1-\nu)}\frac{r}{a^2+b^2}\left(1-\frac{b^2}{r^2}\right)\left(\frac{a^2}{r^2}-1\right)(C\cos\theta + D\sin\theta)$$
$$\tau_{r\theta} = \frac{-E\alpha}{2(1-\nu)}\frac{r}{a^2+b^2}\left(1-\frac{b^2}{r^2}\right)\left(\frac{a^2}{r^2}-1\right)(C\sin\theta - D\cos\theta) \qquad (43.75)$$
$$\sigma_\theta = \frac{E\alpha}{2(1-\nu)}\frac{r}{a^2+b^2}\left(3-\frac{a^2+b^2}{r^2}-\frac{a^2b^2}{r^4}\right)(C\cos\theta + D\sin\theta)$$

where $C = \frac{a^2b^2}{a^2-b^2}\left(\frac{A_1}{b}-\frac{A_1'}{a}\right) \qquad D = \frac{a^2b^2}{a^2-b^2}\left(\frac{B_1}{b}-\frac{B_1'}{a}\right)$

The harmonics in the temperature expansion beyond the first produce only axial stress:

$$\sigma_z = E\alpha\left[\frac{\nu}{1-\nu}\frac{r}{a^2+b^2}\left(2-\frac{a^2+b^2}{r^2}\right)(C\cos\theta + D\sin\theta) - T\right] \qquad (43.76)$$

T is given by

$$T = \sum_{n=1}^{\infty} [(a_nr^n + b_nr^{-n})\cos n\theta + (c_nr^n + d_nr^{-n})\sin n\theta] \qquad (43.77)$$

and the constants $a_n \cdot \cdot \cdot d_n$ follow from the boundary conditions $T = T_b - A_0$ on $r = b$ and $T = T_a - A_0'$ on $r = a$. No resultant axial force is produced by the stresses σ_z.

The preceding results correspond to a tube with free ends. If the axial expansion is completely prevented, σ_z has to be replaced by $\sigma_z = \nu(\sigma_r + \sigma_\theta) - E\alpha T$ (see p. 43–13).

h. Solid Sphere (Radius *a*)

The temperature is assumed to be symmetrical with respect to the center, $T = T(r)$. Introducing spherical coordinates r, ϕ, θ, we obtain, from Eqs. (43.9) and (43.12), for the radial and tangential stress components,

$$\sigma_r = \frac{2E\alpha}{1-\nu}\left(\frac{1}{a^3}\int_0^a Tr^2\,dr - \frac{1}{r^3}\int_0^r Tr^2\,dr\right)$$
$$\sigma_\phi = \sigma_\theta = \frac{E\alpha}{1-\nu}\left(\frac{2}{a^3}\int_0^a Tr^2\,dr + \frac{1}{r^3}\int_0^r Tr^2\,dr - T\right) \qquad (43.78)$$

i. Thin Circular Cylindrical Shell

If the temperature in the shell varies only across its thickness, and if the shell is free to expand, then $\chi = 0$, $Q_\phi = 0$, and hence, from Eqs. (43.59),

$$M_x = M_\theta = (1+\nu)K\alpha\Theta \qquad (43.79a)$$

From Eqs. (43.52*a*,*b*), the radial displacement w is found to be

$$w = \frac{dv}{dx} = a\alpha T_m \qquad (43.79b)$$

At a free end $x = 0$ of the shell, the bending moments M_x must be zero. A second solution corresponding to the homogeneous equations (43.58) has then to be superposed:

$$\chi' = \frac{(1+\nu)\alpha\Theta}{\lambda}e^{-\lambda x}\cos\lambda x \qquad Q_x' = 2(1+\nu)\alpha K\Theta\lambda e^{-\lambda x}\sin\lambda x$$

where $\lambda = [3(1-\nu^2)/a^2t^2]^{1/4}$

The corresponding stress resultants are, from Eqs. (43.59),

$$N'_\theta = -a\frac{dQ'_x}{dx} \qquad M'_x = K\frac{d\chi'}{dx} \qquad M_\theta = \nu M'_x$$

Because of the factor $e^{-\lambda x}$, displacements and stresses connected with the additional solution decrease rapidly with increasing distance from the free end. The formulas may also be used for a free end at $x = l$, by replacing x by $(l - x)$, and changing the sign of Q'_x and χ'.

j. Thin Spherical Shell

A closed spherical shell is exposed to a temperature symmetrical with respect to the center, and varying only across the thickness of the shell. Then, for reasons of symmetry, $\chi = 0$, $N_\phi = N_\theta = 0$, and hence, from Eqs. (43.52*c,d*), $M_\phi = M_\theta = (1 + \nu)\alpha K\Theta$. The radial displacement $w = a\alpha T_m$ [Eqs. (43.52*a,b*)] is entirely due to the mean-temperature change T_m. The bending moments do not produce any deformations.

k. Line Source in a Semi-infinite Solid

In the semi-infinite body $x \geq 0$, heat is generated at constant rate H per unit of time and unit of length along a straight line parallel to the z axis, a distance $x = a$ below the surface $x = 0$ of the body. The surface is kept at zero temperature. The body is assumed to be in a state of plane strain.

Combining solutions of Eqs. (43.27) and (43.28), we obtain, for the state of stress,

$$\begin{aligned}\sigma_x &= M\left[2ax\frac{(x+a)^2 - y^2}{r_2^4} - \ln\frac{r_2}{r_1} - y^2\left(\frac{1}{r_2^2} - \frac{1}{r_1^2}\right)\right]\\ \sigma_y &= M\left[\frac{r_2^2(x+2a) + 2xy^2}{r_2^4} - \ln\frac{r_2}{r_1} - \frac{(x+a)^2}{r_2^2} + \frac{(x-a)^2}{r_1^2}\right]\\ \tau_{xy} &= My\left[\frac{x+a}{r_2^2} - \frac{x-a}{r_1^2} - \frac{2a}{r_2^4}(a^2 - x^2 + y^2)\right]\end{aligned} \tag{43.80}$$

where $M = \dfrac{E\alpha H}{4(1-\nu)\lambda\pi} \qquad r_1 = [(x-a)^2 + y^2]^{1/2} \qquad r_2 = [(x+a)^2 + y^2]^{1/2}$

The constant λ represents the conductivity of the body (cf. Sec. 43.1).

l. Point Source on the Surface of a Semi-infinite Solid

Heat is generated at constant rate Q per unit of time at a point on the surface $x = 0$ of the semi-infinite body $x \geq 0$. Everywhere else the surface is perfectly insulated, and there is no loss of heat to the adjacent medium.

In terms of cylindrical coordinates, with the origin placed at the point source, the temperature distribution is

$$T = \frac{Q}{4\pi\lambda R}$$

where $R = (r^2 + z^2)^{1/2}$. Combining solutions of Eqs. (43.9) and (43.48), we get the following stress distribution:

$$\sigma_r = \frac{-E\alpha Q}{4\pi\lambda}\,\frac{1}{R+z} \qquad \sigma_\theta = \frac{E\alpha Q}{4\pi\lambda}\left(\frac{1}{R+z} - \frac{1}{R}\right) \tag{43.81}$$

$$\tau_{rz} = \sigma_z = 0$$

The displacements u in the radial and w in the axial direction are

$$u = (1+\nu)\frac{\alpha Q}{4\pi\lambda}\,\frac{r}{R+z} \qquad w = (1+\nu)\frac{\alpha Q}{4\pi\lambda}\ln(R+z) \tag{43.82}$$

m. Inclusions in Solids

An otherwise homogeneous body contains an inclusion that possesses the same elastic properties as the surrounding material, but is assumed to have a different coefficient of thermal expansion. Then, if a uniform temperature increase T_0 is imposed on the body and the inclusion, stresses will develop.

It is a general property of such states of stress that there is a discontinuity present in the stress distribution along the surface of the inclusion. If z designates the direction of the normal to the surface, the stress components σ_z, τ_{xz}, and τ_{yz} are continuous when the surface is crossed in the direction of z, but the stress components σ_x and σ_y exhibit a discontinuity as given by

$$(\sigma_x)_i - (\sigma_x)_e = (\sigma_y)_i - (\sigma_y)_e = -\frac{E\eta T_0}{1-\nu} \tag{43.83}$$

where $\eta = \alpha_i - \alpha_e$ is the difference between the coefficients of thermal expansion of the inclusion and the surrounding body, and the indices i and e refer to the space internal and external of the inclusion, respectively.

A similar situation arises if a region in a homogeneous body is heated to a temperature T_0 while the remaining body is kept at zero temperature. In that case, η has to be replaced by α in the formulas of this section.

As an example, consider a spherical inclusion of radius a in an infinite body. Using the thermoelastic potential [Eq. (43.9)], we obtain at once the solution

$$\begin{aligned} &2Gu = Cr \qquad \sigma_r = \sigma_\phi = \sigma_\theta = -2C \qquad r < a \\ &2Gu = Ca^3/r^2 \qquad \sigma_r = -2Ca^3/r^3 \qquad \sigma_\phi = \sigma_\theta = Ca^3/r^3 \qquad r > a \end{aligned} \tag{43.84}$$

where
$$C = \frac{E}{1-\nu}\,\frac{\eta T_0}{3}$$

The discontinuity [Eq. (43.83)] in σ_ϕ and σ_θ along $r = a$ is easily observed.

A second example is provided by considering a circular hot spot of radius a and uniform temperature T_0 in an otherwise cold plate. In this case, from Eq. (43.35),

$$\begin{aligned} &\sigma_r = \sigma_\theta = -E\alpha T_0/2 \qquad \tau_{r\theta} = 0 \qquad r < a \\ &\sigma_r = -\sigma_\theta = -E\alpha T_0 a^2/2r^2 \qquad \tau_{r\theta} = 0 \qquad r > a \end{aligned} \tag{43.85}$$

The factor $1/(1-\nu)$ in the discontinuity of σ_θ is missing here. This is due to the fact that the formulas of plane stress have been used, which represent only approximations to the exact solutions of the equations of elasticity,

43.10. CHANGE IN RIGIDITY AND THERMAL BUCKLING

If the temperature change and, hence, the corresponding thermal stresses in a body are sufficiently large, equilibrium may become unstable, and buckling may occur. There are two causes producing this effect. The first is the appearance, in addition to the loads, of external compressive forces, due to restrictions imposed by external constraints on the free expansion of the body. The second originates from a change in rigidity, due to internal self-equilibrating thermal stress systems, and a reduction in Young's modulus E with elevated temperature. In general, both effects will be present. The change in stiffness may be an increase, raising the critical loads, or a decrease, lowering them.

In the extreme case where the stiffness drops to zero, there will be danger of buckling without any external forces being present at all.

A reduction in stiffness also reduces, and an increase raises, the corresponding natural frequencies of the vibrating body. A shift in resonance frequencies or critical speeds may, therefore, take place.

The changes in stiffness as well as buckling are due to nonlinear, second-order effects. The linearized equations of elasticity, therefore, cannot be used.

A few simple cases are discussed in the following. Modulus E is assumed to be independent of the temperature.

Straight Bar Subject to Uniform Temperature Increase

There is no change in the longitudinal stiffness of the bar. However, if the axial expansion is restricted to a value Δl, then, from Eq. (43.21), a compressive force $N = -EA(\alpha T - \epsilon_0)$ is created, where $\epsilon_0 = \Delta l/l$. The bar will buckle if

$$N = N_{cr} = -\frac{\pi^2 EI}{l_{eq}{}^2}$$

l is the length of the bar, and l_{eq} is its equivalent length (see p. 44–23). $\epsilon_0 = 0$ if the ends of the bar are held fixed.

Flat Plate. Temperature Constant over Thickness

Let the temperature distribution be given by $T = T(x,y)$. Increasing T now to a value $\lambda T(x,y)$, the plate will buckle if the coefficient λ corresponds to an eigenvalue of the homogeneous equation (43.43b),

$$K\,\nabla^2\nabla^2 w = \lambda\left(\frac{\partial^2 F}{\partial y^2}\frac{\partial^2 w}{\partial x^2} - 2\frac{\partial^2 F}{\partial x\,\partial y}\frac{\partial^2 w}{\partial x\,\partial y} + \frac{\partial^2 F}{\partial x^2}\frac{\partial^2 w}{\partial y^2}\right) \tag{43.86}$$

Airy's stress function F is associated by Eqs. (43.42) with the stress resultants N_{ij}, due to the temperature $T(x,y)$. It is obtained as a solution of Eq. (43.47a), with $T_m = T$.

Only the smallest eigenvalue of Eq. (43.86) is, in general, of practical interest.

As an example, consider a rectangular plate of lengths L in the y direction and L/n in the x direction, respectively, where $n = 1, 2, 3, \ldots$. The plate is simply supported along its four edges, and the temperature is assumed to be a function of y only [$T = T(y)$]; see p. 43–11. Let the thermal stress distribution be given by

$$N_{xy} = N_y = 0 \qquad N_x = \frac{-\pi^2 K}{L^2}\sum_{m=0}^{\infty} p_m \cos\frac{m\pi y}{L}$$

Then, upon expanding $w(x,y)$ in a Fourier series,

$$w = \sin\frac{n\pi x}{L}\sum_{k=1}^{\infty} a_k \sin\frac{k\pi y}{L}$$

and, substituting in Eq. (43.86), an infinite, homogeneous set of linear equations in the coefficients a_k is obtained. Its determinant must vanish:

$$\begin{vmatrix} 2(p_0 - k_1/\lambda) - p_2 & p_1 - p_3 & p_2 - p_4 & \cdots \\ p_1 - p_3 & 2(p_0 - k_2/\lambda) - p_4 & p_1 - p_3 & \cdots \\ p_2 - p_4 & p_1 - p_3 & 2(p_0 - k_3/\lambda) - p_6 & \cdots \\ \cdots & \cdots & \cdots & \cdots \end{vmatrix} = 0$$

where
$$k_s = \left(\frac{n^2 + s^2}{n}\right)^2$$

The smallest positive root of this equation provides an upper bound for the smallest critical value of λ.

It is sufficient, in general, to retain only the first two or three rows in the determinant.

More general cases can be treated by using the Ritz-Galerkin method. A finite series

$$w(x,y) = \sum_{j=1}^{N} A_j W_j(x,y) \tag{43.87}$$

is assumed for the deflection w where the suitably chosen functions $W_j(x,y)$ satisfy the kinematic boundary conditions. The coefficients A_j are solutions of the N linear and homogeneous equations:

$$\iint pW_j\,dx\,dy - \oint M_n \frac{\partial W_j}{\partial n}\,ds + \oint \bar{q} W_j\,ds = 0 \tag{43.88}$$

where

$$p = K\nabla^2\nabla^2 w - \lambda\left(N_x \frac{\partial^2 w}{\partial x^2} + 2N_{xy}\frac{\partial^2 w}{\partial x\,\partial y} + N_y \frac{\partial^2 w}{\partial y^2}\right)$$

$$M_n = -K\left(\frac{\partial^2 w}{\partial n^2} + \nu \frac{\partial^2 w}{\partial s^2}\right) \tag{43.89}$$

$$\bar{q} = N_n \frac{\partial w}{\partial n} + N_{ns}\frac{\partial w}{\partial s} - K\left[\frac{\partial^3 w}{\partial n^3} + (2-\nu)\frac{\partial^3 w}{\partial n\,\partial s^2}\right]$$

s is the arc length along the boundary of the plate, taken in the counterclockwise direction, and n is the direction of the normal to the boundary, positive outward.

Putting the determinant of these equations equal to zero yields approximations for the desired eignevalues λ.

Laterally Loaded Flat Plate

In order to study the postbuckling behavior of the plate, or to obtain the influence of thermal stresses on the bending stiffness of a plate subject to lateral loads $p(x,y)$, Eqs. (43.43*a,b*) have to be solved simultaneously. Even approximate solutions of the problem require laborious computations.

Torsional Rigidity of a Straight Bar

A straight bar of constant cross section A and length l is subject to self-equilibrating thermal stresses $\sigma_x(y,z)$, x being in the direction of the axis of the bar. A torque M_T is applied. Then, from the principle of energy, the increase U in strain energy equals the work W done by the torque. Assuming Saint-Venant's theory of torsion to apply (see p. 36–2), U and W are given by

$$U = \tfrac{1}{2} GJ_T\theta^2 l + \iint_A \Delta l\,\sigma_x\,dA \qquad W = \tfrac{1}{2} M_T\theta l$$

Here θ is the twist, GJ_T the torsional rigidity of the bar, and Δl denotes the increase in length due to torsion of a longitudinal fiber. The length of the axis of rotation remains unchanged. Then, with

$$\Delta l = \tfrac{1}{2}(y^2 + z^2)\theta^2 l$$

we obtain

$$M_T = \left[GJ_T + \iint_A (y^2 + z^2)\sigma_x\,dA\right]\theta$$

The effective torsional rigidity of the bar is, therefore,

$$(GJ_T)_{\text{eff}} = GJ_T + \iint_A (y^2 + z^2)\sigma_x\,dA \tag{43.90}$$

Since the thermal stress system is self-equilibrating,

$$\iint_A \sigma_x \, dA = 0 \qquad \iint_A y\sigma_x \, dA = \iint_A z\sigma_x \, dA = 0$$

The integral in Eq. (43.90) is independent of the choice of the axis of rotation.

43.11. QUASI-STATIC NONSTEADY PROBLEMS

If the temperature field, or the stress-strain relations, or both are time-dependent, displacements and stresses will be functions of time. In principle, the problem is then one of dynamics, and no longer one of statics. The propagation of heat in a body is, however, a relatively slow process, and dynamic effects may therefore, in general, be neglected (Duhamel's hypothesis). The corresponding solutions are then termed "quasi-static" (cf. Sec. 43.12).

The formulas and results presented in Secs. 43.3 to 43.10, with the exception of Arts. 43.9*k* and *l*, still remain applicable. The only difference that appears is in the temperature distribution, since T now has to satisfy Eq. (43.13) rather than Laplace's equation $\nabla^2 T = 0$. This fact may be used to obtain a solution of Eq. (43.9) for the thermoelastic potential. Neglecting dynamic terms, i.e., putting $\rho = 0$, we have, after differentiation with respect to t and introduction of Eq. (43.13),

$$\nabla^2 \frac{\partial \Phi}{\partial t} = \frac{\kappa^2 E\alpha}{1 - \nu} \nabla^2 T$$

and, hence,

$$\Phi = \frac{\kappa^2 E\alpha}{1 - \nu} \int_0^t T \, dt + \Phi_0 + t\Phi_1 \qquad (43.91)$$

where Φ_1 is an arbitrary harmonic function, and Φ_0 represents the thermoelastic potential at $t = 0$.

The local instantaneous quasi-static stress produced on the surface of a body by an instantaneous local increase T_0 of the surface temperature is

$$\sigma = -E\alpha T_0/(1 - \nu) \qquad (43.92)$$

43.12. DYNAMIC EFFECTS

General Remarks

By exposing the surface of a body to a very hot or very cold medium, sudden changes in the surface temperature may be produced. They are known as *thermal shock*. We might expect dynamic effects to become of influence in that case. The quasi-static solution, as given in the preceding Secs. 43.3 to 43.11, would then have to be replaced by the corresponding solutions of the equations of motion (cf. Sec. 43.1). An example is given below. It is, however, important to note that mathematically discontinuous changes of the surface temperature, as assumed in this example, cannot occur in reality.

Semi-infinite Elastic Body Exposed to Thermal Shock

The semi-infinite body $x \geq 0$ is at zero temperature. At time $t = 0$, its surface temperature is raised to the temperature T_0, and then held constant.

From Eq. (43.13), the corresponding temperature distribution is

$$T = T_0 \operatorname{erfc}\left(\frac{x}{2\kappa\sqrt{t}}\right)$$

If the lateral expansion of the body is completely prevented ($\epsilon_y = \epsilon_z = 0$), then, from Eqs. (43.2),

$$\sigma_y = \sigma_z = \frac{\nu}{1-\nu}\sigma_x - \frac{E\alpha T}{1-\nu}$$

The equations of motion may now be reduced to a single equation in σ_x:

$$\frac{\partial^2 \sigma_x}{\partial x^2} - \frac{1-2\nu}{1-\nu}\frac{\rho}{2G}\frac{\partial^2 \sigma_x}{\partial t^2} = \frac{1+\nu}{1-\nu}\alpha\rho\frac{\partial^2 T}{\partial t^2} \tag{43.93}$$

On the boundary $x = 0$, the condition $\sigma_x = 0$ has to be satisfied.

The solution of this equation is easily obtained by applying a Laplace transform. We get

$$\sigma_x = \frac{E\alpha T_0}{1-2\nu}\left\{F(x,t) - \tfrac{1}{2}\exp\frac{t}{\kappa^2\omega^2}\left[\exp\frac{x}{\kappa^2\omega}\operatorname{erfc}\left(\frac{x}{2\kappa\sqrt{t}} + \frac{\sqrt{t}}{\kappa\omega}\right) + \exp\frac{-x}{\kappa^2\omega}\operatorname{erfc}\left(\frac{x}{2\kappa\sqrt{t}} - \frac{\sqrt{t}}{\kappa\omega}\right)\right]\right\} \tag{43.94}$$

where $\omega = \left[\dfrac{1-2\nu}{2(1-\nu)}\dfrac{\rho}{G}\right]^{1/2}$ and $F(x,t) = \begin{cases} 0 & \text{if } t < x\omega \\ \exp\dfrac{t-\omega x}{(\kappa\omega)^2} & \text{if } t > x\omega \end{cases}$

We observe that there is a stress wave traveling from the surface into the body of amplitude $\sigma_x = E\alpha T_0/(1-2\nu)$. The stress component σ_x is therefore of the same order of magnitude as σ_y and σ_z, in contrast to the quasi-static solution $\sigma_x = 0$, $\sigma_y = \sigma_z = -E\alpha T/(1-\nu)$. It is, however, important to note that the two solutions give identical results on the surface of the body.

43.13. REFERENCES

[1] H. S. Carslaw, J. C. Jaeger: "Conduction of Heat in Solids," 2d ed., Oxford Univ. Press, New York, 1959.
[2] S. P. Timoshenko, J. N. Goodier: "Theory of Elasticity," 2d ed., McGraw-Hill, New York, 1951.
[3] E. Melan, H. Parkus: "Wärmespannungen," Springer, Vienna, 1953.
[4] H. Parkus: "Instationäre Wärmespannungen," Springer, Vienna, 1959.
[5] B. E. Gatewood: "Thermal Stresses," McGraw-Hill, New York, 1957.
[6] J. N. Goodier: Thermal stresses and deformation, *J. Appl. Mechanics*, **24** (1957), 467–474.
[7] "Behaviour of Metals at Elevated Temperatures," Iliffe, London, 1957.
[8] N. J. Hoff (ed.): "High Temperature Effects in Aircraft Structures," Pergamon, London, 1958.
[9] B. A. Boley, J. H. Weiner: "Theory of Thermal Stresses," Wiley, New York, 1960.

CHAPTER 44

ELASTIC STABILITY

BY

C. LIBOVE, Syracuse, N.Y.

44.1. INTRODUCTION

Definitions

Many structures, when loaded to a critical state, will undergo a marked change in the character of their deformation which is not the result of any failure of the material or alteration of its mechanical properties. Such a change is known as *buckling*. The term *elastic buckling* usually means the occurrence of this phenomenon in structures which obey Hooke's law. Buckling generally occurs because one mode of deformation becomes unstable and the structure seeks another, stable, mode. Therefore, the terms *elastic stability* and *elastic instability* are frequently used to designate the field of elastic buckling problems. The buckling load of a structure is not necessarily the maximum load which the structure is capable of supporting, although in some cases the two loads are practically the same.

Kinds of Buckling

We can distinguish several different kinds of buckling, depending on the type of structure and the manner of loading. Three kinds will be described here.

Classical Buckling. The type of buckling which has received the greatest amount of study is the so-called *classical buckling,* which occurs, for example, in a centrally loaded straight column, in a circular ring under uniform radial compression, and in other structures possessing symmetry of form and loading. This type of buckling is characterized by the fact that, as the loading passes through its critical stage, the structure passes from its unbuckled shape to an *infinitesimally close* buckled shape. The buckled shape looks very much different from the unbuckled one, involving, for example, lateral deflections in the case of the column and circumferential waves in the case of the ring.

It is theoretically possible, from the point of view of equilibrium alone, for the structure to keep deforming in its prebuckling mode even beyond the critical loading. Thus, the perfectly straight column *could* remain straight indefinitely and merely continue to shorten; the ring *could* remain circular indefinitely with a constantly diminishing radius. However, beyond the critical load, these unbuckled shapes are unstable, and cannot be maintained if there is the slightest external force tending to disturb them. Furthermore, if we analyze not perfect structures but slightly imperfect ones, it is found that, as the imperfection becomes vanishingly small, the history of load vs. deformation approaches that of the buckled rather than the unbuckled perfect structure. Hence, the practical assumption can be made that, at the critical loading, the structure will always choose the buckled path of deformation. This existence of two different equilibrium paths of deformation, both emanating from the same state of loading, is referred to as *bifurcation* of the equilibrium deformations, and is characteristic of classical buckling.

As a specific illustration of classical buckling, Fig. 44.1*a* shows the theoretical graph of load vs. shortening for a flat rectangular elastic plate compressed between two rigid blocks, with all four edges kept straight and in their original plane. Line OB represents the plate behaving as a compression specimen before buckling. P_{cr} is the buckling (or critical) load beyond which the flat shape (represented by the dashed extension of line OB) becomes unstable. From that point on, the plate develops lateral deflections, and the load-shortening history is represented by curve BC.

Finite-disturbance Buckling. In the case of the flat plate just discussed, there are no buckled forms of equilibrium at loading states below the critical; i.e., curve BC never drops below point B. For other structures—notably the axially compressed thin-walled cylinder—the loss of stiffness after buckling is so great that the buckled shapes can be maintained in static equilibrium only by a return to an earlier state of loading. This phenomenon is illustrated by the load-shortening curve for a cylinder in Fig. 44.1*b*

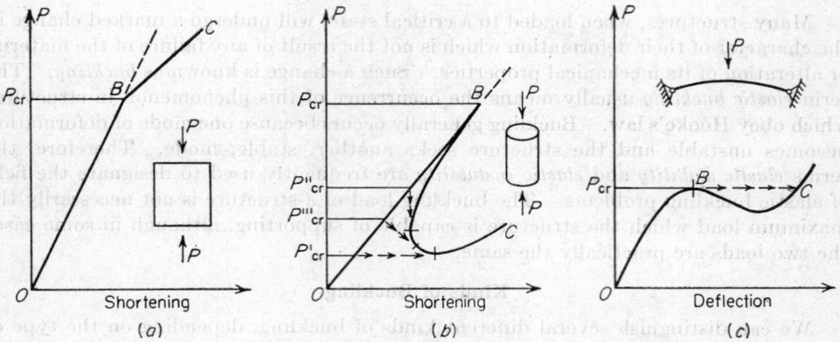

FIG. 44.1. Examples of three kinds of elastic buckling: (*a*) Classical buckling of a flat plate with supported edges; (*b*) finite-disturbance buckling of a cylindrical shell; (*c*) snap-through buckling of a shallow arch with restrained ends.

In structures such as these, another kind of buckling is possible (in fact, probable), for, with a sufficiently large finite disturbance, the structure can pass from an unbuckled equilibrium state to a *nonadjacent* buckled equilibrium state before the classical buckling load (P_{cr} in Fig. 44.1*b*) is reached. We will here refer to such behavior as finite-disturbance buckling.

The arrows in Fig. 44.1*b* indicate three possible occurrences of finite-disturbance buckling for the cylinder. Under dead-weight loading, P'_{cr} is evidently the first load at which this buckling is possible. If the shortening, rather than the load, is kept constant during the buckling process, then P''_{cr} represents the lowest possible buckling load. In an actual testing machine, buckling can first occur at a load P'''_{cr} intermediate between the other two, with a load-shortening history whose slope depends on the stiffness of the testing machine. P'_{cr}, P''_{cr}, and P'''_{cr} are *minimum* loads at which finite-disturbance buckling *can* occur. The question of where buckling *will* occur is still partly a matter of conjecture. It has been suggested that finite-disturbance buckling will be delayed at least to the point where it can occur with no net change in energy for the system made up of the structure and its loading device [6],[7].

Snap-through Buckling. A third kind of buckling is the so-called snap-through (or Durchschlag) buckling, which occurs in structures whose deformation under a load is of such a nature as to lower their stiffness toward further increments of this load. Eventually a point is reached at which the stiffness of the structure becomes zero, and then negative. At this point the structure is unstable under dead-weight loading and snaps into a nonadjacent stable shape.

Figure 44.1*c* illustrates the snap-through buckling of a shallow arch of certain proportions whose ends are restrained from moving apart. Portion *OB* of the load-deflection curve represents the regime of constantly decreasing stiffness; point *B* is the point of zero stiffness; at that point the arch under dead-weight loading will snap into the shape corresponding to point *C*. Snap-through buckling occurs also in shallow spherical domes such as the bottom of an oilcan, for which reason it is also referred to as "oilcanning." The kinking of a venetian blind slat and the flattening instability of curved tubes in bending are additional examples of snap-through buckling.

Because of the shape of the load-flattening curve, it is theoretically possible for the arch of Fig. 44.1*c* to undergo finite-disturbance buckling before point *B* is reached. Figure 44.1*c* represents an always symmetrical deformation. For arches of certain proportions an antimetrical two-lobe type of classical buckling will occur before point *B* is reached.

44.2. THEORETICAL FORMULATIONS OF A CLASSICAL BUCKLING PROBLEM

Some of the equations and theorems of classical buckling, and the methods of analysis which they give rise to, will now be discussed. For the sake of concreteness a specific structure will be considered, one which is simple enough not to require lengthy mathematical developments but yet is sufficiently general to be of some use as a guide for other problems.

The structure selected is a column with one end clamped and the other end hinged, as shown in Fig. 44.2. It is loaded with compressive forces P_0 at the left end and P_1 at the right end, with $P_0 > P_1$ and the difference between them compensated by a leftward-pointing distributed loading whose intensity p per unit length may vary with position along the column.

It is assumed that, during load application, the ratio P_1/P_0 and the shape of the distributed-loading function $p(x)$ remain constant, so that the state of loading can be specified by means of the single parameter P_0. The compressive thrust on any interior cross section can then be expressed as $P(x) = P_0 f(x)$. The functions $P(x)$ and $p(x)$ have the relationship $dP/dx = -p$.

The cross-section centroids lie on a straight line, which is also the line of action of the applied loads before buckling. The principal axes of maximum moment of inertia of the cross sections are assumed to form a single plane, represented by the plane of the paper in Fig. 44.2. During buckling, each element of the distributed loading is assumed to remain constant, but to shift laterally with the deflecting column. Finally, the usual assumptions of beam theory are made. They include, among others, the assumption that during buckling the cross sections remain plane and perpendicular to the (now curved) axis of centroids of the column.

Differential Equation of Equilibrium

One basic method of arriving at a mathematical formulation of a buckling problem is to assume that the structure is buckled and write the necessary equations to make the buckled shape compatible with static equilibrium, material-property, and support requirements. The solution of these equations will indicate the states of loading for which the assumed buckling can exist.[1] This procedure will now be illustrated for the problem at hand.

[1] For a discussion of the shortcomings of this static-equilibrium approach see [8].

Let Fig. 44.2 represent the column just at the point of buckling. Any cross section can be specified by means of its x coordinate in this prebuckling state. Let Fig. 44.3 represent the infinitesimally buckled column. During buckling, the centroid of any cross section is displaced from its prebuckling position by the infinitesimal amounts $u(x)$ parallel to the x axis and $w(x)$ parallel to the z axis. The end forces become P_0' and P_1', and the compressive thrusts on intermediate cross sections become $P'(x)$, where the primed quantities differ infinitesimally from the corresponding unprimed

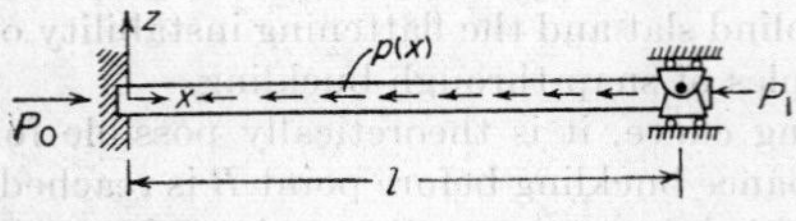

FIG. 44.2. A column-buckling problem.

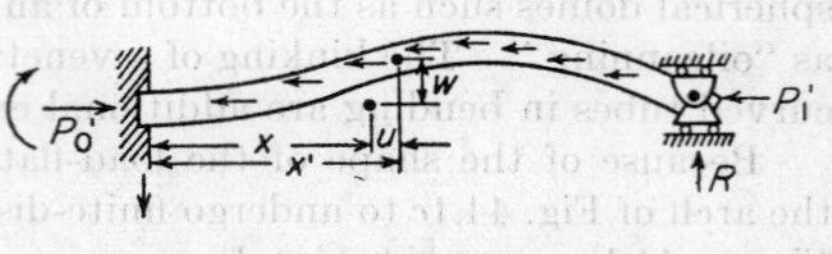

FIG. 44.3. The buckled column.

ones. In addition to the thrusts $P'(x)$, the intermediate cross sections of the buckled columns have on them the infinitesimal bending moments $M(x)$ and shears $V(x)$. Each centroid has the new coordinates $x'(x) = x + u(x)$ and $w(x)$.

Figure 44.4 shows an arbitrarily small element of the column in its prebuckling and buckled positions. If the load application and buckling occur at a vanishingly low rate, as is generally assumed, then each such element must be in static equilibrium. The requirements of equilibrium, correct to infinitesimals of the first order, are

$$V = \frac{dM}{dx} \qquad \text{and} \qquad \frac{dV}{dx} + \frac{d}{dx}\left(P\,\frac{dw}{dx}\right) = 0$$

for equilibrium of moments about an axis perpendicular to the plane of buckling and for equilibrium of forces in the z direction, respectively. (The neglect of higher-order infinitesimals is implied in the second equation, for example, by the use of dx instead of dx', P instead of P', and V in place of its projection. This neglect is legitimate for the analysis of incipient buckling.) Elimination of V between these two equations leads to the following equilibrium equation relating M and w:

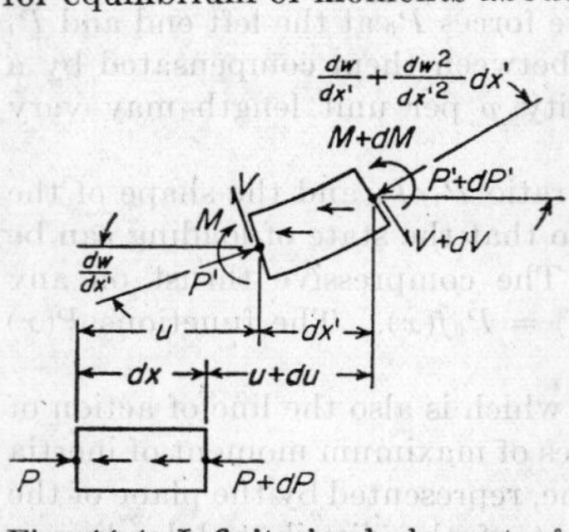

Fig. 44.4. Infinitesimal element of column in prebuckling and during-buckling positions.

$$\frac{d^2M}{dx^2} + \frac{d}{dx}\left(P\,\frac{dw}{dx}\right) = 0 \tag{44.1}$$

In addition to equilibrium, there is the requirement that the internal forces and deformations of the buckled structure be consistent with the stress-strain law of the material. For the element of Fig. 44.4, this requirement is expressed by the beam-flexure equation,

$$M = EI\,\frac{d^2w}{dx^2} \tag{44.2}$$

which, in this form, is also correct to terms of the first order in the buckling displacements.

Elimination of M between Eqs. (44.1) and (44.2) leads to

$$\frac{d^2}{dx^2}\left(EI\,\frac{d^2w}{dx^2}\right) + \frac{d}{dx}\left(P\,\frac{dw}{dx}\right) = 0 \tag{44.3}$$

which is a differential equation for w incorporating the requirements of equilibrium and compatibility of stress and strain.

There are additional restrictions on the buckled shape arising from the boundary conditions. These are of two kinds: *geometric boundary conditions*, which describe any rigid end constraints, and *static boundary conditions*, which state that the internal forces and moments at the column ends must equal any prescribed external forces and moments. In the present example, there are three geometric boundary conditions, namely,

$$w(0) = \frac{dw}{dx}(0) = w(l) = 0 \tag{44.4}$$

and one static boundary condition,

$$M(l) = 0 \tag{44.5}$$

In view of the stress-strain relation [Eq. (44.2)], the static boundary condition may be rewritten as

$$\left(EI\frac{d^2w}{dx^2}\right)_{x\to l} = 0 \tag{44.6}$$

An exact solution of the buckling problem consists of solving Eq. (44.3) for w subject to the boundary conditions (44.4) and (44.6). The form of these equations is characteristic of those obtained for all classical buckling problems: They are linear and homogeneous, and the load parameter appears in the coefficient of a lower-order term of the differential equation. The study of such equations constitutes that branch of mathematics known as *eigenvalue theory*, and much information is available about their properties and methods of solution (see Chap. 18 in this handbook). It is found that only the trivial solution $w = 0$ exists, representing the unbuckled state of equilibrium, unless P_0 takes on one of an infinity of discrete values (eigenvalues). Each nontrivial solution (eigenfunction) for w represents a buckling mode or shape, and the associated value of P_0 specifies the loading state which will maintain that mode in equilibrium with an infinitesimal amplitude. In general, only the lowest of the critical values of P_0 is of practical interest. Because of the homogeneous nature of Eqs. (44.3), (44.4), and (44.6), the amplitude of any buckling mode is indeterminate; i.e., if $w(x)$ is a solution of these equations, then any constant times $w(x)$ is also a solution.

For later use, some important relationships among the buckling modes will now be developed. Let $w_m(x)$ and $w_n(x)$ represent the mth and nth buckling modes and P_{0m} and P_{0n} the respective loading states. The mode w_m satisfies Eq. (44.3) with P replaced by $P_{0m}f(x)$; that is,

$$\frac{d^2}{dx^2}\left(EI\frac{d^2w_m}{dx^2}\right) + P_{0m}\frac{d}{dx}\left(f(x)\frac{dw_m}{dx}\right) = 0$$

Multiplying through by $w_n\,dx$ and integrating from $x = 0$ to $x = l$ gives

$$\int_0^l w_n\frac{d^2}{dx^2}\left(EI\frac{d^2w_m}{dx^2}\right)dx + P_{0m}\int_0^l w_n\frac{d}{dx}\left(f(x)\frac{dw_m}{dx}\right)dx = 0$$

Integrating each term by parts and making use of the boundary conditions on w_m and w_n leads to

$$\int_0^l EI\frac{d^2w_m}{dx^2}\frac{d^2w_n}{dx^2}\,dx - P_{0m}\int_0^l f(x)\frac{dw_m}{dx}\frac{dw_n}{dx}\,dx = 0 \tag{44.7}$$

If now the same steps are repeated but with the roles of w_m and w_n interchanged, the following result is obtained:

$$\int_0^l EI\frac{d^2w_n}{dx^2}\frac{d^2w_m}{dx^2}\,dx - P_{0n}\int_0^l f(x)\frac{dw_n}{dx}\frac{dw_m}{dx}\,dx = 0 \tag{44.8}$$

Subtracting Eq. (44.8) from Eq. (44.7) gives, for $P_{0m} \neq P_{0n}$,

$$\int_0^l f(x) \frac{dw_m}{dx} \frac{dw_n}{dx}\, dx = 0 \tag{44.9}$$

which result, substituted back into either of the two equations, yields

$$\int_0^l EI \frac{d^2w_m}{dx^2} \frac{d^2w_n}{dx^2}\, dx = 0 \tag{44.10}$$

Equations (44.9) and (44.10) are referred to as *orthogonality relationships* among the different buckling modes. Letting $m = n$ in Eq. (44.7) gives, for any one mode, the relationship

$$\int_0^l EI \left(\frac{d^2w_m}{dx^2}\right)^2 dx = P_{0m} \int_0^l f(x) \left(\frac{dw_m}{dx}\right)^2 dx \tag{44.11}$$

Relationship between Elastic Stability Theory and Elasticity Theory. The essential difference between linear (classical) stability theory and the conventional linear elasticity theory (stress analysis) can be described in the following way: In elasticity theory, we ignore the distortions (curvatures, etc.) of the basic element when writing its equilibrium equations, whereas, in stability theory, the equilibrium equations must be written for the *deformed* geometry of the element. Only in this way do products of the applied load and the deflection appear as linear terms in the equations and create an eigenvalue problem.

Had the curvature of the beam element been ignored in the problem just considered, the term $\frac{d}{dx}\left(P\frac{dw}{dx}\right)$ would not have appeared, and Eq. (44.3) would have degenerated into

$$\frac{d^2}{dx^2}\left(EI \frac{d^2w}{dx^2}\right) = 0$$

This equation, in conjunction with the same boundary conditions, has no solution other than $w \equiv 0$, regardless of the magnitude of P.

Potential-energy Methods

In the preceding article, the buckling problem has been represented by a differential equation and boundary conditions. Alternative formulations in terms of certain energy integrals will now be presented.

Consider first the so-called potential-energy integral

$$T = \int_0^l \left[\frac{1}{2} EI \left(\frac{d^2w}{dx^2}\right)^2 - \frac{1}{2} P \left(\frac{dw}{dx}\right)^2\right] dx \tag{44.12}$$

It has a number of important properties which will be stated in the form of theorems. After each theorem, the method or methods of solution which follow from it will be described.

Theorem 1. If in Eq. (44.12) $w(x)$ and $P(x)$ represent a solution to the buckling problem, i.e., if they satisfy Eqs. (44.3), (44.4), and (44.6), then $T = 0$.

This theorem is verified by integrating the right-hand side of Eq. (44.12) by parts (twice in the EI term and once in the P term) to obtain

$$T = \int_0^l \frac{1}{2}\left[\frac{d^2}{dx^2}\left(EI \frac{d^2w}{dx^2}\right) + \frac{d}{dx}\left(P\frac{dw}{dx}\right)\right] w\, dx + \frac{1}{2}\left[EI \frac{d^2w}{dx^2}\frac{dw}{dx}\right]_0^l - \frac{1}{2}\left[\left\{\frac{d}{dx}\left(EI \frac{d^2w}{dx^2}\right) + P\frac{dw}{dx}\right\} w\right]_0^l \tag{44.13}$$

which is evidently zero if w and P satisfy the stated conditions.

Theorem 1 has a simple physical meaning. It is a statement that, during buckling, there is conservation of energy. We need only note that, correct to terms of the lowest order in the buckling displacements, the work of the applied loads during buckling is

$$W = P_0u(0) - P_1u(l) - \int_0^l up\,dx$$
$$= P_0u(0) - P_1u(l) + \int_0^l u\,\frac{dP}{dx}\,dx = -\int_0^l P\,\frac{du}{dx}\,dx$$

the strain energy of bending is

$$U_b = \int_0^l \frac{1}{2}\,EI\left(\frac{d^2w}{dx^2}\right)^2 dx$$

and the increase in extensional strain energy is

$$U_e = -\int_0^l P\left[\frac{du}{dx} + \frac{1}{2}\left(\frac{dw}{dx}\right)^2\right] dx$$

With these expressions the conservation-of-energy condition, namely, $W = U_b + U_e$, is identical with the condition $T = 0$.

There is a still deeper interpretation of this theorem: The statement that $W = U_b + U_e$ means that the column passes from its straight shape to its bent shape without any work by forces external to the system (any strain energy is acquired at the expense of work done by the applied loads of the system). Consequently, the straight column carrying a critical loading is in a state of *neutral* equilibrium. On the other hand, it is known that, for sufficiently small loads (certainly for no load at all), the straight form represents *stable* equilibrium—i.e., positive work would have to be done by forces external to the system in order to bend the column. The plausible inference can therefore be made that one of the critical loadings (the one with the smallest value of P_0) represents the transition point at which the straight form passes from stable to neutral equilibrium and beyond which the straight form represents unstable equilibrium.

Rayleigh Method. If P is replaced by its equivalent form $P_0f(x)$, the equation $T = 0$ gives a formula for any buckling load P_0 in terms of the associated buckling mode $w(x)$. In particular, for the lowest buckling load P_{01} and the corresponding mode $w_1(x)$, the following relationship is obtained:

$$P_{01} = \frac{\int_0^l EI\left(\frac{d^2w_1}{dx^2}\right)^2 dx}{\int_0^l f(x)\left(\frac{dw_1}{dx}\right)^2 dx} \tag{44.14}$$

Equation (44.14) suggests a procedure, known as Rayleigh's method [9], for computing an approximation to the lowest buckling load. The procedure consists in evaluating the right-hand side of Eq. (44.14), not for the true shape $w_1(x)$, which is, of course, generally unknown at the start of a solution, but for a reasonable guess $w_1^*(x)$ which satisfies at least the same geometric boundary conditions as $w_1(x)$, namely, Eqs. (44.4). This results in the following approximation P_{01}^* to the correct buckling load:

$$P_{01}^* = \frac{\int_0^l EI\left(\frac{d^2w_1^*}{dx^2}\right)^2 dx}{\int_0^l f(x)\left(\frac{dw_1^*}{dx}\right)^2 dx} \tag{44.15}$$

The approximation P_{01}^* obtained by this method is never less than the true value of P_{01}. A simple proof of this statement is possible if we assume the validity of the so-called expansion theorem, namely, that any continuous function, and, in particular, $w_1^*(x)$, can be expressed as a linear combination of all the buckling modes. Then

$$w_1^*(x) = a_1 w_1(x) + a_2 w_2(x) + \cdots \tag{44.16}$$

where $w_1, w_2, \ldots$ are the true buckling shapes, assumed, for convenience, to be so normalized that

$$\int_0^l f(x) \left(\frac{dw_n}{dx} \right)^2 dx = 1$$

with which are associated the true buckling loads $P_{01}, P_{02}, \ldots$ $(P_{01} < P_{02} < \cdots)$. Substituting w_1^* from Eq. (44.16) into Eq. (44.15), and making use of the orthogonality relationships (44.9) and (44.10), Eq. (44.11), and the above normalizing condition, we find that

$$P_{01}^* = P_{01} \left[\frac{1 + (a_2/a_1)^2 (P_{02}/P_{01}) + (a_3/a_1)^2 (P_{03}/P_{01}) + \cdots}{1 + (a_2/a_1)^2 + (a_3/a_1)^2 + \cdots} \right] \tag{44.17}$$

The term in brackets is never less than one, and therefore P_{01}^* is never less[1] than P_{01}.

Theorem 1 says that, with the column carrying its critical loading, T is stationary (at the value zero) with respect to lateral deflections which put the straight column into its buckled shape. The following theorem extends this to a much larger class of lateral deflections.

Theorem 2. If $w(x)$ and $P(x)$ represent a nontrivial solution to the buckling problem, then T is stationary with respect to small variatons δw in w, where δw is a function of x satisfying the same geometric boundary conditions as w.

To verify this theorem, let w in the right-hand side of Eq. (44.12) be replaced by $(w + \delta w)$. The first-order change in the value of T produced by this substitution is

$$\delta T = \int_0^l \left(EI \frac{d^2 w}{dx^2} \frac{d^2\, \delta w}{dx^2} - P \frac{dw}{dx} \frac{d\, \delta w}{dx} \right) dx \tag{44.18}$$

or, after integrating by parts,

$$\delta T = \int_0^l \left[\frac{d^2}{dx^2} \left(EI \frac{d^2 w}{dx^2} \right) + \frac{d}{dx} \left(P \frac{dw}{dx} \right) \right] \delta w \, dx + \left[EI \frac{d^2 w}{dx^2} \frac{d\, \delta w}{dx} \right]_0^l - \left[\left\{ \frac{d}{dx} \left(EI \frac{d^2 w}{dx^2} \right) + P \frac{dw}{dx} \right\} \delta w \right]_0^l \tag{44.19}$$

If now w and P satisfy the differential equation (44.3), the bracketed term in the integrand will vanish; if furthermore w satisfies all the boundary conditions (44.4) and (44.6), and δw the geometric boundary conditions (44.4), the remaining terms will vanish. Thus $\delta T = 0$ and, therefore, T is stationary.

Theorem 3. If a $w(x)$ and $P(x)$ can be found such that $w(x)$ satisfies the geometric boundary conditions and T is stationary with respect to small arbitrary variations δw in w also satisfying the geometric boundary conditions, then this $w(x)$ and $P(x)$ represent a solution to the buckling problem—i.e., they satisfy the differential equation and the static boundary condition in addition to the geometric boundary conditions.

This is the converse of Theorem 2. Since δw satisfies the geometric boundary conditions (44.4), the first boundary term at the lower limit and the entire second boundary term in Eq. (44.19) are automatically zero. Since δw is otherwise arbitrary,

[1] We are assuming here that all the critical values of P_0 are positive. It is evident, on physical grounds, that this assumption will be correct if $f(x)$ is never negative—i.e., if the column is in compression along its entire length, and incorrect if $f(x)$ changes sign. Some problems in which the buckling loads are both positive and negative are dealt with by R. V. Southwell [10].

the vanishing of δT implies that

$$\frac{d^2}{dx^2}\left(EI\frac{d^2w}{dx^2}\right) + \frac{d}{dx}\left(P\frac{dw}{dx}\right) = 0$$

and that

$$\left(EI\frac{d^2w}{dx^2}\right)_{x\to l} = 0$$

These are the differential equation and the static boundary condition, thus proving the theorem.

Ritz Method. Theorems 2 and 3 form the basis of a powerful method of solution known as the Ritz method [11]. Instead of solving the differential equation of buckling directly, the Ritz method seeks all values of P_0 and associated shapes $w(x)$ which satisfy the geometric boundary conditions and for which T is stationary with respect to arbitrary δw also satisfying these conditions. By virtue of Theorem 3, these P_0s are buckling loads (of which only the lowest is usually of interest), and the ws are the corresponding buckling modes. Theorem 2 ensures that no buckling loads are missed by this procedure.

In the usual applications of the Ritz method, w is sought in the form of an infinite series

$$w(x) = a_1W_1(x) + a_2W_2(x) + \cdot\cdot\cdot = \Sigma a_nW_n(x) \tag{44.20}$$

where the W_n are specific functions forming a complete set and individually satisfying the geometric boundary conditions, and the a_n are undetermined functions whose relative values are to be selected, in conjunction with P_0, such as to make T stationary. The series for w is substituted into the equation for T[Eq. (44.12) or (44.13)], whose right-hand side then becomes a quadratic function of the a_n. Any arbitrary variation δw consistent with the geometric boundary conditions can be produced in w by giving the a_n in Eq. (44.20) arbitrary variations δa_n. The resulting first-order change in T would be

$$\delta T = \frac{\partial T}{\partial a_1}\,\delta a_1 + \frac{\partial T}{\partial a_2}\,\delta a_2 + \cdot\cdot\cdot$$

Thus, the necessary and sufficient conditions for δT to be zero under all possible small variations in w are

$$\frac{\partial T}{\partial a_1} = 0, \quad \frac{\partial T}{\partial a_2} = 0, \quad \cdot\cdot\cdot \tag{44.21}$$

Equations (44.21) represent an infinite set of linear homogeneous equations in the a_n, with the parameter P_0 appearing in the coefficients of the a_n. These equations have solutions other than the trivial one ($a_1 = a_2 = \cdot\cdot\cdot = 0$), which represents the unbuckled column, only if the determinant of the coefficients of the a_n vanishes. The values of P_0 for which this vanishing occurs are the buckling or critical loads. The buckling mode corresponding to any of the critical values of P_0 can generally be obtained by substituting the value of that P_0 back into all but one of Eqs. (44.21) and solving for the relative values of the a_n.

The calculation of the critical values of P_0, just described in principle, is based on an infinite system of equations, and, therefore, its practical execution, in general, can be accomplished only by means of successive approximation in the following way: Only a finite number of the terms, say the first N, are used in the series representation of w; that is, a_{N+1}, a_{N+2}, etc., are assumed to be zero. Then T becomes a function only of $a_1, a_2, \ldots, a_N$, and the system of Eqs. (44.21) reduces to a finite one. The vanishing of the determinant of this finite system yields approximate buckling loads $\tilde{P}_{01}, \tilde{P}_{02}, \ldots, \tilde{P}_{0N}$. If this calculation is repeated with successively larger values of N, the approximate buckling loads will converge to the true ones.

Each of the approximate buckling loads, $\tilde{P}_{01}, \tilde{P}_{02}, \ldots, \tilde{P}_{0N}$, obtained by using an incomplete series representation for w, is never less than the corresponding one of the first N true buckling loads, $P_{01}, P_{02}, \ldots, P_{0N}$. A proof of this statement as it pertains to the lowest buckling load follows. Let

$$w^*(x) = a_1W_1(x) + a_2W_2(x) + \cdots + a_NW_N(x) \tag{44.22}$$

be the finite-series assumption for the buckling shape. With this series substituted for w in Eq. (44.12) or (44.13), T becomes a function of the N variables: $a_1, a_2, \ldots, a_N$. When the conditions

$$\frac{\partial T}{\partial a_1} = 0, \quad \frac{\partial T}{\partial a_2} = 0, \quad \ldots, \quad \frac{\partial T}{\partial a_N} = 0 \tag{44.23}$$

are imposed, T is made stationary, not with respect to all small δw satisfying the geometric boundary conditions, but only with respect to those δw obtainable by varying the coefficients $a_1, a_2, \ldots, a_N$ in Eq. (44.22). Let $\tilde{P}_{01}$ represent the lowest value of P_0 for which this determinant vanishes, and $w_1^*(x)$ the corresponding shape obtained by solving Eqs. (44.23) for the relative values of the a_n. With P_0 replaced by $\tilde{P}_{01}$, and w replaced by w_1^*, then T as defined by Eq. (44.12) takes on the value

$$T = \int_0^l \left[\frac{1}{2} EI \left(\frac{d^2w_1^*}{dx^2}\right)^2 - \frac{1}{2}\tilde{P}_{01}f(x)\left(\frac{dw_1^*}{dx}\right)^2\right] dx \equiv T^* \tag{44.24}$$

and, by virtue of Eq. (44.23), is stationary at this value with respect to the constrained δw described above. We now note that the stationary value of T, namely, T^*, is zero. This becomes evident when P_0 in Eq. (44.12) is replaced by $\tilde{P}_{01}$ and w by $w_1^* + kw_1^* = (1 + k)w_1^*$, where k is a nonzero constant, small by comparison with 1; then T becomes

$$T = (1 + k)^2T^* = T^* + 2kT^* + k^2T^* \tag{44.25}$$

But kw_1^* is a variation in w obtainable through alteration of the coefficients a_n in Eq. (44.23). Therefore, T is stationary with respect to it, and the first-order change in T, namely, $2kT^*$, must vanish. Since $k \neq 0$, it follows that $T^* = 0$.

As a consequence of the fact that $T^* = 0$, Eq. (44.24) gives the following relationship between $\tilde{P}_{01}$ and w_1^*:

$$\tilde{P}_{01} = \frac{\displaystyle\int_0^l \tfrac{1}{2}EI\left(\frac{d^2w_1^*}{dx^2}\right)^2 dx}{\displaystyle\int_0^l \tfrac{1}{2}f(x)\left(\frac{dw_1^*}{dx}\right)^2 dx} \tag{44.26}$$

Thus $\tilde{P}_{01}$ is identical with the approximate buckling load that would result from the Rayleigh method with w_1^* as the assumed buckling shape [see Eq. (44.15)]. Therefore, $\tilde{P}_{01} \geqq P_{01}$.

Galerkin Method. In using the Ritz method, it is certainly permissible, although not necessary, to seek the buckling modes among functions which satisfy all the boundary conditions, not merely the geometric ones. It would be sufficient then to consider only those variations δw which also satisfy all the boundary conditions. More restrictive forms of Theorems 2 and 3 can readily by devised to cover such a modification of the Ritz method.

The restrictions discussed above can be introduced into the Ritz method by selecting for the $W_n(x)$ in Eq. (44.20) functions which satisfy the static boundary conditions of Eq. (44.6), as well as the geometric ones of Eq. (44.4), and by taking for δw all variations in $w(x)$ obtainable through variations δa_n of the coefficients a_n in Eq. (44.20). Then, in Eq. (44.19), w may be replaced by Σa_nW_n, and δw by

$W_1\delta a_1 + W_2\delta a_2 + \cdot\cdot\cdot$. Furthermore, all the boundary terms in Eq. (44.19) will now vanish, because of either the geometric boundary conditions satisfied by δw or the static boundary conditions satisfied by w. The stationarity condition $\delta T = 0$ therefore becomes

$$\int_0^l \left[\frac{d^2}{dx^2}\left(EI\,\frac{d^2\Sigma a_n W_n}{dx^2}\right) + \frac{d}{dx}\left(P\,\frac{d\Sigma a_n W_n}{dx}\right)\right](W_1\,\delta a_1 + W_2\,\delta a_2 + \cdot\cdot\cdot)\,dx = 0 \tag{44.27}$$

Since Eq. (44.27) must be satisfied for all possible values of $\delta a_1, \delta a_2, \ldots$, it follows that the necessary and sufficient conditions for its satisfaction are

$$\int_0^l \left[\frac{d^2}{dx^2}\left(EI\,\frac{d^2\Sigma a_n W_n}{dx^2}\right) + \frac{d}{dx}\left(P\,\frac{d\Sigma a_n W_n}{dx}\right)\right] W_m\,dx = 0 \qquad m = 1, 2, \ldots \tag{44.28}$$

Equations (44.28) represent a system of homogeneous linear equations in the a_n. The vanishing of the determinant of this system defines the buckling loads. Equations (44.28) will be identical with Eqs. (44.23) if both are based on the same series with each term satisfying all the boundary conditions.

Examination of Eqs. (44.28) shows that they can be regarded as a set of operations applied to the left-hand side of the differential equation (44.3). This set of operations is a method in its own right for the solution of differential equations and is known as the Galerkin method [12]. Here we have chosen to present it as a special form of the Ritz method.

Lagrangian-multiplier Method. The requirement, in the usual application of the Ritz method, that the $W_n(x)$ individually satisfy the geometric boundary conditions is unnecessarily restrictive from the point of view of Theorem 3. It is sufficient that the *sum* of all the $a_n W_n$ satisfy these conditions. Replacing the former requirement by the latter, we arrive at another variation of the Ritz method, proposed by Trefftz [13],[14] and further developed by Budiansky, Hu, and Connor [15],[16]. It is known as the Lagrangian-multiplier method. The greater freedom of choice in the selection of the W_n often makes this method simpler to use than the original Ritz procedure.

In the Lagrangian-multiplier method the series assumed for $w(x)$ is substituted into each geometric boundary condition equation which is not identically satisfied by the individual terms. This substitution furnishes *constraining relationships* among the a_n. We shall symbolize these by

$$g_1(a_1,a_2, \ldots) = 0 \qquad g_2(a_1,a_2, \ldots) = 0 \qquad \text{etc.} \tag{44.29}$$

(In the present problem, there can be, at most, three such constraining relationships; in plate problems, there may be an infinite number of them, inasmuch as there is an infinity of points along each edge of a plate.) With the series for w substituted into the expression for T, the latter becomes a quadratic function of the a_n, and must be made stationary with respect to them. However, the a_n and their variations are no longer independent, but are bound by Eqs. (44.29). Lagrange's method may therefore be used to effect the stationarization. It consists of forming the function

$$F(a_1,a_2, \ldots;\lambda_1,\lambda_2, \ldots) \equiv T - \lambda_1 g_1 - \lambda_2 g_2 - \cdot\cdot\cdot \tag{44.30}$$

where the λs are undetermined constants called Lagrangian multipliers, and making *it* stationary with respect to the constrained a_n by means of the conditions

$$\begin{aligned} \frac{\partial F}{\partial a_1} &= 0, \quad \frac{\partial F}{\partial a_2} = 0, \quad \cdots \\ \frac{\partial F}{\partial \lambda_1} &= 0, \quad \frac{\partial F}{\partial \lambda_2} = 0, \quad \cdots \end{aligned} \tag{44.31}$$

[Note that the second group of Eqs. (44.31) are merely the constraining relationships.] Equations (44.31) constitute a system of homogeneous linear equations in the a_n and the λ_i. The buckling loads are those values of P_0 for which the determinant of this system vanishes, making possible nontrivial solutions for the a_n and λ_i.

The omission of a constraining relationship represents a relaxation of a geometric boundary condition and is therefore equivalent to replacing the actual structure by one with a lower buckling load. Thus, in many applications of the Lagrangian-multiplier method, a lower-limit approximation to the buckling load can be obtained by equating some of the λs to zero, but using all the terms of a complete set in the series for w. If the series for w is truncated, the solution obtained is an upper-limit approximation to the buckling load of a less stiff structure; the relationship between the approximate buckling load and the true one sought is then in doubt.

Complementary-energy Method

The integral

$$C \equiv \int_0^l \left[\frac{1}{2} \frac{M^2}{EI} - \frac{1}{2} P \left(\frac{dw}{dx} \right)^2 \right] dx \tag{44.32}$$

which we shall here call the complementary-energy integral, has properties analogous to those of the potential-energy integral T. These properties are described in the following theorems.

Theorem 4. If $w(x)$, $M(x)$, and $P(x)$ represent a solution to the buckling problem, i.e., if they satisfy Eqs. (44.1), (44.2), (44.4), and (44.5), then $C = 0$.

The simplest verification of this theorem comes from noting that, under the stated conditions, $M = EI(d^2w/dx^2)$; therefore, C is identical with T and zero by Theorem 1.

Theorem 5. If $w(x)$, $M(x)$, and $P(x)$ represent a solution to the buckling problem, then C is stationary with respect to small simultaneous variations δw in w and δM in M which satisfy (1) the equilibrium condition (44.1), and (2) the static boundary condition (44.5).

To verify this theorem let w and M in the definition of C be replaced by $(w + \delta w)$ and $(M + \delta M)$, where δw and δM are functions of x. The first-order change produced thereby in the value of C is

$$\delta C = \int_0^l \left(\frac{M}{EI} \, \delta M - P \frac{dw}{dx} \frac{d\, \delta w}{dx} \right) dx \tag{44.33}$$

or, after integration by parts,

$$\delta C = \int_0^l \frac{M}{EI} \, \delta M \, dx - \left[P \frac{d\, \delta w}{dx} \, w \right]_0^l + \int_0^l w \frac{d}{dx} \left(P \frac{d\, \delta w}{dx} \right) dx \tag{44.34}$$

If, now, condition 1 of the theorem holds, we may replace the term $\dfrac{d}{dx}\left(P \dfrac{d\, \delta w}{dx} \right)$ of the last integral by $-\dfrac{d^2\, \delta M}{dx^2}$. That integral can then be integrated by parts twice, to give the following form for δC:

$$\delta C = \int_0^l \left(\frac{M}{EI} - \frac{d^2 w}{dx^2} \right) \delta M \, dx - \left[\left(P \frac{d\, \delta w}{dx} + \frac{d\, \delta M}{dx} \right) w \right]_0^l + \left[\delta M \, \frac{dw}{dx} \right]_0^l \tag{44.35}$$

By virtue of condition 2, $\delta M(l) = 0$, and the last term vanishes for the upper limit. If, furthermore, M, w, and P represent an exact solution to the buckling problem, they satisfy the stress-strain relation (44.2) and the geometric boundary conditions (44.4). Consequently, the remaining terms on the right-hand side of Eq. (44.35) are identically zero. The vanishing of δC implies the stationarity of C.

Theorem 6. If a $w(x)$, $M(x)$, and $P(x)$ can be found[1] which satisfy the equilibrium equation (44.1) and the static boundary condition (44.5), and for which C is stationary with respect to all small simultaneous variations δw in w and δM in M also satisfying these conditions, then these $w(x)$, $M(x)$, and $P(x)$ represent a solution to the buckling problem—i.e., they fulfill not only the stated conditions, but also the stress-strain law (44.2) and the geometric boundary conditions (44.4).

This is the converse of Theorem 5. Because of Eq. (44.1), we may take Eq. (44.35) as the definition of δC and, because of Eq. (44.5), the last term of that equation has zero for its upper-limit value. Since δM and δw are otherwise arbitrary, the vanishing of δC means that

$$\frac{M}{EI} - \frac{d^2w}{dx^2} = 0 \quad \text{and} \quad w(0) = \frac{dw}{dx}(0) = w(l) = 0$$

which proves the theorem.

Theorems 5 and 6 provide the basis of a method of solution analogous to the Ritz method. $M(x)$ and $w(x)$ can be assumed in the form of series which satisfy identically the equilibrium equation $d^2M/dx^2 + d(P\,dw/dx)/dx = 0$ and the static boundary condition $M(l) = 0$, and which contain undetermined parameters $a_1, a_2, \ldots$. These series can be substituted into the expression for C, whose value is then made stationary by means of the conditions

$$\frac{\partial C}{\partial a_1} = 0, \quad \frac{\partial C}{\partial a_2} = 0, \quad \ldots \tag{44.36}$$

Equations (44.36) will constitute a system of homogeneous linear equations in $a_1, a_2, \ldots$, which can have nontrivial solutions only for such values of P_0 as make their determinant vanish. The lowest nonzero value of P_0 for which the determinant vanishes is the buckling load of practical interest. If the series for $w(x)$ and $M(x)$ are sufficiently complete, the exact buckling load will be obtained; if they are not, an approximate value of the buckling load will result.

The duality of the method of stationary complementary energy just described and the method of stationary potential energy (Ritz method) is brought out in the following parallel statements of the theoretical bases of the two methods (the upper phrase of each pair of braces refers to the potential-energy method, the lower phrase to the complementary-energy method):

Of all nontrivial functions $w(x)$, $M(x)$, and $P(x)$ which satisfy

$$\begin{Bmatrix}\text{Stress-strain relation (44.2)}\\ \text{Equilibrium requirement (44.1)}\end{Bmatrix} \text{ and } \begin{Bmatrix}\text{geometric boundary conditions (44.4)}\\ \text{static boundary condition (44.5)}\end{Bmatrix}$$

those making $\int_0^l \left[\frac{1}{2}\frac{M^2}{EI} - \frac{1}{2}P\left(\frac{dw}{dx}\right)^2\right] dx$ stationary with respect to any small variations δw and δM which also satisfy these conditions will represent solutions to the buckling problem—i.e., they will satisfy not only the stated conditions but also

$$\begin{Bmatrix}\text{Equilibrium requirement (44.1)}\\ \text{Stress-strain relation (44.2)}\end{Bmatrix} \text{ and } \begin{Bmatrix}\text{static boundary condition (44.5)}\\ \text{geometric boundary conditions (44.4)}\end{Bmatrix}$$

Generalization

The $w(x)$, $M(x)$, and $P(x)$ of an exact solution satisfy (1) the stress-strain law and geometric boundary conditions and (2) the equilibrium and static boundary conditions. The methods of stationary potential and stationary complementary energy each satisfy exactly at the outset one of these two pairs of requirements, but leave the

[1] The trivial combination $M(x) \equiv 0$, $P(x) \equiv 0$ is excepted.

other pair to be satisfied indirectly through the stationarity condition. If the stationarity condition is not completely fulfilled, because of incompleteness of the series assumptions for $w(x)$ and $M(x)$, then one pair of requirements will be satisfied only approximately.

One can state a variational theorem which gives preference to none of the requirements of the exact solution and which includes the potential- and complementary-energy theorems as special cases. Such a variational theorem has been presented by E. Reissner [17] for the general boundary-value problem of finite elastic deformations. Using Reissner's work as a guide, we can develop a corresponding theorem for the present buckling problem.[1] It involves, in place of T or C, the quantity

$$R \equiv \int_0^l \left[M \frac{d^2w}{dx^2} - \frac{1}{2}\frac{M^2}{EI} - \frac{1}{2} P \left(\frac{dw}{dx}\right)^2 \right] dx + \left[\left(\frac{dM}{dx} + P \frac{dw}{dx} \right) w \right]_0^l + \left[M \frac{dw}{dx} \right]_{x=0} \tag{44.37}$$

whose first variation, resulting from the variations δw in w and δM in M, is

$$\delta R = \int_0^l \left(M \frac{d^2\,\delta w}{dx^2} + \delta M \frac{d^2w}{dx^2} - \frac{M}{EI}\,\delta M - P \frac{dw}{dx}\frac{d\,\delta w}{dx} \right) dx + \left[\left(\frac{dM}{dx} + P\frac{dw}{dx} \right) \delta w \right]_0^l + \left[\left(\frac{d\,\delta M}{dx} + P \frac{d\,\delta w}{dx} \right) w \right]_0^l + \left[M \frac{d\,\delta w}{dx} + \delta M \frac{dw}{dx} \right]_{x=0}$$

After integration by parts, this becomes

$$\delta R = \int_0^l \left(\frac{d^2w}{dx^2} - \frac{M}{EI} \right) \delta M\, dx + \int_0^l \left[\frac{d^2M}{dx^2} + \frac{d}{dx}\left(P \frac{dw}{dx} \right) \right] \delta w\, dx + \left[\left(\frac{d\,\delta M}{dx} + P \frac{d\,\delta w}{dx} \right) w \right]_0^l + \left[\delta M \frac{dw}{dx} \right]_{x=0} + \left[M \frac{d\,\delta w}{dx} \right]_{x=l} \tag{44.38}$$

From Eq. (44.38) the following stationarity properties of R, analogous to those of C and T, are evident: If $w(x)$, $M(x)$, and $P(x)$ represent an exact solution to the buckling problem—i.e., satisfy Eqs. (44.1), (44.2), (44.4), and (44.5)—then δR is zero (R is stationary) for any δM and δw whatsoever. Conversely, if a $w(x)$, $M(x)$, and $P(x)$ can be found such that $\delta R = 0$ for any δM and δw whatsoever, then these $w(x)$, $M(x)$, and $P(x)$ represent an exact solution to the buckling problem.

With these properties R could, in principle, be used as the basis of a calculation procedure. Note that the functions among which $w(x)$ and $M(x)$ are sought need not satisfy in advance any of the requirements of the exact solution; these are all automatically fulfilled if the stationarity condition is satisfied. On the other hand, if w and M and their variations are constrained to satisfy requirements 1 above, then R reduces to T, and the condition $\delta R = 0$ becomes identical with $\delta T = 0$; if w and M and their variations satisfy instead conditions 2, then R reduces to $-C$, and the condition $\delta R = 0$ becomes equivalent to $\delta C = 0$.

44.3. ILLUSTRATIVE SOLUTIONS

The various methods described in the previous section will now be illustrated by application to a specific problem. The problem selected is that shown in Fig. 44.2 but without the distributed loading and with the flexural stiffness EI constant. In the absence of distributed loading, $f(x) = 1$, and P_0, P_1, and $P(x)$ are all equal and will be designated by the common symbol P. The value of P at which buckling occurs will be represented by the symbol P_{cr}.

[1] An extension of Reissner's principle to buckling problems has also been made by K. Washizu [18].

Solutions Based on Differential Equation

With $P(x)$ and EI constant, Eq. (44.3) reduces to

$$EI \frac{d^4w}{dx^4} + P \frac{d^2w}{dx^2} = 0 \tag{44.39}$$

and has the general solution

$$w = c_1 + c_2x + c_3 \sin kx + c_4 \cos kx \tag{44.40}$$

where

$$k = \sqrt{P/EI} \tag{44.41}$$

and c_1, c_2, c_3, and c_4 are arbitrary constants whose relative values must be such that the general solution satisfies the boundary conditions (44.4) and (44.6). Substitution of the general solution into these boundary-condition equations leads to the following four relationships among the constants:

$$c_1 + c_4 = 0 \qquad c_2 + c_3k = 0$$
$$c_1 + c_2l + c_3 \sin kl + c_4 \cos kl = 0 \qquad c_3 \sin kl + c_4 \cos kl = 0$$

Using the first two equations to eliminate c_1 and c_2 in the last two gives

$$\begin{aligned} c_3\, kl \quad &+ c_4 = 0 \\ c_3 \sin kl + c_4 \cos kl &= 0 \end{aligned} \tag{44.42}$$

Because the right-hand sides of Eqs. (44.42) are zero, these equations will have the trivial solution $c_3 = c_4 = 0$ (and, therefore, $c_1 = c_2 = 0$), leading to $w = 0$ and representing the straight form of equilibrium. Solutions not having c_3 and c_4 both zero are possible if the determinant of the coefficients of c_3 and c_4 in Eqs. (44.42) vanishes, i.e., if

$$\begin{vmatrix} kl & 1 \\ \sin kl & \cos kl \end{vmatrix} = 0$$

or if

$$kl - \tan kl = 0 \tag{44.43}$$

The lowest nonzero root of Eq. (44.43) is $kl = 4.4934$, which, when solved for P, gives the buckling load

$$P_{cr} = 20.19EI/l^2 \tag{44.44}$$

The differential equation can also be solved by means of Fourier series. It will be convenient first to introduce the dimensionless variable

$$\xi = \pi x/2l \tag{44.45}$$

in terms of which the ends of the column are represented by $\xi = 0$ and $\xi = \pi/2$, the differential equation (44.37) by

$$\frac{d^4w}{d\xi^4} + \left(\frac{2kl}{\pi}\right)^2 \frac{d^2w}{d\xi^2} = 0 \tag{44.46}$$

and the boundary conditions (44.4) and (44.6) by

$$w(0) = \frac{dw}{d\xi}(0) = w\left(\frac{\pi}{2}\right) = \frac{d^2w}{d\xi^2}\left(\frac{\pi}{2}\right) = 0 \tag{44.47}$$

The functions $\cos \xi$, $\cos 3\xi$, $\cos 5\xi$, . . . form a complete set for the interval $0 < \xi < \pi/2$. We may, therefore, seek the solution of Eq. (44.46) in the form of the

Fourier series

$$w = a_1 \cos \xi + a_3 \cos 3\xi + a_5 \cos 5\xi + \cdots \tag{44.48}$$

where the a_n are as yet undetermined coefficients related to the buckled shape through the formula

$$a_n = \frac{4}{\pi} \int_0^{\pi/2} w \cos n\xi \, d\xi \tag{44.49}$$

This series satisfies, term by term, all the boundary conditions except $w(0) = 0$. The latter condition yields the following relationship among the Fourier coefficients:

$$a_1 + a_3 + a_5 + \cdots = 0 \tag{44.50}$$

Additional information about the coefficients is obtained from the requirement that the assumed shape, as represented by Eq. (44.48), must satisfy the differential equation. This information, however, cannot be obtained by direct substitution of the series in place of w in Eq. (44.46), for it is by no means certain that the derivatives of w required in that equation are correctly represented by the term-by-term derivatives of the series for w. Therefore an independent series will be assumed for each derivative, namely,

$$dw/d\xi = b_1 \sin \xi + b_3 \sin 3\xi + b_5 \sin 5\xi + \cdots \tag{44.51}$$
$$d^2w/d\xi^2 = c_1 \cos \xi + c_3 \cos 3\xi + c_5 \cos 5\xi + \cdots \tag{44.52}$$
$$d^3w/d\xi^3 = d_1 \sin \xi + d_3 \sin 3\xi + d_5 \sin 5\xi + \cdots \tag{44.53}$$
$$d^4w/d\xi^4 = e_1 \cos \xi + e_3 \cos 3\xi + e_5 \cos 5\xi + \cdots \tag{44.54}$$

and, by means of Stokes' transformation,[1] the coefficients b_n, c_n, d_n, and e_n of these series will be expressed in terms of the coefficients a_n of the series for w. Stokes' transformation consists in integrating by parts in the basic definitions of the b_n, c_n, etc., namely,

$$b_n = \frac{4}{\pi} \int_0^{\pi/2} \frac{dw}{d\xi} \sin n\xi \, d\xi \qquad c_n = \frac{4}{\pi} \int_0^{\pi/2} \frac{d^2w}{d\xi^2} \cos n\xi \, d\xi$$

etc. For example, integrating by parts and making use of the boundary conditions converts the above expression for b_n into

$$\begin{aligned} b_n &= \frac{4}{\pi} \left\{ [w(\xi) \sin n\xi]_0^{\pi/2} - n \int_0^{\pi/2} w \cos n\xi \, d\xi \right\} \\ &= 0 - n \frac{4}{\pi} \int_0^{\pi/2} w \cos n\xi \, d\xi \\ &= -na_n \end{aligned}$$

Operating in the same way with the remaining expressions and making use of boundary conditions, wherever possible, to eliminate boundary terms gives

$$c_n = nb_n = -n^2a_n \qquad d_n = -nc_n = n^3a_n$$
$$e_n = -\frac{4}{\pi} \left(\frac{d^3w}{d\xi^3} \right)_{\xi=0} + nd_n = S + n^4a_n$$

with $$S = -(4/\pi)[d^3w/d\xi^3]_{\xi=0}$$

From these it follows that

$$\frac{d^2w}{d\xi^2} = -\sum n^2a_n \cos n\xi \qquad \frac{d^4w}{d\xi^4} = \sum (S + n^4a_n) \cos n\xi \qquad (44.55), (44.56)$$

[1] See T. J. I'A. Bromwich, "Theory of Infinite Series," 2d ed., Macmillan, London, 1926. Application of Stokes' transformation to the two-dimensional problems of elastic plates was made by A. E. Green [19].

where the symbol Σ here and later is meant to indicate summation only over all the positive odd values of n. These results show that the second derivative of w could have been obtained through term-by-term differentiation of the series of w, but not the fourth derivative.

Substituting the above series into Eq. (44.46), we obtain

$$\sum \left[(S + n^4 a_n) - \left(\frac{2kl}{\pi}\right)^2 n^2 a_n \right] \cos n\xi = 0$$

or

$$a_n = -\frac{S}{n^2[n^2 - (2kl/\pi)^2]} \qquad n = 1, 3, 5, \ldots \qquad (44.57)$$

Using Eq. (44.57) to eliminate the a_n in Eq. (44.50), and noting that $S \neq 0$ for a nontrivial solution, we obtain

$$\sum_{n=1,3,\ldots}^{\infty} \frac{1}{n^2[n^2 - (2kl/\pi)^2]} = 0 \qquad (44.58)$$

as the buckling criterion. The equivalence of Eqs. (44.58) and (44.43) is readily established through the following identities:

$$\frac{1}{n^2[n^2 - (2kl/\pi)^2]} = \left(\frac{\pi}{2kl}\right)^2 \left[\frac{1}{n^2 - (2kl/\pi)^2} - \frac{1}{n^2}\right]$$

$$\sum_{n=1,3,\ldots}^{\infty} \frac{1}{n^2 - (2kl/\pi)^2} = \frac{\pi^2}{8}\frac{\tan kl}{kl}$$

and

$$\sum_{n=1,3,\ldots}^{\infty} \frac{1}{n^2} = \frac{\pi^2}{8}$$

Rayleigh Solutions

For the sake of a simple illustration of the Rayleigh method, let the lowest buckling mode be approximated by the simplest polynomial satisfying the geometric boundary conditions; i.e.,

$$w_1^* = a_1[(x/l)^2 - (x/l)^3] \qquad (44.59)$$

where a_1 is an undetermined amplitude coefficient which will play no role in the solution. Substituting this shape into Eq. (44.15), with EI = Const and $f(x) = 1$, gives the approximate buckling load

$$P_{\text{cr}}^* = \frac{EI}{l^2} \frac{\int_0^1 [4 - 24x/l + 36(x/l)^2]\, d(x/l)}{\int_0^1 [4(x/l)^2 - 12(x/l)^3 + 9(x/l)^4]\, d(x/l)} = 30\frac{EI}{l^2}$$

This result is considerably greater than the correct value of $20.19EI/l^2$. The better result of $21EI/l^2$ is obtained by using, in place of Eq. (44.59), the shape

$$w_1^* = a_1[(x/l)^2 - (x/l)^3] - \tfrac{2}{3}a_1[(x/l)^3 - (x/l)^4]$$

which satisfies *all* the boundary conditions. The coefficient $-\tfrac{2}{3}a_1$ in this expression was selected for the added term in order to make w_1^* satisfy the static boundary condition

$$(d^2w/dx^2)_{x=l} = 0$$

If $-\frac{2}{3}a_1$ is replaced by a coefficient a_2 whose value is permitted to vary so as to minimize P^*_{cr}, the still better approximation $20.92EI/l^2$ is obtained. When the Rayleigh method is used in this fashion, it is equivalent to the Ritz method.

The shape $w^*_1 = a_1(\cos\xi - \cos 3\xi)$, with ξ defined by Eq. (44.45), is an especially good representation of the buckling mode, leading to the result $P^*_{cr} = 20.23EI/l^2$ when substituted in Eq. (44.15). This exceeds the correct value by 0.2%.

Ritz Solution

The series

$$w = a_1(\cos\xi - \cos 3\xi) + a_3(\cos 3\xi - \cos 5\xi) + \cdots \tag{44.60}$$

satisfies, term by term, the geometric boundary conditions, and is therefore an admissible deflection function for use with the Ritz method. Substituting this series for w into Eq. (44.12) gives, for the potential energy,

$$T = \frac{\pi^4 EI}{64l^3}\sum_{n=1,3,\ldots}^{\infty}(\{a_n{}^2[n^4 + (n+2)^4] - 2a_na_{n+2}(n+2)^4\} - L\{a_n{}^2[n^2 + (n+2)^2] - 2a_na_{n+2}(n+2)^2\}) \tag{44.61}$$

where $L \equiv (2kl/\pi)^2$ and k is given by Eq. (44.41). The stationarity conditions

$$\frac{\partial T}{\partial a_m} = 0 \qquad m = 1, 3, 5, \ldots \tag{44.62}$$

lead to the infinite system of equations

$$a_1(82 - 10L) - a_3(81 - 9L) = 0 \tag{44.63}$$

$$-a_{m-2}(m^4 - m^2L) + a_m\{m^4 + (m+2)^4 - [m^2 + (m+2)^2]L\} - a_{m+2}[(m+2)^4 - (m+2)^2L] = 0 \qquad m = 3, 5, 7, \ldots \tag{44.64}$$

In accordance with the earlier discussion of the Ritz method, the buckling loads are defined by the values of L for which the determinant of the coefficients of these equations vanishes, i.e., by the roots of the equation

$$\begin{vmatrix} 82 - 10L & -81 + 9L & 0 & 0 & \cdots \\ -81 + 9L & 706 - 34L & -625 + 25L & 0 & \cdots \\ 0 & -625 + 25L & 3026 - 74L & -2401 + 49L & \cdots \\ 0 & 0 & -2401 + 49L & 8962 - 130L & \cdots \\ \cdots & \cdots & \cdots & \cdots & \cdots \end{vmatrix} = 0 \tag{44.65}$$

Successive approximations of this equation, in the sense discussed earlier, are:

$$82 - 10L = 0 \qquad \begin{vmatrix} 82 - 10L & -81 + 9L \\ -81 + 9L & 706 - 34L \end{vmatrix} = 0$$

$$\begin{vmatrix} 82 - 10L & -81 + 9L & 0 \\ -81 + 9L & 706 - 34L & -625 + 25L \\ 0 & -625 + 25L & 3026 - 74L \end{vmatrix} = 0 \qquad \text{etc.}$$

The first approximation corresponds to the Rayleigh method and gives again the result $L = 8.2$ or $P_{cr} = 20.23EI/l^2$. The second approximation has, for its lowest root, $L = 8.187$ or $P_{cr} = 20.202EI/l^2$. The third gives $L = 8.185$, $P_{cr} = 20.195EI/l^2$.

Galerkin Solution

Each term of the series in Eq. (44.60) satisfies the static as well as the geometric boundary conditions. That series may therefore be used as the basis of a solution by the Galerkin method. For the present application [$P(x)$ and EI constant], the Galerkin equations (44.28) become

$$\int_0^{\pi/2} \left[\frac{d^4}{d\xi^4} (\Sigma a_n W_n) + L \frac{d^2}{d\xi^2} (\Sigma a_n W_n) \right] W_m \, d\xi = 0 \qquad m = 1, 3, 5, \ldots \tag{44.66}$$

where the summation sign Σ represents summation over the positive odd values of n. ξ and L are defined as before, and $W_i \equiv \cos i\xi - \cos (i + 2)\xi$. With W_m, W_n replaced by their appropriate expressions, Eqs. (44.66) become

$$\int_0^{\pi/2} \left\{ \left(\frac{d^4}{d\xi^4} + L \frac{d^2}{d\xi^2} \right) \Sigma a_n [\cos n\xi - \cos (n + 2)\xi] \right\} [\cos m\xi - \cos (m + 2)\xi] \, d\xi = 0$$

$$m = 1, 3, 5, \ldots$$

Carrying out the indicated differentiations and integrations, we obtain precisely the same system of equations, namely, Eqs. (44.63) and (44.64), which the Ritz method gave with the same series.

Lagrangian-multiplier Solution

The series (44.48) is general enough to represent the buckled shape and satisfies all the geometric boundary conditions except $w(0) = 0$. It can therefore be used in connection with a stationary–potential-energy solution only if the constraining relationship (44.50) is imposed on the coefficients. With w replaced by the above series, the potential energy, as defined by Eq. (44.12) with $P(x)$ and EI constant, becomes[1]

$$T = \frac{\pi^4 EI}{64 l^3} \sum a_n^2 \left[n^4 - \left(\frac{2kl}{\pi} \right)^2 n^2 \right] \tag{44.67}$$

The method of stationary potential energy requires that T now be made stationary with respect to the a_n. Because the a_n cannot vary independently but are constrained to satisfy Eq. (44.50) at all times, this stationarity can be accomplished by forming the function

$$F \equiv \frac{\pi^4 EI}{64 l^3} \sum a_n^2 \left[n^4 - \left(\frac{2kl}{\pi} \right)^2 n^2 \right] - \lambda \sum a_n \tag{44.68}$$

and imposing the conditions

$$\frac{\partial F}{\partial \lambda} = 0 \qquad \frac{\partial F}{\partial a_m} = 0 \qquad m = 1, 3, 5, \ldots \tag{44.69}$$

The first of these is merely a restatement of the constraining relationship [Eq. (44.50)]. The rest are

$$\frac{\pi^4 EI}{32 l^3} a_m \left[m^4 - \left(\frac{2kl}{\pi} \right)^2 m^2 \right] - \lambda = 0 \qquad m = 1, 3, 5, \ldots \tag{44.70}$$

whence

$$a_m = \frac{32 l^3 \lambda}{\pi^4 EI} \frac{1}{m^4 - (2kl/\pi)^2 m^2} \qquad m = 1, 3, 5, \ldots \tag{44.71}$$

[1] Here and in the following equations, Σ should be understood to mean summation with respect to all odd values of n only.

Using this result to eliminate the a_n in Eq. (44.50) and dividing through by a constant gives

$$\sum_{m=1,3,\ldots}^{\infty} \frac{1}{m^2[m^2 - (2kl/\pi)^2]} = 0 \tag{44.72}$$

which is the same buckling criterion obtained previously [Eq. (44.58)], by means of a series solution of the differential equation, and is merely another form of the exact solution [Eq. (44.43)].

Solutions by the Method of Stationary Complementary Energy

An approximate solution by means of the complementary-energy method will be obtained by assuming the deflection of the buckled column to be

$$w = a_1[(x/l)^2 - (x/l)^3] \tag{44.73}$$

and the bending moment to be

$$M = R(l - x) - Pa_1[(x/l)^2 - (x/l)^3] \tag{44.74}$$

where R represents the upward reaction at the right end of the column. The first term on the right-hand side of Eq. (44.74) represents the bending moment due to R, and the second term the bending moment produced by the axial force in conjunction with the lateral deflection. As required by the complementary-energy method, the assumed M satisfies the static boundary condition $M(l) = 0$, and the assumed M and w together satisfy the equilibrium equation (44.1).

In terms of the assumed deflection w and bending moment M, the complementary-energy expression [Eq. (44.32)] becomes

$$C = \frac{105EI}{2l}\left\{\alpha^2\left[\left(\frac{Pl^2}{EI}\right)^2 - 14\frac{Pl^2}{EI}\right] - 7\alpha\beta\frac{Pl^2}{EI} + 35\beta^2\right\} \tag{44.75}$$

where $\alpha \equiv a_1/l$ and $\beta \equiv Rl^2/EI$. For C to be stationary with respect to the free constants a_1 and R, it is necessary that

$$\frac{\partial C}{\partial \alpha} = 0 \quad \text{and} \quad \frac{\partial C}{\partial \beta} = 0$$

or

$$\alpha\left(4 - \frac{2}{7}\frac{Pl^2}{EI}\right) + \quad \beta = 0$$

$$\alpha\left(\frac{Pl^2}{EI}\right) \quad - 10\beta = 0$$

These equations can be satisfied in a nontrivial way only if their determinant vanishes. Equating the determinant to zero and solving for Pl^2/EI gives, for the lowest root,

$$\frac{Pl^2}{EI} = \frac{280}{13} \quad \text{or} \quad P_{cr} = 21.54\frac{EI}{l^2}$$

This result is much closer to the exact solution $20.19EI/l^2$ than is the result $30EI/l^2$ obtained by the Rayleigh method with the same deflection function.

In the foregoing example, the assumed deflection shape [Eq. (44.73)] satisfied the geometric boundary conditions. This is not required by the complementary-energy method. In order to illustrate the freedom allowed by the method, let us take for the buckled shape the approximation

$$w = a_1[1 - (x/l)^2] + a_2[1 - (x/l)^3] \tag{44.76}$$

which does not satisfy the geometric boundary condition $w(0) = 0$. An assumed bending moment consistent (from the point of view of equilibrium) with this deflection, and satisfying the static boundary condition $M(l) = 0$, is

$$M = R(l - x) - P\{a_1[1 - (x/l)^2] + a_2[1 - (x/l)^3]\} \tag{44.77}$$

Substituting these assumptions into Eq. (44.32) and making C stationary through the conditions

$$\frac{\partial C}{\partial a_1} = 0 \qquad \frac{\partial C}{\partial a_2} = 0 \qquad \text{and} \qquad \frac{\partial C}{\partial R} = 0$$

leads to the approximation $P_{cr} = 20.53EI/l^2$, which differs from the correct solution by less than 2%.

The complementary-energy method will also yield the exact result if the buckled shape is represented by the Fourier series

$$w = a_1 \cos \xi + a_3 \cos 3\xi + \cdot \cdot \cdot = \Sigma a_n \cos n\xi \tag{44.78}$$

and the bending moment by

$$\begin{aligned} M &= R(l - x) - P \sum a_n \cos n\xi \\ &= \frac{2lR}{\pi}\left(\frac{\pi}{2} - \xi\right) - P \sum a_n \cos n\xi \\ &= \frac{2lR}{\pi} \sum \frac{4}{\pi n^2} \cos n\xi - P \sum a_n \cos n\xi \\ &= \sum \left(\frac{8lR}{\pi^2 n^2} - Pa_n\right) \cos n\xi \end{aligned} \tag{44.79}$$

When substituted into Eq. (44.32), these lead to

$$C = \frac{l}{4EI} \sum \left[\left(\frac{8lR}{\pi^2 n^2} - Pa_n\right)^2 - \frac{PEI\pi^2}{4l^2} a_n{}^2 n^2\right] \tag{44.80}$$

The conditions $\partial C/\partial R = 0$ and $\partial C/\partial a_m = 0$ ($m = 1,3,5, \ldots$) give, respectively,

$$\sum \frac{1}{n^2}\left(\frac{8lR}{\pi^2 n^2} - Pa_n\right) = 0 \tag{44.81}$$

and
$$\frac{8lR}{\pi^2 m^2} - Pa_m + \frac{\pi^2 EI}{4l^2} a_m m^2 = 0 \qquad m = 1, 3, 5, \ldots \tag{44.82}$$

Solving Eqs. (44.82) for the a_m and using the result to eliminate the a_n in Eq. (44.81) gives

$$\sum_{n=1,3,\ldots}^{\infty} \frac{1}{n^2[n^2 - (2kl/\pi)^2]} = 0 \tag{44.83}$$

which, as indicated previously, is equivalent to the exact buckling criterion

$$kl - \tan kl = 0$$

44.4. NUMERICAL DATA

The following tables give numerical data on the buckling strengths of various structural elements. After each table are given the sources for the data presented and, in some cases, sources of information on related problems. These tables are intended as a representative, not an exhaustive, survey of problems solved, For more complete surveys references [1]-[5] on p. 44–42 should be consulted.

Symbols Used in Tables 44.1 to 44.7

A	cross-sectional area
a	(1) rise of unloaded arch axis (Table 44.4) (2) length of plate in direction of applied load (Table 44.5) (3) longer dimension of plate (Tables 44.6, 44.7) (4) base radius of spherical dome (Table 44.7)
b	(1) length of plate in transverse direction (Table 44.5) (2) shorter dimension of plate (Tables 44.6, 44.7)
cr	subscript on loading symbol to designate value of loading at which buckling occurs
E	Young's modulus
G	shear modulus
Γ	warping constant, referred to shear center in Table 44.2
I	moment of inertia of cross section with respect to centroidal axis perpendicular to plane of (1) buckling (Table 44.1); (2) ring or arch (Table 44.4)
I_2	moment of inertia of cross section with respect to axis 2-2
I_y, I_z	moments of inertia of cross section relative to y and z axes, respectively
i_p	polar radius of gyration of cross section with respect to shear center
J	Saint-Venant torsion constant of cross section
K	plate flexural stiffness $= Et^3/[12(1-\nu^2)]$
k	foundation modulus, i.e., stiffness of foundation per unit length of column, force per unit deflection per unit length
l	length of (1) column or beam (Tables 44.1, 44.3); (2) cylinder (Table 44.7); (3) horizontal span of arch (Table 44.4)
m_1, m_2	rotational spring constants, moment per radian
ν	Poisson's ratio
P	applied load
P_{cr}	buckling load
$\bar{P}$	average value of cross-sectional compressive force for columns with distributed axial loading
$\bar{P}_{cr}$	value of $\bar{P}$ at buckling
p_{cr}	critical external pressure
r	(1) mean radius of ring; mean radius of curvature of circular arch (Table 44.4) (2) radius of sphere or cylinder; radius of curvature of cylindrically curved plate (Table 44.7)
α^2	$= E\Gamma/GJl^2$ (Table 44.3B)
σ	applied compressive stress
σ_{cr}	value of σ at buckling
t	thickness of plate or shell
τ	applied shear stress
y, z	principal centroidal axes of cross section
y_S, z_S	absolute values of coordinates of shear center S relative to y and z axes

Table 44.1. Flexural Buckling of Uniform Columns

A. Columns with Simple End Conditions under Uniform Compression*

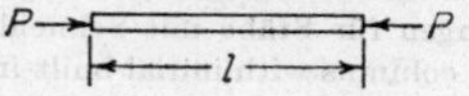

Assumptions: Infinite shear modulus, infinite torsional stiffness (precluding torsional buckling), no local buckling. Load applied through centroids of end cross sections; direction of load unchanged during buckling. $I_z \geq I_y$.

Case	End conditions: Left	End conditions: Right	Buckling mode	Buckling load, P_{cr}
1	Ball bearing or hinge along y axis	Ball bearing or hinge along y axis		$P_{cr} = \frac{\pi^2 EI_y}{l^2}$
2	Clamped	Clamped		$P_{cr} = \frac{4\pi^2 EI_y}{l^2}$
3	Clamped	Ball bearing or hinge along y axis		$P_{cr} = \frac{20.19 EI_y}{l^2}$
4	Clamped	Free		$P_{cr} = \frac{\pi^2 EI_y}{4l^2}$
5	Hinge along z axis	Hinge along z axis	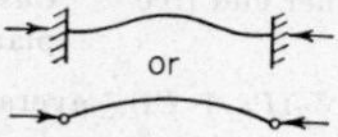	P_{cr} = smaller of $\frac{4\pi^2 EI_y}{l^2}$ or $\frac{\pi^2 EI_z}{l^2}$
6	Hinge along z axis	Hinge along y axis		$P_{cr} = \frac{20.19 EI_y}{l^2}$
7	Hinge in y,z plane, making an angle α with y axis	Hinge parallel to the one at the left end	Space curve	$P_{cr} = \frac{c\pi^2 EI_y}{l^2}$ where c depends on α and I_z/I_y. The variation of $\sqrt{c}$ with α for several values of I_z/I_y is shown in the graph on p. 44-24

* Flexural buckling data are frequently given in terms of an effective length l_e, defined as the length of a hinged-hinged column whose buckling load would be the same as that of the actual column; i.e., l_e and P_{cr} are related through the formula (see case 1) $P_{cr} = \pi^2 EI/l_e^2$, where I is either I_y or I_z as appropriate. Effective-length formulas for various end conditions are readily obtained by comparing this equation with the ones in the last column of the table. Thus, for example, the effective length for a column whose end conditions are those of case 2 is $l_e = l/2$; for case 7, the effective length is $l_e = l/\sqrt{c}$.

Table 44.1. Flexural Buckling of Uniform Columns (*Continued*)

SOURCES: The graph for case 7 is taken from H. Ziegler, Die Knickung des schief gelagerten Stabes, *Schweiz. Bauztg.*, **66** (1948), 87–89. Columns with nonparallel end hinges are treated by P. Fillunger, Über die Eulerschen Knickbedingungen für Stäbe mit Schneidenlagerung, *Z. angew. Math. Mech*, **6** (1926), 294–308. Buckling loads for columns with initial built-in twist are given by H. Ziegler [8].

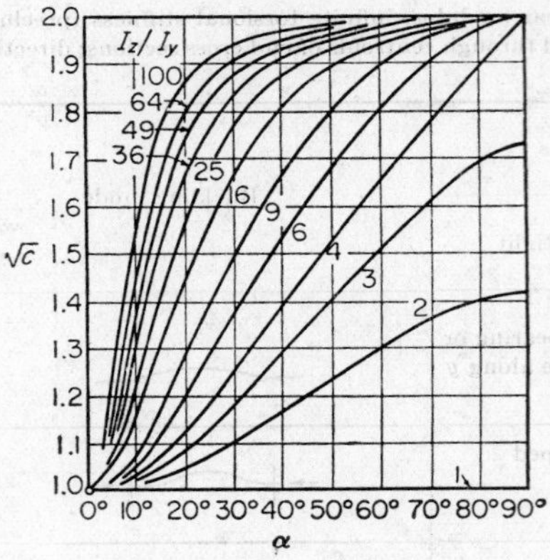

Graph showing $\sqrt{c}$ for case 7, p. 44–23.

B. COLUMNS WITH UNIFORMLY DISTRIBUTED AXIAL LOADING AND END LOADS*

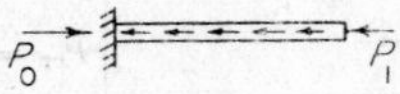

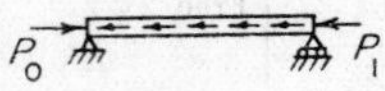

Case 1. One end clamped, other end free

Case 2. Both ends have hinges perpendicular to plane of buckling

$P_0 \geq P_1$ $\quad \bar{P} = \frac{1}{2}(P_0 + P_1)$ average compressive force in column

Additional Assumption: Column constrained to buckle in a single plane (actual constraint may not be necessary if plane of buckling is a principal plane).

P_1/P_0		1.0	0.9	0.8	0.7	0.6	0.5	0.4	0.3	0.2	0.1	0	−0.1	−0.2	−0.3
$\dfrac{\bar{P}_{cr}l^2}{EI}$	Case 1	$\pi^2/4$	2.52	2.58	2.64	2.73	2.84	2.98	3.17	3.38	3.61	3.93	4.37		
	Case 2	1	1.00	0.999	0.998	0.996	0.993	0.989	0.983	0.973	0.959	0.940	0.915	0.884	0.84

* Table shows variation of $\bar{P}_{cr}l^2/EI$ with P_1/P_0 for each case illustrated. Negative values of the ratio P_1/P_0 correspond to negative (or tensile) values of P_1.

SOURCES: The data for case 1 are adapted from [5], p. 104, where the solution is credited to N. Grishcoff. The data for case 2 are adapted from C. Libove, S. Ferdman, and J. J. Reusch, Elastic buckling of a simply supported plate under a compressive stress that varies linearly in the direction of loading, *NACA Tech. Note* 1891 (1949).

Table 44.1. Flexural Buckling of Uniform Columns (*Continued*)

C. Infinitely Long Columns on Elastic Foundation

Additional Assumption: The foundation modulus k is a constant, independent of the buckle pattern of the column, as though the foundation consisted of infinitesimally close independent springs in the plane of buckling.

Case	Description		Buckling load
1	Ends hinged or clamped	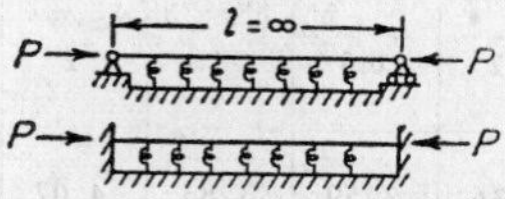	$P_{cr} = 2\sqrt{kEI}$
2	One end free, other end hinged or clamped	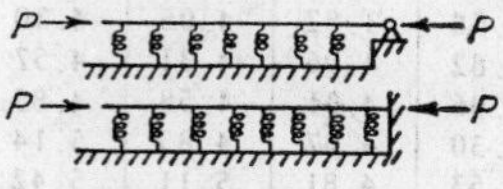	$P_{cr} = \sqrt{kEI}$
3	Ends hinged; foundation attached to rigid slab which is free to translate vertically		$P_{cr} = \sqrt{kEI}$
4	Column with a free cantilever portion	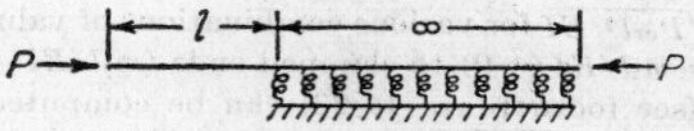	$P_{cr} = c\pi^2 EI/l^2$, where c and k are related by $\sqrt{\dfrac{kl^4}{EI}} = \dfrac{c\pi^2}{1 - \sin\sqrt{c\pi^2}}$
5	Column with a free middle span	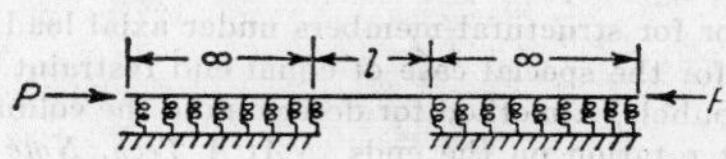	$P_{cr} = c\pi^2 EI/l^2$, where c and k are related by $\sqrt{\dfrac{kl^4}{EI}} = \dfrac{c\pi^2}{1 + \cos\sqrt{c\pi^2/4}}$

Sources: All the above data except for case 3 are taken from M. Hetényi, "Beams on Elastic Foundation," Univ. of Michigan Press, Ann Arbor, 1946. The buckling load for case 3 is from J. N. Goodier, Some observations on elastic stability, *Proc. 1st U.S. Natl. Congr. Appl. Mechanics*, 1952, pp. 193–202.

For data on finite-length columns on elastic foundation see the above-cited work by M. Hetényi.

Table 44.1. Flexural Buckling of Uniform Columns (*Continued*)

D. Columns with Ends Elastically Restrained against Rotation*

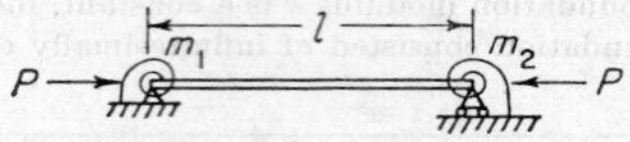

m_1l/EI	m_2l/EI: 0	0.1	0.2	0.5	1	2	5	10	20	50	∞
0	π	3.26	3.36	3.59	3.83	4.07	4.29	4.39	4.44	4.47	4.49
0.1	3.26	3.38	3.48	3.71	3.95	4.19	4.42	4.52	4.57	4.60	4.62
0.2	3.36	3.48	3.58	3.82	4.06	4.30	4.53	4.63	4.68	4.72	4.74
0.5	3.59	3.71	3.82	4.06	4.31	4.57	4.81	4.91	4.96	5.00	5.02
1	3.83	3.95	4.06	4.31	4.58	4.85	5.11	5.21	5.27	5.30	5.33
2	4.07	4.19	4.30	4.57	4.85	5.14	5.42	5.54	5.60	5.64	5.66
5	4.29	4.42	4.53	4.81	5.11	5.42	5.73	5.86	5.92	5.96	5.99
10	4.39	4.52	4.63	4.91	5.21	5.54	5.86	5.99	6.06	6.10	6.13
20	4.44	4.57	4.68	4.96	5.27	5.60	5.92	6.06	6.13	6.18	6.21
50	4.47	4.60	4.72	5.00	5.30	5.64	5.96	6.10	6.18	6.22	6.25
∞	4.49	4.62	4.74	5.02	5.33	5.66	5.99	6.13	6.21	6.25	π

* Table gives values of $\sqrt{P_{cr}l^2/EI}$ for various combinations of values of m_1l/EI and m_2l/EI, ranging from hinged ends ($m_1l/EI = m_2l/EI = 0$) to clamped ends ($m_1l/EI = m_2l/EI = \infty$).

The effective lengths l_e (see footnote, p. 44-23) can be computed from the values of $\sqrt{P_{cr}l^2/EI}$ through the relationship $l_e/l = \pi/\sqrt{P_{cr}l^2/EI}$.

Sources: The above data are adapted from E. E. Lundquist and W. D. Kroll, Extended tables of stiffness and carry-over factor for structural members under axial load, *NACA, Wartime Rept.* L-255 (1944). More precise data for the special case of equal end restraint are given by E. E. Lundquist, C. A. Rossman, and J. C. Houbolt, A method for determining the column curve from tests of columns with equal restraints against rotation on the ends, *NACA Tech. Note* 903 (1943). The influence of rigid-end portions on the buckling strength of elastically restrained columns is considered by W. R. Osgood, Column strength of tubes elastically restrained against rotation at the ends, *NACA Rept.* 615 (1938).

Columns with *intermediate* elastic springs—both deflectional and rotational—are considered by B. Budiansky, P. Seide, and R. A. Weinberger, The buckling of a column on equally spaced deflectional and rotational springs, *NACA Tech. Note* 1519 (1948). A case in which the intermediate springs (deflectional only) are themselves resting on an elastic medium is treated by P. Seide and J. F. Eppler, The buckling of parallel simply supported tension and compression members connected by elastic deflectional springs, *NACA Tech. Note* 1823 (1949).

Table 44.2. Flexural-Torsional Buckling of Thin-walled Columns*

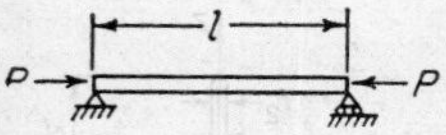

Case 1. End cross sections restrained against rotation about the longitudinal axis but otherwise free to rotate and warp

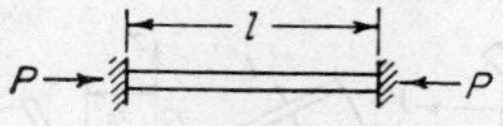

Case 2. End cross sections fully clamped against rotation and warping

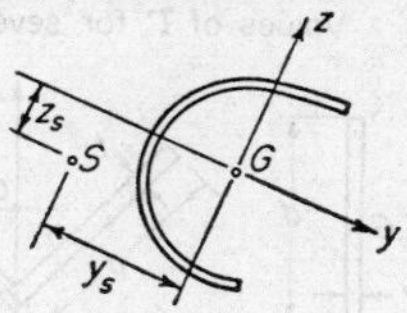

Cross section
G = centroid
S = shear center

Assumptions: No local buckling; centroid of applied load coincides with centroid of end sections, column free to twist about an axis of its own choosing.

Case	Symmetry conditions	Buckling load
a	y and z are both axes of symmetry (y_S, z_S = 0)	P_{cr} = smallest of P_y, P_z, and P_T. Buckling is either purely flexural with no twist, or purely twist about the shear centers
b	z axis is an axis of symmetry ($y_S = 0$, $z_S \neq 0$)	$P_{cr} = P_y$ or smallest root of equation $\left(1 - \frac{P_z}{P_{cr}}\right)\left(1 - \frac{P_T}{P_{cr}}\right) = \left(\frac{z_S}{i_p}\right)^2$, whichever is lower
c	Neither y nor z is an axis of symmetry	P_{cr} = smallest root of equation $\left(1 - \frac{P_y}{P_{cr}}\right)\left(1 - \frac{P_z}{P_{cr}}\right)\left(1 - \frac{P_T}{P_{cr}}\right) - \left(1 - \frac{P_y}{P_{cr}}\right)\left(\frac{z_S}{i_p}\right)^2 - \left(1 - \frac{P_z}{P_{cr}}\right)\left(\frac{y_S}{i_p}\right)^2 = 0$

* Buckling loads for flexure without twist:

$$P_y = \frac{\pi^2 EI_y}{l^2} \quad \text{for case 1} \qquad P_z = \frac{\pi^2 EI_z}{l^2} \quad \text{for case 1}$$

$$= \frac{4\pi^2 EI_y}{l^2} \quad \text{for case 2} \qquad = \frac{4\pi^2 EI_z}{l^2} \quad \text{for case 2}$$

Critical load for torsional buckling about axis of shear centers:

$$P_T = \frac{1}{i_p{}^2}\left(GJ + \frac{\pi^2 E\Gamma}{l^2}\right) \quad \text{for case 1} \qquad P_T = \frac{1}{i_p{}^2}\left(GJ + \frac{4\pi^2 E\Gamma}{l^2}\right) \quad \text{for case 2}$$

Sources: The buckling-load formulas presented above are taken from [1]. The formulas for Γ for all the sections except the zee are from [2]. Values of Γ for many other types of cross section can be found in both these books. An especially clear presentation of the formulas and procedure for the calculation of Γ for arbitrary thin-walled sections is given in [1]. For a discussion of torsional buckling with an enforced axis of rotation—as when the column is attached to a sheet—the reader is referred to the afore-mentioned works.

Table 44.2. Flexural-Torsional Buckling of Thin-walled Columns *(Continued)*

Values of Γ for several cross sections:

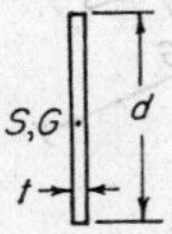

$$\Gamma = \frac{t^3 d^3}{144}$$

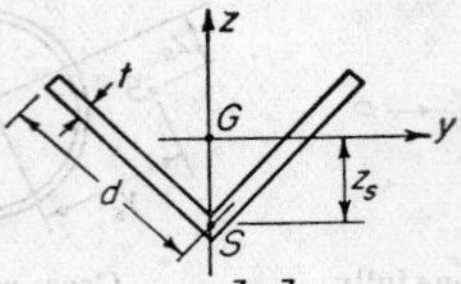

$$\Gamma = \frac{t^3 d^3}{18}$$

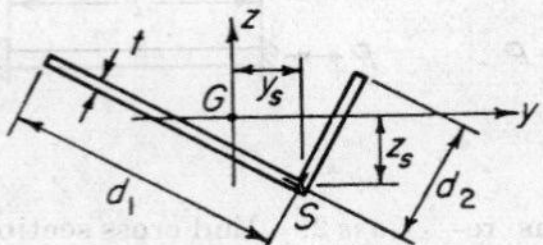

$$\Gamma = \frac{t^3}{36}\left(d_1^3 + d_2^3\right)$$

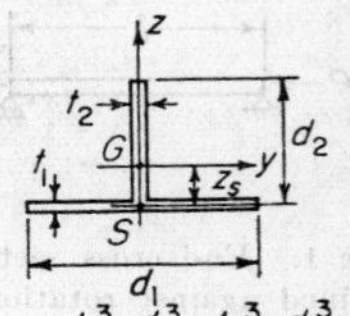

$$\Gamma = \frac{t_1^3 d_1^3}{144} + \frac{t_2^3 d_2^3}{36}$$

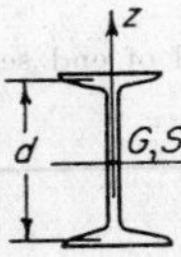

$$\Gamma = \frac{d^2 I_y}{4}$$

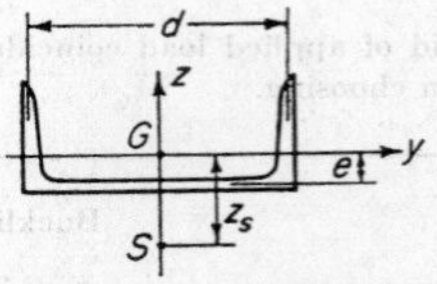

$$\Gamma = \frac{d^2}{4}\left[I_y + e^2 A\left(1 - \frac{d^2 A}{4 I_z}\right)\right]$$

where

A = cross-sectional area

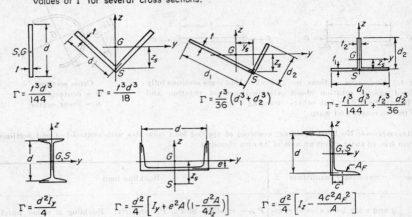

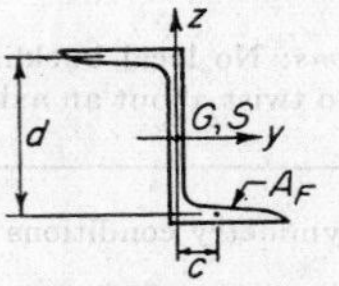

$$\Gamma = \frac{d^2}{4}\left[I_z - \frac{4c^2 A_F^{\,2}}{A}\right]$$

where

A_F = cross-sectional area of one flange

A = total cross-sectional area

c = distance between flange centroid and web centerline

Table 44.3. Lateral Buckling of Deep Beams

A. Narrow Rectangular Cross Sections

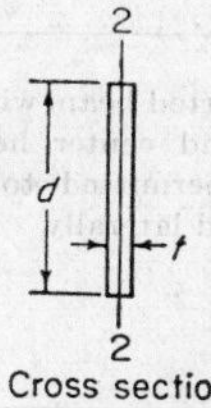

Cross section

Case	Description	Buckling-load formula*
1	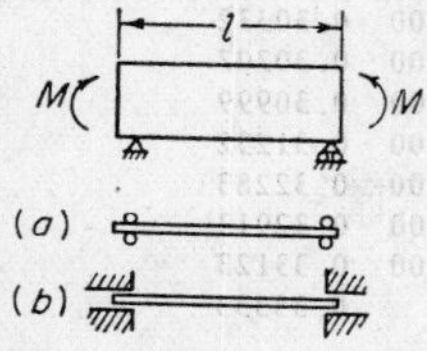 Simply supported beam under constant bending moment M. Ends held in a vertical plane, permitted to rotate longitudinally and (*a*) permitted to rotate laterally (*b*) clamped against lateral rotation	(*a*) $M_{cr} = \pi \sqrt{EI_2GJ}/l$ (*b*) $M_{cr} = 2\pi \sqrt{EI_2GJ}/l$
2	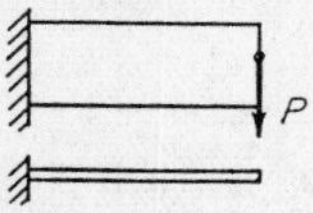 Cantilever with tip load P applied at mid-height	$P_{cr} = 4 \sqrt{EI_2GJ}/l^2$
3	 Cantilever with uniform load p along the length of the beam at its mid-height	$p_{cr} = 12.9 \sqrt{EI_2GJ}/l^3$
4	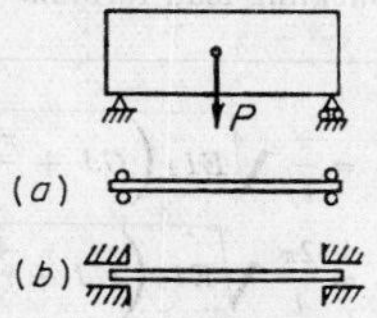 Simply supported beam under central load P at mid-height; ends held as in case 1	(*a*) $P_{cr} = 16.9 \sqrt{EI_2GJ}/l^2$ (*b*) $P_{cr} = 25.9 \sqrt{EI_2GJ}/l^2$
5	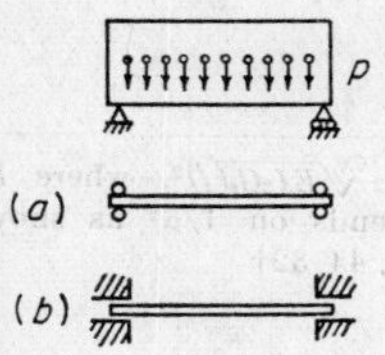 Simply supported beam with uniform load p along length of beam at mid-height. Ends held as in case 1	(*a*) $p_{cr} = 28.3 \sqrt{EI_2GJ}/l^3$ (*b*) $p_{cr} = 43.3 \sqrt{EI_2GJ}/l^3$

Table 44.3. Lateral Buckling of Deep Beams (*Continued*)

Case		Description	Buckling-load formula*
6	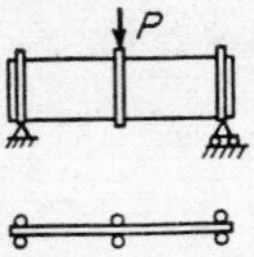	Simply supported beam with central load P. Ends and center held in vertical plane but permitted to rotate longitudinally and laterally	$P_{cr} = 44.5\sqrt{EI_2GJ}/l^2$

* $J = \beta dt^3$, where β depends on d/t as follows:

1.00	0.14058	1.60	0.20374	4.00	0.28081
1.05	0.14744	1.70	0.21093	4.50	0.28665
1.10	0.15398	1.75	0.21428	5.00	0.29135
1.15	0.16021	1.80	0.21743	6.00	0.29832
1.20	0.16612	1.90	0.22332	7.00	0.30332
1.25	0.17173	2.00	0.22868	8.00	0.30707
1.30	0.17707	2.25	0.24012	9.00	0.30999
1.35	0.18211	2.50	0.24936	10.00	0.31232
1.40	0.18690	2.75	0.25696	20.00	0.32283
1.45	0.19145	3.00	0.26332	50.00	0.32913
1.50	0.19576	3.50	0.27331	100.00	0.33123
				∞	0.33333

B. Symmetrical I Beams

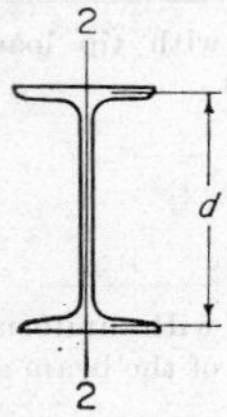

Cross section

$\alpha^2 = E\Gamma/GJl^2$

Case		Description	Buckling-load formula*
1	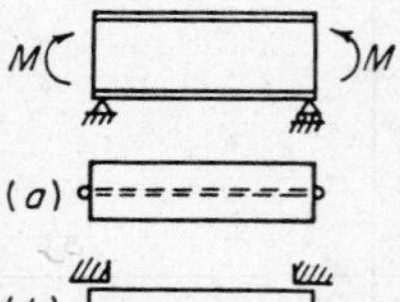	Simply supported beam under constant bending moment M. Ends held in a vertical plane, permitted to rotate longitudinally, and (*a*) permitted to rotate laterally and warp, (*b*) clamped against warping and lateral rotation	(*a*) $M_{cr} = \dfrac{\pi}{l}\sqrt{EI_2\left(GJ + \dfrac{\pi^2 E\Gamma}{l^2}\right)}$ (*b*) $M_{cr} = \dfrac{2\pi}{l}\sqrt{EI_2\left(GJ + \dfrac{4\pi^2 E\Gamma}{l^2}\right)}$
2	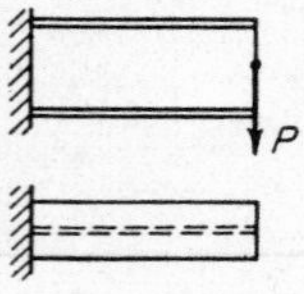	Cantilever with tip load P applied at mid-height	$P_{cr} = F_2\sqrt{EI_2GJ}/l^2$, where F_2 depends on $1/\alpha^2$ as shown on p. 44–32†

Table 44.3. Lateral Buckling of Deep Beams (*Continued*)

Case	Description	Buckling-load formula
3 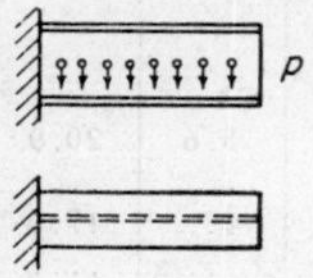	Cantilever with load p distributed uniformly along the length of the beam at its mid-height	$p_{cr} = F_3 \sqrt{EI_2GJ}/l^3$ where F_3 depends on $1/\alpha^2$ as shown on p. 44–32†
4 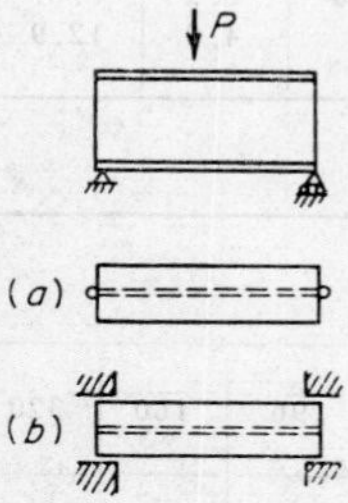	Simply supported beam under central load P. Ends held in a vertical plane, permitted to rotate longitudinally and (*a*) permitted to rotate laterally and warp, (*b*) clamped against lateral rotation and warping	$P_{cr} = F_4 \sqrt{EI_2GJ}/l^2$, where F_4 depends on $1/\alpha^2$ and on the vertical location of P as shown on p 44–32‡
5 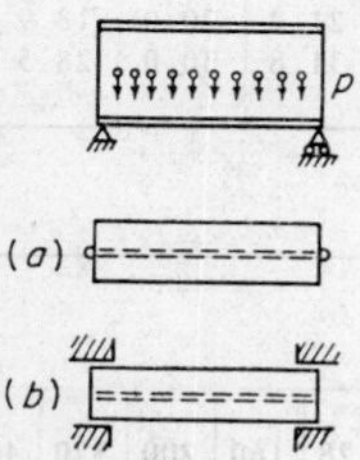	Same as case 4 except that load is p uniformly distributed along length of beam	$p_{cr} = F_5 \sqrt{EI_2GJ}/l^3$ where F_5 depends on $1/\alpha^2$ and on the vertical location of p as shown on p. 44–32§
6 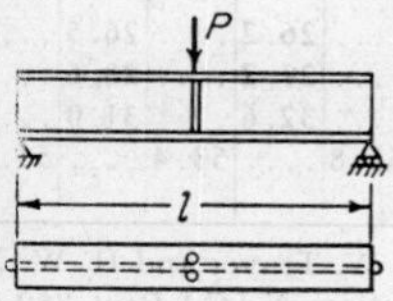	Simply supported beam with central load P. Ends and center section held in a vertical plane but permitted to rotate longitudinally and laterally and to warp	$P_{cr} = F_6 \sqrt{EI_2GJ}/l^2$, where F_6 depends on $1/\alpha^2$ as shown on p. 44–32†

* Values of J for I beams are given by I. Lyse and B. G. Johnston, Structural beams in torsion, *Trans. ASCE*, **101** (1936), 857–896, and by G. W. Trayer and H. W. March, The torsion of members having sections common in aircraft construction, *NACA Tech. Rept.* 334 (1929). $\Gamma = \frac{1}{4} d^2 I_2$ (see Table 44.2).

Table 44.3. Lateral Buckling of Deep Beams (*Continued*)

† Values of F_2, F_3, and F_6 are as follows:

$1/\alpha^2$	F_2	F_3	F_6	$1/\alpha^2$	F_2	F_3	F_6	$1/\alpha^2$	F_2	F_3	F_6
0.1	44.3										
0.4	....	....	466	8	8.0	31.9	114				
1	15.7	66.9		10	...	29.8		40	5.6	20.9	
2	12.2	50.9		12	7.2	28.3		96	...	....	54.2
3	....	44.0		14	...	27.1		100	...	17.6	
4	9.8	39.8	154	16	6.7	26.0	86.4	128	...	....	52.4
6	....	34.9		24	6.2	23.5		200	...	....	49.8
				32	5.9	21.9	69.2	400	...	....	47.4
								∞	4.0	12.9	44.5

‡ Values of F_4 are as follows:

Case	Location of P	$1/\alpha^2$									
		0.4	4	8	16	32	64	96	160	320	∞
a	Top flange	51.4	20.2	17.0	15.4	14.9	14.9	15.0	15.4	15.7	16.9
	Mid-height	86.4	31.9	25.6	21.8	19.4	18.3	17.9	17.5	17.2	16.9
	Bottom flange	145.6	50.0	38.2	30.5	25.5	22.4	21.2	20.0	18.9	16.9
b	Mid-height	268	88.8	65.5	50.2	40.2	34.2	31.8	30.0	28.5	25.9

§ Values of F_5 are as follows:

Case	Location of W	$1/\alpha^2$													
		0.4	4	8	16	32	48	64	96	128	160	200	320	400	∞
a	Top flange	92.8	36.3	30.4	27.4	26.2	26.2	25.8	26.0	....	26.2	....	26.5	....	28.3
	Mid-height	143.2	53.0	42.6	36.3	32.6	31.5	30.5	29.8	....	29.2	....	28.6	....	28.3
	Bottom flange	221.6	78.2	59.4	43.1	40.7	38.1	36.0	34.4	....	32.6	....	31.0	....	28.3
b	Mid-height	488	160.8	119.2	91.2	73.0	....	....	58.0	55.8	....	53.4	....	51.2	43.3

SOURCES: All the above, except case 3 under I beams, is taken from G. W. Trayer and H. W. March, Elastic instability of members having sections common in aircraft construction, *NACA Rept.* 382 (1931). The information for case 3 under I beams is from S. Poley, Lateral buckling of cantilevered I-beams under uniform load, *Trans. ASCE*, **121** (1956), 786–790.

Additional data are contained in the work by Trayer and March and in [5]. Lateral buckling under the combined action of end moments (not necessarily equal) and axial load is considered by M. G. Salvadori, Lateral buckling of eccentrically loaded I-columns, *Trans. ASCE*, **121** (1956), 1163; and by F. DiMaggio, A. Gomza, W. E. Thomas, and M. G. Salvadori, Lateral buckling of beams in bending and compression, *J. Aeronaut. Sci.*, **19** (1952), 574.

Table 44.4. Rings and Arches

Assumptions: Radial depth of ring or arch cross section is very small compared to r. Ring or arch is constrained (if necessary) to buckle in its own plane.

Case	Description	Buckling load
1 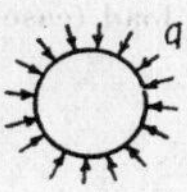	Circular ring under uniform radial loading of q per unit of circumferential length	$q_{cr} = \frac{3EI}{r^3}$
2 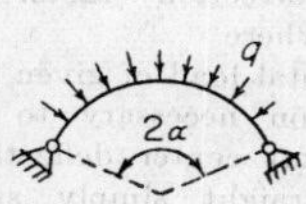	High circular arch with hinged ends and central angle 2α, subjected to uniform radial loading of q per unit of circumferential length	$q_{cr} = \left(\frac{\pi^2}{\alpha^2} - 1\right)\frac{EI}{r^3}$
3 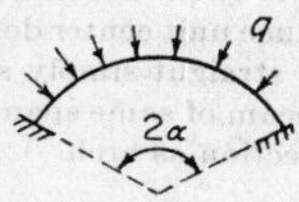	High circular arch with clamped ends and central angle 2α, subjected to uniform radial loading of q per unit circumferential length	$q_{cr} = (k^2 - 1)\frac{EI}{r^3}$ where k is defined by the equation $k \tan \alpha \cot k\alpha = 1$ Values of k corresponding to selected values of α are given on p. 44–34*
4 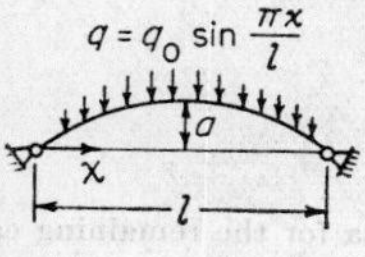	Shallow sinusoidal arch under sinusoidal loading. Ends hinged. Load q per unit horizontal run = $q_0 \sin (\pi x/l)$. Total load $P = 2q_0 l/\pi$ $\lambda \equiv \frac{a}{2}\sqrt{\frac{A}{I}}$ (dimensionless rise parameter) $R \equiv \frac{q_0 l^4}{2\pi^4 EI}\sqrt{\frac{A}{I}}$ (dimensionless load parameter)	For $\lambda \leq 1$: No buckling; arch bends continuously, similar to a straight beam For $1 < \lambda \geq \sqrt{5.5}$: $R_{cr} = \lambda + \sqrt{\frac{4}{27}(\lambda^2 - 1)^3}$ Buckling is symmetrical snap-through type illustrated in Fig. 44.1c For $\lambda \geq \sqrt{5.5}$: $R_{cr} = \lambda + 3\sqrt{\lambda^2 - 4}$ Buckling is antimetrical two-lobe type similar to that of high arches
5 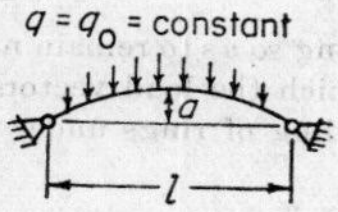	Shallow sinusoidal arch under uniform loading of q_0 per unit horizontal run. Ends hinged. Total load $P = q_0 l$ $\lambda \equiv \frac{a}{2}\sqrt{\frac{A}{I}}$ (dimensionless rise parameter) $R \equiv \frac{q_0 l^4}{2\pi^4 EI}\sqrt{\frac{A}{I}}$ (dimensionless load parameter)	$P_{cr} = \frac{\pi^2}{8}$ times P_{cr} for same arch with sinusoidally distributed load (case 4). Error less than ½%. Consequently, For $\lambda \geq 1$: No buckling For $1 < \lambda \leq \sqrt{5.5}$: $R_{cr} \approx \frac{\pi}{4}\left(\lambda + \sqrt{\frac{4}{27}(\lambda^2 - 1)^3}\right)$ For $\lambda \geq \sqrt{5.5}$: $R_{cr} \approx \frac{\pi}{4}(\lambda + 3\sqrt{\lambda^2 - 4})$

Table 44.4. Rings and Arches (*Continued*)

Case	Description	Buckling load
6	Shallow sinusoidal arch under central concentrated load P. Ends hinged 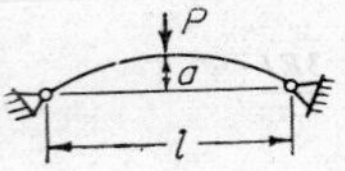	$P_{cr} \approx \frac{\pi}{4}$ times P_{cr} for same arch with sinusoidally distributed load (case 4)
7	Shallow sinusoidal arch under any symmetrically distributed loading of total magnitude P 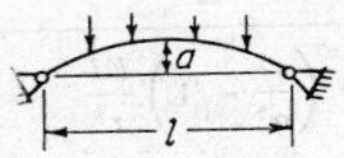	$P_{cr} \approx P_{cr}$ for same arch sinusoidally loaded (case 4) times correction factor P'/P'', where $P' \equiv$ total load of given distribution necessary to produce unit center deflection of a straight simply supported beam of same span and cross section as arch $P'' \equiv$ total load of sinusoidal distribution necessary to produce unit center deflection of a straight simply supported beam of same span and cross section as arch

* Values of k corresponding to selected values of α:

α	k	α	k
30°	8.621	120°	2.364
60°	4.375	150°	2.066
90°	3	180°	2

SOURCES: The data for cases 1, 2, and 3 are taken from [5]. The data for the remaining cases are from Y. C. Fung and A. Kaplan, Buckling of low arches or curved beams of small curvature, *NACA Tech. Note* 2840 (1952). In the latter reference are also considered the effects of initial horizontal-reaction thrusts, the effects of small deviations from symmetry of loading or arch shape, and the buckling of shallow circular arches.

The buckling data for cases 1, 2, and 3 represent classical-type buckling; for the remaining cases the data are for snap-through or classical-type buckling, whichever governs. Some data on finite-disturbance buckling of low arches will be found in the above-cited work by Fung and Kaplan. Experimental data reported therein indicate that the finite-disturbance buckling does not govern for shallow arches except perhaps for $\lambda \ll 3$.

In cases 1, 2, and 3 the load vectors are assumed to rotate during buckling so as to remain normal to the ring or arch, as in the case of fluid-pressure loading. The case in which the load vectors do not rotate is considered by A. P. Boresi, A refinement of the theory of buckling of rings under uniform pressure, *J. Appl. Mechanics*, **22** (1955), 95.

Table 44.5. Flat Rectangular Plates in Compression*

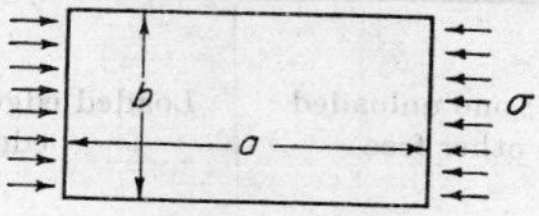

The following symbols are used to designate support conditions:

C clamped edge   S simply supported edge   F free edge

Case 1 All edges simply supported		Case 2 Loaded edges simply supported, unloaded edges clamped		Case 3 Loaded edges simply supported; one unloaded edge simply supported, the other clamped		Case 4 Loaded edges clamped, unloaded edges simply supported		Case 5 All edges clamped	
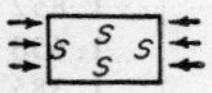		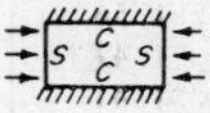		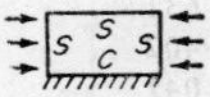		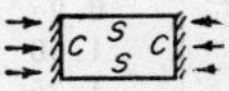		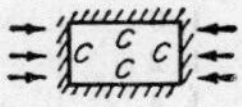	
$\frac{a}{b}$	$\frac{\sigma_{cr}a^2t}{\pi^2K}$	$\frac{a}{b}$	$\frac{\sigma_{cr}a^2t}{\pi^2K}$	$\frac{a}{b}$	$\frac{\sigma_{cr}a^2t}{\pi^2K}$	$\frac{a}{b}$	$\frac{\sigma_{cr}a^2t}{\pi^2K}$	$\frac{a}{b}$	$\frac{\sigma_{cr}a^2t}{\pi^2K}$
0	1.000	0	1.000	0	1.00	0	4.00	0	4.00
0.4	1.346	0.4	1.512	0.5	1.71	0.6	4.82	0.50	4.84
0.6	1.850	0.6	2.541	0.6	2.13	0.8	5.59	0.75	6.56
0.8	2.690	0.8	4.675	0.8	3.46	1.0	6.74	1.00	10.07
1.0	4.000	0.94	7.16*	1.0	5.74	1.2	8.41		
$\frac{a}{b}$	$\frac{\sigma_{cr}b^2t}{\pi^2K}$	$\frac{a}{b}$	$\frac{\sigma_{cr}b^2t}{\pi^2K}$	$\frac{a}{b}$	$\frac{\sigma_{cr}b^2t}{\pi^2K}$	$\frac{a}{b}$	$\frac{\sigma_{cr}b^2t}{\pi^2K}$	$\frac{a}{b}$	$\frac{\sigma_{cr}b^2t}{\pi^2K}$
0.4	8.410	0.4	9.448	0.5	6.85	0.6	13.38	0.75	11.66
0.6	5.138	0.6	7.059	0.6	5.92	0.8	8.73	1.00	10.07
0.8	4.203	0.666	6.98	0.8	5.41	1.0	6.74	1.25	9.25
1.0	4.000	0.8	7.304	1.0	5.74	1.2	5.84	1.50	8.33
1.2	4.134	0.94	8.10*	1.127	6.18*	1.4	5.45	1.75	8.11
1.4	4.470	1.0	7.692	1.2	5.92	1.6	5.34	2.00	7.88
1.414	4.50*	1.2	7.059	1.4	5.51	1.7	5.33	2.50	7.57
1.6	4.203	1.4	6.997	1.6	5.41	1.73	5.33*	3.00	7.37
1.8	4.045	1.61	7.34*	1.8	5.50	1.8	5.18	3.25	7.35
2.0	4.000	2.00	6.98	1.95	5.67*	2.0	4.85	3.50	7.27
2.4	4.134	2.29	7.15*	2.4	5.41	2.5	4.52	3.75	7.24
2.45	4.17*	2.66	6.98	2.76	5.54*	2.83	4.50*	4.00	7.23
3.0	4.000	2.96	7.07*	3.2	5.41	3.0	4.41	∞	6.98
3.44	4.07	3.5	6.997	3.56	5.48*	∞	4.00		
∞	4.000	∞	6.98	∞	5.41				

* An asterisk indicates a known point of slope discontinuity in the graph of $\sigma_{cr}a^2t/\pi^2K$ or $\sigma_{cr}b^2t/\pi^2K$ versus a/b. Such points should not be used as intermediate points in interpolation.

Table 44.5. Flat Rectangular Plates in Compression (*Continued*)

Case 6

Loaded edges simply supported; one unloaded edge simply supported, the other free

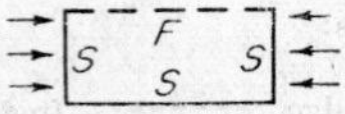

$\nu = 0.25$: $\frac{a}{b}$	$\nu = 0.25$: $\frac{\sigma_{cr} b^2 t}{\pi^2 K}$	$\nu = 0.30$: $\frac{a}{b}$	$\nu = 0.30$: $\frac{\sigma_{cr} b^2 t}{\pi^2 K}$
0.5	4.40	0.8	1.954
1.0	1.440	0.9	1.631
1.2	1.35	1.00	1.402
1.4	0.952	1.25	1.047
1.6	0.835	1.50	0.858
1.8	0.755	1.75	0.742
2.0	0.698	2.00	0.669
2.5	0.610	2.50	0.582
3.0	0.564	3.00	0.533
4.0	0.516	3.50	0.505
5.0	0.506	4.00	0.486
∞	0.456	5.00	0.464
		6.00	0.451
		7.0	0.445
		9.0	0.438
		11.0	0.434
		15.0	0.428
		25.0	0.426
		50.0	0.425
		∞	0.425

Case 7

Loaded edges simply supported; one unloaded edge clamped, the other free

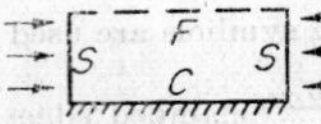

$\nu = 0.25$: $\frac{a}{b}$	$\nu = 0.25$: $\frac{\sigma_{cr} b^2 t}{\pi^2 K}$	$\nu = 0.30$: $\frac{a}{b}$	$\nu = 0.30$: $\frac{\sigma_{cr} b^2 t}{\pi^2 K}$
1.0	1.70	0.8	2.150
1.1	1.56	0.9	1.852
1.2	1.47	1.0	1.658
1.3	1.41	1.25	1.385
1.4	1.36	1.5	1.289
1.5	1.34	1.645	1.281
1.6	1.33	1.75	1.287
1.7	1.33	2.0	1.337
1.8	1.34	2.25	1.424
1.9	1.36	2.42	1.500*
2.0	1.38	2.5	1.385
2.2	1.45	3.0	1.289
∞	1.328	3.290	1.281
		3.5	1.287
		4.0	1.337*
		4.5	1.289
		4.935	1.281
		5.25	1.287
		5.70	1.308*
		6.0	1.289
		∞	1.281

* An asterisk indicates a known point of slope discontinuity in the graph of $\sigma_{cr} a^2 t/\pi^2 K$ or $\sigma_{cr} b^2 t/\pi^2 K$ versus a/b. Such points should not be used as intermediate points in interpolation.

Table 44.5. Flat Rectangular Plates in Compression (*Continued*)

Case 8
Loaded edges simply supported, unloaded edges free

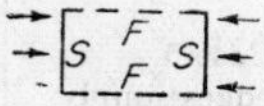

$\frac{a}{b}$	$\sigma_{cr} a^2 t/\pi^2 K$, $\nu = 0.25$	$\nu = 0.30$	$\nu = 0.35$
0	......	0.9962	
0.05	0.998	0.995	0.993
0.10	0.997	0.993	0.990
0.20	0.993	0.987	0.983
0.333	0.989	0.985	0.980
0.50	0.983	0.973	0.961
0.667	0.977	0.965	0.951
1.00	0.968	0.952	0.933
1.25	0.963	0.943	0.922
2.00	0.953	0.929	0.903
2.50	0.949	0.924	0.895
3.33	0.945	0.919	0.889
5.00	0.941	0.914	0.884
10.00	0.938	0.911	0.879
∞	0.9375	0.910	0.8775

Case 9
Loaded edges clamped, unloaded edges free

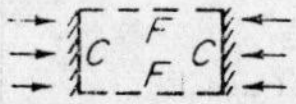

$\frac{a}{2b}$	$\sigma_{cr} a^2 t/4\pi^2 K$, $\nu = 0.30$
0.05	0.998
0.10	0.996
0.20	0.993
0.333	0.988
0.50	0.982
0.667	0.977
1.00	0.967
1.25	0.961
2.00	0.948
2.50	0.940
3.33	0.933
5.00	0.925
10.00	0.917
∞	0.910

SOURCES: Cases 1 and 2: E. E. Lundquist and E. Z. Stowell, Critical compressive stress for flat rectangular plates supported along all edges and elastically restrained against rotation along the unloaded edges, *NACA Rept.* 733 (1942). Case 3 and parts of case 6: Adapted from W. D. Kroll, Tables of stiffness and carry-over factor for flat rectangular plates under compression, *NACA Wartime Rept.* L-398. Case 4 and cases 6 and 7 with $\nu = 0.25$, S. Timoshenko [5]. Case 5: S. Levy, Buckling of rectangular plates with built-in edges, *J. Appl. Mechanics*, **9** (1942), 171–174. Cases 6 and 7; $\nu = 0.30$: E. E. Lundquist and E. Z. Stowell, Critical compressive stress for outstanding flanges, *NACA Rept.* 734 (1942). Some data for case 6 also taken from W. D. Kroll (see earlier citation). Cases 8 and 9: J. C. Houbolt and E. Z. Stowell, Critical stress of plate columns, *NACA Tech. Note* 2163 (1950).

Table 44.6. Flat Rectangular Plates in Shear*

Case 1 All edges simply supported		Case 2 All edges clamped		Case 3 Short edges simply supported, long edges clamped		Case 4 Short edges simply supported or clamped; one long edge simply supported, the other clamped	
$\frac{a}{b}$	$\frac{\tau_{cr} b^2 t}{\pi^2 K}$	$\frac{a}{b}$	$\frac{\tau_{cr} b^2 t}{\pi^2 K}$	$\frac{a}{b}$	$\frac{\tau_{cr} b^2 t}{\pi^2 K}$	$\frac{a}{b}$	$\frac{\tau_{cr} b^2 t}{\pi^2 K}$
1.0	9.35	1.00	14.71	1.0	12.28	∞	6.63
1.2	8.00	1.25	12.48	1.5	11.12		
1.5	7.07	1.50	11.50	2.0	10.21		
2.0	6.59	1.61	11.26*	2.5	9.81		
2.06	6.48*	2.00	10.34	3.0	9.61		
2.5	6.06	2.50	9.82	∞	8.98		
3.0	5.89	2.89	9.69*				
3.5	5.81*	3.00	9.62				
4.0	5.77	5.00	9.21				
∞	5.33	∞	8.98				

* An asterisk denotes a known point of slope discontinuity in the graph $\tau_{cr} b^2 t/\pi^2 K$ versus a/b.

SOURCES: Case 1: M. Stein and J. Neff, Buckling stresses of simply supported rectangular flat plates in shear, *NACA Tech. Note* 1222 (1947). Case 2: B. Budiansky and R. W. Connor, Buckling stresses of clamped rectangular flat plates in shear, *NACA Tech. Note* 1559 (1948). Case 3: S. Iguchi, Die Knickung der rechteckigen Platte durch Schubkräfte, *Ing.-Arch.*, **9** (1938), 1–12 (as cited by F. Bleich [2]). Case 4: A. E. Johnson, Jr., and K. P. Buchert, Critical combinations of bending, shear, and transverse compressive stresses for buckling of infinitely long flat plates, *NACA Tech. Note* 2536 (1951).

AXISYMMETRIC BUCKLING DATA FOR $\lambda > 2.08$ (SEE TABLE 44.7, CASE 2)

Theoretical data		Experimental data			
λ	R_{cr}	λ	R_{cr}	λ	R_{cr}
2.08	2.46	5	33	8	178
3	4.1	6	77	9	250
3.5	8	7	122	10	347
4	14				

Table 44.7. Curved Plates and Shells*

Assumption: $t \ll r$

Case	Description	Buckling-load formula
1	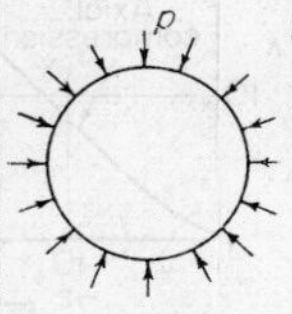 **Spherical shell under external hydrostatic pressure p, producing compressive stress $\sigma = pr/2t$ in the wall**	$p_{cr} = \frac{2}{\sqrt{3(1-\nu^2)}} E \left(\frac{t}{r}\right)^2$ or $\sigma_{cr} = \frac{1}{\sqrt{3(1-\nu^2)}} E \frac{t}{r}$ Critical load for classical buckling. In practice, buckling is usually the finite-disturbance type, occurring at a lower stress
2	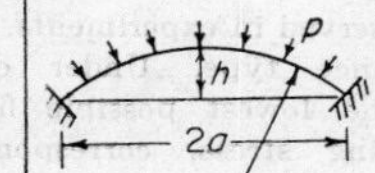 **Shallow spherical dome with fully clamped edge, subjected to uniform radial external pressure p. Limitation: a/h greater than about 8** $\lambda^2 = \frac{a^2}{tr}\sqrt{12(1-\nu^2)} \approx \frac{4h}{t}\sqrt{3(1-\nu^2)}$ (geometrical parameter) $R = \frac{p}{E}\left(\frac{a}{t}\right)^4 (1-\nu^2)$ (load parameter)	For $\lambda < 2.08$: No buckling; shell deforms continuously For $\lambda > 2.08$: Axisymmetric buckling; of the snap-through type for λ up to about 6, probably of the finite-disturbance type for λ greater than about 6. Approximate data are given on p. 44–38.
3	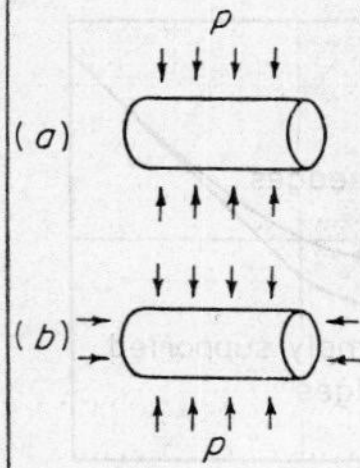 **Cylinder with simply supported edges, subjected to (a) external lateral pressure p applied to curved walls only, (b) external hydrostatic pressure p applied to curved walls and closed ends.** **State of uniform stress assumed in wall.** **Limitation: $Z < 5\left(\frac{r}{t}\right)^2 (1-\nu^2)$**	$p_{cr} = k\frac{\pi^2 K}{l^2 r}$ for classical-type buckling, where k is taken from the appropriate curve below. For large Z (>100), $k = 1.04\sqrt{Z}$ 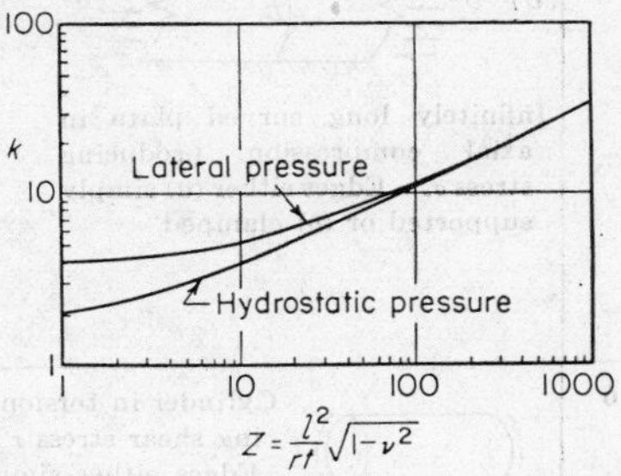 Theoretical calculations for case b (see source below) indicate that, in the geometrical range $0.00289 < rt/l^2 < 0.189$, finite-disturbance buckling can occur at a stress which may be as much as 3% below the classical buckling stress. Experiments show an even greater reduction

Table 44.7. Curved Plates and Shells* (*Continued*)

Case	Description	Buckling-load formula
4	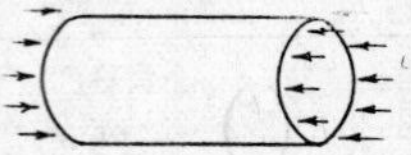 Cylinder with simply supported edges, subjected to axial compression producing stress σ in wall. Limitation: $Z < 6\left(\frac{r}{t}\right)^2 (1 - \nu^2)$	$\sigma_{cr} = k(\pi^2 K/l^2 t)$ for classical buckling, where k is taken from the graph shown. For large Z (>3), $k = 0.702Z$ or $\sigma_{cr} = \frac{0.577}{\sqrt{1-\nu^2}} E \frac{t}{r}$ 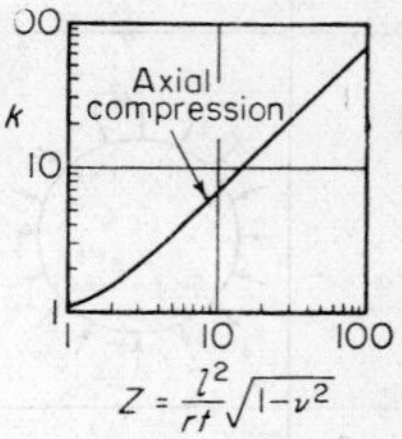 Actual buckling, observed in experiments, is of the finite-disturbance type. Under dead-weight loading, the lowest possible finite-disturbance buckling stress, corresponding to P'_{cr} in Fig. 44.1*b*, is theoretically $\sigma_{cr} = 0.182\ Et/r$ for $\nu = 0.3$. Similarly, under constant shortening, this type of buckling cannot occur below the strain $\epsilon_{cr} = 0.307\ t/r$ or stress $\sigma_{cr} = 0.307\ Et/r$, corresponding to P''_{cr} in Fig. 44.1*b*.
5	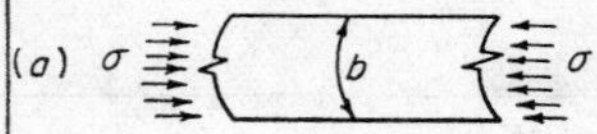 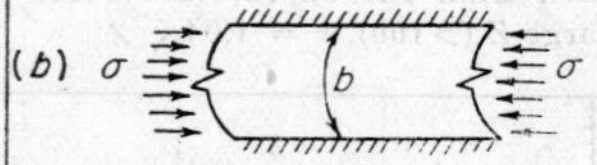 Infinitely long curved plate in axial compression producing stress σ. Edges either (*a*) simply supported or (*b*) clamped	$\sigma_{cr} = k(\pi^2 K/b^2 t)$, where k is taken from the appropriate curve below. (For classical-type buckling) 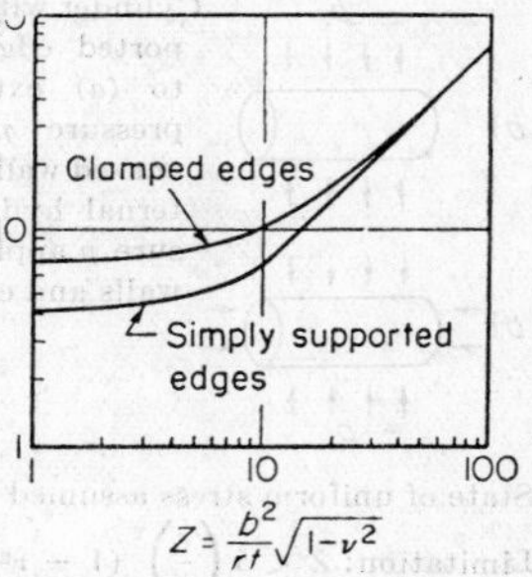
6	 Cylinder in torsion producing shear stress τ in wall. Edges either simply supported or clamped. Limitation: $Z < 10\left(\frac{r}{t}\right)^2 (1 - \nu^2)$	$\tau_{cr} = k(\pi^2 K/l^2 t)$, where k is taken from appropriate curve below. (Classical type buckling) 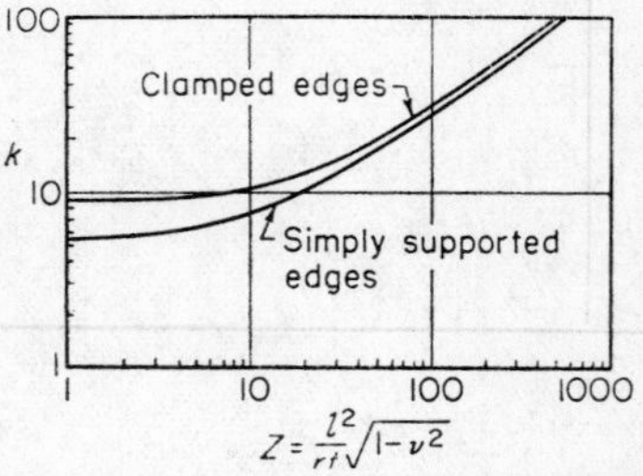 Theory and experiment indicate that slightly lower buckling loads of the finite-disturbance type are possible for this case. See reference cited on p. 44–41

Table 44.7. Curved Plates and Shells* *(Continued)*

Case	Description	Buckling-load formula
7	Cylindrically curved rectangular plates with shear stress τ. Edges simply supported. Case a: circumferential dimension greater than axial dimension; case b: axial dimension greater than circumferential dimension 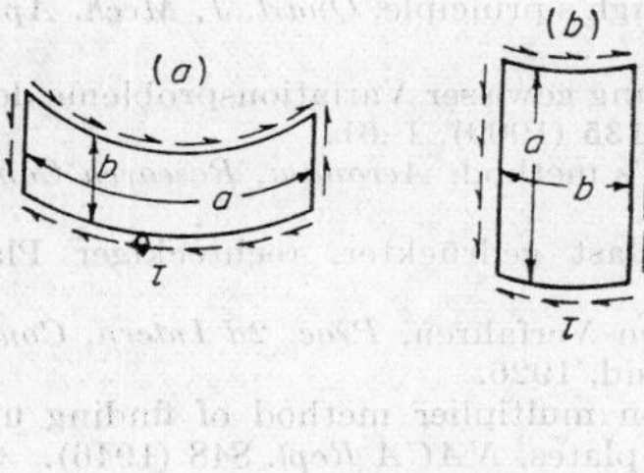	$\tau_{cr} = k(\pi^2 K/b^2 t)$, where k is taken from appropriate graph below. (Classical-type buckling) 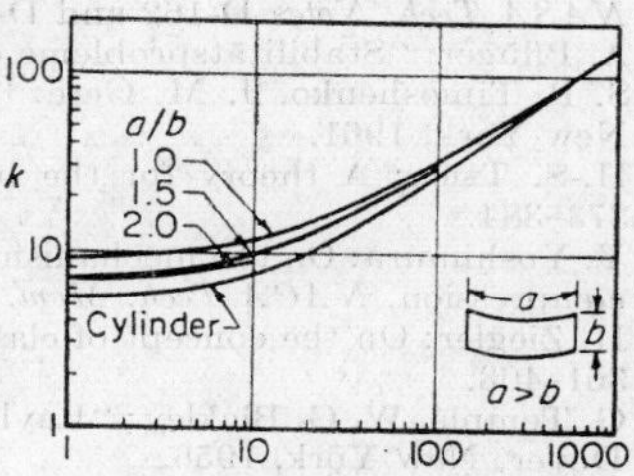 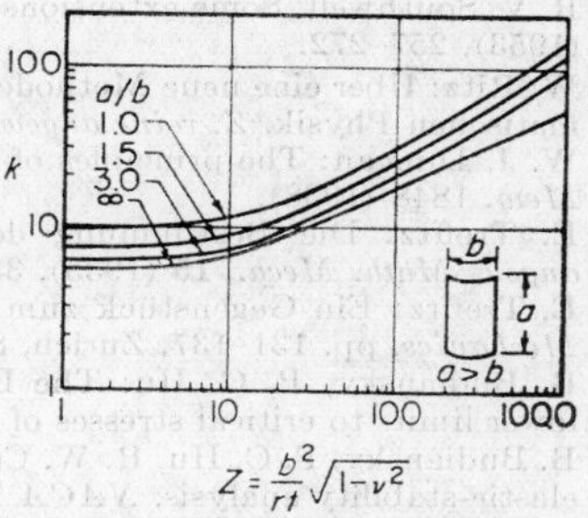
8	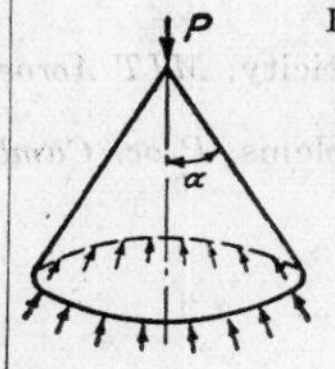 Long conical shell with semi-vertex-angle α under axial vertex load P. Membrane stresses only assumed in prebuckled shell	$P_{cr} \approx \frac{2\pi E t^2 \cos^2 \alpha}{\sqrt{3(1-\nu^2)}}$ Approximate classical buckling load for axisymmetric buckling

* Curvature parameters:

$$Z = \frac{l^2}{rt}\sqrt{1-\nu^2} \quad \text{for cylinders} \qquad Z = \frac{b^2}{rt}\sqrt{1-\nu^2} \quad \text{for curved rectangular plates}$$

Sources: The formula for case 1, from A. van der Neut, is in [5]. Data for case 2 are scaled from graphs of A. Kaplan and Y. C. Fung, A nonlinear theory of bending and buckling of thin elastic shallow spherical shells, *NACA Tech. Note* 3212 (1954). The classical-buckling data for cases 3, 4, 5, 6, and 7 are from S. B. Batdorf, A simplified method of elastic-stability analysis for thin cylindrical shells, *NACA Rept.* 874 (1947). The information on finite-disturbance buckling for cases 3 and 4 is from the following sources: J. Kempner, Postbuckling behavior of axially compressed circular cylindrical shells, *J. Aeronaut. Sci.*, **21** (1954), 329–335, 342; J. Kempner, Recent results in the theory of large deflections of cylindrical shells, *Polytech. Inst. Brooklyn PIBAL Rept.* 360 (August, 1956); J. Kempner, K. A. V. Pandalai, S. A. Patel, and J. Crouzet-Pascal, Postbuckling behavior of circular cylindrical shells under hydrostatic pressure, *J. Aeronaut. Sci.*, **24** (1957), 253–264. Some information on finite-disturbance buckling for case 6 can be found in Tsu-Tao Loo, Effects of large deflections and imperfections on the elastic buckling of cylinders under torsion and axial compression, *Proc. 2d U.S. Natl. Congr. Appl. Mechanics*, 1955, pp. 345–357. The formula for case 8 is due to P. Seide, Axisymmetric buckling of circular cones under axial compression, *J. Appl. Mechanics*, **23** (1956), 625–628. Extensive test data on buckling of conical shells under various types of loading are given by P. Seide, V. I. Weingarten, E. J. Morgan. Final report on the development of design criteria for elastic stability of thin shell structures, *Space Technol. Labs. Rept.* STL/TR-60-0000-19425 (1960).

44.5. REFERENCES

References [1] through [5] contain extensive compilations of numerical data on buckling loads.

[1] J. H. Argyris, P. C. Dunne: "Structural Principles and Data," part 2, structural analysis, 4th ed., Pitman, New York, 1952.
[2] Friedrich Bleich: "Buckling Strength of Metal Structures," McGraw-Hill, New York, 1952.
[3] G. Gerard, H. Becker: Handbook of structural stability, *NACA Tech. Notes* 3781–3786; *NASA Tech. Notes* D-162 and D-163 (1957–1959).
[4] A. Pflüger: "Stabilitätsprobleme der Elastostatik," Springer, Berlin, 1950.
[5] S. P. Timoshenko, J. M. Gere: "Theory of Elastic Stability," 2d ed., McGraw-Hill, New York, 1961.
[6] H.-S. Tsien: A theory for the buckling of thin shells, *J. Aeronaut. Sci.*, **9** (1942), 373–384.
[7] Y. Yoshimura: On the mechanism of buckling of a circular cylindrical shell under axial compression, *NACA Tech. Mem.* 1390 (1955).
[8] H. Ziegler: On the concept of elastic stability, *Advances in Appl. Mechanics*, **4** (1956), 351–403.
[9] G. Temple, W. G. Bickley: "Rayleigh's Principle and Its Applications to Engineering," Dover, New York, 1956.
[10] R. V. Southwell: Some extensions of Rayleigh's principle, *Quart. J. Mech. Appl. Math.* (1953), 257–272.
[11] W. Ritz: Über eine neue Methode zur Lösung gewisser Variationsprobleme der mathematischen Physik, *Z. reine angew. Math.*, **135** (1909), 1–61.
[12] W. J. Duncan: The principles of Galerkin's method, *Aeronaut. Research Comm. Rept. Mem.* 1848 (1938).
[13] E. Trefftz: Die Bestimmung der Knicklast gedrückter, rechteckiger Platten, *Z. angew. Math. Mech.*, **15** (1935), 339.
[14] E. Trefftz: Ein Gegenstück zum Ritzschen Verfahren, *Proc. 2d Intern. Congr. Appl. Mechanics*, pp. 131–137, Zurich, Switzerland, 1926.
[15] B. Budiansky, P. C. Hu: The Lagrangian multiplier method of finding upper and lower limits to critical stresses of clamped plates, *NACA Rept.* 848 (1946).
[16] B. Budiansky, P. C. Hu, R. W. Connor: Notes on the Lagrangian multiplier method in elastic-stability analysis, *NACA Tech. Note* 1558 (1948).
[17] E. Reissner: On a variational theorem for finite elastic deformations, *J. Math. Phys.*, **32** (1953), 129–135.
[18] K. Washizu: On the variational principles of elasticity and plasticity, *MIT Aeroelastic and Structures Research Lab. Tech. Rept.* 25-18 (1955).
[19] A. E. Green: Double Fourier series and boundary value problems, *Proc. Cambridge Phil. Soc.*, **40** (1944), 222–228.

CHAPTER 45

NONLINEAR PROBLEMS

BY

D. G. ASHWELL, Ph.D., Cardiff, Wales, United Kingdom

45.1. CAUSES OF NONLINEARITY

The relationship between the loads applied to a deformable body or structure and the resulting deformations may be nonlinear for one or more of four causes.

1. The interaction between two parts of a structure may depend on the deformations, as, for example, in the case of the beam considered in Sec. 45.2.

2. The strains suffered by elements of the material may become large—i.e., not negligible when compared with unity. With the exception of rubber, engineering materials are not usually subjected to *finite strains,* and the theories developed for such behavior will not be considered here. Standard textbooks are Green and Zerna [1] and Murnaghan [2]; Treloar [3] gives a very brief introduction, and Doyle and Ericksen [4] include a bibliography of recent work.

3. The stress-strain relationship of the material may be nonlinear. Such nonlinearity is not important for problems in which the strains are very small, except for materials exhibiting plasticity; this subject is considered in Part 5 of this handbook. This chapter considers only materials which obey Hooke's law.

4. The deformation of the body may become large—i.e., not negligible when compared with its initial dimensions. Thus, each successive increment of load is applied to a body whose shape has been significantly changed by previous increments. This chapter is concerned largely with problems of this type.

Novozhilov [5] investigates the manner in which large deformations introduce nonlinearity into the equations of strain geometry and force equilibrium. The equations are found to contain terms representing the (small) strains of an element, and the angles through which it rotates in passing from the unstrained to the strained condition. Nonlinearity occurs unless squares and products of the rotations can be neglected when compared with the strains—it is not sufficient that the rotations are small compared with unity. If a body is *massive* i.e., if all three of its dimensions are of the same order of magnitude, the condition of small strains implies that the condition of small products of rotations is satisfied. But, if a body is *flexible,* i.e., if one or two of its dimensions are small compared with the remainder, then the rotations of elements may considerably exceed the strains, and the behavior of the body may become nonlinear. Thus, small-strain nonlinear elasticity is concerned with flexible bodies such as rods, membranes, plates, and shells. In this chapter, examples are given of such behavior and of methods for calculating it.

45.2. BEAM ON ELASTIC FOUNDATION WITH LOCAL LIFTING OFF

Beams on elastic foundations are considered in Chap. 31. If the beam is not attached to the foundation, local lifting off may occur. If the weight of the beam can be neglected, and it is subjected only to a set of loads whose values are proportional to a single parameter, the problem is linear. But, if the beam is subjected to independent loading systems, the length lifting off varies with the magnitudes of the loads, and the problem is nonlinear. This is an example of nonlinearity arising from cause (1) of Sec. 45.1.

As an example, consider the infinite beam shown in Fig. 45.1. It is loaded with a constant, uniform downward load p per unit length, and a varying concentrated upward load P. p, which might be the self-weight of the beam, causes a uniform

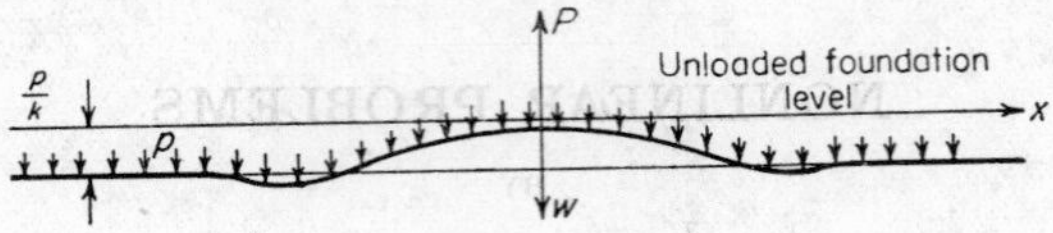

FIG. 45.1

vertical deflection p/k, where k is the foundation modulus. P causes a deflection of the type shown, superimposed on the deflection p/k. The additional deflection, δ_1, caused by P at its point of application, is, from Eq. (31.6a),

$$\delta_1 = -P/8\lambda^3 EI \tag{45.1}$$

where

$$4\lambda^4 = k/EI$$

This linear relationship is true provided $p/k + \delta_1 > 0$ but, if this condition is not satisfied, lifting off will occur on each side of P, and the differential equation (31.3) will no longer apply there.

To investigate the deformation of the beam under these conditions, it is necessary to assume that a length a, say, has lifted off on each side of P, and to consider separately these lengths and the remainder. Figure 45.2 shows the two parts of the beam

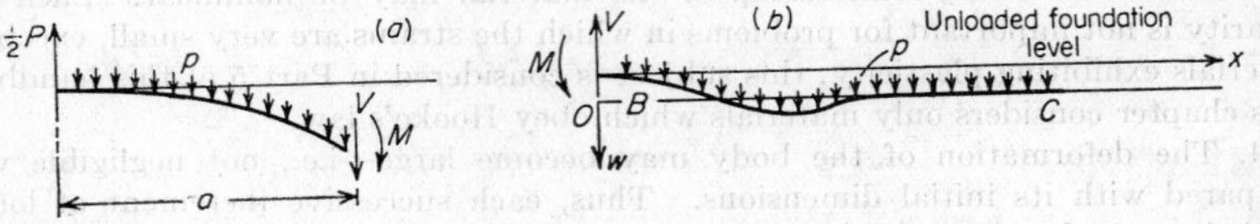

FIG. 45.2

to the right of P, with the forces acting on them. The conditions to be satisfied at $x = a$ provide two equations for V and M: the shearing force and bending moment there. Thus, continuity of slope gives

$$\frac{pa^3}{6EI} + \frac{Va^2}{2EI} + \frac{Ma}{EI} = \frac{1}{2\lambda^2 EI}(V - 2\lambda M) \tag{45.2}$$

where the left-hand side is the slope at the end of the cantilever AB, and the right-hand side is $[dw/dx]_{x=a}$ for the part BC. Also, at $x = a$, the deflection is $w = -p/k$; thus,

$$-\frac{1}{2\lambda^3 EI}(V - \lambda M) = -\frac{p}{k} \tag{45.3}$$

The slope and deflection at $x = a$, for BC, are obtained from the theory of Chap. 31.

The deflection of A above O, and load P corresponding to the assumed free length a, can be obtained by considering the deflection and equilibrium of the cantilever AB. Thus,

$$\delta = \frac{pa^4}{8EI} + \frac{Va^3}{3EI} + \frac{Ma^2}{2EI} + \frac{p}{k} \tag{45.4}$$

and

$$P = 2pa + 2V \tag{45.5}$$

The need for assigning numerical values to the symbols E, I, p, k, λ can be avoided if dimensionless equivalents to the variables P, δ, a, V, M are introduced. The dimensionless forms of these variables will be distinguished by primes:

$$P' = P\lambda/a \qquad \delta' = \delta k/p \qquad a' = \lambda a \qquad V' = V\lambda/p \qquad M' = M\lambda^2/p$$

Equations (45.2) to (45.5) then become, on solving (45.2) and (45.3) for V' and M',

$$V'[\tfrac{1}{2}(a'^2 - 1) + a' + 1] = \tfrac{1}{2}(a' + 1) - \tfrac{1}{6}a'^3 \qquad M' = V' - \tfrac{1}{2}$$
$$\delta' = \tfrac{1}{2}a'^4 + \tfrac{4}{3}a'^3V' + 2a'^2M' + 1 \qquad P' = 2a' + 2V'$$

V' and M' can now be computed for a number of values of a', and the corresponding values of δ' and P' can be calculated. In this way, Fig. 45.3 has been plotted. The curve relating P' and δ' is linear until lifting off occurs at $\delta' = 1$, $P' = 2$. The stiffness $dP'/d\delta'$ then decreases as the free length increases. It is possible, though unlikely, that, for large values of P', further lifting off may occur elsewhere along the beam; this has not been investigated.

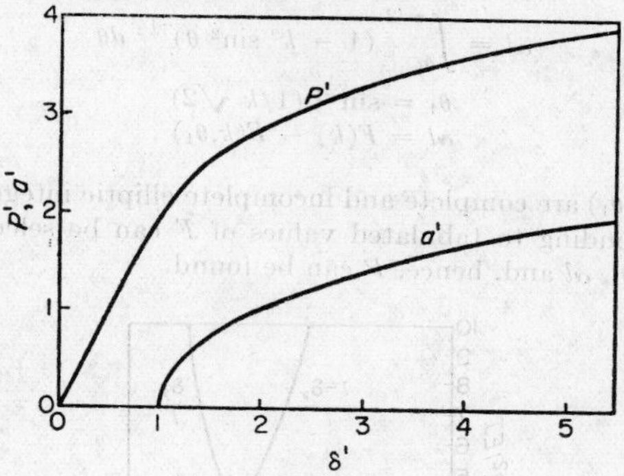

FIG. 45.3

45.3. THE PROBLEM OF THE ELASTICA

The linear theory for determining the deflections of initially straight beams and struts, initially parallel to the x axis, requires that $(dw/dx)^2$ shall be negligible compared with unity. This limitation simplifies analysis in two ways. It allows the expression for curvature of the mid-line to be approximated to d^2w/dx^2; and it allows relative movements in the x direction of the points of applications of the loads to be neglected. If this limitation is removed, so that the deformations of the (flexible) beam or strut can be of any magnitude that does not overstrain its material, the problem becomes that of the *elastica*, or elastic line. In the examples given below, it is assumed that the member is uniform, and that changes of length due to tension or compression can be neglected.

Large Deflections of Beams

In Fig. 45.4 is shown a cantilever, initially horizontal, with a vertical load P at its end. Its behavior for large deflections has been investigated by Bisshop and Drucker [6]. The relationship between curvature and bending moment gives

$$EI\,\frac{d\phi}{ds} = P(l - x - \delta_x) \tag{45.6}$$

or

$$\frac{d^2\phi}{ds^2} = -\alpha^2\,\frac{dx}{ds} = -\alpha^2\cos\phi$$

where

$$\alpha^2 = P/EI$$

Thus,

$$\frac{d\phi}{ds}\,\frac{d^2\phi}{ds^2} = -\alpha^2\cos\phi\,\frac{d\phi}{ds}$$

Integrating with respect to s, and putting $\phi = \phi_l$, $d\phi/ds = 0$, at $s = l$,

$$\frac{1}{2}\left(\frac{d\phi}{ds}\right)^2 = -\alpha^2 \sin\phi + \alpha^2 \sin\phi_l$$

Thus,

$$\frac{d\phi}{ds} = \pm\sqrt{2}\,\alpha(\sin\phi_l - \sin\phi)^{1/2} \tag{45.7}$$

where the positive sign is taken since $d\phi/ds$ is positive. Then

$$\sqrt{2}\,\alpha\int_0^l ds = \sqrt{2}\,\alpha l = \int_0^{\phi_l}(\sin\phi_l - \sin\phi)^{-1/2}\,d\phi \tag{45.8}$$

This integral may be converted to an elliptic integral by the substitutions

$$1 + \sin\phi = (1 + \sin\phi_l)\sin^2\theta \qquad 1 + \sin\phi_l = 2k^2 \tag{45.9}$$

Then

$$\alpha l = \int_{\theta_1}^{\pi/2}(1 - k^2\sin^2\theta)^{-1/2}\,d\theta$$

where

$$\theta_1 = \sin^{-1}(1/k\sqrt{2})$$

Thus,

$$\alpha l = F(k) - F(k,\theta_1)$$

where $F(k)$ and $F(k,\theta_1)$ are complete and incomplete elliptic integrals of the first kind. Values of k corresponding to tabulated values of F can be selected, and the corresponding values of θ_1, αl and, hence, P can be found.

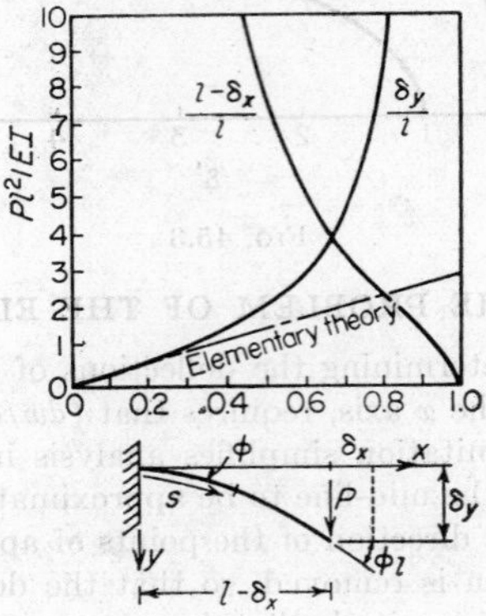

FIG. 45.4

To determine the deflection δ_y, write

$$\frac{dy}{d\phi}\frac{d\phi}{ds} = \frac{dy}{ds} = \sin\phi$$

Thus, from (45.7),

$$\frac{dy}{d\phi}\sqrt{2}\,\alpha(\sin\phi_l - \sin\phi)^{1/2} = \sin\phi$$

and

$$\delta_y = \int_0^{\delta_y} dy = \frac{1}{\sqrt{2}\,\alpha}\int_0^{\phi_l}(\sin\phi_l - \sin\phi)^{-1/2}\sin\phi\,d\phi$$

On making the substitutions as before,

$$\frac{\delta_y}{l} = \frac{1}{\alpha l}\int_{\theta_1}^{\pi/2}\frac{2k^2\sin^2\theta - 1}{(1 - k^2\sin^2\theta)^{1/2}}\,d\theta$$

This may be expressed in terms of elliptic integrals of the first and second kinds. Thus,

$$\frac{\delta_y}{l} = \frac{1}{\alpha l}[F(k) - F(k,\theta_1) - 2E(k) + 2E(k,\theta_1)] \tag{45.10}$$

For δ_x, from (45.6) and (45.7), putting $x = \phi = 0$,

$$P(l - \delta_x) = \sqrt{2}\,EI\alpha(\sin\phi_l)^{1/2}$$

or

$$\frac{l - \delta_x}{l} = \frac{\sqrt{2}}{\alpha l}(\sin\phi_l)^{1/2}$$

where ϕ_l is given, for the selected values of k, by (45.9).

The results of these calculations are plotted on Fig. 45.4.

Corresponding problems of a cantilever with a uniformly distributed load and a centrally loaded beam simply supported on smooth pegs have been solved, the first

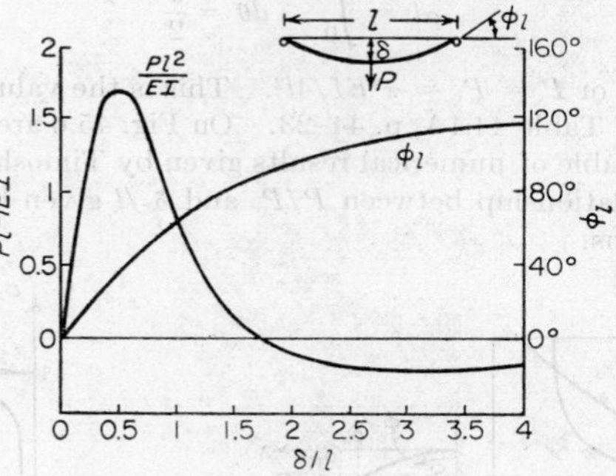

FIG. 45.5

by Rhode [9], and the second by Freeman [10]. Freeman's results are plotted in Fig. 45.5.

Large Deflections of Struts

If the load applied to the end of a cantilever is along its axis, the problem is that of a strut. The solution is similar to that of the laterally loaded cantilever. Proceeding as before, $d\phi/ds$ is given by

$$\frac{d\phi}{ds} = \sqrt{2}\,\alpha(\cos\phi - \cos\phi_l)^{1/2} \tag{45.11}$$

instead of by (45.7). Then

$$\sqrt{2}\,\alpha l = \int_0^{\phi_l}(\cos\phi - \cos\phi_l)^{-1/2}\,d\phi = \frac{1}{2}\sqrt{2}\int_0^{\phi_l}\left(\sin^2\frac{\phi_l}{2} - \sin^2\frac{\phi}{2}\right)^{-1/2}d\phi \tag{45.12}$$

This may be converted to an elliptic integral by the substitutions

$$\sin(\phi_l/2) = k \quad \text{and} \quad \sin(\phi/2) = k\sin\theta$$

and becomes

$$\alpha l = \int_0^{\pi/2}(1 - k^2\sin^2\theta)^{1/2}\,d\theta = F(k) \tag{45.13}$$

For δ_y, write

$$\delta_y = \int_0^{\delta_y} dy = \int_0^l \sin\phi\,ds = \int_0^{\phi_l}\frac{\sin\phi\,d\phi}{\sqrt{2}\,\alpha(\cos\phi - \cos\phi_l)^{1/2}}$$

On making the substitutions as before,

$$\alpha \delta_y = 2k \int_0^{\pi/2} \sin \theta \, d\theta = 2k$$

Finally, for δ_x,

$$l - \delta_x = \int_0^{l-\delta_x} dx = \int_0^l \cos \phi \, ds = \int_0^{\phi_l} \frac{\cos \phi \, d\phi}{\sqrt{2}\, \alpha(\cos \phi - \cos \phi_l)^{1/2}}$$

On making the substitutions as before,

$$\alpha(l - \delta_x) = 2E(k) - F(k)$$

As in the case of the cantilever with a lateral load, values of k can be chosen, and the corresponding values of δ_y, δ_x, and αl, and hence Pl^2/EI, calculated. When ϕ_l, and hence k, is very small, (45.13) can be written as

$$\alpha l = \int_0^{\pi/2} d\theta = \frac{\pi}{2}$$

giving $Pl^2/EI = \pi^2/4$ or $P = P_e = \pi^2 EI/4l^2$. This is the value for the Euler critical load given on line 4 of Table 44.1A, p. 44–23. On Fig. 45.6 are plotted P/P_e and δ_y/l against δ_x/l, from a table of numerical results given by Timoshenko [9]. The broken line represents the relationship between P/P_e and δ_x/l given by the Euler theory for infinitesimal deflections.

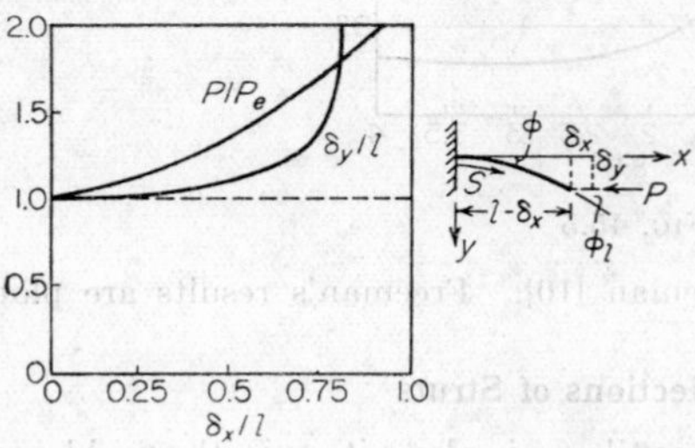

FIG. 45.6

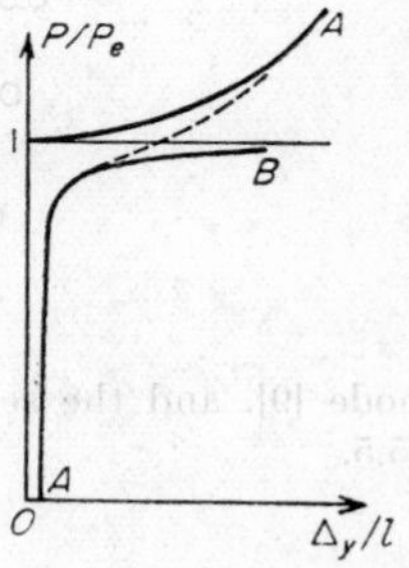

FIG. 45.7

According to the Euler theory, the straight configuration of the strut is unstable for loads above P_e. A strut loaded with P_e is in neutral equilibrium, and there is no configuration which is stable for higher loads. According to the theory of the elastica, however, the load-deflection curve rises indefinitely with a slope that is always positive, and this means that there is a stable configuration for all values of P. Nevertheless, slender struts are found by experiment to fail at loads close to P_e. The reason for this has been discussed by Southwell. All practical struts have slight imperfections, whose effect may be regarded as that of a slight initial lack of straightness. The behavior for such struts can be considered with the help of Fig. 45.7, which shows the P versus δ_y curve predicted by the exact theory for a straight strut (curve A), together with the curve predicted by the small deflection theory for a strut with a slight initial lack of straightness (curve B). When the deflections are small, such a strut will follow curve B, but, as they increase and become large compared with the initial deviations from straightness, the strut will follow curve A. Thus, the curve to be expected is the broken curve.

These curves are for struts infinitely elastic. Practical struts will suffer stress failure when the stress becomes high. Thus, since the stress in a slender strut is primarily due to bending, and since the greatest bending moment in the strut is $P\ \delta_y$, stress failure may be expected when δ_y first becomes large. This, from Fig. 45.7, is seen to occur at a load very close to P_e, the critical load predicted by the Euler theory for a perfect strut. This analysis is important, for theory based on infinitesimal deflections is used for many widely different problems of elastic instability. Where such theory is supported by experiment, the agreement is presumably due to reasons similar to those discussed here.

45.4. BUCKLING OF LOW ARCHES AND SLIGHTLY CURVED BEAMS

A slender arch rib may become unstable under its loads. If the central rise is of the same order of magnitude as the span, the deformation is inextensional, and the magnitude of the buckling load can be found by small deflection theory. But, if the arch is low, compression of the rib becomes important, deflections may be comparable with the initial rise, and the problem is nonlinear. The most complete investigation seems to be that of Fung and Kaplan [10], who include a review of previous work. They consider arches of sinusoidal and other shapes, with and without constraint against spread, and include sinusoidal, uniformly distributed, and concentrated central loads, and the effects of initial axial compression and vertical elastic constraints.

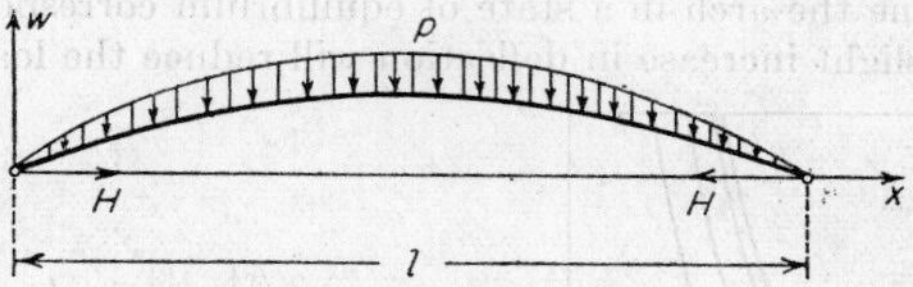

FIG. 45.8

Consider the arch of span l shown in Fig. 45.8. It is sufficiently flat for $(dw/dx)^2$ to be neglected compared with unity. Its load and initial and final shapes are:

$$p = p_0 \sin \frac{\pi x}{l} \qquad w_0 = a_1 l \sin \frac{\pi x}{l} \qquad w = \sum b_m l \sin \frac{m\pi x}{l} \tag{45.14}$$

These, and the horizontal thrust H are connected by the usual equation for a beam with end load H and lateral load p,

$$EI \frac{d^4(w - w_0)}{dx^4} + H \frac{d^2 w}{dx^2} = -p \tag{45.15}$$

The shortening of the arch rib is given by

$$\frac{Hl}{EA} = \frac{1}{2} \int_0^l \left[\left(\frac{dw_0}{dx} \right)^2 - \left(\frac{dw}{dx} \right)^2 \right] dx$$

from which

$$H = \frac{\pi^2 EA}{4} \left(a_1^2 - \sum_m m^2 b_m^2 \right)$$

Substituting in (45.15) for w, w_0, p, and H, and equating coefficients of corresponding terms, the following equations for the Bs are obtained:

$$\begin{aligned} B_1(\Sigma n^2 B_n^2 - \lambda_1^2 + 1) &= -R + \lambda_1 \\ B_2(\Sigma n^2 B_n^2 - \lambda_1^2 + 4) &= 0 \\ \cdots\cdots\cdots\cdots & \\ B_m(\Sigma n^2 B_n^2 - \lambda_1^2 + m^2) &= 0 \\ \cdots\cdots\cdots\cdots & \end{aligned} \tag{45.16}$$

where $B_m = b_m l/2k$, $\lambda_1 = a_1 l/2k$, $R = p_0 l^4/2\pi^4 EIk$, and $k^2 = I/A$.

These equations have two possible solutions. The first is $B_2 = B_3 = \cdots = 0$,

$$B_1^3 - B_1(\lambda_1^2 - 1) = \lambda_1 - R$$

In this equation, λ_1 is proportional to the initial central rise, R to the load, and B_1 to the central rise when loaded. Figure 45.9 shows the relationship graphically. When the initial central rise is given by $\lambda_1 < 1$, the deflection increases steadily with increasing load. But, if $\lambda_1 = 1$, the curve has a horizontal tangent, and, for $\lambda_1 > 1$, the curve has a maximum and a minimum. If an experiment is performed in which the load is applied by steadily increasing dead weights, the load deflection will increase steadily until the maximum is reached, at the point M on the curve for $\lambda_1 = 1.5$, for example. Any further slight increment of load cannot be supported by the arch with any configuration in the neighborhood of M, and the deflection will increase until the point N is reached, the arch having snapped through into a configuration in which its curvature is reversed. This phenomenon is variously called "snapping through," "oil canning," or "Durchschlagen" (see p. 44–2). If the load is then reduced, a similar snapping action will occur from R to S.

It is impossible to plot the part of the curve MR experimentally, using dead-weight loading. For imagine the arch in a state of equilibrium corresponding to a point on the curve MR. A slight increase in deflection will reduce the load which the arch is

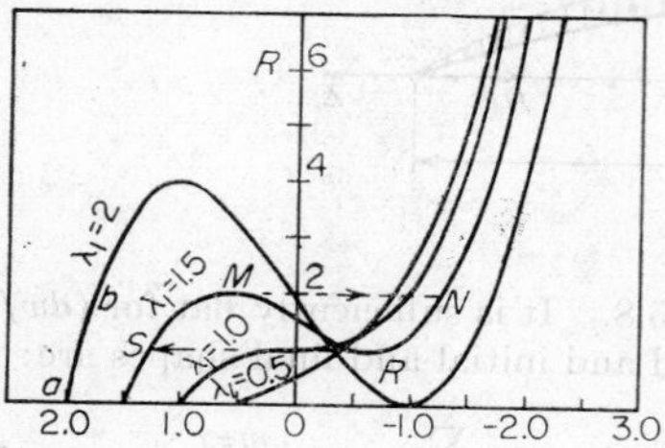

FIG. 45.9

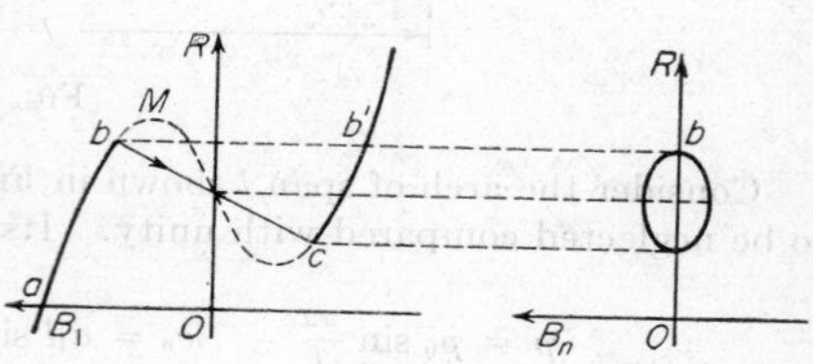

FIG. 45.10

able to support, and the deflection will be increased further, until a condition of equilibrium is found on the rising part of the curve NR. Similarly, a slight decrease in the deflection will *increase* the load which the arch is able to support, and the arch will rise to the condition of equilibrium on the curve SM. Thus, the equilibrium states of the arch and its load represented by MR, are *unstable*. If the arch is made to deform by a device which controls its deflection everywhere, and which measures its reaction, snapping through will not occur, and the whole curve, including the part MR, can be plotted by experiment. The load corresponding to the point M, at which snapping occurs for increasing load, is given by

$$R_{cr} = \lambda_1 + [\tfrac{4}{27}(\lambda_1^2 - 1)^3]^{1/2} \tag{45.17}$$

The second possible solution to Eqs. (45.16) is

$$B_1 = (R - \lambda_1)/(n^2 - 1)$$
$$n^2 B_n^2 = \lambda_1^2 - n^2 - [(R - \lambda_1)/(n^2 - 1)]^2 \tag{45.18}$$

where n has one value, and all other Bs are zero. From (45.18), B_n can have a real value only within a certain range of values of R. On Fig. 45.10, R is plotted against B_1 and B_n. This nth mode first occurs at b, for which

$$B_1 = (\lambda_1^2 - n^2)^{1/2} \qquad B_n = 0 \qquad R_{cr} = \lambda_1 + (n^2 - 1)(\lambda_1^2 - n^2)^{1/2} \tag{45.19}$$

The dashed curve bMc is replaced by the straight line bc, and the load R_{cr} at which the arch collapses will correspond to b, provided that (1) R_{cr}, B_1, and B_n are real; (2) R_{cr} given by (45.19) is less than R_1 given by (45.17); (3) b lies to the left of M; (4) n is chosen for R_{cr} to be as small as possible. These conditions are satisfied if

$$n = 2 \quad \text{and} \quad \lambda_1 > \sqrt{5.5}$$

Thus, for $\lambda_1 < 1$, there is no buckling; for $1 < \lambda_1 \leq \sqrt{5.5}$,

$$R_{cr} = \lambda_1 + [4/27(\lambda_1^2 - 1)^3]^{1/2}$$

and for $\lambda_1 \geq \sqrt{5.5}$,

$$R_{cr} = \lambda_1 + 3(\lambda_1^2 - 4)^{1/2}$$

The critical values for a total uniformly distributed load and a central point load are very closely $\pi^2/8$ and $\pi/4$ times the corresponding total critical sinusoidal load. These values of the critical loads are very sensitive to slight lack of symmetry in the initial shape of the arch.

45.5. EXTERNAL AND INTERNAL INSTABILITY

Instability of the type occurring at the point M on Fig. 45.9 is instability of the system consisting of the arch and the device which applies the load. If the loading device controls the deflection rather than the load, instability will not occur at M. This type of instability might be called *external* instability.

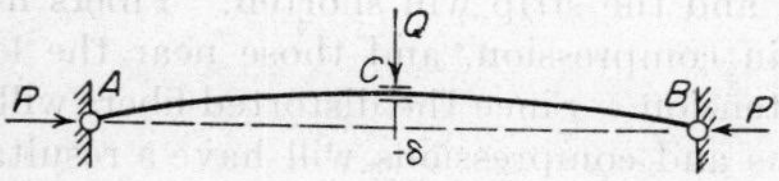

Fig. 45.11

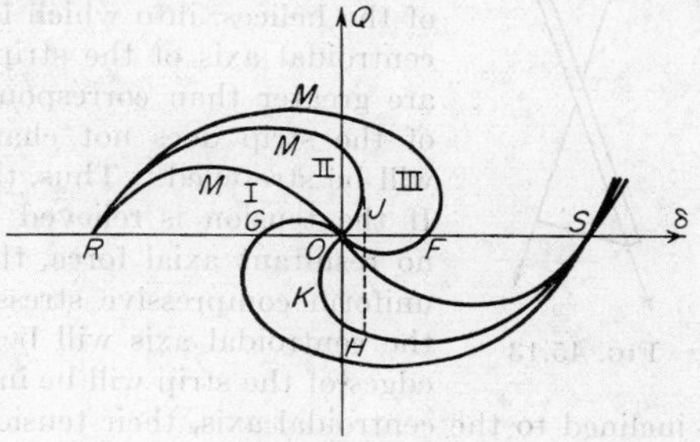

Fig. 45.12

Consider now the structure shown in Fig. 45.11. This consists of a naturally straight uniform rod, held between two fixed pins A and B. The length of the rod slightly exceeds the distance AB. If this excess length is great enough for the compression in the rod, when held straight, to exceed P_e, in the absence of constraint the rod will spring sideways as shown. The rod is loaded laterally at the center C as shown, and a constraint prevents rotation of the tangent there. Assuming $(dw/dx)^2$ to be negligible compared with unity, and symmetrical deformation of the rod, the Q,δ curve may be plotted. When $Q = 0$, the compression in the rod (to the accuracy of the Euler theory) is P_e. As Q and δ increase, the compression in the rod also increases. If this compression, when the rod is held straight, is less than $8P_e$, the Q,δ curve is of the type I in Fig. 45.12; the rod is then quite straight when ACB is a straight line. But, if the excess length is such that the compression in the straight rod slightly exceeds approximately $8P_e$, then the rod will be unstable and will spring into a deformed state with ACB straight, and either a positive or a negative Q will be required to maintain the deflection of C. The Q,δ curve is now of type II in Fig. 45.12. For a certain range of values of δ (those lying between K and J), there are three possible configurations for each value of δ, of which two are stable and one is unstable. If the condition of the rod corresponds to a point on JOK, it will be unstable and will spring to one of the two other conditions having the same value of δ. Thus, even if δ

is increased by a device which controls deflection independently of the reaction of the rod, instability will occur at J, and the rod will snap into condition H.

This type of instability is different in *kind* from the stability occurring at M, for it is quite independent of the nature of the loading device. It might be called *internal* instability. H. L. Cox has pointed out that, if such a Q,δ curve is used to estimate the energy released on snapping, then energy equal to the area $JOKH$ is lost, as the rod travels from state J to state H without change of δ. A further change in the Q,δ curve occurs if the excess length is such that, when held straight, the rod has a compression greater than $9P_e$. In this case, the curve is of type III in Fig. 45.12. The states F and G correspond to buckling of the rod into three half waves with $Q = 0$.

45.6. TORSION OF THIN SECTIONS

When a straight uniform bar is slightly twisted without end constraint, the torque is resisted by shearing stresses distributed over its cross section, in accordance with the Saint-Venant theory of torsion. But, if the twist becomes appreciable, the distortion of longitudinal fibers into helices will set up tensions and compressions in them, and these tensions and compressions, having components at right angles to the axis of the rod, will contribute to the torque exerted by the rod. This effect is nonlinear.

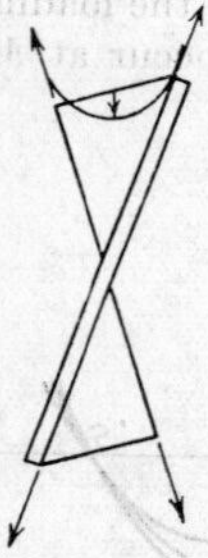

FIG. 45.13

Consider a thin strip, of rectangular cross section, twisted to an appreciable extent (Fig. 45.13). By symmetry, the common axis of the helices, into which longitudinal fibers are distorted, will be the centroidal axis of the strip. Since distances measured along a helix are greater than corresponding distances along its axis, if the length of the strip does not change, all fibers except the centroidal fiber will be stretched. Thus, the strip will experience a resultant tension. If the tension is relieved by an equal compression, so that there is no resultant axial force, this compression will produce an additional uniform compressive stress, and the strip will shorten. Fibers near the centroidal axis will be in compression, and those near the long edges of the strip will be in tension. Since the distorted fibers will be inclined to the centroidal axis, their tensions and compressions will have a resultant torque about the axis, to be added to the Saint-Venant torque.

Let the twist be θ, and the strain of the central fiber be ϵ_0. Then it can be shown, from the geometry of the helix, that a filament distant r from the axis has a strain of $\epsilon_0 + 1 - \cos\phi$, where $\phi = \tan^{-1}(r\theta)$. If $(r\theta)^2$ may be neglected compared with unity, this strain may be taken as $\epsilon_0 + \frac{1}{2}(r\theta)^2$. The axial component of the stress in the fiber is $E[\epsilon_0 + \frac{1}{2}(r\theta)^2]$ to the same accuracy, and, if the thickness t is small compared with the breadth b, the resultant axial tension will be

$$2\int_0^{b/2} E[\epsilon_0 + \tfrac{1}{2}(r\theta)^2]t\,dr = 0$$

from which

$$\epsilon_0 = -b^2\theta^2/24$$

Thus, the fiber tensile stress σ is

$$\sigma = E\theta^2(12r^2 - b^2)/24$$

It can be shown, again from the geometry of the helix, that the component of this tension perpendicular to the centroidal axis is $\sigma \sin\phi$ or $\sigma r\theta$ to the accuracy being used. This has a moment about the axis of $\sigma r^2\theta$, and the resultant torque due to the fiber tensions is

$$M_k' = 2\int_0^{b/2} \frac{E\theta^3}{24}(12r^4 + b^2r^2)t\,dr = \frac{E\theta^3b^5t}{360}$$

Adding this to the Saint-Venant torque [Eq. (36.33)], the total torque is

$$M_k = \tfrac{1}{3}bt^3G\theta\left(1 - \frac{Eb^4\theta^2}{120Gt^2}\right)$$

This analysis ignores the local stress distribution at the ends of the strip, necessary to free the ends from longitudinal stress.

Cullimore [11] obtains similar expressions for I and **Z** sections. For such sections, having two axes of symmetry, or skew symmetry, the position of the fiber which remains straight, and from which r is measured, is obvious. For other sections, it is necessary to obtain the position of this fiber, which is not the usual axis of twist. Methods are given by Cullimore [11] and Ashwell [12].

45.7. TRANSITION BETWEEN BEAM AND PLATE

A beam of compact rectangular section bent about an axis parallel to one face, to a curvature $1/R$, exhibits an *anticlastic* curvature of magnitude ν/R. A thin rectangular strip bent about an axis parallel to the major axis of its cross section, to any appreciable degree, does not exhibit anticlastic curvature. The reason for this concerns the longitudinal stresses present. If anticlastic curvature did occur, longitudinal fibers near the edges would be farther from the axis of curvature than those near the center of the strip. Thus, in the absence of any resultant longitudinal force, outer fibers would be stretched and inner fibers compressed. Since the fibers are curved about the axis of curvature, the resultant longitudinal tensions and compressions would have components forcing the outer filaments toward the axis of curvature, and the inner ones away from it. Thus, they would tend to destroy the anticlastic curvature, and flatten the strip, as the Brazier effect tends to flatten bent tubes (Sec. 45.10). The behavior underlying the transition between the beam and the plate is nonlinear.

If x is measured along the axis of the beam or plate, and y transversely, the deformation of the mid-line of the cross section is found to obey the equation

$$\frac{d^4w}{dy^4} + \frac{12(1-\nu^2)}{R^2t^2}w = 0 \tag{45.20}$$

where t is the thickness of the plate. Thus, a transverse strip across the plate behaves as a beam on an elastic foundation (see Chap. 31). The boundary conditions are $d^2w/dy^2 = -v/R$ and $d^3w/dy^3 = 0$ at $y = \pm b/2$, where b is the breadth of the plate. Hence,

$$w/t = A \operatorname{Cosh} \alpha y \cos \alpha y + B \operatorname{Sinh} \alpha y \sin \alpha y$$

where

$$A,B = \frac{\nu}{[3(1-\nu^2)]^{1/2}} \frac{\operatorname{Cosh}(\alpha b/2)\sin(\alpha b/2) \mp \operatorname{Sinh}(\alpha b/2)\cos(\alpha b/2)}{\operatorname{Sinh}\alpha b + \sin \alpha b}$$

and

$$4\alpha^4 = 12(1-\nu^2)/R^2t^2$$

w/t is plotted against y/b, for $\nu = \frac{1}{3}$ and various values of $C = b^2/Rt$, in Fig. 45.14 (from Ashwell [13]). For values of $C < 1$, the anticlastic curvature takes place effectively unhindered, but, as C increases, flattening occurs, and the distortion is progressively confined to the edges of the plate. For values of $C > 100$, the width of the distorted region is given approximately by $3b/C^{1/2}$. The maximum deflection w at the edges is effectively independent of C, if $C > 16$, and equals $t\nu/2[3(1-\nu^2)]^{1/2}$. For $\nu = \frac{1}{3}$, this is $0.102t$.

The bending moment to be applied to the strip is given by

$$M = -\int_{-b/2}^{+b/2} N_x w \, dy + \int_{-b/2}^{+b/2} M_x \, dy$$

where N_x the longitudinal stress resultant per unit width, and M_x the moment per unit width, are given by

$$N_x = -\frac{Etw}{R} \quad \text{and} \quad M_x = K\left(\frac{1}{R} + \nu\frac{d^2w}{dy^2}\right)$$

If M is expressed as

$$M = \phi EI/R$$

ϕ depends on $(1-\nu^2)^{1/2}C$. For $C = 0$, $\phi = 1$; for $C \to \infty$, $\phi \to 1/(1-\nu^2)$. The coefficient ϕ is plotted against C, for $\nu = 1/3$, in Fig. 45.15 (Ashwell [13]).

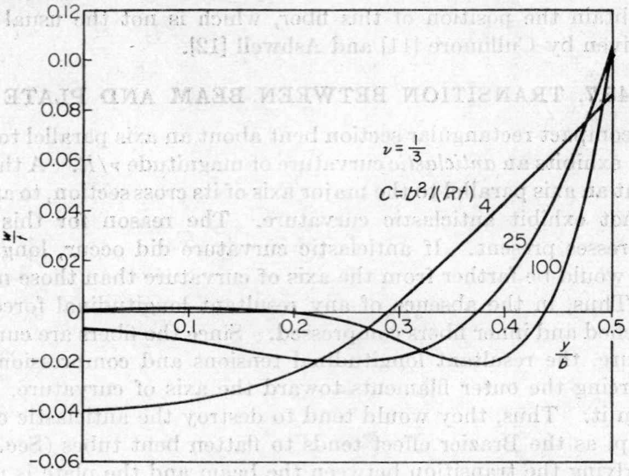

FIG. 45.14

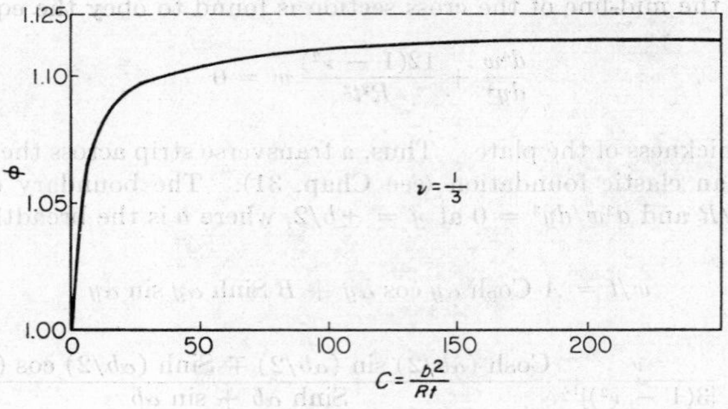

FIG. 45.15

In this paragraph and the previous one were considered the finite twisting and bending of strips. The behavior of a strip both twisted and bent has been considered by Reissner [14].

45.8. DEFORMATION OF PLATES INTO DEVELOPABLE SURFACES

The behavior of the plate considered above is typical of a certain class of problems in the large deflections of plates. For small deflections ($w \ll t$), a plate deforms according to the theory given in Chap. 39, principal curvatures having similar magnitudes, and the effects of membrane forces on the deformations being negligible. As deflections are increased, however, the membrane forces tend to modify the deformed shape, and, if it is possible for a *developable* surface to be found into which the plate can deform under the imposed loads and displacements, the modification is toward such a surface. At each point, one of the principal curvatures is suppressed as the anticlastic curvature of the bent strip was suppressed, deviations from the developable surface being confined to regions close to the free edges. The form of the developable surface, in particular cases, has been investigated by Müller-Magyari [15] and Mansfield [16]. The equilibrium equations for the developable surface are given by Ashwell [17], and for the edge region, where deviations occur, by Fung and Wittrick [18].

45.9. SNAPPING OF A THIN STRIP WITH TRANSVERSE CURVATURE

The analysis of Sec. 45.7 can be applied to the bending of a thin strip with a slight initial transverse curvature of radius R_0. w is now the total (initial plus elastic) displacement from the straight y axis, and the boundary conditions are now $d^2w/dy^2 = (1/R_0) - (\nu/R)$ and $d^3w/dy^3 = 0$, at $y = \pm b/2$. Also, M_x is now given by

$$M_x = K\left(\frac{1}{R} + \nu\frac{d^2w}{dy^2} - \frac{\nu}{R_0}\right)$$

It is found that as the curvature ratio C increases the cross section is rapidly flattened by the longitudinal stresses. The bending moment at first rises with increasing C,

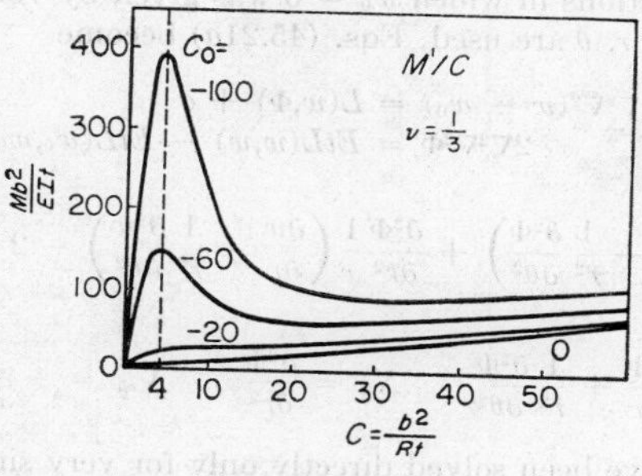

FIG. 45.16

reaches a maximum, falls, and then rises again. On Fig. 45.16, Mb^2/EIt is plotted against C for various negative values of $C_0 = b^2/R_0t$—i.e., for strips with initial and applied curvatures of opposite signs. External instability occurs at $C \approx 4$, and $Mb^2/EIt = 4.155 - 0.166C_0 + 0.0371C_0^2$ (Ashwell, [19]).

45.10. THE BRAZIER EFFECT

The flattening effect of longitudinal stresses on the cross sections of bent thin-walled members was investigated by Brazier [20]. An initially circular tube becomes oval if it is bent, and its bending stiffness $dM/d\kappa$ (where κ is the curvature of its axis) decreases, becoming zero when

$$M_{cr} = \frac{2\sqrt{2}}{9}\frac{E\pi rt^2}{1-\nu^2}$$

where r is the initial mean radius. When this occurs, the section of the tube has become oval, the minimum diameter being $1.56r$. This type of failure occurs in comparatively thick tubes made of material having a low modulus of elasticity, such as rubber. Thin metal tubes in bending become unstable by local buckling on the compression side.

45.11. LARGE DEFLECTIONS OF PLATES

The deformation of a plate into a developable surface has been mentioned above. If a plate is prevented by constraints from deforming into such a surface, increasing deflections lead to increasing membrane forces distributed over the whole plate, and the effect of these is again nonlinear. Rectangular and circular plates whose boundaries are constrained to remain in a plane are examples. The following equations for such plates are given by Marguerre [21]. They are based on the assumption that, though the deflections are large enough for the membrane forces to be important, $(\partial w/\partial x)^2$ and $(\partial w/\partial y)^2$ can be neglected compared with unity.

$$K\,\nabla^2\nabla^2(w - w_0) = \frac{\partial^2\Phi}{\partial x^2}\frac{\partial^2 w}{\partial y^2} + \frac{\partial^2\Phi}{\partial y^2}\frac{\partial^2 w}{\partial x^2} - 2\frac{\partial^2\Phi}{\partial x\,\partial y}\frac{\partial^2 w}{\partial x\,\partial y} + q$$
$$\nabla^2\nabla^2\Phi = Et\left[\left(\frac{\partial^2 w}{\partial x\,\partial y}\right)^2 - \left(\frac{\partial^2 w_0}{\partial x\,\partial y}\right)^2 - \left(\frac{\partial^2 w}{\partial x^2}\frac{\partial^2 w}{\partial y^2} - \frac{\partial^2 w_0}{\partial x^2}\frac{\partial^2 w_0}{\partial y^2}\right)\right] \qquad (45.21a)$$

where the notations are the same as in Chap. 39, and w_0 defines the deviation of the unstressed plate from a plane. Φ is a stress function such that the membrane forces per unit length are

$$N_x = \frac{\partial^2\Phi}{\partial y^2} \qquad N_y = \frac{\partial^2\Phi}{\partial x^2} \qquad N_{xy} = -\frac{\partial^2\Phi}{\partial x\,\partial y}$$

The form of these equations in which $w_0 = 0$ was given by von Kármán.

If polar coordinates r, θ are used, Eqs. (45.21a) become

$$K\nabla^2\nabla^2(w - w_0) = L(w,\Phi) + q$$
$$2\nabla^2\nabla^2\Phi = EtL(w,w) - EtL(w_0,w_0) \qquad (45.21b)$$

where

$$L(w,\Phi) = \frac{\partial^2 w}{\partial r^2}\left(\frac{1}{r}\frac{\partial\Phi}{\partial r} + \frac{1}{r^2}\frac{\partial^2\Phi}{\partial\theta^2}\right) + \frac{\partial^2\Phi}{\partial r^2}\frac{1}{r}\left(\frac{\partial w}{\partial r} + \frac{1}{r^2}\frac{\partial^2 w}{\partial\theta^2}\right) - 2\frac{\partial}{\partial r}\left(\frac{1}{r}\frac{\partial w}{\partial\theta}\right)\frac{\partial}{\partial r}\left(\frac{1}{r}\frac{\partial\Phi}{\partial\theta}\right)$$

and

$$N_r = \frac{1}{r}\frac{\partial\Phi}{\partial r} + \frac{1}{r^2}\frac{\partial^2\Phi}{\partial\theta^2} \qquad N_\theta = \frac{\partial^2\Phi}{\partial r^2} \qquad N_{r\theta} = -\frac{\partial}{\partial r}\left(\frac{1}{r}\frac{\partial\Phi}{\partial\theta}\right)$$

These equations have been solved directly only for very simple cases. The most generally useful method of treating this type of problem is an approximate one using *energy*. If U is the strain energy stored in the plate, expressed in terms of the displacements u, v, w, the principle of virtual displacements gives

$$\delta U - \delta\iint qw\,dx\,dy = 0 \qquad (45.22)$$

for any small variations of u, v, w. U is the sum of the energy of bending U_b, and the energy of the membrane stresses U_m, where, if $w_0 = 0$, from Timoshenko [22],

$$2U_b = K\iint[(w_{xx} + w_{yy})^2 - 2(1-\nu)(w_{xx}w_{yy} - w_{xy}^2)]\,dx\,dy$$
$$2(1-\nu^2)U_m = Et\iint[u_x^2 + u_xw_x^2 + v_y^2 + v_yw_y^2 + \tfrac{1}{4}(w_x^2 + w_y^2)^2 + 2\nu(u_xv_y + \tfrac{1}{2}v_yw_x^2 + \tfrac{1}{2}u_xw_y^2) + \tfrac{1}{2}(1-\nu)(u_y^2 + 2u_yv_x + v_x^2 + 2u_yw_xw_y + 2v_xw_xw_y)]\,dx\,dy$$

A suitable *assumed* form, satisfying the geometrical boundary conditions of a particular problem, and containing undetermined constants a, b, c, . . . , is substituted into (45.22) with

$$\delta = \frac{\partial}{\partial a}, \frac{\partial}{\partial b}, \frac{\partial}{\partial c}, \ldots$$

and the resulting equations are solved for a, b, c, ... Such a solution, for a uniformly loaded plate with clamped edges is given by Timoshenko [22], quoting Way.

45.12. CIRCULAR PLATES WITH AXIAL SYMMETRY

Other methods have been used for circular plates with axial symmetry. For example, Timoshenko [22], quoting Nadai, gives the following equations relating w and the radial displacement u:

$$\frac{\partial^2 u}{\partial r^2} + \frac{1}{r}\frac{\partial u}{\partial r} - \frac{u}{r^2} = -\frac{1-\nu}{2r}\left(\frac{\partial w}{\partial r}\right)^2 - \frac{\partial w}{\partial r}\frac{\partial^2 w}{\partial r^2}$$
$$t^2\left(\frac{\partial^3 w}{\partial r^3} + \frac{1}{r}\frac{\partial^2 w}{\partial r^2} - \frac{1}{r^2}\frac{\partial w}{\partial r}\right) = 12\frac{\partial w}{\partial r}\left[\frac{\partial u}{\partial r} + \frac{\nu}{r}u + \frac{1}{2}\left(\frac{\partial w}{\partial r}\right)^2\right] + \frac{t^2}{Kr}\int_0^r qr\,dr \qquad (45.23)$$

For a plate with edges built in, assume that

$$\frac{dw}{dr} = C\left[\frac{r}{a} - \left(\frac{r}{a}\right)^2\right]$$

which satisfies the boundary conditions for w at $r = a$. This is substituted into the first equation, which is then solved for u. u and dw/dr are then substituted into the second equation, which is solved for q. C and n are then adjusted to make q as nearly uniform as possible.

Two types of *series* solution have been used. The first (Timoshenko [22], quoting Way) expresses the wanted functions as power series in r, with undetermined coefficients. On substituting the series into a suitable form of the differential equations, and equating coefficients of equal powers of r, equations are obtained from which the coefficients can be determined.

The second type of series solution is known as the *perturbation method.* The wanted functions are expressed as power series of the type

$$R_0 + R_1 w_0 + R_2 w_0^2 + \cdots$$

where w_0 is some particular (for example, the central) deflection, and $R_0, R_1, R_2, \ldots$ are functions of r. On substituting the series into a suitable form of the differential equations, and equating coefficients of equal powers of w_0, differential equations for the functions R are obtained. The method is dependent on w_0 being small enough for the series to converge rapidly. Kaplan and Fung [23] have used it to investigate the behavior of shallow spherical domes under uniform pressure.

45.13. RELAXATION METHOD FOR LATERALLY LOADED MEMBRANES

If w_0 and K are put equal to zero in (45.21), Föppl's equations are obtained for deflections of initially flat elastic membranes. In addition to the methods already mentioned for plates, *relaxation methods* have been used for membranes. Shaw and Perrone [24] have solved a rectangular membrane with uniform lateral load. They first transform Föppl's two equations, of which one is of the fourth order, into the following three second-order equations for u, v, and w:

$$\nabla^2 u + u_{xx} - \nu u_{yy} + (1+\nu)v_{xy} = -[(1+\nu)w_y w_{xy} + 2w_x w_{xx} + (1-\nu)w_x w_{yy}]$$
$$\nabla^2 v + v_{yy} - \nu v_{xx} + (1+\nu)u_{xy} = -[(1+\nu)w_x w_{xy} + 2w_y w_{yy} + (1-\nu)w_y w_{xx}]$$
$$w_{xx}(3w_x^2 + w_y^2) + w_{yy}(w_x^2 + 3w_y^2) + 4w_x w_y w_{xy} = -[w_x(2u_{xx} + u_{yy} + v_{xy}) + w_y(2v_{yy} + v_{xx} + u_{xy}) + \nu w_x(v_{xy} - u_{yy}) + \nu w_y(u_{xy} - v_{xx})] - 2[w_{xx}(u_x + \nu v_y) + (1-\nu)w_{xy}(u_y + v_x) + w_{yy}(v_y + \nu u_x)] - \frac{2q}{D} \qquad (45.24a\text{–}c)$$

Here subscripts indicate partial differentiation, ∇^2 is the two-dimensional Laplace operator, and D is the extensional rigidity of the plate [see Eqs. (43.40)].

First u and v are put equal to zero, and Eq. (45.24c) is expressed as a finite-difference equation in terms of the values w at nine points of a square mesh. The changes ΔR of the residuals for a change $\Delta w = 1$ depend on the values w in the neighborhood, and must be recalculated frequently as the relaxation proceeds. This is the essential difference between the relaxation of nonlinear and of linear systems—for linear systems, the operations table is calculated only once, and does not change thereafter.

When a set of values of w has been found in this way, this set, together with $v = 0$, is introduced into a finite-difference form of Eq. (45.24a), and a second relaxation process gives a set of values of u. These values, and those of w already found are substituted in Eq. (45.24b), and a third relaxation process gives a set of values of v. The cycle is started again, after substituting into Eq. (45.24c) the values thus found for u and v, and is repeated until all residuals are sufficiently liquidated. The process is lengthy, but not, apparently, prohibitively so. In a typical case, Shaw and Perrone found four such cycles necessary.

45.14. TYPICAL RESULTS FOR PLATES AND MEMBRANES

The following results are given by Roark [30], who gives references to their derivation. w_0 is the maximum deflection, and σ is the sum of the bending and membrane stresses.

For circular plates with uniform loads, radius a

$$\frac{qa^4}{Et^4} = K_1 \frac{w_0}{t} + K_2 \left(\frac{w_0}{t}\right)^3 \qquad \frac{\sigma a^2}{Et^2} = L_1 \frac{w_0}{t} + L_2 \left(\frac{w_0}{t}\right)^2$$

where

1. Edge simply supported, no edge tension:

$$K_1 = \frac{64}{63(1-\nu)} \qquad K_2 = 0.376 \qquad L_1 = \frac{1.238}{1-\nu} \qquad L_2 = 0.294$$

2. Slope and tension zero at edge:

$$K_1 = \frac{16}{3(1-\nu^2)} \qquad K_2 = 6/7 \qquad L_1\,(\text{edge}) = \frac{4}{1-\nu^2} \qquad L_2\,(\text{edge}) = 0$$

$$L_1\,(\text{center}) = \frac{2}{1-\nu} \qquad L_2\,(\text{center}) = 0.5$$

3. Slope and radial displacement zero at edge:

$$K_1 = \frac{16}{3(1-\nu^2)} \qquad K_2 = \frac{2.603}{1-\nu^2}$$

The membrane stress at the edge is $0.467E(w_0/a)^2$; the bending stress there can be estimated by treating the plate as case 2. The membrane stress at the center is $0.976E(w_0/a)^2$; the *total* stress there can be estimated by treating the plate as case 2.

For rectangular plates with uniform loads, and simply supported edges having $u = v = 0$, and sides a and b $(a < b)$,

$$\frac{16q(1-\nu^2)}{\pi^6 Et^4} = \frac{1}{12}\left(\frac{1}{a^2} + \frac{1}{b^2}\right)^2 \frac{w_0}{t} + \frac{1}{16}\left[\frac{4\nu}{a^2b^2} + (3-\nu^2)\left(\frac{1}{a^4} + \frac{1}{b^4}\right)\right]\left(\frac{w_0}{t}\right)^3$$

at center, parallel to shorter sides:

$$\sigma = \frac{\pi^2 E w_0}{8(1-\nu^2)}\left[\frac{(2-\nu^2)w_0 + 4t}{a^2} + \frac{\nu(w_0 + 4t)}{b^2}\right]$$

For circular membranes, uniform loads, edges held,

$$w_0 = 0.662a\left(\frac{qa}{Et}\right)^{1/3} \qquad \sigma\,(\text{center}) = 0.423\left(\frac{Eq^2a^2}{t^2}\right)^{1/3} \qquad \sigma\,(\text{edge}) = 0.328\left(\frac{Eq^2a^2}{t^2}\right)^{1/3}$$

From Timoshenko [22], for square membranes, uniform loads, edges held, side $2a$,

$$w_0 = 0.802a\left(\frac{qa}{Et}\right)^{1/3} \qquad \sigma\,(\text{center}) = 0.396\left(\frac{Eq^2a^2}{t^2}\right)^{1/3}$$

45.15. POSTBUCKLING BEHAVIOR OF PLATES AND SHELLS

In Chap. 44 are given values obtained by small-deflection theory, for the critical stresses of flat and curved panels in compression. Those for flat panels are found to agree satisfactorily with experiment, but curved panels and thin-walled tubes are found to buckle at stresses very much less than those predicted. This difference in behavior is due to the nonlinear deflections occurring after buckling. A flat panel with its longitudinal edges constrained to remain straight, continues to withstand increasing loads after buckling. Near such edges, the plate is not in a buckled condition; moreover, a restraining transverse tension develops as deflections increase. Thus the

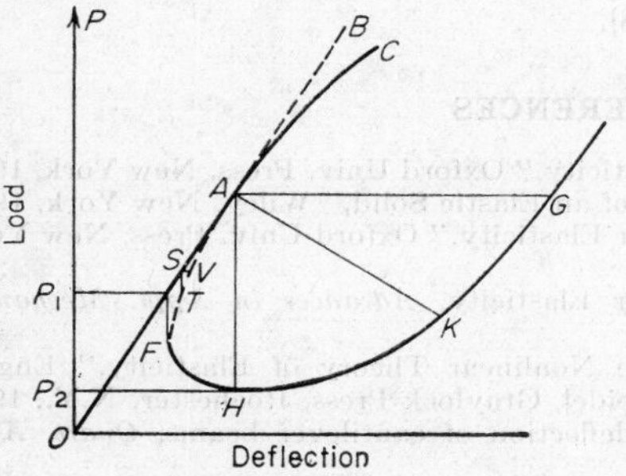

FIG. 45.17

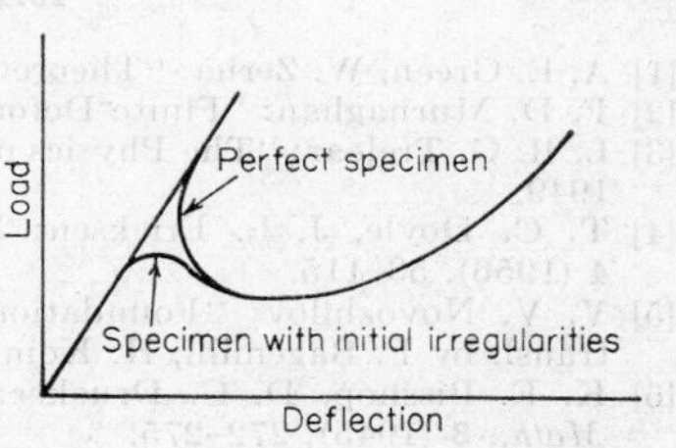

FIG. 45.18

stiffness after buckling is a substantial proportion (between approximately one-half to seven-eights depending on the type and degree of fixing at the edges) of the prebuckling stiffness.

A curved panel, or circular tube behaves differently. If outward buckling were to occur, the transverse curvature increasing, a restraining transverse tension would maintain an appreciable stiffness, as in the case of a flat panel; but, if inward buckling occurs, a corresponding compression arises. This compression causes the distorted form to become unstable (internal instability). In Fig. 45.17 is shown diagrammatically the type of load-deflection curve obtained, the deflection being the average over-all deflection in the direction of the load. The straight line OAB gives the behavior for no lateral deflection. A corresponds to the critical stress given by the small deflection theory, AC to *outward* buckling, AFG to *inward* buckling. The broken curves correspond to states of internal instability. A perfectly formed specimen, loaded under perfect conditions, would be expected to follow OA, becoming internally unstable at A. Subsequent behavior would depend on the loading conditions. In a deflection-controlled experiment, violent inward buckling to state H would occur, with the release of energy equal to the area AFH. In a load-controlled experiment, buckling to state G would occur, again violently, with the release of energy AFG. If the testing machine had finite stiffness, buckling to state K would occur, where the slope of AK was numerically equal to the stiffness of the machine.

Experiments confirm the violent nature of the buckling, and the fall in load to be expected, but there is a scatter of values of the buckling load, which is found to be about half that given by the point A. Two effects have been suggested to explain the lowness of experimental buckling loads.

The first is the effect of initial deviations from the true cylindrical form of the specimen. If the analysis is performed on a curved plate or tube which, when unstressed, has a periodic radial deviation of amplitude w_0, from the cylindrical shape, the predicted load-deflection curve changes in the manner shown in Fig. 45.18. The order of magnitude of w_0/t, required to produce a reduction in maximum load of the order observed, is 0.1 (Cox and Pribram [25]).

The second effect adduced to explain the lowness of experimental buckling loads concerns the narrowness of the elongated cusp OAF of Fig. 45.17. Since the energy required to change the specimen from state S to the unstable state T or V is the *small* quantity SAT or SAV, it is suggested that small disturbances in the form of random impulses in the loading process may supply it. Thus, a lower bound for the buckling load in a deflection-controlled experiment is P_1, obtained by drawing a vertical tangent from F to OA. P_2 is the corresponding load in a load-controlled experiment (Tsien [26]).

Data and bibliographies on the buckling of curved elements are given by Argyris and Dunne [27] and Gerard and Becker [28].

45.16. REFERENCES

[1] A. E. Green, W. Zerna: "Theoretical Elasticity," Oxford Univ. Press, New York, 1954.
[2] F. D. Murnaghan: "Finite Deformation of an Elastic Solid," Wiley, New York, 1951.
[3] L. R. G. Treloar: "The Physics of Rubber Elasticity," Oxford Univ. Press, New York, 1949.
[4] T. C. Doyle, J. L. Ericksen: Nonlinear Elasticity, *Advances in Appl. Mechanics*, **4** (1956), 53–115.
[5] V. V. Novozhilov: "Foundations of the Nonlinear Theory of Elasticity," English transl. by F. Bagemihl, H. Komm, W. Seidel, Graylock Press, Rochester, N.Y., 1953.
[6] K. E. Bisshop, D. C. Drucker: Large deflection of cantilever beams, *Quart. Appl. Math.*, **3** (1945), 272–275.
[7] F. V. Rohde: Large deflection of cantilever beam with uniformly distributed load, *Quart. Appl. Math.*, **11** (1953), 337–338.
[8] J. G. Freeman: Mathematical theory of deflection of beam, *Phil. Mag.*, ser. 7, **37** (1946), 855–862.
[9] S. P. Timoshenko, J. M. Gere: "Theory of Elastic Stability," 2d ed., McGraw-Hill, New York, 1961.
[10] Y. C. Fung, A. Kaplan: Buckling of low arches or curved beams of small curvature, *NACA Tech. Note* 2840 (1952).
[11] M. S. G. Cullimore: The shortening effect: a non-linear feature of pure torsion. *Eng. Structures*, suppl. to *Research (London)*, **2** (1949).
[12] D. G. Ashwell: The two axes of twist in elastic prisms, *Engineering*, **176** (1953), 769–771.
[13] D. G. Ashwell: The anticlastic curvature of rectangular beams and plates, *J. Roy. Aeronaut. Soc.*, **54** (1950), 708–715.
[14] E. Reissner: Finite twisting and bending of thin rectangular plates, *J. Appl. Mechanics*. **24** (1957), 391–396.
[15] F. Müller-Magyari: Endliche Deformtionen dünner Plattenstreifen mit freien Längsrändern, *Österr. Ing.-Arch.*, **7** (1953), 319–328.
[16] E. H. Mansfield: The inextensional theory for thin flat plates, *Quart. J. Mech. Appl. Math.*, **8** (1955), 338–352.
[17] D. G. Ashwell: The equilibrium equations of the inextensional theory for thin flat plates, *Quart. J. Mech. Appl. Math.*, **10** (1957), 169–182.
[18] Y. C. Fung, W. H. Wittrick: A boundary layer phenomenon in the large deflexion of thin plates, *Quart. J. Mech. Appl. Math.*, **8** (1955), 191–210.
[19] D. G. Ashwell: A characteristic type of instability in the large deflexions of elastic plates: I, Curved rectangular plates bent about one axis, *Proc. Roy. Soc. (London)*, **A,214** (1952), 98–118.

[20] L. G. Brazier: On the flexure of thin cylindrical shells and other "thin" sections, *Proc. Roy. Soc. (London)*, **A,116** (1927), 104–114.
[21] K. Marguerre: Zur Theorie der gekrümmten Platte grosser Formänderung, *Proc. 5th Intern. Congr. Appl. Mechanics*, pp. 93–101, Cambridge, Mass., 1938.
[22] S. P. Timoshenko, S. Woinowsky-Krieger: "Theory of Plates and Shells," 2d ed., McGraw-Hill, New York, 1959.
[23] A. Kaplan, Y. C. Fung: A nonlinear theory of bending and buckling of thin elastic shallow spherical shells, *NACA Tech. Note* 3212 (1954).
[24] F. S. Shaw, N. Perrone: A numerical solution for the nonlinear deflection of membranes, *J. Appl. Mechanics*, **21** (1954), 117–128.
[25] H. L. Cox, E. Pribram: The elements of the buckling of curved plates, *J. Roy. Aeronaut. Soc.*, **52** (1948), 551–565.
[26] H.-S. Tsien: Lower buckling load in the nonlinear buckling theory for thin shells, *Quart. Appl. Math.*, **5** (1947), 236–237.
[27] J. H. Argyris, P. C. Dunne: "Structural Principles and Data," part 2, Structural Analysis, 4th ed., Pitman, London, 1952.
[28] G. Gerard, H. Becker: "Handbook of Structural Stability," part 3, Buckling of curved plates and shells, *NACA Tech. Note* 3783 (1957).
[29] P. M. Naghdi: A survey of recent progress in the theory of elastic shells, *Appl. Mechanics Revs.*, **9** (1956), 365–368.
[30] R. J. Roark: "Formulas for Stress and Strain," 3d ed., McGraw-Hill, New York, 1954.

Part 5

PLASTICITY AND VISCOELASTICITY

CHAPTER 46

BASIC CONCEPTS

BY

D. C. DRUCKER, Ph.D., Providence, R.I.

46.1. INTRODUCTION

The purpose of this chapter is to bring together and to analyze, in general rather than in specific terms, some of the available experimental information on structural materials at ambient temperature and the mathematical stress-strain relations which have been proposed for them.

A clear distinction should be made between the mathematical theories of inelastic behavior to be presented and the physical theories on which so many distinguished investigators are working. A mathematical theory is essential for the solution of problems in stress analysis, and also for the correlation of experimental data. However, it is only a formalization of known experimental results, and does not inquire very deeply into their physical and chemical basis. Physical theories attempt to answer the question of *why* things happen, mathematical theories the question of *what* does happen with a given history of loading. The two approaches will be synthesized ultimately, but they are still far apart.

The most useful stress-strain relation is the simplest one which will work for the problem at hand. That it may have little or no generality beyond the immediate application is of no consequence. In fact, as may be seen by comparison of theory and experimental data, the more information to be taken into account, the greater the generality needed in the stress-strain relation. The more complicated the path of loading, the more elaborate the form of the relation must be. An equation of state does not exist. Solutions to problems of stress analysis, on the contrary, become exceedingly difficult to obtain if the stress-strain relation is too involved. A compromise must be made between convenience and physical reality. The basis for a reasonable choice will be described, but the solution of problems is left for other chapters.

46.2. SOME FUNDAMENTAL ASPECTS OF INELASTIC BEHAVIOR

Structural Metals in Tension and Compression

A standard simple tension test at room temperature gives the well-known load-elongation or nominal stress-strain diagrams of Fig. 46.1. The initial elastic region appears as a straight line, extending to quite high values of stress for structural metals. Hot-rolled mild steel exhibits an upper yield point U and a lower yield point L (Fig. 46.1a) under careful central loading. For most metals, however, no sharp yield point is discernible, and a yield strength must be defined arbitrarily. An offset yield σ_{ys} as shown in Fig. 46.1b is customary. Above the yield point, the response of the metal is both elastic and plastic. Unloading at any stage in the deformation reduces the strain along an almost completely elastic unloading line. Reloading retraces the unloading line with relatively minor deviation, and then produces further plastic deformation approximately when the previous maximum stress is exceeded. The extent of the rounding and the hysteresis loop are very much a property of the individual piece of metal and its prior history. Both are remarkably small for the common steels and structural aluminum alloys, far less than shown in Fig. 46.1*b*. To a rather close approximation, then, the strain is separable directly into an elastic or recoverable part, given by Hooke's law, and a plastic part or permanent set.

It is important to note that the flat portion of the mild steel curve, shown dashed in Fig. 46.1*a*, does not really represent a stress-strain curve for the material. Instead it shows an average of a series of unstable jumps from U to L in regions distributed along the gage length of the specimen. The process is inhomogeneous, with plastic regions separated by completely elastic regions.

When the strains are small, it is of no consequence whether the stress is computed from the original cross-sectional area or the actual area, or the strain from the original length or the actual length. When the strains are large, the difference is appreciable. The term *true stress* is employed for load divided by current area, and the term *true strain* for the sum of the strain increments, each computed from the current length.

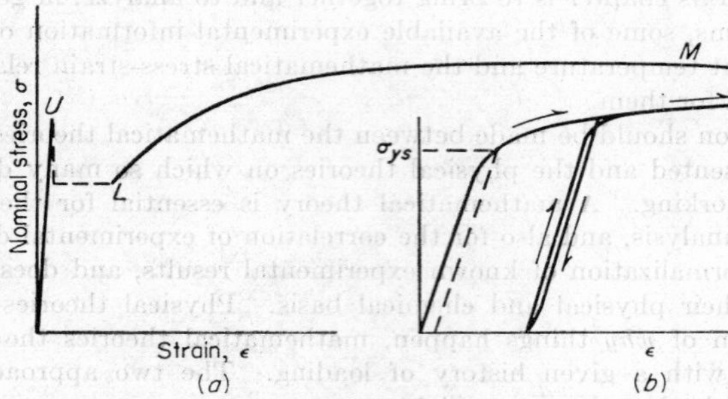

FIG. 46.1. Load-elongation or nominal stress-strain curves, elastic response exaggerated.

True strain, therefore, is the natural logarithm of 1 + the conventional or nominal strain.

The nominal curve dips down beyond the ultimate strength point M because the test bar necks (reduces in area in a small region), and is not able to carry as much load. However, the true stress does increase. The region beyond the nominal ultimate is of great interest in manufacturing processes, but not in the usual stress analysis.

Curves similar to Fig. 46.1, in the moderate strain range, are obtained on the as-received metal in simple compression or shear. However, after prestraining in tension, the stress-strain curve in compression differs from the curve which would be obtained on reloading in tension, or on loading an undisturbed specimen in compression. As illustrated in Fig. 46.2*a*, there is an unequal raising of the new yield point, or often an actual lowering when the stress is reversed. A model which exhibits this Bauschinger effect is shown in Fig. 46.2*b*. Each pair of elements in parallel which carries the load P has a different yield strength, but may itself be without Bauschinger effect, in the sense that a raised yield point in tension carries over in compression (Fig. 46.2*c*). The assemblage of bars of different yield strengths corresponds, qualitatively, to slip planes of different strengths in the actual specimen. Upon unloading, there will be a residual tension in the high-yield-strength bars 3 and 4 and compression in the low-yield-strength 1 and 2. On reversal of load, the low-yield-strength bars yield much earlier than they would on reloading in tension, because they are already in compression.

A complex pattern of internal stresses and strains will exist in a structural metal because of its heating and cooling history and mechanical treatment. Cold-rolled sheets are generally anisotropic, with different tensile stress-strain curves for different directions of the axis of the test specimen with respect to the axis of rolling. Sheets rolled in two perpendicular directions may show reasonable isotropy in the plane, but the properties at right angles are likely to differ considerably from those in the plane. A sheet which is prestrained in tension in one direction will show cross effects which are allied to the Bauschinger effect. Tensile specimens cut at various angles to the direction of prestrain will have different stress-strain curves, even for initially isotropic sheets.

In the discussion of uniaxial testing, so far there has been the tacit assumption that there is a single stress-strain curve for tension or compression, independent of the rate of straining. This is reasonably true for structural metals at room temperature, tested at the ordinary slow speeds of standard testing machines. It is easily observed with a straining machine, however, that the load does drop off when the strain is held constant (*relaxation*). If the load is kept constant, the strain does continue to increase (*creep*). At elevated temperatures, creep and relaxation are very

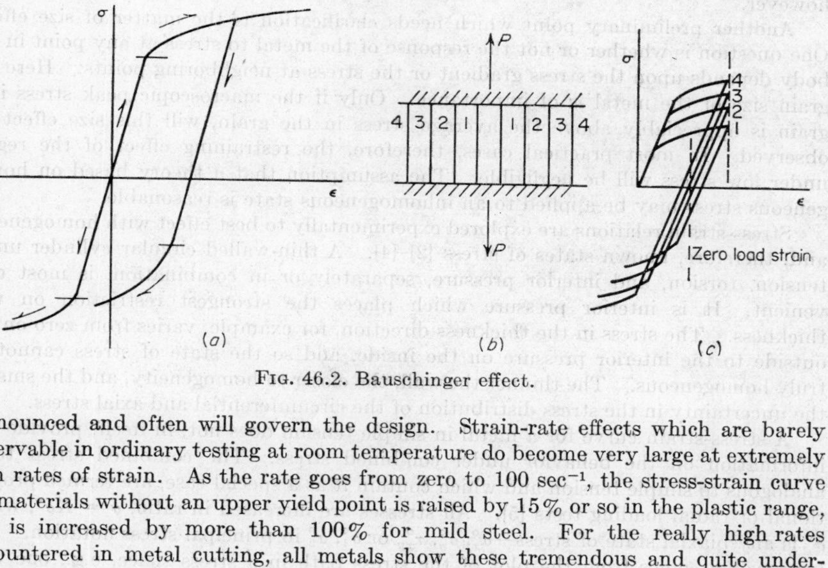

FIG. 46.2. Bauschinger effect.

pronounced and often will govern the design. Strain-rate effects which are barely observable in ordinary testing at room temperature do become very large at extremely high rates of strain. As the rate goes from zero to 100 sec^{-1}, the stress-strain curve for materials without an upper yield point is raised by 15% or so in the plastic range, but is increased by more than 100% for mild steel. For the really high rates encountered in metal cutting, all metals show these tremendous and quite understandable increases.

In a crude way it may be said that it takes time to mobilize weaknesses, or to get dislocations to move and to be generated. The less the time, the higher the stress needed. The limit is the intrinsic strength of perfect material which is never reached. Yielding, therefore, is not instantaneous, and the time delay is a function of stress [1]. Strain-rate effects and delay time for yielding are two aspects of the same physical phenomenon. The higher the stress, the smaller the delay time. The less the time available, the higher must be the stress to produce plastic deformation.

Structural Metals under Combined Stress

The complete behavior of metals under combined stress is exceedingly complex, and every possible simplification is of great value. Two related experimental facts are most helpful. One, shown by Bridgman and others, is that the influence of hydrostatic pressure on yielding and subsequent deformation is not appreciable until very high pressures are reached. Suppose that a tensile test or a torsion test is run at atmospheric pressure in the standard manner, with load and deformation recorded. Immersion of the whole setup in a chamber under high pressure will hardly alter the stress-strain curve obtained in the moderately small strain range. The major effect is to increase the ductility of the material greatly, and so permit much larger deformation prior to fracture.

The related fact is that, even for large plastic deformation, as by repeated torsion and reverse torsion of wires and bars far into the plastic range, the density alters very little. Unless internal cracks are formed, the plastic or permanent change in volume is very small.

The absence of a hydrostatic pressure effect means that shear stresses alone are significant for plastic deformation. The constancy of volume indicates that plastic deformation is a combination of shear strains. That shear stresses and shear strains, in the general sense, should be the significant variables in polycrystalline structural metal is no surprise, because it is known to be true for single crystals. The direct transition from the single crystal to polycrystalline material is never a certainty, however.

Another preliminary point which needs clarification is the matter of size effect. One question is whether or not the response of the metal to stress at any point in the body depends upon the stress gradient or the stress at neighboring points. Here the grain size of the metal is of importance. Only if the macroscopic peak stress in a grain is appreciably above the average stress in the grain, will this size effect be observed. In most practical cases, therefore, the restraining effect of the region under low stress will be negligible. The assumption that a theory based on homogeneous stress may be applied to an inhomogeneous state is reasonable.

Stress-strain relations are explored experimentally to best effect with homogeneous and, therefore, known states of stress [2]–[4]. A thin-walled circular cylinder under tension, torsion, and interior pressure, separately or in combination, is most convenient. It is interior pressure which places the strongest restriction on wall thickness. The stress in the thickness direction, for example, varies from zero on the outside to the interior pressure on the inside, and so the state of stress cannot be truly homogeneous. The thinner the wall, the closer to homogeneity, and the smaller the uncertainty in the stress distribution of the circumferential and axial stress.

A stress-strain curve for a metal in simple tension does not, in itself, provide any information on the behavior under combined stress. The combined stress tests, analogous to simple tension and which contain it as a special case, are termed proportional or radial loading tests [5]. All stresses are increased in ratio, $\sigma = K\sigma^\circ$, where σ° is any biaxial state of stress—σ_x°, σ_y°, τ_{xy}° or σ_1°, σ_2° in principal stress notation. The term *radial* comes from the plot of the stress path in a stress space, e.g., one with coordinate axes σ_x, σ_y, τ_{xy}. Use of σ_1, σ_2 as coordinates in Fig. 46.3 rather than σ_x, σ_y, τ_{xy}, which are in fixed directions with respect to the test specimen, implies an absence of orientation effects or the assumption of isotropy.

The knowledge that shearing stresses and strains are the important factors in the plastic deformation of metals suggests several possibilities for cross-checking and plotting results of radial loading tests. Of these, the simplest for isotropic materials is to consider maximum shearing stress and maximum shearing strain as the significant variables. Plots of test data in this manner are remarkably successful. Curves obtained from tension tests, torsion tests, interior pressure tests, combined pressure and tension tests, all lie within a $\pm 10\%$ band or better when the metal is reasonably isotropic. A simple tension test can thus be used to predict some elements of behavior under combined stress. The maximum shearing stress is, however, only one of infinitely many shearing-stress criteria. It does not reflect any influence of the intermediate principal stress, an influence which Lode showed to be significant [6]. Another popular choice is the octahedral shearing stress or distortional energy criterion or root mean square of the principal shearing stresses [7]. Labeling the principal stresses $\sigma_1, \sigma_2, \sigma_3$ where algebraically $\sigma_1 \geq \sigma_2 \geq \sigma_3$, the maximum shear-stress criterion, which is due to Tresca, considers $\frac{1}{2}(\sigma_1 - \sigma_3)$ as the significant variable. The second criterion, associated with Nadai, Huber, Hencky, and von Mises, considers the effective shearing stress as proportional to $[(\sigma_1 - \sigma_2)^2 + (\sigma_2 - \sigma_3)^2 + (\sigma_1 - \sigma_3)^2]^{1/2}$.

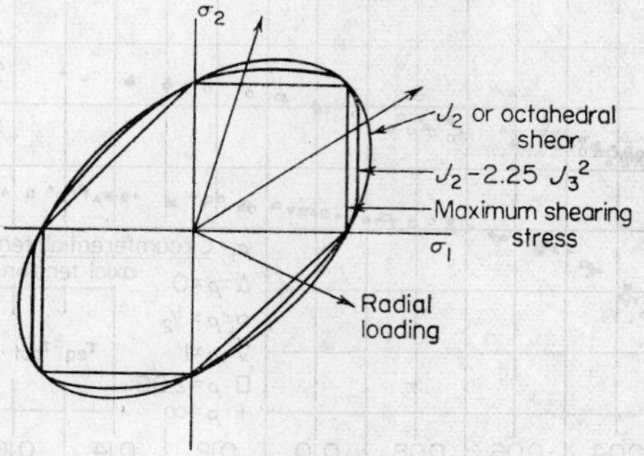

FIG. 46.3. Radial loading paths and three shearing-stress criteria or yield curves.

More generally, if shearing stress is the significant factor, it should be extracted from the state of stress σ_{ij},

$$\begin{matrix} \sigma_x & \tau_{xy} & \tau_{xz} \\ \tau_{yx} & \sigma_y & \tau_{yz} \\ \tau_{zx} & \tau_{zy} & \sigma_z \end{matrix}$$

and examined. The mean normal stress $\sigma_m = \frac{1}{3}(\sigma_x + \sigma_y + \sigma_z)$ must be subtracted from the state of stress, leaving the stress deviation $s_{ij} = \sigma_{ij} - \sigma_m \delta_{ij}$,

$$\begin{matrix} s_x & \tau_{xy} & \tau_{xz} \\ \tau_{yx} & s_y & \tau_{yz} \\ \tau_{zx} & \tau_{zy} & s_z \end{matrix}$$

where $s_x = \sigma_x - \sigma_m$, $s_y = \sigma_y - \sigma_m$, $s_z = \sigma_z - \sigma_m$, and $s_x + s_y + s_z = 0$. The five independent components of the stress deviation describe the state of shear completely, and all five are needed for general anisotropic materials. In the very special isotropic case, it is not important which direction has what label. Two of the three principal stress deviations, s_1, s_2, s_3, thus are sufficient to describe the state of shear. More useful invariants of the stress-deviation tensor are the second invariant,

$$\begin{aligned} J_2 &= \tfrac{1}{2}\Sigma s_{ij}^2 = \tfrac{1}{2}(s_x^2 + s_y^2 + s_z^2) + \tau_{xy}^2 + \tau_{yz}^2 + \tau_{zx}^2 \\ &= \tfrac{1}{6}[(\sigma_1 - \sigma_2)^2 + (\sigma_2 - \sigma_3)^2 + (\sigma_1 - \sigma_3)^2] \end{aligned}$$

and the third invariant,

$$J_3 = \tfrac{1}{3}\Sigma s_{ij}s_{jk}s_{ki} = \tfrac{1}{3}s_1s_2s_3$$

Any function of J_2 and J_3 can be a significant or an effective stress variable for an isotropic material governed by shear. The truly remarkable correlation of Osgood's tests [8] on thin-wall aluminum tubes by $(J_2{}^3 - 2.25J_3{}^2)$ is shown in Fig. 46.4 [5]. As illustrated in Fig. 46.3, this function is about halfway between the maximum shear-stress criterion and the octahedral or shear-strain energy criterion.

If mean normal stress should be important, as it is for soils under usual conditions, and perhaps also for metals at high pressure, $J_1 = \sigma_1 + \sigma_2 + \sigma_3 = 3\sigma_m$, in conjunction with J_2 and J_3, provides a general description [9],[10].

Radial or proportional loading tests, although fundamental, are far too restricted to explore most of the plastic behavior of metals. Another and highly significant basic test for which thin-walled tubes are suitable is the addition of torsion to uniaxial

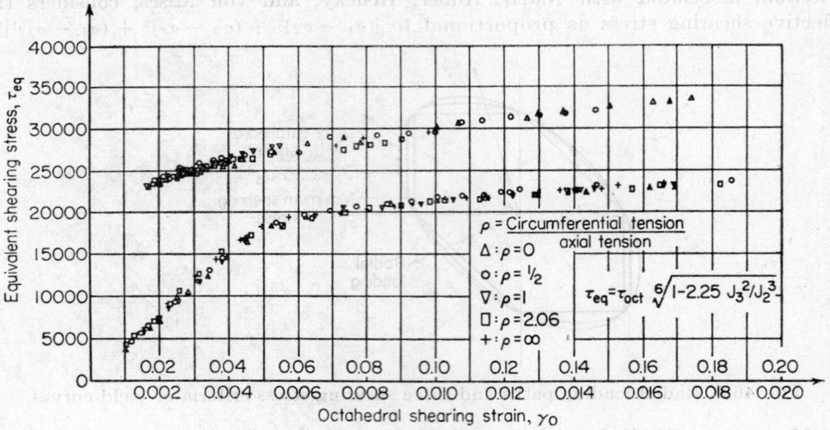

Fig. 46.4. Osgood's test on thin-walled tubes under tension and interior pressure [3],[6].

tension or compression. A very interesting result has been found [11]–[14]. The initial response to torsion is purely elastic, even when the axial stress is far above yield. The initial shear modulus is elastic whether the axial stress is increased, decreased, or kept constant.

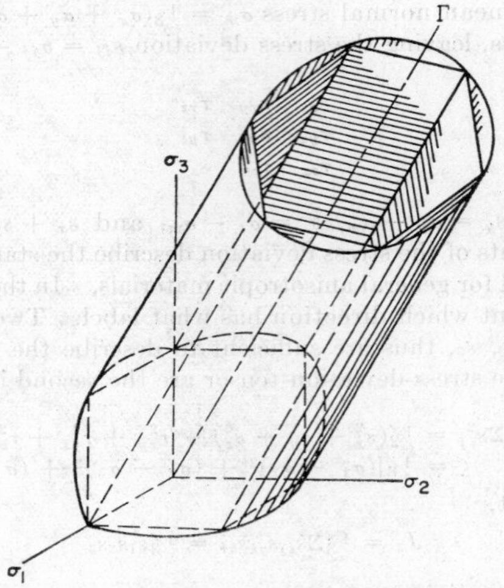

Fig. 46.5. Yield surfaces in principal stress space. (*Courtesy of Oscar Hoffman and George Sachs, "Introduction to the Theory of Plasticity," McGraw-Hill, New York,* 1953.)

Yield Criteria and Subsequent Loading Functions

With some degree of idealization, the structural metals behave purely elastically, as shown in Fig. 46.1, until the yield stress or proportional limit is reached. For the more general radial loading, the limit of purely elastic action defines a yield curve, as shown in Fig. 46.3, or surface, as shown in Fig. 46.5. More generally still, for anisotropic materials, the surface must be drawn in a stress space with more than the three coordinates $\sigma_1, \sigma_2, \sigma_3$. If the state of stress is changed so that the stress point representing it follows a path inside the yield surface, the behavior of the material is elastic, and the paths are reversible. Increasing the stress beyond the yield surface, into the work-hardening range, produces both plastic and elastic action. At each stage of plastic deformation, a new yield surface, called a *loading surface,* is established. Now, if the state of stress is changed so that the stress point representing it moves about inside the new yield surface, no plastic deformation will take place.

46.3. STRESS-STRAIN RELATIONS BASED ON J_2, J_3

The most popular and the simplest self-consistent stress-strain relations are based upon the octahedral shearing stress or second invariant of the stress deviation J_2. The initial yield surface is given by $J_2 = k_0^2$. Subsequent loading surfaces for a work-hardening material are given by $J_2 = k^2$, where k continuously increases as plastic

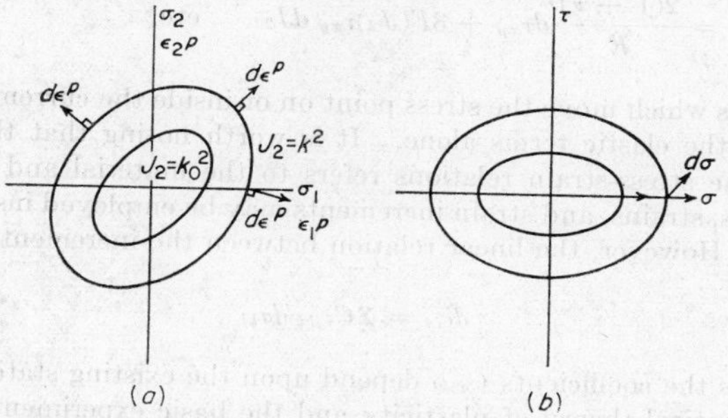

FIG. 46.6. Successive yield or loading curves $J_2 = k^2$. The plastic strain increment vector is normal to the loading surface.

deformation proceeds (Fig. 46.6). If plastic strain coordinates are superposed on the stress coordinates, and plastic deformation is taking place, the plastic strain increment vector is normal to the loading surface.

$$d\epsilon_{ij}^p = d\lambda\, \partial J_2/\partial\sigma_{ij} = d\lambda\, s_{ij} \tag{46.1}$$

in indicial notation. In the usual engineering notation,

$$d\epsilon_x^p = \tfrac{2}{3}\, d\lambda\, [\sigma_x - \tfrac{1}{2}(\sigma_y + \sigma_z)] \qquad d\gamma_{xy}^p = 2\, d\lambda\, \tau_{xy} \tag{46.2}$$

or, eliminating $d\lambda$,

$$\frac{d\epsilon_x^p}{\sigma_x - \tfrac{1}{2}(\sigma_y + \sigma_z)} = \frac{d\epsilon_y^p}{\sigma_y - \tfrac{1}{2}(\sigma_z + \sigma_x)} = \frac{d\epsilon_z^p}{\sigma_z - \tfrac{1}{2}(\sigma_x + \sigma_y)} = \frac{d\gamma_{xy}^p}{3\tau_{xy}}$$
$$= \frac{d\gamma_{yz}}{3\tau_{yz}} = \frac{d\gamma_{zx}}{3\tau_{zx}} \tag{46.3}$$

At first sight, the stress-strain relation (46.2) looks similar to an elastic one with Poisson's ratio of ½ and a plastic modulus replacing Young's modulus. The integrated form for radial loading is, in fact, exactly of this form:

$$\epsilon_x^p = \frac{1}{C}[\sigma_x - \tfrac{1}{2}(\sigma_y + \sigma_z)] \qquad \gamma_{xy}^p = \frac{3}{C}\tau_{xy} \qquad \text{etc.} \tag{46.4}$$

It holds, however, only for radial loading and not for general loading. The incremental form (46.1) is basic, and $d\lambda$ is an incremental quantity which is zero unless plastic deformation is taking place. In the Prager-Mises form of stress-strain relation for a work-hardening material, as J_2 is increased beyond any previously established maximum value, $d\lambda$ is a function of J_2 multiplied by the increment of J_2,

$$d\lambda = \tfrac{3}{2}F(J_2)\,dJ_2 = \tfrac{3}{2}F(J_2)\Sigma s_{mn}\,ds_{mn} \tag{46.5}$$

where the $\frac{3}{2}$ is inserted for convenience in subsequent equations.

The complete stress-strain relation requires the addition of the elastic strain increment. Assuming elastic as well as plastic isotropy, the Prager-Prandtl-Reuss type of equations are obtained:

$$\begin{aligned} d\epsilon_{ij} &= d\epsilon^{e}_{ij} + d\epsilon^{p}_{ij} \\ \text{or} \quad d\epsilon_x &= \frac{1}{E}[d\sigma_x - \nu(d\sigma_y + d\sigma_z)] + F(J_2)[\sigma_x - \tfrac{1}{2}(\sigma_y + \sigma_z)]\,dJ_2 \\ d\gamma_{xy} &= \frac{2(1+\nu)}{E}\,d\tau_{xy} + 3F(J_2)\tau_{xy}\,dJ_2 \qquad \text{etc.} \end{aligned} \tag{46.6}$$

Changes in stress which move the stress point on or inside the current yield or loading surface call for the elastic terms alone. It is worth noting that the term *isotropic* as applied to the stress-strain relations refers to the material and to the fact that principal stresses, strains, and strain increments may be employed instead of, x, y, and z components. However, the linear relation between the increments of stress and of strain

$$d\epsilon_{ij} = \Sigma C_{ijkl}\,d\sigma_{kl} \tag{46.7}$$

is anisotropic, as the coefficients C_{ijkl} depend upon the existing state of stress.

The mathematical theory of plasticity and the basic experimental investigations of the plastic properties of materials have often proceeded independently. It is necessary to ask, therefore, how well (46.6) describes real materials. As already stated, it is permissible to consider the response of structural metals at ambient temperatures as a sum of an elastic and a plastic response. The question then is how well the plastic behavior is described. The computed plastic or permanent change of volume ($d\epsilon^p_x + d\epsilon^p_y + d\epsilon^p_z$) is zero, in agreement with the observation of almost negligible density changes. Of great importance is the fact that the main requirement of consistency is met because of the incremental nature of the stress-strain relation. A loading path on the yield or loading surface produces no additional plastic deformation; one infinitesimally beyond produces infinitesimal plastic strains. As shown in Fig. 46.6*b* for combined tension and shear, as in an isotropic thin-wall tube pulled into the plastic range and then twisted, the normal to the yield and subsequent loading curves is in the axial direction when the shear stress is zero. The initial plastic strain increment vector $d\epsilon^p_{ij}$, therefore, has no component $d\gamma^p$. Shear response is purely elastic, and the initial shear modulus is the elastic shear modulus G, in agreement with experiment.

Another strong point in favor of J_2 as a yield and loading criterion is that the yield stress in shear is $\frac{1}{3}\sqrt{3} = 0.577$ times the yield stress in tension. This value is quite satisfactory for many metals. On the debit side, there is the experimental evidence of a number of investigators that Lode's variables,

$$\mu = \frac{2\sigma_2 - \sigma_1 - \sigma_3}{\sigma_1 - \sigma_3} = \frac{3s_2}{s_1 - s_3} \quad \text{and} \quad \nu^p = \frac{2d\epsilon^p_2 - d\epsilon^p_1 - d\epsilon^p_3}{d\epsilon^p_1 - d\epsilon^p_3} = \frac{3d\epsilon^p_2}{d\epsilon^p_1 - d\epsilon^p_3}$$

are not equal as predicted by J_2. Use of a loading function $f(J_2,J_3)$, based on the third as well as the second invariant of the stress deviation, can fit such experimental data in a consistent manner. The stress-strain law, as discussed by Handelman, Lin, Hodge, and Prager [15], then is

$$d\epsilon_{ij}^p = d\lambda \frac{\partial f}{\partial \sigma_{ij}} = G(J_2, J_3)\left(\frac{\partial f}{\partial J_2} s_{ij} + \frac{\partial f}{\partial J_3} t_{ij}\right) df \tag{46.8}$$

for $df \geq 0$. Here

$$t_{ij} = \sum_k s_{ik}s_{kj} - \tfrac{2}{3}J_2\delta_{ij}$$

is the deviator of the square of the stress deviator s_{ij}. For example,

$$t_{11} \equiv t_x = \tfrac{1}{9}[(\sigma_x - \sigma_y)^2 + (\sigma_x - \sigma_z)^2 - 2(\sigma_y - \sigma_z)^2 + 3\tau_{xy}^2 + 3\tau_{xz}^2 - 6\tau_{yz}^2]$$

and $$t_{12} \equiv t_{xy} = \tfrac{1}{3}[3\tau_{xz}\tau_{yz} - \tau_{xy}(2\sigma_z - \sigma_x - \sigma_y)]$$

The simplest form of f which satisfies the reported experimental condition of constant ratios of the components of the plastic strain increments for radial loading is $f = J_2^3 - cJ_3^2$.

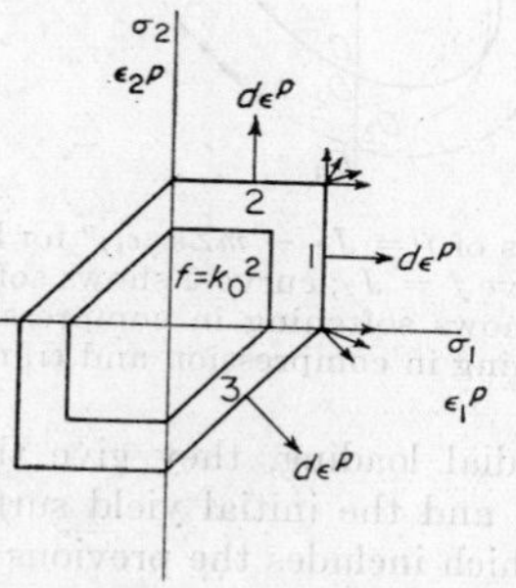

FIG. 46.7. Plastic strain increments associated with maximum shear criterion of Tresca.

A far more elaborate form is required to describe the maximum shear-stress criterion of loading. It is easier instead to think in pictorial terms, and to use the normality condition directly from the picture. At a stress point such as point 1 of Fig. 46.7, the plastic strain increment has no component in the direction of the 2 axis $d\epsilon_3^p = -d\epsilon_1^p$ and $d\epsilon_2^p = 0$. At point 3, $d\epsilon_2^p = -d\epsilon_1^p$ and $d\epsilon_3^p = 0$.

If the loading path remains within part of a quadrant of the two-dimensional plot or an octant of the three-dimensional stress space, the isotropic stress forms $f(J_2,J_3)$ may be good enough for practical use. Corresponding to an effective stress defined in terms of f, there will be an effective strain. For most purposes, the octahedral shearing-strain form $\int \sqrt{\Sigma d\epsilon_{ij} d\epsilon_{ij}}$ will be satisfactory [16].

However, the concept of any nested set of loading surfaces, each corresponding to a numerically higher value of the shear-stress function, clearly cannot agree with many of the known experimental data. Such isotropic stress hardening shows neither Bauschinger nor cross effects. As described previously, raising the yield point in tension does not raise it equally in compression. One direction in the material is no longer the same as any other after plastic deformation has occurred.

46.4. MORE GENERAL LOADING FUNCTIONS

A more realistic set of possible pictures of two successive loading surfaces for a biaxial state of stress [17] is given in Fig. 46.8. The axes are labeled σ_x and σ_y, not

σ_1 and σ_2, because directions in the material are important. It is the term

$$\Sigma s_{ij}\epsilon_{ij}^p = \Sigma\sigma_{ij}\epsilon_{ij}^p \qquad \text{in} \qquad f = J_2 - m\Sigma s_{ij}\epsilon_{ij}^p$$

which introduces the Bauschinger effect and takes into account the actual direction in which the straining occurs. A slightly more complicated loading function

$$f = J_2^{7/2} - 5.5 \times 10^{29}\Sigma\sigma_{ij}\epsilon_{ij}^p$$

proved very successful in correlating some test data of Sachs and Liu on Bauschinger effect in a structural aluminum alloy. Such forms are not general enough for many

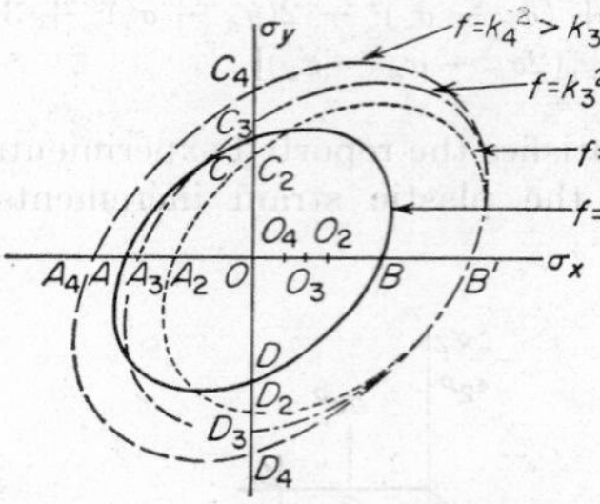

FIG. 46.8. Possible yield curves of $f = J_2 - m\Sigma s_{ij}\epsilon_{ij}{}^p$ for loading in axial tension (σ_x) to B'. Curve 1 is the initial yield curve $f = J_2$; curve 2 shows softening in compression and transverse (σ_y) tension; curve 3 shows softening in compression and hardening in transverse tension; curve 4 shows hardening in compression and transverse tension.

purposes because, under radial loading, they give the same result as J_2. Lode's variables are equal ($\mu = \nu$), and the initial yield surface is the Mises ellipse. The still more elaborate form, which includes the previous ones as special cases,

$$f = F(J_2,J_3) - m\Sigma s_{ij}\epsilon_{ij}^p$$

gives any desired μ versus ν curve and a simple Bauschinger effect. Mathematical expressions can be written to correlate any finite amount of test data of much greater complexity, but there is not yet sufficient experimental evidence to warrant further development here along such lines.

An alternative approach offers more promise because of the possibility of achieving a closer tie to the underlying physical behavior of metals. Koiter, Sanders, and Handelman [18],[19] have suggested considering the yield surface and subsequent loading surfaces as composed of combinations of plane loading surfaces. Each plane may move independently of its neighbors, or planes may influence each other. As shown schematically in Fig. 46.9a, if the Mises ellipse is considered to be the envelope of infinitely many planes, each of which acts independently of the other, simple tension in the x direction produces a new yield surface with a corner. If a few planes only are chosen, as suggested by the Tresca hexagon, the picture is simpler (Fig. 46.9b). By appropriate assumption about the interaction of the plane loading surfaces, Bauschinger and cross effects of various types can be introduced. In a very rough way, the individual loading planes can portray the physical behavior of the slip planes. The experimental investigation of corners in loading surfaces indicates that the loading path tends to produce a fairly sharp but still rounded nose at the stress point [11]–[14],[20],[21]. Both the theories which give rise to corners and those which do not may, therefore, prove useful in practice.

Prager has employed translating yield surfaces (Fig. 46.10), which preserve their shape and orientation in stress space [22],[23]. The kinematics of the motion of the yield surface is determined by thinking of the stress point as a frictionless roller, pressing against the inside of the yield surface, which in turn is guided to move freely in any coordinate direction but which is completely restrained against rotation.

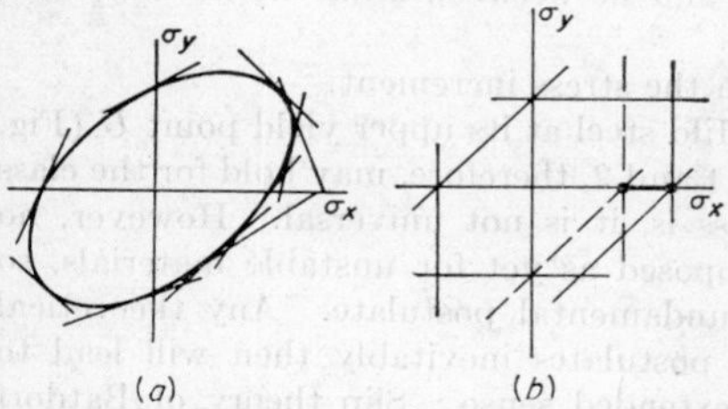

FIG. 46.9. The yield surface considered as formed by independent loading planes—a corner is produced.

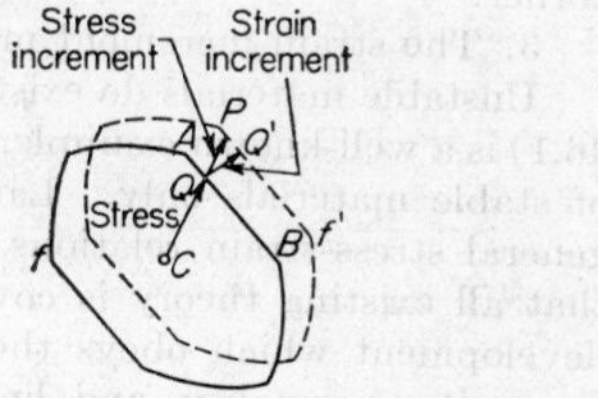

FIG. 46.10. Translating yield surfaces [14].

Successive loading surfaces have the merit of simplicity, and do show Bauschinger and cross effects.

46.5. A FUNDAMENTAL APPROACH TO PLASTICITY AND GENERAL INELASTICITY

Clearly there are limitless specialized yield functions, loading functions, and stress-strain relations which may be devised. The question arises whether there is a unified approach to these basic features of plasticity. One such approach which immediately suggests itself is the application of the principles of thermodynamics. Little is learned from classical thermodynamics except the obvious fact that work must be done on the material to produce plastic deformation. What is needed is a postulate or working hypothesis which goes one step beyond the classical treatment. This is provided by the concept of a stable plastic material in a field of constant temperature at each point [24]. Here it should be recalled that a plastic material, by definition, is independent of strain rate. A stable plastic material is defined in the following manner.

Energy cannot be extracted from a body of the material and the system of forces acting upon it. An alternative and completely equivalent definition may be phrased in terms of an external agency which adds load to the already loaded body.

Positive work is done *by the external* agency during the application of the added loads. The net work performed *by the external agency* over a cycle of application and removal of the added loads is positive if plastic deformation has occurred in the cycle.

As mentioned previously, all materials are actually time-dependent. A postulate of a general stable inelastic material [25] which includes the stable plastic material as a special case is

The work done by the external agency on the *change* in displacements it produces is positive.

It is necessary to speak of the change in displacements produced because, for visco-elastic, viscoplastic, and general inelastic bodies, the existing loads will produce displacements without the addition of an external agency. In a plastic work-hardening material under constant load, by definition there will be no continuing deformation. All the displacement which then occurs is due to the external agency, and the last postulate reduces to the previous one.

It might be thought at first that the above definitions are necessarily valid for all materials, and that their implications, therefore, are manifestations of natural laws. These implications are very far-reaching indeed. For a work-hardening material:

1. The yield surface and all subsequent loading surfaces must be convex.
2. The plastic strain increment vector must be normal to the yield or loading surface at a smooth point $d\epsilon_{ij}^p = d\lambda\, \partial f/\partial\sigma_{ij}$ and lie between adjacent normals at a corner.
3. The strain increment must be linear in the stress increment.

Unstable materials do exist, however. Mild steel at its upper yield point U (Fig. 46.1) is a well-known example. Conclusions 1 and 2, therefore, may hold for the class of stable materials only. Large as this class is, it is not universal. However, no general stress-strain relations have been proposed as yet for unstable materials, so that all existing theory is covered by the fundamental postulate. Any theoretical development which obeys the fundamental postulates inevitably then will lead to convexity, normality, and linearity in the extended sense. Slip theory of Batdorf and Budiansky [26] provides a very good example.

46.6. IDEALIZATIONS AND SIMPLIFICATIONS

Perfect or Ideal Plasticity

The limiting case of no work hardening is termed perfect or ideal plasticity (Fig. 46.11). Unlimited ductility is implied, and it is the true stress which is supposed to remain constant. Successive yield surfaces, as in Figs. 46.6 and 46.7, have approached the initial yield surface and coincide with it. Plastic deformation now occurs when

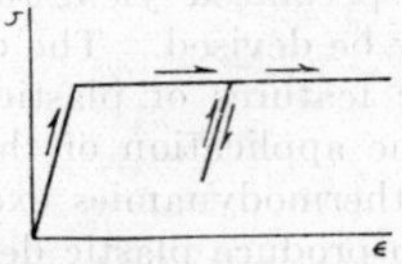

FIG. 46.11. Perfect or ideal plasticity.

the stress point is on the yield surface and the stress point cannot go outside. As for a work-hardening material, the postulate of stability requires convexity of the yield surface and normality of the plastic strain increment vector (Figs. 46.6 and 46.7). Incremental linearity cannot apply, because there is no limit to the plastic deformation for yield at constant stress. The incremental stress-strain relation is of the form

$$d\epsilon_{ij}^p = d\lambda\, \partial f/\partial\sigma_{ij} \tag{46.9}$$

for a smooth yield surface, $f(\sigma_{ij}) = \text{const}$. For $f = J_2$, Eqs. (46.1) to (46.3) then are valid.

Plastic-Rigid

Another idealization is to consider the elastic response as negligible in comparison with the plastic. This is a reasonable approximation in the large strain problems of forging, rolling, extruding, and drawing. A rigid–work-hardening or a rigid–perfectly plastic material may be visualized.

Total or Deformation Stress-strain Relations

When the path of loading is radial, the incremental stress-strain relations such as (46.6) or (46.8) may be integrated to give a relation between stress and total strain such as (46.4). These total or deformation type of expressions which are equivalent to those of nonlinear elasticity then may be taken as valid stress-strain relations, and problems may be solved with them. The answers obtained will be a satisfactory approximation if the path of loading everywhere in the body where the stress is high is nearly radial.

46.7. REFERENCES

[1] D. S. Clark: The plastic behavior of metals under dynamic loading, *Trans. ASM*, **46** (1954), 34–62.

[2] W. Prager, P. G. Hodge, Jr.: "Theory of Perfectly Plastic Solids," Wiley, New York, 1951.

[3] R. Hill: "The Mathematical Theory of Plasticity," Oxford Univ. Press, New York, 1950.

[4] D. C. Drucker: Stress-strain relations in the plastic range of metals: Experiments and basic concepts, chap. 4, pp. 97–119, in "Rheology Theory and Applications," vol. 1 of "Source Book of Rheology," Academic Press, New York, 1956.

[5] D. C. Drucker: The relation of experiments to mathematical theories of plasticity, *J. Appl. Mechanics*, **16** (1949), 349–357.

[6] W. Lode: Versuche über den Einfluss der mittleren Hauptspannung auf das Fliessen der Metalle Eisen, Kupfer und Nickel, *Z. Physik.*, **36** (1926), 913–939.

[7] A. Nadai: "Theory of the Flow and Fracture of Solids," vol. 1, 2d ed., McGraw-Hill, New York, 1950.

[8] W. R. Osgood: Combined stress tests on 24S-T aluminum alloy tubes, *J. Appl. Mechanics*, **14** (1947), 147–153.

[9] D. C. Drucker, W. Prager: Soil mechanics and plastic analysis or limit design, *Quart. Appl. Math.*, **10** (1952), 157–165.

[10] D. C. Drucker, R. E. Gibson, D. J. Henkel: Soil mechanics and work-hardening theories of plasticity, *Proc. ASCE*, **81** (1955), separate 798; *Trans. ASCE*, **122** (1957), 338–346.

[11] B. Budiansky, N. F. Dow, R. W. Peters, R. P. Shepherd: Experimental studies of polyaxial stress-strain laws of plasticity, *Proc. 1st U.S. Natl. Congr. Appl. Mechanics*, pp. 503–512, Chicago, 1951.

[12] J. Marin, H. B. Wiseman: Plastic stress-strain relations for aluminum alloy 14S-T6 subjected to combined tension and torsion, *J. Metals, Trans. AIME*, 1953, pp. 1181–1190.

[13] J. L. Morrison, W. M. Shepherd: An experimental investigation of plastic stress-strain relations, *Proc. Inst. Mech. Engrs. London*, **163** (1950), 1–9.

[14] P. M. Naghdi, F. Essenburg, W. Koff: An experimental study of initial and subsequent yield surfaces in plasticity, *J. Appl. Mechanics*, **25** (1958), 201–209.

[15] G. H. Handelman, C. C. Lin, W. Prager: On the mechanical behavior of metals in the strain-hardening range, *Quart. Appl. Math.*, **4** (1947), 397–407.

[16] G. N. White, D. C. Drucker: Effective stress and effective strain in relation to stress theories of plasticity, *J. Appl. Phys.*, **21** (1950), 1013–1021.

[17] F. Edelman, D. C. Drucker: Some extensions of elementary plasticity theory, *J. Franklin Inst.*, **251** (1951), 581–605.

[18] W. Koiter: Stress-strain relations, uniqueness and variational theorems for elastic-plastic materials with a singular yield surface, *Quart. Appl. Math.*, **11** (1953), 350–353.

[19] J. L. Sanders, Jr.: Plastic stress-strain relations based on infinitely many plane loading surfaces, *Proc. 2d U.S. Natl. Congr. Appl. Mechanics*, pp. 455–460, Ann Arbor, Mich., 1954.

[20] F. D. Stockton, D. C. Drucker: Instrumentation and fundamental experiments in plasticity, *Proc. Soc. Exptl. Stress Anal.* **10, 2** (1951), 127–142.

[21] A. Phillips, L. Kaechele: Combined stress tests in plasticity, *J. Appl. Mechanics*, **23** (1956), 43–48.

[22] W. Prager: The theory of plasticity: A survey of recent achievements, *Proc. Inst. Mech. Engrs. London*, **169** (1955), 41–57.

[23] W. Prager: "Probleme der Plastizitätstheorie," Birkhäuser, Basel, Switzerland, 1955.

[24] D. C. Drucker: A more fundamental approach to stress-strain relations, *Proc. 1st U.S. Natl. Congr. Appl. Mechanics*, pp. 487–491, Chicago, 1951.

[25] D. C. Drucker: A definition of stable inelastic material, *J. Appl. Mechanics*, **26** (1959), 101–106.

[26] S. B. Batdorf, B. Budiansky: A mathematical theory of plasticity based on the concept of slip, *NACA Tech. Note* 1871 (1949).

[27] E. H. Lee, P. S. Symonds (eds.): "Plasticity," Pergamon, London, 1960.

[28] D. C. Drucker: Plasticity, in "Structural Mechanics," Pergamon, London, 1960, pp. 407–488.

[29] W. Prager: "Introduction to Plasticity," Addison-Wesley, Reading, Mass., 1959.

CHAPTER 47

BENDING OF BEAMS

BY

A. PHILLIPS, Dr.-Ing. New Haven, Conn.

47.1. BASIC ASSUMPTIONS

In the theory of plastic bending of beams, the following four assumptions are made:

1. Cross sections that were originally plane and perpendicular to the longitudinal fibers remain plane and normal to the deformed fibers after bending has taken place.
2. The relation between stress and strain for a longitudinal fiber of a beam is the same as between simple tension and compression.
3. The distances between the longitudinal fibers of the beam do not change during bending.
4. The shear stresses are determined from equilibrium considerations, after the normal stresses have been found by means of the three first assumptions.

These four assumptions are the same as the ones used in the elastic theory of beams. For pure bending, with and without axial force, the validity of assumption 1 follows from symmetry considerations. Assumption 2 has been shown to be valid by experiments [1],[6]. Assumption 3 means that the lateral contraction is neglected, which is justified for small strains.

For bending with shear all four assumptions are approximations. Experiments [7], however, indicate that the theory gives acceptable results when the shearing force is not large.

47.2. THE GENERAL EQUATIONS OF BENDING

From assumption 1, it follows that the strain ϵ of fibers parallel to the beam axis is a linear function of the coordinates y and z in the cross section [see Eq. (35.13)]:

$$\epsilon = \alpha + \beta y + \gamma z \tag{47.1}$$

In the plane of each cross section, there exists a straight line $\epsilon = 0$, the neutral axis. Figure 47.1 shows how the intercepts of this line on the coordinate axes are related to α, β and γ.

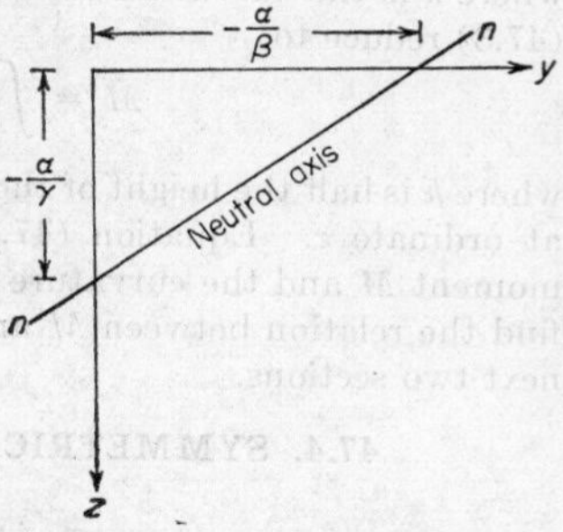

Fig. 47.1

Let $\sigma = f(\epsilon)$ be the stress-strain relation for tension and compression for the material of the beam. It follows from assumption 2 that the stress distribution is given by

$$\sigma = f(\alpha + \beta y + \gamma z) \tag{47.2}$$

If N, M_y, and M_z are the axial force and the bending moments with respect to the y and z axes, respectively, then the equations of equilibrium give:

$$N = \int_A f(\alpha + \beta y + \gamma z)\, dA$$
$$M_y = \int_A f(\alpha + \beta y + \gamma z) z\, dA \qquad M_z = -\int_A f(\alpha + \beta y + \gamma z) y\, dA \tag{47.3}$$

From these three equations, we can find the values of N, M_y, and M_z which correspond to the stress distribution (47.2), when we know the values α, β, and γ. The inverse problem of finding the stress distribution from the values of N, M_y, and M_z is much more difficult.

47.3. THE POSITION OF THE NEUTRAL AXIS

In general, the neutral axis will move when the external forces change. There are cases, however, in which the neutral axis has a constant position, despite the change in the external forces.

Suppose, for example, that the stress-strain relation is given by

$$\sigma = C\epsilon^n \tag{47.4}$$

Then Eqs. (47.3) become

$$N = C\alpha^n \int_A \left(1 + \frac{\beta}{\alpha} y + \frac{\gamma}{\alpha} z\right)^n dA \tag{47.5}$$

and two more equations of the same form for M_y and M_z. But the position of the neutral axis is known when the two ratios β/α, γ/α are given. These ratios remain constant when $N:M_y:M_z$ remains constant. Therefore, when N, M_y, and M_z change proportionally to one another, the neutral axis does not move. For symmetrical bending without axial force, we have $N = M_z = 0$; hence, $N:M_y:M_z = \text{const}$, and the position of the neutral axis remains fixed.

Another case in which the neutral axis does not move is symmetrical bending without axial force, when the cross section has two axes of symmetry and the stress-strain relation for compression is identical with that for tension. In this case, the neutral axis coincides with the axis of symmetry perpendicular to the plane of the external forces. When the y axis coincides with the neutral axis, Eq. (47.1) simplifies to

$$\epsilon = \kappa z \tag{47.6}$$

where κ is the curvature of the axis of the beam [see Eq. (35.8)]. In this case, Eqs. (47.3) reduce to

$$M = \int_A f(\kappa z) z \, dA = 2 \int_0^h f(\kappa z) B(z) z \, dz \tag{47.7}$$

where h is half the height of the cross section, and $B(z)$ is the width of the cross section at ordinate z. Equation (47.7) can be used to find the relationship between the moment M and the curvature κ. Because of the relationship $\epsilon_{\max} = \kappa h$, we can then find the relation between M and $\epsilon_{\max}$. The use of Eq. (47.7) will be discussed in the next two sections.

47.4. SYMMETRICAL BENDING WITHOUT AXIAL FORCE

Power Law

It is assumed that the cross section has two axes of symmetry, and that the stress-strain law for tension and for compression is given by Eq. (47.4), where σ and ϵ stand for the absolute values of stress and strain. The plane of bending is assumed to include the z axis, and, because no axial force is applied, the y axis is the neutral axis. Equations (47.4) and (47.7) give

$$M = 2C\kappa^n \int_0^h z^{n+1} B(z) \, dz = 2C\kappa^n I_{n+1} \tag{47.8}$$

where

$$I_{n+1} = \int_0^h z^{n+1} B(z) \, dz \tag{47.9}$$

Expressions for I_{n+1} are given in Table 47.1. For $n = 1$ (linear elasticity), $2I_2$ is the moment of inertia of the cross section.

Table 47.1[1]

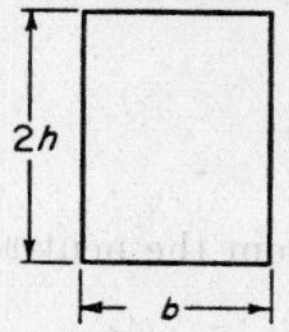	$I_{n+1} = \dfrac{bh^{n+2}}{n+2}$
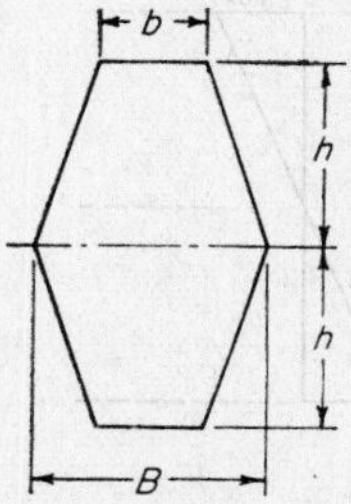	$I_{n+1} = \dfrac{h^{n+2}}{(n+2)(n+3)}(B + 2b + nb)$
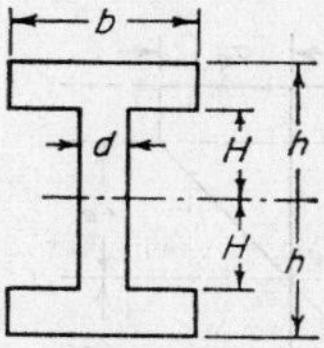	$I_{n+1} = \dfrac{dH^{n+2} + b(h^{n+2} - H^{n+2})}{n+2}$
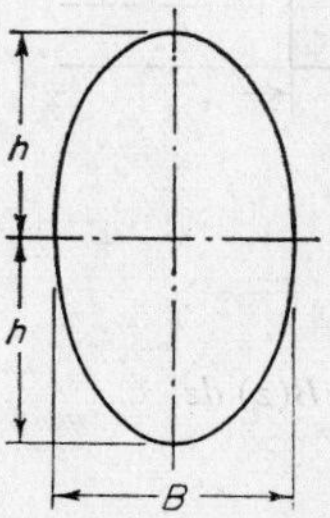	$I_{n+1} = 0.4431 \dfrac{\Gamma(1 + n/2)}{\Gamma[1 + (n+3)/2]} Bh^{n+2}$ where Γ = gamma function

[1] From Aris Phillips, "Introduction to Plasticity," Ronald, New York, 1956.

Perfectly Plastic Material

We consider a cross section with two axes of symmetry and bending in the x,z plane. The stress-strain relation is the same for tension and compression, and is given by Fig. 47.2 (perfect plasticity). As no axial force is applied, the y axis is the neutral axis. The stress and strain distributions are shown in Fig. 47.3. For $\epsilon_{max} \leq \epsilon_e$, we have

$$M = EI\kappa \qquad (47.10)$$

which, for $\epsilon_{max} = \epsilon_e$, becomes

$$M_e = EI\kappa_e \qquad (47.11)$$

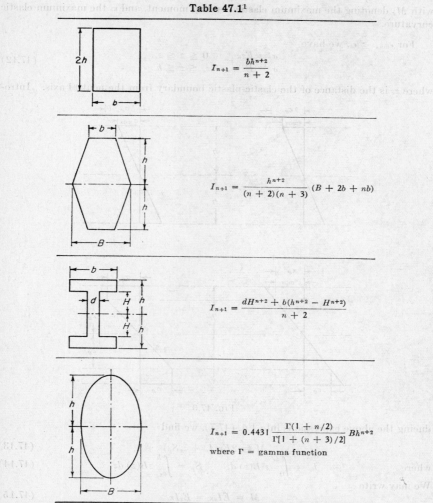

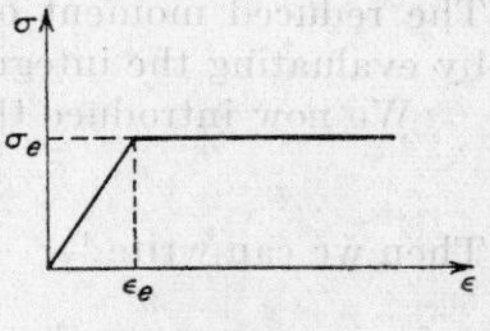

FIG. 47.2

with M_e denoting the maximum elastic bending moment, and κ_e the maximum elastic curvature.

For $\epsilon_{\max} \geqq \epsilon_e$, we have

$$\begin{aligned} \sigma &= Ez\kappa \qquad 0 \leqq z \leqq z_e \\ \sigma &= Ez_e\kappa \qquad z_e \leqq z \leqq h \end{aligned} \tag{47.12}$$

where z_e is the distance of the elastic-plastic boundary from the neutral axis. Intro-

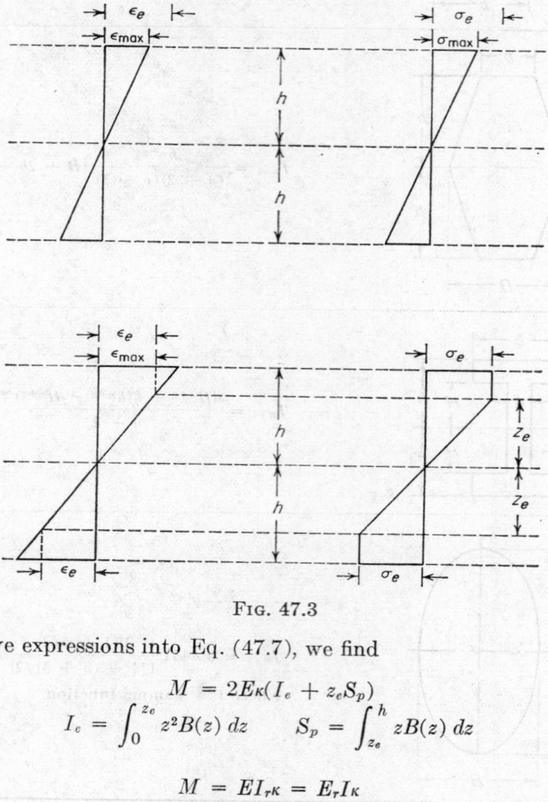

FIG. 47.3

ducing the above expressions into Eq. (47.7), we find

$$M = 2E\kappa(I_e + z_eS_p) \tag{47.13}$$

where

$$I_e = \int_0^{z_e} z^2B(z)\,dz \qquad S_p = \int_{z_e}^{h} zB(z)\,dz \tag{47.14}$$

We may write

$$M = EI_r\kappa = E_rI\kappa \tag{47.15}$$

where

$$I_r = 2I_e + 2z_eS_p \tag{47.16}$$

is the *reduced moment of inertia*, and

$$E_r = E\frac{I_r}{I} \tag{47.17}$$

is the *reduced modulus*. Expressions (47.15) are of the same form as Eq. (47.10). The reduced moment of inertia and the reduced modulus can be calculated easily by evaluating the integrals (47.14).

We now introduce the nondimensional factor t defined by

$$t = I_r/I = E_r/E \tag{47.18}$$

Then we can write

$$M = EI\kappa t \tag{47.19}$$

In the elastic range, we have $t = 1$.

Table 47.2[1]

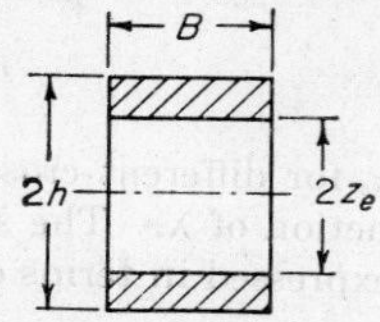

$$t = \frac{3}{2\lambda}\left(1 - \frac{1}{3\lambda^2}\right)$$

$$M_e = 2Bh^2\sigma_e/3$$

$$M_p = Bh^2\sigma_e$$

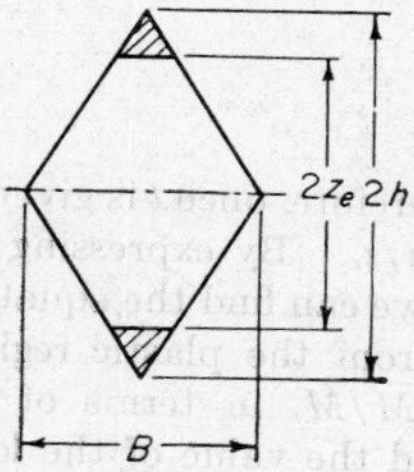

$$t = \frac{2\lambda^3 - 2\lambda + 1}{\lambda^4}$$

$$M_e = Bh^2\sigma_e/6$$

$$M_p = Bh^2\sigma_e/3$$

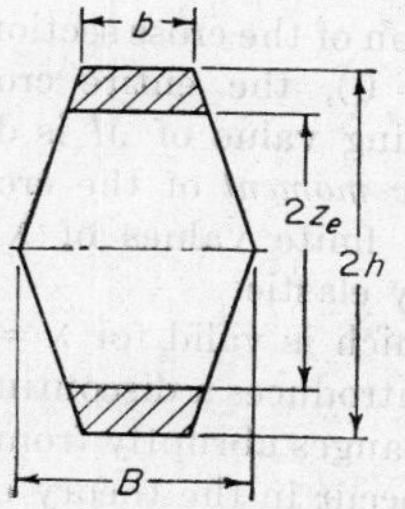

$$t = \frac{2B\lambda(3\lambda^2 - 1) - (B - b)(4\lambda^3 - 1)}{\lambda^4(B + 3b)}$$

$$M_e = (B + 3b)h^2\sigma_e/6$$

$$M_p = (B + 2b)h^2\sigma_e/3$$

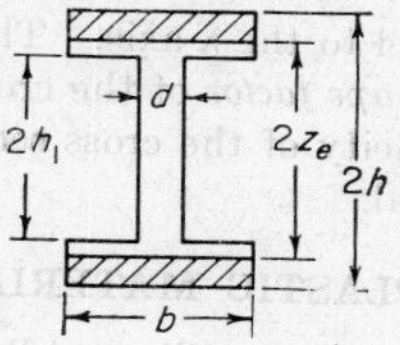

$z_e \geq h_1$

$$t = \frac{(3\lambda^2 - 1) - 2\lambda^3 \dfrac{h_1^3}{h^3}\left(1 - \dfrac{d}{b}\right)}{2\lambda^3\left[1 - \dfrac{h_1^3}{h^3}\left(1 - \dfrac{d}{b}\right)\right]}$$

$$M_e = \tfrac{2}{3}[bh^3 - (b - d)h_1^3]\frac{\sigma_e}{h}$$

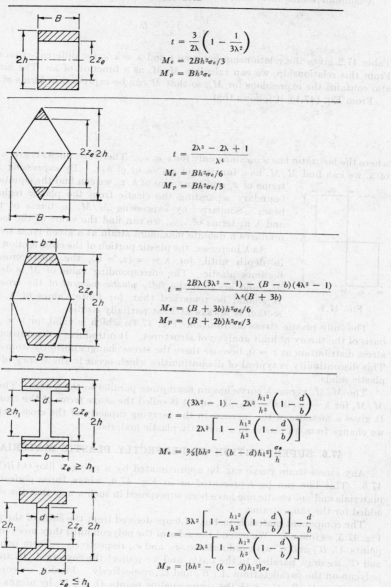

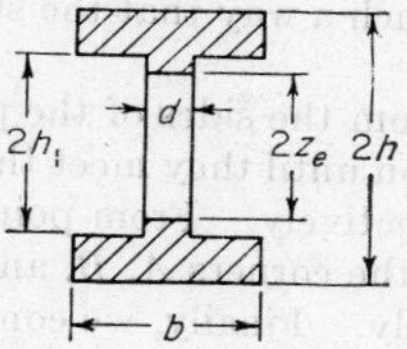

$z_e \leq h_1$

$$t = \frac{3\lambda^2\left[1 - \dfrac{h_1^3}{h^3}\left(1 - \dfrac{d}{b}\right)\right] - \dfrac{d}{b}}{2\lambda^3\left[1 - \dfrac{h_1^3}{h^3}\left(1 - \dfrac{d}{b}\right)\right]}$$

$$M_p = [bh^2 - (b - d)h_1^2]\sigma_e$$

[1] From Aris Phillips, "Introduction to Plasticity," Ronald, New York, 1956.

A nondimensional form of Eq. (47.19) is found from Eqs. (47.11) and (47.19):

$$\frac{M}{M_e} = t\frac{\kappa}{\kappa_e} = t\lambda \tag{47.20}$$

Table 47.2 gives the relationship between t and $\lambda = \kappa/\kappa_e$ for different cross sections. From this relationship, we can calculate M/M_e as a function of λ. The same table also contains the expressions for M_e, so that M can be expressed in terms of κ.

From Eq. (47.1), it follows that

$$\frac{\kappa}{\kappa_e} = \frac{\epsilon_{\max}}{\epsilon_e} = \frac{h}{z_e} \tag{47.21}$$

where the last ratio has a meaning only for $\kappa \geqq \kappa_e$. Therefore, once t is given in terms of λ, we can find M/M_e as a function of $\epsilon_{\max}/\epsilon_e$ or of h/z_e. By expressing M/M_e in terms of x, and λ in terms of h/z_e, we can find the equation of the boundary separating the elastic from the plastic region of the beam. Similarly, by expressing M/M_e in terms of the load, and λ in terms of $\epsilon_{\max}$, we can find the value of the load which produces a definite maximum strain at a given cross section.

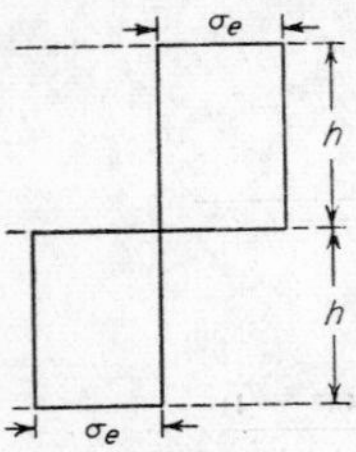

FIG. 47.4

As λ increases, the plastic portion of the cross section increases in depth until, for $\lambda = \infty$ ($z_e = 0$), the entire cross section becomes plastic. The corresponding value of M is denoted by M_p, and is called the *fully plastic moment* of the cross section. It should be remarked that, for finite values of λ, the cross section is always at least partially elastic.

The fully plastic stress distribution (Fig. 47.4), which is valid for $\lambda = \infty$, is the basis of the theory of limit analysis of structures. It introduces a discontinuity in the stress distribution at $z = 0$, because there the stress changes abruptly from σ_e to $-\sigma_e$. This discontinuity is typical of discontinuities which occur in the theory of perfectly plastic solids.

The M/M_e versus λ curve has an asymptote parallel to the λ axis. The value of M/M_e for $\lambda = \infty$ is $s = M_p/M_e$, and it is called the *shape factor* of the cross section. It gives a measure of the increase in the carrying capacity of the cross section when we change from an elastic to a perfectly plastic material.

47.5. SUPERPOSITION OF PERFECTLY PLASTIC MATERIALS

Any stress-strain curve can be approximated by a polygon like $OABCD$ in Fig. 47.5. This line can be represented as in Fig. 47.6, where three perfectly plastic materials and one elastic one have been superposed in such a way that the stresses are added for the same strains.

The component materials in Fig. 47.6 are derived from the sides of the polygon in Fig. 47.5, as follows. We extend the sides of the polygon until they meet the σ axis at points A', B', and C' with ordinates σ_1, σ_2, and σ_3, respectively. From points A', B', and C', we draw parallels to the ϵ axis, and we project the corners A, B, and C of the polygon on the parallels from A', B', and C', respectively. Finally, we connect these projections, A_1, B_1, C_1, with the corresponding points, O, A', B', by means of straight lines. The component materials will be the perfectly plastic materials OA_1A'', $A''A'B_1B''$, $B''B'C_1C''$, and the elastic material $C''C'D$.

Let E_1, E_2, . . . be the slopes $d\sigma/d\epsilon$ of the sides of the polygon. Then the elastic moduli of the component materials are $(E_1 - E_2)$, $(E_2 - E_3)$, $(E_3 - E_4)$, and E_4.

The procedure of getting the representation in Fig. 47.6 from Fig. 47.5 is independent of the number of sides of the polygon in Fig. 47.5.

It is assumed now that (1) the cross section has two axes of symmetry, one of them in the plane of bending, (2) no axial force is acting, and (3) the stress-strain relations

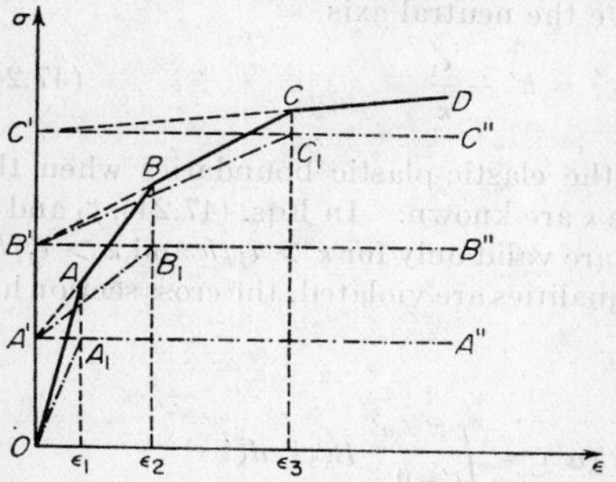

FIG. 47.5

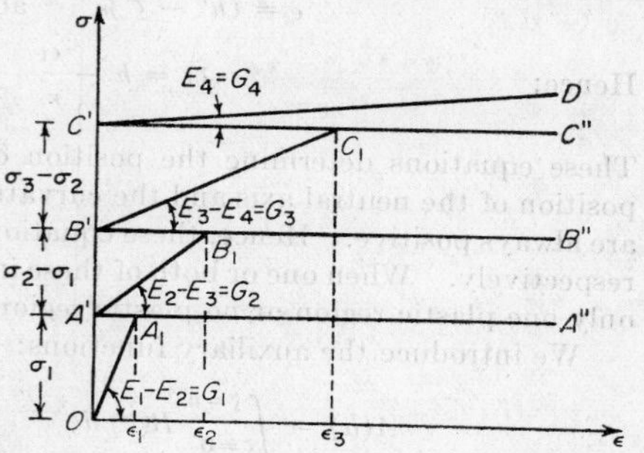

FIG. 47.6

for tension and compression are identical. Then, by using the superposition concept, we conclude that

$$M = \sum_{i=1}^{i=n} M_i$$

From Eq. (47.19), we conclude that

$$M = \kappa I \sum_{i=1}^{i=n} G_i t_i \tag{47.22}$$

where G_i is the elastic modulus of the ith material. The sum in Eq. (47.22) is the *reduced modulus* for the material with the broken line.

47.6. THE NEUTRAL AXIS HAS NO FIXED POSITION; BENDING WITH AXIAL FORCE

The method explained here has been developed in [1]. For an older alternate, see [2] and [3]. We restrict our considerations to beams with a longitudinal plane of

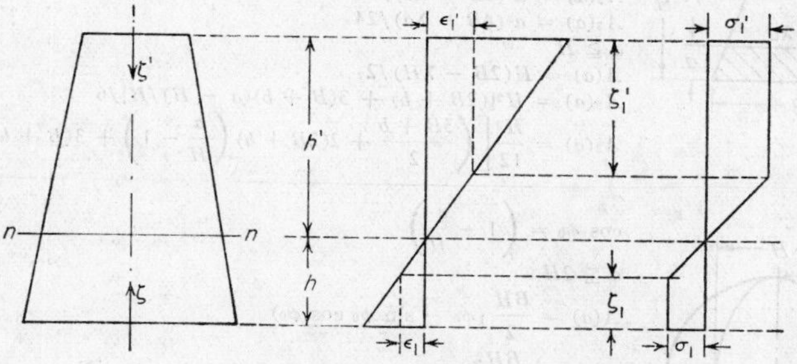

FIG. 47.7

symmetry under the influence of axial and transverse forces acting in this plane. The material of the beam is assumed to be perfectly plastic, but the stress-strain diagram in tension need not be the same as that in compression.

Because the neutral axis does not have a fixed position, it will be useful to express the position of a fiber by its distance from the upper or the lower boundary of the cross section (Fig. 47.7).

With the notations indicated in this figure, the strain is

$$\begin{aligned} \epsilon &= (h - \zeta)\kappa \qquad \text{below the neutral axis} \\ \epsilon &= (h' - \zeta')\kappa \qquad \text{above the neutral axis} \end{aligned} \tag{47.23}$$

Hence:

$$\zeta_1 = h - \frac{\epsilon_1}{\kappa} \qquad \zeta_1' = h' - \frac{\epsilon_1'}{\kappa} \tag{47.24}$$

These equations determine the position of the elastic-plastic boundaries when the position of the neutral axis and the curvature κ are known. In Eqs. (47.24), ζ_1 and ζ_1' are always positive. Hence, these equations are valid only for $\kappa > \epsilon_1/h$ and $\kappa > \epsilon_1'/h'$, respectively. When one or both of these inequalities are violated, the cross section has only one plastic region or no plastic region.

We introduce the auxiliary functions:

$$\begin{aligned} A(a) &= \int_{\zeta=0}^{\zeta=a} B(\zeta)\,d\zeta \qquad & A'(a') &= \int_{\zeta'=0}^{\zeta'=a'} B(\zeta')\,d\zeta' \\ A_2(a) &= \int_{\zeta=0}^{\zeta=a} A(\zeta)\,d\zeta \qquad & A_2'(a') &= \int_{\zeta'=0}^{\zeta'=a'} A'(\zeta')\,d\zeta' \\ A_3(a) &= \int_{\zeta=0}^{\zeta=a} A_2(\zeta)\,d\zeta \qquad & A_3'(a') &= \int_{\zeta'=0}^{\zeta'=a'} A_2'(\zeta')\,d\zeta' \end{aligned} \tag{47.25}$$

Explicit expressions for these functions are given in Table 47.3 for some cross sections.

Table 47.3[1]

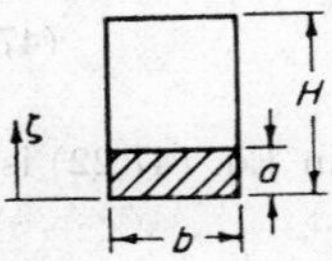 	$a \leqq H$ $A(a) = ba$ $A_2(a) = ba^2/2$ $A_3(a) = ba^3/6$ $a \geqq H$ $A(a) = bH$ $A_2(a) = bH(2a - H)/2$ $A_3(a) = bH(H^2 - 3Ha + 3a^2)/6$
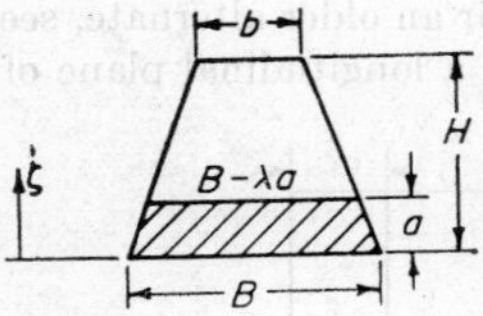 	$a \leqq H$ $\lambda = \dfrac{B - b}{H}$ $A(a) = a(2B - \lambda a)/2$ $A_2(a) = a^2(3B - \lambda a)/6$ $A_3(a) = a^3(4B - \lambda a)/24$ $a \geqq H$ $A(a) = H(2B - \lambda H)/2$ $A_2(a) = H^2[(2B + b) + 3(B + b)(a - H)/H]/6$ $A_3(a) = \dfrac{H^3}{12}\left[\left(\dfrac{3B + b}{2}\right) + 2(2B + b)\left(\dfrac{a}{H} - 1\right) + 3(B + b)\left(\dfrac{a}{H} - 1\right)^2\right]$
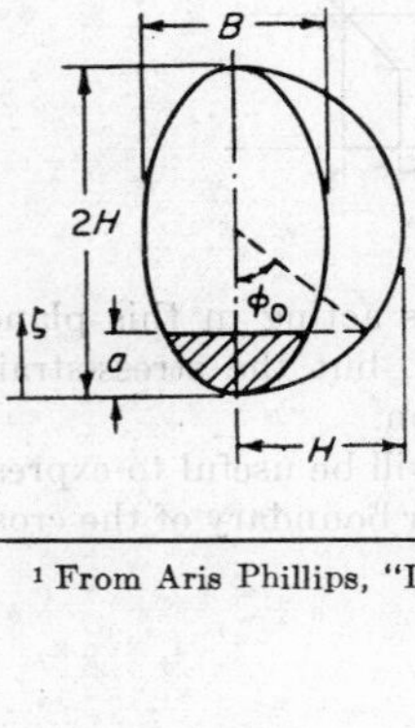 	$\cos\phi_0 = \left(1 - \dfrac{a}{H}\right)$ $a \leqq 2H$ $A(a) = \dfrac{BH}{2}(\phi_0 - \sin\phi_0\cos\phi_0)$ $A_2(a) = \dfrac{BH^2}{6}[-3\phi_0\cos\phi_0 + \sin\phi_0(2 + \cos^2\phi_0)]$ $A_3(a) = \dfrac{BH^3}{48}[3\phi_0(4\cos^2\phi_0 + 1) - \sin\phi_0\cos\phi_0(2\cos^2\phi_0 + 13)]$ $a \geqq 2H$ $A(a) = \pi BH/2$ $A_2(a) = \pi BH(a - H)/2$ $A_3(a) = \pi BH(4a^2 + 5H^2 - 8aH)/16$

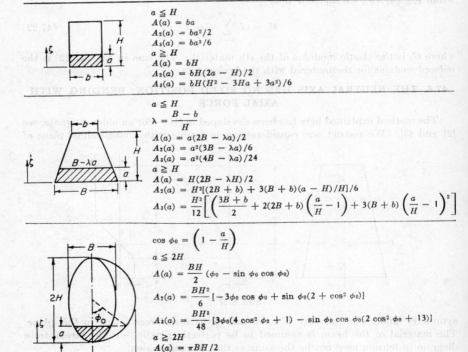

[1] From Aris Phillips, "Introduction to Plasticity," Ronald, New York, 1956.

In [1], it is shown that, to a position of the neutral axis defined by h and h', and to a curvature κ, the axial force transmitted by the cross section is

$$N = \kappa\{E_1[A_2(h) - A_2(\zeta_1)] - E_1'[A_2'(h') - A_2'(\zeta_1')]\} \tag{47.26}$$

while the bending moment with respect to the neutral axis transmitted by the cross section is

$$M = \kappa E_1[2A_3(h) - (h - \zeta_1)A_2(\zeta_1) - 2A_3(\zeta_1)] + \kappa E_1'[2A_3'(h') - (h' - \zeta_1')A_2'(\zeta_1') - 2A_3'(\zeta_1')] \tag{47.27}$$

Equations (47.26) and (47.27) are used as follows. A position of the neutral axis and a value of the curvature are selected (h,h',κ). Then the distances ζ_1 and ζ_1' are determined from Eqs. (47.24), and the values of the auxiliary functions are obtained from Eqs. (47.25) and Table 47.3. Finally, Eqs. (47.26) and (47.27) give the values of N and M which correspond to the selected position of the neutral axis and of the curvature. Equations (47.26) and (47.27) can be used also for finding *interaction formulas*, that is, equations connecting the bending moment and the axial force transmitted by the cross section, for which some other quantity, for example, the depth of the plastic region, remains constant (see [5], chap. 18).

The method of superposition can be used here also. Suppose that n component materials are contributing below the neutral axis, and that n' component materials contribute above the neutral axis. Then we have

$$\begin{aligned} \zeta_i &= h - \frac{\epsilon_i}{\kappa} \qquad i = 1, 2, 3, \ldots, n \\ \zeta_i' &= h' - \frac{\epsilon_i'}{\kappa} \qquad i = 1, 2, 3, \ldots, n' \end{aligned} \tag{47.28}$$

for the position of the elastic-plastic boundary for each material, and

$$\begin{aligned} N &= \kappa\left\{\sum_{i=1}^{i=n} G_i[A_2(h) - A_2(\zeta_i)] - \sum_{i=1}^{i=n'} G_i'[A_2'(h') - A_2'(\zeta_i')]\right\} \\ M &= \kappa\sum_{i=1}^{i=n} G_i[2A_3(h) - \{h - \zeta_i\}A_2(\zeta_i) - 2A_3(\zeta_i)] \\ &\quad + \kappa\sum_{i=1}^{i=n'} G_i'[2A_3'(h') - \{h' - \zeta_i'\}A_2'(\zeta_i') - 2A_3'(\zeta_i')] \end{aligned} \tag{47.29}$$

for the axial force N and the bending moment M transmitted by the cross section.

47.7. DEFLECTIONS

Since the relation

$$\kappa = -\frac{d^2w}{dx^2} \tag{47.30}$$

is of purely geometrical origin, it is valid for any stress-strain relation. It should, however, be remembered that it is based on the assumption of small deflections. For large deflections, it must be replaced by Eq. (35.40). From previous considerations, we already know the relation between bending moment and curvature:

$$\kappa = \psi(M) \tag{47.31}$$

Hence, Eq. (47.30) gives

$$-\frac{d^2w}{dx^2} = \psi(M) \tag{47.32}$$

Table 47.4[1]

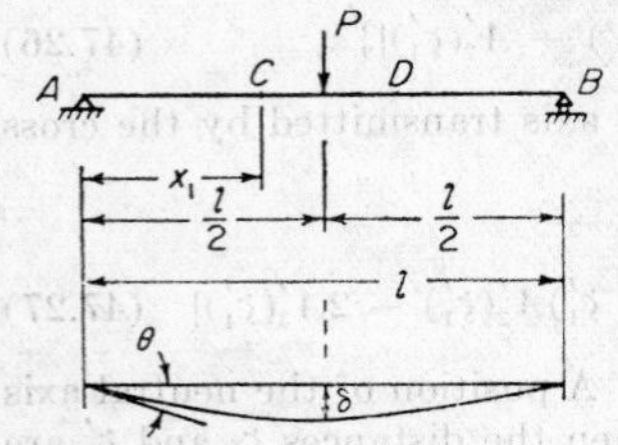

$$x_1 = 2M_e/P$$

$$\theta = \frac{M_e^2}{PEI}\left(3 - 2\sqrt{3 - \frac{Pl}{2M_e}}\right)$$

$$\delta = \frac{M_e^3}{3P^2EI}\left[20 - \left(12 + \frac{Pl}{M_e}\right)\sqrt{3 - \frac{Pl}{2M_e}}\right]$$

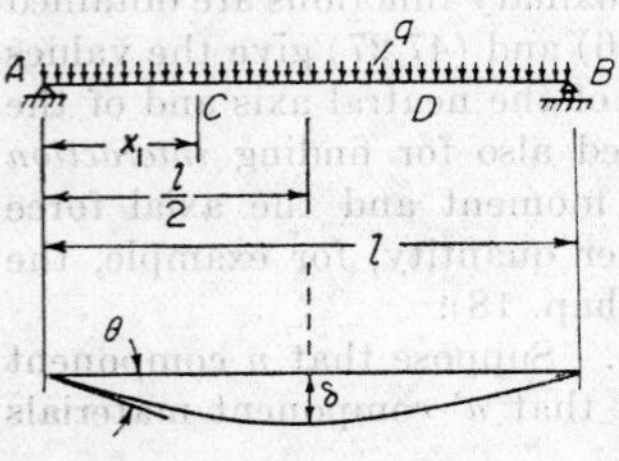

$$x_1 = \frac{l}{2} - \sqrt{\frac{l^2}{4} - \frac{2M_e}{q}}$$

$$\theta = \frac{1}{EI}\left[\frac{qx_1^2}{12}(3l - 2x_1) + M_e\sqrt{\frac{M_e}{q}}\ln\frac{\sqrt{3 - ql^2/4M_e}}{1 - \sqrt{ql^2/4M_e - 2}}\right]$$

$$\delta = \frac{1}{EI}\left\{\frac{qx_1^3}{24}(4l - 3x_1) + M_e\sqrt{\frac{M_e}{q}}\left[\sqrt{\frac{M_e}{q}}\left(\sqrt{3 - \frac{ql^2}{4M_e}} - 1\right) + \frac{l}{2}\ln\frac{\sqrt{3 - ql^2/4M_e}}{1 - \sqrt{ql^2/4M_e - 2}}\right]\right\}$$

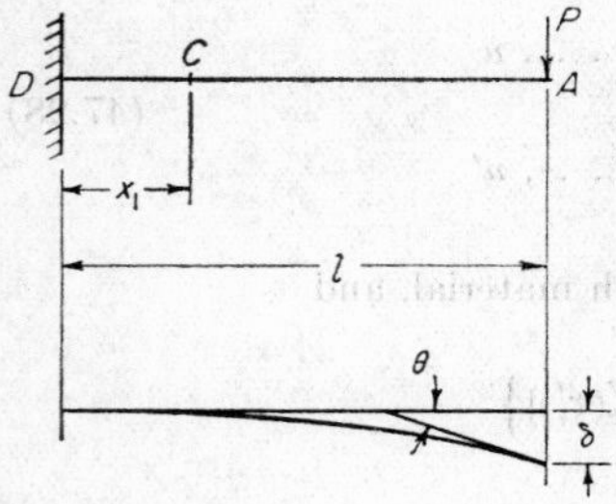

$$x_1 = l - \frac{M_e}{P}$$

$$\theta = \frac{M_e^2}{2PEI}\left(3 - 2\sqrt{3 - \frac{2Pl}{M_e}}\right)$$

$$\delta = \frac{M_e^3}{3P^2EI}\left[5 - \left(3 + \frac{Pl}{M_e}\right)\sqrt{3 - \frac{2Pl}{M_e}}\right]$$

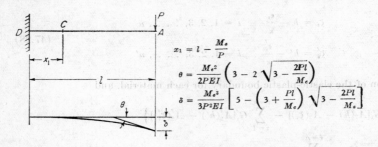

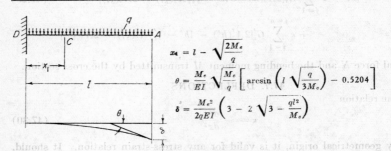

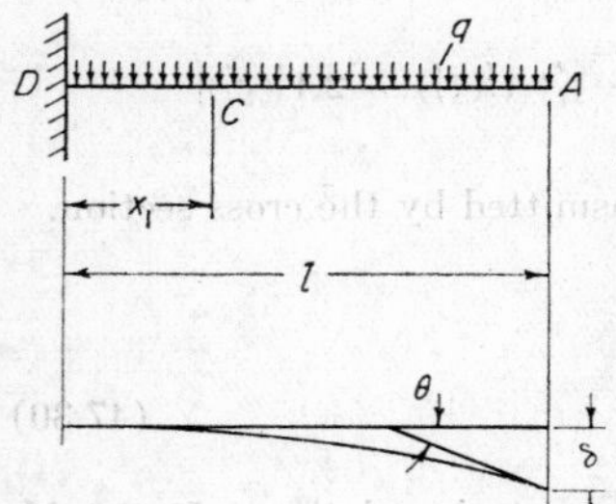

$$x_1 = l - \sqrt{\frac{2M_e}{q}}$$

$$\theta = \frac{M_e}{EI}\sqrt{\frac{M_e}{q}}\left[\arcsin\left(l\sqrt{\frac{q}{3M_e}}\right) - 0.5204\right]$$

$$\delta = \frac{M_e^2}{2qEI}\left(3 - 2\sqrt{3 - \frac{ql^2}{M_e}}\right)$$

[1] From Aris Phillips, "Introduction to Plasticity," Ronald, New York, 1956.

AC = elastic
CD = elastic plastic
DB = elastic

from which we can find the deflections and slopes. Table 47.4 gives the deflections for a rectangular cross section and a perfectly plastic material.

The deflections may also be calculated by means of the area-moment method [3], the virtual-work method [1], and energy methods [8],[9].

47.8. SHEAR STRESSES IN BENDING OF BEAMS

We shall restrict our considerations to rectangular cross sections. Once the normal stresses have been determined, the shear stresses can be found by integrating the differential equation of equilibrium,

$$\frac{\partial \sigma}{\partial x} + \frac{\partial \tau}{\partial z} = 0 \tag{47.33}$$

For the *power law* (47.6), Eqs. (47.1) and (47.8) yield

$$\sigma = Cz^n\kappa^n = \frac{Mz^n}{2I_{n+1}} \tag{47.34}$$

Equation (47.33) then gives

$$\frac{\partial \tau}{\partial z} = -\frac{\partial \sigma}{\partial x} = -\frac{z^n}{2I_{n+1}}\frac{dM}{dx} = -\frac{z^nV}{2I_{n+1}} \tag{47.35}$$

where V is the shear force.
Integrating, and taking into account that, for $z = h$, we have $\tau = 0$, we find

$$\tau = \frac{V}{2(n+1)I_{n+1}}(h^{n+1} - z^{n+1}) \tag{47.36}$$

with

$$\tau_{\max} = \frac{Vh^{n+1}}{2(n+1)I_{n+1}}$$

at $z = 0$.

For a *perfectly plastic material*, the normal stress σ is constant throughout the plastic region; hence,

$$\frac{\partial \tau}{\partial z} = 0$$

and τ is constant as we move in the z direction. Since $\tau = 0$ at the boundary, the shear stress is zero over the entire plastic region.

The shear stresses in the elastic region can be determined by using Eq. (47.36), in which we introduce $n = 1$ and z_e instead of h:

$$\tau = \frac{V}{4I_2}(z_e^2 - z^2) \tag{47.37}$$

For a rectangular cross section, $I_2 = bz_e^3/3$. Hence, Eq. (47.37) becomes

$$\tau = \frac{3V}{4bz_e^3}(z_e^2 - z^2) \qquad \text{and} \qquad \tau_{\max} = \frac{3V}{4bz_e}$$

The above theory gives also a satisfactory approximation for cross sections which have a form slightly different from a rectangular one.

More information about shear stresses in bending can be found in a paper by Horne [11]; and in the books of Prager and Hodge [12]; Baker, Horne, and Heyman [13]; and Neal [14].

47.9. REFERENCES

[1] A. Phillips: "Introduction to Plasticity," Ronald, New York, 1956.
[2] A. Nadai: "Theory of Flow and Fracture of Solids," vol. 1, 2d ed., McGraw-Hill, New York, 1950.
[3] S. Timoshenko: "Strength of Materials," 3d ed., part 2, Van Nostrand, Princeton, N.J., 1956.
[4] A. Freudenthal: "The Inelastic Behavior of Engineering Materials and Structures," Wiley, New York, 1950.
[5] F. B. Seely and J. O. Smith: "Advanced Mechanics of Materials," 2d ed., Wiley, New York, 1952.
[6] D. Morkovin, O. Sidebottom: The effect of non-uniform distribution of stress in the yield strength of steel, *Univ. Illinois Bull.* **45** (1947), no. 26.
[7] J. F. Baker, J. Roderick: Further tests on beams and portals, *Trans. Inst. Welding, London,* **3** (1940), p. 83.
[8] H. Kauderer: "Nichtlineare Mechanik," Springer, Berlin, 1958.
[9] H. Westergaard: On the method of complementary energy and its application to structures stressed beyond the proportional limit, *Trans. Am. Soc. Civil Engrs.*, **107** (1942), 765–793.
[10] J. Argyris: Energy theorems and structural analysis, *Aircraft Eng.* **26** (1954), 347–356, 383–387, 394, **27** (1955), 42–58, 80–94, 125–134, 145–158.
[11] M. R. Horne: The plastic theory of bending of mild steel beams with particular reference to the effect of shear forces, *Proc. Roy. Soc. London,* **A,207** (1951), 216–228.
[12] W. Prager, P. G. Hodge, Jr.: "Theory of Perfectly Plastic Solids," Wiley, New York, 1951.
[13] J. F. Baker, M. R. Horne, J. Heyman: "The Steel Skeleton," vol. 2, Cambridge Univ. Press, New York, 1956.
[14] B. G. Neal: "The Plastic Methods of Structural Analysis," Wiley, New York; Chapman & Hall, London, 1957.

CHAPTER 48

TORSION

BY

W. F. FREIBERGER, Ph.D., Providence, R.I.

48.1. FUNDAMENTAL EQUATIONS

Consider a cylinder of arbitrary cross section, of length l, with one of its bases fixed in the y,z plane, and the other acted on by a torque M_T whose moment vector lies along the x axis. The corresponding stress problem is treated in Sec. 36.2 under the assumption that Hooke's law is valid throughout. When the elastic limit of the material is reached and a plastic zone develops, many of the statements made and equations used in the theory of elastic torsion still apply.

Again the stress system consists of shear stresses τ_{xy}, τ_{xz} only; and these must satisfy the equilibrium condition (36.8). It may still be solved with the help of a stress function ϕ according to Eq. (36.9), but the differential equation (36.10) applies in the elastic region only, since it contains Hooke's law for the shear stresses. The boundary condition (36.11) for the shear stress is still valid, and leads to the same statements (36.13) for the boundary values of the stress function. Consequently, the torque M_T is correctly given by Eqs. (36.14) and (36.16), the integrals being extended over elastic and plastic regions alike and over holes, as explained on p. 36–5.

Equations (36.5) still describe the displacements v and w in terms of the twist θ, and Eqs. (36.6) can be written in the form

$$\gamma_{xy} = \frac{\partial u}{\partial y} + \frac{\partial v}{\partial x} = \frac{\partial u}{\partial y} - \theta z \qquad \gamma_{xz} = \frac{\partial u}{\partial z} + \frac{\partial w}{\partial x} = \frac{\partial u}{\partial z} + \theta y \tag{48.1}$$

but it is no longer possible to write the warping displacement u as $\theta u_1(y,z)$, since the problem of plastic torsion is nonlinear, and the warping is not proportional to the twist. After elimination of u, Eqs. (48.1) yield the relation

$$\frac{\partial \gamma_{xz}}{\partial y} - \frac{\partial \gamma_{xy}}{\partial z} = 2\theta \tag{48.2}$$

This equation and (36.8) are two equations for the four unknowns: τ_{xy}, τ_{xz}, γ_{xy}, γ_{xz}. In the elastic domain, the stress-strain relations connecting τ_{xy} with γ_{xy} and τ_{xz} with γ_{xz} provide the two additional equations required for the determination of the four functions. In the plastic domain, the yield condition

$$\tau_{xy}^2 + \tau_{xz}^2 = k^2(y,z) \tag{48.3}$$

where $k(y,z)$ is the yield stress in pure shear, determines, with (36.8), the two stress components. The plastic torsion problem is thus statically determinate. The function $k(y,z)$ will be a constant k unless the material has been inhomogeneously hardened before torsion or later strain hardening is to be taken into account.

Upon substitution of (36.9) into (48.3), it is seen that, in the plastic domain, ϕ has to satisfy the equation

$$(\partial\phi/\partial y)^2 + (\partial\phi/\partial z)^2 = k^2(y,z) \tag{48.4}$$

whereas, in the elastic domain, where the elastic stress-strain law

$$\tau_{xy} = G\gamma_{xy} \qquad \tau_{xz} = G\gamma_{xz} \tag{48.5}$$

holds, Eqs. (48.2) and (48.5) show that ϕ must satisfy Poisson's equation,

$$\nabla^2\phi = \partial^2\phi/\partial y^2 + \partial^2\phi/\partial z^2 = -2G\theta \tag{48.6}$$

48.2. SAND-HILL ANALOGY

It will be assumed in this section that the yield stress in pure shear

$$k(y,z) = k = \text{const}$$

that is, the material is non-strain-hardening, or perfectly plastic, with a well-defined yield point. The uniform cross section of the bar will be taken as simply connected.

For an elastic bar, it was shown in Sec. 36.3 how Prandtl's membrane analogy, representing the stress function ϕ experimentally, provides a basis for the formation of an intuitive picture of how the solution to a torsion problem should look.

It has been shown [4] that a similar analogy exists for the torsion of perfectly plastic bars. In the plastic region, Eq. (48.4) must be satisfied, which, for constant k, can be written as

$$|\text{grad } \phi| = k \tag{48.7}$$

The meaning of this equation is that the surface

$$\phi = \phi(y,z) \tag{48.8}$$

has, at every one of its points, a steepest slope, equal to the value of the yield shear stress; (48.8) is thus a surface of constant slope, and it follows, from the boundary condition $\phi = 0$, that it will have zero height on the boundary C. Such a surface can be obtained experimentally by piling sand on a horizontal base of shape identical with that of the cross section of the bar; the height of the sand hill will, at any point, be proportional to the value of the stress function at the point, the constant of proportionality depending on k and on the properties of the sand. It follows from (36.16) that the volume of sand going into the sand hill is proportional to the torque required to cause plastic flow over the entire cross section.

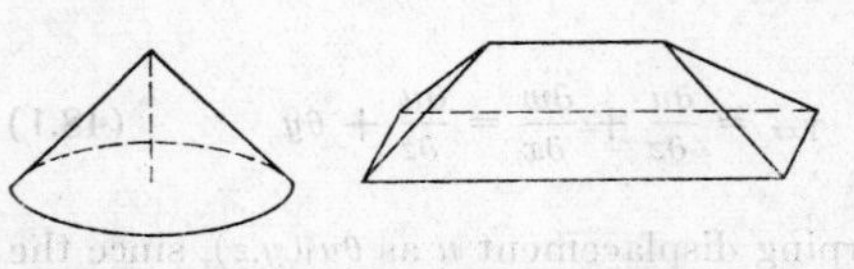

FIG. 48.1. Roofs for circular and rectangular cross sections (sand-hill analogy).

Clearly, the projections, onto the base, of the lines of equal height of the sand hill are the trajectories of the stress vector in the cross section. Figure 48.1 shows the shape of the sand hill for a circular and a rectangular cross section, when each has become fully plastic.

The state of stress in a bar whose cross section has only partly yielded plastically, and still encompasses elastic regions, can be represented experimentally by a combination of the membrane and the sand-hill analogies. A transparent material is used in constructing a surface of constant slope over a base of the same shape as the bar cross section; a rubber membrane is then stretched across the base, and the membrane loaded with a uniformly distributed pressure from the side opposite that of the "roof." For small values of the pressure, corresponding to purely elastic behavior of the bar, the motion of the membrane will not be affected by the presence of the roof.

As these values increase, the membrane will be pressed against the roof, and the projection, onto the base, of the contact areas between membrane and roof corresponds to the plastic regions of the bar. Clearly, these regions will first develop near the boundary, where points of greatest stress occur. Eventually, the membrane will adhere completely to the roof, corresponding to the fully plastic state of the bar.

The stress conditions across the elastic-plastic boundary L can also be gathered from this analogy. This boundary corresponds to the projection, onto the base, of the contact lines between the free (nonadhering) portion of the membrane and the roof; clearly, the roof will be tangent to the membrane there, and thus ϕ, $\partial\phi/\partial y$, and $\partial\phi/\partial z$ continuous across it. This corresponds, from (36.9), to continuous τ_{xy} and τ_{xz}; the following conditions across L are thus satisfied:

$$\tau_{xy}^{e} = \tau_{xy}^{p} \qquad \tau_{xz}^{e} = \tau_{xz}^{p} \tag{48.9}$$

where e denotes elastic and p plastic regions. Photographs showing results of actual experiments with the elastic-plastic analogy are given in [4].

An important property of stress distributions in plastic torsion is represented by the ridges in the sand-hill analogy. Take, for simplicity, a ridge whose projection onto the y,z plane is parallel to the y axis; clearly, across this ridge, the value of $\partial\phi/\partial y$ is continuous, and that of $\partial\phi/\partial z$ is discontinuous. Across the projection of this ridge onto the y,z plane τ_{xy} is thus discontinuous, and τ_{xz} continuous. Such lines of stress discontinuity are of considerable importance in plastic theory in general [13], and torsion in particular. In fact, the component τ_{nx} normal to a line of discontinuity (such as τ_{xz} above) must be continuous across it from continuity considerations; the tangential stress τ_{tx} is permitted a jump in value. Generally, a convex corner in the cross-section contour will give rise to a discontinuity line, starting at that corner; a reentrant corner, on the other hand, is always the apex of a conical portion of the sand hill. These points will be discussed analytically in the next section and they are exemplified by the L-shaped cross section in Fig. 48.2.

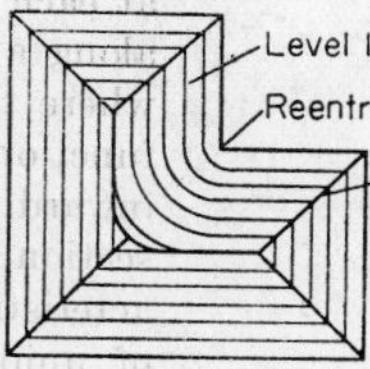

FIG. 48.2. Roof for L-shaped cross section (sand-hill analogy).

48.3. METHOD OF CHARACTERISTICS: UNIFORM BAR

Stresses

The yield stress in pure shear will, in this section, be taken as $k(y,z)$, a function of position in the cross section.

To solve Eq. (48.4),

$$(\partial\phi/\partial y)^2 + (\partial\phi/\partial z)^2 = k^2(y,z) \tag{48.10}$$

in the plastic region, the function $\psi(y,z)$ will be introduced by the equations

$$\tau_{xy} = \partial\phi/\partial z = k(y,z)\cos\psi(y,z) \qquad \tau_{xz} = -\partial\phi/\partial y = k(y,z)\sin\psi(y,z) \tag{48.11}$$

Hence,

$$\tau_{xz}/\tau_{xy} = \tan\psi$$

and ψ is clearly the inclination of the shear-stress vector τ to the y axis; the yield condition (48.3), with $k = k(y,z)$, is thus identically satisfied, and the equilibrium equation (36.8) becomes

$$-\sin\psi\frac{\partial\psi}{\partial y} + \cos\psi\frac{\partial\psi}{\partial z} + \frac{1}{k}\cos\psi\frac{\partial k}{\partial y} + \frac{1}{k}\sin\psi\frac{\partial k}{\partial z} = 0 \tag{48.12}$$

Let ds be a line element in the direction of τ, and dc be a line element obtained by rotating ds through 90° anticlockwise (see Fig. 48.3); Eq. (48.12) can then be written [15] as

$$\frac{\partial \psi}{\partial c} + \frac{1}{k}\frac{\partial k}{\partial s} = 0 \tag{48.13}$$

since

$$\frac{\partial}{\partial c} \equiv -\sin\psi\frac{\partial}{\partial y} + \cos\psi\frac{\partial}{\partial z} \qquad \frac{\partial}{\partial s} \equiv \cos\psi\frac{\partial}{\partial y} + \sin\psi\frac{\partial}{\partial z} \tag{48.14}$$

The direction of dc is the direction of the characteristics of (48.12); for a discussion of characteristics, see pp. 11–8 and 11–20. The equation of the characteristics can be written as

$$dz/dy = -\cot\psi \tag{48.15}$$

and hence, integrating,

$$z + y\cot\psi = f(\psi) \tag{48.16}$$

where $f(\psi)$ is an arbitrary function, determined by the boundary conditions, as shown below. For constant k, it follows from (48.15) that ψ is constant along each characteristic, which is thus a straight line in the y,z plane. For $k = k(y,z)$, the curvature of each characteristic is determined by the value of k at each point; in practice, we would integrate (48.13) along a characteristic, starting from a boundary point where the value of the stress, and thence of the stress function, is known. The characteristics thus continued inward from the boundary will either pass across the section to meet the boundary or intersect other characteristics. In the latter case, there must, because of uniqueness of the stresses, exist a line on which characteristics from different parts of the boundary intersect and terminate: such a line is a line of discontinuity Γ, and its defining property is that the stress component τ_{nx} normal to it be, by equilibrium, continuous across it. The value τ_{tx} of the tangential component will be different on the two sides of Γ.

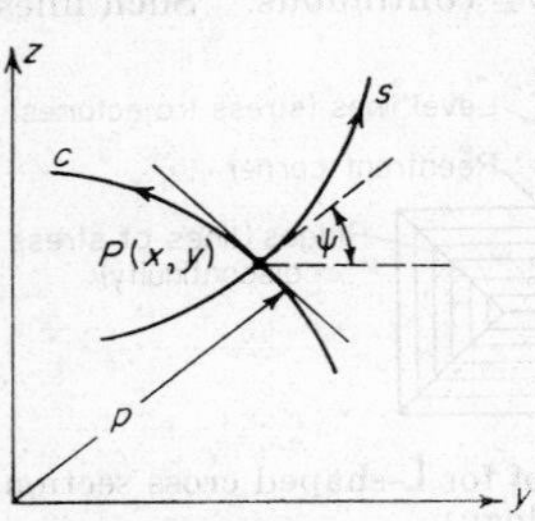

FIG. 48.3. Stress trajectory and characteristic. s = stress-vector trajectory; c = characteristic.

The value of the jump can be determined as follows: Let α be the inclination of Γ to the y axis at any point. Then

$$\tau_{tx} = \tau_{xy}\cos\alpha + \tau_{xz}\sin\alpha \qquad \tau_{nx} = -\tau_{xy}\sin\alpha + \tau_{xz}\cos\alpha$$

and hence, from (48.11),

$$\tau_{tx} = k\cos(\psi - \alpha) \qquad \tau_{nx} = k\sin(\psi - \alpha)$$

If I and II denote the regions on the two sides of Γ, respectively, continuity of τ_{nx} across it requires $\alpha = \frac{1}{2}(\psi_{\mathrm{I}} + \psi_{\mathrm{II}}) - \pi/2$, and the differential equation of Γ becomes

$$dz/dy = -\cot\tfrac{1}{2}(\psi_{\mathrm{I}} + \psi_{\mathrm{II}})$$

Thus, Γ bisects the characteristics meeting on it. Usually Γ has several branches over the cross section.

Let the boundary of the cylindrical bar be given by

$$y = y(s) \qquad z = z(s) \tag{48.17}$$

where s is the boundary parameter. The boundary will be taken free from stress; that is, the stress vector τ must be tangential to the curve (48.17), and hence

$$\frac{\tau_{xz}}{\tau_{xy}} = \frac{dz/ds}{dy/ds} \tag{48.18}$$

that is, from (48.11),

$$\frac{dz/ds}{dy/ds} = \tan \psi \tag{48.19}$$

Thus, the characteristics meet the contour normally, which is indeed obvious from their definition as orthogonal trajectories to the shear-stress vector.

The function $f(\psi)$ in (48.16) may be determined in terms of the boundary parameter s by substituting (48.17) into (48.16), to give the following equation for the characteristic:

$$[y - y(s)] \cos \psi + [z - z(s)] \sin \psi = 0 \tag{48.20}$$

It follows, from the normality of the characteristics to the boundary, that a projecting corner is always the origin of a line of discontinuity which halves the angle between the two parts of the boundary, meeting at that corner. A reentrant corner, on the other hand, is the origin of a fan of characteristics. For k = const, that is, straight-line characteristics, such a fan corresponds to the simple wave well known in fluid dynamics; these points are illustrated in Fig. 48.4, which represents the characteristics field of an L-shaped cross section; the stress field is shown in Fig. 48.2.

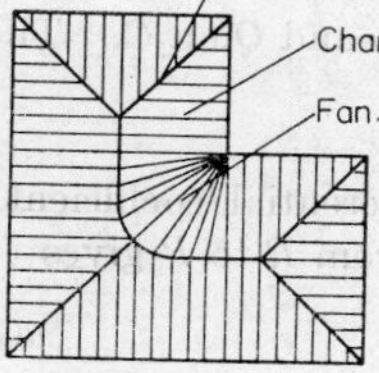

FIG. 48.4. L-shaped cross section.

The above discussion has been for a torque sufficient to make the bar fully plastic. The characteristics are determined by inward integration from the boundary C of the cross section and are, therefore, independent of the elastic-plastic boundary L: they may be computed without reference to it. However, the elastic-plastic problem consists of a matching of the elastic and the plastic solutions such that Eqs. (48.9) are satisfied along the boundary separating these two solutions. Since the elastic torsion problem is not, like the plastic, statically determinate, the displacement solution will be required in the elastic region to determine τ_{xy}^e and on τ_{xz}^e on L. Across L, we must also have continuity in axial displacement:

$$u^e = u^p$$

The torque will be given by

$$M_T = \iint_{R^e} (y\tau_{xz}^e - z\tau_{xy}^e)\, dy\, dz + \iint_{R^p} (y\tau_{xz}^p - z\tau_{xy}^p)\, dy\, dz \tag{48.21}$$

where R^e and R^p denote the elastic and the plastic regions of the cross section.

An example of a semi-inverse method for solving the elastic-plastic problem is given on p. 48–8.

Warping

The flow laws of plasticity theory require that, at every point of the cross section, the stress components be proportional to the components of the rate of strain, the factor of proportionality being a function of position. In the torsion problem, strains may be taken in place of rates of strain, without loss of generality. From (48.1), strain components are

$$-\theta z + \partial u/\partial y \propto \tau_{xy} \qquad \theta y + \partial u/\partial z \propto \tau_{xz}$$

and, by virtue of (48.11), these can be written as

$$-\theta z + \partial u/\partial y = \lambda(y,z)\cos\psi \qquad \theta y + \partial u/\partial z = \lambda(y,z)\sin\psi \tag{48.22}$$

where the variable $k(y,z)$ has been absorbed into $\lambda(y,z)$. Eliminating λ, (48.22) becomes

$$\sin\psi\,\partial u/\partial y - \cos\psi\,\partial u/\partial z = \theta(y\cos\psi + z\sin\psi) \tag{48.23}$$

Equation (48.23) for the warping has the same characteristics as (48.12) for the stress function; it will be written in the form

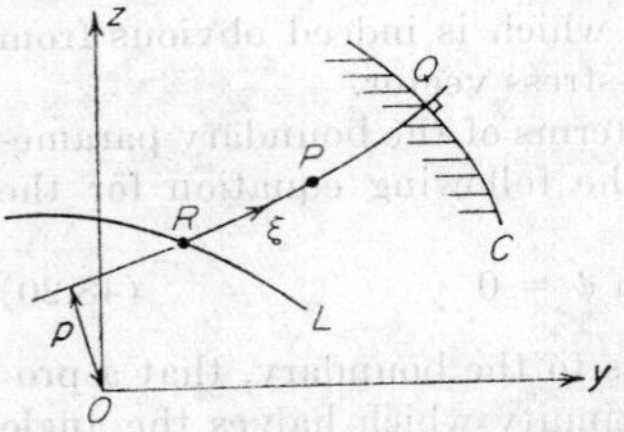

FIG. 48.5. Determination of the warping. L = elastic-plastic boundary; C = cross-section contour; RPQ = characteristic.

$$\partial u/\partial c + \theta p = 0 \tag{48.24}$$

where $\partial/\partial c$ is defined by Eqs. (48.14), and $p = |y\cos\psi + z\sin\psi|$ is the length of the perpendicular from the origin to the tangent to the characteristics at the generic point considered; for constant k, p is simply the distance of the straight-line characteristic from the origin.

The following construction for the determination of the warping at any point is based on Eq. (48.24) (see Fig. 48.5).

Let P be a generic point in the plastic region, L a segment of the elastic-plastic boundary, C of the cross-sectional boundary, RPQ of the characteristic through the points R on L and Q on C. Then, along RPQ,

$$du/dc = \theta p \tag{48.25}$$

the total differential coefficient is used for differentiation along the characteristic. Integrating from R to P gives

$$u(P) = u(R) + \int_R^P \theta p\,dc \tag{48.26}$$

If k = const, so that RPQ is a straight line with $RP = \xi$,

$$u(P) = u(R) + \xi\theta p \tag{48.27}$$

Once the warping function has been determined in the elastic region, its continuity across L provides the starting value $u(R)$ for Eqs. (48.25) and (48.26). When there is no elastic region, as in the solution of the fully plastic problem, the above considerations give no starting point for the integration of the warping function, since it is unknown on the boundary of the cross section. In that case, the lines of discontinuity provide the boundary conditions, and it is necessary to investigate the variation of u on a discontinuity Γ. The normal component of shear stress is continuous across Γ, and the tangential component reverses its direction. Considering Γ as the limit of the elastic region—a thin area in which the stress varies rapidly—the value of the stress in this region must clearly be $\tau < k$. Accordingly, the rate of shear strain must vanish along Γ; that is, from (48.1) and (48.22),

$$\partial u/\partial y = \theta z \qquad \partial u/\partial z = -\theta y$$

and hence, if η denotes variation along Γ,

$$\frac{du}{d\eta} = \frac{\partial u}{\partial y}\frac{dy}{d\eta} + \frac{\partial u}{\partial z}\frac{dz}{d\eta} = \theta\left(z\frac{dy}{d\eta} - y\frac{dz}{d\eta}\right)$$

or

$$\frac{du}{d\eta} = \theta(-z\sin\alpha - y\cos\alpha) \tag{48.28}$$

As before, α is the inclination of Γ to the y axis. Hence, comparing (48.28) with (48.23) and (48.24), it is seen that the equation

$$\frac{du}{d\eta} + \theta p = 0 \tag{48.29}$$

is valid on the characteristics as well as on the lines of discontinuity, provided ξ and p are interpreted as referring to one or the other, as the case may be.

48.4. EXAMPLE: THE SOLID ELLIPTICAL SECTION

Fully Plastic Solution

Consider a cylindrical bar of elliptic cross section, and let the boundary of the cross section be given by

$$y = a \cos s \qquad z = b \sin s \tag{48.30}$$

where s is the boundary parameter [cf. Eq. (48.17)]. Substituting (48.30) into (48.19) and (48.20) gives

$$-(b/a) \cot s = \tan \psi \tag{48.31}$$

and

$$(z - b \sin s) \sin \psi + (y - a \cos s) \cos \psi = 0 \tag{48.32}$$

Eliminating s,

$$z \sin \psi + y \cos \psi - \frac{(a^2 - b^2) \sin \psi \cos \psi}{\sqrt{a^2 \sin^2 \psi + b^2 \cos^2 \psi}} = 0 \tag{48.33}$$

This is the equation of the family of characteristics in terms of the parameter ψ: each value of ψ corresponds to a member of the family. The values of the stress components for any point (y,z) in the plastic region are found by solving (48.33) for $\psi(y,z)$

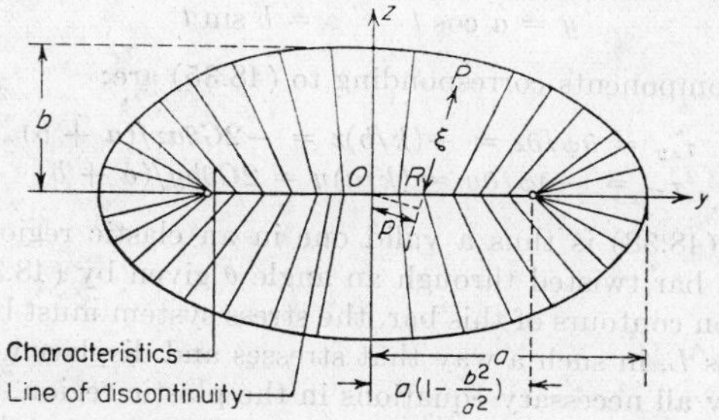

FIG. 48.6. Oval cross section, fully plastic solution.

and substituting its value into (48.11). For a constant yield stress $k(y,z) = k$, the straight-line characteristics intersect in a segment of the y axis. The end points of this segment are the centers of curvature for the end points of the major axis (see Fig. 48.5):

$$y = \pm a(1 - b^2/a^2) \qquad z = 0$$

The warping function is determined from Eq. (48.27)

$$u(P) = u(R) + \theta \xi p \tag{48.34}$$

where P is a generic point of the cross section, R is the point of intersection of the line of discontinuity and the characteristic through P, ξ is the length PR, and p is the perpendicular from the origin to PR. Taking the value of u zero at the center, it is seen from (48.29) that it will be zero along Γ; it will, in fact, vanish along both axes of the ellipse, and increase linearly along the characteristics emanating from Γ.

Elastic-Plastic Solution, Inverse Method

The characteristics were, in the fully plastic case, determined by inward integration, analytically or numerically, from the boundary of the cross section, where the absence of tangential stress provides the necessary boundary condition. Consequently, the characteristics can, in the plastic region, be constructed equally well when there is an elastic core still present. The warping function is found by outward integration along characteristics previously determined; in the fully plastic case, the conditions along the line of discontinuity provided the necessary boundary condition; in the elastic-plastic case, this has to be supplied along the elastic-plastic boundary. Hence, analysis of the elastic region is required for this purpose, and continuity of stress and displacement across the boundary separating it from the plastic region define the position of this boundary.

The analytical determination of the elastic-plastic boundary is a matter of great complexity. Relaxation methods have been used for numerical solution in a number of cases [8]. Here, a semi-inverse method [5], will be explained, for constant $k(y,z) = k$, and applied to the solution of the torsion problem for a bar of uniform oval cross section.

Consider the stress function

$$\phi = -\tfrac{1}{2}k(y^2/a^2 + z^2/b^2) \tag{48.35}$$

which satisfies Poisson's equation (48.6) in the elastic region if

$$k(1/a + 1/b) = 2G\theta \tag{48.36}$$

Along the elastic-plastic boundary L, Eq. (48.7) must hold; the stress function (48.35) will satisfy this equation along the ellipse

$$y = a \cos t \qquad z = b \sin t \tag{48.37}$$

The elastic stress components corresponding to (48.35) are:

$$\begin{aligned} \tau_{xy} &= \partial\phi/\partial z = -(k/b)z = -2G\theta az/(a + b) \\ \tau_{xz} &= -\partial\phi/\partial y = (k/a)y = 2G\theta by/(a + b) \end{aligned} \tag{48.38}$$

The stress system (48.38) is thus a valid one in an elastic region, bounded by the ellipse (48.37), of a bar twisted through an angle θ given by (48.36). To determine possible cross-section contours of this bar, the stress system must be extended into the plastic region across L, in such a way that stresses and displacements are continuous across L and satisfy all necessary equations in the plastic region.

Approaching L from the elastic side, the stress components take, on L, the values

$$\tau_{xy} = -k \sin t \qquad \tau_{xz} = k \cos t \tag{48.39}$$

On the plastic side, they must have, from (48.11), the values

$$\tau_{xy} = k \cos \psi \qquad \tau_{xz} = k \sin \psi \tag{48.40}$$

so that for continuity, on L:

$$\sin t = -\cos \psi \qquad \cos t = \sin \psi \tag{48.41}$$

The equation of the characteristics in the plastic region is now given by (48.16):

$$z + y \cot \psi = f(\psi,\theta) \tag{48.42}$$

where $f(\psi,\theta)$ must be determined from (48.41) so that the stress components are continuous across L. Substituting (48.37) and (48.41) into (48.42) gives

$$f(\psi,\theta) = (a - b) \cos \psi \tag{48.43}$$

and the equation of the characteristics is

$$z = -y \cot \psi + (b - a) \cos \psi \tag{48.44}$$

Since the characteristics must be normal to the cross-section contour C, the determination of C consists in finding orthogonal trajectories to (48.44), that is, solutions of

$$dz/dy = \tan \psi \tag{48.45}$$

where ψ is given by (48.44). Solving (48.44) for ψ, substituting it into (48.45), and solving the resulting equation leads to the following parametric representation of C:

$$y = (B + 3A - A \sin^2 \psi) \sin \psi \qquad z = -(B + A \cos^2 \psi) \cos \psi \tag{48.46}$$

where

$$A = \tfrac{1}{2}(a - b) \tag{48.47}$$

and B must be chosen such that L lies inside C in its entirety. This solution gives no information for the onset of plastic flow, that is, for values of the torque intermediate between that for which plastic flow commences at some point or points on C and that value for which L lies just inside C. Comparing the parametric representations (48.37) for L and (48.46) for C, it is seen that L will lie inside C if

$$B > b \tag{48.48}$$

Solving (48.36) and (48.47) for a and b gives

$$\begin{aligned} a &= k/2G\theta + A + \sqrt{A^2 + (k/2G\theta)^2} \\ b &= k/2G\theta - A + \sqrt{A^2 + (k/2G\theta)^2} \end{aligned} \tag{48.49}$$

The inequality (48.48) becomes, with (48.49) and after solving for θ,

$$\theta > k(A + B)/GB(2A + B) \tag{48.50}$$

Given, then, a bar of uniform cross-section contour (48.46), the elastic-plastic boundary for twists satisfying (48.50) is given by the ellipse (48.37), with axes a and b given by (48.49). The contour (48.46) is oval, approximating an ellipse of semiaxes $(A + B)$ and $(2A + B)$.

The success of the method is due to the fact that, to successive twists θ satisfying (48.50), and a given contour C, values a and b, and thus an elastic-plastic boundary L, can be found. This is possible because only the difference between a and b, and not their separate values, enters into the equations (48.46) for the cross-section contour. Figure 48.7 illustrates these results.

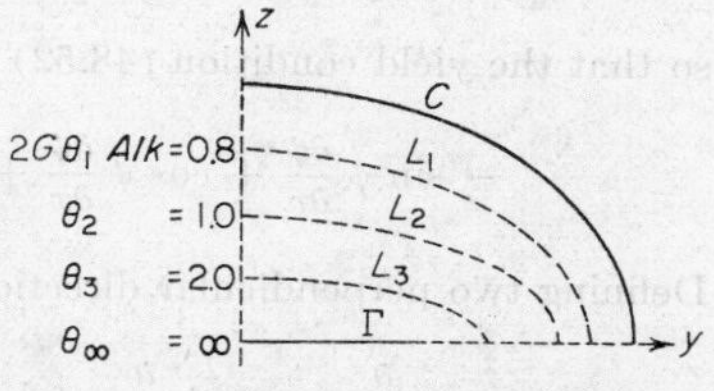

FIG. 48.7. Oval cross section, elastic-plastic boundaries for different values of the twist θ.

The warping function in the elastic region is found by substituting (48.38) into (48.5), and the result into (48.1), and integrating:

$$u^e = -yz\theta(a - b)/(a + b)$$

The warping function for the plastic region now follows from (48.27), with

$$\rho = (a - b) \sin \psi \cos \psi$$

48.5. CIRCULAR RING SECTORS AND NONUNIFORM CIRCULAR SHAFTS

Consider a stress system in cylindrical polar coordinates r, β, and x, in which the only nonvanishing stress components are the shearing stresses transmitted across meridional planes (see Fig. 48.8). These components, $\tau_{r\beta}$ and $\tau_{x\beta}$, are to be functions of r and x alone and to be independent of β. Such systems occur in the torsion of a circular ring sector [16]–[18] (torus), a problem important in the theory of closely

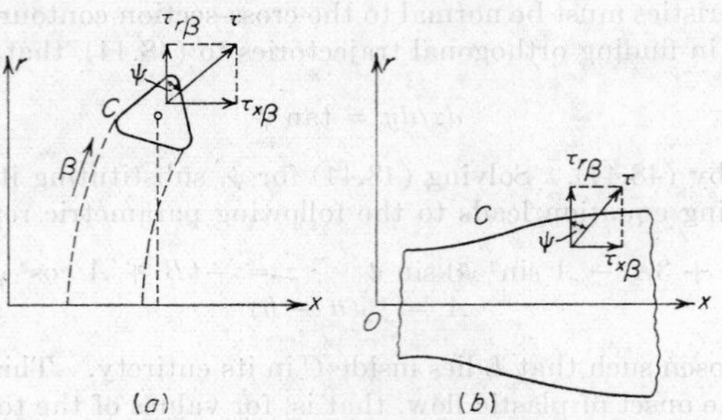

FIG. 48.8. Axisymmetric stress systems in (a) circular ring sector, (b) nonuniform bar.

coiled helical springs, and in the torsion of a shaft of circular cross section and variable diameter [10]. The equations of equilibrium reduce to

$$\frac{\partial}{\partial r}(r^2\tau_{r\beta}) + \frac{\partial}{\partial x}(r^2\tau_{x\beta}) = 0$$

that is,
$$\frac{\partial \tau_{r\beta}}{\partial r} + \frac{\partial \tau_{x\beta}}{\partial x} + 2\frac{\tau_{r\beta}}{r} = 0 \tag{48.51}$$

The plastic yield condition is

$$\tau_{r\beta}^2 + \tau_{\beta x}^2 = k^2(x,r) \tag{48.52}$$

A stress function $\psi(x,r)$ is introduced by the equations

$$\tau_{r\beta} = k(x,r)\cos\psi \qquad \tau_{x\beta} = k(x,r)\sin\psi \tag{48.53}$$

so that the yield condition (48.52) is satisfied identically, and (48.51) becomes

$$-\sin\psi\frac{\partial\psi}{\partial r} + \cos\psi\frac{\partial\psi}{\partial x} + \frac{1}{k}\left(\cos\psi\frac{\partial k}{\partial r} + \sin\psi\frac{\partial k}{\partial x}\right) = -\frac{2}{r}\cos\psi \tag{48.54}$$

Defining two perpendicular directions dc and ds by the identities

$$\frac{\partial}{\partial c} \equiv \cos\psi\frac{\partial}{\partial x} - \sin\psi\frac{\partial}{\partial r} \qquad \frac{\partial}{\partial s} \equiv \sin\psi\frac{\partial}{\partial x} + \cos\psi\frac{\partial}{\partial r} \tag{48.55}$$

Eq. (48.54) can be written as

$$\frac{\partial\psi}{\partial c} + \frac{1}{k}\frac{\partial k}{\partial s} = -\frac{2}{r}\cos\psi \tag{48.56}$$

Except for its right-hand side, this equation is of the same form as (48.13) for the straight bar, where $\partial/\partial c$ and $\partial/\partial s$ were defined by (48.14). The direction of dc is again the direction of the characteristics of (48.56) and (48.54); this direction forms an angle $-\psi$ with the x axis. For constant k, Eqs. (48.54) and (48.56) become, respectively,

$$-\sin\psi\frac{\partial\psi}{\partial r} + \cos\psi\frac{\partial\psi}{\partial x} = -\frac{2}{r}\cos\psi \qquad \frac{\partial\psi}{\partial c} = -\frac{2}{r}\cos\psi \tag{48.57),(48.58}$$

For the remainder of this section, the value of k will be assumed constant.

The equation of the characteristics of (48.57) and (48.58) can be written

$$\frac{dx}{\cos \psi} = \frac{dr}{\sin \psi} = \frac{r\,d\psi}{2 \cos \psi} \tag{48.59}$$

or

$$\frac{dr}{dc} = \sin \psi \qquad \frac{dx}{dc} = -\cos \psi \qquad \frac{d\psi}{dc} = \frac{2}{r} \cos \psi \tag{48.60a–c}$$

Integration of (48.60*a*,*c*) shows that

$$r^2 \cos \psi = \text{const.} \tag{48.61}$$

along each characteristic, the value of the constant being the parameter of the characteristic, determined by the boundary conditions on the stress-free surface. Moreover, since

$$dr/dx = -\tan \psi \qquad \tau_{r\beta}/\tau_{x\beta} = \cot \psi$$

the direction of the stress vector at any point is orthogonal to the direction of the characteristic through that point; the direction of ds in Eq. (48.58) is thus the direction of the stress vector. In particular, the characteristics are perpendicular to the stress-free boundary C. It should be noted that the r,x plane is, for the torus, the plane of a generic cross section of the torus, and for the nonuniform bar, a generic section through its axis. In neither case are the characteristics, in general, straight lines, even for constant k.

From Eq. (48.60*c*), it can be seen that the radius of curvature of a characteristic at any point is given by

$$\rho = \frac{dc}{d\psi} = \frac{r}{2 \cos \psi} \tag{48.62}$$

Since analytical solutions are cumbersome for other than simple boundaries, a graphical construction [16] of the characteristic field can be based on Eq. (48.62). The only boundaries which have been discussed analytically are (1) the circular cross section for the torus [17], and (2) the right circular cone and the stepped shaft for the nonuniform shaft [10].

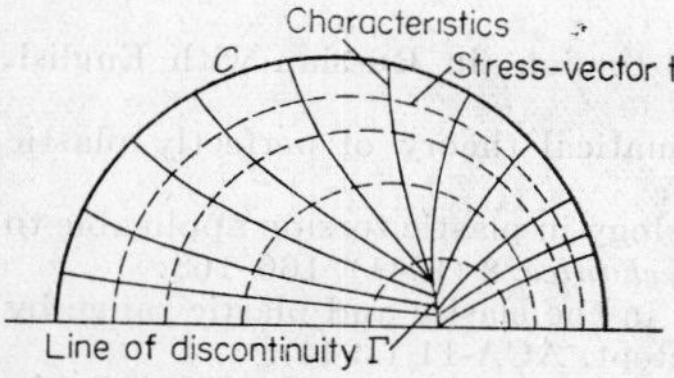

Fig. 48.9. Ring sector, circular cross section.

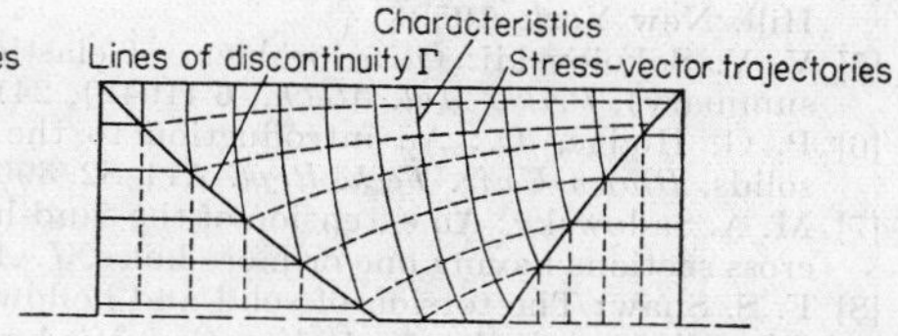

Fig. 48.10. Ring sector, rectangular cross section.

Figures 48.9 and 48.10 show the nature of the stress and characteristics fields for circular ring sectors of circular and rectangular cross sections under fully plastic torsion. It is seen that lines of stress discontinuity occur, in the rectangular case emanating from the corners, in the circular case lying inside the cross section.

For a nonuniform bar, the characteristics form an envelope, and therefore, for single-valued stress, the characteristics cannot be continued beyond that envelope. Consequently, the shaft will, in general, not become fully plastic, no matter how great a torque is applied. An example is shown in Fig. 48.11: For a right circular cone of half angle 15°, a region of $\pm 7°$ around the axis of the shaft will certainly never become plastic. It should be noted that the envelope is not the elastic-plastic boundary.

The elastic-plastic stress distributions for the problems under consideration have been determined for only few cases; analytically, an inverse method such as described for the elliptical cross section of a straight bar has been devised for a torus of circular cross section [17]. Relaxation methods have been used to determine the elastic-plastic stress distribution of stepped shafts [11].

No reference has been made in this section to the determination of the velocity pattern. It can be shown that the warping function for the torus problem satisfies a partial differential equation, which has the same characteristics as the stress equation (48.58); after determination of the stress distribution, the warping function can thus be found by integration along the characteristics from the elastic-plastic boundary or the lines of discontinuity, as the case may be. Similar considerations apply to bars of nonuniform cross section, but much work remains yet to be done in this field.

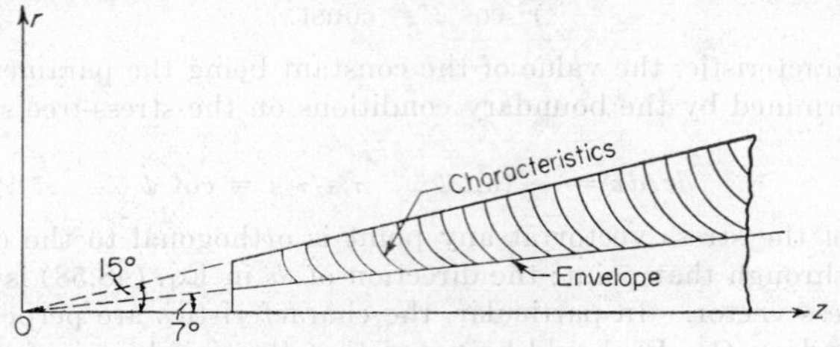

FIG. 48.11. Conical bar.

Multiply connected cross sections, that is, cross sections with holes, can be discussed by the methods developed in this chapter [18]. The only complicating consideration is the nonuniqueness of the displacement pattern; the problem of choice of displacement mode is discussed in [15].

48.6. REFERENCES

[1] W. Prager, P. G. Hodge, Jr.: "Theory of Perfectly Plastic Solids," chap. 3, Wiley, New York, 1951.

[2] R. Hill: "The Mathematical Theory of Plasticity," chap. 4, secs. 8 and 9, Oxford Univ. Press, New York, 1950.

[3] V. V. Sokolovskij: "Theorie der Plastizität" (German transl. of Russian 2d ed.), chap. 4, Verlag Technik, Berlin, 1955.

[4] A. Nadai: "Theory of Flow and Fracture of Solids," vol. 1, 2d ed., chap. 35, McGraw-Hill, New York, 1950.

[5] V. V. Sokolovskij: On a problem of elastic-plastic torsion (in Russian with English summary), *Prikl. Mat. Mekh.*, **6** (1942), 241–246.

[6] P. G. Hodge, Jr.: An introduction to the mathematical theory of perfectly plastic solids, *Brown Univ. Tech. Rept.* A11-S2/396 (1950).

[7] M. A. Sadowsky: An extension of the sand-heap analogy in plastic torsion applicable to cross sections having one or more holes, *J. Appl. Mechanics,* **8** (1941), 166–168.

[8] F. S. Shaw: The torsion of solid and hollow prisms in the elastic and plastic range by relaxation methods, *Australian Council Aeronaut.* Rept. ACA-11 (1944).

[9] J. Mandel: Sur les déformations de la torsion plastique, *Compt. rend. Acad. Sci. Paris,* **222** (1946), 1205–1207.

[10] V. V. Sokolovskij: Plastic torsion of a shaft of circular cross section and variable diameter (in Russian with English summary), *Prikl. Mat. Mekh.*, **1** (1945), 343–346.

[11] R. P. Eddy, F. S. Shaw: Numerical solution of elastoplastic torsion of a shaft of rotational symmetry, *J. Appl. Mechanics,* **16** (1949), 139–148.

[12] L. A. Galin: The elastic-plastic torsion of prismatic bars, *Prikl. Mat. Mekh.*, **13** (1949), 285–296.

[13] W. Prager: Discontinuous fields of plastic stress and flow, *Proc. 2d U.S. Natl. Congr. Appl. Mechanics*, pp. 21–32, Ann Arbor, Mich., 1954.

[14] R. Hill: Plastic torsion of anisotropic bars, *J. Mech. Phys. Solids,* **2** (1954), 87–91.

[15] R. Hill: On the problem of uniqueness in the theory of a rigid-plastic solid, II, *J. Mech. Phys. Solids,* **5** (1957), 153–161.

[16] A. J. Wang, W. Prager: Plastic twisting of a circular ring sector, *J. Mech. Phys. Solids,* **3** (1955), 169–175.

[17] W. Freiberger: Elastic-plastic torsion of circular ring sectors, *Quart. Appl. Math.*, **14** (1956), 259–265; also *Proc. 8th Intern. Congr. Theoret. Appl. Mechanics,* Istanbul, 1955.

[18] W. Freiberger, W. Prager: Plastic twisting of thick-walled circular ring sectors, *J. Appl. Mechanics,* **23** (1956), 461–463.

CHAPTER 49

LIMIT ANALYSIS

BY

P. S. SYMONDS, Ph.D., Providence, R.I.

49.1. INTRODUCTION

The load on a body or structure will be taken to mean a system of surface tractions or forces whose relative proportions are prescribed so that a single parameter specifies the load intensity. Throughout this section, the term *load magnitude* or simply *load* will have this sense unless special qualifications are attached.

The methods of limit analysis furnish estimates of a critical load called the *plastic collapse load.* This load is observed in tests on certain types of structures (notably mild steel beams and rigid-jointed frames) as the load at which the increment of deflection per unit load increment becomes very large. In some cases, the load hardly changes, while the deflections increase to magnitudes several times larger than the maximum deflections in the elastic range. This behavior is caused by the development of plastic flow in the structure, to such an extent that the remaining elastic material plays a relatively insignificant role in sustaining the load. It has been termed *uncontained plastic flow,* to distinguish it from *contained flow,* in which elastic action still plays a major role: here we will use the term *plastic collapse* or simply *collapse.* When this occurs, it may constitute a real failure, in the sense that the structure then can no longer fulfill its intended function. This would be the case for most building structures, for example. Then the plastic collapse load can be used as a realistic basis for design; an efficiently designed structure will be proportioned so that the working load would have to be increased by a specified factor (the *factor of safety*) in order to produce failure. Quite different types of failure might, of course, govern the design as, for example, fatigue, brittle fracture, or buckling; in some cases, the magnitude of deflection (elastic or plastic) is itself the criterion, rather than the imminence of large plastic deflections. Failure by plastic collapse is, however, the governing condition in so many structures that the development of efficient methods for computing the collapse load, or for the direct design of a structure on the basis of this type of failure, has in recent years been of intense practical interest to engineers.

It is essential to distinguish the practical phenomenon of plastic collapse, as it occurs in real structures or bodies, from the special meaning of collapse which will be used for purposes of mathematical analysis. We will use, for mathematical purposes, the concept of plastic collapse of an ideal structure, namely, the condition in which *deflections can increase without limit while the load is held constant.* This rarely (if ever) happens in real bodies or structures, and, hence, the calculation applies strictly, not to the real structure, but to a hypothetical one, in which neither work hardening nor significant shape changes occur. Nevertheless a load computed on the basis of this definition, termed the *limit load,* may give a good approximation to the physical plastic collapse load. Figure 49.1*a* shows a load-deflection curve as it might be measured for a certain frame structure. The load P_c is the limit load. The curve for the real structure has an elastic portion, a region of transition from mainly elastic to mainly plastic behavior; a plastic region, in which the load increases very little while the deflection increases manyfold; and, finally, a region in which either work hardening of the metal or the changes in geometry of the structure (or both) effectively stiffen the structure. In a case such as this (which is actually quite typical of many published test results on beams, continuous frames, and some other bodies), the calculated limit load is a good approximation of the physical collapse load, and has obvious practical importance. Figure 49.1*b*,*c* shows instances in which the structure actually has no plastic collapse load. The curve in Fig. 49.1*b* is the sort that is observed either when the metal has a high degree of work hardening or when there are shape changes that enable the structure to continue to carry load increments, even though the metal is highly ductile and a large amount of plastic flow occurs. Ultimately, there might be a fracture or some other sort of unpleasant effect. The curve in Fig. 49.1*c* is typical of cases in which geometry changes are harmful instead of helpful; because of them, after the load reaches a maximum value, it can then sharply decrease while the structure deflects by large amounts. This might be a case of failure by buckling, and might be a mainly elastic or a mainly plastic phenomenon. In either of these cases, (*b*) or (*c*), the limit load could be computed, but would probably have little physical significance or practical interest.

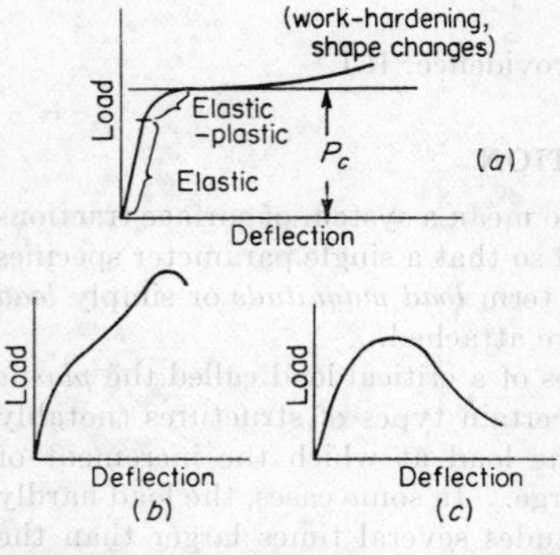

FIG. 49.1 Load-deflection curves: (*a*) shows typical plastic collapse phenomenon.

As has been mentioned, the methods of limit analysis adopt a definition of collapse which is artificial, but which turns out to be enormously fruitful. This is so, not only because the computed load is, in many cases, a good approximation to a physical failure load, but also because the basic theorems of limit analysis are conceptually simple, and often enable useful results to be achieved with remarkably simple and quick calculations. In some cases, in fact, the methods of limit analysis have furnished results which could only have been reached with the greatest difficulty, if at all, by the previously available methods, because of the severe mathematical difficulties that arise.

The general subject of limit analysis can be divided into two parts. One part is concerned with general three-dimensional bodies, involves distributions of stresses, strain rates, and velocities, and is based on such properties as the yield function and the stress-strain rate relations (plastic potential). The other part has to do with *engineering structures*, such as trusses, beams, rigid-jointed frames, arches, grillages, plates, and shells. Here we deal with stress resultants (for example, bending moment or normal force on a beam section), and with corresponding generalized strains and displacements (such as curvature, centroidal strain, and center-line displacement). A yield function and generalized stress-strain rate relations special to each type of problem are used, involving the quantities that appear relevant.

The methods of limit analysis for continuous beams and frames have arrived at a stage of maturity far exceeding that of methods for other types of structures. The basic physical concepts for built-in beams were stated and applied in 1914 by Kazinczy in Hungary, and this was apparently the beginning of the whole subject. In the subsequent decades, the basic concepts and methods for redundant beams became familiar to many engineers in Holland, Germany, and elsewhere on the European continent. Since 1940, there has been a remarkable acceleration of this development. Analytical and experimental investigations have led to practical design methods. This development has been important not only in itself but because it stimulated the development of general theorems and methods for three-dimensional bodies of arbitrary shape, and for more general types of engineering structures. Excellent treatises [3],[4] on the limit analysis of beams and frames are now available. These give details of experimental studies, design aspects, etc., as well as of the relevant theory. These should be consulted also for historical sketches.

The basic theorems of limit analysis concern lower and upper bounds on the limit load, and hence provide a value which, in general, is not exact but approximate, with known limits of error. In view of the uncertainties inherent in all engineering problems, and the essential role of judgment in their solution, it is clear that this is no basic handicap. The real difficulties are those of relating the limit load to the behavior of the physical structure, which always exhibits some degree of work hardening, and which may be sensitive to shape changes. These difficulties will be illustrated.

Although beams and frames are the most important field of application of limit analysis, it is appropriate here to begin with a discussion of the theorems in general form for three-dimensional bodies, rather than the special forms suitable for beams and frames. Although the theorems for these and other special classes of structures can be established independently of the general theorems, they require assumptions and approximations beyond those required for the general theorems. Their validity can best be understood against the background of the general theorems. Apart from this, the potential range of usefulness of the general theorems holds promise of being enormous, since they apply to bodies of arbitrary shape, and not only to metallic structures but also to any other materials (such as soils) whose behavior can, at least approximately, be described in terms of the properties postulated in the following section.

49.2. BASIC CONCEPTS

We will use the term limit load to mean the load at which plastic deformations of arbitrarily large magnitude would take place under constant load if the body or structure possessed the following properties:

1. The material exhibits *perfect* or *ideal* plasticity; i.e., work hardening does not occur. Figure 49.2 shows the stress-strain diagram of a perfectly plastic material.

2. Changes in geometry of the body or structure that occur at the limit load are insignificant; hence, the geometrical description of the structure remains unchanged during deformation at the limit load.

In other words, the limit load is defined as the plastic collapse load of a hypothetical body or structure replacing the actual one, and having the ideal properties listed above.

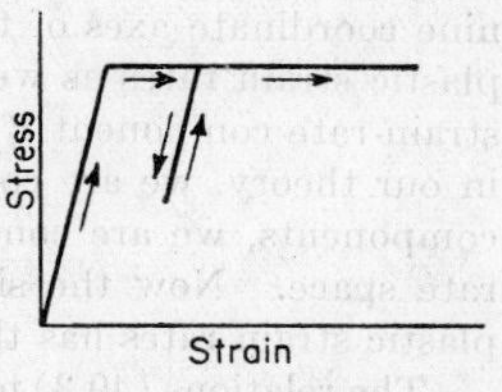

FIG. 49.2. Stress-strain diagram for elastic-ideally plastic material.

We give now a summary of the basic properties of plasticity required in limit analysis of general bodies [1],[2],[5].

Each element of the body is assumed to be governed by a yield function f. For a perfectly plastic (ideally plastic) material, f depends on the set of stress components[1] σ_{ij}, but not on the strain components[1] ϵ_{ij}, although f may be taken as varying from point to point, either because of prior plastic straining or because the material was "born" inhomogeneous. It is a positive-definite function, and may be defined so that plastic flow can occur only when the yield condition

[1] Here the standard tensor suffix notation for stress and strain components is used: i, j take any of the values 1, 2, or 3 corresponding to cartesian axes x_1, x_2, and x_3. Quantities such as σ_{12}, σ_{21} are treated as formally independent variables. In the applications, the usual engineering notation for stress and strain components will be used: σ_x, $\tau_{xy} = \tau_{yx}$, σ_y, etc.; ϵ_x, γ_{xy}, etc., where $\gamma_{xy} = 2\epsilon_{xy}$.

is satisfied,

$$f(\sigma_{ij}) = k^2 \tag{49.1}$$

where k^2 is a constant for a given point. Stress states for which $f > k^2$ are excluded, and $f < k^2$ corresponds to elastic behavior.

It is helpful to visualize a state of stress as a point in a nine-dimensional Euclidean hyperspace, whose coordinate axes are the stress components σ_{ij}. The yield condition (49.1) then defines a surface in this space, called the *yield surface.* We will assume that the yield surface is *convex.* This is discussed later. A state of stress can also be pictured as a vector whose components are the nine σ_{ij}. It will usually be obvious from the context whether σ_{ij} means a stress component or a full set of components; occasionally (σ_{ij}) will be used to denote a stress-state point or vector.

The theorems of limit analysis are also concerned with strain rates $\dot{\epsilon}_{ij}$, generally composed of elastic and plastic parts, $\dot{\epsilon}_{ij} = \dot{\epsilon}^e_{ij} + \dot{\epsilon}^p_{ij}$. The $\dot{\epsilon}^e_{ij}$ are related to the $\dot{\sigma}_{ij}$ through Hooke's law. The $\dot{\epsilon}^p_{ij}$ depend on the state of stress through appropriate *flow rules.* For most metals, the condition of incompressibility

$$\dot{\epsilon}^p_{11} + \dot{\epsilon}^p_{22} + \dot{\epsilon}^p_{33} = 0 \tag{49.2}$$

is very closely obeyed, but this assumption is not required for our purposes. We do assume that the $\dot{\epsilon}^p_{ij}$ are derivable from a *plastic potential* or *loading function,* which coincides with the yield function f, so that the flow rules are

$$\dot{\epsilon}^p_{ij} = \lambda \frac{\partial f}{\partial \sigma_{ij}} \tag{49.3}$$

where $\lambda > 0$ is a scalar proportionality factor; the relations (49.3) determine only the ratios of the strain-rate components, not their magnitudes. This is so, despite our use of strain rates, because viscosity and inertial effects are actually ignored; the time scale is entirely arbitrary. In limit analysis, as in the theory of perfectly plastic materials generally, we are really concerned with small increments of strain. We often deal with strain rates instead of infinitesimal increments as a matter of convenience, but time in fact never enters explicitly.

The geometrical significance of Eq. (49.3) is of particular interest. Suppose the nine coordinate axes of the stress space already referred to represent simultaneously plastic strain rates as well as stresses, each axis σ_{ij} being an axis of the corresponding strain-rate component $\dot{\epsilon}^p_{ij}$. Thus, a point specifies a plastic strain-rate state. Since, in our theory, we are concerned only with the ratios between the plastic strain-rate components, we are concerned with directions, i.e., with rays, in the plastic strain-rate space. Now the significance of Eq. (49.3) is clear: The ray representing the plastic strain rates has the direction of the outward normal to the yield surface.

The relations (49.3) must be supplemented if the yield function is such that derivatives do not exist at all points. The yield surface may have corners or vertices where there is not a continuously turning tangent plane. (The Tresca yield condition requires such corners.) For such cases, it is sufficient to add to Eq. (49.3) the statement that, when the stress-state point lies at a corner or vertex of the yield surface, the plastic strain-rate vector may have any direction within the fan or cone defined by the normals of the contiguous surfaces. This will be exemplified for the Tresca yield condition. The lack of a unique plastic strain rate is found to be no handicap. Figure 49.3 shows the flow relations pictorially. In (a), the yield surface has a continuously turning tangent plane, and the ray representing the plastic strain rate is parallel to the normal to the yield surface at a point σ_{ij}. In (b), there is a corner at A, and the plastic strain-rate ray can have any direction between the normals n_1 and n_2 defined by the adjacent surfaces. Figure 49.3a also illustrates the convexity property. Suppose that the stress state σ_{ij} is a point on the yield surface, while σ''_{ij} is any other point on or within the yield surface. The vectors representing these stress states are shown, together with the vector $(\sigma_{ij} - \sigma''_{ij})$. It is seen that, for a convex yield surface, the scalar product of the vector $(\sigma_{ij} - \sigma''_{ij})$ with the normal vector $\partial f/\partial\sigma_{ij}$ cannot be negative; hence, the convexity property together with the flow rule (49.3) requires[1]

$$(\sigma_{ij} - \sigma''_{ij})\dot{\epsilon}^p_{ij} \geq 0 \qquad (49.4)$$

If the state σ''_{ij} is inside the yield surface, $f(\sigma''_{ij}) < k^2$, then the inequality sign holds in Eq. (49.4).

The relations given above [Eq. (49.4)] express a principle of maximum local energy dissipation due to von Mises: The rate at which plastic work is done on a given plastic strain-rate system has a maximum value for the actual stress state. This is a stationary maximum if the yield surface has a uniquely defined normal everywhere.

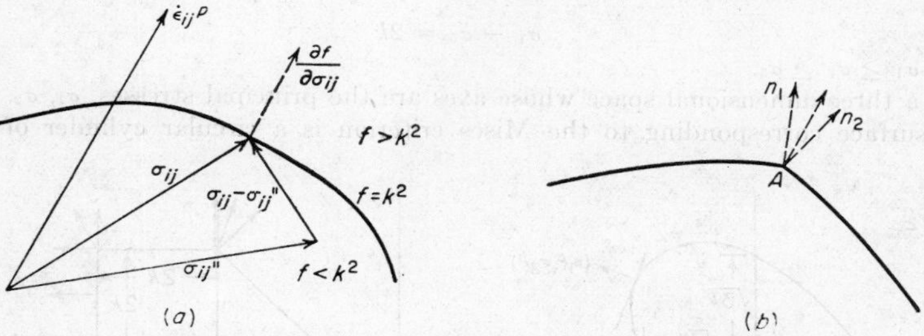

FIG. 49.3. Pictorial representation of yield surface and flow rules: (b) shows yield surface with corner at A.

A continuous field of velocity in a body is related to strain rates through the relation

$$2\dot{\epsilon}_{ij} = \frac{\partial v_i}{\partial x_j} + \frac{\partial v_j}{\partial x_i} \qquad (49.5)$$

where v_i ($\equiv v_1,v_2,v_3$) is the velocity, and x_i the position vector of an element. If T_i, F_i are surface tractions and body forces, respectively, then, when a state of stress σ_{ij} is in equilibrium with these, it satisfies the equations

At surface points $\qquad T_i = n_j\sigma_{ji}$

At interior points $\qquad F_i + \dfrac{\partial\sigma_{ji}}{\partial x_j} = 0 \qquad \sigma_{ji} = \sigma_{ij} \qquad (49.6a,b)$

where n_i is the outward-drawn unit normal vector to a surface element.

In proving the theorems and applying them, an invaluable tool is the theorem of virtual work. For future reference, we give here a suitable form of this theorem. Let σ'_{ij} stand for stresses in equilibrium with surface tractions T'_i and body forces F'_i, Eqs. (49.6) being obeyed. A system of continuous velocities v^*_i and strain rates $\dot{\epsilon}^*_{ij}$ obeys the relations (49.5), but need not be related in any way to the stress system σ'_{ij}. If there is a surface of discontinuity S_D across which, for example, the velocity component tangential to the surface changes abruptly by Δv^*, the shear stress τ' parallel to this surface is assumed continuous. The theorem states that

$$\int T'_i v^*_i \, dS + \int F'_i v^*_i \, dV = \int \sigma'_{ij}\dot{\epsilon}^*_{ij} \, dV + \int \tau' \, \Delta v^* \, dS_D \qquad (49.7)$$

[1] Here the summation convention for repeated Latin subscripts is used (see p. 7-2).

where dS, dV, dS_D are volume and surface elements, as usual. [This is easily derived by an application of the divergence theorem, making use of Eqs. (49.5) and (49.6).]

The two most important yield conditions for isotropic metals are those of von Mises and of Tresca. These are as follows:

von Mises

$$(\sigma_1 - \sigma_2)^2 + (\sigma_2 - \sigma_3)^2 + (\sigma_3 - \sigma_1)^2 = 6k^2$$

or $$(\sigma_x - \sigma_y)^2 + (\sigma_y - \sigma_z)^2 + (\sigma_z - \sigma_x)^2 + 6\tau_{xy}^2 + 6\tau_{yz}^2 + 6\tau_{zx}^2 = 6k^2 \qquad (49.8a,b)$$

where σ_1, σ_2, σ_3, are principal stresses, and k is the yield stress in pure shear. The more fundamental form is

$$f = J_2(\sigma_{ij}) \equiv \tfrac{1}{2} s_{ij} s_{ij} = k^2 \qquad (49.8c)$$

where the s_{ij} are the *stress-deviation* components $s_{ij} \equiv \sigma_{ij} - \tfrac{1}{3}\,\delta_{ij}\sigma_{pp}$, and J_2 is the *second invariant* of the stress-deviation tensor (see p. 46–7).

Tresca (Maximum Shear Stress)

$$\sigma_1 - \sigma_3 = 2k \qquad (49.9)$$

where $\sigma_1 \geq \sigma_2 \geq \sigma_3$.

In a three-dimensional space whose axes are the principal stresses, σ_1, σ_2, σ_3, the yield surface corresponding to the Mises criterion is a circular cylinder of radius

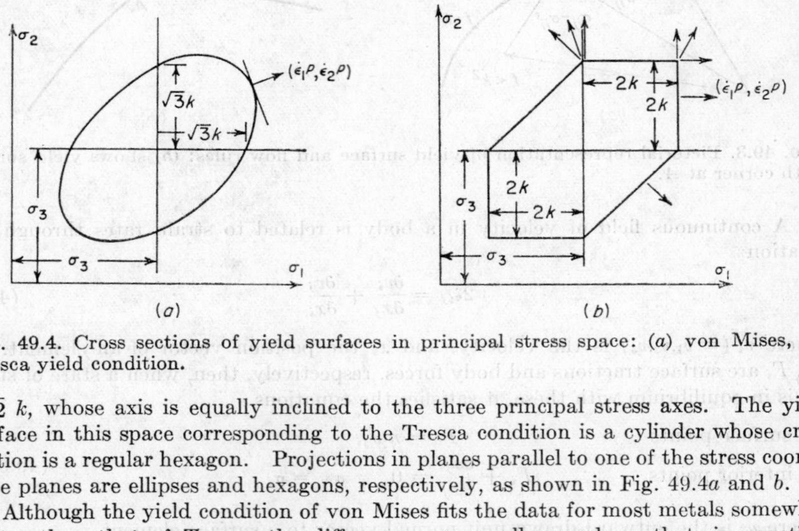

FIG. 49.4. Cross sections of yield surfaces in principal stress space: (*a*) von Mises, (*b*) Tresca yield condition.

$\sqrt{2}\ k$, whose axis is equally inclined to the three principal stress axes. The yield surface in this space corresponding to the Tresca condition is a cylinder whose cross section is a regular hexagon. Projections in planes parallel to one of the stress coordinate planes are ellipses and hexagons, respectively, as shown in Fig. 49.4*a* and *b*.

Although the yield condition of von Mises fits the data for most metals somewhat better than that of Tresca, the differences are small enough so that, in practical problems, the choice is a matter of convenience. Hill [6] showed that the difference between the limit-load predictions of the two yield conditions can be kept less than 8% by suitable choice of k; often it is much less than this.

The Mises and Tresca conditions apply to isotropic metals, for which, to a close approximation, the plastic strain increments obey the incompressibility condition (49.2), and the mean normal stress, $\tfrac{1}{3}(\sigma_1 + \sigma_2 + \sigma_3) \equiv \tfrac{1}{3}J_1(\sigma_{ij})$, is without influence on the plastic flow. As an example of a yield condition intended for different materials, we note the function used by Drucker and Prager [7] in applying the limit theorems to perfectly plastic soils, namely,

$$f = \alpha J_1 + J_2^{1/2} = k \qquad (49.10)$$

where α and k are positive constants of the material. This is a generalization of Coulomb's law for soils, according to which the yield stress in shear depends linearly on the mean normal stress. Because of the presence of J_1 in Eq. (49.10), the plastic dilatation rate is not zero.

49.3. THEOREMS OF LIMIT ANALYSIS

In general, the load system consists of body forces F_i (per unit volume) and surface tractions T_i (per unit area). Each component T_i is specified, except when the corresponding component of displacement is prescribed to be zero. For simplicity, we will let P designate the magnitude of a given load system (T_i, F_i). (P might be a particular force or surface stress, for example.)

There are three fundamental theorems governing the limit load, as follows [8]:

Theorem 1. All stresses remain constant while plastic deformations take place at the limit load; only plastic (not elastic) increments of strain occur.

PROOF. In general, if arbitrary small increments of the applied loads are applied, there are increments of stress and of strain. Expressing increments as rates $\dot{F}_i$, $\dot{T}_i$, etc., the equation of virtual work of the load rates is

$$\int \dot{F}_i v_i \, dV + \int \dot{T}_i v_i \, dS = \int \dot{\sigma}_{ij} \dot{\epsilon}_{ij} \, dV \tag{49.11}$$

if the velocities v_i are continuous. Now, at the limit load, the left-hand side of Eq. (49.11) vanishes, by our definition; $\dot{F}_i = 0$ everywhere, and $\dot{T}_i = 0$ wherever $v_i \neq 0$. Therefore,

$$\int \dot{\sigma}_{ij} (\dot{\epsilon}^e_{ij} + \dot{\epsilon}^p_{ij}) \, dV = 0$$

But it follows from the flow rules (49.3) that $\dot{\sigma}_{ij} \dot{\epsilon}^p_{ij} = 0$, since the vector $\dot{\sigma}_{ij}$ is tangential to the yield surface wherever plastic strains are occurring. Therefore,

$$\int \dot{\sigma}_{ij} \dot{\epsilon}^e_{ij} \, dV = 0 \tag{49.12}$$

A well-known theorem for linear elastic or stable nonlinear elastic materials states that the work (per unit volume) done by any system of stresses during the resulting elastic deformation is a positive-definite quantity (see [5]). Hence the vanishing of the integral in Eq. (49.11) requires that $\dot{\sigma}_{ij} = 0$ throughout the body.

Theorem 2. If a system of stresses σ'_{ij} can be found which satisfies the equations of equilibrium with a load P', that is, with body forces F'_i at all interior points and with surface tractions T'_i at all surface points where these are specified; and which nowhere violates the yield condition (49.1); then the load P' cannot exceed the limit load P_c.

PROOF. The limit load P_c comprises surface tractions T^c_i and body forces F^c_i. Suppose that $P' > P_c$; we show that this leads to a contradiction. Let σ^c_{ij}, v^c_i, $\dot{\epsilon}^{pc}_{ij}$ denote actual stresses, velocities, and strain rates at collapse. The load at collapse T^c_i, F^c_i will be in equilibrium with stresses $\alpha\sigma'_{ij}$, where α is a factor such that $\alpha P' = P_c$; by supposition, $\alpha < 1$. By virtual work,

$$\int T^c_i v^c_i \, dS + \int F^c_i v^c_i \, dV = \int \sigma^c_{ij} \dot{\epsilon}^{pc}_{ij} \, dV = \int \alpha \sigma'_{ij} \dot{\epsilon}^{pc}_{ij} \, dV$$

Hence,

$$\int (\sigma^c_{ij} \dot{\epsilon}^{pc}_{ij} - \alpha \sigma'_{ij} \dot{\epsilon}^{pc}_{ij}) \, dV = 0 \tag{49.13}$$

But if $\alpha < 1$, then $\sigma^c_{ij} \dot{\epsilon}^{pc}_{ij} > \alpha \sigma'_{ij} \dot{\epsilon}^{pc}_{ij}$, since $f(\sigma'_{ij}) \leq k^2$, in view of the convexity property, Eq. (49.14); see Fig. 49.3a. A sum of positive terms cannot vanish. Hence, the supposition $P' > P_c$ is false, and the theorem is proved: $P' \leq P_c$.

Theorem 3. Consider a system of velocities v^*_i vanishing at surface points wherever the corresponding traction is not prescribed, together with strain rates $\dot{\epsilon}^*_{ij} = \dot{\epsilon}^{p*}_{ij}$, regarded as wholly plastic or zero, and satisfying the compatibility relations Eq. (49.5). Let the load $P^* (\equiv T_i^*, F_i^*)$ be computed from the virtual-work equation

$$\int T_i^* v_i^* \, dS + \int F_i^* v_i^* \, dV = \int D(\dot{\epsilon}_{ij}^{p*}) \, dV \qquad (49.14a)$$

where $D(\dot{\epsilon}_{ij}^{p*})$ is the rate of dissipation of energy per unit volume for the plastic strain rates $\dot{\epsilon}_{ij}^{p*}$,

$$D(\dot{\epsilon}_{ij}^{p*}) \equiv \sigma_{ij}^* \dot{\epsilon}_{ij}^{p*} \qquad (49.14b)$$

and σ_{ij}^* are defined wherever $\dot{\epsilon}_{ij}^{p*} \neq 0$ as obeying the yield condition $f(\sigma_{ij}^*) = k^2$, and are related to the $\dot{\epsilon}_{ij}^{p*}$ through the flow rules (49.3). Then P^* cannot be less than the limit load P_c.

PROOF. Suppose that $P^* < P_c$; we show that this leads to a contradiction. As before, σ_{ij}^c, v_i^c, $\dot{\epsilon}_{ij}^{pc}$ denote actual collapse systems. Let $mP^* = P_c$, where $m > 1$ by supposition. Then T_i^*, F_i^* are in equilibrium with $(1/m)\sigma_{ij}^c$. By virtual work,

$$\int T_i^* v_i^* \, dS + \int F_i^* v_i^* \, dV = \int \frac{1}{m} \sigma_{ij}^c \dot{\epsilon}_{ij}^{p*} \, dV$$

Hence,

$$\int \left[D(\dot{\epsilon}_{ij}^{p*}) - \frac{1}{m} \sigma_{ij}^c \dot{\epsilon}_{ij}^{p*} \right] dV = 0 \qquad (49.15)$$

But, if $m > 1$, $D(\dot{\epsilon}_{ij}^{p*}) \equiv \sigma_{ij}^* \dot{\epsilon}_{ij}^{p*} > \frac{1}{m} \sigma_{ij}^c \dot{\epsilon}_{ij}^{p*}$, by the convexity property [Eq. (49.4)]. (In Fig. 49.3a, the stress state σ_{ij} can be identified as σ_{ij}^*, and σ_{ij}'' as σ_{ij}^c). Again, a sum of positive terms is required to vanish by our supposition, which must, therefore, be false; thus, the theorem $P^* \geq P_c$ is proved.

Expressions for the rate of energy dissipation D per unit volume, needed in applying Theorem 3, will now be given for the yield conditions of von Mises [Eqs. (49.8)], and of Tresca [Eq. (49.9)], when the strain rates are continuous. For plastic strain rates $\dot{\epsilon}_{ij}^p$, we wish to compute

$$D = \sigma_{ij} \dot{\epsilon}_{ij}^p$$

where

$$f(\sigma_{ij}) = k^2$$

If the Mises yield condition is used, we have uniquely defined plastic strain rates obeying the flow rule (49.3) so that, by Euler's theorem for homogeneous functions,

$$D = \lambda \sigma_{ij} \frac{\partial f}{\partial \sigma_{ij}} = 2\lambda f = 2\lambda k^2$$

since $f = \frac{1}{2} s_{ij} s_{ij}$ is homogeneous of second degree in the stress components. For this yield condition, $\dot{\epsilon}_{ij}^p = \lambda s_{ij}$, and, hence, $2\lambda k = \sqrt{2\dot{\epsilon}_{ij}^p \dot{\epsilon}_{ij}^p}$. Therefore, the required formula is

$$D(\dot{\epsilon}_{ij}^p) = k \sqrt{2\dot{\epsilon}_{ij}^p \dot{\epsilon}_{ij}^p} \qquad (49.16)$$

If the Tresca yield condition is used, it can easily be shown [19], by considering typical points on the yield surface (Fig. 49.4b), together with the extended form of the flow rule (49.3) and the incompressibility condition (49.2), that the appropriate formula is

$$D(\dot{\epsilon}_{ij}^p) = 2k \max |\dot{\epsilon}^p| \qquad (49.17)$$

where max $|\dot{\epsilon}^p|$ denotes the absolute value of the numerically largest component of the plastic strain rate.

49.4. REMARKS ON THE THEOREMS

The second and third theorems are the basis for the computation of lower bounds and upper bounds, respectively, on the limit load P_c of a body. To calculate a lower bound, we must find a stress system that satisfies equilibrium and yield conditions throughout the body; no attention need be paid to velocities or strains. To calculate an upper bound, we must consider systems of velocities and compatible plastic strains; no attention need be paid to stresses outside the regions where plastic strains occur. The actual limit load has the property that both sets of conditions are satisfied throughout the body; only at the limit load is there a stress system satisfying equilibrium and yield conditions, and, at the same time, related to a system of velocities and compatible plastic strains through the flow rule. Clearly, the limit load P_c is unique.

The first theorem is essential in making clear the role of elastic deformations. It shows that elastic characteristics play no part in the collapse at the limit load *of the ideal structure* (no work hardening or geometry changes during collapse). The geometry of the structure is usually taken as the initial geometry in the unloaded state. However, this is not essential; the geometrical description could be that reached after elastic or plastic deformations have occurred.

Hill [9] has emphasized that the limit load is the yield-point load of a body composed of plastic-rigid material, i.e., a material whose elastic moduli are infinitely large. Then the yield-point load has precise significance as the load at which local distortion first occurs under a given loading program; work hardening is immaterial. However, it is correct to identify this load with the limit load, as defined in the present section, only when the geometry used in the computation of the limit load is taken as the initial geometry of the unloaded and undeformed structure. When this is done, Theorems 2 and 3 become formally identical with theorems governing the yield-point load of a plastic-rigid body, which may be derived from Hill's principle of maximum work and his complementary minimum principle [9].

Although Hill's viewpoint affords a simpler mathematical concept, the advantage of the viewpoint of the theorems as stated above is that, in them, we explicitly face the questions of elastic deformations and of geometry changes, elastic and plastic. This viewpoint does not enable us always to overcome the difficulties involved in taking these effects into account. On the other hand, nothing is gained by, so to speak, sweeping the difficulties under the rug, as in the plastic-rigid point of view. According to both viewpoints, the calculation is made for an ideal body replacing the real one; both viewpoints leave unanswered the vital question of how closely, if at all, the calculated load corresponds to a collapse load of a physical structure. The writer prefers the viewpoint presented here because he thinks it appears more natural to structural engineers, who are, of course, well accustomed to regarding the geometry of their structures as unchanging for specific purposes of calculation.

Some remarks concerning the generality of the theorems are in order.

Regarding the loading, it has been emphasized that all components of a load system at the limit load have prescribed ratios to one another. However, the actual loading program need not be one in which all load components have constant ratios, as their magnitudes are gradually increased from zero to the limit load. On the contrary, only the load proportions at the limit load are prescribed. Any program of loading could be followed that brings the components up to their final proportions and magnitudes at the limit load, provided only that the limit load corresponding to some other proportions is not exceeded on the way.

Clearly, also, prior plastic straining and resulting residual stresses have no effect on the limit load, except as they may influence the yield function, which is regarded as known. The same remark applies to thermal strains and stresses.

It has already been mentioned that discontinuous fields of stress and velocity may be used in applying the theorems. The proofs given above used continuous systems, satisfying Eqs. (49.5) and (49.6*b*) only for brevity. Discontinuous stress fields are actually very useful in deriving lower bounds. Surfaces of stress discontinuity are clearly possible, provided the equilibrium equations of the type of Eqs.

(49.6*a*) are satisfied at all points of these surfaces. Surfaces of velocity discontinuity can also be admitted, provided the energy dissipation is properly computed. These should be regarded as the limiting case of continuous velocity fields, in which one or more velocity components change very rapidly across a thin transition layer, which is replaced by a discontinuity surface as a matter of convenience. The proofs of the limit theorems are easily carried through. The introduction of discontinuous stress fields satisfying equilibrium conditions does not affect the lower-bound Theorem 2. When surfaces of velocity discontinuity are considered in Theorem 3, an appropriate form of the virtual-work theorem, such as that of Eq. (49.7), is used in the proof.

The simplest velocity discontinuity surface is one across which the tangential velocity changes, say by Δv. Since this is regarded as the limiting case of a rapidly varying velocity with large shear-strain rate, the tangential stress at the discontinuity surface is the yield stress in pure shear k. The rate of energy dissipation D per unit area of discontinuity surface is computed from

$$D = k\,\Delta v \tag{49.18}$$

The total energy dissipated in this way is added to that computed from the continuous velocity distribution. Hill [10] has considered a more general type of discontinuity, which can occur, for example, in a thin sheet loaded in tension, where there may be a discontinuity in the normal as well as the tangential component of velocity. Expressions for the rate of dissipation of energy are given by Hill. Finally, we note that a yield function for a perfectly plastic soil, of the type of Eq. (49.10), always requires a volume expansion accompanying a shear deformation. In plane strain, for example, shear in a thin strip must be accompanied by a proportional transverse expansion. These are conveniently treated as discontinuities, as discussed by Drucker and Prager [7].

The most important applications of the theorems are in the numerical calculation of limit loads. Examples of this will be given shortly. They also have clarified the meaning of certain types of solutions, which had been presented previously in the literature of theory of plasticity [11]–[13].

49.5. APPLICATIONS OF THE GENERAL THEOREMS

The examples that follow illustrate the application of the theorems without requiring knowledge of theory of plasticity beyond that outlined above. By the nature of the theorems, familiarity with such techniques of mathematical plasticity as the methods for constructing slip-line fields and velocity fields in plane strain [1],[2],[15] are of great help in achieving accurate results. Here we attempt mainly to show what can be done by applying the ideas of the theorems aided only by a knowledge of ordinary mechanics of materials. Sometimes remarkably good results can be obtained by very elementary methods.

As a first example, consider a long prismatic bar of rectangular cross section, which has one or more cylindrical holes with axis perpendicular to one pair of faces, and is subjected to a force P parallel to the edges. We adopt the Tresca yield condition, so that the yield stress in simple tension $\sigma_0 = 2k$. If the bar has one hole, the limit theorems show that the limit load is $\sigma_0 A'$, where A' is the net area of the cross section, i.e., the cross-sectional area of the bar less the maximum area of the hole in a transverse plane. This is true for a hole of any shape located anywhere in the bar (the line of action of P not being specified), and even holds for any cross section of the bar, not merely a rectangular one. Consider, for simplicity, the rectangular bar with central circular hole. Figure 49.5*a* shows a stress field consisting of three longitudinal strips, two of which have simple tension σ_0, while the one containing the hole has no stress. A lower bound is $P' = \sigma_0(b - d)t = \sigma_0 A'$, t being the thickness.

Figure 49.5b shows a velocity field in which the upper and lower parts of the bar move as rigid bodies relative to each other by sliding along the planes AB and CD, perpendicular to the face of the bar and making the angle α as shown. If the velocity of separation is v, the relative tangential velocity at the plane of sliding is $v/\sin \alpha$. The rate of energy dissipation over the whole sliding surface is $kv(b - d)t/\sin \alpha \cos \alpha$. An upper bound is obtained by setting this equal to the rate at which external work is done, P^*v, and, hence, $P^* = 2k(b - d)t/\sin 2\alpha$. The least upper bound is $\sigma_0(b - d)t$, by the Tresca condition, and, since this is identical with the lower bound just obtained, it is the limit load P_c exactly. Clearly, the result $P_c = \sigma_0 A'$ can be shown to apply to any shapes of bar cross section; the hole could be rectangular, for example.

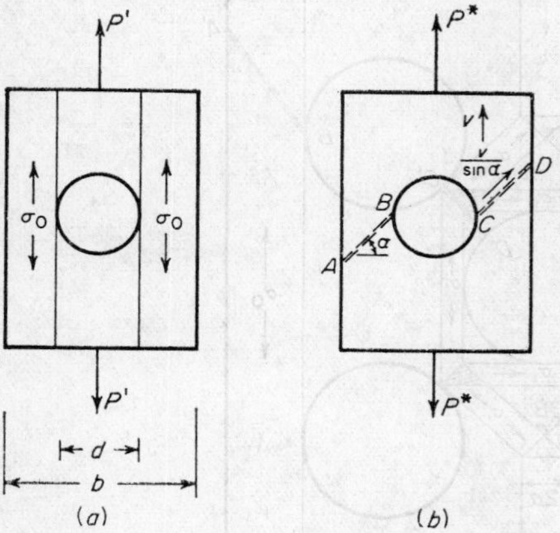

FIG. 49.5. Bar with hole: (a) Stress system for lower bound; (b) velocities for upper bound; sliding occurs along $ABCD$.

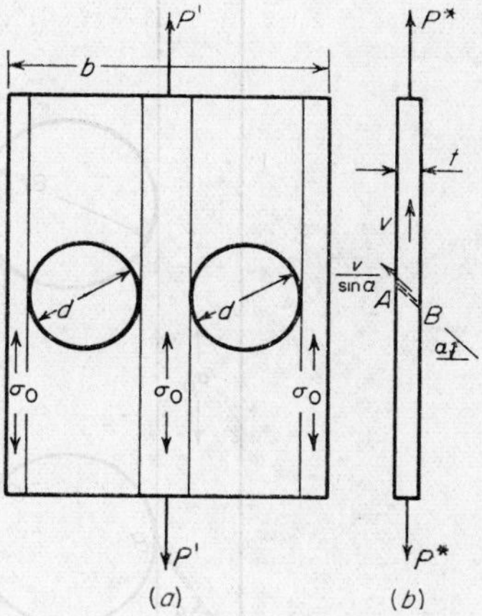

FIG. 49.6. Thin plate with two holes: (a) Stress system; (b) velocities; sliding occurs along AB out of plane of plate.

No such general exact solution is immediately obtainable if the bar has several holes. However, if the bar is a thin plate, with thickness t small compared to the hole dimensions, good solutions may sometimes be quickly found. Consider a plate with two holes with centers on a transverse line, as shown in Fig. 49.6a. Obviously a lower bound P' is $\sigma_0 A'$. To obtain an upper bound, we use a velocity field with sliding out of the plane of the plate, along a plane whose trace is AB in the edge view (Fig. 49.6b). If the rate of separation of the parts above and below the sliding plane is v, the rate of dissipation of energy at the sliding surface approaches $\sigma_0(b - 2d)tv$, as t becomes small compared to d. Hence, for a thin plate with holes in this configuration, we again have $P_c = \sigma_0 A'$.

Finally, consider a thin plate with five holes arranged as in Fig. 49.7. Since this will be treated numerically, we take definite proportions in terms of the thickness t as shown. A field of stresses, in equilibrium with a pulling force P', is indicated in Fig. 49.7. The outer strips are in simple tension at the yield stress σ_0. The central strip of width $2a$ has the tensile stress σ, which is carried around the central hole in the manner indicated. The two triangles above and below the central hole have equal biaxial stress σ_1, while the stress in the four shaded triangles is obtained by superimposing the simple compression σ_1 in the horizontal strip on the tension σ_1 in the strip of width b inclined at the angle β, as shown. The maximum shear stress in the shaded triangles is $\sigma_1 \sin \beta$. Tresca's yield condition is satisfied if the load carried by the central strip is $2a\sigma_1 = 2ak/\sin \beta = \sigma_0 b$. The maximum size of b, for the particular pattern of holes, may be found by trial. In this case, we find $b = 1.7t$, and, hence, a lower bound $P' = 9.7\sigma_0 t$.

A pattern of velocities, also indicated in Fig. 49.7, consists of sliding on the parallel plane surfaces AB, CD, $D'C'$, $B'A'$. These planes are perpendicular to the plate surface; sliding is parallel to the plate surface. If the velocity of separation of the

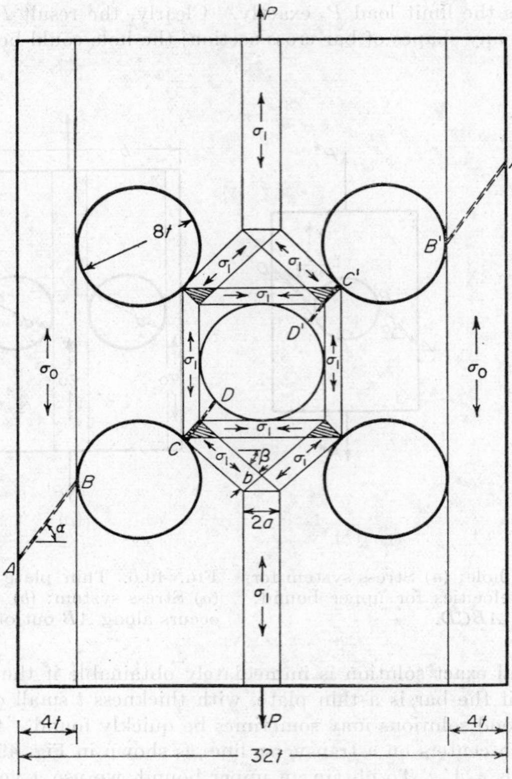

FIG. 49.7. Thin plate with array of holes. Velocity pattern for upper bound is rigid body sliding along $ABCDD'C'B'A'$.

parts above and below the sliding surface is v, the rate of external work is P^*v, and the rate of energy dissipation is $kl'tv/\sin \alpha$, where l' is the total length of the trace of the sliding plane. By trial and adjustment of α, we find $P^* = 12.7\sigma_0 t$. Although this differs considerably from the lower bound, $9.7\sigma_0 t$, the value $11.0\sigma_0 t$ cannot differ from the true limit load by more than about 13%, an accuracy that would be adequate for many purposes.

This problem of the weakening effect of holes in tension sheets is obviously an important practical one. Brady and Drucker [14] have investigated plates with several rows of holes in various configurations, and have compared the estimated limit loads with the observed behavior in tests of hot-rolled steel and 61S-T6 aluminum alloy specimens. The test results are typical of many practical cases, in that the load-deflection curves have different shapes under different circumstances. Brady and Drucker give illuminating discussions of the differences in behavior for various hole configurations, of the scatter in test results due to inhomogeneity of material, and of the interpretation of the limit-analysis results for practical design purposes.

The methods of limit analysis have been applied to a wide variety of problems (apart from the special classes of engineering structures to be discussed). Without attempting a complete list, the following are cited as notable examples: flexure of a wide notched bar [16]; biaxial stressing of a square plate with a central circular hole [17]; reinforcement of holes in plates [18]; indentation by a rigid punch of a plane surface of a semi-infinite body [19] and of a thin layer [20]. Problems of soil mechanics have also been treated [7].

49.6. LIMIT ANALYSIS OF ENGINEERING STRUCTURES

As already mentioned, by this term is meant structures for whose analysis it is convenient and permissible, for practical purposes, to use certain stress resultants and corresponding generalized strains, rather than the general stress and strain tensors. For example, beams are treated in terms of bending moment, normal force, and shear force; plates in terms of bending and twisting moments and transverse shear forces per unit arc length; etc.

The limit analysis of structures requires, first, a choice of the types of plastic deformation to be considered in a given kind of structure. Here we must be guided by experience, as in elastic analysis. For example, in the limit analysis of beams and plates, we commonly ignore plastic shearing deformations. This is justified, as in the comparable elastic treatments, only by reference to test results, or to a more complete limit analysis, in which shear deformation is taken into account. In the limit analysis of a given structure, use is made of a yield condition involving certain of the stress resultants; the decision as to which need to be included is equivalent to the choice of plastic deformation types. The form of the yield condition may then be found by applying the general theorems to a typical element of the structure acted on by the relevant stress resultants.

As in the general theory, the yield function defines a convex surface in an appropriate stress-resultant space [21],[22]. The generalized plastic strain rates are again assumed to be derived from a plastic potential, taken as identical with the yield function. Thus, the flow rule again corresponds to the maximum work principle of von Mises. Regarding the stress space simultaneously as a strain rate space, each coordinate axis corresponds both to one of the relevant stress resultants and to the corresponding generalized plastic strain rate. The flow rule requires the generalized plastic strain-rate vector to have the direction of the normal to the yield surface at all points where this is uniquely defined. At corners or vertices where the yield surface does not have a continuously turning tangent plane, the generalized strain-rate vector may have any direction within the fan or cone defined by the normals of the contiguous surfaces.

Let $Q_1, Q_2, \ldots, Q_n$ denote the relevant stress resultants, and $\dot{q}_1, \dot{q}_2, \ldots, \dot{q}_n$ the corresponding generalized plastic strain rates. The yield condition can be written as

$$F(Q_1, Q_2, \ldots, Q_n) = K^2 \tag{49.19}$$

where K is an appropriate constant for a given point. States for which $F > K^2$ are excluded. If $F < K^2$, behavior is governed by elastic relations. Wherever the yield surface has a unique normal,

$$\dot{q}_k = \Lambda \frac{\partial F}{\partial Q_k} \tag{49.20}$$

where Λ is a positive factor of proportionality.

The three theorems of limit analysis, as stated for general bodies, furnish the corresponding theorems for any special class of engineering structure, when the appropriate stress resultants are inserted in place of the general stress components, and the corresponding generalized strain rates in place of the strain-rate components. Proofs for any class of structures can be given analogous to those outlined above for the general theorems.

Thus, a lower bound will be furnished by showing that a set of stress resultants, $Q'_1, Q'_2, \ldots, Q'_n$, nowhere violates the appropriate yield condition [Eq. (49.19)], and satifies the equations of equilibrium throughout the structure with a load P'; these may be algebraic or differential equations. An upper bound will be obtained by considering any set of velocities agreeing with the given constraints (zero where the corresponding load component is not specified). The assumed velocities determine generalized plastic strain rates throughout the structure, which, in turn, determine [by the appropriate flow rules, Eqs. (49.20)] the rate of energy dissipation. An upper bound P^* is found by equating the rate of work of the external loads to the rate of total energy dissipation.

Discontinuities are permitted in the velocity fields, either of velocities or of their spatial derivatives. These are, again, regarded as limiting cases of continuous but very rapidly changing velocity fields in a narrow strip or segment. In beams and frames, such discontinuities (in angular velocity) are the main form of energy dissipation, as will be seen.

49.7. CONTINUOUS BEAMS AND FRAMES OF DUCTILE MATERIAL

The Limit Moment

A beam cross section is assumed to have an axis of symmetry which is in the plane of loading; a rigid-jointed frame is assumed loaded in the plane of the frame. It is customary to neglect plastic deformations due to shear and normal forces on cross-section planes. The yield condition then concerns only the bending moment and can be written as

$$|M| = M_0 \tag{49.21}$$

where M_0 is the *limit moment* of the cross section. Elastic bending occurs if $|M| < M_0$, while $|M| > M_0$ is assumed impossible. The quantity M_0 depends on the yield stress values in simple tension and compression, assumed to be equal and denoted by σ_0, and on the geometry of a given cross section. Consider a beam element (Fig. 49.8*a*) which is shown subjected to the limit moment M_0. We deduce the value of M_0 by applying the general theorems of Sec. 49.3 to the element. A suitable stress field is shown in Fig. 49.8*c*. Above the plane nn', the material is in simple compression, and, below this plane, in simple tension, in both cases at the yield stress σ_0. The fact that the normal force is zero and that the bending moment is M_0 requires

$$A_1 = A_2 \qquad M_0 = \int_A y\sigma_x \, dA = \sigma_0 \int_{A_1} y \, dA - \sigma_0 \int_{A_2} y \, dA \tag{49.22a,b}$$

with the notation shown in Fig. 49.8*b*. In any case, Eq. (49.22*b*) gives a lower bound, but this is shown to be also an upper bound by considering a field of linear displacement and strain increments as shown in Fig. 49.8*d*, the fibers at the plane nn' being unstrained. Integration over the slice gives the total plastic energy dissipation, and equating this to the external work $M_0 \, d\phi$ gives the result Eq. (49.22*b*). This is, therefore, the correct limit moment, on the assumption that the element is subjected to a pure bending moment.

The terms *fully plastic moment* or *capacity moment* are also often used instead of *limit moment*. For brevity, we have not distinguished between the limit moments for bending in positive and negative senses. The two magnitudes are identical if the tension and compression yield stresses are equal *or* if the cross section has an axis of symmetry perpendicular to the plane of loading. For the great majority of practical cases, either or both of these conditions are satisfied. In commercial rolled beams, there are considerable variations in yield stress, owing to the differences in heat treatment undergone by different elements [23]. It is desirable to determine M_0 by direct test in bending, when possible. Otherwise, σ_0 must be estimated as well as possible from typical stress-strain diagrams.

For a rectangular section with depth h and width b, we have $M_0 = \sigma_0 bh^2/4$. It is useful to compare M_0 with the *yield moment* M_e, meaning the moment at which wholly

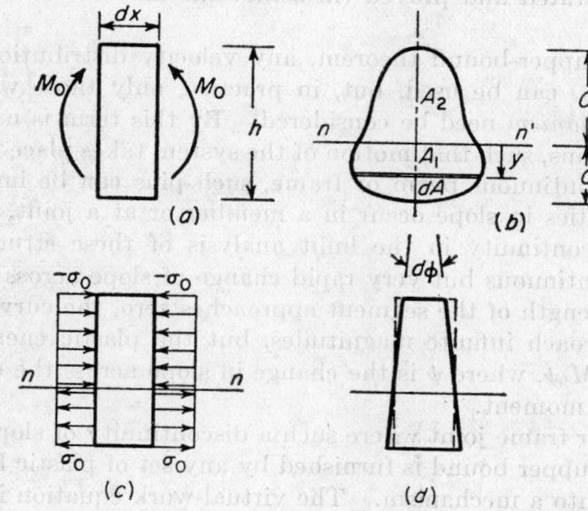

Fig. 49.8. Limit moment M_0, stress and velocity patterns.

elastic action just ceases. Since $M_e = \sigma_0 Z$, where Z is the section modulus of elastic-beam theory, the ratio M_0/M_e for the rectangular beam is

$$\frac{\sigma_0 bh^2/4}{\sigma_0 bh^2/6} = 1.5$$

This ratio, called the *shape factor* and denoted here by α, gives a convenient means of estimating M_0 for I sections, since, for American standard I sections, α is consistently in the range of about 1.10 to 1.25, while, for I sections of wide-flange type, α lies between about 1.05 and 1.15. The section modulus Z is tabulated for commercial beam sections, and $\alpha\sigma_0 Z$ gives a good estimate of M_0 when α and σ_0 are suitably chosen. Tables of M_0/σ_0 for British rolled-steel joist sections are given in [3] and [4]. The shape factor for a solid circular section is $16/3\pi$; that for a very thin-walled hollow circular tube is $4/\pi$.

Limit Theorems

It is reemphasized that the limit load is defined as the load at which plastic collapse occurs of the ideal structure (having the dimensions of the actual structure, but with no work hardening or sensitivity to shape changes). The limit theorems in forms appropriate to beams and frames are as follows:

1. During collapse at the limit load, all bending moments remain constant; only plastic, not elastic, increments of curvature take place.

2. If a system of bending moments M' nowhere violates the yield condition (49.21), and obeys the equations of static equilibrium with a load magnitude P', then $P' \leq P_c$.

3. A system of center-line velocities of beam or frame members which satisfy the prescribed displacement or slope conditions (vanishing wherever corresponding load components are not specified), furnishes a load P^* by equating the rate of work of the external loads to the rate of energy dissipation in the frame; then $P^* \geq P_c$.

Proofs of these theorems can be given, following closely the steps of the proofs stated above for the general theorems. The moment M and rate of plastic curvature $\dot{\kappa}^p$ satisfy relations corresponding to the convexity property,

$$\dot{M}\dot{\kappa}^p = 0 \qquad M\dot{\kappa}^p \geq M''\dot{\kappa}^p$$

where M'' is any moment not violating the yield condition (49.21). The theorems were first precisely stated and proved (in somewhat different forms) by Greenberg and Prager [24].

In applying the upper-bound theorem, any velocity distribution (satisfying the constraint conditions) can be used, but, in practice, only those which convert the structure into a *mechanism* need be considered. By this term is meant a system of rigid bars linked by pins, such that motion of the system takes place through rotations at the pins. In a continuous beam or frame, such pins can be imagined as placed wherever discontinuities in slope occur in a member or at a joint. This is the distinctive type of discontinuity in the limit analysis of these structures. It is the limiting case of a continuous but very rapid change of slope across a small segment. In the limit, as the length of the segment approaches zero, the curvature and strains in the segment approach infinite magnitudes, but the plastic energy dissipation is finite and given by $M_0\dot{\psi}$, where $\dot{\psi}$ is the change in slope across the discontinuity and M_0 is the local limit moment.

A beam section or frame joint where such a discontinuity of slope occurs is called a *plastic hinge.* An upper bound is furnished by any set of plastic hinges which converts the structure into a mechanism. The virtual-work equation is

$$\dot{W}(P^*) = \Sigma M_{0i}\dot{\psi}_i \tag{49.23}$$

where $\dot{W}(P^*)$ is the work rate of external loads P^*, $\dot{\psi}_i$ is the rate of rotation at the ith hinge, M_{0i} the limit moment at the hinge section, and the sum includes all plastic hinges. Each term in the sum is, of course, positive.

The special feature of beams and frames which makes the application of the limit theorems easier and more precise than for other types of structure is the fact that this simple type of velocity distribution, corresponding to certain plastic-hinge locations, actually furnishes the correct limit load of the structure. In other words, there is a set of plastic hinges converting the structure (or some part of it) into a mechanism, and also enabling the plastic-moment condition and the equilibrium equations to be satisfied at all intermediate sections, as well as the hinge sections. Since conditions of both the upper-bound and lower-bound theorems are then satisfied, the load determined is the limit load P_c.

The concept of the plastic hinge is a very old one, going back at least to the pioneer work of Kazinczy [3],[4]. While discontinuities of slope cannot occur in physical beams, short segments of severe curvature are observed in tests of beams and rigid frames of ductile metal, and the actual mechanism of collapse approximates closely that envisaged in the theory.

As a first very simple example, consider the uniform beam of Fig. 49.9*a*, supported at A, built in at B, and with total load P uniformly distributed over the beam. The limit load could be found by calculating the behavior in a series of steps as the load magnitude is gradually increased from zero. The wholly elastic range ends when the greatest moment magnitude, at B, reaches the yield moment M_e. With a further increase in load, the magnitude here reaches M_0. Further load increases can be supported, the moment magnitude at B now remaining constant. In this range, the beam effectively has a plastic hinge at B; finite rotations of the beam tangent at B occur which can be calculated by treating the beam as simply supported. Finally, at a load P_c, the maximum moment at an interior section reaches the value M_0. Above this load, static equilibrium cannot be maintained, and plastic collapse occurs.

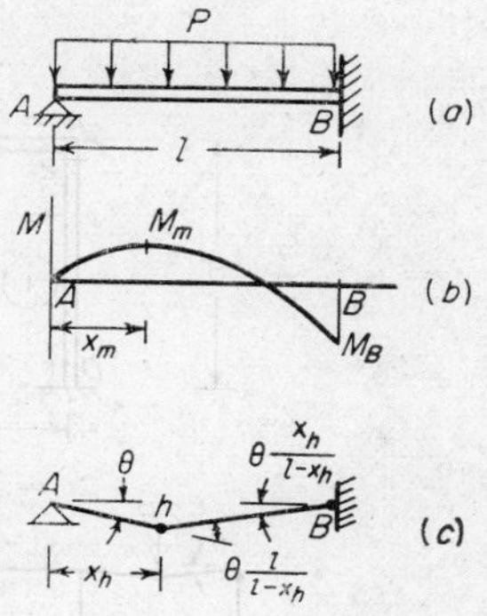

FIG. 49.9. (a) Supported-clamped beam; (b) equilibrium-moment diagram, (c) displacement increment pattern. Black circles show plastic hinges.

This sort of stepwise analysis corresponds quite closely to the observed behavior in tests. While it can always be carried out, for more complex frames it becomes tedious, and fortunately is never necessary. The limit load will now be found by applying the limit-analysis theorems.

We first apply the lower-bound theorem, taking account of equilibrium and plasticity conditions. Equilibrium requires the moment diagram to have the general shape shown in Fig. 49.9*b*. The conditions of the theorem will be met if $M_B \geq -M_0$ and $M_m \leq M_0$, where m is the point of relative maximum moment. The moment at a typical section is

$$M(x) = \frac{Pl}{2}\frac{x}{l}\left(1 - \frac{x}{l}\right) + \frac{x}{l}M_B \qquad (49.24)$$

Taking $M_B = -M_0$, maximizing $M(x)$, and putting $M_m = M_0$, the result obtained is $P_c = 11.65M_0/l$. The maximum moment is at $x_m = 0.414l$.

The upper-bound theorem is next applied. The beam becomes a mechanism when plastic hinges are placed at the built-in end B and at a point h at distance x_h from A, as indicated in Fig. 49.9*c*. If the segment Ah rotates through a small angle θ, the virtual-work equation is

$$\tfrac{1}{2}P^*x_h\theta = M_0\frac{l + x_h}{l - x_h}\theta \qquad (49.25)$$

If P^* is minimized with respect to x_h, we find $x_h = 0.414l$, and, putting this in Eq. (49.25), the result $P_c = 11.65M_0/l$ is again obtained. It may be noted that, for consistency, we should have written a rate-of-work equation, with an angular velocity $\dot{\theta}$. The equation so written would be identical with Eq. (49.25) if the deflection at h is neglected, i.e., if a small motion from the initial unloaded configuration is used. This is generally done in treating beam and frame problems, and the type of notation of Fig. 49.9*c* is convenient. However, finite initial deflections can be taken into account. This will be illustrated later. Note also that good approximate answers for the beam with distributed load are easily obtained by putting the interior plastic hinge at any reasonable location. The guess $x_h = 0.5l$ gives $P^* = 12M_0/l$, less than 3% too high. A second guess $x_h = 0.4l$ gives $P^* = 11.67M_0/l$.

If the load had been concentrated instead of distributed, the solution by either method would have been simpler, since the interior hinge would then necessarily be located at the load point. For a force at distance a from A, the result is

$$P_c = \frac{M_0\,(l + a)}{a(l - a)}$$

Method of Combining Mechanisms

Based directly on the theorems, techniques have been developed which enable the limit load to be computed for complex frames with remarkable rapidity. The *method of combining mechanisms* [25] will next be discussed with reference to the two-bay portal frame of Fig. 49.10*a*. This is taken as a problem of analysis, the limit moments of the five members being given by the numbers in small circles. The result for a frame with definite numerical values is easily translated to the more general form

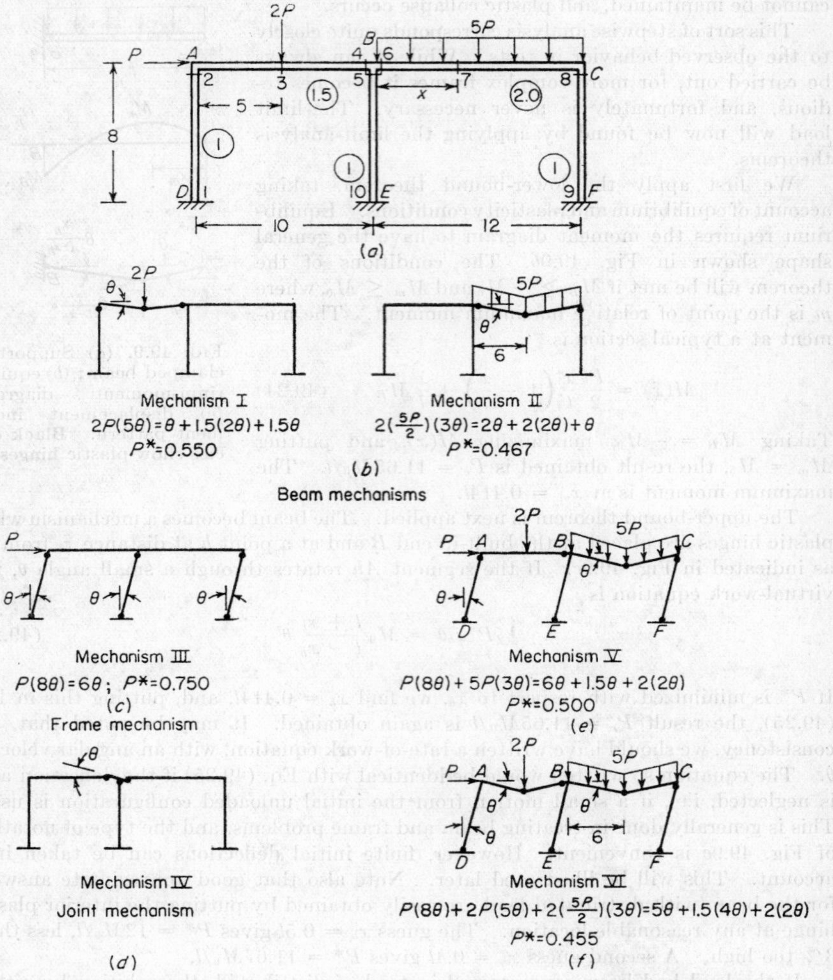

FIG. 49.10. (*a*) Rigid-jointed frame. Numbers in circles are assumed limit moments. Sections marked 1 through 10 show possible plastic hinge positions. (*b*) through (*d*) show four independent mechanisms. (*e*) and (*f*) show mechanisms obtained by combining independent mechanisms. In each case, θ is an infinitesimal arbitrary rotation; black circles show plastic hinges.

$P_c = \lambda M_0/l$, where M_0 and l are a representative limit moment and length, respectively, and λ is a numerical coefficient.

The method of combining mechanisms is based on the upper-bound theorem, and makes use of certain *independent mechanisms* of the structure and their combinations

For the illustrative problem, the three types of independent mechanisms are as shown in Fig. 49.10*b* through *d*, called *beam, frame,* and *joint* mechanisms, respectively. At a joint where there is a choice between two members for a hinge, it is put in the weaker of the two. The interior hinge in the distributed load case, II, is placed at the mid-point. For each of these independent mechanisms the virtual-work equation is written in the figure, with the corresponding upper-bound load. The correct limit load is either the load corresponding to one of the independent mechanisms or that corresponding to some combination of them. Figure 49.10*e* shows a combination of three independent mechanisms; first, mechanisms II and III are combined, and then the joint mechanism IV is added, so as to leave only one hinge at *B*. Figure 49.10*f* shows the result of a further combination of this with the beam mechanism I, so as to eliminate the common hinge at *A*.

This mechanism can be shown to be the correct one except for the location of the interior hinge in beam *BC*, which was put arbitrarily at the mid-point. The adjustment of this hinge to its correct position will be shown later to make a trivial difference; the upper bound $P^* = 0.455$ exceeds the correct limit load by less than 0.5%.

The above procedure of combining mechanisms must be explained and justified. First, it can readily be seen that it will only be advantageous to combine two or more mechanisms when a plastic hinge common to them is thereby eliminated; this is a necessary (but not a sufficient) condition for the load corresponding to the combination to be lower than that of either of the mechanisms being combined. For example, the mechanism V of Fig. 49.10*e* combines with the beam mechanism I (Fig. 49.10*b*) so as to eliminate the common hinge at *A*, and the load 0.455 is less than either 0.500 or 0.550. Moreover, it is usually very easy for the analyst to see when all the possible advantageous combinations have been tried. This is the case, for example, when the mechanism of Fig. 49.10*f* has been reached. At this point the analyst will feel sure that he has the right mechanism (except for locating the hinge under the distributed load). As a check, he should compute the moments at *A* and at *B* where hinges are not present; if these are less in magnitude than the M_0 value of the respective member, then the upper-bound load is the limit load, since the requirements of the lower-bound theorem are satisfied as well.

Second, the procedure of combining mechanisms should be recognized as equivalent to one of linear combination of equations of equilibrium, written in terms of the bending moments at cross sections where plastic hinges may occur. In the illustrative problem, there are 10 such sections, identified by numbers in Fig. 49.10*a*. The moments at these cross sections are governed by four independent equations of equilibrium. These can be written in many different ways: for example, as two beam equations, one equation relating horizontal load to end moments in the columns, and one relating the three end moments at joint *B*. (Equations relating two end moments at *A* and *C* are dispensed with, since only the moment in the weaker section is considered.) Each of the four listed equations of equilibrium corresponds to one of the independent mechanisms that were enumerated. For example, the equation of horizontal equilibrium can be written[1]

$$8P = -M_{AD} - M_{DA} - M_{BE} - M_{EB} - M_{CF} - M_{FC}$$

The largest possible value of P permitted by this equation is $P = 6/8$, obtained when each end moment has the value -1. Evidently, each upper bound furnished by a mechanism could equally well be derived by writing the equation of equilibrium relating moments at sections where plastic hinges occur; the upper bound is the load above which the equilibrium equation could not be satisfied without violating a plastic moment condition. The superposition of two mechanisms is equivalent to adding the corresponding equations, multiplied by suitable factors. The number of independent

[1] M_{AB} denotes the clockwise end moment acting on member *AB* at end *A*.

mechanisms is the number of independent equations of equilibrium; this is the number of moments concerned, less the number of redundancies. In the frame of Fig. 49.10*a*, there are 6 redundant moments; hence, $10 - 6 = 4$ independent mechanisms.

The success of the method, in practice, is really due to two facts: first, that all possible hinge locations are easily recognized; and, second, that only those combinations of mechanisms need be considered which enable a common hinge to be eliminated. This rules out all but a relatively small number of the possible combinations. After listing the independent mechanisms and their corresponding upper bounds, we would use first the mechanism giving the lowest upper bound, trying combinations with this and any other mechanisms that allow a common hinge to be eliminated. Any combination that does lower the upper bound is then examined for further advantageous combinations. Evidently the method is not completely systematic, but the sort of judgment required seems to be easily picked up by engineers. It works particularly well for rectangular-grid-type frames, but has been applied to many other types, such as multibay pitched-roof frames [3]. The method of combining mechanisms is a valuable tool also in the analysis of failure under variable, repeated loads [26].

To finish the illustrative problem, suppose the hinge in beam BC is at distance x from B, in mechanism VI of Fig. 49.10*f*. The virtual-work equation gives the result

$$P^* = \frac{144 - 9x}{(18 + 2.5x)(12 - x)}$$

When this is minimized with respect to x, the results found are $x = 6.4$ and $P_c = 0.453$. As usual, the approximation obtained by putting the hinge at mid-span under a distributed load is amply good enough for practical purposes. Then, to check the final result, moments throughout the frame are computed as follows: $M_{BC} = -1.83$, $M_{BE} = 0.33$, $M_{AD} = 0.05$. In these computations the method of virtual work is a very convenient way of deriving equations containing a single unknown quantity.

Method of Plastic-moment Distribution

Methods have also been based on the lower-bound theorem. The most useful appears to be the method of plastic-moment distribution, [27],[28], so called because

Table 49.1

Line	M_{AD}	M_{AB}	M_G	M_{BA}	M_{BE}	M_{BC}	M_H	M_{CB}	M_{CF}
a	−1.33	−2.50	2.50	2.50	−1.33	−3.75	3.75	3.75	−1.33
b		3.83	1.92				1.21	−2.42	
c					2.58				
	−1.33	*1.33*	*4.42*	*2.50*	*1.25*	*−3.75*	*4.96*	*1.33*	*−1.33*
d	1.14	−1.14	−0.57						
	−0.19	*0.19*	*3.85*						
e			−0.40	0.80		−0.80	0.40		
			3.45	*3.30*		*−4.55*	*4.56*		
f			0.15	−0.31		0.31	−0.31	0.93	−0.93
			3.60	*2.99*		*−4.24*	*4.25*	*2.26*	*−2.26*
g			−0.17	0.34		−0.34	−0.17		
	−0.19	*0.19*	*3.43*	*3.33*	*1.25*	*−4.58*	*4.08*	*2.26*	*−2.26*

of its resemblance to Hardy Cross's method of moment distribution for elastic frames. The two methods differ essentially in that, while the elastic method uses distribution and carry-over factors derived from conditions of slope continuity, the plastic-moment distribution method has no such factors, because no conditions of slope continuity are assumed to exist. Moment increments are governed entirely by equilibrium. At any stage, a lower-bound load is one that makes the ratio $|M|/M_0 \leq 1$ for all members.

To illustrate the method, we consider again the problem of Fig. 49.10*a*. The clockwise-positive convention for end moments acting on members is especially convenient, because, at any joint, the sum of the moments is required by equilibrium to be zero in this notation. We consider, for simplicity only, the moment at the mid-section of the beam carrying distributed load. The moments M_G, M_H at the mid-sections of beams AB, BC, respectively, are positive in the sense corresponding to positive loads on the simply supported members. Equations of equilibrium for the beam loads and for the horizontal load are:

$$-M_{AB} + 2M_G + M_{BA} = 10P \qquad -M_{BC} + 2M_H + M_{CB} = 15P$$
$$M_{AD} + M_{DA} + M_{BE} + M_{EB} + M_{CF} + M_{FC} = -8P \qquad (49.26a\text{–}c)$$

We begin the calculation by writing the smallest moments required in all members when they are isolated from one another by an external agency that prevents rotations at joints A, B, and C. Equal-moment magnitudes at the beam mid-section and end sections are required: $10P/4$ in beam AB, and $15P/4$ in BC. End moments in the columns are each $8P/6$. These starting moments are written in line a of Table 49.1, in each case with the factor P omitted. The object of further operations is first to balance moments at the joints and then to adjust moments so as to raise the lower bound.

The numbers in Table 49.1 show one way (not especially systematic) of reaching a satisfactory lower bound. The numbers in lines below line a are increments in moments. These are governed only by the need to satisfy Eqs. (49.26). Since the

Moments for $P = 1$

Line	M_{DA}		M_{EB}		M_{FC}
a	−1.33		−1.33		−1.33
b					
c	−0.86		−0.86		−0.86
	−2.19		−2.19		−2.19
d	−0.38		−0.38		−0.38
	−2.57		−2.57		−2.57
e					
f	0.31		0.31		0.31
	−2.26		−2.26		−2.26
g					
	−2.26		−2.26		−2.26

moments in line a satisfy these equations (for $P = 1$), increments must satisfy the system

$$-\Delta M_{AB} + 2\Delta M_G + \Delta M_{BA} = 0 \qquad -\Delta M_{BC} + 2\Delta M_A + \Delta M_{CB} = 0$$
$$\Delta M_{AD} + \Delta M_{DA} + \Delta M_{BE} + \Delta M_{EB} + \Delta M_{CF} + \Delta M_{FC} = 0 \qquad (49.27a\text{–}c)$$

In addition, at all joints the sum of moments must be zero. In making adjustments, it is often convenient to change only two at a time of the three moments in a beam; Eqs. (49.27a,b) furnish the relations between the two increments. Line b shows increments restoring equilibrium to joints A and C. Line c then restores equilibrium at joint B, adding increments at the feet of the columns so as to satisfy Eq. (49.27c). Table 49.2 shows the values of P allowed by the various members at this stage of the

Table 49.2. $M_0/\max|M|$ from Table 49.1

Line	Beam AB	Beam BC	Columns AD, BE, CF	P'
c	0.340	0.404	0.456	0.340
d	0.390	0.404	0.390	0.390
e	0.435	0.440	0.390	0.390
f	0.417	0.470	0.443	0.417
g	0.438	0.438	0.443	0.438

calculations; for example, $1.5/4.42 = 0.340$. The smallest of these is a lower bound P'. Lines d, e, f, and g show increments whose effect is to reduce the maximum moments in the members and, hence, to raise P'. The table should be self-explanatory. The reader will probably enjoy more constructing a similar table of his own.

Limit Design

A few words should be said about design: Given the loads and general dimensions of a structure, what strengths should the members have? This is, of course, the real purpose of the methods of analysis that have been outlined. Without further refinement, they already provide a major advance over the design efficiency possible with conventional elastic methods. This is so because the plastic method (assuming plastic collapse is an appropriate form of failure and that the limit load accurately gives the plastic collapse load) enables the design to be based on a definite factor of safety with respect to a real failure of the structure; when the structure is subjected to the working load multiplied by the factor of safety, failure occurs. No such statement can be made when conventional elastic methods, using a maximum allowable (elastic) stress under working loads, are applied to redundant structures. Except for statically determinate structures, the elastic designer has no knowledge of the real failure load and, hence, of the corresponding factor of safety.

Apart from these facts, there have been developed theorems and techniques for determining the optimum design of a frame, of all those that enable a given load to be carried with a specified load factor. *Optimum* is used here in the sense of minimum weight of material, and hence leaves out of account the many other items that affect cost. It is assumed that there is a definite relation between limit moment and weight per unit length. In some limited range, the actual relation may be replaced by a linear one, so that, if w_i is the weight per unit length of a member,

$$w_i = a + bM_i \qquad (49.28)$$

where M_i is the limit moment of the member, and a and b are constants. (Only beams and frames having members of uniform section are considered.) The total weight is

$$\text{Weight} = a\Sigma l_i + b\Sigma M_i l_i \tag{49.29}$$

where l_i is the total length of members having limit moment M_i. The optimum frame is, therefore, one that minimizes the *weight function* G defined as

$$G = \Sigma M_i l_i \tag{49.30}$$

A valuable approach has been suggested [29],[30], using geometrical representations of mechanisms in a hyperspace whose coordinate axes are the limit-moment magnitudes to be chosen, and Foulkes [30] has proved general theorems. We cannot discuss these, but, in order to show how easily the limit analysis is adapted to design, we treat a design problem by elementary methods. Suppose that the loads on the frames are specified as in Fig. 49.11, and that limit moments M_1 of the three columns,

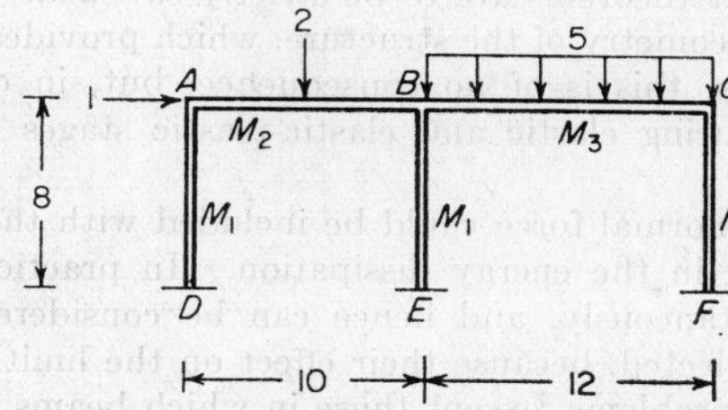

FIG. 49.11. Design problem. Factored load magnitudes (limit loads) are specified. Limit moments M_1, M_2, M_3 are to be selected.

M_2 of beam AB, and M_3 of beam BC are to be selected. The weight function $G = 24M_1 + 10M_2 + 12M_3$. The loads listed are limit loads: working loads multiplied by factor of safety. The moment systems listed in Table 49.1 immediately give safe designs if the limit moment of each section is taken equal to the maximum moment magnitude in the section. The first five lines of Table 49.3 show these

Table 49.3

Line		M_1	M_2	M_3	β	G
1	Table 49.1, line *c*	2.19	4.42	4.96		156
2	Table 49.1, line *d*	2.57	3.85	4.96		160
3	Table 49.1, line *e*	2.57	3.45	4.56		151
4	Table 49.1, line *f*	2.26	3.60	4.25		141
5	Table 49.1, line *g*	2.26	3.43	4.58		143
6	Ratios as in line 4	2.26β	3.60β	4.25β	0.967	136
7	Arbitrary ratios	1.00β	1.50β	2.00β	2.20	138
8	Ratios of minimum sections	1.33β	2.50β	3.75β	1.36	139
9	Minimum weight	1.70	3.91	4.42		133

designs and the corresponding values of G. These are not economical since the given load is not the limit load for these designs. The best of them (line 4) may be modified so as to be a collapse design by multiplying each limit moment by a factor β. Line 6 of Table 49.3 gives the required β and corresponding value of G (collapse occurs in mechanism V of Fig. 49.10). Lines 7 to 9 of the table show some other designs for comparison. Line 7 shows results obtained when the ratios are assumed "arbitrarily," as for the analysis problem of Fig. 49.10*a*. In line 8, the ratios of M_1, M_2, and M_3 are assumed to be those of the minimum values of each section, as previously obtained by supposing the joints are prevented from rotating so that the columns and the two beams are effectively isolated. This simple method often gives very good designs. Finally, line 9 shows the true minimum weight design. The special feature of this is that there are three *alternative* mechanisms of collapse, namely, mechanisms II, III, and VI of Fig. 49.10.

Refinement of the Methods

The methods of analysis and design are remarkably complete and convenient, and many experiments [4],[23] have shown that, for a wide range of types of beams and frames, there is a genuine plastic collapse load, which is closely estimated by the limit load as computed by the methods outlined. Nevertheless, sight must not be lost of their basic limitations. These are of two kinds. One is the consideration only of bending moments in the generalized yield condition, [Eq. (49.21)]. Normal and shear forces are generally present, and should be taken into account if the conditions of the general limit theorems are to be strictly satisfied. The other limitation is the use of the initial geometry of the structure, which provides an enormous simplification. In most cases, this is of no consequence, but, in certain problems, the changes in geometry during elastic and elastic-plastic stages of loading may have drastic significance.

Both a shear and a normal force could be included with the bending moment in the yield condition and in the energy dissipation. In practice, however, they are rarely important simultaneously, and hence can be considered separately. Shear forces can be safely neglected, because their effect on the limit load does not exceed a few per cent in most problems, except those in which beams of I section or of box section have very short span, of the order of five times the beam depth, or less. The importance of shear depends on the manner of loading over the span, not merely on the local ratio of bending moment to shear force. Studies of shear have been critically reviewed recently by Drucker [31].

The inclusion of the normal force and axial strain is of more fundamental concern, not only because these affect the computed limit loads more strongly, but also because they determine whether geometry change effects are minor or decisively alter the response of the structure. We consider, for simplicity, only a cross section with two axes of symmetry, and with equal yield stress in simple tension and compression. The general theorems of limit analysis are applied to a beam element. A system of stresses satisfying yield conditions and equilibrium with bending moment M and normal force N is indicated in Fig. 49.12*c*. This has simple tension below and simple compression above a plane nn', located a distance e above the transverse axis of symmetry, and

$$N = 2\sigma_0 \int_0^e b\,dy \qquad M = 2\sigma_0 \int_e^{h/2} yb\,dy \tag{49.31}$$

These are easily shown to be upper-bound as well as lower-bound values for N and M by considering the pattern of velocities shown in Fig. 49.12*d*, in which one face of the element rotates $d\phi$ with respect to the other, and the strain is zero at the axis nn'. Hence, the yield condition of the form

$$F(M,N) = K^2 \tag{49.32}$$

is obtained by eliminating e in Eqs. (49.31). The rate of dissipation of energy per unit length is

$$D = M\kappa + N\epsilon \tag{49.33}$$

where $\dot{\kappa} = d\dot{\phi}/dx$ is the curvature rate, and $\dot{\epsilon}$ is the strain rate at the center. According to the velocity pattern of Fig. 49.12*d*,

$$\dot{\epsilon} = \dot{\kappa} e \tag{49.34}$$

The relation between $\dot{\kappa}/\dot{\epsilon}$ and M/N is obtained from Eq. (49.34) with the use of Eqs. (49.31), since these determine M/N as a function of e and the cross-section shape.

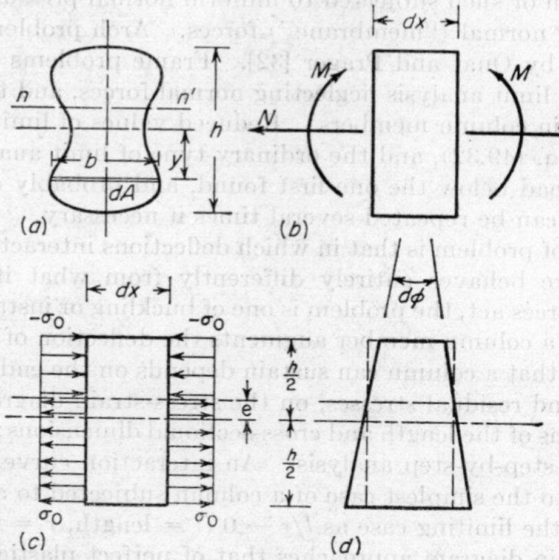

FIG. 49.12. Bending-moment–normal-force interaction: (*c*) and (*d*) show stress and velocity patterns.

It is easily checked that this corresponds to the general flow rule [Eqs. (49.20)], which requires

$$\frac{\dot{\kappa}}{\dot{\epsilon}} = \frac{\partial F/\partial M}{\partial F/\partial N} = -\frac{dN}{dM} = \frac{1}{e}$$

For a rectangular cross section (width b, depth h), the yield condition (49.32) takes the simple form

$$\frac{M}{M_0} + \frac{N^2}{N_0{}^2} = 1 \tag{49.35}$$

where $M_0 = \sigma_0 bh^2/4$, $N_0 = \sigma_0 bh$ are limit moment (pure bending) and limit force (simple tension or compression), respectively. The flow rule is $\dot{\kappa}/\dot{\epsilon} = -dN/dM = N_0{}^2/2NM_0$. It is convenient to use M/M_0 and N/N_0 as nondimensional stress resultants, to which the corresponding generalized plastic strain rates are $M_0\dot{\kappa}$ and $N_0\dot{\epsilon}$, respectively. The solid curve in Fig. 49.13 shows the yield condition for the rectangle. The generalized plastic strain-rate vector $(M_0\dot{\kappa}, N_0\dot{\epsilon})$ is also shown normal to the yield curve. The flow rule is

$$\frac{M_0\dot{\kappa}}{N_0\dot{\epsilon}} = \frac{N_0}{2N} \tag{49.36}$$

This applies at all points except the vertices $M/M_0 = 0$, $N/N_0 = \pm 1$. At these points, the plastic strain-rate vector may have any direction between the normals defined by the adjacent elements of the curves.

For a typical I section, the yield curve is like the dashed curve in Fig. 49.13. The extreme case of an I section with indefinitely thin web and flanges is represented by the square yield curve, with straight lines joining points ± 1 on the axes. The curve for a circular cross section lies outside but very close to that for the rectangle, M/M_0 differing at most by 3%.

Two types of problems involve normal forces. In the first type, the initial geometry of the structure and the manner of loading are such that normal forces are necessary for equilibrium, and may be important irrespective of deflections. This is the case, for example, in frames carrying mainly vertical loads and in arches. An extreme case would be an arch or shell subjected to uniform normal pressure, so that the load is carried entirely by normal ("membrane") forces. Arch problems of several kinds have been discussed by Onat and Prager [32]. Frame problems may be treated by first carrying out the limit analysis neglecting normal forces, and then computing the magnitudes of these in column members. Reduced values of limit moment can then be computed from Eq. (49.32), and the ordinary type of limit analysis repeated; this will furnish a limit load below the one first found, and probably conservative. The cycle of calculations can be repeated several times if necessary.

The second type of problem is that in which deflections interact with normal forces so that the structure behaves entirely differently from what it otherwise would. When compressive forces act, the problem is one of buckling or instability; the moment of the axial force on a column member augments the deflection of the member. The maximum axial load that a column can sustain depends on the end constraints, on the initial straightness and residual stresses, on the stress-strain diagram of the material, and on the proportions of the length and cross-sectional dimensions; it must, in general, be worked out by a step-by-step analysis. An interaction curve of the type of Eq. (49.35) corresponds to the simplest case of a column subjected to axial force and pure bending moment in the limiting case as $l/r \to 0$ (l = length, r = radius of gyration). When the stress-strain diagram approaches that of perfect plasticity, the load on an "ideal" column (no initial stresses, curvature, etc.) reaches a maximum, and then falls off very sharply. With actual columns, the load reaches a lower maximum, and falls off more gradually (Fig. 49.14). Surveys of recent work on the plastic buckling of columns and frames have been presented in recent symposia [33].

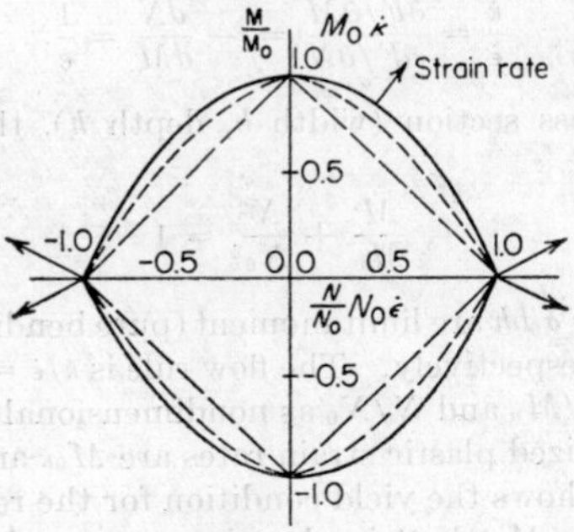

FIG. 49.13 Yield curves for bending moment–normal force. Solid curve shows yield curve for rectangular section; dashed curve represents typical I section. Square with broken lines is for limiting case of I section with vanishing web and flange thickness.

A quite different phenomenon is that in which constraints cause normal *tensile* forces to appear as soon as deflections become finite. Haythornthwaite [34] has studied the case of a beam with ends both clamped and prevented from moving axially. For a central force, the limit load of the ordinary theory is $P_c = 8M_0/l$; this corresponds either to absence of axial constraints or to use of the intial geometry in the presence of axial constraints. We may consider a deflected shape as the initial configuration, and write the equation of virtual work for a small displacement from this, as indicated by the inset figure in Fig. 49.15:

$$P\dot{\delta} = \Sigma M_i \psi_i + \Sigma N_i \xi_i \qquad (49.37)$$

where M_i and N_i are the bending moment and normal force at the three plastic hinges, ψ_i the rates of rotation, and ξ_i the rates of elongation at the three hinge sections. If the deflection $\delta \ll l$, then N can be taken as constant over the length, and it is easy to show, by use of the yield condition and

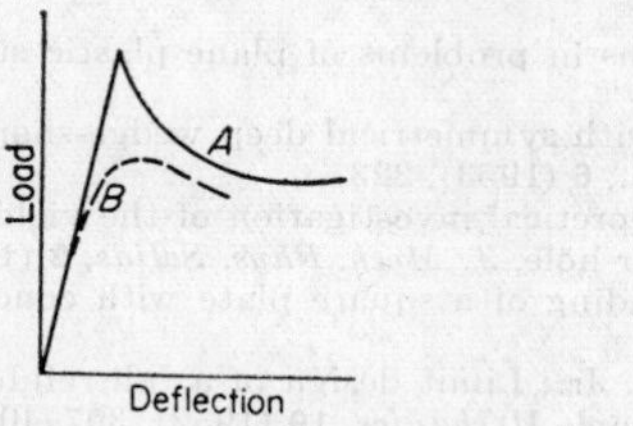

FIG. 49.14. Plastic buckling effects: A is for "perfect" specimen of ideally plastic material; B is for ordinary specimen.

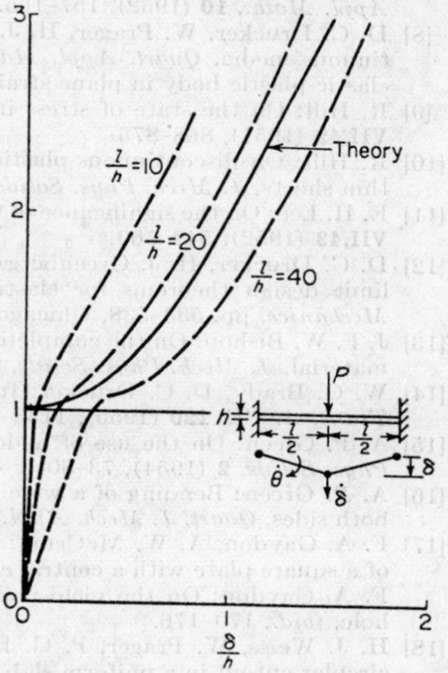

FIG. 49.15. Beam with fully constrained ends [34]. Broken line curves are test results.

flow rule [Eqs. (49.35) and (49.36)] for a rectangular section of depth h, that

For $\delta \leq h$ $\qquad \dfrac{N}{N_0} = \dfrac{\delta}{h} \qquad \dfrac{P}{P_c} = 1 + \dfrac{N^2}{N_0{}^2} = 1 + \dfrac{\delta^2}{h^2}$

For $\delta \geq h$ $\qquad \dfrac{N}{N_0} = 1 \qquad M = 0 \qquad \dfrac{P}{P_c} = \dfrac{2\delta}{h}$

Haythornthwaite's test results (Fig. 49.15) confirm that, when the ends are fully constrained, the load rises rapidly above P_c as soon as δ exceeds h. The limit load, as ordinarily computed, is highly conservative under these circumstances.

Quite apart from the development of normal forces as the result of deflections, exceptional cases occur [35] in which deflections grow to such large magnitudes in the elastic or elastic-plastic phases that the limit load computed from the initial geometry either gives an inaccurate estimate of the collapse load or else loses all significance. Simple methods are available for estimating the deflection of a beam or frame just as the limit load is reached [3],[4].

49.8. REFERENCES

[1] R. Hill: "The Mathematical Theory of Plasticity" Oxford, Univ. Press, New York, 1950.

[2] W. Prager, P. G. Hodge, Jr.: "Theory of Perfectly Plastic Solids," Wiley, New York, 1951.

[3] B. G. Neal: "The Plastic Methods of Structural Analysis," Wiley, New York; Chapman & Hall, London, 1956.

[4] J. F. Baker, M. R. Horne, J. Heyman: "The Steel Skeleton," vol. 2, Cambridge Univ. Press, New York, 1956.

[5] D. C. Drucker: Stress-strain relations in the plastic range of metals: Experiments and basic concepts, chap. 4, pp. 97–119, in "Rheology Theory and Applications," vol. 1 of "Source Book of Rheology," Academic Press, New York, 1956.

[6] R. Hill: A note on estimating yield point loads in a plastic-rigid body, *Phil. Mag.*. **43** (1952), 353–355.

[7] D. C. Drucker, W. Prager: Soil mechanics and plastic analysis or limit design, *Quart. Appl. Math.*, **10** (1952), 157–165.

[8] D. C. Drucker, W. Prager, H. J. Greenberg: Extended limit design theorems for continuous media, *Quart. Appl. Math.*, **9** (1952), 381–389; also: The safety factor of an elastic-plastic body in plane strain, *J. Appl. Mechanics*, **18** (1951), 371–378.

[9] R. Hill: On the state of stress in a plastic-rigid body at the yield point, *Phil. Mag.*, **VII,42** (1951), 868–875.

[10] R. Hill: On discontinuous plastic states, with special reference to localized necking in thin sheets, *J. Mech. Phys. Solids*, **1** (1952), 19–30.

[11] E. H. Lee: On the significance of the limit theorems for a plastic-rigid body, *Phil. Mag.*, **VII,43** (1952), 549–560.

[12] D. C. Drucker, H. J. Greenberg, E. H. Lee, W. Prager: On plastic-rigid solutions and limit design theorems for elastic-plastic bodies, *Proc. 1st U.S. Natl. Congr. Appl. Mechanics*, pp. 533–538, Chicago, 1951.

[13] J. F. W. Bishop: On the complete solution to problems of deformation of a plastic-rigid material, *J. Mech. Phys. Solids*, **2** (1953), 43–53.

[14] W. G. Brady, D. C. Drucker: Investigation and limit analysis of net area in tension, *Trans. ASCE*, **120** (1955), 1133–1164.

[15] A. P. Green: On the use of hodographs in problems of plane plastic strain, *J. Mech. Phys. Solids*, **2** (1954), 73–80.

[16] A. P. Green: Bending of a wide bar with symmetrical deep wedge-shaped notches on both sides, *Quart. J. Mech. Appl. Math.*, **6** (1953), 223.

[17] F. A. Gaydon, A. W. McCrum: A theoretical investigation of the yield point loading of a square plate with a central circular hole, *J. Mech. Phys. Solids*, **2** (1954), 156–169; F. A. Gaydon: On the yield-point loading of a square plate with concentric circular hole, *ibid.*, 170–176.

[18] H. J. Weiss, W. Prager, P. G. Hodge, Jr.: Limit design of a full reinforcement for a circular cutout in a uniform slab, *J. Appl. Mechanics*, **19** (1952), 397–401.

[19] R. T. Shield, D. C. Drucker: The application of limit analysis to punch-indentation problems, *J. Appl. Mechanics*, **20** (1953), 453–461.

[20] R. T. Shield: The plastic indentation of a layer by a flat punch, *Quart. Appl. Math.*, **13** (1955), 27.

[21] W. Prager: "Probleme der Plastizitätstheorie," chap. 3, pp. 34–70, Birkhäuser, Basel, Switzerland, 1955.

[22] W. Prager: The general theory of limit design, *Proc. 8th Intern. Congr. Theoret. Appl. Mechanics*, Istanbul, 1955.

[23] B. G. Johnston, C. H. Yang, L. S. Beedle: An evaluation of plastic analysis as applied to structural design, *Welding J.*, **32** (1953), *Research Suppl.*, 224–239.

[24] H. J. Greenberg, W. Prager: Limit design of beams and frames, *Trans. ASCE*, **117** (1952), 447–484.

[25] B. G. Neal, P. S. Symonds: The rapid calculation of the plastic collapse load for a framed structure, *Proc. Inst. Civil Engrs. London*, **1**, pt. 3 (1952), 58–71.

[26] P. S. Symonds, B. G. Neal: Recent progress in the plastic methods of structural analysis, *J. Franklin Inst.*, **252** (1951), 383–407, 469–492.

[27] M. R. Horne: A moment-distribution method for the analysis and design of structures by the plastic theory, *Proc. Inst. Civil Engrs. London*, **3**, pt. 3 (1954), 51–76.

[28] J. M. English: Design of frames by relaxation of yield hinges, *Trans. ASCE*, **119** (1954), 1143–1153.

[29] W. Prager: Minimum weight design of a portal frame. *J. Eng. Mech. Div., Proc. ASCE*, **82** (1956), paper 1073.

[30] J. Foulkes: The minimum weight design of structural frames, *Proc. Roy. Soc. London* **A,223** (1954), 482–494.

[31] D. C. Drucker: The effect of shear on the plastic bending of beams, *J. Appl. Mechanics*, **23** (1956), 509–514.

[32] E. T. Onat, W. Prager: Limit analysis of arches, *J. Mech. Phys. Solids*, **1** (1953), 77–89.

[33] Symposium on the plastic theory of structures, Cambridge Univ. Eng. Dept., *Brit. Welding J.*, **3** (1956), 331–378, **4** (1957), 1-38; Symposium on the plastic strength of structural members, sponsored by Eng. Mech. Div., *Trans. ASCE.*, **120** (1955), 1019–1164.

[34] R. M. Haythornthwaite: Beams with full end fixity, *Engineering*, **183** (1957), 110–112.

[35] P. S. Symonds, B. G. Neal: The interpretation of failure loads in the plastic theory of continuous beams and frames, *J. Aeronaut. Sci.*, **19** (1952), 15–22.

CHAPTER 50

TWO-DIMENSIONAL PROBLEMS

BY

A. P. GREEN, Hinxton Hall, England

50.1. INTRODUCTION

Many important engineering problems involving plastic deformation are approximately two-dimensional, in the sense that the distribution of stress and strain can be represented by a single typical transverse section of the body, which we take to be the x,y or r,θ plane. In this chapter, we consider such problems for which $\tau_{zx} = \tau_{yz} = 0$, and σ_z is, therefore, a principal stress. The material is assumed to be isotropic, with plastic properties dependent neither on the hydrostatic part of the stress, nor on the strain rate. The theory is well supported by experiment, much of which has been mentioned by Hill [1], or summarized by Hundy [5].

There are three basic types of problem, solutions for which will be discussed in the following three sections. First they are defined, and examples are given of practical conditions under which they are approximated, and of simplifying assumptions made in order to obtain solutions.

Strain ϵ_z Uniform but Not Zero

The dimensions of the body and the stress and strain distribution in it are independent of the z coordinate, and the longitudinal strain ϵ_z is the same for all elements at any stage of the deformation, though, in general, changing as deformation proceeds. Since, in addition, $\tau_{zx} = \tau_{yz} = 0$, by symmetry any originally plane section remains plane.

These conditions are rather restrictive, and are rarely met in practice. The only problem in which accurate account has been taken of a nonzero ϵ_z is the autofrettage of a cylindrical pressure vessel or gun barrel by internal pressure. In practice, the ends are either closed, in which case it is subjected to equal and opposite forces directed along the axis, or sealed by floating pistons, which leave the ends free of load. Provided that the tube is sufficiently long in comparison to its mean diameter, stress and strain are effectively independent of z, except near the ends. Also, by symmetry, originally plane transverse sections remain plane and, hence, ϵ_z is uniform, though not specified in advance. The engineer is interested in accurate information about stresses and strains when the tube is partly elastic and partly plastic. There is no general method for solving such elastic-plastic problems. However, because of the cylindrical symmetry of this problem, the shape of the elastic-plastic boundary is known in advance. It is this simplifying feature which has enabled solutions to be obtained, in spite of other difficulties inherent in elastic-plastic problems.

Plane Strain, $\epsilon_z = 0$

The conditions are the same as before, except that $\epsilon_z = 0$ at all stages of deformation. More concisely, the state of plane strain may be defined by the two conditions, that the flow is everywhere parallel to the x,y plane, and independent of z.

ACKNOWLEDGMENT: This chapter is published by permission of the chairman of Tube Investments, Ltd.

If the hydrostatic component of stress is compressive, a state of plane strain can be achieved artificially by confining the specimen between effectively smooth rigid blocks, with faces parallel to the planes of flow, and applying forces or displacements independent of the z coordinate. In practice, approximate states of plane strain are found in two situations. The most common arises when the dimensions of the plastic region in the plane of flow are small compared with its width, either due to loading being applied over only a restricted region of the body, as in rolling or indentation of a strip between flat dies, or due to the shape of the specimen, as in the stretching of a wide notched bar. The elastic material on either side of the plastic region then severely restricts the lateral strain, except in narrow zones near the edges. Alternatively, in a cylindrical tube whose wall thickness is small in comparison with its diameter, the circumferential strain is negligible if internal contraction is prevented by supporting the bore on a closely fitting mandrel, as in the ironing of thin-walled cups.

In most problems of practical interest, the shape of the elastic-plastic boundary is not known in advance, and we are compelled by mathematical difficulties to neglect the elastic component of strain, and to regard the material as rigid-plastic. The solutions thus obtained provide only partial information. With careful interpretation, however, such information is valuable in many applications, particularly in estimating the loads and the over-all mode of deformation, either at the stage when large plastic strains can first develop without constraint by surrounding elastic material, or when the over-all plastic deformation is large, as in many forming processes. The loading at the former stage, at which the over-all permanent deformation is still of an elastic order of magnitude only, roughly corresponds to the loading at the precisely defined yield point of a similar rigid-plastic body, before which no deformation occurs.

Plane Stress, $\sigma_z = 0$

The state of stress is two-dimensional, all the components σ_z, τ_{zx}, τ_{yz} in the z direction being zero

The only nontrivial problems arise when a plane sheet or plate is loaded along its edge by forces acting in its plane, e.g., the yielding of a notched strip in tension. Unless the sheet bends out of its plane, the stress distribution should be very nearly plane except near the edges, even if the local thickness t varies, provided that dt/ds is small compared with unity, where s is the distance in any direction parallel to the surface.

As in plane strain, the material is commonly assumed to be rigid-plastic for the sake of mathematical simplicity.

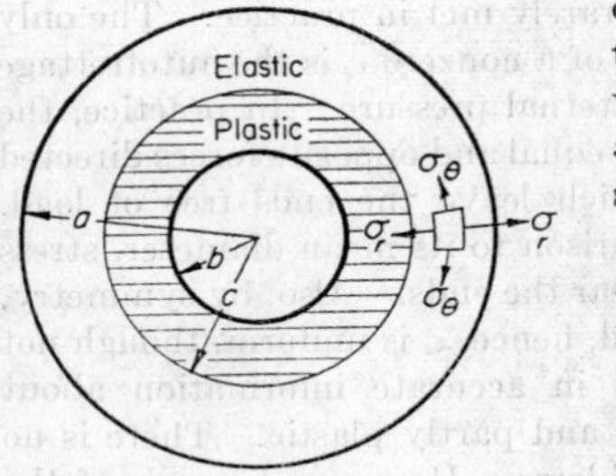

FIG. 50.1. Transverse section of a partially yielded tube.

50.2. THE EXPANSION OF A CYLINDRICAL TUBE

The problems have been defined on p. 50-1, and the notation for stresses and dimensions is shown in Fig. 50.1. We consider in detail only the closed-end condition, for which the axial end load is $\pi b^2 p$, where p is the internal pressure. Work on this problem for various end conditions has been summarized by Hill ([1], Chap. 5) and by Steele [6].

Yield Criterion

Whether the Mises or Tresca yield criterion is more appropriate depends on the tube material. Here we adopt the Tresca criterion since it is simpler and, in any case, provides a very good approximation to the Mises criterion for the closed-end condition. Assuming that there is no work hardening, and that σ_z is the intermediate principal stress (which is confirmed a posteriori), the Tresca criterion gives

$$\sigma_\theta - \sigma_r = \sigma_e \tag{50.1}$$

in the plastic region, where σ_e is the yield stress in uniaxial tension.

Statically Determined Stresses

When the tube is stressed elastically, the well known Lamé solution [Eq. (37.54)] holds, and it is readily shown that, for all end conditions, provided that the tube is initially free of residual stresses, yielding begins on the internal surface, at the pressure

$$p_0 = \tfrac{1}{2}\sigma_e\left(1 - \frac{b_0^2}{a_0^2}\right) \tag{50.2}$$

where a_0 and b_0 are the initial values of a and b, the current external and internal radii. This pressure is, however, very sensitive to initial residual stresses at the bore [7]. As the pressure is increased, the elastic-plastic boundary $r = c$ spreads outward, and the effect of any residual stresses decreases, being virtually eliminated when the tube is entirely plastic. Ignoring any such effect, the boundary conditions for the outer elastic region are $\sigma_r = 0$ at $r = a$, and Eq. (50.1) at $r = c$, and Lamé's solution gives:

$$\begin{aligned} \frac{\sigma_r}{\sigma_e} &= -\frac{c^2}{2a_0^2}\left(\frac{a_0^2}{r^2} - 1\right) \\ \frac{\sigma_\theta}{\sigma_e} &= \frac{c^2}{2a_0^2}\left(\frac{a_0^2}{r^2} + 1\right) \qquad c \le r \le b \\ \frac{\sigma_z}{\sigma_e} &= \frac{E\epsilon_z}{\sigma_e} + \nu\frac{c^2}{a_0^2} \end{aligned} \tag{50.3}$$

and

$$\frac{E}{\sigma_e}\frac{u}{r} = \frac{E}{\sigma_e}\epsilon_\theta = \frac{1}{\sigma_e}[\sigma_\theta - \nu(\sigma_r + \sigma_z)] \tag{50.4}$$

where u is the radial displacement.

In the plastic region, the equation of equilibrium, combined with the yield criterion, leads to

$$\frac{\partial\sigma_r}{\partial_r} = \frac{\sigma_\theta - \sigma_r}{r} = \frac{\sigma_e}{r} \tag{50.5}$$

Integrating, and using the condition for continuity of σ_r across $r = c$,

$$\begin{aligned} \frac{\sigma_r}{\sigma_e} &= -\ln\frac{c}{r} - \frac{1}{2}\left(1 - \frac{c^2}{a_0^2}\right) \\ \frac{\sigma_\theta}{\sigma_e} &= -\ln\frac{c}{r} + \frac{1}{2}\left(1 + \frac{c^2}{a_0^2}\right) \qquad a \le r \le c \end{aligned} \tag{50.6}$$

so that the internal pressure is given by

$$\frac{p}{\sigma_e} = \ln\frac{c}{b} + \frac{1}{2}\left(1 - \frac{c^2}{a_0^2}\right) \tag{50.7}$$

Assuming that the tube is only moderately thick ($a_0/b_0 < 4$, say), the strains remain of an elastic order of magnitude as long as the tube is only partly plastic, and dimensional changes can be neglected. Then σ_r and σ_θ throughout the tube are statically determined by Eqs. (50.3), (50.6), and (50.7), in terms of the pressure p through the parameter c, independently of the end conditions and the stress-strain relations. The same is not true for the Mises criterion since it involves σ_z, which greatly complicates the problem.

Other Quantities

The strain distribution and σ_z can only be determined accurately by following the history of the deformation, making use of some forms of incremental stress-strain relations in the plastic region. With the Reuss relations, a solution in closed form is impossible, but a numerical method of solution has been devised by Hill, Lee, and Tupper [8], and calculated for $a/b = 2$ and $\nu = 0.3$. This solution is valuable, as it enables us to deduce approximate expressions for some quantities, and also to assess the accuracy of various other treatments of the problem. The most useful result is that the quantity

$$q = \frac{1}{\sigma_e} [\sigma_z - \tfrac{1}{2}(\sigma_r + \sigma_\theta)] \tag{50.8}$$

is found to vary only within the narrow limits ± 0.075, and tends to zero with increasing strain. Thus, the assumption that σ_z is the intermediate principal stress is correct, and a rough estimate of σ_z is obtained from Eq. (50.8) by assuming that $q = 0$. Furthermore, the Mises criterion can be written in the form

$$\sigma_\theta - \sigma_r = \frac{2\sigma_e}{\sqrt{3}} (1 - q^2)^{1/2} \tag{50.9}$$

and can therefore be approximated to within 0.3% by writing $2\sigma_e/\sqrt{3}$ instead of σ_e in Eq. (50.1) and elsewhere. The state of stress is everywhere approximately pure shear plus a hydrostatic pressure. This would strictly be true only if the material were incompressible, in which case $q = 0$ and the axial extension would be zero. The accurate theory predicts that the tube will extend but that the amount is limited, and that, after becoming fully plastic, it will deform under approximately plane strain conditions.

A good approximation to the hoop strain at the outer surface, so long as it is elastic, is obtained by eliminating σ_z between Eqs. (50.4) and (50.8), and substituting from Eq. (50.3) for σ_r and σ_θ, to give

$$\left(\frac{E\epsilon_\theta}{\sigma_e}\right)_{r=a} = \tfrac{1}{2}(2 - \nu)\frac{c^2}{a^2} - \nu(q)_{r=a} \approx \tfrac{1}{2}(2 - \nu)\frac{c^2}{a^2} \tag{50.10}$$

which underestimates the true value by less than 3% when $a/b = 2$. No such simple and reasonably accurate expressions for ϵ_z, or ϵ_θ in the bore after yielding begins, have been found.

Koiter [9] has obtained a much simpler solution in closed form by using the Tresca yield criterion and its associated incremental stress-strain relations ([1], p. 52). Comparison with the calculations of Hill, Lee, and Tupper [8] for $a/b = 2$ shows that both ϵ_z and ϵ_θ in the bore are underestimated, the errors increasing to 25 and 17% at a given pressure when the tube is nearly all plastic. The errors in σ_z are even larger.

Other treatments for the closed-end conditions, based on the Hencky relations between stress and total strain, are also simpler, since the difficulty of following the strain history is avoided. In general, these relations are incorrect, but, in this problem, it so happens that reasonable approximations to some of the quantities are obtained (see [1], chap. 5). Allen and Sopwith's [10] solution, which takes account of compressibility, is the most accurate for both σ_z and ϵ_z, though the error in ϵ_z rapidly becomes very large once the tube is fully plastic. The expansion of the bore is badly underestimated well before this, by a greater amount than by Koiter. Steele [6] has achieved a simpler solution in closed form by assuming incompressibility, but his predictions of both ϵ_z and the expansion of the bore are considerably worse than those of Allen and Sopwith.

Provided that the final internal pressure $p_F \leq 2p_0$ [Eq. (50.2)], all elements of the tube recover elastically during unloading. The residual stresses after unloading are found by subtracting the elastic stresses given by Lamé's equations for an internal pressure p_F from the stresses before unloading.

The effect of work hardening on the relation between pressure and outside expansion can be estimated by a method due to Nadai [2], which has been briefly summarized by Hill ([1], chap. 5).

50.3. THEORY OF PLANE PLASTIC STRAIN

The theory of plane plastic strain has been set out in great detail by Hill ([1], chap. 6), who simplified and improved the original presentation by Hencky [11], and Geiringer [12] (see also Prager and Hodge [3]). A summarized version is given here, including mention of work done since Hill's account was written.

Assumptions

The theory is based on the assumptions that the material is (1) rigid-plastic, (2) isotropic with no Bauschinger effect, (3) not influenced in its yielding by hydrostatic pressure, and (4) nonhardening.

From assumption 1, it follows that the volume of an element is constant and, therefore, that any strain increment in plane strain has principal components $d\epsilon$, $-d\epsilon$, 0, which constitute a pure shear. Hence, from assumption 2, it can be shown that the state of stress at each point in the deformed region has principal components of the form $(-p+\tau)$, $(-p-\tau)$, $-p$, where the stress σ_z normal to the planes of flow is equal to $-p$, the hydrostatic component, and so

$$\sigma_z = \tfrac{1}{2}(\sigma_x + \sigma_y) \tag{50.11}$$

From assumptions 3 and 4, it follows that τ must be a constant, k say, and the theory for any yield criterion takes the same form. Only the value of k depends on the yield criterion; for the Tresca and Mises criteria, k is equal to $\sigma_e/2$ and $\sigma_e/\sqrt{3}$, respectively.

Basic Equations Referred to the Slip-line Field

It can be shown that, in order to solve specific problems, the fundamental unknown element which must be determined in the deforming region is the network made up of the two orthogonal families of curves, whose directions at every point coincide with those of the maximum shear strain rate. These lines are known as slip lines and the network is called the slip-line field.[1] Since the material is isotropic, the slip lines coincide with the trajectories of maximum shear stress. The two families of slip lines are labeled α and β lines, respectively, and Fig. 50.2 shows the stress components acting on a curvilinear element bounded by two pairs of neighboring slip lines. By convention, if the α and β lines are regarded as a pair of right-handed curvilinear axes, the line of action of the greatest principal stress falls in the first and third quadrants. If ϕ is the anticlockwise angular rotation of the α line from the x axis, it may be shown, from Mohr's circle of stress, radius k, that

$$\sigma_x = -p - k \sin 2\phi \qquad \sigma_y = -p + k \sin 2\phi$$
$$\tau_{xy} = k \cos 2\phi \tag{50.12}$$

Thus the state of stress depends only on p and ϕ, which, in general, vary throughout the deforming region.

[1] In mathematical terminology, the reason for the basic importance of the slip lines is that they are the real characteristics of the basic equations which are hyperbolic. If the solution is known within an area bounded by characteristics, it cannot be extended outside without additional information, because the solution within does not define all the stress and velocity gradients normal to the characteristics. No other curves have this property.

The basic equations are most usefully expressed in the form of relations along the slip lines. The yield criterion is automatically satisfied by giving k its appropriate value. The equilibrium equations are completely equivalent to the relations

$$\begin{aligned} p + 2k\phi &= \text{const} \qquad \text{on an } \alpha \text{ line} \\ p - 2k\phi &= \text{const} \qquad \text{on a } \beta \text{ line} \end{aligned} \tag{50.13}$$

Deformation in the plane of flow is most conveniently expressed in terms of velocity components u and v along the α and β lines, respectively. Since these lines coincide with the directions of maximum shear strain rate, the condition for zero volume change

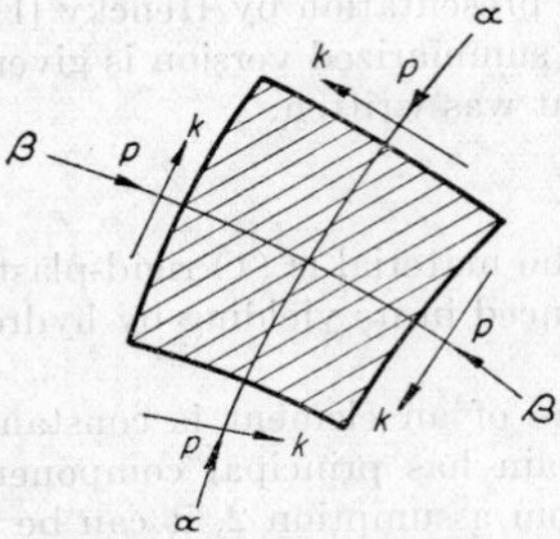

FIG. 50.2. Stress on a small curvilinear element bounded by slip lines.

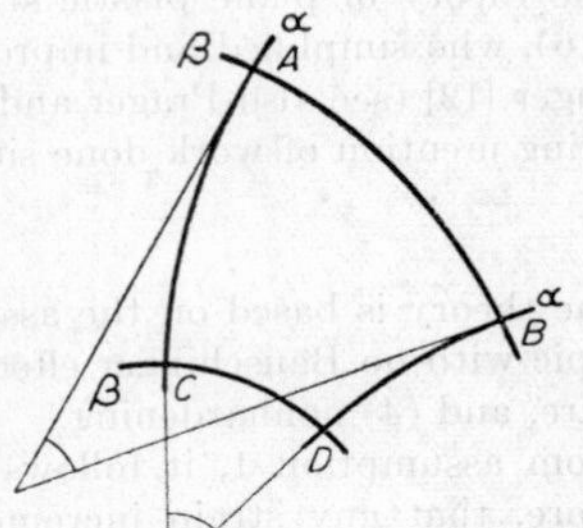

FIG. 50.3. Typical slip-line field demonstrating its geometrical property.

implies that the rate of extension along each slip line is zero. This is equivalent to the relations

$$\begin{aligned} du - v\,d\phi &= 0 \qquad \text{along an } \alpha \text{ line} \\ dv + u\,d\phi &= 0 \qquad \text{along a } \beta \text{ line} \end{aligned} \tag{50.14}$$

It should be emphasized that the stresses are independent of the rate of strain. The velocities do not really involve the element of time, but can be taken as rates of displacement with respect to any monotonically varying quantity, such as an angle or a characteristic length.

Geometry of the Slip-line Field

An important geometrical property of slip-line fields follows from Eqs. (50.13). Consider a curvilinear quadrilateral $ABDC$, bounded by two α lines AC and BD and two β lines AB and CD (Fig. 50.3). Integrating Eqs. (50.13) along ABD and ACD to obtain the pressure difference between A and D, the answers must be the same, and it follows that

$$\phi_D - \phi_C = \phi_B - \phi_A \tag{50.15}$$

In particular, if a segment AB of a β line is straight, then so is the corresponding section CD of any other β line cut off by the α lines through A and B, and, from Eqs. (50.13) and (50.14), p and v are both constant along each of the straight segments. An important special case is a network consisting of straight lines through a point B and concentric circular arcs (for example, BEC in Fig. 50.8a); point B is a stress singularity.

Lines of Discontinuity

The velocity may be discontinuous only across a slip line. For continuity of the material, only the tangential component (u or v) may be discontinuous, and, from Eqs. (50.14), it follows that the jump in u or v is constant along the slip line. Such discontinuities are peculiar to a rigid-plastic material, and correspond in a real elastic-plastic material to a more or less narrow transition region, where the shear strain rate is large; the transition will tend to be less sharp in a rapidly hardening material. A small element crossing a velocity discontinuity abruptly changes its direction, and undergoes a sudden finite shear parallel to the discontinuity.

Stress discontinuities are also possible in a rigid-plastic material across any curve in the plastic region except a slip line. They correspond in a real elastic-plastic material to a narrow transition region of elastic material through which the stress changes rapidly. The velocity must be continuous across the discontinuity, and the rate of extension along it must be zero.

Solving Problems

Consider a particular problem, in which a rigid-plastic body is deformed plastically under given stress and velocity boundary conditions. In general, only part of the body is deforming, and the remainder is rigid. The boundaries between the two parts are slip lines. The solution consists of constructing a slip-line field and an associated velocity distribution in the deforming region, which satisfy all the stress and velocity conditions that directly concern the region, as well as the condition that the rate of plastic work must be everywhere positive. It must also be shown that there exists an associated stress distribution in the rigid regions, which is in equilibrium and nowhere violates the yield criterion.

Hill [13] has shown that, in the deforming regions of such a solution, the stresses, but not necessarily the deformation mode, are unique and independent of initial residual stresses. In the rigid regions, the stresses are not unique, and, hence, though parts may be plastic in any particular solution, the elastic-plastic boundary is not thereby defined. The deforming regions of a rigid-plastic body roughly correspond to the regions in a similar elastic-plastic body, where large plastic strains are possible without constraint from surrounding elastic material. When the deformation mode is not unique, Hill [14] has shown that, to select the correct one, account must be taken of the rate of change of the applied loads, of the shape, and of the hardness, though the maximum possible extent of the deforming region may be found without these considerations [15].

There is, at present, no explicit step-by-step method of finding solutions. Each problem has its peculiarities, which must be appraised qualitatively. The location of the deforming zone is not known in advance, and its approximate position must first be arrived at by physical intuition, aided when possible by experiment. Generally, even if the given boundary conditions involve only the stresses, the problem is not statically determinate (in the sense that a stress distribution can be constructed independent of velocity considerations), because velocities at the boundary of the deforming zone must be compatible with rigid motion outside.

Constructing Slip-line Fields

Certain distinct types of construction, summarized below, frequently recur, and are used to build up successive domains of a slip-line field, across the boundaries of which the curvature of the slip lines often changes abruptly. Only graphical methods of construction are described. These are the simplest and quickest, though, for greater accuracy, numerical methods have often been used, and have been checked in special simple cases by exact analysis ([1], chap. 6).

1. Extending the field beyond two known intersecting slip lines OA and OB (Fig. 50.4): The field is uniquely determined within $OACB$, where AC and BC are the, as yet, unknown slip lines through A and B. The task consists of finding the nodal points $P(m,n)$ of a network formed by the slip lines passing through selected base points $(m,0)$ and $(0,n)$ on OA and OB. From Eq. (50.15), ϕ is determined at all points by the

relation

$$\phi_{m,n} = \phi_{m,0} + \phi_{0,n} - \phi_{0,0}$$

It is often convenient, though not essential, to choose the base points a constant angular distance $\Delta\phi$ apart; the net is then equiangular, all adjacent nodal points being the same angular distance apart.

To find the point (1,1), the two small arcs of which it is the intersection are replaced graphically by chords, whose slopes are the mean of their known terminal slopes. This process is repeated for (2,1), (1,2), etc., until all nodal points are found.

An important special case occurs if O is a singularity, so that one of the given slip lines, OB say, is effectively an arc of zero radius but finite angular span. The method of construction is essentially unchanged.

2. Constructing the field adjacent to a curved segment AB, along which p and ϕ are known (Fig. 50.5): Most commonly, AB is part of the external surface of a specimen, on which given external forces are acting. Then the normal and shear stresses acting on the surface are known, and two values of the third stress component can be found to satisfy the yield criterion (e.g., a free surface may be in compression or

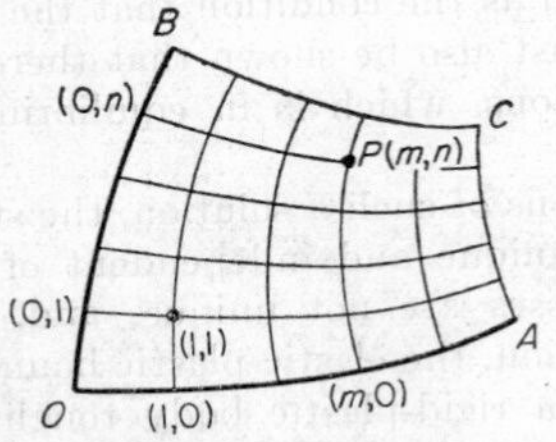

FIG. 50.4. Constructing a field, given two slip lines OA and OB.

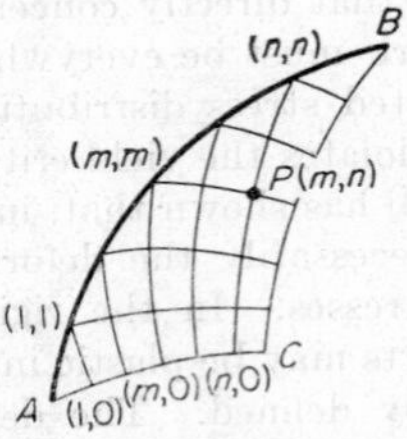

FIG. 50.5. Constructing a field, given p and ϕ on AB.

tension); the correct alternative depends upon the problem as a whole, and this determines which is the α and which the β family of slip lines. The field is uniquely defined within ACB bounded by the, as yet, unknown slip lines through A and B. A network is formed in this area by the slip lines passing through selected base points (m,m) and (n,n) on AB. From Eqs. (50.13), it is found that

$$\phi_{m,n} = \pm \frac{1}{4k}(p_{m,m} - p_{n,n}) + \tfrac{1}{2}(\phi_{m,m} + \phi_{n,n})$$

the $+$ or $-$ sign being taken according to whether AC is an α or a β line. Since ϕ is known at all points, the network can be constructed, point by point, exactly as described in method 1.

The base points on AB can only be chosen so that the net is equiangular if ϕ or p is constant along AB. The former is rare, but the latter occurs when AB is a free surface. If the stress components on AB are such that it is a slip line, the triangular region degenerates to the line AB, and additional information is needed to extend the field. This is the distinguishing feature of a slip line.

3. Constructing the field in the angle ($<\pi$) between a known slip line OA and a curve OB, along which the inclination of the slip lines ϕ is known (Fig. 50.6): Let OA be an α line, then provided that the α direction on OB at O falls between OA and OB, and all α lines cut OB at an acute angle, the field is uniquely defined within OAB bounded by the, as yet, unknown slip line AB. First, if necessary, the field is continued around O as a singularity to OA', the α line which coincides with the α direction on OB at O; this type of construction was described in method 1. Base points $(m,0)$ on OA' define a network with nodal points (m,m) on OB. Unless ϕ is constant on OB, the latter points must be found by a method of successive approximation, since ϕ is

not known in advance. For example, (1,1) is first approximated by the intersection P' of OB with a line through (1,0) normal to OA'; for the next approximation, the line is drawn with a slope corresponding to the mean value of ϕ at P' and (1,0) to give P'', and so on. Other points in the network are plotted successively in the normal way.

When OB is straight, ϕ is constant if the shear stress acting along OB is constant. In particular, if the specimen is in contact with a smooth flat rigid surface, the slip lines meet it at 45°, and the problem is identical with that of finding the field defined by OA and its mirror image in OB.

A similar, but more general, problem arises when the specimen is in contact with a rigid surface, along which the coefficient of friction is constant; p and ϕ are then related, and the successive approximation involves p as well as ϕ.

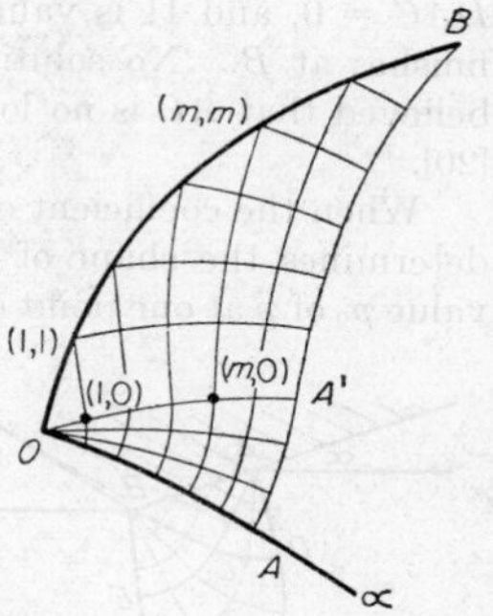

Fig. 50.6. Constructing a field, given one slip line OA and ϕ on the curve OB.

Constructing Velocity Fields

A velocity distribution can be found only on the basis of a known or postulated slip-line field. It is obtained either by means of Eqs. (50.14) or their finite-difference equivalents, or by the graphical construction of a hodograph [16],[17]. Only the latter method will be described here. Before doing so, we summarize the typical boundary-value problems analogous to those already described for the stresses.

1. When normal velocities are given along intersecting slip lines OA and OB (Fig. 50.4), the velocities are uniquely defined in $OACB$.

2. When u and v are given along an arc AB which is not a slip line and which is not cut twice by the same slip line (Fig. 50.5), the velocities are uniquely defined in ACB.

3. When the normal component of velocity is given along a slip line OA', and a boundary condition $f(u,v) = 0$ is given along an intersecting curve OB (Fig. 50.6), the velocities are uniquely defined in OBA', provided there is no intervening slip-line field corresponding to a singularity at O. For example, OB might be a fixed rigid surface so that the velocity component normal to it is zero.

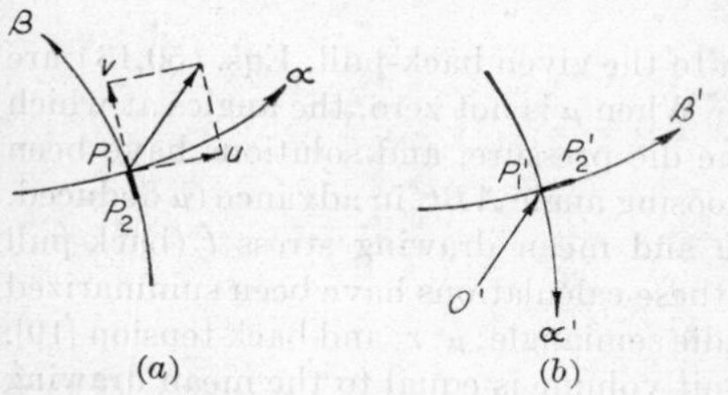

Fig. 50.7. Corresponding points in (a) a slip-line field and (b) a hodograph.

The resultant velocity of a point P_1 in the slip-line field is represented in the hodograph by the vector $O'P'_1$, where O' is fixed and the vector is equal in magnitude and parallel in direction to the velocity at P_1 (Fig. 50.7). Since the rate of extension is zero along a slip-line direction, the vector $P'_1P'_2$ in the hodograph, representing the increments in velocity between two adjacent points P_1 and P_2 on the slip line, is perpendicular to P_1P_2; that is, each slip line is perpendicular to its image in the hodograph, and the angular variations between the corresponding points are identical in the two networks. Thus, the hodograph has the distinctive geometrical property Eq. (50.15)] of a slip-line field, and can be constructed in a similar manner.

Slip-line Field Solutions

The following brief summaries of slip-line field solutions give some idea of the types of problem which have been solved and the methods employed. Velocity discontinuities are indicated in the figures by arrows on the slip lines across which they occur.

Sheet Drawing through a Wedge-shaped Die ([1], Chap. 7). This is a steady-motion problem, in which the stress and velocity do not vary at a fixed point. It so happens that the velocity conditions for rigid motion in and out of a deforming region between the dies are satisfied by making AC straight in the fields I and II shown in Fig. 50.8*a,b*. As the reduction r is increased, field I gives way to II when angle $DAC = 0$, and II is valid so long as the velocity discontinuity transmitted from H finishes at B. No solutions have been calculated for higher reductions, when it is believed that AC is no longer straight and solution is, therefore, much more difficult [20].

When the coefficient of friction $\mu = 0$, all slip lines meet the die face at 45°, which determines the shape of the field, its extent being governed by the reduction. The value p_0 of p at one point on the entry slip lines is obtained by equating the longitudinal

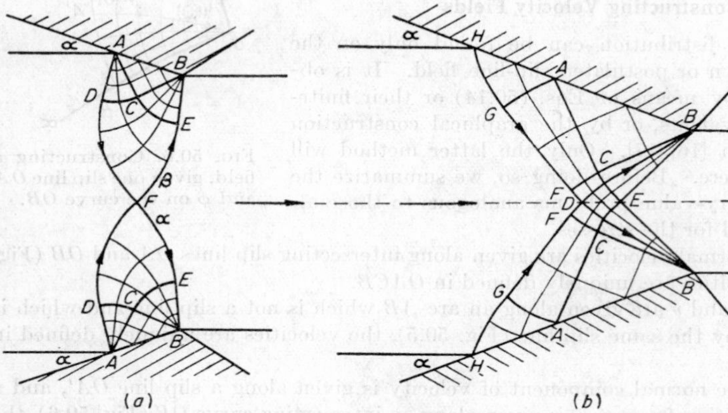

FIG. 50.8. Slip-line fields for sheet drawing: (*a*) Moderate reductions; (*b*) somewhat larger reductions.

component of the total force acting over them to the given back-pull; Eqs. (50.13) are used to express p at any point in terms of p_0. When μ is not zero, the angles at which the slip lines meet the die face depend on the die pressure, and solutions have been obtained either of type I for zero back-pull, choosing angle ABC in advance (μ deduced, not chosen) [18], or of type II for chosen μ and mean drawing stress t (back-pull deduced, not chosen) [19]. The results of all these calculations have been summarized in an empirical formula for t in terms of α, the die semiangle; μ; r; and back tension [19].

When $\mu = 0$, the mean plastic work per unit volume is equal to the mean drawing stress t_0, and, hence, for a nonhardening material a mean effective strain $\bar{\epsilon}$ can be defined by putting $t_0 = \sigma_e \bar{\epsilon}$, where t_0/σ_e is known from the preceding theory. Assuming, to the first approximation, that $\bar{\epsilon}$ is independent of μ and of the shape of the true stress-strain curve σ,ϵ, the drawing stress for a hardening material is obtained by using the mean value $\int_0^{\bar{\epsilon}} \sigma \frac{d\epsilon}{\bar{\epsilon}}$ for σ_e in the formula for t.

Orthogonal Indentation of a Plane Surface by a Smooth Wedge ([1], Chap. 8). The entire configuration remains geometrically similar as indentation proceeds up to the stage at which deformation extends to the base or the side of the block. The general form of the slip-line field is determined by the shape of BD (Fig. 50.9), and the conditions that the slip lines meet the smooth wedge surface at 45° and that AC is a stress-free surface. Fortunately, geometrical similarity can be preserved with BD straight; for then AC is straight, the velocities along one family of slip lines are zero, and along the others meeting AC they are uniform. Hence, AC preserves its shape and direction by moving parallel to itself. The slope of AC depends upon the wedge angle, and is determined together with its length by the conditions that $AC = AB$, and that the volume of material displaced by the wedge must be equal to the volume in the lip formed above the original surface. The stress distribution and, hence, the load on the wedge follow from the fact that $p = k$ on the free surface AC.

Compression between Smooth Flat Dies. The solution depends on t/b, the ratio of the thickness of the block or sheet to the breadth of the dies, and three ranges must be distinguished. Slip lines meet the smooth dies at 45° in all solutions.

When $t/b > 8.74$, the material is forced out of the surfaces adjacent to the dies, and the field I (Fig. 50.10*a*) is the extreme case for that of wedge indentation (wedge angle = 180°), except that it can be extended further. There are an infinite number of possible deformation modes involving positive plastic work and compatible with the motion of the rigid dies.

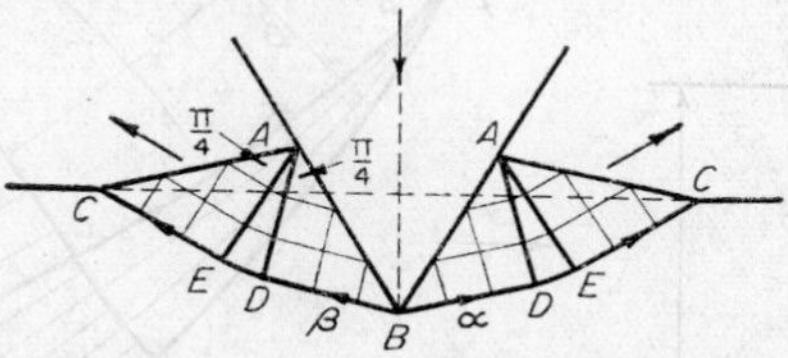

FIG. 50.9. Slip-line field for wedge indentation.

When $8.74 > t/b \geq 1$, field II (Fig. 50.10*b*) gives a lower die pressure ([1], chap. 9) and is therefore to be preferred (see Limit Analysis, p. 50–13). Two regions of deformation join at a point midway between the dies, and the rigid ends are forced apart without the surfaces being deformed. The field and boundary conditions are clearly analogous for those for sheet drawing at lower reductions (for example, Fig. 50.8*a*).

For integral values of $b/t \geq 1$, all the slip lines are straight, but such a field is not valid for nonintegral values, since the velocity discontinuities initiated at the corners would terminate on the exit slip line, which is incompatible with the rigid-body motion of the two ends. In fact, the exit slip lines are in part curved, and cannot be guessed correctly in advance. However, a fortuitous simplification, which has made solution possible, is that the slip-line field and hodograph are identical [20]. This is because their boundary conditions are geometrically identical (with various boundaries interchanged as indicated in Fig. 50.11*a,b*), and the angular variations in the two fields are the same. This is also true for fields I and II, but is not true in general. Thus, a hodograph field constructed to satisfy only the velocity boundary conditions automatically provides a slip-line field satisfying the stress-boundary conditions. Even so, a lengthy numerical procedure had to be adopted to calculate the field. The two types of fields (III and IV in Fig. 50.11*b,c*) constructed in the ranges $1 \leq b/t \leq \sqrt{2}$ and $\sqrt{2} \leq b/t \leq 2$, respectively, are distinguished by the positions of the velocity discontinuities and corresponding regions where sections of one family of slip lines are straight, both of which vanish at the change-over value of $b/t = \sqrt{2}$.

For field I, $p = k$ on the free surfaces, and the uniform die pressure is $2k(1 + \pi/2)$. For fields II, III, and IV, p at a chosen point on an exit slip line was deduced from the condition of zero end thrust on the ends of the block. The results for the die load L are shown in Fig. 50.12.

Yielding of Notched Bars. We consider a bar symmetrically notched on opposite sides. Provided that the notches are sufficiently deep, yielding is confined to a region

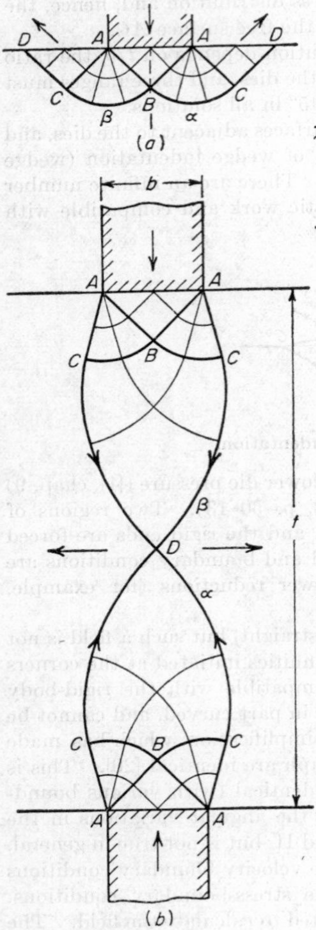

FIG. 50.10. Slip-line fields for compression between smooth flat dies: (*a*) $t/b > 8.74$, showing one die only; (*b*) $8.74 \geq t/b \geq 1$.

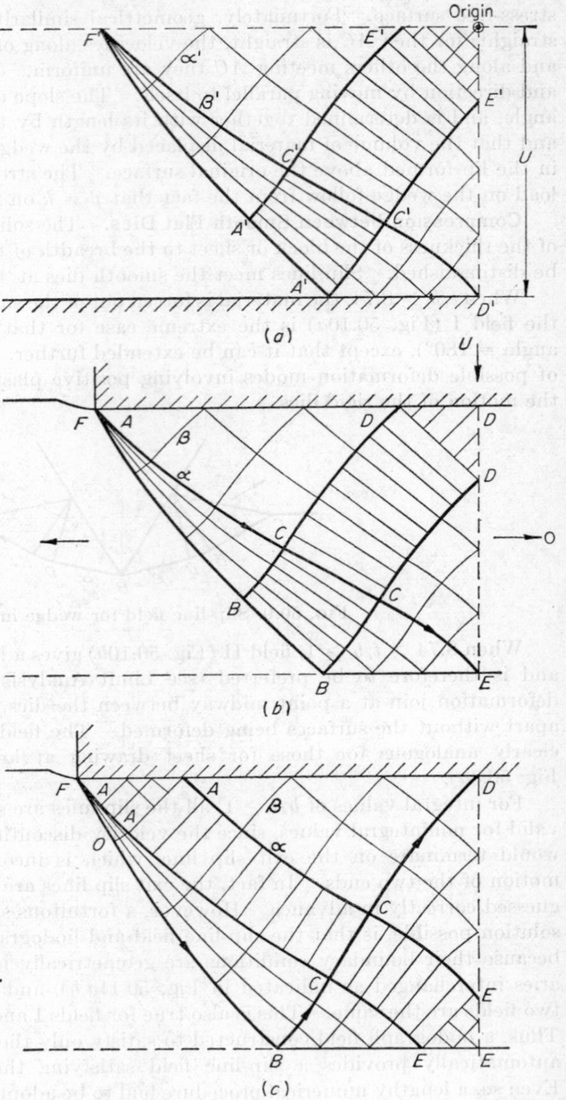

FIG. 50.11. Compression between smooth flat dies (one quadrant of each field is shown to scale): (*a*) Hodograph for $b/t = 1.39$; (*b*) slip-line field for $b/t = 1.39$; (*c*) slip-line field for $b/t = 1.44$.

between them. The shape of the stress-free surfaces of the notches defines the slip-line fields in their vicinity.

If the bar is stretched under tension ([1], chap. 9) $p = -k$ on both notch surfaces, and the solution is obtained by extending the fields symmetrically from the surfaces to meet at a point N, across which p is continuous (Fig. 50.13*a*).

If the bar is bent by pure couples [21], the same form of field may be valid, but now $p = +k$ on one notch surface, and $p = -k$ on the other. There is a stress singularity at N, across which there is a jump in p. Each end of the bar rotates about N, and there are no velocity discontinuities. This solution, however, breaks down for sharper notches if the jump across N is greater than $k\pi$, because then it is impossible to find a stress distribution in the rigid corners at N which does not violate

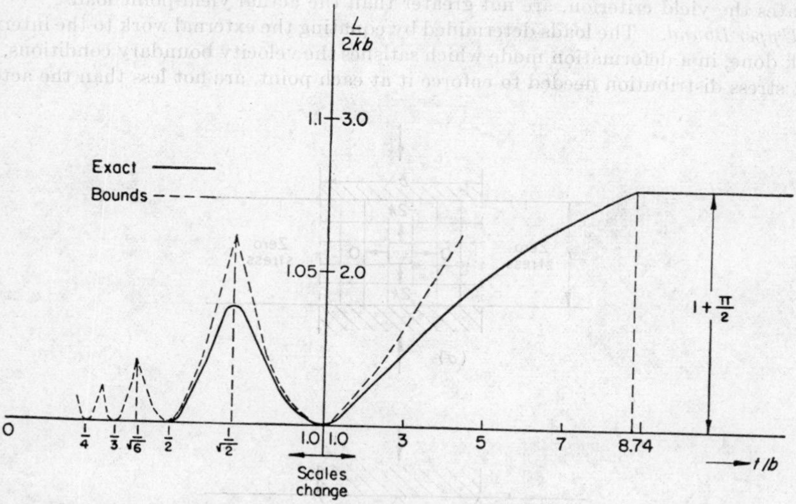

FIG. 50.12. Accurate and approximate calculations of the relation between die load and geometry for compression between smooth flat dies.

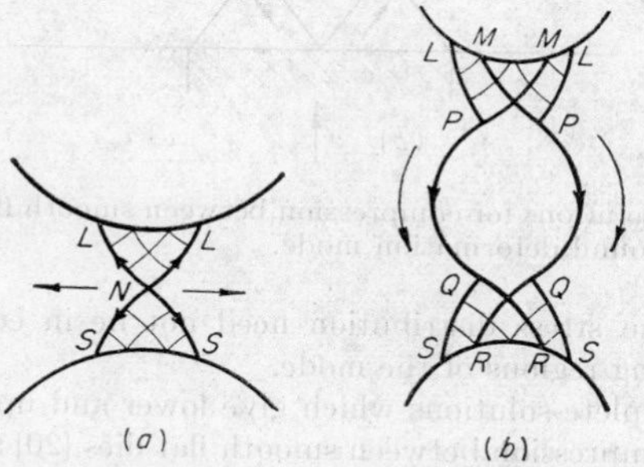

FIG. 50.13. Slip-line fields for symmetrically notched bars: (a) In tension; (b) in pure bending (sharp notches).

the yield criterion [22]. This difficulty is avoided if the two regions are joined by two circular slip lines PQ (Fig. 50.13b), leaving a rigid pivot in the center, about which the two ends rotate. Points P and Q are determined by the symmetry of the problem and the condition that p changes from $+k$ to $-k$ along $MPQR$.

Approximate Methods

Limit Analysis. Limit theorems [23] enable upper and lower bounds to the loads deforming a rigid-plastic body to be obtained under certain conditions without a complete solution. This applies to all types of problems, not merely to plane strain (see chap. 49).

For example, when each load is applied to a smooth plane rigid die (as in indentation), or through a rigid part of the body (as in notched-bar tests), and the ratios of the loads are known in advance, except possibly at smooth fixed constraints, the following two kinds of incomplete solutions provide such bounds:

Lower Bound. The loads determined from an equilibrium stress distribution covering the whole body, which satisfies the stress boundary conditions, and nowhere violates the yield criterion, are not greater than the actual yield-point loads.

Upper Bound. The loads determined by equating the external work to the internal work done, in a deformation mode which satisfies the velocity boundary conditions, by that stress distribution needed to enforce it at each point, are not less than the actual

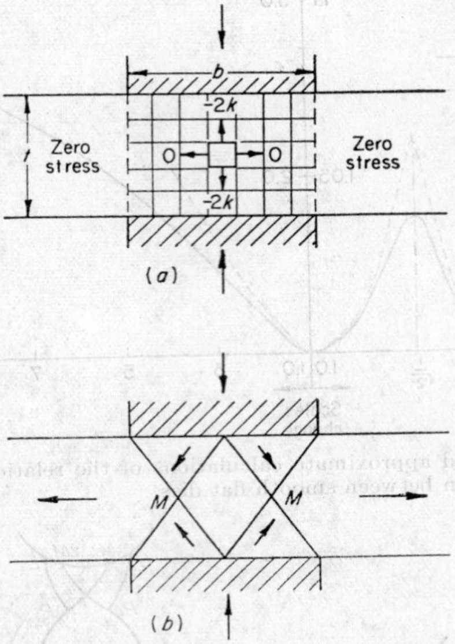

FIG. 50.14. Incomplete solutions for compression between smooth flat dies: (*a*) Lower-bound stress field, (*b*) upper-bound deformation mode.

yield-point loads. The stress distribution need not be in equilibrium, and is only defined in the deforming regions of the mode.

Examples of incomplete solutions which give lower and upper bounds to the load L in the problem of compression between smooth flat dies [20] are shown in Fig. 50.14. In the deformation mode, the number m of intersections M of velocity discontinuities between the rigid regions is chosen to give the least upper bound. The rate of work done at a velocity discontinuity Δv is $k\ \Delta v$ per unit length for unit width of specimen. The resultant bounds,

$$1 \leq \frac{L}{2kb} \leq \frac{1}{2}\left(\frac{mt}{b} + \frac{b}{mt}\right)$$

where $m(m-1) \leq b/t \leq m(m+1)$, are shown as dashed lines in Fig. 50.12.

A complete solution must satisfy both conditions 1 and 2. Most slip-line field solutions are not complete, because, although all conditions pertaining to the deforming region are satisfied, it has not been demonstrated that any stress distribution satisfying conditions 1 exists in the rigid region. Hence, these are strictly only upper-bound solutions, though, in many cases, it seems most likely that they could be completed. Hill [22] has given a general method for determining whether or not the vertex of a wedge of rigid material in any solution can withstand the stresses on it. Unfortunately, no general methods exist for completing solutions, though Bishop [24] has suggested a particular approach involving the construction of fully plastic stress fields in parts of the rigid regions, which is suitable for many problems in plane strain and stress. Such stress fields, and those for lower bounds, frequently involve curves across which the stress is discontinuous.

Other Methods. For certain problems of great practical interest, it has not yet been possible to construct slip-line fields or even reasonably accurate upper- and lower-bound solutions. The latter method is, in any case, not applicable to any process in which energy is dissipated by friction. Recourse then has to be made to ad hoc assumptions to simplify the problem.

The most important example of this is the problem of strip rolling [11],[25]. The main theory, due to von Kármán [26], is based on the assumption that, at any vertical section in the region between the rolls, the normal pressure p on the rolls and the mean horizontal tensile stress q acting over the section are related by the equation

$$p + q = 2k \tag{50.16}$$

This, together with the equation of equilibrium for the vertical slice, gives a linear differential equation for p and q in terms of x, the distance from the exit plane. Various approximate solutions of this equation have been proposed, one of the simplest and most accurate being due to Bland and Ford [27].

Equation (50.16) assumes, in effect, that no redundant work is expended in non-uniform distortion. Orowan [28] has formulated an approximate method of allowing for redundant work, which indicates that the correction is very small when the arc of contact is greater than the mean thickness of the strip, as it is in most industrial cold-rolling practice. Sachs's [4] theories of sheet drawing, of various tube-forming processes, and of wire drawing, are based on a similar assumption, with a similar restriction on the geometry, for which they are reasonably accurate.

50.4. THEORY OF PLANE STRESS

Basic Equations ([1], Chap. 11)

The theory is based on the same assumptions as those for plane strain (p. 50–5), and buckling is not considered. In the plastic region, the set of stress equations (yield criterion and two equilibrium equations) and the set of velocity equations (stress–strain-rate relations) each has a pair of real characteristics only for certain ranges of states of stress and strain rate, respectively.

The velocity characteristics are the two directions of zero rates of extension (cf. slip lines in plane strain), and coincide with the stress characteristics only if the yield function and plastic potential are identical. Denoting this single function by f, the yield criterion and stress–strain-rate relations, in terms of the principal components in the plane of the sheet, are

$$f(\sigma_1,\sigma_2) = \text{const} \qquad \text{and} \qquad \frac{\partial f/\partial \sigma_1}{\partial f/\partial \sigma_2} = \frac{\dot\epsilon_1}{\dot\epsilon_2} \tag{50.17}$$

where the constant depends on the hardness. The two directions of zero rate of extension are each inclined at an angle $(\frac{1}{4}\pi + \frac{1}{2}\psi)$ to the σ_1 direction $(\sigma_1 > \sigma_2)$, where

$$\sin\psi = \frac{\dot\epsilon_1 + \dot\epsilon_2}{\dot\epsilon_1 - \dot\epsilon_2} = \frac{\partial f/\partial\sigma_1 + \partial f/\partial\sigma_2}{\partial f/\partial\sigma_1 - \partial f/\partial\sigma_2} \tag{50.18}$$

Hence, these characteristics are real only if $-\partial f/\partial\sigma_1 \le \partial f/\partial\sigma_2 \le 0$. For example, when the yield function is that of von Mises, then

$$f = \sigma_1^2 - \sigma_1\sigma_2 + \sigma_2^2 = 3k^2 = \sigma_e^2 \qquad \text{and} \qquad \sin\psi = \frac{\sigma_1 + \sigma_2}{3(\sigma_1 - \sigma_2)} \tag{50.19}$$

and, hence,

$$\sigma_1 = 2k \sin(\gamma + \pi/6) \qquad \sigma_2 = 2k \sin(\gamma - \pi/6) \tag{50.20}$$

where

$$\tan\gamma = \sqrt{3}\sin\psi$$

The yield criterion is represented by an ellipse in the σ_1,σ_2 plane (Fig. 50.15), and ψ and the characteristics are real provided $|\sigma_1 + \sigma_2| \leq 3|\sigma_1 - \sigma_2|$, which corresponds to the thickened arc of the ellipse.

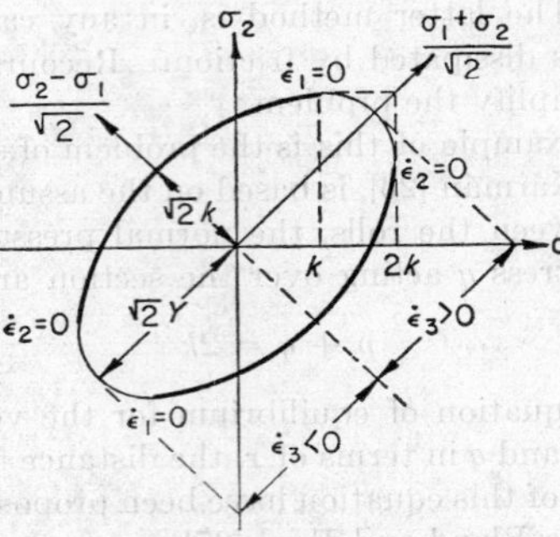

FIG. 50.15. Mises yield criterion in plane stress.

The velocity relations along the characteristics have been derived for any potential and hardness distribution, and the stress relations for any yield function and thickness variation when the hardness is uniform ([1], chap. 11). These relations are more complex than the analogous ones in plane strain. No general methods of integration are available for the parts of the plastic region where the characteristics are unreal.

Lines of Discontinuity [29]

Localized necking through the thickness of the sheet is the most important type of velocity discontinuity, though simple shear discontinuities can occur. The former is

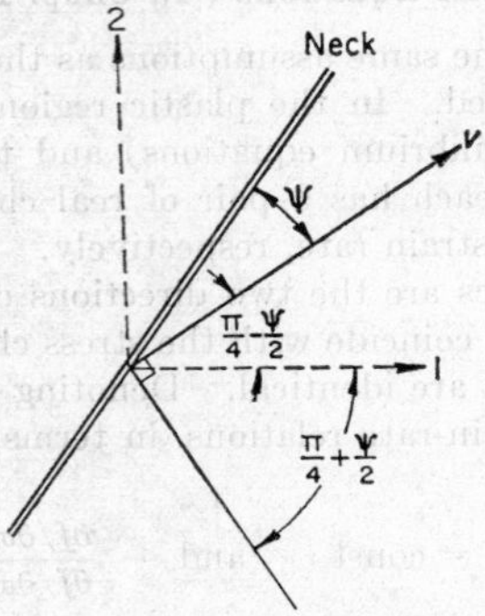

FIG. 50.16. Velocity discontinuity for localized necking.

defined by the relative velocity v between material on either side of the neck (Fig. 50.16). Since one side of the neck, at least, must unload and is therefore rigid, the direction of zero rate of extension at each point must be along the neck, and perpendicular to the local direction of v. These are the directions of the velocity characteristics, and, hence, v makes an angle ψ with the neck.

Necking cannot occur when the characteristics are unreal, or if the material hardens so rapidly that adjacent material cannot unload. For a Mises material in a state of uniform stress, necking is possible only when

$$r/\sqrt{1 - r^2} \leq \sqrt{3} \sin \psi \tag{50.21}$$

where the rate of hardening r is the gradient of the uniaxial stress-strain curve divided by the current yield stress. Thus, in uniaxial tension, where $\dot{\epsilon}_1/\dot{\epsilon}_2 = -2$ and $\psi = \sin^{-1} \frac{1}{3}$, localized necking can occur only when $r \leq \frac{1}{2}$, and the neck is then inclined at $\frac{1}{4}\pi + \frac{1}{2}\psi = 54°44'$ to the tension axis, while v is inclined at $\frac{1}{4}\pi - \frac{1}{2}\psi = 35°16'$. If, however, $r > \frac{1}{2}$, the strip either stretches uniformly or undergoes diffuse necking, symmetrical about the tension axis, according as $r >$ or <1, respectively.

Discontinuities of stress have also been discussed by Hill [29].

Solving Problems

Complete solutions to problems involving a finite amount of distortions, other than those in which the state of stress is uniform, have only been obtained when the configurations are radially symmetric. These are the expansions of a circular hole in an infinite sheet by internal pressure ([1], chap. 11); the contraction of an annular slab by uniform internal tension (as occurs in the deep drawing of a circular blank) ([1], chap. 11); the expansion of an annular slab by uniform external tension [30],[31]; and the bending of a sheet in its plane [32]. The Tresca yield function and the Mises or Tresca potential were adopted in these solutions. When strains are large, accurate solution necessitates taking into account the movement of elements, and numerical methods of integration ([1], chap. 11), solution by successive approximation [32], or a perturbation method [30],[31] were used.

For problems of initial yielding, only two nontrivial types of characteristic field have been constructed (see below), and both these are for a uniformly hard and thick sheet, obeying the Mises yield criterion and potential. The upper- and lower-bound method of approximation (p. 50–14) has been used in a few problems, for example, the yielding of plane slabs with reinforced cutouts [33].

Yielding of a Symmetrically Notched Strip in Tension [29]

Deep V Notch. The characteristic field (Fig. 50.17) is similar to that in the analogous plane-strain problem. In OAB, the stress state is uniaxial tension, the characteristics are straight and, as has been shown above, are inclined at $\lambda = 54°44'$ to the surface. Hence, in OBC the characteristics through O are straight; and the yield criterion, the equilibrium equation, and the tensile state of stress on OB are satisfied by the polar components

$$\sigma_r = k \cos \theta \qquad \sigma_\theta = 2k \cos \theta \qquad \tau_{r\theta} = k \sin \theta$$

where θ is measured from the base line OL, inclined at 2λ to OA. From Eq. (50.19), $\tan \psi = \frac{1}{2} \cot \theta$, and, hence, the curved characteristics have the equation

$$r^2 \sin \theta = \text{const.}$$

They are inflected where $\theta = \frac{1}{2}\pi - \lambda$, and, if continued, would approach OL asymptotically, which is therefore the natural limit of the field. The angle θ_0, corresponding to the innermost characteristic OC through O, is determined by the condition that there should be no shear stress along OO. It is found that $\theta_0 = 0$ when the notch angle $2\alpha = 141°4'$, and the field then shrinks to a coincident pair of characteristics along OO with $\sigma_\theta = 2k$ and $\sigma_r = k$. This solution is also valid for all sharper notches.

There are various deformation modes compatible with the field of Fig. 50.17. The one involving local necking along two characteristics through the centroid is possible only when the hardening rate $r \leq \frac{1}{2}$. An equilibrium stress field not violating the yield criterion has been constructed in the rigid regions [24] to complete this solution (see p. 50–14).

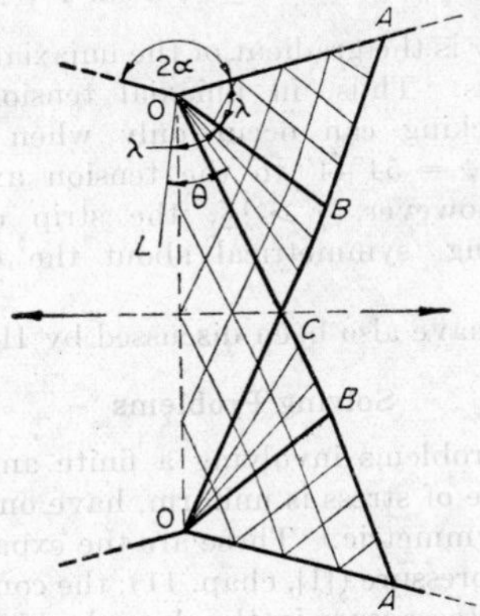

FIG. 50.17. Characteristic field for a symmetrically V-notched strip in tension ($2\alpha > 141°4'$).

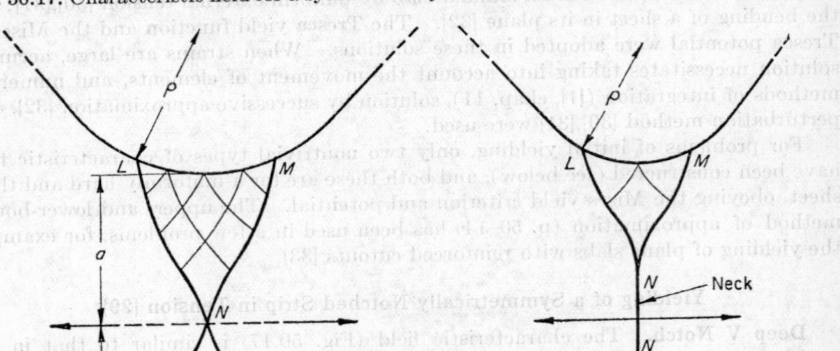

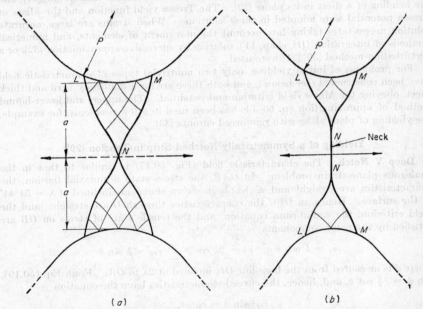

FIG. 50.18. Characteristic fields for a symmetrically notched strip in tension; circular notch root: (a) $a/\rho \leq 1.071$; (b) $a/\rho > 1.071$.

Deep Notch with a Circular Root. Let $2a$ be the distance between the notches, and ρ their radius (Fig. 50.18). For small values of a/ρ, the half of the field LMN defined by the shape of each notch (Fig. 50.18a) is a portion of a radially symmetric field such as occurs in an annular slab of internal radius ρ, yielding under external tension. Hence, in each half of the field, the principal stresses are circumferential and radial, σ_θ and σ_r, and are given by Eqs. (50.20) for σ_1 and σ_2, respectively. From the equilibrium equation $d\sigma_r/dr = (\sigma_\theta - \sigma_r)/r$, and the fact that $\sigma_r = 0$ on the notch surface, it follows that

$$\frac{r^2}{\rho^2} = \frac{\sqrt{3}}{2} \sec \gamma \exp \left[\sqrt{3}\left(\gamma - \tfrac{1}{6}\pi\right)\right] \qquad \tfrac{1}{6}\pi \leq \gamma \leq \tfrac{1}{3}\pi \tag{50.22}$$

The characteristics coincide when $\gamma = \frac{1}{3}\pi$ and $r/\rho = 2.071$. For $a/\rho > 1.071$, these coinciding characteristics are continued straight across the central part of the transverse axis (Fig. 50.18*b*).

No justifying stress field has been constructed in the remainder of the bar. The only deformation mode involves local necking along the outer characteristics, and a diffuse mode for $r > \frac{1}{2}$ is not known.

50.5. REFERENCES

[1] R. Hill: "The Mathematical Theory of Plasticity," Oxford Univ. Press, New York, 1950.

[2] A. Nadai: "The Theory of Flow and Fracture of Solids," vol. 1, 2d ed., chap. 31, McGraw-Hill, New York, 1950.

[3] W. Prager, P. G. Hodge, Jr.: "Theory of Perfectly Plastic Solids," chap. 5, Wiley, New York, 1951.

[4] O. Hoffman, G. Sachs: "Introduction to the Theory of Plasticity," McGraw-Hill, New York, 1953.

[5] B. B. Hundy: Plane plasticity, *Metallurgia*, **49** (1954), 109–118.

[6] M. C. Steele: Partially plastic thick-walled cylinder theory, *J. Appl. Mechanics*, **19** (1952), 133–140.

[7] J. H. Faupel, A. R. Furbeck: Influence of residual stress on behavior of thick-walled closed end cylinders, *Trans. ASME*, **75** (1953), 345–354.

[8] R. Hill, E. H. Lee, S. J. Tupper: Plastic flow in a closed ended tube with internal pressure, *Proc. 1st U.S. Natl. Congr. Appl. Mechanics*, pp. 561–567, Chicago, 1951.

[9] W. T. Koiter: On partially plastic thick-walled tubes, *Biezeno Anniv. Vol.*, p. 232, Stam, Haarlem, 1953.

[10] D. N. de G. Allen, D. G. Sopwith: The stress and strain in a partly plastic thick tube under internal pressure and end-load, *Proc. Roy. Soc. London*, **A,205** (1951), 69–83.

[11] H. Hencky: Über einige statisch bestimmte Fälle des Gleichgewichts in plastischen Körpern, *Z. angew. Math. Mech.*, **3** (1923), 241–251.

[12] H. Geiringer: Fondements mathématiques de la théorie des corps plastiques isotropes, *Mém. sci. math.*, **86,** Gauthier-Villars, Paris, 1937.

[13] R. Hill: On the state of stress in a plastic-rigid body at the yield point, *Phil. Mag.*, **VII,42** (1951), 868–875.

[14] R. Hill: On the problem of uniqueness in the theory of the rigid-plastic solid, *J. Mech. Phys. Solids*, **4** (1956), 247–255; **5** (1957), 153–161.

[15] J. F. W. Bishop, A. P. Green, R. Hill: A note on the deformable region in a rigid-plastic body, *J. Mech. Phys. Solids*, **4** (1956), 256–258.

[16] W. Prager: A geometrical discussion of the slip line field in plane plastic flow, *Trans. Roy. Inst. Technol. Stockholm*, **65** (1953).

[17] A. P. Green: On the use of hodographs in problems of plane plastic strain, *J. Mech. Phys. Solids*, **2** (1954), 73–80, 296.

[18] A. P. Green, R. Hill: Calculations on the influence of friction and die geometry in sheet drawing, *J. Mech. Phys. Solids*, **1** (1952), 31–36.

[19] J. F. W. Bishop: Calculations on sheet drawing under back tension through a rough wedge-shaped die, *J. Mech. Phys. Solids*, **2** (1953), 39–42.

[20] A. P. Green: A theoretical investigation of the compression of a ductile material between smooth flat dies, *Phil. Mag.*, **42** (1951), 900–918.

[21] A. P. Green: The plastic yielding of notched bars due to bending, *Quart. J. Mech. Appl. Math.*, **6** (1953), 223–239.

[22] R. Hill: On the limits set by plastic yielding to the intensity of singularities of stress, *J. Mech. Phys. Solids*, **2** (1954), 278–285.

[23] D. C. Drucker, W. Prager, H. J. Greenberg: Extended limit design theorems for continuous media, *Quart. Appl. Math.*, **9** (1952), 381–389.

[24] J. F. W. Bishop: On the complete solution to problems of deformation of a plastic-rigid material, *J. Mech. Phys. Solids*, **2** (1953), 43–53.

[25] H. Ford: The theory of rolling, *Met. Rev.*, **2** (1957), 1–28.
[26] T. v. Kármán: Beitrag zur Theorie des Walzvorganges, *Z. angew. Math. Mech.*, **5** (1925), 139–142.
[27] D. R. Bland, H. Ford: The calculation of roll force and torque in cold strip rolling with tensions, *Proc. Inst. Mech. Engrs. London*, **159** (1948), 144–153.
[28] E. Orowan: The calculation of roll pressure in hot and cold flat rolling, *Proc. Inst. Mech. Engrs. London*, **150** (1943), 140–167.
[29] R. Hill: On discontinuous plastic states, with special reference to localized necking in thin sheets, *J. Mech. Phys. Solids*, **1** (1952), 19–30.
[30] P. G. Hodge: On the plastic strains in slabs with cutouts, *J. Appl. Mechanics*, **20** (1953), 183–188.
[31] P. G. Hodge: The effect of strain hardening in an annular slab, *J. Appl. Mechanics*, **20** (1953), 530–536.
[32] F. A. Gaydon: An analysis of the plastic bending of a thin strip in its plane, *J. Mech. Phys. Solids*, **1** (1953), 103–112.
[33] P. G. Hodge, N. Perrone: Yield loads of slabs with reinforced cutouts, *J. Appl. Mechanics*, **24** (1957), 85–92.

CHAPTER 51

PLATES AND SHELLS

BY

P. G. HODGE, Jr., Ph.D., Chicago, Ill.

51.1. INTRODUCTION

Most of the available results in plastic plate and shell theory are applicable only to symmetrically loaded circular plates and circular cylindrical shells made of a perfectly plastic material. The mathematics of these two classes of problems are quite similar. Therefore, we will devote most of our attention to the former problem, and briefly indicate the extension to cylindrical shells. Owing to limitations of space, we omit all mention of strain-hardened plates [1]–[3], [27], [28] and shells [4]–[6], [29], noncircular plates [7]–[9], general rotationally symmetric shells [9]–[11], [30]–[33], finite bending of plates [33], [34], or plate and shell problems according to the Mises yield criterion [35]–[38].

51.2. RIGID–PERFECTLY PLASTIC CIRCULAR PLATES [12],[13]

The stress distribution in a circular plate under rotationally symmetric loading is characterized by the radial and circumferential bending-moment resultants M_r and M_θ. These must satisfy an equilibrium equation,

$$\frac{d(rM_r)}{dr} - M_\theta = -\int_0^r rq(r)\,dr \qquad (51.1)$$

where $q(r)$ is the pressure distribution.

We assume that the plate material satisfies Tresca's yield condition of maximum shearing stress; the resultants must then satisfy the similar restriction (Fig. 51.1):

$$\begin{aligned} |M_r| &\leq M_0 \\ |M_\theta| &\leq M_0 \qquad (51.2) \\ |M_r - M_\theta| &\leq M_0 \end{aligned}$$

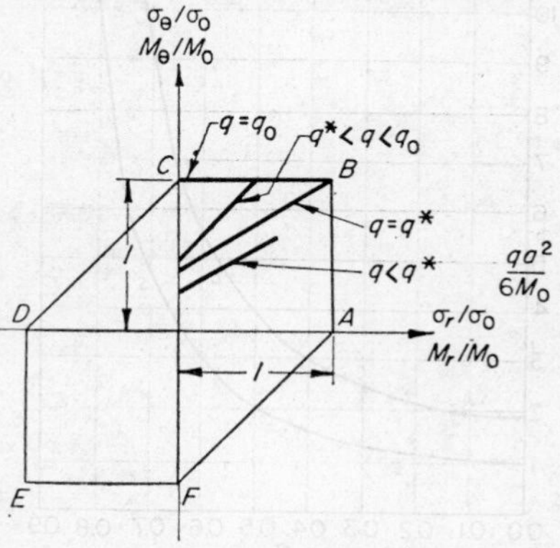

FIG. 51.1. Yield condition and elastic-plastic stress profiles for simply supported circular plates.

If σ_0 is the tensile yield stress, and $2h$ the plate thickness, the maximum moment per unit length is $M_0 = \sigma_0 h^2$.

We consider first a simply supported circular plate of radius a, loaded with a constant pressure q_1 from $r = 0$ to $r = \beta a$, and with a constant pressure q_2 from $r = \beta a$ to $r = a$ $(0 \leq \beta \leq 1)$. We seek the maximum combinations of q_1 and q_2 for which an equilibrium distribution of moments satisfying (51.2) can be found. The general procedure to be followed is first to make an assumption as to which of (51.2) will be equalities. We then solve the problem according to our assumption, and finally check the assumption. For the present problem, M_r must satisfy the boundary conditions

$$M_r(0) = M_\theta(0) \qquad M_r(a) = 0 \tag{51.3}$$

and M_θ is obviously positive. Thus, it appears reasonable to assume that the entire stress profile of the plate is on side BC ($M_\theta = M_0$). Under this assumption, the solution of Eq. (51.1) satisfying the boundary conditions (51.3) and being continuous at $r = a$ is

$$M_r = \begin{cases} M_0 - (q_1/6)r^2 & 0 < r < \beta a \\ M_0 - (q_2/6)r^2 + (q_2 - q_1)(a^2\beta^2/6)(3 - 2a\beta/r), & \beta a < r < a \end{cases} \tag{51.4a}$$

where q_1 and q_2 are related by

$$(1 - \beta)^2 (1 + 2\beta)q_2 + \beta^2(3 - 2\beta)q_1 = 6M_0/a^2 \tag{51.4b}$$

We must first check that this solution lies on BC, rather than on BC extended, i.e., that the remaining inequalities (51.2) are satisfied. In the case where q_1 and q_2 are both positive, it is obvious that M_r is a monotonically decreasing function of r; hence, this condition is everywhere satisfied. Next, we must show that an incipient velocity field can be associated with Eq. (51.4). To this end, we denote the downward displacement of the plate by w; the principal curvatures are then

$$\kappa_r = -d^2w/dr^2 \qquad \kappa_\theta = -(1/r)\,dw/dr \tag{51.5}$$

If the bending moments M_r and M_θ are taken as generalized stresses, the curvatures κ_r and κ_θ are the corresponding generalized strains [14]. The plastic potential strain-rate law then states that the strain-rate vector $\dot{\mathbf{E}} = (\dot\kappa_r, \dot\kappa_\theta)$ must be directed along the outward normal to the moment yield curve at the stress point. Thus, on side BC, $\dot\kappa_r = 0$, $\dot\kappa_\theta \geq 0$. It follows from the above that the velocity field for a simply supported plate is

$$\dot w = \dot w_0(1 - r/a) \tag{51.6}$$

that is, the plate deforms into a cone. The slope discontinuity (hinge) at $r = 0$ is permissible, since $M_r = M_0$ there. Equations (51.4) and (51.6) constitute a complete solution of the problem of the simply supported plate.

If $q_1 = 0$, the solution for $0 \leq r \leq \beta a$ reduces to $M_r = M_0$; hence, the entire stress profile for this region is at the corner B. Figure 51.2 shows the yield-point load for this case. The velocity distribution (51.6) is still a possible one, but it is no longer unique. If $q_2 = 0$, the total load P on the plate is

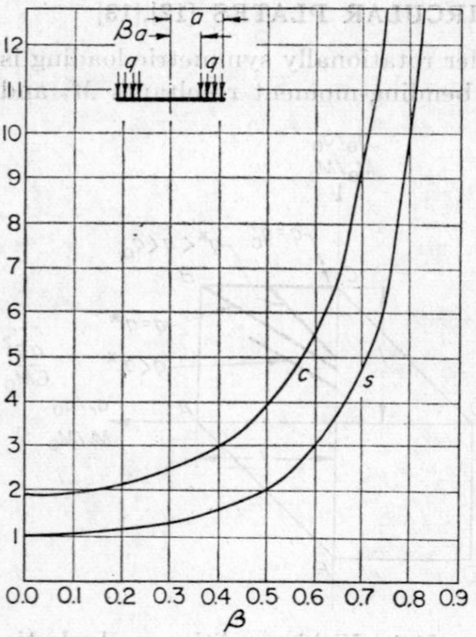

FIG. 51.2. Circular plate with annular loading. s = simply supported plate; c = clamped plate.

$$P = \frac{2\pi M_0}{1 - 2\beta/3} \tag{51.7}$$

In particular, if β tends to zero while P remains finite, we obtain a concentrated load at the center of the plate. The yield-point load for this case is $2\pi M_0$ independently of the radius of the plate.

Similar results may be obtained for the clamped plate. In this case, a hinge circle must develop at the clamped edge with $M_r = -M_0$. The central region is still in regime BC, but the outer region is now in regime CD. For the particular case $q_1 = 0$, the radial moment is given by

$$\frac{M_r}{M_0} = \begin{cases} 1 & 0 < r < \beta a \\ 1 - \dfrac{qa^2}{6M_0}\left(\dfrac{r^2}{a^2} - 3\beta^2 + \dfrac{a\beta^3}{r}\right) & \beta a < r < \rho a \\ \left(1 + \dfrac{qa^2\beta^2}{2M_0}\right)\ln\dfrac{r}{a} - \dfrac{qa^2}{4M_0}\left(\dfrac{r^2}{a^2} - 1\right) - 1 & \rho a < r < a \end{cases} \tag{51.8}$$

where $a\rho$ is the radius of the boundary circle between regimes BC and CD. At $r = a\rho$, M_r must equal zero from either side. This results in two equations, which may be numerically solved for $qa^2/6M_0$ and ρ in terms of β:

$$\frac{qa^2}{6M_0} = \frac{1}{(\rho^2 - \beta^2)(1 + 2\beta/\rho)} = \frac{2(1 - \ln \rho)}{3(1 - \rho^2 + 2\beta^2 \ln \rho)} \tag{51.9}$$

Figure 51.2 shows the collapse load as a function of β.

51.3. ELASTIC–PERFECTLY PLASTIC CIRCULAR PLATE [15]

For simplicity of exposition, we consider an ideal sandwich plate made up of two sheets of thickness h', separated by a core of thickness $2H$. The sheets carry tensile stresses only, which do not vary over the thickness; they have a Young's modulus E' and yield according to Tresca's yield condition with tensile yield stress σ_0'. The core carries only shear and does not yield. This plate may be expected to approximate a uniform plate, provided the constants are chosen so that the yield moment M_0, yield force N_0, and flexural rigidity K agree. This requirement leads to

$$\sigma_0'h' = \sigma_0 h \qquad H = \tfrac{1}{2}h \qquad E'/\sigma_0' = 4E/3\sigma_0 \tag{51.10}$$

Evidently a given element of the plate will be either elastic or plastic. In the former case, the displacement must satisfy the third-order equation, which may be obtained by integrating Eq. (39.93),

$$w''' + \frac{w''}{r} - \frac{w'}{r^2} = \frac{1}{r}\int_0^r rq\,dr \tag{51.11}$$

For sufficiently small values of the load, the entire plate is elastic. For the particular case of a constant load on a simply supported plate, the complete solution is [see Eqs. (39.99)]:

$$\begin{gathered} w = \frac{q}{64K}\left(\frac{5+\nu}{1+\nu}a^4 - 2\frac{3+\nu}{1+\nu}a^2r^2 + r^4\right) \\ M_r = \frac{3+\nu}{16}q(a^2 - r^2) \qquad M_\theta = \frac{3+\nu}{16}q\left(a^2 - \frac{1+3\nu}{3+\nu}r^2\right) \end{gathered} \tag{51.12}$$

The elastic stress distribution may be represented in a stress plane by a straight line (Fig. 51.1). The elastic solution remains valid until the stress profile touches the yield hexagon. Obviously, this will occur at the center of the plate at a load

$$q^* = \frac{16M_0}{(3+\nu)a^2} \tag{51.13}$$

Knowing the stress profiles for $q = q^*$ and the yield-point load $q = q_0$, it is reasonable to expect the stress profile for some intermediate value to be partly in the interior and partly on side BC (Fig. 51.1). Therefore, the elastic solution is to be used for $a\rho < r < 1$, and a plastic solution on side BC for $0 < r < a\rho$, where ρ is to be determined.

In the central plastic region, the stresses are evidently given by

$$M_r = M_0 - qr^2/6 \qquad M_\theta = M_0 \qquad 0 < r < a\rho \tag{51.14}$$

The stress-strain law for this region must include both elastic and plastic effects. The curvature rates will be the sum of an elastic part given by the differentiated form of Hooke's law and a plastic part normal to the yield curve. On side BC, the plastic strain rate is purely circumferential so that the entire radial strain rate is given by Hooke's law, and, hence, the strain is the same as in elasticity. Therefore, it follows from Eqs. (51.5) and (51.14) that

$$w'' = -\frac{M_r - \nu M_\theta}{(1-\nu^2)K} = -\frac{(1-\nu)M_0 - qr^2/6}{(1-\nu^2)K} \tag{51.15}$$

Equation (51.15) is to be integrated in the plastic region, and Eq. (51.11) in the elastic region. The results will contain five integration constants, two of which may be evaluated from the conditions $w(a) = M_r(a) = 0$. Thus,

$$\begin{aligned} Kw &= a^4 B \ln \frac{r}{a} - \left(\frac{B}{2}\frac{1-\nu}{1+\nu} - \frac{q}{32}\frac{3+\nu}{1+\nu}\right)a^2(a^2 - r^2) - \frac{q}{64}(a^4 - r^4) \\ M_r &= \frac{(1-\nu)Ba^4}{r^2} - Ba^2(1-\nu) + \frac{3+\nu}{16}q(a^2 - r^2) \\ M_\theta &= -\frac{(1-\nu)Ba^4}{r^2} - Ba^2(1-\nu) + \frac{3+\nu}{16}q\left(a^2 - \frac{1+3\nu}{3+\nu}r^2\right) \qquad a\rho < r < a \\ Kw &= \frac{1}{1-\nu^2}\left(Aa^4 + Ca^3 r - \frac{1}{2}\nu a^2 r^2 + \frac{q}{72}r^4\right) \qquad 0 < r < a\rho \end{aligned} \tag{51.16}$$

The constants A, B, C and the elastic-plastic interface ρ are determined from the continuity of w, w', M_r, and M_θ at $r = a\rho$. For the particular case $\nu = 1/3$, the results are:

$$\begin{aligned} q &= 24M_0/a^2\Delta \qquad B = -3\rho^4 M_0/2a^2\Delta \qquad C = -8\rho^3 M_0/3a^2\Delta \\ A &= M_0(4 + 7\rho^4 - 4\rho^4 \ln \rho)/3a^2\Delta \qquad \Delta = 5 - 2\rho^2 + \rho^4 \end{aligned} \tag{51.17}$$

Equations (51.14), (51.16), and (51.17) define the complete elastic-plastic solution in terms of the parameter ρ. For $\rho = 0$, we obtain the fully elastic pressure q^* [Eq. (51.13)]; for $\rho = 1$, we regain the yield-point load $q_0 = 6M_0/a^2$. Figure 51.3 shows the displacement at the plate center as a function of the load.

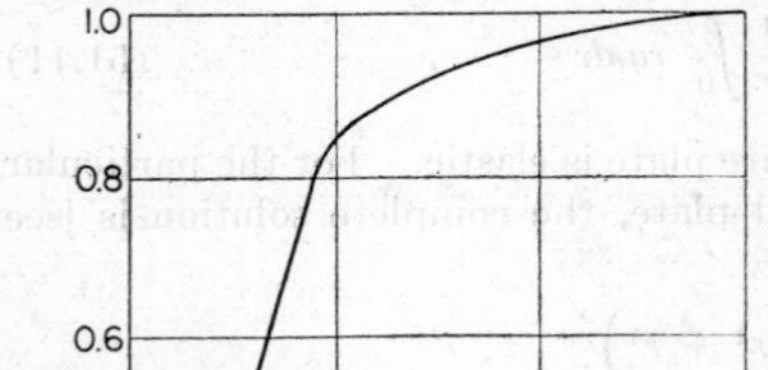

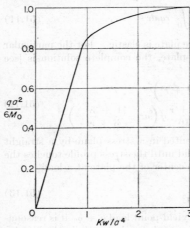

FIG. 51.3. Maximum displacement as a function of load for circular plate.

51.4. DYNAMIC LOADING OF CIRCULAR PLATES [16]

If a rigid, perfectly plastic circular plate is subjected to a pressure which is greater than the yield-point load, but which acts for a very short length of time, the motion will be resisted by inertia forces. In this case, the final deformations may remain within tolerable limits. The equilibrium equation (51.1) must now be modified by the inclusion of the inertia term. Thus,

$$(rM_r)' - M_\theta = -\int_0^r (q - \mu\ddot{w})r\,dr \tag{51.18}$$

where μ is the mass per unit area of middle surface.

We consider a simply supported plate, subject to a load which is constant at any instant and which is monotonically nonincreasing in time. As a hypothesis, we assume that the profile in Fig. 51.1 is everywhere on side BC. It then follows from the flow law that $\dot{w}'' = 0$; hence,

$$w = \phi(t)(1 - r/a) \tag{51.19}$$

where $\phi(t)$ is to be determined. Next we substitute Eq. (51.19) into Eq. (51.18), set $M_\theta = M_0$, and solve for M_r. The result is

$$M_r = M_0 - \tfrac{1}{6}qr^2 + \tfrac{1}{12}\mu\ddot{\phi}(2r^2 - r^3/a) \tag{51.20}$$

Finally, we determine $\ddot{\phi}$ from the condition that M_r vanish at $r = a$. Introducing the yield-point load $q_0 = 6M_0/a^2$, we obtain

$$\ddot{\phi} = (2/\mu)(q - q_0) \tag{51.21}$$

Therefore, the complete solution becomes

$$M_r = \frac{1}{6}\left(1 - \frac{r}{a}\right)[qr^2 + q_0(a^2 + ar - r^2)] \qquad M_\theta = M_0$$
$$w = \frac{2}{\mu}\left(1 - \frac{r}{a}\right)\int_0^t\int_0^t (q - q_0)\,dt^2 \tag{51.22}$$

Before accepting this solution, we must verify that the stress profile is actually on BC, that is, that $0 \le M_r \le M_0$. It is readily verified that this will be true provided $-q_0 \le q \le 2q_0$. If q does not lie within these limits, a different initial hypothesis must be used. An example of this is given in [16].

In addition to the restriction on stress, the velocity $\dot{w}$ must always be positive; hence,

$$\int_0^t (q - q_0)\,dt > 0 \tag{51.23}$$

The pressure q is initially greater than q_0, and decays with time. Thus inequality (51.23) is valid for t sufficiently small, but does not hold for $t > t_2$, where t_2 is defined by

$$\int_0^{t_2} (q - q_0)\,dt = 0 \tag{51.24}$$

At $t = t_2$, the load has dropped below the yield-point load, and the motion stops. Figure 51.4 shows the final deformation and the deformation when q drops to q_0 for a pressure pulse of the form $q = \bar{q}e^{-\beta t}$.

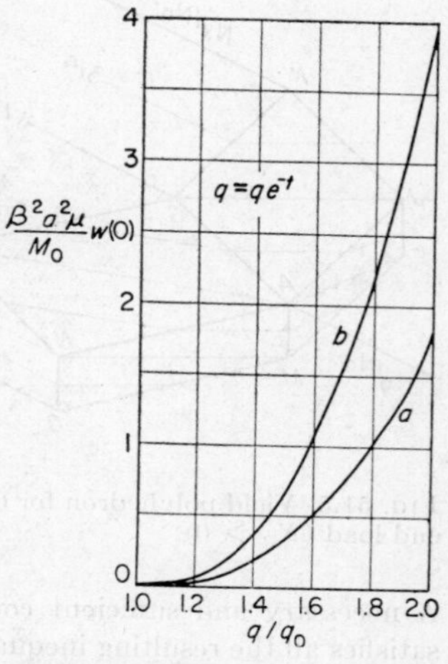

FIG. 51.4. Displacement of plate center as a function of initial load: (a) Displacement at $q = q_0$; (b) final displacement.

51.5. CIRCULAR CYLINDRICAL SHELLS

The stress state of a circular cylindrical shell under axially symmetric loading is characterized by two bending moments, one shear resultant, and two stress resultants: N_x, N_ϕ, M_x, M_ϕ, and Q_x. As in elasticity theory, we assume that the change in circumferential curvature and the shear strain are negligible, so that M_ϕ and Q_x do no work. Then there are three generalized stresses, N_x, N_ϕ, and M_x. These must satisfy two equilibrium equations,

$$N'_x = 0 \qquad M''_x + N_\phi/a + q = 0 \tag{51.25a,b}$$

where primes indicate differentiation with respect to x.

If the shell material satisfies Tresca's yield condition, the yield condition for the stress resultants may be established by a consideration of stress distributions [17],[18] or strain distributions [17],[19] through the shell thickness. The resulting surface in three-dimensional stress space is curved, and, hence, the equations are nonlinear. In most applications, the surface is first approximated by a linear one. A linear approximation can be derived directly by considering the ideal sandwich shell [17]. To this end, let σ_x', σ_ϕ' and σ_x'', σ_ϕ'' be the stresses in the top and bottom sheets, respectively. Then the resultants are:

$$\begin{aligned} N_x &= (\sigma_x' + \sigma_x'')h' \qquad & M_x &= (\sigma_x'' - \sigma_x')h'H \\ N_\phi &= (\sigma_\phi' + \sigma_\phi'')h' \qquad & M_\phi &= (\sigma_\phi'' - \sigma_\phi')h'H \end{aligned} \tag{51.26}$$

Solving Eqs. (51.26) for the four stresses, and substituting into Tresca's yield condition for the top and bottom sheets, we see that the stress resultants must satisfy the 12 inequalities:

$$\left|\frac{N_x}{N_0} \pm \frac{M_x}{M_0}\right| \le 1 \qquad \left|\frac{N_\phi}{N_0} \pm \frac{M_\phi}{M_0}\right| \le 1 \qquad \left|\frac{N_x - N_\phi}{N_0} \pm \frac{M_x - M_\phi}{M_0}\right| \le 1 \tag{51.27}$$

Now, since M_ϕ is not a generalized stress, we can eliminate it from inequalities (51.27). To this end, we solve those inequalities which contain M_ϕ/M_0 for M_ϕ/M_0.

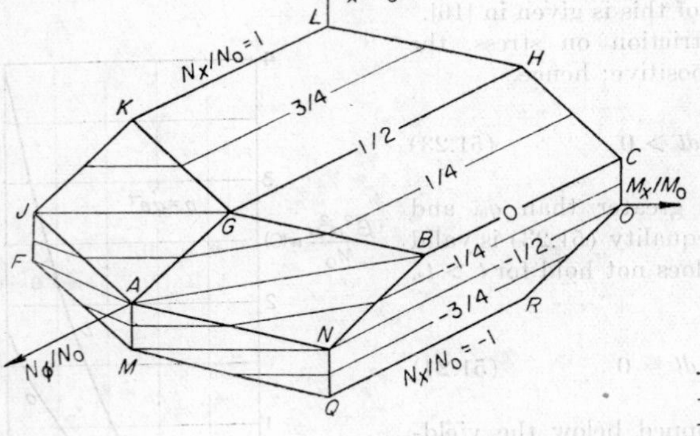

FIG. 51.5. Yield polyhedron for circular cylindrical shell with end load ($N_\phi > 0$).

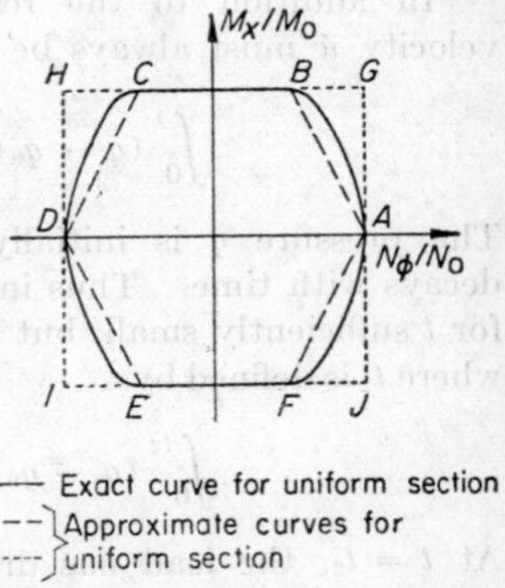

FIG. 51.6. Yield curve for circular cylindrical shell.

A necessary and sufficient condition that there exist some value of M_ϕ/M_0 which satisfies all the resulting inequalities is that each left-hand side be less than each right-hand side. Eliminating redundant inequalities, we finally find that N_x, N_ϕ, and M_x must satisfy the 12 inequalities,

$$\begin{aligned} -1 \le N_\phi/N_0 \le 1 \qquad & -1 \le N_x/2N_0 - N_\phi/N_0 \pm M_x/2M_0 \le 1 \\ -1 \le N_x/N_0 \pm M_x/N_0 \le 1 \qquad & -1 \le N_x/N_0 - N_\phi/N_0 \le 1 \end{aligned} \tag{51.28}$$

Figure 51.5 shows the resulting polyhedron in stress space.

The generalized strains for this problem are the axial and circumferential extensions of the middle surface, and the axial curvature. In terms of the axial displacement u and inward radial displacement w, these are

$$\epsilon_x = u' \qquad \epsilon_\phi = -w/a \qquad \kappa_x = -w'' \tag{51.29}$$

The flow rule states that the strain-rate vector $\dot{\mathbf{E}} = (\dot{\epsilon}_x, \dot{\epsilon}_\phi, \dot{\kappa}_x)$ is normal to the appropriate side of the yield polyhedron.

The solution of problems is now carried out in a manner analogous to that used in circular plate problems. If there is no end load, then only two stress resultants need be considered, and the hexagonal yield curve of Fig. 51.6 may be used. For comparison, the exact nonlinear yield curve for a uniform shell is shown as the solid curve in Fig. 51.6. To further simplify the mathematics, some shell problems have been solved, using the square interaction curve shown as a dotted line in Fig. 51.6. When the end load is different from zero, then Eq. (51.25*a*) shows that N_x is constant throughout the shell for any given load; hence, the analysis is not complicated unduly by the additional stress variable.

51.6. CIRCULAR CYLINDRICAL SHELL PROBLEMS

The mathematical analysis of shell problems closely parallels that of circular plates. Therefore, we will not present any detailed solutions, but refer the reader to some representative literature. In addition to the original results, a survey of some of the aspects may be found in [20], and a textbook treatment is given in Chap. 11 of [9].

We consider first rigid–perfectly plastic shells. The yield condition, in the absence of end load, is derived in [18] and applied to an infinite shell under a ring load. The results are extended to finite shells in [21]. A shell subjected to a uniform radial pressure, with and without the end load produced by a hydrostatic pressure field is considered in [17]. Other combinations of uniform radial pressure and end load are taken up in [19] and [38].

The elastic-plastic analysis of a circular cylindrical shell, subjected to radial pressure only, was carried out in [22]. The effect of end load, including the so-called beam-column effect, i.e., the axial moment produced by the axial force and the radial displacement, was added in [23].

Shells loaded by dynamic radial pressures have been considered in [24], using the square yield condition. This has been compared with the hexagonal condition in [25]; the approximation was found to be a reasonable one. Dynamic loading of an infinite shell by a ring load, both stationary and moving, has been discussed in [26].

51.7. REFERENCES

[1] W. Prager: A new method of analyzing stresses and strains in work-hardening plastic solids, *J. Appl. Mechanics*, **23** (1956), 493–496.

[2] W. Boyce: The bending of a work-hardening circular plate by a uniform transverse load, *Quart. Appl. Math.*, **14** (1956), 277–288.

[3] P. G. Hodge, Jr.: Plasticity, in J. N. Goodier and P. G. Hodge, Jr., "The Mathematical Theory of Elasticity and Plasticity," vol. 1 of Midwest Research Institute, "Surveys in Applied Mathematics," Wiley, New York, 1958.

[4] P. G. Hodge, Jr., F. A. Romano: Deformation of an elastic-plastic cylindrical shell with linear strain hardening, *J. Mech. Phys. Solids*, **4** (1956), 145–161.

[5] P. G. Hodge, Jr.: Piecewise linear isotropic plasticity applied to a circular cylindrical shell with symmetrical radial loading, *J. Franklin Inst.*, **263** (1957), 13–34.

[6] P. G. Hodge, Jr., S. V. Nardo: Carrying capacity of an elastic-plastic cylindrical shell with linear strain-hardening, *J. Appl. Mechanics*, **25** (1958), 79–85.

[7] W. Schumann: On limit analysis of plates, *Quart. Appl. Math.*, **16** (1958), 61–71.

[8] M. Zaid: On the carrying capacity of plates of arbitrary shape and variable fixity under a concentrated load, *J. Appl. Mechanics*, **25** (1958), 598–602.

[9] P. G. Hodge, Jr.: "Plastic Analysis of Structures," McGraw-Hill, New York, 1959.

[10] E. T. Onat, W. Prager: Limit analysis of shells of revolution, *Proc. Roy. Netherlands Acad. Sci.*, **B,57** (1954), 534–548.

[11] D. C. Drucker, R. T. Shield: Limit analysis of symmetrically loaded thin shells of revolution, *J. Appl. Mechanics*, **26** (1959), 61–68.

[12] H. G. Hopkins, W. Prager: The load carrying capacities of circular plates, *J. Mech. Phys. Solids*, **2** (1953), 1–13.

[13] D. C. Drucker, H. G. Hopkins: Combined concentrated and distributed load on ideally-plastic circular plates, *Proc. 2d U.S. Natl. Congr. Appl. Mechanics*, 1954, pp. 517–520.
[14] W. Prager: The general theory of limit design, *Proc. 8th Intern. Congr. Appl. Mechanics*, Istanbul, 1952, vol. 2, pp. 65–72.
[15] R. M. Haythornthwaite: The deflection of plates in the elastic-plastic range, *Proc. 2d U.S. Natl. Congr. Appl. Mechanics*, 1954, pp. 521–526.
[16] H. G. Hopkins, W. Prager: On the dynamics of plastic circular plates, *Z. angew. Math. Phys.*, **5** (1954), 317–330.
[17] P. G. Hodge, Jr.: The rigid-plastic analysis of symmetrically loaded cylindrical shells, *J. Appl. Mechanics*, **21** (1954), 336–342.
[18] D. C. Drucker: Limit analysis of cylindrical shells under axially symmetric loading, *Proc. 1st Midwestern Conf. Solid Mechanics*, pp. 158–163, Urbana, Ill., 1953.
[19] E. T. Onat: Plastic collapse of cylindrical shells under axially symmetrical loading, *Quart. Appl. Math.*, **13** (1955), 63–72.
[20] P. G. Hodge, Jr.: The theory of piecewise linear isotropic plasticity, *Proc. Colloq. Deformation and Flow of Solids*, pp. 147–168, Madrid, 1955.
[21] G. Eason, R. T. Shield: The influence of free ends on the load-carrying capacities of cylindrical shells, *J. Mech. Phys. Solids*, **4** (1955), 17–27.
[22] P. G. Hodge, Jr.: Displacements in an elastic-plastic cylindrical shell, *J. Appl. Mechanics*, **23** (1956), 73–79.
[23] B. Paul, P. G. Hodge, Jr.: Carrying capacity of elastic-plastic shells under hydrostatic pressure, *Proc. 3d U.S. Natl. Congr. Appl. Mechanics*, 1958, pp. 631–640.
[24] P. G. Hodge, Jr.: The influence of blast characteristics on the final deformation of circular cylindrical shells, *J. Appl. Mechanics*, **23** (1956), 617–624.
[25] P. G. Hodge, Jr., B. Paul: Approximate yield conditions in dynamic plasticity, *Proc. 3d Midwestern Conf. Solid Mechanics*, pp. 29–47, 1957.
[26] G. Eason, R. T. Shield: Dynamic loading of rigid-plastic cylindrical shells, *J. Mech. Phys. Solids*, **4** (1956), 53–71.
[27] N. Perrone, P. G. Hodge, Jr.: Strain hardening solutions to plate problems, *J. Appl. Mechanics*, **26** (1959), 276–284.
[28] P. G. Hodge, Jr.: Plastic bending of an annular plate, *J. Math. Phys.*, **36** (1957), 130–137.
[29] N. Perrone, P. G. Hodge, Jr.: On strain hardened circular cylindrical shells, *J. Appl. Mechanics*, **27** (1960), 489–495.
[30] P. G. Hodge, Jr.: The collapse load of a spherical cap, *Proc. 4th Midwestern Conf. Solid Mechanics*, pp. 108–126, Austin, Tex., 1959.
[31] P. G. Hodge, Jr.: Yield conditions for rotationally symmetric shells under axially symmetric loading, *J. Appl. Mechanics*, **27** (1960), 323–331.
[32] P. G. Hodge, Jr.: Plastic analysis of circular conical shells, *J. Appl. Mechanics*, **27** (1960), 696–700.
[33] E. T. Onat: On the plastic analysis of shallow conical shells, *Brown Univ. DAM Rept. DA*-4795/3 (1959).
[34] E. T. Onat, R. M. Haythornthwaite: Load carrying capacity of circular plates at large deflection, *J. Appl. Mechanics*, **23** (1956), 49–55.
[35] H. G. Hopkins and A. J. Wang: Load carrying capacities for circular plates of perfectly-plastic material with arbitrary yield condition, *J. Mech. Phys. Solids*, **3** (1955), 117–129.
[36] P. G. Hodge, Jr.: The Mises yield condition for rotationally symmetric shells, *Quart. Appl. Math.*, **18** (1961), 305–311.
[37] P. G. Hodge, Jr.: A comparison of yield conditions in the theory of plastic shells, in "Problems of Continuum Mechanics," Soc. Indust. Appl. Math., Philadelphia, 1961, pp. 165–177.
[38] P. G. Hodge, Jr., J. Panarelli: Interaction curves for circular cylindrical shells according to the Mises or Tresca yield criterion, *J. Appl. Mechanics* (in press).

CHAPTER 52

INELASTIC BUCKLING

BY

J. E. DUBERG, Ph.D., Newport News, Va.

52.1. INTRODUCTION

The theoretical understanding of the failure of columns in the inelastic range had generally been considered to have reached completion when von Kármán in 1910 [1] further developed the reduced-modulus theory of Considère and Engesser, by establishing the modulus for the rectangular section and the idealized **H** section and comparing the theory with tests. In spite of the acceptance of the reduced-modulus theory as the basic explanation of column strength, subsequent experimental investigators could find little justification for favoring it over the tangent-modulus load as a means of predicting actual column strength [2]. The reduced-modulus concept, however, was required, in order to explain the deflection behavior of columns as influenced by the shape of the cross section and certain secondary maximum strengths indicated in tests of actual steel columns.

Partly on the basis of expediency, but mostly on the basis of extensive tests of columns made of the high-strength aluminum alloys [3], which have more gradually turning stress-strain curves and no anomalies during plastic straining, aircraft designers had accepted the tangent-modulus load as the failing load for columns.

This seeming inconsistency of theory and practice was resolved in 1947 when Shanley [4] was able to show that a simple model of an inelastic column could deflect to a position of equilibrium at the tangent-modulus load if the load were permitted to increase during bending. If the load were further increased, deflections would grow, and the column could, for a constant tangent modulus in the inelastic range, support, at most, the reduced-modulus load. The subsequent analysis demonstrates, in greater detail, the relation of the tangent-modulus and reduced-modulus loads to column behavior, and is based on [5] and [6].

52.2. EQUILIBRIUM OF THE DEFLECTED COLUMN

Equilibrium and Kinematic Relation

The structural element analyzed is the idealized **H**-section column, simply supported at its ends, which is shown in Fig. 52.1. It is composed of two flanges of equal area A, separated by a web of width b that is rigid in shear but has negligible resistance in extension. Static equilibrium of the forces in the left- and right-hand flanges with the end load P requires that

$$\sigma_{L,R} = \frac{P}{2A}\left(1 \pm 2\,\frac{w + w_0}{b}\right) \tag{52.1}$$

in which L and R refer to the left and right flanges, respectively; σ is the compressive stress; w_0 the initial lack of straightness; and w the deflection due to load. The relationship between the displacements and the strains in the flanges can, for small displacements, be expressed as

$$d^2w/dx^2 = -(\epsilon_L - \epsilon_R)/b \tag{52.2}$$

where ϵ is the compressive strain.

Stress-Strain Relations

The essential difficulty in the analysis of inelastic column behavior lies in the lack of uniqueness between stress and strain, which requires that the history of straining at each cross section of the column must be known, in order to determine the appropriate

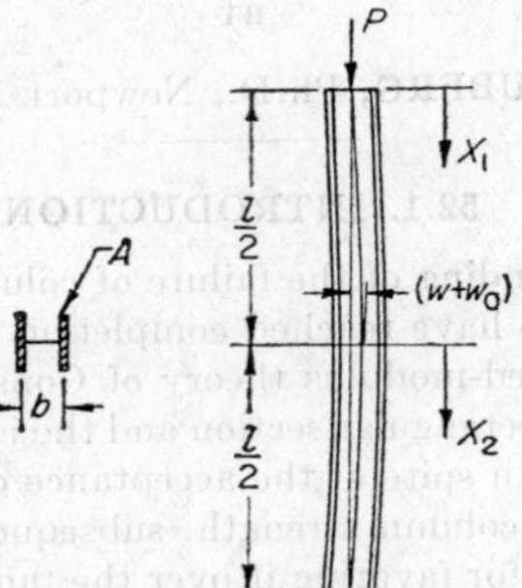

FIG. 52.1. H-section column.

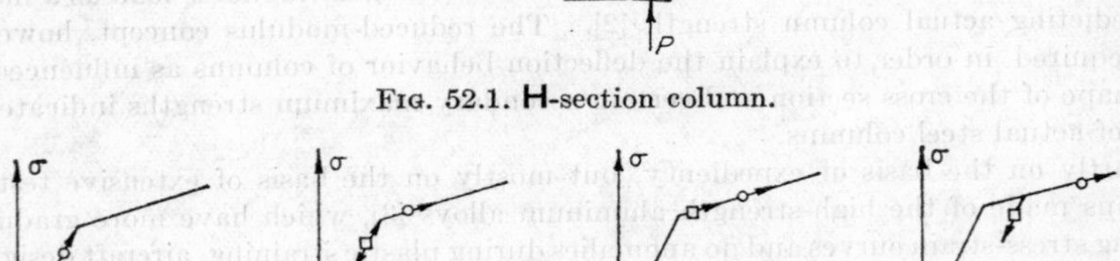

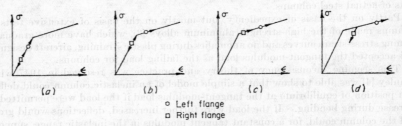

FIG. 52.2. Four stress-strain combinations for the column flanges: (a) Elastic; (b) elastic-plastic with plastic loading; (c) plastic; (d) plastic with elastic unloading.

relationship to use at a given instant. For the simple H-section column loaded to failure, there are four combinations of straining which are of interest. These four are illustrated in Fig. 52.2, and are: both flanges elastic; one elastic, the other plastic and loading; both plastic and loading; one plastic and loading, the other having been plastic and now elastically unloading. The stress-strain properties of the material are assumed to be represented by two straight lines, whose slopes correspond to a Young's modulus of E, and a tangent modulus in the plastic range of E_T.

The four combinations may be expressed analytically as follows:

Elastic

$$\sigma_L = E\epsilon_L \qquad \sigma_R = E\epsilon_R \tag{52.3a}$$

Elastic-plastic with plastic loading

$$\sigma_L = \sigma_1 + E_T(\epsilon_L - \epsilon_1) \qquad \sigma_R = E\epsilon_R \tag{52.3b}$$

Plastic

$$\sigma_L = \sigma_1 + E_T(\epsilon_L - \epsilon_1) \qquad \sigma_R = \sigma_1 + E_T(\epsilon_R - \epsilon_1) \tag{52.3c}$$

Plastic with elastic unloading

$$\sigma_L = \sigma_1 + E_T(\epsilon_L - \epsilon_1) \qquad \sigma_R = \sigma_{\text{rev}} + E(\epsilon_R - \epsilon_{\text{rev}}) \tag{52.3d}$$

The notations σ_1 and ϵ_1 refer to the yield stress and strain of the material, and the notations σ_{rev} and ϵ_{rev} are respectively the stress and strain at which reversal of straining took place at a given section.

Differential Equations

When Eqs. (52.1), (52.2), and one of Eqs. (52.3) are combined, the following differential equations for the deflection are found. Each corresponds to a different type of straining. They are:

Elastic

$$\frac{d^2w}{dx^2} + \frac{2P}{EAb^2}w = -\frac{2P}{EAb^2}w_0 \tag{52.4a}$$

Elastic-plastic with plastic loading

$$\frac{d^2w}{dx^2} + \frac{P}{Ab^2}\left(\frac{1}{E} + \frac{1}{E_T}\right)w = -\frac{P}{Ab^2}\left(\frac{1}{E} + \frac{1}{E_T}\right)w_0 + \frac{P_1 - P}{2Ab}\left(\frac{1}{E_T} - \frac{1}{E}\right) \tag{52.4b}$$

Plastic

$$\frac{d^2w}{dx^2} + \frac{2P}{E_TAb^2}w = -\frac{2P}{E_TAb^2}w_0 \tag{52.4c}$$

Plastic with elastic unloading

$$\frac{d^2w}{dx^2} + \frac{P}{Ab^2}\left(\frac{1}{E_T} + \frac{1}{E}\right)w = -\frac{P}{Ab^2}\left(\frac{1}{E_T} + \frac{1}{E}\right)w_0 - \frac{P_{\text{rev}}}{Ab^2}\left(\frac{1}{E_T} - \frac{1}{E}\right)(w_0 + w_{\text{rev}}) - \frac{P - P_{\text{rev}}}{2Ab}\left(\frac{1}{E_T} - \frac{1}{E}\right) \tag{52.4d}$$

In the above equations, $P_1 = 2A\sigma_1$, w_{rev} = deflection when strain in flange is reversed, and P_{rev} = column load when strain in flange is reversed.

These form the basic set of equilibrium equations for the analysis of the column made of a material with a constant tangent modulus in the plastic range.

52.3. INITIALLY STRAIGHT COLUMN

Infinitesimal Deflection Analysis

The equilibrium paths associated with the infinitesimal deflection of an initially straight column can be found from analysis of Eqs. (52.4*c*,*d*). Let $w_0 = 0$, and let a deflection Δw occur as the load, which is greater than P_1, *increases* from P to $(P + \Delta P)$. Then, neglecting higher-order terms, Eq. (52.4*c*) becomes

$$\frac{d^2\,\Delta w}{dx^2} + \frac{2P}{E_TAb^2}\Delta w = 0$$

and is appropriate to a region in which plastic loading occurs.

Similarly, Eq. (52.4*d*) becomes

$$\frac{d^2\,\Delta w}{dx^2} + \frac{P}{Ab^2}\left(\frac{1}{E_T} + \frac{1}{E}\right)\Delta w = \frac{\Delta P}{2Ab}\left(\frac{1}{E} - \frac{1}{E_T}\right)$$

and is appropriate to a region in which plastic loading occurs in one flange and elastic unloading in the other. Solutions to these two equations are:

$$\Delta w_1 = A_1 \sin\left(\frac{x_1}{b}\sqrt{\frac{2P}{E_T A}}\right) + B_1 \cos\left(\frac{x_1}{b}\sqrt{\frac{2P}{E_T A}}\right)$$

$$\Delta w_2 = A_2 \sin\left[\frac{x_2}{b}\sqrt{\frac{P}{A}\left(\frac{1}{E}+\frac{1}{E_T}\right)}\right] + B_2 \cos\left[\frac{x_2}{b}\sqrt{\frac{P}{A}\left(\frac{1}{E}+\frac{1}{E_T}\right)}\right] - \frac{\Delta P}{P}\frac{b}{2}\frac{E-E_T}{E+E_T}$$

If region x_1 is assumed to exist at the ends of the column with its origin at the support and region x_2 is assumed to exist over the middle part of the column with its origin at the center of the column, then B_1 and A_2 are zero. A set of homogeneous conditions exist for the coefficients A_1, B_2 and the as yet undetermined change of load ΔP, such that, at the junction of the two regions,

$$x_1 = kl \qquad x_2 = (k - \tfrac{1}{2})l$$

$$\Delta w_1 = \Delta w_2 = 0 \qquad \frac{d\Delta w_1}{dx_1} - \frac{d\Delta w_2}{dx_2} = 0 \qquad \Delta\epsilon_R = 0$$

These conditions establish the following equations:

$$A_1 \sin\left(\frac{kl}{b}\sqrt{\frac{2P}{E_T A}}\right) - B_2 \cos\left[\frac{(2k-1)l}{2b}\sqrt{\frac{P}{A}\left(\frac{1}{E_T}+\frac{1}{E}\right)}\right] + \frac{b\,\Delta P}{2P}\frac{E-E_T}{E+E_T} = 0$$

$$A_1 \cos\left(\frac{kl}{b}\sqrt{\frac{2P}{E_T A}}\right) + B_2\sqrt{\frac{E+E_T}{2E}}\sin\left[\frac{(2k-1)l}{2b}\sqrt{\frac{P}{A}\left(\frac{1}{E_T}+\frac{1}{E}\right)}\right] = 0$$

$$A_1 \sin\left(\frac{kl}{b}\sqrt{\frac{2P}{E_T A}}\right) - \frac{b\,\Delta P}{2P} = 0$$

Deflections exist for values of the load and for regions of reversed strains, $(1 - 2k)l$, which satisfy the equation

$$\tan\left(\frac{kl}{b}\sqrt{\frac{2P}{E_T A}}\right)\tan\left[\frac{(1-2k)l}{2b}\sqrt{\frac{P}{A}\left(\frac{1}{E_T}+\frac{1}{E}\right)}\right] = \sqrt{\frac{E+E_L}{2E}}$$

The roots of this equation yield values of loads between the tangent-modulus load

$$P_T = \frac{E_T A b^2 \pi^2}{2l^2}$$

and the reduced-modulus load

$$P_{RM} = \frac{E E_T A b^2 \pi^2}{(E+E_T)l^2}$$

At the tangent-modulus load, the ratio of the increment of load to the increment of center deflection

$$\Delta P/\Delta w_c = 2P_T/b$$

and, at the reduced-modulus load,

$$\Delta P/\Delta w_c = 0$$

The pattern of incremental strain associated with these displacements and loadings is such that, at the tangent-modulus load, all strains continue to increase except that strain at the center of the convex side, and it remains stationary. At the reduced-modulus load, the concave flange continues to load, and the whole convex flange unloads. The deflected shape at both these loads is sinusoidal, which is consistent with a uniform bending stiffness.

A similar argument can be developed for the initially straight column, for which it is assumed that the load may *decrease* as the column bends. It can then be shown that equilibrium paths exist for initial loads lying between the reduced-modulus load and the Euler load $P_E = \pi^2 E A b^2/2l^2$, based on the elastic modulus. For this range of loads, elastic unloading occurs over the whole convex side at the reduced-modulus load, and over increasing lengths of the concave side, such that, at the Euler load, all strains in the column decrease except the strain at the center of the concave side, and it remains stationary. At the Euler load, the ratio of incremental load to incremental center deflection,

$$\Delta P/\Delta w_c = -2P_E/b$$

These ratios of incremental load to incremental deflection can be generalized to any symmetrical prismatic column by requiring that the increment in load be such as to

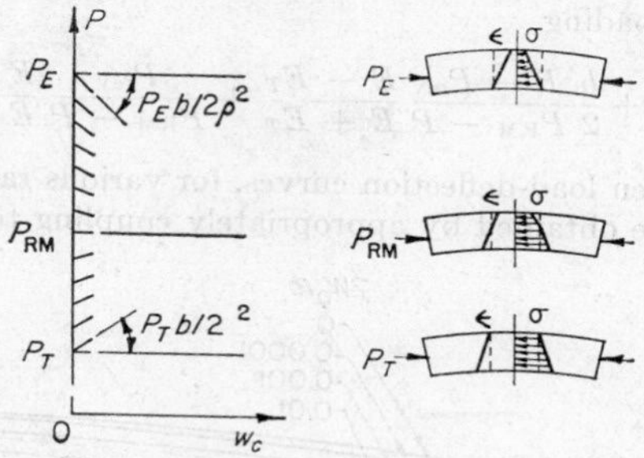

FIG. 52.3. Load-deflection relations and stress and strain increments that are possible for a perfect straight column.

maintain the required strain relations. This yields, at the tangent-modulus load,

$$\Delta P/\Delta w_c = P_T(b/2\rho^2)$$

and, at the Euler load,

$$\Delta P/\Delta w_c = P_E(b/2\rho^2)$$

in which ρ is the radius of gyration of the cross section.

A summary of these slopes and of the stress and strain increments is given in Fig. 52.3.

52.4. COLUMN WITH INITIAL IMPERFECTIONS

Tangent Modulus and Reduced Modulus

It is clear, from the previous analysis, that relaxing the usual requirements of no change in the load at buckling reveals a range of loads at which an initially straight column can deflect. Among these loads is the reduced-modulus load, which corresponds to no change in the load. The lowest load is the tangent-modulus load, and it is associated with the largest rate of change of the load during initial bending.

A column with initial imperfection, and assumed to be made of a material whose stress-strain relationship is represented by two straight lines, is a more realistic representation of a column. An analysis of its behavior, when loaded to failure, can reveal the significance of the reduced-modulus and tangent-modulus loads.

If the H-section column is assumed to be initially bent in the form

$$w_0 \sin(\pi x/l)$$

and to have an additional deflection under load of

$$w \sin(\pi x/l)$$

then the differential equations (52.4) can be solved quite accurately by collocating the equations at the center of the column. The result is a set of algebraic equations which are appropriate to the various regions of straining:

Elastic

$$w = w_0 \frac{P}{P_E - P}$$

Elastic-plastic with no unloading

$$w = w_0 \frac{P}{P_{\mathrm{RM}} - P} + \frac{b}{2} \frac{P - P_1}{P_{\mathrm{RM}} - P} \frac{E - E_T}{E + E_T}$$

Plastic with no unloading

$$w = w_0 \frac{P}{P_T - P}$$

Plastic with elastic unloading

$$w = w_0 \frac{P}{P_{\mathrm{RM}} - P} + \frac{b}{2} \frac{P - P_{\mathrm{rev}}}{P_{\mathrm{RM}} - P} \frac{E - E_T}{E + E_T} + \frac{P_{\mathrm{rev}}}{P_{\mathrm{RM}} - P} \frac{E - E_T}{E + E_T} (w_0 + w_{\mathrm{rev}})$$

In Fig. 52.4 are given load-deflection curves, for various ratios of initial imperfection, $2w_0/b$, which were obtained by appropriately coupling together the solutions of

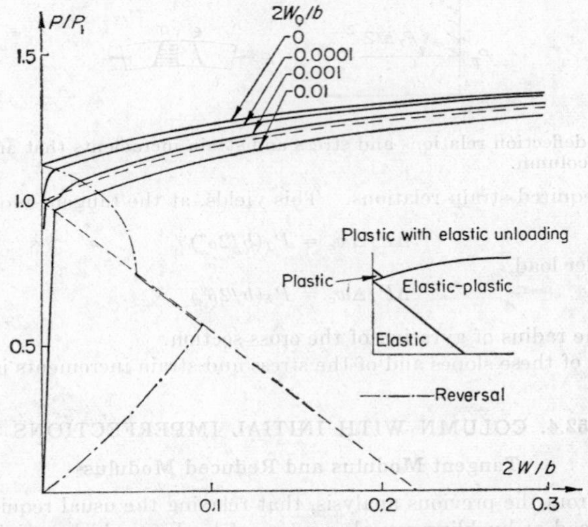

FIG. 52.4. Load-deflection curves for a column with initial deflection and constant tangent modulus in the plastic range.

the above equations. The column and its material are characterized by the following dimensionless ratios:

$$P_T/P_1 = 1.125 \qquad P_{\mathrm{RM}}/P = 1.500$$

The material of the column has a ratio of moduli

$$E_T/E = 0.500$$

There is also included in Fig. 52.4 the boundaries of the several regions, and the loci of points at which the strain at the center of the convex side reverses its direction of straining. Strain reversal occurs for all columns loaded above the tangent-modulus load.

These results indicate that, as the initial imperfections tend to zero, the limit of column behavior is for deflection to start at the tangent-modulus load, and, with the stress-strain curve used. for the maximum load to approach the reduced-modulus load at very large deflections. The reduced-modulus load, at least for small deflections, does not correspond, in this analysis, to a realistic load-deflection point.

Realistic Stress-Strain Relation

The maximum load which a column, made of a hypothetical material with a constant tangent modulus in the plastic range, can support has been shown to be the reduced-modulus load. Real materials have constantly decreasing tangent moduli. Columns made of such materials, even for very small initial curvatures, can, therefore, expect to have maximum loads below the reduced-modulus load. This can readily be seen from the following argument. The incremental strain behavior at maximum load must follow exactly the reduced-modulus concept, except that the column is displaced. If the maximum load were equal to the reduced-modulus load, then, by virtue of the fact that the column is deflected, the compressive strains on the concave side would have to be in excess of those that would exist for the straight column.

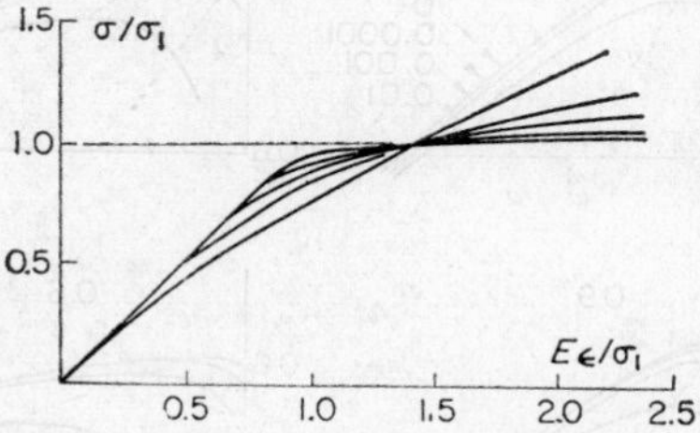

FIG. 52.5. Ramberg-Osgood stress-strain curves.

Such strains are consistent with lower moduli than would exist for the straight column. The bending stiffness is less, and, therefore, it is not possible for the maximum load to be so large as the reduced-modulus load. The maximum load must be some lower load, that is, the Euler load for the instantaneous column stiffness, and is, therefore, a function of shape of the stress-strain curve. A numerical study of the effect of both the shape of the stress-strain curve and initial imperfections was made in [6], and is summarized herein. The static and the strain-displacement relationships are as before.

Stress-Strain Relations. A convenient analytical stress-strain relationship is that proposed by Ramberg and Osgood [7]. It is shown graphically in Fig. 52.5, and in one form can be expressed as

$$\frac{\epsilon}{\epsilon_1} = \frac{\sigma}{\sigma_1} + \frac{3}{7}\left(\frac{\sigma}{\sigma_1}\right)^n$$

The subscript 1 corresponds to the yield stress and strain associated with the $0.7E$ secant. The quantity n determines the rapidity with which the tangent turns. Two stress-strain relations are required. The first is for loading of the flanges, for which

$$\frac{\epsilon_{L,R}}{\epsilon_1} = \frac{\sigma_{L,R}}{\sigma_1} + \frac{3}{7}\left(\frac{\sigma_{L,R}}{\sigma_1}\right)^n$$

The second relationship is required when the strain in the right flange reverses. Then

$$\frac{\epsilon_R}{\epsilon_1} = \frac{\sigma_R}{\sigma_1} + \frac{3}{7}\left(\frac{\sigma_{R\text{rev}}}{\sigma_1}\right)^n$$

Differential Equations. The differential equation obtained with the above stress-strain relationships for both flanges loading is

$$\frac{d^2w}{dx^2} = -\frac{\epsilon_1}{b}\left\{\frac{4P}{P_1}\frac{w + w_0}{b} + \frac{3}{7}\left(\frac{P}{P_1}\right)^n\left[\left(1 + 2\frac{w + w_0}{b}\right)^n - \left(1 - 2\frac{w + w_0}{b}\right)^n\right]\right\}$$

The equation obtained when the right flange reverses direction of straining is

$$\frac{d^2w}{dx^2} = -\frac{\epsilon_1}{b}\left[\frac{4P}{P_1}\frac{w + w_0}{b} + \frac{3}{7}\left(\frac{P}{P_1}\right)^n\left(1 + 2\frac{w + w_0}{b}\right)^n - \frac{3}{7}\left(\frac{P_{\text{rev}}}{P_1}\right)^n\left(1 - 2\frac{w_{\text{rev}} + w_0}{b}\right)^n\right]$$

Numerical Results. With the use of an assumed shape, a half-sine curve for both the initial deflection and the deflection under load, consistent solutions of the differential equations can be obtained by collocation at the center of the column. The

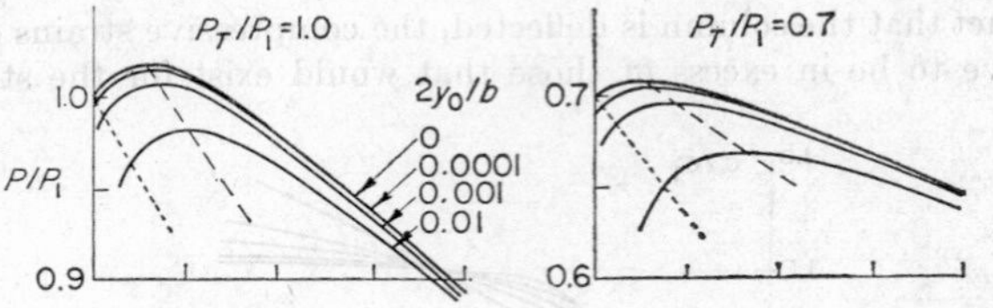

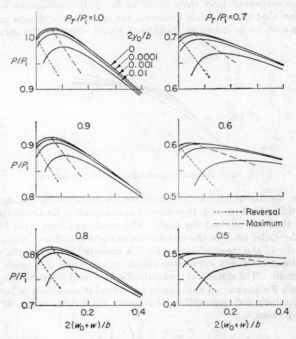

FIG. 52.6. Load-deflection curves for initially curved **H**-section columns having various tangent-modulus loads ($n = 10$).

result is a nonlinear algebraic equation that can be solved without difficulty. Transfer of the solution from the first to the second equation is made when the flange strain reverses.

A summary of the load-deflection curves is given in Fig. 52.6. The results, which show the variation of the maximum load with initial imperfection and tangent-modulus load, are given in Fig. 52.7.

Concluding Remarks. The behavior of the inelastic column may be summarized as follows:

1. There is an infinity of loads at which a perfectly straight column can deflect in the inelastic range. The loads range from the tangent-modulus load to the Euler

load. Those associated with an increase in load lie between the tangent-modulus load and the reduced-modulus load. Those associated with a decrease in load lie between the reduced-modulus load and the Euler load.

2. If a slightly bent column is regarded as a more realistic representation of a true column, its behavior for decreasing amounts of initial imperfection approaches that of a perfect column which deflects at the tangent-modulus load. The tangent-modulus load is the largest load that a column can support without strain reversal.

3. The maximum load supported by a perfectly straight column which starts to deflect at the tangent-modulus load depends on the shape of the stress-strain curve. It cannot exceed the reduced-modulus load of the straight column.

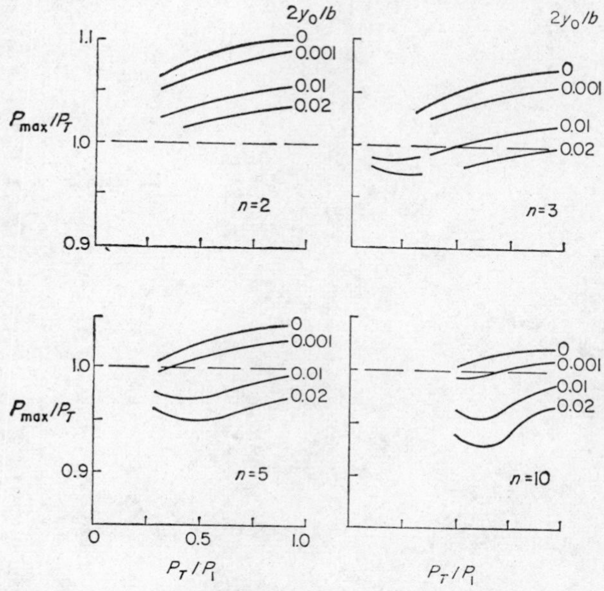

Fig. 52.7. Comparison between maximum load and tangent-modulus load for H-section columns with various amounts of initial deflection.

4. Initial imperfections of the order of those that occur in real columns reduce the maximum load below that of a straight column, and such reductions tend to make the tangent-modulus load a good approximation of the true maximum load.

52.5. REFERENCES

[1] Th. von Kármán: Untersuchungen über Knickfestigkeit, *Mitt. Forschungsarb.*, **81** (1910); also in "Collected Works," vol. 1, pp. 90–140, Butterworth, London, 1956.
[2] O. H. Basquin: The tangent modulus and the strength of steel column in tests, *Natl. Bur. Standards Technol. Papers*, **263** (1924), 381–442.
[3] W. R. Osgood, M. Holt: The column strength of two extruded aluminum alloy H-sections, *NACA Rept.* 656 (1938).
[4] F. R. Shanley: Inelastic column theory, *J. Aeronaut. Sci.*, **14** (1947), 261–267.
[5] J. E. Duberg, T. W. Wilder III: Inelastic column behavior, *NACA Rept.* 1072 (1952).
[6] T. W. Wilder III, W. A. Brooks, Jr., E. E. Mathauser: The effect of initial curvature on the strength of an inelastic column, *NACA Tech. Note* 2872 (1953).
[7] W. Ramberg, W. R. Osgood: Description of stress-strain curves by three parameters, *NACA Tech. Note* 902 (1943).

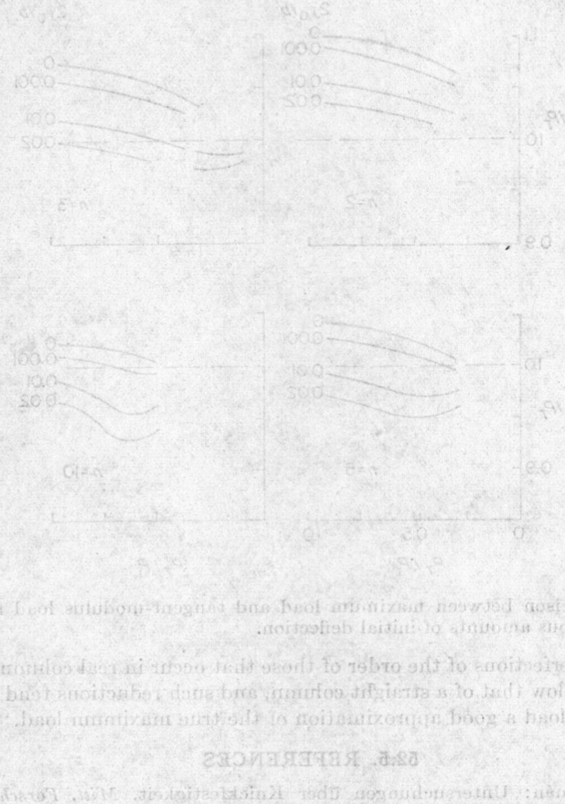

CHAPTER 53

VISCOELASTICITY

BY

E. H. LEE, Ph.D., Providence, R.I.

53.1. INTRODUCTION

Methods of stress analysis have been most completely developed for linear elastic materials subjected to infinitesimal strains. The increasing use of structurally more complex materials such as plastics, and of materials beyond their elastic limits, has emphasized the need for the development of theories for analyzing structures composed of materials exhibiting more complicated stress-strain relations. The classical theory of plasticity considers nonlinear irreversible stress-strain relations, but does not introduce the influence of rate effects. Viscoelastic materials comprise those in which time and rate effects play a dominant role in their response to stress.

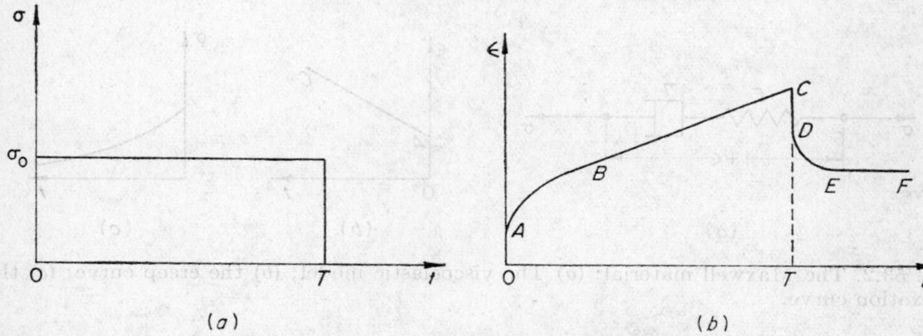

FIG. 53.1. Creep and recovery for a viscoelastic material: (*a*) Stress variation with time; (*b*) resulting strain variation with time.

Figure 53.1 shows the variations of stress and strain with time when a pulse of constant stress of magnitude σ_0 and duration T is applied to a specimen of viscoelastic material. A simple form of loading given by one stress component, σ, is assumed, such as simple tension or shear, and ϵ represents the corresponding strain component. The strain variation exhibits instantaneous elastic response OA in Fig. 53.1*b*, followed by delayed elastic response and viscous flow AB, with continued viscous flow BC. On unloading at the time T, the instantaneous elastic response is recovered immediately, CD; the delayed elastic response recovers gradually, DE; leaving the permanent strain, represented by EF, which is associated with the viscous flow. This complicated response to stress is in marked contrast to that of an elastic material, for which the strain variation would simply reproduce the stress variation to some appropriate scale.

To illustrate the basic components of viscoelastic response to stress, it is helpful to study the simplest laws of this type. For simple stress systems, represented by a single component of stress, σ, and a corresponding strain component ϵ, viscoelasticity is expressed by a relation of the type

$$P[\sigma(t)] = Q[\epsilon(t)] \tag{53.1}$$

where P and Q are rate operators, which may involve differentiation with respect to time, or integration as detailed in Sec. 53.2. The simplest group of such relations which prescribe behavior of the type shown in Fig. 53.1 arises when the operators are linear. The material is then said to be linearly viscoelastic, and methods of stress analysis have been much more completely developed for such materials than for nonlinear materials. While general nonlinear theories have been developed (see, for example, [1]), only very special stress-analysis problems governed largely by symmetry requirements can yet be solved on that basis, and attention here will therefore be directed to linear theory. It has been found to apply, to a satisfactory degree of accuracy, for a number of materials up to certain limiting stress values (see, for example, [2], pp. 20–21).

The simplest linear viscoelastic laws correspond to (53.1) when P and Q are low-order differential operators. For example, the Maxwell material is given by

$$\frac{\dot{\sigma}}{E} + \frac{\sigma}{\eta} = \dot{\epsilon} \tag{53.2}$$

where the dot signifies differentiation with respect to time, and E and η are material constants. The constants are written in this way since it is often convenient to represent the relation (53.2) by a model of a spring and dashpot, as shown in Fig. 53.2*a*.

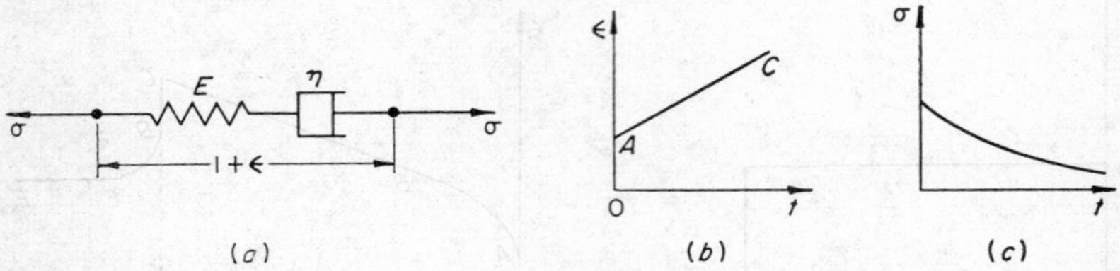

Fig. 53.2. The Maxwell material: (*a*) The viscoelastic model; (*b*) the creep curve; (*c*) the relaxation curve.

If the model is considered to be loaded with the force σ, and E is the spring modulus, and η the viscosity coefficient of the dashpot, then (53.2) gives the extension ϵ of the model. Such a model is often convenient for visualizing the response of a material to stress, although it does not add anything new to the law as expressed by (53.2). Such models do not necessarily have a direct physical significance, with a particular physical mechanism corresponding to the spring, and another to the dashpot, but, nevertheless, they can be useful, as mentioned above.

The creep curve for a Maxwell body, the strain variation associated with a constant suddenly applied stress σ_0, is shown in Fig. 53.2*b*. As is clear from the model (Fig. 53.2*a*), there is an instantaneous strain OA of magnitude σ_0/E associated with the extension of the spring, followed by viscous flow with constant strain rate of magnitude σ_0/η. For the Maxwell material, there is no delayed elastic component of the strain variation. Alternatively, the creep curve (Fig. 53.2*b*) can be deduced from (53.2) by substituting

$$\sigma = \sigma_0\,\Delta(t) \tag{53.3}$$

where $\Delta(t)$ is the Heaviside step function (see p. 19–2). The discontinuity in stress at zero time gives the initial strain response, and, thereafter, the substitution of constant stress σ_0 into (53.2) determines the constant strain rate.

Another test in which viscoelasticity is exhibited is the relaxation test, for which the stress variation is measured corresponding to a suddenly applied constant strain, ϵ_0. The relaxation curve for a Maxwell material is given by substituting $\epsilon = \epsilon_0 \Delta(t)$ into (53.2), with the solution

$$\sigma = E\epsilon_0 e^{-t/\tau} \Delta(t) \tag{53.4}$$

where $\tau = \eta/E$ is called the relaxation time. Just as does the creep curve, the relaxation curve expresses the viscoelastic properties of the material, and we shall see on p. 53–10 how these are related for a linear viscoelastic material. For an elastic material subjected to constant strain, the stress remains constant at the value reached instantaneously, and no relaxation of stress occurs.

The simplest viscoelastic material exhibiting delayed elasticity is the Kelvin or Voigt material, represented by

$$\sigma = E_1\epsilon + \eta_1\dot{\epsilon} \tag{53.5}$$

and the viscoelastic model shown in Fig. 53.3. The creep curve is obtained by substituting the stress variation $\sigma_0 \Delta(t)$ in (53.5) and solving for ϵ, giving

$$\epsilon = \frac{\sigma_0}{E}(1 - e^{-t/\tau_1}) \Delta(t) \tag{53.6}$$

where $\tau_1 = \eta_1/E_1$. *ABG* in Fig. 53.3*b* illustrates this solution. It is clear from Fig. 53.3*a*, which demands that the extension of the spring and dashpot are the same at all

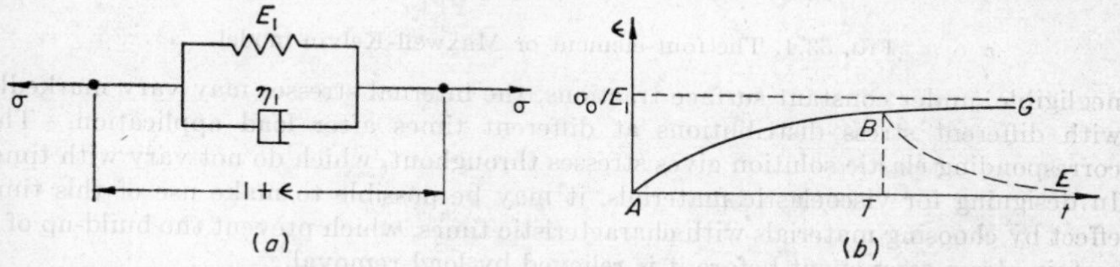

FIG. 53.3. The Kelvin material: (*a*) The viscoelastic model; (*b*) the creep and recovery curve.

times, that the initial strain is zero, since, for finite stress, and thus finite strain rate, the dashpot cannot develop an instantaneous component of strain. With this initial condition, and $\sigma = \sigma_0$, (53.5) is a simple differential equation with the solution (53.6). It is seen that the strain approaches the value σ_0/E_1, asymptotically with a characteristic time given by τ_1, which is known as the delay time.

If the stress is removed at time T, the recovery curve, *BE* in Fig. 53.3*b*, is obtained. This is an exponential curve obtained from (53.5) with zero stress, and it approaches zero strain asymptotically. For long times, the creep and recovery curve shown in Fig. 53.3*b* exhibits elastic behavior with modulus E_1, but this elastic response is approached gradually with a characteristic time τ_1. This behavior is known as delayed elasticity.

The simplest viscoelastic material which exhibits all three types of viscoelastic response—instantaneous elasticity, delayed elasticity, and viscous flow—is represented by the four-element model shown in Fig. 53.4, and is sometimes termed the *Maxwell-Kelvin material*. The corresponding creep and recovery curve is of the type illustrated in Fig. 53.1*b*. The differential operator relation corresponding to the model of Fig. 53.4 is of the form

$$(p_2D^2 + p_1D + p_0)\sigma = (D^2 + q_1D)\epsilon \tag{53.7}$$

where D is the differential operator d/dt, and p_2, p_1, p_0, and q_1 are material constants, which can be expressed in terms of the constants characterizing the model elements.

Special cases of this material are sometimes used, for which either the series spring or the dashpot is removed from the model. A material represented by a Kelvin unit with a spring in series exhibits instantaneous and delayed elasticity but no viscous flow, and is sometimes referred to as the standard linear solid or the material represented by the three-element elastic model. A Kelvin unit in series with a dashpot represents a material which exhibits delayed elasticity and flow, with no instantaneous response to stress. This is known as the *three-element viscous* or *Newton-Kelvin model.* These three-element models provide less restricted deformation characteristics than the Maxwell and Kelvin materials, and the corresponding mathematical representations are sometimes easier to handle in stress-analysis problems than relation (53.7) corresponding to the four-element model, or higher-order relations.

The rate influences which arise from the stress-strain-time relations (53.1), (53.2), (53.5), (53.7) and from other viscoelastic relations determine stress-distribution solutions for a loaded body which exhibit properties in marked contrast to elastic stress analysis. For example, in quasi-static analysis, for which inertia forces are

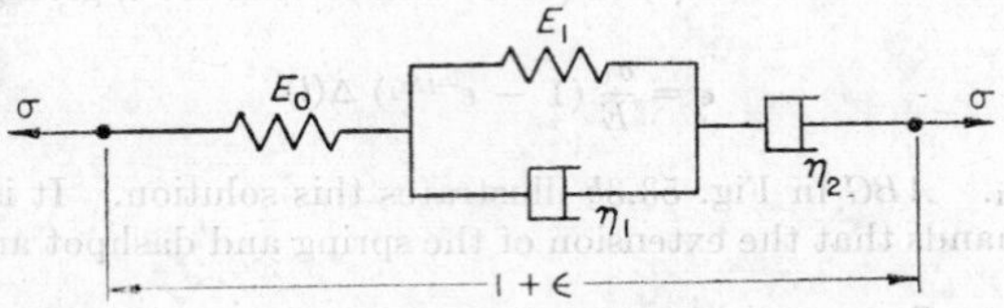

FIG. 53.4. The four-element or Maxwell-Kelvin model.

negligible, under constant surface tractions, the internal stresses may vary markedly with different stress distributions at different times after load application. The corresponding elastic solution gives stresses throughout, which do not vary with time. In designing for viscoelastic materials, it may be possible to make use of this time effect by choosing materials with characteristic times, which prevent the build-up of a certain stress component before it is relieved by load removal.

In applying viscoelastic stress analysis, three main aspects must be considered:

1. Measurement of material properties, including a test of linearity
2. Interpretation of such measurements in a form suitable for stress analysis, such as the determination of the corresponding viscoelastic model
3. Determination of stress distributions

The connections between the various methods of expressing viscoelastic relations are needed for step 2, and are considered in Sec. 53.2. Aspect 3 is discussed in Sec. 53.3, and, making use of the theory developed, the measurement of viscoelastic properties is considered in Sec. 53.4.

53.2. STRESS-STRAIN-TIME RELATIONS OF LINEAR VISCOELASTICITY

In stress-analysis problems, combined stresses will normally be encountered, so that viscoelastic laws will be needed which relate the nine components of stress and of strain. The situation is analogous to that in linear elasticity, with the difference that pairs of rate operators of the type discussed in the previous section replace elastic constants. For isotropic materials, two independent pairs of operators are required to express viscoelastic properties, by analogy with the two independent elastic constants needed for an elastic material. The fact that two are sufficient can be proved by demanding that the response to an applied stress is not changed if the material is rotated arbitrarily before load application, just as for the deduction that two independent constants represent isotropic elasticity. Consideration will be limited to isotropic materials, since this covers most work published in this field, although the extension to anisotropic materials follows that in elasticity according to the analogy mentioned above.

It is particularly significant, in viscoelasticity, to separate shear and dilatational effects, since viscoelastic influences are commonly more marked in response to shear stress. Thus corresponding to the elastic relations

$$\tau_{xy} = G\gamma_{xy}, \quad \ldots, \quad (\sigma_x - \sigma_y) = 2G(\epsilon_x - \epsilon_y), \quad \ldots \tag{53.8}$$

where the dots represent the additional relations obtained by permuting suffixes over the axes x, y, and z, the viscoelastic relations for shear effects are

$$P[\tau_{xy}(t)] = Q[\gamma_{xy}(t)], \quad \ldots, \quad P[\sigma_x(t) - \sigma_y(t)] = 2Q[\epsilon_x(t) - \epsilon_y(t)] \tag{53.9}$$

where P and Q are rate operators of the type discussed in the previous section.

Corresponding to the elastic compressibility relation,

$$\tfrac{1}{3}(\sigma_x + \sigma_y + \sigma_z) = K(\epsilon_x + \epsilon_y + \epsilon_z) \tag{53.10}$$

where K is the modulus of compressibility, isotropic linear viscoelastic behavior requires that

$$\tfrac{1}{3}P'[\sigma_x(t) + \sigma_y(t) + \sigma_z(t)] = Q'[\epsilon_x(t) + \epsilon_y(t) + \epsilon_z(t)] \tag{53.11}$$

where P' and Q' are a pair of rate operators which may be independent of P and Q.

Since a linear operator applied to another linear operator acting on a function gives a product linear operator, as, for example, is clearly valid for the differential operators discussed in the previous section, the basic operator pairs P, Q and P', Q' can be combined as are elastic constants, to determine single operator relations for response to particular stress systems, such as, for example, simple tension. Formally we can use the equivalence

$$Q/P = G \qquad Q'/P' = K \tag{53.12}$$

to obtain combined viscoelastic operators, using the formulas for the relations between elastic constants. This follows since the process of determining the response to special stress systems is by algebraic elimination of unwanted components from (53.8) and (53.10), or equivalent expressions, which carries through in an analogous manner for the linear operator relations (53.9) and (53.11). For example, Young's modulus,

$$E = \frac{\sigma}{\epsilon} \tag{53.13}$$

for simple tension is given in terms of G and K by

$$E = \frac{9KG}{3K + G} \tag{53.14}$$

and, for a linear viscoelastic material prescribed by (53.9) and (53.11), this becomes formally

$$\frac{\sigma(t)}{\epsilon(t)} = \frac{9\,\dfrac{Q'}{P'}\dfrac{Q}{P}}{3\,\dfrac{Q'}{P'} + \dfrac{Q}{P}} \tag{53.15}$$

or, clearing the denominator operators by cross multiplication,

$$[3PQ' + P'Q]\sigma(t) = [9Q'Q]\epsilon(t) \tag{53.16}$$

This is the viscoelastic operator relation for simple tension in terms of the basic operator pairs. If, for example, a body is elastic for dilatational response and behaves as a Maxwell material in shear, (53.10) and the form of Maxwell operator (53.2) give

$$\frac{Q'}{P'} = K \quad \text{and} \quad \frac{Q}{P} = \frac{D}{AD + B} \tag{53.17}$$

where K, A, and B are material constants, and D is the rate operator d/dt. Then (53.16) becomes

$$[(3KA + 1)D + 3KB]\sigma(t) = 9KD\epsilon(t) \tag{53.18}$$

or

$$\left(\frac{3KA + 1}{9K} D + \frac{B}{3}\right)\sigma(t) = D\epsilon(t) \tag{53.18a}$$

which is seen to be a Maxwell relation of type (53.2), with different constants from the shear Maxwell relation (53.17). Thus, the body defined by (53.17) would behave as a Maxwell material in a simple tension test, (53.18*a*), with different constants, and a different relaxation time from those corresponding to the basic shear Maxwell operator.

In a similar manner, an operator relation could be obtained for the relation between the longitudinal and lateral strain in a tension test, which would correspond to the elastic constant, Poisson's ratio.

Thus, in considering the relationships between the various methods of representing viscoelastic operators dealt with below, these can be viewed as applying to either of the basic operator pairs, or to other viscoelastic operators, such as that corresponding to simple tension. The stress variable σ and strain variable ϵ used below may, therefore, be considered to represent shear, volume effects, or simple tension, according to whether (53.9), (53.11), or (53.16) is being considered, or some other related stress and strain components.

A direct generalization of the particular viscoelastic laws, (53.2), (53.5), and (53.7), discussed in the previous section is the differential operator form of (53.1), for which

$$P = \sum_{r=0}^{p} p_r D^r \qquad Q = \sum_{r=0}^{q} q_r D^r \tag{53.19}$$

where p, p_r, q, and q_r are material constants. It can readily be seen that, by adding the strain contribution of the individual Kelvin elements of the model shown in Fig.

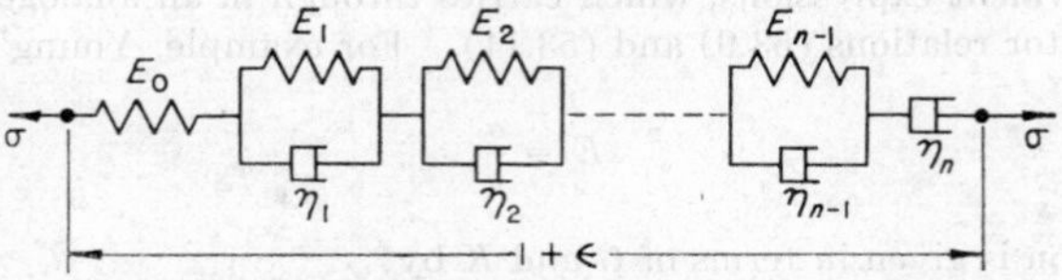

FIG. 53.5. The generalized Kelvin model.

53.5, using relations of the form (53.5), the viscoelastic differential operator relation corresponding to this model is given formally by

$$\epsilon = \left(\frac{1}{E_0} + \sum_{1}^{n-1} \frac{1}{E_r + \eta_r D} + \frac{1}{\eta_n D}\right)\sigma \tag{53.20}$$

Clearing the differential operator D from the denominators by multiplying both sides by

$$\eta_n D \prod_{1}^{n-1} (E_r + \eta_r D)$$

gives a relation of the form (53.19), with the orders p, q of the differential operators P, Q both equal to n. The Maxwell material governed by (53.2) and the model shown in Fig. 53.2*a* corresponds to $n = 1$, and the four-element model [(53.7) and Fig. 53.4] to $n = 2$. For a creep curve corresponding to constant applied stress σ_0, each Kelvin element in Fig. 53.5 gives an exponentially varying strain contribution with a different delay time, if $\tau_r = \eta_r/E_r$ is chosen to be different for each element. This offers the possibility of fitting a creep curve of the type shown in Fig. 53.1*b* by the sum of exponential terms and a linear term, in place of the single exponential and linear term corresponding to the four-element model [Fig. 53.4 and relation (53.7)]. In this way, any measured creep curve of the form of Fig. 53.1*b* can be fitted to any required degree of approximation, a more accurate fit, in general, requiring more exponential terms for an arbitrary creep curve. Thus, greater accuracy calls for a more complicated generalized Kelvin model, of the type shown in Fig. 53.5, with more elements and so higher-order differential operators.

It is convenient to consider the spring E_0 to correspond to a Kelvin element with zero delay time, and the dashpot η_n one with infinite delay time, and to order the rest in a sequence of increasing delay times. Approximations can then be achieved by lumping together Kelvin elements over a band of delay times into a single element

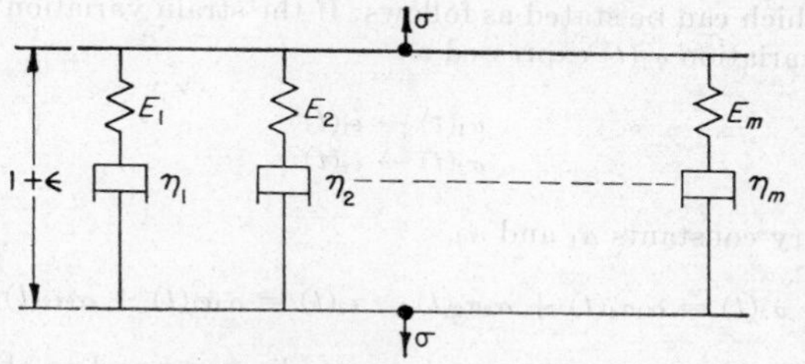

FIG. 53.6. The generalized Maxwell model.

with an average delay time. If, in a particular problem, the response to stress is of interest for only a particular time duration, elements with short delay times compared to this can be considered to have deformed fully, and can be lumped with the spring E_0 in Fig. 53.5, while elements with long delay times compared to the time range of interest will still be deforming mainly under the influence of the dashpots, and can be lumped with the dashpot η_n. Elements with delay times in the time range of interest can be lumped as a single Kelvin element, and the four-element model (Fig. 53.4) is obtained. Such simple models should be considered from this standpoint, since actual materials normally are not represented exactly by them. It will be observed that the model constants depend on the time range of interest of the problem considered, and the best fit of a prescribed model to a particular material will vary according to the loading history to be analyzed.

The generalized Kelvin model, as shown in Fig. 53.5 and expressed in (53.20), is particularly convenient if the stress history is prescribed, and the strain response is to be found, for the same stress acts on each of the Kelvin elements, and the total strain is the sum of the individual contributions. To find the stress associated with a prescribed strain variation, the viscoelastic law associated with the generalized Maxwell model (Fig. 53.6) is more easily handled, since the prescribed strain variation is applied to each Maxwell element individually, and the resulting stress is the sum of the individual contributions. The corresponding differential operator is deduced from (53.2):

$$\sigma = \left(\sum_{r=1}^{m} \frac{D}{D/E_r + 1/\eta_r} \right) \epsilon \tag{53.21}$$

The denominators containing the differential operator D are removed by multiplying both sides by

$$\prod_{1}^{m} \left(\frac{D}{E_r} + \frac{1}{\eta_r} \right)$$

and (53.21) then takes the form (53.19), with the orders of P and Q both equal to m.

A particular pair of operators of the same order of type (53.19) can thus be represented by models in the form of either the generalized Kelvin model (Fig. 53.5) or the generalized Maxwell model (Fig. 53.6). The determination of the model consists of formally dividing through by one of the operators, and representing the quotient by partial fractions. Dividing through by P gives (53.21) and the generalized Maxwell model, and dividing through by Q gives (53.20) and the generalized Kelvin model. It will be noticed that, for operators of a given order, $n = m$ in (53.20) and (53.21), and the same number of material constants is needed to express the equivalent model representations.

Linearity of the viscoelastic operators (53.1) implies the validity of the superposition principle, which can be stated as follows: If the strain variation $\epsilon_1(t)$ is associated with the stress variation $\sigma_1(t)$ expressed as

$$\sigma_1(t) \rightarrow \epsilon_1(t) \tag{53.22}$$

and

$$\sigma_2(t) \rightarrow \epsilon_2(t) \tag{53.23}$$

then, for arbitrary constants α_1 and α_2,

$$\sigma_3(t) = \alpha_1\sigma_1(t) + \alpha_2\sigma_2(t) \rightarrow \epsilon_3(t) = \alpha_1\epsilon_1(t) + \alpha_2\epsilon_2(t) \tag{53.24}$$

This relation is used both to test a specimen for linearity and to obtain the response for arbitrary loading situations from that for particular load variations, such as the creep curve.

Linearity is most simply checked by taking $\alpha_2 = 0$, and generating the load variations

$$\sigma_3(t) = \alpha_1\sigma_1(t) \tag{53.25}$$

for a range of values of α_1, and a particular variation $\sigma_1(t)$. For linearity,

$$\epsilon_3(t)/\alpha_1 = \epsilon_1(t) \tag{53.26}$$

is independent of α_1, so that, if a series of values of $\epsilon_3(t)$ is measured for a range of α_1, the repetition of the same curve for $\epsilon_3(t)/\alpha_1$, plotted against t, will indicate linearity. A common check is to take $\sigma_1(t)$ constant, a creep test, and then the ordinates of the creep curves at any time instant will be proportional to stress for a linear material.

Consider the stress variation shown in Fig. 53.7. It can be approximated by the sum of a series of step functions as indicated in the figure, which corresponds to a series of creep loadings. The creep compliance $J_c(t)$ is the creep strain variation for unit stress applied at $t = 0$, so that the response to the variation depicted in Fig. 53.7 can be represented, making use of the superposition principle, by the integral

$$\epsilon(t_1) = \int_{-\infty}^{t_1} \frac{d\sigma(\tau)}{d\tau} J_c(t_1 - \tau)\, d\tau \tag{53.27}$$

Such an integral is known in the mathematical literature as a *Duhamel integral*. The lower limit is taken to be $-\infty$ to allow for the initiation of load at any time, but, if the origin of t is taken to denote the commencement of loading, with the material previously undisturbed, the lower limit can be replaced by zero.

The representation of viscoelasticity by a relation of the form (53.27) was first introduced by V. Volterra under the terminology hereditary material, since the kernel of the integral, $J_c(t_1 - \tau)$, can be considered as a memory function, transforming the influence of the pulse of load applied at $t = \tau$ to the instant $t = t_1$. Relation (53.27) is equivalent to the differential operator form (53.1) and (53.19), since $J_c(t)$ is then a particular solution of the differential equation for $\epsilon(t)$ corresponding to a step function of stress. It is related to the Green's function of the differential operator Q.

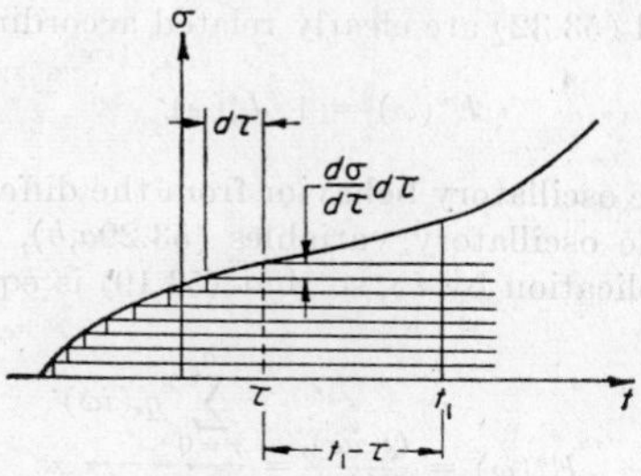

FIG. 53.7. Superposition of creep responses.

A relation similar to (53.27) can be obtained in terms of the relaxation modulus $E_R(t)$, which is the relaxation stress per unit applied strain,

$$\sigma(t_1) = \int_{-\infty}^{t_1} \frac{d\epsilon(\tau)}{d\tau} E_R(t_1 - \tau)\, d\tau \tag{53.28}$$

Another common method of expressing viscoelastic properties is the response to steady-state oscillatory stress. For a linear material, steady-state response has the same frequency as the excitation, and it is convenient to use the complex notation common in alternating-current theory:

$$\sigma = \Re\,(\sigma_0 e^{i\omega t}) \qquad \epsilon = \Re\,(\epsilon_0 e^{i\omega t}) \tag{53.29a,b}$$

For a linear material, the ratio of stress to strain is a complex number, the complex modulus E^*, since, for a viscoelastic body, the stress and strain will not, in general, be in phase:

$$\frac{\sigma}{\epsilon} = \frac{\sigma_0}{\epsilon_0} = E^*(\omega) = E_1(\omega) + iE_2(\omega) \tag{53.30}$$

$E_1(\omega)$ is called the real part of the complex modulus, or the storage modulus since it is associated with the storage of elastic strain energy in the body. $E_2(\omega)$, the imaginary component of the complex modulus, is sometimes referred to as the loss modulus since it represents the dissipation of energy during cyclic loading,

$$E_2/E_1 = \tan\delta \tag{53.31}$$

where δ is the phase angle between the variation of stress and the variation of strain.

It is sometimes convenient to use the reciprocal of (53.30), involving the complex compliance $J^*(\omega)$:

$$\frac{\epsilon}{\sigma} = \frac{\epsilon_0}{\sigma_0} = J^*(\omega) = J_1(\omega) - iJ_2(\omega) \tag{53.32}$$

The various methods of expressing viscoelastic properties, (53.19), (53.27), (53.28), (53.30), and (53.32), can all be considered as variations of (53.1) with the P

and Q taking on the form of differential (53.19), integral (53.27) and (53.28), and complex algebraic (53.30) and (53.32) operators, respectively. For a particular viscoelastic material, they are all equivalent, so that each can be deduced from any of the others by purely mathematical manipulation. The ease with which this can be done varies greatly between the various operator pairs. In practice, it is important to be able to make the change from one form of operator to another, since it may be convenient to measure viscoelastic properties in one form, and yet another form may be more convenient for stress analysis.

Equations (53.30) and (53.32) are clearly related according to

$$E^*(\omega) = 1/J^*(\omega) \tag{53.33}$$

It is also simple to deduce oscillatory behavior from the differential operator relations (53.19). For steady-state oscillatory variables (53.29a,b), the differential operator D is equivalent to multiplication by $i\omega$, so that (53.19) is equivalent to

$$E^*(\omega) = \frac{Q(i\omega)}{P(i\omega)} = \frac{\sum_{r=0}^{q} q_r(i\omega)^r}{\sum_{r=0}^{p} p_r(i\omega)^r} \tag{53.34}$$

This relation shows that (53.19), with finite-order differential operators, is a more restricted relation than (53.30), since it corresponds to E^* in the form of a rational function of $(i\omega)$, whereas E^* need not be so restricted. In a similar way, (53.27) and (53.28), for arbitrary functions $J_c(t)$ and $E_R(t)$, are also more general than (53.19) with finite p and q. Finite differential operators give these functions as solutions of differential equations with constant coefficients, so that they take the form of finite sums of exponential terms. From a practical standpoint, this restriction is not significant, since general functions can be represented by rational functions or sums of exponentials to any required degree of accuracy. Thus finite differential operators, and so finite viscoelastic models, can supply a satisfactory approximation for arbitrary linear viscoelastic behavior.

General relations among all the forms of operators have been discussed by Gross [3]. In many instances, these are expressed in terms of the theory of analytic functions, and apply for operator functions known analytically. If measured values are to be utilized, such as measured creep or relaxation functions, some difficulty arises in using these analytic procedures. Methods are now being developed to solve this problem for measured functional values. Roesler and Pearson [4] and Roesler and Twyman [5] have made use of Fourier series representations for this purpose; Schwarzl [6] and Leaderman [7] of asymptotic representations of the analytic forms; and Hopkins and Hamming [8] have used numerical integration of the integral equations (53.27) and (53.28); Bland and Lee [9] have given a graphical method for assessing the accuracy to which a measured variation of complex compliance with frequency can be reproduced by the response of a four-element model of the type shown in Fig. 53.4. This has proved satisfactory for a frequency range of one or two decades, which is often adequate for stress-analysis problems. It may be possible to obtain a satisfactory representation of a transient loading pulse by a Fourier series with between 10 and 100 terms, so that the frequency band mentioned above would be adequate for analyzing the response of the material to this load. Such problems contrast with the model representation of viscoelastic properties over very wide frequency ranges, which are often needed in the physicochemical studies of materials, and, in these cases, a more elaborate model would be needed.

Figure 53.4 corresponds to the complex compliance expression,

$$J_1(\omega) - iJ_2(\omega) = \frac{1}{E_0} + \frac{1}{E_1 + \eta_1\omega i} + \frac{1}{\eta_2\omega i} \tag{53.35}$$

as can be deduced from (53.20), with $n = 2$, and the operator D replaced by $i\omega$. This implies the linear relation

$$\frac{E_1}{\eta_1} J_1(\omega) + \omega J_2(\omega) = \frac{E_1}{E_0\eta_1} + \frac{1}{\eta_2} + \frac{1}{\eta_1} \tag{53.36}$$

between $J_1(\omega)$ and $\omega J_2(\omega)$. Experimental values of $J_1(\omega)$ and $\omega J_2(\omega)$ over a band of frequency can thus be plotted, and an approximate linear result implies the satisfactory application of an equivalent four-element viscoelastic model or quadratic operator relation. The gradient and intercept of the straight line give

$$\frac{E_1}{\eta_1} \qquad \text{and} \qquad \frac{E_1}{E_0\eta_1} + \frac{1}{\eta_2} + \frac{1}{\eta_1}$$

Equation (53.35) also implies the relation

$$J_1(\omega) = \frac{1}{E_0} + \frac{1}{E_1}\left[1 + \left(\frac{\eta_1}{E_1}\right)^2 \omega^2\right]^{-1} \tag{53.37}$$

The term $[1 + (\eta_1/E_1)^2\omega^2]^{-1}$ can be obtained for varying ω from the gradient of the previous line (53.36), so that a plot of $J_1(\omega)$ against this will give $1/E_1$ and $1/E_0$ from the gradient and intercept of the straight line (53.37). All constants can then be determined. An assessment of the accuracy of the model is shown by the degree to which the two plots give approximations to straight lines over the frequency range of interest.

This method is useful in considering the response of a body to a short pulse of stress, since, for the high frequencies involved, oscillatory tests are the most convenient method of measuring viscoelastic properties. Since the differential operator has particular advantages for certain stress-analysis problems, the transformation of compliance variation to an equivalent differential operator is needed.

53.3. STRESS ANALYSIS

Figure 53.8 illustrates a viscoelastic body V subjected to surface tractions $[T_x(x,y,z,t), \ldots]$ over the part S_1 of the boundary S, where, as in the previous section, the dots stand for the other components in the system of axes (x,y,z). Over the part S_2 of the boundary, the displacements $[u(x,y,z,t), \ldots]$ are prescribed, and this part could include a support where the displacements are zero. A body force $[f_x(x,y,z,t), \ldots]$ per unit volume may also be acting throughout the body V. The present section is concerned with the determination of the variation of stress and strain distributions within the body. It will be observed that, in contrast to quasi-static elasticity problems, time appears in the boundary conditions and body force, and plays an important role in the analysis. It will be seen later that, even for constant boundary values, the internal stress distribution will, in general, vary with time for a viscoelastic body.

The analysis presented below is limited to infinitesimal displacements and strains. Thus, the displacements and strain components are connected by linear relations [see Eqs. (33.2a–f)]:

$$\epsilon_x = \frac{\partial u}{\partial x}, \quad \ldots, \quad \gamma_{xy} = \frac{\partial u}{\partial y} + \frac{\partial v}{\partial x}, \quad \ldots \tag{53.38}$$

Boundary conditions will, in general, be considered to be satisfied on the undisturbed boundaries of V, as is common for infinitesimal elasticity theory. Extensions to such situations as instability problems, for which boundary motion must be taken into account, can be made as in elasticity theory. The particular case of contact stresses corresponding to the Hertz solution in elasticity (see p. 42–2), which also requires the consideration of boundary motion, is discussed below.

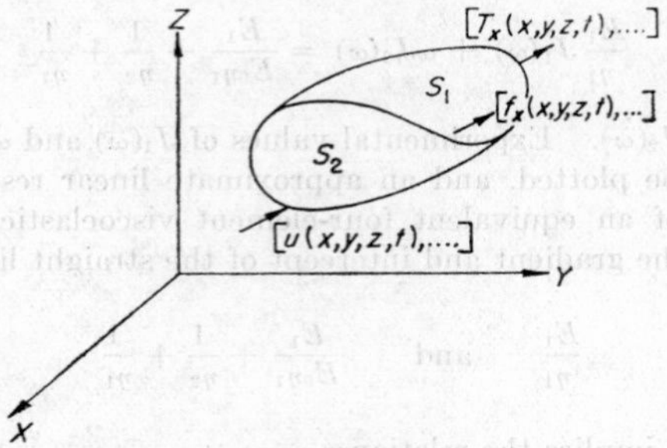

FIG. 53.8. The stress-analysis problem.

The stress and strain components at each point in the body must satisfy the viscoelastic stress-strain relations (53.9) and (53.11), where the components are functions of x, y, z, and t:

$$\sigma_x = \sigma_x(x,y,z,t), \quad . \; . \; . \tag{53.39}$$

so that derivatives with respect to time now become partial derivatives. The pairs of operators P, Q and P', Q' can take on any of the forms discussed in the previous subsection, with all derivatives and integrals with respect to time t occurring at constant values of (x,y,z).

The equations of motion are:

$$\frac{\partial \sigma_x}{\partial x} + \frac{\partial \tau_{xy}}{\partial y} + \frac{\partial \tau_{xz}}{\partial z} + f_x = \rho \frac{\partial^2 u}{\partial t^2}, \quad . \; . \; . \tag{53.40}$$

where ρ is the density.

Stress and strain distributions must satisfy (53.9), (53.11), (53.38), and (53.40) throughout V, and the boundary conditions

$$T_x(x,y,z,t) = \sigma_x n_x + \tau_{xy} n_y + \tau_{xz} n_z, \quad . \; . \; . \tag{53.41}$$

on S_1, where n_x, n_y, and n_z are the direction cosines of the outward normal, and

$$u(x,y,z,t), \quad . \; . \; . \quad \text{prescribed} \tag{53.42}$$

on S_2. As in elasticity, it is also possible to have mixed conditions of prescribed displacement and traction components, but these will not be detailed here.

Equations (53.9), (53.11), (53.38), and (53.40) form a system of 15 partial differential equations for the six independent stress components, six strain components, and three displacement components. The six independent equations (53.9) and (53.11) contain only time derivatives, or operators; the six equations (53.38) contain only space derivatives; and the three equations (53.40) are mixed. For quasi-static problems for which the inertia terms in (53.40) are negligible, these equations also contain only space derivatives.

Because of the time derivatives, initial conditions will generally also be needed, and these will depend on the viscoelastic operators. For the common case of an initially undisturbed body, the initial conditions will be that stress and strain, and their derivatives with respect to time to all orders needed, will be equal to zero.

In certain circumstances, the time and space operations in the system of equations and boundary conditions can be separated, and solutions obtained by integrating separately with respect to the space variables and the time. In such cases, the integration with respect to time reduces to the analysis of a viscoelastic body for homogeneous stress and strain, as discussed in the previous section, and the space integration reduces to an equivalent elastic problem. A common example of such behavior arises for steady-state oscillatory boundary conditions. If, apart from the oscillatory variation of surface tractions and prescribed displacement, the boundary conditions are independent of time, for example, the boundary regions S_1 and S_2 in Fig. 53.8 do not change with time, then all dependent variables, stress components, strain components, and displacements can be expressed in the form

$$f(x,y,z,t) = \Re\,[f^*(x,y,z)e^{i\omega t}] \tag{53.43}$$

Equations (53.9) and (53.11) are then most conveniently expressed in terms of the complex shear and bulk moduli:

$$Q/P = G^*(\omega) \qquad Q'/P' = K^*(\omega) \tag{53.44}$$

The time factor $e^{i\omega t}$ cancels from all equations and boundary conditions, and (53.9), (53.11), (53.38), and (53.40) reduce to an elastic problem in $\sigma_x^*, \ldots, \epsilon_x^*, \ldots, u^*, \ldots$, with the additional body force $[\omega^2 \rho u^*, \ldots]$, arising from the inertia forces. Since the associated elastic moduli are complex, stress components given by the elastic solution will, in general, also be complex, which means, according to (53.43), that they will oscillate with different phases varying with position.

As an example, consider an oscillatory point load,

$$P(t) = P_0 e^{i\omega t} \tag{53.45}$$

acting on the surface of a semi-infinite viscoelastic solid, and directed normally to the surface toward the interior of the solid. In (53.45), according to the usual convention, the specification, real part, has been omitted. We will consider no prescribed body force, and a frequency sufficiently low so that the additional inertial body force $[\omega^2 \rho u^*, \ldots]$ is negligible. The solution of the associated elastic problem for a load P_0 is well known [see Eqs. (41.61)], and, for example, the radial stress in cylindrical coordinates r, z, θ is given by

$$\sigma_r^* = \frac{P_0}{2\pi}\left\{(1 - 2\nu^*)\left[\frac{1}{r^2} - \frac{z}{r^2}(r^2 + z^2)^{-1/2}\right] - 3r^2 z(r^2 + z^2)^{-5/2}\right\} \tag{53.46}$$

ν^* is the complex Poisson's ratio, associated with the complex moduli through the equivalent of the usual elastic relation

$$1 - 2\nu^* = \frac{3G^*}{3K^* + G^*} \tag{53.47}$$

The two terms in (53.46), one having the complex coefficient (53.47), show that the radial stress oscillates with different phases at different points.

In some cases, for example, plane problems with prescribed surface tractions independently in equilibrium on the separate boundaries for multiconnected regions, the stresses in the plane for an elastic body are not dependent on the elastic moduli. In the equivalent viscoelastic case for steady-state oscillatory loading, the corresponding separated stress components will be real and equal to those for the elastic body with the same loading, so that the stress throughout will oscillate in phase with the applied tractions. Lamé's solution for a cylinder with applied internal and external pressures is an example of this behavior.

A transient load can sometimes be conveniently represented by a Fourier integral or Fourier series, and the principle of superposition enables each component to be analyzed as described above, and the resulting solutions superposed. The Fourier series is mentioned because, although it involves periodic behavior, it can sometimes be used to represent a transient loading for viscoelastic stress analysis. Response to stress tends to be damped out with time, apart from viscous flow, which can be treated separately, so that the influence of a single pulse in a periodic sequence of pulses may be dissipated before the next pulse arrives. The advantages of using a Fourier series rather than a Fourier integral, in such cases, are associated with the ease of carrying a finite number of terms for superposition, rather than the limit of an infinite sum in the form of an integral, and the fact that a narrow frequency band may be utilized, with the consequent use of a low-order differential operator viscoelastic relation. In a particular case, the relaxation curve indicates the time interval over which transient pulses will be effectively dissipated, and the choice of periodic time for the Fourier series should be of this magnitude. A glance at the evaluated stress and strain variations will show whether there is interaction between the separate pulses.

A wider class of viscoelastic stress-analysis problems can be treated by removing the time dependence in the system of equations and boundary conditions by applying the Laplace transform with respect to time:

$$\bar{f}(x,y,z,s) = \int_0^\infty f(x,y,z,t)e^{-st}\,dt \tag{53.48}$$

This method can be used directly if the boundary conditions at each surface point retain the same form throughout the period for which the stress analysis is required; otherwise, for example, at a particular point, tractions might be prescribed during part of the time, and surface displacements at later times; the Laplace transform of neither the surface traction nor the displacement could then be evaluated. Thus, for this method, S_1 and S_2 in Fig. 53.8 must not change, nor must the form of the body itself.

Application of the Laplace transform replaces the operator D in (53.19) by the transform parameter s for zero initial conditions, so that the general differential operator stress-strain relation takes the form

$$\left(\sum_{r=0}^{p} p_r s^r\right)\bar{\sigma} = \left(\sum_{r=0}^{q} q_r s^r\right)\bar{\epsilon} \tag{53.49}$$

with some additional polynomial terms in s for nonzero conditions. The transforms of the integral operators (53.27) and (53.28) for zero initial conditions are given by the convolution theorem [see Eq. (19.17)]:

$$\bar{\epsilon} = \bar{J}_c(s)s\bar{\sigma} \qquad \bar{\sigma} = \bar{E}_R(s)s\bar{\epsilon} \tag{53.50a,b}$$

Thus, for transformed coordinates, an elastic stress-strain relation applies, with the modulus a function of the transform parameter s. The transforms of the equations of viscoelastic stress analysis, (53.9), (53.11), (53.38), (53.40), (53.41), and (53.42), then determine an elastic problem, known as the associated elastic problem. In addition to its appearance in the elastic moduli, the parameter s also appears in the boundary conditions, and from the inertia term in the equations of motion. When the associated elastic problem can be solved, the stresses in the viscoelastic case are given by the inversion of the stresses for the associated elastic solution. This result permits the techniques and special solutions of linear elasticity to be utilized directly for corresponding viscoelastic problems.

As an example, consider the force $P(t)$ applied on the surface of a semi-infinite viscoelastic body, and acting normally to the surface. Again, making use of the elastic solution for point loading [see Eq. (53.46)], the transform of the radial stress component is given by

$$\bar{\sigma}_r(r,z,s) = \frac{\bar{P}(s)}{2\pi}\left\{\frac{3Q/P}{3Q'/P' + Q/P}\left[\frac{1}{r^2} - \frac{z}{r^2}(r^2 + z^2)^{-1/2}\right] - 3r^2z(r^2 + z^2)^{-5/2}\right\} \tag{53.51}$$

As a particular case, suppose that the body behaves as a Kelvin material for shear response, and is elastic under hydrostatic compression. Equations (53.9) and (53.11) then become

$$\tau_{xy} = (AD + B)\gamma_{xy}, \quad \ldots \tag{53.52}$$

$$(\sigma_x + \sigma_y + \sigma_z) = 3K(\epsilon_x + \epsilon_y + \epsilon_z) \tag{53.53}$$

and (53.51) takes the form

$$\bar{\sigma}_r(r,z,s) = \frac{\bar{P}}{2\pi}\left\{\frac{3(As + B)}{3K + As + B}\left[\frac{1}{r^2} - \frac{z}{r^2}(r^2 + z^2)^{-1/2}\right] - 3r^2z(r^2 + z^2)^{-5/2}\right\} \tag{53.54}$$

For a constant force P_0 suddenly applied and maintained, this can be inverted to give

$$\sigma_r(r,z,t) = \frac{P_0}{2\pi}\left(\frac{3}{3K + B}\left\{B + 3K\exp\left[-(3K + B)\frac{t}{A}\right]\right\}\left[\frac{1}{r^2} - \frac{z}{r^2}(r^2 + z^2)^{-1/2}\right] - 3r^2z(r^2 + z^2)^{-5/2}\right) \tag{53.55}$$

showing, in contrast to the behavior of elastic stress analyses, that the stress components vary with time under constant applied load. For differential operators of finite order, inversion of the Laplace transform can, in general, be achieved as above by means of the method of partial fractions associated with a rational function of s. The case for variable loading $P(t)$ can be inverted, using the convolution theorem, giving, for the particular material considered above,

$$\sigma_r = \frac{1}{2\pi}\left(\int_0^t 3\left\{\delta(\tau) - \frac{3K}{A}\exp\left[-(3K + B)\frac{\tau}{A}\right]\Delta(\tau)\right\}P(t - \tau)\,d\tau \times \left[\frac{1}{r^2} - \frac{z}{r^2}(r^2 + z^2)^{-1/2}\right] - P(t)3r^2z(r^2 + z^2)^{-5/2}\right) \tag{53.56}$$

where $\delta(\tau)$ is the Dirac delta function.

Using the associated elastic solution method and the Laplace transform, Woodward and Radok [10] have determined the varying stress distribution for a viscoelastic cylinder with elastic reinforcement on the outside and subjected to constant internal pressure. This stress distribution can be obtained directly from the Lamé solution for an elastic cylinder. For the particular case of a material which behaves as an elastic body under hydrostatic compression, and as a Maxwell body in shear, Fig. 53.9 shows the variations of radial and circumferential compressive stress for a particular geometry and reinforcement strength. Under the initial load, tensile circumferential stresses are produced adjacent to the internal cavity, but, as the Maxwell response in shear generates viscous flow, the circumferential stress becomes increasingly compressive. Finally, the viscoelastic body acts effectively as a compressible liquid, the shear stresses having relaxed to zero, and the full internal pressure is transmitted to the reinforcing ring. The wide variation of circumferential stress under constant load and the associated gradual build-up of the pressure on the reinforcement indicated by the external radial stress illustrate the possibility of utilizing time effects in design to prevent build-up of a certain stress component during the time of loading by choosing the characteristic time of the material appropriately.

It was pointed out above that only special problems, for which the type of boundary condition at each surface point does not change with time, can be treated by means of the Laplace transform. Essentially equivalent results can be obtained using the Fourier transform [11],[12] under the same restrictions. For the more general type of problem, it is sometimes possible to integrate the system of equations (53.9), (53.11), (53.38), and (53.40) separately with respect to time and to the space variables for

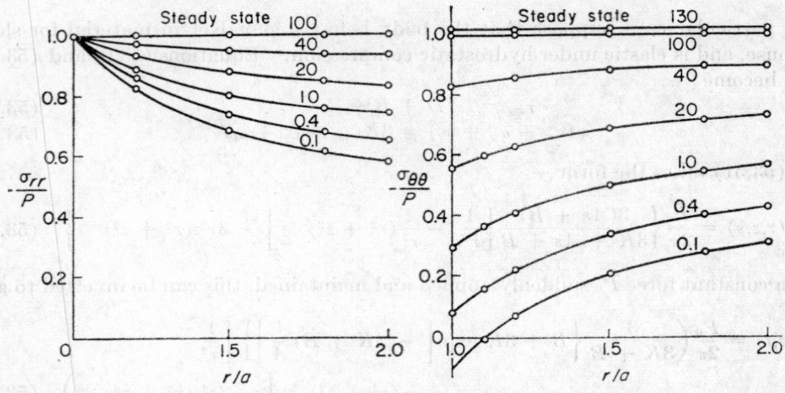

FIG. 53.9. Variation of radial and circumferential stress distribution for a reinforced cylinder subjected to constant internal pressure. Numbers on curves give time elapsed after loading.

quasi-static problems, for which, as we have seen, the individual equations of the system contain only time derivatives or only space derivatives. Because of the need to carry arbitrary functions when such integrations are performed, this procedure becomes very cumbersome, and Radok [13] has suggested a simpler approach, which applies for some problems not amenable to transform analysis.

Radok suggests considering the elastic constants in elastic solutions as time operators, and interpreting the expressions for elastic stress components accordingly. For example, if the expression for a stress component contains rational functions of the elastic constants, this expression will become a differential equation for that stress component in the viscoelastic case, if the differential form of the viscoelastic relation is used. This method differs from the separate integration of the equations mentioned above, since the elastic solution is obtained without the elaboration of time operators. This neglect can lead to difficulties, as detailed in a particular case below, but it is always possible to check the solution by transform methods, once the surface tractions or displacements have been obtained over the whole surface.

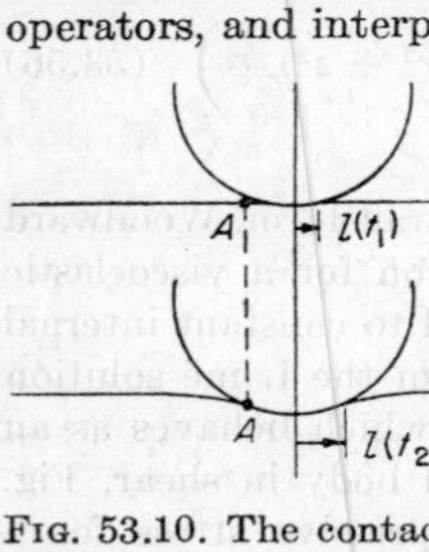

FIG. 53.10. The contact problem.

The problem of contact stresses, analogous to the Hertz solution in elasticity, is one for which the type of boundary conditions changes with time at a fixed surface point. Thus, in Fig. 53.10 point A is outside the surface of contact at $t = t_1$, and zero surface traction is prescribed. At $t = t_2$, A falls in the contact region, and the displacement is prescribed. Thus, the Laplace transform method cannot be used directly for this problem. Radok and Lee [14] have shown that the viscoelastic contact problem can be solved with certain restrictions by an appropriate interpretation of the elastic solution.

For convenience of illustration, consider the case of a rigid spherical indenter of radius R_2, pressing against a semi-infinite incompressible viscoelastic body. In the elastic case [see Eqs. (42.1) and (42.2)], the expression for the radius of contact $l(t)$, in terms of the total contact force $P(t)$, is

$$P(t) = \frac{8}{3}\frac{2G}{R_2}\, l(t)^3 \tag{53.57}$$

and the contact pressure is given by

$$p(r,t) = \frac{4}{\pi}\frac{2G}{R_2}\sqrt{l(t)^2 - r^2} \tag{53.58}$$

To interpret this solution for a viscoelastic body, the shear modulus G must be replaced by a viscoelastic operator. It is convenient, in this case, to use the shear creep compliance operator (53.27), and then (53.57) and (53.58) become

$$\int_{-\infty}^{t} J_c(t-\tau)\frac{dP(\tau)}{d\tau}\,d\tau = \frac{8}{3}\frac{2}{R_2}\, l(t)^3 \tag{53.59}$$

and

$$\int_{-\infty}^{t} J_c(t-\tau)\frac{\partial p(r,\tau)}{\partial\tau}\,d\tau = \frac{4}{\pi}\frac{2}{R_2}\sqrt{l(t)^2 - r^2} \tag{53.60}$$

For step-function loading,

$$P(t) = P_0\,\Delta(t) \tag{53.61}$$

Eq. (53.59) becomes

$$P_0 J_c(t) = \frac{8}{3}\frac{2}{R_2}\, l(t)^3 \tag{53.62}$$

In this case, because we have considered the material to be incompressible, only one time operator occurs in the stress-analysis problem, and it is possible to interpret the solution for a general creep compliance. An equivalent differential operator could have been used, and (53.59) and (53.60) would have taken the form of differential equations for the radius of contact and contact pressure.

As we would expect, (53.62) shows that a body which creeps indefinitely under constant stress, such as a Maxwell body, will determine a continuously increasing contact area under constant load. However, a body exhibiting restricted creep, such as the Kelvin body, will determine a limiting maximum radius of contact under constant load.

No difficulties arise in interpreting the elastic stress distribution to solve the viscoelastic problem unless the area of contact decreases. In this case, it is found that the solution gives a residual tensile traction on the part of the surface formerly in contact, and this violates the condition for contact for which nonzero surface traction occurs only in the region of contact. This difficulty arises since the zero pressure in the elastic case corresponds to a homogeneous operator equation in the case of the viscoelastic body. Thus, if differential operators are used, the zero traction in the elastic case is replaced by the complementary function of the differential equation in the viscoelastic case, and this leads to nonzero tractions outside the region of contact. The solution deduced still represents a satisfactory viscoelastic problem, but not the contact problem for which the pressure must immediately fall to zero outside the region of contact. Difficulties of this type must be anticipated in interpreting elastic solutions as viscoelastic operator relations, but they can always be checked by computing the tractions over the surface, and using the solution obtained by the transform method to ensure that the desired boundary conditions are satisfied.

The above discussion has been confined to quasi-static problems for which the inertia terms in (53.40) can be neglected. For steady-state oscillations of a viscoelastic body, the inertia terms take the form of a body force depending on the displacement, and the methods discussed above can be used, although the associated elastic problem becomes much more difficult to solve. Adler, Sawyer, and Ferry [15] and Oestreicher [16] have presented examples of such problems. In the transient case, wave propagation occurs, and the evaluation of solutions is much more difficult. Investigations have, in the main, been confined to the propagation of waves in one dimension. This theory can be interpreted to cover longitudinal waves in rods or shear waves or longitudinal waves in bodies of large lateral dimensions. A single component of stress and strain will appear in the analysis in all these cases, and it is simply necessary to make the appropriate choice of stress and displacement components for each application.

The viscoelastic stress-strain relation is given by (53.1), and the equation of motion in one dimension, in the absence of body forces, takes the form

$$\frac{\partial \sigma}{\partial x} = \rho \frac{\partial^2 u}{\partial t^2} \tag{53.63}$$

When (53.1) is taken in the form of differential operators (53.19), this with (53.63) comprises a pair of partial differential equations in the dependent variables σ and u. Either variable can be eliminated, giving a single partial differential equation, the order of which increases with the order of the viscoelastic differential operators.

For the Maxwell material, (53.2) and (53.63) lead to

$$\frac{\partial^2 \sigma}{\partial x^2} - \frac{1}{c^2}\frac{\partial^2 \sigma}{\partial t^2} - \frac{\rho}{\eta}\frac{\partial \sigma}{\partial t} = 0 \tag{53.64}$$

where $c^2 = E/\rho$ is the square of the elastic-wave velocity corresponding to infinite viscosity of the dashpot. The same partial differential equation governs the velocity, strain, and displacement, as can be seen by eliminating the appropriate variables in each case. Equation (53.64) is the same as the telegraph equation which governs the propagation of electricity along a cable, and particular solutions are available in the literature. For a semi-infinite rod subjected to a constant velocity impact of magnitude V, the resulting variation of the stress distribution is given, according to [17], by

$$\sigma = -\rho c V e^{-Et/2\eta} I_0 \left[\frac{E}{2\eta}\left(t^2 - \frac{x^2}{c^2}\right)^{1/2} \right] \Delta\left(t - \frac{x}{c}\right) \tag{53.65}$$

where I_0 is the modified Bessel function of zero order. Figure 53.11 shows, in dimensionless coordinates, the stress distributions at successive times according to this solution. For comparison, the corresponding solution for elastic impact is shown by the broken lines. It is seen that propagation in the Maxwell body exhibits a sharp wavefront moving with speed c, but that, in contrast to the corresponding elastic problem, the stress amplitude is attenuated.

Wave-propagation solutions for impact on rods of materials corresponding to the four-element model of Fig. 53.4, and simpler two- and three-element models, have been compared by Lee and Morrison [18]. It is found that, if the body exhibits instantaneous elastic response, a definite wavefront is obtained ahead of which no influence of the impact is felt. However, bodies not exhibiting instantaneous elastic response give parabolic- rather than hyperbolic-type equations, and diffusive behavior is obtained, with the impact spreading immediately throughout the bar, but approaching zero amplitude with increasing x. It is clear that, although such solutions may provide useful approximations at later times, they will be in error for large x and small t, because all materials must exhibit some instantaneous elastic response, and so give a definite expanding wavefront, beyond which the influence of the impact is not felt. The Kelvin body (Fig. 53.3) suffers from this limitation.

The principle of superposition can be used to combine elementary solutions in order to satisfy other boundary requirements. For example, just as for elastic waves, the solution for impact on a semi-infinite rod can be used to determine the solution for a finite rod by application of the method of images [17]. This principle is also useful in breaking down quasi-static problems into more elementary components which can be superposed.

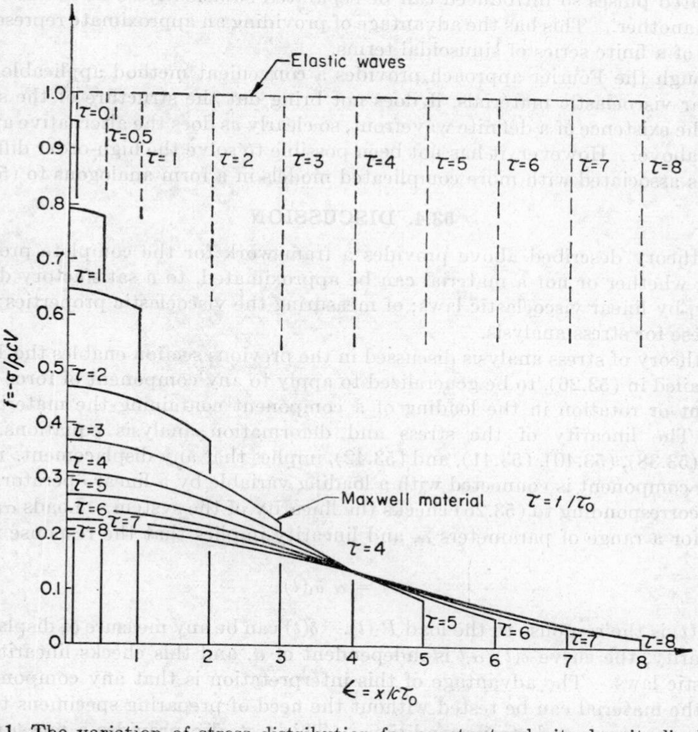

FIG. 53.11. The variation of stress distribution for constant-velocity longitudinal impact on a rod of Maxwell material.

An alternative approach to wave-propagation analysis through Fourier representation has been given by Kolsky [19]. Considering a simple sinusoidal stress wave propagating in the direction of increasing x, the stress can be written in the form

$$\sigma = \sigma_0 e^{i\omega(t-kx)} \tag{53.66}$$

Since, for fixed x, the stress varies sinusoidally with frequency ω, the strain is given in terms of the complex modulus for steady-state conditions:

$$\epsilon = \frac{\sigma_0}{E^*} e^{i\omega(t-kx)} \tag{53.67}$$

Substitution into the equation of motion (53.63) determines the parameter k, which is frequency-dependent:

$$k = \sqrt{\rho/E^*} \tag{53.68}$$

Since, for a viscoelastic material, E^* is a complex number, k is also complex; the real part is the reciprocal of the wave velocity, and the complex part gives the attenuation of the wave. This solution requires only the variation of complex modulus with frequency, and can be applied independently of the complexity of the viscoelastic model representation.

The propagation of a transient is obtained by representing it as a Fourier integral, and considering the propagation of each component wave as detailed above. For practical purposes the Fourier integral can often be replaced by a Fourier series, since the repeated pulses so introduced can be separated sufficiently so as not to interfere with one another. This has the advantage of providing an approximate representation in terms of a finite series of sinusoidal terms.

Although the Fourier approach provides a convenient method applicable to general linear viscoelastic materials, it does not bring out the structure of the solution, such as the existence of a definite wavefront, so clearly as does the alternative approach detailed above. However, it has not been possible to solve the high-order differential equations associated with more complicated models in a form analogous to (53.65).

53.4. DISCUSSION

The theory described above provides a framework for the complete problem of checking whether or not a material can be approximated, to a satisfactory degree of accuracy, by linear viscoelastic laws; of measuring the viscoelastic properties; and of using these for stress analysis.

The theory of stress analysis discussed in the previous section enables the linearity test, detailed in (53.26), to be generalized to apply to any component of force and displacement or rotation in the loading of a component containing the material to be tested. The linearity of the stress and deformation analysis equations, (53.9), (53.11), (53.38), (53.40), (53.41), and (53.42), implies that any displacement, rotation, or strain component is connected with a loading variable by a linear operator, so that the test corresponding to (53.26) checks the linearity of the system. Loads $\alpha P_1(t)$ are applied for a range of parameters α, and linearity implies that the response is of the form

$$\delta(t) = \alpha\ \delta_1(t) \tag{53.69}$$

where $\delta_1(t)$ is the response to the load $P_1(t)$. $\delta(t)$ can be any measure of displacement. For linearity, the curve $\delta(t)/\alpha,t$ is independent of α, and this checks linearity of the viscoelastic laws. The advantage of this interpretation is that any component containing the material can be tested without the need of preparing specimens to obtain homogeneous stress and strain conditions. This test also provides a measure of the influence of viscoelasticity for cases in which inertia forces are negligible. For an elastic material, $\delta_1(t)$ would then be proportional to $P_1(t)$, so that, if linearity of the operators is exhibited, curvature of the $P_1(t),\delta_1(t)$ relation indicates the influence of viscoelasticity. Inertia effects produce a qualitatively similar influence, so that they must be negligible for this test to be used to distinguish viscoelastic from elastic materials, although this limitation is not necessary for the check of linearity.

Since two basic linear operators are needed to prescribe linear viscoelastic behavior, it is strictly necessary to make a check which tests both of these. However, in most stress-analysis solutions, a complicated combined operator arises, such as that detailed in (53.51), so that a check of linearity effectively tests both basic operators. For a homogeneous body subjected to hydrostatic compression, pure dilatation results with no shear strain, and linearity deduced from such loading, would apply only to P', Q', [Eq. (53.11)], and not to P, Q, [Eq. (53.9)], since the arguments of both sides of (53.9) are identically zero, and this relation does not influence the deformation. Such a situation, however, is extremely unlikely to arise in a more complicated loading situation.

Once linear viscoelastic behavior has been established, measurements must be made on specimens carefully designed to give homogeneous stress and strain, in order to determine the basic viscoelastic operators. Tests for both shear and dilatational modes of deformation are needed. At the present time, this forms one of the most difficult aspects of the whole problem, particularly the accurate measurement of dilatational viscoelastic response.

For times of the order of minutes or longer, creep and relaxation tests are the most convenient means of measuring viscoelastic properties. If measurements over shorter times are needed, the sudden loading associated with creep and relaxation tests causes initial vibrations in the stress and strain, which disturb the requirements of constant stress and strain, respectively. For such short times, it is easier to generate steady oscillatory motion, and to measure the amplitude and phase of the stress and strain variations. This gives the complex modulus, or compliance, and variation of either of these with frequency defines the viscoelastic properties of the material. For frequencies higher than about ten cycles per second, a vibration test is most easily carried out by constructing a dynamic system containing the specimen which can be put into forced oscillation. The equations of motion of the system then enable the complex modulus to be determined from the amplitude and phase of the response. The vibration of a reed of the material to be tested, which is clamped in an oscillator at one end and free at the other, is a common method of measurement in the frequency range 10 to 500 cycles per second. Measurement of the amplitude and phase of the free-end motion, in comparison with the forced motion of the clamped end, determines the complex modulus associated with the tensile stress induced by bending. The complete analysis is complicated [20], but it is possible to deduce the complex modulus from amplitude measurements only in the region of a resonance peak. Torsional oscillation of a rod would determine shear response, and would be easier to interpret because of the greater simplicity of the torsional equations compared with the bending equations. In many cases, an analysis based on an equivalent single-degree-of-freedom system can be used with considerable saving of complexity as described by Nolle [21].

For dilatational response, McKinney, Edelman, and Marvin [22] have used a sample of the material submerged in a liquid in a cavity in which dilatational oscillations are generated. Only preliminary results have yet been announced, since considerable difficulties, such as the influence of air bubbles, affect the accuracy of the measurement.

Few checks of the complete process, from the measurement of material properties to the determination of stress and displacement distributions, have been made, mainly because of the difficulty of measurement of viscoelastic properties for dilatational strain. Particular aspects have been checked, such as the deduction of oscillatory behavior from creep measurements [5]. Kolsky [19] has presented confirmation of the application of the theory to wave propagation in polyethylene rods. Remarkably good agreement was obtained between the varying shape of a stress pulse as it propagated down a rod and the deduction of this through Fourier analysis and the principle of superposition based on complex modulus measurements.

53.5. REFERENCES

[1] A. E. Green, R. S. Rivlin: The mechanics of nonlinear materials with memory, *Arch. Rat. Mech. Anal.*, **1** (1957), 1–21.
[2] H. A. Stuart (ed.): "Die Physik der Hochpolymeren," vol. 4, Springer, Berlin, 1956.
[3] B. Gross: "Mathematical Structure of the Theories of Viscoelasticity," Hermann, Paris, 1953.
[4] F. C. Roesler, J. R. A. Pearson: Determination of relaxation spectra from damping measurements, *Proc. Phys. Soc. London*, **B,67** (1954), 338–347.
[5] F. C. Roesler, W. A. Twyman: An iteration method for the determination of relaxation spectra, *Proc. Phys. Soc. London*, **B,68** (1955), 97–105.

[6] F. Schwarzl: Näherungsmethoden in der Theorie des viscoelastischen Verhaltens, I, *Physica*, **17** (1951), 830–840.

[7] H. Leaderman: Rheology of polyisobutylene, IV: Calculation of the retardation time function and dynamic response from creep data, *Proc. 2d Intern. Congr. Rheol.*, pp. 203–212, 1954.

[8] I. L. Hopkins, R. W. Hamming: On creep and relaxation, *J. Appl. Phys.*, **28** (1957), 906–909.

[9] D. R. Bland, E. H. Lee: On the determination of a viscoelastic model for stress analysis of plastics, *J. Appl. Mechanics*, **23** (1956), 416–420.

[10] W. B. Woodward, J. R. M. Radok: Stress distribution in a reinforced hollow viscoelastic cylinder subjected to time dependent internal pressure, *Brown Univ. Tech. Rept.* PA-TR/14 (1955), 1–20.

[11] W. T. Read: Stress analysis for compressible viscoelastic materials, *J. Appl. Phys.*, **21** (1950), 671–674.

[12] D. R. Bland: Application of the one-sided Fourier transform to the stress analysis of linear viscoelastic materials, *Proc. Conf. Properties of Materials at High Rates of Strain*, pp. 156–163, Inst. Mech. Engrs., London, 1957.

[13] J. R. M. Radok: Viscoelastic stress analysis, *Quart. Appl. Math.*, **15** (1957), 198–200.

[14] J. R. M. Radok, E. H. Lee: Stress analysis in linearly viscoelastic materials, *Proc. 9th Intern. Congr. Appl. Mechanics*, **5** (1957) pp. 321–329.

[15] F. T. Adler, W. M. Sawyer, J. D. Ferry: Propagation of transverse waves in viscoelastic media, *J. Appl. Phys.*, **20** (1949), 1036–1041.

[16] H. L. Oestreicher: Field and impedance of an oscillating sphere in a viscoelastic medium with an application to biophysics, *J. Acoust. Soc. Am.*, **23** (1951), 707–714.

[17] E. H. Lee, I. Kanter: Wave propagation in finite rods of viscoelastic material, *J. Appl. Phys.*, **24** (1953), 1115–1122.

[18] E. H. Lee, J. A. Morrison: A comparison of the propagation of longitudinal waves in rods of viscoelastic materials, *J. Polymer Sci.*, **19** (1956), 93–110.

[19] H. Kolsky: The propagation of stress pulses in viscoelastic solids, *Phil. Mag.*, **1** (1956), 693–710.

[20] D. R. Bland, E. H. Lee: Calculation of the complex modulus of linear viscoelastic materials from vibrating reed measurements, *J. Appl. Phys.*, **26** (1955), 1497–1503.

[21] A. W. Nolle: Methods of measuring dynamic mechanical properties of rubber-like materials, *J. Appl. Phys.*, **19** (1948), 753–774.

[22] J. S. McKinney, S. Edelman, R. S. Marvin: Apparatus for the direct determination of the dynamic bulk modulus, *J. Appl. Phys.*, **27** (1956), 425–430.

CHAPTER 54

VISCOELASTIC BUCKLING

BY

J. KEMPNER, Ph.D., Brooklyn, N.Y.

54.1. INTRODUCTION

Whenever a structural member is required to resist compressive stresses, the possibility of an instability type of failure must be considered. This is particularly true if the dimension of at least one of the characteristic lengths of the region of compression is small compared to that of one of the other two. If the material of which the member is fabricated is viscoelastic, it is to be expected that a time-dependent instability phenomenon will be obtained. On the other hand, if the material is represented by a system of inertialess springs and dashpots, the instantaneous response of the member to small instantaneous disturbances, applied at any time, will be independent of the past loading history and, hence, of time. Either purely elastic deformations or none at all will result, depending on the particular array of springs and dashpots required to represent the mechanical properties of the material. Thus, the buckling load of a perfectly straight column whose material is Maxwellian will not be influenced by the viscous component which can deform only with time. It will be equal to the corresponding Euler load of a column whose modulus of elasticity is that of the spring component. If the spring is itself nonlinear elastic or inelastic, then a corresponding effective modulus can be introduced. When the material of the column is represented by a retarded elastic element (Kelvin model), the buckling load is indefinitely large.

However, if a viscoelastic column is acted upon by a lateral load, simultaneously with an axial compressive load whose magnitude is less than the buckling load, it is to be expected that deflections will develop with time and that these deflections will be very sensitive to the magnitude of the axial load. Similarly, if the column is not perfectly straight before the application of the compressive end load, the column will deform with time, the action being triggered by the bending moments resulting from the simultaneous existence of initial deformation and end load. This behavior suggests that, in addition to the instantaneous buckling load, the design of a viscoelastic column must be based upon a concept involving the time required for the development of a prescribed deflection or bending moment, or upon some quantity related to these parameters. The loading time associated with such a condition can then be considered as the useful life of the column.

Thus, whenever the mechanical behavior of the material of a compression member is assumed to be analogous to the behavior of a system of linear or nonlinear springs and dashpots whose individual physical constants are independent of time, the buckling load, in the classical sense, is a well-defined quantity, whereas the design load must depend upon a prescribed column lifetime. In general, it can be expected that, for every load less than the classical buckling load, a time exists such that a maximum permissible deflection or bending moment, etc., is reached.

54.2. DEFLECTION-TIME CHARACTERISTICS OF LINEARLY VISCOELASTIC COLUMNS

In order to determine the essential features of the load-deflection characteristics of linearly viscoelastic columns or beam-columns, the mechanical behavior of the material of which the column is fabricated may be assumed to be analogous to that of a system consisting of a Maxwell element connected in series to a Kelvin element [1]. Thus, the material will be capable of exhibiting instantaneous elastic response, as well as retarded elasticity (transient creep), and viscous flow (steady creep). The corresponding stress-strain-time law is given by the relation [see Eq. (53.7)]

$$\frac{1}{E_1}\ddot{\sigma} + \left(\frac{E_2}{E_1\eta_2} + \frac{1}{\eta_1} + \frac{1}{\eta_2}\right)\dot{\sigma} + \frac{E_2}{\eta_1\eta_2}\sigma = \ddot{\epsilon} + \frac{E_2}{\eta_2}\dot{\epsilon} \tag{54.1}$$

in which E_1 and η_1, respectively, are the elasticity and viscosity coefficients associated with the Maxwell element, and E_2 and η_2 are the corresponding parameters of the Kelvin element.

Solutions to Eq. (54.1) are subject to the conditions resulting from the assumption that all loads are applied instantaneously. If time is measured from the instant of load application, the initial conditions are, at $t = 0$,

$$\epsilon(0) = \frac{1}{E_1}\sigma(0) \qquad \frac{1}{E_1}\dot{\sigma}(0) + \left(\frac{1}{\eta_1} + \frac{1}{\eta_2}\right)\sigma(0) = \dot{\epsilon}(0) \qquad (54.1a,b)$$

These relations correspond to the condition that, while the elastic component of the Maxwell element shows immediate response to the loading, the corresponding viscous component, as well as the entire Kelvin element, remain rigid. Thus, for a rod under a constant uniform axial stress σ_0,

$$\frac{\epsilon}{\epsilon_0} = \frac{E_1}{E_2}[1 - e^{-(E_2/\eta_2)t}] + \frac{E_1}{\eta_1}t + 1 \qquad (54.1c)$$

in which $\epsilon_0 = \sigma_0/E_1$.

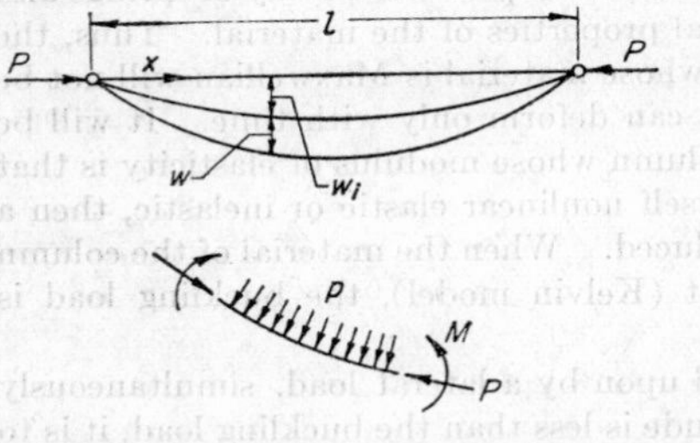

Fig. 54.1. Beam-column with initial deformation.

Now, consider a beam-column loaded in such a manner that the applied bending moment M acts in a plane containing the beam axis and a principal centroidal axis of the cross section (see Fig. 54.1). Assuming that plane cross sections remain plane (see p. 35–5), we find that the strain along the axis is given by Eq. (54.1c) with $\epsilon_0 = -P/E_1A$, where P is the applied axially compressive load, and A is the cross-sectional area. For small lateral deflections w, it furthermore follows that these and M are related by the differential equation

$$-E_1I\left(\ddot{w} + \frac{E_2}{\eta_2}\dot{w}\right)_{,xx} = \ddot{M} + \left(\frac{E_2}{\eta_2} + \frac{E_1}{\eta_1} + \frac{E_1}{\eta_2}\right)\dot{M} + \frac{E_1E_2}{\eta_1\eta_2}M \tag{54.2}$$

where I is the appropriate moment of inertia of the cross section, and subscripts preceded by a comma indicate differentiation (see Fig. 54.1).

The bending moment can be expressed as

$$M = P(w + w_i) + M^*(x) \tag{54.3}$$

in which w_i represents an initial deviation from straightness of the original beam axis, before loading (see Fig. 54.1), and M^* is the moment due to statically applied lateral loads. Equations (54.2) and (54.3) yield

$$E_1I\left(\ddot{w}+\frac{E_2}{\eta_2}\dot{w}\right)_{,xx}+P\left[\ddot{w}+\left(\frac{E_2}{\eta_2}+\frac{E_1}{\eta_1}+\frac{E_1}{\eta_2}\right)\dot{w}+\frac{E_1E_2}{\eta_1\eta_2}w\right]=-\frac{E_1E_2}{\eta_1\eta_2}Pw_i-\frac{E_1E_2}{\eta_1\eta_2}M^* \quad (54.4)$$

Furthermore, if $p(x)$ is an applied distributed lateral load, Eq. (54.4) becomes

$$\left[E_1I\left(\ddot{w}+\frac{E_2}{\eta_2}\dot{w}\right)_{,xx}\right]_{,xx}+\left\{P\left[\ddot{w}+\left(\frac{E_2}{\eta_2}+\frac{E_1}{\eta_1}+\frac{E_1}{\eta_2}\right)\dot{w}+\frac{E_1E_2}{\eta_1\eta_2}w\right]\right\}_{,xx}=-\left(\frac{E_1E_2}{\eta_1\eta_2}Pw_i\right)_{,xx}+\frac{E_1E_2}{\eta_1\eta_2}p \quad (54.5)$$

Equation (54.5) is the equation of bending with small deflections of an inertialess beam-column, whose material behaves in a manner analogous to a Maxwell-Kelvin model, shear deformation of the beam being neglected.

For a uniform beam under the action of equal and opposite constant couples M_0 applied at the ends $x = 0$ and l, Eqs. (54.1*a*,*b*) and (54.2) yield

$$\frac{w}{w_0}=\frac{E_1}{E_2}(1-e^{-(E_2/\eta_2)t})+\frac{E_1}{\eta_1}t+1 \quad (54.6)$$

in which $w_0 = (M_0/2E_1I)(lx - x^2)$ is the purely elastic deflection obtained at $t = 0$.

Because of the interaction between end load and deflection, the action of a beam-column is somewhat more involved than that of a simple beam. In the case of a uniform column whose axis initially deviates from a straight line, and upon which only compressive end loads are applied (see Fig. 54.1), Eq. (54.5) reduces to

$$E_1I\left(\ddot{w}+\frac{E_2}{\eta_2}\dot{w}\right)_{,xxxx}+P\left[\ddot{w}+\left(\frac{E_2}{\eta_2}+\frac{E_1}{\eta_1}+\frac{E_1}{\eta_2}\right)\dot{w}+\frac{E_1E_2}{\eta_1\eta_2}w\right]_{,xx}=-\frac{E_1E_2}{\eta_1\eta_2}Pw_{i,xx} \quad (54.7)$$

For a column simply supported at the ends ($x = 0$ and l), and with an initial deviation from straightness taken as

$$w_i = a_1\sin\frac{\pi x}{l} \quad (54.8)$$

the solution of Eq. (54.7), subject to the initial conditions described by Eqs. (54.1*a*,*b*), can be expressed as

$$\frac{w+w_i}{w_0+w_i}=\frac{1}{n_1-n_2}[(\beta-n_2)e^{n_1t}-(\beta-n_1)e^{n_2t}] \quad (54.9)$$

provided that $\alpha = P/P_E < 1$, where the Euler buckling load $P_E = \pi^2E_1I/l^2$, and

$$\beta=\frac{\alpha E_1}{1-\alpha}\left(\frac{1}{\eta_1}+\frac{1}{\eta_2}\right)\qquad 2n_{1,2}=\beta-\frac{E_2}{\eta_2}\pm\left[\left(\beta-\frac{E_2}{\eta_2}\right)^2+\frac{4\alpha}{1-\alpha}\frac{E_1E_2}{\eta_1\eta_2}\right]^{1/2}$$

In these relations, n_1 is real and positive, and n_2 is real and negative for nonvanishing, finite values of E and η. The parameter w_0 represents the increment in deflection at $t = 0$ due to the application of the end load. It is given by the well-known expression for an elastic column with modulus E_1 as

$$w_0=\frac{\alpha}{1-\alpha}a_1\sin\frac{\pi x}{l} \quad (54.10)$$

Equation (54.9) shows that, for $0 < (E,\eta) < \infty$, the deflections of a Maxwell-Kelvin column increase with time, and become indefinitely large only as t approaches infinity for any load $P < P_E$. Thus, the viscoelastic deformations do not affect the critical load under which sudden buckling may occur but, in effect, simply result in a time-dependent magnification of the initial deformations. Results of computations based upon Eqs. (54.9) are shown in Fig. 54.2 for three values of P/P_E, for a material whose elasticity and viscosity coefficients are given in the figure [1]. Examination of Eq. (54.9) reveals that, except when $\alpha \ll 1$, the deflections are very much more sensitive to changes in load than to changes in the amplitude of initial deformations.

In the absence of any special criterion, any compressive load can be considered to lead eventually to collapse of the column. However, if the column design is associated with allowable deflections, moments, strains, or stresses, then Eq. (54.9) or (54.3) can be applied to determine the corresponding lifetime. Results so obtained are, of course, valid only if the slope at any position along the column axis is small compared to unity.

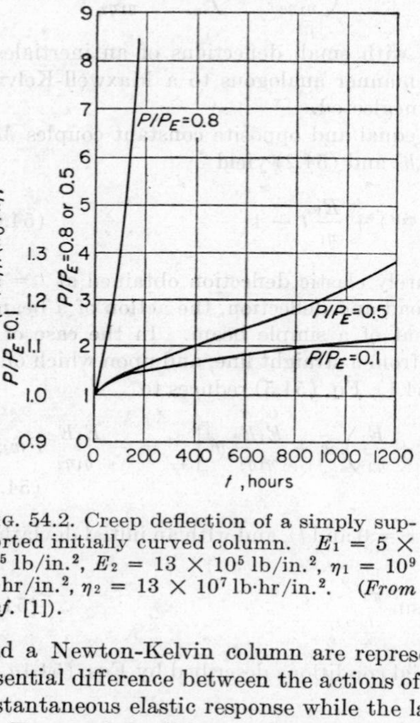

FIG. 54.2. Creep deflection of a simply supported initially curved column. $E_1 = 5 \times 10^5$ lb/in.², $E_2 = 13 \times 10^5$ lb/in.², $\eta_1 = 10^9$ lb·hr/in.², $\eta_2 = 13 \times 10^7$ lb·hr/in.². (*From Ref.* [1]).

The deflection-time characteristics of columns whose viscoelastic model configurations are degenerate forms of the four-parameter Maxwell-Kelvin model can be readily obtained from the foregoing results. For example, if the complete model consists of a Kelvin element connected in series to a viscous element (Newton-Kelvin model), then it corresponds to the Maxwell-Kelvin model with $E_1 \to \infty$. In such a case, $\alpha \to 0$, while $\alpha E_1 = Pl^2/\pi^2 I$ remains finite. Thus, deflections of a column with an initial sinusoidal deviation from straightness are given by Eq. (54.9) with $w_0 = 0$. In Fig. 54.3, the deflection-time characteristics of both a Maxwell-Kelvin column and a Newton-Kelvin column are represented schematically by curve 1. The only essential difference between the actions of the two columns is that the former exhibits instantaneous elastic response while the latter does not (for example, $w_0 = 0$).

For a standard linear solid, which can be represented by a spring in series with a Kelvin element, Eq. (54.9) with $\eta_1 \to \infty$ reduces to

$$\frac{w + w_i}{w_0 + w_i} = \frac{\alpha E_1}{\alpha E_1 - (1 - \alpha)E_2}\left(\exp\left\{\left[\frac{\alpha E_1}{(1 - \alpha)E_2} - 1\right]\frac{E_2}{\eta_2}t\right\} - 1\right) + 1 \quad (54.11)$$

In Fig. 54.3 the deflection-time characteristics in accordance with Eq. (54.11) for nonvanishing finite values of E_1, E_2, and η_2 are shown schematically by curves 2, 3, and 4. Curve 2 indicates an exponential and curve 3 a linear increase of the deflections with time for $\alpha < 1$. Thus, for the conditions stipulated, the deflections increase indefinitely for any compressive load, no matter how small. Curve 4 shows that, when $\alpha E_1/(1 - \alpha)E_2 < 1$, the deflections remain finite for all time, and approach a limiting value as $t \to \infty$ corresponding to $(w + w_i)/(w_0 + w_i) = \delta$, where

$$\delta = 1 + \frac{\alpha E_1}{(1 - \alpha)E_2 - \alpha E_1} \quad (54.12)$$

In this case, it is possible to find a limiting value of the applied load such that the limiting deflections tend toward infinity. Such a limiting load corresponds to

$$\alpha E_1/(1-\alpha)E_2 = 1 \tag{54.12a}$$

and is analogous to the Euler load when the latter is determined on the basis of a column with initial imperfections. It is also seen that it is always possible to select a load for an allowable deflection such that the column has an indefinitely large lifetime.

Equation (54.12*a*) can also be written as

$$P = \frac{E_2}{E_1+E_2} P_E = \frac{\pi^2 I}{l^2}\,\frac{E_1E_2}{E_1+E_2} \tag{54.12b}$$

and thus the limiting load corresponds to the buckling load of a column whose material model array consists of two series-connected springs, one with modulus E_1, the other with modulus E_2. This result reflects the fact that, once the rate of deformation becomes zero, the viscous component of the Kelvin element of the standard linear solid supports no load. The limiting deflection for $\alpha E_1/(1-\alpha)E_2 < 1$ is found to be

$$w = \frac{\alpha(E_1+E_2)/E_2}{1-\alpha(E_1+E_2)/E_2}\, a_1 \sin \frac{\pi x}{l} \tag{54.12c}$$

Comparison of this expression with Eq. (54.10) shows that Eq. (54.12*c*) yields the deflection due to loads of a column whose Euler load is given by Eq. (54.12*b*).

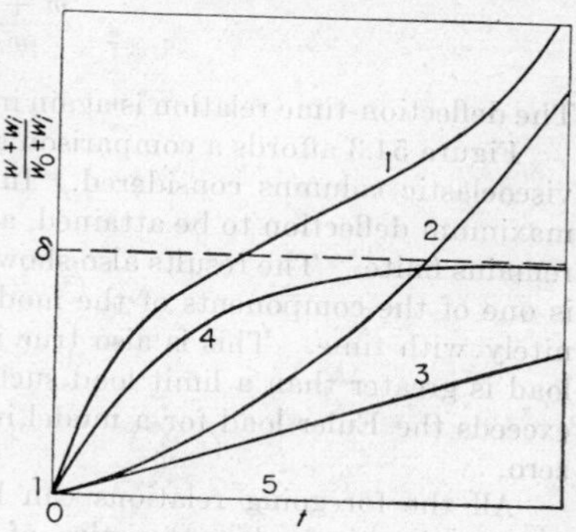

FIG. 54.3. Deflection-time curves for viscoelastic columns with initial deformations ($\alpha < 1$): 1, Maxwell-Kelvin model ($0 < (\eta, E) < \infty$); Newton-Kelvin model ($w_0 = 0$). 2, Standard linear solid ($\alpha E_1/(1-\alpha)E_2 > 1$); Kelvin model ($w_0 = 0$, $\alpha' > 1$); Maxwell model; viscous model ($w_0 = 0$). 3, Standard linear solid ($\alpha E_1/(1-\alpha)E_2 = 1$); Kelvin model ($w_0 = 0$, $\alpha' = 1$). 4, Standard linear solid ($\alpha E_1/(1-\alpha)E_2 < 1$); Kelvin model ($w_0 = 0$, $\alpha' < 1$). 5, Elastic solid ($w = w_0$).

A column of a Kelvin material corresponds to the condition $E_1 \to \infty$ in addition to $\eta_1 \to \infty$, and, hence,

$$\frac{w+w_i}{w_i} = \frac{\alpha'}{\alpha'-1}\left(e^{(\alpha'-1)(E_2/\eta_2)t} - 1\right) + 1 \tag{54.13}$$

in which $\alpha' = P/P'_E$, where $P'_E = \pi^2 E_2 I/l^2$. With the exception that there are no purely elastic deformations (for example, $w_0 = 0$), the behavior of the column of Kelvin material is analogous to that of the standard linear solid. Curves 2, 3, and 4 of Fig. 54.3 now correspond to $\alpha' > 1$, $= 1$, < 1, respectively, with the limiting deflection depicted for curve 4 corresponding to

$$\delta = 1 + \frac{\alpha'}{1-\alpha'} \tag{54.14}$$

Thus, the limiting value of the applied load for this case is given by $P = P'_E$, where P'_E is the Euler load of a purely elastic column having a modulus E_2, and the limiting deflection for $P < P'_E$ is $w = [\alpha'/(1-\alpha')]a_1 \sin(\pi x/l)$.

The relations governing the behavior of a column of Maxwell material are obtained from those of the Maxwell-Kelvin column with the condition $\eta_2 \to \infty$. Hence,

$$\frac{w+w_i}{w_0+w_i} = \exp \frac{\alpha E_1 t}{(1-\alpha)\eta_1} \tag{54.15}$$

The deflection-time diagram corresponding to Eq. (54.15) is indicated by curve 2 in Fig. 54.3. This result, which shows that the deflections increase indefinitely with time, appears to have been obtained first by Freudenthal [3],[4]. An analogous behavior was found for an eccentric column (w_i = const) in [5].

For a purely viscous column $E_1, \eta_2 \to \infty$, and, hence, Eq. (54.9) or (54.15) yields

$$\frac{w + w_i}{w_i} = \exp \frac{Pl^2t}{\pi^2 I \eta_1} \tag{54.16}$$

The deflection-time relation is again indicated by curve 2 of Fig. 54.3, but with $w_0 = 0$.

Figure 54.3 affords a comparison among the deflection-time relations of the several viscoelastic columns considered. In each case, an infinite time is required for the maximum deflection to be attained, although, under certain conditions, this deflection remains finite. The results also show that, whenever an unrestrained viscous element is one of the components of the model array selected, the deflections increase indefinitely with time. This is also true in a generalized sense, that is, when the applied load is greater than a limit load such as that given by Eq. (54.12*b*), that is, when it exceeds the Euler load for a model in which all viscosity coefficients are set equal to zero.

All the foregoing relations can be readily applied to the determination of the deformation-time characteristics of simply supported columns with an arbitrary initial deviation from straightness. In such a case, w_i and w_0, respectively, can be replaced by the Fourier series

$$\sum_{m=1}^{\infty} w_{im} = \sum_{m=1}^{\infty} a_m \sin \frac{m\pi x}{l} \quad \text{and} \quad \sum_{m=1}^{\infty} w_{0m} = \sum_{m=1}^{\infty} \left[\frac{\alpha}{m^2} \Big/ \left(1 - \frac{\alpha}{m^2}\right) \right] a_m \sin \frac{m\pi x}{l}$$

Equation (54.9) and the special solutions stemming therefrom then determine the mth harmonic component of the deflection, that is, $(w_m + w_{im})/(w_{0m} + w_{im})$, provided that, in Eq. (54.9), the column length l is replaced by l/m, wherever it appears.

The previous analysis can readily be extended to include viscoelastic model representations involving more general arrays than that considered. With the aid of the viscoelastic analog, Hilton [6] has presented an analysis in which the simply supported column has both arbitrary initial deviations from straightness and arbitrary viscoelastic model configuration. A modified form of this analysis follows:

For one-dimensional systems of stress and strain, the stress-strain-time law governing the behavior of an arbitrary viscoelastic material can be expressed in the form [see [2] and Eqs. (53.1), (53.19)]

$$\sum_{0}^{p} p_r \frac{\partial^r \sigma}{\partial t^r} = \sum_{0}^{q} q_r \frac{\partial^r \epsilon}{\partial t^r} \tag{54.17}$$

in which p_r and q_r are material constants, and $\partial^\circ/\partial t^\circ = 1$. For small displacements of an elastic beam, $M = -EIw_{,xx}$ and, hence, from the viscoelastic analog, Eq. (54.17) transforms into

$$\sum_{0}^{p} p_r \frac{\partial^r M}{\partial t^r} = -I \sum_{0}^{q} q_r \frac{\partial^r w_{,xx}}{\partial t^r} \tag{54.18}$$

For a simply supported, initially deformed column, Eq. (54.18) becomes

$$P \sum_{0}^{p} p_r \frac{\partial^r (w + w_i)}{\partial t^r} = -I \sum_{0}^{q} q_r \frac{\partial^r w_{,xx}}{\partial t^r} \tag{54.19}$$

The results obtained from the solution of the Maxwell-Kelvin column [see Eq. (54.9)] and from the generalization to include an arbitrary initial deformation suggest that a suitable solution of Eq. (54.19) can be found in the form

$$w(x,t) = \sum_{m=1}^{\infty} w_m(x,t) \tag{54.20}$$

where

$$w_m = [w_{0m}(x) + w_{im}(x)]T_m(t) - w_{im}(x)$$

$$w_{0m} = \frac{\alpha/m^2}{1-(\alpha/m^2)}\, a_m \sin\frac{m\pi x}{l} \qquad w_{im} = a_m \sin\frac{m\pi x}{l} \qquad T_m(0) = 1 \tag{54.21a–d}$$

Equation (54.21*d*) corresponds to the condition $w(x,0) = \Sigma w_{0m}$, where the latter quantity represents the deflection due to loads at $t = 0$.

Equations (54.19) to (54.21*c*) yield the following equation for determination of T_m:

$$\frac{P(l/m)^2}{\pi^2 I} \sum_0^p p_r \frac{d^r T_m}{dt^r} - \sum_0^q q_r \frac{d^r T_m}{dt^r} = -\frac{w_{im}}{w_{0m} + w_{im}}\, q_0 \tag{54.22}$$

The solution of Eq. (54.22), together with Eq. (54.21*d*) and additional initial conditions appropriate to the particular viscoelastic law chosen, serves to determine the deflection $w(x,t)$. The additional initial conditions, such as that stemming from Eq. (54.1*b*), are required only when $r > 2$.

An alternative solution to the Maxwell-Kelvin column problem has been presented by Lin [7]. In this analysis, the viscoelastic law was expressed in terms of a memory function, and the Laplace transformation of the law was applied in the formulation of the governing equations. In this connection, Lin refers to the work of Brull [8] on the development of the general three-dimensional viscoelastic equations for bodies not subjected to buckling.

Thus, instead of the form given in Eq. (54.1) or (54.17), the viscoelastic law can be written as

$$\sigma = E_1[\epsilon - \int_0^t \phi(t - t')\epsilon(t')\, dt'] \tag{54.23}$$

where ϕ is the memory function associated with the particular viscoelastic model considered. The Laplace transformation with respect to time of Eq. (54.23) is

$$\bar{\sigma} = E_1(1 - \bar{\phi})\bar{\epsilon} \tag{54.24}$$

where a bar over a symbol denotes the transformed quantity. For a Maxwell-Kelvin model, Eqs. (54.1) and (54.1*a*,*b*) afford the determination of a relation between $\bar{\sigma}$ and $\bar{\epsilon}$ which, together with Eq. (54.24), yields

$$1 - \bar{\phi} = \frac{(s + E_2/\eta_2)s}{s^2 + E_1(E_2/E_1\eta_2 + 1/\eta_1 + 1/\eta_2)s + E_1E_2/\eta_1\eta_2} \tag{54.25}$$

where s is the argument of the Laplace transformation with respect to time.

For a simply supported column with arbitrary initial deviation from straightness, we obtain

$$E_1 I(1 - \bar{\phi})\bar{w}_{,xx} + P\bar{w} = -\frac{Pw_i}{s} \tag{54.26}$$

When Eq. (54.25) is applied, the Laplace transformations of Eqs. (54.20) to (54.21*c*) yield a solution to Eq. (54.26), whose inverse is identical with the displacement function described previously for a Maxwell-Kelvin column. A completely analogous procedure can be applied to the analysis of columns having more general viscoelastic properties than those considered in the determination of Eq. (54.25). Alternatively, Eq. (54.25) can be specialized to correspond to the previously described degenerate cases to yield:

Newton-Kelvin model $$E_1(1 - \bar{\phi}) = \frac{(s + E_2/\eta_2)s}{(1/\eta_1 + 1/\eta_2)s + E_2/\eta_1\eta_2}$$

Standard linear solid $$1 - \bar{\phi} = \frac{s + E_2/\eta_2}{s + (1/\eta_2)(E_1 + E_2)}$$

Kelvin model $$E_1(1 - \bar{\phi}) = \eta_2(s + E_2/\eta_2)$$

Maxwell model $$1 - \bar{\phi} = \frac{s}{s + E_1/\eta_1}$$

Viscous model $$E_1(1 - \bar{\phi}) = \eta_1 s$$

The foregoing analysis of the behavior of linearly viscoelastic columns was generalized by Lin [9] to apply to homogeneous isotropic viscoelastic plates.

54.3. DEFLECTION-TIME CHARACTERISTICS OF NONLINEARLY VISCOELASTIC COLUMNS

When the material of compression members cannot be characterized by the behavior of systems of Hookean and Newtonian elements, the foregoing analyses cannot be expected to predict their response to loading realistically. Such is generally the case with metallic columns. Here the material does not behave according to a linear law of the type represented by either Eq. (54.17) or (54.24); rather, uniaxial creep tests reveal that the strain is dependent upon the stress in a highly nonlinear manner. Although, as yet, no stress-strain-time law has been revealed which accounts for the many factors (e.g., loading history, structural changes of material, etc.) which influence creep, several column creep-buckling theories have been developed (e.g., [10]–[32]). These theories are based upon empirical descriptions of the results of uniaxial creep tests. One of the simplest of such column-creep analyses proposed is based on a generalization of the Maxwell model, in the form of a linear spring in series with a nonlinear dashpot, whose strain rate is proportional to a power of the applied stress [16]. The corresponding stess-strain-time relation can be adapted to approximate the creep characteristics of structural materials, such as aluminum or steel, by assuming that the usual creep curve for constant stress and temperature can be represented by a straight line in the ϵ,t plane. This line is chosen to coincide with the secondary or steady stage of creep, and to intersect the $t = 0$ axis. Hence, the idealized creep curve takes into account the actual secondary state (steady rate) of creep, approximates the initial elastic or elastoplastic and primary (transient) states, and ignores the final stage.

Stress-strain-time relations corresponding to the assumptions described are given by the expressions

$$\dot{\epsilon} = \frac{\dot{\sigma}}{E_t'} + \left(\frac{\sigma}{\lambda_t}\right)^{n_t} \qquad \sigma > 0 \tag{54.27a}$$

$$\dot{\epsilon} = \frac{\dot{\sigma}}{E_c'} - \left(-\frac{\sigma}{\lambda_c}\right)^{n_c} \qquad \sigma < 0 \tag{54.27b}$$

in which E' is an effective elastic modulus, and n and λ are positive constants defining the viscous behavior of the material; the subscripts c and t distinguish between constants determined from compressive and tensile creep tests, respectively. Equation (54.27b) reduces to Eq. (54.27a) when the creep properties are identical for compression and tension and n is an odd integer.

The creep parameters E', λ, and n are considered constant for a given temperature, and can be evaluated from constant-stress, uniaxial creep tests. Thus, for a tensile test, if the strain ϵ_0 at $t = 0$ is taken as σ_0/E'_t, the relationship among stress, strain, and time becomes

$$\frac{\epsilon}{\epsilon_0} = \frac{E'_t}{\lambda^{n_t}} \sigma_0^{n_t-1} t + 1 \tag{54.28}$$

In Ref. [16], Eqs. (54.27*a*,*b*) were applied in the derivation of the differential equations of bending of an idealized H-section beam-column (see Fig. 54.4). In this case, the strain rates $\dot{\epsilon}_1$ and $\dot{\epsilon}_2$, respectively, on the concave and convex sides of the beam of depth h are, from Eq. (54.27*b*),

$$\dot{\epsilon}_1 = -\frac{1}{E'_c}\left(\frac{\dot{P}}{A} + \frac{2\dot{M}}{Ah}\right) - \left[\frac{1}{\lambda_c}\left(\frac{P}{A} + \frac{2M}{Ah}\right)\right]^{n_c}$$
$$\dot{\epsilon}_2 = -\frac{1}{E'_c}\left(\frac{\dot{P}}{A} - \frac{2\dot{M}}{Ah}\right) - \left[\frac{1}{\lambda_c}\left(\frac{P}{A} - \frac{2M}{Ah}\right)\right]^{n_c} \tag{54.29a}$$

when both flanges, each of area $A/2$, are in compression (that is, $P/A \geq 2M/Ah$). When the concave flange is in compression while the convex flange is in tension ($P/A \leq 2M/Ah$), Eq. (54.27*a*) must be considered, and, hence, the second of Eqs. (54.29*a*) is replaced by the relation

$$\dot{\epsilon}_2 = \frac{1}{E'_t}\left(\frac{2\dot{M}}{Ah} - \frac{\dot{P}}{A}\right) + \left[\frac{1}{\lambda_t}\left(\frac{2M}{Ah} - \frac{P}{A}\right)\right]^{n_t} \tag{54.29b}$$

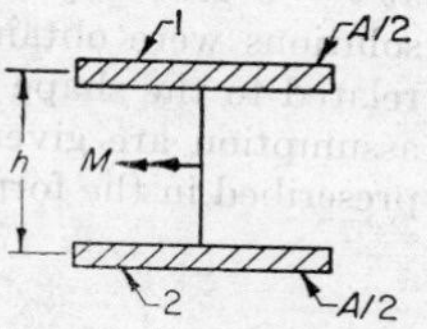

FIG. 54.4. Idealized H section.

It should be noted that, for arbitrary distribution of the applied loads, both tension and compression can exist simultaneously in different regions of each flange, and, hence, in such a case, appropriate combinations of Eqs. (54.29*a*,*b*) must be used.

For small displacements,

$$\epsilon_1 - \epsilon_2 = hw_{,xx} \tag{54.30}$$

and, hence, if $P/A \geq 2M/Ah$, this relation, together with Eqs. (54.29*a*), yields

$$h\dot{w}_{,xx} = -\frac{4}{E'_c Ah}\dot{M} - \left(\frac{1}{\lambda_c}\right)^{n_c}\left[\left(\frac{P}{A} + \frac{2M}{Ah}\right)^{n_c} - \left(\frac{P}{A} - \frac{2M}{Ah}\right)^{n_c}\right] \tag{54.31a}$$

If $P/A \leq 2M/Ah$, consideration of Eq. (54.29*b*) leads to the alternate relation,

$$h\dot{w}_{,xx} = -\frac{1}{E'_c}\left(\frac{2\dot{M}}{Ah} + \frac{\dot{P}}{A}\right) - \frac{1}{E'_t}\left(\frac{2\dot{M}}{Ah} - \frac{\dot{P}}{A}\right) - \left[\frac{1}{\lambda_c}\left(\frac{2M}{Ah} + \frac{P}{A}\right)\right]^{n_c} - \left[\frac{1}{\lambda_t}\left(\frac{2M}{Ah} - \frac{P}{A}\right)\right]^{n_t} \tag{54.31b}$$

Thus Eqs. (54.31*a*,*b*) are the differential equations applicable to the analysis of the creep behavior of an idealized H-section beam-column, the material of which is characterized by Eqs. (54.27*a*,*b*).

For an initially deformed, simply supported, massless, H-section column, the stresses in concave and convex regions of the flanges, respectively, are:

$$\sigma_1 = -\bar{\sigma}\left[1 + \frac{2}{h}(w + w_i)\right] \qquad \sigma_2 = -\bar{\sigma}\left[1 - \frac{2}{h}(w + w_i)\right] \tag{54.32}$$

where $\bar{\sigma} = P/A$. Hence, if the creep constants are assumed to be the same for tension and compression, Eqs. (54.31*a,b*) yield

$$h\dot{w}_{,xx} + \frac{4\bar{\sigma}}{hE'}\dot{w} + \left(\frac{\bar{\sigma}}{\lambda}\right)^n \left\{\left[1 + \frac{2}{h}(w + w_i)\right]^n - \left[1 - \frac{2}{h}(w + w_i)\right]^n\right\} = 0 \quad (54.33a)$$

provided $P/A \geq 2M/Ah$, and, hence, when $w + w_i \leq h/2$, and

$$h\dot{w}_{,xx} + \frac{4\bar{\sigma}}{hE'}\dot{w} + \left(\frac{\bar{\sigma}}{\lambda}\right)^n \left\{\left[\frac{2}{h}(w + w_i) + 1\right]^n + \left[\frac{2}{h}(w + w_i) - 1\right]^n\right\} = 0 \quad (54.33b)$$

when $w + w_i \geq h/2$. Except when n is an odd integer, Eq. (54.33*a*) is valid only in those regions in which the total deformation is sufficiently small so that both flanges are in compression, while Eq. (54.33*b*) is applicable in regions in which the concave and convex flanges are in compression and tension, respectively. When n is an odd integer, Eqs. (54.33*a,b*) are identical, and, when $n = 1$, they reduce to the equation for a column of Maxwell material.

In Ref. [16], Eqs. (54.33*a,b*) were applied to a simply supported column with an initial deviation from straightness given by Eq. (54.8) and, hence, with a deflection at $t = 0$ given by Eq. (54.10) with E' replacing E_1. In this analysis, approximate solutions were obtained with the assumption that the deflected shape was affinely related to the shape obtained upon the instant of loading. Justifications of such an assumption are given in [18] and [28]. Thus, the assumed deflected shape can be prescribed in the form

$$w(x,t) = \left[ha_c(t) + \frac{\alpha}{1 - \alpha}a_1\right] \sin\frac{\pi x}{l} \quad (54.34)$$

in which $a_c(t)$ is the nondimensional mid-span deflection due to creep alone $[a_c(0) = 0]$. Upon the application of the collocation method, in which Eq. (54.34) is introduced into Eq. (54.33*a*), and the latter is enforced at $x = l/2$, the following relation is obtained:

$$\tau = \int_{a'_1}^{a_c + a'_1} \frac{dz}{(\frac{1}{2} + z)^n - (\frac{1}{2} - z)^n} \quad (54.35a)$$

where $a'_1 = a_1/h(1 - \alpha)$ represents the nondimensional total mid-span deflection of the loaded column at $t = 0$, and the nondimensional time variable τ is defined as

$$\tau = 2^{n-2}\frac{E'}{\sigma_E - \bar{\sigma}}\left(\frac{\bar{\sigma}}{\lambda}\right)^n t \quad (54.36)$$

where $\sigma_E = \bar{\sigma}/\alpha$ is the Euler buckling stress. Equation (54.35*a*) is limited to the case in which both flanges are completely in compression, i.e., when $a'_1 \leq a_c + a'_1 \leq \frac{1}{2}$, which, in turn, corresponds to $0 \leq \tau \leq \tau_{\text{lim}}$, where τ_{lim} is proportional to the time at which the convex flange becomes unstressed at its mid-span. Thus, when $a'_1 \leq \frac{1}{2}$,

$$\tau_{\text{lim}} = \int_{a'_1}^{1/2} \frac{dz}{(\frac{1}{2} + z)^n - (\frac{1}{2} - z)^n} \quad (54.35b)$$

In accordance with the present solution, once τ exceeds τ_{lim}, Eq. (54.33*b*) applies. Hence, for $\frac{1}{2} \leq a_c + a'_1 \leq \infty$ and, therefore, for $\tau_{\text{lim}} \leq \tau \leq \tau_{\text{cr}}$, where τ_{cr} is defined as the critical time parameter corresponding to infinite deflections,

$$\tau = \tau_{\text{lim}} + \int_{1/2}^{a_c + a'_1} \frac{dz}{(z + \frac{1}{2})^n + (z - \frac{1}{2})^n} \quad (54.37a)$$

For $a_1 \leq \frac{1}{2}$, the critical time is determined by the relation

$$\tau_{cr} = \tau_{lim} + \int_{\frac{1}{2}}^{\infty} \frac{dz}{(z + \frac{1}{2})^n + (z - \frac{1}{2})^n} \tag{54.37b}$$

If $a'_1 \geq \frac{1}{2}$, only one flange is completely in compression for the entire duration of loading, and, hence, Eqs. (54.37*a*,*b*) are replaced by the expressions

$$\tau = \int_{a'_1}^{a_c + a'_1} \frac{dz}{(z + \frac{1}{2})^n + (z - \frac{1}{2})^n} \tag{54.38a}$$

and

$$\tau_{cr} = \int_{a'_1}^{\infty} \frac{dz}{(z + \frac{1}{2})^n + (z - \frac{1}{2})^n} \tag{54.38b}$$

The deflection-time relations represented by Eqs. (54.35*a*) to (54.38*a*) do not reflect the fact that, even when $\tau \geq \tau_{lim}$, a considerable spanwise region of the convex flange can be in compression. Thus, when $\tau \geq \tau_{lim}$, the solution presented predicts a greater rate of increase in the curvature than would actually occur [see Eq. (54.30)], and, hence, it can be expected that the critical time computed with either Eq. (54.37*b*) or Eq. (54.38*b*) will be conservative. The restriction on the degrees of freedom of the deformation, which would tend to result in unconservative predictions for the critical time, has been shown in [28] to have a negligible effect, at least for the case $n = 3$.

In Ref. [16] the integrations indicated in the foregoing relations for τ, τ_{lim}, and τ_{cr} were performed for $n = 1$, 2, 3, 4, and 5. For odd-integral values of n, only Eqs. (54.38*a*,*b*) require consideration, regardless of the magnitude of a'_1. When $n = 3$, for example, the deflection due to creep is related to the time in the following manner:

$$\frac{a_c}{a'_1} = \left[\frac{3e^{3\tau}}{4a_1'^2(1 - e^{3\tau}) + 3}\right]^{\frac{1}{2}} - 1 \tag{54.39}$$

and the critical-time parameter is given by the expression

$$\tau_{cr} = \tfrac{1}{3} \ln\left(1 + \frac{3}{4a_1'^2}\right) \tag{54.40}$$

Typical creep deflection-time curves for several values of the nondimensional total deflection a'_1 (obtained at $t = 0$) and for $n = 1$, 2, and 3 are given in Fig. 54.5. In Ref. [17] the integrals appearing in the several expressions for τ were evaluated in closed form for any integral value of n. With the exception of the linearly viscoelastic column ($n = 1$), finite critical times were found to exist provided $\alpha \leq 1$ (that is, $\bar{\sigma} \leq \sigma_E$) for all $a_1 < \infty$. Some of the results obtained in [17] are presented in Fig. 54.6. The curves show that, for prescribed values of E', λ, n, σ_E, and $\bar{\sigma}$, the critical time decreases with increasing amplitude of the initial deformation. However, the critical time is much more sensitive to changes in the applied stress, $\bar{\sigma} = P/A$, than to changes in the initial deformation, especially when the value of the applied load is close to that of the Euler load.

Since the foregoing analysis is based upon small-deflection theory, the critical time determined by the relations presented is a nominal quantity which can, at best, serve as a guide in predicting the lifetime of a column. If, as in the case of linearly viscoelastic columns, the column design is associated with allowable deflections, moments, strains, or stresses, then Eq. (54.35*a*), (54.37*a*), or (54.38*a*) can serve as a basis for useful-lifetime predictions. In such a case, the critical time computed with either Eq. (54.37*b*) or Eq. (54.38*b*) will represent an upper limit. One of the major shortcomings of the creep law represented by Eqs. (54.27*a*,*b*) is the inadequacy of the effective elastic modulus E' for characterizing accurately material properties in the primary creep range, and when stresses have exceeded the proportional limit. Even for very slender columns, it is to be expected that, because of the magnification of initial deformations (and, hence, of stresses) with time, the central region of each flange will not respond to changes in stress in the manner assumed in Eqs. (54.27*a*,*b*). Furthermore, since E' is smaller than the actual modulus governing instantaneous elastic response, the buckling stress σ_E is always less than the actual instantaneous

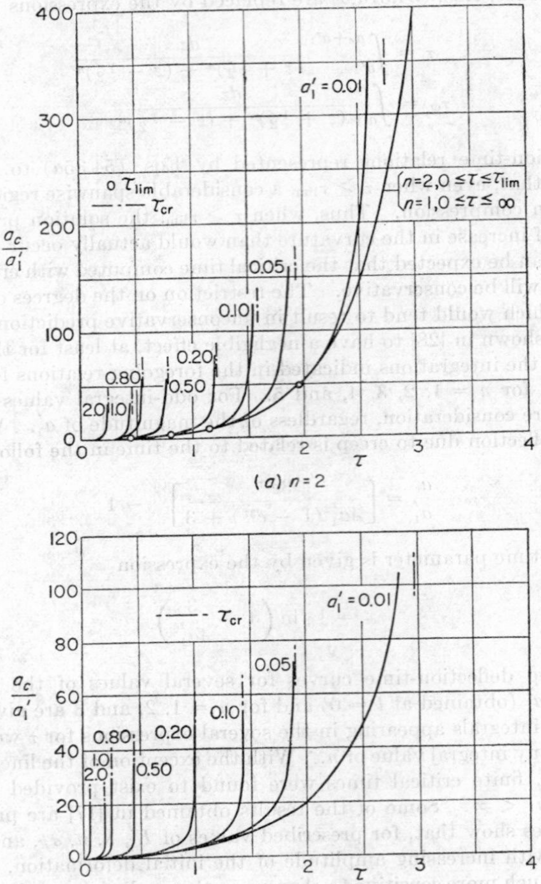

FIG. 54.5. Variation of deflection accrued for $t > 0$ with time parameter τ for $n = 2$ and $n = 3$ and several values of a_1'. (*From Ref.* [16].)

elastic buckling stress. Hence, for slender columns, the computed zero-lifetime stress ($\bar{\sigma} = \sigma_E$) is always less than the corresponding actual stress. Thus, deflection-time characteristics obtained on the basis of Eqs. (54.27*a*,*b*) are realistic only if primary creep strain is negligible, and if deformations are limited to stresses within the proportional limit of the material.

In order to obtain a more accurate analysis of creep buckling, Hoff [20] extended the previous work to include the effects of instantaneous plastic deformations and primary creep, in a more realistic manner than that afforded by the use of Eqs.

(54.27*a*,*b*). Thus, these relations were replaced by the creep law,

$$\dot{\epsilon} = \frac{\dot{\sigma}}{E} + k_1 \left(\frac{\sigma}{\mu}\right)^m \left(\frac{\dot{\sigma}}{\mu}\right) + k_2 \left(\frac{\sigma}{\lambda}\right)^n \tag{54.41}$$

in which E, λ, μ, m, and n are positive constants for a given material and temperature,

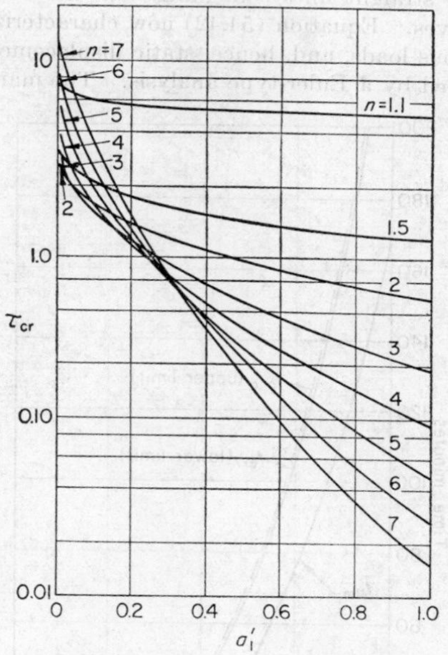

FIG. 54.6. Critical time parameter τ_{cr} for $1.1 \leq n \leq 7$ and $0.01 \leq a_1' \leq 1.0$. (*From Ref.* [17].)

with m and n assumed to be integers; k_1 and k_2 depend on the conditions of loading as follows:

When $\sigma > 0, \dot{\sigma} > 0$ (loading in tension): $k_1 = 1$

$\sigma > 0, \dot{\sigma} < 0$ (unloading in tension): $k_1 = 0$

$\sigma < 0, \dot{\sigma} < 0$ (loading in compression): $k_1 = (-1)^m$

$\sigma < 0, \dot{\sigma} > 0$ (unloading in compression): $k_1 = 0$

$\sigma > 0$ (tension): $k_2 = 1$

$\sigma < 0$ (compression): $k_2 = (-1)^{n+1}$

In Eq. (54.41), the first term on the right-hand side corresponds to the actual instantaneous elastic response, and, hence, E is the modulus of elasticity; the last term corresponds to purely secondary creep as in Eqs. (54.27*a*,*b*) (for a creep-buckling theory based on a creep law consisting of this term alone, see [18]); and the remaining term approximates the net effect of both instantaneous plastic deformation and primary creep. The form of the latter term was suggested by Odqvist [19], where he extended Hoff's [18] purely viscous (nonlinear) column analysis to include primary creep.

When the stress is constant, the creep curve stemming from Eq. (54.41) is once again a straight line in the ϵ,t plane; n and λ are selected, as previously discussed, on the basis of the secondary stages of conventionally obtained creep curves. The $t = 0$

intercept is determined from integration of the first two terms on the right-hand side of Eq. (54.41) to be

$$\epsilon_o = \frac{\sigma_0}{E} + \frac{k_1}{m+1}\left(\frac{\sigma_0}{\mu}\right)^{m+1} \tag{54.42}$$

With E known, m and μ can be chosen in a manner such that ϵ_0 approximates the $t = 0$ intercepts of the straight lines coinciding with the secondary-creep stages of conventional creep curves. Equation (54.42) now characterizes the response of the material to instantaneous loads, and, hence, static displacements and buckling loads are no longer determined by a Euler-type analysis. The manner in which the static

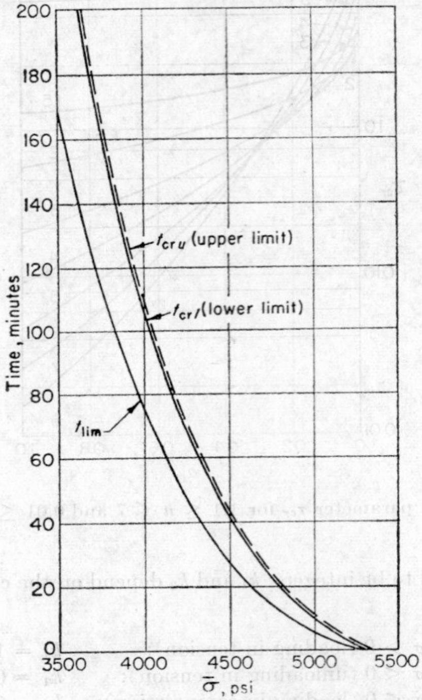

Fig. 54.7. Effect of load on critical time in creep buckling. 24S-T4 aluminum alloy at 600°F: Amplitude of initial deflection = $0.03h$, $2l/h = 111$. (*By permission from Ref.* [21].)

deflections and buckling load of an initially deformed idealized H-section column whose stress-strain law is governed by Eq. (54.42) are calculated was described by Hoff [20],[21]. These quantities now depend on the parameters E, μ, and m rather than on E'.

The application of Eqs. (54.41) and (54.42) to the analysis of the creep buckling of an H-section column is given in [20]. Comparison of the creep-buckling theory, which stems from consideration of the creep law expressed by Eqs. (54.27*a,b*), with the theory based upon the law given in Eq. (54.41), shows that whereas in the former analysis the buckling mechanism is associated with indefinitely large deflections, in the latter buckling occurs at a finite-displacement amplitude for which the rate of deflection becomes infinite. Furthermore, buckling of a given column under a prescribed load takes place at a fixed value of the amplitude of the total displacement

$(w + w_i)$, whether the deflections are due to static loading or to creep [25],[29]. This essential difference in the behavior predicted by the two theories is readily associated with the difference in response to instantaneous loading of the materials characterized by Eqs. (54.27*a*,*b*) and Eq. (54.41). The latter equation includes the effect of the decreasing slope of the instantaneous stress-strain relation and the concomitant loss of rigidity, while the former equations do not. In conformance with the latter theory, creep-buckling experiments with rectangular section columns have indicated that, immediately before collapse, the deflections are relatively small [30]. Such results also afford a justification of the use of small-deflection theory.

The theory presented in [20] leads to the determination of upper and lower bounds on the critical time. Some of the results obtained from the analysis of a simply supported 24S-T4 aluminum alloy **H**-section column are presented in Fig. 54.7 [21], where t_{cru} and t_{crl}, respectively, are the upper and lower bounds on the critical time and t_{lim} is the time at which the mid-span deflection equals one-half the depth of the column. For the particular example considered, the difference between the upper and lower bounds on the critical time is quite small. Once again it is seen that the critical time is extremely sensitive to changes in the applied stress.

54.4. REFERENCES

[1] J. Kempner: Creep bending and buckling of linear viscoelastic columns, *NACA Tech. Note* 3136 (1954); also *Polytech. Inst. Brooklyn PIBAL Rept.* 195 (1952).

[2] T. Alfrey, Jr.: "Mechanical Behavior of High Polymers," Interscience, New York, 1948.

[3] A. M. Freudenthal: Some time effects in structural analysis (unpublished), 6th Intern. Congr. Appl. Mechanics, Paris, 1946.

[4] A. M. Freudenthal: "The Inelastic Behavior of Engineering Materials and Structures," Wiley, New York, 1950.

[5] J. Kempner, F. V. Pohle: On the nonexistence of a finite critical time for linear viscoelastic columns, *J. Aeronaut. Sci.*, **20** (1953), 572–573.

[6] H. H. Hilton: Creep collapse of viscoelastic columns with initial curvatures, *J. Aeronaut. Sci.*, **19** (1952), 844–846.

[7] T. H. Lin: Stresses in columns with time dependent elasticity, *Proc. 1st Midwestern Conf. Solid Mechanics*, pp. 196–199, Urbana, Ill., 1953.

[8] M. A. Brull: A structural theory incorporating the effect of time-dependent elasticity, *Proc. 1st Midwestern Conf. Solid Mechanics*, pp. 141–147, Urbana, Ill., 1953.

[9] T. H. Lin: Creep deflection of viscoelastic plate under uniform edge compression, *J. Aeronaut. Sci.*, **23** (1956), 883–887.

[10] J. Marin: Creep deflections in columns, *J. Appl. Phys.*, **18** (1947), 103–109.

[11] D. Rosenthal, H. W. Baer: An elementary theory of creep buckling of columns, *Proc. 1st U.S. Natl. Congr. Appl. Mechanics*, pp. 603–611, Chicago, 1951.

[12] F. R. Shanley: "Weight-Strength Analysis of Aircraft Structures," McGraw-Hill, New York, 1952.

[13] T. P. Higgins, Jr.: Effect of creep on column deflection, chap. 20, pp. 359–385, in Ref. [12].

[14] C. Libove: Creep buckling of columns, *J. Aeronaut. Sci.*, **19** (1952), 459–467.

[15] C. Libove: Creep buckling analysis of rectangular-section columns, *NACA Tech. Note* 2956 (1953).

[16] J. Kempner: Creep bending and buckling of nonlinear viscoelastic columns, *NACA Tech. Note* 3137 (1954); also *PIBAL Rept.* 200 (1952).

[17] J. Kempner, S. A. Patel: Creep buckling of columns, *NACA Tech. Note* 3138 (1954); also *PIBAL Rept.* 205 (1952).

[18] N. J. Hoff: Buckling and stability, *J. Roy. Aeronaut. Soc.*, **58** (1954), 1–52.

[19] F. K. G. Odqvist: Influence of primary creep on column buckling, *J. Appl. Mechanics*, **21** (1954), 295.

[20] N. J. Hoff: Creep buckling, *Aero Quart.*, **7** (1956), 1–20; also *PIBAL Rept.* 252 (1954).

[21] N. J. Hoff: Rapid creep in structures, *J. Aeronaut. Sci.*, **22** (1955), 661–672, 700.

[22] S. A. Patel: Buckling of columns in the presence of creep, *Aero Quart.*, **7** (1956), 125–134; also *PIBAL Rept.* 292 (1955).

[23] L. W. Hu, N. H. Triner: Bending creep and its application to beam columns, *J. Appl. Mechanics*, **23** (1956), 35–42.

[24] T. H. Lin: Creep stresses and deflections of columns, *J. Appl. Mechanics*, **23** (1956), 214–218.

[25] N. J. Hoff: Buckling at high temperature, *J. Roy Aeronaut. Soc.*, **61** (1957), 756–774, also *PIBAL Rept.* 356 (1956).

[26] J. L. Sanders, Jr., H. G. McComb, F. R. Schlechte: A variational theorem for creep with applications to plates and columns, *NACA Tech. Note* 4003 (1957).

[27] J. C. Chapman, B. Erickson, N. J. Hoff: A theoretical and experimental investigation of creep buckling, *PIBAL Rept.* 406 (1957).

[28] S. A. Patel, J. Kempner: Effect of higher-harmonic deflection components on the creep buckling of columns, *Aero Quart.*, **8** (1957), 215–225; also *PIBAL Rept.* 290 (1955).

[29] B. Fraeijs de Veubeke: Creep buckling, chap. 11 in N. J. Hoff (ed.), "High Temperature Effects in Aircraft Structures," Pergamon, London, 1958.

[30] S. A. Patel, J. Kempner, B. Erickson, A. H. Mobassery: Correlation of creep-buckling tests with theory, *NACA Research Mem.* 56C20 (1956); also *PIBAL Rept.* 285 (1955).

[31] G. N. Rabotnov, S. A. Shesterikov: Creep stability of columns and plates, *J. Mech. Phys. Solids*, **6** (1957), 27–34.

[32] N. J. Hoff: A survey of the theories of creep buckling, *Proc. 3d U.S. Natl. Congr. Appl. Mechanics*, pp. 29–49, Providence, R.I., 1958.

Part 6

VIBRATIONS

CHAPTER 55

KINEMATICS OF VIBRATIONS

BY

R. S. AYRE, Ph.D., New Haven, Conn.

55.1. DEFINITIONS

A periodic (or cyclic) vibration (Fig. 55.1) repeats itself in all respects in a constant interval of time (the duration of one cycle) called the period T. The frequency f of vibration is the reciprocal of the period $f = 1/T$. Usually the period is measured in seconds, and the frequency in cycles per second.

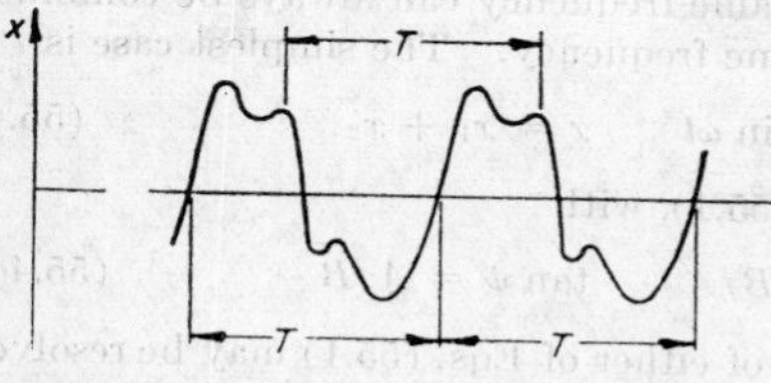

FIG. 55.1. Periodic vibration of general form.

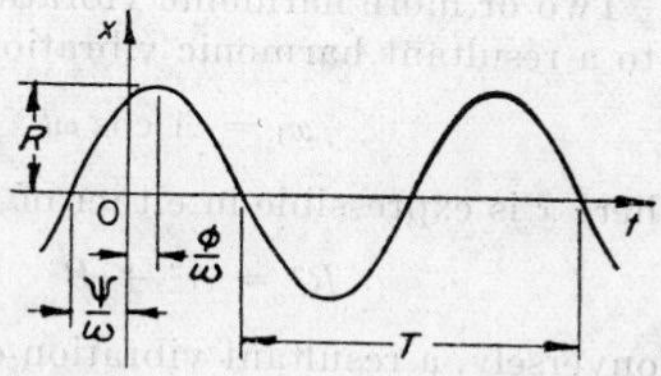

FIG. 55.2. Harmonic (sinusoidal) vibration.

The simplest periodic vibration is harmonic (sinusoidal) (Fig. 55.2). It may be described by either of the following expressions:

$$x = R \cos(\omega t - \phi) \quad \text{or} \quad x = R \sin(\omega t + \psi) \tag{55.1}$$

where R is the amplitude of vibration, ω the circular frequency, and ϕ and ψ are phase angles which give the phase of the vibration with respect to the time origin. Since $\omega T = 2\pi$, the following relationships among ω, T, and f may be written:

$$\omega = \frac{2\pi}{T} = 2\pi f \tag{55.2}$$

The time derivatives of a harmonic vibration are also harmonic, for example:

$$x = R \cos(\omega t - \phi) \qquad \dot{x} = -R\omega \sin(\omega t - \phi) \qquad \ddot{x} = -R\omega^2 \cos(\omega t - \phi) \tag{55.3}$$

55.2. VECTORIAL REPRESENTATION

The representation of harmonic vibration in vector form is often advantageous. For example, Eqs. (55.3) may be represented vectorially as in Fig. 55.3. The three radius vectors of lengths R, $R\omega$, and $R\omega^2$, which are equal in length to the maximum displacement (amplitude), maximum velocity, and maximum acceleration, maintain a constant relationship to one another, and rotate in the counterclockwise direction with an angular velocity ω.

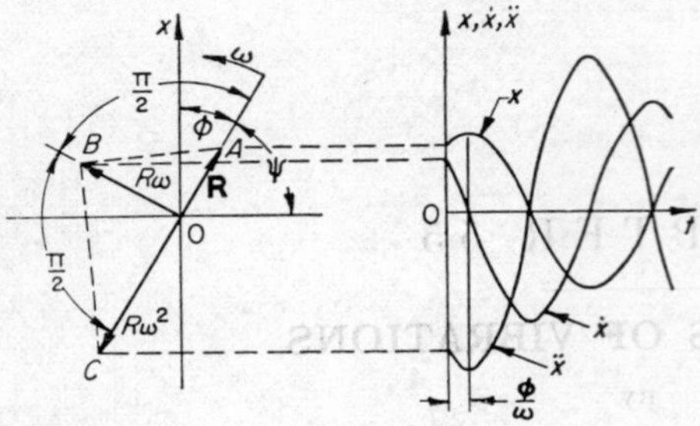

Fig. 55.3

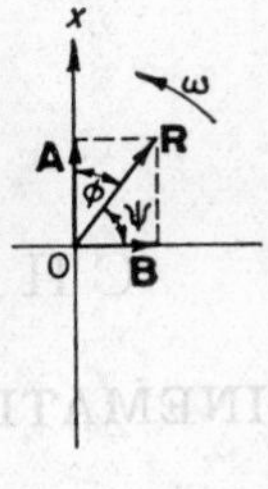

Fig. 55.4

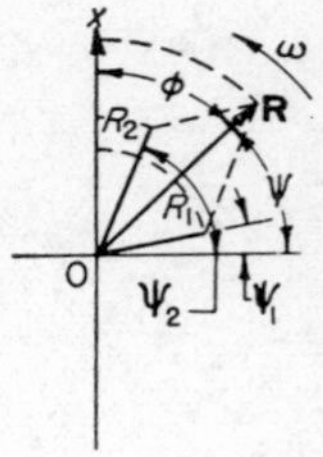

Fig. 55.5

Fig. 55.3. Vectorial representation of harmonic displacement, velocity, and acceleration.
Fig. 55.4. Vectorial representation of two harmonic motions of the same frequency having a phase difference of $\pi/2$ radians.
Fig. 55.5. Vectorial representation of two harmonic motions of the same frequency having a general phase relationship.

55.3. COMBINATION OF HARMONIC VIBRATIONS OF THE SAME FREQUENCY

Two or more harmonic vibrations of the same frequency can always be combined into a resultant harmonic vibration of the same frequency. The simplest case is

$$x_1 = A \cos \omega t \qquad x_2 = B \sin \omega t \qquad x = x_1 + x_2 \tag{55.4}$$

where x is expressible in either of the forms (55.1), with

$$R^2 = A^2 + B^2 \qquad \tan \phi = B/A \qquad \tan \psi = A/B \tag{55.4a}$$

Conversely, a resultant vibration of the form of either of Eqs. (55.1) may be resolved into components [Eq. (55.4)] by use of the relations

$$A = R \cos \phi \qquad B = R \sin \phi$$
or
$$A = R \sin \psi \qquad B = R \cos \psi \tag{55.4b}$$

These formulas indicate that arrows **A** and **B** as in Fig. 55.4 can be added according to the parallelogram rule, i.e., that they are vectors.

In the more general case (Fig. 55.5) of arbitrary amplitudes and phase angles,

$$x_1 = R_1 \sin (\omega t + \psi_1) \qquad x_2 = R_2 \sin (\omega t + \psi_2) \tag{55.5}$$

the resultant vibration $x = x_1 + x_2$ is given by Eqs. (55.1) with

$$R^2 = R_1{}^2 + R_2{}^2 + 2R_1R_2 \cos (\psi_1 - \psi_2)$$
and
$$\tan \psi = \frac{R_1 \sin \psi_1 + R_2 \sin \psi_2}{R_1 \cos \psi_1 + R_2 \cos \psi_2} \tag{55.5a}$$

Any number of collinear, harmonic, component vibrations of the same frequency may be compounded in the following manner:

Let
$$x = \Sigma R_j \sin (\omega t + \psi_j) \tag{55.6}$$
Then
$$x = \sin \omega t(\Sigma R_j \cos \psi_j) + \cos \omega t(\Sigma R_j \sin \psi_j) = R \sin (\omega t + \psi) \tag{55.6a}$$
where
$$R^2 = (\Sigma R_j \cos \psi_j)^2 + (\Sigma R_j \sin \psi_j)^2$$
$$\tan \psi = \frac{\Sigma R_j \sin \psi_j}{\Sigma R_j \cos \psi_j} \tag{55.6b}$$

55.4. NONHARMONIC PERIODIC MOTIONS

An example of a nonharmonic periodic vibration has been shown in Fig. 55.1. Periodic motions composed of two or more harmonic motions of different frequency

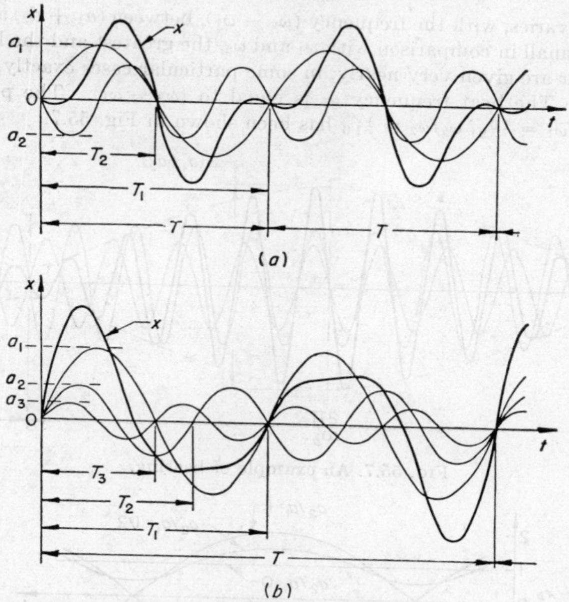

FIG. 55.6. Combination of harmonic motions of different frequency: (*a*) Two component motions; (*b*) three component motions.

often occur. In the examples shown in Fig. 55.6, the periods of the component harmonic motions are in the ratios of whole numbers, and the resultant motion repeats itself; that is, it is periodic, in a period equal to the least common multiple of the component periods. If the component periods are not commensurable, the resultant motion will not be exactly periodic.

The equation for x in Fig. 55.6*a* is

$$x = a_1 \sin \frac{2\pi t}{T_1} + a_2 \sin \frac{2\pi t}{T_2} \tag{55.7a}$$

and, for the particular case shown, $T_1 = 2T_2 = T$. For Fig. 55.6*b*, the motion is given by

$$x = a_1 \sin \frac{2\pi t}{T_1} + a_2 \sin \frac{2\pi t}{T_2} + a_3 \sin \frac{2\pi t}{T_3} \tag{55.7b}$$

and the period relationships are $2T_1 = 3T_2 = 4T_3 = T$.

The representation of a general periodic function by the use of Fourier series has been described on p. 15–2 and in Table 20.3.

55.5. BEATING

If the difference between the frequencies of two component vibrations of harmonic form is small, the resultant vibration takes a form that is called *beating*. For example, let

$$x = a_1 \sin \omega_1 t + a_2 \sin \omega_2 t \tag{55.8a}$$

This can be rewritten in the combined form,

$$x = [a_1^2 + a_2^2 + 2a_1a_2 \cos (\omega_2 - \omega_1)t]^{1/2} \sin (\omega_1 t + \psi)$$

where $$\tan \psi = \frac{a_2 \sin (\omega_2 - \omega_1)t}{a_1 + a_2 \cos (\omega_2 - \omega_1)t} \tag{55.8b}$$

The radical varies, with the frequency $(\omega_2 - \omega_1)$, between $(a_1 + a_2)$ and $(a_1 - a_2)$. If $(\omega_2 - \omega_1)$ is small in comparison with ω_1 and ω_2, the greatest and the least values of the maxima of x are given very nearly, in some particular cases exactly, by $(a_1 + a_2)$ and $(a_1 - a_2)$. The beat frequency ω_b is equal to $(\omega_2 - \omega_1)$. The particular case $a_2/a_1 = 4/3$, $\omega_2/\omega_1 = 10/9$, $\omega_b/\omega_2 = 1/10$ has been shown in Fig. 55.7.

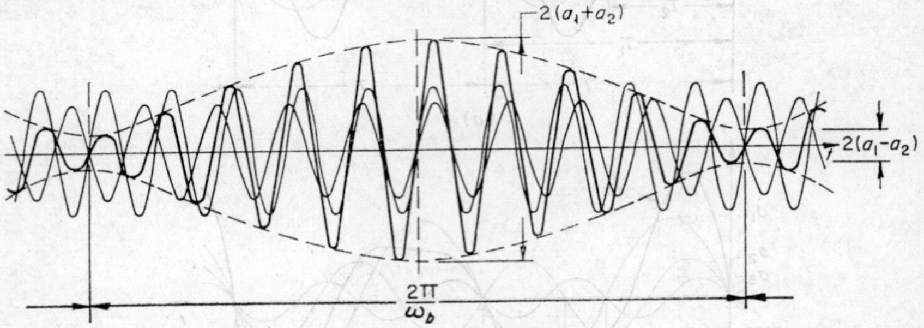

FIG. 55.7. An example of beating.

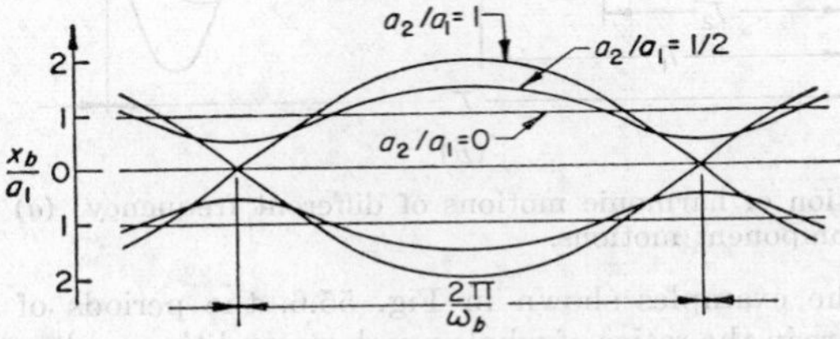

FIG. 55.8. Envelopes of beat vibration.

If the radical in Eq. (55.8*b*) is put in dimensionless form,

$$\frac{x_b}{a_1} = \left[1 + \left(\frac{a_2}{a_1}\right)^2 + 2\frac{a_2}{a_1}\cos\omega_b t\right]^{1/2} \tag{55.8c}$$

a family of envelopes of beat vibration may be drawn as in Fig. 55.8, showing the effect of variation in the ratio a_2/a_1 of the component amplitudes.

55.6. COMBINATION OF TWO HARMONIC MOTIONS OCCURRING AT RIGHT ANGLES TO EACH OTHER; LISSAJOUS FIGURES

The plane motion of a particle, having two harmonic-motion components perpendicular to each other, may be described by

$$x = a_x \cos(\omega_x t - \phi) \qquad y = a_y \cos \omega_y t \tag{55.9}$$

Lissajous figures are curves showing the path of the particle in the x,y plane. In all cases, the path is contained within the rectangle having sides of lengths $2a_x$ and $2a_y$.

If the two frequencies are equal, the path is, in general, an ellipse (Fig. 55.9), its equation being given by

$$\frac{x^2}{a_x^2} + \frac{y^2}{a_y^2} - \frac{2xy}{a_x a_y}\cos\phi - \sin^2\phi = 0 \tag{55.10}$$

The phase angle ϕ may be determined from the following relationships: At the point of tangency of the ellipse with the line $y = a_y$, $\cos\phi = x/a_x$. Similarly, when $x = a_x$, $\cos\phi = y/a_y$.

If the two frequencies are not equal, the path may be very complicated. If the frequencies are in the ratio of two whole numbers, the trace will repeat itself. If the ratio is of two small whole numbers, the trace may be reasonably simple, as shown in the example of Fig. 55.10.

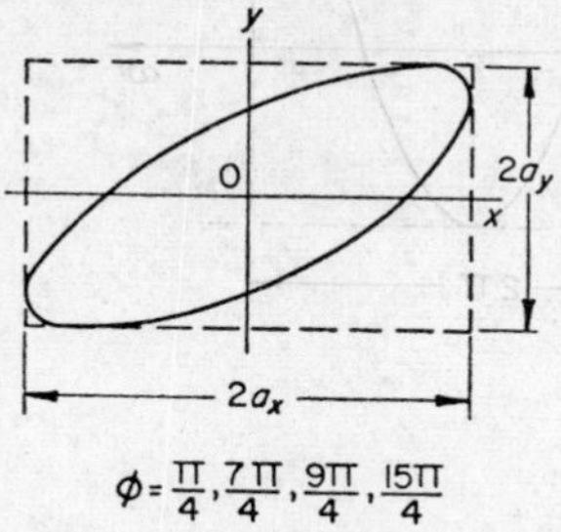

FIG. 55.9. A Lissajous figure for the case $\omega_x = \omega_y$.

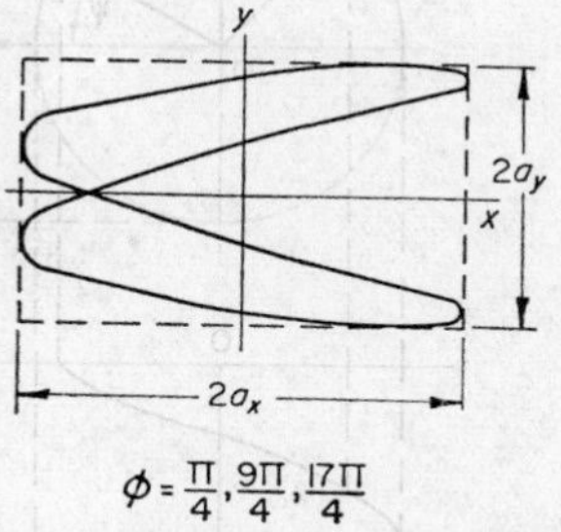

FIG. 55.10. A Lissajous figure for the case $\omega_x = 2\omega_y$.

55.7. PHASE-PLANE REPRESENTATION OF VIBRATION

The phase plane, commonly the velocity-displacement plane, provides a convenient means of representing one-dimensional vibration. In the case of harmonic motion about $x = 0$ as the equilibrium position, the displacement and velocity may be given by

$$x = R \sin (\omega t + \psi) \qquad \dot{x} = R\omega \cos (\omega t + \psi) \tag{55.11}$$

Eliminating t, we obtain the equation

$$(\dot{x}/\omega)^2 + x^2 = R^2 \tag{55.11a}$$

which represents a circle if x and $\dot{x}/\omega$ are chosen as coordinates. This is the phase-plane trajectory for this particular case.

The trajectory (Fig. 55.11) is generated by the counterclockwise revolution of a radius vector of length R rotating with the angular velocity ω. Its zero-time position is given by the phase angle ψ, and the time corresponding to any given location on the trajectory by the angle ωt. Curves in the time-displacement and time-velocity planes may be constructed by direct projection from the phase plane.

If the motion is about a displaced position of equilibrium located a distance δ from the origin, measured along the x axis, the displacement and velocity may be described by

$$x = \delta + R \sin (\omega t + \psi) \qquad \dot{x} = R\omega \cos (\omega t + \psi) \tag{55.12}$$

The equation of the phase-plane trajectory is then given by

$$(\dot{x}/\omega)^2 + (x - \delta)^2 = R^2 \tag{55.12a}$$

which, as shown in Fig. 55.12, is the equation of a circle with its center at $(0,\delta)$.

The concepts just shown provide the basis for a stepwise graphical method of solution [4]–[6] of transient forced-vibration problems for which the forcing function, dissipation function, and restoring force function may be represented adequately by the stepwise variation of δ. In effect, the motion is represented by a connected sequence of harmonic motions, each of short duration. The method is applicable to nonlinear as well as to linear systems.

Figure 55.13 shows phase trajectories for several types of response of a linear oscillator: (1) damped free vibration resulting from the initial conditions $\dot{x} = 0$, $x = R$, with damping of Coulomb type; (2) resonant, sinusoidally forced vibration,

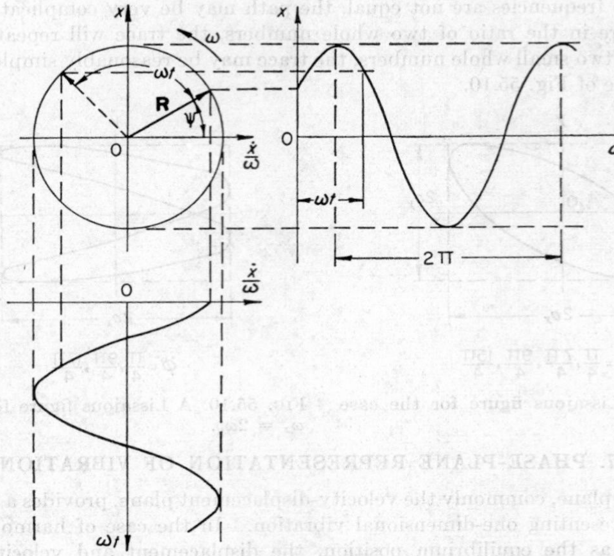

FIG. 55.11. Phase-plane representation of harmonic motion.

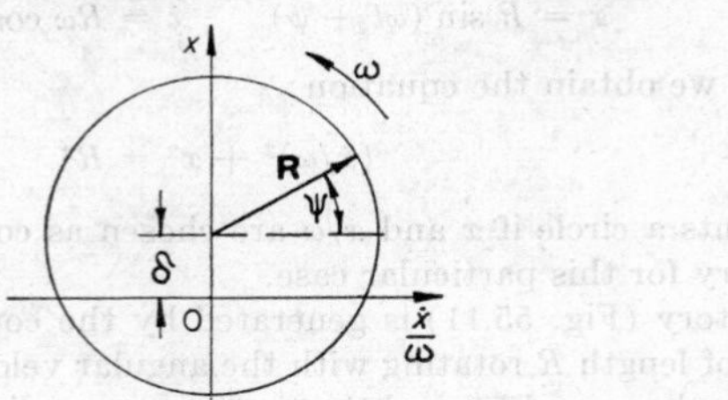

FIG. 55.12. Phase-plane representation of harmonic motion about a displaced position of equilibrium.

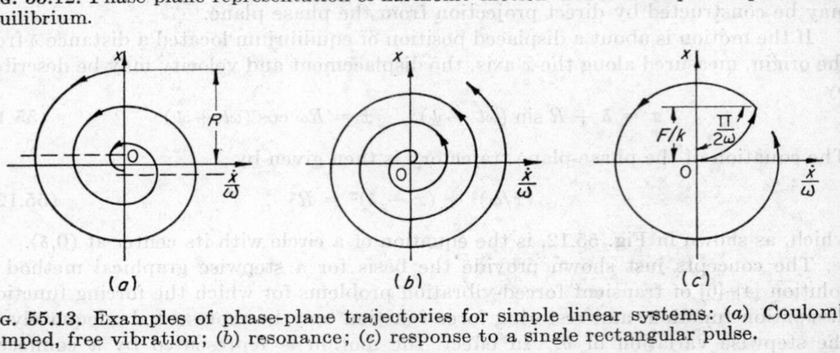

FIG. 55.13. Examples of phase-plane trajectories for simple linear systems: (*a*) Coulomb damped, free vibration; (*b*) resonance; (*c*) response to a single rectangular pulse.

without damping; (3) transient response of the undamped system to a *rectangular* force pulse of magnitude F and duration $\pi/2\omega$.

The most important use of the phase plane is in the topological study of the stability, instability, and other properties of nonlinear vibrating systems. This topic may be found in Chap. 65, Nonlinear Vibrations.

55.8. AMPLITUDE MODULATION AND FREQUENCY MODULATION

When either the amplitude R or the frequency ω is a function of time, we have a vibration with amplitude modulation

$$x = R(t) \sin (\omega_0 t + \psi_0) \tag{55.13a}$$

or with frequency modulation

$$x = R_0 \sin [\omega(t) \cdot t + \psi_0] \tag{55.13b}$$

A simple example of the amplitude modulation (Fig. 55.14) is given by

$$x = [R_0 + R_m \sin (\omega_m t + \psi_m)] \sin (\omega_0 t + \psi_0) \tag{55.14}$$

The modulation function (in brackets) is sinusoidal; the frequency of modulation is

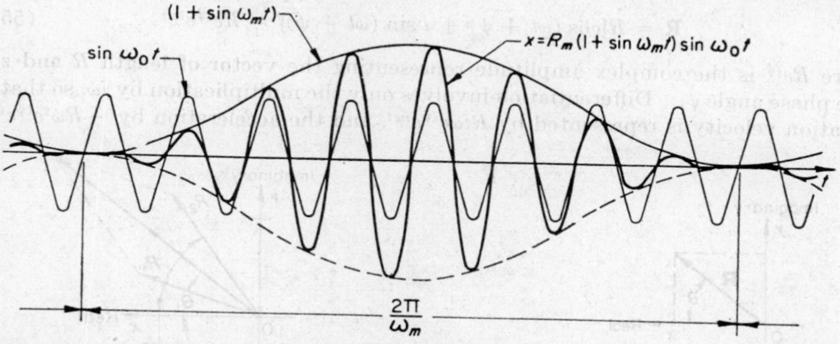

FIG. 55.14. Example of sinusoidal amplitude modulation: $\psi_0 = \psi_m = 0$, $R_0 = R_m$, $\omega_0 = 8\omega_m$.

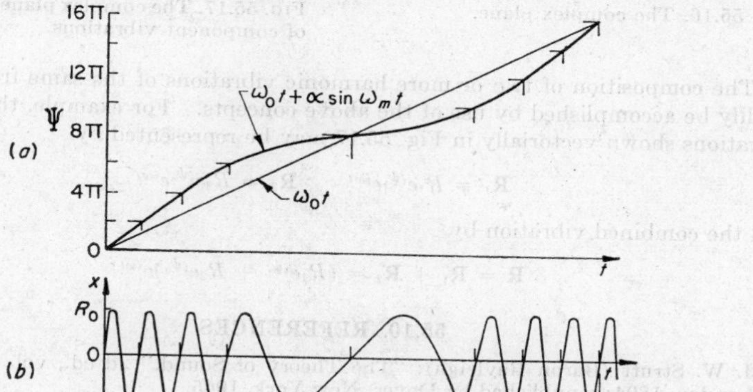

FIG. 55.15. Example of sinusoidal frequency modulation: $\psi_0 = \psi_m = 0$, $\alpha = 2\pi$, $\omega_0 = 8\omega_m$. (*a*) Determination of times for $x = 0$; (*b*) the modulated waveform.

ω_m. Equation (55.14) may be rewritten in the frequencies ω_0, $(\omega_0 - \omega_m)$, and $(\omega_0 + \omega_m)$ as follows:

$$x = R_0 \sin (\omega_0 t + \psi_0) + \frac{R_m}{2} \cos [(\omega_0 - \omega_m)t + (\psi_0 - \psi_m)] - \frac{R_m}{2} \cos [(\omega_0 + \omega_m)t + (\psi_0 + \psi_m)] \tag{55.14a}$$

A simple case of frequency modulation is given by

$$x = R_0 \sin [\omega_0 t + \alpha \sin (\omega_m t + \psi_m) + \psi_0] = R_0 \sin \Psi(t) \tag{55.15}$$

where $x = 0$ when $\Psi = n\pi$, n being an integer. If $\psi_0 = \psi_m = 0$, then $x = 0$ when $(\omega_0 t + \alpha \sin \omega_m t) = n\pi$ (Fig. 55.15).

55.9. COMPLEX NUMBERS AND EXPONENTIALS

The rotating vector **R** of Fig. 55.3 may be expressed in complex form as

$$\mathbf{R} = x + iy = R(\cos \theta + i \sin \theta) \tag{55.16}$$

with

$$R^2 = x^2 + y^2 \qquad \tan \theta = y/x$$

as shown in Fig. 55.16. The vibration displacement is represented by the real part of Eq. (55.16*a*). Let $\theta = \omega t + \psi$. Then,

$$\mathbf{R} = R[\cos (\omega t + \psi) + i \sin (\omega t + \psi)] = Re^{i\psi}e^{i\omega t} \tag{55.17}$$

where $Re^{i\psi}$ is the complex amplitude representing the vector of length R and zero-time phase angle ψ. Differentiation involves only the multiplication by $i\omega$, so that the vibration velocity is represented by $Ri\omega e^{i\psi}e^{i\omega t}$, and the acceleration by $-R\omega^2 e^{i\psi}e^{i\omega t}$.

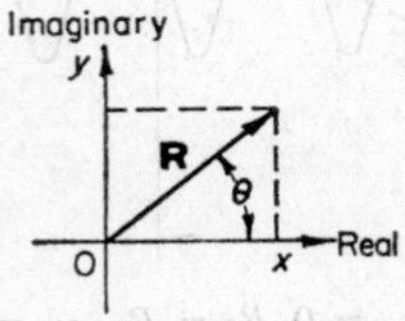

FIG. 55.16. The complex plane.

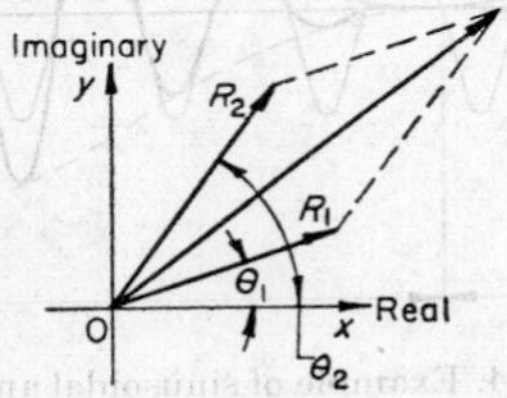

FIG. 55.17. The complex plane; combination of component vibrations.

The composition of two or more harmonic vibrations of the same frequency may readily be accomplished by use of the above concepts. For example, the component vibrations shown vectorially in Fig. 55.17 may be represented by

$$\mathbf{R}_1 = R_1 e^{i\psi_1}e^{i\omega t} \qquad \mathbf{R}_2 = R_2 e^{i\psi_2}e^{i\omega t} \tag{55.18a}$$

and the combined vibration by

$$\mathbf{R} = \mathbf{R}_1 + \mathbf{R}_2 = (R_1 e^{i\psi_1} + R_2 e^{i\psi_2})e^{i\omega t} \tag{55.18b}$$

55.10. REFERENCES

[1] J. W. Strutt (Baron Rayleigh): "The Theory of Sound," 2d ed., vol. 1, Macmillan, London, 1894; republished by Dover, New York, 1945.
[2] K. Klotter: "Technische Schwingungslehre," vol. 1, 2d ed., Springer, Berlin, 1951.
[3] J. P. Den Hartog: "Mechanical Vibrations," 4th ed., McGraw-Hill, New York, 1956.
[4] L. S. Jacobsen, R. S. Ayre: "Engineering Vibrations," McGraw-Hill, New York, 1958.
[5] J. Lamoen: Étude graphique des vibrations de systèmes à un seul degré de liberté, *Rev. universelle mines*, [8]**11** (1935), no. 7, pp. 3–16.
[6] L. S. Jacobsen: On a general method of solving second-order ordinary differential equations by phase-plane displacements, *J. Appl. Mechanics*, **19** (1952), 543–553.

Special acknowledgment is due to Refs. [1] and [2], particularly in regard to the combination of harmonic motions.

CHAPTER 56

SYSTEMS OF ONE DEGREE OF FREEDOM

BY

W. T. THOMSON, Ph.D., Los Angeles, Calif.

56.1. UNDAMPED FREE VIBRATION

The simplest oscillatory system is one of a single degree of freedom, the motion of which can be described by a single coordinate x. A mass m suspended from a spring of negligible mass as shown in Fig. 56.1 represents such a system.

Free vibration takes place when a system in the absence of impressed forces is disturbed from its equilibrium position. The system under free vibration will oscillate at its natural frequency and its amplitude will gradually diminish, owing to energy dissipated by the motion. The natural frequency is dependent mainly on the mass and stiffness of the system, and in most cases is only slightly affected by damping. Damping, therefore, is generally neglected in the calculation of the natural frequency.

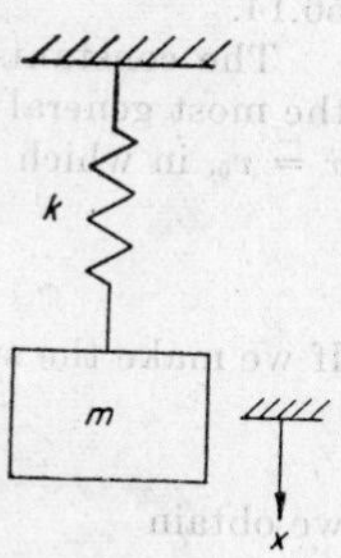

FIG. 56.1. Undamped spring-mass system.

The behavior of a single-degree-of-freedom system can be discussed by studying the spring-mass system of Fig. 56.1. With the mass m constrained to move along a straight line, its position is completely specified by the value of the displacement x. We take as reference the statical equilibrium position of the mass. In this position, the gravitational force W acting on the mass is balanced by the spring force $k\delta$, where k is the spring stiffness, defined as the force required per unit extension or compression, and δ is the statical deflection of the spring due to the weight $W = mg$. It is evident then that, by choosing the origin of the coordinate x at the static equilibrium position, only forces due to displacement from this position need be considered.

Letting all vector quantities in the downward direction be positive, the unbalanced force acting on the mass m in position x is $-kx$, and its motion is described by Newton's second law to be

$$m\ddot{x} = -kx \tag{56.1}$$

By rearranging terms and introducing $\omega_n^2 = k/m$, this equation can be rewritten as

$$\ddot{x} + \omega_n^2 x = 0 \tag{56.2}$$

This is a homogeneous second-order differential equation, and has the general solution

$$x = A \sin \omega_n t + B \cos \omega_n t \tag{56.3}$$

describing a motion with the period

$$\tau = \frac{2\pi}{\omega_n} = 2\pi \sqrt{\frac{m}{k}} \tag{56.4}$$

the frequency

$$f = \frac{1}{2\pi}\sqrt{\frac{k}{m}} \tag{56.5}$$

and the circular frequency

$$\omega_n = \sqrt{\frac{k}{m}} \tag{56.6}$$

Equation (56.5) may be expressed in terms of the statical deflection δ by noting that $k\delta = W = mg$, from which we obtain

$$f = \frac{1}{2\pi}\sqrt{\frac{g}{\delta}} \tag{56.7}$$

Thus the frequency is found to be a function of the statical deflection of the system. Using $g = 386$ in./sec² and δ in inches, the relationship between the two quantities becomes

$$f = \frac{3.127}{\sqrt{\delta}} \text{ cycles/sec} = \frac{187.6}{\sqrt{\delta}} \text{ cycles/min} \tag{56.8}$$

A plot of this equation is presented by the line marked "natural frequency" in Fig. 56.14.

The constants A and B in Eq. (56.3) are evaluated from initial conditions. In the most general case, the system may be started from position $x = x_0$ with velocity $\dot{x} = v_0$, in which case the general solution is

$$x = \frac{v_0}{\omega_n}\sin \omega_n t + x_0 \cos \omega_n t \tag{56.9}$$

If we make the substitutions

$$x_0 = X \sin \phi \qquad \frac{v_0}{\omega_n} = X \cos \phi$$

we obtain

$$X = \sqrt{x_0{}^2 + \left(\frac{v_0}{\omega_n}\right)^2} \qquad \tan \phi = \frac{\omega_n x_0}{v_0} \tag{56.10}$$

and

$$x = X \sin (\omega_n t + \phi) \tag{56.11}$$

which is another form of the general solution.

In an alternative approach we can arrive at the equation for the natural frequency ω_n by making use of the principle of conservation of energy. Thus, in the vibration of any nondissipative system, the maximum kinetic energy

$$T_{\max} = \tfrac{1}{2}m\dot{x}^2_{\max} = \tfrac{1}{2}m\omega_n{}^2X^2$$

must equal the maximum potential energy

$$U_{\max} = \tfrac{1}{2}kx^2_{\max} = \tfrac{1}{2}kX^2$$

When these are equated, the amplitude X cancels, and Eq. (56.6) results.

The differential equation (56.1) and its solution (56.3) with the circular frequency (56.6) may also be applied to a system consisting of a rigid body which is able to rotate about a fixed axis and is held in a position of equilibrium by a spring. In this case m must be replaced by the mass moment of inertia J of the body, and x by its angular displacement θ, and the spring constant k must be defined as the moment per unit of θ.

The single-degree-of-freedom system is an idealization which in many cases is

Table of Spring Stiffness

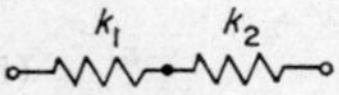	$k = \dfrac{1}{1/k_1 + 1/k_2}$	
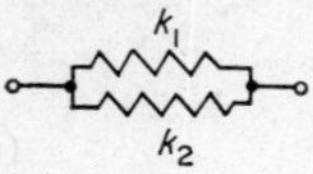	$k = k_1 + k_2$	
	$k = \dfrac{EI}{l}$	I = moment of inertia of cross-sectional area; l = total length
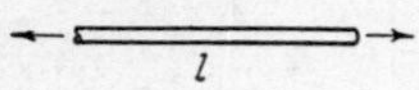	$k = \dfrac{EA}{l}$	A = cross-sectional area
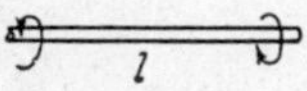	$k = \dfrac{GJ}{l}$	J = torsion constant of cross section [see Eq. (36.25)]
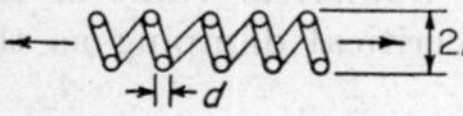	$k = \dfrac{Gd^4}{64nR^3}$	n = number of turns
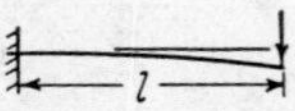	$k = \dfrac{3EI}{l^3}$	
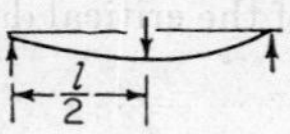	$k = \dfrac{48EI}{l^3}$	
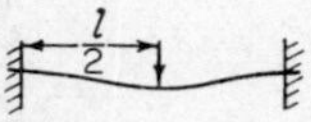	$k = \dfrac{192EI}{l^3}$	
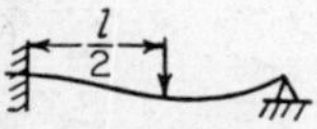	$k = \dfrac{768EI}{7l^3}$	
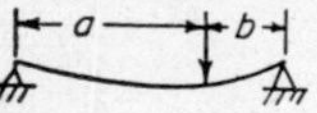	$k = \dfrac{3EIl}{a^2b^2}$	

justified. Such approximations give excellent results when the mass of the elastic element is small compared to the lumped mass which it supports, and when the deflection of the elastic element is large compared to all other deformations occurring in the system. The fundamental frequency of the system can then be determined with good accuracy from Eq. (56.5), provided the proper stiffness k is used. The table presents the stiffness of various types of springs.

56.2. DAMPED FREE VIBRATION

Energy dissipation takes place in all vibrating systems, and the force associated with the dissipation is called *damping force.* There are several different types of damping forces, and the one which is the simplest to consider is viscous damping proportional to the velocity.

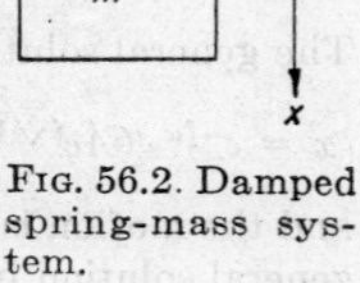

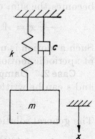

FIG. 56.2. Damped spring-mass system.

Figure 56.2 is a schematic diagram for a spring-mass system with viscous damping. The equation of motion for the damped free vibration is

$$m\ddot{x} + c\dot{x} + kx = 0 \tag{56.12}$$

Equation (56.12) is a homogeneous second-order differential equation which can be solved by assuming a function of the form

$$x = e^{st} \tag{56.13}$$

Upon substitution, we find that s must satisfy the equation

$$s^2 + \frac{c}{m} s + \frac{k}{m} = 0 \tag{56.14}$$

which has the two roots

$$s_{1,2} = -\frac{c}{2m} \pm \sqrt{\left(\frac{c}{2m}\right)^2 - \frac{k}{m}} \tag{56.15}$$

Hence, the general solution becomes

$$x = Ae^{s_1 t} + Be^{s_2 t} \tag{56.16}$$

where A and B must be evaluated from initial conditions.

Critical Damping

The free motion of the damped system depends on the numerical value of the radical of Eq. (56.15). As a reference quantity we define critical damping as the value of c which reduces this radical to zero, or

$$c_c = 2\sqrt{km} = 2m\omega_n \tag{56.17}$$

The actual damping of the system can then be specified in terms of the critical damping by the damping factor, which is

$$\zeta = \frac{c}{c_c} \tag{56.18}$$

and the quantity $c/2m$ becomes

$$\frac{c}{2m} = \zeta \frac{c_c}{2m} = \zeta\omega_n \tag{56.19}$$

Equation (56.15) can now be rewritten as

$$s_{1,2} = (-\zeta \pm \sqrt{\zeta^2 - 1})\omega_n \tag{56.20}$$

and the behavior of the system is dependent on ζ.

Case 1. Damping Greater than Critical ($\zeta > 1$). The radical in this case is real, and is always less than ζ, so that s_1 and s_2 are negative. The displacement x then becomes the sum of two decaying exponentials:

$$x = A \exp\left(-\zeta + \sqrt{\zeta^2 - 1}\right)\omega_n t + B \exp\left(-\zeta - \sqrt{\zeta^2 - 1}\right)\omega_n t \tag{56.21}$$

Such a motion is nonoscillatory and is called *aperiodic*. Figure 56.3 shows some cases of aperiodic motion.

Case 2. Damping Less than Critical ($\zeta < 1$). The radical in this case is imaginary, and s can be written as

$$s_{1,2} = (-\zeta \pm i\sqrt{1 - \zeta^2})\omega_n \tag{56.22}$$

The general solution then becomes

$$x = e^{-\zeta\omega_n t}\left(Ae^{i\sqrt{1-\zeta^2}\omega_n t} + Be^{-i\sqrt{1-\zeta^2}\omega_n t}\right) = Xe^{-\zeta\omega_n t} \sin\left(\sqrt{1 - \zeta^2}\,\omega_n t + \phi\right) \tag{56.23}$$

and the motion is oscillatory with diminishing amplitude, as shown in Fig. 56.4. The general solution in terms of the initial displacement x_0 and initial velocity v_0 is

$$x = e^{-\zeta\omega_n t}\left[x_0 \cos \sqrt{1 - \zeta^2}\,\omega_n t + \left(\frac{v_0}{\omega_n\sqrt{1 - \zeta^2}} + \zeta x_0\right) \sin \sqrt{1 - \zeta^2}\,\omega_n t\right] \tag{56.24}$$

Case 3. Critical Damping ($\zeta = 1$). $\zeta = 1$ represents a transition between oscillatory and nonoscillatory conditions. The roots s_1 and s_2 approach each other, and become equal to $-\omega_n$. The general solution then has the form (see p. 10–8)

$$x = (A + Bt)e^{-\omega_n t} \tag{56.25}$$

which retains the necessary number of arbitrary constants in conformity with the order of the differential equation. For a motion initiated with velocity v_0 from position x_0, the solution becomes

$$x = \left[x_0 + \left(\frac{v_0}{\omega_n} + x_0\right)\omega_n t\right] e^{-\omega_n t} \tag{56.26}$$

which is illustrated in Fig. 56.5. The motion is similar to the aperiodic motion of case 1; however, it has the smallest damping possible for aperiodic motion, and hence returns to the equilibrium position in the shortest time. The moving parts of many

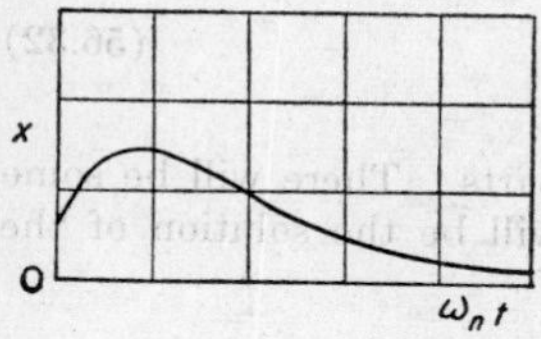

FIG. 56.3. Aperiodic motion $\zeta > 1$.

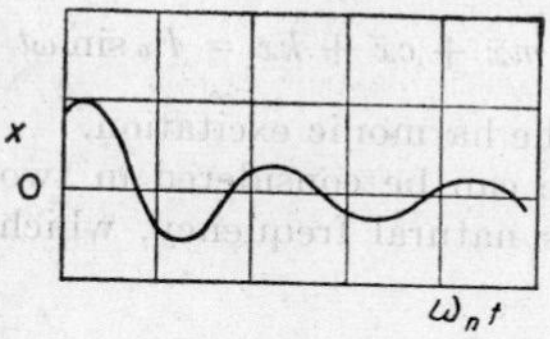

FIG. 56.4. Damped oscillations $\zeta < 1$.

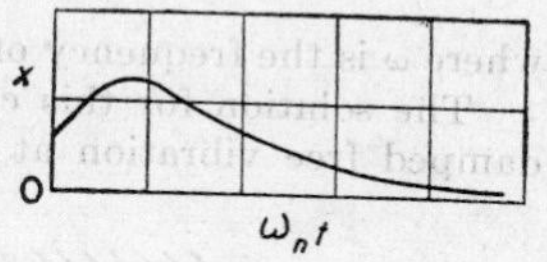

FIG. 56.5. Critically damped motion $\zeta = 1$.

electric meters and instruments are critically damped to take advantage of this property.

Logarithmic Decrement

The amount of damping present in an oscillatory system can be determined by measuring the rate of decay of oscillation. For this purpose, we define the logarithmic decrement, which is the natural logarithm of the ratio of any two successive maxima x_i, x_{i+1} of the displacement x. If we let the damped oscillation of Fig. 56.4 be described by Eq. (56.23), the logarithmic decrement defined by the equation

$$\delta = \ln \frac{x_i}{x_{i+1}} \tag{56.27}$$

can be determined in terms of the damping factor.

When $\sin(\sqrt{1-\zeta^2}\,\omega_n t + \phi) = 1$, the curve is tangent to the exponential envelope $Xe^{-\zeta\omega_n t}$, which is slightly to the right of the point of maximum amplitude. Generally this discrepancy is negligible, and the amplitude at the point of tangency may be taken equal to the maximum amplitude. The logarithmic decrement δ is then expressed as

$$\delta = \ln \frac{e^{-\zeta\omega_n t_1}}{e^{-\zeta\omega_n(t_1+\tau)}} = \ln e^{\zeta\omega_n\tau} = \zeta\omega_n\tau \tag{56.28}$$

Since the period of damped oscillation is equal to

$$\tau = \frac{2\pi}{\omega_n\sqrt{1-\zeta^2}} \tag{56.29}$$

the logarithmic decrement in terms of ζ becomes

$$\delta = \frac{2\pi\zeta}{\sqrt{1-\zeta^2}} \approx 2\pi\zeta \tag{56.30}$$

A convenient procedure to establish ζ is to determine the number of cycles of oscillation required for the amplitude to diminish to half its value. Letting x_0 be the initial amplitude and x_n the amplitude after n oscillations, the following relationship can be derived:

$$\delta = \frac{1}{n} \ln \frac{x_0}{x_n} = 2\pi\zeta$$

Thus, for $x_n/x_0 = \frac{1}{2}$, we arrive at the result

$$n\zeta = 0.110 \tag{56.31}$$

56.3. FORCED VIBRATION WITH HARMONIC EXCITATION

We consider here a viscously damped spring-mass system excited by a harmonic force, as shown in Fig. 56.6. The differential equation of motion can then be written as

$$m\ddot{x} + c\dot{x} + kx = F_0 \sin \omega t \tag{56.32}$$

where ω is the frequency of the harmonic excitation.

The solution for this case can be considered in two parts. There will be some damped free vibration at its natural frequency, which will be the solution of the

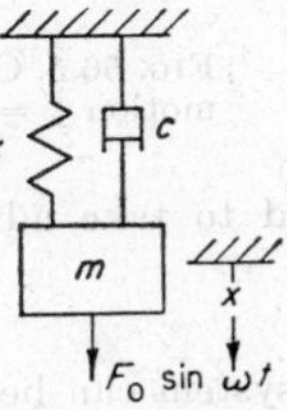

FIG. 56.6. Damped spring-mass system harmonically excited.

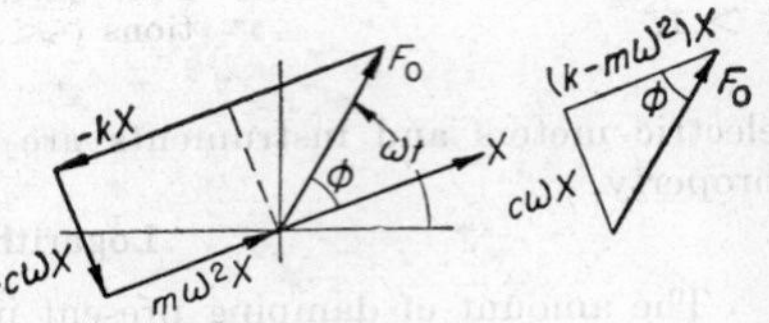

FIG. 56.7. Vector relationship of Eq. (56.34).

homogeneous equation and which has already been discussed in the previous section, and a particular solution, which is a steady oscillation at the frequency of the exciting force. The free vibration is introduced as a transient by the condition at $t = 0$, and will gradually disappear because of damping. The steady-state oscillation at the excitation frequency will be harmonic, with the displacement lagging the force vector by some phase angle ϕ. We can therefore assume the particular solution to be

$$x = X \sin (\omega t - \phi) \tag{56.33}$$

The amplitude X and the phase angle ϕ are found by substitution into the differential equation:

$$m\omega^2 X \sin (\omega t - \phi) - c\omega X \sin \left(\omega t - \phi + \frac{\pi}{2}\right) - kX \sin (\omega t - \phi) + F_0 \sin \omega t = 0 \tag{56.34}$$

The individual terms of this equation can be represented graphically as in Fig. 56.7, which clearly shows the phase relationship of each force with respect to each other. From this diagram, it is also possible to deduce the equations for the amplitude and phase angle, which are

$$X = \frac{F_0}{\sqrt{(k - m\omega^2)^2 + (c\omega)^2}} \qquad \tan \phi = \frac{c\omega}{k - m\omega^2} \tag{56.35}$$

Thus, the solution to the forced-vibration problem including the free vibration is

$$x = Xe^{-\zeta\omega_n t} \sin (\sqrt{1 - \zeta^2}\, \omega_n t - \phi_1) + \frac{F_0 \sin (\omega t - \phi)}{\sqrt{(k - m\omega^2)^2 + (c\omega)^2}} \tag{56.36}$$

For convenience of presentation and discussion, the steady-state solution is generally reduced to nondimensional form, which can be plotted as in Fig. 56.8 (response

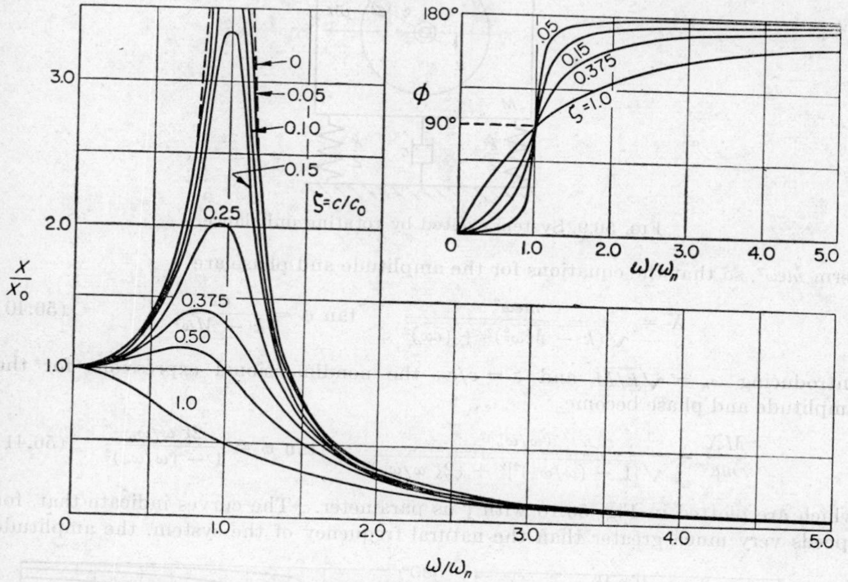

FIG. 56.8. Amplitude and phase [Eq. (56.37)] for excitation $F_0 \sin \omega t$.

curve). These equations are

$$\frac{X}{X_0} = \frac{1}{\sqrt{[1 - (\omega/\omega_n)^2]^2 + (2\zeta\, \omega/\omega_n)^2}} \qquad \tan \phi = \frac{2\zeta\, \omega/\omega_n}{1 - (\omega/\omega_n)^2} \tag{56.37}$$

where $X_0 = F_0/k$ = zero frequency deflection
$\omega_n = \sqrt{k/m}$ = natural frequency of undamped system
$\zeta = c/c_c$ = damping factor

At resonance, i.e., when the system is excited at its natural frequency ($\omega = \omega_n$), Eqs. (56.37) yield $x/x_0 = 1/2\zeta$ and $\phi = \pi/2$.

Rotating Unbalance

The unbalance of rotating machinery is a very common source of vibration excitation. Referring to Fig. 56.9, let me be the product of the unbalanced mass and eccentricity of a machine part rotating with angular speed ω. The displacement of m is

$$x + e \sin \omega t \tag{56.38}$$

where x is the displacement of the nonrotating mass $M - m$. Applying Newton's second law, we now have

$$(M - m)\ddot{x} + m \frac{d^2}{dt^2} (x + e \sin \omega t) = -kx - c\dot{x}$$

which rearranges to

$$M\ddot{x} + c\dot{x} + kx = (me\omega^2) \sin \omega t \tag{56.39}$$

It is evident then that the excitation force F of the previous article is replaced by the

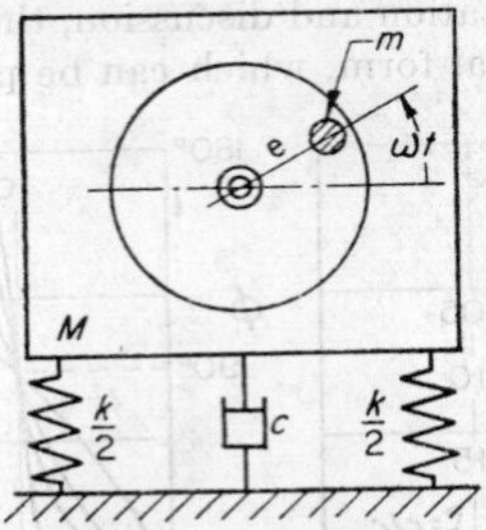

FIG. 56.9. System excited by rotating unbalance.

term $me\omega^2$, so that the equations for the amplitude and phase are

$$X = \frac{me\omega^2}{\sqrt{(k - M\omega^2)^2 + (c\omega)^2}} \qquad \tan \phi = \frac{c\omega}{k - M\omega^2} \tag{56.40}$$

Introducing $\omega_n = \sqrt{k/M}$ and $\zeta = c/c_c$, the nondimensional expressions for the amplitude and phase become

$$\frac{MX}{me} = \frac{(\omega/\omega_n)^2}{\sqrt{[1 - (\omega/\omega_n)^2]^2 + (2\zeta\, \omega/\omega_n)^2}} \qquad \tan \phi = \frac{2\zeta\, \omega/\omega_n}{1 - (\omega/\omega_n)^2} \tag{56.41}$$

which are plotted in Fig. 56.10 with ζ as parameter. The curves indicate that, for speeds very much greater than the natural frequency of the system, the amplitude

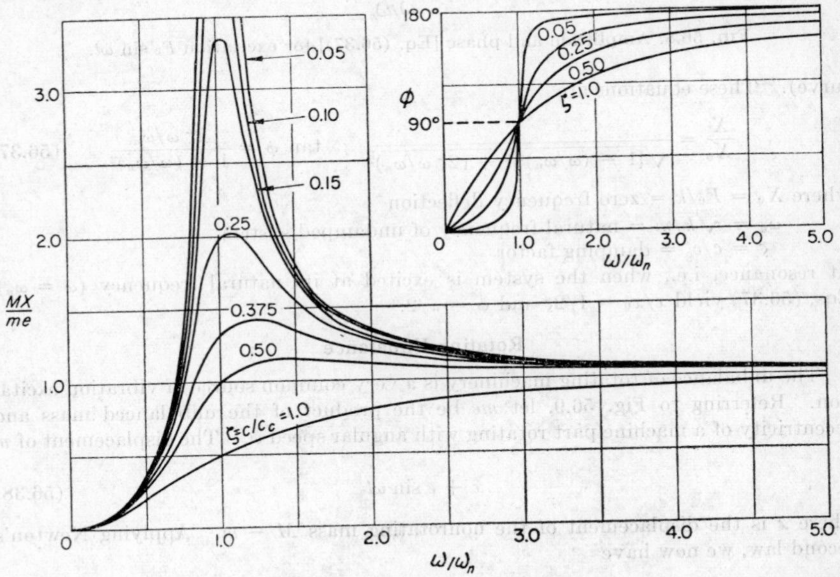

FIG. 56.10. Amplitude and phase [Eq. (56.41)] for rotating unbalance, or relative base motion [Eq. (56.52)].

becomes $X = (m/M)e$; thus, for a given unbalance me, X can be kept small by a large mass M.

Displacement Excitation

A vibratory system is sometimes excited by a prescribed motion of some point in the system. If we let y be the motion of the support point of the system of Fig. 56.11,

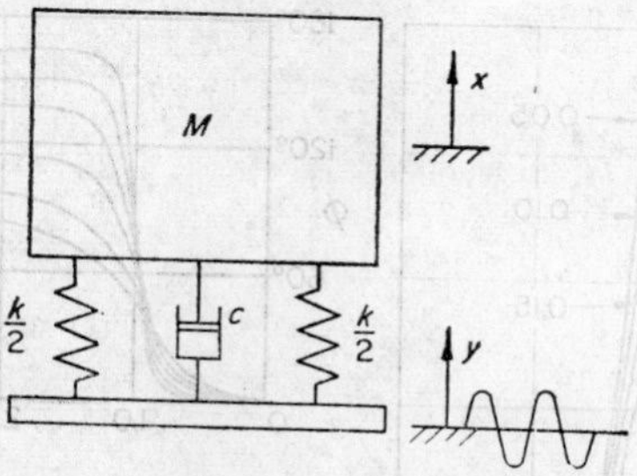

FIG. 56.11. System excited by base motion.

the differential equation becomes

$$M\ddot{x} = -k(x - y) - c(\dot{x} - \dot{y})$$

which may be rearranged to

$$M\ddot{x} + c\dot{x} + kx = ky + c\dot{y} \tag{56.42}$$

We will introduce now the method of complex algebra, which often simplifies the task of arriving at the steady-state solution. This procedure recognizes the fact that, in the forced-vibration problems where the impressed excitation is harmonic, the steady-state solution is also harmonic, of the same frequency but differing in phase. Thus, the excitation and amplitude are rotating vectors, differing in phase by the angle ϕ and expressible by the exponential forms

$$y = Ye^{i\omega t} \qquad x = Xe^{i(\omega t - \phi)} = Xe^{-i\phi}e^{i\omega t} \tag{56.43}$$

Substituting these into the differential equation, we obtain

$$(-m\omega^2 + i\omega c + k)Xe^{-i\phi} = (k + i\omega c)Y \tag{56.44}$$

from which the amplitude ratio is

$$\frac{Xe^{-i\phi}}{Y} = \frac{k + i\omega c}{(k - m\omega^2) + i\omega c} \tag{56.45}$$

The absolute value of the amplitude ratio is then

$$\left|\frac{X}{Y}\right| = \sqrt{\frac{k^2 + (c\omega)^2}{(k - m\omega^2)^2 + (c\omega)^2}} = \sqrt{\frac{1 + (2\zeta\,\omega/\omega_n)^2}{[1 - (\omega/\omega_n)^2]^2 + (2\zeta\,\omega/\omega_n)^2}} \tag{56.46}$$

To find the phase angle ϕ, we put $e^{-i\phi} = \cos\phi - i\sin\phi$, and equate the real and imaginary parts of Eq. (56.44) to determine $\sin\phi$ and $\cos\phi$. The ratio then results in the equation for the phase angle, which is

$$\tan\phi = \frac{mc\omega^3}{k^2[1 - (\omega/\omega_n)^2] + (c\omega)^2} = \frac{2\zeta(\omega/\omega_n)^3}{1 - (\omega/\omega_n)^2 + (2\zeta\,\omega/\omega_n)^2} \tag{56.47}$$

Equations (56.46) and (56.47) are plotted in Fig. 56.12.

Vibration Isolation

Vibratory forces generated by machines and engines are often unavoidable; however, they can be reduced substantially by properly designed springs, which we can refer to as *isolators.*

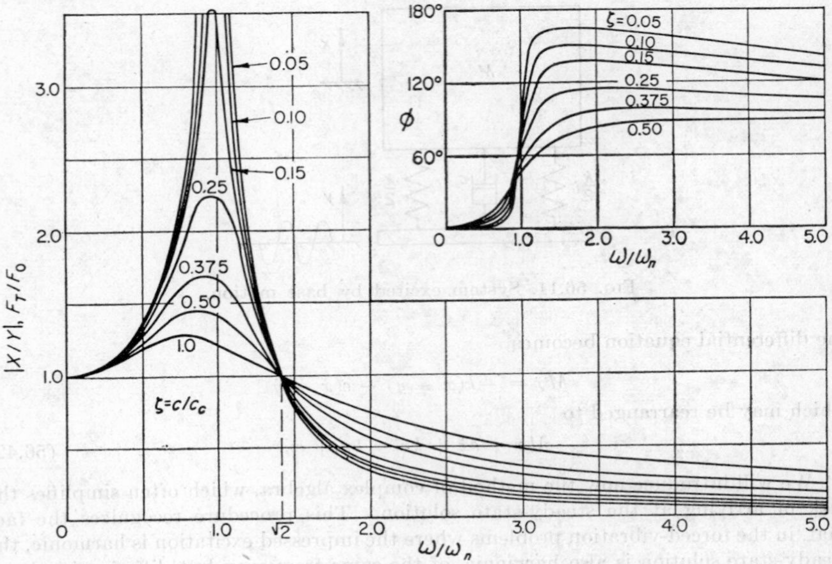

Fig. 56.12. Amplitude [Eq. (56.46)] and phase [Eq. (56.47)] for base excitation, or force transmissibility [Eq. (56.49)].

In Fig. 56.13, let F be the excitation source to be isolated by the spring k. The transmitted force through the spring and damper is

$$F_T = \sqrt{(kX)^2 + (c\omega X)^2} = kX\sqrt{1 + \left(\frac{c\omega}{k}\right)^2} \tag{56.48}$$

Since the amplitude X developed under the force $F = F_0 \sin \omega t$ is given by Eq. (56.35), the above equation reduces to

$$F_T = \frac{F_0\sqrt{1 + (c\omega/k)^2}}{\sqrt{[1 - m\omega^2/k]^2 + (c\omega/k)^2}} = \frac{F_0\sqrt{1 + (2\zeta\,\omega/\omega_n)^2}}{\sqrt{[1 - (\omega/\omega_n)^2]^2 + (2\zeta\,\omega/\omega_n)^2}} \tag{56.49}$$

Comparison of Eqs. (56.46) and (56.49) indicates that the ratio of the transmitted force to the exciting force is identical with the ratio of the transmitted displacement to the exciting displacement. Each of these ratios is referred to as *transmissibility*, and is plotted in Fig. 56.12 for various values of damping ζ. These curves show that the transmissibility is less than unity only for $\omega/\omega_n > \sqrt{2}$, thereby establishing the fact that vibration isolation is possible only for $\omega/\omega_n > \sqrt{2}$. The results also indicate that an undamped spring is superior to a damped spring in reducing the transmissibility; however, some damping may be desirable when it is necessary for ω to pass through the resonant region.

Figure 56.14 gives the percentage of isolation possible at any excitation frequency and statical deflection for zero damping. The curves are obtained from the undamped-transmissibility equation

$$T = \frac{1}{(\omega/\omega_n)^2 - 1} = \frac{1}{(2\pi f)^2\,\delta/g - 1} \tag{56.50}$$

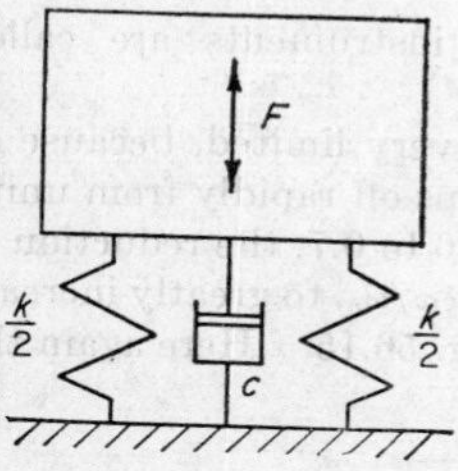

FIG. 56.13. Vibration isolation.

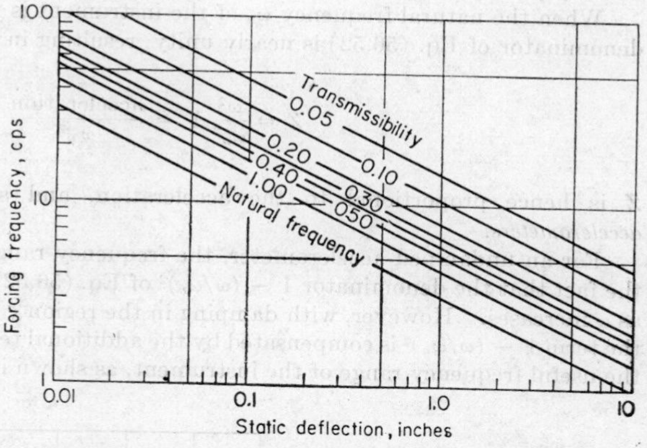

FIG. 56.14. Undamped transmissibility [Eq. (56.50)].

Relative Motion

In many cases we are interested only in the relative motion of the mass with respect to the supporting base of the spring and damper. In such cases, we again start from the differential equation (56.42) where $y = Y \sin \omega t$ is the motion of the base.

Introducing the relative displacement $z = Z \sin \omega t = x - y$, Eq. (56.42) can be written as

$$M\ddot{z} + c\dot{z} + kz = M\omega^2 Y \sin \omega t \tag{56.51}$$

which is similar to Eq. (56.39), with $M\omega^2 Y$ replacing $me\omega^2$. The steady-state solution can hence be written as

$$\frac{Z}{Y} = \frac{(\omega/\omega_n)^2}{\sqrt{[1 - (\omega/\omega_n)^2]^2 + (2\zeta\,\omega/\omega_n)^2}} \qquad \tan \phi = \frac{2\zeta\omega/\omega_n}{1 - (\omega/\omega_n)^2} \tag{56.52}$$

and the curves of Fig. 56.10 are applicable by replacing MX/me by Z/Y.

Theory of Seismic Instruments

The spring-supported mass of Fig. 56.11 forms the basic element of all seismic instruments. The motion to be measured is established from the relative motion between the mass and the base, and Eq. (56.52) is applicable. The characteristics of the instrument are determined mainly by the natural frequency of the seismic mass, which may be low or high compared to the frequency of the vibration to be measured.

When the natural frequency of the instrument is low, $\omega/\omega_n \gg 1$, and Eq. (56.52) reduces to $Z = Y$. Thus, the displacement Y to be measured is equal to the relative displacement of the seismic mass, which may be recorded by a stylus on a rotating drum. It is also obvious here that Y to be measured is limited by the size of the instrument.

Instead of recording the relative displacement Z mechanically, the motion may be converted into electric voltage by means of a coil of wire attached to the seismic mass, and moving relative to a magnet fixed to the base or vice versa. Since the voltage generated in a coil is proportional to the rate of cutting of the magnetic field, such an instrument will measure velocity. Accordingly, such instruments are referred to as velocity pickups.

When the natural frequency ω_n of the instrument is very high, $\omega/\omega_n \ll 1$, and the denominator of Eq. (56.52) is nearly unity, resulting in the approximate relation

$$Z = \frac{\omega^2}{\omega_n{}^2} Y = \frac{\text{acceleration}}{\omega_n{}^2} \tag{56.53}$$

Z is hence proportional to the acceleration, and such instruments are called *accelerometers*.

For an undamped accelerometer, the frequency range is very limited, because of the fact that the denominator $1 - (\omega/\omega_n)^2$ of Eq. (56.52) drops off rapidly from unity as ω increases. However, with damping in the region $\zeta = 0.6$ to 0.7, the reduction in the term $1 - (\omega/\omega_n)^2$ is compensated by the additional term $2\zeta\omega/\omega_n$, to greatly increase the useful frequency range of the instrument, as shown in Fig. 56.15. Here again the

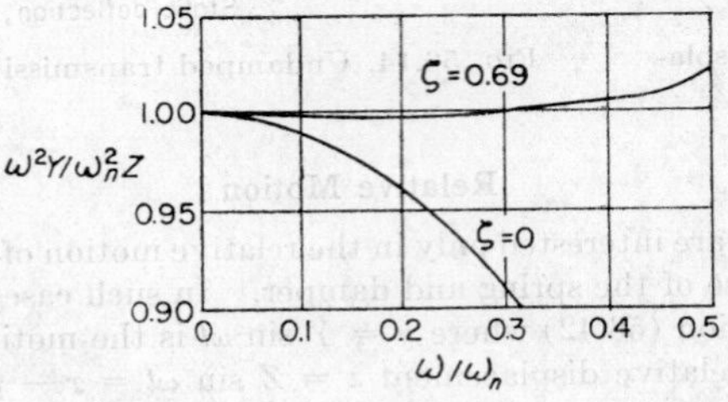

FIG. 56.15. Effect of damping on useful range of accelerometer.

relative motion Z may be converted to electric voltage by means of a coil-magnet arrangement mentioned previously.

Phase Distortion in Instruments

In order that a complex wave be reproduced without a change in shape, it is necessary that the phase of all harmonic components be shifted equally along the time axis. This can be accomplished if the phase angle ϕ increases linearly with frequency. For instance, if $\phi = (\pi/2)\ \omega/\omega_n$, which is very nearly satisfied when $\zeta = 0.70$, a complex acceleration wave of frequency ω_1 and ω_2

$$\ddot{y} = -\omega_1{}^2 Y_1 \sin \omega_1 t - \omega_2{}^2 Y_2 \sin \omega_2 t$$

will result in the following relative displacement,

$$z = \left(\frac{\omega_1}{\omega_n}\right)^2 Y_1 \sin \omega_1 \left(t - \frac{\pi}{2\omega_n}\right) + \left(\frac{\omega_2}{\omega_n}\right)^2 Y_2 \sin \omega_2 \left(t - \frac{\pi}{2\omega_n}\right)$$

Since both harmonic components are shifted along the time axis by the same amount $\pi/2\omega_n$, the original waveshape of acceleration is retained, and no phase distortion results.

Energy Dissipated by Damping

Consider the general case of a harmonic displacement lagging the force by an angle ϕ as in a forced vibration. Letting these two quantities be expressed by the equations

$$F = F_0 \sin \omega t \qquad x = X \sin (\omega t - \phi) \tag{56.54}$$

the work done in a cycle of motion is

$$\begin{aligned} W &= \int F\, dx = \int F \frac{dx}{dt}\, dt \\ &= \omega F_0 X \int_0^{2\pi/\omega} \sin \omega t \cos (\omega t - \phi)\, dt \\ &= \pi F_0 X \sin \phi \end{aligned} \tag{56.55}$$

It is evident from this equation that, for a given amplitude, the maximum energy dissipation takes place when $\phi = 90°$. In forced vibration, this is the condition at resonance. Substituting $\phi = 90°$, and F_0 from the resonant amplitude

$$X = \frac{F_0}{2\zeta k} \tag{56.56}$$

we arrive at the work done per cycle at resonance to be

$$W = 2\zeta\pi k X^2 = \pi c \omega X^2 \tag{56.57}$$

The energy dissipated per cycle by the damping force can be represented graphically as follows. Letting the displacement be given by Eq. (56.54), the velocity is

$$\dot{x} = \omega X \cos (\omega t - \phi) = \pm \omega \sqrt{X^2 - x^2}$$

and the viscous damping force is represented by the ellipse

$$F_d = c\dot{x} = \pm c\omega \sqrt{X^2 - x^2}$$

The work done per cycle is then the area enclosed by the ellipse shown in Fig. 56.16. The diagram also shows the total force of the spring and the damper.

$$F_s + F_d = kx \pm c\omega \sqrt{X^2 - x^2} \tag{56.58}$$

Solid Damping

In the vibration of any elastic element, a damping force independent of frequency and proportional to the amplitude of vibration is always encountered. Thus, in the cyclic variation of load on most metals, the stress-strain curve will enclose a finite area, which is shown in Fig. 56.17 and is referred to as *hysteresis loop*. The area

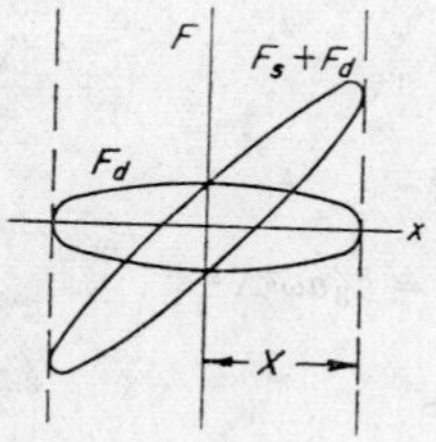

FIG. 56.16. Energy dissipated by viscous damping.

FIG. 56.17. Hysteresis loop for solid damping.

enclosed by the hysteresis curve is proportional to the stress amplitude and represents energy dissipated per cycle.

To simulate the damping from this source of energy dissipation, a damping force proportional to the elastic spring force but out of phase with the velocity is introduced in vibration. Thus, in vibration analysis, the solid damping force is approximated by the vector equation

$$\mathbf{F}_d = -i\gamma k\mathbf{x} \tag{56.59}$$

where γ is the solid damping factor which, in general, is less than 0.05 for most structures. Letting $x = X \cos(\omega t - \phi)$ be the real part of the vector $\mathbf{x} = Xe^{i(\omega t - \phi)}$, the damping force F_d is found from the real part of $\mathbf{F}_d$ to be

$$F_d = \gamma kX \sin(\omega t - \phi) = \pm\gamma k\sqrt{X^2 - x^2} \tag{56.60}$$

and the same type of ellipse is obtained as in the viscous case where γk replaces $c\omega$.

Equivalent Viscous Damping

In an actual vibratory system, many different types of damping may be encountered simultaneously. The differential equation including all these damping forces will be complicated and, in general, beyond the possibility of solution.

As shown in Fig. 56.8, damping is of importance in forced vibration only at resonance when the amplitude, assuming viscous damping, is

$$X = \frac{F_0}{c\omega_n} \tag{56.61}$$

For other types of damping, the resonant amplitude can be estimated with good accuracy by using in the above equation the equivalent viscous damping, determined by equating the total energy dissipated W to that of viscous damping as follows.

$$\pi c_{eq}\omega X^2 = 2\zeta\pi kX^2 = W \qquad c_{eq} = \frac{W}{\pi\omega X^2} \qquad \zeta_{eq} = \frac{W}{2\pi kX^2} \tag{56.62}$$

The dissipated energy W for different types of damping will, in general, be some function of the amplitude and frequency. Thus, for solid damping $W = \pi\gamma kX^2$, and the equivalent viscous damping is

$$\zeta_{eq} = \frac{\gamma}{2} \tag{56.63}$$

As another example, consider a damping force proportional to the square of the velocity

$$F_d = \pm a\dot{x}^2 \tag{56.64}$$

The energy dissipated per cycle is then

$$W = 2\int_{-X}^{X} a\dot{x}^2\,dx$$

Letting $x = -X \cos \omega t$, this integral is evaluated as

$$W = 2a\omega^2X^3\int_0^{\pi} \sin^3 \omega t\, d(\omega t) = \tfrac{8}{3}a\omega^2X^3 \tag{56.65}$$

from which the equivalent viscous damping becomes

$$c_{eq} = \frac{8}{3\pi}a\omega X \tag{56.66}$$

Upon substitution into Eq. (56.61), the resonant amplitude is found as

$$X = \sqrt{\frac{3\pi F_0}{8a\omega_n^2}} \tag{56.67}$$

Sharpness of Resonance

In forced vibration there is a quantity Q related to damping which is a measure of the sharpness of resonance. To determine its equation, we will assume viscous damping, and start with Eq. (56.37).

When $\omega/\omega_n = 1$, the resonant amplitude $X/X_0 = 1/2\zeta$, as shown in Fig. 56.18. We now seek the two frequencies on either side of resonance (often referred to as

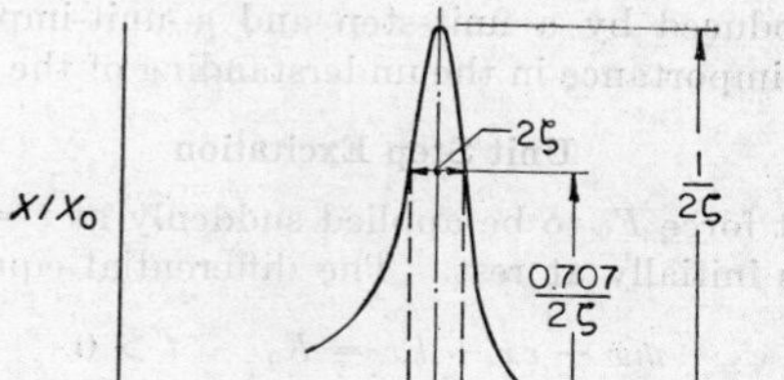

FIG. 56.18. Width of resonance curve at half-power points.

sidebands), where X/X_0 is $\frac{1}{2}\sqrt{2}\,(1/2\zeta)$. These points are also referred to as the *half-power points*.

Substituting into Eq. (56.37), we obtain

$$\frac{1}{2}\left(\frac{1}{2\zeta}\right)^2 = \frac{1}{[1-(\omega/\omega_n)^2]^2 + (2\zeta\,\omega/\omega_n)^2}$$

which yields the equation

$$\left(\frac{\omega}{\omega_n}\right)^4 - 2(1-2\zeta^2)\left(\frac{\omega}{\omega_n}\right)^2 + (1-8\zeta^2) = 0$$

Solving, we obtain

$$\left(\frac{\omega}{\omega_n}\right)^2 = (1-2\zeta^2) \pm 2\zeta\sqrt{1+\zeta^2}$$

If we now assume that $\zeta \ll 1$, and neglect higher-order terms, we arrive at the result

$$\left(\frac{\omega}{\omega_n}\right)^2 = 1 \pm 2\zeta$$

Letting the two frequencies corresponding to the roots of this equation be ω_2 and ω_1, we obtain

$$4\zeta = \frac{\omega_2^2 - \omega_1^2}{\omega_n^2} \cong 2\,\frac{\omega_2 - \omega_1}{\omega_n}$$

The quantity Q is then defined as

$$Q = \frac{\omega_n}{\omega_2 - \omega_1} = \frac{f_n}{f_2 - f_1} = \frac{1}{2\zeta} \tag{56.68}$$

Here again equivalent damping can be used to define Q for systems with other forms of damping. Thus, for solid damping, Q is equal to

$$Q = \frac{f_n}{f_2 - f_1} = \frac{1}{\gamma} \tag{56.69}$$

56.4. FORCED VIBRATION WITH NONPERIODIC EXCITATION

In this section, we will consider the response of a spring-mass system to a suddenly applied nonperiodic excitation. The responses to such excitation are called *transients*, since steady-state oscillations are generally not produced. The excitation may be a force, displacement, velocity, or an acceleration which is prescribed as a function of time.

Before discussing the case of arbitrary excitation, we will consider two special types of response produced by a unit-step and a unit-impulse excitation. These two basic cases are of importance in the understanding of the more general case.

Unit Step Excitation

Assume a constant force F_0 to be applied suddenly at $t = 0$ to a damped spring-mass system which is initially at rest. The differential equation of motion is then

$$m\ddot{x} + c\dot{x} + kx = F_0 \qquad t > 0 \tag{56.70}$$

The solution will here be the sum of the solution to the homogeneous equation and a particular solution which, in this case, is F_0/k. We have already discussed the homogeneous equation under free vibration, and have found that, for the oscillatory case, the solution is given by Eq. (56.23). Thus the complete solution is

$$x = \frac{F_0}{k} + Xe^{-\zeta\omega_n t} \sin\left(\sqrt{1 - \zeta^2}\,\omega_n t + \phi\right) \tag{56.71}$$

It must satisfy the initial conditions $x = \dot{x} = 0$ at $t = 0$, which lead to

$$\tan\phi = \frac{\sqrt{1 - \zeta^2}}{\zeta} \qquad X = -\frac{F_0}{k\sqrt{1 - \zeta^2}}$$

and, hence,

$$x = \frac{F_0}{k}\left[1 - \frac{e^{-\zeta\omega_n t}}{\sqrt{1 - \zeta^2}} \sin\left(\sqrt{1 - \zeta^2}\,\omega_n t + \phi\right)\right] \tag{56.72}$$

An alternative form is

$$x = \frac{F_0}{k}\left[1 - \frac{e^{-\zeta\omega_n t}}{\sqrt{1 - \zeta^2}} \cos\left(\sqrt{1 - \zeta^2}\,\omega_n t - \phi'\right)\right] \tag{56.73}$$

with $\phi + \phi' = \pi/2$. A plot of this equation is given in Fig. 56.19 for various values of ζ.

The velocity is

$$\dot{x} = \frac{F_0}{\sqrt{km(1 - \zeta^2)}} e^{-\zeta\omega_n t} \sin \sqrt{1 - \zeta^2}\,\omega_n t \tag{56.74}$$

and, by equating $\dot{x}$ to zero, the time corresponding to the peak amplitude is found to be

$$\omega_n t_p = \frac{\pi}{\sqrt{1 - \zeta^2}} \tag{56.75}$$

Substituting this quantity into Eq. (56.72), its maximum value is

$$\left(\frac{xk}{F_0}\right)_{\max} = n = 1 + \exp\left(-\frac{\zeta\pi}{\sqrt{1 - \zeta^2}}\right) \tag{56.76}$$

which is also plotted in Fig. 56.19 as a function of ζ. Since $F_0/k = \delta$ is the static displacement, the quotient n in Eq. (56.76) is called the dynamic response factor.

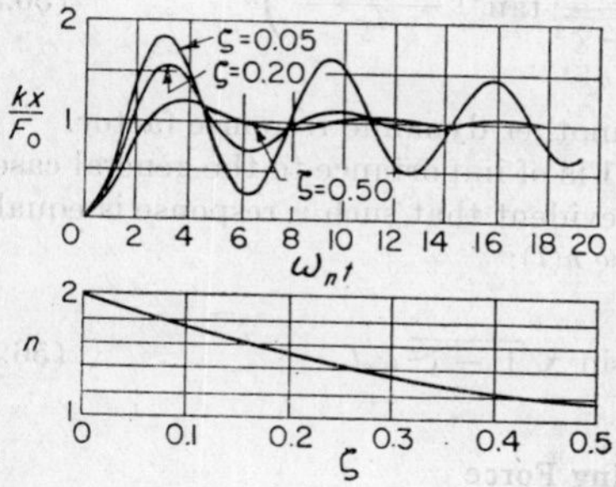

FIG. 56.19. Response [Eq. (56.73)] and dynamic response factor [Eq. (56.76)] for step excitation.

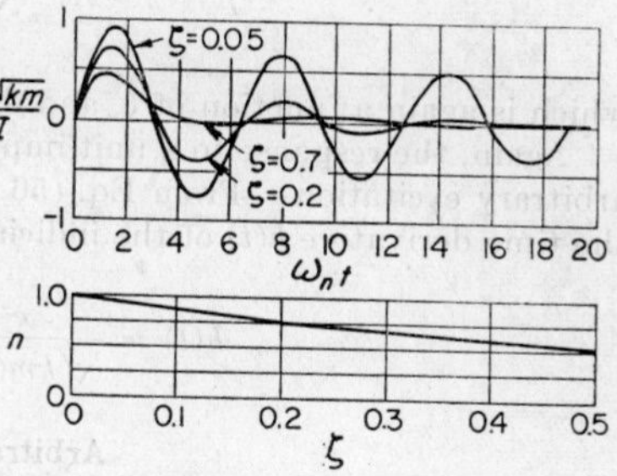

FIG. 56.20. Response [Eq. (56.80)] and dynamic response factor [Eq. (56.82)] for impulse excitation.

The response to a unit step force $F_0 = 1$ is of special importance and is referred to as the *indicial response* $h(t)$.

$$h(t) = \frac{1}{k}\left[1 - \frac{e^{-\zeta\omega_n t}}{\sqrt{1-\zeta^2}}\cos\left(\sqrt{1-\zeta^2}\,\omega_n t - \phi'\right)\right] \tag{56.77}$$

We will have occasion to refer to this quantity later.

Unit Impulse

An impulse I is here defined as a very large force acting for a very short time such that the time integral of the force is I. In the limiting case, the impulsive force is

$$\lim \frac{I}{\Delta t} \to \infty \qquad \Delta t \to 0 \tag{56.78}$$

and we can ignore all other forces of nonimpulsive nature while I is acting.

The initial effect of such a limiting impulse is to produce an initial velocity equal to $v_0 = I/m$, with initial displacement of zero. Since the differential equation for $t > 0$ is the homogeneous equation with the right side equal to zero, the solution

$$x = Xe^{-\zeta\omega_n t}\sin\left(\sqrt{1-\zeta^2}\,\omega_n t - \phi\right)$$

must be fitted to $x = 0$ and $\dot{x} = I/m$ at $t = 0$. This yields

$$X = \frac{I}{m\omega_n\sqrt{1-\zeta^2}} \qquad \phi = 0 \tag{56.79}$$

The displacement at any time t is then

$$x = \frac{I}{\sqrt{km(1-\zeta^2)}}\,e^{-\zeta\omega_n t}\sin\sqrt{1-\zeta^2}\,\omega_n t \tag{56.80}$$

which is plotted in Fig. 56.20 with ζ as parameter.

The time corresponding to the peak displacement is found from the equation

$$\tan\sqrt{1-\zeta^2}\,\omega_n t_p = \frac{\sqrt{1-\zeta^2}}{\zeta} \tag{56.81}$$

and the equation for the peak displacement becomes

$$\frac{x_{\max}\sqrt{km}}{I} = \exp\left(-\frac{\zeta}{\sqrt{1-\zeta^2}}\tan^{-1}\frac{\sqrt{1-\zeta^2}}{\zeta}\right) \tag{56.82}$$

which is again a function of ζ, and describes another dynamic response factor.

Again, the response to a unit impulse $I = 1$ is of importance to the general case of arbitrary excitation. From Eq. (56.74), it is evident that such a response is equal to the time derivative $\dot{h}(t)$ of the indicial response $h(t)$:

$$\dot{h}(t) = \frac{e^{-\zeta\omega_n t}}{\sqrt{km(1-\zeta^2)}} \sin\sqrt{1-\zeta^2}\,\omega_n t \tag{56.83}$$

Arbitrary Exciting Force

We consider next the damped spring-mass system excited by a force which is some arbitrary function of time, as shown in Fig. 56.21. The increment of force applied in

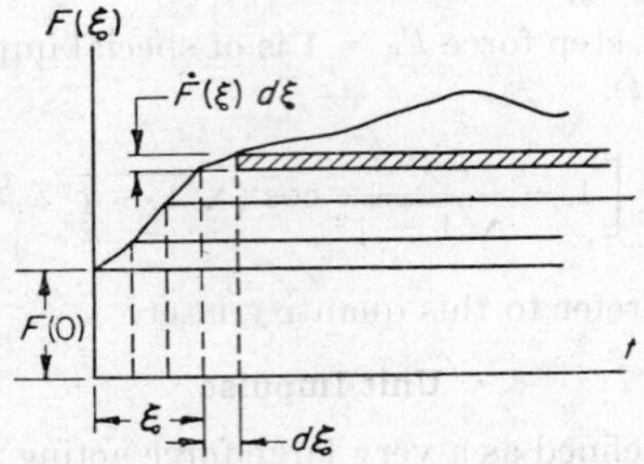

FIG. 56.21. Arbitrary excitation as series of step functions.

any time interval $d\xi$ is a step function force of magnitude $\dot{F}\,d\xi$, to which the solution (56.77) is applicable. Since this force is applied at time ξ, it produces, at time $t > \xi$, the displacement

$$dx = \dot{F}(\xi)h(t-\xi)\,d\xi$$

Summing all such contributions, the response at t is found to be

$$x(t) = F(0)h(t) + \int_0^t \dot{F}(\xi)h(t-\xi)\,d\xi \tag{56.84}$$

This equation is known as *Duhamel's integral*, and it gives the response of a system to any arbitrary excitation when the indicial response $h(t)$ is known. Another form of this equation is obtained by integrating the last term by parts:

$$x(t) = F(t)h(0) + \int_0^t F(\xi)\dot{h}(t-\xi)\,d\xi \tag{56.85}$$

For any mechanical system with mass, $h(0)$ is zero, and the first term of Eq. (56.85) should be deleted.

Rectangular Pulse Input

The rectangular pulse input of Fig. 56.22 can be considered to be the superposition of two step functions, the negative step function being delayed by t_0. The solution can thus be obtained by the superposition of Eq. (56.73) as follows:

$$x = \frac{F_0}{k}\left[1 - \frac{e^{-\zeta\omega_n t}}{\sqrt{1-\zeta^2}} \cos\left(\sqrt{1-\zeta^2}\,\omega_n t - \phi'\right)\right] \qquad 0 < t < t_0 \tag{56.86a}$$

$$x = \frac{F_0}{k}\left(\left[1 - \frac{e^{-\zeta\omega_n t}}{\sqrt{1-\zeta^2}} \cos\left(\sqrt{1-\zeta^2}\,\omega_n t - \phi'\right)\right] - \left\{1 - \frac{e^{-\zeta\omega_n (t-t_0)}}{\sqrt{1-\zeta^2}} \cos\left[\sqrt{1-\zeta^2}\,\omega_n (t-t_0) - \phi'\right]\right\}\right) \qquad t > t_0 \tag{56.86b}$$

If the peak displacement occurs before the end of the pulse, the peak time and the peak amplitude are the same as those for the step function.

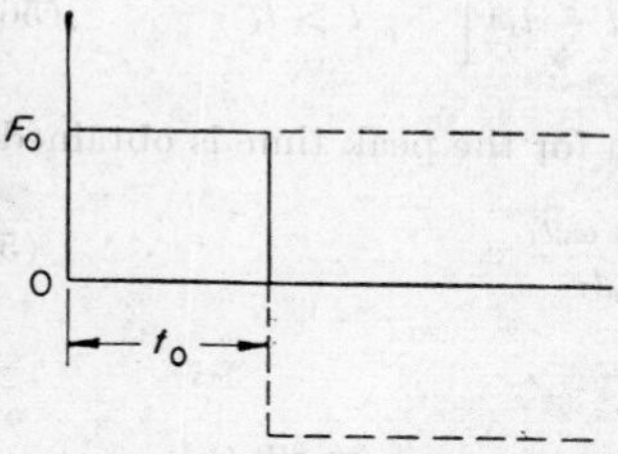

FIG. 56.22. Rectangular pulse as superposition of two step functions.

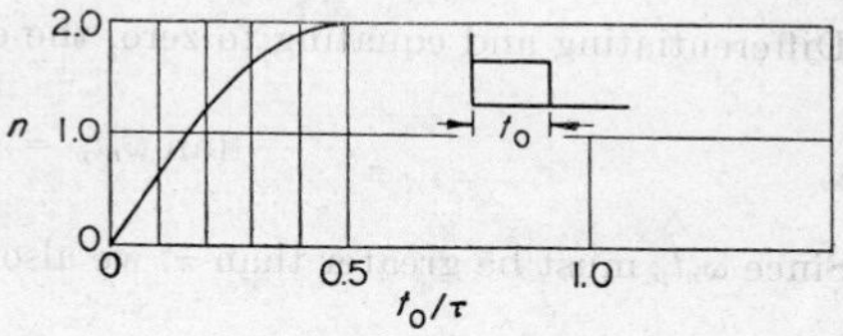

FIG. 56.23. Dynamic response factor for rectangular pulse [Eq. (56.89)].

If the peak displacement occurs when $t > t_0$, the velocity is given by

$$\dot{x} = \frac{F_0}{\sqrt{km(1-\zeta^2)}} e^{-\zeta\omega_n t} \left[\sin\sqrt{1-\zeta^2}\,\omega_n t - e^{\zeta\omega_n t_0} \sin\sqrt{1-\zeta^2}\,\omega_n (t-t_0)\right]$$

from which the equation for the peak time becomes

$$\tan\sqrt{1-\zeta^2}\,\omega_n t_p = \frac{-e^{\zeta\omega_n t_0}\sin\sqrt{1-\zeta^2}\,\omega_n t_0}{1 - e^{\zeta\omega_n t_0}\cos\sqrt{1-\zeta^2}\,\omega_n t_0} \tag{56.87}$$

Since $\omega_n t_0 = 2\pi t_0/\tau$, where τ is the natural period of the system, the peak time is a function of both t_0/τ and ζ.

When $t_0 \to 0$, the above equation reduces to Eq. (56.81) for the peak time of the limiting impulsive load.

For $\zeta = 0$, the peak time is given by

$$\tan\frac{2\pi t_p}{\tau} = \frac{-\sin(2\pi t_0/\tau)}{1 - \cos(2\pi t_0/\tau)} = -\cot\frac{\pi t_0}{\tau} \tag{56.88}$$

and the peak amplitude or the dynamic response factor becomes

$$n = \frac{x_p k}{F_0} = 2\sin\frac{\pi t_0}{\tau} \tag{56.89}$$

which is plotted in Fig. 56.23. For $t_0/\tau > 0.5$, the peak response will occur in the region $t < t_0$, and the dynamic response factor will be $n = 2.0$.

The Effect of Rise Time t_1

In general, there is a rise time associated with the applied load, as shown in Fig. 56.24. We use Duhamel's integral (56.84), and assume that $\zeta = 0$, in which case we will have, for $t < t_1$,

$$F(t) = \frac{F_0 t}{t_1} \qquad h(t) = \frac{1}{k}(1 - \cos \omega_n t)$$

$$x(t) = \frac{F_0}{kt_1}\int_0^t [1 - \cos \omega_n(t - \xi)]\, d\xi = \frac{F_0}{kt_1}\left(t - \frac{1}{\omega_n}\sin \omega_n t\right) \qquad t < t_1 \tag{56.90a}$$

We now superimpose on this the negative of the same solution delayed by t_1, to obtain the solution for $t > t_1$:

$$x(t) = \frac{F_0}{kt_1}\left[\left(t - \frac{1}{\omega_n}\sin \omega_n t\right) - \left((t - t_1) - \frac{1}{\omega_n}\sin \omega_n(t - t_1)\right)\right]$$

$$= \frac{F_0}{kt_1}\left[t_1 - \frac{1}{\omega_n}\sin \omega_n t + \frac{1}{\omega_n}\sin \omega_n(t - t_1)\right] \qquad t > t_1 \tag{56.90b}$$

Differentiating and equating to zero, the equation for the peak time is obtained as

$$\tan \omega_n t_p = \frac{1 - \cos \omega_n t_1}{\sin \omega_n t_1} \tag{56.91}$$

Since $\omega_n t_p$ must be greater than π, we also obtain

$$\sin \omega_n t_p = -\sqrt{\tfrac{1}{2}(1 - \cos \omega_n t_1)} \qquad \cos \omega_n t_p = \frac{-\sin \omega_n t_1}{\sqrt{2(1 - \cos \omega_n t_1)}}$$

Substituting these quantities into x, we obtain the maximum amplitude for $t > t_1$ to be

$$n = \left(\frac{xk}{F_0}\right)_{\max} = 1 + \frac{1}{2\pi\, t_1/\tau}\sqrt{2\left(1 - \cos 2\frac{\pi t_1}{\tau}\right)} \tag{56.92}$$

The dynamic response factor given by this equation is plotted in Fig. 56.25.

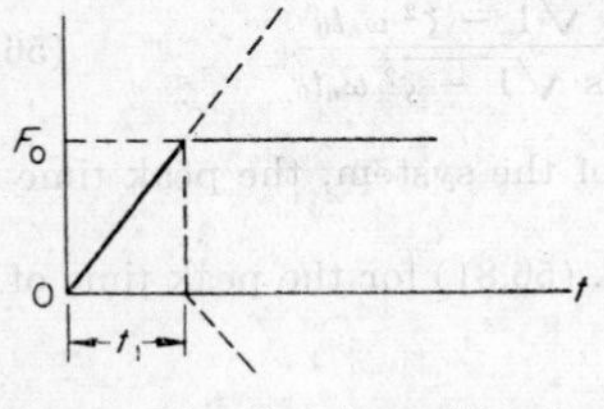

FIG. 56.24. Rise time in exciting force as superposition of two lines.

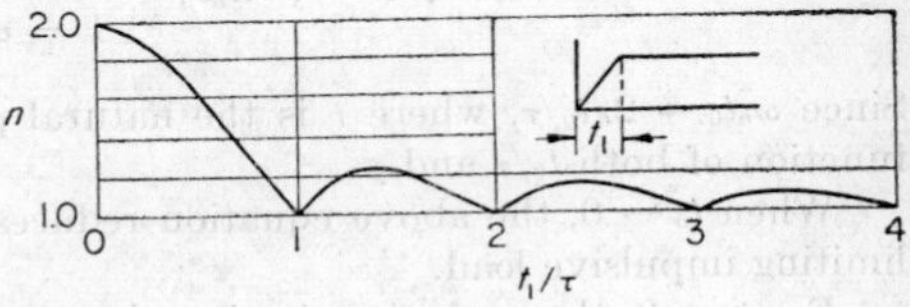

FIG. 56.25. Dynamic response factor for constant force with rise time [Eq. (56.92)].

Sinusoidal Pulse

In impact loading the force-time relationship is often approximated by the sinusoidal pulse. The duration of the pulse in relation to the natural period of the system is of importance here in determining the dynamic response factor.

In deriving the equations of motion, the sinusoidal pulse of Fig. 56.26 can be considered to be a superposition of two continuous sine waves, the second one starting at $t = t_1$ to cancel the excitation after the first pulse. Assuming an undamped system, the general solution, consisting of the homogeneous and the particular solutions, is

$$x = A \sin \omega_n t + B \cos \omega_n t + \frac{F_0 \sin (\pi t/t_1)}{m\omega_n^2[1 - (\pi/\omega_n t_1)^2]} \tag{56.93}$$

Introducing the initial conditions $x = \dot{x} = 0$ and replacing ω_n by $2\pi/\tau$, the solution to the continuous sine-wave excitation started at $t = 0$ becomes

$$x = \frac{F_0}{k(\tau/2t_1 - 2t_1/\tau)}\left(\sin\frac{2\pi t}{\tau} - \frac{2t_1}{\tau}\sin\frac{\pi t}{t_1}\right) \qquad t < t_1 \tag{56.94a}$$

The solution for $t > t_1$ can be obtained by adding to Eq. (56.94a) the same solution, with the time t replaced by $(t - t_1)$.

$$x = \frac{F_0}{k(\tau/2t_1 - 2t_1/\tau)}\left[\left(\sin 2\pi\frac{t}{\tau} - \frac{2t_1}{\tau}\sin\frac{\pi t}{t_1}\right) + \left(\sin 2\pi\frac{t - t_1}{\tau} - \frac{2t_1}{\tau}\sin\frac{\pi}{t_1}(t - t_1)\right)\right] \qquad t > t_1 \tag{56.94b}$$

In determining the peak response, it is necessary to assign a numerical value to the parameter t_1/τ and note whether the peak occurs in the region $t > t_1$ or $t < t_1$. It is evident that for small values of t_1/τ, the peak response occurs in the region $t > t_1$ and Eq. (56.94b) must be used. The dynamic response factor $n = kx_{max}/F_0$ will, in this case, increase with t_1/τ, as shown in the first part of Fig. 56.27.

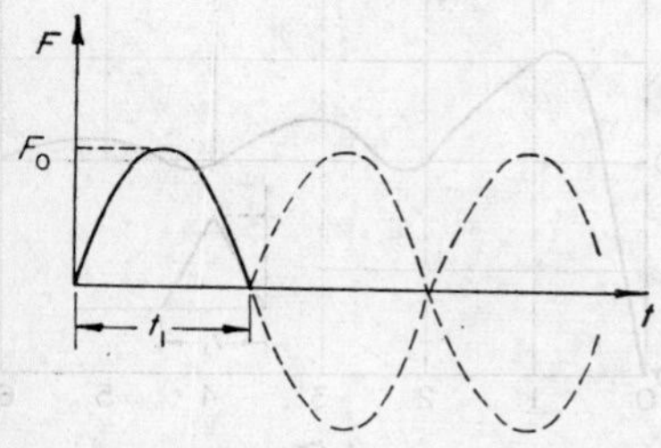

FIG. 56.26. Sinusoidal pulse as superposition of shifted sine waves.

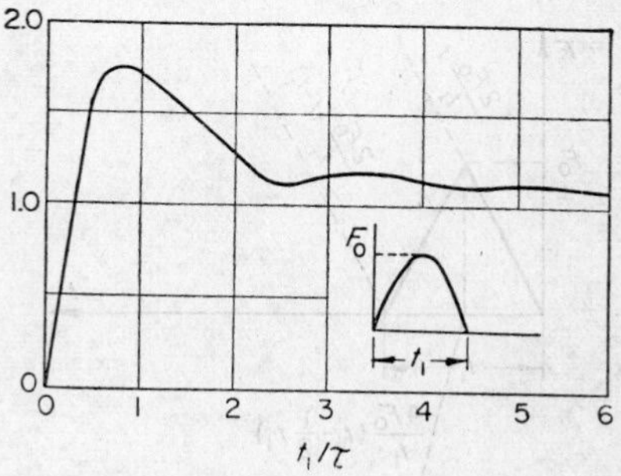

FIG. 56.27. Dynamic response factor for sinusoidal pulse.

As t_1/τ continues to increase, the peak response will occur in the region $t < t_1$, and Eq. (56.94a) must be used. Differentiating it and equating to zero, the time t_p corresponding to the peak response is given by solving the equation

$$\cos\frac{2\pi t_p}{\tau} = \cos\frac{\pi t_p}{t_1} \tag{56.95}$$

It is evident from this equation that the value of t_1/τ which produces a peak response at $t_p = t_1$ is $t_1/\tau = \frac{1}{2}$. Thus, for all $t_1/\tau > \frac{1}{2}$, the peak response will occur in the interval $t_p < t_1$, where Eq. (56.94a) must be used. As an example, when $t_1/\tau = 2.0$, $\pi t_p/t_1 = 72°$, and its substitution into Eq. (56.94a) results in $n = 1.27$.

Triangular Pulse

The triangular pulse of Fig. 56.28 is another approximation often used for simulating impact loading. For the derivation of the equation for response, the triangular pulse can be considered to be the superposition of three straight lines shown dotted in Fig. 56.28.

In the region $0 < t < \frac{1}{2}t_1$, the response is due to the excitation $2F_0t/t_1$, the equation for the displacement being

$$x = \frac{2F_0}{k}\left(\frac{t}{t_1} - \frac{\tau}{2\pi t_1}\sin 2\pi\frac{t}{\tau}\right) \qquad 0 < t < \tfrac{1}{2}t_1 \qquad (56.96a)$$

In the second region $\frac{1}{2}t_1 < t < t_1$, the response is due to the excitation $2F_0t/t_1$ started at $t = 0$, and a second excitation $-(4F_0/t_1)(t - \frac{1}{2}t_1)$ started at $t = \frac{1}{2}t_1$. Thus, by superimposing the above solution, we arrive at the equation

$$x = \frac{2F_0}{k}\left\{1 - \frac{t}{t_1} + \frac{\tau}{2\pi t_1}\left[2\sin\frac{2\pi}{\tau}(t - \tfrac{1}{2}t_1) - \sin 2\pi\frac{t}{\tau}\right]\right\} \qquad \tfrac{1}{2}t_1 < t < t_1 \qquad (56.96b)$$

In the third region $t > t_1$, we add to Eq. (56.96*b*) the response due to the excitation $(2F_0/t_1)(t - t_1)$ started at $t = t_1$, the result being

$$x = \frac{2F_0}{k}\left\{\frac{\tau}{2\pi t_1}\left[2\sin\frac{2\pi}{\tau}(t - \tfrac{1}{2}t_1) - \sin\frac{2\pi}{\tau}(t - t_1) - \sin\frac{2\pi}{\tau}t\right]\right\} \qquad t > t_1 \qquad (56.96c)$$

In all these equations, the variable t/τ can be written as $\frac{t_1}{\tau}\frac{t}{t_1}$, so that t_1/τ becomes the parameter governing the response. The dynamic response factor $n = kx_{\max}/F_0$ is

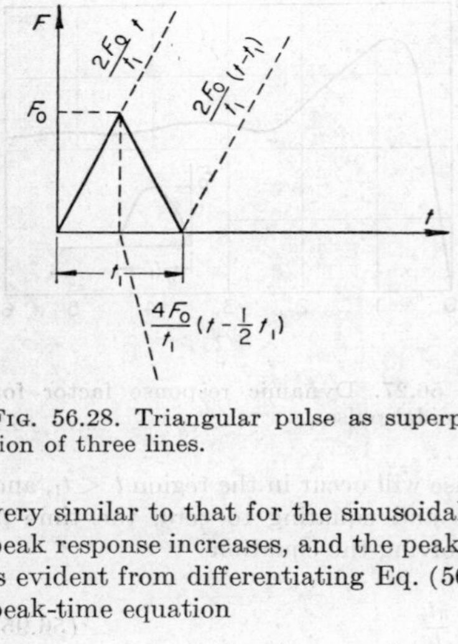

FIG. 56.28. Triangular pulse as superposition of three lines.

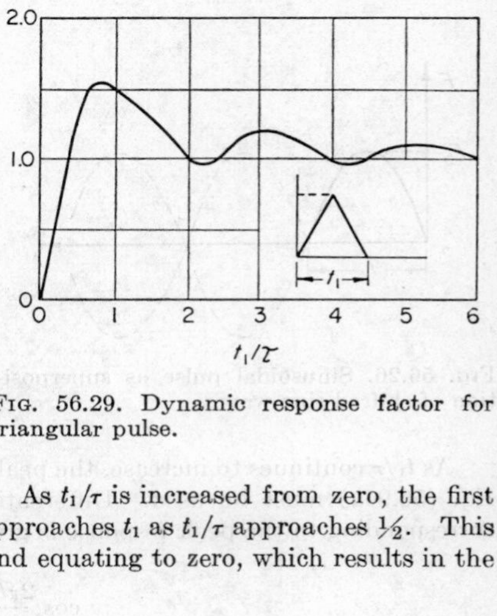

FIG. 56.29. Dynamic response factor for triangular pulse.

very similar to that for the sinusoidal pulse. As t_1/τ is increased from zero, the first peak response increases, and the peak time approaches t_1 as t_1/τ approaches $\frac{1}{2}$. This is evident from differentiating Eq. (56.96*c*) and equating to zero, which results in the peak-time equation

$$2\cos\frac{2\pi t_1}{\tau}\left(\frac{t_p}{t_1} - 0.5\right) - \cos 2\pi\frac{t_1}{\tau}\left(\frac{t_p}{t_1} - 1\right) - \cos\frac{2\pi t_1}{\tau}\frac{t_p}{t_1} = 0 \qquad (56.97a)$$

If we let $t_p/t_1 = 1.0$, the parameter t_1/τ becomes $\frac{1}{2}$, indicating that the first peak response occurs in the region $t_p > t_1$ for $t_1/\tau < \frac{1}{2}$.

To find the value of t_1/τ corresponding to a peak response occurring at $t_p = \frac{1}{2}t_1$, we can differentiate Eq. (56.96*a*) and obtain the equation

$$1 - \cos 2\pi\frac{t_1}{\tau}\frac{t_p}{t_1} = 0 \qquad (56.97b)$$

If t_p/t_1 is set equal to $\frac{1}{2}$, the parameter $t_1/\tau = 2$. Thus, for the range of values $\frac{1}{2} < t_1/\tau < 2$, the peak response occurs in the region $\frac{1}{2}t_1 < t < t_1$, and Eq. (56.96*b*) must be used.

For $t_1/\tau > 2$, the peak response will occur in the region $t < \frac{1}{2}t_1$, for which Eq. (56.96*a*) is applicable. By assigning values for $t_1/\tau > 2$, t_p/t_1 corresponding to the peak is found, which, substituted into Eq. (56.96*a*), results in the dynamic response factor n. n as a function of t_1/τ is plotted in Fig. 56.29 for each of the three regions.

Shock Spectrum

A shock spectrum is a plot of the maximum peak response of a single-degree-of-freedom oscillatory system to a specified shock excitation, as a function of the natural frequency of the oscillatory system. Thus, the curves for n versus t_1/τ of the previous articles constitute shock spectra for the specified excitations. The present article gives a general discussion of the shock spectrum for a single pulse excitation.

Referring to Duhamel's integral with the initial conditions $x(0) = \dot{x}(0) = 0$, the response to an arbitrary pulse excitation $F(t)$ is given by Duhamel's integral (56.85), where $h(t)$ from Eq. (56.83) is rewritten as

$$h(t) = \frac{\omega_n}{k\sqrt{1-\zeta^2}} e^{-\zeta\omega_n t} \sin \sqrt{1-\zeta^2}\,\omega_n t$$

Letting $F(\xi) = F_0 f(\xi)$, where F_0 is the peak value of the excitation, the equation for the dynamic response factor n becomes

$$n = \left(\frac{kx}{F_0}\right)_{\max} = \max_{0<t<\infty} \left[\frac{\omega_n}{\sqrt{1-\zeta^2}} \int_0^t f(\xi) e^{-\zeta\omega_n(t-\xi)} \sin \sqrt{1-\zeta^2}\,\omega_n(t-\xi)\,d\xi\right]$$

Factoring out terms involving t from the integral,

$$n = \max_{0<t<\infty} \frac{\omega_n e^{-\zeta\omega_n t}}{\sqrt{1-\zeta^2}} \left[\sin \sqrt{1-\zeta^2}\,\omega_n t \int_0^t f(\xi) e^{\zeta\omega_n\xi} \cos \sqrt{1-\zeta^2}\,\omega_n\xi\,d\xi \right.$$

$$\left. - \cos \sqrt{1-\zeta^2}\,\omega_n t \int_0^t f(\xi) e^{\zeta\omega_n\xi} \sin \sqrt{1-\zeta^2}\,\omega_n\xi\,d\xi\right] \qquad (56.98)$$

If the natural period τ of the system is large compared to the pulse duration t_1, the maximum peak response will occur in the region $t > t_1$, and the integrals of Eq. (56.98) will become constants. Making the following substitution,

$$A \cos \phi = \frac{\omega_n}{\sqrt{1-\zeta^2}} \int_0^{t_1} f(\xi) e^{\zeta\omega_n\xi} \cos \sqrt{1-\zeta^2}\,\omega_n\xi\,d\xi \qquad (56.99a)$$

$$A \sin \phi = \frac{\omega_n}{\sqrt{1-\zeta^2}} \int_0^{t_1} f(\xi) e^{\zeta\omega_n\xi} \sin \sqrt{1-\zeta^2}\,\omega_n\xi\,d\xi \qquad (56.99b)$$

the response for this condition becomes

$$\frac{kx}{F_0} = A e^{-\zeta\omega_n t} \sin (\sqrt{1-\zeta^2}\,\omega_n t - \phi) \qquad (56.100)$$

which is a damped harmonic oscillation. The dynamic response factor is then

$$n = A e^{-\zeta\omega_n t_p} \qquad (56.101)$$

where t_p is the time corresponding to the peak response. When damping is negligible, the following simpler equations can be written for this case,

$$n = A = \left|\omega_n \int_0^\infty f(\xi) e^{i\omega_n\xi}\,d\xi\right| \qquad (56.102)$$

where the upper limit of the integral is changed to ∞ without its numerical value being altered since $f(\xi) \equiv 0$ for $t > t_1$.

For very large τ or when $\omega_n \to 0$, the equation for $\tan \phi$ obtained by dividing Eq. (56.99*b*) by Eq. (56.99*a*) indicates that $\phi \to 0$, and the peak time from Eq. (56.100) approaches the value

$$\omega_n t_p \to \frac{\pi}{2\sqrt{1-\zeta^2}} \tag{56.103}$$

We also have, for this limiting case,

$$n \to \omega_n \int_0^{t_1} f(\xi)\, d\xi \tag{56.104}$$

The slope of the shock spectrum at the origin when n is plotted as a function of t_1/τ is then

$$\frac{dn}{d(t_1/\tau)} = \frac{2\pi}{t_1} \int_0^{t_1} f(\xi)\, d\xi \tag{56.105}$$

For large values of ω_n or small natural period τ, the excitation pulse variation will be slow in comparison to the natural oscillations of the system, and the response will differ only slightly from its static result, which is equal to that of the pulse. Thus, for this case, the response is

$$\frac{kx}{F_0} = f(\xi) \qquad \omega_n \to \infty \tag{56.106}$$

and the dynamic response factor n approaches unity.

56.5. REFERENCES

[1] S. P. Timoshenko: "Vibration Problems in Engineering," 3d ed., Van Nostrand, Princeton, N.J., 1955.

[2] J. P. Den Hartog: "Mechanical Vibrations," 4th ed., McGraw-Hill, New York, 1956.

[3] N. O. Myklestad: "Fundamentals of Vibration Analysis," McGraw-Hill, New York, 1956.

[4] C. E. Crede: "Vibration and Shock Isolation," Wiley, New York, 1951.

[5] W. T. Thomson: "Mechanical Vibrations," 2d ed., Prentice-Hall, Englewood Cliffs, N.J., 1953.

[6] J. M. Frankland: Effect of impact on simple structures, *Proc. Soc. Exptl. Stress Anal.*, **6** (1948), 7–27.

[7] Y. C. Fung, M. V. Barton: Some characteristics and uses of shock spectra, *Ramo-Wooldrich Corp. Rept.* AM No. 6-14, Oct. 15, 1956.

[8] Y. C. Fung: Some general properties of the dynamic amplification spectra, *J. Aeronaut. Sci.*, **24** (1957), 547–549.

[9] Y. C. Fung, M. V. Barton: Some shock spectra characteristics and uses, *J. Appl. Mechanics*, **25** (1958), 365–372.

[10] L. S. Jacobsen, R. S. Ayre: "Engineering Vibrations," McGraw-Hill, New York, 1958.

CHAPTER 57

SYSTEMS OF SEVERAL DEGREES OF FREEDOM

BY

F. R. E. CROSSLEY, D. Eng., New Haven, Conn.

57.1. NUMBER OF DEGREES OF FREEDOM

The number of degrees of freedom in a system capable of vibration is defined by the minimum number of independent variables, representing various displacements (linear or angular), that are needed to specify the state of the system with respect to some fixed geometrical (usually static equilibrium) configuration.

Usually only a few of the possible motions of a system are of interest: for the purposes of analysis, certain hypothetical restraints may be assumed to be imposed on a mechanical system, thus reducing the number of degrees of freedom. For instance, if a weight is suspended by a helical spring, the weight may move up and down, or laterally in two senses; it may also twist about the vertical axis of the spring, or even about its two horizontal axes, and then in the spring itself ripples can run up and down. But some degree of independence among these effects is usually tacitly assumed (and often, but not always, this is mathematically justifiable), so that, for the purpose of study, the weight may be taken to move only vertically, or only torsionally. Similarly, a turbine wheel may oscillate as a unit, either torsionally or transversely, on its shaft, and then it may flex as a drum membrane, or the buckets on its periphery may vibrate. It is permissible though to consider that it is only able to oscillate torsionally on its shaft, thus imposing unreal constraints which simplify the analysis because of the directed interest. It must always be remembered, however, that, by imposing these simplifying restraints, certain phenomena that occur in practice may be inexplicable by the simplified theory.

When the decision has then been made as to what is the system to be analyzed, and how many motions are of interest, the number of degrees of freedom will correspond to the minimum number of coordinates needed to specify the instantaneous configuration of the system. The system with more than one degree of freedom will be either (1) one or more masses, for which more than one motion is possible, or (2) several masses, each capable of moving in a similar manner but independently of each other.

An example of system 1 is the simple spherical pendulum, a bob suspended by a string from the ceiling: Here the bob may move in either the north-south vertical plane or the east-west vertical plane. Any position will need two coordinates for specification; the system has two degrees of freedom.

An example of system 2 is the multistage turbine in torsional oscillation: Each wheel needs one angular coordinate, and there are, therefore, as many degrees of freedom as wheels.

57.2. FROM PHYSICAL TO MATHEMATICAL FORM

The initial task is to write in mathematical form the equations that govern the physical problem. There will usually be as many separate equations as there are degrees of freedom. Two approaches are possible: The equations may be discovered by applying the principles of dynamical equilibrium (d'Alembert's principle) to each motion of each mass; or the principles of work and energy may be applied. In the latter method however, a statement of the total kinetic and potential energy of the system will provide only one equation, which is not enough. For the application of the energy method, Lagrange's equations have been developed (see p. 23–24), and their use is explained on p. 57–4.

Some examples of the differential equations governing typical systems as derived by appling d'Alembert's principle follow.

Example 1. Figure 57.1 represents a three-stage axial compressor. Its oscillation is not affected by its rotation, so that it may be considered not to be revolving; but each wheel is oscillating torsionally. The angular positions of the three wheels

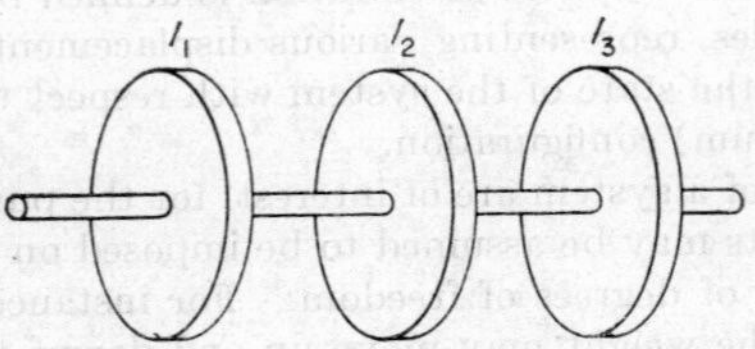

FIG. 57.1

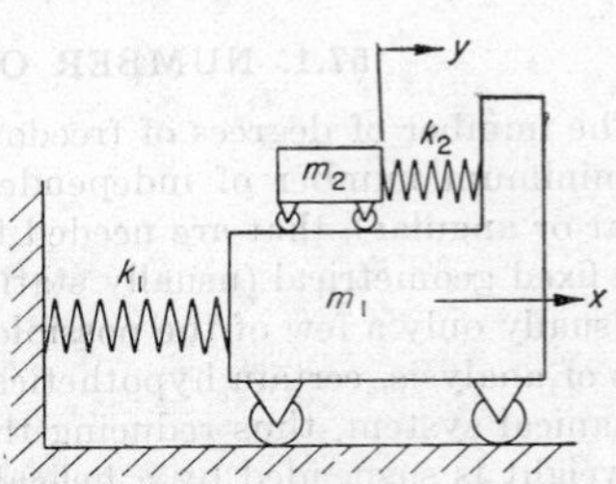

FIG. 57.2

are respectively θ_1, θ_2, and θ_3; their moments of inertia are I_1, I_2, and I_3. The torque needed to twist the shaft between I_1 and I_2 through a unit angle is k_{12}, and for the shaft between I_2 and I_3 it is k_{23}. Then, without externally applied torque, assume that the wheels got into positions θ_1, θ_2, and θ_3 (unspecified), and the motion of the wheels must be governed by

$$\begin{aligned} I_1\ddot{\theta}_1 + k_{12}(\theta_1 - \theta_2) &= 0 \\ I_2\ddot{\theta}_2 + k_{12}(\theta_2 - \theta_1) + k_{23}(\theta_2 - \theta_3) &= 0 \\ I_3\ddot{\theta}_3 + k_{23}(\theta_3 - \theta_2) &= 0 \end{aligned} \tag{57.1}$$

Example 2. Figure 57.2 shows a large body m_1, able to move horizontally with motion x, but restrained by a spring of stiffness k_1. Connected to m_1 there is another body m_2, which moves with relative motion y, so that the force in the connecting spring is k_2y. To the body m_2 an external independent force $F_0 \sin \omega t$ is applied horizontally. Then

$$m_1\ddot{x}_1 + k_1x - k_2y = 0 \qquad m_2(\ddot{x} + \ddot{y}) + k_2y = F_0 \sin \omega t \tag{57.2a,b}$$

To understand the latter, note that the absolute acceleration of m_2 is

$$(d^2/dt^2)(x + y) = \ddot{x} + \ddot{y}$$

since y is only relative to m_1.

Example 3. Figure 57.3 shows an unbalanced rotor m_2 on a flexible shaft. We presume to consider only vertical motions induced by the force $m_2r\omega^2 \sin \omega t$, which is the vertical component of the force of unbalance $m_2r\omega^2$. The machine of mass m_1 containing the rotor is mounted on the floor by isolators of combined stiffness k_1. The rotor's flexible shaft bends an amount y relative to its bearings, and its resistance to bending is measured by the stiffness k_2. The equations of this system are the same as Eqs. (57.2), with the substitution of $m_2r\omega^2$ for F_0.

Example 4. In Fig. 57.4, the position of the mass m of the spherical pendulum is measured by the angle ϕ of inclination of its string to the vertical and by the angle of longitude θ. The string is of constant length l. To write the equations in two senses on the surface of the sphere, the bob must be considered at a general position (ϕ,θ) and

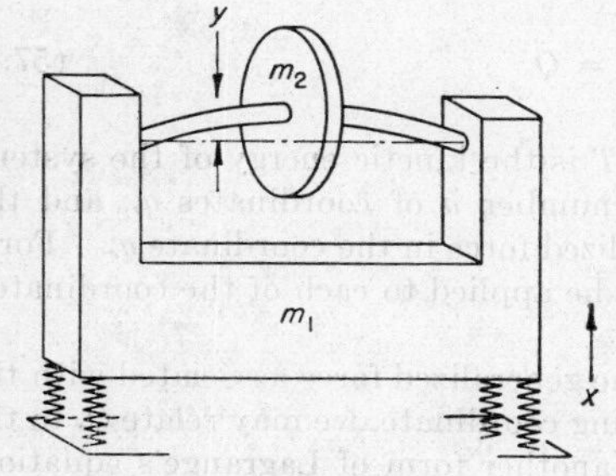

FIG. 57.3

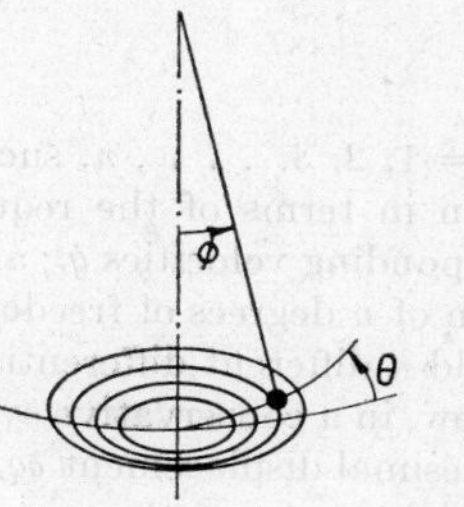

FIG. 57.4

having all velocity and acceleration components due to $\dot\phi$, $\ddot\phi$, $\dot\theta$, and $\ddot\theta$. The equations are, therefore,

$$
\begin{aligned}
ml^2\ddot\phi - m(l\dot\theta)^2 \sin\phi\cos\phi + mgl\sin\phi &= 0 \\
m(l\sin\phi)^2\ddot\theta + m(2l\dot\phi\dot\theta\cos\phi)l\sin\phi &= 0
\end{aligned}
\qquad (57.3)
$$

57.3. DAMPING

In all problems, damping is normally present. Mathematically, it is only possible to take care of it if it be regarded as proportional to velocity (viscous damping), or if it is a constant with periodically reversing sign (Coulomb or solid friction). (See p. 57–18.) These terms should then properly be included in the equations such as have been derived.

However, usually only three types of problem are paramount in mechanical engineering:

1. To find the natural frequencies of a system, in order to know the spectrum of resonances
2. To find the forced steady-state amplitudes of any part of a system at frequencies well divorced from any resonance
3. To investigate transient effects thoroughly when a system is subjected to sudden shock or impulse

In calculations of type 1, the damping has no effect, and, in type 2, it has little effect; so in both these the damping may be ignored. In type 3, of course, the effects of damping must always be reckoned with. Occasionally, also, it is necessary that a machine contain enough damping, when acted upon by a variable-speed vibration source, that it may pass through a resonance without damage. And, again, damping may be the cause of the instability of a system, as is discussed on p. 57–19. So for the sake of these, damping will be fully discussed later, in Sec. 57.14 and those sections following it.

But first there is sufficient reason to study undamped free oscillations thoroughly.

57.4. LAGRANGIAN METHOD

There are two methods available for obtaining the equations of motion of a system. The first is by the application of the principle of d'Alembert, considering forces, masses, and accelerations, and this has been described. The second utilizes Lagrange's equations.

These equations are a concise statement, derived from consideration of forces and energy, of the condition of any Newtonian system when displaced from its position of static equilibrium. Their derivation and an explanation of the slightly different forms in which they may be written are to be found on p. 23–24.

The basic form of Lagrange's equation is

$$\frac{d}{dt}\left(\frac{\partial T}{\partial \dot{q}_r}\right) - \frac{\partial T}{\partial q_r} = Q_r \tag{57.4}$$

for $r = 1, 2, 3, \ldots, n$, successively; in this, T is the kinetic energy of the system, written in terms of the required (holonomic) number n of coordinates q_r, and the corresponding velocities $\dot{q}_r$; and Q_r is the generalized force in the coordinate q_r. For a system of n degrees of freedom, Eq. (57.4) may be applied to each of the coordinates, to yield n different differential equations.

Now, in a conservative system, since Q_r is the generalized force associated with the infinitesimal displacement δq_r of the corresponding coordinate, we may relate Q_r to the potential energy of the system, and so obtain another form of Lagrange's equation. For, if

$$Q_r\,\delta q_r = \delta W$$

that is, if the work done by the force Q_r in moving its coordinate a small amount is δW, then this work must indicate an equal loss in the potential energy V of the system, or

$$\delta W = -\frac{\partial V}{\partial q_r}\delta q_r$$

When this is incorporated in Eq. (57.4), we have, for a holonomic conservative system: (see Refs. [21], p. 203, and [31], p. 34):

$$\frac{d}{dt}\left(\frac{\partial T}{\partial \dot{q}_r}\right) - \frac{\partial T}{\partial q_r} + \frac{\partial V}{\partial q_r} = 0 \qquad r = 1, 2, 3, \ldots, n \tag{57.5}$$

This equation may be used on the same physical examples, for which the equations of motion were just obtained by the d'Alembertian method, as follows. But proof of and further work with Lagrangian methods may be found in Refs. [5],[19],[21],[25]–[27],[29]–[31].

Example 5. With the two-mass system in Fig. 57.2, the two coordinates $q_1 = x$ and $q_2 = y$ are chosen. Then

$$T = \tfrac{1}{2}m_1\dot{x}^2 + \tfrac{1}{2}m_2(\dot{x} + \dot{y})^2 \qquad V = \tfrac{1}{2}k_1x^2 + \tfrac{1}{2}k_2y^2$$

$$\frac{\partial T}{\partial \dot{x}} = m_1\dot{x} + m_2(\dot{x} + \dot{y}) \qquad \frac{\partial T}{\partial x} = 0 \qquad \frac{\partial V}{\partial x} = k_1x$$

So, applying Lagrange's equation (57.5) for a conservative system,

$$m_1\ddot{x} + m_2(\ddot{x} + \ddot{y}) + k_1x = 0 \tag{57.6a}$$

Again, with respect to y,

$$\frac{\partial T}{\partial \dot{y}} = m_2(\dot{x} + \dot{y}) \qquad \frac{\partial T}{\partial y} = 0 \qquad \frac{\partial V}{\partial y} = k_2y$$

so that, reapplying Lagrange's equation (57.5),

$$m_2(\ddot{x} + \ddot{y}) + k_2y = 0 \tag{57.6b}$$

It is easily seen that Eqs. (57.6) and (57.2), the latter with $F_0 = 0$, are linear combinations of each other.

Example 6. With the pendulum of Fig. 57.4, let the two coordinates be θ and ϕ, as shown. Then

$$T = \tfrac{1}{2}m(l \sin \phi)^2\dot{\theta}^2 + \tfrac{1}{2}m(l\dot{\phi})^2 \qquad V = mgl(1 - \cos \phi)$$

$$\frac{\partial T}{\partial \dot{\theta}} = ml^2\dot{\theta} \sin^2 \phi \qquad \frac{\partial T}{\partial \theta} = 0 \qquad \frac{\partial V}{\partial \theta} = 0$$

and so

$$ml^2\ddot{\theta} \sin^2 \phi + 2ml^2\dot{\theta}\dot{\phi} \sin \phi \cos \phi = 0$$

or, integrated,

$$ml^2\dot{\theta} \sin^2 \phi = \text{const}$$

$$\frac{\partial T}{\partial \dot{\phi}} = ml^2\dot{\phi} \qquad \frac{\partial T}{\partial \phi} = ml^2\dot{\theta}^2 \sin \phi \cos \phi \qquad \frac{\partial V}{\partial \phi} = mgl \sin \phi$$

Thus,

$$ml^2\ddot{\phi} - ml^2\dot{\theta}^2 \sin \phi \cos \phi + mgl \sin \phi = 0$$

In this case, when $(\partial V/\partial \theta - \partial T/\partial \theta)$ vanishes, and Lagrange's equation appears in the highly simplified form

$$\frac{d}{dt}\left(\frac{\partial T}{\partial \dot{\theta}}\right) = 0$$

which is immediately integrable, θ is called an *ignorable coordinate.*

57.5. GENERAL EQUATIONS AND SOLUTION: FREE VIBRATION OF SMALL AMPLITUDE

To find the frequencies and modes of small oscillation of a system with n degrees of freedom about a configuration of stable equilibrium, it is not necessary to consider any damping. The equations will always appear as an array of n linear equations of second order as follows:

$$\begin{array}{l} m_{11}\ddot{x}_1 + m_{12}\ddot{x}_2 + \cdots + k_{11}x_1 + k_{12}x_2 + \cdots + k_{1n}x_n = 0 \\ m_{21}\ddot{x}_1 + m_{22}\ddot{x}_2 + \cdots + k_{21}x_1 + k_{22}x_2 + \cdots + k_{2n}x_n = 0 \\ \cdots\cdots\cdots\cdots\cdots\cdots\cdots\cdots\cdots\cdots \\ m_{n1}\ddot{x}_1 + m_{n2}\ddot{x}_2 + \cdots + k_{n1}x_1 + k_{n2}x_2 + \cdots + k_{nn}x_n = 0 \end{array} \tag{57.7}$$

Of course, in most practical cases, a number of the above coefficients may be expected to vanish. The coefficients m_{ij} (if they exist, and $i \neq j$) are known as *dynamic coupling* coefficients; and the coefficients k_{ij} (for $i \neq j$) are termed coefficients of *static coupling.*

Solutions of the above set of equations are in the form

$$\begin{array}{l} x_1 = a_1 \cos (\omega t + \phi) \\ x_2 = a_2 \cos (\omega t + \phi) \\ \cdots\cdots\cdots\cdots \\ x_n = a_n \cos (\omega t + \phi) \end{array} \tag{57.8}$$

in which the circular frequency ω and phase angle ϕ are common to all xs. Substitution verifies this and produces a new similar array of algebraic homogeneous equations:

$$\begin{array}{l} (k_{11} - m_{11}\omega^2)a_1 + (k_{12} - m_{12}\omega^2)a_2 + \cdots + (k_{1n} - m_{1n}\omega^2)a_n = 0 \\ (k_{21} - m_{21}\omega^2)a_1 + (k_{22} - m_{22}\omega^2)a_2 + \cdots + (k_{2n} - m_{2n}\omega^2)a_n = 0 \\ \cdots\cdots\cdots\cdots\cdots\cdots\cdots\cdots\cdots\cdots \\ (k_{n1} - m_{n1}\omega^2)a_1 + (k_{n2} - m_{n2}\omega^2)a_2 + \cdots + (k_{nn} - m_{nn}\omega^2)a_n = 0 \end{array} \tag{57.9}$$

If the amplitudes $a_1, a_2, a_3, \ldots, a_n$ are not all to be zero, the coefficient determinant of these equations must vanish:

$$\begin{vmatrix} (k_{11} - m_{11}\omega^2) & (k_{12} - m_{12}\omega^2) & \cdots & (k_{1n} - m_{1n}\omega^2) \\ (k_{21} - m_{21}\omega^2) & \cdots & \cdots & \cdots \\ \cdots & (k_{ij} - m_{ij}\omega^2) & & \cdots \\ (k_{n1} - m_{n1}\omega^2) & \cdots & \cdots & (k_{nn} - m_{nn}\omega^2) \end{vmatrix} = 0 \qquad (57.10)$$

This determinant, which is called the *Lagrangian determinant,* may then be expanded, and yields a polynomial of nth degree in ω^2, called the *frequency equation.* The n roots are the n values of ω for which the solutions in Eq. (57.8) are acceptable.

These n roots are called *eigenvalues* or *characteristic values,* because it is only when ω has these particular values that all the coordinates a_i do not vanish.

Equation (57.10) has been derived from the set of equations (57.9), n in number: there remain $(n - 1)$ independent equations (57.9) which establish the relative magnitudes of the coordinates a_i. For example, suppose a_1 is chosen as the one arbitrary constant: then the $(n - 1)$ ratios $a_2/a_1, a_3/a_1, \ldots, a_i/a_1$ may be determined from the equations (57.9), after substitution of each value of ω^2; and they will be different for each different value of ω^2.

The n eigenvalues of ω are the n natural frequencies of the system. They are usually thought to be arranged in order of ascending magnitude, the lowest being called the *first natural frequency,* etc. At each such natural frequency, the geometrical picture presented by the system with each part moving simultaneously to its maximum in the ratios a_2/a_1, a_3/a_1, etc., appropriate to that frequency is called a *mode of vibration.* Thus, with the lowest natural frequency, the system oscillates in its first mode; with the second natural frequency, it oscillates in the second mode; and so forth.

The most general motion of which a system is capable is not restricted to any one mode, but is the sum of all possible modes superimposed. This general solution supersedes Eqs. (57.8) and appears as

$$x_i = \sum_{k=1}^{n} a_{i(k)} \cos (\omega_k t + \phi_k) \qquad i = 1, 2, \ldots, n \qquad (57.11)$$

That this is so, is called Daniel Bernoulli's *principle of superposition.*

57.6. OBTAINING THE ROOTS OF A POLYNOMIAL

The eigenvalues, i.e., the natural frequencies of a system, are found as the roots of an algebraic polynomial expression. For quadratics, this presents no problem, but from expressions of higher degree it is more difficult to extract roots. The matter is dealt with in Chap. 2; it is generally necessary to be satisfied with approximate roots, for which the accuracy may be improved by successive iterations.

57.7. MODES OF FREE VIBRATION OF SMALL AMPLITUDE

Once the values of the natural frequencies that correspond to the roots of the polynomial of degree $2n$ obtained from the Lagrangian determinant (57.10) have been found, Eqs. (57.9) may be used to determine the relative amplitudes a_i of each of the n coordinates.

Example 7. Consider again Fig. 57.2, for which the equations were stated as (57.2) and again derived by Lagrange's equations in Sec. 57.4. Suppose that $F_0 = 0$, $m_2 = \frac{1}{2}m_1$, $k_1/m_1 = k_2/m_2 = \Omega^2$. The general equations (57.9) assume, then, the form

$$(k_1 - m_1\omega^2)a_1 - k_2a_2 = 0 \qquad (-m_2\omega^2)a_1 + (k_2 - m_2\omega^2)a_2 = 0 \qquad (57.12)$$

The determinant expanded, therefore, is

$$(k_1 - m_1\omega^2)(k_2 - m_2\omega^2) - k_2 m_2 \omega^2 = 0$$

or, dividing by k_1k_2,

$$1 - \frac{5}{2}\frac{\omega^2}{\Omega^2} + \frac{\omega^4}{\Omega^4} = 0 \qquad (57.13)$$

This biquadratic has roots given by

$$\frac{\omega^2}{\Omega^2} = 5/4 \pm \sqrt{(5/4)^2 - 1}$$

that is, $\omega_1{}^2 = \Omega^2/2$, and $\omega_2{}^2 = 2\Omega^2$. From either of Eqs. (57.12), obtain

$$a_2 = 2\left(1 - \frac{\omega^2}{\Omega^2}\right) a_1 \qquad (57.14)$$

Thus, when the system is oscillating in the first mode, with natural frequency ω_1, then $a_1 = a_2$; alternatively, when it is oscillating in the second mode, with natural frequency ω_2, then $a_2 = -2a_1$.

These should be visualized: In the first mode, while the first mass m_1 moves to the right a distance a_1, so does the second mass in its motion relative to the first: thus the second moves in phase with the first but with twice the absolute motion. In contrast to this, the masses are moving in their second natural mode when they move antiphase to each other, with equal absolute amplitudes. In this second mode, there exists a node or point that remains stationary with respect to the oscillatory motion of all the other points: the nodal point is obviously (from the geometry of this mode) halfway along the spring k_2.

Whenever there is a node, it is a simple matter to check the arithmetic that led to this conclusion: Thus mass m_2 is shown to be oscillating on half the length of spring k_2; its frequency should then be given by $\omega^2 = (2k_2)/m_2$, and this is $2\Omega^2$. Secondly, the mass m_1 is moving relative to the fixed frame because of two springs, of stiffness k_1 and $2k_2$; its frequency should then be $\omega^2 = (k_1 + 2k_2)/m_1 = 2\Omega^2$, and this checks also.

Example 8. A Degenerate Case. The example of Fig. 57.1 shows a system with three degrees of freedom, but with the third mode degenerating so that the system has only two modes of oscillation. This fact will appear while doing the analysis; no special prescience is expected. When the substitutions of the form $\theta_i = a_i \sin(\omega t + \phi)$ are made, the three equations of torque (57.1) become

$$\begin{aligned} (k_{12} - I_1\omega^2)a_1 \quad k_{12}a_2 &= 0 \\ -k_{12}a_1 + (k_{12} + k_{23} - I_2\omega^2)a_2 - k_{23}a_3 &= 0 \\ -k_{23}a_2 + (k_{23} - I_3\omega^2)a_3 &= 0 \end{aligned} \qquad (57.15)$$

Suppose then, for example, that $k_{12} = k_{23} = k$ and $I_1 = I_2 = I_3 = I$. Let $k/I = \Omega^2$ and $\omega/\Omega = \rho$. Then the above three equations appear as

$$\begin{aligned} (1 - \rho^2)a_1 - a_2 &= 0 \\ -a_1 + (2 - \rho^2)a_2 - a_3 &= 0 \\ -a_2 + (1 - \rho^2)a_3 &= 0 \end{aligned} \qquad (57.16)$$

and the frequency equation obtained from the Lagrangian determinant is

$$(1 - \rho^2)^2(2 - \rho^2) - 2(1 - \rho^2) = 0$$

which factors to

$$\rho^2(1 - \rho^2)(\rho^2 - 3) = 0 \qquad (57.17)$$

The first factor gives the root $\rho^2 = 0$; hence, $\omega = 0$ and Eqs. (57.16) yield for this mode $a_1 = a_2 = a_3$. With $\phi = 0$, Eqs. (57.8) describe a static rigid-body displacement, whereas, with $\phi = 90°$, the trigonometric function degenerates into a linear function and describes a rigid-body rotation of the entire shaft at an arbitrary but constant angular velocity. This is the degenerate part of the solution. The two nonvanishing natural frequencies are found from $\rho^2 = 1$ and $\rho^2 = 3$.

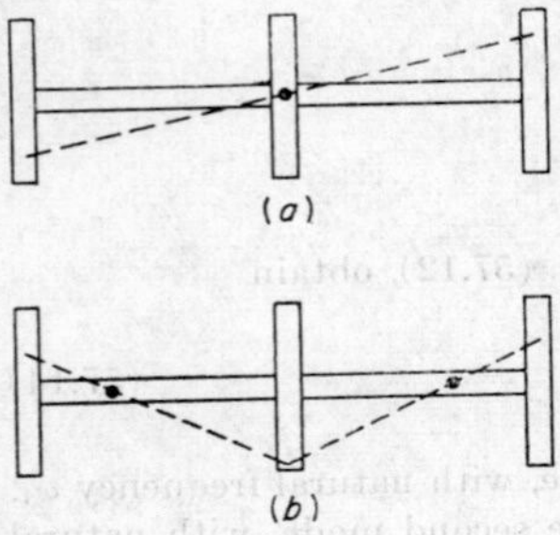

FIG. 57.5

In the first mode, $\omega^2 = \Omega^2 = k/I$, and, from (57.15), it appears that $a_2 = 0$ while $a_1 = -a_3$. This mode is illustrated in Fig. 57.5*a*, in which the dashed line represents a graph of the torsional angle θ of displacement at every section along the shaft. There is a node at the central mass here because of the symmetry.

In the second mode, $\omega^2 = 3\Omega^2$ and $a_1 = a_3 = -\frac{1}{2}a_2$. This is shown by the dashed line in Fig. 57.5*b*. The two nodes lie one-third of the distance along the span between the wheels measured from the outer ends.

It will be noted, in general, that each successive normal mode has one more node than the preceding mode.

The dashed lines in Fig. 57.5 illustrating the deformation in the modes are often called the *normal elastic curves*.

The general solution to the equations of motion of this system is then expressed by the superposition of the three possible motions of the three coordinates:

$$\begin{aligned}\theta_1 &= A_1 \sin(\omega_1 t + \phi_1) + B_1 \sin(\omega_2 t + \phi_2) + C_1 + Dt \\ \theta_2 &= -2B_1 \sin(\omega_2 t + \phi_2) + C_1 + Dt \\ \theta_3 &= -A_1 \sin(\omega_1 t + \phi_1) + B_1 \sin(\omega_2 t + \phi_2) + C_1 + Dt\end{aligned}$$

in which A_1 corresponds to the amplitude a_1 in the first mode and B_1 to the same in the second mode; C_1 is then that arbitrary motion of zero frequency in the degenerate mode, in which all participate equally if at all. Six initial conditions, being the values of θ_1, θ_2, θ_3, $\dot\theta_1$, $\dot\theta_2$, and $\dot\theta_3$ at $t = 0$, must, however, be freely substituted, and this requires a sixth arbitrary constant D, beyond the five already reckoned (A_1, B_1, C_1, ϕ_1, ϕ_2). So the term Dt is added, denoting a possible constant rotation of the system.

Before leaving this example, note that such degenerating cases are those for which a set of coordinates may be chosen, of which one will be an ignorable coordinate.

Choose, for the coordinates of the example, the angles of twist in the two sections of shaft, which are relative angles, and, for reference to the fixed frame, the angular position of the first wheel. Then, in terms of the θ's previously used,

$$\psi_1 = \theta_1 \qquad \psi_2 = \theta_2 - \theta_1 \qquad \psi_3 = \theta_3 - \theta_2 \tag{57.18}$$

Thus, the potential energy is very simply written as

$$V = \tfrac{1}{2}k\psi_2^2 + \tfrac{1}{2}k\psi_3^2 \tag{57.19}$$

The kinetic energy is

$$T = \tfrac{1}{2}I_1\dot\psi_1^2 + \tfrac{1}{2}I_2(\dot\psi_1 + \dot\psi_2)^2 + \tfrac{1}{2}I_3(\dot\psi_1 + \dot\psi_2 + \dot\psi_3)^2 \tag{57.20}$$

Thus, ψ_1 is an ignorable coordinate, for $\partial V/\psi_1 = 0$, and $\partial T/\partial\psi_1 = 0$. The application of the Lagrange equation (57.5) gives, for this coordinate,

$$I_1\dot\psi_1 + I_2(\dot\psi_1 + \dot\psi_2) + I_3(\dot\psi_1 + \dot\psi_2 + \dot\psi_3) = \text{const}$$

which is a statement of the conservation of angular momentum of the system in the absence of any external forces.

57.8. WANDERING OF THE ENERGY

In a double-mode system (that is, one with two degrees of freedom), the general motion of free oscillation is expressed as

$$\begin{aligned} q_1 &= a_1 \sin(\omega_1 t + \phi_1) + b_1 \sin(\omega_2 t + \phi_2) \\ q_2 &= a_2 \sin(\omega_1 t + \phi_1) + b_2 \sin(\omega_2 t + \phi_2) \end{aligned} \qquad (57.21)$$

There will, moreover, be a relationship between a_1 and a_2, and between b_1 and b_2, which appears from the analysis. Suppose then that it so happens that the two natural frequencies ω_1 and ω_2 are only slightly different: this will be the case if two similar single-mode systems, such as two equal pendulums, are coupled by a light spring as in Fig. 57.6 (see also p. 55–5). The motion q_1 may be regarded as the

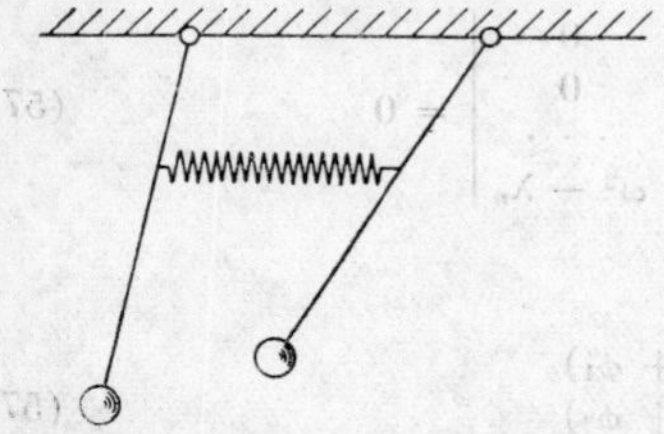

FIG. 57.6

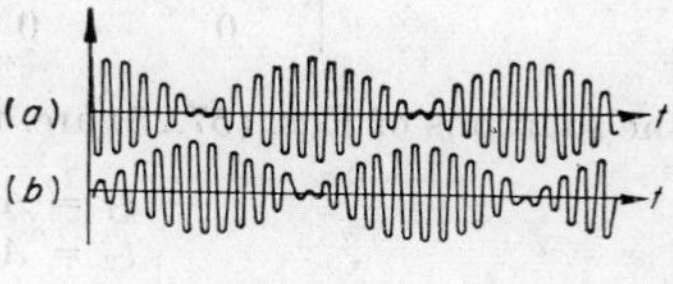

FIG. 57.7

projection of the sum of two vectors rotating at slightly different rates. At times they will be nearly in phase, so that the amplitude of the motion is about $(a_1 + b_1)$; at other times, intermediate between the former, they will be out of phase, so that the amplitude is $(a_1 - b_1)$.

The motion q_1 is then represented by a wave having beats as in Fig. 57.7*a*. Now, if the system is conservative, the amplitude and kinetic energy of the motion q_1 can increase only at the expense of the motion q_2. This latter must therefore also have a motion, of which the phase of the beats is opposite to the former.

If the two masses of the pendulums, for example, started with $\dot{\theta}_1 = \dot{\theta}_2 = 0$ and $\theta_2 = 0$, Eqs. (57.21) become

$$\theta_1 = a \cos \omega_1 t + a \cos \omega_2 t \qquad \theta_2 = a \cos \omega_1 t - a \cos \omega_2 t \qquad (57.22)$$

Each pendulum bob periodically and alternately comes instantaneously to rest while the other reaches its maximum amplitude $2a$. The energy is seen to wander from one to the other and back again.

57.9. NORMAL COORDINATES

The expressions for the kinetic and potential energies of a system with n degrees of freedom, which is either wholly linear or engaged in oscillations of small amplitude only about a state of stable equilibrium, are written in terms of the generalized coordinates q as

$$\begin{aligned} T &= \tfrac{1}{2}(m_{11}\dot{q}_1^2 + m_{22}\dot{q}_2^2 + \cdots + 2m_{12}\dot{q}_1\dot{q}_2 + 2m_{13}\dot{q}_1\dot{q}_3 + \cdots) \\ V &= \tfrac{1}{2}(k_{11}q_1^2 + k_{22}q_2^2 + \cdots + 2k_{12}q_1q_2 + 2k_{13}q_1q_3 + \cdots) \end{aligned} \qquad (57.22a,b)$$

Now Sylvester shows that there exists a new set of coordinates ξ, in terms of which the kinetic and potential energies may alternatively be expressed (see [31], pp. 178–185) as

$$T = \tfrac{1}{2}(\xi_1^2 + \xi_2^2 + \cdots + \xi_n^2) \qquad V = \tfrac{1}{2}(\lambda_1\xi_1^2 + \lambda_2\xi_2^2 + \cdots + \lambda_n\xi_n^2) \qquad (57.23a,b)$$

where the coefficients λ are all constants. In these, it is to be noted that no product terms appear. The theorem proves that this new set of coordinates ξ may be found from the set of coordinates q by means of a real linear transformation:

$$\begin{aligned} \xi_1 &= h_{11}q_1 + h_{12}q_2 + h_{13}q_3 + \cdots + h_{1n}q_n \\ \xi_2 &= h_{21}q_1 + h_{22}q_2 + \cdots + \cdots + h_{2n}q_n \\ &\cdots\cdots\cdots\cdots\cdots\cdots \end{aligned} \tag{57.24}$$

In any conservative system, if Lagrange's equation (57.5) is applied to the expressions (57.23) with respect to the n coordinates ξ_i, there will be obtained n equations, each of which appears very simply as

$$\ddot{\xi}_i + \lambda_i \xi_i = 0 \qquad i = 1, 2, \ldots, n \tag{57.25}$$

For such an array of equations, the Lagrangian determinant is reduced to a diagonal of terms:

$$\begin{vmatrix} \omega^2 - \lambda_1 & 0 & \cdots & 0 \\ 0 & \omega^2 - \lambda_2 & \cdots & 0 \\ \cdots & \cdots & \cdots & \cdots \\ 0 & 0 & \cdots & \omega^2 - \lambda_n \end{vmatrix} = 0 \tag{57.26}$$

and the solutions of Eqs. (57.25) are, therefore,

$$\begin{aligned} \xi_1 &= A_1 \sin(\omega_1 t + \phi_1) \\ \xi_2 &= A_2 \sin(\omega_2 t + \phi_2) \\ &\cdots\cdots\cdots\cdots \end{aligned} \tag{57.27}$$

in which the terms A_i and ϕ_i are $2n$ arbitrary constants dependent on the initial conditions. Each coordinate ξ represents the motion and the relative amplitudes of all different parts of the system oscillating in a given mode. The motions in each of the several modes are completely uncoupled. These coordinates ξ are called the *normal coordinates*, or *principal coordinates*. It is quite simple (see [20], p. 195, and [21], p. 25) to find the normal coordinates of a given system, and the coefficients of the transformation (57.24). If Eqs. (57.27) are compared with Eqs. (57.11) in terms of q, we need only substitute $\xi_1, \xi_2, \ldots$ for $\sin(\omega_1 t + \phi_1)$, $\sin(\omega_2 t + \phi_2)$, etc., to obtain the inverse form of Eqs. (57.24).

This may be stated also in matrix notation: If the n equations represented by

$$\begin{bmatrix} m_{11} & m_{12} & \cdots \\ m_{21} & \cdots & \cdots \\ \cdots & \cdots & m_{nn} \end{bmatrix} \begin{bmatrix} \ddot{q}_1 \\ \cdots \\ \ddot{q}_n \end{bmatrix} + \begin{bmatrix} k_{11} & k_{12} & \cdots \\ k_{21} & \cdots & \cdots \\ \cdots & \cdots & k_{nn} \end{bmatrix} \begin{bmatrix} q_1 \\ \cdots \\ q_n \end{bmatrix} = \begin{bmatrix} 0 \\ 0 \\ 0 \end{bmatrix}$$

are written concisely as

$$M\{\ddot{q}\} + K\{q\} = \{0\} \tag{57.28}$$

and changed to

$$\{\ddot{q}\} + \Omega\{q\} = \{0\} \tag{57.29}$$

where $\Omega = M^{-1}K$, then a real linear transformation (57.24) can be found which changes Ω to a diagonal matrix Λ.

Example 9. As shown in Fig. 57.8, a rectangular body with mass m and moment of inertia I is held by two springs; the first of stiffness k_1 is at one end; the second is centrally placed, and its stiffness $k_2 = 3k_1$. The two natural frequencies are desired for small motions in the plane without lateral motions of the springs.

The two coordinates most readily chosen are the vertical displacement x of the center of the body, and its angular motion θ about the same point. θ is taken to be positive when clockwise, x positive when downward.

Then the energies are

$$T = \tfrac{1}{2}m\dot{x}^2 + \tfrac{1}{2}I\dot{\theta}^2 \qquad V = \tfrac{1}{2}k_1(x - a\theta)^2 + \tfrac{1}{2}k_2x^2 \tag{57.30}$$

and so, applying the Lagrange equation,

$$m\ddot{x} + (k_1 + k_2)x - k_1a\theta = 0 \qquad I\ddot{\theta} - k_1ax + k_1a^2\theta = 0 \tag{57.31}$$

With this particular set of coordinates, the system exhibits static coupling but not

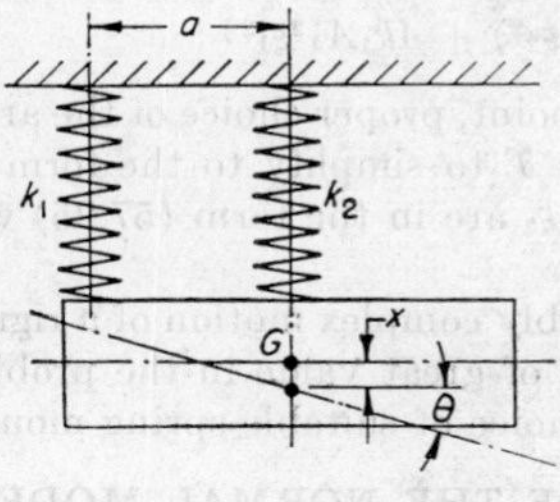

FIG. 57.8

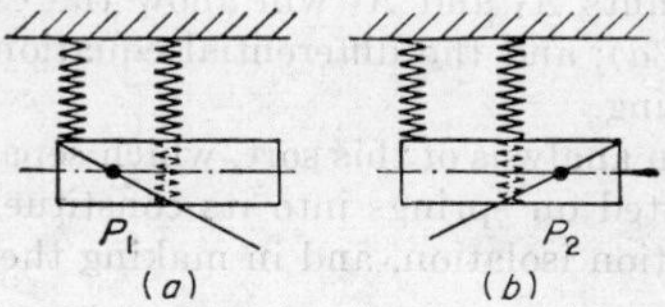

FIG. 57.9

dynamic coupling.

Substitution of $x = x_0 \sin(\omega t + \phi)$ $\quad \theta = \theta_0 \sin(\omega t + \phi)$

gives the equations

$$(k_1 + k_2 - m\omega^2)x_0 - k_1a\theta_0 = 0$$
$$-k_1ax_0 + (k_1a^2 - I\omega^2)\theta_0 = 0$$

Here we may substitute $p_1{}^2 = (k_1 + k_2)/m = 4k_1/m$ and $p_2{}^2 = k_1a^2/I$, to get

$$\begin{aligned}(1 - \omega^2/p_1{}^2)x_0 - \tfrac{1}{4}a\theta_0 &= 0\\ -x_0 + (1 - \omega^2/p_2{}^2)a\theta_0 &= 0\end{aligned} \tag{57.32}$$

The determinant yields the quadratic

$$\omega^4 - (p_1{}^2 + p_2{}^2)\omega^2 + \tfrac{3}{4}p_1{}^2p_2{}^2 = 0$$

Supposing now also that $p_1{}^2 = p_2{}^2 = p^2$, which is to say that $I = ma^2/4$:

$$\omega^2 = (1 \pm \tfrac{1}{2})p^2 \qquad \text{or} \qquad \omega_1{}^2 = p^2/2 \qquad \omega_2{}^2 = 3p^2/2$$

For the amplitude relationships of the mode, substitute these in (57.32).

$$\begin{aligned}&\text{With } \omega_1, & a\theta_0 &= 2x_0\\ &\text{With } \omega_2, & a\theta_0 &= -2x_0\end{aligned} \tag{57.33}$$

The general solutions are, therefore,

$$\begin{aligned}x &= A_1 \sin(\omega_1t + \phi_1) + A_2 \sin(\omega_2t + \phi_2)\\ a\theta &= 2A_1 \sin(\omega_1t + \phi_1) - 2A_2 \sin(\omega_2t + \phi_2)\end{aligned} \tag{57.34}$$

in which A_1, A_2, ϕ_1, and ϕ_2 are arbitrary constants.

To find the *normal coordinates* of the system, substitute ξ_1 and ξ_2 for the sine terms,

$$x = A_1\xi_1 + A_2\xi_2 \qquad \tfrac{1}{2}a\theta = A_1\xi_1 - A_2\xi_2 \tag{57.35}$$

which invert to

$$\xi_1 = B_1(x + \tfrac{1}{2}a\theta) \qquad \xi_2 = B_2(x - \tfrac{1}{2}a\theta)$$

A study of conditions (57.33) shows that the modes of oscillation appear as in Fig. 57.9. In the first mode, with frequency ω_1 there is rotation about a node P_1; in the second mode, about node P_2. Displacement of the nodal point P_2 can occur only in the first mode; it is independent of the second-mode motion. This displacement is, therefore, proportional to ξ_1; similarly, the displacement of node P_1 is proportional to ξ_2.

Substitution of the variables ξ_1 and ξ_2 into expressions (57.30) for the energies will be found to give

$$T = \tfrac{1}{2}\left(m + \frac{4I}{a^2}\right)(A_1{}^2\dot{\xi}_1{}^2 + A_2{}^2\dot{\xi}_2{}^2)$$
$$V = \tfrac{1}{2}(k_1 + k_2)(A_1{}^2\xi_1{}^2 + A_2{}^2\xi_2{}^2) + 4k_1A_1{}^2\xi_1{}^2$$

from which all product terms are absent. At this point, proper choice of the arbitrary constants A_1 and A_2 will allow the expression for T to simplify to the form of Eq. (57.23*a*); and the differential equations in ξ_1 and ξ_2 are in the form (57.25) with no coupling.

An analysis of this sort, which separates a possibly complex motion of a rigid body mounted on springs into its constituent modes, is of great value in the problems of vibration isolation, and in making the optimum choice of suitable spring mounts.

57.10. THE STATIONARY PROPERTY OF THE NORMAL MODES

The effects are considered by Whittaker [31], p. 192, and by Ramsey [21], p. 268, of adding geometrical frictionless constraints to a conservative system, thereby reducing its n degrees of freedom to just one degree. If $q_1, \ldots, q_n$ are the generalized coordinates, then, for small oscillations, the normal coordinates $\xi_1, \ldots, \xi_n$ are obtained by n linear relations of the form

$$\xi_i = h_{i1}q_1 + h_{i2}q_2 + \cdots \qquad i = 1, 2, \ldots, n$$

For such small oscillations, any imposed geometrical constraints can be written by the linear relations between the normal coordinates:

$$\xi_1 = \mu_1 x, \quad \xi_2 = \mu_2 x, \quad \ldots, \quad \xi_n = \mu_n x \tag{57.36}$$

where the μs are constants, and x is any chosen one of the ξs or of the qs. When the expressions for the kinetic and potential energies of the multiple-mode system are written in terms of the normal coordinates, as

$$T = \tfrac{1}{2}(\dot{\xi}_1{}^2 + \dot{\xi}_2{}^2 + \cdots + \dot{\xi}_n{}^2)$$
$$V = \tfrac{1}{2}(\omega_1{}^2\xi_1{}^2 + \omega_2{}^2\xi_2{}^2 + \cdots \omega_n{}^2\xi_n{}^2) \tag{57.37}$$

the coefficients ω_i are the n natural frequencies of the unconstrained system. When the constraints (57.36) are added,

$$T = \tfrac{1}{2}(\mu_1{}^2 + \mu_2{}^2 + \cdots + \mu_n{}^2)\dot{x}^2$$
$$V = \tfrac{1}{2}(\mu_1{}^2\omega_1{}^2 + \mu_2{}^2\omega_2{}^2 + \cdots + \mu_n{}^2\omega_n{}^2)x^2 \tag{57.38}$$

and the circular frequency in this constrained mode is given by

$$\omega^2 = \frac{\mu_1{}^2\omega_1{}^2 + \mu_2{}^2\omega_2{}^2 + \cdots + \mu_n{}^2\omega_n{}^2}{\mu_1{}^2 + \mu_2{}^2 + \cdots + \mu_n{}^2} \tag{57.39}$$

Consider now the effect of varying the constraints. Suppose, for instance, that $(n - 1)$ of the quantities are very small in comparison with μ_i: then, since the constraints are insisting that the system move approximately in the ith mode, by Eq. (57.39) $\omega \approx \omega_i$ very closely. This is the basis of the Rayleigh method, which assumes that an elastic body maintains a geometrical shape while vibrating, so as to obtain the frequencies of beams, membranes, plates, etc. (q.v.). It also shows that, since ω^2 clearly lies between the values of the least and greatest of the normal mode frequencies

$\omega_1^2, \omega_2^2, \ldots, \omega_n^2$, then, if further constraints μ are imposed to approximate the lowest natural mode shape, the approximate frequency ω as calculated can only be higher than (or perhaps equal to) ω_1, but not lower.

When all the quantities μ become very small except μ_i, the value of ω^2 will differ from ω_i^2 by small quantities of the second order. Thus, the frequency of the constrained system "has a stationary value for those constraints which make the vibration a normal vibration of the unconstrained system" (Whittaker).

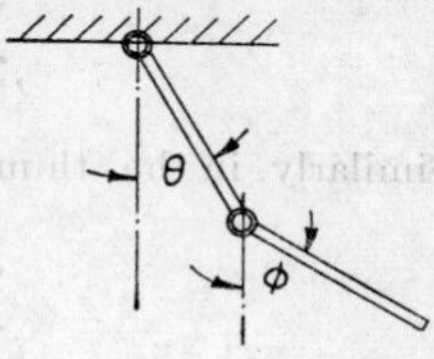

FIG. 57.10

An example of the use of this feature to find the mode and frequencies of a system with two degrees of freedom follows.

Example 10. Two equal uniform rods of length l are hinged together to form a double pendulum as in Fig. 57.10. The angles that each rod makes with the vertical are respectively θ and ϕ. Then, for small values of these angles,

$$\begin{aligned} T &= \tfrac{1}{2}(\tfrac{1}{3}ml^2)\dot{\theta}^2 + \tfrac{1}{2}m(l\dot{\theta} + \tfrac{1}{2}l\dot{\phi})^2 + \tfrac{1}{2}ml^2\dot{\phi}^2/12 \\ &= \tfrac{1}{2}ml^2(\tfrac{4}{3}\dot{\theta}^2 + \dot{\theta}\dot{\phi} + \tfrac{1}{3}\dot{\phi}^2) \\ V &= \tfrac{1}{2}mgl[\tfrac{1}{2}\theta^2 + (\theta^2 + \tfrac{1}{2}\phi^2)] \end{aligned} \tag{57.40}$$

Let the constraint be $\phi = \mu\theta$, but in this μ remains an unspecified parameter. Then, let θ be the chosen coordinate in which T and V are written

$$T = \tfrac{1}{2}ml^2(\tfrac{4}{3} + \mu + \tfrac{1}{3}\mu^2)\dot{\theta}^2 \qquad V = \tfrac{1}{2}mgl(\tfrac{3}{2} + \tfrac{1}{2}\mu^2)\theta^2$$

so that

$$\omega^2 = \frac{3g}{2l}\,\frac{3 + \mu^2}{4 + 3\mu + \mu^2} \tag{57.41}$$

For μ to reflect a principal mode, ω^2 will have a stationary value, or

$$\frac{d\omega^2}{d\mu} = \frac{3g}{2l}\,\frac{2\mu(4 + 3\mu + \mu^2) - (3 + \mu^2)(3 + 2\mu)}{(4 + 3\mu + \mu^2)^2} = 0$$

which means that $3\mu^2 + 2\mu - 9 = 0$, whence

$$\mu = \tfrac{1}{3}(-1 \pm \sqrt{28}) \qquad \text{i.e., } 1.43 \text{ or } -2.10$$

Substitution of these values of μ into Eq. (57.41) will give the two natural frequencies,

$$\omega^2 = \frac{3g}{l}\left(1 \pm \frac{2}{\sqrt{7}}\right)$$

for the lower of which the mode is pictured with the two parts swinging in the ratio $\phi = 1.43\theta$, and for the higher of which the two parts swing against each other with $\phi = -2.10\theta$.

57.11. ORTHOGONALITY OF THE NORMAL MODES

Suppose first that a certain multiple-mode linear system has only static coupling. Its equations of free oscillation will be

$$\begin{aligned} m_1\ddot{q}_1 + k_{11}q_1 + k_{12}q_2 + \cdots + k_{1n}q_n &= 0 \\ m_2\ddot{q}_2 + k_{21}q_1 + k_{22}q_2 + \cdots + \cdots &= 0 \\ \cdots\cdots\cdots\cdots\cdots\cdots\cdots\cdots & \end{aligned} \tag{57.42}$$

in which $k_{ij} = k_{ji}$. Suppose that the solution of these has yielded the n eigenvalues, of which any two are ω_r and ω_s. Then, application of these to the above equations will give the relationship between the amplitudes a_{ir} of the displacements q_i in the rth mode,

$$\sum_{j=1}^{n} k_{ij}a_{jr} = \omega_r^2 m_i a_{ir} \qquad i = 1, 2, \ldots, n \tag{57.43a}$$

Similarly, in the sth mode,

$$\sum_{i=1}^{n} k_{ji}a_{is} = \omega_s^2 m_j a_{js} \qquad j = 1, 2, \ldots, n \tag{57.43b}$$

If we multiply Eq. (57.43*a*) by a_{is} and then take the sum over all the is, and if we multiply Eq. (57.43*b*) by a_{jr} and take the sum over all the js, we obtain

$$\begin{aligned}\sum_{i=1}^{n}\sum_{j=1}^{n} k_{ij}a_{jr}a_{is} &= \omega_r^2 \sum_{i=1}^{n} m_i a_{ir}a_{is}\\ \sum_{i=1}^{n}\sum_{j=1}^{n} k_{ji}a_{is}a_{jr} &= \omega_s^2 \sum_{j=1}^{n} m_j a_{js}a_{jr}\end{aligned} \tag{57.44a,b}$$

Because of $k_{ij} = k_{ji}$, the left-hand sides of these equations are equal, and the sums to the right differ only by the notation used for the summation index. Hence, by subtraction,

$$(\omega_r^2 - \omega_s^2) \sum_{i=1}^{n} m_i a_{ir}a_{is} = 0$$

and, therefore,

$$\sum_{i=1}^{n} m_i a_{ir}a_{is} = 0 \qquad \text{for } \omega_r \neq \omega_s \tag{57.45}$$

This relation is known as the *orthogonality relation* for the principal modes of oscillation. It may be used to check solutions found by other methods, or used directly to obtain solutions beyond the first.

That the two principal motions of a sprung mass free to oscillate in a plane are geometrically orthogonal is shown by Hansen and Chenea [5], pp. 114–116. For the extension to three or more dimensions, see [29], pp. 289–291; [30], pp. 172–175; and [25], p. 129.

Example 11. The system shown in Fig. 57.8 has static coupling. Equations (57.30) show the coefficients to be $m_1 = m$, $m_2 = I = ma^2/4$. The amplitudes for the two modes are given in Eqs. (57.33), and it follows from Eq. (57.45) that

$$mx_{01}x_{02} + I\frac{2x_{01}}{a}\frac{-2x_{02}}{a} = 0$$

which agrees with the results obtained.

The notion of orthogonality may be extended to include systems having dynamic coupling, though not in the form (57.45). The array of equations for free vibrations of small amplitude of a system with n degrees of freedom is

$$\begin{aligned} m_{11}\ddot{q}_1 + m_{12}\ddot{q}_2 + \cdots + k_{11}q_1 + k_{12}q_2 + \cdots &= 0\\ m_{21}\ddot{q}_1 + m_{22}\ddot{q}_2 + \cdots + k_{21}q_1 + k_{22}q_2 + \cdots &= 0\end{aligned} \tag{57.46}$$

These may be derived as Lagrange's equations from expressions (57.22*a*,*b*), for T and V. Consequently, $m_{ij} = m_{ji}$ and $k_{ij} = k_{ji}$, since each pair stems from one coefficient in T or V, respectively. By following a procedure similar to that explained before, the following, instead of Eqs. (57.44), may be obtained:

$$\sum_{i=1}^{n}\sum_{j=1}^{n} k_{ij}a_{jr}a_{is} = \omega_r^2 \sum_{i=1}^{n}\sum_{j=1}^{n} m_{ij}a_{jr}a_{is}$$
$$\sum_{i=1}^{n}\sum_{j=1}^{n} k_{ji}a_{is}a_{jr} = \omega_s^2 \sum_{i=1}^{n}\sum_{j=1}^{n} m_{ji}a_{is}a_{jr} \qquad (57.47)$$

Again the left-hand sides are the same, and so are the double sums on the right-hand sides. This leads to

$$\sum_{i=1}^{n}\sum_{j=1}^{n} m_{ij}a_{is}a_{jr} = 0 \quad \text{for } \omega_r \neq \omega_s \qquad (57.48)$$

and this will hold for all systems with both static and dynamic coupling. For example, it may be applied to check the example of Fig. 57.10; in this, $m_{12} = \frac{1}{2}ml^2$, which is ascertained from the expression (57.40) for the kinetic energy.

57.12. A MATRIX ITERATION METHOD

The problem of finding the eigenvalues buried within a determinant for a system with very many degrees of freedom calls for some special method. One such is an iterative method by which, after several repetitions, a close approximation to certain eigenvalues may be obtained, and at the same time an approximation to the corresponding mode shape (see p. 18–6).

For a demonstration of the method, consider a conservative system having only static coupling. After substitution of the usual assumption that each coordinate has a harmonic solution

$$q_i = a_i \sin(\omega t + \phi)$$

we are faced with a set of equations,

$$k_{j1}a_1 + k_{j2}a_2 + \cdots + k_{jn}a_n = m_j\omega^2 a_j \qquad j = 1, 2, \ldots, n \qquad (57.49)$$

If each of these is divided by its own coefficient m_j, all the terms on the right side have the common factor ω^2; and the array is written

$$\begin{bmatrix} \lambda_{11} & \lambda_{12} & \cdots & \lambda_{1n} \\ \lambda_{21} & \lambda_{22} & \cdots & \lambda_{2n} \\ \cdots & \cdots & \cdots & \cdots \\ \lambda_{n1} & \lambda_{n2} & \cdots & \lambda_{nn} \end{bmatrix} \begin{bmatrix} a_1 \\ a_2 \\ \cdot \\ a_n \end{bmatrix} = \omega^2 \begin{bmatrix} a_1 \\ a_2 \\ \cdots \\ a_n \end{bmatrix} \qquad (57.50)$$

or, more concisely,

$$\Lambda a = \omega^2 a \qquad (57.51)$$

In this, the value of each term $\lambda_{ij} = k_{ij}/m_i$ is known numerically, and the problem is to find the vector a for which the effect of multiplication by the matrix operator Λ is merely to lengthen it by a scalar factor ω^2.

The procedure is as follows: Guess a series of amplitudes ${}_0A_1, {}_0A_2, \ldots, {}_0A_n$ (probably a series of unit values) to be the zero-order approximations for $a_1, a_2, \ldots, a_n$. Here the prefix subscript refers to the order of the approximation. Substitute these values, premultiply by Λ, and find the column ${}_1B$; take the largest value among its elements ${}_1B_j$, and factor it out to obtain the column ${}_1A$:

$$\Lambda \begin{bmatrix} {}_0A_1 \\ {}_0A_2 \\ \cdot \\ {}_0A_n \end{bmatrix} = \begin{bmatrix} {}_1B_1 \\ {}_1B_2 \\ \cdot \\ {}_1B_n \end{bmatrix} = {}_1\omega^2 \begin{bmatrix} {}_1A_1 \\ {}_1A_2 \\ \cdot \\ {}_1A_n \end{bmatrix} \qquad (57.52)$$

This largest value ${}_1B_j = {}_1\omega^2$ will serve as a first approximation to one of the values ω^2, and the remaining vector components ${}_1A_j$ will be first approximations to the relative amplitudes, or improvements on the values of the guesses ${}_0A_j$.

The process is then repeated by putting the new ${}_1A_j$ into the left-hand side of Eq. (57.50) thus,

$$\Lambda\, {}_1A_j = {}_2B_j = {}_2\omega^2\, {}_2A_j$$

and a second improved set of values for A and ω^2 is obtained. When, after p such iterations, the new set of values ${}_pA_j$ agrees with the previous set ${}_{(p-1)}A_j$ to the desired number of decimal places, the values both of a natural frequency and of the relative amplitudes of the coordinates for that frequency are obtained.

The process will converge toward the highest natural frequency of the system (see, for example, [7], p. 68, and [24], p. 181).

Example 12. A statically coupled system with three degrees of freedom has the following equations relating the amplitudes a_i and the frequency ω, when harmonic motion is assumed:

$$\begin{aligned} (16 - 3\omega^2)a_1 \quad - 4a_2 \qquad\qquad &= 0 \\ -4a_1 + (2 - \omega^2)a_2 \quad - 3a_3 &= 0 \\ -3a_2 + (16 - 3\omega^2)a_3 &= 0 \end{aligned} \qquad (57.53a\text{–}c)$$

In order to demonstrate the efficacy of the iteration process and the nearness of the approximation, we shall do this by the determinant method first. From Lagrange's determinant, the frequency equation is found to be

$$(16 - 3\omega^2)[(16 - 3\omega^2)(2 - \omega^2) - 25] = 0$$

for which the roots are

$$\omega^2 = \tfrac{1}{3},\ \tfrac{16}{3},\ \text{or } 7$$

For the third mode, Eqs. (57.53) give the amplitudes

$$a_1 = -0.80a_2; \quad a_2; \quad a_3 = -0.60a_2 \qquad (57.54)$$

Now, commencing the matrix iteration method, we have, for Eq. (57.52),

$$\Lambda\{A\} \equiv \begin{bmatrix} \tfrac{16}{3} & -\tfrac{4}{3} & 0 \\ -4 & 2 & -3 \\ 0 & -1 & \tfrac{16}{3} \end{bmatrix} \begin{bmatrix} {}_0A_1 \\ {}_0A_2 \\ {}_0A_3 \end{bmatrix} = \begin{bmatrix} {}_1B_1 \\ {}_1B_2 \\ {}_1B_3 \end{bmatrix} \qquad (57.55)$$

For the highest mode, two nodes are expected, and so we guess

$$\{{}_0A_1, {}_0A_2, {}_0A_3\} = \{1, -1, 1\}$$

This makes $\{{}_1B_1, {}_1B_2, {}_1B_3\} = \{\tfrac{20}{3}, -9, \tfrac{19}{3}\}$. Dividing by the largest (that is, 9), we obtain ${}_1\omega^2 = 9$ and $\{{}_1A_1, {}_1A_2, {}_1A_3\} = \{\tfrac{20}{27}, -1, \tfrac{19}{27}\}$. Then, by proceeding to operate again as described, the values shown below are reached in succession:

	Iteration number				
	0	1	2	3	4
a_1	1	+0.741	+0.747	+0.760	0.769
a_2	−1	−1.000	−1.000	−1.000	−1.000
a_3	1	+0.704	+0.672	+0.654	0.641
ω^2	—	9.000	7.074	7.0035	7.0002

At this stage, ω^2 appears to have reached its asymptote more nearly, and use of the separate equations (57.53) might help to reach the results (57.54) more quickly.

Once the value of one natural frequency has been obtained, the equations of motion, the coefficient relations (57.53), and the Λ matrix (57.55) may be reduced by one degree of freedom by use of the principle of orthogonality. For, by Eq. (57.45),

$$m_1 a_{1(3)} a_1 + m_2 a_{2(3)} a_2 + m_3 a_{3(3)} a_3 = 0 \tag{57.56}$$

where a_1, a_2, a_3 without a second subscript apply to any one mode, not the third (about which we already have found the relations). Substituting these, Eq. (57.56) then reads

$$3(0.8)a_1 - a_2 + 3(0.6)a_3 = 0 \tag{57.57}$$

This may be regarded then as an equation for a_3 in terms of a_1 and a_2, by means of which a_3 can be eliminated from Eq. (57.53*b*). Thus, for modes other than the third,

$$(16 - 3\omega^2)a_1 - 4a_2 = 0 \qquad -4a_1 + (2 - \omega^2)a_2 + (4a_1 - 5\!/\!_3 a_2) = 0$$

and from this pair of equations the two other roots ω^2 and the corresponding relative values of a_1 and a_2 may be found. For the values of a_3, recourse is taken to Eq. (57.57), or back to Eq. (57.53).

The matrix iteration method may be adjusted a little, so that it first yields the value of the lowest natural frequency of the system, instead of the highest. This is achieved by inverting the system of equations (57.51) so that the factor is the reciprocal $1/\omega^2$.

Let Eqs. (57.49) be inverted; define some terms F_j by

$$\begin{aligned} k_{11}a_1 + k_{12}a_2 + \cdots + k_{1n}a_n &= F_1 \\ k_{21}a_1 + k_{22}a_2 + \cdots + k_{2n}a_n &= F_2 \\ \cdots\cdots\cdots\cdots\cdots\cdots & \quad \text{etc.} \end{aligned} \tag{57.58}$$

and solve for a_1, a_2, obtaining

$$\begin{aligned} a_1 &= \nu_{11}F_1 + \nu_{12}F_2 + \cdots + \nu_{1n}F_n \\ a_2 &= \nu_{21}F_1 + \nu_{22}F_2 + \cdots + \nu_{2n}F_n \\ \cdots\cdots\cdots\cdots\cdots\cdots & \quad \text{etc.,} \end{aligned}$$

where the ν's are a new set of constants. Then, since $F_j = m_j\omega^2 a_j$, write the matrix equation

$$N\{a_j\} = \frac{1}{\omega^2}\{a_j\} \tag{57.59}$$

where the element of N in the ith row and jth column is $\nu_{ij}m_j$. From here on, the procedure is exactly as it was when using Eq. (57.51), starting with a good guess for the values ${}_0A_1$, ${}_0A_2$, . . . , and finding a convergence toward the maximum reciprocal eigenvalue $1/\omega^2$.

Equation (57.59) may also be applied directly for the solution of systems with dynamic instead of static coupling.

Other matrix methods are available for finding the natural frequencies of systems with a large number of degrees of freedom. Two that may be mentioned are the Hestenes-Stiefel method of *conjugate gradients* [6] and Lanczos' method [12],[13]. The Lanczos method has one great advantage over the method described in the previous section, in that, for a system having n degrees of freedom, precisely n iterations will produce an exact solution; it does not approach the solution asymptotically but absolutely.

Indications of a new method of tackling a very large or complex system, by which the system may be torn apart, the pieces solved more simply and then interconnected, appear in an article by Kron [11].

The great interest in matrix methods for the solution of problems of this sort derives from the ease with which they may be assigned to an electronic computer.

57.13. SYSTEMS OF CHAIN FORM

In a system of chain form, each mass is coupled statically only with its two neighbors. In their Lagrangian determinant, the principal diagonal and the two adjacent diagonals contain nonvanishing elements, but all remaining elements are zero.

The best-known method of solution for this type is the Holzer tabular method (see p. 58–9). Moreover, this method has been elaborated by Myklestad and Prohl (see p. 61–31) to make it applicable to continuous beams and airplane wing structures.

A more rapid method of finding the eigenvalues of such a system is, however, furnished by the algorithm of Crossley and Germen [32], which yields the coefficients of the frequency equation. The masses and spring constants are, beginning at one end, denoted by k_{01}, m_1, k_{12}, m_2, k_{23}, ... In free-ended chains one or both of k_{01} and $k_{n(n+1)}$ are absent. These numbers are processed as shown in the following example.

Assume $m_1 = 2$, $m_2 = 1$, $m_3 = 3$, $m_4 = 4$, and $k_{01} = 10$, $k_{12} = 2$, $k_{23} = 3$, $k_{34} = 12$, $k_{45} = 0$. Calculate the values k_{01}/m_1, k_{12}/m_1, k_{12}/m_2, k_{23}/m_2, . . . , arrange them in the first column of the following array, and call their sum A.

5	13	65	40	200	6	30
1	11	11	24	24		$D = 30$
2	8	16	3	6		
3	7	21		$C = 230$		
1	3	3				
4		$B = 116$				
3						
$A = 19$						

For a second column, add the terms of the first column starting from the bottom, placing the accumulating sums two rows higher as each term is added. For the third column, multiply first- and second-column terms, add and call the sum B. For the fourth (and every even-numbered) column, add the terms of the preceding column from the bottom, moving again two rows higher. For the fifth (and every odd-numbered) column, multiply the preceding and the first column. Continue until no terms are left. The polynomial expansion of the Lagrangian determinant is then

$$\omega^8 - A\omega^6 + B\omega^4 - C\omega^2 + D = 0$$

57.14. DAMPED FREE OSCILLATIONS OF SYSTEMS

The preceding several sections have been describing theory and methods of analysis applicable when damping may be ignored, for the reasons outlined in Sec. 57.3. However, it may be desirable to include a consideration of damping, if it is thought that its presence will shift the values of the natural vibrational frequencies considerably, or if it may cause instability.

In a system having only a single degree of freedom, the equation governing its damped free oscillation is

$$(mD^2 + cD + k)q = 0$$

in which m, c, and k are constants; D is the differential operator with respect to time; and it is assumed that damping occurs only in proportion to the velocity (any other damping law will render the system nonlinear).

For a system with n degrees of freedom, there will be as many equations as coordinates; and possibly every coordinate q with both its first and second derivatives may appear in each equation. It is thus convenient to introduce the differential operator,

$$e_{ij} = m_{ij}D^2 + c_{ij}D + k_{ij} \tag{57.60}$$

in which m_{ij}, c_{ij}, k_{ij} are respectively the coefficients of dynamic, friction, and static coupling when $i \neq j$, and the array of n equations will appear as

$$\begin{aligned} e_{11}q_1 + e_{12}q_2 + e_{13}q_3 + \cdots + e_{1n}q_n &= 0 \\ e_{21}q_1 + e_{22}q_2 + \cdots + \quad\quad + e_{2n}q_n &= 0 \\ \cdots\cdots\cdots\cdots\cdots\cdots\cdots\cdots &\quad \text{etc.} \end{aligned} \tag{57.61}$$

or, in matrix notation,

$$E\{q\} = \{0\} \tag{57.62}$$

This is a linear system of homogeneous equations, and, although the solution can no longer be written as a sine or cosine as in the undamped system, it is known to be an exponential. A solution such as

$$q_i = a_i e^{\lambda t} \qquad i = 1, 2, \ldots, n \tag{57.63}$$

is therefore assumed, and substitution converts Eqs. (57.61) into an array of linear algebraic equations in the n different amplitudes a_i. Each equation is also a quadratic in λ. The Lagrange determinant then is

$$L = \begin{vmatrix} \lambda_{11} & \lambda_{12} & \cdots & \lambda_{1n} \\ \lambda_{21} & \lambda_{22} & \cdots & \cdots \\ \cdots & \cdots & \cdots & \cdots \\ \lambda_{n1} & \lambda_{n2} & \cdots & \lambda_{nn} \end{vmatrix} = 0 \tag{57.64}$$

for which every element may be found directly from the corresponding element of the matrix E by changing D in Eq. (57.60) into λ, that is,

$$\lambda_{ij} = m_{ij}\lambda^2 + c_{ij}\lambda + k_{ij} \tag{57.65}$$

and, since λ is expected to be complex, every term of L is complex.

Expanding this into the characteristic equation, then, will produce a polynomial in λ of degree $2n$, and in this it differs from the undamped case, in which a polynomial of degree n in λ^2 arose. The problem of determining the roots of such a polynomial has already been discussed. Since the general form of each of the roots is complex, they will occur in conjugate pairs, such as

$$\lambda_{p,q} = r \pm i\omega$$

which means that the coordinate q_i will have the form

$$q_i = e^{rt}(a_p \sin \omega t + a_q \cos \omega t)$$

Thus, if the real part r of any complex root of λ should be positive, or when a real root should occur, if that real root should be positive, q_i would have a solution which would increase exponentially with time. This indicates an unstable condition. The conditions which preclude the appearance of such instability are called the *Routh-Hurwitz criteria.*

57.15. THE ROUTH-HURWITZ STABILITY CRITERIA

The determinant L of Eq. (57.64), when expanded, becomes a polynomial

$$P(\lambda) = a_0\lambda^n + a_1\lambda^{n-1} + a_2\lambda^{n-2} + \cdots + a_{n-1}\lambda + a_n \tag{57.66}$$

It has been noted that there will be stability of the system if, and only if, all the real roots and the real part of all complex roots of $P(\lambda)$ are negative, i.e., when all the zeros of $P(\lambda)$ lie in the left half of the complex (Argand) plane. A polynomial satisfying this condition is known as a *Hurwitz polynomial.*

The necessary and sufficient conditions that a polynomial P should be a Hurwitz polynomial are called, in dynamics, Routh's stability criteria [23], but, in electrical work, the Hurwitz criteria [8].

1. A necessary condition for such stability is that every coefficient a should be positive; none can be zero.

2. The necessary and sufficient conditions for the polynomial to be a Hurwitz polynomial are that all those determinants Δ should be positive, which are formed as follows:

$$\Delta_1 = a_1 \qquad \Delta_2 = \begin{vmatrix} a_1 & a_0 \\ a_3 & a_2 \end{vmatrix} \qquad \Delta_3 = \begin{vmatrix} a_1 & a_0 & 0 \\ a_3 & a_2 & a_1 \\ a_5 & a_4 & a_3 \end{vmatrix}$$

$$\Delta_r = \begin{vmatrix} a_1 & a_0 & 0 & 0 & 0 \\ a_3 & a_2 & a_1 & a_0 & 0 \\ a_5 & a_4 & a_3 & a_2 & a_1 & \cdots & 0 \\ \cdots & \cdots & \cdots & \cdots & \cdots & \cdots & \cdots \\ a_{2r-1} & a_{2r-2} & a_{2r-3} & a_{2r-4} & & \cdots & a_r \end{vmatrix} \tag{57.67}$$

for $r = 1, 2, \ldots, n$

In writing the general rth determinant, it is assumed (1) that Δ_r has only r columns and rows; (2) that, wherever an element a_k is called for where $k > n$, a zero is written in Δ.

Proofs may be found in the original references, or in [4], pp. 396–408. Conditions when the coefficients a are complex numbers, which happens in flutter analysis, are outlined [24], p. 295, and [3], pp. 151–155.

Example 13. Determine whether the system is stable which has the characteristic equation

$$\omega^6 + 2\omega^5 + 5\omega^4 + 7\omega^3 + 6\omega^2 + 4\omega + 3 = 0$$

Then

$$\Delta_1 = 2 \qquad \Delta_2 = \begin{vmatrix} 2 & 1 \\ 7 & 5 \end{vmatrix} = +3 \qquad \Delta_3 = \begin{vmatrix} 2 & 1 & 0 \\ 7 & 5 & 2 \\ 4 & 6 & 7 \end{vmatrix} = +5$$

$$\Delta_4 = \begin{vmatrix} 2 & 1 & 0 & 0 \\ 7 & 5 & 2 & 1 \\ 4 & 6 & 7 & 5 \\ 0 & 3 & 4 & 6 \end{vmatrix} = +20 \qquad \Delta_5 = \begin{vmatrix} 2 & 1 & 0 & 0 & 0 \\ 7 & 5 & 2 & 1 & 0 \\ 4 & 6 & 7 & 5 & 2 \\ 0 & 3 & 4 & 6 & 7 \\ 0 & 0 & 0 & 3 & 4 \end{vmatrix} = -25$$

Thus, with Δ_5 negative, the system is unstable.

A dynamically unstable system is subject to oscillations which appear and build up, as though in a single-mode system there were a negative damping term. Such motions are called *self-excited* and *relaxation* oscillations, and are dealt with elsewhere: two commonly occurring types are airplane flutter (Chap. 63) and the hunting of a servomechanism.

57.16. LAGRANGE'S EQUATION FOR THE NONCONSERVATIVE CASE

The equations of a multiple-mode system that is forced and damped may be obtained from a form of Lagrange's equation that is a modification of those obtained in Sec. 57.4.

If the generalized force Q_r in Eq. (57.4) can be derived from a potential, the

Lagrange equation assumes the form (57.5). If Q_r is the sum of several forces, it may happen that only some of them can be derived from a potential. In this case, Eq. (57.4) may be written as

$$\frac{d}{dt}\left(\frac{\partial T}{\partial \dot{q}_r}\right) - \frac{\partial T}{\partial q_r} + \frac{\partial V}{\partial q_r} = Q_r \tag{57.68}$$

where Q_r refers to all those forces different from potential forces. Such a force might be a sinusoidal exciting force in an x direction, so that

$$Q_x = F_0 \sin \omega t$$

When damping is to be reckoned with, and if damping is proportional to velocity, then a force such as

$$F_r = -\sum_s c_{rs}\dot{q}_s$$

must be included in Eq. (57.68) as part of Q_r. Frictional forces of this type may be derived from a function $\mathfrak{F}$, known as *Rayleigh's dissipation function*,[1] which is defined as

$$\mathfrak{F} = \tfrac{1}{2}\sum_r \sum_s c_{rs}\dot{q}_r\dot{q}_s \tag{57.69}$$

and yields

$$F_r = -\frac{\partial \mathfrak{F}}{\partial \dot{q}_r} \tag{57.70}$$

It is plain that $\mathfrak{F}$, like the kinetic energy T, is a positive-definite and homogeneous quadratic function of the velocities. Now, peeling off, as it were, these frictional forces from the generalized force Q_r in Eq. (57.68), Lagrange's equation becomes

$$\frac{d}{dt}\left(\frac{\partial T}{\partial \dot{q}_r}\right) - \frac{\partial T}{\partial q_r} + \frac{\partial V}{\partial q_r} + \frac{\partial \mathfrak{F}}{\partial \dot{q}_r} = Q_r \tag{57.71}$$

and this may be applied to each coordinate successively.

57.17. RESPONSE OF A DOUBLY FREE SYSTEM TO HARMONIC STIMULUS

The oscillations of a system having only two degrees of freedom illustrate a number of concepts which are most easily exemplified here before taking up the system with a larger number of coordinates.

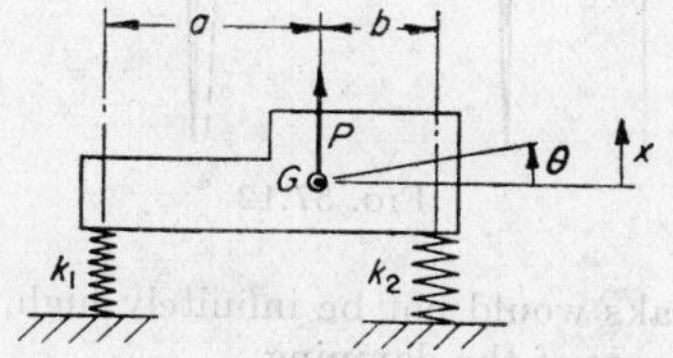

FIG. 57.11

The equations of motion for a typical undamped system such as shown in Fig. 57.11 are:

$$\begin{aligned} m\ddot{x} + (k_1 + k_2)x - (k_1a - k_2b)\theta &= P\cos pt \\ I\ddot{\theta} - (k_1a - k_2b)x + (k_1a^2 + k_2b^2)\theta &= 0 \end{aligned} \tag{57.72}$$

[1] Ref. [22], pp. 102, 120; also [29], p. 281; [30], p. 219; [31], p. 232.

In this, it is presumed that the center of mass is at G, located a distance a from the vertical line of the left spring (stiffness k_1), and a distance b from the other spring (stiffness k_2). The exciting force P is applied to G vertically.

As with a system having a single degree of freedom, the steady-state forced vibration will have a circular frequency agreeing with the excitation; therefore, assume that

$$x = a_1 \cos pt \qquad \theta = a_2 \cos pt \tag{57.73}$$

and solve for the amplitudes a_1 and a_2. If, first, Eqs. (57.72) are written here as

$$mx'' + k_{11}x + k_{12}\theta = P \cos pt \qquad I\theta'' + k_{12}x + k_{22}\theta = 0 \tag{57.74}$$

for the sake of the abbreviated form, then the substitution of (57.73) gives the nonhomogeneous algebraic equations,

$$(k_{11} - mp^2)a_1 + k_{12}a_2 = P \qquad k_{12}a_1 + (k_{22} - Ip^2)a_2 = 0 \tag{57.75}$$

whence

$$a_1 = \frac{P(k_{22} - Ip^2)}{\Delta} \qquad a_2 = -\frac{k_{12}P}{\Delta} \tag{57.76a,b}$$

where Δ is the Lagrangian determinant of the free system, that is,

$$\Delta = \begin{vmatrix} (k_{11} - mp^2) & k_{12} \\ k_{12} & (k_{22} - Ip^2) \end{vmatrix}$$

But it is known that this determinant vanishes when p equals either ω_1 or ω_2, the two natural frequencies; and, at these two frequencies, both a_1 and a_2 become infinite; that is, there are two resonances.

The plot of the amplitudes of a_1 and a_2 against the variable frequency p shows the response of the system. It will have the form of Fig. 57.12. If there were damping

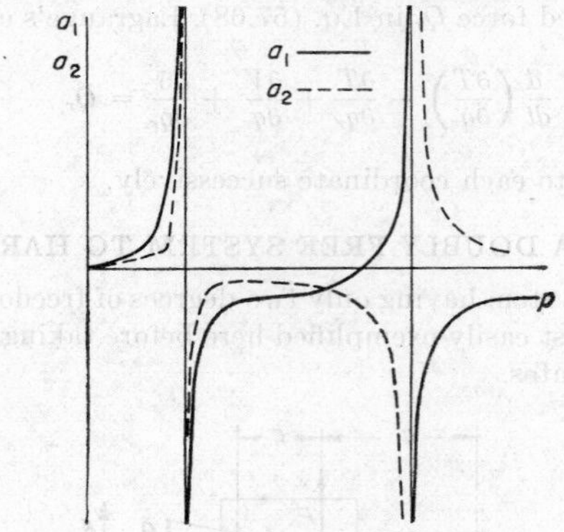

FIG. 57.12

in the system, the two peaks would not be infinitely high, but would reach summits depending on the effectiveness of the damping.

Equations for a typical damped and forced system with two degrees of freedom are:

$$\begin{aligned} m_1\ddot{x} + c_{11}\dot{x} + k_{11}x + c_{12}\dot{y} + k_{12}y &= P \cos pt \\ m_2\ddot{y} + c_{12}\dot{x} + k_{12}x + c_{22}\dot{y} + k_{22}y &= 0 \end{aligned} \tag{57.77}$$

For this it is not practical to assume solutions of the form (57.73), because the damping will cause a phase difference between P and the motions. Solutions such as

$$x = a_1 \cos pt + b_1 \sin pt \qquad y = a_2 \cos pt + b_2 \sin pt \tag{57.78}$$

may be taken, but, as a result, four algebraic equations for the coefficients a_1, a_2, b_1, and b_2 must be solved.

A recommended alternative utilizes complex notation. To start this method, the right-hand side of Eqs. (57.77) is changed to

$$P \cos pt = \Re(Pe^{ipt})$$

and then, of course, the solutions represented by Eqs. (57.78) may be expected in the form $x = \Re(Xe^{ipt})$, $y = \Re(Ye^{ipt})$, where X and Y are complex numbers. The real and imaginary parts of these *complex amplitudes* will then represent the components of the amplitudes of x and y that are respectively in phase and 90° out of phase with F; that is, for x,

$$x = \Re(Xe^{ipt}) = a_1 \cos pt + b_1 \sin pt$$

with $X = a_1 - ib_1$. The total amplitude of x is $\sqrt{a_1^2 + b_1^2}$. Substituting these assumed solutions into Eqs. (57.77),

$$\begin{aligned} [(k_{11} - m_1p^2) + ic_{11}p]X + (k_{12} + ic_{12}p)Y &= P \\ (k_{12} + ic_{12}p)X + [(k_{22} - m_2p^2) + ic_{22}p]Y &= 0 \end{aligned} \qquad (57.79)$$

whence X and Y may be solved for very simply. In a complex plane, Xe^{ipt} and Ye^{ipt} are vectors rotating with angular velocity p, and their projection on the real axis at any time gives the values of x and y, respectively.

57.18. THE DYNAMIC ABSORBER

The dynamic absorber, the tuned damper, and the Frahm damper are various names, more or less synonymous, for a device by which a certain body, which tends to vibrate severely because of an exciting force, can be most effectively reduced to near rest.

The way in which the undamped absorber works may be seen quite clearly from the equations of the preceding section. Because its action depends upon the choice of its natural frequency, and because no damping exists, it should hardly be called a damper.

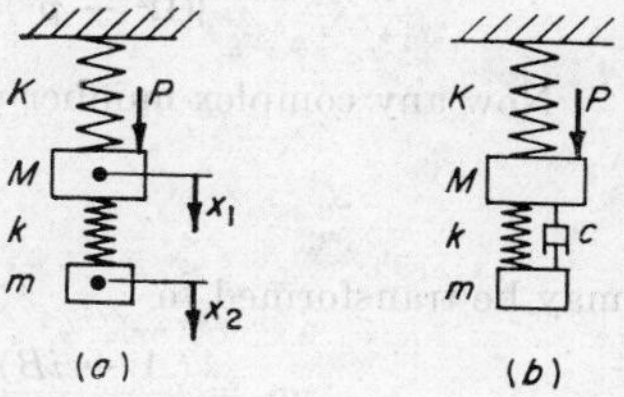

FIG. 57.13

Suppose that, in the system of Fig. 57.13a, the force $P \cos pt$ acts upon the combination K, M, with a frequency so nearly resonant that the response of mass M is objectionably large. The extra spring and mass, k, m, are added as a Frahm absorber. The equations of the resulting response will be found from the equations of motion:

$$M\ddot{x}_1 + Kx_1 + k(x_1 - x_2) = P \cos pt \qquad m\ddot{x}_2 + k(x_2 - x_1) = 0 \qquad (57.80)$$

Compare these with Eqs. (57.74) of the preceding section, and it follows immediately, from the resulting equation (57.76a) of that argument, that the amplitude of x_1 in this case is

$$a_1 = \frac{P(k - mp^2)}{\Delta} \qquad (57.81)$$

Thus, the response of the principal mass M may be reduced to zero, so that it does not move at all, at the one previously troublesome frequency $(K/M)^{1/2}$, by choosing the spring and mass of the absorber so that $k/m = K/M$.

This is very valuable when the impressed frequency p is invariant. But, if p is expected to change, damping is better added between the principal body and the absorber.

The analysis presented here of the damped system as shown in Fig. 57.13b follows that of Den Hartog [2], pp. 93–102. The governing equations are similar, of course, to Eq. (57.80), being

$$M\ddot{x}_1 + Kx_1 + c(\dot{x}_1 - \dot{x}_2) + k(x_1 - x_2) = P\cos pt$$
$$m\ddot{x}_2 + c(\dot{x}_2 - \dot{x}_1) + k(x_2 - x_1) = 0 \tag{57.82}$$

The amplitude of x_1 is of principal interest, and this may be obtained by the process, described in the last section, of using complex assumed results, after changing $P\cos pt$ to the complex vector form Pe^{ipt}. Thus, if a_1 and a_2 are respectively the complex amplitudes of x_1 and x_2, with circular frequency p,

$$[(K + k - Mp^2) + ipc]a_1 - (k + ipc)a_2 = P$$
$$-(k + ipc)a_1 + [(k - mp^2) + ipc]a_2 = 0 \tag{57.83}$$

So

$$a_1 = \frac{[(k - mp^2) + ipc]P}{\Delta} \tag{57.84}$$

when

$$\Delta = [(K + k - Mp^2) + ipc][(k - mp^2) + ipc] - (k + ipc)^2$$
$$= [(K - Mp^2)(k - mp^2) - mkp^2] + ipc[K - (M + m)p^2] \tag{57.85}$$

The expressions (57.84) and (57.85) are simplified by the use of the following parameters. Let

$$\Omega^2 = K/M \qquad \omega^2 = k/m$$
$$c_c = 2m\Omega \qquad c/c_c = \gamma$$
$$\mu = m/M \qquad x_{st} = P/K$$

Then

$$a_1 = x_{st}\Omega^2 \frac{(\omega^2 - p^2) + 2i\gamma p\Omega}{[(\Omega^2 - p^2)(\omega^2 - p^2) - \mu p^2\omega^2] + 2i\gamma p\Omega[\Omega^2 - (1 + \mu)p^2]} \tag{57.86}$$

Now any complex number which is expressed as the ratio

$$a_1 = \frac{A + iB}{C + iD}$$

may be transformed to

$$a_1 = \frac{(A + iB)(C + iD)}{C^2 + D^2} = \frac{AC + BD}{C^2 + D^2} + i\frac{BC - AD}{C^2 + D^2}$$

and then the length of this vector on the complex plane is found from the root of the sums of the squares of these components as

$$X_1 = \sqrt{\frac{A^2 + B^2}{C^2 + D^2}}$$

That being so, from Eq. (57.86) the absolute value of amplitude a_1 is

$$X_1 = x_{st}\Omega^2\sqrt{\frac{(\omega^2 - p^2)^2 + 4\gamma^2p^2\Omega^2}{[(\Omega^2 - p^2)(\omega^2 - p^2) - \mu p^2\omega^2]^2 + 4\gamma^2p^2\Omega^2[\Omega^2 - (1 + \mu)p^2]^2}}$$

and we now set $f = \omega/\Omega$, for the internal frequency ratio of the system, and $g = p/\Omega$, for the ratio of frequency of excitation to frequency of basic system, then

$$\frac{X_1}{x_{st}} = \sqrt{\frac{(g^2 - f^2)^2 + (2\gamma g)^2}{[\mu f^2g^2 - (g^2 - 1)(g^2 - f^2)]^2 + (2\gamma g)^2[(1 + \mu)g^2 - 1]^2}} \tag{57.87}$$

When certain numerical values are chosen for the parameters of the system, such as that

$$f = 1 \qquad \text{and} \qquad \mu = 1/20$$

then the response of the amplitude X_1 may be plotted for varying frequencies of excitation, for any chosen amount of damping given by the parameter γ. Such a graph appears as Fig. 57.14.

It is to be noted that all curves pass through two certain points, labeled P and Q

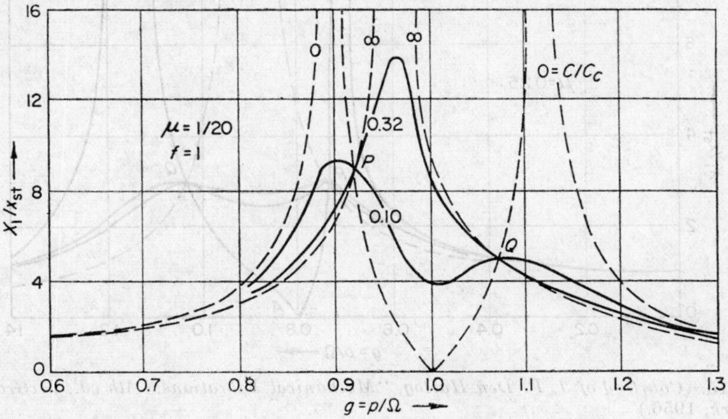

FIG. 57.14. (*Courtesy of J. P. Den Hartog, "Mechanical Vibrations," 4th ed., McGraw-Hill, New York*, 1956.)

in the figure. We proceed now to prove what is referred to as the *fixed-point theorem:* this is to say that there exist certain values of the abscissa g for which the value of the ordinate X_1/x_{st} is independent of the damping parameter γ, in Eq. (57.87).

This is plain enough; because, for any chosen value of g, Eq. (57.87) is seen to have the form

$$\left(\frac{X_1}{x_{st}}\right)^2 = \frac{A + B\gamma^2}{C + D\gamma^2}$$

where A, B, C, D are constants, and this will have a value independent of γ^2 provided that

$$A/C = B/D$$

When this requirement is expanded, it appears either that $g = 0$, which is trivial, or that

$$g^4 - 2g^2 \frac{1 + (1 + \mu)f^2}{2 + \mu} + \frac{2f^2}{2 + \mu} = 0 \tag{57.88}$$

There are two real roots, and therefore two points through which all curves in Fig. 57.14 must pass.

We turn then to consider optimum conditions: What is the best combination of frequency ω and damping c in designing an absorber for a given application?

The first thought is that, if every curve for varying c (or γ) must pass through two points such as P and Q, the best value of c is that which makes either point P or point Q a maximum. Let the curve have zero slope through the higher point, if this is possible.

The second thought comes from noting that the points P and Q must always lie on the lines of the particular curve $c/c_c = \gamma = 0$, and that the sum of the two roots in g^2 is found from the coefficient of the middle term of Eq. (57.88). Trial will show that, as P is lowered, Q is raised, by changing values of f. The optimum tuning of the absorber will occur when P and Q have equal ordinates.

Den Hartog [2] shows that, when P and Q have equal heights, as in Fig. 57.15, it is not actually possible to satisfy the first stipulation above; however, the curve plotted

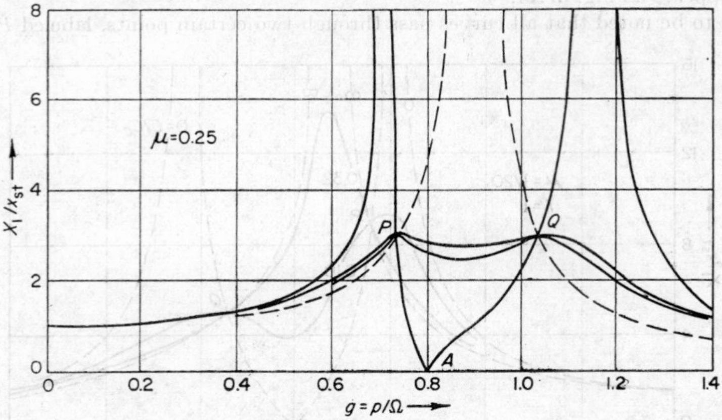

FIG. 57.15. (*Courtesy of J. P. Den Hartog, "Mechanical Vibrations," 4th ed., McGraw-Hill, New York,* 1956.)

shows that the maximum ordinate is very little greater than those of the fixed points. At the latter, the ordinate is

$$\frac{X_1}{x_{st}} = \sqrt{1 + \frac{2}{\mu}}$$

when the tuning of the absorber system is given by

$$f = \frac{1}{1 + \mu}$$

57.19. EXTENSION OF FIXED-POINT THEOREM TO MULTIPLE-COORDINATE SYSTEMS

It is shown by Lewis [15] that the theorem of the last section, by which the existence of two points P and Q in Fig. 57.14 is established, may be extended to systems with more than two degrees of freedom.

Consider an undamped system of $(n - 1)$ coordinates to which is added a damped absorber with characteristics k, c, m. For the special case with $c = 0$, there is no damping in the system at all, and the response curve must have n infinitely high peaks. Then, for another special case, let $c = \infty$, and thus the absorber mass is effectively glued to its neighboring mass: as a result, there are only $(n - 1)$ coordinates in independent motion, still with no damping, and the response curve will have $(n - 1)$ peaks, infinitely high.

Where these two response curves intersect, there are established $2(n - 1)$ points such as P, Q, . . . , through which every curve for $0 \leq c < \infty$ must pass.

57.20. GENERAL EQUATIONS OF FORCED OSCILLATIONS OF A LINEAR SYSTEM

The general equations of forced vibration of a system having n degrees of freedom may be taken from the equations of free oscillation of Sec. 57.14, and appear concisely in the form

$$[M]\{\ddot{q}\} + [C]\{\dot{q}\} + [K]\{q\} = \{F(t)\} \tag{57.89}$$

in which { } are column matrices, and [ ] are square matrices; respectively, $[M]$ is the inertia matrix, $[C]$ the damping matrix, and $[K]$ the stiffness matrix. This expression can be further shortened, as indicated in Sec. 57.14, to

$$[E]\{q\} = \{F(t)\} \tag{57.90}$$

in which the elements of the matrix E are differential operators.

The situation most commonly under study is one in which most of the forces $F(t)$ are zero but some are harmonically varying with time. Suppose then that

$$\begin{aligned} F_1(t) &= A \sin \omega_1 t + B \sin \omega_2 t \\ F_2(t) &= 0 \\ F_3(t) &= C \cos \omega_3 t \\ F_r(t) &= 0 \qquad \text{for } r = 4, 5, \ldots, n \end{aligned} \tag{57.91}$$

From the general theory of linear differential equations, it is known that the solution of an equation such as

$$[E]\{q\} = \{F(t)\} + \{G(t)\} + \{H(t)\} + \cdots$$

is the sum of the solutions of each equation such as (57.90), in which the functions on the right-hand side are taken one by one. Then the column matrix corresponding to the functions of the example (57.91) can be written as

$$F(t) = \begin{bmatrix} A \sin \omega_1 t \\ 0 \\ 0 \\ 0 \\ \cdots \\ 0 \end{bmatrix} + \begin{bmatrix} B \sin \omega_2 t \\ 0 \\ 0 \\ 0 \\ \cdots \\ 0 \end{bmatrix} + \begin{bmatrix} 0 \\ 0 \\ C \cos \omega_3 t \\ 0 \\ \cdots \\ 0 \end{bmatrix} + \cdots$$

and the most general problem is thereby reduced to one in which all functions $F(t)$ are zero except one. Moreover, because all periodic functions may be written as a Fourier series, all such excitations, including square-wave pulses, are soluble when the system

$$\begin{aligned} m_{11}\ddot{q}_1 + m_{12}\ddot{q}_2 + \cdots + c_{11}\dot{q}_1 + c_{12}\dot{q}_2 + \cdots + k_{11}q_1 + k_{12}q_2 + \cdots &= F \cos \omega t \\ m_{21}\ddot{q}_1 + m_{22}\ddot{q}_2 + \cdots + c_{21}\dot{q}_1 + c_{22}\dot{q}_2 + \cdots + k_{21}q_1 + k_{22}q_2 + \cdots &= 0 \end{aligned} \tag{57.92}$$

is soluble. Particular solutions may be found by substituting harmonics into this, but the complex-variable method introduced in Sec. 57.17 is easier. Change $F \cos \omega t$ to $Fe^{i\omega t}$, and it may be assumed that every q will have the form

$$q_r = A_r e^{i\omega t} \qquad r = 1,2, \ldots, n \tag{57.93}$$

in which the complex amplitude A_r will be given by

$$A_r = a_r e^{-i\phi_r}$$

Here a_r is the absolute value of the amplitude of q_r, and ϕ_r is its phase angle with respect to the force F. Then substitution of these will change the system of differential equations (57.89) or (57.92) into the nonhomogeneous algebraic equations:

$$([K] - \omega^2[M] + i\omega[C])\{A\} = \begin{bmatrix} F \\ 0 \\ 0 \\ \cdots \\ 0 \end{bmatrix} \tag{57.94}$$

When we write

$$[Z] = [K] - \omega^2[M] + i\omega[C]$$

the matrix Z is called the *mechanical impedance matrix* of the system; its elements are identical with those of the Lagrangian determinant of the system, given by Eq. (57.64), when $i\omega$ is written in place of λ. The value of every element of the matrix is, of course, variable in ω, the applied frequency.

So Eq. (57.94) is now written

$$[Z]\{A\} = \{F\} \tag{57.95}$$

Our purpose is to obtain the amplitudes A_r, and, therefore, the most direct method logically is to consider that the matrix Z is inverted, so that we have

$$\{A\} = [Y]\{F\} \tag{57.96}$$

in which $[Y]$ is called the *admittance matrix* of the system, and then, from each row of this when expanded, the values of A_r are obtained. Actually the process of inversion of a large matrix is best left to a computer [10], for which standard programs are available. For a small number of coordinates, the simultaneous algebraic equations (57.95) are solved, as discussed below.

All this will yield the steady-state solutions, that is, the particular integrals, which satisfy the equations of forced motion given by Eqs. (57.93). For the general solution, each of these solutions must be added to the n damped sine waves with n different natural frequencies and n phase angles that are the free transient vibrations of the system. The $2n$ arbitrary constants appearing in this part of the solution are found by consideration of the initial conditions of displacement and velocity of each of the n coordinates q_r.

Further study of Eq. (57.95) for the steady-state forced vibration shows that, if it is written

$$\begin{aligned} z_{11}A_1 + z_{12}A_2 + \cdots &= F_1 \\ z_{21}A_1 + z_{22}A_2 + \cdots &= 0 \end{aligned} \tag{57.97}$$

in which the terms z are known complex numbers involving the frequency ω of F_1, the solution for any amplitude A_r is given by

$$A_r = \frac{F_1 M_{1r}}{\Delta(z)} \tag{57.98}$$

in which M_{1r} is the minor of the term z_{1r} (together with its proper sign) in the determinant of the coefficients z, which is $\Delta(z)$. Thus, of course, are found the elements of the admittance matrix $[Y]$ in Eq. (57.96).

An aspect of the Maxwell reciprocal relationship is apparent here, in that, at any frequency ω, $M_{rs} = M_{sr}$; hence,

$$A_r = \frac{F_s M_{sr}}{\Delta} = A_s = \frac{F_r M_{rs}}{\Delta}$$

Therefore, the following theorem may be stated: If an oscillating force of unit amplitude, applied to any point s of a system, produces a certain steady-state displacement at another point r, the unit force applied instead to point r, at the same frequency, will produce a displacement of equal amplitude at s.

Again, if we step back to Eq. (57.90), from which all subsequent equations were derived, then, instead of the array of algebraic equations as (57.97), we have

$$\begin{aligned} e_{11}q_1 + e_{12}q_2 + \cdots &= F_1 e^{i\omega t} \\ e_{21}q_1 + e_{22}q_2 + \cdots &= 0 \\ \cdots\cdots\cdots\cdots\cdots \end{aligned} \tag{57.99}$$

in which the coefficients e are differential operators. Now, by eliminating all but one coordinate q_r, there arises an equation of similar form to Eq. (57.98), or

$$\Delta(e) \cdot q_r = F_1 M_{1r} e^{i\omega t} \tag{57.100}$$

Both Δ and M are polynomial functions of the differential operator D, of degree $2n$ and $(2n - 2)$, respectively. This equation then compares exactly with the type encountered in analyzing the response of a servomechanism, that is,

$$f_1(D)\theta_o = f_2(D)\theta_i$$

for θ_i input, θ_o output (response), and f_1 and f_2 polynomials in D.

57.21. TRANSIENT MOTIONS AND SHOCK ISOLATION

The transient responses of a system with a single degree of freedom to a pulse have been discussed on p. 56–16; and the analysis of the response of systems with many degrees of freedom to the same sort of input is, in general, merely an extension of the same methods. It has been found that there are several important types of pulse or shock, which, in order to be expressed mathematically, must be idealized somewhat; among these there are, in particular: (1) the unit step function of Heaviside, (2) the velocity shock, (3) the square-wave pulse, (4) the half-sine pulse, (5) the exponentially increasing or decaying pulse.

Physically type (1) is associated with a sudden displacement such as a rolling wheel meeting a curbstone; type (2) arises from sudden momentum change, either hammer blows or a falling body meeting the ground. Type (3) is a double form of the unit step, and type (4) is mathematically a rather easier form to consider. Of the last two, the exponentially increasing form modifies the suddenness of the unit step, thus simulating some damping in its application; the exponentially decreasing pulse is similar in form to that received from an explosion.

To find the transient response of any linear system to these various forms of shock, the procedure for analysis is just the same as is needed to find the complete solution to a sinusoidal excitation. Thus, if the motion of the system is to be represented by the set of linear equations [compare Eq. (57.90)]

$$[E]\{x\} = \{F(t)\} \tag{57.101}$$

where $\{F(t)\}$ is a column matrix of zeros and one force $F(t)$ of a type such as those listed above, then the solution in any x consists of the complementary solution of $[E]\{x\} = 0$ added to a particular solution due to $F(t)$.

The response to any unit step function is thus found to consist of a number of sinusoids, and a constant displacement. The values of the latter for each coordinate are those changes which will bring the system back into static equilibrium after the step has been taken. For a system of n degrees of freedom, there will be $2n$ arbitrary constant coefficients to the sine and cosine waves, which are found by consideration of the initial conditions of displacement and velocity.

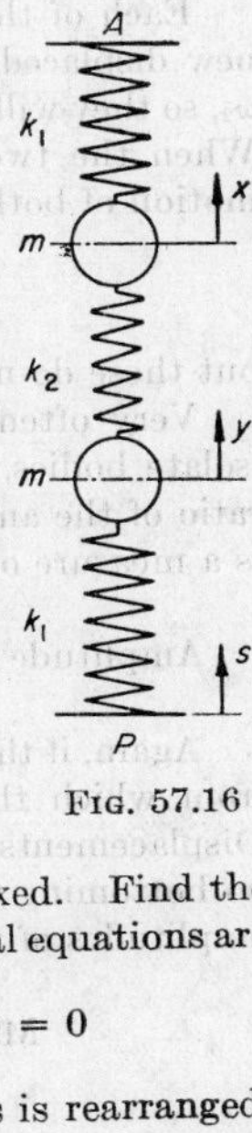

FIG. 57.16

Example 14. In the system of Fig. 57.16, two equal springs k_1 and two equal masses m are joined by a spring k_2. The point P at the bottom is suddenly raised a distance s; the top, however, remains fixed. Find the motion of the upper mass, if damping may be ignored. The differential equations are

$$m\ddot{x} + k_1 x + k_2(x - y) = 0 \qquad m\ddot{y} + k_2(y - x) + k_1(y - s) = 0$$

for which the initial conditions are $x = y = 0$ and $\dot{x} = \dot{y} = 0$. This is rearranged, by putting $(k_1 + k_2)/m = \alpha^2$ and $k_2/m = \beta^2$, and becomes

$$\ddot{x} + \alpha^2 x - \beta^2 y = 0 \qquad \ddot{y} - \beta^2 x + \alpha^2 y = sk_1/m$$

For the homogeneous equations, the constant term on the right-hand side is dropped, and the two frequencies $\omega_1 = \sqrt{k_1/m}$ and $\omega_2 = \sqrt{(k_1 + 2k_2)/m}$ are found from the Lagrangian determinant. Thus, the complementary solutions are:

$$\begin{aligned} x &= A \sin \omega_1 t + B \cos \omega_1 t + C \sin \omega_2 t + D \cos \omega_2 t \\ y &= A \sin \omega_1 t + B \cos \omega_1 t - C \sin \omega_2 t - D \cos \omega_2 t \end{aligned} \tag{57.102}$$

The particular integrals for x and y are both constants; if we let $x = x_0$ and $y = y_0$ represent these, then

$$\alpha^2 x_0 - \beta^2 y_0 = 0 \qquad \alpha^2 y_0 - \beta^2 x_0 = sk_1/m$$

which is to say that

$$x_0 = \frac{k_2 s}{k_1 + 2k_2} \qquad y_0 = \frac{(k_1 + k_2)s}{k_1 + 2k_2} \tag{57.103}$$

Now adding x_0 and y_0, respectively, to the expressions (57.102), and using the initial conditions, it will be found that

$$A = C = 0 \qquad B + D + x_0 = 0 \qquad B - D + y_0 = 0$$

Thus,

$$B = -\frac{s}{2} \qquad D = \frac{k_1}{k_1 + 2k_2} \cdot \frac{s}{2}$$

so that the transient response is given by

$$\begin{aligned} x &= \frac{s}{2}\left[(1 - \cos \omega_1 t) - \frac{k_1}{k_1 + 2k_2}(1 - \cos \omega_2 t)\right] \\ y &= \frac{s}{2}\left[(1 - \cos \omega_1 t) + \frac{k_1}{k_1 + 2k_2}(1 - \cos \omega_2 t)\right] \end{aligned} \tag{57.104}$$

Each of these is shown to be a pair of waves superimposed, oscillating about the new displaced equilibrium positions (x_0 and y_0); and, according to the ratio of ω_1 to ω_2, so they will appear either as a long wave with a ripple or as an oscillation with beats. When the two component waves are instantaneously in phase, amplitudes of the motion of both masses are

$$X_{\max} = Y_{\max} = \frac{s}{2}\left(1 + \frac{k_1}{k_1 + 2k_2}\right) = y_0$$

but these do not occur simultaneously, of course.

Very often the analysis of a problem of this sort is occasioned by the need to isolate bodies, such as these masses m, from neighboring shocks. In this case, the ratio of the amplitude of the mass to the amplitude of the step pulse which caused it is a measure of the isolating properties of the mounting. Here then

$$\text{Amplitude reduction ratio} = \frac{X_{\max}}{s} = \frac{Y_{\max}}{s} = \frac{k_1 + k_2}{k_1 + 2k_2} = \frac{1}{2}\left(1 + \frac{\omega_1^2}{\omega_2^2}\right) \tag{57.105}$$

Again, if the masses m represent delicate instruments, it may be acceleration forces from which the spring mounting is intended to protect them, not displacements. Displacements may be of interest only to the extent that it is determined that there is no bottoming of the springs. By differentiating Eqs. (57.104) twice, the sum of the amplitudes of the two acceleration component waves gives

$$\text{Maximum acceleration of } m = \omega_1^2 + \omega_2^2\left(\frac{k_1}{k_1 + 2k_2}\right) = 2\omega_1^2 \tag{57.106}$$

In this case there is no ratio, for the step input is discontinuous, and its acceleration infinite. Note that the value of the maximum is independent of the stiffness k_2 of the middle spring.

A variation on the above problem would consider that the top anchor A (in Fig. 57.16) moved with the bottom one P. In this case, it should be observed that the solutions would be the same if the step impulse were omitted and both the masses reckoned to have an initial displacement of the amount $-s$.

A similar form of analytic procedure is followed when a velocity shock is applied to a system: the particular integrals are easily found. And again, in some cases, an alternative method is available, in which all masses have initial velocity: for consider a crate full of instruments carefully packed, but dropped off the tail gate of a truck; the vibrations started by the velocities acquired in falling and the sudden stop of the exterior crate as it hits the road are the same as when all elements are stationary and the crate is given a sudden velocity shock, as it might if it were encased within a rocket.

For the two shocks of types (3) and (4), two stages of analysis are needed: For the duration (say t_0) of the pulse, the problem is one of forced oscillation, and, at time t_0, the displacement and velocity of each mass must be calculated. These then are the starting conditions for the second stage.

If the pulse is of either sinusoidal or exponential form, and the system is linear, the particular integrals for the forced oscillation must also be of exponential form. The procedure of the last section may then be followed.

For all problems of transient response, the method of the Laplace transform (see chap. 19) is a very powerful alternative that is preferred by many.

Further studies in these matters will be helped by consulting [9] and [17].

57.22. VIBRATION ISOLATION IN MULTIPLE-MODE SYSTEMS

The purposes of isolation are either of two: to protect a delicate object from vibrations that may be present in the base to which the object is attached, or to prevent a large proportion of the vibratory forces that may occur in a machine from reaching the floor.

In general, it has been found, in studying the isolation of systems having a single degree of freedom, that the most effective isolator is the soft cushion. Mathematically, this is to say that the natural frequency of the machine or system on its mounts must be low in comparison to the frequency of the vibrations to be excluded or contained.

A rigid body such as a machine is capable of motion in six coordinates, three linear and three angular. A major purpose in successful isolator design is so to locate the necessary spring mounts, and choose their relative stiffness, that the natural frequencies in these six modes are bunched together into a narrow range. Then this narrow range is put as low in the frequency spectrum as possible. The alternative is to have the six natural frequencies with some low and some medium speeds, with the likelihood of trouble at the higher speeds if the modes are coupled.

Since the designer of such mounts for a machine is, at the same time, making some compromise between the horns of the dilemma—that on the one side the provision of soft springs will mean that the machine will seem to float like a boat, and on the other side the preference of most machine operators is for a perfectly rigid machine—so the successful solution of a machine isolator installation requires considerable skill. It needs true engineering art.

Certain analytic studies have been made; but most simplify the basic problem by considering that the "box" to be mounted has one or more planes of symmetry. References [1], [16], and [17] should be consulted.

57.23. REFERENCES

[1] C. E. Crede: "Vibration and Shock Isolation," Wiley, New York, 1951.
[2] J. P. Den Hartog: "Mechanical Vibrations," 4th ed., McGraw-Hill, New York, 1956.
[3] R. A. Frazer, W. J. Duncan, A. R. Collar: "Elementary Matrices," Macmillan, New York, 1947.
[4] E. A. Guillemin: "The Mathematics of Circuit Analysis," Wiley, New York, 1949.
[5] H. M. Hansen, P. F. Chenea: "Mechanics of Vibration," Wiley, New York, 1952.
[6] M. R. Hestenes, E. S. Stiefel: Method of conjugate gradients for solving linear systems, *J. Research Natl. Bur. Standards*, **49** (1952), 409–436.
[7] F. B. Hildebrand: "Methods of Applied Mathematics," Prentice-Hall, Englewood Cliffs, N.J., 1952.
[8] A. Hurwitz: Über die Bedingungen, unter welchen eine Gleichung nur Wurzeln mit negativen reellen Theilen besitzt, *Math. Ann.*, **46** (1894), 273–284.
[9] L. S. Jacobsen, R. S. Ayre: "Engineering Vibrations," McGraw-Hill, New York, 1958.
[10] G. Kron: Inverting a 256 × 256 matrix, *Engineering*, **179** (1955), 309–312.
[11] G. Kron: Tearing, tensors and topological models, *Am. Scient.* **45** (1957), 401–413.
[12] C. Lanczos: Solutions of systems of linear equations by minimized iterations, *J. Research Natl. Bur. Standards*. **49** (1952), 33–53.
[13] C. Lanczos: Chebyshev polynomials in the solution of large-scale linear systems, *Proc. Assoc. Computing Machines*, 1952, pp. 124–133.
[14] D. F. Lawden: "Mathematics of Engineering Systems," Wiley, New York, 1955.
[15] F. M. Lewis: The extended theory of the viscous vibration damper, *J. Appl. Mechanics*, **22** (1955), 377–382.
[16] R. C. Lewis, K. Unholtz: A simplified method for the design of vibration-isolating suspensions, *Trans. ASME*, **69** (1947), 813–820.
[17] J. N. MacDuff, J. R. Curreri: "Vibration Control," McGraw-Hill, New York, 1958.
[18] R. D. Mindlin: Dynamics of package cushioning, *Bell System Tech. J.*, **24** (1945), 353–461.
[19] N. O. Myklestad: "Fundamentals of Vibration Analysis," McGraw-Hill, New York, 1956.
[20] L. A. Pipes: "Applied Mathematics for Engineers and Physicists," 2d ed., McGraw-Hill, New York, 1958.
[21] A. S. Ramsey: "Dynamics," part II, Cambridge Univ. Press, New York, 1947.
[22] Lord Rayleigh: "The Theory of Sound," 2d ed., Dover, New York, 1945.
[23] E. J. Routh: "Dynamics of a System of Rigid Bodies," part 2, 5th ed., Macmillan, London, 1892.
[24] R. H. Scanlan, R. Rosenbaum: "Introduction to the Study of Aircraft Vibration and Flutter," Macmillan, New York, 1951.
[25] J. C. Slater, N. H. Frank: "Mechanics," McGraw-Hill, New York, 1947.
[26] A. Sommerfeld: "Mechanics," Academic Press, New York, 1952.
[27] J. L. Synge, B. A. Griffith: "Principles of Mechanics," 3d ed., McGraw-Hill, New York, 1959.
[28] S. P. Timoshenko: "Vibration Problems in Engineering," 3d ed., Van Nostrand, Princeton, N.J., 1955.
[29] S. P. Timoshenko, D. H. Young: "Advanced Dynamics," McGraw-Hill, New York, 1948.
[30] T. v. Kármán, M. A. Biot: "Mathematical Methods in Engineering," McGraw-Hill, New York, 1940.
[31] E. T. Whittaker: "A Treatise on the Analytical Dynamics of Particles and Rigid Bodies," 4th ed., Dover, New York, 1944.
[32] F. R. E. Crossley, Ü. Germen: A method of numerical evaluation of a large determinant, *J. Appl. Mechanics*, **27** (1960), 350–351.

CHAPTER 58

ROTATING AND RECIPROCATING MACHINES

BY

S. H. CRANDALL, Ph.D., Cambridge, Mass.

58.1. INTRODUCTION

A machine which imparts power to or receives power from a rotating shaft is usually either a *reciprocating* machine, in which pistons are connected to a rotating crankshaft by connecting rods, or a *rotating* machine, in which a rotor revolves within a stator. The aim of this chapter is to survey the practically important vibrations of such machines.

Vibrations usually represent parasitic oscillations, involving very small amounts of power in comparison to the total power handled by the machine. Indeed, if any appreciable fraction of the available energy is funneled into a vibration, there is sure to be a failure due to excessive stress or excessive deformation. For example, each blade of the transonic compressor in the propulsion wind tunnel at Tullahoma is designed to develop 2500 hp. In a vibratory fatigue test, it was found that less than 2.4 hp input to a shaker was required to cause blade failure within 24 hours when the blade was vibrated at resonance.

Analytical work on the vibrations of machinery is always performed on an idealized model of a real situation. There are many stages of refinement in the models employed. The simplest model is that in which all deformability is neglected. The machine is considered to be made up of rigid masses in motion. Most balancing arguments are made in terms of this model. More complicated models take into account the elasticity of various parts: e.g., the drive shaft, the bearings, the blades, and the disks of a turbine. Usually the shaft is the most flexible member, and the most important natural vibrations of the machine involve modes in which the shaft is either bent or twisted. A straight shaft which bends away from its initial position while rotating is said to undergo *whirling*. A shaft in which the relative twist angles fluctuate with time is said to undergo *torsional vibration*.

The rotational frequency of a machine at which whirling or torsional vibration occurs, or potentially could occur, is called a *critical* speed. The whirling or torsional vibration which takes place does not necessarily have to synchronize with the rotational frequency.

58.2. BALANCING OF ROTORS

Rigid Rotors

Figure 58.1 shows a rigid body B constrained to rotate about an axis CC. The x, y, z axes are fixed in space, with the center of mass G of the body moving in the y,z plane. The ξ, η, ζ, axes are fixed in the body with the ξ axis coinciding with the x axis, and the η,ζ plane containing the center of mass G. If the eccentricity of G is ϵ with components ϵ_η and ϵ_ζ, and the mass of the body is m, the external force on B must be

$$F_\eta = -m\epsilon_\eta\omega^2 - m\epsilon_\zeta\dot{\omega} \qquad F_\zeta = -m\epsilon_\zeta\omega^2 + m\epsilon_\eta\dot{\omega} \qquad (58.1)$$

where $\omega = \dot{\theta}$. The external moment on B obtained from Euler's angular momentum equations must be

$$M_\xi = I_\xi \dot{\omega} \qquad M_\eta = -I_{\xi\eta}\dot{\omega} + I_{\xi\zeta}\omega^2 \qquad M_\zeta = -I_{\xi\zeta}\dot{\omega} - I_{\xi\eta}\omega^2 \tag{58.2}$$

where the I_{jk} are the *products* of inertia of B with respect to the j and k axes, and I_j is the *moment* of inertia with respect to the j axis. The forces F_η and F_ζ and the torques M_η and M_ζ revolve with the body, and give rise to rotating bearing reactions. The torque M_ξ is supplied by the drive shaft.

If $\epsilon = 0$, the forces F_η and F_ζ vanish, and the body B is said to have *static* balance. If the axis of rotation is a principal axis of inertia, then

$$I_{\xi\eta} = I_{\xi\zeta} = 0 \quad \text{and} \quad M_\eta = M_\zeta = 0$$

for all ω and all $\dot{\omega}$

If $\epsilon = 0$ and $I_{\xi\eta} = I_{\xi\zeta} = 0$, then the body B is said to have *dynamic* balance. In this case, the rotating bearing reactions vanish entirely. The torque M_ξ vanishes when the rotor revolves at constant speed.

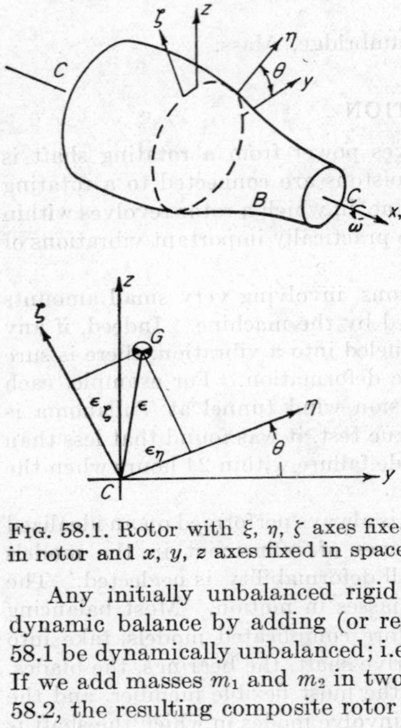

FIG. 58.1. Rotor with ξ, η, ζ axes fixed in rotor and x, y, z axes fixed in space.

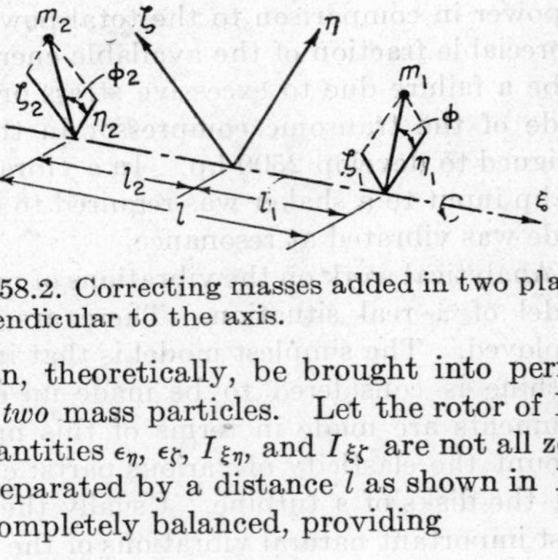

FIG. 58.2. Correcting masses added in two planes perpendicular to the axis.

Any initially unbalanced rigid rotor can, theoretically, be brought into perfect dynamic balance by adding (or removing) *two* mass particles. Let the rotor of Fig. 58.1 be dynamically unbalanced; i.e., the quantities ϵ_η, ϵ_ζ, $I_{\xi\eta}$, and $I_{\xi\zeta}$ are not all zero. If we add masses m_1 and m_2 in two planes separated by a distance l as shown in Fig. 58.2, the resulting composite rotor will be completely balanced, providing

$$m_1\eta_1 = \frac{-I_{\xi\eta} - m\epsilon_\eta l_2}{l} \qquad m_1\zeta_1 = \frac{-I_{\xi\zeta} - m\epsilon_\zeta l_2}{l}$$
$$m_2\eta_2 = \frac{I_{\xi\eta} - m\epsilon_\eta l_1}{l} \qquad m_2\zeta_2 = \frac{I_{\xi\zeta} - m\epsilon_\zeta l_1}{l} \tag{58.3}$$

The left-hand sides of (58.3) may be represented by a pair of radial vectors at stations 1 and 2, having magnitudes m_1r_1 and m_2r_2 and angles ϕ_1 and ϕ_2. These vectors are sometimes called the *balance* vectors. It is often convenient to visualize the original state of unbalance in terms of a pair of *unbalance* vectors, which are equal and opposite to the balance vectors. The configuration of most rotors severely limits the choice of the values l_1, l_2, r_1, and r_2. The balance conditions (58.3) are then met by adjusting the angles ϕ_1 and ϕ_2 and the amounts of mass m_1 and m_2 added (or removed). There are a large number of specialized techniques (see [1], secs. 6.3 to 6.5) employed in balancing, but, basically, they all involve measuring the existing unbalance and determining the magnitude and location of the correcting masses which satisfy (58.3).

The existing unbalance in a rotor is usually obtained indirectly by measuring the amplitude and phase of the steady-state vibration at two different stations along the axis (usually at the bearings), when the rotor revolves steadily in flexible supports. These vibrations may be represented by a pair of rotating vectors. If the system is linear, it can be shown that, for any fixed speed of rotation, the vibration vectors are linear vector functions of the unbalance vectors. The coefficients of these functions can be determined experimentally by adding known amounts of unbalance to the rotor and observing the resulting changes in the vibration vectors. It is then possible to compute the magnitude and location of the correction masses required to annul the original unbalance.

Flexible Rotors

If an absolutely rigid rotor is in perfect dynamic balance, then it is balanced for all speeds of a rotation. A flexible rotor, on the other hand, may be dynamically balanced when the rotor is undeformed, but, if there are compensating local unbalances, the distribution of centrifugal forces when rotating may bend the rotor, which will shift the entire mass distribution. This gives rise to an over-all unbalance which depends on the speed. In order for a flexible rotor to be in perfect balance at all speeds, every axial section $d\xi$ in Fig. 58.1 must itself be dynamically balanced.

An alternative description can be given in terms of the natural modes of the flexible rotor as a free-free beam. If the inertia distribution is such that the centrifugal force distribution along the axis of the undeformed rotor is *orthogonal* to the rigid-body modes and to the successive free-free bending modes, then the rotor can revolve about its axis with no tendency for transverse motion, and, hence, no rotating bearing reactions are required. The only centrifugal force distribution which is orthogonal to *all* modes is that which results when every axial section of Fig. 58.1 is balanced.

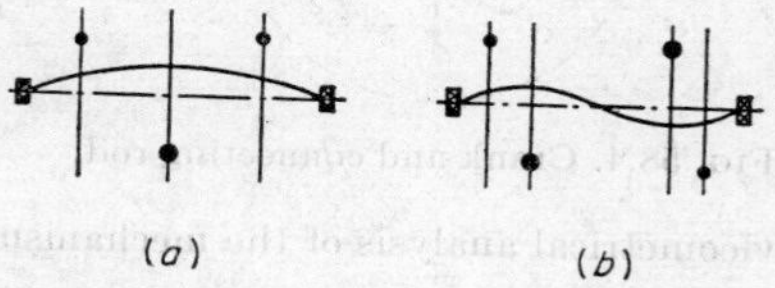

FIG. 58.3. Pattern of correcting masses required to balance against (a) whirling in first bending mode, (b) whirling in second bending mode.

In balancing flexible rotors, it is usual to employ several balancing planes but only two vibration pickups. The process is as follows. At relatively low speed, the rotor is balanced in the usual manner as though it were rigid. Then the speed is increased to near the first bending critical for the given bearing configuration. If the rotor runs rough, it must be balanced; but it must be balanced in such a way that the low-speed balance is not altered. This is accomplished by adding a pattern of balancing masses, as shown in Fig. 58.3*a*. This is a distribution which is *orthogonal* to the rigid-body modes, but does have an effective component in the first in-bearings mode. When the amplitude and phase of this pattern have been adjusted, it should now be possible to run smoothly through the first critical and up to the vicinity of the second critical. If the rotor requires further balancing, a pattern such as that in Fig. 58.3*b* must be employed. This is a distribution which is orthogonal to the rigid-body modes *and* the first bending mode. Continuing in this manner, it is possible to obtain smooth operation throughout the speed range of the machine. Note that it is necessary to know in advance the successive bending modes of the rotor, in order to construct patterns of balancing masses which will be orthogonal to the lower modes.

58.3. BALANCING OF RECIPROCATING MACHINES

The fluctuating momenta of the moving parts in a piston machine do not, in general, balance completely, and require either compensating momentum from the frame of the machine if the frame is suspended in a soft mounting, or fluctuating external reaction forces if the frame is rigidly mounted. One aspect of reciprocating-machine design concerns selecting the geometrical configuration of the machine and proportioning the mass distributions of the moving parts so as to minimize the vibratory motions of the frame or the vibratory reaction forces developed. Another aspect has to do with the fluctuating drive-shaft torque, which acts as excitation for forced torsional vibrations.

Single Cylinder

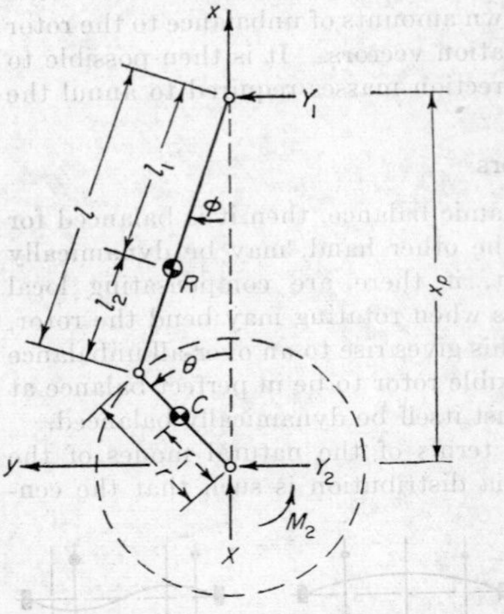

FIG. 58.4. Crank and connecting rod.

Referring to Fig. 58.4, let C and R be the centers of mass of the crank and connecting rod, respectively, and let

m_p = mass of piston
m_r = mass of connecting rod
m_c = mass of crank
$m_{\text{rot}} = \frac{r_1}{r} m_c + \frac{l_1}{l} m_r$ = equivalent rotating mass
$m_{\text{rec}} = m_p + \frac{l_2}{l} m_r$ = equivalent reciprocating mass
k_r = radius of gyration of connecting rod about R
$I_{\text{red}} = m_r(k_r^2 - l_1 l_2)$ = reduced moment of inertia of connecting rod
$\lambda = \frac{r}{l}$ = crank-connecting rod ratio

Geometrical analysis of the mechanism yields

$$\frac{x_p}{r} = \left(\frac{1}{\lambda} - \frac{\lambda}{4} - \frac{3\lambda^3}{64} - \frac{5\lambda^5}{256} - \cdots\right) + \cos\theta + \left(\frac{\lambda}{4} + \frac{\lambda^3}{64} + \frac{15\lambda^5}{512} + \cdots\right)\cos 2\theta - \left(\frac{\lambda^3}{64} + \frac{3\lambda^5}{256} + \cdots\right)\cos 4\theta + \left(\frac{\lambda^5}{512} + \cdots\right)\cos 6\theta \cdots \tag{58.4}$$

$$\phi = \left(\lambda + \frac{\lambda^3}{8} + \frac{3\lambda^5}{64} + \cdots\right)\sin\theta - \left(\frac{\lambda^3}{24} + \frac{3\lambda^5}{128} + \cdots\right)\sin 3\theta + \left(\frac{3\lambda^5}{640} + \cdots\right)\sin 5\theta \cdots \tag{58.5}$$

Dynamic analysis under the assumptions that the crank revolves at a steady rate ω, and that there are no friction and no gas pressure, yields the following inertia forces and moments

$$X = r\omega^2 \left\{ m_{\text{rot}} \cos\theta + m_{\text{rec}} \left[\cos\theta + \left(\lambda + \frac{\lambda^3}{4} + \frac{15\lambda^5}{128} + \cdots\right)\cos 2\theta - \left(\frac{\lambda^3}{4} + \frac{3\lambda^5}{16} + \cdots\right)\cos 4\theta + \left(\frac{9\lambda^5}{128} + \cdots\right)\cos 6\theta - \cdots \right] \right\} \tag{58.6}$$

$$Y = Y_1 + Y_2 = r\omega^2 m_{\text{rot}} \sin\theta \tag{58.7}$$

$$M_z = m_{\text{rec}} r^2 \omega^2 \left[\left(\frac{\lambda}{4} + \frac{\lambda^3}{16} + \frac{15\lambda^5}{512} + \cdots\right)\sin\theta - \left(\frac{1}{2} + \frac{\lambda^4}{32} + \cdots\right)\sin 2\theta - \left(\frac{3\lambda}{4} + \frac{9\lambda^3}{32} + \frac{81\lambda^5}{512} + \cdots\right)\sin 3\theta - \left(\frac{\lambda^2}{4} + \frac{7\lambda^4}{32}\right)\sin 4\theta + \left(\frac{5\lambda^3}{32} + \frac{75\lambda^5}{512} + \cdots\right)\sin 5\theta + \left(\frac{3\lambda^4}{32} + \cdots\right)\sin 6\theta \left(\frac{21\lambda^5}{512} + \cdots\right)\sin 7\theta + \cdots \right] + I_{\text{red}}\omega^2 \left[\left(\frac{\lambda^2}{2} - \cdots\right)\sin 2\theta - \left(\frac{\lambda^4}{4} + \cdots\right)\sin 4\theta + \cdots \right] \tag{58.8}$$

$$
\begin{aligned}
Y_1x_p = & -m_{\text{rec}}r^2\omega^2 \left[\left(\frac{\lambda}{4} + \frac{\lambda^3}{16} + \frac{15\lambda^5}{512} + \cdots \right) \sin \theta - \left(\frac{1}{2} + \frac{\lambda^4}{32} + \cdots \right) \sin 2\theta \right. \\
& - \left(\frac{3\lambda}{4} + \frac{9\lambda^3}{32} + \frac{81\lambda^5}{512} + \cdots \right) \sin 3\theta - \left(\frac{\lambda^2}{4} + \frac{7\lambda^4}{32} + \cdots \right) \sin 4\theta \\
& + \left(\frac{5\lambda^3}{32} + \frac{75\lambda^5}{512} + \cdots \right) \sin 5\theta + \left(\frac{3\lambda^4}{32} + \cdots \right) \sin 6\theta \\
& \left. - \left(\frac{21\lambda^5}{512} + \cdots \right) \sin 7\theta - \cdots \right] - I_{\text{red}}\omega^2 \left[\left(\lambda + \frac{\lambda^3}{8} + \frac{3\lambda^5}{64} \right. \right. \\
& \left. + \cdots \right) \sin \theta + \left(\frac{\lambda^2}{2} - \cdots \right) \sin 2\theta - \left(\frac{3\lambda^3}{8} + \frac{27\lambda^5}{128} + \cdots \right) \sin 3\theta \\
& \left. - \left(\frac{\lambda^4}{4} + \cdots \right) \sin 4\theta + \left(\frac{15\lambda^5}{128} + \cdots \right) \sin 5\theta + \cdots \right]
\end{aligned} \tag{58.9}
$$

These inertia forces and moments act on the frame and crankshaft in the directions shown in Fig. 58.4. Under the assumption that the crankshaft bearings and the cylinder walls are integral parts of a rigid frame, the frame is acted on by the resultant of X, Y_1, and Y_2. The crankshaft, which usually has no torque return to the frame, is acted on by the inertia torque M_z. The inertia torque acting on the frame about the crankshaft axis is Y_1x_p. In Eqs. (58.4) to (58.9), all powers of λ up to λ^5 have been included. Accurate values for the coefficients in (58.6) have been tabulated (see [2], p. 7) for several values of λ and a more extensive table [3] of the coefficients in (58.8) has been published.

If there is a *gas pressure* p in the cylinder, there will be a force pA, where A is the area of the piston, acting upward on the frame, and a force pA acting downward on the piston. As a consequence, forces $X = -pA$, $Y_2 = -Y_1 = pA \tan \phi$ and a torque $M_z = pAx_p \tan \phi$ will be delivered to the frame and the crankshaft. Assuming the crankshaft bearings and the cylinder to be parts of a single rigid frame, there will be no net force on the frame, but there will be the torque M_z acting on the crankshaft, and an equal and opposite torque reaction on the frame.

If the pressure variation throughout the cycle of the machine is known, it is possible to evaluate the gas-pressure torque as a function of crank angle θ. It is usually convenient to make a harmonic analysis of this function in the following form:

$$
\begin{aligned}
M_z = b_o + a_{1/2} \sin \frac{\theta}{2} + b_{1/2} \cos \frac{\theta}{2} + a_1 \sin \theta + b_1 \cos \theta + a_{3/2} \sin \frac{3\theta}{2} + b_{3/2} \cos \frac{3\theta}{2} \\
+ a_2 \sin 2\theta + b_2 \cos 2\theta + \cdots
\end{aligned} \tag{58.10}
$$

In a two-stroke machine, the cycle is complete in a single revolution, and only integer orders occur. In a four-stroke machine, the cycle requires two revolutions, and, in general, half-integer as well as integer orders will occur. The coefficients of all orders up to $j = 18$ for a number of representative engine cycles have been tabulated [3].

Neglecting gravity, the total torque delivered to the crankshaft is then the sum of the inertia torque (58.8) and the gas-pressure torque (58.10). The total torque on the frame is that due to inertia (58.9) plus the negative of (58.10) due to gas pressure. The total forces on the frame are (58.6) and (58.7) due only to inertia. See [3] for the inclusion of gravity.

In principle, it would be possible to balance all the inertia forces and moments by making $m_{\text{rec}} = m_{\text{rot}} = I_{\text{red}} = 0$. It is not too difficult to make $m_{\text{rot}} = I_{\text{red}} = 0$, but to make $m_{\text{rec}} = 0$ would require an overhung connecting rod (a negative value of l_2 in Fig. 58.4). This has so many drawbacks (increased crankcase size, increased weight of connecting rod, etc.) that it has never been adopted in practice. The gas-pressure torque is inherently pulsating, and cannot possibly be balanced in a single cylinder.

Multicylinder Machines

By combining several cylinders acting on the same drive shaft into a single rigid frame, it is possible to balance out some of the important harmonics in the forces and moments of the individual cylinders. Although many configurations are possible [2], we will here discuss only the *in-line* machine in which n identical cylinders are equally spaced along a straight line, as shown in Fig. 58.5. Let the crank offset angles be θ_j $(j = 2, \ldots, n)$ with respect to the first crank, as shown. Then the forces and moments exerted on the frame and crankshaft by the jth cylinder are given by Eqs.

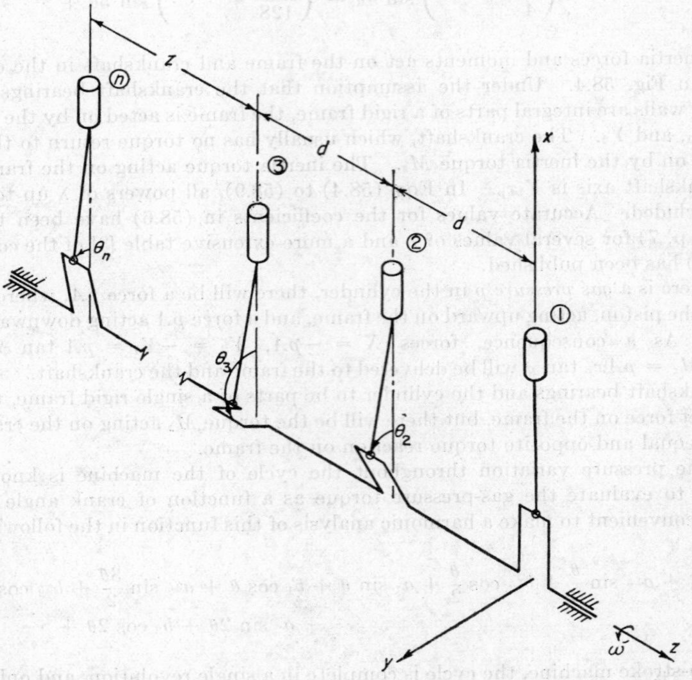

FIG. 58.5. In-line machine with n equally spaced cylinders.

(58.6) through (58.10), with θ replaced by $\omega t + \theta_j$ (in a four-stroke cycle machine it is necessary to consider certain of the offset angles in the form $2\pi + \theta_j$, depending on the firing order employed). The resultant force acting on the frame has components

$$X = (m_{\text{rot}} + m_{\text{rec}})r\omega^2\mathcal{R}[(1 + e^{i\theta_2} + \cdots e^{i\theta_n})e^{i\omega t}] + m_{\text{rec}}r\omega^2\mathcal{R}\left[(1 + e^{i2\theta_2} + \cdots + e^{i2\theta_n})\left(\lambda + \frac{\lambda^3}{4} + \frac{15\lambda^5}{128} + \cdots\right)e^{i2\omega t} - \cdots\right] \tag{58.11}$$

$$Y = m_{\text{rot}}r\omega^2\mathcal{R}[(1 + e^{i\theta_2} + \cdots + e^{i\theta_n})e^{i\omega t}] \tag{58.12}$$

In general, there is a yawing moment

$$M_x = m_{\text{rot}}r\omega^2 d\mathcal{R}[(e^{i\theta_2} + 2e^{i\theta_3} + \cdots + (n-1)e^{i\theta_n})e^{i\omega t}] \tag{58.13}$$

and a pitching moment

$$M_y = -(m_{\text{rot}} + m_{\text{rec}})r\omega^2 d\Re[(e^{i\theta_2} + 2e^{i\theta_3} + \cdots + (n-1)e^{i\theta_n})e^{i\omega t}] - m_{\text{rec}}r\omega^2 d\Re\left[(e^{i2\theta_2} + 2e^{i2\theta_3} + \cdots + (n-1)e^{i2\theta_n})\left(\lambda + \frac{\lambda^3}{4} + \frac{15\lambda^5}{128} + \cdots\right)e^{i2\omega t} + \cdots\right] \quad (58.14)$$

The total rolling moment acting on the frame due to inertia (58.9) and gas pressure (58.10) is

$$Y_1x_p = -nb_0 - \mathcal{I}[a_{1/2}(1 + e^{i\theta_2/2} + \cdots + e^{i\theta_n/2})e^{i\omega t/2}] - \Re[b_{1/2}(1 + e^{i\theta_2/2} + \cdots + e^{i\theta_n/2})e^{i\omega t/2}] - \mathcal{I}\left\{\left[m_{\text{rec}}r^2\omega^2\left(\frac{\lambda}{4} + \frac{\lambda^3}{16} + \frac{15\lambda^5}{512} + \cdots\right) + I_{\text{red}}\,\omega^2\left(\lambda + \frac{\lambda^3}{8} + \frac{3\lambda^5}{64} + \cdots\right) + a_1\right][1 + e^{i\theta_2} + \cdots + e^{i\theta_n}]e^{i\omega t}\right\} - \Re[b_1(1 + r^{i\theta_2} + \cdots + e^{i\theta_n})e^{i\omega t}] - \cdots \quad (58.15)$$

For certain arrangements of cylinders, crank angles, and firing orders, these forces and moments can be quite small. For example, a six-cylinder four-stroke cycle engine with crank offset angles of 0, 120, 240, 120, and 0°, whose cylinders fire in the order 142635 has $\theta_1 = 0$, $\theta_2 = 480°$, $\theta_3 = 240°$, $\theta_4 = 600°$, $\theta_5 = 120°$, and $\theta_6 = 360°$. On inserting these angles in the above formulas, it is seen that $Y = M_x = 0$, and that the first nonvanishing harmonic in X and M_y is the sixth. The resultant of these is the force

$$m_{\text{rec}}r\omega^2 6\left(\frac{9\lambda^5}{128} + \cdots\right)\cos 6\omega t \quad (58.16)$$

acting upward at the center of the engine frame. The other nonvanishing harmonics in X are the 12th, 18th, etc. The rolling moment on the frame consists of the average reaction torque, $-6b_0$, plus harmonics of order 3, 6, 9, 12, etc. The third harmonic is

$$\left[6m_{\text{rec}}r^2\omega^2\left(\frac{3\lambda}{4} + \frac{9\lambda^3}{32} + \frac{81\lambda^5}{512} + \cdots\right) + 6I_{\text{red}}\omega^2\left(\frac{3\lambda^3}{8} + \frac{27\lambda^5}{128} + \cdots\right) - 6a_3\right]\sin 3\omega t - 6b_3\cos 3\omega t \quad (58.17)$$

For a more detailed discussion of balancing problems in engine design, see [4].

58.4. TORSIONAL VIBRATIONS

Consider the three-inertia torsional-vibration system shown in Fig. 58.6. The governing equations (57.15) were given on p. 57–7. The natural frequencies are $\omega_0 = 0$, ω_1, and ω_2, where the latter two are given by

$$\omega_{1,2}^2 = \frac{k_1(I_1 + I_2)}{2I_1I_2} + \frac{k_2(I_2 + I_3)}{2I_2I_3} \mp \sqrt{\left[\frac{k_1(I_1 + I_2)}{2I_1I_2} - \frac{k_2(I_2 + I_3)}{2I_2I_3}\right]^2 + \frac{k_1k_2}{I_2{}^2}} \quad (58.18)$$

The amplitude ratios for the corresponding natural modes are given by

$$x_1\left(1 - \frac{I_1\omega^2}{k_1}\right) = x_2 = x_3\left(1 - \frac{I_3\omega^2}{k_2}\right) \quad (58.19)$$

when the appropriate natural frequency is substituted for ω.

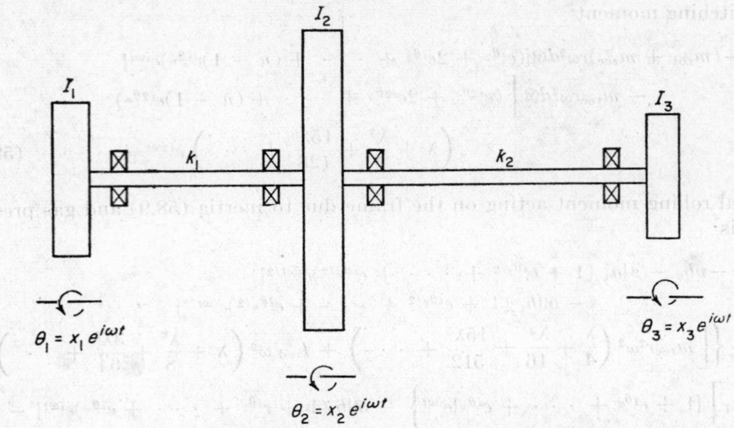

FIG. 58.6. Natural mode of vibration of a three-inertia torsional system.

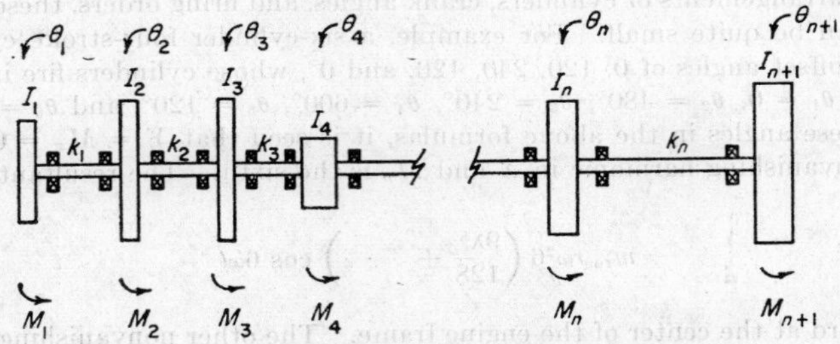

FIG. 58.7. General single-shaft torsional-vibration system.

To extend this analysis to an $(n + 1)$ inertia system as shown in Fig. 58.7, it is convenient to define the stiffness matrix K and the inertia matrix I as

$$K = \begin{bmatrix} k_1 & -k_1 & 0 & \cdots \\ -k_1 & k_1 + k_2 & -k_2 & \cdots \\ 0 & -k_2 & k_2 + k_3 & \cdots \\ \cdots & \cdots & \cdots & \cdots \end{bmatrix} \qquad I = \begin{bmatrix} I_1 & 0 & 0 & \cdots \\ 0 & I_2 & 0 & \cdots \\ 0 & 0 & I_3 & \cdots \\ \cdots & \cdots & \cdots & \cdots \end{bmatrix} \tag{58.20}$$

and the column matrices or vectors

$$\theta = \begin{bmatrix} \theta_1 \\ \theta_2 \\ \theta_3 \\ \cdot \\ \cdot \\ \cdot \end{bmatrix} \qquad M = \begin{bmatrix} M_1 \\ M_2 \\ M_3 \\ \cdot \\ \cdot \\ \cdot \end{bmatrix} \tag{58.21}$$

for the displacements and external twisting moments. Note that each matrix has $(n + 1)$ rows. In terms of these, the differential equations of motion are

$$K\theta + I\ddot{\theta} = M \tag{58.22}$$

By setting $M_j = 0$ and $\theta_j = x_j e^{i\omega t}$ for $j = 1, \ldots, n + 1$, we obtain the following matrix eigenvalue problem for the natural modes and natural frequencies,

$$KX = \omega^2 IX \tag{58.23}$$

where X is the column vector whose elements are x_j.

It can be shown that, when all the I_i and k_i are positive, the natural frequencies of (58.23) are $\omega_0 = 0$ (for the rigid-body mode) and n nonrepeated roots $0 < \omega_1 < \omega_2 < \cdots < \omega_n$. The natural mode corresponding to ω_m has exactly m nodes. The rigid-body mode $x_1 = x_2 = \cdots = x_{n+1}$ has no node.

The natural modes and frequencies of (58.23) may be solved by methods given in chap. 18. One caution is that, because of the rigid-body mode with $\omega_0 = 0$, any procedure which attempts to invert the stiffness matrix will fail. The most straightforward process is to evaluate the *characteristic* polynomial numerically for several trial values of ω, and locate the roots by interpolation. A widely used hand computational procedure for doing this utilizes the *Holzer table* (see [1], p. 189), presented as Table 58.1. One such table is required for each trial value of ω. Beginning with an arbitrary value of x_1, say $x_1 = 1$, the table works out the forced-vibration mode resulting from a single shaking torque applied to I_{n+1}. The entries in column 4 give the inertia torque amplitudes to the left of a given section, and thus represent the twisting torque amplitude in the shaft. The twist angle in this section is given in column 7. The entry in column 3 in any line is obtained by subtracting the column 7 entry from the column 3 entry in the previous line: for example, $x_2 = 1.00 - I_1\omega^2/k_1$ in the table. Note that this is equivalent to evaluating x_2 from the first row of the matrix equation (58.23). The final entry in column 5 is the resultant inertia torque amplitude M_{res} of the entire system, and is equal and opposite to the external shaking torque amplitude which would be required to maintain the motion. The form of the plot of resultant torque vs. trial frequency is shown in Fig. 58.8. It can be shown [5] that, in systems of this kind if the mode shape in column 3 of Table 58.1 has m nodes, then the trial ω must lie between ω_{m-1} and ω_{m+1}. Furthermore, the algebraic sign of the remainder torque can be used with Fig. 58.8 to decide whether the trial ω is larger or smaller than ω_m.

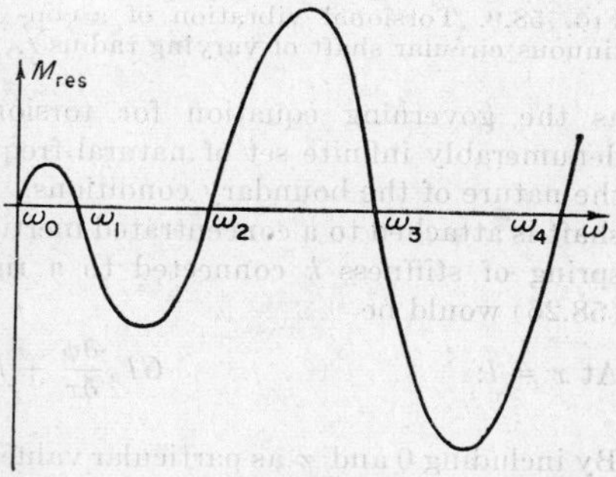

FIG. 58.8. The resultant inertia torque vanishes at the natural frequencies and is alternately positive and negative between the natural frequencies.

Table 58.1. Holzer Table

(1)	(2)	(3)	(4)	(5)	(6)	(7)
I	$I\omega^2$	x	$I\omega^2 x$	$\Sigma I\omega^2 x$	k	$\frac{1}{k}\sum I\omega^2 x$
I_1	$I_1\omega^2$	1.000	$I_1\omega^2$	$I_1\omega^2$	k_1	$I_1\omega^2/k_1$
I_2	$I_2\omega^2$	x_2	. . .	. . .	k_2	. . .
⋮	⋮	⋮	⋮	⋮	⋮	⋮
I_{n+1}	$I_{n+1}{}^2$	x_{n+1}	$I_n{}^2 x_n$	M_{res}		

A *continuous* shaft with distributed stiffness and inertia is shown in Fig. 58.9. If the variation in $r(x)$ is not too severe, the twisting moment M_T is related to the twisting angle by

$$M_T = GI_p \frac{\partial \phi}{\partial x} \tag{58.24}$$

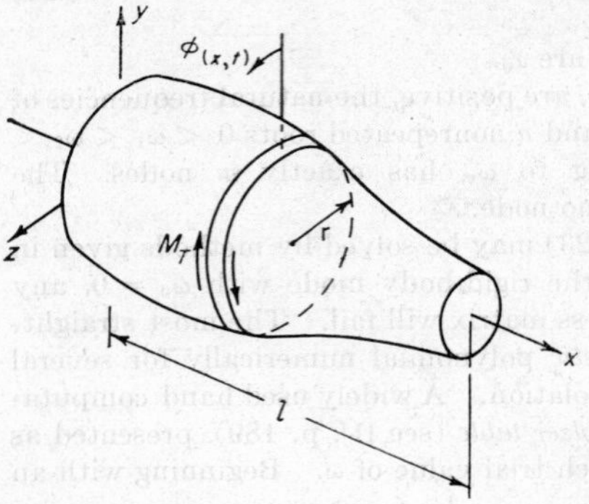

FIG. 58.9. Torsional vibration of a continuous circular shaft of varying radius r.

where $I_p(x) = \pi r^4/2$ is the polar moment of inertia of the cross section, and G is the shear modulus. The torque–angular acceleration relation is

$$\frac{\partial M_T}{\partial x} = \rho I_p \frac{\partial^2 \phi}{\partial t^2} \tag{58.25}$$

where ρ is the mass density. Eliminating M_T between (58.24) and (58.25) leads to

$$\frac{\partial}{\partial x}\left(GI_p \frac{\partial \phi}{\partial x} \right) = \rho I_p \frac{\partial^2 \phi}{\partial t^2} \tag{58.26}$$

as the governing equation for torsional vibration. Any finite length l has a denumerably infinite set of natural frequencies and natural modes which depend on the nature of the boundary conditions. For example, if at the right end ($x = l$), the shaft is attached to a concentrated inertia I which is restrained by a massless torsional spring of stiffness k connected to a rigid foundation, the boundary condition for (58.26) would be

At $x = l$:
$$GI_p \frac{\partial \phi}{\partial x} + k\phi + I \frac{\partial^2 \phi}{\partial t^2} = 0 \tag{58.27}$$

By including 0 and ∞ as particular values of I and k in (58.27), a wide range of possible end conditions can be represented. Natural modes and frequencies are obtained by substituting

$$\phi(x,t) = \psi(x) e^{i\omega t} \tag{58.28}$$

in (58.26) and (58.27). Results for a few special cases of *uniform* shafts are quoted below.

1. Shaft of length l, clamped at $x = 0$ and free at $x = l$:

$$\psi_m = \sin \frac{(2m-1)\pi x}{2l} \qquad \omega_m = \sqrt{\frac{G}{\rho}}\frac{(2m-1)\pi}{2l} \qquad m = 1, 2, \ldots \tag{58.29}$$

2. Shaft of length l, free at $x = 0$ and free at $x = l$:

$$\psi_m = \cos \frac{m\pi x}{l} \qquad \omega_m = \sqrt{\frac{G}{\rho}}\frac{m\pi}{l} \qquad m = 0, 1, 2, \ldots \tag{58.30}$$

3. Shaft of length l, clamped at $x = 0$ and clamped at $x = l$:

$$\psi_m = \sin \frac{m\pi x}{l} \qquad \omega_m = \sqrt{\frac{G}{\rho}}\frac{m\pi}{l} \qquad m = 1, 2, \ldots \tag{58.31}$$

For *forced* vibrations of torsional systems, we return to (58.22), where the $M_j(t)$ are prescribed excitation torques. In principle, solutions may be obtained for any sort of excitation, transient or steady. In the design of machinery, the most important considerations are the periodic torques inherently involved in normal operation.

These are such torques as the inertia torque (58.8) and the gas-pressure torque (58.10) of a piston machine, or the fluctuating electromagnetic torques in electric motors and generators. In analyzing these, it is convenient to make Fourier expansions such as (58.10), and to consider each harmonic component separately. In single-shaft reciprocating machines, it is customary to refer the order m of the harmonic to the rotational speed of the shaft. This means that, for two-stroke cycle machines, there will be only integer-order numbers, while, for four-stroke cycle machines, there will be half-integer-order numbers.

A complete analysis of the forced torsional vibrations of a machine throughout its operating speed range is almost never attempted. It is, however, customary to consider all resonance possibilities, and to estimate the severity of resonances. Any speed ω for which the frequency $m\omega$ of some order m coincides with one of the torsional natural frequencies is called a *critical speed.* When a machine must operate over a wide speed range, and there are many Fourier components in the excitation and many natural frequencies, there may be a large number of different combinations yielding critical speeds. At a critical speed, resonance is possible. Whether or not a large vibration actually occurs depends on

1. The magnitudes of the Fourier components of the particular order in question that exist in the exciting torques
2. The relative phasing of these components at the various points of excitation
3. The mode shape of the natural vibration in question
4. The amount of damping available in that mode

In this paragraph, a method of estimating the amplitude of a resonance in the pth mode of vibration with a critical speed of order m is outlined. From a Fourier analysis, we obtain the magnitude M_{jm} of the mth-order excitation torque acting at each point j of the system. We also determine the relative phase angles θ_j of these excitations. The excitation may then be represented as a row matrix:

$$M_m = [M_{1m}e^{im\theta_1}, M_{2m}e^{im\theta_2}, \; . \; . \; . \; , M_{n+1,m}e^{im\theta_{n+1}}] \tag{58.32}$$

This torque can be represented as a linear combination of the natural-mode inertia torques as follows:

$$M_m = \sum_i c_i \omega_i{}^2 X_i{}^T I \tag{58.33}$$

where the constants c_i are called the *mode participation factors.* To determine a particular mode participation factor c_p, it is only necessary to know the corresponding natural frequency ω_p and mode X_p. We postmultiply (58.33) by X_p and use the orthogonality principle (p. 18–17) to obtain

$$c_p = \frac{M_m X_p}{\omega_p{}^2 X_p{}^T I X_p} \tag{58.34}$$

This factor also gives the ratio of the static displacement angles due to the pth mode component of M_m as compared with the angles X_p. Finally, it is necessary to estimate the *dynamic amplification factor* Q_p, which gives the ratio of the amplitudes at resonance to the static displacements. Empirical correlation with existing machinery is the best guide here. In systems with no obvious damping devices, the value of Q_p is of the order of 25 to 50. This corresponds to an equivalent damping ratio of 0.02 to 0.01. The displacements and twisting moments at resonance are then obtained by multiplying the displacements and twisting moments in the Holzer table by the factor

$$Q_p c_p = Q_p \frac{M_m X_p}{\omega_p{}^2 X_p{}^T I X_p} \tag{58.35}$$

For a particular natural mode (p fixed), the *relative* severity of the criticals of various order m can be ascertained by examining the scalar product $M_m X_p$. For in-line reciprocating machines with identical cylinders, the magnitudes M_{jm} will equal M_m, say, at all cylinders, and we have only the relative phases to consider in evaluating $M_m X_p$. This may be done by laying out a *star diagram*, in which the jth element of X_p is placed at the angle $m\theta_j$. The vector sum of the star, when multiplied by M_m, gives the product $M_m X_p$. In this way, the most severe critical speed in the operating speed range for the pth mode can be determined.

An alternative method of estimating the amplitudes at resonance is based on an energy balance. This has the advantage that it may be used in systems in which known damping devices have been installed. It is, however, assumed that the damping is sufficiently small that, at resonance, the mode shape is still the same as that of the undamped system. Under steady motion at the natural frequency, the work done per cycle by the excitation torques equals the energy dissipated in the damping devices. By evaluating these two energies for the mode ϵX_p, where ϵ is an unknown scalar multiple, it is possible to evaluate ϵ, since the input work will be proportional to ϵ, while the dissipation will be proportional to ϵ^2 for linear viscous damping.

It is sometimes of interest to evaluate a nonresonant forced vibration. For example, if there is a strong critical speed somewhat above the top speed of the system, it would be important to estimate the forced vibration at top speed. Away from resonance, it is usually justifiable to neglect damping in estimating response. We thus return to (58.22). Near a resonance of an order m, a useful approximation to the total response could be obtained by considering only the component of that order in the excitation. In this case, setting $\theta_j = x_j e^{im\omega t}$ in (58.22) leads to the following linear algebraic equation for the amplitudes X:

$$KX - m^2\omega^2 IX = M_m \tag{58.36}$$

The Holzer table may be adapted to evaluating (58.36) by inserting additional columns between columns 5 and 6, to add in the excitation elements M_{jm}. To solve (58.36), the initial amplitude is taken as x_1 (unknown) instead of unity, and the table is completed in terms of x_1. At the end, x_1 is solved for, by using the boundary condition at $n + 1$; for example, the resultant torque should vanish if there is no excitation at the final station. In general, the elements of M_m are complex numbers; see (58.32). This requires complex arithmetic.

The Holzer table may also be adapted to analyzing forced vibrations of systems with viscous dampers installed. The damping need not necessarily be small. A viscous element with damping coefficient c between two inertias is treated as a torsional spring with the imaginary stiffness $ic\omega$.

Torsional Vibrations of Crankshafts

In analyzing crankshaft systems such as that shown in Fig. 58.10*a*, it is customary to replace them by idealized systems like that of Fig. 58.10*b*. The torsional spring constants k_j are supposed to represent the equivalent torsional stiffness of the complex crankshaft structure when supported in its bearings. Exact analysis is almost hopeless here. Several empirical formulas have been proposed [6]. The moments of inertia I_j are supposed to represent the equivalent inertia of all the rotating and reciprocating parts associated with the jth cylinder. The usual approximation is to take

$$I_{\text{eq}} = I_{\text{rot}} + \tfrac{1}{2} m_{\text{rec}} r^2 \tag{58.37}$$

although other choices have been suggested (see [2], p. 167).

Equivalent systems like that of Fig. 58.10 are usually satisfactory for representing the first two or three natural modes of crankshaft systems. This is generally all that is required in practice. Refinements on this theory can be made (see [2], secs. III-8 and III-9) by modeling the stiffness of the crankshaft more faithfully and by accounting for the variable inertia effects of the reciprocating masses.

The element of the stiffness matrix of (58.20) in the ith row and jth column can be interpreted as the torque at the station i required to maintain a unit angular deflection at the station j. The triple diagonal form of the matrix implies that torques need only be applied at stations $(j - 1)$, j, and $(j + 1)$ in order to keep the system from deflecting anywhere except at j. Though this is true for the equivalent system of Fig. 58.10*b*, it is not true, in general, for a system like that of Fig. 58.10*a*. Here there is a carry-over effect, much like that for a continuous beam on many supports. In order to hold zero deflection at all points but j, it is necessary for small torques to be applied at $(j - 2)$ and $(j + 2)$, and even smaller torques at $(j - 3)$ and $(j + 3)$, etc. These additional torques are said to constitute *induced* torsion. A qualitative description of induced torsion can be obtained by idealizing the crankshaft into a system of beams and shafts supported in bearings with clearances; however, for realistic crankshaft dimensions, it does not seem possible to predict the magnitudes of induced torsion analytically. One of the few cases which have been computed, using a stiffness matrix which included induced torsion, led to essentially no change in the lowest natural frequency, and gradually increasing discrepancies in the higher natural frequencies.

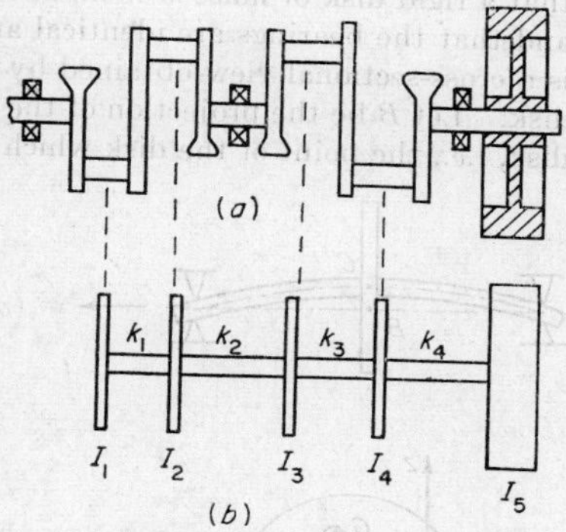

FIG. 58.10. Idealized equivalent torsional system for a crankshaft.

The approximation (58.37) of a cylinder's inertia effect by a constant moment of inertia can be improved by using a variable moment of inertia

$$I = I_{\text{rot}} + \tfrac{1}{2}m_{\text{rec}}r^2\,(1 - \cos 2\omega t) \tag{58.38}$$

When this is inserted in (58.22), we have a system of linear differential equations with variable coefficients of the form known as Hill's equation [7]. By perturbation methods, it can be shown (see [2], p. 248) that the effect of the variable inertia is to convert each natural frequency ω_p of the system with constant inertia into a narrow range of frequencies (including ω_p), in which the system is unstable and executes self-excited vibrations (in the absence of damping). Experimental confirmation of this phenomenon has yet to be obtained.

58.5. WHIRLING OF SHAFTS

Under certain conditions, some of the power transmitted by a rotating shaft can be diverted into building up and maintaining transverse vibrations of the shaft. Analysis of this problem is one of the most fascinating chapters in dynamics. The characteristics of the vibratory system are influenced by the rotation (for example, the natural frequencies are affected by gyroscopic phenomena), and the mechanisms by which energy is supplied to the vibration are extremely diverse. The most common excitation is unbalance in the rotating system, and the most common whirling phenomenon can be viewed as resonance between the rotational speed of the unbalance and a transverse natural frequency of the system. There are many other secondary phenomena in which the vibration does not synchronize with the rotational speed. The nature of the damping of a transverse vibration of a shaft which is simultaneously rotating is imperfectly understood. Internal hysteresis, windage, and friction at the bearings, either dry or lubricated, undoubtedly all contribute, but, under certain circumstances, *every one* (see [1], pp. 292, 295, and 320) of these mechanisms can be guilty of supplying *negative* damping, which can build up severe self-excited vibrations. Here we will discuss the basic case of synchronous whirl, with a brief survey of some of the secondary phenomena. See [11] for an experimental study of these phenomena.

Symmetrical Single-disk System

Consider a flexible shaft, mounted in bearings as shown in Fig. 58.11*a*. We assume that a rigid disk of mass m is mounted in the center of a uniform and massless shaft, and that the bearings are identical and rigidly fixed in inertial space. Figure 58.11*b* is a cross-sectional view obtained by passing a plane P through the mid-plane of the disk. Let B be the projection of the bearing centers, and let C be the "center" of the disk, i.e., the point of the disk which coincides with B when the shaft is straight and motionless. For simplicity, we will neglect gravity. If the shaft bends during rotation, we use the displacements v and w to locate C with respect to B. The center of mass G of the disk will not, in general, coincide with C. Let the eccentricity be ϵ. The line CG is fixed in the disk, and may be used to indicate its rotation.

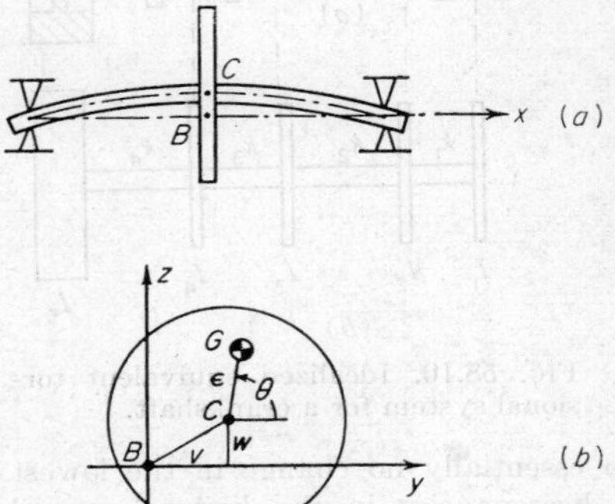

FIG. 58.11. Whirling of a symmetrical single-disk system.

Our assumptions about the symmetry of the system imply that the disk does not tip out of the plane P when the shaft bends. The dynamic system thus has three degrees of freedom, measured by v, w, and θ. We will consider that an external agent inexorably maintains a uniform rotation $\theta = \omega t$, and investigate the remaining two-degree-of-freedom system.

The mass of the disk is subjected to elastic and damping forces. Let the effective stiffnesses at C be $k_y = k_z = k$, and let the effective linear damping coefficients at C be $c_y = c_z = c$. The equations of motion are:

$$kv + c\dot{v} + m(\ddot{v} - \epsilon\omega^2 \cos \omega t) = 0$$
$$kw + c\dot{w} + m(\ddot{w} - \epsilon\omega^2 \sin \omega t) = 0 \qquad (58.39)$$

The complete solutions [see Eq. (56.36)] can be written in the following form:

$$v = A_1 e^{-\zeta\omega_n t} \cos(\sqrt{1 - \zeta^2}\, \omega_n t + \phi_1) + \frac{\epsilon\omega^2}{\sqrt{(\omega_n{}^2 - \omega^2)^2 + 4\zeta^2\omega_n{}^2\omega^2}} \cos(\omega t - \phi)$$

$$(58.40)$$

$$w = A_2 e^{-\zeta\omega_n t} \cos(\sqrt{1 - \zeta^2}\, \omega_n t + \phi_2) + \frac{\epsilon\omega^2}{\sqrt{(\omega_n{}^2 - \omega^2)^2 + 4\zeta^2\omega_n{}^2\omega^2}} \sin(\omega t - \phi)$$

where A_1, A_2, ϕ_1, and ϕ_2 are constants of integration; we have employed the notation

$$\omega_n{}^2 = \frac{k}{m} \qquad \zeta^2 = \frac{c^2}{4km} \qquad (58.41)$$

for natural frequency and damping ratio; and the phase angle ϕ is given by

$$\tan \phi = \frac{2\zeta\omega\omega_n}{\omega_n{}^2 - \omega^2} \qquad (58.42)$$

The solutions (58.40) may be interpreted as free vibrations plus forced vibrations.

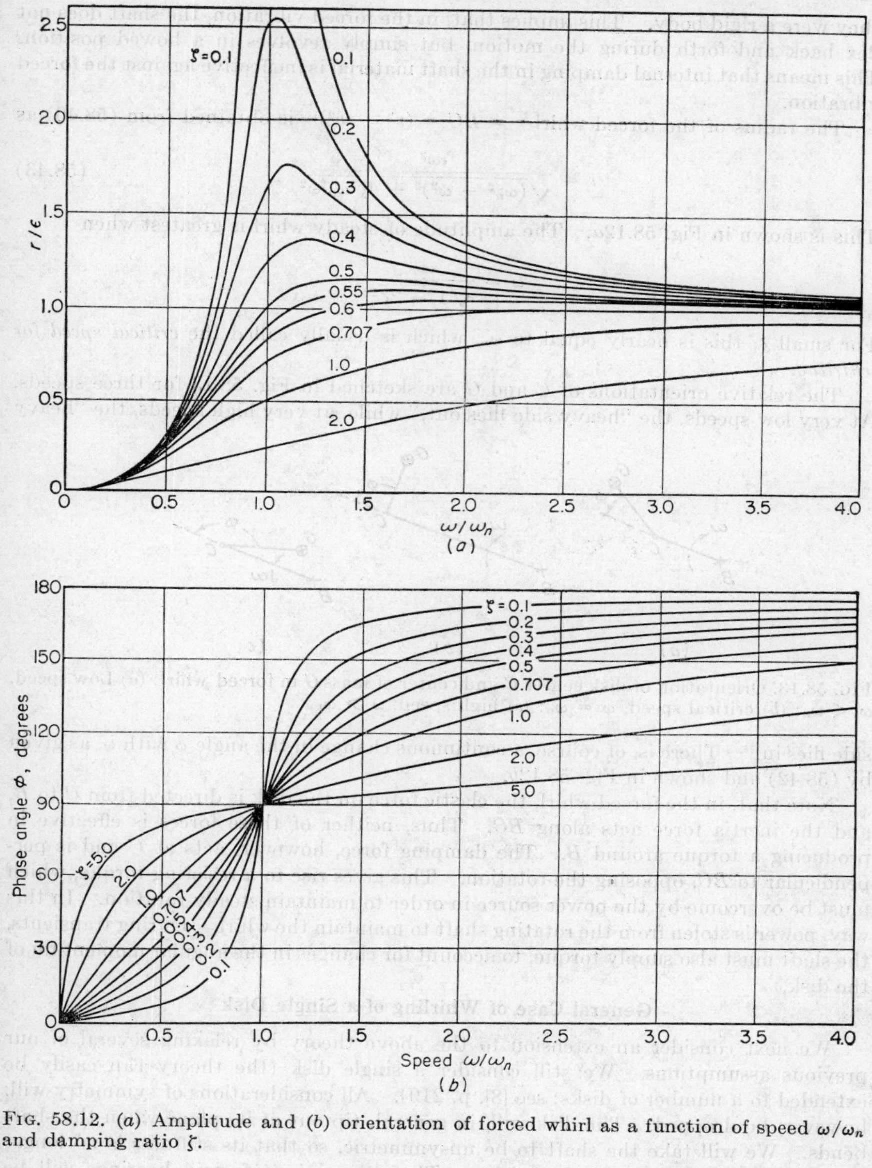

FIG. 58.12. (*a*) Amplitude and (*b*) orientation of forced whirl as a function of speed ω/ω_n and damping ratio ζ.

The free vibrations depend on the initial conditions and decay with time, and the v and w motions are independent of each other with regard to amplitude and phase. In the forced vibration, however, the x and y motions are so related that the path of C is a circle; i.e., the motion is *circularly polarized.*

Furthermore, as C travels in a circle at frequency ω, line CG rotates at the same frequency, so that, in the forced vibration, the line segments BC and CG in Fig. 58.11 have no relative motion with respect to each other. They rotate about B as though they were a rigid body. This implies that, in the forced vibration, the shaft does not flex back and forth during the motion, but simply revolves in a bowed position. This means that internal damping in the shaft material is ineffective against the forced vibration.

The radius of the forced whirl $r = BC = (v^2 + w^2)^{1/2}$ is obtained from (58.40) as

$$r = \frac{\epsilon\omega^2}{\sqrt{(\omega_n^2 - \omega^2)^2 + 4\zeta^2\omega_n^2\omega^2}} \tag{58.43}$$

This is shown in Fig. 58.12*a*. The amplitude of steady whirl is greatest when

$$\omega = \frac{\omega_n}{\sqrt{1 - 2\zeta^2}}$$

For small ζ, this is nearly equal to ω_n, which is usually called the *critical speed for whirling.*

The relative orientations of C and G are sketched in Fig. 58.13 for three speeds. At very low speeds, the "heavy side flies out," while, at very high speeds, the "heavy

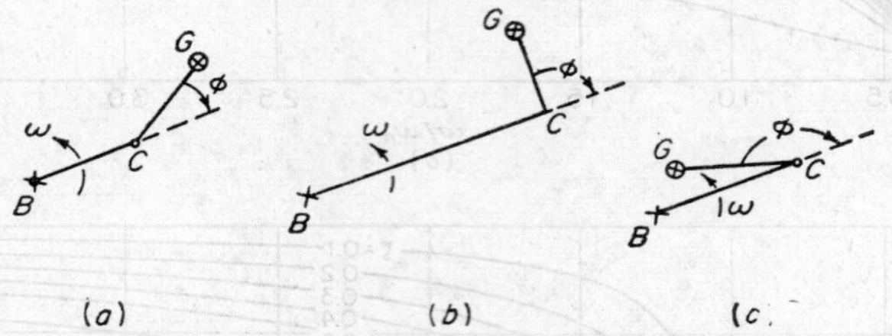

FIG. 58.13. Orientation of disk center C and center of mass G in forced whirl: (*a*) Low speed, $\omega < \omega_n$; (*b*) critical speed, $\omega = \omega_n$; (*c*) high speed, $\omega > \omega_n$.

side flies in." There is, of course, a continuous change in the angle ϕ with ω, as given by (58.42) and shown in Fig. 58.12*b*.

Note that, in the forced whirl, the elastic force on the disk is directed from C to B, and the inertia force acts along BC. Thus, neither of these forces is effective in producing a torque around B. The damping force, however, acts at C and is perpendicular to BC, opposing the rotation. This gives rise to a resisting torque, which must be overcome by the power source in order to maintain steady rotation. In this way, power is stolen from the rotating shaft to maintain the whirl. During transients, the shaft must also supply torque, to account for changes in the angular momentum of the disk.

General Case of Whirling of a Single Disk

We next consider an extension to the above theory by relaxing several of our previous assumptions. We still consider a single disk (the theory can easily be extended to a number of disks; see [8], p. 219). All considerations of symmetry will, however, be dropped. The disk will, in general, tip out of its plane when the shaft bends. We will take the shaft to be unsymmetric, so that its stiffness is no longer independent of its rotational position. The assumption of rigid bearings will be relaxed, by assuming unsymmetric bearing compliances. The inertia distribution of the disk will be taken as unsymmetrical, so that, not only is there an eccentricity of the mass center, but also the principal axes of inertia through the mass center will not, in general, be aligned with the axis and diameters of the disk. Gravity will be included, assuming that the shaft is horizontal. We will obtain generalized equations of motion including all these effects. Then, by specializing these equations, we will discuss several particular cases.

In Fig. 58.14 we assume that when the system is motionless and gravity is not acting, the shaft is straight, with C coinciding with B, and with both bearings B_1 and B_2 centered. The mass center G is displaced a distance ϵ from C. The axes x_1, y_1, z_1 are centered at G, and always remain parallel to the fixed x, y, z axes centered at B. The ξ, η, ζ axes centered at G are fixed in the disk; η and ζ are diameters, and ξ is the disk axis which coincides with x_1 when the system is undeformed. As a rigid body, the disk has six degrees of freedom. We will assume that motion of G parallel to x is negligible, and that the angular motion $\theta = \omega t$ is forced by an external agent. This leaves four degrees of freedom, which we take as v and w, the transverse displacements of C, and ϕ_y and ϕ_z, the angular displacements of the disk with respect to the x_1, y_1, z_1 axes. To obtain a linear theory, these four displacements are all taken as first-order small quantities, i.e., squares and products of v, w, ϕ_y, ϕ_z, and their derivatives will be neglected. Note that the y_2, z_2 axes are a pair of diametral axes which are obtained by

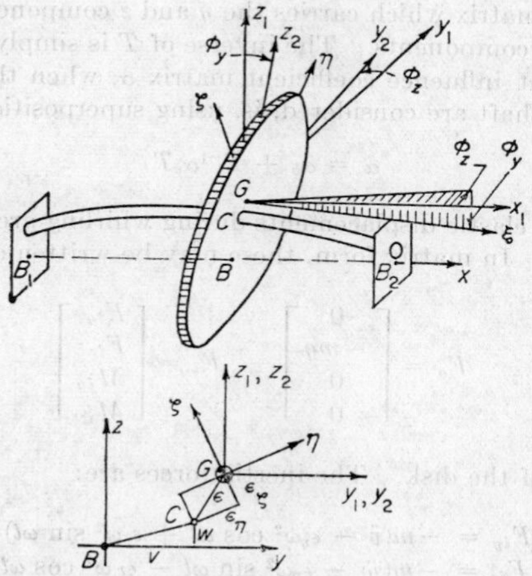

FIG. 58.14. Unsymmetrical disk and shaft on unsymmetrical bearings.

rotating the η and ζ axes about ξ back through the angle θ. The triad (ξ, y_2, z_2) does not partake in rotation about x_1, but does make the displacements ϕ_y and ϕ_z.

To describe the elasticity of the system, we must consider the flexibility of the shaft and the compliance of the bearings. Taking the bearings first, let us assume that, for a *rigid* shaft, the displacements and forces at the disk are related by

$$X = \alpha_b F \tag{58.44}$$

where

$$X = \begin{bmatrix} y \\ z \\ \phi_y \\ \phi_z \end{bmatrix} \qquad F = \begin{bmatrix} F_y \\ F_z \\ M_y \\ M_z \end{bmatrix} \tag{58.45}$$

are column matrices of the disk displacements and corresponding forces, and α_b is a four-by-four matrix of influence coefficients. For bearings which are equally compliant in all directions, there would be no coupling between (y, ϕ_z) displacements and (z, ϕ_y) displacements. We will assume that all couplings are present. The matrix α_b is symmetric and positive-definite.

For the shaft, the situation is complicated by the rotation of the unsymmetrical shaft. Let us assume that, if the bearings were completely *rigid*, and the disk was in the position $\theta = 0$, then the displacements and forces would be related by

$$X = \alpha_s F \tag{58.46}$$

where α_s is a matrix of influence coefficients for the shaft. Again we assume general coupling. Now, when $\theta = \omega t$, we must transform the y and z components of F into η and ζ, before using (58.46) to get the η and ζ components of displacement. These must, in turn, be transformed back to y and z to obtain X. We write

$$X = T^{-1}\alpha_s T F \tag{58.47}$$

where
$$T = \begin{bmatrix} \cos \omega t & \sin \omega t & 0 & 0 \\ -\sin \omega t & \cos \omega t & 0 & 0 \\ 0 & 0 & \cos \omega t & \sin \omega t \\ 0 & 0 & -\sin \omega t & \cos \omega t \end{bmatrix} \tag{58.48}$$

is the transformation matrix which carries the y and z components of F or X into the corresponding η and ζ components. The inverse of T is simply its transpose (see [9], p. 55). The resultant influence coefficient matrix α, when the compliances of *both* the bearings and the shaft are considered, is, using superposition,

$$\alpha = \alpha_b + T^{-1}\alpha_s T \tag{58.49}$$

The forces causing elastic displacements during whirling are due to gravity and to the inertial reactions. In matrix form, these may be written as

$$F_g = \begin{bmatrix} 0 \\ -mg \\ 0 \\ 0 \end{bmatrix} \qquad F_i = \begin{bmatrix} F_{iy} \\ F_{iz} \\ M_{iy} \\ M_{iz} \end{bmatrix} \tag{58.50}$$

where m is the mass of the disk. The inertia forces are:

$$\begin{aligned} F_{iy} &= -m(\ddot{v} - \epsilon_\eta \omega^2 \cos \omega t + \epsilon_\zeta \omega^2 \sin \omega t) \\ F_{iz} &= -m(\ddot{w} - \epsilon_\eta \omega^2 \sin \omega t - \epsilon_\zeta \omega^2 \cos \omega t) \end{aligned} \tag{58.51}$$

To obtain the inertia moments, we write the angular momentum of the disk about G in ξ, η, ζ components [see Eq. (23.47) and the footnote on p. 23–12]:

$$\begin{bmatrix} H_\xi \\ H_\eta \\ H_\zeta \end{bmatrix} = \begin{bmatrix} I_\xi & -I_{\xi\eta} & -I_{\xi\zeta} \\ -I_{\xi\eta} & I_\eta & -I_{\eta\zeta} \\ -I_{\xi\zeta} & -I_{\eta\zeta} & I_\zeta \end{bmatrix} \begin{bmatrix} \omega \\ \omega_\eta \\ \omega_\zeta \end{bmatrix} \tag{58.52}$$

where we have taken a completely general inertia distribution with respect to the ξ, η, ζ axes fixed in the disk. Then, transforming to ξ, y_2, z_2 components, we have

$$\begin{bmatrix} H \\ H_{y_2} \\ H_{z_2} \end{bmatrix} = \begin{bmatrix} 1 & 0 & 0 \\ 0 & \cos \omega t & -\sin \omega t \\ 0 & \sin \omega t & \cos \omega t \end{bmatrix} \begin{bmatrix} I_\xi & -I_{\xi\eta} & -I_{\xi\zeta} \\ -I_{\xi\eta} & I_\eta & -I_{\eta\zeta} \\ -I_{\xi\zeta} & -I_{\eta\zeta} & I_\zeta \end{bmatrix} \begin{bmatrix} 1 & 0 & 0 \\ 0 & \cos \omega t & \sin \omega t \\ 0 & -\sin \omega t & \cos \omega t \end{bmatrix} \begin{bmatrix} \omega \\ \omega_{y_2} \\ \omega_{z_2} \end{bmatrix} \tag{58.53}$$

Now, since the reference frame (ξ,y_2,z_2) undergoes a rotation $\boldsymbol{\omega}_{\text{frame}}$ (with components ω_{y_2} and ω_{z_2}) the absolute rate of change of H is the relative rate of change of **H** with respect to the frame plus $\boldsymbol{\omega}_{\text{frame}} \times \mathbf{H}$ due to the rotation of the frame [see Eq. (25.13)]:

$$\left(\frac{d\mathbf{H}}{dt}\right)_{\text{abs}} = \left(\frac{d\mathbf{H}}{dt}\right)_{\text{rel to frame}} + \boldsymbol{\omega}_{\text{frame}} \times \mathbf{H} \tag{58.54}$$

Applying (58.54) to (58.53) yields the time rate of change of the disk's angular momentum. The y and z components of this are the negatives of the inertia moments M_{iy} and M_{iz} required in (58.50). To carry out (58.54), we neglect squares and products of ω_{y_2} and ω_{z_2}. This leads to ξ, y_2, and z_2 components of $d\mathbf{H}/dt$, which are all of first order in ω_{y_2} and ω_{z_2}. Now, since the angles between the (ξ,y_2,z_2) triad and the (x_1,y_1,z_1) triad are also of the first order, it is correct to first order to take the x_1, y_1, z_1 components of $d\mathbf{H}/dt$ as being identical with the ξ, y_2, z_2 components. Furthermore, we can, to the same order, make the approximations

$$\begin{aligned} \omega_{y_2} &\approx \dot{\phi}_y & \dot{\omega}_{y_2} &= \ddot{\phi}_y \\ \omega_{z_2} &\approx \dot{\phi}_z & \dot{\omega}_{z_2} &= \ddot{\phi}_z \end{aligned} \tag{58.55}$$

in our final expression. It would, of course, be incorrect to make these approximations before applying (58.54). Carrying out this program leads to

$$-\begin{bmatrix} M_{iy} \\ M_{iz} \end{bmatrix} = \begin{bmatrix} I_y & -I_{yz} \\ -I_{yz} & I_z \end{bmatrix} \begin{bmatrix} \ddot{\phi}_y \\ \ddot{\phi}_z \end{bmatrix} + \omega \begin{bmatrix} 2I_{yz} & I_\xi + (I_y - I_z) \\ -I_\xi + (I_y - I_z) & -2I_{yz} \end{bmatrix} \begin{bmatrix} \dot{\phi}_y \\ \dot{\phi}_z \end{bmatrix} + \omega^2 \begin{bmatrix} I_{\xi\eta} \sin \omega t + I_{\xi\zeta} \cos \omega t \\ -I_{\xi\eta} \cos \omega t + I_{\xi\zeta} \sin \omega t \end{bmatrix} \tag{58.56}$$

where we have used the following abbreviations:

$$\begin{aligned} I_y &= \frac{I_\eta + I_\zeta}{2} + \frac{I_\eta - I_\zeta}{2} \cos 2\omega t + I_{\eta\zeta} \sin 2\omega t \\ I_z &= \frac{I_\eta + I_\zeta}{2} - \frac{I_\eta - I_\zeta}{2} \cos 2\omega t - I_{\eta\zeta} \sin 2\omega t \\ I_{yz} &= \qquad\quad -\frac{I_\eta - I_\zeta}{2} \sin 2\omega t + I_{\eta\zeta} \cos 2\omega t \end{aligned} \tag{58.57}$$

Now, finally, assembling (58.49), (58.50), (58.51), and (58.56), we have the general equation of motion of an unsymmetrical disk on an unsymmetrical shaft in unsymmetrical bearings

$$X = (\alpha_b + T^{-1}\alpha_s T)(F_g - \omega^2 U - \omega A\dot{X} - B\ddot{X}) \tag{58.58}$$

where U is the unbalance column,

$$U = \begin{bmatrix} -m\epsilon_\eta \cos \omega t + m\epsilon_\zeta \sin \omega t \\ -m\epsilon_\eta \sin \omega t - m\epsilon_\zeta \cos \omega t \\ I_{\xi\eta} \sin \omega t + I_{\xi\zeta} \cos \omega t \\ -I_{\xi\eta} \cos \omega t + I_{\eta\zeta} \sin \omega t \end{bmatrix} \tag{58.59}$$

and A and B are the following inertia matrices with variable coefficients:

$$A = \begin{bmatrix} 0 & 0 & 0 & 0 \\ 0 & 0 & 0 & 0 \\ 0 & 0 & 2I_{yz} & (I_\xi + I_y - I_z) \\ 0 & 0 & (-I_\xi + I_y - I_z) & -2I_{yz} \end{bmatrix} \tag{58.60}$$

$$B = \begin{bmatrix} m & 0 & 0 & 0 \\ 0 & m & 0 & 0 \\ 0 & 0 & I_y & -I_{yz} \\ 0 & 0 & -I_{yz} & I_z \end{bmatrix} \tag{58.61}$$

Equation (58.58) represents four simultaneous linear second-order differential equations with periodic coefficients. The excitation is represented by the nonrotating constant F_g and the unbalance U which rotates at the shaft speed ω. The general solution of (58.58) is unknown. We consider several special cases to give an indication of the range of possible phenomena.

First of all, we can see that the variable coefficients in the inertia matrices will disappear if the disk has equal diametral moments of inertia for all diameters, and the variable coefficients in the compliance matrix will disappear if the shaft is symmetric. In this symmetric situation, (58.58) has constant coefficients, and the motion is independent of how the total compliance is divided between the bearings and the shaft. The effect of gravity, in this case, is only in fixing the equilibrium configuration about which whirling takes place. The gyroscopic phenomena still remain, however, to make this symmetric system different from the simple case of Fig. 58.11.

Gyroscopic Effect in a Symmetrical System

Neglecting gravity and unbalance, and calling the diametral moment of inertia I_d, (58.58) becomes

$$\begin{bmatrix} v \\ w \\ \phi_y \\ \phi_z \end{bmatrix} = \begin{bmatrix} \alpha_{11} & 0 & 0 & \alpha_{12} \\ 0 & \alpha_{11} & -\alpha_{12} & 0 \\ 0 & -\alpha_{12} & \alpha_{22} & 0 \\ \alpha_{12} & 0 & 0 & \alpha_{22} \end{bmatrix}$$

$$\times \left\{ -\omega \begin{bmatrix} 0 & 0 & 0 & 0 \\ 0 & 0 & 0 & 0 \\ 0 & 0 & 0 & I_\xi \\ 0 & 0 & -I_\xi & 0 \end{bmatrix} \begin{bmatrix} \dot{v} \\ \dot{w} \\ \dot{\phi}_y \\ \dot{\phi}_z \end{bmatrix} - \begin{bmatrix} m & 0 & 0 & 0 \\ 0 & m & 0 & 0 \\ 0 & 0 & I_d & 0 \\ 0 & 0 & 0 & I_d \end{bmatrix} \begin{bmatrix} \ddot{v} \\ \ddot{w} \\ \ddot{\phi}_y \\ \ddot{\phi}_z \end{bmatrix} \right\} \tag{58.62}$$

which describes the free motions of a symmetric gyroscopic disk, with symmetric elastic compliance driven at the constant speed ω. Taking advantage of the symmetry, we introduce the complex displacements,

$$r = u + iv \qquad \phi = -i\phi_y + \phi_z \tag{58.63}$$

into (58.62) to obtain the simpler formulation

$$\begin{bmatrix} r \\ \phi \end{bmatrix} = -\begin{bmatrix} \alpha_{11} & \alpha_{12} \\ \alpha_{12} & \alpha_{22} \end{bmatrix} \left\{ \begin{bmatrix} m & 0 \\ 0 & I_d \end{bmatrix} \begin{bmatrix} \ddot{r} \\ \ddot{\phi} \end{bmatrix} + i\omega \begin{bmatrix} 0 & 0 \\ 0 & I_\xi \end{bmatrix} \begin{bmatrix} \dot{r} \\ \dot{\phi} \end{bmatrix} \right\} \tag{58.64}$$

The natural modes and frequencies can be derived from (58.64) by making the substitution

$$r = Re^{i\omega_n t} \qquad \phi = \Phi e^{i\omega_n t} \tag{58.65}$$

which, together with (58.63), implies that the natural modes of the symmetric system are circularly polarized; i.e., the natural modes are configurations which rotate at frequency ω_n without change of shape. The two natural frequencies obtained as the roots of the characteristic equation of (58.64) are given by

$$\omega_n^2 = \frac{\alpha_{11}m + \alpha_{22}I_{eq} \pm \sqrt{(\alpha_{11}m - \alpha_{22}I_{eq})^2 + 4\alpha_{12}^2 m I_{eq}}}{2(\alpha_{11}\alpha_{22} - \alpha_{12}^2)mI_{eq}} \tag{58.66}$$

where

$$I_{eq} = I_d - \frac{\omega}{\omega_n} I_\xi \tag{58.67}$$

Note that (58.66) is an *implicit* equation for ω_n, since ω_n appears in I_{eq}. In practice, it is easiest to insert fixed values for the ratio ω/ω_n in (58.67) and (58.66), and plot the results in a graph of ω_n versus ω as shown in Fig. 58.15. In this diagram, positive and negative natural frequencies are plotted against positive shaft rotational frequency. The actual directions here are immaterial: What is important is the relative direction of the rotation of the circularly polarized natural motion and the rotation of the shaft. For any given shaft speed, there are four natural modes, two forward and two backward, with four natural frequencies.

When the shaft is not rotating ($\omega = 0$), the four natural frequencies become a pair of double roots:

$$\omega_1^2,\ \omega_2^2 = \frac{\alpha_{11}m + \alpha_{22}I_d \pm \sqrt{(\alpha_{11}m - \alpha_{22}I_d)^2 + 4\alpha_{12}^2 mI_d}}{2(\alpha_{11}\alpha_{22} - \alpha_{12}^2)mI_d} \tag{58.68}$$

There are still four natural modes. The forward and backward circularly polarized modes at the same frequency may be alternately represented as independent bending modes in the x and y direction.

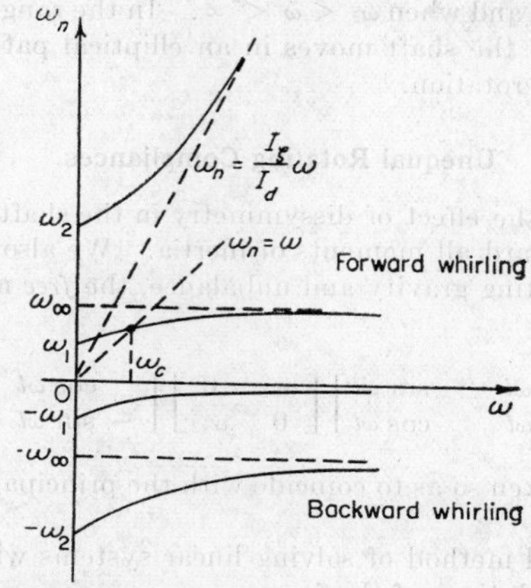

FIG. 58.15. Natural frequencies of a symmetrical whirling system as functions of shaft rotation frequency, showing gyroscopic effects.

When the shaft rotates very rapidly ($\omega \to \infty$), the axial angular momentum becomes so large that the disk cannot be tipped out of its own plane; that is, ϕ remains zero during whirling. The natural frequency, under these circumstances, is

$$\omega_\infty^2 = \frac{\alpha_{22}}{m(\alpha_{11}\alpha_{22} - \alpha_{12}^2)} \tag{58.69}$$

The interaction of rotating unbalance with a symmetrical gyroscopic system can be investigated by including the matrix U of (58.58) and (58.59) in (58.62). Since the unbalance rotates at the shaft speed, there is a resonance at the *critical speed* ω_c where $\omega_n = \omega$. From Fig. 58.15, it is clear that there is one critical speed between ω_1 and ω_∞. For disklike inertias, this is the only critical speed, since the asymptote to the second forward whirl has a slope I_ξ/I_d of about 2. For inertias with considerable axial length, I_ξ/I_d may be less than unity, in which case there is a second critical speed.

Unequal Nonrotating Compliances

Returning to (58.58), we investigate the effect of dissymmetry in the bearing compliances. To see the phenomenon in its simplest guise, we will disregard all moments of inertia. Neglecting gravity, the equations of motion of the system are

$$\begin{bmatrix} v \\ w \end{bmatrix} = -m\left\{\begin{bmatrix} \alpha_{b1} & 0 \\ 0 & \alpha_{b2} \end{bmatrix} + \begin{bmatrix} \alpha_s & 0 \\ 0 & \alpha_s \end{bmatrix}\right\}\begin{bmatrix} \ddot{v} - \epsilon\omega^2 \cos \omega t \\ \ddot{w} - \epsilon\omega^2 \sin \omega t \end{bmatrix} \tag{58.70}$$

if the y and z directions are taken to coincide with the principal directions of the bearing compliance and if the η and ζ axes are taken so that $\epsilon_\eta = \epsilon$ and $\epsilon_\zeta = 0$. The system (58.70) is *uncoupled.* The natural frequency for y motion is given by $\omega_y{}^2 = 1/[m(\alpha_{b1} + \alpha_s)]$, and the natural frequency for z motion is given by $\omega_z{}^2 = 1/[m(\alpha_{b2} + \alpha_s)]$. The steady-state forced motion is

$$v = \epsilon \frac{\omega^2}{\omega_y{}^2 - \omega^2} \cos \omega t \qquad w = \epsilon \frac{\omega^2}{\omega_z{}^2 - \omega^2} \sin \omega t \tag{58.71}$$

which, in general, represents elliptical paths for the shaft center. Resonance of the y motion occurs when $\omega = \omega_y$, and resonance of the z motion occurs when $\omega = \omega_z$. Assuming $\omega_y < \omega_z$, the elliptical paths are traced in the *same* sense as the shaft rotates when $0 < \omega < \omega_y$ and when $\omega_z < \omega < \infty$. In the range $\omega_y < \omega < \omega_z$ between resonances, the center of the shaft moves in an elliptical path in the *opposite* sense with respect to the shaft rotation.

Unequal Rotating Compliances

We next investigate the effect of dissymmetry in the shaft flexibility. Again, for simplicity, we will disregard all moments of inertia. We also assume no compliance in the bearings. Neglecting gravity and unbalance, the *free* motion of the system is described by

$$\begin{bmatrix} v \\ w \end{bmatrix} = -m \begin{bmatrix} \cos \omega t & -\sin \omega t \\ \sin \omega t & \cos \omega t \end{bmatrix} \begin{bmatrix} \alpha_{s1} & 0 \\ 0 & \alpha_{s2} \end{bmatrix} \begin{bmatrix} \cos \omega t & \sin \omega t \\ -\sin \omega t & \cos \omega t \end{bmatrix} \begin{bmatrix} \ddot{v} \\ \ddot{w} \end{bmatrix} \tag{58.72}$$

if the η and ζ axes are taken so as to coincide with the principal directions of the shaft compliance.

Following the general method of solving linear systems with periodic coefficients [10], we would assume solutions of the form

$$v = e^{\mu t} \sum_{-\infty}^{\infty} A_n e^{in\omega t} \qquad w = e^{\mu t} \sum_{-\infty}^{\infty} B_n e^{in\omega t} \tag{58.73}$$

and obtain an infinite algebraic system for μ and the coefficients A_n and B_n. In this particular case, we find that the infinite system collapses, and that only the terms corresponding to $n = -1$ and $n = 1$ are required. This fortunate circumstance is clarified if we reconsider (58.72) in terms of the rotating displacements (u_η, u_ζ) given by

$$\begin{bmatrix} u_\eta \\ u_\zeta \end{bmatrix} = \begin{bmatrix} \cos \omega t & \sin \omega t \\ -\sin \omega t & \cos \omega t \end{bmatrix} \begin{bmatrix} v \\ w \end{bmatrix} \tag{58.74}$$

The equations of motion then become linear equations with *constant coefficients:*

$$\begin{bmatrix} u_\eta \\ u_\zeta \end{bmatrix} = -m \begin{bmatrix} \alpha_{s1} & 0 \\ 0 & \alpha_{s2} \end{bmatrix} \left\{ \begin{bmatrix} \ddot{u}_\eta \\ \ddot{u}_\zeta \end{bmatrix} + 2\omega \begin{bmatrix} 0 & -1 \\ 1 & 0 \end{bmatrix} \begin{bmatrix} \dot{u}_\eta \\ \dot{u}_\zeta \end{bmatrix} - \omega^2 \begin{bmatrix} u_\eta \\ u_\zeta \end{bmatrix} \right\} \tag{58.75}$$

Solutions corresponding to (58.73) can be obtained by inserting

$$u_\eta = ae^{\mu t} \qquad u_\zeta = be^{\mu t} \tag{58.76}$$

in (58.75). The roots of the characteristic equation are

$$\mu^2 = -\left(\omega^2 + \frac{\omega_1{}^2 + \omega_2{}^2}{2}\right) \pm \sqrt{\left(\omega^2 + \frac{\omega_1{}^2 + \omega_2{}^2}{2}\right)^2 - (\omega^2 - \omega_1{}^2)(\omega^2 - \omega_2{}^2)} \tag{58.77}$$

where $\omega_1^2 = 1/(\alpha_{s1}m)$ and $\omega_2^2 = 1/(\alpha_{s2}m)$ give the two nonrotating natural frequencies of the mass on the unsymmetrical shaft. When μ^2 is negative, the solution (58.76) is bounded, but, when μ^2 is positive, initial displacements will grow without limit. Study of (58.77) reveals that this latter case occurs when $\omega_1 < \omega < \omega_2$. Thus, unequal rotating flexibilities provide a mechanism for *self-excited vibrations.* When $(\omega_2 - \omega_1)$ is small compared to $\omega_0 = \frac{1}{2}(\omega_2 + \omega_1)$, the maximum value of μ occurs at $\omega \approx \omega_0$ with $\mu \approx \frac{1}{2}(\omega_2 - \omega_1)$. Assuming that the effect of linear damping is still the same as that in (58.40), we would expect that a damping ratio ζ of magnitude

$$\zeta = \frac{1}{2}\frac{\omega_2 - \omega_1}{\omega_0} \tag{58.78}$$

would be required to annul the self-excited vibration.

Outside the range of self-excited vibrations, the values of μ given by (58.77) are imaginary, and the free motions consist of ellipses traced out at frequencies $\omega_n = i\mu$

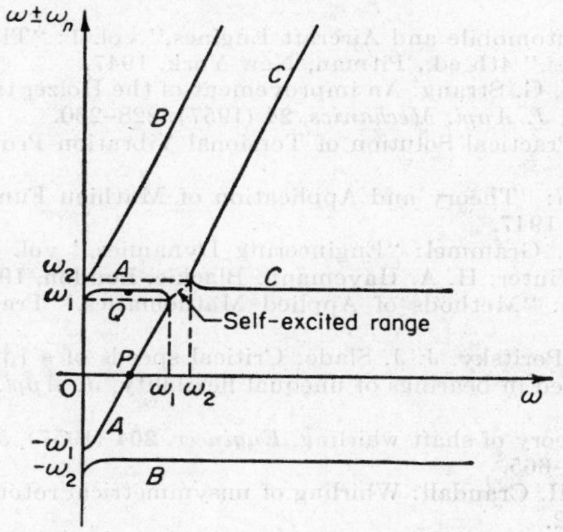

FIG. 58.16. Frequencies of circularly polarized components of free motion of system with unequal rotational flexibility, as seen from nonrotating reference frame.

with respect to the rotating coordinate system. In order to describe these in terms of the nonrotating coordinate system, it is convenient to visualize each elliptical motion as the superposition of two circularly polarized motions, one rotating in the same sense as the ellipse with amplitude $(a + b)/2$, and the other rotating in the opposite sense with amplitude $(a - b)/2$, where the semimajor and semiminor axes of the ellipse are a and b, respectively. In a nonrotating reference frame, these circularly polarized motions remain circularly polarized, but now have frequencies of $(\omega + \omega_n)$ and $(\omega - \omega_n)$. These resultant frequencies are sketched in Fig. 58.16. The curves are labeled AA, BB, and CC, to indicate that a component on one branch of the B curve, say, is invariably accompanied by a component on the other B branch. When $(\omega_2 - \omega_1)$ is small, the major response consists of the forward and backward whirls at frequencies between ω_1 and ω_2. There are, however, the branches whose slopes are approximately two plus the small self-excited range, which mark the special character of a system with unequal rotational stiffness.

The frequencies in Fig. 58.16 occur in free motions of the system. It is to be expected that, if a circularly polarized excitation were to have a frequency coinciding with one of these, there would be a *resonance* of the forced motion. For the case of unbalance, it is easy to show that, without damping, there is infinite response when

$\omega = \omega_1$ and $\omega = \omega_2$. In between these frequencies, the forced response is finite, but, as we have already seen, the free motion is unbounded.

An interesting phenomenon is associated with the point P in Fig. 58.16. Here, where ω is approximately *one-half* of $\omega_0 = \frac{1}{2}(\omega_1 + \omega_2)$, there is a free-motion component which *does not rotate.* The accompanying component, indicated by point Q, is a circularly polarized whirl of frequency approximately equal to ω_0. When $(\omega_2 - \omega_1)$ is small, the ratio of the amplitude associated with P to that associated with Q is approximately $(\omega_2 - \omega_1)/\omega_0$; that is, the whirl is large compared with the nonrotating deflection. We would expect that a nonrotating excitation such as gravity could cause a resonance of this motion. This is, in fact, the case (see [1], p. 338).

58.6. REFERENCES

[1] J. P. Den Hartog: "Mechanical Vibrations," 4th ed., McGraw-Hill, New York, 1956.

[2] C. B. Biezeno, R. Grammel: "Engineering Dynamics," vol. 4: "Internal Combustion Engines," transl. by M. B. White, Blackie, London, 1954.

[3] F. P. Porter: Harmonic coefficients of engine torque curves, *J. Appl. Mechanics,* **10** (1943), 33–48.

[4] A. W. Judge: "Automobile and Aircraft Engines," vol. 1: "The Mechanics of Petrol and Diesel Engines," 4th ed., Pitman, New York, 1947.

[5] S. H. Crandall, W. G. Strang: An improvement of the Holzer table based on a suggestion of Rayleigh's, *J. Appl. Mechanics,* **24** (1957), 228–230.

[6] W. K. Wilson: "Practical Solution of Torsional Vibration Problems," 3d ed., Wiley, New York, 1956.

[7] N. W. McLachlan: "Theory and Application of Mathieu Functions," Oxford Univ. Press. New York, 1947.

[8] C. B. Biezeno, R. Grammel: "Engineering Dynamics," vol. 3: "Steam Turbines," transl. by E. F. Winter, H. A. Havemann, Blackie, London, 1954.

[9] F. B. Hildebrand: "Methods of Applied Mathematics," Prentice-Hall, Englewood Cliffs, N.J., 1952.

[10] W. R. Foote, H. Poritsky, J. J. Slade: Critical speeds of a rotor with unequal shaft flexibilities mounted in bearings of unequal flexibility, *J. Appl. Mechanics,* **10** (1943), 77–87.

[11] E. Downham: Theory of shaft whirling, *Engineer,* **204** (1957), 518–522, 552–555, 588–591, 624–628, 660–665.

[12] P. J. Brozens, S. H. Crandall: Whirling of unsymmetrical rotors, *J. Appl. Mechanics,* **28** (1961), 355–362.

CHAPTER 59

SERVOMECHANISMS, AUTOMATIC CONTROL

BY

G. A. SMITH, Ph.D., Palo Alto, Calif.

59.1. INTRODUCTION

The word *servomechanism* refers to a mechanism, usually a group of mechanical, electrical, and electronic components, whose output is constrained to follow its input by the use of feedback, and which utilizes an external source of energy so that the power available at the output can be much greater than that supplied to the input. The most common type of servomechanism accepts a varying positional command input at low power level, and delivers a nearly exact replica of this position as an output. The words "nearly exact replica" are responsible for the existence of the considerable

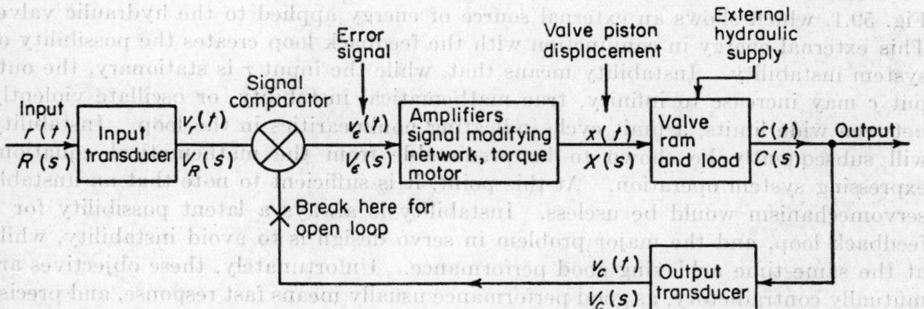

FIG. 59.1. Block diagram of a typical servomechanism.

body of servo theory which is presented here in greatly abbreviated form. If a servomechanism could produce an exact replica, the need for the highly specialized design methods required for a servo capable of approaching the ideal would vanish.

A block diagram of a typical positional servomechanism is shown in Fig. 59.1. The input r is the angular position of a dial, which can be varied arbitrarily by hand. The input transducer is an electromechanical device, which produces an electric voltage v_r proportional to r. The controlled member might be a large antiaircraft searchlight, whose horizontal angular position is c. The output transducer produces a voltage v_c proportional to c; it thus measures output position. The signal comparator compares the voltages v_r and v_c, and produces a voltage v_ϵ equal to their difference and thus proportional to the error between input and output. An amplifier increases the magnitude of this error signal voltage, and a signal-modifying network introduces dynamic lead and lag compensation to the signal. Then a power amplifier drives a spring-restrained torque motor, which controls the position x of the piston of the hydraulic valve. The valve controls a source of external hydraulic power, which it applies to the hydraulic ram. The ram positions the load.

Block diagrams such as Fig. 59.1 are used extensively in the servo field. The lines connecting the blocks indicate the flow of signals, while the blocks themselves represent actual physical components whose inputs and outputs are connected in the sequence indicated by the diagram. For example, the input to the amplifier is an electric voltage, and its output is also a voltage; the input to the transducer is a mechanical angle, and its output is a voltage; the input to the torque motor is a voltage, and its output is a mechanical position.

One fundamental characteristic of a servomechanism is indicated in Fig. 59.1, where it is noted that a signal v_c proportional to the output position c is fed back (right to left along the lower connecting line) to the signal comparator, where it is subtracted from the input signal v_r to produce the error signal voltage v_ϵ. It is this error signal which is then used to control the output position. Since a continuous signal path exists from the comparator through the amplifier, valve, ram, and output transducer, back to the comparator, this is usually referred to as a *feedback loop*. Since a number of separate components are involved in a servomechanism, the complete assemblage is often referred to as a *servomechanism system*. The study of servomechanism systems is based on two relations apparent from Fig. 59.1: the relation among input, output, and error signals, $v_\epsilon = v_r - v_c$; and a relation expressing the output as a function of the error signal, $c = F(v_\epsilon)$. Actually only a relation between output and input is of practical interest such as $c = F(r)$, the error signal being of no concern. However, enormous simplifications in the theoretical analysis and synthesis of servomechanism systems are realized if the error is considered.

The other fundamental characteristic of a servo system is also apparent from Fig. 59.1, which shows an external source of energy applied to the hydraulic valve. This external energy in conjunction with the feedback loop creates the possibility of system instability. Instability means that, while the input r is stationary, the output c may increase to infinity, true mathematical instability, or oscillate violently between wide limits, a limit cycle indicating nonlinearities in the loop. Instability will subsequently be shown to be predictable from the mathematical equations expressing system operation. At this point, it is sufficient to note that an unstable servomechanism would be useless. Instability is always a latent possibility for a feedback loop, and the major problem in servo design is to avoid instability, while at the same time achieving good performance. Unfortunately, these objectives are mutually contradictory, as good performance usually means fast response, and precise dynamic correspondence between input and output. This implies high gain in the amplifier and the hydraulic valve, which then causes the feedback loop to tend toward instability.

Although an all mechanical-hydraulic or pneumatic servomechanism is entirely possible, the great majority of servomechanisms include some electrical and electronic components. A servomechanism may be used to control quantities other than position, such as temperature, pressure, flow, or speed. Many regulators are servomechanisms with steady or quasi-steady desired outputs.

59.2. MATHEMATICAL DESCRIPTION OF A SERVO SYSTEM

This treatment of servomechanisms is limited to linear systems, which comprise the majority of practical systems. It is therefore assumed that performance of each element of the system, each block in Fig. 59.1, may be described by an ordinary linear differential equation with constant coefficients. When the differential equation for each block is subjected to the Laplace transformation, the ratio of the output transform to the input transform is found as a function of the Laplace variable s. This ratio is called the *transfer function* of the block. The blocks are chosen to be unilateral; that is, their input is not affected by their output, and so the transfer function of a sequence of blocks is the product of their individual transfer functions.

Actual physical quantities in Fig. 59.1 are designated by lower-case letters as functions of time $r(t)$ or $v(t)$. Their Laplace transforms are designated by upper-case letters as functions of the Laplace variable s, like $R(s)$ or $V(s)$.

Figure 59.2 is a conventional simplified block diagram of a servomechanism, in which the transfer functions of all the separate elements of Fig. 59.1 have been combined to give a single equivalent block, whose transfer function is designated $KG(s)$, where K is a constant and $G(s)$ is a rational function of s.

It is often convenient to consider the open-loop performance with the feedback path broken as at X, in Fig. 59.1 or 59.2. Then an input to the system $R(s)$ is also the input to the block $KG(s)$, since no feedback signal has been subtracted from it, and the output is given by $C(s) = R(s)KG(s)$. Because of this relation, $KG(s)$ is called the *open-loop* transfer function. If now the feedback loop is considered to be closed at X, as it normally is, we have the relations $E(s) = R(s) - C(s)$ and $C(s) = E(s)KG(s)$ which, on elimination of $E(s)$, yield $C/R = KG/(1 + KG)$, which is called the *closed-loop* transfer function. Because a servomechanism always operates with the loop closed, only the closed-loop transfer function is of practical interest since it gives the Laplace transform of the output $C(s)$ in terms of the Laplace transform of the input $R(s)$. However, from an analytical and theoretical standpoint, a great deal of attention centers on the open-loop transfer function $KG(s)$, which gives the Laplace transform of the output in terms of the Laplace transform of the error.

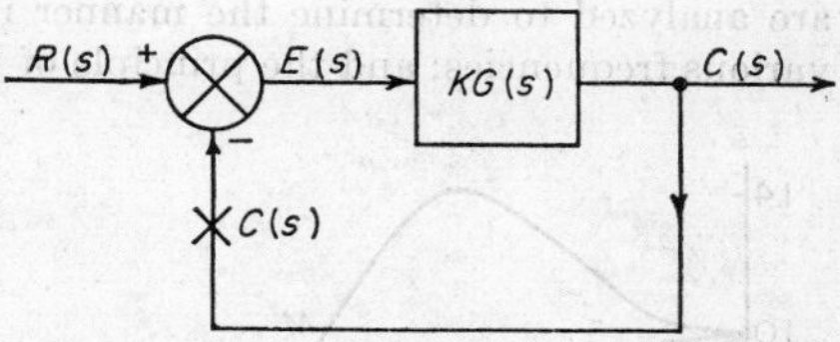

Fig. 59.2. Conventional simplified block diagram of a servomechanism.

If one possesses detailed knowledge of the physical performance of each element of a servo system, one can write differential equations for each element, take the Laplace transforms of these equations, and combine the resulting individual transfer functions to obtain the over-all closed-loop performance $C(s) = KG(s)R(s)/[1 + KG(s)]$. For any input as a function of time $r(t)$, one can substitute its Laplace transform $R(s)$ in the above relation and arrive at an expression for $C(s)$, the Laplace transform of the output. It is now merely necessary to take the inverse Laplace transform of both sides of this equation to obtain $c(t)$, the actual time response of the output. Unfortunately, there are two major disadvantages to this seemingly straightforward procedure; considerable effort may be required to find the inverse transform, and, even when found, it provides very little really useful information about how to improve the system performance. In connection with the first difficulty, it should be recalled that, in order to find the inverse Laplace transform of $KGR/(1 + KG)$, it must be put into a form such that recognizable transform pairs from a table can be used. This usually means a partial fraction expansion of the expression, and this, in turn, requires a determination of the roots of the denominator. Since this denominator is frequently of fourth or higher degree, considerable labor is involved. The second difficulty arises because the system parameters, such as amplifier gain or modifying network transfer function, which can be adjusted to provide a desirable system response, are imbedded in the expression $KG/(1 + KG)$, and, if we wish to determine the effect of changing one of these parameters, a completely new inverse Laplace transform must be found, and the roots of the new denominator determined. What we really want is a simple and rapid method that will enable the effects of changing parameters to be clearly seen, so that they may be adjusted to give the most desirable performance.

Two different approaches to servomechanism theory which more or less achieve these objectives are widely used, and they will be described in the following sections. These methods both have the same objective, of determining a system design that will be stable and will have a high degree of static and dynamic accuracy. These approaches are:

1. The frequency-response method, using the Nyquist stability criterion
2. The root-locus method, displaying stability directly by pole positions in the complex plane

59.3. FREQUENCY-RESPONSE APPROACH TO SERVO DESIGN

The frequency-response approach to servomechanisms is a direct application and extension of electrical communication circuit theory where random time-function waveshapes are analyzed in terms of their Fourier frequency components; networks are analyzed to determine the manner in which they transmit sinusoidal signals of various frequencies; and the principle of linear superposition is employed to establish the output due to the transmission of the various frequency components of the input signal through the network.

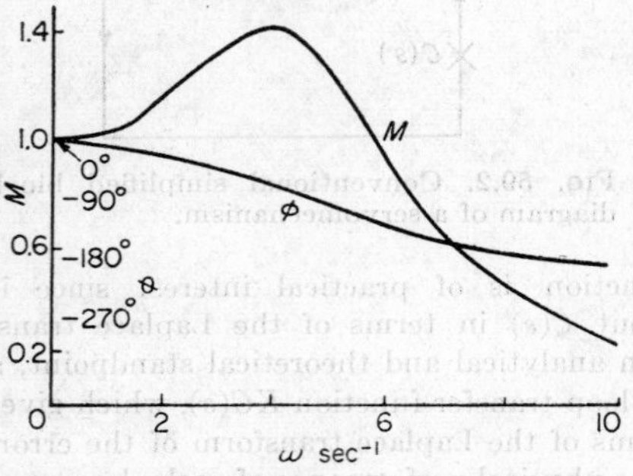

FIG. 59.3. Frequency response of a servomechanism, linear plot.

The inputs applied to a servo are usually of a transient character, and their analysis typically discloses that they contain a band of frequencies extending from zero upward, the magnitudes of the frequency components falling off rapidly as frequency increases. A servomechanism can be analyzed to determine its frequency response, which is then plotted as in Fig. 59.3 for a typical case, where the amplitude ratio M and phase ϕ of the output with respect to a steady-state sinusoidal input are plotted against frequency. The low-frequency components of an input signal will be transmitted to the output quite faithfully since M is unity and ϕ is relatively linear in this region. The medium-frequency components will be somewhat exaggerated in the output, as indicated by the hump in the M curve; the higher-frequency components will be drastically reduced. The typical servo thus has a frequency reponse like a low-pass filter. When the frequency response of a servo has been obtained, it could be used in connection with a Fourier analysis of an input signal to compute the resulting output. Usually the amplitude ratio curve is merely used to estimate the output; a servo whose amplitude response does not fall off until a higher frequency will more faithfully reproduce an input with sharp breaks and high rates of change, indicating high-frequency components in the input, than will a servo whose M curve falls off at a lower frequency. On the other hand, a servo whose M curve falls off at a relatively low frequency will pass less extraneous noise from the input to the output.

The frequency response of a servomechanism can be obtained relatively easily, either experimentally or by calculation. An experimental frequency response can be obtained by applying a sinusoidal signal to the input, $r(t) = A \sin \omega_1 t$, and measuring the resulting output, $c(t) = B \sin (\omega_1 t + \phi)$. This gives one point on the amplitude ratio response, $M = B/A$ at a frequency ω_1, and one point on the phase curve, ϕ at ω_1. This procedure is repeated for a range of input frequencies, to obtain the other points on the curves. It should be noted that the assumption of linear elements yields a sinusoidal output for a sinusoidal input. If practical system nonlinearities produce a somewhat nonsinusoidal output, a satisfactory procedure is to estimate the amplitude ratio and phase shift of the fundamental component.

The closed-loop frequency response can be calculated directly (Fig. 59.2) from the closed-loop transfer function, $C(s)/R(s) = KG(s)/[1 + KG(s)]$, by merely replacing the Laplace variable s by $i\omega$ to give $C(i\omega)/R(i\omega) = KG(i\omega)/[1 + KG(i\omega)]$. This expression, which has real and imaginary parts, can be used to calculate the magnitude M and phase ϕ for steady-state sinusoidal excitation at any frequency, if it is remembered that K is real and $G(i\omega)$ is complex, with a magnitude $|G(i\omega)|$ and a phase $\theta = \arg G(i\omega)$, or, in rectangular form, $G(i\omega) = X + iY$, where X and Y are real, so that $|G(i\omega)| = \sqrt{X^2 + Y^2}$ and $\theta = \arg G(i\omega) = \tan^{-1}(Y/X)$.

Although the closed-loop frequency response,

$$\frac{C(i\omega)}{R(i\omega)} = \frac{KG(i\omega)}{1 + KG(i\omega)}$$

is of interest in servomechanism evaluation, the open-loop frequency response, $C(i\omega)/E(i\omega) = KG(i\omega)$, is of much greater theoretical interest. The open-loop frequency response is obtained from the open-loop transfer function,

$$\frac{C(s)}{E(s)} = KG(s)$$

of Fig. 59.2, by replacing s by $i\omega$ to get $C(i\omega)/E(i\omega) = KG(i\omega)$. In terms of the previous symbols, the open-loop frequency response is a complex quantity with magnitude $K|G(i\omega)|$ and phase θ. This open-loop frequency response can be used to solve the all-important problem of determining closed-loop stability of the actual operating closed-loop system. The method of using the open-loop frequency response to determine closed-loop stability utilizes the Nyquist criterion.

The Nyquist Stability Criterion

A dynamic system is unstable if the denominator of its transfer function has any zeros with positive real parts. The closed-loop transfer function of a simple servomechanism, as given in Fig. 59.2, is $C/R = KG/(1 + KG)$.

If $G(s)$ is written in terms of its numerator and denominator $N(s)$ and $D(s)$, which are polynomials in s, we have $G = N/D$ and $C/R = KN/(KN + D)$. Stability of the servomechanism thus depends upon the zeros of $(KN + D)$. However, $1 + KG = (KN + D)/D$ has the same zeros, in addition to poles, when $D = 0$. Since KG is usually known in factored form, it is much more convenient to determine whether any of the zeros of $(1 + KG)$ have positive real parts than it is to check the zeros of $(KN + D)$.

The Nyquist criterion utilizes the fact that the Laplace variable s is a complex variable, $s = \sigma + i\omega$, so that $KG(s)$ describes a conformal mapping of the s plane onto the KG plane.

The Nyquist criterion states that, if the contour of KG is drawn on the KG plane as s encircles the entire right-half s plane clockwise, then the KG contour will encircle the point $KG = -1$ in the clockwise direction $Z - P$ times. Z is the number of zeros of $(1 + KG)$ with positive real parts, and P is the number of poles of $(1 + KG)$ with positive real parts. The system is stable only if $Z = 0$.

This test does not immediately specify the value of Z since P must first be determined. However, P is usually zero, since the poles of $(1 + KG)$ which occur at $D = 0$ cannot have positive real parts unless the system is unstable with the loop open. This never occurs for the simple single-loop systems of Fig. 59.1 or 59.2, since each block is itself stable. However, a term $1 - Ts$ sometimes occurs in the denominator of an aircraft transfer function or might arise in a multiloop system where a minor loop of stable elements could be unstable with the minor loop closed. If such a possibility exists, it must be checked by an initial Nyquist or Routh test of the minor loop.

The Nyquist test rests on a theorem of complex function theory that the contour integral of the logarithmic derivative of $f(x)$ is $2\pi i$ times the number of poles less the number of zeros of $f(x)$ within the contour, that is,

$$\oint \frac{f'(x)}{f(x)}\,dx = 2\pi i(P - Z)$$

and the fact that the contour for $(1 + KG)$ can be obtained by shifting the contour for KG one unit to the right so that, if the contour of KG encircles the -1 point, the contour of $(1 + KG)$ will encircle the origin.

Conformal Mapping Involved in Applying the Nyquist Criterion

Figure 59.4*a* illustrates the contour followed by s in encircling the right-half s plane, and Fig. 59.4*b* illustrates the conformal mapping of this contour onto the KG

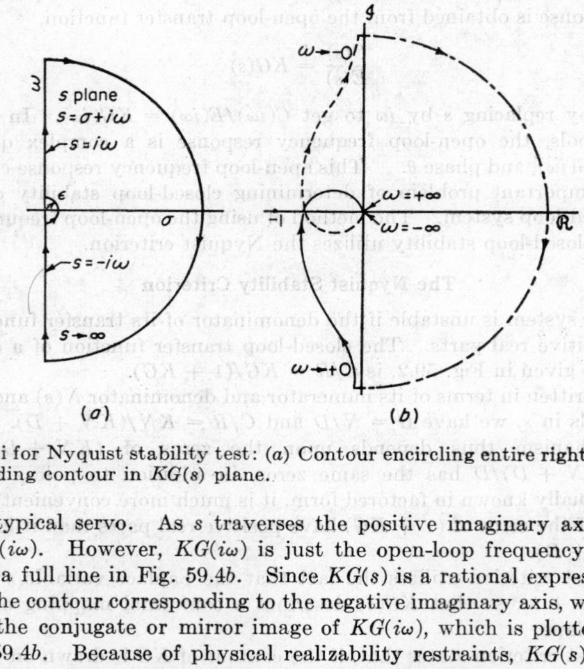

FIG. 59.4. Loci for Nyquist stability test: (*a*) Contour encircling entire right half of s plane; (*b*) corresponding contour in $KG(s)$ plane.

plane for a typical servo. As s traverses the positive imaginary axis, $s = i\omega$ and $KG(s) = KG(i\omega)$. However, $KG(i\omega)$ is just the open-loop frequency response. It is plotted as a full line in Fig. 59.4*b*. Since $KG(s)$ is a rational expression with real coefficients, the contour corresponding to the negative imaginary axis, where $s = -i\omega$, is $KG(-i\omega)$, the conjugate or mirror image of $KG(i\omega)$, which is plotted as a dotted line in Fig. 59.4*b*. Because of physical realizability restraints, $KG(s)$ approaches a constant, usually zero, as s approaches infinity, and so the KG locus remains at the origin as s traverses the infinite-radius semicircle clockwise. The open-loop transfer function $KG(s)$ normally contains a pole at $s = 0$, and so the contour on the s plane must avoid the origin by traversing a small semicircular indentation into the right half plane along which $s = \epsilon e^{i\theta}$ and $KG(s) \to K/s$ as $\epsilon \to 0$. So the KG contour describes a semicircle of infinite radius clockwise, shown by the dot-dash line in Fig. 59.4*b*, as s traverses its small semicircle counterclockwise around the origin from $s = \epsilon e^{-i\pi/2}$ to $s = \epsilon e^{i\pi/2}$.

After the KG locus has been constructed, it remains to determine whether the point $KG = -1$ is encircled. If it is not obvious, it can be checked by considering a vector from -1 to a point on the locus, and observing whether this vector makes any net revolutions as the point traverses the entire locus. If it does, the point $KG = -1$ is encircled, and the system is unstable. Figure 59.4 is the locus of a stable system.

The KG locus can be computed if an analytic expression for KG exists, since, for the path along the imaginary axis, $KG(s) = KG(i\omega)$ is the frequency response. However, it is equally valid to determine the frequency response experimentally, plot the locus from amplitude ratio and phase data, and sketch in the obvious parts as ω becomes very large or small. The possibility of using experimental frequency-response data when an analytic expression is not available is one of the most important advantages of the frequency-response method.

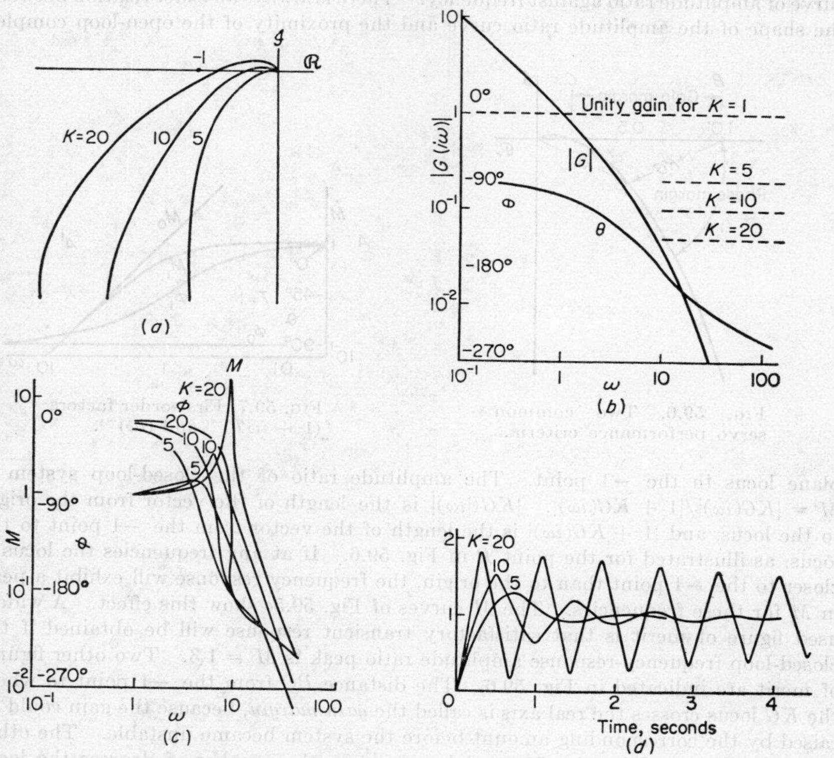

FIG. 59.5. Response of a servomechanism as gain is increased:

$$KG(i\omega) = K/i\omega(1 + 0.2i\omega)(1 + 0.06i\omega).$$

(a) $KG(i\omega)$ loci as K increases; (b) $KG(i\omega)$ amplitude ratio and phase, open loop; (c) $KG(i\omega)/[1 + KG(i\omega)]$ amplitude ratio and phase, closed loop; (d) time responses to step inputs.

The majority of frequency-response theory deals with simple systems such as Fig. 59.4, with a single loop and a first-order pole at $s = 0$. For such systems, the frequency response discloses a great deal of information, in addition to just whether the system is stable or not. Attention is centered on the portion of the positive frequency locus near the origin as in Fig. 59.5a. If the -1 point lies to the left of the point where the locus crosses the negative real axis, the system is stable. Furthermore, the degree of stability is related to the proximity of the locus to the -1 point. In general, the closer the locus approaches the point $KG = -1$, the less stable the system is. The degree of system stability may be conveniently indicated by the time response of the system to a step input.

Figure 59.5 shows a typical example of a system whose KG locus is modified only by increasing K; this does not affect the shape of the locus, but merely expands the scale. The corresponding time responses of the actual closed-loop servo (Fig. 59.5*d*) to a step input indicate a larger overshoot and more oscillation as the KG locus approaches the -1 point. The correlation can be expressed exactly for a second-order characteristic equation, and it follows approximately for higher-order systems. A further correlation is seen if the closed-loop frequency response is displayed in the curve of amplitude ratio against frequency. There is always an exact relation between the shape of the amplitude ratio curve and the proximity of the open-loop complex

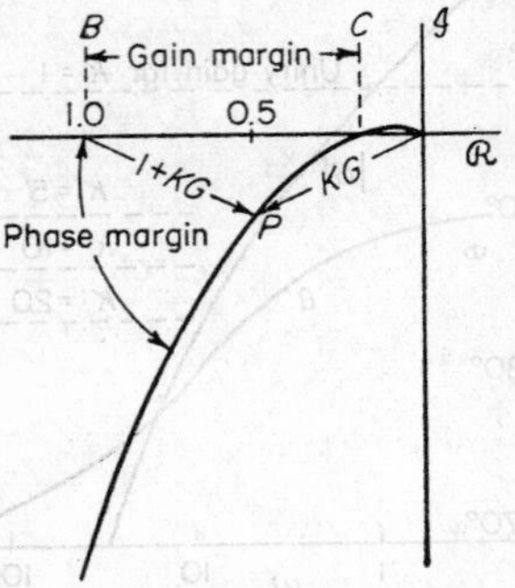

FIG. 59.6. Two common servo performance criteria.

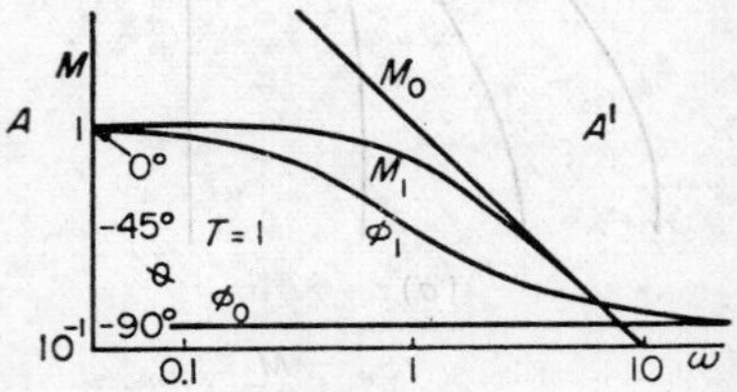

FIG. 59.7. First-order factors $(1 + i\omega T)^{-1}$ and $(i\omega)^{-1}$.

plane locus to the -1 point. The amplitude ratio of the closed-loop system is $M = |KG(i\omega)|/|1 + KG(i\omega)|$. $|KG(|i\omega)|$ is the length of the vector from the origin to the locus, and $|1 + KG(i\omega)|$ is the length of the vector from the -1 point to the locus, as illustrated for the point P in Fig. 59.6. If at any frequencies the locus is closer to the -1 point than to the origin, the frequency response will exhibit a peak in M for these frequencies. The M curves of Fig. 59.5*c* show this effect. A widely used figure of merit is that satisfactory transient response will be obtained if the closed-loop frequency-response amplitude ratio peak is $M \approx 1.3$. Two other figures of merit are indicated in Fig. 59.6. The distance BC from the -1 point to where the KG locus crosses the real axis is called the *gain margin,* because the gain could be raised by the corresponding amount before the system became unstable. The other criterion illustrated in Fig. 59.6 is *phase margin,* the number of degrees the locus lacks being on the negative real axis (180°) when the open-loop amplitude ratio, length of vector from the origin to the locus, is one. For Fig. 59.6, which is an expanded plot of the data for Fig. 59.5*a* with $K = 5$, the phase margin is about 52°, and the gain margin is about 0.8. Satisfactory margins frequently cited are 0.6 to 0.9 gain margin and 30 to 40° phase margin.

The Bode Plot

The frequency-response information contained in the Nyquist plot (Fig. 59.5*a*) can also be represented by plotting the open-loop amplitude response $|C/E|$ and the phase $\theta = \arg (C/E)$ separately, as functions of the frequency. In the Bode plot, this is done using logarithmic scales for ω and $|KG|$ and a linear scale for θ (Fig. 59.5*b*). The system is stable if the phase shift is less than 180° at the frequency for which the amplitude response equals unity. Where the phase is 180°, the amplitude-response curve indicates the gain margin.

The advantage of the Bode plots over the Nyquist plot lies in the rapidity with which they can be constructed and used, to quickly show the effect of added gain or changes in the signal modifier on the open-loop frequency response. These effects on the open-loop frequency response are used to predict the effects on the closed-loop frequency response, through the correlations between open-loop proximity to the -1 point, gain margin, phase margin, and the closed-loop M peak and transient response.

The analytical expression for $KG(s)$ is a ratio of two polynomials in s. These polynomials can be broken up into first- and second-order factors. In fact, $KG(s)$

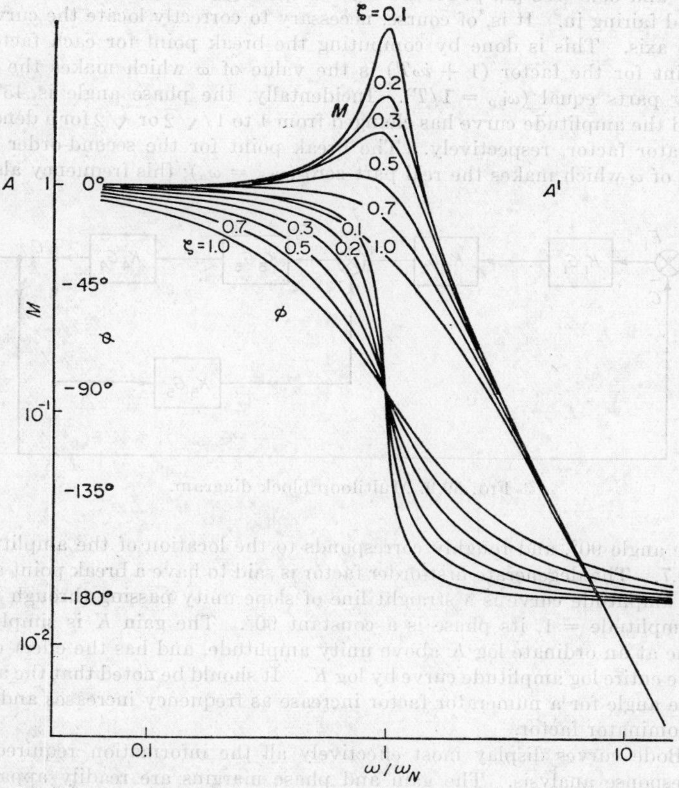

FIG. 59.8. Second-order factors $[1 - (\omega/\omega_n)^2 + 2i\zeta\omega/\omega_n]^{-1}$.

is usually computed in factored form when combining the transfer functions of the elements in the loop. For frequency-response calculations, s is replaced by $i\omega$ in each of the factors. The standard form of the first-order factor is $(1 + i\omega T)$; T is called the time constant. A degenerate form of first-order factor usually occurring in the denominator is $i\omega$. Figure 59.7 is a plot of the phase and log amplitude of these factors versus log ω for the usual case, in which they occur as denominator factors. The phase and amplitude for $1/(1 + i\omega T)$ are designated as ϕ_1 and M_1, and for $1/i\omega$ as ϕ_0 and M_0. These curves should be reflected about the horizontal axis AA' to represent numerator factors. It can be assumed that $1/(1 + i\omega T)$ is plotted for $T = 1$, or else the abscissa for this curve can be designated as ωT. The standard form of second-order factor $1 + 2\zeta s/\omega_n + s^2/\omega_n^2$ becomes $1 - \omega^2/\omega_n^2 +$

$i2\zeta\omega/\omega_n$ upon the substitution of $i\omega$ for s, where ζ is the damping ratio, and ω_n is the natural frequency. Figure 59.8 is a plot of the phase and log amplitude of this factor versus log ω for several values of ζ, and for the usual case of a denominator factor. These curves should be reflected about the horizontal AA' axis to represent a numerator factor.

When plotted in this fashion against log ω, the shape of the factors is not affected by their location along the log ω axis. It is, therefore, possible to draw the amplitude and phase curves for each individual factor by using templates of the curves of Figs. 59.7 and 59.8 (see [2], p. 605), or by transposing a few points from the standard curves and fairing in. It is, of course, necessary to correctly locate the curves along the log ω axis. This is done by computing the break point for each factor. The break point for the factor $(1 + i\omega T)$ is the value of ω which makes the real and imaginary parts equal ($\omega_{bp} = 1/T$). Incidentally, the phase angle is 45° at this point, and the amplitude curve has changed from 1 to $1/\sqrt{2}$ or $\sqrt{2}$ for a denominator or numerator factor, respectively. The break point for the second-order factor is the value of ω which makes the real part zero ($\omega_{bp} = \omega_n$); this frequency also makes

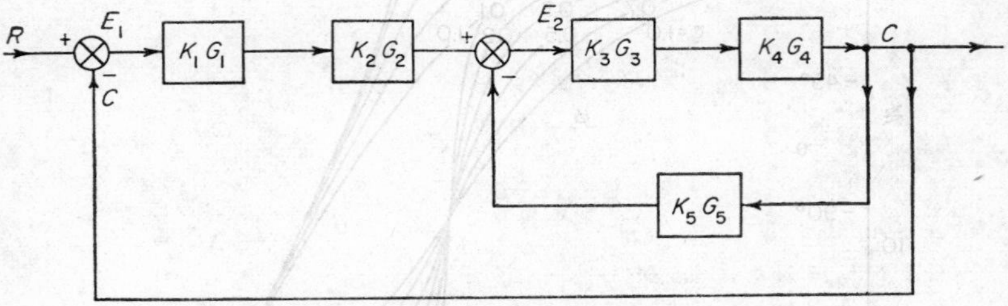

FIG. 59.9. Multiloop block diagram.

the phase angle 90°, and roughly corresponds to the location of the amplitude peak for $\zeta < 0.7$. The degenerate first-order factor is said to have a break point at $\omega = 1$; since the amplitude curve is a straight line of slope unity passing through the point $\omega = 1$, amplitude $= 1$, its phase is a constant 90°. The gain K is simply a horizontal line at an ordinate log K above unity amplitude, and has the effect of merely raising the entire log amplitude curve by log K. It should be noted that the amplitude and phase angle for a numerator factor increase as frequency increases and decrease for a denominator factor.

The Bode curves display most effectively all the information required for frequency-response analysis. The gain and phase margins are readily apparent, the maximum K for stability is evident, the effect of the modifying network in introducing phase lead is clearly shown, and one can quickly visualize what the effect would be of shifting the critical frequencies originally chosen for the modifying network. If the open-loop frequency response is first plotted without a modifying network, an intelligent choice of network can be made from inspection of the Bode curves, which indicate a desirable fashion in which to modify the original curves to permit a maximum K, and still leave an acceptable gain and phase margin.

After the summation curves for phase and amplitude, including the appropriate value of K, have been drawn, the closed-loop frequency response can be computed from the relation $C/R = KG/(1 + KG)$. This is also presented as curves of phase and log amplitude versus log ω. The computations involve converting $KG(i\omega)$ from polar to rectangular form, so that the 1 can be added, and then reconverting to polar form to give closed-loop amplitude ratio and phase.

An example of the block-diagram algebra used for multiloop systems is illustrated in Fig. 59.9, where an auxiliary feedback loop K_5G_5, usually velocity feedback, is employed, in addition to the normal position loop feedback. The frequency-response analysis of this system is concerned with the denominator of the transfer function, that is, $1 + K_3K_4K_5G_3G_4G_5 + K_1K_2K_3K_4G_1G_2G_3G_4$, and is correspondingly more lengthy than the analysis of the single-loop system previously discussed, though the same general principles are involved. In exchange for the added complexity, more freedom in system performance optimization exists, since both the series compensation K_1G_1 and the auxiliary feedback K_5G_5 can be manipulated. Figure 59.10 illustrates the block-diagram algebra involved for an external torque load U acting on the output. It will be noted that the output is the result of both the input R and

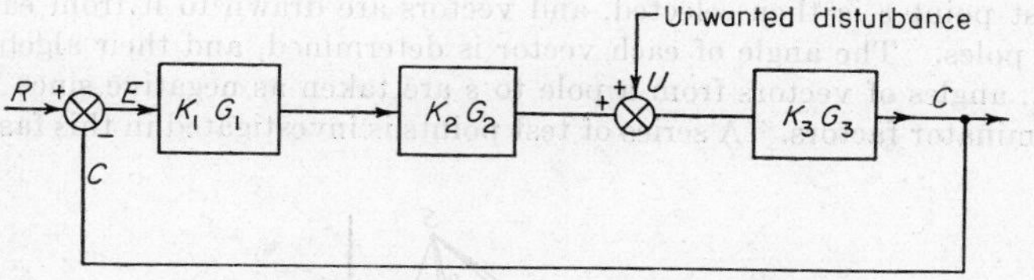

FIG. 59.10. Load-disturbance block diagram.

the load U. It is seen that, in order to minimize the effects of U, the transfer function between input and the point where U is applied ($K_1G_1K_2G_2$) should be made as large as possible.

59.4. ROOT-LOCUS METHOD

The closed-loop transfer function from Fig. 59.2 becomes $C/R = KN/(KN + D)$ when the open-loop transfer function is written in terms of its numerator and denominator polynomials ($KG = KN/D$). The root-locus method developed by W. R. Evans requires an analytical expression for the open-loop transfer function, from which the roots of N and D can be determined; it then employs a graphical technique to determine the location of the roots of $(KN + D)$ on the complex s plane, as K is varied from zero to infinity. These roots lie on continuous curves, which constitute the root locus.

The characteristic equation of the servomechanism is $KN + D = 0$, which, in factored form, is $(s - r_1)(s - r_2)(s - r_3) \cdots = 0$, where $r_1, r_2, r_3, \ldots$ are the roots. The transient response of the system is given by the sum of terms of the form $A_1e^{r_1t} + A_2e^{r_2t} + A_3e^{r_3t} + \cdots$. For negative real roots, these terms represent a sum of decaying exponentials with characteristic times equal to the reciprocals of the roots and magnitudes A equal to the residues of $C = [KN/(KN + D)]R$ at the roots. Complex roots occur in conjugate pairs ($\sigma \pm i\omega$), and the corresponding transient-response terms are $(Ae^{(\sigma+i\omega)t} + \bar{A}e^{(\sigma-i\omega)t})$, where the residues A and $\bar{A}$ are conjugate also. These terms combine to a single decaying sinusoid $2|A|e^{\sigma t} \cos (\omega t + B)$ where B is the angle of A. System time response can thus be computed in detail when the roots are known. The root-locus method utilizes a graphical procedure to avoid the labor that would be involved in the analytical solutions for the roots of the characteristic equation as K is varied.

The expression $1 + KG = (KN + D)/D$ is zero when $KG = -1$, which can occur only when $G(s)$ has a phase angle of $-180°$. The phase angle of

$$G(s) = \frac{(s - r_a)(s - r_b)(s - r_c) \cdots}{(s - r_d)(s - r_e)(s - r_f) \cdots}$$

is the algebraic sum of the individual phase angles contributed by each factor. A factor $(s - r_a)$ may be represented on the complex s plane by a straight line from r_a to s, and its phase angle is measured by the angle between it and the real axis. The root-locus method essentially tests points of the complex s plane, to see whether the total phase angle of $G(s)$ is $-180°$ at the test point. The totality of all points in the s plane, where $G(s)$ has $-180°$ phase shift, constitutes the root locus, since any such value of s makes $KN + D = 0$ for the value of K which makes the magnitude of KG equal one.

The graphical procedure for checking whether a particular point s lies on the root locus is illustrated in Fig. 59.11. The zeros and poles of the open-loop transfer function $KG(s) = KN/D$ are first plotted on the s plane, indicated by ○ and ×, respectively. They remain fixed during the investigation, as they do not depend on K. A test point s is then selected, and vectors are drawn to it from each of these zeros and poles. The angle of each vector is determined, and their algebraic sum is computed; angles of vectors from a pole to s are taken as negative since they represent denominator factors. A series of test points is investigated in this fashion; those

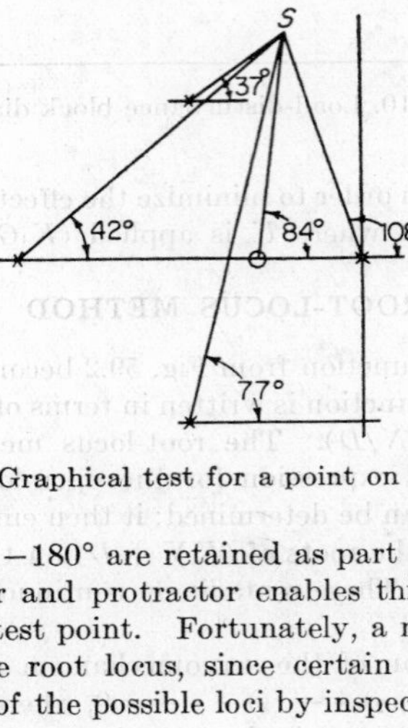

Fig. 59.11. Graphical test for a point on the root locus.

whose total phase angle is $-180°$ are retained as part of the root locus. A systematized procedure with ruler and protractor enables this graphical test to be carried out quite rapidly for each test point. Fortunately, a relatively few points suffice to establish the curves of the root locus, since certain relationships can be used to indicate the general paths of the possible loci by inspection of the original plot of the open-loop zeros and poles.

Figure 59.12 is a root-locus plot of the data from Fig. 59.5. Values of K are marked along the loci. The system becomes unstable for values of K greater than $K = 21.7$, since two of the roots then have positive real parts as the locus crosses into the right half plane. For $K = 10$, the complex roots represent a damped sinusoidal term with frequency $\omega = 6.4 \text{ sec}^{-1}$ and a damping ratio $\zeta = 0.15$ in response to a step input. The corresponding negative real root represents an exponentially decaying term with characteristic time $\tau = 0.05$ sec. The complete time response is shown by the curve for $K = 10$ in Fig. 59.5*d*.

Interest usually centers on the roots nearest the right half plane, since further increase in K will move them into the right half plane and cause system instability; also they usually dominate the transient response. The largest value of K is usually selected which will still produce adequate damping for the transient terms from these critical roots in response to a step input. The actual magnitude and phase of the residues are seldom computed, except for the final configuration, although a relatively simple graphical method for evaluating the residues exists.

The actual frequency ω and damping σ in the transient response due to a pair of conjugate roots are related to the natural frequency ω_n and damping ratio ζ of the standard second-order factor, obtained from the product of the factors by the relations $(s + \sigma + i\omega)(s + \sigma - i\omega) = s^2 + 2\sigma s + \sigma^2 + \omega^2 = s^2 + 2\zeta\omega_n s + \omega_n^2$, so that $\sigma = \zeta\omega_n$ and $\omega = \omega_n\sqrt{1 - \zeta^2}$. Figure 59.13 illustrates that ω_n is the length of the vector from the origin to the root, and the angle made by this vector with the negative real axis is $\cos^{-1}\zeta$.

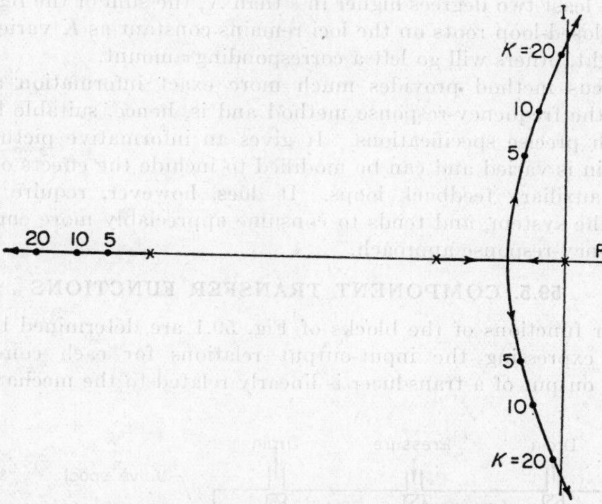

Fig. 59.12. Root-locus plot, data of Fig. 59.5.

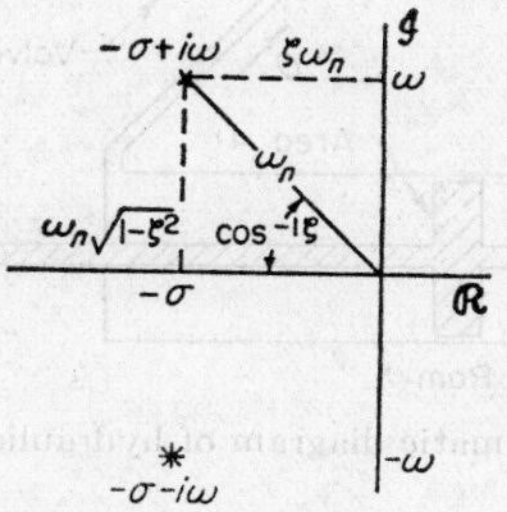

Fig. 59.13. Complex-plane plot of poles of the standard second-order term.

Aids in Sketching the Root Locus

1. The loci start at the poles of N/D for $K = 0$, and proceed along smooth curves, approaching the zeros of N/D as $K \rightarrow \infty$. Zeros at infinity must be included so that there are an equal number of poles and zeros, and one locus for each pair.
2. Sections of the real axis to the left of an odd number of critical points of N/D (finite roots or poles) are part of the root locus.
3. The loci occur in conjugate pairs.
4. The loci approach the n zeros at infinity along asymptotes spaced at n equal angles around the origin, the first angle being $180°/n$. The extensions of these asymptotes meet at a point on the real axis whose abscissa is the sum of the abscissas

of the poles, minus the sum of the abscissas of the zeros of N/D, divided by the number of poles, minus the number of finite zeros.

5. If the open-loop frequency response is written in real and imaginary parts, that is, $KG(i\omega) = (R_N + iI_N)/(R_D + iI_D)$, the root locus crosses the imaginary axis at values of $s = i\omega$ which make $R_N I_D = R_D I_N$, and the corresponding gain is $K = -R_D/R_N$.

6. If D is at least two degrees higher in s than N, the sum of the real parts of the instantaneous closed-loop roots on the loci remains constant as K varies. Hence, as some loci go right, others will go left a corresponding amount.

The root-locus method provides much more exact information about system behavior than the frequency-response method and is, hence, suitable for the design of systems with precise specifications. It gives an informative picture of system behavior, as gain is varied and can be modified to include the effects of stabilization networks and auxiliary feedback loops. It does, however, require an analytical expression for the system, and tends to consume appreciably more engineering time than the frequency-response approach.

59.5. COMPONENT TRANSFER FUNCTIONS

The transfer functions of the blocks of Fig. 59.1 are determined by considering the equations expressing the input-output relations for each component. The electric voltage output of a transducer is linearly related to the mechanical input; so

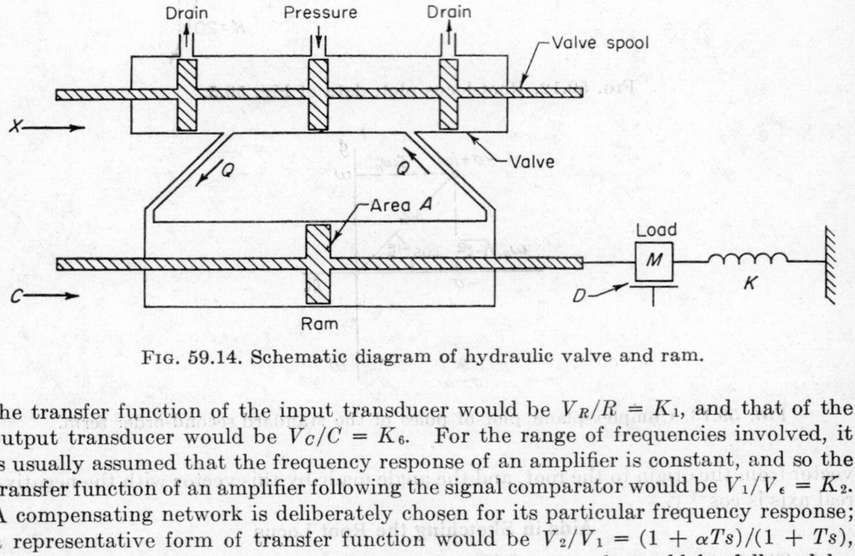

FIG. 59.14. Schematic diagram of hydraulic valve and ram.

the transfer function of the input transducer would be $V_R/R = K_1$, and that of the output transducer would be $V_C/C = K_6$. For the range of frequencies involved, it is usually assumed that the frequency response of an amplifier is constant, and so the transfer function of an amplifier following the signal comparator would be $V_1/V_\epsilon = K_2$. A compensating network is deliberately chosen for its particular frequency response; a representative form of transfer function would be $V_2/V_1 = (1 + \alpha Ts)/(1 + Ts)$, which, for $\alpha > 1$, represents a lead network. This network would be followed by another amplifier of transfer function $V_3/V_2 = K_3$, which would drive a spring-restrained torque motor as a linear function of the voltage to give a transfer function $X/V_3 = K_4$. The complete transfer function of the block between the error and the valve position X would then be $X/V_\epsilon = K_2K_3K_4(1 + \alpha Ts)/(1 + Ts)$ where K has been used to designate a constant, and V is the Laplace transform of a voltage.

The hydraulic valve and ram of Fig. 59.1 represent a relatively complicated subsystem. Flow of hydraulic fluid through an orifice is actually nonlinear ($Q = C\sqrt{P}$), where Q is the rate of flow, P the pressure drop, and C a constant. However, a linearizing assumption which corresponds rather well with experimental pressure-flow data for actual servo valves can be made, in order to permit a linear analysis [4]. Figure 59.14 is a schematic diagram of a hydraulic valve and ram, driving a load consisting of mass, viscous damping, and spring. It is assumed that sufficient force is available to control the valve spool, so that its position x is not influenced by the flow.

The curves of valve flow vs. pressure drop across the ram for underlapped servo valves are approximately linear over a considerable range of operation, so that $Q = K_1X - C_1P$, where Q is total valve flow, P is pressure drop across the ram, and X is valve opening. This valve flow is distributed among leakage flow C_2P, compressibility flow $K_3\, dP/dt$, and ram volume flow $A\, dC/dt$. Application of the Laplace transform gives the equation $K_1X - C_1P = C_2P + K_3sP + AsC$. The load-force balance equation is $AP = Ms^2C + DsC + KC$. Pressure may be eliminated from these two equations to give

$$\frac{C}{X} = \frac{AK_1}{s^3K_3M + s^2(MC_1 + MC_2 + K_3D) + s(DC_1 + DC_2 + K_3K + A^2) + K(C_1 + C_2)}$$

The coefficients K_1 and C_1 are obtained from the valve pressure-flow curves, C_2 is a leakage coefficient, and K_3 is the compressibility coefficient given by $K_3 = V/B$, where V is the volume of oil under compression, and B is the bulk modulus.

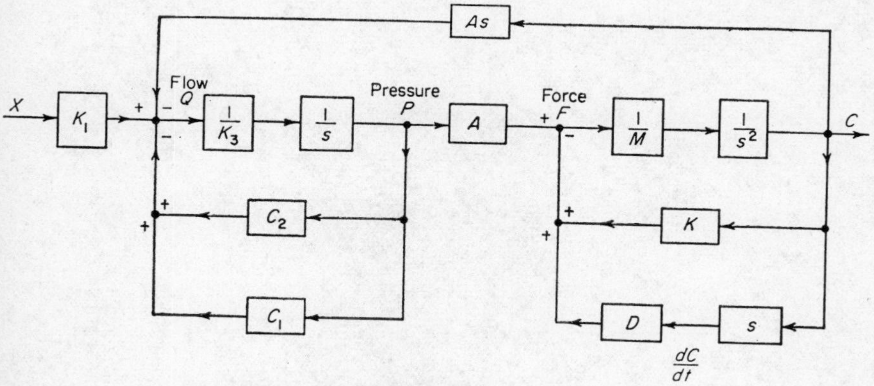

FIG. 59.15. Signal flow diagram of hydraulic valve ram and load.

The two basic equations of flow and force can be represented by the signal flow diagram of Fig. 59.15. The blocks of a signal flow diagram represent mathematical operations rather than specific pieces of physical equipment. Since Laplace transforms are assumed, the output signal of each block is the product of the input signal and the transfer function represented by the block. A set of differential equations may be represented by many different signal flow diagrams.

If the load is taken merely as the mass M, the transfer function reduces to

$$\frac{C}{X} = \frac{AK_1}{s^3K_3M + s^2(MC_1 + MC_2) + sA^2} = \frac{K_1/A}{s\{1 + s[M(C_1 + C_2)/A^2] + s^2(K_3M/A^2)\}}$$

$$= \frac{K_5}{s[1 + (2\zeta/\omega_n)s + s^2/\omega_n^2]}$$

where $\omega_n = \dfrac{A}{\sqrt{K_3M}}$ and $\zeta = \dfrac{(C_1 + C_2)\sqrt{M}}{2A\sqrt{K_3}}$

59.6. REFERENCES

[1] H. Chestnut, R. W. Mayer: "Servomechanisms and Regulating System Design," 2d ed., vol. 1, Wiley, New York, 1959.

[2] C. S. Draper, W. McKay, S. Lees: "Instrument Engineering," vol. 2, McGraw-Hill, New York, 1953.

[3] J. G. Truxal: "Automatic Feedback Control System Synthesis," McGraw-Hill, New York, 1955.

[4] J. L. Shearer: Dynamic characteristics of valve controlled hydraulic servomotors, *Trans. ASME*, **76** (1954), 895–903.

CHAPTER 60

SURGE TANKS

BY

C. JAEGER, Dr. ès Sc. techn., Rugby, England

60.1. INTRODUCTION

Let us consider the elementary hydraulic system consisting of reservoir, pressure tunnel, surge tank, pressure pipeline, or penstock and control valve, near the turbine. Any time the demand for power on the electric grid varies, the turbine governor automatically opens or closes the control valve or turbine wicket gates. The steady flow

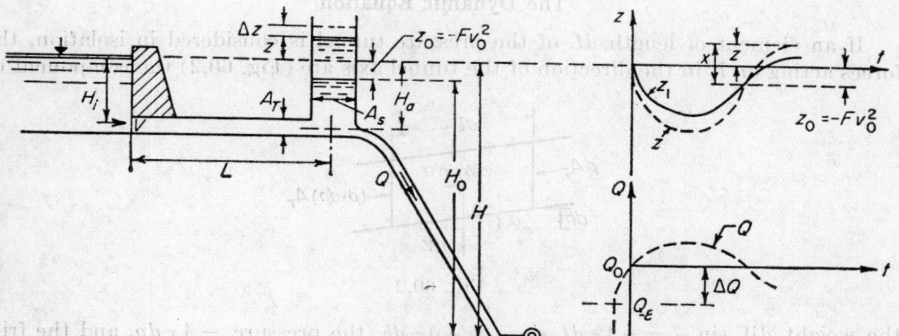

FIG. 60.1

of water in the hydraulic pressure system is disturbed. Masses of water have to be suddenly accelerated or decelerated, which causes considerable forces to develop. Elastic pressure waves, called *water hammer*, develop in the pipeline or penstock. The pressure tunnel is sensitive to pressure variations, and the purpose of the surge tank is to protect the tunnel from sudden high-pressure variations.

The water in the tunnel too has to be accelerated or decelerated. This causes surges to develop in surge tanks, which can be compared to mass oscillations in communicating vessels.

The period of the water-hammer waves is short: a fraction of a second to a maximum of a few seconds. The surge oscillations are slower, the period ranging from 100 to 400 or 500 sec.

The theory of surges in surge tanks considers mass displacements of water in the pressure tunnel and in the surge tank, assuming the water not to be elastic. Friction losses in the tunnel are the main damping force.

Notation (See Fig. 60.1)

H = gross head
H_0 = net head available at surge tank, in steady state, losses in the pipeline being neglected
$\pm Fv^2$ = head loss in tunnel, positive when $v > 0$
$z_0 = -Fv_0^2$ = head loss in tunnel in steady state, $H_0 = H - Fv_0^2$
L = length of tunnel
A_T = area of cross section of tunnel, $A_T = \pi D_T^2/4$
A_S = horizontal area of cross section of surge tank
v = velocity of flow in tunnel, positive for flow from reservoir
Q = instantaneous flow required by turbines
z = water level in surge tank above reservoir level
u = dz/dt
τ = closing or opening time of valve or wicket gate
γ = specific weight of water

60.2. THE BASIC EQUATIONS OF MASS OSCILLATIONS IN THE SIMPLE SURGE TANK

We define the "simple" surge tank as a tank or shaft of constant horizontal cross-sectional area A_S.

The Dynamic Equation

If an element of length dL of the pressure tunnel is considered in isolation, the forces acting on it in the direction of the tunnel axis are (Fig. 60.2) the component of

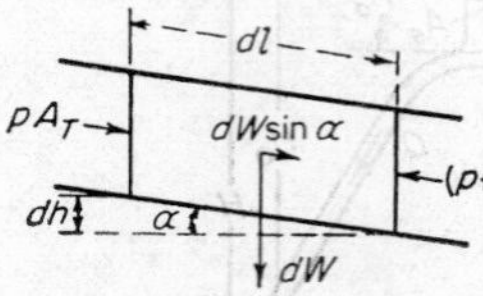

FIG. 60.2

the weight $dW \sin\alpha = \gamma A_T\, dL \sin\alpha = \gamma A_T\, dh$, the pressure $-A_T\, dp$, and the frictional resistance $-\gamma A_T i\, dL$, where i is the energy gradient. As the mass of the element of water in the tunnel is $(\gamma/g)A_T\, dL$, the dynamic equation for this element is

$$(\gamma/g)A_T\, dL\, \partial v/\partial t = \gamma A_T\, dh - A_T\, dp - \gamma A_T i\, dL$$

In order to make possible the integration of these equations from 0 to L, the following assumptions are necessary: (1) Both the tunnel and the water are incompressible, and $\partial v/\partial t = dv/dt$. (2) As a first approximation, the velocity head $v^2/2g$ is negligible. (3) The value for the frictional resistance substituted in the equations is the value for steady flow at any one time t when, instantaneously, the energy gradient i obtains. The head loss iL is proportional to v^2 or $iL = Fv^2$. (4) The mass of the water in the surge shaft, in a first approximation, is negligible.

The dynamic equation can then be integrated, and becomes (see Fig. 60.1)

$$\frac{1}{g}\frac{dv}{dt}\int_0^L dL = \int_{H_i}^{H_a} dh - \int_{H_i}^{H_a+z}\frac{dp}{\gamma} - i\int_0^L dL$$

or, after simplification,

$$\frac{L}{g}\frac{dv}{dt} + z \pm Fv^2 = 0 \qquad (60.1)$$

The positive sign is to be used when the direction of flow is from the reservoir to the surge tank.

Equation of Continuity

Using the nomenclature of Fig. 60.1, we obtain

$$vA_T = A_S \frac{dz}{dt} + Q \tag{60.2}$$

Direct integration of Eqs. (60.1) and (60.2) is possible in a limited number of cases. Step-by-step integration or graphical methods are being used in others.

60.3. SOLUTION OF EQS. (60.1) AND (60.2) NEGLECTING TUNNEL FRICTION

Sudden Closure of Turbine Valve

The case of sudden rejection of load (sudden closure of the turbine valve) can be solved by direct integration when friction losses in the tunnel are neglected. At the time $t = 0 - \epsilon$, $Q = Q_0$, and, at the time $t = 0 + \epsilon$, $Q = 0$. Also, $A_T v = A_S\, dz/dt$, and $dv/dt = (A_S/A_T)(d^2z/dt^2)$. As $F = 0$,

$$\frac{L}{g}\frac{dv}{dt} + z = 0$$

or

$$\frac{L}{g}\frac{A_S}{A_T}\frac{d^2z}{dt^2} + z = 0$$

The general solution of the equation is

$$z = C_1 \cos\frac{2\pi t}{T} + C_2 \sin\frac{2\pi t}{T} \qquad T = 2\pi\sqrt{\frac{LA_S}{gA_T}} \tag{60.3}$$

When $t = 0$, $z = 0$, and, consequently, $C_1 = 0$. Putting $C_2 = Z_*$, we obtain the equations

$$z = Z_* \sin\frac{2\pi t}{T} \qquad \text{and} \qquad v = v_0 \cos\frac{2\pi t}{T} \tag{60.4}$$

whence

$$Z_* = v_0\sqrt{\frac{LA_T}{gA_S}} \tag{60.5}$$

In the case of partial closure, reducing the flow from Q_0 to Q_1,

$$z = (v_0 - v_1)\sqrt{\frac{LA_T}{gA_S}}\sin\frac{2\pi t}{T}$$

Sudden Opening of Turbine Valve or Sudden Increase in Load

If the flow Q_1 is suddenly increased to Q_0, with corresponding steady tunnel velocities v_1 and v_0, we obtain the equation

$$z = (v_1 - v_0)\sqrt{\frac{LA_T}{gA_S}}\sin\frac{2\pi t}{T}$$

where $v_1 < v_0$.

Linear Rate of Change of Load

If the friction in the tunnel is neglected, this case also is capable of rigorous analysis.

60.4. CALCULATION OF WATER-LEVEL OSCILLATIONS INCLUDING TUNNEL FRICTION

Sudden Complete Closure of Turbine Valve

From the basic equations, the following relationship between v and z may be derived:

$$\left(\frac{v}{v_0}\right)^2 = -\frac{z}{Fv_0^2} + \frac{LA_Tv_0^2}{2gA_S(Fv_0^2)^2}\left[1 - \exp\left(-\frac{2gA_SFv_0^2(z + Fv_0^2)}{LA_Tv_0^2}\right)\right] \qquad (60.6)$$

The maximum surge $z_{\max}$ corresponds to $v = 0$.

This is the only case which can be dealt with by direct integration when friction is not neglected.

Step-by-step Integration

The differential equations (60.1) and (60.2) can be replaced by difference equations wherein the infinitesimally small time interval dt is replaced by a small, but finite, interval Δt. The difference equations

$$\frac{L}{g}\frac{\Delta v}{\Delta t} + z_m \pm Fv_m^2 = 0 \qquad v_mA_T = Q_m + A_{Sm}\frac{\Delta z}{\Delta t} \qquad (60.7a,b)$$

where

$$t_{i+1} = t_i + \Delta t \qquad z_m = z_i + \frac{\Delta z}{2} \qquad Q_m = ½(Q_i + Q_{i+1})$$

$$v_m = ½(v_i + v_{i+1}) = v_i + ½\Delta v$$

can then be integrated step by step. Δv can be calculated from the equation

$$\pm ¼F(\Delta v)^2 + \left(\frac{L}{g\,\Delta t} + \frac{A_T}{4A_{Sm}}\Delta t \pm Fv_i\right)\Delta v + z_i + \frac{A_T}{2A_{Sm}}v_i\,\Delta t - \frac{Q_m}{2A_{Sm}}\Delta t \pm Fv_i^2 = 0$$

In recent years, the difference equations have been used for extensive surge-tank investigations carried out on digital computers.

60.5. INTRODUCTION OF RATIOS INTO THE CALCULATION OF SURGE TANKS

Dimensionless parameters of great significance may be introduced into the discussion of surge-tank problems because of the existence of categories of surge tanks with similar oscillations: If the oscillations of one surge tank are known, we are able to predict the behavior of all other tanks belonging to the same category. The parameters to be used are those introduced by Calame and Gaden [5]. Using Z_* and T from Eqs. (60.5) and (60.3) as the units of surge and of time, we introduce

$$z_r = z/Z_* \qquad F_{r0} = Fv_0^2/Z_* \qquad F_r = Fv^2/Z_* = F_{r0}(v/v_0)^2 \qquad t' = t/T$$

$$w = Q/A_S \qquad u_0 = w_0 = Q_0/A_S \qquad v_0 = Q_0/A_T$$

$$u_r = u/u_0 \qquad v_r = v/v_0 \qquad w_r = w/w_0$$

The dimensionless velocities u_r, v_r, w_r are indicative of the water velocities in the surge tank, the tunnel, and the pipeline, respectively.

After the substitution

$$\frac{dv}{dt} = \frac{A_S}{A_T}\frac{du}{dz}\frac{dz}{dt} + \frac{A_S}{A_T}\frac{dw}{dt}$$

the dynamic equation (60.1) may be brought into the form

$$u_r\frac{du_r}{dz_r} + \frac{1}{2\pi}\frac{dw_r}{dt'} + z_r \pm F_r = 0 \qquad (60.8)$$

and the continuity equation (60.2) becomes

$$v_r = u_r + w_r \tag{60.9}$$

Application to the Calculation of Instantaneous Total Closure (Instantaneous Rejection of Load)

In this case,

$$w_r = 0 \qquad \frac{dw_r}{dt'} = 0 \qquad v_r = u_r$$

which leads to

$$F_r = F_{r0}v_r^2 = F_{r0}u_r^2$$

The dynamic equation may be rewritten in the form

$$u_r \frac{du_r}{dz_r} + z_r + F_{r0}u_r^2 = 0 \qquad \text{or} \qquad \frac{d(u_r^2)}{dz_r} + 2F_{r0}u_r^2 = -2z_r$$

Its solution for the initial conditions $u_r = 1$, $z_r = F_{r0}$ at $t' = 0$ is

$$u_r^2 = \frac{1}{2F_{r0}^2}\{1 - 2F_{r0}z_r - \exp[-2F_{r0}(z + F_{r0})]\} \tag{60.10}$$

It is identical with Eq. (60.6). The highest level is reached when $dz/dt = 0$ or when $u_r = 0$; therefore,

$$1 - 2F_{r0}z_r - \exp[-2F_{r0}(z_r + F_{r0})] = 0 \tag{60.11}$$

yields the first maximum surge z_{r1} in relative values.

An approximative solution was given by Eydoux:

$$z_{r1} = 1 - \tfrac{2}{3}F_{r0} + \frac{1}{9}F_{r0}^2 - \cdots \tag{60.12}$$

Numerically Eq. (60.11) yields the following table:

$F_{r0} =$	0	0.1	0.2	0.4	0.6	0.8	1.0
$z_{r1} =$	1	0.933	0.875	0.76	0.65	0.555	0.475

The most important conclusion of this discussion is that the ratio z_{r1} is solely a function of $F_{r0} = Fv_0^2/Z_*$. A *single curve*, instead of the family of curves required for dimensional variables, connects the two dimensionless numbers z_{r1} and F_{r0}. All surge tanks whose parameters F_{r0} are equal have similar surges.

Other Important Results

The first draw-down after closure is given by

$$z_{r2} = -1 + 2F_{r0} \tag{60.13}$$

The first downsurge after sudden opening from rest is equal to

$$z_r' = -1 - 0.125F_{r0} \qquad \text{for } F_{r0} < 0.8 \tag{60.14}$$

60.6. THE GRAPHICAL METHOD OF SCHOKLITSCH (FIG. 60.3)

Direct integration is limited to the few cases mentioned. Numerical methods of integration are tedious, unless they are made on the digital computer. Since the early days, therefore, graphical methods have been developed and found both speedy and useful to the practicing engineer.

The graphical method, whose very simple principles are due to Schoklitsch, may be adapted to the analysis of a great variety of types of surge systems. The main graph consists of a curve relating z to v. The time t is treated as a variable parameter, so that the curve relating z to t follows on easily.

In the main diagram, with v as abscissa and z as ordinate, two auxiliary curves are plotted first to suitable scales (see Fig. 60.3): Curve 1, the curve of head losses $\Delta H_e = \pm Fv^2$, with v as abscissa and ΔH_e as ordinate, where ΔH_e is plotted downwards if $v > 0$; curve 2, the straight line representing the equation $\Delta v = -(g/L)\,\Delta t\, z$, which also passes through the origin. It is here assumed that the time increment Δt, on which choice there is no restriction, is constant.

Two further auxiliary curves are required, of cubic content and quantity of water, and for these, too, suitable scales have to be chosen. Line 3 connects v as abscissa

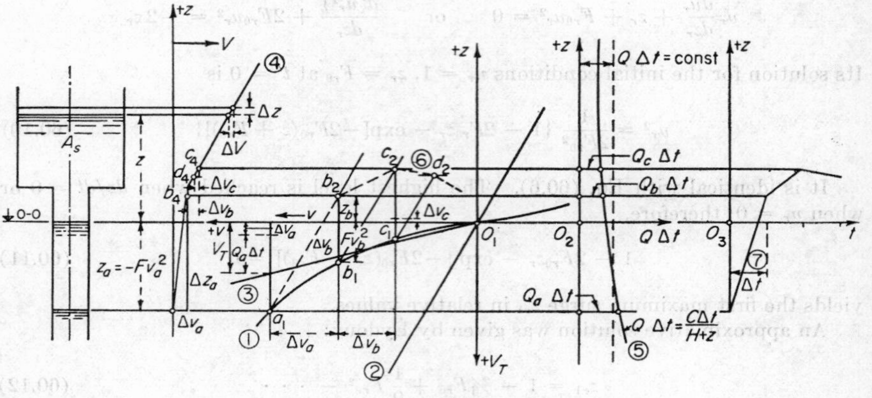

FIG. 60.3

with V_T as ordinate, where $V_T = vA_T\,\Delta t$ is the volume of water flowing through the tunnel in the time Δt. This equation is represented by a straight line, also passing through the origin. It is convenient to plot V_T positive downward, for $v > 0$. Auxiliary curve 4 relates the cubic content V of the surge tank to z. The elevation z is plotted as ordinate, and V as abscissa. The origin of curve 4, which represents the equation $\Delta V = A_S \Delta z$, may be chosen anywhere.

The dynamic equation (60.1) may be written in the form

$$\Delta v = -(g/L)\,\Delta t\,(z + Fv^2) \tag{60.15}$$

It represents a straight line parallel to auxiliary line 2. The values $z = -Fv^2$ and $\Delta v = 0$ will always satisfy Eq. (60.15), and the lines representing it will pass through points a_1, b_1, c_1, . . . on curve 1, whose coordinates are $v = v_a, v_b, v_c,$ . . . and $z = -Fv_a^2, -Fv_b^2, -Fv_c^2,$ . . . .

The equation of continuity (60.2) is

$$V_T = vA_T\,\Delta t = Q\,\Delta t \pm \Delta V \tag{60.16}$$

where $\Delta V = A_S\,\Delta z$ is the increase in storage volume in the surge tank due to a rise Δz in water level. In curve 4, ΔV is measured horizontally. $V_T = vA_T\,\Delta t$ is the ordinate of line 3.

$Q\,\Delta t$ represents the quantity of water flowing to the turbine during the time interval Δt. $Q\,\Delta t$ is plotted against z on auxiliary curve 5. If, for example, Q = const (assuming that the flow to the turbine is kept constant), curve 5 will be a straight line, parallel to the vertical axis and at a distance $Q\,\Delta t$ from it. If $Q\,\Delta t$ varies with z, curve 5 must show their relationship. If, for instance, the turbine valves are set at a given opening A_a (perhaps the maximum opening), then $Q\,\Delta t = A_a\sqrt{2g(H+z)}\,\Delta t$ yields a parabola; if $Q\,\Delta t = \eta_0 Q_0 H_0\,\Delta t/\eta(H+z)$ (governing equation), the corresponding curve will be similar to a hyperbola.

At time $t = t_a$, point a_1 on curve 1 is known ($v = v_a$ and $z = z_a = -Fv_a^2$). A straight line a_1b_2 is drawn through point a_1 on curve 1, parallel to line 2. $Q_a\,\Delta t$ and V_{Ta} are known. Graphical construction yields ΔV_a, which is plotted as abscissa on curve 4, giving b_4, and, on a horizontal line the point b_2. The curve representing the oscillation may be drawn through points $a_1b_2c_2d_2$ . . . .

60.7. THE STABILITY OF SINGLE AND MULTIPLE SURGE TANKS

Some surge tanks have proved to be "unstable." Under certain circumstances the surges are of the *forced oscillation type*. D. Thoma was the first to show that, with automatically governed turbines, the surge tank will only be (hydraulically) stable if the horizontal cross-sectional area of the tank A_S exceeds a certain minimum value A_{Th} (Thoma's area).

The possibility of this kind of instability may, with reference to Fig. 60.1, be deduced from simple physical considerations. At the instant $t = -\epsilon$, the flow through the pressure tunnel and pipeline is at the uniform rate $Q_{-\epsilon}$. In the short interval of time ϵ, the rate of flow $Q_{-\epsilon}$ to the turbine is increased by an amount ΔQ to $Q_0 = Q_{-\epsilon} + \Delta Q$. It is now possible to calculate the surges that could result if the governor ensured constant discharge Q_0. The resultant oscillation in the surge tank has been plotted as z_1 in Fig. 60.1.

In practice, the governor on a turbine does not ensure constant discharge but constant power, that is,

$$N = \eta_0 H_0 Q_0 \gamma = \eta Q(H+z)\gamma \tag{60.17}$$

As $(H+z)$ varies with z, then Q will also vary, but in the opposite direction. As shown in Fig. 60.1, the graph of Q plotted against time t is not a horizontal straight line, but a curve which oscillates in phase with z, against t. A more precise calculation will now yield a new curve z, showing increased oscillations. It is not difficult to see that oscillations of this type may, under certain circumstances, become unstable. There must be a limiting value A_{Th} for A_S, which will depend on the ratio z_{max}/H_0.

A simplified analytical treatment may start from the dynamic equation (60.8) and the equation of continuity (60.9) written in relative values. Instead of the surges Z or $z = Z/Z_*$ being measured from the static level, they may be measured from the dynamic level, so that $X = Z + Fv_0^2$ or, in relative values, $x = z + F_r$, with $dx = dz$. Let us assume the oscillation to be very small (x very small). The governing condition (60.17) yields, in relative values,

$$w = \frac{Q}{Q_0} = \frac{H_0}{H_0 + Fv_0^2 + Z} = \frac{h_0}{h_0 + F_r + z} = \frac{h_0}{h_0 + x} \approx 1 - \frac{x}{h_0}$$

For small surges, the pressure loss, in relative values, can be written as

$$F_r = F_{r0}v^2 = F_{r0}(u+w)^2 \approx F_{r0}\left(u + 1 - \frac{x}{h_0}\right)^2 \approx F_{r0}\left(1 + 2u - \frac{2x}{h_0}\right)$$

In addition,

$$u = \frac{1}{2\pi}\frac{dx}{dt'} \qquad u\frac{du}{dz} = u\frac{du}{dx} = \frac{1}{2\pi}\frac{du}{dt'} = \left(\frac{1}{2\pi}\right)^2\frac{d^2x}{dt'^2}$$

Introducing these values in the dynamic equation (60.8), we finally get

$$\frac{d^2x}{dt'^2} + 2\pi\left(2F_{r0} - \frac{1}{h_0}\right)\frac{dx}{dt'} + 4\pi^2\left(1 - \frac{2F_{r0}}{h_0}\right)x = 0 \tag{60.18}$$

an equation of the type

$$\frac{d^2x}{dt'^2} + 2a\frac{dx}{dt'} + bx = 0 \tag{60.19}$$

Conditions for damped oscillations are that $a > 0$ and $b > 0$. The condition $b > 0$ implies that

$$\frac{F_{r0}}{h_0} < \frac{1}{2} \qquad \text{or} \qquad Fv_0^2 < \frac{H_0}{2}$$

a condition which is usually satisfied on economical grounds.

The condition $a \geq 0$ leads to

$$2F_{r0} \geq \frac{1}{h_0} \qquad \text{or} \qquad \frac{Fv_0^2}{Z_*} \geq \frac{Z_*}{2H_0} \qquad \text{where} \qquad Z_*^{\,2} = v_0\frac{L}{g}\frac{A_T}{A_S}$$

and, therefore,

$$A_S \geq \frac{v_0^2}{2g}\frac{LA_T}{Fv_0^2H_0} = A_{\text{Th}} \qquad \text{or} \qquad A_S = nA_{\text{Th}} \qquad n \geq 1 \tag{60.20}$$

This is Thoma's condition for stability of single surge tanks with oscillations of small amplitude, and A_{Th} is Thoma's least area of cross section of a surge tank.

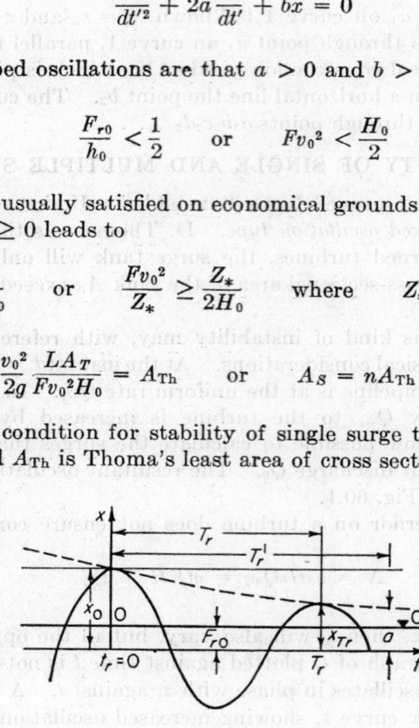

FIG. 60.4

If the Manning-Strickler formula is used to determine the friction loss Fv_0^2, with $k_s = 85$ (metric units) or $M = 125$ (foot units), a convenient formula is obtained for the horizontal dimensions of a surge tank or shaft:

$$A_S = \frac{nk_s^2\pi D_T^{10/3}}{2^{17/3}gH_0} \approx 45n\frac{D_T^{10/3}}{H_0} \qquad \text{m}$$

$$A_S = \frac{nM^2\pi D_T^{10/3}}{2^{17/3}gH_0} \approx 30n\frac{D_T^{10/3}}{H_0} \qquad \text{ft}$$

This formula must apply to every horizontal cross section of the surge tank that the water surface may occupy, including the narrow shaft between expansion chambers.

Estimate of the Factor n and Decrement Δ

The oscillation is damped provided $a > 0$ and $n > 1$. Referring to Fig. 60.4, we define a damping factor or decrement Δ as

$$\Delta = \frac{x_{t'+T_r}}{x_{t'}}$$

where T_r is the period, in relative time t'. The relation of the decrement Δ to the factor n is to be investigated. The integral of Eq. (60.19) is known to be

$$x = \frac{A_0}{Z_*} e^{-at'} \sin (\beta t' - \mu) \qquad (60.21)$$

where

$$\beta = \sqrt{b - a^2} = 2\pi \sqrt{1 - \frac{1}{4}\left(\frac{1}{h_0} + 2F_{r0}\right)^2}$$

If $t' = 0$ when $x = A_0/Z_*$, then $\mu = -\frac{1}{2}\pi$. By definition, the relative period is

$$T_r = \frac{2\pi}{\beta} = \left[1 - \frac{1}{4}\left(\frac{1}{h_0} + 2F_{r0}\right)^2\right]^{-1/2}$$

and the true period of oscillation is

$$T^* = T_r T = 2\pi T_r \sqrt{\frac{LA_S}{gA_T}}$$

The decrement Δ becomes

$$\Delta = \frac{\exp [\pi(1/h_0 - 2F_{r0})(t' + T_r)]}{\exp [\pi(1/h_0 - 2F_{r0})t']} \cdot \frac{\sin [\beta(t' + T_r) - \mu]}{\sin (\beta t' - \mu)} = \exp [\pi(1/h_0 - 2F_{r0})T_r]$$

For $A_S = A_{\text{Th}}$ and $n = 1$, $a = 0$. If

$$Z'_* = v_0 \sqrt{\frac{LA_T}{gA_{\text{Th}}}} \qquad \text{and} \qquad T' = 2\pi \sqrt{\frac{LA_{\text{Th}}}{gA_T}}$$

new relative values, $h'_0 = H_0/Z'_*$, $F'_{r0} = Fv_0^2/Z'_*$, are introduced, and $1/h'_0 = 2F'_{r0}$ because $a = 0$. For $n \neq 1$:

$$Z_* = Z'_* \sqrt{\frac{1}{n}} \qquad h_0 = \sqrt{n}\, h'_0 \qquad F_{r0} = \sqrt{n}\, F'_{r0}$$

$$T_r = \left[1 - \frac{1}{4}\left(\frac{2}{\sqrt{n}} F'_{r0} + 2\sqrt{n}\, F'_{r0}\right)^2\right]^{-1/2} = \left[1 - (F'_{r0})^2 \frac{(1+n)^2}{n}\right]^{-1/2}$$

and

$$\Delta = \exp\left[\pi\left(\frac{1}{h'_0\sqrt{n}} - 2F'_{r0}\sqrt{n}\right) T_r\right] = \exp\left[2\pi F'_{r0}\left(\frac{1}{\sqrt{n}} - \sqrt{n}\right) T_r\right] \qquad (60.22)$$

For $n > 1$, the argument of the exponential function is clearly negative, as required. Δ decreases with increasing n.

Stability if Amplitudes Are Large

In order to get a linear differential equation of the type (60.19) with constant coefficients a and b, the oscillations have to be assumed to be very small. For large oscillations, the differential equation would be of the form

$$\frac{d^2x}{dt^2} + \phi(t)\frac{dx}{dt} + \psi(t)x = 0 \qquad (60.23)$$

$\phi(t)$ and $\psi(t)$ being functions of the time t.

Equation (60.23) can be dealt with in an approximate manner. Remarking that, if all the terms of Eq. (60.19) are being multiplied by $(dz/dt)\,dt$ and integrated between 0 and T, we get

$$\int_0^t \frac{d^2x}{dt^2}\frac{dx}{dt}\,dt + b\int_0^t x\frac{dx}{dt}\,dt = -2a\int_0^t \left(\frac{dx}{dt}\right)^2 dt$$

or

$$\left[\frac{1}{2}\left(\frac{dx}{dt}\right)^2\right]_0^t + b\left(\frac{x^2}{2}\right) = -2a\int_0^t \left(\frac{dx}{dt}\right)^2 dt$$

The left-hand part of this equation represents the variation of the sum of both kinetic and potential energies of the oscillating system during the time interval 0 to t. If a and b are positive, the right-hand term shows that this sum must always be negative: the total energy of a stable oscillation decreases. Equation (60.23) may be treated in the same manner. If the values of $\phi(t)$ and $\psi(t)$ do not vary greatly with time, two mean values ϕ_m and ψ_m may be defined so that the equation

$$\left[\frac{1}{2}\left(\frac{dx}{dt}\right)^2\right]_0^t + \psi_m\left[\frac{x^2}{2}\right]_0^t = -\phi_m\int_0^t\left(\frac{dx}{dt}\right)^2 dt$$

represents stable oscillations if both $\phi_m > 0$ and $\psi_m > 0$. For oscillations which approach the limit of stability, this approximate calculation leads to

$$n^* = \frac{A_S}{A_{\text{Th}}} = 1 + 0.48\,\frac{Z_*}{H + z_0} \tag{60.24}$$

Stability of Systems of Surge Tanks

The analysis of stability conditions can be extended to the systems represented in Fig. 60.5a,b. In both cases, the oscillations are represented by equations of the form

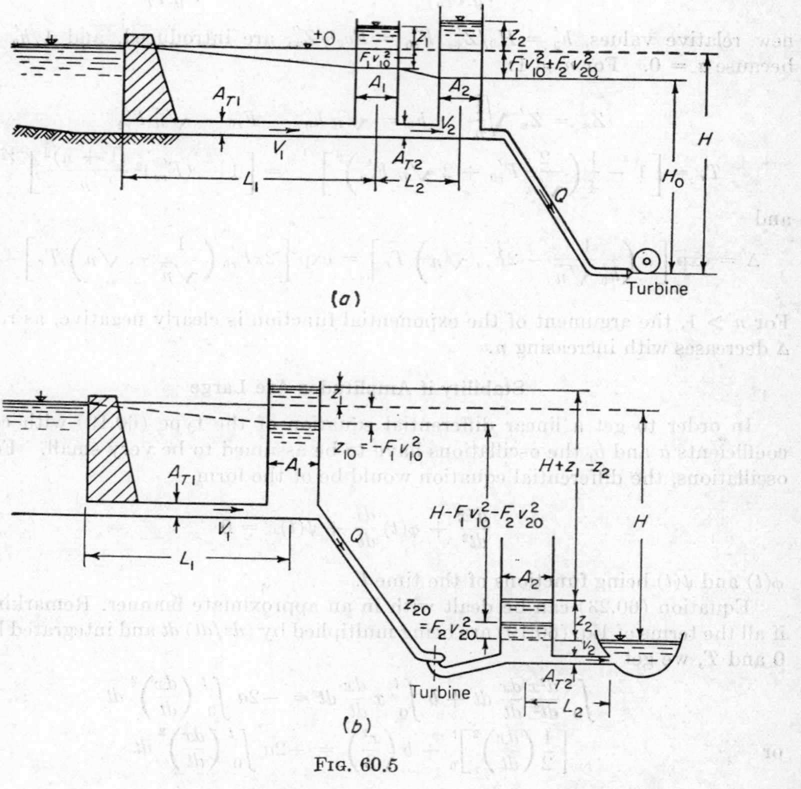

FIG. 60.5

(in relative values)

$$\frac{d^2x_1}{dt^2} + a_1\frac{dx_1}{dt} + b_1x_1 = A_1\frac{d^2x_2}{dt^2} + B_1\frac{dx_2}{dt} + C_1x_2$$
$$\frac{d^2x_2}{dt^2} + a_2\frac{dx_2}{dt} + b_2x_2 = A_2\frac{d^2x_1}{dt^2} + B_2\frac{dx_1}{dt} + C_2x_1$$

suffix 1 referring to the upstream tank, suffix 2 to the other. The use of the Laplace transformation leads to the calculation of the roots of an equation of the type

$$\Delta(p) \equiv \begin{vmatrix} p^2 + a_1p + b_1 & -(A_1p^2 + B_1p + C_1) \\ -(A_2p^2 + B_2p + C_2) & p^2 + a_2p + b_2 \end{vmatrix} = 0$$

or

$$p^4 + \alpha_1p^3 + \alpha_2p^2 + \alpha_3p + \alpha_4 = 0$$

The stability conditions are then $\alpha_1 > 0$, $\alpha_2 > 0$, $\alpha_3 > 0$, $\alpha_4 > 0$, and

$$\Gamma = \alpha_3^2 - \alpha_1(\alpha_2\alpha_3 - \alpha_1\alpha_4) < 0$$

For the further treatment of the problem, the areas

$$A_{\text{Th}1} = \frac{1}{2g}\left(\frac{Q_0}{A_{T1}}\right)^2 \frac{L_1A_{T1}}{F_1v_{10}^2H_0} \qquad \text{and} \qquad A_{\text{Th}2} = \frac{1}{2g}\left(\frac{Q_0}{A_{T2}}\right)^2 \frac{L_2A_{T2}}{F_2v_{20}^2H_0}$$

are introduced and all relative values defined in functions of these. For stability just to be reached, we should have at least

$$A_1 = n_1A_{\text{Th}1} \qquad \text{and} \qquad A_2 = n_2A_{\text{Th}2}$$

It can be shown that n_1 and n_2 depend only on the three parameters

$$\bar{m}^* = \sqrt{\frac{L_2}{L_1}\frac{A_{\text{Th}2}A_{T1}}{A_{\text{Th}1}A_{T2}}} \qquad F'_{r1} = \frac{F_1v_{10}^2}{Z'_{*1}} \qquad F'_{r2} = \frac{F_2v_{20}^2}{Z'_{*2}}$$

where $F_1v_{10}^2$ and $F_2v_{20}^2$ are the losses in the tunnels L_1 and L_2 (Fig. 60.5*a*,*b*) and

$$Z'_{*1} = Q_0\sqrt{\frac{L_1}{gA_{T1}A_{\text{Th}1}}} \qquad Z'_{*2} = Q_0\sqrt{\frac{L_2}{gA_{T2}A_{\text{Th}2}}}$$

60.8. REFERENCES

[1] R. D. Johnson: The surge tank in water power plants, *Trans. ASME*, **30** (1908), 443–501.

[2] F. Prášil: Wasserschlossprobleme. *Schweiz. Bauztg.*, **52** (1908), 271–277, 301–306, 317–320, 333–336.

[3] A. Schoklitsch: Spiegelbewegung in Wasserschlössern, *Schweiz. Bauztg.*, **81** (1923), 129–131, 146–149.

[4] D. Thoma: "Beiträge zur Theorie des Wasserschlosses bei selbsttätig geregelten Turbinenanlagen," Munich, 1910.

[5] Calame, Gaden: "Théorie des chambres d'équilibre," Paris and Lausanne, 1926.

[6] J. S. McNown: Surges and water hammer, chap. 7 in H. Rouse (ed.), "Engineering Hydraulics," Wiley, New York, 1950.

[7] Ch. Jaeger: "Engineering Fluid Mechanics," Blackie, London, 1956.

[8] A. Gardel: "Chambres d'équilibre," Lausanne, 1956.

[9] Ch. Jaeger: Present trends in surge tank design, *Proc. Inst. Mech. Engrs. London*, **168** (1954), 91–124.

[10] Ch. Jaeger: The double surge tank system; general discussion of the stability problem, *Water Power*, **9** (1957), 253–258, 301–305, and *Trans. ASME*, **80** (1958), 1574–1584.

[11] Ch. Jaeger: A review of surge-tank stability criteria, *J. Basic Eng.* **82** (1960), 765–783.

CHAPTER 61

CONTINUOUS SYSTEMS

BY

D. YOUNG, Ph.D., New Haven, Conn.

61.1. DEFINITIONS. GENERAL PROCEDURES

Assuming small oscillations about an equilibrium position, the free vibrations of a conservative elastic system are governed by a linear differential equation and linear homogeneous boundary conditions. The motion can be analyzed into an infinite number of periodic motions, in each of which the motion of every particle is simple harmonic, with the same frequency and phase for all the particles, and the displacement of any particle bears a definite ratio to the displacement of any chosen reference particle. When the system is oscillating in one of these ways, it is said to be vibrating in a *normal* (or *principal*) mode.

Consider, for example, a one-dimensional system which can vibrate in only the x direction. If $u(x,t)$ is the displacement and ω is the circular frequency, the motion in a normal mode has the form

$$u(x,t) = U(x) \sin (\omega t + \epsilon)$$

Substituting this into the governing differential equation and boundary conditions leads to an equation for ω which is, in general, transcendental and is called the *frequency equation.* The frequency equation has an infinite number of real roots, ω_1, ω_2, ω_3, ... Corresponding to these roots, the solution of the differential equation determines the functions $U_1(x)$, $U_2(x)$, . . . , respectively. The functions $U_n(x)$ are called the *eigenfunctions* (or *normal* functions or *characteristic* functions) of the system. The frequencies are generally assumed to be numbered in order of increasing magnitude so that $\omega_1 < \omega_2 < \omega_3$, etc. The first or lowest frequency ω_1 is called the *fundamental* frequency.

The most general free vibration of the system can be represented by superposing the motions in the different normal modes, that is,

$$u(x,t) = \sum_{n=1}^{\infty} A_n U_n(x) \sin (\omega_n t + \epsilon_n) \tag{61.1}$$

where A_n and ϵ_n are constants which depend upon the initial conditions.

If the frequencies and eigenfunctions of a system are known, problems in forced vibrations can be solved in the following manner. Take the displacement in the form

$$u(x,t) = \sum_{n=1}^{\infty} U_n(x) q_n(t) \tag{61.2}$$

The quantities $q_n(t)$ are called the *normal coordinates* of the system. It can be shown that the kinetic energy T and the potential energy V of the system reduce to sums of squares of the normal coordinates and have the following forms

$$T = \tfrac{1}{2} \sum_{n=1}^{\infty} m_n \dot{q}_n{}^2 \qquad V = \tfrac{1}{2} \sum_{n=1}^{\infty} m_n \omega_n{}^2 q_n{}^2 \tag{61.3a,b}$$

where m_n, which is called the *generalized mass*, is a quantity that can be defined in terms of the parameters of any given system.

Lagrange's equation of motion is

$$\frac{d}{dt}\frac{\partial T}{\partial \dot{q}_n} - \frac{\partial T}{\partial q_n} + \frac{\partial V}{\partial q_n} = Q_n \tag{61.4}$$

where Q_n is the generalized force. Substituting the series expressions for T and V into Lagrange's equation gives

$$m_n \ddot{q}_n + m_n \omega_n{}^2 q_n = Q_n \tag{61.5}$$

from which q_n may be determined.

As an alternative to the above procedure, forced-vibration problems can be solved by means of the Laplace transform [43],[44] or other mathematical techniques.

The general procedures outlined above are specialized to particular systems in the following sections of this chapter. Because of the technical importance of beams, and to serve as an illustration of methods of analysis, the treatment of lateral vibrations of beams is given in more detail than the other problems.

The methods and data presented herein, as well as the types of problems covered, are necessarily limited. Additional details and analyses of more complicated systems are covered in the references listed on pp. 61–33 to 61–34. This is a selected bibliography which is, with few exceptions, restricted to readily available references in the English language.

16.2. LONGITUDINAL VIBRATIONS OF UNIFORM BARS

Free Vibrations

The elementary theory of longitudinal vibrations of bars is based on the assumptions that cross sections remain plane, that stress is proportional to strain and is uniformly distributed over each cross section, and that lateral deformations can be neglected.

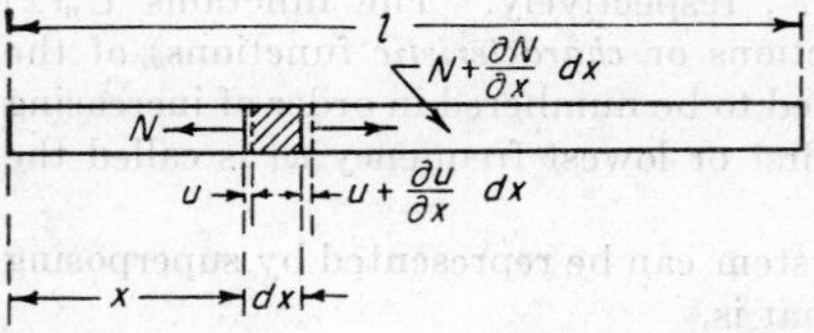

FIG. 61.1. Longitudinal vibrations of a bar.

Referring to Fig. 61.1, $u(x,t)$ is the longitudinal displacement of a cross section during vibration, and N is the longitudinal force at a cross section. The equation of motion for an element of the bar is $N' = \rho A \ddot{u}$. The strain is u' and, hence, $N = EAu'$. Substituting this expression for N into the equation of motion gives

$$(EAu')' = \rho A \ddot{u} \tag{61.6}$$

This is the differential equation that governs longitudinal vibrations. For a uniform bar of constant cross section, the equation reduces to

$$c_0{}^2 u'' = \ddot{u} \tag{61.7}$$

where $c_0 = \sqrt{E/\rho}$.

To determine the frequencies and eigenfunctions, take $u(x,t) = U(x) \sin (\omega t + \epsilon)$.

Substituting into (61.7) gives

$$U'' + (\omega^2/c_0^2)U = 0$$

for which the general solution is

$$U(x) = C_1 \cos(\omega x/c_0) + C_2 \sin(\omega x/c_0) \tag{61.8}$$

As shown in the examples below, the frequencies ω_n and the eigenfunctions $U_n(x)$ are determined by substituting (61.8) into the boundary conditions. The general solution for free vibrations is

$$u(x,t) = \sum_{n=1}^{\infty} U_n(x)(A_n \cos \omega_n t + B_n \sin \omega_n t) \tag{61.9}$$

The constants A_n and B_n can be determined from the initial conditions, as explained on p. 61–4 in connection with forced vibrations.

Example 1. Bar Fixed at Both Ends. The boundary conditions are $U(0) = 0$ and $U(l) = 0$. To satisfy the first condition, $C_1 = 0$. To satisfy the second condition, $C_2 \sin(\omega l/c_0) = 0$, which has nontrivial solutions when $\sin(\omega l/c_0) = 0$. This is the frequency equation for the problem. Its roots determine the frequencies

$$\omega_n = n\pi c_0/l$$

where $n = 1, 2, 3, \ldots$. The corresponding eigenfunctions are

$$U_n(x) = C_n \sin(\omega_n x/c_0) = C_n \sin(n\pi x/l)$$

where C_n is an undetermined constant, which may be chosen equal to unity or any other convenient value.

Example 2. Bar Fixed at $x = 0$ and Free at $x = l$. The boundary conditions are $U(0) = 0$ and $U'(l) = 0$. The frequency equation is $\cos(\omega l/c_0) = 0$, and the frequencies and eigenfunctions are

$$\omega_n = \frac{(2n-1)\pi c_0}{2l} \qquad U_n(x) = C_n \sin \frac{(2n-1)\pi x}{2l}$$

Example 3. Bar Free at Both Ends. The boundary conditions are $U'(0) = 0$ and $U'(l) = 0$. The frequency equation is $\sin(\omega l/c_0) = 0$, and the frequencies and eigenfunctions are

$$\omega_n = n\pi c_0/l \qquad U_n(x) = C_n \cos(n\pi x/l)$$

where $n = 0, 1, 2, 3, \ldots$. For $n = 0$, $\omega_0 = 0$ and $U_0 = \text{const}$; this represents a rigid-body translation of the bar.

Forced Vibrations

Suppose that the bar is acted upon by a distributed force $f(x,t)$ and one or more concentrated forces $F_i(t)$ applied at sections which are at distances x_i from the origin. All the forces are assumed to act in the direction of the bar axis and through the centroid of the cross section.

Assuming that the eigenfunctions and frequencies are known, the response to the impressed forces can be found by using normal coordinates. Take

$$u(x,t) = \sum_{n=1}^{\infty} U_n(x) q_n(t) \tag{61.10}$$

For a bar with any combination of free and fixed ends, such as all those in the preceding examples, it can be verified that the eigenfunctions satisfy the following orthogonality relations:

$$\int_0^l U_n U_m \, dx = 0 \qquad \int_0^l U'_n U'_m \, dx = 0 \qquad m \neq n \tag{61.11}$$

Using these relations, it can be verified that the expressions for the kinetic energy T and the potential energy V, in terms of the normal coordinates, reduce to

$$T = \tfrac{1}{2} \int_0^l A\rho \dot{u}^2 \, dx = \tfrac{1}{2} \sum_{n=1}^{\infty} m_n \dot{q}_n^2 \tag{61.12}$$

$$V = \tfrac{1}{2} \int_0^l EA(u')^2 \, dx = \tfrac{1}{2} \sum_{n=1}^{\infty} m_n \omega_n^2 q_n^2 \tag{61.13}$$

$$m_n = \int_0^l \rho A U_n^2 \, dx \tag{61.14}$$

It should be noted that, for bars with certain other types of boundary conditions, such as an end mass or elastic restraint, the orthogonality relations (61.11) and the expression (61.14) for the generalized mass must be modified. An example of such a case is given on p. 61–5.

Substituting (61.12) and (61.13) into the Lagrangian equation (61.4) gives

$$m_n \ddot{q}_n + m_n \omega_n^2 q_n = Q_n(t) \tag{61.15}$$

where Q_n is the generalized force corresponding to the nth coordinate, and is found in the usual manner by considering the work done in a virtual displacement $U_n(x)\,\delta q_n$; thus,

$$Q_n(t) = \int_0^l f(x,t) U_n(x) \, dx + \sum_i F_i(t) U_n(x_i) \tag{61.16}$$

The solution of (61.15) can be written in terms of the free-vibration solution plus a convolution integral. Substituting this solution into (61.10) gives the general solution of the forced-vibration problem in the following form:

$$u(x,t) = \sum_{n=1}^{\infty} U_n(x) \left[A_n \cos \omega_n t + B_n \sin \omega_n t + \frac{1}{m_n \omega_n} \int_0^t Q_n(\tau) \sin \omega_n (t - \tau) \, d\tau \right] \tag{61.17}$$

The constants A_n and B_n can be determined from the initial conditions in the following manner. Let $u_0(x)$ be the initial displacement and $\dot{u}_0(x)$ the initial velocity. To satisfy these conditions, A_n and B_n must be chosen so that

$$\sum_{n=1}^{\infty} A_n U_n(x) = u_0(x) \qquad \sum_{n=1}^{\infty} \omega_n B_n U_n(x) = \dot{u}_0(x)$$

Multiplying each of these expressions by $U_m(x)\,dx$, integrating over the length, and using the orthogonality relations (61.11) yields the following formulas for determining the constants:

$$A_n = \frac{\rho A}{m_n} \int_0^l u_0(x) U_n(x) \, dx \qquad B_n = \frac{\rho A}{m_n \omega_n} \int_0^l \dot{u}_0(x) U_n(x) \, dx \tag{61.18}$$

Bar With Concentrated Mass at End

Consider a uniform bar, which is fixed at $x = 0$ and carries a concentrated mass m_c at $x = l$. The boundary conditions are $U(0) = 0$ and $EAU'(l) = m_c\omega^2 U(l)$. Substituting (61.8) into these conditions gives $\omega_n = \beta_n c_0/l$, where the β_n are the roots of the frequency equation $\beta \tan \beta = \rho A l/m_c$. The corresponding eigenfunctions are

$$U_n(x) = \sin(\beta_n x/l) \tag{61.19}$$

The orthogonality relations for this case can be shown to be, for $n \neq m$,

$$A\rho \int_0^l U_n U_m \, dx + m_c U_n(l) U_m(l) = 0 \qquad \int_0^l U'_n U'_m \, dx = 0 \tag{61.20a,b}$$

The energy expressions reduce to the same form as (61.12) and (61.13), except, for this case, the generalized mass is

$$m_n = \rho A \int_0^l U_n{}^2 \, dx + m_c U_n{}^2(l) \tag{61.21}$$

The general solution of a forced-vibration problem is given by (61.17), using (61.19) for $U_n(x)$ and (61.21) for m_n. The formulas for determining the integration constants become

$$m_n A_n = \rho A \int_0^l u_0(x) U_n(x) \, dx + m_c u_0(l) U_n(l)$$

$$\omega_n m_n B_n = \rho A \int_0^l \dot{u}_0(x) U_n(x) \, dx + m_c \dot{u}_0(l) U_n(l)$$

61.3. STRINGS, SHAFTS, HELICAL SPRINGS, FLUID COLUMNS

The free vibrations of a number of different systems are governed by differential equations having the same form as that for the longitudinal vibrations of a uniform bar, namely, Eq. (61.7). The method and results given in Sec. 61.2, therefore, apply directly to these other systems. Of course, the variables and parameters have different physical significance in the different systems. The equations for a number of such systems are given in the following articles.

Transverse Vibrations of an Elastic String

Consider an elastic string of negligible bending stiffness which is stretched between two points with a tension N. Let $u(x,t)$ be the transverse deflection and μ be the mass per unit length. Assume that deflections are small so that N may be considered to be constant during vibrations. The transverse component of the tensile force at a section is Nu' and the resultant of the transverse forces acting at the two ends of an element of length dx is $Nu''\,dx$. Equating this resultant force to the mass times the acceleration of the element gives

$$Nu'' = \mu\ddot{u} \tag{61.22}$$

For a uniform string, this is the same as (61.17) with $c_0 = \sqrt{N/\mu}$.

Torsional Vibrations of a Uniform Shaft

Let M_T be the twisting moment and θ the twist at a section. For a circular shaft, $M_T = GI_p\theta'$, where I_p is the polar moment of inertia of the cross section. The resultant twisting moment acting on an element of length dx is $M'\,dx$, which equals $GI_p\theta''\,dx$. The mass moment of inertia of the element is $\rho I_p\,dx$. The equation of rotation of the element is, therefore,

$$GI_p\theta'' = \rho I_p\ddot{\theta} \tag{61.23}$$

This is the same as (61.7), with $c_0 = \sqrt{G/\rho}$. For a noncircular shaft, the torsional stiffness term GI_p on the left-hand side of (61.23) must be replaced by the actual torsional stiffness of the specified cross section (see p. 36–2)

Helical Spring

The longitudinal vibrations of a uniform helical spring are directly analogous to the longitudinal vibrations of a bar. Let $u(x,t)$ be the longitudinal displacement of a point on the spring axis, and μ be the mass per unit length of the spring. The axial force at any section is $k_1 u'$, where k_1 is the spring constant for a unit length of the spring. If the length of the spring is l, and k is the over-all spring constant, then $k_1 = kl$. The differential equation is

$$k_1 u'' = \mu \ddot{u} \tag{61.24}$$

This is the same as (61.7), with $c_0 = \sqrt{k_1/\mu}$.

Fluid in a Pipe

The longitudinal vibrations of a liquid or gas in a uniform pipe are governed by (61.7), provided c_0 is taken equal to the velocity of sound in the fluid: that is, $c_0 = \sqrt{K/\rho}$, where K is the bulk modulus, and ρ the density of the fluid. If the expansion of the pipe is taken into consideration, $c_0 = \sqrt{K/\rho}\,\sqrt{1/[1 + (2RK/E_1 h)]}$, where R is the pipe radius, h is the pipe wall thickness, and E_1 is the modulus of elasticity of the pipe material.

61.4. LATERAL VIBRATIONS OF BEAMS: CLASSICAL THEORY

Free Vibrations

The classical theory of beam vibrations, often called the Bernoulli-Euler theory, is based on the same assumptions as elementary flexure theory, namely, that plane sections remain plane, stress is proportional to strain, bending is in a principal plane, slopes of the deflection curve are small compared to unity, deflection due to shear is negligible. In this treatment, the only inertia forces considered are those due to the lateral translation of a beam element, as shown in Fig. 61.2. Actually, during vibration, each element of the beam rotates through an angle $\partial w/\partial x$, and there is thus a distributed inertia couple equal to $-I\rho\, \partial^3 w/\partial x\, \partial t^2$ per unit length, which is called the *rotatory inertia effect*. Correction for this rotatory inertia becomes important when the cross-sectional dimensions of the beam are not small compared with the beam length. The effect of rotatory inertia is included in the Timoshenko theory, which is treated in Sec. 61.5.

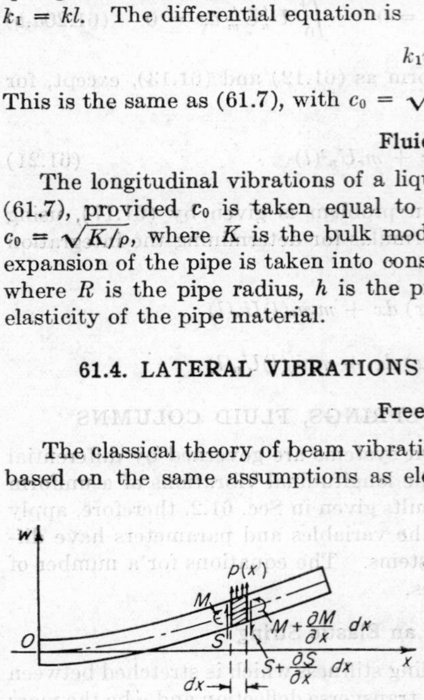

FIG. 61.2. Coordinate system and forces acting on a beam element.

Let $w(x,t)$ be the lateral deflection of the neutral axis, and $\mu(x)$ be the mass per unit length. Referring to Fig. 61.2, the deflection due to a load $p(x)$ is governed by the equation $(EIw'')'' = p(x)$. For a vibrating beam, the inertia force per unit length is $-\mu\ddot{w}$. Substituting this for $p(x,t)$ gives the following for the partial differential equation of free vibrations:

$$(EIw'')'' = -\mu\ddot{w} \tag{61.25}$$

For vibrations in a normal mode, take

$$w(x,t) = \phi(x) \sin(\omega t + \epsilon) \tag{61.26}$$

Substitution into (61.25) yields the ordinary differential equation,

$$(EI\phi'')'' = \omega^2\mu\phi \tag{61.27}$$

This equation, plus the boundary conditions, constitutes the eigenvalue problem whose solution yields the frequencies ω_n and the eigenfunctions $\phi_n(x)$.

For variable-section beams, relatively few exact solutions are known. Hence, for such problems, it is generally necessary to use some approximate procedure, such as the Rayleigh-Ritz method, the Stodola method, matrix iteration, or the Myklestad-Prohl method (see Sec. 61.10). For uniform beams, exact solutions can be obtained as shown below.

Free Vibrations of Uniform Beams

For a uniform beam, (61.27) becomes

$$\phi^{\text{iv}} = (\mu\omega^2/EI)\phi \tag{61.28}$$

for which the general solution is

$$\phi(x) = C_1 \operatorname{Cosh}(\lambda x/l) + C_2 \operatorname{Sinh}(\lambda x/l) + C_3 \cos(\lambda x/l) + C_4 \sin(\lambda x/l) \tag{61.29}$$

where

$$\lambda = l(\mu\omega^2/EI)^{1/4} \tag{61.30}$$

Consider, for example, a clamped-free beam for which the boundary conditions are $\phi(0) = \phi'(0) = \phi''(l) = \phi'''(l) = 0$. From the first two conditions, it is found that

$$C_3 = -C_1 \qquad C_4 = -C_2 \tag{61.31}$$

Substituting these relations into (61.29), and then applying the two conditions at $x = l$, gives

$$(\operatorname{Cosh}\lambda + \cos\lambda)C_1 + (\operatorname{Sinh}\lambda + \sin\lambda)C_2 = 0 \tag{61.32}$$

$$(\operatorname{Sinh}\lambda - \sin\lambda)C_1 + (\operatorname{Cosh}\lambda + \cos\lambda)C_2 = 0 \tag{61.33}$$

Equating to zero the determinant of these equations yields the frequency equation,

$$\operatorname{Cosh}\lambda \cos\lambda + 1 = 0 \tag{61.34}$$

The roots of (61.34) may be found by numerical trial-and-error calculations. Thus, it may be shown that the first five roots are $\lambda_n = 1.875, 4.694, 7.855, 10.966, 14.137$, and that, for higher values of n, the roots are approximately $\lambda_n \approx (2n - 1)\pi/2$. From (61.30), the frequencies are

$$\omega_n = (\lambda_n^2/l^2)\sqrt{EI/\mu} \tag{61.35}$$

The eigenfunction $\phi_n(x)$ for the nth mode is found by substituting λ_n into (61.29) and using the relations (61.31) and either (61.32) or (61.33). Thus,

$$\phi_n(x) = \operatorname{Cosh}(\lambda_n x/l) - \cos(\lambda_n x/l) - \alpha_n[\operatorname{Sinh}(\lambda_n x/l) - \sin(\lambda_n x/l)] \tag{61.36}$$

where $\alpha_n = -(C_2/C_1)$ and, using (61.32), is given by

$$\alpha_n = (\operatorname{Cosh}\lambda_n + \cos\lambda_n)/(\operatorname{Sinh}\lambda_n + \sin\lambda_n)$$

The right-hand side of (61.36) can be multiplied by any desired constant factor. For example, it is sometimes convenient to choose the factor so that, for each mode,

$$\int_0^l \mu\phi_n^2\,dx = m$$

where m is the total mass of the beam. In other problems it may be convenient to choose the factor so that $\phi_n(l) = 1$ for all modes.

The above solution for a clamped-free beam is given to illustrate the method of solution. Solutions for beams with other boundary conditions are found in a similar

Table 61.1. Frequencies and Eigenfunctions for Uniform Beams

Type	Boundary conditions	Frequency equation	Eigenfunction $\phi_n(x)$	Roots of frequency equation λ_n
Clamped-clamped 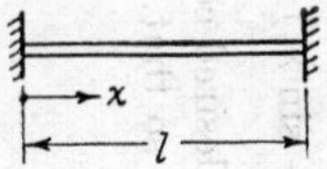	$\phi(0) = \phi'(0) = 0$ $\phi(l) = \phi'(l) = 0$	$\cos\lambda \operatorname{Cosh}\lambda = 1$	$J\left(\frac{\lambda_n x}{l}\right) - \frac{J(\lambda_n)}{H(\lambda_n)} H\left(\frac{\lambda_n x}{l}\right)$	$\lambda_1 = 4.7300$ $\lambda_2 = 7.8532$ $\lambda_3 = 10.9956$ $\lambda_4 = 14.1372$ For n large, $\lambda_n \approx (2n + 1)\pi/2$
Clamped-hinged 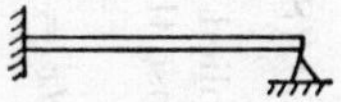	$\phi(0) = \phi'(0) = 0$ $\phi(l) = \phi''(l) = 0$	$\tan\lambda = \operatorname{Tanh}\lambda$	$J\left(\frac{\lambda_n x}{l}\right) - \frac{J(\lambda_n)}{H(\lambda_n)} H\left(\frac{\lambda_n x}{l}\right)$	$\lambda_1 = 3.9266$ $\lambda_2 = 7.0686$ $\lambda_3 = 10.2102$ $\lambda_4 = 13.3518$ For n large, $\lambda_n \approx (4n + 1)\pi/4$
Clamped-free 	$\phi(0) = \phi'(0) = 0$ $\phi''(l) = \phi'''(l) = 0$	$\cos\lambda \operatorname{Cosh}\lambda = -1$	$J\left(\frac{\lambda_n x}{l}\right) - \frac{G(\lambda_n)}{F(\lambda_n)} H\left(\frac{\lambda_n x}{l}\right)$	$\lambda_1 = 1.8751$ $\lambda_2 = 4.6941$ $\lambda_3 = 7.8548$ $\lambda_4 = 10.9955$ For n large, $\lambda_n \approx (2n - 1)\pi/2$
Clamped-guided 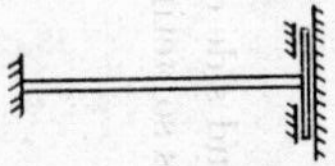	$\phi(0) = \phi'(0) = 0$ $\phi'(l) = \phi'''(l) = 0$	$\tan\lambda = -\operatorname{Tanh}\lambda$	$J\left(\frac{\lambda_n x}{l}\right) - \frac{F(\lambda_n)}{J(\lambda_n)} H\left(\frac{\lambda_n x}{l}\right)$	$\lambda_1 = 2.3650$ $\lambda_2 = 5.4978$ $\lambda_3 = 8.6394$ $\lambda_4 = 11.7810$ For n large, $\lambda_n \approx (4n - 1)\pi/4$
Hinged-hinged 	$\phi(0) = \phi''(0) = 0$ $\phi(l) = \phi''(l) = 0$	$\sin\lambda = 0$	$\sin\frac{n\pi x}{l}$	$\lambda_n = n\pi$

Table 61.1. Frequencies and Eigenfunctions for Uniform Beams (*Continued*)

Type	Boundary conditions	Frequency equation	Eigenfunction $\phi_n(x)$	Roots of frequency equation λ_n
Hinged-guided 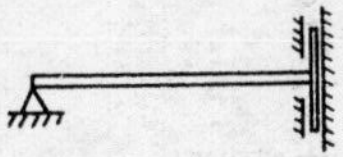	$\phi(0) = \phi''(0) = 0$ $\phi'(l) = \phi'''(l) = 0$	$\cos \lambda = 0$	$\sin \frac{(2n-1)\pi x}{2l}$	$\lambda_n = (2n-1)\pi/2$
Guided-guided 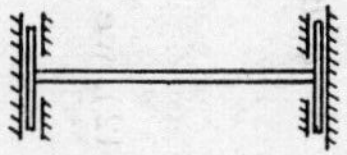	$\phi'(0) = \phi'''(0) = 0$ $\phi'(l) = \phi'''(l) = 0$	$\sin \lambda = 0$	$\cos \frac{n\pi x}{l}$	$\lambda_n = n\pi$
Free-free 	$\phi''(0) = \phi'''(0) = 0$ $\phi''(l) = \phi'''(l) = 0$	$\cos \lambda \operatorname{Cosh} \lambda = 1$	$G\left(\frac{\lambda_n x}{l}\right) - \frac{J(\lambda_n)}{H(\lambda_n)} F\left(\frac{\lambda_n x}{l}\right)$	Same as for clamped-clamped beam
Free-hinged 	$\phi''(0) = \phi'''(0) = 0$ $\phi(l) = \phi''(l) = 0$	$\tan \lambda = \operatorname{Tanh} \lambda$	$G\left(\frac{\lambda_n x}{l}\right) - \frac{G(\lambda_n)}{F(\lambda_n)} F\left(\frac{\lambda_n x}{l}\right)$	Same as for clamped-hinged beam
Free-guided 	$\phi''(0) = \phi'''(0) = 0$ $\phi'(l) = \phi'''(l) = 0$	$\tan \lambda = -\operatorname{Tanh} \lambda$	$G\left(\frac{\lambda_n x}{l}\right) - \frac{H(\lambda_n)}{F(\lambda_n)} F\left(\frac{\lambda_n x}{l}\right)$	Same as for clamped-guided beam

1. The circular frequency is

$$\omega_n = \frac{\lambda_n^2}{l^2} \sqrt{\frac{EI}{\mu}}$$

where

EI = bending stiffness
μ = mass per unit length
l = length of the beam

2. Notation used in expressions for the eigenfunctions:

$$F(u) = \operatorname{Sinh} u + \sin u$$
$$G(u) = \operatorname{Cosh} u + \cos u$$
$$H(u) = \operatorname{Sinh} u - \sin u$$
$$J(u) = \operatorname{Cosh} u - \cos u$$

manner. Table 61.1 gives a summary of the frequencies and eigenfunctions for a number of common cases.

The general solution of a free-vibration problem is obtained by superposing the solution for each mode. This gives

$$w(x,t) = \sum_{n=1}^{\infty} \phi_n(x)(A_n \cos \omega_n t + B_n \sin \omega_n t) \tag{61.37}$$

Using the orthogonality relations derived in the following article, it can be shown that the integration constants A_n and B_n can be calculated by the formulas

$$A_n = \frac{1}{m_n}\int_0^l w_0 \mu \phi_n \, dx \qquad B_n = \frac{1}{m_n \omega_n}\int_0^l \dot{w}_0 \mu \phi_n \, dx \tag{61.38}$$

where $w_0(x)$ and $\dot{w}_0(x)$ are the initial displacement and velocity, respectively, and m_n is defined by (61.48).

Mathematical Properties of the Eigenfunctions

The eigenfunctions satisfy certain mathematical relations, which are important in various applications. Considering the general case of a variable-section beam, the eigenfunction ϕ_n satisfies the differential equation (61.27), with ω replaced by ω_n; that is,

$$(EI\phi_n'')'' = \omega_n^2 \mu \phi_n \tag{61.39}$$

It also satisfies the boundary conditions of the particular beam under consideration.

Let ϕ_n and ϕ_m be any two of the eigenfunctions of a given beam, and let ω_n and ω_m be the corresponding frequencies. It is assumed that $\omega_m \neq \omega_n$. Multiplying (61.39) by ϕ_m and integrating over the length gives

$$\omega_n^2 \int_0^l \mu \phi_n \phi_m \, dx = \int_0^l \phi_m (EI\phi_n'')'' \, dx \tag{61.40}$$

Integrating the right-hand side by parts twice gives

$$\omega_n^2 \int_0^l \mu \phi_n \phi_m \, dx = \int_0^l EI\phi_n''\phi_m'' \, dx + [\phi_m(EI\phi_n'')' - \phi_m' EI\phi_n'']_0^l \tag{61.41}$$

The same relation holds if the subscripts are interchanged; that is,

$$\omega_m^2 \int_0^l \mu \phi_m \phi_n \, dx = \int_0^l EI\phi_m''\phi_n'' \, dx + [\phi_n(EI\phi_m'')' - \phi_n' EI\phi_m'']_0^l \tag{61.42}$$

Subtracting (61.42) from (61.41) yields

$$\int_0^l \mu \phi_n \phi_m \, dx + \frac{1}{\omega_n^2 - \omega_m^2}[\phi_n(EI\phi_m'')' - \phi_m(EI\phi_n'')' - \phi_n' EI\phi_m'' + \phi_m' EI\phi_n'']_0^l = 0 \tag{61.43}$$

This is the orthogonality relation of the first type.

Eliminating $\int_0^l \mu \phi_n \phi_m \, dx$ from (61.41) and (61.42), we get

$$\int_0^l EI\phi_n''\phi_m'' \, dx + \frac{1}{\omega_n^2 - \omega_m^2}[\omega_n^2\phi_n(EI\phi_m'')' - \omega_m^2\phi_m(EI\phi_n'')' - \omega_n^2\phi_n' EI\phi_m'' + \omega_m^2\phi_m' EI\phi_n'']_0^l = 0 \tag{61.44}$$

This is the orthogonality relation of the second type.

For $n = m$, Eq. (61.41) gives the following relation:

$$\int_0^l EI(\phi_n'')^2\,dx = \omega_n^2 \int_0^l \mu\phi_n^2\,dx + [\phi_n' EI\phi_n'' - \phi_n(EI\phi_n'')']_0^l \tag{61.45}$$

For a beam with any combination of clamped, free, hinged, and guided ends, the boundary values inside the brackets in (61.43), (61.44), and (61.45) vanish, and the orthogonality relations reduce to the following simple forms:

$$\begin{aligned}\int_0^l \mu\phi_n\phi_m\,dx &= 0 && \text{for } m \neq n\\ &= m_n && \text{for } m = n\end{aligned} \tag{61.46}$$

$$\begin{aligned}\int_0^l EI\phi_n''\phi_m''\,dx &= 0 && \text{for } m \neq n\\ &= m_n\omega_n^2 && \text{for } m = n\end{aligned} \tag{61.47}$$

where

$$m_n = \int_0^l \mu\phi_n^2\,dx \tag{61.48}$$

Normal Coordinates

Any deflection of a beam can be represented by an infinite series of its eigenfunctions by taking

$$w(x,t) = \sum_{n=1}^{\infty} \phi_n(x)q_n(t) \tag{61.49}$$

where the functions $q_n(t)$ are the normal coordinates.

In terms of the normal coordinates, the kinetic and potential energies reduce to sums of squares of $\dot{q}_n$ and q_n, respectively. Consider a beam with any combination of hinged, clamped, free, and guided ends. The kinetic and potential energies are

$$T = \tfrac{1}{2}\int_0^l \mu\dot{w}^2\,dx \qquad V = \tfrac{1}{2}\int_0^l EI(w'')^2\,dx \tag{61.50}$$

Substituting (61.49) into (61.50), and making use of the orthogonality relations (61.46) through (61.48), it is found that

$$T = \tfrac{1}{2}\sum_{n=1}^{\infty} m_n\dot{q}_n^2 \qquad V = \tfrac{1}{2}\sum_{n=1}^{\infty} m_n\omega_n^2 q_n^2 \tag{61.51}$$

where

$$m_n = \int_0^l \mu\phi_n^2\,dx \tag{61.52}$$

and is called the *generalized mass.*

Forced Vibrations without Damping

Consider a beam acted upon by a distributed force $p(x,t)$ and one or more concentrated forces $P_i(t)$ applied at x_i, as shown in Fig. 61.3. Assuming that the eigenfunctions and frequencies are known, normal coordinates can be used. In such a case, the deflection is given by (61.49), and the kinetic and potential energies by (61.51).

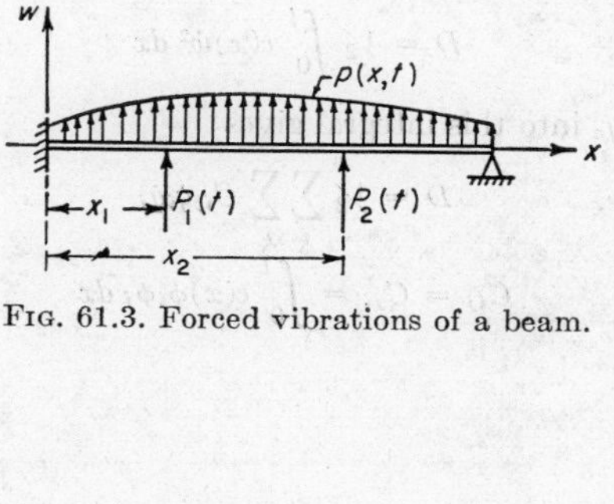

FIG. 61.3. Forced vibrations of a beam.

Substituting into the Lagrangian equation (61.4) gives

$$m_n\ddot{q}_n + m_n\omega_n^2 q_n = Q_n(t) \qquad n = 1, 2, 3, \ldots \tag{61.53}$$

where the generalized force Q_n is given by

$$Q_n(t) = \int_0^l p(x,t)\phi_n(x)\,dx + \sum_i P_i(t)\phi_n(x_i) \tag{61.54}$$

The solution of (61.53) can be written as the sum of the associated homogeneous equation and a convolution integral. Substituting this solution into (61.49) gives the following for the general solution of the forced-vibration problem:

$$w(x,t) = \sum_{n=1}^{\infty} \phi_n(x)\left[A_n \cos\omega_n t + B_n \sin\omega_n t + \frac{1}{m_n\omega_n}\int_0^t Q_n(\tau)\sin\omega_n(t-\tau)\,d\tau\right] \tag{61.55}$$

The constants A_n and B_n depend upon the initial conditions, and can be calculated using formulas (61.38).

Example 4. Harmonic Force. Consider a beam acted upon by a concentrated harmonic force $P_0 \sin \omega t$ at $x = a$. The generalized force becomes $Q_n = \phi_n(a)P_0 \sin \omega t$. Using this, the general solution is found to be

$$w(x,t) = P_0 \sin\omega t \sum_{n=1}^{\infty} \frac{\phi_n(a)\phi_n(x)}{m_n(\omega_n^2 - \omega^2)} + \sum_{n=1}^{\infty} \phi_n(x)(C_n \cos\omega_n t + D_n \sin\omega_n t)$$

where C_n and D_n can be determined from the initial conditions.

Example 5. Step-function Force. Let P_0 be a suddenly applied force acting at $x = a$. Assume that the beam is at rest initially. The generalized force for $t > 0$ is $P_0\phi_n(a)$. Using (61.55), the complete solution is found to be

$$w(x,t) = P_0 \sum_{n=1}^{\infty} \frac{\phi_n(a)\phi_n(x)}{m_n\omega_n^2}(1 - \cos\omega_n t)$$

Forced Vibrations with Damping

For the beam shown in Fig. 61.3, assume that, in addition to the applied forces, there is a distributed damping force per unit length equal to $c(x) \cdot \dot{w}$, where $c(x)$ is a viscous damping coefficient which, in general, is a function of x. For systems with viscous damping, Lagrange's equation can be written as

$$\frac{d}{dt}\frac{\partial T}{\partial \dot{q}_n} - \frac{\partial T}{\partial q_n} + \frac{\partial V}{\partial q_n} + \frac{\partial D}{\partial \dot{q}_n} = Q_n \tag{61.56}$$

where D is the Rayleigh dissipation function and represents one-half the time rate at which energy is dissipated by the damping. For this problem

$$D = \tfrac{1}{2}\int_0^l c(x)\dot{w}^2\,dx$$

Substituting $w = \Sigma\phi_n q_n$ into this integral gives

$$D = \tfrac{1}{2}\sum_i\sum_j C_{ij}\dot{q}_i\dot{q}_j \tag{61.57}$$

where

$$C_{ij} = C_{ji} = \int_0^l c(x)\phi_i\phi_j\,dx \tag{61.58}$$

The expressions for the kinetic and potential energies and for the generalized force are given by (61.51) and (61.54). Substituting these expressions and (61.57) into (61.56) gives

$$m_n\ddot{q}_n + m_n\omega_n^2 q_n + \sum_{i=1}^{\infty} C_{in}\dot{q}_i = Q_n \qquad n = 1, 2, 3, \ldots \tag{61.59}$$

which represents an infinite set of velocity-coupled equations.

In certain special cases, the coupling terms vanish, and the equation for each q_n reduces to that of a one-degree-of-freedom system with viscous damping. One important case in which this occurs is when the distribution of the damping coefficient is proportional to the distribution of the beam mass, that is, when

$$c(x) = \beta\mu(x) \tag{61.60}$$

where β is a positive constant. In this case, by virtue of the orthogonality relation (61.46),

$$\begin{aligned} C_{jn} = \beta \int_0^l \mu\phi_j\phi_n\, dx &= 0 \qquad \text{for } j \neq n \\ &= \beta m_n \qquad \text{for } j = n \end{aligned}$$

and, hence, (61.59) reduces to

$$\ddot{q}_n + \beta\dot{q}_n + \omega_n^2 q_n = Q_n/m_n \tag{61.61}$$

It will be observed that (61.60) is fulfilled for a uniform beam if $c(x)$ is a constant.

The solution of (61.61) can be written as the sum of the solution of the associated homogeneous equation and a convolution integral. This leads to the following general solution,

$$w(x,t) = \sum_{n=1}^{\infty} \phi_n(x) \left[e^{-\beta t/2}(A_n \cos \bar{\omega}_n t + B_n \sin \bar{\omega}_n t) + \frac{1}{m_n\bar{\omega}_n} \int_0^l Q_n(\tau) e^{-(\beta/2)(t-\tau)} \sin \bar{\omega}_n(t - \tau)\, d\tau \right] \tag{61.62}$$

where $\bar{\omega}_n = \sqrt{\omega_n^2 - (\beta/2)^2}$. If $w_0(x)$ and $\dot{w}_0(x)$ are the initial displacement and velocity, it can be shown, with the help of the orthogonality relations, that the integration constants A_n and B_n can be calculated by the following formulas:

$$A_n = \frac{1}{m_n}\int_0^l w_0\mu\phi_n\, dx \qquad B_n = \frac{1}{m_n\bar{\omega}_n}\int_0^l \dot{w}_0\mu\phi_n\, dx + \frac{\beta A_n}{2\bar{\omega}_n} \tag{61.63}$$

In problems involving harmonic vibrations, structural damping is sometimes used in place of viscous damping. *Structural damping* (also called solid damping) is defined as being proportional to the displacement and in phase with the velocity. For example, consider the response of a beam to a harmonic concentrated force $P_0e^{i\omega t}$ applied at $x = a$. The force is taken in complex exponential form for convenience in the subsequent analysis. Structural damping is introduced by replacing the viscous damping term $\beta\dot{q}_n$ in (61.61) with the term $i\gamma\omega_n^2 q_n$, where γ is the structural damping constant and $i = \sqrt{-1}$. The generalized force for this example is $\phi_n(a) P_0e^{i\omega t}$. With these changes, (61.61) becomes

$$\ddot{q}_n + \omega_n^2(1 + \gamma i)q_n = \frac{1}{m_n}\phi_n(a)P_0e^{i\omega t} \tag{61.64}$$

Omitting consideration of the starting transient, the steady-state solution of (61.64) is

$$w_{ss} = \sum_{n=1}^{\infty} \frac{\phi_n(a)\phi_n(x)P_0 e^{i(\omega t - \epsilon_n)}}{m_n\omega_n^2 \sqrt{(1 - \omega^2/\omega_n^2)^2 + \gamma^2}} \tag{61.65}$$

where $\epsilon_n = \tan^{-1} \dfrac{\gamma}{1 - \omega^2/\omega_n^2}$

For comparison, if the same harmonic force is applied to a beam with viscous damping for which (61.61) is valid, the steady-state solution is

$$w_{ss} = \sum_{n=1}^{\infty} \frac{\phi_n(a)\phi_n(x)P_0 e^{i(\omega t - \epsilon_n)}}{m_n\omega_n^2 \sqrt{(1 - \omega^2/\omega_n^2)^2 + (\beta\omega/\omega_n^2)^2}} \tag{61.66}$$

where $\epsilon_n = \tan^{-1} \dfrac{\beta\omega/\omega_n^2}{1 - \omega^2/\omega_n^2}$

Additional Problems in Beams Vibrations

The procedures given above must be modified and extended to the analysis of other problems such as (1) the effect of axial forces on lateral vibrations, (2) beams continuous over several supports, (3) beams on elastic foundations, (4) vibrations due to moving loads crossing a beam, (5) rotating beams, including vibrations of turbine blades and propellers, and (6) beam vibrations with time-dependent boundary conditions.

For methods of analysis and data relative to these additional types of problems, see Timoshenko [3] and also Refs. [10] through [19].

61.5. LATERAL VIBRATIONS OF BEAMS: TIMOSHENKO THEORY

Free-vibration Equations

The Timoshenko theory of beam vibrations takes into account the effect of shearing deflections and of rotatory inertia. Other assumptions are the same as in the Bernoulli-Euler theory (see p. 61–6). Let

w = total deflection due to both bending and shear
Ψ = slope due to bending
J = mass moment of inertia per unit length
$r = \sqrt{J/\mu}$ = radius of gyration of a beam element
k = shear-deflection coefficient
S = transverse shearing force

Referring to Fig. 61.4, the equations of motion for rotation and translation, respectively, of a beam element are

$$M' - S = J\ddot{\Psi} \qquad S' = -\mu\ddot{w} \tag{61.67}$$

The moment and shear are related to the deflection by the two elastic relations,

$$M = EI\Psi' \qquad S = -kGA(w' - \Psi) \tag{61.68}$$

Eliminating M and S from the four relations above gives the pair of coupled equations:

$$(EI\Psi')' + kGA(w' - \Psi) - J\ddot{\Psi} = 0 \qquad \mu\ddot{w} - [kGA(w' - \Psi)]' = 0 \tag{61.69a,b}$$

The boundary conditions for the four common types of end support are: (1) Hinged end: $w = 0$, $\Psi' = 0$; (2) clamped end: $w = 0$, $\Psi = 0$; (3) free end: $w' - \Psi = 0$, $\Psi' = 0$; (4) guided end: $w' = 0$, $\Psi = 0$.

In the general case of a variable-section beam, it is not possible to solve (61.69) analytically, and such problems must, therefore, be solved by numerical methods. However, for uniform beams, analytical solutions can be obtained.

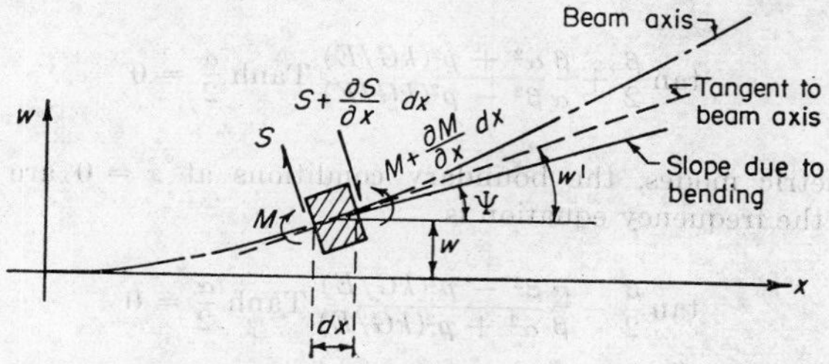

FIG. 61.4. Notation and geometry of an element of a Timoshenko beam.

Free Vibrations of Uniform Beams

Assume a uniform homogeneous beam, for which $\mu = A\rho$, $J = I\rho$, and $I = Ar^2$. Equations (61.69) become

$$\Psi'' - \frac{1}{c_0^2}\ddot{\Psi} - \frac{1}{r^2}\frac{kG}{E}\Psi + \frac{1}{r^2}\frac{kG}{E}w' = 0$$
$$\ddot{w} - c_0^2\frac{kG}{E}w'' + c_0^2\frac{kG}{E}\Psi' = 0 \qquad (61.70a,b)$$

where $c_0 = \sqrt{E/\rho}$. For harmonic vibrations, take

$$w(x,t) = \phi(x)\sin(\omega t + \epsilon) \qquad \Psi(x,t) = \psi(x)\sin(\omega t + \epsilon) \qquad (61.71)$$

Substituting into (61.70) and solving the resulting pair of ordinary differential equations leads to the following general expressions for the eigenfunctions ϕ and ψ,

$$\phi(x) = C_1\cosh\frac{\alpha x}{l} + C_2\sinh\frac{\alpha x}{l} + C_3\cos\frac{\beta x}{l} + C_4\sin\frac{\beta x}{l}$$
$$\psi(x) = \frac{(kG/E)p^2 + \alpha^2}{\alpha l}\left(C_2\cosh\frac{\alpha x}{l} + C_1\sinh\frac{\alpha x}{l}\right) \qquad (61.72a,b)$$
$$+ \frac{(kG/E)p^2 - \beta^2}{\beta l}\left(-C_4\cos\frac{\beta x}{l} + C_3\sin\frac{\beta x}{l}\right)$$

where

$$p^2 = \omega^2 l^2/c_0^2$$
$$\left.\begin{matrix}\alpha^2\\ \beta^2\end{matrix}\right\} = \frac{p^2}{2}\left[\mp\left(1 + \frac{E}{kG}\right) + \sqrt{\left(1 - \frac{E}{kG}\right)^2 + \frac{4l^2}{p^2r^2}}\right] \qquad (61.73a,b)$$

Substituting (61.72) into the four boundary conditions gives a set of four homogeneous algebraic equations in the unknowns C_1, C_2, C_3, C_4. The frequency equation is obtained by equating to zero the determinant of the set. Corresponding to any root ω_n of the frequency equation, the set of algebraic equations can be solved to obtain the relative magnitudes of C_1, C_2, C_3, C_4, and these magnitudes, when substituted back into (61.72), determine the eigenfunctions $\phi_n(x)$ and $\psi_n(x)$, except for an arbitrary multiplicative factor.

Example 6. Clamped-Free Beam. The boundary conditions are $\phi(0) = 0$, $\psi(0) = 0$, $\phi'(l) - \psi(l) = 0$, $\psi'(l) = 0$. The frequency equation is

$$2 + \frac{\alpha^2 - \beta^2}{\alpha\beta}\sinh\alpha\sin\beta + \left(\frac{\beta^2 - p^2}{\alpha^2 + p^2} + \frac{\alpha^2 + p^2}{\beta^2 - p^2}\right)\cosh\alpha\cos\beta = 0$$

Example 7. Clamped-Clamped Beam. Take the origin at the mid-point of the beam, with the clamped ends at $x = \pm l/2$. The boundary conditions at $x = l/2$ are $\phi(l/2) = 0$ and $\psi(l/2) = 0$. For the symmetric modes, the boundary conditions at $x = 0$ are $\psi(0) = 0$ and $\phi'(0) - \psi(0) = 0$, and the frequency equation is

$$\tan\frac{\beta}{2} + \frac{\beta}{\alpha}\frac{\alpha^2 + p^2(kG/E)}{\beta^2 - p^2(kG/E)} \operatorname{Tanh}\frac{\alpha}{2} = 0$$

For the antimetric modes, the boundary conditions at $x = 0$ are $\phi(0) = 0$ and $\psi'(0) = 0$, and the frequency equation is

$$\tan\frac{\beta}{2} - \frac{\alpha}{\beta}\frac{\beta^2 - p^2(kG/E)}{\alpha^2 + p^2(kG/E)} \operatorname{Tanh}\frac{\alpha}{2} = 0$$

Example 8. Free-Free Beam. Take the origin at the mid-point of the beam with the free ends at $x = \pm l/2$. The boundary conditions at $x = l/2$ are $\psi'(l/2) = 0$ and $\phi'(l/2) - \psi(l/2) = 0$. The boundary conditions at $x = 0$ are the same as in Example 7. For the symmetric modes, the frequency equation is

$$\tan\frac{\beta}{2} + \frac{\beta}{\alpha}\frac{\alpha^2 + p^2}{\beta^2 - p^2} \operatorname{Tanh}\frac{\alpha}{2} = 0$$

For the antimetric modes, the frequency equation is

$$\tan\frac{\beta}{2} - \frac{\alpha}{\beta}\frac{\beta^2 - p^2}{\alpha^2 + p^2} \operatorname{Tanh}\frac{\alpha}{2} = 0$$

Example 9. Hinged-Hinged Beam. For a beam which is hinged at $x = 0$ and $x = l$, it may be verified that the eigenfunctions have the simple form

$$\phi_n(x) = A_n \sin(n\pi x/l) \qquad \psi_n(x) = B_n \cos(n\pi x/l) \tag{61.74}$$

where $n = 1, 2, 3, \ldots$. Making use of this fact, a direct solution for the frequencies can be obtained in the following manner. Substituting (61.74) into (61.70) gives

$$\begin{aligned} \left(\frac{kG}{E}\frac{l}{n\pi r^2}\right) A_n + \left(\lambda^2 - 1 - \frac{kG}{E}\frac{l}{n^2\pi^2 r^2}\right) B_n &= 0 \\ \left(\lambda^2 - \frac{kG}{E}\right) A_n + \left(\frac{kG}{E}\frac{l}{n\pi}\right) B_n &= 0 \end{aligned}$$

where $\lambda = \omega l/nc_0\pi$. Equating the determinant to zero and solving for λ^2,

$$\left.\begin{matrix}\lambda_{n1}^2 \\ \lambda_{n2}^2\end{matrix}\right\} = \frac{1}{2}\left[1 + \frac{kG}{E} + \frac{kG}{E}\frac{l^2}{n^2\pi^2 r^2} \mp \sqrt{\left(1 + \frac{kG}{E} + \frac{kG}{E}\frac{l}{n^2\pi^2 r^2}\right)^2 - 4\frac{kG}{E}}\right] \tag{61.75}$$

Thus, for each n, there are two frequencies:

$$\omega_{n1} = n\pi c_0 \lambda_{n1}/l \qquad \omega_{n2} = n\pi c_0 \lambda_{n2}/l$$

The corresponding amplitude ratios are

$$\left(\frac{B_n}{A_n}\right)_s = \frac{n\pi}{l}\left(1 - \frac{E}{kG}\lambda_{ns}^2\right) \qquad s = 1, 2$$

Using these, the general solution of the free-vibration problem may be written as follows:

$$w(x,t) = \sum_{n=1}^{\infty} \sin\frac{n\pi x}{l}\,[A_{n1}\sin(\omega_{n1}t + \epsilon_{n1}) + A_{n2}\sin(\omega_{n2}t + \epsilon_{n2})]$$

$$\Psi(x,t) = \sum_{n=1}^{\infty} \frac{n\pi}{l}\cos\frac{n\pi x}{l}\left[\left(1 - \frac{E}{kG}\lambda_{n1}^2\right)A_{n1}\sin(\omega_{n1}t + \epsilon_{n1}) + \left(1 - \frac{E}{kG}\lambda_{n2}^2\right)A_{n2}\sin(\omega_{n2}t + \epsilon_{n2})\right]$$

Orthogonality Relations. Normal Coordinates

The orthogonality relations for the eigenfunctions of a Timoshenko beam can be developed in a manner analogous to that used on p. 61–10 (see [14]). Thus it can be shown that, for a beam with any combination of clamped, free, hinged, and guided ends, the eigenfunctions satisfy the relations

$$\int_0^l (\phi_n\phi_m + r^2\psi_n\psi_m)\mu\,dx = 0 \qquad \text{for } m \neq n$$
$$= m_n \qquad \text{for } m = n \tag{61.76}$$

$$\int_0^l EI\psi_n'\psi_m'\,dx + \int_0^l kAG(\phi_n' - \psi_n)(\phi_m' - \psi_m)\,dx = 0 \qquad \text{for } m \neq n$$
$$= m_n\omega_n^2 \qquad \text{for } m = n \tag{61.77}$$

where
$$m_n = \int_0^l (\phi_n^2 + r^2\psi_n^2)\mu\,dx \tag{61.78}$$

Any deflection of a Timoshenko beam can be represented by a pair of infinite series of the eigenfunctions; thus,

$$w(x,t) = \sum_{n=1}^{\infty} \phi_n(x)q_n(t) \qquad \Psi(x,t) = \sum_{n=1}^{\infty} \psi_n(x)q_n(t) \tag{61.79a,b}$$

where the functions $q_n(t)$ are called the *normal coordinates*.

Substituting these series into the general expressions for the kinetic and potential energies, and making use of the orthogonality relations, it can be shown that the energy expressions reduce to

$$T = \tfrac{1}{2}\sum_{n=1}^{\infty} m_n\dot{q}_n^2 \qquad V = \tfrac{1}{2}\sum_{n=1}^{\infty} m_n\omega_n^2 q_n^2 \tag{61.80a,b}$$

where m_n is defined by (61.78).

Forced Vibrations

Using normal coordinates, the solution of a forced-vibration problem is similar to that given on p. 61–11. The general solution for $w(x,t)$ has the same form as (61.55), with the appropriate changes in the definitions of $\phi_n(x)$ and m_n. The general solution for $\Psi(x,t)$ is similar to that for $w(x,t)$, except that $\psi_n(x)$ replaces $\phi_n(x)$.

As shown in Example 9, the eigenfunctions for a uniform hinged-hinged beam are proportional to $\sin(n\pi x/l)$ and $\cos(n\pi x/l)$, and, corresponding to each n, there are two frequencies ω_{n1} and ω_{n2}. In this case, it is convenient to specialize the analysis by taking

$$w(x,t) = \sum_{n=1}^{\infty} \sin\left(\frac{n\pi x}{l}\right) [q_{n1}(t) + q_{n2}(t)]$$

$$\Psi(x,t) = \sum_{n=1}^{\infty} \cos\left(\frac{n\pi x}{l}\right) \left[\frac{k_{n1}}{l} q_{n1}(t) + \frac{k_{n2}}{l} q_{n2}(t)\right] \tag{61.81a,b}$$

where
$$k_{ns} = n\pi\left(1 - \frac{E}{kG}\lambda_{ns}^2\right) \qquad s = 1, 2 \tag{61.82a,b}$$
$$\lambda_{ns} = \omega_{ns}\, l/n\pi c_0$$

In these equations, q_{n1} and q_{n2} are the normal coordinates corresponding to ω_{n1} and ω_{n2}, respectively.

The kinetic and potential energies are

$$T = \tfrac{1}{2} \sum_{n=1}^{\infty} (m_{n1}\dot{q}_{n1}^2 + m_{n2}\dot{q}_{n2}^2) \qquad V = \tfrac{1}{2} \sum_{n=1}^{\infty} (m_{n1}\omega_{n1}^2 q_{n1}^2 + m_{n2}\omega_{n2}^2 q_{n2}^2) \tag{61.83a,b}$$

where
$$m_{ns} = \int_0^l \left(\sin^2 \frac{n\pi x}{l} + \frac{k_{ns} r^2}{l^2} \cos^2 \frac{n\pi x}{l}\right) \mu\, dx = \frac{\mu l}{2}\left(1 + k_{ns}\frac{r^2}{l^2}\right) \tag{61.84}$$

Example 10. Hinged-Hinged Beam. Find the bending moment in a uniform hinged-hinged beam subjected to a suddenly applied constant force P_0 acting at $x = a$, assuming that the beam is initially at rest. Substituting T and V into Lagrange's equation gives $\ddot{q}_{ns} + \omega_{ns}^2 q_{ns} = Q_{ns}/m_{ns}$, where the generalized force for this example is $Q_{ns} = P_0 \sin(n\pi a/l)$. The solution corresponding to the specified initial conditions is

$$q_{ns} = \frac{P_0 \sin(n\pi a/l)}{m_{ns}\omega_{ns}^2}(1 - \cos\omega_{ns}t) \qquad s = 1, 2;\ n = 1, 2, 3, \ldots \tag{61.85}$$

The completed solutions for $w(x,t)$ and $\Psi(x,t)$ are obtained by substituting (61.85) into (61.81*a,b*).

The bending moment is $M = EI\Psi'$. Substituting (61.85) into (61.81*b*) and differentiating, we obtain, after considerable algebraic manipulation,

$$M = -\frac{2P_0 l}{\pi^2} \sum_{n=1}^{\infty} \frac{1}{n^2} \sin\frac{n\pi a}{l} \sin\frac{n\pi x}{l} \left[\frac{\lambda_{n2}^2}{\lambda_{n2}^2 - \lambda_{n1}^2}(1 - \cos\omega_{n1}t) - \frac{\lambda_{n1}^2}{\lambda_{n2}^2 - \lambda_{n1}^2}(1 - \cos\omega_{n2}t)\right] \tag{61.86}$$

61.6. CIRCULAR RINGS

The data given herein are for a complete circular ring of constant cross section. It is assumed that the cross-sectional dimensions of the ring are small compared to the radius of the ring center line, and that one of the principal axes of the cross section is in the plane of the ring. Derivations of the equations of vibrations are given in references such as Love [2] and Timoshenko [3].

In this section, R = radius of the ring center line; μ = mass per unit length of the ring; I = moment of inertia of the cross section about an axis perpendicular to the plane of the ring; u, v = radial and tangential displacements, respectively, of a point on the center line; and θ = angle from a reference radius to a point on the ring.

Flexural Vibrations in the Plane of the Ring

Neglecting rotatory inertia and extension of the center line, the differential equation of free vibrations which governs the radial displacement is

$$\frac{\partial^2}{\partial t^2}\left(u - \frac{\partial^2 u}{\partial \theta^2}\right) = \frac{EI}{\mu}\left(\frac{\partial^6 u}{\partial \theta^6} + 2\frac{\partial^4 u}{\partial \theta^4} + \frac{\partial^2 u}{\partial \theta^2}\right) \tag{61.87}$$

This is satisfied by taking

$$u = C_n \cos n\theta \sin (\omega_n t + \epsilon_n) \tag{61.88}$$

The frequency is

$$\omega_n{}^2 = \frac{n^2(n^2-1)^2}{n^2+1}\frac{EI}{\mu R^4} \tag{61.89}$$

where $n = 1, 2, 3, \ldots$ = number of complete wavelengths in a circumference.

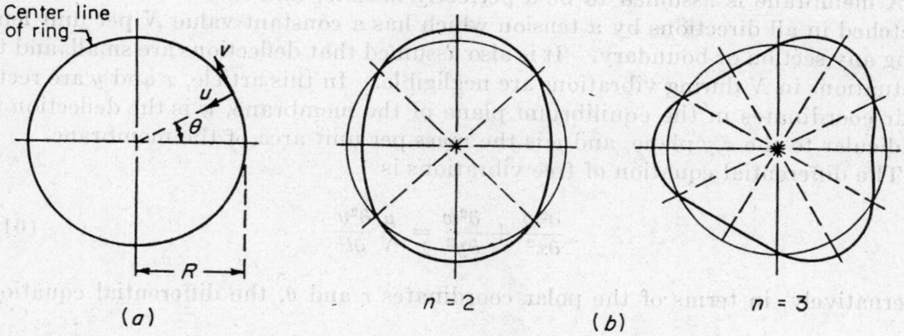

FIG. 61.5. Circular ring: (*a*) Geometry of ring; (*b*) normal mode shapes.

When $n = 1$, the frequency vanishes, and the displacement is as a rigid body. The ring geometry and typical displacement shapes are shown in Fig. 61.5.

Flexural Vibrations Perpendicular to Plane of Ring

Vibrations of this type involve flexure perpendicular to the plane of the ring, coupled with torsion. For a *circular* cross section, the frequency is

$$\omega_n{}^2 = \frac{n^2(n^2-1)^2}{n^2+1+\nu}\frac{EI}{\mu R^4} \tag{61.90}$$

where ν is Poisson's ratio, and $n = 1, 2, 3, \ldots$. Thus, for a circular cross section, these frequencies are nearly equal to those in (61.89).

Extensional Vibrations

Vibrations of this type are analogous to the longitudinal vibrations of a bar. The frequency is

$$\omega_n{}^2 = (n^2+1)E/\rho R^2 \tag{61.91}$$

where $n = 0, 1, 2, 3, \ldots$. The displacements have the form

$$u = C_n \cos n\theta \sin (\omega_n t + \epsilon_n) \qquad v = -nC_n \sin n\theta \sin (\omega_n t + \epsilon_n) \tag{61.92a,b}$$

When $n = 0$, the displacement component v vanishes, and the ring vibrates with a pure radial motion in which u is independent of θ.

Torsional Vibrations

Vibrations of this type are analogous to the torsional vibrations of shafts or prismatic bars. For a *circular* cross section, the frequency is

$$\omega_n^2 = (n^2 + 1 + \nu)G/\rho R^2 \tag{61.93}$$

where $n = 0, 1, 2, 3, \ldots$.

Circular Arcs

For circular arcs which are only a part of a complete circular ring, it is generally necessary to resort to approximate methods. Frequency data determined by Rayleigh's method for a number of cases are given by Den Hartog [4].

61.7. MEMBRANES

Free-vibration Equations

A membrane is assumed to be a perfectly flexible, thin sheet, which is uniformly stretched in all directions by a tension which has a constant value N per unit length along any section or boundary. It is also assumed that deflections are small, and that fluctuations in N during vibrations are negligible. In this article, x and y are rectangular coordinates in the equilibrium plane of the membrane, w is the deflection perpendicular to the x,y plane, and μ is the mass per unit area of the membrane.

The differential equation of free vibrations is

$$\frac{\partial^2 w}{\partial x^2} + \frac{\partial^2 w}{\partial y^2} = \frac{\mu}{N}\frac{\partial^2 w}{\partial t^2} \tag{61.94}$$

Alternatively, in terms of the polar coordinates r and θ, the differential equation is

$$\frac{1}{r}\frac{\partial}{\partial r}\left(r\frac{\partial w}{\partial r}\right) + \frac{1}{r^2}\frac{\partial^2 w}{\partial \theta^2} = \frac{\mu}{N}\frac{\partial^2 w}{\partial t^2} \tag{61.95}$$

The kinetic and potential energies are

$$T = \frac{1}{2}\iint \mu\left(\frac{\partial w}{\partial t}\right)^2 dx\,dy \qquad V = \frac{N}{2}\iint\left[\left(\frac{\partial w}{\partial x}\right)^2 + \left(\frac{\partial w}{\partial y}\right)^2\right] dx\,dy \tag{61.96a,b}$$

Rectangular Membrane

Let a and b be the length of the sides of a rectangular membrane, which is fixed along the four edges at $x = 0$, $x = a$, $y = 0$, and $y = b$. The frequency is

$$\omega_{mn}^2 = \pi^2\left(\frac{m^2}{a^2} + \frac{n^2}{b^2}\right)\frac{N}{\mu} \tag{61.97}$$

and the corresponding eigenfunction is

$$W_{mn} = C_{mn}\sin\frac{m\pi x}{a}\sin\frac{n\pi y}{b} \tag{61.98}$$

where $m, n = 1, 2, 3, \ldots$. The lowest frequency is obtained by taking $m = n = 1$. It is seen that m is the number of half wavelengths in the x direction of the deflection shape, and that n is the same in the y direction.

Circular Membrane

For a circular membrane of radius a which is fixed at its boundary, the frequency equation is

$$J_n(ka) = 0 \tag{61.99}$$

where $k = \omega \sqrt{\mu/N}$ and J_n is the Bessel function of the first kind and nth integral order. The corresponding eigenfunction is

$$w_{ns} = C_{ns}J_n(kr) \cos n\theta \tag{61.100}$$

The frequencies obtained from the roots of (61.99) have the form

$$\omega_{ns} = \frac{\alpha_{ns}}{a} \sqrt{\frac{N}{\mu}} \tag{61.101}$$

where α_{ns} is a constant. The subscript n indicates the number of nodal diameters, and the subscript s indicates the number of nodal circles including the boundary circle. The lowest frequency is obtained when $n = 0$ and $s = 1$, in which case $\alpha_{01} = 2.404$. Other values of α_{ns} are given in Table 61.2.

Table 61.2

	$n = 0$	$n = 1$	$n = 2$	$n = 3$
$s = 1$	2.404	3.832	5.135	6.379
$s = 2$	5.520	7.016	8.417	9.760
$s = 3$	8.654	10.173	11.620	13.017
$s = 4$	11.792	13.323	14.796	16.224

References

For more complete discussions of membrane vibrations, see Rayleigh [1], Timoshenko [3], and Morse [5].

61.8. PLATES

Equations of Free Vibrations

The classical theory of plate vibrations assumes that (1) the material is elastic, homogeneous, and isotropic; (2) the plate thickness h is small compared to the other dimensions; (3) deflections are small, so that stretching of the middle surface is negligible; and (4) a normal to the initially plane middle surface remains normal to this surface during bending. Effects of rotatory inertia and shear are neglected. The analysis and data given herein are based on these assumptions and, in addition, are restricted to plates of uniform thickness. Discussions of more complicated cases are given in [1]–[3] and [20]–[29].

In the following, x and y are cartesian coordinates in the plane of the middle surface, $\mu = \rho h$ is the mass of the plate per unit area, w is the transverse deflection, and $K = Eh^3/12(1 - \nu^2)$ is the bending stiffness. The differential equation of free vibrations is

$$\nabla^2\nabla^2 w = -\frac{\mu}{K}\frac{\partial^2 w}{\partial t^2} \tag{61.102}$$

where ∇^2 is the Laplacian operator [Eq. (11.64)]. For harmonic vibrations, take $w(x,y,t) = W(x,y) \sin \omega t$. Substitution into (61.102) yields

$$\nabla^2\nabla^2 W = \gamma^2 W \tag{61.103}$$

where

$$\gamma^2 = \mu\omega^2/K \tag{61.104}$$

The boundary conditions for a plate are (see p. 39–4)

Supported (hinged) edge: $w = 0 \qquad \frac{\partial^2 w}{\partial \xi^2} + \nu \frac{\partial^2 w}{\partial \eta^2} = 0$ (61.105)

Clamped edge: $w = 0 \qquad \frac{\partial w}{\partial \xi} = 0$ (61.106)

Free edge: $\frac{\partial^2 w}{\partial \xi^2} + \nu \frac{\partial^2 w}{\partial \eta^2} = 0 \qquad \frac{\partial}{\partial \xi}\left[\frac{\partial^2 w}{\partial \xi^2} + (2 - \nu)\frac{\partial^2 w}{\partial \eta^2}\right] = 0$ (61.107)

In these expressions, ξ is the coordinate normal to the boundary, and η the coordinate tangential to the boundary.

For use in energy methods, the expressions for the kinetic and potential energies of a plate are needed:

$$T = \frac{\mu}{2}\iint \left(\frac{\partial w}{\partial t}\right)^2 dx\, dy$$

$$V = \frac{K}{2}\iint \left\{\left(\frac{\partial^2 w}{\partial x^2} + \frac{\partial^2 w}{\partial y^2}\right)^2 - 2(1-\nu)\left[\frac{\partial^2 w}{\partial x^2}\frac{\partial^2 w}{\partial y^2} - \left(\frac{\partial^2 w}{\partial x\, \partial y}\right)^2\right]\right\} dx\, dy \qquad (61.108a,b)$$

where the integration is over the area of the plate. The corresponding expressions in polar coordinates are

$$T = \frac{\mu}{2}\iint \left(\frac{\partial w}{\partial t}\right)^2 r\, d\theta\, dr$$

$$V = \frac{K}{2}\iint \left(\left(\frac{\partial^2 w}{\partial r^2} + \frac{1}{r}\frac{\partial w}{\partial r} + \frac{1}{r^2}\frac{\partial^2 w}{\partial \theta^2}\right)^2 - 2(1-\nu)\left\{\frac{\partial^2 w}{\partial r^2}\left(\frac{1}{r}\frac{\partial w}{\partial r} + \frac{1}{r^2}\frac{\partial^2 w}{\partial \theta^2}\right) - \left[\frac{\partial}{\partial r}\left(\frac{1}{r}\frac{\partial w}{\partial \theta}\right)\right]^2\right\}\right) r\, d\theta\, dr \qquad (61.109a,b)$$

Rectangular Plates

Let the length of the sides be denoted by a and b, and assume that the boundaries are at $x = 0$, $x = a$, $y = 0$, and $y = b$.

For a plate which is simply supported at all four edges, it may be verified that the differential equation (61.103) and boundary conditions (61.105) are satisfied by taking the frequency and eigenfunction as follows:

$$\omega_{mn} = \pi^2\left(\frac{m^2}{a^2} + \frac{n^2}{b^2}\right)\sqrt{\frac{K}{\mu}} \qquad m, n = 1, 2, 3, \ldots \qquad (61.110)$$

$$W_{mn} = \sin\frac{m\pi x}{a}\sin\frac{n\pi y}{b} \qquad (61.111)$$

For a plate which is simply supported at two opposite edges, and has any boundary conditions at the other two edges, a solution can be obtained in the following manner. Suppose that the two simply supported edges are at $x = 0$ and $x = a$. Take the eigenfunction in the form

$$W(x,y) = Y(y)\sin\frac{m\pi x}{a} \qquad m = 1, 2, 3, \ldots \qquad (61.112)$$

Substituting into (61.103), the general solution for Y is found to be

$$Y = C_1 \operatorname{Cosh} \alpha y + C_2 \operatorname{Sinh} \alpha y + C_3 \cos \beta y + C_4 \sin \beta y \qquad (61.113)$$

where

$$\alpha = \sqrt{\gamma + \frac{m^2\pi^2}{a^2}} \qquad \beta = \sqrt{\gamma - \frac{m^2\pi^2}{a^2}} \qquad (61.114)$$

The frequency equation is determined by substituting (61.113) into the boundary conditions. For example, assume that the plate is clamped at $y = 0$ and $y = b$. The boundary conditions are

$$Y(0) = 0 \qquad Y'(0) = 0 \qquad Y(b) = 0 \qquad Y'(b) = 0$$

Applying (61.113) to each of these conditions, in turn, gives the following four equations:

$$\begin{aligned} 0 &= C_1 + C_3 \\ 0 &= \alpha C_2 + \beta C_4 \\ 0 &= C_1 \operatorname{Cosh} \alpha b + C_2 \operatorname{Sinh} \alpha b + C_3 \cos \beta b + C_4 \sin \beta b \\ 0 &= \alpha C_1 \operatorname{Sinh} \alpha b + \alpha C_2 \operatorname{Cosh} \alpha b - \beta C_3 \sin \beta b + \beta C_4 \cos \beta b \end{aligned}$$

Table 61.3. Frequencies of Rectangular Plates

$$\omega = \frac{\lambda}{a^2}\sqrt{\frac{K}{\mu}}$$

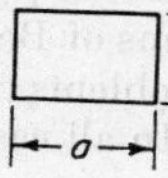

Boundary conditions:

S = supported (hinged) edge, C = clamped edge, F = free edge

Edge conditions	$\frac{b}{a}$	Frequency parameter λ		
		Lowest frequency	Second frequency	Third frequency
S S S S		See Eq. (61.110)		
F F F F	1	14.1	20.5	23.9
C F F C	1	6.96	24.1	26.8
C F F F	2 1 0.5	3.51 3.49 3.47	5.37 8.55 14.9	22.0 21.4 21.6
C S S S	2 1 0.5	17.3 23.7 51.7	51.7	58.7
C S C S	2 1 0.5	23.8 29.0 54.8	54.8	69.3
C C C C	1 2 3 ∞	36.0 24.6 23.2 22.4	73.4	108.3

From the condition that the determinant of the system vanishes, we obtain the following frequency equation

$$2 - 2\cos\beta b \operatorname{Cosh}\alpha b + \left(\frac{\alpha}{\beta} - \frac{\beta}{\alpha}\right)\sin\beta b \operatorname{Sinh}\alpha b = 0$$

This may be solved by trial and error, to obtain the frequency parameter λ.

For rectangular plates without two opposite edges simply supported, exact solutions of the differential equation are not known. Such cases may be solved approximately by the Rayleigh-Ritz method or other approximate methods. See, for example, [3] and [29].

Frequency data are given in Table 61.3 for rectangular plates with a number of different edge conditions.

Circular Plates

The exact solution of the problem of circular plates with clamped or free boundaries can be obtained in terms of Bessel functions, as shown by Rayleigh [1]. Solutions for other circular-plate problems can be obtained by the Rayleigh-Ritz method, as shown by Timoshenko [3]. In all cases, the frequency has the form

$$\omega_{ns} = \frac{\lambda_{ns}}{a^2}\sqrt{\frac{K}{\mu}} \tag{61.115}$$

where a is the radius of the plate boundary. Numerical values of λ_{ns} for circular plates with clamped and with free edges are given in Table 61.4.

Table 61.4. Frequencies of Circular Plates

$$\omega_{ns} = \frac{\lambda_{ns}}{a^2}\sqrt{\frac{K}{\mu}}$$

s	Values of parameter λ_{ns}						
	Clamped plate			Free plate			
	$n = 0$	$n = 1$	$n = 2$	$n = 0$	$n = 1$	$n = 2$	$n = 3$
0	10.2	21.2	34.8	.....	....	5.25	12.2
1	39.8	60.8	88.4	9.08	20.5	35.2	52.9
2	88.9	120.2	157.9	38.5	59.9	88	120

n = number of nodal diameters
s = number of nodal circles, excluding the boundary in the case of the clamped plate
Nodal pattern example:

$n = 1$
$s = 2$

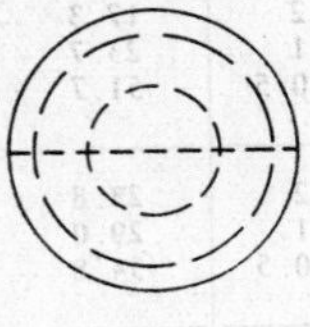

Rotating Disks

For a disk which is rotating with angular velocity Ω, the potential energy due to flexure is composed of two parts, namely, (1) that due to strain energy of bending, which is given by (61.109b), and (2) that due to the centrifugal forces, which is proportional to Ω^2. The kinetic energy expression is the same as for a stationary plate (61.109a). As a consequence, the frequency of lateral vibrations can be expressed approximately by the relation

$$\omega^2 = \omega_{st}^2 + B\Omega^2$$

where ω_{st} is the frequency of the disk when it is stationary, and B is a numerical factor, which has different values for different modes of vibration. Values of B for a complete circular disk of uniform thickness are given in [3]. For variable-thickness disks and disks with central hubs, it is generally necessary to determine B experimentally or by approximate methods.

For more complete discussions of the rotating-disk problem, see [3], [24], [27], and [39].

61.9. THIN SHELLS

For convenience in analysis, vibrations of thin shells are considered to be of two classes, namely, extensional and inextensional, depending on whether the middle surface does or does not undergo extension. Actually both types of deformation occur together, and it is only in limiting cases that there is absolute separation into these two classes. However, as a first approximation, it is generally satisfactory to treat the two classes of vibrations independently.

In the following it is assumed that the shell thickness is constant and small compared to the principal radii of curvature of the middle surface. The shell thickness is denoted by h, and the mass of the shell per unit area by μ.

Cylindrical Shells: Extensional Vibrations

Using cylindrical coordinates r, θ, and x, as shown in Fig. 61.6, let

a = mean radius of the cylindrical shell
L = length of longitudinal half wave
n = 0, 1, 2, 3, . . . = number of complete circumferential waves
u, v, w = eigenfunctions representing displacements in the longitudinal, tangential, and radial directions, respectively
U_n, V_n, W_n = constants

Consider an infinitely long cylindrical shell. Assuming the equations of motion to have the form given by Rayleigh [1], it can be shown that the eigenfunctions are

$$\begin{aligned} u &= U_n \cos n\theta \cos (\pi x/L) \\ v &= V_n \sin n\theta \sin (\pi x/L) \\ w &= W_n \cos n\theta \sin (\pi x/L) \end{aligned} \tag{61.116}$$

and that the frequency equation is

$$\left(\lambda^2 - \frac{\pi^2a^2}{L^2} + n^2\right)\left\{(1-\nu)\lambda^2\left[\lambda^2 - 2(1-\nu)\left(\frac{\pi^2a^2}{L^2} + n^2 + 1\right)\right] + (1+\nu)\frac{4\pi^2a^2}{L^2}\right\} + (1+\nu)\frac{4\pi^2a^2n^2}{L^2} = 0 \tag{61.117}$$

where λ is a frequency parameter, which is defined in terms of the circular frequency ω as follows:

$$\omega = \frac{\lambda}{a}\sqrt{\frac{Gh}{\mu}} \tag{61.118}$$

For a given n and a/L, there are three roots, $\lambda_n^{(1)}$, $\lambda_n^{(2)}$, $\lambda_n^{(3)}$, of (61.117) and, hence, three frequencies, $\omega_n^{(1)}$, $\omega_n^{(2)}$, $\omega_n^{(3)}$. Numerical values of λ for a few values of n and

a/L are given in Table 61.5. More extensive tabular data are given by Baron and Bleich [32]. This reference also gives correction factors for the effect of inextensional deformations when the shell bending stiffness is not negligible. For additional data

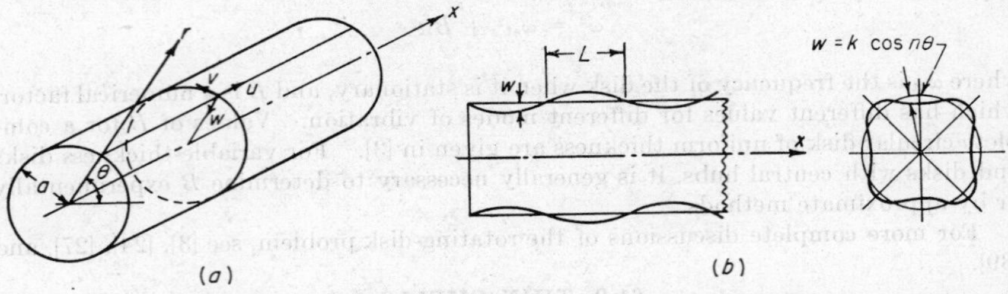

FIG. 61.6. Cylindrical shell: (*a*) Notation and geometry; (*b*) modal shape geometry.

on cylindrical shells, see Arnold and Warburton [30],[31]. The effect of internal pressure on the frequencies is analyzed by Fung, Sechler, and Kaplan [34].

Table 61.5. Frequencies of Infinitely Long Cylindrical Shells

n	L/a	$\lambda_n^{(1)}$	$\lambda_n^{(2)}$	$\lambda_n^{(3)}$
0	1	3.142	1.604	5.338
	2	1.571	1.569	2.729
	4	0.785	1.201	1.783
	8	0.393	0.628	1.704
	∞	0	0	1.691
1	1	1.428	2.784	5.611
	2	0.968	2.109	3.294
	4	0.434	1.501	2.579
	8	0.148	1.169	2.434
	∞	0	1.000	2.391
2	1	1.102	3.840	6.357
	2	0.553	2.709	4.491
	4	0.190	2.231	3.956
	8	0.053	2.064	3.924
	∞	0	2.000	3.780
3	1	0.813	4.458	7.425
	2	0.328	3.477	5.908
	4	0.098	3.135	5.489
	8	0.026	3.035	5.382
	∞	0	3.000	5.346

Cylindrical Shells: Inextensional Vibrations

As shown by Love [2], there are two types of purely inextensional vibrations. The first type is two-dimensional, for which the eigenfunctions are independent of x, and have the form

$$u = 0 \qquad v = V_n \cos n\theta \qquad w = -nV_n \sin n\theta \tag{61.119}$$

The corresponding frequencies are

$$\omega_n{}^2 = \frac{Eh^3}{12(1-\nu)^2\mu a^4}\,\frac{n^2(n^2-1)^2}{n^2+1} \tag{61.120}$$

The second type is three-dimensional, for which the eigenfunctions are

$$u = -\frac{a}{n} V_n \sin n\theta \qquad v = xV_n \cos n\theta \qquad w = -nxV_n \sin n\theta \tag{61.121}$$

For a cylinder with ends at $x = \pm l$, the corresponding frequencies are

$$\omega_n^2 = \frac{Eh^3}{12(1-\nu^2)\mu a^4} \frac{n^2(n^2-1)^2}{n^2+1} \frac{1+(1-\nu)6a^2/n^2l^2}{1+3a^2/(n^4+n^2)l^2} \qquad (61.122)$$

Spherical Shells

For a complete spherical shell, there are no inextensional vibrations, and all modes of vibration are extensional. These extensional modes are of two types, characterized respectively by (1) wholly tangential motion, and (2) partly radial and partly tangential motion. As shown in Love ([2], art. 335) the frequency of modes of the first type is

$$\omega_n^2 = (n-1)(n-2)Gh/\mu a^2 \qquad (61.123)$$

where n is an integer, and a is the mean radius of the spherical shell.

For vibrations of the second type, the frequency equation is

$$\lambda^4 - \lambda^2\left[(n^2+n+4)\frac{1+\nu}{1-\nu} + (n^2+n-2)\right] + 4(n^2+n-2)\frac{1+\nu}{1-\nu} = 0 \qquad (61.124)$$

where λ is defined by (61.118), in which a is the radius of the sphere. For $n > 1$, there are two frequencies for each n. The lowest frequency is for $n = 2$, and is $\omega = 1.176a^{-1}\sqrt{Gh/\mu}$ when $\nu = 0.25$. For purely radial vibrations ($n = 0$), the frequency is

$$\omega = \frac{2}{a}\sqrt{\frac{Gh}{\mu}\frac{1+\nu}{1-\nu}}$$

Purely inextensional vibrations of a segment of a spherical shell of which the edge line is a circle are discussed by Rayleigh ([1], chap. Xa). In that reference it is shown that, for a hemispherical shell, the three lowest frequencies are

$$\omega_1 = 2.14\frac{h}{a^2}\sqrt{\frac{G}{\rho}} \qquad \omega_2 = 6.01\frac{h}{a^2}\sqrt{\frac{G}{\rho}} \qquad \omega_3 = 11.6\frac{h}{a^2}\sqrt{\frac{G}{\rho}}$$

For a "saucer" of 120° total subtended angle, the two lowest frequencies are

$$\omega_1 = 3.27\frac{h}{a^2}\sqrt{\frac{G}{\rho}} \qquad \omega_2 = 8.55\frac{h}{a^2}\sqrt{\frac{G}{\rho}}$$

An analysis and data for shallow spherical shell segments are given by E. Reissner [36].

61.10. SPECIAL METHODS

Rayleigh's Method

This is a procedure for determining an approximate value for the lowest frequency of a conservative elastic system. For vibration in a normal mode with frequency ω, the displacement from the static equilibrium position is $w = W \sin \omega t$, where W is a function of the space coordinates. The maximum kinetic energy is

$$T_{\max} = \omega^2 T^* \quad \text{where} \quad T^* = \tfrac{1}{2}\int \rho W^2\,dv$$

Here dv is a volume element, and the integration is over the volume of the body. The maximum potential energy $V_{\max}$ is a function of W and its space derivatives. Assuming that the zero datum for V is taken at the equilibrium position, conservation of energy requires that $T_{\max} = V_{\max}$, and, hence,

$$\lambda = V_{\max}/T^* \qquad (61.125)$$

where $\lambda = \omega^2$.

The procedure is to assume a function W, and substitute it into (61.125). The calculated value of λ will be greater than or equal to ω_1^2 (the lowest natural frequency squared of the system), provided the assumed function W satisfies the so-called "essential" boundary conditions. *Essential* boundary conditions are those involving only the function and its first $(m - 1)$ derivatives, where $2m$ is the order of the governing differential equation. The practicability of the procedure rests upon the empirical fact that the values of λ are not sensitive to the choice of W so long as it is a reasonable approximation to the true first eigenfunction.

For a more detailed discussion of Rayleigh's method, see [1],[3],[4],[6], and [7].

Example 11. Calculate the lowest frequency of a clamped-free beam with a concentrated mass m_1 at its free end. The mass per unit length of the beam is μ. For the deflection function, assume that

$$W = W_0\left(\frac{3x^2}{2l^2} - \frac{x^3}{2l^3}\right)$$

This is an arbitrary choice; it happens to be the static deflection of the beam due to a concentrated load at its end. Using this, the calculated values are:

$$T^* = \tfrac{1}{2}\mu \int_0^l W^2\,dx + \tfrac{1}{2}m_1W_0^2 = \tfrac{1}{2}W_0^2(\tfrac{33}{140}\mu l + m_1)$$

$$V_{\max} = \tfrac{1}{2}\,EI \int_0^l (W'')^2\,dx = \frac{3EIW_0^2}{2l^3}$$

$$\omega_1^2 \le \frac{V_{\max}}{T^*} = \frac{3EI/l^3}{m_1 + 33\mu l/140}$$

Example 12. Calculate the lowest frequency of a uniform square plate which is clamped at all four edges. Assume that the clamped edges are at $x = \pm b$ and $y = \pm b$ For the deflection function, take

$$W = (x^2 - b^2)^2(y^2 - b^2)^2$$

which is an arbitrary choice that satisfies the boundary conditions. Using this,

$$T^* = \tfrac{1}{2}\mu \int_{-h}^{h}\int_{-h}^{h} W^2\,dx\,dy = \frac{\mu}{2}\left(\frac{256}{315}\right)^2 b^{18}$$

Substituting W into the general expression for V of a plate [Eq. (61.108*b*)],

$$V_{\max} = \tfrac{1}{2}Kb^{14}[81(\tfrac{256}{315})^2]$$

and, hence,

$$\omega_1^2 \le V_{\max}/T^* = 81/K\mu b^4$$

If $a = 2b$ is the length of each side of the square plate, the above gives

$$\omega_1 \le 36\sqrt{K/\mu a^4}$$

This is unusually close to the more exact value, which has been found by S. Tomotika to be $\omega_1 = 35.987\sqrt{K/\mu a^4}$.

The Ritz Method

This is an extension of Rayleigh's method, which provides a means of obtaining a more accurate value of the lowest frequency, and also gives approximations to higher frequencies and eigenfunctions. Choose W as a series of n terms,

$$W = \sum_{i=1}^{n} A_iW_i \qquad (61.126)$$

where the A_i are constants, and the W_i are functions which satisfy at least the essential boundary conditions. Substitute (61.126) into (61.125), and minimize by putting $(\partial/\partial A_i)(V_{max}/T^*) = 0$. Differentiating V_{max}/T^* as a quotient and using (61.125), this minimizing relation becomes

$$\frac{\partial V_{max}}{\partial A_i} - \lambda \frac{\partial T^*}{\partial A_i} = 0 \qquad i = 1, 2, 3, \ldots, n \tag{61.127}$$

The partial derivatives in this expression are linear functions of the A_i, and hence, (61.127) represents a set of n homogeneous equations in the A_i. Setting the determinant equal to zero gives the frequency equation, which has n real roots, $\lambda_1, \lambda_2, \ldots, \lambda_n$. Assuming that these roots are ranked in order of increasing numerical magnitude, it can be shown [7] that these are upper-bound approximations to the first n frequencies squared: that is $\lambda_1 \geq \omega_1^2$, $\lambda_2 \geq \omega_2^2$, etc. Substituting any root λ_j back into (61.127), the equations can be solved to determine the relative magnitudes of the A_i corresponding to the jth mode. Using these values in (61.126), the resulting series defines approximately the jth eigenfunction.

In general, the accuracy of the calculated data is best for the first mode, and degenerates progressively for the higher modes. The highest root λ_n is likely to be a very inaccurate approximation to ω_n^2.

For a more detailed discussion of the Ritz method, see [3],[7],[8], and [29].

Example 13. Consider a clamped-free beam of constant depth, and a width which varies linearly from a maximum at the fixed end to zero at the free end. Taking the origin of coordinates at the fixed end, $I = I_0(1 - x/l)$ and $\mu = \mu_0(1 - x/l)$, where I_0 and μ_0 are respectively the moment of inertia and mass per unit length at the fixed end. Choose a two-term series,

$$W = A_1x^2 + A_2x^3$$

The kinetic and potential energies are:

$$T^* = \tfrac{1}{2} \int_0^l \mu W^2\,dx = \frac{\mu_0}{2} \int_0^l \left(1 - \frac{x}{l}\right) (A_1x^2 + A_2x^3)^2\,dx$$
$$V_{max} = \tfrac{1}{2} \int_0^l EI(W'')^2\,dx = \frac{EI_0}{2} \int_0^l \left(1 - \frac{x}{l}\right) (2A_1 + 6A_2x)^2\,dx$$

Integrating, taking the partial derivatives, and substituting into (61.127) gives the pair of equations,

$$(2 - \beta/30)A_1 + (2 - \beta/42)A_2l = 0$$
$$(2 - \beta/42)A_1 + (3 - \beta/56)A_2l = 0 \tag{61.128a,b}$$

where $\beta = \mu_0 l^4 \lambda / EI_0$. The roots of the frequency equation are $\beta_1 = 51.25$ and $\beta_2 = 1377$, and, hence, the first two frequencies are $\omega_1^2 \leq 51.25EI_0/\mu_0l^4$ and $\omega_2^2 \leq 1377EI_0/\mu_0l^4$

From (61.128a), the amplitude ratio is $lA_2/A_1 = -(2 - \beta/30)/(2 - \beta/42)$. Substituting β_1, and β_2 into this, in turn, gives -0.374 and -1.426, respectively. The approximate eigenfunctions for the first two modes are, thus,

$$(W)_1 = x^2 - 0.374x^3/l \qquad (W)_2 = x^2 - 1.426x^3/l$$

Stodola-Vianello Method

This is an iteration method for determining the frequencies and eigenfunctions of a conservative system. While the procedure converges to the exact solution, in actual applications, where only a limited number of iteration steps are carried out, the results are approximate.

Suppose that the free-vibration problem is governed by the differential equation $M[w] = \omega^2 N[w]$ and the linear homogeneous boundary conditions $B_i[w] = 0$, where M, N, and B_i are linear differential operators and $i = 1, 2, 3, \ldots, m$. The iteration scheme is as follows. Starting with any assumed function $w_0(x)$, solve in succession for $w_1(x)$, $w_2()x$, . . . , using the equation $M[w_k] = N[w_{k-1}]$ and the boundary conditions $B_i[w_k] = 0$. Let

$$\bar{\omega}^2 = \frac{w_{k-1}(x)}{w_k(x)} \tag{61.129}$$

It can be shown [7] that, in the limit as $k \to \infty$, $\bar{\omega}$ approaches the first natural frequency of the system, and w_k approaches the first eigenfunction.

For a finite number of iterations, w_{k-1}, and w_k will not be exactly proportional, and, hence, $\bar{\omega}$ will be a function of x. In such cases, a practical procedure is to use the average values; that is, take

$$\bar{\omega}^2 = \frac{\int_0^l w_{k-1}\,dx}{\int_0^l w_k\,dx} \tag{61.130}$$

The integrations required in solving $M[w_k] = N[w_{k-1}]$ can be carried through graphically in many problems, as shown in [4] and [7]. It is possible to extend the method so as to determine the frequencies and eigenfunctions of higher modes, but the calculations involved are cumbersome and generally impractical.

For more detailed discussion of this method, see [4],[6],[7], and [39].

Example 14. Consider a free-clamped beam for which I and μ vary linearly from zero at the free end to I_0 and μ_0, respectively, at the fixed end. Taking the origin of coordinates at the free end, $I = I_0x/l$ and $\mu = \mu_0 x/l$. The differential equation is $(xw'')'' = \beta^2 xw$, where $\beta^2 = \mu_0\omega^2/EI_0$, and hence, $M[w] = (xw'')''$ and $N[w] = xw$. The iteration equation is $(xw_k'')'' = xw_{k-1}$. The boundary conditions $B_i[w_k] = 0$, where $i = 1, 2, 3, 4$, are $(xw_k'')'_{x=0} = 0$, $(xw_k'')_{x=0} = 0$, $w_k(l) = 0$, and $w_k'(l) = 0$.

To start, choose arbitrarily $w_0 = 1$. The equation for determining w_1 is $(xw_1'')'' = x$. Integrating this equation and satisfying the boundary conditions gives

$$w_1 = \tfrac{1}{72}(x^4 - 4l^3x + 3l^4)$$

The equation for determining w_2 is $(xw_2'')'' = xw_1$. Substituting w_1 into the right-hand side and integrating gives

$$w_2 = \frac{1}{72 \times 11{,}760}(5x^8 - 196l^3x^5 + 490l^4x^4 - 1{,}020l^7x + 721l^8)$$

Stopping the iteration procedure at this stage and substituting w_1 and w_2 into (61.130) gives $\beta^2 = 50.97/l^4$ or $\omega_1^2 \approx 50.97EI_0/\mu_0 l^4$.

An alternative scheme is to substitute w_k into the Rayleigh quotient (61.125) and calculate $\lambda \approx \omega_1^2$ therefrom.

Matrix Iteration

This is a numerical procedure for calculating the frequencies and eigenvectors of a lumped system. To apply it to a continuous system, the system must be approximated by an equivalent lumped system.

Assume a lumped system with s degrees of freedom and static coupling only. Using d'Alembert's principle, the equations of motion are

$$w_i = -\sum_{j=1}^{s} a_{ij}m_j\ddot{w}_j \qquad \text{where } i = 1, 2, \ldots, s$$

In this expression w_j is the displacement of the mass m_j, and a_{ij} is the influence coefficient, which gives the displacement of m_i due to a unit load at m_j. Taking $w_i = A_i \sin \omega t$, the equations of motion reduce to

$$A_i = \omega^2 \sum_{j=1}^{\infty} a_{ij} m_j A_j$$

In matrix notation, this can be written as

$$\{A_i\} = \omega^2[a_{ij}][m_{ii}]\{A_i\} = \omega^2[b_{ij}]A_i \tag{61.131}$$

where $\{A_i\}$ is the column matrix of the A_i, $[a_{ij}]$ is the square matrix of the influence coefficients, $[m_{ii}]$ is the diagonal inertia matrix, and $[b_{ij}] = [a_{ij}][m_{ii}]$ is the dynamic matrix.

The iteration scheme is as follows. Starting with any assumed set of amplitudes $\{(A_i)_0\}$, calculate successively the sets $\{(A_i)_1\}$, $\{(A_i)_2\}$, . . . , using the equation

$$\{(A_i)_k\} = [b_{ij}]\{(A_i)_{k-1}\} \tag{61.132}$$

This iteration scheme is equivalent to the repeated premultiplication of an arbitrary column matrix $\{(A_i)_0\}$ by the dynamic matrix $[b_{ij}]$.

It can be shown (see [6] or [40]) that, in the limit as k approaches infinity,

$$\lim_{k\to\infty} \frac{(A_i)_{k-1}}{(A_i)_k} = \omega_1^2 \qquad \text{and} \qquad \lim_{k\to\infty} (A_i)_k = cA_i^{(1)} \tag{61.133}$$

where ω_1 is the lowest frequency of the lumped system, the $A_i^{(1)}$ are the components of the eigenvector of the first mode, and c is an arbitrary constant factor.

After the frequency and eigenvector components of the first mode are determined, the corresponding results for the second mode can be calculated. This is done by using the orthogonality relation to reduce by one the order of the dynamic matrix. Continuing in the same manner, the third and higher modes can be determined.

For a more detailed discussion of the matrix iteration method, and numerical examples, see [7] and [40].

Myklestad-Prohl Method

This is a numerical method for calculating the frequencies and eigenfunctions of lateral vibrations of a beam. The beam is represented by a lumped system, consisting of N masses concentrated at stations 1, 2, . . . , N along the beam axis. Let

m_n = mass at station n
M_n = bending moment at station n (positive as shown in Fig. 61.7)
S_n = shear just to left of mass m_n (positive as shown in Fig. 61.7)
w_n, θ_n = amplitude of deflection and slope, respectively, at station n
h_n = length of beam segment between stations n and $(n + 1)$
α_n, η_n = angular displacement at station $(n + 1)$ relative to tangent at station n due respectively to unit shear and unit moment at station $(n + 1)$
δ_n, γ_n = linear displacement at station $(n + 1)$ relative to tangent at station n due respectively to unit shear and unit moment at station $(n + 1)$

For example, if the bending stiffness E_nI_n in segment h_n is constant,

$$\alpha_n = \gamma_n = h_n^2/2E_nI_n \qquad \eta_n = h_n/E_nI_n \qquad \delta_n = h_n^3/3E_nI_n$$

The inertia force acting on m_n is $m_n\omega^2 w_n$. Referring to Fig. 61.7, the quantities at station $(n + 1)$ can be found from those at station n by the following relations:

$$
\begin{aligned}
S_{n+1} &= S_n - m_n\omega^2 w_n \\
M_{n+1} &= M_n - h_n S_{n+1} \\
\theta_{n+1} &= \theta_n + \alpha_n S_{n+1} + \eta_n M_{n+1} \\
w_{n+1} &= w_n + h_n\theta_n + \delta_n S_{n+1} + \gamma_n M_{n+1}
\end{aligned}
\qquad (61.134a\text{–}d)
$$

To explain the calculation procedure, consider, for example, a cantilever beam which is free at station 1 and fixed at station N. The boundary conditions are

$$S_1 = M_1 = 0 \qquad \text{and} \qquad w_N = \theta_N = 0$$

Assume a frequency ω. Starting with $w_1 = w_1$, $\theta_1 = \theta_1$, $S_1 = 0$, and $M_1 = 0$, apply (61.134), and calculate successively the quantities at stations 1, 2, and so on, to

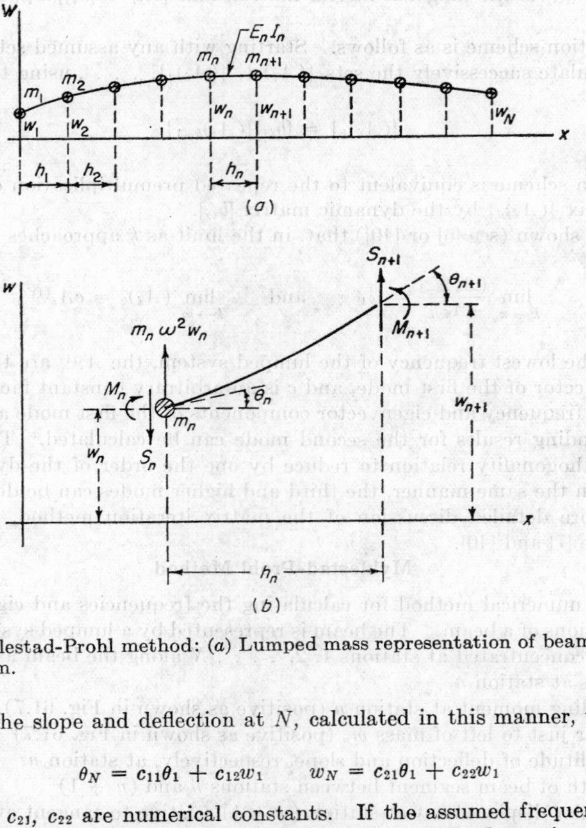

FIG. 61.7. Myklestad-Prohl method: (a) Lumped mass representation of beam; (b) segment of lumped beam.

station N. The slope and deflection at N, calculated in this manner, will have the form

$$\theta_N = c_{11}\theta_1 + c_{12}w_1 \qquad w_N = c_{21}\theta_1 + c_{22}w_1$$

where c_{11}, c_{12}, c_{21}, c_{22} are numerical constants. If the assumed frequency were one of the natural frequencies, θ_N and w_N would be zero, and the determinant

$$\Delta = c_{11}c_{22} - c_{12}c_{21} \qquad (61.135)$$

would vanish. The process is repeated with different values of ω, and the determinant Δ is calculated for each. The natural frequencies are those for which the curve of Δ versus ω passes through zero.

At a natural frequency, $\theta_1 = -c_{12}w_1/c_{11} = -c_{22}w_1/c_{21}$. Substituting this value of θ_1 into the calculations of (61.134) gives the deflections at each station relative to w_1, and, hence, defines the eigenfunction.

For beams with boundary conditions other than those in the above example, the calculations are handled in a similar manner, with obvious changes in the details at each end station, to account for the particular set of boundary conditions.

It will be observed that one of the characteristics of this method is that the calculations for each mode are independent of those for the other modes. Several alternative forms of (61.134) are used in practice.

The rotatory inertia of the lumped masses may be taken into account by adding the term $-J_n\omega^2\theta_n$ to Eq. (61.134*b*), so that it becomes

$$M_{n+1} = M_n - h_nS_{n+1} - J_n\omega^2\theta_n \tag{61.136}$$

where J_n is the mass moment of inertia of m_n.

For more detailed discussion of the Myklestad-Prohl method, see [4], p. 229; [6], p. 163; and [41],[42], and [46].

61.11. REFERENCES

General References

[1] Lord Rayleigh: "The Theory of Sound," 2d ed., Dover, New York, 1945.
[2] A. E. H. Love: "A Treatise on the Mathematical Theory of Elasticity," Dover, New York, 1944.
[3] S. P. Timoshenko: "Vibration Problems in Engineering," 3d ed., Van Nostrand, Princeton, N.J., 1955.
[4] J. P. Den Hartog: "Mechanical Vibrations," 4th ed., McGraw-Hill, New York, 1956.
[5] P. M. Morse: "Vibration and Sound," 2d ed., McGraw-Hill, New York, 1948.
[6] R. H. Scanlon, R. Rosenbaum: "Introduction to the Study of Aircraft Vibration and Flutter," Macmillan, New York, 1951.
[7] L. Collatz: "Eigenwertproblems," Chelsea, New York, 1948.
[8] C. B. Biezeno, R. Grammel: "Engineering Dynamics," transl. from 2d German ed., Van Nostrand, Princeton, N.J., 1954–1956.
[9] J. N. MacDuff, R. P. Felgar: Vibration design charts, *Trans. ASME*, **79** (1957), 1459–1475.
[9*a*] L. S. Jacobsen, R. S. Ayre: "Engineering Vibrations," McGraw-Hill, New York, 1958.
[9*b*] K. N. Tong: "Theory of Mechanical Vibrations," Wiley, New York, 1960.
[9*c*] R. E. D. Bishop, D. C. Johnson: "The Mechanics of Vibrations," Cambridge Univ. Press, New York, 1960.

Beams

[10] R. D. Mindlin, L. E. Goodman: Beam vibrations with time-dependent boundary conditions, *J. Appl. Mechanics*, **17** (1950), 377–380.
[11] E. T. Cranch, A. A. Adler: Bending vibrations of variable section beams, *J. Appl. Mechanics*, **23** (1956), 103–108.
[12] E. T. Kruszewski: Effect of transverse shear and rotary inertia on the natural frequencies of a uniform beam, *NACA Tech. Note* 1909 (1949).
[13] R. A. Anderson: Flexural vibrations in uniform beams according to the Timoshenko theory, *J. Appl. Mechanics*, **20** (1953), 504–510.
[14] G. Herrmann: Forced motions of Timoshenko beams, *J. Appl. Mechanics*, **22** (1955), 53–56.
[15] R. D. Mindlin, H. Deresiewicz: Timoshenko's shear coefficient for flexural vibrations of beams, *Proc. 2d U.S. Natl. Congr. Appl. Mechanics*, 1954, pp. 175–178.
[16] R. T. Yntema: Simplified procedures and charts for the rapid estimation of bending frequencies of rotating beams, *NACA Tech. Note* 3459 (1955).
[17] R. S. Ayre, L. S. Jacobsen: Natural frequencies of continuous beams of uniform span length, *J. Appl. Mechanics*, **17** (1950), 391–395.
[18] R. S. Ayre, L. S. Jacobsen, C. S. Hsu: Transverse vibrations of one or two span beams under the action of a moving mass load, *Proc. 1st U.S. Natl. Congr. Appl. Mechanics*, 1951, pp. 81–90.
[19] D. Young, R. P. Felgar: Tables of characteristic functions representing the normal modes of vibration of a beam, *Univ. Texas Eng. Research Ser.*, **44** (1949).

Plates

[20] B. W. Anderson: Vibration of triangular cantilever plates by the Ritz method, *J. Appl. Mechanics*, **21** (1955), 365–370.
[21] M. V. Barton: Vibration of rectangular and skew cantilever plates, *J. Appl. Mechanics*, **18** (1954), 129–134.
[22] R. F. S. Hearmon: Frequency of vibration of rectangular isotropic plates, *J. Appl. Mechanics*, **19** (1952), 402–403.
[23] N. J. Huffington, W. H. Hoppman: On the transverse vibrations of rectangular orthotropic plates, *J. Appl. Mechanics*, **25** (1958), 389–395.
[24] H. Lamb, R. V. Southwell: The vibrations of a spinning disk, *Proc. Roy. Soc. London*, **A,99** (1921), 272.
[25] R. D. Mindlin: Influence of rotatory inertia and shear on flexural motions of isotropic, elastic plates, *J. Appl. Mechanics*, **18** (1951), 31–38.
[26] R. D. Mindlin, A. Schacknow, H. Deresiewicz: Flexural vibrations of rectangular plates, *J. Appl. Mechanics*, **23** (1956), 430–436.
[27] J. Prescott: "Applied Elasticity," chap. 18, Vibrations of rotating disks, Dover, New York, 1946.
[28] G. B. Warburton: The vibration of rectangular plates, *Proc. Inst. Mech. Engrs. London*, **168** (1954), 371–384.
[29] D. Young: Vibration of rectangular plates by the Ritz method, *J. Appl. Mechanics*, **17** (1950), 448–453.

Shells

[30] R. N. Arnold, G. B. Warburton: Flexural vibrations of the walls of thin cylindrical shells having freely supported ends, *Proc. Roy. Soc. London*, **A,197** (1949), 238.
[31] R. N. Arnold, G. B. Warburton: The flexural vibrations of thin cylinders, *Proc. Inst. Mech. Engrs. London*, **167** (1953), 62–74.
[32] M. L. Baron, H. H. Bleich: Tables for frequencies and modes of free vibrations of infinitely long cylindrical shells, *J. Appl. Mechanics*, **21** (1954), 178–184.
[33] W. Flügge: "Statik und Dynamik der Schalen," pp. 263–272, 2d ed., Springer, Berlin, 1957.
[34] Y. C. Fung, E. E. Sechler, A. Kaplan: On the vibration of thin cylindrical shells under internal pressure, *J. Aeronaut. Sci.*, **24** (1957), 650–660.
[35] P. M. Naghdi, J. G. Berry: On the equations of motion of cylindrical shells, *J. Appl. Mechanics*, **21** (1954), 160–166.
[36] E. Reissner: On transverse vibrations of thin shallow elastic shells, *Quart. Appl. Math.*, **13** (1955), 169–176.
[37] E. Reissner: On axi-symmetrical vibrations of shallow spherical shells, *Quart. Appl. Math.*, **13** (1955), 279–290.
[38] Y. Y. Yu: Free vibrations of thin cylindrical shells having finite lengths with freely supported and clamped edges, *J. Appl. Mechanics*, **22** (1955), 547–552.

Special Methods

[39] A. Stodola: "Steam Turbines," transl. by L. C. Loewenstein, McGraw-Hill, New York, 1945.
[40] R. A. Frazer, W. J. Duncan, A. R. Collar: "Elementary Matrices," Macmillan, New York, 1947.
[41] N. O. Myklestad: "Fundamentals of Vibration Analysis," McGraw-Hill, New York, 1956.
[42] W. T. Thomson: Matrix solution for the vibration of nonuniform beams, *J. Appl. Mechanics*, **17** (1950), 337–339.
[43] R. V. Churchill: "Operational Mathematics," 2d ed., McGraw-Hill, New York, 1958.
[44] W. T. Thomson: "Laplace Transformation," Prentice-Hall, Englewood Cliffs, N.J., 1950.
[45] J. C. Houbolt, R. A. Anderson: Calculation of uncoupled modes and frequencies in bending or torsion of nonuniform beams, *NACA Tech. Note* 1522 (1948).
[46] L. S. Dzung: Influence matrix of beam vibration, *J. Aeronaut. Sci.*, **20** (1953), 437.

CHAPTER 62

DYNAMIC BUCKLING

BY

E. METTLER, Dr. rer. techn., Karlsruhe, Germany

62.1. INTRODUCTION

It is well known that, under a static load, a thin elastic body, such as a column, a plate, or a shell subjected to direct stresses, may deflect laterally when a condition of neutral elastic equilibrium has been reached. When the load is further increased, the elastic body leaves its former, now unstable position, in order to assume a new, supercritical position. The stability problems connected with this phenomenon are of two distinctly different types: bifurcation problems and snap-through problems (see p. 44–2). Although the critical load of bifurcation problems (e.g., the Euler load of a column) is an eigenvalue of a linear differential equation, the study of snap-through problems and of the supercritical behavior in bifurcation problems requires the use of nonlinear theory.

The problem of dynamic buckling is obtained by simply replacing a static load with a time-dependent one. In doing this, we assume, in agreement with nearly all cases so far treated, that the load varies harmonically. Then the forced vibrations of the elastic body take the place of the static equilibrium displacements. Neutral points and supercritical conditions occur in analogy to the static phenomena, although naturally the relations become more complicated, since time enters as an additional variable. In particular, in the dynamic case, the supercritical phenomena, in the form of nonlinear vibrations, play a more important role than in the static case.

62.2. DYNAMIC STABILITY OF A BEAM UNDER PLANE MOTION

Equations of Motion

As the first and simplest problem, we consider the plane motion of a simply supported bar of constant cross section (Fig. 62.1). In order to derive the two possible

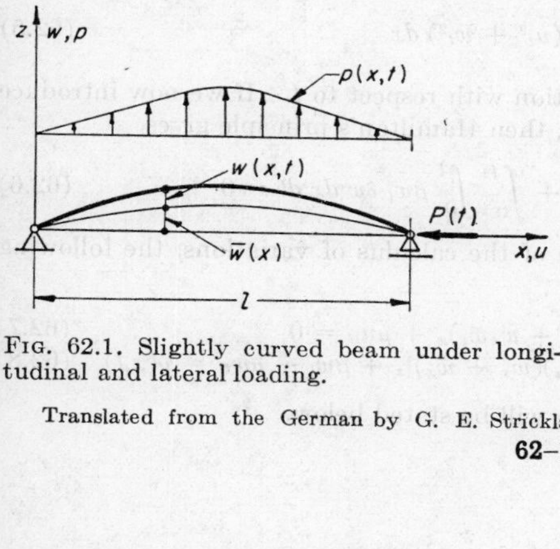

FIG. 62.1. Slightly curved beam under longitudinal and lateral loading.

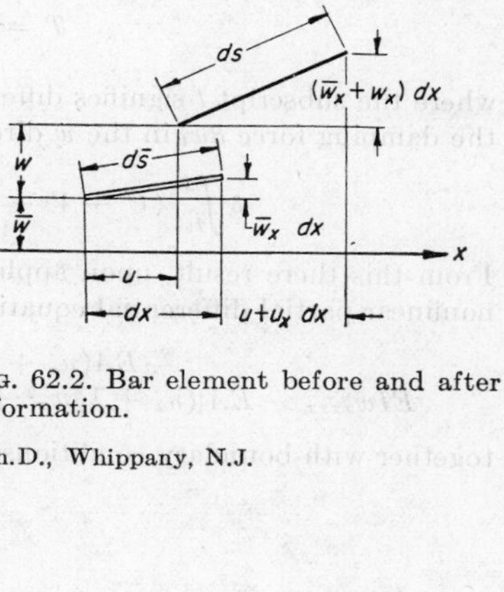

FIG. 62.2. Bar element before and after deformation.

Translated from the German by G. E. Strickland, Ph.D., Whippany, N.J.

cases of dynamic instability from the same equations, we assume that the axis of the unloaded beam has a small deflection $\bar{w}(x)$, on which the dynamic deflection $w(x,t)$ is superposed. In addition, we assume that there is a lateral loading $p(x,t)$, as well as an axial load $P(t)$. For this problem, the differential equations of motion will be derived from Hamilton's principle, incorporating the most important nonlinear terms of the elastic equations [3].

Before deformation, a line element of the axis of the bar (Fig. 62.2) has the length

$$d\bar{s} = (1 + \bar{w}_x^2)^{1/2}\, dx$$

When the bar has been deformed, the length is

$$ds = [(1 + u_x)^2 + (\bar{w}_x + w_x)^2]^{1/2}\, dx$$

Here a subscript x indicates a derivative with respect to x.

The extension of the bar axis is

$$\epsilon_0 = (ds - d\bar{s})/d\bar{s}$$

or, after an expansion up to second-order terms,

$$\epsilon_0 = u_x + \tfrac{1}{2}w_x^2 + w_x\bar{w}_x \tag{62.1}$$

The additional extension of a fiber a distance z from the axis is, as in the usual theory of the bending of beams, $\epsilon_1 = -zw_{xx}$, and the total extension is, therefore,

$$\epsilon = \epsilon_0 + \epsilon_1 \tag{62.2}$$

Under the simplifying assumptions of the usual beam bending theory, which considers only extensions in the x direction, and which takes Hooke's law as a basis, the strain energy becomes

$$U = \frac{E}{2}\int_V \epsilon^2\, dV$$

where we integrate over the volume V of the beam. Upon the insertion of (62.2) and the execution of partial integration over the cross section of the beam, we have

$$U = \frac{EA}{2}\int_0^l (u_x + \tfrac{1}{2}w_x^2 + w_x\bar{w}_x)^2\, dx + \frac{EI}{2}\int_0^l w_{xx}^2\, dx \tag{62.3}$$

The potential energy of the external forces is

$$V = -\int_0^l pw\, dx + Pu(l,t) \tag{62.4}$$

where P is taken as positive when compressive, and the kinetic energy is

$$T = \frac{\mu}{2}\int_0^l (u_t^2 + w_t^2)\, dx \tag{62.5}$$

where the subscript t signifies differentiation with respect to t. If we now introduce the damping force βw_t in the w direction, then Hamilton's principle gives

$$\delta \int_{t_1}^{t_2} (U + V - T)\, dt + \int_{t_1}^{t_2}\int_0^l \beta w_t\, \delta w\, dx\, dt = 0 \tag{62.6}$$

From this there result, upon application of the calculus of variations, the following nonlinear partial differential equations:

$$-EA(u_x + \tfrac{1}{2}w_x^2 + w_x\bar{w}_x)_x + \mu u_{tt} = 0 \tag{62.7}$$

$$EIw_{xxxx} - EA[(u_x + \tfrac{1}{2}w_x^2 + w_x\bar{w}_x)(w_x + \bar{w}_x)]_x + \beta w_t + \mu w_{tt} = p(x,t) \tag{62.8}$$

together with boundary conditions which will be stated below.

Snap-through of the Slightly Curved, Transversely Loaded Beam

The dynamic analog to the static snap-through phenomenon of a slightly curved beam is obtained [7], [11] if the ends of the beam are pinned at $x = 0$ and at $x = l$. Then only the transverse load $p(x,t)$ comes into play, while the P term in (62.4) vanishes. The boundary conditions for Eqs. (62.7) and (62.8) then become

$$u = w = w_{xx} = 0 \qquad \text{for } x = 0 \text{ and } x = l \tag{62.9}$$

To simplify the calculation, we now assume that ϵ_0 does not depend on x, an assumption which may easily be justified from the result, with the help of Eq. (62.7). It then follows, from (62.1) and (62.9), that

$$\epsilon_0 = \frac{1}{l}\int_0^l \epsilon_0\, dx = \frac{1}{l}\int_0^l (\tfrac{1}{2}w_x{}^2 + w_x\bar{w}_x)\, dx$$

and, when this and Eq. (62.1) are inserted into Eq. (62.8), the following integro-differential equation for w is obtained:

$$EIw_{xxxx} - \frac{EA}{l}\left[\int_0^l (\tfrac{1}{2}w_x{}^2 + w_x\bar{w}_x)\, dx\right](w_{xx} + \bar{w}_{xx}) + \mu w_{tt} = p \tag{62.10}$$

Here, for the sake of brevity, the damping term has been dropped ($\beta = 0$).

The solution of (62.10) will be explained here for the simple case that

$$\bar{w} = w_0 \sin\frac{\pi x}{l} \qquad p = p_0 \sin\frac{\pi x}{l}\cos \omega t \tag{62.11}$$

with constant w_0 and p_0. For functions other than (62.11), a solution is possible, but more complicated [11]. With the assumption

$$w = q(t)\sin\frac{\pi x}{l} \tag{62.12}$$

Eq. (62.10) is reduced to an ordinary differential equation for $q(t)$:

$$\ddot{q} + \omega_1{}^2 q + aq^3 + bq^2 = \frac{p_0}{\mu}\cos \omega t \tag{62.13}$$

where

$$\omega_1{}^2 = \frac{1}{\mu}\left(\frac{\pi}{l}\right)^4 (EI + \tfrac{1}{2}EAw_0{}^2) \qquad a = \frac{1}{\mu}\left(\frac{\pi}{l}\right)^4 \frac{EA}{4} \qquad b = 3aw_0$$

Following Duffing, an approximate solution may be assumed in the form

$$q = S\cos \omega t + S_1$$

Introducing this into (62.13), and equating to zero the constant term and the coefficient of $\cos \omega t$, we obtain an approximate relation among S, p_0, and ω:

$$S(\omega^2 - \omega_1{}^2) + cS^3 = -p_0/\mu \tag{62.14}$$

Here the constant

$$c = \frac{b^2}{\omega_1{}^2} - \tfrac{3}{4}a$$

is employed, where $c \gtreqless 0$ for $w_0 \gtreqless i_y/\sqrt{2.5}$, and i_y is the radius of gyration of the cross section.

Equation (62.14) may be represented as a surface in the coordinate system ($S, p_0, \omega^2/\omega_1{}^2$). A section $p_0 = \text{const}$ represents the amplitude-frequency curve of an ordinary nonlinear oscillator. Figure 62.3*a* shows the curve for the case $c > 0$, which leads to the characteristic for a softening restoring force.[1] The relationship is understandable, for (62.13) is indeed the differential equation of a nonlinear vibration, but with the peculiarity that both a cubic and a quadratic restoring force occur. Hence, we also know that, in Fig. 62.3*a*, stable and unstable branches are present. At the place where the tangent is perpendicular, there lies a neutral point, and, when S and ω reach this point from small values, the well-known *jump phenomenon* occurs, in which S jumps with a change of sign to a greater magnitude.

When we set ω = const, Eq. (62.14) yields a curve in the (S,p_0) system, as shown in Fig. 62.3*b*, which is directly analogous to the load-displacement curve of the static snap-through problem: When p_0 and S increase from zero, a neutral point with horizontal tangent is reached, from which the unstable branch proceeds downward, while the vibration S jumps over (snaps through) to a stable value, as indicated by the arrow.

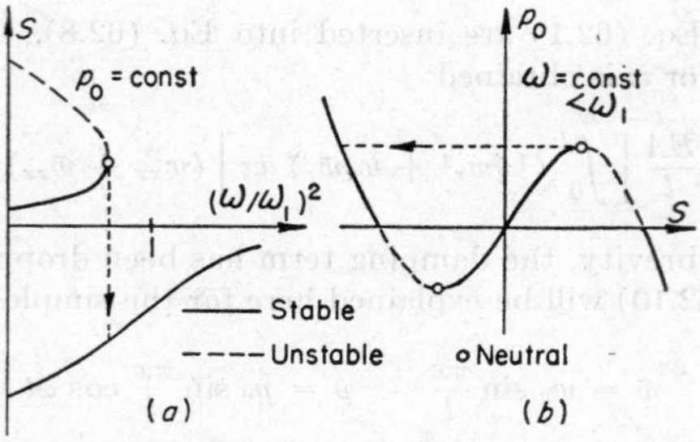

FIG. 62.3. Snap-through of a beam: (*a*) Amplitude-frequency diagram; (*b*) amplitude-load diagram, shown for $c > 0$.

It is noteworthy that the static snap-through phenomenon occurs only for $w_0 \neq 0$, while the dynamic snap-through phenomenon is still obtainable as a special case for $w_0 = 0$ (a straight beam). On the other hand, the dynamic phenomenon disappears for $w_0 = (2.5)^{-1/2} i_y$ in our approximation.

Buckling of a Beam under Periodic Axial Forces; the Linearized Bifurcation Problem

The problem of simple dynamic beam buckling occurs when we put $\bar{w} = 0$ (straight beam) and $p = 0$ in (62.7) and (62.8). It is now assumed that $P(t) \neq 0$, and that the support at $x = l$ is axially unrestrained. The boundary conditions for (62.7) and (62.8) are now

$$\begin{aligned} u = w = w_{xx} &= 0 \qquad \text{for } x = 0 \\ EA(u_x + \tfrac{1}{2}w_x^2) + P = w = w_{xx} &= 0 \qquad \text{for } x = l \end{aligned} \tag{62.15}$$

It can be seen immediately that the straight beam can theoretically remain straight and, hence, perform purely axial vibrations. However, these do not always turn out to be stable. On the contrary, the axial force can, under certain conditions, produce *transverse* vibrations, which correspond to the deflected equilibrium positions of static buckling.

As long as we are interested only in the stability of the longitudinal vibrations [1],[2],[4],[9], we may linearize the basic equations (62.7) and (62.8), just as is done in the static buckling problem. In (62.7) and (62.15), we strike out the nonlinear term $w_x^2/2$, and obtain a linear differential equation with linear boundary conditions for the function $u(x,t)$. The longitudinal motion thereby determined we call the *primary*

[1] For $c < 0$, the characteristic for a hardening restoring force results.

motion. Hence, corresponding to the usual procedure in stability theory, it is calculated on a linearized basis. In (62.8), we likewise strike out the term $w_x^2/2$, but we retain the important quadratic term $u_x w_x$, and insert therein for u_x the solution already obtained. Then (62.8) is an equation for w alone, a linear differential equation with time-dependent coefficients. Its solution is called the *secondary motion*.

To simplify the problem, we assume that

$$P(t) = P_0 + P_1 \cos \omega t \qquad \text{with constant } P_0, P_1 \tag{62.16}$$

$$\omega \ll \text{fundamental frequency of longitudinal vibration} \tag{62.17}$$

Assumption (62.17) implies that we can make the approximation

$$u = -\frac{P(t)}{EA} x$$

an approximation whose accuracy is controllable. Equation (62.8) now becomes

$$EIw_{xxxx} + P(t)w_{xx} + \beta w_t + \mu w_{tt} = 0 \tag{62.18}$$

and transforms under the assumption

$$w = q(t) \sin \frac{k\pi x}{l} e^{-\beta t/2} \qquad k = 1, 2, \ldots \tag{62.19}$$

into an ordinary differential equation for $q(t)$:

$$\ddot{q} + \omega_k^2 \left(1 - \frac{\beta^2}{4\omega_k^2} - \alpha \cos \omega t\right) q = 0 \tag{62.20}$$

where

$$\omega_k = \frac{k\pi}{l} \left[\frac{1}{\mu}(k^2 P_E - P_0)\right]^{1/2} \tag{62.20a}$$

are the frequencies of the lateral vibration,

$$\alpha = P_1/(k^2 P_E - P_0) \tag{62.20b}$$

is a measure of the amplitude of the exciting force, and $P_E = (\pi/l)^2 EI$ is the Euler load. It is assumed that $\beta^2 \ll \omega_k^2$.

Equation (62.20) is the Mathieu differential equation, whose mathematical theory is well known. The solution of (62.20) has, in general, the form

$$q = c_1 e^{\kappa\omega t}\phi_1(t) + c_2 e^{-\kappa\omega t}\phi_2(t) \tag{62.21}$$

where c_1 and c_2 are constants of integration, and ϕ_1 and ϕ_2 are functions with period $2\pi/\omega$. The *characteristic exponent* κ is, in general, a complex number, whose real part $\mathcal{R}\,\kappa$ determines the stability conditions; for $\omega|\mathcal{R}\,\kappa| - \beta/2 > 0$ (<0), the disturbed motion grows (dies out), and the primary motion u, that is, the forced axial vibration of the beam, is unstable (stable).

We shall not enter here into the various mathematical methods for the determination of the characteristic exponent and of the stability conditions for the solution. Only the most important results will be set forth briefly. These results can be surveyed best by means of the Strutt diagram (Fig. 62.4). There the regions in the plane of the coordinates ω/ω_k and α, in which instability occurs, are indicated by shading. The heavy lines bound the regions for zero damping ($\beta = 0$). We see that the tips of these instability regions impinge upon the (ω/ω_k) axis at the points

$$\omega/\omega_k = 2/n \qquad n = 1, 2, \ldots \tag{62.22}$$

The lightly drawn lines give the boundaries of the regions for a certain $\beta \neq 0$. Damping, therefore, makes the instability regions smaller and, in particular, elevates their lowest points to certain distances α_0 above the ω/ω_k axis. These α_0 are called the threshold values.

Outside the instability regions, the bending vibrations die out, while, within, they increase; on the boundary curves, a strictly periodic bending vibration w is possible. The frequency of this vibration is equal to $\omega/2$ for regions with odd n [compare (62.22)], and is equal to ω for regions with even n. The points of the boundary curves, therefore, have the significance of bifurcation points, where the purely axial vibration may be accompanied by a secondary lateral vibration. Details on this follow in the next section.

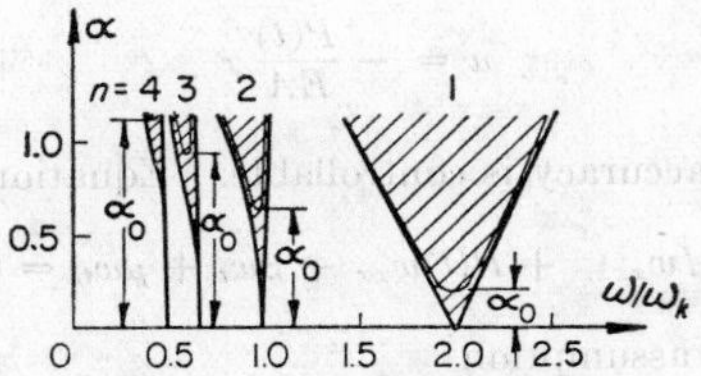

FIG. 62.4. Strutt diagram: instability regions (shaded) for the column with oscillating axial load, shown up to $n = 4$.

The following are the most important quantitative facts. For small α (in the case of zero damping, $\beta = 0$), the boundary curves of the first two regions ($n = 1$ and $n = 2$) have the equations

$$\frac{\omega}{\omega_k} = 2 \pm \frac{\alpha}{2} - \frac{\alpha^2}{32} + O(\alpha^3) \qquad \text{for } n = 1$$

$$\frac{\omega}{\omega_k} = 1 + \frac{\alpha^2}{24} + O(\alpha^3) \qquad \frac{\omega}{\omega_k} = 1 - \tfrac{5}{24}\alpha^2 + O(\alpha^3) \qquad \text{for } n = 2$$

With damping ($\beta \neq 0$), the first and most important region ($n = 1$) has, somewhat less accurately, the boundary equation

$$\left(\frac{\omega}{\omega_k}\right)^2 = 4 \pm 2\left[\alpha^2 - \left(\frac{\beta}{\mu\omega_k}\right)^2\left(\frac{\omega}{\omega_k}\right)^2\right]^{1/2} \tag{62.23}$$

The threshold values, which demonstrate the effect of damping, are

$$\alpha_0 = 2\left(\frac{\beta}{\mu\omega_k}\right)^{1/n}$$

We can calculate the characteristic exponent κ from (62.36) when we put $N = 1$ there.

Buckling of a Column under Oscillating Axial Load, Nonlinear Lateral Vibrations

The linear theory thus far presented yields stability rules and bifurcation points. In the case of instability, however, it leads to lateral vibrations, whose amplitude increases beyond all bounds. Since such vibrations cannot really occur, we seek stationary bending vibrations of finite amplitude. These are the analogs to the supercritical equilibrium conditions in static instability, and can be derived only with a consideration of nonlinear terms [1],[8],[10].

For that reason, we return to the nonlinear equations (62.7) and (62.8) with $\bar{w} = 0$ and $p = 0$, and apply the previously used assumption that ϵ_0 can be taken approximately independent of x. This is obviously correct when assumption (62.17) is correct. We must now distinguish between two cases whose distinction was not necessary in the linear theory. We prescribe, at $x = l$, either (1) the end displacement,

$$u(l,t) = u_0 + u_1 \cos \omega t \tag{62.24}$$

or (2) the end force,

$$P(t) = P_0 + P_1 \cos \omega t \tag{62.16}$$

Case 1 will be treated first. It may be, at least approximately, satisfied in many practical cases, as, for example, when the bar is part of a larger vibrating structure. The boundary conditions then read

$$u = w = w_{xx} = 0 \qquad \text{at } x = 0$$
$$w = w_{xx} = 0 \qquad u = u(l,t) \qquad \text{at } x = l$$

From integrating (62.1), there follows

$$\int_0^l \epsilon_0 \, dx = \epsilon_0 l = u(l,t) + \tfrac{1}{2} \int_0^l w_x{}^2 \, dx \tag{62.25}$$

and, with (62.1) and (62.25), there follows, from (62.8),

$$EIw_{xxxx} - \frac{EA}{l} \left[u(l,t) + \frac{1}{2} \int_0^l w_x{}^2 \, dx \right] w_{xx} + \beta w_t + \mu w_{tt} = 0 \tag{62.26}$$

We solve this integrodifferential equation for the fundamental mode of the beam through assumption (62.12), and so obtain the differential equation for $q(t)$:

$$\ddot{q} + \frac{\beta}{\mu} \dot{q} + \omega_1{}^2 q(1 - \alpha \cos \omega t + \gamma q^2) = 0 \tag{62.27}$$

with

$$\gamma = \frac{A}{4I(1 - P_0/P_E)} \qquad P_0 = -\frac{EAu_0}{l} \qquad P_1 = -\frac{EAu_1}{l}$$

ω_1 and α are given by (62.20a,b) with $k = 1$.

The periodic solution of Eq. (62.27) can be obtained in various ways, for example, with the help of the method of slowly varying phase and amplitude. For the first and most important instability region ($n = 1$), we try

$$q = S_1 \cos \frac{\omega t}{2} + S_2 \sin \frac{\omega t}{2} \tag{62.28}$$

as an approximation for the solution, since the lateral vibration has half the frequency of the excitation. There then follows a relation involving the amplitude

$$S = (S_1{}^2 + S_2{}^2)^{1/2}$$

the excitation frequency ω, and the parameter α, which we recall is a measure of the load:

$$3\gamma S^2 = \left(\frac{\omega}{\omega_1}\right)^2 - 4 \pm 2 \left[\alpha^2 - \left(\frac{\beta}{\mu\omega_1}\right)^2 \left(\frac{\omega}{\omega_1}\right)^2 \right]^{1/2} \tag{62.29}$$

For a given α and ω, it determines the amplitude. For $S = 0$, this equation reduces, as it must, to Eq. (62.23) for the boundary of the first instability region.

For $S \neq 0$ and $\alpha = \text{const}$, Eq. (62.29) is represented graphically in Fig. 62.5 for various values of the damping constant β. The curves should be interpreted as response curves for the lateral vibration. As a more exact analysis shows, the solid lines correspond to stable, and the broken lines to unstable bending vibrations. The

ω axis ($S = 0$) designates the straight beam, and, hence, the primary motion, which is stable on the heavily drawn portion of the axis, that is, outside the instability region. On the boundary of the region lie bifurcation points where the amplitude curves of the bending vibration branch off, stable to the left and unstable to the right. The bending vibrations are known as secondary vibrations, and, in particular, as such of the first kind, since they are simply periodic. With another terminology we may speak of "parametrically excited" vibrations in connection with Eq. (62.27) from which they are derived.

The derived secondary vibrations differ from linear forced vibrations in two respects: According to (62.28), they have a frequency different from—in fact one-half of—the exciting frequency, and their response curve has an entirely different shape.

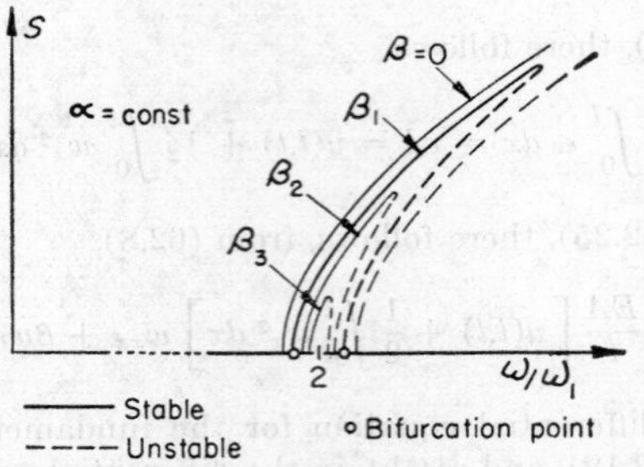

FIG. 62.5. Response curves for the lateral vibrations of a beam with prescribed axially oscillating end displacements and with various damping constants: $0 < \beta_1 < \beta_2 < \beta_3$.

For the second instability region ($n = 2$), similar, though more complicated relations and a diagram corresponding to Fig. 62.5 may be obtained, with a solution of the form

$$q = S_0 + S_1 \cos \omega t + S_2 \sin \omega t$$

In case 2, the end force $P(t)$ is prescribed according to (62.16), and so the boundary conditions (62.14) apply, and the differential equation (62.8) transforms, under the assumption $\epsilon_0 = \epsilon_0(t)$, into the previously treated equation (62.18). Thus, in this case, our considerations do not lead us beyond the linear theory.

In any event, assumption 2 is scarcely realizable in this form. If we wish, say, in an experiment (such as has actually been carried out in various places, see [1] and [9]), to impose a definite force on the end of the beam, then we always need an end mass m (Fig. 62.6) to transmit the force. The situation is similar in practice.

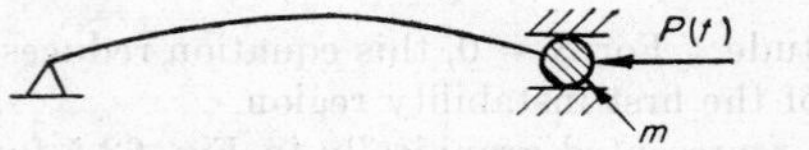

FIG. 62.6. Bar with end mass m under oscillating axial load.

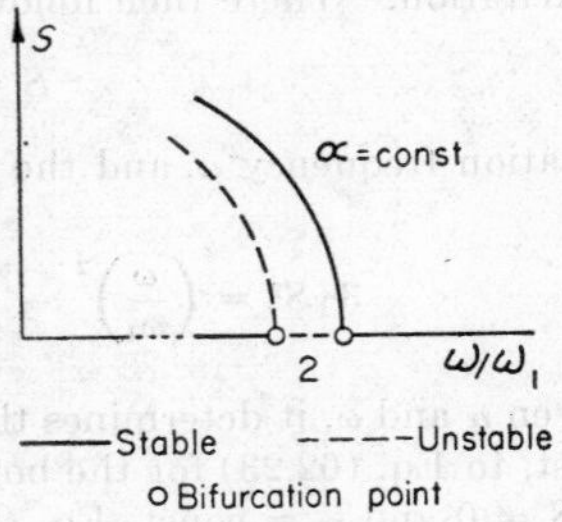

FIG. 62.7. Response curve for the bending vibration of a beam with end mass and prescribed oscillating axial load, but without damping.

Mechanically, however, that means an actual alteration of the problem, since the boundary conditions (62.15) for $x = l$ must now be replaced by

$$EA(u_x + \tfrac{1}{2}w_x^2) + P(t) + mu_{tt} = w = w_{xx} = 0$$

The further development of the theory of case 2 leads to the result that the secondary bending vibration has a sublinear character. For vanishing damping ($\beta = 0$), assumption (62.12), together with $q = S \cos(\omega t/2)$ or $q = S \sin(\omega t/2)$, leads to the relation

$$\frac{ml}{4\mu}\left(\frac{\pi}{l}\right)^4 S^2 = -\frac{\omega}{\omega_1} + 2 \pm \frac{\alpha}{2}$$

which is represented in Fig. 62.7 for α = const.

62.3. ADDITIONAL TOPICS

General Theory of Dynamic Stability[1]

The following discussion applies quite generally to a three-dimensional finite deformable body of isotropic elastic material, which is in a state of forced vibrations. The principal subject of the theory is the stability of these vibrations [1],[5],[6].

We describe the deformation of the body according to the Lagrangian method in a cartesian coordinate system: Before the deformation, a point has the coordinates x_i; afterward, $(x_i + u_i)$, where u_i is the displacement vector; and we use the x_i as independent variables. We also define the body force X_i, the external surface force p_i, and the density ρ, all with respect to the undeformed body. Damping is assumed to be absent. Then Hamilton's principle (62.6) reads

$$\delta \int_{t_1}^{t_2} (U + V - T)\, dt = 0 \qquad (62.30)$$

with the kinetic energy

$$T = \frac{1}{2} \iiint \rho \frac{\partial u_i}{\partial t} \frac{\partial u_i}{\partial t}\, dx_1\, dx_2\, dx_3$$

the potential of the applied forces

$$V = -\textstyle\iiint X_i u_i\, dx_1\, dx_2\, dx_3 - \iint p_i u_i\, dS \qquad (62.31)$$

(dS = an element of the outer surface), and the strain energy

$$U = \textstyle\iiint \bar{U}(\epsilon_{ij})\, dx_1\, dx_2\, dx_3 \qquad (62.32)$$

The volume integrals are to be taken over the entire volume, and the surface integral in (62.31) is to be taken over those parts of the bounding surface on which forces are given. In (62.32), $\bar{U}(\epsilon_{ij})$ is a function of the components of the strain tensor

$$\epsilon_{ij} = \tfrac{1}{2}(u_{i,j} + u_{j,i} + u_{r,i}u_{r,j})$$

and represents, in a reversible adiabatic deformation, the internal-energy density; in an isothermal deformation, the free-energy density.

We expand $\bar{U}$ as a power series in ϵ_{ij}:

$$\bar{U} = \tfrac{1}{2} c_{ijkl}\epsilon_{ij}\epsilon_{kl} + \tfrac{1}{3} c_{ijklmn}\epsilon_{ij}\epsilon_{kl}\epsilon_{mn} + \cdots \qquad (62.33)$$

in which the coefficients of the quadratic terms follow from Hooke's law as

$$c_{ijkl} = G\left(\delta_{ik}\,\delta_{jl} + \delta_{jk}\,\delta_{il} + \frac{2\nu}{1 - 2\nu}\,\delta_{ij}\,\delta_{kl}\right)$$

[1] In this section, cartesian tensor notation (see Chap. 7) is used. All Latin subscripts run from 1 to 3.

Let the applied loads be

$$X_i = X_i' + \alpha X_i'' \cos \omega t \qquad p_i = p_i' + \alpha p_i'' \cos \omega t$$

where the X_i', X_i'', p_i', p_i'' are functions of position only. These loads induce certain displacements $\mathring{u}_i$ in the elastic body, and we assume that they can be calculated from the linearized theory of elasticity. They can be written in the form

$$\mathring{u}_i = u_i' + u_i'' \alpha \cos \omega t$$

and represent the primary vibration whose stability is to be investigated. We now consider the displacement

$$u_i = \mathring{u}_i + \bar{u}_i$$

where now $\bar{u}_i$ denotes the secondary motion, and we pose the problem of calculating $\bar{u}_i$. This is made possible by the mixed Ritz expression

$$\bar{u}_i = \sum_{\xi=1}^{N} \overset{\xi}{U}_i(x_1,x_2,x_3)q_\xi(t)$$

in which the $\overset{\xi}{U}_i$ represent the N first fundamental modes of vibration of the body under the given static load. The Ritz expression is inserted into the variational equation (62.30), where, however, only the quadratic terms are retained because of the smallness of $\bar{u}_i$. Under the simplifying assumption of the classical basic theory, and after carrying out the volume integration and the variation, we obtain a system of differential equations for the time functions $q_\xi(t)$:

$$\ddot{q}_\xi + \omega_\xi{}^2 q_\xi + \alpha \cos \omega t \sum_{\eta=1}^{N} F_{\xi\eta}q_\eta = 0 \qquad \xi = 1, 2, \ldots, N \tag{62.34}$$

The constants $F_{\xi\eta}$ are

$$F_{\xi\eta} = \iiint d_{ijkl}\overset{\xi}{U}_{j,i}\overset{\eta}{U}_{k,l}\, dx_1\, dx_2\, dx_3$$

where $$d_{ijkl} = c_{ijkl} + (2c_{ijklmn} + \delta_{jk}c_{ilmn})u''_{m,n}$$

ω_ξ denotes the ξth eigenfrequency of the statically loaded body.

The system of equations (62.34) is a generalization of Eq. (62.20). According to the Floquet theory of linear differential equations with periodic coefficients (see p. 10–16), it has the general solution

$$q_\xi = \sum_{\eta=1}^{N} [c_\eta e^{\kappa_\eta \omega t}\phi_{\xi\eta}(t) + c_{\eta+N}e^{-\kappa_\eta \omega t}\phi_{\xi\eta}(-t)] \tag{62.35}$$

Only in special cases which are not of interest here does the solution have a different form. The c_η are integration constants; the $\phi_{\xi\eta}(t)$, periodic functions of t; and the κ_η, the characteristic exponents. The κ_η obviously determine, by their real parts, the stability or instability of the primary motion, and must be calculated first of all. That can be done on the basis of the following formulas. Let the functions $q_{\xi\eta}(t)$ be n particular solutions of (62.34) which satisfy, for $t = 0$, the initial conditions,

$$q_{\xi\eta}(0) = \begin{cases} 1 \text{ for } \xi = \eta \\ 0 \text{ for } \xi \neq \eta \end{cases} \qquad \frac{1}{\omega}\dot{q}_{\xi\eta}(0) = 0 \qquad \xi, \eta = 1, \ldots, N$$

Furthermore, let us put

$$q_{\xi\eta}\left(\frac{2\pi}{\omega}\right) = h_{\xi\eta} \quad \text{and} \quad z_\eta = \operatorname{Cosh} 2\pi\kappa_\eta$$

Then the z_η are the roots of the equation

$$\begin{vmatrix} h_{11} - z_\eta & \cdots & h_{1N} \\ \cdots & \cdots & \cdots \\ h_{N1} & \cdots & h_{NN} - z_\eta \end{vmatrix} = 0 \tag{62.36}$$

Further analysis leads again to a Strutt diagram, that is, to a set of instability regions in the ω,α plane. The peaks of these regions are at the points

$$\omega = (\omega_\xi + \omega_\eta)/n \qquad \xi,\eta = 1, 2, \ldots, N$$

of the ω axis. When $\xi = \eta$, the instability is said to be of the first kind. It is the only type occurring in the simplest case of Mathieu's equation (see p. 62–5). When $\xi \neq \eta$, the instability is said to be of the second kind. In this case, the secondary motion has remarkably different properties [5],[7]. The most important fact is that a periodic excitation does not produce a periodic secondary motion.

Additional Problems

In addition to the bifurcation problems presented in Sec. 62.2, many other problems have been treated, in particular, straight beams, curved beams, circular rings, frames, plates, and shells. In all these, one considers a kinetic analogy to the static stability problem, in which one adds a pulsating load (usually with the time varying harmonically) to the static one. In all the special cases, the phenomena are essentially those which have been described here for the general case, and for the example of the straight column. Of course, each of the above-mentioned special problems requires many special mathematical and mechanical considerations.

There is only one comprehensive book on the subject of dynamic buckling, namely, the work of V. V. Bolotin [1]. It contains much additional material, and also a survey of the very extensive Russian literature. In other countries, only scattered papers on this subject have so far appeared. A few of these are cited in the following bibliography. Additional references will be found in the reports [6],[7]. The whole field of dynamic buckling is still very much under development.

62.4. REFERENCES

[1] V. V. Bolotin: "The Dynamic Stability of Elastic Systems," (in Russian) Moscow, 1956. German translation "Kinetische Stabilität elastischer Systeme," Deutscher Verlag d. Wissenschaft, Berlin, 1961.
[2] S. Lubkin, J. J. Stoker: Stability of columns and strings under periodically varying forces, *Quart. Appl. Math.*, **1** (1943), 215.
[3] K. Marguerre: Über die Anwendung der energetischen Methode auf Stabilitätsprobleme, *Jahrb.* 1938 *deut. Luftfahrt-Forsch.*, p. I 433.
[4] E. Mettler: Über die Stabilität erzwungener Schwingungen elastischer Körper, *Ing.-Arch.*, **13** (1942), 97.
[5] E. Mettler: Allgemeine Theorie der Stabilität erzwungener Schwingungen elastischer Körper, *Ing.-Arch.*, **17** (1949), 418.
[6] E. Mettler: Zum Problem der nicht-linearen Schwingungen elastischer Körper, *Publs. sci. tech. Ministère Air France*, **281** (1953), 77.
[7] E. Mettler: Erzwungene nichtlineare Schwingungen elastischer Körper, *Proc. 9th Intern. Congr. Appl. Mechanics*, **5** (1956) 5, Brussels.
[8] E. Mettler, F. Weidenhammer: Der axial pulsierend belastete Stab mit Endmasse, *Z. angew. Math. Mech.*, **36** (1956), 284.
[9] I. Utida, K. Sezawa: Dynamical stability of a column under periodic longitudinal forces, *Tokyo Imp. Univ. Aeronaut. Research Inst. Rept.*, **15** (1940), 193.
[10] F. Weidenhammer: Das Stabilitätsverhalten der nichtlinearen Biegeschwingungen des axial pulsierend belasteten Stabes, *Ing.-Arch.*, **24** (1956), 53.
[11] E. Mettler, F. Weidenhammer: Kinetisches Durchschlagen des schwach gekrümmten Stabes, *Ing.-Arch.* **29** (1960), 301.

CHAPTER 63

FLUTTER

BY

Y. C. FUNG, Ph.D., Pasadena, Calif.

63.1. INTRODUCTION: RELATIONSHIP OF FLUTTER TO OTHER AEROELASTIC PROBLEMS

Flutter, a special problem in *aeroelasticity,* is generally described as a self-excited oscillation, with a sustained, or divergent amplitude, which occurs when a structure is placed in a uniform flow. The aerodynamic forces that are responsible for the flutter motion are caused principally by the elastic deformation of the structure. If the flow remains laminar at all times, the problem lies in the domain of classical theory of aerodynamics and elasticity, and is often called a *classical flutter*. If the flow becomes *separated* over part of the body or over part of the cycle, the resulting motion is called *stall flutter*.

Structural oscillation may also be excited by the turbulences in a flow. Such a motion is often irregular, and is called *buffeting.* Buffeting over a part of an airplane may be caused by a flow separation over another part of the airplane, as, for example, tail buffeting due to a separation at the wing-fuselage junction. Sometimes it may be difficult to distinguish buffeting from stall flutter. For example, a stalled wing creates a turbulent wake, but the wing may induce in itself a motion that borders between buffeting and stall flutter. Sometimes the motion of the wing is quite random, sometimes it is quite regular, and sometimes it is a mixture of the two: e.g., random in bending, while more or less a regular sinusoidal oscillation in torsion. The last feature is described by the term *buffeting-flutter.*

The limiting case of flutter at vanishing frequency is called *divergence.* At the critical-divergence speed of flow, a structure becomes neutrally unstable: under an arbitrary, small deflection, the structure will have no tendency to return to its original position.

The line of demarcation between flutter and other aeroelastic problems is not always clear. For example, in an automatically controlled airplane, flutter and dynamic stability of the airplane may merge into a single problem. Also, the analysis of gust response has many features in common with the flutter analysis.

Classical flutter is important to aircraft design, because flutter is to be avoided in the flight speed range. Stall flutter is important in the design of rotating machineries, such as propellers, turbine blades, and compressors, which have to operate at higher angles of incidence. Buffeting limits the operational range of Mach number and lift coefficient of an airplane, and is of decisive importance in high-altitude flight. Buffeting-flutter of a wing is a problem of great concern to transonic flight. In the following, however, only the classical flutter will be discussed. An extensive bibliography can be found in the references listed at the end of this chapter.

The basic cause of flutter is the extraction of energy from a flow by a moving

This chapter deals with flutter and aeroelasticity with emphasis on the structural aspects; see Chap. 80 for the aerodynamic aspects.

object, and feeding of this energy into some particular mode of oscillation. The mechanism of flutter is, therefore, a mechanism of adjusting the relationship between the transient aerodynamic pressure and the transient wing displacement, in such a way that, in certain specific modes of oscillation, the work done by the air balances or exceeds the energy that is dissipated by other sources of damping. It is important that the energy balance mentioned above is related to a flutter mode. Thus, consider a bending-torsion flutter of a cantilever wing. When the flight speed slightly exceeds the first critical flutter speed, the energy exchange between the wing and the flow becomes such that the amplitude of the bending-torsion flutter oscillation increases with time. But, in the meantime, the wing drag is increased so that more power is required to maintain the flight.

In a harmonic motion, it is the phase shift between the force and the displacement that provides a mechanism for the energy exchange. In-phase (or directly out-of-phase) force and displacement have no net energy change in any complete cycle of harmonic motion. Thus, the inertia forces and the elastic forces are conservative.

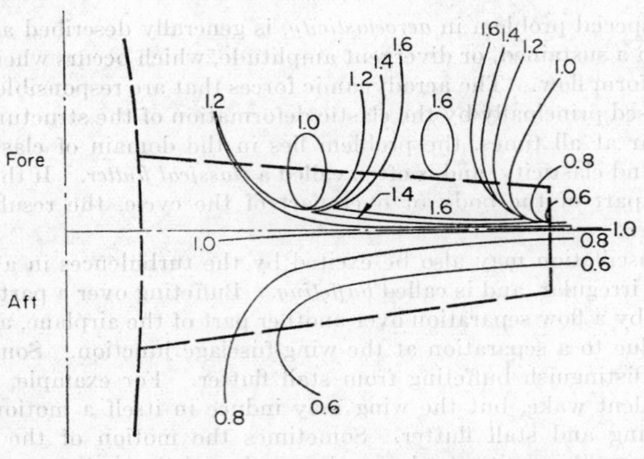

FIG. 63.1 Contours of constant flutter speed for unswept tapered wing, showing effect of concentrated mass location. Symmetric flutter. (*From Wilts* [6], *courtesy NACA*.)

But the aerodynamic force has, in general, a component 90° out of phase with the displacement, and thus is directly responsible for any possible flutter.

The aerodynamic response to a wing motion depends on the dynamic pressure of the flow, the wing geometry, the amplitude of the motion, the mode of motion (i.e., the deflection surface), and the *reduced frequency* of the oscillation. It is through these dependences that the inertial and elastic characteristics of the structure exercise their influence, for the mode and frequency of motion are determined through the interaction of the inertial, elastic, and aerodynamic forces.

Such a simple observation implies some deductions which may appear surprising: for example, that the addition of a mechanical damper to a structure, say, by dry friction or by viscous damping, is not necessarily beneficial to the prevention of flutter. The removal of some rigid or elastic constraint does not necessary lower the critical flutter speed. The reason is that constraints, dampers, mechanical or electromagnetic exciters, etc., may affect the phase shift between forces and displacements, and their effect may be quite subtle. Whereas, in the theory of mechanical vibrations, we have such beautifully simple theorems as a small increment of mass lowers the frequency, an increment in stiffness raises the frequency, etc.; in the flutter theory, we are not at all certain whether an increment in mass or stiffness will raise or reduce the flutter speed. Adding a degree of freedom may or may not cause a stable wing to flutter (which implies the importance of the rigid-body degrees of freedom of a model in a wind-tunnel testing). It can also be seen that an often-suggested method of flight-flutter testing is without foundation: that the power output of a vibration exciter which excites a wing oscillation of certain amplitude be measured, and that the decrease of this power output be used as an indication of approaching flutter.

In modern airplane design, a flutter analysis is usually started at an early stage, so that certain decisions, such as the location of the engines and fuel tanks, etc., can be made. That such an analysis is quite necessary can be shown by an example. Figure 63.1 presents the results of a systematic study of the effect of engine locations on the flutter speed by means of an electric analog computation. A point on the contour curves of constant flutter speed represents the actual location of the center of gravity of the added mass, and the numbers shown refer to unity based on the flutter speed of the bare wing, without any added mass. Thus an engine whose center of gravity is located at a point on a contour labeled 1.4 will improve the flutter characteristics of the bare wing by raising its flutter speed by a factor of 1.4. A knowledge of such trend curves is of great value to the designer; but such accurate information certainly cannot be given by any simple criterion.

63.2. MATHEMATICAL FORMULATION OF THE FLUTTER PROBLEM

Flutter is a result of the interplay of the aerodynamic force, the elastic force, and the inertia force. As prerequisites for a flutter analysis, we must know (1) the deformation pattern of the structure under an arbitrarily distributed system of

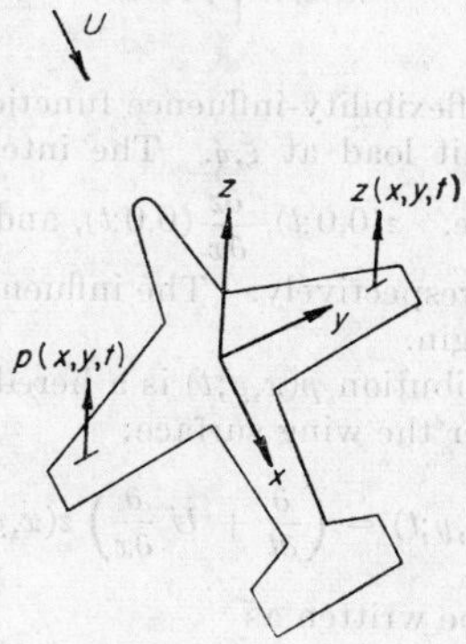

FIG. 63.2

transient loading, and (2) the aerodynamic pressure distribution, induced by an arbitrary transient deformation pattern of the structure. In an ordinary aircraft structure, the anelasticity (elastic hysteresis, damping) of the structure is small, and the elastic property of the structure can be described concisely by the static influence functions (more precisely, the static, structural-flexibility influence functions), which give the deformation of the structure under a unit static load acting at an arbitrary point. The influence functions are found by solving the partial differential equations of elasticity, under appropriate boundary conditions. Use of influence functions in the flutter analysis avoids these partial differential equations. In a similar manner, the partial differential equations of fluid mechanics are avoided by expressing the relationship between the downwash distribution over the structure and the aerodynamic pressure distribution, by means of an integral equation, the kernel of which may be interpreted as an *aerodynamic influence function*.

To describe the mathematical character of the flutter problem, let us consider a *small* perturbation of a steady, rectilinear, uniform flight of an airplane wing. Assume that the airplane is very thin, lying approximately in the x,y plane. Let the origin of the reference coordinates be chosen at a fixed point in the airplane, with the x axis pointing in the direction of the undisturbed flow (Fig. 63.2). The x, y, z axes form a rectangular cartesian system. The wing will be assumed to be infinitely rigid in its own plane; hence, only the loads and deflections normal to the wing surface are of significance.

Let $z(x,y;t)$ denote the deflection of the wing from the steady flight condition. Let $\mu(x,y)$ denote the mass per unit projected area of the wing, and let $p(x,y;t)$ denote the aerodynamic lift per unit area, in the z direction, caused by the deflection $z(x,y;t)$. Then an application of d'Alembert's principle yields the following equations of motion:

$$\iint_S \mu(\xi,\eta) \frac{\partial^2 z}{\partial t^2}(\xi,\eta;t)\, d\xi\, d\eta = \iint_S p(\xi,\eta;t)\, d\xi\, d\eta \tag{63.1}$$

$$\iint_S \xi\mu(\xi,\eta) \frac{\partial^2 z}{\partial t^2}(\xi,\eta;t)\, d\xi\, d\eta = \iint_S \xi p(\xi,\eta;t)\, d\xi\, d\eta \tag{63.2}$$

$$\iint_S \eta\mu(\xi,\eta) \frac{\partial^2 z}{\partial t^2}(\xi,\eta;t)\, d\xi\, d\eta = \iint_S \eta p(\xi,\eta;t)\, d\xi\, d\eta \tag{63.3}$$

$$\begin{aligned} z(x,y;t) - z(0,0;t) - x\frac{\partial z}{\partial x}(0,0;t) - y\frac{\partial z}{\partial y}(0,0;t) \\ = \iint_S C(x,y;\xi,\eta)\left[p(\xi,\eta;t) - \mu(\xi,\eta)\frac{\partial^2 z}{\partial t^2}(\xi,\eta;t)\right] d\xi\, d\eta \end{aligned} \tag{63.4}$$

where $C(x,y;\xi,\eta)$ is the static flexibility-influence function of the wing describing the deflection at x,y due to a unit load at ξ,η. The integrals in the above equations extend over the entire airplane. $z(0,0;t)$, $\frac{\partial z}{\partial x}(0,0;t)$, and $\frac{\partial z}{\partial y}(0,0;t)$ refer to the deflection and slopes at the origin, respectively. The influence function is measured when the airplane is fixed at the origin.

The aerodynamic lift distribution $p(x,y;t)$ is a hereditary functional of the downwash distribution $w(x,y;t)$ over the wing surface:

$$w(x,y;t) = \left(\frac{\partial}{\partial t} + U\frac{\partial}{\partial x}\right) z(x,y;t) \tag{63.5}$$

The general relationship can be written as

$$p(x,y;t) = \int_{-\infty}^{t} \iint_S H(x, y; \xi, \eta; t - \tau) \frac{\partial w}{\partial \tau}(\xi,\eta;\tau)\, d\xi\, d\eta\, d\tau \tag{63.6}$$

where $H(x, y; \xi, \eta; t - \tau)$ is the indicial response function describing the lift per unit area at x,y, and time t due to a unit-step change in downwash $w(\xi,\eta)$ at time τ. The lower limit of the Duhamel integral in (63.6) is set at $-\infty$ for convenience of writing. If the motion commences at time 0, with a sudden initial value of w_0, the discontinuity at $\tau = 0$ should be accounted for, and is understood to mean

$$\int_{-\infty}^{t} H(t-\tau)\frac{\partial w(\tau)}{\partial \tau}\, d\tau = H(t)w_0 + \int_0^t H(t-\tau)\frac{\partial w}{\partial \tau}(\tau)\, d\tau \tag{63.7}$$

Any other discontinuous change in w at some instant of time τ must be accounted for in a similar manner.

If the functions C and H were known, the above equations describe the wing motion after some small initial disturbances. If there were any other external excitations, the external forces should be added to the loading term in the above equations. We shall consider, however, that no other forces act on the wing.

The system of Eqs. (63.1) to (63.6) is linear and homogeneous in $z(x,y;t)$. It is possible to separate the variables x and y from t so that

$$z(x,y;t) = Z(x,y)e^{i\omega t} \qquad w(x,y;t) = W(x,y)e^{i\omega t} \tag{63.8}$$

If ω is a real number, the real part of (63.8) represents a sinusoidal oscillation. If we let ω approach a real number from the negative half plane (i.e., with a negative imaginary part that tends to zero), then the limiting value of $p(x,y;t)$ from Eq. (63.6) assumes the form

$$p(x,y;t) = P(x,y)e^{i\omega t} \tag{63.9}$$

The time factor $e^{i\omega t}$ can then be eliminated from all the equations. The resulting equations are:

$$-\omega^2 \iint_S \mu(\xi,\eta)Z(\xi,\eta)\,d\xi\,d\eta = \iint_S P(\xi,\eta)\,d\xi\,d\eta \tag{63.10}$$

$$-\omega^2 \iint_S \xi\mu(\xi,\eta)Z(\xi,\eta)\,d\xi\,d\eta = \iint_S \xi P(\xi,\eta)\,d\xi\,d\eta \tag{63.11}$$

$$-\omega^2 \iint_S \eta\mu(\xi,\eta)Z(\xi,\eta)\,d\xi\,d\eta = \iint_S \eta P(\xi,\eta)\,d\xi\,d\eta \tag{63.12}$$

$$Z(x,y) - Z(0,0) - x\frac{\partial Z}{\partial x}(0,0) - y\frac{\partial Z}{\partial y}(0,0) = \iint_S C(x,y;\xi,\eta)[P(\xi,\eta) + \omega^2\mu(\xi,\eta)Z(\xi,\eta)]\,d\xi\,d\eta \tag{63.13}$$

$$P(x,y) = \iint_S \mathcal{K}(x,y;\xi,\eta)\left(i\omega + U\frac{\partial}{\partial\xi}\right)Z(\xi,\eta)\,d\xi\,d\eta \tag{63.14}$$

where $\mathcal{K}(x,y;\xi,\eta)$ may be called a *lift kernel* or an *aerodynamic-lift influence function*. The explicit form of the function $\mathcal{K}$ is known for several cases [see Eqs. (80.14), (80.29), (80.39)].

Further simplification can be obtained if the origin is chosen at the center of mass of the airplane. Then, multiplying (63.13) successively by $\mu(x,y)$, $x\mu(x,y)$, $y\mu(x,y)$; integrating throughout the surface S; and using (63.10) to (63.12) to eliminate $Z(0,0)$, etc., we obtain

$$Z(x,y) + \frac{1}{\omega^2}\iint_S\left(\frac{1}{M} + \frac{x\xi}{I_y} + \frac{y\eta}{I_x}\right)P(\xi,\eta)\,d\xi\,d\eta - \iint_S C'(x,y;\xi,\eta)[P(\xi,\eta) + \omega^2\mu(\xi,\eta)Z(\xi,\eta)]\,d\xi\,d\eta = 0 \tag{63.15}$$

where

$$C'(x,y;\xi,\eta) = C(x,y;\xi,\eta) - \iint_S\left(\frac{1}{M} + \frac{xr}{I_y} + \frac{ys}{I_x}\right)\mu(r,s)C(r,s;\xi,\eta)\,dr\,ds \tag{63.16}$$

and M, I_x, I_y are the total mass and the central mass moments of inertia about the x and y axes, respectively:

$$M = \iint_S \mu(x,y)\,dx\,dy$$
$$I_x = \iint_S y^2\mu(x,y)\,dx\,dy \qquad I_y = \iint_S x^2\mu(x,y)\,dx\,dy \tag{63.17}$$

Equations (63.14) and (63.15) or (63.10) to (63.14) define an eigenvalue problem, because nontrivial solutions exist only for special values of some parameters. For example, the parameter ω may be chosen. It is easily recognized that, if the density of air vanishes, the functions $\mathcal{K}$ and P will vanish and Eq. (63.15) is reduced to an equation for the free vibration of the wing in a vacuum, in which case the eigenvalue ω and the eigenfunction $Z(x,y)$ are real-valued. If the density of air is finite, aerodynamic forces exist. The lift kernel $\mathcal{K}(x,y;\xi,y)$ is complex-valued and nonsymmetric. The eigenvalue ω, if it exists, is, in general, complex-valued. But, from the physical meaning of the parameter ω as given by Eq. (63.8) it is seen that flutter (a divergent motion) exists only if the imaginary part of ω is negative or zero, the critical condition being a real-valued eigenvalue ω. Hence, it becomes clear that a second parameter, say, the air density ρ, must be allowed to vary; and the dependence of the eigenvalue ω on ρ must be searched in such a manner that the intervals of ρ in which ω has a nonpositive imaginary part be determined. When all such intervals of ρ are determined, the boundary values defining these intervals will give the critical flutter conditions, and the corresponding ω and z will give the critical flutter frequency and deflection mode, respectively.

63.3. EXAMPLE

To illustrate the mathematical problem, let us consider the panel flutter which may occur in a supersonic flow. Consider an infinite strip of flat plate of uniform thickness, one side of which is exposed to a uniform supersonic flow in the direction of the chord of

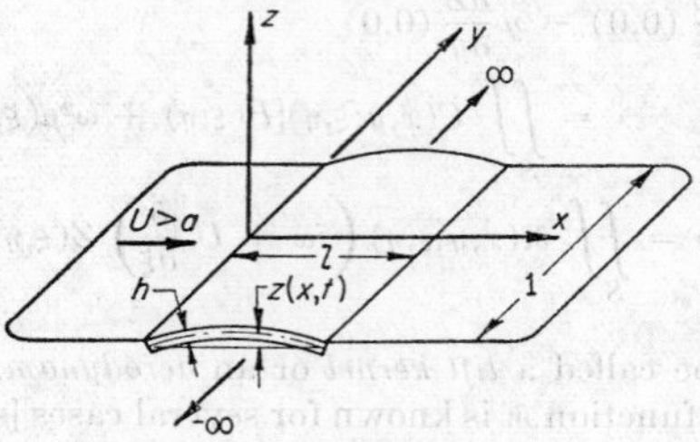

FIG. 63.3

the strip, whereas the other side is exposed to stagnant air. When not deflected, the plate lies in the x,y plane. Flat solid walls, parallel to the uniform flow, extend to infinity both in front and in rear of the panel (Fig. 63.3). We shall assume that the amplitude of flutter, if it exists, is so small that the elastic and aerodynamic actions can be described by linearized equations. The acoustic pressure beneath the plate will be neglected. The flow condition and deflection surface will be assumed to be two-dimensional; hence, the coordinate y plays no role in the problem. The equations can be simply written for a strip of unit span and of chord length l. Let $c(x,\xi)$ be the influence function describing the deflection z at x due to a unit load at ξ. The coordinates are chosen as in Fig. 63.3. Since no rigid-body motion is possible, Eqs. (63.10) to (63.12) are meaningless, and the terms $Z(0)$, $\dfrac{dZ}{dx}(0)$ may be omitted. Equations (63.13), (63.14) will be rewritten in dimensionless form. Let ρ_0 be the density of the air at infinity and ω_0 the fundamental frequency of the plate, and replace x by lx, ξ by $l\xi$, Z by lZ, $C(x,\xi)$ by $C(x,\xi)/\mu l\omega_0^2$, $P(x)$ by $\frac{1}{2}\rho_0 U^2 P(x)$, $W(x)$ by $UW(x)$; then the new quantities x, ξ, C, P, W are dimensionless. Furthermore, we introduce the dimensionless parameters:

$$
\begin{aligned}
\gamma &= \mu/\rho_0 l && \text{density ratio plate to air} \\
k &= \omega l/U && \text{reduced frequency} \\
k_0 &= \omega_0 l/U && \text{a structural rigidity parameter}
\end{aligned}
\tag{63.18}
$$

Then Eq. (63.13) assumes the form

$$Z(x) = \frac{k^2}{k_0^2}\int_0^1 C(x,\xi)Z(\xi)\,d\xi + \frac{1}{2\gamma k_0^2}\int_0^1 C(x,\xi)P(\xi)\,d\xi \tag{63.19}$$

The aerodynamic load is induced by the downwash distribution

$$W(x) = \left(\frac{\partial}{\partial x} + ik\right) Z(x) \tag{63.20}$$

The kernel function for the present case is known [see Eq. (80.29)]:

$$P(\xi) = -\frac{2}{\sqrt{M^2-1}}\left[W(0)G(\xi) + \int_0^\xi G(\xi - v)\left(\frac{d}{dv} + ik\right) W(v)\,dv\right] \tag{63.21}$$

where M is the Mach number of the flow and

$$G(x) = e^{-i\kappa Mx}J_0(\kappa x) \qquad \kappa = \frac{Mk}{M^2-1} \tag{63.22}$$

Combining (63.21) with (63.19), we can show that, for plates with hinged or clamped edges, for which

$$Z(0) = Z(1) = C(x,1) = 0$$

the above equations can be reduced to

$$Z(x) = -\frac{1}{\gamma k_0^2\sqrt{M^2-1}}\int_0^1 Z(v)\left[\left(-\frac{\partial}{\partial v} + ik - \sqrt{M^2-1}\,k^2\gamma\right)C(x,v) + \int_v^1 g(\xi - v)\left(-\frac{\partial}{\partial \xi} + ik\right)C(x,\xi)\,d\xi\right] dv \tag{63.23}$$

where

$$g(x) \equiv \left(ik + \frac{d}{dx}\right)G(x) = -\kappa e^{-i\kappa Mx}\left[J_1(\kappa x) + \frac{i}{M}J_0(\kappa x)\right] \tag{63.24}$$

The influence function $C(x,\xi)$ is known. For example, for a plate clamped at both leading and trailing edges,

$$\begin{aligned} C(x,\xi) &= \frac{\mu l^4\omega_0^2}{6K}\,x^2(1-\xi)^2[3\xi - x(1+2\xi)] && \text{for } x \le \xi \\ &= \frac{\mu l^4\omega_0^2}{6K}\,\xi^2(1-x)^2[3x - \xi(1+2x)] && \text{for } x \ge \xi \end{aligned} \tag{63.25}$$

where K is the flexural rigidity of the plate [see Eq. (39.4)].

The fundamental frequency for the clamped plate is

$$\omega_0 = \frac{(1.51\pi)^2}{l^2}\sqrt{\frac{K}{\mu}} \tag{63.26}$$

It is most convenient to regard (63.23) as an eigenvalue problem for the function $Z(x)$ and the eigenvalue k_0^2. k is real-valued since harmonic motion is assumed. For each set of specific real-valued k, γ, and M, the kernel of the integral equation is complex-valued. It is necessary to vary k, γ, M systematically, in order to locate a *real-valued* eigenvalue k_0^2 (a complex-valued k_0^2 has no physical significance).

63.4. METHODS OF SOLUTION

The method of iteration can be applied to the complex eigenvalue problem such as Eq. (63.23). An arbitrary function $Z_0(x)$ is assumed. We substitute Z_0 into the integral on the right-hand side, evaluate the integral, and collocate with Z_0 on the left-hand side at some point to obtain an approximate eigenvalue k_0^2. The right-hand side can then be designated as Z_1, and the process repeated until it converges. Wielandt [9] has proved that, if an integral equation of the form (63.23) has an eigenvalue, that eigenvalue can be obtained by such an iteration procedure, and the iterated eigenvalue k_0^2 always converges to the one with the largest modulus. Biorthogonality relationships may be used in the iteration procedure to derive successively smaller eigenvalues, and modified formulas can be devised for two or more eigenvalues having the same modulus.

A second method of solution is to represent the integral equation as a matrix equation, through the use of the method of finite differences, and then solve the eigenvalue problem for the matrix. Take Eq. (63.23) as an example. Let us represent $Z(x)$ by a column matrix of the values of Z, evaluated at a set of selected points, $x_1, x_2, \ldots, x_n$.

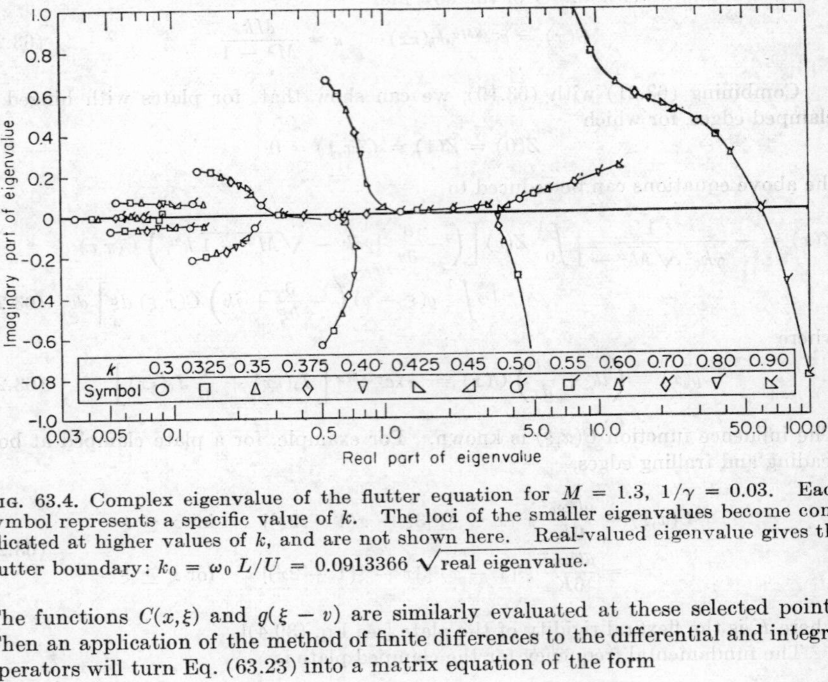

FIG. 63.4. Complex eigenvalue of the flutter equation for $M = 1.3$, $1/\gamma = 0.03$. Each symbol represents a specific value of k. The loci of the smaller eigenvalues become complicated at higher values of k, and are not shown here. Real-valued eigenvalue gives the flutter boundary: $k_0 = \omega_0 L/U = 0.0913366 \sqrt{\text{real eigenvalue}}$.

The functions $C(x,\xi)$ and $g(\xi - v)$ are similarly evaluated at these selected points. Then an application of the method of finite differences to the differential and integral operators will turn Eq. (63.23) into a matrix equation of the form

$$\{Z\} = \frac{1}{k_0^2}[A]\{Z\} \tag{63.27}$$

The square matrix $[A]$ is complex-valued and nonhermitian. Several practical methods of solution are available, the best known is Lanczos' method of *minimized iteration* [10]. The actual calculation with complex numbers is usually too difficult for hand computation, but modern high-speed electronic computers have been used with success.

These two methods, when applied to Eq. (63.23), give good agreement in the calculated eigenvalues. When 10 equal divisions are taken across the chord, a ninth-order matrix equation is obtained. For fixed values of Mach number M and density ratio γ, the variation of the eigenvalues k_0^2 with the reduced frequency k is shown in

Fig. 63.4 for the case $M = 1.3, 1/\gamma = 0.03$. It is necessary to repeat such calculations for other combinations of M and γ in order to obtain a complete picture of the flutter characteristics.

For a complete theoretical treatment, one would have to study the convergence of the eigenvalues as the number of points $x_1, x_2, \ldots, x_n$ tends to infinity, and the distances between neighboring points tend to zero. For the relatively simple problem of Eq. (63.23), a solution taking 20 equal divisions in the interval $0 \leq x \leq 1$ gives the largest real eigenvalue k_0^2, which does not differ much from that of the 10 division case. The complete program of convergence study, however, has never been attempted so far. The lack of an existence theorem for real-valued eigenvalues of nonhermitian matrices or of integral equations with asymmetric complex-valued kernels is the most important unsolved mathematical difficulty in the theory of flutter.

63.5. DISCRETE MASS APPROXIMATION

In the example above, the integral equation is approximated by a matrix equation. In many cases, much practical advantage can be achieved by regarding the final matrix equation as a numerical approximation of the integral equation, because mathematical manipulations such as change of variables and integration by parts can be performed, and greater accuracy can be obtained from the numerical viewpoint.

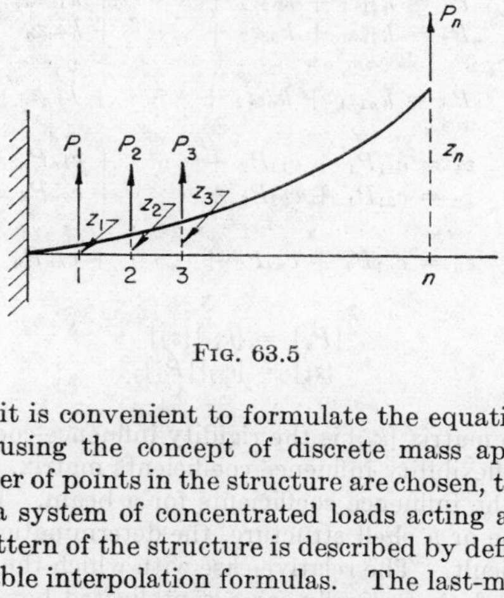

FIG. 63.5

On the other hand, it is convenient to formulate the equations of motion directly as a matrix equation, using the concept of discrete mass approximation. In this approach, a finite number of points in the structure are chosen, the distributed external loading is replaced by a system of concentrated loads acting at these chosen points, and the deformation pattern of the structure is described by deflections at these points through the use of suitable interpolation formulas. The last-mentioned interpolation procedure is very important in flutter work, for the aerodynamic load depends on the slope of the deflection surface as well as on the deflection itself. Rauscher's station-function method [11] is a well-known example.

The structural flexibility-influence functions are determined by numerical calculations; and the results are expressed as a matrix of influence coefficients. In practice, it is easier to calculate the inverse of the flexibility matrix: the matrix of the *rigidity influence coefficients*, which give the forces that are required to maintain a specific deformation pattern.

As an illustration, consider a cantilever beam whose deflection pattern is to be specified by the displacements $z_1, z_2, \ldots, z_n$ at a set of n stations, $1, 2, 3, \ldots, n$; with the actual mass and loading distributions represented as closely as possible by n discrete masses, $\mu_1, \mu_2, \ldots, \mu_n$; and n concentrated forces $P_1, P_2, \ldots, P_n$. Figure 63.5 shows such a beam. Consider the static equilibrium of the beam under the set of forces. The strain energy V can be expressed either in terms of the deflections z

alone or in terms of the forces P alone. The first gives an expression

$$V_z = \tfrac{1}{2}(k_{11}z_1^2 + k_{22}z_2^2 + \cdots + k_{nn}z_n^2) + (k_{12}z_1z_2 + \cdots + k_{rs}z_rz_s + \cdots) \tag{63.28}$$

and the second gives

$$V_P = \tfrac{1}{2}(c_{11}P_1^2 + c_{22}P_2^2 + \cdots + c_{nn}P_n^2) + (c_{12}P_1P_2 + \cdots + c_{rs}P_rP_s + \cdots) \tag{63.29}$$

These are homogeneous quadratic functions in which the reciprocal relations

$$k_{rs} = k_{sr} \qquad c_{rs} = c_{sr} \tag{63.30}$$

hold, under the assumptions of linearity between load and displacement. According to Castigliano's theorems, we have

$$\frac{\partial V_z}{\partial z_r} = P_r \qquad \frac{\partial V_P}{\partial P_r} = z_r \tag{63.31a,b}$$

that is, in expanded form,

$$\begin{aligned} P_1 &= k_{11}z_1 + k_{12}z_2 + \cdots + k_{1n}z_n \\ P_2 &= k_{21}z_1 + k_{22}z_2 + \cdots + k_{2n}z_n \\ &\cdots\cdots\cdots\cdots\cdots\cdots \\ P_n &= k_{n1}z_1 + k_{n2}z_2 + \cdots + k_{nn}z_n \end{aligned} \tag{63.32}$$

$$\begin{aligned} z_1 &= c_{11}P_1 + c_{12}P_2 + \cdots + c_{1n}P_n \\ z_2 &= c_{21}P_1 + c_{22}P_2 + \cdots + c_{2n}P_n \\ &\cdots\cdots\cdots\cdots\cdots\cdots \\ z_n &= c_{n1}P_1 + c_{n2}P_2 + \cdots + c_{nn}P_n \end{aligned} \tag{63.33}$$

or, in matrix form,

$$\{P_i\} = [k_{ij}]\{z_j\} \tag{63.34}$$

$$\{z_i\} = [c_{ij}]\{P_j\} \tag{63.35}$$

The symmetric square matrix $[k_{ij}]$ is the rigidity-influence-coefficients matrix, and its reciprocal, $[c_{ij}]$, is the flexibility-influence-coefficients matrix.

It is easy to find the influence coefficients for a beam. However, for a built-up structure or for a plate or a shell structure, the determination of the influence coefficients can be very difficult. The relative ease with which the rigidity influence coefficients can be calculated theoretically for a complicated box-beam type of structure has been discussed by D. Williams [12]. When an approximate matrix $[k_{ij}]$ is found, it can be inverted to obtain an approximate $[c_{ij}]$. The matrix $[c_{ij}]$, however, is nearly singular (although it is never exactly so), and a slight error in the elements c_{ij} will cause large errors in $[k_{ij}]$. Thus, the inversion of the matrix $[c_{ij}]$ should be avoided in any practical calculations, especially when the order of the matrix is large.

To illustrate the use of influence coefficients, let us consider a fluttering airplane which is restrained from plunging and rotation at the origin of the reference axes, with respect to which the influence coefficients are measured. Since Hooke's law is assumed, it is evident that Eqs. (63.34) and (63.35) hold as well when P_i, z_i are functions of time, $P_i = \mathcal{P}_i e^{i\omega t}$, $z_i = Z_i e^{i\omega t}$, where ω is the flutter frequency. All equations will be written for harmonic oscillations. The aerodynamic influence coefficients q_{ij} will be defined by

$$\{\mathcal{P}_i\} = \tfrac{1}{2}\rho_0 U^2 [q_{ij}]\{Z_j\} \tag{63.36}$$

where $\mathcal{P}_i$ represents the amplitude of the resultant aerodynamic force at the point i, and Z_i the amplitude of the deflection at the same point. ρ_0 and U are the air density and speed at infinity. The wing deflection under the influence of the inertia and aerodynamic load is therefore, according to Eq. (63.35),

$$\{Z_i\} = \omega^2[c_{ij}][\mu_j]\{Z_j\} + \tfrac{1}{2}\rho_0 U^2[c_{ij}][q_{jl}]\{Z_l\} \tag{63.37}$$

where the matrix $[\mu_j]$ is a diagonal matrix of the mass associated with the point j.

An eigensolution may exist and flutter may occur when the determinantal equation is satisfied:

$$|[I] - \omega^2[c][\mu] - \tfrac{1}{2}\rho_0 U^2[c][q]| = 0 \tag{63.38}$$

where $[I]$ is the unit matrix.

It is also interesting to write the eigenvalue problem in terms of the aerodynamic load distribution $\mathcal{P}_i$. Let the relationship between the downwash and deflection, Eq. (63.5), be integrated with proper boundary conditions, and let the result be written as a matrix equation:

$$\{Z_i\} = [\alpha_{ij}]\{W_j\} \tag{63.39}$$

The relationship between the aerodynamic load and the downwash can be expressed as

$$\{W_i\} = [\gamma_{ij}]\{\mathcal{P}_j\} \tag{63.40}$$

The elastic forces corresponding to a given deflection pattern are expressed in terms of the rigidity influence coefficients k_{ij} as in Eq. (63.32). The balance of the aerodynamic, inertial, and elastic forces, therefore, gives the equation of motion:

$$\{\mathcal{P}_i\} + \omega^2[\mu_i]\{Z_i\} + [k_{ij}]\{Z_j\} = 0 \tag{63.41}$$

Combining (63.39), (63.40), and (63.41), we obtain:

$$\{\mathcal{P}_i\} + \omega^2[\mu_i][\alpha_{ij}][\gamma_{jl}]\{\mathcal{P}_l\} + [k_{ij}][\alpha_{jl}][\gamma_{lm}]\{\mathcal{P}_m\} = 0 \tag{63.42}$$

The flutter determinant is now

$$|[I] + \omega^2[\mu][\alpha][\gamma] + [k][\alpha][\gamma]| = 0 \tag{63.43}$$

It should be pointed out that the aerodynamic-downwash-influence coefficients $[\gamma_{ij}]$ can be derived from elementary solutions, and are known for arbitrary wing planform and Mach number. On the other hand, the aerodynamic lift-influence function $\mathcal{K}$ of Eq. (63.14) is known only for a few special cases [see Eqs. (80.14), (80.29), (80.39)], and, therefore, the coefficients q_{ij} of Eq. (63.36) are not easy to obtain.

63.6. STRUCTURAL DAMPING

So far, the material damping or any other source of damping has not been considered. For aircraft structures, the anelasticity is small, and a simplified concept of structural damping is often used. It is assumed that the damping force is proportional to the elastic restoring force in magnitude, and directly opposes the velocity of motion. For small damping and for a harmonic oscillation, one modifies the stiffness influence coefficients k_{lm} by a factor $(1 + ig)$ into $(1 + ig)k_{lm}$; or the flexibility influence coefficients c_{lm} into $(1 - ig)c_{lm}$, where g is assumed to be a small number (of order 0.01 to 0.05) and is called the *structural damping factor*.

63.7. MODAL APPROACH. GALERKIN'S METHOD OF SOLUTION

Galerkin's method is often used to obtain approximate solutions of the integro-differential equations of the preceding sections or their matrix approximations. In this approach, the deformation is assumed to be of the form

$$z(x,y;t) = a_1(t)Z_1(x,y) + a_2(t)Z_2(x,y) + \cdot \cdot \cdot + a_n(t)Z_n(x,y) \qquad (63.44)$$

where the a_i are functions of time, and the $Z_i(x,y)$ are the assumed modes, which satisfy the boundary constraints. In a complex representation of the harmonic motion, all a_i are proportional to the time factor $e^{i\omega t}$ at the critical flutter condition. Then the time factor can be removed from the governing equations, and the a_i may be regarded as complex-valued constants. The modes Z_i can be either the normal modes of free vibration of the whole structure or the *uncoupled* vibration modes of the component structures, or some convenient set of analytic functions or interpolation formulas. The proper choice of the modes Z_i is of utmost importance if an efficient and accurate solution is to be obtained. The principle is that the flutter mode must be capable of being well represented by the chosen finite series.

When the series (63.44) is substituted into equations such as (63.14), (63.15), or (63.37), and the coefficients a_i are so chosen that these equations are satisfied with the minimum mean-squares error, an approximate solution in the sense of Galerkin is obtained.

The a_i of Eq. (63.44) are called the *generalized displacements*. The equation governing a_i can also be obtained from the energy principle. The resulting Lagrange's equations of motion are

$$\frac{d}{dt}\left(\frac{\partial T}{\partial \dot{a}_i}\right) + \frac{\partial V}{\partial a_i} = Q_i \qquad i = 1, 2, \ldots, n \qquad (63.45)$$

where T is the kinetic energy, V is the potential energy, and Q_i is the generalized aerodynamic force arising from the work done by the aerodynamic loading in the Z_i mode. T and V are determined on the basis of Eq. (63.44) and are, in general, quadratic forms in the a_i and their time derivatives. These equations can be represented schematically by

$$[\mu]\{a_i\} + [E]\{a_i\} = [A]\{a_i\} \qquad (63.46)$$

where $[\mu]$, $[E]$, and $[A]$ are square matrices, denoting the inertia, elastic, and aerodynamic matrices, respectively, and a_i is a column matrix. The elements of the square matrices are functions of the reduced frequency of the harmonic motion, and are integrated quantities weighted by the chosen mode shapes. The critical flutter condition is the determinantal equation

$$|[\mu] + [E] - [A]| = 0 \qquad (63.47)$$

The order of the determinant is n. Since the real and imaginary parts of the above determinant must vanish separately, two independent characteristic equations are obtained, from which two real-valued eigenvalues, or the real and imaginary part of one complex-valued eigenvalue, can be determined. The selection of which physical parameters are to be the eigenvalues is a matter of individual judgment, depending on the nature of the particular problem, the scope of calculation, and the answers desired.

63.8. ELECTRIC ANALOG METHODS

A powerful tool for flutter analysis is the electric analog method [7]. In this method, the aeroelastic system is assumed to be linear, and a linear electric network is constructed, whose electrical behavior approximates the dynamic behavior of the linearized structure. An example of such an analog sets up the following correspondence:

Capacitors—concentrated or lumped inertia properties
Inductors—lumped flexibility properties
Transformers—geometric properties
Voltage—velocities
Current—forces

In such an analog, electronic equipment is used to produce currents which depend on voltages in the electric system, in the same manner in which aerodynamic forces depend on the velocities of the airfoil.

The electric analog of the aircraft can be regarded as a model whose properties can be altered easily. Thus, a parametric study of the aircraft can be done with great rapidity. The analysis consists in observing the behavior of the model in flight. That behavior which is most readily observed is the transient response to a sudden disturbance. The logarithmic decrement of the response and the frequency of oscillation can be easily obtained. Flutter speed is found by computing the logarithmic decrement for specific values of flight speed, and interpolating to find the speed at which the logarithmic decrement is zero. It is a particular advantage of the electric analog method that tuned pulses may be used to separate two or more nearly unstable modes of motion.

Whereas the most convenient form of aerodynamic information used in the influence-functions formulation presented above is the frequency response, in the analog method indicial responses are used: e.g., the lift due to a sudden change of angle of attack, the moment due to a sudden aileron deflection, etc. They can be approximated by a finite sum of exponential functions. The electric analog for the aerodynamic system can then be recognized relatively easily by examining the Laplace transforms of the indicial responses.

63.9. CONCLUDING REMARKS

Only a bird's-eye view of the flutter problem is presented above. Practical details of an actual analysis involve questions such as the calculation of inertial and rigidity matrices, the estimation of the finite-differences errors, the incorporation of finite-aspect ratio effect in aerodynamic forces, etc. Various components of the airplane have different flutter characteristics. Various Mach-number ranges may have their own crop of problems, such as the *aileron buzz* at the transonic flow speeds. Mass balancing techniques for the control of control-surface flutter, pressurization of the structure for the prevention of panel flutter, model and flight-test techniques, etc., are special problems which constitute the bulk of literature on the subject. Bibliographies can be found in Refs. [1]–[3] listed below.

63.10. REFERENCES

[1] R. L. Bisplinghoff, H. Ashley, R. L. Halfman: "Aeroelasticity," Addison-Wesley, Reading, Mass., 1955.

[2] Y. C. Fung: "An Introduction to the Theory of Aeroelasticity," Wiley, New York, 1955.

[3] R. H. Scanlan, R. Rosenbaum: "Introduction to the Study of Aircraft Vibration and Flutter," Macmillan, New York, 1951.

[4] H. Templeton: "Massbalancing of Aircraft Control Surfaces," Chapman & Hall, London, 1954.

[5] R. L. Bisplinghoff: Some structural and aeroelastic considerations of high-speed flight; 19th Wright Brothers Lecture, *J. Aeronaut. Sci.*, **23** (1956), 289–329.

[6] C. H. Wilts: Incompressible flutter characteristics of representative aircraft wings, *NACA Tech. Note* 3780 (1957).

[7] R. H. MacNeal, G. D. McCann, C. H. Wilts: The solution of aeroelastic problems by means of electrical analogies, *J. Aeronaut. Sci.*, **18** (1951), 777–789.

[8] I. E. Garrick: Aerodynamic theory and its application to flutter, paper prepared for Structures and Materials Panel of Advisory Group for Aeronautical Research and Development, 1956.

[9] H. Wielandt: Solution of general linear eigenvalue problems, *AVA Göttingen Rept.* AVA 44/J/19–20; Translation: TS-976, ATI 18804, Air Materiel Command, Wright Air Force Base, Dayton, Ohio.

[10] C. Lanczos: An iteration method for the solution of the eigenvalue problem of linear differential and integral operators, *J. Research Natl. Bur. Standards*, **45** (1950), 255.

[11] M. Rauscher: Station functions and air density variations in flutter analysis, *J. Aeronaut. Sci.*, **16** (1949), 345.

[12] D. Williams: Recent developments in the structural approach to aeroelastic problems, *J. Roy. Aeronaut. Soc.*, **58** (1954), 403.

CHAPTER 64

PROPAGATION OF ELASTIC WAVES

BY

E. E. ZAJAC, Ph.D., Murray Hill, N.J.

64.1. INTRODUCTION

Broadly speaking, the subject of wave propagation in an elastic continuum has evolved from two directions. On the one hand, so-called "approximate" or "elementary" theories have developed. Such theories have taken an approximate description of the continuum as their starting point, for example, the approximation that plane cross sections of a bar remain plane. These theories arrive at comparatively simple but approximate differential equations for describing the propagation phenomena. Often the equations can be solved for a wide variety of boundary conditions, but the detailed accuracy of the resulting solutions is always in doubt because of the assumptions made at the outset.

On the other hand, more precise descriptions of the continuum, for example, the equations of classical infinitesimal elasticity, have been used as a starting point in so-called "exact" theories. However, the solution of these equations for meaningful boundary conditions can be formidable. In specific problems, approximations must be often made to render the mathematics of the exact theories tractable, and, in a sense, the distinction between *approximate* and *exact* theories is artificial. However, this distinction is usually made in the literature, and, for convenience, we shall make it also.

Although, in many cases, neither an approximate nor an exact theory gives so complete a description of elastic wave propagation as would be desired, both approaches, nevertheless, have fruitful applications, and are powerful tools for the analysis of wave propagation. In ordinary structural problems, where the nature of the material and loads is not precisely known, the approximate theories to be described generally give results which are as accurate as those obtained in static structural and stress analyses. Their comparative mathematical simplicity allows one to obtain approximate solutions for problems which, by an exact theory, would be out of reach: for example, the axial impact of a finite-length bar. By the same token, in other applications, for example, in the field of ultrasonics, one is sometimes interested in details of the propagation phenomena which are outside the scope of the approximate theories. Often, an insight into these details can be obtained from such things as the phase and group velocities and mode shapes which are given in an exact theory, even when solutions, according to this theory, which satisfy all boundary and initial conditions are difficult to obtain.

Comprehensive bibliographies of both approximate and exact treatments of elastic wave propagation are given by Davies [1] and Ewing, Jardetzky, and Press [2]. The particular problems of elastic waves in rods and beams have been surveyed in detail by Abramson, Plass, and Ripperger [12]. For other expository treatments the reader is referred to Kolsky [3], Love [4], and Pfeiffer [5].

64.2. SMALL-AMPLITUDE TRANSVERSE WAVES IN A FLEXIBLE STRING

Let w be the transverse displacement of a point on a flexible string, which is under tension T (Fig. 64.1). For small displacements w, we may assume that T remains constant, and we find then for dynamic equilibrium of an element of the string that

$$\frac{\partial}{\partial x}\left(T\frac{\partial w}{\partial x}\right) = \mu\frac{\partial^2 w}{\partial t^2} \tag{64.1}$$

where μ is the mass per unit length of the string. For convenience, we rewrite Eq. (64.1) as

$$\frac{\partial^2 w}{\partial x^2} - \frac{1}{c_0{}^2}\frac{\partial^2 w}{\partial t^2} = 0 \tag{64.2}$$

where $c_0{}^2 = T/\mu$. Equation (64.2) is the simplest mathematical representation of a wave-propagation phenomenon. We shall call it the *one-dimensional wave equation*

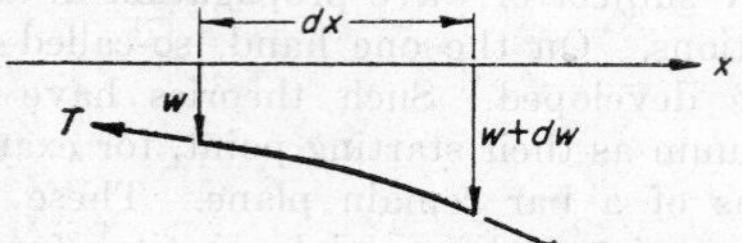

FIG. 64.1. Force diagram of a differential string element.

or simply *wave equation*, although there exist many other equations which have wave-type solutions.

By writing Eq. (64.2) as

$$\left(\frac{\partial}{\partial x} - \frac{1}{c_0}\frac{\partial}{\partial t}\right)\left(\frac{\partial}{\partial x} + \frac{1}{c_0}\frac{\partial}{\partial t}\right) w = 0$$

it is seen that its solution is of the general form

$$w = f(x - c_0 t) + g(x + c_0 t) \tag{64.3}$$

Consider the function $f(x - c_0 t)$. At a given time t, $f(x - c_0 t)$ will represent a function of x alone. At Δt later, $f(x - c_0 t)$ will represent the same function at values of x displaced by the amount $\Delta x = c_0 \, \Delta t$. Thus, $f(x - c_0 t)$ represents a wave traveling in the direction of positive x at velocity $\Delta x/\Delta t = c_0$. Similarly, $g(x + c_0 t)$ represents a wave traveling at velocity $-c_0$ in the negative x direction. Any solution of Eq. (64.2) can, therefore, be represented as the sum of two waves traveling in opposite directions. We note, further, that waves $f(x - c_0 t)$ and $g(x + c_0 t)$ preserve their shape as they travel along the string. The fact that waves described by the wave equation propagate without distortion is one of their most important properties.

Solutions of Eq. (64.2) which explicitly show the wave-propagation phenomena usually can be more conveniently obtained by use of the Laplace transform than from Eq. (64.3). We illustrate this by an example.

Example. One end of a string of length l is excited by a small transverse motion $F(t)$. Find the response of the string if the other end is fixed and the string is initially at rest.

The initial and boundary conditions in this problem are

$$w(x,0) = \frac{\partial w}{\partial t}(x,0) = 0 \tag{64.4}$$

$$w(0,t) = F(t) \qquad w(l,t) = 0 \tag{64.5}$$

Denoting the Laplace transform of a function with respect to time by a bar, we obtain from Eq. (64.2), for initial conditions (64.4),

$$\frac{d^2\bar{w}}{dx^2} - \frac{s^2}{c_0{}^2}\bar{w} = 0 \tag{64.6}$$

Furthermore, the boundary conditions (64.5) transform to

$$\bar{w}(0,s) = \bar{F}(s) \qquad \bar{w}(l,s) = 0 \tag{64.7}$$

The solution of Eq. (64.6) for the boundary conditions (64.7) is

$$\bar{w}(x,s) = \frac{\bar{F}(s)}{1 - e^{-2sl/c_0}} (e^{-sx/c_0} - e^{-2sl/c_0}e^{sx/c_0})$$

Utilizing the series expansion for $(1 - e^{-2sl/c})^{-1}$, we may write this in the form

$$\bar{w}(x,s) = \sum_{n=0}^{\infty} \left(\bar{F}(s) \exp\left[-\frac{s}{c_0}(2nl + x) \right] - \bar{F}(s) \exp\left[-\frac{s}{c_0}[(2n+1)l - x] \right] \right)$$

For convenience, we introduce here an *initiating operator*, which we define for an arbitrary function $g(\xi)$ with argument ξ as follows:

$$\begin{aligned} g\{\xi\} &= g(\xi) \qquad &\text{when } \xi > 0 \\ &= 0 &\text{when } \xi < 0 \end{aligned} \tag{64.8}$$

The shifting theorem [Eq. (19.14)] can then be written as

$$L^{-1}[e^{-sa}\bar{g}(s)] = g\{t - a\}$$

By the use of this theorem, the expression for $\bar{w}(x,s)$ can be inverted, term by term:

$$w(x,t) = \sum_{n=0}^{\infty} F\left\{ t - \frac{1}{c_0}(2nl + x) \right\} - \sum_{n=0}^{\infty} F\left\{ t - \frac{1}{c_0}[2(n+1)l - x] \right\} \tag{64.9}$$

For $t < l/c_0$, we get

$$w(x,t) = F\{t - x/c_0\}$$

Thus, at a station x along the string, the end motion is initially reproduced undistorted, but, starting at the time $t = x/c_0$ instead of $t = 0$. Consider now the nth terms in the series expansions contained in Eq. (64.9). The nth term of the first series represents a wave propagating in the positive x direction, which, after it starts at $t = 2nl/c_0$ at the excited end, is identical with the initial wave. The nth term of the second series represents a wave starting at the fixed end at time $t = 2(n + 1)l/c_0$, and propagating in the negative x direction, with the same form as the initial wave but with opposite sign. The quantity l/c_0 is the time it takes a wave to propagate the length of the string. Hence, the sequence of events described by Eq. (64.9) is the following. The initial wave propagates to the fixed end of the string, where it is reflected with a change of sign. The reflected wave, propagating in the positive x direction, superposes on the initial wave, which continues to propagate in the positive x direction. When the reflected wave reaches the excited end, it itself is reflected with a change of sign. This reflection is superposed on the initial wave, and the first reflected wave, and so forth. The sequence is illustrated in Fig. 64.2 for $F(t)$ in the form of a ramp function in time.

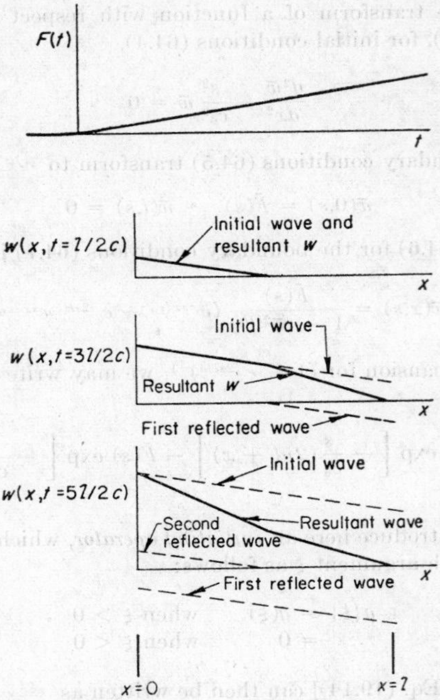

FIG. 64.2. Resultant string displacements. One end fixed, the other end displaced by $F(t)$.

64.3. LONGITUDINAL WAVES IN A BAR OR STRING

We assume that initially plane cross sections of a uniform bar remain plane when the bar is in motion, and that the normal stress across the bar is uniformly distributed.

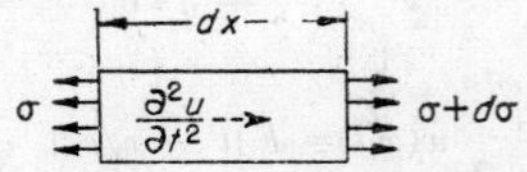

FIG. 64.3. Stresses and acceleration of a differential element of a bar.

If u is the longitudinal displacement of a point x along the bar, then, for dynamic equilibrium (Fig. 64.3), we have

$$A\,\frac{\partial \sigma}{\partial x} = \mu\,\frac{\partial^2 u}{\partial t^2}$$

where A is the cross-sectional area of the bar, and μ is its mass density per unit length. Hooke's law gives, in turn,

$$\sigma = E\,\frac{\partial u}{\partial x} \tag{64.10}$$

Combining the above equations, we arrive at

$$\frac{\partial^2 u}{\partial x^2} - \frac{1}{c_B{}^2}\,\frac{\partial^2 u}{\partial t^2} = 0 \tag{64.11}$$

where the *bar velocity* c_B is given by

$$c_B^2 = \frac{EA}{\mu}$$

For a homogeneous bar, $\mu = \rho A$, where ρ is the mass density, and $c_B^2 = E/\rho$. We recognize that Eq. (64.11) is again the wave equation. However, here the stress σ rather than the displacement u is likely to be of interest. The connection between σ and u is given by Eq. (64.10).

By techniques similar to those used in the previous section, the solution to a wide range of problems relating to longitudinal wave propagation in a bar or string can be obtained. Table 64.1 lists the solutions to commonly occurring problems of this kind. These solutions are easily applied to transverse wave propagation in strings, and, as will be seen in the next section, to torsional wave propagation in bars as well.

Table 64.1. Solutions of Common Axial Impact Problems

$$G(t) = u_0(t) = \frac{c_B}{E}\int_0^t \sigma_0(t)\,dt$$

$$c_B^2 = \frac{EA}{\mu},\ \theta_n = t - \frac{1}{c_B}(2nl + x),\ \phi_n = t - \frac{1}{c_B}(2nl - x),\ G'(\xi) = \frac{dG}{d\xi},$$

$$L_n(\xi) = \frac{1}{n!}\left[\frac{d}{d\xi} - 1\right]^n \xi^n \text{ (Laguerre polynomial see p. 15-4)}$$

$\{\ \}$ = initiating operator, i.e., $G\{\theta_n\} = G(\theta_n)$ when $\theta_n > 0$
$= 0$ when $\theta_n < 0$

1. Semi-infinite 	$u(x,t) = G\{\theta_0\}$ $\sigma(x,t) = -\frac{E}{c_B}G'\{\theta_0\}$
2. Free-ended 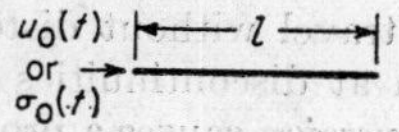	$u(x,t) = \sum_{n=0}^{\infty} (-)^n[G\{\theta_n\} + G\{\phi_{n+1}\}]$ $\sigma(x,t) = -\frac{E}{c_B}\sum_{n=0}^{\infty} (-)^n[G'\{\theta_n\} - G'\{\phi_{n+1}\}]$
3. Fixed-ended 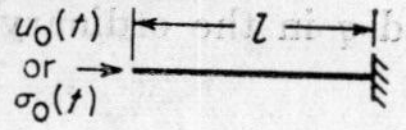	$u(x,t) = \sum_{n=0}^{\infty} [G\{\theta_n\} - G\{\phi_{n+1}\}]$ $\sigma(x,t) = -\frac{E}{c_B}\sum_{n=0}^{\infty} [G'\{\theta_n\} + G'\{\phi_{n+1}\}]$
4. Viscously damped end 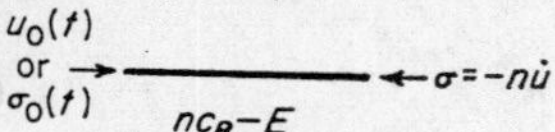 $\gamma = \frac{nc_B - E}{nc_B + E}$	$u(x,t) = \sum_{n=0}^{\infty} [\gamma^n G\{\theta_n\} - \gamma^{n+1} G\{\phi_{n+1}\}]$ $\sigma(x,t) = -\frac{E}{c_B}\sum_{n=0}^{\infty} [\gamma^n G'\{\theta_n\} + \gamma^{n+1} G'\{\phi_{n+1}\}]$

64.4. TORSIONAL WAVES IN A ROD OF CIRCULAR CROSS SECTION

Assume that, in the twisting of a rod of circular cross section, initially radial lines remain radial. Then dynamic equilibrium (Fig. 64.4) and elementary theory give

$$\frac{\partial M_T}{\partial x} = \rho I_p \frac{\partial^2 \phi}{\partial t^2} \qquad M_T = G I_p \frac{\partial \phi}{\partial x} \tag{64.12a,b}$$

where ϕ is the angle through which a radial line rotates, and ρ is the mass density of the rod. The above equations combine to yield

$$\frac{\partial^2 \phi}{\partial x^2} - \frac{1}{c_T^2}\frac{\partial^2 \phi}{\partial t^2} = 0 \tag{64.13}$$

where $c_T^2 = G/\rho$. Equation (64.13) is again the wave equation. In addition to the twist $\theta = \partial\phi/\partial x$, the maximum shearing stress is often of interest. This is given by

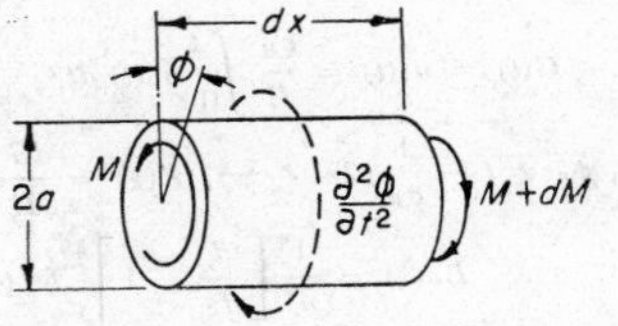

FIG. 64.4. Moments and angular acceleration of a rod of circular cross section in torsion.

$\tau_{\max} = Ga\ \partial\phi/\partial x$. Table 64.1 can be applied directly to torsional wave-propagation problems by means of the following correspondences:

Longitudinal waves	*Torsional waves*
u	$a\phi$
σ	$\tau_{\max}$
E	G

64.5. WAVES IN DISPERSIVE MEDIA—FLEXURAL WAVES IN A BEAM (BERNOULLI-EULER THEORY)

Elastic waves described by the wave equation (64.2) travel without distortion, their form being changed only by reflection or refraction at discontinuities in the medium. However, in most media, the phenomenon of *dispersion* causes a propagating disturbance to distort as it travels. We illustrate this phenomenon by considering flexural waves in a beam as described by elementary theory, due originally to Bernoulli and Euler.

The equation of beam motion, according to this theory, can be obtained by considering the inertia force $-\mu\ \partial^2 w/\partial t^2$ as a distributed load q in the ordinary beam equation. This gives

$$EI\frac{\partial^4 w}{\partial x^4} + \mu\frac{\partial^2 w}{\partial t^2} = 0 \tag{64.14}$$

For convenience, we introduce dimensionless space and time variables, $\xi = x/I^{1/4}$, $\tau = t/\sqrt{\mu/E}$, and write Eq. (64.14) in the form

$$\frac{\partial^4 w}{\partial \xi^4} + \frac{\partial^2 w}{\partial \tau^2} = 0 \tag{64.15}$$

We ask: Under what conditions can a harmonic wave train $w = w_0 e^{i\kappa(\xi - c\tau)}$ exist in the beam? Here c will be the propagation velocity of the harmonic wave train, and κ its *wave number*, defined by $\kappa = 2\pi/\lambda$, where λ is the wavelength. For a given κ and c, a given value of the *phase* or argument $\phi(\kappa) = \kappa(\xi - c\tau)$ propagates at the velocity c. Hence, c is called the *phase velocity*. By substitution into Eq. (64.15), we find that a harmonic wave train can exist if $c = \kappa$. Thus, the velocity of the harmonic wave train depends on its wavelength. A propagating disturbance can, by Fourier theory, be represented as the sum of elemental harmonic wave trains. In media such as the Bernoulli-Euler beam, where the velocity of a harmonic wave train depends on its wavelength, component elemental waves travel at different velocities and thus *disperse*, resulting in a distortion of the initial disturbance. In media governed by the wave equation (64.2), the wave velocity is independent of wavelength. In these cases, no dispersion and, hence, no distortion occurs.

In dispersive media, the idea of a definite velocity for the whole of a propagating disturbance becomes vague. However, one can often associate definite velocities with particular features of the disturbance. For example, the discontinuities in a disturbance may travel at distinct, characteristic velocities, even though the entire form of the disturbance is not preserved as it propagates. Still other aspects of a disturbance propagate at the so-called *group velocity*. To illustrate the group-velocity concept, we consider first a group of two harmonic wave trains of wave numbers κ and κ_1 and of unit amplitude. We write

$$w = \cos \kappa(\xi - c\tau) + \cos \kappa_1(\xi - c_1\tau) \tag{64.16}$$

If κ and κ_1 differ only slightly, say $\kappa_1 = \kappa + \Delta\kappa$, we may neglect higher-order terms and write Eq. (64.16) as

$$w = 2 \cos \frac{\Delta\kappa}{2} \left[\xi - \frac{\Delta(\kappa c)}{\Delta\kappa} \tau \right] \cos \kappa(\xi - c\tau)$$

At a given time, the two harmonic wave trains reinforce and interfere with each other along the space coordinate ξ, to give rise to the familiar beat phenomenon. Thus we have a slowly varying envelope of wavelength $4\pi/\Delta\kappa$, forming periodically spaced "groups" or "beats," by enveloping or modulating a rapidly varying component with wavelength $2\pi/\kappa$. The envelope propagates with velocity $\Delta(\kappa c)/\Delta\kappa$, while the individual harmonic waves propagate at the phase velocities c and $(c + \Delta c)$. The limiting value of the envelope velocity for $\Delta\kappa \to 0$ we call the *group velocity* c_g, that is, $c_g = d(\kappa c)/d\kappa$. We note that c_g can be alternatively represented in terms of the frequency $\omega = \kappa c$ or the wavelength $\lambda = 2\pi/\kappa$:

$$c_g = \frac{d\omega}{d\kappa} \qquad c_g = c - \lambda \frac{dc}{d\lambda}$$

In the case of the Bernoulli-Euler beam, $c = \kappa$ and $c_g = 2\kappa = 2c$.

The above results for a simple group of two harmonic wave trains can be extended to groups composed of a continuous spectrum of harmonic waves. By Fourier theory, a propagating disturbance can be compounded of elemental harmonic wave trains of amplitudes $A(\kappa)\, d\kappa$:

$$w(\xi,\tau) = \frac{1}{2\pi} \int_{-\infty}^{\infty} A(\kappa) e^{i\kappa(\xi - c\tau)}\, d\kappa \tag{64.17}$$

If $A(\kappa)$ is negligible outside of a narrow band of wave numbers $(\kappa_0 - \epsilon \le \kappa \le \kappa_0 + \epsilon)$, then we expand $\kappa(\xi - c\tau)$ in a Taylor series about κ_0,

$$\kappa(\xi - c\tau) = \kappa_0(\xi - c_0\tau) + (\kappa - \kappa_0)(\xi - c_g(\kappa_0)\tau) + \cdots$$

where $c_0 = c(\kappa_0)$. If the higher terms in this expansion can be neglected, we substitute into Eq. (64.17) to get

$$w(\xi,\tau) = \alpha(\xi,\tau)e^{i\kappa_0(\xi - c_0\tau)} \tag{64.18}$$

where

$$\alpha(\xi,\tau) = \frac{1}{2\pi}\int_{\kappa_0-\epsilon}^{\kappa_0+\epsilon} A(\kappa)\exp\left[i(\kappa - \kappa_0)(\xi - c_g(\kappa_0)\tau)\right] d\kappa$$

The result is again an amplitude function $\alpha(\xi,\tau)$ modulating the harmonic wave train $e^{i\kappa_0(\xi - c_0\tau)}$. As before, constant values of $\alpha(\xi,\tau)$, and, hence, of the amplitude of $w(\xi,\tau)$, propagate at the group velocity $c_g(\kappa_0)$. It is important to note that a sharp clustering of frequencies in the Fourier representation generally means a propagating disturbance which is widely spread in space and time. For example, a disturbance concentrated at a single frequency corresponds to a harmonic wave train of infinite extent in both time and space.

Likewise, a propagating disturbance which is sharp in the time or space domain will be generally spread over a wide range in the frequency domain, and the above evaluation will not apply. However, if the amplitude function of the disturbance varies slowly in the frequency domain, the propagating disturbance may often be evaluated by the *method of stationary phase*, due to Kelvin and Stokes. More precisely, consider an amplitude function $A(\kappa)$, which changes little during a change in 2π of the phase $\phi(\kappa) = \kappa(\xi - c\tau)$. We have $e^{i(\phi+\pi)} = -e^{i\phi}$. Hence, over a range of κ for which $d\phi/d\kappa \neq 0$, the elemental waves corresponding to the first π increment in a 2π change in $\phi(\kappa)$ are largely annulled by those corresponding to the second π increment. However, in the neighborhood of κ for which ϕ is stationary ($d\phi/d\kappa = 0$), the elemental waves reinforce one another. Hence, if the amplitude function $A(\kappa)$ is slowly varying, the integral of Eq. (64.17) can be approximately evaluated by considering only the group of waves clustered around stationary phase values of κ. These are given by

$$\frac{d\phi}{d\kappa} = \xi - \frac{d(\kappa c)}{d\kappa}\tau = \xi - c_g\tau = 0$$

or by the value κ for which the corresponding group velocity is $c_g = \xi/\tau$. Thus, in the case of the Bernoulli-Euler beam, for which $c_g = 2\kappa$, a group of harmonic wave trains of wave number near $\kappa = \xi/2\tau$ make up the main contribution to the propagating disturbance.

To carry out the approximation of the integral[1] in Eq. (64.17), we let κ_0 represent the stationary phase value of κ, and denote by a zero subscript values of functions of κ evaluated at κ_0. Expansion of the phase about κ_0 gives

$$\phi(\kappa) = \kappa(\xi - ct) = \kappa_0(\xi - c_0\tau) - \frac{(\kappa - \kappa_0)^2}{2}\left(\frac{dc_g}{d\kappa}\right)_0 \tau + \cdots$$

where we have made use of $[d\phi/d\kappa]_0 = 0$. Hence, we have

$$w(\xi,\tau) \approx \frac{A(\kappa_0)}{2\pi} e^{i\kappa_0(\xi - c_0\tau)} \int_{-\infty}^{\infty} \exp\left[-i\frac{(\kappa - \kappa_0)^2}{2}\left(\frac{dc_g}{d\kappa}\right)_0 \tau\right] d\kappa$$

Using the known result,

$$\int_{-\infty}^{\infty} e^{\pm im^2\eta^2}\, d\eta = \frac{\sqrt{\pi}}{m}\, e^{\pm i\pi/4}$$

we get

$$w(\xi,\tau) \approx A(\kappa_0)\left[2\pi\left|\left(\frac{dc_g}{d\kappa}\right)_0\right|\tau\right]^{-1/2} \exp i\left[\kappa_0(\xi - c_0\tau) \pm \frac{\pi}{4}\right]$$

[1] For a more complete discussion of the evaluation of integrals of this type by the method of stationary phase, together with a list of references on the subject, the reader is referred to [2].

where the sign is to be chosen opposite the sign of $(dc_g/d\kappa)_0$. For example, if a bending-moment impulse of unit strength, that is, $M(\xi) = \delta(\xi)$, is applied to a Bernoulli-Euler beam initially at rest, the resulting bending moment in the beam can be shown to be

$$M(\xi,\tau) = \Re\left(\frac{1}{2\pi}\int_{-\infty}^{\infty} e^{i\kappa(\xi - c\tau)}\,d\kappa\right)$$

In this case $A(\kappa_0) = 1$, $\kappa_0 = c_0 = \xi/2\tau$, $(dc_g/d\kappa)_0 = 2$, and the stationary phase evaluation is

$$M(\xi,\tau) = (4\pi\tau)^{-1/2}\cos\left(\frac{\xi^2}{4\tau} - \frac{\pi}{4}\right) \tag{64.19}$$

It can be verified that, in this particular example, the method of stationary phase gives, in fact, the exact result.

Still another aspect of wave propagation which travels at the group velocity is the energy of a harmonic wave train. To keep physical quantities in focus, we return to the Bernoulli-Euler equation of motion in terms of x and t [Eq. (64.14)]. We denote the corresponding phase velocities, group velocities, and wave numbers by bars: $\bar{c}$, $\bar{c}_g$, $\bar{\kappa}$. In a harmonic wave train $w = w_0 \cos \bar{\kappa}(x - \bar{c}t)$, we have $\bar{c} = \bar{\kappa}(EI/\mu)^{1/2}$. The average energy $\mathcal{E}_{av}$ per unit length in a length of beam equal to one wavelength is

$$\mathcal{E}_{av} = \frac{\bar{\kappa}}{2\pi}\int_0^{2\pi/\bar{\kappa}} \left[\frac{EI}{2}\left(\frac{\partial^2 w}{\partial x^2}\right)^2 + \frac{\mu}{2}\left(\frac{\partial w}{\partial t}\right)^2\right] dx = \frac{EI}{2}\,\bar{\kappa}^4 w_0^2 \tag{64.20}$$

We note that $\mathcal{E}_{av}$ is independent of x and t. The wave train continuously transports energy as it propagates, the rate of transport into one end of a section of the beam being, on the average, equal to the rate out of the other end. This rate of transport, or energy flux, is given by the rate of work of the forces acting on a beam cross section. Over one period $T = 2\pi/\bar{\kappa}\bar{c}$ the average energy flux $\mathcal{G}_{av}$ is given by

$$\begin{aligned}\mathcal{G}_{av} &= \frac{1}{T}\int_0^T \left(M\,\frac{\partial^2 w}{\partial x\,\partial t} + V\,\frac{\partial w}{\partial t}\right) dt \\ &= \frac{1}{T}\int_0^T EI\left(-\frac{\partial^2 w}{\partial x^2}\frac{\partial^2 w}{\partial x\,\partial t} + \frac{\partial^3 w}{\partial x^3}\frac{\partial w}{\partial t}\right) dt \\ &= EI\bar{\kappa}^4\bar{c}w_0^2\end{aligned}$$

We have then, from the above and Eq. (64.20),

$$\frac{\mathcal{G}_{av}}{\mathcal{E}_{av}} = \bar{c}_g \tag{64.21}$$

Thus the rate at which energy is transported is here proportional to the group velocity. Equation (64.21) has been shown by Biot [6] to hold very generally for conservative linear media.

For short wavelengths ($\bar{\kappa} \to \infty$), the group velocity $\bar{c}_g = 2\bar{\kappa}(EI/\mu)^{1/2}$ approaches infinity. As is shown by the solution to the moment-impulse problem [Eq. (64.19)], this means that a suddenly applied disturbance is immediately felt along the beam. However, one would expect the first effects of a disturbance, no matter how sudden, to propagate at a finite velocity. This anomaly is the result of the approximations made at the outset in the Bernoulli-Euler theory. A refinement of the Bernoulli-Euler theory that does not contain this deficiency is the *Timoshenko theory*, discussed in the next section.

64.6. FLEXURAL WAVES IN A BEAM (TIMOSHENKO THEORY)

In the Timoshenko beam theory [7], it is assumed, as in the Bernoulli-Euler theory, that originally plane cross sections of the beam remain plane. However, the plane cross sections are not assumed to remain at right angles to the neutral axis of the loaded beam. Figure 64.5 shows the forces and deflections associated with a loaded element of a beam, according to the Timoshenko theory. Newton's law gives

$$\frac{\partial V}{\partial x} = \mu \frac{\partial^2 w}{\partial t^2} \qquad V - \frac{\partial M}{\partial x} = \rho I \frac{\partial^2 \psi}{\partial t^2} \tag{64.22a,b}$$

Here ψ is the rotation of an originally vertical cross section, and ρI is the rotatory inertia per unit length of the beam. From Hooke's law and kinematic considerations, the following are obtained:

$$-EI \frac{\partial \psi}{\partial x} = M \qquad V = -GA_s \left(\psi - \frac{\partial w}{\partial x} \right) \tag{64.23a,b}$$

It is assumed in Eq. (64.23b) that the angle change $(\psi - \partial w/\partial x)$ is caused by the total shear force V acting on an *effective shear area* A_S, which, in general, will differ from the actual cross-sectional area. Similarly, the area moment of inertia used to compute EI may differ from the mass moment of inertia used to determine ρI.

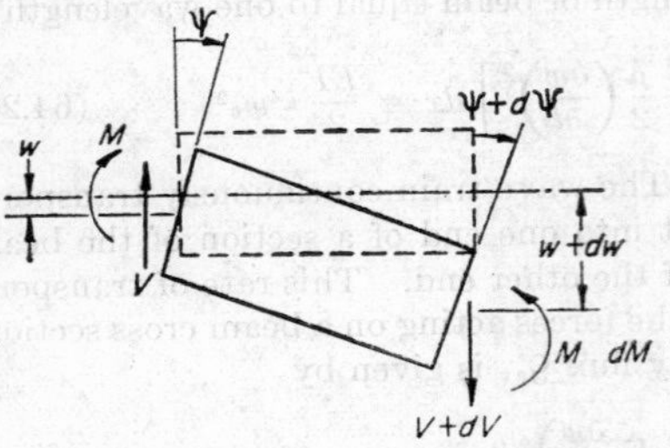

FIG. 64.5. Force on a differential element in the Timoshenko theory.

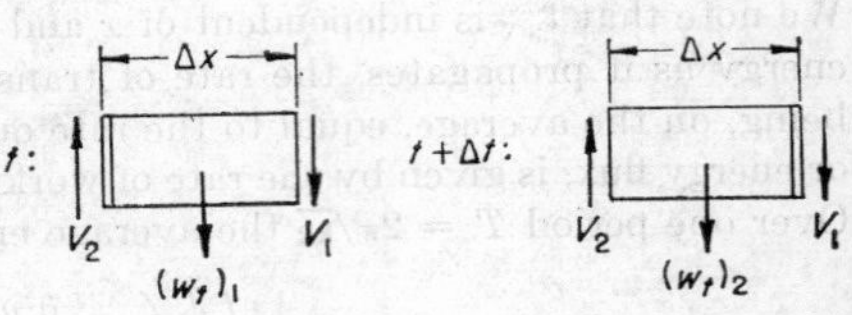

FIG. 64.6. Shear wave passing through a differential element.

By differentiation and elimination, we can reduce the four above equations to a single equation of fourth order in w:

$$\frac{\partial^4 w}{\partial x^4} - \left(\frac{1}{c_M{}^2} + \frac{1}{c_V{}^2} \right) \frac{\partial^4 w}{\partial x^2\, \partial t^2} + \frac{1}{c_M{}^2 c_V{}^2} \frac{\partial^4 w}{\partial t^4} + K \frac{\partial^2 w}{\partial t^2} = 0 \tag{64.24}$$

where $\qquad c_M{}^2 = \dfrac{EI}{\rho I} \qquad c_V{}^2 = \dfrac{GA_s}{\mu} \qquad K = \dfrac{\mu}{EI}$

In a homogeneous beam, $EI/\rho I = E/\rho$, and $c_M = c_B$, the bar velocity.

Like the theories governed by the basic wave equation, the Timoshenko theory predicts that discontinuities will propagate at definite, finite velocities. This was first shown by Flügge [8] as follows. It is observed at the outset that there can be no discontinuities in w and ψ without tearing, and these can, therefore, be taken as continuous. Assume then that a jump in V is traveling in the beam at a constant velocity. Let $V_1(t)$ and $V_2(t)$ be the shearing forces on the right and left sides of an element Δx, through which the jump propagates during a time Δt (Fig. 64.6). We denote quantities to the right and left of the jump by the subscripts 1 and 2, respectively, and define the jump operator δ by $\delta(\ \ ) = (\ \ )_1 - (\ \ )_2$. From the impulse-momentum law,

$$\int_t^{t+\Delta t} \delta V\, dt = -\mu\, \Delta x\, \delta \left(\frac{\partial w}{\partial t} \right)$$

Dividing by Δt and letting Δt approach zero, we have

$$\delta V = -\mu c\, \delta \left(\frac{\partial w}{\partial t}\right) \tag{64.25}$$

Hence, a jump in V is accompanied by a jump in $\partial w/\partial t$, both traveling at the as yet unknown velocity c. To evaluate c, we note that $\delta(\partial w/\partial t) + c\,\delta(\partial w/\partial x) = 0$ because of the continuity of w. Hence, by Eq. (64.25), $\delta V = \mu c^2\,\delta(\partial w/\partial x)$. But, from Eq. (64.23*b*) (Hooke's law), $\delta V = GA_S[\delta(\partial w/\partial x) - \delta\psi] = GA_S\,\delta(\partial w/\partial x)$, since $\delta\psi = 0$ by continuity. Hence,

$$\mu c^2\,\delta\left(\frac{\partial w}{\partial x}\right) = GA_S\,\delta\left(\frac{\partial w}{\partial x}\right)$$

and, for $\delta(\partial w/\partial x) \neq 0$, we arrive at $c^2 = GA_S/\mu = c_V{}^2$ as the velocity of propagation of a discontinuity in V, $\partial w/\partial t$, or $\partial w/\partial x$. By similar reasoning, we find that the propagation velocity of a jump in M, $\partial\psi/\partial t$, or $\partial\psi/\partial x$ is $c_M{}^2 = EI/\rho I$. Thus, in the case of the Timoshenko beam, discontinuities can, in general, propagate at two distinct velocities c_M and c_V, called the *bending wave* and *shear wave* velocities, respectively. Furthermore, it can be shown that, in a uniform bar, the magnitude of a jump discontinuity remains *constant* as it propagates.

The substitution of $w = e^{i\kappa(x-ct)}$ into the equation of motion (64.24) shows that the phase velocity c, and hence, also the group velocity c_g, is a double-valued function of the wave number κ. This means that there are two modes of propagation possible. The dispersion curves for these two modes for a circular bar of radius a and Poisson's ratio $\nu = 0.29$ are shown in Fig. 64.9. Here we have followed Mindlin's [9] suggestion, and have chosen A_S in Eq. (64.23*b*) such that c_V is equal to the Rayleigh velocity (see Sec. 64.9). In the lower mode, denoted by the subscript I, the slope $\partial w/\partial x$ and the rotation ψ are in phase, while, in the upper mode, denoted by the subscript II, they are 180° out of phase. We observe, for both the phase and group velocities, that the lower-mode asymptote for $\kappa \to \infty$ is the shear wave velocity c_V and that the upper-mode asymptote is the bending wave velocity c_M.

Recently, theories similar to the Timoshenko theory have been developed and applied to a wide variety of elastic bodies such as plates, shells, crystals, etc., especially by R. D. Mindlin [9] and his coworkers.

64.7. WAVES IN AN UNBOUNDED ISOTROPIC MEDIUM

To obtain the equations of motion of classical elasticity, we need only to consider the inertia forces $-\rho\,\partial^2u/\partial t^2$, $-\rho\,\partial^2v/\partial t^2$, $-\rho\,\partial^2w/\partial t^2$ as body forces. Then, if no other body forces are acting, the equations of equilibrium (33.56) in rectangular cartesian coordinates yield directly

$$\begin{aligned}
\frac{G}{1-2\nu}\frac{\partial e}{\partial x} + G\,\nabla^2 u &= \rho\frac{\partial^2 u}{\partial t^2}\\
\frac{G}{1-2\nu}\frac{\partial e}{\partial y} + G\,\nabla^2 v &= \rho\frac{\partial^2 v}{\partial t^2}\\
\frac{G}{1-2\nu}\frac{\partial e}{\partial z} + G\,\nabla^2 w &= \rho\frac{\partial^2 w}{\partial t^2}
\end{aligned} \tag{64.26a–c}$$

If these equations are differentiated with respect to x, y, and z, respectively, and added, the result is

$$\nabla^2 e = \frac{1}{c_1{}^2}\frac{\partial^2 e}{\partial t^2} \tag{64.27}$$

where $c_1{}^2 = 2(1-\nu)G/(1-2\nu)\rho$. This is the wave equation in three dimensions. If the dilatation e is a function only of the radius r from some point in space, a general solution of Eq. (64.27) is the sum, multiplied by $1/r$, of a wave propagating in the positive r direction and one propagating in the negative r direction:

$$e = \frac{1}{r}[F(r - c_1t) + G(r + c_1t)] \tag{64.28}$$

The velocity c_1 is called the *dilatational* velocity.

Likewise, if we differentiate Eq. (64.26*b*) with respect to z and Eq. (64.26*c*) with respect to y and subtract, we find that the x component of the rotation $\omega_x = \frac{1}{2}(\partial w/\partial y - \partial v/\partial z)$ satisfies

$$\nabla^2 \omega_x = \frac{1}{c_2^2}\frac{\partial^2 \omega_x}{\partial t^2} \tag{64.29}$$

where $c_2^2 = G/\rho$. Similar equations obtain for the other components of the rotation ω_y, ω_z. Hence, the rectangular cartesian components of the rotation are governed by a wave equation with the propagation constant c_2. This velocity is variously known as the *distortional, rotational, equivoluminal,* or *shear* velocity.

The above may be extended to arbitrary coordinate systems by working with the equations of motion (64.26) in vector form,

$$\rho\frac{\partial^2 \mathbf{u}}{\partial t^2} = \frac{G}{1 - 2\nu}\operatorname{grad}\operatorname{div}\mathbf{u} + G\,\nabla^2\mathbf{u} \tag{64.30}$$

where $\mathbf{u}$ is the displacement vector, and ∇^2 is the generalized vector form of the Laplacian operator, given by Eq. (6.23*b*). Equation (64.30) follows immediately from (64.26) if it is noted that the dilatation is div $\mathbf{u}$. Taking the divergence of both sides of Eq. (64.30), we get Eq. (64.27) directly. Noting Eqs. (6.14*a*) and (6.23*c*), we find, on taking the curl of both sides of Eq. (64.30), that

$$\frac{\partial^2 \boldsymbol{\omega}}{\partial t^2} = c_2^2\,\nabla^2\boldsymbol{\omega} \tag{64.31}$$

where $\boldsymbol{\omega} = \frac{1}{2}$ curl $\mathbf{u}$. Equation (64.31) is a generalized form of Eq. (64.29). We note that the components of the rotation vector will not, in general, individually satisfy the scalar wave equation.

The equations of motion of classical elasticity may thus be equivalently written as a scalar wave equation governing the dilatation and a vector wave equation governing the components of the rotation. In two special cases, the components of the displacements in rectangular cartesian coordinates are themselves governed by the wave equation.

If the dilatation e is zero, Eqs. (64.26) yield

$$\nabla^2 u = \frac{1}{c_2^2}\frac{\partial^2 u}{\partial t^2} \tag{64.32}$$

with similar equations for v and w. Zero dilatation implies no volume change. Hence, these waves are called *equivoluminal waves,* or *waves of distortion.*

Likewise, if the rotation is zero, the rectangular cartesian displacements satisfy the wave equation. To show this, we note the identity

$$\nabla^2 u = \frac{\partial e}{\partial x} - \frac{\partial}{\partial y}\left(\frac{\partial v}{\partial x} - \frac{\partial u}{\partial y}\right) + \frac{\partial}{\partial z}\left(\frac{\partial u}{\partial z} - \frac{\partial w}{\partial x}\right)$$

which, in the case of vanishing rotation, gives, upon substitution into Eq. (64.26*a*),

$$\nabla^2 u = \frac{1}{c_1^2}\frac{\partial^2 u}{\partial t^2}$$

In a similar fashion, the same equation can be shown to govern v and w. These waves are called *irrotational waves or waves of dilatation*.

We can again write the equations corresponding to the vanishing dilatation and rotation cases for a general coordinate system in vector notation. For vanishing dilatation we have div $\mathbf{u} = 0$, which gives, from Eq. (64.30),

$$\nabla^2\mathbf{u} = \frac{1}{c_2{}^2}\frac{\partial^2\mathbf{u}}{\partial t^2}$$

Similarly, if the rotation vanishes, we have curl $\mathbf{u} = 0$. The vector identity (6.23*b*), together with Eq. (64.30), then gives

$$\nabla^2\mathbf{u} = \frac{1}{c_1{}^2}\frac{\partial^2\mathbf{u}}{\partial t^2}$$

Any plane wave must travel at either the dilatational or the distortional velocity in an unbounded elastic medium. For, if we take the x axis as the direction of propagation of such a wave, then u, v, and w are of the form

$$u = u(x - ct) \qquad v = v(x - ct) \qquad w = w(x - ct)$$

If differentiation with respect to argument is denoted by a prime, Eqs. (64.26) become in this case

$$\frac{2(1-\nu)}{1-2\nu}Gu'' = \rho c^2 u'' \qquad Gv'' = \rho c^2 v'' \qquad Gw'' = \rho c^2 w''$$

These equations, in turn, can be satisfied only if $c = c_1$ or $c = c_2$. In the former case, the particle motion is longitudinal or along the direction of propagation; in the latter case, it is transverse or perpendicular to the propagation direction.

64.8. REFLECTION OF PLANE WAVES AT A PLANE BOUNDARY

Of interest is how plane waves reflect at a plane, stress-free boundary. To illustrate the reflection phenomenon, we consider a harmonic plane wave whose component displacements are of the form

$$(u,v,w) = (u_0,v_0,w_0)e^{i(\alpha x+\gamma z-pt)} \tag{64.33}$$

in the half space $z < 0$, with $z = 0$ the stress-free boundary. In the planes $\alpha x + \gamma z =$ const, the motion is identical. If $p > 0$, a vector normal to these planes

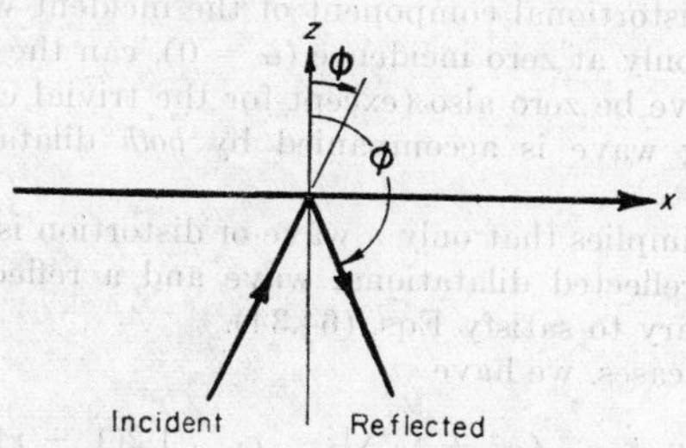

FIG. 64.7. Definition of incident and reflected plane waves.

and pointing in the direction of propagation of the wave makes the angle ϕ with the z axis, where $\sin\phi = \alpha/(\alpha^2 + \gamma^2)^{1/2}$, $\cos\phi = \gamma/(\alpha^2 + \gamma^2)^{1/2}$, and ϕ is clockwise positive (Fig. 64.7). We call the wave an incident wave when $0 \le \phi < \pi/2$ ($\alpha \ge 0$, $\gamma > 0$) and a reflected wave when $\pi/2 < \phi \le \pi$ ($\alpha \ge 0, \gamma < 0$).

Consider, first, transverse waves in which the particle motion is parallel to the y axis; that is, $u = w = 0$. These are waves of shear or distortion for which, by the previous result, $c = c_2$. A solution of the equations of motion representing an incident and a reflected wave with equal angles of incidence and reflection is, thus,

$$v = v_{01}e^{i(\alpha x + \gamma z - pt)} + v_{02}e^{i(\alpha x - \gamma z - pt)}$$

where α, γ, and p are now all positive numbers. It is easily verified that the boundary conditions,

$$\sigma_z = \frac{2\nu}{1 - 2\nu} Ge + 2G\epsilon_z = 0 \qquad \tau_{xz} = G\left(\frac{\partial u}{\partial z} + \frac{\partial w}{\partial x}\right) = 0 \qquad \tau_{yz} = G\left(\frac{\partial v}{\partial z} + \frac{\partial w}{\partial y}\right) = 0 \tag{64.34}$$

at $z = 0$ imply $v_{01} = -v_{02}$. Therefore, in this case, we have ordinary Snell's law reflection.

Consider next the case where $v = 0$, that is, no motion parallel to the y axis. The particle motion can be resolved into a longitudinal component along the propagation direction and a transverse component normal to it. The former will propagate at the dilatational velocity $c = c_1$ and have u_0, w_0 in the ratio $u_0/w_0 = \alpha/\gamma$; the latter will propagate at the shear velocity $c = c_2$ and have u_0, w_0 in the ratio $u_0/w_0 = -\gamma/\alpha$. However, it is found, in general, in this case, that an incident and a reflected dilatational wave by themselves cannot satisfy the conditions of no stress on the boundary $z = 0$, nor can an incident and reflected distortional wave alone satisfy these conditions. On the other hand, a superposition of incident and reflected dilatational *and* distortional waves can satisfy the stress-free condition. The superposed waves can be written in the form

$$\begin{aligned} u &= \alpha_1(u_{01}e^{i\gamma_1 z} + u_{02}e^{-i\gamma_1 z})e^{i(\alpha_1 x - p_1 t)} + \gamma_2(u_{03}e^{i\gamma_2 z} + u_{04}e^{-i\gamma_2 z})e^{i(\alpha_2 x - p_2 t)} \\ w &= \gamma_1(u_{01}e^{i\gamma_1 z} - u_{02}e^{-i\gamma_1 z})e^{i(\alpha_1 x - p_1 t)} - \alpha_2(u_{03}e^{i\gamma_2 z} - u_{04}e^{-i\gamma_2 z})e^{i(\alpha_2 x - p_2 t)} \end{aligned} \tag{64.35}$$

where $\alpha_1^2 + \gamma_1^2 = p^2/c_1^2$, $\alpha_2^2 + \gamma_2^2 = p^2/c_2^2$. Substituting from Eqs. (64.35) into the boundary conditions (64.34), it is found that

$$\alpha_1 = \alpha_2 = \alpha \qquad p_1 = p_2 = p$$

and that

$$\begin{aligned} (u_{01} + u_{02})(\gamma_2^2 - \alpha^2) - (u_{03} + u_{04})2\alpha\gamma_2 &= 0 \\ (u_{01} - u_{02})2\alpha\gamma_1 + (u_{03} - u_{04})(\gamma_2^2 - \alpha^2) &= 0 \end{aligned} \tag{64.36}$$

If $u_{03} = 0$, then the distortional component of the incident wave is zero. But Eqs. (64.36) indicate that, only at zero incidence ($\alpha = 0$), can the distortional component u_{04} of the reflected wave be zero also (except for the trivial case $u = w = 0$), and, in general, an impinging wave is accompanied by *both* dilatational *and* distortional reflected waves.

Similarly, $u_{01} = 0$ implies that only a wave of distortion is incident on the boundary. Again, both a reflected dilatational wave and a reflected wave of distortion are, in general, necessary to satisfy Eqs. (64.34).

In both the above cases, we have

$$\frac{\sin\phi_1}{\sin\phi_2} = \left(\frac{\alpha^2 + \lambda_2^2}{\alpha^2 + \lambda_1^2}\right)^{1/2} = \frac{c_1}{c_2} = \left[\frac{2(1 - \nu)}{1 - 2\nu}\right]^{1/2}$$

where ϕ_1 and ϕ_2 refer to the waves of dilatation and distortion, respectively. Thus, if the incident wave is purely dilatational, ϕ_1 is its angle of incidence, and ϕ_2 is the angle of reflection of the resulting distortional wave. For a purely distortional incident wave, ϕ_2 is the angle of incidence and ϕ_1 is the angle of reflection of the dilatational wave. In the latter case, there is no reflected dilatational wave if ϕ_2 exceeds a critical angle defined by $\sin\phi_2 = c_2/c_1$. If it is assumed that Poisson's ratio ν varies from

0 to ½, this critical angle varies from a maximum value of 45° at $\nu = 0$ to a minimum value of zero at $\nu = \frac{1}{2}$.

Grazing incidence has been studied by Goodier and Bishop [10]. By means of a limiting process, they have shown that both grazing dilatational and grazing shear waves give rise to reflections which contain a term whose amplitude is linear in z.

64.9. RAYLEIGH SURFACE WAVES

The last section has shown that the presence of a boundary changes the wave-propagation phenomenon from that which occurs in an unbounded medium. Rayleigh found, for example, that waves could propagate without dispersion in a half space ($z < 0$). These waves, which are of considerable seismic importance, decay away from the surface exponentially with the depth z, and are called *Rayleigh surface waves*.

We take the x direction as the direction of propagation, and assume that $v = 0$ and that

$$u = \frac{\partial \phi}{\partial x} - \frac{\partial \psi}{\partial z} \qquad w = \frac{\partial \phi}{\partial z} + \frac{\partial \psi}{\partial x} \tag{64.37}$$

Then the dilatation $e = \partial u/\partial x + \partial w/\partial z$ is given by $\nabla^2\phi$, and the y component of the rotation by $\omega_y = -\frac{1}{2}\nabla^2\psi$. From our previous results [Eqs. (64.31) and (64.32)], the dilatation and the components of rotation in rectangular cartesian coordinates satisfy wave equations, and we have

$$\frac{\partial^2(\nabla^2\phi)}{\partial t^2} = c_1^2\nabla^2(\nabla^2\phi) \qquad \frac{\partial^2(\nabla^2\psi)}{\partial t^2} = c_2^2\nabla^2(\nabla^2\psi) \tag{64.38}$$

Clearly solutions of

$$\frac{\partial^2\phi}{\partial t^2} = c_1^2\nabla^2\phi \qquad \frac{\partial^2\psi}{\partial t^2} = c_2^2\nabla^2\psi \tag{64.39}$$

are solutions of Eqs. (64.38) as well. As solutions of Eqs. (64.39), we take, in turn, ϕ and ψ of the forms

$$\begin{aligned} \phi &= A \exp[-qz + i\kappa(x - c_3t)] \\ \psi &= B \exp[-sz + i\kappa(x - c_3t)] \end{aligned} \tag{64.40}$$

where κ is the wave number, and c_3 is the Rayleigh wave velocity. It is found, on substitution into Eqs. (64.39), that

$$q^2 = \kappa^2\left(1 - \frac{c_3^2}{c_1^2}\right) \qquad s^2 = \kappa^2\left(1 - \frac{c_3^2}{c_2^2}\right) \tag{64.41}$$

We note that we must have $c_3/c_2 < 1$ if the waves are to have exponential decay with depth.

The boundary conditions $\sigma_z = \tau_{xz} = \tau_{yz} = 0$ on $z = 0$ yield two homogeneous equations for A and B:

$$2iq\kappa A + [2 - (c_3^2/c_2^2)]\kappa^2 B = 0 \qquad [2 - (c_3^2/c_2^2)]\kappa^2 A - 2is\kappa B = 0$$

The resulting determinantal equation can be put in the form

$$\left(2 - \frac{c_3^2}{c_2^2}\right)^4 = 16\left(1 - \frac{c_3^2}{c_1^2}\right)\left(1 - \frac{c_3^2}{c_2^2}\right) \tag{64.42}$$

which is a cubic for c_3^2/c_2^2. Since $c_2^2/c_1^2 = (1 - 2\nu)/2(1 - \nu)$, the roots of this cubic depend only on Poisson's ratio ν. Hence, the wave velocity c_3 is independent of wavelength, and the waves are nondispersive.

In general, only one of the roots of Eq. (64.42) is real and less than unity, and gives a wave with an exponential decrease with z. The value of this root increases monotonically from $c_3/c_2 = 0.8740$ at $\nu = 0$ to $c_3/c_2 = 0.9553$ at $\nu = \frac{1}{2}$. For the value $\nu = 0.29$, which is commonly used for steel, the root is $c_3/c_2 = 0.9258$.

64.10. WAVES IN AN INFINITE SLAB

By extending the development for Rayleigh waves, we may investigate wave propagation in an infinite slab, that is, the plane-strain analog of a beam. We take the x,y plane as the median plane of a slab, which is bounded by stress-free surfaces at $z = \pm d$, and assume wave propagation in the positive x direction. Equations (64.37) through (64.40), which yield a solution of the equations of motion in terms of two functions ϕ and ψ, follow as before—except that now, for ϕ and ψ, we use the form

$$\begin{aligned} \phi &= (A_1 \operatorname{Cosh} qz + A_2 \operatorname{Sinh} qz)e^{i\kappa(x-ct)} \\ \psi &= (B_1 \operatorname{Cosh} sz + B_2 \operatorname{Sinh} sz)e^{i\kappa(x-ct)} \end{aligned} \tag{64.43}$$

where c is the phase velocity corresponding to the wave number κ, and

$$q^2 = \kappa^2(1 - c^2/c_1^2) \qquad s^2 = \kappa^2(1 - c^2/c_2^2) \tag{64.44}$$

We ask for free-vibration solutions of the above type which satisfy the boundary conditions $\sigma_z = \tau_{xz} = \tau_{yz} = 0$ on $z = \pm d$. These conditions yield four homogeneous equations for A_1, A_2, B_1, and B_2, which split into two sets:

$$\text{(I)} \qquad \begin{aligned} \kappa^2\left(2 - \frac{c^2}{c_2^2}\right) A_1 \operatorname{Cosh} qd + 2is\kappa B_2 \operatorname{Cosh} sd &= 0 \\ 2iq\kappa A_1 \operatorname{Sinh} qd - \kappa^2\left(2 - \frac{c^2}{c_2^2}\right) B_2 \operatorname{Sinh} sd &= 0 \end{aligned}$$

$$\text{(II)} \qquad \begin{aligned} \kappa^2\left(2 - \frac{c^2}{c_2^2}\right) A_2 \operatorname{Sinh} qd + 2is\kappa B_1 \operatorname{Sinh} sd &= 0 \\ 2iq\kappa A_2 \operatorname{Cosh} qd - \kappa^2\left(2 - \frac{c^2}{c_2^2}\right) B_1 \operatorname{Cosh} sd &= 0 \end{aligned}$$

From Eqs. (64.37) it is verified that set I represents *longitudinal* propagation modes, in which u is even and w odd about $z = 0$, and set II represents *transverse* modes, for which u is odd and w is even about $z = 0$.

The determinantal equations for these two sets are

$$\text{(I)} \qquad \frac{\operatorname{Tanh} sd}{\operatorname{Tanh} qd} = \frac{4sq\kappa^2}{(\kappa^2 + s^2)^2}$$

$$\text{(II)} \qquad \frac{\operatorname{Tanh} qd}{\operatorname{Tanh} sd} = \frac{4sq\kappa^2}{(\kappa^2 + s^2)^2}$$

These equations, together with Eqs. (64.44), determine the phase velocity c as a multivalued function of the wave number κ. For Poisson's ratio $\nu = 0.29$, the first two branches of this relationship, corresponding to the two lowest modes of propagation, are plotted in Figs. 64.8 and 64.9 for longitudinal (case I) and transverse (case II) propagation, respectively. The main features of the behavior of the phase velocity may be summarized as follows.

1. For the lowest mode of both longitudinal and transverse propagation, the limiting phase velocity for short wavelengths ($\kappa \to \infty$) is the Rayleigh wave velocity.

2. For all higher modes, the limiting phase velocity, in both cases I and II, is the distortional velocity c_2.

3. For first-mode longitudinal waves (Fig. 64.8) the limiting phase velocity for long waves ($\kappa \to 0$) is $c = E/\rho(1 - \nu^2)$. This is the bar velocity c_B with the additional factor $(1 - \nu^2)$ caused by the plane-strain constraint $v = 0$.

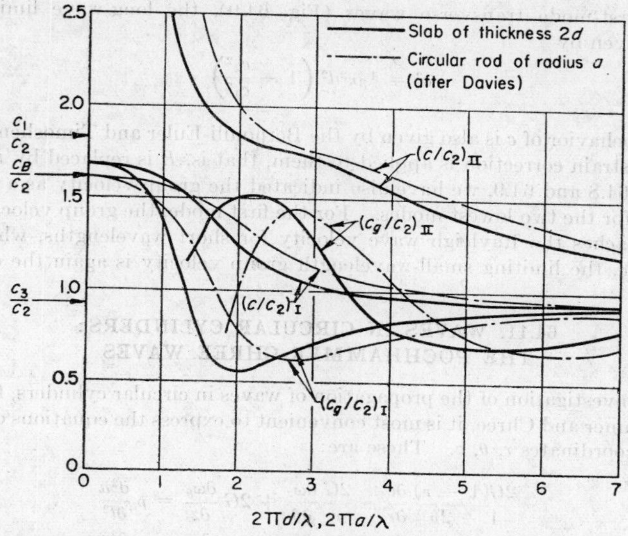

FIG. 64.8. Dispersion curves for the two lowest longitudinal modes in a slab and in a circular rod (Poisson's ratio $\nu = 0.29$).

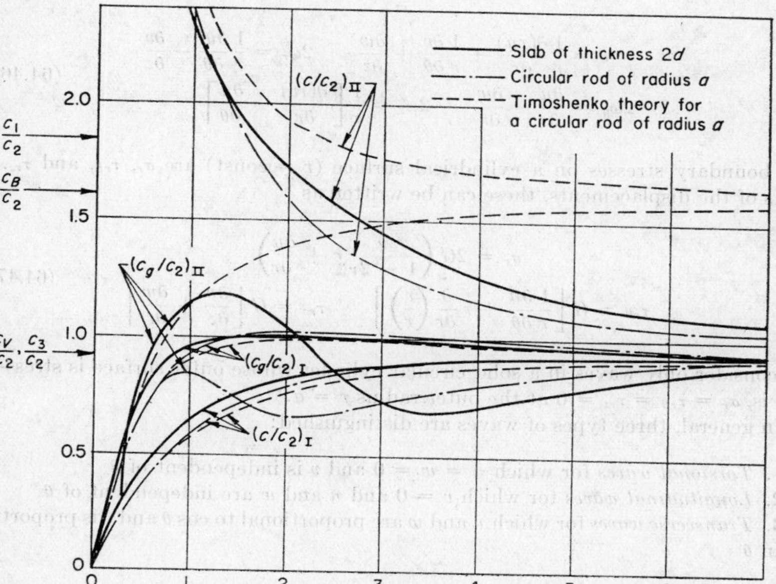

FIG. 64.9. Dispersion curves for the Timoshenko theory and for the two lowest transverse modes in a slab and in a circular rod (Poisson's ratio $\nu = 0.29$).

4. For first-mode transverse waves (Fig. 64.9), the long-wave limiting phase velocity is given by

$$c^2 = \tfrac{4}{3}\kappa^2 d^2 \left(1 - \frac{c_2^2}{c_1^2}\right)$$

This initial behavior of c is also given by the Bernoulli-Euler and Timoshenko theories if the plane-strain correction is applied to them, that is, E is replaced by $E/(1 - \nu^2)$.

In Figs. 64.8 and 64.9, we have also indicated the group velocity as a function of wavelength for the two lowest modes. For the first mode, the group velocity, in both cases, approaches the Rayleigh wave velocity for short wavelengths, while, for the second mode, the limiting small-wavelength group velocity is again the distortional velocity c_2.

64.11. WAVES IN CIRCULAR CYLINDERS: THE POCHHAMMER-CHREE WAVES

For the investigation of the propagation of waves in circular cylinders, first studied by Pochhammer and Chree, it is most convenient to express the equations of motion in cylindrical coordinates r, θ, z. These are:

$$\begin{aligned}
\frac{2G(1-\nu)}{1-2\nu}\frac{\partial e}{\partial r} - \frac{2G}{r}\frac{\partial \omega_z}{\partial \theta} + 2G\frac{\partial \omega_\theta}{\partial z} &= \rho\frac{\partial^2 u}{\partial t^2} \\
\frac{2G(1-\nu)}{1-2\nu}\frac{1}{r}\frac{\partial e}{\partial \theta} - 2G\frac{\partial \omega_r}{\partial z} + 2G\frac{\partial \omega_z}{\partial r} &= \rho\frac{\partial^2 v}{\partial t^2} \\
\frac{2G(1-\nu)}{1-2\nu}\frac{\partial e}{\partial z} - \frac{2G}{r}\frac{\partial (r\omega_\theta)}{\partial r} + \frac{2G}{r}\frac{\partial \omega_r}{\partial \theta} &= \rho\frac{\partial^2 w}{\partial t^2}
\end{aligned} \qquad (64.45a\text{–}c)$$

where the dilatation e and the components of the rotation ω_r, ω_θ, ω_z are, in turn, given by

$$\begin{aligned}
e &= \frac{1}{r}\frac{\partial (ru)}{\partial r} + \frac{1}{r}\frac{\partial v}{\partial \theta} + \frac{\partial w}{\partial z} \qquad & 2\omega_r &= \frac{1}{r}\frac{\partial w}{\partial \theta} - \frac{\partial v}{\partial z} \\
2\omega_\theta &= \frac{\partial u}{\partial z} - \frac{\partial w}{\partial r} \qquad & 2\omega_z &= \frac{1}{r}\left[\frac{\partial (rv)}{\partial r} - \frac{\partial u}{\partial \theta}\right]
\end{aligned} \qquad (64.46a\text{–}d)$$

The boundary stresses on a cylindrical surface (r = const) are σ_r, $\tau_{r\theta}$, and τ_{rz}. In terms of the displacements, these can be written as

$$\begin{aligned}
\sigma_r &= 2G\left(\frac{\nu}{1-2\nu}e + \frac{\partial u}{\partial r}\right) \\
\tau_{r\theta} &= G\left[\frac{1}{r}\frac{\partial u}{\partial \theta} + r\frac{\partial}{\partial r}\left(\frac{v}{r}\right)\right] \qquad \tau_{rz} = G\left[\frac{\partial u}{\partial z} + \frac{\partial w}{\partial r}\right]
\end{aligned} \qquad (64.47a\text{–}c)$$

We consider only waves in a solid circular cylinder whose outer surface is stress-free; that is, $\sigma_r = \tau_{r\theta} = \tau_{rz} = 0$ at the outer radius $r = a$.

In general, three types of waves are distinguished:

1. *Torsional waves* for which $u = w = 0$ and v is independent of θ
2. *Longitudinal waves* for which $v = 0$ and u and w are independent of θ
3. *Transverse waves* for which u and w are proportional to $\cos\theta$ and v is proportional to $\sin\theta$

We consider each of these in turn with the substitution

$$u = U(r,\theta)e^{i\kappa(z-ct)} \qquad v = V(r,\theta)e^{i\kappa(z-ct)} \qquad w = W(r,\theta)e^{i\kappa(z-ct)} \qquad (64.48)$$

Torsional Waves

In this case, $e = \omega_\theta = 0$, ω_r and ω_z are independent of z, and only the second of Eqs. (64.45) is not identically zero. Its solution and the stress-free condition at the boundary lead to

$$\frac{\partial}{\partial r}\left(\frac{J_1(kr)}{r}\right)_{r=a} = 0 \qquad k^2 = -s^2 = \kappa^2\left(\frac{c^2}{c_2^2} - 1\right) \tag{64.49a,b}$$

One of the roots of this equation is $k = 0$, which yields $c^2 = c_2^2$ independent of κ. Furthermore, the corresponding form of V is $V = \text{const} \times r$. Hence, both the wave velocity and the mode of propagation are identical with those obtained in Sec. 64.4, and the approximate theory described there is, in fact, the first mode of the exact theory.

The exact theory, however, yields, in addition, higher modes obtained from the roots of Eq. (64.49a). Unlike the fundamental mode, these modes are dispersive, with the phase velocity c approaching the shear velocity c_2 for waves of short wavelength ($\kappa \to \infty$).

Longitudinal Waves

In this case, $\omega_r = \omega_z = 0$. The solution of the equations of motion can be shown to be [4]

$$U = A_1 \frac{\partial}{\partial r} J_0(hr) + A_2 \kappa J_1(kr)$$

$$W = A_1 i\kappa J_0(hr) + \frac{iA_2}{r}\frac{\partial}{\partial r}[rJ_1(kr)] \qquad h^2 = -q^2 = \kappa^2\left(\frac{c^2}{c_1^2} - 1\right) \tag{64.50}$$

where A_1 and A_2 are constants, and k is given by Eq. (64.49b).

The stress-free boundary condition gives two homogeneous equations for A_1 and A_2, whose determinantal equation yields the dispersion curves of the phase velocity c as a function of wave number κ. For Poisson's ratio $\nu = 0.29$, these curves are plotted in Fig. 64.8 from curves given by Davies [11]. It is noted that, in agreement with the approximate theory (Sec. 64.3), long waves of the lowest mode propagate with the bar velocity $c_B = (E/\rho)^{1/2}$.

Transverse Waves

We assume that $U = U^* \cos\theta$, $V = V^* \sin\theta$, and $W = W^* \cos\theta$. The dilatation and all components of the rotation are, in this case, nonvanishing, and the solution of the equations of motion (64.45) can be shown to be [4]

$$U^* = A\frac{\partial J_1(hr)}{\partial r} + B\kappa\frac{\partial J_1(kr)}{\partial r} + C\frac{J_1(kr)}{r}$$

$$V^* = -A\frac{J_1(hr)}{r} - B\frac{\kappa J_1(kr)}{r} - C\frac{\partial J_1(kr)}{\partial r}$$

$$W^* = iA\kappa J_1(hr) - iBk^2 J_1(kr)$$

This solution, when substituted into the boundary conditions $\sigma_r = \tau_{rz} = \tau_{r\theta} = 0$ at $r = a$, yields the dispersion equation, which, in this case, is complicated. In Fig. 64.9, we show, for $\nu = 0.29$, the phase and group velocities for the lowest two modes.[1] We mention, in passing, that it is sometimes erroneously stated that the dispersion curves, in this case, contain only their lowest branch.

[1] The author is indebted to M. C. Gray and G. R. Nelson of the Bell Telephone Laboratories for these curves.

64.12. REFERENCES

[1] R. M. Davies: Stress waves in solids, *Appl. Mechanics Revs.*, **6** (1953), 1–3.
[2] W. M. Ewing, W. S. Jardetzky, F. Press: "Elastic Waves in Layered Media," McGraw-Hill, New York, 1957.
[3] H. Kolsky: "Stress Waves in Solids," Oxford Univ. Press, New York, 1953.
[4] A. E. H. Love: "A Treatise on the Mathematical Theory of Elasticity," Dover, New York, 1944.
[5] F. Pfeiffer: Elastokinetik, in H. Geiger, K. Scheel (eds.), "Handbuch der Physik," vol. 6, pp. 309–403, Springer, Berlin, 1928.
[6] M. A. Biot: General theorems of equivalence of group velocity and energy transport, *Phys. Rev.*, **105** (1957), 1129–1137.
[7] S. P. Timoshenko: (*a*) On the correction for shear of the differential equation for transverse vibrations of prismatic bars, *Phil. Mag.*, **VI,41** (1921), 744–746; (*b*) On the transverse vibrations of bars of uniform cross-section, *Phil. Mag.*, **VI,43** (1922), 125–131.
[8] W. Flügge: Die Ausbreitung von Biegungswellen in Stäben, *Z. angew. Math. Mech.*, **22** (1942), 312–318.
[9] R. D. Mindlin: Influence of rotatory inertia and shear on flexural motions of isotropic, elastic plates, *J. Appl. Mechanics*, **18** (1951), 31–38.
[10] J. N. Goodier, R. E. D. Bishop: A note on critical reflections of elastic waves at free surfaces, *J. Appl. Phys.*, **23** (1952), 124–126.
[11] R. M. Davis: A critical study of the Hopkinson pressure bar, *Phil. Trans. Roy. Soc. London*, **240** (1946–1948), 375–457.
[12] H. N. Abramson, H. J. Plass, E. A. Ripperger: Stress propagation in rods and beams, *Advances Appl. Mechanics*, **5** (1958), 111–194.

CHAPTER 65

NONLINEAR VIBRATIONS

BY

K. KLOTTER, Dr.-Ing., Darmstadt, Germany

65.1. FREE OSCILLATIONS IN SYSTEMS NOT CAPABLE OF AUTO-OSCILLATIONS

Undamped Systems; Exact Integration

Characteristics Allowing Direct Integration. Not many nonlinear differential equations allow exact integration. The phenomena considered here do.

The systems under consideration obey the differential equation

$$\ddot{q} + \kappa^2 f(q) = 0 \tag{65.1}$$

We will require that dim f = dim q; hence, dim $\kappa = T^{-1}$; furthermore $f(0) = 0$.

$f(q)$ is a function describing the behavior of the *restoring force*. The f,q diagram frequently is referred to as the "characteristic" of the system.

If $f'(q)$ is monotonously increasing–decreasing, the restoring force is called hardening–softening (Fig. 65.1). Alternate designations are overlinear–underlinear.

In systems for which

$$\lim_{q \to 0} \left[\frac{f(q)}{q} \right] = 1 \tag{65.2}$$

holds, the parameter κ may be interpreted as the (circular) natural frequency of small oscillations.

If $f(q) = q$, the characteristic is a straight line; the differential equation is linear, and its solution has the familiar form

$$q = A \cos \kappa t + B \sin \kappa t$$

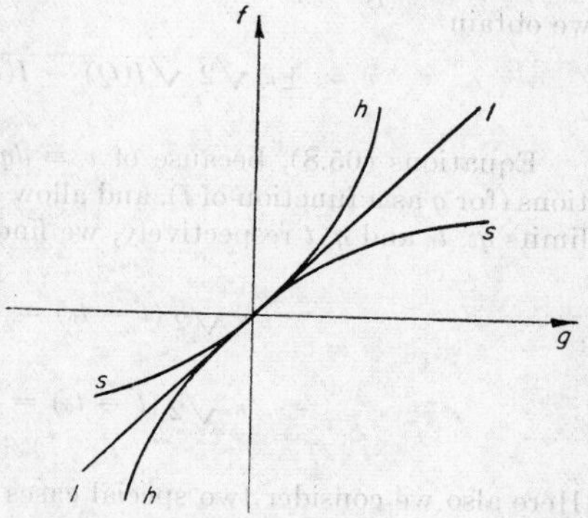

FIG. 65.1. Characteristic lines: h = hardening, s = softening, l = linear system.

In regard to nonlinear functions $f(q)$, we start our considerations with odd functions; the characteristic line, then, is point-symmetrical with respect to the origin. Later we will admit more general functions f.

Oddness of $f(q)$ does not necessarily exclude the appearance of even powers in the functional relationship. The most general odd function involving a power is

$$f(q) = (\text{sgn } q)|q|^n \tag{65.3a}$$

with n being any real number. Of course, sums of functions (65.3a) also qualify. For odd integers n, Eq. (65.3a) may be replaced by the simpler form

$$f(q) = q^n \tag{65.3b}$$

For the purpose of integration we first transform Eq. (65.1). Putting $\dot{q} = v$, and making use of the identity

$$\ddot{q} \equiv \frac{1}{2}\frac{d(v^2)}{dq} \tag{65.4}$$

we find, from (65.1),

$$\frac{d(v^2)}{dq} + 2\kappa^2 f(q) = 0 \tag{65.5}$$

This is a first-order differential equation in the dependent variable v^2, with q as the independent variable. It allows separation of variables immediately. Introducing

$$I(q) = \int_0^q f(x)\,dx \tag{65.6}$$

and integrating (65.5) between limits v_1, q_1 and v, q, respectively, we obtain

$$v^2 - v_1^2 = 2\kappa^2[I(q_1) - I(q)] \tag{65.7}$$

or, equivalently,

$$v(q) = \pm\kappa\sqrt{2}\sqrt{\frac{v_1^2}{2\kappa^2} + I(q_1) - I(q)} \tag{65.7'}$$

as the result of the first step of integration. The upper sign corresponds to a motion with positive, the lower one to a motion with negative velocity.

We note two special cases of initial conditions. Putting into evidence the

	Maximum displacement Q	Maximum velocity v_m	
	$q_1 = Q$ for $v_1 = 0$	$q_1 = 0$ for $v_1 = v_m$	
we obtain			
	$v = \pm\kappa\sqrt{2}\sqrt{I(Q) - I(q)}$	$v = \pm\kappa\sqrt{2}\sqrt{\frac{v_m^2}{2\kappa^2} - I(q)}$	(65.8a,b)

Equations (65.8), because of $v \equiv dq/dt$, in turn are first-order differential equations (for q as a function of t), and allow separation of variables. Integrating between limits q_0, t_0 and q, t respectively, we find

$$\kappa\sqrt{2}\,(t - t_0) = \pm\int_{q_0}^{q}\frac{dx}{\sqrt{I(Q) - I(x)}}$$

$$\kappa\sqrt{2}\,(t - t_0) = \pm\int_{q_0}^{q}\frac{dx}{\sqrt{v_m^2/2\kappa^2 - I(x)}} \tag{65.9a,b}$$

Here also we consider two special cases of initial conditions (for either case a or b):

(α)	$t_0 = 0$	for $q_0 = 0$
(β)	$t_0 = 0$	for $q_0 = Q$

Thus, we obtain [choosing appropriate signs in (65.9) and distinguishing between t_α and t_β]

$$\kappa t_\alpha = \tfrac{1}{2}\sqrt{2}\int_0^q\frac{dx}{\sqrt{I(Q) - I(x)}} \qquad \kappa t_\alpha = \tfrac{1}{2}\sqrt{2}\int_0^q\frac{dx}{\sqrt{v_m^2/2\kappa^2 - I(x)}}$$

$$\kappa t_\beta = \tfrac{1}{2}\sqrt{2}\int_q^Q\frac{dx}{\sqrt{I(Q) - I(x)}} \qquad \kappa t_\beta = \tfrac{1}{2}\sqrt{2}\int_q^Q\frac{dx}{\sqrt{v_m^2/2\kappa^2 - I(x)}} \tag{65.10a–d}$$

From Eqs. (65.7) and (65.9), it can be seen readily that the time spent on the way between coordinates q_0 and q is the same for motions in either direction. Furthermore, we confirm that the time from zero displacement to displacement q is the same for motion to either side. Hence, the time spent between $q = 0$ and $q = Q$ is a quarter of the full period T,

$$\kappa \frac{T}{4} = \tfrac{1}{2}\sqrt{2}\int_0^Q \frac{dx}{\sqrt{I(Q) - I(x)}} \qquad \kappa \frac{T}{4} = \tfrac{1}{2}\sqrt{2}\int_0^Q \frac{dx}{\sqrt{v_m{}^2/2\kappa^2 - I(x)}} \tag{65.11a,b}$$

and we find, for both cases a and b,

$$t_\alpha + t_\beta = T/4 \tag{65.12}$$

Equations (65.9) and (65.10) describe, for certain types of initial conditions, the solutions of Eq. (65.1). The solutions appear in the form of quadratures. Of particular interest are those cases where the quadratures can be expressed in terms of functions with known properties; such cases allow easy discussions, in particular, in regard to the influences of parameters involved. Therefore, we investigate the results (65.9) and (65.10) in this respect.

Assuming that the characteristic line may be described by a polynomial,

$$f(q) = (\operatorname{sgn} q) \sum_{\nu=0}^{n} \mu_\nu{}^{\nu-1}|q|^\nu \tag{65.13}$$

with integer values ν (where the coefficients μ_ν have dimensions $\dim \mu_\nu = \dim q^{-1}$, and where $\mu_1{}^0 = 1$), we find that expressions (65.9), (65.10), and (65.11) are algebraic functions if $n = 0$, circular functions (arcsin, arccos) if $n = 1$, elliptic integrals if $n = 2$ or $n = 3$, and hyperelliptic integrals if $n > 3$.

Because of the symmetry of the characteristic line, and the resulting properties of the integrals (65.9), (65.10), (65.11), it will suffice to consider the first quadrant of the f,q diagram, where both f and q are positive. Hence, for our present purposes, it suffices to use

$$f(q) = \sum_{\nu=0}^{n} \mu_\nu{}^{\nu-1}q^\nu \tag{65.13a}$$

instead of the complete form (65.13).

First we treat the cases of *pure powers,* where the polynomials (65.13a) reduce to a single term $f = \mu^{n-1}q^n$. We introduce the abbreviations

$$i_\alpha(n,q/Q) = \int_0^{q/Q} \frac{dx}{\sqrt{1 - x^{n+1}}} \qquad i_\beta(n,q/Q) = \int_{q/Q}^{1} \frac{dx}{\sqrt{1 - x^{n+1}}}$$
$$i(n) = \int_0^1 \frac{dx}{\sqrt{1 - x^{n+1}}} \tag{65.14a}$$

for which the relation

$$i_\alpha(n,q/Q) + i_\beta(n,q/Q) = i(n) \tag{65.14b}$$

holds. After some (easily verifiable) transformations, we obtain from Eqs. (65.10) the expressions

$$\kappa t_{\alpha,\beta} = \frac{1}{(\mu Q)^{(n-1)/2}} \sqrt{\frac{n+1}{2}}\, i_{\alpha,\beta}(n,q/Q) \tag{65.15}$$

and

$$\kappa \frac{T}{4} = \frac{1}{(\mu Q)^{(n-1)/2}} \sqrt{\frac{n+1}{2}}\, i(n) = \frac{1}{(\mu Q)^{(n-1)/2}}\, \psi(n) \tag{65.15a}$$

Furthermore,

$$\frac{t_{\alpha,\beta}}{T/4} = \frac{i_{\alpha,\beta}(n,q/Q)}{i(n)} \tag{65.16}$$

Equation (65.16) expresses a noteworthy property of *pure power characteristics.* For a given value of the exponent n the *relative time* t/T is a function of the *relative displacement* q/Q alone, whatever the value Q may be. The q,t diagrams have the same shape (the oscillations, hence, the same content of harmonics), whatever the maximum displacement Q may be. The systems with pure power characteristics are the only ones possessing that property.

In Table 65.1 there are listed, for $n = 0, 1, 2, 3, 5$, the functions $i_\alpha(n,q/Q)$, $i_\beta(n,q/Q)$, $i(n)$, and $\psi(n)$ needed to calculate t_α, t_β, $T/4$. The elliptic integrals are reduced to their Legendre standard forms.

Table 65.1.*

n	$i_\alpha(n,q/Q)$	$i_\beta(n,q/Q)$	$i(n)$	$\psi(n)$
0	$2(1 - \sqrt{1 - q/Q})$	$2\sqrt{1 - q/Q}$	2	$\sqrt{2} = 1.414$
1	$\arcsin(q/Q)$	$\arccos(q/Q)$	$\pi/2$	$\pi/2 = 1.571$
2	$\frac{1}{\sqrt[4]{3}}[F(75°,74°.5) - F(75°,\phi_2)]$ $= 0.760[1.848 - F(75°,\phi_2)]$	$0.760F(75°,\phi_2)$	1.402	1.720
	with $\phi_2 = \arccos \frac{\sqrt{3} - 1 + q/Q}{\sqrt{3} + 1 - q/Q}$			
3	$\frac{1}{\sqrt{2}}\{K(45°) - F[45°, \arccos(q/Q)]\}$	$\frac{1}{\sqrt{2}}F[45°, \arccos(q/Q)]$	1.311	1.854
5	$\frac{1}{2\sqrt[4]{3}}F(15°,\phi_5)$	$0.380[F(15°,\pi) - F(15°,\phi_5)]$	1.214	2.104
	with $\phi_5 = \arccos \frac{1 - (\sqrt{3} + 1)(q/Q)^2}{1 + (\sqrt{3} - 1)(q/Q)^2}$			

* Functions $f(q)$ of fifth degree principally lead to hyperelliptic integrals. The special case considered here (pure power) happens to allow reduction to elliptic integrals [2],[3].

In addition to the pure power characteristics, we consider briefly those of polynomial type. If the degree of the polynomial is two or three, expressions (65.9), (65.10), or (65.11) are elliptic integrals.

For systems (65.1) having

$$f(q) = q + \mu^2 q^3 \quad \Big| \quad f(q) = q - \mu^2 q^3$$

by making use of the abbreviation $\theta = \mu^2 Q^2$, we obtain

$$\kappa t_\alpha(q) = \sqrt{\frac{1}{1 + \theta}}\, F(k,\phi) \quad \Big| \quad \kappa t_\alpha(q) = \sqrt{\frac{2}{\theta}}\, F(k,\phi) \tag{65.17}$$

and

$$\kappa \frac{T}{4} = \sqrt{\frac{1}{1 + \theta}}\, K(k) \quad \Big| \quad \kappa \frac{T}{4} = \sqrt{\frac{2}{\theta}}\, K(k) \tag{65.18}$$

where

$$k^2 = \frac{\theta}{2(1 + \theta)} \quad \Big| \quad k^2 = \frac{\theta}{2 - \theta}$$

and

$$\sin^2 \phi = \frac{2(q/Q)^2(1 + 1/\theta)}{1 + (q/Q)^2 + 2/\theta} \quad \Big| \quad \sin \phi = q/Q$$

In the last column, obviously $\theta < 1$ is required.

Another important characteristic is $f(q) = \sin q$, which applies to all types of pendulums. This case too leads to elliptic integrals for expressions (65.9), (65.10), and (65.11).

From $f(q) = \sin q$, we find $I(q) = 1 - \cos q = 2\sin^2(q/2)$, and hence, for Eq. (65.10a),

$$\kappa t_\alpha = \int_0^{q/2} \frac{d(\eta/2)}{\sqrt{\sin^2(Q/2) - \sin^2(\eta/2)}} = \int_0^{\phi} \frac{du}{\sqrt{1 - \sin^2(Q/2)\sin^2 u}} = F(k,\phi) \tag{65.19}$$

with $k = \sin(Q/2)$ and $\phi = \sin^{-1}[\sin(q/2)/\sin(Q/2)]$. For the quarter period $\phi = \pi/2$, and, therefore,

$$\kappa \frac{T}{4} = K(k) \tag{65.20}$$

Case (65.10c), because of (65.12), leads to

$$\kappa t_\beta = K(k) - F(k,\phi) \tag{65.21}$$

We compare now the period T of the pendulum having maximum displacement Q with the period $T_0 = 2\pi/\kappa$ of its small oscillations. From (65.20), we find

$$\frac{T}{T_0} = \frac{2}{\pi} K \tag{65.22}$$

Calling $\Delta = (T - T_0)/T_0$, we find

$$\Delta = \frac{2}{\pi} K - 1 \tag{65.23}$$

and, from the expansion for $K(Q)$,

$$\Delta = \tfrac{1}{4}Q^2 + \tfrac{9}{64}Q^4 + \tfrac{25}{256}Q^6 + \cdots \tag{65.24}$$

Equation (65.23) leads to Table 65.2 and Fig. 65.2.

Table 65.2

Q, degrees	Δ	*Q, degrees*	Δ
0	0.0000	60	0.0732
10	0.0019	90	0.1804
20	0.0076	120	0.3739
30	0.0174	150	0.7622

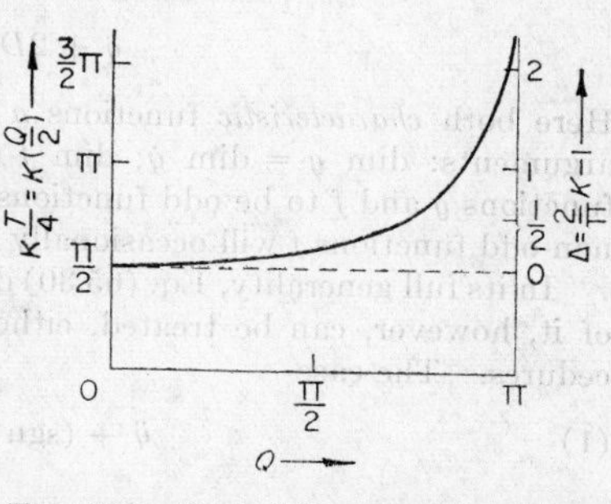

FIG. 65.2. Period T versus maximum displacement ("amplitude") Q of pendulum.

For the non-odd functions $f(q)$, the fact still remains that the time t spent between displacements q_1 and q_2 is the same, irrespective of the direction of the motion. However, the behavior of the oscillations is different for positive and for negative displacements. The period, hence, is composed of

$$T = 2t_1 + 2t_2 \tag{65.25}$$

if t_1 and t_2 are the times spent between $q = 0$ and $q = Q_1 > 0$ or $q = Q_2 < 0$, respectively. The relationship between the maximum displacements Q_1 and Q_2 can be found by equating the potential energies in either position, i.e., from

$$I(Q_1) = I(Q_2) \tag{65.26}$$

For the example $f = q + q^2$, Eq. (65.26) leads to

$$3Q_1^2 + 2Q_1^3 = 3Q_2^2 - 2Q_2^3$$

Except for the fact that the motions to the positive side and to the negative side have to be treated separately, the statements and discussions of the earlier case (odd functions f) can be applied to the present case also.

Characteristics Composed of Straight-line Segments [16],[17]. Only in comparatively rare cases, the quadratures in (65.9), (65.10), and (65.11) can be reduced to known functions. When they cannot, the properties of the motion may be deduced from the quadratures themselves. This, though principally possible, is a tedious procedure, and not cherished by those interested in ready results. Resort to numerical computation is open too, of course. Sometimes another procedure is indicated, however: replacement of the given characteristic $f(q)$ by a sequence of straight-line segments. If this replacement is done in such a way that the new characteristic $\bar{f}$ satisfies either the condition

$$\bar{f}(q) \geqq f(q) \qquad \text{or} \qquad \bar{f}(q) \leqq f(q) \tag{65.27a,b}$$

we are assured of either

$$\bar{t}(q) < t(q) \qquad \text{or} \qquad \bar{t}(q) > t(q) \tag{65.28a,b}$$

respectively (unless the equality signs in (65.27) hold throughout the range). By *bracketing* the original characteristic, we are able to obtain lower and upper bounds for $t(q)$.

Occasionally systems immediately exhibit characteristics composed of straight-line segments, e.g., systems with straight-line characteristics but having *free play* (backlash) or prestressed springs [16],[17].

Damped Systems

The damped oscillations of a completely linear system are described by the differential equation

$$\ddot{q} + 2D\kappa\dot{q} + \kappa^2 q = 0 \tag{65.29}$$

If both damping and restoring forces are nonlinear, we write

$$\ddot{q} + 2D\kappa g(\dot{q}) + \kappa^2 f(q) = 0 \tag{65.30}$$

Here both *characteristic* functions g and f have the dimensions of their respective arguments: dim g = dim $\dot{q}$; dim f = dim q. For what follows, we assume both functions g and f to be odd functions of their arguments. Some remarks concerning non-odd functions f will occasionally be inserted.

In its full generality, Eq. (65.30) does not allow explicit integration. Special cases of it, however, can be treated, either by exact integration or by approximate procedures. The case

(1) $$\ddot{q} + (\operatorname{sgn} \dot{q}) \frac{\delta}{2} \dot{q}^2 + \kappa^2 f(q) = 0 \tag{65.31}$$

allows an exact first integration; the three cases (with linear restoring forces)

$$\begin{aligned}
&\text{(2A)} \qquad \ddot{q} + \kappa^2 q + \alpha_n|\dot{q}|^n(\operatorname{sgn} \dot{q}) = 0\\
&\text{(2B)} \qquad \ddot{q} + \kappa^2 q + \beta_n|q|^n(\operatorname{sgn} \dot{q}) = 0\\
&\text{(2C)} \qquad \ddot{q} + \kappa^2 q + \gamma_n Q^n(\operatorname{sgn} \dot{q}) = 0
\end{aligned} \tag{65.32a,b,c}$$

can be treated by the *method of equivalent linearization* (see p. 65–32). Here we shall use the first version of that method.

Arbitrary Restoring Forces, Quadratic Damping Forces [18]. By virtue of the identity (with $V \equiv \dot{q}^2$),

$$\ddot{q} = \frac{1}{2}\frac{dV}{dq} \tag{65.33}$$

Eq. (65.31) can be written as

$$\frac{dV}{dq} + (\operatorname{sgn} \dot{q})\, \delta V = -2\kappa^2 f(q) \tag{65.34}$$

This constitutes a linear differential equation (or rather a pair of such) for V as a function of q.

We begin by considering a motion which starts from rest at a positive displacement Q_1, therefore having negative velocity $\dot{q}$. Hence, (65.34) becomes

$$\frac{dV}{dq} - \delta V = -2\kappa^2 f(q) \tag{65.34a}$$

Its solution, fitted to the initial condition, is

$$V(q) = 2\kappa^2 e^{\delta q} \int_q^{Q_1} f(\sigma) e^{-\delta\sigma}\, d\sigma \tag{65.35}$$

Making use of the abbreviation

$$N(q) = \int_0^q f(\sigma) e^{-\delta\sigma}\, d\sigma \tag{65.36}$$

we obtain for (65.35) the expression

$$V(q) = 2\kappa^2 e^{\delta q}[N(Q_1) - N(q)] \tag{65.37}$$

Equation (65.37) describes the motion as long as Eq. (65.34*a*) holds, that is, until the next *maximum* displacement Q_2 is reached. This second maximum displacement Q_2 is determined by the subsequent zero of (65.37), hence by

$$N(Q_1) = N(Q_2) \qquad Q_2 < 0 \tag{65.38}$$

Equation (65.38) gives the (transcendental) relationship between a maximum displacement Q_1 on the positive side and the subsequent maximum displacement Q_2 on the negative side. In what follows, we shall speak of *amplitudes* instead of *maximum displacements;* hence we shall refer to (65.38) as the *amplitude relationship.*

It can be realized readily (by just reverting the direction of q) that, when f is odd, (65.38) describes also the relationship between Q_2 and the subsequent amplitude Q_3.

If f is non-odd, however, and composed of odd (f_o) and even (f_e) parts, $f = f_o + f_e$, for the "return journey," f is to be replaced by $(f_o - f_e)$ in (65.38) and (65.36).

We proceed by showing the use of (65.38) for special functions f.

First, we consider

$$f(q) = \mu^{n-1} q^n \tag{65.39}$$

N in Eq. (65.36) then becomes

$$N(q) = \frac{\mu^{n-1}}{\delta^{n+1}} \{n! - e^{-\delta q}[(\delta q)^n + n(\delta q)^{n-1} + n(n-1)(\delta q)^{n-2} + \cdots + n!]\} \tag{65.40}$$

For the case of linear restoring forces ($n = 1$), this leads to

$$N(q) = \frac{1}{\delta^2}[1 - e^{-\delta q}(\delta q + 1)] \tag{65.41}$$

Instead of using N as it stands, we deal with the function $y = -\ln[-(\delta^2 N - 1)]$. Writing x for δq, we obtain

$$y = x - \ln(x + 1) \tag{65.42a}$$

Two successive amplitudes Q_1 and Q_2 are such values for which the arguments $x_1 = \delta Q_1$ and $x_2 = \delta Q_2$ render the same value to the function y. Figure 65.3*a*,*b* shows the plot of $y(x)$ given by (65.42a). Each diagram exhibits a pair of abscissas x_1 and x_2 belonging to one swing.

An important feature may be especially noted: The presence of an asymptote to the diagram at $x = -1$ means that the absolute value of the second amplitude Q_2, because of $|x_2| < 1$, never exceeds $1/\delta$, however large the starting amplitude Q_1 may have been. This fact is in strict contrast to the behavior of a linearly damped system, where any amplitude has a fixed ratio to the preceding one, $|Q_2|/|Q_1| = \exp(\theta/2)$, with θ denoting the logarithmic decrement of the system.

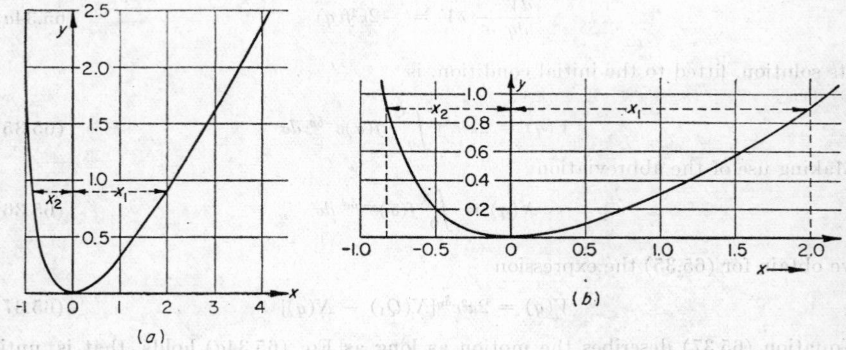

FIG. 65.3. Amplitude relationship [Eq. (65.42a)], linear restoring force: (a) $x \leq 4$; $x \leq 2$.

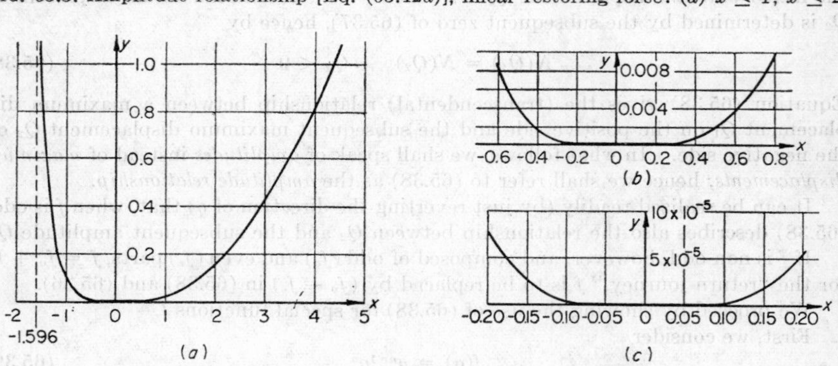

FIG. 65.4. Amplitude relationship [Eq. (65.42b)], cubic restoring force.

For the case $n = 3$ in (65.39), Eq. (65.40) becomes (with $x = \delta Q$)

$$N(x) = (\mu^2/\delta^4)[6 - e^{-x}(x^3 + 3x^2 + 6x + 6)]$$

Replacing N by

$$y = -\ln[-(\delta^4/\mu^2)N + 6] + \ln 6$$

we obtain

$$y = x - \ln\left(\frac{x^3}{6} + \frac{x^2}{2} + x + 1\right) \tag{65.42b}$$

Figure 65.4 shows the plot for function (65.42b). The same general features appear as in the former case, 2A. In particular, an asymptote exists which puts an upper bound to any second amplitude; the abscissa of the asymptote is found to be $x_a = -1.596$.

For the case $n = 5$, a pair of amplitudes is determined by a pair of abscissas x, rendering the same value to the function

$$y = x - \ln\left(\frac{x^5}{120} + \frac{x^4}{24} + \frac{x^3}{6} + x + 1\right) \tag{65.42c}$$

The asymptote is found at $x_a = -2.180$.

In the general case n, we find the expression for y to be

$$y = x - \ln\left(1 + x + \frac{x^2}{2!} + \cdots + \frac{x^n}{n!}\right) \tag{65.42d}$$

The abscissa x_a of the asymptote to the curve is given by the first negative zero of the polynomial in the argument of the logarithm.

For use in the vicinity of the origin, the amplitude relationships given by Eqs. (65.42) will profitably be replaced by their respective power-series expansions [1].

If (65.39) is replaced by a polynomial, similar considerations apply. For the case $f(q) = q + \mu^2 q^3$ (with $\mu^2 > 0$), the function y corresponding to the functions (65.42) reads (with $p^2 = \mu^2/\delta^2$)

$$y = x - \ln\left[(x + 1) + p^2(x^3 + 3x^2 + 6x + 6)\right] + \ln(1 + 6p^2) \tag{65.43}$$

Finally we consider $f(q) = \sin q$. The function N in Eq. (65.36) becomes

$$N(Q) = \frac{1}{1 + \delta^2}\left[1 - e^{-\delta Q}(\delta Q + \cos Q)\right]$$

From it we do not derive a function y involving logarithms, but simply

$$Y = 1 - (1 + \delta^2)N$$

This gives

$$Y = e^{-\delta Q}(\delta Q + \cos Q) \tag{65.44}$$

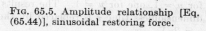

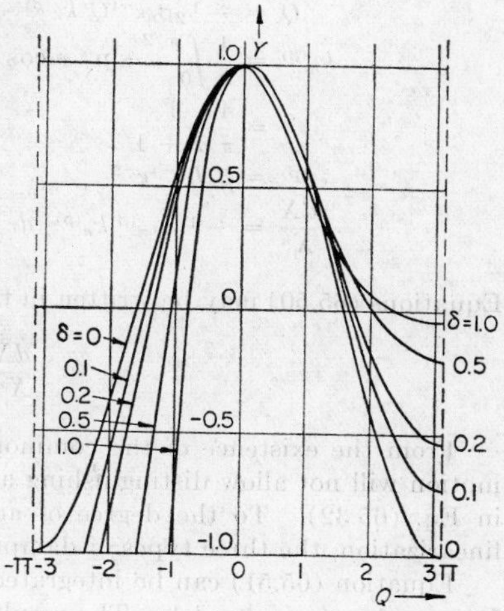

FIG. 65.5. Amplitude relationship [Eq. (65.44)], sinusoidal restoring force.

Plots for (65.44) are shown in Fig. 65.5.

Cases with non-odd restoring forces can be treated in a similar manner [1].

Linear Restoring Forces, Damping Forces Proportional to a Power of the Velocity or of the Displacement [19]. These cases are described by Eq. (65.32). They are special cases of

$$\ddot{q} + \kappa^2 q + \epsilon\kappa^2 h(q,\dot{q}) = 0 \tag{65.45}$$

Hence, they are amenable to the procedure of equivalent linearization, as described on p. 65–32. Of the two variants of that method, we shall use here the first one, which culminates in Eq. (65.152). It can be readily verified that, for all three cases A, B, C of Eq. (65.32), the change of frequency $\dot{\theta}$ according to Eq. (65.152b) vanishes. We are able to restrict our attention to the rate of attenuation $\dot{Q}$ according to Eq. (65.152a).

Case A. From Eq. (65.152a), we find

$$\dot{Q} = -\tfrac{1}{2}\alpha_n\kappa^{n-1}Q^n \frac{4}{\pi}\int_0^{\pi/2} \cos^{n+1} x\, dx \tag{65.46a}$$

Replacing the *amplitude* Q and the time t by the dimensionless variables (L being a representative length)

$$X = Q/L \qquad \tau = \kappa t \tag{65.47}$$

respectively, using the abbreviation

$$k_n{}^{(\alpha)} = \frac{4}{\pi}\int_0^{\pi/2} \cos^{n+1} x\, dx \equiv \phi(n) \tag{65.48a}$$

(see p. 65–35), and denoting

$$\epsilon_n{}^{(\alpha)} = \alpha_n L^{n-1}\kappa^{n-2} \tag{65.49a}$$

we find the differential equation for the attenuation,

$$\frac{dX}{X^n} = -\tfrac{1}{2}(\epsilon_n{}^{(\alpha)} k_n{}^{(\alpha)})\, d\tau \tag{65.50a}$$

Cases B and C. The corresponding formulas read

$$\dot{Q} = -\tfrac{1}{2}\beta_n\kappa^{-1}Q^n k_n{}^{(\beta)} \qquad \dot{Q} = -\tfrac{1}{2}\gamma_n\kappa^{-1}Q^n k_n{}^{(\gamma)}$$

$$k_n{}^{(\beta)} = \frac{4}{\pi}\int_0^{\pi/2} \sin^n x \cos x\, dx \qquad k_n{}^{(\gamma)} = \frac{4}{\pi}\int_0^{\pi/2} \cos x\, dx \tag{65.46b,c}$$

$$= \frac{4}{\pi}\frac{1}{n+1} \qquad = \frac{4}{\pi} \tag{65.48b,c}$$

$$\epsilon_n{}^{(\beta)} = \beta_n L^{n-1}\kappa^{-2} \qquad \epsilon_n{}^{(\gamma)} = \gamma_n L^{n-1}\kappa^{-2} \tag{65.49b,c}$$

$$\frac{dX}{X^n} = -\tfrac{1}{2}(\epsilon_n{}^{(\beta)} k_n{}^{(\beta)})\, d\tau \qquad \frac{dX}{X^n} = -\tfrac{1}{2}(\epsilon_n{}^{(\gamma)} k_n{}^{(\gamma)})\, d\tau \tag{65.50b,c}$$

Equations (65.50) may be written in the common form

$$\frac{dX}{X^n} = -\rho_n\, d\tau \tag{65.51}$$

From the existence of the common form, we may conclude that the record of a motion will not allow distinguishing among the three damping expressions appearing in Eq. (65.32). To the degree of accuracy inherent in the method of equivalent linearization, the three types of damping laws are equivalent.

Equation (65.51) can be integrated readily. Two different formulas appear: one for $n = 1$, one for $n \neq 1$. The results are listed in Table 65.3. In that table, the dimensionless time τ is replaced by the number N of oscillations performed, where $\tau = 2\pi N$.

Table 65.3

$n = 1$:	$\ln X = \ln X_0 - 2\pi\rho_1 N$
$n \neq 1$:	$\dfrac{1}{X^{n-1}} = \dfrac{1}{X_0{}^{n-1}} + 2\pi(n-1)\rho_n N$
$n = 0$:	$X = X_0 - 2\pi\rho_0 N$
$n = 2$:	$\dfrac{1}{X} = \dfrac{1}{X_0} + 2\pi\rho_2 N$
$n = 3$:	$\dfrac{1}{X^2} = \dfrac{1}{X_0{}^2} + 4\pi\rho_3 N$
$n = 4$:	$\dfrac{1}{X^3} = \dfrac{1}{X_0{}^3} + 6\pi\rho_4 N$
$n > 1$:	Asymptotically for $2\pi(n-1)\rho_n N \gg X_0{}^{-n+1}$ $\ln X = \dfrac{1}{n-1}\ln\dfrac{1}{2\pi(n-1)\rho_n} - \dfrac{1}{n-1}\ln N$

From the formulas of Table 65.3, procedures can be developed which allow the damping law (exponent n) to be determined from the record of an oscillation [19].

The formulas developed for the single-term damping laws (65.32) may be extended to polynomial-type expressions, as, for example,

$$\ddot{q} + \kappa^2 q + (\operatorname{sgn} \dot{q}) \sum_k \alpha_k |\dot{q}|^k = 0 \tag{65.52}$$

Details are given in [19].

65.2. SELF-SUSTAINED OSCILLATIONS (AUTO-OSCILLATIONS)

The Phase Plane

General Discussions. Self-sustained oscillations are described by autonomous differential equations,

$$\ddot{q} + \kappa^2 h(q,\dot{q}) = 0 \tag{65.53}$$

Not all functions h in (65.53) lead to self-sustained oscillations, of course; free damped oscillations, for instance (see p. 65–6), are another special class comprised in (65.53).

An appropriate tool for investigating and representing autonomous phenomena in general and, hence, self-sustained oscillations in particular, is the phase plane. As coordinates x and y in the phase plane, we shall use

$$x = q \qquad \text{and} \qquad y = \dot{q}/\kappa \tag{65.54}$$

By making use of (65.54) and of the identity $\ddot{q} = \dot{q}(d\dot{q}/dq)$, Eq. (65.53) can be transformed into

$$\frac{dy}{dx} = -\frac{h(x,\kappa y)}{y} \tag{65.55}$$

a first-order differential equation. Hence, the dynamical problem described by the second-order autonomous differential equation (65.53) has been transformed into the purely geometrical problem of finding a curve $y(x)$ described by the first-order differential equation (65.55).

A curve $y = y(x)$ in the phase plane is known as a phase-plane trajectory, and the family of trajectories representing all solutions of Eq. (65.55) as the *phase portrait* of the motion. Periodic motions are represented by closed curves; conversely, closed curves (unless they pass through singular points) correspond to periodic motions.

By transforming the dynamical problem (65.53) into the geometrical problem (65.55), the order of the differential equation is reduced from two to one. This reduction is indicative of the fact that the dynamical problem is not completely described by (65.55). What is missing is the relationship to time. In order to find the time involved in the motion, we shall have to perform a second integration. This part of the problem will be treated on p. 65–14.

Subsequently, we shall not deal with the most general form (65.53), or equivalently (65.55), of the differential equation but with a more specific form, for example,

$$\ddot{q} + 2D\kappa g(q,\dot{q}) + \kappa^2 q = 0 \tag{65.56}$$

Equation (65.56) comprises the following two classical examples:

van der Pol's differential equation: $\ddot{q} - 2D\kappa\dot{q}(1 - \alpha^2 q^2) + \kappa^2 q = 0$ (65.57)

Rayleigh's differential equation: $\ddot{q} - 2D\kappa\dot{q}(1 - \beta^2 \dot{q}^2) + \kappa^2 q = 0$ (65.58)

These two equations are equivalent; (65.57) follows from (65.58) if this equation is differentiated, with $\dot{q} = q$, and $\alpha^2 = 3\beta^2$.

Methods of Graphical Integration. The simplest way of discussing (and graphically producing) the integral curves of Eq. (65.55) is by way of the isoclines,

$$y\sigma + h(x,\kappa y) = 0 \tag{65.59}$$

with σ denoting the slope of the integral curves.

Another way is given by Liénard's construction. In phase-plane coordinates x and y, Eq. (65.56) reads

$$\frac{dy}{dx} = -\frac{2D\frac{1}{\kappa}g(x,\kappa y) + x}{y} \tag{65.60}$$

Liénard's construction deals with that special case of (65.56) or (65.60) where, instead of a function $g(q,\dot{q})$, there appears a function $g(\dot{q})$ of $\dot{q}$ alone

$$\frac{dy}{dx} = -\frac{2D\frac{1}{\kappa}g(\kappa y) + x}{y} \tag{65.61}$$

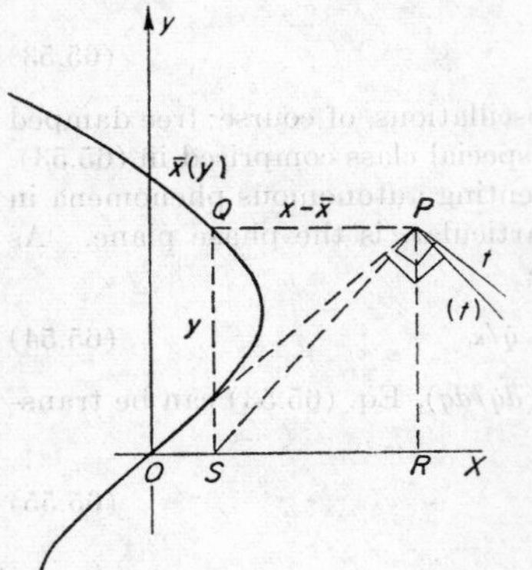

FIG. 65.6. Liénard's construction.

By putting

$$\bar{x}(y) = -2D\frac{1}{\kappa}g(\kappa y) \tag{65.62}$$

Eq. (65.61) may be written as

$$\frac{dy}{dx} = -\frac{x - \bar{x}}{y} \tag{65.63}$$

Hence, if we plot the curve $x = \bar{x}(y)$ in the phase plane, Eq. (65.63) immediately suggests the following procedure: In order to find the slope of the integral curve at an arbitrarily chosen point P, we draw a horizontal line through P to the intersection Q with $\bar{x}(y)$. The ordinate in Q has foot point S. S is to be connected with P. The line Pt, perpendicular to SP, has the direction of the tangent to the integral curve (Fig. 65.6).

In Fig. 65.6 the line OP and its perpendicular direction $P(t)$ are indicated. Remembering that $P(t)$ is the direction of the tangent to the circle (with O as center) through P and that this circle represents the undamped motion, $D = 0$, of (65.61), we will be able to derive all the essential properties (regions of "increase," of "decrease") of the integral curve of (65.61) from the sketch.

Figures 65.7 and 65.8 show an integration performed for Rayleigh's differential equation (65.58) with (intermediate) parameter values $2D = 0.2$ and $2D = 2$, respectively.

It is easily realized that, for small values of the parameter ($D \ll 1$), the steady-state shape of the trajectory, the limit cycle, will closely resemble a circle. The oscillations performed will be nearly sinusoidal.

In case $D \gg 1$, curve (65.62) would spread far to the left and right. Therefore, instead of coordinate $x \equiv q$, we use

$$\xi = q/2D \qquad (\text{or } x = 2D\xi) \tag{65.64}$$

Equation (65.63), then, turns into

$$\frac{dy}{d\xi} = -(2D)^2\frac{\xi - \bar{\xi}}{y} \tag{65.65}$$

where

$$\bar{\xi} = -\frac{1}{\kappa}g(\kappa y) \tag{65.66}$$

(free from $2D$) now is the auxiliary curve to be plotted.

Equation (65.65) shows clearly that, if $2D \gg 1$, the slope of the integral curve has high absolute values everywhere except in the close vicinity of the curve $\xi = \bar{\xi}(y)$; the integral curve, hence, ascends or descends steeply. On the curve $\xi = \bar{\xi}(y)$, the slopes are zero; in close proximity of it (but only there) they may have finite values. From this discussion if follows (see Fig. 65.9) that, after a short "starting branch" (ABF' in Fig. 65.9), the trajectory consists of the limit cycle, and that this limit cycle is composed of two (nearly) vertical lines CD and EF and of the parts DE and FC of the

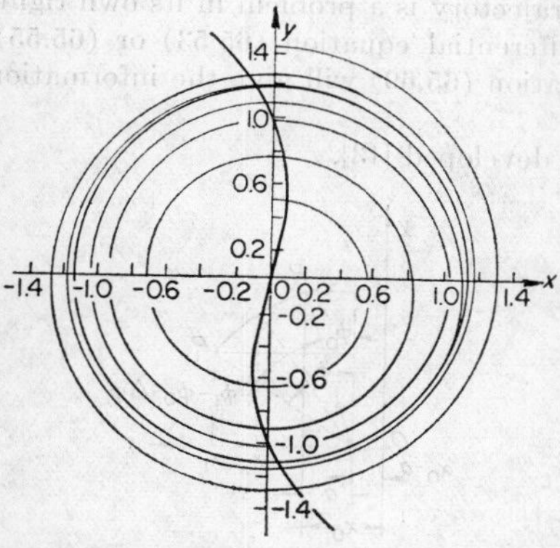

FIG. 65.7. Integration of Rayleigh's differential equation (65.58); $D = 0.1$.

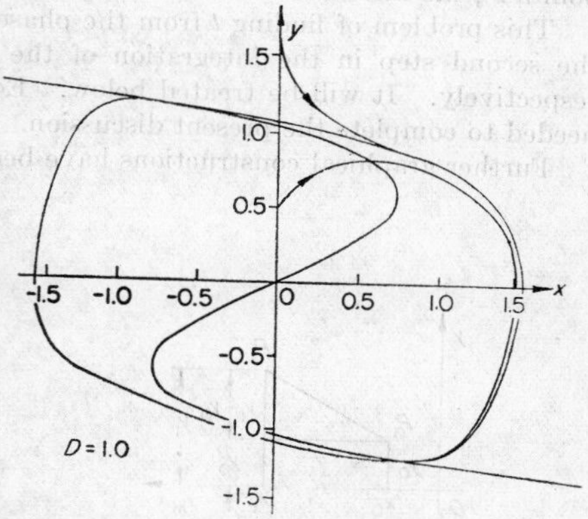

FIG. 65.8. Integration of Rayleigh's differential equation (65.58); $D = 1.0$.

auxiliary curve $\xi = \bar{\xi}(y)$. From these facts it is seen that the q,t and $\dot{q},t$ diagrams represent rather jerky oscillations.

$2D \ll 1$ and $2D \gg 1$ clearly exhibit the features of the two limiting cases of auto-oscillations, which occasionally are referred to as (1) "dedamped oscillations" (energy exchanged per period is small compared to energy content of the oscillation; energy input simply serves to replace the energy losses; oscillation is nearly sinusoidal; frequency is the natural frequency of the free oscillations) and (2) "relaxation oscillations" (energy exchanged per period is large compared to energy content; oscillation is

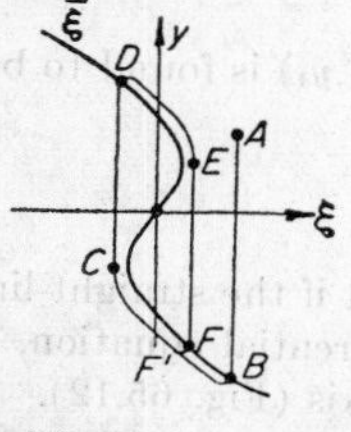

FIG. 65.9. Limiting case: $D \gg 1$; relaxation oscillation.

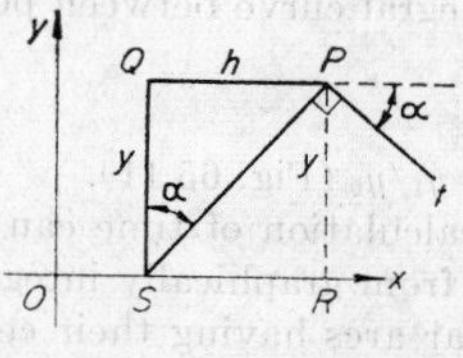

FIG. 65.10. Generalization of Liénard's construction.

jerky, far from sinusoidal; frequency is decidedly determined by the mechanism of exchange).

The Liénard construction, which is linked to Eq. (65.61), can be generalized to fit Eq. (65.55) (Fig. 65.10): If the quantity (length) h is calculated from the coordinates x and y of point P, and then h is plotted from P to give Q, by finding the foot point S, connecting S with P and drawing the perpendicular to SP, we obtain in Pt the direction (tangent; line element) of the integral curve. The proof is obvious from (65.55) and the geometry of the figure.

The procedures may be extended to the nonautonomous equation

$$\ddot{q} + \kappa^2 k(q,\dot{q},\kappa t) = 0 \tag{65.67a}$$

equivalently,

$$\frac{dy}{dx} = -\frac{k(x,\kappa y,\kappa t)}{y} \tag{65.67b}$$

The arguments in k, then, will not be completely furnished by the coordinates of the points P; the increments of time between successive points P will have to be known.

This problem of finding t from the phase trajectory is a problem in its own right, the second step in the integration of the differential equation (65.53) or (65.55), respectively. It will be treated below. Equation (65.69) will give the information needed to complete the present discussion.

Further graphical constructions have been developed [13].

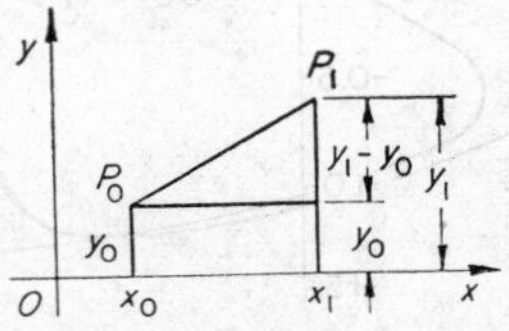

FIG. 65.11. Straight segment of trajectory.

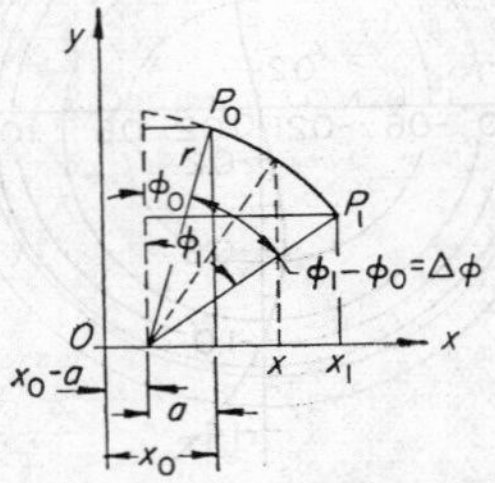

FIG. 65.12. Circular segment of trajectory.

Time Involved in Motion (Second Step of Integration). To calculate the time involved in the motion, we use the second of Eqs. (65.54) and obtain

$$\kappa\, dt = dx/y \tag{65.68}$$

hence,

$$\kappa(t - t_0) = \int_{x_0}^{x} \frac{dx}{y} \quad \text{and} \quad \kappa T = \oint \frac{dx}{y} \tag{65.68a}$$

From Eq. (65.68), the increment of time Δt associated with a straight-line segment,

$$y = y_0 + \frac{y_1 - y_0}{x_1 - x_0}(x - x_0)$$

of the integral curve between points $P_0(x_0,y_0)$ and $P_1(x_1,y_1)$ is found to be

$$\kappa\, \Delta t = \frac{\Delta x}{y_0}\frac{\ln \eta}{\eta - 1} \tag{65.69a}$$

with $\eta = y_1/y_0$ (Fig. 65.11).

The calculation of time can be somewhat simplified if the straight-line segments, resulting from graphically integrating a first-order differential equation, are replaced by circular arcs having their centers on the abscissa axis (Fig. 65.12). The time is then found very simply from

$$\kappa \Delta t = \Delta\phi \tag{65.69b}$$

with $\Delta\phi$ denoting the central angle (in radians) of the arc [20]. Such an approximation by a circular arc, however, contains a dose of arbitrariness; it prejudices the curvature of the integral curve, of which nothing is known from the differential equation.

Topological Considerations. Equation (65.55) is a special case of the equation

$$\frac{dy}{dx} = \frac{P(x,y)}{Q(x,y)} \tag{65.70}$$

which has been investigated thoroughly. Hence, the whole arsenal of tools developed in the theory of differential equations, for treating Eq. (65.70), is available for discussing (65.55). The equations

$$P(x,y) = 0 \quad \text{and} \quad Q(x,y) = 0 \tag{65.71}$$

represent loci of *horizontal* and *vertical tangency*, respectively; the intersections of these loci are the *singular points*. Knowledge about distribution and nature of singular points gives important clues to the behavior of the integral curves. We refer the reader to the literature [9],[21],[22].

Analytical Treatment

Radius of Limit Cycle. A quantity of primary importance is the size (in the case of a circle: the radius) of the limit cycle. The phase-plane methods discussed above do not provide ready means for determining that size. Here the analytical methods prove advantageous. The method which is best suited to the present objectives is the first version of the method of equivalent linearization discussed on p. 65–32. It applies to Eq. (65.147), or equivalently Eq. (65.56), if ϵ or $2D$ are small quantities. We confine our attention to Eqs. (65.57) and (65.58). For both cases, Eq. (65.152*b*) leads to $\theta \equiv 0$, so that only Eq. (65.152*a*) remains to be considered. From Eq. (65.152*a*), in the case of Eq. (65.57), because of $\epsilon\kappa h = -2D\dot{q}(1 - \alpha^2 q^2)$, we find

$$\dot{Q} = D\kappa Q\left[1 - \left(\frac{\alpha Q}{2}\right)^2\right] \tag{65.72}$$

in the case of Eq. (65.58), because of $\epsilon\kappa h = -2D\dot{q}(1 - \beta^2\dot{q}^2)$, we find

$$\dot{Q} = D\kappa Q[1 - \tfrac{3}{4}\beta^2 Q^2 \kappa^2] \tag{65.73}$$

Both equations determine the rate of change of the *amplitude* $Q(t)$. The first-order differential equations allow integration by separation of variables. Calling the *starting amplitude* Q_0 (for $t = 0$), we obtain

$$Q = Q_0 \frac{e^{D\kappa t}}{\sqrt{1 + (\alpha Q_0/2)^2(e^{2D\kappa t} - 1)}} \tag{65.74}$$

and

$$q = Q \sin(\kappa t + \theta_0) \tag{65.74a}$$

It can easily be verified that, from any starting value Q_0, the amplitude Q tends toward the asymptotic value,

$$Q_L = \frac{2}{\alpha} \tag{65.75}$$

Q_L represents the radius of the limit cycle; it is independent of Q_0, κ, D and depends on α only.

Equations (65.74) and (65.74*a*) describe the transitions (though not the "transients") from the starting conditions into the steady state (limit cycle) of the self-sustained oscillation. The time necessary for approaching the steady state (to within a certain ratio Q/Q_L) depends on D decisively. From Eq. (65.74), by linearization, we derive the relationship

$$2D\kappa t = \frac{1 - (Q/Q_0)^2}{(Q/Q_L)^2 - 1} \tag{65.76}$$

On the left-hand side the number of oscillations N may be introduced through

$$\kappa t = 2\pi N$$

From (65.76), all pertinent information may be obtained.

Impulse Systems. Oscillations frequently are sustained by mechanisms in which the energy is replaced in a pulselike fashion. As one example, we cite the pendulum of a clock; as a second example, a system with relay excitation. We shall treat these two examples by analytical methods (of approximate integration; although both cases could be handled exactly by fitting together solutions available in different regions). The method used will be the second version of the method of equivalent linearization, described on p. 65–33 [23].

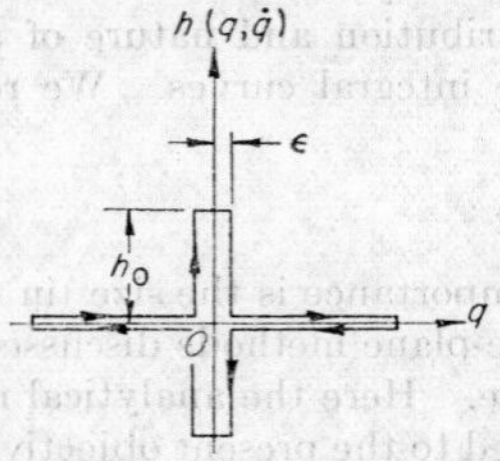

FIG. 65.13. Diagram for function $h(q,\dot{q})$ in Eq. (65.77).

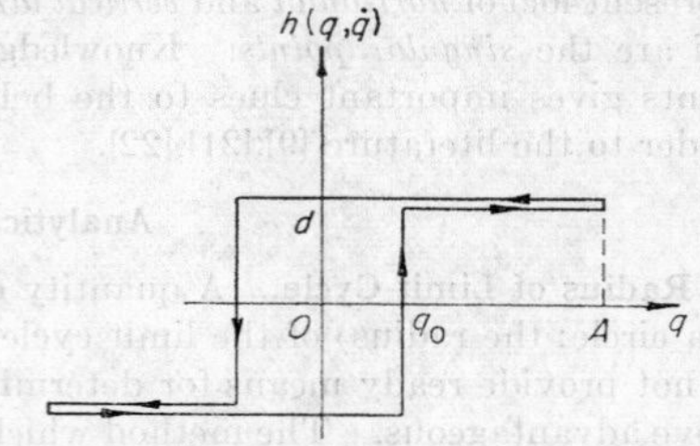

FIG. 65.14. Diagram for function $h(q,\dot{q})$ in Eq. (65.85).

PENDULUM OF A CLOCK. The differential equation of motion is assumed to be

$$\ddot{q} + 2D\kappa\dot{q} + \kappa^2 q = \kappa^2 h(q,\dot{q}) \tag{65.77}$$

with the function $h(q,\dot{q})$ given by the diagram of Fig. 65.13. For the height h_0 and the width 2ϵ of the pulse, we assume the following relations to exist:

$$v_0 = \kappa^2 h_0 \,\Delta t \tag{65.78}$$

is the increase in velocity during the duration Δt of the pulse, whereas the width 2ϵ (displacement) is determined by

$$2\epsilon = \dot{q}(0)\,\Delta t = A\kappa\,\Delta t \tag{65.79}$$

Considering Eqs. (65.154), (65.155), (65.164), (65.168), and (65.170), we find the *equivalence*

$$-\kappa^2 h(q,\dot{q}) \rightarrow -\kappa^2 \frac{F}{\pi A^2 \kappa}\,\dot{q} \tag{65.80}$$

with

$$F = 4\epsilon h_0 \tag{65.81}$$

(see Fig. 65.13). Using (65.78) and (65.79), we transform the right-hand side of (65.80):

$$-\kappa^2 \frac{F}{\pi A^2 \kappa}\,\dot{q} = -\frac{2v_0}{\pi A}\,\dot{q} \tag{65.82}$$

With this value the linearized version of Eq. (65.77) becomes

$$\ddot{q} + \dot{q}\left(2D\kappa - \frac{2v_0}{\pi A}\right) + \kappa^2 q = 0 \tag{65.83}$$

From this Eq. (65.83), it is seen that sustained oscillations will have amplitudes

$$A_s = \frac{v_0}{\pi D \kappa} \tag{65.84}$$

RELAY SYSTEM. We write the governing differential equation in the form

$$T\ddot{q} + \dot{q} + h(q,\dot{q}) = 0 \tag{65.85}$$

with h given by the diagram of Fig. 65.14. The process of equivalent linearization, according to p. 65–32, leads to the equivalence

$$h(q,\dot{q}) \rightarrow K_s(h)q + K_c(h)\dot{q} \tag{65.86}$$

For oscillations having amplitude $A > q_0$, we find

$$K_s(h) = \frac{4d}{\pi A} \sqrt{1 - \left(\frac{q_0}{A}\right)^2} \qquad K_c(h) = -\frac{4dq_0}{\pi A^2 \omega} \tag{65.87a,b}$$

Hence, the equivalent linearized differential equation reads

$$T\ddot{q} + \left(1 - \frac{4dq_0}{\pi A^2 \omega}\right)\dot{q} + \frac{4d}{\pi A}\sqrt{1 - \left(\frac{q_0}{A}\right)^2} = 0 \tag{65.88}$$

From (65.88), we obtain the circular frequency ω according to

$$\omega^2 = \frac{4d}{\pi A T}\sqrt{1 - \left(\frac{q_0}{A}\right)^2} \tag{65.89}$$

whereas the amplitudes A for steady-state oscillations follow from

$$\pi A^3 \sqrt{1 - \left(\frac{q_0}{A}\right)^2} = 4dq_0{}^2 T \tag{65.90}$$

Systems with Dead Time. So far, all phenomena investigated could be described by differential equations. Occasionally systems have to be considered where certain terms have arguments which differ from t by a *delay time* or *dead time* t_ν. Phenomena in such systems are described by difference-differential equations (d.d.e.).

We consider the set of equations

$$\dot{x}_i = \sum_{\nu=1}^{n} f_{i\nu}[x(t - t_\nu)] \qquad i = 1, 2, \ldots, n \tag{65.91}$$

Equations (65.91) may be looked upon as a special case of Eq. (65.161). Applying the second version of the method of equivalent linearization, we find, as an equivalent set of linearized differential equations [24],

$$\dot{x}_i = \sum_{\nu=1}^{n} (\bar{a}_{i\nu} x_\nu + \bar{a}^*_{i\nu} \dot{x}_\nu) \tag{65.92}$$

with coefficients $\bar{a}_{i\nu}$ and $\bar{a}^*_{i\nu}$ following from Eq. (65.162),

$$\begin{aligned} \bar{a}_{i\nu} &= \frac{1}{\pi A_\nu}\int_0^{2\pi} f_{i\nu}[A_\nu \sin a(t - t_\nu)] \sin \omega t \, d\omega t \\ \bar{a}^*_{i\nu} &= \frac{1}{\pi A_\nu \omega}\int_0^{2\pi} f_{i\nu}[A_\nu \sin \omega(t - t_\nu)] \cos \omega t \, d\omega t \end{aligned} \tag{65.93}$$

Putting $\sin \omega t = \sin [\omega(t - t_\nu) + \omega t_\nu]$, expanding according to the addition theorem, and making use of the definitions of the integral transforms (65.165), we find, from Eqs. (65.93),

$$\begin{aligned} \bar{a}_{i\nu} &= \cos \omega t_\nu \, K_s\{f_{i\nu}\} + \omega \sin \omega t_\nu \, K_c\{f_{i\nu}\} \\ \bar{a}^*_{i\nu} &= \cos \omega t_\nu \, K_c\{f_{i\nu}\} - \frac{1}{\omega} \sin \omega t_\nu \, K_s\{f_{i\nu}\} \end{aligned} \tag{65.94}$$

As an example, we treat the d.d.e. of second order,

$$T\ddot{q} + \dot{q} + c \operatorname{sgn} [q(t - t_1)] = 0 \tag{65.95}$$

The equivalent linearized differential equation is found to be

$$T\ddot{q} + \dot{q}\left[1 - \frac{4c}{\pi A \omega} \sin \omega t_1\right] + \left(\frac{4c}{\pi A} \cos \omega t_1\right) q = 0 \tag{65.96}$$

Steady-state oscillations require the bracketed term to be zero,

$$1 - \frac{4c}{\pi A\omega} \sin \omega t_1 = 0 \tag{65.97a}$$

Their frequency is determined by the coefficients of the last and the first terms:

$$\omega^2 = \frac{4c}{\pi AT} \cos \omega t_1 \tag{65.97b}$$

By eliminating the amplitude A, the frequency condition is found to be

$$\tan \omega t_1 = \frac{1}{\omega t_1} \tag{65.98}$$

Having obtained ω, we find the amplitude from (65.97a) as

$$A = \frac{4c}{\pi\omega} \sin \omega t_1 \tag{65.99}$$

65.3. FORCED OSCILLATIONS

Systems Not Capable of Auto-oscillations. Response Curves

For forced oscillations, Eq. (65.1) or (65.30) is to be supplemented by a *driving term.* This term we assume to be of the form $p \sin \Omega t$. However, it should be kept in mind that results obtained with sinusoidal driving forces do not possess the same degree of generality for nonlinear systems as they do for linear ones. Because superposition does not apply, the effect of periodic driving forces cannot be calculated from the effects of their sinusoidal components.

The differential equation to be treated reads

$$E[q] \equiv \ddot{q} + 2D\kappa g(\dot{q}) + \kappa^2 f(q) - p \sin \Omega t = 0 \tag{65.100}$$

In regard to the functions f and g, the same remarks apply as for Eq. (65.30). Equation (65.100) does not allow an exact integration. We shall have to resort to an approximate method of integration. Various such methods may be and have been applied to Eq. (65.100): for example, the method of perturbations, iteration procedures, the method of equivalent linearization (see p. 65–32).

This author is of the opinion that the Ritz-Galerkin method (see p. 65–32) is best suited to the present purposes. This method allows us to find (approximations to) the steady state of the resulting motion. For both functions f and g, any single-valued and integrable function of the respective arguments is admitted. The functions may possess discontinuities; different expressions in various regions may be used. Certainly, the functions need not be *quasi-linear.* Additionally, the method has the advantage of allowing a treatment for unspecified expressions f and g (having the above properties) [25],[26].

One-frequency Approximations (Fundamental Harmonics). Although the method applies to any type of function (within the limits mentioned), for the sake of easy presentation we start by requiring that both f and g be odd functions. In this case, the resulting steady-state motion has the mean value zero. Therefore, an appropriate assumption (65.141), if we content ourselves with terms of a single frequency, may be (with $\Omega t = \tau$)

$$\tilde{q} = Q \sin (\tau - \epsilon) = A \sin \tau - B \cos \tau \tag{65.101}$$

where

$$A = Q \cos \epsilon \qquad B = Q \sin \epsilon \tag{65.101a}$$

The Ritz-Galerkin conditions (65.144), where

$$\psi_1 = \sin\tau \qquad \psi_2 = \cos\tau \tag{65.102}$$

finally lead to

$$[F(Q) - \eta^2]^2 + 4D^2G^2(\Omega,Q) = (s/Q)^2 \tag{65.103a}$$

and

$$\tan\epsilon = \frac{2DG(\Omega,Q)}{F(Q) - \eta^2} \tag{65.103b}$$

Here the following abbreviations have been used:

$$F(Q) = \frac{4}{\pi Q}\int_0^{\pi/2} f(Q\cos\sigma)\cos\sigma\, d\sigma = \frac{4}{\pi Q}\int_0^{\pi/2} f(Q\sin\sigma)\sin\sigma\, d\sigma$$

$$G(\Omega,Q) = \frac{4}{\kappa\pi Q}\int_0^{\pi/2} g(\Omega Q\sin\sigma)\sin\sigma\, d\sigma = \frac{4}{\kappa\pi Q}\int_0^{\pi/2} g(\Omega Q\cos\sigma)\cos\sigma\, d\sigma \tag{65.104a,b}$$

$$\eta^2 = \Omega^2/\kappa^2 \qquad s = p/\kappa^2 \tag{65.104c}$$

The functions $F(Q)$ and $G(\Omega,Q)$, defined by Eqs. (65.104a,b), are certain integral transformations of $f(q)$ and $g(\dot q)$, respectively. For the frequent cases where $f(q)$ or $g(\dot q)$ is described by a polynomial (within the first quadrant),

$$f(q) = \sum_{k=1}^{m} \mu_k{}^{k-1}q^k \qquad g(\dot q) = \sum_{k=1}^{n} \nu_k{}^{k-1}\dot q^k \tag{65.105}$$

$$(\text{with } \mu_1^0 = 1) \qquad (\text{with } \nu_1^0 = 1)$$

by making use of the functions $\phi(n)$, defined and described on p. 65–35, we obtain

$$F(Q) = \sum_{k=1}^{m} \phi(k)(\mu_k Q)^{k-1} \qquad G(\Omega,Q) = \eta \sum_{k=1}^{n} \phi(k)(\nu_k\kappa\eta Q)^{k-1} \tag{65.106}$$

We list separately special cases, frequently encountered:

$$\begin{array}{ll} f = q & F = 1 \\ f = q \pm \mu^2 q^3 & F(Q) = 1 \pm \tfrac{3}{4}\mu^2Q^2 \\ f = \sin q & F(Q) = \dfrac{1}{Q}J_1(Q) \\ f = \operatorname{Sinh} q & F(Q) = \dfrac{1}{Q}I_1(Q) \\ g = \dot q & G(\Omega,Q) = \eta \\ g = (\operatorname{sgn}\dot q)\nu\dot q^2 & G(\Omega,Q) = \tfrac{2}{3}\eta^2\nu\kappa Q \end{array} \tag{65.106a–f}$$

In Eqs. (65.106c,d), J_1 and I_1 represent the Bessel function and the modified Bessel function.

Equation (65.103a) is the expression for the *response curves*, usually given in Q,η^2 diagrams, with s as parameter.

If the driving term has the form $p\cos\tau$ instead of $p\sin\tau$, we would use $\tilde q = Q\cos(\tau - \epsilon)$ instead of (65.101). The resulting equations (65.103), however, remain unaffected by these replacements.

The most serious deficiency, from the mathematical point of view, of the Ritz-Galerkin method is the absence of estimates for the errors involved. Although such estimates are missing, the usefulness of the results is well established. The method, of course, allows refinements, taking more terms, either of higher frequencies (superharmonics) or of lower frequencies (subharmonics). We shall deal with these refinements on p. 65–24.

For *undamped systems* ($D = 0$), Eq. (65.103*a*) simplifies to

$$\eta^2 = F(Q) - \frac{s}{Q} \qquad (65.107)$$

The equation

$$\eta^2 = F(Q) \qquad (65.107a)$$

which is obtained by putting $s = 0$, applies to the free vibrations; it shows their amplitude-frequency relationship. The corresponding curve customarily is referred to as the *backbone curve* of the set (65.107).

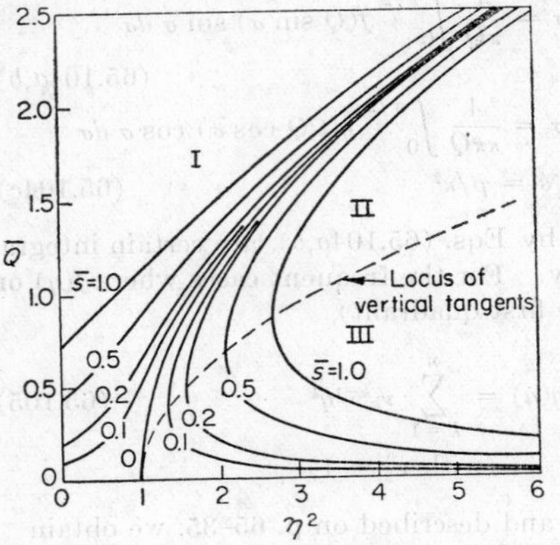

FIG. 65.15. Response curves $|\bar{Q}|$ versus η^2 for undamped system obeying Eq. (65.108), hardening springs.

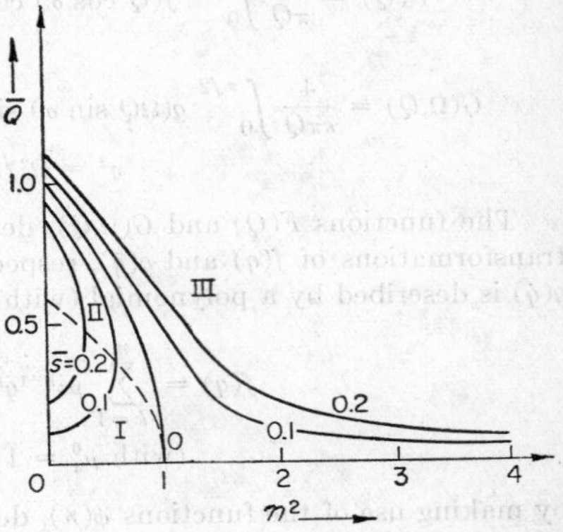

FIG. 65.16. Same as Fig. 65.15, but softening springs.

Figures 65.15 and 65.16 show two representative examples of response curves for undamped systems. Both pertain to the *Duffing equation*

$$\ddot{q} + \kappa^2 q(1 \pm \mu^2 q^2) = p \sin \Omega t \qquad (65.108)$$

the first one to a system possessing a *hardening* spring [upper sign in (65.108)], the second one to a system possessing a *softening* spring [lower sign in (65.108)]. Equation (65.107) becomes

$$\eta^2 = 1 \pm \tfrac{3}{4}\mu^2 Q^2 - \frac{s}{Q} \qquad (65.109)$$

respectively. By use of the dimensionless quantities $\bar{Q} = \sqrt{3/4}\,\mu Q$ and $\bar{s} = \sqrt{3/4}\,\mu s$, they may be written as

$$\eta^2 = 1 \pm \bar{Q}^2 - \bar{s}/\bar{Q} \qquad (65.109a)$$

In the diagrams, these dimensionless quantities $\bar{Q}$ (as ordinates) and $\bar{s}$ (as parameters) are used.

Figure 65.17*a* shows the response curves of a system whose characteristic line is given by Fig. 65.17*b*. Here the ordinates are $\bar{Q} = Q/q_1$, the parameters $\bar{s} = s/q_1$. Similar response curves for numerous other systems having characteristic lines composed of straight-line segments may be found in [25].

From Eqs. (65.105), (65.106), (65.107) in general and Eqs. (65.109) in particular, it is seen that, for systems with a hardening or softening characteristic, the backbone curve in the response diagram leans to the right or to the left, respectively. Figures 65.15 and 65.16 show that, in certain ranges of the abscissa values η^2, there exist several values of ordinates $\bar{Q}$. In regions I and III the sequence of the response curves is such that $\partial\bar{Q}/\partial\bar{s} > 0$ (natural behavior), in II $\partial\bar{Q}/\partial\bar{s} < 0$ (unnatural behavior). The separatrices between the regions are either the backbone curves $\eta^2 = F(Q)$ or the *loci of vertical tangents* $\eta^2 = F(\bar{Q}) + \bar{Q}\partial F/\partial\bar{Q}$; for the examples (65.108), $\eta^2 = 1 \pm \bar{Q}^2$ and $\eta^2 = 1 \pm 3\bar{Q}^2$.

It is important to note that the conditions $\partial\bar{Q}/\partial\bar{s} \gtrless 0$ which describe a natural–unnatural behavior can be shown to express the pertinent stability criteria: The upper sign indicates stability, the lower one instability (see p. 65–29). Some consequences of these stability conditions will be discussed in the following.

When damping forces are present, we have to consider the complete equations (65.103). Equation (65.103*a*) is the equation of the response curves; its degree (or transcendentality) depends on the types of functions f and g present in (65.100) [cf. the examples (65.106)]. In contrast to the results for undamped systems (where the

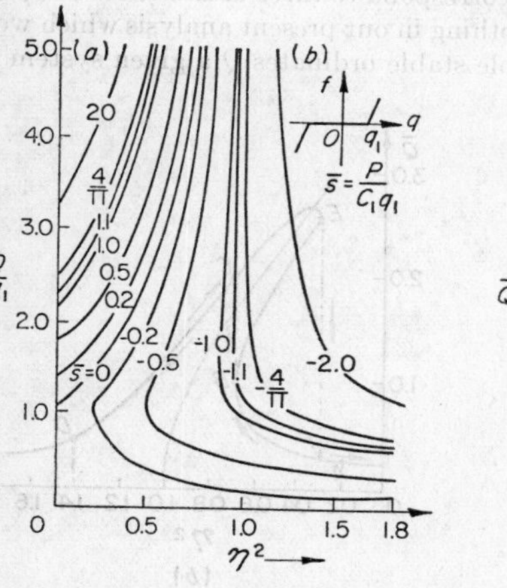

FIG. 65.17. (*a*) Response curves for undamped system having characteristic shown in (*b*).

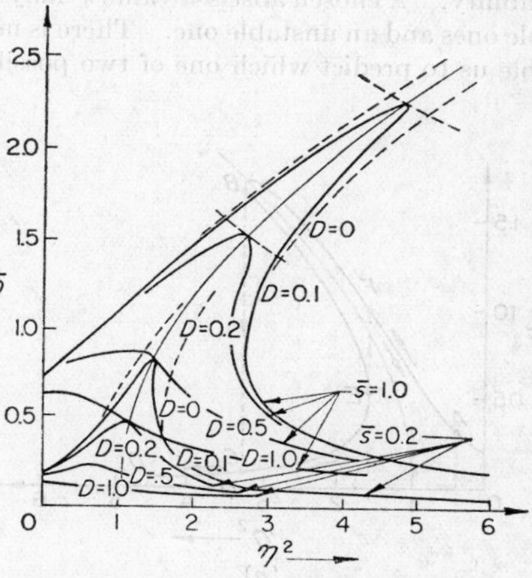

FIG. 65.18. Response curves of system used in Fig. 65.15 with linear damping forces.

phase angle ϵ may assume values 0 or π only), ϵ for damped systems may have any value; here, both Q and s will have to be considered as positive throughout.

Figure 65.18 shows representative examples of response curves. The response curves in the upper regions do not accompany the backbone curve indefinitely (as they do for undamped systems), but they cross it. The crossing point can easily be determined: Because on the backbone curve (65.107*a*) the first term in (65.103*a*) vanishes, we obtain

$$2DG = s/Q \qquad \text{equivalently} \qquad 2DG = \bar{s}/\bar{Q} \tag{65.110}$$

Hence, in the case of linear damping forces,

$$2D\eta = s/Q \qquad \text{or} \qquad \eta Q = s/2D \tag{65.110a}$$

of quadratic damping forces

$$\tfrac{4}{3}D\eta^2\nu\kappa Q = s/Q \qquad \text{or} \qquad \eta^2 Q^2 = \frac{3s}{4D\nu\kappa} \tag{65.110b}$$

The second forms of (65.110*a*,*b*) describe the loci of crossing points (for given parameters s, D, etc.) for any backbone curve.

The exact shape of the response curve of a damped system is given by Eq. (65.103a). Its detailed calculation only rarely will be required. For many purposes, it will suffice to produce, first, the *undamped* response curve (by which the *damped* one is bracketed) and its backbone from the much simpler Eq. (65.107) and, second, the crossing points with the backbone by means of the loci (65.110). It should be noted that the crossing point, because of (65.103b), corresponds to a phase angle $\epsilon = \pi/2$.

In systems with softening characteristics, where the backbone leans toward the left, more than one crossing point, and, hence, more than one branch of the response curve, may be found. In those regions, extreme conditions prevail, and the accuracy of the Ritz method becomes doubtful; special considerations will be necessary under those circumstances.

In Fig. 65.18, for each curve, two regions may be distinguished. In region I, there is $\partial Q/\partial s > 0$; hence, stability exists (see p. 65–29) in region II, $\partial Q/\partial s < 0$, indicating instability. A chosen abscissa value η^2 may correspond to three ordinate values Q, two stable ones and an unstable one. There is nothing in our present analysis which would enable us to predict which one of two possible stable ordinates Q a given system will

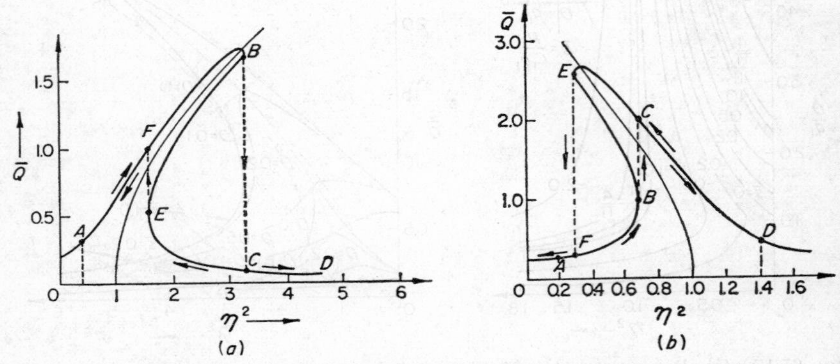

FIG. 65.19. Jump phenomenon: (*a*) Hardening springs; (*b*) softening springs.

actually assume. The selection will depend on the initial conditions or the "history" of the motion (see p. 65–26).

One line of reasoning, however, is very instructive, and is well confirmed by experimental evidence. Although the curves in our response diagrams (as in any response diagram, whether the system be linear or not) are to be interpreted strictly *pointwise* (each point corresponding to a steady-state condition) and, principally, cannot be *traversed*, very slow changes in the parameters of the steady state will result in the representative point traveling slowly along the curves. Small changes in parameters will result in small changes of the response coordinates; hence, the nearest of two *possible* representative points will be the actual one. In such cases, we know the condition the system will assume.

Following this plan, we may perform the following thought experiment with a view to Fig. 65.19a. Assuming that a steady state exists corresponding to point A, we let the driving frequency Ω, hence η^2, be increased slowly. The representative point will move upward on the response curve, as indicated by the arrow above the line. This process may continue until point B is reached. A further slight increase of η^2 allows no other points but those on the lower branch: The representative point will drop ("jump") from B to C. From C, it may move continuously on that branch to the right. If, however, the starting state corresponds to point D and the frequency η^2

is decreased, the point moves from D via C continuously until E is reached. From this position, it jumps to F, and then it continues its steady journey down that branch.

The two jumps, of course, will be accompanied by transient effects (of which we do not learn from our present analysis). The jumps occur at different frequencies as η^2 increases or decreases; and the differences in the ordinates, in general, will not be equal either.

What has been described in view of Fig. 65.19*a* (system with hardening characteristic) applies similarly to Fig. 65.19*b* (system with softening characteristic).

The jump phenomenon is an essentially nonlinear feature. It is a consequence of the fact that the pertinent nonlinear differential equations do not possess unique solutions.

Finally, we shed the restriction imposed on functions f and g on p. 65-18, and admit non-odd functions also. In such cases, the resulting motion will not have zero mean displacement, but will show a bias. We have to include a mean value M in the assumption for Q. Equation (65.101), therefore, will be replaced by

$$\tilde{q} = M + Q \sin(\tau - \epsilon) = M + A \sin \tau - B \cos \tau \qquad (65.111)$$

This expression contains three open parameters, M, A, B or M, Q, ϵ, respectively. Consequently, there will be three Ritz conditions for determining those three parameters. These conditions are:

$$\int_0^{2\pi} E[\tilde{q}]\, d\tau = 0 \qquad \int_0^{2\pi} E[\tilde{q}] \sin \tau\, d\tau = 0 \qquad \int_0^{2\pi} E[\tilde{q}] \cos \tau\, d\tau = 0 \qquad (65.112)$$

By introducing new integral transforms, F_0, F_s, F_c, G_0, G_s, G_c, analogous to functions F and G [Eq. (65.104*a*,*b*)], three equations for the three parameters M, Q, ϵ will be obtained from (65.112) in terms of the F_i and G_i. They can be treated similarly to Eqs. (65.103*a*,*b*) for the case of odd functions.

FIG. 65.20. Element with input x and output y.

The results presented on the preceding pages allow interpretation in still other terms and applications to other fields [27],[28]. In linear *system analysis* (e.g., of feedback systems), the concept of frequency response in the form of (direct or inverse) transfer functions has been found a very useful tool. An element (Fig. 65.20) may possess an input $x(t)$ and an output $y(t)$; its *behavior*, the relationship between input and output, may be described by the linear differential equation

$$E[y] \equiv M[y] - x(t) = 0 \qquad (65.113)$$

If the input is supposed to be sinusoidal,

$$x = X \sin \Omega t \qquad (65.114a)$$

the output will be sinusoidal too,

$$y = Y \sin(\Omega t - \epsilon) \qquad (65.114b)$$

The ratio of the complex amplitude of the output $Y^* = Ye^{-i\epsilon}$ to the input amplitude X is known as the *direct* transfer function $(Y/X)e^{-i\epsilon}$; its reciprocal as the *inverse* transfer function $(X/Y)e^{i\epsilon}$.

If the behavior of the element is described by a nonlinear differential equation (65.113), and if the same input (65.114*a*) is applied, the output will be nonsinusoidal, although periodic. Using the first harmonic Y_1 of the periodic output, we may construct counterparts to the two transfer functions, namely, $(Y_1/X)e^{-i\epsilon}$ and $(X/Y_1)e^{i\epsilon}$, respectively. These expressions may be called *equivalent* (direct or inverse) transfer functions. For these equivalent transfer functions, the term *describing functions* has been introduced and has become customary.

It is obvious that the results presented above may be interpreted in these terms of systems analysis. If we replace q by y, hence Q by Y_1, and $s\kappa^2 \sin \Omega t$ by $x = X\kappa^2 \sin \Omega t$, Eq. (65.103*a*) is the equation for the amount $|X/Y_1|$; Eq. (65.103*b*) for the phase shift ϵ of the inverse equivalent transfer function $(X/Y_1)e^{i\epsilon}$ (describing function) of the nonlinear element.

Not many systems of more than one degree of freedom have been investigated up to date. Among those dealt with there are the forced oscillations treated by the Ritz method with sinusoidal approximations [29],[30],[31].

Bifrequency Approximations; Superharmonics, Subharmonics. Whereas, in a linear system, the motion resulting from a sinusoidal driving force will again be sinusoidal and will possess the same frequency as the driving term, in a nonlinear system, the resulting motion will be nonsinusoidal, although periodic. This statement is equivalent to saying that higher harmonics (*superharmonics*) will appear in the solution—provided the period T of the resulting motion is the same as the period T of the driving force. Under certain conditions, however, nonlinear systems are capable of forced motions whose period is an integer multiple of the driving period. In such a case, terms will appear in the solution whose frequencies are submultiples of the driving frequency (*subharmonics*). In general, under such circumstances, the frequencies may have any rational relationship to the driving frequency (*super-subharmonics*). In free oscillations, no subharmonics appear; superharmonics, however, are an intrinsic part of free oscillations in nonlinear systems.

As an example, we consider the (undamped) Duffing equation (65.108) in regard to superharmonics of order 3 and to subharmonics of order ⅓. In either case, two frequencies will appear in the form $\tilde{q}$ (*bifrequency approximation*). As a means of investigation, we continue to use the Ritz method [32].

We start our considerations with the free oscillations. The differential equation to be discussed is

$$E[q] \equiv \ddot{q} + \kappa^2 q(1 + \mu^2 q^2) = 0 \tag{65.115}$$

Its harmonic (one-frequency) approximation has been treated on pp. 65–18ff. With the bifrequency approximation (denoting $Q_3 = \alpha Q_1$),

$$q = Q_1 \cos \omega t + \alpha Q_1 \cos 3\omega t \tag{65.116}$$

the Ritz conditions (65.144) yield (if the earlier abbreviations and notations are taken over)

$$\begin{aligned} \eta^2 - 1 - \bar{Q}_1^2(1 + \alpha + 2\alpha^2) &= 0 \\ 9\alpha\eta^2 - \alpha - \bar{Q}_1^2(1/3 + 2\alpha + \alpha^3) &= 0 \end{aligned} \tag{65.117}$$

Eliminating η^2 between these two equations results in

$$\bar{Q}_1^2 = \frac{24\alpha}{1 - 21\alpha - 27\alpha^2 - 51\alpha^3} \tag{65.118}$$

Real values for $\bar{Q}_1$ exist only for

$$0 \leqq \alpha \leqq 0.044818 \tag{65.119}$$

The smallness of the quantity α justifies neglecting higher powers of α in (65.115) (*linearization in* α). Thus, we find

$$\eta^2 = 1 + \bar{Q}_1^2(1 + \alpha) \tag{65.120a}$$

and

$$\alpha = \frac{\bar{Q}_1^2}{21\bar{Q}_1^2 + 24} \tag{65.120b}$$

Now α is bracketed by

$$0 \leqq \alpha \leqq 0.047619 \qquad (65.119')$$

The present example, because it allows exact integration, also may well serve as an illustrative case of the sizes of errors involved in the various approximations. Figure 65.21 tells the story. $\bar{Q}_m$ is an abbreviation for $\bar{Q}_m = \bar{Q}_1(1 + \alpha)$.

Now we turn to the forced oscillations (65.108). First, we consider the superharmonics of order 3; that is, we make the assumption (65.116) for $\tilde{q}$. Thus, we obtain

$$\eta^2 = 1 + \bar{Q}_1{}^2(1 + \alpha + 2\alpha^2) - \frac{\bar{s}}{\bar{Q}_1} \qquad (65.121)$$
$$9\alpha\eta^2 = \alpha + \bar{Q}_1{}^2(1/3 + 2\alpha + \alpha^3)$$

and, from these equations (after linearization in α),

$$\alpha = \frac{\bar{Q}_1{}^2}{21\bar{Q}_1{}^2 + 24 + 27(\bar{s}/\bar{Q}_1)} \qquad (65.121a)$$

From (65.121) and (65.121a), we conclude that, if $|\alpha|$ is to be $\leqq 1/10$, the value $\eta^2 \gtrsim 0.6$. For the frequency range above $\eta^2 \approx 0.6$, the first harmonic is a good approximation to

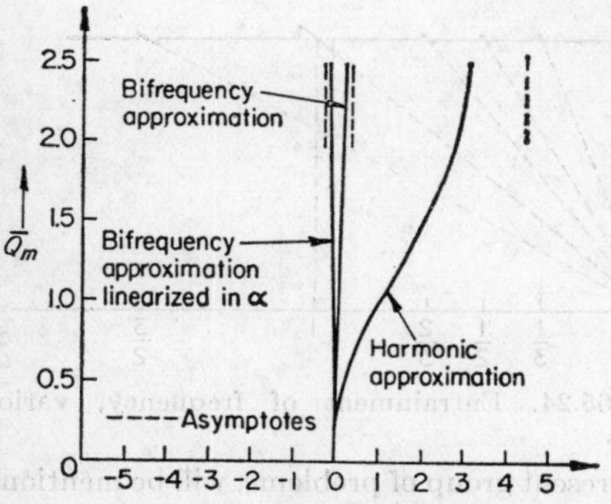

FIG. 65.21. Free undamped Duffing equation, percentage error in η^2 for various Ritz approximations.

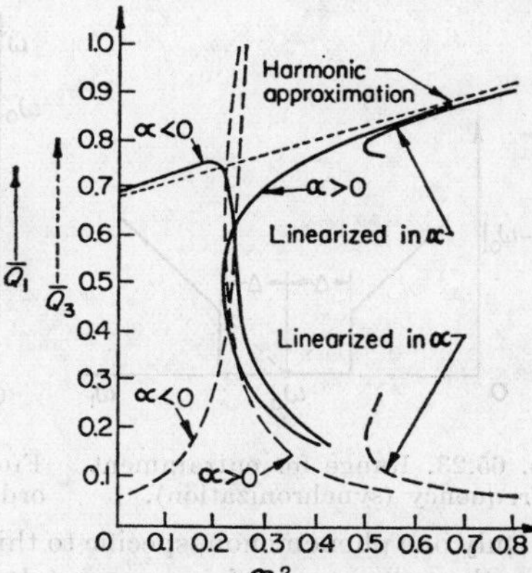

FIG. 65.22. Duffing equation, Ritz bifrequency approximation: (a) Response curves for $\bar{Q}_1$; (b) response curves for $\bar{Q}_3$.

the nonsinusoidal solution. In other ranges, the more complete equations (65.121) have to be used. From a discussion of these equations, Fig. 65.22 has been developed (as a representative example).

In order to discuss the subharmonics of order 1/3, we put

$$\tilde{q} = \beta Q_1 \cos(\omega t/3) + Q_1 \cos \omega t \qquad (65.122)$$

The Ritz conditions yield

$$\beta\eta^2 = 9\beta[1 + \bar{Q}_1^2(2 + \beta + \beta^2)]$$
$$\eta^2 = 1 + \bar{Q}_1^2[1 + 2\beta^2 + 1/3\beta^3] - \frac{\bar{s}}{\bar{Q}_1} \qquad (65.123)$$

If we compare Eqs. (65.121) (for the case of superharmonics) with (65.123) (for the case of subharmonics), we find that Eqs. (65.123) admit $\beta = 0$ as a possible solution, whereas Eqs. (65.121) do not allow $\alpha = 0$. This observation points to an essential feature: Superharmonics must appear, subharmonics may or may not appear (in the undamped system). Detailed investigations show that, furthermore, the subharmonics are extremely sensitive to damping. Comparatively small amounts of damping destroy the conditions for the appearance of subharmonics altogether. For the system considered here with $\bar{s} = 1.0$, the critical value D (for linear damping) was found to be $D = 0.0244$.

Transition into Steady State. So far, only the steady states of the forced oscillations have been considered. Nothing has been said yet about how the system behaves between its initial state and the finally achieved steady state. This question profitably can be treated by a variant of the method based on the concept of variation of parameters, shown on p. 65–32. That variant of the method originally was used by B. van der Pol [33]; later it has been applied sucessfully by many Russian authors; a detailed presentation has been given by Ch. Hayashi [34]. We refer the reader to this last source.

Systems Capable of Auto-oscillations

On the mathematical side, there exists an extensive body of papers, the majority of them contributed by J. E. Littlewood and M. L. Cartwright. A representative account of this work has been given by Miss Cartwright [35].

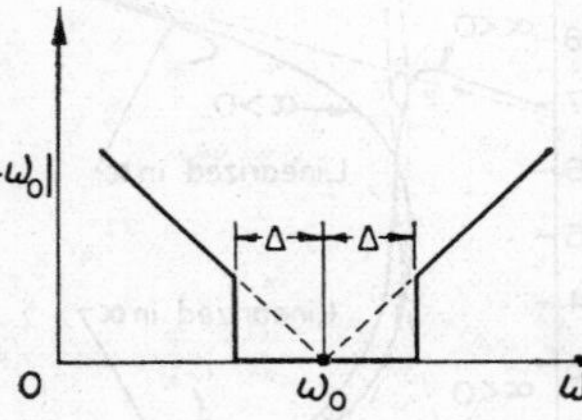

FIG. 65.23. Range for entrainment of frequency (synchronization).

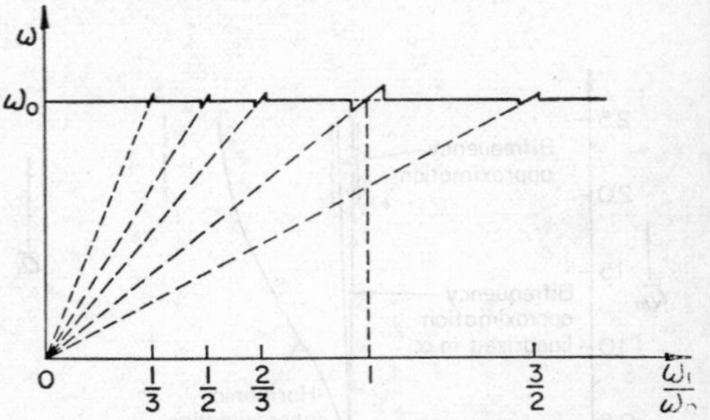

FIG. 65.24. Entrainment of frequency, various orders.

Only one phenomenon, specific to this present group of problems, will be mentioned here: the *entrainment* of frequency (also called *synchronization*). In a linear system, when oscillations of two different frequencies occur, these oscillations perform as if present alone, they do not influence each other. Superposition of two sinusoidal oscillations with closely neighboring frequencies leads to the familiar phenomenon of beats. The beats become the more pronounced, the smaller the difference in frequencies gets. In a system capable of auto-oscillations, this phenomenon changes radically. Let ω_0 be the *natural* (self-sustained) frequency of the system, and ω_1 the driving frequency, and let ω_1 have slowly increasing values. As soon as the difference $|\omega_1 - \omega_0|$ becomes smaller than a certain value, say Δ, the oscillations of frequency ω_0 cease to exist altogether, and only those of frequency ω_1 persist (see Fig. 65.23). The self-sustained oscillations are *entrained* by the driving frequency; they are *synchronized.*

The phenomenon of entrainment occurs not only if $\omega_1 : \omega_0$ is close to $1:1$, it also takes place near $p\omega_0 = q\omega_1$, with p and q being integers. The phenomenon, however, becomes weak when the sum $(p + q)$ gets large, and finally disappears altogether. During entrainment, the frequency ω follows from $p\omega = q\omega_1$. One refers to the case $p/q = 1$ as *ordinary* synchronization, to the case $p/q > 1$ as *subharmonic* synchronization (frequency demultiplication), and to $p/q < 1$ as *ultrasubharmonic* (if p/q is not an integer) or as *ultraharmonic* (if p/q is an integer) synchronization (frequency multiplication). Figure 65.24 gives a survey of these facts [36].

65.4. STABILITY CONSIDERATIONS

Stability of Autonomous Equivalent Linear Systems

For deciding on the stability of autonomous (homogeneous) linear systems, a number of useful criteria have been developed (Routh, Hurwitz, Nyquist, and others). Although their appearances, and also the procedures for applying them, differ greatly, all these criteria grow from the same basic ideas.

One of the methods for treating autonomous nonlinear systems is the method of *equivalent linearization.* (Second version, p. 65–33.) By that method, from the nonlinear system an equivalent linear system is produced, and to this system the stability criteria for linear systems may be applied. Here we follow a presentation by K. Magnus [23] (the source to which the non-Russian reader may turn for further information). Of the multitude of versions for the stability criteria, we select the Hurwitz version as the one best suited to our purposes.

The equivalent linear set of differential equations is given by Eq. (65.155):

$$\frac{dx_i}{dt} = \sum_{\nu=1}^{n} \left(\bar{a}_{i\nu} x_\nu + \bar{a}^*_{i\nu} \frac{dx_\nu}{dt} \right) \tag{65.124}$$

From it, by calling the complex exponent of the e function λ and letting $\delta_{i\nu}$ stand for the Kronecker symbol, the characteristic equation follows in the form

$$|\bar{a}_{i\nu} + \bar{a}^*_{i\nu}\lambda - \delta_{i\nu}\lambda| = 0 \tag{65.125a}$$

or, by putting the coefficients into evidence,

$$\sum_{\nu=0}^{n} c_{n-\nu}\lambda^\nu = 0 \tag{65.125b}$$

The Hurwitz criteria tell us that stability is assured if (1) all the coefficients c_i have the same sign, and (2) the n Hurwitz determinants H_ν have positive signs (see p. 57–20). It can be shown [37] that, among these conditions, the two,

$$c_n > 0 \qquad \text{and} \qquad H_{n-1} > 0 \tag{65.126a}$$

are the most stringent ones. The stability region proper, hence, is bounded by

$$c_n = 0 \qquad H_{n-1} = 0 \tag{65.126b}$$

On the first boundary, the characteristic equation has one root, $\lambda = 0$; on the second boundary, there exists a pair of imaginary roots, signifying that undamped oscillations will occur. We concern ourselves with only the latter part of the stability boundary, and, of this boundary, only that part on which all other stability criteria are satisfied simultaneously. This boundary we denote by

$$(H_{n-1}) = 0 \tag{65.127}$$

Equation (65.127) describes a surface in the space of parameters $\bar{a}$ and $\bar{a}^*$. Frequently, only a selected group of these parameters will be of interest; the parameter space considered, then, may have fewer dimensions. The frequency ω of oscillations occurring on the boundary, by the way, is given by

$$\omega^2 = \frac{c_n H_{n-3}}{H_{n-2}} \tag{65.128}$$

if $n \geqq 3$ and $H_0 = 1$.

For investigating the stability of a linear system, we must decide whether the representative point is situated inside $[(H_{n-1}) < 0]$ or outside $[(H_{n-1}) > 0]$ the stability region. In dealing with equivalent linear systems, we must remember that the values of the parameters (coordinates) $\bar{a}_{i\nu}$ and $\bar{a}^*_{i\nu}$ depend on the (reference) amplitude A. Hence, the representative point moves on a curve, which we call the *A curve.* The representative position of this A curve and the boundary $(H_{n-1}) = 0$ gives all the answers in regard to stability.

Figure 65.25 shows some of the possible relative positions (for the case of two significant parameters $\bar{a}_1$ and $\bar{a}_2$): Curve I is situated entirely inside the stability region; the system is stable for any amplitude A. The converse is true for curve II, situated entirely outside the stability region. Curve III crosses the boundary from the outside to the inside; the amplitude at the crossing point is A'. This means that small oscillations tend to increase, and large ones tend to decrease, to the stable steady-state value A' (limit cycle). The converse is true for curve IV; the steady-state value A' is unstable in this case. Of course, there may be more than one crossing point, corresponding to values A' and A'', respectively. It is readily realized

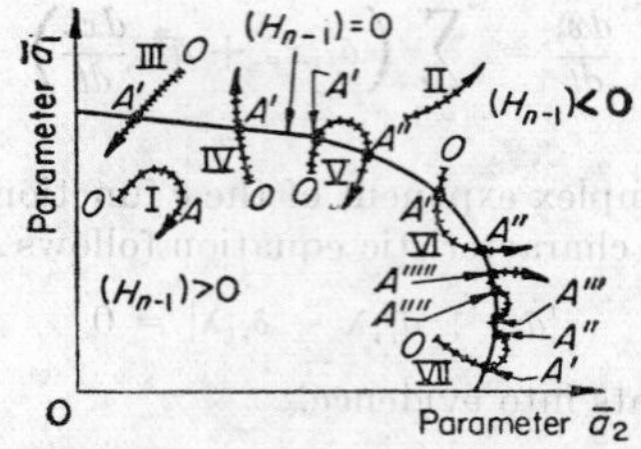

FIG. 65.25. Boundary of stability $(H_{n-1}) = 0$ with various types of A curves.

that the system of curve V will exhibit an unstable limit cycle of radius A' and a stable one of radius A''; the system of curve VI will possess a stable limit cycle of radius A' and an unstable one of radius A''.

In inspecting the curves, the reader should keep in mind that, inside the stability region, the amplitudes tend to decrease; outside to increase.

Of course, there may appear more complicated cases (curve VII), and also special ones (for example, A curves tangent to the boundary, etc.). However, the examples given should suffice to convey the general idea of how to deal with the stability problem for the equivalent linear systems.

Stability of Steady-state Forced Oscillations

On p. 65–18 we discussed the response curves of nonlinear systems which are subjected to sinusoidal driving forces. We found that (under certain conditions) more than one solution is possible. We touched briefly on the problem of stability, and discussed the associated phenomenon of jumps. Here we shall take up once more the problem of stability; we shall treat a representative example rather than develop a general theory. The example will be the Duffing equation (65.108) discussed on p. 65–20. The response curves of that system are described by Eqs. (65.109) and (65.109*a*); the curves proper are shown in Figs. 65.15 and 65.16. Here we shall confine our attention to the case of the hardening spring (Fig. 65.15).

Incremental Differential Equations. The method to be presented is known as the method of the incremental differential equation (sometimes also called *variational differential equation*). The underlying concept is to replace the variable q by $(q_0 + \delta)$ with $\delta \ll q_0$, to develop a linearized differential equation for δ by assuming q_0 to obey the original differential equation [for the example (65.108)], and to investigate the stability of that linear incremental differential equation for δ.

By introducing $(q_0 + \delta)$ for q in (65.108), where $q_0 = Q \sin \omega t$, by taking (65.109) and the abbreviation $\bar{Q}$ into account, and by making use of the identity

$$\sin^2 x = \tfrac{1}{2}(1 - \cos 2x)$$

we find

$$\ddot{\delta} + \kappa^2 \delta[(1 + 2\bar{Q}^2) - 2\bar{Q}^2 \cos 2\omega t] = 0 \tag{65.129}$$

Writing τ for ωt, replacing the derivatives with respect to t by those with respect to τ (denoted by primes), we obtain

$$\delta'' + \frac{1}{\eta^2} \delta[(1 + 2\bar{Q}^2) - 2\bar{Q}^2 \cos 2\tau] = 0 \tag{65.130}$$

Equation (65.130) is a Mathieu differential equation, whose standard form is

$$\delta'' + \delta(\lambda - \gamma \cos 2\tau) = 0 \tag{65.131}$$

Hence, the parameters λ and γ read

$$\lambda = \frac{1}{\eta^2}(1 + 2\bar{Q}^2) \qquad \gamma = \frac{1}{\eta^2} 2\bar{Q}^2 \tag{65.132}$$

The behavior of the solutions of (65.131) is well known. Regions of stability (bounded solutions) and of instability (unbounded solutions), in the plane of the parameters λ and γ, are indicated by the stability chart (also called Ince-Strutt chart) (see, for example, [38]). Of importance is the first region of instability near $\lambda = 1$. The boundary curves of that region have tangents at the crossing point described by

$$\lambda = 1 \pm \frac{\gamma}{2} \tag{65.133}$$

By introducing the values (65.132), we obtain

$$\eta^2 = 1 + \bar{Q}^2 \qquad \eta^2 = 1 + 3\bar{Q}^2 \tag{65.134}$$

respectively. The (tangents of the) boundary curves in the λ,γ plane, hence, map into the backbone curve and the curve of vertical tangency of Fig. 65.15, respectively, the two curves which constitute the boundaries of region II of that diagram. Points (solutions) in that region, hence, are to be considered unstable.

In a similar fashion, other nonlinear differential equations may be treated. Because of the periodicity of q_0, all the linear incremental differential equations will have periodic coefficients; hence, they will be of the type of Hill's differential equation. The stability theory for Hill's differential equations has been developed, in general. However, not many numerical results (as for the special case of Mathieu's equation) are available. Hence, the various cases will have to be investigated individually.

A More Comprehensive Criterion. In order to avoid the necessity for such individual investigations, an attempt has been made to prove that the descriptive statement for regions I and III of the response diagrams on the one hand and region II on the other, namely, $\partial Q/\partial s \gtrless 0$, does constitute the stability criterion proper. The attempt has proved successful for the large class of differential equations

$$\ddot{q} + \kappa^2 q + \epsilon h(q,\dot{q}) = p \sin \Omega t \tag{65.135}$$

with $\epsilon \ll 1$. The method employed was that described on p. 65–32. For details, see [39].

Lyapunov's Method

Whereas the methods described so far in this chapter in one way or another reduced the considerations to those for linear systems and equations, a method based on ideas of A. Lyapunov [5] truly deals with the nonlinear equations. Here we follow a presentation by W. Hahn [40], based on Russian publications. For a survey of the state of the art in regard to stability problems for nonlinear systems, see another paper by W. Hahn [41].

In order to present the basic ideas of the *Lyapunov method,* we choose a simple example,

$$\dot{x} = y - x(x^2 + y^2) \qquad \dot{y} = -x - y(x^2 + y^2) \tag{65.136}$$

without regard to possible physical interpretations. We want to learn about the behavior of the functions $x(t)$ and $y(t)$ for large values of the independent variable t; we are particularly interested to find out whether or not the functions $x(t)$ and $y(t)$ tend toward zero as t increases indefinitely. In case they do, we speak of *asymptotic stability* of the system.

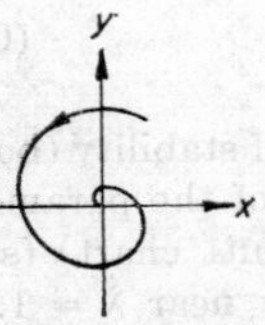

FIG. 65.26. Curve of state for asymptotically stable motion.

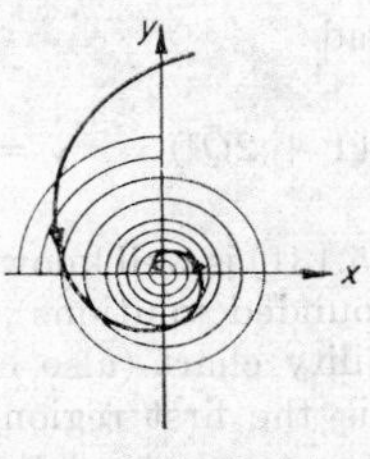

FIG. 65.27. Curve of state with reference curves (circles).

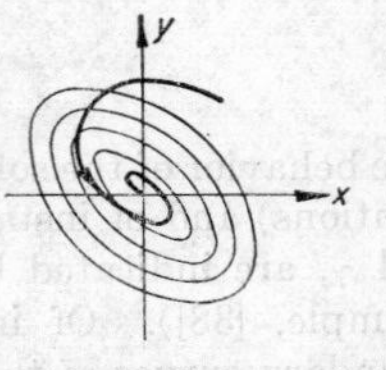

FIG. 65.28. Curve of state with reference curves (ellipses).

We start by assuming that somehow we know a solution $\bar{x}(t),\bar{y}(t)$. We interpret x and y as rectangular coordinates in a *plane of state.* (If $y = \dot{x}$, the plane of state is identical with the phase plane.) The solution $\bar{x}(t),\bar{y}(t)$ represents a curve in that plane, a *curve of state.* The point $\bar{x}(0),\bar{y}(0)$ selects a special curve. If $\bar{x}(t),\bar{y}(t)$ tends toward zero, the curve approaches the origin (Fig. 65.26).

Next we imagine the plane of state covered with a family of closed curves, the *reference curves.* For the present example, we select concentric circles,

$$x^2 + y^2 = V \tag{65.137}$$

as reference curves (Fig. 65.27). A larger value V corresponds to a larger radius of the circle. The reference curves are *numbered,* with numbers increasing from inner to outer circles. The left-hand side of Eq. (65.137) will be called the *reference function.* Let us now consider V at points situated on a chosen curve of state $\bar{x}(t),\bar{y}(t)$. In this case $V(\bar{x},\bar{y}) = V(\bar{x}(t),\bar{y}(t))$ becomes a function of t. With this fact in mind, the behavior of the curves of state mentioned above can easily be expressed analytically: A curve which approaches the origin arbitrarily closely must penetrate the reference circles from the exterior to the interior—at least for large values of t. This statement is equivalent to saying that the value V in (65.137) must decrease, or, in other words,

$$\dot{V} = \frac{\partial V}{\partial x}\dot{x} + \frac{\partial V}{\partial y}\dot{y} \tag{65.138}$$

must be negative along the curve of state. For our example (65.136) we find

$$\dot{V} = 2x\dot{x} + 2y\dot{y} = 2[xy - x^2(x^2 + y^2) - xy \; - y^2(x^2 + y^2)] = -2(x^2 + y^2)^2 \tag{65.138a}$$

Hence, $\dot{V}$ is negative everywhere outside the origin. This is true for any curve of state, because the initial values do not appear in expression (65.138). We conclude that *all* solutions $x(t),y(t)$ approach zero, as t grows indefinitely.

For the example considered, the reference curves are very simple curves. It is readily realized that circles will not be the appropriate reference curves under all circumstances. In case the curves of state have a shape as indicated in Fig. 65.28, circles will not be helpful; however, we may try to work with ellipses. The nature of the reference curves has no significance whatever. The only important facts are (1) that exactly one reference curve passes through each point of the plane of state, and (2) that the curves of state cross the reference curves from the exterior to the interior. It will suffice if these two conditions are satisfied in a certain vicinity of the origin.

Similar considerations apply to sets of three equations of first order. Under such circumstances, we have to work with *reference surfaces* in a *space of state*. Finally, the considerations may be generalized to spaces of n dimensions.

We have to be content with these hints and add just one more remark: The Lyapunov method, in requiring $V > 0$, $\dot{V} < 0$, yields *sufficient* conditions. If V proves to be indefinite (capable of either sign), no conclusion can be drawn. If, however, $V > 0$ and $\dot{V} > 0$, all curves of state cross the reference surfaces from the interior to the exterior, and the system certainly is unstable.

65.5. SPECIAL METHODS

The Ritz Method

The Ritz method is one of the so-called "direct" methods for solving problems in calculus of variations.

If

$$E[q(t)] = 0 \tag{65.139}$$

is the (second-order) differential equation of a phenomenon (motion), this equation may be considered the Euler equation of a related variational problem,

$$I = \int_{t_1}^{t_2} F(q,\dot{q},t)\,dt = \text{stat} \tag{65.140}$$

Instead of solving (65.139), we may solve (65.140). The direct methods allow us to do so without resorting to (65.139).

The First Ritz Method (Ritz Minimizing Method) [42]. According to the idea of W. Ritz, the function $q(t)$ in (65.140) is expanded into a series, in terms of a given set of (orthogonal) functions $\psi_k(t)$, retaining only a finite number of terms, thus replacing $q(t)$ by the polynomial

$$\tilde{q}(t) = \sum_{k=1}^{n} a_k \psi_k(t) \tag{65.141}$$

By introducing (65.141) into (65.140), the integral I becomes a function of the parameters a_k. These parameters then are determined by the conditions

$$\frac{\partial I}{\partial a_k} = 0 \qquad k = 1, 2, \ldots, n \tag{65.142}$$

If, as it is frequently the case, problem (65.140) is a minimum problem, the function $\tilde{q}(t)$ with coefficients determined by (65.142) renders a *relative* minimum to I, whereas the exact solution $q(t)$ would produce an *absolute* minimum.

The Second Ritz Method (Ritz Averaging Method or Ritz-Galerkin Method) [42]–[44]. It can be shown quite generally that the set of n conditions (65.142) for the coefficients a_k is equivalent to the set of equations

$$\int_{t_1}^{t_2} E[\tilde{q}]\psi_k \, dt + \left[\frac{\partial F}{\partial \dot{\tilde{q}}} \psi_k\right]_{t_1}^{t_2} = 0 \qquad k = 1, 2, \ldots, n \tag{65.143}$$

If the integrated parts vanish or can be made to vanish, these conditions simplify to

$$\int_{t_1}^{t_2} E[\tilde{q}]\psi_k \, dt = 0 \qquad k = 1, 2, \ldots, n \tag{65.144}$$

We will call Eqs. (65.144) the Ritz-Galerkin conditions.

For all our applications of the method, we shall choose the $\psi_k(t)$ as periodic functions having period T. Because it is possible to identify t_2 with $(t_1 + T)$, the conditions for the vanishing of the second term in (65.143) are satisfied: We shall be able to use conditions (65.144). For applications, see p. 65–18.

Occasionally advantages can be gained from using an assumption $\tilde{q}$ of a type more general than (65.141). If

$$\tilde{q} = \sum_{i=1}^{n} \tilde{q}_i(\gamma_{i1}, \gamma_{i2}, \ldots, \gamma_{in}, t) \tag{65.145}$$

is used instead of (65.141), the conditions of stationarity are

$$\frac{\partial I}{\partial \gamma_{ik}} \equiv \int_{t_1}^{t_2} E[\tilde{q}] \frac{\partial \tilde{q}}{\partial \gamma_{ik}} \, dt + \left[\frac{\partial F}{\partial \dot{\tilde{q}}} \frac{\partial \tilde{q}}{\partial \gamma_{ik}}\right]_{t_1}^{t_2} = 0 \qquad i = 1, 2, \ldots, n; \; k = 1, 2, \ldots, n \tag{65.146}$$

Details of procedure, examples, and discussions of benefits may be found in [44].

The Methods of Equivalent Linearization

The First Version [45]. The method deals with autonomous equations of the form

$$\ddot{q} + \kappa^2 q + \epsilon\kappa^2 h(q, \dot{q}) = 0 \qquad \epsilon \ll 1 \tag{65.147}$$

Equation (65.147) describes *quasi-linear* systems; the nonlinearities, lumped in the last term, are supposed to be small. If $\epsilon = 0$, the solution of (65.147) would be

$$q = Q \sin(\kappa t + \theta) \tag{65.148}$$

with Q and θ denoting the parameters of integration. For $\epsilon \neq 0$ in the spirit of the *method of varying the parameters*, the solution is assumed to have the form

$$q = Q(t) \sin[\kappa t + \theta(t)] \equiv Q(t) \sin \alpha(t) \tag{65.149}$$

In other words: Instead of $q(t)$, two new unknown functions $Q(t)$ and $\theta(t)$ are introduced, and differential equations are established for these new functions.

Differentiating (65.149) and requiring

$$Q\dot{\theta} \cos \alpha + \dot{Q} \sin \alpha = 0 \tag{65.150a}$$

after substituting q, $\dot{q}$ into (65.147) we find

$$-Q\kappa\dot{\theta} \sin \alpha + \dot{Q}\kappa \cos \alpha = \epsilon\kappa^2 h(Q \sin \alpha, \kappa Q \cos \alpha) \tag{65.150b}$$

Equations (65.150a,b) lead to

$$\dot{Q} = -\epsilon\kappa h \cos \alpha \qquad Q\dot{\theta} = +\epsilon\kappa h \sin \alpha \tag{65.151a,b}$$

The right-hand sides in (65.151*a,b*) are periodic functions in α with periods 2π. They can be expanded into Fourier series. By retaining only the constant terms of these expansions, we perform processes of *averaging* over the period. By replacing the exact values for $\dot{Q}$ and $\dot{\theta}$ by those average values, we obtain

$$\begin{aligned} \dot{Q} &= -\epsilon\kappa \frac{1}{2\pi} \int_0^{2\pi} h(Q \sin \alpha, \kappa Q \cos \alpha) \cos \alpha \, d\alpha \\ \dot{\theta} &= +\epsilon\kappa \frac{1}{Q} \frac{1}{2\pi} \int_0^{2\pi} h(Q \sin \alpha, \kappa Q \cos \alpha) \sin \alpha \, d\alpha \end{aligned} \tag{65.152a,b}$$

Equations (65.152*a,b*) are first-order differential equations for the newly introduced functions $Q(t)$ and $\theta(t)$.

If the function h splits into

$$h(q,\dot{q}) = h_0(q) + h_1(\dot{q}) \tag{65.153}$$

then $\dot{Q}$ is unaffected by h_0, and $\dot{\theta}$ is unaffected by h_1.

The Second Version [23]. This version is suited to deal with (autonomous) systems of any order (sets of differential equations). In its most general form, it can be presented in this way: Given the set

$$\frac{dx_i}{dt} = \sum_{\nu=1}^{n} a_{i\nu}x_\nu + \mu_i f_i(x_1,x_2, \ldots ,x_n) \qquad i = 1, 2, \ldots , n \tag{65.154}$$

we wish to find the *equivalent linear set*

$$\frac{dx_i}{dt} = \sum_{\nu=1}^{n} \left(\bar{a}_{i\nu}x_\nu + \bar{a}^*_{i\nu} \frac{dx_\nu}{dt} \right) \tag{65.155}$$

The *equivalence* is understood in the following sense: The coefficients $\bar{a}_{i\nu}$ and $\bar{a}^*_{i\nu}$ are to be chosen in such a way that the amplitudes A_ν of the sinusoidally varying quantities x_ν in (65.155) for all values of those amplitudes coincide with the first harmonics of the corresponding periodic solutions of (65.154).

We put

$$x_\nu = A_\nu \sin \alpha_\nu = A\gamma_\nu \sin \alpha_\nu \tag{65.156}$$

where

$$\alpha_\nu = \omega t + \theta_\nu \qquad \gamma_\nu = A_\nu/A \tag{65.156a}$$

and A denotes any reference amplitude, γ_ν the amplitude ratio.

If the variables x_ν [Eq. (65.156)] are introduced into (65.154), the functions f_i also become periodic functions with period $T = 2\pi/\omega$. It is assumed that the functions f_i can be expanded into Fourier series (not containing constant terms). The fundamental harmonics of those functions can be written as

$$C_i \cos \alpha_1 + D_i \sin \alpha_1 \tag{65.157}$$

with

$$\begin{aligned} C_i &= \frac{1}{\pi} \int_0^{2\pi} f_i(A\gamma_1 \sin \alpha_1, A\gamma_2 \sin \alpha_2, \ldots , A\gamma_n \sin \alpha_n) \cos \alpha_1 \, d\alpha_1 \\ D_i &= \frac{1}{\pi} \int_0^{2\pi} f_i(A\gamma_1 \sin \alpha_1, A\gamma_2 \sin \alpha_2, \ldots , A\gamma_n \sin \alpha_n) \sin \alpha_1 \, d\alpha_1 \end{aligned} \tag{65.158}$$

(Here the fundamental harmonics of the functions f_i are resolved according to the components of the first variable x_1. In a similar manner, any other variable x_ν could be chosen.)

Comparing the coefficients of (65.155) and of (65.154), produced by introducing (65.156) with attention to (65.158), we find

For $\nu \neq 1$: $\bar{a}_{i\nu} = a_{i\nu} \qquad \bar{a}^*_{i\nu} = 0$

For $\nu = 1$:

$$\bar{a}_{i1} = a_{i1} + \frac{\mu_i}{\pi A \gamma_1} \int_0^{2\pi} f_i(\cdot \cdot \cdot) \sin \alpha_1 \, d\alpha_1 \tag{65.159}$$

$$a^*_{i1} = 0 + \frac{\mu_i}{\omega \pi A \gamma_1} \int_0^{2\pi} f_i(\cdot \cdot \cdot) \cos \alpha_1 \, d\alpha_1$$

Matrix $\bar{a}_{i\nu}$ differs only in its first column from matrix $a_{i\nu}$; matrix $\bar{a}^*_{i\nu}$ contains elements in its first column exclusively.

Equations (65.159) determine matrices $\bar{a}_{i\nu}$ and $\bar{a}^*_{i\nu}$ in general. An important and frequently occurring special case allows considerable simplifications. If the functions f_i can be split according to

$$f_i(x_1, x_2, \ldots, x_n) = \sum_{\nu=1}^{n} f_{i\nu}(x_\nu) \tag{65.160}$$

the original set of differential equations (65.154) reads

$$\frac{dx_i}{dt} = \sum_{\nu=1}^{n} [a_{i\nu} x_\nu + \mu_i f_{i\nu}(x_\nu)] \qquad i = 1, 2, \ldots, n \tag{65.161}$$

In this case, after forms (65.156) are introduced, the functions $f_{i\nu}$ preferably are resolved, not according to the components $\cos \alpha_1$ and $\sin \alpha_1$, but individually according to $\cos \alpha_\nu$ and $\sin \alpha_\nu$. Thus, instead of (65.159), the equations

$$\bar{a}_{i\nu} = a_{i\nu} + \frac{\mu_i}{\pi A \gamma_\nu} \int_0^{2\pi} f_{i\nu}(A\gamma_\nu \sin \alpha_\nu) \sin \alpha_\nu \, d\alpha_\nu \tag{65.162}$$

$$\bar{a}^*_{i\nu} = 0 + \frac{\mu_i}{\omega \pi A \gamma_\nu} \int_0^{2\pi} f_{i\nu}(A\gamma_\nu \sin \alpha_\nu) \cos \alpha_\nu \, d\alpha_\nu$$

result. Now matrix $\bar{a}_{i\nu}$ differs in all places from $a_{i\nu}$, and matrix $\bar{a}_{i\nu}^*$ has elements in all columns. However, the computation of the integrals in (65.162) is greatly facilitated compared to those in (65.159).

Using the abbreviations (integral transformations)

$$K_s\{f_{i\nu}\} = \frac{1}{\pi A \gamma_\nu} \int_0^{2\pi} f_{i\nu}(A\gamma_\nu \sin \alpha_\nu) \sin \alpha_\nu \, d\alpha_\nu \tag{65.163}$$

$$K_c\{f_{i\nu}\} = \frac{1}{\omega \pi A \gamma_\nu} \int_0^{2\pi} f_{i\nu}(A\gamma_\nu \sin \alpha_\nu) \cos \alpha_\nu \, d\alpha_\nu$$

(which are very similar to those used on p. 65–19), we may write Eqs. (65.162) as

$$\bar{a}_{i\nu} = a_{i\nu} + \mu_i K_s\{f_{i\nu}\} \qquad \bar{a}_{i\nu}{}^* = 0 + \mu_i K_c\{f_{i\nu}\} \tag{65.164}$$

For ready reference we mention a few of the more important properties of the integral transforms:

$$K_s\{f\} = \frac{1}{\pi A} \int_0^{2\pi} f(A \sin \phi) \sin \phi \, d\phi \tag{65.165}$$

$$K_c\{f\} = \frac{1}{\omega \pi A} \int_0^{2\pi} f(A \sin \phi) \cos \phi \, d\phi$$

First, we find, for either transform,

$$K\{f_1 + f_2\} = K\{f_1\} + K\{f_2\} \qquad (65.166)$$

and

$$K\{cf\} = cK\{f\} \qquad (65.167)$$

Second, for all even functions $f_e(x)$, there is

$$K_s\{f_e\} = 0 \qquad (65.168)$$

Third, for all unique functions $f(x)$, there is

$$K_c\{f\} = 0 \qquad (65.169)$$

Fourth, for nonunique functions $f(x)$, there is

$$K_c\{f\} = \frac{F}{\omega A^2 \pi} \qquad (65.170)$$

where F denotes the area of the pertinent f,x diagram circumscribed during the integration. $F \gtrless 0$, if the sense of direction is clockwise–counterclockwise.

The Auxiliary Functions $\phi(n)$

Both the *method of equivalent linearization* and the *Ritz-Galerkin method* frequently lead to integrals over powers of the functions sin x and cos x. It has been found desirable to deal with those cases by means of a special auxiliary function.

We define the function (of positive argument)

$$\phi(n) = \frac{4}{\pi}\int_0^{\pi/2} \sin^{n+1} x\, dx = \frac{4}{\pi}\int_0^{\pi/2} \cos^{n+1} x\, dx \qquad (65.171)$$

For integer arguments n, the functional values can be expressed as

$$\phi(0) = 4/\pi$$

$$\phi(1) = 1$$

$$\phi(n) = \frac{n}{n+1}\,\frac{n-2}{n-1}\cdots\frac{2}{3}\cdot\frac{4}{\pi} \qquad n \text{ even, } n \geqq 2$$

$$\phi(n) = \frac{n}{n+1}\,\frac{n-2}{n-1}\cdots\frac{3}{4} \qquad n \text{ odd, } n \geqq 3$$

For noninteger (and integer) arguments, $\phi(n)$ can be expressed (in various ways) in terms of the gamma function. Two possible forms are

$$\phi(n) = \frac{1}{2^n}\,\frac{\Gamma(n+2)}{\left[\Gamma\left(\frac{n+3}{2}\right)\right]^2} \qquad (65.172)$$

or

$$\phi(n) = \frac{4}{n+2}\,\frac{1}{\sqrt{\pi}}\,\frac{\Gamma\left(\frac{n+4}{2}\right)}{\Gamma\left(\frac{n+3}{2}\right)} \qquad (65.173)$$

A short table of numerical values is added:

n	$\phi(n)$	n	$\phi(n)$
0	1.2732	4	0.6791
1	1.0000	5	0.6250
2	0.8488	6	0.5820
3	0.7500	7	0.5469

65.6. REFERENCES

Classics

[1] Lord Rayleigh: On maintained vibrations, *Phil. Mag.*, **V,15** (1883), 229–235.
[2] H. Poincaré: "Les Méthodes nouvelles de la mécanique céleste," vol. 1, Gauthier-Villars, Paris, 1892.
[3] H. Poincaré: Sur les courbes définies par une équation différentielle, *J. math. pures*, (3), **7** (1881), 375; **8** (1882), 251; and "Oeuvres," vol. 1, Gauthier-Villars, Paris, 1892.
[4] A. Lindstedt: Differentialgleichungen der Störungstheorie, *Mem. Acad. Sci. St. Petersbourg*, **31** (1883).
[5] A. Liapunov: Problème général de la stabilité du mouvement, *Ann. fac. sci. univ. Toulouse*, **2** (1907); reproduced in *Ann. Math. Studies*, **17**, Princeton Univ. Press, Princeton, N.J., 1949. (Transl. of Russian original of 1892.)
[6] B. van der Pol: A theory of the amplitude of free and forced triode vibrations, *Radio Rev.*, **1** (1920), 701–710, 754–762.
[7] G. Duffing: Erzwungene Schwingungen bei veränderlicher Eigenfrequenz, Vieweg, Brunswick, Germany, 1918.

Textbooks, Surveys

[8] N. Minorsky: "Introduction to Non-linear Mechanics," Edwards, Ann Arbor, 1947.
[9] J. J. Stoker: "Nonlinear Vibrations," Interscience, New York, 1950.
[10] W. C. Cunningham: "Introduction to Nonlinear Analysis," McGraw-Hill, New York, 1958.
[11] Y. H. Ku: "Analysis and Control of Nonlinear Systems," Ronald, New York, 1958.
[12] H. Kauderer: "Nichtlineare Mechanik," Springer, Berlin, 1958.
[13] K. Klotter: Neuere Methoden und Ergebnisse auf dem Gebiet nicht-linearer Schwingungen, *VDI-Ber.*, **4** (1955), 35–46.
[14] K. Klotter: Non-linear oscillations, *Appl. Mechanics Revs.*, **10** (1957), 495–498.

Sec. **65.1**

[15] P. F. Byrd, M. D. Friedman: "Handbook of Elliptic Integrals," p. 256, Springer, Berlin, 1954.
[16] K. Klotter: Über die freien Bewegungen einfacher Schwinger mit nicht gerader Kennlinie, *Ing.-Arch.*, **7** (1936), 87–99.
[17] K. Klotter: "Technische Schwingungslehre," vol. 1, 2d ed. p. 153, Springer, Berlin, 1951.
[18] K. Klotter: Free oscillations of systems having quadratic damping and arbitrary restoring forces, *J. Appl. Mechanics*, **22** (1955), 493–499.
[19] K. Klotter: The attenuation of damped free vibrations and the derivation of the damping law from recorded data, *2d U.S. Natl. Congr. Appl. Mechanics*, 1954, pp. 85–93.

Sec. **65.2**

[20] L. S. Jacobsen: On a general method for solving second order differential equations by phase-plane displacements, *J. Appl. Mechanics*, **19** (1952), 543–553.
[21] A. A. Andronov, C. E. Chaikin (English ed. by S. Lefschetz): "Theory of Oscillations," Princeton Univ. Press, Princeton, N.J., 1949.
[22] J. Kestin, S. K. Zaremba: Geometrical methods in the analysis of ordinary differential equations, *Appl. Sci. Research*, **B,3** (1954), 149–189.
[23] K. Magnus: Über ein Verfahren zur Untersuchung nichtlinearer Schwingungs- und Regelungs-Systeme, *VDI Forschungsheft*, no. 451 (1955). (Excerpts with the permission of the copyright owner.)
[24] K. Magnus: Stationäre Schwingungen in nicht-linearen dynamischen Systemen mit Totzeiten, *Ing.-Arch.*, **24** (1956), 341–350.

Sec. **65.3**

[25] K. Klotter: Non-linear vibration problems treated by the averaging method of W. Ritz, *Proc. 1st U.S. Natl. Congr. Appl. Mechanics*, 1951, pp. 125–131.
[26] K. Klotter: Steady state vibrations in systems having arbitrary restoring and arbitrary damping forces, *Proc. Symposium Nonlinear Circuit Anal.*, New York, 1953, pp. 234–257.

[27] K. Klotter: How to obtain describing functions for nonlinear feedback systems, *Trans. ASME*, **79** (1957), 509–512.
[28] K. Klotter: An extension of the conventional concept of the describing function, *Proc. Symposium Nonlinear Circuit Anal.*, New York, 1956, pp. 151–162.
[29] F. R. Arnold: Steady state oscillations in non-linear systems of two degrees of freedom, Ph.D. diss., Stanford Univ., 1953.
[30] F. R. Arnold: Steady state behavior of systems provided with non-linear dynamic vibration absorbers, *J. Appl. Mechanics*, **22** (1955), 487–492.
[31] K. Klotter: Steady state oscillations in non-linear multi-loop circuits, *Trans. IRE*, **CT-1** (1954), no. 4, pp. 13–18.
[32] J. C. Burgess: Harmonic, superharmonic and subharmonic response for single degree of freedom systems of the Duffing type, Ph.D. diss., Stanford Univ., 1954.
[33] B. van der Pol: The nonlinear theory of electric oscillations, *Proc. IRE*, **22** (1934), 1051–1086.
[34] Ch. Hayashi: "Forced Oscillations in Non-linear Systems," Nippon Print. and Publ. Co., Osaka, Japan, 1953.
[35] M. L. Cartwright: Forced oscillations in non-linear systems, "Contributions to the Theory of Nonlinear Oscillations," vol. 1, p. 149, Princeton Univ. Press, Princeton, N.J., 1950.
[36] J. S. Schaffner: Almost sinusoidal oscillations in nonlinear systems, part II: Synchronization, *Univ. Illinois Eng. Expt. Sta. Bull.* 400 (1952).

Sec. 65.4

[37] L. Cremer: Die Verringerung der Zahl der Stabilitätskriterien bei Voraussetzung positiver Koeffizienten der charakteristischen Gleichung, *Z. angew. Math. Mech.* **33** (1953), 221–227.
[38] N. W. McLachlan: "Theory and Application of Mathieu Functions," Oxford Univ. Press, New York, 1947.
[39] K. Klotter, E. Pinney: A comprehensive stability criterion for forced vibrations in nonlinear systems, *J. Appl. Mechanics*, **20** (1953), 9–12.
[40] W. Hahn: Behandlung von Stabilitätsproblemen mit der zweiten Methode von Liapunov, pp. 51–66 in: "Nichtlineare Regelungsvorgänge, Beihefte zur Regelungstechnik," Oldenbourg, Munich, 1956. (Excerpts with permission of copyright owner.)
[41] W. Hahn: Probleme und Methoden der modernen Stabilitäts-theorie, *MTW-Mitt.* (*Tech. Hochsch. Wien*), no. IV/5 (1957), 119–134.

Sec. 65.5

[42] W. Ritz: Über eine neue Methode zur Lösung gewisser Variationsprobleme der mathematischen Physik, *Crelles J. reine u. angew. Math.*, **135** (1909), 1–61.
[43] I. S. Sokolnikoff: "Mathematical Theory of Elasticity," 2d ed., pp. 404–411, McGraw-Hill, New York, 1956.
[44] K. Klotter, Ph. R. Cobb: The use of non-sinusoidal approximating functions for nonlinear oscillation problems, *J. Appl. Mechanics*, **27** (1960), 579–583.
[45] N. Kryloff, N. Bogoliuboff (English ed. by S. Lefschetz): "Introduction to Non-linear Mechanics," Princeton Univ. Press, Princeton, N.J., 1949.

CHAPTER 66

STOCHASTIC LOADS

BY

A. C. ERINGEN, Ph.D., Lafayette, Ind.

66.1. INTRODUCTION

Basic Concepts

In order to determine a physical quantity x, an experimentalist performs many seemingly identical experiments. Let us imagine that all these experiments are performed simultaneously, and the outcome x_i $(i = 1, 2, \ldots, N)$ is tabulated. If $x_1 = x_2 = \cdots = x_N = x$ for large N, we are certain that the outcome is x. However, often we find that these outcomes differ from one another. Suppose that x_1 occurred n_1 times, x_2 occurred n_2 times, and x_M occurred n_M times, x_i being distinct; we are then tempted to take the numerical average,

$$\bar{x} = \frac{1}{N}(x_1 n_1 + x_2 n_2 + \cdots + x_M n_M) \tag{66.1}$$

where

$$N = n_1 + n_2 + \cdots + n_M \tag{66.2}$$

as the outcome x for our experiment. Passing to the limit $N \to \infty$, and defining the probable occurrence of random variable x_k by

$$p(x_k) = n_k/N \tag{66.3}$$

we have

$$E\{x\} = \bar{x} = \sum_{k=1}^{M} x_k p(x_k) \tag{66.4}$$

where $E\{x\}$ represents the expected value or the mean of x_k. Sometimes $\langle x \rangle$ is used to denote the expectation.[1] For a continuous random variable, the corresponding definition is

$$E\{x\} = \bar{x} = \int_{-\infty}^{\infty} x p(x)\, dx \tag{66.5}$$

We notice, from (66.2) and (66.3), that we must have

$$\sum_{k=1}^{N} p(x_k) = \sum_{k=1}^{N} \frac{n_k}{N} = 1 \qquad \text{or} \qquad \int_{-\infty}^{\infty} p(x)\, dx = 1 \tag{66.6a,b}$$

a condition that the probability density function (p.d.f.) $p(x)$ must satisfy. This function is non-negative.

[1] Generally, angular brackets $\langle\ \rangle$ are used to denote the time averages. When the system has the ergodic property, the distinction between $\langle\ \rangle$ and $E\{\ \}$ disappears. Roughly speaking, a system is said to have the ergodic property if the time and the system averages are the same.

Similar to (66.5), the expectation of a function $f(x)$ of a random variable x is defined by[1]

$$E\{f(x)\} = \int_{-\infty}^{\infty} f(x)p(x)\,dx \tag{66.7}$$

Of special interest is the nth moment of the density function,

$$m_n = E\{x^n\} = \int_{-\infty}^{\infty} x^n p(x)\,dx \tag{66.8}$$

which for $n = 1$ gives $\bar{x}$, and for $n = 2$ the mean square.

All the foregoing is valid for a random function $x(t)$ at a given time t. Here t is treated as a fixed parameter. For different t, the statistical character of x may be different. For example, at two distinct instants, t_1 and t_2, the random variables $x(t_1)$ and $x(t_2)$ may have different statistical moments. Thus, treating $x_1 \equiv x(t_1)$, $x_2 \equiv x(t_2)$ as two random variables, we may inquire about their dependence on each other. A second-order probability density function $p_2(x_1,x_2)$ is just what is needed here. In order to indicate the dependence on t_1 and t_2, we must actually write p_2 as $p_2(x_1,x_2;t_1,t_2)$. This being understood, we can calculate the expected value of a function by

$$E\{f(x_1,x_2)\} = \iint_{-\infty}^{\infty} f(x_1,x_2)p_2(x_1,x_2)\,dx_1\,dx_2 \tag{66.9}$$

Note that p_2 is again a non-negative function, and must satisfy[2]

$$\iint_{-\infty}^{\infty} p(x_1,x_2)\,dx_1\,dx_2 = 1 \tag{66.10}$$

Being in possession of p_2, we can calculate second-order moments:

$$m_{mn} = E\{x_1{}^m x_2{}^n\} = \iint_{-\infty}^{\infty} x_1{}^m x_2{}^n p_2(x_1,x_2)\,dx_1\,dx_2 \tag{66.11}$$

Of particular importance is the autocorrelation function $m_{11} \equiv R(t_1,t_2)$ obtained from (66.11) by setting $m = n = 1$. This function for $t_1 = t_2$ gives the mean square of x, which in many instances in the physical system is related to the energy involved,

$$R(t_1,t_2) = E\{x_1x_2\} = \iint_{-\infty}^{\infty} x_1x_2p_2(x_1,x_2)\,dx_1\,dx_2 \tag{66.12}$$

It is now simple to see how this can be extended to higher-order statistics in which the third, fourth, etc., p.d.f. are defined.

Roughly speaking, a process in which various probability density functions are defined at given times is called a *stochastic process*. In physical systems, the ultimate desire is to calculate the p.d.f. of the output when those of the input are given. In the

[1] In this article, we consider only continuous random variables, since they are of greater interest in physical systems.

[2] For a useful account on probability theory, the reader is referred to Chap. 3 on probability and error and to Refs. [2] and [3].

theory of elasticity, statistical fog occurs because of the presence of any combination of:

1. External load
2. The initial and boundary conditions
3. The media

Below, we first give a simple example, which illustrates a method of attack. The following sections are concerned with category 1, for which some feeble attempts toward a solution have been made.

Damped Simple Harmonic Oscillator

Consider a single-degree-of-freedom system, consisting of a mass m, a linear spring having the spring constant k, and a dashpot with a damping constant c, subject to a load $F(t)$. The differential equation of motion is

$$m \frac{d^2x}{dt^2} + c \frac{dx}{dt} + kx = F(t) \tag{66.13}$$

where x is the distance of the mass from its equilibrium position. The character of the general solution of this equation depends on the discriminant:

$$\Delta_0 = c^2 - 4km \tag{66.14}$$

[see Eq. (56.15)]. The underdamped, critically damped, and overdamped cases are distinguished by $\Delta_0 < 0$, $\Delta_0 = 0$, and $\Delta_0 > 0$, respectively. We consider here the

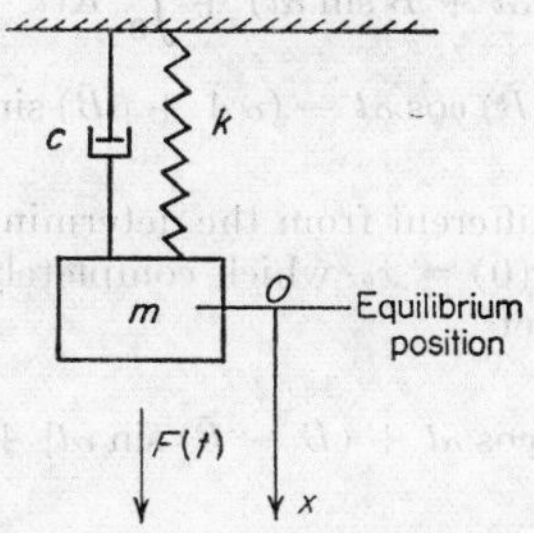

FIG. 66.1. Harmonic oscillator.

interesting case of $\Delta_0 < 0$, the underdamped oscillator. The other two cases are similarly treated. In this case, the general solution of (66.13) is

$$x(t) = e^{-\beta t}(A \cos \alpha t + B \sin \alpha t) + \int_0^t K(t - \tau)F(\tau)\, d\tau \tag{66.15}$$

where

$$K(t) = \frac{1}{m\alpha} e^{-\beta t} \sin \alpha t$$

$$\beta = \frac{c}{2m} \qquad \alpha = \frac{1}{2m}\sqrt{4km - c^2} = \frac{1}{2m}\sqrt{-\Delta_0} \tag{66.16}$$

$$\omega^2 = \beta^2 + \alpha^2 = \frac{k}{m}$$

and A and B are constants which are generally determined from the initial conditions:

$$x(0) = x_0 \qquad \dot{x}(0) = \dot{x}_0 \tag{66.17}$$

In this way the solution of the initial-value problem is completed. Let us now see what happens if some randomness is introduced into this system without destroying the structure of the system. This may be done either through the load $F(t)$ or through the initial conditions (66.17). Thus, in the first case, we may be observing the behavior of a collection of damped harmonic systems, all identical, but, for some reason, the applied load is not controllable, so that, at a given time $t = t_1$, $F_i(t_1)$ applied to the ith oscillator may be different from that applied to the jth oscillator, $F_j(t_1)$. We may know the system average of $\bar{F}(t)$, defined as in (66.1), or various order moments, defined as in (66.8).

If more is known about the external loads separated at the two different times t_1 and t_2, writing $F_1 \equiv F(t_1)$, $F_2 \equiv F(t_2)$, we may possess the knowledge of the second probability density $p_2(F_1,F_2;t_1,t_2)$. In this case, we can calculate such cross moments as are defined in (66.11).

Ideally, given the p.d.f. of some order for $F_i \equiv F(t_i)$, we would like to determine the same for the response $x_i \equiv x(t_i)$. Once this is known, various statistical averages can be calculated. This problem remains unsolved. If, however, we are modest enough to wish simply to find the mean-square displacements, given the mean-square forces, we can achieve some success. Siegert [4] has shown that, for linear systems, when the input is a gaussian process, the output is also a gaussian process. Since a gaussian process is completely determined by prescribing the mean and the correlation functions of the process, then, for this case, we solve the problem completely by determining these functions alone.

Let us first find the mean of $x(t)$ and $dx/dt \equiv \dot{x}(t)$. Since the expectation operator E is exchangeable with other linear operators, we have

$$E\{x(t)\} = \bar{x} = e^{-\beta t}(\bar{A}\cos\alpha t + \bar{B}\sin\alpha t) + \int_0^t K(t-\tau)\bar{F}(\tau)\,d\tau$$

$$E\{\dot{x}(t)\} = e^{-\beta t}[(-\beta\bar{A} + \alpha\bar{B})\cos\alpha t - (\alpha\bar{A} + \beta\bar{B})\sin\alpha t] + \int_0^t \dot{K}(t-\tau)\bar{F}(\tau)\,d\tau$$

We notice that these are no different from the deterministic case. In fact, in view of (66.17), we have $\bar{x}(0) = x_0$, $\dot{\bar{x}}(0) = \dot{x}_0$, which completely determine $\bar{A}$ and $\bar{B}$. If we subtract $\bar{x}$ from (66.15), we get

$$x(t) - \bar{x}(t) = e^{-\beta t}[(A-\bar{A})\cos\alpha t + (B-\bar{B})\sin\alpha t] + \int_0^t K(t-\tau)[F(\tau) - \bar{F}(\tau)]\,d\tau$$

To focus our attention on the effect of the random force, rather than on that of the initial conditions, let us assume that the initial data are deterministic. Hence, $A = \bar{A}$, $B = \bar{B}$, and we have

$$y(t) = \int_0^t K(t-\tau)G(\tau)\,d\tau \tag{66.18}$$

where

$$y(t) = x(t) - \bar{x}(t) \qquad G(\tau) = F(\tau) - \bar{F}(\tau) \tag{66.19}$$

We now proceed to determine the correlation function. To this end, we form

$$E\{y(t_1)y(t_2)\} = \int_0^{t_1}\int_0^{t_2} K(t_1-\tau_1)K(t_2-\tau_2)E\{G(\tau_1)G(\tau_2)\}\,d\tau_1\,d\tau_2 \qquad t_2 \geq t_1 \tag{66.20}$$

which, for $t_1 = t_2$, gives the mean square $E\{y^2\}$. When the correlation function of the external force is given, we have

$$\begin{aligned} E\{G(\tau_1)G(\tau_2)\} &= E\{(F_1-\bar{F}_1)(F_2-\bar{F}_2)\} = E\{F_1F_2\} - \bar{F}_1\bar{F}_2 \\ E\{y(t_1)y(t_2)\} &= E\{(x_1-\bar{x}_1)(x_2-\bar{x}_2)\} = E\{x_1x_2\} - \bar{x}_1\bar{x}_2 \end{aligned} \tag{66.21}$$

Hence, (66.20) and (66.21) determine the correlation function $E\{x_1x_2\}$ in terms of that of the forces. Higher-order moments are similarly determined.

If the initial conditions are not important, that is, the system at $t = -\infty$ starts from zero displacement and velocity, we are facing a steady-state condition. In this case, a certain simplicity is achieved. Writing (66.18) as

$$y(t) = \int_{-\infty}^{t} K(t-\tau)G(\tau)\,d\tau = \int_0^\infty K(s)G(t-s)\,ds \tag{66.22}$$

where we have written $t - \tau = s$, corresponds to the Fourier integral solution of (66.13). Now, the mean and the correlation functions become

$$\begin{aligned} E\{y(t)\} &= \int_0^\infty K(s)E\{G(t-s)\}\,ds \\ E\{y_1y_2\} &= \int_0^\infty \int_0^\infty K(s_1)K(s_2)E\{G(t_1-s_1)G(t_2-s_2)\}\,ds_1\,ds_2 \end{aligned} \tag{66.23}$$

An ideal case is a gaussian process in which

$$E\{u\} = 0 \qquad E\{G(u_1)G(u_2)\} = D\,\delta(u_2 - u_1) \tag{66.24}$$

where $\delta(u)$ is the Dirac's delta function, and D is a constant, the spectral density of the correlation function. This is known as the *purely random gaussian noise.* In this case, the first of (66.23) gives a zero mean for $y(t)$, and the second reduces to

$$E\{y(t)y(t+\tau)\} = D\int_0^\infty K(s)K(\tau+s)\,ds = \frac{D}{2ck}e^{-\beta\tau}\left(\cos\alpha\tau + \frac{\beta}{\alpha}\sin\alpha\tau\right) \tag{66.25}$$

where we have written $t_2 - t_1 = \tau$. Similarly, we may calculate

$$\begin{aligned} E\{\dot{y}(t)y(t+\tau)\} &= \frac{D}{4\alpha\beta m^2}e^{-\beta\tau}\sin\alpha\tau \\ E\{\dot{y}(t)\dot{y}(t+\tau)\} &= \frac{D}{4\beta m^2}e^{-\beta\tau}\left(\cos\alpha\tau - \frac{\beta}{\alpha}\sin\alpha\tau\right) \end{aligned} \tag{66.26}$$

Since G is gaussian, it is now possible to write various probability density functions. For example,

$$p_2(y_1,y_2;t_1,t_2) = \frac{1}{2\pi\sigma_1\sigma_2(1-\rho^2)^{1/2}}\exp\left[-\frac{1}{2(1-\rho^2)}\left(\frac{y_1^2}{\sigma_1^2} + \frac{y_2^2}{\sigma_2^2} - \frac{2\rho}{\sigma_1\sigma_2}y_1y_2\right)\right] \tag{66.27}$$

where

$$\sigma_1^2 = E\{y^2(t_1)\} \qquad \sigma_2^2 = E\{y^2(t_2)\} \qquad \rho\sigma_1\sigma_2 = E\{y(t_1)y(t_2)\} \tag{66.28}$$

all of which are given by (66.25).

We notice that the result, in this case, is independent of t_2 and t_1; it depends only on the difference $\tau = t_2 - t_1$. Such a process is known as a *stationary* or *homogeneous* process. Based on (66.25) and (66.26), we may, in fact, write the fourth-order probability density function by recalling that

$$p_s(y_1, \ldots, y_s) = \frac{1}{(2\pi)^{s/2}\sqrt{\Delta}}\exp\left(-\frac{1}{2\Delta}\sum_{k,l=1}^{s}\Delta_{kl}y_ky_l\right) \tag{66.29}$$

where $\Delta = \det(R_{kl})$, Δ_{kl} = autocovariance, i.e., cofactor of R_{kl} in Δ, and R_{kl} is the covariance or the correlation functions $R_{kl} = E\{y_ky_l\}$. In the present problem,

$$\begin{aligned} R_{11} &= E\{y(t)y(t+\tau)\} & R_{12} &= E\{y(t)\dot{y}(t+\tau)\} \\ R_{21} &= E\{y(t+\tau)\dot{y}(t)\} & R_{22} &= E\{\dot{y}(t)\dot{y}(t+\tau)\} \end{aligned} \tag{66.30}$$

all of which are known through (66.25) and (66.26). Thus, using (66.29) for $s = 4$, we can obtain the four-dimensional probability density function for $y(t)$, $\dot{y}(t)$ at t_1 and t_2 if we desire.

The treatment of the problem with randomness emanating from the initial conditions is similar to and, in fact, somewhat simpler than the foregoing analysis.

66.2. DIFFERENTIAL EQUATIONS OF DISTRIBUTION AND CORRELATION FUNCTIONS

Fokker-Planck Equation

Differential equations of motion of a physical system, together with a given process for the external loads, may be used to obtain the differential equations of the probability density function (p.d.f.) or the characteristic function (c.f.) of the output. For the independent, gaussian, and a few other very special processes, this has been possible. For a simple Gauss-Markoff process input E_{ij}, the output $y_j(t)$ of the linear system,

$$\sum_{j=1}^{n} \left(L_{ij} \frac{d^2 y_j}{dt^2} + R_{ij} \frac{dy_j}{dt} + G_{ij} y_j \right) = \sum_{j=1}^{n} E_{ij}(t) \qquad i = 1, 2, \ldots, n \tag{66.31}$$

will be a gaussian random process [4],[5]; here L_{ij}, R_{ij}, and G_{ij} are constants. We assume that

$$\begin{aligned} E\{E_{ij}\} &= 0 \qquad E\{E_{ij}E_{kl}\} = 0 \\ E\{E_{ij}(t_1)E_{ji}(t_2)\} &= 2R_{ij}k_1T\,\delta(t_2 - t_1) \\ E\{E_{ij}(t_1)E_{ij}(t_2)\} &= 2|R_{ij}|k_1T\,\delta(t_2 - t_1) \end{aligned} \tag{66.32}$$

where $\delta(t)$ is the Dirac delta function, and k_1 and T are constants. Under the assumptions of (66.32), we find that the $2n$ variables, $y_j(t)$, dy_j/dt, will form a $2n$-dimensional Markoff process governed by the *Fokker-Planck* equation for the probability density p:

$$\frac{\partial p}{\partial t} = -\sum_{i=1}^{2n} \frac{\partial}{\partial x_i}(A_i p) + \frac{1}{2} \sum_{i,j=1}^{2n} \frac{\partial^2}{\partial x_i\, \partial x_j}(D_{ij} p) \tag{66.33}$$

where $x_1, \ldots, x_{2n}$ denote the variables $y_1, \ldots, y_n, dy_1/dt, \ldots, dy_n/dt$,

$$A_i = \sum_k a_{ik} x_k \tag{66.34}$$

and the $2n$-by-$2n$ matrices $\mathbf{a}$ and $\mathbf{D}$ are given by

$$\mathbf{a} = \begin{bmatrix} \mathbf{0} & \mathbf{I} \\ -\mathbf{L}^{-1}\mathbf{G} & -\mathbf{L}^{-1}\mathbf{R} \end{bmatrix} \qquad \mathbf{D} = \begin{bmatrix} \mathbf{0} & \mathbf{0} \\ \mathbf{0} & 2k_1T\mathbf{L}^{-1}\mathbf{R}\mathbf{L}^{-1} \end{bmatrix} \tag{66.35}$$

where $\mathbf{L}$, $\mathbf{R}$, and $\mathbf{G}$ are the matrices of the coefficients L_{ij}, R_{ij}, and G_{ij}, respectively, and $\mathbf{I}$ is the unit matrix. For further detail and for other cases, see [3] and [5].

Let us now apply this method to the simple harmonic oscillator. In this case, $L_{ij} = R_{ij} = G_{ij} = E_{ij} = 0$ for all i, j except $i = 1, j = 1$, for which we have $L_{11} = m$, $R_{11} = c$, $G_{11} = k$, $E_{11} = F(t)$. Thus, according to (66.34) and (66.35), we have

$$\mathbf{a} = \begin{bmatrix} 0 & 1 \\ -\omega^2 & -2\beta \end{bmatrix} \qquad \mathbf{D} = \begin{bmatrix} 0 & 0 \\ 0 & D \end{bmatrix} \tag{66.36}$$

$$A_1 = \dot{x} \qquad A_2 = -\omega^2 x - 2\beta\dot{x} \qquad 2k_1Tc/m^2 = D$$

where ω^2, β are given by (66.16). Hence, (66.33) reads

$$\frac{\partial p}{\partial t} = -\frac{\partial}{\partial x}(\dot{x}p) + \frac{\partial}{\partial \dot{x}}[(\omega^2 x + 2\beta\dot{x})p] + \frac{1}{2}\frac{\partial^2}{\partial \dot{x}^2}(Dp) \tag{66.37}$$

which has to be solved under the condition that, at $t = 0$,

$$p(x,\dot{x},0) = \delta(x - x_0)\,\delta(\dot{x} - \dot{x}_0) \tag{66.38}$$

It is simpler to work with the variables

$$z_1 = \dot{x} + a_1 x \qquad z_2 = \dot{x} + a_2 x \tag{66.39}$$

where

$$a_1 = \beta + i\alpha \qquad a_2 = \beta - i\alpha \tag{66.40}$$

with α given by (66.16). Equation (66.37) now reduces to

$$\frac{\partial p}{\partial t} = a_2\frac{\partial}{\partial z_1}(z_1 p) + a_1\frac{\partial}{\partial z_2}(z_2 p) + D\left(\frac{\partial}{\partial z_1} + \frac{\partial}{\partial z_2}\right)^2 p \tag{66.41}$$

We find [5] that the fundamental solution of (66.41) is a two-dimensional gaussian distribution in z_1 and z_2 with the averages and variances

$$\begin{aligned} \bar{z}_i = z_{i0}e^{-a_i t} \qquad i = 1, 2 \qquad E\{(z_1 - \bar{z}_1)^2\} &= (D/a_1)(1 - e^{-a_1 t}) \\ E\{(z_2 - \bar{z}_2)^2\} &= (D/a_2)(1 - e^{-a_2 t}) \\ E\{(z_1 - \bar{z}_1)(z_2 - \bar{z}_2)\} &= \frac{2D}{a_1 + a_2}[1 - e^{-(a_1+a_2)t}] \end{aligned} \tag{66.42}$$

where $\bar{z}_{i0}$ are the initial values of z_1 and z_2 corresponding to x_0 and $\dot{x}_0$. This result can be shown to agree with the results obtained on pp. 66–4 and 66–5.

Differential Equations of Mean-square Output

Often the calculation of the mean-square value of the random variable is sufficient for engineering purposes. Here, we give the result for the nth-order differential equation,

$$\frac{d^n y}{dt^n} + a_1(t)\frac{d^{n-1}y}{dt^{n-1}} + \cdots + a_n(t)y = b_0(t)\frac{d^n x}{dt^n} + b_1(t)\frac{d^{n-1}x}{dt^{n-1}} + \cdots + b_n(t)x \tag{66.43}$$

where $a_j(t)$ and $b_j(t)$ are given, and $x(t)$ is a purely random gaussian process specified by the correlation function

$$E\{x(t_1)x(t_2)\} = \delta(t_1 - t_2) \tag{66.44}$$

It is convenient to write (66.43) in the form

$$\begin{aligned} y &= y_1 + F_0(t)x \\ \frac{dy_j}{dt} &= y_{j+1} + F_j(t)x \qquad j = 1, 2, \ldots, n-1 \\ \frac{dy_n}{dt} &= -a_1(t)y_n - a_2(t)y_{n-1} - \cdots - a_n(t)y_1 + F_n(t)x \end{aligned} \tag{66.45}$$

so that

$$F_0(t) = b_0(t)$$

$$F_i(t) = b_i(t) - \sum_{k=0}^{i-1}\sum_{s=0}^{i-k}\binom{n+s-i}{n-i}a_{i-k-s}(t)\frac{d^s F_k}{dt^s} \qquad i = 1, 2, \ldots, n \tag{66.46}$$

where $\binom{n}{r}$ is the binomial coefficient, Eq. (1.20). If we set

$$\theta_{i,j}(t) = E\{y_i(t)y_j(t)\} \tag{66.47}$$

we find that $\theta_{i,j}$ must satisfy the system

$$\begin{aligned} \frac{d\theta_{i,j}}{dt} &= \theta_{i,j+1} + \theta_{i+1,j} + F_iF_j \qquad i, j = 1, 2, \ldots, n-1 \\ \frac{d\theta_{i,n}}{dt} &= \theta_{i+1,n} + a_1\theta_{i,n} - a_2\theta_{i,n-1} - \cdots - a_n\theta_{i,1} + F_iF_n \\ &\qquad\qquad i = 1, 2, \ldots, n-1 \\ \frac{d\theta_{n,n}}{dt} &= F_n^2 - 2(a_1\theta_{n,n} + a_2\theta_{n,n-1} + \cdots + a_n\theta_{n,1}) \end{aligned} \tag{66.48}$$

subject to the initial conditions $\theta_{i,j}(0) = 0$ when the system is initially at rest. After solving for $\theta_{1,1}$, we can calculate

$$E\{y^2(t)\} = E\{[y_1(t) + F_0(t)x(t)]^2\} = \theta_{1,1}(t) + F_0(t)F_1(t) + F_0(t)\,\delta(0) \tag{66.49}$$

which indicates that, when $F_0(t) \neq 0$, the differential operator on the left of (66.43) is of the same order as the operator on the right, and that $y(t)$ may contain white noise since $\delta(0) = \infty$ [6]. Since it is possible to write a first-order system of equations in the form (66.45), these results apply to a linear system with a purely random input. It is left to the reader to apply the present method to the linear oscillators.

Correlation Analysis

An alternative way of calculating the correlation functions of a linear system of the form

$$\frac{dy_i}{dt} = \sum_{k=1}^{n} a_{ik}(t)y_k(t) + x_i(t) \qquad i = 1, 2, \ldots, n \tag{66.50}$$

with $a_{ik}(t)$ deterministic and $x_i(t)$ a random input, is through the solution

$$y_i(t) = F_i(t) + \sum_{k=1}^{n} \int_{-\infty}^{t} W_{ik}(t,\tau)x_k(\tau)\,d\tau \tag{66.51}$$

where $F_i(t)$ is the general solution of the homogeneous equation, that is, (66.50) with $x_i(t) \equiv 0$. The integration constants are to be determined from

$$E\{y_i(t)\}_{t=0} = \langle y_i(0)\rangle \tag{66.52}$$

Hence, they may be looked upon as nonstatistical quantities. The weighting functions $W_{ik}(t,\tau)$ for a fixed τ satisfy the differential system

$$\frac{dW_{ik}(t,\tau)}{dt} = \sum_{j=1}^{n} a_{ij}(t)W_{jk}(t,\tau) \tag{66.53}$$

subject to

$$W_{ik}(t,t) = \delta_{ik} \tag{66.54}$$

at $t = \tau$. Once the solution (66.51) is known, we can calculate the mean and the correlation functions of $y_i(t)$ in a fashion similar to the method used on p. 66–4.

For example,

$$E\{y_i(t)\} \equiv m_i(t) = F_i(t) + \sum_{k=1}^{n} \int_{-\infty}^{t} W_{ik}(t,\tau)E\{x_k(\tau)\}\,d\tau$$

$$\times E\{[y_i(t_1) - m_i(t_1)][y_j(t_2) - m_j(t_2)]\} = \sum_{k=1}^{n} \sum_{l=1}^{n} \int_{-\infty}^{t_1} \int_{-\infty}^{t_2} W_{ik}(t_1,\tau_1) \cdot W_{jl}(t_2,\tau_2)$$
$$\times E\{[x_k(\tau_1) - \bar{m}_k(\tau_1)][x_l(\tau_2) - \bar{m}_l(\tau_2)]\}\,d\tau_1\,d\tau_2 \qquad (66.55)$$

where $\bar{m}_k(\tau) \equiv E\{x_k(\tau)\}$. If, furthermore, $x_k(t)$ is a gaussian white noise with a zero mean, that is,

$$E\{x_k(t_1)x_l(t_2)\} = \delta_{kl}\delta(t_1 - t_2) \qquad E\{x_k(t)\} = 0 \qquad (66.56)$$

then (66.55) reduces to

$$E\{y_i(t_1)y_j(t_2)\} = \sum_{k=1}^{n} \int_{\infty}^{t_1} W_{ik}(t_1,\tau)W_{jk}(t_2,\tau)\,d\tau \qquad (66.57)$$

In the following, we shall be concerned mainly with the last method of computation, for, once the product moments are known, we can calculate the *covariance function*, $\mu_{ij}^{(r)}$ for the rth component $y_r(t)$ by

$$\mu_{ij}^{(r)} = E\{[y_r(t_i) - m_r(t_i)][y_r(t_j) - m_r(t_j)]\} = E\{y_r(t_i)y_r(t_j)\} - m_r(t_i)m_r(t_j) \qquad (66.58)$$

This, in turn, determines the probability density function $p_n[y_r(t_1),y_r(t_2), \ldots ,y_r(t_n)]$ whenever the output is gaussian:

$$p_n[y_r(t_1),y_r(t_2), \ldots ,y_r(t_n)] = (2\pi)^{-n/2}\Delta^{-1/2} \exp\left(-\frac{1}{2\Delta}\sum_{i,j} \Delta_{ij}\mu_{ij}^{(r)}\right) \qquad (66.59)$$

where Δ_{ij} is the cofactor of $\mu_{ij}^{(r)}$ in $\Delta = \det \mu_{ij}^{(r)}$. If we wish to determine the p.d.f. for the joint distribution of $y_1(t)$, $y_2(t)$, . . . , $y_n(t)$, at time t, we may use

$$p_n(y_1,y_2, \ldots ,y_n) = (2\pi)^{-n/2}\Delta^{-1/2} \exp\left(-\frac{1}{2\Delta}\sum_{i,j} \Delta_{ij}\mu_{ij}\right)$$
$$\mu_{ij} = E\{y_iy_j\} - m_im_j \qquad (66.60)$$

66.3. CORRELATION ANALYSIS OF VISCOELASTIC SYSTEMS

The differential equations of linear viscoelasticity theory for a homogeneous isotropic medium are

$$Mu_{i,jj} + (L + M)u_{j,ij} - \rho\ddot{u}_i = -f_i \qquad (66.61)$$

where $u_i(x_1,x_2,x_3,t) \equiv u_i(x,t)$ is the displacement vector, and ρ and f_i are the mass density and the body force per unit volume, respectively. Indices after a comma represent partial differentiation with respect to space coordinates, and repeated indices indicate summation (see pp. 7-2 and 7-7). Linear time operators L and M are defined by

$$LF \equiv \lambda F + \int_{-\infty}^{t} \lambda'(t - s)F(s)\,ds \qquad MF \equiv \mu F + \int_{-\infty}^{t} \mu'(t - s)F(s)\,ds \qquad (66.62)$$

where λ and μ are Lamé constants, and λ' and μ' are the functions of internal damping (heredity functions).

The stress tensor σ_{ij} and the displacement vector u_i are related to each other by

$$\sigma_{ij} = Lu_{k,k}\delta_{ij} + M(u_{i,j} + u_{j,i}) \tag{66.63}$$

If we extend the definition of λ' and μ' to the range $(-\infty, \infty)$ by assuming that

$$\lambda'(t - s) = \mu'(t - s) = 0 \qquad \text{for } s > t$$

we may take Fourier transforms of (66.61) and (66.63) to obtain

$$\bar{M}\bar{u}_{i,jj} + (\bar{L} + \bar{M})\bar{u}_{j,ji} + \rho\zeta^2\bar{u}_i = -\bar{f}_i \tag{66.64}$$

where a superposed bar represents the Fourier transforms with respect to time, and ζ is the argument of the transform. It is now clear that Eq. (66.64) in form is identical with the equations of elasticity without internal damping, except that $\bar{L}$ and $\bar{M}$ replace λ and μ. Hence, if we obtain the solution of the transformed equations of elasticity without internal damping, and replace λ and μ by $\bar{L}$ and $\bar{M}$, respectively, we have the Fourier transform of the solution of (66.61). Two simple cases of internal friction have been amenable to analytical treatment. These are Kelvin (Voigt) and Maxwell materials. For these, we have respectively

$$\begin{array}{lll} \bar{L} = \lambda - i\lambda'\zeta & \bar{M} = \mu - i\mu'\zeta & \text{Kelvin (Voigt) material} \\ \bar{L} = (1 + ik'/\zeta)^{-1}\lambda & \bar{M} = (1 + ik'/\zeta)^{-1}\mu & \text{Maxwell material} \end{array} \tag{66.65}$$

where k' is a constant. Let $U_k(x,\zeta)$ be solutions of (66.64) with $\bar{f}_i \equiv 0$, where x stands for x_1, x_2, x_3. Then a general solution of (66.64) may be written as

$$2\pi u_j(x,t) = \int_{-\infty}^{\infty} A_{jk}(\zeta)U_k(x,\zeta)e^{-i\zeta t}\,d\zeta - \int_{-\infty}^{\infty}\int_{V(\xi)} G_{jk}(x,\xi,\zeta)f_k(\xi,\zeta)e^{-i\zeta t}\,dV(\xi)\,d\zeta \tag{66.66}$$

where $A_{jk}(\zeta)$ are arbitrary functions to be determined from the initial conditions, and G_{jk} is the Green's function for the problem. If the initial and the boundary conditions are not statistical in nature, then the mean of (66.66) is calculated from

$$2\pi\langle u_j(x,t)\rangle = \int_{-\infty}^{\infty} A_{jk}(\zeta)U_k(x,\zeta)e^{-i\zeta t}\,d\zeta - \int_{-\infty}^{\infty}\int_{V(\xi)} G_{jk}(x,\xi,\zeta)\langle f_k(\xi,\zeta)\rangle e^{-i\zeta t}\,dV(\xi)\,d\zeta \tag{66.67}$$

For convenience, we may define

$$v_j(x,t) = u_j(x,t) - \langle u_j(x,t)\rangle \qquad g_j(x,t) = f_j(x,t) - \langle f_j(x,t)\rangle \tag{66.68}$$

and note that the means of v_j and g_j are zero. Thus, subtracting (66.67) from (66.66), we obtain

$$2\pi v_j(x,t) = -\int_{-\infty}^{\infty}\int_{V(\xi)} G_{jk}(x,\xi,\zeta)g_k(\xi,\zeta)e^{-i\zeta t}\,dV(\xi)\,d\zeta \tag{66.69}$$

The correlation function for v_j is obtained by taking the product of v_j by v_m and averaging

$$\begin{aligned} 4\pi^2\langle v_j(x,t)v_m(y,\tau)\rangle = \iint_{-\infty}^{\infty}\int_{V(\xi)}\int_{V(\eta)} & G_{jk}(x,\xi,\zeta)G_{mr}(y,\eta,\omega) \\ & \times \langle g_k(\xi,\zeta)g_r(\eta,\omega)\rangle \exp\,[-i(\xi t + \omega\tau)]\,dV(\xi)\,dV(\eta)\,d\zeta\,d\omega \end{aligned} \tag{66.70}$$

If the body force is a white-noise process, we have

$$\langle g_k(\xi,\zeta)g_r(\eta,\omega)\rangle = h_{kr}(\xi,\eta)\,\delta(\zeta + \omega) \tag{66.71}$$

which, upon substitution into (66.70), gives

$$4\pi^2\langle v_j(x,t)v_m(y,\tau)\rangle = \int_{-\infty}^{\infty}\int_{V(\xi)}\int_{V(\eta)} G_{jk}(x,\xi,\zeta)G_{mr}(y,\eta,\zeta) \times h_{kr}(\xi,\eta)\exp[-i\xi(t-\tau)]\,dV(\xi)\,dV(\eta)\,d\zeta \quad (66.72)$$

Further simplification would be obtained only if the space distribution of the body force is white noise in one, two, or in all directions. For each direction, we can carry out one integration similar to the foregoing formalism. Let us assume that we have white noise in all directions; then formally

$$h_{kr}(\xi,\eta) = H\delta(\xi - \eta)\delta_{kr} \quad (66.73)$$

which simplifies (66.72) to

$$4\pi^2\langle v_j(x,t)v_m(y,\tau)\rangle = H\int_{-\infty}^{\infty}\int_{V(\xi)} G_{jk}(x,\xi,\zeta)G_{mk}(y,\xi,\zeta)\exp[-i\zeta(t-\tau)]\,dV(\xi)\,d\zeta \quad (66.74)$$

where H is the spectral density of the white noise in space and time. No further simplification of (66.74) is possible unless the Green's function is specifically known. From the knowledge of the correlation function for v_j, we can compute the same for u_j, since

$$\langle u_j u_m\rangle = \langle v_j v_m\rangle + \langle u_j\rangle\langle u_m\rangle \qquad \langle f_j f_m\rangle = \langle g_j g_m\rangle + \langle g_j\rangle\langle g_m\rangle \quad (66.75)$$

Mean square, $\langle u_j^2\rangle$, follows from this by setting $j = m$ (without summing). If the solution of a body-force problem is known, then the mean $\langle u_j\rangle$ will be known. Thus, we may disregard the determination of the mean in the statistical computations, since we may also measure the body forces from their mean. This is valid for all linear systems. We shall, henceforth, assume that the means of applied forces are zero. The determination of the correlation tensor thus rests on finding the Green's function G_{jk} for a given problem. We remember that G_{jk} must satisfy (66.64) with $f_i = 4\pi\delta_{ij}\delta(x_j - \xi_j)$ and homogeneous boundary conditions.

We shall not pursue this subject in this generality any further, but give some examples.

66.4. RANDOM LOADS ON INFINITE AND SEMI-INFINITE MEDIA

Infinite Medium

Consider the case of a static force distribution $f_i(\xi)$, which has zero mean and is applied to the points ξ_i. In this case, we have the Green's function [see Eqs. (41.66), (41.67), and (41.5)]

$$\begin{aligned} G_{ij}(x,\xi) &= ABr^{-1}\delta_{ij} - A(r^{-1})_{,i}(x_j - \xi_j) \\ r &= [(x_1-\xi_1)^2 + (x_2-\xi_2)^2 + (x_3-\xi_3)^2]^{1/2} \\ A &= (\lambda+\mu)/8\pi\mu(\lambda+2\mu) \qquad B = (\lambda+3\mu)/(\lambda+\mu) \end{aligned} \quad (66.76)$$

Hence, the correlation function is given by

$$\langle v_j(x)v_m(y)\rangle = H\iiint_{V(\xi)} [ABr^{-1}\delta_{jk} - A(r^{-1})_{,j}(x_k - \xi_k)] \times [ABr_1^{-1}\delta_{mk} - A(r_1^{-1})_{,m}(y_k - \eta_k)]\,d\xi_1\,d\xi_2\,d\xi_3 \quad (66.77)$$

where r_1 is obtained from r by replacing x_i by y_i. If we set $x = y$ in (66.77), we obtain

$$\langle v_j(x)v_m(x)\rangle = H\iiint_{V(\xi)} A^2r^{-2}[B^2\delta_{jm} + (1+2B)r_{,j}r_{,m}]\,dV(\xi) \quad (66.78)$$

This result is simplified further by taking $j = m$ and summing over j:

$$\langle v_j(x)v_j(x)\rangle = HA^2(3B^2 + 2B + 1) \iiint_{V(\xi)} r^{-2}\, dV(\xi) \tag{66.79}$$

If the region in which the body forces act is a sphere of radius R, this integral can be carried out, leading to

$$\langle v_j(x)v_j(x)\rangle = 4\pi HA^2(3B^2 + 2B + 1)R \tag{66.80}$$

We notice that, when the body forces are distributed in a purely random fashion throughout the infinite medium, the mean-square displacement becomes infinite.

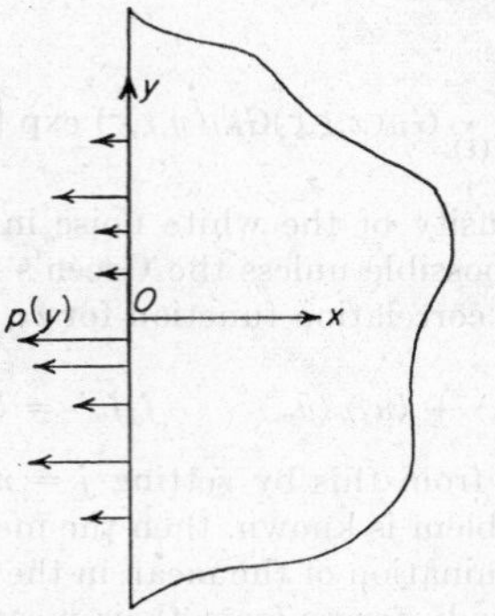

FIG. 66.2. Semi-infinite plane.

Semi-infinite Medium

As an example here, we select the case of normal random forces $p(y)$ applied to the boundary $x = 0$ of a semi-infinite medium $x > 0$. In this case, the stress components σ_{jk} are given by [8]

$$\sigma_{jk}(x,y) = \frac{1}{2\pi}\int_{-\infty}^{\infty} G_{jk}(x,\xi)\bar{p}(\xi)e^{-i\xi y}\, d\xi$$
$$G_{11} = (1 + |\xi|x)e^{-|\xi|x} \qquad G_{22} = (1 - |\xi|x)e^{-|\xi|x} \qquad G_{12} = i\xi xe^{-|\xi|x} \tag{66.81}$$

For white noise, we have

$$\langle p(x)p(y)\rangle = H\delta(x - y) \tag{66.82}$$

This, in turn, gives

$$\langle \bar{p}(\xi)\bar{p}(\eta)\rangle = 2\pi H\delta(\xi + \eta) \tag{66.83}$$

Hence, from (66.81) we find

$$\langle \sigma_{jk}(x,y)\sigma_{mn}(x_1,y_1)\rangle = \frac{H}{2\pi}\int_{-\infty}^{\infty} G_{jk}(x,\xi)G_{mn}(x_1,-\xi)\exp\,[-i\xi(y - y_1)]\, d\xi \tag{66.84}$$

Upon substituting G_{ij} given by (66.81) and integrating, we obtain

$$\begin{aligned}
\langle \sigma_{11}(x,y)\sigma_{11}(x_1,y_1)\rangle &= \pi^{-1}a^{-1}H(A + B + xx_1a^{-2}C)\\
\langle \sigma_{22}(x,y)\sigma_{22}(x_1,y_1)\rangle &= \pi^{-1}a^{-1}H(A - B + xx_1a^{-2}C)\\
\langle \sigma_{12}(x,y)\sigma_{12}(x_1,y_1)\rangle &= \pi^{-1}a^{-3}Hxx_1C\\
\langle \sigma_{11}(x,y)\sigma_{22}(x_1,y_1)\rangle &= \pi^{-1}a^{-1}H[A + (x - x_1)a^{-1}B - xx_1a^{-2}C]\\
\langle \sigma_{11}(x,y)\sigma_{12}(x_1,y_1)\rangle &= -\pi^{-1}a^{-1}H(x_1a^{-1}B' + xx_1a^{-2}C')\\
\langle \sigma_{22}(x,y)\sigma_{12}(x_1,y_1)\rangle &= -\pi^{-1}a^{-1}H(x_1a^{-1}B' - xx_1a^{-2}C')
\end{aligned} \tag{66.85}$$

where

$$A = [1 + (h/a)^2]^{-1} \qquad B = [1 - (h/a)^2]A^2 \qquad C = 2[1 - 3(h/a)^2]A^3$$
$$B' = 2(h/a)A^2 \qquad C' = 2(h/a)[3 - (h/a)^2]A^3 \tag{66.86}$$
$$a = x + x_1 \qquad h = y - y_1$$

We notice that the stress correlations are functions of $(y - y_1)$ as well as of x and x_1. Hence, the output process is *homogeneous* in y. Mean-square stresses are obtained from the above by taking $x = x_1$ and $y = y_1$. This gives

$$\langle \sigma_{11}{}^2 \rangle = 5H/4\pi x \qquad \langle \sigma_{22}{}^2 \rangle = \langle \sigma_{12}{}^2 \rangle = \langle \sigma_{11}\sigma_{22} \rangle = H/4\pi x$$
$$\langle \sigma_{11}\sigma_{12} \rangle = \langle \sigma_{22}\sigma_{12} \rangle = 0 \tag{66.87}$$

which reveals that $\langle \sigma_{11}{}^2 \rangle$ enjoys the privilege of being five times as large as the remaining mean squares and that the shearing stress σ_{12} is uncorrelated with the normal stresses [18].

66.5. BROWNIAN MOTION OF BARS AND PLATES

Ornstein [9] and Houdijk [10] gave a treatment for the Brownian motion of strings. Van Lear and Uhlenbeck [11] treated the case of elastic rods. Eringen [12] gave a method of solution for the problems of random motions of beams and plates subject to random loads. Here we present the result of the latter two works.

The classical differential equation of vibrating damped plates is

$$\nabla^4 w + \frac{\partial^2 w}{\partial t^2} + 2\beta \frac{\partial w}{\partial t} = p(x,y,t) \tag{66.88}$$

Here $w(x,y,t)$ is the displacement as a function of rectangular coordinates x, y and reduced time t; β and p are the reduced-velocity damping coefficient and the transverse pressure. These are related to damping coefficient β_0, pressure p_0, and time t_0 by

$$t = (K/\rho)^{1/2} t_0 \qquad \beta = \tfrac{1}{2}\beta_0/(\rho K)^{1/2}$$
$$p = p_0/K \qquad K = Eh^3/12(1 - \nu^2) \tag{66.89}$$

where E is Young's modulus, ν is Poisson's ratio, h is the thickness, ρ is the mass density per unit area. If we take $EI \equiv K$, $y \equiv \partial/\partial y \equiv 0$, we obtain equations of elastic bars. If $p(x,y,t)$ is a purely random gaussian process in t, we have

$$\langle p(x,y,t + \tau)p(\xi,\eta,t) \rangle = H(x,y,\xi,\eta)\ \delta(\tau)$$

The method of the previous section in this case leads to

$$\langle w(x,y,t)w(\xi, \eta, t + \tau) \rangle = \tfrac{1}{2} \sum_{mnrs} T_{mnrs}(\tau) V_{mn}(x,y) V_{rs}(\xi,\eta) \overline{Q_{mn}Q_{rs}} \tag{66.90}$$

where

$$T_{mnrs}(\tau) = \frac{2\pi}{\omega_{mn}} e^{-\beta\tau} \frac{(\lambda_{rs}{}^4 - \lambda_{mn}{}^4 + 4\beta^2) \sin \omega_{mn}\tau + 4\beta\omega_{mn} \cos \omega_{mn}\tau}{(\lambda_{rs}{}^4 - \lambda_{mn}{}^4 + 4\beta^2)^2 + (4\beta\omega_{mn})^2} \tag{66.91}$$
$$\omega_{mn} = (\lambda_{mn}{}^4 - \beta^2)^{1/2} \qquad \tau \geq 0 \qquad \lambda_{mn}{}^4 > \beta^2$$
$$\overline{Q_{mn}Q_{rs}} = \frac{1}{\pi b_{mn} b_{rs}} \int_A \int_A H(x,y,\xi,\eta) V_{mn}(x,y) V_{rs}(\xi,\eta)\, dA(x,y)\, dA(\xi,\eta)$$
$$\int_A V_{mn}(x,y) V_{rs}(x,y)\, dA = \begin{cases} b_{mn} & \text{for } r = m,\ n = s \\ 0 & \text{for } r \neq m \text{ or } n \neq s \end{cases} \tag{66.92}$$

λ_{mn} and V_{mn} are respectively the eigenvalues and the eigenfunctions, which depend on the boundary conditions. The above integrations are to be taken over the area of the plate. Below, we give various cases for a purely random process, in which

$$H(x,y,\xi,\eta) = H_0 \delta(x - \xi)\delta(y - \eta) \tag{66.93}$$

Some other cases are discussed in [12].

Simply Supported Beam of Span l

$$\langle ww \rangle = \frac{H_0}{\pi l} \sum_n T_{nn} \sin \frac{n\pi x}{l} \sin \frac{n\pi \xi}{l}$$

$$T_{nn} = \frac{l^4 e^{-\beta\tau}}{2\pi^3 n^4} \left(\frac{\sin \omega_n \tau}{\omega_n} + \frac{\cos \omega_n \tau}{\beta} \right) \tag{66.94}$$

$$\omega_n = (n^4 \pi^4 l^{-4} - \beta^2)^{1/2} \qquad \tau \geq 0$$

For $\tau = 0$, we can sum the series in (66.94),

$$\langle ww \rangle = (H_0 l^3/24\beta)[l^{-4}(x^3\xi + x\xi^3) - l^{-3}(3x^3 + \xi^3) + 2l^{-2}x\xi] \tag{66.95}$$

valid for $0 \leq x + \xi \leq 2l$, $0 \leq x - \xi \leq 2l$. For $x = \xi$, this gives

$$\langle w^2 \rangle = \frac{H_0 l^3}{12\beta} \left(\frac{x}{l} \right)^2 \left(1 - \frac{x}{l} \right)^2 \qquad 0 \leq x \leq l, \tau = 0, x = \xi \tag{66.96}$$

For the mid-point $x = \xi = l/2$ but arbitrary τ, we find

$$\left\langle w\left(\frac{l}{2}, t\right) w\left(\frac{l}{2}, t + \tau\right) \right\rangle = \frac{H_0 l^3}{2\beta\pi^4} e^{-\beta\tau} \sum_{n=1,3,\ldots} n^{-4}(\cos \omega_n \tau + \beta\omega_n^{-1} \sin \omega_n \tau) \qquad \tau \geq 0 \tag{66.97}$$

Cantilever Beam of Span l Clamped at $x = 0$

$$\langle ww \rangle = \frac{H_0}{2\pi} \sum_n b_n^{-1} T_{nn}(\tau) V_n(x) V_n(\xi)$$

$$V_n(x) = \frac{\operatorname{Cosh} \lambda_n x - \cos \lambda_n x}{\operatorname{Cosh} \lambda_n l - \cos \lambda_n l} - \frac{\operatorname{Sinh} \lambda_n x - \sin \lambda_n x}{\operatorname{Sinh} \lambda_n l - \sin \lambda_n l} \tag{66.98}$$

$$b_n = (l/4)[V_n{}^2(x)]_{x=l}$$

where T_{nn} is identical with that of (66.94), and λ_n is the nth root of

$$\cos \lambda l \operatorname{Cosh} \lambda l + 1 = 0 \tag{66.99}$$

When $\tau = 0$ and $x = \xi = l$, (66.98) gives

$$\langle w^2(l,t) \rangle = H_0 l^3/12\beta \tag{66.100}$$

The case of $x = \xi = l$, but τ not necessarily zero, may be approximated very well by

$$\langle w(l, t + \tau) w(l,t) \rangle \simeq (H_0 l^3/12\beta) e^{-\beta\tau} (\cos \omega_1 \tau + \beta\omega_1^{-1} \sin \omega_1 \tau) \tag{66.101}$$

Circular Plate of Radius a Clamped at the Outer Edge

$$\langle ww \rangle = \frac{H_0}{4\beta} e^{-\beta\tau} \sum_{m,n} b_{mn}{}^{-1} \lambda_{mn}{}^{-4} (\cos \omega_{mn}\tau + \beta\omega_{mn}{}^{-1} \sin \omega_{mn}\tau) V_{mn}(r,\theta) V_{mn}(\rho,\phi)$$

$$V_{mn}(r,\theta) = [J_n(\lambda_{mn} r) + B I_n(\lambda_{mn} r)] \sin (n\theta + \gamma) \tag{66.102}$$

$$B = -J_n(\lambda_{mn} a)/I_n(\lambda_{mn} a) \qquad b_{mn} = 2\pi a^2 J_n{}^2(\lambda_{mn} a)$$

where J_n and I_n are the Bessel functions, and λ_{mn} are the roots of

$$\frac{J_{n+1}(\lambda a)}{J_n(\lambda a)} + \frac{I_{n+1}(\lambda a)}{I_n(\lambda a)} = 0 \tag{66.103}$$

At the center of the plate ($r = 0$), a good approximation to (66.102) is

$$\langle w(0,\theta,t+\tau)w(0,\phi,t)\rangle = [H_0a^2 \sin^2 \gamma_0/833.6\pi\beta]e^{-\beta\tau} \times \left[\frac{1}{J_0(\lambda_{00}a)} - \frac{1}{I_0(\lambda_{00}a)}\right]^2 (\cos \omega_{00}\tau + \beta\omega_{00}^{-1} \sin \omega_{00}\tau) \quad (66.104)$$

Rectangular Plate, Simply Supported at All Edges (Edge Lengths a, b)

$$\langle ww\rangle = \frac{H_0}{ab\beta} e^{-\beta\tau} \sum_{m,n} \left[\left(\frac{m\pi}{a}\right)^2 + \left(\frac{n\pi}{b}\right)^2\right]^{-2} \sin\frac{m\pi x}{a} \sin\frac{n\pi y}{b} \times \sin\frac{m\pi\xi}{a} \sin\frac{n\pi\eta}{b} (\cos \omega_{mn}\tau + \omega_{mn}^{-1} \sin \omega_{mn}\tau)$$

$$\omega_{mn} = \{[(m\pi/a)^2 + (n\pi/b)^2]^2 - \beta^2\}^{1/2} \qquad \tau \geq 0 \quad (66.105)$$

Bending stress correlations in all cases diverge. Thus, a more improved theory must be used. Calculations based on Timoshenko beam theory for a simply supported bar having translatory and rotatory velocity damping show that the bending stress correlation is convergent [13]. A few results of both theories are shown in Figs. 66.3 and 66.4.

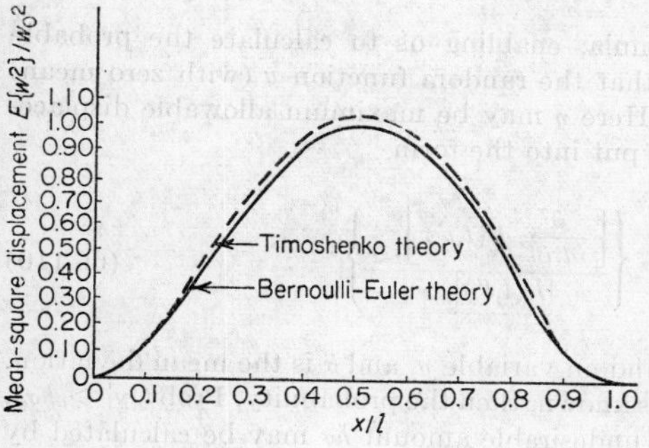

Fig. 66.3. Mean-square displacement along beam. Simply supported steel bar: $Al^2/I = 400$, $w_0^2 = H_0l^3/2\pi^4\beta$.

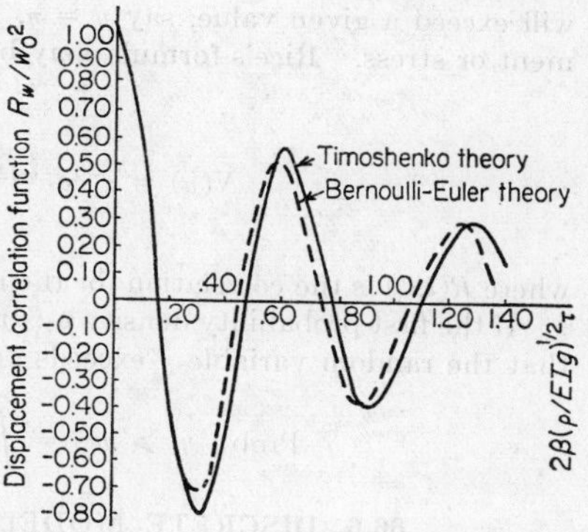

Fig. 66.4. Displacement correlation function (unnormalized): R_w/w_0^2 versus correlation interval $2\beta\tau(\rho/EIg)^{1/2}$. Simply supported steel bar: $Al^2/I = 400$.

The foregoing analysis gives the steady-state response. It does not show how the system reaches this state. This, of course, is also clear from the fact that we have employed the generalized Fourier analysis. Van Lear and Uhlenbeck [11], for a cantilever bar initially at rest, have calculated the mean-square displacement as a function of time t. In our notation, this reads

$$\langle w^2(x,t)\rangle = \frac{H_0}{\beta l} \sum_n \frac{V_n^2(x)}{V_n^2(l)\lambda_n^4} \{1 - \beta\omega_n^{-2}e^{-2\beta t} \times [(\lambda_n^4 + \beta^2)\beta^{-1} + \beta \cos 2\omega_n t + \omega_n \sin 2\omega_n t]\} \quad (66.106)$$

where we replaced $4kT\beta/K$ by our H_0. Here $kT/2$ is the thermal equivalent of the mean potential energy (kinetic energy) for each eigenvibration (the mean being taken over a canonical ensemble) in the statistical mechanics. This equation for $t \to \infty$

gives our result (for $\tau = 0$), indicating, with its form in time dependence, how the steady-state condition is reached. For $x = l$ and $\tau = 0$, it gives $\langle w^2(l) \rangle = kTl^3/3K$, which was already discovered by Houdijk [10].

Van Lear and Uhlenbeck have also given $\langle w^2(l,t) \rangle$ for a given initial displacement $w_0(x)$ but arbitrary initial velocities. In our notation, this reads

$$\langle w^2(l,t) \rangle \approx (H_0 l^3/12\beta) + (w_0{}^2 - H_0 l^3/12\beta) e^{-2\beta t} (\cos \omega_1 t + \beta \omega_1{}^{-1} \sin \omega_1 t)^2 \tag{66.107}$$

If the distribution function for the external loads is gaussian, then the mean and the correlation functions of the outputs are sufficient to determine the first probability distribution of the output. To this end, we use (66.59). If the p.d.f. for the external loads is not gaussian but symmetrical, then we may use Gauss' inequality for an estimate:

$$\text{Prob} (|w - \langle w \rangle| \geq k\sigma) \leq 4/gk^2 \qquad \sigma = \langle w^2 \rangle - \langle w \rangle^2 \tag{66.108}$$

Often the p.d.f. can be approximated by a gaussian one,

$$\text{Prob} (|w - \langle w \rangle| \geq k\sigma) \approx e^{-k^2/2}/k(2\pi)^{1/2} \qquad k \gg 1 \tag{66.109}$$

Another useful result is Rice's formula, enabling us to calculate the probable number of times, $N_0(\eta)$, per unit time that the random function y (with zero mean) will exceed a given value, say $y = \eta$. Here η may be maximum allowable displacement or stress. Rice's formula may be put into the form

$$N(\eta) = \pi^{-1} e^{-\eta^2/2\sigma^2} \left\{ \frac{\left[\dfrac{\partial^2}{\partial t\, \partial t'} R(t,t') \right]_{t=t'}}{[R(t,t')]_{t=t'}} \right\}^{1/2} \tag{66.110}$$

where $R(t,t')$ is the correlation for the random variable y, and σ is the mean deviation.

If the first probability density $p_1(y)$ is known, then the probability, Prob $(|y| > h\sigma)$, that the random variable y exceeds an undesirable amount $h\sigma$ may be calculated by

$$\text{Prob} (|y| > h\sigma) = \int_{-\infty}^{-h\sigma} p_1(y)\, dy + \int_{h\sigma}^{\infty} p_1(y)\, dy \tag{66.111}$$

66.6. DISCRETE MODELS FOR CONTINUOUS SYSTEMS

Continuous elastic structures may sometimes be conveniently discretized in the form of masses, springs, and dashpots, thus simplifying the mathematical problem a great deal. For example, the problem of random motions of a tall building subject to earthquake or wind loading may be made tractable by concentrating the masses at the floor levels, considering the structural and external damping in the form of suitable dashpots, and replacing the wall by springs (Fig. 66.5).

In the case of constant mass m (for each floor), interfloor damping c, and external damping d, the differential equations of an N-story building, subject to random earthquake displacement $f(t)$ at the foundation, take the form

$$\begin{aligned} m\ddot{y}_j + c(2\dot{y}_j - \dot{y}_{j+1} - \dot{y}_{j-1}) + d\dot{y}_j + k(2y_j - y_{j+1} - y_{j-1}) &= 0 \\ m\ddot{y}_N + c(\dot{y}_N - \dot{y}_{N-1}) + d\dot{y}_N + k(y_N - y_{N-1}) &= 0 \\ y_0 = f(t) \qquad j = 1, 2, \ldots, N-1 & \end{aligned} \tag{66.112a–c}$$

Here Eqs. (66.112*b*,*c*) are the boundary conditions. In this case, we may use the result of p. 66–6, when $f(t)$ is a Gauss-Markoff process, to obtain the Fokker-Plank equation for the transition probabilities p. If we are interested only in the correlation functions of the mean-square displacements, the method of p. 66–9 is most convenient. For white-noise input $f(t)$, the result [14] is

$$R_{jk}(\tau) = E\{y_j(t)y_k(t+\tau)\} = 4\pi^{-2}f_0^2(2N+1)^{-2}\sum_{n=1}^{N}\sum_{p=1}^{N}(-)^{n+p}\sin\lambda_n$$
$$\times \sin\lambda_p \cos\lambda_n(N+\tfrac{1}{2}-j)\cos\lambda_p(N+\tfrac{1}{2}-k)e^{-\alpha_p\tau}(C_{np}\cos\omega_p\tau + D_{np}\sin\omega_p\tau) \quad (66.113)$$

where f_0^2 is the spectral density of the input, and

$$\begin{aligned} C_{np} &= -2\kappa[(2\epsilon\gamma^2-\kappa)(\alpha_n+\alpha_p) - 4\gamma\alpha_n\alpha_p]/\Delta_{np} \\ D_{np} &= \{-4\gamma^2\alpha_p\omega_p(\alpha_n+\alpha_p) + \omega_p^{-1}[(\alpha_n+\alpha_p)^2+\omega_n^2-\omega_p^2][\gamma^2(\omega_p^2-\alpha_p^2)+\kappa^2]\}/\Delta_{np} \\ \Delta_{np} &= [(\alpha_n+\alpha_p)^2+(\omega_n-\omega_p)^2][(\alpha_n+\alpha_p)^2+(\omega_n+\omega_p)^2] \end{aligned} \quad (66.114)$$

$$\alpha_n = \gamma(1+\epsilon-\cos\lambda_n) \qquad \omega_n = [2\kappa(1-\cos\lambda_n)-\alpha_n^2]^{1/2}$$
$$\lambda_n = \pi(2n-1)/(2N+1) \qquad \gamma = c/m \qquad \epsilon = d/2c \qquad \kappa = k/m$$

For a gaussian white-noise input, the cross-correlation function (66.113) is sufficient to determine the probability density $p[y_j(t_1), y_j(t_2), \ldots, y_j(t_n)]$ with

$$t_1 = t,\ t_k = t + (k-1)\tau,$$

and $p(y_1, y_2, \ldots, y_N)$ with $\tau = 0$. To this end, use Eqs. (66.59) and (66.60), respectively.

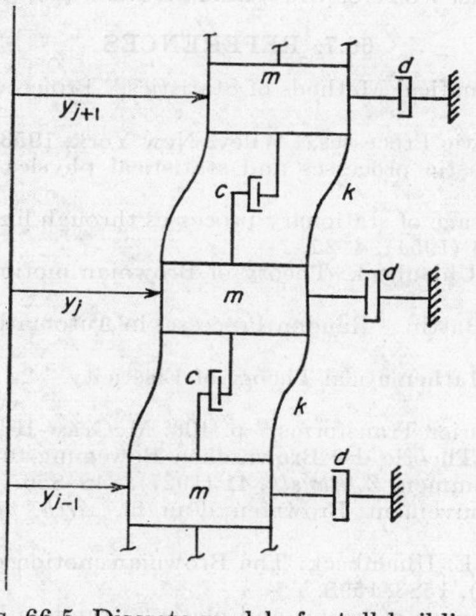

FIG. 66.5. Discrete model of a tall building.

A continuous body, such as an elastic string or a bar, may be considered as a structure consisting of N discrete masses, attached by series of springs and dashpots. This, in turn, reduces the partial differential equations into a set of ordinary differential equations such as (66.112). By use of such a model, Spiegel [15] treats the random motion of a string.

Differential equations for the discretized string with no internal damping ($c \equiv 0$) have the same form as (66.112), except that there is a force $F_j(t)$ on the right of (66.112), and $k = \tau/d$ is the ratio of tension τ in the springs to distance d between particles. The Fokker-Plank equation, when solved for an n-particle system with the end particles fixed (boundary conditions), gives a $2n$-dimensional gaussian distribution. With the help of this density function, Spiegel calculates mean-square displacements

$$E\{(y_j - m_j)^2\} = \frac{E\mu}{4La^2\beta} \sum_{r=1}^{n} \frac{\sin^2 [jr\pi/(n+1)]}{\sin^2 [r\pi/(2n+2)]} (1 - U_r) \tag{66.115}$$

where

$$U_r = \frac{\beta e^{-\beta t}}{4\omega_r^2} \left[2\omega_r \sin 2\omega_r t - \beta \cos 2\omega_r t + 16a^2 \sin^2 \left(\frac{r\pi}{2n+2} \right) \right]$$

$$E = \frac{2\beta kT}{\mu} \qquad a^2 = \frac{\tau(n+1)^2}{\mu L^2} \qquad \mu = \frac{m}{d} \qquad \beta = \frac{d}{m} \tag{66.116}$$

$$\omega_r = \left[\frac{4\tau(n+1)^2}{\mu L^2} \sin^2 \frac{r\pi}{2n+2} - \frac{\beta^2}{4} \right]^{1/2}$$

Here L is the total length of the string, k is Boltzmann's constant, and T is the temperature of the medium in which vibration is taking place. According to (66.32), kT is related to the statistics of the external forces $F_j(t)$. In fact, $kT/2$ is the mean square of the potential energy of the nth eigenvibration as $t \to \infty$ [11]. As $n \to \infty$ and $t \to \infty$, the mean-square displacement approaches

$$E\{y^2\} = \frac{kLT}{\tau} \left[\frac{x}{L} - \left(\frac{x}{L} \right)^2 \right] \tag{66.117}$$

where x and y are the coordinates of a point on the string. For further discussion of this and related problems, we refer the reader to Refs. [15],[16], and [17].

66.7. REFERENCES

[1] H. Cramér: "Mathematical Methods of Statistics," Princeton Univ. Press, Princeton, N.J., 1946.

[2] J. L. Doob: "Stochastic Processes," Wiley, New York, 1953.

[3] J. E. Moyal: Stochastic processes and statistical physics, *J. Roy. Stat. Soc.*, **B,11** (1949), 150–210.

[4] A. J. F. Siegert: Passage of stationary processes through linear and nonlinear devices, *Trans. IRE*, **PGIT-3** (1954), 4–25.

[5] M. C. Wong, G. E. Uhlenbeck: Theory of Brownian motion, *Revs. Modern Phys.*, **17** (1945), 323–342.

[6] J. H. Laning, R. H. Battin: "Randon Processes in Automatic Control," McGraw-Hill, New York, 1956.

[7] I. S. Sokolnikoff: "Mathematical Theory of Elasticity," 2d ed., p. 336, McGraw-Hill, New York, 1956.

[8] I. N. Sneddon: "Fourier Transforms," p. 406, McGraw-Hill, New York, 1951.

[9] L. S. Ornstein: Zur Theorie der Brownschen Bewegung für Systeme, worin mehrere Temperaturen vorkommen, *Z. Physik*, **41** (1927), 848–856.

[10] A. Houdijk: Le mouvement Brownien d'un fil, *Arch. nèerl. sci.*, **III A,11** (1928), 212–277.

[11] G. A. Van Lear, G. E. Uhlenbeck: The Brownian motion of strings and elastic rods, *Phys. Rev.*, **38** (1931), 1583–1598.

[12] A. C. Eringen: Response of beams and plates to random loads, *J. Appl. Mechanics*, **24** (1957), 46–52.

[13] J. C. Samuels, A. C. Eringen: *Office Naval Research*, *Tech. Rept.* 10, Purdue Univ., 1957.

[14] A. C. Eringen: Response of tall buildings to random earthquakes, *Proc. 3d U.S. Natl. Congr. Appl. Mechanics*, 1958, pp. 141–151.

[15] M. R. Spiegel: The random vibrations of a string, *Quart. Appl. Math.*, **10** (1952), 25–33.

[16] R. H. Lyon: Response of strings to random noise fields, *J. Acoust. Soc. Am.*, **28** (1956), 391–398.

[17] A. Powell: On the response of structures to random pressures and to jet noise in particular, Chap. 8 in S. H. Crandall (ed.), "Random Vibrations," Technology Press, MIT, Cambridge, Mass., Wiley, New York, 1959.

[18] A. C. Eringen, J. W. Dunkin: The elastic half plane subjected to surface tractions with random magnitude or separation, *J. Appl. Mechanics*, **27** (1960), 701–709.

CHAPTER 67

ACOUSTICS

BY

O. K. MAWARDI, Ph.D., Cleveland, Ohio

67.1. THE SOUND FIELD

The Character of Plane Waves

An acoustic disturbance propagates in a compressible fluid of unlimited extent as a traveling wave. The simplest type of such a wave is that which results from an initial disturbance, uniformly distributed over an infinite plane. Practically, a plane wave is reasonably well approximated by the propagation of an acoustic wave along the inside of a tube of uniform cross section.

To study the properties of such a wave, consider the fluid element originally contained in a length dx of a tube of unit cross section. In equilibrium, it has the mass $\rho_0\, dx$. At some time t, its terminal cross sections have undergone displacements ξ and $\xi + (\partial\xi/\partial x)\, dx$, respectively. Its length is then $(1 + \partial\xi/\partial x)\, dx$, and the density is $(\rho_0 + \rho)$; hence, the mass

$$(\rho_0 + \rho)\left(1 + \frac{\partial\xi}{\partial x}\right) dx$$

The conservation of mass requires that this be the same as $\rho_0\, dx$, whence

$$\rho = -\rho_0 \frac{\partial\xi}{\partial x} \tag{67.1}$$

When the fluid element is in accelerated motion, the difference between the pressure forces at both ends must equal the inertial force; hence,

$$\rho_0 \frac{\partial^2\xi}{\partial t^2} = -\frac{\partial p}{\partial x} \tag{67.2}$$

These equations have been linearized by neglecting all terms that are quadratic in the small quantities, ρ, ξ, p.

If the fluid is a gas, satisfying the adiabatic equation of state,

$$P\rho_t^{-\gamma} = \text{const} \tag{67.3}$$

where $P = P_0 + p$, $\rho_t = \rho_0 - \rho$, and γ is the ratio of specific heats. Combining (67.1), (67.2), and (67.3), we find that p, ρ, and ξ satisfy an equation of the form

$$\frac{\partial^2\phi}{\partial x^2} - \frac{1}{c^2}\frac{\partial^2\phi}{\partial t^2} = 0 \tag{67.4}$$

where ϕ stands for p, ρ, or ξ, and $c = \sqrt{\gamma P_0/\rho_0}$ is the velocity of propagation of sound waves.

The noises of interest to the acoustical engineer often have a random spectrum.

In the study of plane waves, it is simpler, however, to restrict the discussion to the behavior of waves of single frequencies. By means of the Fourier integral theorem [1], we can then reconstruct the behavior for an arbitrary complex noise. Let us accordingly assume that the time dependence of the acoustic quantities is of the form $e^{i\omega t}$, where ω is the radian frequency of the sound. It is known that Eq. (67.4) is satisfied by two solutions of the form $e^{i(\omega/c)(x+ct)}$ and $e^{i(\omega/c)(x-ct)}$. These represent waves traveling to the left and right, respectively. Suppose we consider a particular solution,

$$p = Ae^{i(\omega/c)(x-ct)} \tag{67.5}$$

where A is an arbitrary constant denoting the amplitude of the fluctuating pressure wave traveling to the right. By means of (67.1), (67.2), and (67.3), we find that

$$\xi = \frac{i}{\rho_0 c\omega} Ae^{i(\omega/c)(x-ct)} \tag{67.6}$$

and

$$\rho = \frac{A}{c^2} e^{i(\omega/c)(x-ct)} \tag{67.7}$$

It is very useful to introduce the local fluctuating velocity $v = \partial\xi/\partial t$, because v is easier to measure than ξ. Accordingly, it will be convenient to express the results in functions of v. Using (67.6), we obtain

$$v = -i\omega\xi = \frac{A}{\rho_0 c} e^{i(\omega/c)(x-ct)} \tag{67.8}$$

The quantity $e^{i\omega t}$ appears in all acoustical terms. To remove the necessity of working with instantaneous quantities, we make use of the concept of acoustic impedance Z, which is, by definition, the ratio of the pressure to the particle velocity. In a plane traveling wave, this is a constant, independent of position. From (67.5) and (67.6), we find the specific acoustic impedance,

$$Z = \frac{p}{v}\ \rho_0 c \tag{67.9}$$

Typical values of this important constant for different materials are given in Table 67.1

Table 67.1

Substance	ρ_0, g/cm³	c, m/sec	$\rho_0 c$, g/cm² sec
Air (20°C)	0.0012	344	41.4
Water (13°C)	1.0	1441	14.4×10^4
Rubber (soft)	0.95	70	0.67×10^4
Rubber (hard)	1.1	1400	15×10^4
Brick	1.8	3700	67×10^4
Concrete	2.6	3100	81×10^4
Glass	2.4	5000	120×10^4
Steel	7.7	5000	390×10^4
Lead	11.3	1200	130×10^4

Energy in a Plane Wave. The rate at which energy is being transmitted along the wave per unit area of wavefront is called the *intensity* I of the sound wave. From elementary considerations, I is the average of the product pv, taken over the period $2\pi/\omega$ of one oscillation,

$$I = \overline{pv} = \frac{\omega}{2\pi}\frac{A^2}{\rho_0 c}\int_0^{\pi/\omega} \cos^2 \frac{\omega}{c}(x-ct)\, dt = \frac{A^2}{2\rho_0 c} \tag{67.10}$$

The *energy density* at a point is the energy per unit volume in a sound wave. The energy density E bears a simple relation to I for a *plane* wave; indeed,

$$E = \frac{I}{c} \tag{67.11}$$

The Concept of Level. Although we deal with alternating quantities which are periodic functions of the time of the form $e^{i\omega t}$, $\sin \omega t$, $\cos \omega t$, etc., they are cumbersome. To avoid the necessity of explicitly writing the functional time dependence, the acoustic quantities are expressed in terms of the peak amplitude or of the root-mean-square (rms) value. For instance, the amplitude of the fluctuating pressure,

$$p = p_+ e^{i(\omega/c)(x-ct)} \tag{67.12}$$

is p_+, while, by definition, its rms value is

$$p_{\text{rms}} = \sqrt{\frac{\omega}{2\pi}\int_0^{2\pi/\omega}\left[p_+ \cos\frac{\omega}{c}(x-ct)\right]^2 dt} = \frac{p_+}{\sqrt{2}} \tag{67.13}$$

which indicates that the amplitude is $\sqrt{2}$ times the rms value for a *sinusoidal* time dependence. Henceforth, unless indicated differently, the time dependence will be dropped, and rms values will be used throughout, the notation $p(x)$ standing for the rms value of the pressure at the point x. It then follows from (67.8) and (67.10) that

$$I = \frac{p^2}{\rho_0 c} = \rho_0 c v^2 \tag{67.14}$$

for *plane waves.*

The range of pressure fluctuations encountered in acoustic work extends from pressure as low as 10^{-4} dyne/cm² to fluctuations of the order of an atmosphere (10^6 dynes/cm²). To easily span this large range of variations, and also because the human ear, under certain conditions, responds linearly to logarithmic changes of the sound pressure, acoustic quantities are expressed in decibel levels.

By *definition* [2], the *sound-pressure level* in decibels is 20 times the logarithm to the base 10 of the ratio of the pressure of this sound to a reference pressure of 0.0002 dyne/cm². Similarly, the *intensity level* in decibels, by means of (67.14), is 10 times the logarithm to the base 10 of the intensity I of this sound to the reference intensity I_0 corresponding to the reference pressure mentioned above. For noise work, we generally take this reference intensity to be approximately 10^{-16} watt/cm².

Reaction on a Piston

Suppose we have a tube of uniform cross-sectional area S, starting at the origin, and extending to the right along the x axis. Let a flat-topped piston be fitted at the end of the tube at $x = 0$. If the piston is imparted with an oscillating motion, a plane wave is generated inside the tube. There are three extreme cases that can be considered:

1. The tube is infinitely long. The wave produced by the piston will travel unimpeded along the tube, and equations similar to those derived in the previous article will apply. The force required to drive the piston is

$$F = pS \tag{67.15}$$

The average power expended to drive the piston is

$$Fv = (pv)S = Sv^2\rho_0 c \tag{67.16}$$

It becomes clear that energy continuously "flows" down the pipe, as though the piston moving with a velocity v is driven against a resistance $R = S\rho_0 c$.

2. The tube is of finite length l, and is closed at one end. In this case, a plane wave produced by the piston will be reflected at the end of the tube, and the reaction on the piston will be due to the system of incident plus reflected wave. To find this reaction, we write that the pressure p is

$$p = A_{-}e^{-i(\omega/c)x} + A_{+}e^{i(\omega/c)x} \tag{67.17}$$

that is, the system of incident and reflected wave will set up a standing wave. To evaluate the arbitrary coefficients A_{-}, A_{+}, we use the boundary conditions that, at $x = l$ where the tube is closed with a rigid partition, the velocity vanishes, while, at the origin, the velocity is that of the piston assumed to be v_0. Combining these conditions together with Eqs. (67.1) and (67.17), we find that

$$v = \frac{v_0 \sin (\omega/c)(l - x)}{\sin (\omega l/c)} \tag{67.18}$$

$$p = -i\rho_0 c v_0 \frac{\cos (\omega/c)(l - x)}{\sin (\omega l/c)} \tag{67.19}$$

Because of the presence of i in the expression for the pressure, the average power expended to drive the piston is zero. Indeed, this means that the impedance sensed by the piston is

$$p/v = -i\rho_0 c \cot \frac{\omega l}{c} \tag{67.20}$$

When cot $(\omega l/c)$ is negative, the piston behaves as though it was working against an inertial load. On the other hand, the reaction on the piston is similar to that of a spring if the sign of cot $(\omega l/c)$ is positive. The transition in sign occurs at the place where the cotangent becomes infinite. This, however, is an ideal case, because losses have been neglected in the discussion. In the practical case, losses are always present. As a result, the impedance becomes large but always remains finite.

When $\omega l \ll c$, cot $(\omega l/c) \approx c/\omega l$, and the impedance is $\rho_0 c^2/i\omega l$ per unit area. Hence, a small cavity acts as a spring to the piston. If the cross section of the tube is S, the acoustic impedance becomes, for this case,

$$Z = \frac{p}{vS} = \frac{\rho_0 c^2}{i\omega(lS)} = \frac{\rho_0 c^2}{i\omega V} \tag{67.21}$$

where V is the volume of the cavity. We will see an important application of this result.

3. The tube is of finite length, and is open at one end. By repeating a discussion similar to case 2, with the boundary condition that the pressure p vanishes at $x = l$, we find [3] that the acoustic impedance

$$Z = \frac{p}{vS} = \frac{i\rho_0 c}{S} \tan \frac{\omega l}{c} \tag{67.22}$$

Here also we notice that the presence of i will indicate that the system is reactive, no power being received on the average. The sign of the impedance, however, is such that, for a short tube ($l\omega \ll c$), the acoustic impedance is

$$Z = i(\rho_0 l S) \frac{\omega}{S^2} = im \frac{\omega}{S^2} \tag{67.23}$$

where $m = \rho_0 l S$ is the mass of air in the short tube. The impedance is then purely due to the inertia of the air.

Resonators and Mufflers

An air cavity terminated with a small opening constitutes an acoustic resonator. From the above discussion, the cavity provides the stiffness, and the opening, the inertia, so that the combination will be similar to a mass-loaded spring that will resonate at a given frequency. The resulting large amplitude will produce large losses at the resonance frequency. The arrangement shown in Fig. 67.1 is a practical way of attenuating undesirable sounds of frequencies close to the resonance frequency.

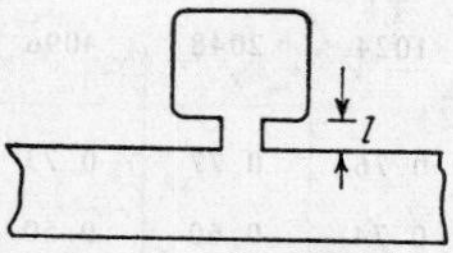

FIG. 67.1. Resonator attached to duct: l = length of neck, V = volume.

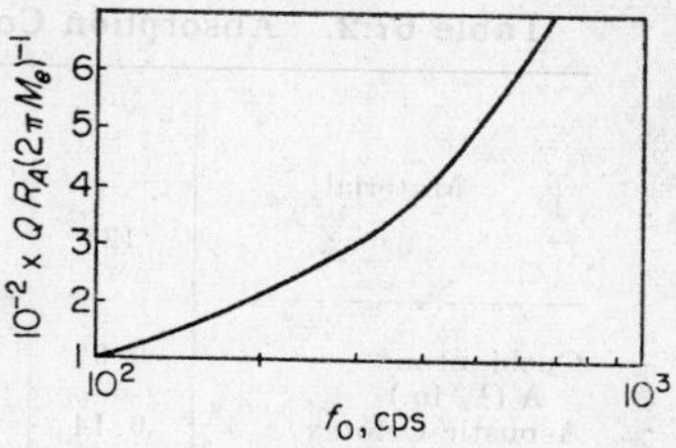

FIG. 67.2. Universal curve for resonators.

The effectiveness of the device of Fig. 67.1 is measured by the attenuation, i.e., by 20 log (p_1/p_2), that is, by the relative levels of the pressure at both ends of the duct. As is well known from vibration theory, losses cause the sharpness of resonance to be finite.

Reverberation in Enclosures

When a noise source is placed within an enclosure, the sound field from the source will be drastically affected by the presence of the walls. It can be shown [5] that any source configuration may be represented by a large number of plane waves of different intensities, and propagating at different angles to one another. This representation is very useful in understanding the effect of the walls on the sound field. If the enclosure is large, any one *component* plane wave starting from the source will suffer several reflections from the walls. As a result of the large number of standing waves that are thus formed, the acoustic energy may be considered uniformly diffused in the enclosure. Since each reflection must be accompanied by a loss in the intensity of the reflected waves, the sound energy density inside the room does not increase indefinitely, but stabilizes at the equilibrium condition for which the energy lost through the walls per unit time is equal to the power of the source.

We now introduce a *coefficient of absorption* α, which is defined as the ratio of the energy of the reflected wave to that of a plane wave incident at right angles to the wall. If the walls of the enclosure are coated with different materials of coefficients $\alpha_1, \alpha_2, \ldots$, which cover areas S_1, S_2, etc., respectively, the average coefficient of absorption

$$\bar{\alpha} = \frac{S_1\alpha_1 + S_2\alpha_2 + \cdots}{S_1 + S_2 + \cdots} \tag{67.24}$$

The absorption coefficients for some usual materials are given in Table 67.2. We notice that α is a function of the frequency. It can be shown [6] that the average acoustic energy density E_0 due to a source of power Π placed in an enclosure, is

$$E_0 = \frac{4\Pi}{\bar{\alpha}c} \tag{67.25}$$

On the other hand, if the source of sound Π is suddenly switched off, the sound will decay exponentially, according to the formula

$$E = E_0 \exp\left(-\frac{Sc\bar{\alpha}t}{4V}\right) \tag{67.26}$$

where $S = S_1 + S_2 + \cdots$ is the total surface of the enclosure, and V its volume. By definition [7], the time taken for the density to decay by 60 db is called the reverberation time.

Table 67.2. Absorption Coefficients for Some Usual Materials[1]

Material	Cycles per second					
	128	256	512	1024	2048	4096
Cushiontone A (½ in.)	0.05	0.18	0.56	0.76	0.77	0.73
Acoustic-Celotex C_2 (1¼ in.)	0.14	0.42	0.99	0.74	0.60	0.50
Sanacoustic KK pad (1¼ in.)	0.25	0.58	0.96	0.97	0.85	0.72
Acoustifibre (⅝ in.)	0.10	0.16	0.62	0.97	0.81	0.73
Fiberglas Acoustic tile perforated (¾ in.)	0.02	0.16	0.76	0.99	0.63	0.44

[1] From *Acoust. Materials Assoc. Bull.* **XI** (1949).

Spherical Diverging Waves

If outgoing plane waves, of all possible inclinations and all of the same amplitude, are compounded together, the result is a spherical wave. This is the type of wave that is generated by a sphere which is allowed to pulsate, its radius changing its magnitude in a sinusoidal fashion with the time.

It can be shown ([8], p. 236) that, at large distances r from the center of the sphere,

$$p = \frac{A_0}{r} e^{-i\omega r/c} \qquad v \approx \frac{p}{\rho_0 c} \tag{67.27a,b}$$

It may be seen from these equations that the intensity of the sound falls off inversely with the square of the distance, being in fact given by

$$I = \frac{A_0{}^2}{r^2 \rho_0 c} \tag{67.28}$$

It is also deduced that the radiation impedance at a large distance from the source is

$$R = \rho_0 c \tag{67.29}$$

which is similar to that of a plane propagating wave.

67.2. MECHANISMS OF SOUND GENERATION

Simple Sources

One of the earliest methods of sound production to be understood is that analogous to the flat piston at one end of a tube (p. 67–3), or to the pulsating sphere just described. Both these sources set up periodic fluctuations of the density which propagate as waves. Such sources are referred to as simple sources, and their functions can be idealized by devices that inject—or extract—mass in the medium.

In order to derive the acoustic power of a simple source, it is convenient to introduce the concept of the strength of a source. By definition, the strength of the source is the volume of fluid injected by the source in—or extracted from—the medium per unit time. For instance, the strength of a flat piston oscillating at one end of a tube of a radius R_0 and with an excursion ξ is $\omega\xi(\pi R_0^2)$. In a similar fashion, the strength of a spherical source of radius R_0 pulsating with amplitude ξ at a frequency ω is $(4\pi R_0^2)\omega\xi$.

The power of the source can now be simply expressed in terms of the strength of the source and of the impedance presented to the source. In fact, it can be shown that the power P radiated from the simple source is

$$P = (A^2 R_e)/S \tag{67.30}$$

where A is the strength of the source, S its area, and R_e the real part of the specific impedance presented to the source.

Driving-point and Mechanical Impedance

The driving-point impedance presented to the source is the specific impedance sensed by the source at a given point. This concept is useful only in dealing with point sources. Otherwise, it is more meaningful to resort to the mechanical impedance. By definition, the mechanical impedance is the ratio of the fluctuating force to the acoustic velocity. The velocity is related to the pressure ([8], p. 8) by the equation

$$v = -\frac{\text{grad } p}{i\omega\rho_0} \tag{67.31}$$

which, in the one-dimensional case, reduces to Eq. (67.2).

As a first example of an impedance calculation, consider the sound field radiated from a pulsating sphere. The pressure is given by Eq. (67.27a), and, from it and Eq. (67.31), the velocity is found to be

$$v = \frac{A_0}{i\omega\rho_0}\left(\frac{i\omega}{rc} + \frac{1}{r^2}\right)e^{-i\omega r/c} \tag{67.32}$$

At large distances, the term containing $1/r^2$ may be neglected. Combining (67.32) and (67.27a), the driving-point (specific) impedance Z is found to be

$$Z = \rho_0 c\,\frac{\omega^2 r^2 + i\omega rc}{\omega^2 r^2 + c^2} \tag{67.33}$$

At large distances, the impedance tends to $\rho_0 c$, which checks the relation given in (67.27b). The above relation indicates that, at $\omega r \gg c$, the sound waves, except for the divergence factor $1/r$, radiate like plane waves. On the other hand, close to the source, the impedance is a complex quantity, and the positive sign of the imaginary part indicates that the medium presents an inertial as well as a resistive load. The power radiated from the source is given by

$$P = A_0^2\rho_0\omega^2/\pi c^2 \tag{67.34}$$

in accordance with Eq. (67.30) for the limiting case of r tending to zero.

The Piston in a Baffle. The next example of impedance calculation is for a piston mounted at one end of a tube, which has already been obtained on p. 67–4.

To calculate the radiation impedance of the piston, it is necessary, first, to find the solution for the sound field configuration around the piston. To obtain this configuration, we imagine the piston to be replaced by a distribution of point sources spread over the surface of the piston. The calculations are straightforward, but tedious. We find that the acoustic impedance is given by [9]

$$Z = \rho_0 c(\theta + i\beta) \tag{67.35}$$

The two functions θ and β are

$$\theta = 1 - 2J_1\left(\frac{2\omega a}{c}\right) \tag{67.36}$$

and
$$\beta = \frac{4}{\pi}\int_0^{\pi/2} \sin\left(\frac{2\omega a}{c}\cos\phi\right)\sin^2\phi\, d\phi \tag{67.37}$$

and a is the radius of the piston.

Here J_1 stands for the Bessel function of the first kind and of order one. θ and β are shown in graphical form in Fig. 67.3 as a function of the parameter $\omega a/c$. We

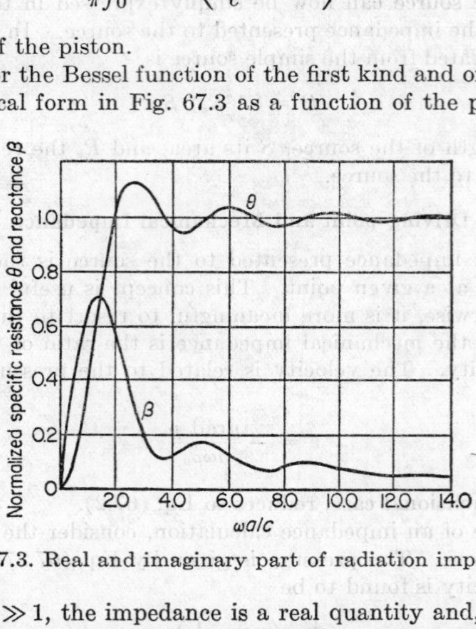

FIG. 67.3. Real and imaginary part of radiation impedance.

notice that, as $\omega a/c \gg 1$, the impedance is a real quantity and, in fact, is similar to that of a piston oscillating at the end of a long tube. When this happens, the sound energy radiates from the piston in the form of a beam, the area of which is that of the piston area, and hardly any energy is radiated in an off-axis direction. At lower frequency, however, the fluid is set into motion over a larger angle, but less energy is radiated, the motion being confined in the vicinity of the piston; hence, the substantial contribution of the inertia term β of (67.35).

If the piston moves with a velocity v_0, then its strength is $A = \pi a^2 v_0$. Hence, by (67.30), the power radiated is

$$P = \frac{A^2 \rho_0 c\theta}{\pi a^2} \tag{67.38}$$

which reduces to $(\pi a^2)v_0{}^2\rho_0 c$ at high frequencies. This result was already obtained in Eq. (67.16), where $S = \pi a^2$.

Radiation from Dipoles

The production of sound by means of a tuning fork cannot be explained by appeal to the mechanism of sound radiation from simple sources. It is clear that the sound emanates from the end of the vibrating prongs of the fork. When the prongs oscillate, the air is swished *around* the profile of the prongs, and it is not reasonable to imagine the density of the medium in contact with the prongs to be affected. Consideration of this situation leads one to deduce that, as the prong oscillates, a periodic *force* is applied to the fluid, and this will give rise to the sound. It can be shown [10], however, that, from the point of view of sound generation, a periodic force is equivalent to a pair of sources of opposite sign, very close to each other and working in opposite phase. Such a pair of sources is called a *dipole.*

The sound field due to a dipole can be computed by calculating the contribution of two simple sources of opposite sign (i.e., a source and a sink) a distance l apart, and then making l vanishingly small. We find [11] in this way that the pressure field at a point Q is given by

$$p = i\omega\rho_0 A_1 \frac{1 + ikr}{r^2} e^{-ikr} \cos\theta \tag{67.39}$$

Here

$$A_1 = \lim_{l\to 0} (A_0 l)$$

where A_0 is the strength of either source, $k = \omega/c$, and θ is the angle between the axis of the dipole (the line joining the centers of the two sources) and a line joining Q to either one of the sources. It is conventional to refer to A_1 as the strength of the dipole.

The velocity field due to the dipole can be derived by using Eq. (67.31). We find in this way that, at a large distance from the dipole, the velocity is mainly in a radial direction, and is found to be

$$v = \left(\frac{i}{\omega\rho_0} \frac{\partial p}{\partial r}\right)_{r\to\infty} = \frac{A_1\omega^2}{c^2} \frac{\cos\theta}{r} e^{-ikr} \tag{67.40}$$

To estimate the total power radiated from the dipole, we cannot use the simple relation (67.28), because here the net volume flow from the dipoles is zero. So, instead, we calculate the real value of the product pv over a large sphere of radius r. We obtain in this manner that

$$P = \frac{\rho A_1{}^2\omega^4}{3c^3} \tag{67.41}$$

This clearly shows that the dipole source is less efficient than the simple source.

Quadrupole Source

There is another mechanism of sound production which is still different from those described so far. In fact, if we can cause a change in the rate of momentum flux across a surface, then we can, in principle, generate sound. This process had been speculated upon by Stokes ([8], p. 246), who actually showed that, if the surface of a sphere vibrates in such a manner that its volume and the position of its centroid remain fixed, then the sphere would radiate sound. But this is a very inefficient mechanism. If the sphere had been pulsating with the same maximum amplitude, it can be shown that it would radiate approximately 1000 times more acoustic energy.

A simple way of constructing a sound source of the kind described above is by considering the limiting case of two sources and two sinks of equal strength and infinitely close together. It is seen that such a source, called *quadrupole,* is also the limit of a source composed of two dipoles. From the way such a source is constructed, we find two possible groups of combinations that lead to a lateral quadrupole or to a longitudinal quadrupole. The combinations correspond to those indicated in Fig. 67.4.

The field due to quadrupole sources can be estimated by a procedure similar to the one used for the piston in a wall. Here, however, the situation is complicated by the fact that there are five possible combinations for the sources as shown in Fig.

67.4. The general expression for the pressure due to a quadrupole is given by [12]

$$p \approx \frac{A_2}{r^3} [a_{20}P_2{}^0(\cos\theta) + (a_{21}\cos\psi + b_{21}\sin\psi)P_2{}^1(\cos\theta) + (a_{22}\cos 2\psi + b_{22}\sin 2\psi)P_2{}^2(\cos\theta)]e^{i\omega r/c} \qquad (67.42)$$

In the above relation, A_2 is the strength of the quadrupole, and a_{20}, a_{21}, a_{22}, b_{21}, b_{22} are five arbitrary constants, indicating the relative strengths of the five arrangements shown in Fig. 67.4. The angles θ and ψ are the zenith and azimuth angles in a

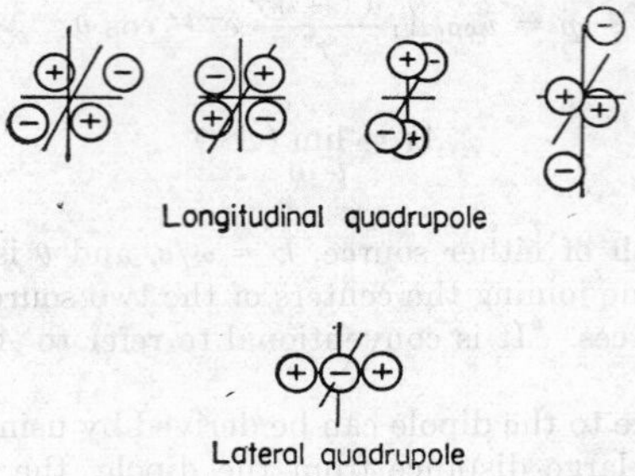

FIG. 67.4. Combinations of four simple sources that lead to a quadrupole.

spherical coordinate system, and $P_2{}^0$, $P_2{}^1$, and $P_2{}^2$ are associated Legendre functions of the first kind. It is seen from (67.42) that, when all constants except a_{20} vanish, the source configuration becomes that of a lateral quadrupole. On the other hand, when a_{20} vanishes, the source corresponds to longitudinal quadrupoles, and these may be expressed by spherical harmonics of second order and of zero order, that is, by a combination of a lateral quadrupole and simple sources.

67.3. REFERENCES

[1] R. V. Churchill: "Fourier Series and Boundary Value Problems," pp. 78ff., McGraw-Hill, New York, 1941.
[2] American Standards Association Bulletin on Acoustical Terminology.
[3] H. F. Olson: "Elements of Acoustical Engineering," 2d ed., p. 72, Van Nostrand, Princeton, N.J., 1947.
[4] L. L. Beranek: "Acoustics," pp. 62ff., McGraw-Hill, New York, 1954.
[5] A. Sommerfeld: "Partial Differential Equations," p. 286, Dieterich, Wiesbaden, 1947.
[6] V. O. Knudsen: "Architectural Acoustics," pp. 120ff., Wiley, New York, 1932.
[7] W. C. Sabine: "Collected Papers on Acoustics," Harvard Univ. Press, Cambridge, Mass., 1927.
[8] Lord Rayleigh: "The Theory of Sound," 2d ed., vol. 2, Macmillan, London, 1940.
[9] P. M. Morse: "Vibration and Sound," 2d ed., p. 246, McGraw-Hill, New York, 1948.
[10] H. Lamb: "Hydrodynamics," 6th ed., p. 502, Dover, New York, 1945.
[11] H. Lamb: "The Dynamical Theory of Sound," 2d ed., pp. 231ff., Arnold, London, 1931.
[12] J. A. Stratton: "Electromagnetic Theory," p. 182, McGraw-Hill, New York, 1941.

Part 7

FLUID MECHANICS

CHAPTER 68

BASIC CONCEPTS

BY

C. E. BROWN, Yorktown, Va.

68.1. PROPERTIES OF FLUIDS

A fluid is most often thought of as material having the ability to continually alter its form under applied forces and, for example, take on the shape of a container into which it is placed. It will be noticed that such a definition leaves open the question of how long the material will be allowed to reach such a state of conformity; hence, the definition includes such fluids as the glasses and other plastics which contain no crystalline structure and which, when given time, will ultimately flow to relieve the applied stresses. It is seen then that some materials coming under the definition of a fluid will have a shear modulus of elasticity, and it is known that even liquids such as water may have a shear modulus. Measurements of such a quantity are carried out by subjecting the fluid to high-frequency strain, in which the strain reaches a peak before the action of viscosity can relieve it. Consideration of such phenomena and, generally, the study of the flow of materials having very high viscosity constitute part of the field of rheology (see Chaps. 53 and 54); the field of fluid mechanics concerns itself mostly with flow of fluids having small viscosity, such as water and air. There is inevitably an overlap, however.

According to the kinetic theory of matter, what appears to be a smooth continuous distribution of material is, in reality, a vibrating and moving mass of discrete molecules, each molecule sharing with all the others a common law of force interaction, which arises from the electrical structure of the molecular particles. These laws of force, involving attraction at long ranges and repulsion at short ranges, provide the basis for a statistical explanation of the various states of a homogeneous substance. The simplest state is that of a gas. Here the molecules are, on the average, so widely spaced that they lie outside the effective influence of the molecular forces. Only when two molecules approach close to each other, do the forces come into play, and it is these "collisions" that the kinetic theory of gases must accurately take into account. In the liquid state, the molecules have, on the average, insufficient kinetic energy to prevent the attractive forces from exercising a dominant effect on their motion. Hence, in a manner of speaking, the molecules are within one another's effective sphere of influence at all times. As cooling takes place, the molecules loose kinetic energy, and thus become more and more closely bound by the attractive force fields; therefore, the random wandering motions of the molecules become smaller. Finally, but still gradually, the material assumes the apparently solid state. In crystalline materials, of course, a sharp transition to a favored grouping occurs, which produces a highly stable and elastic structure. The liquids, however, manifest no such preferred arrangements as required by their definition.

The transport properties of fluids, viscosity, heat conduction, and diffusion may all be explained and calculated by means of kinetic theory [1],[2]; however, it is fortunate that the concept of discrete particles need not generally be carried over to the analysis of flow problems, for the complexity of even simple flows is extremely great. Instead of a discrete-particle approach, it is possible to assume that the fluid is a continuous distribution of matter which possesses the various static and transport properties so clearly explained by kinetic theory. It should be clear then that the assumption of a continuum is valid as long as any characteristic length of importance in the flow to be described is orders of magnitude larger than the intermolecular distances. Fortunately, the bulk of problems of practical interest to date have been concerned with fluid states in which the assumption of a continuum is completely justified. In the consideration of the mechanics of the fluid continuum in the following paragraphs, recourse has been made to the extensive body of literature presently available, of which only a few of the outstanding sources are listed as [3] through [5].

68.2. DYNAMICS OF FLUIDS

Fluid Stress

When a stress is suddenly applied to a fluid element, considered infinitesimal with respect to a characteristic length of the macroscopic flow field and yet large compared to the molecular spacing, a reorientation of the equilibrium molecular state takes place. For liquids which may have a certain shear elasticity, the first motion of the molecules is one within their former molecular binding ranges, and this initial dislocation produces the same elastic type of stress usually associated with solids and crystalline materials. With continued application of the stress, a molecule leaves its local bounds, and drifts along (on the average) in response to the shearing action. It will be seen, then, that the magnitude of the effects mentioned should be influenced by the rate of change of stress, and by the rate of molecular diffusion of the liquid. Since, for *gases*, the molecules are largely out of one another's binding force fields, the elastic shear effects should always be zero. For mobile liquids, such as water or even lubricating oils, the rate of change of stress needed to produce a sensible elastic shear effect is extremely high—of the order of 10^{11} and 10^8 cycles per second, respectively. It is seen, then, that there exists only a narrow class of fluid flows in which such effects must be considered, and we shall henceforth assume the fluid shear elasticity to be negligible.

The determination of stress at a point in a fluid requires a knowledge of the normal and shearing pressures acting at the point on any three mutually perpendicular planes passed through the point. It is seen from Eqs. (33.31) that these requirements are sufficient, since the stresses for any other plane depend on them. We shall use here the same notation for the stresses as explained in Fig. 33.3, that is, σ_x, . . . for the normal stresses, positive when a tension, and τ_{xy}, . . . for the shear stresses, with $\tau_{xy} = \tau_{yx}$. For a given state of stress, a change in the orientation of the axes results in changes in the components of the stress tensor. There exists one axis orientation for which all shear stresses vanish simultaneously (see p. 33–13). These special axes for which only normal stresses are present are called the principal axes of stress.

Fluid Rate of Strain

As discussed in the introductory paragraphs, a steady fluid stress cannot be produced by displacement of the elements of the fluid particle, but rather by the relative motion of the elements with respect to one another. There is here an exact analogy with the case of a solid undergoing displacement of its elements, and the necessary relations may be obtained from the theory of elasticity by replacing the displacements in the solid by the relative velocities in the fluid.

The velocity field in the immediate vicinity of a point (x,y,z) may be written as

$$\begin{aligned} u &= u_0 + \frac{\partial u}{\partial x}\,dx + \frac{\partial u}{\partial y}\,dy + \frac{\partial u}{\partial z}\,dz \\ v &= v_0 + \frac{\partial v}{\partial x}\,dx + \frac{\partial v}{\partial y}\,dy + \frac{\partial v}{\partial z}\,dz \\ w &= w_0 + \frac{\partial w}{\partial x}\,dx + \frac{\partial w}{\partial y}\,dy + \frac{\partial w}{\partial z}\,dz \end{aligned} \tag{68.1}$$

where u_0, v_0, and w_0 are the components of the fluid velocity at the point (x,y,z), and the differentials are infinitesimal distances from the point (x,y,z).

Since we are interested in the rate of strain components which produce the stress, it is clear that here, as in the case of an elastic solid, the rotation of the elements about the particle must be eliminated from consideration. In the fluid, the *rotation* is, of course, a rate of rotation in which the fluid elements may be considered to be rotating like a wheel. As given by Eqs. (33.3), the component rotations of the fluid at a point are:

$$\omega_x = \frac{1}{2}\left(\frac{\partial w}{\partial y} - \frac{\partial v}{\partial z}\right) \qquad \omega_y = \frac{1}{2}\left(\frac{\partial u}{\partial z} - \frac{\partial w}{\partial x}\right) \qquad \omega_z = \frac{1}{2}\left(\frac{\partial v}{\partial x} - \frac{\partial u}{\partial y}\right) \tag{68.2}$$

Many problems of practical interest involve flows for which the rotation is everywhere zero; such *irrotational* flows usually afford great simplification in so far as their mathematical solution is concerned. Introducing the above expressions for the rotations in Eqs. (68.1), the velocity field may be expressed in the following form:

$$\begin{aligned} u &= u_0 + \omega_y\,dz - \omega_z\,dy + \frac{\partial u}{\partial x}\,dx + \frac{1}{2}\left(\frac{\partial v}{\partial x} + \frac{\partial u}{\partial y}\right) dy + \frac{1}{2}\left(\frac{\partial w}{\partial x} + \frac{\partial u}{\partial z}\right) dz \\ v &= v_0 + \omega_z\,dx - \omega_x\,dz + \frac{\partial v}{\partial y}\,dy + \frac{1}{2}\left(\frac{\partial v}{\partial x} + \frac{\partial u}{\partial y}\right) dx + \frac{1}{2}\left(\frac{\partial w}{\partial y} + \frac{\partial v}{\partial z}\right) dz \\ w &= w_0 + \omega_x\,dy - \omega_y\,dx + \frac{\partial w}{\partial z}\,dz + \frac{1}{2}\left(\frac{\partial w}{\partial x} + \frac{\partial u}{\partial z}\right) dx + \frac{1}{2}\left(\frac{\partial w}{\partial y} + \frac{\partial v}{\partial z}\right) dy \end{aligned} \tag{68.3}$$

The motion expressed above is now in the form of a translation (u_0,v_0,w_0), a mean rotation $(\omega_x,\omega_y,\omega_z)$, and a residual distortion characterized by the six terms

$$\frac{\partial u}{\partial x}, \qquad \frac{\partial v}{\partial y}, \qquad \frac{\partial w}{\partial z}, \qquad \frac{1}{2}\left(\frac{\partial u}{\partial y} + \frac{\partial v}{\partial x}\right), \qquad \frac{1}{2}\left(\frac{\partial u}{\partial z} + \frac{\partial w}{\partial x}\right), \qquad \frac{1}{2}\left(\frac{\partial v}{\partial z} + \frac{\partial w}{\partial y}\right)$$

Clearly the stress must be related only to these six quantities.

Stress–Rate-of-strain Relationship

We consider next the relationship which must connect the six quantities of stress and the six pertinent rate-of-strain quantities. From the assumption of a homogeneous fluid, it is necessary to consider only the two types of stress, normal and shear; and the three normal stress relations must exhibit a symmetry or lack of preference for a particular coordinate axis. Take first the normal-stress-to-rate-of-strain relationship. For complete generality, we must admit that it is possible for all six previously mentioned rate-of-strain terms to contribute to the normal stress. We may, however, attempt to write a very general relationship by expressing a stress component by means of a Taylor series expansion in terms of the six pertinent rate-of-strain components derived previously. Thus the expression might be written as

$$\begin{aligned} \sigma_x = -p_s + a\,\frac{\partial u}{\partial x} + b\,\frac{\partial v}{\partial y} + c\,\frac{\partial w}{\partial z} + d\,\left(\frac{\partial u}{\partial y} + \frac{\partial v}{\partial x}\right) \\ + e\left(\frac{\partial u}{\partial z} + \frac{\partial w}{\partial x}\right) + f\left(\frac{\partial v}{\partial z} + \frac{\partial w}{\partial y}\right) + \cdots \end{aligned} \tag{68.4}$$

and succeeding terms would involve, of course, powers and cross products of the six rate-of-strain quantities. The terms p_s, a, b, c, etc., are all functions of the fluid state. The quantity p_s we define as the *state pressure*, that is, the pressure usually associated with the equilibrium state equation $p = f(\rho, T)$, where ρ is the fluid density, and T the temperature. Equation (68.4) thus represents the normal stress as a perturbation from the equilibrium state, in which no rate of strain exists, and for which then all three normal stresses would be equal to p_s. Determination of the coefficients $a, b, c, \ldots$ must be carried out either by experiment or by recourse to the kinetic theory. In either event, determination of the coefficients of the higher-order terms is extremely difficult, and of questionable importance since for vanishing strains, the linear terms must dominate, but for large strains, the theory is already compromised by the assumption of a continuum. In any case, the first-order result is all that is available to us without excessively complicating the analysis; hence, the hypothesis is put forward that there exists a linear dependence of stress on strain! In (68.4) we may put $d = e = f = 0$ on the ground that a reversal of the sign of the strain would reverse the sign of the normal stress contribution, thus requiring a directional preference, which is not possible in the homogeneous medium.

The coefficients b and c must be equal to each other for reasons of symmetry, and hence the relationship between normal stress and strain becomes

$$\sigma_x = -p_s + a\frac{\partial u}{\partial x} + b\left(\frac{\partial v}{\partial y} + \frac{\partial w}{\partial z}\right)$$

With slight manipulation, we may write

$$\sigma_x = -p_s + (a - b)\frac{\partial u}{\partial x} + b\left(\frac{\partial u}{\partial x} + \frac{\partial v}{\partial y} + \frac{\partial w}{\partial z}\right) \qquad (68.5a)$$

and the homogeneity of the medium then dictates that

$$\begin{aligned} \sigma_y &= -p_s + (a - b)\frac{\partial v}{\partial y} + b\left(\frac{\partial u}{\partial x} + \frac{\partial v}{\partial y} + \frac{\partial w}{\partial z}\right) \\ \sigma_z &= -p_s + (a - b)\frac{\partial w}{\partial z} + b\left(\frac{\partial u}{\partial x} + \frac{\partial v}{\partial y} + \frac{\partial w}{\partial z}\right) \end{aligned} \qquad (68.5b,c)$$

The shear-stress terms will also be connected linearly to the rate-of-strain terms; however, it is evident from the symmetry of the motions that the linear dilatation terms involving $\partial u/\partial x$, $\partial v/\partial y$, $\partial w/\partial z$ cannot contribute to the shear stress. Furthermore, for reasons of symmetry, the shear motion $\partial u/\partial y$ cannot conceivably produce a shear stress in the z direction, etc. Thus, it becomes evident that the shear-stress relations to strain must be

$$\tau_{xy} = \mu\left(\frac{\partial u}{\partial y} + \frac{\partial v}{\partial x}\right) \qquad \tau_{yz} = \mu\left(\frac{\partial v}{\partial z} + \frac{\partial w}{\partial y}\right) \qquad \tau_{xz} = \mu\left(\frac{\partial w}{\partial x} + \frac{\partial u}{\partial z}\right) \qquad (68.6a\text{–}c)$$

where μ is a coefficient depending on the fluid state at the particle center.

The fluid stress-strain relations contain three coefficients: a, b, and μ. It is possible, however, to show that a unique relationship exists between μ and $(a - b)$, thus leaving only two coefficients to be determined experimentally.

To obtain the $\mu(a - b)$ relationship, we transform both stress and rate of strain to principal axes. Let l, m, and n be the direction cosines of the direction x_1 of the principal stress σ_1; then, from Eq. (33.31a),

$$\sigma_1 = l^2\sigma_x + m^2\sigma_y + n^2\sigma_z + 2lm\tau_{xy} + 2mn\tau_{yz} + 2nl\tau_{xz} \qquad (68.7)$$

The velocity component u_1 in the direction x_1 is

$$u_1 = ul + vm + wn$$

Hence,

$$\frac{\partial u_1}{\partial x_1} = \frac{\partial u}{\partial x} l^2 + \frac{\partial v}{\partial y} m^2 + \frac{\partial w}{\partial z} n^2 + \left(\frac{\partial u}{\partial y} + \frac{\partial v}{\partial x}\right) lm + \left(\frac{\partial v}{\partial z} + \frac{\partial w}{\partial y}\right) mn + \left(\frac{\partial w}{\partial x} + \frac{\partial u}{\partial z}\right) nl \tag{68.8}$$

On the other hand, Eqs. (68.5) must hold for any spatial direction, and we may apply it to σ_1. Before writing it, however, it is necessary to note that the quantity $(\partial u/\partial x + \partial v/\partial y + \partial w/\partial z)$ expresses the *divergence* of the fluid velocity at the particle and, as is derived in the next section, it expresses the percentage-time–rate-of-density change at the point and may be written

$$\frac{\partial u}{\partial x} + \frac{\partial v}{\partial y} + \frac{\partial w}{\partial z} = -\frac{1}{\rho}\frac{D\rho}{Dt} \tag{68.9}$$

where ρ is the fluid density, t is time, and the symbol D/Dt expresses the time variation relative to a given fluid particle. The normal stress may now be written as

$$\sigma_1 = -p_s + (a - b)\frac{\partial u_1}{\partial x_1} - \frac{b}{\rho}\frac{D\rho}{Dt} \tag{68.10}$$

If we introduce (68.5) and (68.6) into (68.7), and then (68.7) and (68.8) into (68.10), the simple result is obtained that

$$a - b = 2\mu \tag{68.11}$$

The normal stress equation can now be written as

$$\sigma_x = -p_s + 2\mu\frac{\partial u}{\partial x} + b\left(\frac{\partial u}{\partial x} + \frac{\partial v}{\partial y} + \frac{\partial w}{\partial z}\right) \tag{68.12}$$

The coefficient μ is called the *coefficient of viscosity*, and has been the subject of extensive measurement. Kinetic theory calculations have also been carried to great lengths [2], and the values of this coefficient are tabulated for various fluids in engineering and technical handbooks. The coefficient b is called the *bulk viscosity*, and its value is of importance only for compressible fluids since for incompressible fluids, $\partial u/\partial x + \partial v/\partial y + \partial w/\partial z \equiv 0$. For monatomic gases, kinetic theory gives the result that $b = -\tfrac{2}{3}\mu$.

For diatomic gases and gases having more complicated molecules, an analysis by Busemann [5] has suggested the relation

$$b = (1 - \gamma)\mu \tag{68.13}$$

where γ is the ratio of specific heats of the gas. Unfortunately, accurate experimental determination of b is very difficult since the quantities to be measured are invariably of second order with respect to the pressure; hence, no reliable data are available. It is doubtful, however, that many cases will arise in which the value of b will be required with great precision. It is probable that the relation given in Eq. (68.13) or even the monatomic value of $-\tfrac{2}{3}\mu$ is adequate. It is of interest to compare the values of p_s and the mean value of the normal stresses, which will be defined as $-p_m$. Using Eq. (68.12) and the corresponding equations in the other coordinate directions, we obtain

$$-p_m = \tfrac{1}{3}(\sigma_x + \sigma_y + \sigma_z) = -p_s + (\tfrac{2}{3}\mu + b)\left(\frac{\partial u}{\partial x} + \frac{\partial v}{\partial y} + \frac{\partial w}{\partial z}\right)$$

It is seen that p_s and p_m are equal when the fluid is incompressible [see Eq. (68.9)], or when the fluid is a monatomic gas ($b = -\frac{2}{3}\mu$).

Equation of Continuity

In order to apply the conservation laws of mass, momentum, and energy to a continuous fluid, we consider the subsequent motion of a mass of fluid which at some arbitrary time, t_0, is contained within an infinitesimal cubical volume centered at the arbitrary location (x,y,z). Such a fluid particle is shown in Fig. 68.1. During an infinitesimal time lapse dt, the fluid element moves to a new position and undergoes changes in shape, stress, and energy. By keeping track of these changes, we can express the fundamental laws as they apply to any fluid element.

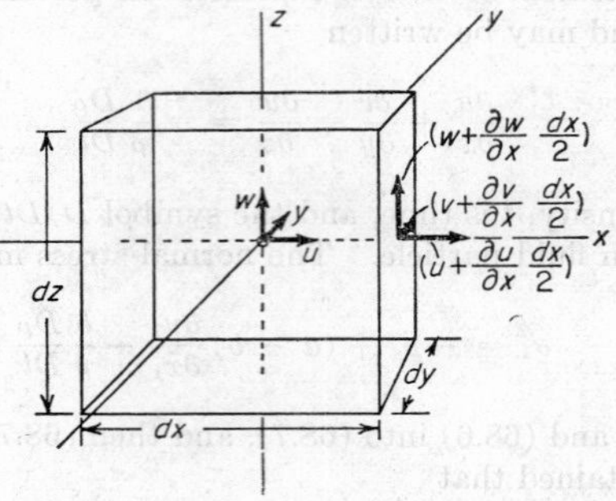

FIG. 68.1

The conservation of mass is expressed by the statement that for a given infinitesimal element, the product of density and volume must remain constant. For the cubic element of Fig. 68.1, the time variation of this product must be zero; hence,

$$\frac{D(\rho Q)}{Dt} = \rho\frac{DQ}{Dt} + Q\frac{D\rho}{Dt} = 0 \tag{68.14}$$

where $Q = dx\,dy\,dz$ is the elemental volume, and the operation D/Dt denotes the time rate of change as observed by an observer on the particle as distinct from $\partial/\partial t$ denoting time rate of change at a fixed point (x,y,z). The changes occurring following the particle may be composed of the sum of the temporal and spatial changes as

$$\frac{D}{Dt} = \frac{\partial}{\partial t} + u\frac{\partial}{\partial x} + v\frac{\partial}{\partial y} + w\frac{\partial}{\partial z} \tag{68.15}$$

The volume Q of the cubical element after the time dt may be found to first order by the product of its mean dimensions in the x, y, and z directions. The new length of the particle in the x direction is

$$dx\left(1 + \frac{\partial u}{\partial x}dt\right)$$

Hence, the volume at $(t + dt)$ is to the first order

$$dx\,dy\,dz\left(1 + \frac{\partial u}{\partial x}dt + \frac{\partial v}{\partial y}dt + \frac{\partial w}{\partial z}dt\right)$$

The initial volume is $dx\,dy\,dz$, and, therefore, the quantity

$$\frac{DQ}{Dt} = \left(\frac{\partial u}{\partial x} + \frac{\partial v}{\partial y} + \frac{\partial w}{\partial z}\right)dx\,dy\,dz$$

Upon substitution of the above in (68.14), we obtain

$$\frac{\partial u}{\partial x} + \frac{\partial v}{\partial y} + \frac{\partial w}{\partial z} = -\frac{1}{\rho}\frac{D\rho}{Dt} \qquad (68.16)$$

which shows that the quantity on the left-hand side is a scalar quantity and, therefore, independent of the choice of coordinates, a fact which was used in the preceding analysis of fluid stress. Equation (68.16) may be put in a somewhat different form by the use of (68.15); thus,

$$\frac{\partial \rho}{\partial t} + \frac{\partial(\rho u)}{\partial x} + \frac{\partial(\rho v)}{\partial y} + \frac{\partial(\rho w)}{\partial z} = 0 \qquad (68.17)$$

Equations of Motion

We now proceed to make a force and momentum balance for the elemental cube of fluid. For this we observe that the acceleration of the particle must be produced by the forces due to the variation of stress across the particle and the extraneous forces of gravitational or electromagnetic origin. The magnitudes of these *body forces* per unit mass in the x, y, and z directions, respectively, are denoted by the symbols X, Y, and Z. The force acting in the x direction is thus $\rho X\,dx\,dy\,dz$. The force in the x direction produced by the stresses arises from the difference in the stresses on opposite faces of the cube. Thus the x-wise force is

$$\left(\frac{\partial \sigma_x}{\partial x} + \frac{\partial \tau_{xy}}{\partial y} + \frac{\partial \tau_{xz}}{\partial z}\right) dx\,dy\,dz$$

The complete equation for the x direction may now be written as

$$\rho\frac{Du}{Dt} = \rho X + \frac{\partial \sigma_x}{\partial x} + \frac{\partial \tau_{xy}}{\partial y} + \frac{\partial \tau_{xz}}{\partial z} \qquad (68.18a)$$

The corresponding equations in y and z may then be written as

$$\rho\frac{Dv}{Dt} = \rho Y + \frac{\partial \tau_{xy}}{\partial x} + \frac{\partial \sigma_y}{\partial y} + \frac{\partial \tau_{yz}}{\partial z} \qquad \rho\frac{Dw}{Dt} = \rho Z + \frac{\partial \tau_{xz}}{\partial x} + \frac{\partial \tau_{yz}}{\partial y} + \frac{\partial \sigma_z}{\partial z} \qquad (68.18b,c)$$

Substitution of Eqs. (68.6) and (68.12) into Eqs. (68.18) yields, for the force relationship in terms of strain,

$$\begin{aligned}
\rho\frac{Du}{Dt} &= \rho X - \frac{\partial p_s}{\partial x} + \frac{\partial}{\partial x}\left[2\mu\frac{\partial u}{\partial x} + b\left(\frac{\partial u}{\partial x} + \frac{\partial v}{\partial y} + \frac{\partial w}{\partial z}\right)\right] \\
&\quad + \frac{\partial}{\partial y}\left[\mu\left(\frac{\partial u}{\partial y} + \frac{\partial v}{\partial x}\right)\right] + \frac{\partial}{\partial z}\left[\mu\left(\frac{\partial u}{\partial z} + \frac{\partial w}{\partial x}\right)\right] \\
\rho\frac{Dv}{Dt} &= \rho Y - \frac{\partial p_s}{\partial y} + \frac{\partial}{\partial y}\left[2\mu\frac{\partial v}{\partial y} + b\left(\frac{\partial u}{\partial x} + \frac{\partial v}{\partial y} + \frac{\partial w}{\partial z}\right)\right] \\
&\quad + \frac{\partial}{\partial z}\left[\mu\left(\frac{\partial v}{\partial z} + \frac{\partial w}{\partial y}\right)\right] + \frac{\partial}{\partial x}\left[\mu\left(\frac{\partial v}{\partial x} + \frac{\partial u}{\partial y}\right)\right] \qquad (68.19) \\
\rho\frac{Dw}{Dt} &= \rho Z - \frac{\partial p_s}{\partial z} + \frac{\partial}{\partial z}\left[2\mu\frac{\partial w}{\partial z} + b\left(\frac{\partial u}{\partial x} + \frac{\partial v}{\partial y} + \frac{\partial w}{\partial z}\right)\right] \\
&\quad + \frac{\partial}{\partial x}\left[\mu\left(\frac{\partial w}{\partial x} + \frac{\partial u}{\partial z}\right)\right] + \frac{\partial}{\partial y}\left[\mu\left(\frac{\partial w}{\partial y} + \frac{\partial v}{\partial z}\right)\right]
\end{aligned}$$

These are the *Navier-Stokes equations*, the momentum equations which govern the flow of a compressible, viscous fluid. If, as is often true, the viscosities μ and b are negligibly small, each equation reduces to only the first three terms. In this reduced form, they are known as *Euler's equations*, [see Eqs. (71.1)].

Energy Equation

The requirement of the conservation law for energy is that the gain in total energy of the fluid particle in time dt must result from the work done on the particle by the action of external forces and the conduction of heat energy into the particle from the external fluid. The total energy of the fluid is completely specified by its kinetic and internal energy. The internal energy denoted by the symbol E contains, in the most general case, the chemical energy of fluid, as well as that part represented by the molecular kinetic energy. In cases of chemical reaction, dissociation, or variation in the specific heat with temperature, the fluid may not be considered strictly homogeneous; however, in practical cases, this approximation is often valid. The rate of change of internal and kinetic energy is given by the expression

$$\rho \frac{D}{Dt} [\tfrac{1}{2}(u^2 + v^2 + w^2) + E] \, dx \, dy \, dz$$

The work done by action of the stress on any face of the cube is equal to the force produced by the stress times the distance in the direction of force that the particular cube face moves. Thus, the contribution of the normal stress σ_x on the face $dy\,dz$ at $x + \tfrac{1}{2}dx$ is, over a time dt,

$$\left[\sigma_x + \left(\frac{\partial \sigma_x}{\partial x}\right)\frac{dx}{2}\right]\left(u + \frac{\partial u}{\partial x}\frac{dx}{2}\right) dt \, dy \, dz$$

Including the shear-stress terms, and summing up over all six faces, we obtain for the rate of energy input to the cube by the stresses

$$\left[\frac{\partial}{\partial x}(\sigma_x u + \tau_{xy} v + \tau_{xz} w) + \frac{\partial}{\partial y}(\tau_{xy} u + \sigma_y v + \tau_{yz} w) + \frac{\partial}{\partial z}(\tau_{xz} u + \tau_{yz} v + \sigma_z w)\right] dx \, dy \, dz$$

The rate of energy input by the body forces is simply

$$\rho(Xu + Yv + Zw) \, dx \, dy \, dz$$

Heat conduction into the cube provides the energy flux,

$$\left[\frac{\partial}{\partial x}\left(k\frac{\partial T}{\partial x}\right) + \frac{\partial}{\partial y}\left(k\frac{\partial T}{\partial y}\right) + \frac{\partial}{\partial z}\left(k\frac{\partial T}{\partial z}\right)\right] dx \, dy \, dz$$

where k is the coefficient of heat conduction[1] assumed to be a scalar function of the fluid state. It is undoubtedly true that a state of stress renders the medium inhomogeneous with respect to heat conduction, in that heat would diffuse more rapidly along certain preferred axes. This is also true of the diffusion of momentum when a temperature variation is present in the fluid producing a slight inhomogeneity. Both these effects are, of course, neglected in this development, on the basis that their over-all influence on the fluid flow is negligible. There may be cases, however, in which considerations of this sort must be taken into account.

We now write the complete energy equation by summing up the various energy contributions.

$$\begin{aligned}\rho \frac{D}{Dt} [\tfrac{1}{2}(u^2 + v^2 + w^2) + E] &= \rho(Xu + Yv + Zw) \\ &+ \left[\frac{\partial}{\partial x}\left(k\frac{\partial T}{\partial x}\right) + \frac{\partial}{\partial y}\left(k\frac{\partial T}{\partial y}\right) + \frac{\partial}{\partial z}\left(k\frac{\partial T}{\partial z}\right)\right] + \frac{\partial}{\partial x}(\sigma_x u + \tau_{xy} v + \tau_{xz} w) \\ &+ \frac{\partial}{\partial y}(\tau_{xy} u + \sigma_y v + \tau_{yz} w) + \frac{\partial}{\partial z}(\tau_{xz} u + \tau_{yz} v + \sigma_z w)]\end{aligned} \quad (68.20)$$

[1] Note that k must be expressed in units consistent with the dimensions of energy being used.

In its present form, the energy equation (68.20) merely gives a balance of inputs and outputs; however, it is possible to learn something more about the distribution of energy.

In (68.20), the internal energy term represents the rate of heat increase in the particle. We may inquire as to what part of this heat is produced by internal friction, and what part is produced in a reversible manner by compression and heat conduction. To this end, we make use of the momentum equations (68.18) and multiply the first by u, the second by v, the third by w, and form the sum

$$\rho \frac{D}{Dt}[\tfrac{1}{2}(u^2+v^2+w^2)] = \rho(Xu+Yv+Zw) + u\left(\frac{\partial \sigma_x}{\partial x}+\frac{\partial \tau_{xy}}{\partial y}+\frac{\partial \tau_{xz}}{\partial z}\right) + v\left(\frac{\partial \tau_{xy}}{\partial x}+\frac{\partial \sigma_y}{\partial y}+\frac{\partial \tau_{yz}}{\partial z}\right) + w\left(\frac{\partial \tau_{xz}}{\partial x}+\frac{\partial \tau_{yz}}{\partial y}+\frac{\partial \sigma_z}{\partial z}\right) \tag{68.21}$$

Subtracting (68.21) from (68.20) gives

$$\rho \frac{DE}{Dt} = \left[\frac{\partial}{\partial x}\left(k\frac{\partial T}{\partial x}\right)+\frac{\partial}{\partial y}\left(k\frac{\partial T}{\partial y}\right)+\frac{\partial}{\partial z}\left(k\frac{\partial T}{\partial z}\right)\right] + \sigma_x\frac{\partial u}{\partial x}+\sigma_y\frac{\partial v}{\partial y} + \sigma_z\frac{\partial w}{\partial z}+\tau_{xy}\left(\frac{\partial u}{\partial y}+\frac{\partial v}{\partial x}\right)+\tau_{yz}\left(\frac{\partial v}{\partial z}+\frac{\partial w}{\partial y}\right)+\tau_{xz}\left(\frac{\partial w}{\partial x}+\frac{\partial u}{\partial z}\right)$$

If the stress–rate-of-strain relations are now introduced from Eqs. (68.6) and (68.12), then

$$\rho \frac{DE}{Dt} = \left[\frac{\partial}{\partial x}\left(k\frac{\partial T}{\partial x}\right)+\frac{\partial}{\partial y}\left(k\frac{\partial T}{\partial y}\right)+\frac{\partial}{\partial z}\left(k\frac{\partial T}{\partial z}\right)\right] - p_s\left(\frac{\partial u}{\partial x}+\frac{\partial v}{\partial y}+\frac{\partial w}{\partial z}\right) + b\left(\frac{\partial u}{\partial x}+\frac{\partial v}{\partial y}+\frac{\partial w}{\partial z}\right)^2 + 2\mu\left[\left(\frac{\partial u}{\partial x}\right)^2+\left(\frac{\partial v}{\partial y}\right)^2+\left(\frac{\partial w}{\partial z}\right)^2\right] + \mu\left[\left(\frac{\partial u}{\partial y}+\frac{\partial v}{\partial x}\right)^2+\left(\frac{\partial v}{\partial z}+\frac{\partial w}{\partial y}\right)^2+\left(\frac{\partial w}{\partial x}+\frac{\partial u}{\partial z}\right)^2\right] \tag{68.22}$$

Equation (68.22) expresses the result that the increase in internal energy may be broken into three parts: first, the heat conducted into the element; second, the work done by the pressure in changing the volume; and third, an irreversible contribution arising from the state of fluid stress. The last three terms of (68.22) constitute an irreversible production of heat since the squared velocity gradients would not change sign if the motion were reversed. This dissipation of mechanical energy into heat is characterized by the symbol Φ, which is known as the *dissipation function*. From Eq. (68.22), we therefore define Φ thus:

$$\Phi = b\left(\frac{\partial u}{\partial x}+\frac{\partial v}{\partial y}+\frac{\partial w}{\partial z}\right)^2 + 2\mu\left[\left(\frac{\partial u}{\partial x}\right)^2+\left(\frac{\partial v}{\partial y}\right)^2+\left(\frac{\partial w}{\partial z}\right)^2\right] + \mu\left[\left(\frac{\partial u}{\partial y}+\frac{\partial v}{\partial x}\right)^2+\left(\frac{\partial v}{\partial z}+\frac{\partial w}{\partial y}\right)^2+\left(\frac{\partial w}{\partial x}+\frac{\partial u}{\partial z}\right)^2\right] \tag{68.23}$$

It will be noted that Φ has the dimension of energy dissipated per unit time per unit volume. We may inquire as to the possibility of a dissipation-free flow; it is evident from the foregoing, however, that in such a flow there can be no relative motion between fluid particles. Hence, dissipation-free flows are restricted to motions in which the fluid moves in uniform translational or rotational motion. Also, certain accelerated motions are possible under the action of body forces. Fortunately, in many practical fluid-flow problems, the dissipation is negligible by virtue of a combination of small viscosities and weak velocity gradients.

Integration of the Equations for Reversible Flows

In many fluids of importance, the coefficients of viscosity and heat conduction are very small, and may be neglected in the general equations, without loss of the important features of the flow. Thus, the energy equation (68.20) becomes

$$\rho \frac{D}{Dt}[\tfrac{1}{2}(u^2+v^2+w^2)+E] = \rho(Xu+Yv+Zw) - \left(u\frac{\partial p_s}{\partial x}+v\frac{\partial p_s}{\partial y}+w\frac{\partial p_s}{\partial z}\right) \tag{68.24}$$

If the body forces arise from a potential Ω (as is the usual case) such that

$$X = -\frac{\partial \Omega}{\partial x} \qquad Y = -\frac{\partial \Omega}{\partial y} \qquad Z = -\frac{\partial \Omega}{\partial z} \tag{68.25}$$

and, if we introduce the enthalpy

$$h = E + \frac{p_s}{\rho} \tag{68.26}$$

we obtain for Eq. (68.20), making use of (68.15),

$$\frac{D}{Dt}[\tfrac{1}{2}(u^2+v^2+w^2)+h+\Omega] = \frac{\partial \Omega}{\partial t}+\frac{1}{\rho}\frac{\partial p_s}{\partial t} \tag{68.27}$$

When the motion is steady, the terms on the right vanish, and the equation may be integrated; thus we obtain, for steady flow,

$$\tfrac{1}{2}(u^2+v^2+w^2)+h+\Omega = \text{const} \tag{68.28}$$

We see then that the sum of the particle kinetic, heat, and potential energy is conserved and, furthermore, since every particle along any given streamline passes through a common set of points, the sum of the energies shown must be conserved all along the streamline. When all the streamlines of a given flow originate at locations of common pressure and density, we observe that the constant of (68.28) is common to the entire flow field.

If, in addition to the above restrictions, the fluid is taken to be incompressible, as is quite nearly the case for water, oil, and liquids in general, (68.22) shows the change in internal energy to be zero for all particles, and Eq. (68.28) may be written as

$$p_s + \rho\Omega = -\frac{\rho}{2}(u^2+v^2+w^2) + \text{const} \tag{68.29}$$

which is known as *Bernoulli's equation.* For cases where the body forces are small, we obtain the *Pitot equation*

$$p_t - p_s = \frac{\rho}{2}(u^2+v^2+w^2) \tag{68.30}$$

where p_t is the stagnation pressure or total head. This equation has been utilized in a wide number of practical applications for the measurement of fluid velocity.

68.3. BOUNDARY CONDITIONS

General Considerations

The equations of motion derived in the preceding paragraphs represent the application of the conservation laws to an arbitrary fluid particle. They appear as nonlinear partial differential equations of second order, involving the fluid velocities, density, pressure, and temperature as unknowns, with time and three position coordinates as independent variables. Use of the equation of state allows the replacement of pressure, density, or temperature with a function of the remaining two. Since the equations govern all possible fluid motions, it is necessary to specify sufficient additional "boundary conditions" to obtain a unique solution of the mathematically posed problem. A unique solution requires, of course, that the unknown quantities be specified at every point of space and time.

The boundary conditions are, in no sense, mathematical conditions, but must be obtained from the physical requirements of the flow problem. In a properly posed problem, the number of boundary conditions available is just sufficient to permit solution. There are cases, however, in which the initial conditions are not known, and only a statistical type of solution is desired. For more information on this type of problem, the reader is referred to the section on turbulence. Once a solution has been obtained through the use of boundary conditions, there still remains the question of whether or not the solution is unique. In most physical problems, the question of uniqueness is settled by physical intuition, since uniqueness studies of the solution to each important fluid-flow problem would increase the labor required many times over and, in a majority of cases, would delay the final acceptance indefinitely. Since the equations of motion are *deterministic*, that is, they arise from the assumption of Newton's laws, together with the conservation of energy and mass, the intuition referred to above would imply that, for a set of properly given boundary conditions, there would exist only one physical solution. Hence, when any solution is found, it should be the only one and the proper one. This reasoning is quite probably correct, but one continually encounters so-called "paradoxes," in addition to problems to which there would appear to be many solutions. It is generally true that when a paradox appears, it arises because the complete equations have been found too difficult to solve, and an approximate set of equations has been substituted. To cite an example, we have the famous d'Alembert paradox, in which the mathematical solution gives the unphysical result that the resistance of a body moving in a liquid is zero. In this case, the complete equations of motion had been approximated by omitting the effects of fluid viscosity, with the result that a dissipation-free flow had been predicted. Had the full equations (Navier-Stokes equations) been solved, the resulting flow would have been found completely altered in character from that of the approximate analysis. It should be remarked, however, that in many cases the solution found would probably not resemble the physical flow at all. The reason for this remark is that, in formulating the boundary conditions, the analyst must anticipate what the physical conditions will be. That is, he must decide, a priori, whether there will be random disturbances introduced during the flow development, whether the fluid is truly homogeneous, etc. In short, he must accurately know the initial conditions and subsequent disturbance inputs. In any real experiment, these initial conditions will be known only approximately; hence, the correlation between real flow and calculated flow will depend critically on the sensitivity of the solution to small variations in the assumed inputs. If an infinitesimal variation in the initial conditions produces a large variation in the flow developed, then the flow is termed unstable, and the possibility arises of having a large number of physically possible flows, each depending critically on unobservable initial disturbances. In many important cases, the final flow field exhibits a cyclic regularity. Such a flow is illustrated by the well-known *Kármán vortex street*, in which cyclic production of large vortices takes place behind a cylinder placed in a fluid flowing in a direction perpendicular to the cylinder axis. In such flows it is probable, though conjectural, that a small disturbance of almost any type would lead to the observed flow. In cases where the flow produced is not cyclic, it is an open question as to the influence of the infinitesimal initiating disturbances. The development of high-speed digital computing machines has led to the hope that some of these difficulties might be resolved; however, the complexity of the general equations of motion is so great, even in the case of incompressible fluids, that further development of the machines is required.

Physical Considerations

Most fluid-flow problems involve contact of the fluid with a solid surface. That no fluid may pass into the solid is obviously a fundamental boundary condition, but this requires only the vanishing of the normal component of the velocity. The tangential component at the interface must also vanish, although it is not immediately so evident. There have been attempts to find surfaces which would provide some slip; however, to date there have been no data brought forward to indicate such a result. It is worthy of note that, in certain cases, the attainment of a reasonable slip at the boundary would provide a reduction of heat transfer as well as frictional resistance. In addition to solid boundaries, a physical problem may involve an interface between two fluids such as air and water, or a liquid and its vapor. When such a contact surface exists, it cannot sustain force and, hence, the pressure must be continuous across it. Here a fundamental difficulty arises; the location of the surface is an unknown in the problem! We see, therefore, that while the boundary condition is specified, the boundary is not, and the resulting problem is extremely difficult mathematically. Typical problems in this category are those of finite waves on liquids having a free surface, and cavitation on hydrofoils and in water turbines. Cavitation occurs when the local pressure in a liquid becomes low enough at the existing temperature to permit boiling to occur. A region filled with vapor is thus produced, and it is necessary to specify the boundary condition on the liquid-vapor interface. Because of the very great difference in density between the gas and liquid, it is often possible to assume that the gas pressure is constant along the interface. Such an approximation leads to great mathematical simplification, and characterizes the so-called "free streamline flows."

In many cases, it is convenient and necessary to substitute boundary conditions at infinity in place of the real conditions. For example, even though an airfoil may be flying in a limited atmosphere having a finite pressure gradient, the airfoil's disturbance is sensible only in a limited region around and behind the airfoil. It then becomes permissible to substitute for the real atmosphere a fictitious one spreading to infinity in all directions. The condition required at infinity is merely that the fluid there is quiescent, and is introducing no outside disturbances into the flow field.

In the present state of the art, it is generally necessary to resort to some simplification of the general equations and boundary conditions which still leave hope for obtaining the salient features of a given flow problem. Such modifications vary, of course, with the individual problem, and the following chapters of Part 7, Fluid Mechanics, will deal, in the main, with the methods of solution specialized for the various types of flow problems.

68.4. REFERENCES

[1] H. S. Green: "The Molecular Theory of Fluids," Interscience, New York, 1952.
[2] S. Chapman, T. G. Cowling: "The Mathematical Theory of Non-uniform Gases," Cambridge Univ. Press, New York, 1939.
[3] H. Lamb: "Hydrodynamics," 6th ed., Dover, New York, 1945.
[4] M. Planck: "The Mechanics of Deformable Bodies," Macmillan, London, 1932.
[5] A. Busemann: Gasdynamik, in "Handbuch der Experimentalphysik," vol. 4, part 1, pp. 343–460, Akad.-Verlag., Leipzig, 1931.

CHAPTER 69

DIMENSIONLESS PARAMETERS

BY

D. C. BAXTER, Ph.D., Ottawa, Ont., Canada

There are certain dimensionless groups of variables which play a frequent role in fluid mechanics studies. They are useful because they allow convenient and systematic manipulation of the governing quantities, and lend themselves to generalizations and correlations of results. Their magnitudes indicate the relative importance of various types of physical phenomena, and so also serve to orient the particular problem at hand for the investigator.

Some of these parameters can be thought of as ratios of forces acting in the problem; others as ratios of energy transfers which are taking place. Still others are geometric factors, but these need not be considered here. Since the feature common to fluid mechanics problems is the presence of a *flowing* medium, it is not unnatural that the force against which others are measured should be an inertia force, formed dimensionally with a mass ρL^3 times an acceleration V^2/L, or, in other words, $\rho V^2 L^2$. Here ρ is the mass density, V the velocity, and L some characteristic length dimension.

Perhaps the most widely occurring of the force-ratio parameters is the *Reynolds number*, the ratio of inertia to viscous forces. The latter exist because of the shear deformation of fluid elements, or, more picturesquely, because of the drag of one fluid layer on another. The resulting shear stress is proportional to the velocity gradient, with the constant of proportionality being the viscosity coefficient μ. This viscous shear force is thus of the form μVL, and so the Reynolds number is

$$\text{Re} = \rho V^2 L^2/\mu VL = \rho VL/\mu \tag{69.1}$$

Superficially, at least, the larger the Reynolds number, the less important are viscosity effects to the problem. Since, in many phases of fluid mechanics, the Reynolds number plays a particularly conspicuous role, the fluid properties which affect it are often combined into a single quantity, the kinematic viscosity

$$\nu = \mu/\rho \tag{69.2}$$

Gravity forces can be characterized by a quantity of the form $\rho L^3 g$, where g is the acceleration due to gravity. The *Froude number* denotes the ratio of the inertia force to this gravity force, and so is

$$\text{Fr} = V^2/Lg \tag{69.3}$$

Gravity may, in addition, play a part in some problems, through the buoyant forces produced in a fluid due to density changes resulting from heating. Such a force is characterized by $\rho g \beta \,\Delta T\, L^3$, in which β is the temperature coefficient of volumetric expansion and, ΔT is the driving temperature difference. Since the velocity in such problems is usually small, this force is related to others in such a way as to be independent of velocity. In fact, the parameter used is the *Grashoff number*

$$\mathrm{Gr} = g\beta\,\Delta T\,L^3/\nu^2 \tag{69.4}$$

the ratio of the buoyant forces to viscous forces multiplied by the Reynolds number.

In problems with a free surface, such as between a liquid and a gas, surface tension forces may have to be considered. If the kinematic surface tension (i.e., the ratio of the surface tension to the density) is σ, the ratio of inertia to surface tension, called the *Weber number*, is defined by

$$W = V^2L/\sigma \tag{69.5}$$

A small value of the Weber number indicates that surface tension is relatively important.

The pressure forces acting on a system involve some pressure difference, Δp. The ratio of the force ΔpL^2 to the inertia force leads to the formation of a *pressure coefficient*, $C_p = \Delta p/\rho V^2$, or more conventionally

$$C_p = \Delta p/\tfrac{1}{2}\rho V^2 \tag{69.6}$$

Resultant forces in various problems are also measured relative to the inertia force, actually to half of it again, and so there are found force coefficients such as the lift or drag coefficients of aeronautical engineering. These are of the form

$$C_F = F/\tfrac{1}{2}\rho V^2L^2 \tag{69.7}$$

in which F is any of the forces involved.

As the velocities of the problem approach the speed of sound, the compressive force generated becomes important. Its ratio to the inertia force is the square of the *Mach number*

$$M = V/c \tag{69.8}$$

where c is the speed of sound. In some problems, such as those in which a liquid is being considered, the Mach number is, for practical purposes, zero and the fluid behaves as if it were incompressible. On the other hand there are situations in which the Mach number is large, and then, in addition to compressibility, even elastic effects, such as shock waves, will have to be considered.

The parameters considered above have all been related to forces and, in some sense, are related to momentum transfers. In addition, there are energy transfers, which play a part in many situations, in particular those associated with heat-transfer phenomena. Two of the common parameters here can be formed by considering two modes of heat transfer and one form of energy storage. The first mode is conduction, with the conduction heat-transfer rate proportional to the temperature gradient and with the proportionality constant being the thermal conductivity k. The next is convection, in which the heat-transfer rate is proportional to a temperature difference, and the proportionality constant is the heat-transfer coefficient h. Finally thermal energy can be stored in the flowing medium itself, with the specific heat c_p representing the energy stored in unit weight at unit temperature above datum. Thus, the rates of thermal energy flow represented by three processes are characterized dimensionally by $k\,\Delta T\,L$, $h\,\Delta T\,L^2$, and $\rho Vc_p\,\Delta T\,L^2$, respectively. The ratio of convection to conduction rates leads to the *Nusselt number*,

$$\mathrm{Nu} = hL/k \tag{69.9}$$

while the ratio of the convection rate to the flowing storage effect gives the *Stanton number*

$$\mathrm{St} = h/\rho Vc_p \tag{69.10}$$

The transport properties of the medium can be combined into the *Prandtl number*

$$\mathrm{Pr} = c_p\mu/k \tag{69.11}$$

the ratio of momentum to thermal diffusivities. Low-Prandtl-number fluids such as liquid metals have, loosely speaking, a high thermal conductivity compared to their viscosity. The opposite is true of oils where the Prandtl number is high, while for gases the Prandtl number is approximately unity.

CHAPTER 70

TWO-DIMENSIONAL IDEAL FLUID FLOW

BY

V. L. STREETER, Sc.D., Ann Arbor, Mich.

70.1. DEFINITIONS

In two-dimensional flow, the fluid particles move in plane paths in which all such planes are parallel and streamline patterns are identical in each plane. An ideal fluid is one that is frictionless and incompressible. With an additional requirement, that the flow be irrotational (or that a velocity potential exist), the concept of the flow net may be developed as an aid in analyzing two-dimensional flow cases. The *Prandtl hypothesis* states that, for fluids of low viscosity, the effects of viscosity are appreciable only in a narrow region surrounding the fluid boundaries. For flow situations in which the boundary layer remains thin, ideal fluid results may be applied to flow of a real fluid to a satisfactory degree of approximation. Converging or accelerating flow situations generally have thin boundary layers, but decelerating flow may have separation of the boundary layer and development of a wake that is difficult to predict analytically.

An ideal fluid must satisfy the following requirements:

1. The continuity equation
2. Newton's second law of motion, at every point at every instant
3. Neither penetration of fluid into, nor gaps between fluid and boundary at, any solid boundary

If, in addition to requirements 1, 2, and 3, the assumption of irrotational flow is made, the resulting fluid motion closely resembles real fluid motion for fluids of low viscosity, outside the boundary layer.

By using the above conditions, the application of Newton's second law to a fluid particle leads to Euler's equations, which, together with the assumption of irrotational flow, may be integrated to obtain the Bernoulli equation. The unknowns in a fluid-flow situation with given boundaries are the pressure and the velocity components at each point. The continuity equation (1) and Newton's second law (2) produce three partial differential equations, and condition (3) yields the necessary boundary conditions. Unfortunately, in most cases, it is impossible to proceed directly from the given boundary conditions to equations for velocity and pressure distribution.

70.2. EULER'S EQUATIONS

Newton's second law states that the resultant force on any fluid particle must always equal the product of the mass of the particle and its acceleration, and that the acceleration is in the direction of the resultant force. By taking component forces in the x and y directions, scalar equations of motion are obtained. The forces acting are either surface forces or body forces. Since the fluid is frictionless, all surface forces are normal to the particle surface.

If X and Y are the body-force components per unit mass, the equations are

$$X - \frac{1}{\rho}\frac{\partial p}{\partial x} = u\frac{\partial u}{\partial x} + v\frac{\partial u}{\partial y} + \frac{\partial u}{\partial t} \qquad Y - \frac{1}{\rho}\frac{\partial p}{\partial y} = u\frac{\partial v}{\partial x} + v\frac{\partial v}{\partial y} + \frac{\partial v}{\partial t} \quad (70.1a,b)$$

These are Euler's equations. The first two terms on the right-hand side are the *convective* acceleration, i.e., the acceleration resulting from changes of velocity in space. The final term is the *local* acceleration, i.e., the acceleration arising because the flow is *unsteady* or changing at a point with time.

Natural Coordinates

A curve which is everywhere tangent to the resultant velocity q is a *streamline.* We rewrite Eqs. (70.1) for a coordinate system (s,n) (Fig. 70.1) whose s axis is a

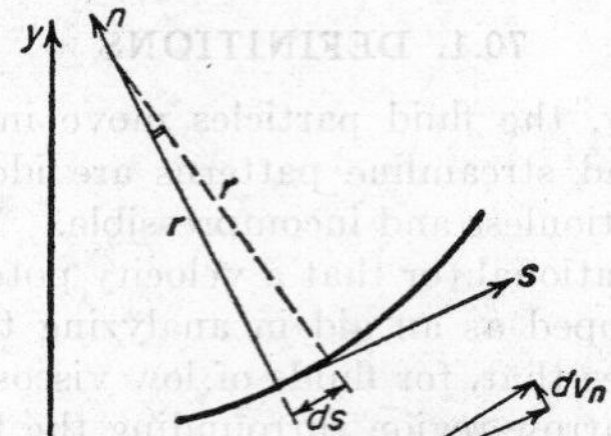

FIG. 70.1. Natural coordinates.

tangent to a streamline. Then u and v must be replaced by $v_s = q$ and v_n. At the origin of the (s,n) system, $v_n = 0$, but, at $s = ds$, it equals

$$dv_n = \frac{\partial v_n}{\partial s}\,ds = \frac{q}{r}\,ds$$

If we assume additionally that gravity $Y = -g$ is the only body force, then X, Y in (70.1) must be replaced by the forces $-g\,\partial y/\partial s$, $-g\,\partial y/\partial n$ in directions s, n, and the Euler equations are

$$-\frac{1}{\rho}\frac{\partial}{\partial s}(p + \gamma y) = q\frac{\partial q}{\partial s} + \frac{\partial q}{\partial t} \qquad -\frac{1}{\rho}\frac{\partial}{\partial n}(p + \gamma y) = \frac{q^2}{r} + \frac{\partial v_n}{\partial t} \quad (70.2a,b)$$

Here $\gamma = \rho g$ is the (constant) specific weight of the fluid. For steady flow, the last term in each equation vanishes. Equation (70.2*a*) may then be integrated with respect to s to produce the Bernoulli equation for rotational flow, in which the constant of integration varies with n, or the particular streamline. Equation (70.2*b*) shows how the pressure varies across streamlines, and may be integrated if v_s and r are known functions of n.

70.3. CONTINUITY EQUATION

The continuity equation states that the net inflow per unit time into any small fixed volume must be zero. It takes the form

$$\frac{\partial u}{\partial x} + \frac{\partial v}{\partial y} = 0 \quad (70.3)$$

This equation holds for both steady and unsteady flow at every point except singular points where the partial derivatives are not defined. When the continuity equation is satisfied by expressions for u and v as functions of x, y, and t, no gaps can occur in a fluid, and it cannot pile up at any point.

70.4. BOUNDARY CONDITIONS

The kinematical conditions that must be satisfied at every point on a solid-fluid boundary are that the fluid does not penetrate the boundary and that no gaps occur between fluid and boundary. If the boundary is *stationary*, this may be stated as

$$u \cos \theta + v \sin \theta = 0 \tag{70.4}$$

in which θ is the angle that the normal from the boundary makes with the $+x$ axis. In other words, the fluid velocity component normal to the boundary at the boundary must be zero.

When the boundary *itself* is in motion, then the fluid velocity component normal to the boundary must equal the velocity of the boundary normal to itself. If ν is allowed to be this velocity of the boundary normal to itself, the boundary condition becomes

$$u \cos \theta + v \sin \theta = \nu \tag{70.5}$$

for every point on the boundary. No restrictions are placed on the tangential component of fluid velocity at a boundary when ideal fluids are considered.

If the boundary surface is expressed by

$$F(x,y,t) = 0 \tag{70.6}$$

it may be proved that

$$u \frac{\partial F}{\partial x} + v \frac{\partial F}{\partial y} + \frac{\partial F}{\partial t} = 0 \tag{70.7}$$

is the desired boundary condition.

Where two different fluids are in contact, the pressure must vary continuously across the boundary.

In dealing with motion of bodies in an infinite fluid, it is assumed, in general, that the motion starts from rest; i.e., initially both fluid and bodies are stationary. When forces are applied to the bodies to set them in motion, a necessary condition is that the fluid at infinity remain at rest. If this were not so, it would imply that the finite forces acting on the bodies had imparted to the fluid infinite kinetic energy in finite time, which violates the principles of work and energy.

70.5. IRROTATIONAL FLOW. VELOCITY POTENTIAL

The individual particles of a frictionless incompressible fluid initially at rest cannot be caused to rotate. This may be visualized by considering a small free body of fluid the shape of a sphere. Surface forces must act normal to the surface since the fluid is frictionless, and therefore act through the center of the sphere. Similarly, body forces act through the mass center. Hence, no couple can be exerted on the sphere, and it remains without rotation. Likewise, once an ideal fluid has rotation, there is no way of altering it.

Rotation at a point may be defined as the average angular velocity through the point of any two infinitesimal linear elements which are perpendicular to each other and the axis. These two lines may conveniently be taken parallel to the x and y axes (Fig. 70.2), although any other two perpendicular lines in the plane through the point would give the same result. The particle is at P and has the velocity components u, v. The angular velocity of the line element dx is $\partial v/\partial x$ and the angular velocity of dy is $-\partial u/\partial y$ if counterclockwise is positive. Hence, the average rotation is $\frac{1}{2}(\partial v/\partial x - \partial u/\partial y)$, and, for irrotational flow, the condition

$$\frac{\partial v}{\partial x} = \frac{\partial u}{\partial y} \tag{70.8}$$

must be satisfied at every point. The double of the rotation, that is,

$$\omega = \frac{\partial v}{\partial x} - \frac{\partial u}{\partial y} \tag{70.9}$$

is called the *vorticity*. It is the absolute value of the curl of the velocity vector.[1]

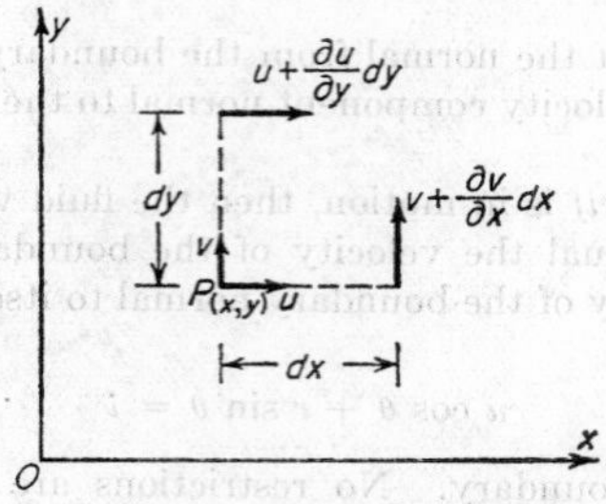

FIG. 70.2. Rotation of an element.

In an actual fluid having small viscosity, such as air or water, irrotational flow occurs, for all practical purposes, for a short time after motion starts.

Where the flow is irrotational, a *velocity potential* exists. It is defined as a scalar function of space and time such that its derivative with respect to any direction is the velocity in that direction, in particular,

$$u = \frac{\partial \phi}{\partial x} \qquad v = \frac{\partial \phi}{\partial y} \tag{70.10a,b}$$

Differentiation of Eqs. (70.10*a*) with respect to y and (70.10*b*) with respect to x yields Eq. (70.8). Thus, it is seen that the existence of a velocity potential implies irrotational flow.

70.6. INTEGRATION OF EULER'S EQUATIONS. BERNOULLI EQUATION

Equations (70.1*a,b*) can be integrated by assuming that

1. The extraneous force components are derivable from a potential Ω, that is,

$$X = -\frac{\partial \Omega}{\partial x} \qquad Y = -\frac{\partial \Omega}{\partial y} \tag{70.11a,b}$$

Gravity is given by $\Omega = gy$.

2. The fluid is incompressible, or ρ = const.
3. The flow is irrotational.

When these substitutions are made, Eqs. (70.1) become respectively

$$\frac{\partial}{\partial x}\left(\Omega + \frac{p}{\rho} + \frac{q^2}{2} + \frac{\partial \phi}{\partial t}\right) = 0 \qquad \frac{\partial}{\partial y}\left(\Omega + \frac{p}{\rho} + \frac{q^2}{2} + \frac{\partial \phi}{\partial t}\right) = 0$$

in which $q^2 = u^2 + v^2$. Since the term in parentheses is not a function of either x or y, at most it is a function of time

$$\Omega + \frac{p}{\rho} + \frac{q^2}{2} + \frac{\partial \phi}{\partial t} = F(t) \tag{70.12}$$

[1] Note that in many books and also in other parts of this handbook, ω is used to denote the average rotation.

This is the energy equation for unsteady, irrotational flow of an ideal fluid. In steady flow, $\partial\phi/\partial t = 0$, and $F(t)$ becomes a constant E,

$$\Omega + \frac{p}{\rho} + \frac{q^2}{2} = E \tag{70.13}$$

This equation shows that the energy is everywhere constant throughout the fluid. By considering gravity the only extraneous force,

$$gy + \frac{p}{\rho} + \frac{q^2}{2} = E \tag{70.14}$$

The pressure term may be separated into two parts, the hydrostatic pressure p_s and the dynamic pressure p_d, so that $p = p_s + p_d$. Equation (70.14) then contains the terms $(\gamma y + p_s)/\rho$, which, together, are a constant and may be included in E. Dropping the subscript on the dynamic pressure,

$$\frac{p}{\rho} + \frac{q^2}{2} = E \tag{70.15}$$

This simple equation permits the variation of dynamic pressure to be determined if the velocity distribution is known, or *vice versa.*

70.7. LAPLACE EQUATION. EQUIPOTENTIAL LINES

When the velocity potential ϕ from Eqs. (70.10) is introduced into the continuity equation (70.3), the Laplace equation results:

$$\frac{\partial^2\phi}{\partial x^2} + \frac{\partial^2\phi}{\partial y^2} \equiv \nabla^2\phi = 0 \tag{70.16}$$

Any solution ϕ of this equation represents a possible flow pattern, and the corresponding pressure field may be calculated from Eq. (70.15). One fruitful method of approach is to examine solutions of the Laplace equation in order to determine which particular boundary conditions they fulfill.

When the boundary conditions are specified in advance, it may be very difficult to find the proper function ϕ to satisfy them as well as the Laplace equation. Several methods of attack are available other than the direct method of finding ϕ from the boundary conditions. These include the electric analogy method, the Hele Shaw method, and graphical trial-and-error methods.

Equipotential lines are lines ϕ = const. It follows from Eqs. (70.10) that the velocity component tangential to the line vanishes. Equipotential lines and streamlines are orthogonal to each other.

70.8. KINETIC ENERGY THEOREM

The kinetic energy of the fluid is given by the area integral

$$T = \tfrac{1}{2}\rho\iint q^2\,dx\,dy$$

to be evaluated over the entire region of the flow. For irrotational flow, Green's theorem [see Eq. (6.19) with $\psi = \phi$ and div grad $\phi = 0$ (continuity)] yields

$$T = -\tfrac{1}{2}\rho\int \phi\frac{\partial\phi}{\partial n}\,ds \tag{70.17}$$

This integral is to be evaluated over the entire boundary (including the boundary at infinity, if the flow field extends that far), and $\partial\phi/\partial n$ is the derivative of ϕ with respect to the normal to the boundary, drawn positive into the fluid.

Equation (70.17) may also be shown to hold for an infinite expanse of fluid, at rest at infinity.

70.9. UNIQUENESS THEOREMS

Several uniqueness theorems may be proved as a consequence of Eq. (70.17). They are limited to ideal fluid-flow cases where the velocity potential is single-valued.

1. Irrotational motion of a fluid is impossible if the boundaries are fixed.
2. Irrotational motion of a fluid will cease when the boundaries are brought to rest.
3. Irrotational motion of a fluid that satisfies the Laplace equation and prescribed boundary conditions is uniquely determined by the motion of the boundaries.
4. Irrotational motion of a fluid at rest at infinity is impossible if the interior boundaries are at rest.
5. Irrotational motion of a fluid at rest at infinity is uniquely determined by the motion of the interior solid boundaries.

Statements 1, 2, and 3 apply to a finite region.

70.10. STREAM FUNCTION

Let A and P represent two points in the plane of flow (Fig. 70.3), and consider that the flow has unit thickness. The rate of flow across any two lines ACP, ABP must be the same as a consequence of continuity. Now consider A fixed in the x, y plane and P a movable point. The flow rate across any line connecting the two points is a function of the position of P and of the time. Let this function be ψ, and take as sign convention that it denotes the flow rate from left to right across any line connecting A and P, when the observer is at A looking along the line toward P; thus,

$$\psi = \psi(x,y,t)$$

is defined as the stream function.

Let ψ and ψ' represent the values of the stream function at points P and P', respectively, in Fig. 70.3. Then $(\psi' - \psi)$ is the flow across PP' from left to right. If some other point besides A were taken as fixed point, say O, then the values of ψ and ψ' would be increased by the same amount, viz., the flow across OA. ψ is then indeterminate to the extent of an arbitrary constant.

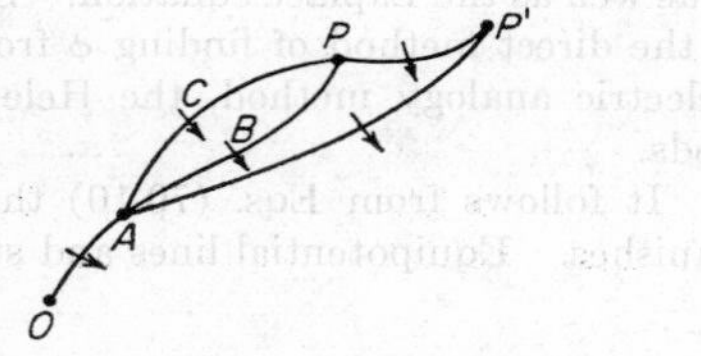

FIG. 70.3. Flow between points in a fluid region.

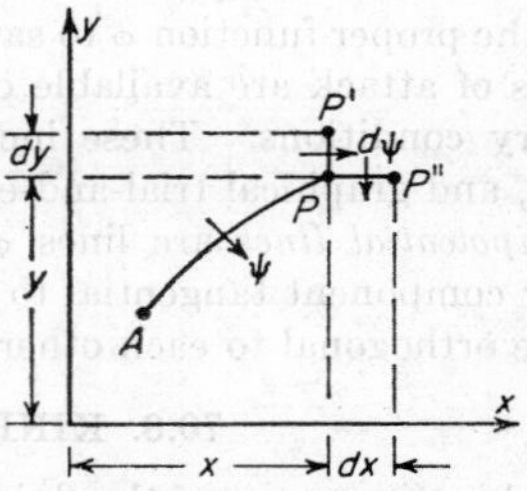

FIG. 70.4. Displacement of P to show the relation between ψ, u, and v.

Let ψ be the stream function at P (Fig. 70.4). The flow across the line element $PP' = dy$ is then

$$u\, dy = \frac{\partial \psi}{\partial y}\, dy$$

Similarly, for an element $PP'' = dx$, the flow is

$$-v\,dx = \frac{\partial\psi}{\partial x}\,dx$$

Hence,

$$u = \frac{\partial\psi}{\partial y} \qquad v = -\frac{\partial\psi}{\partial x} \tag{70.18}$$

In words, the partial derivative of the stream function with respect to any direction gives the velocity component 90° clockwise to that direction.

Equations (70.18) are true whether the flow has rotation or not. For irrotational flow, however, Eq. (70.8) applies. Substituting Eqs. (70.18) into this expression produces

$$\frac{\partial^2\psi}{\partial x^2} + \frac{\partial^2\psi}{\partial y^2} = \nabla^2\psi = 0 \tag{70.19}$$

showing that ψ may be construed as a velocity potential for some other flow. The stream function has the dimensions L^2T^{-1}, the same as the velocity potential.

When the two points P, P' of Fig. 70.3 lie on the same streamline, the rate of flow across AP and AP' is the same, and $\psi = \psi'$. Therefore, streamlines are lines $\psi =$ const. The flow rate between any two streamlines (per unit width) is given by the difference of the values of the stream function. Relations between stream function and velocity potential are found by equating the expressions for velocity components:

$$\frac{\partial\phi}{\partial x} = \frac{\partial\psi}{\partial y} \qquad \frac{\partial\phi}{\partial y} = -\frac{\partial\psi}{\partial x} \tag{70.20}$$

70.11. CIRCULATION

The line integral of the velocity vector taken around a closed curve C (Fig. 70.5) is said to be the *circulation:*

$$\Gamma = \oint q\cos\alpha\,ds = \oint(u\,dx + v\,dy) \tag{70.21}$$

By use of Stokes' theorem [Eq. (6.16)],

$$\Gamma = \iint\left(\frac{\partial v}{\partial x} - \frac{\partial u}{\partial y}\right)dx\,dy \tag{70.22}$$

in which the integration is performed over the area enclosed by the boundary C. A restriction on Eq. (70.22) is that no singular points may exist within the curve C.

The integrand of Eq. (70.22) is ω from Eq. (70.9); hence, for an element of area $dS = dx\,dy$

$$d\Gamma = \omega\,dS \tag{70.23}$$

As any surface may be divided into surface elements dS, in words: The circulation about any closed curve is equal to twice the surface integral of the rotation over the surface enclosed by the curve. A restriction on this statement is that ω be continuous and defined at all points on the surface.

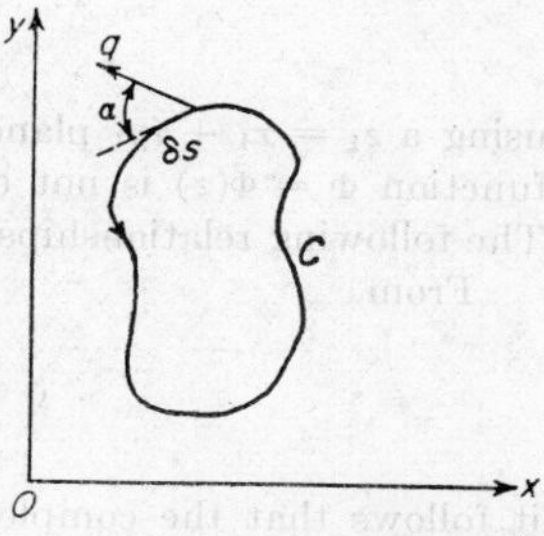

FIG. 70.5. Notation for circulation around a curve.

When the flow is irrotational, the right-hand side of Eq. (70.23) is zero; hence, there is no circulation about any closed curve in the flow not enclosing a singular point. Conversely, if the circulation about every closed curve in a fluid is zero, when the curve contains no singular points, the flow is irrotational. It is, therefore, necessary that the rotation be everywhere zero for the circulation to be zero. Circulation has the dimension L^2T^{-1}.

70.12. COMPLEX VARIABLE, CONFORMAL MAPPING

Let $\Phi = \phi + i\psi$ be an analytic function of the complex variable $z = x + iy$. Then the real and imaginary parts, ϕ and ψ, each satisfy the Laplace equation [see Eqs. (9.16)], and Eqs. (70.20) are the Cauchy-Riemann equations (9.14) (note that u, v, w there are identical with ϕ, ψ, Φ here). Therefore, the real and imaginary parts $\phi(x,y)$ and $\psi(x,y)$ of any analytic function of z may be interpreted as the potential and the stream function, respectively, of a two-dimensional flow. The complex-valued function $\Phi(x + iy)$ is called the complex potential of the flow.

From Eqs. (9.14) and (9.15) it follows that, in the present notation,

$$\frac{d\Phi}{dz} = \frac{\partial \phi}{\partial x} - i\frac{d\phi}{\partial y} = u - iv = w \tag{70.24}$$

The derivative $w = d\Phi/dz$ is called the complex velocity. Its absolute value is equal to the actual speed q of the flow,

$$|w| = \sqrt{u^2 + v^2} = q \tag{70.25}$$

but its vertical component is opposed in direction to the velocity component v. Points where $w = 0$ and, hence, $q = 0$ are called *stagnation points*.

A complex function $\Phi(z)$ represents a coordinate transformation and, as shown on p. 9–6, may be interpreted as a conformal mapping of the Φ plane into the z plane or vice versa. The equipotential lines $\phi = \text{const}$ and the streamlines $\psi = \text{const}$ in the Φ plane describe a uniform parallel flow in the direction of the ϕ axis. Through the function $\Phi(z)$, this flow pattern is mapped into another one in the z plane, the lines $\phi(x,y) = \text{const}$ and $\psi(x,y) = \text{const}$ being its equipotential lines and its streamlines, respectively. At points where $d\Phi/dz = 0$ (stagnation points) or $= \infty$, the transformation is not conformal. At such singular points, streamlines and equipotential lines do not intersect at right angles. Singular points may easily be recognized in the flow patterns of Figs. 70.6, 70.7, 70.8, and 70.9.

It is often convenient to consider a complicated function as a sequence of several conformal mappings. The function $\Phi = \sin e^z$ may, for example, be represented by the two consecutive mappings

$$\Phi = \sin z_1 \qquad z_1 = e^z$$

using a $z_1 = x_1 + iy_1$ plane as an intermediate plane. It also may happen that the function $\Phi = \Phi(z)$ is not explicitly known, but that the inverse $z = z(\Phi)$ is given. The following relationships are useful in this case.

From

$$\zeta = \frac{dz}{d\Phi} = \frac{1}{d\Phi/dz} = \frac{1}{u - iv} = \frac{u + iv}{q^2}$$

it follows that the complex number ζ has the same argument (represents a vector of the same direction) as the velocity, but has a modulus which is the reciprocal of the speed $1/q$.

Writing the Cauchy-Riemann equations [see Eqs. (9.14)] for $z(\Phi)$ yields

$$\frac{\partial x}{\partial \phi} = \frac{\partial y}{\partial \psi} \qquad \frac{\partial y}{\partial \phi} = -\frac{\partial x}{\partial \psi} \tag{70.26}$$

and, applying again Eq. (9.15), we have

$$\frac{dz}{d\Phi} = \frac{\partial x}{\partial \phi} + i\,\frac{\partial y}{\partial \phi}$$

whence

$$\left|\frac{dz}{d\Phi}\right| = \frac{1}{q} = \sqrt{\left(\frac{\partial x}{\partial \phi}\right)^2 + \left(\frac{\partial y}{\partial \phi}\right)^2}$$

70.13. LINEAR OPERATIONS

When in a complex potential $\Phi = \Phi(z)$ the independent variable z is replaced by $z - c = z - a - ib$ (with real a, b), this amounts to shifting the corresponding flow field by amounts a and b in the x and y directions, respectively. *Example:* The complex potential $\Phi = (M/2\pi) \ln z$ represents the flow from a source located at the origin $z = 0$ [see Eq. (70.27)]. Then $\Phi = (M/2\pi) \ln (z - a)$ represents the flow from a source located at $x = a$, $y = 0$.

When the variable z in $\Phi(z)$ is replaced by $ze^{i\beta}$, this corresponds to a clockwise rotation of the flow pattern by an angle β. *Example:* Dipole flow (p. 70–12).

Let $\Phi_1(z)$ and $\Phi_2(z)$ be two complex potentials with velocity fields u_1,v_1 and u_2,v_2, respectively. Then $\Phi = \Phi_1 + \Phi_2$ represents a flow field with velocity components $u = u_1 + u_2$, $v = v_1 + v_2$; that is, complex potentials and velocities obey the law of linear superposition. The pressure p, however, cannot be obtained as the sum of the pressures of the component flow fields, but must be calculated from Eq. (70.15) for the resultant velocity.

70.14. SIMPLE CONFORMAL TRANSFORMATIONS

Any streamline in a steady-flow field may be replaced by a solid boundary, as on it the condition is satisfied that the velocity component normal to it be zero. Using this idea, it is possible to find solutions of two-dimensional steady-flow problems, by inspecting the flow fields generated by various functions $\Phi(z)$ and finding out whether they contain streamlines which might represent boundaries of practical interest.

Powers of z

A number of useful flow fields can be derived from $\Phi = Az^n$ by specifying various real, complex, or pure imaginary values of A and real values of n.

For $n = 1$ and a complex $A = a + ib$, the complex velocity $w = A$ is a constant. The flow is a uniform parallel flow: $u = a$, $v = -b$.

For $n = 2$ and real A, there is

$$\phi = A(x^2 - y^2) \qquad \psi = 2Axy \qquad w = d\Phi/dz = 2Az$$

Figure 70.6 shows the flow lines in the z plane. The vertical lines $\phi = \text{const}$ in the Φ plane become a family of hyperbolas $x^2 - y^2 = \text{const}$, having axes coincident with the x and y axes and asymptotes $y = \pm x$. The horizontal lines $\psi = \text{const}$ of the Φ plane map into the rectangular hyperbolas $xy = \text{const}$, having axes $y = \pm x$ and with the coordinate axes as asymptotes. One of the streamlines consists of the positive branches of the x and y axes combined. When this is taken as a solid boundary, the right upper part of Fig. 70.6 represents the flow in a rectangular corner. At the origin, $w = 0$. This is a singular point, and here an equipotential line and a streamline meet at angles of 45°.

For $n = -1$ and real A, that is, for $\Phi = A/z$, the real and imaginary parts of Φ are

$$\phi = \frac{Ax}{x^2 + y^2} \qquad \psi = -\frac{Ay}{x^2 + y^2}$$

The equipotential lines $\phi = \text{const}$ in the z plane are circles through the origin with centers on the x axis, and the streamlines are circles through the origin with centers on the y axis (Fig. 70.7). The flow field is the limiting case of Fig. 70.12 when the two

FIG. 70.6. Flow net for $\Phi = Az^2$.

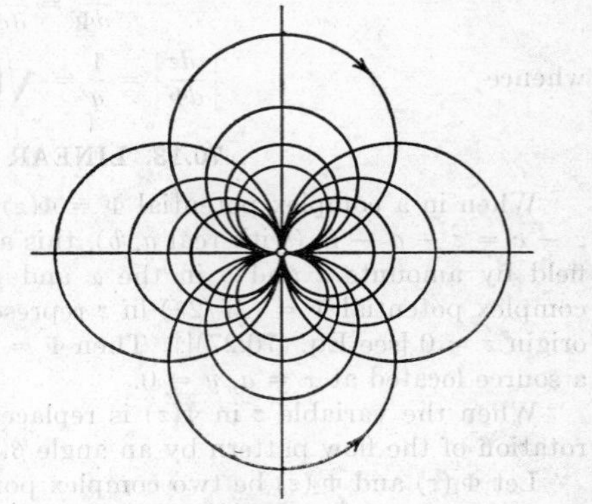

FIG. 70.7. Flow net for $\Phi = A/z$.

singular points coincide (doublet). When the origin is approached along any streamline it is found that $u \to -\infty$, $v = 0$.

For $n = \pi/\alpha$ with $0 < \alpha < 2\pi$ and real A, the flow field is best examined using polar coordinates r and θ in the x,y plane. Then

$$\Phi = Az^{\pi/\alpha} = Ar^{\pi/\alpha}\left(\cos\frac{\pi\theta}{\alpha} + i\sin\frac{\pi\theta}{\alpha}\right)$$

from which

$$\phi = Ar^{\pi/\alpha}\cos\frac{\pi\theta}{\alpha} \qquad \psi = Ar^{\pi/\alpha}\sin\frac{\pi\theta}{\alpha} \qquad w = A\,\frac{\pi}{\alpha}\,z^{\pi/\alpha-1}$$

The streamline $\psi = 0$ is given by $\theta = 0$ and $\theta = \alpha$. Hence, the uniform flow in the Φ plane is transformed into flow between two plane boundaries at an angle α. At the origin, w becomes zero if $\alpha < \pi$, and becomes infinite if $\alpha > \pi$. Figure 70.8 shows

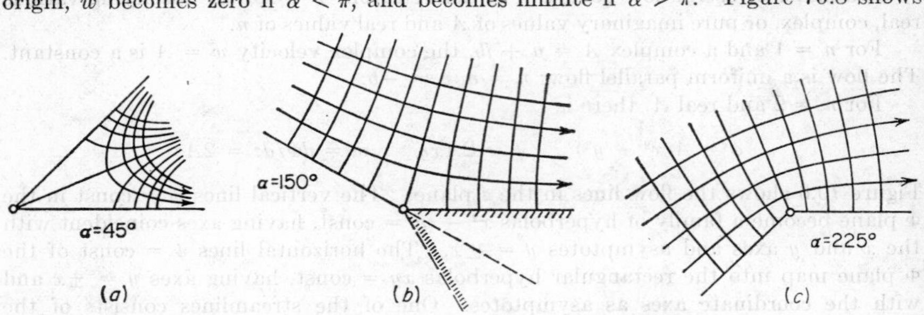

FIG. 70.8. Flow along two inclined plane surfaces.

the flow net for three cases: $\alpha = 45°$, $150°$, and $225°$. In Fig. 70.8*b* it has been indicated how the flow net may be interpreted as representing one-half the flow against a 60° wedge.

Source and Vortex

These are derived from the complex potential

$$\Phi = A\ln z = A\ln(re^{i\theta}) \tag{70.27}$$

When $A = M/2\pi$, then

$$\phi = \frac{M}{2\pi} \ln r \qquad \psi = \frac{M}{2\pi} \theta \qquad w = \frac{M}{2\pi z}$$

The equipotential lines are concentric circles about the origin, and the streamlines are radii, as shown in Fig. 70.9. On any circle of radius r, the velocity is directed radially outward, and has the magnitude $v_r = \partial\phi/\partial r = M/2\pi r$. Integration over the circle shows that the total outflow (discharge) is M, and this amount of liquid must be

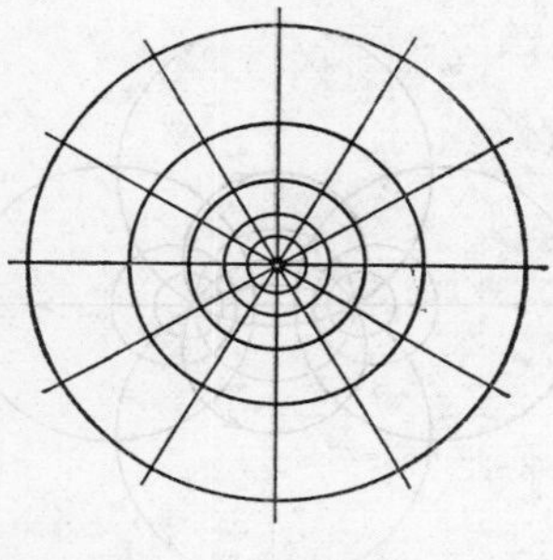

FIG. 70.9. Flow net for source or vortex.

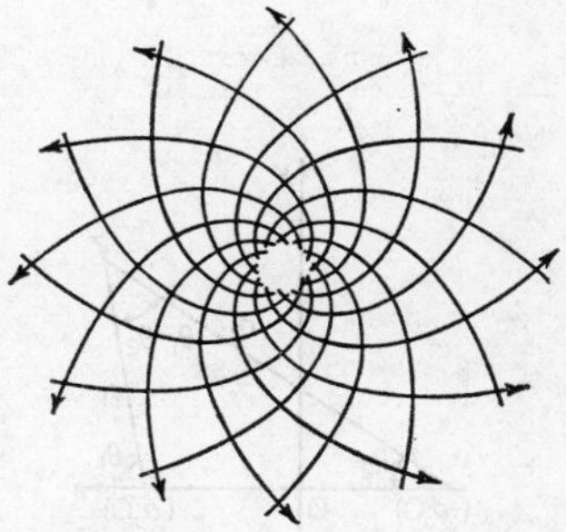

FIG. 70.10. Combination of source and vortex.

generated per unit of time at the singular point $r = 0$. This singularity is called a *source* of strength M. A source of negative strength is called a *sink*.

When $A = -i\Gamma/2\pi$ in Eq. (70.27), a quite different flow results. In this case,

$$\phi = \frac{\Gamma}{2\pi} \theta \qquad \psi = -\frac{\Gamma}{2\pi} \ln r \qquad w = -\frac{i\Gamma}{2\pi z}$$

Figure 70.9 is the flow pattern for this case also, except that the equipotential lines are now the radii, and the streamlines are the circles. The flow is a counterclockwise vortex flow, as described on p. 70–22, and Γ is the strength of the vortex.

Figure 70.10 shows the flow field resulting from the superposition of a source of strength M and a vortex of strength $\Gamma = M$. Both equipotential lines and streamlines are logarithmic spirals.

Source and Sink of Equal Strength

The complex potential for a source and a sink of equal strength M, located at $(a,0)$ and $(-a,0)$, respectively, is

$$\Phi = \frac{M}{2\pi} \ln \frac{z-a}{z+a} \tag{70.28}$$

By using the notation of Fig. 70.11, in which

$$z - a = r_1 e^{i\theta_1} \qquad z + a = r_2 e^{i\theta_2}$$

for any point z, then

$$\Phi = \frac{M}{2\pi}\left[\ln \frac{r_1}{r_2} + i(\theta_1 - \theta_2)\right]$$

and

$$\phi = \frac{M}{2\pi} \ln \frac{r_1}{r_2} \qquad \psi = \frac{M}{2\pi}(\theta_1 - \theta_2) \tag{70.29a,b}$$

$$w = \frac{Ma}{\pi(z-a)(z+a)}$$

By introducing coordinates x, y instead of r_1, r_2, θ_1, θ_2, Eqs. (70.29a,b) may be brought into the form

$$\left(x + a \operatorname{Coth} \frac{2\pi\phi}{M}\right)^2 + y^2 = \left(a \operatorname{Csch} \frac{2\pi\phi}{M}\right)^2$$
$$x^2 + \left(y - a \cot \frac{2\pi\psi}{M}\right)^2 = \left(a \csc \frac{2\pi\psi}{M}\right)^2$$

which shows that equipotential lines ϕ = const are circles of radius a Csch $2\pi\phi/M$

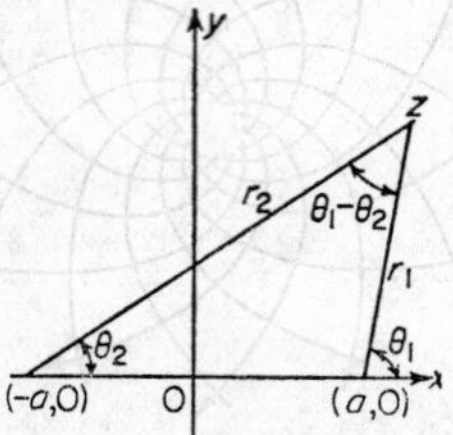

FIG. 70.11. Notation for source and sink.

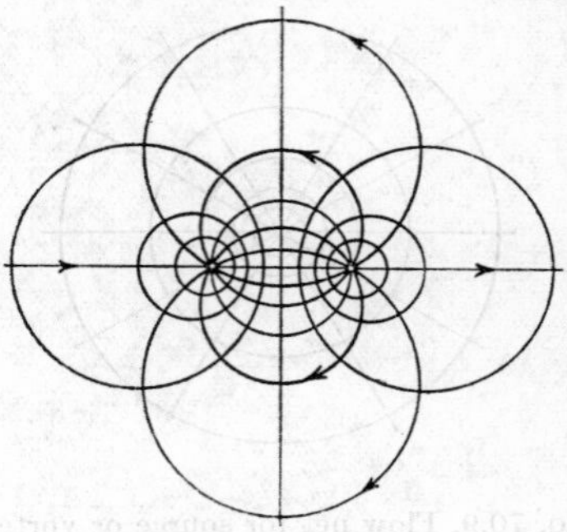

FIG. 70.12. Flow pattern for source and sink of equal strength or for two equal vortices of opposite sign.

with centers at ($-a$ Coth $2\pi\phi/M$, 0), while streamlines ψ = const are circles of radius a csc $2\pi\psi/M$ having centers at (0, a cot $2\pi\psi/M$). The flow pattern is given in Fig. 70.12.

Doublets

The doublet or dipole is defined as the limiting case of a source and sink of equal strength which approach each other such that the product of their strength by the distance between them remains constant. The complex potential for source and sink of equal strength [Eq. (70.28)] becomes

$$\Phi = \frac{M}{2\pi}\left[\ln (z - a) - \ln (z + a)\right] = \frac{M}{2\pi}\left[\ln\left(1 - \frac{a}{z}\right) - \ln\left(1 + \frac{a}{z}\right)\right]$$

By expanding the logarithmic terms in series and by rearranging,

$$\Phi = \frac{M}{2\pi}\left[-\frac{a}{z} - \frac{a^2}{2z^2} - \frac{a^3}{3z^3} - \cdots - \left(+\frac{a}{z} - \frac{a^2}{2z^2} + \frac{a^3}{3z^3} - \cdots\right)\right]$$
$$= -\frac{Ma}{\pi}\left[\frac{1}{z} + \frac{a^2}{3z^3} + \cdots\right]$$

By placing $2aM = m$, the strength of the doublet, and by taking the limit as a approaches zero,

$$\Phi = -\frac{m}{2\pi z} \tag{70.30}$$

This is the equation for a doublet with axis (from sink drawn toward source) in the $+x$ direction.

When the same limiting process is applied to a vortex Γ at $z = ia$ and a vortex $-\Gamma$ at $z = -ia$, the potential (70.30) is again obtained, if m is defined as $m = 2a\Gamma$. This indicates another physical interpretation of the doublet flow.

Flow Due to a Series of Equal and Equidistant Sources along the y Axis

The complex potential,

$$\Phi = \frac{M}{2\pi} \ln \operatorname{Sinh} \frac{\pi z}{a} \tag{70.31}$$

may be written as

$$\Phi = \frac{M}{2\pi} \ln \left[\frac{\pi z}{a} \prod_{n=1}^{\infty} \left(1 + \frac{z^2}{n^2 a^2} \right) \right]$$

By expanding,

$$\Phi = \frac{M}{2\pi} [\ln z + \ln (z - ia) + \ln (z + ia) + \ln (z - i2a) + \ln (z + i2a) + \cdots] + \text{const}$$

In this form the complex potential is seen to be that due to an infinite series of sources, all of strength M, located at the points (0,0), $(0, \pm a)$, $(0, \pm 2a)$, ... The constant part of the series can be dropped, as it contributes nothing to the flow pattern.

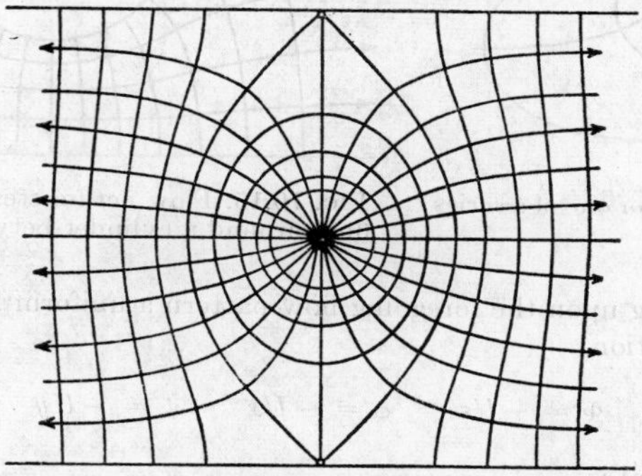

FIG. 70.13. Flow net for one of a series of equal and equidistant sources.

By returning to the first form of the complex potential, and by separating into real and pure imaginary parts,

$$\phi = \frac{M}{4\pi} \ln \frac{1}{2} \left(\operatorname{Cosh} \frac{2\pi x}{a} - \cos \frac{2\pi y}{a} \right)$$

$$\psi = \frac{M}{4\pi} \tan^{-1} \frac{\tan (\pi y/a)}{\operatorname{Tanh} (\pi x/a)}$$

From the symmetry of the sources, there is no flow across the lines:

$$y = 0, \pm \frac{a}{2}, \pm a, \pm \frac{3a}{2}, \cdots$$

The formulas may apply to a source midway between two fixed parallel boundaries. The flow pattern is shown in Fig. 70.13.

Infinite Number of Doublets along the y Axis

Differentiation of the complex potential (70.31) produces the complex potential for a series of doublets with axes in the $+x$ direction (if C' is negative)

$$\Phi = \frac{d}{dz} \left(\frac{M}{2\pi} \ln \operatorname{Sinh} \frac{\pi z}{a} \right) = C' \operatorname{Coth} \frac{\pi z}{a} \tag{70.32}$$

which can be proved by differentiation of the infinite series expression. By separating into real and pure imaginary parts,

$$\phi = C' \frac{\text{Sinh } (2\pi x/a)}{\text{Cosh } (2\pi x/a) - \cos (2\pi y/a)}$$

$$\psi = -C' \frac{\sin (2\pi y/a)}{\text{Cosh } (2\pi x/a) - \cos (2\pi y/a)}$$

from which the streamlines and equipotential lines of Fig. 70.14 are plotted.

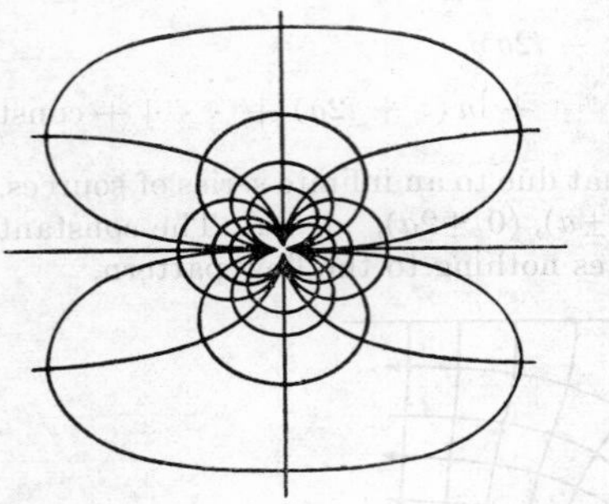

FIG. 70.14. Flow net for one of a series of doublets.

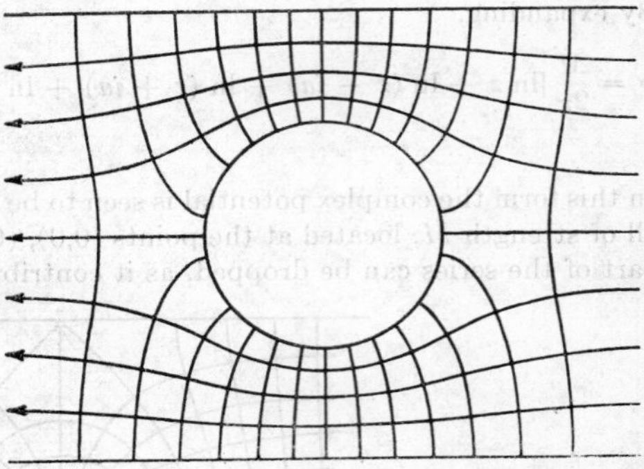

FIG. 70.15. Flow net for steady flow of an infinite fluid around a cylinder between parallel walls.

After superposing upon the foregoing flow pattern a uniform flow of velocity U in the negative x direction,

$$\Phi = -Uz \qquad \phi = -Ux \qquad \psi = -Uy$$

the complex potential is

$$\Phi = -Uz + C' \text{ Coth } \frac{\pi z}{a} \tag{70.33}$$

The velocity potential and stream function are

$$\phi = -Ux + C' \frac{\text{Sinh } (2\pi x/a)}{\text{Cosh } (2\pi x/a) - \cos (2\pi y/a)}$$

$$\psi = -Uy - C' \frac{\sin (2\pi y/a)}{\text{Cosh } (2\pi x/a) - \cos (2\pi y/a)}$$

The streamline $\psi = 0$ may be shown to be the boundary of a cylinder which encloses the doublet (Fig. 70.15). The smaller that C' is taken, the more nearly the cylinder becomes a circle.

70.15. INVERSE TRANSFORMATIONS. ELLIPTIC-HYPERBOLIC NET

Several flow patterns may be derived from the function

$$z = c \text{ Cosh } \Phi$$

If c is real, then separating real and imaginary parts yields

$$x = c \text{ Cosh } \phi \cos \psi \qquad y = c \text{ Sinh } \phi \sin \psi$$

from which, by eliminating first ψ, then ϕ,

$$\frac{x^2}{c^2 \operatorname{Cosh}^2 \phi} + \frac{y^2}{c^2 \operatorname{Sinh}^2 \phi} = 1 \qquad \frac{x^2}{c^2 \cos^2 \psi} - \frac{y^2}{c^2 \sin^2 \psi} = 1$$

This shows that the equipotential lines in the z plane are confocal ellipses, with foci at $(c,0)$ and $(-c,0)$, while the streamlines are confocal hyperbolas with the same foci. The streamlines $\psi = 0, \pi$ represent the parts $x^2 > c^2$ of the x axis (Fig. 70.16). When they are replaced by a solid boundary, the flow pattern represents flow through an aperture of width $2c$ in a thin plate.

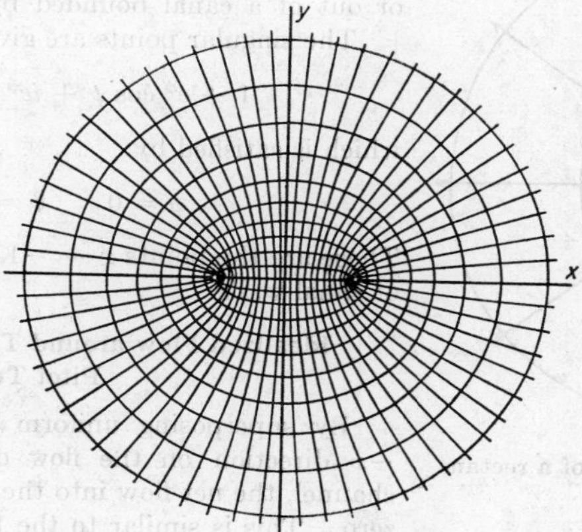

FIG. 70.16. Circulation about elliptic cylinder. Flow through a rectangular slot.

If c is purely imaginary, ϕ and ψ change places, and the ellipses become the streamlines, portraying circulation around an elliptic cylinder or, in the limit, circulation around the rectangular lamina extending from $x = -c$ to $x = +c$ on the x axis.

Flow Out of a Rectangular Channel ($z = \Phi + e^{\Phi}$)

The advantages of the special relations developed for inverse transformations are apparent from this function, which cannot be solved explicitly for Φ. By separating into its real and pure imaginary parts,

$$x + iy = \phi + i\psi + e^{\phi} \cos \psi + ie^{\phi} \sin \psi$$

from which

$$x = \phi + e^{\phi} \cos \psi \qquad y = \psi + e^{\phi} \sin \psi \qquad \frac{dz}{d\Phi} = 1 + e^{\Phi}$$

The streamline $\psi = 0$ is given by

$$x = \phi + e^{\phi} \qquad y = 0$$

As ϕ varies from $-\infty$ to $+\infty$, x also varies from $-\infty$ to $+\infty$; hence, the x axis is the streamline $\psi = 0$, and flow is in the $+x$ direction. The streamline $\psi = \pi$ is given by

$$x = \phi - e^{\phi} \qquad y = \pi$$

As ϕ varies from $-\infty$ to zero, x varies from $-\infty$ to -1. As ϕ varies from $+\infty$ to zero, x again varies from $-\infty$ to -1. Therefore, the line

$$y = \pi \qquad -\infty < x \leq -1$$

is the streamline $\psi = \pi$. This line may be considered as bent back upon itself. Similarly, the streamline $\psi = -\pi$ is

$$y = -\pi \qquad -\infty < x \leq -1$$

The streamlines lying between $\psi = -\pi$ and $\psi = \pi$ are all contained within the two straight lines $y = \pm\pi$ $(-\infty < x \leq -1)$ for negative values of ϕ. For positive values of ϕ, the intermediate streamlines fan out and cover the complete z plane. The flow net is shown in Fig. 70.17. It represents the flow into or out of a canal bounded by two parallel walls.

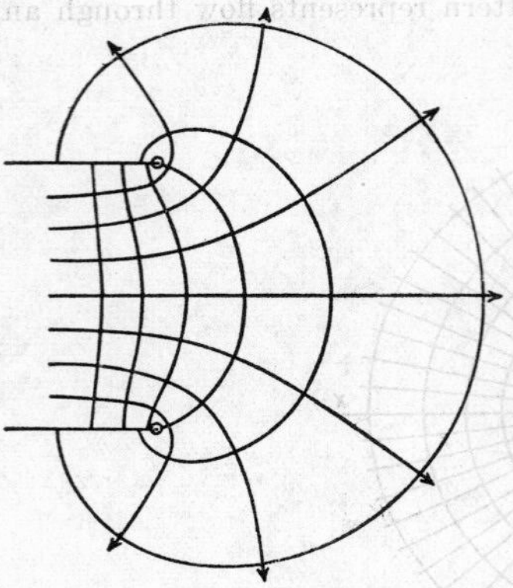

FIG. 70.17. Flow out of a rectangular channel.

The singular points are given by

$$1 + e^{\phi} \cos \psi + ie^{\phi} \sin \psi = 0$$

which is satisfied by

$$\phi = 0 \qquad \psi = \pm\pi$$

These are the points $x = -1$, $y = \pm\pi$, or the end of canal walls.

Idealized Flow around Two-dimensional Pitot Tube

By superposing uniform unit velocity in the $-x$ direction on the flow out of a rectangular channel, the net flow into the channel is reduced to zero. This is similar to the flow around an idealized pitot tube except that, being two-dimensional, the parallel walls take the place of a cylindrical tube.

Unit velocity in the $-x$ direction is given by $\Phi'' = -z$, $\phi'' = -x$, $\psi'' = -y$, in which the primes are used to distinguish the function from the final complex potential that is sought. The complex potential for flow out of the channel is given by

$$z = \Phi' + e^{\Phi'} \qquad \Phi' = \phi' + i\psi'$$

or

$$x = \phi' + e^{\phi'} \cos \psi' \qquad y = \psi' + e^{\phi'} \sin \psi'$$

The resultant complex potential is

$$\Phi = \phi + i\psi = \Phi' + \Phi'' = \phi' + i\psi' + \phi'' + i\psi''$$
$$= \phi' - x + i(\psi' - y)$$

or

$$\phi' = \phi + x \qquad \psi' = \psi + y \qquad \Phi' = \Phi + z$$

By eliminating the primed quantities,

$$\Phi + e^{\Phi + z} = 0$$

$$\phi + e^{\phi + x} \cos(\psi + y) = 0 \qquad \psi + e^{\phi + x} \sin(\psi + y) = 0$$

Now, by eliminating first y and then x,

$$x = -\phi + \ln \sqrt{\phi^2 + \psi^2} \qquad y = -\psi + \tan^{-1} \frac{\psi}{\phi}$$

By combining these expressions,

$$z = -\Phi + \ln \Phi$$

is the inverse function. Streamlines for this case are shown in Fig. 70.18.

Boundary Conditions for the Translation of Any Cylinder through an Infinite Fluid

The boundary condition for translation of a cylinder through an infinite fluid is simply expressed in terms of the stream function. Let the cylinder have the velocity U in the x direction, as in Fig. 70.19. The boundary condition states that the velocity of fluid normal to the surface at the boundary equals the velocity of the surface normal to itself. Taking s positive in the direction shown and the normal positive when directed into the fluid, then (s,n) forms a right-handed system similar to (x,y). The

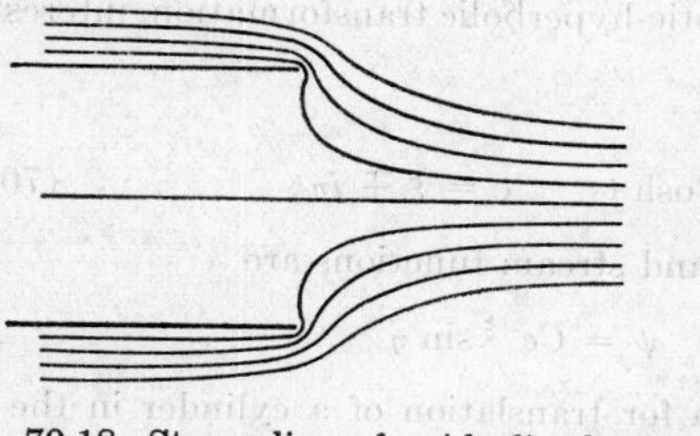

FIG. 70.18. Streamlines for idealized two-dimensional pitot tube.

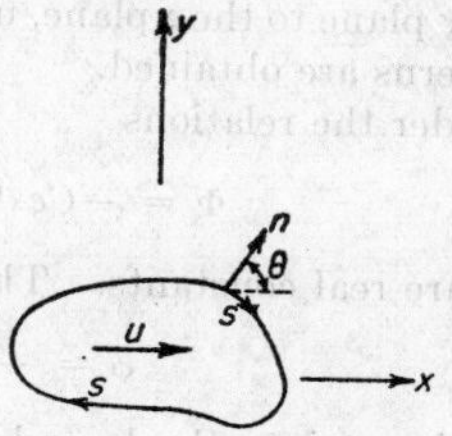

FIG. 70.19. Translation of cylinder.

fluid velocity in the n direction is $-\partial\psi/\partial s$. The velocity of the boundary normal to itself is $U \cos \theta$. However, $\cos \theta$ may be expressed as $-dy/ds$. Hence, the boundary condition in differential form is

$$-\frac{\partial \psi}{\partial s} = -U \frac{dy}{ds}$$

After integrating around the periphery

$$\psi = Uy + \text{const}$$

Similarly,

$$\psi = -Vx + \text{const}$$

is the boundary condition for translation of a cylinder with velocity V in the $+y$ direction. The two conditions may be combined for translation of a cylinder in any direction:

$$\psi = -Vx + Uy + \text{const} \tag{70.34}$$

The function

$$\psi = Uy$$

satisfies the boundary condition identically for any shape of body translating in the x direction. It is the case of uniform fluid motion with constant velocity $u = U$. It must then be the case of fluid contained within a cylinder which is in translation. The complex potential is $\Phi = Uz$.

By expressing the boundary condition for translation in the x direction in polar coordinates,

$$\psi = Ur \sin \theta + \text{const}$$

By selecting the function

$$\psi = \frac{AU}{r} \sin \theta$$

and by substituting into the boundary condition, it is satisfied if $r = \sqrt{A} = a$, which is a circular cylinder. Hence,

$$\psi = \frac{Ua^2}{r} \sin \theta$$

is the stream function for translation of a circular cylinder through an infinite fluid otherwise at rest. The complex potential is

$$\Phi = \frac{Ua^2}{z}$$

The flow net is given by Fig. 70.7 if a circle is drawn in with center at the origin.

Translation of an Elliptic Cylinder

By making use of two transformations, first from Φ to an auxiliary ζ plane, then from the ζ plane to the z plane, using the elliptic-hyperbolic transformation, interesting flow patterns are obtained.

Consider the relations

$$\Phi = -Ce^{-\zeta} \qquad z = c \operatorname{Cosh} \zeta \qquad \zeta = \xi + i\eta \tag{70.35}$$

C and c are real constants. The potential and stream functions are

$$\phi = -Ce^{-\xi} \cos \eta \qquad \psi = Ce^{-\xi} \sin \eta$$

Substituting ψ into the boundary condition for translation of a cylinder in the $+x$ direction,

$$\begin{aligned} Ce^{-\xi} \sin \eta &= Uy + \text{const} \\ &= Uc \operatorname{Sinh} \xi \sin \eta + \text{const} \end{aligned}$$

By letting the constant be zero, the boundary condition is satisfied by one value of ξ, say ξ_0, determined by

$$Ce^{-\xi_0} = Uc \operatorname{Sinh} \xi_0$$

$\xi = \xi_0$ is the equation of an elliptic cylinder, with semimajor and semiminor axes a and b, respectively

$$a = c \operatorname{Cosh} \xi_0 \qquad b = c \operatorname{Sinh} \xi_0 \qquad c = \sqrt{a^2 - b^2}$$

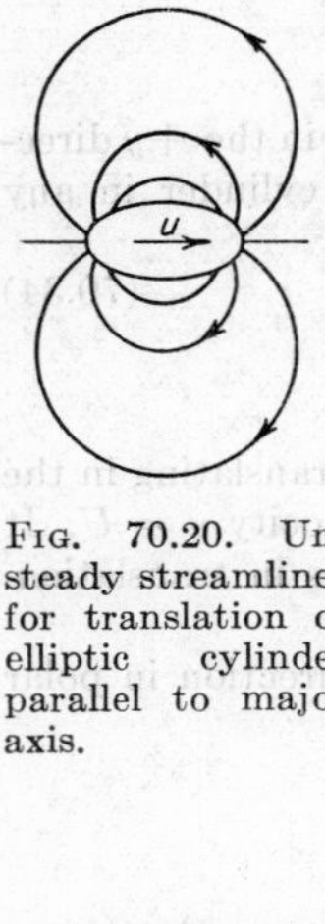

Fig. 70.20. Unsteady streamlines for translation of elliptic cylinder parallel to major axis.

After solving for C,

$$C = e^{\xi_0} \qquad Uc \operatorname{Sinh} \xi_0 = \sqrt{\frac{a+b}{a-b}}\, Ub$$

since

$$e^{\xi_0} = \sqrt{\frac{a+b}{a-b}}$$

By substituting back,

$$\psi = Ub \sqrt{\frac{a+b}{a-b}}\, e^{-\xi} \sin \eta$$

is the stream function for an elliptic cylinder of semiaxes a and b, moving parallel to the major axis with velocity U in an infinite fluid, otherwise at rest. A few streamlines are shown in Fig. 70.20.

Steady Flow around an Elliptic Cylinder

The preceding case of translation of an elliptic cylinder can be reduced to a steady-flow case by addition of a uniform velocity U in the $-x$ direction; thus,

$$\Phi = -Ub \sqrt{\frac{a+b}{a-b}}\, e^{-\zeta} - Uc \operatorname{Cosh} \zeta$$

Steady flow of fluid around the same elliptic cylinder, approaching broadside, or parallel to the minor axis, is given by

$$\Phi = -iVa \sqrt{\frac{a+b}{a-b}}\, e^{-\zeta} + iVc \operatorname{Cosh} \zeta$$

The two cases may be added to produce flow around the elliptic cylinder for an arbitrary approach velocity given by the components U, V

$$\Phi = -(Ub + iVa) \sqrt{\frac{a+b}{a-b}}\, e^{-\zeta} - (U - iV)c \operatorname{Cosh} \zeta \tag{70.36}$$

The streamline pattern for $U = V$ is shown in Fig. 70.21.

Boundary Condition for Rotation of Any Cylinder in an Infinite Fluid

The boundary condition for rotation of a rigid cylinder about an axis through the origin is obtained from the original definition: The fluid velocity component at the

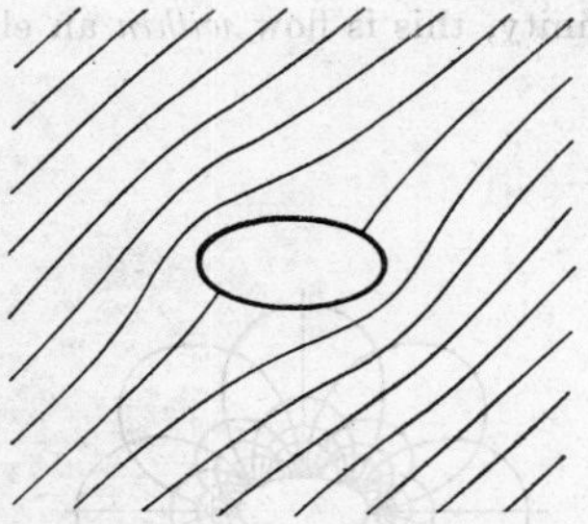

FIG. 70.21. Steady flow around an elliptic cylinder ($U = V$).

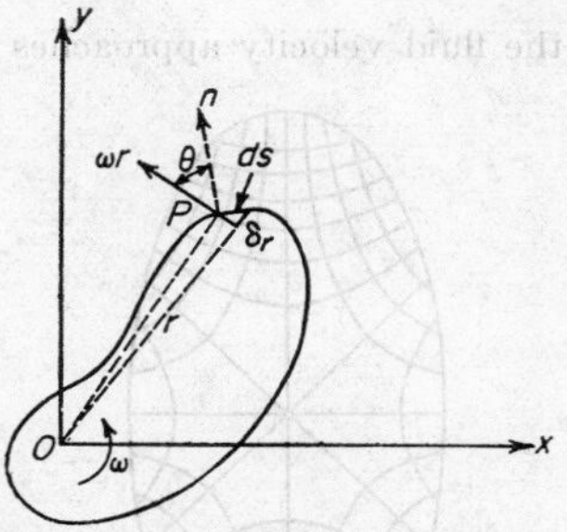

FIG. 70.22. Rotation of cylinder about origin.

boundary normal to the boundary must equal the velocity of the boundary normal to itself. In Fig. 70.22, let ds denote an element of the surface of the cylinder. The positive direction of s is as shown, and the positive normal to the surface is drawn into the fluid. The fluid velocity normal to the surface is $-\partial\psi/\partial s$. The velocity of a point P on the surface is ωr, and the velocity normal to the surface is $\omega r \cos\theta$. Since $\cos\theta = dr/ds$, the differential form of the boundary condition is

$$-\frac{\partial\psi}{\partial s} = \omega r \frac{dr}{ds}$$

which applies equally to external or internal boundaries.

After integrating,

$$\psi = -\tfrac{1}{2}\omega r^2 + \text{const} \tag{70.37}$$

in which the constant is arbitrary.

Fluid Contained within a Rotating Elliptic Cylinder

The complex potential

$$\Phi = -iAz^2$$

has the potential and stream functions

$$\phi = 2Axy \qquad \psi = -A(x^2 - y^2)$$

By substituting ψ into the boundary condition,

$$A(x^2 - y^2) = \tfrac{1}{2}\omega(x^2 + y^2) - C$$

After rearranging,

$$\frac{x^2}{C/(\frac{1}{2}\omega - A)} + \frac{y^2}{C/(\frac{1}{2}\omega + A)} = 1$$

which is an ellipse when $A < \frac{1}{2}\omega$. Denoting the semimajor and semiminor axes by a and b, respectively,

$$a^2 = \frac{C}{\frac{1}{2}\omega - A} \qquad b^2 = \frac{C}{\frac{1}{2}\omega + A}$$

After solving for A by eliminating C,

$$A = \frac{\omega}{2}\frac{a^2 - b^2}{a^2 + b^2}$$

The stream function is

$$\psi = \frac{\omega}{2}\frac{a^2 - b^2}{a^2 + b^2}(x^2 - y^2)$$

Since the fluid velocity approaches infinity at infinity, this is flow *within* an elliptic cylinder. The flow net is shown in Fig. 70.23.

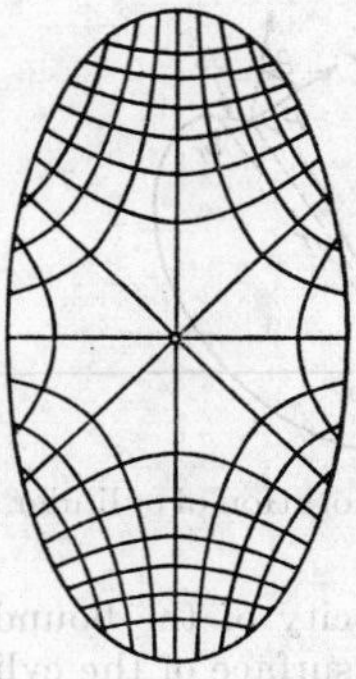

FIG. 70.23. Flow net for fluid within a rotating elliptic cylinder.

FIG. 70.24. Flow net for rotation of elliptic cylinder about its axis in an infinite fluid.

Rotation of an Elliptic Cylinder in an Infinite Fluid

The fluid motion due to the rotation of an elliptic cylinder about its axis in an infinite fluid is given by

$$\Phi = -iCe^{-2\zeta} \qquad z = c \operatorname{Cosh} \zeta \tag{70.38}$$

The potential and stream functions are

$$\phi = -Ce^{-2\xi} \sin 2\eta \qquad \psi = -Ce^{-2\xi} \cos 2\eta$$

By expressing the boundary condition in terms of ξ and η

$$\psi = -\frac{1}{4}c^2\omega (\operatorname{Cosh} 2\xi + \cos 2\eta) + D$$

in which D is the arbitrary constant. After substituting in the stream function,

$$Ce^{-2\xi} \cos 2\eta = \frac{1}{4}c^2\omega (\operatorname{Cosh} 2\xi + \cos 2\eta) - D$$

This equation is satisfied by $\xi = \xi_0$, provided

$$Ce^{-2\xi_0} = \frac{1}{4}c^2\omega \qquad \frac{1}{4}c^2\omega \operatorname{Cosh} 2\xi_0 - D = 0$$

By letting a and b be the semimajor and semiminor axes of the elliptic cylinder $\xi = \xi_0$,

$$a = c \operatorname{Cosh} \xi_0 \qquad b = c \operatorname{Sinh} \xi_0$$

and, since

$$e^{\xi_0} = \operatorname{Sinh} \xi_0 + \operatorname{Cosh} \xi_0 = \frac{a+b}{c}$$

then

$$c = \tfrac{1}{4}\omega(a+b)^2$$

The flow net is given in Fig. 70.24. By setting $b = 0$, the case of rotation of a rectangular lamina in an infinite fluid is obtained.

70.16. BLASIUS' THEOREM

The Blasius theorem provides formulas for determination of resultant fluid forces and moments on cylinders. If X is the force component in the $+x$ direction, Y is the force component in the $+y$ direction, and M is the moment about the origin, the Blasius theorem takes the form

$$X - iY = \frac{i}{2}\rho \oint \left(\frac{d\Phi}{dz}\right)^2 dz$$

and

$$M + iN = -\frac{\rho}{2} \oint z \left(\frac{d\Phi}{dz}\right)^2 dz \qquad (70.39a,b)$$

in which N is the pure imaginary part of the second equation and has no physical significance. Φ is the complex potential of the irrotational flow, and the integrals are carried out over the periphery of the cylinder.

By means of the Cauchy integral theorem, the integration may be carried out around indefinitely large circular cylinders, in place of the actual cylinder, provided no singular points occur between the cylinder and the large circle. By expressing $(d\Phi/dz)^2$ as a series,

$$\left(\frac{d\Phi}{dz}\right)^2 = A_0 + \frac{A_1}{z} + \frac{A_2}{z^2} + \cdots \qquad (70.40)$$

in which the coefficients A_0, A_1, A_2, . . . are usually complex, the integrals are evaluated; thus,

$$X - iY = -\pi\rho A_1 \qquad M + iN = -i\pi\rho A_2 \qquad (70.41a,b)$$

70.17. FLOW AROUND CIRCULAR CYLINDERS

The complex potential

$$\Phi = U\left(z + \frac{a^2}{z}\right) = U\left(re^{i\theta} + \frac{a^2}{r}e^{-i\theta}\right) \qquad (70.42)$$

in the superposition of the potentials of a uniform flow U in x direction and of a doublet flow [Eq. (70.30)]. Separation of real and imaginary parts yields

$$\phi = U\left(r + \frac{a^2}{r}\right)\cos\theta \qquad \psi = U\left(r - \frac{a^2}{r}\right)\sin\theta$$

The streamline $\psi = 0$ is given by $\theta = 0, \pi$, and by $r = a$, that is, by the x axis and by the cylinder $r = a$.

The complex velocity is

$$\frac{d\Phi}{dz} = U - \frac{a^2U}{z^2}$$

showing the uniform velocity $u = U$ at great distances from the cylinder. To find A_1, A_2,

$$\left(\frac{d\Phi}{dz}\right)^2 = U^2 - \frac{2a^2U^2}{z^2} + \frac{a^4U^2}{z^4}$$

from which $A_1 = 0$, $A_2 = -2a^2U^2$. Hence, the resultant force and moment on the cylinder are zero. Stagnation points occur at $x = \pm a$, $y = 0$. The flow net is shown in Fig. 70.25.

The complex potential

$$\Phi = \frac{i\Gamma}{2\pi} \ln z = \frac{i\Gamma}{2\pi} \ln (re^{i\theta}) \tag{70.43}$$

is for circulation Γ about the origin in the clockwise direction (p. 70–11). Separating into real and pure imaginary parts,

$$\phi = -\frac{\Gamma\theta}{2\pi} \qquad \psi = \frac{\Gamma}{2\pi} \ln r$$

from which the streamlines are seen to be circles concentric with the origin.

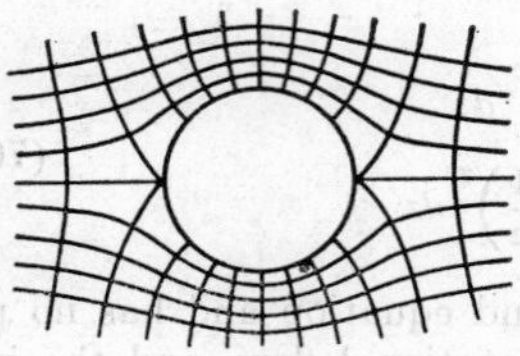

FIG. 70.25. Flow pattern for uniform flow around a circular cylinder without circulation.

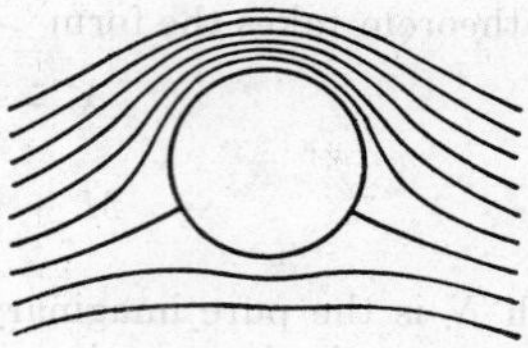

FIG. 70.26. Streamlines for uniform flow around a circular cylinder with circulation.

By superposing this flow upon the steady flow around a circular cylinder,

$$\Phi = U\left(z + \frac{a^2}{z}\right) + \frac{i\Gamma}{2\pi} \ln z \tag{70.44}$$

$$\frac{d\Phi}{dz} = U\left(1 - \frac{a^2}{z^2}\right) + \frac{i\Gamma}{2z}$$

The potential and stream functions are

$$\phi = U\left(r + \frac{a^2}{r}\right) \cos\theta - \frac{\Gamma\theta}{2\pi} \qquad \psi = U\left(r - \frac{a^2}{r}\right) \sin\theta + \frac{\Gamma}{2\pi} \ln r$$

The streamline $\psi = (\Gamma/2\pi) \ln a$ is the circular cylinder $r = a$, which shows that this is still the case of uniform flow around a circular cylinder.

To evaluate the fluid force and moment about the origin, the coefficients A_1 and A_2 are determined from $(d\Phi/dz)^2$,

$$\left(\frac{d\Phi}{dz}\right)^2 = U^2 + i\frac{U\Gamma}{\pi z} - \left(2U^2a^2 + \frac{\Gamma^2}{4\pi^2}\right)\frac{1}{z^2} - i\frac{Ua^2\Gamma}{z^3} + \frac{U^2a^4}{z^4}$$

from which

$$A_1 = i\frac{U\Gamma}{\pi} \qquad A_2 = -\left(2U^2a^2 + \frac{\Gamma^2}{4\pi^2}\right)$$

Hence, the resultant force is

$$X = 0 \qquad Y = \rho U\Gamma$$

and the resultant moment is zero because A_2 is real. There is no drag force on the cylinder (in the direction of the uniform flow), but there is a *lift* force $\rho U\Gamma$. This thrust is referred to as *Magnus effect*. Streamlines for this case are shown in Fig. 70.26.

70.18. PRINCIPLES OF FREE-STREAMLINE FLOW

The boundary conditions imposed on the flow cases in the preceding examples do not permit separation of flow from a boundary. The assumption of *free streamlines* permits separation of the flow to take place at those sudden changes in direction of boundaries which cause infinite velocities. At separation points in steady flow of a fluid around a body, the streamlines leave the body. These dividing streamlines are called free streamlines in two-dimensional flow, and the fluid in contact with the body downstream from the separation points and separated from the main flow by the free streamlines is known as the *wake*. The fluid in the wake is assumed to be at rest in steady-flow problems.

The Blasius theorem does not apply to cases where there is separation. The method of free streamlines provides a drag in irrotational flow of a frictionless fluid around bodies. Because of this and the avoidance of points of infinite velocity, this method permits the solution of many problems that conform closely to similar problems with actual fluids. The assumption that the fluid is at rest in the wake is considerably in error for actual fluids, and frequently leads to a theoretical drag that is much less than in an actual case. When the wake contains another fluid of much less density, the theory should give results comparing favorably with experiment. An example would be the discharge of water out of a slot into air, or the flow around a cavitation bubble (see p. 87–7).

In the following treatment, the effects of gravity are neglected. The pressure in the wake is, therefore, constant, since it is at rest, and the pressure along the free streamline must be constant. According to the Bernoulli equation, the velocity of the free streamline must also be constant. A streamline in contact with the boundary upstream from the separation point is referred to as a *bounding streamline*. Since the resultant drag on a body is the same, whether viewed by an observer as steady or unsteady motion, examples are considered for steady flow to take advantage of the simpler form of the Bernoulli equation.

When the bounding streamlines are straight, the shape of the free streamlines *in two-dimensional flow* can be found by use of several conformal transformations, including the *hodograph* plane and the Schwarz-Christoffel theorem.

Transformations Used in Free-streamline Problems

The transformations required in free-streamline problems are conformal, and provide a means of mapping the uniform flow of the Φ plane into the free-streamline pattern of the z plane. There are several ways to define a hodograph plane. It may be the Argand plane for the complex velocity $w = u - iv$, for its conjugate $\bar{w}$ or for its reciprocal $\zeta = 1/w$. The velocity of the free streamlines is assumed to be unity.

The general form of the free streamlines may be shown on the z plane, along with the bounding streamlines. One hodograph transformation, from the z plane to another plane, say the ζ plane, is

$$\zeta = \frac{dz}{d\Phi}$$

When ζ is known or can be expressed as a function of z and Φ, the problem is solved. Expressing $dz/d\Phi$ in terms of the complex velocity,

$$\zeta = \frac{1}{d\Phi/dz} = \frac{1}{u - iv} = \frac{1}{q}\left(\frac{u}{q} + i\frac{v}{q}\right) = \left|\frac{1}{q}\right| e^{i\theta} \qquad (70.45)$$

in which u, v are component velocities in the x, y directions and q is the speed. The velocity vector has the direction θ at a point. $\left|\frac{1}{q}\right|$ is the modulus of ζ. Since

$$\left(\frac{u}{q}\right)^2 + \left(\frac{v}{q}\right)^2 = 1$$

it is evident that ζ is a complex number having a modulus that is the reciprocal of the speed and an argument that is the angle the velocity vector makes with the $+x$ axis.

The next transformation is usually to the ζ' plane, with

$$\zeta' = \ln \zeta = \ln \left|\frac{1}{q}\right| + i\theta$$

This transformation causes the bounding streamlines and free streamlines to form a straight-sided polygon, usually with some vertices at infinity.

The Schwarz-Christoffel theorem provides a method for mapping the inside of any straight-sided simply connected polygon into the upper half of the plane, and, by being applied one or more times, finally converts the bounding and free streamlines into horizontal lines that satisfy conditions for the Φ plane.

Borda's Mouthpiece in Two-dimensional Flow

Borda's mouthpiece, in two dimensions, is a reentrant slot in a large container, as shown in the z plane of Fig. 70.27. The bounding streamlines AB_∞, $A'B'_\infty$ are assumed

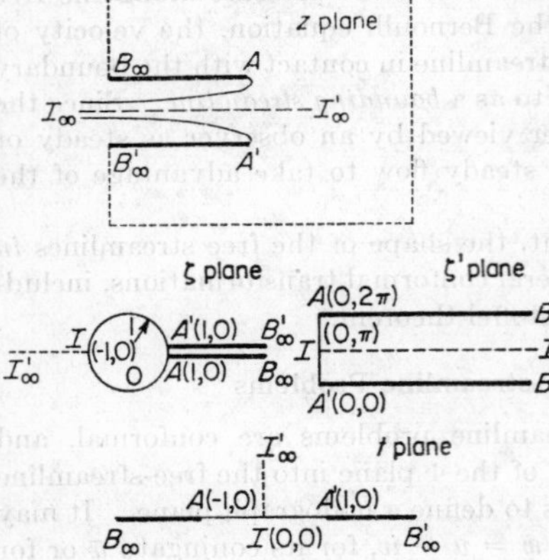

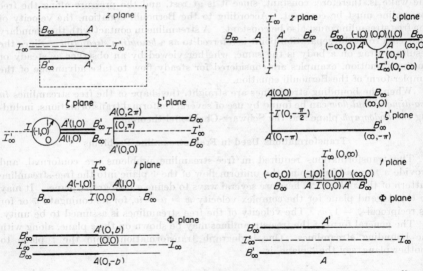

FIG. 70.27. Mapping planes for Borda's mouthpiece in two dimensions.

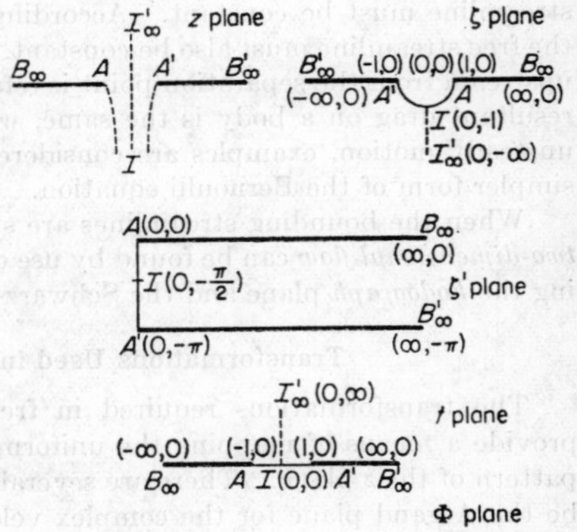

FIG. 70.28. Mapping planes for a two-dimensional orifice.

to be so long that the velocity at B_∞ and B'_∞ is zero. The point I'_∞ is in the tank along the line of symmetry, and is sufficiently removed from the entrance so that the velocity is also zero; I_∞ is a point on the jet sufficiently far downstream so that no more contraction takes place; $I'_\infty I_\infty$ is a streamline.

The first two transformations, to the ζ and ζ' planes have been discussed. The third transformation is

$$t = \text{Cosh}\,\frac{\zeta'}{2}$$

The final transformation is

$$\Phi = -\frac{2b}{\pi} \ln t + ib$$

as shown in Fig. 70.27. $2b$ is the discharge per unit length of slot.

The free streamlines are plotted by expressing x and y as parameters of θ,

$$x = \frac{2b}{\pi}\left(\sin^2\frac{\theta}{2} - \ln \sec\frac{\theta}{2}\right) \qquad 0 \le \theta \le \pi$$

$$y = \frac{b}{\pi}(\theta - \sin\theta)$$

in which the origin of the (x,y) system is at A'. Since the asymptotic value of y is b, the distance between walls is $4b$, and the coefficient of contraction is 0.50, in agreement with Borda's theory.

Flow Out of a Two-dimensional Orifice

The transformations are similar to the preceding example, except for constants. They are

$$\zeta = \frac{dz}{d\Phi} = \left|\frac{1}{q}\right| e^{i\theta} \qquad \zeta' = \ln \zeta = \ln\left|\frac{1}{q}\right| + i\theta$$

$$\zeta' = \text{Cosh}^{-1}\, t - i\pi \qquad \Phi = -\frac{2b}{\pi}\ln t + ib$$

in which $2b$ is again the final thickness of jet. The planes are shown in Fig. 70.28. By taking the origin at A, in the z plane, the equations of the free streamline are

$$x = \frac{4b}{\pi}\sin^2\frac{\theta}{2} \qquad -\frac{\pi}{2} \le \theta \le 0$$

$$y = \frac{2b}{\pi}\left[\ln\tan\left(\frac{\pi}{4} + \frac{\theta}{2}\right) - \sin\theta\right]$$

The asymptotic value of x is $2b/\pi$; hence, the total width of slot is $(4b/\pi + 2b)$. The coefficient of contraction is

$$\frac{2b}{4b/\pi + 2b} = \frac{\pi}{\pi + 2} = 0.611$$

Infinite Stream Impinging upon a Fixed Plane Lamina

An infinite stream is divided by a rectangular fixed plane lamina of breadth l at right angles to the stream. The flow separates into two portions, divided internally by two free streamlines, as shown in Fig. 70.29. The dividing streamline is taken as

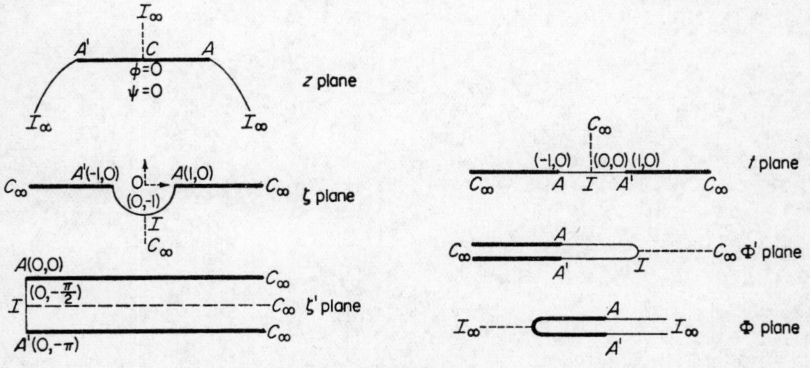

FIG. 70.29. Mapping planes for an infinite stream impinging on a lamina.

$\psi = 0$, and the equipotential line through the stagnation point C as $\phi = 0$. The transformations are:

$$\zeta = \frac{dz}{d\Phi} = \left|\frac{1}{q}\right| e^{i\theta} \qquad \zeta' = \ln \zeta = \ln \left|\frac{1}{q}\right| + i\theta$$

$$\zeta' = \text{Cosh}^{-1}\, t - i\pi \qquad \Phi = -\frac{\pi + 4}{l} t^2$$

and

$$\Phi' = -\frac{1}{\Phi}$$

The equations for the free streamline are

$$x = \frac{2l}{\pi + 4}\left(\sec \theta + \frac{\pi}{4}\right)$$

$$y = \frac{l}{\pi + 4}\left[\sec \theta \tan \theta - \ln \tan \left(\frac{\pi}{4} + \frac{\theta}{2}\right)\right] \qquad -\frac{\pi}{4} \le \theta \le 0$$

with the origin taken as the center of the lamina. The resultant force on the lamina per unit length is

$$\frac{\pi}{\pi + 4} \rho q_0{}^2 l$$

in which q_0 is the undisturbed stream velocity.

70.19. REFERENCES

[1] H. Lamb: "Hydrodynamics," 6th ed., Dover, New York, 1945.
[2] L. M. Milne-Thomson: "Theoretical Hydrodynamics," 3d ed., Macmillan, London, 1955.
[3] L. Prandtl, O. G. Tietjens: "Fundamentals of Hydro- and Aeromechanics," McGraw-Hill, New York, 1934.
[4] A. S. Ramsey: "A Treatise on Hydromechanics," G. Bell, London, 1942.
[5] H. Rouse: "Fluid Mechanics for Hydraulic Engineers," McGraw-Hill, New York, 1938.
[6] V. L. Streeter: "Fluid Dynamics," McGraw-Hill, New York, 1948.

CHAPTER 71

THREE-DIMENSIONAL IDEAL FLUID FLOW

BY

I. FLÜGGE-LOTZ, Dr.-Ing., Los Altos, Calif.

71.1. BASIC EQUATIONS

The concept of ideal flow is discussed in Chap. 70. The notation used here is based on that employed in the preceding chapter. Definitions are given here only when the necessary extension to three dimensions is not self-explanatory.

Throughout this section, the coordinate systems shown in Fig. 71.1 are used. The notations for the cylindrical and spherical coordinates are different from those used elsewhere in this book.

General Equations

The flow is governed generally by the Euler equations,

$$\begin{aligned}\frac{\partial u}{\partial t} + u\frac{\partial u}{\partial x} + v\frac{\partial u}{\partial y} + w\frac{\partial u}{\partial z} &= X - \frac{1}{\rho}\frac{\partial p}{\partial x}\\ \frac{\partial v}{\partial t} + u\frac{\partial v}{\partial x} + v\frac{\partial v}{\partial y} + w\frac{\partial v}{\partial z} &= Y - \frac{1}{\rho}\frac{\partial p}{\partial y}\\ \frac{\partial w}{\partial t} + u\frac{\partial w}{\partial x} + v\frac{\partial w}{\partial y} + w\frac{\partial w}{\partial z} &= Z - \frac{1}{\rho}\frac{\partial p}{\partial z}\end{aligned} \tag{71.1a–c}$$

and the continuity equation,

$$\frac{\partial u}{\partial x} + \frac{\partial v}{\partial y} + \frac{\partial w}{\partial z} = 0 \tag{71.2}$$

A particular flow is given by adding to these equations the necessary boundary conditions (e.g., statements about the velocity components) on all surfaces limiting the fluid-filled space under consideration. These may be, for instance, the surface of an obstacle and infinity, represented by a sphere of infinite radius; or they may be the surface of an obstacle and the walls of a tunnel in which this obstacle is placed. In the latter case, the velocity at the open-end sections of the tunnel must also be given.

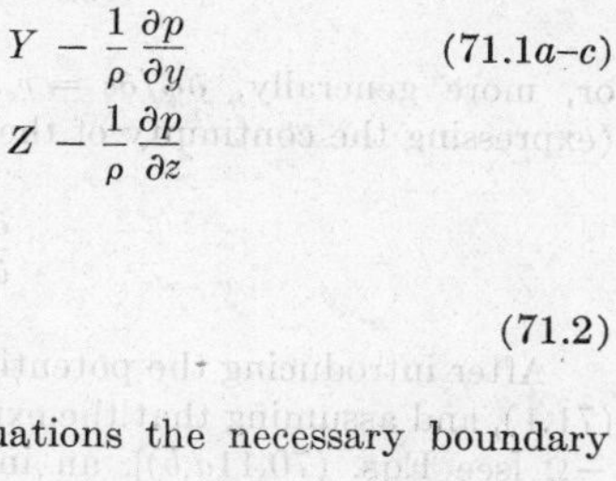

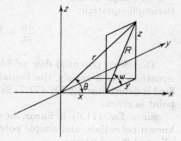

FIG. 71.1. Coordinates.

In some cases, an obstacle is placed in a limited space of special form; e.g., an obstacle flies over ground which is assumed to be a plane (say, the x,y plane). Instead of considering the flow in the upper half space, we may use an image of the obstacle in the lower half space for satisfying the boundary conditions on the ground. In this way, the problem is reduced to one in the unlimited space. This *method of images* can be used for more complicated cases, for instance, for the flow along an obstacle in a tunnel with rectangular cross section, etc. It can also be adapted to satisfying the condition of constant pressure on a free surface.

Irrotational Flow

For irrotational flow, the vorticity components

$$\omega_x = \frac{\partial w}{\partial y} - \frac{\partial v}{\partial z} \qquad \omega_y = \frac{\partial u}{\partial z} - \frac{\partial w}{\partial x} \qquad \omega_z = \frac{\partial v}{\partial x} - \frac{\partial u}{\partial y} \tag{71.3}$$

are zero at all regular points. In such flows, the circulation $\oint \mathbf{q} \cdot d\mathbf{s}$ along a closed path is zero because of Stokes' theorem (6.16):

$$\begin{aligned}\oint \mathbf{q} \cdot d\mathbf{s} &= \oint (u\,dx + v\,dy + w\,dz) \\ &= \iint_S \left(\frac{\partial w}{\partial y} - \frac{\partial v}{\partial z}\right) dy\,dz + \iint_S \left(\frac{\partial u}{\partial z} - \frac{\partial w}{\partial x}\right) dx\,dz \\ &\qquad + \iint_S \left(\frac{\partial v}{\partial x} - \frac{\partial u}{\partial y}\right) dx\,dy\end{aligned} \tag{71.4}$$

For the surface integrals, any surface S may be chosen as long as it lies entirely in the fluid-filled space and its border line is identical with the path along which the line integral is taken.

In irrotational flows, a potential ϕ can be assumed with

$$\frac{\partial \phi}{\partial x} = u \qquad \frac{\partial \phi}{\partial y} = v \qquad \frac{\partial \phi}{\partial z} = w \tag{71.5}$$

or, more generally, $\partial\phi/\partial s = v_s$. The potential ϕ satisfies the Laplace equation (expressing the continuity of the flow):

$$\frac{\partial^2 \phi}{\partial x^2} + \frac{\partial^2 \phi}{\partial y^2} + \frac{\partial^2 \phi}{\partial z^2} = 0 \tag{71.6}$$

After introducing the potential ϕ into the system of the nonlinear Euler equations (71.1), and assuming that the external-force components are derivatives of a potential $-\Omega$ [see Eqs. (70.11a,b)], an integration can be performed. It yields the general Bernoulli equation:

$$\frac{\partial \phi}{\partial t} + \frac{q^2}{2} = -\Omega - \frac{1}{\rho} p + F(t) \tag{71.7}$$

Therefore, potential flow problems require that solutions be found of the Laplace equation which satisfy the boundary condition of a particular flow. The pressure field is determined by Eq. (71.7) as soon as ϕ is known and the pressure at one reference point is given.

Since Eq. (71.6) is linear, new potentials can be found by superposing various known potentials, and simple potential flows may be used for building up more complicated flow patterns.

In flows with axial symmetry, it is convenient to use cylindrical coordinates x and R (Fig. 71.1). The differential equation for the potential in these coordinates is given by

$$\frac{\partial^2 \phi}{\partial x^2} + \frac{\partial^2 \phi}{\partial R^2} + \frac{1}{R}\frac{\partial \phi}{\partial R} = 0 \tag{71.8}$$

Stream Function

Streamlines are determined by the relations

$$\frac{dx}{u} = \frac{dy}{v} = \frac{dz}{w} \tag{71.9}$$

The entity of all streamlines intersecting a fixed curve form a stream surface. When the curve is closed, the stream surface encloses a stream tube. In axisymmetric flow the surfaces defined by circles R = const for x = const are of particular interest. They are a set of nonintersecting surfaces of revolution (see Fig. 71.2*a*,*b*). In this

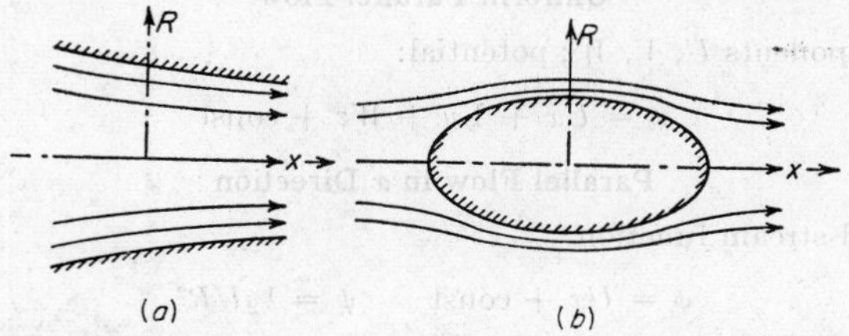

FIG. 71.2. Axisymmetric flows: (*a*) Interior flow; (*b*) exterior flow.

case, a stream function ψ^* (Stokes' stream function) may be defined as the total flux inside the stream tube. Evidently

$$\frac{\partial \psi^*}{\partial R} = 2\pi R u \qquad -\frac{\partial \psi^*}{\partial x} = 2\pi R v_R \tag{71.10a,b}$$

where v_R denotes the velocity component normal to the x axis. Often $\psi = \psi^*/2\pi$ is introduced, resulting in the relations

$$\frac{1}{R}\frac{\partial \psi}{\partial R} = u \qquad -\frac{1}{R}\frac{\partial \psi}{\partial x} = v_R \tag{71.10c,d}$$

The concept of a stream function is not restricted to flow with axial symmetry; however, in other cases, it usually does not offer mathematical and computational advantages.

The condition of zero vorticity in flow with axial symmetry is expressed by

$$\frac{\partial u}{\partial R} - \frac{\partial v_R}{\partial x} = 0 \tag{71.11}$$

From this relation, the differential equation for the stream function is obtained,

$$\frac{\partial^2 \psi}{\partial x^2} + \frac{\partial^2 \psi}{\partial R^2} - \frac{1}{R}\frac{\partial \psi}{\partial R} = 0 \tag{71.12}$$

which is similar in form to Eq. (71.8) for the potential ϕ in flow with axial symmetry.

The surfaces ϕ = const and ψ = const intersect at right angles at regular flow points, and the following differential relations exist:

$$\frac{\partial \phi}{\partial x} = \frac{1}{R}\frac{\partial \psi}{\partial R} \qquad \frac{\partial \phi}{\partial R} = -\frac{1}{R}\frac{\partial \psi}{\partial x} \tag{71.13a,b}$$

General Remarks about Solutions

There are two methods of finding solutions of Eqs. (71.6), (71.8), and (71.12): the method of separation of variables, and the singularity method.

In the first case, we assume the solution to be a product of functions of one variable, for example, $\phi = f_1(x)f_2(y)f_3(z)$ or $\phi = g_1(R)g_2(x)$, $\psi = h_1(R)h_2(x)$. There are many possibilities of separating variables in the three-dimensional Laplace equation. They are well described in Ref. [1], pp. 1252–1309 and 656–664. An example of this method is given in Sec. 71.3, where the flow along ellipsoids of revolution is treated.

In the following, the very powerful singularity method is mostly used. In Sec. 71.2, fields of important singularities are described. By superposition of such fields, very general flow fields can be generated.

71.2. SIMPLE FLOW FIELDS

Uniform Parallel Flow

Velocity components U, V, W; potential:

$$\phi = Ux + Vy + Wz + \text{const} \tag{71.14}$$

Parallel Flow in x Direction

Potential and stream function:

$$\phi = Ux + \text{const} \qquad \psi = \tfrac{1}{2}UR^2 \tag{71.15a,b}$$

Point Source

The strength M of the source is defined as the fluid volume produced per unit of time. For a source located at the origin, the radial velocity is

$$v_r = \frac{d\phi}{dr} = \frac{M}{4\pi r^2}$$

and the potential is

$$\phi = -\frac{M}{4\pi r} + \text{const} = -\frac{M}{4\pi}\frac{1}{\sqrt{x^2 + R^2}} + \text{const} \tag{71.16a}$$

The stream function follows from Eqs. (71.13),

$$\psi = -\frac{M}{4\pi}(1 + \cos\theta) = \frac{M}{4\pi}\left(1 + \frac{x}{\sqrt{x^2 + R^2}}\right) \tag{71.16b}$$

if $\psi = 0$ is desired for $R = 0$ and $x < 0$ (see Fig. 71.3). This way of numbering the stream function is advantageous when the source flow is to be superposed on a parallel flow.

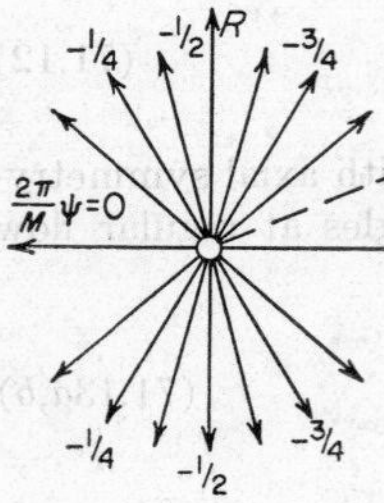

FIG. 71.3. Source flow.

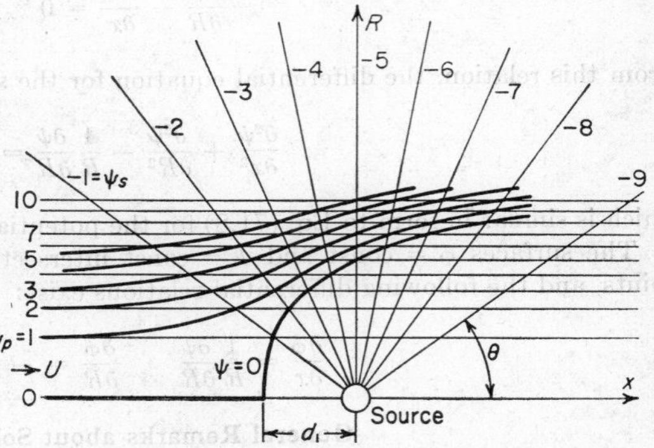

FIG. 71.4. Flow around a half body. Thin lines: stream lines for a parallel flow and for a source flow; heavy lines: stream lines for the flow around the half body.

Source in a Parallel Flow

Figure 71.4 shows, in thin lines, the streamline patterns of the two component flows and, in heavy lines, the result of the superposition. The flow outside the streamline $\psi = 0$ may be considered as the flow about a half body, which for $x \to \infty$ approaches the cross section:

$$\pi R_\infty{}^2 = M/U \tag{71.17}$$

The apex of the body coincides with the stagnation point of the flow, and is at the distance

$$d_1 = \sqrt{\frac{M}{4\pi U}} \tag{71.18}$$

to the left of the source. The tangential velocity along a meridian of the half body increases from zero at the stagnation point to a value larger than U at $\theta \approx 70°$, and then decreases toward U, which is reached for $x \to \infty$.

Source and Sink of Equal Strength

In a parallel flow, this phenomenon produces the flow shown in Fig. 71.5. The flow outside $\psi = 0$ can be considered as the flow around a body of revolution of finite

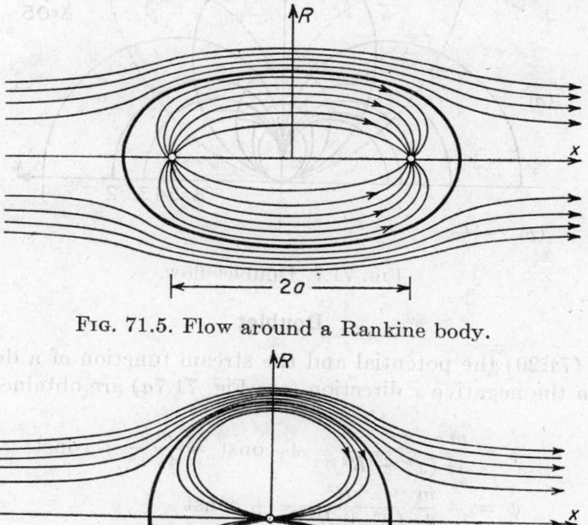

Fig. 71.5. Flow around a Rankine body.

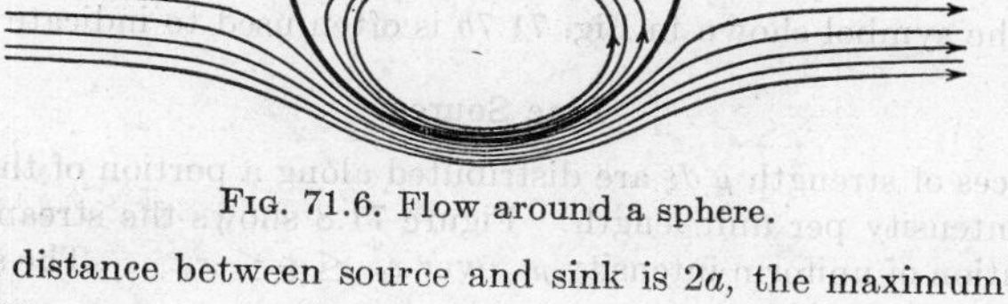

Fig. 71.6. Flow around a sphere.

length. If the distance between source and sink is $2a$, the maximum radius of this body is given by

$$R_{\max}^2 \sqrt{1 + \frac{R_{\max}^2}{a^2}} = \frac{M}{\pi U} \tag{71.19}$$

If the distance $2a$ is decreased and the product $2aM = m$ is kept constant, the body becomes less elongated. In the limiting case $M \to \infty$ and $a \to 0$, there results the flow around a sphere (Fig. 71.6). The singularity which is produced by a source and a sink of infinite strength at zero distance is called a *doublet*. The strength of the doublet m is related to the radius r_0 of the sphere by

$$r_0^3 = m/2\pi U$$

The potential of the flow about the sphere is

$$\phi = Ux\left(1 + \frac{1}{2}\frac{r_0^3}{(x^2 + R^2)^{3/2}}\right) + \text{const} \tag{71.20a}$$

and the stream function is

$$\psi = U\frac{R^2}{2}\left(1 - \frac{r_0^3}{(x^2 + R^2)^{3/2}}\right) \tag{71.20b}$$

The bodies formed by superposing source and sink of equal strength at distance $2a$ to a parallel flow are called *Rankine bodies*. Their nose and tail are rather blunt. By choosing the distance $2a$ and the ratio $M/4\pi U a^2$, the size and the slenderness ratio of a Rankine body are determined.

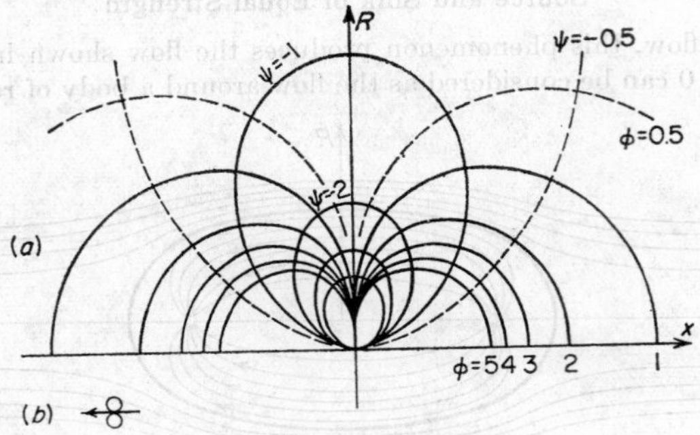

FIG. 71.7. Doublet flow.

Doublet

From Eqs. (71.20) the potential and the stream function of a doublet m with its axis pointing in the negative x direction (see Fig. 71.7a) are obtained as

$$\begin{aligned} \phi &= \frac{m}{4\pi}\frac{x}{(x^2 + R^2)^{3/2}} + \text{const} = \frac{m}{4\pi}\frac{x}{r^3} + \text{const} \\ \psi &= -\frac{m}{4\pi}\frac{R^2}{(x^2 + R^2)^{3/2}} + \text{const} \end{aligned} \tag{71.21a,b}$$

In diagrams, the symbol shown in Fig. 71.7b is often used to indicate a doublet.

Line Source

When sources of strength $\mu\, d\xi$ are distributed along a portion of the x axis, then μ is the source intensity per unit length. Figure 71.8 shows the streamline field for a source distribution of uniform intensity μ_m over $x_m \le \xi \le x_{m+1}$. The stream function is found from Eq. (71.16b):

$$\begin{aligned} \psi &= \int_{x_m}^{x_{m+1}} d\psi = -\frac{\mu_m}{4\pi}\int_{x_m}^{x_{m+1}}\left(1 + \frac{x - \xi}{[(x - \xi)^2 + R^2]^{1/2}}\right) d\xi \\ &= -\frac{\mu_m}{4\pi}\left[(x_{m+1} - x_m) - (r_{m+1} - r_m)\right] \end{aligned} \tag{71.22}$$

with

$$r_m^2 = (x - x_m)^2 + R^2 \tag{71.23}$$

Source Ring

Point sources of intensity $\mu\,ds$ are distributed along a circle of radius R' lying in the y,z plane (see Fig. 71.9). The resulting flow field has axial symmetry. Since

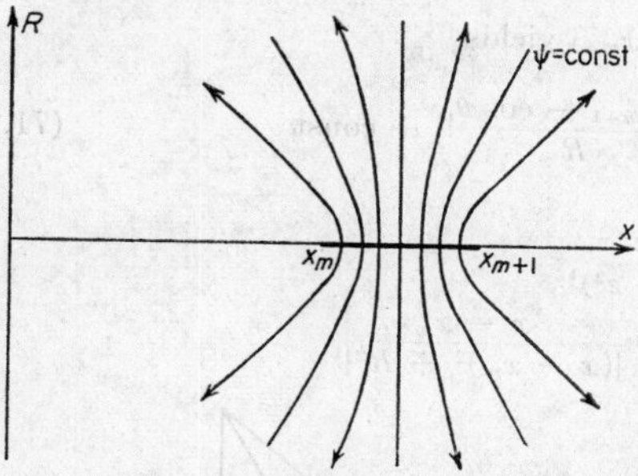

FIG. 71.8. Flow from a line source.

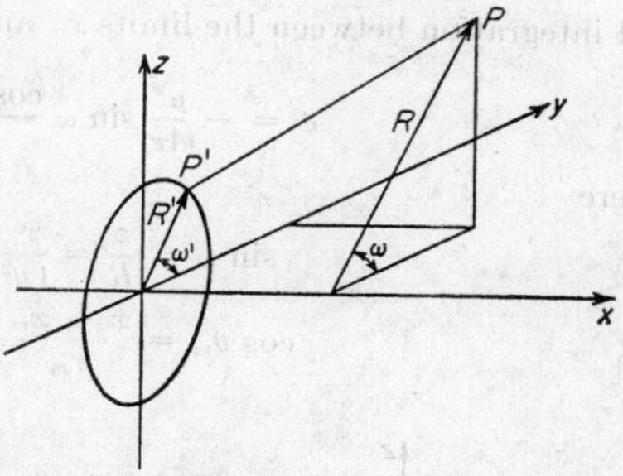

FIG. 71.9. Line doublet.

the velocity components u and v_R at P are independent of ω, we can choose $\omega = \pi/2$ for computing them. With $x/R' = \bar{x}$ and $R/R' = \bar{R}$, we obtain

$$\begin{aligned} u(x,R) &= \frac{\mu\bar{x}}{4\pi R'}\int_0^{2\pi}\frac{d\omega'}{(\bar{x}^2+\bar{R}^2+1-2\bar{R}\sin\omega')^{3/2}} \\ v_R(x,R) &= \frac{\mu}{4\pi R'}\int_0^{2\pi}\frac{(R-\sin\omega')\,d\omega'}{(\bar{x}^2+\bar{R}^2+1-2\bar{R}\sin\omega')^{3/2}} \end{aligned} \qquad (71.24a,b)$$

These integrals can be reduced to complete elliptic integrals with the modulus k [see Eqs. (15.118)], where

$$k^2 = \frac{4\bar{R}}{\bar{x}^2+(\bar{R}+1)^2}$$

Tables for

$$u(x,R) = \frac{\mu}{2\pi R'}\,\frac{2\bar{x}}{[\bar{x}^2+(\bar{R}+1)^2]^{1/2}[\bar{x}^2+(\bar{R}-1)^2]}\,E(k) \qquad (71.24c)$$

and

$$v_R(x,R) = \frac{\mu}{2\pi R'}\,\frac{1}{\bar{R}[\bar{x}^2+(\bar{R}+1)^2]^{1/2}}\left\{K(k)-\left[1-\frac{2\bar{R}(\bar{R}-1)}{\bar{x}^2+(\bar{R}-1)^2}\right]E(k)\right\} \qquad (71.24d)$$

are given in [2], pp. 313–314 (notation slightly different!). A table for the stream function of the source ring, defined by

$$\psi_\mu(x,R) = -R\int_0^x v_R\,dx^* + \int_0^R R^*u\,dR^* + \text{const} \qquad (71.25)$$

is given in [2], p. 315.

Line Doublets

Consider doublets distributed continuously along some part of the x axis, the intensity being μ^*. When the doublets point in the positive x direction, they are equivalent to a source distribution of intensity $\mu = -d\mu^*/dx$ plus a source at one end and a sink at the other.

When the axes of the doublets point in the negative z direction (Fig. 71.10), the potential of a doublet $\mu^* \, d\xi$ at $x = \xi$ is

$$d\phi = \frac{\mu^* \, d\xi}{4\pi} \frac{z}{[(x - \xi)^2 + R^2]^{3/2}}$$

and integration between the limits x_m and x_{m+1} yields

$$\phi = -\frac{\mu^*}{4\pi} \sin \omega \frac{\cos \theta_{m+1} - \cos \theta_m}{R} + \text{const} \tag{71.26}$$

where

$$\sin \omega = \frac{z}{R} = \frac{z}{(y^2 + z^2)^{1/2}}$$

$$\cos \theta_m = \frac{x - x_m}{r_m} = \frac{x - x_m}{[(x - x_m)^2 + R^2]^{1/2}}$$

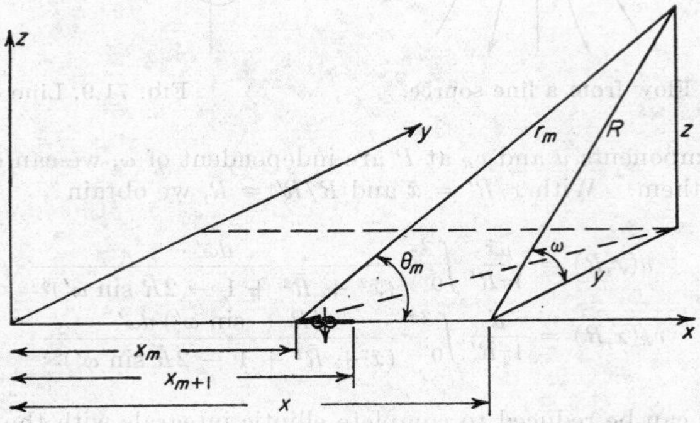

FIG. 71.10. Line doublet.

71.3. ELLIPSOIDS OF REVOLUTION

Elliptic-Hyperbolic Coordinates

For the computation of the flow around ellipsoids, new coordinates μ, ζ (Fig. 71.11) are introduced by the transformation

$$\frac{x}{c} = \mu\zeta \qquad \frac{R}{c} = \sqrt{1 - \mu^2}\,\sqrt{\zeta^2 - 1} \qquad \omega = \omega \tag{71.27}$$

Evidently, μ = const yields hyperbolas in the x,R plane,

$$\frac{(x/c)^2}{\mu^2} - \frac{(R/c)^2}{1 - \mu^2} = 1 \tag{71.28}$$

and ζ = const yields ellipses in the x,R plane,

$$\frac{(x/c)^2}{\zeta^2} + \frac{(R/c)^2}{\zeta^2 - 1} = 1 \tag{71.29}$$

These ellipses and hyperbolas are confocal. For $\zeta \to 1$, the ellipses become very elongated, and, for $\zeta \to \infty$, the ellipses approach circular form. The coordinate μ varies between 0 and 1.

The partial differential equation for the potential in the new coordinates is given by

$$\frac{\partial}{\partial \mu}\left[(1 - \mu^2)\frac{\partial \phi}{\partial \mu}\right] + \frac{\partial}{\partial \zeta}\left[(\zeta^2 - 1)\frac{\partial \phi}{\partial \zeta}\right] + \frac{\zeta^2 - \mu^2}{(1 - \mu^2)(\zeta^2 - 1)}\frac{\partial^2 \phi}{\partial \omega^2} = 0 \tag{71.30}$$

Ellipsoid of Revolution Moving with Constant Speed in the Direction of Its Axis

In this case, Eq. (71.30) for the potential is reduced to

$$\frac{\partial}{\partial \mu}\left[(1-\mu^2)\frac{\partial \phi}{\partial \mu}\right] + \frac{\partial}{\partial \zeta}\left[(\zeta^2-1)\frac{\partial \phi}{\partial \zeta}\right] = 0 \tag{71.31}$$

At the surface S of the ellipsoid of revolution moving with constant speed U through air at rest, the velocity v_ν normal to the surface is

$$v_\nu = \frac{1}{c}\sqrt{\frac{\zeta^2-1}{\zeta^2-\mu^2}}\frac{\partial \phi}{\partial \zeta} = -U\cos(x,\nu)$$

or

$$\frac{\partial \phi}{\partial \zeta} = -U\mu c \tag{71.32}$$

At infinity, the velocity components must vanish.

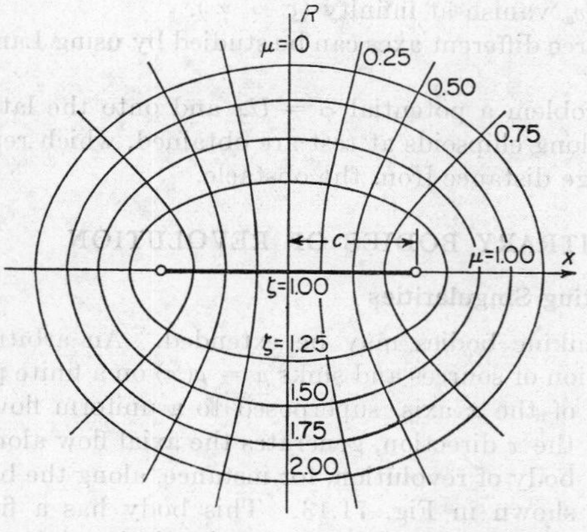

FIG. 71.11. Elliptic-hyperbolic coordinates.

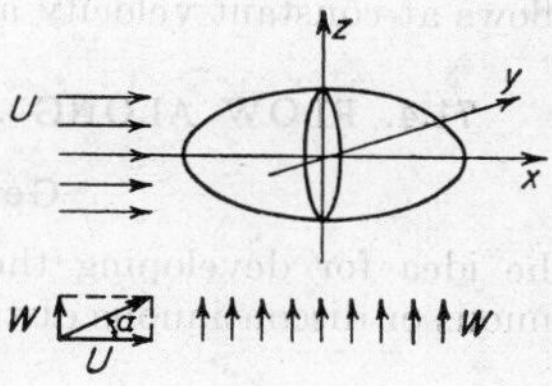

FIG. 71.12. Ellipsoid of revolution.

The differential equation (71.32) can be satisfied by a product of a function of μ only and a function of ζ only (see [3], p. 282)

$$\phi = f_1(\mu) \cdot f_2(\zeta)$$

In particular,

$$\phi = A\mu\left(\tfrac{1}{2}\zeta \ln\frac{\zeta+1}{\zeta-1} - 1\right) \tag{71.33}$$

is a solution which satisfies the boundary conditions of this problem if the constant A is correctly chosen.

The surface of the moving ellipsoid is characterized by a particular value $\zeta_0 = a/c$. For this value ζ_0, Eq. (71.32) must be satisfied. This yields

$$A = -Uc\left(\tfrac{1}{2}\ln\frac{\zeta_0+1}{\zeta_0-1} - \frac{\zeta_0}{\zeta_0^2-1}\right)^{-1} \tag{71.34}$$

At infinity, that is, for $\zeta \to \infty$, the velocity components vanish.

The Lateral Flow around an Ellipsoid of Revolution

Since the motion of an ellipsoid in a direction which is inclined to its axis of symmetry can be decomposed into a motion in the direction of the axis and one normal to it, the lateral motion (see Fig. 71.12) of an ellipsoid is important. This flow around the ellipsoid has no longer axial symmetry, and the potential, therefore, has to satisfy the general equation (71.30). The boundary condition on the surface is

$$v_\nu = -W\cos(z,\nu) = \frac{1}{c}\sqrt{\frac{\zeta^2-1}{\zeta^2-\mu^2}}\frac{\partial\phi}{\partial\zeta} \quad \text{or} \quad \frac{\partial\phi}{\partial\zeta} = -W\frac{\partial z}{\partial\zeta}$$

Since $z = R\sin\omega$, we obtain $\partial\phi/\partial\zeta = -W(\partial R/\partial\zeta)\sin\omega$. The solution of Eq. (71.30) is a product of three functions of μ, ζ, ω, respectively:

$$\phi = B\sqrt{1-\mu^2}\sqrt{\zeta^2-1}\left[\tfrac{1}{2}\ln\frac{\zeta+1}{\zeta-1} - \frac{\zeta}{\zeta^2-1}\right]\sin\omega \tag{71.35}$$

with

$$B = -\left[\tfrac{1}{2}\ln\frac{\zeta_0+1}{\zeta_0-1} - \frac{\zeta_0{}^2-2}{\zeta_0(\zeta_0{}^2-1)}\right]^{-1}Wc \tag{71.36}$$

All three velocity components v_μ, v_ζ, v_ω vanish at infinity ($\zeta\to\infty$).

The motion of an ellipsoid with three different axes can be studied by using Lamé's functions (see [4], sec. 110).

By superposing onto the axial problem a potential $\phi = Ux$ and onto the lateral problem a potential $\phi = Wz$, flows along ellipsoids at rest are obtained, which represent flows at constant velocity at large distance from the obstacle.

71.4. FLOW ALONG ARBITRARY BODIES OF REVOLUTION

Generating Singularities

The idea for developing the Rankine bodies may be extended. An arbitrary continuous or discontinuous distribution of sources and sinks $\mu = \mu(x)$ on a finite part of the x axis, superposed to a uniform flow in the x direction, generates the axial flow along a body of revolution, for instance, along the body shown in Fig. 71.13. This body has a finite length if $\int\mu\,dx = 0$, that is, if the sinks just absorb all the fluid produced by the sources. Similarly, a distribution of doublets (see Fig. 71.10), together with a uniform flow in the z direction generates the transverse flow around a body of revolution.

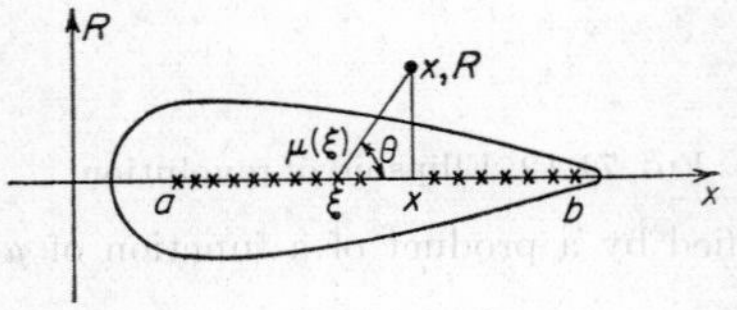

FIG. 71.13. Body of revolution.

The procedure is applicable only when the actual flow field permits an analytic continuation into the interior of the body, with no singular points except on the body axis. This is the case for slender bodies only. No simple mathematical criterion is available which states which shapes can be treated in this way.

Figure 71.14 shows a body with very sudden changes of diameter. It is known that, in such a case, singularities on rings of finite diameter will be needed, in addition to those on the axis. Therefore, in case of bodies with very blunt nose or sudden changes of diameter, it is recommended that surface layers of singularities be used. This is always possible when the body shape has a well-defined tangent plane in each surface point; i.e., bodies with conical nose and tail have to be excluded. However, a very small rounding at the vertex of the cone (such as practically always exists) will suffice to handle the problem.

The intensity of the source layer on a surface S of a body is determined by an integral equation [5]. This equation expresses the fact that a desired velocity component $v_\nu(x,y,z)$ has to be produced by all sources spread over the body surface. The source intensity is called $f\,d\sigma$, with $d\sigma$ being the surface element. For a body of revolution (see Fig. 71.15), the integral equation is given by

$$2\pi f(x,y,z) + \int_S f(\xi,\eta,\zeta)\frac{\cos\psi}{r^2}\,d\sigma = 4\pi v_\nu(x,y,z) \tag{71.37}$$

where ξ, η, ζ are the coordinates of the element $d\sigma$,

$$r^2 = (x-\xi)^2 + (y-\eta)^2 + (z-\zeta)^2$$

and ψ is the angle between r and the surface normal at the point (x,y,z) of the surface. For more details and special problems treated in this way, see [5] and [6].

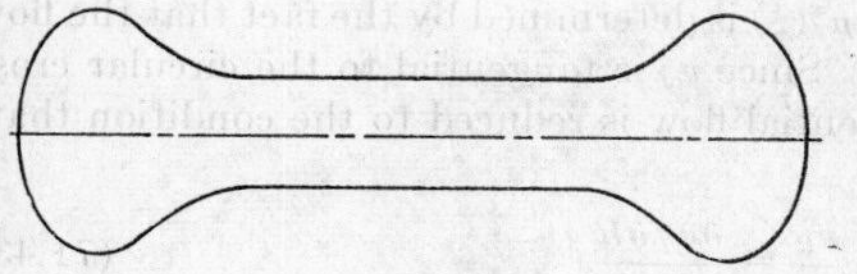

FIG. 71.14. Body of revolution with blunt bow and stern and strong diameter changes.

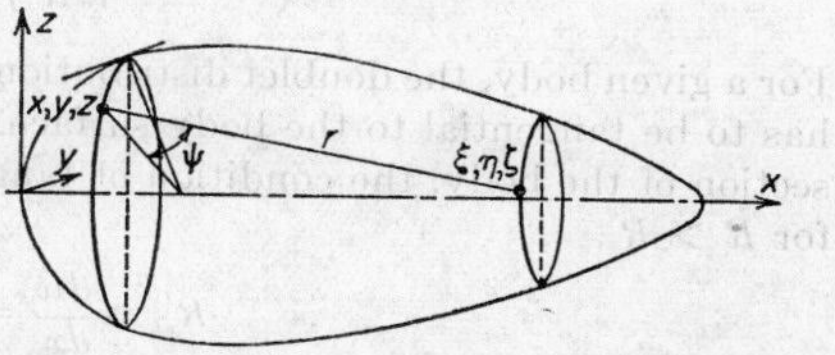

FIG. 71.15. Surface layer on a body of revolution.

Axial Flow along Bodies of Revolution

If a continuous distribution of sources along the major portion of the body axis, from $x = a$ to $x = b$ (Fig. 71.13), is assumed, an integral equation has to be solved for determining the distribution. For closed bodies, the integral over the source distribution has to be zero:

$$\int_a^b \mu(\xi)\,d\xi = 0$$

The total stream function [see Eqs. (71.15b) and (71.16b)] is given by

$$\begin{aligned}\psi &= \tfrac{1}{2}UR^2 - \frac{1}{4\pi}\int_a^b \mu(\xi)\frac{x-\xi}{\sqrt{(x-\xi)^2+R^2}}\,d\xi \\ &= \tfrac{1}{2}UR^2 - \frac{1}{4\pi}\int_a^b \mu(\xi)\cos\theta\,d\xi\end{aligned} \tag{71.38}$$

where the coordinates are the same as in Fig. 71.10. Since the stream function $\psi = 0$ on the body surface $R = R_s(x)$, the source distribution can be obtained from the integral equation:

$$\tfrac{1}{2}UR^2 = \frac{1}{4\pi}\int_a^b \mu(\xi)\cos\theta\,d\xi \tag{71.39}$$

For very slender bodies,

$$\mu(x) = U\frac{d(R_s{}^2\pi)}{dx} \tag{71.40}$$

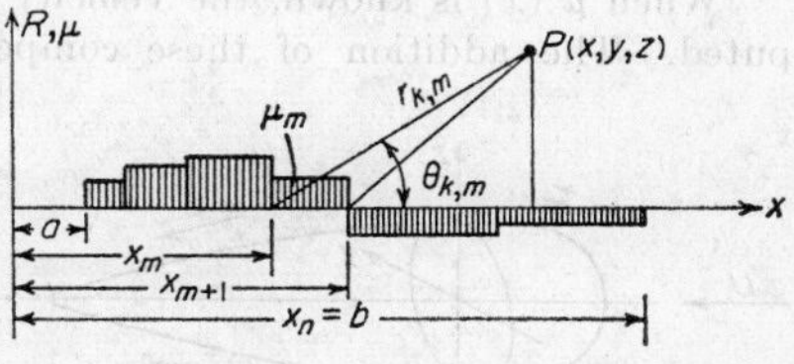

FIG. 71.16. Distribution of line sources.

is a good approximation for the source distribution, with the restriction that this relation is not valid near the body ends. If the source distribution is approximated by n line sources (see Fig. 71.16), the intensity of these line sources is given by a system of linear equations [see Eq. (71.22)],

$$\frac{1}{2\pi}\sum_{m=0}^{n-1}\mu_m(r_{k,m+1} - r_{k,m}) = -U_\infty R_k{}^2 \tag{71.41}$$

and $\psi = 0$ can be specified at n points R_k $(k = 1,2, \ldots, n)$. (For details, see [7], pp. 253–273.)

Lateral Flow about a Slender Body of Revolution

Upon arranging doublets of intensity $\mu^* \, d\xi$ on the axis, we obtain, with the notations of Fig. 71.13, the potential [see Eq. (71.21*a*)]

$$\begin{aligned} \phi &= Wz + \frac{1}{4\pi} \int_a^b \mu^*(\xi) \frac{z}{[(x-\xi)^2 + R^2]^{3/2}} \, d\xi + \text{const} \\ &= \sin \omega \left(WR + \frac{1}{4\pi R} \int_{\theta_a}^{\theta_b} \mu^* \sin \theta \, d\theta \right) + \text{const} \end{aligned} \tag{71.42}$$

For a given body, the doublet distribution $\mu^*(\xi)$ is determined by the fact that the flow has to be tangential to the body surface. Since v_ω is tangential to the circular cross section of the body, the condition of tangential flow is reduced to the condition that, for $R > R_s$,

$$R_s' \equiv \frac{dR}{dx} = \frac{v_R}{u} = \frac{\partial \phi / \partial R}{\partial \phi / \partial x} \tag{71.43}$$

The doublet distribution is determined by the integral equation:

$$-3R_s' \int_{\theta_a}^{\theta_b} \mu^* \sin^2 \theta \cos \theta \, d\theta + \int_{\theta_a}^{\theta_b} \mu^*(2 - 3 \cos^2 \theta) \sin \theta \, d\theta = 4\pi W R_s^2 \tag{71.44}$$

If the doublet distribution is approximated by a finite number of line doublets, we obtain a system of linear equations for the intensities of the n line doublets:

$$\begin{aligned} R_k' \sum_{m=0}^{n-1} \mu_m^*(\sin^3 \theta_{k,m} - \sin^3 \theta_{k,m+1}) + \sum_{m=0}^{n-1} \mu_m^*[2(\cos \theta_{k,m} - \cos \theta_{k,m+1}) \\ - (\cos^3 \theta_{k,m} - \cos^3 \theta_{k,m+1})] = 4\pi W R_k^2 \end{aligned} \tag{71.45}$$

This system enforces tangential flow at n surface rings with the radii $R_k(x)$ $(k = 1,2, \ldots, n)$. For details, see [7].

For very slender elongated bodies (slow change of R with x), the doublet intensity $\mu^*(x)$ is approximately that needed for the two-dimensional flow around a cylinder of radius $R_s(x)$:

$$\mu^*(x) = 2\pi R_s^2(x) W \tag{71.46}$$

This relation is not valid near the ends of the body.

When $\mu^*(x)$ is known, the velocity components for the lateral flow can be computed. The addition of these components to the components of the axial flow permits the computation of the resultant flow field around the body flying under any angle of attack (Fig. 71.12).

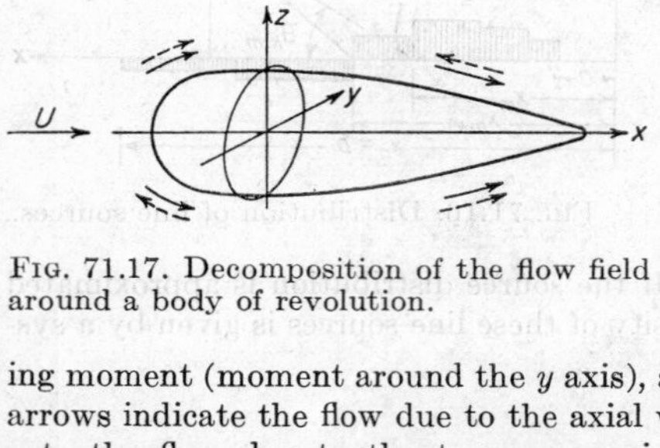

FIG. 71.17. Decomposition of the flow field around a body of revolution.

Moment on an Elongated Body of Revolution

By applying the momentum theorem, we find that, in inviscid fluid, the lift and drag of a body flying under an angle of attack are zero; however, there is a pitching moment (moment around the y axis), as can be seen easily (Fig. 71.17). The solid arrows indicate the flow due to the axial wind component U; the dotted arrows indicate the flow due to the transverse wind component W. The moment tends to increase the angle of attack. This means that a body without tailplane is unstable.

For elongated bodies (Fig. 71.18), a simple approximate calculation of the pitching moment is possible. For each ring of the body surface, we calculate the z component of the force and neglect the influence of the x component. The velocity components

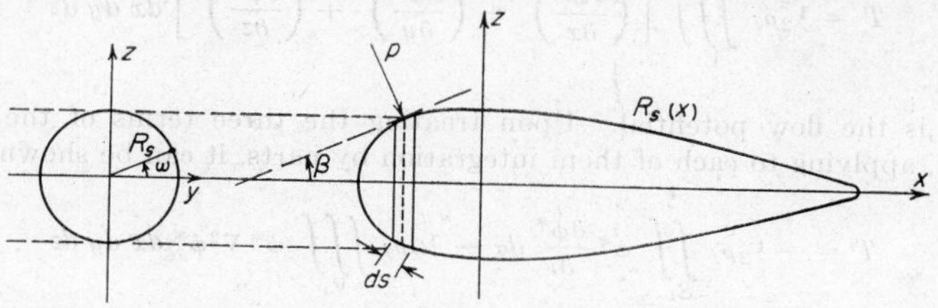

FIG. 71.18. Computation of the transverse force on a ring.

u_1, v_{R1} for axial flow and u_2, v_{R2}, v_ω for lateral flow are known. The pressure p is given by the Bernoulli equation, with p_∞ being the pressure at infinity:

$$p - p_\infty = \tfrac{1}{2}\rho\{U^2 + W^2 - [(u_1 + u_2)^2 + (v_{R1} + v_{R2})^2 + v_\omega^2]\} \qquad (71.47)$$

The force on the ring is then given by

$$dF_z = -\int_0^{2\pi} pR_s\, d\omega\, (ds \cos\beta) \sin\omega$$
$$= -\int_0^{2\pi} (p - p_\infty)R_s\, d\omega\, dx \sin\omega \qquad (71.48)$$

When introducing $(p - p_\infty)$ from Eq. (71.47), it is important for the integration to express the dependency of the velocity components on ω explicitly:

$$u_1 = u_1(R) \qquad u_2 = \bar{u}_2(R) \sin\omega$$
$$v_{R1} = v_{R1}(R) \qquad v_{R2} = \bar{v}_{R2}(R) \sin\omega \qquad (71.49)$$
$$v_\omega = \bar{v}_\omega(R) \cos\omega$$

Thus, the integration can be performed easily; it yields, for the distribution of the ring forces,

$$\frac{dF_z}{dx} = \rho\pi R_s(1 + R_s'^2)u_{1s}\bar{u}_{2s} \qquad (71.50)$$

71.5. KINETIC ENERGY OF THE FLUID FLOW; APPARENT MASS

Translatory Motion

Consider a body with surface S starting from rest and moving through a fluid of density ρ_f. Assume that the fluid also was originally at rest. In order to pass through the fluid, the body has to push fluid particles out of their places, to which they will return after the body has passed. The kinetic energy of the moving particles is given by

$$T = +\tfrac{1}{2}\rho_f \iiint q^2\, dx\, dy\, dz \qquad (71.51)$$

The integral has to be taken over the entire fluid-filled space, i.e., over the space between the surfaces S and Σ in Fig. 71.19. When the body has obtained a certain speed in one direction, a certain amount of energy has been put into the surrounding fluid, which does not change if the body moves on with this speed, and which is proportional to the square of this speed.

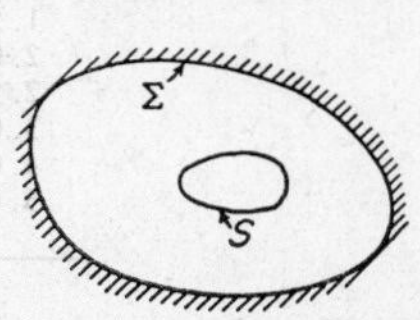

FIG. 71.19. Body with surface S in a fluid-filled space, outer boundary Σ.

In order to calculate the kinetic energy in the surrounding fluid, Green's theorem is applied. Consider some region V_1 of the fluid space, enclosed by a surface S_1, and let

ν be the normal to S_1, pointing inward. Then the kinetic energy of a potential flow in V_1 is

$$T = \tfrac{1}{2}\rho_f \iiint_{V_1} \left[\left(\frac{\partial \phi^*}{\partial x}\right)^2 + \left(\frac{\partial \phi^*}{\partial y}\right)^2 + \left(\frac{\partial \phi^*}{\partial z}\right)^2 \right] dx\,dy\,dz \qquad (71.52)$$

where ϕ^* is the flow potential. Upon treating the three terms of the integrand separately, applying to each of them integration by parts, it can be shown that

$$T = -\tfrac{1}{2}\rho_f \iint_{S_1} \phi^* \frac{\partial \phi^*}{\partial \nu}\,d\sigma - \tfrac{1}{2}\rho_f \iiint_{V_1} \phi^* \nabla^2 \phi^*\,dx\,dy\,dz \qquad (71.53)$$

Since $\nabla^2\phi^* = 0$, the volume integral vanishes, and the kinetic energy is given by

$$T = -\tfrac{1}{2}\rho_f \iint_{S_1} \phi^* \frac{\partial \phi^*}{\partial \nu}\,d\sigma \qquad (71.54)$$

For applying this theorem to the space around a moving body, we imagine the mass of fluid particles in motion enclosed by two surfaces: the body surface S, and a surface of a very big sphere Σ whose center lies somewhere in the interior of the body (compare Fig. 71.19). If we allow the radius of this sphere to grow toward infinity, we will obtain the energy of the fluid surrounding the body moving in unlimited space. For a straight motion of the body with constant speed U_0, the potential ϕ^* is proportional to this speed U_0, and the energy T can be written as

$$T = \tfrac{1}{2}\rho_f \textstyle\iiint q^2\,dx\,dy\,dz = \tfrac{1}{2}\rho_f U_0^2 k V_{\text{body}} \qquad (71.55)$$

The coefficient k depends on the shape of the body and the direction of its motion. It is called the coefficient of additional apparent mass, because, when Newton's law for the acceleration of the body is written down, the mass $\rho_f k V_{\text{body}}$ must be added to the actual mass of the body to represent the inertia of the surrounding fluid.

For *prolate ellipsoids of revolution* (a = axis in the x direction, b = axis normal to the x direction), the coefficients k_1 for axial motion and k_2 for transverse motion are given in Table 71.1.

Table 71.1

a/b	k_1	k_2	a/b	k_1	k_2
1.00	0.500	0.500	6.0	0.045	0.918
1.5	0.305	0.621	7.0	0.036	0.933
2.0	0.209	0.702	8.0	0.029	0.945
2.5	0.157	0.762	9.0	0.024	0.954
3.0	0.121	0.804	10.0	0.021	0.960
4.0	0.082	0.860	∞	0	1.000
5.0	0.059	0.895			

For the *circular disk* moving normal to its plane (Fig. 71.20), the kinetic energy is given by

$$T = \tfrac{1}{2}\rho_f U_0^2 (\tfrac{8}{3} b^3) \qquad (71.56)$$

and, for a *plate strip* (two-dimensional case, Fig. 71.21), we obtain

$$dT = \tfrac{1}{2}\rho_f U_0^2 d^2 \pi\,dz \qquad (71.57)$$

For bodies of general shape, the apparent mass can be computed easily, if the system of singularities is known which generate the body in a particular flow. The

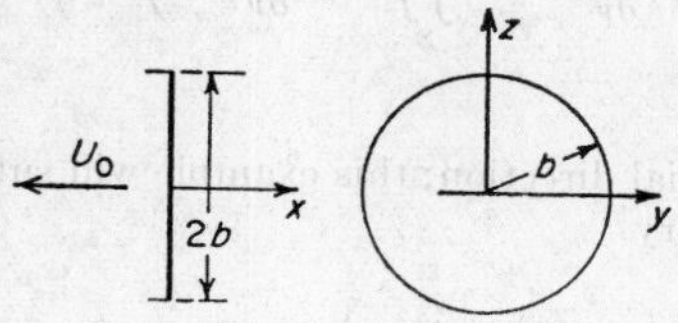

FIG. 71.20. Circular disk.

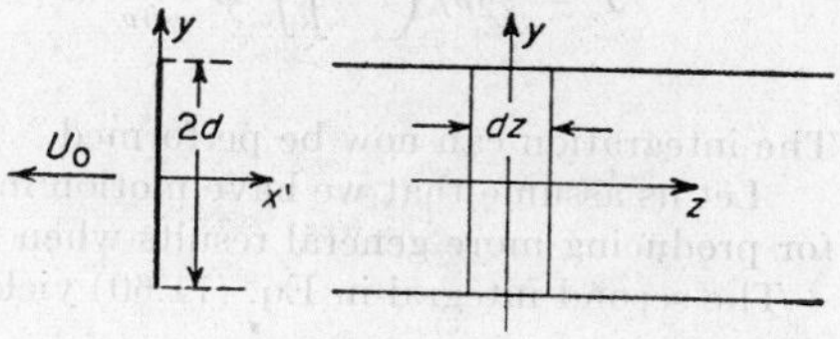

FIG. 71.21. Plate strip.

energy follows from Eq. (71.54). The integration has to be extended over the surface S of the moving body and the surface Σ of the sphere with an infinitely large radius (Fig. 71.19):

$$T = -\tfrac{1}{2}\rho_f \left[\iint_S \phi^* \frac{\partial \phi^*}{\partial \nu}\, d\sigma + \iint_\Sigma \phi^* \frac{\partial \phi^*}{\partial \nu}\, d\sigma \right] \tag{71.58}$$

The singularities substituting the body are either sources and sinks or doublets. Since, for closed bodies, the sum over all sources and sinks has to be zero, it is possible to combine sources and sinks in groups (see Fig. 71.22). If $\mu_1 \, \Delta x_1 = -\mu_2 \, \Delta x_2$, these line singularities combined will influence the flow at far-distant points like a doublet of intensity $M = \mu_1 \, \Delta x_1 \cdot d$. Therefore, we can consider doublets as the decisive singularities when evaluating the energy integral. Since the potential due to doublets decreases with the square of the distance, and the surface of a sphere increases with the square of the radius, the integral over Σ vanishes, and

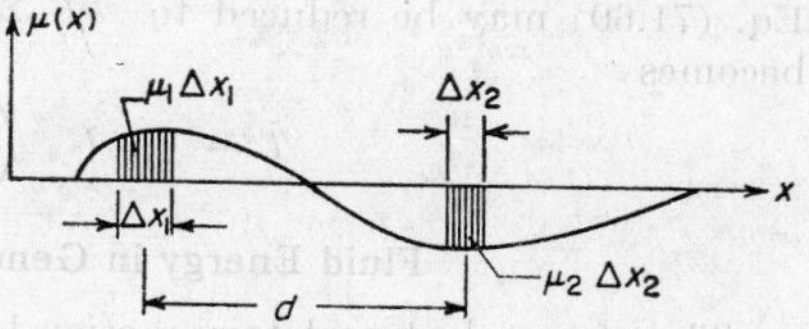

FIG. 71.22. Substitution of doublets for a source distribution.

$$T = -\tfrac{1}{2}\rho_f \iint_S \phi^* \frac{\partial \phi^*}{\partial \nu}\, d\sigma \tag{71.59}$$

The potential ϕ^* of the moving body can be expressed by the potential ϕ of the body at rest in a parallel flow and the potential ϕ_p of a parallel flow:

$$\phi^* = \phi - \phi_p$$

Therefore,

$$T = -\tfrac{1}{2}\rho_f \iint_S \phi^* \left(\frac{\partial \phi}{\partial \nu} - \frac{\partial \phi_p}{\partial \nu} \right) d\sigma = \tfrac{1}{2}\rho_f \iint_S \phi^* \frac{\partial \phi_p}{\partial \nu}\, d\sigma$$

because $\partial\phi/\partial\nu = 0$ at the body surface. Since the flow does not have singular points in the region between S and Σ, Green's theorem (6.20) gives the following relation,

$$\iint_S \phi^* \frac{\partial \phi_p}{\partial \nu}\, d\sigma + \iint_\Sigma \phi^* \frac{\partial \phi_p}{\partial \nu}\, d\sigma = \iint_S \phi_p \frac{\partial \phi^*}{\partial \nu}\, d\sigma + \iint_\Sigma \phi_p \frac{\partial \phi^*}{\partial \nu}\, d\sigma$$

$$= -\iint_S \phi_p \frac{\partial \phi_p}{\partial \nu}\, d\sigma + \iint_\Sigma \phi_p \frac{\partial \phi^*}{\partial \nu}\, d\sigma$$

and the kinetic energy becomes

$$T = \tfrac{1}{2}\rho_f \left(-\iint_\Sigma \phi^* \frac{\partial \phi_p}{\partial \nu}\, d\sigma - \iint_S \phi_p \frac{\partial \phi_p}{\partial \nu}\, d\sigma + \iint_\Sigma \phi_p \frac{\partial \phi^*}{\partial \nu}\, d\sigma \right) \tag{71.60}$$

The integration can now be performed.

Let us assume that we have motion in the axial direction; this example will suffice for producing more general results when necessary.

The second integral in Eq. (71.60) yields

$$\iint_S \phi_p \frac{\partial \phi_p}{\partial \nu}\, d\sigma = U_0{}^2 V_{\text{body}}$$

The first and the third integrals depend on the form of ϕ^*. Since only values for large r are needed, ϕ^* may be written as the sum of the potentials of doublets m_i, and it may be assumed that all these doublets are located at the origin. Hence,

$$\phi^* = \frac{1}{4\pi} \sum_i m_i \frac{\cos\theta}{r^2}$$

With $x = r\cos\theta$ and $d\sigma = 2\pi r^2 \sin\theta\, d\theta$, the sum of the first and third integrals of Eq. (71.60) may be reduced to $-U_0 \Sigma m_i$, and the kinetic energy of the flow field becomes

$$T = \tfrac{1}{2}\rho_f U_0{}^2 \left(\frac{1}{U_0} \sum m_i - V_{\text{body}} \right) \tag{71.61}$$

Fluid Energy in General Motion of an Obstacle

Thus far, only translatory motion has been considered, and a coefficient of additional apparent mass of a body has been defined. For a pure rotation, a coefficient of additional apparent moment of inertia can be defined. In a general motion composed of translations in the x, y, and z directions with the velocities U, V, W, and of rotation around these axes with the angular velocities p, q, r, the total velocity of the body is a linear combination of these six values. To accelerate a body in ideal fluid to a steady-state motion of the described form means generating in the surrounding fluid an energy content which depends on U, V, W, p, q, r. The velocity of each fluid particle is proportional to the total velocity of the body. The computation of the energy content of the fluid surrounding the body, therefore, demands evaluation of 21 integrals corresponding to six squares U^2, V^2, . . . and 15 products UV, . . . , pq, . . . , Up, ... In such a general motion, we could define 21 coefficients of additional inertia. Fortunately, most bodies of technical interest have some symmetry, and this fact reduces the number of the coefficients.

When the motion of a body in a fluid is studied, the fluid energy can be neglected if the ratio of fluid density to body density is small. However, if this ratio is not small, or if the body consists only of a solid shell filled with a light medium (e.g., airships, submarines), the neglect of additional apparent inertia would lead to serious mistakes.

71.6. THREE-DIMENSIONAL FLOW WITH VORTICITY IN LIMITED REGIONS

Vortex Line or Filament, Vortex Tube

Let us assume that, in a general motion of nonviscous incompressible fluid, the vorticity ω given by its three components ω_x, ω_y, ω_z [Eqs. (71.3)] is different from zero.

A line whose tangent at every point lies in the direction of the vorticity vector is called a vortex line. The differential equations

$$\frac{dx}{\omega_x} = \frac{dy}{\omega_y} = \frac{dz}{\omega_z} \tag{71.62}$$

determine the vortex lines. The circulation around a bundle of vortex lines through a cross section S' (see Fig. 71.23) is given by

$$\oint \mathbf{q} \cdot d\mathbf{s} = \oint (u\,dx + v\,dy + w\,dz)$$
$$= \iint_{S'} (\omega_x\,dy\,dz + \omega_y\,dz\,dx + \omega_z\,dx\,dy) \tag{71.63}$$

There exist motions which have vorticity only in very limited regions, for instance, the so-called potential vortex flow (see p. 70–22). This is a two-dimensional motion, say, around the y axis. The streamlines are circles in the x,z plane and the planes parallel to the x,z plane. The circulation around the y axis is finite and constant. The y axis is a vortex line.

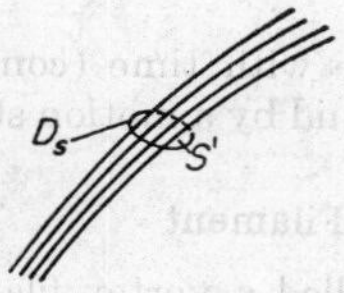

FIG. 71.23. Fascicle of vortex lines.

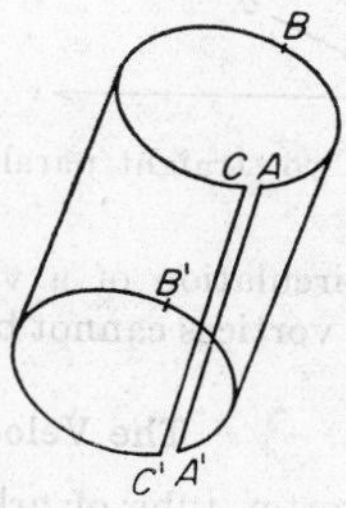

FIG. 71.24. Part of the surface of a vortex tube.

A vortex tube is formed by the vortex lines through a small area enclosed by a curve D_s (see Fig. 71.23). The circulation along all circuits around a vortex tube is the same. To prove this theorem, the circuit $ABCC'B'A'A$ is drawn on a vortex tube (see Fig. 71.24). CC' and AA' are apart in the figure, but are imagined to approach each other closely. The circuit encloses a part S of the surface of the vortex tube. If l, m, n are the direction cosines of a unit vector $\boldsymbol{\nu}$ normal to the surface of the vortex tube, then

$$J \equiv \int_S \boldsymbol{\omega} \cdot \boldsymbol{\nu}\, d\sigma = \int_S (\omega_x l + \omega_y m + \omega_z n)\, d\sigma = 0$$

Stokes' theorem (6.16) allows this surface integral to be transformed into a line integral along the path which borders the surface.

$$J = \int_{ABC} \mathbf{q} \cdot d\mathbf{s} + \int_{CC'} \mathbf{q} \cdot d\mathbf{s} + \int_{C'B'A'} \mathbf{q} \cdot d\mathbf{s} + \int_{A'A} \mathbf{q} \cdot d\mathbf{s}$$

Since the integrals over the parts CC' and $A'A$ cancel, it follows that

$$\Gamma = \int_{ABC} \mathbf{q} \cdot d\mathbf{s} = \int_{A'B'C'} \mathbf{q} \cdot d\mathbf{s}$$

that is, the circulation Γ along the circuits ABC and $A'B'C'$ is the same. This yields:

Theorem 1. The circulation does not change along a vortex tube. A consequence of this fact is:

Theorem 2. Vortex lines either are closed curves or start and end at the boundary of the fluid-filled space, which may be infinity (Helmholtz's theorem).

With the help of the equation of motion (Euler equation), it can be shown ([8], p. 184) that:

Theorem 3. The vortex lines move with the fluid.

As an example, consider two straight potential vortices of circulations Γ_1, Γ_2 with axes parallel to the y axis. At time $t = 0$, the centers of these vortices have the positions indicated in Fig. 71.25. Vortex Γ_2 produces, at the center of vortex Γ_1, the velocity $\Gamma_2/2\pi d$. The center of vortex Γ_1 will move with this speed. Vortex Γ_2 will move under the influence of Γ_1. It is obvious how the path of each vortex center can be computed step by step.

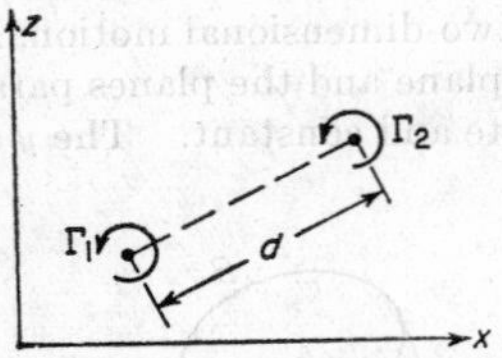

FIG. 71.25. Two straight parallel vortices.

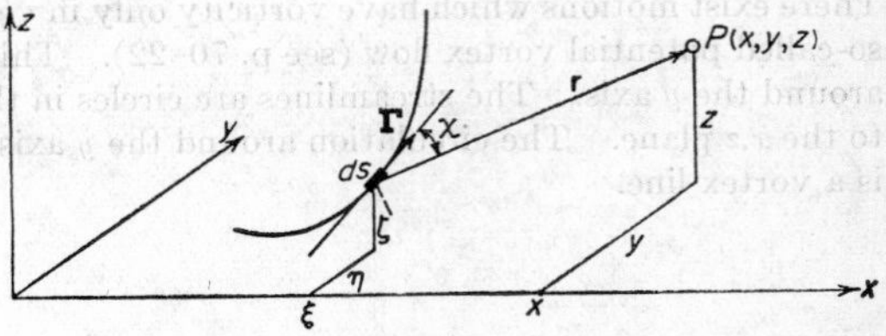

FIG. 71.26. Biot-Savart's law.

As the circulation of a vortex line does not change with time (consequence of Theorem 3), vortices cannot be created in a nonviscous fluid by a motion starting from rest.

The Velocity Field around a Vortex Filament

A thin vortex tube of arbitrary shape is usually called a vortex filament. The straight vortex is a very special case; the velocity field around it is well known. In the general case, the velocity field is given by an integral over the elements of the vortex filament. Consider a line element ds of the vortex filament (Fig. 71.26), with the vorticity $\boldsymbol{\omega}$ (components ω_x, ω_y, ω_z). The circulation $\boldsymbol{\Gamma}$ around this vortex element is $\boldsymbol{\Gamma} = \int \boldsymbol{\omega}\, d\sigma$, integrated over the small, but finite cross section of the vortex filament. Like $\boldsymbol{\omega}$, it is a vector pointing in the direction of the tangent of the axis of the filament.

The vortex element ds at ξ, η, ζ produces at the point $P(x,y,z)$ the velocity

$$d\mathbf{q} = \frac{\boldsymbol{\Gamma} \times \mathbf{r}\, ds}{4\pi r^3} \qquad dq = \frac{\Gamma\, ds}{4\pi}\frac{\sin\chi}{r^2} \tag{71.64}$$

where $r = [(x - \xi)^2 + (y - \eta)^2 + (z - \zeta)^2]^{1/2}$ is the distance between the vortex element and P, and χ is the angle between the vectors $\boldsymbol{\Gamma}\, ds$ and $\mathbf{r}$. The velocity $d\mathbf{q}$ is perpendicular to the plane through ds and $P(x,y,z)$. Equation (71.64) is often called Biot-Savart's theorem.[1] There is a complete analogy to the theorem in electrodynamics which describes the magnetic field vector produced by an electric current.

It is obvious that the computation of the velocity field around an arbitrarily shaped vortex filament can be rather troublesome. The velocity field around filaments which consist of portions of straight lines or which form a circular ring is easier.

Horseshoe Vortex

The horseshoe vortex (see Fig. 71.27) is a vortex filament that has found ample use in fluid dynamics, essentially in wing theory. The velocity at an arbitrary point P is

[1] For a derivation of Eq. (71.64), see [8], pp. 185–191; [4], pp. 208–211; and [9], pp. 201–206.

composed of three parts due to the center part I (span $b = 2s$) and the parts II and III (trailing vortices). Positive circulation is indicated by the circular arrows. If the elements ds of a filament lie on a straight line, a modified form of Eq. (71.64) is

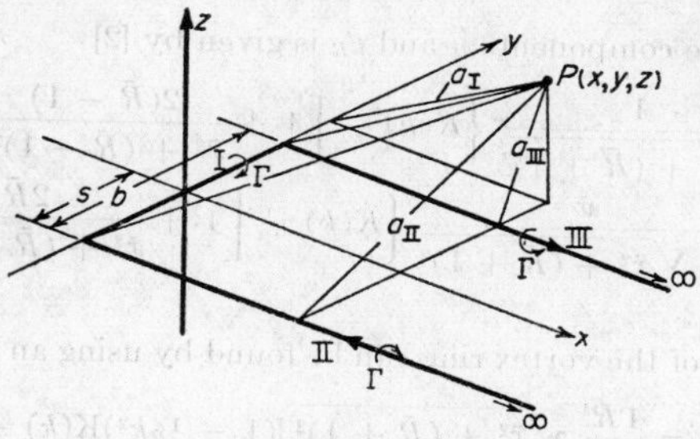

Fig. 71.27. Horseshoe vortex.

convenient. Upon introducing the shortest distance a between the point, at which the velocity is desired, and the straight line (see Fig. 71.27), we obtain

$$dq = \frac{\Gamma}{4\pi} \frac{\sin \chi \, d\chi}{a} \tag{71.65}$$

Upon using this form, the necessary integrations become extremely simple for the horseshoe vortex. The velocities and their components due to parts I, II, III are

$$q_{I} = \frac{\Gamma}{4\pi} \frac{1}{\sqrt{x^2 + z^2}} \left(\frac{y + s}{\sqrt{x^2 + z^2 + (y + s)^2}} - \frac{y - s}{\sqrt{x^2 + z^2 + (y - s)^2}} \right) \tag{71.66}$$

with

$$u_{I} = q_{I} \frac{z}{\sqrt{x^2 + z^2}} \qquad v_{I} = 0 \qquad w_{I} = -q_{I} \frac{x}{\sqrt{x^2 + z^2}}$$

$$q_{II} = \frac{\Gamma}{4\pi} \frac{1}{\sqrt{z^2 + (y + s)^2}} \left(1 + \frac{x}{\sqrt{x^2 + z^2 + (y + s)^2}} \right) \tag{71.67}$$

with

$$u_{II} = 0 \qquad v_{II} = q_{II} \frac{z}{\sqrt{z^2 + (y + s)^2}} \qquad w_{II} = -q_{II} \frac{y + s}{\sqrt{z^2 + (y + s)^2}}$$

$$q_{III} = \frac{\Gamma}{4\pi} \frac{1}{\sqrt{z^2 + (y - s)^2}} \left(1 + \frac{x}{\sqrt{x^2 + z^2 + (y - s)^2}} \right) \tag{71.68}$$

with

$$u_{III} = 0 \qquad v_{III} = -q_{III} \frac{z}{\sqrt{x^2 + (y - s)^2}} \qquad w_{III} = q_{III} \frac{y - s}{\sqrt{z^2 + (y - s)^2}}$$

Vortex Ring

A vortex filament forms a circular ring in the y,z plane (see Fig. 71.9). The velocity components at a point $P(x,R,\omega)$ are independent of ω, since the field has axial symmetry. We choose a point P with $\omega = 0$ for computing u and v_R. With the abbreviations $x/R' = \bar{x}$ and $R/R' = \bar{R}$, the following integrals are obtained:

$$u(x,R) = -\frac{\Gamma}{4\pi R'} \int_0^{2\pi} \frac{(\bar{R} \cos \omega' - 1) \, d\omega'}{(\bar{x}^2 + \bar{R}^2 + 1 - 2\bar{R} \cos \omega')^{3/2}}$$

$$v_R(x,R) = \frac{\Gamma}{4\pi R'} \int_0^{2\pi} \frac{\bar{x} \cos \omega' \, d\omega'}{(\bar{x}^2 + \bar{R}^2 + 1 - 2\bar{R} \cos \omega')^{3/2}} \tag{71.69a,b}$$

These integrals can be evaluated by using complete elliptic integrals of modulus k where

$$k^2 = \frac{4\bar{R}}{\bar{x}^2 + (\bar{R}+1)^2} \tag{71.70}$$

The final form for the components u and v_R is given by [2]

$$u(x,R) = \frac{\Gamma}{2\pi R'} \frac{1}{\sqrt{\bar{x}^2 + (\bar{R}+1)^2}} \left\{K(k) - \left[1 + \frac{2(\bar{R}-1)}{\bar{x}^2 + (\bar{R}-1)^2}\right] E(k)\right\}$$

$$v_R(x,R) = -\frac{\Gamma}{2\pi R'} \frac{\bar{x}}{\bar{R}\sqrt{\bar{x}^2 + (\bar{R}+1)^2}} \left\{K(k) - \left[1 + \frac{2\bar{R}}{\bar{x}^2 + (\bar{R}-1)^2}\right] E(k)\right\} \tag{71.71a,b}$$

The stream function of the vortex ring can be found by using an equation like (71.25):

$$\psi(x,R) = \frac{\Gamma R'}{2\pi} \sqrt{\bar{x}^2 + (\bar{R}+1)^2}\,[(1 - \tfrac{1}{2}k^2)\mathrm{K}(k) - E(k)] \tag{71.72}$$

The velocity components and the stream function are tabulated in [2], pp. 307, 309, and 311 (notation slightly different).

The Use of Vortex Filaments for Producing Resultant Fluid Forces on Three-dimensional Obstacles

As is shown in the chapter on airfoil theory, the horseshoe vortex can be used to generate lift in a section of a finite wing (Fig. 71.28). In this case, part I is called the "bound" portion of the filament and located in the wing, while II and III are the "free" trailing vortices.

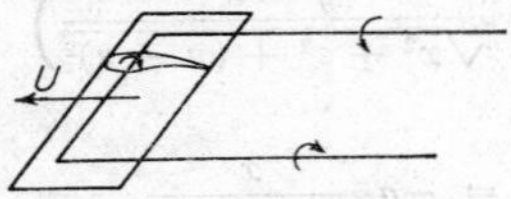

FIG. 71.28. Finite wing.

FIG. 71.29. Doubly connected body.

The fact that vortices cannot end in the fluid-filled space makes it impossible to create circulation around a section of a simply connected three-dimensional body without having free vortices in the fluid field. For bodies of the topologically next higher form (see Fig. 71.29), it is possible to arrange circulation around sections without the need of having free ends of the vortex filament floating in the surrounding fluid. In such a doubly connected body, a closed vortex filament can be arranged in the interior.

71.7. REFERENCES

[1] P. M. Morse, H. Feshbach: "Methods of Theoretical Physics," McGraw-Hill, New York, 1953.

[2] D. Küchemann, J. Weber: "Aerodynamics of Propulsion," McGraw-Hill, New York, 1953.

[3] M. M. Munk: Fluid mechanics, in W. F. Durand (ed.), "Aerodynamic Theory," vol. 1, pp. 224–304, Springer, Berlin, 1934.

[4] H. Lamb: "Hydrodynamics," 6th ed., Dover, New York, 1945.

[5] I. Lotz: Calculation of potential flow past arbitrary bodies in yaw, *NACA Tech. Mem.* 675 (1932), transl. from *Ing.-Arch.*, **2** (1931), 507–527.

[6] A. M. O. Smith, J. Pierce: Exact solutions of the Neumann problem; calculation of noncirculatory plane and axially symmetric flows about or within arbitrary boundaries, *Proc. 3d U.S. Natl. Congr. Appl. Mechanics*, Providence, 1958, pp. 807–815.

[7] Th. von Kármán: "Collected Works," vol. 2 (1914–1932), pp. 253–273, Butterworth, London, 1956; English transl. in *NACA Tech. Mem.* 574 (1930).

[8] V. L. Streeter: "Fluid Dynamics," McGraw-Hill, New York, 1948.

[9] L. Prandtl, O. G. Tietjens: "Fundamentals of Hydro- and Aeromechanics," McGraw-Hill, New York, 1934.

CHAPTER 72

AIRFOIL THEORY

BY

A. ROBINSON, Ph.D., D.Sc., Jerusalem, Israel[1]

72.1. INTRODUCTORY REMARKS

Wing theory, or airfoil theory, is concerned with the aerodynamic forces which act on a lifting surface in flight. The information it provides is required for performance and structural strength calculations or, in the unsteady case, for purposes of stability and control. The present chapter will be concerned only with steady conditions. Moreover, it will be assumed that the fluid medium through which the wing moves (for most practical purposes, air) is incompressible and inviscid. This assumption implies that the forward speed of the wing is considerably smaller than the velocity of sound. As the forward speed increases, the results calculated by the methods of the present chapter become gradually less reliable, until a period is reached when the effects of compressibility become critical.

Viscosity affects the aerodynamic forces and their calculations in a rather different way. Suppose that a wing of finite dimensions moves with constant speed and direction through an inviscid fluid. It is natural to suppose that the motion of the wing must have started from rest at some previous moment, however remote. Standard hydrodynamics leads us to conclude that the fluid is then irrotational. However, in that case, it can be proved that the resultant forces on the wing vanish. This conclusion, which is contrary to experience is known as *d'Alembert's paradox*. In actual fact, the action of the viscosity, however small, produces regions of rotational flow (vorticity). Though the presence of viscosity can be neglected for many practical purposes, the vorticity and circulation, which are ultimately due to viscosity, must be taken into account in any wing theory. Theoretical arguments and the interpretation of empirical evidence then lead to the result that these effects can be taken into account in an inviscid theory by the introduction of the following two conditions:

1. The Joukowski Condition. The velocity field remains finite in the neighborhood of the wing trailing edge, even though the latter is assumed to be sharp.

2. The Vortex Wake. The free vorticity in the fluid is concentrated in a layer which extends downstream from the trailing edge. For practical purposes, it may be assumed that this layer consists of vortex lines which are parallel to the direction of the free stream (in a system of reference which is at rest relative to the wing.)

A more detailed theory brings out the fact that, even if the effects of viscosity are taken into account directly, it is sufficient to suppose that they are confined to a thin layer surrounding, and aft of, the wing, the so-called boundary layer, but that now shear stresses must be taken into account, which act on the surface of the wing (and which are, of course, not obtained in an inviscid theory). It follows that the resultant forces and moments which are due to the normal pressures, such as the lift and the pitching moment, are (except for terms of a higher order of smallness) furnished correctly by the theory of the present chapter, whereas the forces and moments to

[1] The author is indebted to Miss Naomi Salinger for preparing the diagrams of this chapter.

which the tangential stresses make an important contribution, such as the drag and the yawing moment, can be obtained by means of it only in part. In the latter cases, the results obtained in this chapter have to be supplemented, even in the first approximation, by considerations of boundary-layer theory.

Throughout this chapter, we shall use a frame of reference which is fixed relative to the wing. We shall denote the velocity vector by $\mathbf{q} = (u,v,w)$, where u, v, and w are the velocity components in the direction of the x, y, and z axes, respectively. Since the flow is supposed to be irrotational (except in the infinitely thin vortex wake), there exists a velocity potential ϕ such that

$$\mathbf{q} = \text{grad}\ \phi \qquad u = \frac{\partial \phi}{\partial x} \qquad v = \frac{\partial \phi}{\partial y} \qquad w = \frac{\partial \phi}{\partial z} \tag{72.1}$$

Also, the equation of continuity is, for steady incompressible flow,

$$\text{div}\ \mathbf{q} = \frac{\partial u}{\partial x} + \frac{\partial v}{\partial y} + \frac{\partial w}{\partial z} = 0$$

or

$$\text{div grad}\ \phi = \frac{\partial^2 \phi}{\partial x^2} + \frac{\partial^2 \phi}{\partial y^2} + \frac{\partial^2 \phi}{\partial z^2} = 0 \tag{72.2}$$

which is Laplace's equation.

Let (u_0,v_0,w_0) be the free-stream velocity; then

$$\mathbf{q}' = (u',v',w') = (u - u_0,\ v - v_0,\ w - w_0)$$

is said to be the induced velocity. $\mathbf{q}'$ is equal to the gradient of the induced velocity potential ϕ', where

$$\phi' = \phi - u_0 x - v_0 y - w_0 z$$

Both ϕ and ϕ' are unique except for the addition of an arbitrary constant, and ϕ' also is a solution of Laplace's equation.

The pressure at any point of the field of flow is given by Bernoulli's equation, which reads, for incompressible flow,

$$p + \tfrac{1}{2}\rho q^2 = p + \tfrac{1}{2}\rho(u^2 + v^2 + w^2) = H \tag{72.3}$$

where p is the pressure, ρ the density, and H a constant, which is the same for the entire field.

A topic which will not be considered in the present chapter is the theory of wind-tunnel interference. This theory suggests corrections which are to be made to the results of wind-tunnel measurements, in order to yield a better approximation to free-flight conditions (see [1], pp. 280–297).

72.2. TWO-DIMENSIONAL WING THEORY

Consider a wing of infinite aspect ratio, whose generators are parallel to the z axis. The cross section of the wing is the same in all planes parallel to the x,y plane, and it will, therefore, be sufficient to consider conditions in that plane alone. The velocity potential now depends only on x and y, and, being a solution of Laplace's equation, it may be regarded as the real part of an analytic function Φ of the complex variable $(x + iy)$. Since the third spatial dimension no longer intervenes in the calculations, we may write $z = x + iy$, $w = u - iv$, so that

$$\phi(x,y) = \Re\,\Phi(z) \qquad w(z) = \frac{d\Phi}{dz} \tag{72.4}$$

Moreover, by symmetry, a vortex wake such as described by condition 2, p. 72–1, must vanish under two-dimensional conditions, so that $w(z)$ is regular except possibly at or inside the contour of the wing. It can therefore be expanded into a Laurent series:

$$w(z) = A_0 + \frac{A_1}{z} + \frac{A_2}{z^2} + \cdots + \frac{A_n}{z^n} + \cdots \tag{72.5}$$

A_0 is the complex velocity at infinity, $A_0 = Ue^{-i\alpha}$ say, where U is the magnitude of the free-stream velocity and $u_0 = U \cos \alpha$, $v_0 = U \sin \alpha$ are its components.

The calculation of the force and moment which act on the wing is carried out most conveniently by means of the following formulas, which are due to Blasius [2].

Let S be a simple closed contour, which surrounds the trace of the wing in the x,y plane, W (Fig. 72.1). The wing being supposed fixed and impermeable, there is no flux of momentum across W. It follows that the rate of change of the total linear momentum $\mathbf{I}$ in the volume V (of unit thickness) between W and S is equal to the resultant pressure across S, $\mathbf{T}$ say, together with the force exerted on the fluid by the wing, $-\mathbf{R}$, and the rate of flow, or flux, of momentum across S into V, $\mathbf{f}$. But $d\mathbf{I}/dt = 0$, since we are dealing with steady conditions, and so

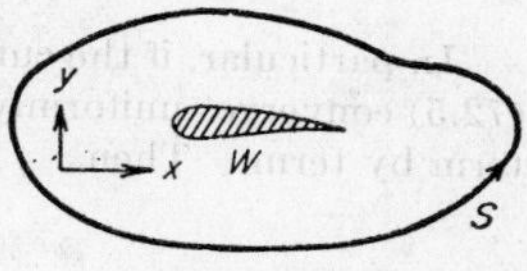

FIG. 72.1

$$\mathbf{R} = \mathbf{T} + \mathbf{f} \tag{72.6}$$

where $\mathbf{R}$ is the force (per unit span) exerted by the fluid on the wing.

Now,

$$\mathbf{T} = -\int_S p(dy, -dx)$$

and

$$\mathbf{f} = \rho \int_S (u,v)(-u\,dy + v\,dx)$$

Hence,

$$\mathbf{R} = \int_S [\rho(-dy,dx) + (u,v)(-u\,dy + v\,dx)] \tag{72.7}$$

Note that, here and elsewhere, all integrals will be taken in counterclockwise direction.

Also, by Bernoulli's equation (72.3).

$$p = H - \tfrac{1}{2}\rho(u^2 + v^2)$$

Substituting in (72.7) and simplifying, we obtain

$$\mathbf{R} = (X,Y) = \tfrac{1}{2}\rho \int_S \{(u^2 + v^2)(dy,-dx) + 2(u,v)(-u\,dy + v\,dx)\}$$

or, in complex notation,

$$X - iY = \tfrac{1}{2}\rho i \int_S w^2\,dz \tag{72.8}$$

This is the first of Blasius' formulas. There exists a similar formula for the moment M of the aerodynamic forces on the wing about a point (x_0,y_0). (Note that a nose-down moment is positive.) It is based on the fact that M is equal to the moment about (x_0,y_0) exerted by the pressure on V across S, together with the flux of angular momentum about (x_0,y_0) across S into V; in symbols:

$$M = \int_S p[(x - x_0)\,dx + (y - y_0)\,dy] + \rho \int_S [(x - x_0)v - (y - y_0)u](v\,dx - u\,dy)$$

Again making use of Bernoulli's equation, we obtain, after some algebra,

$$M = -\tfrac{1}{2}\rho\,\mathcal{R} \int_S w^2(z - z_0)\,dz \tag{72.9}$$

where

$$z_0 = x_0 + iy_0$$

Still referring to the same contour S, we recall that the circulation around S is given by

$$\Gamma = \int_S (u\,dx + v\,dy)$$

while the flow across S is

$$\Delta = \int_S (u\,dy - v\,dx)$$

Hence,

$$\Gamma + i\Delta = \int_S (u - iv)(dx + i\,dy) = \int_S w\,dz \tag{72.10}$$

In particular, if the curve S is sufficiently far from the origin so that the expansion (72.5) converges uniformly on S, we may substitute (72.5) into (72.10) and integrate, term by term. Then

$$\Gamma + i\Delta = 2\pi i A_1$$

But, if the wing is impermeable, then the flow across S must vanish ($\Delta = 0$), and, even with suction present, Δ can usually be neglected. Then

$$A_1 = -\frac{i}{2\pi}\Gamma \tag{72.11}$$

that is, A_1 is pure imaginary. Then,

$$w(z) = Ue^{-i\alpha} - \frac{i\Gamma}{2\pi z} + \frac{A_2}{z^2} + \cdots \tag{72.12}$$

Substituting this expression for w in (72.7) and integrating term by term, we obtain

$$X - iY = \rho i U e^{-i\alpha}\Gamma$$

Hence

$$X + iY = \rho U \Gamma e^{i(\alpha - \pi/2)} \tag{72.13}$$

$$|X + iY| = \rho U|\Gamma| \tag{72.14}$$

Equation (72.13) shows that the direction of the force (X,Y) is perpendicular to the direction of the free stream. Together with (72.14), which yields the magnitude of (X,Y), this constitutes the law of Kutta-Joukowski [3],[4]. If we define the drag D as the component of the aerodynamic force in the direction of the main flow, and the lift L as the component in the normal direction, then the law of Kutta-Joukowski may be restated in the form

$$L = -\rho U \Gamma \qquad D = 0 \tag{72.15}$$

In order to obtain the flow round a wing section, we may write down the complete potential of flow round a circular cylinder (i.e., a circle) in an auxiliary ζ plane, and then try to find a suitable complex function which maps the region outside the circle on the region outside a wing-shaped contour. The Joukowski condition (p. 72–1) requires that the wing have a sharp trailing edge. Since the contour is the image of a circle, it follows that the mapping must be singular at that point. A more detailed analysis shows that we must have $dz/d\zeta = 0$ at the trailing edge.

The general flow around a circular cylinder of radius a and center ζ_0 in the auxiliary ζ plane is given by the complex potential [see Eq. (70.44)]:

$$\Phi(\zeta) = Ue^{-i\alpha}(\zeta - \zeta_0) - \frac{i\Gamma}{2\pi}\ln(\zeta - \zeta_0) + \frac{Ue^{i\alpha}a^2}{\zeta - \zeta_0} \tag{72.16}$$

The corresponding complex velocity in the ζ plane is

$$\frac{d\Phi}{d\zeta} = U\left[e^{-i\alpha} - \frac{e^{i\alpha}a^2}{(\zeta - \zeta_0)^2}\right] - \frac{i\Gamma}{2\pi(\zeta - \zeta_0)} \tag{72.17}$$

Now let $z = f(\zeta)$ be a function which maps the region outside a certain airfoil in the z plane on the region outside the circle just considered, in such a way that the point at infinity in the ζ plane is mapped on the point at infinity in the z plane (Fig. 72.2). According to the theory of conformal mapping, $f(\zeta)$ is rendered unique by imposing the additional condition that a specified direction at infinity in the ζ plane be trans-

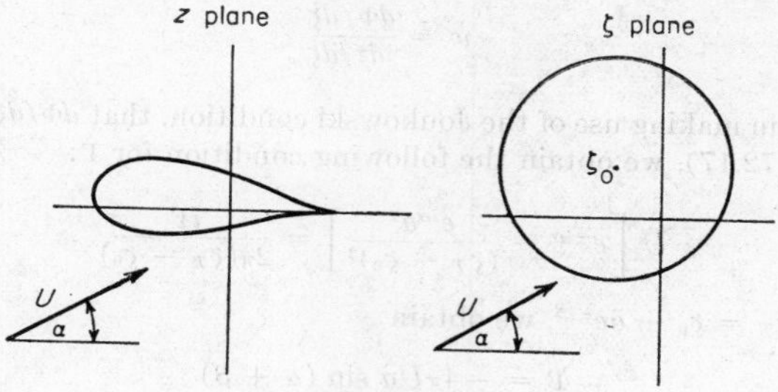

FIG. 72.2

formed into a specified direction in the z plane. For sufficiently large ζ, we may write $f(\zeta)$ in the form of a convergent Laurent series:

$$z = f(\zeta) = C_{-1}(\zeta - \zeta_0) + \sum_{n=0}^{\infty} \frac{C_n}{(\zeta - \zeta_0)^n} \tag{72.18}$$

Then $$\frac{dz}{d\zeta} = f'(\zeta) = C_{-1} - \frac{C_1}{(\zeta - \zeta_0)^2} - \frac{2C_2}{(\zeta - \zeta_0)^3} - \cdots$$

Hence, if $C_{-1} = 1$, then the corresponding directions at infinity in the ζ and z planes coincide. Also, the circulation round the wing is the same as that round the circular cylinder, Γ. It then follows from the law of Kutta-Joukowski that, for $C_{-1} = 1$, the forces which act on the wing and the cylinder are equal and parallel.

In order to calculate the moment on the wing in terms of the potential of flow around the cylinder and in terms of the mapping function $f(\zeta)$, we proceed as follows. According to the second Blasius formula (72.9), we have

$$M = -\tfrac{1}{2}\rho \Re \int w^2(z - z_0)\, dz$$

$$= -\tfrac{1}{2}\rho \Re \int w^2 \left[(\zeta - \zeta_0) + \sum_{n=1}^{\infty} \frac{C_n}{(\zeta - \zeta_0)^n} \right] \frac{dz}{d\zeta}\, d\zeta - \tfrac{1}{2}\rho \Re \left[(C_0 - z_0) \int w^2\, dz \right]$$

where we have assumed again that $C_{-1} = 1$, and where

$$w = \frac{d\Phi}{dz} = \frac{d\Phi}{d\zeta}\frac{d\zeta}{dz}$$

is the complex velocity in the (physical) z plane. Now the last term on the right-hand side may be written in the form

$$\Re[-i(C_0 - z_0)\tfrac{1}{2}\rho i \textstyle\int \omega^2\, dz] = \mathscr{I}[(C_0 - z_0)(X - iY)]$$

Also, after some algebra, the first term is found to be equal to $2\pi\rho U^2 \mathscr{I}(C_1 e^{-2i\alpha})$. Hence,

$$M = 2\pi\rho U^2 \mathscr{I}(C_1 e^{-2i\alpha}) - \mathscr{I}[(C_0 - z_0)(X - iY)] \tag{72.19}$$

As stated earlier, the Joukowski condition implies that

$$f'(\zeta) = \frac{dz}{d\zeta} = 0$$

at the trailing edge, for $\zeta = \zeta_T$. Since the complex velocity w in the z plane is given by

$$w = \frac{d\Phi/d\zeta}{dz/d\zeta}$$

we conclude, again making use of the Joukowski condition, that $d\Phi/d\zeta = 0$ for $\zeta = \zeta_T$. Substituting in (72.17), we obtain the following condition for Γ:

$$U\left[e^{-i\alpha} - \frac{e^{i\alpha}a^2}{(\zeta_T - \zeta_0)^2}\right] = \frac{i\Gamma}{2\pi(\zeta_T - \zeta_0)}$$

Hence, writing $\zeta_T = \zeta_0 + ae^{-i\beta}$, we obtain

$$\Gamma = -4\pi Ua \sin(\alpha + \beta) \tag{72.20}$$

This yields the following expression for the lift:

$$L = 4\pi\rho U^2 a \sin(\alpha + \beta) \tag{72.21}$$

Again, putting

$$C_0 - z_0 = re^{-i\psi} \qquad C_1 = d^2 e^{2i\gamma}$$

and substituting in (72.19), we obtain

$$M = 2\pi\rho U^2[2ar \sin(\alpha + \beta) \cos(\psi - \alpha) + d^2 \sin 2(\gamma - \alpha)] \tag{72.22}$$

Note that α is the angle of attack measured in terms of the inclination of the flow relative to the x axis. For $\alpha = -\beta$, the lift vanishes so that $\alpha = -\beta$ is the *angle of zero lift*, and $(\alpha + \beta)$ measures the deviation from the angle of zero lift. $(\alpha + \beta)$ is called the *effective angle of attack*. Also, for $\alpha = -\beta$, the moment becomes

$$M = M_0 = 2\pi\rho U^2 d^2 \sin 2(\gamma + \beta)$$

and a simple calculation then shows that, in general,

$$M = M_0 + 4\pi\rho U^2 \sin(\alpha + \beta)[ar \cos(\psi - \alpha) - d^2 \cos(2\gamma + \beta - \alpha)] \tag{72.23}$$

We may ask whether there exists a point about which the moment is independent of the angle of attack. This will be the case if the expression in the brackets on the right-hand side of (72.23) vanishes, yielding the condition

$$r = \frac{d^2}{a} \qquad \psi = 2\gamma + \beta$$

and so

$$z_0 = C_0 - \frac{d^2}{a} e^{i(2\gamma+\beta)} \tag{72.24}$$

z_0 is called the *aerodynamic center of the wing*.

Besides formulas for the over-all force and moment on the wing, it is also important to have an expression for the pressure distribution. This is done by the use of Bernoulli's equation, which may be written in the form

$$p = p_0 + \tfrac{1}{2}\rho(U^2 - q^2)$$

where p_0 is the pressure at infinity. $q = |w|$ can be expressed conveniently in terms of the polar coordinate θ, which varies from 0 to 2π along the auxiliary circle, so that

$$\zeta = \zeta_0 + ae^{i\theta}$$

The result is

$$q = |w| = 4U \left| \cos\left(\alpha + \frac{\beta - \theta}{2}\right) \sin\frac{\beta + \theta}{2} \right| \left| \frac{d\zeta}{dz} \right| \qquad (72.25)$$

Studying this expression in the region of the trailing edge, we find that the velocity vanishes at the trailing edge except when the latter is a cusp (i.e., when the tangents to the top and bottom surfaces of the wing are parallel at that point).

72.3. JOUKOWSKI AIRFOILS [5]

Following a course of development similar to that found in other branches of engineering science, the solution of the problem of finding the pressure distribution around wings of arbitrary shape was preceded by the solution of the problem for a particular family of shapes, which was selected precisely for the ease with which the problem under consideration can be solved. The study of the family of wings in question—the so-called Joukowski airfoils—is still of considerable interest, since these shapes display many properties which apply, exactly or approximately, to wings in general. The point of the method is that, instead of specifying a wing and trying to find the corresponding transformation to a circle—a problem to which an exact solution is usually not to be found, and to which even an approximate solution can be determined only with some difficulty—we choose a simple mapping function, and then try to find circles in the ζ plane which are transformed into winglike shapes in the z plane. The transformation in question is

$$z = \zeta + \frac{d^2}{\zeta} \qquad d > 0 \qquad (72.26)$$

and is known as the *Joukowski transformation.* It is, in the order of nonvanishing C_n which appear in (72.18), the simplest possible transformation which does not map all circles on circles. Note that

$$\frac{dz}{d\zeta} = 1 - \frac{d^2}{\zeta^2} \qquad (72.27)$$

so that the transformation is singular for $\zeta = 0$, where the derivative becomes infinite, and for $\zeta = \pm d$, where it vanishes.

When choosing a circle in the ζ plane, as explained, we must make the trailing edge correspond to a singular point of the transformation, so as to satisfy the Joukowski condition (for, otherwise, the smooth outline of the circle yields a blunt trailing edge). Accordingly, we choose $\zeta_T = d$. It can be shown that every wing section obtained on this basis has a cusp at the trailing edge. It remains for us only to select the center of the circle, ζ_0.

We take first $\zeta_0 = 0$, so that $a = d$ (Fig. 72.3). This yields, for points $\zeta = de^{i\theta}$ on the circle,

$$z = 2d \cos \theta$$

that is, as ζ varies round the circle, z varies along the straight segment $(-2d \leq z \leq 2d)$ from the trailing edge to the leading edge and back. The chord length, i.e., the distance from leading edge to trailing edge, is $c = 4d = 4a$, and the formula for the lift is [compare (72.21)]

$$L = \pi\rho U^2 c \sin \alpha \approx \pi\rho U^2 c\alpha \qquad (72.28)$$

The lift coefficient C_L is defined by $C_L = L/\tfrac{1}{2}\rho U^2 c$, and so

$$C_L = 2\pi \sin \alpha \approx 2\pi\alpha \qquad (72.29)$$

and

$$\frac{dC_L}{d\alpha} \approx 2\pi \qquad (72.30)$$

the approximate relations being valid for small angles α.

The position of the aerodynamic center [see (72.25)] is given by $z_0 = -d = -\frac{1}{4}c$; that is, the aerodynamic center is located one-quarter of the chord aft of the leading edge, at the so-called quarter-chord point. Also, at zero incidence, the moment acting on the wing (about a given point) vanishes, since the pressure is constant throughout, and so the moment about the aerodynamic center vanishes at all incidences. Thus, the force on the wing always acts through the aerodynamic center.

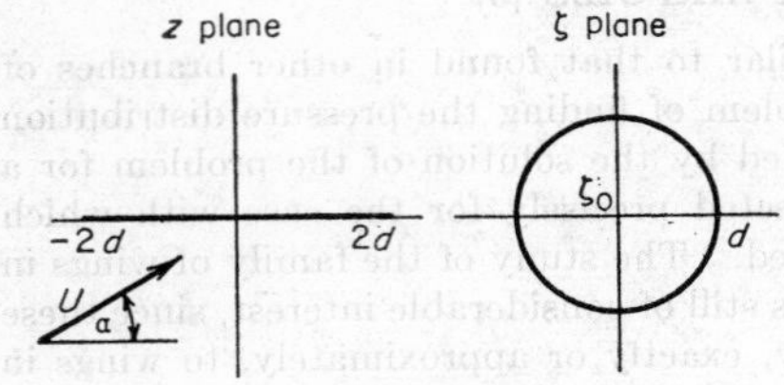

FIG. 72.3

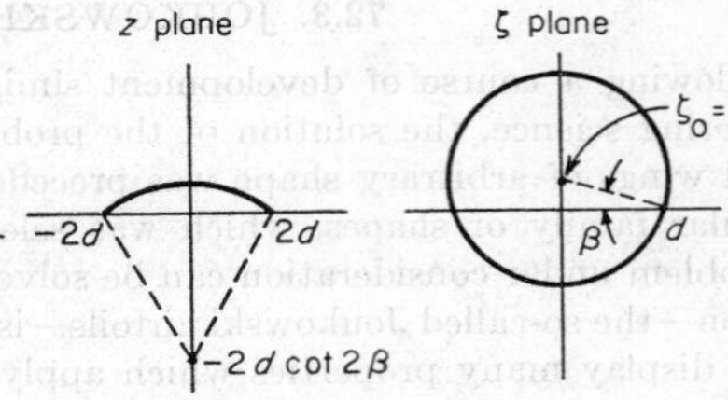

FIG. 72.4

The moment around any other point, e.g., around the mid-chord, can then be found from statics or from one of our formulas:

$$M = -\tfrac{1}{8}\pi\rho U^2 c^2 \sin 2\alpha \approx -\tfrac{1}{4}\pi\rho U^2 c^2 \alpha$$

The corresponding moment coefficient C_M is defined by $C_M = M/\frac{1}{2}\rho U^2 c^2$, and so, in the present case,

$$C_M \approx -\tfrac{1}{2}\pi\alpha \qquad \frac{dC_M}{d\alpha} \approx -\frac{\pi}{2} \tag{72.31}$$

The velocity component at the wing and tangential to it is given by

$$u = U\frac{\cos(\alpha - \frac{1}{2}\theta)}{\cos\frac{1}{2}\theta}$$

and, since the normal component must vanish, we have

$$q = |u| = U\frac{\cos(\alpha - \frac{1}{2}\theta)}{\cos\frac{1}{2}\theta} \tag{72.32}$$

From this formula, the pressure distribution is calculated by means of Bernoulli's equation. The induced velocity component in the direction of the x axis is

$$u' = u - U\cos\alpha = -U\sin\alpha\tan\frac{\theta}{2}$$

Thus, the induced tangential velocity is infinite at the leading edge, and vanishes at the trailing edge.

Next, we take ζ_0 on the positive imaginary axis (Fig. 72.4). Then $\zeta_0 = id\tan\beta$ and $a = d\sec\beta$, since the circle passes through $\zeta_T = d$. Some algebra shows that the corresponding contour in the z plane is given by

$$x^2 + (y + 2d\cot 2\beta)^2 = 4d^2\csc^2 2\beta$$

This is a circular arc with center $y = -2d\cot 2\beta$ on the y axis and radius $2d\csc 2\beta$. The leading and trailing edges of the wing are on the x axis and are given by $x = \pm 2d$, as before, and the maximum ordinate of the wing is on the y axis and is given by $y = 2d\tan\beta$.

The following general definitions have found wide acceptance. The *leading edge* is the point of maximum curvature near the nose of the airfoil. This agrees with our previous terminology, for the case of a sharp nose. The *chord* is the straight segment from leading to trailing edge. The *mean camber line* is the locus of mid-points of segments cut out by the wing section on the straight lines perpendicular to the chord. For example, suppose that the chord is parallel to the x axis, and that the equations of the upper and lower surfaces of the wing are $y = y_u(x)$ and $y = y_l(x)$. Then the equation of the mean camber line is

$$y = y_c(x) = \tfrac{1}{2}[y_u(x) + y_l(x)]$$

The *camber* χ of the wing is defined as the ratio of the maximum distance of the mean camber line from the chord and the chord length c.

Whenever the wing section is infinitely thin, as is the case for the circular arc, it coincides with its mean camber line. Thus, in the concrete case now under consideration, the camber is given by

$$\chi = \frac{2d \tan \beta}{4d} \approx \tfrac{1}{2}\beta$$

The expression for the lift now is

$$L = \pi \rho U^2 c \frac{\sin (\alpha + \beta)}{\cos \beta}$$

and so, for small β,

$$\frac{dC_L}{d\alpha} \approx 2\pi \qquad \text{as before}$$

Similarly, we find that, to the first approximation, the aerodynamic center is still located at the quarter chord. The moment about the aerodynamic center is

$$M = \tfrac{1}{8}\pi \rho U^2 c^2 \sin 2\beta \approx \tfrac{1}{2}\pi \rho U^2 c^2 \chi$$

and the corresponding moment coefficient is $C_M \approx \pi\chi$.

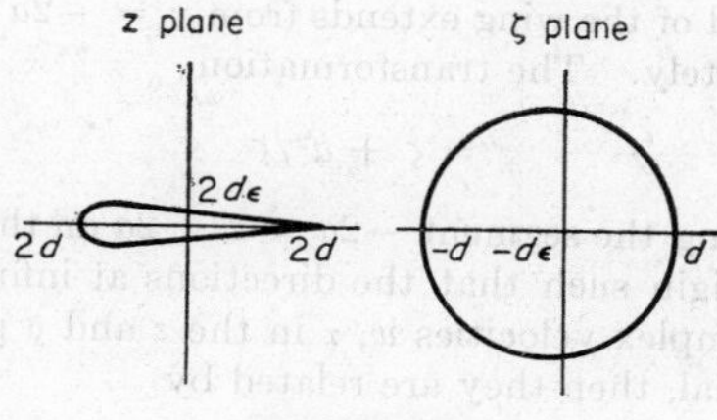

FIG. 72.5

Next we wish to investigate the effect of thickness. For this purpose, we choose $\zeta_0 = -\epsilon d$, where ϵ is small and positive (Fig. 72.5). In this case, the leading edge is at

$$z_L \approx -2d(1 + \epsilon^2) \approx -2d$$

except for terms of the second order of smallness in ϵ. To the same degree of accuracy, the chord is still given by $c = 4d$, and a parametric equation of the airfoil is

$$x = 2d \cos \theta \qquad y = d\epsilon(2 \sin \theta - \sin 2\theta) = 2d\epsilon \sin \theta(1 - \cos \theta)$$

Differentiation shows that y possesses a single maximum, at $y_0 = \tfrac{3}{2}\sqrt{3}\, d\epsilon$, for $\theta = 120°$. Thus, the maximum thickness of the airfoil is $2y_0$, and its maximum thickness-chord ratio is

$$\left(\frac{t}{c}\right)_0 = \frac{2y_0}{c} \approx 1.30\epsilon$$

Except for terms of the second or higher orders of smallness in ϵ, the aerodynamic center is still at the quarter chord. The lift is given by

$$L = \pi\rho U^2 c(1 + \epsilon) \sin \alpha$$

indicating an increase of lift with thickness. Owing to the presence of viscosity (boundary layer) effects, this result disagrees with experience.

The general wing of the Joukowski family is obtained by taking ζ_0 in a general position, thus obtaining a wing with both thickness and camber. In order to obtain more general families of wings, it is natural to consider, in place of the Joukowski transformation, somewhat more general transformations, such as

$$z = f(\zeta) = \zeta + \frac{C_2}{\zeta} + \frac{C_3}{\zeta^2} + \cdots + \frac{C_{n+1}}{\zeta^n}$$

or

$$\frac{z - 2d}{z + 2d} = \left(\frac{\zeta - d}{\zeta + d}\right)^\nu$$

The first transformation has been considered in this connection by von Mises [6]; the second by von Kármán and Trefftz [7]. It is found that the first transformation still leads to cusped trailing edges, while the second yields a trailing edge angle ϵ_T (that is, the angle between the tangents to the upper and lower surfaces of the wing at the trailing edge) for

$$\nu = 2 - \epsilon_T/\pi$$

72.4. THIN AIRFOIL THEORY

We now turn to the more difficult problem of determining the pressure distribution on a wing of given shape. The theory which we proceed to describe provides an approximate solution to the problem. It has the advantage of relative simplicity coupled, nevertheless, with wide applicability.

Suppose that the chord of the wing extends from $x = -2a$ to $x = 2a$ on the x axis in the z plane, approximately. The transformation

$$z = \zeta + a^2/\zeta$$

maps the z plane, cut along the segment $-2a < z < 2a$ on the region outside a circle of radius a round the origin such that the directions at infinity remain unchanged. We recall that, if two complex velocities w, τ in the z and ζ planes are derived from the same complex potential, then they are related by

$$\tau = \frac{d\phi}{d\zeta} = \frac{d\phi}{dz}\frac{dz}{d\zeta} = w\frac{dz}{d\zeta}$$

Similarly, defining the induced complex velocity w' by $w' = w - Ue^{-i\alpha}$, we may introduce the transformed induced velocity by

$$\tau' = w'\frac{dz}{d\zeta} = w'\left(1 - \frac{a^2}{\zeta^2}\right) \tag{72.33}$$

Suppose that τ' is given by a Laurent series, which converges for $|\zeta| \geq a$:

$$\tau' = \frac{T_1}{\zeta} + \frac{T_2}{\zeta^2} + \cdots + \frac{T_n}{\zeta^n} + \cdots \tag{72.34}$$

Write

$$\frac{T_n}{a^n} = \tau_n + i\sigma_n \qquad n = 1, 2, 3, \ldots$$

and suppose that the wing neither absorbs nor emits fluid (or that the amount of fluid emitted or absorbed is negligible). Then T_1 is pure imaginary, $\tau_1 = 0$.

Making use of (72.33), we then obtain the following expressions for the velocity components u and v for points of the segment $(-2a,2a)$:

$$u = U\cos\alpha - \frac{1}{2\sin\theta}(-\sigma_1 + \tau_2\sin\theta - \sigma_2\cos\theta + \tau_3\sin 2\theta - \sigma_3\cos 2\theta + \cdots) \tag{72.35}$$

$$v = U\sin\alpha + \frac{1}{2\sin\theta}(\tau_2\cos\theta + \sigma_2\sin\theta + \tau_3\cos 2\theta + \sigma_3\sin 2\theta + \cdots) \tag{72.36}$$

Let β be the slope at any point of the wing. Then the boundary condition is, at any such point,

$$u/v = \beta \tag{72.37}$$

We now introduce the following simplifying assumptions. First, we make use of the fact that the longitudinal velocity is, on the whole, small compared with the free-stream velocity. Accordingly, we replace u in the denominator of (72.37) by $U\cos\alpha$. Second, we apply the boundary conditions at the normal projection of a point of the wing on the segment $(-2a,2a)$ rather than at the point itself. It is then convenient to express the slope as a function of the corresponding coordinate θ $[\beta = \beta(\theta)]$. Equation (72.37) now becomes

$$v = U\beta(\theta)\cos\alpha \tag{72.38}$$

Substituting the right-hand side of (72.36) for v in (72.38), we may, after a slight modification, determine $\sigma_2, \tau_2, \sigma_3, \tau_3, \ldots$ as the coefficients of the Fourier series for

$$2U\sin\theta[\beta(\theta)\cos\alpha - \sin\alpha]$$

σ_1 does not appear in (72.36), but is obtained by applying the Joukowski condition to (72.35). For u to remain finite at the trailing edge, the expression in parentheses on the right-hand side of Eq. (72.35) must vanish, hence;

$$\sigma_1 = -\sigma_2 - \sigma_3 - \cdots$$

The final result is expressed conveniently in terms of the integrals

$$\alpha_n = \frac{\cos\alpha}{\pi}\int_0^{2\pi}\beta(\theta)\cos n\theta\,d\theta \qquad \beta_n = \frac{\cos\alpha}{\pi}\int_0^{2\pi}\beta(\theta)\sin n\theta\,d\theta \qquad n = 1, 2, \ldots \tag{72.39}$$

and

$$\alpha_0 = \frac{\cos\alpha}{\pi}\int_0^{2\pi}\beta(\theta)\,d\theta - 2\sin\alpha$$

which are related to σ_n, τ_n by

$$\sigma_n = -U(\alpha_n - \alpha_{n-2}) \qquad \tau_n = U(\beta_n - \beta_{n-2}) \qquad n = 2, 3, \ldots$$

We obtain for u at the wing [i.e., at the segment $(-2a,2a)$]

$$u = U\left[\cos\alpha - \tfrac{1}{2}\alpha_0\tan\frac{\theta}{2} + \sum_{n=1}^{\infty}(\beta_n\cos n\theta - \alpha_n\sin n\theta)\right] \tag{72.40}$$

while the expression for v is, in terms of the same coefficients,

$$v = U\left[\sin\alpha + \tfrac{1}{2}\alpha_0 + \sum_{n=1}(\alpha_n\cos n\theta + \beta_n\sin n\theta)\right] \tag{72.41}$$

The condition that the wing close up implies that $\beta_1 = 0$.

It will be seen that the infinite series in (72.40) is conjugate to the infinite series in (72.41). However, in addition, Eq. (72.40) contains the term $-\frac{1}{2}\alpha_0 \tan \frac{1}{2}\theta$, which corresponds to $\frac{1}{2}\alpha_0$ in the expression for v. This term—which yields the solution for a flat plate—becomes infinite at the leading edge, and it is because of its presence that, in general, u does not possess a Fourier expansion at the wing.

Having found u for given $\beta(\theta)$, we may calculate the pressure distribution over the wing. In keeping with the general quality of the approximation, we use the linearized form of Bernoulli's equation, and obtain

$$p - p_0 = -\rho U(u' \cos \alpha + v' \sin \alpha)$$

where $u' = u - U \cos \alpha$, $v' = v - U \sin \alpha$, or, even more simply,

$$p - p_0 = -Uu' \tag{72.42}$$

Also, using (72.15) and (72.19), and introducing the usual approximations, we obtain, for the lift and moment coefficients,

$$C_L = 2\pi(\alpha + \epsilon_0) \qquad C_{M0} = \tfrac{1}{2}\pi\epsilon_0 - 2\mu_0$$

where

$$\epsilon_0 = \frac{2}{\pi}\int_0^\pi \frac{y_m}{c}\frac{d\theta}{1 - \cos\theta} \qquad \mu_0 = -\int_0^\pi \frac{y_m}{c}\cos\theta\, d\theta$$

and where

$$y = y_m(x) = y_m(2a \cos \theta)$$

is the equation of the mean camber line.

The drag on the wing is still equal to zero, in agreement with a general result obtained previously. However, if, instead of using (72.15), we integrate the pressure over the wing surface, then the resulting force component in the direction of the stream is, in general, different from zero, and this is true even if we make use of the exact Bernoulli equation. This is obvious in the case of a flat plate at incidence where the resultant force, as calculated from the pressure, will, by necessity, be normal to the wing and not to the direction of flow. The discrepancy is due to the fact that the induced velocity may become infinite at the leading edge. This implies the presence of a concentrated force at the leading edge, the so-called suction force. The component of that force in the direction of flow precisely balances the component of the pressure integral over the wing in that direction, while the component of the force in the direction of the lift is negligible. Lighthill [8] has modified the theory so as to eliminate the infinity at the leading edge.

The above analysis is not completely general, since we had to assume that v' (though not u) possesses a Fourier expansion, in order to determine the coefficients α_n, β_n. However, for a specified pressure distribution, the possibility of a rounded leading edge (yielding infinite slope) cannot be ruled out, and, in order to take it into account, it is convenient to derive complementary formulas, as follows. Suppose that it is desired to design a wing section to a given pressure distribution. Accepting (72.42), we may then determine u'. Suppose that

$$u' = U\left[\tfrac{1}{2}\gamma_0 + \sum_{n=1}^{\infty} (\gamma_n \cos n\theta + \delta_n \sin n\theta)\right]$$

Then the corresponding induced velocity component in the y direction is, by a similar analysis as before (see [1], pp. 127–129):

$$v' = U\left[\tfrac{1}{2}\gamma_0 \tan\frac{\theta}{2} + \sum_{n=1}^{\infty} (\gamma_n \sin n\theta - \delta_n \cos n\theta)\right]$$

and this determines the shape of the airfoil, by (72.38). Note that only the terms involving γ_0 are new, compared with Eqs. (72.40) and (72.41).

In order to simplify the analysis, it may be convenient to consider the effects of camber and thickness separately. This is possible in view of the linearization of the boundary conditions. For, if we put

$$y_m(x) = \tfrac{1}{2}[y_u(x) + y_l(x)] \qquad y_t(x) = \tfrac{1}{2}[y_u(x) - y_l(x)]$$

where $y = y_u(x)$ and $y = y_l(x)$ are the equations of the upper and lower surfaces of the wing, respectively, then the flow round the actual wing is obtained by the superposition of the flow over a very thin wing, which is given by $y = y_m(x)$, that is, which coincides with the mean camber line of the original wing, and of the flow over a thick wing without camber, whose upper and lower surfaces are given by $y = \pm\tfrac{1}{2}y_t(x)$. The procedure can be simplified further by taking the free-stream direction parallel to the x axis, so that the effect of incidence is expressed by the position of the mean camber line. Taking the two cases separately, suppose first that we are given an infinitely thin wing, such that the slope is the same at corresponding points of the upper and lower surfaces, while the free-stream velocity is parallel to the x axis. Considerations of symmetry then show that v is the same at points (x,y) and $(x,-y)$, whereas the corresponding x components of the induced velocity u' are of equal magnitude and opposite sign at such points. Moreover, v is continuous across the wing, while u' is discontinuous across it.

According to standard hydrodynamics, a discontinuity of tangential velocity across a curve is equivalent to a distribution of vortices along the curve, the strength of the vortices γ being equal to the discontinuity. The earlier approach to the problem of a very thin wing is based on this fact ([9], p. 91). It provides an interesting physical interpretation of the effect of the presence of the wing, but, in other respects, it is more arbitrary and complicated than the complex-variable approach presented above. We note that for the flat plate at incidence, we obtain from Eq. (72.40), putting $\beta(\theta) = 0$ so that $\alpha_0 = -2 \sin \alpha$, $\alpha_n = \beta_n = 0$, $n > 0$,

$$\gamma = 2u' = 2U \sin\alpha \tan\frac{\theta}{2} \qquad 0 \leq \theta < \pi$$

Similarly, consider the flow around a symmetrical wing section at zero incidence, so that the contour of the wing is given by $y = \pm\tfrac{1}{2}y_t(x)$, while the free-stream velocity is still parallel to the x axis. In this case, u and u' are the same at points (x,y) and $(x,-y)$, while $v(x,-y) = -v(x,y)$. Also, u and u' are now continuous across the wing, while v is discontinuous across it. Thus, the wing now has the effect of a source and sink distribution along the chord, the intensity of the sources or sinks being proportional to the discontinuity in v, and hence to the slope. This can again be made the basis of the analysis, in place of the complex-variable approach.

72.5. THEODORSEN'S METHOD [10]

There exist more exact methods for the calculation of the pressure distribution round a given wing, or for the design of a wing section to a given pressure distribution. For each of these problems, we shall sketch one convenient method of solution.

As stated earlier, the problem of finding the pressure distribution over a given wing is solved if we can find a function which maps the region outside the wing on the region outside a circle. To find such a function, we first select a point z_L within the wing contour and near its leading edge, and we choose our system of coordinates in such a way that $z_L = -2d$ on the x axis, while the trailing edge is given by $z_T = 2d$. Then the transformation $z = z' + d^2/z'$ maps the z plane cut from $-2d$ to $2d$ on the region outside a circle C of radius d round the origin of an auxiliary z' plane. At the same time, the wing contour is transformed into a curve W, which may be expected not to differ greatly from C.

Next, we wish to map the region outside the curve W in the z' plane on the region outside a circle S in the ζ plane, in such a way that the points at infinity correspond, and such that $dz'/d\zeta = 1$ at infinity. These conditions determine the (unknown) radius a of the circle S uniquely.

Write $z' = de^{\psi + i\omega}$ and $\zeta = de^{\chi + i\theta}$, and define χ_0 by $a = de^{\chi_0}$. Then

$$\ln \frac{z'}{\zeta} = \ln \frac{de^{\psi+i\omega}}{de^{\chi+i\theta}} = \psi - \chi + i(\omega - \theta)$$

$$= \sum_{n=1}^{\infty} \frac{\alpha_n + i\beta_n}{\zeta^n} = \sum_{n=1}^{\infty} \frac{\alpha_n + i\beta_n}{d^n} e^{-n\chi}(\cos n\theta - i \sin n\theta)$$

where the coefficients α_n, β_n are unknown to begin with. For points on the circle $S(\chi = \chi_0)$, we then have, separating real and imaginary parts,

$$\psi = \chi_0 + \sum_{n=1}^{\infty} (\alpha^n d^{-n} e^{-n\chi_0} \cos n\theta + \beta_n d^{-n} e^{-n\chi_0} \sin n\theta) \tag{72.43}$$

and

$$\omega - \theta = \sum_{n=1}^{\infty} (\beta_n d^{-n} e^{-n\chi_0} \cos n\theta - \alpha_n d^{-n} e^{-n\chi_0} \sin n\theta) \tag{72.44}$$

The right-hand side of (72.43) is the Fourier expansion of the left-hand side, and so

$$\chi_0 = \frac{1}{2\pi} \int_0^{2\pi} \psi \, d\theta \tag{72.45}$$

$$\left.\begin{matrix}\alpha_n \\ \beta_n\end{matrix}\right\} = \frac{d^n e^{n\chi_0}}{\pi} \int_0^{2\pi} \psi \left\{\begin{matrix}\cos n\theta \\ \sin n\theta\end{matrix}\right\} d\theta \tag{72.46}$$

Moreover, the quantity $(\omega - \theta)$ is conjugate to ψ, by (72.43) and (72.44). Hence, using Poisson's integral formula,

$$\delta = \omega - \theta = \frac{1}{2\pi} \int_0^{2\pi} \psi \cot \frac{\theta_1 - \theta}{2} \, d\theta_1 \tag{72.47}$$

where the principal value of the integral is to be taken on the right-hand side.

We can find the functional connection between ψ and ω since it describes the curve W. However, in order to evaluate the above integrals, we require to know ψ as a function of θ. Since ω may be expected to differ but little from the corresponding θ, we set $\psi = g(\theta)$ in the first approximation. This enables us to compute $\delta = \omega - \theta$ from (72.47). A second approximation to the dependence between ψ and θ is then given by $\psi = g(\theta + \delta)$. Substituting back into (72.47) we obtain a second approximation for δ, etc. In practice, the first or second approximations appear to be adequate. Substituting either $g(\theta)$ or $g(\theta + \delta)$ for ψ in (72.45) and (72.46), we then determine χ_0, α_n, β_n, and hence the mapping function from the ζ plane to the z' plane. The subsequent calculation of the pressure distribution, though still quite complicated, does not present any essential difficulties.

There is a variant of Theodorsen's method, in which some account is taken of the experimental results for the over-all lift coefficient. Other methods, which may be said to be intermediate between Theodorsen's method and the method of thin airfoil theory, are due to S. Goldstein [11].

72.6. LIGHTHILL'S METHOD [12]

The problem of designing a wing to a given pressure distribution is equivalent, by Bernoulli's theorem, to designing a wing to a given variation of the magnitude of the velocity q. However, to begin with, we have to agree in terms of what the pressure or velocity distribution is to be specified. Clearly, a direct reference to the points of the wing surface is out of place, in view of the fact that the wing still has to be determined. The most obvious answer would seem to be that we may wish to specify the pressure along the chord, and this is the procedure followed by thin airfoil theory. However, the point of Lighthill's method is precisely that we specify the velocity distribution along the circle into which the wing is transformed by the mapping function $z = f(\zeta)$, under which points at infinity correspond, with $dz/d\zeta = 1$ at infinity. As we have seen, the appropriate transformation for a straight segment carries the segment into a circle whose diameter equals one-half the length of the segment. By supposing that the circle in question is the unit circle, we only affect the scale of the wing. Finally, we shall assume that the trailing edge corresponds to $\zeta = 1$, and that $U = 1$. Then (72.25) becomes

$$q = 4|\cos(\alpha - \tfrac{1}{2}\theta)\sin\tfrac{1}{2}\theta|\left|\frac{d\zeta}{dz}\right| \tag{72.48}$$

Introducing the element of length of the wing ds, and its local direction $e^{i\omega}$, we have, taking into account that $d\zeta = ie^{i\theta}\,d\theta$ at the unit circle,

$$q = 4\left|\cos(\alpha - \tfrac{1}{2}\theta)\sin\tfrac{1}{2}\theta\frac{d\theta}{ds}\right| = 2\frac{\left|\dfrac{\cos(\alpha - \frac{1}{2}\theta)}{\cos\frac{1}{2}\theta}\right|}{\left|\dfrac{ds}{d(\cos\theta)}\right|} \tag{72.49}$$

In particular, if $\alpha = 0$, then

$$q = q_0 = \frac{2}{\left|\dfrac{ds}{d(\cos\theta)}\right|} \tag{72.50}$$

It follows that we may write, for arbitrary incidence,

$$q = q_0\left|\frac{\cos(\alpha - \frac{1}{2}\theta)}{\cos\frac{1}{2}\theta}\right| \tag{72.51}$$

Since the velocity must be tangential to the surface at any point of the wing, we have $w = qe^{-i\omega}$. In particular, at zero incidence, $\alpha = 0$, we write $w_0 = q_0e^{-i\omega}$. A simple argument then shows that the Laurent series for $\ln w_0$ begins with the second-order term, and so

$$\ln w_0 = \ln q_0 - i\omega = \frac{A_2}{\zeta^2} + \frac{A_3}{\zeta^3} + \cdots \tag{72.52}$$

Thus, at the unit circle, we may expand $\ln q_0$ and $-\omega$ into conjugate Fourier series; so

$$\ln q_0 = \alpha_2\cos 2\theta + \beta_2\sin 2\theta + \alpha_3\cos 3\theta + \beta_3\sin 3\theta + \cdots \tag{72.53}$$

$$-\omega = \beta_2\cos 2\theta - \alpha_2\sin 2\theta + \beta_3\cos 3\theta - \alpha_3\sin 3\theta + \cdots \tag{72.54}$$

Given q as a function of θ for a given design incidence α, we may find q_0 from (72.51). Equation (72.53) shows that we can expect a solution only if q_0 is such that the first three Fourier coefficients of $\ln q_0$ vanish:

$$\int_0^{2\pi} \ln q_0 \, d\theta = \int_0^{2\pi} \ln q_0 \cos \theta \, d\theta = \int_0^{2\pi} \ln q_0 \sin \theta \, d\theta = 0 \tag{72.55}$$

Using Poisson's formula, we may then calculate ω for any particular θ_0,

$$\omega(\theta_0) = -\frac{1}{2\pi} \int_0^{2\pi} \ln q_0(\theta) \cot \frac{\theta - \theta_0}{2} \, d\theta \tag{72.56}$$

Also,

$$\frac{dz}{d(\cos \theta)} = \frac{ds}{d(\cos \theta)} e^{i\omega} = \pm \frac{2}{q_0} e^{i\omega}$$

by (72.50), and so, writing $z = x + iy$ and separating real and imaginary parts,

$$x = \int \pm \frac{2}{q} \cos \omega \, d(\cos \theta) \qquad y = \int \pm \sin \omega \, d(\cos \theta) \tag{72.57}$$

In particular, if q_0 is an even function of θ, $q_0(-\theta) = q_0(\theta)$, then the direction of flow always coincides with the direction of increasing $\cos \theta$. Hence,

$$x = -\int \frac{2}{q_0} \cos \omega \sin \theta \, d\theta \qquad y = -\int \frac{2}{q_0} \sin \omega \sin \theta \, d\theta \tag{72.58}$$

Although the details given so far settle the problem in principle, the practical application of the method still requires considerable skill. An arbitrary variation of the pressure distribution may lead to intersecting top and bottom surfaces, which is inadmissible. Also, the choice of a suitable pressure distribution (possibly involving the use of suction) depends, to a large extent, on boundary-layer theory, and is, therefore, beyond the scope of the present chapter.

72.7. LIFTING-LINE THEORY

For wings of finite span, three-dimensional effects have to be taken into account. This can be done satisfactorily by means of lifting-line theory, so long as the wing is of sufficiently large aspect ratio (= span²/area), not less than 5, say, with little sweepback or none, and such that the wing section does not vary too rapidly along the span. The physical picture on which the method is based was understood by Lanchester [13], but the development of an effective mathematical analysis is due to Prandtl [14].

For wings of finite span, a sheet of trailing vortices extends downstream from the trailing edge. This sheet may be supposed to consist of vortex lines, which are parallel to the direction of flow (Fig. 72.6). Further analysis shows that the principal three-dimensional effect that has to be taken into account now is the presence of a velocity field induced at the wing by the trailing vortices. Neglecting the chordwise and vertical dimensions of the wing in the calculation of this effect, we may suppose that the wing extends from $-s$ to s along the y axis. We shall suppose that the free-stream velocity is parallel to the x axis, while the z axis points upward. Then the velocity induced by the trailing vortex sheet at the wing is parallel to the z axis, and we denote it by $w = w(y)$ (the "downwash"). In general, $w(y)$ is small compared with the free-stream velocity U.

According to two-dimensional theory, the circulation is related to the angle of attack α, now measured from zero lift, by the formula

$$\Gamma = \tfrac{1}{2} a_0 U c \alpha \tag{72.59}$$

[Compare (72.20) with α for sin $(\alpha + \beta)$. The sign is reversed, owing to the use of a different system of coordinates.] U is the free-stream velocity, and c is the chord. Theoretically, $a_0 = 2\pi$, but, in connection with the present theory, any other value of a_0 which has better experimental justification may be used as well. Note that a_0 as well as c and α may depend on y.

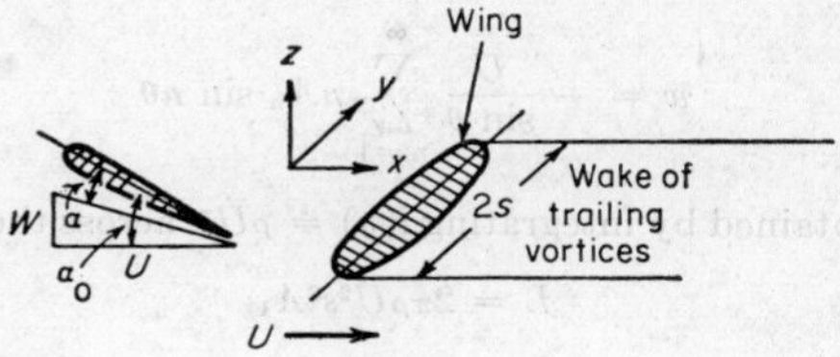

FIG. 72.6

Using the law of Kutta-Joukowski (72.14), we now obtain, for the lift per unit span $l(y)$,

$$l(y) = \rho U \Gamma = \tfrac{1}{2} a_0 \rho U^2 c \alpha \tag{72.60}$$

Now the induced velocity w deflects the free-stream velocity by an angle w/U. The angle of attack is modified by this amount, and we now have, putting $\alpha = \alpha_0 + w/U$ in (72.59),

$$\Gamma = \tfrac{1}{2} a_0 U c \left(d_0 + \frac{w}{U} \right) = \tfrac{1}{2} a_0 c (U \alpha_0 + w) \tag{72.61}$$

where α_0 is the angle of attack relative to the free stream (the so-called geometrical angle of attack).

Let $\gamma(y)$ be the strength of the trailing vortices $(-s < y < s)$. By the law of Biot-Savart, the induced velocity $w(y)$ is then given by

$$w(y) = -\frac{1}{4\pi} \int_{-s}^{s} \frac{\gamma(\eta)}{\eta - y} \, d\eta \tag{72.62}$$

where the principal value is to be taken in the integral on the right-hand side.

We still require an expression for $\gamma(y)$ in terms of the circulation $\Gamma(y)$. Now one of Helmholtz's vortex laws shows that, on any small segment dy along the span,

$$\gamma(y)\, dy + \Gamma(y + dy) - \Gamma(y) = 0$$

Hence,

$$\gamma(y) = -\frac{d\Gamma}{dy} = -\Gamma'(y) \tag{72.63}$$

Combining Eqs. (72.61) to (72.63), we finally obtain Prandtl's equation,

$$\Gamma(y) = \tfrac{1}{2} a_0 c \left[U \alpha_0(y) + \frac{1}{4\pi} \int_{-s}^{s} \frac{\Gamma'(\eta)}{\eta - y} \, d\eta \right] \tag{72.64}$$

which is an integrodifferential equation for Γ.

We introduce the variable θ by

$$y = s \cos \theta \qquad 0 \leq \theta \leq \pi$$

and express Γ as a Fourier series,

$$\Gamma = 4sU(A_1 \sin \theta + A_2 \sin 2\theta + A_3 \sin 3\theta + \cdots) \tag{72.65}$$

Differentiating with respect to y and substituting the result in (72.64), we obtain, after some rearrangement,

$$\sum_{n=1}^{\infty} A_n \sin n\theta(n\mu + \sin\theta) = \mu\alpha \sin\theta \qquad 0 < \theta < \pi \tag{72.66}$$

where $\mu = a_0 c/8s$. In particular, the expression for w alone is

$$w = -\frac{U}{\sin\theta} \sum_{n=1}^{\infty} nA_n \sin n\theta \tag{72.67}$$

The total lift is obtained by integrating $l(y) = \rho U\Gamma$ across the span. This leads to

$$L = 2\pi\rho U^2 s^2 A_1 \tag{72.68}$$

Also, since the downwash $w(y)$ deflects the main stream (usually in a downward direction), the force on the given section is tilted by an angle w/U, and this yields a force component

$$-l(y)\frac{w}{U} = -\rho w\Gamma$$

in the direction of the x axis. Substituting from (72.65) and (72.67), and integrating across the span, we obtain

$$D_i = 2\pi\rho U^2 s^2 \sum_{n=1}^{\infty} nA_n^2 \tag{72.69}$$

D_i, which evidently is always positive, is called the *induced drag*. The relation between lift and induced drag is one of the starting points of the theory of aircraft performance.

Keeping U constant, we may say that, for given lift, i.e., for given A_1, D_i clearly is a minimum if $A_2 = A_3 = \cdots = 0$. For constant (geometrical) angle of attack α_0 and constant a_0, this implies that the chord $c = c(y)$ is given by a formula

$$c(y) = \text{const} \cdot \sqrt{s^2 - y^2}$$

which is satisfied by wings of elliptic planform. At the same time, (72.65) and (72.67) show that the curve $l(y)$ versus y is a semiellipse, while w and hence α also are constant across the span.

It may be mentioned here that there is an alternative method for the derivation of (72.66), which is due to Trefftz. The method also provides a more rigorous justification of the expression for the induced drag.

Various methods (see [9], chap. 9; [15],[16],[1], pp. 183–203, and further references mentioned in this source) have been proposed for the solution of (72.66). Thus, Glauert retains only a finite number (for example, $n = 8$) in the infinite series on the left-hand side of (72.66), and then tries to satisfy the equation at a finite number of points along the span. The method can be simplified by considering separately conditions which are symmetrical and those which are antimetrical across the span. In the former case, the A_k with even subscripts vanish automatically; in the latter, the same applies to the coefficients with odd subscripts. Having found the coefficients A_1, . . . , A_k, we may then compute the lift, the drag, and the lift distribution across the span. In case it is the latter function which is of particular interest, it may be advantageous to use Multhopp's method, which is based on certain trigonometrical interpolation formulas. For given n, define the angles θ_k by

$$\theta_k = \frac{k\pi}{n+1} \qquad k = 1, \ldots, n$$

Denote by μ_k, α_k, Γ_k, the values of μ, α_0, and Γ for $\theta = \theta_k$. Then Multhopp shows that the Γ_k can be obtained as the solution of a system of linear equations,

$$b_k\Gamma_k = \alpha_k + \sum_{1 \le j \le n}^{j \ne k} b_{jk}\Gamma_j \qquad k = 1, \ldots, n$$

where

$$b_k = \frac{1}{2\mu_k} + \frac{n+1}{4 \sin \theta_k}$$
$$b_{kj} = 0 \qquad \text{for } |k - j| = 2, 4, 6, \ldots$$
$$b_{kj} = \frac{\sin \theta_j}{(n+1)(\cos \theta_j - \cos \theta_k)^2} \qquad \text{for } |k - j| = 1, 3, 5, \ldots$$

It may be noted that, except for computational inaccuracies, the same results are obtained by Glauert's and Multhopp's methods if the equations are satisfied for the same θ_k.

72.8. SMALL-ASPECT-RATIO THEORY

There is another class of wings for which the pressure distribution and aerodynamic forces can be calculated with relative ease. These are the wings of small aspect ratio whose planform has a pointed leading edge. An outstanding example of this class is the triangular (delta) wing of small aspect ratio. The theory was initiated by R. T. Jones [17], who adapted a method used by Munk [18] in the calculation of aerodynamic forces on airships. The theory can be based on the idea, justifiable in greater detail, that, for the class of wings under consideration, the velocity potential varies only slowly in chordwise direction, and, accordingly, may be supposed to satisfy Laplace's equation in two dimensions,

$$\frac{\partial^2 \phi}{\partial y^2} + \frac{\partial^2 \phi}{\partial z^2} = 0$$

where the coordinates are defined as in Fig. 72.6. The same then applies to the induced velocity potential ϕ'. We confine ourselves to the case that the wing is very thin, so that $z = f(x,y)$ is the equation of the wing at both its upper and lower surfaces. The boundary condition at the wing, with the same approximation as in Sec. 72.4, is

$$w = U \frac{\partial f}{\partial x} = -U\alpha(x,y) \tag{72.70}$$

where α is the local incidence, chosen with the appropriate sign. We shall assume, for simplicity, that the z,x plane is a plane of symmetry for the wing planform, although the analysis can be carried out also for more general cases. We shall also suppose that the trailing edge is straight, at $x = x_T$, while the nose is pointed, at $x = x_L$, and approximately in the x,y plane. For any x between x_L and x_T, suppose that the wing extends from $-s(x)$ to $s(x)$, where $s(x)$ is an increasing function of x. For any such x, we map the plane normal to the x axis at that chord station on the region outside the unit circle in an auxiliary ζ plane, by means of the familiar transformation,

$$y + iz = \frac{s}{2}\left(\zeta + \frac{1}{\zeta}\right)$$

Putting $\zeta = re^{i\theta}$, we then find if convenient (see [1], pp. 269–280) to express the induced velocity potential ϕ', for given x, in the form

$$\phi' = \sum_{n=1}^{\infty} C_n r^{-n} \sin n\theta \tag{72.71}$$

This leads to the following expression for the velocity component w at the wing [i.e., at the segment $(-s,s)$, corresponding to $r = 1$],

$$w = -\frac{1}{s \sin \theta} \sum_{n=1}^{\infty} nC_n \sin n\theta \tag{72.72}$$

Also, at the wing $y = s \cos \theta$, and so the boundary condition (72.70) may be written in the form

$$U\alpha(x, s \cos \theta) s \sin \theta = \sum_{n=1}^{\infty} nC_n \sin n\theta \tag{72.73}$$

In other words, the right-hand side is the Fourier sine series of the left-hand side, which may be supposed known. This yields

$$C_n = \frac{2Us}{\pi n} \int_0^{\pi} \alpha(x, s \cos \theta) \sin \theta \sin n\theta \, d\theta \tag{72.74}$$

By Bernoulli's equation, the pressure difference between the lower and upper surfaces of the wing is given by

$$p_l - p_u \equiv \Delta p = 2\rho U \frac{\partial \phi'}{\partial x}$$

where $\partial\phi'/\partial x$ is calculated at the upper surface of the wing. But, at the wing, $r = 1$, and so

$$\frac{\partial \phi'}{\partial x} = \sum_{n=1}^{\infty} \frac{dC_n}{dx} \sin n\theta + nC_n \frac{d\theta}{dx} \cos n\theta$$

In order to find $d\theta/dx$, we differentiate the relation $y = s \cos \theta$, keeping y constant. This yields

$$\frac{d\theta}{dx} = \cot \theta \frac{ds}{dx}$$

At the same time, the derivatives dC_n/dx are obtained by differentiating (72.74).

The total lift is found by integrating the pressure difference over the wing, and turns out to be equal to

$$L = \pi\rho U s C_1$$

where s and C_1 are taken at the trailing edge.

The following results are obtained for a plane delta wing at incidence α. Suppose that the apex of the wing is at the origin and its trailing edge at $x = c$, and let γ be the semiapex angle. Then the pressure difference at a point (x,y) of the wing is given by

$$\Delta p = 2\rho U^2 \alpha \frac{x \tan^2 \gamma}{\sqrt{x^2 \tan^2 \gamma - y^2}}$$

This shows that the pressure difference becomes infinite at the leading edges.

The spanwise lift distribution $l(y)$ is given by

$$l(y) = 2\rho U^2 \alpha \sqrt{c^2 \tan^2 \gamma - y^2}$$

and is, therefore, elliptic.

The total lift is given by

$$L = \pi\rho U^2 \alpha c^2 \tan^2 \gamma = \tfrac{1}{4}\pi\rho U^2 S \alpha \mathit{A\!R}$$

where $S = c^2 \tan \gamma$ is the area of the wing, and Æ is the aspect ratio

$$Æ = \frac{4c^2 \tan^2 \gamma}{S} = 4 \tan \gamma$$

This yields

$$\frac{dC_L}{d\alpha} = \frac{\pi}{2} Æ$$

Finally, it can be shown that the induced drag is equal to

$$D_i = \tfrac{1}{2} L\alpha$$

The theory can be extended so as to cover trailing edges which are curved or consist of polygonal lines. However, in general, the results for the pressure break down in the immediate neighborhood of the trailing edge.

72.9. GENERAL THREE-DIMENSIONAL THEORY

For moderately small aspect ratios, $2 < Æ < 5$, say, neither the lifting-line theory nor the low-aspect-ratio theory of the preceding sections are adequate. The same may be true even for larger aspect ratios, e.g., if there is a considerable degree of sweepback. Unfortunately the methods which are available for dealing with this range are relatively complicated, and they do not lead to compact results. We can do no more than sketch a number of them.

First of all, following the procedure given at the end of Sec. 72.4, we may split the general case into two parts.

Thus, we take first the case of a wing with symmetrical cross sections at zero incidence relative to the free stream (which is parallel to the x axis). Then, the upper and lower surfaces of the wing are given by a pair of equations:

$$z = \pm f(x,y)$$

This yields the boundary condition,

$$w = \frac{\partial \phi'}{\partial z} = \pm U \frac{\partial f}{\partial x} = \pm \beta(x,y) \tag{72.75}$$

where ϕ' is the induced velocity potential, and $\beta(x,y)$ is the local slope.

To find the pressure increment $(p - p_0)$, we use the linearized form of Bernoulli's equation,

$$p - p_0 = -\rho U u' = -\rho U \frac{\partial \phi'}{\partial x}$$

This formula may become inadequate near the leading edge, where the presence of a suction force may have to be taken into account.

In order to determine the induced velocity potential for this case, we represent it by a distribution of sources and sinks over the wing, or, rather, over the projection of the wing onto the x,y plane; so

$$\phi'(x,y,z) = -\frac{1}{4\pi} \iint_S \frac{\sigma(x_0,y_0)\, dx_0\, dy_0}{\sqrt{(x - x_0)^2 + (y - y_0)^2 + z^2}} \tag{72.76}$$

The source strength σ is related directly to the local slope (of the upper surface, say) by the relation

$$\sigma(x,y) = 2U\beta(x,y) = 2U \frac{\partial f}{\partial x}$$

and so the expression for ϕ' becomes

$$\phi'(x,y,z) = -\frac{U}{2\pi}\iint_S \frac{\beta(x_0,y_0)\,dx_0\,dy_0}{\sqrt{(x-x_0)^2+(y-y_0)^2+z^2}} \tag{72.77}$$

The induced velocity u' and, hence, the pressure can now be found by differentiation, and this would appear to settle the problem. In actual fact, however, it is sometimes not easy to carry out the integrations and differentiations in question, and, for this reason, the effective application of the method to particular cases may still require considerable effort.

Consider next the antimetrical case. It is now assumed that the wing is very thin, so that its top and bottom surfaces are given by the same equation, $z = f(x,y)$.

In this case, the induced velocity potential can be represented by a distribution of doublets. However, in view of the presence of the vortex wake, we have to take the doublet distribution over the wake, as well as over the wing. The expression for the contribution of the wake is simplified by the fact that the latter consists (approximately) of straight vortices which extend downstream from the trailing edge in the direction of the free stream. It follows that the induced velocity potential is constant along the wake in chordwise direction. Suppose then that the wing extends from $-s$ at port to s at starboard, so that s is the semispan and the trailing edge (situated approximately in the x,y plane) is given by

$$x = t(y) \qquad -s \leq y \leq s$$

Then the induced velocity potential may be written as follows, in terms of a doublet distribution function $\tau(x,y)$ over the projection of the wing onto the x,y plane, W:

$$\phi'(x,y,z) = \frac{1}{4\pi}\frac{\partial}{\partial z}\iint_W \frac{\tau(x_0,y_0)\,dx_0\,dy_0}{\sqrt{(x-x_0)^2+(y-y_0)^2+z^2}} + \frac{z}{4\pi}\int_{-s}^{s}\frac{\tau(t(y_0),y_0)}{(y-y_0)^2+z^2}\left\{1+\frac{x-t(y_0)}{\sqrt{[x-t(y_0)]^2+(y-y_0)^2+z^2}}\right\}dy_0 \tag{72.78}$$

The corresponding z component of the velocity at the wing, $w(x,y,0)$, is obtained by differentiating ϕ' with respect to z and passing to the limit. For a given wing, $\tau(x,y)$ is then again determined by the boundary condition,

$$\left(\frac{\partial\phi'}{\partial z}\right)_{z=0} = w = U\beta(x,y)$$

Unfortunately, this is a very complicated equation, which cannot be solved exactly in most cases. We therefore have to fall back upon one of a number of approximate methods. Thus, we may assume a suitable expansion for $\tau(x,y)$, and we may then determine the coefficients of a finite (small) number of terms of the expansion by satisfying the boundary condition at a well-chosen set of points. In the choice of such particular solutions, we are helped by previous experience with two-dimensional theory, and by the fact that $\tau(x,y)$ is equal to the discontinuity of the velocity potential across the wing. The reader is referred to a paper by W. P. Jones [19] for the details of such a method.

Alternatively, the induced velocity field may be thought of as being due to a distribution of vortices over wing and wake. This leads to a result which is mathematically equivalent to the formulas for a doublet distribution given above. However, from a physical point of view, it gives a different, and in some ways more intuitive, picture of the situation, and suggests different approximate methods. Thus,

Falkner [20] makes use of a system of "horseshoe vortices," which consist of a segment parallel to the y axis (the "bound vortex"), together with the trailing vortices which issue from ends of the segment in downstream direction.

For sweptback wings, Weissinger [21] uses systems of vortices in which the bound part is itself swept back. Thus, in the simplest version, assume that the locus of the quarter-chord points on a sweptback wing is (approximately) straight, except for the break at the centerline. Weissinger replaces the system of vortices over the wing by a single vortex line of variable strength along the quarter-chord line (together with the usual system of trailing vortices emanating from it in the downstream direction). He then calculates w, and he determines the bound vorticity distribution, $\nu(y)$, and hence the corresponding lift distribution, by satisfying the boundary condition at a certain number of points along the three-quarter chord.

The expression obtained for $w(x,y,0)$ is

$$w = \frac{1}{4\pi}\int_{-s}^{0} \Gamma'(\eta)\left\{\frac{1}{\eta - y}\left[1 + \frac{\sqrt{(x + \eta\tan\phi)^2 + (y-\eta)^2}}{x + y\tan\phi}\right] - 2\tan\phi\,\frac{\sqrt{x^2+y^2}}{x^2 - y^2\tan^2\phi}\right\} d\eta + \frac{1}{4\pi}\int_0^s \frac{\Gamma'(\eta)}{\eta - y}\left[1 + \frac{\sqrt{(x - \eta\tan\phi)^2 + (y-\eta)^2}}{x - y\tan\phi}\right] d\eta \qquad (72.79)$$

A method similar to Multhopp's (Sec. 72.7) can be applied in the final stages of the procedure.

Finally, we may mention two papers on the subject by Küchemann [22] and Lawrence [23]. Küchemann's method provides a rational interpolation between the region of low aspect ratio which is covered by the method of R. T. Jones and the region of high aspect ratio to which lifting-line theory is applicable. The idea of Lawrence's paper is to extend the scope of Jones's method in the direction of increasing aspect ratio. It is based on an analysis of Reissner's [24] (applying also to unsteady conditions), from which it may be concluded that the methods of Prandtl and Jones provide genuine limiting values for extreme aspect ratios.

Another method which is effective in some cases is based on Prandtl's acceleration potential[1] (see [1], pp. 249–268, where further references are given). This method makes use of the fact that since ϕ' is constant along the wake in the direction of the x axis, $\partial\phi'/\partial x$ vanishes at the wake and, hence, is continuous across it. Thus, $\partial\phi'/\partial x$ is a solution of Laplace's equation, which is continuous except at the wing. In the case where the wing is circular or, more generally, elliptical, we may therefore use, as particular solutions for ϕ', the normal solutions provided by potential theory in terms of particular systems of coordinates. The theory thus provides some interesting additional information for a very limited class of wings.

72.10. OTHER PROBLEMS AND METHODS IN THREE-DIMENSIONAL THEORY

In this account, we have not been able to include descriptions of the theory of biplanes, or of compound lifting systems such as tail units, and the reader is referred to the literature (for example, [1], pp. 213–216 and 203–213). Nor have we expounded Munk's interesting variational analysis [25] of lifting systems. One of Munk's conclusions is the so-called "stagger theorem," which reads as follows:

If we displace the individual components of a system of lifting units (wings) in the direction of the free stream, and at the same time adjust the incidence of the

[1] The expression *acceleration potential method* is due to the fact that $U\,\partial\phi'/\partial x$ (which is proportional to $\partial\phi'/\partial x$) is the linearized expression for the ratio of pressure and density, from which the acceleration is obtained by taking the gradient. It is in this form that the concept can be extended without difficulty to unsteady motion.

individual units, in such a way that the lift distribution over the various components remains unchanged, then the total induced drag (the drag associated with the lift) remains unchanged.

Another consequence of Munk's general analysis is that, at a lifting line with minimum drag for given total lift, the induced normal velocity (downwash) must be constant along the span. This fact has been verified already by the methods of Sec. 72.7.

72.11. REFERENCES

[1] A. Robinson, J. A. Laurmann: "Wing Theory," Cambridge Univ. Press, New York, 1956.

[2] H. Blasius: Funktionentheoretische Methoden in der Hydrodynamik, *Z. Math. Phys.*, **58** (1910), 90.

[3] W. M. Kutta: Auftriebskräfte in strömenden Flüssigkeiten, *Illustr. Aeron. Mitt.*, July, 1902, p. 133.

[4] N. E. Joukowski: Sur les tourbillons adjoints, *Trans. Phys. Sec., Imp. Soc. Friends Natl. Sci. Moscow*, **23** (1906).

[5] N. E. Joukowski: Über die Konturen der Tragflächen der Drachenflieger, *Z. Flugtech.*, **22** (1910).

[6] R. von Mises: Zur Theorie des Tragflächenauftriebs, *Z. Flugtech.*, **11** (1920), 68.

[7] Th. von Kármán, E. Trefftz: Potentialströmung um gegebene Tragflächenquerschnitte, *Z. Flugtech.*, **9** (1918), 111.

[8] M. J. Lighthill: A new approach to thin aerofoil theory, *Aeronaut. Quart.*, **3** (1951), 193.

[9] H. Glauert: "The Elements of Aerofoil and Airscrew Theory," 2d ed., Cambridge Univ. Press, New York, 1957.

[10] Th. Theodorsen: Theory of wing sections of arbitrary shape, *NACA Rept.* 411 (1932).

[11] S. Goldstein: A theory of aerofoils of small thickness, parts 1–6, *Aeronaut. Research Comm. Current Papers*, 68–73 (1952).

[12] M. J. Lighthill: A new method of two-dimensional aerodynamic design, *Aeronaut. Research Comm. Rept. and Mem.* 2112 (1945).

[13] F. W. Lanchester: "Aerodynamics," Constable, London, 1907.

[14] L. Prandtl: Tragflügeltheorie, *Nachr. Ges. Wiss. Göttingen Math.-physik Kl.*, 1918, pp. 451–477.

[15] H. Multhopp: Die Berechnung der Auftriebsverteilung von Tragflügeln, *Luftfahrt-Forsch.*, **15** (1938), 153–169.

[16] I. Lotz: Berechnung der Auftriebsverteilung beliebig geformter Flügel, *Z. Flugtech.*, **22** (1931), 189–195.

[17] R. T. Jones: Properties of low-aspect-ratio pointed wings at speeds below and above the speed of sound, *NACA Rept.* 835 (1946).

[18] M. M. Munk: The aerodynamic forces on airship hulls, *NACA Rept.* 184 (1924).

[19] W. P. Jones: Theoretical determination of the pressure distribution on a finite wing in steady motion, *Aeronaut. Research Comm. Rept. and Mem.* 2145 (1943).

[20] V. M. Falkner: The calculation of aerodynamic loading on surfaces of any shape, *Aeronaut. Research Comm., Rep. and Mem.* 1910 (1943).

[21] J. Weissinger: The lift distribution of swept-back wings, *NACA Tech. Mem.* 1120 (1947) (transl. of German report of 1942).

[22] D. Küchemann: A simple method for calculating the span and chordwise loading on straight and swept back wings of any given aspect ratio at subsonic speeds, *Aeronaut. Research Comm. Rept. and Mem.* 2935 (1952).

[23] H. R. Lawrence: The lift distribution on low aspect ratio wings at subsonic speeds, *J. Aeronaut. Sci.*, **18** (1951), 683–695.

[24] E. Reissner: On the general theory of thin airfoils for non-uniform motion, *NACA Tech. Note* 946 (1944).

[25] M. Munk: Isoperimetrische Aufgaben aus der Theorie des Fluges, Diss., Göttingen, 1919.

CHAPTER 73

GENERAL THERMODYNAMICS

BY

K. OSWATITSCH, Dr. phil., Aachen, Germany

73.1. FUNDAMENTAL CONCEPTS

Most of the statements made in this chapter apply to a fluid of given composition, whose properties are isotropic. Such a fluid is then physically homogeneous. A common example is air, which, at room temperature, is a mixture of nitrogen (78% by volume), oxygen (21%), argon, etc. At very high temperatures, the oxygen and nitrogen molecules dissociate into atoms; at still higher temperatures, ionization occurs, yielding electrons and positive ions (plasmas). At chemical equilibrium, the relative proportions of molecules, atoms, and ions are fixed by the absolute temperature T, the density ρ, and the pressure p. As long as the composition can be specified in this manner, the requirements of homogeneity can be met, and the results of this section are rigorously applicable.

A suspension of water droplets in air, such as a mist, may also be regarded as a homogeneous system, as long as the portion existing as liquid is fixed by equilibrium conditions and the dimensions of the system are sufficiently large in comparison with the droplets. In meteorology, clouds are considered to be homogeneous if the drops are small enough to remain in suspension. When rain occurs, however, the liquid content of the air changes at constant temperature and pressure, a typical example of inhomogeneity.

There are situations when a homogeneous system may be assumed although complete chemical equilibrium does not exist. For example, it is not necessary that a cloud be continually saturated with water vapor. Also, when the state of the system changes so rapidly that its components do not have time to readjust (a situation of complete relaxation), the system can be treated as a homogeneous one, because its composition can be specified.

A physically homogeneous system can be assumed for the majority of the flow problems encountered, including processes with combustion and other chemical transformations. In such cases, the mixture of the reacting substances and the mixture of the end products must each be treated as single substances. Examples of systems which may not be treated in this manner are boundary layers through which the diffusion of matter is occurring and those systems in which relaxation phenomena must be taken into account.

73.2. EQUATION OF STATE

Properties whose values depend on the state of a system are called *state properties*. In addition to the properties ρ, p, and T mentioned above, other state properties are the internal energy u, the enthalpy h, and the entropy s, all referring to a unit mass of the given substance. In a homogeneous system, only two state properties may be

Translated from the German by G. A. Agoston, Sc. D., Menlo Park, Calif.

selected as independent variables; the remaining ones are dependent. Since analytical expressions relating these properties are generally complicated, diagrams are often employed to demonstrate simultaneously the effects of changes of several variables. The equation relating the state properties p, ρ, and T is called the *equation of state*. At low densities, the pressure is substantially proportional to the density and the absolute temperature. Then the equation of state takes the simple form

$$p = R\rho T \tag{73.1}$$

This relation is obeyed rigorously only in the limit of vanishing density and, hence, is commonly called the equation of state of an ideal gas or, simply, the *perfect-gas law*. The factor R is a constant for a gas of given composition; it is inversely proportional to the molecular weight m. The product Rm is the universal gas constant, whose dimensions are set by the units selected for p, ρ, and T. For example, when p is expressed as pounds (weight) per square foot, $1/\rho$ as cubic feet per pound (mass) mole, and T as degrees Rankine, then

$$Rm = 49{,}700 \frac{\text{lb wt}}{\text{ft}^2} \frac{\text{ft}^3}{\text{lb mole}} \frac{1}{°\text{R}} = 49{,}700 \frac{\text{ft}^2}{°\text{R} \cdot \text{sec}^2}$$

This form of the universal constant is useful in dynamics calculations, such as the speed of sound. On the other hand, if this value is divided by the gravitational constant g, another universal constant is obtained, which bears energy units. Because of these units, the following forms of the universal constant are used extensively in thermal calculations [for example, Eq. (73.22)]:

$$\begin{aligned} Rm &= 1545 \text{ ft-lb}/°\text{R} \cdot \text{lb mole} \\ &= 1.986 \text{ Btu}/°\text{R} \cdot \text{lb mole} \\ &= 8.315 \text{ joule}/°\text{K} \cdot \text{g mole} \\ &= 1.986 \text{ cal}/°\text{K} \cdot \text{g mole} \end{aligned} \tag{73.2}$$

Here g mole represents one molecular weight of gas in grams mass; lb mole represents one molecular weight of gas in pounds mass. In Eq. (73.1), temperature T is the absolute temperature measured above absolute zero. At the freezing point of water, $T = 273.15°\text{K} = 459.69°\text{R}$. In the first of these scales (degrees Kelvin), centigrade units are used; in the second (degrees Rankine), Fahrenheit units are used.

In this chapter, all relative quantities are referred to the unit of mass. If, instead, the unit of weight is used (with a standard or conventional value of the gravitational constant g), the dimensions of R, c_p, c_v, etc., change, but the numerical values are the

Table 73.1. Some Properties of Gases at 0°C and Low Pressure

Gas	Symbol	Molecular weight, m	Specific heat, Btu/lb · °F		γ	n	Reaction	Latent heat l_r, Btu/lb
			c_p	c_v				
Oxygen	O_2	32.00	0.219	0.157	1.40	5	$O_2 \rightarrow 2O$	6.592
Nitrogen	N_2	28.016	0.248	0.177	1.40	5	$N_2 \rightarrow 2N$	14.46*
Air		29.0	0.240	0.171	1.40	5		
Hydrogen	H_2	2.016	3.400	2.415	1.41	5	$H_2 \rightarrow 2H$	92.95
Steam	H_2O	18.016	0.444	0.334	1.33	6		
Argon	A	39.944	0.125	0.076	1.65	3		

* Uncertain.

same when gram and pound mass are replaced by gram and pound weight. It is then, of course, necessary to replace the density ρ in all equations by the specific weight.

Equation (73.1) is not limited to ideal gases. It applies as well to gases in high-temperature regions where dissociation or ionization occur, with necessary changes in molecular weight or R. The equilibrium composition of molecules and atoms, or atoms and ions, depends not only on temperature but also on density. Therefore, the molecular weight is dependent on pressure and temperature $m(T,p)$ (Fig. 73.1), and the direct proportionality between p and ρT which characterizes ideal-gas behavior

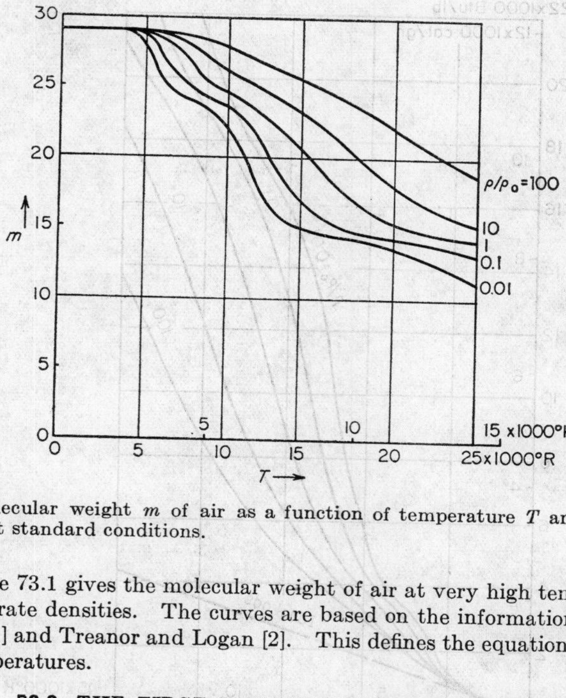

FIG. 73.1. Molecular weight m of air as a function of temperature T and density ρ [1]. ρ_0 = density at standard conditions.

is lost. Figure 73.1 gives the molecular weight of air at very high temperatures and low and moderate densities. The curves are based on the information of Hilsenrath and Beckett [1] and Treanor and Logan [2]. This defines the equation of state for air at higher temperatures.

73.3. THE FIRST LAW OF THERMODYNAMICS

If energy is introduced into a system in the form of heat, electricity, or mechanical energy (for example, by a stirring device such as a paddle), then the total internal energy of the system increases in accordance with the energy conservation law. Part of this energy is in the form of kinetic energy, which is associated with the translational, rotational, and vibrational motions of the molecules. The remaining part exists as potential energy, which, in solids and liquids, holds the substance together. In gases, the potential energy is limited to energy holding the atoms together as molecules and, indeed, the electrons and ions, as atoms.

Clearly the internal energy cannot change if the state does not change; otherwise, the law of energy conservation would not hold. Therefore, the internal energy u is a state property and, in a homogeneous medium, is dependent on only two state variables, for example, T and p. The relation

$$u = u(T,p) \tag{73.3}$$

is sometimes called the caloric equation of state.

For an ideal gas, the internal energy u depends only on T, a fact which becomes apparent below in the discussion of the second law of thermodynamics. The dependency on temperature alone becomes intuitively reasonable if we consider the energetics of molecules in the gas phase. In low-density gases, the attractive forces between the molecules are negligible and, hence, cannot contribute to the internal energy. The bonding forces within molecules and atoms depend on the temperature only. Thus, the internal energy of such a gas is dependent only on temperature. Of course, if dissociation or ionization occurs, causing the molecular weight to change, then clearly

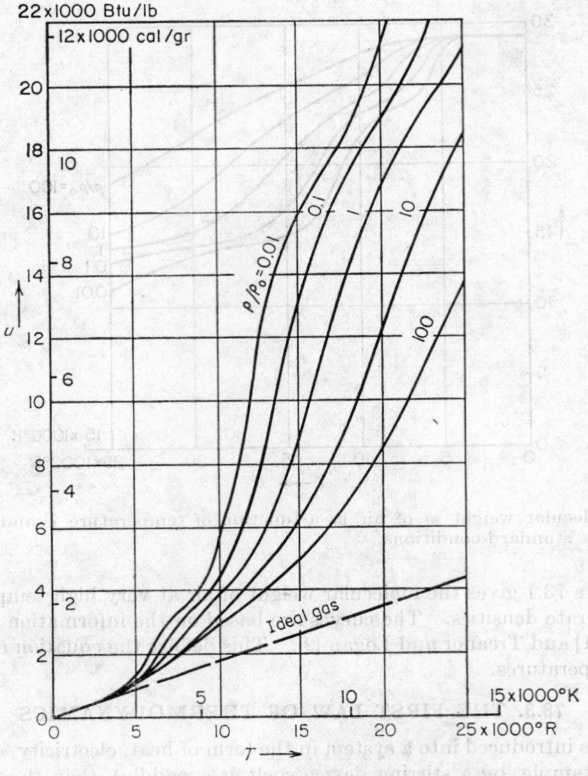

FIG. 73.2. Internal energy u of air as a function of temperature T and density ρ [1]. ρ_0 = density at standard conditions.

u becomes a function of pressure and density as well. Figure 73.2 gives the internal energy of air as a function of T and ρ, based on information from Ref. [1]. It is clear that, at certain temperature levels, the heat required for dissociation is several times as large as that required in the form of sensible heat. The thermal properties of air at high temperatures are thus described by the equation of state (73.1) and Figs. 73.1 and 73.2. At normal conditions, the internal energy of liquids and solids also depends only on temperature.

With the use of a paddle stirring mechanism, it is possible to find how much mechanical energy is required to raise the temperature of a fluid a definite amount. In thermodynamics, the unit of energy most commonly employed is the *kilocalorie* (1 Cal = 1000 cal). This is defined as the amount of heat required to raise the temperature of one kilogram mass of water from 14.5 to 15.5°C. In the English system the corresponding unit is the Btu (*British thermal unit*). The following relationship (called the mechanical equivalent of heat) exists between the mechanical and thermal units:

$$1 \text{ joule} = 0.2389 \text{ cal} = 0.9478 \times 10^{-3} \text{ Btu} \tag{73.4}$$

This relation is applied in Eq. (73.2).

The expansion of a thermally insulated gas without work output represents one of the simplest tests of the energy conservation law. Such a change of state without heat exchange with the surroundings is called *adiabatic*. If a gas is released from a thermally insulated pressure vessel through a valve into a larger evacuated, insulated container and equilibrium is attained, then the density and the pressure change in an easily predicted manner. Since no energy exchange occurs with the surroundings, and no work is performed by the system, the internal energy in the final state (subscript 2) and in the initial state (subscript 1) must be the same:

$$u(T_2,p_2) = u(T_1,p_1) \tag{73.5}$$

Since the internal energy of ideal gases depends only on T, the following applies for an adiabatic expansion without work output:

$$T_2 = T_1 \tag{73.6}$$

The throttling process (Fig. 73.3) is similar to the one just described. A gas at state T_1,p_1 in an entrance duct flows continually through a throttle valve into an exit duct at state T_2,p_2. By the use of entrance and exit ducts with sufficiently large cross-sectional areas, the influence of kinetic energy can be made negligible. Thus, in addition to the change in internal energy in the throttling process, only the flow work required to force the gas continually through the valve needs to be considered.

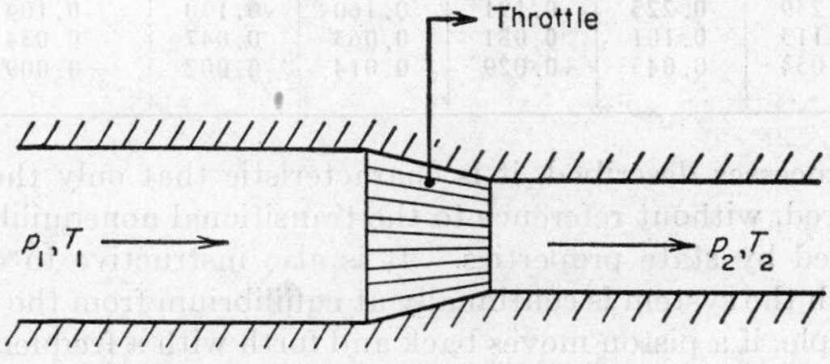

FIG. 73.3. Throttling process.

At the entry, the volume of a unit mass of gas at constant pressure p_1 is $1/\rho_1$. With A_1 as the cross-sectional area, and l_1 as the length of a unit mass of gas, $1/\rho_1 = A_1 l_1$. The force applied is $p_1 A_1$, and the work performed on forcing the gas into the constriction is

$$p_1 A_1 l_1 = p_1/\rho_1 \tag{73.7}$$

Similarly, the work yielded on expanding from the constriction is then p_2/ρ_2. The difference between these flow-work quantities must equal the increase in internal energy:

$$p_1/\rho_1 - p_2/\rho_2 = u_2 - u_1 \tag{73.8}$$

If now a new state property called *enthalpy* h is introduced such that

$$h = u + p/\rho \tag{73.9}$$

then, in consideration of Eq. (73.8), the throttling process is represented by

$$h(T_2,p_2) = h(T_1,p_1) \tag{73.10}$$

Equations (73.1) and (73.9) show that the enthalpy of an ideal gas also depends only on temperature; therefore, Eq. (73.6) applies to the throttling process for an ideal gas, just as it does to the adiabatic expansion without work.

In a real gas, on the other hand, the temperature changes in the throttling process, a phenomenon called the *Joule-Thomson effect*. Since T is a function of two independent variables, for example, p and h in a homogeneous system,

$$dT = \left(\frac{\partial T}{\partial p}\right)_h dp + \left(\frac{\partial T}{\partial h}\right)_p dh \tag{73.11}$$

For the throttling process, $dh = 0$, causing the second term on the right-hand side of Eq. (73.11) to drop out. The remaining term $(\partial T/\partial p)_h$ is called the *Joule-Thomson coefficient* (Table. 73.2). At average temperature levels, the temperature falls with pressure; the cooling effect obtainable at high pressures is significant only at low temperature levels. At both high and very low temperatures, heating can occur on expansion.

Table 73.2. Joule-Thomson Coefficients for Air at 0°C [4]
Values of $(\partial T/\partial p)_h$ are in degrees Fahrenheit per atmosphere

T, °F	p, atm						
	1	20	60	100	140	180	220
−148	1.037	1.012	0.850	0.511	0.256	0.135	0.056
32	0.479	0.448	0.385	0.320	0.261	0.203	0.146
212	0.239	0.223	0.191	0.160	0.130	0.104	0.081
392	0.113	0.101	0.081	0.063	0.047	0.034	0.023
536	0.054	0.045	0.029	0.014	0.002	−0.009	−0.020

In the two processes described, it is characteristic that only the initial and final states are compared, without reference to the transitional nonequilibrium step, which cannot be specified by state properties. It is also instructive to consider idealized processes, in which the system is continually at equilibrium from the initial to the final state. For example, if a piston moves back and forth with a frequency f in a cylinder of length l, then pressure waves originating at the piston face spread out continually, through the chamber with the speed of sound a. Continuous equilibrium in the moving gas can be attained only if the pressure waves bounce between the piston face and the cylinder wall a great number of times during the period $1/f$. This means that the time between bounces l/a is very much smaller than $1/f$:

$$lf/a \ll 1 \tag{73.12}$$

For a frequency $f = 150\ \text{sec}^{-1}$, a cylinder length $l = 4$ in., and an average sound velocity $a = 1300$ ft/sec, $lf/a = 0.038$. Thus, the change of state can occur rather swiftly without the homogeneity of the system being disturbed significantly. A process in which the physical and chemical equilibrium is continuously preserved is called *quasi-static* or infinitely slow. This is an idealization which is valid in many practical situations, possibly even when high rates of change are encountered.

The work which a unit mass of gas performs quasi-statically against the pressure p in an expansion is equal to the product of pressure and change in volume, $p \cdot d(1/\rho)$

(Fig. 73.4). If dq is the heat transferred during a small change of state, then the energy conservation law requires that dq be equal to the increase in internal energy plus the work performed by the gas. This is an expression of the first law of thermodynamics as it applies to quasi-static processes:

$$dq = du + p\,d(1/\rho) \tag{73.13}$$

By introducing h [see Eq. (73.9)], Eq. (73.13) can also be written in the form

$$dq = dh - \frac{1}{\rho}\,dp \tag{73.14}$$

If a quasi-static process is adiabatic ($dq = 0$), the internal energy must decrease during expansion in accordance with Eq. (73.13), because the energy for expanding can be drawn only from this source. For this reason, on quasi-static adiabatic expansion a gas cools, and on compression it becomes heated.

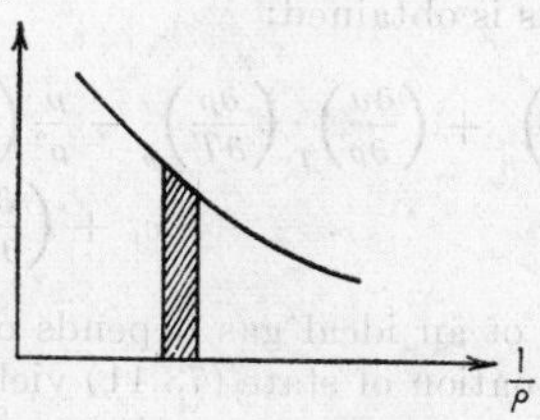

FIG. 73.4. Performance of work by a gas.

Under conditions of heat addition to a closed container, the entire energy added is equal to the increase in internal energy, because the density is constant [Eq. (73.13)]. On the other hand, under constant-pressure conditions, the added heat is equal to the increase in enthalpy [Eq. (73.14)]. Constant-pressure processes are of great practical interest, for example, in combustion chambers.

73.4. SPECIFIC HEATS, HEATS OF TRANSITION

The quantity of heat which must be added to a unit mass of a substance in order to raise its temperature one degree is called the specific heat. Although the dimensions differ, depending on whether mass or weight units are employed, the numerical values are the same. The dimensions of specific heat are Cal/kg · °C or Btu/lb · °F on one hand, and Cal/kg wt · °C or Btu/lb wt · °F on the other, where the terms kg and lb refer to mass units and kg wt and lb wt refer to weight units.

For a gas, it is important to know under what conditions heat addition occurs. If it occurs at constant volume or at constant density, then the entire energy input is transformed into internal energy. The specific heat at constant volume c_v is expressed as follows:

$$c_v = \left(\frac{\partial u}{\partial T}\right)_\rho \tag{73.15}$$

The internal energy of an ideal gas does not depend on ρ; hence,

$$du = c_v\,dT \tag{73.16}$$

For an ideal gas of constant specific heat, closely typified by air at ordinary pressures and temperatures, Eq. (73.16) may be integrated, yielding a linear relation between

u and T:

$$u = c_v T + \text{const} \tag{73.17}$$

When heat addition occurs at constant pressure, a definite portion of the energy is expended in the expansion of the gas. On the basis of the first law of thermodynamics [Eq. (73.14)], the specific heat at constant pressure is given by

$$c_p = \left(\frac{\partial q}{\partial T}\right)_p = \left(\frac{\partial h}{\partial T}\right)_p \tag{73.18}$$

Thus, for an ideal gas [compare Eq. (73.16)],

$$dh = c_p\, dT \tag{73.19}$$

and, for an ideal gas with constant specific heat,

$$h = c_p T + \text{const} \tag{73.20}$$

With the use of Eq. (73.9) and the chain rule of differentiation, the following relation between the two specific heats is obtained:

$$c_p = \left(\frac{\partial(u + p/\rho)}{\partial T}\right)_p = \left(\frac{\partial u}{\partial T}\right)_\rho + \left(\frac{\partial u}{\partial \rho}\right)_T \left(\frac{\partial \rho}{\partial T}\right)_p - \frac{p}{\rho^2}\left(\frac{\partial \rho}{\partial T}\right)_p$$

$$= c_v + \left(\frac{\partial \rho}{\partial T}\right)_p \left[\left(\frac{\partial u}{\partial \rho}\right)_T - \frac{p}{\rho^2}\right] \tag{73.21}$$

Since the internal energy of an ideal gas depends only on the temperature, Eq. (73.21) combined with the equation of state (73.11) yields

$$c_p - c_v = R \tag{73.22}$$

Although both c_p and c_v for ideal gases may vary with temperature, the difference in their values is simply the gas constant R, or the universal gas constant divided by the molecular weight. The ratio of specific heats $\gamma = c_p/c_v$ plays an important role in aerodynamics. With the aid of Eq. (73.22), this ratio may be calculated if either c_p or c_v is known. In Table 73.3, c_p and γ are presented for air and water vapor at normal pressure conditions. At low temperatures, particularly below 0°C, the numerical values are practically constant. At higher temperatures, the temperature dependence becomes appreciable. Furthermore, at high pressures, above 50 atm, deviations from the ideal-gas state become significant. Extensive data for air up to dissociation temperatures and to 100 atm pressure are given in [3].

Table 73.3. c_p and γ for Air and Steam at Low Pressure and Temperatures Up to 3632°F

Temperature		Air		H_2O	
$T - 460$°F, °F	$T - 273$°C, °C	γ	c_p, Btu/lb · °F	γ	c_p, Btu/lb · °F
32	0	1.401	0.240	1.331	0.444
392	200	1.390	0.244	1.313	0.464
752	400	1.368	0.255	1.292	0.489
1112	600	1.349	0.266	1.265	0.527
1472	800	1.332	0.276	1.244	0.563
1832	1000	1.320	0.283	1.228	0.596
2732	1500	1.303	0.295	1.203	0.659
3632	2000	1.294	0.302	1.187	0.701

In the dissociation regime (up to 15,000°K), c_v may be evaluated from Fig. 73.2 by means of Eq. (73.15). Equation (73.21) may then be employed to evaluate c_p. For this purpose, the following relation is useful, which is based on Eqs. (73.1) and (73.2):

$$-\frac{T}{\rho}\left(\frac{\partial \rho}{\partial T}\right)_p = 1 - \frac{T}{m}\left(\frac{\partial m}{\partial T}\right)_p \tag{73.23}$$

Both c_p and c_v in Fig. 73.5 were determined in this manner.

Definite quantities of heat become bound or liberated with each change of phase or change in composition (such as in dissociation or ionization reactions). It is important to know whether such transitions occur at constant volume and density, as in a closed container, or at constant pressure, as in a vessel open to the atmosphere. In the present discussion, consideration is given only to the isobaric conversions. The cases considered are those where two states of a substance exist, as ice and water, water and steam, or oxygen molecules and oxygen atoms. Furthermore, the initial substances entering a reaction or the end products may be a mixture of various molecular species.

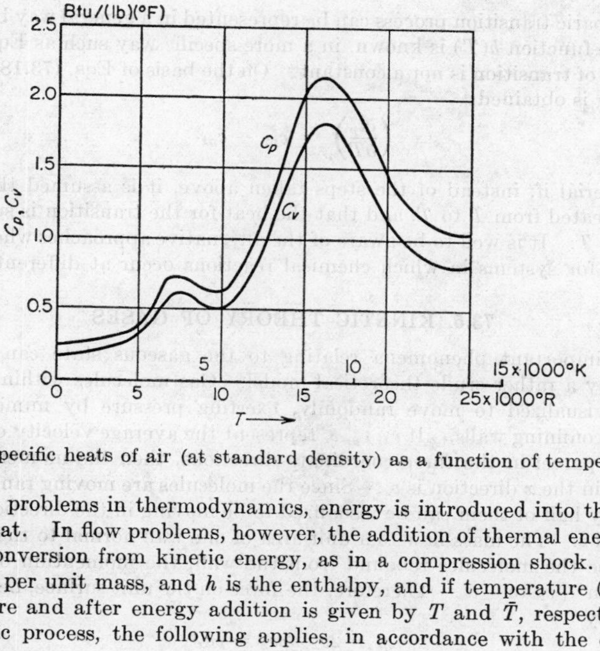

FIG. 73.5. Specific heats of air (at standard density) as a function of temperature [1].

Usually in problems in thermodynamics, energy is introduced into the system in the form of heat. In flow problems, however, the addition of thermal energy can also result from conversion from kinetic energy, as in a compression shock. If q is the added energy per unit mass, and h is the enthalpy, and if temperature (uniform) of the fluid before and after energy addition is given by T and $\bar{T}$, respectively, then, for an isobaric process, the following applies, in accordance with the energy conservation law:

$$q = h(\bar{T}) - h(T) \tag{73.24}$$

In Eq. (73.24), the components present during a transition occurring when the temperature varies from T to $\bar{T}$ are not designated specifically, because the enthalpy terms refer to mixtures. If, on the other hand, the concentrations of the individual components are given by c_1 and c_2 before energy addition, and by $\bar{c}_1$ and $\bar{c}_2$ after energy addition, and the enthalpies are given by h_1 and h_2, then

$$\begin{aligned} h(T) &= c_1 h_1(T) + c_2 h_2(T) \qquad & c_1 + c_2 &= 1 \\ h(\bar{T}) &= \bar{c}_1 h_1(\bar{T}) + \bar{c}_2 h_2(\bar{T}) \qquad & \bar{c}_1 + \bar{c}_2 &= 1 \end{aligned} \tag{73.25}$$

With the aid of Eq. (73.25), Eq. (73.24) may be written as

$$\begin{aligned} q &= (\bar{c}_1 - c_2)h_1(T) + (\bar{c}_2 - c_2)h_2(T) \\ &\quad + \bar{c}_1[h_1(\bar{T}) - h_1(T)] + \bar{c}_2[h_2(\bar{T}) - h_2(T)] \\ &= (\bar{c}_2 - c_2)[h_2(T) - h_1(T)] + \bar{c}_1[h_1(\bar{T}) - h_1(T)] + \bar{c}_2[h_2(\bar{T}) - h_2(T)] \end{aligned} \tag{73.26}$$

It may be assumed that the transition occurs at temperature T, and that the resulting final mixture is heated to temperature $\bar{T}$; then the energy q has been used for the following processes: (1) conversion of the mixture from c_2 to $\bar{c}_2$ at T, and (2) isobaric heating of products ($\bar{c}_1$ and $\bar{c}_2$) from T to $\bar{T}$. The last two terms of Eq. (73.26) refer to the isobaric heating step. The heat of transition (e.g., the heat of reaction, when chemical conversions occur) at constant pressure is given by

$$l_p(T,p) = h_2(T,p) - h_1(T,p) \tag{73.27}$$

Equation (73.26) may be simplified to the following form for components whose specific heats are essentially constant within the temperature range of interest:

$$q = (\bar{c}_2 - c_2)l_p(T) + (\bar{c}_1 c_{p1} + \bar{c}_2 c_{p2})(\bar{T} - T) \tag{73.28}$$

Thus, an isobaric transition process can be represented in a general way by Eq. (73.24) or, when the function $h(T)$ is known, in a more specific way such as Eq. (73.28).

The heat of transition is not a constant. On the basis of Eqs. (73.18) and (73.27), the following is obtained:

$$\left(\frac{\partial l_p}{\partial T}\right)_p = c_{p2} - c_{p1} \tag{73.29}$$

It is immaterial if, instead of the steps taken above, it is assumed that the initial mixture is heated from T to $\bar{T}$, and that the heat for the transition is supplied at the temperature $\bar{T}$. It is well to be aware of the alternative approaches when performing calculations for systems in which chemical reactions occur at different temperature levels.

73.5. KINETIC THEORY OF GASES

Several important phenomena relating to the gaseous state can be described effectively by a rather crude theoretical model. Gas molecules within a closed container are visualized to move randomly, exerting pressure by numerous impacts against the confining walls. If v_x, v_y, v_z represent the average velocity components of a small gas parcel in the three coordinate directions, then the momentum per unit mass of gas in the x direction is v_x. Since the molecules are moving randomly, at any moment only half of them can be visualized to be moving in the directions of increasing x, y, and z. The mass flow to a unit area of one wall normal to the x direction is $\frac{1}{2}\rho v_x$. Since the molecules rebound from the wall, the momentum of a unit mass changes from $+v_x$ to $-v_x$. Therefore, the force on the unit surface, i.e., the pressure p, is given by

$$p = 2v_x \tfrac{1}{2}\rho v_x = \rho v_x^2$$

an expression which is related to the momentum conservation principle of aerodynamics. If the mean molecular velocity is c, and if the difference between the square of a mean and the mean of a square is disregarded, then

$$c^2 = v_x^2 + v_y^2 + v_z^2$$

Since none of the directions of motion is preferred, $v_x^2 = c^2/3$. Hence,

$$p = \tfrac{1}{3}\rho c^2 \tag{73.30}$$

From Eq. (73.30) and the equation of state (73.1), the following relation between the absolute temperature and the molecular velocity is obtained:

$$c^2 = 3RT \tag{73.31}$$

In the equation for the speed of sound a (that is, $a^2 = \gamma RT$), the factor γ appears in place of the number 3. This shows that the average velocity of gas molecules exceeds somewhat the speed of sound in the medium.

In the kinetic theory of gases, the internal energy of a substance is looked upon as the energy of motion of the molecules. Here, of course, such complicating factors as ionization are not considered. The simplest case is that of a unimolecular gas which has only translational molecular motion in the three directions. Thus, the internal energy being equal to the kinetic energy can be expressed as follows:

$$u = \tfrac{1}{2}c^2 = \tfrac{3}{2}RT \tag{73.32}$$

Thus, the specific heat at constant volume for a unimolecular gas is, according to Eq. (73.15),

$$c_v = \left(\frac{\partial u}{\partial T}\right)_\rho = \tfrac{3}{2}R \tag{73.33}$$

This constant value for the specific heat has been confirmed in the case of the rare gases (see Argon in Table 73.1).

The kinetic energy of polyatomic molecules includes rotational and vibrational as well as the three translational forms. The amount of internal energy apportioned to translational motion in one direction is clearly one-third the amount given by Eq. (73.32). If it is assumed that the internal energy is apportioned equally to the n different forms of motion of a molecule (i.e., its n degrees of freedom), then the internal energy is given by

$$u = \frac{n}{2}RT \tag{73.34}$$

From Eqs. (73.15) and (73.22), the specific heats (for constant molecular weights) may be obtained:

$$c_v = \frac{n}{2}R \qquad c_p = \left(1 + \frac{n}{2}\right)R \tag{73.35}$$

The ratio of specific heats γ and n are related as follows:

$$\gamma = \frac{c_p}{c_v} = \frac{2+n}{n} \qquad n = \frac{2}{\gamma - 1} \tag{73.36}$$

According to the principle of the equipartition of energy, the kinetic energy is distributed equally among the degrees of freedom corresponding to the various forms of motion. Thus, atoms in the gaseous state possess three degrees of freedom ($n = 3$). A diatomic molecule without vibrational motion (for example, O_2 or N_2, at low or ambient temperatures) possesses five degrees of freedom ($n = 5$). In this case, the molecule is conceived to be rod-shaped, with both atoms rigidly bound. In addition to the three translational forms of motion always present, there is rotational motion independently about two axes. For molecules containing a larger number of atoms, the question arises whether the structure is one-dimensional (rodlike), two-dimensional (platelike), or three-dimensional. In the two- and three-dimensional cases, there are three forms of rotational motion. Table 73.1 presents data which confirm the values predictable by Eqs. (73.35) and (73.36).

An oscillation with one degree of freedom possesses twice the unit amount of energy per mole, that is, not $\tfrac{1}{2}RT$ but instead RT, because both the potential energy of the elastic bond forces and the kinetic energy of motion are involved.

The number n increases by 2 with each kind of molecular vibration. But, according to quantum theory, the classical partition of energy applies only to the so-called fully excited vibrations. The product of frequency and Planck's constant must be sufficiently small in comparison with the product of the Boltzmann constant and the absolute temperature. For this reason, full conformity with the classical laws is not always obtained in the case of rod-shaped molecules rotating about their axes. For example, at very low temperatures, all rotational motion of the hydrogen molecule ceases. The calculations of u (Fig. 73.2) are therefore rather complex, because full excitation of the vibrational motion and the bond forces does not occur and because, at high temperatures, dissociation and, finally, ionization must be considered. From this, it is clear that n from Eq. (73.36) represents only an approximation of the number of degrees of freedom.

73.6. THE SECOND LAW OF THERMODYNAMICS

The amount of heat which passes into a body is not a state property. A system can be transformed into various states without heat addition or removal, for example, by adiabatic expansion without the performance of work or by the quasi-static adiabatic change of state. Nevertheless, the state property called entropy s is closely related to the quasi-static heat addition [Eq. (73.13)], as is evident from its definition:

$$ds = \frac{dq}{T} = \frac{du + p\, d(1/\rho)}{T} \tag{73.37}$$

For an ideal gas with constant specific heat, this equation can be transformed into the following by substituting Eqs. (73.1), (73.16), and (73.22):

$$ds = c_v \frac{dT}{T} + (c_p - c_v)\rho\, d\left(\frac{1}{\rho}\right)$$

It is obvious from this equation that ds is an exact differential (and, hence, that s is a state property), because $(c_p - c_v)$ is a constant for the gas in question, and c_v (or u) is dependent on T only, a fact which is discussed below. The above equation may be integrated and converted by substitution of Eq. (73.1) to the following forms, where s_0 corresponds to some arbitrary state given by p_0, ρ_0, T_0:

$$\begin{aligned} s - s_0 &= c_v \ln \frac{T}{T_0} - (c_p - c_v) \ln \frac{\rho}{\rho_0} \\ &= c_p \ln \frac{T}{T_0} - (c_p - c_v) \ln \frac{p}{p_0} \\ &= c_v \ln \frac{p}{p_0} - c_p \ln \frac{\rho}{\rho_0} \end{aligned} \tag{73.38}$$

Equations (73.38) are useful in the many instances where the assumption of an ideal gas of constant specific heat is acceptable.

The quasi-static adiabatic change of state is given by Eq. (73.13) when $dq = 0$. In the light of Eq. (73.37), this means that, for the ideal gas with constant specific heat (and, generally, for all substances, as is shown below), a quasi-static adiabatic change is characterized by no change in entropy. Quasi-static adiabatic (or constant entropy) changes are therefore called *isentropic*, just as state changes at constant temperature, pressure, or density are called *isothermal*, *isobaric*, or *isochoric*, respectively. For an isentropic change ($s = s_0$) of an ideal gas of constant specific heat, Eqs. (73.36) and (73.38) lead to the following:

$$\left(\frac{T}{T_0}\right)^{1/(\gamma-1)} = \frac{\rho}{\rho_0} \qquad \left(\frac{T}{T_0}\right)^{\gamma/(\gamma-1)} = \frac{p}{p_0} = \left(\frac{\rho}{\rho_0}\right)^{\gamma} \tag{73.39}$$

Various state changes may occur in sequence, to form a cyclic process, as in the heat engine. The working fluid (a gas, for example), initially in a compressed state, is heated at constant volume with a corresponding increase in internal energy. Then the heated fluid is allowed to expand isentropically; the temperature and internal energy decrease. In the expanded state, the fluid is allowed to cool to such a point that it can be isentropically compressed to the initial state. The compression work performed is less than the work output. In accordance with the energy conservation law, the net work produced is equal to the difference between the added and abstracted heat.

The concept of the Carnot cycle has considerable significance in thermodynamics. In the Carnot cycle, heat addition and abstraction occur isothermally. In many practical cases, however, such as in gas-turbine operation and in the use of heat exchangers, it is more appropriate to consider heat addition and abstraction to occur isobarically. Figure 73.6 shows an isobaric process in the form of a $1/\rho$, p diagram.

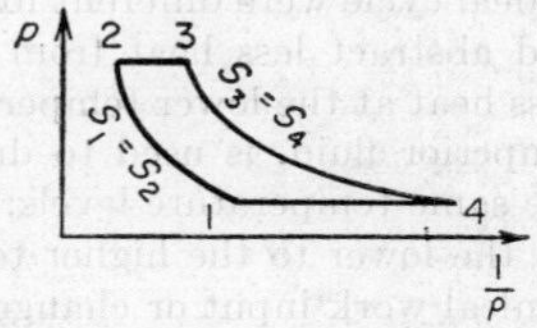

FIG. 73.6. Cyclic process with heat exchange at constant pressure.

Point 1 represents the state at the beginning of isentropic compression. Between points 2 and 3, isobaric heat addition occurs, then isentropic expansion occurs between points 3 and 4, and between 4 and 1 isobaric heat removal takes place. Work is performed not only along the isentropes but also along the isobars. If $q_{2,3}$ is the heat addition along path 2 to 3 and $q_{4,1}$ is the heat addition along 4 to 1 per unit mass of working fluid, then, for an ideal gas with constant specific heat,

$$q_{2,3} = c_p(T_3 - T_2) \qquad q_{4,1} = c_p(T_1 - T_4) \tag{73.40}$$

Since $T_1 < T_4$, then $q_{4,1} < 0$, implying that heat is being taken away at pressure $p_4 = p_1$. Since $p_3/p_4 = p_2/p_1$, isentropic relations (73.39) lead to

$$\frac{T_1}{T_2} = \frac{T_4}{T_3} \tag{73.41}$$

The efficiency η of the process is defined as the ratio of the net work output to the heat input. According to the energy conservation law, the net work output is equal to the sum of the positive and negative quantities of heat transferred. Hence, on the basis of Eqs. (73.40) and (73.41), the efficiency is expressed as follows:

$$\eta = \frac{q_{2,3} + q_{4,1}}{q_{2,3}} = \frac{T_3 - T_2 + T_1 - T_4}{T_3 - T_2} = \frac{T_2 - T_1}{T_2} \tag{73.42}$$

Under normal circumstances, the efficiency of this ideal cyclic process is considerably less than unity. For the compression ratio $p_2/p_1 = 5$ (corresponding to practical gas-turbine operation) and a specific heat ratio of $\gamma = 1.40$, the temperature ratio yielded by Eq. (73.39) is $T_2/T_1 = 1.58$, and, therefore, the efficiency is $\eta = 0.37$. An ideal cyclic process can also be operated in the reversed direction. Then mechanical work must be performed, with the result that heat is added at a lower temperature and discarded at a higher temperature. In this case, the cyclic process functions as a refrigerator or heat pump.

It is a well-established fact that, without the exertion of work, heat cannot flow from a lower to a higher temperature level. The process of heat transfer cannot be reversed. It is also impossible to reverse the action of friction, which causes the dissipation of part of the energy as heat when uesful mechanical work is performed. The throttling process and the adiabatic expansion of a gas without work output are other examples in which irreversibility occurs. If the irreversibility of such basic processes is accepted as an irrefutable experimental fact, then the efficiency of a reversible cycle, regardless of the working fluid, cannot differ from the efficiency calculable from Eq. (73.24) for an ideal gas with a constant specific heat. This statement, which is demonstrated below, is an expression of the second law of thermodynamics.

According to the first law of thermodynamics, the energy conservation law, the difference between the added and abstracted heat is equal to the net work. Thus, for a given net work, the numerator $(q_{2,3} + q_{4,1})$ of Eq. (73.42) is fixed for an ideal cycle. Now, if the efficiency of this ideal cycle were different for two different working fluids, then the superior fluid would abstract less heat from the higher-temperature heat reservoir, and would reject less heat at the lower temperature. Then, if an ideal heat engine, operating with the superior fluid, is used to drive a heat pump, employing the inferior fluid, between the same temperature levels, it is clear that heat would be transported continually from the lower to the higher temperature reservoir, without the need of additional mechanical work input or changes in the surroundings.

This illustration leads to another statement of the second law of thermodynamics: The efficiency of an ideal reversible cyclic process is the same for all working fluids. Another rather well-known statement is: No heat engine can convert into work all the heat it receives (a violation of this statement is called perpetual motion of the second kind).

From Eq. (73.37), it can be seen that the following expression applies to a cyclic process in which the working fluid is an ideal gas with constant specific heat:

$$\oint \frac{dq}{T} = 0 \tag{73.43}$$

Actually Eq. (73.43) applies as well to any real fluid in a reversible cyclic process. If, at some temperature T, the added heat dq were different for two gases, then a reversal of heat flow or a perpetual motion of the second kind would be possible. Since Eq. (73.43) applies to every substance and for every reversible process, the integral between the two limits must be independent of the path (i.e., the sequence of events between the two limits). This is possible only when dq/T is an exact differential. Equation (73.37) is, therefore, perfectly general.

A reversible process takes place quasi-statically at all moments over its full course. Such a process can rigorously be taken to be comprised of a series of infinitesimal steps. Any departure from equilibrium, in pressure or temperature, for example, results in irreversibility. It is not surprising that, in nature, all processes display at least a small amount of irreversibility; in most processes encountered, an appreciable amount of irreversibility is found.

Whenever the infinitesimal elemental steps of an ideal process proceed reversibly, i.e., quasi-statically, then the sum of the entropies of all substances involved in the process does not change. Any heat exchange which occurs between substances in the system must take place at temperature equilibrium. According to Eq. (73.37), this means that one substance increases in entropy and some other decreases in entropy such that the total remains unaltered.

In irreversible processes, the entropy increases. Thus, entropy is a measure of irreversibility. An increase in entropy can be demonstrated easily in the throttling process. From Eqs. (73.37) and (73.14), the following is obtained:

$$T\,ds = dh - \frac{1}{\rho}dp \tag{73.44}$$

Since, in a throttling process, the pressure decreases ($dp < 0$) and the enthalpy remains constant ($dh = 0$), it is clear from Eq. (73.44) that the fluid leaves the system at a higher entropy. The total entropy of the system increases in the process. This observation leads to an analytical or quantitative statement of the second law of thermodynamics: The sum of the entropies of all of the substances in a process cannot decrease.

Some of the most widely applied relations in aerodynamics are deduced on the basis of Eq. (73.44). Since various pairs of independent variables are commonly employed in thermodynamics, for example, p, ρ; p, T; T, s or h, s, care must always be taken to indicate which variables are held constant during differentiation. This is done by placing a subscript after the parenthesis. Thus $(\partial s/\partial h)_p$ denotes the slope of the curve of s versus h when p is held constant. The following relations are based on Eq. (73.44):

$$\left(\frac{\partial p}{\partial h}\right)_s = \rho = \frac{1}{(\partial h/\partial p)_s} \qquad \left(\frac{\partial p}{\partial s}\right)_h = -\rho T = \frac{1}{(\partial s/\partial p)_h} \qquad \left(\frac{\partial h}{\partial s}\right)_p = T = \frac{1}{(\partial s/\partial h)_p} \tag{73.45}$$

73.7. SOME CONSEQUENCES OF THE SECOND LAW OF THERMODYNAMICS

The second law is one of the most remarkable laws of exact science. This is demonstrated by the fact that its consequences have frequently no apparent connection with the law itself. This is evident in the following examples.

In this discussion, T and ρ are chosen as the independent variables. In conformity with practice, subscripts after the parentheses are employed, although this is actually unnecessary whenever two independent variables are used exclusively. Thus, since u can be taken to be a function of T and ρ,

$$du = \left(\frac{\partial u}{\partial T}\right)_\rho dT + \left(\frac{\partial u}{\partial \rho}\right)_T d\rho$$

Then substitution of du into Eq. (73.37) yields

$$T\,ds = \left(\frac{\partial u}{\partial T}\right)_\rho dT + \left[\left(\frac{\partial u}{\partial \rho}\right)_T - \frac{p}{\rho^2}\right] d\rho \tag{73.46}$$

From this, the following derivatives are obtained:

$$\frac{\partial s}{\partial T} = \frac{1}{T}\left(\frac{\partial u}{\partial T}\right)_\rho \qquad \frac{\partial s}{\partial \rho} = \frac{1}{T}\left[\left(\frac{\partial u}{\partial \rho}\right)_T - \frac{p}{\rho^2}\right]$$

If appropriate second derivatives are taken,

$$\frac{\partial^2 s}{\partial T\,\partial \rho} = \frac{1}{T}\frac{\partial^2 u}{\partial \rho\,\partial T} = -\frac{1}{T^2}\left(\frac{\partial u}{\partial \rho}\right)_T + \frac{1}{T}\frac{\partial^2 u}{\partial \rho\,\partial T} - \frac{1}{\rho^2}\frac{\partial}{\partial T}\left(\frac{p}{T}\right)$$

then, by canceling the term $\frac{1}{T}\frac{\partial^2 u}{\partial \rho\,\partial T}$, a relation for $(\partial u/\partial \rho)_T$ is found:

$$\left(\frac{\partial u}{\partial \rho}\right)_T = -\frac{T^2}{\rho}\left[\frac{\partial}{\partial T}\left(\frac{p}{\rho T}\right)\right]_\rho \tag{73.47}$$

Equation (73.47) is valid without any restrictions. If the equation of state (73.1) is introduced, then the following is obtained for the case of variable molecular weight:

$$\left(\frac{\partial u}{\partial \rho}\right)_T = \frac{R}{m}\frac{T^2}{\rho}\left(\frac{\partial m}{\partial T}\right)_\rho \tag{73.48}$$

As mentioned earlier, u for an ideal gas (m = const) is clearly a function of T only.

Another example concerns dissociation, which occurs in high-temperature reaction zones in combustion chambers and in aerodynamic-flow regions at high velocities. As an illustration, consideration is given here to a mixture of a substance of molecular weight m_1, and its dissociation product of molecular weight $m_2 = \frac{1}{2}m_1$, such as molecular nitrogen N_2 and atomic nitrogen N, or, in the case of ionization, oxygen atoms O and the mixture of oxygen ions O^+ and electrons (a mixture which has a molecular weight half the atomic weight of O). Furthermore, the gas mixture is held at a total pressure p and is thermally insulated. Now, when the mixture of component 1 and its dissociation product 2 remains for a sufficient time in the container, equilibrium is finally established. In the period when equilibrium is being approached, the entropy of the whole system increases; when equilibrium is attained, the entropy is at a maximum value, and no further change occurs. If both components are regarded as ideal gases with mass concentrations c_1 and c_2, partial pressures p_1 and p_2, and densities ρ_1 and ρ_2, then

$$c_1 + c_2 = 1 \qquad c_1 = \rho_1/\rho \qquad c_2 = \rho_2/\rho \qquad p = p_1 + p_2 \tag{73.49}$$

The energy conservation law requires that the total enthalpy in an isobaric process without heat addition be constant:

$$h = c_1h_1 + c_2h_2 = \text{const} \tag{73.50}$$

Since the entropy

$$s = c_1s_1 + c_2s_2 \tag{73.51}$$

possesses a maximum at equilibrium, its differential must vanish [see Eq. (73.44)]:

$$ds = 0 = \frac{c_1}{T}\left(dh_1 - \frac{1}{\rho_1}\,dp_1\right) + \frac{c_2}{T}\left(dh_2 - \frac{1}{\rho_2}\,dp_2\right) + s_1\,dc_1 + s_2\,dc_2$$

This equation, together with relations obtained from Eqs. (73.49) and (73.50),

$$dc_1 = -dc_2 \qquad c_1/\rho_1 = c_2/\rho_2 \qquad dp_1 = -dp_2$$
$$0 = c_1\,dh_1 + c_2\,dh_2 + h_1\,dc_1 + h_2\,dc_2$$

yields the equilibrium relation,

$$h_1 - Ts_1 = h_2 - Ts_2 \tag{73.52}$$

Now, according to Eq. (73.27), the enthalpy difference is the heat of reaction l_p. Equation (73.52) shows that the entropy of each component increases when the enthalpy increases, just as it does when heat is added to the system.

The influence of the heat of reaction is further demonstrated if it is expressed in terms of the variables p and T or ρ and T. For substances each possessing a constant specific heat in the region of interest, Eqs. (73.38) and (73.52) yield

$$\frac{l_p}{T} = c_{v2}\ln T - R_2\ln\rho_2 - c_{v1}\ln T + R_1\ln\rho_1 + \text{const} \tag{73.53}$$

The constant includes all the terms which are not dependent upon the state. It can be shown that, in Eq. (73.53), it is immaterial whether the heat of reaction is at constant pressure l_p or at constant density l_ρ and, furthermore, whether the heat of reaction l is taken as a function of temperature $l(T)$ or at some standard condition. For an ideal gas with constant specific heat, all these quantities differ by an added term, which is porportional to T. This results simply in a change in the added constant and, in Eq. (73.54), in a change in the multiplication constant A.

After introducing R_1 and R_2 [see Eq. (73.2)] and the mass concentrations c_1 and c_2 from Eqs. (73.49), then Eq. (73.53) becomes

$$\frac{c_2^2}{c_1} = Ae^{-l/R_1T}T^{(c_{v2}-c_{v1})/R_1}\rho^{-1} \tag{73.54}$$

By introducing the gas kinetic relations (73.35), the exponent containing the specific heats may be written as follows:

$$\frac{c_{v2} - c_{v1}}{R_1} = \frac{1}{R_1}\left(\frac{n_2}{2}R_2 - \frac{n_1}{2}R_1\right) = n_2 - \tfrac{1}{2}n_1$$

In general, the coefficient n_2 representing the degrees of freedom in the dissociated state is more than half that for the undissociated molecules, n_1. Therefore, the exponent of T is usually positive. Nevertheless, the controlling influence of temperature resides almost entirely in the exponential term. Accordingly, the concentration of the dissociated or ionized species is found to increase exponentially with temperature. For the same reason, it increases exponentially with l/R_1, or lm_1 [see Eq. (73.2)]. Since, for the dissociation reaction $O_2 \rightarrow 2O$, $lm = 117$ Cal/mole (see Table 73.1), and, for $N_2 \rightarrow N$, $lm = 225$ Cal/mole, and since, for ionization $O \rightarrow O^+ +$ electron, $lm = 313$ Cal/mole and, for $N \rightarrow N^+ +$ electron, $lm = 343$ Cal/mole, it is clear why oxygen dissociates at a lower temperature than nitrogen, and why ionization of atoms becomes prevalent only at much higher temperatures. Furthermore, Eqs. (73.54) and (73.2) show that, at constant temperature, dissociation increases (and, hence, molecular weight decreases), with increase in the total density of the system. Equation (73.54) is a special form of the law of mass action.

73.8. REFERENCES

Tables

[1] J. Hilsenrath, W. Beckett: Tables of thermodynamic properties of argon-free air to 15,000°K, *AEDC Tech. Note* 56–12 (September, 1956); *ASTIA Doc.* AD-98974.
[2] C. E. Treanor, J. G. Logan: Tables of thermodynamic properties of air from 3000°K to 10,000°K, *Cornell Aeronaut. Lab. Rept.* AD-1052-A-2.
[3] "Handbook of Supersonic Aerodynamics," vol. 5, a Bureau of Ordnance publication, *NAVORD Rept.* 1488 (August, 1953).
[4] J. D'Ans, E. Lax: "Taschenbuch für Chemiker und Physiker," Springer, Berlin, 1949.

Books

[5] F. D. Rossini (ed.): "Thermodynamics and Physics of Matter," vol. 1 of "High Speed Aerodynamics and Jet Propulsion," Princeton Univ. Press, 1955.
[6] J. K. Roberts, A. R. Miller: "Heat and Thermodynamics," Blackie, London and Glasgow, 1955.
[7] M. W. Zemansky: "Heat and Thermodynamics," 4th ed., McGraw-Hill, New York, 1957.
[8] J. H. Keenan: "Thermodynamics," Wiley, New York, 1941.

CHAPTER 74

THERMODYNAMICS OF GAS FLOW

BY

A. ROSHKO, Ph.D., Pasadena, Calif.

74.1. COMPARISON WITH THE STATIC CASE

Classical thermodynamics (Chap. 73) deals with a uniform, static system whose condition is defined by a few properties of state, and which can exchange energy with its surroundings. In applying these concepts to a flowing fluid, we take as the system a fluid particle, arbitrarily small so that it may be assumed to be uniform. This assumption actually has a large area of validity; we shall discuss on p. 74–11 the conditions under which it breaks down.

Thermodynamic properties of state (T, ρ, p, h, u, s, etc.) are associated with the system (fluid particle) exactly as in the static case; they are related by the equations of state. Their values, which would be measured by an observer moving with the particle, that is, having no motion relative to the fluid, are called the *static* values, as distinguished from the *total* values, to be discussed presently.

A flowing fluid possesses an additional form of energy, namely, the kinetic energy, which it does not have in the static form. One of the main extensions of classical thermodynamics to a flowing fluid concerns the exchanges that occur among the kinetic energy, the internal energy, and the work performed by the pressure. A further difference is that the element of work performed by the flowing fluid is $d(p/\rho)$ as compared to $p\,d(1/\rho)$ in the static case. The difference between these, dp/ρ, is the portion of the work that is related to the fluid motion, and does not depend on a density change. This *flow work* was derived on p. 73–5 in connection with the *throttling process* [see Eq. (73.8)].

We shall consider only *steady* flow, for which the conditions at each point of the flow, from a suitable frame of reference, are invariant with time. The velocity field of such a flow is defined by a stationary set of streamlines, or stream tubes. A fluid particle moves along a definite streamline. In general, conditions vary along a streamline, and the properties of a particle change accordingly, as it moves along. From the thermodynamic point of view, it is natural to describe the changes occurring in the particle (the system), whereas, for the flow-field description, it is preferable to describe conditions along the streamline. For steady flow, the two points of view are equivalent, since a particle passes through every point on the streamline; that is, the changing properties of one are related to the variable conditions along the other. The relations between different streamlines are taken up on p. 74–12.

74.2. THE ENERGY EQUATION

The throttling equation (73.8) accounts for the internal energy and flow work of a flowing fluid, but assumes the kinetic energy to be negligible. In problems of gas dynamics, this assumption is generally not valid. When the kinetic energy is included, the throttling equation generalizes to the important equation

$$h_2 + \tfrac{1}{2}v_2^2 = h_1 + \tfrac{1}{2}v_1^2 \tag{74.1}$$

The enthalpy h and kinetic energy $\frac{1}{2}v^2$ are for unit mass of fluid. The relation expressed here is an *adiabatic* one; that is, no energy is transferred into or out of the stream tube, between the two reference sections, 1 and 2. The equation is very general, in that no further assumption is made about the flow at any points between 1 and 2. For instance, it applies if there is a region of dissipation, as at a resistance screen, between 1 and 2.

If an amount of heat q per unit mass is added to the flow between 1 and 2, for example, by a radiator or by an electric coil, then we have the *diabatic* relation

$$h_2 + \tfrac{1}{2}v_2^2 = h_1 + \tfrac{1}{2}v_1^2 + q \tag{74.2}$$

As another example, if an amount of energy w per unit mass is extracted from the flow, e.g., by a turbine, then

$$h_2 + \tfrac{1}{2}v_2^2 = h_1 + \tfrac{1}{2}v_1^2 - w \tag{74.3}$$

The external energy supplied or extracted appears as an additional term. Energy that is not exchanged with the surroundings is all accounted for in the enthalpy. Such is the case with the so-called transformation energy or transition heat, such as heat of vaporization, heat of dissociation, and "chemical" energy in general. It is permissible, of course, to account for such transformation energy separately, and this is often done, for example, in detonation and deflagration theory (cf. also the remarks on p. 73–10 for isobaric processes).

If an adiabatic flow is uniform at each section of the stream tube, that is, h and v can be defined for each section, then Eq. (74.1) may be written as

$$h + \tfrac{1}{2}v^2 = \text{const} \tag{74.4}$$

If it can be applied continuously along the flow, we may use the differential form,

$$dh + v\,dv = 0 \tag{74.4a}$$

Equation (74.4) shows that a decrease in v implies an increase in h, and vice versa. In particular, if $v = 0$, h has a maximum value h_t, which is called the *total* enthalpy. Thus, if the flow is accelerated from a reservoir, e.g., in a wind tunnel or a rocket nozzle, the fluid in the reservoir is practically at rest, and so the enthalpy there is h_t; hence, the alternative name, *reservoir* enthalpy. The term *stagnation* enthalpy is also used. Equation (74.4) may now be written as

$$h + \tfrac{1}{2}v^2 = h_t \tag{74.5}$$

That is, the total enthalpy, which may be considered to be a flow property of the fluid at any point, is the sum of the static enthalpy and the kinetic energy. In adiabatic flow it remains constant.

The maximum speed to which a fluid may be accelerated is determined by its total enthalpy. Equation (74.5) shows that the maximum speed v_m would be attained if the static enthalpy became zero; that is,

$$\tfrac{1}{2}v_m^2 = h_t \tag{74.6}$$

Thus, the maximum exhaust velocity attainable by a rocket fuel depends on its total enthalpy.

74.3. AREA-VELOCITY RELATIONS; COMPRESSIBILITY

The relationship between the flow variables and the stream-tube geometry cannot be obtained from the energy equation alone; the continuity equation is required. For steady flow, this states that the product of the density ρ, the velocity v, and the cross-sectional tube area A—that is, the mass flow m—is the same for every section of the tube,

$$m = \rho v A = \text{const} \tag{74.7}$$

This could now be combined with the energy equation, which is the procedure we shall follow in Sec. 74.7. At this point, it is instructive to study the *differential* relations

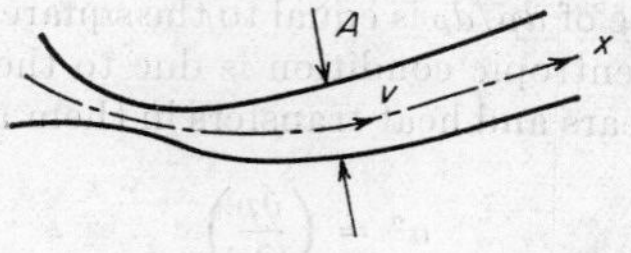

FIG. 74.1. Stream tube.

between the area and the flow variables. Accordingly, we write Eq. (74.7) in differential form:

$$\frac{d\rho}{\rho} + \frac{dv}{v} + \frac{dA}{A} = 0 \tag{74.7a}$$

If we consider the flow to be frictionless, the adiabatic energy equation is equivalent to the momentum equation (70.1):

$$v\frac{dv}{dx} = -\frac{1}{\rho}\frac{dp}{dx} \qquad \text{or} \qquad v\,dv = -\frac{1}{\rho}\,dp \tag{74.8}$$

The reason for the equivalence is as follows. In Eq. (74.8), the acceleration $v\,dv/dx$ is due to pressure forces only. The assumption that friction is absent is equivalent to the assumption that the flow is reversible. Since it is also adiabatic, then it must be *isentropic,* according to a general result of thermodynamics (see p. 73–12); that is,

$$s = \text{const} \qquad \text{or} \qquad ds = 0$$

If we now compare the momentum equation (74.8) with the differential form of the energy equation (74.4*a*),

$$dh + v\,dv \equiv du + p\,d\left(\frac{1}{\rho}\right) + \frac{1}{\rho}\,dp + v\,dv = 0 \tag{74.9}$$

remembering that $du + p\,d(1/\rho) = 0$ in an isentropic process [see Eq. (73.37)], we see that the two are equivalent. In frictionless adiabatic flow, the energy equation splits into a "thermodynamic part," which is identical with that in the static case,

$$du = -p\,d\left(\frac{1}{\rho}\right)$$

and a "mechanical part,"

$$d\left(\frac{v^2}{2}\right) = -\frac{1}{\rho}\,dp$$

To study the interplay between the continuity equation (74.7*a*) and the momentum equation (74.8), we eliminate the pressure by writing it in the form

$$dp = \left(\frac{\partial p}{\partial \rho}\right)_s d\rho + \left(\frac{\partial p}{\partial s}\right)_\rho ds \tag{74.10}$$

This simply expresses the fact that p is related to two other independent variables of state, which we take to be ρ and s.

The coefficient $dp/d\rho$ is related to the bulk coefficient of elasticity, that is, the pressure required to produce unit change of volume,

$$E = -\frac{1}{\rho}\frac{dp}{d(1/\rho)} = \rho\frac{dp}{d\rho}$$

Since the speed of elastic waves in a medium is $\sqrt{E/\rho}$, we see that $dp/d\rho$ is the square of a wave speed.

Now the isentropic value of $dp/d\rho$ is equal to the square of the acoustic wave speed, or speed of sound a; the isentropic condition is due to the fact that sound waves are so weak that dissipative shears and heat transfers in them are extremely small. Thus,

$$a^2 = \left(\frac{\partial p}{\partial \rho}\right)_s \tag{74.11}$$

The role of the speed of sound as a measure of fluid elasticity, or, inversely, of its compressibility, is now clear. Changes of flow speed are determined by pressure changes, through the momentum equation, and thus are related to density changes, through a^2. That is, Eq. (74.8) may be written as

$$v\,dv = -a^2\frac{d\rho}{\rho} \tag{74.12}$$

A more useful measure of the influence of compressibility is obtained by introducing the *Mach number*,

$$M = \frac{v}{a} \tag{74.13}$$

which, in Eq. (74.12), gives

$$\frac{d\rho}{\rho} = -M^2\frac{dv}{v} \tag{74.14}$$

This relation shows that if the flow speed is very small compared to the speed of sound ($M \ll 1$), then a change in speed produces a very small change in density; that is, the flow may be considered to be *incompressible*. At higher Mach numbers, the density changes become more pronounced. At *subsonic* speed ($M < 1$), the relative density change is smaller than the velocity change, while, at *supersonic* speed ($M > 1$), it is larger. This has a remarkable effect on the relation between velocity and area, as may be seen by putting Eq. (74.14) into Eq. (74.7*a*). The result is

$$\frac{dA}{A} = (M^2 - 1)\frac{dv}{v} \tag{74.15}$$

which states that, in subsonic flow, an increase in speed is produced by a *decrease* in area, while, at supersonic speed, an increase in speed requires an *increase* in area. At supersonic speed, the density is changing faster than the speed, requiring an increasing area to satisfy the continuity equation.

It follows that acceleration of a fluid from subsonic to supersonic speed requires that the flow pass through a *throat*, where the stream changes from converging to diverging, and conversely for deceleration from supersonic to subsonic speed. In these cases, the throat is *sonic* ($M = 1$); only at a throat can sonic conditions be attained. (However, the converse is not necessary; a throat need not be sonic.) The sonic condition provides another convenient reference condition (cf. total conditions): thus, sonic enthalpy h^*, sonic speed v^*, and the sonic value of the speed of sound a^*. The latter two are equal,

$$v^* = a^*$$

since the sonic value of Mach number is unity. The constant in Eq. (74.4) may now be evaluated in another way:

$$h + \tfrac{1}{2}v^2 = h_t = h^* + \tfrac{1}{2}a^{*2} \tag{74.16}$$

74.4. PROCEDURE FOR A GENERAL EQUATION OF STATE

The relations obtained so far for isentropic flow in a stream tube are quite general, and apply to any fluid, since no assumption has been made about the equation of state. In the general case, it is not possible to express the equation of state in simple analytical form, particularly if the fluid is a mixture of components and a large temperature range is to be covered. Tabulated values or a chart, such as the Mollier chart, must be used [7],[9]. With h and s as coordinates on the Mollier chart, other state variables appear as parameters. For solving the equations of isentropic flow, it may be seen, from the previous section, that convenient parameters are ρ and a. [The speed of sound a may evidently be treated as a state variable, since it is related to p, ρ, and s through Eq. (74.11).]

For example, given the reservoir conditions h_t and a_t or s_t, what are the sonic conditions, h^* and a^*? On a Mollier h,s chart, in which lines of constant a are plotted, the reservoir condition (h_t,a_t) is located. The trajectory

$$h^* = h_t - \tfrac{1}{2}a^{*2}$$

is plotted. Its intersection with the line $s = \text{const} = s_t$ determines (h^*,a^*), as well as any other state variables that may be plotted on the chart.

The solution for any other point in the stream tube may be similarly found from the basic equations:

$$\rho v A = m = \rho^* a^* A^* \qquad h + \tfrac{1}{2}v^2 = h^* + \tfrac{1}{2}a^{*2} = h_t \qquad s = \text{const} \tag{74.17}$$

For example, elimination of v from the first two equations gives

$$h - h^* = \tfrac{1}{2}a^{*2}\left[1 - \left(\frac{A^*}{A}\right)^2\left(\frac{\rho^*}{\rho}\right)^2\right]$$

The condition for given A is found at the intersection of this equation with the line $s = \text{const}$. In general, there will be two solutions, a subsonic and a supersonic one.

74.5. EQUATIONS FOR A PERFECT GAS

Given simple analytical forms of the equations of state, explicit relations between all the flow variables may be found. Here we consider only the case of a gas which is *thermally* perfect, that is, satisfies the perfect-gas law,

$$p = R\rho T \tag{74.18}$$

and *calorically* perfect, that is, has constant specific heat,

$$h = c_p T \tag{74.19}$$

These relations are particularly simple, and they have a wide range of usefulness. They have to be modified only at temperatures so low that van der Waals effects become important, or at temperatures so high that molecular vibration, dissociation, or ionization effects appear. In aerodynamics, the low-temperature effects may be encountered in high-speed wind tunnels, where the gas may be expanded to extremely low temperatures; the high-temperature effects become important at hypersonic flight speeds. Van der Waals effects may also be important in high-pressure reservoirs.

For the perfect-gas law, we can now calculate

$$a^2 = \left(\frac{\partial p}{\partial \rho}\right)_s = \gamma \left(\frac{\partial p}{\partial \rho}\right)_T = \gamma RT = \gamma \frac{p}{\rho} \tag{74.20}$$

Also, for constant specific heat,

$$h = c_p T = \frac{c_p}{R}\frac{p}{\rho} = \frac{\gamma}{\gamma - 1}\frac{p}{\rho} = \frac{a^2}{\gamma - 1} \tag{74.21}$$

The energy equation may now be rewritten in different forms; putting these various expressions for h in Eq. (74.5), we have the following adiabatic relations:

$$c_p T + \tfrac{1}{2}v^2 = c_p T_t \qquad \frac{\gamma}{\gamma - 1}\frac{p}{\rho} + \frac{v^2}{2} = \frac{\gamma}{\gamma - 1}\frac{p_t}{\rho_t} \qquad (74.22a\text{–}c)$$

$$\frac{a^2}{\gamma - 1} + \frac{v^2}{2} = \frac{a_t^2}{\gamma - 1} = \frac{v_m^2}{2} = \frac{1}{2}\frac{\gamma + 1}{\gamma - 1}a^{*2}$$

The right-hand sides are all equivalent, being simply different expressions for the total enthalpy h_t.

Dividing Eq. (74.22c) by $a^2/(\gamma - 1)$, remembering that $M = v/a$, we have

$$\left(\frac{a}{a_t}\right)^2 = \frac{T}{T_t} = \frac{h}{h_t} = \left(1 + \frac{\gamma - 1}{2}M^2\right)^{-1} \tag{74.23}$$

From this, it follows that

$$\left(\frac{v}{a_t}\right)^2 = M^2\left(1 + \frac{\gamma - 1}{2}M^2\right)^{-1} = \frac{2}{\gamma + 1}\left(\frac{v}{a^*}\right)^2 = \frac{2}{\gamma - 1}\left(\frac{v}{v_m}\right)^2 \tag{74.24}$$

or

$$\begin{aligned} M^2 &= \left(\frac{v}{a_t}\right)^2\left[1 - \frac{\gamma - 1}{2}\left(\frac{v}{a_t}\right)^2\right]^{-1} \\ &= \frac{2}{\gamma + 1}\left(\frac{v}{a^*}\right)^2\left[1 - \frac{\gamma - 1}{\gamma + 1}\left(\frac{v}{a^*}\right)^2\right]^{-1} \qquad (74.25a\text{–}c) \\ &= \frac{2}{\gamma - 1}\left(\frac{v}{v_m}\right)^2\left[1 - \left(\frac{v}{v_m}\right)^2\right]^{-1} \end{aligned}$$

For the preceding relations [Eqs. (74.22) to (74.25)] to be valid, it is sufficient for the flow to be adiabatic. If it is also isentropic, along the stream tube, then the isentropic relations for a perfect gas [Eqs. (73.39)] may be introduced:

$$\frac{p}{p_t} = \left(\frac{\rho}{\rho_t}\right)^\gamma = \left(\frac{T}{T_t}\right)^{\gamma/(\gamma-1)} \tag{74.26}$$

From Eq. (74.23), we then have

$$\frac{p}{p_t} = \left(1 + \frac{\gamma - 1}{2}M^2\right)^{-\gamma/(\gamma-1)} \qquad \frac{\rho}{\rho_t} = \left(1 + \frac{\gamma - 1}{2}M^2\right)^{-1/(\gamma-1)} \qquad (74.27a,b)$$

The relation between pressure and speed, in isentropic flow, is called the *Bernoulli equation.* This may be found by using the isentropic relation to eliminate ρ from Eq. (74.22b), obtaining

$$\tfrac{1}{2}v^2 + \frac{\gamma}{\gamma - 1}\frac{p_t}{\rho_t}\left(\frac{p}{p_t}\right)^{(\gamma-1)/\gamma} = \frac{\gamma}{\gamma - 1}\frac{p_t}{\rho_t} \tag{74.28}$$

Many more such useful relations between various ratios may be found from the energy equation; an extensive list of these is given in [5]. Numerical tables and charts may be found in this and other references.

74.6. THE CONSTANTS OF THE ENERGY EQUATION

In an adiabatic flow, the total enthalpy h_t is constant throughout, and (for a perfect gas) so are the related values T_t and a_t. On the other hand, the total pressure p_t and the total density ρ_t are invariant only if the flow is isentropic, that is, no losses occur. If losses occur, the entropy s increases, and p_t and ρ_t decrease.

Where p_t and ρ_t appear in any of the forms of the adiabatic energy equation, we may take either of two points of view: (1) The equation applies to an isentropic flow, in which p_t and ρ_t are invariant; (2) the flow is not necessarily isentropic, but the equation applies *locally*, giving a relation, for example, Eq. (74.27), between the local static pressure and local total pressure. The *local* total values at any point are defined as those which would be attained if the fluid there were brought to rest isentropically. Evidently then there is no distinction between static entropy and total entropy ($s = s_t$). Furthermore, the total values give a measure of the entropy, through Eqs. (73.38).

Similar remarks may be made about the other reference values. Thus h^*, T^*, a^{*2}, and $v_m{}^2$, which are related to h_t, are constant in an adiabatic flow, while p^* and ρ^*, which are related to p_t and ρ_t, are constant only if the flow is isentropic. The latter relations may be obtained by putting $M = 1$ into Eqs. (74.27):

$$\frac{p^*}{p_t} = \left(\frac{2}{\gamma + 1}\right)^{\gamma/(\gamma-1)} \qquad \frac{\rho^*}{\rho_t} = \left(\frac{2}{\gamma + 1}\right)^{1/(\gamma-1)} \tag{74.29a,b}$$

The above remarks can evidently be generalized to diabatic flow, in which case h_t may also change.

74.7. THE AREA RELATION FOR A PERFECT GAS

The area-velocity relation (74.15) can be integrated, first eliminating M or v by means of the energy equation. But more simply, we have

$$m = \rho v A = \rho^* a^* A^* = \left(\frac{2}{\gamma + 1}\right)^{\frac{1}{2}(\gamma+1)/(\gamma-1)} \rho_t a_t A^*$$
$$= \left(\frac{2}{\gamma + 1}\right)^{\frac{1}{2}(\gamma+1)/(\gamma-1)} \sqrt{\frac{\gamma}{RT_t}}\, p_t A^* \tag{74.30}$$

Then

$$\frac{A}{A^*} = \frac{\rho^* a^*}{\rho v} = \left(\frac{2}{\gamma + 1}\right)^{\frac{1}{2}(\gamma+1)/(\gamma-1)} \frac{\rho_t}{\rho}\frac{a_t}{a}\frac{a}{v}$$
$$= \frac{1}{M}\left[\frac{2}{\gamma + 1}\left(1 + \frac{\gamma - 1}{2} M^2\right)\right]^{\frac{1}{2}(\gamma+1)/(\gamma-1)} \tag{74.31}$$

This is an isentropic relation, since Eq. (74.27*b*) was used for ρ_t/ρ. [Equation (74.23) was used for a_t/a.] The function A/A^* has a minimum value equal to 1 at $M = 1$. Every other (higher) value of A/A^* corresponds to two values of M, one subsonic and one supersonic. This agrees with our previous conclusions, that the flow can pass through sonic speed only at a throat.

It is not necessary that a sonic throat actually occur in the flow; Eq. (74.31) may still be used to relate A and M at different points of an isentropic flow; A^* appears then simply as a flow parameter. M may be replaced by any other flow variable to which it is related through the energy equation.

The right-hand side of Eq. (74.30) gives the mass flow through a sonic throat A^*. This sonic condition may not actually exist in a given flow; A^* may be merely a parameter. If, however, an actual throat A_0 exists, not necessarily sonic, then we may rewrite Eq. (74.30):

$$m = \left(\frac{2}{\gamma+1}\right)^{\frac{1}{2}(\gamma+1)/(\gamma-1)} \sqrt{\frac{\gamma}{RT_t}}\, p_t \frac{A^*}{A_0} A_0 \tag{74.32}$$

Now the factor A^*/A_0 [Eq. (74.31)] is always less than unity, unless A_0 is sonic. Thus, for given p_t, T_t, the mass flow through a throat becomes a maximum when the throat becomes sonic, and is then given by Eq. (74.30). The increase of mass flow in a nozzle, for given p_t, T_t, is increased by decreasing the pressure at its exit. Once the throat becomes sonic, the flow upstream of the throat becomes independent of downstream conditions, since signals cannot be propagated upstream against a flow that is sonic or supersonic. Then it is governed only by the reservoir (total) conditions and the throat area, according to Eq. (74.30).

74.8. SHOCK-WAVE EQUATIONS

A shock wave is a surface defining the points in a flow field at which discontinuous changes occur in all the flow variables. It is distinguished from streamline (slip) surfaces and contact surfaces in that the shock wave cuts *across* streamlines, so that fluid crosses it, and undergoes a change in pressure and velocity. A shock wave may be stationary, in a steady-flow field, or it may be propagating, even accelerating; it may

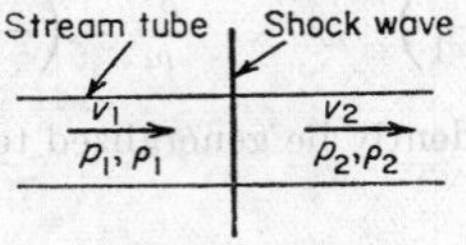

FIG. 74.2. Stationary shock wave.

cross streamlines obliquely, and it may be curved. Whatever the configuration, most of the pertinent relations may be determined by considering the simple case of a stationary shock wave normal to a stream tube, as shown in Fig. 74.2. The pertinent relations are the changes or "jumps" in the flow variables from condition 1 to condition 2. These depend on the normal relative velocity v_1 between the shock wave and the fluid ahead of it. This may always be determined by a suitable velocity transformation, and the shock wave observed in a coordinate system which is attached to it, as in Fig. 74.2. An example of such a transformation, applied to oblique shocks, is described on p. 77–2.

This generality in the treatment of a shock wave is due to the fact that, so long as the wave may be regarded as a discontinuity in the flow field, the two reference points 1 and 2, on either side of it, are arbitrarily close together, making all other length and time scales irrelevant, so that a purely local analysis applies. In a real shock wave, the flow properties cannot actually change discontinuously; but the changes occur in so small a distance that the representation as a discontinuity is very accurate for most purposes. Only for small body dimensions and extremely low densities (and for some cases where chemical reaction rates are involved) will the relative shock thickness become large enough to be taken into account in practical problems.

A shock wave exists only for supersonic relative velocity v_1, that is, for $v_1/a_1 > 1$. If fluid is entering a stationary shock, it must be flowing at supersonic speed; conversely, the speed of a wave moving into stationary fluid is sonic or greater. The higher the relative speed, the stronger is the wave; i.e., the larger is the jump in pressure, etc. The lower limit ($v = a_1$) is the speed of a vanishingly weak wave, i.e., a sound wave.

The conservation equations which relate conditions at 1 and 2 are independent of the details of the flow within the shock wave, i.e., the actual shock structure. They are sometimes called *jump* equations. Two of them, the continuity and energy equations (74.7) and (74.1) have been discussed in this chapter. The third one, the momentum equation, may be obtained from Eqs. (70.1). Since the two sections on either side of the shock wave are arbitrarily close together, and since the stream tube cannot change its area abruptly, we have $A_2 = A_1$. Thus, the conservation equations for mass, momentum, and energy are

$$\rho_2 v_2 = \rho_1 v_1 = \frac{m}{A} \tag{74.33}$$

$$p_2 + \rho_2 v_2^2 = p_1 + \rho_1 v_1^2 \tag{74.34}$$

$$h_2 + \tfrac{1}{2} v_2^2 = h_1 + \tfrac{1}{2} v_1^2 = h_t \tag{74.35}$$

together with an equation of state,

$$h = h(p,\rho) \tag{74.36}$$

Given four of the quantities, usually the four upstream values h_1, p_1, ρ_1, u_1, these four equations determine the remaining four quantities. In general, the equation of state is not in simple analytic form, and the set of equations must be solved numerically; this is facilitated by using the continuity equation to eliminate v_2 from the momentum and energy equations, which then take the form

$$p_2 = p_1 + \rho_1 v_1^2 \left(1 - \frac{\rho_1}{\rho_2}\right) \qquad h_2 = h_1 + \tfrac{1}{2} v_1^2 \left(1 - \frac{\rho_1^2}{\rho_2^2}\right) \tag{74.37a,b}$$

These can be solved by iteration: Guessing a value for ρ_2, compute p_2 and h_2; with these, find ρ_2 from the equation of state; repeat. The process converges especially quickly for strong shocks, since in that case ρ_1/ρ_2 is small.

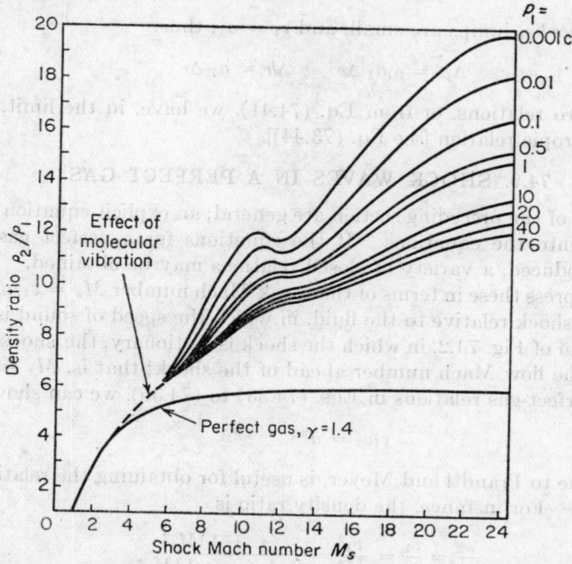

FIG. 74.3. Density ratio across a normal shock wave in air. $T_1 = 300°K$. *(From Ref.* [7], *by courtesy of AVCO Research Laboratory.)*

The behavior of the density ratio in air, calculated with the real equations of state for air, is shown in Fig. 74.3. The curve for a perfect gas [Eq. (74.45)], with $\gamma = 1.4$, is shown for comparison.

Equations (74.37) can be recast in another useful form in terms of the shock Mach number $M_s = v_1/a_1$. The thermodynamic variables in the fluid ahead of the shock satisfy Eqs. (74.21): $a_1^2 = \gamma_1 p_1/\rho_1 = h_1(\gamma_1 - 1)$. With these, we find

$$\frac{p_2}{p_1} = 1 + \gamma_1 M_s^2 \left(1 - \frac{\rho_1}{\rho_2}\right) \qquad \frac{h_2}{h_1} = 1 + \frac{\gamma_1 - 1}{2} M_s^2 \left(1 - \frac{\rho_1^2}{\rho_2^2}\right) \tag{74.38a,b}$$

Alternative Forms of the Jump Equations

Equations (74.33) to (74.35) may also be rewritten in the following forms,

$$\begin{aligned} p_2 - p_1 &= \rho_1 v_1 (v_1 - v_2) = \rho_1 v_1 \, \Delta v \\ h_2 - h_1 &= \tfrac{1}{2}(v_1 + v_2)(v_1 - v_2) = (v_1 - \tfrac{1}{2}\Delta v)\, \Delta v \end{aligned} \tag{74.39a,b}$$

where Δ denotes the absolute value of a jump. Also we have

$$\begin{aligned} p_2 - p_1 &= \rho_1^2 v_1^2 \left(\frac{1}{\rho_1} - \frac{1}{\rho_2}\right) = \rho_1^2 v_1^2 \, \Delta\left(\frac{1}{\rho}\right) \\ h_2 - h_1 &= \tfrac{1}{2}\rho_1^2 v_1^2 \left(\frac{1}{\rho_1^2} - \frac{1}{\rho_2^2}\right) = \tfrac{1}{2}\, \rho_1^2 v_1^2 \, \Delta\left(\frac{1}{\rho^2}\right) \end{aligned} \tag{74.40a,b}$$

and, from the last two, the Hugoniot relation, which involves only the thermodynamic variables,

$$\frac{h_2 - h_1}{p_2 - p_1} = \frac{1}{2}\left(\frac{1}{\rho_1} + \frac{1}{\rho_2}\right) \tag{74.41}$$

For very *strong* shocks, $v_1 \gg v_2$, and then Eqs. (74.39a,b) give

$$\Delta p \approx \rho_1 v_1^2 \qquad \Delta h \approx \tfrac{1}{2} v_1^2 \tag{74.42a,b}$$

For very *weak* shocks, jumps are small, and $v_1 \approx a_1$; thus,

$$\Delta p \approx \rho_1 a_1 \, \Delta v \qquad \Delta h \approx a_1 \, \Delta v \tag{74.43a,b}$$

From the last two relations, or from Eq. (74.41), we have, in the limit, $dp/dh = \rho$, which is an isentropic relation [see Eq. (73.44)].

74.9. SHOCK WAVES IN A PERFECT GAS

The relations of the preceding section are general; an explicit equation of state was not introduced into the equations. If the equations for a perfect gas, (74.18) to (74.21), are introduced, a variety of closed relations may be obtained. It is usually convenient to express these in terms of the shock Mach number $M_s = v_1/a_1$, where v_1 is the speed of the shock relative to the fluid, in which the speed of sound is a_1. In the coordinate system of Fig. 74.2, in which the shock is stationary, the shock Mach number is equal to the flow Mach number ahead of the shock; that is, $M_s = M_1$.

Using the perfect-gas relations in Eqs. (74.33) to (74.35), we can show that

$$v_1 v_2 = a^{*2} \tag{74.44}$$

This relation, due to Prandtl and Meyer, is useful for obtaining the relations between other quantities. For instance, the density ratio is

$$\frac{\rho_2}{\rho_1} = \frac{v_1}{v_2} = \frac{v_1^2}{a^{*2}} = \frac{(\gamma + 1) M_s^2}{2 + (\gamma - 1) M_s^2} \tag{74.45}$$

the last equality being obtained from Eq. (74.24).

The pressure ratio is obtained by dividing Eq. (74.39a) by p_1:

$$\frac{p_2 - p_1}{p_1} = \frac{p_1 v_1^2}{p_1}\left(1 - \frac{v_2}{v_1}\right) = \frac{2\gamma}{\gamma + 1}(M_s^2 - 1) \tag{74.46}$$

the last step being worked out with the help of Eqs. (74.20) and (74.45).

The temperature ratio is obtained from Eqs. (74.45) and (74.46) and the perfect-gas law.

Writing the Prandtl relation in the form $v_2/a^* = a^*/v_1$ and remembering that a^* is the same on both sides of the stationary shock, we can use Eq. (74.24) to relate the downstream flow Mach number, $M_2 = v_2/a_2$, to the upstream Mach number M_1:

$$\frac{(\gamma + 1)M_2^2}{2 + (\gamma - 1)M_2^2} = \frac{2 + (\gamma - 1)M_1^2}{(\gamma + 1)M_1^2} \tag{74.47}$$

Since $M_1 > 1$, it follows that $M_2 < 1$.

From Eq. (73.38), we can obtain a relation between the entropy change and the changes in the other thermodynamic variables. This is

$$\begin{aligned} s_2 - s_1 &= c_v \ln \frac{p_2}{p_1} - c_p \ln \frac{\rho_2}{\rho_1} = c_p \ln \frac{T_2}{T_1} - R \ln \frac{p_2}{p_1} \\ &= c_v \ln \left[1 + \frac{2\gamma}{\gamma + 1}(M_s^2 - 1)\right] - c_p \ln \frac{(\gamma + 1)M_s^2}{2 + (\gamma - 1)M_s^2} \end{aligned} \tag{74.48a,b}$$

From this expression it follows that $s_2 > s_1$, if $M_s > 1$ and $s_2 < s_1$ if $M_s < 1$. It follows then that shock Mach numbers less than unity are not possible, since that would require an adiabatic decrease of entropy for every fluid particle passing through the shock, and this is not possible according to the second law of thermodynamics.

A shock wave is an entropy-producing mechanism. The dissipation which accounts for the entropy production is connected with the large stresses and heat transfer (and mass diffusion in case of chemical reaction) that occur within the shock, where the gradients of the flow variables are very high (rather than infinite, as they would be in the ideal discontinuity). The shock thickness is, in fact, determined by the entropy jump and the coefficients of viscosity, conductivity, diffusion, since the gradients must adjust themselves to produce just the right amount of entropy, as required by the jump equations. We might note here that the rate of entropy production is made up of terms containing squares of the gradients of velocity, temperature, chemical concentration, etc., each multiplied by the corresponding (diffusion) coefficient [2].

Inside a shock wave, and in other regions (e.g., boundary layers) where gradients are very high, it is not possible to assume that a particle is a thermodynamically uniform system. To describe the state of affairs inside such regions, it is necessary to employ flow equations, e.g., the Navier-Stokes equations, which take into account the viscous stresses, heat transfers, and other diffusive terms which are present. Thus, the limit on the assumption of uniformity, discussed at the very beginning of this chapter, depends on whether the gradients become large enough to produce terms which must be taken into account, compared to the usual ones.

Equation (74.48b) is for a perfect gas, but the general remarks concerning entropy increase are valid for any fluid.

Since total entropy and static entropy are one and the same (p. 74–3), we can use total values instead of static values on the right-hand sides of Eq. (74.48a). The total temperature is the same on both sides of the shock, and so the second of these equations gives

$$s_2 - s_1 = -R \ln \frac{p_{2t}}{p_{1t}} \tag{74.49}$$

that is, corresponding to the increase of entropy, there is a decrease in total pressure across the shock (and a corresponding decrease in total density).

For very weak shocks ($M_s^2 - 1 \ll 1$), Eq. (74.48*b*) may be reduced to

$$s_2 - s_1 = \tfrac{2}{3} c_v \frac{\gamma(\gamma - 1)}{(\gamma + 1)^2} (M_s^2 - 1)^3$$

Comparing with Eq. (74.46), we see that, for weak shocks, the entropy increase is vanishingly small, to the third order in the shock strength.

In the preceding discussion, we have considered the shock to be stationary, in the coordinate system of Fig. 74.2. But the relations between the thermodynamic ratios and the shock Mach number are valid in any system, so long as M_s is formed from the normal relative velocity and the value a_1 ahead of the shock. Another useful relation may be obtained by inverting each term in Eq. (74.45) and subtracting it from unity. This gives

$$\frac{\Delta v}{a_1} = \frac{v_1 - v_2}{a_1} = \frac{2}{\gamma + 1}\left(M_s - \frac{1}{M_s}\right) \tag{74.50}$$

which is a relation among the velocity jump, the shock Mach number, and a_1. Since M_s is related to the thermodynamic ratios, the latter are thus related to the velocity jump. This is useful in cases where both fluid and shock are moving, from the point of view of the observer.

74.10. THERMODYNAMICS OF A FLOW FIELD

Throughout this chapter we have, in effect, been studying the thermodynamics of a fluid particle as it moves along a streamline (or stream tube). Since the flow is steady, this is also the "thermodynamics" of the streamline, as pointed out on p. 74–1. To extend this to a general flow field, we have to know how the thermodynamic constants may vary from one streamline to another.

The thermodynamic *constants* are the total values, p_t, ρ_t, h_t, s, etc. As we have seen, p_t, ρ_t, s may be changed along a streamline in an adiabatic flow, by the effects of dissipative mechanisms, e.g., shock waves. In diabatic flow, these as well as h_t may be changed by energy transported into the stream tube. Now evidently different streamlines may carry different total values. However, these cannot be assigned arbitrarily, or, conversely, streamlines cannot be patched together arbitrarily, since there is a dynamical constraint through the momentum equation. We have already made use of the momentum equation component along the streamline,

$$\frac{\partial p}{\partial x} = -\rho v \frac{\partial v}{\partial x}$$

[Eq. (74.8)]. The constraint is due to the other component, which, in the plane of curvature of the streamline, is

$$\frac{\partial p}{\partial r} = \frac{\rho v^2}{r} \tag{74.51}$$

where r is the radius of curvature. Now the pressure variation is related to the enthalpy and entropy variation through Eq. (73.44), and the enthalpy variation can be related to the variation of total enthalpy through the energy relation (74.4), so that

$$\frac{dp}{\rho} = dh_t - T\,ds - v\,dv$$

Putting this into Eq. (74.51) gives

$$\frac{\partial h_t}{\partial r} - T\frac{\partial s}{\partial r} = v\left(\frac{v}{r} + \frac{\partial v}{\partial r}\right) = v\zeta \tag{74.52}$$

Since $s \equiv s_t$, we can write $T_t\,ds = dh_t - dp_t/\rho_t$ and rewrite Eq. (74.52) in the form

$$\left(1 - \frac{T}{T_t}\right)\frac{\partial h_t}{\partial r} + \frac{1}{\rho_t}\frac{T}{T_t}\frac{\partial p_t}{\partial r} = v\zeta \tag{74.53}$$

Equation (74.52), known as *Crocco's relation*, shows that the variation of total enthalpy and entropy across streamlines is related to the kinematic quantity $v\zeta$. The quantity ζ, which is twice the angular velocity of the particle, is called the *vorticity*. A flow which has vorticity is also called *rotational*. An example of a rotational flow field is the flow downstream of a curved shock wave, such as the detached bow wave on a blunt body. Here the inclination of the shock relative the oncoming flow determines its strength (see Chap. 77) and entropy change, and thus there is a continual variation of s from streamline to streamline, on the downstream side. In this example, h_t is constant throughout. A case of varying h_t would be obtained for fluid which had passed through a nonuniform cooler or heater or flame front.

In the solution of flow problems, one often works with the velocity field, and then the relation between the thermodynamic parameters and the vorticity is very useful.

74.11. REFERENCES

Textbooks

[1] A. B. Cambel, B. H. Jennings: "Gas Dynamics," McGraw-Hill, New York, 1958.
[2] H. W. Liepmann, A. Roshko: "Elements of Gasdynamics," Wiley, New York, 1957.
[3] K. Oswatitsch: "Gas Dynamics," Academic Press, New York, 1956.
[4] A. H. Shapiro: "The Dynamics and Thermodynamics of Compressible Fluid Flow," 2 vols., Ronald, New York, 1953.

Tables, Charts, Summaries

[5] Ames Research Staff: Equations, tables and charts for compressible flow, *NACA Rept.* 1135 (1953). (Extensive tables and charts for isentropic flow and shock waves for $\gamma = 1.4$.)
[6] H. W. Emmons: "Gas Dynamics Tables for Air," Dover, New York, 1947.
[7] S. Feldmann: Hypersonic gas dynamic charts for equilibrium air, AVCO Research Laboratory, Everett, Mass., 1957. (Contains a Mollier chart and shock charts for air, up to 13,000°K.)
[8] N. A. Hall: "Thermodynamics of Fluid Flow," Prentice-Hall, Englewood Cliffs, N.J., 1951.
[9] I. Korobkin, S. M. Hastings: Mollier chart for air in dissociated equilibrium at temperatures of 2000°K to 15,000°K, *U.S. Naval Ordnance Lab. NAVORD Rept.* 4446, 1957.

Encyclopedic Works

[10] H. W. Emmons (ed.): "Fundamentals of Gas Dynamics," vol. 3 of "High Speed Aerodynamics and Jet Propulsion," Princeton Univ. Press, Princeton, N.J., 1958.
[11] L. Howarth (ed.): "Modern Developments in Fluid Dynamics: High Speed Flow," vols. 1 and 2, Oxford Univ. Press, New York, 1953.
[12] W. R. Sears (ed.): "General Theory of High Speed Aerodynamics," vol. 6 of "High Speed Aerodynamics and Jet Propulsion," Princeton Univ. Press, Princeton, N.J., 1954.

Classical Handbook Articles

[13] J. Ackeret: Gasdynamik, in "Handbuch der Physik," vol. 7, Springer, Berlin, 1927.
[14] A. Busemann: Gasdynamik, in "Handbuch der Experimentalphysik," vol. 4, part 1, Akad. Verlag, Leipzig, 1931.
[15] G. I. Taylor, J. W. Macoll: The mechanics of compressible fluids, in W. F. Durand (ed.), "Aerodynamic Theory," vol. 3, Calif. Inst. of Technol., Pasadena, Calif., 1943.

CHAPTER 75

SUBSONIC FLOW

BY

N. ROTT, Dr. sc. techn., Los Angeles, Calif.

75.1. INTRODUCTION

Definitions

The term *subsonic flow* will be applied, following common usage, to a steady-flow field of a compressible medium (gas), in which the velocity nowhere exceeds the speed of sound, i.e., where the local Mach number never reaches 1. Subsonic flow problems of great practical interest arise in connection with the consideration of flight of bodies in the unlimited atmosphere at subsonic speeds. In a body-fixed system, the Mach number at infinity will be referred to as the *flow Mach number*. The *critical Mach number* is the flow Mach number for which the local speed of sound is first reached at a point in the flow field. If the flow Mach number exceeds the critical Mach number, a local supersonic region will appear embedded in the subsonic flow field; the treatment of these *mixed-flow* problems is found in Chap. 76.

The critical Mach number and the methods of its prediction from subsonic flow theories will be included in this chapter, although some aspects of this problem can be understood only from transonic flow theory.

Limited regions of subsonic flow are found embedded in many supersonic flow fields (e.g., behind shocks). For this type of mixed-flow problems, Chaps. 77 and 78 should be consulted.

The present chapter is restricted to the treatment of the *subcritical* subsonic flow of a frictionless gas. Effects of viscosity and heat conduction are treated in Chap. 83.

Basic Equations

Let the flow field be described by the velocity vector $\mathbf{q}(u,v,w)$, the density ρ, and the pressure p; other functions of state, such as the temperature T and the enthalpy h, follow by the equation of state of the gas. In the cartesian coordinate system (x,y,z), the notion of the *substantial derivative* with respect to the time t will be employed:

$$\frac{D}{Dt} = \frac{\partial}{\partial t} + u\frac{\partial}{\partial x} + v\frac{\partial}{\partial y} + w\frac{\partial}{\partial z}$$

The basic equations of the flow are, as derived in Chap. 68, the continuity equation,

$$\frac{D\rho}{Dt} + \rho \operatorname{div} \mathbf{q} = 0 \tag{75.1}$$

and the momentum (Eulerian) equation,

$$\rho\frac{D\mathbf{q}}{Dt} + \operatorname{grad} p = 0 \tag{75.2}$$

As a third equation, in the particular case of a gas without friction and heat conduction, it suffices to state that the entropy S for each flow element is constant:

$$\frac{DS}{Dt} = 0 \tag{75.3}$$

[Implied is also the absence of relaxation phenomena, and, more generally, of all reactions involving irreversibilities; also, the absence of discontinuities (shocks), naturally fulfilled in subsonic flow.] From Eqs. (75.1), (75.2), and (75.3), the usual energy equation follows: By Eq. (73.44), Eq. (75.3) is equivalent to

$$\rho \frac{Dh}{Dt} - \frac{Dp}{Dt} = 0 \tag{75.4}$$

Scalar multiplication of Eq. (75.2) by **q** yields the "steady" part of Dp/Dt, which, introduced into Eq. (75.4) gives the important equation,

$$\rho \frac{DH}{Dt} - \frac{\partial p}{\partial t} = 0 \tag{75.5}$$

where

$$H = h + \tfrac{1}{2}q^2 \tag{75.6}$$

is the *kinetic enthalpy*.

In the particular case of steady flow, Eqs. (75.3) and (75.5) state that

$$S = \text{const} \qquad H = h + \tfrac{1}{2}q^2 = \text{const} \tag{75.7a,b}$$

along a streamline. [Note that Eq. (75.7*b*) also holds in cases where Eq. (75.7*a*) is not fulfilled, namely, across shocks, and generally for stream tubes of adiabatic though not necessarily irreversible flow. These cases are not considered here.] By the pair of equations (75.7), the magnitude q of the velocity vector **q** is connected to any function of state along a stream tube. Equations (74.22) are explicit relations valid for a perfect gas. For any equation of state, an expansion in powers of the enthalpy change Δh (for constant entropy) is always possible, e.g., for the pressure

$$p = p_0 + \left(\frac{\partial p}{\partial h}\right)_{s0} \Delta h + \cdots$$

Noting that $(\partial p/\partial h)_s = \rho$, and

$$\left(\frac{\partial p}{\partial \rho}\right)_s = a^2 \tag{75.8}$$

where a is the velocity of sound, this expansion takes the form

$$p - p_0 = \rho_0 \,\Delta h + \frac{\rho_0}{a_0^2}\frac{\Delta h^2}{2!} + \frac{\rho_0}{a_0^4}\left[1 - 2\left(\frac{\partial \ln a}{\partial \ln \rho}\right)_s\right]\frac{\Delta h^3}{3!} + \cdots \tag{75.9}$$

If the reference state (subscript 0) is the stagnation point $q = 0$, $h_0 = H$, $\Delta h = h - H$, we have

$$p - p_0 = -\tfrac{1}{2}\rho_0 q^2 + \frac{1}{8}\frac{\rho_0}{a_0^2}q^4 - \cdots \tag{75.10}$$

that is, the first term yields Bernoulli's equation as known for incompressible flow, and the subsequent terms represent a series in growing powers of the square of the Mach number $M = q/a_0$. It is seen that incompressible flow ($a_0 \to \infty$) is a limiting case equivalent to low-speed flow ($q \ll a_0$).

For a perfect gas with constant specific heats, the expression appearing in the third term of Eq. (75.9) can be evaluated:

$$\left(\frac{\partial \ln a}{\partial \ln \rho}\right)_s = \frac{\gamma - 1}{2} \tag{75.11}$$

Conversely, this equation defines an "equivalent" γ for an imperfect gas, for procedures of successive approximation. The same expression arises if the velocity of sound is found by a power series:

$$a^2 = a_0^2 + 2\left(\frac{\partial \ln a}{\partial \ln \rho}\right)_s \Delta h + \cdots \tag{75.12}$$

For a perfect gas with $\gamma = \text{const}$,

$$a^2 = a_0^2 - \frac{\gamma - 1}{2} q^2 \tag{75.13}$$

and all other functions of state can be given explicitly.

Potential Flow

If

$$\mathrm{q} = \operatorname{grad} \phi \tag{75.14}$$

several simplifications are found. In steady flow, if the stagnation enthalpy has the same value for every streamline, i.e., if the flow is homenergic, then a potential flow is also homentropic, that is, $S = \text{const}$ everywhere (Crocco's theorem, see Chap. 74). Equations (75.7*a*,*b*) are valid throughout the flow field.

In unsteady homentropic flow, from the Euler equation (75.2) and from (75.14), the Bernoulli equation can be derived:

$$\frac{\partial \phi}{\partial t} + \tfrac{1}{2}q^2 + h = C(t) \tag{75.15}$$

where $C(t)$ is a function of time, which, however can be "absorbed" in the definition of the potential, so that the right-hand side of Eq. (75.15) becomes a constant independent of time. The equation for the potential ϕ is then

$$\phi_{xx}(a^2 - \phi_x^2) + \phi_{yy}(a^2 - \phi_y^2) + \phi_{zz}(a^2 - \phi_z^2) - 2\phi_x\phi_y\phi_{xy} - 2\phi_y\phi_z\phi_{yz} \\ - 2\phi_z\phi_x\phi_{zx} - 2\phi_x\phi_{xt} - 2\phi_y\phi_{yt} - 2\phi_z\phi_{zt} - \phi_{tt} = 0 \tag{75.16}$$

The physical conditions for the existence of subsonic (subcritical) potential flow are the same as in the incompressible case (Chap. 70). In particular, subsonic flows past bodies which were brought into motion in a homogeneous frictionless gas at rest are potential flows; this type of problem is extensively treated in the present chapter. Important cases of subsonic flow regions where a potential does not exist are found behind curved shocks, in supersonic flow. In these cases, a stream function can be used in steady plane and axisymmetric flow; its equations are given in Chaps. 70 and 71.

General Remarks on the Methods of Solution

For compressible flow, even the potential flow problem leads to a nonlinear equation, so that approximate methods have to be used. The character of Eq. (75.16), in the case of steady subsonic flow, is still related to the incompressible flow case inasmuch as the equation is still *elliptic*, a notion explained on p. 11–5. It is therefore to be expected that the first step in practically all methods of solution is an approximate reduction of the problem to the incompressible case.

The first method to be described is the *linearized theory* or method of small disturbances, which has important aeronautical applications, as ideal designs do cause small disturbances. Next, methods for obtaining higher approximations will be discussed, based either on the linearized theory or on incompressible flow solutions. Finally, the *hodograph method* for plane flow will be treated, which uses the fact that if the velocity components are employed as independent variables, the potential fulfills a linear equation.

The similarity of solutions for steady-flow problems in an inviscid gas depends on two parameters: the Mach number M and the ratio of specific heats γ; the latter characterizes the properties of a perfect gas. More complicated similarity conditions exist for real gases; for subsonic flow, as only a small change of state of the gas is considered, this is of little practical importance. Moreover, linearized theory (valid for thin bodies) is independent of γ, and so is the first approximation describing the departure from incompressible flow (see p. 75–18), which is valid for all thicknesses. The next approximation, in both cases, will contain γ, but it can be concluded that this parameter has only a "weak" influence in subsonic flow.

75.2. LINEARIZED THEORY OF SUBSONIC FLOW

The Linearized Equation for Potential Flow

Let the potential ϕ be replaced by the expression

$$\phi = Ux + \varphi(x,y,z) \tag{75.17}$$

The first term represents the potential of a *basic* flow of constant speed U in positive x direction, while φ is the *disturbance* potential, steady in the x,y,z space. If expression (75.17) is introduced into Eq. (75.16), and only terms *linear* in φ are kept, then φ fulfills the equation

$$\beta^2\varphi_{xx} + \varphi_{yy} + \varphi_{zz} = 0 \tag{75.18}$$

where

$$\beta^2 = 1 - M^2 \tag{75.19}$$

and $M = U/a_\infty$ is the flow Mach number (with a_∞ being the undisturbed sound speed).

Equation (75.18) is the *Prandtl-Glauert equation* [4],[5] which, *for subsonic flow*, is reduced by an affine transformation to Laplace's equation, and thereby to the incompressible case. Using the new variables,

$$\xi = x \qquad \eta = \beta y \qquad \zeta = \beta z \tag{75.20}$$

Eq. (75.18) is transformed into

$$\varphi_{\xi\xi} + \varphi_{\eta\eta} + \varphi_{\zeta\zeta} = 0 \tag{75.21}$$

It is seen that the *Prandtl-Glauert transformation* (75.20) consists of the introduction of a new coordinate system, which is "shrunk" in the direction transverse to the free stream by the factor β. (A "stretching" in stream direction by the factor β^{-1} is equivalent; see p. 75–7.)

The exact energy equation in case of a potential of the form (75.17) is

$$h - h_\infty = \Delta h = -U\varphi_x - \tfrac{1}{2}(\varphi_x^2 + \varphi_y^2 + \varphi_z^2) \tag{75.22}$$

The pressure follows from Eq. (75.9); the following expression includes all quadratic terms:

$$p - p_\infty = -\rho_\infty U\varphi_x - \frac{\rho_\infty}{2}(\beta^2\varphi_x^2 + \varphi_y^2 + \varphi_z^2) \tag{75.23}$$

Consistent linearization means the use of only the first term in Eqs. (75.22) and (75.23). The quadratic terms are needed, however, in connection with some linearized solutions; this will be discussed on p. 75–17.

Singular Solutions of the Prandtl-Glauert Equation

Singular solutions for linearized subsonic flow are found as the generalization of singular solutions of Laplace's equation (75.21) by application of transformation (75.20). The fact that valid linearization breaks down when the singularity is approached does not detract from the importance of the singular solutions, which are, just as for incompressible flow, the elements of meaningful solutions obtained by proper superposition.

Example 1. Point Source. The potential is

$$\varphi = -\frac{A}{4\pi\sqrt{x^2 + \beta^2(y^2 + z^2)}} \tag{75.24}$$

The constant A is a measure of the source strength, determined by calculation of the flux of the mass flow vector. Note that the x component of the mass flow vector, $(U + \varphi_x)(\rho_\infty + \Delta\rho)$, contains two first-order terms: $(\rho_\infty\varphi_x + U\Delta\rho)$. According to Eq. (75.23),

$$U\,\Delta\rho = U\,\Delta p/a_\infty{}^2 = -\rho_\infty M^2\varphi_x$$

so that the mass flow vector has the components $\rho_\infty\beta^2\varphi_x$, $\rho_\infty\varphi_y$, and $\rho_\infty\varphi_z$. [Equation (75.18) states that its divergence vanishes.] The total mass flux through any surface surrounding the source is found to be $\rho_\infty A$.

Example 2. Line Source (on the z Axis)

$$\varphi = \frac{B}{4\pi}\ln(x^2 + \beta^2 y^2) \tag{75.25}$$

The mass flux (per unit height in z direction) is $\rho_\infty\beta B$.

Example 3. Line Vortex

$$\varphi = -\frac{\Gamma}{2\pi}\tan^{-1}\frac{\beta y}{x} \tag{75.26}$$

The circulation is evidently Γ, as the change in potential for any counterclockwise circumvention of the origin is $-\Gamma$ (Γ shall be counted positive clockwise). A momentum survey shows that the Kutta-Joukowski theorem holds for linearized subsonic flow: A *lift* (in the y direction) of the magnitude $\rho_\infty U\Gamma$ is found, while the *drag* vanishes. Actually this result holds exactly (not linearized) for plane inviscid subsonic irrotational-isentropic circulatory flow past any body shape: it can be shown that far away from the body the leading term of the disturbance potential will always have the form (75.26). Thus, a momentum balance at infinity gives always the same result.

Wavy Wall in Plane Flow

The following example is due to Ackeret [4]. The plane subsonic flow (in the upper half-space) is to be determined, limited by a "wavy wall," with the ordinate

$$y_0 = h\sin(2\pi x/\lambda)$$

For the solution of this problem in linearized theory, it is consistent to fulfill boundary conditions in linearized approximation only. The linearization of the boundary conditions was discussed on p. 72–11. Accordingly, the boundary condition can be stated for the velocity component v on the x axis ($y = 0$):

$$v = U\frac{dy_0}{dx} = \frac{2\pi Uh}{\lambda}\cos\frac{2\pi x}{\lambda}$$

The problem is solved by the following potential:

$$\varphi = Ce^{-2\pi\beta y/\lambda}\cos\frac{2\pi x}{\lambda}$$

where the constant C has to be adjusted to the boundary condition above; this done, the velocity components become

$$u = -\frac{2\pi Uh}{\beta\lambda}e^{-2\pi\beta y/\lambda}\cos\frac{2\pi x}{\lambda} \qquad v = \frac{2\pi Uh}{\lambda}e^{-2\pi\beta y/\lambda}\sin\frac{2\pi x}{\lambda} \tag{75.27}$$

The result differs from the incompressible case (included above for the special value $\beta = 1$) in two ways: (1) The x component for $y = 0$ is increased by the factor β^{-1}; the same factor applies for the linearized pressure; (2) the exponential decay of the disturbance in y direction is made weaker, or the decay length is increased compared to the wavelength λ by the factor β^{-1}. Both the increasing pressures and the strengthened "transverse" effect of disturbances are characteristic subsonic compressibility effects.

In this connection, it is also of interest to consider the solution

$$\varphi = Ce^{-2\pi x/\beta\lambda} \cos (2\pi y/\lambda)$$

which represents a disturbance periodic in y direction with a given wavelength λ, as it may be found behind cascades or other obstacles in periodic arrangement with the spacing λ. Compressibility means here a stronger decay in the x direction, or a reduction of the streamwise decay length by the factor β.

Solution of Boundary-value Problems

Boundary-value problems in linearized approximation are posed such that, along certain boundaries, the value of the *crosswise* flow components v and w are prescribed. It will be shown that, after solving such a problem for incompressible flow, its generalization to subsonic linearized flow can be accomplished by a simple rule.

First, let the Prandtl-Glauert transformation (75.20) be generalized in an apparently trivial way by introducing two factors κ and k, which are left undetermined for the moment:

$$\xi = \kappa x \qquad \eta = \kappa\beta y \qquad \zeta = \kappa\beta z \tag{75.28}$$

and

$$\varphi = k\varphi_i(\kappa x, \kappa\beta y, \kappa\beta z) \tag{75.29}$$

where φ_i is the incompressible solution, which fulfills Laplace's equation in the ξ,η,ζ space, while φ is the potential to be determined for the compressible case. For the velocity components $u = \partial\varphi/\partial x$, etc., and $u_i = \partial\varphi_i/\partial\xi$, etc., the following rules are obtained:

$$u = k\kappa u_i \qquad v = k\kappa\beta v_i \qquad w = k\kappa\beta w_i \tag{75.30}$$

Note that the following ratios are transformed independently of the "trivial" constants k and κ:

$$\frac{y}{x} = \frac{1}{\beta}\frac{\eta}{\xi} \qquad \frac{z}{x} = \frac{1}{\beta}\frac{\zeta}{\xi} \tag{75.31}$$

$$\frac{v}{u} = \beta\frac{v_i}{u_i} \qquad \frac{w}{u} = \beta\frac{w_i}{u_i} \tag{75.32}$$

It is significant that the factors β^{-1} and β are found in corresponding places in Eqs. (75.31) and (75.32), expressing the fact that the gradient vector changes in a way "opposite" to (or contravariant to) the space vector.

In linearized boundary-value problems, the prescribed values of v and w (or v_i and w_i) are simply proportional to slopes dy/dx and dz/dx (or $d\eta/d\xi$ and $d\zeta/d\xi$) on the thin bodies. It will be required now that this proportionality shall be preserved by the transformation, so that the properly (i.e., pointwise) transformed bodies, put in the undisturbed flow of velocity U in both the x,y,z and the ξ,η,ζ space, shall generate linearized boundary-value problems, i.e., crosswise flow components, which are related in the same way as by Eq. (75.30). This will be accomplished by putting

$$v = \frac{1}{\beta} v_i \qquad w = \frac{1}{\beta} w_i \tag{75.33}$$

as the slopes naturally transform according to (75.31); Eqs. (75.32) gives

$$u = \frac{1}{\beta^2} u_i \tag{75.34}$$

and, in comparison with Eqs. (75.30), leads to the condition

$$k\kappa = \beta^{-2} \tag{75.35}$$

The particular requirement expressed above and its consequences [Eqs. (75.33) to (75.35)] are due (in this generality) to Göthert [6.] It may be noted that there are many ways, identical in principle, to achieve the same results, but Göthert's rule has been found particularly useful because of its simplicity and straightforwardness. It helps to eliminate errors, which can be found in the older literature on this subject, in spite of the inherent simplicity of the problem.

The condition (75.35) on the product $k\kappa$ still leaves a certain freedom, although the choice is of little consequence. The most common conventions are $k = \beta^{-1}$ and $\kappa = \beta^{-1}$ or $k = \beta^{-2}$ and $\kappa = 1$. Here, the latter values will be chosen, and the Göthert rule will be summed up as follows: Given a boundary-value problem for compressible subsonic flow, with the linearized solution to be determined in the x,y,z space. Transform this space, including the given body, according to Eqs. (75.20), leading to a body which is made "more slender" by the factor β, as compared to the original. If the solution of the incompressible flow problem is $\varphi_i(\xi,\eta,\zeta)$, the original problem is solved by

$$\varphi = \beta^{-2}\varphi_i(x,\beta y,\beta z) \tag{75.36}$$

In particular, the u component and thereby the linearized pressure coefficient transform according to Eq. (75.34); in words: The pressure coefficient in the original flow is increased by the factor β^{-2} in relation to the incompressible solution for the more slender body.

Solution for Thin Airfoils

As a first application of Göthert's rule, consider the problem of the flow past a thin airfoil. Let τ be the percentage thickness of the airfoil. In imcompressible flow, the flow potential for a given shape is a known function of ξ and η, and its dependence on the thickness parameter τ is also known; thus, suppose that the function

$$\varphi_i = \varphi_i(\xi,\eta;\tau) \tag{75.37}$$

is given. Consider now the profile of thickness τ in the compressible flow case. According to the rule, the incompressible flow has to be found for the airfoil with the thickness $\beta\tau$; the second step is the application of the transformation (75.36), so that the final result is

$$\varphi = \beta^{-2}\varphi_i(x,\beta y;\beta\tau) \tag{75.38}$$

Now, for thin airfoils, the parametric dependence of φ_i on τ is known to be a simple proportionality; thus, put

$$\varphi_i(\xi,\eta;\tau) = \tau\varphi_i'(\xi,\eta) \tag{75.39}$$

and Eq. (75.38) becomes

$$\varphi = \beta^{-1}\tau\varphi_i'(x,\beta y) \tag{75.40}$$

or

$$\varphi = \beta^{-1}\varphi_i(x,\beta y;\tau) \tag{75.41}$$

This result contains a simple rule connecting the solutions for the same thickness τ in compressible and incompressible flow; applied to the velocity component u on the profile surface ($y \approx 0$), it states that

$$u(x;\tau) = \beta^{-1}u_i(x;\tau)$$

and the same holds for the linearized pressure coefficient,

$$\frac{\Delta p}{\frac{1}{2}\rho_\infty U^2} = C_p(x;\tau) = \beta^{-1} C_{pi}(x;\tau) \tag{75.42}$$

This is *Prandtl's rule*, stating that, for a given airfoil, when the Mach number is increased, the pressure coefficient grows with β^{-1} (see Fig. 75.1). The same result can be applied for the part of the solution which is due to the angle of attack α, for which evidently α plays the same role as τ; thus, for the slope of the lift coefficient curve, it follows that

$$\frac{\partial C_L}{\partial \alpha} = \beta^{-1}\left(\frac{\partial C_L}{\partial \alpha}\right)_i = 2\pi\beta^{-1} \tag{75.43}$$

[The value $(\partial C_L/\partial\alpha)_i = 2\pi$ for all thin airfoils is derived on p. 72–12.] It should be reemphasized that Eq. (75.42) is a consequence of the behavior of thin bodies in two-dimensional incompressible flow as expressed by Eq. (75.39). Limitation of these results and comparison with other theories and experiment will be discussed on p. 75–23, etc.

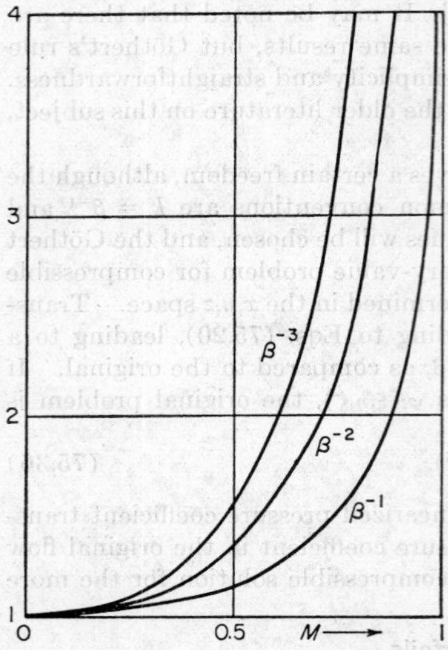

FIG. 75.1. Correction factors for linearized subsonic flow theory.

Solution for Bodies of Revolution

The arguments and the results of the preceding paragraph remain unchanged up to Eq. (75.38); applied to a slender three-dimensional affine family of bodies with thickness τ, Eq. (75.38) is simply generalized:

$$\varphi = \beta^{-2}\varphi_i(x,\beta y,\beta z;\beta\tau) \tag{75.44}$$

The dependence of φ_i on τ is, however, quite different from the two-dimensional case. For bodies of revolution at zero angle of attack, the results for small τ obtained in Chap. 71 [Eqs. (71.16*a*) and (71.40)] can be expressed as follows:[1]

$$\varphi_i = \tau^2 \varphi_i'(x,r) \tag{75.45}$$

where $r = \sqrt{x^2 + y^2}$. Application of the Göthert rule [Eq. (75.44)] leads to the following result:

$$\varphi = \tau^2\varphi_i'(x,\beta r) = \varphi_i(x,\beta r;\tau) \tag{75.46}$$

When this result is applied on the surface of the body, it must be remembered that φ_i becomes singular for $r \to 0$, so that, in contrast to the two-dimensional case, all quantities will be calculated on the actual body radius R, and not for $r = 0$. The actual limiting behavior of φ_i for small r leads to

$$\varphi_i(x,r) = \frac{U}{2\pi}\left[\frac{dS}{dx}\ln r + g(x)\right] \tag{75.47}$$

where $S = \pi R^2$ is the cross section of the body, and $g(x)$ is a function of x along the body which does not need to be specified here. Equation (75.47) gives, when intro-

[1] This formulation is correct only if the radius r under consideration is independent of τ; on the body, for example, as the body radius depends on τ, we have $\varphi_i = \tau^2\varphi_i'(x,\tau R)$; that is, φ_i' still has a parametric dependence on τ. Application to Eq. (75.47) gives a typical example.

duced into Eq. (75.46),

$$\varphi(x,r) = \varphi_i(x,r) + \frac{U}{2\pi}\frac{dS}{dx} \ln \beta \tag{75.48}$$

and, for the pressure coefficient, it is found that

$$C_p = C_{pi} - \frac{d^2(R^2)}{dx^2} \ln \beta \tag{75.49}$$

It is seen that this compressibility correction has a completely different character from the corresponding results in the two-dimensional case: Here, an *additive* term, depending on the body shape, expresses the compressibility effect in linearized approximation. Although the different character of the effects makes a good comparison difficult, it can be said (somewhat vaguely) that three-dimensional compressibility corrections are "smaller" or "less severe" than the two-dimensional ones, a statement which will be made more precise by discussions on p. 75–25.

The effects of angle of attack are fully discussed in Chap. 79. The important result is that the flow field is independent of β near the body, so that no compressibility corrections are needed on the body surface. The incompressible solution for an angle of attack α (in the x,z plane) is, for small values of the radius, given by the *crossflow potential*

$$\varphi_i = -U\alpha R^2 \frac{z}{y^2 + z^2} \tag{75.50}$$

This expression is found to be invariant with respect to the operations involved in the Göthert rule.

Applications to Wing Theory

The approximations used in the wing theory for incompressible flow, as discussed in Chap. 72, can be introduced in several steps. The first step involves the linearization of the boundary conditions, and leads to the *lifting-surface* problem, which can be separated from the thickness problem; the latter will not be considered here. Practically all methods dealing directly with the lifting-surface problem are approximate and very tedious. A further simplification is introduced when the lifting surface is replaced by the *lifting line,* leading to Prandtl's classical theory, valid for straight wings, with large aspect ratios. For swept wings, other simplified theories (notably Weissinger's) are used.

Proceeding now to the case of subsonic flow, first, the lifting-surface problem will be formulated, without being concerned with methods of solution. Let the wing surface S be situated in the x,y plane, and let T be the domain occupied by the trailing vortices. In linearized theory, T is a half-infinite strip of width b (= wing span) in the plane behind the wing, extending in the positive x direction. Owing to the particular symmetries of the lifting case, the boundary-value problem can be formulated for the half space $z \geq 0$ (say), with the following boundary conditions to be fulfilled in the plane $z = 0$:

$$\begin{aligned} \varphi_z &= w_0 && \text{given in } S \\ \varphi_x &= 0 && \text{in } T \\ \varphi &= 0 && \text{in the remainder of the } x,y \text{ plane} \end{aligned}$$

In addition, the Kutta condition requires that φ_x be regular on the border between S and T.

According to the Göthert rule, the following problem has to be solved in incompressible flow:

$$\begin{aligned} \varphi_{i\zeta} &= \beta w_0 && \text{given in } S' \\ \varphi_{i\xi} &= 0 && \text{in } T' \\ \varphi_i &= 0 && \text{in the remainder of the } \xi,\eta \text{ plane} \end{aligned}$$

where S' and T' are surfaces obtained by affine reduction of the y coordinates of S and T, applying the factor β (see Fig. 75.2). In other words, the aspect ratio is changed from Æ to $\beta\text{Æ}$.

The lifting-surface problem in the ξ,η,ζ space can be treated either directly or by a further simplification to the lifting-line method. The latter step is valid only if $\beta\text{Æ} \gg 1$. Thus, a lifting-line theory for subsonic flow has not only the *aspect-ratio* limitation as in incompressible flow, but also an upper limit for the Mach number, or a lower limit in β.

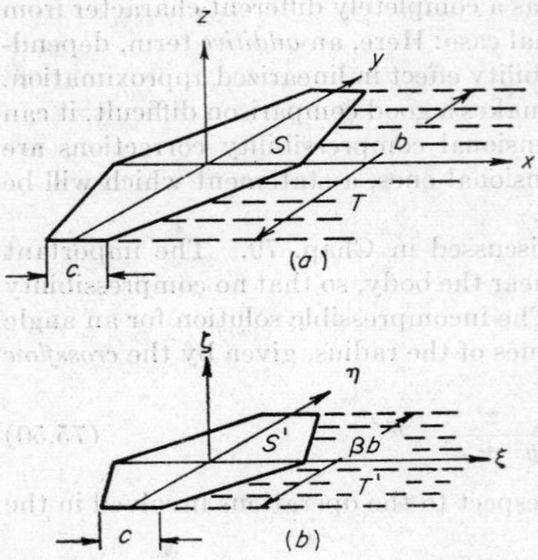

FIG. 75.2. The transformation of a lifting-surface problem according to the Göthert rule.

As an example, consider the lift coefficient, calculated for an affine family of planforms in incompressible flow; it will have the form

$$C_{Li} = C_{L\alpha i}(\text{Æ}) \cdot \alpha \qquad (75.51)$$

or the derivative $C_{L\alpha i}$ will be some function of Æ. According to the Göthert rule,

$$C_L = \beta^{-2} C_{L\alpha i}(\beta\,\text{Æ})\beta\alpha$$

$$\text{or} \quad C_{L\alpha} = \beta^{-1} C_{L\alpha i}(\beta\,\text{Æ}) \qquad (75.52)$$

For an elliptic planform, the lifting-line theory result is

$$C_{L\alpha i} = \frac{m}{1 + m/\pi\,\text{Æ}} \qquad (75.53)$$

where m is the two-dimensional profile derivative. Equation (75.52) yields

$$C_{L\alpha} = \frac{m}{\beta + m/\pi\,\text{Æ}} \qquad (75.54)$$

If $M \to 1$, the limiting value is $C_{L\alpha} = \pi\,\text{Æ}$, which is "off" by a factor ½, as, for $\beta \to 0$, $\text{Æ} \to 0$, slender-wing theory leads to $C_{L\alpha} = \pi\,\text{Æ}/2$. Nevertheless, for a high enough aspect ratio, Eq. (75.54) will be valid up to the "transonic" limit of the flow past the wing.

[*Note:* By eliminating m in Eqs. (75.53) and (75.54), the formula

$$C_{L\alpha} = \frac{C_{L\alpha i}}{\beta(1 - C_{L\alpha i}/\pi\,\text{Æ}) + C_{L\alpha i}/\pi\,\text{Æ}}$$

is obtained, which can be used as a rule of thumb for a quick estimate of the compressibility effect also for nonelliptic planforms, if $C_{L\alpha i}$ is known, but the explicit dependence on Æ is not available.]

The consideration of the lifting-surface problem is imperative in subsonic-wing theory, even if it serves only for an estimate of the errors committed in further simplifications. An accidental advantage is that the Göthert rule is easily applied to this boundary-value problem. The explicit use of the Biot-Savart law can also be extended to subsonic flow, but this is a more tedious procedure.

The knowledge of the downwash distribution behind the wing is important for the estimate of the tail effectiveness in longitudinal stability. It is possible to construct a *Prandtl-Glauert model* of a whole airplane, by an affine reduction of all lateral dimensions by the factor β. The distance between wing and tail remains unchanged, and becomes thereby a β^{-1} times bigger multiple of the wing span βb of the model. Thus, for subsonic flow, the tail appears to be further behind the wing, as M increases. Methods and results for the downwash calculation in incompressible flow can be used, if this rule is followed.

For swept wings, the Göthert rule requires increased sweep for the "model" in incompressible flow. This should be kept in mind for the application of the Weissinger method (p. 72–23).

As an example for the application of the method to the case of swept wings, consider the thickness effect for an infinite swept wing of constant chord. Let C_{pi0} be the pressure coefficient for a straight (two-dimensional) wing in incompressible flow. Let the wing be "swept" or "sheared" back such that its chord in the x direction remains unchanged, but its leading edge will form an angle Λ with the flow normal (i.e., with the unyawed leading edge in the former case). The corresponding pressure coefficient, still in incompressible flow, is easily found to be

$$C_{pi} = C_{pi0} \cos \Lambda = C_{pi0}(1 + \tan^2 \Lambda)^{-\frac{1}{2}} \tag{75.55}$$

Consider now the wing, swept back by an angle Λ, in compressible flow. The transformed wing will be swept back by an angle Λ' such that

$$\tan \Lambda' = \beta^{-1} \tan \Lambda \tag{75.56}$$

and the pressure coefficient is

$$C_p = \beta^{-2} C_{pi}(\Lambda', \beta\tau)$$

or, as C_{pi} is proportional to the thickness ratio,

$$C_p = \beta^{-1} C_{pi}(\Lambda', \tau)$$

According to Eqs. (75.55) and (75.56), this equals

$$C_p = \beta^{-1} C_{pi0}(1 + \beta^{-2} \tan^2 \Lambda)^{-\frac{1}{2}}$$

which is easily transformed to

$$C_p = C_{pi}(1 - M^2 \cos^2 \Lambda)^{-\frac{1}{2}} \tag{75.57}$$

It is found that the correction follows the Prandtl rule, calculated with the Mach-number component perpendicular to the infinite wing. This result is really obvious, but the calculation illustrates the procedure to be applied in more complicated cases.

Note on Cascades and Wind-tunnel Effects

The problem of cascades (in two-dimensional flow) is, in practice, rarely tractable by linearized theory, which would be valid only for very low deflections of the flow. In cases when the Prandtl-Glauert theory is applied, the change in stagger angle caused by the affine transformation has to be noted.

A related problem occurs when wind-tunnel corrections are considered. For example, the case of a nonlifting nonstaggered plane cascade is identical with the case of the two-dimensional flow past a profile symmetrically arranged between two walls. In incompressible flow, the leading term of the wind-tunnel corrections is found by replacing all the "image" members of the cascade by doublets; at the profile, an excess velocity will be found proportional to the profile thickness and inversely proportional to the square of the spacing (wind-tunnel height). The Göthert rule gives immediately that, for a fixed geometry, the wind-tunnel correction becomes proportional to β^{-3}: a strong compressibility effect. The application of this result is limited by a phenomenon beyond the scope of linearized theory: choking. This effect will be discussed briefly on p. 75–26.

The wind-tunnel corrections due to the displacement (thickness) effect of bodies of revolution are also found to vary proportional to β^{-3}.

Note on Unsteady Flow

If in Eq. (75.17) the disturbance potential is replaced by an unsteady one:

$$\phi = Ux + \varphi(x,y,z,t) \tag{75.58}$$

the linearization of Eq. (75.16) leads to the following equation for φ:

$$(1 - M^2)\varphi_{xx} + \varphi_{yy} + \varphi_{zz} - \frac{2M}{a}\varphi_{xt} - \frac{1}{a^2}\varphi_{tt} = 0 \tag{75.59}$$

The variables

$$X = x - Mat \qquad Y = y \qquad Z = z \qquad T = t \tag{75.60}$$

transform Eq. (75.59) into the classical wave equation of acoustics:

$$\varphi_{XX} + \varphi_{YY} + \varphi_{ZZ} = \frac{1}{a^2}\varphi_{TT} \tag{75.61}$$

showing the identity between linearized aerodynamic theory and acoustics. Also, the set of variables

$$\xi = x \qquad \eta = \beta y \qquad \zeta = \beta z \qquad \tau = \beta^2 t + \frac{M}{a}x \tag{75.62}$$

transforms Eq. (75.59) into one which has a form identical with Eq. (75.61):

$$\varphi_{\xi\xi} + \varphi_{\eta\eta} + \varphi_{\zeta\zeta} = \frac{1}{a^2}\varphi_{\tau\tau} \tag{75.63}$$

In Eq. (75.62), the Prandtl-Glauert transformation appears as a special case of a more general set, which is applicable to the acoustics of moving sound waves as well as to the aerodynamics of unsteady compressible flow. The transformation between X, Y, Z, T and ξ, η, ζ, τ, obtained by elimination of x, y, z, t in Eqs. (75.60) and (75.62), is formally identifiable with the Lorentz transformation. (It may be noted that the applications quoted above, which naturally do not leave the framework of classical mechanics, were already given by Voigt in 1887.)

75.3. HIGHER APPROXIMATIONS

Introductory Remarks

The application of the method of successive approximations to flow problems presupposes that a family of problems will be considered depending on a parameter; the solution will be sought—a classical procedure—in the form of a power series in terms of the chosen parameter. In one application, the parameter is the thickness ratio (τ, say) of an affine family of thin bodies. The first (or lowest) approximation is often referred to as the linearized theory, although this is not always justified. The lowest-order theory may lead to a nonlinear differential equation, as in cases of transonic flow. For subsonic flow, however, all thin-body potential problems apparently lead to the linearized (Prandtl-Glauert) equation.

A method for a first-order solution of a family of flow problems must specify, for the potential: (1) the differential equation to be fulfilled, (2) the boundary conditions, and (3) the evaluation of the pressure (and other functions of state) from the velocity components. In each of these steps, linearity means simplification, but the importance of the simplification is vastly different in each step. Most important naturally is the linearity of the differential equation; this is justified for the first-order solution in subsonic flow. Boundary conditions can be fulfilled "in principle" to any degree of accuracy, but, because of the considerable labor involved, it is important to simplify the boundary conditions such that only the requirements significant in the first-order theory are retained. The linearization of the boundary conditions involves the same steps for subsonic flow as for the small perturbation theory of incompressible flow. Finally, for the evaluation of the pressure from the velocity components, there is only a trivial difference between the effort needed for a linearized formula and for the quadratic expression given by Eq. (75.23), or even for the use of the exact (isentropic) relations between velocity and pressure.

A *first-order method* is defined as the sum of the rules involved in the three steps above. In particular, it will lead to a limit (often in closed analytic form) for all flow quantities as the thickness τ tends to zero. Such a well-defined mathematical limit is unambiguous, whether the proposed expansion is analytic or asymptotic.

If a prediction is desired from a first-order theory for a particular value of τ, or a comparison with experiment is made, the afore-mentioned limit is not necessarily the best prediction to which the theory is capable. True that it is the only *fully assured part* of the theory. But every numerical prediction includes a guess of the higher-order terms, and "zero" is a guess, which is not superior to any other value. Thus, an improvement in the fulfillment of the boundary conditions, which is actually a *higher-order* effect, or, even more trivially, an improved evaluation of the pressure, can have practical importance for the applicability of the first-order theory. Such steps are proposed mostly on semiempirical grounds, as comparisons with known exact solutions or experiment. If too much importance is attached to the prediction of such a "hybrid" theory, in contrast to the "pure" first-order limit, a full investigation of the higher-order theory is indicated. This, however, is generally a major undertaking.

It is also appropriate to point out in advance a few limitations of the higher-order theories. Suppose the first-order theory "fails" in the vicinity of some points in the flow field, where it yields a "singular" result. *Example:* the stagnation point of thin bodies in subsonic flow. Here, a higher-order theory does not mean an improvement, as it will lead to "worse" (i.e., higher order) singularities, unless special precautions are taken. The first general treatment of this problem is due to Lighthill (1949), who proposed a technique [19] for eliminating the growing "divergence" in singular regions. Here, it can only be mentioned that Lighthill's technique obtains the new "degree of freedom" needed for this procedure by the expansion of the independent (space) variables simultaneously with the usual expansion of the dependent variable, i.e., the potential. Research in recent years has shown that Lighthill's technique is no panacea, but this work has given, by showing the crux of the problem, invaluable guidance to research in the field of higher approximations.

A word of caution is also necessary in connection with the question of convergence; its existence and its limits are at best a matter of conjecture. Thus, a second approximation defines rather than extends the limits of the first approximation, unless either the limits of convergence can be reasonably argued or a third approximation is known at least. A second-order calculation may yield essentially only a clarification and limitation of the first approximation; this is naturally a very valuable result.

Second-order Calculation of Plane Subsonic Flow past Airfoils

A second approximation to plane subsonic flow problems, based on the Prandtl-Glauert first-order solution, was carried out first by Görtler [7] for the wavy wall, by Hantzsche and Wendt [8] for Joukowski profiles, and by Kaplan [10] for a smooth "bump" in a wall. Important progress was made when Hayes [11] (starting from a result of Imai [12]) gave an explicit solution of the second-order airfoil problem in subsonic flow, in terms of the first- and second-order solutions in incompressible flow. Hayes's results were clarified and given a poignant formulation by Van Dyke [13], who also has cast the method into a ready-to-use form [14]. The following presentation outlines the derivation of Hayes's solution.

Putting again $\phi = Ux + \varphi$, now the equation for φ is needed, including all quadratic terms in the disturbance potential; introduction into the basic equation (75.16), together with (75.13), gives

$$\beta^2 \varphi_{xx} + \varphi_{yy} = \frac{M}{a} [(\gamma - 1)(\varphi_{xx} + \varphi_{yy})\varphi_x + (\varphi_x{}^2 + \varphi_y{}^2)_x] \tag{75.64}$$

neglecting cubes in φ. An expansion of φ will be sought in terms of a parameter τ, the (percentage) thickness of the airfoil (or body) under consideration:

$$\varphi = U\left(\frac{\tau}{\beta}\varphi_1 + \frac{\tau^2}{\beta^2}\varphi_2 + \cdots\right) \tag{75.65}$$

The parameter τ/β was chosen as this form ensures that the first-order solution φ_1 becomes (on the profile surface) independent of both τ and β (see p. 75–7). Introduction of (75.65) into (75.64) gives, by gathering equal powers of τ, the following set of equations:

$$\beta^2 \varphi_{1xx} + \varphi_{1yy} = 0 \tag{75.66}$$

$$\beta^2 \varphi_{2xx} + \varphi_{2yy} = \frac{\gamma + 1}{2} M^4(\varphi_{1x}{}^2)_x + M^2(\beta^2\varphi_{1x}{}^2 + \varphi_{1y}{}^2)_x \tag{75.67}$$

Next, the boundary conditions will be considered. At infinity, $\varphi \to 0$, and also $\varphi_1 \to 0$, $\varphi_2 \to 0$. The systematic expansion of the boundary conditions in terms of τ follows the procedure for incompressible flow. The exact relations will be formulated on the profile with the ordinate $y_p(x) = \tau f(x)$ by using a Taylor series starting from the chord line $y = 0$:

$$\frac{dy_p}{dx} = \tau f' = \frac{v + v_y \tau f + \frac{1}{2} v_{yy} \tau^2 f^2 + \cdots}{u + u_y \tau f + \frac{1}{2} u_{yy} \tau^2 f^2 + \cdots}$$

where

$$v = U\frac{\tau}{\beta}\varphi_{1y} + U\frac{\tau^2}{\beta^2}\varphi_{2y} + \cdots \qquad u = U + U\frac{\tau}{\beta}\varphi_{1x} + \cdots$$

Expansion in powers of τ yields

$$\varphi_{1y} = \beta f'(x) \qquad y = 0 \tag{75.68}$$

or

$$\beta f(x) = \int_{x_0} \varphi_{1y}\, dx \qquad y = 0 \tag{75.69}$$

and, by use of Eq. (75.69),

$$\varphi_{2y} = \varphi_{1x}\varphi_{1y} - \varphi_{1yy} \int_{x_0} \varphi_{1y}\, dx \qquad y = 0 \tag{75.70}$$

The next step is the application of the Prandtl-Glauert transformation

$$\xi = x \qquad \eta = \beta y \tag{75.71}$$

Equations (75.66) to (75.70) take the form

$$\nabla^2 \varphi_1 \equiv \varphi_{1\xi\xi} + \varphi_{1\eta\eta} = 0 \tag{75.66a}$$

$$\nabla^2 \varphi_2 = \frac{\gamma + 1}{2}\frac{M^4}{\beta^2}(\varphi_{1\xi}{}^2)_\xi + M^2(\varphi_{1\xi}{}^2 + \varphi_{1\eta}{}^2)_\xi \tag{75.67a}$$

$$\varphi_{1\eta} = f'(\xi) \qquad \eta = 0 \tag{75.68a}$$

$$f(\xi) = \int_{\xi_0} \varphi_{1\eta}\, d\xi \qquad \eta = 0 \tag{75.69a}$$

$$\varphi_{2\eta} = \varphi_{1\xi}\varphi_{1\eta} - \beta^2 \varphi_{1\eta\eta} \int_{\xi_0} \varphi_{1\eta}\, d\xi \qquad \eta = 0 \tag{75.70a}$$

As φ_1 is a harmonic function in the ξ,η plane, it is possible to construct the conjugate harmonic, i.e., the stream function ψ_1, fulfilling the relations

$$\varphi_{1\eta} = -\psi_{1\xi} \qquad \varphi_{1\xi} = \psi_{1\eta} \tag{75.72}$$

With the help of ψ_1, the boundary condition (75.70a) can be written as

$$\varphi_{2\eta} = \varphi_{1\xi}\varphi_{1\eta} + \beta^2\psi_1\varphi_{1\eta\eta} \tag{75.70b}$$

Both the integral in Eq. (75.70a) and the stream function in Eq. (75.70b) contain an undetermined constant, which has to be adjusted such that the *total source strength* calculated from the distribution $\varphi_{2\eta}$ along the chord vanishes. (This is the same as requiring the closure of the profile.)

For later use, the purely formal rewriting of Eq. (75.67a), namely,

$$\nabla^2\varphi_2 = (K-1)(\varphi_{1\xi}{}^2 - \varphi_{1\eta}{}^2)_\xi + (K-\beta^2)(\varphi_{1\xi}{}^2 + \varphi_{1\eta}{}^2)_\xi \tag{75.67b}$$

will be advantageous; here,

$$K = 1 + \frac{\gamma+1}{4}\frac{M^4}{\beta^2} \tag{75.73}$$

The second-order problem is now clearly formulated: to solve Eq. (75.67b), a Poisson-type equation for φ_2, with the boundary conditions (75.70b).

For the solution, consider first the incompressible case in the ξ,η plane, putting

$$\varphi_i = U(\tau\varphi_{1i} + \tau^2\varphi_{2i} + \cdots) \tag{75.74}$$

where φ_{1i} and φ_{2i} are the special solutions of Eqs. (75.66a) to (75.70a) in the case $M = 0$, $\beta = 1$. It is clear from Eqs. (75.66a) and (75.68a) that

$$\varphi_{1i}(\xi,\eta) = \varphi_1(\xi,\eta) \tag{75.75}$$

a result merely stating what was known from the Prandtl-Glauert theory, namely, that putting the disturbance potential into the form (75.65), together with the transformation (75.71), reduces the compressible problem exactly to the incompressible one, in the first approximation. Correspondingly, the subscript i will not be needed in addition to the subscript 1.

For the second-order solutions, Eqs. (75.67b) and (75.70b) simplify to

$$\nabla^2\varphi_{2i} = 0 \tag{75.67c}$$

$$\varphi_{2i\eta} = \varphi_{1\xi}\varphi_{1\eta} + \psi_1\varphi_{1\eta\eta} \qquad (\eta = 0) \tag{75.70c}$$

In order to solve Eq. (75.67b), consider the function

$$\varphi^* = \eta\,\varphi_{1\xi}\varphi_{1\eta} \tag{75.76}$$

It is easy to prove [using $\nabla^2(\varphi_{1\xi}\varphi_{1\eta}) = 0$] that

$$\nabla^2\varphi^* = -(\varphi_{1\xi}{}^2 - \varphi_{1\eta}{}^2)_\xi \tag{75.77}$$

Also, consider the function

$$\varphi^{**} = \int_0 \psi_1\varphi_{1\eta\eta}\,d_\eta \tag{75.78}$$

Upon partial integration, it can be shown that

$$\varphi^{**}_\xi = \psi_1\varphi_{1\xi\eta} - \tfrac{1}{2}(\varphi_{1\eta}{}^2 + \varphi_{1\xi}{}^2) \tag{75.79}$$

and

$$\nabla^2\varphi^{**} = -(\varphi_{1\eta}{}^2 + \varphi_{1\xi}{}^2)_\xi \tag{75.80}$$

Now, the final result can be obtained as the function φ_2, fulfilling Eq. (75.67b), is found to be

$$\varphi_2 = K\varphi_{2i} - (K-1)\varphi^* - (K-\beta^2)\varphi^{**} \tag{75.81}$$

as is immediately clear from Eqs. (75.67c), (75.77), and (75.80). Also, the boundary conditions (75.70b) are fulfilled, as can be seen from Eq. (75.70c) and the definitions (75.76) and (75.78). Thus, the second-order solution φ_2 is given by the incompressible second-order solution φ_{2i}, and the particular solutions φ^* and φ^{**}, derived from the first-order solution. The latter can fulfill the prescribed boundary conditions only on the line $y = 0$, so that solution (75.81) is complete and valid only for airfoils in an infinite stream, or for other problems containing fixed boundaries only near the axis $y = 0$.

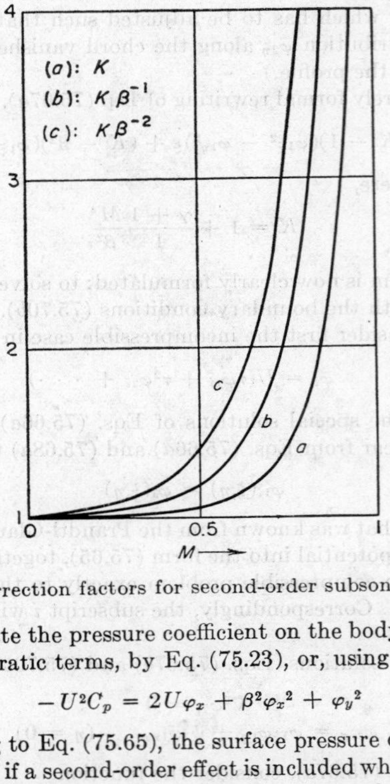

FIG. 75.3. Correction factors for second-order subsonic flow theory.

It remains to calculate the pressure coefficient on the body surface. The pressure is given, including quadratic terms, by Eq. (75.23), or, using a pressure coefficient C_p,

$$-U^2C_p = 2U\varphi_x + \beta^2\varphi_x^2 + \varphi_y^2 \tag{75.82}$$

Introducing φ according to Eq. (75.65), the surface pressure coefficient is obtained by going to the limit $y = 0$, if a second-order effect is included which is due to the fact that the first-order pressure has to be evaluated at $y = y_p$ instead of $y = 0$, a result accomplished again by a series expansion:

$$-C_p = 2\frac{\tau}{\beta}(\varphi_{1x} + \varphi_{1xy}\tau f) + \frac{\tau^2}{\beta^2}(2\varphi_{2x} + \beta^2\varphi_{1x}^2 + \varphi_{1y}^2)$$

Replacing f by Eq. (75.69) and using the variables ξ and η, this transforms into

$$-C_p = 2\frac{\tau}{\beta}\varphi_{1\xi} + \frac{\tau^2}{\beta^2}[2\varphi_{2\xi} - \beta^2(2\psi_1\varphi_{1\xi\eta} - \varphi_{1\xi}^2 - \varphi_{1\eta}^2)] \tag{75.83}$$

Introducing φ_2 for $\eta \to 0$ and making use of Eq. (75.79), the final result is

$$-\tfrac{1}{2}C_p = \frac{\tau}{\beta}\varphi_{1\xi} + \frac{K\tau^2}{\beta^2}(\varphi_{2i\xi} - \varphi_\xi^{**}) \tag{75.84}$$

In the particular case $M = 0$,

$$-\tfrac{1}{2}C_{pi} = \tau\varphi_{1\xi} + \tau^2(\varphi_{2i\xi} - \varphi_\xi^{**}) \tag{75.85}$$

Comparison of Eqs. (75.84) and (75.85) yields the following rule (restricted to the pressure coefficient on the body surface): If the first- and second-order terms are known in incompressible flow, the values for compressible flow are obtained by multiplying the first-order term by β^{-1} and the second-order term by $K\beta^{-2}$, where K is given by Eq. (75.73); see Fig. 75.3. It is remarkable that this coefficient is the same as Busemann's second-order coefficient for supersonic flow, discussed on p. 77–10.

The remaining problems are (1) to find a practical method for the second-order solution in incompressible flow, and (2) to apply special consideration to stagnation regions, in order to avoid the growing divergence discussed before. Both questions are treated by Van Dyke [14] who uses Lighthill's treatment for the second-order thin airfoil theory in incompressible flow (see p. 72–15).

Example: It is known that, for incompressible symmetric flow past an ellipse of the thickness ratio τ, the maximum velocity is exactly $U(1 + \tau)$, so that the minimum pressure coefficient is

$$-C_{pi} = 2\tau + \tau^2 \tag{75.86}$$

For compressible flow, the Hayes–Van Dyke rule gives

$$-C_p = 2\tau\beta^{-1} + \frac{4\beta^2 + (\gamma + 1)M^4}{4\beta^4}\tau^2 \tag{75.87}$$

a result already obtained for this case by Hantzsche and Wendt [8]. It may be noted here that Hantzsche [9] calculated further approximations for this example and found, in compressible flow, additional terms of order τ^3 and $\tau^4 \ln \tau$.

Bodies of Revolution

For bodies of revolution, the first approximation already needs clarification and proper limitation by second-order considerations. The problem is the following: On the surface of a body of thickness τ, linearized theory yields v/U and w/U of order τ, and u/U of order $\tau^2 \ln \tau$. The pressure coefficient has a $\tau^2 \ln \tau$ term from its linear part, while the quadratic terms contribute to an expression proportional to τ^2. Now, for cases of practical interest, the theory will be applied up to thicknesses where the orders $\tau^2 \ln \tau$ and τ^2 can hardly be distinguished. The question arises whether it is consistent to calculate the pressure from a formula containing quadratic terms together with the solution of a linearized equation. The answer is naturally yes in the particular case of incompressible flow, where the exact equation is linear, and it can be shown that the answer is yes also for supersonic flow, where the second-order theory of bodies of revolution was particularly successful (see p. 77–25). It is therefore generally conjectured that the same answer is obtained for subsonic flow. Practically, this means the following: Take the first-order rule obtained on p. 75–9, namely, Eq. (75.49). Here it appears permissible to use the "exact" (or rather, second-order) value for C_{pi}.

The second-order solution for an ellipsoid of revolution was calculated by Schmieden and Kawalki [20], who showed that second-order effects are of the order $\tau^4 (\ln \tau)^2$, $\tau^4 \ln \tau$, and τ^4. It happens that their solution is slightly incorrect because of some difficulties associated with the transformation of the Stokes stream function. This error was corrected and a theory for general body shapes was given by Van Dyke [21]. According to his work, the excess speed at the center of an ellipsoid of thickness ratio τ at zero angle of attack in subsonic flow is

$$\frac{u}{U} = \tau^2 \left(\ln \frac{2}{\beta\tau} - 1\right) + \tau^4\beta^2 \left(\ln \frac{2}{\beta\tau}\right)^2 + \tau^4(M^2 - \tfrac{1}{2}) \ln \frac{2}{\beta\tau} + \tau^4 \left[\frac{\gamma + 1}{4}\frac{M^4}{\beta^2} - \tfrac{1}{4} + M^2(\tfrac{1}{2} \ln 2 - \tfrac{3}{4})\right]$$

The terms in τ^2 are the usual slender-body solution. The second-order terms in τ^4

agree with the result of Schmieden and Kawalki except for the coefficient of $M^2\tau^4$, which they give as $-5/4$ instead of $\frac{1}{2}\ln 2 - \frac{3}{4} = -0.4034$.

The Janzen-Rayleigh Approximation

The earliest proposal for a method of successive approximations was made by Janzen [16] and Rayleigh [17]. In their method, the first term is the (exact) solution for incompressible flow past the given body, and the subsonic flow is found by successive approximations arising from a power development in terms of the free-stream Mach number M. It is easy to show that such an expansion will contain only even powers of M, so that a potential ϕ will be assumed having the form

$$\phi = U(\phi_0 + M^2\phi_1 + M^4\phi_2 + \cdots) \tag{75.88}$$

(which explains the name *M^2 expansion method*). Here, ϕ_0 fulfills Laplace's equation,

$$\nabla^2\phi_0 \equiv \phi_{0xx} + \phi_{0yy} + \phi_{0zz} = 0 \tag{75.89}$$

while substitution of series (75.88) into Eqs. (75.16) and (75.13) yields

$$\nabla^2\phi_1 = \nabla\phi_0 \cdot \nabla[\tfrac{1}{2}(\nabla\phi_0)^2] \tag{75.90}$$

for ϕ_1, and further equations of the Poisson type with increasingly complicated "inhomogeneous" terms for ϕ_2 etc.

The practical value of the M^2 expansion method is limited if it is desired (as it is in practice) to consider flows near the critical Mach number. It has been often argued that the first appearance of sonic velocity in the field determines the limit of convergence of the M^2 method; based on this plausible assumption, it may be expected that many terms of expansion (75.88) would be needed for accurate predictions near the critical value of M. It is true that the Prandtl-Glauert approximation also breaks down in the "transonic" limit, but there is no doubt that, for thin bodies, approximations valid for near-critical subsonic Mach numbers are obtained already in the first step.

On the other hand, the Janzen-Rayleigh method is not limited to thin bodies. At first glance, this appears to be of little practical importance. However, independently of the body thickness, this method is free of the afore-mentioned growing divergence difficulties near the stagnation point, a fact which offers vital help in eliminating these difficulties in the thickness-expansion method discussed above. These arguments were used first by Van Dyke [14],[15], who considered the conclusions that can be drawn from a simultaneous expansion in both M^2 and τ.

In summary, up until now the Janzen-Rayleigh solutions were more important for the general theory of subsonic flow (where they play the role of solutions which are "exact" in principle) than in the everyday applications of aeronautical practice. A recent publication by Imai [18] contains a fast method for obtaining the M^2 term for airfoils and cascades and the M^4 term for airfoils.

75.4. THE HODOGRAPH METHOD

Basic Equations

It was discovered by Molenbroeck [22] and Chaplygin [23] that the problem of the plane steady potential flow of a gas becomes linear when the velocity components are used as independent variables. With the advance of high-speed aerodynamics, the use of the hodograph plane (where the velocity components u, v are cartesian coordinates) proved to be extremely useful for the description and treatment of flow problems although in many cases the basic advantage of linearity is partially offset by the practical difficulty of fulfilling boundary conditions given in the physical plane.

The transformation to the hodograph plane is best accomplished by considering the potential φ and the stream function ψ simultaneously; in the physical plane x,y they are defined by

$$u = \frac{\partial \phi}{\partial x} = \frac{\rho_0}{\rho}\frac{\partial \psi}{\partial y} \qquad v = \frac{\partial \phi}{\partial y} = -\frac{\rho_0}{\rho}\frac{\partial \psi}{\partial x} \tag{75.91}$$

(The stagnation density ρ_0 is used as a normalizing factor.) In the hodograph plane, polar coordinates q, θ (velocity magnitude and direction) will be used:

$$u = q\cos\theta \qquad v = q\sin\theta \tag{75.92}$$

The differentials of ϕ and ψ are

$$d\phi = u\,dx + v\,dy \qquad \rho_0\,d\psi = -\rho v\,dx + \rho u\,dy \tag{75.93}$$

Solution of Eqs. (75.93) for dx and dy gives

$$dx = \frac{1}{q}\left(\cos\theta\,d\phi - \frac{\rho_0}{\rho}\sin\theta\,d\psi\right) \qquad dy = \frac{1}{q}\left(\sin\theta\,d\phi + \frac{\rho_0}{\rho}\cos\theta\,d\psi\right) \tag{75.94}$$

Now we desire to consider ϕ and ψ as functions of q and θ:

$$d\phi = \phi_q\,dq + \phi_\theta\,d\theta \qquad d\psi = \psi_q\,dq + \psi_\theta\,d\theta \tag{75.95}$$

so that from Eqs. (75.94), the following partial derivatives are known:

$$\begin{aligned} x_q &= \frac{1}{q}\left[(\cos\theta)\,\phi_q - \frac{\rho_0}{\rho}(\sin\theta)\,\psi_q\right] \qquad & x_\theta &= \frac{1}{q}\left[(\cos\theta)\phi_\theta - \frac{\rho_0}{\rho}(\sin\theta)\,\psi_\theta\right] \\ y_q &= \frac{1}{q}\left[(\sin\theta)\,\phi_q + \frac{\rho_0}{\rho}(\cos\theta)\,\psi_q\right] \qquad & y_\theta &= \frac{1}{q}\left[(\sin\theta)\phi_\theta + \frac{\rho_0}{\rho}(\cos\theta)\,\psi_\theta\right] \end{aligned} \tag{75.96}$$

By making use of Bernoulli's equation in computing

$$\frac{d}{dq}\left(\frac{\rho_0}{\rho}\right) = -\frac{\rho_0}{\rho^2}\frac{d\rho}{dp}\frac{dp}{dq} = \frac{\rho_0}{\rho}\frac{q}{a^2} \tag{75.97}$$

cross differentiation of Eqs. (75.96) leads to the following set:

$$\phi_q = -\frac{\rho_0}{\rho}(1 - M^2)\frac{1}{q}\psi_\theta \qquad \frac{1}{q}\phi_\theta = \frac{\rho_0}{\rho}\psi_q \tag{75.98}$$

(Here, $M = q/a$ is again the local Mach number.) These are the fundamental equations for ϕ and ψ in the hodograph plane. The sound speed a and the density ρ are in homenergic-homentropic flow functions of q; in particular, for the perfect gas with constant specific heat, the following relations are recalled:

$$2a^2 + (\gamma - 1)q^2 = 2a_0^2 \tag{75.99}$$

$$\frac{\rho_0}{\rho} = \left(1 + \frac{\gamma - 1}{2}M^2\right)^{1/(\gamma-1)} \tag{75.100}$$

It is often advantageous to change the independent variable q in Eqs. (75.98) to some other function of q. For instance, consider the quantity λ given by

$$d\lambda = \sqrt{1 - M^2}\,\frac{dq}{q} = \beta\,\frac{dq}{q} \tag{75.101}$$

(whereby a "local" value of β is defined). By using λ, Eqs. (75.98) transform into the more symmetrical form:

$$\phi_\lambda = -\frac{\rho_0}{\rho}\beta\psi_\theta \qquad \phi_\theta = \frac{\rho_0}{\rho}\beta\psi_\lambda \tag{75.102}$$

Finally, cross differentiation in Eqs. (75.98) [or in any modified form, as Eqs. (75.102)] eliminates either ϕ or ψ. From the set (75.98), the following equations can be obtained:

$$\phi_{qq} + \frac{1 + \gamma M^4}{1 - M^2} \frac{1}{q} \phi_q + (1 - M^2) \frac{1}{q^2} \phi_{\theta\theta} = 0 \tag{75.103}$$

$$\psi_{qq} + (1 + M^2) \frac{1}{q} \psi_q + (1 - M^2) \frac{1}{q^2} \psi_{\theta\theta} = 0 \tag{75.104}$$

A function derived from the potential by the so-called Legendre transformation, namely,

$$\chi = x\phi_x + y\phi_y - \phi$$

also fulfills a simple equation in the hodograph plane. χ has the property that

$$\chi_u = x \qquad \chi_v = y$$

so that

$$\phi = u\chi_u + v\chi_v - \chi$$

It can be shown that the equation for χ is

$$\chi_{qq} + (1 - M^2) \frac{1}{q} \chi_q + (1 - M^2) \frac{1}{q^2} \chi_{\theta\theta} = 0$$

Chaplygin's Gas and the Kármán-Tsien Method

Many important applications of the hodograph equations are found in the field of "mixed" or transonic plane flow, p. 76–8 (where the use of the hodograph variables leads to the only possible linearization of the problem), and for the characteristics method in supersonic flow. Both applications are discussed in the corresponding chapters, whereas here the following presentation is restricted to the only known typically (and exclusively) subsonic case.

In the limit of incompressible flow ($M \to 0$, $\beta = 1$, $\rho = \rho_0$), Eqs. (75.102) reduce to the Cauchy-Riemann relations (9.14); the same would hold for all M in a hypothetical gas for which

$$\frac{\rho}{\rho_0} = \beta = \sqrt{1 - M^2} \tag{75.105}$$

Comparison with Eq. (75.100) shows that this relation would be true for a gas with $\gamma = -1$; that is, for a substance with an isentropic pressure–density relation of the form

$$p = \frac{\text{const}}{\rho} \tag{75.106}$$

The plane potential flow of this hypothetical substance was first investigated by Chaplygin [23]. The speed of sound is, according to Eq. (75.99), given for $\gamma = -1$ by

$$a^2 - q^2 = a_0{}^2 \tag{75.107}$$

so that

$$M = \frac{q}{\sqrt{q^2 + a_0{}^2}} \qquad \beta = \frac{a_0}{\sqrt{q^2 + a_0{}^2}} \tag{75.108}$$

showing that Chaplygin's gas is capable of subsonic flow only. (Replacing the constant $a_0{}^2$ by $-a_0{}^2$ restricts q to supersonic values everywhere.)

Although Chaplygin's gas is clearly a physically impossible substance (all real gases have $\gamma > 1$), its properties can approximate, in limited regions, the behavior of real gases. The two equations (75.105) and (75.100) do coincide, for all values of γ, to the first approximation in M^2. Thus, calculations based on Chaplygin's gas exhibit the correct M^2-order compressibility effects, equivalent to the M^2 term of the Janzen-Rayleigh series. In other words, curve (75.106) will approximate the true gas behavior if it is adjusted so that it is tangent to the true pressure-density relationship at a certain point. In Chaplygin's original work, this point was chosen to represent the state at stagnation conditions. According to a method due to von Kármán [24] and Tsien [25], the point of tangency represents the state of the undisturbed flow (at infinity), so that Eq. (75.106) will be replaced by

$$p - p_\infty = \left(\frac{dp}{d(1/\rho)}\right)_\infty \left(\frac{1}{\rho} - \frac{1}{\rho_\infty}\right) = -a_\infty^2 \rho_\infty^2 \left(\frac{1}{\rho} - \frac{1}{\rho_\infty}\right) \tag{75.109}$$

For the further development of the Kármán-Tsien method, the differential equation (75.101) for the variable λ can be integrated upon introduction of M from Eq. (75.108), yielding

$$\lambda = \tfrac{1}{2} \ln \frac{\sqrt{a_0^2 + q^2} - a_0}{\sqrt{a_0^2 + q^2} + a_0} \tag{75.110}$$

As ϕ and ψ fulfill the Cauchy-Riemann relations in λ and θ, the complex potential $F = \phi + i\psi$ is an analytic function of $\lambda - i\theta$. The difficulty is, as always, to adjust the solution to boundary conditions given in the physical plane. However, if one solution were known, a whole family could be derived easily by changing the parameter a_0 in Eq. (75.110), which means (as a_0 is the stagnation sound speed) a change of the flow Mach number; in particular, the incompressible limit will be a member of the family in the special case $a_0 \to \infty$. Therefore, it will be possible to start from a known solution in incompressible flow and derive a whole family of compressible flows of the Chaplygin gas. It has to be emphasized that the flow in the physical plane, i.e., the body contour in the flow, does not remain constant for the members of the family, but undergoes a continuous deformation in function of the parameter a_0, as will be calculated presently. However, it is clear that the shapes will be "related" to the original incompressible case, and that, in principle, it should be possible to choose a shape in incompressible flow such that a prescribed shape for compressible flow is realized. Particularly simple results can be expected if the shape change turns out to be negligibly small.

First, consider the limiting solution in the incompressible case. Let Λ and Q be the limits of λ and q for $a_0 \to \infty$, according to Eq. (75.110):

$$\Lambda = \tfrac{1}{2} \ln \frac{Q^2}{4a_0^2}$$

so that, for $\lambda = \Lambda$,

$$\frac{Q^2}{4a_0^2} = \frac{\sqrt{a_0^2 + q^2} - a_0}{\sqrt{a_0^2 + q^2} + a_0} = \frac{1 - \beta}{1 + \beta} \tag{75.111}$$

These relations can be inverted to give

$$\beta = \frac{\rho}{\rho_0} = \frac{4a_0^2 - Q^2}{4a_0^2 + Q^2} \tag{75.112}$$

$$\frac{q}{a_0} = \frac{4a_0 Q}{4a_0^2 - Q^2} \tag{75.113}$$

Equations (75.111) to (75.113) give the "generalization" of the velocity magnitude from its incompressible value Q to the general quantity q. In particular, the connection between the corresponding pressure coefficients is of interest, referred to the free-stream speeds Q_∞ and q_∞, respectively. In the compressible case, from Eq. (75.109), the pressure coefficient is

$$C_p = \frac{2}{M_\infty^2}\left(1 - \frac{\rho_\infty}{\rho}\right) = \frac{2}{M_\infty^2}\left(1 - \frac{\beta_\infty}{\beta}\right) \tag{75.114}$$

while the pressure coefficient in incompressible flow is defined as

$$C_{pi} = 1 - \frac{Q^2}{Q_\infty^2} \tag{75.115}$$

If β and β_∞ are introduced by (75.111) into Eq. (75.115), and finally β is eliminated between (75.114) and (75.115), the well-known Kármán-Tsien formula is obtained:

$$C_p = \frac{C_{pi}}{\beta_\infty + (1 - \beta_\infty)(C_{pi}/2)} \tag{75.116}$$

It was proposed by von Kármán and Tsien to use Eq. (75.116) (which, for small C_{pi}, approaches the Prandtl-Glauert rule) as an engineering approximation to the pressure coefficient calculated in function of M_∞ for the same profile, thereby neglecting the shape changes still to be determined, which are inherent to the original method. In this form the Kármán-Tsien rule is particularly simple; although its errors and limitations are very hard to predict from theory (little more than speculations exist), the results compare favorably both with experiments and with results of second-order calculations according to the more elaborate methods of p. 75–13. This point will be discussed again on p. 75–23; here the conclusion is mentioned that the success of the Kármán-Tsien formula may be accidental in some respects, but it is so far well documented.

It remains to calculate the change in the shape, to be determined by Eqs. (75.94), which can be written by Eqs. (75.112) and (75.113) in the form

$$\begin{aligned} dx &= \frac{1}{Q}\left[\left(1 - \frac{Q^2}{4a_0^2}\right)\cos\theta\, d\phi - \left(1 + \frac{Q^2}{4a_0^2}\right)\sin\theta\, d\psi\right] \\ dy &= \frac{1}{Q}\left[\left(1 - \frac{Q^2}{4a_0^2}\right)\sin\theta\, d\phi + \left(1 + \frac{Q^2}{4a_0^2}\right)\cos\theta\, d\psi\right] \end{aligned} \tag{75.117}$$

Put

$$dx + i\,dy = dz \qquad d\phi + i\,d\psi = dF \tag{75.118}$$

and let $Z = X + iY$ be the incompressible plane, for which the complex velocity is given by

$$\frac{dF}{dZ} = Qe^{-i\theta} \tag{75.119}$$

With the help of Eqs. (75.118) and (75.119), it is possible to express and integrate (75.117) in complex form as follows:

$$z = Z - \frac{1}{a_0^2}\overline{\int \left(\frac{dF}{dZ}\right)^2 dZ} \tag{75.120}$$

The second term here expresses the distortion or change from the particular *shape* Z in incompressible flow to the general value z. Actually this formula can be used only if z is single-valued, i.e., if the integral in (75.120) vanishes if evaluated following a path enclosing the profile. It is easy to realize that this is true if F represents a complex potential without circulation (i.e., logarithmic term). In the presence of circulation, instead of Eq. (75.119), the more general case

$$\frac{dF}{dZ} = Qe^{-i\theta}G(Z;M_\infty) \tag{75.121}$$

has to be considered, where G is an analytic function of Z. depending on the parameter M_∞ such that $G(Z;0) = 1$, and to be determined such that z becomes a single-valued function of Z. Many solutions to this problem were found by different authors; so far, however, the Kármán-Tsien method with distortions included has been seldom used in practice. More detailed presentations are found in [1], [2], and [3].

75.5 CONCLUSIONS AND DISCUSSION OF THE CRITICAL MACH NUMBER

Plane Flow

Before a critical comparison of the different methods for subsonic flow (and the experimental results) is made, a peculiarity of all these discussions has to be emphasized, namely, that the greatest interest (both theoretically and practically) is concentrated on the problem of the limit of the subsonic region, i.e., the critical Mach number. For low subsonic (that is, definitely subcritical) Mach numbers, the mutual agreement between theoretical predictions and their agreement with experiment can be taken for granted. When the critical Mach number is approached, the difference between predictions from available theories becomes experimentally noticeable. Also, the limit of validity for all theories discussed in this chapter is, in some unknown way, connected with, and certainly near to, the critical Mach number. Thus, it is understandable that discussions are limited to the critical region. However, certain fortuitous agreements may result from this restriction, as will be seen later.

The connection between the critical Mach number and the minimum (i.e., largest negative) pressure coefficient at the *sonic point* follows easily from the isentropic-pressure–Mach-number relationship; it is

$$C_{p\,\min} = \frac{2}{\gamma M_{cr}^2}\left[\left(\frac{2+(\gamma-1)M_{cr}^2}{\gamma+1}\right)^{\gamma/(\gamma-1)} - 1\right] \qquad (75.122)$$

Values of $C_{p\,\min}$, calculated in function of M by different theories, will be inserted in Eq. (75.122) in order to determine M_{cr} (numerically or graphically).

For plane subsonic flow, three methods will be compared: the Prandtl-Glauert method, the Kármán-Tsien rule, and the second-order theory in the Imai-Hayes form, or the Hayes–Van Dyke rule. (The M^2 method approaches the critical region too slowly and will not be considered.) The minimum pressure coefficient is, as in incompressible flow, to be found on the body surface. The Prandtl-Glauert rule, given by Eq. (75.42), will be applied, as is mostly and preferably done in practice, with C_{pi} meaning the *exact* pressure coefficient in incompressible flow. To be inserted in Eq. (75.122) is

$$C_{p\,\min} = \frac{C_{pi\,\min}}{\sqrt{1-M^2}} \qquad (75.123)$$

so that a "universal" relation between $C_{pi\,\min}$ and M_{cr} is obtained. The same kind of relationship follows from the Kármán-Tsien formula (75.116), that is,

$$C_{p\,\min} = \frac{C_{pi\,\min}}{\sqrt{1-M^2} + (1-\sqrt{1-M^2})(C_{pi\,\min}/2)} \qquad (75.124)$$

The well-known effect of the use of the Kármán-Tsien formula is a systematic reduction of M_{cr}, as compared to the Prandtl-Glauert result; the effect is also shown in Fig. 75.4.

The Hayes–Van Dyke rule, as given by Eq. (75.84), does not permit such a simple plot. Practically, the rule can be formulated as follows: take the exact value C_{pi} and subdivide it into two parts, $C_{pi}^{(1)}$ (say) and $(C_{pi} - C_{pi}^{(1)})$. The first-order part $C_{pi}^{(1)}$, (which is generally the "bulk" part) is multiplied by the Prandtl-Glauert factor, while the "remainder" gets a stronger amplification, according to the formula:

$$C_p = \frac{1}{\beta} C_{pi}^{(1)} + \frac{1}{\beta^2}\left(1 + \frac{\gamma+1}{4}\frac{M^4}{\beta^2}\right)(C_{pi} - C_{pi}^{(1)}) \qquad (75.125)$$

The difficulty is that the *subdivision* between C_{pi} and $C_{pi}^{(1)}$ depends on the thickness and the shape of the airfoil, as well as on the position on the surface.

As an example, consider $C_{pi\,\min}$ for elliptical profiles, where [as already stated in Eq. (75.86)]

$$C_{pi\,\min} = -2\tau - \tau^2$$

and

$$C^{(1)}_{pi\ \min} = -2\tau$$

so that, in this particular case, elimination of the thickness τ between these two equations gives

$$C^{(1)}_{pi\ \min} = 2(1 - \sqrt{1 - C_{pi\,\min}}) \tag{75.126}$$

It is proposed here to use Eq. (75.126) as a *rule of subdivision* between C_{pi} and $C_{pi}{}^{(1)}$ in cases where $C_{pi}{}^{(1)}$ is not known, with the justification that (1) Eq. (75.126) is correct in a *typical* case, and (2) Eq. (75.125) is not too sensitive to the proper subdivision. Introducing Eq. (75.126) into (75.125) yields again a pressure coefficient for compressible flow depending on C_{pi} and M only.

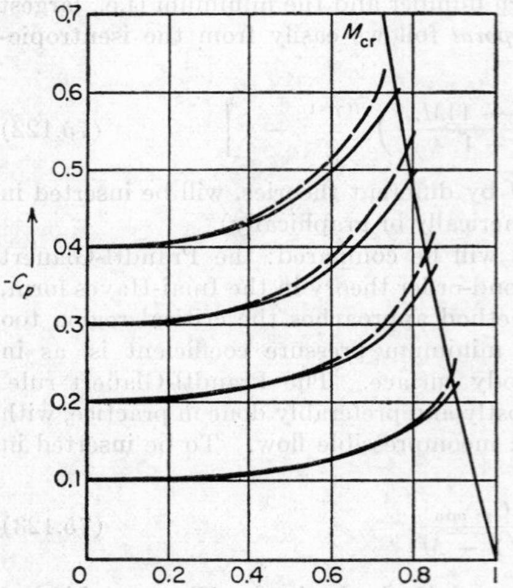

FIG. 75.4. The pressure coefficient in plane subsonic flow in function of the Mach number. Solid line: first-order theory; broken line: second-order theory. The latter result is exact only for the middle point of elliptic profiles, but it can be used approximately for the minimum C_p of any profile shape.

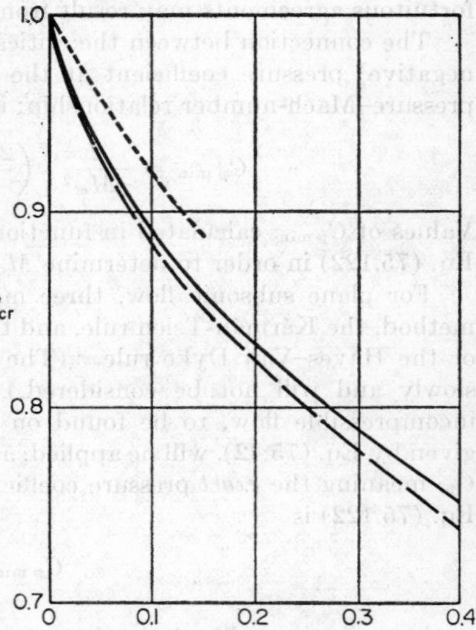

FIG. 75.5 The critical Mach number against the minimum pressure coefficient in incompressible flow, according to first-order theory (solid line) and second-order theory (broken line) for plane flow. (See Fig. 75.4 for restrictions on the latter theory.) Dotted line applies to ellipsoids of revolution, calculated from the first-order theory of axisymmetric flow.

The result turns out to be almost indistinguishable from the Kármán-Tsien rule, in particular around M_{cr}; this behavior supports the Hayes–Van Dyke rule, as the Kármán-Tsien formula is well supported by experimental investigations in the critical region. Nevertheless, the agreement is fortuitous in the sense that it is not possible to find a term-by-term correspondence between the Kármán-Tsien rule and the second-order theories over the full range of subsonic flow.

Figure 75.4 shows the results of the Prandtl-Glauert rule and of the Hayes–Van Dyke rule as described above; Fig. 75.5 gives a cross plot of M_{cr} against the minimum pressure coefficient in incompressible flow. It should be reemphasized that these second-order results hold exactly only for ellipses, in which case they agree with the Kármán-Tsien rule so closely that a plot of the latter theory could not be included in the figures.

Concluding the discussion of plane flow, it can be pointed out that all calculations of M_{cr} comply in the limit $C_p \to 0$, $\tau \to 0$, $\beta_{cr} \to 0$ with the rules of transonic similarity to be discussed (see p. 76–3). In this limit, Eq. (75.122) can be replaced by

$$C_{p\,\min} = -\frac{2}{\gamma+1}\beta_{cr}^2 \tag{75.127}$$

Combined with the Prandtl-Glauert rule, this gives

$$\beta_{cr}^3 = \frac{\gamma+1}{2}C_{pi\,\min} \tag{75.128}$$

As $C_{pi\,\min}$ is proportional to τ, Eq. (75.128) states that M_{cr} occurs at a given value of Kármán's similarity parameter

$$\mathcal{K} = \frac{\gamma+1}{2}\frac{\tau}{\beta^3} \tag{75.129}$$

Calculations using the Imai–Hayes–Van Dyke method also follow the transonic similarity rule. The ratio between the second-order and the first-order terms changes with the Mach number and thickness with the factor $K\tau/\beta$, and in the limit $\beta \to 0$,

$$\frac{K\tau}{\beta} \to \frac{\gamma+1}{4}\frac{\tau}{\beta^3} = \frac{1}{2}\mathcal{K} \tag{75.130}$$

It is to be expected that the ratio between the second-order and first-order terms—which is a measure of the convergence of the series—follows transonic similitude. (Actually, Imai [12] originally investigated the second-order term in order to obtain a parameter K which possibly would give a better correlation of transonic data than $\mathcal{K}$, for not so small values of β. This idea was not entirely successful; for other modifications of $\mathcal{K}$, see p. 76–3.) The use of the approximate equations (75.127) and (75.130) strongly simplifies the calculation of M_{cr}, but the loss in numerical accuracy is significant except for the thinnest profiles.

For an infinite swept wing, it makes sense to define M_{cr} as the Mach number for which the velocity component normal to the wing reaches sonic speed, because this is the limit for the first possible appearance of a shock wave which has to be swept (i.e., oblique) too. Critical conditions are easily obtained by reformulating the problem with complete disregard of the (constant) spanwise velocity component; the case is thereby reduced to the plane unswept wing problem. This procedure is not permissible for finite swept wings, owing to "end" and "middle" effects.

Bodies of Revolution

For bodies of revolution, the use of Eq. (75.49) together with (75.122) gives results which cannot be expressed as a universal relationship between the minimum C_{pi} and M_{cr}. This makes the comparison between the two- and three-dimensional cases difficult. It is well known that, for the same thickness ratio, bodies of revolution give much lower excess speeds, and lead to considerably higher values of M_{cr} than the same profiles in plane flow. A more interesting basis for a comparison can be obtained by considering a particular family of bodies, e.g., ellipsoids of revolution, and replotting the results in a diagram showing $C_{pi\,\min}$ against M_{cr}. Results of such a calculation (after Sears [1]) are included in Fig. 75.5, together with results for plane flow. It is seen that, even in this plot, bodies of revolution have an advantage over wing profiles: M_{cr} is higher for the same value of $C_{pi\,min}$. It has to be noted, however, that ellipsoids of revolution are probably the "best" in this respect, and that the dependence on body shape is not negligible in the three-dimensional case. As an extreme example, consider a "bulge" on a long body of revolution: its contribution to C_p will be transformed essentially according to the rules governing plane flow. Thus, the curves in Fig. 75.4 may be considered as *extremes* which probably will "bracket" a wide range of cases of practical interest.

Wind-tunnel Effects

On p. 75–11 wind-tunnel corrections varying with β^{-3} were found. For any arrangement of practical interest, wind-tunnel corrections for incompressible flow must be very small; their growth with β^{-3} indicates, however, their importance when sonic speed is approached. Nevertheless, these corrections calculated from linearized theory are less important than an effect which is beyond the scope of the linear theories, namely, choking. Theoretical considerations (as well as experiments) show that a closed wind tunnel with a model in it can be considered as a "channel" in the hydraulic sense; i.e., the velocity is determined by the available cross section, using a constant average speed value, for the purpose of choking calculations. The Mach number M_{ch} for choking is obtained from the formulas for isentropic flow in the convergent-divergent nozzle (p. 74–7) having sonic speed at the "throat," i.e., the minimum cross section. An experiment in a closed tunnel is meaningful only if the critical Mach number (calculated without wind-tunnel effects) is definitely lower than the Mach number for choking.

As an example, consider the plane flow past a profile of thickness t, having a chord c in a tunnel with parallel walls at a distance h. The critical Mach number will be obtained at a certain value of the parameter $\mathcal{K}$, given by Eq. (75.129), so that

$$1 - M_{cr} \approx A\left(\frac{\gamma+1}{2}\frac{t}{c}\right)^{2/3} \tag{75.131}$$

Here, A is a constant (depending on the profile shape) of order 1.

On the other hand, M_{ch} can be obtained by expanding the formulas for isentropic channel flow, near the throat, for small changes in the cross section. The result is

$$1 - M_{ch} = \sqrt{\frac{\gamma+1}{2}\frac{t}{h}} \tag{75.132}$$

The condition for a meaningful experiment is $M_{cr} < M_{ch}$, which leads to

$$\frac{h}{c} > \frac{1}{A^2}\left(\frac{2}{\gamma+1}\frac{c}{t}\right)^{1/3} \tag{75.133}$$

This result is in accordance with a transonic similarity rule, due to Spreiter, discussed on p. 76–3.

Now, if M_{cr} is calculated with $C_{p\,min}$ including the effects of wind-tunnel corrections from linearized theory, i.e., using a form

$$C_{p\,min} \sim \frac{\tau}{\beta} + B\frac{\tau}{\beta^3}\frac{c^2}{h^2}$$

(where B is a number depending on geometry), the calculation of β_{cr} follows von Kármán's transonic similarity if c/h varies according to Eq. (75.133).

It seems paradoxical that, for a fixed chord, the thinner profiles require, according to Eq. (75.133), greater tunnel height. This means, however, simply that the thinner profiles have higher critical Mach numbers, so that the choking problem becomes more severe.

Practically, choking difficulties are overcome by using test sections which have partly closed and partly "free" boundaries; the latter permit an effect which may properly be called unchoking. A successful design principle for these facilities is to apply the condition that wind-tunnel corrections should vanish for incompressible flow.

In conclusion, it can be said that the transonic limit is the most severe difficulty encountered in subsonic flow, both for experiment and theory, and that many results in Chap. 76, Transonic Flow, are significant in connection with the problems discussed here.

75.6. REFERENCES

Textbooks

[1] W. R. Sears (ed.): "General Theory of High Speed Aerodynamics," vol. 6 of "High Speed Aerodynamics and Jet Propulsion," Princeton Univ. Press, Princeton, N.J., 1954: chap. C: W. R. Sears, Small perturbation theory; chap. E: M. J. Lighthill, Higher approximations; chap. F: Y. H. Kuo, W. R. Sears, Plane subsonic and transonic potential flows.

[2] M. J. Lighthill: The hodograph transformation, chap. 7 in vol. 1 of L. Howarth (ed.), "Modern Developments in Fluid Dynamics: High Speed Flow," Oxford Univ. Press, New York, 1953.

[3] A. Robinson, J. A. Laurmann: "Wing Theory," Cambridge Univ. Press, New York, 1956.

Linearized Theory

[4] J. Ackeret: Über die Luftkräfte bei sehr grossen Geschwindigkeiten, insbesondere bei ebenen Strömungen, *Helv. Phys. Acta,* **1** (1928), 301–322.

[5] H. Glauert: The effect of compressibility on the lift of an aerofoil, *Proc. Roy. Soc. London,* **A,118** (1928), 113–119.

[6] B. Göthert: Ebene und räumliche Strömung bei hohen Unterschallgeschwindigkeiten, *Jahrb.* 1941 *deut. Luftfahrt-Forsch.*, pp. I, 156–I, 157, and *Ber. Lilienthal-Ges. Luftfahrt-Forsch.*, no. 127, pp. 97–101 (1940).

Second-order Theory

[7] H. Görtler: Gasströmungen mit Übergang von Unterschall zu Überschallgeschwindigkeiten, *Z. angew. Math. Mech.* **20** (1940), 254–262.

[8] W. Hantzsche, H. Wendt: Der Kompressibilitätseinfluss für dünne, wenig gekrümmte Profile bei Unterschallgeschwindigkeit, *Z. angew. Math. Mech.*, **22** (1942), 72–86.

[9] W. Hantsche: Die Prandtl-Glauertsche Näherung als Grundlage für ein Iterationsverfahren zur Berechnung kompressibler Unterschallströmungen, *Z. angew. Math. Mech.*, **23** (1943), 185–199.

[10] C. Kaplan: The flow of a compressible fluid past a curved surface, *NACA Rept.* 768 (1943).

[11] W. D. Hayes: Second-order pressure law for two-dimensional compressible flow, *J. Aeronaut. Sci.*, **22** (1955), 284–286. Originally published as *Aeronaut. Research Comm. Paper* 15 (1953), 722 (F.M. 1877).

[12] I. Imai: Extension of von Kármán's transonic similarity rule, *J. Phys. Soc. Japan,* **9** (1954), 103–108.

[13] M. D. Van Dyke: The second-order compressibility rule for airfoils, *J. Aeronaut. Sci.*, **21** (1954), 647–648.

[14] M. D. Van Dyke: Second-order subsonic airfoil theory including edge effects, *NACA Rept.* 1274 (1956).

[15] M. D. Van Dyke: The similarity rules for second-order subsonic and supersonic flow, *NACA Tech. Note* 3875 (1957).

[16] O. Janzen: Beitrag zu einer Theorie der stationären Strömung kompressibler Flüssigkeiten, *Z. Physik,* **14** (1913), 639–643.

[17] Lord Rayleigh: On the flow of compressible fluid past an obstacle, *Phil. Mag.*, **32** (1916), 1–6.

[18] I. Imai: Application of the M^2-expansion method to compressible flow past isolated and lattice airfoils, *J. Phys. Soc. Japan,* **12** (1957), 58–67.

[19] M. J. Lighthill: A technique for rendering approximate solutions to physical problems uniformly valid, *Phil. Mag.*, **40** (1949), 1179–1201.

[20] C. Schmieden and K. Kawalki: Beiträge zum Umströmungsproblem bei hohen

Geschwindigkeiten, *Ber.* S 13/1 *der Lilienthal-Ges.* (1942). (Also, *NACA Tech. Mem.* 1233.)

[21] M. D. Van Dyke: Second-order slender-body theory—Axisymmetric flow, *NACA Tech. Note* 4281 (1958).

Hodograph Method

[22] P. Molenbroeck: Über einige Bewegungen eines Gases bei Annahme eines Geschwindigkeitspotentials, *Arch. Math. Phys.*, **II,9** (1890), 157–195.

[23] S. A. Chaplygin: On gas jets, *Sci. Ann. Imp. Univ. Moscow, Phys.-Math. Div.*, **21** (1904), 1–121; also as *NACA Tech. Mem.* 1063 (1944).

[24] Th. von Kármán: Compressibility effects in aerodynamics, *J. Aeronaut. Sci.*, **8** (1941), 337–356.

[25] H.-S. Tsien: Two-dimensional subsonic flow of compressible fluids, *J. Aeronaut. Sci.*, **6** (1939), 399–407.

CHAPTER 76

TRANSONIC FLOW

BY

J. R. SPREITER, Ph.D., Los Altos, Calif.

76.1. INTRODUCTION

The term *transonic* refers to flows in which both subsonic and supersonic regions are present. It is also usually implied that the free-stream Mach number M_∞ is not too far removed from 1. The need for a special theory for transonic flow arises from the fact that most of the standard methods for the solution of the equations of compressible flow fail when applied to transonic flows. The difficulties are associated, fundamentally, with a lack of knowledge of the properties of partial differential equations that are both nonlinear and of mixed elliptic-hyperbolic type. Even such questions as existence and uniqueness of solutions remain largely unanswered at the present time [1],[2].

The small disturbance theory of transonic flow has grown out of efforts directed originally toward providing a useful first approximation for the aerodynamics of thin airfoils and slender bodies in inviscid flows with $M_\infty \approx 1$. It has evolved into a unified theory for subsonic, transonic, and supersonic flow, however, and is the simplest theory proposed to date that is capable of yielding reliable results throughout this range of speeds. The fundamental equations of this theory differ from those of linear theory by the addition of one nonlinear term in the differential equation for the perturbation potential and in the shock relation. The following discussion is based, almost entirely, on this theory.

More complete accounts of many of the following topics, as well as extensive lists of references, can be found in [3] through [8].

76.2. FUNDAMENTAL EQUATIONS

Let the free-stream velocity be U_∞, introduce a cartesian coordinate system with the x axis extending in the direction of the free stream, and denote the perturbation velocity components parallel to the x, y, and z axes by u, v, and w. They are the gradient of a perturbation potential ϕ that satisfies the following nonlinear partial differential equation,

$$(1 - M_\infty^2)\phi_{xx} + \phi_{yy} + \phi_{zz} = M_\infty^2 \frac{\gamma + 1}{U_\infty} \phi_x \phi_{xx} \equiv k\phi_x\phi_{xx} \tag{76.1}$$

where

$$\phi_x = u \qquad \phi_y = v \qquad \phi_z = w$$

and γ is the ratio of specific heats. The type is dependent on the sign of $(1 - M_\infty^2 - ku)$ as follows:

$$1 - M_\infty^2 - ku \begin{cases} > 0 & \text{elliptic} \quad \text{(subsonic)} \\ < 0 & \text{hyperbolic} \quad \text{(supersonic)} \end{cases} \tag{76.2}$$

The quantity $(1 - M_\infty^2 - ku)$ is equivalent, in transonic flow theory, to $(1 - M^2)$, where M is the local Mach number. The critical value for u associated with $M = 1$

is thus

$$\frac{u_{cr}}{U_\infty} = \frac{1 - M_\infty^2}{M_\infty^2(\gamma + 1)} \tag{76.3}$$

Shock waves are a prominent feature of most transonic flows, and additional relations are needed for the transition through the shock. They are that ϕ is continuous across a shock wave, and that u, v, and w are discontinuous in such a manner that the values immediately ahead and behind the shock wave are related according to

$$(1 - M_\infty^2)(u_a - u_b)^2 + (v_a - v_b)^2 + (w_a - w_b)^2 = k\,\frac{u_a + u_b}{2}\,(u_a - u_b)^2 \tag{76.4}$$

The shock wave is oriented so that the following relations hold among the direction cosines of the normal n to the downstream face of the shock:

$$\cos(n,x):\cos(n,y):\cos(n,z) = (u_b - u_a):(v_b - v_a):(w_b - w_a) \tag{76.5}$$

Conditions associated with a normal shock follow directly upon equating v_a, v_b, w_a, and w_b to zero, whence

$$u_b = 2\,\frac{M_\infty^2 - 1}{k} - u_a \qquad \cos(n,x) = 1 \qquad \cos(n,y) = \cos(n,z) = 0 \tag{76.6}$$

Conditions associated with an oblique shock in planar flow in, say, the x,z plane follow similarly, upon equating v_a and v_b to zero. The angle σ between the tangent to the shock wave and the x axis is related to u and w according to

$$\tan\sigma = -\frac{\cos(n,x)}{\cos(n,z)} = -\frac{u_b - u_a}{w_b - w_a} \tag{76.7}$$

Sonic velocity occurs behind the shock wave if

$$u_b = u_{cr} = -\frac{M_\infty^2 - 1}{k} \qquad |w_b - w_a| = \frac{(M_\infty^2 - 1 + ku_a)^{3/2}}{\sqrt{2}\,k} \tag{76.8}$$

The conditions for shock detachment are

$$|w_b - w_a|_{\max} = \frac{4}{3\sqrt{3}}\,\frac{(M_\infty^2 - 1 + ku_a)^{3/2}}{k} \qquad u_b = -\frac{4}{3}\,\frac{M_\infty^2 - 1}{k} - \frac{u_a}{3} \tag{76.9}$$

The boundary conditions require that ϕ is constant far ahead of the body, and that the flow must be tangential to the body surface. The latter condition, together with the expression relating the pressure coefficient C_p to the velocity, is approximated in the same way as in linear theory. For example, the appropriate relations for thin wings having ordinates given by $Z(x,y)$ are

$$\left(\frac{\partial\phi}{\partial z}\right)_{z=0} = U_\infty\,\frac{\partial Z}{\partial x} \qquad C_p = -2\,\frac{\phi_x}{U_\infty} \tag{76.10}$$

and those for axisymmetric flow past slender bodies of revolution having ordinates and cross-section area given by $R(x)$ and $S(x)$ are

$$(r\phi_r)_{r=0} = U_\infty R\,\frac{dR}{dx} = \frac{U_\infty}{2\pi}\,\frac{dS}{dx} \qquad C_p = -2\,\frac{\phi_x}{U_\infty} - \frac{\phi_r^2}{U_\infty^2} \tag{76.11}$$

where $r = \sqrt{y^2 + z^2}$. It is also presumed necessary to prescribe that the Kutta condition applies whenever the component of the velocity normal to a sharp trailing edge is subsonic, and that the direct influence of a disturbance in the supersonic region proceeds only in the downstream direction.

Many investigations of transonic flows have been based on equations in which the coefficient $k = M_\infty{}^2(\gamma + 1)/U_\infty$ is replaced by $(\gamma + 1)U_\infty$. The use of the latter expression is not recommended because considerable loss of accuracy results at M_∞ removed from 1 [9]. Results derived from consideration of this or any other expression for k can be readily converted so as to be consistent with the present formulation of transonic flow theory, by expressing the result in terms of k, and then substituting $M_\infty{}^2(\gamma + 1)/U_\infty$ for k.

76.3. SIMILARITY RULES, WINGS

The equations of transonic flow theory contain similarity rules that relate the aerodynamic properties of families of wings of affinely related geometry [9]. The ordinates of all members of such a family are given by

$$\frac{Z}{c} = \tau\left[\pm f\left(\frac{x}{c}, \frac{y}{b}\right) - \frac{\alpha}{\tau}\frac{x}{c}\right] \tag{76.12}$$

where b and c are the span and chord, which are arbitrary; τ and α are the thickness ratio and angle of attack, which must be small but are otherwise arbitrary; and $f(x/c,y/b)$ is the thickness distribution function, which must be the same for all members of the family.

The similarity rule states that the pressures on the wings of this family are related in such a way that the following functional dependence holds:

$$\bar{C}_p = \frac{(U_\infty k)^{1/3}}{\tau^{2/3}} C_p = \mathcal{P}\left(\xi_\infty, \bar{A}, \bar{\alpha}; \frac{x}{c}, \frac{y}{b}\right) \tag{76.13}$$

where ξ_∞, $\bar{A}$, and $\bar{\alpha}$ are similarity parameters defined by

$$\xi_\infty = \frac{M_\infty{}^2 - 1}{(U_\infty k\tau)^{2/3}} \qquad \bar{A} = (U_\infty k\tau)^{1/3} Æ \qquad \bar{\alpha} = \frac{\alpha}{\tau} \tag{76.14}$$

where Æ refers to the aspect ratio. The corresponding relations for the lift and drag coefficients C_L and C_D are

$$\left(\frac{\bar{C}_L}{\alpha}\right) = (U_\infty k\tau)^{1/3}\left(\frac{C_L}{\alpha}\right) = \mathcal{L}(\xi_\infty, \bar{A}, \bar{\alpha}) \qquad \bar{C}_D = \frac{(U_\infty k)^{1/3}}{\tau^{5/3}} C_D = \mathcal{D}(\xi_\infty, \bar{A}, \bar{\alpha}) \tag{76.15}$$

It should be pointed out that the similarity parameters can be combined in an arbitrary manner, and that it is sometimes convenient to replace $\bar{A}$ by $\sqrt{|\xi_\infty|}$ Æ or $\sqrt{|1 - M_\infty{}^2|}$ Æ.

Simplifications occur if attention is confined to cases such as $M_\infty = 1$ ($\xi_\infty = 0$), nonlifting wings ($\bar{\alpha} = 0$), two-dimensional flows ($\bar{A} = \infty$), or low-aspect-ratio wings ($\bar{A} \to 0$). Such considerations lead, for instance, to the conclusion that the pressure distribution and drag of nonlifting airfoil sections ($\bar{\alpha} = 0$, $\bar{A} = \infty$) at $M_\infty = 1$ are proportional to $\tau^{2/3}$ and $\tau^{5/3}$, respectively. It can be deduced further that the section lift-curve slope $dc_l/d\alpha$ at $\alpha = 0$ at $M_\infty = 1$ is inversely proportional to $\tau^{1/3}$. If τ vanishes and the airfoil is merely an inclined flat plate ($\bar{\alpha} = \infty$), the lift is proportional to $\alpha^{2/3}$.

76.4. SIMILARITY RULES, BODIES OF REVOLUTION

The similarity rules for bodies of revolution relate the areodynamic properties of families of bodies of affinely related geometry in axisymmetric flow [10]. The ordinates of all members of such a family of bodies are given by

$$\frac{R}{l} = \tau f\left(\frac{x}{l}\right) \tag{76.16}$$

where l is the length, which is arbitrary; τ is the thickness ratio, which must be small but is otherwise arbitrary; and $f(x/l)$ is the thickness distribution function, which must be the same for all members of the family.

If the subscripts 1 and 2 refer to different members of a given family, the similarity rule states that the values for C_p at corresponding points defined by

$$\frac{x_2}{l_2} = \frac{x_1}{l_1} \qquad \sqrt{U_{\infty 2}k_2}\,\tau_2\,\frac{r_2}{l_2} = \sqrt{U_{\infty 1}k_1}\,\tau_1\,\frac{r_1}{l_1} \tag{76.17}$$

are related according to

$$\frac{C_{p2}}{\tau_2{}^2} = \frac{C_{p1}}{\tau_1{}^2} \tag{76.18}$$

provided

$$\xi_{\infty 2} \equiv \frac{M_{\infty 2}{}^2 - 1}{U_{\infty 2}k_2\tau_2{}^2} = \frac{M_{\infty 1}{}^2 - 1}{U_{\infty 1}k_1\tau_1{}^2} \equiv \xi_{\infty 1} \tag{76.19}$$

This rule cannot be used to relate C_p on the surfaces of bodies of different τ because the ordinates of surface points are not related to one another in conformity with Eq. (76.17).

An approximate similarity rule for the surface pressures can be established if it can be assumed that $r\phi_r$ is independent of r, not only between the axis and the surface of the body, but also beyond the surface to a distance r_2 given by

$$\frac{r_2}{R_2} = \sqrt{\frac{U_{\infty 1}k_1}{U_{\infty 2}k_2}\left(\frac{\tau_1}{\tau_2}\right)^2} \tag{76.20}$$

This rule states that the surface pressures on an affinely related family of bodies are related in such a way that

$$C_p + \frac{1}{\pi}\frac{d^2S}{dx^2}\ln\left(\tau^2\sqrt{U_\infty k}\right) = \tau^2\mathcal{P}\left(\xi_\infty; \frac{x}{l}\right) \tag{76.21}$$

The corresponding relation for drag D is

$$\frac{D}{\frac{1}{2}\rho_\infty U_\infty{}^2 l^2} + \frac{1}{2\pi l^2}\left[\frac{dS}{dx}(l)\right]^2 \ln\left(\tau^2\sqrt{U_\infty k}\right) = \tau^4\mathcal{D}(\xi_\infty) \tag{76.22}$$

where $dS/dx(l)$ represents the value for dS/dx at the base of the body ($x = l$). The latter relation simplifies if the body is either cylindrical or pointed at the base so that $dS/dx(l) = 0$.

It is pointed out in [11] that the similarity rule for axisymmetric flow is useful for Mach numbers in the neighborhood of unity, and also for Mach numbers that are sufficiently small or large that the flow is either purely subsonic or supersonic, but not for Mach numbers near the bounds of the transonic range. This is evident from the fact that the expression for ξ_∞ given by Eq. (76.19) indicates in certain instances that a transonic flow is to be related to a subsonic flow, or to a supersonic flow; whereas it is clear from elementary considerations of the differences between such flows, particularly with regard to the properties of the shock system and the associated drag, that the results indicated by the simple relations given by the similarity rules are unacceptable physically. It is conjectured that the source of this difficulty is associated with an insufficient attention, in the derivation of the similarity rule for axisymmetric flow, to the conditions of dependence and influence in regions of supersonic flow.

76.5. MACH NUMBER FREEZE AT $M_\infty = 1$

An important property of considerable generality is that the local Mach number M at an arbitrary point in the vicinity of a wing or body is independent of M_∞ for small changes in M_∞ near $M_\infty = 1$; thus [4]

$$\left(\frac{dM}{dM_\infty}\right)_{M_\infty=1} = 0 \tag{76.23}$$

The conditions under which this result holds have not been established in all entirety, but it has been shown that the leading term in the expansion of the deviation of M from its value for $M_\infty = 1$ is proportional to $(M_\infty - 1)^3$ for planar flows and $(M_\infty - 1)^{5/3}$ for axisymmetric flows [3].

Combination of Eq. (76.23) with the exact isentropic relation for C_p in terms of M and M_∞ yields

$$\left(\frac{dC_p}{dM_\infty}\right)_{M_\infty=1} = \frac{4}{\gamma+1} - \frac{2}{\gamma+1}(C_p)_{M_\infty=1} \tag{76.24}$$

The approximate relation in transonic flow theory that corresponds to (76.23) is

$$\left(\frac{d\xi}{d\xi_\infty}\right)_{\xi_\infty=0} = 0 \qquad \xi = \frac{M_\infty{}^2 - 1 + ku}{(U_\infty k\tau)^{2/3}} \tag{76.25}$$

Combination with Eqs. (76.10) and (76.13) yields [9]

$$\left(\frac{dC_p}{d\xi_\infty}\right)_{\xi_\infty=0} = 2$$

or

$$\left(\frac{dC_p}{dM_\infty}\right)_{M_\infty=1} = \frac{4}{\gamma+1} - \tfrac{2}{3}(C_p)_{M_\infty=1} \tag{76.26}$$

76.6. TRANSONIC EQUIVALENCE RULE

The transonic equivalence rule relates the flow around a slender body of arbitrary cross section to that around an *equivalent* nonlifting body of revolution having the same longitudinal distribution of cross-section area $S(x)$ [12]. This rule is closely related to one of the simplest results of transonic slender-body theory, namely [13], that the expression for ϕ in the vicinity of a slender body is approximately of the form

$$\phi = \phi_2(x;y,z) + g(x) \tag{76.27}$$

where ϕ_2 is the solution of

$$\phi_{yy} + \phi_{zz} = 0 \tag{76.28}$$

for the given boundary conditions in the y,z plane at each x station, and $g(x)$ is an additional contribution dependent upon M_∞ and $S(x)$. Thus, for a nonlifting body of revolution,

$$\phi_2 = \frac{U_\infty}{2\pi}\frac{dS}{dx}\ln r \tag{76.29}$$

The corresponding expression for a thin nonlifting wing having a planform given by $y = \pm s(x)$ is

$$\phi_2 + \frac{U_\infty}{\pi}\int_{-s(x)}^{+s(x)} \frac{\partial[\Delta Z(x,\eta)]}{\partial x} \ln[(y-\eta)^2 + z^2]\,d\eta \tag{76.30}$$

where $\Delta Z(x,y)$ represents the difference between the ordinates of the upper and lower surfaces. Other cases may be more complicated, but many methods are available for the determination of ϕ_2, since the problem is mathematically equivalent to a classical problem of potential theory. The equivalence rule follows immediately if we write Eq. (76.27) once for a body of arbitrary cross section (subscript W), once for the equivalent body (subscript B), and subtract one from the other. Thus

$$\phi_W = \phi_{2W} - \phi_{2B} + \phi_B \tag{76.31}$$

The order of the error in Eq. (76.31) has been established for the case where the quantities with subscript W refer to a thin wing of aspect ratio $\mathit{A\!R}$ and thickness ratio τ [13]. Thus, the magnitude of the quantities $\phi_W/U_\infty c$ retained in Eq. (76.31) is $O(\mathit{A\!R}\tau \ln \mathit{A\!R})$, whereas that of the quantities discarded in the derivation for $M_\infty = 1$ is $O(\mathit{A\!R}^4\tau^2 \ln \mathit{A\!R})$. Since the magnitude of the quantities discarded in the derivation of the same result in linearized theory is $O(\mathit{A\!R}^3\tau \ln \mathit{A\!R})$, it follows that Eq. (76.31) can be applied to wings of greater aspect ratio at $M_\infty = 1$ than at other Mach numbers.

Several important relations for the pressures and forces on slender bodies follow directly from the equivalence rule [13]. A key relation in the derivation of many of these is the following relation between $g'(x)$ and C_{pB} on the surface of a nonlifting body of revolution:

$$g'(x) = -\frac{U_\infty}{2}\left[C_{pB} + \frac{S''(x)}{2\pi}\ln\frac{S}{\pi} + \frac{S'^2(x)}{4\pi S}\right] \tag{76.32}$$

where the prime denotes differentiation with respect to x. This relation, together with Eq. (76.27) and the appropriate expression for C_{pW}, permits the determination of the pressure distribution on a slender body of arbitrary cross section, provided the pressure distribution on the equivalent body is known. If, for instance, the arbitrary body is a thin finite cone of elliptic cross section, the following relation is obtained between the values for C_{pW} on the elliptic cone and C_{pB} on the equivalent circular cone

$$C_{pW} = C_{pB} - \frac{\mathit{A\!R}\tau}{8}\left(1 + \ln\frac{\mathit{A\!R}}{8\tau}\right) \mp \frac{\alpha\,\mathit{A\!R}}{2\sqrt{1 - y^2/s^2}} \tag{76.33}$$

where the minus sign is to be used for the upper surface and the plus sign for the lower surface.

The lift and other lateral forces on slender wings and bodies arise entirely from ϕ_2. They are thus independent of M_∞, and are the same as indicated by linearized slender-body theory for subsonic or supersonic flow.

Equation (76.31) leads to the following relationship for drag [13]:

$$D_W = D_B - \frac{\rho_\infty}{2}\left(\int_{C_W}\phi_{2W}\frac{\partial\phi_{2W}}{\partial n}\,d\sigma_C - \int_{C_B}\phi_{2B}\frac{\partial\phi_{2B}}{\partial n}\,d\sigma_C\right) \tag{76.34}$$

where each of the integrals is a line integral along a curve C, situated in a plane perpendicular to the x axis, that goes around the base of the body and also encloses any vortex wake that may be present. The difference $(D_W - D_B)$ is thus independent of M_∞. If the arbitrary body is inclined at an angle α, the first integral provides a contribution to the drag that is proportional to α^2. This quantity is exactly the vortex drag. If attention is confined to nonlifting cases, there exist several classes of shapes for which the contributions of the two integrals cancel and $D_W = D_B$. One such class includes shapes that taper to a point at the rear, since then both integrals vanish. Another includes shapes that are cylindrical at the base, since then $\partial\phi_{2W}/\partial n = \partial\phi_{2B}/\partial n = 0$. Still another includes bodies that have the same shape and surface slopes at the base, since then both integrals are carried out over the same curve C, along which $\phi_{2W} = \phi_{2B}$ and $\partial\phi_{2W}/\partial n = \partial\phi_{2B}/\partial n$, and the integrals again cancel. These and many other cases constitute the class of shapes for which the transonic area rule applies. This rule is generally quoted as stating that "near the speed of sound, the zero-lift drag rise of thin low-aspect-ratio wing-body combinations is primarily dependent on the axial distribution of cross-sectional area normal to the airstream" [14]. It is important to recognize, however, that D_W is not always equal to D_B. A simple example is furnished by a cone-cylinder of elliptic cross section for which

$$D_W = D_B - \frac{\pi}{16}\frac{\rho_\infty U_\infty{}^2}{2}\frac{a^2b^2}{l^2}\ln\left[\frac{a}{4b}\left(1+\frac{b}{a}\right)^2\right] \tag{76.35}$$

where l refers to the length of the cone, and a and b refer to the major and minor axes of the elliptic cross section at the base of the cone.

76.7. SELF-SIMILAR SOLUTIONS FOR $M_\infty = 1$

Important properties of flows with $M_\infty = 1$ are revealed from consideration of exact solutions of the self-similar type, in which the variation of ϕ with x is similar at all z. Since it is convenient to combine the discussion of planar and axisymmetric flows, consider Eq. (76.1) rewritten as follows,

$$\phi_{zz} + \frac{\sigma}{z}\phi_z = k\phi_x\phi_{xx} \tag{76.36}$$

where $\sigma = 0$ for planar, and $\sigma = 1$ and $z = r$ for axisymmetric flows. If solutions are sought having the form

$$\phi = \frac{U_\infty}{z^m}f(\zeta) \qquad (kU_\infty)^{1/3}\zeta = \frac{x}{z^n} \tag{76.37}$$

the condition $m + 3n = 2$ must apply if the resulting differential equation is to involve only ζ and $f(\zeta)$. The perturbation velocity components then become

$$\frac{u}{U_\infty} = \frac{\zeta}{xz^{2-3n}}f'(\zeta) \qquad \frac{w}{U_\infty} = -\frac{2-3n}{z^{3(1-n)}}\left[f(\zeta) + \frac{n}{2-3n}\zeta f'(\zeta)\right] \tag{76.38}$$

The ordinary differential equation for $f(\zeta)$ is

$$f'f'' = [3(1-n)-\sigma](2-3n)f + [5(1-n)-\sigma]n\zeta f' + n^2\zeta^2 f'' \tag{76.39}$$

The asymptotic behavior of the flow at great distances from a body that is pointed at the nose can be calculated by consideration of the above equations with $n = 4/5$ for planar flows and $n = 4/7$ for axisymmetric flows [3]. The determination of u and w requires integration of Eq. (76.39), but knowledge of the value for n is sufficient to indicate that, as one proceeds away from the origin along lines of ζ = const, u and w decrease as $z^{-2/5}$ and $z^{-3/5}$ for planar flows, and as $z^{-6/7}$ and $z^{-9/7}$ for axisymmetric flows. The results also show that a shock wave must occur, and that it is situated along a line ζ = const at great distances from the body.

The use of other values for n leads to solutions that can be associated with flows about bodies that extend to infinity or to flows in nozzles [15]. A relatively simple case in which a first integral of Eq. (76.39) can be written immediately follows if n is equated to either $1/2$ or $(3-\sigma)/4$. The first integral is

$$f'^2 = 2n^2\frac{d}{d\zeta}(\zeta^2 f) + 2n(5-9n-\sigma)\zeta f + \text{const} \tag{76.40}$$

If the flow is undisturbed for $x < 0$, the constant of integration is zero, and the resulting equation is homogeneous and can be integrated. Two families of exact solutions appear. The first represents purely supersonic flows. It is, in the planar case, the transonic equivalent of the Prandtl-Meyer expansion around a convex corner of a flow that is initially parallel to the x axis, and leads to the following expression for C_p:

$$C_p = \frac{2}{M_\infty^2(\gamma+1)}\left\{M_\infty^2 - 1 - \left[(M_\infty^2-1)^{3/2} - \tfrac{3}{2}M_\infty^2(\gamma+1)\frac{dZ}{dx}\right]^{2/3}\right\} \qquad \frac{dZ}{dx} < 0 \tag{76.41}$$

where M_∞ is the Mach number upstream of the corner and dZ/dx is the slope of the surface downstream of the corner. The companion solution in axisymmetric flow corresponds to a conical flow which coalesces onto the axis. The second family of solutions represents subsonic flows. It is possible, in either the planar or the axisymmetric case, to join the two solutions. The transition involves a single shock of parabolic shape. These solutions simulate the flow over the rear half of an infinite body.

76.8. HODOGRAPH METHOD

The initial step in many investigations of planar flows is to linearize the differential equation by use of the hodograph transformation in which the dependent and independent variables are interchanged [2],[3],[6]. Either of the two classical transformations of compressible flow theory, the Molenbroek transformation or the Legendre transformation, may be used. In either case, linearization is achieved with no loss of accuracy.

A simple way to transform Eq. (76.1) in two dimensions into hodograph variables is to rewrite it as the following pair of first-order partial differential equations,

$$(1 - M_\infty^2)\frac{\partial u}{\partial x} + \frac{\partial w}{\partial z} = ku\frac{\partial u}{\partial x} \qquad \frac{\partial u}{\partial z} = \frac{\partial w}{\partial x} \tag{76.42}$$

and to consider the spatial coordinates x and z as functions of the velocity components so that $x = x(u,w)$ and $z = z(u,w)$. The equations describing the transformation process can be derived from the fundamental relations for du and dw,

$$du = u_x\,dx + u_z\,dz \qquad dw = w_x\,dx + w_z\,dz \tag{76.43}$$

by solving for dx and dz,

$$dx = (w_z\,du - u_z\,dw)/J \qquad dz = (-w_x\,du + u_x\,dw)/J \tag{76.44}$$

where

$$J = u_xw_z - u_zw_x = u_x^2(M_\infty^2 - 1 + ku) - u_z^2 \tag{76.45}$$

is the Jacobian of the transformation; and comparing with the fundamental relations for dx and dz,

$$dx = x_u\,du + x_w\,dw \qquad dz = z_u\,du + z_w\,dw \tag{76.46}$$

Thus,

$$\begin{aligned} x_u &= w_z/J & x_w &= -u_z/J \\ z_u &= -w_x/J & z_w &= u_x/J \end{aligned} \tag{76.47}$$

Substitution of these relations in each term of Eq. (76.42) leads directly to the transonic hodograph equations, provided the Jacobian is finite and different from zero,

$$(1 - M_\infty^2)\frac{\partial z}{\partial w} + \frac{\partial x}{\partial u} = ku\frac{\partial z}{\partial w} \qquad \frac{\partial x}{\partial w} = \frac{\partial z}{\partial u} \tag{76.48}$$

or, upon elimination of x, to

$$(1 - M_\infty^2)\frac{\partial^2 z}{\partial w^2} + \frac{\partial^2 z}{\partial u^2} = ku\frac{\partial^2 z}{\partial w^2} \qquad \frac{\partial x}{\partial w} = \frac{\partial z}{\partial u} \tag{76.49}$$

The Jacobian cannot vanish in regions of subsonic flow, but can and frequently does vanish in regions of supersonic flow. As a result, application of the hodograph method to transonic flows is usually confined to the subsonic region and a limited part of the supersonic region. The remainder of the solution is then calculated by some other method, such as characteristics, which is more appropriate for supersonic flow.

Introduction of the new variables

$$\eta = (M_\infty{}^2 - 1 + ku)/(U_\infty k)^{2/3} \qquad \theta = w/U_\infty \tag{76.50}$$

leads to Tricomi's equation,

$$\frac{\partial^2 z}{\partial \eta^2} = \eta \frac{\partial^2 z}{\partial \theta^2} \qquad \frac{\partial x}{\partial \theta} = (U_\infty k)^{1/3} \frac{\partial z}{\partial \eta} \tag{76.51}$$

The pair of relations given in Eq. (76.49) can be expressed alternatively as a single second-order differential equation by the introduction of a potential Φ such that

$$x = \Phi_u \qquad z = \Phi_w \tag{76.52}$$

whence

$$(1 - M_\infty{}^2)\Phi_{ww} + \Phi_{uu} = ku\Phi_{ww} \tag{76.53}$$

The potential Φ is known as the Legendre potential and is related to the perturbation potential ϕ according to

$$\Phi = xu + zw - \phi \tag{76.54}$$

Introduction of the variables defined in Eq. (76.50) leads again to Tricomi's equation,

$$\Phi_{\eta\eta} = \eta\Phi_{\theta\theta} \tag{76.55}$$

Introduction of the stream function ψ leads similarly to

$$\psi_{\eta\eta} = \eta\psi_{\theta\theta} \tag{76.56}$$

Simple solutions of Tricomi's equation can be found in a variety of ways [3]. Application of the usual procedures of separation of variables leads, for instance, to solutions of the form

$$\Phi = e^{\pm i\lambda\theta} \sqrt{\eta}\, Z_{1/3}(\tfrac{2}{3}\lambda\eta^{3/2}) \tag{76.57}$$

where $Z_{1/3}$ is any linear combination of Bessel functions of order $\frac{1}{3}$, and λ may be real or imaginary. There are also self-similar solutions of the form

$$\Phi = |\eta|^n f_n(\zeta) \qquad \zeta = \frac{9}{4}\frac{\theta^2}{\eta^3} \tag{76.58}$$

where n is an arbitrary number, and the function $f_n(\zeta)$ satisfies the hypergeometric differential equation,

$$\zeta(\zeta - 1)\frac{d^2 f_n}{d\zeta^2} + [(\tfrac{4}{3} - \tfrac{2}{3}n)\zeta - \tfrac{1}{2}]\frac{df_n}{d\zeta} + \frac{n(n-1)}{9} f_n = 0 \tag{76.59}$$

The solution for $n = -\frac{5}{2}$ has particular importance in flows about nonlifting airfoils at $M_\infty = 1$, because it represents the singularity in the hodograph plane corresponding to the velocity at infinity. A solution for flow with $M_\infty = 1$ past the forepart of a particular airfoil is determined in [3] by superposition of solutions for $n = -\frac{5}{2}$ and $n = \frac{1}{2}$.

Even with knowledge of simple solutions such as given above and the availability of the principle of superposition, the solution of boundary-value problems associated with transonic flow around airfoils remains a complex matter. The difficulties are diminished considerably when the airfoil is described by straight lines, and the hodograph method has been applied with considerable success in the study of transonic flow around wedge and flat-plate airfoils [3]. Even greater simplifications occur in some applications, if the flow is considered to separate from the airfoil surface and follow a free-stream line [7],[16],[17]. Both analytical and numerical (relaxation) methods have been employed in these investigations. Extension of these methods to permit calculation of transonic flows around airfoils with curved boundaries or around three-dimensional bodies appears, however, to be a very difficult task.

76.9. METHOD OF SUCCESSIVE APPROXIMATION

Approximate solutions of the equations of transonic flow theory can be obtained by application of the method of successive approximation, in which the solution is expressed in a power series in terms of the thickness ratio τ [7]. In this method, the first approximation $\phi^{(1)}$ is obtained by solving the linear equation derived from equating the right-hand sides of Eqs. (76.1) and (76.4) to zero. The result is precisely that of linear theory, to which the solutions of transonic flow theory converge as M_∞ departs far from unity in either direction. The second approximation is determined by solving the equation, again linear, obtained by using the results of the first approximation to evaluate the right-hand sides of Eqs. (76.1) and (76.4); that is, $k\phi_x\phi_{xx}$ is replaced by $k\phi_x{}^{(1)}\phi_{xx}{}^{(1)}$, etc. Higher approximations follow in a similar manner, but the difficulties of integration are considerable. This method is very useful for obtaining results of high accuracy for M_∞ well removed from 1. It is now generally believed, however, that the series expression for the result fails to converge if the flow is transonic.

76.10. INTEGRAL EQUATION METHOD

The basic equations of transonic flow theory can be recast into the form of a nonlinear integral equation by application of Green's theorem to appropriate regions surrounding the body, its wake, and the associated shock waves. There are many forms of Green's theorem, and it is possible to derive many different integral equations in this way. Various of these have particular merit in the analysis of certain facets of transonic flow theory.

In actual application, this method is closely related to the method of successive approximation, and can perhaps be best described by consideration of the following integral equation [5], valid for two-dimensional shock-free flow around an arbitrary nonlifting airfoil with $M_\infty < 1$:

$$\bar{u} = \bar{u}_L + \frac{1}{2\pi} \iint_{-\infty}^{+\infty} \frac{\partial}{\partial \xi}\left(\frac{\bar{u}^2}{2}\right) \frac{\bar{x} - \xi}{(\bar{x} - \xi)^2 + (\bar{z} - \zeta)^2}\, d\xi\, d\zeta \tag{76.60}$$

where

$$\bar{u} = \frac{u}{u_{cr}} = \frac{ku}{1 - M_\infty{}^2} \qquad \bar{x} = x \qquad \bar{z} = \sqrt{1 - M_\infty{}^2}\, z$$

and $\bar{u}_L$ represents the expression for $\bar{u}$ given by linear theory. The solution of this equation can be sought by equating $\bar{u}$ to $\bar{u}_L$ as a first approximation, then substituting $\bar{u}_L$ for $\bar{u}$ in the integral and integrating for the second approximation, etc. The result is precisely that of the method of successive approximation described above, and the same difficulties are experienced with the occurrence of transonic flow.

The integral equation that corresponds to Eq. (76.60) for flows containing shock waves includes additional line integrals extending along the shock waves. Partial integration with respect to ξ of the double integral yields additional line integrals, canceling those previously present, and the following integral equation results [13]:

$$\bar{u} - \frac{\bar{u}^2}{2} = \bar{u}_L - \frac{1}{2\pi} \iint_{-\infty}^{+\infty} \frac{\bar{u}^2}{2} \frac{(\bar{x} - \xi)^2 - (\bar{z} - \zeta)^2}{[(\bar{x} - \xi)^2 + (\bar{z} - \zeta)^2]^2}\, d\xi\, d\zeta \tag{76.61}$$

Discussion of the properties of Eq. (76.61) is facilitated by introduction of the abbreviation $I/2$ for the integral, and rewriting Eq. (76.61) in the form

$$\bar{u} = 1 \pm \sqrt{I - L}$$

where $L = 2\bar{u}_L - 1$ [18]. If follows immediately that $I \geq L$. The choice of plus or minus sign determines whether the flow is subsonic or supersonic. A change of sign at a point where $I = L$ corresponds to a smooth transition through sonic velocity. A change in sign at a point where $I > L$ results in a discontinuous jump in velocity and corresponds physically to a shock wave. If I and L are continuous at such a point, the discontinuity corresponds to a normal shock wave. Such discontinuities are permissible when they proceed from greater to lesser velocities (compression shock), but are inadmissible when they proceed in the opposite direction.

It is necessary, at the present time, to introduce approximations to simplify Eq. (76.61) before results can be calculated for specific applications. One simplification that has proved useful for the calculation of $\bar{u}$ at the surface of a number of convex airfoils is to approximate $\bar{u}(\xi,\zeta)$ in the integral by

$$\bar{u}(\xi,\zeta) = \frac{\bar{u}(\xi,0)}{[1 + (\zeta/b)]^2} \qquad b = -2\,\frac{(1 - M_\infty{}^2)^{3/2}}{kU_\infty}\,\frac{\bar{u}(\xi,0)}{d^2Z/d\bar{x}^2} \tag{76.62}$$

thereby permitting integration with respect to ζ. The single integral equation that results can be solved by a procedure involving numerical and iteration techniques [18].

Theoretical results for transonic flows with free-stream Mach numbers less than unity do not agree with experimental results as well as do those for other Mach numbers because the shock wave frequently induces the flow to separate over the rear of the airfoil.

The integral equations of the corresponding theory for three-dimensional flows or for flows with $M_\infty > 1$ can be established in a similar manner, but neither the exact nor the approximate solution of these equations has yet been accomplished.

76.11. METHOD OF LOCAL LINEARIZATION

A simple method for the approximate solution of the equations of transonic flow theory is based on the idea of linearizing the equations in a small region by replacing part of the nonlinear term by a constant λ, and then introducing different values for λ for different points in the field [11],[19]. Results obtained by such a procedure depend, of course, on the choice of λ, and must be assembled in order to determine the final results. This step is accomplished by putting the results in such a form that a first-order nonlinear ordinary differential equation is obtained after λ is replaced by the quantity it originally represented. In many cases, this equation is of sufficiently simple form so that it can be integrated analytically and the solution expressed in closed analytic form. In other cases, the integration must be performed numerically, but the equation is of such a form that standard methods can be applied. Although the mathematical basis of the method is not established in all details, the utility of the method has been amply demonstrated by numerous comparisons with results indicated by other theoretical methods or by experiment. A simple account of the method follows:

Two-dimensional Flows

Subsonic Flows. Consider, first, two-dimensional subsonic flow past a nonlifting airfoil, the upper surface of which has ordinates given by $Z(x)$. Introduce the symbol λ as an abbreviation for $(1 - M_\infty{}^2 - ku)$, and rewrite Eq. (76.1) in the form

$$\lambda\phi_{xx} + \phi_{zz} = 0 \qquad \lambda = 1 - M_\infty{}^2 - ku > 0 \tag{76.63}$$

It is now assumed that λ varies sufficiently slowly that it can be considered as a constant in the initial stages of the analysis. The solution u_E at the airfoil surface is

$$u_E = \frac{U_\infty}{\pi\sqrt{\lambda}} \int_0^c \frac{dZ/d\xi}{x - \xi}\, d\xi = \frac{u_i}{\sqrt{\lambda}} \tag{76.64}$$

where the subscript i refers to the values for $M_\infty = 0$. Differentiation and subsequent replacement of λ by $(1 - M_\infty{}^2 - ku)$, so that the local value for λ is used at each point, leads to the following nonlinear ordinary differential equation for u:

$$\frac{du}{dx} = \frac{1}{\sqrt{1 - M_\infty{}^2 - ku}} \frac{du_i}{dx} \tag{76.65}$$

This equation can be solved by separation of variables. The result, expressed in terms of C_p rather than u, is

$$C_p = -\frac{2}{M_\infty{}^2(\gamma + 1)} \{(1 - M_\infty{}^2) - [C + \tfrac{3}{4}M_\infty{}^2(\gamma + 1)C_{pi}]^{2/3}\} \tag{76.66}$$

where $C_{pi} = -2u_i/U_\infty$, and C is a constant of integration. If the flow is subsonic everywhere, C can be evaluated simply by use of the result suggested by Eq. (76.64) that $u = 0$ where $u_i = 0$. This procedure leads to the following relation between C_p and C_{pi}:

$$C_p = -\frac{2}{M_\infty{}^2(\gamma + 1)} \{(1 - M_\infty{}^2) - [(1 - M_\infty{}^2)^{3/2} + \tfrac{3}{4}M_\infty{}^2(\gamma + 1)C_{pi}]^{2/3}\} \tag{76.67}$$

This relation can also be derived by application of the WKB method [22] of approximation [7] to the equations of transonic flow theory, as expressed in hodograph variables [17].

Supersonic Flows. Application of the same procedures to supersonic flow, for which $\lambda < 0$, leads to the following results in place of Eqs. (76.66) and (76.67):

$$C_p = \frac{2}{M_\infty{}^2(\gamma + 1)} \left\{(M_\infty{}^2 - 1) - \left[C - \tfrac{3}{2}M_\infty{}^2(\gamma + 1)\frac{dZ}{dx}\right]^{2/3}\right\} \tag{76.68}$$

$$C_p = \frac{2}{M_\infty{}^2(\gamma + 1)} \left\{(M_\infty{}^2 - 1) - \left[(M_\infty{}^2 - 1)^{3/2} - \tfrac{3}{2}M_\infty{}^2(\gamma + 1)\frac{dZ}{dx}\right]^{2/3}\right\} \tag{76.69}$$

The latter relation is the exact equivalent, in transonic flow theory, to that given by simple wave theory.

Flows with $M_\infty = 1$. An essential feature of most flows with $M_\infty = 1$ is the transition from subsonic to supersonic flow somewhere along the chord. If the sonic point is situated at a point where the airfoil surface is smoothly curved, the transition is accomplished with a finite acceleration, and additional considerations are necessary because the foregoing results lead to infinite accelerations at the sonic point. The technique adopted is to apply the procedures described above to the relations that follow upon introduction of the symbol λ as an abbreviation for $k\, \partial u/\partial x$ in Eq. (76.1); thus

$$\phi_{zz} = \lambda\phi_x \qquad \lambda = k\frac{\partial u}{\partial x} > 0 \tag{76.70}$$

It is assumed, again, that λ varies sufficiently slowly that it can be considered as a constant in the initial stages of the analysis. The solution at the airfoil surface is

$$u_p = -\frac{U_\infty}{\sqrt{\pi\lambda}} \frac{d}{dx} \int_0^x \frac{dZ/d\xi}{\sqrt{x - \xi}}\, d\xi \tag{76.71}$$

If $k\,du/dx$ is now restored in place of λ, a nonlinear ordinary differential equation is obtained that can be solved by separation of variables. The result, expressed in terms of C_p is

$$C_p = -2\left[\frac{3}{\pi(\gamma+1)}\int_{x^*}^{x}\left(\frac{d}{dx_1}\int_0^{x_1}\frac{dZ/d\xi}{\sqrt{x_1-\xi}}\,d\xi\right)^2 dx_1\right]^{1/3} \tag{76.72}$$

where x^* is the value for x at which the local velocity is sonic.

Application to a single-wedge profile of thickness t and chord $c/2$, for which the sonic point is known, from a priori considerations, to be situated at the shoulder, yields the following expressions for the pressure distribution and drag:

$$\begin{aligned}\bar{C}_p &= \frac{(\gamma+1)^{1/3}}{\tau^{2/3}}\,C_p = -2\left(\frac{3}{\pi}\ln\frac{x}{c/2}\right)^{1/3}\\ \bar{c}_d &= \frac{\gamma+1}{\tau^{5/3}}\,c_d = 2\left(\frac{3}{\pi}\right)^{1/3}\Gamma\left(\tfrac{4}{3}\right) = 1.758\end{aligned} \tag{76.73}$$

where $\tau = t/c$.

Application to smoothly curved airfoils requires the determination of the location of the sonic point. This can be readily accomplished, however, since the condition that du/dx be finite at the sonic point requires that $u = 0$ where $u_p = 0$. Thus x^* is the value of x for which

$$\frac{d}{dx}\int_0^x \frac{dZ/d\xi}{\sqrt{x-\xi}}\,d\xi = 0 \tag{76.74}$$

Application of Eqs. (76.72) and (76.74) to a parabolic-arc airfoil of thickness ratio τ and chord c having ordinates given by

$$\frac{Z}{c} = 2\tau\left[\frac{x}{c} - \left(\frac{x}{c}\right)^2\right] \tag{76.75}$$

leads to the following expressions for $\bar{C}_p$ and $\bar{c}_d$:

$$\bar{C}_p = -2\left\{\frac{12}{\pi}\left[\ln\left(4\frac{x}{c}\right) - 8\frac{x}{c} + 8\left(\frac{x}{c}\right)^2 + \frac{3}{2}\right]\right\}^{1/3} \qquad \bar{c}_d = 4.77 \tag{76.76}$$

An important problem in many applications is that of designing an airfoil to have a given pressure distribution. The solution of this problem requires the determination of an expression for the ordinates Z in terms of C_p, and can be accomplished for flows with $M_\infty = 1$ by inversion of the relations given above for C_p in terms of Z. The result is

$$Z = \frac{1}{2}\sqrt{\frac{\gamma+1}{2\pi}}\int_0^x dx_1\int_0^{x_1} C_p\sqrt{\frac{-dC_p}{d\xi}}\,\frac{d\xi}{\sqrt{x_1-\xi}} \tag{76.77}$$

Situations are frequently encountered, particularly in the analysis of flows past an airfoil having a cusped trailing edge, in which du/dx is positive over the front of the airfoil and negative over the rear. The expressions derived in this section are applicable to the front of such an airfoil, but not to the rear. It is possible to calculate the pressures on the rear by use of Eq. (76.68), however, if the constant of integration is selected so that the values for C_p match at the point where the two results are joined together. The selection of a point for joining the solutions can be made in many cases requiring that dC_p/dx, as well as C_p, match.

Axisymmetric Flows

An analysis similar to that given for planar flows can likewise be given for axisymmetric flows. These procedures lead directly to the following nonlinear ordinary differential equation for u on the surface of a slender body of revolution,

$$\frac{d(u/U_\infty)}{dx} = \frac{S'''(x)}{4\pi} \ln (1 - M_\infty{}^2 - ku) + \frac{d}{dx}\left[\frac{S''(x)}{4\pi} \ln \frac{S}{4\pi x(l-x)} + \frac{1}{4\pi}\int_0^l \frac{S''(x) - S''(\xi)}{|x-\xi|}\,d\xi\right] \tag{76.78}$$

for subsonic flows, to

$$\frac{d(u/U_\infty)}{dx} = \frac{S'''(x)}{4\pi} \ln (M_\infty{}^2 - 1 + ku) + \frac{d}{dx}\left[\frac{S''(x)}{4\pi} \ln \frac{S}{4\pi x^2} + \frac{1}{2\pi}\int_0^x \frac{S''(x) - S''(\xi)}{x-\xi}\,d\xi\right] \tag{76.79}$$

for supersonic flows, and to

$$\frac{u}{U_\infty} = \frac{S''(x)}{4\pi} \ln \left\{\left[\frac{d}{dx}\left(\frac{u}{U_\infty}\right) - \frac{S'S''}{4\pi S}\right]\left(\frac{U_\infty k S e^C}{4\pi x}\right)\right\} + \frac{1}{4\pi}\int_0^x \frac{S''(x) - S''(\xi)}{x-\xi}\,d\xi \tag{76.80}$$

for flows with $M_\infty = 1$. The primes indicate differentiation with respect to x, the symbol C represents Euler's constant = 0.5772, and

$$\frac{d}{dx}\left(\frac{u}{U_\infty}\right) - \frac{S'S''}{4\pi S} = \frac{\partial u}{\partial x}$$

Application of Eq. (76.80) to flow with $M_\infty = 1$ past a finite cone of length $l/2$ and maximum diameter $\tau l/2$, for which the point of sonic velocity ($u = 0$) is known to be situated at the shoulder, leads to the following expression for the pressure distribution along the surface of the cone:

$$C_p = -2\tau^2 \ln \frac{\tau x}{l/2} + \tau^2 \ln \left\{\tau^2 + \frac{4\left[1 - \left(\frac{x}{l/2}\right)^2\right]}{e^C \tau^2(\gamma+1)}\right\} - \tau^2 \tag{76.81}$$

The corresponding expression for the drag is

$$D = \frac{\rho_\infty}{2} U_\infty{}^2 \pi \tau^4 \left(\frac{l}{2}\right)^2 \left[-1 + \ln \frac{4}{e^C \tau^4(\gamma+1)}\right] \tag{76.82}$$

Application of Eq. (76.78), (76.79), or (76.80) to the flow past smoothly curved bodies of revolution requires the evaluation of a constant of integration. A simple procedure that leads to satisfactory results in applications of Eq. (76.78) to purely subsonic flows or of Eq. (76.79) to purely supersonic flows is to specify that u be equal to the value indicated by the solution of the linearized partial differential equation at the point where $S''(x) = 0$. This procedure, which is analogous to that used in the analysis of planar flows, does not suffice in the case of Eq. (76.80). The reason is that this equation is singular at the point where $S''(x) = 0$, and infinitely many integral curves pass through this value for u at the point where $S''(x)$ vanishes. Of all these curves, only one is analytic (all derivatives finite) at this point, and selection of it suffices to determine a unique solution, which is in good agreement with experimental data. This procedure assures that the solution for u can be expanded in a Taylor series in the neighborhood of the point where $S''(x)$ vanishes. The remainder of the solution can be determined by application of standard numerical methods. In some cases with $M_\infty = 1$, such as flow past the front half of a parabolic-arc body, the values for u

indicated by Eq. (76.80) are so slightly affected by the value of $\partial u/\partial x$ that good results can be obtained by simply replacing $\partial u/\partial x$ by some reasonable constant [20].

Since the flow is subsonic at the tail of a body that closes to a point, it follows that the calculation of the pressure distribution on a complete body of revolution usually requires consideration of flows that accelerate over the forward portion of the body and decelerate over the rearward portion. Satisfactory results can be obtained for such cases by joining solutions in the manner described in the preceding discussion of planar flows. Results can also be calculated for the supersonic region by application of a characteristics method for axisymmetric flow, derived from the equations of transonic flow theory [21].

76.12. REFERENCES

[1] G. Guderley: On the presence of shocks in mixed subsonic-supersonic flow patterns, *Advances in Appl. Mechanics*, **3** (1953), 145–184.

[2] L. Bers: "Mathematical Aspects of Subsonic and Transonic Gas Dynamics," Wiley, New York, 1958.

[3] K. G. Guderley: "Theorie schallnaher Strömungen," Springer, Berlin, 1957.

[4] H. W. Liepmann, A. Roshko: "Elements of Gasdynamics," Wiley, New York, 1957.

[5] K. Oswatitsch: "Gas Dynamics," Academic Press, New York, 1956.

[6] W. R. Sears (ed.): "General Theory of High Speed Aerodynamics," vol. 6 of "High Speed Aerodynamics and Jet Propulsion," Princeton Univ. Press, Princeton, N.J., 1954.

[7] I. Imai: Approximation methods in compressible fluid dynamics, *Inst. Fluid Dynamics Appl. Math. Univ. Maryland Tech. Note* BN-95 (1957).

[8] J. R. Spreiter: Aerodynamics of wings and bodies at transonic speeds, *J. Aero Space Sci.*, **26** (1959), 465–486.

[9] J. R. Spreiter: On the application of transonic similarity rules to wings of finite span, *NACA Rept.* 1153 (1953).

[10] K. Oswatitsch, S. B. Berndt: Aerodynamic similarity at axisymmetric transonic flow around slender bodies, *Roy. Inst. Technol. Stockholm, Swed., Div. Aeronaut. KTH-Aero Tech. Note* 15 (1950).

[11] J. R. Spreiter, A. Y. Alksne: Slender body theory based on approximate solution of the transonic flow equation, *NASA Rept.* 2 (1959).

[12] K. Oswatitsch: The area rule, *Appl. Mechanics Revs.*, **10** (1957), 543–545.

[13] M. A. Heaslet, J. R. Spreiter: Three-dimensional transonic flow theory applied to slender wings and bodies, *NACA Rept.* 1319 (1957).

[14] R. T. Whitcomb: A study of the zero-lift drag-rise characteristics of wing-body combinations near the speed of sound, *NACA Rept.* 1273 (1956).

[15] M. A. Heaslet, F. B. Fuller: Particular solutions for flows at Mach number 1, *NACA Tech. Note* 3868 (1956).

[16] J. B. Helliwell, A. G. Mackie: Two-dimensional subsonic and sonic flow past thin bodies, *J. Fluid Mechanics*, **3** (1957), 93–109.

[17] K.-I. Kusukawa: On the two-dimensional compressible flow over a thin symmetrical obstacle with sharp shoulders placed in an unbounded fluid and in a choked wind tunnel, *J. Phys. Soc. Japan*, **12** (1957), 1031–1041.

[18] J. R. Spreiter, A. Y. Alksne, B. J. Hyett: Theoretical pressure distributions for several related nonlifting airfoils at high subsonic speeds, *NACA Tech. Note* 4148 (1958).

[19] J. R. Spreiter, A. Y. Alksne: Thin airfoil theory based on approximate solution of the transonic flow equation, *NACA Rept.* 1359 (1958).

[20] K. Oswatitsch, F. Keune: The flow around bodies of revolution at Mach number 1, *Proc. Conf. High-speed Aeronaut.*, Brooklyn, 1955, pp. 113–131.

[21] K. Oswatitsch: Die Berechnung wirbelfreier, achsensymmetrischer Überschallfelder, *Österr. Ing.-Arch.*, **10** (1956), 359–382.

[22] P. M. Morse, H. Feshbach: "Methods of Theoretical Physics," vol. 2, pp. 1092–1106, McGraw-Hill, New York, 1953.

CHAPTER 77

SUPERSONIC FLOW

BY

M. D. VAN DYKE, Ph.D., Stanford, Calif.

77.1. INTRODUCTION

This chapter is devoted to steady flow of a uniform supersonic stream of gas past a solid body. The flow field is assumed to be supersonic throughout. Otherwise, if local subsonic regions appear, the flow is *transonic* (Chap. 76). The gas is assumed to be thermally and calorically perfect; significant imperfections appear only in *hypersonic* flow (Chap. 78). Viscosity and heat conduction are ignored.

Characteristics

Supersonic flow is often simpler than subsonic because of the *rule of forbidden signals*. Disturbances propagate only within a downstream *region of influence*, which is the envelope of a nearly spherical sound wave that expands as it is swept downstream (Fig. 77.1). Conversely, conditions at a point depend only upon changes

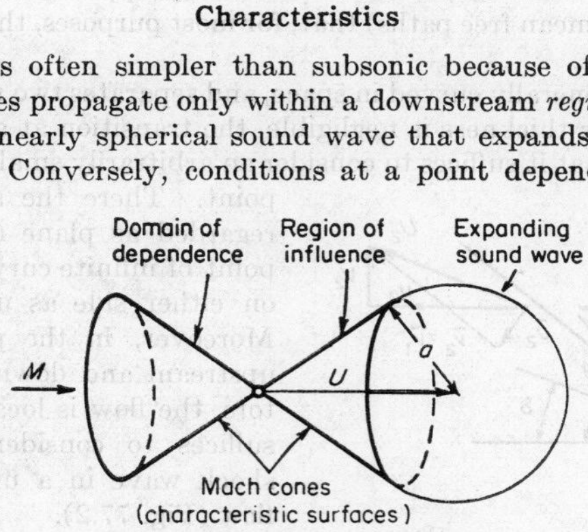

FIG. 77.1. Characteristic regions for point in supersonic flow.

inside an upstream *domain of dependence*. For infinitesimal disturbances, these two complementary regions are bounded by the downstream and upstream *Mach cones* (Mach lines in plane flow). Their surface makes the local *Mach angle*

$$\mu = \sin^{-1}\frac{U}{a} = \sin^{-1}\frac{1}{M}$$

with the streamline. In two dimensions, a point disturbance propagates only along the downstream Mach lines; in three dimensions, it affects the whole interior of the Mach cone.

Mathematically, the Mach cones are the *characteristic surfaces* of the differential equations of motion (p. 11–8). First and higher derivatives of the velocity, pressure, etc., can be discontinuous across characteristic surfaces (lines in two dimensions). This property is the basis of Massau's *numerical method of characteristics*, which permits a plane or axisymmetric supersonic flow field to be constructed step by step downstream [5].

Shock Waves and Drag

Supersonic flows almost always involve *shock waves,* across which the velocity jumps discontinuously and pressure, density, and temperature rise. Entropy also rises discontinuously, but is elsewhere constant along streamlines. Associated with the entropy increase at shock waves is *wave drag,* which has no counterpart in subsonic flow, and can be avoided only with exceptional nonlifting configurations (p. 77–11).

A curved shock wave in a uniform stream produces entropy gradients across streamlines, and vorticity, so that a velocity potential (which implies irrotational flow) does not exist. The total enthalpy is unaffected by a shock wave; a uniform stream remains *isoenergetic.*

77.2. OBLIQUE SHOCK WAVES

Nature of Shock Waves

A *bow shock wave* forms near the nose of any body in supersonic flight. The bow shock is *detached* and stands ahead of a blunt-nosed body (see Fig. 78.9), may be attached to the tip of a sharp one (p. 77–6), and forms behind and away from a cusped nose. A closed body generates at least one additional shock wave, at or near its tail.

The detailed structure of a shock wave involves consideration of viscosity, heat conduction, etc. [6]. However, shock waves of any appreciable strength are so thin (of the order of a few mean free paths) that, for most purposes, they can be regarded as having zero thickness.

A shock wave is generally curved in space, and separates two regions of nonuniform flow. However, if its thickness is negligible, the transition at any point takes place instantaneously, so that it suffices to consider an arbitrarily small neighborhood of the point. There the shock wave may be regarded as plane (except at a singular point of infinite curvature), and the flows on either side as uniform and parallel. Moreover, in the plane containing the upstream and downstream velocity vectors, the flow is locally plane. Hence, it suffices to consider a straight oblique shock wave in a uniform parallel-plane flow (Fig. 77.2).

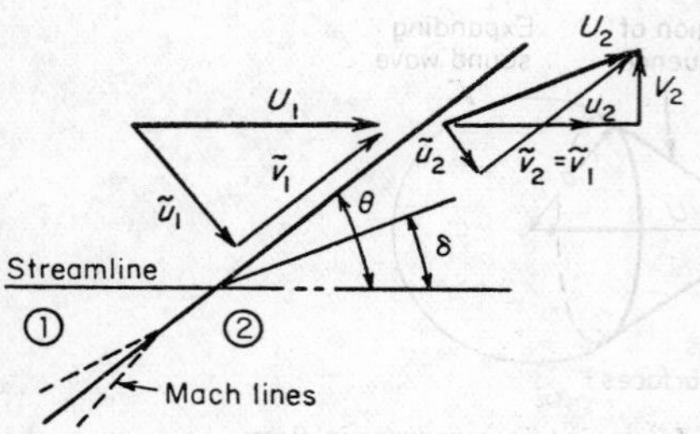

Fig. 77.2. Notations for oblique shock wave.

The velocity vector is turned toward the shock wave in passing through, because its normal component is reduced. The slope of the shock wave is slightly greater than that of the Mach waves upstream, less than that downstream. A weak shock wave bisects those Mach directions to second order.

Relations for Oblique Shock Waves

Conservation of mass and tangential momentum across an oblique shock wave show that the tangential velocity component $\tilde{v}$ is unchanged. To an observer moving along the shock wave with velocity $\tilde{v}$, it therefore appears to be a normal shock. Hence, the relations of Chap. 74 (except those involving stagnation pressure) apply to oblique shocks if the upstream and downstream Mach numbers are replaced by their normal components. This leads to the following useful relations:

Given M_1 and θ:

$$\frac{p_2}{p_1} = \frac{2\gamma M_1{}^2 \sin^2 \theta - (\gamma - 1)}{\gamma + 1} \qquad \frac{p_2}{\rho_1 U_1{}^2} = \frac{2\gamma M_1{}^2 \sin^2 \theta - (\gamma - 1)}{\gamma(\gamma + 1)M_1{}^2} \qquad (77.1a,b)$$

$$\frac{\rho_2}{\rho_1} = \frac{\tilde{u}_1}{\tilde{u}_2} = \frac{(\gamma + 1)M_1^2 \sin^2\theta}{(\gamma - 1)M_1^2 \sin^2\theta + 2} \tag{77.2}$$

$$\frac{T_2}{T_1} = \frac{a_2^2}{a_1^2} = \frac{[2\gamma M_1^2 \sin^2\theta - (\gamma - 1)][(\gamma - 1)M_1^2 \sin^2\theta + 2]}{(\gamma + 1)^2 M_1^2 \sin^2\theta} \tag{77.3}$$

$$M_2^2 = \frac{(\gamma + 1)^2 M_1^4 \sin^2\theta - 4(M_1^2 \sin^2\theta - 1)(\gamma M_1^2 \sin^2\theta + 2)}{[2\gamma M_1^2 \sin^2\theta - (\gamma - 1)][(\gamma - 1)M_1^2 \sin^2\theta + 2]} \tag{77.4}$$

$$\frac{\tilde{u}_2}{U_1} = \frac{(\gamma - 1)M_1^2 \sin^2\theta + 2}{(\gamma + 1)M_1^2 \sin\theta} \qquad \frac{\tilde{v}_1}{U_1} = \frac{\tilde{v}_2}{U_1} = \cos\theta \tag{77.5a,b}$$

$$\frac{u_2}{U_1} = 1 - \frac{2(M_1^2 \sin^2\theta - 1)}{(\gamma + 1)M_1^2} \qquad \frac{v_2}{U_1} = \frac{2(M_1^2 \sin^2\theta - 1)}{(\gamma + 1)M_1^2} \cot\theta \tag{77.6a,b}$$

$$\frac{v_2^2}{U_1^2} = 1 - 4\,\frac{(M_1^2 \sin^2\theta - 1)(\gamma M_1^2 \sin^2\theta + 1)}{(\gamma + 1)^2 M_1^4 \sin^2\theta} \tag{77.7}$$

$$\cot\delta = \left[\frac{(\gamma + 1)M_1^2}{2(M_1^2 \sin^2\theta - 1)} - 1\right]\tan\theta \qquad \tan\delta = \frac{2(M_1^2 \sin^2\theta - 1)\cot\theta}{2 + M_1^2(\gamma + 1 - 2\sin^2\theta)} \tag{77.8a,b}$$

$$\frac{p_2}{p_{t_1}} = \frac{2\gamma M_1^2 \sin^2\theta - (\gamma - 1)}{\gamma + 1}\left[\frac{2}{(\gamma - 1)M_1^2 + 2}\right]^{\gamma/\gamma-1} \tag{77.9}$$

$$\frac{p_2}{p_{t_2}} = \left\{2\,\frac{[2\gamma M_1^2 \sin^2\theta - (\gamma - 1)][(\gamma - 1)M_1^2 \sin^2\theta + 2]}{(\gamma + 1)^2 M_1^2 \sin^2\theta[(\gamma - 1)M_1^2 + 2]}\right\}^{\gamma/(\gamma-1)} \tag{77.10}$$

$$\frac{p_{t_2}}{p_{t_1}} = \frac{\rho_{t_2}}{\rho_{t_1}} = e^{-\Delta s/R} = \left[\frac{(\gamma + 1)M_1^2 \sin^2\theta}{(\gamma - 1)M_1^2 \sin^2\theta + 2}\right]^{\gamma/(\gamma-1)} \times \left[\frac{\gamma + 1}{2\gamma M_1^2 \sin^2\theta - (\gamma - 1)}\right]^{1/(\gamma-1)} \tag{77.11}$$

$$\frac{p_{t_2}}{p_1} = \left[\frac{\gamma + 1}{2\gamma M_1^2 \sin^2\theta - (\gamma - 1)}\right]^{1/(\gamma-1)} \times \left\{\frac{(\gamma + 1)M_1^2 \sin^2\theta[(\gamma - 1)M_1^2 + 2]}{2[(\gamma - 1)M_1^2 \sin^2\theta + 2]}\right\}^{\gamma/(\gamma-1)} \tag{77.12}$$

$$\frac{\Delta s}{c_v} = (\gamma - 1)\frac{\Delta s}{R} = -(\gamma - 1)\ln\frac{p_{t_2}}{p_{t_1}} = \ln\frac{2\gamma M_1^2 \sin^2\theta - (\gamma - 1)}{\gamma + 1} - \gamma\ln\frac{(\gamma + 1)M_1^2 \sin^2\theta}{(\gamma - 1)M_1^2 \sin^2\theta + 2} \tag{77.13}$$

$$\frac{p_2 - p_1}{\frac{1}{2}\rho_1 U_1^2} = \frac{4(M_1^2 \sin^2\theta - 1)}{(\gamma + 1)M_1^2} \tag{77.14}$$

For weak shock waves ($M_1 \sin\theta$ slightly greater than 1),

$$\frac{p_{t_2}}{p_{t_1}} = 1 - \frac{2\gamma}{3(\gamma + 1)^2}(M_1^2 \sin^2\theta - 1)^3 + \frac{2\gamma^2}{(\gamma + 1)^3}(M_1^2 \sin^2\theta - 1)^4 + \cdots \tag{77.15}$$

$$\frac{\Delta s}{R} = \frac{1}{\gamma - 1}\frac{\Delta s}{c_v} = \frac{2\gamma}{3(\gamma + 1)^2}(M_1^2 \sin^2\theta - 1)^3 - \frac{2\gamma^2}{(\gamma + 1)^3}(M_1^2 \sin^2\theta - 1)^4 + \cdots \tag{77.16}$$

Given θ and δ:

$$\frac{1}{M_1^2} = \sin^2\theta - \frac{\gamma + 1}{2}\frac{\sin\theta\sin\delta}{\cos(\theta - \delta)} = \sin^2\theta - \frac{\gamma + 1}{2}\frac{\tan\delta}{\tan\delta + \cot\theta} \tag{77.17}$$

$$M_1^2 = \frac{2(\cot\theta + \tan\delta)}{\sin 2\theta - (\gamma + \cos 2\theta)\tan\delta} \tag{77.18}$$

$$\frac{p_2 - p_1}{\frac{1}{2}\rho_1 U_1^2} = 2\,\frac{\sin\theta\sin\delta}{\cos(\theta - \delta)} = \frac{2\tan\delta}{\tan\delta + \cot\theta} = \frac{2\tan\theta}{\tan\theta + \cot\delta} \tag{77.19}$$

$$\frac{\rho_2 - \rho_1}{\rho_2} = \frac{\sin\delta}{\sin\theta\cos(\theta - \delta)} \tag{77.20}$$

Given M_1 and δ: No convenient explicit relations exist. The value of $\sin^2\theta$ is found from the cubic

$$\sin^6\theta - \left(1 + \frac{2}{M_1^2} + \gamma\sin^2\delta\right)\sin^4\theta + \left\{\frac{2M_1^2+1}{M_1^4} + \left[\left(\frac{\gamma+1}{2}\right)^2 + \frac{\gamma-1}{M_1^2}\right]\sin^2\delta\right\}\sin^2\theta = \frac{\cos^2\delta}{M_1^4} \quad (77.21)$$

The smallest of the three roots corresponds to a decrease in entropy and should be disregarded. For weak shock waves (δ in radians),

$$\frac{p_2}{p_1} = 1 + \frac{\gamma M_1^2}{(M_1^2-1)^{1/2}}\,\delta + \gamma M_1^2\,\frac{(\gamma+1)M_1^4 - 4(M_1^2-1)}{4(M_1^2-1)^2}\,\delta^2 + \cdots \quad (77.22)$$

$$\frac{p_2 - p_1}{\frac{1}{2}\rho_1 U_1^2} = \frac{2}{(M_1^2-1)^{1/2}}\,\delta + \frac{(\gamma+1)M_1^4 - 4(M_1^2-1)}{2(M_1^2-1)^2}\,\delta^2 + \cdots \quad (77.23)$$

$$\frac{\rho_2}{\rho_1} = 1 + \frac{M_1^2}{(M_1^2-1)^{1/2}}\,\delta + M_1^2\,\frac{(3-\gamma)M_1^2(M_1^2-2)+4}{4(M_1^2-1)^2}\,\delta^2 + \cdots \quad (77.24)$$

$$\frac{T_2}{T_1} = 1 + \frac{(\gamma-1)M_1^2}{(M_1^2-1)^{1/2}}\,\delta + (\gamma-1)M_1^2\,\frac{(\gamma+1)M_1^4 - 2(M_1^2+2)(M_1^2-1)}{4(M_1^2-1)^2}\,\delta^2 + \cdots \quad (77.25)$$

Large-scale charts for oblique shock waves with $\gamma = 7/5$ are given in [7].

Busemann's Shock Polar

The velocities associated with an oblique shock wave are conveniently represented in the velocity-vector (hodograph) plane. For a given upstream Mach number, all

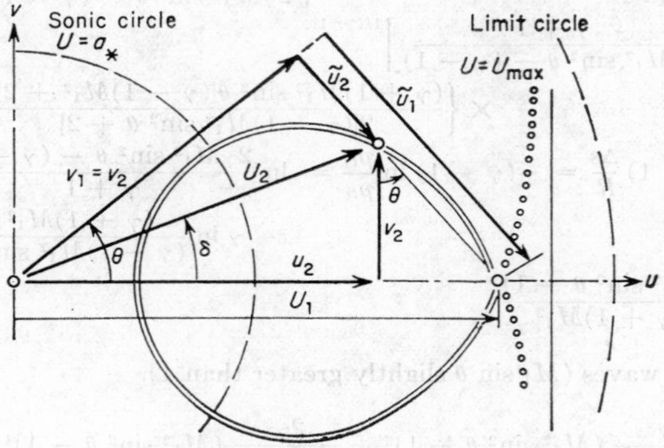

FIG. 77.3. Shock polar associated with oblique shock wave of Fig. 77.2. ($M_1 = 3$, $u_2/U_1 = 0.75$, $\gamma = 1.4$).

possible velocity vectors behind the shock lie on a single curve (Fig. 77.3). Only the closed loop represents real shock waves with nondecreasing entropy. Its equation is

$$v_2^2 = (U_1 - u_2)^2\,\frac{u_2 - a_*^2/U_1}{2U_1/(\gamma+1) + a_*^2/U_1 - u_2} \quad (77.26)$$

Other forms convenient for computation are:

$$\left(\frac{v_2}{U_1}\right)^2 = \left(1 - \frac{u_2}{U_1}\right)^2 \frac{(M_1^2-1) - \frac{1}{2}(\gamma+1)M_1^2(1 - u_2/U_1)}{1 + \frac{1}{2}(\gamma+1)M_1^2(1-u_2/U_1)} \quad (77.27)$$

$$\left(\frac{v_2}{a_*}\right)^2 = \left(\frac{U_1}{a_*} - \frac{u_2}{a_*}\right)^2 \frac{(U_1/a_*)(u_2/a_*) - 1}{1 + 2(U_1/a_*)^2/(\gamma+1) - (U_1/a_*)(u_2/a_*)} \quad (77.28)$$

$$\left(\frac{v_2}{U_{max}}\right)^2 = \left(\frac{U_1}{U_{max}} - \frac{u_2}{U_{max}}\right)^2 \frac{(\gamma+1)(U_1/U_{max})(u_2/U_{max}) - (\gamma-1)}{2(U_1/U_{max})^2 + (\gamma-1) - (\gamma+1)(U_1/U_{max})(u_2/U_{max})} \quad (77.29)$$

The shock-wave angle for sonic flow downstream is given by

$$\sin^2 \theta_* = \frac{1}{4\gamma M_1^2}\{(\gamma+1)M_1^2 - (3-\gamma) + \sqrt{(\gamma+1)[(\gamma+1)M_1^4 - 2(3-\gamma)M_1^2 + (\gamma+9)]}\} \quad (77.30)$$

and that for maximum stream deflection by

$$\sin^2 \theta_{\delta\,\max} = \frac{1}{4\gamma M_1^2}\{(\gamma+1)M_1^2 - 4 + \sqrt{(\gamma+1)[(\gamma+1)M_1^4 + 8(\gamma-1)M_1^2 + 16]}\} \quad (77.31)$$

The flow downstream is subsonic in the latter case. For small deflections (hence, Mach numbers close to unity), these angles (in radians) are given approximately by

$$\delta_* = \frac{1}{\sqrt{2}\,(\gamma+1)}\frac{(M_1^2-1)^{3/2}}{M_1^2} \qquad \delta_{\max} = \frac{4}{3\sqrt{3}\,(\gamma+1)}\frac{(M_1^2-1)^{3/2}}{M_1^2} \quad (77.32)$$

All the above relations are valid across an unsteady shock wave if instantaneous velocities are used, measured relative to the shock. Reference [1] gives a thorough treatment of shock waves in both steady and unsteady flow.

77.3. PRANDTL-MEYER EXPANSION

There is a useful exact solution for isentropic expansion of a uniform supersonic stream over a single convex plane wall. The relations are conveniently given for the special case of a stream with initial Mach number 1 flowing over a sharp corner (Fig. 77.4). Upstream and downstream regions of uniform parallel flow are connected by a *Prandtl-Meyer fan*. All flow quantities are constant along rays of the fan, which are Mach lines. The angle ν, through which the stream turns in expanding from $M = 1$ to a supersonic Mach number M, is

$$\nu = \sqrt{\frac{\gamma+1}{\gamma-1}}\tan^{-1}\sqrt{\frac{\gamma-1}{\gamma+1}(M^2-1)} - (90° - \mu)$$

where

$$(90° - \mu) = \cos^{-1}\frac{1}{M} = \tan^{-1}\sqrt{M^2-1} \quad (77.33)$$

The corresponding ratio of static to total pressure is given by

$$\left(\frac{p}{p_t}\right)^{(\gamma-1)/\gamma} = \frac{1}{\gamma+1}\left\{1 + \cos\left[2\sqrt{\frac{\gamma-1}{\gamma+1}}(\nu + 90° - \mu)\right]\right\} \quad (77.34)$$

Other flow quantities are given by the isentropic streamtube relations of Chap. 74. The expansion angle cannot exceed a certain maximum value

$$\nu_{\max} = \left(\sqrt{\frac{\gamma+1}{\gamma-1}} - 1\right)90° = 130.45° \quad \text{for } \gamma = 7/5 \quad (77.35)$$

at which the flow has expanded to a vacuum, with $M = \infty$.

For expansions through small angles $\Delta\nu$, the ratio of final to initial static pressures is given by

$$\frac{p_2}{p_1} = 1 - \frac{\gamma M_1^2}{(M_1^2-1)^{1/2}}(\Delta\nu) + \gamma M_1^2\frac{(\gamma+1)M_1^4 - 4(M_1^2-1)}{4(M_1^2-1)^2}(\Delta\nu)^2 + \cdots \quad (77.36)$$

Up to the term in $(\Delta\nu)^2$ this coincides with the series for compression through an oblique shock, [Eq. (77.22)]. Entropy changes affect only terms in $(\Delta\nu)^3$.

These relations apply to expansion of a uniform supersonic stream over any single convex wall (Fig. 77.4). The nonuniform *simple wave* joining two uniform streams corresponds to a sector of the full Prandtl-Meyer fan. If flow quantities are known at one point, those at another can be read from a table by identifying the change in flow

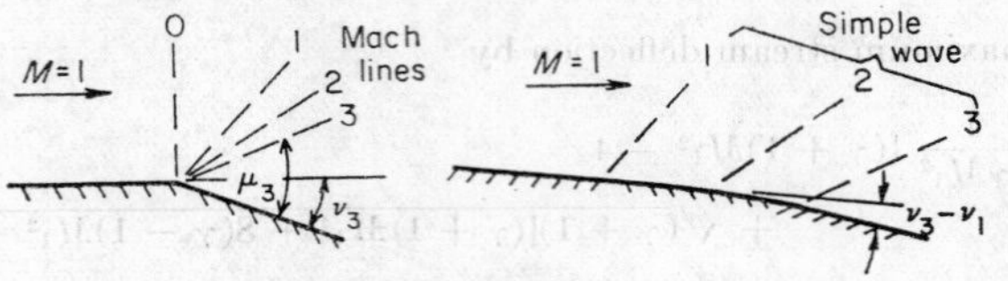

FIG. 77.4. Prandtl-Meyer expansion around sharp corner and convex wall.

angle with $\Delta\nu$. Compression of a uniform supersonic stream by a smooth concave wall follows a reversed Prandtl-Meyer expansion up to the region of influence of the shock wave that forms away from the wall.

Table 77.1. Values of ν in Degrees for $\gamma = 1.4$

M	ν	M	ν	M	ν
1.00	0	1.50	11.90	2.50	39.12
1.05	0.49	1.60	14.86	2.75	44.69
1.10	1.34	1.70	17.81	3.00	49.76
1.15	2.38	1.80	20.72	3.50	58.53
1.20	3.56	1.90	23.59	4.00	65.78
1.25	4.83	2.00	26.38	6.00	84.96
1.30	6.17	2.10	29.10	8.00	95.62
1.35	7.56	2.20	31.73	10.00	102.32
1.40	8.99	2.30	34.28	20.00	116.20
1.45	10.44	2.40	36.75	∞	130.45

Reference [7] gives a more extensive table for $\gamma = 1.4$.

77.4. WEDGE AND CIRCULAR CONE

Conical Flow

The wedge and cone in purely supersonic flow are simple examples of *conical flow*. (Prandtl-Meyer expansion around a sharp corner is another.) The base does not affect the flow upstream of it, which, therefore, contains no characteristic length. Hence, the velocity, pressure, etc., are constant on each ray through the vertex. All shock waves must also be conical.

Flow past a Wedge

The bow shock wave cannot be attached, no matter how high the Mach number, to a wedge of semivertex angle greater than $\sin^{-1}(1/\gamma)$, which is 45.6° for $\gamma = 7/5$. For a thinner wedge, the bow wave is attached above a certain Mach number (Fig. 77.5), and the flow is purely supersonic above a slightly higher one. For thin wedges, the values are given approximately by Eq. (77.32).

The bow wave is straight if the flow is purely supersonic. The subsequent flow consists of uniform streams parallel to the faces of the wedge (Fig. 77.6). The flow over either face may be regarded as obtained from the oblique shock pattern (Fig. 77.2). Equations (77.1) to (77.32) give all flow quantities.

The shock polar (Fig. 77.3) shows that two different shock waves and flow patterns are possible for a given wedge and Mach number. However, only the weaker of the two shock waves (smaller θ) gives supersonic flow downstream (neither does near detachment), and it alone can, therefore, occur on an isolated finite wedge.

An inclined wedge can be treated similarly if neither face exceeds the angle for purely supersonic flow (Fig. 77.5). The flows over the two faces are then independ-

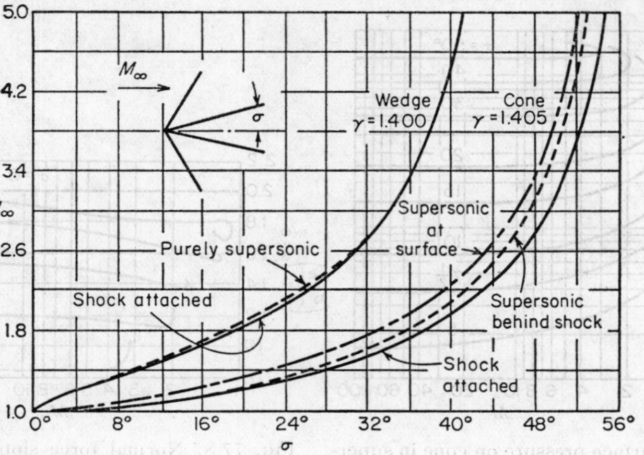

FIG. 77.5. Critical Mach numbers for wedge and cone in supersonic flow.

ent, and involve unequal oblique shock waves. At angles of attack exceeding the semivertex angle, the flow over the leeward surface is given instead by a Prandtl-Meyer expansion.

Circular Cone at Zero Incidence

As for wedges, the bow wave is attached above a certain Mach number for a cone of moderate thickness (semivertex angle less than 57.5° for $\gamma = 1.405$), and the flow becomes purely supersonic at a higher speed. Because the flow varies between

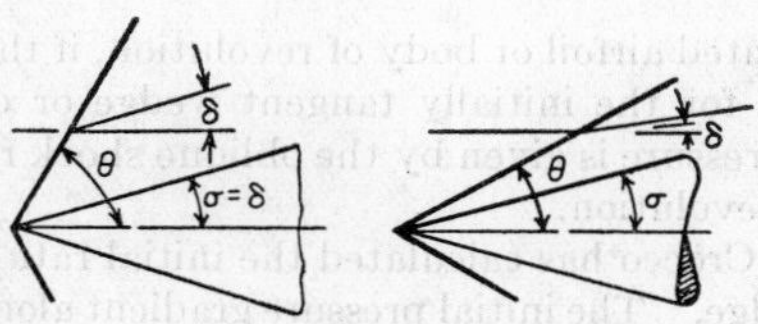

FIG. 77.6. Flow past wedge and cone.

the shock wave and the body, there is an intermediate speed at which sonic flow occurs just behind the bow wave (Fig. 77.5). For slender cones (hence, Mach numbers close to unity), the first and third of these values are given approximately by

$$\sigma_{\max} = 0.86 \sqrt{\frac{M_\infty^2 - 1}{\gamma + 1}} \qquad \sigma_* = 0.723 \sqrt{\frac{M_\infty^2 - 1}{\gamma + 1}} \tag{77.37}$$

The bow wave is itself a circular cone. Flow quantities are constant on concentric cones between the shock and the body, and so depend on only one space variable. The transition across the bow wave is governed by the oblique shock relations. It is followed by continuous isentropic compression to the surface, which can be calculated numerically. Extensive tables of flow quantities are given in Ref. [8*a*] for $\gamma = 1.405$ and $\frac{4}{3}$. As for wedges, two solutions exist for each cone and Mach number, but only the weaker shock wave can occur on an isolated finite cone. Figure 77.7 shows the surface pressure. Values on the lower branches (weak family of shocks) have been often well confirmed by experiment.

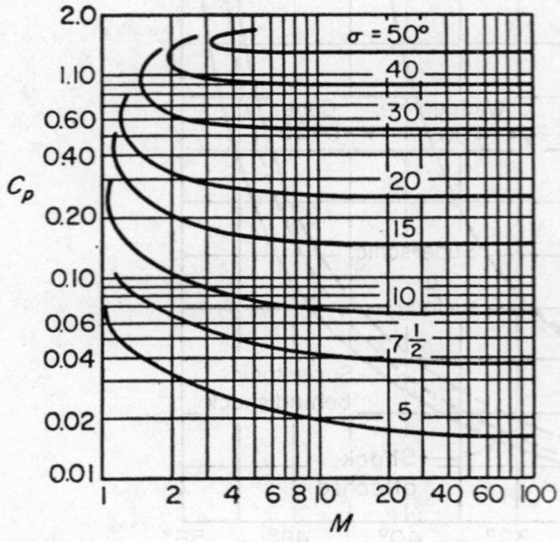

FIG. 77.7. Surface pressure on cone in supersonic flow ($\gamma = 1.405$).

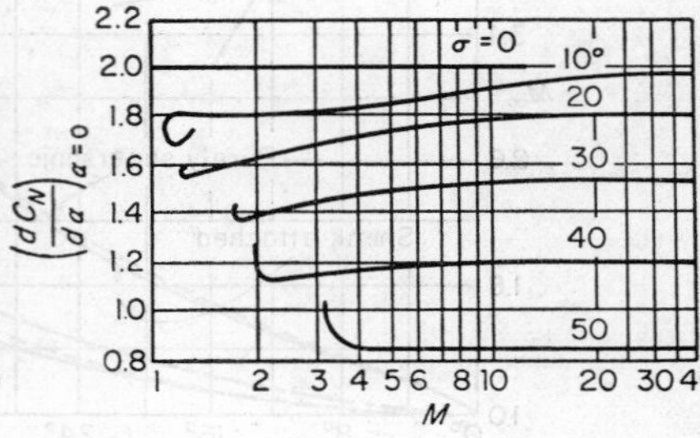

FIG. 77.8. Normal force slope for cone in supersonic flow ($\gamma = 1.405$).

Inclined Circular Cone

The effects of slightly inclining a circular cone have been calculated. Tables of flow quantities are given in [8*b*]. However, care must be taken in interpreting those values (see Ref. [9]). Figure 77.8 shows the normal force.

Second-order effects in angle of attack have also been calculated, but are not free from error. In any case, experiments [10] show that the first-order theory, properly interpreted, is accurate up to angles at which effects of crossflow separation become important.

Curved Tips

At the tip of any pointed airfoil or body of revolution, if the bow wave is attached, the flow is the same as for the initially tangent wedge or cone. Hence, Fig. 77.5 applies, and the initial pressure is given by the oblique shock relations for an airfoil, or Fig. 77.7 for a body of revolution.

For a curved airfoil, Crocco has calculated the initial rate of departure of the flow field from that of the wedge. The initial pressure gradient along the surface is given by

$$\frac{1}{p}\frac{\partial p}{\partial s} = \gamma \frac{1 + \frac{1}{2}\left[\frac{\cos^2 \mu_2}{\cos^2(\theta - \delta)} + \frac{1}{M_\infty^2 \sin^2 \theta}\right]}{\frac{\tan^2(\theta - \delta)}{\tan^2 \mu_2} + \frac{1}{2}\left[\frac{\cos^2 \mu_2}{\cos^2(\theta - \delta)} + \frac{1}{M_\infty^2 \sin^2 \theta}\right]} \frac{\tan(\theta - \delta)}{\sin^2 \mu_2} \kappa \qquad (77.38)$$

where μ_2 is the Mach angle behind the shock wave and κ the wall curvature. Reference [11] gives tables of numerical values.

77.5. AIRFOIL SECTION THEORY

Supersonic airfoil section theory may be exact over considerable portions of a finite wing, in contrast to the subsonic case. On straight untapered wings of constant section, the flow is accurately two-dimensional, except in the regions of influence of the tips. On swept untapered wings, the region affected by the root juncture must also be excluded, and then section theory can be applied, using the sweepback principle (p. 77–16).

a. Exact and Shock-expansion Methods

Supersonic flow past a sharp-nosed airfoil can be calculated with any desired accuracy by the numerical method of characteristics [5]. Polygonal approximations to the two families of characteristic lines and the bow shock wave are constructed step by step downstream (Fig. 77.9). Conditions at points 1 and 2 are used to find those at 3, etc. At high Mach numbers, entropy gradients must be included; then streamlines are a third set of characteristics.

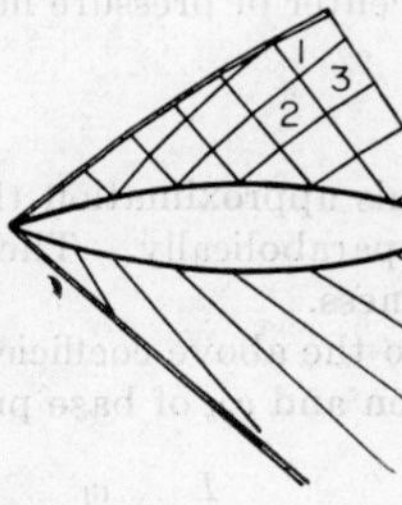

FIG. 77.9. Flow pattern about sharp airfoil showing numerical characteristics mesh (above), and simple wave of shock-expansion method (below).

A less laborious approximation is the *shock-expansion method*. Exact conditions at the tip are found from the oblique shock relations (Sec. 77.2), or the Prandtl-Meyer expansion (Sec. 77.3) if one face is inclined away from the wind. The subsequent flow over the surface is approximated by a Prandtl-Meyer expansion. This neglects the interaction of the outgoing simple wave with the bow shock, and subsequent reflection back to the surface (Fig. 77.9). The reflection coefficient is small except when the bow wave is about to detach [11]; so the method is remarkably accurate up to arbitrarily high Mach numbers. For airfoils having a wedge nose of finite length, the reflected interaction may pass behind the trailing edge, so that the shock-expansion pressures are exact.

b. Ackeret's Linear Theory

Very simple results are given by linearized small-disturbance theory (Sec. 77.6). To first order, the pressure coefficient on a single thin airfoil at small angle of attack is

$$C_p \equiv \frac{p - p_\infty}{\frac{1}{2}\rho_\infty U^2} = \frac{2\delta}{\sqrt{M_\infty^2 - 1}} \tag{77.39}$$

where δ is the local inclination of the surface from the free-stream direction, positive into the wind.

Consider an airfoil with sharp leading edge and possibly blunt base (Fig. 77.10). Measure the angle of attack α from a chord line passing through the nose and bisecting

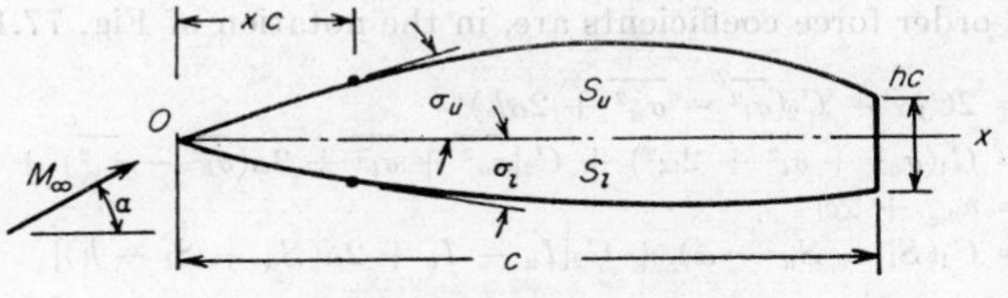

FIG. 77.10. Notation for thin airfoil.

the base. Then the first-order section coefficients of lift, wave drag, and pitching moment (positive nose up) about a point a fraction x of the chord behind the nose are

$$c_l = \frac{4\alpha}{\sqrt{M_\infty^2 - 1}} \qquad c_d = \frac{2}{\sqrt{M_\infty^2 - 1}}\left(\overline{\sigma_u^2} + \overline{\sigma_l^2} + 2\alpha^2\right) \tag{77.40}$$

$$c_{m_x} = \frac{2}{\sqrt{M_\infty^2 - 1}}\left[S_l - S_u - \alpha(1 - 2x)\right] \tag{77.41}$$

Here σ_u and σ_l are the angles (or slopes) of the upper and lower surfaces, measured positive as in Fig. 77.10, and

$$\overline{\sigma_u^2} = \int_0^1 \sigma_u^2\,dx \qquad \overline{\sigma_l^2} = \int_0^1 \sigma_l^2\,dx \tag{77.42}$$

S_u and S_l are the areas above and below the chord line, made dimensionless with c^2. The center of pressure lies at

$$x_{cp} = \frac{1}{2} + \frac{S_u - S_l}{2\alpha} \tag{77.43}$$

To this approximation the lift and moment vary linearly with angle of attack; the drag parabolically. The minimum drag is proportional to the square of the airfoil thickness.

To the above coefficient of wave drag must be added the contributions c_{d_f} of skin friction and c_{d_b} of base pressure for a blunt trailing edge. Then the lift-drag ratio is

$$\frac{L}{D} = \frac{c_l}{c_d} = \frac{\alpha}{\alpha^2 + \frac{1}{2}(\overline{\sigma_u^2} + \overline{\sigma_l^2}) + \frac{1}{4}\sqrt{M_\infty^2 - 1}\,(c_{d_f} + c_{d_b})} \tag{77.44}$$

The maximum, occurring at

$$\alpha_{\max\,L/D} = [\tfrac{1}{2}(\overline{\sigma_u^2} + \overline{\sigma_l^2}) + \tfrac{1}{4}\sqrt{M_\infty^2 - 1}\,(c_{df} + c_{db})]^{1/2} \tag{77.45}$$

is $(L/D_{\max}) = (2\alpha_{\max\,L/D})^{-1}$.

c. Busemann's Second-order Theory

The series for the pressure changes through a weak oblique shock wave [Eq. (77.22)] and a Prandtl-Meyer expansion [Eq. (77.36)] are identical up to the square of the deflection angle. Hence, to second order, the pressure on a single wall depends only on the local inclination from the stream, independent of how it was achieved. The surface pressure coefficient is

$$C_p = C_1\delta + C_2\delta^2 \qquad C_1 = \frac{2}{\sqrt{M_\infty^2 - 1}} \qquad C_2 = \frac{(\gamma + 1)M_\infty^4 - 4(M_\infty^2 - 1)}{2(M_\infty^2 - 1)^2} \tag{77.46}$$

where δ is the slope from the stream direction into the wind. The first term gives Ackeret's linear theory. The nonlinear term intensifies compressions and weakens expansions.

The second-order force coefficients are, in the notation of Fig. 77.10:

$$c_l = 2C_1\alpha + C_2(\overline{\sigma_l^2} - \overline{\sigma_u^2} + 2\alpha h) \tag{77.47}$$

$$c_d = C_1(\overline{\sigma_u^2} + \overline{\sigma_l^2} + 2\alpha^2) + C_2[\overline{\sigma_u^3} + \overline{\sigma_l^3} + 3\alpha(\overline{\sigma_l^2} - \overline{\sigma_u^2}) + 3h\alpha^2] \tag{77.48}$$

$$c_{m_x} = c_{m_0} + xc_l \tag{77.49}$$

$$c_{m_0} = C_1(S_l - S_u - \alpha) + C_2[I_u - I_l + 2\alpha(S_u + S_l - h)] \tag{77.50}$$

Here the σ^2 are defined by Eq. (77.42), and

$$\overline{\sigma_{u,l}^3} = \int_0^1 \sigma_{u,l}^3\,dx \qquad I_{u,l} = \int_0^1 x\sigma_{u,l}^2\,dx \tag{77.51}$$

For symmetrical (uncambered) profiles the center of pressure is

$$x_{cp} = \frac{1}{2} - \frac{C_2}{C_1}(S - h) \tag{77.52}$$

where S is the total area of the section. Skin friction and base drag must be added to the above wave drag.

d. Third- and Fourth-order Theory

Entropy changes through a shock wave are proportional to the cube of its strength [Eq. (77.16)]. Hence, further terms in Eq. (77.46) are affected by shock waves upstream. A sharp-nosed convex airfoil has at most one shock wave, at the nose, ahead of its trailing edge. Then third-order terms in pressure involve only the initial slope of the airfoil, and fourth-order terms also the initial curvature. The surface pressure coefficient [12] is given by

$$C_p = C_1\delta + C_2\delta^2 + (C_3\delta^3 + D_1\delta_0^3) + (C_4\delta^4 + D_2\delta_0^4 + D_3\delta_0^3\delta + D_4\delta_0^3\delta_0'x) \quad (77.53)$$

where

$$C_1 = 2\beta^{-1} \qquad C_2 = (M^2N - 2)\beta^{-2} \qquad \beta = \sqrt{M_\infty^2 - 1} \qquad N = \tfrac{1}{2}(\gamma + 1)M_\infty^2\beta^{-2}$$
$$C_3 = \tfrac{1}{3}[4M_\infty^2N^2 + M_\infty^2(M_\infty^2 - 10)N + 2(M_\infty^2 + 2)]\beta^{-3}$$
$$C_4 = \tfrac{1}{12}[4M_\infty^2(M_\infty^2 + 3)N^3 - M_\infty^2(M_\infty^4 + 4M_\infty^2 + 72)N^2 + 2M_\infty^2(M_\infty^4 - 3M_\infty^2 + 38)N + 8(2M_\infty^2 + 1)]\beta^{-4}$$
$$D_1 = \tfrac{1}{12}M_\infty^2N[(3M_\infty^2 - 4)N - 4\beta^2]\beta^{-3}$$
$$D_2 = -\tfrac{1}{4}M_\infty^2N(M_\infty^2 - 2)[(M_\infty^2 - 2)N^2 - (3M_\infty^2 - 4)N + 2\beta^2]\beta^{-4}$$
$$D_3 = \tfrac{1}{6}M_\infty^4N[2(3M_\infty^2 - 1)N^2 - (4M_\infty^2 - 5)N + 2\beta^2]\beta^{-4}$$
$$D_4 = \tfrac{1}{16}M_\infty^4N^2[(3M_\infty^2 - 4)N - 4\beta^2]\beta^{-4}$$

Here δ is the local slope into the free stream; and δ_0 and δ_0' are the slope and its derivative (or the curvature) at the leading edge, which are to be replaced by zero if δ_0 is negative (since there is then no bow shock).

e. Optimum Airfoil Shapes

The calculus of variations has been applied to several of the preceding theories to find airfoils having minimum pressure drag under various structural criteria [13]. In linearized theory, the symmetrical double-wedge and biconvex profiles are optimum for given thickness and area, respectively, if base pressure is disregarded. The trailing edge should be blunt if the base-pressure parameter $C_{pb}\sqrt{M^2 - 1}/(t/c)$ exceeds 4 and 8, respectively (where t/c is the thickness ratio), with intermediate values for other structural criteria [13*a*]. The optimum shapes predicted by shock-expansion theory differ little from those of linearized theory [13*b*].

f. Interacting Waves; Busemann Biplane

The previous approximations apply only to single airfoils, from which waves travel out primarily in one direction. According to linearized theory, momentum is carried to infinity along Mach waves. More accurately, shock waves form that ultimately recede from the Mach lines as the square root of the distance from the airfoil. From either viewpoint, a single airfoil necessarily experiences wave drag.

Two or more interacting airfoils involve waves of both families that cross and modify each other. Busemann showed that a nonlifting supersonic biplane can be designed to cancel all external waves so that the wave drag is zero [14]. The leading and trailing edges must be cusped to avoid shock waves, and the outer surfaces must be flat and parallel to the stream.

g. Accuracy of Airfoil Theory

Linearized theory (p. 77–9) is a good approximation to the inviscid flow (and second-order theory is better) if the second term in Eq. (77.46) is small compared with the first. This requires that the parameters

$$\frac{[M_\infty^2(\gamma + 1)\delta]^{2/3}}{M_\infty^2 - 1} \qquad \text{and} \qquad M_\infty\delta$$

of transonic [Eq. (76.14)] and hypersonic flow be small. Otherwise, nonlinearity is dominant, and the series fail to converge. Linearized theory is least reliable for the center of pressure, because the error is of the order of the first power of thickness, whereas, in lift and moment, it is of the second power, and, in drag, the third.

Viscous effects may be significant near the trailing edge. The boundary layer and trailing shock wave interact to raise the pressure, and so reduce the drag. Figure 77.11 compares various approximations with experiment.

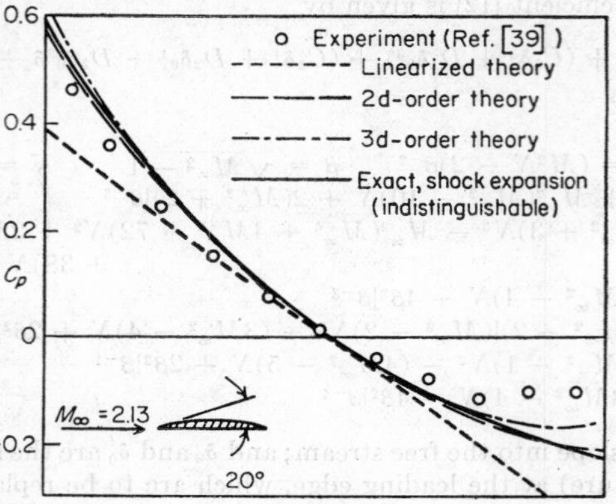

FIG. 77.11. Pressure distribution on half-circular-arc airfoil.

77.6. LINEARIZED THEORY

Efficient aerodynamic shapes usually disturb the air only slightly. This is the basis of *small-disturbance theory:* in particular, the first approximation of *linearized theory*. For moderate supersonic speeds, linearized theory is the only feasible means of treating most three-dimensional shapes, and it has been extensively developed.

Equations of Motion

Shock waves almost invariably appear in supersonic flow. However, their effects are, at most, of third order in the deflection angle (cf. p. 77–11). Hence, to second order, a uniform stream remains irrotational and isentropic, so that a velocity potential Φ exists (p. 75–3). As in subsonic flow (p. 75–4), it is convenient to introduce a *perturbation velocity potential* ϕ according to

$$\Phi = U(x + \phi) \tag{77.54}$$

where U is the speed of the free stream in the x direction.[1] The velocity is given by $\mathbf{q} = U(\mathbf{i}_x + \operatorname{grad} \phi)$; in cartesian coordinates,

$$\frac{u}{U} = 1 + \phi_x \qquad \frac{v}{U} = \phi_y \qquad \frac{w}{U} = \phi_z \tag{77.55}$$

Introducing the perturbation potential into the equations (75.16) of steady isentropic flow yields the quasi-linear second-order partial differential equation:

$$\begin{aligned} \phi_{yy} + \phi_{zz} - (M_\infty{}^2 - 1)\phi_{xx} \\ = M_\infty{}^2 \Big[\frac{\gamma - 1}{2} (2\phi_x + \phi_x{}^2 + \phi_y{}^2 + \phi_z{}^2)(\phi_{xx} + \phi_{yy} + \phi_{zz}) + 2\phi_x\phi_{xx} \\ + \phi_x{}^2\phi_{xx} + \phi_y{}^2\phi_{yy} + \phi_z{}^2\phi_{zz} + 2\phi_y\phi_z\phi_{yz} + 2(1 + \phi_x)\phi_z\phi_{xz} \\ + 2(1 + \phi_x)\phi_y\phi_{xy} \Big] \end{aligned} \tag{77.56}$$

[1] Note the difference in the notations of Chap. 75 and the present chapter, as represented by Eqs. (75.17) and (77.54).

Neglecting the nonlinear terms on the right gives the *linearized* equation,

$$\phi_{yy} + \phi_{zz} - (M_\infty{}^2 - 1)\phi_{xx} = 0 \tag{77.57a}$$

or, in cylindrical polar coordinates,

$$\phi_{rr} + \frac{\phi_r}{r} + \frac{\phi_{\theta\theta}}{r^2} - (M_\infty{}^2 - 1)\phi_{xx} = 0 \tag{77.57b}$$

This is formally identical with the linearized equation (75.18) of subsonic flow. However, it is *hyperbolic* rather than *elliptic* (p. 11–5), and is, in fact, the classical wave equation. It has two sets of real characteristic surfaces, the right circular *Mach cones* (p. 77–1) described by

$$x \pm \sqrt{M_\infty{}^2 - 1}\, r = \text{const} \tag{77.58}$$

whose semivertex angles are the free-stream Mach angle.

The linearization is justified if (cf. p. 77–11)

$$\frac{1}{M_\infty{}^2 - 1}\left|\frac{u - U}{U}\right| \ll 1 \quad \text{and} \quad M_\infty\left|\frac{v}{U}\right|, M_\infty\left|\frac{w}{U}\right| \ll 1$$

The first condition excludes transonic flow, the second hypersonic flow.

Boundary Conditions

For a body described by $S(x,y,z) = 0$, the exact condition of tangent flow at the surface is

$$\text{grad}\,\Phi \cdot \text{grad}\,S = 0 \qquad \text{at } S = 0 \tag{77.59}$$

The linearized form, which is adequate whenever Eq. (77.57) applies, is

$$\phi_y S_y + \phi_z S_z + S_x = 0 \qquad \text{at } S = 0 \tag{77.60}$$

For *quasi-cylindrical* bodies that lie everywhere near some streamwise cylinder (e.g., flat wing, circular tube of nearly constant radius), the tangency condition may, to first order, be imposed at the reference cylinder.

By the rule of forbidden signals (p. 77–1), the perturbation potential must vanish ahead of the body. In linearized theory, it actually vanishes on the bow Mach cone. For additional boundary conditions on thin wings, see p. 77–16.

Pressure Relation

The exact isentropic relation between pressure coefficient and speed q is

$$C_p = \frac{2}{\gamma M_\infty{}^2}\left\{\left[1 + \frac{\gamma - 1}{2} M_\infty{}^2\left(1 - \frac{q^2}{U^2}\right)\right]^{\gamma/(\gamma-1)} - 1\right\} \tag{77.61}$$

To the order of accuracy of linearized theory, this becomes

$$C_p = -2\frac{u - U}{U} - \frac{v^2 + w^2}{U^2} = -2\phi_x - (\phi_y{}^2 + \phi_z{}^2) \tag{77.62}$$

The squared terms are significant for slender fusiform objects, but not for wide flat wings or other quasi-cylinders.

Similarity Rule

Consider an affinely related family of bodies derived from one another by expanding or contracting all dimensions normal to the stream. Let τ be some characteristic slope or thickness ratio. Then linearized flow field depends, not on M and τ

separately, but only on the combination

$$\sqrt{M_\infty^2 - 1}\,\tau = \beta\tau \qquad \textit{supersonic similarity parameter}$$

For a fixed value of this parameter, all streamline and surface slopes must vary in proportion to the slope of the Mach cone, as must lateral separations of compound bodies (biplane, body in wind tunnel). Then the supersonic similarity rule for the perturbation velocity potential is, in alternative forms,

$$\phi(x,y,z;M,\tau) = \beta^{-2}fn(x,\beta y,\beta z;\beta\tau) = (\tau/\beta)fn(\ \ ) = \tau^2 fn(\ \ ) \tag{77.63}$$

where the fn are (different) general functions of the *reduced coordinates* x, βy, βz and the similarity parameter. The rule for pressure coefficient has the same forms. The rules for force coefficients are

$$\begin{aligned} C_D &= \tau^{2+k}fn(\beta\tau) = (\tau^{1+k}/\beta)fn(\beta\tau) = (\tau^k/\beta^2)fn(\beta\tau) \\ C_L &= \tau^{1+k}fn(\beta\tau) = (\tau^k/\beta)fn(\beta\tau) \end{aligned} \tag{77.64a,b}$$

where k is 0 or 1 according to whether coefficients are referred to a base or a planform area. The rule for moment is implicit in the statement that center of pressure in reduced coordinates depends only on $\beta\tau$. Analogous, but more complex rules exist for second-order theory.

These rules happen to coincide with those for hypersonic flow, and so represent combined supersonic-hypersonic rules valid at all Mach numbers above the transonic range.

General Solution for Plane Flow

For motions independent of z, the linearized equation (77.57a) reduces to

$$\phi_{yy} - \beta^2\phi_{xx} = 0 \qquad \beta = \sqrt{M_\infty^2 - 1} \tag{77.65}$$

Its general solution

$$\phi(x,y) = f(x - \beta y) + g(x + \beta y) \tag{77.66}$$

involves two arbitrary functions, representing waves traveling along each of the characteristic (Mach) directions. The first term applies above an isolated airfoil, the second below; and imposing the condition of tangent flow leads to Ackeret's pressure formula (p. 77–9). Between two bodies, both families of waves appear (p. 77–11).

Elementary Solutions

The three-dimensional subsonic source has as its counterpart the *supersonic point source* located at the point (x_0,y_0,z_0) with the potential

$$\phi = -\frac{1}{2\pi}\frac{1}{\sqrt{(x - x_0)^2 - \beta^2[(y - y_0)^2 + (z - z_0)^2]}} \tag{77.67}$$

This implies real velocities inside the upstream as well as the downstream Mach cones (Fig. 77.1). The former are to be disregarded to satisfy the rule of forbidden signals. Hence, the strength is double that of Eq. (75.24) to preserve unit flux.

From this basic solution, the following elementary solutions are derived by shifting the origin, integrating, and differentiating:

Point doublet (dipole) at origin, axis in x, y, or z direction

$$\phi = \frac{1}{2\pi}\frac{(-x),\ \beta y,\ \text{or}\ \beta z}{[x^2 - \beta^2(y^2 + z^2)]^{3/2}} \tag{77.68}$$

Integrating this over a line or surface yields a divergent integral (which can be interpreted according to its "finite part" [2a]. However, for a distribution along a streamwise line of doublets with axis normal to the stream, this complication is avoided by using instead the

Elementary horseshoe vortex at origin, lift in y or z direction

$$\phi = \frac{1}{2\pi\beta} \frac{xy \text{ or } xz}{(y^2 + z^2)\sqrt{x^2 - \beta^2(y^2 + z^2)}} \tag{77.69}$$

Semi-infinite line source of constant strength downstream from origin

$$\phi = -\frac{1}{2\pi} \operatorname{Cosh}^{-1} \frac{x}{\beta\sqrt{y^2 + z^2}} \tag{77.70}$$

Semi-infinite line source of constant strength across wind from origin in y or z direction

$$\phi = \frac{1}{2\pi} \cos^{-1} \frac{\beta y}{\sqrt{x^2 - \beta^2 z^2}} \quad \text{or} \quad \frac{1}{2\pi} \cos^{-1} \frac{\beta z}{\sqrt{x^2 - \beta^2 y^2}} \tag{77.71}$$

Semi-infinite conical line source (linearly increasing strength) *downstream from origin*

$$\phi = -\frac{1}{2\pi}\left[x \operatorname{Cosh}^{-1} \frac{x}{\beta\sqrt{y^2 + z^2}} - \sqrt{x^2 - \beta^2(y^2 + z^2)}\right] \tag{77.72}$$

Semi-infinite line doublet of constant strength downstream from origin

$$\phi = -\frac{1}{2\pi\beta} \frac{\sqrt{x^2 - \beta^2(y^2 + z^2)}}{y^2 + z^2} (y \text{ or } z) \tag{77.73}$$

Semi-infinite conical doublet line downstream from origin

$$\phi = \frac{\beta}{4\pi}\left[\operatorname{Cosh}^{-1} \frac{x}{\beta\sqrt{y^2 + z^2}} - \frac{x}{\beta^2(y^2 + z^2)}\sqrt{x^2 - \beta^2(y^2 + z^2)}\right](y \text{ or } z) \tag{77.74}$$

Solutions for oblique source and doublet lines lying in the plane $z = 0$ along $x = \lambda\beta y$ (inside the Mach cone) are obtained from Eqs. (77.70) and (77.72) to (77.74), by making the *oblique transformation,*

$$\begin{aligned} x &= \bar{x} - \lambda\beta\bar{y} \\ \beta y &= \beta\bar{y} - \lambda\bar{x} \qquad \lambda^2 < 1 \\ \beta z &= \sqrt{1 - \lambda^2}\, \beta\bar{z} \end{aligned} \tag{77.75}$$

which leaves the linearized equation (77.57) unchanged.

If ϕ is a solution of the linearized equation, so is ϕ_x (and ϕ_y, ϕ_z). Hence, all the above can be taken as elementary solutions for $(u - U)/U$ or (for thin flat wings or other quasi-cylinders) C_p. Hayes [15] has given a comprehensive treatment of elementary solutions of various types and degrees of homogeneity in the coordinates.

Perturbation of Other Basic Flows

Conventional linearized theory involves disturbances about a uniform stream. We can likewise consider small perturbations about another known flow. The resulting linearized equation is more complicated, because its coefficients are not constants but functions of the basic flow. Thus Stone perturbed the axisymmetric flow past a cone to find the effects of incidence (p. 77–8). Ferri [16] has discussed the method in general. Extensive tables for calculating supersonic flow about any cone that departs only slightly from a circular cone (e.g., an elliptical cone) are given in [17].

77.7. THIN WINGS

Sweep

In inviscid flow past an infinite swept wing of constant section (Fig. 77.12), the spanwise component of free-stream velocity is unaffected, and can be ignored. The pressure distribution is that on a profile having the section normal to the leading edge, in a stream of speed $U \cos \Lambda$ and Mach number $M \cos \Lambda$. If this effective Mach number exceeds unity, the methods of Sec. 77.5 apply. The effective angle of attack is $\alpha_{\text{eff}} = \alpha / \cos \Lambda$, where α is the actual angle in the streamwise direction. Thus, according to linearized theory (p. 77–9), the surface pressure coefficient is

$$C_p = \frac{2\delta \cos \Lambda}{\sqrt{M_\infty^2 \cos^2 \Lambda - 1}} = \frac{2\delta \cot \Lambda}{\sqrt{\beta^2 \cot^2 \Lambda - 1}} \qquad \beta^2 = M_\infty^2 - 1 \tag{77.76}$$

where δ is the streamwise slope of the surface into the wind.

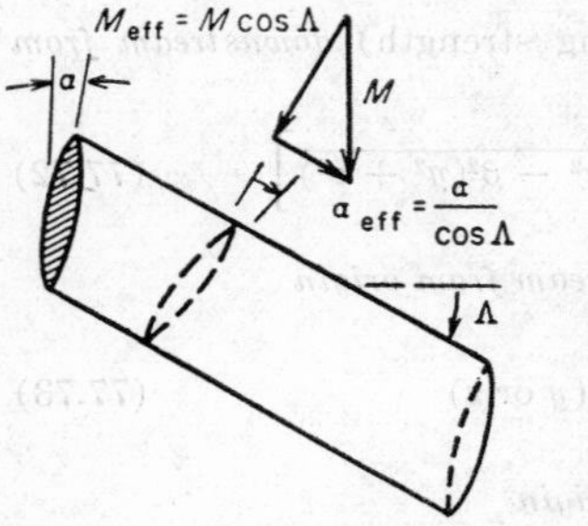

FIG. 77.12. Swept wing.

This simple sweep principle applies also within triangular regions of finite wings having straight leading edges swept at less than the Mach angle (as in sketches in Table 77.2).

Separation of Thickness and Lift

In linearized theory, a thin flat wing is a *planar system*—a special quasi-cylindrical body (p. 77–13) for which the tangency condition can, for simplicity, be imposed at a mean plane (say $z = 0$) rather than at the actual surface. As a consequence, pressures due to thickness, camber, and angle of attack are additive, and can be calculated separately. Lift and moments are also additive; drag is not. The perturbation potential ϕ, streamwise velocity increment, pressure coefficient, and sidewash are even functions of z in thickness problems, and odd ones in lifting problems; the downwash velocity has the opposite parity. Off the wing, the pressure must be continuous, so that it vanishes off a lifting wing in the plane.

Planform Edges

Any portion of the periphery of a wing planform is classified as a *side edge* if it lies in the streamwise direction; otherwise as a *leading* or *trailing edge;* and as *subsonic* or *supersonic* according to whether it is swept at more or less than the Mach angle. At a subsonic trailing edge, the Kutta condition of noninfinite (actually zero) pressure difference must be enforced in the lifting case for uniqueness.

At subsonic leading edges, singularities in pressure characteristic of incompressible flow must be required for uniqueness. These may contribute to the drag. Hence, the theoretical drag is given by the integral over the wing of pressure times slope, plus an additional *leading-edge thrust* in the lifting example and *leading-edge drag* in the thickness example, for a round edge. The lifting pressure will be infinite like

$$C_p \sim -2 \frac{k}{\sqrt{x - x_{\text{LE}}}} = -2k \sqrt{\frac{\cos \Lambda}{n}} \tag{77.77}$$

where n is the normal distance into the edge, and Λ the local angle of sweep. Then the leading-edge thrust normal to the edge, per unit length of edge, is

$$T_n = \pi \rho_\infty U^2 \sqrt{\sec^2 \Lambda - M_\infty^2}\, k^2 = \pi \rho_\infty U^2 \sqrt{\tan^2 \Lambda - \beta^2}\, k^2 \tag{77.78}$$

Similarly, at a round edge in the thickness solution, the normal drag per unit length of edge is

$$D_n = \tfrac{1}{2}\rho_\infty(U \cos \Lambda)^2 \frac{\pi R}{\sqrt{1 - M_\infty^2 \cos^2 \Lambda}} \tag{77.79}$$

where R is the leading-edge radius. At a sharp subsonic edge of finite angle in the thickness problem, the pressure is logarithmically infinite, but contributes nothing to drag.

Wing of Given Thickness

The thickness problem is easier than the lifting one for wings of given shape. (Conversely, the lifting problem is easier if the pressure is prescribed.) The wing can be represented by a distribution of supersonic sources [Eq. (77.67)] over the planform in the plane of the wing. The source strength is proportional to the local streamwise slope $\partial Z/\partial y$ if the surface is $z = \pm Z(x,y)$. The pressure at any point

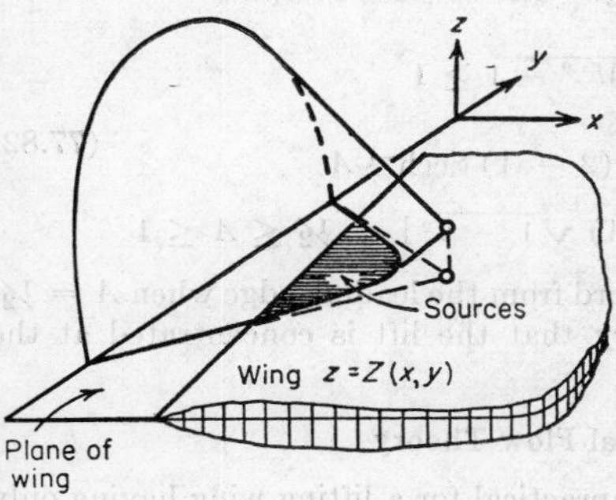

FIG. 77.13. Mach forecone for thin wing.

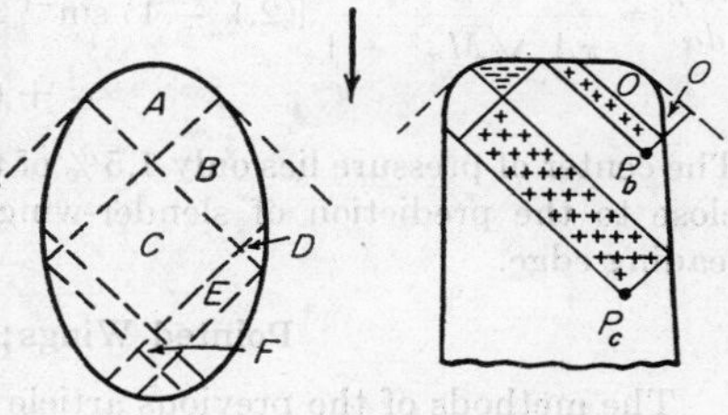

FIG. 77.14. Regions on lifting wing.

depends only upon the sources inside the hyperbolic intersection of the forward-going Mach cone with the plane of the wing (Fig. 77.13). Hence,

$$\phi(x,y,z) = -\frac{1}{\pi}\iint \frac{Z_y(\xi,\eta)\, d\xi\, d\eta}{\sqrt{(x-\xi)^2 - (M_\infty^2 - 1)\{(y-\eta)^2 + z^2\}}} \tag{77.80}$$

The integration extends over the region where the denominator is real (and the numerator nonzero). On the wing itself,

$$C_p(x,y) = \frac{2}{\pi}\frac{\partial}{\partial x}\iint \frac{Z_y(\xi,\eta)\, d\xi\, d\eta}{\sqrt{(x-\xi)^2 - (M_\infty^2 - 1)(y-\eta)^2}} \tag{77.81}$$

the region of integration being bounded by the two forward Mach lines and the leading edge.

For simple sections (prismatic, biconvex) and simple planforms (usually polygonal), the integration can be carried out in closed form. The results typically involve inverse trigonometric, hyperbolic, and elliptic functions. Table 77.2 gives a short guide to existing solutions.

Lifting Wings; Evvard-Krasilshchikova Method

Consider a lifting wing having some extent of supersonic leading edge. Calculation of lifting pressure grows increasingly difficult in the successive regions of Fig. 77.14.

Region A: The Mach forecone intersects only the supersonic edges. The upper and lower surfaces act independently, and so the distinction between the lifting and thickness problems disappears. Equation (77.81) applies.

Region B: The forecone intersects one supersonic and one subsonic leading (or side) edge. Evvard [18] and Krasilshchikova showed that regions marked O in Fig. 77.14 have no net effect. Pressure at P_B is given by Eq. (77.81) with integration extended only over area shaded $+$.

Region C: The forecone intersects two subsonic leading (or side) edges. Repeated application of the rule for region B shows that pressure at P_C is given by Eq. (77.81) with integration extended over the parallelogram shaded $+$ in Fig. 77.14 and, with negative sign, over area shaded $-$ (if it exists). Further reflections on low-aspect-ratio wings can be treated by methods of lift cancellation [19],[20].

Regions D and E: The forecone intersects one or two subsonic trailing edges. The Kutta condition must be imposed. For details, see [18].

Example 1. Rectangular Wing. The flat rectangular lifting wing of aspect ratio $\mathcal{R}$ has been solved [21] for $\mathcal{R}\sqrt{M_\infty^2 - 1} \geq \frac{1}{2}$. The lift-curve slope is

$$\frac{dC_L}{d\alpha} = \frac{4}{\sqrt{M_\infty^2 - 1}}\left(1 - \frac{1}{2A}\right) \qquad A \equiv \mathcal{R}\sqrt{M_\infty^2 - 1} \geq 1$$

$$\frac{dC_L}{d\alpha} = \frac{4}{\pi A\sqrt{M_\infty^2 - 1}}\,[(2A - 1)\sin^{-1} A - A(2 - A)\,\mathrm{Sech}^{-1} A + (1 + A)\sqrt{1 - A^2}\,] \qquad \tfrac{1}{2} \leq A \leq 1 \tag{77.82}$$

The center of pressure lies only 1.5% of the chord from the leading edge when $A = \frac{1}{2}$, close to the prediction of slender-wing theory that the lift is concentrated at the leading edge.

Pointed Wings; Conical Flow Theory

The methods of the previous article are impractical for a lifting wing having only subsonic leading edges. At least near its vertex, a flat wing of such planform corresponds to a flat triangular wing swept inside the Mach cone. The flow field ahead

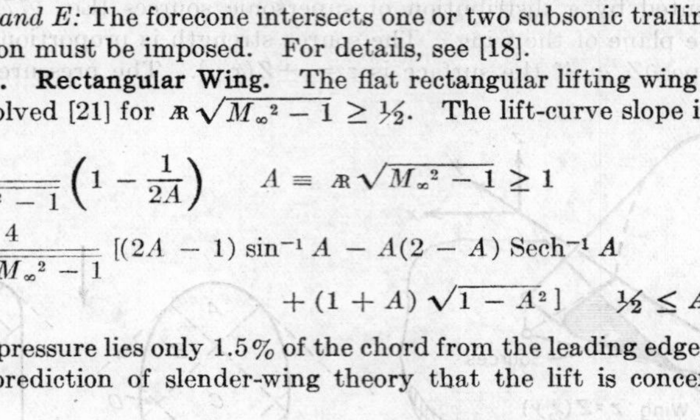

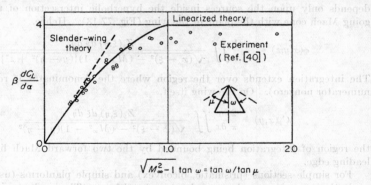

FIG. 77.15. Lift of flat triangular wing.

of the trailing edge contains no characteristic length. Hence, it is a *conical flow* (p. 77–6). Flow quantities are constant along rays through the vertex, and so depend upon only two space variables. Busemann showed how they can be determined by two-dimensional potential theory, using the complex variable [22]. References [19] and [23] give detailed applications. Reference [2*b*] gives a useful tabulation of conical-flow solutions for thickness as well as lift.

The surface loading on the unyawed flat lifting triangle with subsonic edges (Fig. 77.15) is

$$\frac{\Delta p}{\frac{1}{2}\rho_\infty U^2} = \frac{4\alpha}{\beta E'(k)} \frac{\beta y/x}{\sqrt{k^2 - (\beta y/x)^2}} \qquad k = \beta \tan \omega \tag{77.83}$$

The lift coefficient is

$$C_L = \frac{2\pi \tan \omega}{E'(k)} \alpha \tag{77.84}$$

where $E'(k)$ is the complete elliptic integral of the second kind of modulus $(1 - k^2)$. The drag coefficient is

$$C_D = \alpha C_L - \epsilon \frac{C_L^2 \sqrt{1 - \beta^2 \tan^2 \omega}}{4\pi \tan \omega} \tag{77.85}$$

where ϵ varies from 0 to 1 according to whether the leading-edge thrust is lost or fully realized (p. 77–22). When the edges become supersonic, the lift has the two-dimensional value of Eq. (77.40), and $C_D = \alpha C_L$ because no leading-edge thrust is possible. The lift coefficient is compared in Fig. 77.15 with slender-wing theory (Chap. 79) and experiment.

Camber and twist proportional to powers of distance from the vertex can be treated by an extension of conical-flow theory [23].

Guide to Linearized Solutions for Thin Flat Wings

Table 77.2 summarizes most of the existing solutions. The references (numbers in brackets) are to the most useful rather than the first treatments. In some cases, detailed design charts are available; in others, only formulas.

The drag and lift-curve slope of sweptback wings apply also to their swept-forward counterparts. For purely supersonic edges (planforms 2, 4*a*, etc.) the thickness profile need not be symmetric, since the upper and lower surfaces are independent. For a general survey, see [2*b*], for comprehensive charts, [24].

Reverse Flow and Reciprocity Relations

For steady flow past thin flat wings, there is a general reciprocity relation [26], analogous to those familiar in other branches of mechanics (for example, p. 33–22). Consider a uniform stream of gas flowing past a thin wing of given planform S, the normal velocity at the surface being $v_1(x,y)$ and the pressure $p_1(x,y)$. Let the direction of the stream be reversed, and the surface slopes changed arbitrarily so that the normal velocity and pressure are $v_2(x,y)$ and $p_2(x,y)$. Then, in the approximation of linearized (subsonic or supersonic) theory,

$$\iint_S v_1(x,y) p_2(x,y)\, dx\, dy = \iint_S v_2(x,y) p_1(x,y)\, dx\, dy \tag{77.86}$$

where the integration extends over both the upper and the lower surfaces of the wing. For definiteness, it may be supposed that the Kutta condition is imposed at subsonic trailing edges (though it is only required that the nature of edge singularities be known).

Applications are of two types [26],[2*a*]. First, *reversal theorems* are found by considering identical wings in forward and reverse flight:

1. The pressure drag of symmetrical nonlifting wings is the same in forward and reverse flight.

Table 77.2

No.	Planform and Mach lines	Thickness solution For streamwise section shown (of constant thickness ratio if not specified)	Lifting solution Flat plate if not specified
1	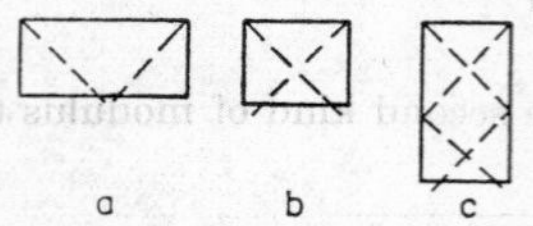 a b c	[47] [46] a Varying thickness [72]	[21] Yaw [56] Twist [86] Camber [73]
2	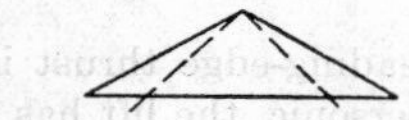	[67] [78] Varying thickness [63] Parabolic hyperboloid [82]	[67] Yaw [56] Twist and camber [74]
3	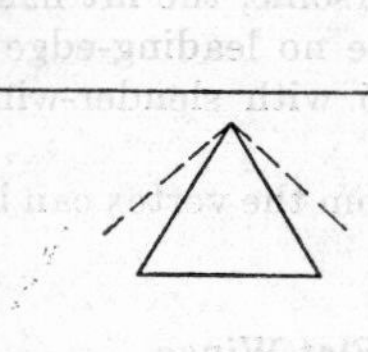	[67] [78] Varying thickness [63] Parabolic hyperboloid [82] Elliptic across stream [88]	[87] Yaw [44] Camber and twist [51], [61], [76], [77] Uniform load [43]
4	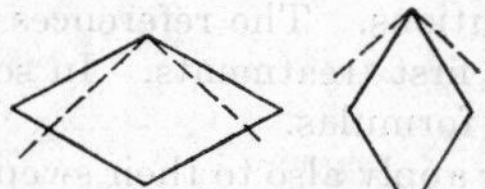 a	[47] a [70]	a [70]
5	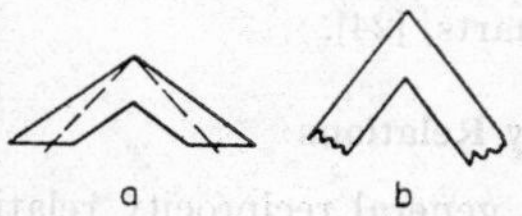 a b	b [41]	a [53] a Yaw [56] Uniform load [43]
6	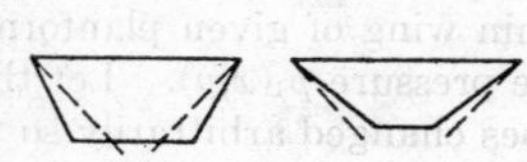 a b		a [53] b 2-dimensional (sec. 5-2) Yaw [56]
7	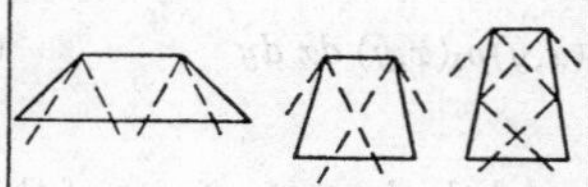 a b c		a,b [53] a,b Yaw [56] c [89] or [2b]
8	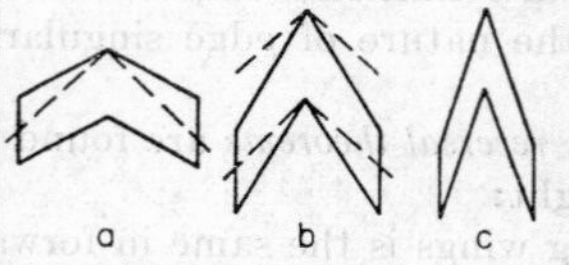 a b c	[45] [48] a [46] b [42]	a [54]

Table 77.2. (*Continued*)

9	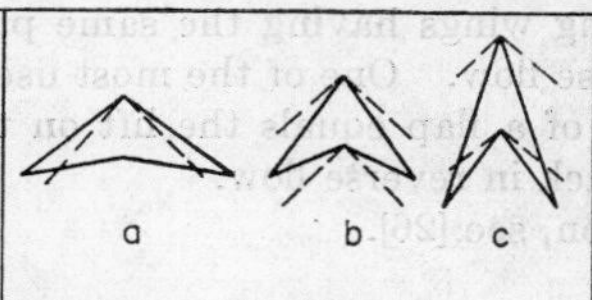	[68] [80] 3rd degree surface [82] Parabolic paraboloid [81]	a [68] b [68], [53] b Yaw [49], [56] a Twist [71] a, Twist and camber [83]
10	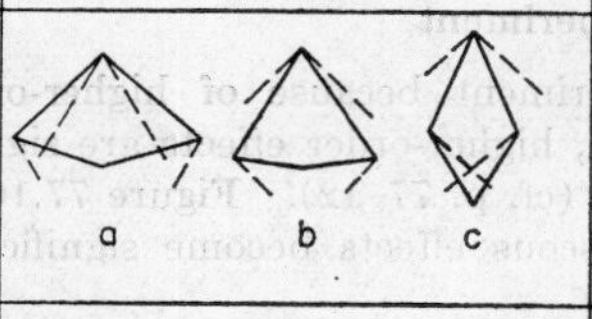	[68]	a [68] b Yaw [49] b Twist [75]
11	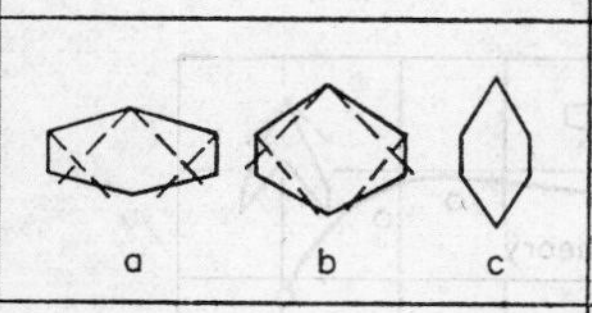	[47], [84] b Varying thickness [66]	a,b [84] a,b Twist [85]
12	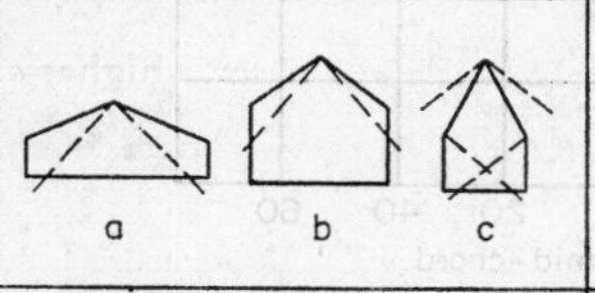	[47] Parabolic hyperboloid [81]	a,b [69] c [79]
13	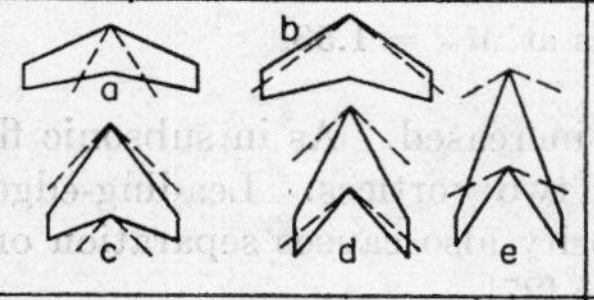	Ridge supersonic [50] " subsonic [45] (1 case) [57] (special cases) [70] Parabolic hyperboloid [81]	a,b [59], [62] b,c,d,e [55] a,b,c,d Yaw [64]
14	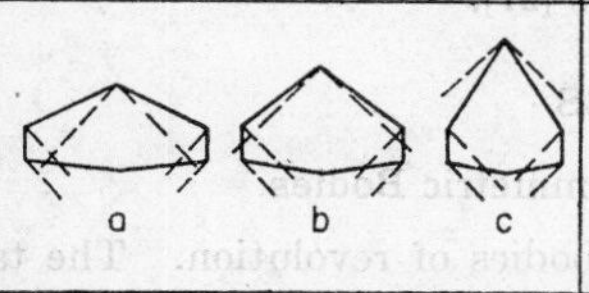	a Ridge subsonic [45]	a,b [59] c [52] d [60]
15	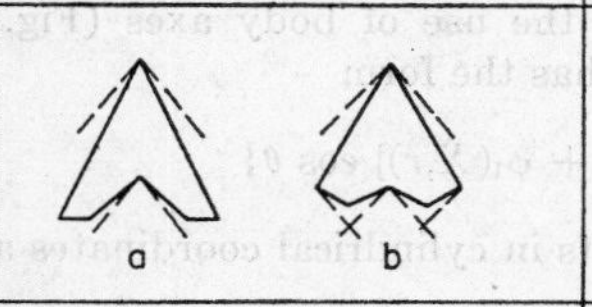		[58]
16	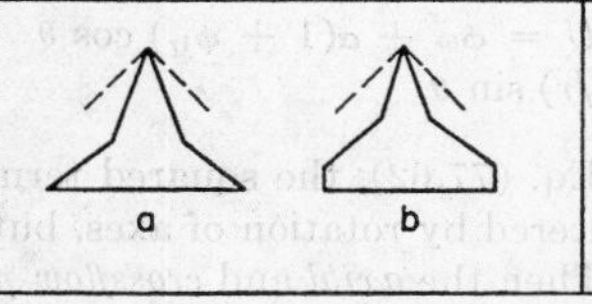		[65]

2. The lift-curve slope is the same in forward and reverse flight.

Second, *reciprocity rules* are found by considering wings having the same planform but different surface slopes in forward and reverse flow. One of the most useful is

3. The lift on a wing due to unit deflection of a flap equals the lift on the flap portion of the wing at uniform unit angle of attack in reverse flow.

For others, and generalization to unsteady motion, see [26].

Comparison with Experiment

Linearized theory may disagree with experiment because of higher-order or viscous effects. For wings of practical thinness, higher-order effects are significant only in making the center of pressure uncertain (cf. p. 77–12). Figure 77.16 shows the accuracy of linearized theory for lift. Viscous effects become significant for

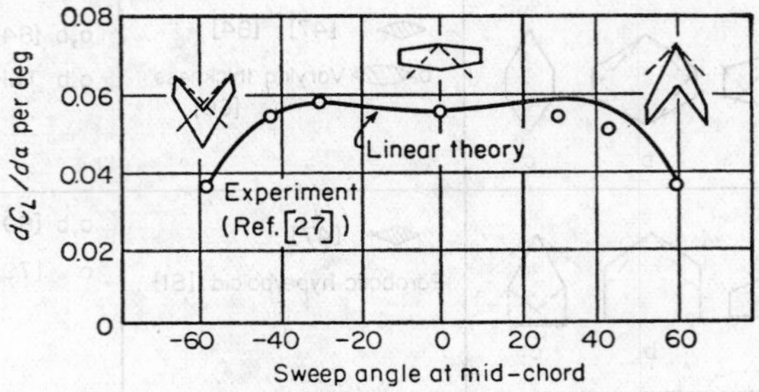

FIG. 77.16. Lift on thin wings at $M_\infty = 1.53$.

low-aspect-ratio wings as the angle of attack is increased. As in subsonic flow, the flow separates from the leading edges, forming two vortices. Leading-edge thrust is lost, but the lift curve remains linear. Viscosity also causes separation or transition to turbulent flow on highly swept planforms [27].

77.8. BODIES

Linearized Flow past Axisymmetric Bodies

The simplest three-dimensional shapes are bodies of revolution. The tangency condition is simplified at angle of attack by the use of body axes (Fig. 77.17). Then, for small angles α, the velocity potential has the form

$$\Phi = U\{[X + \phi_0(X,r)] + \alpha[r + \phi_1(X,r)] \cos \theta\} \tag{77.87}$$

instead of Eq. (77.54). The velocity components in cylindrical coordinates are

$$\begin{gathered} u/U = (1 + \phi_{0x}) + \alpha\phi_{1x} \cos \theta \qquad v/U = \phi_{0r} + \alpha(1 + \phi_{1r}) \cos \theta \\ w/U = -\alpha(1 + \phi_1/r) \sin \theta \end{gathered} \tag{77.88}$$

and the pressure is given in terms of these by Eq. (77.62), the squared terms being essential. The linearized equation is actually altered by rotation of axes, but should be taken as unchanged for best accuracy [28]. Then the *axial* and *crossflow perturbation potentials* ϕ_0 and ϕ_1 satisfy

$$\phi_{0rr} + \phi_{0r}/r - \beta^2\phi_{0XX} = 0 \qquad \phi_{1rr} + \phi_{1r}/r - \phi_{1\theta\theta}/r^2 - \beta^2\phi_{1XX} = 0 \tag{77.89a}$$

$$\phi_{0r} = R' \qquad \phi_{1r} = -1 \qquad \text{at } r = R(X) \tag{77.89b}$$

and vanish at the bow Mach cone. The general solutions are given by arbitrary distributions of point sources [Eq. (77.67)] and doublets [Eq. (77.69)] along the axis:

$$\phi_0 = -\int_0^{X-\beta r} \frac{f(\xi)\,d\xi}{\sqrt{(X-\xi)^2 - \beta^2 r^2}} = -\int_0^{\operatorname{Cosh}^{-1} X/\beta r} f(X - \beta r \operatorname{Cosh} \xi)\,d\xi$$

$$\phi_1 = \frac{1}{\beta r}\int_0^{X-\beta r} \frac{(X-\xi)g(\xi)\,d\xi}{\sqrt{(X-\xi)^2 - \beta^2 r^2}} = \int_0^{\operatorname{Cosh}^{-1} X/\beta r} \operatorname{Sinh} \xi\, g(X - \beta r \operatorname{Cosh} \xi)\,d\xi \qquad (77.90a,b)$$

the second forms being useful for differentiation. The tangency conditions (77.89*b*) give Volterra integral equations for $f(X)$, $g(X)$ that cannot be solved explicitly.

Linearized Supersonic Flow past Cone

Taking $f(X)$, $g(X)$ proportional to X gives the conical source [Eq. (77.72)] and doublet [Eq. (77.74)]. Choosing constants to satisfy tangency on a cone of semi-angle σ gives, with $t = \beta r/x$,

$$\frac{\Phi}{U} = X\left[1 - \frac{\sigma^2}{\sqrt{1-\beta^2\sigma^2}}\left(\operatorname{Sech}^{-1} t - \sqrt{1-t^2}\right)\right]$$
$$+ \alpha r\left[1 + \frac{\beta^2\sigma^2}{\sqrt{1-\beta^2\sigma^2} + \beta^2\sigma^2 \operatorname{Sech}^{-1}\beta\sigma}\left(\frac{\sqrt{1-t^2}}{t^2} - \operatorname{Sech}^{-1} t\right)\right]\cos\theta \qquad (77.91)$$

The surface pressure coefficient is

$$C_p = \sigma^2\left(2\frac{\operatorname{Sech}^{-1}\beta\sigma}{\sqrt{1-\beta^2\sigma^2}} - 1\right) - \frac{4\sigma\sqrt{1-\beta^2\sigma^2}}{\sqrt{1-\beta^2\sigma^2} + \beta^2\sigma^2 \operatorname{Sech}^{-1}\beta\sigma}\,\alpha\cos\theta \qquad (77.92)$$

and the lift coefficient

$$C_L = \frac{2\sqrt{1-\beta^2\sigma^2}}{\sqrt{1-\beta^2\sigma^2} + \beta^2\sigma^2 \operatorname{Sech}^{-1}\beta\sigma}\,\alpha \qquad (77.93)$$

The accuracy of the lift can be considerably increased [28] by using the linearized velocities in the exact isentropic pressure relation of Eq. (77.61).

Numerical Method for General Axisymmetric Bodies

The linearized solution for other bodies of revolution can be approximated by superposing conical solutions, starting at points O, X_1, X_2, . . . , on the axis (Fig.

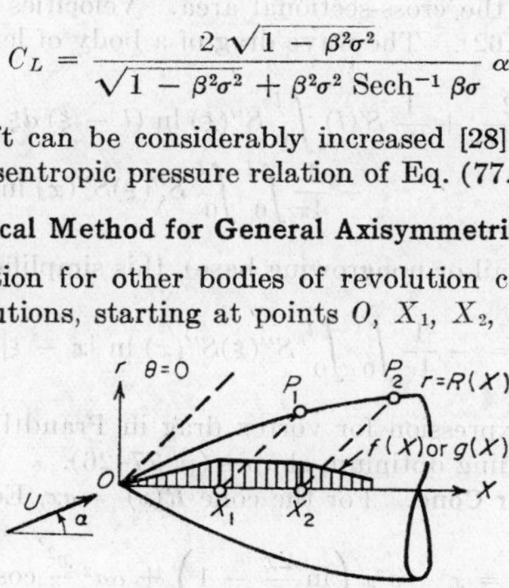

FIG. 77.17. Polygonal approximation to source or dipole distribution.

77.17). Thus, the unknown source and dipole functions are replaced by broken lines. The conical solution starting at X_n gives velocities

$$\frac{u-U}{U} = -C_n \operatorname{Sech}^{-1} t_n + 2D_n \frac{\sqrt{1-t_n^2}}{t_n}\,\alpha\cos\theta \qquad t_n \equiv \frac{\beta r}{X - X_n}$$

$$\frac{v}{U} = \beta C_n \frac{\sqrt{1-t_n^2}}{t_n} - \beta D_n\left(\frac{\sqrt{1-t_n^2}}{t_n^2} + \operatorname{Sech}^{-1} t_n\right)\alpha\cos\theta \qquad (77.94)$$

$$\frac{w}{U} = -\beta D_n\left(\frac{\sqrt{1-t_n^2}}{t_n^2} - \operatorname{Sech}^{-1} t_n\right)\alpha\sin\theta$$

The constants C_0 and D_0 are determined by imposing tangency [Eq. (77.89b)] at point P_1 (which is affected only by the first conical solution), then C_1 and D_1 from tangency at P_2 (which is affected by the first two solutions), etc. This procedure applies also to purely supersonic flow outside open-nosed bodies.

If a discontinuity in slope occurs at point P_m, the constant C_m is determined from the slope just ahead of the corner, and tangency for ϕ_0 just behind is satisfied by adding a proper multiple of the axisymmetric *corner solution*, starting at X_m [28], with velocities

$$\frac{u-U}{U} = \frac{1}{\sqrt{r}}\sqrt{\frac{2t_m}{1+t_m}}\,K\left(\frac{1-t_m}{1+t_m}\right) \qquad t_m \equiv \frac{\beta r}{X-X_m}$$
$$\frac{v}{U} = -\frac{\beta}{\sqrt{r}}\sqrt{\frac{2t_m}{1+t_m}}\left[\frac{1+t_m}{t_m}\,E\left(\frac{1-t_m}{1+t_m}\right) - K\left(\frac{1-t_m}{1+t_m}\right)\right] \tag{77.95}$$

where $K(k^2)$ and $E(k^2)$ are the complete elliptic integrals of the first and second kinds.

Slender-body Solution for Axisymmetric Shapes

As discussed on p. 79–3, the slender-body potential is a solution of Laplace's equation in the cross plane. For a body of revolution described by $r = R(x)$ at angle of attack α, it is found by simplifying the linearized solution [29]. This gives, for the velocity potential in wind axes,

$$\frac{\Phi}{U} = x + \frac{1}{2\pi}\left[S'(x)\ln\frac{\beta r}{2x} + \int_0^x \frac{S'(x)-S'(\xi)}{x-\xi}\,d\xi\right] + \alpha\frac{R^2(x)}{r}\cos\theta \tag{77.96}$$

where $S(x) = \pi R^2(s)$ is the cross-sectional area. Velocities and pressure are given by Eqs. (77.55) and (77.62). The wave drag of a body of length l is

$$\frac{D}{\rho_\infty U^2} = \frac{1}{4\pi}[S'(l)]^2\ln\frac{2}{\beta R(l)} + \frac{1}{2\pi}S'(l)\int_0^l S''(\xi)\ln(l-\xi)\,d\xi - \frac{1}{4\pi}\int_0^l\int_0^l S''(\xi)S''(x)\ln|x-\xi|\,dx\,d\xi \tag{77.97a}$$

For $S'(l) = 0$ (pointed tail or nongrowing base), this simplifies to

$$\frac{D}{\rho_\infty U^2} = -\frac{1}{4\pi}\int_0^l\int_0^l S''(\xi)S''(x)\ln|x-\xi|\,dx\,d\xi \tag{77.97b}$$

This is similar to the expression for vortex drag in Prandtl's lifting-line theory, an analogy exploited in finding optimum shapes (p. 77–26).

Example 2. Circular Cone. For the cone $R(x) = \sigma x$, Eq. (77.96) gives

$$\frac{\Phi}{U} = x - \sigma^2 x\left(\ln\frac{2x}{\beta r} - 1\right) + \alpha\sigma^2\frac{x^2}{r}\cos\theta \tag{77.98}$$

and the surface pressure coefficient becomes

$$C_p = \sigma^2\left(2\ln\frac{2}{\beta\sigma} - 1\right) - 2\alpha\sigma\cos\theta - \alpha^2 \tag{77.99}$$

These are the expansions for small σ and t of the linearized results [Eqs. (77.91) and (77.92)].

Slender-body theory approximates surface pressures uniformly if the meridian curvature $R''(x)$ is continuous, (and integrated forces if the slope R' is continuous). Then the relative error for a body of thickness ratio τ is $O(\tau^2 \ln \tau)$, which is the same as in linearized theory. However, linearized theory exceeds the slender-body approximation in actual numerical accuracy for Mach numbers appreciably greater than 2 (cf. Fig. 77.18).

Slender-body Solution for General Shapes

For any fusiform object, the thickness effect in the slender-body approximation is that for a body of revolution having the same distribution of cross-sectional area $S(x)$. Hence,

$$\phi(x,y,z) = \phi_2(y,z) + \frac{1}{2\pi}\left[S'(x)\ln\frac{\beta}{2x} + \int_0^x \frac{S'(x) - S'(\xi)}{x - \xi}\,d\xi\right] \qquad (77.100)$$

Here ϕ_2 is a solution of Laplace's equation in the y,z cross plane (see Chap. 79), determined so as to satisfy the tangency condition and (in wind axes) vanish as $r \to \infty$, except for a term $(2\pi)^{-1}S(x)\ln r$. For details, see [30].

Thus, the solution for a slender elliptic cone is derived in [31]. At zero incidence, the drag coefficient (referred to base area) is

$$C_D = ab\left[2\ln\frac{4}{\beta(a+b)} - 1\right] \qquad (77.101)$$

where a and b are the side and top slopes. For given cross-sectional area, flattening reduces the drag.

Quasi-cylindrical Theory

If a body deviates only slightly from a streamwise circular cylinder, the tangency condition can be transferred to that cylinder (quasi-cylindrical approximation, p. 77–13). Then surface pressures can be expressed in terms of universal functions. The first is the result of a discontinuity in slope on an axisymmetric body; by superposition, it yields approximately the pressures over any body of revolution ([2*a*], p. 453). Higher functions vary as $\cos n\theta$ about the streamwise axis. They are extensively tabulated and applications are discussed in [32].

Exact, Higher-order Methods for Axisymmetric Flow

The numerical method of characteristics yields accurate solutions for bodies of revolution without incidence [5]. Hand computation requires 20 hours in a typical case. References [33] and [34] give reliable families of solutions for ogival noses and cone-cylinders. The axisymmetric solution can, in principle, be perturbed to calculate the effects of incidence, but no reliable results exist (except for the cone,

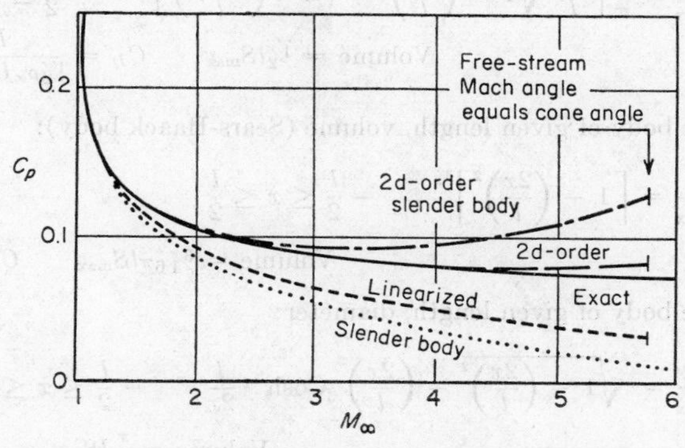

FIG. 77.18. Pressure on circular cone of 10° half angle.

p. 77–8), owing to practical difficulties. A numerical method for three dimensions is not yet fully developed.

The counterpart for axisymmetric flow of Busemann's second-order airfoil theory (p. 77–10) is given in [35], where the cone is solved explicitly. In slender-body theory (p. 77–24), the second approximation for pressure on the cone is

$$C_p = \sigma^2\left(2 \ln \frac{2}{\beta\sigma} - 1\right) + \sigma^4\left[3\beta^2\left(\ln \frac{2}{\beta\sigma}\right)^2 - (5M_\infty{}^2 - 1)\ln\frac{2}{\beta\sigma} + (\gamma + 1)M_\infty{}^2/\beta^2 + 1\tfrac{3}{4}M_\infty{}^2 + \tfrac{1}{2}\right] \quad (77.102)$$

Figure 77.18 shows the relative accuracies of the linearized and second-order solutions and their slender-body counterparts.

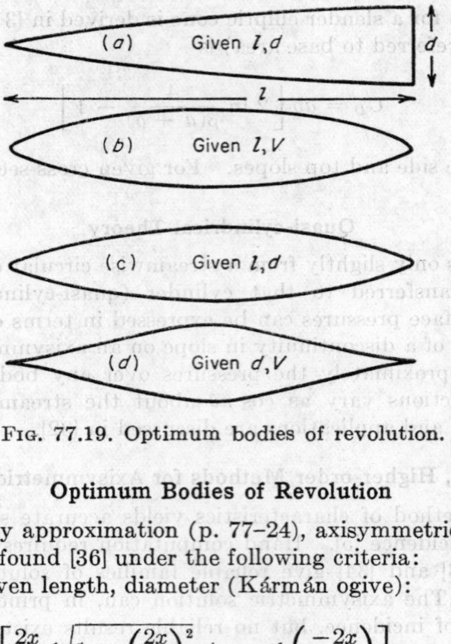

Fig. 77.19. Optimum bodies of revolution.

Optimum Bodies of Revolution

In the slender-body approximation (p. 77–24), axisymmetric shapes of minimum wave drag have been found [36] under the following criteria:

1. Half body of given length, diameter (Kármán ogive):

$$\left(\frac{R}{R_{\max}}\right)^2 = \frac{S}{S_{\max}} = \frac{1}{\pi}\left[\frac{2x}{l}\sqrt{1 - \left(\frac{2x}{l}\right)^2} + \cos^{-1}\left(\frac{-2x}{l}\right)\right] \qquad -\frac{l}{2} \le x \le \frac{l}{2}$$

$$\text{Volume} = \tfrac{1}{2} l S_{\max} \qquad C_D = \frac{D}{\tfrac{1}{2}\rho_\infty U^2 S_{\max}} = \left(\frac{d}{l}\right)^2$$

2. Complete body of given length, volume (Sears-Haack body):

$$\left(\frac{R}{R_{\max}}\right)^2 = \frac{S}{S_{\max}} = \left[1 - \left(\frac{2x}{l}\right)^2\right]^{3/2} \qquad -\frac{l}{2} \le x \le \frac{l}{2}$$

$$\text{Volume} = \tfrac{3}{16}\pi l S_{\max} \qquad C_D = \tfrac{9}{8}\pi^2\left(\frac{d}{l}\right)^2$$

3. Complete body of given length, diameter:

$$\left(\frac{R}{R_{\max}}\right)^2 = \frac{S}{S_{\max}} = \sqrt{1 - \left(\frac{2x}{l}\right)^2} - \left(\frac{2x}{l}\right)^2 \operatorname{Cosh}^{-1}\frac{l}{2x} \qquad -\frac{l}{2} \le x \le \frac{l}{2}$$

$$\text{Volume} = \frac{\pi}{6} l S_{\max} \qquad C_D = \pi^2\left(\frac{d}{l}\right)^2$$

4. Complete body of given diameter, volume:

$$\left(\frac{R}{R_{\max}}\right)^2 = \frac{S}{S_{\max}} = 3\sqrt{1-\left(\frac{2x}{l}\right)^2} - 2\left[1-\left(\frac{2x}{l}\right)^2\right]^{3/2} - 3\left(\frac{2x}{l}\right)^2 \operatorname{Cosh}^{-1}\frac{l}{2x}$$

$$\text{Volume} = \frac{\pi}{8} lS_{\max} \qquad C_D = 3/2\pi^2 \left(\frac{d}{l}\right)^2$$

These shapes are shown in Fig. 77.19. The latter three are symmetric fore and aft. Further reduction of pressure drag by blunting their bases (cf. p. 77–11) has not yet been studied.

Optimum bodies of revolution have also been investigated, using linearized theory [37] and full nonlinear isentropic theory [38].

77.9. REFERENCES

General References

[1] R. Courant, K. O. Friedrichs: "Supersonic Flow and Shock Waves," Interscience, New York, 1948.

[2] "High Speed Aerodynamics and Jet Propulsion": (*a*) vol. 6: W. R. Sears (ed.): "General Theory of High Speed Aerodynamics," (*b*) vol. 7: A. F. Donovan, H. R. Lawrence (eds.): "Aerodynamic Components of Aircraft at High Speeds," Princeton Univ. Press, Princeton, N.J., 1954, 1957.

[3] H. W. Liepmann, A. Roshko: "Elements of Gasdynamics," Wiley, New York, 1957.

[4] K. Oswatitsch: "Gas Dynamics," Academic Press, New York, 1956.

Specific References

[5] J. S. Isenberg, C. C. Lin: The method of characteristics in compressible flow, part I, Steady supersonic flow, *Air Materiel Command, Tech. Rept.* F-TR-1173A-ND (6DAM A9-M11/1), Wright-Patterson Air Force Base, Ohio, 1947.

[6] D. Gilbarg, D. Paolucci: The structure of shock waves in the continuum theory of fluids, *J. Rat. Mechanics*, **2** (1953), 617–642.

[7] Ames Research Staff: Equations, tables, and charts for compressible flow, *NACA Rept.* 1135 (1953).

[8] Staff of the Computing Section, Center of Analysis (under direction of Z. Kopal): (*a*) Tables of supersonic flow around cones, (*b*) Tables of supersonic flow around yawing cones, *MIT Tech. Repts.* 1 and 3, (1947).

[9] R. C. Roberts, J. D. Riley: A guide to the use of the MIT cone tables, *J. Aeronaut. Sci.*, **21** (1954), 336–342.

[10] M. Holt, J. Blackie: Experiments on circular cones at yaw in supersonic flow, *J. Aeronaut. Sci.*, **23** (1956), 931–936.

[11] A. J. Eggers, Jr., C. A. Syvertson, S. Kraus: A study of inviscid flow about airfoils at high supersonic speeds, *NACA Rept.* 1123 (1953).

[12] A. E. Donov: A flat wing with sharp edges in a supersonic stream, *NACA Tech. Mem.* 1394 (1956); transl. from *Izvest. Akad. Nauk SSSR*, **1939**, 603–626.

[13] D. R. Chapman: Airfoil profiles for minimum pressure drag at supersonic velocities: (*a*) *NACA Rept.* 1063 (1952) and (*b*) *Tech. Note* 2787 (1952).

[14] A. Busemann: Aerodynamischer Auftrieb bei Überschallgeschwindigkeit, *Luftfahrt-Forsch.*, **12** (1935), 210–220; reprinted in G. F. Carrier: "Foundations of High-Speed Aerodynamics," pp. 131–141, Dover, New York, 1951.

[15] W. D. Hayes: Linearized supersonic flow with axial symmetry, *Quart. Appl. Math.*, **4** (1946), 255–261.

[16] A. Ferri: The linearized characteristics method and its applications to practical nonlinear supersonic problems, *NACA Rept.* 1102 (1952).

[17] N. Ness, T. T. Kaplita: Tabulated values of linearized conical flow solutions for solution of supersonic conical flows without axial symmetry, *Polytech. Inst. Brooklyn PIBAL Rept.* 220 (1954).

[18] J. C. Evvard: Use of source distributions for evaluating theoretical aerodynamics of thin finite wings at supersonic speeds, *NACA Rept.* 951 (1950).

[19] P. A. Lagerstrom: Linearized supersonic theory of conical wings, *NACA Tech. Note* 1685 (1950).

[20] H. Mirels: A lift-cancellation technique in linearized supersonic wing theory, *NACA Rept.* 1004 (1951).

[21] P. A. Lagerstrom, M. E. Graham: Low aspect ratio rectangular wings in supersonic flow, *Douglas Aircraft Co. Rept.* SM-13110 (1947).
[22] A. Busemann: Infinitesimal conical supersonic flow, *NACA Tech. Mem.* 1100 (1957); transl. of Infinitesimale kegelige Überschallströmung, *Jahrb. deut. Akad. Luftfahrt-Forsch.*, 1942–43, p. 455.
[23] P. Germain: General theory of conical flows and its application to supersonic aerodynamics, *NACA Tech. Mem.* 1354 (1955); transl. of La théorie générale des mouvements coniques et ses applications à l'aérodynamique supersonique, *ONERA Rept.* 34 (1949).
[24] E. Lapin: Charts for the computation of lift and drag of finite wings at supersonic speeds, *Douglas Aircraft Co. Rept.* SM-13480 (1949).
[25] F. Ursell, G. N. Ward: On some general theorems in the linearized theory of compressible flow, *Quart. J. Mech. Appl. Math.*, **3** (1950), 326–348.
[26] M. A. Heaslet, J. R. Spreiter: Reciprocity relations in aerodynamics, *NACA Rept.* 1119 (1953).
[27] W. G. Vincenti: Comparison between theory and experiment for wings at supersonic speeds, *NACA Rept.* 1033 (1951).
[28] M. D. Van Dyke: First- and second-order theory of supersonic flow past bodies of revolution, *J. Aeronaut. Sci.*, **18** (1951), 161–178.
[29] M. A. Heaslet, H. Lomax: The calculation of pressure on slender airplanes in subsonic and supersonic flow, *NACA Tech. Note* 2900 (1953).
[30] G. N. Ward: Supersonic flow past slender pointed bodies, *Quart. J. Mech. Appl. Math.*, **3** (1949), 75–97.
[31] L. E. Fraenkel: Supersonic flow past slender bodies of elliptic cross-section, *Aeronaut. Research Council Rept. and Mem.* 2954 (1955).
[32] J. N. Nielsen: Tables of characteristic functions for solving boundary-value problems of the wave equation with application to supersonic interference, *NACA Tech. Note* 3873 (1957).
[33] V. J. Rossow: Applicability of the hypersonic similarity rule to pressure distributions which include the effects of rotation for bodies of revolution at zero angle of attack, *NACA Tech. Note* 2399 (1951).
[34] R. F. Clippinger, J. H. Giese, W. C. Carter: Tables of supersonic flows about cone cylinders, parts I and II, *Ballistic Research Lab. Repts.*, 729 and 730 (1950).
[35] M. D. Van Dyke: Practical calculation of second-order supersonic flow past non-lifting bodies of revolution, *NACA Tech. Note* 2744 (1952).
[36] W. R. Sears: On projectiles of minimum wave drag, *Quart. Appl. Math.*, **6** (1947), 361–366.
[37] H. M. Parker: (*a*) Minimum drag ducted and pointed bodies of revolution based on linearized supersonic theory, *NACA Rept.* 1213 (1955); (*b*) Minimum drag ducted and closed three-point body of revolution based on linearized supersonic theory, *NACA Tech. Note* 3704 (1956).
[38] Yu. D. Shmyglevskii: Some variational problems of gas dynamics for axisymmetric supersonic flows (in Russian), *Prikl. Mat. Mekh.*, **21** (1957), 195–206.
[39] A. Ferri: Experimental results with airfoils tested in the high-speed tunnel at Guidonia, *NACA Tech. Mem.* 946 (1940); transl. of Alcuni risultati sperimentali riguardanti profili alari provati alla galleria ultrasonora di Guidonia, *Atti Guidonia*, no. 17 (1939).
[40] E. S. Love: Investigations at supersonic speeds of 22 triangular wings representing two airfoil sections for each of 11 apex angles, *NACA Rept.* 1238 (1955).

References for Table 77.2

[41] R. T. Jones: *NACA Rept.* 851 (1946).
[42] S. M. Harmon, M. D. Swanson: *NACA Tech. Note* 1319 (1947).
[43] R. T. Jones: *NACA Tech. Note* 1350 (1947).
[44] M. A. Heaslet, H. Lomax, A. L. Jones: *NACA Rept.* 889 (1947).
[45] K. Margolis: *NACA Tech. Note* 1448 (1947).
[46] S. M. Harmon: *NACA Tech. Note* 1449 (1947).
[47] J. N. Nielsen: *NACA Tech. Note* 1487 (1947).
[48] K. Margolis: *NACA Tech. Note* 1543 (1948).
[49] W. E. Moeckel: *NACA Tech. Note* 1549 (1948).
[50] K. Margolis: *NACA Tech. Note* 1672 (1948).
[51] B. S. Baldwin: *NACA Tech. Note* 1816 (1949).
[52] F. S. Malvestuto, K. Margolis, H. S. Ribner: *NACA Tech. Note* 1860 (1949).
[53] A. L. Jones, R. M. Sorenson, E. E. Lindler: *NACA Tech. Note* 1916 (1949).
[54] R. O. Piland: *NACA Tech. Note* 1977 (1949).
[55] D. Cohen: *NACA Rept.* 1059 (1951).

[56] A. L. Jones, A. Alksne: *NACA Tech. Note* 2007 (1950).
[57] J. H. Kainer: *NACA Tech. Note* 2009 (1950).
[58] K. Margolis: *NACA Tech. Note* 2048 (1950).
[59] S. M. Harmon, I. Jeffreys: *NACA Tech. Note* 2114 (1950).
[60] F. S. Malvestuto, D. B. Hoover: *NACA Tech. Note* 2294 (1951).
[61] H. Lomax, M. A. Heaslet: *NACA Tech. Note* 2497 (1951).
[62] J. C. Martin, K. Margolis, I. Jeffreys: *NACA Tech. Note* 2699 (1952).
[63] A. Henderson: *NACA Tech. Note* 2858 (1952).
[64] K. Margolis, W. L. Sherman, M. E. Hannah: *NACA Tech. Note* 2989 (1953).
[65] D. Cohen, M. D. Friedman: *NACA Tech. Note* 2959 (1953).
[66] A. Henderson, J. M. Goodwin: *NACA Tech. Note* 3418 (1955).
[67] A. E. Puckett: *J. Aeronaut. Sci.*, **13** (1946), 475.
[68] A. E. Puckett, H. J. Stewart: *J. Aeronaut. Sci.*, **14** (1947), 567.
[69] E. Lapin: *J. Aeronaut. Sci.*, **16** (1949), 381.
[70] B. Beane: *J. Aeronaut. Sci.*, **18** (1951), 7.
[71] J. H. Kainer: *J. Aeronaut. Sci.*, **20** (1953), 469.
[72] K. Walker: *J. Aeronaut. Sci.*, **20** (1953), 509.
[73] N. Rott: *J. Aeronaut. Sci.*, **20** (1953), 642.
[74] H. K. Zienkiewicz: *J. Aeronaut. Sci.*, **21** (1954), 421.
[75] R. N. Haskell, J. J. Hosek, W. S. Johnson: *J. Aeronaut. Sci.*, **22** (1955), 274.
[76] P. Kafka: *J. Aeronaut. Sci.*, **22** (1955), 725.
[77] H.-S. Tsien: *J. Aeronaut. Sci.*, **22** (1955), 805.
[78] M. Fenain: *Recherche aéronaut.*, **16** (1950), 27.
[79] A. Gilles: *Recherche aéronaut.*, **19** (1951), 3.
[80] M. Bismut: *Recherche aéronaut.*, **21** (1951), 9.
[81] M. Fenain: *Recherche aéronaut.*, **24** (1951), 25.
[82] M. Fenain, D. Vallée: *Recherche aéronaut.*, **44** (1955), 9.
[83] M. Fenain, D. Vallée: *Recherche aéronaut.*, **50** (1956), 17.
[84] P. A. Lagerstrom, D. Wall, M. E. Graham: *Douglas Aircraft Co. Rept.* SM-11901 (1946).
[85] P. A. Lagerstrom, M. E. Graham: *Douglas Aircraft Co. Rept.* SM-13200 (1948).
[86] M. E. Graham: *Douglas Aircraft Co. Rept.* SM-13303 (1948).
[87] H. J. Stewart: *Quart. Appl. Math.*, **4** (1946), 246.
[88] H. B. Squire: *Brit. Aeronaut. Research Comm. Rept. and Mem.* 2549 (1951).
[89] H. Behrbohm: *SAAB Aircraft Co. (Sweden) Tech. Note* 10 (1952).

CHAPTER 78

HYPERSONIC FLOW

BY
A. J. EGGERS, Jr., Ph.D., Los Altos, Calif.
C. A. SYVERTSON, Saratoga, Calif.

78.1. INTRODUCTION

This chapter treats steady inviscid flow, generated by a solid body immersed in a stream of gas, which is uniform and has a velocity large compared to the speed of sound ahead of the body. Thus,

$$M_\infty \gg 1$$

and such flows are described as being hypersonic. A hypersonic flow may contain local supersonic or transonic regions adjacent to the body.

Inertia forces frequently predominate over elastic forces in hypersonic flow, because even normally slender bodies produce strong shock waves. Thus, the simplifying assumption of potential flow is not usually permissible (see p. 77–12). Strong shock waves may cause temperatures to exceed values for which the gas behaves ideally, and such real-gas effects as dissociation may have to be considered. Attention herein is focused primarily on hypersonic flow of an ideal gas.

78.2. EXACT SOLUTIONS

Hypersonic flow through oblique shock waves, in Prandtl-Meyer expansions, and about wedges and circular cones is defined by the same "exact" solutions which arise in supersonic flow (see pp. 77–2, 77–5, 77–7). Hypersonic flow is also susceptible to solution by the method of characteristics. It is primarily in the selection of suitable approximate methods of solution that the treatment of hypersonic flows differs from that of supersonic flows.

78.3. APPROXIMATE SOLUTIONS

Newtonian Impact Theory

The simplest approximate theory for hypersonic flow about a solid body is obtained when inertia forces are large compared to elastic forces in the gas. This theory [1] is usually termed Newtonian or impact theory, and it yields surface pressure coefficients in the form

$$\begin{aligned} C_p &= 2\sin^2\delta \qquad && \delta \geq 0 \\ C_p &= 0 && \delta \leq 0 \end{aligned} \tag{78.1}$$

where δ is the angle through which the stream is deflected by the surface. Note that only forward-facing surfaces experience pressure; any surface on which the free stream cannot impinge directly is at zero pressure. Impact theory is especially useful in predicting the pressures on bodies of revolution [2] and for calculating minimum-drag bodies [3] at hypersonic speeds. The theory can be deduced from the oblique shock relations (p. 77–2) to be strictly applicable only in the limiting case of $M_\infty \to \infty$

and $\gamma \to 1$. Under these circumstances, the shock wave generated by a body coincides with the body surface. Impact theory neglects the centrifugal forces in the gas when it flows over a surface curved in the stream direction. Busemann [4] considered these forces and arrived at a modified impact theory for two-dimensional and axially symmetric flows, which yields

$$C_{p_i} = 2 \sin \delta_i \left[\sin \delta_i + \frac{1}{y_i{}^k} \left(\frac{d\delta}{dy}\right)_i \int_0^{y_i} y^k \cos \delta \, dy \right] \tag{78.2}$$

where the integration is performed along the surface from the beginning of a streamline to the point i, where pressure is desired. The constants $k = 0$ for planar flow and $k = 1$ for axially symmetric flow. Impact theory is most accurate for bodies of revolution, and Busemann theory does not become useful until γ approaches 1.

Shock-expansion Theory

This theory [5] is frequently more accurate than impact theory for hypersonic flow of air about sharp-nosed objects. The application of the theory to two-dimensional airfoils is described in the treatment of supersonic flow (p. 77–9). The

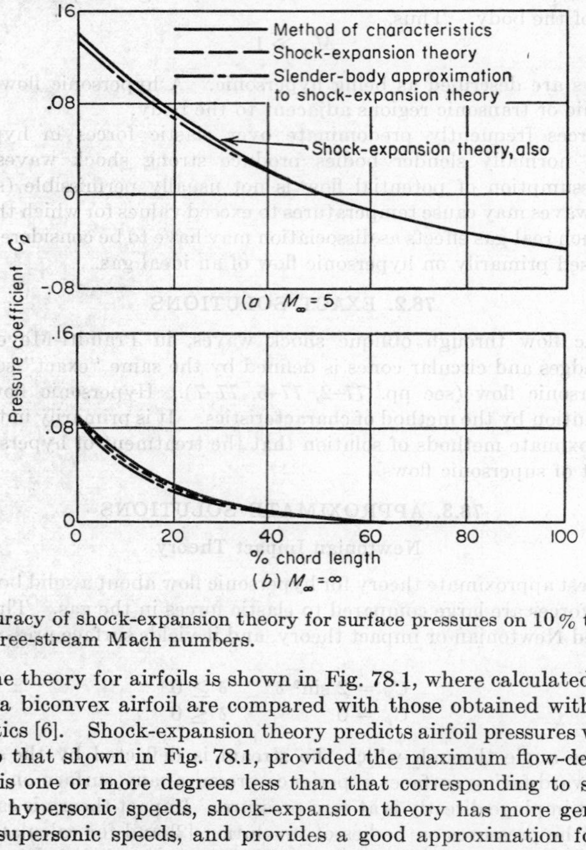

FIG. 78.1. Accuracy of shock-expansion theory for surface pressures on 10% thick biconvex airfoil at two free-stream Mach numbers.

accuracy of the theory for airfoils is shown in Fig. 78.1, where calculated pressure distributions on a biconvex airfoil are compared with those obtained with the method of characteristics [6]. Shock-expansion theory predicts airfoil pressures with accuracy comparable to that shown in Fig. 78.1, provided the maximum flow-deflection angle of the airfoil is one or more degrees less than that corresponding to shock detachment [6]. At hypersonic speeds, shock-expansion theory has more general application than at supersonic speeds, and provides a good approximation for flow about three-dimensional shapes such as bodies of revolution [7]–[9]. To apply the theory, the appropriate solution for the flow at the vertex or leading edge is combined with the Prandtl-Meyer expansion equations to obtain the flow downstream along streamlines. Consistent with the approximations of the method, surface streamlines are taken as geodesics [7]. For bodies of revolution, the vertex solution is that for a circular cone tangent to the body at the vertex (p. 77–7), and the appropriate surface geodesics are meridian lines. An example of the application of the method to an inclined tangent ogive is shown in Fig. 78.2. The method applies

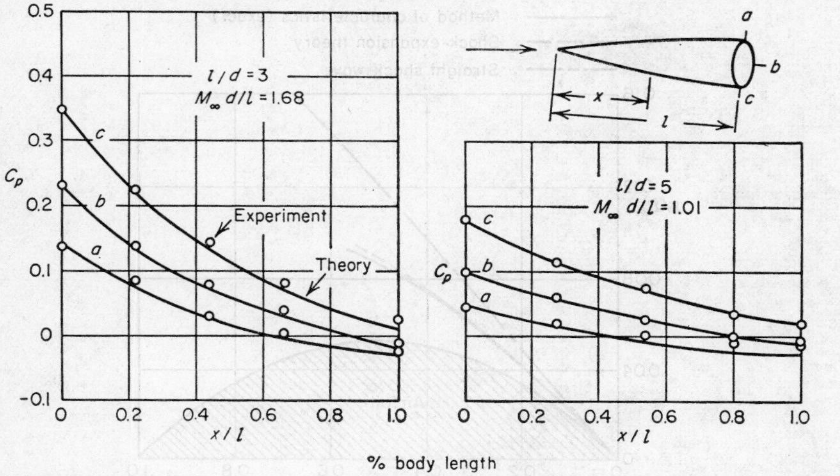

FIG. 78.2. Accuracy of shock-expansion theory for surface pressures on two inclined tangent ogives at $M_\infty = 5.05$ and $\alpha = 15°$ angle of attack.

with good accuracy to bodies of revolution at hypersonic speeds, so long as the hypersonic similarity parameter, $M_\infty d/l$ (where d is the maximum diameter of the body, and l is the total length), is greater than about 1. The method in any application requires that the flow be everywhere supersonic, and thus it cannot be applied in the region of a blunt nose. Shock-expansion theory can be employed to treat hypersonic flows of imperfect (i.e., nonideal) gases [6], provided the imperfections do not reduce the ratio of specific heats γ to less than about 1.3.

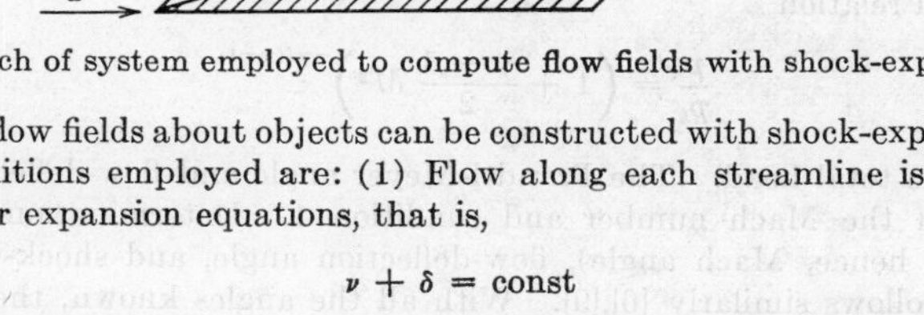

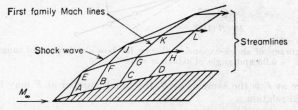

FIG. 78.3. Sketch of system employed to compute flow fields with shock-expansion theory.

Complete flow fields about objects can be constructed with shock-expansion theory. The two conditions employed are: (1) Flow along each streamline is given by the Prandtl-Meyer expansion equations, that is,

$$\nu + \delta = \text{const}$$

where ν is the Prandtl-Meyer expansion angle [see Eq. (77.33)], and (2) the pressure is constant along first family Mach lines (i.e., Mach lines positively inclined with respect to the local velocity vector). A schematic diagram of a flow-field construction is shown in Fig. 78.3. Points along the surface such as A, B, C, D are selected, and the Mach number, flow-deflection angle, and pressure are determined at each point with the aid of condition 1. From condition 2, the pressure at point E is the same as at A. The shock-wave angle, Mach number, flow-deflection angle, and total head at point E are then determined from the oblique-shock relations (p. 77–2). Since the total head is constant along streamlines downstream of the shock, and the

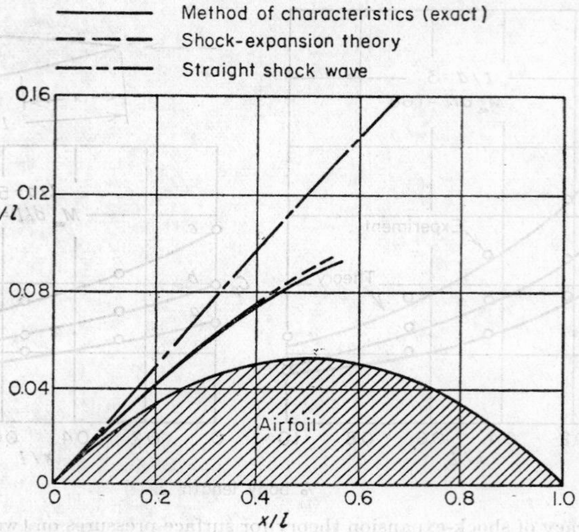

FIG. 78.4. Accuracy of shock-expansion theory for bow shock wave of noninclined 10% thick biconvex airfoil at $M_\infty = \infty$.

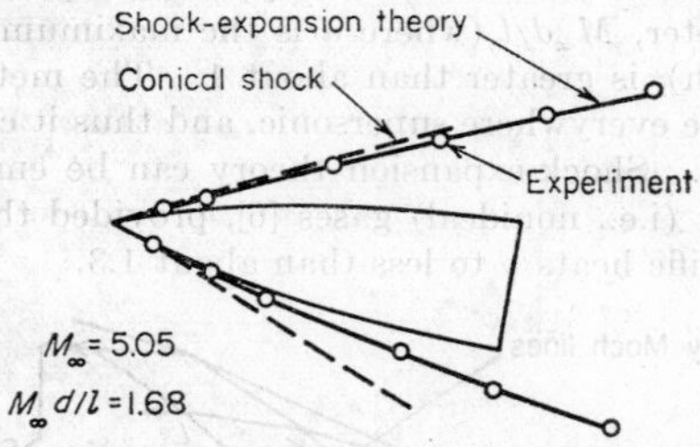

FIG. 78.5. Accuracy of shock-expansion theory for bow shock wave of tangent ogive with $l/d = 3$ at $M_\infty = 5.05$ and angle of attack $\alpha = 10°$.

static pressure at F is the same as at B, the Mach number at F may be determined from the usual relation

$$\frac{p}{p_t} = \left(1 + \frac{\gamma - 1}{2} M^2\right)^{-\gamma/(\gamma-1)} \tag{78.3}$$

where p_t is the total head. The Prandtl-Meyer angle and flow-deflection angle are obtained from the Mach number and condition 1. Determination of the Mach number (and, hence, Mach angle), flow-deflection angle, and shock-wave angle at other points follows similarly [6],[9]. With all the angles known, the slopes of the various segments of Mach lines, streamlines, and shock wave can be calculated by averaging the values at each end. With the slopes known, the lines themselves can be constructed graphically or numerically. The procedure is similar to, although markedly simpler than, the use of the method of characteristics. Shock waves determined with shock-expansion theory for an airfoil and a body of revolution are shown in Figs. 78.4 and 78.5. In both cases, it is seen that shock-expansion theory predicts the shock-wave shape with good accuracy.

Slender-body Shock-expansion Theory

Under the assumption that $\delta \ll 1$ and, thus, $M \gg 1$, shock-expansion theory can be reduced to a very simple form [6]–[8],[10]. For airfoils, the surface pressure coefficient is given by

$$C_p = \frac{2}{\gamma M_\infty^2}\left\{\frac{p_N}{p_\infty}\left[1 - \frac{\gamma - 1}{2} M_N\delta_N\left(1 - \frac{\delta}{\delta_N}\right)\right]^{2/(\gamma-1)} - 1\right\} \tag{78.4}$$

where

$$\frac{p_N}{p_\infty} = \frac{2\gamma M_\infty^2\sigma_N^2 - (\gamma - 1)}{\gamma + 1}$$

$$M_N\delta_N = \frac{M_\infty^2\sigma_N^2 - 1}{\sqrt{\{1 + [(\gamma - 1)/2]M_\infty^2\sigma_N^2\}[\gamma M_\infty^2\sigma_N^2 - (\gamma - 1)/2]}}$$

and

$$M_\infty\sigma_N = \frac{\gamma + 1}{4} M_\infty\delta_N + \sqrt{1 + \left(\frac{\gamma + 1}{4} M_\infty\delta_N\right)^2}$$

In these expressions, p_N, M_N, δ_N, and σ_N are the pressure, Mach number, flow-deflection angle, and shock-wave angle at the leading edge, respectively, and δ is the flow-deflection angle at the point where the pressure coefficient is desired. If $\delta_N \leq 0$, no shock wave is generated at the surface, and

$$C_p = \frac{2}{\gamma M_\infty^2}\left[\left(1 + \frac{\gamma - 1}{2} M_\infty\delta\right)^{2\gamma/(\gamma-1)} - 1\right] \tag{78.5}$$

As shown in Fig. 78.1, the slender-body approximation to shock-expansion theory provides accurate results. For noninclined bodies of revolution, Eq. (78.4) still applies, however, in this case,

$$\frac{p_N}{p_\infty} = \frac{p_S}{p_\infty}\frac{p_N}{p_S} \tag{78.6}$$

where

$$\frac{p_S}{p_\infty} = 1 + \gamma(M_\infty\delta_N)^2 \qquad \frac{p_N}{p_S} = \left(\frac{M_S}{M_N}\right)^{2\gamma(\gamma-1)}$$

$$\left(\frac{M_S}{M_N}\right)^2 = 1 + \frac{\gamma - 1}{2}\left(\frac{M_S}{M_\infty}\right)^2(M_\infty\delta_N)^2\left[1 + \ln\left(\frac{\delta_S}{\delta_N}\right)^2 - \left(\frac{\delta_S}{\delta_N}\right)^2\right]$$

$$\left(\frac{M_S}{M_\infty}\right)^2 = \frac{1 + [(\gamma + 1)/2](M_\infty\delta_N)^2}{[1 + \gamma(M_\infty\delta_N)^2][1 + [(\gamma - 1)/2](M_\infty\delta_N)^2]}$$

and

$$\left(\frac{\delta_S}{\delta_N}\right)^2 = \frac{(M_\infty\delta_N)^2}{1 + [(\gamma + 1)/2](M_\infty\delta_N)^2}$$

and the term $M_N\delta_N$ is determined from

$$(M_N\delta_N)^2 = (M_\infty\delta_N)^2\left(\frac{M_N}{M_S}\right)^2\left(\frac{M_S}{M_\infty}\right)^2$$

These equations can also be applied to inclined bodies of revolution in the manner shown in [8].

A Second-order Shock-expansion Theory

As previously noted, the shock-expansion theory gives accurate results for bodies of revolution when $M_\infty d/l$ is greater than about 1. For lower values of this parameter, a second-order shock-expansion theory is available. To apply this theory to a body of revolution, the surface of the body is approximated by a series of tangent conical frustums, as shown in Fig. 78.6. About each corner joining two frustums, the Prandtl-Meyer expansion equations give the exact change in pressure. Along the surface of the frustum, the pressure is approximated by

$$p = p_c - (p_c - p_2)e^{-\eta} \tag{78.7}$$

where
$$\eta = \left(\frac{\partial p}{\partial s}\right)_2 \frac{x - x_2}{(p_c - p_2)\cos\delta_2}$$

and
$$\left(\frac{\partial p}{\partial s}\right)_2 = \frac{\psi_2}{r_2}\left(\frac{\Omega_1}{\Omega_2}\sin\delta_1 - \sin\delta_2\right) + \frac{\psi_2}{\psi_1}\frac{\Omega_1}{\Omega_2}\left(\frac{\partial p}{\partial S}\right)_1$$

$$\psi = \frac{\gamma p M}{2(M^2-1)} \qquad \Omega = \frac{1}{M}\left\{\frac{1 + [(\gamma-1)/2]M^2}{(\gamma+1)/2}\right\}^{(\gamma+1)/2(\gamma-1)}$$

In these equations, p_c is the pressure on a cone of the same semivertex angle as the frustum (that is, δ_2), and r is the radial ordinate of the body. The subscript 1 refers

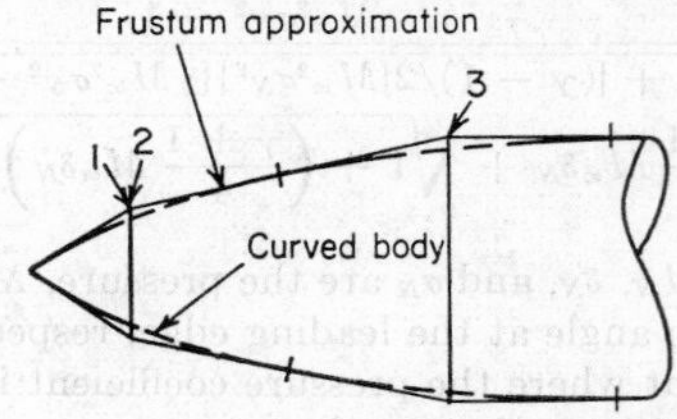

FIG. 78.6. Use of conical frustums to approximate curved body for second-order shock expansion method.

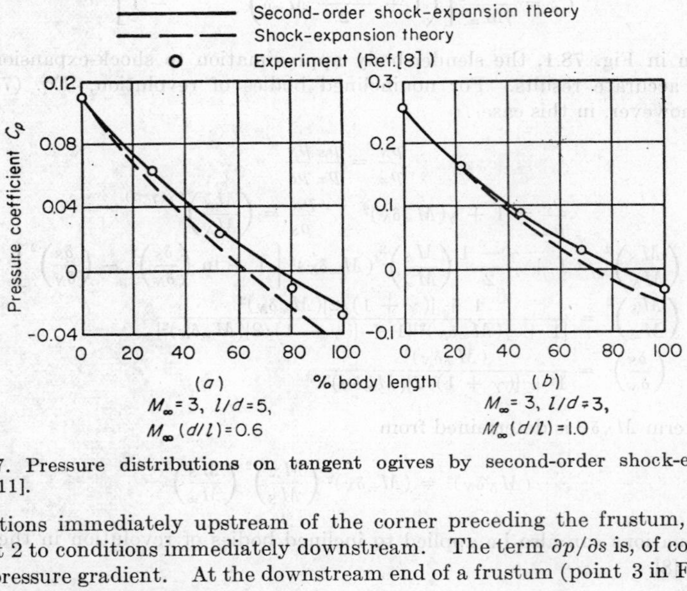

FIG. 78.7. Pressure distributions on tangent ogives by second-order shock-expansion method [11].

to conditions immediately upstream of the corner preceding the frustum, and the subscript 2 to conditions immediately downstream. The term $\partial p/\partial s$ is, of course, the surface pressure gradient. At the downstream end of a frustum (point 3 in Fig. 78.6), the pressure gradient is given by

$$\left(\frac{\partial p}{\partial s}\right)_3 = \left(\frac{\partial p}{\partial s}\right)_2 e^{-\eta_3} \tag{78.8}$$

where η_3 is the value of η for $x = x_3$, and, thus, with the conditions known at the end of a frustum, it is possible to proceed to the next frustum. In using this method, the pressure at the tangent point for each frustum is calculated and applied to the original body. The accuracy of this method is demonstrated in Fig. 78.7 for $M_\infty d/l = 0.6$ and 1.0. It is seen that the second-order theory offers some improvement over the shock-expansion theory itself for values of $M_\infty d/l$ of the order of and less than one. The second-order theory gives good accuracy for values of $M_\infty d/l$ as low as 0.5. It can be applied so long as $\eta \geq 0$, which is usually the case if $M_\infty d/l < 2.5$. This theory can also be used to determine the normal-force and pitching-moment derivatives for bodies of revolution at zero angle of attack [11].

Hypersonic Similarity Rule

A useful tool for correlating experimental data and enlarging the utility of hypersonic theory of flow about slender objects is the hypersonic similarity rule [12]–[15].

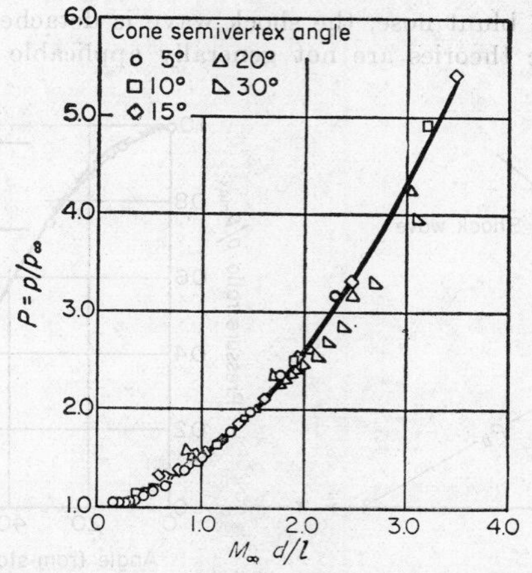

FIG. 78.8. Correlation of surface pressures on cones by hypersonic similarity rule.

This rule permits the aerodynamic characteristics of two slender configurations to be related, provided their shapes are related geometrically by an affine transformation, and provided the similarity parameters

$$K_t = M_\infty(t/c) \qquad K_b = M_\infty(b/c)$$
$$K_\alpha = M_\infty\alpha \qquad K_\beta = M_\infty\beta \qquad K_\varphi = \varphi$$
$$K_p = M_\infty\left(\frac{pb}{V_\infty}\right) \qquad K_q = M_\infty\left(\frac{qc}{V_\infty}\right) \qquad K_r = M_\infty\left(\frac{rc}{V_\infty}\right)$$

are the same for both configurations. In these expressions, t is some characteristic thickness, c is some characteristic length, and b is some characteristic lateral dimension of the configuration. In addition α is the angle of attack; β is the sideslip angle; φ is the angle of roll; p, q, and r are the angular velocities of roll, pitch, and yaw, respectively; and V_∞ is the free-stream velocity. Consideration of the differential equations of motion and boundary conditions [15] leads to the conclusion that the following parameters will be the same for related configurations:

$$\tilde{C}_L = M_\infty{}^{1+m}C_L \qquad \tilde{C}_D = M_\infty{}^{2+m}C_D \qquad \tilde{C}_Y = M_\infty{}^{1+m}C_Y$$
$$\tilde{C}_m = M_\infty{}^{1+m}C_m \qquad \tilde{C}_n = M_\infty{}^{m}C_n \qquad \tilde{C}_l = M_\infty{}^{1+m}C_l$$

where $m = 0$ if the coefficients are referenced to base or frontal area, and $m = 1$ if they are referenced to wing or plan area. Local pressures correlated according to

$$\tilde{C}_p = M_\infty{}^2 C_p \qquad \text{or} \qquad \tilde{p} = p/p_\infty$$

also satisfy the rule. To demonstrate the accuracy of this aspect of the rule, the correlation achieved for the pressures on noninclined cones [16] is shown in Fig. 78.8. It should be noted that the similarity rule applies only to pressure forces, and, thus, the drag coefficient employed in determining $\tilde{C}_D$ must not include the friction drag. If in all the similarity relations, the Mach number M_∞ is replaced by $B_\infty = \sqrt{M_\infty{}^2 - 1}$, a combined hypersonic-supersonic similarity rule results [17], although, in this case, pressures must be correlated with $\tilde{C}_p$. At hypersonic speeds, the difference between M and B is, of course, small.

Effects of Nose Bluntness

If an object has a blunt nose, the shock wave is detached, and the previously described approximate theories are not generally applicable to the calculation of

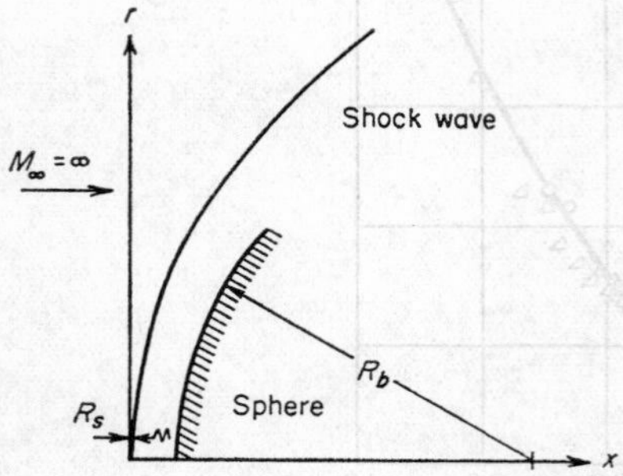

FIG. 78.9. Shock wave generated by sphere at $M_\infty = \infty$ as determined by numerical method of Ref. [18].

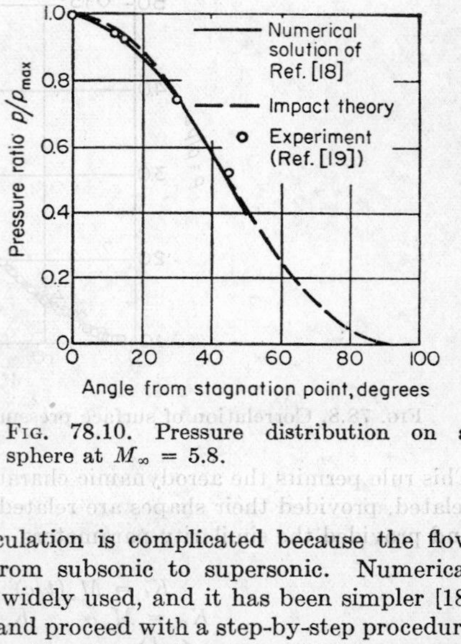

FIG. 78.10. Pressure distribution on a sphere at $M_\infty = 5.8$.

flow in the region of the nose. This calculation is complicated because the flow between the shock and the nose varies from subsonic to supersonic. Numerical methods of calculating the flow have been widely used, and it has been simpler [18] to assume the shape of the detached shock, and proceed with a step-by-step procedure to determine the shape of the surface generating the shock. Thus, if the assumed shock wave is described by a prolate ellipsoid of revolution of axis ratio $b/a = 1/\sqrt{2}$, so that (see Fig. 78.9)

$$r^2 = 2R_S x - \tfrac{1}{2}x^2$$

then the nose generating the shock is spherical to within less than 0.2% of its radius R_b at $M_\infty = \infty$, $\gamma = 1.4$. The surface pressures, according to the numerical calculations [18] for $M_\infty = 5.8$, are compared in Fig. 78.10 with experiment [19]. Theory and experiment are in good agreement. Interestingly enough, simple impact theory also predicts the variation of surface pressures away from the stagnation point with good accuracy [19], and, for many applications, this theory may suffice. Bluntness also affects surface pressures and shock-wave shapes in the flow downstream of the nose [20], and these effects must be included in any calculation of the flow if the bluntness is large (that is, the body fineness ratio is of order 1).

78.4. REAL-GAS AND VISCOUS EFFECTS

Real-gas effects have a relatively minor influence on the inviscid flow about slender configurations at Mach numbers up to the highest interesting values (about 35) [6]. Their influence on the inviscid flow about blunt configurations is primarily to cause the flow to approximate more nearly Newtonian flow [20]. In hypersonic boundary-layer flows, the real-gas effects can be more significant [21]–[24]. Further complication can arise at hypersonic speeds if, in addition, the Reynolds number is low. In this case, the boundary layer is very thick, and can extend through the entire disturbed flow field about a body out to the bow shock wave. Under these conditions, the usual separation of the inviscid and viscous flow is not possible, and the analysis of the disturbed flow is relatively involved [25]–[29].

78.5. REFERENCES

[1] I. Newton: "Principia. Motte's Translation Revised," Univ. of California Press, Berkeley, Calif., 1946.

[2] G. Grimminger, E. P. Williams, G. B. W. Young: Lift on inclined bodies of revolution in hypersonic flow, *J. Aeronaut. Sci.*, **17** (1950), 675–690.

[3] A. J. Eggers Jr., M. M. Resnikoff, D. H. Dennis: Bodies of revolution having minimum drag at high supersonic airspeeds, *NACA Rept.* 1306 (1957).

[4] A. Busemann: Flüssigkeits- und Gasbewegung, in "Handwörterbuch der Naturwissenschaften," 2d ed., pp. 244–279, G. Fischer, Jena, Germany, 1933.

[5] P. S. Epstein: On the air resistance of projectiles, *Proc. Natl. Acad. Sci.*, **17** (1931), 532–547.

[6] A. J. Eggers Jr., C. A. Syvertson, S. Kraus: A study of inviscid flow about airfoils at high supersonic speeds, *NACA Rept.*, 1123 (1953).

[7] A. J. Eggers Jr., R. C. Savin: A unified two-dimensional approach to the calculation of three-dimensional hypersonic flows, with applications to bodies of revolution, *NACA Rept.* 1249 (1955).

[8] R. C. Savin: Application of the generalized shock-expansion method to inclined bodies of revolution traveling at high supersonic airspeeds, *NACA Tech. Note* 3349 (1955).

[9] A. J. Eggers Jr., R. C. Savin, C. A. Syvertson: The generalized shock-expansion method and its application to bodies traveling at high supersonic airspeeds, *J. Aeronaut. Sci.*, **22** (1955), 231.

[10] R. D. Linnell: Two-dimensional airfoils in hypersonic flows, *J. Aeronaut. Sci.*, **16** (1949), 22–30.

[11] C. A. Syvertson, D. H. Dennis: A second-order shock-expansion method applicable to bodies of revolution near zero lift, *NACA Rept.* 1328 (1958).

[12] H.-S. Tsien: Similarity laws of hypersonic flows, *J. Math. Phys.*, **25** (1946), 247–251.

[13] W. D. Hayes: On hypersonic similitude, *Quart. Appl. Math.*, **5** (1947), 105–106.

[14] K. Oswatitsch: Similarity laws for hypersonic flow, *Roy. Inst. Technol. Stockholm, Swed., Div. Aeronaut. KTH-Aero Tech. Note* 16 (1950).

[15] F. M. Hamaker, S. E. Neice, T. J. Wong: The similarity law for hypersonic flow and requirements for dynamic similarity of related bodies in free flight, *NACA Rept.* 1147 (1953).

[16] D. M. Ehret, V. J. Rossow, V. I. Stevens: An analysis of the applicability of the hypersonic similarity law to the study of flow about bodies of revolution at zero angle of attack, *NACA Tech. Note* 2250 (1950).

[17] M. D. Van Dyke: The combined supersonic-hypersonic similarity rule, *J. Aeronaut. Sci.*, **18** (1951), 499–500.

[18] M. D. Van Dyke: The supersonic blunt-body problem: Review and extension, *J. Aeronaut. Sci.*, **25** (1958), 485–496.

[19] R. E. Oliver: An experimental investigation of flow about simple blunt bodies at a nominal Mach number of 5.8, *J. Aeronaut. Sci.*, **23** (1956), 177–179.

[20] L. Lees: Recent developments in hypersonic flow, *Jet Propulsion*, **27** (1957), 1162–1178.

[21] Y. H. Kuo: Dissociation effects in hypersonic viscous flows, *J. Aeronaut. Sci.*, **24** (1957), 345–350.

[22] A. J. Eggers Jr., C. F. Hansen, B. E. Cunningham: Stagnation point heat transfer to blunt shapes in hypersonic flight, including effects of yaw, *NACA Tech. Note* 4229 (1958).

[23] J. A. Fay, F. R. Riddell: Theory of stagnation point heat transfer in dissociated air, *AVCO Research Lab., Rept.* 1 (1957).

[24] L. Lees: Laminar heat transfer over blunt-nosed bodies at hypersonic flight speeds, *Jet Propulsion*, **26** (1956), 259–269.
[25] H. T. Nagamatsu, T.-Y. Li: Shock wave effects on the laminar skin friction of an insulated flat plate at hypersonic speeds, *J. Aeronaut. Sci.*, **20** (1953), 345–355.
[26] L. Lees: Influence of the leading-edge shock wave on the laminar boundary layer at hypersonic speeds, *GALCIT Tech. Rept.* 1 (1954).
[27] L. Lees: Hypersonic viscous flow over an inclined wedge, *J. Aeronaut. Sci.*, **20** (1953), 794–795.
[28] M. H. Bertram: Viscous and leading-edge thickness effects on the pressures on the surface of a flat plate in hypersonic flow, *J. Aeronaut. Sci.*, **21** (1954), 430–431.
[29] A. G. Hammitt, S. M. Bogdonoff: A study of the flow about simple bodies at Mach numbers from 11 to 15, *WADC Tech. Rept.* 54–257 (1954).

CHAPTER 79

SLENDER-BODY THEORY

BY

A. E. BRYSON, Jr., Ph.D., Cambridge, Mass.

79.1. INTRODUCTION

Slender-body theory can be used to find approximate steady or unsteady aerodynamic forces on wings, bodies, and complete configurations, whose lateral dimensions change slowly with distance parallel to the flow direction. If the configuration is very slender, the theory applies approximately over the speed range from zero to moderate supersonic Mach numbers. If the configuration is only moderately slender, the theory is a good approximation only near Mach number one; for this reason, it is sometimes regarded as a transonic theory. It was originated by Munk [1] in connection with airship design, but not until Tsien [2], Jones [3], and Spreiter [4] proposed

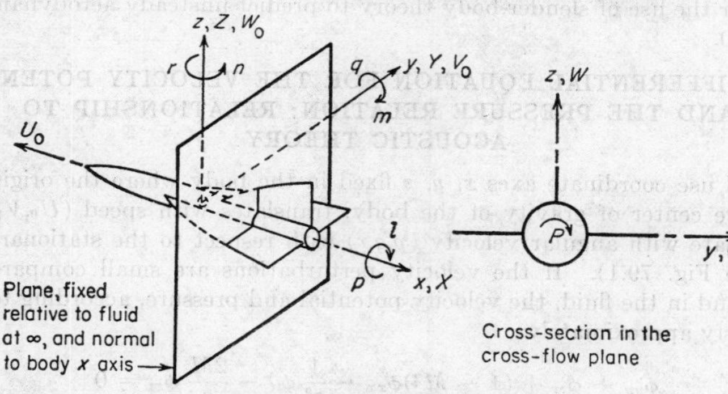

FIG. 79.1. Slender body, reference axes.

its use at high speeds and for wings and wing-body combinations, did its generality become evident.

The basic idea is as follows: Imagine a plane fixed with respect to a large mass of stationary fluid, and consider a slender body penetrating this plane with its long axis instantaneously perpendicular to the plane (see Fig. 79.1). The "hole" in the plane being made by the body is changing slowly in shape and moving laterally. In the vicinity of the body cross section, the fluid velocities in the plane are very nearly those corresponding to the two-dimensional unsteady flow created by an infinite cylinder whose cross section is changing shape and moving laterally like the cross section of the three-dimensional body as it penetrates the plane.

The relationship between distance along the body in the three-dimensional flow and time in the related two-dimensional flow is simply

$$dt = \frac{dx}{U_0}$$

where U_0 is the *axial* velocity of body, i.e., the velocity of the body perpendicular to the fixed plane.

The related two-dimensional unsteady flow is simpler to analyze than the three-dimensional flow, particularly if all the velocities normal to the surface of the cross section in the crossflow plane are small compared to the speed of sound in the fluid, for, in that case, the flow can be treated as incompressible. We have then only to solve Laplace's equation in two dimensions to get the velocity potential, and we have available for this the powerful methods of complex-variable theory.

A difficulty arises in this simple approach when we consider slender bodies with thickness. However, as first shown by Ward [5], the main effect of thickness near the body is simply to change the level of the pressure at each cross section. Clearly this is important in calculating drag (it explains the transonic area rule), but it does not affect the lateral force at each cross section. We shall concern ourselves in this chapter only with the lateral forces and moments; drag of slender bodies is considered in other chapters, and, unlike the lateral forces and moments, it shows a dependence on flight Mach number.

When the velocities normal to the surface of the cross section become comparable to the speed of sound, we can no longer use an "incompressible" crossflow. We may, however, still retain the approximation of a two-dimensional unsteady crossflow. This is the basis of the hypersonic small-disturbance theory where nonlinear compressible-flow relations are used. The linearized equations for the two-dimensional unsteady crossflow (acoustic equation) may be used for high-frequency oscillations or indicial motions of slender bodies at subsonic to moderately supersonic flight speeds. For the use of slender-body theory to predict unsteady aerodynamic forces, see Chap. 80.

79.2. DIFFERENTIAL EQUATION FOR THE VELOCITY POTENTIAL AND THE PRESSURE RELATION: RELATIONSHIP TO ACOUSTIC THEORY

We shall use coordinate axes x, y, z fixed in the body where the origin, usually taken at the center of gravity of the body, translates with speed (U_0, V_0, W_0) and the axes rotate with angular velocity (p,q,r) with respect to the stationary fluid at infinity (see Fig. 79.1). If the velocity perturbations are small compared to the speed of sound in the fluid, the velocity potential and pressure, according to acoustic theory, satisfy approximately

$$\phi_{yy} + \phi_{zz} + (1 - M^2)\phi_{xx} - \frac{1}{a^2}\phi_{tt} - \frac{2M}{a}\phi_{xt} = 0 \tag{79.1}$$

$$\frac{p_1 - p_\infty}{p_\infty} = -\phi_t - U_0\phi_x + (V + pz)\phi_y + (W - py)\phi_z - \tfrac{1}{2}[(1 - M^2)\phi_x^2 + \phi_y^2 + \phi_z^2] \quad \text{(near body)} \tag{79.2}$$

where $M = U_0/a$
a = speed of sound in undisturbed fluid
$V = V_0 + rx$
$W = W_0 - qx$
$ry + qz,\ V_0,\ W_0 \ll U_0$

For slender configurations, ϕ_t and ϕ_x are smaller in magnitude than the crossflow velocities ϕ_y and ϕ_z, and, for nonplanar bodies particularly, even the squares of ϕ_y and ϕ_z may be comparable in size to ϕ_t and ϕ_x (see Lighthill in [6]), so that the only term we might neglect in (79.2) would be the term $(1 - M^2)\phi_x^2$. In (79.1), however, we can neglect all but the first two terms for slender bodies. For steady flows $(\partial\phi/\partial t = 0)$ near $M = 1$, even for bodies that are not so slender, the same approximations can be made.

Thus we have, approximately, in body axes,

$$\phi_{yy} + \phi_{zz} = 0 \tag{79.3}$$

$$\frac{p_1 - p_\infty}{p_\infty} = -\phi_t - U_0\phi_x + (V + pz)\phi_y + (W - py)\phi_z - \tfrac{1}{2}(\phi_y^2 + \phi_z^2) \tag{79.4}$$

Solutions to this two-dimensional Laplace equation are of the form

$$\phi = \Re F(\zeta;x,t) + g(x,t) \qquad \text{where } \zeta = y + iz \tag{79.5}$$

and g is an arbitrary function. The determination of g cannot be made within the framework of slender-body theory, but must come from three-dimensional considerations [5]. F is the complex potential for the crossflow, and x and t occur in it only

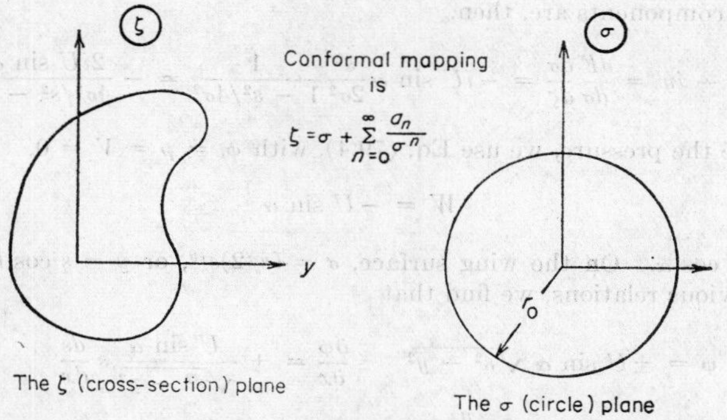

FIG. 79.2. Conformal mapping of exterior of cross section onto exterior of a circle with no distortion at infinity.

as parameters. If we map the exterior region of the cross section (in the $\zeta = y + iz$ plane) conformally onto the exterior of a circle of radius r_0 in the σ plane, with no distortion at infinity (see Fig. 79.2), we know that F can be expressed as

$$F = \frac{Q(x,t) - i\Gamma(x,t)}{2\pi} \ln \sigma + \sum_{n=0}^{\infty} \frac{A_n(x,t)}{\sigma^n} \tag{79.6}$$

(where Q = net source strength within the body, Γ = circulation about the body), since the velocities

$$\phi_y - i\phi_z = \frac{dF}{d\sigma}\frac{d\sigma}{d\zeta}$$

must vanish as $\sigma \to \infty$. The infinite (Laurent) series converges for $\sigma > r_0$; so, if we consider the pressure on $\sigma = (r_0 + \epsilon)e^{i\theta}$, where $\epsilon \ll r_0$, we may use the real part of this expression in our pressure equation. In most cases, Γ_t, Γ_x, and Q_t are zero, but Q_x will not be zero if the body cross-section area changes with x; in this case, the ϕ_x term of the pressure contains a term

$$\frac{U_0 Q_x}{2\pi} \ln |\sigma| + U_0 g_x \tag{79.7}$$

but near the body surface, $|\sigma| = r_0 + \epsilon$, this is a constant pressure and does not contribute to the lateral forces or rolling moment on the body cross section. The source term does, however, enter significantly into the quadratic terms in the pressure equation.

79.3. LIFT OF FLAT WINGS AND BODIES OF REVOLUTION

Flat-plate Wing with Unswept Trailing Edge at Angle of Attack

This simple example will serve to show how the theory is used. As the wing pierces the crossflow plane, its trace is a slit of width $2s$, increasing in width with time, and moving down with speed $U \sin \alpha$. The complex potential for this two-dimensional incompressible flow is

$$F = -iU \sin \alpha(\sigma - \zeta) + iU \sin \alpha \frac{s^2}{4\sigma} = iU \sin \alpha \frac{s^2}{2\sigma} \tag{79.8}$$

where $\zeta = \sigma + s^2/4\sigma$ is the conformal mapping that maps the exterior of the slit onto the exterior of a circle in the σ plane, with no distortion at infinity ($r_0 = s/2$). The velocity components are, then,

$$v - iw = \frac{dF}{d\sigma}\frac{d\sigma}{d\zeta} = -iU \sin \alpha \frac{s^2}{2\sigma^2}\frac{1}{1 - s^2/4\sigma^2} = -\frac{2iU \sin \alpha}{4\sigma^2/s^2 - 1} \tag{79.9}$$

To determine the pressure, we use Eq. (79.4), with $\phi_t = p = V = 0$,

$$W = -U \sin \alpha$$

and $U_0 = U \cos \alpha$. On the wing surface, $\sigma = (s/2)e^{i\theta}$, or $y = s \cos \theta$, $z = 0$, and, from the previous relations, we find that

$$\phi = \pm U \sin \alpha \sqrt{s^2 - y^2} \qquad \frac{\partial \phi}{\partial x} = \pm \frac{U \sin \alpha}{\sqrt{s^2 - y^2}} s \frac{ds}{dx}$$

$$v = \mp U \sin \alpha \frac{y}{\sqrt{s^2 - y^2}} \qquad w = -U \sin \alpha \tag{79.10}$$

$$C_p = \frac{p_1 - p_\infty}{q} = \mp \frac{\sin 2\alpha}{\sqrt{s^2 - y^2}} s \frac{ds}{dx} + \sin^2 \alpha \left(1 - \frac{y^2}{s^2 - y^2}\right)$$

where the upper and lower signs pertain to the top and the bottom of the wing, respectively, and $q = \frac{1}{2}\rho U^2$ is the dynamic pressure. The normal force per unit length of the wing in the x direction is, then,

$$\frac{d}{dx}\left(\frac{N}{q}\right) = 2\int_0^s (C_{pl} - C_{pu})\, dy = 4s\frac{ds}{dx} \sin 2\alpha \int_0^s \frac{dy}{\sqrt{s^2 - y^2}} = 2\pi s \frac{ds}{dx} \sin 2\alpha \tag{79.11}$$

Note there is no normal force per unit length predicted on parts of the wing where the span is constant.

The total normal force is obtained by integration with respect to x,

$$\frac{N}{q} = 2\pi \sin 2\alpha \int_0^{x_0} ss'\, dx = \pi s_0^2 \sin 2\alpha \tag{79.12}$$

where we have assumed that $s(0) = 0$; that is, the wing is pointed, and s_0 = maximum semispan. Note that the plan form of the wing is irrelevant; all that matters is its *maximum span*. The normal force coefficient is

$$C_N = \frac{N}{qS} = \frac{\pi}{2} \mathcal{R} \frac{\sin 2\alpha}{2} \tag{79.13}$$

where S = plan form area of the wing, and $\mathcal{R} = 4s_0^2/S$ is the aspect ratio of the wing. The moment coefficient about the apex of the wing is given by

$$C_m = \frac{m}{qS(2s_0)} = \frac{\pi}{Ss_0} \sin 2\alpha \int_0^{x_0} xs \frac{ds}{dx} dx \tag{79.14}$$

Alternative Approach, Using the Concept of Apparent Additional Mass

The aerodynamic force on a two-dimensional body moving in an infinite incompressible fluid can be accounted for by simply adding a small mass m, known as the apparent additional mass, to the actual mass M of the body in Newton's law of motion (see, for example, [18], Chap. VI):

$$\frac{d}{dt}[(M + m)w] = F \qquad \text{or} \qquad \frac{d}{dt}(Mw) = -\frac{d}{dt}(mw) + F \tag{79.15}$$

where F is the external force (other than aerodynamic), and $-d(mw)/dt$ is the aerodynamic force.

In the case just considered, it is easy to show that

$$m = \pi\rho s^2 \tag{79.16}$$

that is, the apparent additional mass per unit length of a slit is the mass of air in a circle of diameter equal to the width of the slit. The normal force per unit length is then

$$\frac{dN}{dx} = U \cos\alpha \frac{d}{dx}[(\pi\rho s^2)(U \sin\alpha)] \qquad \text{since} \quad \frac{d}{dt} = U\cos\alpha\frac{d}{dx} \quad \text{and} \quad w = U \sin\alpha$$

or

$$\frac{d}{dx}\left(\frac{N}{q}\right) = 2\pi s \frac{ds}{dx} \sin 2\alpha \tag{79.17}$$

which is, of course, the same as the result obtained previously [Eq. (79.11)].

Body of Revolution at Angle of Attack α

The apparent additional mass per unit length of a circular cylinder is

$$m = \pi\rho a^2 \tag{79.18}$$

where a is the radius of the cylinder (see [7], p. 228). Hence, the normal force per unit length of a body of revolution is simply

$$\frac{d}{dx}\left(\frac{N}{q}\right) = 2\pi a \frac{da}{dx} \sin 2\alpha \tag{79.19}$$

Thus slender-body theory predicts the same lift on a body of revolution as on a flat wing if they have the same plan form. The normal-force coefficient is

$$C_N = \frac{N}{q(\pi a_0^2)} = \sin 2\alpha \tag{79.20}$$

where a_0 is the radius of the base of the body. The moment coefficient about the nose of the body is given by

$$C_m = \frac{m}{q(\pi a_0^2)(2a_0)} = \frac{\sin 2\alpha}{a_0^3} \int_0^{x_0} xa \frac{da}{dx} dx \tag{79.21}$$

79.4. THE VORTEX WAKE BEHIND SLENDER CONFIGURATIONS

Planar Wings

For a slender wing with a straight trailing edge perpendicular to the flight direction (see Fig. 79.3 with trailing edge a), there is no influence of the trailing edge on the pressure distribution on the wing according to slender-body theory. From irrotationality, the sidewash just behind the trailing edge is the same as that just ahead, and, furthermore, it is discontinuous (antimetric) through the plane of the wing; this means that there is a vortex sheet shed from the trailing edge. The downwash behind the wing can be calculated from slender-body theory, by utilizing the fact that there is no pressure jump across the vortex sheet leaving the trailing edge. If the span is still increasing at the trailing edge, there will be a discontinuity in pressure; at subsonic speeds, this cannot really happen, and we interpret this result to

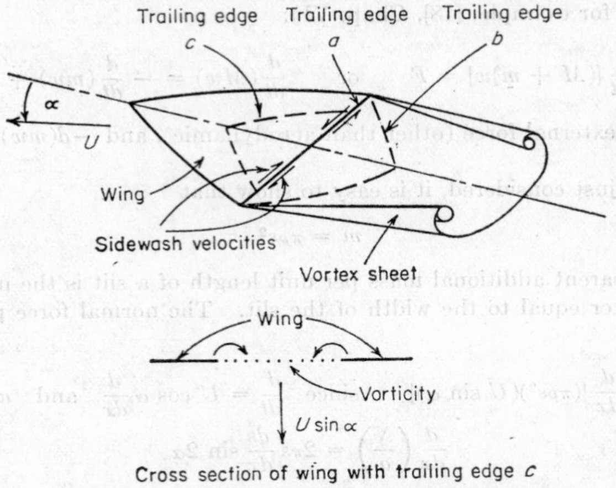

FIG. 79.3. The vortex wake behind planar wings.

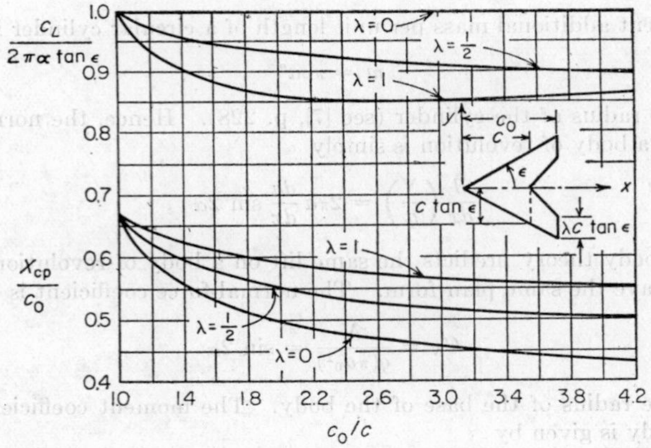

FIG. 79.4. Lift and center of pressure for slender swept wing at angle of attack: $C_L = L/qA$, where A = wing planform area.

mean a "rapid" pressure change near the trailing edge. Within the framework of slender-body theory, the vortex sheet creates a downwash at the sheet exactly equal to the downwash at the wing trailing edge. Thus, it is exactly as though the wing were "continued" downstream, with span and downwash distribution the same as that of the trailing edge. If the wing had a trailing edge like b, as shown in Fig. 79.3, the theory indicates that there would be no lift on the portion aft of the maximum span, since the vortex sheet shed from the trailing edges would make it appear as though the wing had been continued with span equal to the maximum span aft of the maximum span point.

The situation is more complicated if the trailing edges are *reentrant*, like trailing edge c in Fig. 79.3. The calculation of the pressure distribution on the wing, aft of the forward point of the trailing edge, involves the solution of an integral equation, since the vorticity in the sheet between the two parts of wing at any section (like that shown in Fig. 79.3) depends on the lift distribution ahead. Also the Kutta condition must be used at the trailing edge. Figure 79.4 shows the lift and center of pressure for some sweptback wings with straight edges, taken from [8] and [9], where the theory is given in detail.

Roll-up of the Trailing Vortex Sheet

It is well known that the vortex sheet behind a lifting surface is *unstable*, in the sense that a small disturbance will cause it to start "rolling up" at the edges. This roll-up process can be followed by approximating the continuous sheet by a number of concentrated vortices (for example, [10]). Theory and experiment show that the distance d behind the trailing edge for the sheet to effectively roll up into two concentrated regions of vorticity is given very roughly by

$$\frac{d}{b} = 0.28\,\frac{C_L}{Æ} \qquad (79.22)$$

where b = wing span, $Æ$ = wing aspect ratio, and C_L = lift coefficient based on wing plan-form area. For low-aspect-ratio wings, Eq. (79.13) gives $C_L = (\pi/2)Æ\alpha$, which indicates, again very roughly, that

$$\frac{d}{b} = 0.44\alpha \qquad \text{as } Æ \to 0 \qquad (79.23)$$

Thus, roll-up occurs in a very short distance behind a low-aspect-ratio wing; often the approximation can be made that roll-up occurs right at the trailing edge, in order to simplify the analysis of the motion of the vortex wake downstream. The forces induced on a downstream lifting surface by a vortex sheet are often little different from those induced by the rolled-up version of the sheet.

Leading-edge Separation

For low-aspect-ratio wings, the vortex sheet beings to "peel off" the leading edges, and rolls up into regions of concentrated vorticity, as shown in Fig. 79.5.

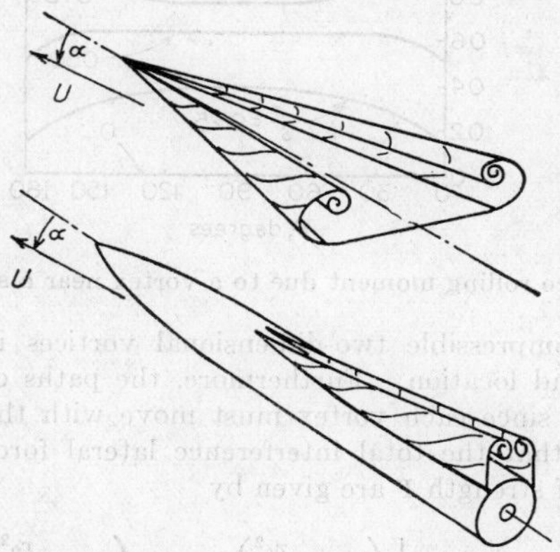

FIG. 79.5. Leading-edge separation and body vortices.

The high suction under these vortices increases the wing lift well above that predicted by linear theory. An approximate nonlinear slender-body theory for a delta wing shows that [11], [12]

$$C_L = \frac{\pi}{2} A\!\!R\,\alpha + \pi A\!\!R^{1/3}\alpha^{5/3} + \cdots \tag{79.24}$$

Reasonable agreement with experiment is obtained.

Body Vortices

In a manner very similar to leading-edge separation, an elongated slender body sheds vorticity on its lee side which again increases the lift above that predicted by linear theory (see Fig. 79.5 and [13],[14]). This phenomenon is very similar to that which occurs when a two-dimensional cylinder is started impulsively from rest in an incompressible fluid [15]. First, two symmetric vortices appear and move downstream; then, depending on the crossflow Reynolds number, the symmetry disappears and a *vortex street* appears, finally developing into a turbulent wake.

Vortex Interference

The vorticity shed from forward lifting components creates flow disturbances downstream, which induce nonlinear interference forces and moments on the tail and afterbody. In leading-edge separation and body crossflow separation, the nonlinear forces are induced on the component shedding the vorticity. Slender-body theory provides an approximate method for calculating these interference forces

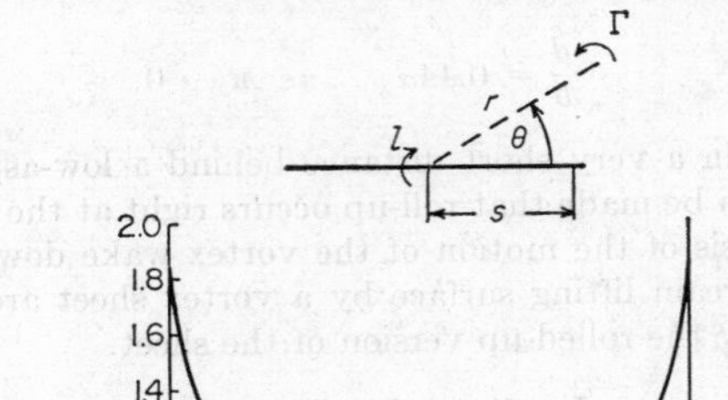

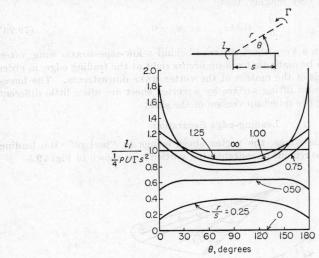

FIG. 79.6. Interference rolling moment due to a vortex near a slender planar wing.

by simply adding incompressible two-dimensional vortices in the crossflow with appropriate strength and location. Furthermore, the paths of the vortices can be traced in this manner, since each vortex must move with the local fluid velocity. Sacks [16] has shown that the total interference lateral forces in steady straight flight due to a vortex of strength Γ are given by

$$Y_I + iZ_I = i\rho U_0 \Gamma \left[\left(\sigma_1 - \frac{r_0^2}{\bar{\sigma}_1} \right)_{x=x_b} - \left(\sigma_1 - \frac{r_0^2}{\bar{\sigma}_1} \right)_{x=x_s} \right] \tag{79.25}$$

where $(\sigma_1 - r_0^2/\bar{\sigma}_1)$ is the distance between the vortex and its image in the circle in the σ plane, x_s is the station where the vortices were shed, and x_b is the station of the base plane. The interference rolling moment for steady straight flight is

$$l_l = \tfrac{1}{2} U_0 \rho \Re \left\{ \left(\oint_C \zeta\bar{\zeta}\, dF' + \Gamma \zeta_1 \bar{\zeta}_1 \right)_{x=x_b} - \left(\oint_C \zeta\bar{\zeta}\, dF + \Gamma \zeta_1 \bar{\zeta}_1 \right)_{x=x_s} + 2\Gamma \frac{(V_0 - iW_0)}{U_0} \int_{x=x_s}^{x_b} \left(\sigma_1 - \frac{r_0^2}{\bar{\sigma}_1} \right) dx \right\} \qquad (79.26)$$

where F' is the complex potential due to the vortex and $\oint_C$ is the contour integral around the body cross section. Figure 79.6 shows the first-order interference rolling moment due to a single vortex near a planar wing when $V_0 = W_0 = 0$.

79.5. MODIFIED SLENDER-BODY THEORY

Many results of slender-body theory do not check experiment well enough for engineering purposes, except for very slender configurations. In particular, no variation of aerodynamic force coefficients with Mach number is predicted by slender-body theory. Nielsen et al. [17] have proposed a modification of slender-body theory

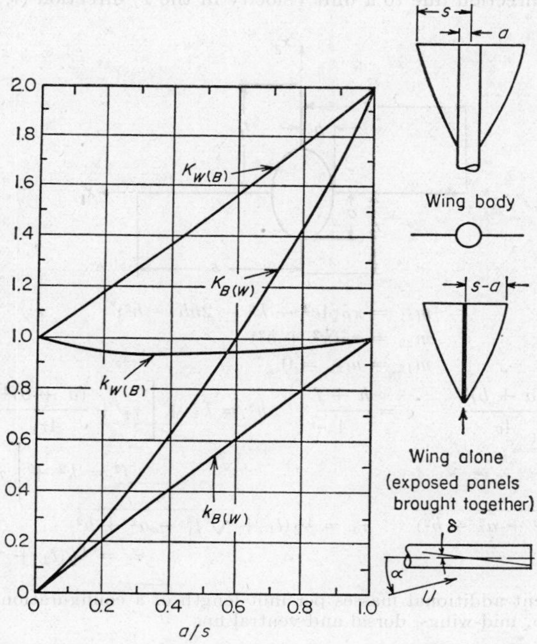

FIG. 79.7. Slender-body interference factors for planar wing–circular body: $C_{LW(B)} = (K_{W(B)}\alpha + k_{W(B)}\delta)(C_{L\alpha})_W$, $C_{LB(W)} = (K_{B(W)}\alpha + k_{B(W)}\delta)(C_{L\alpha})_W$.

that improves agreement with experiment markedly. They argue that *linearized theory*[1] for planar wings and bodies of revolution obviously checks experiment better than slender-body theory, and is tractable enough so that results are known for many plan forms and shapes; however, nonplanar configurations (other than the body of revolution) are nearly impossible to analyze by linearized theory, whereas, quite

[1] By this, we mean acoustic theory, using the complete equation (79.1).

often, they can be analyzed by slender-body theory. By taking the *ratios* of the forces and moments on such nonplanar configurations to the forces and moments on an appropriate "wing alone," *interference factors* are formed, which are then applied to the forces and moments on the wing alone as predicted by linearized theory; thus, a dependence on Mach number is introduced into the force and moment coefficients. Figure 79.7 is an example of this technique applied to the problem of predicting wing-body interference on a cylindrical fuselage with a wing at incidence δ, relative to the fuselage axis, and fuselage at angle of attack α. $L_{W(B)}$ is the lift on the wing in the presence of the body, $L_{B(W)}$ is the lift on the body in the presence of the wing, and L_W is the lift on the wing alone. Of course, still a further improvement is to apply the interference factors to experimental values for the wing alone.

79.6. QUASI-STEADY AERODYNAMIC FORCES ON SLENDER AIRPLANES AND MISSILES

All the aerodynamic forces except drag can be determined for quasi-steady motions of a slender configuration by slender-body theory (for example, see [19]–[21]). Figure 79.8 is an example of the degree of configuration complexity that can be readily handled by this technique. Note that m_{ij} is the apparent additional mass per unit length in the x_i direction due to a unit velocity in the x_j direction ($i, j = 1, 2$).

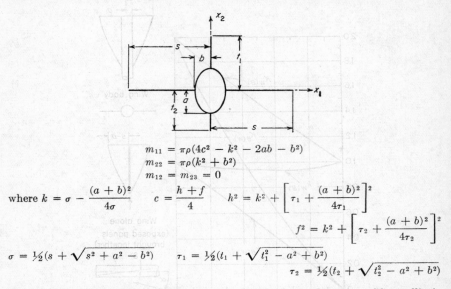

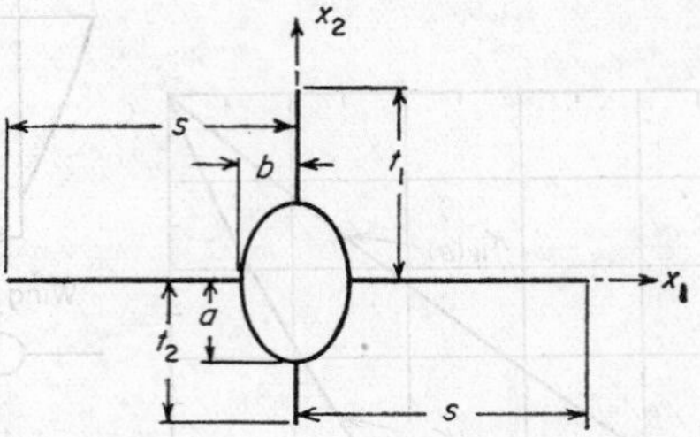

$$m_{11} = \pi\rho(4c^2 - k^2 - 2ab - b^2)$$
$$m_{22} = \pi\rho(k^2 + b^2)$$
$$m_{12} = m_{23} = 0$$

$$\text{where } k = \sigma - \frac{(a+b)^2}{4\sigma} \qquad c = \frac{h+f}{4} \qquad h^2 = k^2 + \left[\tau_1 + \frac{(a+b)^2}{4\tau_1}\right]^2$$
$$f^2 = k^2 + \left[\tau_2 + \frac{(a+b)^2}{4\tau_2}\right]^2$$
$$\sigma = \tfrac{1}{2}(s + \sqrt{s^2 + a^2 - b^2}) \qquad \tau_1 = \tfrac{1}{2}(t_1 + \sqrt{t_1^2 - a^2 + b^2})$$
$$\tau_2 = \tfrac{1}{2}(t_2 + \sqrt{t_2^2 - a^2 + b^2})$$

FIG. 79.8. Apparent additional masses per unit length of a configuration with an elliptic body cross section, mid-wings, dorsal and ventral fins.

79.7. REFERENCES

[1] M. M. Munk: The aerodynamic forces on airship hulls, *NACA Rept.* 184 (1924).
[2] H.-S. Tsien: Supersonic flow over an inclined body of revolution, *J. Aeronaut. Sci.*, **5** (1938), 480–483.
[3] R. T. Jones: Properties of low-aspect-ratio pointed wings at speeds below and above the speed of sound, *NACA Rept.* 835 (1946).
[4] J. R. Spreiter: Aerodynamic properties of slender wing-body combinations at subsonic, transonic, and supersonic speeds, *NACA Tech. Note* 1662 (1948).
[5] G. N. Ward: Supersonic flow past slender pointed bodies, *Quart. J. Mech. Appl. Math.*, **2** (1949), 75–98.

[6] W. R. Sears (ed.): "General Theory of High Speed Aerodynamics," vol. 6 of "High Speed Aerodynamics and Jet Propulsion," pp. 96–100, 240–275, 451–476, Princeton Univ. Press, Princeton, N.J., 1954.
[7] L. M. Milne-Thompson: "Theoretical Hydrodynamics," 3d ed., Macmillan, London, 1955.
[8] K. W. Mangler: Calculation of the pressure distribution over a wing at sonic speeds, *Aeronaut. Research Comm. Rept. and Mem.* 2888 (1955).
[9] H. Mirels: Aerodynamics of slender wings and wing-body combinations having swept trailing edges, *NACA Tech. Note* 3105 (1954).
[10] J. R. Spreiter, A. H. Sacks: The rolling up of the trailing vortex sheet and its effect on the downwash behind wings, *J. Aeronaut. Sci.*, **18** (1951), 21–32.
[11] R. H. Edwards: Leading edge separation from delta wings at supersonic speeds, *J. Aeronaut. Sci.*, **20** (1953), 430.
[12] C. E. Brown, W. H. Michael: On slender delta wings with leading edge separation, *NACA Tech. Note* 3430 (1955).
[13] H. J. Allen, E. W. Perkins: A study of effects of viscosity on flow over slender inclined bodies of revolution, *NACA Rept.* 1048 (1951).
[14] H. R. Kelly: The estimation of normal-force, drag, and pitching-moment for blunt-based bodies of revolution of large angles of attack, *J. Aeronaut. Sci.*, **21** (1954), 549–555.
[15] Rubach: Göttingen Diss., 1914, see S. Goldstein (ed.), "Modern Developments in Fluid Mechanics," vol. 1, pp. 62–65, Oxford Univ. Press, New York, 1938.
[16] A. H. Sacks: Aerodynamic interference of slender wing-tail combinations, *NACA Tech. Note* 3725 (1957).
[17] J. N. Nielsen, E. D. Katzen, K. K. Tang: Lift and pitching moment interference between a pointed cylindrical body and triangular wings of various aspect ratios at $M = 1.50$ and $M = 2.02$, *NACA Tech. Note* 3795 (1956).
[18] H. Lamb: "Hydrodynamics," 6th ed., Dover, New York, 1945.
[19] A. E. Bryson: Evaluation of the inertia coefficients of the cross-section of a slender body, *J. Aeronaut. Sci.*, **21** (1954), 424–426.
[20] A. E. Bryson: Stability derivatives for a slender missile with application to a wing-body-vertical tail configuration, *J. Aeronaut. Sci.*, **20** (1953), 297–306.
[21] A. H. Sacks: Aerodynamic forces, moments, and stability derivatives for slender bodies of general cross-section, *NACA Tech. Note* 3283 (1954).

Other general sources of slender-body theory:

[22] A. F. Donovan, H. R. Lawrence (eds.): "Aerodynamic Components of Aircraft at High Speeds," vol. 7 of "High Speed Aerodynamics and Jet Propulsion," pp. 97–109, 214–217, 343–360, 748–752, 817, Princeton Univ. Press, Princeton, N.J., 1957.
[23] H. W. Liepmann, A. Roskho: "Elements of Gasdynamics," chap. 9, Wiley, New York, 1957.
[24] G. N. Ward: "Linearized Theory of Steady High-speed Flow," Cambridge Univ. Press, New York, 1955.

[6] W. R. Sears (ed.): "General Theory of High Speed Aerodynamics," vol. 6 of "High Speed Aerodynamics and Jet Propulsion," pp. 90–100, 240–275, 451–476, Princeton Univ. Press, Princeton, N.J., 1954.
[7] L. M. Milne-Thomson: "Theoretical Hydrodynamics," 3d ed., Macmillan, London, 1955.
[8] K. W. Mangler: Calculation of the pressure distribution over a wing at sonic speeds, Aeronaut. Research Council Rept. and Mem. 2888 (1955).
[9] H. Mirels: Aerodynamics of slender wings and wing-body combinations having swept trailing edges, NACA Tech. Note 3105 (1954).
[10] J. R. Spreiter, A. H. Sacks: The rolling up of the trailing vortex sheet and its effect on the downwash behind wings, J. Aeronaut. Sci. 18 (1951), 21–32.
[11] R. H. Edwards: Leading edge separation from delta wings at supersonic speeds, J. Aeronaut. Sci. 20 (1953), 430.
[12] C. E. Brown, W. H. Michael: On slender delta wings with leading edge separation, NACA Tech. Note 3430 (1955).
[13] H. J. Allen, E. W. Perkins: A study of effects of viscosity on flow over slender inclined bodies of revolution, NACA Rept. 1048 (1951).
[14] H. R. Kelly: The estimation of normal-force, drag, and pitching-moment for blunt-based bodies of revolution at large angles of attack, J. Aeronaut. Sci. 21 (1954), 549–555.
[15] Rubach: Göttingen Diss., 1914, see S. Goldstein (ed.): "Modern Developments in Fluid Mechanics," vol. 1, pp. 62–65, Oxford Univ. Press, New York, 1938.
[16] A. H. Sacks: Aerodynamic interference of slender wing-tail combinations, NACA Tech. Note 3725 (1957).
[17] J. N. Nielsen, E. D. Katzen, K. K. Tang: Lift and pitching moment interference between a pointed cylindrical body and triangular wings of various aspect ratios at $M = 1.50$ and $M = 2.02$, NACA Tech. Note 3795 (1956).
[18] H. Lamb: "Hydrodynamics," 6th ed., Dover, New York, 1945.
[19] A. E. Bryson: Evaluation of the inertia coefficients of the cross-section of a slender body, J. Aeronaut. Sci. 21 (1954), 424–427.
[20] A. E. Bryson: Stability derivatives for a slender missile with application to a wing-body-vertical tail configuration, J. Aeronaut. Sci. 20 (1953), 297–308.
[21] A. H. Sacks: Aerodynamic forces, moments, and stability derivatives for slender bodies of general cross-section, NACA Tech. Note 3283 (1954).

Other general sources of slender-body theory:

[22] A. F. Donovan, H. R. Lawrence (eds.): "Aerodynamic Components of Aircraft at High Speeds," vol. 7 of "High Speed Aerodynamics and Jet Propulsion," pp. 97–109, 214–217, 348–360, 748–752, 817, Princeton Univ. Press, Princeton, N.J., 1957.
[23] H. W. Liepmann, A. Roshko: "Elements of Gasdynamics," chap. 9, Wiley, New York, 1957.
[24] G. N. Ward: "Linearized Theory of Steady High-speed Flow," Cambridge Univ. Press, New York, 1955.

CHAPTER 80

FLUTTER

BY

I. E. GARRICK, Langley Field, Va.

80.1. INTRODUCTORY REMARKS

A vibrating aircraft structure in a uniform flow is ordinarily damped by the flow; that is, energy is taken from the vibration; however, for a sufficiently high-speed flow, a condition of self-maintained oscillations termed *flutter*, wherein energy is extracted from the flow, may occur and become oscillatory-divergent and violent. Flutter of an aircraft or its components must be avoided in design and practice. The determination of flutter characteristics is an increasingly important engineering art and science.

At an *initial* flutter condition, there is mechanical equilibrium among the inertia forces associated with the kinetic energy of the vibrating system, the interior forces associated with the elastic-strain potential energy and the internal energy of damping (or of other sources of internal energy), and the unsteady oscillatory aerodynamic forces.

There are, thus, three distinct aspects to flutter analysis: (1) structural or elastic aspects, which require determination of quantities such as the inertial and stiffness distributions, stiffness or flexibility influence functions, vibration modes and frequencies, and internal damping or impedances (see other chapters of this handbook); (2) aerodynamic aspects requiring nonsteady air forces such as the aerodynamic loading of a lifting surface vibrating harmonically with small amplitudes at a given frequency in a definite mode; and (3) aeroelastic aspects which combine 1 and 2 in a dynamic analysis and develop the eigenvalue (instability) or response problems. Many experimental and theoretical skills and techniques may be required for determining the primary data, particularly in cases wherein the margins between flight speeds and flutter speeds may become small. Flutter, although strictly a stability or rather an instability problem, is representative of the general dynamic aeroelastic problem, and hence may be regarded as including the static aeroelastic problems such as determination of the steady air load distributions of a deformable wing, control effectiveness and control reversal problems, wing divergence which may be termed zero-frequency flutter, and the dynamic response problems such as penetration into gusts which require considerations of transient flows. It is the purpose of this section to indicate and to consolidate some of the well-developed theoretical aspects of the aerodynamics of flutter.

This chapter deals with flutter and aeroelasticity, with emphasis on the aerodynamic aspects; see Chap. 63 for the structural aspects.

The author of this section is indebted to Dr. John Dugundji of the Aeronautical Engineering Department, Massachusetts Institute of Technology, for helpful suggestions and critical reading of the manuscript.

80.2. NONSTEADY POTENTIAL FLOW

Consideration here is restricted to potential flows, thus putting aside various classes of phenomena such as stall flutter, buffeting, and control-surface buzz, which largely require empirical developments and experimental measurement.

The governing partial differential equation for potential-flow aerodynamics may be shown to be, in general, a nonlinear form of the wave equation ([5], p. 660), where the disturbance represented by the velocity potential is convected by the local fluid velocity, and is propagated as a wave which spreads at a rate equal to the local speed of sound. When this local wave equation for the velocity potential is linearized and considered for small disturbances from the main flow with respect to space-fixed coordinates, the classical equation of wave propagation arises; and, with respect to coordinate systems moving uniformly with the mainstream velocity, a modified form of the wave equation (80.1) occurs. Both forms of the wave equation reduce to Laplace's equation for an incompressible flow, on allowing the velocity of sound to be infinite or the Mach number zero. The building blocks for solutions of the wave equation are various elementary generalizations of sources and doublets, including also source pulses, the latter of which are sources which act only for a short time. The linearized relation for the pressure [Eq. (80.2)] also satisfies the wave equation, and, hence, the sources and doublets may also be used to represent pressure singularities, in addition to their more customary use in conjunction with the velocity potential for representing mass flow or vorticity distributions. The so-called acceleration potential, which has been employed with some success in linearized nonsteady aerodynamics, corresponds to the use of the pressure singularities to synthesize the flow.

The boundary conditions are usually stated to apply at the projection of the lifting surface onto the mean planar boundary surface, and require mainly that the resultant flow be unseparated and tangential to the boundary surfaces. In addition, it is required that the trailing-edge condition of smooth flow be satisfied for subsonic edges, and that the wake vorticity float freely without mutual interaction. Also, proper behavior near sharp leading edges is postulated. Linearization of the boundary conditions leads to separation of problems of camber and of thickness.

The customary linearized differential equation for the velocity potential, with respect to cartesian coordinates moving with the main-stream velocity U in the negative x direction, is

$$(1 - M^2)\phi_{xx} + \phi_{yy} + \phi_{zz} - \frac{2M^2}{U}\phi_{xt} - \frac{M^2}{U^2}\phi_{tt} = 0 \tag{80.1}$$

where M is the Mach number U/a_∞, and a_∞ is the sound speed in the undisturbed flow. The linearized pressure relation is

$$p - p_\infty = -\rho_\infty(\phi_t + U\phi_x) \tag{80.2}$$

where p is the local pressure and p_∞, ρ_∞ the pressure and density of the undisturbed stream.

The equations are valid for small perturbations of the main stream whether at subsonic ($M < 1$) or supersonic ($M > 1$) speeds; for flow conditions in which the perturbations may be large, they should be supplemented by the addition of nonlinear terms and more complete relations. For flutter, the aerodynamical forces for simple harmonic wing motions of small amplitudes are particularly important, and in the next few sections there is a presentation of appropriate unsteady flow aerodynamics for idealized types of motions in two- and three-dimensional flows. The main nondimensional parameters of the dynamical and aerodynamical problems will be indicated.

80.3. DYNAMICS AND AERODYNAMICS OF IDEALIZED TWO-DIMENSIONAL CASE: TRANSLATION AND ROTATION OF A HARMONICALLY OSCILLATING WING SECTION

Consider as a simplest illustrative case—a two-degree-of-freedom system—a uniform wing, capable of rotation as a rigid body about an axis $x = x_a$ with angular motion $\alpha = \alpha_0 e^{i\omega t}$, and of displacement in the $-z$ direction of $h = h_0 e^{i\omega t}$. (See Fig. 80.1 for positive directions; note that x is here nondimensional, and bx is the actual distance.) The circular frequency is ω, and the amplitudes α_0 and h_0 are

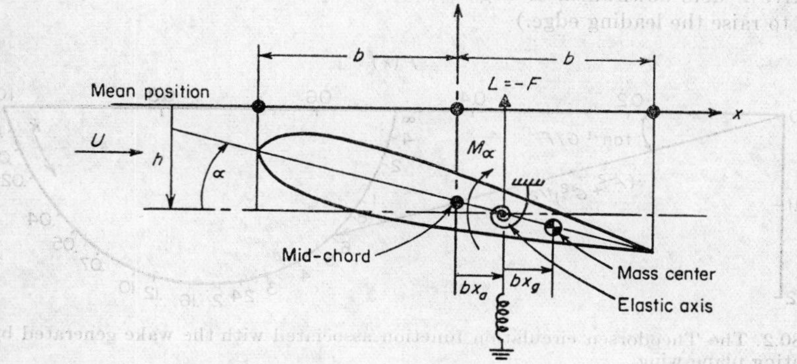

FIG. 80.1. Coordinates and dynamical parameters for bending-torsion flutter of two-dimensional wing system.

complex numbers, whose magnitudes represent maximum amplitudes, and whose arguments determine the relative phase. The equations of mechanical equilibrium for vertical forces and moments about $x = x_a$ are

$$m\ddot{h} + S_\alpha\ddot{\alpha} + m\omega_h^2 h = F \qquad S_\alpha\ddot{h} + I_\alpha\ddot{\alpha} + I_\alpha\omega_\alpha^2\alpha = M_\alpha \tag{80.3}$$

where m = mass per unit span

I_α = mass moment of inertia per unit span about $x = x_a$

S_α = static mass moment per unit span about $x = x_a$, positive when the center of gravity is aft

$\omega_h = \sqrt{C_h/m}$, uncoupled natural bending frequency

$\omega_\alpha = \sqrt{C_\alpha/I_\alpha}$, uncoupled natural torsional frequency

C_h, C_α = effective spring constants per unit span

The air force and moment per unit span length are defined by

$$F = b\int_{-1}^{1} P\,dx \qquad M_\alpha = b^2\int_{-1}^{1}(x - x_a)P\,dx \tag{80.4a,b}$$

where the pressure difference between upper and lower surfaces is

$$P = p_u - p_l \tag{80.5}$$

Expressions for the air force and moment per unit span length in Eqs. (80.3) and (80.4) for incompressible flow ($M = 0$) are

$$\begin{aligned} F &= -2\pi\rho UbC(k)[\dot{h} + U\alpha + (\tfrac{1}{2} - x_a)b\dot{\alpha}] - \rho b^2(\pi\ddot{h} + U\pi\dot{\alpha} - \pi b x_a\ddot{\alpha}) \\ M_\alpha &= 2\pi\rho Ub^2(\tfrac{1}{2} + x_a)C(k)[\dot{h} + U\alpha + (\tfrac{1}{2} - x_a)b\dot{\alpha}] \\ &\quad - \rho b^2[-\pi b x_a\ddot{h} + \pi(\tfrac{1}{2} - x_a)Ub\dot{\alpha} + \pi b^2(\tfrac{1}{8} + x_a^2)\ddot{\alpha}] \end{aligned} \tag{80.6a,b}$$

where b = reference length, the half chord, for which $c/2$ will also be used

$k = \omega c/2U = \omega b/U$, reduced frequency (number of wing oscillations made during a forward travel of 2π half chords)

U = forward speed in $-x$ direction

$C(k) = \dfrac{H_1^{(2)}(k)}{H_1^{(2)}(k) + iH_0^{(2)}(k)} = F(k) + iG(k)$, Theodorsen's function representing the wake effects (Fig. 80.2)

ρ = density of medium

$\dot{h} = i\omega h$; $\ddot{h} = -\omega^2 h$; etc.

x_a = nondimensional abscissa of axis of rotation (or elastic axis)

(Positive F acts downward, or negative F corresponds to lift L, while positive M_α tends to raise the leading edge.)

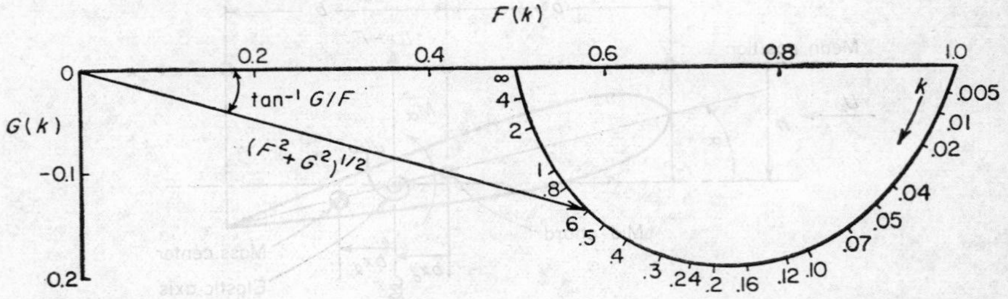

FIG. 80.2. The Theodorsen circulation function associated with the wake generated by an oscillating plane wing.

80.4. AERODYNAMIC COEFFICIENTS FOR THE HARMONIC CASE

In coefficient form, we have

$$-F = L = \tfrac{1}{2}\rho U^2 c c_l \qquad M_\alpha = \tfrac{1}{2}\rho U^2 c^2 c_m \tag{80.7a,b}$$

where (with $b = c/2$)

$$c_l = (c_{l,h})\frac{h}{b} + (c_{l,\alpha})\alpha \qquad c_m = (c_{m,h})\frac{h}{b} + (c_{m,\alpha})\alpha$$

and where the complex coefficients are defined by

$$\frac{1}{\pi} c_{l,h} = -k^2 + 2ikC(k) \qquad \frac{1}{\pi} c_{l,\alpha} = ik + x_a k^2 + 2C(k)[1 + (\tfrac{1}{2} - x_a)ik]$$

$$-\frac{2}{\pi} c_{m,h} = +x_a k^2 - 2(\tfrac{1}{2} + x_a)ikC(k) \tag{80.8}$$

$$-\frac{2}{\pi} c_{m,\alpha} = (\tfrac{1}{2} - x_a)ik - (\tfrac{1}{8} + x_a^2)k^2 - 2(\tfrac{1}{2} + x_a)C(k)[1 + (\tfrac{1}{2} - x_a)ik]$$

Each coefficient may be separated into real and imaginary parts, that is, into parts in phase and in quadrature with the respective displacement terms being modified; e.g., the lift coefficient for the α term is

$$c_{l,\alpha} = (c_{l,\alpha})_r + i(c_{l,\alpha})_i \tag{80.9}$$

where

$$(c_{l,\alpha})_r = \pi x_a k^2 + 2\pi F(k) - 2\pi(\tfrac{1}{2} - x_a)kG(k)$$

$$(c_{l,\alpha})_i = \pi k + 2\pi G(k) + 2\pi(\tfrac{1}{2} - x_a)kF(k)$$

Another way of representing this term appears in the literature in the type form

$$l_\alpha \alpha + l_{\dot{\alpha}}\dot{\alpha} + l_{\ddot{\alpha}}\ddot{\alpha}$$

where the various coefficients are termed aerodynamic derivatives (respectively stiffness, damping, and inertia derivatives) and, in this example, may be identified with the corresponding α terms in F in Eq. (80.6).

It may be observed that F in Eq. (80.6) may be identified as composed of three forces: (1) the wake or circulation term with factor $C(k)$ which acts at the one-quarter chord position, (2) the angular velocity term with factor $\dot{\alpha}$ which acts at the three-quarter chord position, and (3) the apparent mass terms with the factors $\ddot{h}$ and $\ddot{\alpha}$ which act at the mid-chord position [that with $\ddot{\alpha}$ involves also a pure couple $M = -(\pi/8)\rho b^4\ddot{\alpha}$].

80.5. FLUTTER DETERMINANT

When a parameter such as the flight speed is varied, a damped vibration may become a self-excited one. In between there is a border case where an undamped harmonic vibration is possible. This flutter condition is determined by substitution of the harmonic forms for the vibration modes and the requirement that a solution, which is more than merely a state of rest, exists for the homogeneous equations of the aeroelastic system, and this corresponds to the vanishing of a determinant of the coefficients of the amplitudes. The solution of this determinantal equation will yield the critical flutter speed and frequency and, moreover, the relative amplitudes as well as the differences in phase of the modes that exist at flutter.

The equations for the bending-torsion system (h,α) considered in Eq.(80.6) are

$$A_{11}\frac{h_0}{b} + A_{12}\alpha_0 = 0 \qquad A_{21}\frac{h_0}{b} + A_{22}\alpha_0 = 0$$

and the flutter determinant is

$$\begin{vmatrix} A_{11} & A_{12} \\ A_{21} & A_{22} \end{vmatrix} = 0 \tag{80.10}$$

where

$$A_{11} = \mu\left[-k^2 + \left(\frac{\omega_h}{\omega_\alpha}\right)^2 k_\alpha{}^2\right] + \frac{1}{\pi}c_{l,h} \qquad A_{12} = -\mu x_g k^2 + \frac{1}{\pi}c_{l,\alpha}$$

$$A_{21} = -\mu x_g k^2 - \frac{2}{\pi}c_{m,h} \qquad A_{22} = \mu r_\alpha{}^2(-k^2 + k_\alpha{}^2) - \frac{2}{\pi}c_{m,\alpha}$$

Here the aerodynamic coefficients are given by Eqs. (80.8), and the other dynamical parameters are as follows:

$\mu = m/\pi\rho b^2$ = mass-density parameter,

$x_g = S_\alpha/mb$ = nondimensional distance of center of gravity from axis of rotation (or elastic axis). [The center of gravity is thus located at $(1 + x_a + x_g)$ half chords from the leading edge.]

ω_h/ω_α = ratio of uncoupled bending to torsion frequencies

$r_\alpha = \sqrt{I_\alpha/mb^2}$ = nondimensional radius of gyration with respect to axis at $x = x_a$

$k = \omega b/U;\ k_\alpha = \omega_\alpha b/U$

The determinantal equation (80.10) may be solved for the values of k and k_α, and these will yield both the flutter speed and flutter frequency; afterward the ratio of the complex amplitudes $h_0/b\alpha_0$ may be determined, giving the proportion of bending and torsion, and their separation in phase in the flutter condition. Sometimes the determinant is multiplied through by $1/k^2$, and the variables chosen for evaluation may be k and ω_α/ω, or often in trend studies it may be convenient to solve for some other combination of parameters. Also, it is often very useful to introduce so-called structural damping terms in such a way that complex factors of the type $(1 + ig)$ occur, and multiply the elastic restoring terms. The g factors, when small, can be interpreted as energy-dissipation factors. The order of the determinant in the example is two; in general, it corresponds to the number of degrees of freedom chosen to represent the system.

An approximate expression for the flutter speed for the foregoing two-dimensional incompressible-flow case that is useful when the frequency ratio of bending to torsion is small ($\omega_h/\omega_\alpha \ll 1$) and the mass parameter large ($\mu > \sim 5.0$) is

$$\nu = b\omega_\alpha \sqrt{\mu} \sqrt{\frac{r_\alpha^2}{2(\frac{1}{2} + x_a + x_g)}}$$

In this formula $(\frac{1}{2} + x_a + x_g)b$ is the distance from the center of pressure (the quarter-chord point) to the center of gravity. For a summary of how various flutter experiences on simple configurations relate to this expression and to certain modifications of it, see [9].

80.6. PRESSURE FORMULA FOR GENERAL HARMONIC MOTION (TWO-DIMENSIONAL FLOW AT $M = 0$)

In the preceding example (see Fig. 80.1), the deflection in the z direction and the normal velocity for the rigid-section case may be expressed as

$$Z = -[h + b(x - x_a)\alpha] \qquad W = -[\dot{h} + U\alpha + b(x - x_a)\dot{\alpha}] \tag{80.11}$$

In order to obtain general air forces for *arbitrary* small displacements Z of the two-dimensional section, such as would be necessary for camber modes, or control surface motions, we may write

$$W(x,t) = \frac{\partial Z}{\partial t} + U\frac{\partial Z}{\partial(bx)} \tag{80.12}$$

There follows a classical result (see [5], p. 688) in integral form and in an evaluated form for the pressure difference $P(x)$ [see Eq. (80.5)] in incompressible flow, associated with the general oscillating normal velocity,

$$W(\xi,t) = w(\xi)e^{i\omega t} \tag{80.13}$$

namely,

$$\begin{aligned} P(x) = &-\frac{b\rho}{\pi}\oint_{-1}^{1} \ln\left|\frac{1 - x\xi - \sqrt{1 - x^2}\sqrt{1 - \xi^2}}{x - \xi}\right| \frac{\partial w(\xi)}{\partial t}\, d\xi \\ &+ \frac{2\rho U}{\pi}\sqrt{\frac{1 - x}{1 + x}}\oint_{-1}^{1}\sqrt{\frac{1 + \xi}{1 - \xi}}\frac{1}{\xi - x}\, w(\xi)\, d\xi \\ &+ \frac{2\rho U}{\pi}\sqrt{\frac{1 - x}{1 + x}}\,[C(k) - 1]\int_{-1}^{1}\sqrt{\frac{1 + \xi}{1 - \xi}}\, w(\xi)\, d\xi \end{aligned} \tag{80.14}$$

where the three terms in Eq. (80.14) represent respectively (1) apparent mass effect, (2) quasi-steady effect, i.e., the result that would be obtained by thin airfoil theory if the instantaneous velocity were frozen in time, and (3) effect of the harmonic wake extending to infinity. Cauchy principal parts are implied in the integrations where necessary.

A convenient *explicit* result, in which the above integrations are performed, is obtained as follows, by writing $x = -\cos\theta$ and $\xi = -\cos\eta$. Let the normal velocity distribution (which represents the boundary condition of the mean camber surface and is an even function of θ or η) be expressed as a Fourier series in the form

$$w = U\left(A_0 + 2\sum_{1}^{\infty} A_n \cos n\eta\right) \tag{80.15}$$

Then the pressure difference [Eq. (80.5)] is, in general,

$$P = \rho U^2 e^{i\omega t}\left(2a_0 \cot\frac{\theta}{2} + 4\sum_{1}^{\infty} a_n \frac{\sin n\theta}{n}\right) \tag{80.16}$$

where

$$a_0 = C(k)(A_0 - A_1) + A_1 \qquad a_n = \frac{ik}{2}A_{n-1} - nA_n - \frac{ik}{2}A_{n+1} \qquad n \geqq 1 \tag{80.17}$$

The total section force and moment about the axis at $x = x_a$ per unit span may be written on the basis of the form chosen for the pressure distribution in Eq. (80.16):

$$\begin{aligned} L &= -(\tfrac{1}{2}\rho U^2 c)e^{i\omega t}2\pi(a_0 + a_1) \\ M_\alpha &= -(\tfrac{1}{2}\rho U^2 c^2)e^{i\omega t}\frac{\pi}{2}\left[(1 + 2x_a)a_0 + 2x_a a_1 + \frac{a_2}{2}\right] \end{aligned} \tag{80.18}$$

The moment about the one-quarter chord axis ($x_a = -\tfrac{1}{2}$) simplifies to

$$M_{1/4} = (\tfrac{1}{2}\rho U^2 c^2)e^{i\omega t}\frac{\pi}{2}\left(a_1 - \frac{a_2}{2}\right) \tag{80.19}$$

Partial forces and moments or aileron and tab forces may be determined from Eq. (80.16) or Eq. (80.14) as required, or found in the literature. The results given in Eqs. (80.6) to (80.9) for the rigid wing section are obtained by the use of Eqs. (80.11) and (80.15) where $A_0 = -(\dot{h} + U\alpha - bx_a\dot{\alpha})/U$ and $A_1 = b\dot{\alpha}/2U$.

80.7. GENERALIZED AERODYNAMIC FORCES

Often in a flutter analysis, in order to go beyond the two-dimensional analysis, the wing's motion is assumed to be synthesized by a combination of rigid modes, such as translation, pitch, and roll, and a finite number of selected elastic modes of vibration termed semirigid. These may be of several types: (1) calculated natural or normal modes, which are orthogonal modes involving mechanical coupling of flexure and torsion, (2) uncoupled normal modes, for example, obtained if it is feasible to decouple bending and torsion motions by use of an elastic axis, (3) arbitrary modes defined mathematically, e.g., by polynomials in spanwise and chordwise coordinates, and (4) measured ground modes. The choice depends on a number of motivating factors, such as convenience, the stage of the design, or experience. It is not demanded nor necessary, in practice, that the elastic modes satisfy all boundary constraints. The aerodynamic coefficients that appear in the analysis depend on the choice of modes and the method of solution. If, as is commonly done, energy methods are employed for setting up the dynamical problem, and Lagrange's equations of motion are used, the aerodynamic forces appear *generalized*, and are related to the virtual aerodynamic work in the chosen mode. For example, let the deflection be given, relative to a system of axes fixed in space in the uniform stream (i.e., relative to a system of axes moving uniformly and attached to an origin in the mean planar surface), as

$$Z(x,y,t) = \sum_{1}^{n} q_j(t)Z_j(x,y) \tag{80.20}$$

where the q_j are generalized coordinates to be assumed as harmonic functions of time, and the Z_j are modal types of deflection, which may include rigid-body modes. The kinetic energy T and potential energy U appearing in Lagrange's equations,

$$\frac{d}{dt}\frac{\partial T}{\partial \dot{q}_j} - \frac{\partial T}{\partial q_j} + \frac{\partial U}{\partial q_j} = Q_j \tag{80.21}$$

are generally (at least for small perturbations from equilibrium) quadratic forms in q_j and its time derivatives, and are here assumed to be known. Their expressions will involve dynamical parameters similar to those in the simple example given earlier, but weighted by appropriate integrations over the plan form and involving the mode shapes. For a choice of normal modes, both T and U will contain no cross-product terms in the q_js; for uncoupled modes, this holds for U but not necessarily for T. The generalized aerodynamic force Q_j depends on the work of the air forces for virtual displacement δq_j in a particular mode $Z_j(x,y)$; thus, the total work is (see [1], p. 556)

$$W = \sum_{1}^{n} Q_j \, \delta q_j$$

and

$$Q_j = \frac{\partial W}{\partial q_j} = \sum_{k=1}^{n} Q_{j,k} q_k = \iint F_z(x,y,t) Z_j(x,y) \, dx \, dy \tag{80.22}$$

where F_z is the force per unit area in the z direction corresponding to $-P$, where P is the pressure difference, and where the integration is taken over appropriate surfaces of the aircraft wherein the integrand is of significance. Thus, Q_j may be separated into terms associated with the partial work done by air forces arising from mode k in moving through the virtual displacement δq_j, and pressure differences are required for each chosen mode of vibration. In general, this is a large undertaking and nearly always greatly simplified.

One such simplification that has proved useful for large-aspect-ratio lifting surfaces is the *strip analysis* method. This assumes each section to behave aerodynamically as though two-dimensional and, therefore, utilizes aerodynamics such as given for $M = 0$ in the preceding pages, and for other values of the Mach number subsequently. In general, the local chord for each strip should be retained dimensionally as a function of the span variables, for it appears in various powers in the spanwise integrations. An over-all reference length b_r may be chosen conveniently, such as the half chord at the root or at the three-quarter span position.

80.8. ILLUSTRATIVE EXAMPLE FOR CANTILEVER WING

Consider two degrees of freedom represented by a mode in bending and one in torsion of an unswept cantilever wing. Let these be

$$h(y,t) = h(t)f(y) \qquad \alpha(y,t) = \alpha(t)\phi(y)$$

where $h(t)$ and $\alpha(t)$ are the generalized coordinates (q_1 and q_2), having harmonic time variation $h = h_0 e^{i\omega t}$, $\alpha = \alpha_0 e^{i\omega t}$; and $f(y)$ and $\phi(y)$ define the modal shapes. (It is convenient, though not necessary, to define each of these to be unity at the tip.) The quadratic forms for the kinetic and potential energies may be written as

$$T = \tfrac{1}{2}\bar{m}\dot{h}^2 + \bar{S}_\alpha \dot{h}\dot{\alpha} + \tfrac{1}{2}\bar{I}_\alpha \dot{\alpha}^2 \qquad U = \tfrac{1}{2}\bar{m}\omega_h^2 h^2 + \tfrac{1}{2}\bar{I}_\alpha \omega_\alpha^2 \alpha^2$$

and the equations of motion corresponding to Eq. (80.3) are

$$\bar{m}\ddot{h} + \bar{S}_\alpha \ddot{\alpha} + \bar{m}\omega_h^2 h = Q_1 \qquad \bar{S}_\alpha \ddot{h} + \bar{I}_\alpha \ddot{\alpha} + \bar{I}_\alpha \omega_\alpha^2 \alpha = Q_2$$

where the generalized dynamical parameters are

$$\bar{m} = \int_0^l m(y) f^2(y) \, dy$$

$$\bar{S}_\alpha = \int_0^l S_\alpha(y) f(y)\phi(y) \, dy = b_r \int_0^l m x_g \frac{b}{b_r} f(y)\phi(y) \, dy = b_r \bar{m} \bar{x}_g$$

$$\bar{I}_\alpha = \int_0^l I_\alpha \phi^2(y) \, dy = b_r^2 \int_0^l m r_\alpha^2 \left(\frac{b}{b_r}\right)^2 \phi^2(y) \, dy = b_r^2 \bar{m} \bar{r}_\alpha^2$$

$$\bar{m}\omega_h^2 = \bar{C}_h = \int_0^l EI \left[\frac{d^2}{dy^2} f(y)\right]^2 dy \qquad \bar{I}_\alpha \omega_\alpha^2 = \bar{C}_\alpha = \int_0^l GJ \left[\frac{d}{dy}\phi(y)\right]^2 dy$$

and where l is the span length.

The generalized aerodynamic forces are given, for harmonic time variation, by [see Eqs. (80.7) and (80.8)]

$$Q_1 = -\tfrac{1}{2}\rho U^2 c_r \bar{c}_l \qquad Q_2 = \tfrac{1}{2}\rho U^2 c_r^2 \bar{c}_m$$

where

$$\bar{c}_l = \bar{c}_{l,h}\frac{h}{b_r} + \bar{c}_{l,\alpha}\alpha \qquad \bar{c}_m = \bar{c}_{m,h}\frac{h}{b_r} + \bar{c}_{m,\alpha}\alpha$$

and where

$$\bar{c}_{l,h} = \int_0^l c_{l,h} f^2(y)\,dy \qquad \bar{c}_{l,\alpha} = \int_0^l \frac{c}{c_r} c_{l,\alpha} f(y)\phi(y)\,dy$$

$$\bar{c}_{m,h} = \int_0^l \frac{c}{c_r} c_{m,h} f(y)\phi(y)\,dy \qquad \bar{c}_{m,\alpha} = \int_0^l \left(\frac{c}{c_r}\right)^2 c_{m,\alpha}\phi^2(y)\,dy$$

Wherever $k = \omega b/U$ appears, it may be conveniently replaced by $k_r c/c_r$ where $k_r = \omega b_r/U$ and where $c_r = 2b_r$.

The determinantal flutter equation corresponding to Eq. (80.10) may be expressed as

$$\begin{vmatrix} \bar{A}_{11} & \bar{A}_{12} \\ \bar{A}_{21} & \bar{A}_{22} \end{vmatrix} = 0$$

where

$$\bar{A}_{11} = \bar{\mu}\left[-k_r^2 + \left(\frac{\omega_h}{\omega_\alpha}\right)^2 k_\alpha^2\right] + \frac{1}{\pi}\bar{c}_{l,h} \qquad \bar{A}_{12} = -\bar{\mu}\bar{x}_g k_r^2 + \frac{1}{\pi}\bar{c}_{l,\alpha}$$

$$\bar{A}_{21} = -\bar{\mu}\bar{x}_g k_r^2 - \frac{2}{\pi}\bar{c}_{m,h} \qquad \bar{A}_{22} = \bar{\mu}\bar{r}_\alpha^2(-k_r^2 + k_\alpha^2) - \frac{2}{\pi}\bar{c}_{m,\alpha}$$

and where

$$\bar{\mu} = \frac{\bar{m}}{\pi\rho b_r^2} \qquad b_r = \text{reference half chord} \qquad k_r = \frac{\omega b_r}{U} \qquad k_\alpha = \frac{\omega_\alpha b_r}{U}$$

and all barred terms have previously been defined as generalized quantities. The foregoing modal method illustrated for two modes may utilize n degrees of freedom, and is sometimes referred to as a Rayleigh-Ritz-type flutter analysis. An equivalent variation of it is Galerkin's method, (see p. 63–11).

When the reduced frequency is small, it is sometimes feasible to introduce over-all factors from steady or quasi-steady aerodynamics to modify the strip-analysis results, as the measured slope of the lift curve, for example. Other three-dimensional flow considerations for including finite-span or aspect-ratio effects are mentioned below.

Discrete mass methods, which are alternative ones for the modal methods, have long been used in the field of elasticity for determining vibration modes, for example, of a segmented wing. These methods lend themselves to systematic matrix formulations for the inertia matrix and for the flexibility influence coefficients, and are particularly valuable when high-speed computing machinery is available, and iterative procedures may be used. The aerodynamic matrix which must be included for the aeroelastic problems represents a set of influence coefficients appropriate to the chosen segments or boxes, and is usually synthesized from results for the continuous-mode methods or by strip analysis. Though many special analytic (and electric analog) devices exist, practical procedures are tailored to each case, and cannot be prescribed in general. A typical formulation for harmonic oscillations leads to relations

$$\{Z\} = \omega^2[G][M + \rho A]\{Z\}$$

where $\{Z\}$ is a column matrix of deflections, $[G]$ a square matrix of flexibility influence coefficients, $[M]$ a mass matrix, and $[\rho\omega^2 A]$ an aerodynamic matrix. When the aerodynamic matrix is absent, the relations correspond to the mechanical-vibration case. The matrix $[G]$ is the inverse to the stiffness influence coefficient matrix, and the product $[G][M]$ is sometimes referred to as the dynamical matrix (see also Sec. 63.5).

80.9. SUPERSONIC FLOW (TWO-DIMENSIONAL, NONSTEADY)

The linearized methods for this case make use of the concept of independence of the upper and lower surfaces of the plan form, i.e., no interaction of the flows over the separate surfaces; hence, distributions of sources or source pulses may synthesize the flow, and explicit solutions of relatively simple forms are obtained.

The general surface velocity potential at a point x (dimensional) for an arbitrary normal velocity distribution $W(\xi,t)$ is [5]

$$\Phi(x,t) = -\frac{1}{\pi\beta}\int_0^x\int_0^\pi W\left(\xi,t - \frac{x-\xi}{U}\frac{M^2}{\beta^2} - \frac{x-\xi}{U}\frac{M}{\beta^2}\cos\theta\right) d\theta\, d\xi \tag{80.23}$$

where $M = U/a$ the Mach number, and $\beta = |M^2 - 1|^{1/2}$, and the leading edge has been designated by ξ or $x = 0$. For example, for the harmonically oscillating case, let

$$W(\xi,t) = w(\xi)e^{i\omega t} \qquad \Phi(x,t) = \phi(x)e^{i\omega t} \tag{80.24}$$

Then

$$\phi(x) = -\frac{1}{\beta}\int_0^x w(\xi)G(x-\xi)\,d\xi \tag{80.25}$$

where

$$G(x) = e^{-i\omega M^2 x/U\beta^2} J_0\left(\frac{\omega}{U}\frac{M}{\beta^2}x\right) \tag{80.26}$$

The linearized pressure difference associated with Eq. (80.23) is, in general,

$$P(x,t) = -2\rho\left(\frac{\partial\Phi}{\partial t} + U\frac{\partial\Phi}{\partial x}\right) \tag{80.27}$$

and, for the oscillating case, by writing

$$P(x,t) = P(x)e^{i\omega t} \tag{80.28}$$

there is obtained

$$P(x) = \frac{2\rho U}{\beta}\left[w(0)G(x) + \int_0^x G(x-\xi)\left(\frac{\partial}{\partial\xi} + \frac{i\omega}{U}\right)w(\xi)\,d\xi\right] \tag{80.29}$$

For illustration, taking the case of translation and rotation of a rigid section, as developed for $M = 0$ in Eq. (80.6); it results for $M > 1$ that

$$\begin{aligned} F &= -\frac{2\rho c}{\beta}\left[Ur_1\dot{h} + cr_2\ddot{h} + U^2 r_1\alpha + (-Ucx_0r_1 + 2Ucr_2)\dot{\alpha} + \frac{cr_3}{2}\ddot{\alpha}\right] \\ M_0 &= -\frac{2\rho c^2}{\beta}\left[Uq_1\dot{h} + \frac{cq_2}{2}\ddot{h} + U^2q_1\alpha + (-Ucx_0q_1 + Ucq_2)\dot{\alpha} + c^2\frac{q_3}{6}\ddot{\alpha}\right] - cx_0F \end{aligned} \tag{80.30}$$

where x_0 is the location of the axis of rotation as a fraction of the chord c from the leading edge ($x_0 = (1 + x_a)/2$), and the coefficients $r_1, r_2, \ldots q_3$ are expressible in closed form in terms of Bessel functions $J_0(\bar{\omega}/M)$ and $J_1(\bar{\omega}/M)$ and the following basic integral

$$f_0(M,\bar{\omega}) = \frac{1}{\bar{\omega}}\int_0^{\bar{\omega}} e^{-iu}J_0\left(\frac{u}{M}\right)du \tag{80.31}$$

The argument $\bar{\omega}$ is defined as

$$\bar{\omega} = \frac{c\omega}{U}\frac{M^2}{\beta^2} = \frac{2kM^2}{\beta^2} \tag{80.32}$$

Tables and expansions in powers of $\bar{\omega}$ exist for all functions involved. (It may be noted that the parameter $\bar{\omega}$ becomes very large for a given value of k as $M \to 1$).

For small values of the frequency parameter $\bar{\omega}$, the functions are approximately

$$
\begin{aligned}
r_1 &\approx 1 - i\frac{\bar{\omega}}{2} & q_1 &\approx \tfrac{1}{2} - i\frac{\bar{\omega}}{3} \\
r_2 &\approx \tfrac{1}{2} - i\frac{\bar{\omega}}{6} & q_2 &\approx \tfrac{2}{3} - i\frac{\bar{\omega}}{4} \\
r_3 &\approx \tfrac{1}{3} - i\frac{\bar{\omega}}{12} & q_3 &\approx \tfrac{3}{4} - i\frac{\bar{\omega}}{5}
\end{aligned}
\tag{80.33}
$$

80.10. HIGH MACH-NUMBER FLOWS

The limit approached by the linearized two-dimensional results for the perturbation pressure at any surface point for high (infinite) Mach numbers is

$$p = \frac{\rho_\infty U}{\beta} W(x,t) \approx \frac{\rho_\infty U}{M} W(x,t) \approx \rho_\infty a_\infty W(x,t) \tag{80.34}$$

where $\beta = |M^2 - 1|^{1/2} \approx M$, and subscript ∞ is employed to denote the undisturbed stream values. Equation (80.34) corresponds to the basic steady-flow linearized result of Ackeret; hence, it indicates the valid use of quasi-steady results that employ the local values of the normal velocity W. The result also corresponds to the relation between perturbation pressure and particle velocity in a plane acoustic wave. An immediate generalization extends the result to the *simple* wave theory which employs the local values of density and sound speed instead of the stream values. It leads to the *piston theory* results, as follows,

$$dp = \rho a\, dW \qquad p/p_\infty = (\rho/\rho_\infty)^\gamma \qquad dp/d\rho = a^2 \tag{80.35}$$

where γ is the adiabatic index. By integration, the local pressure is given as a function of the local value of W and of the main-stream conditions as

$$p_l(W) = p_\infty \left(1 + \frac{\gamma - 1}{2a_\infty} W\right)^{2\gamma/(\gamma-1)} \tag{80.36}$$

or, expanded to the third power in W (consideration of compression shocks influences this term in any case), the perturbation pressure is

$$p = p_l - p_\infty = \rho_\infty a_\infty^2 \left[\frac{W}{a_\infty} + \frac{\gamma+1}{4}\left(\frac{W}{a_\infty}\right)^2 + \frac{\gamma+1}{12}\left(\frac{W}{a_\infty}\right)^3 + \cdots\right] \tag{80.37}$$

If $W(x,t)$ is separated into

$$W = W_c \pm W_t \tag{80.38}$$

where W_c is a part due to the mean camber or mean angle of attack and W_t is a part due to the thickness, and where, applied to the upper surface, the $+$ sign, and, for the lower surface, the $-$ sign is used, it results that the pressure difference is

$$P = -\rho_\infty a_\infty^2 \left[2 + (\gamma+1)\frac{W_t}{a_\infty} + \frac{\gamma+1}{2}\left(\frac{W_t}{a_\infty}\right)^2\right]\frac{W_c}{a_\infty} - \tfrac{1}{6}(\gamma+1)\left(\frac{W_c}{a_\infty}\right)^3 \tag{80.39}$$

The main effect of thickness is already contained in the first term with factor W_t, while the term with factor 2 corresponds to the linear result indicated in Eq. (80.34). In order to apply Eq. (80.39) to lower supersonic Mach numbers, it may be proposed to replace the term with factor 2 by the linearized theory results obtained by use of Eqs. (80.23) to (80.29) [or by use of Eq. (80.40)], and to add the thickness term as a correction. (This procedure should correspond, in the large, to results obtained by more elaborate methods of iteration, as by Van Dyke [10].)

The foregoing results, though nonlinear, are of sufficiently simple form to be readily applied. Moreover, there is no need to limit considerations to the border harmonic solution, so that complex values of the frequency term ω may be directly employed. Such applications have been found useful in panel-flutter analysis for supersonic flows (see p. 63–6).

80.11. SUPERSONIC FLOW (THREE-DIMENSIONAL, NONSTEADY)

Figure 80.3 indicates that, for supersonic flow, several types of regions must be analyzed. If the given plan form is circumscribed by the forward- and rearward-facing Mach lines corresponding to the stream Mach number under consideration, a number of regions of arc segments are defined by their points of tangency, A, E, B, D, and by those of the outermost stream lines, C, F. Segment AB is a supersonic

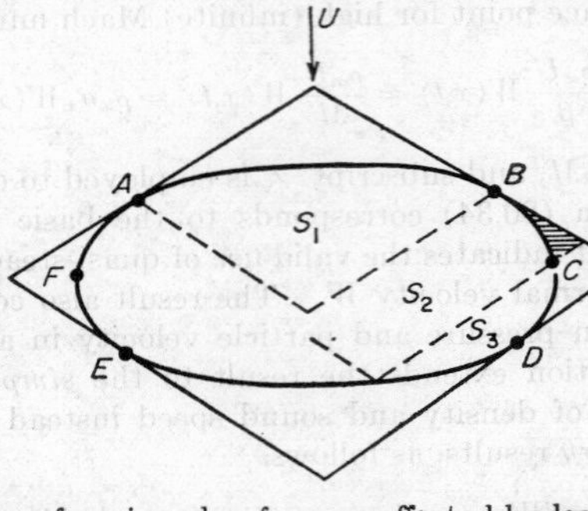

FIG. 80.3. Various regions of a wing plan form as affected by leading and trailing edges in a supersonic flow.

leading edge characterized by the fact that the component of the main-stream velocity normal to the edge remains supersonic; similarly, DE is a supersonic trailing edge; segments BC and AF are subsonic leading edges; and segments CD and FE are subsonic trailing edges. It is observed that the proportions of the regions such as S_1, S_2, and S_3 influenced by these edges change with the stream Mach number. For high values of M, type S_1 mainly occurs, whereas, for values of M near unity, types S_2 and S_3 may predominate.

The flow problem for the neighborhood of any point on the plan form influenced only by regions such as S_1 bounded by the supersonic edge may be given a general explicit solution as follows. The linearized perturbation velocity potential may be expressed as a definite integration of sources taken over the plan-form surface as

$$\phi(x,y,z,t) = -\frac{1}{2\pi}\iint \frac{W(\xi,\eta,0,t-\tau_1) + W(\xi,\eta,0,t-\tau_2)}{R}\,d\xi\,d\eta \qquad (80.40)$$

where

$$\tau_1 = [M(x-\xi) - R]/a\beta^2 \qquad \tau_2 = [M(x-\xi) + R]/a\beta^2$$

$$R = \sqrt{(x-\xi)^2 - \beta^2(y-\eta)^2 - \beta^2 z^2} \qquad \beta = |M^2 - 1|^{1/2} \text{ and } M = U/a$$

The normal velocity W is arbitrary: the result may be applied to obtain local superstream pressures independently on the separate surfaces of the plan form, and then the difference may be taken. A significant part of the nonlinear effects associated with thickness will be included, in this way, by employing the general pressure formula, which may be expressed as

$$C_p = \frac{p_l - p_\infty}{\frac{1}{2}\rho_\infty U^2} = \frac{2}{\gamma M^2}\left\{\left[1 + \frac{\gamma - 1}{2}M^2\left(1 - \frac{(\bar{U} + \nabla\phi)\cdot(\bar{U} + \nabla\phi) + 2\phi_t}{U^2}\right)\right]^{\gamma/(\gamma-1)} - 1\right\} \qquad (80.41)$$

where U is the main-stream velocity and $\nabla\phi$ is the disturbance velocity. The first few terms may be written as

$$C_p \approx -\frac{2}{U^2}(\phi_t + U\phi_x) + \frac{\beta^2}{U^2}\phi_x{}^2 - \frac{1}{U^2}\phi_y{}^2 + \frac{2a^2}{U}\phi_x\phi_t + \frac{a^2}{U^2}\phi_t{}^2$$

where partial derivatives have been denoted by subscripts and where the first term on the right side is the linear-pressure term.

For the harmonic case, Eq. (80.40) may be written as

$$\phi(x,y,z,t) = \phi(x,y,z)e^{i\omega t}$$
$$= -\frac{e^{i\omega t}}{\pi}\iint W(\xi,\eta)e^{i\mu(x-\xi)}\frac{\cos\kappa R}{R}\,d\xi\,d\eta \qquad (80.40a)$$

where $\kappa = \omega/a\beta^2$, $\mu = \omega M/a\beta^2$, and R is given in (80.40). Evaluation of Eq. (80.40) or (80.40*a*) has been proposed by covering the plan form of type S_1 with an array of small rectangular boxes and performing the integrations approximately by a combination of analytic, numerical, and sequential treatments. Similar but more elaborate numerical methods have been advanced for regions of type S_2, while general methods for regions of type S_3 are also feasible, in principle, but presently are not well-developed. An analytic development of air forces on wings having triangular and related plan forms with subsonic leading edges, which assumes that the mode shapes are given as polynomials in powers of x and y and develops the results in powers of the frequency, has been given in [11]–[13].

80.12. OSCILLATING AIRFOIL IN SUBSONIC COMPRESSIBLE FLOW (TWO-DIMENSIONAL AND THREE-DIMENSIONAL METHODS)

For M less than about 0.5, it is easier and perhaps just as satisfactory, for most purposes, to employ the methods for the incompressible flow case ($M = 0$). For M greater than about 0.75, there may arise doubts about the validity of the shock-free nonseparated flow solutions. For such reasons, theoretical methods for subsonic flows have been rather infrequently used, and empirical corrections have been sought. Nevertheless, a brief classification of the existing methods is desirable, as trend studies with change in Mach numbers have been useful.

The earliest method for two-dimensional flow is based on *Possio's integral equation* which may be put into the following form

$$W(x) = \frac{1}{4\pi\rho}\int_{-1}^{1} L(\xi)K(x-\xi)\,d\xi \qquad (80.42)$$

where $W(x)$ is known from the harmonic motion of the section $W(x,t) = W(x)e^{i\omega t}$
$L(\xi)$ is the distribution of lifting-pressure difference to be determined
$K(x-\xi)$ is a known kernel function

The kernel K is obtained by starting with an isolated pressure jump or doublet singularity of the two-dimensional wave equation for the pressure, and inverting Eq. (80.27) to yield the velocity field induced by this singularity. The function is analytically known, and can be tabulated.

The simplest method in principle of obtaining the loading $L(\xi)$ is by collocation; a series of basic pressure distributions with unknown coefficients is assumed for L such as

$$L(x) = a_0\sqrt{\frac{1-x}{1+x}} + a_1\sqrt{1-x^2} + a_2x\sqrt{1-x^2} + \cdots \qquad (80.43)$$

A number of downwash points are picked, at which W is determined; a set of simultaneous equations is developed by Eq. (80.42) for the unknown coefficients a_n which permit these to be found. (The method corresponds in the steady-flow case to the Prandtl-Glauert correction factor for the incompressible-flow pressure distribution, and may be interpreted, when the frequency $\omega \to 0$ or when the reduced frequency k is small, to lead to a correction factor $(1 - M^2)^{1/2}$ for the dynamic pressure at flutter or to a correction factor $(1 - M^2)^{1/4}$ for the flutter speed as compared to incompressible flow results.)

Another method (Dietze's) is an iteration procedure, which employs the known result [Eqs. (80.15) to (80.17)] for the incompressible flow as a basis. Still another method (Fettis) approximates K in an appropriate way to permit analytic inversion of Eq. (80.42). Finally, a classical method should be mentioned, which employs appropriate elliptic coordinates and separation of variables to yield product-type solutions in terms of Mathieu functions. Numerical tables by this method and series expansions in powers of the frequency are indicated, for example, in [14].

For three-dimensional subsonic flow, a generalization of the collocation method has been proposed. The integral equation is of similar form to Eq. (80.42), where the integration is to be taken over the plan form. The kernel is the downwash function appropriate to a doublet singularity in three-dimensional space, and the pressure-distribution functions as in Eq. (80.43) contain also chosen factors of the span variable. (The appropriate forms of the distribution functions will vary if sonic and supersonic speed ranges are also to be included.) The three-dimensional methods require high-speed machine computing and are in the process of being systematized for routine use.

80.13. REMARKS ON INTERMEDIATE METHODS, SMALL-ASPECT-RATIO WINGS, AND SWEPTBACK WINGS

There are a large number of methods which attempt to improve on the two-dimensional results for incompressible flow by moving in the direction of lifting-surface procedures, but which do not go all the way. The methods (for example, of Cicala, Küssner, and Reissner) each lead to a single integral equation in the span variable, one which reduces to the classical integral equation of Prandtl's lifting-line theory in the case of steady flow, and which is made to contain the two-dimensional unsteady-flow case for infinite aspect ratios. A representative full development with applications to unswept wings is to be found in [15] and [16]. These single lifting-line methods have not been generally adopted, since they frequently have led to unconservative flutter results. A variation, in which the chord variable instead of the span variable is employed, has been developed by Lawrence and Gerber [17] to apply more suitably to low-aspect-ratio wings. The limiting case of vanishingly small aspect ratio has been given a full development by generalization of low aspect ratio and slender-body theory of R. T. Jones. (For example, see [5], p. 748.) General methods for treating swept wings in unsteady flow would depend on lifting-surface procedures; however, the much simpler methods of two-dimensional strip analysis are usually employed. The procedures involve two main concepts: (1) recognition that a swept elastic axis (angle of sweep Λ) undergoing bending and torsion leads to additional terms in the normal velocity distribution, depending on the local bending slope of the axis and the local change in twist, and on the component of the main stream along the span $U \sin \Lambda$, and (2) the assumption that the component of the main stream normal to the leading edge of the plan form, $U \cos \Lambda$, is the aerodynamically effective component [18]. Alternatively, an approximate analysis based on strips taken in the streamwise direction may also be employed, usually in conjunction with coupled modes (see [6], p. 351).

80.14. INDICIAL FUNCTIONS ASSOCIATED WITH TRANSIENT FLOWS (GROWTH OF LIFT AND GUST FUNCTIONS); THE SUPERPOSITION PRINCIPLE

Since, according to linearized theory, solutions are additive, it is possible to relate the results for the harmonically oscillating wing to those for the indicial case of a unit-step time history, either by Fourier integral or by Laplace transform methods. In principle, this may be done for a given Mach number for *any* chosen mode of normal velocity $w(x,y)$, to obtain either local results as pressure distributions, or integrated results as lift or moment. The principle of superposition permits generalizing the indicial results for the chosen mode to an arbitrary time history. For example, consider the oscillatory case of uniform vertical translation as given in Eq. (80.6) with $\alpha \equiv 0$, for which

$$w(x,y)e^{i\omega t} = \dot{h} = w_0 e^{i\omega t}$$

where w_0 is a constant having the dimension of a velocity. The oscillatory lift (denoted by subscript 0) per unit span is given by

$$L_0 = \left[\pi\rho U c C(k) + \frac{\pi\rho c^2}{4}\frac{\partial}{\partial t}\right] w_0 e^{i\omega t} \tag{80.44}$$

Then the indicial lift (denoted by subscript 1) for the unit-step time function $w_0\Delta(t)$, where $\Delta(t) = 1$ for $t > 0$, and $\Delta(t) = 0$ for $t < 0$, is

$$L_1 = \left[\pi\rho U c k_1(\sigma) + \frac{\pi\rho c^2}{4}\,\delta(\sigma)\right] w_0\Delta(\sigma) \tag{80.45}$$

where it is convenient to introduce the variable σ, distance traversed in half chords, instead of t as follows,

$$\omega t = \left(\frac{\omega c}{2U}\right)\left(\frac{Ut}{c/2}\right) = k\sigma \tag{80.46}$$

and where, in Eq. (80.45), $\delta(\sigma)$ is the Dirac impulse function [the derivative of the unit step, $\Delta'(\sigma)$], associated with the fact that an incompressible fluid has been assumed and the apparent mass effects act instantaneously; and $k_1(\sigma)$ is known as Wagner's growth of lift function. To express the relations between $k_1(\sigma)$ and $C(k)$, consider the classical transforms [Eqs. (19.1) and (19.22)]:

$$\frac{f(p)}{p} = \int_0^\infty e^{-pt}A(t)\,dt \qquad A(t) = \frac{1}{2\pi i}\int_{c-i\infty}^{c+i\infty}\frac{f(p)}{p}e^{pt}\,dp$$

These may be recognized from operational calculus as the tie between the indicial time function $A(t)$ and the complex admittance $f(p) = 1/Z(p)$. By introducing the variables σ for t, ik for p, where $\sigma = Ut/b$ and $k = \omega b/U$, there may be obtained the relations

$$\frac{C(k)}{ik} = \int_0^\infty k_1(\sigma)e^{-ik\sigma}\,d\sigma \tag{80.47}$$

or also, in inverted form, writing $C(k) = F(k) + iG(k)$,

$$k_1(\sigma) = F(0) + \frac{2}{\pi}\int_0^\infty \frac{F(k) - F(0)}{k}\sin k\sigma\,dk$$

or

$$k_1(\sigma) = F(0) + \frac{2}{\pi}\int_0^\infty \frac{G(k)}{k}\cos k\sigma\,dk \tag{80.48}$$

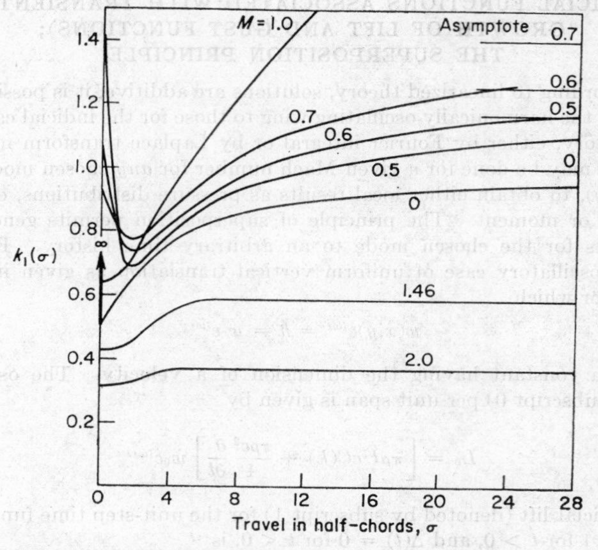

FIG. 80.4. Indicial growth-of-lift function $k_1(\sigma)$ for two-dimensional flows.

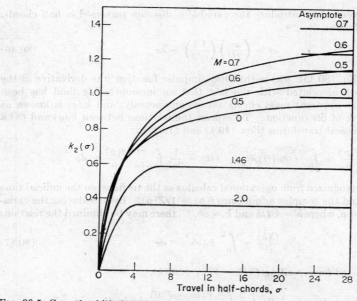

FIG. 80.5. Growth-of-lift functions $k_2(\sigma)$ for gust penetration for two-dimensional flows.

By introducing *generalized* functions $\bar{C}(k)$ and $\bar{k}_1(\sigma)$, the results in Eqs. (80.44) and (80.45) may be expressed as

$$L_0 = \pi\rho U c\bar{C}(k)w_0e^{i\omega t} \tag{80.44a}$$

and

$$L_1 = \pi\rho U c\bar{k}_1(\sigma)w_0\Delta(\sigma) \tag{80.45a}$$

where

$$\bar{k}_1(\sigma) = k_1(\sigma) + \frac{c/2}{U}\frac{\delta(\sigma)}{2}$$

Relations between $\bar{C}(k) = \bar{F}(k) + i\bar{G}(k)$ and $k_1(\sigma)$ are of the same type as given in Eqs. (80.47) and (80.48).

For an *arbitrary* time history of vertical motion expressed as $w(\sigma)$, use of the superposition principle in the form of Duhamel's integral yields

$$L_w = w(0)k_1(\sigma) + \int_0^\sigma k_1(\lambda - \sigma)\frac{d}{d\lambda}w(\lambda)\,d\lambda \tag{80.49}$$

As indicated, similar considerations may be applied to determine growth of lift response for any functions $w(x,y)f(t)$ or $w(x,y;\sigma)$. In this connection, it is often desirable to express the indicial functions in a form convenient for operational or transform methods, or for electric analog methods. For example, Wagner's function $k_1(\sigma)$ may be approximated by

$$k_1(\sigma) \approx 1 - 0.165e^{-0.0455\sigma} - 0.335e^{-0.3\sigma} \tag{80.50}$$

The asymptotic value as $\sigma \to \infty$ is $k_1(\sigma) = 1$, and corresponds to the slope of the lift curve for two-dimensional steady flow. Similar forms may be utilized for finite-span cases, wherein the asymptotic value may correspond to steady flow as measured or

Table 80.1. Growth-of-Lift Indicial Function $k_1(\sigma)$ and Gust Function $k_2(\sigma)$

$k(\sigma) = (C_L/2\pi)(b_0 + b_1e^{-\beta_1\sigma} + b_2e^{-\beta_2\sigma} + b_3e^{-\beta_3\sigma})$

A. Two-dimensional Wing

Indicial function	Mach number	$C_L/2\pi$	b_0	b_1	b_2	b_3	β_1	β_2	β_3
$k_1(\sigma)$	0	1.00	1.00	−0.165	−0.335	0	0.0455	0.300	
	0.5	1.155	1.00	−0.352	−0.216	0.669	0.0754	0.372	1.890
	0.6	1.250	1.00	−0.362	−0.504	0.714	0.0646	0.481	0.958
	0.7	1.400	1.00	−0.364	−0.405	−0.419	0.0536	0.357	0.902
$k_2(\sigma)$	0	1.00	1.00	−0.236	−0.513	−0.171	0.058	0.364	2.42
	0.5	1.155	1.00	−0.390	−0.407	−0.203	0.0716	0.374	2.165
	0.6	1.250	1.00	−0.328	−0.430	−0.242	0.0545	0.257	1.461
	0.7	1.400	1.00	−0.402	−0.461	−0.137	0.0542	0.3125	1.474

B. Elliptic Wing at $M = 0$

Indicial function	Aspect ratio	$C_L/2\pi$	b_0	b_1	b_2	b_3	β_1	β_2	β_3
$k_1(\sigma)$	∞	1.00	1.00	−0.165	−0.335	0	0.0455	0.300	
	6	0.75	1.00	−0.361	0	0	0.381		
	3	0.600	1.00	−0.283	0	0	0.540		
$k_2(\sigma)$	∞	1.00	1.00	−0.236	−0.513	−0.171	0.058	0.364	2.42
	6	0.75	1.00	−0.448	−0.272	−0.193	0.290	0.725	3.00
	3	0.600	1.00	−0.678	−0.227	0	0.558	3.20	

calculated. Table 80.1 presents a few such results for the growth of the lift indicial function $k_1(\sigma)$, resulting from a sudden acquisition of uniform downwash, and the function $k_2(\sigma)$, which represents the growth of lift after penetration into a normal indicial gust field. The two-dimensional cases (part A of Table 80.1) are illustrated for a sequence of Mach numbers from $M = 0$ to $M = 2$ in Figs. 80.4 and 80.5. Results for wings of elliptical plan form for aspect ratios ∞, 6, and 3 are given in part B of Table 80.1.

80.15. CONCLUDING REMARKS

The theoretical aspects of the aerodynamics of flutter must often be supplemented by (1) measurement of aerodynamic forces and moments over appropriate ranges of frequency, Reynolds numbers, angles of attack; (2) flutter experiments in subsonic, transonic, and supersonic wind tunnels; (3) use of scaled flutter models for prototype testing and research trends; and (4) techniques of vibration testing in flight. Various simple rules or design criteria relating to principles of mass balance, increase in torsional stiffness, avoidance of frequencies nearly alike in two coupled modes, etc., may be found in the general literature. Very few of these rules are always valid or generally foolproof.

80.16. REFERENCES

[1] R. L. Bisplinghoff, H. Ashley, R. L. Halfman: "Aeroelasticity," Addison-Wesley, Reading, Mass., 1955.

[2] Civil Air Regulations, "Airplane Airworthiness," part 4.

[3] A. H. Flax: Aeroelasticity and flutter, sec. C in vol. 8 of "High Speed Aerodynamics and Jet Propulsion," Princeton Univ. Press, Princeton, N.J., 1961.

[4] Y. C. Fung: "An Introduction to the Theory of Aeroelasticity," Wiley, New York, 1955.

[5] I. E. Garrick: Nonsteady wing characteristics, sec. F in A. F. Donovan, H. R. Lawrence (eds.), "Aerodynamic Components of Aircraft at High Speeds," vol. 7, of "High Speed Aerodynamics and Jet Propulsion," Princeton Univ. Press, Princeton, N.J., 1957.

[6] R. H. Scanlan, R. Rosenbaum: "Introduction to Study of Aircraft and Flutter," Macmillan, New York, 1951.

[7] Th. Theodorsen: General theory of aerodynamic instability and the mechanism of flutter, *NACA Rept.* 496 (1935).

[8] *U.S. Air Force Spec.* 1817: Prevention of flutter, divergence, and reversal in aircraft.

[9] D. J. Martin: Summary of flutter experiences as a guide to the preliminary design of lifting surfaces on missiles, *NACA Tech. Note* 4197 (1958).

[10] M. D. Van Dyke: Supersonic flow past oscillating airfoils including nonlinear thickness effects, *NACA Rept.* 1183 (1954).

[11] C. E. Watkins, J. H. Berman: Air forces and moments on triangular and related wings with subsonic leading edges oscillating in supersonic potential flow, *NACA Rept.* 1099 (1952).

[12] C. E. Watkins, J. H. Berman: Velocity potential and air forces associated with a triangular wing in supersonic flow, with subsonic edges, and deforming harmonically according to a general quadratic equation, *NACA Tech. Note* 3009 (1953).

[13] H. J. Cunningham: Total lift and pitching moment on thin arrowhead wings oscillating in supersonic potential flow, *NACA Tech. Note* 3433 (1955).

[14] R. Timman, A. I. van de Vooren, J. H. Greidanus: Aerodynamic coefficients of an oscillating airfoil in two-dimensional subsonic flow, *J. Aeronaut. Sci.*, **21** (1954), 499–501.

[15] E. Reissner: Effect of finite span on the airload distributions for oscillating wings, I, *NACA Tech. Note* 1194 (1947).

[16] E. Reissner, J. E. Stevens: Effect of finite span on the airload distributions for oscillating wings, II, *NACA Tech. Note* 1195 (1947).

[17] H. R. Lawrence, E. H. Gerber: The aerodynamic forces on low aspect ratio wings oscillating in an incompressible flow, *J. Aeronaut. Sci.*, **19** (1952), 769–781.

[18] J. G. Barmby, H. J. Cunningham: Study of effects of sweep on the flutter of cantilever wings, *NACA Rept.* 1014 (1951).

CHAPTER 81

FLOW AT LOW REYNOLDS NUMBERS

BY

P. A. LAGERSTROM, Ph.D., Pasadena, Calif.
I. D. CHANG, Ph.D., Palo Alto, Calif.

81.1. INTRODUCTION

Equations and Boundary Conditions. Dimensional Analysis

Equations. The Navier-Stokes equations for incompressible viscous flow are:

Momentum equation $$\frac{D\mathbf{q}}{Dt} + \frac{1}{\rho}\operatorname{grad} p = \nu \nabla^2 \mathbf{q} \tag{81.1a}$$

Continuity equation $$\operatorname{div} \mathbf{q} = 0 \tag{81.1b}$$

where μ = viscosity, and $\nu = \mu/\rho$ = kinematic viscosity. These equations are obtained from Eqs. (68.19) and (68.16), by neglecting compressibility and body forces and rewriting the result in vector form. If body forces such as gravity are present, a term **f** should be added to the right-hand side of Eq. (81.1*a*), where **f** = force per unit mass. The influence of such terms will be neglected in the present chapter except in the discussion of fundamental solutions of the Oseen equation (p. 81–18). The corresponding velocity components will be denoted by u, v, w or by u_1, u_2, u_3. Equations (81.1) may then be written as

$$\frac{\partial u_i}{\partial t} + \sum_{j=1}^{3} u_j \frac{\partial u_i}{\partial x_j} + \frac{1}{\rho}\frac{\partial p}{\partial x_i} = \nu \sum_{j=1}^{3} \frac{\partial^2 u_i}{\partial x_j{}^2} \tag{81.2a}$$

$$\sum_{j=1}^{3} \frac{\partial u_j}{\partial x_j} = 0 \tag{81.2b}$$

Boundary Conditions. At a surface where the fluid considered is in contact with a solid or with another fluid, a condition on the velocity of the fluid is prescribed. The condition generally assumed is the no-slip condition. If the fluid is in contact with a boundary at point P, then the velocity of the fluid at point P is equal to the velocity of boundary point P. The most important case of this is flow past a stationary solid. In this case, $\mathbf{q} = 0$ at the boundary of the solid. A more complicated example occurs when the fluid is a gas and drops of liquid move in this gas. At points of contact, the velocity of the gas is equal to the velocity of the liquid. In this case, there may be fluid motion inside the drops of liquid. The liquid may then also be considered as a fluid obeying Eqs. (81.1).

In addition, the pressure must be prescribed at some point, and, for unbounded fluids, the velocity at large distances must be given. For flow of an unbounded fluid past an object, one prescribes U, the velocity at infinity, and p_∞, the pressure at infinity.

Parameters. Dimensional Analysis. The equations contain two parameters: ρ and ν. In addition, the boundary conditions may contain parameters, say U, the free-stream velocity; p_∞, the free-stream pressure; and l, a characteristic length of the object. If these are the only parameters, dimensional analysis may be used to reduce the number of parameters to one, namely, the Reynolds number Re, defined by

$$\mathrm{Re} = \frac{lU}{\nu} \tag{81.3}$$

To show this, we shall write the Navier-Stokes equation in nondimensional form. For the length scale, we may use either the geometrical length l or the viscous length ν/U; the significance of the two are, however, different. When $\mathbf{q}/U = \mathbf{f}(\mathbf{x}/l)$, the flows have the property of geometric similarity, in other words, the flow pattern is similar for similar bodies. On the other hand, $\mathbf{q}/U = \mathbf{f}(\mathbf{x}U/\nu)$ implies a physical similarity which means that the similarity exists between flows of different μ or U. In general, a flow does not possess either of the two similarities: i.e., change of one scale will necessarily cause the change in flow pattern unless the other length scale varies simultaneously. A few exceptional cases will be mentioned on p. 81–4.

The nondimensional variables may be formed in various ways. The most important such variables are defined below.

Stokes variables $$x_i^* = \frac{x_i}{l} \qquad t^* = \frac{Ut}{l} \tag{81.4}$$

Oseen variables $$\bar{x}_i = \frac{Ux_i}{\nu} = \mathrm{Re}\, x_i^* \qquad \bar{t} = \frac{U^2 t}{\nu} = \mathrm{Re}\, t^* \tag{81.5}$$

The justification of the names will be apparent later. Since both ρU^2 and $\mu U/l$ have the dimensions force per unit area, the nondimensional pressure may be defined by

$$p^+ = \frac{l(p - p_\infty)}{\mu U} \tag{81.6}$$

This variable is convenient when Stokes variables are used. When Oseen variables are used, the classical pressure coefficient is more convenient:

$$p^* = \frac{p - p_\infty}{\rho U^2} = \frac{p^+}{\mathrm{Re}} \tag{81.7}$$

(Note that ρU^2 rather than $\frac{1}{2}\rho U^2$ is used in the denominator.) The nondimensional velocity is defined by

$$\mathbf{q}^* = \frac{\mathbf{q}}{U} \qquad u_i^* = \frac{u_i}{U} \qquad \text{etc.} \tag{81.8}$$

The Navier-Stokes equations may then be written in the following nondimensional form, using Stokes variables:

$$\mathrm{Re}\frac{D\mathbf{q}^*}{Dt^*} + \mathrm{grad}\, p^+ = \nabla^2\mathbf{q}^* \qquad \mathrm{div}\, \mathbf{q}^* = 0 \tag{81.9a,b}$$

An alternative form of the equations, using Oseen variables, is

$$\frac{D\mathbf{q}^*}{D\bar{t}} + \mathrm{grad}\, p^* = \nabla^2\mathbf{q}^* \qquad \mathrm{div}\, \mathbf{q}^* = 0 \tag{81.10a,b}$$

In Eqs. (81.9) and (81.10), the vector operators grad, div, etc., are formed in terms of Stokes and Oseen variables, respectively. As an example, the momentum equation

(81.2*a*) in cartesian coordinates is

$$\mathrm{Re}\left(\frac{\partial u_i^*}{\partial t^*} + \sum_{j=1}^{3} u_j^* \frac{\partial u_i^*}{\partial x_j^*}\right) + \frac{\partial p^+}{\partial x_i^*} = \sum_{j=1}^{3} \frac{\partial^2 u_i^*}{\partial x_j^{*2}} \tag{81.11}$$

in Stokes variables, and in Oseen variables

$$\frac{\partial u^*}{\partial \tilde{t}} + \sum_{j=1}^{3} u_j^* \frac{\partial u_i^*}{\partial \tilde{x}_j} + \frac{\partial p^*}{\partial \tilde{x}_i} = \sum_{j=1}^{3} \frac{\partial^2 u_i^*}{\partial \tilde{x}_j^2} \tag{81.12}$$

The only parameter in Eqs. (81.9) is Re. In general, this parameter is absent from the boundary conditions. If, for example, the solid is a sphere of diameter l, then the boundary conditions are imposed at $r^* = \frac{1}{2}$, where $r^* = r/l$, and

$$r^2 = r_1{}^2 + r_2{}^2 + r_3{}^2$$

In Eqs. (81.10), all parameters have been eliminated. The parameter Re will, however, in general reappear in the boundary conditions. For flow past a sphere, the boundary conditions are imposed at $\tilde{r} = \mathrm{Re}/2$, where $\tilde{r} = Ur/\nu$.

Forces and Torque on a Solid

Local Forces. We consider an infinitesimal part of a solid area dS (in two dimensions, *area* means linear measure). This area element may be represented by an infinitesimal vector $d\mathbf{n}$ whose magnitude is dS and whose direction is that of the outward normal. The force $d\mathbf{f}$ exerted by the fluid on this infinitesimal part of the solid has then the components

$$df_i = -p\,dn_i + \sum_{j=0}^{3} \tau_{ij}\,dn_j \tag{81.13a}$$

Here τ_{ij} is the viscous stress tensor whose components in cartesian coordinates are

$$[\tau_{ij}] = \mu \begin{bmatrix} 2\dfrac{\partial u}{\partial x} & \dfrac{\partial u}{\partial y} + \dfrac{\partial v}{\partial x} & \dfrac{\partial u}{\partial z} + \dfrac{\partial w}{\partial x} \\ \dfrac{\partial u}{\partial y} + \dfrac{\partial v}{\partial x} & 2\dfrac{\partial v}{\partial y} & \dfrac{\partial v}{\partial z} + \dfrac{\partial w}{\partial y} \\ \dfrac{\partial u}{\partial z} + \dfrac{\partial w}{\partial x} & \dfrac{\partial v}{\partial z} + \dfrac{\partial w}{\partial y} & 2\dfrac{\partial w}{\partial z} \end{bmatrix} \tag{81.13b}$$

If the surface element considered has zero velocity, i.e. if $\mathbf{q} = 0$ there, it may be shown that Eq. (81.13) reduces to the more convenient form

$$d\mathbf{f} = -p\,d\mathbf{n} + \mu\,\mathrm{curl}\,\mathbf{q} \times d\mathbf{n} \tag{81.14}$$

Total Force. The total force on a solid may, of course, be obtained by integrating the local forces. It is sometimes convenient to use the method of momentum integrals. Consider stationary flow of an unbounded fluid past a finite solid. Let the velocity of the fluid at infinity be $U\mathbf{i}_x$, where $\mathbf{i}_x$ is the unit vector in the x direction. It can then be shown that the drag of the solid, i.e., the total force in the x direction is

$$D = \iint_{x=x_1} [-p' - \rho u'(U + u')]\,dy\,dz \tag{81.15a}$$

Here p' and u' are the perturbation quantities,

$$p' = p - p_\infty \qquad u' = u - U \tag{81.15b}$$

The integral is taken over a plane $x = x_1 =$ const, with the only restriction that this plane be downstream of the solid. In the limiting case of a plane very far downstream, the quadratic term $\rho(u')^2$ may be neglected, and we find

$$D = \iint_{x=\infty} (-p' - \rho U u')\, dy\, dz \tag{81.16}$$

The force in the y direction, which we shall arbitrarily call the lift L, is

$$L = \rho U \int_{z=-\infty}^{z=\infty} \Gamma(z)\, dz \tag{81.17}$$

where $\Gamma(z)$ is the clockwise circulation around a very large contour in the plane $z =$ const. Equation (81.17) is the familiar Joukowski formula for the lift. However, for viscous flow, the formula is true only if the circulation is evaluated over a contour that is infinitely far away in the limit.

Types of Flows

In discussing the role of the Reynolds number, it is, for our purpose, convenient to distinguish the following cases:

Case 1. No Reynolds Number Can Be Defined. A typical example is a uniform flow past a semi-infinite plate, where no characteristic length can be specified. Obviously Oseen variables should be used (unless a completely artificial length l is introduced). If the plate is located along the positive x axis, the boundary conditions are:

$$\begin{aligned} \mathbf{q}^* &= 0 \qquad \text{for } \tilde{y} = 0,\ \tilde{x} > 0 \\ \mathbf{q}^* &= \mathbf{i}_x \qquad \text{for } \tilde{y} \to \infty \end{aligned} \tag{81.18a,b}$$

A similar situation occurs for flow past a body without a length parameter, i.e., a conical body. Clearly, for such a problem, it does not make sense to talk about flow at low Reynolds numbers. The flows exhibit complete physical similarity; i.e., the flow field is a function of $\tilde{x}_i = Ux_i/\nu$ only. A change in viscosity coefficient μ or free-stream velocity U is equivalent to a uniform stretching of the flow field.

Case 2. Reynolds Number Can Be Defined but Does not Appear in Equations or Boundary Conditions. Consider, for example, plane Couette flow where a Reynolds number Ul/ν may be defined (see p. 81–5). Owing to similarity, the transport term $D\mathbf{q}/Dt$ is zero. If the equations are written in Stokes variables, no Reynolds number occurs in the boundary conditions or in the equations. Hence, the solution is independent of Re. The distinction between flow at large and small Reynolds numbers is then unimportant in this case. A characteristic of such flows is the kinematic similarity of the flow pattern; i.e., the flow field is similar for similar bodies; the viscosity coefficient μ usually modifies only the magnitude of flow pressure [cf. Eq. (81.6)].

Case 3. Reynolds Number Is Defined and Cannot Be Eliminated. This is the case for which fundamental distinction can be made between flows of high Reynolds number and flows of low Reynolds number. The main part of this chapter will be devoted to the study of flows of low Reynolds number (slow flows). This occurs when the flow is very slow, the object is very small, or the viscosity coefficient of the fluid is very large. It is clear from Eq. (81.9) that, for sufficiently small Re, the viscous effect must, at least in the immediate neighborhood of the body, outweight the transport effects. In the theory of flow at low Reynolds numbers, this is the basic idea utilized for obtaining approximate solutions to the Navier-Stokes equations.

81.2. SOME SIMPLE EXACT SOLUTIONS OF THE NAVIER-STOKES EQUATIONS

The principal mathematical difficulty in solving the Navier-Stokes equations arises from the fact that the transport, or acceleration term, $D\mathbf{q}/Dt$, is nonlinear [cf. Eq. (81.2*a*)]. Approximate solutions for low Reynolds numbers are made possible by omitting or linearizing this term (methods of Stokes and Oseen, respectively). These approximate methods will be discussed in subsequent sections. In the present section, we shall consider some flow problems with very simple boundary conditions for which exact solutions may actually be obtained.

In the first example, that of plane stagnation flow, no over-all Reynolds number can be defined (case 1). Owing to the simple nature of the boundary conditions, the Navier-Stokes equations may be reduced to ordinary differential equations, which may be integrated in spite of the presence of nonlinear terms.

The other examples represent various forms of Couette flow and Poiseuille flow. In these cases, the symmetry of the boundary conditions implies that the transport terms are either identically zero or at least very simple, for any Reynolds number. The simplest case is that of parallel flow; i.e., all velocity vectors have the same direction. If this direction is taken as the x direction, then $v = w = 0$. Furthermore, we assume that u is independent of x. The transport term is then identically zero, and the Reynolds number is absent from the problem (case 2). As will be seen below, certain simple boundary conditions are consistent with these assumptions. The two most important cases are plane Couette flow and flow in pipes (Poiseuille flow). A simple generalization of plane Couette flow is circular Couette flow. In this case, the streamlines are concentric circles rather than parallel lines. Since a particle moving on a circle has a centripetal acceleration, the transport terms are not identically zero. Owing to the symmetry of the problem, the transport terms will, however, be greatly simplified, and solutions can be obtained easily.

Plane Stagnation Flow

There exist solutions of the Navier-Stokes equations which describe flow toward an infinite plane. These solutions satisfy the equations exactly, and certain boundary conditions, to be described in detail below, are also satisfied exactly. The same solutions are also used to describe the viscous flow in the neighborhood of the forward stagnation point on a blunt body; in this case, the boundary conditions are satisfied only approximately. Only the two-dimensional case will be discussed here; the case of rotational symmetry may be treated by the same method.

More generally, we may consider two-dimensional flow toward an infinite wedge. A wedge is a two-dimensional conical body since it has no length scale. Consequently, no Reynolds number can be defined for such a body (cf. p. 81–4). Because of the absence of a geometrical length, the boundary-layer equations for wedges may be reduced to an ordinary differential equation containing one parameter, which depends on the opening angle of the wedge. This equation is known as the *Falkner-Skan equation*, and its solution has been studied extensively. For zero opening angle, i.e., for the case of a semi-infinite flat plate parallel to the flow, this equation reduces to the Blasius equation. In general, a solution of the Falkner-Skan equation does not satisfy the Navier-Stokes equations exactly, and has approximately validity only far downstream. However, for the total opening angle equal to π, i.e., for flow toward a plane, the Falkner-Skan solution satisfies the Navier-Stokes equations exactly, and is valid in the whole flow field.

For potential flow past a wedge whose opening angle is greater than zero, the velocity at infinity is infinite. Accordingly, it may not be required that the corresponding viscous flow have constant velocity at upstream infinity; instead, it is required that the viscous flow agree with the potential flow at upstream infinity, in the sense described by (81.21b,c).

We shall choose our coordinate system in such a way that the line $y = 0$ represents the plane, and the half plane $y > 0$ represents the flow region. The line of symmetry for the flow will be taken as the positive y axis. The potential flow, denoted by subscript e, is then

$$u_e = ax \qquad v_e = -ay \qquad a = \text{const} \tag{81.19a}$$

$$p_e - p_0 = -\frac{\rho}{2}(u_e^2 + v_e^2) = -\frac{\rho a^2}{2}(x^2 + y^2) \tag{81.19b}$$

Here a is a constant whose reciprocal has the dimension time. We note that the case $a > 0$ represents flow toward the wall, and that $a < 0$ is for flow away from the wall. The pressure is, of course, determined only within an arbitrary constant; we denote by p_0 the value of p_e at the stagnation point $x = 0$, $y = 0$. It may be shown that no solution exists for viscous flow away from the wall. An intuitive explanation of this is that, for flow away from the wall, vorticity generated at the wall will be transported by the fluid large distances from the wall. For this reason, the viscous flow field for large values of y cannot agree with the potential flow field described by (81.19a,b). Hence, we shall assume from now on that a is positive.

We define the following nondimensional variables:

$$\bar{x} = \sqrt{\frac{a}{\nu}}\,x \qquad \bar{y} = \sqrt{\frac{a}{\nu}}\,y \qquad \mathbf{q}^* = \frac{\mathbf{q}}{\sqrt{a\nu}} \qquad p^* = \frac{p - p_0}{\rho a \nu} \tag{81.20a-d}$$

The boundary conditions for the viscous flow are then

$$\begin{aligned} \mathbf{q}^* &= 0 && \text{at } \bar{y} = 0 \\ u^* \to \bar{x},\ v^* &\to -\bar{y} && \text{as } \bar{y} \to \infty \\ p^* &\to -\tfrac{1}{2}(\bar{x}^2 + \bar{y}^2) && \text{as } \bar{y} \to \infty \end{aligned} \tag{81.21a-c}$$

The corresponding equation is Eq. (81.12) with $\partial/\partial\bar{t} = 0$, $\partial/\partial\bar{x}_3 = 0$, and $\bar{x}_i$, u_i^* replaced by $\bar{x}$, $\bar{y}$, u^*, and v^*, respectively. This equation may be simplified considerably by assuming the following functional form of the dependent variables:

$$u^* = \bar{x}f'(\bar{y}) \qquad v^* = -f(\bar{y}) \tag{81.22a}$$

and

$$p^* = -\tfrac{1}{2}[\bar{x}^2 + g(\bar{y})] \tag{81.22b}$$

Here $f(\bar{y})$ and $g(\bar{y})$ are functions to be determined. The prime denotes differentiation with respect to $\bar{y}$. After substitution of expressions (81.22) into Eq. (81.12), the following set of ordinary differential equations results:

$$\begin{gathered} f''' + ff'' + (1 - f'^2) = 0 \\ f(0) = f'(0) = 0 \qquad f'(\infty) = 1 \end{gathered} \tag{81.23a-b}$$

and

$$\begin{gathered} g' = 2(ff' + f'') \\ \lim_{\bar{y}\to\infty} \frac{g(\bar{y})}{\bar{y}^2} = 1 \end{gathered} \tag{81.24a-b}$$

Equations (81.23) have been integrated numerically (see, for example, [5],[7],[26]). Figure 81.1 gives a plot of f, f', and f''.

The values of u and v in dimensional form are

$$\frac{u}{u_e} = f'(\bar{y}) \qquad \frac{v}{v_e} = \frac{1}{\bar{y}}f(\bar{y}) \tag{81.25a,b}$$

where $\tilde{y} = \sqrt{a/\nu}\, y$. From Fig. 81.1, we see that $u/u_e = 0.99$ at $\tilde{y} = 2.4$; that is, the flow velocity reaches the corresponding inviscid solution within 1% at $\tilde{y} = 2.4$.

Equations (81.24) are integrated to

$$g = f^2 + 2f' + \text{const} \qquad (81.26)$$

If it is required that $p = p_0$ at the potential stagnation point, the constant must be put to zero. Note that then $g = 0$ at the wall. The pressure distribution on the

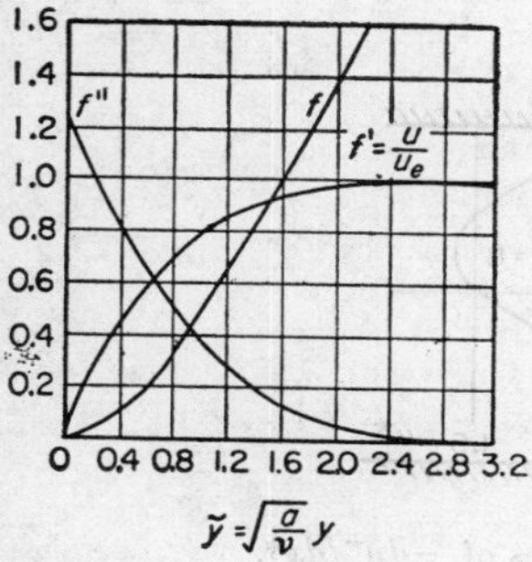

FIG. 81.1. Velocity distribution of plane flow at stagnation point [26].

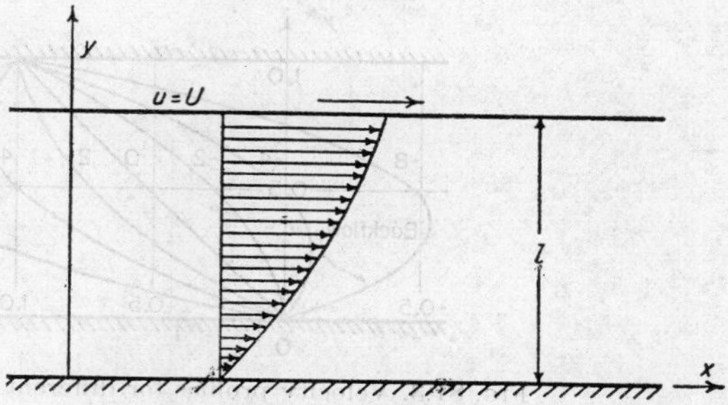

FIG. 81.2. Plane parallel flow.

wall is then the same as that given by the potential solution, namely,

$$p - p_0 = -\frac{\rho a^2 x^2}{2} \qquad (81.27)$$

Plane Parallel Shear Flow

General Solution. We consider two-dimensional steady flow generated by the motion of two parallel walls. The distance l between the planes is unchanged during the motion. By using a Galilean transformation, we may always introduce a system of coordinates such that the velocity of one wall is zero. Without loss of generality, we may, therefore, assume the boundary conditions shown in Fig. 81.2.

The conditions on the pressure are, for the moment, left unspecified, except that we assume that the pressure distribution is such that u is independent of x. Since $\partial u/\partial x = 0$, the continuity equation tells us that $\partial v/\partial y = 0$. From the boundary conditions at the lower wall, it then follows that $v = 0$. The momentum equations are then

$$\frac{\partial p}{\partial x} = \mu \frac{\partial^2 u}{\partial y^2} \qquad \frac{\partial p}{\partial y} = 0 \qquad (81.28a,b)$$

The second equation states that p is independent of y. Hence, in the first equation, the left-hand side may depend on x only and, by the assumption on u, the right-hand side may depend on y only. The two terms of this equation are thus constant. Integrating and using the boundary conditions, we find the solution

$$\frac{u}{U} = \frac{y}{l}\left[1 - \frac{l^2}{2\mu U}\frac{dp}{dx}\left(1 - \frac{y}{l}\right)\right] \qquad \frac{dp}{dx} = \text{const} \qquad (81.29a)$$

or, in nondimensional form,

$$u^* = y^*\left[1 - \frac{1}{2}\frac{dp^+}{dx^*}(1 - y^*)\right] \qquad (81.29b)$$

The only nondimensional parameter is thus

$$\frac{dp^+}{dx^*} \equiv \frac{l^2}{\mu U}\frac{dp}{dx} \tag{81.30}$$

It measures the ratio between the pressure gradient and an average viscous stress $\mu U/l$. A Reynolds number Ul/ν may be defined. However, it does not enter the solution (unless the solution is written in terms of $p^* = p^+/\mathrm{Re}$ instead of p^+). The

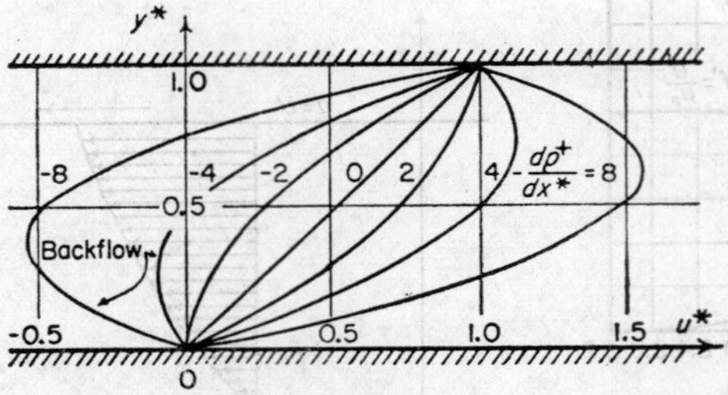

FIG. 81.3. Velocity profiles for different values of $-dp^+/dx^*$.

velocity profiles for different pressure gradients are shown in Fig. 81.3. When the pressure is adverse and strong enough, backflow may occur near the wall.

The local viscous shear is

$$\tau = \mu\frac{\partial u}{\partial y} = \frac{dp}{dx}y + \tau_0 \tag{81.31a}$$

where

$$\tau_0 = \frac{\mu U}{l} - \frac{1}{2}\frac{dp}{dx}l = \text{shear on stationary wall} \tag{81.31b}$$

Note that the shear on the upper wall is given by

$$\tau = \frac{\mu U}{l} + \frac{1}{2}\frac{dp}{dx}l \qquad \text{at } y = l \tag{81.31c}$$

Plane Couette Flow. This is defined as the case for which the driving force is supplied by the motion of the upper wall alone; that is, $dp/dx = 0$. Then

$$\frac{u}{U} = \frac{y}{l} \qquad \tau = \tau_0 = \frac{\mu U}{l} \tag{81.32}$$

Plane Poiseuille Flow (Channel Flow). In this case, both walls are stationary, and the driving force is provided only by a linearly decreasing pressure. Putting $U = 0$ in Eq. (81.29), we find that

$$u = \frac{l^2}{2\mu}\left(-\frac{dp}{dx}\right)\frac{y}{l}\left(1 - \frac{y}{l}\right) \tag{81.33}$$

The profile is parabolic. The maximum velocity occurs at $y = l/2$ and is

$$U_{\max} = \frac{l^2}{8\mu}\left(-\frac{dp}{dx}\right) \tag{81.34a}$$

The mean velocity $\bar{U}$ is

$$\bar{U} = \frac{1}{l}\int_0^l u\,dy = \frac{l^2}{12\mu}\left(-\frac{dp}{dx}\right) = \tfrac{2}{3}U_{\max} \tag{81.34b}$$

The total mass flow per unit time is

$$Q = \int_0^l \rho u \, dy = \rho \bar{U} l = \frac{\rho l^3}{12\mu}\left(-\frac{dp}{dx}\right) \tag{81.34c}$$

The dimensional parameters of the problem are ρ, μ, l, and dp/dx. A nondimensional combination of these parameters is $(l^3/\rho\nu^2)(dp/dx)$. Within a multiplicative constant, this parameter may be interpreted as a Reynolds number based on maximum velocity,

$$\frac{U_{\max} l}{\nu} = \frac{l^3}{8\mu\nu}\left(-\frac{dp}{dx}\right) \tag{81.35a}$$

or on mass flow or mean velocity,

$$\frac{Q}{\mu} = \frac{\bar{U} l}{\nu} = \frac{l^3}{12\mu\nu}\left(-\frac{dp}{dx}\right) \tag{81.35b}$$

In the present case there is only one independent nondimensional parameter.

A nondimensional form of Eq. (81.33) is

$$\frac{u}{\bar{U}} = 6y^*(1 - y^*) \tag{81.36}$$

Again we see that the solution is independent of Reynolds number. No parameter occurs in Eq. (81.36) and, hence, all plane Poiseuille flows are similar.

The shearing stress is a linear function antimetric about the axis of the channel,

$$\tau = \frac{dp}{dx}\left(y - \frac{l}{2}\right) \tag{81.37}$$

Pipe Flow (Three-dimensional Poiseuille Flow)

We consider steady flow through a cylindrical pipe. The axis of the cylinder will be taken as the z axis. The velocity at the wall of the cylinder is zero. We shall assume parallel flow,

$$u = v = 0 \qquad \frac{\partial w}{\partial z} = 0 \tag{81.38}$$

This implies that p depends on z only and is linear in z,

$$\frac{\partial p}{\partial x} = \frac{\partial p}{\partial y} = 0 \qquad \frac{\partial p}{\partial z} = \text{const} \tag{81.39}$$

The continuity equation and the first two momentum equations are then automatically satisfied. The momentum equation in the z direction is a simple two-dimensional Poisson equation:

$$\frac{\partial^2 w}{\partial x^2} + \frac{\partial^2 w}{\partial y^2} = -F \qquad F = -\frac{1}{\mu}\frac{dp}{dz} = \text{const} \tag{81.40}$$

The boundary condition is that $w = 0$ at the wall of the pipe. This equation may, of course, be solved in principle for any cross-sectional shape. The solutions for some simple cases are given in Table 81.1.

For the case of circular cross section we find

Maximum velocity $$w_{\max} = \frac{R^2F}{4} \tag{81.41a}$$

Mean velocity $$\bar{w} = \frac{1}{\pi R^2}\iint w\,dy\,dz = \frac{R^2F}{8} = \frac{w_{\max}}{2} \tag{81.41b}$$

Skin friction $$\tau = -\frac{dp}{dx}\frac{R}{2} \tag{81.41c}$$

For further solutions of pipe flow, reference is made to [1].

Table 81.1. Pipe-flow Solutions for Various Cross Sections

Cross-sectional shape	Velocity w	Mass flow per unit time Q
Circle of radius R.	$\frac{R^2F}{4}\left(1 - \frac{r^2}{R^2}\right)$	$\frac{\pi\rho R^4F}{8}$
Ellipse with semiaxes of length a and b.	$\frac{a^2b^2F}{2(a^2+b^2)}\left(1 - \frac{x^2}{a^2} - \frac{y^2}{b^2}\right)$	$\frac{\pi\rho a^3b^3F}{4(a^2+b^2)}$
Equilateral triangle, sides $y = \pm x\sqrt{3}$, $y = a$.	$\frac{F}{4a}(y-a)(3x^2-y^2)$	$\frac{\rho a^4F}{60\sqrt{3}}$

Analogy with Problem in Elasticity. Equation (81.40) is identical with the differential equation of the torsion problem [Eq. (36.10)], and also the boundary condition is the same in both cases. The torsion stress function corresponds to the velocity w of the pipe flow, and the torque M_T from Eq. (36.16) corresponds to $2Q/\rho$.

Circular Couette Flow

Consider the motion between two rotating coaxial cylinders of infinite length Let the radius of the cylinders be R_1 and R_2, respectively, and their angular velocities be ω_1 and ω_2, respectively. We denote the tangential velocity by u and the radial velocity by v. The boundary conditions are shown in Fig. 81.4. The continuity equation shows that $v \equiv 0$; that is, the streamlines are circles. For steady flow, the radial and tangential momentum equations are

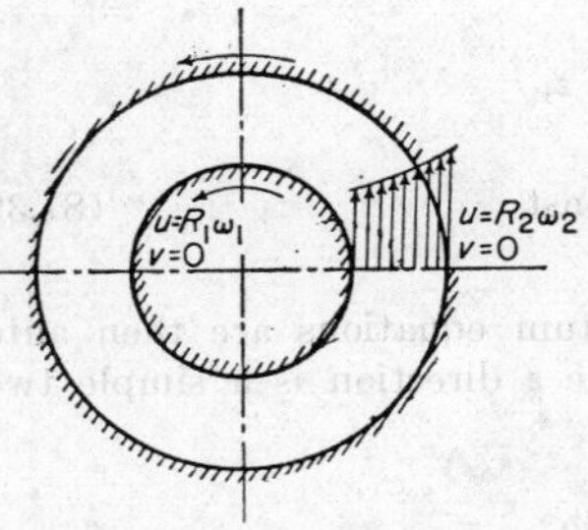

FIG. 81.4. Circular Couette flow.

$$\frac{\rho u^2}{r} = \frac{dp}{dr} \qquad \frac{d^2u}{dr^2} + \frac{1}{r}\frac{du}{dr} - \frac{u}{r^2} = 0 \tag{81.42a,b}$$

Equation (81.42b) states that the tangential viscous shear is zero. The tangential pressure gradient and the tangential acceleration vanish by symmetry. The left-hand side of Eq. (81.42a) contains the remnant of the transport term; it is the centripetal acceleration. However, this causes no mathematical difficulty. The solution for u is obtained from Eq. (81.42b) alone; Eq. (81.42a) may then be used for computing the pressure.

The general solution for u is

$$u = \frac{(R_2/r - r/R_2)R_1\omega_1 + (r/R_1 - R_1/r)R_2\omega_2}{R_2/R_1 - R_1/R_2} \tag{81.43}$$

When the inner cylinder is absent ($R_1 = 0$), we find that

$$u = \omega_2 r \tag{81.44}$$

In this case, the fluid rotates like a solid. The angular velocity is everywhere equal to ω_2.

When the outer cylinder is absent ($R_2 = \infty$, $\omega_2 = 0$), we find

$$u = \frac{\omega_1 R_1^2}{r} \tag{81.45}$$

This is the solution for a potential vortex [see Eq. (70.43)]. The general solution [Eq. (81.43)] has the form

$$u = Ar + \frac{B}{r} \tag{81.46}$$

and is hence a superposition of a solid-body rotation and a potential vortex. The two constants A and B are to be chosen so as to satisfy the conditions at the two cylinders. When one cylinder is absent, one of the constants must be equal to zero, in order to avoid a singularity at infinity or at zero.

Note that the viscosity does not occur in the solution. Hence, if Eq. (81.43) is written in nondimensional form, the Reynolds number would not occur.

If the viscous stress tensor (81.13*b*) is expressed in polar coordinates and the symmetries of the present problem are used, we find that the viscous stress is

$$\tau = \mu \left(\frac{\partial u}{\partial r} - \frac{u}{r} \right) \tag{81.47}$$

At the inner cylinder, a positive value of τ means a force in the positive θ direction; at the outer cylinder, a positive value of τ means a force in the negative θ direction. A comparison with Eqs. (81.46) and (81.43) shows that

$$\tau = -\frac{2\mu B}{r^2} = \frac{2\mu}{r^2} \frac{R_1^2 R_2^2(\omega_2 - \omega_1)}{R_2^2 - R_1^2} \tag{81.48}$$

The torque M is then

$$M = 2\pi r^2 \tau = 4\pi\mu \frac{R_1^2 R_2^2(\omega_2 - \omega_1)}{R_2^2 - R_1^2} \tag{81.49}$$

The torques on the inner and outer cylinders are then equal in magnitude and of opposite sign. This may, of course, be concluded a priori: since the total angular momentum of the fluid remains constant in time, the total torque exerted by the walls on the fluid must be zero.

Equation (81.49) may be used for experimental determination of the viscosity of a fluid.

81.3. STOKES EQUATIONS

Basic Equations

As pointed out in Sec. 81.1, with the exception of some cases with very simple boundary conditions, we may define a Reynolds number for a flow problem, and this parameter cannot be eliminated from the system. Solutions for a general value of Re are, in general, impossible to find. In solving the Navier-Stokes equations, we must, therefore, either make numerical calculations for specific values of Re or try to obtain approximate methods for the cases of very low or very high Reynolds numbers. Prandtl's boundary-layer theory is an attempt to deal with flow at very high Reynolds numbers. In the present chapter we are concerned with the case of very low Reynolds numbers. A method for dealing with this problem was proposed by Stokes in 1845 [2]. His method actually yields the first approximation for small values of the Reynolds numbers, with the exception of the case of two-dimensional flow of an unbounded fluid past a solid. The failure of Stokes' method for the exceptional case is referred to as the *Stokes paradox*. The basic nature of the Stokes paradox was discovered by Oseen [3],[4]. Oseen proposed a modification of the Stokes equations. His equations yield answers for the two-dimensional case and increased accuracy, relative to the Stokes equations, in the three-dimensional case [4]. The general nature of the asymptotic expansions of the Navier-Stokes equation has been clarified only recently by the work of Saul Kaplun (cf. Sec. 81.5). Kaplun's investigations show the meaning of the Stokes and Oseen approximations, and also give methods for obtaining higher-order approximations.

In the present section, we shall discuss the Stokes approximation. Using the notation on p. 81–2, we may express the Stokes method as follows. If the Navier-Stokes equations are written in Stokes variables [Eqs. (81.9)], we see that the parameter Re does not occur in the boundary conditions. In the equations, it occurs only as a multiplicative factor of the transport term. It is then natural to assume that we obtain equations approximately valid for low Re by putting Re = 0 in Eqs. (81.9), that is, by neglecting the transport terms in the Navier-Stokes equations. We then obtain the Stokes equations,

$$\operatorname{grad} p^{+} = \nabla^2 \mathbf{q}^* \qquad \operatorname{div} \mathbf{q}^* = 0 \tag{81.50a,b}$$

or, in dimensional form,

$$\operatorname{grad} p = \mu \nabla^2 \mathbf{q} \qquad \operatorname{div} \mathbf{q} = 0 \tag{81.51a,b}$$

The vorticity $\mathbf{\Omega}$, defined by

$$\mathbf{\Omega} = \operatorname{curl} \mathbf{q} \tag{81.52}$$

and the pressure p both satisfy the Laplace equation,

$$\nabla^2 \mathbf{\Omega} = 0 \qquad \nabla^2 p = 0 \tag{81.53a,b}$$

In two-dimensional flow, a stream function ψ may be introduced:

$$u = \frac{\partial \psi}{\partial y} \qquad v = -\frac{\partial \psi}{\partial x} \tag{81.54a}$$

Then

$$\nabla^2 \psi = -\Omega \qquad \text{where } \mathbf{\Omega} = \Omega \mathbf{i}_z \tag{81.54b}$$

and

$$\nabla^2 \nabla^2 \psi = 0 \tag{81.54c}$$

Hence the stream function ψ obeys the biharmonic equation. Since

$$\operatorname{grad} p = \mu \nabla^2 \mathbf{q} = -\mu \operatorname{curl} \operatorname{curl} \mathbf{q} = -\mu \operatorname{curl} (\mathbf{i}_z \Omega)$$

we also have

$$\frac{1}{\mu}\frac{\partial p}{\partial x} = -\frac{\partial \Omega}{\partial y} \qquad \frac{1}{\mu}\frac{\partial p}{\partial y} = \frac{\partial \Omega}{\partial x} \tag{81.55}$$

Equation (81.55) shows that p/μ and $-\Omega$ are harmonic conjugates, and hence equipressure and equivorticity lines are orthogonal.

The following characteristic properties of the Stokes equations are apparent:

1. The transport effects are neglected completely, and Re disappears from the equations and boundary conditions. It follows then that, for similar bodies, the flow pattern is geometrically similar, independent of Re.

2. For similar bodies, p^+ are the same at corresponding points with the same nondimensional coordinates. Hence, by Eq. (81.6),

$$p - p_\infty \sim \frac{U}{l}\mu \qquad \text{Drag } D \sim lU\mu \qquad C_D \equiv \frac{D}{\frac{1}{2}\rho U^2 l^2} \sim \frac{1}{\text{Re}} \qquad (81.56a\text{–}c)$$

Equation (81.56a) shows that, if U and l are fixed, $(p - p_\infty)$ is directly proportional to μ, an important result in lubrication theory. Equation (81.56b) shows that the dimensional force D itself is proportional to the free-stream velocity U.

Three-dimensional Flow Past a Solid

Flow Past a Sphere. For a uniform flow $U\mathbf{i}_x$ past a sphere of radius a, centered at the origin, the velocity field may be written as the sum of an irrotational part $\mathbf{q}_1$,

$$\mathbf{q}_1 = \frac{U}{4} \text{ grad} \left[a^3 \frac{\partial}{\partial x}\left(\frac{1}{r}\right) + 3a\left(\frac{x}{r}\right) + 4x \right] \qquad (81.57a)$$

and a rotational part which has only an x component,

$$\mathbf{q}_2 = -\frac{3}{2}\frac{Ua}{r}\mathbf{i}_x \qquad (81.57b)$$

where

$$r^2 = x^2 + y^2 + z^2$$

The pressure is

$$p - p_\infty = -\frac{3}{2}\frac{\mu U a x}{r^3} \qquad (81.58)$$

The force on the sphere consists of a viscous part and a pressure part. Let $\mathbf{r}$ be the radius vector from the origin to a point on the surface of a sphere, θ the polar angle between $\mathbf{r}$ and the positive x axis, and $\mathbf{i}_n$ a unit vector normal to the x axis and contained in the plane spanned by $\mathbf{r}$ and the x axis. Using Eq. (81.14), we find

$$\begin{aligned} \text{Viscous stress} &= \mu(\text{curl } \mathbf{q}) \times \mathbf{n} \\ &= \frac{3\mu U}{2a} \sin\theta(\mathbf{i}_x \sin\theta - \mathbf{i}_n \cos\theta) \\ \text{Pressure} &= \frac{3\mu U}{2a} \cos\theta(\mathbf{i}_x \cos\theta + \mathbf{i}_n \sin\theta) \end{aligned}$$

The total force per unit area of the sphere is then $(3\mu U/2a)\mathbf{i}_x$, directed along the x axis. The total drag is then given by the Stokes formula [2],

$$D = \frac{3\mu U}{2a} 4\pi a^2 = 6\pi\mu U a \qquad (81.59)$$

of which the viscous and pressure drags contribute in the ratio 2:1. The drag coefficient based on the front area πa^2 is

$$C_D \equiv \frac{D}{\frac{1}{2}\rho U^2 \pi a^2} = \frac{24}{\text{Re}} \qquad \text{Re} = \frac{2Ua}{\nu} \qquad (81.60)$$

The streamlines around the sphere are shown in Fig. 81.5.

The flow pattern is symmetric with respect to the plane $x = 0$ and, hence, exhibits no wake behind the sphere; this is a typical characteristic of Stokes solutions.

Other Solutions. The motion of a sphere was solved by Stokes in 1850 [2]. He also solved the problems of spherical and cylindrical pendulums which are performing small oscillations in a straight line. The torsional oscillations of spheres and cylinders were discussed by Helmholtz and Piotrowski, and Oberbeck has attained the solution for the steady motion of an ellipsoid which moves parallel to an axis. Various other solutions are discussed in [5]–[11] and, in particular, in [1] and [4], where further references are given.

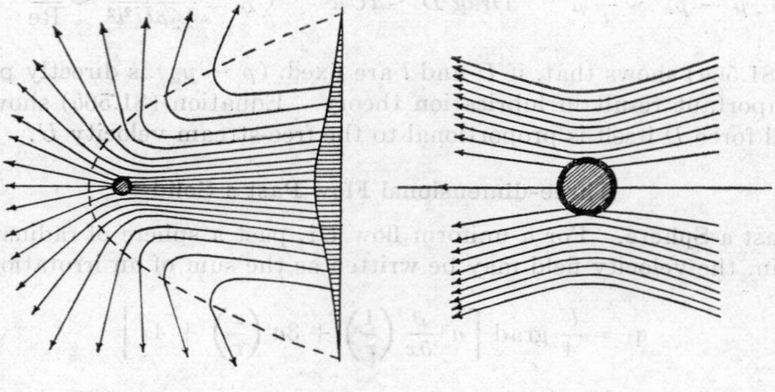

FIG. 81.5. Flow around sphere [7]. (Sphere moving to the left with constant velocity; fluid at rest at infinity.)

Insensitivity of Drag to the Shape of Bodies. In Stokes flow, the drag coefficient is relatively insensitive to the shape and the orientation of the body. For example, the drag of a circular disk of radius a placed perpendicular in the stream is, according to the computation of Oberbeck [4],

$$D = 16\mu U a$$

If the same disk is placed edgewise into the flow, the result is

$$D = {}^{32}\!/_{3}\mu U a$$

Hence, even in these two extreme cases, the drag differs from that of a sphere of radius a only by the ratio of $6\pi : 16$ and $6\pi : {}^{32}\!/_{3}$, respectively.

Strong Effect of Solid Walls. The Stokes formula for the drag of a sphere has been applied to many physical occasions. One obvious application is the determination of the viscosity coefficient of a fluid by measuring the drag of a small sphere which is dropping through a column of liquid to be investigated.

This problem was first studied by H. A. Lorentz, who determined the influence of an infinite plane wall on the motion of a sphere. The resistance experienced by a sphere when moving along the axis of an unlimited cylindrical tube was studied by Ladenburg. According to his result, the drag on the sphere is increased in comparison with the usual formula of Stokes by the factor $(1 + 2.4a/b)$, where a is the radius of the sphere, and b is the radius of the tube [1],[12].

Stokes Flow in Two Dimensions. Stokes Paradox

Method of Complex Variables. The Stokes equations for two dimensions are given by Eqs. (81.54). Thus, in particular, we may obtain solutions by solving the biharmonic equation for the stream function. Another method of obtaining solutions is the following. It may be shown that the general solution may be written as

$$u + iv = \overline{\frac{dF}{dz}} - \frac{1}{4}\left(G - z\,\overline{\frac{dG}{dz}}\right) \tag{81.61}$$

where F and G are analytic functions of the complex variable $z = x + iy$. A bar over functions indicates the complex conjugate.

The vorticity and the pressure are then given by

$$\frac{p}{\mu} - i\Omega = \frac{dG}{dz} \tag{81.62}$$

This method is discussed [13] and [14].

The Stokes Paradox. Many solutions for two-dimensional flow may be obtained, using the above method. However, for two-dimensional flow of an unbounded fluid past a solid, solutions of the Stokes equations, satisfying the proper conditions, do not exist. This fact, referred to as the Stokes paradox, was discovered by Stokes himself in 1850 for the special case of a circular cylinder [2]. Rigorous mathematical proofs of the Stokes paradox can be given (see, for example, [14]–[16]).

The explanation of this paradox was first given by Oseen. His reasoning was essentially as follows: The basic assumption in deriving the Stokes equations was that, for Re small, the term Re $D\mathbf{q}^*/Dt$ is negligible compared to the $\nabla^2\mathbf{q}^*$, that is, that viscous effects outweigh transport (inertia) effects. However, in the Stokes solution for the sphere, discussed on p. 81-13, the term $\nabla^2 u^*$ is of the order $1/r^{*3}$, and the transport term Re $u^* \, \partial u^*/\partial x^*$ is of the order Re$/r^{*2}$. The ratio of Re $u^* \, \partial u^*/\partial x^*$ to $\nabla^2 u^*$ is of the order Re r^*. Thus, as Re tends to zero, the transport term Re $(D\mathbf{q}/Dt)$ becomes negligible compared to the viscous term $\nabla^2\mathbf{q}^*$ but not uniformly in space. Within a given radius r_0, the first term may be made an arbitrarily small fraction of the second by choosing Re sufficiently small. But, for this same Reynolds number, the transport terms may be made much larger than the viscous terms by choosing $r \gg r_0$. This indicates that, for points which are at a large distance from the sphere, the Stokes approximation is not valid. It is inconsistent in the sense that the solution does not satisfy the assumptions on which the approximation is based. In a sense, the real paradox is not that a solution does not exist in two dimensions, but rather that Stokes' method gives a good approximation in three dimensions. In three dimensions, the difficulties due to nonuniformity do not appear until we try to find higher-order approximations. To remedy the difficulties, Oseen proposed a new set of equations, which will be discussed in Sec. 81.4.

81.4. OSEEN EQUATIONS

Equations and Boundary Conditions

The Oseen equations are obtained by linearizing the Navier-Stokes equations (81.1a,b) about the free-stream velocity $\mathbf{q} = U\mathbf{i}_x$, that is, by neglecting terms which are quadratic in $\mathbf{q}' = \mathbf{q} - U\mathbf{i}_x$. The resulting equations are:

$$\frac{\partial \mathbf{q}}{\partial t} + U\frac{\partial \mathbf{q}}{\partial x} + \frac{1}{\rho}\operatorname{grad} p = \nu\,\nabla^2\mathbf{q} \qquad \operatorname{div}\mathbf{q} = 0 \tag{81.63a,b}$$

The corresponding equation for $\mathbf{\Omega}$ is

$$\frac{\partial \mathbf{\Omega}}{\partial t} + U\frac{\partial \mathbf{\Omega}}{\partial x} = \nu\,\nabla^2\mathbf{\Omega} \tag{81.64}$$

When the flow is two-dimensional, a stream function ψ[Eq. (81.54a)] may be introduced, which will then satisfy the equation

$$\frac{\partial}{\partial t}(\nabla^2\psi) + U\frac{\partial}{\partial x}(\nabla^2\psi) = \nu\,\nabla^2\,\nabla^2\psi \tag{81.65}$$

Since only its derivatives appear, $\mathbf{q}$ may be interpreted either as the perturbation velocity $\mathbf{q}'$ or as the full velocity $(\mathbf{q}' + U\mathbf{i}_x)$. In the former interpretation, the boundary conditions for flow past a solid at rest are

$$\mathbf{q} = -U\mathbf{i}_x \quad \text{at solid} \qquad \mathbf{q} \to 0 \quad \text{at infinity} \tag{81.66a,b}$$

In the latter interpretation,

$$\mathbf{q} = 0 \quad \text{at solid} \qquad \mathbf{q} \to U\mathbf{i}_x \quad \text{at infinity} \tag{81.67a,b}$$

The free-stream velocity may, in particular, have the value $U = 0$. In this case, the only acceleration term occurring in the momentum equation (81.63a) is $\partial\mathbf{q}/\partial t$. The resulting equation is, in a sense, as general as the original equation (81.63a), since we may always make $U = 0$ with the aid of a Galilean transformation.

The Oseen equations (81.63) differ from the Stokes equations (81.51) by the inclusion of the linearized transport term $U\ \partial\mathbf{q}/\partial x$ (for convenience, we consider the stationary flow past a finite body). Since $\mathbf{q}' \ll U\mathbf{i}_x$ at large distances from the body, it is to be expected that the Oseen equations are approximately valid at large distances from the solid for any Reynolds numbers. A priori, we are tempted to use a similar argument to conclude that the Stokes equations are better than the Oseen equations near the body: One way of obtaining the Stokes equations is to linearize the value of the velocity at the surface of the body; that is, $\mathbf{q} = 0$. This point of view is, however, not very fruitful. The Stokes equations are of importance only for flow at low Reynolds numbers. If the Oseen equations are written in Stokes variables [cf. Eqs. (81.9)], we find that the first term is Re $(\partial\mathbf{q}/\partial x^*)$. For low Reynolds numbers, the inclusion of this term is harmless in the region near the body. Actually the Oseen equations, unlike the Stokes equations, are uniformly valid for small values of Re, and a solution of the Oseen equations is as good as or better than the corresponding solution of the Stokes equations. The arguments just given are, of course, not conclusive, but the statements made are justified by Kaplun's more systematic study of the problems (see Sec. 81.5). Circumstantial evidence is also provided by actual comparison of corresponding Stokes and Oseen solutions.

Propagation of Vorticity. According to the full Navier-Stokes equations, vorticity is transported with the fluid, and, at the same time, it is conducted like heat from streamline to streamline. The rate of this conduction, or diffusion, is proportional to viscosity. (In addition, certain three-dimensional effects influence the propagation of vorticity; these will not be discussed here.) On the other hand, Eq. (81.53a) shows that, in the Stokes approximation, only the diffusion effect remains. Vorticity, which is generated at the boundary of a solid, spreads uniformly in all directions; the transport effect is neglected. This may be contrasted with the laws of propagation, according to the Euler equations for inviscid flow. In this approximation, the diffusion is completely neglected; transport with the fluid (see Theorem 3, p. 71–18), together with the three-dimensional effect mentioned above, account solely for the propagation of vorticity. Finally, Eq. (81.64) shows that, according to the Oseen approximation, vorticity is conducted like heat, while at the same time being transported with the streamlines of the undisturbed flow. Thus, in the neighborhood of the solid, vorticity is transported through the solid with the undisturbed streamlines rather than around the solid with the true streamlines. This apparently paradoxical result shows that the Oseen equations do not give a good approximation near the solid, except for very low Reynolds numbers. In this case, diffusion of vorticity is the dominant mode of propagation, and the physically incorrect assumption regarding its transport does not introduce a large error. The mathematical reason for this was pointed out earlier. In the nondimensional form of Eq. (81.64), the factor Re multiplies the transport term. Hence, these terms, which are incorrect near the body, are rendered harmless if Re is small.

Longitudinal and Transversal Waves. Splitting

Many linear equations have the property that an arbitrary solution of these equations may be split into a finite sum of solutions of simple type. In particular, if **q** is a solution of the Oseen equations, it may be written as a sum of the longitudinal velocity field (wave) $\mathbf{q}_L$ and a transversal field $\mathbf{q}_T$. There is a corresponding splitting of the pressure fields.

A solution of Eqs. (81.63) is said to represent a *longitudinal wave* if the velocity field is irrotational; that is, curl $\mathbf{q} = 0$. Hence, the viscous term $\nu \nabla^2\mathbf{q} = -\nu$ curl curl **q** disappears in the momentum equation, and Eqs. (81.63) reduce to

$$\frac{\partial \mathbf{q}_L}{\partial t} + U\frac{\partial \mathbf{q}_L}{\partial x} + \frac{1}{\rho}\,\text{grad}\, p = 0 \qquad \text{div}\,\mathbf{q}_L = 0 \qquad \text{curl}\,\mathbf{q}_L = 0 \qquad (81.68a\text{–}c)$$

The last equation implies that $\mathbf{q}_L$ may be derived from a scalar potential ϕ,

$$\mathbf{q}_L = \text{grad}\,\phi \qquad (81.69a)$$

and, by Eq. (81.68*b*),

$$\nabla^2\phi = 0 \qquad (81.69b)$$

Equation (81.68*a*) then gives

$$p' = p - p_\infty = -\rho\left(Uu_L + \frac{\partial \phi}{\partial t}\right) \qquad (81.70)$$

Hence, the velocity components of a longitudinal wave are simply those of a potential flow field. The pressure, however, is evaluated from a linearized Bernoulli's law.

A solution of Eqs. (81.63) is said to represent a *transversal wave* if the velocity field is solenoidal; that is, div $\mathbf{q} = 0$, and the pressure field is identically zero. The equations in such a case reduce to

$$\frac{\partial \mathbf{q}_T}{\partial t} + U\frac{\partial \mathbf{q}_T}{\partial x} = \nu\nabla^2\mathbf{q}_T \qquad \text{div}\,\mathbf{q}_T = 0 \qquad (81.71a,b)$$

Equation (81.71*b*) implies a vector potential **A** such that

$$\mathbf{q}_T = \text{curl}\,\mathbf{A} \qquad (81.72a)$$

and, from Eq. (81.71*a*),

$$\frac{\partial \mathbf{A}}{\partial t} + U\frac{\partial \mathbf{A}}{\partial x} = \nu\nabla^2\mathbf{A} \qquad (81.72b)$$

In obtaining (81.72*b*), a potential term due to integration has been omitted.

The following theorem may then be proved:

Splitting Theorem. Any solution $(\mathbf{q},p')$ of the Oseen equation (81.63) may be decomposed into a longitudinal wave $(\mathbf{q}_L,p')$ and a transversal wave $(\mathbf{q}_T,0)$ where $(\mathbf{q}_L,p')$ satisfies Eqs. (81.68) and $(\mathbf{q}_T,0)$ satisfies Eqs. (81.71).

Mathematical proof for this theorem may be found in [17].

For certain types of steady Oseen flows, an additional splitting of the transversal wave is possible. Let κ be a scalar function, satisfying

$$\frac{\partial \kappa}{\partial t} + U\frac{\partial \kappa}{\partial x} = \nu\nabla^2\kappa \qquad (81.73)$$

and vanishing at infinity. Then it is easily seen that the following vector field is a transversal wave:

$$\mathbf{q}_T = -\frac{1}{2\lambda}\,\text{grad}\left(\frac{\partial \kappa}{\partial x}\right) + \mathbf{i}_x\left(\frac{\partial \kappa}{\partial x} + \frac{1}{U}\frac{\partial \kappa}{\partial t}\right) = \mathbf{q}_1 + \mathbf{q}_2 \qquad \lambda = \frac{U}{2\nu} \qquad (81.74a,b)$$

Note, however, that, in general, div $\mathbf{q}_2$ and, hence, div $\mathbf{q}_1$ are not equal to zero. Hence, in general, neither $\mathbf{q}_1$ nor $\mathbf{q}_2$ satisfies the Oseen equations by itself.

Examples which exhibit such a splitting will be given later. Table 81.2 summarizes the results:

$$\mathbf{q} = \mathbf{q}_L + \mathbf{q}_T \qquad \mathbf{q}_T = \mathbf{q}_1 + \mathbf{q}_2 \tag{81.75}$$

Table 81.2

		Curl	Divergence	Representation
$\mathbf{q}_L$		0	0	$\mathbf{q}_L = \nabla\phi,\ \nabla^2\phi = 0$
$\mathbf{q}_T$	$\mathbf{q}_1$	0	$-\frac{1}{2\lambda}\nabla^2\frac{\partial\kappa}{\partial x}$	$\mathbf{q}_1 = -\frac{1}{2\lambda}\nabla\frac{\partial\kappa}{\partial x},\ \frac{\partial\kappa}{\partial t} + U\frac{\partial\kappa}{\partial x} = \nu\nabla^2\kappa$
	$\mathbf{q}_2$	$0,\ \frac{1}{2\lambda}\frac{\partial}{\partial z}\nabla^2\kappa,\ -\frac{1}{2\lambda}\frac{\partial}{\partial y}\nabla^2\kappa$	$\frac{\partial^2\kappa}{\partial x^2} + \frac{1}{U}\frac{\partial^2\kappa}{\partial x\,\partial t}$	$\mathbf{q}_2 = \mathbf{i}_x\left(\frac{\partial\kappa}{\partial x} + \frac{1}{U}\frac{\partial\kappa}{\partial t}\right),\ \frac{\partial\kappa}{\partial t} + U\frac{\partial\kappa}{\partial x} = \nu\nabla^2\kappa$
	$\mathbf{q}_1 + \mathbf{q}_2$	$0,\ \frac{1}{2\lambda}\frac{\partial}{\partial z}\nabla^2\kappa,\ -\frac{1}{2\lambda}\frac{\partial}{\partial y}\nabla^2\kappa$	0	$\mathbf{q}_T = \text{curl}\ \mathbf{A},\ \frac{\partial\mathbf{A}}{\partial t} + U\frac{\partial\mathbf{A}}{\partial x} = \nu\nabla^2\mathbf{A}$

Fundamental Solutions

The concept of fundamental solutions plays an important role in the theory of linear differential equations. The general theory of fundamental solutions is given in [18], where also many examples are discussed. The fundamental solutions of the Oseen equations are discussed in [4] and [17]. Here we shall discuss the results for two-dimensional and three-dimensional stationary flow. These two cases have many important implications.

Consider the equations

$$U\frac{\partial\mathbf{q}}{\partial x} + \text{grad}\ p = \nu\nabla^2\mathbf{q} + \mathbf{f} \qquad \text{div}\ \mathbf{q} = 0 \tag{81.76a,b}$$

where

$$\mathbf{f} = \frac{1}{\rho}\,\delta(P,Q)\mathbf{a} \qquad |\mathbf{a}| = 1 \tag{81.76c}$$

If $P = x_i$ and $Q = \xi_i$, the delta function $\delta(P,Q)$, regarded as a function of the x_i and the ξ_i as parameters, is zero except when $P = Q$. At this point, its value becomes infinite in such a way that the integral of $\delta(P,Q)$ over any domain including Q is unity. Since Eq. (81.76a) is a momentum equation, the physical significance of the added term $\mathbf{f}$ is the following: At the point Q, a concentrated force $\rho\mathbf{f}$ acts on the fluid. The magnitude of this force is unity, and its direction is that of the unit vector $\mathbf{a}$. Since the Oseen equations are invariant under a translation of the origin, we may, without loss of generality, assume that the force is located at the origin; that is, $Q = \xi_i = 0$. This assumption will be made below. It is furthermore sufficient to study the case for which the force is in the flow direction and the case for which the force is normal to the flow direction.

Note that, because of the special assumptions of $\mathbf{f}$, the solution for $\mathbf{q}$ will have the dimension of velocity per unit force.

Singular Flat Plate. We consider the case of two-dimensional flow, and assume that the force is located at the origin and directed along the x axis. Actually it is convenient to have the force acting in the opposite direction of the flow. We thus put

$$\mathbf{f} = -\frac{\delta(x,y)}{\rho}\mathbf{i}_x \tag{81.77}$$

The solution of Eqs. (81.76) may then be interpreted as the flow field due to a singular flat plate located at the origin. The drag of the flat plate is unity, and its lift is zero. In considering this limiting case, we may first consider a finite flat plate located on the x axis between $x = -x_0$ and $x = x_0$. At the surface of the plate, $\mathbf{q}$ is not zero, but has a value $\mathbf{q}_0$ chosen so as to make its drag unity. In the limit, x_0 tends to zero. In order for the drag to remain unity, the retarding action of the plate must increase; that is, $\mathbf{q}_0 \to -\infty\mathbf{i}_x$ as $x_0 \to 0$.

The solution is then

$$\mathbf{q} = \mathbf{q}_L + \mathbf{q}_T \tag{81.78a}$$

where $$\mathbf{q}_L = \frac{1}{2\pi\rho U}\,\text{grad}\ln r \qquad r^2 = x^2 + y^2 \tag{81.78b}$$

$$\mathbf{q}_T = \frac{1}{2\pi\mu}\left\{\frac{1}{2\lambda}\,\text{grad}\,[e^{\lambda x}K_0(\lambda r)] - e^{\lambda x}K_0(\lambda r)\mathbf{i}_x\right\} \qquad \lambda = \frac{U}{2\nu} \tag{81.78c}$$

$K_0(x)$ being a modified Bessel function. It is seen that if we put

$$\frac{\partial\kappa}{\partial x} = -\frac{1}{2\pi\mu}e^{\lambda x}K_0(\lambda r) \qquad \frac{\partial\kappa}{\partial t} = 0 \tag{81.79}$$

then $\mathbf{q}_T$ is split according to Eq. (81.74). The pressure is

$$p' = -\rho U u_L = -\frac{x}{2\pi r^2} \tag{81.80}$$

The longitudinal wave represents simply potential source flow. The total outflow from the source is $1/\rho U$. The transverse wave has a sink of the same strength, so that the net flow through any closed contour is zero.

For $r \ll 1$, the total flow field is approximately

$$u(x,y) \approx \frac{1}{4\pi\mu}\left(\ln\frac{\gamma_0 Ur}{4\nu} - \frac{x^2}{r^2}\right) \qquad v(x,y) \approx -\frac{1}{4\pi\mu}\left(\frac{xy}{r^2}\right) \qquad \Omega = \frac{\partial v}{\partial x} - \frac{\partial u}{\partial y} \approx -\frac{y}{2\pi\mu r^2} \tag{81.81a–c}$$

where $\gamma_0 = e^\gamma \approx 1.781$, $\gamma =$ Euler's constant ≈ 0.577. Note that the above expressions, together with Eq. (81.80), form a set of exact solutions for the Stokes equations.

For large values of r, the transversal flow field may be given asymptotically as

$$u_T(x,y) \approx -\frac{1}{4\rho}\sqrt{\frac{1}{\pi\nu Ur}}\left(1 + \frac{x}{r}\right)e^{-\lambda(r-x)}$$

$$v_T(x,y) \approx -\frac{1}{4\rho}\sqrt{\frac{1}{\pi\nu Ur}}\left(\frac{y}{r}\right)e^{-\lambda(r-x)} \tag{81.82a–c}$$

$$\Omega \approx -\frac{\sqrt{\lambda}\,y}{\sqrt{\pi}\,\mu(2r)^{3/2}}e^{-\lambda(r-x)}$$

The contours of constant $e^{-\lambda(r-x)}$ are given by $r - x = c$, and are parabolas with foci at the origin and opening downstream. With a sufficiently high value of c, both the transversal wave and the vorticity are then negligible. Consider, for instance, the value of u on the x axis. Upstream ($x = -r$), it dies off exponentially, whereas, downstream of the plate, it decreases only as $1/\sqrt{x}$. Thus, the transversal wave and the vorticity are negligible, except in a parabolic wake.

The existence of the transversal wave is due to the retarding effect of the plate. Vorticity generated at the plate diffuses like heat, and is at the same time transported downstream with the streamlines of the undisturbed flow. This is the reason why the vorticity is very weak upstream, and comparatively strong downstream. The longitudinal wave may be thought of as due to a displacement effect. The slowing down of the fluid at the plate gives the plate an apparent shape. Since the longitudinal wave is a simple source, the apparent shape is that of a semi-infinite body whose thickness downstream approaches a finite value. The effective (or apparent) thickness may be defined as $\int_{-\infty}^{\infty} u_T\,dy$ for a fixed value of x. Roughly speaking, u_T increases as $1/\sqrt{x}$ inside the wake, and is negligible outside. On the other hand, the thickness of the wake increases as $\sqrt{x}$. Hence, the apparent thickness due to the transversal wave is of the order of $(1/\sqrt{x})\sqrt{x} = 1$.

The drag of the flat plate may be evaluated with the aid of the momentum integral (81.16). It is seen that the longitudinal part of u cancels the pressure term. Hence,

$$D = \lim_{x_1 \to \infty} \int_{x=x_1} (-\rho U u_T)\,dy \tag{81.83}$$

Evaluation of this integral shows that the value of D is indeed unity, as originally assumed. A simple way of evaluating (81.83) is to observe that, according to (81.83), $D/\rho U$ is the total sink strength of the transversal wave. This must equal the total source strength of the longitudinal wave, which, as pointed out above, is $1/\rho U$.

Singular Lifting Element (Two Dimensions). We shall now consider the case of a singular element of unit lift located at the origin. If the positive y direction is the direction of lift, the force exerted on the fluid is directed along the negative y axis. The flow field due to the lift is then the solution of Eqs. (81.76) with $\mathbf{a} = -\mathbf{i}_y$. It can be shown that this solution is

$$\mathbf{q} = \mathbf{i}_x \frac{\partial A}{\partial y} - \mathbf{i}_y \frac{\partial A}{\partial x} \tag{81.84a}$$

where $$A = \frac{1}{2\pi\rho U}[\ln r + e^{\lambda x} K_0(\lambda r)] \qquad r^2 = x^2 + y^2 \tag{81.84b}$$

Equations (81.84) can also be written in the following form,

$$\mathbf{q} = \mathbf{q}_L + \mathbf{q}_T \tag{81.85a}$$

where $$\mathbf{q}_L = -\frac{1}{2\pi\rho U}\,\text{grad}\left(\tan^{-1}\frac{y}{x}\right) \qquad \mathbf{q}_T = \text{curl}\left[\mathbf{i}_z \frac{1}{2\pi\rho U} e^{\lambda x} K_0(\lambda r)\right] \tag{81.85b,c}$$

$\mathbf{i}_z$ being the unit vector directed along the z axis. The pressure is

$$p' = -\rho U u_L = -\frac{y}{2\pi r^2} \tag{81.86}$$

The longitudinal wave is the flow field due to a potential vortex whose clockwise circulation $\Gamma = 1/\rho U$. The transversal wave at large distances from the origin is

$$u_T \approx -\frac{1}{4\rho}\sqrt{\frac{1}{\pi U \nu r}}\left(\frac{y}{r}\right)e^{-\lambda(r-x)}$$
$$v_T \approx -\frac{1}{4\rho}\sqrt{\frac{1}{\pi U \nu r}}\left(1 - \frac{\nu}{4Ur} - \frac{x}{r} - \frac{3}{4}\frac{\nu x}{Ur^2}\right)e^{-\lambda(r-x)} \tag{81.87a,b}$$

As before, it may be seen that the transversal wave is exponentially negligible outside the wake. Furthermore, the contribution of the transversal wave to the circulation around a contour at large distances from the origin is negligible. Hence Γ, the clockwise circulation of the total flow field around a very distant contour enclosing the origin, is equal to $1/\rho U$. An application of the momentum integral (81.17) then shows that the magnitude of the lift has the correct value, namely, unity.

Singular Needle. We now consider the three-dimensional analog of the singular flat plate. A force of unit strength, located at the origin, and directed in the negative x direction, is exerted on the fluid. The resulting flow field may then be regarded as the flow field due to a singular needle of unit drag located at the origin. The solution is

$$\mathbf{q} = \mathbf{q}_L + \mathbf{q}_T \tag{81.88a}$$

where

$$\mathbf{q}_L = -\frac{1}{4\pi\rho U}\,\text{grad}\,\frac{1}{r} \qquad r^2 = x^2 + y^2 + z^2$$

$$\mathbf{q}_T = \frac{1}{4\pi\mu}\left(\frac{1}{2\lambda}\,\text{grad}\,\frac{e^{-\lambda(r-x)}}{r} - \frac{e^{-\lambda(r-x)}}{r}\,\mathbf{i}_x\right) \tag{81.88b,c}$$

If we put

$$\frac{\partial\kappa}{\partial x} = -\frac{1}{4\pi\mu}\,\frac{e^{-\lambda(r-x)}}{r} \qquad \frac{\partial\kappa}{\partial t} = 0 \tag{81.89}$$

then $\mathbf{q}_T$ is split according to Eq. (81.74). The pressure is

$$p' = \frac{x}{4\pi r^3} \tag{81.90}$$

The longitudinal wave $\mathbf{q}_L$ represents potential perturbation flow due to a source of strength $1/\rho U$. The transversal wave $\mathbf{q}_T$ vanishes exponentially upstream and decreases as $1/x$ downstream, with a paraboloidal wake of cross-sectional area proportional to x. The transversal wave has a sink at the origin of the same strength as the longitudinal source. The inflow through a plane $x = x_1$ in the limit as $|x_1| \to \infty$ is zero if $x_1 < 0$, and equal to $1/\rho U$ if $x_1 > 0$. Hence, ρU times the transversal inflow in the wake is unity, i.e., equal to the drag [cf. (81.83)].

Singular Lifting Element (Three Dimensions). Finally, we consider the case of three-dimensional flow for which the singular force is normal to the free-stream direction. All such directions are equivalent because of the rotational symmetry of the equations. Hence, it is sufficient to select a special direction. We shall assume that the force is directed along the negative y direction. The resulting flow field is then due to a singular element of unit lift located at the origin. The solution is

$$\mathbf{q} = \mathbf{q}_L + \mathbf{q}_T \tag{81.91a}$$

where

$$\mathbf{q}_L = \frac{1}{4\pi\rho U}\,\text{grad}\left[\frac{\partial}{\partial y}\ln(r - x)\right]$$

$$\mathbf{q}_T = -\frac{1}{4\pi\rho U}\,\text{grad}\left[e^{-\lambda(r-x)}\frac{\partial}{\partial y}\ln(r - x)\right] - \frac{e^{-\lambda(r-x)}}{4\pi\mu r}\,\mathbf{i}_y \tag{81.91b,c}$$

and

$$\lambda = \frac{U}{2\nu} \qquad r^2 = x^2 + y^2 + z^2$$

The pressure is

$$p' = -\rho U u_L = -\frac{y}{4\pi r^3} \tag{81.92}$$

The longitudinal wave is identical with the potential flow field due to the infinitesimal horseshoe vortex of unit lift. All the vorticity is contained in the last term of the transversal wave, and is

$$\Omega_1 = -\frac{\partial v_2}{\partial z} \qquad \Omega_2 = 0 \qquad \Omega_3 = \frac{\partial v_2}{\partial x} \qquad v_2 = -\frac{e^{-\lambda(r-x)}}{4\pi\mu r} \tag{81.93}$$

This function, which is the coefficient of $\mathbf{i}_y$ in the last term in Eq. (81.91c), satisfies the following equation at any point except the origin:

$$\operatorname{div}\,(\operatorname{grad} v_2 - 2\lambda v_2 \mathbf{i}_x) \equiv \nabla^2 v_2 - 2\lambda \frac{\partial v_2}{\partial x} = 0 \qquad (81.94a)$$

The singularity at the origin of the left-hand expression is equal to that of $-(1/4\pi\mu)$ $\nabla^2(1/r)$. By integrating Eq. (81.94a), we then obtain the following equation,

$$\iiint_V \frac{\partial v_2}{\partial x}\, dV = \frac{1}{2\lambda} \iiint_V \operatorname{div} \operatorname{grad} v_2 \, dV - \frac{1}{\rho U} \qquad (81.94b)$$

where V is any region containing the origin. If B is the boundary of V and V increases indefinitely in all directions, then

$$\iint_B (\operatorname{grad} v_2) \cdot d\mathbf{n} \to 0 \qquad (81.94c)$$

A comparison of Eqs. (81.94a), (81.94b), and (81.94c) shows that

$$\iiint \Omega_3 \, dV = -\frac{1}{\rho U} \qquad (81.95)$$

if the integral is taken over the entire space. Since the lift is supposed to be unity, this equation checks the lift formula obtained by the momentum method [Eq. (81.17)].

Methods for deriving the fundamental solutions given, as well as the fundamental solutions for nonstationary flow, are presented in [4] and [17]. The use of the fundamental solutions will be illustrated in the following two articles.

Solutions of Boundary-value Problems of the Oseen Equations

Finite Flat Plate in Two-dimensional Flow. We consider a finite flat plate located on the x axis between $x = -a$ and $x = a$. We may regard this plate as composed of infinitely many singular flat plates, those located between $x = \xi$ and $x = \xi + d\xi$ having the drag $g(\xi)\, d\xi$. Thus, $g(\xi)$ is the local force distribution on the plate. By superposition, we then find that the total flow field due to these plates is

$$\mathbf{q}(x,y) = \int_{-a}^{a} \mathbf{q}_F(x - \xi,\, y) g(\xi)\, d\xi \qquad (81.96)$$

where $\mathbf{q}_F(x,y)$ is the fundamental solution given by Eq. (81.78). In order to determine the function $g(\xi)$, we note that the no-slip condition requires the perturbation velocities u and v to be $-U$ and zero, respectively, on the finite plate. The second condition is identically satisfied. The first condition gives

$$-U = \int_{-a}^{a} u_F(x - \xi,\, 0) g(\xi)\, d\xi \qquad (81.97)$$

where $u_F(x,y)$ is the x component of $\mathbf{q}_F(x,y)$. Equation (81.97) is an integral equation for the unknown function $g(\xi)$.

An approximate solution for low Reynolds numbers will now be given. The Reynolds number of the plate is defined as

$$\text{Re} = \frac{2aU}{\nu} \qquad (81.98)$$

In the integral equation the maximum value of the nondimensional quantity $(x - \xi)U/\nu$ is Re. Hence, if the latter is small, we may use an expression for $\mathbf{q}_F$

valid for small values of $(x - \xi)U/\nu$, such as given by Eq. (81.81*a*). We then obtain the approximate equation

$$U = \frac{1}{4\pi\nu} \int_{-a}^{a} (\ln|x - \xi| + c) g(\xi)\, d\xi \qquad -a \le x \le a \tag{81.99}$$

where

$$c = \ln \frac{\gamma_0 U}{4\nu} - 1$$

The solution of this integral equation is (cf., for example, [19], p. 143)

$$g(x) = \frac{4U\nu}{\ln(\gamma_0 \text{ Re}/16) - 1} \frac{1}{\sqrt{a^2 - x^2}} \tag{81.100}$$

Since the force distribution on the plate is now determined, we find the drag D by integrating:

$$C_D = \frac{D}{\frac{1}{2}\rho U^2 \cdot 2a} = \frac{8\pi}{\text{Re}} \frac{1}{1 - \ln(\gamma_0 \text{ Re}/16)} \tag{81.101}$$

The complete flow field may now be obtained from Eq. (81.96). In this integral, we may use the force distribution as determined above. However, except in the immediate neighborhood of the plate, the values for $\mathbf{q}_F$ appearing explicitly in Eq. (81.96) may not be approximated by the values given in Eq. (81.81). At large distances, Eq. (81.82) should be used. It is easily checked that, at large distances, the dominant part of the flow field created by the finite flat plate is the same as the flow field due to a singular flat plate of drag D. Near the plate, the solution given satisfies the Stokes equation.

Circular Cylinder. We consider a circle of radius a and center at the origin. The diameter $2a$ will be used as the characteristic length. The Reynolds number and Stokes variables [Eq. (81.4)] are then defined by

$$\text{Re} = \frac{2aU}{\nu} \qquad x^* = \frac{x}{2a} \qquad \text{etc.} \tag{81.102}$$

Nondimensional variables will be used. $\mathbf{q}$ will denote the *complete velocity vector*, that is, $U\mathbf{i}_x$ plus perturbation velocity.

The boundary conditions are then

$$\mathbf{q}^* = \mathbf{i}_x \quad \text{at infinity} \tag{81.103a}$$

$$\mathbf{q}^* = 0 \quad \text{at the cylinder, i.e., for } \tilde{r} = \text{Re}/2,\ r^* = \tfrac{1}{2} \tag{81.103b}$$

An approximate solution for low Reynolds number is

$$\mathbf{q}^* = \mathbf{i}_x + 2\epsilon\mathbf{q}^{(1)} - \frac{\epsilon\,\text{Re}^2}{8}\mathbf{q}^2 \tag{81.104a}$$

where

$$\mathbf{q}^{(1)} = \text{grad}\left[\ln\tilde{r} + e^{\tilde{x}/2} K_0\left(\frac{\tilde{r}}{2}\right)\right] - e^{\tilde{x}/2} K_0\left(\frac{\tilde{r}}{2}\right)\mathbf{i}_x$$

$$\mathbf{q}^{(2)} = \text{grad}\frac{\partial}{\partial\tilde{x}}\ln\tilde{r} \qquad \epsilon = \frac{1}{\frac{1}{2} - \ln(\gamma_0 \text{ Re}/8)} \tag{81.104b–d}$$

Here, in accordance with a previous convention,

$$\text{grad} \equiv \mathbf{i}_x \frac{\partial}{\partial\tilde{x}} + \mathbf{i}_y \frac{\partial}{\partial\tilde{y}}$$

The flow field corresponding to $2\epsilon\mathbf{q}^{(1)}$ is the perturbation field due to a singular flat plate of drag $2\epsilon \cdot 2\pi\mu U$, as seen by comparison with Eq. (81.78). The function $\mathbf{q}^{(2)}$ represents a potential flow field, namely, the perturbation field of a dipole directed along the x axis. Its contribution to the drag is zero, as seen, for example, from momentum considerations.

The remarks above imply that $\mathbf{q}^*$ satisfies the Oseen equation and the boundary conditions at infinity. We shall now investigate to what degree the condition at the cylinder is satisfied. Since $\tilde{r} = \text{Re}/2$ at the cylinder and Re is assumed small, we develop $\mathbf{q}^*$ for small values of $\tilde{r}$. Since $\tilde{r} = \text{Re}\, r^*$, this is equivalent to rewriting Eq. (81.104) in Stokes variables and expanding in orders of ϵ. We find

$$\mathbf{q}^* = \epsilon \left[\left(\ln 2r^* + \tfrac{1}{2} - \frac{1}{8r^{*2}} \right) \mathbf{i}_x + \left(-\frac{x^*}{r^*} + \frac{x^*}{4r^{*3}} \right) \text{grad } r^* \right] + O(\text{Re}\, \epsilon) \qquad (81.105)$$

The boundary condition at $r^* = \tfrac{1}{2}$ [Eq. (81.103b)] is thus satisfied to order Re ϵ.

It can be easily verified that $\mathbf{i}_x + 2\epsilon \mathbf{q}^{(1)}$ satisfies Eq. (81.103b) to order unity whenever $\epsilon = 1/(c - \ln \text{Re})$, c = arbitrary constant. If we choose $c = \tfrac{1}{2} - \ln(\gamma_0/8)$ and add a suitable multiple of a potential dipole, Eq. (81.103b) will be satisfied to order Re ϵ, as just seen.

It was remarked above that the total drag is $4\pi\mu U\epsilon$. Hence,

$$C_D = \frac{D}{\tfrac{1}{2}\rho U^2 \cdot 2a} = \frac{8\pi}{\text{Re}} [\epsilon + O(\text{Re}\, \epsilon)] \qquad (81.106)$$

This drag formula is due to Lamb (see, for example, [6], p. 615). The connection with the fundamental solution was pointed out by Oseen [4], p. 177ff.

The general nature of the flow field is easily found from previous comments on $\mathbf{q}^{(1)}$. A viscous wake exists which far downstream is that of a singular flat plate of drag D. Outside the wake, the flow field is essentially potential at large distances from the cylinder. More specifically, it is the flow field due to potential source and potential dipole of appropriate strengths.

In [20], results of numerical computations for Oseen flow past a circular cylinder are given. The computations show two standing eddies on the downstream part of the cylinder. For increasing Re, these eddies become more and more elongated in the downstream direction. It is also proved that, within the Oseen approximation, the pressure drag is exactly equal to the drag due to viscous shear.

Sphere. We shall now discuss low Reynolds number flow past a sphere of radius a. A solution analogous to Eq. (81.104a) has been given by Oseen [4], p. 166ff. It is

$$\mathbf{q}^* = \mathbf{i}_x + \tfrac{3}{4}(1 + \tfrac{3}{16}\,\text{Re})\,\text{Re}\,\mathbf{q}^{(1)} - \tfrac{1}{32}\left(1 + \frac{3\,\text{Re}}{16}\right)\text{Re}^3\,\mathbf{q}^{(2)} \qquad (81.107a)$$

where $$\mathbf{q}^{(1)} = \text{grad}\, \frac{\exp[-(\tilde{r} - \tilde{x})/2] - 1}{\tilde{r}} - \frac{\mathbf{i}_x}{\tilde{r}} \exp\left(-\frac{\tilde{r} - \tilde{x}}{2}\right) \qquad (81.107b)$$

$$\mathbf{q}^{(2)} = -\,\text{grad}\, \frac{\partial}{\partial \tilde{x}} \left(\frac{1}{\tilde{r}}\right) \qquad (81.107c)$$

$$r^2 = x^2 + y^2 + z^2 \qquad \text{Re} = \frac{2aU}{\nu} \qquad \tilde{r} = \frac{Ur}{\nu} \qquad \text{etc.}$$

This solution is obviously analogous to that previously given for the circular cylinder, and the discussion of the present case will, therefore, be very brief. However, some special comments are needed about the boundary condition at the sphere ($\mathbf{q}^* = 0$ for $\tilde{r} = \text{Re}/2$). The term Re $\mathbf{q}^{(1)}$, which is the flow field due to a singular needle of drag $8\pi\mu Ua$ [cf. Eq. (81.88)], may be expanded for small values of $\tilde{r}$ as [cf. Eq. (81.105)]

$$\text{Re}\,\mathbf{q}^{(1)} \approx \left[-\frac{x^*}{2r^{*2}} + \text{Re}\left(\frac{1}{8} - \frac{x^{*2}}{8r^{*2}}\right)\right] \text{grad } r^* + \left[-\frac{1}{2r^*} + \text{Re}\left(\frac{1}{4} - \frac{x^*}{4r^*}\right)\right] \mathbf{i}_x \qquad (81.108a)$$

where $x^* = x/2a$, etc. $\mathrm{Re}^3\, \mathbf{q}^{(2)}$, which represents a potential dipole, is, rewritten in Stokes variables:

$$\mathrm{Re}^3\, \mathbf{q}^{(2)} = \frac{1}{r^{*3}}\, \mathbf{i}_x - \frac{3x^*}{r^{*4}} \operatorname{grad} r^* \tag{81.108b}$$

It is then seen, if the term 3 Re/16 is omitted in the factor $(1 + 3\,\mathrm{Re}/16)$ in Eq. (81.107a), that, $\mathbf{q}^* = O(\mathrm{Re})$ at $r^* = \frac{1}{2}$, so that the boundary conditions are satisfied to order unity. If the full factor $(1 + 3\,\mathrm{Re}/16)$ is retained, all terms of order unity and most terms of order Re cancel out at $r^* = \frac{1}{2}$. The uncanceled terms have the

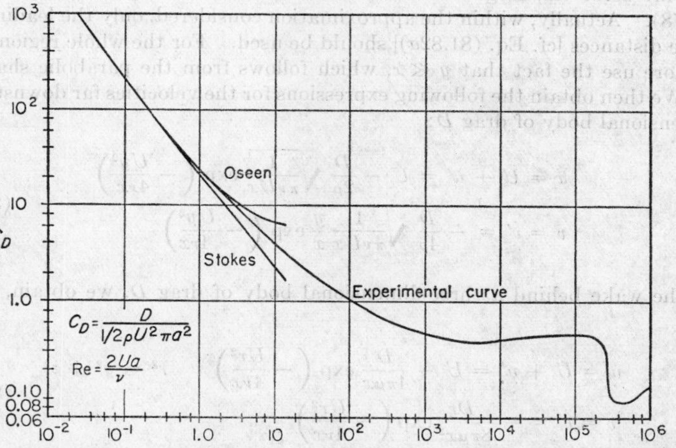

FIG. 81.6. Drag coefficient of smooth spheres. For Stokes solution, see Eq. (81.60); for Oseen solution, see Eq. (81.109); experimental curve from [26].

property that the x components of the velocity are odd functions of x, whereas the y and z components are even. It follows by symmetry that the Oseen solution, which may be added to cancel these terms, contributes to the drag only with terms of the order Re^2 ([4], p. 173ff). (This symmetry argument is based on the fact that, near the cylinder, the unsymmetrical Oseen equation may be replaced by the symmetrical Stokes equation, as explained below.) Hence, the solution given by Eq. (81.107) satisfies the boundary conditions only to order unity, but gives the drag correctly to order Re. The drag comes only from the $\mathbf{q}^{(1)}$ term. It is then (see Fig. 81.6)

$$C_D = \frac{D}{\frac{1}{2}\rho U^2 \pi a^2} = \frac{24}{\mathrm{Re}}\left(1 + \frac{3\,\mathrm{Re}}{16} + O(\mathrm{Re}^2)\right) \tag{81.109}$$

For further discussion of Oseen flow past a sphere, see [20].

Elliptic Cylinders. Solutions of the Oseen equations for flow past elliptic cylinders are discussed in [20] and [21]. The effect of angle of attack is considered there. The finite flat plate is treated as a limiting case of an elliptic cylinder.

Use of the Oseen Fundamental Solutions in Studying Wake Flow

In connection with the solution for the finite flat plate, it was pointed out that, at large distances, the flow field agrees with that of a singular flat plate of drag equal to the drag of the finite plate. We may also verify the corresponding statements for the solutions for the circular cylinder and the sphere. Actually the flow field at large distances from a finite body of drag D and lift L is always given, to the first approximation, by the corresponding fundamental equations for singular drag and lifting elements. The same statement is true for the full Navier-Stokes equations at an arbitrary Reynolds number. The values of drag and lift can then not be calculated from the Oseen equations. However, once the values of D and L are assumed, the flow field at large distances is given by the fundamental Oseen solutions multiplied by D and L, respectively. This is due to the facts that the Oseen equations, being a linearization about the free-stream velocity, are approximately correct at large distances, and that D and L may be evaluated by Eqs. (81.16) and (81.17), respectively.

As an example, consider flow past a two-dimensional body, and let the drag be D and the lift zero. At large distances, the flow field is $D \cdot \mathbf{q}$, where $\mathbf{q}$ is given by Eq. (81.78). Actually, within the approximation considered, only the leading term of $\mathbf{q}$ at large distances [cf. Eq. (81.82a)] should be used. For the whole region, we may furthermore use the fact that $y \ll x$, which follows from the parabolic shape of the wake. We then obtain the following expressions for the velocities far downstream of a two-dimensional body of drag D:

$$\begin{aligned} u &= U + u' = U - \frac{D}{2\rho}\sqrt{\frac{1}{\pi\nu U x}}\exp\left(-\frac{Uy^2}{4\nu x}\right) \\ v &= v' = -\frac{D}{4\rho}\sqrt{\frac{1}{\pi\nu U x}}\,\frac{y}{x}\exp\left(-\frac{Uy^2}{4\nu x}\right) \end{aligned} \tag{81.110a,b}$$

For the wake behind a three-dimensional body of drag D, we obtain, from Eq. (81.88),

$$\begin{aligned} u &= U + u' = U - \frac{D}{4\pi\mu x}\exp\left(-\frac{Ur^2}{4\nu x}\right) \qquad r^2 = y^2 + z^2 \\ v &= v' = -\frac{Dr}{8\pi\mu x^2}\exp\left(-\frac{Ur^2}{4\nu x}\right) \end{aligned} \tag{81.111a,b}$$

These formulas may be obtained directly as follows. Inside the wake, the flow has certain boundary-layer characteristics; the pressure variation $\partial p/\partial x$ and the contribution of $\mu\, \partial^2 u/\partial x^2$ to the viscous force are small, and may therefore be neglected from the first momentum equation. At the same time, the perturbation u' is small compared to U far downstream of the body; hence, the equations can be linearized as in the Oseen flow. The governing equations and boundary conditions for the wake behind a two-dimensional body are, therefore,

$$U\frac{\partial u'}{\partial x} = \nu\frac{\partial^2 u'}{\partial y^2} \qquad \frac{\partial u'}{\partial x} + \frac{\partial v'}{\partial y} = 0 \tag{81.112a,b}$$

$$u'(0,y) = u'(x,\infty) = 0 \qquad \text{and} \qquad \frac{\partial u'}{\partial y} = 0 \qquad \text{at } y = 0 \tag{81.112c}$$

The solutions of these equations are given by Eqs. (81.110) within a multiplicative constant, which is then determined by the momentum theorem for drag [cf. (81.16)].

In three dimensions, Eqs. (81.112) are replaced by

$$U\frac{\partial u'}{\partial x} = \nu\left(\frac{\partial^2 u'}{\partial r^2} + \frac{1}{r}\frac{\partial u'}{\partial r}\right) \qquad r^2 = y^2 + z^2 \tag{81.113a}$$

$$\frac{\partial u'}{\partial x} + \frac{1}{r}\frac{\partial}{\partial r}(rv') = 0 \tag{81.113b}$$

$$u'(0,r) = u'(x,\infty) = 0 \qquad \text{and} \qquad \frac{\partial u'}{\partial r} = 0 \qquad \text{at } r = 0 \tag{81.113c}$$

The solution obtained is then Eqs. (81.111).

It must be pointed out, however, that these results have only limited applicability, because, in a real wake, the flow is usually unsteady, and is very often turbulent instead of laminar as assumed above.

81.5. DISCUSSION OF THEORETICAL RESULTS. HIGHER-ORDER APPROXIMATIONS

We shall now discuss the relations of the Stokes and the Oseen solutions to each other, and to the solutions of the more accurate Navier-Stokes equations. In particular, we shall consider the case of flow of an unbounded fluid past finite bodies.

In the preceding section, the question was raised whether the Stokes solutions give a better approximation near the body than the corresponding Oseen solutions. This is not true, in spite of the fact that the transport terms in the Oseen equations are "wrong" near the body. To fix the ideas, consider the solution for flow past a sphere. An approximate solution of the Oseen equations was given by Eq. (81.107*b*). This solution satisfies the Oseen equation to order unity. Near the body, it satisfies the Stokes equations; in fact, it contains the exact solution of the Stokes equations. In Oseen variables $\tilde{x}_i$, the radius of the body is Re/2. Hence, to study the Oseen solution near the body for small values of Re, we should consider the values of the solution for small values of $\tilde{r}$. From the point of view of a general theory of expansions for low Reynolds numbers, it is convenient to use the following formulation. The Oseen solution is rewritten in Stokes variables. Since $\tilde{x}_i = \text{Re}\, x_i^*$, an expansion of the solution in powers of Re is equivalent to an expansion for small values of $\tilde{x}_i$. This may be achieved with the aid of Eq. (81.108). If only the term of order unity is retained in this expansion, we obtain exact Stokes solutions. Note that the term $-(\text{Re}^3/32)\mathbf{q}^{(2)}$ in the Oseen solution (81.107*a*) is of order unity when rewritten in Stokes variables (81.108*b*). This term is thus of order Re^3 far away from the body, but of order unity near the body. Such terms of apparently high order must be included in the Oseen solution, in order for this solution to be uniformly valid to order unity. We finally observe that the leading term of the Oseen formula for the drag (81.109) agrees with the Stokes formula for the drag (81.60). The Oseen solution gives, in addition, a correction term for the drag of order Re. This term is correct in the sense that a solution of the Navier-Stokes equations gives the same term [cf. Eq. (81.121)]. It is true, in general, that the Oseen equations give the drag correctly to order Re for three-dimensional bodies with fore-aft symmetry (cf. p. 81–25). This result is stated for the sphere by Goldstein [5], p. 492, and proved in general by Kaplun [22], p. 592.

So far, the Stokes and Oseen solutions have been compared. In order to obtain a theory for flow at low Reynolds numbers, the relations of these solutions to the corresponding Navier-Stokes solutions must be clarified. The solution of this problem, for two or three dimensions, has been given by S. Kaplun. His theory can only be indicated very briefly here. The essence of Kaplun's ideas is stated in [23], and they are further amplified in [22] and [24]. These papers also contain some specific applications. A detailed investigation of flow past a sphere is given by Proudman and Pearson [25].

Kaplun observes that the problem of flow at low Reynolds numbers may be treated in analogy with the problem of flow at high Reynolds numbers. He introduces an outer, or Oseen, expansion,

$$\mathbf{q}^* \sim \sum_{j=0} \epsilon_j(\text{Re})\mathbf{g}_j(\tilde{x}_i) \tag{81.114}$$

The functions $\epsilon_j(\text{Re})$ are of increasingly higher order; that is,

$$\epsilon_0 = 1, \quad \lim_{\text{Re}\to 0} \frac{\epsilon_{j+1}}{\epsilon_j} = 0 \tag{81.115}$$

Equation (81.114) gives an asymptotic expansion which is uniformly valid away from the body, that is,

$$\lim_{\text{Re}\to 0} \frac{\mathbf{q}^* - \sum_{j=0} \epsilon_j(\text{Re})\mathbf{g}_j(\tilde{x}_i)}{\epsilon_n(\text{Re})} = 0 \quad \text{uniformly for } \tilde{r} > \text{const} \tag{81.116}$$

Note that, in this limit, the Oseen variables $\tilde{x}_i$ are kept fixed. In the analogy with flow at high Reynolds numbers, the leading terms of the expansion (81.114) correspond to potential flow, flow due to displacement thickness, etc.

The expansion (81.114) is complemented by an inner, or Stokes, expansion, which is valid near the body, i.e., for $r^* <$ const,

$$\mathbf{q}^* \sim \sum_{j=0} \epsilon_j(\text{Re})\mathbf{h}_j(x_i^*) \tag{81.117}$$

The first nonvanishing term of this expansion corresponds to the boundary-layer solution for high Reynolds numbers.

The equations governing the $\mathbf{g}_j$ and $\mathbf{h}_j$ are easily found by inserting expansions (81.114) and (81.117) into the Navier-Stokes equations written in Oseen variables (81.10) and Stokes variables (81.9), respectively. For flow of an unbounded fluid past a finite object in two or three dimensions, it can be seen that $\mathbf{g}_0 = \mathbf{i}_x$. This implies that $\mathbf{g}_1$ obeys the Oseen equations. For $j > 1$, the $\mathbf{g}_j$ then obey Oseen equations, with possibly a known function of the $\tilde{x}_i$, obtained from the lower-order $\mathbf{g}_j$, added as a forcing term. The leading term of expansion (81.117) obeys the Stokes equations, the other terms obey Stokes equations, possibly with a known forcing term added.

The outer expansion should satisfy the boundary conditions at infinity, and the inner expansion the boundary conditions at the body. In order to completely determine the $\mathbf{g}_j$, $\mathbf{h}_j$, and ϵ_j, further conditions are needed. These are provided by conditions of matching between the outer and inner expansions. (Compare the matching of boundary-layer flow with potential flow and flow due to displacement thickness in the theory of high Reynolds numbers.) An example of the matching will be given here (cf. [23], p. 874). Consider flow past a circular cylinder of radius a. A solution of the Stokes equations satisfying the boundary conditions at the body is

$$\mathbf{q}^* = A\mathbf{h} \tag{81.118a}$$

where

$$\mathbf{h} = (\ln 2r^* + \tfrac{1}{2})\mathbf{i}_x - \frac{x^*}{r^*}\,\text{grad}\, r^* - \tfrac{1}{8}\,\text{grad}\,\frac{x^*}{r^{*2}} \tag{81.118b}$$

The constant A is as yet undetermined. This solution may be matched with the leading term of the outer expansion, $\mathbf{g}_0 = \mathbf{i}_x$, by requiring that

$$\lim_{\text{Re}\to 0} (\mathbf{i}_x - A\mathbf{h}) = 0 \quad \tilde{r} \text{ fixed} \tag{81.119}$$

This requirement implies

$$A = -\frac{1}{\ln \text{Re}} + O\left(\frac{1}{\ln \text{Re}}\right) \tag{81.120}$$

Hence, $\mathbf{h}_0 = 0$, $\mathbf{h}_1 = \mathbf{h}$, and $\epsilon_1(\text{Re}) = A$. Thus the method of Kaplun gives a Stokes solution in two dimensions. This solution gives, of course, an infinite velocity at infinity. This is no paradox, since it is required only to satisfy the boundary condition at the sphere. The drag as computed from Eqs. (81.118) and (81.120) agrees to the lowest order with Eq. (81.101), that is, with the approximate solution of the Oseen equation.

The ideas underlying the matching are explained in [22]. In [25], calculations for the sphere are carried through to the order $\mathrm{Re}^2 \ln \mathrm{Re}$, and the following formula is obtained:

$$C_D = \frac{24}{\mathrm{Re}}[1 + 3/16\ \mathrm{Re} + 9/160\ \mathrm{Re}^2 \ln \mathrm{Re} + O(\mathrm{Re}^2)] \tag{81.121}$$

It should be noted that, once the inner and outer expansions have been found to a certain order, it is easy to form a composite expansion which is uniformly valid to the same order, see, for example, [22], p. 592, and [23].

81.6. REFERENCES

[1] H. L. Dryden, F. D. Murnaghan, H. Bateman: "Hydrodynamics," Dover, New York, 1956.

[2] G. G. Stokes: (*a*) On the theories of internal friction of fluids in motion, and of the equilibrium and motion of elastic solids, *Trans. Cambridge Phil. Soc.*, **8** (1845), 287–319, also: "Mathematical and Physical Papers," vol. 1, pp. 75–129, 1880; (*b*) On the effect of the internal friction of fluids on the motion of pendulums, *Trans. Cambridge Phil, Soc.*, **9**, part 2 (1850), 8–106, also: "Mathematical and Physical Papers," vol. 3. pp. 1–141, 1901.

[3] C. W. Oseen: Über die Stokes'sche Formel und über eine verwandte Aufgabe in der Hydrodynamik, *Arkiv. Mat. Astron. Fysik*, **6** (1911), 1–20.

[4] C. W. Oseen: Neuere Methoden und Ergebnisse in der Hydrodynamik, Akad.-Verlag., Leipzig, 1927.

[5] S. Goldstein: "Modern Developments in Fluid Dynamics," Oxford Univ. Press, New York, 1938.

[6] H. Lamb: "Hydrodynamics," 6th ed., Dover, New York, 1945.

[7] L. Prandtl: The mechanics of viscous fluids, in. W. F. Durand (ed.), "Aerodynamic Theory," vol. 3, div. G, Springer, Berlin, 1935.

[8] H. S. Hele-Shaw: (*a*) Investigation of the nature of surface resistance of water and of stream-line motion under certain experimental conditions, *Trans. Inst. Naval Architects*, **40** (1898), 21–40; (*b*) The flow of water, *Nature*, **58** (1898), 34–36; (*c*) The motion of perfect liquid, *Proc. Roy. Inst.*, **16** (1899), 49–64.

[9] G. G. Stokes: Mathematical proof of the identity of the stream lines obtained by means of a viscous film with those of a perfect fluid moving in two dimensions, "Mathematical and Physical Papers," vol. 5, pp. 278–282, 1905.

[10] F. Riegels: Zur Kritik des Hele-Shaw Versuchs, *Z. angew. Math. Mech.*, **18** (1938), 95–106.

[11] W. R. Dean: (*a*) Note on the slow motion of a fluid, *Proc. Cambridge Phil. Soc.*, **32** (1936), 598–613; **36** (1940), 300–313; (*b*) Note on the shearing motion of fluid past a projection, *ibid.*, **40** (1944), 214–222; (*c*) On the steady motion of viscous liquid in a corner, *ibid.*, **45** (1949), 389–394; (*d*) Slow motion of viscous liquid near a half-Pitot tube, *ibid.*, **48** (1952), 149–167.

[12] M. S. Smoluchowski: On the practical applicability of Stokes' law of resistance and the modification of it required in certain cases, *Proc. 5th Intern. Congr. Math.*, Cambridge, 1912, vol. 2, pp. 192–201.

[13] P. M. Morse, H. Feshbach: "Methods of Theoretical Physics," part 1, chaps. 10 and 13, McGraw-Hill, New York, 1953.

[14] M. Krakowski, A. Charnes: Stokes' paradox and biharmonic flows, *Carnegie Inst. Tech. Dept. Math. Tech. Rept.* 37 (1953).

[15] R. Finn, W. Noll: On the uniqueness and non-existence of Stokes flows, *Arch. Rat. Mech. Anal.*, **1** (1957), 97–106.

[16] G. Birkhoff: "Hydrodynamics: A Study in Logic, Fact, and Similitude," p. 33, Dover, New York, 1955.

[17] P. A. Lagerstrom, J. D. Cole, L. Trilling: Problems in the theory of viscous compressible fluids, *Office Naval Research GALCIT Rept.*, 1949.

[18] R. Courant, D. Hilbert: "Methoden der Mathematischen Physik," vol. 2, Interscience, New York, 1937.

[19] W. Magnus, F. Oberhettinger: "Formulas and Theorems for the Special Functions of Mathematical Physics," Chelsea, New York, 1949.

[20] S. Tomotika, T. Aoi: (*a*) The steady flow of viscous fluid past a sphere and circular cylinder at small Reynolds numbers. *Quart. J. Mech. Appl. Math.*, **3** (1950), 140–161; (*b*) An expansion formula for the drag on a circular cylinder through a viscous fluid at small Reynolds numbers, *ibid.*, **4** (1951), 401–406; (*c*) The steady flow of a viscous

fluid past an elliptic cylinder and a flat plate at small Reynolds numbers, *ibid.*, **6** (1953), 290–311.

[21] I. Imai: (*a*) On the asymptotic behaviour of viscous fluid flow at a great distance from a cylindrical body, with special reference to Filon's paradox, *Proc. Roy. Soc. London*, **A,208** (1951), 487–516; (*b*) A new method of solving Oseen's equations and its application to the flow past an inclined elliptic cylinder, *ibid.*, **A,224** (1954), 141–160.

[22] S. Kaplun, P. A. Lagerstrom: Asymptotic expansions of Navier-Stokes solutions for small Reynolds numbers, *J. Math. Mech.*, **6** (1957), 585–593.

[23] P. A. Lagerstrom, J. D. Cole: Examples illustrating expansion procedures for Navier-Stokes equations, *J. Rat. Mech. Anal.*, **4** (1955), 817–882.

[24] S. Kaplun: Low Reynolds number flow past a circular cylinder, *J. Math. Mech.*, **6** (1957), 595–603.

[25] I. Proudman, J. R. A. Pearson: Expansions at small Reynolds numbers for the flow past a sphere and a circular cylinder, *J. Fluid Mechanics*, **2** (1957), 237–262.

[26] H. Schlichting: "Boundary Layer Theory," 2d ed., transl. by J. Kestin, McGraw-Hill, New York, 1960.

CHAPTER 82

INCOMPRESSIBLE BOUNDARY LAYERS

BY

J. KESTIN, Ph.D., (Eng), Providence, R.I.

82.1. OUTLINE OF FLUID MOTION WITH FRICTION; REAL AND PERFECT FLUIDS

Real fluids are viscous and compressible. In a field of flow, they develop tangential stresses in addition to normal forces, and change their density when the pressure is changed. Pressure and density changes are accompanied by temperature changes. An ideal fluid is an approximation in which these two properties are neglected. The theory of perfect fluids (Chaps. 70 to 72) supplies a satisfactory description of a range of phenomena, but fails completely to account for the occurrence of drag on solid bodies placed in a stream. It leads to the result (d'Alembert's paradox) that a body which moves uniformly through an incompressible fluid extending to infinity experiences no drag.

Because of the absence of shear, a perfect fluid will, in general, move with a velocity relative to that of a solid body immersed in the flow, or containing it; on the boundary between a solid body and a perfect fluid, there is *slip*. In a real fluid, molecular interactions between the fluid and the wall cause the fluid to adhere to it, giving rise to the *condition of no slip*, which can be supported by the fluid because of its ability to develop shearing stresses. All real fluids, even those whose viscosity is very small and otherwise negligible, adhere to solid walls, except when their density is so small that the mean free path of a molecule becomes comparable with the linear dimensions of the body. On a microscopic scale, a negligible amount of slip will be found to exist under all conditions.

The essential difference between the mechanics of perfect and real fluids consists in the fact that, in the latter, the condition of no slip and the presence of viscosity are systematically taken into account. The account given in this chapter of the handbook will neglect compressibility, assuming that the equation of state of the fluid is $\rho =$ const, and the presence of temperature fields and questions concerned with the transfer of heat will be ignored.

82.2. THE PHYSICAL CONCEPT OF A BOUNDARY LAYER

Boundary Layer

At high Reynolds numbers, particularly when the measured pressure distribution agrees well with that predicted by perfect-fluid theory, the influence of viscosity is confined to a very thin layer near the solid wall. Because of the no-slip condition, the velocity in that thin layer increases from zero at the wall to a value equal to that in the external frictionless flow. This creates a very large velocity gradient $\partial u/\partial y$ at

This chapter is based on the author's translation of Professor H. Schlichting's treatise on boundary layers. It is mainly a condensation of the book, and has been done with Professor H. Schlichting's permission, for which the author wishes to express his appreciation. The holders of the copyright, Verlag, G. Braun of Karlsruhe, also gave permission to abstract the book, and kindly loaned the original drawings for reproduction. For a more extensive treatment of this subject and for a comprehensive list of bibliographical entries, reference should be made to [1] and [2].

right angles to the wall; the very large velocity gradient, multiplied, even by the very small viscosity of such fluids as air or water, produces large shearing stresses $\tau = \mu(\partial u/\partial y)$, which play an essential part in the balance of forces. They are very small outside this thin layer of fluid, and can be ignored there. This thin region of flow is called the *boundary layer*.

The division of the field of flow into a boundary layer where shearing stresses are essential and into an external region of nearly ideal flow brings with it a considerable simplification of the theory of the motion of fluids of low viscosity, and makes it possible to give it an exact mathematical form.

Separation and Vortex Formation

Because of the cumulative action of viscosity in the boundary layer, progressively larger masses of fluid are retarded in it by the shearing forces, and the thickness δ of the boundary layer increases in the downstream direction. The growth of a boundary layer downstream from the leading edge of an infinitely thin flat plate in a parallel stream is shown sketched in Fig. 82.1. It is seen that the initially uniform velocity distribution is changed along the plate, the zone of retarded flow (boundary layer) growing in the downstream direction. The loss of momentum by individual particles may become so large that, in some cases, they cease to be able to overcome the pressure gradient which is *impressed* on the boundary layer by the external flow if it acts against

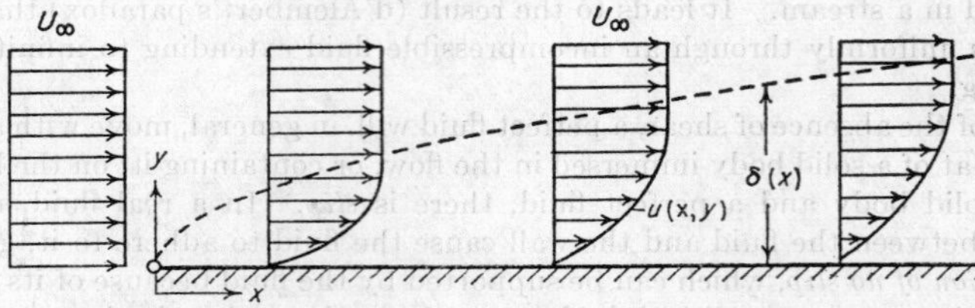

FIG. 82.1. Sketch of boundary layer on a semi-infinite flat plate in parallel flow at zero incidence. Vertical coordinate highly exaggerated. (*From* [2], *p.* 22.)

the flow (adverse pressure gradient). The decelerated fluid particles are first arrested, and then forced to reverse their direction of motion; they then penetrate the region of potential flow, and the boundary layer may thus *separate* from the wall. This phenomenon is always accompanied by the formation of vortices, and by large energy losses in the *wake* of the body.

Boundary-layer separation occurs most readily near blunt bodies, where the external flow impresses large adverse pressure gradients on the boundary layer. The wake behind the separated region considerably modifies the potential pressure distribution by displacing the streamlines so that a large drag force is created. The pressure gradient along an infinitely thin flat plate at zero incidence in a parallel stream is equal to zero, and, in principle, the boundary layer does not separate from it.

The occurrence of separation explains the presence of pressure or form drag. It also explains why the measured pressure distribution agrees so well with the ideal one around a slender streamline body, and why it deviates from it around a blunt body.

The streamlines in the boundary layer near a point of separation are shown sketched in Fig. 82.2; it is seen that the reversal of the flow direction causes a considerable thickening of the boundary layer near the point of separation A. The point of separation divides the region of forward flow, $\partial u/\partial y > 0$, from that of backflow, $\partial u/\partial y < 0$, and is given by

$$\frac{\partial u}{\partial y} = 0 \qquad \text{at wall (separation)} \qquad (82.1)$$

This is equivalent to stating that, at the point of separation, the shearing stress at the wall

$$\tau_0 = \mu \left(\frac{\partial u}{\partial y}\right) = 0 \qquad \text{at wall (separation)} \tag{82.1a}$$

Since separation is associated with an adverse pressure gradient, it is absent in convergent nozzles, but may cause large losses in diffusers; it is also responsible for the phenomenon of *stall* in airfoils and turbine blades.

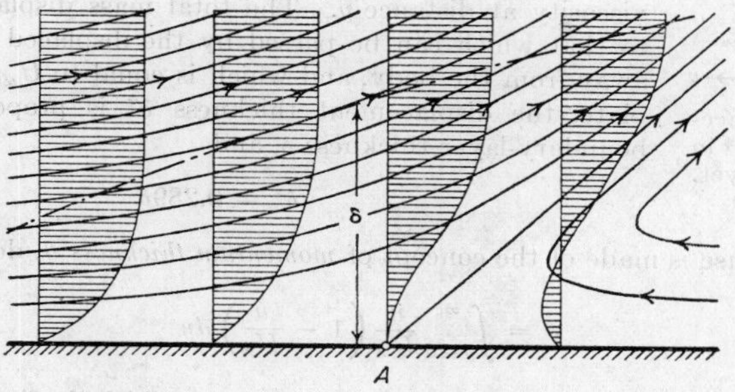

FIG. 82.2. Streamlines in boundary layer near separation. A = point of separation. (*From* [2], *p.* 29.)

Boundary-layer Thickness

The thickness of an unseparated boundary layer can be estimated by equating the inertia and friction forces. These are of order $\rho U_\infty{}^2/l$ and $\mu U_\infty/\delta^2$, respectively, whence

$$\delta = C \sqrt{\frac{\nu l}{U_\infty}} \tag{82.2}$$

where U_∞ denotes the velocity in the free stream, and l is a characteristic length measured along the stream. For a semi-infinite flat plate at zero incidence, the factor $C = 5.0$, so that, for a *laminar boundary layer* on a flat plate, we obtain

$$\frac{\delta}{l} \approx 5.0 \sqrt{\frac{\nu}{U_\infty l}} = \frac{5.0}{\sqrt{\mathrm{Re}_l}} \tag{82.2a}$$

Here Re_l denotes the length Reynolds number. In the limit of $\mathrm{Re}_l \to \infty$ (for example, when $\nu \to 0$), we have $\delta/l \to 0$. By an analogous estimate, we find that the dimensionless shearing stress is

$$\frac{\tau}{\rho U_\infty{}^2} = \frac{C_1}{\sqrt{\mathrm{Re}_l}} \tag{82.3}$$

The total drag is given by

$$D = C_3 b \sqrt{\rho \mu U_\infty{}^3 l} \qquad b = \text{width of plate}$$

and the drag coefficient is

$$c_D = \frac{C_3}{\sqrt{\mathrm{Re}_l}} \qquad C_3 = 1.328 \text{ for flat plate} \tag{82.4}$$

The boundary-layer thickness δ can be defined as that distance from the wall where the velocity reaches a value $u = 0.99U_\infty$. It is somewhat arbitrary, because, in principle, transition to U_∞ is asymptotic. A more meaningful physical quantity is the average distance by which the external streamlines are displaced, owing to the presence of the boundary layer. It is called the *displacement thickness* δ^* and is defined by

$$\delta^* = \frac{1}{U_\infty} \int_{y=0}^{\infty} (U_\infty - u)\, dy \tag{82.5}$$

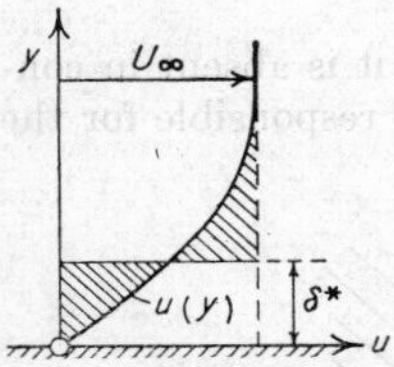

FIG. 82.3. Displacement thickness δ^* in a boundary layer. (*From* [2], *p.* 25.)

It should be noted (Fig. 82.3) that $\rho(U_\infty - u)$ is the mass flow forced out of the boundary layer by the retardation caused by viscosity at distance y. The total mass displaced is equated to that which can be passed by the displaced streamlines far away from the body, and which is equal to $U_\infty \rho \delta^*$. For a flat plate, the displacement thickness δ^* is proportional to the boundary-layer thickness δ, and

$$\delta^* = 0.289\delta$$

Very often use is made of the concept of *momentum thickness* ϑ, defined by

$$\vartheta = \int_{y=0}^{\infty} \frac{u}{U_\infty}\left(1 - \frac{u}{U_\infty}\right) dy \tag{82.5a}$$

which represents that distance in the free stream through which the flux of momentum $\rho U_\infty{}^2 \vartheta$ is equal to the *loss* of momentum in the boundary layer $\rho \int_0^\infty u(U_\infty - u)\, dy$. Here ρu is the mass flow, and $(U_\infty - u)$ is the deficiency in the velocity at a given distance y.

The *energy dissipation thickness* δ^{**} is defined by

$$\delta^{**} = \int_{y=0}^{\infty} \frac{u}{U_\infty}\left[1 - \left(\frac{u}{U_\infty}\right)^2\right] dy \tag{82.5b}$$

It is that distance in the free stream through which the flow of energy $(\frac{1}{2}\rho U_\infty{}^2)U_\infty$ is equal to the flux of energy $\frac{1}{2}\rho \int_0^\infty u(U_\infty{}^2 - u^2)\, dy$ dissipated in the boundary layer.

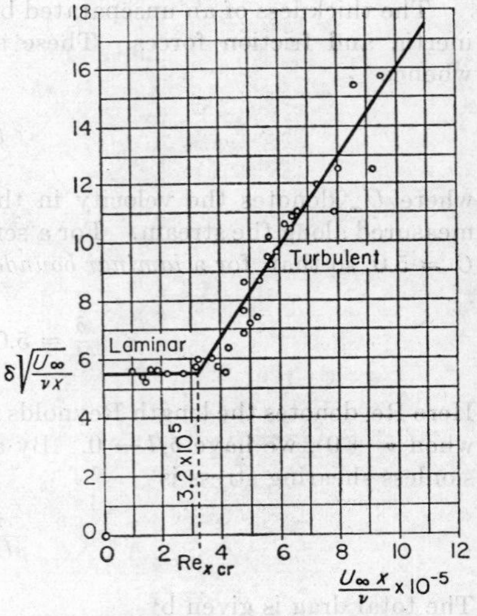

FIG. 82.4. Experimental evidence for the existence of transition and of turbulent boundary layers. Plot of $\delta/\sqrt{\nu x/U_\infty}$ against length Reynolds number $U_\infty x/\nu$. (*From* [2], *p.* 33.)

Turbulent Flow in a Pipe and in a Boundary Layer

Experiments, first performed systematically by O. Reynolds, show that the steady flow in a channel may be of one of two types. At low Reynolds numbers, laminae of fluid slide over one another, and there are no radial velocity components if the pipe is circular and long. The velocity profile is then parabolic, and the pressure drop is proportional to the first power of the mean flow velocity. This region of flow is called *laminar*. At higher Reynolds numbers, the laminar flow pattern becomes unstable, and ceases to be truly steady. On the axial motion, there are now superimposed *random* fluctuations, which exert a mixing effect, and momentum is exchanged across the flow. These fluctuations have a zero time average when the flow is *quasi-steady.* Owing to the exchange of momentum, the average velocity profile becomes nearly uniform at the axis; energy is dissipated, and the losses increase, the pressure drop now being nearly proportional to the square of the average velocity. Such a pattern of flow is termed *turbulent*.

Transition from laminar to turbulent flow occurs (approximately) at a *critical Reynolds number*. The value of the critical Reynolds number strongly depends on the pattern of random disturbances suffered by the fluid before it entered the channel. For a sharp-edged entrance, it is given by

$$\mathrm{Re}_{cr} = \left(\frac{\bar{u}d}{\nu}\right)_{cr} \approx 2300 \tag{82.6}$$

In most engineering appliances, the Reynolds number is sufficiently high for turbulent flow to occur as a rule.

The motion in a boundary layer, like that in a pipe, can also become unstable, and *transition* to turbulent flow may occur in it. It is clearly discernible by a sudden and large increase in the boundary-layer thickness. According to Eq. (82.2), the dimensionless ratio $\delta/\sqrt{\nu x/U_\infty}$ (x = current lengths) is constant for a laminar boundary layer on a flat plate at zero incidence. A plot of this quantity against the length Reynolds number $U_\infty x/\nu$ is shown in Fig. 82.4. The very sharp increase of this function at $\mathrm{Re}_x = 3.2 \times 10^5$, as well as a very sharp increase in the wall shearing stress (not shown), provides an experimental proof for the existence of transition and of turbulent boundary layers. The critical Reynolds number formed with the boundary-layer thickness, $(U_\infty\delta/\nu)_{cr}$, has a value of 2800, which is nearly that for a circular pipe.

Other things being equal, a turbulent boundary layer separates later than a laminar one. This explains the sudden decrease in the drag coefficient of spheres (Fig. 81.6) and cylinders. The Reynolds number at which this occurs is also called *critical;* its value depends on the *intensity of turbulence* [see Eq. (82.89)] of the external stream. When the Reynolds number becomes large enough to cause transition, the point of separation moves downstream, and the wake becomes smaller.

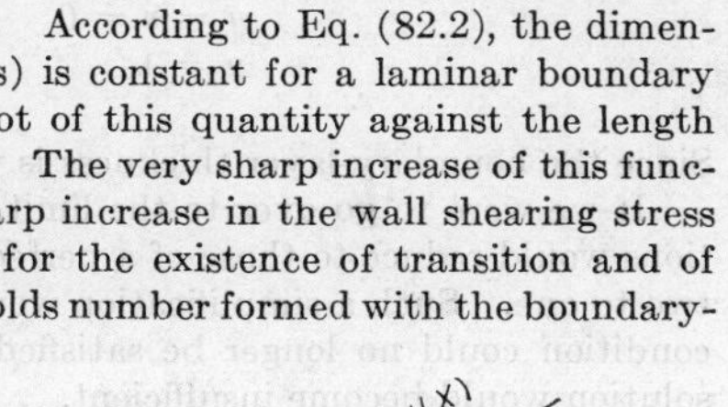

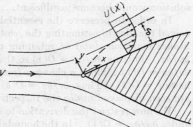

FIG. 82.5. Boundary-layer flow along a wall. (*From* [2], *p.* 94.)

The larger thickness of a turbulent boundary layer on a flat plate at zero incidence is given by Eq. (82.95), instead of the value in Eq. (82.2*a*).

With few exceptions, separation is an undesirable phenomenon, because it is accompanied by large pressure and energy losses. Separation can be prevented artificially, the most practical method being *boundary-layer suction*.

82.3. THE MATHEMATICAL CONCEPT OF A BOUNDARY LAYER

The Boundary-layer Approximation for Two-dimensional Flow along a Flat Plate or along a Curved Wall

We shall now derive a simplified form of the Navier-Stokes equations (68.19) in two dimensions, which will be applicable under the assumption that the Reynolds number is very large.

We begin by rewriting the Navier-Stokes equations in dimensionless form, in order to show explicitly how they depend on the Reynolds number. The notation and system of coordinates are explained in Fig. 82.5. When all velocities are divided by the free-stream velocity V, lengths by a characteristic dimension L, pressures by ρV^2, and time by L/V, we obtain the following equations:

$$\text{Direction } x \qquad \underset{1}{\frac{\partial u}{\partial t}} + \underset{1}{u}\,\underset{1}{\frac{\partial u}{\partial x}} + \underset{\delta}{v}\,\underset{\delta^{-1}}{\frac{\partial u}{\partial y}} = -\underset{(1)}{\frac{\partial p}{\partial x}} + \underset{\delta^2}{\frac{1}{\text{Re}}}\left(\underset{1}{\frac{\partial^2 u}{\partial x^2}} + \underset{\delta^{-2}}{\frac{\partial^2 u}{\partial y^2}}\right) \tag{82.7}$$

$$\text{Direction } y \qquad \underset{\delta}{\frac{\partial v}{\partial t}} + \underset{1}{u}\,\underset{\delta}{\frac{\partial v}{\partial x}} + \underset{\delta}{v}\,\underset{1}{\frac{\partial v}{\partial y}} = -\underset{(\delta)}{\frac{\partial p}{\partial y}} + \underset{\delta^2}{\frac{1}{\text{Re}}}\left(\underset{\delta}{\frac{\partial^2 v}{\partial x^2}} + \underset{\delta^{-1}}{\frac{\partial^2 v}{\partial y^2}}\right) \tag{82.8}$$

$$\text{Continuity} \qquad \underset{1}{\frac{\partial u}{\partial x}} + \underset{1}{\frac{\partial v}{\partial y}} = 0 \tag{82.9}$$

The dimensionless quantities u, v, p, etc., are here denoted by the same symbols as before, a change of notation being superfluous. Furthermore $\text{Re} = VL/\nu$, and the boundary conditions are

$$\begin{aligned} u = v &= 0 \qquad \text{at } y = 0 \qquad \text{no slip} \\ u &= 1 \qquad \text{at } y \to \infty \end{aligned} \tag{82.10}$$

Since the boundary-layer thickness is very small, we now have $\delta \ll 1$.

If we were to go over to the limit $\text{Re} \to \infty$ ($\nu = 0$) directly, the preceding equations would reduce to those of a perfect fluid, and their order would be reduced from two to one. Such a simplification would alter their physical character, as the no-slip condition could no longer be satisfied because the number of free constants in the solution would become insufficient.

In order to preserve the essential physical features of the problem, Prandtl conceived the idea of estimating the *relative order of magnitude* of the terms in the Navier-Stokes equations, and of retaining only those whose order is the same as that of $\partial u/\partial x$. Putting $\partial u/\partial x = O(1)$ so that $\partial^2 u/\partial x^2 = O(1)$, we see, from Eq. (82.9), that $\partial v/\partial y = O(1)$. Since v changes from 0 to its highest value across a distance δ, we have $v = O(\delta)$. Thus, $\partial v/\partial x = O(\delta)$ and $\partial^2 v/\partial x^2 = O(\delta)$. The orders of magnitude are seen displayed under the respective terms in the above equations.

We now restrict the derivation to cases of *moderate acceleration* when $\partial u/\partial t = O(1)$ and $u\,\partial u/\partial x = O(1)$. In the boundary layer at the wall, the dominant viscous forces, i.e., those proportional to $\partial^2 u/\partial y^2$ and $\partial^2 v/\partial y^2$, must be of the same order as the inertia forces, which are proportional to $u(\partial u/\partial x) = O(1)$. Hence, we estimate first $\partial u/\partial y = O(\delta^{-1})$ since u increases from 0 to 1 (in dimensionless terms) across a distance δ, and then $\partial^2 u/\partial y^2 = O(\delta^{-2})$. Consequently, $\text{Re} = O(\delta^{-2})$. This completes the estimate except for the pressure gradient. From Eq. (82.8), we see that $\partial p/\partial y = O(\delta)$ at most, as this is the order of all terms except $(1/\text{Re})(\partial^2 v/\partial x^2) = O(\delta^3)$. The order of magnitude of the pressure gradient $\partial p/\partial x$ can be estimated from the external flow where, in accordance with Euler's equation (in dimensional form), we have

$$\frac{\partial U}{\partial t} + U\frac{\partial U}{\partial x} = -\frac{1}{\rho}\frac{\partial p}{\partial x} \tag{82.11}$$

and where $U = U(x,t)$ [see Eq. (70.1*a*)]. Hence, the dimensionless pressure gradient $\partial p/\partial x = O(1)$. This argument confirms mathematically that the external pressure $p(x,t)$ is *impressed* on the boundary layer. The pressure *across* the boundary layer is constant, and varies *along* it in the same way as in the external flow, so that $p = p(x,t)$ only.

Omitting all terms of order lower than 1, we see that the two equations (82.7) and (82.8) reduce to one, and that Eq. (82.9) remains unaltered. Reverting to dimensional notation, we obtain

$$\frac{\partial u}{\partial t} + u\frac{\partial u}{\partial x} + v\frac{\partial u}{\partial y} = -\frac{1}{\rho}\frac{\partial p}{\partial x} + \nu\frac{\partial^2 u}{\partial y^2} \tag{82.12}$$

$$\frac{\partial U}{\partial t} + U\frac{\partial U}{\partial x} = -\frac{1}{\rho}\frac{\partial p}{\partial x} \tag{82.13}$$

$$\frac{\partial u}{\partial x} + \frac{\partial v}{\partial y} = 0 \tag{82.14}$$

$U(x,t)$ being given. This simplification preserves the order of the equations, and the full boundary conditions, which can all be satisfied, are

$$\begin{aligned} u = v = 0 \quad & \text{at } y = 0 \quad \text{no slip} \\ u = U \quad & \text{at } y \to \infty \end{aligned} \tag{82.15}$$

In order to determine the problem, it is, finally, necessary to prescribe a velocity profile at some cross section, say $u(x_0,y)$ at $x = x_0$.

It should be noted, however, that Prandtl's simplification of the Navier-Stokes equations has changed their type from elliptic to parabolic. The omission of Eq. (82.8) is physically equivalent to assuming that a fluid particle has zero mass and viscosity, as far as its motion in the transverse direction is concerned. This and the other assumptions are satisfied with an increasing degree of accuracy as the Reynolds number increases, and the process of solving the boundary-layer equations amounts to a process of *asymptotic integration*.

In agreement with Eq. (82.2*a*), the analysis confirms that $\delta/L = O(\text{Re}^{-1/2})$. For steady flow, the terms $\partial u/\partial t$ and $\partial U/\partial t$ vanish identically.

The preceding derivation makes it clear that the resulting system of equations is valid under *two additional* restrictive conditions, implied but not yet stated explicitly. First, the equations describe only *laminar motion* in the boundary layer. Second, owing to the boundary condition at $y = 0$, it is seen that when the *external flow* is *turbulent*, the free-stream velocity U is a function of time, even in cases when the flow is quasisteady. Strictly speaking, even when we are to solve a problem involving *steady laminar* flow in the boundary layer, but in the presence of *turbulent* flow in the free stream, we are not justified in deleting $\partial u/\partial t$ from Eq. (82.12), because the boundary condition $u = U(x,t)$ at $y \to \infty$ is time-dependent, the presence of random, time-dependent fluctuations being an essential characteristic of turbulent motion. In practice, it is always assumed that $u = \bar{U}(x)$ where

$$\bar{U}(x) = \frac{1}{t}\int_0^t U(x,t)\,dt$$

is the time average of the fluctuating velocity $U(x,t)$ for t large enough. To date, the implications of this assumption have not been examined in detail, either mathematically or experimentally.

The preceding result can be easily extended to curved walls, introducing a curvilinear system of coordinates, in which x is measured along the curved wall and y at right angles to it, and R denotes the local radius of curvature. In cases when $\delta \ll R$ and when the curvature is gentle, so that $dR/dx = O(1)$, Eqs. (82.12), (82.14), and (82.15) can be retained unchanged. The presence of a centrifugal force changes the transverse pressure gradient to $(1/\rho)(\partial p/\partial y) = u^2/R$, so that $\partial p/\partial y = O(1)$, instead of $O(\delta)$. The pressure *difference* across the boundary layer becomes of the order δ, as against $O(\delta^2)$ previously. Since it is small, it is customary to neglect $\partial p/\partial y$ at the cost of committing an additional error.

Separation

At the point of separation, the boundary layer satisfies the condition in Eq. (82.1). In the separated region, the boundary-layer simplifications (p. 82–7) do not apply, so that the equations can be used only as far as the point of separation. Even this is limited to slender bodies, because blunt bodies have large wakes which, in turn, displace the external streamlines and modify the external flow [Eq. (82.13)]. In such cases, the pressure distribution $p(x,t)$ must be known from experiment.

In steady flow, near the wall, Eq. (82.12) can be reduced to

$$\mu \left(\frac{\partial^2 u}{\partial y^2}\right)_0 = \frac{dp}{dx} \qquad \text{at wall} \tag{82.16}$$

which shows that the curvature of the velocity profile at the wall is of the same sign as the pressure gradient, and changes sign with dp/dx. At $y \to \infty$, we always have $(\partial^2 u/\partial y^2)_\infty = -0$ (Fig. 82.6). Hence for accelerated flow ($dp/dx < 0$), the curvature

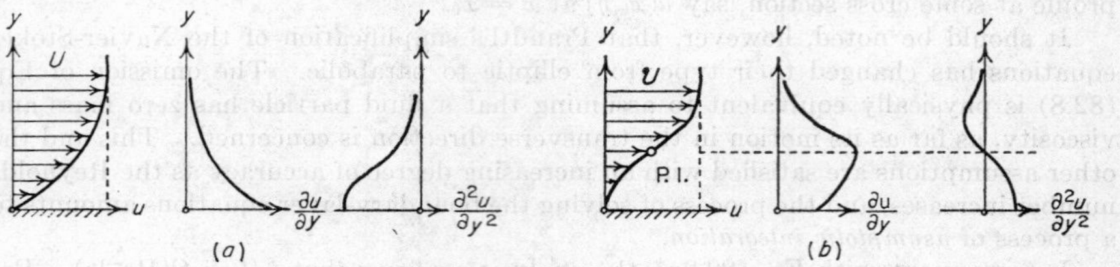

FIG. 82.6. Velocity profile, its gradient and curvature in a boundary layer: (*a*) Pressure decreasing downstream ($dp/dx < 0$); (*b*) pressure increasing downstream ($dp/dx > 0$). P.I. = point of inflection. (*From* [2], *p.* 100.)

is always negative. For decelerated flow, $dp/dx > 0$ implies that $(\partial^2 u/\partial y^2)_0 > 0$, so that there must exist a point of inflection, $\partial^2 u/\partial y^2 = 0$, at some value of y. Such profiles are inherently less stable than profiles displaying no point of inflection (Sec. 82.7). Condition (82.16) is equivalent to

$$\left(\frac{\partial^3 u}{\partial y^3}\right)_0 = 0 \qquad \text{at wall} \tag{82.17}$$

Stream Function

Since the equation of continuity (82.14) is the same as for any two-dimensional incompressible flow [Eq. (70.3)], it can be satisfied by the introduction of a stream function $\psi(x,y,t)$ defined by

$$u = \frac{\partial \psi}{\partial y} \qquad v = -\frac{\partial \psi}{\partial x} \tag{82.18}$$

so that Eq. (82.12) becomes

$$\frac{\partial^2 \psi}{\partial y\,\partial t} + \frac{\partial \psi}{\partial y}\cdot\frac{\partial^2 \psi}{\partial x\,\partial y} - \frac{\partial \psi}{\partial x}\frac{\partial^2 \psi}{\partial y^2} = -\frac{1}{\rho}\frac{\partial p}{\partial x} + \nu\frac{\partial^3 \psi}{\partial y^3} \tag{82.19}$$

and the boundary conditions are

$$\begin{aligned} \frac{\partial \psi}{\partial x} = \frac{\partial \psi}{\partial y} &= 0 \qquad \text{at } y = 0 \qquad \text{no slip} \\ \frac{\partial \psi}{\partial y} &= U \qquad \text{at } y \to \infty \end{aligned} \tag{82.20}$$

Compared with the full Navier-Stokes equations written in terms of a stream function, Eq. (82.19) for ψ is of one order lower.

Skin Friction

The shear stress at the wall is

$$\tau_0 = \mu \left(\frac{\partial u}{\partial y}\right)_0$$

and the skin friction force (viscous drag) (Fig. 82.7) is given by

$$D_f = b \int_0^l \tau_0 \cos \phi \, ds = b \int_0^l \mu \left(\frac{\partial u}{\partial y}\right)_0 dx \tag{82.21}$$

where b denotes the width of the body. Here ds is measured along the curved profile, and $dx = ds \cos \phi$ is measured along the x axis in Fig. 82.7. Equation (82.21) is valid only for unseparated laminar boundary layers.

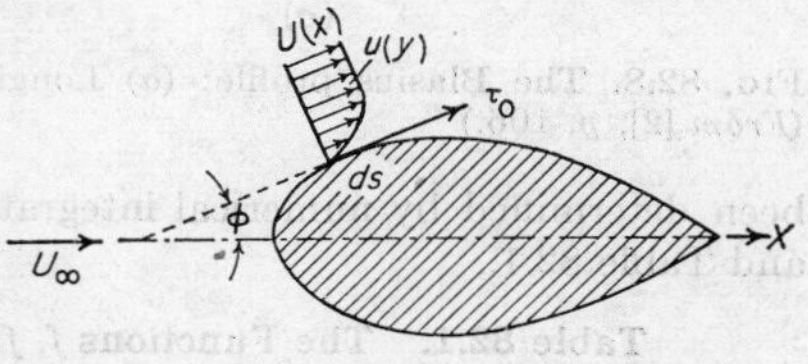

FIG. 82.7. Skin friction. (*From* [2], p. 101.)

The Flat Plate: The Blasius Profile

The case of a flat plate at zero incidence in a parallel stream is the simplest example of a boundary layer. It was first analyzed by H. Blasius, and the resulting profile is known as a *Blasius profile.* Remembering that now $dp/dx = 0$, we are led to the following two equations,

$$u\frac{\partial u}{\partial x} + v\frac{\partial u}{\partial y} = \nu\frac{\partial^2 u}{\partial y^2} \qquad \frac{\partial u}{\partial x} + \frac{\partial v}{\partial y} = 0 \tag{82.22}$$

with the boundary conditions

$$\begin{aligned} u = v = 0 \quad \text{at } y = 0 \qquad & u = U_\infty \quad \text{at } y = \infty \\ u = U_\infty \quad v = 0 \qquad & \text{at } x = 0 \end{aligned} \tag{82.22a}$$

There being no preferred lengths in the system, we suppose that all velocity profiles are similar, or that

$$\frac{u}{U_\infty} = \phi\left(\frac{y}{\delta}\right) \qquad \text{at all } x$$

Since $\delta \propto \sqrt{\nu x/U_\infty}$, we try to solve Eqs. (82.22) by the substitution

$$\eta = y\sqrt{\frac{U_\infty}{\nu x}} \tag{82.23}$$

and by the assumption of the stream function

$$\psi = \sqrt{\nu x U_\infty}\, f(\eta) \tag{82.24}$$

so that

$$u = U_\infty f'(\eta) \qquad v = \frac{1}{2}\sqrt{\frac{\nu U_\infty}{x}}\,(\eta f' - f) \tag{82.25}$$

With these substitutions, the system of equations (82.22) is reduced to one nonlinear ordinary differential equation of the third order, namely,

$$\begin{gathered} 2f''' + ff'' = 0 \\ f = f' = 0 \quad \text{at } \eta = 0 \qquad f' = 1 \quad \text{at } y = \infty \end{gathered} \tag{82.26}$$

The equation has a singularity at the leading edge, $x = 0$ (see [3]); it cannot be solved in closed form, but a solution in the form of a power expansion in η has been obtained.

We shall omit the details of the calculation, and assume that the function $f(\eta)$ has

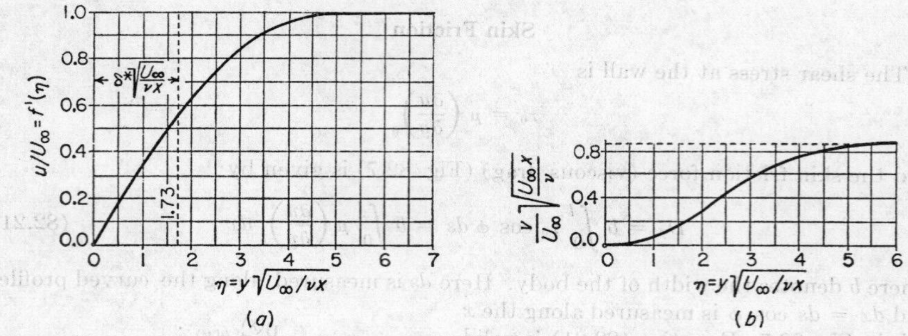

FIG. 82.8. The Blasius profile: (*a*) Longitudinal component; (*b*) transverse component. (*From* [2], *p.* 106.)

been determined by numerical integration, the result being summarized in Fig. 82.8 and Table 82.1.

Table 82.1. The Functions f, f', and f'' for a Blasius Profile from [4]

$\eta = y\sqrt{\dfrac{U_\infty}{\nu x}}$	f	$f' = \dfrac{u}{U_\infty}$	f''
0	0	0	0
0.2	0.00664	0.06641	0.33199
0.4	0.02656	0.13277	0.33147
0.6	0.05974	0.19894	0.33008
0.8	0.10611	0.26471	0.32739
1.0	0.16557	0.32979	0.32301
2.0	0.65003	0.62977	0.26675
3.0	1.39682	0.84505	0.16136
4.0	2.30576	0.95552	0.06424
5.0	3.28329	0.99155	0.01591
6.0	4.27964	0.99898	0.00240
7 0	5.27926	0.99992	0.00022
8.0	6.27923	1.00000	0.00001
8.8	7.07923	1.00000	0.00000

It is worth noting that the Blasius profile possesses a point of inflection at $y = 0$ and a very small curvature near $y = 0$. At infinity, $v_\infty \neq 0$, and is

$$v_\infty = 0.865 U_\infty \sqrt{\frac{\nu}{x U_\infty}} \tag{82.27}$$

The shearing stress at the wall is

$$\tau_0 = \frac{0.332}{\sqrt{\mathrm{Re}_x}}\,\rho U_\infty{}^2 = 0.332\sqrt{\frac{\nu}{U_\infty x}}\,\rho U_\infty{}^2 \tag{82.28}$$

and the skin friction for a plate wetted on *both* sides,

$$\begin{aligned} 2D &= 1.328 b \sqrt{U_\infty{}^3 \mu \rho l} \\ &= 1.328 b \rho U_\infty{}^2 \sqrt{\frac{\nu l}{U_\infty}} \end{aligned} \tag{82.29}$$

so that

$$c_f = \frac{2D}{\frac{1}{2}\rho A U_\infty{}^2} = \frac{1.328}{\sqrt{\mathrm{Re}_x}} \qquad \mathrm{Re}_x = \sqrt{\frac{U_\infty x}{\nu}} \tag{82.30}$$

where $A = 2bx$ is the *total* wetted area. The 99% boundary-layer thickness is

$$\delta = 5.0 \sqrt{\frac{\nu x}{U_\infty}} \tag{82.31}$$

and the displacement thickness is

$$\delta^* = \sqrt{\frac{\nu x}{U_\infty}} [\eta_1 - f(\eta_1)] = 1.72 \sqrt{\frac{\nu x}{U_\infty}} \tag{82.32}$$

Here $\eta_1 - f(\eta_1)$ taken at $\eta = \infty$ is equal to 1.72.

The momentum thickness is

$$\vartheta = \sqrt{\frac{\nu x}{U_\infty}} \int_0^\infty f'(1 - f')\, d\eta = 0.664 \sqrt{\frac{\nu x}{U_\infty}} \tag{82.33}$$

Careful experimental investigations have fully confirmed Blasius' analysis of the problem.

82.4. TRANSFORMATION OF THE BOUNDARY-LAYER EQUATIONS; STEADY, TWO-DIMENSIONAL FLOWS

Elimination of Reynolds Number

Since the Reynolds number constitutes the only parameter in the boundary-layer equations, it can be eliminated by the use of a suitable transformation. We introduce, first, the dimensionless quantities,

$$u' = u/U_\infty \qquad v' = v/U_\infty \qquad x' = x/L \qquad y' = y/L \qquad \mathrm{Re} = U_\infty L/\nu \tag{82.34}$$

and apply the transformation,

$$v'' = v' \sqrt{\mathrm{Re}} \qquad y'' = y' \sqrt{\mathrm{Re}} \tag{82.35}$$

The boundary-layer equations will then assume the form

$$u' \frac{\partial u'}{\partial x'} + v'' \frac{\partial u'}{\partial y''} = U' \frac{dU'}{dx'} + \frac{\partial^2 u'}{\partial y''^2} \qquad \frac{\partial u'}{\partial x'} + \frac{\partial v''}{\partial y''} = 0 \tag{82.36}$$
$$u' = v'' = 0 \quad \text{at } y'' = 0 \qquad u' = U' \quad \text{at } y'' = \infty$$

A solution of the system of equations (82.36) is valid for any Reynolds number, and for a whole class of geometrically similar problems. A change in the Reynolds number is equivalent to a transformation of the transverse coordinate and the transverse velocity component. Hence, the position of the point of separation is seen to be independent of the Reynolds number, as the condition for it [Eq. (82.17)] remains unaffected by the transformation; only the angle of the streamline through the point of separation changes in the ratio $1/\mathrm{Re}^{1/2}$. The existence of a point of separation is unaffected by a passage to the limit $\mathrm{Re} \to \infty$. This feature is due to the fact that here, in contrast with the perfect-fluid theory, the passage to the limit has been effected in the solution and not in the differential equation itself.

Affine Transformation: Wedge Flow

If the longitudinal velocity profiles $u(x,y)$ in a boundary layer can be made congruent for all values of x by applying suitable scale factors to u and y, they can be reduced to one plot in a system of coordinates which have been made dimensionless with the aid of these scale factors. This amounts to a continuous, *affine transformation* of the boundary-layer profiles. The natural scale factor for u is the potential velocity $U(x)$, and the scale factor for y must be a function of x, say $g(x)$. Following Goldstein and Mangler, we can give a general answer as to the class of potential flows for which an affine transformation can be found. We introduce a stream function (p. 82–8), and make use of Eqs. (82.12) and (82.13). Hence, we obtain

$$\frac{\partial\psi}{\partial y}\frac{\partial^2\psi}{\partial x\,\partial y} - \frac{\partial\psi}{\partial x}\frac{\partial^2\psi}{\partial y^2} = U\frac{dU}{dx} + \nu\frac{\partial^3\psi}{\partial y^3}$$
$$\partial\psi/\partial x = \partial\psi/\partial y = 0 \quad \text{at } y = 0 \qquad \partial\psi/\partial x = U \quad \text{at } y = \infty \tag{82.37}$$

Introducing dimensionless quantities, applying the scale factor $g(x)$ to y, and making the stream function dimensionless, we substitute

$$\xi = \frac{x}{L} \qquad \eta = \frac{y\sqrt{\text{Re}}}{L\cdot g(x)} \qquad f(\xi,\eta) = \frac{\psi(x,y)\sqrt{\text{Re}}}{L\cdot U(x)\cdot g(x)} \tag{82.38}$$

into Eq. (82.37). An affine transformation will exist if the dependence of f on ξ can be canceled, making it a function of η only. Thus, the *partial differential equation* for the stream function ψ will be replaced by an *ordinary differential equation* for the function $f(\eta)$. In this manner, we obtain

$$f''' + \alpha ff'' + \beta(1 - f'^2) = \frac{U}{U_\infty}g^2\left(f'\frac{\partial f'}{\partial\xi} - f''\frac{\partial f}{\partial\xi}\right)$$
$$\alpha = \frac{Lg}{U_\infty}\frac{d}{dx}(Ug) \qquad \beta = \frac{L}{U_\infty}g^2U' \tag{82.39}$$

(Here $U' = dU/dx$, $f' = df/d\eta$, $g' = dg/dx$, etc.) An affine transformation will exist if the right-hand side of Eq. (82.39) vanishes and if the coefficients α and β on the left-hand side are independent of x. This furnishes two differential equations for $U(x)$ and $g(x)$.

The original equation is then transformed to

$$f''' + \alpha ff'' + \beta(1 - f'^2) = 0$$
$$f = f' = 0 \quad \text{at } \eta = 0 \qquad f' = 1 \quad \text{at } \eta = \infty \tag{82.40}$$

and the velocity components of the flow are given by the equations

$$u = Uf' \qquad v = -\frac{1}{\sqrt{\text{Re}}}L\left[f\frac{d}{dx}(Ug) - Ug'\eta f'\right] \tag{82.41}$$

Solving the equations for $U(x)$ and $g(x)$, we are led to two cases.

Case 1. $2\alpha - \beta \neq 0$. As long as $\alpha \neq 0$, we may put $\alpha = 1$ and include the ratio α/β in $g(x)$. The results can be summarized in the following equations:

$$\frac{U}{U_\infty} = K^{m+1}\left(\frac{2}{m+1}\frac{x}{L}\right)^m \qquad m = \frac{\beta}{2-\beta} \tag{82.42}$$

Here U is of the form

$$U = Cx^m \qquad C = \text{const}$$

and represents, among others, the flow about a wedge whose included angle is equal to $\beta\pi$ $(0 < \beta < 2,\ m > 0)$; see p. 70–10. Furthermore,

$$g = \sqrt{\frac{2}{m+1}\frac{x}{L}\frac{U_\infty}{U}} \tag{82.42a}$$

and the transformed ordinate is

$$\eta = y\sqrt{\frac{m+1}{2}\frac{U}{\nu x}} \tag{82.42b}$$

The following particular cases may be mentioned: $\alpha = 1, \beta = 1$; hence, $m = 1$; here $U(x) \propto x$ and corresponds to plane stagnation, or *Hiemenz* flow (p. 81–5). $\alpha = 1$, $\beta = 0$; hence, $m = 0$; here $U = U_\infty$ and corresponds to a flat plate at zero incidence (p. 82–9). [Here $\eta = y\sqrt{U_\infty/2\nu x}$ and differs by a factor $\sqrt{2}$ from η in Eq. (82.23).]

For further solutions and the case where $\alpha = 0$, see p. 82–14.

Case 2. $2\alpha = \beta$. In this case, we are led to solutions of the type $U(x) \propto e^{px}$ (p = const), which are of no importance in boundary-layer theory.

The Integral Momentum and Energy Equations

Even the simplified boundary-layer equations present great difficulties when *exact* solutions are sought. It is possible to devise *approximate* solutions by considering only the gross features of the motion and by satisfying the boundary conditions, together with some of the compatibility conditions given on p. 82–18. The differential equations (82.12), (82.13), and (82.14) describe the motion of an elementary particle. By integrating them over the boundary-layer thickness, it is possible to obtain equations which describe the motion of the bulk of the fluid in the boundary layer. Integrating the equations directly, we shall obtain a relation for the momentum of the fluid, i.e., the von Kármán *integral momentum equation.* Integrating these equations after first multiplying them by u, we obtain Wieghardt's *integral energy equation.* The integration is restricted to the case of steady flow only.

In deriving the integral momentum equation, it is convenient to replace $\mu(\partial u/\partial y)_0$ in the first integral of Eq. (82.19) by the shearing stress at the wall τ_0, because then the equation applies to laminar as well as to turbulent flow. The upper limit of integration h is chosen to be *outside* the boundary layer. Hence,

$$\int_0^h \left(u\frac{\partial u}{\partial x} + v\frac{\partial u}{\partial y} - U\frac{dU}{dx}\right) dy = -\frac{\tau_0}{\rho} \qquad h > \delta$$

Replacing v by

$$v = -\int_0^h \frac{\partial u}{\partial x}\, dy$$

from Eq. (82.14) and integrating by parts, we obtain

$$\int_0^h \frac{\partial}{\partial x}\,[u(U-u)]\, dy + \frac{dU}{dx}\int_0^h (U-u)\, dy = \frac{\tau_0}{\rho} \tag{82.43}$$

Since the integrands vanish outside the boundary layer, we may put $h \to \infty$. Introducing the displacement thickness δ^*, and the momentum thickness ϑ, from Eqs. (82.5) and (82.5*a*), we have

$$\frac{\tau_0}{\rho} = \frac{d}{dx}\,(U^2\vartheta) + \delta^* U\frac{dU}{dx} \tag{82.44}$$

Similarly, we can obtain the integral energy equation,

$$\frac{d}{dx}\,(U^3\delta^{**}) = 2\nu\int_0^\infty \left(\frac{\partial u}{\partial y}\right)^2 dy \tag{82.45}$$

The right-hand term in the equation represents the energy dissipated in the boundary layer.

82.5. EXACT SOLUTIONS OF BOUNDARY-LAYER EQUATIONS

In this section, a solution to a boundary layer problem is considered exact if it satisfies the differential equations (82.12) to (82.14), even if it involves numerical or mathematical approximations. The approximate solutions in Sec. 82.6 will be obtained with the aid of the integral momentum equation (p. 82–13). Exact analytical solutions are difficult to obtain, and only a few of them are known.

Wedge Flow

The potential flow past a wedge of included angle $\pi\beta$ $(0 < \beta < 2,\ m > 0)$ (p. 70–10), is given by $U(x) = u_1 x^m$, where x is measured from the stagnation point and $m = \beta/(2 - \beta)$.† The property of affinity possessed by the velocity profiles in wedge flow was discussed on p. 82–11. The solution of the resulting ordinary differential equation (82.40) must be obtained numerically; for a given value of m (or β), the solution can be represented by a single curve of $u/U = f'(\eta)$ plotted against

$$\eta = \frac{m+1}{2}\sqrt{\frac{U}{\nu x}}\, y$$

as seen in Fig. 82.9. For accelerated flow $(m > 0,\ 0 < \beta < 2)$, there is no point of inflection in the profile. For decelerated flows ($m < 0$, $\beta < 0$—these are no longer wedge flows), a point of inflection appears. Separation occurs for $\beta = -0.199$, $m = -0.091$, that is, for a very small deceleration.

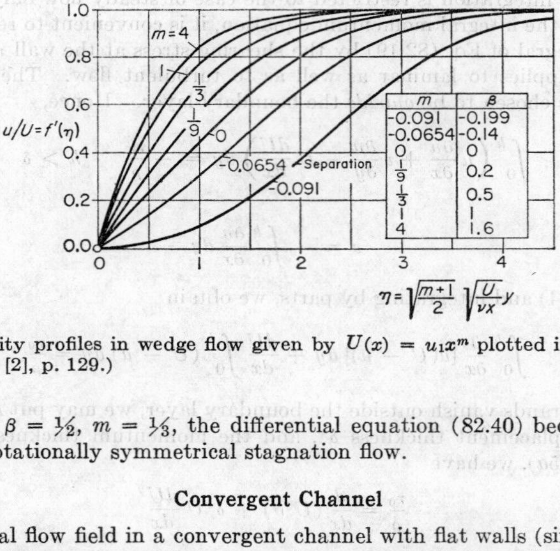

FIG. 82.9. Velocity profiles in wedge flow given by $U(x) = u_1 x^m$ plotted in reduced coordinates. (*From* [2], p. 129.)

In the case $\beta = \frac{1}{2}$, $m = \frac{1}{3}$, the differential equation (82.40) becomes identical with that for rotationally symmetrical stagnation flow.

Convergent Channel

The potential flow field in a convergent channel with flat walls (sink) is given by $U(x) = u_1/(-x)$ $(u_1 > 0)$, and belongs to the category of affine solutions (p. 82–12) with $\alpha = 0$, $\beta = +1$; Eq. (82.40) then becomes

$$f''' - f'^2 + 1 = 0 \tag{82.46}$$

$$f = f' = 0 \quad \text{at } \eta = 0 \qquad \text{and} \qquad f' = 1 \quad \text{at } \eta = \infty$$

with
$$\eta = y\sqrt{\frac{U}{-x\nu}} = \frac{y}{x}\sqrt{\frac{Q}{2\pi\nu}} \tag{82.46a}$$

where $Q = 2\pi u_1$ is the volume of flow for a full opening angle 2π (strength of sink). Exceptionally, this case can be solved in closed form. Introducing the polar angle

† Special cases of viscous wedge flow, namely, $\beta = 0$ (flat plate) and $\beta = 1$ (stagnation flow), have been treated on pp. 82–9 and 81–5, respectively.

$\theta = y/x$ and the radial distance r from the sink by putting $Q = 2\pi r U$, we find

$$\frac{u}{U} = f' = 3\,\mathrm{Tanh}^2\left(\frac{\eta}{\sqrt{2}} + 1.146\right) - 2$$
$$\eta = \theta\sqrt{\frac{Ur}{\nu}} = \frac{y}{x}\sqrt{\frac{Ur}{\nu}} \tag{82.47}$$

At $\eta = 3$, $u/U = 0.9825$, so that $\delta = 3x\sqrt{\nu/Ur}$.

Flow past a Symmetrical Cylinder; The Blasius Series

The general case of flow past a cylinder (symmetrical or not) is solved by series expansions. The potential velocity is assumed to have the form of a power series in x (measured along the contour from the stagnation point). The velocity profile in the boundary layer is also assumed in the form of a power series in x with y-dependent coefficients (Blasius series). By the use of a substitution discovered by L. Howarth [5], the y-dependent coefficients are given in the form of universal (tabulated) functions which are independent of the shape of the cylinder. The usefulness of such a series is severely restricted because, for a slender body, it is necessary to take a very large number of terms in the series. It is seldom possible to reach the point of separation because of the prohibitive number of terms required, but the solution in the neighborhood of the stagnation point can be obtained. It is then possible to continue the solution by numerical methods, e.g., the method given on p. 82–17.

We now restrict the description to the case of symmetrical cylinders with blunt noses and in symmetrical flow. The potential flow is then assumed to be given by

$$U(x) = u_1 x + u_3 x^3 + u_5 x^5 + u_7 x^7 + \cdots \tag{82.48}$$

where the coefficients $u_1, u_3, \ldots$ depend on the shape of the cylinder only. Hence,

$$-\frac{1}{\rho}\frac{dp}{dx} = U\frac{dU}{dx} = u_1^2 x + 4u_1u_3x^3 + (6u_1u_5 + 3u_3^2)x^5 + (8u_1u_7 + 8u_3u_5)x^7 + (10u_1u_9 + 10u_3u_7 + 5u_5^2)x^9 + \cdots \tag{82.48a}$$

The stream function ψ is also assumed in the form of a power series. If the series for $U(x)$ is terminated at x^k, the pressure term in Eq. (82.48a) will end with x^{2k-1}, and the series for ψ must contain as many terms. The independent variable is chosen as

$$\eta = y\sqrt{\frac{u_1}{\nu}} \tag{82.48b}$$

and the stream function is assumed to be of the form

$$\psi = \sqrt{\frac{\nu}{u_1}}\,[u_1xf_1(\eta) + 4u_3x^3f_3(\eta) + 6u_5x^5f_5(\eta) + 8u_7x^7f_7(\eta) + 10u_9x^9f_9(\eta) + \cdots] \tag{82.49}$$

If the functions $f_1(\eta)$, $f_3(\eta)$, etc., could be made independent of the coefficients u_1, u_3, etc., they would thereby become independent of the shape of the cylinder, and, hence, universal. This cannot be achieved directly, but the functions can be split up as follows:

$$f_5 = g_5 + \frac{u_3^2}{u_1u_5}h_5$$
$$f_7 = g_7 + \frac{u_3u_5}{u_1u_7}h_7 + \frac{u_3^3}{u_1^2u_7}k_7 \tag{82.50}$$
$$f_9 = g_9 + \frac{u_3u_7}{u_1u_9}h_9 + \frac{u_5^2}{u_1u_9}k_9 + \frac{u_3^2u_5}{u_1^2u_9}j_9 + \frac{u_3^4}{u_1^3u_9}q_9 \qquad \text{etc.}$$

The scheme leads to a set of ordinary differential equations,

$$\begin{aligned} f_1'^2 - f_1 f_1'' &= 1 + f_1''' \\ 4f_1' f_3' - 3f_1'' f_3 - f_1 f_3'' &= 1 + f_3''' \\ 6f_1' g_5' - 5f_1'' g_5 - f_1 g_5'' &= 1 + g_5''' \\ 6f_1' h_5' - 5f_1'' h_5 - f_1 h_5'' &= \tfrac{1}{2} + h_5''' - 8(f_3'^2 - f_3 f_3'') \qquad \text{etc.} \end{aligned} \tag{82.51}$$

with the boundary conditions

$$\begin{aligned} &f_1 = f_1' = f_3 = f_3' = g_5 = g_5' = h_5 = h_5' = 0 \qquad \text{at } \eta = 0 \\ &f_1' = 1 \qquad f_3' = \tfrac{1}{4} \qquad g_5' = \tfrac{1}{6} \qquad h_5' = 0 \qquad \text{at } \eta = \infty. \end{aligned} \tag{82.51a}$$

The first equation is the only nonlinear equation of the set; it is identical with the Hiemenz equation (81.23a). All the remaining linear equations have functional coefficients which are determined by the preceding equations. They have been tabulated for series up to and including the terms in x^9 (see [2], p. 134). With the aid of these functions, the velocity components can be expressed as

$$\begin{aligned} u &= u_1 x f_1' + 4u_3 x^3 f_3' + 6u_5 x^5 f_5' + 8u_7 x^7 f_7' + 10 u_9 x^9 f_9' + \cdots \\ v &= -\sqrt{\frac{\nu}{u_1}}\,(u_1 f_1 + 12 u_3 x^2 f_3 + 30 u_5 x^4 f_5 + 56 u_7 x^6 f_7 + 90 u_9 x^8 f_9 + \cdots) \end{aligned} \tag{82.52}$$

Carrying out this scheme for a circular cylinder, we can determine the velocity profile as a function of the polar angle ϕ measured from the stagnation point, and hence the point of separation from condition (82.1) or (82.1a). The solution is found to be $\phi_{\text{sep}} = 108.8°$, which is larger than that found by experiment.

There is no assurance that a particular Blasius series converges. In fact, M. Van Dyke has shown [6] that the expression for the skin-friction coefficient for a parabolic cylinder whose radius of curvature at the vertex is equal to a diverges for $s/a > 0.62$, where s is the arc length. Convergence can be improved if the argument is changed to $x/(x + \frac{1}{2}a)$, where x is measured along the axis of the parabola. It is conjectured that the Blasius series diverges at a distance of the order of the nose radius.

A remarkable generalization of the Blasius-series method was given by H. Gőertler [7].

Flow behind the Wake of a Flat Plate at Zero Incidence

The applicability of the boundary-layer equations is not restricted to regions near a solid wall. They can also be applied to regions within the fluid where the action of shear predominates. This occurs, for example, in the regions of shear created by two parallel streams of different velocities as in the wake of a body, or at the boundaries of a jet being discharged through an orifice.

Behind the trailing edge of a body, the two boundary-layer profiles coalesce into one, giving a velocity profile with a local depression (Fig. 82.10). The velocity profile in the wake is closely related to the drag on the body, which can be determined by the application of the momentum equation. For a flat plate of width b, the momentum equation written for the control surface AA_1B_1B (Fig. 82.10) gives

$$2D = 2b\rho \int_0^\infty u(U_\infty - u)\,dy = b\rho U_\infty^2 \theta \tag{82.53}$$

for both sides. Equation (82.53) can be used for a cylindrical body, on the condition that $u(y)$ is measured at a section where the static pressure has its undisturbed value, i.e., sufficiently far downstream. It represents the loss of momentum due to friction between a given section, say at x, and the leading edge of the flat plate. In the case of a cylinder, additional pressure terms must be included if the integral is taken in the wake at a short distance behind the body.

The calculation of the wake of a flat plate is performed in two steps: (1) through an expansion in the downstream direction starting from the trailing edge by a continuation of the Blasius profile, and (2) through an asymptotic integration for a large distance behind the plate (valid for a cylinder of arbitrary shape).

Thus the velocity profile in the wake can be shown to be given approximately by

$$\frac{u_1}{U_\infty} = \frac{0.664}{\sqrt{\pi}} \left(\frac{x}{l}\right)^{-\frac{1}{2}} \exp\left(-\frac{y^2 U_\infty}{4x\nu}\right) \tag{82.54}$$

In practice, Eq. (82.54) can be used from $x = 3$ onward. The velocity distribution in Eq. (82.54) is seen to be gaussian in character.

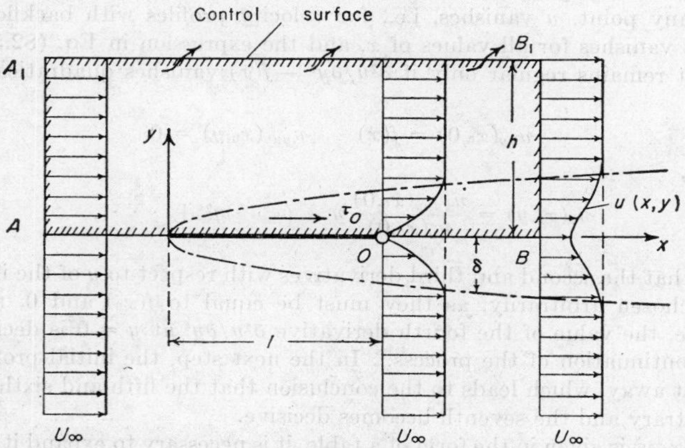

FIG. 82.10. Velocity profile in the wake of a flat plate and calculation of drag. (*From* [2], *p.* 138.)

In itself, Eq. (82.54) is of limited importance because, in most practical cases, the flow in a wake is turbulent, even though the boundary layer up to the trailing edge may have been laminar. This is due to the fact that the velocity profiles in Eq. (82.54) have points of inflection and are unstable (see p. 82–26). The importance of the present case lies in the discovery that, far behind a body, the velocity profile in a wake is given by a gaussian curve.

Continuation of Given Solution. Conditions of Compatibility

It is clear from the preceding examples that analytical methods become impracticable for large values of x because of the slow convergence of the resulting series expansions. It is, therefore, necessary to be in a position to continue a given solution $u(x_0,y)$ beyond the section x_0. This can be done numerically by a step-by-step procedure, which will be outlined for two-dimensional steady flows only. In addition to the profile at x_0, the pressure is prescribed, say, in the form

$$\frac{1}{\rho}\frac{dp}{dx} = \nu f(x) \tag{82.55}$$

From Eq. (82.19) we determine that, at $(x_0 + \Delta x)$, we have

$$u(x_0 + \Delta x, y) = u(x_0,y) + u_x(x_0,y)\,\Delta x \qquad \text{etc.} \tag{82.55a}$$

Making use of Eq. (82.14), we rewrite Eq. (82.12) in the forms

$$u\frac{\partial u}{\partial x} + v\frac{\partial u}{\partial y} = v\frac{\partial u}{\partial y} - u\frac{\partial v}{\partial y} = -u^2\frac{\partial}{\partial y}\left(\frac{v}{u}\right)$$

$$\frac{\partial}{\partial y}\left(\frac{v}{u}\right) = \nu\frac{1}{u^2}\left[f(x)\ \frac{\partial^2 u}{\partial y^2}\right]$$

$$v = \nu u\int_0^y \frac{1}{u^2}\left[f(x) - \frac{\partial^2 u}{\partial y^2}\right]dy$$

Hence,

$$\frac{\partial u}{\partial x} = -\frac{\partial v}{\partial y} = \nu\left\{\frac{\partial u}{\partial y}\int_0^y\left[\frac{\partial^2 u}{\partial y^2} - f(x)\right]dy + \left[\frac{1}{u}\frac{\partial^2 u}{\partial y^2} - f(x)\right]\right\} \tag{82.56}$$

This form allows us to calculate $\partial u/\partial x$ at (x_0,y), enabling us to proceed in the x direction by a distance Δx, provided that the derivative remains bounded. The procedure fails if, at any point, u vanishes, i.e., for velocity profiles with backflow. At the wall itself u vanishes for all values of x, and the expression in Eq. (82.56) becomes singular. It remains regular only if $\partial^2 u/\partial y^2 - f(x)$ vanishes quadratically, and we must have

$$u_{yy}(x_0,0) = f(x) \qquad u_{yyy}(x_0,y) = 0 \tag{82.57}$$

In this case,

$$u_x(x_0,y) = \frac{u_{yyyy}(x_0,0)}{u_y(x_0,0)}\,y + (\cdot\cdot\cdot)y^2 + \cdot\cdot\cdot,$$

and we see that the second and third derivatives with respect to y of the initial profile cannot be chosen arbitrarily, as they must be equal to $f(x_0)$ and 0, respectively. Furthermore, the value of the fourth derivative $\partial^4 u/\partial y^4$ at $y = 0$ is decisive for the successful continuation of the process. In the next step, the initial profile contains $\partial^4 u/\partial y^4$ right away, which leads to the conclusion that the fifth and sixth derivatives are not arbitrary and the seventh becomes decisive.

When $u(x_0,y)$ is given in the form of a table, it is necessary to expand it into a series in y, and to make sure that the preceding *compatibility conditions at the wall* [Eq. (82.57)], as well as the succeeding ones, are satisfied. Putting

$$u(x_0,y) = a_1 y + \frac{a_2}{2!}y^2 + \frac{a_3}{3!}y^3 + \cdot\cdot\cdot \tag{82.58}$$

we find that

$$\begin{aligned} &a_1 \text{ is free} \qquad a_2 = F(x) \qquad a_3 = 0 \qquad a_4 = a_1 a_1' \quad \text{hence free} \\ &a_5 = 2a_1F' \qquad a_6 = 2FF' \qquad a_7 = 4a_1{}^2 a_1'' - a_1 a_1'^2 \quad \text{hence free} \\ &a_8 = 10a_1{}^2F'' - 13a_1a_1'F' + 9(a_1a_1'' + a_1'^2)F \\ &a_9 = 40a_1FF'' - 16a_1F'^2 \quad \text{where } F(x) = \mu f(x) \end{aligned} \tag{82.59}$$

Consequently, only the coefficients a_1, a_4, a_7, a_{10}, . . . are free, the remaining ones being determined by the conditions of compatibility.

A remarkable simplification in the above numerical scheme was described by H. Göertler [8].

The Mangler Transformation. Flow past Bodies of Revolution

For the case when the stream is parallel to the axis of symmetry, it can be shown that compared with two-dimensional flow [Eqs. (82.12), (82.13), and (82.14)], it is only necessary to rewrite the equation of continuity to read

$$\frac{\partial(ur)}{\partial x} + \frac{\partial(vr)}{\partial y} = 0 \tag{82.60}$$

retaining the other two. For a body of revolution, x is measured along its meridian, and y in a direction normal to it, positive outward. When the contour of the body is given by its radius $r(x)$, the approximation is valid if d^2r/dx^2 does not assume very large values. The pressure gradient is of the same order as that on p. 82–7 for a curved cylindrical wall.

The appropriate form of the stream function $\psi(x,y)$ is now such that

$$u = \frac{1}{r}\frac{\partial(\psi r)}{\partial y} = \frac{\partial \psi}{\partial y} \qquad v = -\frac{1}{r}\frac{\partial(\psi r)}{\partial x} = -\frac{\partial \psi}{\partial x} - \frac{1}{r}\psi\frac{dr}{dx} \tag{82.61}$$

The boundary layer on a body of revolution can be determined by the same general method as that given on p. 82–15.

It is instructive to record the principal difference between the equations for the two-dimensional and axially symmetrical cases. In the former, the shape of the body enters the problem only through its influence on the potential velocity distribution as the term $U\,dU/dx$. In the latter, it appears also explicitly as $r(x)$ in Eq. (82.61). It is natural, therefore, to ask whether the axially symmetrical case can be reduced to the two-dimensional case by a suitable transformation. Such a transformation was indicated by W. Mangler [9] for *steady flow*. Making the substitutions

$$\begin{aligned} x^* &= \frac{1}{L^2}\int_0^x r^2(x)\,dx \qquad y^* = \frac{r(x)}{L}\,y \\ u^* &= u \qquad v^* = \frac{L}{r(x)}\left[v + \frac{r'(x)}{r(x)}\,yu\right] \qquad U^* = U \end{aligned} \tag{82.62}$$

where L denotes a constant length, we can verify that the starred quantities satisfy a system of equations which is identical with Eqs. (82.12) to (82.14). It is interesting to note that the Mangler transformation remains valid for compressible boundary layers, and for thermal boundary layers in laminar steady flow.

82.6. APPROXIMATE METHODS FOR THE SOLUTION OF THE BOUNDARY-LAYER EQUATIONS

The approximate method is based principally on the integral momentum equation (82.44), and an idea of the resulting accuracy can be obtained by applying it to a flat plate for which an exact solution (p. 82–9) is known.

Integral Momentum Equation Applied to Flat Plates

Applying Eq. (82.44) to a flat plate, we obtain

$$U_\infty^2 \frac{d\vartheta}{dx} = \frac{\tau_0}{\rho} \tag{82.63}$$

The essence of the method consists in *assuming* a suitable expression for the velocity profile, $u(y)$, and then solving for $\vartheta(x)$. Taking advantage of the similarity of velocity profiles on a flat plate, we put

$$\frac{u}{U_\infty} = f\left[\frac{y}{\delta(x)}\right] = f(\eta) \tag{82.64}$$

It is convenient now to assume that the free-stream velocity U_∞ is attained at a finite distance $\delta(x)$ from the wall, rather than asymptotically, at $y = \infty$. Hence, by Eq. (82.5*a*),

$$U_\infty^2\vartheta = U_\infty^2\,\delta(x)\int_0^1 f(1-f)\,d\eta$$

and

$$\vartheta = \alpha_1\delta \qquad \text{with} \qquad \alpha_1 = \int_0^1 f(1-f)\,d\eta \tag{82.65}$$

Similarly [Eq. (82.5)],

$$\delta^* = \alpha_2\delta \qquad \text{with} \qquad \alpha_2 = \int_0^1 (1-f)\,d\eta \tag{82.66}$$

Furthermore,

$$\frac{\tau_0}{\rho} = \nu \left(\frac{\partial u}{\partial y} \right)_{y=0} = \beta_1 \frac{\nu U_\infty}{\delta} \quad \text{with} \quad \beta_1 = f'(0) \tag{82.67}$$

The differential equation (82.63) now becomes

$$\delta \frac{d\delta}{dx} = \frac{\beta_1}{\alpha_1} \frac{\nu}{U_\infty}$$

which integrates to

$$\delta(x) = \sqrt{\frac{2\beta_1}{\alpha_1}} \sqrt{\frac{\nu x}{U_\infty}} \tag{82.68}$$

so that

$$\tau_0(x) = \sqrt{\frac{\alpha_1 \beta_1}{2}}\, \mu U_\infty \sqrt{\frac{U_\infty}{\nu x}} \tag{82.69}$$

and

$$2D = 2b \sqrt{2\alpha_1\beta_1} \sqrt{\mu \rho l U_\infty^3} \tag{82.70}$$

Finally,

$$\delta^* = \alpha_2 \sqrt{\frac{2\beta_1}{\alpha_1}} \sqrt{\frac{\nu x}{U_\infty}} \tag{82.71}$$

By comparison with p. 82–10, it is seen that all equations of the approximate theory correctly formulate the dependence of the relevant quantities on the free-stream velocity U_∞, as well as on the properties of the fluid, i.e., on μ, ν, and ρ.

In adopting an assumption for $f(\eta)$, it is necessary to satisfy the condition of no slip, $u = 0$ at $y = 0$, and the condition of continuity, $u = U_\infty$ at $y = \delta$. In addition, we may seek to satisfy further conditions, such as the conditions of the continuity of derivatives, for example, $\partial u/\partial y = 0$ at $y = \delta$, $\partial^2 u/\partial y^2 = 0$ at $y = \delta$, etc. For a flat plate, it is important to satisfy the condition $\partial^2 u/\partial y^2 = 0$ at $y = 0$, as seen from the exact solution [Eq. (82.26)].

Since the assumption for $f(\eta)$ is not unique, several possible forms have been compared with the exact solution in Table 82.2. The results are satisfactory and depend little on the form of $f(\eta)$, except when the very crude approximation $f(\eta) = \eta$ is used.

Table 82.2. Comparison of Different Assumptions for the Velocity Profile $f(\eta)$ with the Blasius Solution for a Flat Plate at Zero Incidence

Velocity distribution $u/U_\infty = f(\eta)$	α_1	α_2	β_1	$\delta^* \sqrt{\frac{U_\infty}{\nu x}}$	$\frac{\tau_0}{\mu U_\infty} \sqrt{\frac{\nu x}{U_\infty}}$	$C_f \left(\frac{U_\infty l}{\nu} \right)^{1/2}$	$\frac{\delta^*}{\vartheta}$
1. $f(\eta) = \eta$	$1/6$	$1/2$	1	1.732	0.289	1.155	3.00
2. $f(\eta) = \frac{3}{2}\eta - \frac{1}{2}\eta^3$	$39/280$	$3/8$	$3/2$	1.740	0.323	1.292	2.70
3. $f(\eta) = 2\eta - 2\eta^3 + \eta^4$	$37/315$	$3/10$	2	1.752	0.343	1.372	2.55
4. $f(\eta) = \sin \frac{\pi\eta}{2}$	$\frac{4-\pi}{2\pi}$	$\frac{\pi-2}{\pi}$	$\frac{\pi}{2}$	1.741	0.327	1.310	2.66
5. Exact	—	—	—	1.729	0.332	1.328	2.61

$$\vartheta \sqrt{\frac{U_\infty}{\nu x}} = \frac{2\tau_0}{\mu U_\infty} \sqrt{\frac{\nu x}{U_\infty}} \qquad C_f \left(\frac{U_\infty l}{\nu} \right)^{1/2} = 2\vartheta \sqrt{\frac{U_\infty}{\nu x}}$$

The Kármán-Pohlhausen Method

In the present article, the preceding method will be applied to the general case of two-dimensional, steady flow past a cylinder of *arbitrary* cross section. The calculation is so arranged that the characteristics of a particular problem, i.e., the potential function $U(x)$, need not be substituted into the equations until the last step in the calculation is reached. In this way, a maximum amount of generality is achieved.

The fundamental equation (82.44) is now written as

$$U^2 \frac{d\vartheta}{dx} + (2\vartheta + \delta^*)U \frac{dU}{dx} = \frac{\tau_0}{\rho} \tag{82.72}$$

and we *assume*

$$\frac{u}{U} = f(\eta) = a\eta + b\eta^2 + c\eta^3 + d\eta^4 \qquad 0 \le \eta \le 1$$
$$\frac{u}{U} = 1 \qquad \eta \ge 1 \tag{82.72a}$$

The four constants in $f(\eta)$ are determined from the boundary conditions:

$$y = 0 \qquad u = 0 \qquad \nu \frac{\partial^2 u}{\partial y^2} = \frac{1}{\rho}\frac{dp}{dx} = -U\frac{dU}{dx}$$
$$y = \delta \qquad u = U \qquad \frac{\partial u}{\partial y} = 0 \qquad \frac{\partial^2 u}{\partial y^2} = 0$$

The first condition is of particular importance because it determines the curvature of the velocity profile near the wall, and makes sure that there is no point of inflection in regions of decreasing pressure.

Introducing the *shape factor*,

$$\Lambda = \frac{\delta^2}{\nu}\frac{dU}{dx} = -\frac{dp}{dx}\frac{\delta}{\mu U/\delta} \tag{82.73}$$

that is, the ratio of pressure forces to viscous forces, we obtain

$$\frac{u}{U} = F(\eta) + \Lambda G(\eta) \tag{82.74}$$

where

$$F(\eta) = 1 - (1-\eta)^3(1+\eta)$$
$$G(\eta) = \tfrac{1}{6}\eta(1-\eta)^3 \tag{82.75}$$

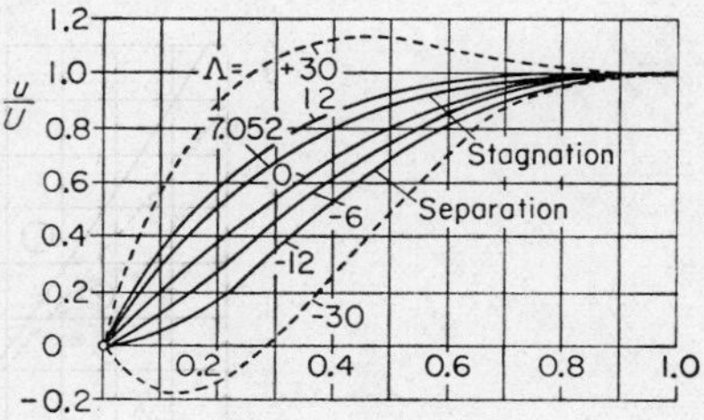

FIG. 82.11. The one-parameter family of velocity profiles from Eq. (82.74). (*From* [2], *p.* 208.)

so that now the velocity profiles form a one-parameter family of curves. A graphical representation of this family of curves is given in Fig. 82.11. The profile for $\Lambda = 0$ corresponds to $dp/dx = 0$, that is, to the flat plate (example 3 in Table 82.2) and to points where the velocity of the potential flow passes through a maximum or a minimum. Separation, $(\partial u/\partial y)_0 = 0$, occurs at $\Lambda = -12$; the profile at a stagnation point corresponds to $\Lambda = 7.052$. Values of $\Lambda > 12$ must be excluded because, for them, u becomes greater than U in the boundary layer. Hence, the range of useful values is $-12 \le \Lambda \le 12$.

It is easy to derive that

$$\frac{\delta^*}{\delta} = \frac{3}{10} - \frac{\Lambda}{120} \qquad \frac{\vartheta}{\delta} = \frac{37}{315} - \frac{\Lambda}{945} - \frac{\Lambda^2}{9072} \qquad \frac{\tau_0\delta}{\mu U} = 2 + \frac{\Lambda}{6} \tag{82.76}$$

Rearranging the integral momentum equation (82.72), we obtain the following differential equation for $\Lambda(x)$:

$$\frac{U\vartheta\vartheta'}{\nu} + \left(2 + \frac{\delta^*}{\vartheta}\right)\frac{U\vartheta^2}{\nu} = \frac{\tau_0\vartheta}{\mu U} \tag{82.77}$$

The equation does not contain the fortuitous quantity δ, but includes the really important physical quantities, δ^*, ϑ, and τ_0, all of them functions of Λ.

It is now convenient to introduce a second shape factor,

$$K = \frac{\vartheta^2}{\nu}\frac{dU}{dx} = Z\frac{dU}{dx} \quad \text{with } Z = \frac{\vartheta^2}{\nu} \tag{82.78}$$

related to Λ by the universal function

$$K = \left(\frac{37}{315} - \frac{\Lambda}{945} - \frac{\Lambda^2}{9072}\right)^2 \Lambda \tag{82.79}$$

Denoting

$$\frac{\delta^*}{\vartheta} = f_1(K) = \frac{3/10 - \Lambda/120}{37/315 - \Lambda/945 - \Lambda^2/9072} \tag{82.79a}$$

$$\frac{\tau_0\vartheta}{\mu U} = f_2(K) = \left(2 + \frac{\Lambda}{6}\right)\left(\frac{37}{315} - \frac{\Lambda}{945} - \frac{\Lambda^2}{9072}\right) \tag{82.79b}$$

and

$$2f_2(K) - 4K - 2Kf_1(K) = F(K) \tag{82.79c}$$

we can reduce Eq. (82.77) to the highly condensed form,

$$\frac{dZ}{dx} = \frac{F(K)}{U} \qquad K = ZU' \tag{82.80}$$

The auxiliary, universal function $F(K)$ is seen plotted in Fig. 82.12.

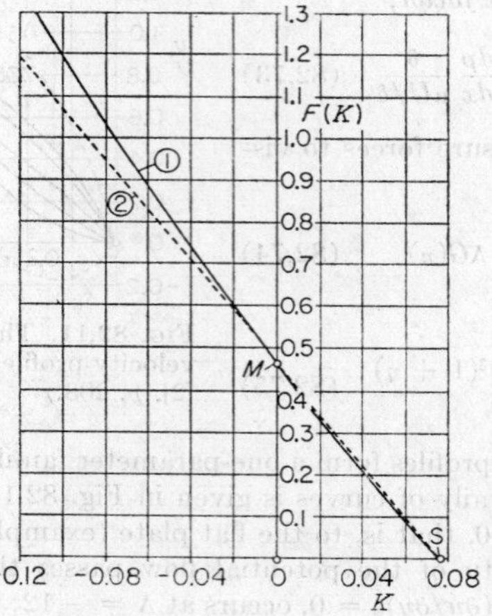

FIG. 82.12. The auxiliary function $F(K)$. Curve 1, for laminar boundary layers; it can be approximated by $F(K) = 0.470 - 6K$, curve 2. S = stagnation point, M = velocity maximum. (*From* [2], *p.* 213.)

The solution is best started at the stagnation point $x = 0$, where $U = 0$ and $dU/dx \neq 0$ and finite, unless the body possesses there a sharp cusped edge. The derivative dZ/dx does not become infinite at the stagnation point, because simultaneously $F(K) = 0$, and the stagnation point is singular. Thus, $F(K) = 0$ for $K = K_0 = 0.0770$, that is, for $\Lambda = \Lambda_0 = 7.052$, and

$$Z_0 = \frac{K_0}{U_0'} = \frac{0.0770}{U_0'} \qquad \left(\frac{dZ}{dx}\right)_0 = -0.0652\,\frac{U_0''}{U_0'^2} \tag{82.81}$$

Now integration can proceed numerically, or graphically by the method of isoclines. It is completed at the point of separation, $\Lambda = -12$, where $K = -0.1567$.

The function $F(K)$ can be approximated by a linear relation so that

$$F(K) = U\frac{dZ}{dx} = a - bK \qquad a = 0.470 \qquad b = 6 \tag{82.82}$$

when we obtain

$$\frac{U\vartheta^2}{\nu} = \frac{0.470}{U^5}\int_0^x U^5\,dx \tag{82.83}$$

In practice, the criterion for separation, $K = -0.1567$, is replaced by $K = -0.082$, which gives better agreement with known exact solutions.

Adverse Pressure Gradient; Separation

Whenever an adverse pressure gradient occurs in a flow (divergent channel, upper portion of airfoil), it is necessary to examine it for possible *separation*, in order to be able to avoid it, because it always causes large energy losses.

An estimate of the permissible magnitude of the adverse pressure gradient for which separation is just prevented can be obtained from Eq. (82.78) by putting $\Lambda = -10 =$ const with $K = -0.1369$ and $F(K) = 1.523$. In this manner, we can derive that separation will not occur if

$$U'' > 11U'^2/U$$

or if

$$U(x) = \left(1 + 100\frac{\nu x}{U_0\delta_0^2}\right)^{-0.1} \qquad \delta(x) = \delta_0\left(1 + 100\frac{\nu x}{U_0\delta_0^2}\right)^{0.55} \tag{82.84}$$

where U_0 and δ_0 are the values at $x = 0$.

The permissible deceleration is very small and proportional to $x^{-0.1}$. Since, for a flat plate, $\delta \propto x^{0.5}$ as against $\delta \propto x^{0.55}$ here, it is realized that the permissible pressure gradient is very small indeed.

82.7. ORIGIN OF TURBULENCE

General Remarks

The flow in a turbulent boundary layer is *never steady*, and only average values of the variables of the flow can be considered. The velocity, pressure, etc., fluctuate with time at any point in the flow in a *random manner;* in the case of vector quantities, a change in magnitude as well as in direction is involved.

Transition from laminar to turbulent flow is accompanied by a profound change in the velocity profile, and the shearing stress is markedly increased, but separation is delayed. Experimental evidence amply demonstrates that transition does not depend on velocity alone, i.e., that it does not occur at a fixed *critical Reynolds number* under given conditions. It is highly influenced by the disturbances carried by the free stream. For example, transition on a flat plate with a sharp leading edge can be delayed from an average value $\mathrm{Re}_{cr} = 3$ to 5×10^5, to as high a value as 3×10^6, if the disturbances are carefully eliminated from the free stream. In a pipe, transition can be delayed from an average value $\mathrm{Re}_{cr} = 2300$ to $\mathrm{Re}_{cr} = 40\,000$. Transition is also markedly affected by the magnitude of the pressure gradient, by the application of suction, by heat transfer, and by roughness.

The Tollmien-Schlichting Stability Theory

A very successful mathematical theory of the origin of turbulence was developed by W. Tollmien and H. Schlichting. The theory assumes that random oscillations are always carried by a fluid stream. They originate at the various solid boundaries and, depending on circumstances, are either damped out in the flow or amplified.

The amplification of random disturbances begins at a well-defined Reynolds number, known as *the limit of stability* (point of instability). Transition occurs at a somewhat larger, but unspecified by the theory, *critical Reynolds* number (point of transition).

A detailed account of the mathematical argument which expresses the preceding ideas can be found in chaps. 16 and 17 of [1] and [2] and in [10]. Only the barest outline can be given here.

In the case of two-dimensional parallel flows, the random, equally two-dimensional, disturbance is assumed to be expanded in a Fourier series, each of whose terms represents a partial harmonic oscillation given by a stream function of the form

$$\psi(x,y,t) = \phi(y) \exp [i(\alpha x - \beta t)] \tag{82.85}$$

that is, a progressing wave of wavelength $\lambda = 2\pi/\alpha$. The complex frequency

$$\beta = \beta_r + i\beta_i$$

contains the circular frequency β_r and the *amplification factor* β_i, whose sign is decisive for stability. The disturbances are damped (stable) if $\beta_i < 0$; they are amplified (unstable) if $\beta_i > 0$, and the limiting case $\beta_i = 0$ corresponds to neutral stability. The sign for β_i for a given α is derived from an eigenvalue problem formulated *on the basis of the boundary-layer equations* for the stream function (82.85).

In the theory, it is sufficient to consider two-dimensional disturbances only, because it has been shown by H. B. Squire (see [2] or [10]) that a two-dimensional flow pattern becomes unstable at a *higher* Reynolds number for a *three-dimensional* disturbance than for one which is assumed to be two-dimensional.

In a typical problem, the equation for the eigenvalues determines a curve of neutral stability, which divides the plane into a region of stable and a region of unstable points. The solution for a flat plate at zero incidence (Blasius profile) is shown represented graphically in Figs. 82.13 and 82.14. Figure 82.13 represents the values of the wavelengths of the disturbance, made dimensionless with reference to the displacement thickness δ^*, and given in the form $\alpha\delta^* = 2\pi\delta^*/\lambda$, which lie on the curve of neutral stability ($\beta_i = 0$). The independent variable is the Reynolds number $\mathrm{Re} = U_\infty\delta^*/\nu$ based on the displacement thickness. For $\mathrm{Re} < \mathrm{Re}_{cr}$ ($\mathrm{Re}_{cr} = 420$, corresponding to $\mathrm{Re}_x = U_\infty x/\nu = 0.82 \times 10^5$), disturbances of any wavelength are damped out; the flow is always stable. At higher Reynolds numbers, a definite *band of disturbances* is amplified, but this is sufficient to promote transition. As $\mathrm{Re} \rightarrow \infty$, the amplified band narrows down to zero. The companion Figure 82.14 shows the frequencies β_r given in the dimensionless form $\beta_r\delta^*/U_\infty$ and the propagation velocities c_r, given by $c = \beta/\alpha = c_r + ic_i$, and made dimensionless in the form c_r/U_∞, of the disturbances which are amplified at various Reynolds numbers.

It is noteworthy that only a comparatively narrow range of wavelengths and frequencies is "dangerous" for stability, and that there exist *upper limits* for the characteristic magnitudes. These are

$$c_r/U_\infty = 0.42 \qquad \alpha\delta^* = 0.36 \qquad \beta_r\delta^*/U_\infty = 0.15 \tag{82.86}$$

The smallest unstable wavelength

$$\lambda = \frac{2\pi}{0.36}\,\delta^* = 17.5\delta^* \approx 6\delta \tag{82.87}$$

It is comparatively large with respect to the boundary-layer thickness.

In the detailed mathematical development of the theory, use is made of two general theorems due to Lord Rayleigh. The second theorem is quoted here in its extended form, given by W. Tollmien, and both are valid only with the reservation that the influence of the viscosity on the disturbance, but not on the profile, has been neglected.

Theorem 1. If the mean (undisturbed) velocity profile $u(y)$ has a curvature which does not change sign ($u'' < 0$ everywhere), there exists at least one neutral

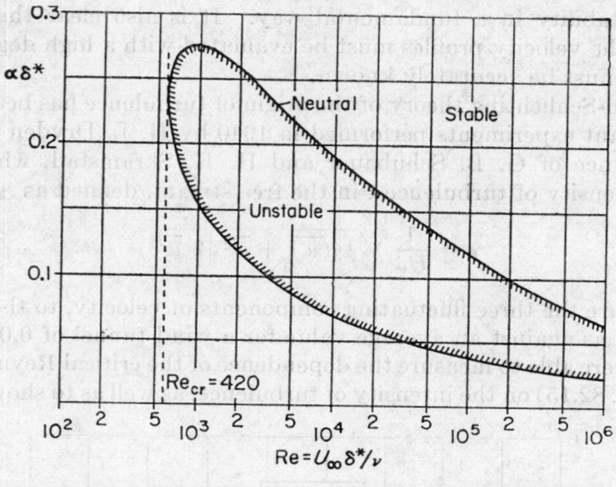

FIG. 82.13. Curve of neutral stability for the wavelengths $\alpha\delta^*$ of the disturbance in terms of the Reynolds number for the boundary layer on a flat plate. (*From* [2], *p.* 329.)

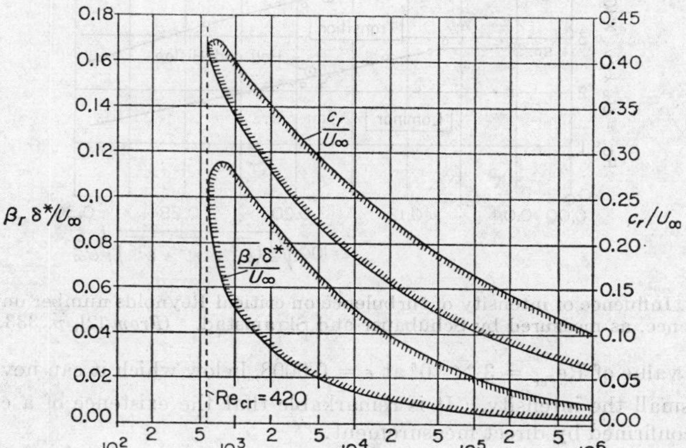

FIG. 82.14. Curve of neutral stability for the frequency β_r of the disturbance and wave-propagation velocity c_r for the boundary layer on a flat plate at zero incidence (Blasius profile). (*From* [2], *p.* 329.)

disturbance ($\beta_i = c_i = 0$) for which the wave-propagation velocity is equal to the mean velocity at some place y_K in the velocity field, so that

$$u - c = 0 \qquad \text{or} \qquad c = c_r = u(y_K) \tag{82.88}$$

In addition, it is shown that, at this critical layer, the oscillating x component, u', of the velocity induced by this disturbance undergoes a jump in phase.

Theorem 2. The existence of a point of inflection in the mean velocity profile ($u'' = 0$) is a necessary and sufficient condition for the amplification of disturbances ($\beta_i > 0,\ c_i > 0$). This means that velocity profiles with a point of inflection are inherently unstable.

The preceding two theorems demonstrate that the curvature of the velocity profile affects stability in a fundamental way. It is also clear that, for stability investigations, the velocity profiles must be evaluated with a high degree of precision and that $u''(y)$ must be accurately known.

The Tollmien-Schlichting theory of the origin of turbulence has been confirmed by a series of brilliant experiments performed in 1940 by H. L. Dryden (see [2] p. 332), with the assistance of G. B. Schubauer and H. K. Skramstad, who succeeded in reducing the intensity of turbulence ϵ in the free stream, defined as

$$\epsilon = \frac{1}{U_\infty} \sqrt{\tfrac{1}{3}(\overline{u'^2} + \overline{v'^2} + \overline{w'^3})} \tag{82.89}$$

where u', v', w' are the three fluctuating components of velocity, to the extremely low value of 0.0002, as against an average value for a wind tunnel of 0.002 to 0.01. In addition, they were able to measure the dependence of the critical Reynolds number on a flat plate (Fig. 82.15) on the intensity of turbulence, as well as to show that it reaches

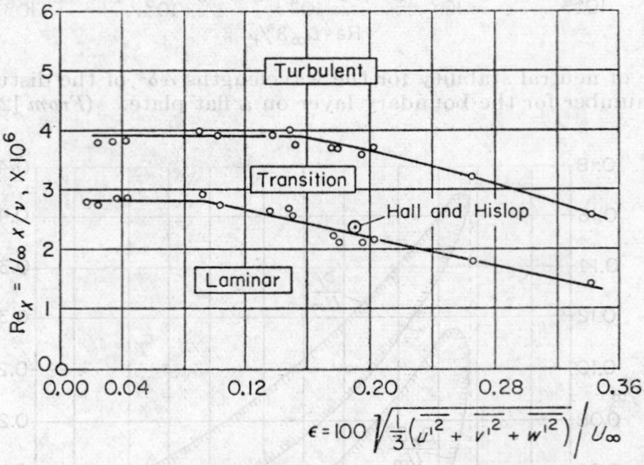

FIG. 82.15. Influence of intensity of turbulence on critical Reynolds number on flat plate at zero incidence, as measured by Schubauer and Skramstad. (*From* [2], *p.* 333.)

a limiting value of $Re_{x_{cr}} = 3 \times 10^6$ at $\epsilon = 0.0008$, below which it can never decrease, however small the intensity. It is remarkable that the existence of a critical layer was also confirmed by direct measurement.

The excellent agreement between the Tollmien-Schlichting theory and experiment leaves no room for doubt that it presents a *possible* mechanism of transition. However, recent experiments show that, in actual flows, transition occurs in a considerably more complex manner than that postulated. Transition is never abrupt and fixed, and takes place in a zone of Reynolds numbers around the critical, its most characteristic feature being intermittency. In the transition zone, at a given point, the flow continually alternates in time between laminar and turbulent. Turbulence is generated in "spots," and propagates itself downstream in a divergent wedge. Such spots occur at irregular intervals of time at apparently randomly placed points. Furthermore, it appears that truly two-dimensional flows and disturbances do not exist even in extremely carefully controlled flows. The third velocity component which is always present can be ignored in purely laminar flow, but comes into play, when the limit of stability has been exceeded, by contributing to the production of longitudinal vortices. Hence, it would appear that transition is essentially a three-dimensional phenomenon. Consequently, the understanding of the instability of three-dimensional flows appears to be essential in the study of transition.

In the case of three-dimensional flows, instability need not necessarily lead to turbulence; it may first lead to a different pattern of laminar flow with a *secondary* vortexlike *flow,* superimposed on the previous pattern. This occurs in the flow between two concentric rotating cylinders (G. I. Taylor type of instability [11]), and in flows along *concave* walls (H. Görtler type of instability [12],[13]), being due to the action of centrifugal forces. The former type of instability (Fig. 82.16) does not

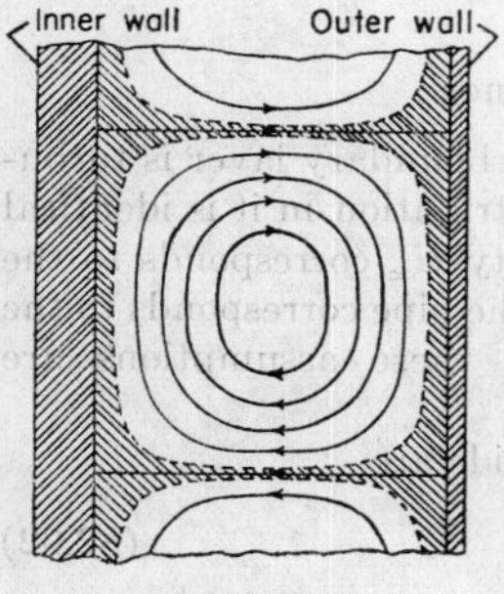

Fig. 82.16

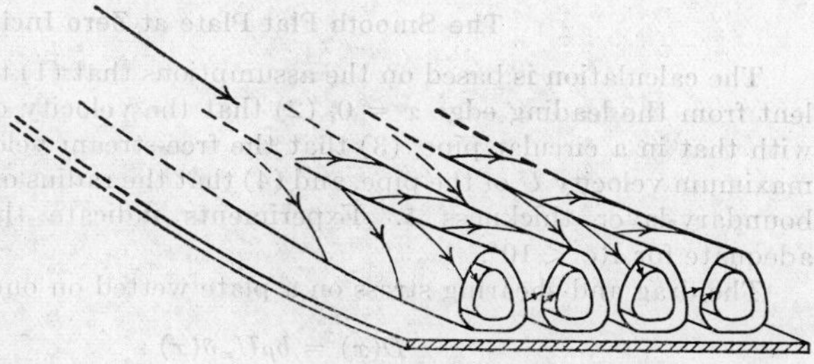

Fig. 82.17

Fig. 82.16. Pattern of streamlines of secondary flow between two concentric rotating cylinders, the outer cylinder being at rest and the inner cylinder rotating, after G. I. Taylor. (*From* [2], *p.* 360.)
Fig. 82.17. Vortices caused by three-dimensional disturbances in the flow along a concave wall, after Görtler. (*From* [2], *p.* 360.)

occur when the outer cylinder rotates and the inner cylinder is at rest. The limit of stability is given by

$$\frac{Uh}{\nu} \geq 41.3 \sqrt{\frac{R}{h}} \tag{82.90}$$

where h = width of gap, R = radius of inner cylinder, and U = peripheral velocity of inner cylinder. In the second case (Fig. 82.17), instability does not occur on convex walls, and the limit of stability for concave walls is given by

$$\frac{U_\infty \vartheta}{\nu} \sqrt{\frac{\vartheta}{R}} \geq 0.58 \tag{82.91}$$

the wavelength of the vortex pattern being $\lambda = 0.45\vartheta = 17\delta^*$. In both cases, as the speed is increased, the vortices eventually break up, and turbulence results. Experiments show that instability occurs when

$$\frac{U_\infty \vartheta}{\nu} \sqrt{\frac{\vartheta}{R}} \geq 6 \text{ to } 9$$

depending on the intensity of the free-stream turbulence.

82.8. TURBULENT FLOW ALONG A FLAT PLATE

General Remarks

The hypotheses which lead to the calculation of shearing stresses in turbulent flow will be discussed on p. 84–3; it might be supposed that calculations related to turbulent boundary layers can be performed with the aid of the equations of motion and by the same general methods as those applied to laminar boundary layers. So far this scheme has met with no success, owing to the fact that nothing is known about the zone of transition from the turbulent boundary layer to the *laminar sublayer*, or about the laws of friction in the latter. From this point of view, conditions are more favorable in motions in which no solid boundaries exist. In all other problems, semi-empirical methods based principally on the integral momentum equation (p. 82–13) must be employed; here we shall only illustrate the principle of the calculation by giving one example. The modern treatment of turbulent flows by statistical methods [14] has not yet reached a point where questions of practical importance can be answered.

The Smooth Flat Plate at Zero Incidence

The calculation is based on the assumptions that (1) the boundary layer is turbulent from the leading edge $x = 0$, (2) that the velocity distribution in it is identical with that in a circular pipe, (3) that the free-stream velocity U_∞ corresponds to the maximum velocity U of the pipe, and (4) that the radius of the pipe corresponds to the boundary-layer thickness δ. Experiments indicate that these assumptions are adequate for $\mathrm{Re} < 10^6$.

The drag and shearing stress on a plate wetted on one side are

$$D(x) = b\rho U_\infty \vartheta(x) \tag{82.92}$$

$$\tau_0(x) = \frac{1}{b}\frac{dD(x)}{dx} = \rho U_\infty{}^2 \frac{d\vartheta}{dx} \tag{82.92a}$$

We assume that velocity profiles are similar and obey a one-seventh power law [Eq. (84.16′)], that is,

$$\frac{u}{U_\infty} = \left(\frac{y}{\delta}\right)^{1/7} \tag{82.93}$$

and that the shearing stress at the wall is equal to that in a pipe, that is,

$$\frac{\tau_0}{\rho U_\infty{}^2} = 0.0225\left(\frac{\nu}{U_\infty \delta}\right)^{1/4} \tag{82.93a}$$

This leads to $\delta^* = \frac{1}{8}\delta$, $\vartheta = \frac{7}{72}\delta$ and, hence, to the differential equation,

$$\frac{7}{72}\frac{d\delta}{dx} = 0.0225\left(\frac{\nu}{U_\infty \delta}\right)^{1/4} \tag{82.94}$$

On integration, we obtain (Prandtl)

$$\delta(x) = 0.37x\,\mathrm{Re}_x{}^{-1/5} \qquad \vartheta(x) = 0.36x\,\mathrm{Re}_x{}^{-1/5} \qquad D(x) = 0.036\,\rho U b x\,\mathrm{Re}_x{}^{-1/5}$$

$$c_f' = \frac{\tau_0}{\frac{1}{2}\rho U_\infty{}^2} = 0.0592\,\mathrm{Re}_x{}^{-1/5} \qquad c_f = \frac{\int_0^l D(x)\,dx}{\frac{1}{2}\rho U_\infty{}^2 bl} = 0.074\,\mathrm{Re}_l{}^{-1/5} \tag{82.95†}$$

$$5 \times 10^5 < \mathrm{Re}_l < 10^7$$

In an actual case, when the boundary layer at the leading edge is laminar, transition occurring at Re_{cr}, it is found that Eqs. (82.95) can still be used, assuming that x is measured *from the leading edge*. However, the integration in the equation for drag must proceed from x_{cr} at the point of transition to the current point $x > x_{cr}$. This

† The constant 0.074 has been adjusted empirically; if it were derived from the preceding data, it would have the value 0.072. Similarly the constant in c_f' would be 0.0576.

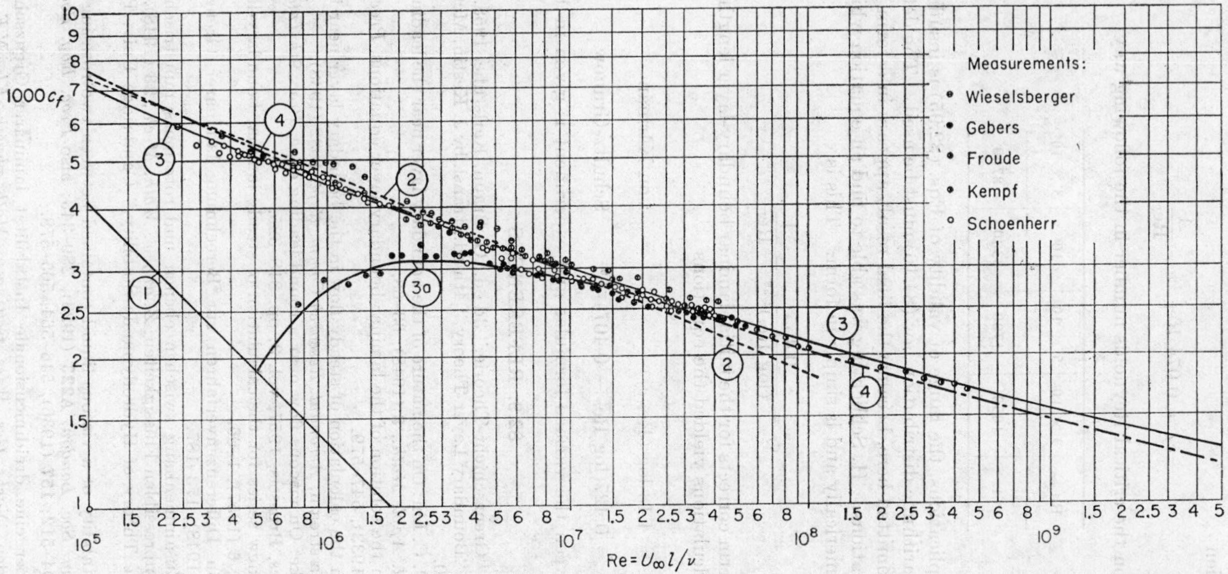

FIG. 82.18. Resistance formula for smooth flat plates at zero incidence: (1) Laminar flow, Eq. (82.30); (2) turbulent flow, Eq. (82.95), Prandtl; (3) turbulent flow, Eq. (82.97), Prandtl-Schlichting; (3*a*) transition; (4) turbulent, refined theory due to Schultz-Grunow. (*From* [2], *p.* 439.)

leads to the equation

$$c_f = 0.074\ \mathrm{Re}_l^{-1/5} - \frac{A}{\mathrm{Re}_l} \tag{82.96}$$

where A depends on the critical Reynolds number in the following way:

$\mathrm{Re}_{x\mathrm{cr}}$	3×10^5	5×10^5	10^6	3×10^6
A	1050	1700	3300	8700

In practical applications, the range of validity of Eqs. (82.95) is insufficient, and the asymptotic logarithmic distribution law (84.16) must be used. The details of the derivation will be omitted here, because the final result appears in a form unsuitable for practical applications. H. Schlichting was able to find an equation which fits the final expression numerically and is simple in form. This is

$$c_f = \frac{0.455}{(\log \mathrm{Re}_l)^{2.58}} - \frac{A}{\mathrm{Re}_l} \tag{82.97}$$

where the second term corrects for the initial laminar boundary-layer length as before.

Very similar calculations yielded the equations

$$\frac{1}{\sqrt{c_f}} = 4.13 \log (\mathrm{Re}_l\, c_f) \qquad \text{von Kármán} \tag{82.97a}$$

$$c_f = 0.427(\log \mathrm{Re}_l - 0.407)^{-2.64} \qquad \text{Schultz-Grunow} \tag{82.97b}$$

A comparison between the various formulas and experiment is given in Fig. 82.18.

82.9. REFERENCES

[1] H. Schlichting: "Grenzschicht-Theorie," 3d ed., G. Braun, Karlsruhe, 1958.
[2] H. Schlichting: "Boundary Layer Theory," 4th ed., transl. by J. Kestin, McGraw-Hill, New York, 1960.
[3] G. F. Carrier, C. C. Lin: On the nature of the boundary layer near the leading edge of a flat plate, *Quart. Appl. Math.*, **8** (1948), 63–68.
[4] L. Howarth: On the solution of the laminar boundary layer equations, *Proc. Roy. Soc. London*, **A164** (1933), 547–579.
[5] L. Howarth: On the calculation of steady flow in the boundary layer near the surface of a cylinder in a stream, *Aeronaut. Research Comm. Rept.* 1632 (1938).
[6] M. D. Van Dyke: On viscous flow near a round leading edge, *Proc. 9th Intern. Congr. Appl. Mechanics*, Brussels, 1957, vol. 3, pp. 318–325.
[7] H. Görtler: A new series for the calculation of steady laminar boundary-layer flows, *J. Math. Mech.*, **6** (1957), 1–66.
[8] H. Görtler: Ein Differenzenverfahren zur Berechnung laminarer Grenzschichten, *Ing.-Arch.*, **16** (1948), 173–187.
[9] W. Mangler: Zusammenhang zwischen ebenen und rotationssymmetrischen Grenzschichten in kompressiblen Flüssigkeiten, *Z. angew. Math. Mech.*, **28** (1948), 97–103.
[10] C. C. Lin: "The Theory of Hydrodynamic Stability," Cambridge Univ. Press, New York, 1955.
[11] G. I. Taylor: Stability of a viscous liquid contained between two rotating cylinders, *Phil. Trans. Roy Soc. London*, **A223** (1923), 289–343; also *Proc. Roy. Soc. London*, **A151** (1935), 494–512; **157** (1936), 546–564, 565–578.
[12] H. Görtler: Über eine dreidimensionale Instabilität laminarer Grenzschichten an konkaven Wänden, *Nachr. Ges. Wiss. Göttingen, Math.-physik. Kl., N.F.*, **2** (1940), 1–26; also *Z. angew. Math. Mech.*, **21** (1941), 250–252.
[13] H. Görtler: Dreidimensionale Instabilität der ebenen Staupunktströmung gegenüber wirbelartigen Störungen, in W. Tollmien, H. Görtler (eds.), "Fifty Years of Boundary Layer Theory," pp. 304–314, Vieweg, Brunswick, Germany, 1955.
[14] G. K. Batchelor: "The Theory of Homogeneous Turbulence," Cambridge Univ. Press, New York, 1953.

CHAPTER 83

COMPRESSIBLE BOUNDARY LAYERS

BY

K. STEWARTSON, Ph.D., Durham, England

83.1. INTRODUCTION

The main differences between incompressible and compressible boundary layers stem from thermal effects. Not only is the velocity gradient large in a compressible boundary layer, but so also are the temperature and density gradients. Furthermore, the physical properties of the fluid, viscosity and diffusivity, which control the behavior of the fluid are now themselves functions of temperature. In consequence, the equations of motion and energy reduce to a pair of interdependent equations whose treatment presents such a formidable problem that the solution can be obtained only with the aid of high-speed calculating machines, or after making certain approximations.

In addition to the calculation of skin friction and separation, which are the principal aims of the theory of incompressible boundary layers, the heat transfer between a solid body and the fluid or, alternatively, the temperature rise at the surface of an insulated wall due to viscous dissipation, is also required from a theory of compressible boundary layers. The major portion of this chapter is concerned with the progress which has been made when the flow is steady, two-dimensional and the main stream is prescribed. There is also a section on the stability of laminar boundary layers. Finally, two problems, the hypersonic boundary layer and the interaction between a shock wave and a boundary layer, which have no counterpart in incompressible flow, are discussed.

83.2. EQUATIONS

Consider a viscous compressible perfect gas streaming steadily along a semi-infinite flat plate. Let x measure the distance along the plate from its leading edge and y the distance normal to the plate, and let u, v be the components of the velocity of the fluid in these directions. Furthermore, let p be the pressure, ρ the density, T the absolute temperature, μ the viscosity, k the thermal conductivity, and c_p the specific heat of the gas at constant pressure. Then, neglecting body forces and heat sources, the Navier-Stokes equations (68.19) for this problem are [2]:

$$\begin{aligned}\rho u \frac{\partial u}{\partial x} + \rho v \frac{\partial u}{\partial y} &= -\frac{\partial p}{\partial x} + \frac{\mu}{3}\frac{\partial}{\partial x}\left(\frac{\partial u}{\partial x} + \frac{\partial v}{\partial y}\right) - \frac{2}{3}\frac{\partial \mu}{\partial x}\left(\frac{\partial u}{\partial x} + \frac{\partial v}{\partial y}\right) \\ &\quad + 2\frac{\partial u}{\partial x}\frac{\partial \mu}{\partial x} + \frac{\partial \mu}{\partial y}\left(\frac{\partial u}{\partial y} + \frac{\partial v}{\partial x}\right) + \mu\left(\frac{\partial^2 u}{\partial x^2} + \frac{\partial^2 u}{\partial y^2}\right) \\ \rho u \frac{\partial v}{\partial x} + \rho v \frac{\partial v}{\partial y} &= -\frac{\partial p}{\partial y} + \frac{\mu}{3}\frac{\partial}{\partial y}\left(\frac{\partial u}{\partial x} + \frac{\partial v}{\partial y}\right) - \frac{2}{3}\frac{\partial \mu}{\partial y}\left(\frac{\partial u}{\partial x} + \frac{\partial v}{\partial y}\right) \\ &\quad + 2\frac{\partial v}{\partial y}\frac{\partial \mu}{\partial y} + \frac{\partial \mu}{\partial x}\left(\frac{\partial u}{\partial y} + \frac{\partial v}{\partial x}\right) + \mu\left(\frac{\partial^2 v}{\partial x^2} + \frac{\partial^2 v}{\partial y^2}\right)\end{aligned} \tag{83.1a,b}$$

The equation of continuity (68.17) is, in this case,

$$\frac{\partial}{\partial x}(\rho u) + \frac{\partial}{\partial y}(\rho v) = 0 \tag{83.2}$$

the equation of energy (68.22) is, with c_p = const,

$$c_p\rho u\frac{\partial T}{\partial x} + c_p\rho v\frac{\partial T}{\partial y} - u\frac{\partial p}{\partial x} - v\frac{\partial p}{\partial y} = \frac{\partial}{\partial x}\left(k\frac{\partial T}{\partial x}\right) + \frac{\partial}{\partial y}\left(k\frac{\partial T}{\partial y}\right) + \Phi \tag{83.3}$$

where Φ is the dissipation function, given by

$$\Phi = 2\mu\left(\frac{\partial u}{\partial x}\right)^2 + 2\mu\left(\frac{\partial v}{\partial y}\right)^2 + \mu\left(\frac{\partial v}{\partial x} + \frac{\partial u}{\partial y}\right)^2 - \frac{2}{3}\mu\left(\frac{\partial u}{\partial x} + \frac{\partial v}{\partial y}\right)^2 \tag{83.4}$$

[see Eq. (68.23)], and the equation of state is

$$p = R\rho T \tag{83.5}$$

where R is the gas constant.

The boundary-layer equations are derived from these equations, on the assumption that the boundary layer is thin and separates the plate from the main stream, in which viscosity and thermal conductivity may be neglected. As a first approximation, therefore, the inviscid equations are solved with appropriate boundary conditions, neglecting the viscous conducting layer. In general, the condition on temperature at the plate and the condition of no slip of gas past the plate will not be satisfied, and a boundary layer is necessary to adjust the inviscid solution to satisfy these conditions on the plate. In this chapter, it will be supposed that the inviscid equations have been solved and that, according to them,

$$u = u_1(x) \qquad T = T_1(x) \qquad p = p_1(x) \qquad \rho = \rho_1(x) \tag{83.6}$$

on the plate. The boundary layer comprises a velocity layer, in which u increases from zero on the plate to u_1 in the main stream, and a thermal layer, in which the temperature changes from its value T_w at the plate to T_1 in the main stream. The thermal layer is also necessary if the plate is thermally insulated because heat is supplied to it via the dissipation function Φ.

Since the boundary layer is thin, it may be assumed that

$$\frac{\partial}{\partial x} \ll \frac{\partial}{\partial y} \tag{83.7}$$

whence $v \ll u$ and Eqs. (83.1) and (83.4) reduce to

$$\rho u\frac{\partial u}{\partial x} + \rho v\frac{\partial u}{\partial y} = -\frac{\partial p}{\partial x} + \frac{\partial}{\partial y}\left(\mu\frac{\partial u}{\partial y}\right) \qquad 0 = -\frac{\partial p}{\partial y} \tag{83.8a,b}$$

$$c_p\rho u\frac{\partial T}{\partial x} + c_p\rho v\frac{\partial T}{\partial y} - u\frac{\partial p}{\partial x} = \frac{\partial}{\partial y}\left(k\frac{\partial T}{\partial y}\right) + \mu\left(\frac{\partial u}{\partial y}\right)^2 \tag{83.9}$$

while the equations of continuity and of state are unaffected. The solution of (83.8*a*) is significant only if the inertia terms on the left-hand side are of the same order as the viscous terms on the right-hand side. Hence, if δ_u is representative of the thickness of the velocity layer, and l is a typical length in the x direction,

$$\frac{\partial}{\partial x} = O(l^{-1}) \qquad \frac{\partial}{\partial y} = O(\delta_u^{-1})$$

in (83.8*a*), whence

$$\delta_u/l,\ v/u_1 = O\left(\frac{\mu}{u_1 l\rho}\right)^{1/2}$$

Similarly, if δ_T is representative of the thickness of the thermal layer so that

$$\partial/\partial y = O(\delta_T^{-1})$$

in (83.9),

$$\delta_T/l = O\left(\frac{k}{u_1 l \rho c_p}\right)^{1/2}$$

Hence, the boundary-layer equations can be regarded as limiting forms of the Navier-Stokes equations when $\mu_1 \to 0$, keeping $y/\mu_1^{1/2}$, k/μ_1 constant. The ratio of the thickness of the two layers is

$$\delta_u/\delta_T = O(\text{Pr}^{1/2})$$

where $\text{Pr} = \mu c_p/k$ is the Prandtl number. If, as is usual, $\text{Pr} = O(1)$, the thermal and velocity layers are of the same order of thickness, but exceptions can occur in which Pr is either very small or very large. The boundary-layer equations can still be valid, but care must be taken in solving them.

The boundary conditions to be assumed in this chapter are first that

$$u = v = 0 \qquad \text{and} \qquad \text{either } \partial T/\partial y = 0 \text{ or } T = T_w \qquad \text{at } y = 0,\ x > 0 \tag{83.10}$$

Second, in view of the limiting procedure $\mu_1 \to 0$, by which the boundary-layer equations may be derived, it follows that their solution as $y \to \infty$ must be matched to the inviscid solution just outside the boundary layer. There is no contradiction involved here since, when y is large on the scale of the boundary layer, it is still only $O(\mu_1^{1/2})$ on the scale of the inviscid flow. Thus, the second set of boundary conditions is that

$$u \to u_1(x) \qquad T \to T_1(x) \qquad p \to p_1(x) \qquad \text{as } y \to \infty,\ x > 0 \tag{83.11}$$

In general, $v(x, \infty)$ cannot be prescribed. Third, the parabolic nature of the equations means that initial conditions must be prescribed along some line $x =$ const, and here

$$u(0,y) = u_1(0),\ T(0,y) = T_1(0) \tag{83.12}$$

will be taken. These boundary conditions (83.10) to (83.12), although not the most general possible, are of frequent occurrence and sufficient to lead to a broad understanding of the nature of the boundary layers.

The second momentum equation (83.8*b*) implies that

$$p(x,y) = p_1(x) \tag{83.13}$$

that is, the pressure does not vary across the boundary layer. Hence,

$$-\frac{\partial p}{\partial x} = \rho_1 u_1 \frac{du_1}{dx} \tag{83.14}$$

Furthermore, if the inviscid flow outside the boundary layer is homenergic,

$$c_p T_1 + \tfrac{1}{2} u_1^2 = \text{const} \tag{83.15}$$

The first momentum equation (83.8*a*) now becomes

$$\rho u \frac{\partial u}{\partial x} + \rho v \frac{\partial u}{\partial y} = \rho_1 u_1 \frac{du_1}{dx} + \frac{\partial}{\partial y}\left(\mu \frac{\partial u}{\partial y}\right) \tag{83.16}$$

the energy equation

$$c_p \rho u \frac{\partial T}{\partial x} + c_p \rho v \frac{\partial T}{\partial y} + \rho_1 u u_1 \frac{du_1}{dx} = \frac{\partial}{\partial y}\left(\frac{c_p \mu}{\sigma}\frac{\partial T}{\partial y}\right) + \mu\left(\frac{\partial u}{\partial y}\right)^2 \tag{83.17}$$

and the equation of state

$$\rho T = \rho_1 T_1 \tag{83.18}$$

It is convenient to define the stream function ψ from the equation of continuity by

$$\rho u = \rho_\infty \frac{\partial \psi}{\partial y} \qquad \rho v = -\rho_\infty \frac{\partial \psi}{\partial x} \tag{83.19}$$

Two of the most important quantities to be derived from the boundary-layer theory are the skin friction τ_w and the heat transfer k_w per unit area and unit time from the plate to the fluid. These are defined by

$$\tau_w = \left(\mu \frac{\partial u}{\partial y}\right)_{y=0} \qquad k_w = -\left(\frac{\mu}{\mathrm{Pr}} c_p \frac{\partial T}{\partial y}\right)_{y=0} \tag{83.20}$$

83.3. PROPERTIES OF Pr AND μ

For most practical purposes, it is sufficient to assume that Pr is a constant, equal to 0.72 for air, and that μ is a function of temperature only. The most adequate description of μ at ordinary temperatures and pressures is given by Sutherland's formula

$$\mu = \frac{BT^{3/2}}{T + D} \tag{83.21}$$

where B and D are constants, D being 114°K for air. Several authors, however, have found it convenient to use an approximate form

$$\mu \propto T^\omega \tag{83.22}$$

where ω is a constant whose chosen value has varied from 0 to 1.25. Even so, using the formulas requires extensive numerical calculations unless $\omega = 1$, when the equations simplify considerably. On the other hand, this law is not realistic. Accordingly, Chapman proposed an average viscosity of the form

$$\mu/\mu_\infty = CT/T_\infty \tag{83.23}$$

The constant C can be chosen to agree with (83.21) at any point, and it is clearly important to have the correct viscosity coefficient near the plate where the shear is greatest. For constant wall temperature, therefore, C will be defined by

$$C = \frac{\mu_w T_\infty}{\mu_\infty T_w} \tag{83.24}$$

The simplification mentioned earlier, which occurs when μ is strictly proportional to T can easily be extended to include $C \neq 1$.

83.4. RANGE OF APPLICATION OF THE EQUATIONS

Although the boundary-layer equations have been derived from the Navier-Stokes equations, which are only valid in the limiting case of a monatomic gas with zero mean free path, in the particular case of flow past a flat plate, they have a much wider application. First, such disagreements between theory and observation as have been reported can only in a very few extreme cases be attributed to the breakdown of the continuum hypothesis. Second, although, in a polyatomic gas, there is a second coefficient of viscosity λ, which does *not* satisfy Stokes' relation $3\lambda + 2\mu = 0$, it disappears when the approximations of boundary-layer theory are applied. Third, if the plate is replaced by a wall of finite curvature κ and x, y measure distance along and normal to the wall, the only difference in the boundary-layer equations is that (83.8*b*) is replaced by

$$\rho \kappa u^2 = \partial p/\partial y \tag{83.25}$$

It follows that the change in pressure across the boundary layer may still be neglected and (83.14) used. The final equations, (83.16) to (83.19), are therefore unchanged.

Finally, the corresponding equations describing the steady flow past axially symmetric bodies of revolution for which the boundary layer is sufficiently thin may also be transformed into (83.16) to (83.19). For let r denote distance from the axis of symmetry and x, y measure distance along and normal to a meridian section of the surface of the body. Then, since the boundary layer is thin, r may be replaced by r_0, the radius of the body at the same value of x, whenever it occurs explicitly. Accordingly, the only modification to (83.16) to (83.19) is in the definition of ψ, which is now

$$\rho_\infty \frac{\partial \psi}{\partial y} = r_0 \rho u \qquad \rho_\infty \frac{\partial \psi}{\partial x} = -r_0 \rho v \tag{83.26}$$

The extra factor r_0 may be removed, using Mangler's transformation, in which

$$\begin{aligned} \bar{x} &= \int_0^x \frac{r_0^2}{l^2}\, dx \qquad \bar{y} = \frac{r_0}{l}\, y \qquad \bar{\psi}(\bar{x},\bar{y}) = \frac{1}{l}\, \psi(x,y) \\ \bar{T} &= T \qquad \bar{\rho} = \rho \qquad \bar{\mu} = \mu \qquad \bar{k} = k \end{aligned} \tag{83.27}$$

On substitution, it follows that the barred quantities satisfy (83.16) to (83.19), thus reducing the problem to the two-dimensional case. It follows from (83.27) that the main-stream velocities just outside the boundary layers are the same at corresponding points. Furthermore, it can be shown that the correlation may be extended as far as the stagnation point of a bluff body, even though $r_0 = 0$ there.

83.5. NONDIMENSIONAL FORM OF THE EQUATIONS

This may be achieved by writing

$$\begin{aligned} x' &= x/l \qquad y' = (y/l)\,\mathrm{Re}^{1/2} \qquad u' = u/u_\infty \qquad v' = \frac{v}{u_\infty}\,\mathrm{Re}^{1/2} \\ T' &= T/T_\infty \qquad \mu' = u/u_\infty \qquad \rho' = \rho/\rho_\infty \end{aligned} \tag{83.28}$$

when (83.16) to (83.19) reduce to

$$\begin{aligned} \rho' u' \frac{\partial u'}{\partial x'} + \rho' v' \frac{\partial u'}{\partial y'} &= \rho_1' u_1' \frac{du_1}{dx'} + \frac{\partial}{\partial y'}\left(\mu' \frac{\partial u'}{\partial y'}\right) \\ \rho' T' = \rho_1' T_1' \qquad \frac{\partial}{\partial x'}(\rho' u') &+ \frac{\partial}{\partial y'}(\rho' v') = 0 \end{aligned} \tag{83.29}$$

and

$$\begin{aligned} \rho' u' \frac{\partial T'}{\partial x'} + \rho' v' \frac{\partial T'}{\partial y'} &+ (\gamma - 1) M_\infty^2 u' \rho_1' u_1' \frac{du_1'}{dx'} \\ &= \frac{1}{\mathrm{Pr}} \frac{\partial}{\partial y'}\left(\mu' \frac{\partial T'}{\partial y'}\right) + (\gamma - 1) M_\infty^2 \mu' \left(\frac{\partial u'}{\partial y'}\right)^2 \end{aligned} \tag{83.30}$$

where M_∞ is the Mach number of the inviscid flow at infinity, and $\mathrm{Re} = u_\infty \rho_\infty l/\mu_\infty$, the Reynolds number.

The boundary layer must be regarded as compressible as soon as temperature changes take place in it as a result of dissipation or through heat sources, at the plate, for example. Thus, even when $M_\infty = 0$, so that the flow outside the boundary layer is incompressible, the temperature and density inside can be made to vary by heating (or cooling) the plate. When $M_\infty \neq 0$, the effect of dissipation can be seen most easily if it is made the only forcing term in (83.30) by assuming that u_1 is constant and that the plate is insulated. It then follows that there is a rise in temperature $O(M_\infty^2)$ in the boundary layer, particularly at the plate. At high values of M_∞ this effect is serious, because it can lead to a structural weakening and even melting of the plate. The rise in temperature is controlled by balancing the frictional heating with the convection and diffusion of heat. The diffusion of heat is inversely proportional to Pr, and, hence, the temperature in the boundary layer must increase with Pr.

The effect of the rise in temperature of the gas is to diminish ρ and increase μ. Both these consequences lead to a thickening of the boundary layer and a decrease in the velocity gradient. The skin friction is, therefore, modified by two partially canceling trends, and changes rather slowly. It is actually independent of M_∞ and the plate temperature if the viscosity is proportional to the temperature and u_1 is constant.

There is no qualitative difference between supersonic and subsonic boundary layers *as such*, because, unlike inviscid flow, the pressure is not the dominant force controlling the flow. Nevertheless, since the Mach number also enters through the boundary conditions, different kinds of problems are likely to occur in the two cases, particularly as regards their interaction with the exterior inviscid flow.

83.6. ALTERNATIVE FORMS OF THE BOUNDARY-LAYER EQUATIONS

Several alternative forms of the differential equations have been used, either to emphasize the role played by certain parameters or for the purposes of numerical calculation. Here two of them are given, one due to von Mises and the other the restricted form of Howarth's transformation.

von Mises Transformation

In this transformation, originally introduced for incompressible flow, the independent variables are taken to be x and ψ instead of x and y. The equations become

$$u\left(\frac{\partial u}{\partial x}\right)_\psi - \frac{\rho_1 u_1}{\rho}\frac{du_1}{dx} = u\frac{\partial}{\partial \psi}\left(\rho u \mu \frac{\partial u}{\partial \psi}\right)_x \tag{83.31}$$

$$c_p\left(\frac{\partial T}{\partial x}\right)_\psi + \frac{\rho_1 u_1}{\rho}\frac{du_1}{dx} = \frac{\partial}{\partial \psi}\left(c_p\frac{\mu\rho u}{\mathrm{Pr}}\frac{\partial T}{\partial \psi}\right)_x + \rho\mu u\left(\frac{\partial u}{\partial \psi}\right)^2 \tag{83.32}$$

Near the plate $u \sim \psi^{1/2}$ so that the differential equations are singular at $\psi = 0$, but, nevertheless, they have proved useful for developing approximate theories and in hypersonic flow.

Howarth's Transformation

In this transformation, the independent variables are taken to be x and η, where

$$\eta = \int_0^y \frac{\rho\, dy}{\rho_\infty} \tag{83.33}$$

The equation of continuity now reduces to

$$\left(\frac{\partial u}{\partial x}\right)_\eta + \left(\frac{\partial \bar{v}}{\partial \eta}\right)_x = 0 \qquad \text{where} \qquad \bar{v} = \frac{\rho}{\rho_\infty}\left(v - u\left(\frac{\partial y}{\partial x}\right)_\eta\right) \tag{83.34}$$

while the equations of motion and energy become

$$u\left(\frac{\partial u}{\partial x}\right)_\eta + \bar{v}\left(\frac{\partial u}{\partial \eta}\right)_x - \frac{\rho_1}{\rho}u_1\frac{du_1}{dx} = \frac{1}{\rho_\infty{}^2}\frac{\partial}{\partial \eta}\left(\mu\rho\frac{\partial u}{\partial \eta}\right)_x \tag{83.35}$$

$$u\left(\frac{\partial T}{\partial x}\right)_\eta + \bar{v}\left(\frac{\partial T}{\partial \eta}\right)_x + \frac{\rho_1}{c_p\rho}uu_1\frac{du_1}{dx} = \frac{1}{\rho_\infty{}^2}\frac{\partial}{\partial \eta}\left(\frac{\mu\rho}{\mathrm{Pr}}\frac{\partial T}{\partial \eta}\right)_x + \frac{\mu\rho}{c_p\rho_\infty{}^2}\left(\frac{\partial u}{\partial \eta}\right)_x^2 \tag{83.36}$$

83.7. INTEGRAL RELATIONS

The momentum integral of the compressible boundary layer can be obtained on integrating (83.16) with respect to y from 0 to ∞, if it is borne in mind that when y is large on the scale of the boundary layer, it can still be very small (and indeed zero in the limiting case $\mu_1 \to 0$) on the scale of the inviscid flow. Defining the displacement thickness δ_1 and the momentum thickness ϑ by

$$\delta_1 = \int_0^\infty \left(1 - \frac{\rho u}{\rho_1 u_1}\right) dy \qquad \vartheta = \int_0^\infty \frac{\rho u}{\rho_1 u_1}\left(1 - \frac{u}{u_1}\right) dy \tag{83.37}$$

it is found that

$$\frac{d\vartheta}{dx} + \frac{1}{u_1}\frac{du_1}{dx}(2\vartheta + \delta_1) + \frac{\vartheta}{\rho_1}\frac{d\rho_1}{dx} = \frac{\tau_w}{\rho_1 u_1^2} \tag{83.38}$$

which is closely related to the momentum integral of the incompressible boundary layer.

Before integrating the energy equation, it is convenient to add to it the momentum integral multiplied by u, obtaining

$$\rho u \frac{\partial}{\partial x}(c_pT + \tfrac{1}{2}u^2) + \rho v \frac{\partial}{\partial y}(c_pT + \tfrac{1}{2}u^2) = \frac{\partial}{\partial y}\left(\frac{\mu}{\mathrm{Pr}}\frac{\partial}{\partial y}\left(c_pT + \frac{\mathrm{Pr}}{2}u^2\right)\right) \tag{83.39}$$

On integration, it follows that

$$\frac{d}{dx}\int_0^\infty \rho u[c_p(T - T_1) + \tfrac{1}{2}(u^2 - u_1^2)]\,dy = -\frac{c_p\mu}{\mathrm{Pr}}\frac{\partial T}{\partial y} \tag{83.40}$$

since $u = 0$ on the plate. In particular, if the plate is insulated, then

$$\int_0^\infty \rho u[c_p(T - T_1) + \tfrac{1}{2}(u^2 - u_1^2)]\,dy = \text{const} \tag{83.41}$$

It is noted that, when $\mathrm{Pr} = 1$, Eq. (83.39) has a simple integral, due to Crocco,

$$c_pT + \tfrac{1}{2}u^2 = \text{const} \tag{83.42}$$

which satisfies all the boundary conditions if the external flow is homenergic and the plate is insulated. The integral is particularly useful in approximate theories for gases such as air in which $\mathrm{Pr} = O(1)$. It may also be written in the form

$$\frac{T}{T_1} = 1 + \frac{\gamma - 1}{2}M_1^2\left(1 - \frac{u^2}{u_1^2}\right) \tag{83.43}$$

83.8. THE BOUNDARY LAYER ON A FLAT PLATE IN A UNIFORM STREAM

This boundary layer, in which the pressure gradient is zero, is, because of its relative simplicity, the natural starting point of an investigation of the general properties of boundary layers. It is also of particular importance because it frequently occurs in practice. In this section, it will be assumed that the temperature of the plate is constant, when the governing partial differential equations may be reduced to a pair of ordinary nonlinear differential equations. It is then possible to study with comparative ease the dependence of skin friction, heat transfer, and recovery temperature on Mach number, viscosity law, Prandtl number, and wall temperature. Furthermore, the properties of this boundary layer may be used to check the accuracy of approximate methods of calculation.

The differential equations for this problem are (83.16) to (83.19), with the simplification that $u_1 = u_\infty$ and is constant. The boundary conditions are that

$$\begin{aligned} &u = v = 0 \quad \text{and} \quad \text{either } T = T_w \quad \text{or } \partial T/\partial y = 0 \quad \text{at } y = 0 \\ &u = u_\infty,\ T = T_\infty \quad \text{at } y = \infty,\ x > 0 \quad \text{and} \quad \text{at } x = 0,\ y > 0 \end{aligned} \tag{83.44}$$

where T_w is constant. If $\partial T/\partial y = 0$ at $y = 0$, the corresponding plate temperature T_r, called the *recovery temperature*, is also constant.

These equations may be reduced to a pair of mutually dependent ordinary nonlinear equations for T and ψ on writing

$$\theta = \eta\left(\frac{u_\infty\rho_\infty}{2\mu_\infty x}\right)^{1/2} \qquad \psi = \left(\frac{2\mu_\infty u_\infty x}{\rho_\infty}\right)^{1/2} f(\theta) \qquad T = T(\theta) \tag{83.45}$$

where f and T are functions of θ only, and η is defined in (83.33). The new equations are

$$f(\theta)f''(\theta) + \left[\frac{\mu\rho}{\mu_\infty\rho_\infty} f''(\theta)\right]' = 0 \qquad (83.46)$$

$$\Pr f(\theta)T'(\theta) + \left[\frac{\mu\rho}{\mu_\infty\rho_\infty} T'(\theta)\right]' + \Pr(\gamma - 1)M_\infty{}^2 \frac{\mu\rho T_\infty}{\mu_\infty\rho_\infty}[f''(\theta)]^2 = 0 \qquad (83.47)$$

with boundary conditions

$$\begin{aligned} f(0) = f'(0) = 0 \quad &\text{and either } T(0) = T_w \quad \text{or } T'(0) = 0 \\ f'(\infty) = 1 \quad &\quad T(\infty) = T_\infty \end{aligned} \qquad (83.48)$$

The skin friction τ_w and heat transfer κ_w from the plate, given by

$$\tau_w = u_\infty \frac{(\mu\rho)_w}{\rho_\infty}\left(\frac{u_\infty\rho_\infty}{2\mu_\infty x}\right)^{1/2} f''(0) \qquad \kappa_w = \frac{c_p(\mu\rho)_w}{\Pr\rho_\infty}\left(\frac{u_\infty\rho_\infty}{2x\mu_\infty}\right)^{1/2} T'(0) \qquad (83.49)$$

are both proportional to $x^{-1/2}$.

Short of a full-scale numerical computation, further progress in solving these equations depends on making some simplifying assumption about μ or Pr. The simplest assumption, which also leads to results of the greatest value, is to take the viscosity proportional to the absolute temperature [that is, Eq. (83.23) with $C = 1$] and $\Pr = 1$. Then, since $\mu\rho = \mu_\infty\rho_\infty$, the equations reduce to

$$ff'' + f''' = 0 \qquad (83.50)$$

$$fT' + T'' + (\gamma - 1)T_\infty M_\infty f''^2 = 0 \qquad (83.51)$$

The momentum equation (83.50) is now explicitly independent of T and is identical with Blasius' equation which occurs in the corresponding incompressible problem. Hence, $f''(0) = 0.470$, and

$$\gamma_w = 0.332 u_\infty \left(\frac{u_\infty\mu_\infty\rho_\infty}{x}\right)^{1/2} \qquad (83.52)$$

independent of Mach number and of the thermal plate condition. Thus, the two ways in which dissipation and heat transfer each modify the skin friction, namely, by thickening the boundary layer and increasing the viscosity, exactly cancel.

The energy equation may be integrated immediately in terms of f, giving

$$\frac{T}{T_\infty} = 1 + \frac{\gamma - 1}{2} M_\infty{}^2(1 - f'^2) + (1 - f')\left(\frac{T_w}{T_\infty} - 1 - \frac{\gamma - 1}{2} M_\infty{}^2\right) \qquad (83.53)$$

when the plate temperature is T_w. If the plate is insulated, the recovery temperature

$$T_r = \left(1 + \frac{\gamma - 1}{2} M_\infty{}^2\right) T_w$$

in agreement with (83.43), while, in general,

$$\kappa_w = \frac{c_p \tau_w (T_w - T_r)}{u_\infty} \qquad (83.54)$$

Having obtained the temperature distribution, the independent variable can be changed back to y and various properties of the compressible boundary layer deduced. As an example, the effect of compressibility on the thickness of the boundary layer is now found. Strictly speaking, the thickness of the boundary layer is infinite, for the viscous stresses are never zero, but they are exponentially small when y is large. It is convenient, therefore, to define the outer limit of the layer at that value of y for which u is within 0.1% of its main-stream value. If the boundary layer is incompressible, its thickness is then

$$\delta_i = \left(\frac{2\mu_\infty x}{u_\infty \rho_\infty}\right)^{1/2} \theta^* \qquad \text{where } f'(\theta^*) = 0.999 \tag{83.55}$$

and, if compressible, its thickness is

$$\delta_c = \left(\frac{2\mu_\infty x}{u_\infty \rho_\infty}\right)^{1/2} \int_0^\theta \frac{\rho_\infty}{\rho}\, d\theta \tag{85.56}$$

In particular, if the plate is insulated,

$$(\delta_c - \delta_i)\left(\frac{u_\infty \rho_\infty}{2\mu_\infty x}\right)^{1/2} = \frac{\gamma - 1}{2} M_\infty{}^2 \int_0^{\theta^*} (1 - f'^2)\, d\theta = 0.85(\gamma - 1)M_\infty{}^2 \tag{83.57}$$

and increases with Mach number. On the other hand, the amount of the fluid in the boundary layer upstream of the line $x = X$ is

$$\int_0^X dx \int_0^{\delta_c} \rho\, dy \tag{83.58}$$

which is independent of Mach number if X is fixed.

The next simplest assumption to consider is that Pr is an arbitrary constant while retaining the linear viscosity law. The momentum equation still reduces to Blasius' equation, but the energy equation is now

$$T'' + \text{Pr}\, fT' + \text{Pr}\,(\gamma - 1)\, M_\infty{}^2 T_\infty f''^2 = 0 \tag{83.59}$$

with solution

$$\frac{T}{T_\infty} = 1 + B \int_\theta^\infty [f''(\phi)]^{\text{Pr}}\, d\phi + \text{Pr}(\gamma - 1)M_\infty{}^2 \int_\theta^\infty [f''(\phi_1)]^{\text{Pr}} \int_0^{\phi_1} [f''(\phi_2)]^{2-\text{Pr}}\, d\phi_2\, d\phi_1 \tag{83.60}$$

where B is determined from the condition at the plate. It is zero if the plate is insulated, and the recovery temperature is then

$$T_r = T_\infty + \text{Pr}\,(\gamma - 1)M_\infty{}^2 T_\infty \int_0^\infty [f''(\phi_1)]^{\text{Pr}} \int_0^{\phi_1} [f''(\phi_2)]^{2-\text{Pr}}\, d\phi_2\, d\phi_1 \tag{83.61}$$

which has been calculated by Pohlhausen [6] for various values of Pr. It is given with good accuracy by

$$T_\infty + \tfrac{1}{2}\sqrt{\text{Pr}}\,(\gamma - 1)M_\infty{}^2 T_\infty \tag{83.62}$$

when Pr $\approx$ 1. If the temperature T_w on the plate is prescribed, then

$$T_w - T_r = BT_\infty \int_0^\infty [f''(\phi)]^{\text{Pr}}\, d\phi \tag{83.63}$$

whence

$$\frac{\text{Pr}\, \kappa_w}{\mu_\infty c_p (T_w - T_r)} = \frac{\tau_w 2^{1/2} a(\text{Pr})}{\mu_\infty u_\infty f''(\text{Pr})} \tag{83.64}$$

where

$$a(\text{Pr}) = \frac{[f''(0)]^{\text{Pr}}}{\sqrt{2} \int_0^\infty [f''(\phi)]^{\text{Pr}}\, d\phi}$$

Pohlhausen has also tabulated $a(\text{Pr})$ and found that, if Pr $\approx$ 1,

$$a(\text{Pr}) \approx 0.332\text{Pr}^{1/3} \tag{83.65}$$

The next logical step is to consider a more general law for the viscosity. If $\Pr = 1$, the energy equation integrates immediately, giving (83.53), but the momentum equation is intractable except after extensive numerical computations. Many of these have been carried out, using various viscosity laws, and extended to cover values of Pr other than unity. The reader is referred to [2] and [4] for a discussion of the results obtained. Unfortunately, these results are, of necessity, restricted to certain sets of prescribed conditions, and any extension to a new set or to a more general situation, for example, a pressure gradient, is not usually immediate and requires additional heavy computation. A simple approximate theory is required, which can readily be generalized, and which agrees closely with the predictions of the numerical computations. This is provided by using Chapman's viscosity law (83.23), it being only necessary to replace μ_∞ by $C\mu_\infty$ wherever it occurs in the theory discussed above for a linear viscosity law, determining C finally from the correct viscosity law and the plate temperature. For example, if the true viscosity law is

$$\mu/\mu_\infty = (T/T_\infty)^\omega$$

$\Pr \doteqdot 1$, and the plate is insulated, then

$$C = (1 + \tfrac{1}{2}\sqrt{\Pr}\,(\gamma - 1)M_\infty{}^2)^{\omega-1} \tag{83.66}$$

from (83.62) and (83.23), so that

$$\tau_w = u_\infty\left(\frac{\mu_\infty\rho_\infty u_\infty}{2x}\right)^{1/2}[1 + \tfrac{1}{2}\sqrt{\Pr}\,(\gamma - 1)M_\infty{}^2]^{(\omega-1)/2}f''(0) \tag{83.67}$$

In particular, if $\omega = 0.76$, $\Pr = 1$, $\gamma = 1.4$, $M_\infty = 10$, the correct and approximate values of τ_w are

$$0.244u_\infty\left(\frac{\mu_\infty\rho_\infty u_\infty}{x}\right)^{1/2} \qquad \text{and} \qquad 0.231u_\infty\left(\frac{\mu_\infty\rho_\infty u_\infty}{x}\right)^{1/2} \tag{83.68}$$

respectively. Of course the more ω and Pr differ from unity, the greater will be the error incurred on using (83.23), but it is clear that, in a wide range of practically important problems, (83.23) can be regarded as a good approximation.

83.9. BOUNDARY LAYERS WITH PRESSURE GRADIENT

The previous section was concerned with boundary layers in a uniform stream and estimated the influence of the other parameters of the flow in determining their properties. Here attention is concentrated on the effect of pressure gradient; it is assumed that the other parameters may be allowed for in the practically important cases by taking $\Pr = 1$ and using Chapman's viscosity law. The appropriate equations are, after applying Howarth's transformations (83.33) and using the equation of state to eliminate ρ,

$$u\frac{\partial u}{\partial x} + \bar{v}\frac{\partial u}{\partial \eta} - \frac{a_\infty{}^2 T}{a_1{}^2 T_\infty}u_1\frac{du_1}{dx} = C\nu_\infty\frac{p_1}{p_\infty}\frac{\partial^2 u}{\partial \eta^2} \tag{83.69}$$

$$\frac{\partial u}{\partial x} + \frac{\partial \bar{v}}{\partial \eta} = 0 \tag{83.70}$$

$$c_p u\frac{\partial T}{\partial x} + c_p\bar{v}\frac{\partial T}{\partial \eta} + \frac{a_\infty{}^2}{a_1{}^2}\frac{T}{T_\infty}uu_1\frac{du_1}{dx} = C\nu_\infty c_p\frac{p_1}{p_\infty}\frac{\partial^2 T}{\partial \eta^2} + C\nu_\infty\frac{p_1}{p_\infty}\left(\frac{\partial u}{\partial \eta}\right)^2 \tag{83.71}$$

where $\nu = \mu/\rho$ is the kinematic viscosity, and $a = (\gamma p/\rho)^{1/2}$, the local speed of sound. The boundary conditions are that

$$\begin{aligned} &u = \bar{v} = 0 \quad \text{and} \quad \text{either } T = T_w(x) \quad \text{or } \partial T/\partial \eta = 0 \quad \text{at } \eta = 0,\ x > 0 \\ &\qquad u = u_1(x),\ T = T_1(x) \quad \text{at } \eta = \infty,\ x > 0 \\ &\qquad u = u_1(0),\ T = T_1(0) \quad \text{at } x = 0,\ \eta > 0 \end{aligned} \tag{83.72}$$

Furthermore, the flow outside the boundary layer is assumed to be homenergic, so that

$$\tfrac{1}{2}u_1^2 + c_pT_1 = \tfrac{1}{2}u_1^2 + \frac{a_1^2}{\gamma - 1} = \text{const} \tag{83.73}$$

Let

$$T = T_\infty\left(1 + \frac{\gamma - 1}{2} M_\infty^2\right) S - \frac{\gamma - 1}{2} T_\infty \frac{u^2}{a_\infty^2} \tag{83.74}$$

which reduces the energy equation to

$$u\frac{\partial S}{\partial x} + \bar{v}\frac{\partial S}{\partial \eta} = C\nu_\infty \frac{p_1}{p_\infty}\frac{\partial^2 S}{\partial \eta^2} \tag{83.75}$$

In particular, if $\partial S/\partial \eta = 0$ at $\eta = 0$,

$$S \equiv 1 \tag{83.76}$$

in agreement with (83.42). In terms of S, the momentum equation is

$$u\frac{\partial u}{\partial x} + \bar{v}\frac{\partial u}{\partial \eta} - \left(\frac{a_\infty}{a_1}\right)^2 \left[\left(1 + \frac{\gamma - 1}{2} M_\infty^2\right) S - \frac{\gamma - 1}{2}\frac{u^2}{a_\infty^2}\right] u_1 \frac{du_1}{dx} = C\nu_\infty \frac{p_1}{p_\infty}\frac{\partial^2 u}{\partial \eta^2} \tag{83.77}$$

The Illingworth-Stewartson transformation may be used to simplify (83.77). First, write

$$Y = \frac{a_1}{a_\infty}\eta \qquad u = \frac{a_1}{a_\infty}\frac{\partial \psi}{\partial Y} \qquad \text{whence} \qquad \bar{v} = -\frac{\partial \psi}{\partial x} - \frac{Y}{a_1}\frac{da_1}{dx}\frac{\partial \psi}{\partial Y} \tag{83.78}$$

and (83.77) becomes

$$\frac{\partial \psi}{\partial Y}\frac{\partial^2 \psi}{\partial Y\,\partial x} - \frac{\partial \psi}{\partial x}\frac{\partial^2 \psi}{\partial Y^2} - \left(\frac{a_\infty}{a_1}\right)^2 S\left(1 + \frac{\gamma - 1}{2} M_\infty^2\right) u_1 \frac{du_1}{dx} = C\nu_\infty \frac{p_1}{p_\infty}\left(\frac{a_\infty}{a_1}\right)\frac{\partial^3 \psi}{\partial Y^3} \tag{83.79}$$

where ψ is now a function of x and Y. Finally, write

$$X = \int_0^x \frac{p_1 a_1}{p_\infty a_\infty}\,dx = \int_0^x \left(\frac{a_1}{a_\infty}\right)^{(3\gamma-1)/(\gamma-1)} dx \tag{83.80}$$

whence

$$\frac{\partial \psi}{\partial Y}\frac{\partial^2 \psi}{\partial Y\,\partial X} - \frac{\partial \psi}{\partial X}\frac{\partial^2 \psi}{\partial Y^2} - a_\infty^2 S M_1 \frac{dM_1}{dX} = C\nu_\infty \frac{\partial^3 \psi}{\partial \eta^3} \tag{83.81}$$

since

$$a_\infty^2 M_1 \frac{dM_1}{dX} = \left(\frac{a_\infty}{a_1}\right)^2 u_1 \frac{du_1}{dX}\left(1 + \frac{\gamma - 1}{2} M_1^2\right) = \left(\frac{a_\infty}{a_1}\right)^4 \left(1 + \frac{\gamma - 1}{2} M_\infty^2\right) u_1 \frac{du_1}{dX} \tag{83.82}$$

In terms of X and Y, S satisfies

$$\frac{\partial \psi}{\partial Y}\frac{\partial S}{\partial X} - \frac{\partial \psi}{\partial X}\frac{\partial S}{\partial Y} = C\nu_\infty \frac{\partial^2 S}{\partial Y^2} \tag{83.83}$$

The appropriate boundary conditions are

$$\psi = \frac{\partial \psi}{\partial Y} = 0 \quad \text{and} \quad \text{either } S = S_\infty(x) \quad \text{or} \quad \partial S/\partial Y = 0 \quad \text{at } Y = 0,\ X > 0$$

$$\partial \psi/\partial Y = a_\infty M_1(x),\ S = 1 \quad \text{at } Y = \infty,\ X > 0 \quad \text{and} \quad \text{at } X = 0,\ Y > 0 \tag{83.84}$$

where

$$T_\infty S_\infty(x)\left(1 + \frac{\gamma - 1}{2} M_\infty^2\right) = T_w(x)$$

The skin-friction and heat-transfer coefficients are now

$$\begin{aligned} \tau_w &= \left(\mu \frac{\partial u}{\partial y}\right)_{y=0} = C\mu_\infty \left(\frac{a_1}{a_\infty}\right)^2 \left(\frac{\partial^2 \psi}{\partial Y^2}\right)_{Y=0} \\ \kappa_w &= -\left(k \frac{\partial T}{\partial y}\right)_{y=0} = -\frac{Ca_1 a_\infty \mu_\infty}{\gamma - 1}\left(1 + \frac{\gamma - 1}{2} M_\infty^2\right)\left(\frac{\partial S}{\partial Y}\right)_{Y=0} \end{aligned} \tag{83.85}$$

As a result of the transformations, the Mach number of the flow only appears in the boundary conditions. A correlation is therefore established between boundary layers in gases for which $M_\infty \neq 0$ and those for which $M_\infty = 0$, although it is perhaps simpler to regard both as equivalent to a boundary layer in an incompressible fluid with a pressure gradient both along and across the layer.

In particular, if the plate is insulated, $S = 1$ and the equations are reduced to those for an incompressible boundary layer, with no heat transfer from the plate and, therefore, no transverse pressure gradient. The whole apparatus of the theory of incompressible boundary layers may, therefore, be taken over into the theory of such compressible boundary layers. It is only necessary to bear in mind that, in effecting the correlation, both the x and y coordinates are distorted, and that the main-stream velocity of the incompressible boundary layer is proportional to the Mach number of the main stream of the related compressible boundary layer. It is difficult to estimate qualitatively the effect of compressibility on the skin friction for it enters in several ways, some of which tend to increase and some to decrease the skin friction. However, the position of separation depends only on the stretching of the x coordinate, and on the modification in the main-stream velocity, both of which augment the effect of the pressure gradient and can be expected to move the point of separation, $\tau_w = 0$, upstream toward the pressure minimum as M_∞ increases.

The effect of heat transfer is to vary the pressure gradient across the layer. When the plate is heated so that $T_w > T_r$, it may be expected that $S \geq 1$, and the main-stream pressure gradient is augmented in the boundary layer. On the other hand, when the plate is cooled, the main-stream pressure gradient is diminished in the boundary layer. Hence, in particular, separation, defined by $\tau_w = 0$, moves forward toward the pressure minimum on heating the plate and backward on cooling it. Confirmation of this conclusion is provided by the numerical solution of (83.81) and (83.83), taking $T_w = \text{const}$ and $M_1 = X^m$, m constant, when they reduce to ordinary differential equations [10].

Very little work has been done on the dependence of skin friction and heat transfer on Pr when $M_\infty > 0$. When $M_\infty = 0$, it appears that the skin friction is almost independent of Pr near Pr = 1, but that the heat transfer varies appreciably with Pr. The approximate formula given in the previous section,

$$\kappa_w(\text{Pr}) = \text{Pr}^{-2/3}\kappa_w(1)$$

expressing the dependence of heat transfer on Pr near Pr = 1 is also fairly accurate here, provided that the pressure gradient is favorable. Theories have also been developed for the special cases Pr large and Pr small; the reader is referred to [1] for a discussion of them.

83.10. STABILITY

The theory of the stability of the compressible laminar boundary layer follows lines similar to the theory when the fluid is incompressible. Thus, perturbations in each of the dependent variables, i.e., velocity components, pressure, density, and temperature, of the form

$$\phi(y) \exp i\alpha(x \cos \theta + z \sin \theta - ct)$$

are assumed, where z measures distance parallel to the plate and across the main stream, and α, θ, c are constants of which α, θ are real. On substituting into the Navier-Stokes equations, a similar but more complicated equation than Orr and Sommerfeld's is obtained. The problem is now to find, for given values of α, θ and Re (the Reynolds number of the undisturbed flow), the characteristic values of c for which solutions satisfying the required conditions at the plate and at the outer edge of the boundary layer can be found. Of particular importance is the critical value Re_c of Re such that, if $Re < Re_c$, *all* characteristic values of c have negative imaginary parts, so that the corresponding solutions are damped in time, while, if $Re > Re_c$, some characteristic values of c have positive imaginary parts, so that the corresponding solutions are amplified. Hence, it is then possible to say that Re_c is the maximum Reynolds number at which the flow is stable. For a fuller discussion of the problems involved, the reader is referred to [11]; here only a brief summary of the conclusions is given.

If the plate is insulated, Re_c decreases steadily as M_∞ increases from a value of about 80,000 at $M_\infty = 0$ to about 1000 at $M_\infty = 3$, when the boundary-layer assumptions can hardly be expected to hold. Heating the plate, in general, is destabilizing, while cooling is stabilizing. In fact, for a range of moderate supersonic speeds, $1 \leq M_\infty \leq M_1$, where M_1 depends on the physical properties of the fluid, two-dimensional disturbances (that is, $\theta = 0$) can be completely stabilized if the plate is cooled sufficiently. However, three-dimensional disturbances (that is, $\theta \neq 0$) are not stabilized completely, and lead to moderate values of Re_c (for example, $Re_c = 1.65 \times 10^6$ when $T_w = 1.073T_\infty$ and $M_\infty = 1.6$). Three-dimensional disturbances are ordinarily not important at subsonic speeds, while in incompressible flow they may be neglected altogether, since they are then equivalent to two-dimensional disturbances at a lower Reynolds number. Finally, the theory is confirmed to some extent by a small body of experimental evidence.

Once the laminar boundary layer is unstable, it can be expected to undergo a transition to turbulence, and, from the theory for an insulated plate, it would appear that transition will tend to occur earlier as M_∞ increases. This is certainly true at moderate supersonic speeds, but, at $M_\infty \approx 3.5$, the trend is reversed, and the transition Reynolds number thereafter rises rapidly, the hypersonic boundary layer being extraordinarily stable. The theory discussed here is strictly not valid when M_∞ is large, and no complete explanation of the phenomenon is known at present.

83.11. INTERACTION BETWEEN SHOCK WAVES AND BOUNDARY LAYERS

When a supersonic stream is flowing past a wall, a shock wave nearly always occurs at some stage. Except when the shock is weak, its interaction with the boundary layer at the wall is spectacular, leading to a secondary shock or shocks being developed, upstream of the main shock, whose properties depend in a complicated way on the curvature of the wall, on the properties of the main stream, and on whether the boundary layer is turbulent or laminar. Here only the simplest problem, although still of great difficulty, will be discussed, in which the main stream is uniform apart from the shock, and the wall is straight. In Figs. 83.1 and 83.2 are shown sketches of the interactions which occur when the boundary layer is laminar and when it is turbulent. In the laminar case, the most interesting feature is the large distance upstream of the point of incidence at which the effect of the main shock is felt, notwithstanding the parabolic nature of the boundary-layer equations and the hyperbolic nature of the equations of supersonic inviscid flow, both of which forbid the propagation of disturbances upstream. The boundary layer is, however, actually lifted away from the wall, and the disturbance caused by the main shock propagates upstream *via* an elongated bubble of relatively slowly moving, circulating fluid, lying between it and the wall. The bubble starts at the separation of the boundary layer from the plate, where, apart from position, conditions are virtually independent of the incident shock which provokes it. The mechanism at separation is known as a free interaction, and operates as follows. There is a compression fan in the main stream, with a small pressure change but a comparatively large pressure gradient. Its effect on the boundary layer is small, except near the plate where the fluid is moving slowly. The stream tubes there thicken, pushing the rest of the boundary layer away from the wall, although not otherwise seriously modifying it. The thickening of the boundary layer means that the stream tubes of the supersonic inviscid flow just above are compressed, producing an adverse pressure gradient.

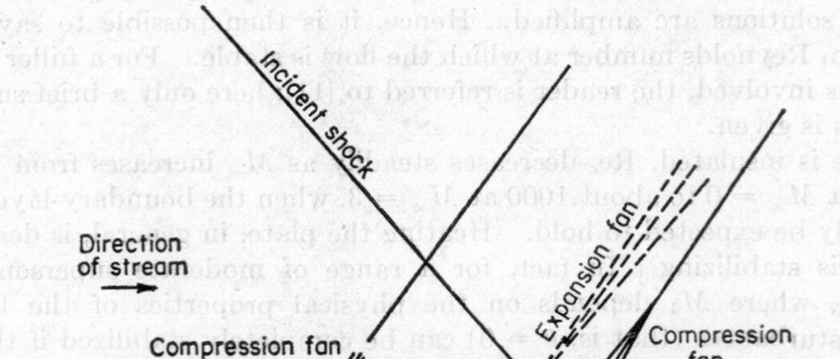

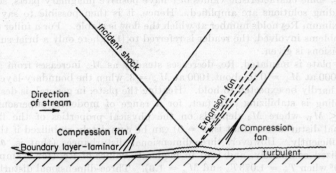

FIG. 83.1. Schematic drawing of the intersection of a shock wave and a boundary layer which is initially laminar and becomes turbulent.

A balance must be set up between the rate of thickening of the boundary layer and of the inviscid compression fan. Chapman has given a simple dimensional argument on these lines, which Gadd has confirmed by obtaining a characteristic solution of the boundary-layer equations appropriate to a free interaction. He finds that the pressure p_s at separation is

$$p_s = p_\infty + \tfrac{1}{2}\rho_\infty U_\infty^2 \frac{1.13}{(M_\infty^2 - 1)^{1/4}\mathrm{Re}_s^{1/4}} \tag{83.86}$$

where Re_s is the Reynolds number of the boundary layer at separation. Although the theory is strictly valid only in the limit $\mathrm{Re}_s \to \infty$, the observed dependence of p_s on M_∞ and Re_s is the same as (83.86), except for the numerical coefficient which is approximately equal to 0.93 instead of 1.13.

Downstream of separation, the pressure continues to rise until the boundary layer is clear of the plate, when it flattens out to a plateau pressure $\approx p_0 + 2(p_s - p_0)$. The boundary layer is thereafter a free jet separating the main stream from the elongated bubble of circulating air near the plate. The plate pressure is nearly constant, because the air in the bubble is moving relatively slowly, until the main shock strikes the detached shear layer, turning it back toward the plate, when it gradually rises toward the value expected on inviscid theory. No wholly satisfactory theory to determine quantitatively the length of the elongated bubble has yet been found; perhaps the most hopeful approach is Chapman's, which has been successful in finding important flow properties when separation occurs at the leading edge.

In general, it is difficult to produce conditions in which the boundary layer remains laminar right through the interaction region until it reattaches. If the boundary layer is laminar at separation, the detached shear layer has a low reserve of stability, and transition to turbulence often starts before reattachment. Qualitatively, the main effect of transition on the plate pressure is to shorten the distance after incidence of the main shock before it reaches the inviscid value.

When the flow is completely turbulent, pressure distributions observed in different laboratories in apparently similar circumstances show marked divergences, which can at present only be ascribed to effects normally neglected. The general features, however, are similar to those when the flow is laminar, the differences being of degree rather than of character. Thus, the turbulent boundary layer is much more reluctant to separate than is a laminar layer. Furthermore, a much greater proportion of the boundary layer is moving supersonically and contracts when slowed down. Hence, to achieve a free interaction layer, the bottom of the turbulent boundary layer must

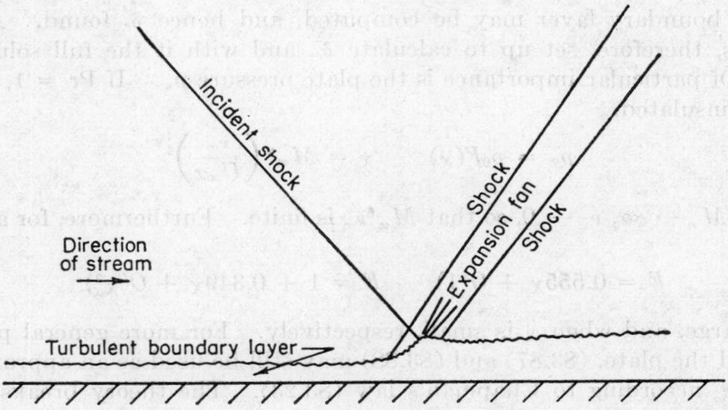

FIG. 83.2. Schematic drawing of the intersection of a shock wave and a turbulent boundary layer.

be lifted more sharply from the wall than in the laminar case. Subsequently, the pressure rises more slowly until the main shock reaches the free jet, and thereafter more rapidly, but a presentation in greater detail must wait until a larger measure of experimental agreement has been obtained.

83.12. HYPERSONIC BOUNDARY LAYER

The nature of the boundary layer on a flat plate in a uniform stream has already been discussed in Sec. 83.8. In hypersonic flow when M_∞ is large, the main characteristics are that it is much thicker, $\delta_c = O(M_\infty{}^2\delta_i)$, and the fluid in it is much lighter, $\rho = O(M_\infty{}^{-2}\rho_0)$, and much hotter, $T = O(M_\infty{}^2 T_0)$. In the limit $M_\infty \to \infty$, two new properties emerge. First, the boundary layer has a sharp edge. Although, in incompressible flow, the edge is blurred [cf. (83.55)], the blurring is almost independent of Mach number and so, in the limit $M_\infty \to \infty$, it is a vanishingly small fraction of the thickness of the layer. Second, this edge is a streamline of the inviscid flow outside. Although this result is true in general, it may be seen most easily in the special case when $\mathrm{Pr} = 1$ and $\mu \propto T$; for then the amount of fluid in the boundary layer is independent of M_∞. Hence, of the fluid originally within a distance δ_c of the line $y = 0$ upstseam of the plate, only a fraction $O(\delta_i/\delta_c) = O(M_\infty{}^{-2})$ has entered the boundary layer at or upstream of the station x. In the limit $M_\infty \to \infty$, therefore, the edge is a streamline of the inviscid flow, although not, of course, of the boundary-layer flow because fluid must continually be drawn into it. These two results derived when the main stream is uniform also apply when there is a pressure gradient.

Hence, when M_∞ is large, the effect of the boundary layer on the main stream is as though the plate were replaced by a "body" of shape $y = \pm\,\delta_c(x)$, whose boundary is a streamline, but whose shape depends on the main stream. As a result, a strong shock wave is set up near the leading edge, gradually weakening until ultimately it degenerates into a Mach wave, to divert the main stream past the body. In the theory, it is assumed that $\delta_c \ll x$, which is satisfied over almost all the plate. The perturbation of the component, parallel to the plate, of the main-stream velocity through the shock is negligible, although the pressure jump may be large. This means that, in the inviscid region behind the shock, the conventional theory of hypersonic flow may be applied. In principle, the method of solution is to assume a general form for δ_c, and calculate the resulting inviscid flow past a body of that shape, determining, in particular, the pressure on the body which is, of course, the pressure in the boundary layer at the same value of x. Once the pressure is known, the flow in the boundary layer may be computed, and hence δ_c found. An iterative procedure is, therefore, set up to calculate δ_c, and with it the full solution of the problem. Of particular importance is the plate pressure p_w. If Pr = 1, $\mu \propto T$, and the plate is insulated,

$$p_w = p_0 F(\chi) \qquad \chi = M_\infty{}^3 \left(\frac{\nu_\infty}{U_\infty x} \right)^{1/2} \tag{83.87}$$

in the limit $M_\infty \to \infty$, $\nu_\infty \to 0$, so that $M_\infty{}^6 \nu_\infty$ is finite. Furthermore, for air, in which $\gamma = 1.4$,

$$F = 0.555\chi + O(1) \qquad F = 1 + 0.349\chi + O(\chi^2) \tag{83.88}$$

when χ is large, and when χ is small, respectively. For more general properties of the fluid and the plate, (83.87) and (83.88) may still be used as an approximation on modifying μ_∞ according to Chapman's law (83.23). The theory breaks down near the leading edge, being invalid if

$$M_\infty \left(\frac{\nu_\infty}{U_\infty x} \right)^{1/2} = O(1)$$

but, when x is that small, the continuum hypothesis is no longer tenable. The theoretical conclusions are verified qualitatively by experiment, but quantitatively it appears that the plate pressure is remarkably sensitive to the degree of bluntness of the leading edge. The observations differ from the predictions by varying factors up to about two, except when the leading edge is extremely sharp.

83.13. REFERENCES

General Information

[1] C. C. Lin (ed.): "High Speed Aerodynamics and Jet Propulsion," vol. 5, Princeton Univ. Press, Princeton, N.J., 1959.

[2] L. Howarth (ed.): "Modern Developments in Fluid Dynamics: High Speed Flow," vols. 1 and 2, Oxford Univ. Press, New York, 1953.

[3] S.-I Pai: "Viscous Flow Theory," vols. 1 and 2, Van Nostrand, Princeton, N.J., 1956, 1957.

[4] G. Kuerti: The laminar boundary layer in compressible flow, *Advances in Appl. Mechanics*, **2** (1951), 21–92.

Uniform Main Stream

[5] L. Crocco: Una caratteristica trasformazione delle equazioni dello strato limite nei gas, *Atti di Guidonia*, **7** (1939), 105–120.

[6] E. Pohlhausen: Der Wärmeaustausch zwischen festen Körpern und Flüssigkeiten mit kleiner Reibung und kleiner Wärmeleitung, *Z. angew. Math. Mech.*, **1** (1921), 115–121.

With Pressure Gradient

[7] D. R. Chapman, M. W. Rubesin: Temperature and velocity profiles in the compressible laminar boundary layer with arbitrary distribution of surface temperature, *J. Aeronaut. Sci.*, **16** (1949), 547–565.

[8] C. R. Illingworth: Steady flow in the laminar boundary layer of a gas, *Proc. Roy. Soc. London*, **A,199** (1949), 533–558.

[9] K. Stewartson: Correlated incompressible and compressible boundary layers, *Proc. Roy. Soc. London*, **A,200** (1949), 84–100.

[10] C. B. Cohen, E. Reshotko: Similar solutions for the compressible laminar boundary layer with heat transfer and pressure gradient, *NACA Tech. Note* 3325 (1955).

Stability

[11] C. C. Lin: "The Theory of Hydrodynamic Stability," Cambridge Univ. Press, New York, 1955.

[12] E. R. Van Driest: Calculation of the stability of the laminar boundary layer in a compressible fluid on a flat plate with heat transfer, *J. Aeronaut. Sci.*, **19** (1952), 801–812.

[13] D. W. Dunn, C. C. Lin: On the role of three-dimensional disturbances in the stability of supersonic boundary layers, *J. Aeronaut. Sci.*, **20** (1953), 577–578.

Interaction with Shock Waves

[14] M. J. Lighthill: On boundary layers and upstream influence, II; Supersonic flows without separation, *Proc. Roy. Soc. London*, **A,214** (1953), 478–507.

[15] G. E. Gadd: A theoretical investigation of laminar separation in supersonic flow, *J. Aeronaut. Sci.*, **24** (1957), 759–771, 784.

[16] D. R. Chapman, D. M. Kuehn, H. K. Larson: Investigation of separated flows in supersonic and subsonic streams with emphasis on the effect of transition, *NACA Tech. Note* 3869 (1957).

Hypersonic Boundary Layer

[17] L. Lees: Hypersonic flow, *Proc. 5th Intern. Conf. Inst. Aeronaut. Sci.*, 1955, pp. 241–276.

[18] W. D. Hayes, R. O. Probstein: "Hypersonic Flow Theory," Academic Press, New York, 1959.

Turbulence

[19] E. R. Van Driest: Turbulent boundary layer on a cone in a supersonic flow at zero angle of attack, *J. Aeronaut. Sci.*, **19** (1952), 55–57.

[20] R. H. Korkegi: Transition studies and skin friction measurements on an insulated flat plate at a Mach number of 5.8, *J. Aeronaut. Sci.*, **23** (1956), 97–107.

CHAPTER 84

TURBULENCE

BY

S. I. PAI, Ph.D., College Park, Md.

84.1. INTRODUCTION

It is common experience that the flows observed in nature, such as rivers and winds, usually differ from the streamline flow or the laminar flow of a viscous fluid. The mean motions of such flows do not satisfy the Navier-Stokes equations (68.19) for a viscous fluid. Such flows, which occur at high Reynolds numbers, are often termed *turbulent* flows. In turbulent flow, the apparently steady motion of the fluid is only steady in so far as the temporal mean values of the velocities and the pressures are concerned, whereas actually both the velocities and the pressures are irregularly fluctuating. The velocity and the pressure distributions in turbulent flows, as well as the energy losses, are determined mainly by the turbulent fluctuations.

The essential characteristic of turbulent flow is that the turbulent fluctuations are random in nature. Hence, the final and logical solution of the turbulence problem requires the application of methods of statistical mechanics. In turbulent flow, the most important problem is the problem of transfer of forces by such random motion.

84.2. REYNOLDS' EXPERIMENTS AND OTHER FACTS ABOUT TURBULENT FLOWS

The first systematic investigation of turbulent flow was made by Reynolds [19]. In his experiments, a filament of colored liquid was introduced into water flowing along a long straight circular tube with a smooth entry. He defined a nondimensional quantity $\mathrm{Re} = Ud/\nu$ (now called the *Reynolds number*), where U = mean velocity of water in the tube, d = diameter of the tube, and ν = coefficient of kinematic viscosity of the water. At small Reynolds numbers, the colored filament is maintained as a straight line, with slight increase in width through molecular diffusion. The flow is laminar, and the pressure drop along the axis of the tube agrees with that given by the Hagen-Poiseuille formula. At large Reynolds numbers, the colored filament no longer maintains its straight-line shape, but is uniformly mixed with the whole fluid after a short distance from the entrance section. The flow is turbulent, and the pressure drop is much larger than that given by the Hagen-Poiseuille formula. Reynolds found that there was a critical Reynolds number $\mathrm{Re}_c \approx 2000$. If the Reynolds number Re of the test is smaller than Re_c, the flow is laminar; if $\mathrm{Re} > \mathrm{Re}_c$, the flow is turbulent. However, it was found later that Re_c depends on the test conditions, particularly the shape of the inlet, and the flow conditions of the supply of the water. If the water is free of disturbance before entering the tube, Re_c may be as high as 24,000. In practice, we are interested in the lowest critical Reynolds number which is obtained for an arbitrary disturbance in the entrance flow. Reynolds found that, in turbulent flow, the pressure drop is proportional to $U^{1.73}$, instead of U as in laminar flow. In the transition region, the turbulent mixing and the laminar motion alternated more or less regularly.

It was found later that the mean velocity distribution of turbulent flow in a tube is more uniform in the central part of the cross section, with a larger gradient near the wall than in the corresponding laminar flow case.

Turbulent flow occurs also in the boundary layer over the surface of a solid body (see pp. 82–23 to 82–30).

84.3. THE TURBULENT FIELD AND MEAN FLOW

In turbulent motion, even though the fluid is regarded as a continuum where the average over molecular motion has been taken, we still have to deal with turbulent velocity fluctuations superimposed on a mean motion. The separation of mean motion and turbulent fluctuation depends mainly on the scale of turbulence. Different scales give different descriptions of turbulent flow. Once the scale of turbulence is chosen, we may write, for instance, the instantaneous velocity component u_i as

$$u_i = \overline{u_i} + u_i' \tag{84.1}$$

where u_i is the ith component of the total fluid velocity, $\overline{u_i}$ is the ith mean velocity component, and u_i' is the ith component of the turbulent fluctuating velocity.

There are four different methods of averaging a turbulent quantity, viz.:

1. Time average in which we take the average at a fixed point in space over a long period of time
2. Space average in which we take the average over all the space at a given time
3. Space-time average in which we take the average over a long period of time and over the space
4. Statistical average in which we take the average over the whole collection of sample turbulent functions for a fixed point in space and at a fixed time

In an experimental investigation, time averages are used almost exclusively, and seldom space averages, but never statistical averages. In the theory, the statistical averages are used almost exclusively. The question is: What is the connection between the time averages and the statistical averages? Are they equal or not? In classical statistical mechanics, the answer is given by the ergodic theorem, which states sufficient conditions for the equality of the two kinds of average for almost all samples. In fluid mechanics, no ergodic theorem connecting these two entirely different types of averages has yet been proved.

84.4. REYNOLDS EQUATIONS AND REYNOLDS STRESSES

In turbulent flow, we usually assume that the instantaneous velocity components satisfy the Navier-Stokes equations. Substituting the expression for the instantaneous velocity components (84.1) into the Navier-Stokes equations of an incompressible fluid, and taking the mean values of the equations, we have the following *Reynolds equations* of motion for the turbulent flow of an incompressible fluid:

$$\begin{aligned}
&\frac{\partial \bar u}{\partial x} + \frac{\partial \bar v}{\partial y} + \frac{\partial \bar w}{\partial z} = 0 \\
&\rho\left(\frac{\partial \bar u}{\partial t} + \bar u\frac{\partial \bar u}{\partial x} + \bar v\frac{\partial \bar u}{\partial y} + \bar w\frac{\partial \bar u}{\partial z}\right) = -\frac{\partial \bar p}{\partial x} + \mu\nabla^2\bar u + \frac{\partial(-\rho\overline{u'^2})}{\partial x} + \frac{\partial(-\rho\overline{u'v'})}{\partial y} + \frac{\partial(-\rho\overline{u'w'})}{\partial z} \\
&\rho\left(\frac{\partial \bar v}{\partial t} + \bar u\frac{\partial \bar v}{\partial x} + \bar v\frac{\partial \bar v}{\partial y} + \bar w\frac{\partial \bar v}{\partial z}\right) = \frac{\partial \bar p}{\partial y} + \mu\nabla^2\bar v + \frac{\partial(-\rho\overline{u'v'})}{\partial x} + \frac{\partial(-\rho\overline{v'^2})}{\partial y} + \frac{\partial(-\rho\overline{v'w'})}{\partial z} \\
&\rho\left(\frac{\partial \bar w}{\partial t} + \bar u\frac{\partial \bar w}{\partial x} + \bar v\frac{\partial \bar w}{\partial y} + \bar w\frac{\partial \bar w}{\partial z}\right) = -\frac{\partial \bar p}{\partial z} + \mu\nabla^2\bar w + \frac{\partial(-\rho\overline{u'w'})}{\partial x} + \frac{\partial(-\rho\overline{v'w'})}{\partial y} + \frac{\partial(-\rho\overline{w'^2})}{\partial z}
\end{aligned} \tag{84.2}$$

The additional terms over the Navier-Stokes equations may be interpreted as stresses, the Reynolds or eddy stresses. For an incompressible fluid, these eddy or turbulent stresses are:

$$\begin{aligned} \sigma_x &= -\rho\overline{u'^2} & \tau_{xy} &= -\rho\overline{u'v'} & \tau_{xz} &= -\rho\overline{u'w'} \\ \tau_{yx} &= -\rho\overline{u'v'} & \sigma_y &= -\rho\overline{v'^2} & \tau_{yz} &= -\rho\overline{v'w'} \\ \tau_{zx} &= -\rho\overline{u'w'} & \tau_{zy} &= -\rho\overline{v'w'} & \sigma_z &= -\rho\overline{w'^2} \end{aligned} \tag{84.3}$$

These stresses form a tensor of second order, and represent the rate of transfer of momentum across the corresponding surfaces due to the turbulent velocity fluctuations.

The solutions of the Reynolds equations will properly represent the turbulent flow. However, there are only four Reynolds equations (84.2) for ten unknowns: the mean pressure, three mean velocity components, and six Reynolds stresses. In general, the Reynolds equations are not sufficient for determining these unknowns. This is one of the main difficulties in the theoretical investigation of turbulent flow.

In a similar manner, the Reynolds equations of motion for the turbulent flow of a compressible fluid may be obtained. But the expressions for the eddy stresses of a compressible fluid are much more complicated, because the fluctuation of density must be considered.

Similar to the Navier-Stokes equations, at the present time, it is not possible to solve the Reynolds equations for any practical flow problem. Additional assumptions and hypotheses are necessary to simplify these equations, in order to obtain some approximate solutions for important practical cases.

84.5. SEMIEMPIRICAL THEORIES OF TURBULENCE

Various semiempirical theories of turbulence have been formulated, based on different hypotheses about the Reynolds stresses.

Boussinesq [10] introduced the turbulent exchange coefficient ϵ such that, for a simple parallel mean flow $\bar{u} = \bar{u}(y)$, $\bar{v} = 0$, $\bar{w} = 0$, he wrote

$$\overline{u'v'} = -\epsilon \frac{d\bar{u}}{dy} \tag{84.4}$$

In actual analysis, one has to adopt further hypotheses about the variation of ϵ. In general, it was found that it is difficult to make a plausible assumption for ϵ in most problems of fluid flow, such as the flow in a pipe in which ϵ is different for different shapes of the pipe. But, for some particular flow problems such as those of atmospheric turbulence and of jet mixing, a simple assumption for ϵ may be made, and, in these cases, the theory has been successfully worked out [3].

Another semiempirical theory of turbulence, which is the most successful and the most well known, is the mixing-length theory originated by Prandtl [17]. In this theory, a mixing length l is introduced similar to the mean free path in the kinetic theory of gases, representing the length through which a certain quantity of the turbulent flow is assumed to be preserved. In two-dimensional parallel mean flow, we have

$$|q'| = l \frac{d\bar{q}}{dy} \tag{84.5}$$

where q is the transferable quantity of the fluid. The main advantage of the mixing-length over Boussinesq's theory is that it is generally easier to make a plausible assumption for l than for ϵ. For simple parallel flow, it may be shown that

$$\epsilon = -\overline{lv'} \tag{84.6}$$

There are many different mixing-length theories based on different quantities being transferred. In Prandtl's momentum transfer theory [17], the momentum of the fluid elements is assumed to be preserved in the mixing process. Prandtl assumed that

$$|v'| \propto |u'| \tag{84.7}$$

By the help of Eq. (84.5), Prandtl's formula for the shearing stress τ of nearly parallel turbulent flow is obtained as follows:

$$\tau = \rho l^2 \left| \frac{d\bar{u}}{dy} \right| \frac{d\bar{u}}{dy} \tag{84.8}$$

This formula has been successfully used in the calculations of many turbulent flow problems such as the flow along a flat plate, in jets or wakes [2],[5],[8], and other flows.

Prandtl's formula (84.8) has been extended to curved flow, divergent flow, and many other cases [5]. These extensions are not so successful as that for nearly parallel flow where Eq. (84.8) is applicable.

In Taylor's vorticity transfer theory [20], the transferable quantity is the vorticity ω (the curl of the velocity). For parallel mean flow, we have

$$|\omega'| = l_1 \frac{d\bar{\omega}}{dy} \tag{84.9}$$

where l_1 is the mixing length in Taylor's vorticity transfer theory, which is different from the mixing length l of Prandtl's momentum theory. The Reynolds equation of motion for this simple case becomes

$$\frac{dp}{dx} = -\rho|v'|l_1 \frac{d\bar{\omega}}{dy} = \rho|v'|l_1 \frac{d^2\bar{u}}{dy^2} \tag{84.10}$$

The corresponding formula of Prandtl's momentum transfer theory is

$$\frac{dp}{dx} = \frac{d\tau}{dy} = \rho \frac{d}{dy} \left[l^2 \left(\frac{d\bar{u}}{dy} \right)^2 \right] \tag{84.11}$$

Only when the mixing lengths l and l_1 are constant and $|v'| = l_1\, d\bar{u}/dy$, do Eqs. (84.10) and (84.11) give the same results with $2l^2 = l_1^2$.

In general, these two theories give different results. Even when the mean velocity distributions given by these two theories are the same, there is a difference between the temperature distributions for the two corresponding cases, e.g. in the two-dimensional turbulent jet-mixing problem. The results of vorticity transport theory give better agreement with experiment.

Von Kármán made a similarity hypothesis [22] to determine the mixing length, so that no special model for a transferable quantity is required. He found that, for parallel flow, the mixing length may be written as

$$l = \kappa|d\bar{u}/dy|/|d^2\bar{u}/dy^2| \tag{84.12}$$

where κ is a universal constant. From experimental results, $\kappa = 0.4$ is found. The shearing stress in two-dimensional parallel mean flow, according to von Kármán's similarity hypothesis, is

$$\tau = \kappa^2\rho(d\bar{u}/dy)^4/(d^2\bar{u}/dy^2) \tag{84.13}$$

84.6. THE LOGARITHMIC VELOCITY PROFILE AND GENERAL RESISTANCE LAW

The most important deduction from von Kármán's similarity hypothesis is the universal velocity distribution for the flow in circular pipes or between parallel plane walls (two-dimensional channel). Von Kármán first pointed out that the ratio between the velocity defect $(U_m - \bar{u})$ and the quantity $\sqrt{\tau_0/\rho}$ is a universal function of the ratio $(y_0 - y)/y_0$, that is,

$$\frac{U_m - \bar{u}}{\sqrt{\tau_0/\rho}} = f\left(\frac{y_0 - y}{y_0}\right) \tag{84.14}$$

where y_0 is the radius of the circular pipe or the half width of the channel, and y is the distance from the wall. U_m is the maximum axial velocity $\bar{u}$ occurring at $y = y_0$, that is, at the center of the pipe or the channel. τ_0 is the shearing stress at the wall $y = 0$.

The universal velocity distribution may be obtained by integrating Eq. (84.13) with replacing the shearing stress τ by its value on the wall τ_0. We then have

$$\bar{u} = \frac{1}{\kappa}\sqrt{\frac{\tau_0}{\rho}}\ln\frac{y + \Delta}{\delta} \tag{84.15}$$

where Δ and δ are constants of integration. Δ is of the same order of magnitude as the thickness of the laminar sublayer. Neglecting Δ with respect to y, we have

$$\bar{u} = \frac{1}{\kappa}\sqrt{\frac{\tau_0}{\rho}}\ln\frac{y}{\delta} \tag{84.16}$$

This is the well-known logarithmic velocity distribution near the wall, which has been verified by experimental results [2],[5],[8]. For a smooth wall, the length δ is determined by the physical parameters ρ and ν of the fluid, while, for a very rough surface, it is determined by the roughness of the wall.

For engineering application, it is customary to introduce a nondimensional pipe resistance coefficient λ to give the pressure drop in a pipe, that is,

$$\frac{dp}{dx} = \frac{p_1 - p_2}{L} = \frac{\lambda}{d}\frac{\rho}{2}u_0^2 \tag{84.17}$$

where $(p_1 - p_2)$ is the pressure drop on a pipe of length L and diameter d, and u_0 is the average mean velocity over a section of the pipe.

For a smooth pipe with logarithmic velocity distribution, it is found that

$$\frac{1}{\sqrt{\lambda}} = 2.035\log\left(\frac{u_0 d}{\nu}\sqrt{\lambda}\right) - 0.91 \tag{84.18}$$

Equation (84.18) has been verified experimentally by Nikuradse [16] with slight modification of the numerical values. Nikuradse's experiments give

$$\frac{1}{\sqrt{\lambda}} = 2.0\log\left(\frac{u_0 d}{\nu}\sqrt{\lambda}\right) - 0.80 \tag{84.19}$$

In the noncircular pipe, the characteristic length is often represented by the hydraulic mean length,

$$L_h = \frac{2S}{L_w} \tag{84.20}$$

where S is the cross-sectional area of the pipe, and L_w is the wetted circumferential length. If the hydraulic mean length is used instead of the radius of the circular pipe, the resistance law (84.19) may be used for noncircular pipes with an accuracy of within a few per cent.

For a pipe of rough surface, the resistance depends on the size, shape, and spacing of the roughness elements. Only for closely packed roughness, the linear dimension h of the roughness alone may be used to describe it. If $\sqrt{\tau_0/\rho} \cdot h/\nu < 4$, the surface may be considered as hydraulically smooth; that is, Eq. (84.19) holds; if $\sqrt{\tau_0/\rho} \cdot h/\nu > 100$, the surface may be considered as completely rough; in the case $4 \leq \sqrt{\tau_0/\rho} \cdot h/\nu \leq 100$, the effects of both roughness and viscosity are of equal importance.

For a completely rough pipe, the resistance law obtained experimentally by Nikuradse is

$$\frac{1}{\sqrt{\lambda}} = 2.00 \log \frac{y}{h} + 1.74 \tag{84.21}$$

When the Reynolds number $\mathrm{Re} = u_0 d/\nu$ is not large, say $\mathrm{Re} \leq 10^5$, the logarithmic law (84.16) may be approximated by a one-seventh power law,

$$\bar{u} = 8.74 \sqrt{\frac{\tau_0}{\rho}} \left(\frac{\sqrt{\tau_0/\rho} \cdot y}{\nu} \right)^{1/7} \tag{84.16$'$}$$

and the corresponding smooth pipe resistance coefficient is

$$\lambda = 0.3164 \left(\frac{u_0 d}{\nu} \right)^{1/4} \tag{84.19$'$}$$

Equation (84.19′) was found empirically by Blasius before the logarithmic velocity distribution was known. For $\mathrm{Re} > 10^5$, a better approximation is obtained by using an exponent $1/8$, $1/9$, or $1/10$, instead of $1/7$. It is now known that these power laws are approximations of the logarithmic law, valid in certain ranges of the Reynolds number.

84.7. TURBULENT JET-MIXING REGIONS AND WAKES

Another type of turbulent flow, in which no solid wall is in the flow field considered, is known as the free turbulence problem. The problems of jet mixing and wakes fall in this category. In the mixing-length theory of a mixing zone, we may write

$$v' \sim u' \sim l \frac{\partial \bar{u}}{\partial y} \tag{84.22}$$

where $\bar{u}$ is the mean velocity in the x direction. The rate of increase of the width of the mixing zone $2b$ with respect to time is comparable to v', the fluctuating velocity component across the flow. Then we have

$$\frac{Db}{Dt} \sim \frac{l}{b} U_{\max} \tag{84.23}$$

where $U_{\max}$ is the maximum mean velocity in the mixing region. Equation (84.23) is obtained from Eq. (84.22) by replacing $\partial\bar{u}/\partial y \sim U_{\max}/b$. In mixing-length theory, we may assume that l/b is a constant across a given section of the mixing region. Then, for a steady flow of a jet in a medium at rest, Eq. (84.23) gives

$$b = \text{const} \times x \tag{84.24}$$

where a proper origin of x is chosen. It can be shown that the corresponding $U_{\max}$ is given by the formula

$$U_{\max} = \text{const} \times \sqrt{\frac{M_0}{\rho} \frac{1}{b^{1+\delta}}} \tag{84.25}$$

where M_0 is the total rate of change of momentum in the x direction, $\delta = 0$ for the two dimensional case, and $\delta = 1$ for the axisymmetric case.

For a wake of a body in a uniform flow of velocity U_0, Eq. (84.23) gives

$$b = \text{const} \times (C_D S_b x)^{1/(2+\delta)} \tag{84.26}$$

where C_D is the drag coefficient of the body, and S_b is the reference area of the body. The maximum value of $u_1 = U_0 - \bar{u}$ is

$$\frac{u_{1\,\max}}{U_0} = \text{const} \times \left(\frac{C_D S_b}{x^{1+\delta}}\right)^{1/(1+\delta)} \tag{84.27}$$

The mean velocity distribution in the mixing region may be calculated from the mixing-length theory, with proper assumptions on the variation of mixing length in the flow field. The theoretical curve agrees well with the experimental results over a major portion of the flow field.

From experimental results, it is found that there is an analogy between the turbulent transfer of free turbulence and the molecular transfer of laminar flow. The essential difference seems to be only quantitative, the turbulent friction being much larger than laminar friction. This led Reichardt to the development of the inductive theory of free turbulence, in which the mean value of the quantities to be transferred, such as longitudinal mean velocity distribution, mean temperature distribution, or mean concentration distribution, can be represented by error functions or error integrals, which closely agree with the experimental results.

The spread of the concentration of the fluid of a jet in the jet-mixing region is usually wider than that of the velocity profile, and is close to that of the temperature profile.

84.8. FUNDAMENTALS OF THE STATISTICAL THEORY OF TURBULENCE

Even though the semiempirical theory had successfully predicted the mean velocity distributions in many practical problems, it is known to have serious limitations and inconsistencies. For an understanding of turbulent flow in general, a study of the mean velocity distribution is not sufficient. The fields of turbulent fluctuations must be studied in detail. Since these fluctuations are random in nature, the logical solution of the turbulence problem requires the application of the methods of statistical mechanics.

Modern statistical theory of turbulence is developed largely from physical intuition. G. I. Taylor [21] first introduced new notions into the study of the statistical theory of turbulence. These have been further developed by von Kármán [23],[24], Kolmogoroff [15], Heisenberg [13], and others. The most important new notions are the correlation function and the spectral function. The mathematical foundation for the theory of the correlation function and the spectral function has been studied extensively by Kampé de Fériet [14], Batchelor [1], Burgers [11], and others.

To define a random function, it is not sufficient to give the correlation function and the spectral function; it is also necessary to prescribe the probability distribution of the random function and the joint probability distribution of the values of the random function at different times and different positions in the space. A complete statistical specification of a turbulent motion thus requires a vast array of information. The general discussion of statistical theory of turbulence is rather complicated.

G. I. Taylor also introduced the concept of statistically isotropic turbulence. Isotropy introduces a great simplification into the theory. Great progress has been made on the statistical theory of isotropic turbulence. Most of the results from the statistical theory of turbulence are concerned with isotropic turbulence.

84.9. TURBULENCE IN COMPRESSIBLE FLUID FLOW

For the turbulent flow of an incompressible fluid, the effect of variation of density in the expression of Reynolds stresses is neglected. This effect is no longer negligible for the turbulent flow of a compressible fluid; i.e., it cannot be neglected for high-speed flow, or flow with large variation of temperature, or both. In the study of the turbulent flow of a compressible fluid, besides the correlation of the velocity components, we must also deal with the correlation of velocity and density, and that of the pressure and velocity. This complicates the analysis very much.

Mixing-length theories may be extended to compressible fluids. For two-dimensional parallel flow with mean flow field

$$\bar{u} = \bar{u}(y) \qquad \bar{v} = 0 \qquad \bar{\rho} = \bar{\rho}(y) \qquad \bar{T} = \bar{T}(y) \tag{84.28}$$

the fluctuations of the velocity component u, the density ρ, and the temperature T of the fluid may be written as

$$|u'| = l\,\frac{d\bar{u}}{dy} \qquad |\rho'| = l_\rho\,\frac{d\rho}{dy} \qquad |T'| = l_T\,\frac{d\bar{T}}{dy} \tag{84.29}$$

where l, l_ρ, and l_T are the corresponding mixing lengths for the velocity, density, and temperature, respectively. It is customary to assume that these mixing lengths are equal, in order to simplify the analysis so that practical problems may be solved. But there is experimental evidence that they are not equal. For instance, in the problem of jet mixing of a compressible fluid [3], the spread of temperature is wider than that of velocity. One way to explain this phenomenon is to assume that l_T is larger than l.

84.10. REFERENCES

Books

[1] G. K. Batchelor: "The Theory of Homogeneous Turbulence," Cambridge Univ. Press, New York, 1953.

[2] S. Goldstein: "Modern Developments in Fluid Dynamics," vols. 1 and 2, Oxford Univ. Press, New York, 1938.

[3] S. I. Pai: "Fluid Dynamics of Jets," Van Nostrand, Princeton, N.J., 1954.

[4] S. I. Pai: "Viscous Flow Theory," vol. 1, "Laminar Flow," Van Nostrand, Princeton, N.J., 1956.

[5] S. I. Pai: "Viscous Flow Theory," vol. 2, "Turbulent Flow," Van Nostrand, Princeton, N.J., 1957.

[6] L. Prandtl: The mechanics of viscous fluids, in W. F. Durand (ed.), "Aerodynamic Theory," vol. 3, div. G, Springer, Berlin, 1935.

[7] L. Prandtl: "Essentials of Fluid Dynamics," Hafner, New York, 1952.

[8] H. Schlichting: "Boundary Layer Theory," 2d ed., transl. by J. Kestin, McGraw-Hill, New York, 1960.

[9] A. A. Townsend: "The Structure of Turbulent Shear Flow," Cambridge, Univ. Press, New York, 1956.

Papers and Reports

[10] T. V. Boussinesq: Théorie de l'écoulement tourbillant, *Mém. prés. par div. savants*, **23** (1877).

[11] J. M. Burgers: On turbulent fluid motion, *Calif. Inst. Technol. Hydrodynamics Lab. Rept.* E-341 (1951).

[12] H. L. Dryden: The turbulence problem today, *Proc. 1st Midwestern Conf. Fluid Mechanics*, 1951, pp. 1–20.

[13] W. Heisenberg: Zur statistischen Theorie der Turbulenz, *Ann. Phys.*, **124** (1948), 628.

[14] J. Kampé de Fériet: Introduction to the statistical theory of turbulence; correlation and spectrum, *J. Soc. Ind. Appl. Math.*, **2** (1954), 1–9, 143–174, 244–271; **3** (1955), 90–117.

[15] A. N. Kolmogoroff: Dissipation of energy in locally isotropic turbulence, *Comp. rend. acad. sci. URSS*, **31** (1941), 538–540; **32** (1941), 16.

[16] J. Nikuradse: Laws of resistance and velocity distribution for turbulent flow of water in smooth and rough tubes, *Proc. 3d Intern. Congr. Appl. Mechanics*, Stockholm, 1930, vol. 1, pp. 239–248.

[17] L. Prandtl: Bericht über Untersuchungen zur ausgebildeten Turbulenz, *Z. angew. Math. Mech.*, **5** (1925), 136–264.

[18] H. Reichardt: Über eine neue Theorie der freien Turbulenz, *Z. angew. Math. Mech.*, **21** (1941), 257–264.

[19] O. Reynolds: An experimental investigation of the circumference which determines whether the motion of water shall be direct or sinuous and of the law of resistance in parallel channels, *Phil. Trans. Roy. Soc. London*, **A,174** (1883), 935–982; "Scientific Papers," vol. 2, p. 51.

[20] G. I. Taylor: The transport of vorticity and heat through fluids in turbulent motion, *Proc. Roy. Soc. London*, **A,135** (1932), pp. 685–702.

[21] G. I. Taylor: Statistical theory of turbulence, *Proc. Roy. Soc. London*, **A,151** (1935), pp. 421–478.

[22] Th. von Kármán: Mechanische Ähnlichkeit und Turbulenz, *Nachr. Ges. Wiss. Göttingen Math.-physik. Kl.*, **5** (1930), 58–76; also transl. as Mechanical similitude and turbulence, *NACA Tech. Mem.* 611 (1931).

[23] Th. von Kármán: The fundamentals of the statistical theory of turbulence, *J. Aeronaut. Sci.*, **4** (1937), 134–138.

[24] Th. von Kármán, D. C. Lin: On the statistical theory of isotropic turbulence, *Advances in Appl. Mechanics*, **2** (1951), 1–19.

CHAPTER 85

LUBRICATION

BY

H. PORITSKY, Ph.D., Schenectady, N.Y.

85.1. SLIDER BEARING

Consider a layer of a viscous liquid, confined between two parallel walls, as shown in Fig. 81.2. Let the distance between the walls be h, and replace the coordinate y by z; then Eq. (81.29a) gives the velocity of the fluid flow as

$$u = \frac{Uz}{h} - \frac{\partial p}{\partial x}\frac{z(h-z)}{2\mu} \tag{85.1}$$

Here μ is the viscosity of the liquid, U the velocity of the upper plate, and the pressure gradient $\partial p/\partial x$ is a constant. The net flow per unit depth (normal to the x,z plane) is

$$Q = \int_0^h u\,dz = \frac{Uh}{2} - \frac{\partial p}{\partial x}\frac{h^3}{12\mu} \tag{85.2}$$

In a plane z = const, the shear stress is

$$\tau_{xz} = \mu\frac{\partial u}{\partial z} = \frac{\mu U}{h} - \frac{\partial p}{\partial x}\left(\frac{h}{2} - z\right) \tag{85.3}$$

Figure 85.1 shows a slider bearing consisting of a "shoe" moving over a plane surface $z = 0$ with velocity U. The simplest theory of this bearing assumes that the

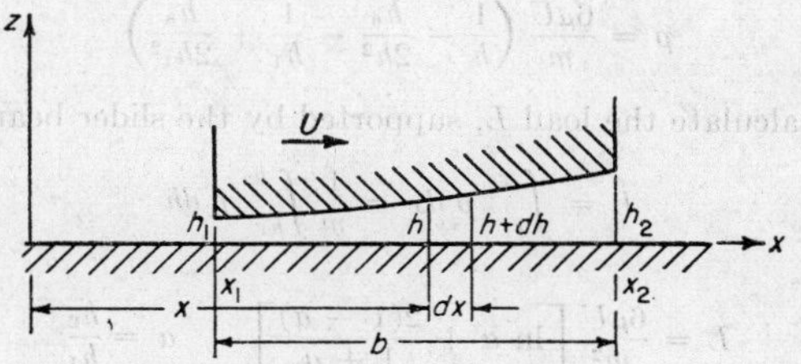

FIG. 85.1

flow is two-dimensional in the x,z plane, and that $h = h(x)$ varies so slowly that, at each x, Eqs. (85.1) and (85.2) may be applied with the local value of h, allowing p and $\partial p/\partial x$ to vary with x only.

The present chapter is a greatly abridged version of the author's original manuscript. The handbook editor takes the responsibility for the changes made. He wishes to express his thanks to Dr. W. A. Gross, San Jose, Calif., for his advice.

The continuity condition for an element of length dx is

$$dQ = U\frac{dh}{dx}\,dx$$

whence

$$Q(x) - h(x)U = \text{const} \tag{85.4}$$

When Q from (85.2) is substituted, there results

$$\frac{Uh}{2} + \frac{\partial p}{\partial x}\frac{h^3}{12\mu} = \text{const} \tag{85.5}$$

and this can be put in the form

$$\frac{\partial p}{\partial x} = 6\mu U\,\frac{h_0 - h}{h^3} \tag{85.6}$$

where h_0 is a proper constant, which agrees with the value of h at the point, if any, where $\partial p/\partial x = 0$. The solution of the differential equation (85.6) contains another constant besides h_0. Both can be determined from the boundary conditions that the pressure vanishes at the bearing ends:

At $x = x_1$, $x = x_2$:

$$p = 0 \tag{85.7}$$

In particular, h_0 follows from the relation

$$h_0 \int_{x_1}^{x_2} \frac{dx}{h^3} = \int_{x_1}^{x_2} \frac{dx}{h^2} \tag{85.8}$$

For the inclined flat plate of slope m, we have $h = h_0 + mx$, and Eq. (85.6) may be written as

$$\frac{dp}{dh} = \frac{6\mu U}{m}\left(\frac{h_0}{h^3} - \frac{1}{h^2}\right) \tag{85.9}$$

Equation (85.8) now yields

$$\frac{1}{h_0} = \frac{1}{2}\left(\frac{1}{h_1} + \frac{1}{h_2}\right) \tag{85.10}$$

Thus, h_0 is the harmonic mean of h_1, h_2, and on Fig. 85.1 will correspond to a location to the left of the center.

Substituting in (85.9), putting $p = 0$ for $h = h_1$, and integrating yields

$$p = \frac{6\mu U}{m}\left(\frac{1}{h} - \frac{h_0}{2h^2} - \frac{1}{h_1} + \frac{h_0}{2h_1^{\,2}}\right) \tag{85.11}$$

We may now calculate the load L, supported by the slider bearing,

$$L = \int_{x_1}^{x_2} p\,dx = \frac{1}{m}\int_{h_1}^{h_2} p\,dh \tag{85.12}$$

There results

$$L = \frac{6\mu U}{m^2}\left[\ln a + \frac{2(1-a)}{1+a}\right] \qquad a = \frac{h_2}{h_1} \tag{85.13}$$

The pressure distribution is lopsided, the amount of lopsidedness varying with a [3]. The velocity profile is linear at $x = x_0$ where $p = p_{\max}$, and elsewhere it is convex to the side of decreasing pressure.

Of practical interest is the center of pressure or pivot point. This is the point $x = x_c$, about which the moment of the pressure forces vanishes. It is found from the equation

$$\int_{x_1}^{x_2} (x - x_c)p\,dx = 0 \tag{85.14}$$

The ratio x_c/b depends on h_1/b and m. For details, see [4], p. 168, where curves for $\mu U/L$ versus h_1/b are given for various m. It will be found that the same load can be carried for different h_1 by shifting the pivot.

To calculate the friction force, we evaluate the resultant shear force

$$F = \mu \int_{x_1}^{x_2} \left(\frac{\partial u}{\partial z}\right)_{z=0} dx \tag{85.15}$$

which the fluid exerts on the plate $z = 0$. Equations (85.3), (85.6) yield, for any slider shape,

$$F = \mu U \int_{x_1}^{x_2} \left(\frac{4}{h} - \frac{3h_0}{h^2}\right) dx \tag{85.16}$$

For the flat plate of slope m, there results

$$F = \frac{-2\mu U}{m}\left(3\,\frac{a-1}{a+1} - 2\ln a\right) \qquad a = \frac{h_2}{h_1} \tag{85.17}$$

The ratio F/L is the *coefficient of friction;* it is independent of μ and U, is proportional to h_1/b, and depends on a.

It is of interest to note that, if the friction is evaluated from the shear stress at the slider plate, a different force is obtained. The explanation of this paradox lies in the added pressure force component mL in the negative x direction, acting on the pad. Though m is small, the pressure resultant L is large. For a graphical representation of the friction force and the coefficient of friction, see [4], p. 171.

85.2. JOURNAL BEARING

Figure 85.2 shows, in a cross section, a journal bearing or rotor shaft, of radius a, revolving in a larger complete 360° circular bearing, of radius $(a + b)$ with $b \ll a$.

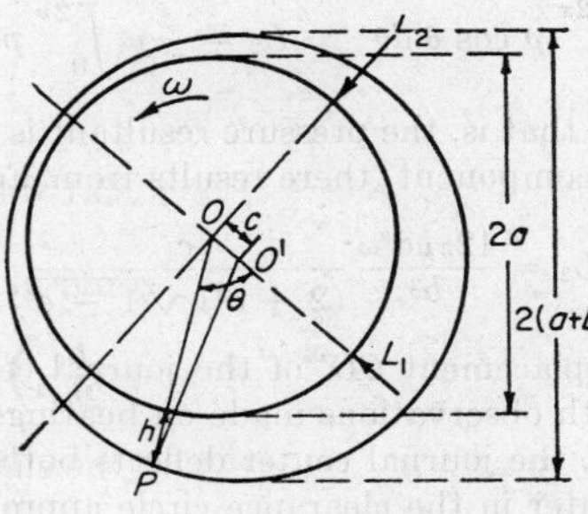

FIG. 85.2

The locus of possible positions of the journal center O is a circle of radius b and center at the bearing center O', known as the clearance circle. The ratio

$$e = c/b \tag{85.18}$$

is called the eccentricity.

Let P be any point on the bearing surface. Since the angle OPO' is small, it will be seen from Fig. 85.2 that very nearly

$$OP - O'P = c\cos\theta \tag{85.19}$$

Since $OP = a + h$, $O'P = a + b$, there results

$$h = b(1 + e\cos\theta) \tag{85.20}$$

Now "straighten" the journal surface so that it becomes the x axis (Fig. 85.3), replacing the bearing surface by a wavy surface,

$$z = h(x) = b(1 + e\cos\theta) \tag{85.21}$$

moving with the velocity $U = a\omega$ relative to the journal, and in the direction shown in Fig. 85.3. Then the slider bearing analysis may be applied. Equation (85.6) yields

$$p(\theta) - p(0) = \frac{6\mu a^2\omega}{b^2}\left(k\int_0^\theta \frac{d\theta}{(1 + e\cos\theta)^3} - \int_0^\theta \frac{d\theta}{(1 + e\cos\theta)^2}\right) \tag{85.22}$$

where $k = h_0/b$ is a proper constant. After integrating and solving for k so that $p(\theta)$ is periodic in θ with period 2π, there results

$$p(\theta) - p(0) = -\frac{6\mu a^2\omega}{b^2}\frac{e(2 + e\cos\theta)\sin\theta}{(2 + e^2)(1 + e\cos\theta)^2} \tag{85.23}$$

The right-hand member of (85.23) is odd in θ; it is positive over the convergent half of the film ($\pi \le \theta \le 2\pi$), negative over the divergent half.

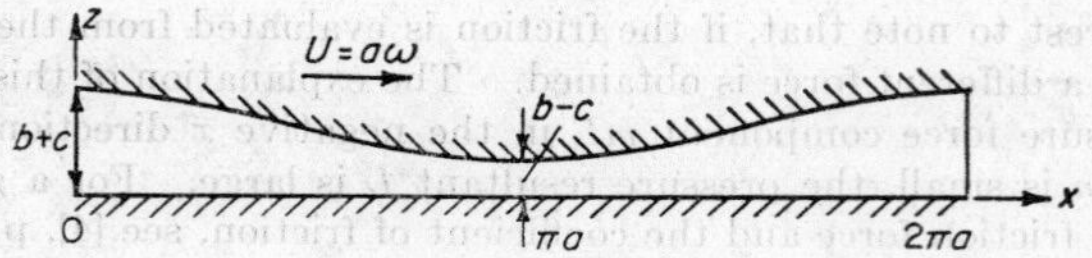

FIG. 85.3

The components L_1 and L_2 of the force acting on the journal, per unit axial length, are

$$L_1 = a\int_0^{2\pi} p\cos\theta\,d\theta \qquad L_2 = -a\int_0^{2\pi} p\sin\theta\,d\theta \tag{85.24}$$

Since p is odd in θ, $L_1 = 0$; that is, the pressure resultant is at right angles to the line of centers. For the other component, there results from Eq. (85.23):

$$L_2 = \frac{12\pi\mu a^3\omega}{b^2}\frac{e}{(2 + e^2)\sqrt{1 - e^2}} \tag{85.25}$$

The result, that the displacement OO' of the journal center is at right angles to the load, does not agree with observations made on bearings of finite length. Corresponding to a vertical load, the journal center deflects both downward and sidewise; the locus of the journal center in the clearance circle approximates half an ellipse of semiaxes 0.775:1, with the major axis vertical and extending from the journal center (at light loads or high speeds) to the lowest point (at heavy loads or low speeds) [12].

The torque on the bearing may be found as in Eq. (85.15) from

$$M_T = Fa = \mu a^2\int_0^{2\pi}\left(\frac{\partial u}{\partial z}\right)_{z=0} d\theta \tag{85.26}$$

and Eq. (85.3).

The dimensionless parameter $S = \mu N/p'$, where $N = \omega/2\pi$ is the journal speed in revolutions per second, and $p' = L_2/2a$, the load per unit projected area, is often used in describing lubrication phenomena and is known as the *Sommerfeld number*. Equation (85.25) can evidently be put in the form

$$\left(\frac{a}{b}\right)^2 S = \frac{(2 + e^2)\sqrt{1 - e^2}}{12\pi^2 e} \tag{85.27}$$

thus expressing the dimenisonless product Sa^2/b^2 in terms of the eccentricity only. Curves of S and the coefficient of friction versus e are given in [4], Figs. 5-23 to 5-27.†

As in Eq. (85.15), there is an even more serious discrepancy between the computed torque on the journal and that on the bearing. This is due to the fact that, in (85.26), the moment of the shear force on the bearing was computed about the bearing center O, rather than the journal center O'.

Further corrections in L_1, L_2, M_T are due to the shear forces and to the moment of the pressure forces. However, only the exact solution of the Navier-Stokes equations subject to the proper boundary conditions can remove these discrepancies completely.

Equation (85.22) and Figs. 85.2 and 85.3 can be applied equally well to *partial* circular journal bearings. Here, in addition to the eccentricity e, the "attitude" or the direction of the line of centers relative to one edge angle of the bearing enters as a further unknown. The constant k in (85.22) is now determined so as to satisfy $p = 0$ at the bearing ends, and h_0 is given by Eq. (85.8).

Similar to the case of the slider bearing, it turns out that the same load can be carried for many positions of the journal center. No condition of zero moment about a pivot now exists, since a rotating shaft is generally supplied with, and transmits, large torques. Therefore, some other criterion must be chosen for deciding which center position is proper. D. G. Christopherson [13] has shown that the proper position of the journal center can be determined by requiring that the pressure gradient vanish at the point where the pressure drops to zero.

Partial bearings have been studied analytically by Waters [9]. Karelitz carried out the integrations (85.22) numerically (see Ref. [10], p. 196). Howarth [11] has studied partial bearings by graphical methods and obtained curves of attitude and eccentricity.

85.3. NEGATIVE PRESSURES

As noted above, the theoretical prediction that the force component L_1 on an eccentric journal vanishes does not agree with observed attitudes of the journal center for complete 360° bearings. Another discrepancy between theory and tests occurs for symmetric and symmetrically placed slider bearings (Fig. 85.4), or between a roller and its races in roller bearings, or between gear teeth. Application of Eq. (85.6), with $x = 0$ placed at the center of symmetry, leads, for any h_0, to a pressure distribution which is odd in x, and, hence, to a vanishing total load.

The explanation of these discrepancies between theory and experiment lies in the assumption that the negative pressures predicted by integrating Eq. (85.6) can actually obtain in the lubricant. Generally, a liquid will not support negative pressure, since any tendency to develop tension will be resisted by cavitation and foaming, and the journal bearing will not "run full" there. Consequently, the above theories have to be corrected to exclude negative pressure areas, replacing p there by $p = 0$. If p is the pressure above atmospheric p_a, then only $p < -p_a$ corresponds to tension.

Applying this to a 360° journal bearing, we might make use of Eq. (85.23) only over the arc where $p > 0$, with $p(0)$ determined from a specified oil pressure at the inlet location. Now L_1 will no longer vanish, and L_2 will differ from its expression (85.25). The performance of the infinite bearing will correspond more nearly to that of a finite bearing. However, Eq. (85.23) was obtained from (85.22) by assuming that p is periodic in θ; with finite boundaries $p = 0$ for the portion of the film running full, the condition of periodicity no longer applies. In fact, many different eccentricity ratios and attitude angles are theoretically possible for the same load on the journal.

† In [4], the left-hand member of Eq. (85.27) is referred to as the Sommerfeld number.

For a symmetric slider bearing as in Fig. 85.4, the choice of various h_0 in (85.6) leads to pressure curves schematically shown in Fig. 85.5. One of the curves is tangent to the x axis, and satisfies the criterion mentioned before. To the left

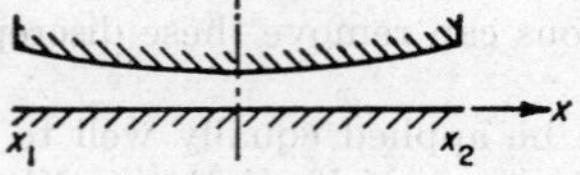

FIG. 85.4

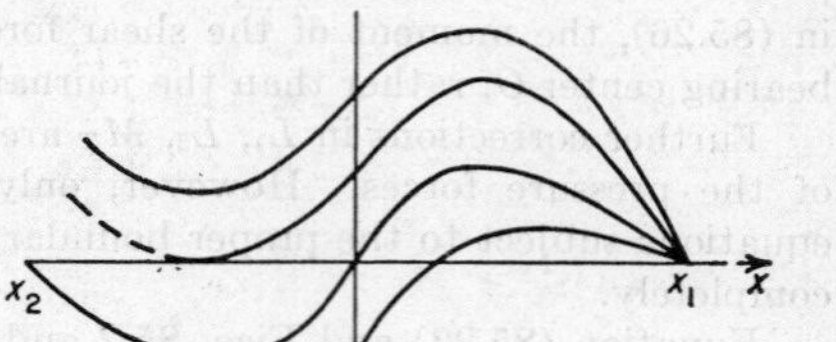

FIG. 85.5

of the contact point, the pressure is replaced by $p = 0$. This leads to the maximum possible L for a given minimum distance of approach.

85.4. END LEAKAGE, REYNOLDS' EQUATION

It is evident that, only for journal bearings of great axial length, is it possible to neglect side leakage and to assume two-dimensional flow. For this reason, we shall now consider the more general case that flow can take place in the two directions x and y perpendicular to the film thickness.

We suppose that the film is bounded by two rigid surfaces, $z = 0$ and $z = h(x,y)$, with $\partial h/\partial x$, $\partial h/\partial y$ small, and that the pressure is independent of z.

The basic unidirectional flow discussed on p. 85–1 can now be adapted to each location of the film, by rotating the solution through an angle ϕ relative to the x axis, and resolving the fluid velocity along the x and y directions. Thus, we obtain from Eq. (85.1), upon replacing $U \cos \phi$ and $U \sin \phi$ by U and V respectively,

$$u = \frac{Uz}{h} - \frac{\partial p}{\partial x}\frac{z(h-z)}{2\mu} \qquad v = \frac{Vz}{h} - \frac{\partial p}{\partial y}\frac{z(h-z)}{2\mu} \tag{85.28}$$

where U and V are the velocities of the upper boundary parallel to the plane of the lower boundary $z = 0$, which is assumed to be stationary. To provide for general situations, we allow for the upper boundary even a normal velocity component W in the z direction; however, it does not affect Eqs. (85.28).

Integrating (85.28) with respect to z yields

$$Q_x = \int_0^h u\,dz = \frac{Uh}{2} - \frac{\partial p}{\partial x}\frac{h^3}{12\mu} \qquad Q_y = \int_0^h v\,dz = \frac{Vh}{2} - \frac{\partial p}{\partial y}\frac{h^3}{12\mu} \tag{85.29a,b}$$

where Q_x represents the volume flow per unit y across the film section x = const, and Q_y the flow per unit x across the film section y = const (Fig. 85.6).

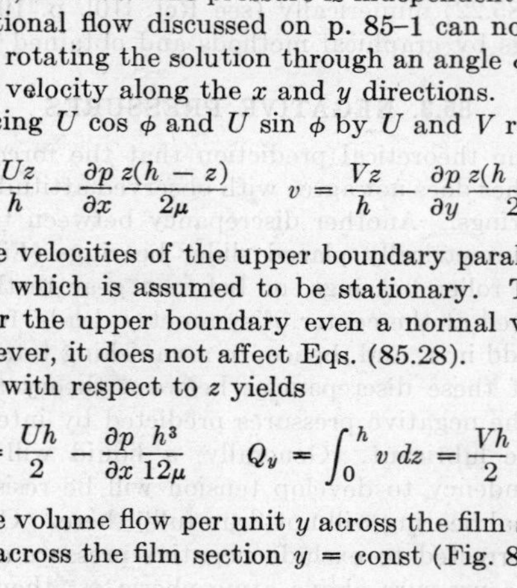

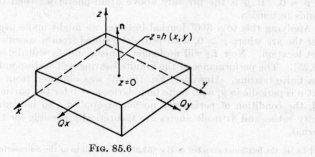

FIG. 85.6

For incompressible flow of the lubricant, the continuity condition is

$$\frac{\partial Q_x}{\partial x} + \frac{\partial Q_y}{\partial y} + \mathbf{U} \cdot \mathbf{n} = 0, \tag{85.30}$$

where $\mathbf{U} = U\mathbf{i}_x + V\mathbf{i}_y + W\mathbf{i}_z$ is the resultant velocity of the upper surface, and $\mathbf{n}$ the unit vector in the direction of the outward normal to the upper film boundary. Under the assumption that the slopes are small,

$$\mathbf{n} = -\frac{\partial h}{\partial x}\mathbf{i}_x - \frac{\partial h}{\partial y}\mathbf{i}_y + \mathbf{i}_z \tag{85.31}$$

and Eq. (85.30) becomes

$$\frac{\partial Q_x}{\partial x} + \frac{\partial Q_y}{\partial y} = U\frac{\partial h}{\partial x} + V\frac{\partial h}{\partial y} - W \tag{85.32}$$

After substitution from (85.29), there results

$$\frac{1}{6\mu}\left[\frac{\partial}{\partial x}\left(h^3\frac{\partial p}{\partial x}\right) + \frac{\partial}{\partial y}\left(h^3\frac{\partial p}{\partial y}\right)\right] = -\left(U\frac{\partial h}{\partial x} + V\frac{\partial h}{\partial y}\right) + h\left(\frac{\partial U}{\partial x} + \frac{\partial V}{\partial y}\right) + 2W \tag{85.33}$$

This equation is known as *Reynolds' equation.* It will suffice for analyzing thin hydrodynamic lubricating films under very general conditions, for instance, for pulsating loads, gears, etc.

When the above analysis is applied to Fig. 85.1, Eqs. (85.4) and (85.6) result as a special case.

Equation (85.33) has to be solved in the bearing area subject to the conditions $p = 0$ over the bearing boundary and $p = p_0$ over the lubricant inlet, where p_0 is the inlet pressure. However, as indicated on p. 85–5, negative pressure areas must be avoided. Potential negative areas are characterized by

$$p = 0 \qquad \partial p/\partial n = 0 \qquad \text{over } C$$

where C is the boundary between such an area and the bearing portion which is running full.

Reynolds' equation may be put in vector form:

$$\nabla^2 p + 3\nabla p \cdot \nabla \ln h = \frac{6\mu}{h^3}[-\nabla \cdot (\mathbf{V}h) + 2h\nabla \cdot \mathbf{V} + 2W] \tag{85.34}$$

where $\quad \mathbf{V} = U\mathbf{i}_x + V\mathbf{i}_y \qquad \nabla = \mathbf{i}_x\frac{\partial}{\partial x} + \mathbf{i}_y\frac{\partial}{\partial y}$

In this form, it may easily be applied to other coordinate systems, for instance, to polar coordinates r, θ (where $x = r\cos\theta$, $y = r\sin\theta$). For a thrust bearing rotating about its axis with velocity ω, we have $V = r\omega\mathbf{i}_\theta$, and hence

$$\frac{\partial^2 p}{\partial r^2} + \frac{1}{r}\frac{\partial p}{\partial r} + \frac{1}{r^2}\frac{\partial^2 p}{\partial \theta^2} + 3\left(\frac{\partial p}{\partial r}\frac{\partial \ln h}{\partial r} + \frac{1}{r^2}\frac{\partial p}{\partial \theta}\frac{\partial \ln h}{\partial \theta}\right) = -\frac{6\mu\omega}{h^3}\frac{\partial h}{\partial \theta} \tag{85.35}$$

85.5. APPLICATION OF REYNOLDS' EQUATION

Consider the lubricant flow in a slider bearing as shown in Fig. 85.1, but of a finite extent c in the y direction. Assume that the shoe has a plane surface,

$$h = h_0 + mx \tag{85.36}$$

Then Eq. (85.33) assumes the form

$$\frac{\partial}{\partial x}\left(\frac{h^3}{6\mu}\frac{\partial p}{\partial x}\right) + \frac{\partial}{\partial y}\left(\frac{h^3}{6\mu}\frac{\partial p}{\partial y}\right) = -Um \tag{85.37}$$

and has to be solved in the area $x_1 \leq x \leq x_2$, $0 \leq y \leq c$, subject to the boundary conditions

$$p = 0 \qquad \text{over} \quad x = x_1,\, x = x_2,\, y = 0,\, y = c \tag{85.38}$$

Even a bearing slanted in both the x and y directions can be handled, replacing (85.36) by a linear expression in x and y.

The solution can be obtained [14],[15] as a sum of products,

$$p = \sum_{n=1,3,\ldots}^{\infty} p_n(x) \sin \frac{n\pi y}{c} \tag{85.39}$$

when the right-hand member of (85.37) is expanded in a similar series (see Table 20.3, Fig. 20.25). Upon substituting (85.39) into (85.37), there results the ordinary differential equation

$$\frac{d}{dx}\left(h^3 \frac{dp_n}{dx}\right) - \frac{n^2\pi^2}{c^2} h^3 p_n = -\frac{24\mu U m}{\pi n} \tag{85.40}$$

which must be solved subject to the boundary condition $p_n = 0$ for $x = x_1$, $x = x_2$.

Upon introduction of new variables,

$$\xi = x + h_0/m \qquad h = m\xi \qquad p_n = q_n/\xi$$

Eq. (85.40) takes the form

$$\frac{d^2q_n}{d\xi^2} + \frac{1}{\xi}\frac{dq_n}{d\xi} - \left(\frac{1}{\xi^2} + \frac{n^2\pi^2}{c^2}\right) q_n = -\frac{24\mu U}{\pi n m^2}\frac{1}{\xi^2} \tag{85.41}$$

A particular solution of this equation can be obtained by assuming that

$$q_n = c_0 + c_2\xi^2 + c_4\xi^4 + \cdots$$

and determining the coefficients by substituting into (85.41). The homogeneous form of Eq. (85.41) can be converted to a Bessel equation and has the solution

$$q_n = AI_1(n\pi\xi/c) + BK_1(n\pi\xi/c) \tag{85.42}$$

with arbitrary constants A and B. For details, see [14].

The effect of leakage is, of course, to decrease the pressure and hence the load, relative to the results presented on p. 85–2, where end leakage is neglected.

Similar solutions of Reynolds' equation (85.33) and Eq. (85.20) for a journal in a cylindrical bearing of finite axial length, including side leakage, have been carried out by Muskat and Morgan [16], Waters [9], and Hays [17].

If the x dimension of the bearing is much larger than the y dimension, and if $\partial h/\partial y \equiv 0$, a fair approximation to the pressure is obtained by assuming it as

$$p = P(x)(c^2 - y^2) \tag{85.43}$$

Equation (85.37) cannot then be satisfied pointwise. However, if it is integrated over $-c \leq y \leq c$ for each x, there results the equation

$$\frac{d}{dx}\left(\frac{h^3}{6\mu}\frac{dP}{dx}\right) - \frac{h^3}{2\mu}\frac{P}{c^2} = -\frac{3}{2}\frac{Um}{c^2} \tag{85.44}$$

Its solution satisfies (85.37) on the average over y for each x. Equation (85.44) is similar to (85.40), and may be solved similarly.

For journal bearings of relatively small width [18], in Eq. (85.29a), the term $\frac{1}{2}Uh$ predominates, while, in (85.29b), $V = 0$. Hence, and because h is independent of y,

Eq. (85.32) yields

$$\frac{h^3}{6\mu}\frac{\partial^2 p}{\partial y^2} = -U\frac{dh}{dx} = -\omega\frac{dh}{d\theta} \tag{85.45}$$

Integration with respect to y, subject to the condition $p = 0$ for $y = \pm c$, yields the result (short-bearing solution):

$$p = 3\mu\omega(c^2 - y^2)\frac{1}{h^3}\frac{dh}{d\theta} \tag{85.46}$$

This is again parabolic in y as in Eq. (85.43), but now with a completely determined coefficient $P(x)$. The film thickness h depends on θ according to Eq. (85.20). Equation (85.46) is applicable only over the convergent half of the film, $\pi \leq \theta \leq 2\pi$, since only here is $p > 0$; to avoid $p < 0$, we put $p = 0$ over $0 \leq \theta \leq \pi$. Then there results [19]

$$p = -\frac{3\mu\omega}{b^2}\frac{e \sin\theta}{(1 + e\cos\theta)^3}(c^2 - y^2) \qquad \pi \leq \theta \leq 2\pi \tag{85.47}$$

$$p = 0 \qquad 0 \leq \theta \leq \pi$$

A more realistic solution uses the condition of Christopherson (p. 85–5), see [13],[20],[21].

Integration according to Eqs. (85.24) yields the components of the resultant force on the journal:

$$L_1 = -\frac{8\mu a\omega c^3}{b^2}\frac{e^2}{(1 - e^2)^2} \qquad L_2 = \frac{2\pi\mu a\omega c^3}{b^2}\frac{e}{(1 - e^2)^{3/2}} \tag{85.48}$$

The eccentricity and attitude may now be calculated relative to the resultant load and its direction.

85.6. HYDROSTATIC BEARINGS

Slider and journal bearings owe their load-carrying capacity to the motion of the slider or journal surface, which "drags" the lubricant into the wedge-shaped space. In a hydrostatic bearing, there is no (or very minor) relative motion of the surfaces of the thin film, and the load-carrying capacity is due to high-pressure oil which is supplied by a pump.

For stationary surfaces $z = 0$, $z = h$, the right-hand member in Eq. (85.37) vanishes. If h is constant, Eq. (85.37) reduces to the Laplace equation:

$$\frac{\partial^2 p}{\partial x^2} + \frac{\partial^2 p}{\partial y^2} = 0 \tag{85.49}$$

As a simple illustration, consider an axially symmetric hydrostatic bearing shown in Fig. 85.7. Oil at pressure p_0 (above atmospheric) is supplied at the center into a

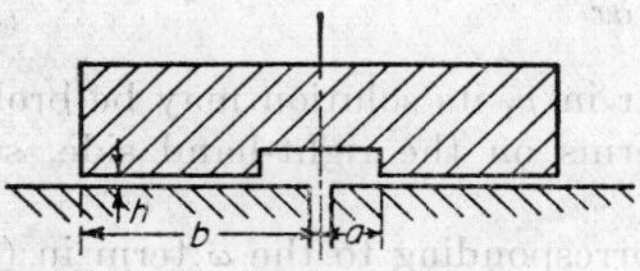

FIG. 85.7

recessed cylinder of radius a, from which it leaks out through a thin film of thickness h to the outer edge of radius b.

When the pressure drop in the recessed space is neglected, the boundary conditions for the solution of (85.49) are:

$$\text{At } r = a\text{: } p = p_0 \qquad \text{at } r = b\text{: } p = 0 \tag{85.50}$$

The solution is given by

$$p = p_0 \frac{\ln (b/r)}{\ln (b/a)} \tag{85.51}$$

The total load carried is

$$L = p_0 \pi a^2 + 2\pi \int_a^b pr\, dr = \frac{\pi p_0 (b^2 - a^2)}{2 \ln (b/a)} \tag{85.52}$$

Of interest is also the amount of oil leakage. The analog of Eqs. (85.29) is

$$Q_r = -\frac{\partial p}{\partial r} \frac{h^3}{12\mu} \qquad Q_\theta = 0 \tag{85.53}$$

and yields, for the total volume flow rate,

$$Q = 2\pi r Q_r = \frac{\pi p_0 h^3}{6\mu \ln (b/a)} \tag{85.54}$$

For further examples and solutions of Eq. (85.49), see [10] and [22], p. 215.

85.7. DYNAMICALLY LOADED BEARING

If the load on a journal or bearing slider varies with time, the bearing is called a dynamically loaded bearing. The fluid flow induced by the varying geometry of the bearing surfaces produces certain new features.

Suppose that, in the journal (Fig. 85.2), the center O is moving in its circle of clearance. Developing the film as in Fig. 85.3, we may describe the motion of the journal bearing by

$$\begin{aligned} U &= a\omega + V_1 \sin\theta + V_2 \cos\theta \qquad V = 0 \\ W &= \frac{\partial h}{\partial t} = -V_1 \cos\theta + V_2 \sin\theta \end{aligned} \tag{85.55}$$

where V_1, V_2 are the components of the velocity of the center O, opposed to the directions of the forces L_1, L_2 shown in Fig. 85.2, and h is again given by Eq. (85.20).

We consider the Reynolds equation (85.33) corresponding to these velocities, neglecting side leakage and overlooking the question of negative pressures. Since the results of Sec. 85.2 do not agree completely with tests, the use of the same assumptions here can only be justified as an attempt to render the problem tractable. Substituting from (85.55) into (85.33) and neglecting the terms involving b/a, h/a, the following equation is obtained:

$$\frac{1}{6\mu} \frac{d}{dx}\left(h^3 \frac{dp}{dx}\right) = -a\omega \frac{\partial h}{\partial x} - 2V_1 \cos\theta + 2V_2 \sin\theta \qquad \frac{\partial h}{\partial x} = -\frac{be}{a} \sin\theta \tag{85.56}$$

Since this equation is linear in p, its solution may be broken up into three separate parts, due to the three terms on the right-hand side, separately, and the results (pressures, forces) added.

A particular solution corresponding to the ω term in (85.56) may be found by a simple integration,

$$\frac{1}{6\mu} h^3 \frac{dp}{dx} = -a\omega h + \text{const} \tag{85.57}$$

in agreement with (85.6) and leading to Eqs. (85.23) and (85.25). Next it will be noted that the first and third terms in Eq. (85.56) can be combined into $(\omega - 2\Omega)be \sin\theta$, where $\Omega = -V_2/be$ is the angular velocity with which the journal center is moving about the bearing center. Hence, no separate calculation is required for the effect of V_2. Integrating (85.56) for the V_1 contribution and adding all the parts of the solution leads to the following expressions for the resultant forces:

$$L_1 = -\frac{12\pi\mu a^3}{b^2}\frac{V_1}{(1-e^2)^{3/2}} \qquad L_2 = \frac{12\pi\mu a^3}{b^2}\frac{e}{(2+e^2)(1-e^2)^{1/2}}(\omega - 2\Omega) \tag{85.58}$$

These equations are due to Harrison [31]. Robertson [32] included the terms involving b/a and h/a. The resulting corrections are small, except possibly for $\Omega = \omega/2$.

Equations (85.58) give the forces when V_1 and Ω are known. To find the movement of the shaft when the forces are known, we put

$$V_1 = -\dot{r} = -b\dot{e} \qquad \Omega = \dot{\psi} \tag{85.59}$$

where r and ψ are polar coordinates of the journal center in its circle of clearance. With this substitution, Eqs. (85.58) take the form

$$\frac{\dot{e}}{(1-e^2)^{3/2}} = \frac{b^2}{12\pi\mu a^3}L_1 \qquad \frac{e(\omega - 2\dot{\psi})}{(2+e^2)(1-e^2)^{1/2}} = \frac{b^2}{12\pi\mu a^3}L_2 \tag{85.60}$$

If the variable load is described by its components L_x and L_y in a fixed coordinate system (x,y), then

$$L_1 = L_x \cos\psi + L_y \sin\psi \qquad L_2 = -L_x \sin\psi + L_y \cos\psi \tag{85.61}$$

Substituting these into (85.60), we obtain two first-order differential equations in r and ψ, whose solution yields the journal center path for a massless shaft.

The solution for a given load depends on the initial journal center position. Harrison [31] studied the possible paths for a constant load. They are closed paths, enclosing the steady-state position of the journal center. Even for no load, by putting $\Omega = \frac{1}{2}\omega$, $\dot{e} = 0$, we may obtain possible circular paths described at half the rotation frequency of the journal.

It is known that, under certain conditions, journal bearings may induce vibrations and instability in high-speed rotors, known as "oil whirl" and "oil whip" [33]. In general, these vibrations occur at speeds exceeding twice the critical rotor speed (as defined by bending vibrations), the shaft "whipping" at the critical frequency.

Robertson [32] has studied the stability of a rotor with journal bearings under the journal forces (85.60). He showed that, at all speeds, the rotor will be unstable, and was not able to explain stability below twice the critical speed. Poritsky [34] has studied the dynamical behavior of a rotor in journal bearings for small eccentricities. While corroborating Robertson for low speeds, he further showed that, if the forces (85.60) be modified by the addition of a radial restoring force, the rotor motion is stable at all speeds below twice the critical, and unstable at all speeds above twice the critical, where the critical is calculated by including the stiffness of the oil film. As pointed out before, Eqs. (85.60) are based on pressure distributions, which may very well become negative. The solutions of the Reynolds equations for which the pressure is positive do exhibit restoring forces; see, for instance, Eqs. (85.48).

Hagg and Warner [35] have studied oil whirl by means of an electric analog.

For small vibrations, at least, the cause of the oil whirl may be shown to be the force L_2 normal to the line of centers. To minimize L_2, the 360° bearing may be replaced by several pivoted bearings. Another remedy consists in increasing the journal eccentricity. This may be accomplished by increasing the load on the journal through properly designed dams in the upper half of the bearing.

85.8. LUBRICATION OF GEARS AND ROLLERS

The following application of Reynolds' equation to the lubrication of gear teeth also applies to the lubrication of rolls or between rollers and races in a roller bearing.

Let $2h_c$ be the distance of closest approach of two mating gear teeth (Fig. 85.8). The line of closest approach, where this distance is measured, will here be referred to as the line of contact. The film thickness at any place is then

$$2h = 2h_c + \frac{1}{2}\left(\frac{1}{R_1} + \frac{1}{R_2}\right) x^2 + \cdot \cdot \cdot \qquad (85.62)$$

where R_1 and R_2 are the radii of curvature of the tooth surfaces. Equation (85.62) is not affected if, as in Fig. 85.9, we replace the two teeth by circular cylindrical surfaces of equal radii R, where

$$\frac{1}{R} = \frac{1}{2}\left(\frac{1}{R_1} + \frac{1}{R_2}\right)$$

Let V_1 and V_2 be the velocities with which the line of contact moves on the mating gear teeth. The relative motion of the teeth consists of a slipping motion or transla-

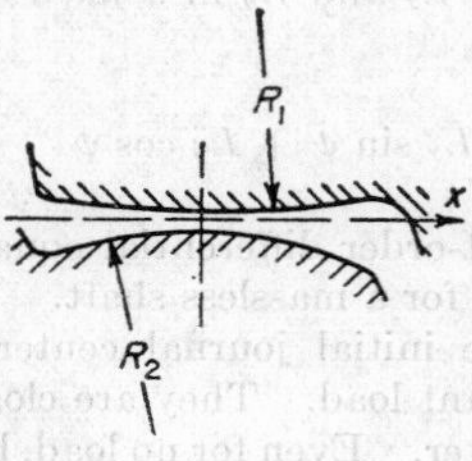

FIG. 85.8

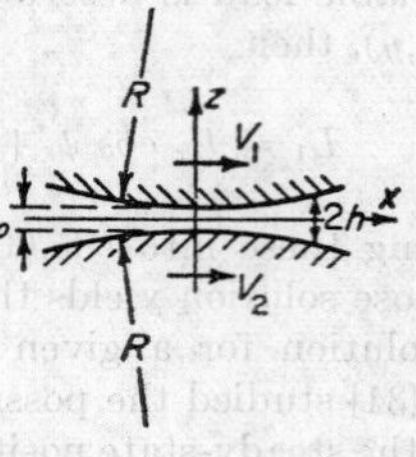

FIG. 85.9

tion of velocity $(V_1 - V_2)$, and a rolling motion with the contact point moving with the same velocity $\frac{1}{2}(V_1 + V_2)$ on each surface.

In Fig. 85.9, we replace the slipping motion by translations of the upper and lower surfaces, of respective velocities $\frac{1}{2}(V_1 - V_2)$ and $\frac{1}{2}(V_2 - V_1)$. This leads to the film velocity

$$u = \tfrac{1}{2}(V_1 - V_2)\frac{z}{h} \qquad (85.63)$$

It is evident, from the analysis given on p. 85-1, that no pressure gradient is induced by this component of the flow. There results, however, a shearing force, which leads to added drag and to losses in the lubricant.

Now consider the rolling motion. By changing to a coordinate system which moves with velocity $U = -\frac{1}{2}(V_1 + V_2)$, the motion of the cylinders is replaced by a rotation of each cylinder about its axis, giving rise to the peripheral velocity U. Equation (85.1) is now replaced by

$$u = U - \frac{\partial p}{\partial x}\frac{h^2 - z^2}{2\mu} \qquad (85.64)$$

while Eq. (85.6) is replaced by

$$U(h - h_0) = \frac{\partial p}{\partial x}\frac{h^3}{3\mu} \qquad (85.65)$$

The integration of (85.62) and (85.65) has been carried out by different authors [36]–[38]. In this integration, we run into the difficulty pointed out on p. 85-5, namely, that if h_0 is picked so as to satisfy the conditions $p = 0$ at $x = -\infty, +\infty$, there results a pressure which is positive over the convergent film half $x > 0$, negative over the divergent half $x < 0$, giving rise to a vanishing net resultant. This difficulty is overcome by disregarding the negative pressure, as discussed in Sec. 85.3.

85.9. REFERENCES

[1] H. Lamb: "Hydrodynamics," 6th ed., Dover, New York, 1945.

[2] O. Reynolds: On the theory of lubrication and its application etc., *Phil. Trans. Roy. Soc.*, **177** (1886), 157–234.

[3] A. E. Norton, J. R. Muenger: "Lubrication," p. 74, McGraw-Hill, New York, 1942.

[4] M. C. Shaw, E. F. Macks: "Analysis and Lubrication of Bearings," pp. 168, 171, McGraw-Hill, New York, 1949.

[5] Lord Rayleigh: Notes on the theory of lubrication, *Phil. Mag.*, **35** (1918), 1–12.

[6] A. Sommerfeld: Zur hydrodynamischen Theorie der Schmiermittelreibung, *Z. Math. Phys.*, **50** (1904), 97–155.

[7] D. D. Fuller: "A Survey of Journal Bearing Literature," Am. Soc. of Lubrication Engrs., Chicago, 1958.

[8] E. B. Sciulli: "A Bibliography on Gas-lubricated Bearings," Franklin Inst., 1957.

[9] E. O. Waters: Characteristics of centrally supported journal bearings, *Trans. ASME*, **64** (1942), 711–719.

[10] D. D. Fuller: "Theory and Practice of Lubrication for Engineers," pp. 19–25, Wiley, New York, 1956.

[11] H. A. S. Howarth: The loading and friction of thrust and journal bearings, *Trans. ASME*, **57** (1935), 169–187.

[12] A. Cameron, W. L. Wood: The full journal bearing, *Proc. Inst. Mech. Engrs. London*, **161** (1949), 59–64.

[13] D. G. Christopherson: Boundary conditions in lubricating films, *Engineer*, **203** (1957), 100–101.

[14] A. G. M. Michell: The lubrication of plane surfaces, *Z. Math. Phys.*, **52** (1905), 123–137.

[15] W. L. Wood: Note on a new form of solution of Reynolds' equation for Michell rectangular and sector-shaped pads, *Phil. Mag.*, **VII, 40** (1949), 220–226.

[16] M. Muskat, F. Morgan: The theory of thick film lubrication of flooded journal bearings and bearings with circumferential grooves, *J. Appl. Phys.*, **10** (1939), 398–407.

[17] D. F. Hays: A variational approach to lubrication problems and the solution of the finite journal bearing, *J. Basic Eng.*, **1** (1959), 13–23.

[18] F. W. Ocvirk: Short-bearing approximation for full journal bearings, *NACA Tech. Note* 2808 (1952).

[19] L. Gümbel, E. Everling: "Reibung und Schmierung im Maschinenbau," Krayn, Berlin 1925.

[20] H. W. Swift: The stability of lubricating films in journal bearings, *Proc. Inst. Civil Engrs. London*, **233** (1932), 267–288.

[21] W. Stieber: Das Schwimmlager, Hydrodynamische Theorie des Gleitlagers, *Forsch.* **4** (1933), 259.

[22] D. F. Wilcock, E. R. Booser: "Bearing Design and Application," McGraw-Hill, New York, 1957.

[23] A. Kingsbury: On problems in the theory of fluid-film lubrication, with experimental method of solution, *Trans. ASME*, **53** (1931), 59–75.

[24] S. J. Needs: Effects of side leakage in 120° centrally supported journal bearings, *Trans. ASME*, **56** (1934), 721–732.

[25] D. G. Christopherson: A new mathematical method for the solution of film lubrication problems, *Proc. Inst. Mech. Engrs.*, **146** (1941), 126–135.

[26] O. Pinkus: Solution of Reynolds' equations for finite journal bearings, *Trans. ASME*, **80** (1958), 858–864.

[27] B. Sternlicht, F. J. Maginnis: Application of digital computers to bearing design, *Trans. ASME*, **79** (1957), 1483–1493.

[28] G. Duffing: Beitrag zur Theorie der Flüssigkeitsbewegung zwischen Zapfen und Lager, *Z. angew. Math. Mech.*, **4** (1924), 296–314.

[29] H. Reissner: Ebene und räumliche Strömung zäher, inkompressibler, trägheitsfreier Flüssigkeiten zwischen exzentrischen, relativ zueinander rotierenden Zylinderflächen, *Z. angew. Math. Mech.*, **15** (1935), 81–87.

[30] G. H. Wannier: A contribution to the hydrodynamics of lubrication, *Quart. Appl. Math.*, **8** (1950), 1–32.

31] W. J. Harrison: The hydrodynamical theory of lubrication with special reference to air as a lubricant, *Trans. Cambridge Phil. Soc.*, **22** (1913), 39–54.
[32] D. Robertson: Whirling of a journal in a sleeve bearing, *Phil. Mag.*, **15** (1933), 113–130.
[33] B. L. Newkirk, H. D. Taylor: Shaft whipping due to oil action in journal bearings, *Gen. Elec. Rev.*, **28** (1925), 559–568.
[34] H. Poritsky: Contribution to the theory of oil whip, *Trans. ASME*, **75** (1953), 1153–1161.
[35] A. C. Hagg, P. C. Warner: Oil whip of flexible rotors, *Trans. ASME*, **75** (1953), 1339–1344.
[36] Martin: Lubrication of gear teeth, *Engineering*, **102** (1916), 119–121.
[37] E. K. Gatcombe: Lubrication characteristics of involute spur gears: a theoretical investigation, *Trans. ASME*, **67** (1945), 177–178.
[38] H. Poritsky: Lubrication of gear teeth, including the effect of elastic displacement, *Proc. Am. Soc. Lubrication Engrs.*, Fundamentals of Lubrication in Engineering, 1952.
[39] *Proc. Conf. on Lubrication and Wear*, Inst. Mech. Engrs., London, 1957.
[40] S. Abramovitz: Turbulence in a tilting-pad thrust bearing, *Trans. ASME*, **78** (1956), 7–11.
[41] G. I. Taylor: Stability of a viscous liquid contained between two rotating cylinders, *Phil. Trans. Roy. Soc. London*, **A,223** (1923), 289–343.
[42] D. F. Wilcock: Turbulence in high speed journal bearings, *Trans. ASME*, **72** (1950), 825–834.

CHAPTER 86

SURFACE WAVES

BY

T. Y. WU, Ph.D., Pasadena, Calif.

86.1. GENERAL FORMULATION OF A SURFACE-WAVE PROBLEM

A liquid, such as water, is assumed to be initially at rest, occupying the space R_0 defined in the cartesian coordinate system as $-\infty < (x,z) < \infty$, $-h(x,z) < y \leq 0$, with the y axis pointing vertically upward and $y = 0$ denoting the undisturbed free surface. The space above the liquid medium is supposed to be filled with a fluid of much smaller density, such as air, and may be assumed to be an empty space. Starting from the time $t = 0$, the liquid is set into motion by some prescribed disturbances, and the problem is to determine the resulting flow, in particular the form of the free surface S_f: $y = \zeta(x,z,t)$. The space occupied by the liquid in the disturbed state will be denoted by R. The liquid medium is assumed to be incompressible and nonviscous, and the resulting flow irrotational; hence, the flow velocity $\mathbf{q} = (u,v,w)$ has a potential $\phi(x,y,z,t)$ such that $\mathbf{q} = \text{grad } \phi$. It then follows from the continuity equation, $\text{div } \mathbf{q} = 0$, that ϕ satisfies [see Eq. (71.6)]

$$\nabla^2\phi = \phi_{xx} + \phi_{yy} + \phi_{zz} = 0 \quad \text{in } R \tag{86.1}$$

where, as throughout this chapter, the letter subscripts denote the partial differentiation. The momentum equation governing the flow can be then integrated to yield the Bernoulli equation,

$$\phi_t + \tfrac{1}{2} (\text{grad } \phi)^2 + p/\rho + gy = 0 \quad \text{in } R \tag{86.2}$$

which gives the relation between ϕ and the pressure p. Here ρ denotes the liquid density, and g the acceleration due to gravity. The boundary conditions of this problem may be stated as follows. On the free surface S_f, denoted by $y = \zeta(x,z,t)$, the kinematic condition that the fluid particles on S_f move with, and hence remain on, S_f is

$$\phi_y = \zeta_t + \phi_x\zeta_x + \phi_z\zeta_z \quad \text{on } S_f \tag{86.3}$$

When the external pressure over S_f is prescribed, say $p_0(x,z,t)$, the dynamic condition, with the surface tension also taken into account, becomes

$$\phi_t + \tfrac{1}{2} (\text{grad } \phi)^2 + g\zeta + \sigma K = -p_0/\rho \quad \text{on } S_f \tag{86.4}$$

where σ is the ratio of the surface tension of the liquid-air interface to the liquid density (σ being often called the kinematic surface tension), and K is the total curvature of S_f, taken to be positive if the curvature center lies inside the liquid. If p_0 is a constant, it can be chosen to be zero. Furthermore, on a solid surface S, such as on the bottom $y = -h(x,z)$, or on some floating or submerged body which may be in motion with velocity $\mathbf{Q}$ relative to the coordinate system, the condition of vanishing normal velocity is imposed that

$$\mathbf{n} \cdot (\text{grad } \phi - \mathbf{Q}) = 0 \quad \text{on } S \tag{86.5}$$

where **n** is the vector normal to S. If the liquid has an infinite depth, $h(x,z) \to \infty$, it is further required that

$$\phi,\ |\text{grad}\ \phi| \to 0 \qquad \text{as} \qquad y \to -\infty \tag{86.6}$$

There should be no particular difficulty in modifying the present general formulation so as to make it applicable to other special cases. For example, if there is a layer of lighter liquid above a heavier one, then the flow in both media should be considered separately, and conditions (86.3) and (86.4) should also be imposed at the interface between the layers. In other problems, particularly those involving moving disturbances, it may become convenient to adopt a coordinate system fixed at the disturbance. The formulation in the latter case can be obtained directly from the above by performing the appropriate transformation of coordinate systems.

The problem is seen to be intrinsically nonlinear because, first, conditions (86.3) and (86.4) are nonlinear, and, second, the free surface S_f is not known a priori. Furthermore, the time dependence of the solution enters into the problem through conditions (86.3) and (86.4); consequently, the solution at time t will depend on its past history, or on some initial conditions. This brief discussion points out the main difficulties usually encountered in the surface-wave problem.

Obviously the mathematical analysis will be facilitated if the problem can be linearized. In most practical cases, the free surface does not deviate appreciably from its undisturbed state, $y = 0$. With the motion assumed to be small, the nonlinear terms in (86.2), (86.3), and (86.4) may be neglected and conditions (86.3) and (86.4) may be applied on $y = 0$ to become

$$\zeta_t = \phi_y(x,0,z,t) \tag{86.7}$$

$$\phi_t(x,0,z,t) + g\zeta - \sigma(\zeta_{xx} + \zeta_{zz}) = -p_0/\rho \tag{86.8}$$

which are to replace conditions (86.3) and (86.4).

The dynamical similarity of this problem is worth mentioning. Suppose that the flow is characterized by a typical velocity U, a typical length L (for example, the extent of a disturbance, or the size of a body, or the water depth), and a typical time T (for example, for an oscillatory motion with frequency f, $T = 1/f$, or else $T = U/g$, or σ/U^3). Then, in terms of the nondimensional quantities,

$$t' = t/T \qquad x' = x/L \qquad \phi' = \phi/UL \qquad p' = p/\rho U^2$$

Eq. (86.8) becomes

$$\left(\frac{L}{UT}\right)\frac{\partial\phi'}{\partial t'} + \left(\frac{gL}{U^2}\right)\zeta' - \left(\frac{\sigma}{LU^2}\right)\left(\frac{\partial^2\zeta'}{\partial x'^2} + \frac{\partial^2\zeta'}{\partial z'^2}\right) = -p_0' \qquad \text{on } y' = 0$$

Thus, three nondimensional parameters of the flow naturally arise:

$$\Omega = \frac{L}{UT} \qquad \text{Fr} = \frac{U^2}{gL} \qquad \text{Wr} = \frac{LU^2}{\sigma} \tag{86.9}$$

which are usually referred to respectively as the generalized reduced frequency, the Froude number, and the Weber number. The value of Ω measures the importance of the time-rate variation of the flow. The value of Fr gives a measure of the relative magnitudes of the typical inertia force ρU^2 and the typical gravitational action ρgL. Lastly, the value of Wr gives a measure between the inertia effect and the effect of surface tension; this effect disappears if the phenomenon of surface tension is absent in the problem. For two geometrically similar flows, their solutions will be identical in terms of the nondimensional quantities, provided the two flows also have the same value of Ω, Fr, and Wr.

86.2. TWO-DIMENSIONAL PROBLEMS

The problems considered in this section will be those in which the flow is independent of z and will thereby be described in an x,y plane.

Standing Waves

Consider the simple solution of (86.1), which is simple harmonic in t and satisfies (86.7), (86.8), and (86.5) with $p_0 = 0$ and the only solid boundary at $y = -h$ (a constant). It is readily seen that the required solution is of the form

$$\zeta = A \sin(\omega t + \gamma) \cos(kx + \beta)$$
$$\phi = A \cos(\omega t + \gamma) \frac{\omega \operatorname{Cosh} k(y + h)}{k \operatorname{Sinh} kh} \cos(kx + \beta) \tag{86.10a}$$

which, as a pair, satisfies (86.1), (86.7), (86.8), and (86.5), provided that

$$\omega^2 = k(g + \sigma k^2) \operatorname{Tanh} kh \tag{86.10b}$$

Here the real coefficient A is the wave amplitude, k the wave number (related to the wavelength λ by $\lambda = 2\pi/k$), and $\omega/2\pi$ the frequency of the wave. Moreover, β and γ are two arbitrary phase angles. For $\beta = 0$, the surface elevation ζ is always zero at the points $kx = (n + \frac{1}{2})\pi$, n being an integer; these points are called *nodes*. The points $kx = n\pi$ correspond to the maximum of $|\zeta|$ at any time and are called *loops*. Furthermore, the expressions for ϕ_x, ϕ_y show that the fluid particles below nodes move horizontally, and those below loops move vertically (see Fig. 86.1). Thus, a wave of this type actually does not propagate, and is called a standing wave. Consequently, the total energy of the wave motion between two loops is always conserved. By rocking a rectangular tank of length L in an appropriate manner, two-dimensional standing waves can be generated in the water contained in the tank. Since ϕ_x must vanish at $x = 0$ and L, these two vertical sides must be loops, and L must be a multiple of $\frac{1}{2}\lambda$, or $kL = n\pi$, $n = 1, 2, \ldots$. With these eigenvalues of k, the frequency of the motion is obtained from (86.10b),

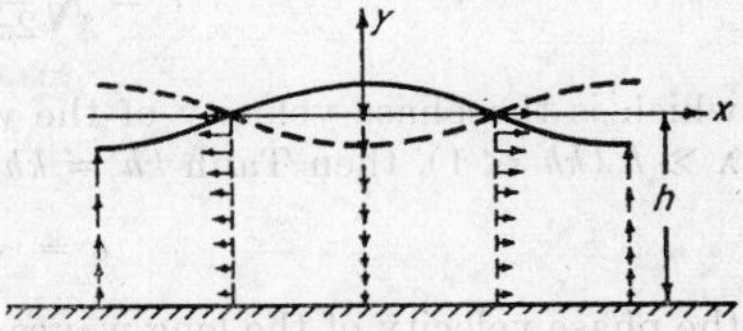

FIG. 86.1. Standing wave.

$$\omega^2 = \frac{n\pi}{L}\left[g + \sigma\left(\frac{n\pi}{L}\right)^2\right] \operatorname{Tanh} \frac{n\pi h}{L} \qquad n = 1, 2, \ldots \tag{86.11}$$

If $\sigma/g \ll (L/\pi)^2$, the capillary effect may be neglected for the slow modes (n small); only for n large may this effect become important. Furthermore, if $h \geq L$, $\operatorname{Tanh}(n\pi h/L) \approx 1$; the motion is then said to be in deep water.

In the general case when the container shape is not simple enough to make the problem readily solvable (such as a circular section), the frequency of the slowest mode, however, can be approximated by using Rayleigh's principle in the theory of vibrations. The formula is

$$\omega^2 = \frac{\int_{S_f} (g\phi_y{}^2 + \sigma\phi_{xy}{}^2)\, dx}{\int_{S_f} \phi\phi_y\, dx} \tag{86.12}$$

where ϕ is the potential of the slowest mode, satisfying (at least to a good approximation) the condition $\partial\phi/\partial n = 0$ on the container wall. Since the time factor cos $(\omega t + \gamma)$ cancels out from (86.12), it may be omitted. If the solution ϕ is exact, (86.12) gives the exact value of ω for all modes. However, if ϕ is approximate, (86.12) gives the lowest ω slightly higher than its exact value; the approximation becomes poorer for the higher modes.

Progressing Free Waves; Phase Velocity and Group Velocity

Within the framework of the linear theory, the standing waves may be superimposed appropriately to yield a train of progressing waves,

$$\zeta = A \cos kx \cos \omega t + A \sin kx \sin \omega t = A \cos (kx - \omega t) \tag{86.13a}$$

The corresponding velocity potential of this flow in water of depth h is

$$\phi = A \frac{\omega \operatorname{Cosh} k(y + h)}{k \operatorname{Sinh} kh} \sin (kx - \omega t) \tag{86.13b}$$

with $\omega(k)$ given by (86.10b). This solution represents a train of progressing waves which advance unchanged toward x positive with constant phase velocity,

$$c = \left(\frac{dx}{dt}\right)_{\varphi = \text{const}} = -\frac{\phi_t}{\phi_x} = \frac{\omega}{k} = \left[\left(\frac{g}{k} + \sigma k\right) \operatorname{Tanh} kh\right]^{1/2} \tag{86.14}$$

For $h > \lambda$, or $kh > 2\pi$, then Tanh $kh \approx 1$; hence,

$$c = \sqrt{\frac{g\lambda}{2\pi} + \frac{2\pi\sigma}{\lambda}} \qquad \text{for } h > \lambda \tag{86.14a}$$

which is the phase velocity of the waves in deep water. On the other hand, when $\lambda \gg h$ $(kh \ll 1)$, then Tanh $kh \approx kh$ and σk may be neglected, giving

$$c = \sqrt{gh} \qquad \text{for } \lambda \gg h \tag{86.14b}$$

the phase velocity of the long waves in shallow water. The fluid particles engaged in the wave motion in deep water each describe a circular path of radius $r_0 = A \exp (ky_0)$ where y_0 is its vertical position when undisturbed. Hence, the motion diminishes rapidly from the surface downward. For example, at $y_0 = -\lambda/2$, $r_0 = 0.043A$. Furthermore, the particle motion is in phase with the wave motion; i.e., the particles at crests move in the direction of the wave, whereas those at troughs move in the opposite direction. When the depth effect is not negligible, the path described by each particle becomes an ellipse.

Equation (86.14a) may also be written as

$$\frac{c}{c_m} = \left[\frac{1}{2}\left(\frac{\lambda}{\lambda_m} + \frac{\lambda_m}{\lambda}\right)\right]^{1/2} \qquad \lambda_m = 2\pi\sqrt{\frac{\sigma}{g}} \qquad c_m = (4g\sigma)^{1/4} \tag{86.15}$$

The curve $c(\lambda)$, as plotted in Fig. 86.2, has a minimum value $c = c_m$ at $\lambda = \lambda_m$. Hence, no surface wave, of any wavelength, can propagate with phase velocity less than c_m. On the other hand, for every given $c > c_m$, there are two permissible wavelengths, $\lambda_1 > \lambda_m > \lambda_2$ (see Fig. 86.2). Following Lord Kelvin, the waves with $\lambda > \lambda_m$ will be called *gravity* waves, and those with $\lambda < \lambda_m$ *capillary* waves, or ripples. For a water-air interface at the normal condition, $\sigma = 74$ cm³/sec², $g = 981$ cm/sec²; hence, $\lambda_m = 1.73$ cm and $c_m = 23.2$ cm/sec.

Since the phase velocity c of water waves, in general, depends on λ [see Eq. (86.14)] superposition of simple waves of different λ gives rise to the phenomenon of dispersion. As each component of a group of waves having different wavelengths propagates with velocity proper to its own wavelength, the group will move with a different velocity, called the group velocity, given by

$$c_g = \frac{d\omega}{dk} = c + k\frac{dc}{dk} = c - \lambda\frac{dc}{d\lambda} \tag{86.16}$$

and the form of the wave group will change continually as it progresses (see also p. 64–7). Equation (86.16) leads to a geometrical construction of c_g. As indicated in

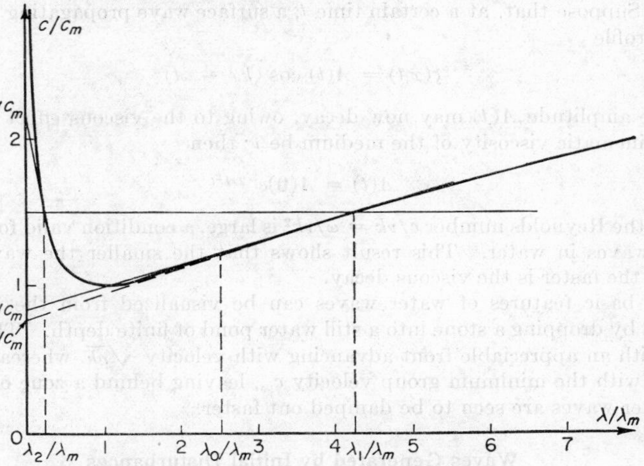

FIG. 86.2. Phase velocity c versus wavelength λ.

Fig. 86.2, c_g is the intercept made by a tangent to the curve $c = c(\lambda)$ on the c axis. For surface waves in deep water, we obtain, from Eqs. (86.15) and (86.16),

$$c_g = \frac{c}{2}\frac{(\lambda/\lambda_m)^2 + 3}{(\lambda/\lambda_m)^2 + 1} \tag{86.17}$$

with c and λ_m given by (86.15). Thus, c_g is greater or less than c, according to whether $\lambda < \lambda_m$ or $\lambda > \lambda_m$. In particular, $c_g \approx c/2$ for the gravity wave with $\lambda \gg \lambda_m$ and $c_g \approx 3c/2$ for the short capillary wave, $\lambda \ll \lambda_m$. In an isolated group of gravity waves (such as that sent out by dropping a stone in still water), the individual waves are observed to move faster than the group as a whole, dying out at the front, while new ones arise at the rear. The individual capillary waves, however, move through their group from the front to the rear. On the other hand, for long waves in shallow water, c becomes independent of λ [see Eq. (86.14*b*)]; hence, it follows from Eq. (86.16) that $c_g = c$, and the phenomenon of dispersion thereby vanishes.

It may be noted that the group velocity of water waves, like the phase velocity, also has a minimum value, c_{g_0}, at $\lambda = \lambda_0$, where

$$\lambda_0/\lambda_m = (3 + 2\sqrt{3})^{1/2} = 2.542 \qquad c_{g_0}/c_m = 0.767 \tag{86.18}$$

For the water-air interface, $\lambda_0 = 4.4$ cm and $c_{g_0} = 17.8$ cm/sec.

One important physical significance of group velocity is its connection with the mean rate of transmission of energy. When the wave group consists of several simple waves of different wavelengths, it can be shown that the energy of the wave motion is transmitted on the average (over a few oscillations of the individual components) with the group velocity c_g relative to the fluid at rest (cf. [7]). It is of interest to note that this physical significance of c_g holds valid even for the monochromatic waves (i.e., an infinite train of one simple wave). For a mathematical proof of this result, see [1], p. 458. Physically, the infinite train of simple waves may be regarded as an idealized limiting case of a single group having an immense group length. As in reality, there always exists an appreciable wavefront of such a group, though at a large distance, and though the waves near the front are no longer simple, the front will nevertheless propagate at the rate c_g (see [8]).

The viscous effect on a free running simple wave (see [1], p. 624) may be stated as follows. Suppose that, at a certain time t, a surface wave propagating in deep water has the profile

$$\zeta(x,t) = A(t) \cos (kx - \omega t) \tag{86.19}$$

where the amplitude $A(t)$ may now decay, owing to the viscous effect of the liquid. Let the kinematic viscosity of the medium be ν; then

$$A(t) = A(0)e^{-2\nu k^2 t} \tag{86.20}$$

provided the Reynolds number $c/\nu k = \omega/\nu k^2$ is large, a condition valid for all practical cases of waves in water. This result shows that the smaller the wavelength (the larger k), the faster is the viscous decay.

These basic features of water waves can be visualized from the wave pattern generated by dropping a stone into a still water pond of finite depth. The long waves appear with an appreciable front advancing with velocity $\sqrt{gh}$, whereas its rear also advances with the minimum group velocity c_{g_0}, leaving behind a zone of calm water. The shorter waves are seen to be damped out faster.

Waves Generated by Initial Disturbances

Suppose that the flow in the region $0 > y > -h$ has been at rest for time $t < 0$, and that, at $t = 0$, the initial disturbance is described as

$$\zeta(x,0) = f(x) \qquad \phi(x,y,0) = 0 \tag{86.21}$$

where the arbitrary function $f(x)$ is assumed to be absolutely integrable for $|x| < \infty$. Then, for $t > 0$ and $y < 0$, the solution is readily verified to be

$$\phi(x,y,t) = -\frac{1}{\pi}\int_0^\infty \frac{\omega \operatorname{Cosh} k(y+h)}{k \operatorname{Sinh} kh} \sin \omega t \, dk \int_{-\infty}^{\infty} f(\xi) \cos k(x-\xi)\, d\xi \tag{86.22}$$

$$\zeta(x,t) = \frac{1}{\pi}\int_0^\infty \cos \omega t \, dk \int_{-\infty}^{\infty} f(\xi) \cos k(x-\xi)\, d\xi \tag{86.23}$$

where $\omega(k)$ is given by (86.10b) for the general case, and reduces to $\omega = \sqrt{gk}$ for the case of gravity waves in deep water. Only for a very limited number of simple forms of $f(x)$ can the above integrals be expressed in terms of known functions. For arbitrary $f(x)$ and sufficiently large t, however, these integrals can be approximated with good accuracy by their asymptotic value. One widely used method of obtaining the asymptotic representation of the wave solution is the principle of stationary phase [9], which can be explained as follows. The product of the trigonometric functions in the integrand of (86.22) and (86.23) may be expressed as the linear combination of cosine or sine functions of argument $(kx \pm \omega t)$; these terms may be regarded as elementary progressive waves. At a given x for a sufficiently large t, only those wave components which nearly have the same phase will reinforce, to produce the predominant part of the result, whereas the other components differ in phase and annul one another because of mutual interference. Hence, for given x and t, the phase will be stationary for those wave components whose wave number k satisfies

$$d(kx - \omega t)/dk = 0 \qquad \text{or} \qquad c_g = d\omega/dk = x/t \tag{86.24}$$

where c_g is the group velocity defined in (86.16). Hence, the local group velocity of the wave passing point x at time t is x/t. By making use of this method, the following typical examples of gravity waves in deep water are presented with their asymptotic solutions, exhibiting the basic features of the water waves generated.

(1) $$f(x) = B\delta(x)$$

where B is a constant, having the dimension of length squared, and $\delta(x)$ is the Dirac

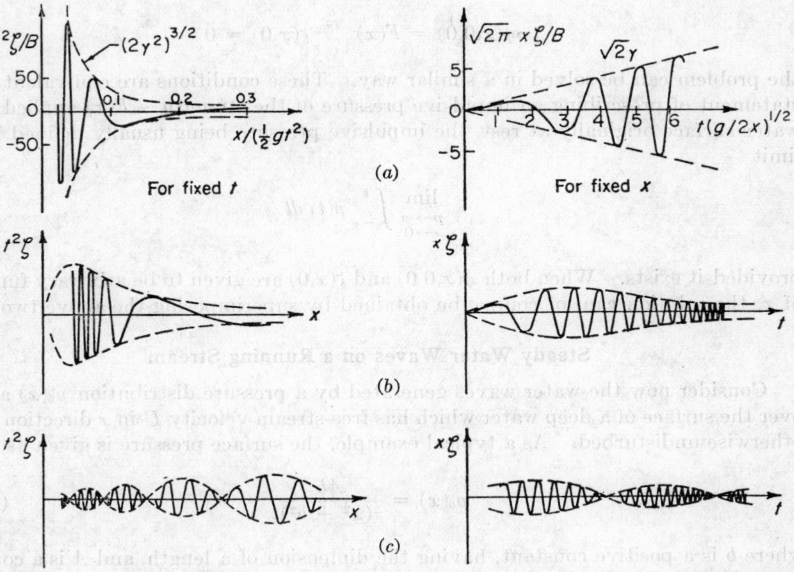

FIG. 86.3. Waves generated by initial disturbances. Left figures: t = const; right figures: x = const. (a) See Eq. (86.25); (b) see Eq. (86.26); (c) see Eq. (86.27).

delta function [see Eq. (19.7)]. This is the classical case first treated by Cauchy and Poisson [1], giving

$$\zeta(x,t) \cong \frac{B\gamma}{\sqrt{\pi}\,|x|}\left[\cos\left(\gamma^2 - \frac{\pi}{4}\right) + O\left(\frac{1}{|x|\gamma^2}\right)\right] \quad \text{for } \gamma^2 = \frac{gt^2}{4|x|} \gg 1 \tag{86.25}$$

(2) $$f(x) = \frac{Bb}{\pi(x^2 + b^2)}$$

The asymptotic solution is

$$\zeta \cong \frac{B\gamma}{\sqrt{\pi}\,|x|}\,e^{-\epsilon\gamma^2}\left[\cos\left(\gamma^2 - \frac{\pi}{4}\right) + O\left(\frac{1}{|x|\gamma^2}\right)\right] \quad \text{for } \gamma^2 = \frac{gt^2}{4|x|} \gg 1,\ \epsilon = \frac{b}{|x|} \ll 1 \tag{86.26}$$

(3) $$\begin{aligned} f(x) &= A \quad \text{for } |x| < b \\ &= 0 \quad \text{for } |x| > b \end{aligned}$$

Then $$\zeta = \frac{2A}{\gamma\sqrt{\pi}}\sin(\epsilon\gamma^2)\cos\left(\gamma^2 - \frac{\pi}{4}\right) \quad \text{for } \gamma^2 = \frac{gt^2}{4|x|} \gg 1,\ \epsilon = \frac{b}{|x|} \ll 1 \tag{86.27}$$

These solutions are depicted qualitatively in Fig. 86.3. It shows that, for given x, the wave motion is established only for moderate and large values of γ (more precisely

$\gamma > 2$, or $t > 4\sqrt{x/g}$). For smaller values of γ (that is, for less time or for farther distance), the disturbance is weaker, and does not bear the feature of a wave motion. Thus, the leading part of the wave pattern may be regarded as the apparent wavefront. In the region where the wave motion has been established, changes in wavelength, period, and phase velocity are all propagated with the group velocity $c_g = x/t$ [Eq. (86.24)]. That is to say, if at x_1 and t_1, λ is found to be λ_1, then $\lambda = \lambda_1$ at $x/t = x_1/t_1$ for other places and other time instants.

When the initial conditions are given by

$$\rho\phi(x,0,0) = F(x) \qquad \zeta(x,0) = 0$$

the problem can be solved in a similar way. These conditions are equivalent to the statement of prescribing an impulsive pressure of the strength $-F(x)$ applied to the water surface originally at rest, the impulsive pressure being usually defined by the limit

$$\lim_{\substack{p\to\infty \\ \epsilon\to 0}} \int_{-\epsilon}^{\epsilon} p(t)\,dt$$

provided it exists. When both $\phi(x,0,0)$ and $\zeta(x,0)$ are given to be arbitrary functions of x, the solution can, of course, be obtained by superimposing the above two cases.

Steady Water Waves on a Running Stream

Consider now the water waves generated by a pressure distribution $p_0(x)$ applied over the surface of a deep water which has free-stream velocity U in x direction and is otherwise undisturbed. As a typical example, the surface pressure is given as

$$p_0(x) = \frac{Ab}{\pi(x^2 + b^2)} \tag{86.28}$$

where b is a positive constant, having the dimension of a length, and A is a constant with the dimension of pressure times length. Suppose that the motion has been maintained for an infinite time under the influence of both gravity and surface tension, then the steady-state surface elevation for $U > c_m$ is

$$\zeta(x) = -\frac{2A}{\rho(U^4 - c_m{}^4)^{1/2}} \left\{ \begin{matrix} e^{-b\kappa_1} \sin \kappa_1 x \\ e^{-b\kappa_2} \sin \kappa_2 x \end{matrix} + L(x) \right\} \qquad \text{for } \left\{ \begin{matrix} x > 0 \\ x < 0 \end{matrix} \right. \tag{86.29}$$

with

$$\left. \begin{matrix} \kappa_1 \\ \kappa_2 \end{matrix} \right\} = \frac{1}{2\sigma} [U^2 \mp (U^4 - c_m{}^4)^{1/2}] \tag{86.30}$$

This result shows that a stationary train of gravity waves of length $\lambda_1 = 2\pi/\kappa_1$ ($\lambda_1 > \lambda_m$) propagates downstream of the origin, and a train of capillary waves of length $\lambda_2 = 2\pi/\kappa_2$ ($\lambda_2 < \lambda_m$) exists upstream. The function $L(x)$ merely represents the elevation near the origin; it becomes negligible when $|x|$ exceeds half the greater wavelength λ_1 ($|L(x)| < 0.007$ for $|\kappa_1 x| > 2\pi$). The exponential factors in (86.29) show that the capillary wave has a smaller amplitude than the gravity wave, since $\kappa_2 > \kappa_1$.

For $U < c_m$, the surface elevation falls off rapidly away from the origin, and thus no surface waves in the ordinary sense can be generated. In this respect, c_m is related to the minimum wind speed required for the onset of surface waves. For $U = c_m$, it can be shown that no steady-state solution exists when the viscosity is neglected [8].

When the viscous effect is also taken into account, then, for large values of the Reynolds number $\mathrm{Re} = (U/\nu)(\sigma/g)^{1/2}$, where ν is the kinematic viscosity of the medium, and for $U > c_m$, the steady solution is obtained as [10]

$$\zeta(x,t) = -\frac{2A}{\rho(U^4 - c_m{}^4)^{1/2}} \begin{cases} \exp(-b\kappa_1 - 4\nu\kappa_1{}^2 x/U) \sin \kappa_1 x \\ \exp(-b\kappa_2 + 4\nu\kappa_2{}^2 x/U) \sin \kappa_2 x \end{cases} \quad \text{for} \begin{cases} x > 0 \\ x < 0 \end{cases} \tag{86.31}$$

In the above, the local effect of the elevation is again omitted. Thus, the space rate of attenuation is inversely proportional to the square of the wavelength; hence, the attenuation of the capillary wave is much more marked than that of the gravity wave, especially when $U \gg c_m$, for then $\kappa_2 \gg \kappa_1$.

It is of interest to consider the corresponding initial-value problem. The uniform stream of velocity U was undisturbed for the time $t < 0$; at $t = 0$, a pressure distribution $p_0(x)$, given by (86.28), is applied on the free surface, maintaining its strength thereafter. Then, for $U \gg c_m$ and $t \gg U/g$, the asymptotic solution can be shown to be [8]

$$\begin{aligned} \zeta(x,t) &\cong -\frac{2A}{\rho U^2} e^{-gb/U^2} \sin \frac{gx}{U^2} \qquad \text{for } 0 \le x < \frac{Ut}{2}\left(1 - \sqrt{\frac{U}{gt}}\right) \\ \zeta(x,t) &\cong -\frac{2A}{\rho U^2} e^{-U^2 b/\sigma} \sin \frac{U^2 x}{\sigma} \qquad \text{for } -\frac{Ut}{2}\left(1 - \sqrt{\frac{\sigma}{U^3 t}}\right) < x \le 0 \end{aligned} \tag{86.32}$$

In the above expression, the local effect near $x = 0$ and the other transient waves which are insignificant for large time are omitted. Furthermore, for $U < c_m$, no surface wave in the usual sense is generated, the disturbance being merely local; and, for $U = c_m$, the amplitude of the wave grows like $t^{1/2}$ as the time increases, implying that no steady-state linear solution exists in this case. The above result for $U \gg c_m$ clearly shows that the wave trains have their apparent wavefronts extending outward from the origin, at the rate equal to their respective group velocity with respect to the fluid at rest, and the apparent wavefronts diffuse, in the meantime, at the rate proportional to $t^{1/2}$. This result is in accordance with the physical principle governing the rate of energy transmission stated in Sec. 86.2. This rate of advance of the wavefront is useful for predicting the growth of waves as the waves generated in a stormy area propagate into the area of calm.

Nonlinear Effects; Waves of Finite Amplitude

Several theories have been developed for deep water waves of finite amplitude. The irrotational gravity waves of arbitrary amplitude have been treated by Stokes [11], Rayleigh [12], and Levi-Civita [13]; the irrotational capillary waves, by Crapper [14], and the rotational trochoidal wave, by Gerstner [15]. The profile of the Stokes waves, when expressed in terms of the coordinates fixed with respect to the wave, is found by successive approximations to be

$$\zeta/a = \cos kx + (\tfrac{1}{2}ka + \tfrac{17}{24}k^3a^3)\cos 2kx + \tfrac{3}{8}k^2a^2 \cos 3kx + \tfrac{1}{3}k^3a^3 \cos 4kx + O(k^4a^4) \tag{86.33}$$

and the wave velocity c is given by

$$c^2 = \frac{g}{k}[1 + k^2a^2 + \tfrac{5}{4}k^4a^4 + O(k^6a^6)] \tag{86.34}$$

The wavelength λ is, of course, $2\pi/k$, whereas the wave height H (i.e., the vertical distance between trough and crest) is related to the quantity a by

$$H = 2a[1 + \tfrac{3}{8}k^2a^2 + O(k^4a^4)] \tag{86.35}$$

Equation (86.33) shows that the wave profile has very nearly the outline of a trochoid, in which the circumference of the rolling circle is $\lambda = 2\pi/k$ and the length of the tracing arm is a. This waveform is sharper near the crests, and flatter in the troughs, than for the simple harmonic waves of infinitesimal amplitude; and these features become prominent as the amplitude is increased (see Fig. 86.4). According to Michell [16], however, the greatest possible height of Stokes waves is $H = 0.142\lambda$, with the corresponding wave velocity $c = 1.12(g/k)^{1/2}$. The mathematical existence of this kind of permanent gravity wave has been shown by Levi-Civita [13]. Although the theory indicates that there is a definite limit to the wave height for a given wavelength, there are no known records of ocean or laboratory waves attaining such

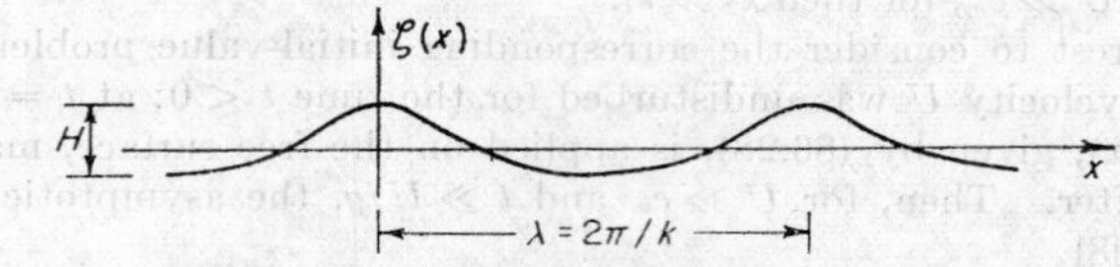

FIG. 86.4. The wave profile given by Eq. (86.33) for $ka = \frac{1}{2}$.

extreme heights. Generally, waves in the ocean are much less steep than this limit. Another feature of Stokes waves is that each fluid particle has a constant orbital velocity, and is not exactly periodic in its position. Upon completion of each nearly circular path, the particles at depth h advance a short distance in the direction of wave propagation at the mean velocity $k^2a^2ce^{-2kh}$.

Recently, the problem of progressing capillary waves of finite amplitude has been investigated by Crapper [14], when only surface tension, and not gravity, is taken into account as the restoring force. The greatest possible wave height H is found to be 0.730λ, at which the wave surface traps air bubbles in the troughs, and is tangent to itself above the bubbles; higher waves become impossible, because the wave surface then crosses itself. The wave velocity is found to be

$$c = (\sigma k)^{1/2}(1 + \tfrac{1}{16}k^2H^2)^{-1/4} \qquad \text{for } H \leq 0.73\lambda \tag{86.36}$$

Unlike the gravity wave, for which the wave velocity is higher than the value given by the linear theory, the velocity of the capillary wave decreases from its value of the linear theory as its amplitude increases.

The gravity waves in deep water considered by Gerstner [15] have a rotational flow field, and have exactly the trochoidal profile for which the equation

$$c = \sqrt{g/k} \tag{86.37}$$

is valid for all possible wave heights without approximation. The greatest possible height of Gerstner waves is $H = \lambda/\pi = 0.318\lambda$, obtained when the trochoids become cycloid. Furthermore, the Gerstner waves are without mass transport velocity.

86.3. THREE-DIMENSIONAL PROBLEMS

Wave Pattern Due to a Moving Surface Disturbance; Theory of Ship Waves

Suppose that a pressure disturbance, applied over a limited area on the free surface (the x,z plane), moves along a curve C which has the parametric representation

$$x = x_0(t) \qquad z = z_0(t) \tag{86.38}$$

for a given interval of the time t. For the purpose of finding the flow at large distances from the disturbance, the moving pressure system may be approximated by a point impulse on the same course C. Let it be given that, at $t = \tau$, the point impulse was moving with velocity $U(\tau)$, tangent to curve C at the point Q (see Fig. 86.5); and, at some later insant t, the impulse has just arrived at the point Q_0. Then, for a sufficiently large time interval $(t - \tau)$, the disturbance generated at some point P

by the original impulse at Q will have had many oscillations. Consequently, according to the principle of stationary phase, only in the region where the phase is nearly stationary, can the resulting disturbance have any significant value, whereas the wave elements in the remaining portions of the free surface differ in phase and cancel each other, leaving very small disturbances there. It can be shown (cf. [1], p. 433) that the waves due to the impulse exerted at point Q, at $t = \tau$, will have their phase stationary at some later instant t, only on the circle which passes through Q and has

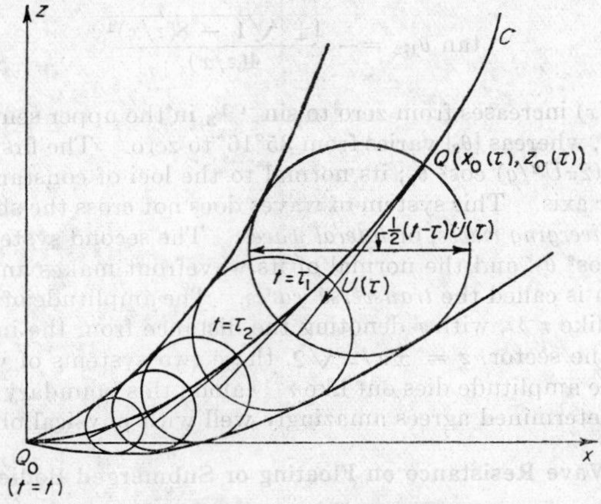

FIG. 86.5. Path of a moving surface disturbance.

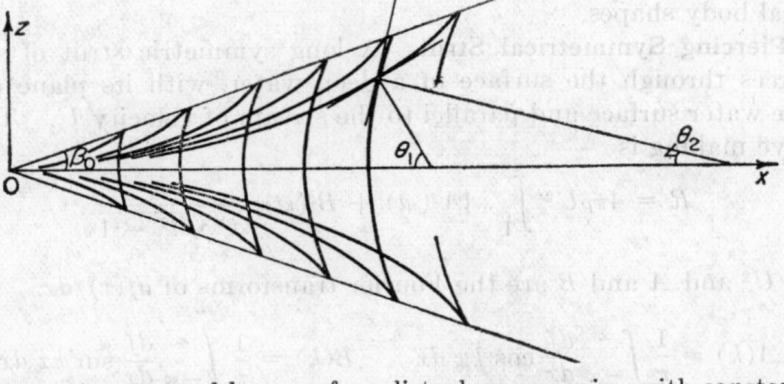

FIG. 86.6. Wave pattern caused by a surface disturbance moving with constant speed from right to left.

a diameter of length $(\frac{1}{2})U(\tau)(t - \tau)$ tangent to the instantaneous velocity of the impulse at τ (see Fig. 86.5). As the impulse moves along curve C, τ varies from zero (say) to t, ending up at Q_0, the position at present. Thus, a family of such circles of stationary phase can be constructed; the region within the two envelopes of these circles then is the region in which the surface elevation becomes the most conspicuous at the instant t; the elevation outside this region falls off more rapidly with distance and becomes much less significant.

In particular, if the impulse moves with constant velocity U on a straight course in the negative x direction, then the noticeable waves are confined to a sector of semiangle $\arcsin \frac{1}{3} = 19°28'$. It is of interest to note that this angle is independent of the speed as long as the speed is constant. Further detailed calculation shows that there are two systems of waves within this sector (see Fig. 86.6), with their surface elevations proportional respectively to

$$\sin\left(\frac{g}{U^2}\,\frac{x\cos\theta_1 + z\sin\theta_1}{\cos^2\theta_1} - \frac{\pi}{4}\right)$$
$$\sin\left(\frac{g}{U^2}\,\frac{x\cos\theta_2 + z\sin\theta_2}{\cos^2\theta_2} + \frac{\pi}{4}\right) \tag{86.39}$$

where θ_1 and θ_2 are related to x and z by

$$\tan\theta_{1,2} = -\frac{1 \pm \sqrt{1 - 8(z/x)^2}}{4(z/x)} \tag{86.40}$$

Thus, as $\tan^{-1}(z/x)$ increases from zero to $\sin^{-1}\frac{1}{3}$ in the upper semisector, $|\theta_1|$ varies from 90° to 35°16′, whereas $|\theta_2|$ varies from 35°16′ to zero. The first wave system has local wavelength $(2\pi U^2/g)\cos^2\theta_1$; its normal to the loci of constant phase makes an angle θ_1 with the x axis. This system of waves does not cross the ship's course, and is often called the *diverging waves*, or *lateral waves*. The second system has local wavelength $(2\pi U^2/g)\cos^2\theta_2$, and the normal of its wavefront makes an angle θ_2 with the x axis; this system is called the *transverse waves*. The amplitude of these two systems of waves falls off like $r^{-1/2}$, with r denoting the distance from the impulse point. On the boundary of the sector, $z = \pm x/2\sqrt{2}$, these two systems of waves meet with a cusp, and the wave amplitude dies out like $r^{-1/3}$ along this boundary. The theoretical wave pattern so determined agrees amazingly well with physical observations.

Wave Resistance on Floating or Submerged Bodies

The wave resistance experienced by an obstacle floating or submerged in a water stream can be calculated from the principle of momentum conservation, or by integrating the pressure over the body surface. The following cases are presented for several typical body shapes.

Surface Piercing Symmetrical Strut. A long symmetric strut of semithickness $z = f(x)$ pierces through the surface of a deep water, with its plane of symmetry normal to the water surface and parallel to the stream of velocity U. The resistance R due to wave making is

$$R = 4\pi\rho U^2 \int_1^\infty [A^2(\kappa t) + B^2(\kappa t)]\,\frac{dt}{t^2\sqrt{t^2-1}} \tag{86.41}$$

where $\kappa = g/U^2$ and A and B are the Fourier transforms of $df(x)/dx$,

$$A(k) = \frac{1}{\pi}\int_{-\infty}^{\infty}\frac{df}{dx}\cos kx\,dx \qquad B(k) = \frac{1}{\pi}\int_{-\infty}^{\infty}\frac{df}{dx}\sin kx\,dx \tag{86.42}$$

Wave Resistance of a Thin Ship. Let the ship hull be given by $z = f(x,y)$, where $z = 0$ is the plane of symmetry from bow to stern, and $y = 0$ is the undisturbed free surface. The ship is said to be thin if, for almost everywhere over the hull, $|f_x| \ll 1$, $|f_y| \ll 1$. Then the wave resistance experienced by the ship is [17]

$$R = \frac{4}{\pi}\rho U^2\kappa^2 \int_1^\infty [M^2(t) + N^2(t)]\,\frac{t^2\,dt}{\sqrt{t^2-1}} \tag{86.43}$$

where $\kappa = g/U^2$ and

$$\left.\begin{matrix} M(t) \\ N(t) \end{matrix}\right\} = \int_{-\infty}^{\infty} dx \int_{-\infty}^{0} e^{\kappa t^2 z} \left\{\begin{matrix}\cos \\ \sin\end{matrix}\right. \kappa t x\,\frac{\partial f}{\partial x}\,dz \tag{86.44}$$

Submerged Solid. When a solid body is submerged in an otherwise uniform stream, with one of the three principal axes (of permanent translation) coinciding with the stream direction, the resistance due to wave making is then given by ([1], p. 437):

$$R = (g^4/\pi\rho U^6)(A + \rho Q)^2 I \tag{86.45}$$

where A denotes the apparent mass along the direction of motion, Q is the volume of the solid, U its velocity, and

$$I(\alpha) = \tfrac{1}{4} e^{-\alpha}\left[K_0(\alpha) + \left(1 + \frac{1}{2\alpha}\right) K_1(\alpha)\right] \tag{86.46}$$

with $K_n(\alpha)$ standing for the modified Bessel function of the second kind ($\alpha = gh/U^2$), h being the depth of submergence.

86.4. REFERENCES

[1] H. Lamb: "Hydrodynamics," 6th ed., Dover, New York, 1945.
[2] J. J. Stoker: "Water Waves," Interscience, New York, 1957.
[3] L. M. Milne-Thomson: "Theoretical Hydrodynamics," 3d ed., Macmillan, London, 1955.
[4] N. J. Kotschin, I. A. Kibel, N. W. Rose: "Theoretische Hydromechanik," vol. 1, Akad.-Verlag, Berlin, 1954.
[5] H. F. Thorade: "Probleme der Wasserwellen," ("Probleme der kosmischen Physik," vols. 13 and 14), Akad. Verl.-Ges., Leipzig, 1931.
[6] H. Rouse: "Engineering Hydraulics," Wiley, New York, 1950.
[7] T. H. Havelock: The Propagation of Disturbances in Dispersive Media, *Cambridge Tracts in Math. and Math. Phys.*, no. 17, Cambridge Univ. Press, New York, 1914.
[8] C. R. De Prima, T. Y. Wu: On the theory of surface waves in water generated by moving disturbances, *Calif. Inst. Technol. Eng. Div. Rept.* 21–23 (1957).
[9] A. Erdélyi: "Asymptotic Expansions," Dover, New York, 1956.
[10] T. Y. Wu, R. E. Messick: Viscous effect on surface waves generated by steady disturbances, *Calif. Inst. Technol. Eng. Div. Rept.* 85–8 (1958).
[11] G. G. Stokes: On the theory of oscillatory waves, *Trans. Cambridge Phil. Soc.*, **8** (1847), 441–455; also "Scientific Papers," vol. 1, pp. 197, 314.
[12] Lord Rayleigh: On waves, *Phil. Mag.*, (5)**1** (1876), 257–279.
[13] T. Levi-Civita: Détermination rigoureuse des ondes permentes d'ampleur finie, *Math. Ann.*, **93** (1925), 264–314.
[14] G. D. Crapper: An exact solution for progressive capillary waves of arbitrary amplitude, *J. Fluid Mechanics*, **2** (1957), 532–540.
[15] Gerstner: Theorie der Wellen, *Abhandl. Böhm. Ges. Wiss.*, 1802.
[16] J. H. Michell: The highest waves in water, *Phil. Mag.*, (5)**36** (1893), 430–437.
[17] J. H. Michell: The wave resistance of a ship, *Phil. Mag.*, (5)**45** (1898), 106–123.

CHAPTER 87

CAVITATION

BY

M. S. PLESSET, Ph.D., Pasadena, Calif.
B. PERRY, Ph.D., Menlo Park, Calif.

87.1. INTRODUCTION

Cavitating flow occurs in a liquid when the pressure is locally low enough to cause the appearance of a gas or vapor phase. Since its occurrence may appreciably change the flow pattern, cavitation may, among other things, influence the drag of submerged bodies and the efficiency of pumps, turbines, and propellers. Moreover, it may cause noise and damage to solid surfaces [1],[2].

It is convenient to distinguish three regimes of flow, occurring in sequence as the velocity increases. *Noncavitating flow* occurs at the lowest speeds. *Incipient cavitation* is characterized by the appearance of small bubbles, which grow and collapse in the flow. Damage and noise are characteristic for this regime. With increasing speed, the number of bubbles grows until they merge into one large vapor cavity. This last regime is called *cavity flow*, or sometimes *supercavitating flow*. In the incipient cavitation regime, the gross velocity and pressure fields do not differ greatly from those in noncavitating flow; in the cavity-flow regime, however, they are essentially different.

This division into flow regimes is, in part, for convenience in analysis, and the transition from one flow condition to another is not always well defined. In particular, the point at which incipient cavitation begins cannot be precisely determined, since this point may have one value if visual means of bubble detection are used, and a different one if the onset of noise is used as the criterion. The transition from incipient to full cavity flow is also poorly defined, since it usually covers a range of flow parameters. Another significant feature of the transition in flow regimes is the hysteresis effect, which often gives a significantly different transition point, according to whether the flow speed increases or decreases.

The basic nondimensional parameter which is used to describe cavitating flow is the *cavitation parameter* K defined by

$$K = \frac{p_\infty - p_v}{\rho q_\infty^2/2} \tag{87.1}$$

where p_∞ is the static pressure of the undisturbed liquid, p_v is the vapor pressure of the liquid, ρ is the liquid density, and q_∞ is the speed of the undisturbed flow. Clearly one cannot expect a single parameter to describe so complex a phenomenon as cavitating flow over, say, for example, a submerged body. For a given body, there is, however, a qualitative correlation between the numerical values of K and the three regimes of flow just described. Sufficiently large positive values of K correspond to noncavitating flow; as K decreases, the flow passes through the incipient to the full cavity regime, in which the body may eventually become enclosed in the cavity. Zero is the limiting value for K since $K = 0$ implies that the liquid is at the boiling point everywhere.

A particular value of the cavitation parameter which is frequently used in technical applications is the *incipient cavitation number* K_i, which, for a fixed body in a flow, is the value of K at which cavitation first appears. Thus, K_i is determined by the pressure distribution over the body. If p is the pressure at a point on the body in noncavitating flow, then the familiar pressure coefficient is

$$C_p = \frac{p - p_\infty}{\rho q_\infty^2/2} \tag{87.2}$$

It might be expected that cavitation would begin at that point on the body at which p is first reduced to p_v, which will occur when the flow velocity has reached some sufficiently high value $q_{\infty i}$. The corresponding minimum value of C_p is

$$C_{p\,\min} = \frac{p_v - p_\infty}{\rho q_\infty{}^2{}_i/2} = -K_i \tag{87.3}$$

If $C_{p\,\min}$ is known, then K_i can be found, and the speed $q_{\infty i}$ at which cavitation will first occur can be estimated from Eq. (87.3) (p_v and p_∞ will usually be known). Cavitation on the surface of the body will not appear until K has fallen *below* K_i. This is a manifestation of *boiling delay,* which refers to the fact that a liquid will withstand some transient tension before boiling occurs. The value of the cavitation parameter for which cavitation first appears on the surface of a smooth streamlined body is only slightly less than K_i, as given by (87.3). For bluff bodies, on the other hand, the points at which minimum pressures occur are often in vortex cores shed by the body. In such cases, incipient cavitation appears not on the body surface but out in the liquid, and the value of K for the onset of cavitation may be much larger than the K_i given by (87.3). This vortex cavitation also occurs in the tip vortices of lifting surfaces, such as hydrofoils and propeller blades.

Cavitation may appear as a consequence of the dynamic pressure reduction in a liquid flow, such as takes place in pipes, through valves, and over submerged bodies. Cavitation may also be generated in a static liquid, by means of acoustic pressure oscillations.

87.2. INCIPIENT CAVITATION

Environmental Effects

Ordinary water and other liquids have tensile strengths which are made evident when they are subjected to tensions of short duration. The theoretical tensile strength of pure water, even when containing dissolved gas, is very high, of the order of 1000 atm. Such a tensile strength is orders of magitude greater than that observed. A pragmatic explanation of this discrepancy is that an ordinary liquid, such as water, has an appreciable number of imperfections or *nuclei,* from which a bubble can begin to grow. Though a nucleus may be of submicroscopic dimensions, it is still presumed to be large enough so that bubble growth starts against moderate surface-tension forces. More specific pictures of these nuclei have been put forward. One conjecture is that a nucleus is a small bubble of undissolved gas. The difficulty with this suggestion is in stabilizing such a bubble so that it does not go into solution because of the surface tension which acts on it. A more tenable view, perhaps, is that the nuclei consist of small solid particles, which may or may not have pockets of gas attached to them. Evidence for the existence of nuclei has been furnished by experiments of Harvey et al. [3],[4]. It was observed that water which had been subjected to high pressures (of the order of 1000 atm) exhibited very large tensile strengths when tested at ordinary atmospheric pressure.

In terms of nuclei, one would expect some correlation between the pressure reduction required for a bubble to grow and the amount of gas dissolved in the liquid.

Experiments have indeed indicated two types of bubble growth processes, depending upon the amount of air dissolved in the water samples. With significant amounts of dissolved air, macroscopic bubbles would form with small pressure reductions. This type of cavitation is called *gaseous cavitation*. With small amounts of dissolved air, noticeably greater pressure reductions are required to produce bubble growth. This condition is called *vaporous cavitation*.

In the experiments above, the cavitation was generated acoustically. The physical situation is somewhat different when we consider the onset of cavitation in the flow over a submerged body. For the flow over a streamlined body, the minimum static pressure appears at the body surface. Water-tunnel studies have been made under such circumstances [5], and it has been observed that the bubbles begin as microscopic bubbles in the boundary layer, where they remain fixed in position for a significant period of time. The liquid in the neighborhood of such a bubble would be supersaturated with dissolved air, even if the liquid in the undisturbed stream were only saturated. These fixed microscopic bubbles, therefore, grow by air diffusion. This growth proceeds until the bubble is larger than the boundary-layer thickness, at which time it will be torn out of the boundary layer. It then moves downstream with the flow, and will be subjected to the pressure variations along the body.

Theory of Growth and Collapse of a Single Cavitation Bubble

The growth and collapse of an incipient cavitation bubble is a central problem in cavitation hydrodynamics. For an analytic treatment, some idealizations are made which simplify the theory but still preserve the general applicability of the results. It will be assumed that the bubble growth and collapse can be treated as if there were a single bubble in the liquid field. This assumption is justified if there are only a few bubbles in the flow, as at the inception of cavitation. The Rayleigh theory for the collapse of a vapor bubble and its extension to growth [6] can be used to obtain a satisfactory general understanding of the problem.

It will be assumed that the bubble is spherical throughout its growth and collapse. Let the radius of the bubble at any time t be R. The origin of coordinates is taken at the bubble center, and r is the radial distance to any point in the liquid which is supposed to be incompressible. With the limitation to irrotational flow, we have the velocity potential

$$\phi = -\frac{\dot{R}R^2}{r} \tag{87.4}$$

Furthermore, the liquid equation of motion gives the Bernoulli integral

$$\frac{\partial \phi}{\partial t} + \tfrac{1}{2}(\nabla\phi)^2 = -\frac{p(r,t)}{\rho} + C(t) \tag{87.5}$$

Evaluation of Eq. (87.5) at $r = R$ gives the equation of motion of the bubble wall,

$$R\ddot{R} + \tfrac{3}{2}\dot{R}^2 = \frac{p(R) - P}{\rho} \tag{87.6}$$

where ρ is the liquid density, $p(R)$ is the pressure in the liquid at the bubble boundary, and P is the external pressure in the liquid at a distance from the bubble. The pressure $p(R)$ is related to the pressure within the bubble p_i by

$$p(R) = p_i - \frac{2\sigma}{R} - 4\mu\frac{\dot{R}}{R} \tag{87.7}$$

where σ is the surface tension constant, and μ is the coefficient of viscosity. The effect of viscosity is important only toward the end of the bubble collapse, when $\dot{R}/R$

becomes large. The behavior near the end of collapse will be discussed below in more detail, and the viscous term in (87.7) will be omitted for the present. It is assumed that the gas in the bubble, over most of the growth-collapse range to be followed, consists of the vapor of the liquid, so that $p_i = p_v$, the liquid vapor pressure. The vapor pressure will be taken to be constant here. If the external pressure P is also constant, say P_0, then the equation of motion (87.6) may be readily integrated. For the case of a growing bubble ($p_i > P_0$), it is convenient to introduce a radius R_0 defined by the relation

$$\frac{2\sigma}{R_0} = p_v - P_0 \tag{87.8}$$

Physically, R_0 is the effective initial radius of the nucleus from which the bubble begins its growth; it represents an extrapolation of the free spherical bubble down to an equilibrium radius for the given initial conditions. It should be noted that a bubble at rest with radius R_0 is in unstable equilibrium. The actual nucleus from which the bubble grows is not necessarily spherical, and its surface energy may be less than $4\pi\sigma R_0^2$ (it may be zero); however; such a nucleus and the free spherical bubble of radius R_0 will have essentially the same growth behavior after their radii have increased beyond two or three times R_0.

In terms of the parameter R_0, the equation of motion (87.6) may be written for a growing bubble as

$$R\ddot{R} + \frac{3}{2}\dot{R}^2 = \frac{2\sigma}{\rho R_0}\left(1 - \frac{R_0}{R}\right)$$

or

$$\frac{1}{2R^2\dot{R}}\frac{d}{dt}(R^3\dot{R}^2) = \frac{2\sigma}{\rho R_0}\left(1 - \frac{R_0}{R}\right) \tag{87.9}$$

Integration gives

$$\dot{R}^2 = \frac{R_0{}^3}{R^3}\dot{R}_0{}^2 + \frac{4\sigma}{3\rho R_0}\left(1 - \frac{R_0{}^3}{R^3}\right) - \frac{2\sigma}{\rho R}\left(1 - \frac{R_0{}^2}{R^2}\right) \tag{87.10}$$

For $R \gg R_0$, Eq. (87.10) gives

$$\dot{R} \approx \sqrt{\frac{4\sigma}{3\rho R_0}} = \sqrt{\frac{2}{3\rho}(p_v - P_0)} \tag{87.11}$$

Thus, the rate of growth of a cavitation bubble in a liquid under constant tension, $(p_v - P_0)$, tends to a constant value.

For a cavitation bubble with initial radius R_m, collapsing from rest under a constant pressure $P_0 - p_v > 0$, we have similarly

$$\dot{R}^2 = \frac{2(P_0 - p_v)}{3\rho}\left(\frac{R_m{}^3}{R^3} - 1\right) + \frac{2\sigma}{\rho R}\left(\frac{R_m{}^2}{R^2} - 1\right) \tag{87.12}$$

so that, as $R \to 0$, the collapse velocity has (with $\sigma = 0$) the behavior

$$\dot{R} \approx \sqrt{\frac{2(P_0 - p_v)}{3\rho}\frac{R_m{}^3}{R^3}} \tag{87.13}$$

The preceding formulas do not describe the history of a cavitation bubble as it appears in a flow. There a nucleus grows into a cavitation bubble, which subsequently collapses because it is subjected to a varying external pressure. Equation (87.6) describes this situation with $P = P(t)$. Under this circumstance, a numerical integration of the equation of motion is required to obtain a solution. Such a procedure has been used, for example, to analyze the experimental observations on bubbles which move in the flow over a body, and are therefore affected by a varying pressure field [6]. In this analysis, the pressure field was calculated from the known pressure distribution over the body in noncavitating flow. Presumably the pressure coefficient is not significantly affected by the small number of cavitation bubbles in the flow. The pressure field so determined, together with the measured bubble position as a function of time, gives the function $P(t)$. Since the initial conditions for the bubble are not known, the integration constants of the equation of motion were fixed by using the observed value for the maximum radius R_m, where $\dot{R}_m = 0$. The agreement between observed radii and calculated values is shown in Fig. 87.1, where it may be observed that the bubble grows during the interval of liquid tension, and later collapses in a region of higher pressure. The discrepancy between the theoretical curve and the observed values in the early stages of growth presumably is due to the effects of permanent gas (air), which is known to be in the bubble. It is also of interest to remark that there is a tendency for the predicted collapse to be somewhat more rapid than actually observed. This effect is explained by the fact that the bubble collapses

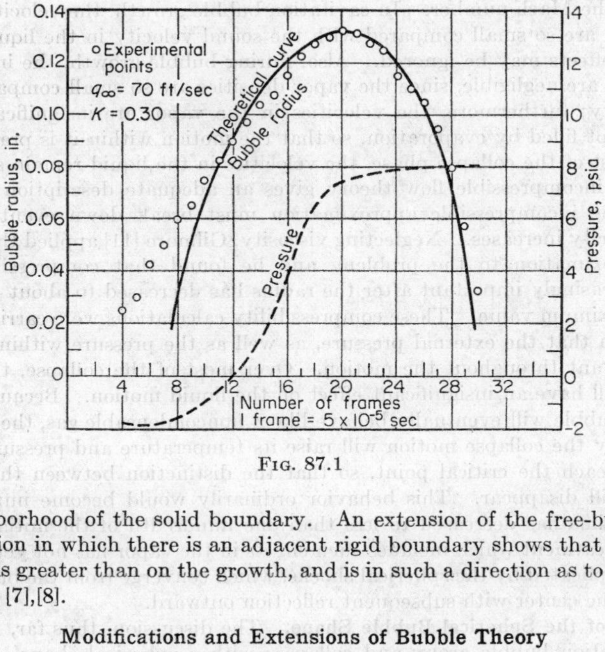

FIG. 87.1

in the neighborhood of the solid boundary. An extension of the free-bubble theory to the situation in which there is an adjacent rigid boundary shows that the effect on the collapse is greater than on the growth, and is in such a direction as to decrease the collapse rate [7],[8].

Modifications and Extensions of Bubble Theory

Thermal Effects. As a vapor bubble grows, the heat required for evaporation produces a cooling effect so that the temperature and vapor pressure p_v decrease. Conversely, during bubble collapse, the heat released by vapor condensation causes a rise in bubble temperature and vapor pressure. The dynamics of a vapor bubble can be strongly affected by these heat flows; however, cavitation bubbles in the usual hydrodynamic situations occur at temperatures appreciably below the ordinary boiling temperature, so that the vapor density and pressure are small, and, under such a condition, the thermal effect for a growing bubble is negligible [9]. Furthermore, under this same condition, the temperature rise in a collapsing bubble is unimportant except near the very end of collapse. One may therefore say, for ordinary cavitation bubbles, that the thermal effects may be neglected over the entire growth phase, and over nearly all of the collapse phase.

In addition to the temperature rise which appears toward the end of the collapse, it may also be remarked that the vapor will cease to behave like a condensable gas, and will begin to behave like a permanent gas, since the rate of condensation is finite and the bubble collapse velocity becomes very large. As will be seen, there are other difficulties with the theoretical analysis of the bubble behavior near the end of collapse.

Viscous Effect. The effect of the liquid viscosity for a spherical bubble appears in the boundary condition given by Eq. (87.7), and it is equivalent to an increase in the surface tension σ by the amount $2\mu\dot{R}$. For a growing bubble, this term makes a negligible contribution. In a liquid such as water, with a small coefficient of viscosity, the viscous effect remains small also during collapse, except toward the end of the collapse, when $\dot{R}$ becomes very large. The viscous term makes no contribution to a gross quantity such as the total time for collapse and will affect the collapse velocity only for very small values of R [10].

Compressibility. A general criterion for the importance of fluid compressibility is given by the Mach number. In cavitation bubble growth, the velocities produced in the liquid are so small compared with the sound velocity in the liquid that compressibility effects may be ignored. Also, during bubble growth, the inertial effects of the vapor are negligible, since the vapor densities are so small compared with the liquid density; furthermore, the velocities in the vapor are insignificant since the bubble is kept filled by evaporation, so that the motion within it is practically zero.

Over most of the collapse phase, the velocities in the liquid remain small enough so that the incompressible flow theory gives an adequate description of the flow. However, the incompressible approximation must break down eventually as the collapse velocity increases. Neglecting viscosity, Gilmore [11] applied the Kirkwood-Bethe approximation to the problem, and he found that compressibility effects become increasingly important after the radius has decreased to about one-tenth its initial or maximum value. These compressibility calculations were carried out under the condition that the external pressure, as well as the pressure within the bubble, remain constant throughout the motion. Over most of the collapse, the inertia of the vapor will have an insignificant effect on the liquid motion. Because the vapor within the bubble will eventually behave like a noncondensable gas, the compression of this gas by the collapse motion will raise its temperature and pressure. The gas will finally reach the critical point, so that the distinction between the liquid and gas phase will disappear. This behavior ordinarily would become important only when the radius has decreased to less than one hundredth of the initial radius. A theoretical treatment which includes such effects in the vapor has not yet been made, and it is complicated by the spherical shocks, which converge from the bubble boundary wall to the center with subsequent reflection outward.

Stability of the Spherical Bubble Shape. The discussion, thus far, has assumed that a cavitation bubble grows and collapses with a spherical shape. Among the conditions which could perturb the spherical shape are the presence of a solid boundary and an external force field such as gravity. Finally, there is the fundamental question of the inherent stability of the growth and collapse motion for a free bubble.

Theoretical study [7],[8] of the growth and collapse of a bubble in the neighborhood of a rigid boundary shows that the bubble will remain essentially spherical throughout its growth and during most of its collapse. Distortion from the spherical shape appears during the later stage of collapse and becomes very pronounced after the bubble radius has decreased to about 0.2 of its initial value.

If a bubble grows relatively slowly in a gravity field, its shape will become somewhat distorted by the translational motion it acquires because of buoyancy. Such a deformation is not too important, and generally the distortions from the spherical shape for a growing bubble are minor when the translational velocity of the bubble relative to the liquid is small. The situation is different for a bubble collapsing in a gravity field toward the end of its collapse. As $R \to 0$, the bubble acquires a large velocity in the direction of the buoyant force [12]. An approximate value for this upward velocity is found by consideration of the impulse which the bubble experiences; in this way, the upward velocity V is found to be

$$V(t) = \frac{2g}{R^3(t)} \int_0^t R^3(t')\, dt'$$

so that $V \to \infty$ as $R \to 0$. The associated large distortions which a bubble undergoes during the collapse have been observed experimentally.

An analysis of the inherent stability of a free spherical bubble in a force-free liquid field [13] shows that an expanding vapor cavity, initially spherical in shape, is stable, in the sense that any deformation remains small in comparison with the bubble radius if the initial amplitude of the deformation is small compared with the initial bubble radius. For a collapsing spherical cavity, on the other hand, the distortion amplitude remains small over the interval $1 \geq R/R_m \geq 0.2$, where R_m is the initial value of the radius. As $R \to 0$, the distortion amplitude increases as $R^{-1/4}$. Thus, a collapsing spherical cavity has an inherent instability in shape toward the end of the collapse. This inherent instability, even for a free bubble, represents an important and serious complication for any theoretical analysis of the last stage of bubble collapse.

Cavitation Damage and Noise

The damage produced by cavitation is a subject of great technical interest and has provided the main incentive for the development of the preceding theory. Quantitative results are, however, still mostly experimental. Accelerated damage tests have been devised, in which cavitation bubbles grow and collapse at ultrasonic frequencies. It has been shown that the damage to a solid occurs when a cavitation bubble collapses near its surface. The direct experimental evidence for this phenomenon has been obtained by Ellis and Sutton [14], who photographed the strain waves developed in a photoelastic plastic when a cavitation bubble collapsed at its surface. The strain wave generated in a solid upon collapse is of very short duration ($\lesssim 0.2$ μsec), and, while it is difficult to determine the magnitude of the stress originating from a bubble collapse, these same experimenters estimate that the stress is greater than 10^5 lb/in.^{2} over a very small area. From the very short duration of the pressure pulse radiated upon bubble collapse, it is clear that the associated noise spectrum would extend to acoustic frequencies of the order of megacycles [15].

Experimental studies on the mechanism of accelerated cavitation damage with an ultrasonic generator [16] show that the cavitation damage begins with the plastic deformation of the exposed solid structure. A solid, therefore, fatigues by this cold-work process. The stresses which produce this fatigue have components of extremely high frequency, so that it cannot be expected that the fatigue properties of solids at low frequencies would necessarily be correlated with their cavitation damage resistance. Since cavitation damage is a fatigue process, it must be expected that it could be accelerated when it takes place in a chemically active environment. There should, therefore, be a corrosive enhancement of cavitation damage, even though the basic physical mechanism is plastic deformation.

87.3. CAVITY FLOW

Assumptions of Cavity-flow Theory

When the cavitation parameter is sufficiently small so that a full cavity flow ensues, the cavity is filled with vapor or gas at nearly uniform pressure. In this flow regime, theory may be used to predict the shape of the cavity and the pressures and forces on the body. It has proved feasible to calculate these quantities for many simple cavity flows by the methods of potential flow theory [17]–[19]. The definition of the cavitation parameter K may now be generalized to include the possibility of a contribution to the pressure of a permanent gas in the cavity. If p_c is the total pressure in the cavity, then we write

$$K = \frac{P_\infty - p_c}{\rho q_\infty^2/2} \tag{87.14}$$

In the special case of no permanent gas in the cavity so that $p_c = p_v$, this definition reduces to Eq. (87.1).

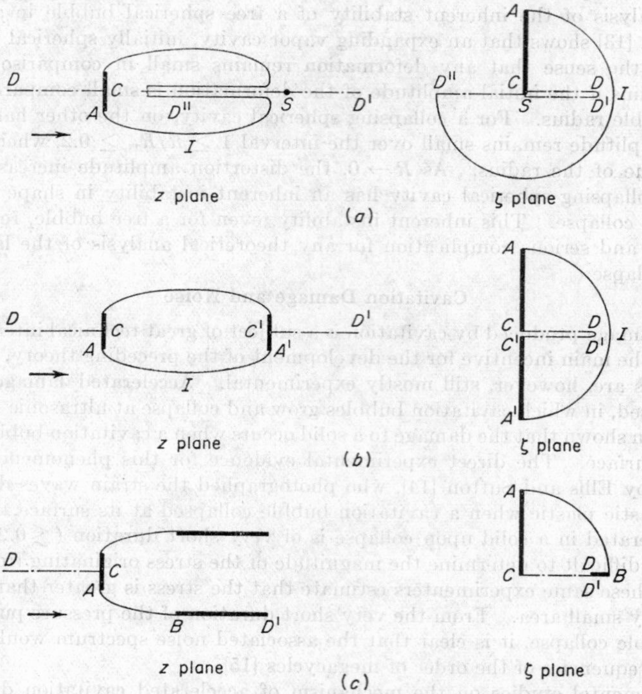

FIG. 87.2

While the assumption of potential flow is reasonable for the flow over the body surface and for the forward part of the cavity, the flow near the rear of the cavity deviates strongly from potential flow, since it contains a region of turbulent mixing. An idealized theory should attempt to account for this deviation in some way which does not disturb the accuracy of the predictions of the physical quantities of major interest, that is, the flow in the forward part of the cavity and over the body itself. When gravity and surface-tension effects are negligible, it has proved satisfactory to use any one of several potential flow models [20], which are shown in Fig. 87.2. Though none of these models actually corresponds to the flow situation at the rear of the cavity, that of Fig. 87.2*a* is probably nearest the physical situation if we imagine that the reentrant jet is broken up by turbulent mixing. The Riabouchinsky flow (Fig. 87.2*b*) is somewhat easier to compute, and gives accurate results for body forces and cavity shape. The Joukowsky flow (Fig. 87.2*c*) is even simpler to analyze and gives adequate results for the forces on the body; it is not very useful for determination of the cavity shape except for the forward portion. None of these models is accurate when the cavity length is very short (K large), since, in this case, the whole region is taken up with the turbulent mixing process. With regard to the flow conditions at the rear of a cavity, it should also be remarked that gravity can greatly alter the physical processes in this region of the flow.

For any of these potential flow models, the formulation of the problem proceeds in a straightforward manner. The flow is assumed to be irrotational and incompressible, so that the velocity potential exists, and satisfies Laplace's equation. The boundary condition on the surface of the body, and on any solid image surfaces introduced by the flow model, is the usual one that the normal component of the fluid velocity is zero. The application of the boundary condition on the free surface representing the cavity wall, on the other hand, presents a difficulty characteristic of these problems. This boundary condition arises from the physical requirement that the pressure on the free surface is constant. The direct application of Bernoulli's equation for steady flow, where gravity and surface tension are neglected, leads to the kinematic boundary condition that the magnitude of the free streamline velocity q_s is constant; that is,

$$q_s = q_\infty \sqrt{1 + K} = \text{const}$$

We do not know, however, where this condition is to be applied since the location of the free surface is itself unknown. Thus, the free surface introduces an inherent nonlinearity into the mathematical problem.

Two-dimensional Cavity-flow Theory

Most of the theoretical results on cavity hydrodynamics refer to the case of plane flow where the powerful methods of complex-variable theory may be applied. If $\phi(x,y)$ is the velocity potential, and $\psi(x,y)$ is the stream function, then the complex potential $W = \phi + i\psi$ has as its derivative the complex velocity

$$\zeta = \frac{dW}{dz} = u - iv$$

where u and v are the x and y components of the fluid velocity. The ζ plane, with coordinates $+u$ and $-v$, is known as the *hodograph plane*. The special importance of the hodograph plane is that the free boundaries of the flow on which the flow speed must be constant correspond to circular segments in this plane. The difficulty of the unknown free boundary in the flow field, or in the *physical plane*, does not occur in the hodograph plane. If the part of the body in contact with the liquid has only straight walls, then the remainder of the hodograph is also very easily constructed, as is illustrated for the example of the flat lamina in Fig. 87.2. If the hodograph plane is now mapped on the complex potential plane, which is also known, the coordinates of the flow are given by a quadrature

$$z = \int \frac{dW}{\zeta}$$

For bodies of simple shape, this program can often be carried through in terms of elementary conformal mappings, although the integration may have to be completed by numerical means. The example of a wedge in cavity flow, which has some practical applications, is shown in the diagram of Fig. 87.3. The potential and hodograph planes are shown together, with some auxiliary planes which are used in the conformal mappings. For this case, the sequence of transformations

$$\eta^{2\beta/\pi} = \zeta \qquad t = \frac{i}{2}\left(\frac{1}{\eta} - \eta\right) \qquad W = \frac{bt}{(\tan^2 \alpha + t^2)^{1/2}}$$

maps the hodograph on the potential plane. The constant b determines the scale for the cavity length, and the parameter α is related to the cavitation parameter as follows:

$$1 + K = \left(\frac{1 + \sin \alpha}{\cos \alpha}\right)^{4\beta/\pi}$$

The coordinates in the physical plane are found by integration, and the remaining flow parameters are then calculated. Such calculations on the cavity flow with wedges give the following limiting behavior for the drag coefficient C_D, and for the length-to-maximum-width ratio, l/d_m, of the cavity:

$$C_D(K) \approx C_D(0)(1 + K) \qquad K \to 0 \tag{87.15}$$

$$\frac{l}{d_m} \approx \frac{2}{K} \qquad K \to 0 \tag{87.16}$$

Equation (87.16) reflects the fact that the over-all shape of the cavity is independent of the specific shape of the body from which it arises when K is small.

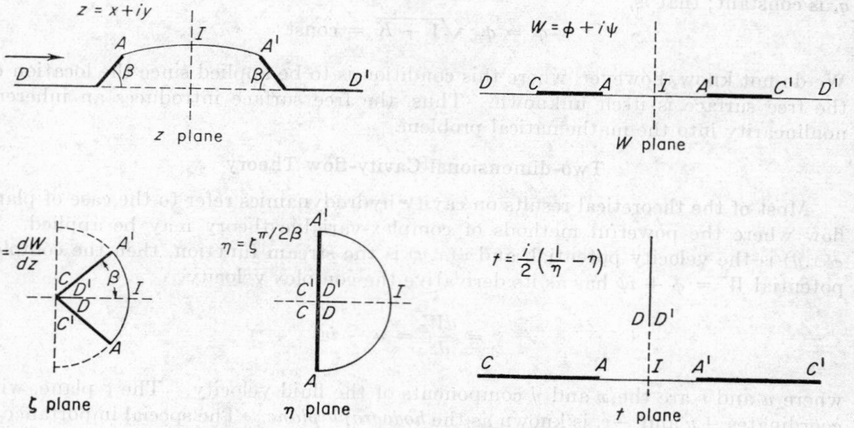

FIG. 87.3

For a curved body, the main ideas just described are applicable, but now the shape of the hodograph boundary, which corresponds to the flow on the surface of the body, is unknown, as is also, in general, the point on the body at which the cavity separation takes place. This problem can be attacked indirectly by an inverse method. The shape of that part of the hodograph boundary which corresponds to the flow over the solid boundary is assumed arbitrarily, and the flow in the physical plane including the form of the boundary is then determined from a calculation which is essentially the same as that described in the paragraphs above. Such an inverse procedure does not, of course, usually lead to the cavity flow with a preassigned solid boundary shape; however, a wide range of boundary possibilities can be explored by adjusting the coefficients in the infinite series which represents the mapping of the arbitrary hodograph on the potential plane. Moreover, a systematic method for determining the series coefficients, so as to obtain a required body shape by successive approximations, has been developed; it has been used to obtain numerical solutions for such cavity flows as circular cylinders ($K \geq 0$) and circular arcs ($K \geq 0$). Calculations of this type have recently been put on a more accurate basis by use of high-speed computing machines, to obtain numerical results [19].

Another approach to the cavity-flow problem has been devised, which uses methods first developed in thin airfoil theory [21],[22]. Supposing that the attack angle is small, it is assumed that the disturbance of the free-stream velocity is everywhere small, and that the boundary conditions need not be applied on the actual free streamline, which has an unknown location, but on a line with the direction of the free flow. The calculations may be carried through with conformal mapping techniques, which are familiar from thin airfoil theory. The linearized results compare favorably with the exact theory for very thin wedges in cavity flow as $K \to 0$. Agreement with the more exact lifting hydrofoil theory at small angles of attack is also reasonable, but fairly large errors are present in some cases. While the linearized theory may be subject to such errors, it has great qualitative value for a wide variety of problems.

Three-dimensional Cavity Flow

An extensive experimental study with bluff bodies (disks and cones) was made by Reichardt [23], who observed the cavity-flow phenomena at extremely small values of the cavitation parameter ($K \leq 0.1$), with both vapor- and air-filled cavities. He also calculated the shape of the axially symmetric cavity as a function of the cavitation parameter, by application of the method of distributed sources and sinks. While this technique is open to objection because it does not properly account for the singularity at the separation point, the over-all cavity shapes so obtained seem to be in satisfactory agreement with the available data. More recently, Garabedian [24] has analyzed a specific example of a finite cavity in axially symmetric flow with the Riabouchinsky model (symmetric fore and aft disks). The improvement from a theoretical point of view is that, in addition to sources and sinks, singularities are included which are known to have the correct behavior near the separation point. To carry through the numerical calculations, an iterative method for adjusting the strength of the singularities is used. In addition, he has devised a new approach for the calculation of axially symmetric cavity flows with $K = 0$. In two dimensions the stream function ψ satisfies Laplace's equation $\nabla^2\psi = 0$, while in axially symmetric flows it satisfies the equation

$$\nabla^2\psi - \frac{1}{y}\frac{\partial\psi}{\partial y} = 0$$

These two problems may be related by considering them as special cases of the more general equation

$$\nabla^2\psi - \frac{\epsilon}{y}\frac{\partial\psi}{\partial y} = 0 \tag{87.17}$$

The parameter ϵ is now allowed to have an arbitrary value, although it has a physical interpretation only for $\epsilon = 0$ (plane flow) and for $\epsilon = 1$ (axially symmetric flow). The properties of the solution of Eq. (87.17) can be analyzed in sufficient detail so that good estimates of flow quantities can be made. Thus, in addition to the known case with $\epsilon = 0$, several degenerate cases can be calculated, and $\partial\psi/\partial\epsilon$ at $\epsilon = 0$ can be found. By using these data, one may interpolate to estimate the physically important case with $\epsilon = 1$. The result for the drag coefficient of a circular disk is $C_D(0) = 0.827$. In an additional calculation, Garabedian shows that the limiting formula of Eq. (87.15) also holds in three dimensions for bodies with fixed separation points, so that the drag of a circular disk for small K is

$$C_D(K) \approx 0.827(1 + K) \tag{87.18}$$

87.4. REFERENCES

[1] Cavitation in Hydrodynamics, *Proc. Symposium Natl. Phys. Lab.*, H. M. Stationery Office, London, 1956.

[2] P. Eisenberg: Modern developments in the mechanics of cavitation, *Appl. Mechanics Revs.*, **10** (1957), 85–89.
[3] E. N. Harvey, W. S. McElroy, A. H. Whitely: On cavity formation in water, *J. Appl. Phys.*, **18** (1947), 162–172.
[4] D. C. Pease, L. R. Blinks: Cavitation from solid surfaces in the absence of gas nuclei, *J. Phys. Colloid Chem.*, **51** (1947), 556–567.
[5] R. W. Kermeen, J. T. McGraw, B. R. Parkin: Mechanism of cavitation inception and the related scale-effects problem, *Trans. ASME*, **77** (1955), 533–541.
[6] M. S. Plesset: The dynamics of cavitation bubbles, *J. Appl. Mechanics*, **16** (1949), 277–282.
[7] M. Rattray: Perturbation effects in cavitation bubble dynamics, Thesis, Calif. Inst. of Technol., 1951.
[8] J. M. Green: The hydrodynamics of spherical cavities in the neighborhood of a rigid plane, *Calif. Inst. Technol. Eng. Div. Rept.* 85-1 (1957).
[9] M. S. Plesset: Physical effects in cavitation and boiling, Symposium on Naval Hydrodynamics, pp. 297–318, *Natl. Acad. Sci. Publ.* 515 (1956).
[10] R. H. Pennington: Surface instabilities on pulsating gas bubbles, *Stanford Univ. Appl. Math. and Statist. Lab. Tech. Rept.* 22 (1954).
[11] F. R. Gilmore: The growth or collapse of a spherical bubble in a viscous compressible liquid, *Calif. Inst. Technol. Hydrodynamics Lab. Rept.* 26-4 (1952).
[12] R. H. Cole: "Underwater Explosions," p. 308, Princeton Univ. Press, Princeton, N.J., 1948.
[13] M. S. Plesset, T. P. Mitchell: On the stability of the spherical shape of a vapor cavity in a liquid, *Quart. Appl. Math.*, **13** (1956), 419–430.
[14] A. T. Ellis: Techniques for pressure pulse measurements and high-speed photography in ultrasonic cavitation, Paper 8 of Ref. [1].
[15] H. M. Fitzpatrick, M. Strasberg: Hydrodynamic sources of sound, Symposium on Naval Hydrodynamics, pp. 241–276, *Natl. Acad. Sci. Publ.* 515 (1956).
[16] M. S. Plesset, A. T. Ellis: On the mechanism of cavitation damage, *Trans. ASME*, **77** (1955), 1055–1064.
[17] D. Gilbarg: Jets and cavities, in S. Flügge (ed.), "Encyclopedia of Physics," vol. 9, Springer, Berlin, 1960.
[18] D. Gilbarg: Free streamline theory, *Appl. Mechanics Revs.*, **10** (1957), 181–184.
[19] G. Birkhoff, E. Zarantonello: "Jets, Wakes and Cavities," Academic Press, New York, 1957.
[20] M. S. Plesset, B. Perry: On the application of free streamline theory to cavity flows, Mémoires sur la méchanique des fluides offerts à M. D. Riabouchinsky à l'occasion de son jubilé scientifique, *Publ. sci. et tech. ministère air*, 1954.
[21] M. P. Tulin: Steady two-dimensional cavity flows about slender bodies, *David Taylor Model Basin Rept.* 843 (1953).
[22] B. R. Parkin: Munk integrals for fully cavitated hydrofoils, *J. Aerospace Sci.* (to be published).
[23] H. Reichardt: The laws of cavitation bubbles at axially symmetrical bodies in a flow, *MAP Rept. and Transl.* 766 (1946), distrib. by Office of Naval Research; originally UM 6620, Göttingen, 1945.
[24] P. R. Garabedian: Calculation of axially symmetric cavities and jets, *Pacific J. Math.*, **6** (1956), 611–684.

CHAPTER 88

FLOW THROUGH POROUS MEDIA

BY

J. S. ARONOFSKY, Ph.D., New York, N.Y.

88.1. INTRODUCTION

The theory of fluid flow in a porous medium has found applications in various fields. It describes the movement of ground water in soil and porous rock, the seepage of water through earth fills and concrete dams and of fluids in filters, and the movement of oil and gas in oil fields. In the last case, gas is dissolved in the oil, and is released where the pressure drops below the saturation value. Such two-phase fluids will not be treated here.

88.2. FLUIDS

The flow of a homogeneous fluid through a porous medium depends on its viscosity μ and its density ρ. For a compressible liquid, we may set

$$\rho = \rho_0 e^{\beta(p-p_0)} \tag{88.1}$$

For most liquids, the compressibility β is a very small quantity, of the order of 10^{-6} lb·in.$^{-2}$. Hence, it is often possible to expand Eq. (88.1) in a Taylor series and to drop the higher terms:

$$\rho = \rho_0[1 + \beta(p - p_0)] \tag{88.2}$$

For a gas, we write

$$\rho = \rho_0(p/p_0)^m \tag{88.3}$$

where m determines the thermodynamic character of the gas expansion: $m = 1$ corresponds to an isothermal expansion, $m = \gamma = c_p/c_v$ to an adiabatic expansion.

88.3. POROUS MEDIUM

A porous medium such as sand or foam rubber contains innumerable voids of varying sizes and shapes. These pores may be isolated from each other, or they may be interconnected to form a network of channels through which a fluid may flow. We are concerned only with the interconnected part of the pore system, the *effective pore space.* The ratio of the effective pore space to the total volume of a certain part of the medium is called the porosity f.

88.4. PERMEABILITY, DARCY'S LAW

The velocity of the flow depends on the pressure gradient. In the simplest case, this relation is linear. Figure 88.1 may serve to define the pertinent quantities. In the apparatus shown, fluid is percolated at the rate Q through a cylindrical sample of the porous material. The flow rate per unit of its cross section A is called the filter velocity or macroscopic velocity $q = Q/A$. An experiment as shown in Fig. 88.1 may yield the relation

$$q = K\frac{h_2 - h_1}{h} \tag{88.4}$$

which is known as *Darcy's law* [1]. The constant K depends on mechanical properties of the fluid and the medium. In underground hydrology it is called the permeability. In other fields Darcy's law is used in the form

$$q = \frac{K}{\rho g}\left(\frac{p_2 - p_1}{h} + \rho g\right) \tag{88.4'}$$

or as

$$q = \frac{k}{\mu}\left(\frac{p_2 - p_1}{h} + \rho g\right) \tag{88.4''}$$

The permeability constant k in Eq. (88.4″) does not depend on the fluid viscosity μ. It has the dimension of a length squared, while K has the dimension of a velocity. Experimental methods for determining porosity and permeability are given in [6] and [7].

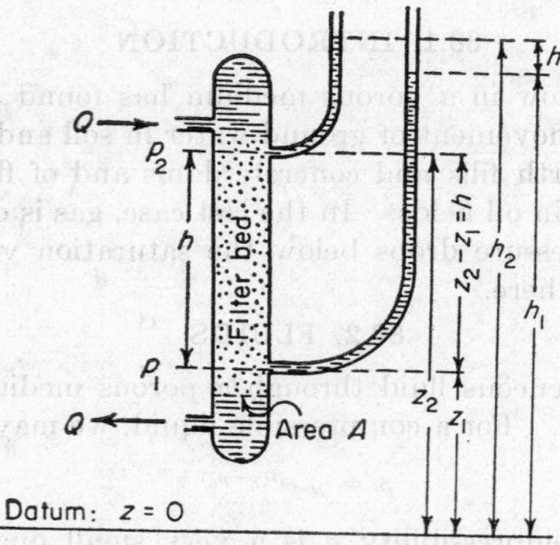

FIG. 88.1. The Darcy experiment. *(By permission from Ref. [7], p. 56.)*

Darcy's law may be extended to three dimensions by replacing the pressure difference in Eqs. (88.4′) and (88.4″) by a pressure gradient. Using a velocity vector **q** and a gravity vector **g** pointing vertically downward, we may write Eq. (88.4″) in the generalized form

$$\mathbf{q} = -(k/\mu)(\text{grad } p - \rho\mathbf{g}) \tag{88.5}$$

When a force potential ϕ is defined as

$$\phi = -gz + \int_{p_0}^{p} \frac{dp}{\rho(p)} \tag{88.6}$$

Eq. (88.5) becomes

$$\mathbf{q} = -\frac{\rho k}{\mu}\text{ grad } \phi \tag{88.7}$$

When written out explicitly for a three-dimensional cartesian coordinate system, Eq. (88.5) takes the following form,

$$u = -\frac{k}{\mu}\frac{\partial p}{\partial x} \qquad v = -\frac{k}{\mu}\frac{\partial p}{\partial y} \qquad w = -\frac{k}{\mu}\left(\frac{\partial p}{\partial z} - \rho g\right) \tag{88.8}$$

where the z axis is directed downward, and u, v, and w are the components in the x, y, and z directions of the vector **q**. There is some controversy as to whether Eq. (88.5) is a suitable generalization of Eq. (88.4″), and alternate equations have been proposed (see [7], p. 60).

The validity of Darcy's law has been tested on many occasions [7]. Equation (88.4) will ultimately break down as the rate of flow is increased. The range of fluid velocities over which Darcy's law is obeyed is termed *viscous* or *laminar*, and the range of higher velocities over which it breaks down as *turbulent*. Reynolds' number is introduced as follows:

$$\mathrm{Re} = \rho q d/\mu$$

where d is a linear dimension measuring the size of the passageway (d is referred to as either grain diameter or pore diameter). For flow through porous media, the transition between viscous and turbulent flow is gradual as the Reynolds number is increased. The gradual transition is undoubtedly due to the spread in sizes of the pores and passageways (pore-size distribution). Some investigators have taken d to be the average grain size and have found that the transition region lies at values of Re of the order of 1 to 10 [2].

88.5. FLUID-FLOW EQUATIONS

Compressible Fluid in an Incompressible Medium

In order to determine the flow pattern in a porous medium for given boundary conditions, the three unknowns $\mathbf{q}$, p, and ρ must be determined. One equation is the differential form of Darcy's law (88.5), another is the pressure-density relationship (88.1), and the third is the continuity equation:

$$\operatorname{div}(\rho \mathbf{q}) = -f \frac{\partial \rho}{\partial t} \tag{88.9}$$

When gravity forces are absent and when Eq. (88.1) is consistently replaced by its linearized form (88.2), then the elimination of all other unknowns leads to the differential equation

$$\operatorname{div}\left(\frac{k}{\mu} \operatorname{grad} p\right) = f\beta \frac{\partial p}{\partial t} \tag{88.10}$$

for the pressure p. This equation has the same form as the heat conduction equation (classical diffusion equation). Thus, the methods used for solving problems in heat conduction are also applicable to the transient flow in porous media [12].

Compressible Fluid in a Compressible Medium

Here the porous medium itself is considered to be compressible. The flow of a compressible fluid through an elastic, porous, and permeable medium is characterized by a coupling between pressure in the fluid and stresses in the solid [14],[16]. The interaction between fluid pressure and stresses in the solid may be greatly simplified by assuming that the porous medium is an elastic body that follows Hooke's law in the form

$$\frac{df}{dp} = \lambda \tag{88.11}$$

where λ is a compressibility coefficient. References [4],[7], and [10] discuss the use of Eq (88.11). The continuity equation becomes

$$\operatorname{div}(\rho \mathbf{q}) = -\left(f + \frac{\lambda}{\beta}\right) \frac{\partial \rho}{\partial t} \tag{88.12}$$

A comparison shows that the porosity f in Eq. (88.9) is replaced by $(f + \lambda/\beta)$ in Eq. (88.12). Otherwise, the two equations are identical. It readily follows that the partial differential equation (88.10) will have the analogous form

$$\operatorname{div}\left(\frac{k}{\mu} \operatorname{grad} p\right) = (f\beta + \lambda) \frac{\partial p}{\partial t} \tag{88.13}$$

which is also the diffusion equation. The analogy between Eqs. (88.13) and (88.10) indicates that the effect of a compressible medium on the flow distribution can be duplicated by assuming a "supercompressibility" of the liquid.

Incompressible Flow

When the fluid and the medium are incompressible, Eqs. (88.6) and (88.7) may be replaced by

$$\varphi = \rho\phi = p - \rho g z \tag{88.14}$$

$$\mathbf{q} = -\frac{k}{\mu}\operatorname{grad}\varphi \tag{88.15}$$

The continuity equation (88.9) reduces to

$$\operatorname{div}\mathbf{q} = 0 \tag{88.16}$$

and Eq. (88.10) assumes the simpler form

$$\operatorname{div}\left(\frac{k}{\mu}\operatorname{grad} p\right) = 0 \tag{88.17}$$

When k and μ are constants, the pressure and the flow potential φ satisfy Laplace's equation:

$$\nabla^2 p = 0 \qquad \nabla^2\varphi = 0 \tag{88.18a,b}$$

The flow field has then the same kinematic properties as that of an ideal fluid flow (see Chaps. 70 and 71). In two dimensions x, y, there exists a complex potential $\Phi = \varphi + i\psi$, whose real part is the potential φ of Eq. (88.14), and which is an analytic function of the complex coordinate $z = x + iy$. The complex velocity $w = u - iv$ is the derivative of the complex potential [see Eq. (70.24)]. All the methods of conformal mapping described on pp. 70-8 to 70-21 may be used to solve two-dimensional flow problems in porous media, except that the factor $-k/\mu$ of Eq. (88.15) enters into all relations between the potential and the velocity.

Anisotropic Permeability

Measurements from layers of porous media have shown that the permeability normal to the bedding plane may be appreciably different from that parallel to the bedding plane. The layered material is considered to be anisotropic with respect to permeability, and a generalized form of Darcy's law becomes

$$\mathbf{q} = -(K/\mu)\operatorname{grad} p \tag{88.19}$$

where K is a symmetric tensor (called the permeability tensor). Two important properties are: (1) The pressure gradient grad p and the filter velocity $\mathbf{q}$ need not act in the same direction. (2) There are three orthogonal axes in space along which the pressure gradient and velocity do have the same direction. They are the principal axes of the permeability tensor.

Let us choose the principal axes as coinciding with the x, y, z coordinate axes (corresponding permeabilities k_x, k_y, k_z). Equation (88.19) can now be written in the form

$$u = -\frac{k_x}{\mu}\frac{\partial p}{\partial x} \qquad v = -\frac{k_y}{\mu}\frac{\partial p}{\partial y} \qquad w = -\frac{k_z}{\mu}\frac{\partial p}{\partial z} \tag{88.20}$$

If the flow is incompressible and if the permeabilities and the viscosity are constants, combination of Eqs. (88.20) with the continuity equation (88.16) leads to

$$k_x\frac{\partial^2 p}{\partial x^2} + k_y\frac{\partial^2 p}{\partial y^2} + k_z\frac{\partial^2 p}{\partial z^2} = 0 \tag{88.21}$$

as the differential equation of the pressure field. The following transformation of the coordinate system,

$$\bar{x} = x/\sqrt{k_x} \qquad \bar{y} = y/\sqrt{k_y} \qquad \bar{z} = z/\sqrt{k_z} \tag{88.22}$$

will reduce Eq. (88.19) to Laplace's equation

$$\frac{\partial^2 p}{\partial \bar{x}^2} + \frac{\partial^2 p}{\partial \bar{y}^2} + \frac{\partial^2 p}{\partial \bar{z}^2} = 0 \tag{88.23}$$

Thus, the effect of anisotropy is reflected by an apparent shrinking or expansion of the coordinates.

88.6. TYPICAL SOLUTIONS

Steady-state Radial Flow

Figure 88.2 shows a porous formation sealed to fluid flow above and below, so that radial flow into the well bore is assured. The flow is assumed to be incompressible, and gravity effects are not present. It can therefore be described by

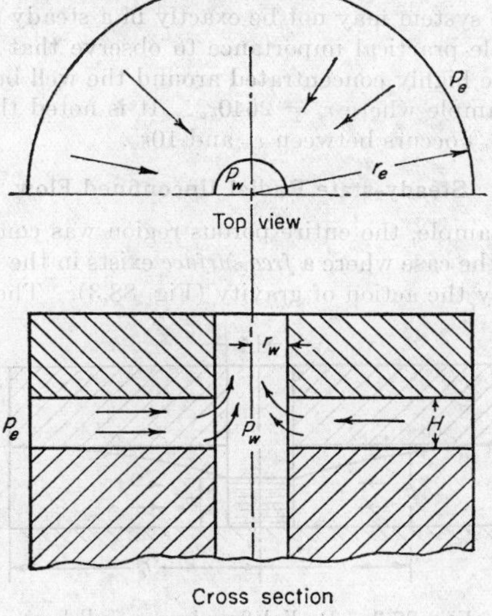

FIG. 88.2. Radial flow into a well bore.

Eq. (88.18*a*) which, for radial symmetry and in polar coordinates, reads

$$\frac{1}{r}\frac{d}{dr}\left(r\frac{dp}{dr}\right) = 0 \tag{88.24}$$

Integrating Eq. (88.24) yields

$$p(r) = c_1 \ln r + c_2 \tag{88.25}$$

where c_1 and c_2 are the constants of integration. Two boundary conditions are required in order to evaluate them. One boundary condition is specified at the well radius ($r = r_w$) when the fluid pressure there is held at the constant value p_w. The other boundary condition is specified for any outer radius r_e where the pressure is assumed to have the value p_e. Hence, the boundary conditions are:

$$\text{At } r = r_w: \qquad p = p_w$$
$$\text{At } r = r_e: \qquad p = p_e$$

and the solution is

$$p(r) = p_w + \frac{p_e - p_w}{\ln (r_e/r_w)} \ln \frac{r}{r_w} \tag{88.26}$$

Equation (88.26) represents the pressure distribution for the radial flow system where the pressure varies logarithmically with increasing radii. The velocity distribution is obtainable by differentiating Eq. (88.26) with respect to r:

$$q_r = \frac{k}{\mu}\frac{\partial p}{\partial r} = -\frac{k}{\mu}\frac{1}{r}\frac{p_e - p_w}{\ln (r_e/r_w)} \tag{88.27}$$

and it is shown that the velocity varies inversely with increasing radii. The total flow rate Q into the well bore is

$$Q = -2\pi r H q_r = \frac{2\pi k H(p_e - p_w)}{\mu \ln (r_e/r_w)} \tag{88.28}$$

The total flow rate Q is directly proportional to the permeability, the thickness, and the pressure drop, and inversely proportional to the viscosity. Extensive use is made of this result whenever fluid is flowing through a porous material into a well bore, even though the system may not be exactly in a steady state.

It is of considerable practical importance to observe that the pressure gradients and flow resistance are highly concentrated around the well bore. To illustrate this point, consider an example where $r_e = 2640 r_w$. It is noted that 29.2% of the total pressure drop $(p_e - p_w)$ occurs between r_w and $10 r_w$.

Steady-state Radial Unconfined Flow

In the previous example, the entire porous region was completely filled with the fluid. Consider now the case where a *free surface* exists in the porous material that is primarily controlled by the action of gravity (Fig. 88.3). There is still a radial flow

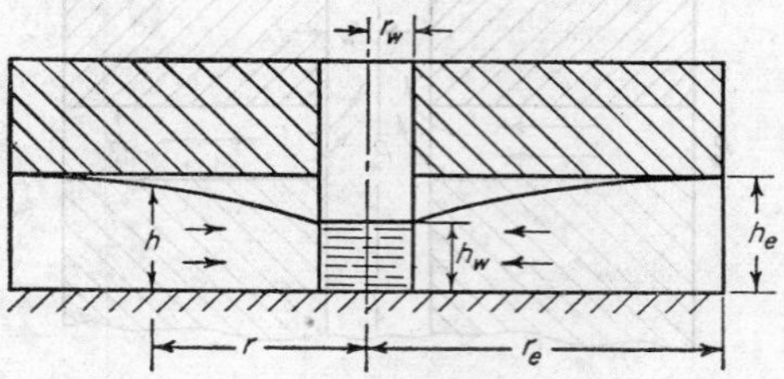

Fig. 88.3. Radial flow into a well bore.

into a well bore. The formation is sealed to fluid flow at the bottom, and the fluid level in the well bore is maintained at h_w, measured from the bottom of the formation. It is assumed that the fluid level at radius r_e remains at h_e, which is the thickness of the formation. Thus, a free surface exists when $h_w < h_e$.

Along the free surface, the pressure is constant, and the intersection of the free surface with any radial plane is a streamline. Problems involving free surfaces are not readily tractable, and approximate methods have been developed. The best-known approximate method is the Dupuit-Forchheimer theory. The assumption is made that, for small inclinations of the free surface, the streamlines can be taken as horizontal, and the velocities are proportional to the slope of the free surface, but are independent of the depth [2],[9]. With a polar coordinate r in a horizontal plane, the radial velocity is given by

$$q_r = -K \frac{\partial h}{\partial r} \tag{88.29}$$

where h represents the height of the free surface above the bottom of the formation. For steady-state flow, the continuity equation becomes

$$\frac{\partial}{\partial x}(hu) + \frac{\partial}{\partial y}(hv) = 0 \tag{88.30}$$

which leads to Forchheimer's result:

$$\nabla^2(h^2) = 0 \tag{88.31}$$

The boundary conditions are:

$$\text{At } r = r_w: \quad h = h_w$$
$$\text{At } r = r_e: \quad h = h_e$$

and the solution of Laplace's equation (88.31) is

$$h^2 = \frac{h_e^2 - h_w^2}{\ln(r_e/r_w)} \ln \frac{r}{r_w} + h_w^2 \tag{88.32}$$

Again, Eq. (88.32) is identical with Eq. (88.26), except that p is replaced by h^2. The total flow rate into the well bore becomes

$$Q = 2\pi Krh \frac{\partial h}{\partial r} = \pi K \frac{h_e^2 - h_w^2}{ln\ (r_e/r_w)} \tag{88.33}$$

It should be pointed out that the assumptions basic to this simple theory of Dupuit and Forchheimer are not quite warranted and, therefore, have been questioned [2]. In spite of its shortcomings, the Dupuit-Forchheimer theory is still widely used because of its simplicity, and because the total flow-rate formula (88.33) compares very favorably with both experiments and other approximate methods of analysis.

Seepage through Dams

Figure 88.4 illustrates a seepage problem of a particular kind, the flow of water with a free surface. If the length of the dam is much larger than its thickness, the flow is two-dimensional. The flow field in the x,y plane is bounded by the line $ABCDE$, which contains segments of different types. Among them is the free-surface segment CD. Different from all the others, its shape is not known, not even the location of its end C on the downstream face of the dam. In return, there are two boundary conditions to be satisfied instead of only one as on the other segments: The pressure is constant (capillarity effects being neglected), and the velocity has no component normal to the boundary.

Problems of this type, like the free-streamline problems of ideal flow (p. 70-23), are best solved by the hodograph method. It consists in using the relations between two Argand planes, the physical plane of Fig. 88.4 (the z plane with $z = x + iy$), and the w plane with $w = u - iv$ (or, instead, a $\bar{w}$ plane with $\bar{w} = u + iv$). To every point P of the z plane, there corresponds a point P' of the w plane, whose coordinates u and $-v$ are the components of the flow velocity at P. Since

$$w = -(k/\mu)\, d\Phi/dz$$

is an analytic function of z, the relation between the two planes is that of a conformal mapping. From this fact, the method for solving the problem is derived.

The first step is to use the boundary conditions of the flow to find the points A', B', . . . of the w plane which correspond to the boundary points of the flow field in the z plane. Then the complex function $w(z)$, which maps the z plane onto the w plane, may be found by analytical or numerical methods. When this has been done, the velocity components for each point of the z plane are known, and the problem is solved. For the actual work, it is often preferred to use the $\bar{w}$ plane, since in it the vector with the components u, v is parallel to the actual velocity vector in the physical plane. In this case, however, the hodograph plane is not a conformal mapping, but the mirror image of a conformal mapping, of the physical plane.

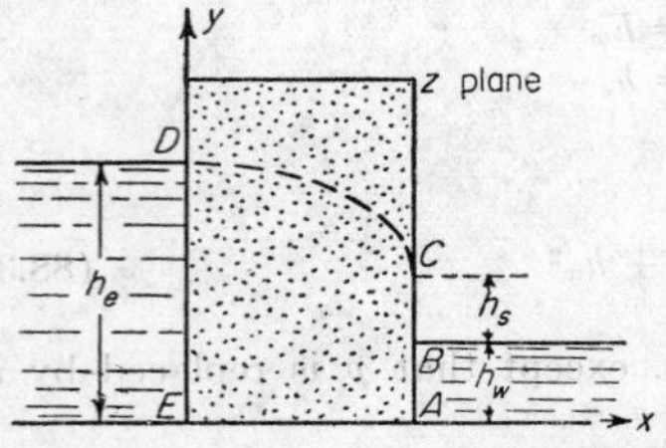

FIG. 88.4. Seepage through a dam with vertical faces. (*By permission from Ref.* [2], *p.* 303.)

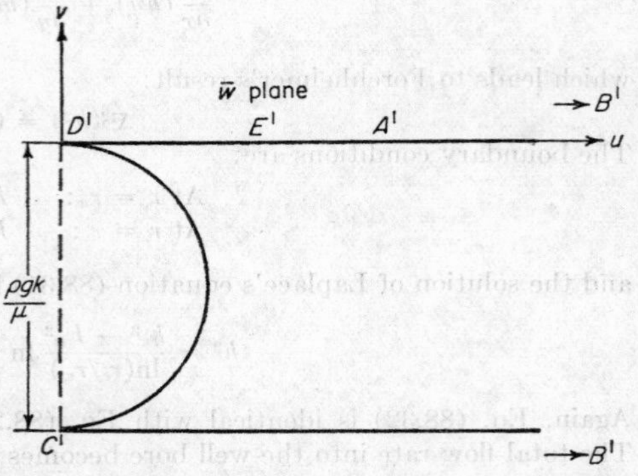

FIG. 88.5. The hodograph plane of Fig. 88.4. (*By permission from Ref.* [2], *p.* 304.)

In the example of Fig. 88.4, the boundary $A'B'C'D'E'$ in the hodograph plane (Fig. 88.5) is found by the following steps:

1. Along DE, the potential φ is constant according to Eq. (88.14), in which the vertical coordinate z must be replaced by $-y$. Consequently, $v = -(k/\mu)\,\partial\varphi/\partial y = 0$, and the points of $D'E'$ are found on the u axis of the hodograph plane. In particular, at D, at the water level, there is no pressure gradient in the horizontal direction and, therefore, also $u = 0$; that is, D' coincides with the origin of the (u,v) system.

2. Along the impermeable boundary EA, there is $v = 0$, and the image $E'A'$ is the horizontal continuation of $D'E'$.

3. The region AB is similar to DE, and $A'B'$ is another part of the u axis.

4. The segment DC of the boundary lies in a free surface. The condition of constant pressure, when introduced into Eq. (88.14) (with y instead of $-z$), yields

$$\varphi - \rho g y = \text{const} \tag{88.34}$$

The second boundary condition, that the total velocity q has the direction of the boundary, may be written in the form

$$q = \sqrt{u^2 + v^2} = -\frac{k}{\mu}\frac{\partial\varphi}{\partial s} \tag{88.35}$$

where s is the arc length measured from D along the boundary. If dx and dy are the components of a line element ds, the vertical component of the velocity is

$$v = -\frac{k}{\mu}\frac{\partial\varphi}{\partial s}\frac{dy}{ds}$$

Differentiating Eq. (88.34) with respect to s and then multiplying by $\partial\varphi/\partial s$ yields the relation

$$\left(\frac{\partial\varphi}{\partial s}\right)^2 = \rho g\frac{dy}{ds}\frac{\partial\phi}{\partial s} = -\rho g\frac{\mu}{k}v$$

whence ultimately

$$u^2 + v^2 + \frac{\rho g k}{\mu}v = 0$$

This is the equation of a circular arc passing through D'. The location of its other end C' follows from the fact that, at C, the vertical component of the velocity is

$$v = -\frac{k}{\mu}\frac{\partial\varphi}{\partial y} = -\frac{k}{\mu}\rho g$$

Therefore, $D'C'$ is the semicircle shown in Fig. 88.5.

5. The segment BC of the boundary lies in a seepage surface. Here the pressure is constant [Eq. (88.34)], but the velocity may have a component normal to the boundary. This means that here water may leave the porous medium and either evaporate or run down on the surface of the dam. The magnitude of the component u is not known, but v may be found from (88.34) to be $v = -\rho gk/\mu$. Therefore, the boundary BC maps on a horizontal line $B'C'$, as shown in Fig. 88.5.

As may be seen from Fig. 88.5, point B' is required to lie on two different horizontal lines. The apparent contradiction is resolved if B' lies at infinity, i.e., if $u = \infty$ at B.

After the image $A'B'C'D'E'$ of the boundary has been found from physical considerations, the solution of the problem proceeds along the same lines as described on p. 70–24. Further details may be found in the book by Muskat [2].

Two-fluid System: Displacement of an Interface

The analysis for single-phase flow in porous media may be made applicable to multiple-phase flow if certain stringent approximations are made. Consider two immiscible fluids that are in motion within a porous medium and in contact along an interface (e.g., oil and water in porous rock). Both fluids are assumed to be incompressible.

Two regions, 1 and 2, are involved. In each, the fluid flow is governed by a potential, φ_1 or φ_2, according to Eq. (88.14). At the interface, the pressure and the normal velocity component must be continuous. These two conditions may be written in the form

$$\varphi_1 + \rho_1 gz = \varphi_2 + \rho_2 gz \qquad \frac{k_1}{\mu_1}\frac{\partial\varphi_1}{\partial n} = \frac{k_2}{\mu_2}\frac{\partial\varphi_2}{\partial n}$$

where n is the normal to the interface.

Let the equation of the interface be

$$F(x,y,z,t) = 0 \tag{88.36}$$

At time $(t + \delta t)$ the coordinate x has changed to $(x + u\,\delta t)$ because the fluid moves with the interface. Equation (88.36) reads then

$$F(x + u\,\delta t,\ y + v\,\delta t,\ z + w\,\delta t,\ t + \delta t) = 0 \tag{88.37}$$

Subtraction of Eq. (88.36) from (88.37) and transition to the limit $\delta t \to 0$ yields the differential equation

$$\frac{\partial F}{\partial t} + \mathbf{q} \cdot \operatorname{grad} F = 0 \tag{88.38}$$

which must be satisfied at the interface. It may be written as

$$\frac{\partial F}{\partial t} - \frac{k}{\mu} \operatorname{grad} \varphi \cdot \operatorname{grad} F = 0 \tag{88.39}$$

and in this equation either the constants k_1, μ_1 and the potential φ_1, or the corresponding quantities for region 2 must be used.

As an example, consider Fig. 88.6. It describes an axisymmetric movement in a thin horizontal layer and toward a well of radius r_w. In this case, the gravity term may be dropped from Eq. (88.14), and the potential $\varphi = p$. At $r = r_w$, the potential $\varphi = \varphi_w$ is given. To make the problem determinate, it is assumed that, on a circular

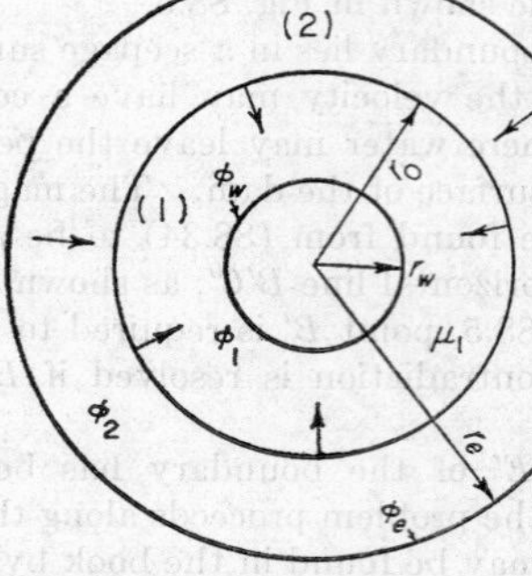

FIG. 88.6. Radial displacement. (*By permission from Ref.* [7], *p.* 169.)

boundary $r = r_e$ of domain 2, the potential $\varphi = \varphi_e$ is also known. The solution of this problem is due to Muskat [12]. He finds that

$$\frac{r_0^2}{r_e^2}\left[\ln\frac{r_e^2}{r_w^2} - (1-\epsilon) + (1+\epsilon)\ln\frac{r_0^2}{r_e^2}\right] = -\frac{4k_1 t(\varphi_e - \varphi_w)}{\mu_1 r_e^2} + \ln\frac{r_e^2}{r_w^2} - (1-\epsilon)$$

where $\epsilon = k_1\mu_2/k_2\mu_1$. This equation describes the variation of the interface radius r_0 as a function of t. From (88.39) grad $\varphi = d\varphi/dr$ may be calculated, and then it is an easy task to use the solution (88.25) in each domain to find φ, and hence p and q.

Nonsteady Motion

Equation (88.10) describes the flow of a compressible fluid in an incompressible medium. Here we are concerned with the transient flow into a well (Fig. 88.2). When we assume that k and μ are constant, Eq. (88.10) reduces to

$$\frac{k}{\mu}\frac{1}{r}\frac{\partial}{\partial r}\left(r\frac{\partial p}{\partial r}\right) = f\beta\frac{\partial p}{\partial t} \tag{88.40}$$

We introduce dimensionless variables,

$$\bar{r} = r/r_w \qquad \bar{p} = p/p_i \qquad \bar{t} = kt/f\beta\mu r_w^2$$

where r_w is the well radius, and p_i the initial pressure in the entire domain. The boundary and initial conditions are specified as follows:

For $\bar{r} = 1$: $\quad Q = \text{const}$
For $\bar{r} = \bar{r}_e$ and all times $t \geq 0$: $\quad p = p_i$
For $t = 0$ at all points: $\quad p = p_i$

The solution for the pressure at $r = r_w$ has been given by several authors [6],[13]. It may be written as

$$p_i - p(\bar{r}_w,\bar{t}) = \frac{\mu Q}{2\pi kH}\left\{\ln\bar{r}_e + 2\sum_n \frac{J_0^2(\lambda_n\bar{r}_e)e^{-\lambda_n^2\bar{t}}}{\lambda_n^2[J_0^2(\lambda_n\bar{r}_e) - J_1^2(\lambda_n)]}\right\} \tag{88.41}$$

where λ_n stands for the solutions of the transcendental equation

$$Y_1(\lambda_n)J_0(\lambda_n\bar{r}_e) - J_1(\lambda_n)Y_0(\lambda_n\bar{r}_e) = 0 \tag{88.42}$$

The pressure drop in Eq. (88.41) approaches asymptotically the steady-state value,

$$p_i - p(\bar{r}_w, \infty) = \frac{\mu Q}{2\pi kH} \ln \frac{r_e}{r_w} = \Delta p_\infty$$

which agrees with Eq. (88.28).

The manner in which the pressure distribution builds up is shown (after Hurst [15]) in Fig. 88.7. This set of curves is, of course, independent of the flow Q.

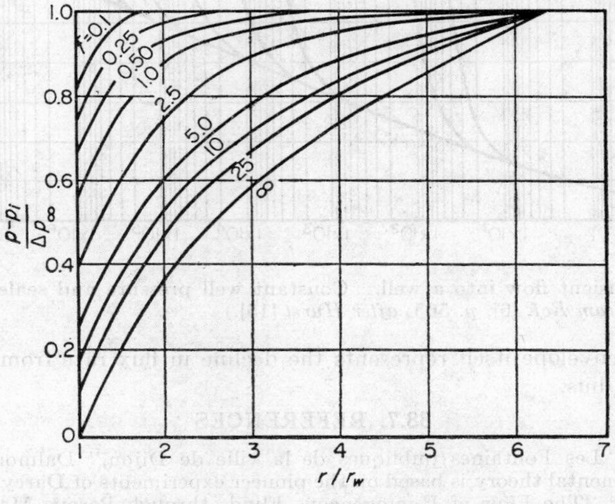

FIG. 88.7. Transient flow into a well. Constant flow rate into well and constant outer pressure. (*By permission from Ref.* [6], *p.* 561; *after Hurst* [15].)

As a second example, consider a sealed system described by the following boundary and initial conditions:

For $\bar{r} = 1$ and $t > 0$:	$p = p_w$
For $r = r_e$ and all times $t \geq 0$:	$\partial p/\partial r = 0$
For $t = 0$ at all points:	$p = p_i$

In this case, the outer radius is sealed, and the system initially contains liquid at the pressure p_i. The pressure at the well bore is suddenly changed to p_w. Solutions to this problem were published by Hurst [15], and the flow per unit thickness at the well $r = r_w$ is

$$Q = \frac{2\pi k(p_i - p_w)}{\mu} \sum_n \frac{J_1{}^2(\lambda_n\bar{r}_e)e^{-\lambda_n{}^2\bar{t}}}{J_0{}^2(\lambda_n) - J_1{}^2(\lambda_n r_e)} \tag{88.43}$$

where the λ_n are the roots of

$$Y_1(\lambda_n\bar{r}_e)J_0(\lambda_n) - J_1(\lambda_n\bar{r}_e)Y_0(\lambda_n) = 0 \tag{88.44}$$

The variation with time of the efflux rate is plotted in Fig. 88.8. For convenience in plotting, the ordinates in this figure were chosen as $k(p_i - p_w)/\mu Q$, i.e., proportional to the reciprocal of Q. As is to be expected, the curves for the different $\bar{r}_e$ coincide initially but successively break away from the common envelope as the effect of the finite radius of the closed system comes into play and the flux rate begins to decline

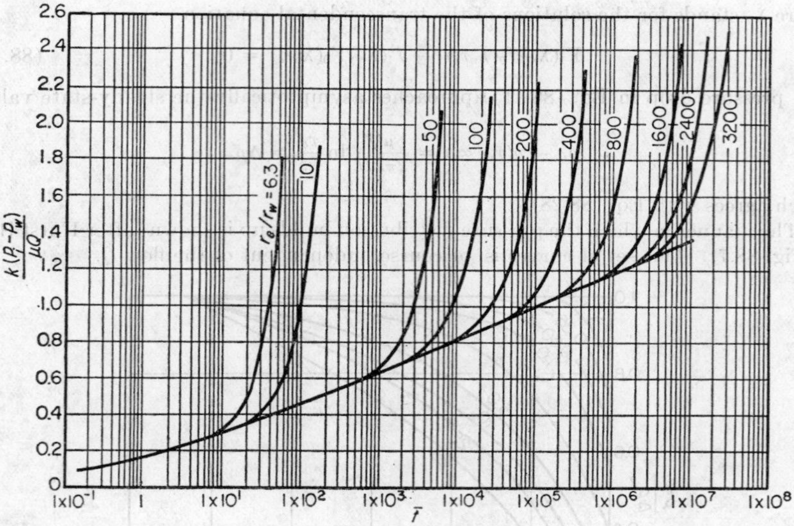

FIG. 88.8. Transient flow into a well. Constant well pressure and sealed outer radius. (*By permission from Ref.* [6], *p.* 565; *after Hurst* [15].)

sharply. The envelope itself represents the decline in flux rate from a reservoir of infinite outer radius.

88.7. REFERENCES

[1] H. Darcy: "Les Fontaines publiques de la ville de Dijon," Dalmont, Paris, 1856. (The fundamental theory is based on the pioneer experiments of Darcy.)

[2] M. Muskat: "The Flow of Homogeneous Fluids through Porous Media," Edwards, Ann Arbor, 1946. (This book is an excellent reference source.)

[3] M. K. Hubbert: The theory of ground water motion, *J. Geol.*, **48** (1940), 785–944. (Presents comprehensive development of field equations.)

[4] L. S. Leibenson: Motion of natural fluids and gases in a porous medium (in Russian), *Gosudarst. Izdat. Tekh.-teoret. Lit.*, Moscow and Leningrad, 1947. (Stresses the physical principles.)

[5] P. Ya. Polubarinova-Kochina: Theory of motion of ground water, *Gosudarst. Izdat. Tekh.-Teoret. Lit.*, Moscow, 1952.

[6] M. Muskat: "Physical Principles of Oil Production," McGraw-Hill, New York, 1949. (Comprehensive review of flow of two-phase fluids with particular emphasis on oil production.)

[7] A. E. Scheidegger: "The Physics of Flow through Porous Media," Univ. of Toronto Press, Toronto, and Macmillan, New York, 1957. (Contains an extensive bibliography for both single-phase and multiphase fluids.)

[8] F. Engelund: On the laminar and turbulent flow of ground water through homogeneous sands, *Trans. Danish Acad. Tech. Sci.*, no. 3, 1953.

[9] C. E. Jacob: Flow of ground water, chap. 5 in H. Rouse (ed.), "Engineering Hydraulics," Wiley, New York, 1950.

[10] A. E. Scheidegger: Hydrodynamics in porous media, in S. Flügge (ed.), "Encyclopedia of Physics," vol. 9, Springer, Berlin, 1960.

[11] P. Ya. Polubarinova-Kochina, S. B. Falkovich: Theory of filtration of liquids in porous media, *Advances in Appl. Mechanics*, **2** (1951), 154–225.

[12] H. S. Carslaw, J. C. Jaeger: "Conduction of Heat in Solids," Oxford Univ. Press, New York, 1947.

[13] G. Hamel: Über Grundwasserströmung, *Z. angew. Math. Mech.*, **14** (1934), 129–157.

[14] M. Biot: General theory of three-dimensional consolidation, *J. Appl. Phys.*, **12** (1944), 155–165.

[15] W. Hurst: Unsteady flow of fluids in oil reservoirs, *Physics*, **5** (1934), 20–30.

[16] K. Terzaghi: "Theoretical Soil Mechanics," Wiley, New York, 1943.

INDEX